Earth and Life

International Year of Planet Earth

Series Editors:

The book series is dedicated to the United Nations International Year of Planet Earth. The aim of the Year is to raise worldwide public and political awareness of the vast (but often under-used) potential of Earth sciences for improving the quality of life and safeguarding the planet. Geoscientific knowledge can save lives and protect property if threatened by natural disasters. Such knowledge is also needed to sustainably satisfy the growing need for Earth's resources by more people. Earth scientists are ready to contribute to a safer, healthier and more prosperous society. IYPE aims to develop a new generation of such experts to find new resources and to develop land more sustainably.

For further volumes:
http://www.springer.com/series/8096

John A. Talent
Editor

Earth and Life

Global Biodiversity, Extinction Intervals and Biogeographic Perturbations Through Time

planetearth
Earth Sciences for Society

United Nations
Educational, Scientific and
Cultural Organization

IUGS
International Union of Geological Sciences

Springer

Editor
Em. Prof. John A. Talent
Earth and Planetary Sciences
Macquarie University 2109
Australia
jatalent32@gmail.com

ISBN 978-94-024-0491-3 ISBN 978-90-481-3428-1 (eBook)
DOI 10.1007/978-90-481-3428-1
Springer Dordrecht Heidelberg London New York

Softcover re-print of the Hardcover 1st edition 2012

Printed on acid-free paper

Springer is part of Springer Science+Business Media (www.springer.com)

Foreword

The International Year of Planet Earth (IYPE) was established as a means of raising worldwide public and political awareness of the vast, though frequently under-used, potential the earth sciences possess for improving the quality of life of the peoples of the world and safeguarding Earth's rich and diverse environments.

The International Year project was jointly initiated in 2000 by the International Union of Geological Sciences (IUGS) and the Earth Science Division of the United Nations Educational, Scientific and Cultural Organisation (UNESCO). IUGS, which is a non-governmental organization, and UNESCO, an inter-governmental organization, already shared a long record of productive cooperation in the natural sciences and their application to societal problems, including the International Geoscience Programme (IGCP) now in its fourth decade.

With its main goals of raising public awareness of and enhancing research in the Earth sciences on a global scale in both the developed and less-developed countries of the world, two operational programmes were demanded. In 2002 and 2003, the series editors together with Dr. Ted Nield and Dr. Henk Schalke (all four being core members of the Management Team at that time) drew up outlines of a science and an outreach programme. In 2005, following the UN proclamation of 2008 as the United Nations International Year of Planet Earth, the "year" grew into a triennium (2007–2009).

The outreach programme, targeting all levels of human society from decision makers to the general public, achieved considerable success in the hands of member states representing over 80% of the global population. The science programme concentrated on bringing together like-minded scientists from around the world to advance collaborative science in a number of areas of global concern. A strong emphasis on enhancing the role of the Earth sciences in building a healthier, safer and wealthier society was adopted – as declared in the Year's logo strap-line "Earth Sciences *for* Society".

The organizational approach adopted by the science programme involved recognition of 10 global themes that embrace a broad range of problems of widespread national and international concern, as follows:

- Human health: this theme involves improving understanding of the processes by which geological materials affect human health as a means identifying and reducing a range of pathological effects.
- Climate: particularly emphasizes improved detail and understanding of the non-human factor in climate change.
- Groundwater: considers the occurrence, quantity and quality of this vital resource for all living things against a background that includes potential political tension between competing neighbour nations.

- Ocean: aims to improve understanding of the processes and environment of the ocean floors with relevance to the history of planet Earth and the potential for improved understanding of life and resources.
- Soils: this thin "skin" on Earth's surface is the vital source of nutrients that sustain life on the world's landmasses, but this living skin is vulnerable to degradation if not used wisely. This theme emphasizes greater use of soil science information in the selection, use and ensuring sustainability of agricultural soils so as to enhance production and diminish soil loss.
- Deep Earth: in view of the fundamental importance of deep the Earth in supplying basic needs, including mitigating the impact of certain natural hazards and controlling environmental degradation, this theme concentrates on developing scientific models that assist in the reconstruction of past processes and the forecasting of future processes that take place in the solid Earth.
- Megacities: this theme is concerned with means of building safer structures and expanding urban areas, including utilization of subsurface space.
- Geohazards: aims to reduce the risks posed to human communities by both natural and human-induced hazards using current knowledge and new information derived from research.
- Resources: involves advancing our knowledge of Earth's natural resources and their sustainable extraction.
- Earth and Life: it is over 2½ billion years since the first effects of life began to affect Earth's atmosphere, oceans and landmasses. Earth's biological "cloak", known as the biosphere, makes our planet unique but it needs to be better known and protected. This theme aims to advance understanding of the dynamic processes of the biosphere and to use that understanding to help keep this global life-support system in good health for the benefit of all living things.

The first task of the leading Earth scientists appointed as theme leaders was the production of a set of theme brochures. Some 3500 of these were published, initially in English only but later translated into Portuguese, Chinese, Hungarian, Vietnamese, Italian, Spanish, Turkish, Lithuanian, Polish, Arabic, Japanese and Greek. Most of these were published in hard copy and all are listed on the IYPE web site.

It is fitting that, as the International Year's triennium terminates at the end of 2009, the more than 100 scientists who participated in the 10 science themes should bring together the results of their wide ranging international deliberations in a series of state-of-the-art volumes that will stand as a legacy of the International Year of Planet Earth. The book series was a direct result of interaction between the International Year and the Springer Verlag Company, a partnership which was formalized in 2008 during the acme of the triennium.

This IYPE-Springer book series contains the latest thinking on the chosen themes by a large number of earth science professionals from around the world. The books are written at the advanced level demanded by a potential readership consisting of Earth science professionals and students. Thus, the series is a legacy of the science programme, but it is also a counterweight to the earth science information in several media formats already delivered by the numerous national committees of the International Year in their pursuit of worldwide popularization under the outreach programme.

The discerning reader will recognize that the books in this series provide not only a comprehensive account of the individual themes but also share much common ground that makes the series greater than the sum of the individual volumes. It is to be hoped that the scientific perspective thus provided will enhance the reader's appreciation of the nature and scale of earth science as well as the guidance it can offer to governments, decision makers and others seeking solutions to national and global problems, thereby improving everyday life for present and future residents of planet Earth.

Eduardo F.J. de Mulder
Executive Director International Secretariat
International Year of Planet Earth

Edward Derbyshire
Goodwill Ambassador
International Year of Planet Earth

Series Preface

This book series is one of the many important results of the International Year of Planet Earth (IYPE), a joint initiative of UNESCO and the International Union of Geological Sciences (IUGS), launched with the aim of ensuring greater and more effective use by society of the knowledge and skills provided by the earth sciences.

It was originally intended that the IYPE would run from the beginning of 2007 until the end of 2009, with the core year of the triennium (2008) being proclaimed as a UN Year by the United Nations General Assembly. During all 3 years, a series of activities included in the IYPE's science and outreach programmes had a strong mobilizing effect around the globe, not only among earth scientists but also within the general public and, especially, among children and young people.

The outreach programme has served to enhance cooperation among earth scientists, administrators, politicians and civil society and to generate public awareness of the wide ranging importance of the geosciences for human life and prosperity. It has also helped to develop a better understanding of Planet Earth and the importance of this knowledge in building of a safer, healthier, and wealthier society.

The scientific programme, focused upon 10 themes of relevance to society, has successfully raised geoscientists' awareness of the need to develop further the international coordination of their activities. The programme has also led to some important updating of the main challenges the geosciences are, and will be confronting within an agenda closely focused on societal benefit.

An important outcome of the work of the IYPE's scientific themes includes this thematic book as one of the volumes making up the IYPE-Springer Series, which was designed to provide an important element of the legacy of the International Year of Planet Earth. Many prestigious scientists, drawn from different disciplines and with a wide range of nationalities, are warmly thanked for their contributions to a series of books that epitomize the most advanced, up-to-date and useful information on evolution and life, water resources, soils, changing climate, deep earth, oceans, non-renewable resources, earth and health, natural hazards, and megacities.

This legacy opens a bridge to the future. It is published in the hope that the core message and the concerted actions of the International Year of Planet Earth throughout the triennium will continue and, ultimately, go some way toward helping to establish an improved equilibrium between human society and its home planet. As stated by the Director General of UNESCO, Koichiro Matsuura, "Our knowledge of the Earth system is our insurance policy for the future of our planet". This book series is an important step in that direction.

R. Missotten
Chief, Global Earth Observation Section
UNESCO

Alberto C. Riccardi
President
IUGS

We cling to a sort of magical thinking about the impending ecological disaster. It is too easy to accept the assertions of the World's professional optimists, or the pronouncements of those who noisily deny scientific facts. If more of us cared to experience the natural world unmediated by anything at all – to simply sit, quietly, open to its sublime beauty – we might take better care of it – our heritage with its almost incomprehensibly long history!

From Sara Maitland in *A Book of Silence*, published by Granta, 2008

This volume is dedicated to my son Ross who, on the point of submitting his PhD, died in a car accident near Allora, southern Queensland, on 21 September 1991. A mathematician, biologist and statistician, his research on statistical modelling of complex ecosystems would have complemented this volume perfectly. He would have relished many of the contributions here.

Preface to this Volume

This volume is one of a series of volumes, Springer's Legacy Series, produced as a tangible by-product of the activities of UNESCO's International Year of Planet Earth. The volume was directed towards considering the broad pattern of increasing biodiversity through time, and recurrent events of minor and major ecosphere reorganization, some regional, such as the swiftly developing collapse of reef systems: or local, exemplified by the 79 AD disaster about Pompei. Intense scrutiny was sought on the pattern of physical (including isotopic), sedimentary and biotic circumstances through the time intervals during which life crises occurred. These events affected terrestrial, lacustrine and estuarine ecosystems, locally and globally, but also affected continental shelf (including intertidal) ecosystems and even deep ocean ecosystems. The pattern of such events is an important part of the backdrop against which modelling the pattern of future environmental change needs to be evaluated.

A broad spectrum of authors was sought but, inevitably, recently published exhaustive surveys resulted in lacunae in this particular volume. Authors who felt they could usefully contribute were encouraged to take liberties with subject, organization and point of view. As a consequence, some of the contributions are highly personal and digressive (such as the concluding chapter of this volume); some are essays, others are reflective overviews of broad topics, and yet others present the detailed results of specialized scholarship undertaken over many years. It is anticipated that all have adroitly balanced scientific data and inference. Writing is often a slippery commodity. An author's voice and the readability of his/her contribution can be easily lost by over-editing. Care has been taken to maintain the personal voice and viewpoints of the authors.

Every fossil, large or small, can be viewed as a time-capsule or as a data bank – as my late friend, Paul Tasch, insisted nearly 40 years ago in his *Paleobiology of the Invertebrates: Data Retrieval from the Fossil Record.* Paul's thesis was that all fossils may be viewed as parts of an infinitely larger data bank of transcendental scale and that the information each fossil can deliver (alone or in conjunction with other fossils) can answer many questions concerning the fossils themselves and the environments in which they lived. There have been countless trillions of such tiny time-capsules. Their stories – as individuals or as members of former communities, supplemented by information about the sediments on or in which they lived and were buried – provide us with grand panoramas of deep time, and provide much information about the context of the recent past, and help explain the present-day ecosphere of our planet. In this volume, 74 authors contributed 36 manuscripts that present 'snapshots' of that grand saga – the vast array of patterns, and a mind-numbing panorama of tiny events leading up to the present day – and have thereby contributed a special standpoint from which to consider our planet's immediate future.

Many scientists have interpreted parts of this saga. Some are painstakingly probing how life could have originated from the simple chemicals present on the early Earth – a long, almost intractable endeavour, developing understanding of how protocells could have formed and then imported RNA building blocks into their structure. Others – the majority – have concentrated on aspects of the 'tree of life' such as the relationships of now-extinct organisms. Though most of the trunk and branches of this 'tree' disappeared long ago, all living species are ultimately twigs of a single tree. Others have focused on reconstructing ancient environments. Such reconstructions, especially their cyclicities, gain significance as we approach the present day and consider the contribution of humankind to atmospheric chemistry, ecosystem degradation and global warming – and how the impact of these might be mollified.

The history of our biosphere, now under serious threat, is a stunningly complex story. Many ancient environments, seemingly little changed through hundreds of millions of years, are still with us. These include the environments of oceanic trenches (hadal regions) and abyssal plains, where oozes slowly form from the tests of microscopic organisms (radiolarians and foraminifers). Other environments, if not being destroyed, are changing rapidly.

If used quantitatively, exhaustive palaeontologic databases can be used not only for deciphering grand-scale patterns in the history of the biosphere, but can also be brought to bear on gross global tectonic problems – especially by quantitative analysis of biogeographic data for inferring the relative disposition and motions of crustal blocks, major and minor, during the past 550 million years or so. This has considerable potential value: being able to use data from the foldbelts of the globe where palaeomagnetic data has lost value from overprinting by subsequent tectonic episodes (intervals of intense earth movements).

Five of the nine chapters at the beginning of this volume present a fresh look at the data and methods of palaeontology (Martin Aberhan and Wolfgang Kiessling), 'coordinated stasis' (Carlton E. Brett), the possible impact of periodic astronomical phenomena on the history of life (Bruce S. Lieberman and Adrian L. Melott), a background chapter on climate change through time (John Dodson) and a major synthesis of much data on the development of intertidal biotas through the last 550 million years (Markes E. Johnson and B. Gudveig Baarli). Two chapters focus on marine encrusting and endolithic communities (sclerobiofacies) (Carlton E. Brett et al., and Bruno Mistiaen et al.) – the first with experimental data from shells positioned at various depths in the deep sea, the second a statistical study of Devonian epibionts. This sequence of nine chapters ends with a contribution on the significance of taphonomy and fish microfossils ('fish-bits') in unravelling the history of Devonian vertebrates (Carole J. Burrow and Susan Turner), and another on one of palaeontology's most spectacular 'windows' on deep time, the Eocene-age Messel Pit near Darmstadt in Germany (Stephan Schaal).

John Dodson maintains that the components and processes of Earth's climate system are subject to the same physical laws as the rest of the Universe. Climate change occurs in response to changes in Earth's orbit about the Sun. The behaviour of the climate system can therefore be understood and predicted through careful, systematic study. On a grand scale, rearrangements of continents through plate-tectonic motions have surely wrought massive changes in global climate during deep time.

The first four of the next nine chapters present a survey of latest Precambrian embryos and the Early Cambrian Chengxiang faunas from China and explain their evolutionary significance (Jun-yuan Chen); grand-scale evolution during the

Ordovician is chronicled by Arnold I. Miller; Jiří Frýda details the revolution in gastropod classification that has come from ontogenies revealed by SEM photography; and Philippe Steemans et al. discuss innovations in the micro- and megafossil plant record, from earliest plant spores to earliest seeds. Three chapters provide a panorama on the appearance and disappearance of a unique group, the tentaculitids (Eberhard Schindler), results of intensive investigation of shell morphology of Palaeozoic ammonoids (Dieter Korn and Christian Klug) and the outcomes of statistical studies of Late Palaeozoic trilobites (Rudy Lerosey-Aubril and Raimund Feist). The sequence ends with an overview of the evolution of cladid crinoids (Gary D. Webster) and an elegantly illustrated synthesis of the aerodynamics and origin of flight during the dinosaur–bird transition (Sankar Chatterjee and R. Jack Templin).

There were several important thresholds in the saga of life. The first of these was the origin of single-cell organisms on the surface of Earth, occurring as much as 3.7 billion years ago. The next great event, about 2.7 billion years ago, was development of photosynthesizing bacteria and production of their waste product, oxygen. Increasing levels of oxygen facilitated development of lowly multicellular organisms and, eventually (around 575 million years ago), the swift rise to prominence of bizarre soft-bodied creatures known as the Ediacara fauna. The name of another old friend, the late Martin Glaessner, is closely associated with elegant, early work on the strange Ediacaran animals. Martin drew analogies to present-day groups including jellyfish, sea pens and flat worms, but these assumptions were eventually, in large measure, discarded when it became clear that most were not related to Cambrian and younger groups. Dolf Seilacher was the great motivator in bringing about this rethinking.

Recent discoveries of phosphatised embryos of more than 100 taxa in the Duoshantou phosphate deposit (580 million years ago) at Weng'an, Guizhou, China, some so beautifully preserved that 'fact maps', the sequence of cell division, can be inferred. They include lineages closely related to living sponges, cnidarians and bilaterians (Jun-Yuan Chen, herein). Their discovery is consistent with the suggestion that several lineages of metazoan life (including the above major lineages), previously thought to be of Cambrian age or younger, flourished before the Cambrian explosion – despite not being known yet from fossilised mature individuals.

Close affinities between many of the appreciably younger Maotianshan Shale biota (c. 530 mya) found in the Chengjiang, Haikou and Anning areas near Kunming in Yunnan, China and the appreciably younger Burgess Shale biota (c. 505 mya) of British Columbia, Canada, accord with these lineages having rapidly become geographically (perhaps globally) widespread and, as suggested by Chen Junyuan, having possibly become less rapidly evolving following the Early Cambrian 'explosion'.

By the beginning of the Cambrian period (542 mya), the acme of the Ediacara fauna had passed, following the first readily recognisable mass extinction (perhaps a suite of associated sub-events) or grand-scale but gradual faunal turnover in Earth's history, referred to by some as the Kotlin Crisis. Though rare elements of the Ediacara fauna survived into the Cambrian, they declined dramatically in size, diversity and abundance of them through the Kotlin Crisis.

It was succeeded by the earliest Cambrian assemblages containing trace fossils and, finally, by the 'Big Bang' in animal evolution: the appearance of almost all groups of bilaterian animals – the groups that constitute nearly all of our present-day faunas. There had been rapid diversification of grazing animals with shells composed of phosphate or calcium carbonate – this is considered to be a response to the appearance of the first 'predators'.

There is evidence from latest Precambrian (Ediacaran) trace fossils (burrows and trails) of organisms ingesting soft sediments and thereby obtaining nourishment from tiny organisms that lived in or very close to the surface of shallow-marine sands and silts. By Middle Cambrian time, tiny animals had begun to utilize the oozes of deep oceanic seafloors – specifically the ultra-fine soft, siliceous sediments on the abyssal plains – as a new food source. The first 60 million years of this saga (not elaborated herein) is presently being elaborated by Yoshitaka Kakuwa of the University of Tokyo.

Though the saga of life displays a broad pattern of increasing biodiversity through time, there have been recurrent events of minor and major ecosphere reorganisation. Scales range from minor and local events, scarcely discriminable from the normal 'background' pattern of extinctions, to the mother of all extinction events at the end of the Permian era when as much as 95% of all taxa became extinct, globally. Most, perhaps all, of these events including the end-Permian life crisis were not instantaneous but involved significant intervals of time with definable sub-events.

Though the Kotlin Crisis appears to have been spread over a substantial period of time, it could be regarded as the first of at least eight (rather than five) major events – if one includes the Hangenberg Extinction (360 mya) in the latest Devonian and the now swiftly developing Anthropogenic Event as well. They have been referred to as First-Order extinction events, but how one draws a line between these and the most devastating among the 85–90 other extinction events (many regional rather than global; compilation made by the late Otto Walliser) lacks consensus. Among these are the complex Ireviken Event (Silurian: Landovery–Wenlock boundary, about 428 mya), the Lau Event (mid-Late Silurian; Lennart Jeppsson et al., herein), the Klonk Event (spanning the Silurian–Devonian boundary), the Taghanic Event (mid-Devonian: late Givetian, about 385 mya; James J. Zambito IV et al., herein) and possibly the Cenomanian–Turonian Event (about 93.5 mya).

These 'Second Order' extinction events as well as the largest of all extinction events about the Permian–Triassic boundary (mentioned above; around 251 mya) were not instantaneous as one might expect had they been generated by an asteroid or comet impact. Rather, they were, 'packages of sub-events'. For example, Lennart Jeppsson, from an enormous database of painstakingly and systematically accumulated conodonts (about a half million of them), from the island of Gotland, has discriminated eight discrete extinction episodes (sub-events) that together constitute the Ireviken Event at the base of the Middle Silurian. At least five of these sub-events can be identified elsewhere around the globe (cf. Peter D. Molloy and Andrew J. Simpson). Lennart and fellow workers have shown that a subsequent (mid-Late Silurian) global extinction event, the Lau Event, not only had a significant impact globally, but also can be shown from intensive sampling to have included sub-events defined by identifiable 'data points'.

Popular science, as noted above, refers to five major extinction events punctuating the broad increase in biodiversity that occurred through Phanerozoic time (the last 550 million years of Earth history) and ignores the rest. Characteristics of the database on which this assertion is based smooth out all but the most dramatic events. But even then, the massive Late Devonian 'extinction event' is a composite of data from three significant events: the Lower Kellwasser (late Frasnian, early in the *linguiformis* Conodont Zone), Upper Kellwasser (at the base of the Famennian) and Hangenberg (latest Famennian) global extinction events.

James J. Zambito IV et al. believe that lesser studied and more frequent events that resulted in faunal restructuring and replacement at regional to global scales may have had a greater aggregate effect on the evolution of life than the 'Big Eight'. This they exemplify by the late Middle Devonian (Givetian) Taghanic Biocrisis in the type area, northern Appalachian North America, where three main bioevents – a series of pulsed biotic transitions and extinctions – ultimately resulting in an end to previous faunal provinciality and appearance of a global cosmopolitan fauna.

Enzo Farabegoli and Cristina Perri, by meticulous observation, have demonstrated the 'mother of all extinction events' at the Permian–Triassic boundary consisted of at least three sub-events.

Gerta Keller insists that the massive Cretaceous–Cenozoic (K/T) Boundary Event (about 65.5 mya) could not have been connected with the Chicxulub bolide on the Yucatan Peninsula, the widely accepted 'culprit'. She argues that it impacted about 300,000 years before the K/T global extinction event and points to the voluminous extrusion of the Deccan Traps volcanics as a more likely driving force behind the devastating K/T extinction event. Jeffrey D. Stilwell and Eckart Håkansson probe extinction/survivorship patterns of some of the biota through the K/T boundary interval, and Vivi Vajda neatly analyses the important role of fungi as a driving force in normalisation of the terrestrial carbon cycle following that massive extinction event.

Six contributions focus on past biogeography: a quantitative analysis of brachiopod biogeography in northern Asia through several time-slices of the Early and Middle Devonian (John A. Talent et al.); the biogeography of Pennsylvanian crinoids and blastoids (Johnny Waters and Gary D. Webster); the biogeography of Late Mesozoic ostracods of western Australia (Michelle Guzel); two contributions that, to a degree, complement one another: one on Mesozoic dinosaurs of the southern hemisphere (Federico Fanti), the other on Mesozoic mammals (Thomas H. Rich and Patricia Vickers-Rich); and extinction patterns in North American Miocene gastropods (Edward J. Petuch). The time is ripe for using large databases! Databases thus generated, coupled with increased knowledge of past climate, are important for modelling the biosphere and its history, specifically past patterns of biogeography and, from these patterns, information on the disposition of crustal blocks (major and minor) during at least the past 550 million years – and of the pattern of extinction events, exemplified earlier, that have affected life regionally and even globally since the appearance of the first metazoan life-forms.

Two contributions round out the volume: environmental shifts during the Cenozoic (Brian McGowran) and a comprehensive survey of the biostratigraphy, phylogenetic palaeoecology and palaeobiogeography of Australian marsupials (Karen H. Black et al.). The latter, serendipitously, draws attention to the impact of climate change on Australia's marsupials over the past 55 million years. Their chapter ends with the words: '. . .it seems probable that the next time the history of these unique marsupials is reviewed there will be considerably less scepticism about the capacity of climate change to drive extinctions'.

Acknowledgements

A genial flock of referees reviewed the many manuscripts: Richard Aldridge (Leicester), William Ausich (Columbus, Ohio), Sara C. Ballent (La Plata, Argentina), Richard Bambach (Washington, DC and Blacksburg, Virginia), James E. Barrick

(Lubbock, Texas), Michael Benton (Bristol), Walter Boles (Sydney), Arthur J. Boucot (Corvallis), Monica J. Campi (Melbourne), Euan Clarkson (Edinburgh), Keith A.W. Crook (Canberra), Thomas A. Darragh (Melbourne), Richard V. Dingle (Rondebosch, Cape Town), Enzo Farabegoli (Bologna), Raimund Feist (Montpellier), Rodney Feldman (Kent), Karl Foellmi (Lausanne); Terry Furey-Grieg (Kandos, Australia), Gavin Hanke (Victoria, British Columbia), Graeme Hirst (Toronto), David Holloway (Melbourne), Yoshitaka Kakuwa (Tokyo), Lennart Jeppsson (Lund), Neil Landman (New York), Alan R. Lord (Frankfurt), Royal Mapes (Athens, Ohio), John Marshall (Southampton), Ruth Mawson (Sydney), William D. MacDonald (Binghampton, New York), Brian McGowran (Adelaide), Arnold I. Miller (Cincinnati, Ohio), Michael Melchin (Antigonish, Nova Scotia); William Miller III (Arcata, California), Ian Percival (Sydney), Maria Cristina Perri (Bologna), Geoffrey Playford (Brisbane), Thomas H. Rich (Melbourne), Matthew Saltzman (Ohio State), Lawrence Sherwin (Orange, Australia), Andrew J. Simpson (Sydney), Steven M. Stanley (Manoa, Hawaii), Noel Tait (Sydney), Nadia Talent (Toronto), Tony Thulborn (Brisbane), Kate Trinajstic (Perth, Australia), Fons VandenBerg (Melbourne), Charles Wellman (Sheffield, UK) and Margaret Yaccobucci (Bowling Green, Ohio).

I am indebted to my Macquarie University Centre for Ecostratigraphy and Palaeobiology friends Ruth Mawson, Peter Cockle, Robyn Isberg, David Mathieson, Colin McKenzie, Janine Miller, Karen Novotny and George A Wilson who, when energies flagged or the going became viscous, cheerfully re-edited slabs of manuscript, checked bibliographies, tracked down obscure references and hunted what eventually proved, in some cases, to be phantom items and, in general, maintained optimism. Deadlines came, and went, whooshing by like owls in the night. Special mention must be made of Ruth Mawson who in various ways energetically propelled the volume towards the finishing line, and Peter Cockle who dealt with myriads of computer problems, small as well as large. All of the above were kind and generous friends who helped enormously in bringing the volume to fruition. I raise my glass to them, to the 48 reviewers (several of whom reviewed more than one manuscript), and to every one of the 74 authors who contributed 36 manuscripts to this book.

Sydney, NSW, Australia John A. Talent

Contents

Contributors

Martin Aberhan Museum für Naturkunde, Leibniz Institute for Research on Evolution and Biodiversity, Humboldt University Berlin, D-10115 Berlin, Germany, martin.aberhan@mfn-berlin.de

Anita Andrew CSIRO Petroleum Exploration, North Ryde, NSW 1670, Australia, a.andrew@dem.csiro.au

Michael Archer University of New South Wales, Sydney, NSW, Australia, m.archer@unsw.edu.au

B. Gudveig Baarli Department of Geosciences, Williams College, Williamstown, MA 01267, USA, gbaarli@williams.edu

Gordon C. Baird Department of Geosciences, SUNY College at Fredonia, Fredonia, NY 14063, USA, Gordon.Baird@fredonia.edu

Karen H. Black Evolution of Earth and Life Sciences Research Group, School of Biological, Earth and Environmental Sciences, University of New South Wales, Sydney, NSW 2052, Australia, k.black@unsw.edu.au

Carlton E. Brett Department of Geology, University of Cincinnati, Cincinnati, OH 45221-0013, USA, carlton.brett@uc.edu

Pierre Breuer Geological Technical Services Division, Dhahran 31311, Saudi Aramco, Saudi Arabia, pierre.breuer@aramco.com

Denise Brice Laboratoire de Paleontologie stratigraphique FLST and ISA, UMR 8014 CNRS, 59046 Lille cedex, France, d.brice@isa-lille.fr

Carole J. Burrow Geosciences, Queensland Museum, Hendra, Qld 4011, Australia, carole.burrow@gmail.com

Monica J. Campi Department of Primary Industries, GeoScience Victoria, Melbourne, Vic 3001, Australia; School of Life and Environmental Sciences, Deakin University, Burwood, Vic 3125, Australia, monica.campi@dpi.vic.gov.au

Sankar Chatterjee Museum of Texas Tech University, Lubbock, TX 79409, USA, SANKAR.CHATTERJEE@ttu.edu

Jun-Yuan Chen LPS Institute of Geology and Palaeontology of Academia Sinica, Nanjing 210008, People's Republic of China; Institute of Evolution and Developmental Biology, Nanjing University, Nanjing 210093, People's Republic of China, chenjunyuan@163.net; chenjy@nju.edu.cn

Carlo Corradini Dipartimento di Scienze della Terra, Università di Cagliari, I-09127 Cagliari, Italy, corradin@unica.it; ccorr@unimo.it

John Dodson Institute for Environmental Research, Australian Nuclear Science and Technology Organisation, Sydney, NSW, Australia, john.dodson@ansto.gov.au

Federico Fanti Dipartimento di Scienze della Terra e Geologico-Ambientali, Alma Mater Studiorum Università di Bologna, 40126 Bologna, Italy, federico.fanti@unibo.it

Enzo Farabegoli Dipartimento di Scienze della Terra e Geologico-Ambientali, Alma Mater Studiorum – Università di Bologna, 40126 Bologna, Italy, enzo.farabegoli@unibo.it

Raimund Feist Laboratoire de Paléontologie, Institut des Sciences de l'Evolution, Université Montpellier II, 34095 Montpellier, France, raimund.feist@univ-montp2.fr

Jiří Frýda Faculty of Environmental Sciences, Czech University of Life Sciences Prague, 165 21 Suchdol, Czech Republic; Czech Geological Survey, 118 21 Prague 1, Czech Republic, bellerophon@seznam.cz

Philippe Gerrienne University of Liège, Paléobotanique, Paléopalynologie et Micropaléontologie, B-4000 Liège 1, Belgium, p.gerrienne@ulg.ac.be

Henk Godthelp Vertebrate Palaeontology Laboratory, School of Biological, Earth and Environmental Sciences, University of New South Wales, Sydney, NSW 2052, Australia, h.godthelp@unsw.edu.au

Rimma T. Gratsianova† Trofimuk Institute of Petroleum Geology and Geophysics (IPGG), Siberian Branch, Russian Academy of Science (SB RAS), Uinversitetskiy Prospect, 3, 630090 Novosibirsk, Russia

Michelle Guzel School of Life and Environmental Sciences, Deakin University, Burwood, Vic 3125, Australia, mguz@deakin.edu.au

Eckart Håkansson School of Earth and Geographical Sciences, University of Western Australia, Crawley, WA 6009, Australia; Geological Institute, University of Copenhagen, DK–1350 Copenhagen K, Denmark, eckart@cyllene.uwa.edu.au

Suzanne J. Hand Vertebrate Palaeontology Laboratory, School of Biological, Earth and Environmental Sciences, University of New South Wales, Sydney, NSW 2052, Australia, s.hand@unsw.edu.au

Chris Holmden Saskatchewan Isotope Laboratory, Department of Geological Sciences, University of Saskatchewan, Saskatoon, SK S7N 5E2, Canada, ceh933@mail.usask.ca

Karla Parsons Hubbard Department of Geology, Oberlin College, Oberlin, OH 44074, USA, karla.hubbard@oberlin.edu

Lennart Jeppsson Department of Geology, University of Lund, SE 223 62 Lund, Sweden, Lennart.Jeppsson@geol.lu.se

Markes E. Johnson Department of Geosciences, Williams College, Williamstown, MA 01267, USA, markes.e.johnson@williams.edu

Gerta Keller Department of Geosciences, Princeton University, Princeton, NJ 08544, USA, gkeller@princeton.edu

Wolfgang Kiessling Museum für Naturkunde, Leibniz Institute for Research on Evolution and Biodiversity, Humboldt University Berlin, D-10115 Berlin, Germany, wolfgang.kiessling@mfn-berlin.de

Tat'yana P. Kipriyanova Trofimuk Institute of Petroleum Geology and Geophysics (IPGG), Siberian Branch, Russian Academy of Science (SB RAS), Uinversitetskiy Prospect, 3, 630090 Novosibirsk, Russia

Christian Klug Palaeontologisches Institut, CH-8006 Zürich, Switzerland, chklug@pim.uzh.ch

Dieter Korn Museum für Naturkunde, Leibniz-Institut für Evolutions und Biodiversitätsforschung an der Humboldt-Universität zu Berlin, D-10115 Berlin, Germany, dieter.korn@museum.hu-berlin.de

Anna Kozłowska Institute of Palaeobiology, Polish Academy of Sciences, 00-818 Warszawa, Poland, akd@twarda.pan.pl

Alfred C. Lenz Department of Earth Sciences, University of Western Ontario, London, ON N6A 5B7, Canada, aclenz@uwo.ca

Rudy Lerosey-Aubril Senckenberg Forschungsinstitut und Naturmuseum Frankfurt, 60325 Frankfurt am Main, Germany, rudy.leroseyaubril@senckenberg.de

Bruce S. Lieberman Department of Geology, and Natural History Museum/Biodiversity Research Center, University of Kansas, Lawrence, KS 66045, USA, blieber@ku.edu

Christian Loones Musée d'Histoire Naturelle et d'Ethnologie, 59000 Lille, France, loones_christian@hotmail.com

Monika Masiak Institute of Geological Sciences, Polish Academy of Sciences, 00-818 Warszawa, Poland, mmasiak@twarda.pan.pl

Paula Mauller-Mendlowicz Department of De Estratigrafia e paleontologia, FGEL, UERJ, Rio de Janeiro, RJ, Brasil, mendlowicz@gmail.com

Ruth Mawson Earth and Planetary Sciences, Macquarie University 2109, Australia, rmawson37@gmail.com

Brian McGowran Earth and Environmental Sciences, The University of Adelaide, Adelaide, SA 5005, Australia, brian.mcgowran@adelaide.edu.au

Adrian L. Melott Department of Physics and Astronomy, University of Kansas, Lawrence, KS 66045, USA, melott@ku.edu

Arnold I. Miller Department of Geology, University of Cincinnati, Cincinnati, OH 45221-0013, USA, arnold.miller@uc.edu

Bruno Mistiaen Laboratoire de Paleontologie stratigraphique FLST and ISA, UMR 8014 CNRS, 59046 Lille cedex, France, b.mistiaen@isa-lille.fr

Peter D. Molloy† Earth and Planetary Sciences, Macquarie University 2109, Australia

Paula J. Noble Department of Geological Sciences and Engineering, University of Nevada, Reno, NV 89557-0138, USA, noblepj@unr.edu

Maria Cristina Perri Dipartimento di Scienze della Terra e Geologico-Ambientali, Alma Mater Studiorum – Università di Bologna, 40126 Bologna, Italy, mariacristina.perri@unibo.it

Edward J. Petuch Department of Geosciences, Florida Atlantic University, Boca Raton, FL 33431, USA, epetuch@fau.edu

Elodie Petus University of Liège, Paléobotanique, Paléopalynologie et Micropaléontologie, B-4000 Liège 1, Belgium, elodiepetus@hotmail.fr

Simon R. Poulson Department of Geological Sciences and Engineering, University of Nevada, Reno, NV 89557-0138, USA, poulson@mines.unr.edu

Thomas H. Rich Museum Victoria, Melbourne, Vic 3001, Australia; School of Geosciences, Monash University, Vic 3800, Australia, trich@museum.vic.gov.au

Stephan Schaal Senckenberg Forschungsinstitut und Naturmuseum Frankfurt, 60325 Frankfurt am Main, Germany, stephan.schaal@senckenberg.de

Eberhard Schindler Senckenberg Forschungsinstitut und Naturmuseum Frankfurt, 60325 Frankfurt am Main, Germany, eberhard.schindler@senckenberg.de

Hans Peter Schönlaub Kötschach 350, A-9640 Kötschach-Mauthen, Austria, hp.schoenlaub@aon.at

Andrew J. Simpson Department of Earth and Planetary Sciences, Museum Studies and Centre for Ecostratigraphy and Palaeobiology (MUCEP), Macquarie University 2109, Australia, andrew.simpson@mq.edu.au

Trisha Smrecak Paleontological Research Institution, Ithaca, NY 14850, USA, trish.smrecak@gmail.com

Philippe Steemans University of Liège, Paléobotanique, Paléopalynologie et Micropaléontologie, B-4000 Liège 1, Belgium, p.steemans@ulg.ac.be

Jeffrey D. Stilwell Applied Palaeontology and Basin Studies Group, School of Geosciences, Monash University, Clayton, Vic 3800, Australia; Centre for Evolutionary Research, Australian Museum, Sydney, NSW, Australia, Jeffrey.Stilwell@sci.monash.edu.au

John A. Talent Earth and Planetary Sciences, Macquarie University 2109, Australia, jatalent32@gmail.com

R. Jack Templin National Aeronautical Establishment (Retired), 2212 Aster Street, Ottawa, ON, Canada, K1H 6R6, jtemplin@rogers.com

Susan Turner School of Geosciences, Queensland Museum Geosciences, Hendra, Qld 4011, Australia; School of Geosciences, Monash University, Melbourne, Vic 3800, Australia, paleodeadfish@yahoo.com; sue.turner@qm.qld.gov.au

Vivi Vajda Department of Earth and Ecosystem Sciences, Lund University, 223 62 Lund, Sweden, vivi.vajda@geol.lu.se

Patricia Vickers-Rich School of Geosciences, Monash University, Melbourne, Vic 3800, Australia; Museum Victoria, Vic 3001, Australia, pat.rich@sci.monash.edu.au

Sally Walker Department of Geology, University of Georgia, Athens, GA 30602, USA, swalker@gly.uga.edu

Johnny Waters Department of Geology, Appalachian State University, Boone, NC 28608, USA, watersja@appstate.edu

Gary D. Webster School of Earth and Environmental Sciences, Washington State University, Pullman, WA 99164-2812, USA, webster@wsu.edu

Jane Wigforss-Lange Department of Geology, University of Lund, SE 223 62, Lund, Sweden, Jane.Wigforss-Lange@geol.lu.se

Yevgeniy A. Yolkin† Trofimuk Institute of Petroleum Geology and Geophysics (IPGG), Siberian Branch, Russian Academy of Science (SB RAS), Uinversitetskiy Prospect, 3, 630090 Novosibirsk, Russia

Valentina N. Yolkina† Trofimuk Institute of Petroleum Geology and Geophysics (IPGG), Siberian Branch, Russian Academy of Science (SB RAS), Uinversitetskiy Prospect, 3, 630090 Novosibirsk, Russia

James J. Zambito IV Department of Earth and Atmospheric Sciences, Central Michigan University, Mt. Pleasant, MI 48859, USA, zambi1jj@cmich.edu

Mikolaj K. Zapalski Faculty of Geology, Warsaw University, 02-089 Warszawa, Poland, m.zapalski@uw.edu.pl

Matthew K. Zimmerman Department of Geological Sciences and Engineering, University of Nevada, Reno, NV 89557-0138, USA, zimmermk2000@yahoo.com

Part I

Articles of a General Nature

Phanerozoic Marine Biodiversity: A Fresh Look at Data, Methods, Patterns and Processes

Martin Aberhan and Wolfgang Kiessling

Abstract

Patterns of Phanerozoic global biodiversity continue to be a major focus of palaeobiological research. Recent advances have been fuelled by the establishment of the Paleobiology Database. This new type of data compilation, based on the actual occurrences of taxa in fossil collections, has entailed the development and application of a whole set of new analytical methods. These allow to account for large-scale biases that affect estimates of palaeodiversity, in particular uneven sampling and differential preservation of rocks and fossils. The new curve of global diversity of marine genera (Alroy et al. 2008) deviates in important aspects from traditional ones. Rather than an exponential post-Palaeozoic increase in diversity, it shows an only modest rise from the mid-Palaeozoic to the Neogene. This rise is paralleled by an increase in within-assemblage (alpha) diversity. Contrasting previous conceptions, between-assemblage (beta) diversity, here presented for the first time by a Phanerozoic beta curve, does not show a long-term trend. We approximate the new global Phanerozoic marine diversity trajectory by two consecutive logistic curves, one for the Palaeozoic and one for the Mesozoic to Cenozoic. This implies two thresholds for global diversity in the Phanerozoic, although these are not far apart. We discuss the concept of a biosphere-wide carrying capacity and argue that competition for limited resources and incumbency effects are constraining the diversity of marine life on macroevolutionary scales.

Keywords

Palaeobiodiversity • Sampling • Biases • Analytical methods • Carrying capacity • Diversity thresholds

Introduction

The entire diversity of life – from the diversity of genes to the diversity of species and ecosystems up to the diversity of biomes – is currently shrinking at unprecedented rates and affects all hierarchical levels.

M. Aberhan (✉)
Museum für Naturkunde, Leibniz Institute for Research on Evolution and Biodiversity, Humboldt University Berlin, D-10115 Berlin, Germany
e-mail: martin.aberhan@mfn-berlin.de

J.A. Talent (ed.), *Earth and Life*, International Year of Planet Earth,
DOI 10.1007/978-90-481-3428-1_1, © Springer Science+Business Media B.V. 2012

In the midst of this biodiversity crisis, we increasingly recognize that diversity matters. Apart from ethical arguments, there are strong practical and economical reasons to conserve biological diversity. As the consequences of biodiversity loss are unclear, a thorough understanding of diversity patterns and controlling factors is not just an academical exercise.

Although not strictly comparable to modern biodiversity, the historical perspective of the fossil record offers the unique opportunity to study the temporal and spatial dynamics of diversity over long intervals of time, in particular the 542 million years of the Phanerozoic eon. In spite of its obvious biases, the fossil record documents how the diversity of life has evolved; it promotes an understanding of how past diversity has shaped present diversity; it highlights the evolutionary consequences of mass extinctions; and it enables us to assess the evolutionary dynamics of the biotic recovery following these major extinctions. Thus, the study of diversity of the geological past can yield valuable insights of relevance and urgency to our society.

At the global scale, diversity is the result of the interplay between the origination and extinction of taxa. At the regional and local scale, the immigration and emigration of taxa play an additional role. Most commonly, diversity is calculated as taxonomic (or taxic) diversity, which measures the number of taxa, for palaeontologists often genera. Another way of expressing diversity is phylogenetic diversity, which refers to the evolutionary history of clades and is a metric for the gain and loss of evolutionary distinctiveness. Morphological diversity (disparity) analyses the occupation of morphospace through time, and functional diversity characterizes the impact of different ecological groups (Erwin 2008). Further measures include architectural and developmental diversity as well as behavioural complexity.

In this chapter we focus on the well-studied taxonomic dimension of Phanerozoic diversity history. We confine ourselves to the marine realm because its record is richest and exhibits high levels of preservational completeness and palaeoecological reliability (e.g. Kidwell 2001, 2005). We evaluate the two main biases affecting palaeodiversity analyses – uneven sampling and differential preservation – and summarize the state-of-the-art methods to overcome these deficiencies. We review diversity patterns at various geographical scales, ranging from the diversity of local assemblages (alpha diversity) to global diversity (gamma diversity). We investigate between-assemblage diversity (beta diversity) and present the first Phanerozoic curve on beta diversity based on the Paleobiology Database (PaleoDB). We evaluate two principal classes of models that have been applied to understand the processes controlling global biodiversity. In the equilibrium model, diversity increase is described as a logistic curve with initial exponential growth that slows down as saturation begins and finally stops (Sepkoski 1978). In contrast, the expansion model allows global diversity to rise without a predictable limit (Benton 2009). Finally, we assess the role of the potential drivers of diversity dynamics: extrinsic abiotic factors stress changes in the physico-chemical environment; extrinsic biotic forcing mechanisms comprise biological interactions such as competition and predation; and intrinsic biological controls revolve around key adaptive innovations and clade-specific rates of origination and extinction.

We conclude that global, marine genus diversity has twice reached an upper limit in the Phanerozoic, one in the Devonian and one in the Late Cretaceous. Logistic growth models driven by intrinsic diversification rates are suitable to capture the rough pattern but global environmental perturbations and large-scale climatic fluctuations need to be evoked to model the second-order pattern.

Estimating Diversity from an Incomplete Record of Rocks and Fossils

For a long time, researchers of palaeobiodiversity took the Phanerozoic fossil record at face value to characterize the dynamics of diversifications, major extinctions and the recovery thereof. This has led to the perception that the diversity of species, genera and families has increased considerably through time (Fig. 1a). More recently, variations in the availability of sedimentary rocks and the intensity with which fossils are collected were recognized as important sources of systematic biases (Raup 1976, 1979; Smith 2001; Peters and Foote 2001). A proper understanding of the trajectory of past diversity in time and space requires identifying and correcting for such biases. Such modified diversity curves look quite different from the previous ones (Fig. 1b and see below), and it appears that much of the

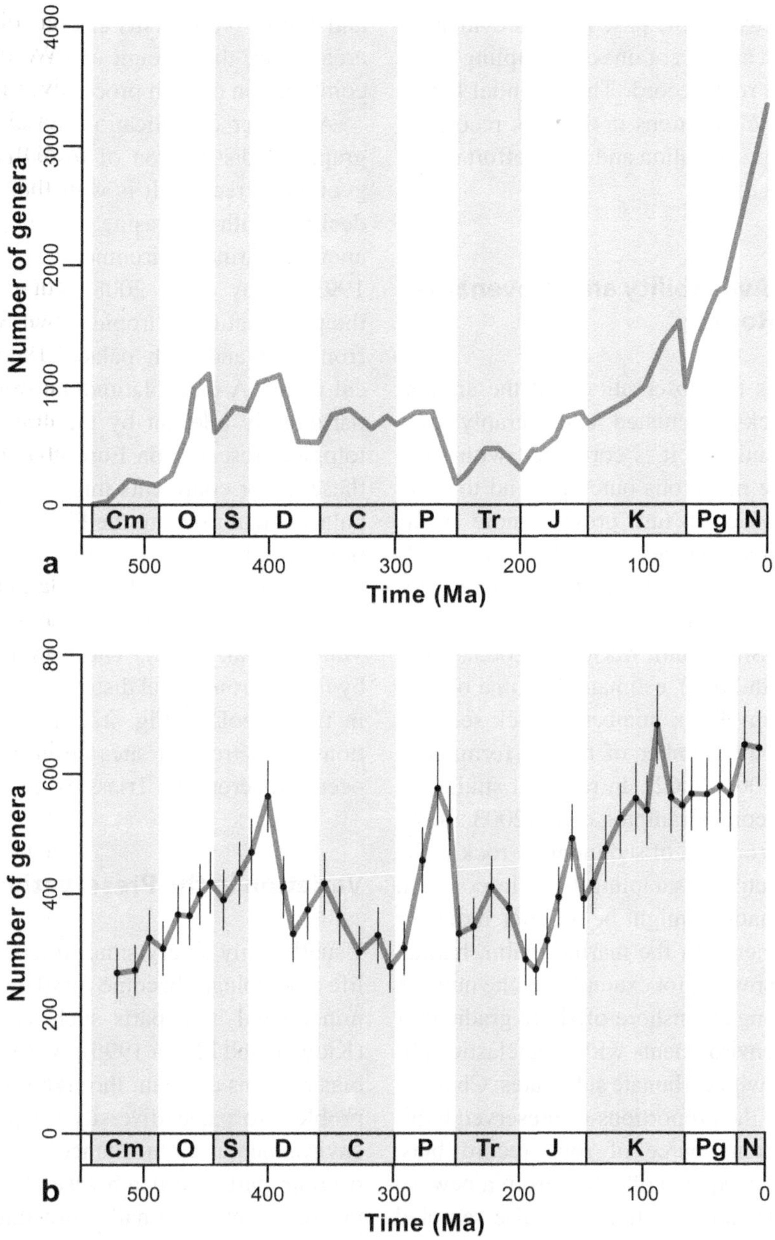

Fig. 1 Global diversity of marine genera (metazoans less tetrapods) during the Phanerozoic. (**a**) Traditional diversity curve utilizing Sepkoski's (2002) compendium of fossil marine animals resolved to the same time intervals as b. The curve is based on the record of a genus' first and last occurrence; each genus is counted at all interval boundaries lying between the times of its first and last occurrence (boundary crossers). It is increasingly recognized that these kinds of curves are severely biased and suffer from the Pull of the Recent effect (see text). (**b**) Diversity curve based on an analysis of the Paleobiology Database. Diversity values rest on the actually recorded occurrence of genera for altogether 48 time intervals of roughly equal duration (ca. 11 Ma). Data are standardized to account for bin-to-bin variations in geographic coverage of samples and sampling intensity. *Vertical lines* denote 95% confidence intervals. For detailed methodology see text and Alroy et al. (2008). Ma, million years ago. Cm, Cambrian; O, Ordovician; S, Silurian; D, Devonian; C, Carboniferous; P, Permian; Tr, Triassic; J, Jurassic; K, Cretaceous; Pg, Paleogene; N, Neogene. Redrawn from Alroy et al. (2008)

steep diversity increase of the past 100 Ma evident in previous curves is an artefact of uneven sampling effort and variation in the rock record. The potential biases arise from systematic variations in the rock record, in the quality of fossil preservation and in the effort raised to sample collections.

Variation in the Availability and Provenance of Fossiliferous Rocks

Of major concern is the observation that the amount of sedimentary rocks fluctuated considerably over Phanerozoic time and that it is correlated with sampled diversity. More numerous outcrops tend to yield more collections, which in turn provide more specimens and finally more species. The close link holds true for several proxies of rock quantity, namely geological map area covered by sedimentary rock (Raup 1976; Smith 2001; Smith and McGowan 2005, 2007; McGowan and Smith 2008), estimated volume of sedimentary rock (Raup 1976), number of rock sections (Peters 2005), and the number of named formations (Peters and Foote 2001, 2002). In regional studies of the New Zealand record, Crampton et al. (2003, 2006) concluded that outcrop area of sedimentary rock is the more reliable predictor of sampling bias. In contrast, the number of formations might be a better indicator of habitat heterogeneity. In the marine realm, habitat heterogeneity is expressed, for example, by the number of environments along an onshore-offshore gradient or the distribution of environments with siliciclastic substrates versus those with carbonate substrates. Changes in the spectrum and the proportions of preserved habitats are an important source of rock record bias. Everything else being equal, collecting from a new, so far unsampled environment is likely to raise sampled diversity at a higher rate than continued sampling of the same environment. Ultimately, the preserved volume and outcrop area of rock as well as the nature of represented environments are a function of tectonically mediated sea-level change (see Smith 2007b and references therein). The flooding of continents increases both the amount of habitable shelf area and the depositional area of shallow marine sediments that becomes part of the rock record. In contrast, regression reduces the area of shallow marine shelf and erosion leads to less marine sediments entering the rock record. Therefore, the correlation between the amount of rock and marine biodiversity can be the result of the species-area effect, the amount of rock that is preserved or a combination of both processes (Smith 2007a).

A further complication arises by the uneven geographical distribution of fossiliferous deposits in the geological record. It is well established that diversity declines with increasing latitude in both modern and ancient marine environments (e.g. Allison and Briggs 1993; Alroy et al. 2008). Changes through time in the preservation of tropical environments versus those from mid and high palaeolatitudes are thus a critical factor. A palaeolatitudinal sampling bias becomes particularly relevant by the strong focus of palaeontological research on Europe and North America. As these major continents migrated northward across the palaeoequator during the Phanerozoic, the representation of high-diversity, tropical shelf area (and the area of shallow-water carbonate accumulation) decreased over the same time interval (Allison and Briggs 1993; Walker et al. 2002). These trends are well reflected by the environmental distribution of marine collections in the PaleoDB (Fig. 2). The major drop of collections from tropical sites and carbonate environments occurred across the Triassic-Jurassic transition.

Variation in the Preservation of Fossils

Usually, only a very small fraction of the members of life assemblages become fossilized, even if they posses mineralized hard parts such as shells and skeletons (Kidwell and Flessa 1996). As long as this taphonomic bias remains constant through time, it does not pose a problem to palaeodiversity trajectories. However, the environmental circumstances under which organisms become part of a death assemblage are embedded in the sediment and finally turn into fossils during diagenesis can vary considerably. An extreme example is conservation Lagerstätten, which exhibit extraordinary completeness including the preservation of soft parts under anoxic conditions. Although they offer a highly complete record of ancient biodiversity at a particular locality, their erratic distribution in time and space introduces a bias. As palaeodiversity studies rely heavily on skeletal organisms, a temporal heterogeneity in the proportions of skeletal versus non-skeletal organisms can also introduce a major bias. Although the current contention is that there was little variation after the Cambro-Ordovician increase of skeletal

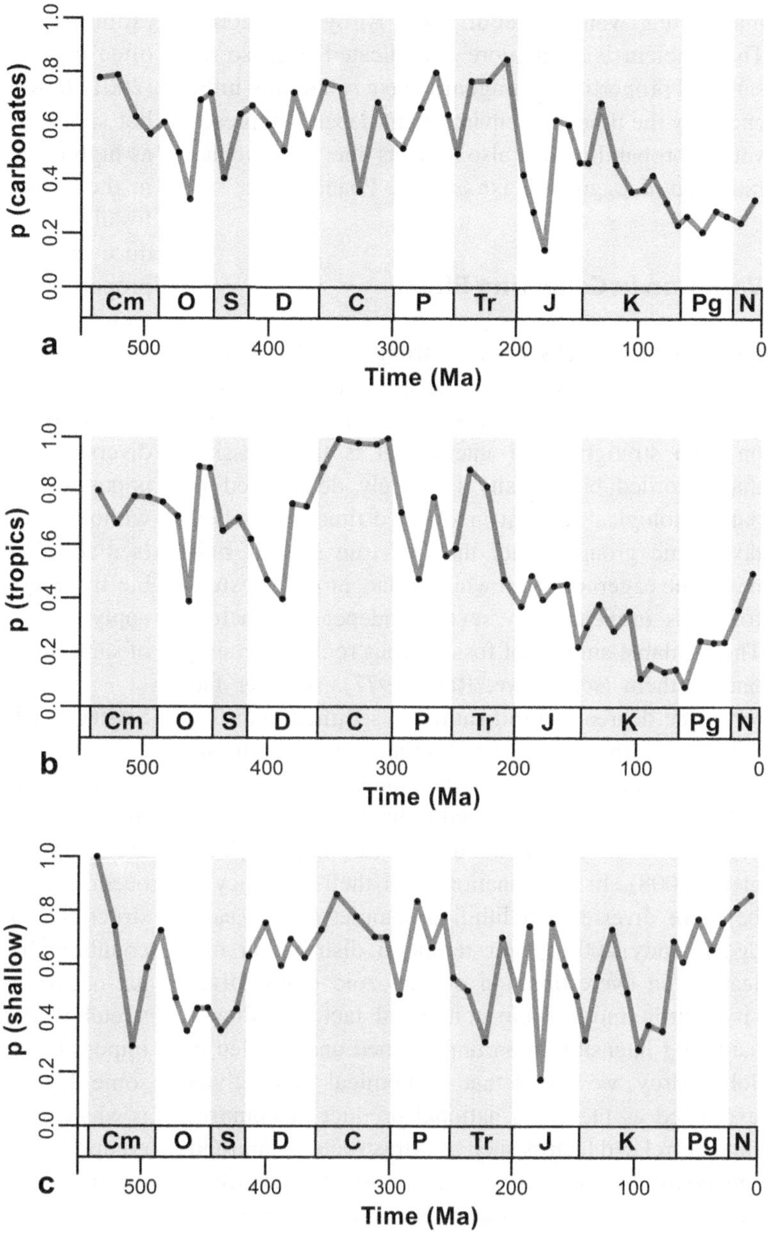

Fig. 2 Properties of the paleobiology database. This database records the occurrences of species and genera for individual fossil collections with exact information on stratigraphic age, geographic coordinates, lithology, depositional setting and many other features. This allows for detailed analysis of taxonomic, geographic, environmental and other subsets of the data. Shown here are the proportions of collections from (**a**) carbonates versus siliciclastic substrates (mixed lithologies excluded); (**b**) tropical versus extra-tropical palaeolatitudes (with a cut-off at 30° N and S); and (**c**) shallow-water (above storm wave base) versus deeper water habitats. For legend see Fig. 1

taxa (Knoll 2003), more quantitative studies are clearly needed on Lagerstätten preserving both skeletal and non-skeletal taxa.

The hard parts of organisms differ markedly in their preservation potential. Calcareous shells of molluscs, for example, can be composed of aragonite, calcite or a combination of both carbonate minerals. Because of the increased solubility of aragonite compared to calcite, shells exclusively made up of aragonite are more strongly affected by the selective dissolution of shells. Indeed, it has been argued that the taphonomic loss of aragonitic molluscs was quite large in at least some Palaeozoic and Mesozoic environments (Wright et al. 2003), whereas Cenozoic collections frequently preserve fossils with original aragonite (Kiessling et al. 2008). While some analyses found no large-scale bias of shell mineralogy on preservation, possibly because the time intervals are long enough to record an adequate range of taphonomic settings (e.g. Kidwell 2005; Cooper et al. 2006), others concluded that this effect matters (e.g. Valentine et al. 2006; Alroy et al. 2008). The problem is even more complicated because the sampled proportion of aragonitic taxa is not only influenced by the time- and environment-dependent preservation probability but also reflects the evolutionary radiation of aragonitic taxa over the Phanerozoic.

Variation in Collecting Effort

Sheehan (1977) observed a strong correlation over Phanerozoic time between the estimated total number of species and the number of researchers working on each stratigraphical interval. It is thus possible that recorded biodiversity is largely determined by palaeontological interest in selected time intervals or taxonomic groups rather than by true species richness. The eagerness with which palaeontologists study fossils is influenced by several independent factors. The available amount of fossiliferous rock is certainly one of them (see above, Raup 1977). Another factor is the degree of lithification of sediment and thus the ease with which even fragile and small fossils can be collected from each interval. Collections from unlithified and poorly lithified sediments are relatively common in the Cenozoic but very rare before (Alroy et al. 2008). In combination with their tendency to be more diverse than lithified samples of the same age (Hendy 2009), their temporal distribution may lead to an overestimation of Cenozoic global diversity. Furthermore, socioeconomical factors influence sampling intensity. In an unpublished analysis led by John Alroy, we found that economical productivity, expressed as the gross national product of countries (and correlated factors such as infrastructure and funding possibilities), is the most important predictor of collection intensity. Similarly, Kiessling (2005b) in an analysis of Phanerozoic reef distribution concluded that reefs are more likely to be sampled in wealthy countries, irrespective of geological and ecological controls.

Appropriate Methods of Global Diversity Analyses

Before we discuss modern methods of biodiversity analyses with fossils, we have to emphasize one crucial point: It is the scope of all analyses to achieve unbiased trajectories of diversity but not to achieve estimates of true diversity in a given time. This point is often overlooked. For example, Jackson and Johnson (2001) have emphasized that some recent biodiversity hot spots in the sea yield generic diversities that are as high or even higher than inferred global diversities in the Palaeozoic and Mesozoic. Even well-sampled fossil sites such as the Pliocene of the Caribbean produce generic diversities that represent a much larger proportion of the inferred global diversity than one would expect from its limited area. These assertions are not conducive because they refer to our limited knowledge of true global diversity at any point in time (including today) but are silent on the changes of biodiversity through time. As we shall detail below, it is possible to infer reliable trajectories of biodiversity without knowing the true total diversity at any time (see also Vermeij and Leighton 2003). This is possible by interpolation rather than extrapolation and by applying appropriate counting methods and methods of sampling standardization.

Sources of Palaeodiversity Data

There are two fundamentally different types of data for assessing palaeodiversity. The most commonly used approach in small-scale studies is to record the occurrences of taxa in sampling intervals and to construct diversity trajectories from these data, usually by counting the number of species. Recording individual occurrences at a global scale and through longer stretches of geological time is of course a huge task, impossible to achieve by a single person if at least some degree of comprehensiveness is requested. This is why earlier accounts of global diversity limited their compilations to the record of a taxon's first and last occurrence. This is the second type of data, which we refer to as compendium data.

The compilation of stratigraphical ranges is still a tremendous undertaking, and only Jack Sepkoski has done this single-handedly for marine orders, families and genera (Sepkoski 1978, 1982, 1993, 2002). The only comparable database is the Fossil Record 2, for which Benton (1993, 1995, see also http://www.fossilrecord.net/fossilrecord/) has recruited a number of experts to compile estimates of family diversity for the marine and terrestrial realm. Most of these datasets are accessible, either online or in print. Sepkoski's (2002) compendium is accompanied by a CD-ROM that contains the stratigraphical ranges of 36,340 marine genera. The dataset and Software to use these data for plotting and analyses have been made available online by Tapanila (2007, http://geology.isu.edu/FossilPlot/). This dataset (and its unpublished precursor) has clearly been the most widely used source for palaeodiversity studies in the last two decades.

We have mentioned above that a global and Phanerozoic-scale coverage of occurrence data is out of reach for an individual worker. However, as will be detailed below, these data are needed to allow the application of modern counting and standardization methods. This is why the Paleobiology Database (PaleoDB, http://paleodb.org) has been founded. Being arguably the largest database on fossils of all taxonomic groups, ages and environments, this database is a major international endeavour. The PaleoDB is in constant development. More occurrences and taxonomic data are entered on a daily basis and new contributors join every 2 weeks or so. Therefore, any numbers reported here will soon be obsolete. At the data of final submission of this typescript there are about 220 contributors from 17 countries who have compiled a database of 30,000 references, being linked to 805,000 taxonomic occurrences of 150,000 taxa. The PaleoDB is rapidly becoming the major source of palaeodiversity data. Comprehensive metadata on age, location, lithology, depositional environment and many other attributes allow for a detailed analysis of taxonomic, geographical, environmental and other subsets of the data (Fig. 2). So far, close to 100 official publications were published in international journals.

Taxic Versus Phylogenetic Approach

All currently available estimates on the trajectories of global diversities are based on a taxic approach, that is, they count names of genera or families irrespective of their phylogenetic relationships. Without phylogenies, the timing of originations can be substantially flawed. Phylogenies often infer long ghost ranges, where the presence of a taxon is predicted from the origination time of its sister group but has not been sampled so far (Smith 1994). Thus, so cladists argue, the ghost-taxon has to be added to diversity estimates. When this is done, diversity patterns may differ substantially from the original patterns (Smith 1994). One major flaw in this assertion is that ranges can only be corrected in one direction, that is, downwards. There is no way to correct for unsampled ranges beyond the last recorded fossil. These zombie lineages are just as likely to occur as the ghost lineages but the phylogenetic method offers no tool to correct for this bias. Simulations have shown that correcting for ghost lineages may produce more accurate patterns when most of the taxa are extant, but the taxic method may be superior when a group is extinct or contains few extant lineages (Lane et al. 2005). Note that this refers to raw patterns of taxic diversity that are derived from the analysis of compendia. Proper sampling standardization will further improve the taxic patterns, while standardization cannot be applied to ghost-lineages.

A further issue with the taxic approach is that true terminations of phylogenetic lineages cannot be separated from pseudo-extinctions. A pseudo-extinction of a species occurs when members of the species become extinct but a daughter species survives. While this affects especially assessments of extinction rates, some metrics of measuring diversity are potentially sensitive to pseudo-extinctions. The important issue is thus whether or not pseudo-extinctions produce a significant bias. A recent study suggests that they do not. Breaking 60% of the "lineages" recorded in Sepkoski's compendium into sub-lineages with random originations and extinctions has no significant effect on extinction rates (Peters 2008).

Counting Methods

How should we measure diversity? This question has bothered ecologists and palaeontologists for long, and major biases can be introduced by inappropriate counting methods. The problem of relative abundances aside, diversity through time can be measured in several ways (Fig. 3). Of course the most straightforward method is to just count everything in a given time interval, be it a bed or a geological stage. This approach called "sampled in bin" (SIB) has the advantage of not making assumptions on what could be present but has not been sampled. However, by not taking unsampled ranges into account (we use this term to avoid confusion with ghost ranges, which mean something else; see above), diversity patterns can become very noisy and it is impossible to address turnover rates. In any case, the SIB method can only be applied to actual occurrence data such as those recorded in field studies or in the PaleoDB.

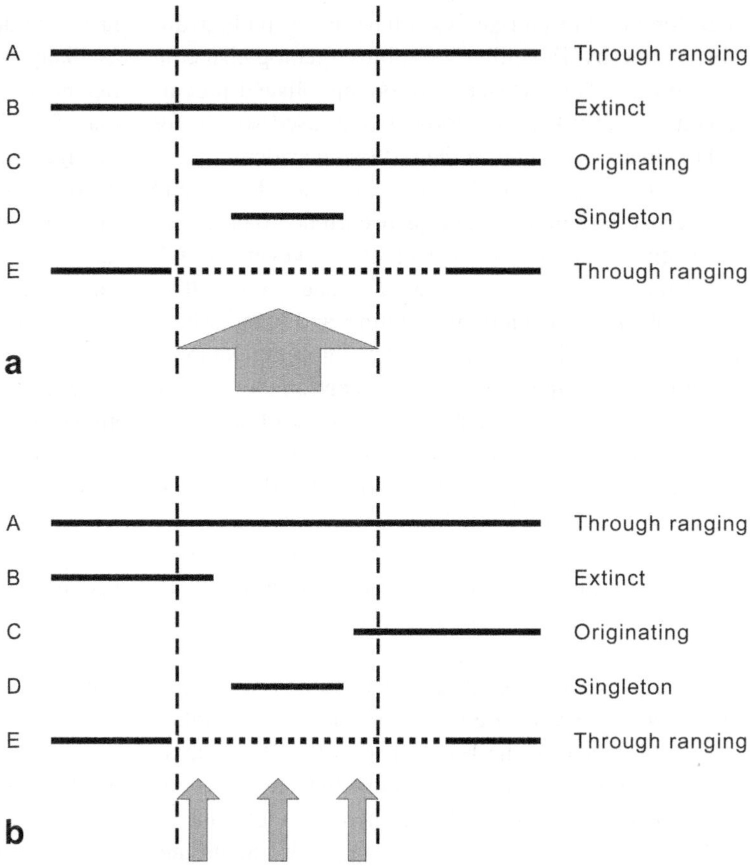

Fig. 3 The various methods of counting diversity. A–E denote different taxa. The *horizontal bars* show their stratigraphic ranges. Stippled segments indicate an interval of time in which the respective taxon was not recorded, but its presence is inferred on the basis of both older and younger records. The illustrations demonstrate the differences between counting diversity within a temporal bin (**a**) and counting diversity at a point in time, usually at the boundary between bins (**b**). For detailed explanations see text

Compendia of stratigraphical ranges have to apply different methods because only the oldest and youngest occurrence of a taxon are recorded. These counting methods are known as range-through. The traditional approach is to estimate diversity from counting everything that is known to occur in an interval (e.g. first and last occurrences) plus everything that is inferred to be present (e.g. recorded before and after an interval) (Fig. 3a). One has to make a decision to include or exclude singletons. Singletons are taxa known from only one bin. These can be either genuinely short-lived taxa or represent monographic or taxonomic artefacts. In the context of global diversity studies, singletons are usually omitted because the artifactual patterns are thought to prevail (Pease 1985; Alroy 1998; Foote 2000). Even if a singleton represented a biological reality its inclusion would tell us little about standing diversity within an interval, that is, the number of taxa that coexisted at any one time (Pease 1985). This is because singletons can be spread through an interval without overlap.

For the same reason, the traditional range-through approach of counting diversity is flawed. Consider the five taxa represented in Fig. 3a. The range-through diversity in the middle interval would be estimated as five with the singleton and four without. Now consider that the extinct taxon would actually become extinct right after crossing the bottom interval and the originating taxon would originate just before the boundary to the next interval (Fig. 3b). The cumulative diversity in the interval would be unchanged but the standing diversity would never exceed three because the ranges of the extinct, singleton and originating taxa do not overlap. To avoid this bias, the metric of boundary crossers is now commonly used. Counting diversity at bin boundaries instead of within bins has the advantage of actually assessing standing diversity and automatically omitting singletons because those do not cross boundaries by definition (Bambach et al. 2004; Kiessling et al. 2007).

A general problem with all range-through counts are the edge effects. Unsampled taxa can only be inferred to be present when they occur before and after a bin. Therefore, diversity at beginnings and ends of time series is artificially depressed. This problem is less obvious in Phanerozoic-scale analyses because the earliest bins are genuinely poor in taxa and the Recent is by far the best sampled. However, shorter time series are severely affected by edge effects and the great sampling of the Recent introduces a bias in its own right: the Pull of the Recent (POR, Raup 1979). The POR refers to the fact that extant taxa will contribute to the standing diversity of all older bins until their oldest record. This applies also to poorly preservable taxa, which would have been recorded as singletons if they would not be extant. Extinct taxa in contrast need both a first and last occurrence in the fossil record to be counted. Thus, range extensions cannot be applied and the diversity of extinct taxa will be dampened. How severe the POR-effect really is has been debated. Using bivalves, Jablonski et al. (2003) argued that the POR accounts for only 5% of the Cenozoic diversity increase, but this was done by specifically targeting Pleistocene records of extant bivalve genera. With the occurrence data in the PaleoDB it can be shown that the POR effect is dramatic and explains most of the steep increase in Mesozoic-Cenozoic diversity derived from compendia (Alroy et al. 2008).

For these reasons, counting methods involving range-through are associated with substantial biases and this is why Alroy et al. (2008) returned to the SIB method of counting diversity. To reduce the short-term variation inherent to SIB counting they used a correction for sampling completeness. These measures of completeness can also be used in conjunction with counts of short-term boundary crossers (so-called two timers and three timers) to address taxonomic turnover rates (Alroy 2008). A prerequisite for SIB methods of counting to be meaningful, however, is sampling standardization.

Sampling Standardisation

That diversity is sampling-dependent is trivial but to correct for sampling biases in diversity curves is not. This is due to the non-linear relationship between the amount of sampling and the number of taxa. Importantly also, sound corrections can only be applied by interpolation, that is, subsampling, and not by extrapolation. Within communities, species richness (S) increases with the number of individuals (N) in a convex upward fashion, that is, first increasing rapidly as more individuals are encountered and then levelling off at higher values of N. When the number of individuals for each species is known, the expected S can be predicted at any given N_S that is below the original N. This method is called rarefaction. One can do rarefaction mathematically by using standard equations (Tipper 1979) or by randomly culling a given number of individuals from the total sample and tabulating the number of encountered species. The latter method is at the core for all modern subsampling techniques. Sampling units are randomly culled until a certain quota is reached. The size of the subsampling quota depends on the bin with the lowest sample size for which subsampling is being conducted. Consider four samples with 30, 80, 100 and 200 individuals. The quota would have to be 30 or lower to allow a rarefied diversity of all samples to be assessed. If this number appears too low, the first sample should be dropped (and can then not be compared to the others) and the subsampling can use a quota of 80 or less. A common feature of all subsampling methods is also that random culls are repeated many times and the diversity estimates are averaged.

Rarefaction can also be applied to regional or even global datasets by drawing taxonomic occurrences rather than individuals. A taxonomic occurrence is the documentation of a species (or higher taxon) in a sample. The sample can range from the collection of fossils from a single bed at one locality to a collection of a whole sequence of beds over a larger area. When subsampling, occurrences of all taxa in a bin are lumped, occurrences are drawn at random until a pre-defined quota is reached and the diversity is calculated. In doing so, Miller and Foote (1996) demonstrated that the Ordovician radiation occurred much earlier than suggested in the raw data of stratigraphical ranges. Rarefaction was also used in conjunction with range-through counting methods by culling the same number of occurrences from each bin and then ranging the encountered first and last occurrences through (Alroy et al. 2001; Kiessling et al. 2007; Kiessling and Aberhan 2007a, b). A problem with classical rarefaction is that taxonomic occurrences are drawn independently from a pooled dataset. More advanced methods of subsampling take a note of the composition of faunal lists (Alroy 2000; Alroy et al. 2001). One such method simply draws entire faunal lists until a quota of such lists is reached (by-list unweighted, UW). This makes the unrealistical assumption of average collection sizes to be constant, while there is a significant trend towards larger collections through time (Alroy et al. 2008). Another method known as occurrences weighted (OW) also draws lists at random but the quota is determined by the cumulative number of occurrences in these lists. In this way, OW is very similar to classical rarefaction and indeed yields almost identical results (Alroy et al. 2001; Bush et al. 2004). Finally, a method known as O^2W squares the number of occurrences in each list and the quota is based on the sum of squared occurrences. This method, originally employed by Alroy (2000), was thought to make more realistical assumptions about the abundance distribution of taxa within collections. However, although perhaps valid for land mammals, the exponent of two was found to be too high for marine faunas and would overcorrect for sample-size bias (Bush et al. 2004). A pitfall in most previous analyses is that they assumed a single exponent for the entire Phanerozoic, when in fact it should be variable depending on the evenness of palaeo-communities (Alroy et al. 2008).

The currently most advanced subsampling protocol was employed by Alroy et al. (2008). These authors computed actual or inferred counts of specimens for each collection and inversely weighted the chance of a collection to be drawn during subsampling. With this protocol they avoided a major pitfall of previous approaches that preferentially drew either long lists or common taxa, thereby underestimating global diversity. In an attempt to standardize for geographical spread, the authors also used a quota of references for each bin. It has to be admitted that even this protocol is not perfect. Pitfalls are as follows: (1) The composition of faunal lists is accepted as a complete record of what can be found at a sampling site and its length is biased by geological factors; (2) the geographical and environmental spread of collections is not really standardized; (3) Lagerstätten effects are not completely omitted; and (4) a single averaged evenness value is assumed for all collections in a bin. Unfortunately, a further refinement of subsampling methods is presently hampered by the available data.

In summary, counting and subsampling methods are both important for recovering unbiased patterns in global Phanerozoic diversity, but for the purpose of identifying accurate long-term trajectories counting methods are probably the most crucial. The Pull of the Recent has clearly introduced the most substantial bias in Phanerozoic diversity trajectories computed from compendia.

Diversity Patterns from Marine Fossils

The major concern of this chapter are patterns of global standing diversity, but an understanding of factors governing global diversity requires insights into two underlying patterns: turnover rates and geographical and environmental subsets. Turnover rates (rates of origination and extinction) are the principal drivers of global diversity. Global diversity will increase when diversification rates (origination minus extinction) are positive and decrease when diversification rates are negative. Geographical and environmental subsets highlight the interplay of local richness and differences among communities. This interplay is crucial to understanding the factors governing major trends in global diversity.

Diversity Dynamics

Mass extinctions are classical examples of drops in diversity, but they are not defined by such drops. Rather, the rate of extinction is significantly above background in a mass extinction (Raup and Sepkoski 1982). How strongly the intensity of a mass extinction is linked to diversity loss depends critically on the origination rates during and immediately after times of mass extinctions (Bambach et al. 2004). The length of sample intervals is also critical in determining how strong a mass extinction event is reflected in standing diversity. Long sampling intervals will dampen the diversity signal of mass extinction events.

Although the patterns in the PaleoDB can be matched with those from Sepkoski's compendium when binning schemes and counting methods are identical, the patterns of extinction and origination are similar even with different methods (Fig. 4). Three observations are especially noteworthy: (1) There is an overall trend towards lower extinction and origination rates from the Cambrian to the Neogene; (2) extinction rates tend to increase in the aftermath of diversity increases; and (3) peak originations are noted in the aftermath of mass extinctions (Raup and Sepkoski 1982; Sepkoski 1998; Alroy 2008). It is obvious that

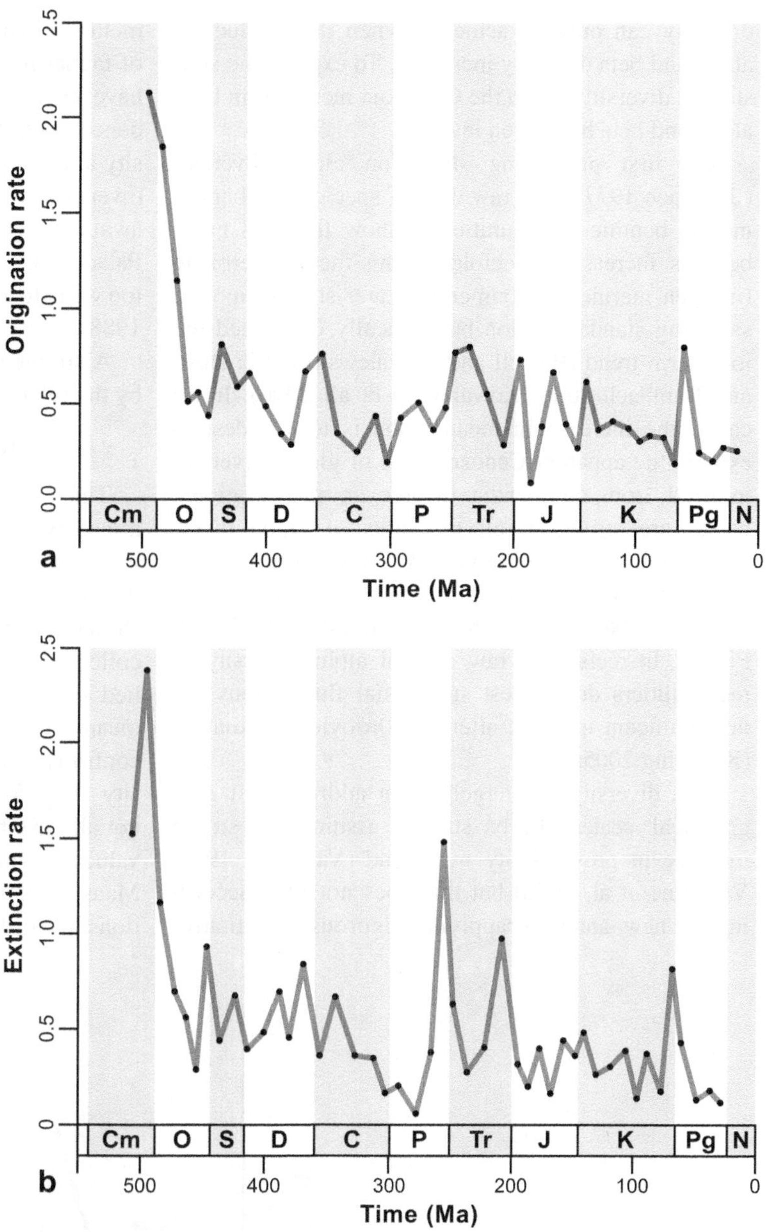

Fig. 4 Origination rates (**a**) and extinction rates (**b**) of marine invertebrate genera during the Phanerozoic. Values are based on sample-standardized data of the Paleobiology Database. Both rates decline through the Phanerozoic. Time intervals are the same as in Fig. 1. Redrawn from Alroy (2008)

the latter two observations have implications for the mechanism by which global diversity could be constrained (see Discussion in section "Mechanisms of Constraints in Global Diversity").

Geographical and Environmental Components

The concept of alpha, beta and gamma diversity is extremely useful for dissecting global diversity patterns. Some pooled diversity of a region (=gamma diversity) consists of several point diversities (alpha diversity) and the taxonomic differences among these (beta diversity). Alpha and gamma diversities can be measured in a straightforward way and beta can be calculated by

$$\text{beta} = \text{gamma/mean alpha}.$$

Using the inverse of traditional similarity indices (such as Jaccard or percent similarity), beta diversity can also be calculated directly between pairs of samples (Miller et al. 2009). Increases of global diversity can only be achieved when the product of alpha and beta diversity increases. To explain the substantial diversity rise in the Cenozoic increases in both alpha and beta have been invoked.

The first pioneering study on alpha diversity (Bambach 1977) used raw data of species numbers in marine benthic communities to show that this number has increased threefold during the Phanerozoic (in open marine environments). Later studies applied sampling standardization but basically confirmed the long term trend (Powell and Kowalewski 2002; Bush and Bambach 2004; Kowalewski et al. 2006). In all cases, the increase in mean alpha is too modest to explain the apparent Cenozoic rise of global diversity inferred from compendia. Moreover, preservational biases are strong in all measures of alpha diversity. Communities in the very well-preserved Cassian Formation (Carnian, Italy) achieve alpha diversities very similar to the Neogene (Kowalewski et al. 2006). Finally, in reefs even raw data of alpha diversity of reef builders do suggest substantial fluctuations but no significant increase after the Ordovician radiation (Kiessling 2005a).

Beta diversity has largely been addressed at geographical scales. Early studies assumed a strong increase in provinciality over time (Valentine 1970; Valentine et al. 1978) but this does not hold according to new analyses applying rigorous quantitative methods (Miller et al. 2009). Geographical disparity of faunal lists within comparable distances appears to have been rather constant and shows no correspondence with large-scale climatic cycles. Beta diversity at environmental scales (e.g. between-assemblage diversity of onshore and offshore faunas) is still awaiting rigorous testing at Phanerozoic scale. In the Palaeozoic, an increase of beta was noted but this was too weak to explain the Ordovician radiation (Sepkoski 1988).

A simple analysis of total beta diversity is possible by the relationship

$$\text{mean alpha} \times \text{beta} = \text{gamma}.$$

Thus, beta for a time interval can be simply determined by dividing the total genus richness observed in an interval by the mean number of genera found in each collection of this bin. Because beta diversity shows a non-linear relationship with the number of collections (Bush et al. 2004), we have randomly tallied 400 collections from each bin and calculated the mean value of beta after 20 iterations. The results confirm that there is no Phanerozoic trend in beta diversity (Fig. 5). There are substantial fluctuations, but not all have biological meaning. For example, the low values during most of the Carboniferous and in the Maastrichtian are due to a strong clustering of collections in North America. A biotic homogenization often

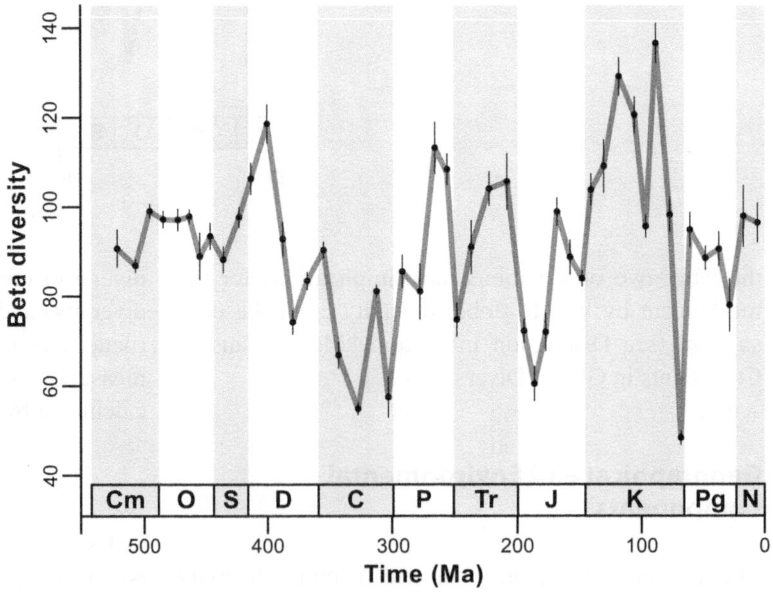

Fig. 5 Phanerozoic marine beta diversity. The curve implies that there is no Phanerozoic trend in beta diversity. Values are calculated by dividing the total sampled genera in a subset of 400 collections from each interval by the mean number of genera in each collection. Time intervals are the same as in Fig. 1. See text for discussion

inferred after mass extinctions and biotic crises (Erwin 1998; McKinney and Lockwood 1999) is only recognizable after the end-Permian and end-Triassic mass extinctions, when beta diversity dropped significantly.

The New Global Marine Diversity Curve

Given the obvious problems of traditional counting methods and the impossibility of sampling standardization, all diversity curves based on simple compendia approaches have to be viewed with extreme caution. Accordingly, we focus our discussion to the new diversity curve of Alroy et al. (2008) (Fig. 1b). In the original paper, the discussion was mostly limited to the key result, which is an only modest rise in global diversity from the mid-Palaeozoic to the Neogene. Some of the additional features, highlighted by Alroy et al. (2008), are (1) a substantial Cambro-Ordovician radiation, also noted in previous studies; (2) a large mid-Devonian drop prior to main Late Devonian extinction phases; (3) a substantial rise in the Permian roughly coinciding with the end of the Permo-Carboniferous ice age; (4) a pronounced effect of, but also a rapid recovery from the end-Permian mass extinction; and (5) a substantial drop after the end-Triassic mass extinction but a lack thereof after the end-Cretaceous mass extinction. An immediately obvious result is also that there is no cyclicity in standing diversity contradicting hints from compendium analyses (Rohde and Muller 2005; see also Omerbashich 2006). Alroy et al. (2008) were careful not to emphasize minor wiggles of the curve because some of them are not robust, that is, they are not seen in different treatments of the data.

However, some features of the diversity curve are robust in this sense, but nevertheless may be somewhat arbitrary. For example, the low Late Carboniferous diversity is largely due to an extreme concentration of collections in Laurentia and thus may be more a regional rather than global signal. Dissecting global diversity patterns can yield important insights as to the controlling factors (Miller 1997). In the above example, however, limiting the other bins to a similar geographical spread would turn the Phanerozoic diversity curve into a trajectory of arbitrary regional patterns. Similarly, standardizing for individual environments rather than grand totals of all recorded environments would introduce large gaps in the record. A disturbing feature of the new diversity curve is also the lack of substantial diversity drops after several of the traditional mass extinctions. In most cases this is due to the long sampling bins chosen by the PaleoDB. On average they are 11 million years in duration and incorporate almost two traditional geological stages. Therefore, the post-extinction diversity incorporates much of the recovery, which has been shown to start sooner after extinctions (Lu et al. 2006) than assumed previously (Kirchner and Weil 2000). It is still strange, however, that the Early Jurassic diversity is lower than the Early Triassic diversity. These issues remind us that the new diversity curve is not perfect and more data and analyses, and a finer temporal resolution, are still needed.

The important message, however, is that all subsampling methods agree in recording a modest Cenozoic diversity rise relative to the Palaeozoic. This modest rise could even be exaggerated given the better preservation of anatomical details in younger rocks. This means that thresholds for global diversity not only existed in the Palaeozoic (Sepkoski 1979, 1988) but probably throughout the Phanerozoic. Our discussion will thus focus on how these thresholds might be achieved.

Mechanisms of Constraints in Global Diversity

Assuming that the sophisticated methods applied by Alroy et al. (2008) led to a more realistic approximation of the shape of the global marine diversity curve than previous attempts, we suggest that global marine diversity was in fact constrained, that is, an upper threshold to global biodiversity was reached before the present day. Previous interpretations of compendia approaches have seen thresholds at the ordinal and family level but not at the level of genera. For marine orders it appears that a single threshold existed for the entire Phanerozoic, which was reached at the end of the Ordovician radiation (Sepkoski 1978). The trajectory of marine ordinal diversity can be modelled by a logistic equation, that is, an initial exponential growth that slows down as the threshold is approached and finally stops, giving rise to an overall sigmoidal shape. At the level of marine families, logistic equations are also behind a more complex model that involves separate intrinsic growth rates and thresholds for each of the three evolutionary faunas (Sepkoski 1981), competition among those faunas, and mass extinctions (Sepkoski 1984). Already noted by Sepkoski (1984) was the poor fit of the logistic model to the observed diversity trajectory in the Mesozoic and Cenozoic. We think the POR is largely responsible for this lack of fit and also for the seemingly unbridled diversity increase at the genus level (Fig. 1a and discussion above). The important message of Sepkoski's analyses is that long-term diversity thresholds are not necessarily fixed but can evolve due to evolutionary innovations.

The new diversity curve (Fig. 1b) suggests two different thresholds for global diversity with two different growth rates: one in the Palaeozoic and one in the Mesozoic-Cenozoic (Fig. 6). The thresholds are not far apart, suggesting that evolutionary innovations in the post-Palaeozoic could not increase the threshold substantially. The Palaeozoic threshold was first reached in the early Devonian but diversity stayed well below this threshold for a long stretch of time ranging from the Middle Devonian to the Early Permian. The post-Palaeozoic threshold appears to have been reached already by the Late Cretaceous.

How are these thresholds brought about? Can we simply scale up from simple ecological models and speak of carrying capacities for the global biosphere on macroevolutionary time scales or are there different causes? Global standing diversity in an interval of geologic time is the result of a variety of ecological and evolutionary processes that occur at multiple spatial scales, ranging from local environments to larger regions and biogeographic provinces and finally to the global scale (e.g. Miller 1998, 2000; Badgley 2003). For example, severe and very widespread environmental perturbations, induced, e.g. by global climate change, volcanism or a catastrophic bolide impact, affect global diversity directly. At the other end of the spatial spectrum are local physical and biological processes, e.g. geographically restricted environmental disturbances and biotic interactions among individuals and populations, which shape small-scale species richness (e.g. Thayer 1983). They contribute to the global Phanerozoic diversity trajectory as the sum of diversity measures for individual localities and habitats. A hierarchical approach, in which processes are nested according to multiple spatial and temporal scales, with different variables emerging at different levels, shows great promise for understanding the multifarious patterns of biodiversity (e.g. Willis and Whittaker 2002).

Carrying Capacity in Population Ecology

In population ecology, the carrying capacity is the population size at which there is no further net growth. The carrying capacity is manifested by resource availability (e.g. food), diseases, competition and predation. These "checks" to population size have long been noted as universally affecting all organisms (Darwin and Wallace 1858) and led to the well-known logistic growth models. The carrying capacity is to be seen

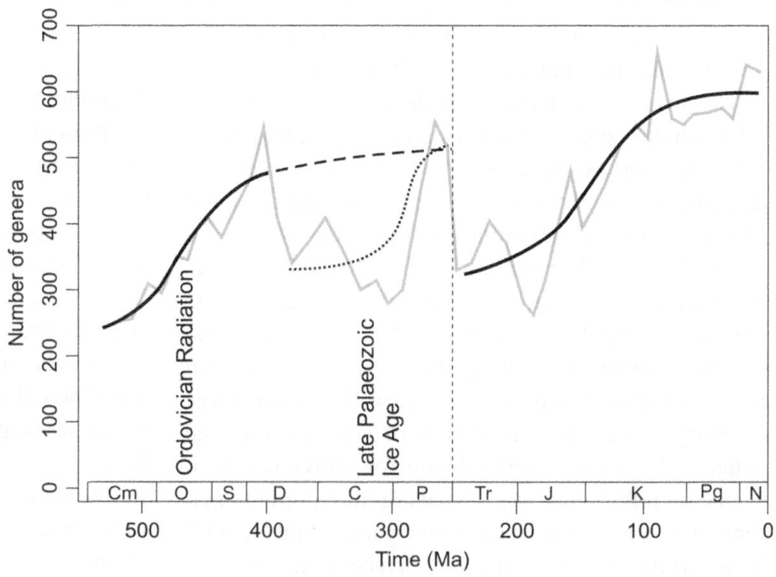

Fig. 6 Simple logistic models of diversity, modelled visually on the new global diversity curve (see Fig. 1b). This interpretation suggests two thresholds, one for the Palaeozoic and a somewhat higher one for the post-Palaeozoic

as an equilibrium point rather than a fixed threshold. Population size can be much lower than the carrying capacity (e.g. due to a regional disaster) and it can be above the carrying capacity for a limited time. As population sizes are largely determined by birth and death rates, these rates seem to be density-dependent. Obviously, decreases in birth rates will have the same net effect as increases in death rates – both factors appear to be equally important.

Another question relates to the constancy of the carrying capacity itself. When a population size exceeds a long-term average, is this due to a short disequilibrium in the birth/death rate ratio leading to overpopulation or has the carrying capacity raised due to, for example, the extinction of a predator or beneficial climatic conditions? This problem is widely discussed in the context of human population size (Seidl and Tisdell 1999) and is especially acute in variable environments (McLeod 1997).

Carrying Capacity in Macroevolution

It is tempting to apply the ecological concepts of carrying capacities to macroevolution: individuals are equivalent to species (or genera), birth and death rates are origination and extinction rates, and resources are resources. But this analogy has some pitfalls especially because density-dependence has a different meaning for individuals and species. Nevertheless, the concept of carrying capacity has been applied at multiple hierarchical levels, comprising populations, communities, ecosystems and the global biosphere (del Monte-Luna et al. 2004), and the equilibrium models elaborated by MacArthur and Wilson (1963) and Sepkoski (1978, 1984) evoke a carrying capacity on evolutionary time scales.

We can reasonably assume that an ecosystem, even if as extensive as all marine habitats taken together, can only support a finite amount of biomass. The view of the oceanic biosphere as an essentially closed system helps to translate the ecological parameters into evolutionary ones. In the global ocean, the potential for biomass production depends primarily on the availability of nutrients and primary productivity. Assuming that global biomass has reached saturation, and unless primary productivity increases or the food requirements of consumers decrease, the addition of new species can only be accomplished by a decrease in the average population size of species (Levinton 1979; Stanley 2007). This compensation is limited to the degree that population sizes become so small and unstable that stochastic events lead to the extinction of species. With this mechanism it is obvious that a threshold to global diversity must exist, although it is still surprising that it was already reached in the Mesozoic.

In his classical description of seafood through time, Bambach (1993) suggested that primary productivity and the supply of food in the oceans have increased during the Phanerozoic. This would provide potential for an increase in marine diversity. On the other hand, the documented Phanerozoic shift towards higher proportions of marine genera with more energetic modes of life, such as predators and organisms with relatively high mobility (Bambach 1993, 1999; Bambach et al. 2002), may have used up much of this plus in nutritional resources.

A second limiting resource is the availability of space. Its relation to diversity is well known as the species-area effect, the tendency for the number of species to increase in a predictable way with area (e.g. Collins et al. 2002). In addition to more space, larger geographical areas also tend to hold a greater diversity of habitats, and environmental heterogeneity in turn promotes biodiversity (Ricklefs 2004; Barnosky et al. 2005). The rate at which diversity rises with increasing area depends on the spatial scale and the overlap in faunal composition of the localities from which a species-area curve is drawn. The area (and water volume) of the marine shelf, which yields the great majority of our marine diversity data, is determined by sea-level fluctuations, degree of continental fragmentation and the hypsometric curve of continents (e.g. Kiessling 2005b). Owing to the similar effects of sampling intensity, the species-area effect has not reliably been demonstrated in the fossil record (Crampton et al. 2006) and in the new sampling-standardized curve there is no significant correlation between changes in diversity and changes in shelf area ($R = 0.22$, $p = 0.23$).

A third phenomenon that is expressed at the global scale is mass extinction events, caused by large-scale perturbations of the physico-chemical environment. Mass extinctions create evolutionary opportunities and have the potential to redirect the course of the history of life (Erwin 2001). However, while new ecological patterns may arise in the aftermath of mass extinctions (e.g. Jablonski 1986; Sepkoski 1996; Bambach et al. 2002; Kiessling et al. 2008), these occasional setbacks appear to have no first-order effect on the long-term trajectory of global diversity (Sepkoski 1984). Therefore, mass extinctions are not directly relevant to carrying capacities, although they may explain why diversity stays substantially below an inferred threshold for some time.

The same is probably true for climatic fluctuations, even at the scale of global greenhouse-icehouse intervals (Veevers 1990). The only exception seems to be the late Palaeozoic ice age, during which diversity and turnover rates were depressed for nearly 50 million years (Stanley and Powell 2003; Powell 2005). An alternate state of the global marine ecosystem has been proposed, brought about by unusually large and stable populations of species with broad ecological niches and large geographical ranges (Stanley and Powell 2003). Although the diversity depression is indeed a remarkable feature also in the new curve (Fig. 1b), one has to be aware of the limited geographic sampling regime during this time (see above). We could also argue that this depression is just a long-term deviation from a stable global carrying capacity. This view is supported by dramatic diversity increase in the Middle and Late Permian.

In sum, it appears that resource limitations and biotic interactions are the most important drivers of a carrying capacity on evolutionary time scales. We may legitimately ask which sort of biotic interactions are the most relevant, although the answer is difficult.

Biotic Interactions

Statistical relationships between metrics of biodiversity dynamics, that is, rates of origination and extinction, can elucidate the role of biotic interactions in shaping diversity patterns (see Section "Diversity Dynamics"). The rapidly increasing origination rates in the aftermath of mass extinctions are well known (Lu et al. 2006). Moreover, considering the Phanerozoic as a whole, there is a strong general cross correlation between extinction and origination rates with a temporal lag (Alroy 2008). Together with the correlation between diversity increases and subsequent increases in extinctions, this clearly indicates that biotic interactions are the principal driver of the carrying capacity in macroevolution. The direct role of biotic interactions at global scales depends on the degree to which existing species (and new species) are able to disperse from their area of distribution (and their point of origination) and interact with other species elsewhere. Biotic interactions also become indirectly relevant to global diversity as the aggregate sum of interactions at smaller, local to regional scales.

The relaxation of niche incumbency and filling of empty niches have long been favoured to explain the rapid rebounds after mass extinctions (see review of Erwin 2001). Although this is clearly an oversimplification because niches have to be realized rather than occupied, the model of incumbency is appealing. A long-established incumbent species with large local population size can prevent invaders from colonization simply by being there first, even if the invaders are competitively superior. As long as existing populations limit the probability of survival of newly evolved species, the cumulative effect could be a constrained diversity globally. The extinction of species can effectually reduce the incumbent fauna below carrying capacity and create opportunities for newly evolving species and immigrants. However, restored diversity after extinctions may be virtually at the same level as before. The persistence in the fossil record of large-scale diversity patterns, such as the latitudinal diversity gradient, partly brought about by migration out of but not into the tropics, suggests that the effects of ecological incumbency have been a general feature (Valentine et al. 2008). A potential problem with the incumbency model is the spread of opportunistic taxa in the aftermath of some mass extinctions (Erwin 1998). These do have large population sizes but nevertheless origination rates increase.

The altogether moderate Phanerozoic rise in species richness of benthic communities in open marine environments (see Section "Geographical and Environmental Components") indicates that alpha diversity is constrained. The increase was explained by the exploitation of new ecospace (reflected in the larger number of guilds in geologically younger communities), an increase in species packing and resource partitioning, and change to a more nutrient-rich sea (Bambach 1983, 1985; Vermeij 1987). With respect to the partitioning of resources within communities, Wagner et al. (2006) have demonstrated that in the Palaeozoic ecologically simple and complex communities are about equally common. In contrast, in the Mesozoic and Cenozoic, complex communities are on average significantly more common. There is no long-term trend within these two intervals. Simple communities are those in which all species utilize similar resources such that the second most common species can only use resources left by the dominant species and so forth. Complex communities are characterized by a more diverse resource utilization, accompanied by ecological feedbacks such as the provision of adaptive opportunities for other species (niche construction). Competition for resources and a limited evolutionary inventiveness to utilize these limited resources in different ways (e.g. by tiering or different feeding modes) would thus be likely mechanisms for achieving a carrying capacity at the community level. It is striking that both global diversity (Fig. 6) as well as ecological patterns of communities (Wagner et al. 2006) show a first-order shift from the Palaeozoic to the Mesozoic-Cenozoic. Therefore, the idea of a global carrying capacity, brought about by the concerted, resource-limited carrying capacities of local communities, is a plausible explanation for constraining global diversity.

Stanley (2007) has criticized the competition hypothesis just as the carrying capacity model. He argued that population sizes in marine communities are governed by predation, physical disturbance and imponderabilities of larval recruitment rather than competition for food or space. There are some experimental data supporting the contention that competitive exclusion fails to reduce diversity at an ecological time scale (cited in Stanley 2007). However, it remains to be tested whether this plays out on macroevolutionary time scales. Processes that govern population sizes are not necessarily identical to those governing diversity at global scales.

With respect to predation, release from heavy predation pressure during mass extinctions would favour the diversification of prey in their aftermath and thus promote the reestablishment of pre-extinction diversity levels. Also, moderate predation pressure facilitates diversity, mostly by hindering the monopolization of communities by a few species (Paine 1966). In contrast, predation can suppress speciation by holding the population sizes of incipient species at low levels so that extinction risk is high (Stanley 2007). Long-term ecological feedback of interactions between predator and prey (escalation, Vermeij 1987) can create or reduce diversity (e.g. Aberhan et al. 2006), but net effects on Phanerozoic global diversity are unclear.

Conclusions

With the production of a new, less-biased curve of Phanerozoic marine diversity, we have to depart from the conventional view of a steep Cretaceous and Cenozoic diversity increase with a seemingly unbridled exponential diversity rise. The challenge seems to be to explain that diversity fluctuated within limits. We think that the finding of constraints to Phanerozoic diversity is well compatible with ecological principles and is justified biologically. We argue that on a global scale a carrying capacity for marine species can not be exceeded, at least not for long. Empirical results show that reductions of global diversity by extinctions are compensated by increasing origination rates in the aftermath. By approaching the carrying capacity, ecological feedback processes limit diversity by increasing extinction rates.

The exact number of species at a given time is the product of a multitude of factors that offer ample opportunity to explain diversity fluctuations through the Phanerozoic. Nevertheless, a few major processes can plausibly explain the overall pattern, in particular the observation that diversity of the global ocean did (and does) not rise infinitely.

Limitations of resources, in particular food, seem to define the overall setting at a global scale. However, species do not compete for global resources. This would require a degree of global connectedness among species that remains unsupported. Rather, the ecological interactions that influence global diversity occur at local to regional scales. They ultimately define the global diversity trajectory as the aggregate sum of biotic interactions at smaller scales. Of these we identify incumbency effects, the competition for limited resources, and possibly predation as being most relevant (see Rosenzweig and McCord 1991).

While within-assemblage (alpha) and global diversity (gamma) both exhibit a slight increase, our analysis of between-assemblage diversity (beta) demonstrates that this fluctuated substantially without an overarching trend. Although analyses at finer geographic and environmental scales are clearly needed, these results suggest that the first-order pattern in global diversity in governed by the number of species that are able to coexist within local or regional communities (see Patzkowsky and Holland 2003; Ricklefs 2004 for criticism of local processes). Future research should investigate the interplay among more resolved spatial and temporal scales, explicitly testing for nestedness (Wright et al. 1998) and integrating empirical and theoretical modelling approaches.

References

Aberhan M, Kiessling W, Fürsich FT (2006) Testing the role of biological interactions in the evolution of mid-Mesozoic marine benthic ecosystems. Paleobiology 32:259–277

Allison PA, Briggs DEG (1993) Paleolatitudinal sampling bias, Phanerozoic species diversity, and the end-Permian extinction. Geology 21:65–68

Alroy J (1998) Equilibrial diversity dynamics in North American mammals. In: McKinney ML, Drake JA (eds) Biodiversity dynamics. Columbia University Press, New York, pp 232–287

Alroy J (2000) New methods for quantifying macroevolutionary patterns and processes. Paleobiology 26:707–733

Alroy J (2008) Dynamics of origination and extinction in the marine fossil record. Proc Natl Acad Sci 105:11536–11542

Alroy J, Aberhan M, Bottjer DJ, Foote M, Fürsich FT, Hendy AJW, Holland SM, Ivany LC, Kiessling W, Kosnik MA, Marshall CR, McGowan AJ, Miller AI, Olszewski TD, Patzkowsky ME, Wagner PJ, Bonuso N, Borkow PS, Brenneis B, Clapham ME, Ferguson CA, Hanson VL, Jamet CM, Krug AZ, Layou KM, Leckey EH, Nürnberg S, Peters SE, Sessa JA, Simpson C, Tomasovych A, Visaggi CC (2008) Phanerozoic trends in the diversity of marine invertebrates. Science 321:97–100

Alroy J, Marshall CR, Bambach RK, Bezusko K, Foote M, Fürsich FT, Hansen TA, Holland SM, Ivany LC, Jablonski D, Jacobs DK, Jones DC, Kosnik MA, Lidgard S, Low S, Miller AI, Novack-Gottshall PM, Olszewski TD, Patzkowsky ME, Raup DM, Roy K, Sepkoski JJ Jr, Sommers MG, Wagner PJ, Webber A (2001) Effects of sampling standardization on estimates of Phanerozoic marine diversification. Proc Natl Acad Sci USA 98:6261–6266

Badgley C (2003) The multiple scales of biodiversity. Paleobiology 29:11–13

Bambach RK (1977) Species richness in marine benthic habitats through the Phanerozoic. Paleobiology 3:152–167

Bambach RK (1983) Ecospace utilization and guilds in marine communities through the Phanerozoic. In: Tevesz MJS, McCall PM (eds) Biotic interactions in recent and fossil benthic communities. Plenum, New York, pp 719–746

Bambach RK (1985) Classes and adaptive variety: the ecology of diversification in marine faunas through the Phanerozoic. In: Valentine JW (ed) Phanerozoic diversity patterns: profiles in macroevolution. Princeton University Press, Princeton, NJ, pp 191–253

Bambach RK (1993) Seafood through time: changes in biomass, energetics, and productivity in the marine ecosystem. Paleobiology 19:372–397

Bambach RK (1999) Energetics in the global marine fauna: a connection between terrestrial diversification and change in the marine biosphere. Geobios 32:131–144

Bambach RK, Knoll AH, Sepkoski JJ Jr (2002) Anatomical and ecological constraints on Phanerozoic animal diversity in the marine realm. Proc Natl Acad Sci 99:6854–6859

Bambach RK, Knoll AH, Wang SC (2004) Origination, extinction, and mass depletions of marine diversity. Paleobiology 30:522–542

Barnosky AD, Carrasco MA, Davis EB (2005) The impact of the species-area relationship on estimates of paleodiversity. PLoS Biol 3:e266

Benton MJ (ed) (1993) The fossil record 2. Chapman and Hall, London

Benton MJ (1995) Diversification and extinction in the history of life. Science 268:52–58

Benton MJ (2009) The Red Queen and the Court Jester: species diversity and the role of biotic and abiotic factors through time. Science 323:728–732

Bush AM, Bambach RK (2004) Did alpha diversity increase during the Phanerozoic? Lifting the veils of taphonomic, latitudinal, and environmental biases. J Geol 112: 625–642

Bush AM, Markey MJ, Marshall CR (2004) Removing bias from diversity curves: the effects of spatially organized biodiversity on sampling-standardization. Paleobiology 30:666–686

Collins MD, Vázquez DP, Sanders NJ (2002) Species-area curves, homogenization and the loss of global diversity. Evol Ecol Res 4:457–464

Cooper RA, Maxwell PA, Crampton JS, Beu AG, Jones CM, Marshall BA (2006) Completeness of the fossil record: estimating losses due to small body size. Geology 34:241–244

Crampton JS, Beu AG, Cooper RA, Jones CM, Marshall B, Maxwell PA (2003) Estimating the rock volume bias in paleobiodiversity studies. Science 301:358–360

Crampton JS, Foote M, Beu AG, Maxwell PA, Cooper RA, Matcham L, Marshall BA, Jones CM (2006) The ark was full! Constant to declining Cenozoic shallow marine biodiversity on an isolated midlatitude continent. Paleobiology 32:509–532

Darwin CR, Wallace AR (1858) On the tendency of species to form varieties; and on the perpetuation of varieties and species by natural means of selection. J Proc Linnean Soc London Zool 3:45–62

Del Monte-Luna P, Brook BW, Zetina-Rejón MJ, Cruz-Escalona VH (2004) The carrying capacity of ecosystems. Global Ecol Biogeogr 13:485–495

Erwin DH (1998) The end and the beginning: recoveries from mass extinctions. Trends Ecol Evol 13:344–349

Erwin, DH (2001) Lessons from the past: biotic recoveries from mass extinctions. Proc Natl Acad Sci 98:5399–5403

Erwin, DH (2008) Macroevolution of ecosystem engineering, niche construction and diversity. Trends Ecol Evol 23:304–310

Foote M (2000) Origination and extinction components of taxonomic diversity: general problems. Paleobiology 26:74–102

Hendy AJW (2009) The influence of lithification on Cenozoic marine biodiversity trends. Paleobiology 35:51–62

Jablonski D (1986) Background and mass extinctions: the alternation of macroevolutionary regimes. Science 231:129–133

Jablonski D, Roy K, Valentine JW, Price RM, Anderson PS (2003) The impact of the Pull of the Recent on the history of marine diversity. Science 300:1133–1135

Jackson JBC, Johnson KG (2001) Measuring past biodiversity. Science 293:2401–2404

Kidwell, SM (2001) Preservation of species abundance in marine death assemblages. Science 294:1091–1094

Kidwell, SM (2005) Shell composition has no net impact on large-scale evolutionary patterns in mollusks. Science 307:914–917

Kidwell SM, Flessa KW (1996) The quality of the fossil record: populations, species, and communities. Annu Rev Earth Planetary Sci 24:433–464

Kiessling W (2005a) Long-term relationships between ecological stability and biodiversity in Phanerozoic reefs. Nature 433:410–413

Kiessling W (2005b) Habitat effects and sampling bias on Phanerozoic reef distribution. Facies 51:27–35

Kiessling W, Aberhan M (2007a) Environmental determinants of marine benthic biodiversity dynamics through Triassic-Jurassic times. Paleobiology 33:414–434

Kiessling W, Aberhan M (2007b) Geographical distribution and extinction risk: lessons from Triassic-Jurassic marine benthic organisms. J Biogeogr 34:1473–1489

Kiessling W, Aberhan M, Brenneis B, Wagner PJ (2007) Extinction trajectories of benthic organisms across the Triassic-Jurassic boundary. Palaeogeogr Palaeoclimatol Palaeoecol 244:201–222

Kiessling W, Aberhan M, Villier L (2008) Phanerozoic trends in skeletal mineralogy driven my mass extinctions. Nat Geosci 1:527–530

Kirchner JW, Weil A (2000) Delayed biological recovery from extinctions throughout the fossil record. Nature 404:177–180

Knoll AH (2003) Biomineralization and evolutionary history. Rev Mineral Geochem 54:329–356

Kowalewski M, Kiessling W, Aberhan M, Fürsich FT, Scarponi D, Wood SLB, Hoffmeister AP (2006) Ecological, taxonomic, and taphonomic components of the post-Paleozoic increase in sample-level species diversity of marine benthos. Paleobiology 32:533–561

Lane A, Janis CM, Sepkoski JJ Jr (2005) Estimating paleodiversities: a test of the taxic and phylogenetic methods. Paleobiology 31:21–34

Levinton JS (1979) A theory of diversity equilibrium and morphological evolution. Science 204:335–336

Lu PJ, Yogo M, Marshall CR (2006) Phanerozoic marine biodiversity dynamics in light of the incompleteness of the fossil record. Proc Natl Acad Sci USA 103:2736–2739

MacArthur RH,s Wilson EO (1963) An equilibrium theory of insular zoogeography. Evolution 17:373–387

McGowan AJ, Smith AB (2008) Are global Phanerozoic marine diversity curves truly global? A study of the relationship between regional rock records and global Phanerozoic marine diversity. Paleobiology 34:80–103

McKinney ML, Lockwood JL (1999) Biotic homogenization: a few winners replacing many losers in the next mass extinction. Trends Ecol Evol 14:450–453

McLeod SR (1997) Is the concept of carrying capacity useful in variable environments? Oikos 79:529–542

Miller AI (1997) Dissecting global diversity patterns: examples from the Ordovician radiation. Annu Rev Ecol Syst 28:85–104

Miller AI (1998) Biotic transitions in global marine diversity. Science 281:1157–1160

Miller AI (2000) Conversations about Phanerozoic global diversity. Paleobiology 26(Supplement):53–73

Miller AI, Foote M (1996) Calibrating the Ordovician radiation of marine life: implications for Phanerozoic diversity trends. Paleobiology 22:304–309

Miller AI, Aberhan M, Buick DP, Bulinski KV, Ferguson CA, Hendy AJW, Kiessling W (2009) Phanerozoic trends in the global geographic disparity of marine biotas. Paleobiology 35:612–630

Omerbashich M (2006) A Gauss–Vanicek spectral analysis of the Sepkoski compendium: no new life cycles. Comput Sci Eng 8:26–30

Paine RT (1966) Food web complexity and species diversity. Am Nat 100:65–75

Patzkowsky ME, Holland SM (2003) Lack of community saturation at the beginning of the Paleozoic plateau: the dominance of regional over local processes. Paleobiology 29:545–560

Pease CM (1985) Biases in the durations and diversities of fossil taxa. Paleobiology 11:272–292

Peters SE (2005) Geologic constraints on the macroevolutionary history of marine animals. Proc Natl Acad Sci USA 102:12326–12331

Peters SE (2008) Environmental determinants of extinction selectivity in the fossil record. Nature 454:626–629

Peters SE, Foote M (2001) Biodiversity in the Phanerozoic: a reinterpretation. Paleobiology 27:583–601

Peters SE, Foote M (2002) Determinants of extinction in the fossil record. Nature 416:420–424

Powell MG (2005) Climatic basis for sluggish macroevolution during the late Paleozoic ice age. Geology 33:381–384

Powell MG, Kowalewski M (2002) Increase in evenness and sampled alpha diversity through the Phanerozoic: comparison of early Paleozoic and Cenozoic marine fossil assemblages. Geology 30:331–334

Raup DM (1976) Species diversity in the Phanerozoic: an interpretation. Paleobiology 2:289–297

Raup DM (1977) Species diversity in the Phanerozoic: systematists follow the fossils Paleobiology 3:328–329

Raup DM (1979) Biases in the fossil record of species and genera. Bull Carnagie Museum Nat Hist 13:85–91

Raup DM, Sepkoski JJ Jr (1982) Mass extinctions in the marine fossil record. Science 215:1501–1503

Ricklefs RE (2004) A comprehensive framework for global patterns in biodiversity. Ecol Lett 7:1–15

Rohde RA, Muller RA (2005) Cycles in fossil diversity. Nature 434:208–210

Rosenzweig ML, McCord RD (1991) Incumbent replacement: evidences for long-term evolutionary progress. Paleobiology 17:202–213

Seidl I, Tisdell CA (1999) Carrying capacity reconsidered: from Malthus' population theory to cultural carrying capacity. Ecol Econ 31:395–408

Sepkoski JJ Jr (1978) A kinetic model of Phanerozoic taxonomic diversity. I. Analysis of marine orders. Paleobiology 4:223–251

Sepkoski JJ Jr (1979) A kinetic model of Phanerozoic taxonomic diversity. II. Early Phanerozoic families and multiple equilibria. Paleobiology 5:222–251

Sepkoski JJ Jr (1981) A factor analytic description of the Phanerozoic marine fossil record. Paleobiology 7:36–53

Sepkoski JJ Jr (1982) A compendium of fossil marine families. Milwaukee Public Museum, Contrib Biol Geol 51:1–125

Sepkoski JJ Jr (1984) A kinetic model of Phanerozoic taxonomic diversity. III. Post-Paleozoic families and mass extinctions. Paleobiology 10:246–267
Sepkoski JJ Jr (1988) Alpha, beta, or gamma: where does all the diversity go? Paleobiology 14:221–234
Sepkoski JJ Jr (1993) Ten years in the library: new data confirm paleontological patterns. Paleobiology 19:43–51
Sepkoski JJ Jr (1996) Patterns of Phanerozoic extinction: a perspective from global data bases. In: Walliser OH (ed) Global events and event stratigraphy in the phanerozoic. Springer, Berlin, pp 35–51
Sepkoski JJ Jr (1998) Rates of speciation in the fossil record. Philos Trans R Soc Lond B Biol Sci 353:315–326
Sepkoski JJ Jr (2002) A compendium of fossil marine animal genera. Bull Am Paleontol 363:1–563
Sheehan PM (1977) Species diversity in the Phanerozoic: a reflection of labor by systematists? Paleobiology 3: 325–329
Smith AB (1994) Systematics and the fossil record: documenting evolutionary patterns. Blackwell, Oxford
Smith AB (2001) Large-scale heterogeneity of the fossil record: implications for Phanerozoic biodiversity studies. Philos Trans R Soc Lond B: Biol Sci 356:351–367
Smith AB (2007a) Marine diversity through the Phanerozoic: problems and prospects. J Geol Soc (Lond) 164: 731–745
Smith AB (2007b) Intrinsic versus extrinsic biases in the fossil record: contrasting the fossil record of echinoids in the Triassic and early Jurassic using sampling data, phylogenetic analysis, and molecular clocks. Paleobiology 33: 310–323
Smith AB, McGowan AJ (2005) Cyclicity in the fossil record mirrors rock outcrop area. Biol Lett 1:443–445
Smith AB, McGowan AJ (2007) The shape of the phanerozoic marine palaeodiversity curve: how much can be predicted from the sedimentary rock record of western Europe? Palaeontology 50:765–774
Stanley SM (2007) An analysis of the history of marine animal diversity. Paleobiology 33:1–55
Stanley SM, Powell MG (2003) Depressed rates of origination and extinction during the late Paleozoic ice age: a new state for the global marine ecosystem. Geology 31:877–880
Tapanila L (2007) FossilPlot, an Excel-based computer application for teaching stratigraphic paleontology using the Sepkoski compendium of fossil marine genera. J Geosci Educ 55:133–137
Thayer CW (1983) Sediment-mediated biological disturbance and the evolution of marine benthos. In: Tevesz MJS, McCall PL (eds) Biotic interactions in recent and fossil benthic communities. Plenum Press, New York, pp 479–625
Tipper JC (1979) Rarefaction and rarefiction – the use and abuse of a method in paleoecology. Paleobiology 5:423–434
Valentine JW (1970) How many marine invertebrate fossil species? A new approximation. J Paleontol 44:410–415
Valentine JW, Foin TC, Peart D (1978) A provincial model of Phanerozoic marine diversity. Paleobiology 4:55–66
Valentine JW, Jablonski D, Kidwell S, Roy K (2006) Assessing the fidelity of the fossil record by using marine bivalves. PNAS 103:6599–6604
Valentine JW, Jablonski D, Krug AZ, Roy K (2008) Incumbency, diversity, and latitudinal gradients. Paleobiology 34:169–178
Veevers JJ (1990) Tectonic-climatic supercycle in the billion-year plate-tectonic eon: Permian Pangean icehouse alternates with Cretaceous dispersed-continents greenhouse. Sedimentary Geol 68:1–16
Vermeij GJ (1987) Evolution and escalation – an ecological history of life. Princeton University Press, Princeton, NJ
Vermeij GJ, Leighton LR (2003) Does global diversity mean anything? Paleobiology 29:3–7
Wagner PJ, Kosnik MA, Lidgard S (2006) Abundance distributions imply elevated complexity of post-Paleozoic marine ecosystems. Science 314:1289–1292
Walker LJ, Wilkinson BH, Ivany LC (2002) Continental drift and Phanerozoic carbonate accumulation in shallow-shelf and deep-marine settings. J Geol 110:75–87
Willis KJ, Whittaker RJ (2002) Species diversity – scale matters. Science 295:1245–1248
Wright DH, Patterson BD, Mikkelson GM, Cutler A, Atmar W (1998) A comparative analysis of nested subset patterns of species composition. Oecologia 113:1–20
Wright P, Cherns L, Hodges P (2003) Missing molluscs: field testing taphonomic loss in the Mesozoic through early large-scale aragonite dissolution. Geology 31:211–214

Coordinated Stasis Reconsidered: A Perspective at Fifteen Years

Carlton E. Brett

Abstract

"Coordinated stasis" concerns larger-scale ecological-evolutionary units (EE = "blocks of coordinated stasis") defined by persistence of lineages and having similar biofacies composition at species level. It encompasses concurrent taxonomic stasis, ecologic stasis, and coordinated biofacies change. This pattern of shared stability and change, introduced by the author, was never intended as a hypothesis about process. It is an extension of the concept of punctuated equilibrium, but implies synchronous stasis and punctuation in numerous lineages rather than within a single lineage. It does not necessarily apply everywhere but enables a systematic approach to identifying cases showing similarities of pattern. It is separate from hypotheses and speculations about processes.

Keywords

Coordinated taxonomic stasis • Ecologic stasis • Biofacies stasis • Recurrent biofacies • Extinction and restructuring • Habitat tracking • Diversity thresholds • Appalachian Basin • Hamilton Group • Regressions • Transgressions • Biogeographic shifts

Introduction

It has now been just over 20 years since the term "coordinated stasis" was first used informally (Brett et al. 1990) and about 15 years since the concept was elaborated (Brett and Baird 1995; Brett et al. 1996). In the ensuing time this concept has engendered substantial response, both positive and negative (for reviews see: Jackson 1994; Miller 1996, 1997; Buzas and Culver 1994, 1998; Roy and Wagner 1995; Baumiller 1996; Stanton and Dodd 1997; Patzkowsky and Holland 1997; Ivany 1996, 1999; Bonuso et al. 2002a, b; DiMichele et al. 2004; Brett et al. 2007a, b; Ivany et al. 2009). It seems an appropriate time to review where the concept of coordinated stasis stands, particularly with regard to its type area, and to consider why it does or does not apply and its implications for larger scale evolution.

C.E. Brett (✉)
Department of Geology, University of Cincinnati, Cincinnati, OH 45221-0013, USA
e-mail: carlton.brett@uc.edu

J.A. Talent (ed.), *Earth and Life*, International Year of Planet Earth,
DOI 10.1007/978-90-481-3428-1_2, © Springer Science+Business Media B.V. 2012

What's in a Name? Semantics of Coordinated Stasis

Discussions with many colleagues have suggested that some of the problems that people had with "coordinated stasis" was the name. In retrospect, the term "coordinated", which was meant in the common sense of "moving together" or "synchronous", was often misconstrued and together with conflation of coordinated stasis with ecological locking, gave the impression of "lock step" or dependent interactions, which was never intended.

In the early formulation of the concepts of shared taxonomic and ecological stasis I cast around for an effective name. Initially, I considered "community stasis" but discussions with colleagues, including Niles Eldredge, indicated that the term "community" was so ambiguous and loaded with meanings that this was a poor term; moreover, I rejected this name because I realized that it might imply strong ecological interactions and burden the concept with unnecessary implications of process. Other possibilities that I toyed with were "concurrent stasis" or "concatenated stasis", as well as "coordinated stasis". A discussion with Steve Gould during a seminar I presented at Harvard in 1991 cemented the term "coordinated stasis", which Gould felt was a strong and evocative term. Also, it suggested (to me at least) the coincidence of taxonomic/morphologic stasis and ecological stasis. In hindsight, "concurrent stasis" might have been the better term as it is less burdened with implications that I had never intended. It was the synchronicity of stasis and change in both lineages and biofacies that I most wanted to emphasize.

Reacting in part to the loaded terminology of "coordinated stasis", Arnie Miller (1997) suggested the term "coincident relative stability"; this rather awkward phrase seemed to me to also give an incorrect implication: the idea that the shared stasis was merely the result of chance or coincidence (see also Roy and Wagner 1995). Yet the existence of repeated EE-subunits with shared stasis of lineages and biofacies and abrupt change suggested the process was not merely a coincidence of conditions of a particular time and place.

Perhaps the term "concurrent stasis" would have been less loaded, but coordinated stasis is in the literature. Furthermore, simply implying that the stasis of many lineages is synchronous does not tell the whole story. There is an ecological aspect to the phenomenon, although not implying tight ecological interactions, but rather that species/lineages maintain about the same ecological relationships through the time span. That is to say that they occupy generally similar environments and co-occur with similar (though not identical) groups of other species along gradients during stable time blocks ("dynasties" or ecological-evolutionary subunits). At the termination of these blocks some species do persist locally but they are in different associations of other species and may even appear to change ecology somewhat, becoming, for example, more eurytopic in certain instances (Holland and Patzkowsky 2004, 2007). Discrete communities may or may not exist, but gradients of faunal composition can be documented and, in some cases, they remain rather consistent for sometimes millions of years (Brett et al. 2007a, b). It is in these senses that the phenomena of taxonomic and morphologic stasis do appear to be coordinated; hence, the term was retained.

Overview of the Concept of "Coordinated Stasis" and Ecological-Evolutionary Subunits

History of the Concepts

We originally proposed the hypothesis of coordinated stasis on the basis of empirical field observations and data compiled from primary literature (Brett and Baird 1995) on the Silurian-Devonian interval of the northern Appalachian Basin region: New York, Pennsylvania, and southern Ontario. We defined ten blocks of relative stability that were termed ecological-evolutionary subunits, taking that term from Boucot (1975, 1990, 1996) who had already discussed larger-scale ecological-evolutionary units in some detail, defining these as intervals bounded by major extinctions, between which general biofacies showed strong similarities at generic and familial levels. We argued that ecologic-evolutionary (=EE) subunits show persistence of lineages and similar biofacies composition at a species level. Overall, Brett and Baird (1995) recognized a series of 11 relatively long intervals, EE-subunits or "faunas", each 1–6 million years in duration, in which a majority species persisted in their favored environments and during which ecological aspects of biofacies remained similar. Each was apparently terminated by a geologically abrupt interval of at least local extinction, immigration/emigration and new speciation. EE subunits were defined as "blocks of coordinated stasis" and characterized by low rates of holdover (species held over from a previous EE subunit) and carryover (species that persisted to the next younger subunit).

To examine the internal patterns of EE subunits, we chose to focus on the best studied of those intervals, the Hamilton fauna, as an exemplar of the pattern (Brett and Baird 1995). In more recent papers (Bonelli et al. 2006; Brett et al. 2007a, b; Ivany et al. 2009) we have further documented and statistically tested for patterns of similarity in species-level composition and biofacies similarity, specifically in the Hamilton fauna. In addition, we have tested for the discreteness of EE-subunits overall and have refined and modified our definitions of Devonian EE-subunits with the addition of two such intervals: Stony Hollow fauna and Lower Tully fauna not previously singled out as discrete entities (Brett et al. 2009).

We attempted to clarify definitions and possible metrics for defining coordinated stasis (Brett et al. 1996, 2007a, b) and to suggest that this pattern, or some aspects of it, *might* be found widely in the fossil record. As stated therein, coordinated stasis is a postulated pattern of shared stability and change in large numbers of independent lineages; it was *not* intended as a hypothesis about *processes* that may regulate evolutionary ecology, although several researchers have conflated pattern with process. Brett et al. (1996) in fact discussed several possible implications of coordinated stasis for evolutionary process, but these are derivations from the pattern, not necessary adjuncts.

What Coordinated Stasis Was Meant to Imply

In brief, the concept of coordinated stasis makes not one, but three distinct, though potentially related, claims about pattern: (a) stasis of lineages; (b) stability of biofacies; and (c) concurrent turnover of many lineages and abrupt change of biofacies. Each of these claims can be examined briefly.

Taxonomic Stasis

First, coordinated stasis implies that large numbers of species (or closely related species groups within lineages) may persist in appropriate facies/environments with effective stasis during approximately synchronous long time intervals (*taxonomic stasis*, Ivany et al. 2009). In this sense, coordinated stasis is an extension of the concept of punctuated equilibrium (Eldredge and Gould 1972) except that, rather than stasis in a single lineage at a time, the concept implies synchronous stasis and punctuation in many, indeed a majority, of lineages. Brett et al. (1996) used a cutoff of >60% to define "majority of persistent lineages" and a cutoff of <30% carryover and holdover to delimit EE subunits. Persistence, as defined, also assumes range-through: counting species as present within the basin in intervals provided they are known from higher and lower samples even if they were not directly observed. This might not be valid if biofacies are reassembled from outside species pools.

Apparent taxonomic stability is predicated on the work of competent systematists, who have recognized particular species at multiple levels. It implies that chronospecies or groups of related populations through time show little or no *net* change (*concurrent morphological stasis*), although, in fact, for most lineages there have not been rigorous morphometric studies (but see Lieberman et al. 1995; Lieberman and Kloc 1997). Recently, Hunt (2008) has distinguished three possible categories of pattern within lineages through time: directional change, stasis, and random walks. In testing actual data against hypothetical patterns, he inferred that directional change was rare but that there were numerous examples of both true stasis and random walks. It is possible that some of the lineages considered to show stasis in fact display random walks, but if so, the amount of morphological change is quite minor. Brett and Baird (1995) argued that because most of the early systematists were inclined to split lineages on slight morphological grounds to refine biostratigraphy, data on distributions of species were, if anything, biased toward short ranges; thus, the reports of long-ranging taxa must represent lineages with very little change.

Ivany et al. (2009) compared the actual pattern of occurrences of 247 species from the diverse coral-brachiopod biofacies among 13 horizons representing the entire Hamilton Group with 1000 mirror datasets with random deletions of taxa from particular horizons to simulate incomplete sampling. They found that the pattern of actual occurrences was statistically indistinguishable from a pattern in which all species persisted through the entire time span, but were incompletely sampled. In fact, not all lineages do persist and few do show possible directional changes, but these are clearly the exceptions in the studied intervals.

The requirement that at least 60% of species in a fauna must persist throughout an EE-subunit to identify cases of coordinated stasis was perhaps extreme; the number was based on well-known Appalachian Basin faunas for which even rare taxa are well documented; and all of the EE subunits identified by Brett and Baird did appear to meet this threshold. In the Michigan Basin Subprovince, the Traverse fauna, coeval with the Hamilton fauna in the Appalachian Basin, shows synchronous and equally major turnovers at its beginning and end. However, the criterion of 60% persistence is not met (Bartholomew and Brett 2007), although >50% of recognized species do persist. In part, this no doubt reflects the limited geographic distribution of samples of any given age (see below); many formations are represented by only a single good outcrop (or none!). This low value of persistence may also result from a high degree of species splitting in the Michigan Basin faunas: nearly each formation and member has different species or subspecies of many brachiopods. In many cases, species are ill defined, overly split, and overall rather subjective. Similar problems may have prevented other faunas similar to the Hamilton from being recognized as examples of coordinated stasis, although the counter claim could also be made that possibly the Hamilton species are overly lumped. For these reasons it has been suggested that genera and subgenera would be better taxonomic units to examine; using this metric the Traverse fauna does meet the 60% persistence threshold.

Ecological Stasis

The second aspect of coordinated stasis is that recurring species associations or biofacies ("communities" sensu lato) and their ecological properties are relatively stable for long time spans (*ecological stasis*; Ivany et al. 2009). A good deal of discussion regarding coordinated stasis involves this point. The obvious question is what is meant by relatively stable. In the original formulation of the model, Brett and Baird (1995) suggested that ecological stability might include similarity of species lists, species richness, and perhaps, though not definitely, rank order and relative abundance. Numerous studies (e.g., Bennington and Bambach 1996; Olszewski and Patzkowsky 2001; Bonelli et al. 2006; Brett et al. 2007a, b) have demonstrated that while species lists and general richness may be similar in recurrent biofacies, rank abundance of dominant taxa, and certainly relative abundance, are rarely if ever the same in two successive samples of a generalized biofacies. This aspect of coordinated stasis did not prove viable even in the type examples. These properties typically vary strongly even among samples from a single locality and horizon (see Bennington and Bambach 1996).

Coordinated Biofacies Change

The third aspect of coordinated stasis is that large numbers of species undergo concurrent change (local extinction/evolution, emigration/immigration) during intervals that are brief (<10%) relative to times of stasis (*coordinated biofacies change*; see Ivany 1996). This pattern can be tested to varying degrees, but only in cases of relatively complete records; in many cases biases of preservation will militate against recovery of taxa, especially rare species at the true first and last appearances in a basin (see discussions in Signor and Lipps 1982; Marshall 1997; Bush et al. 2004). Thus, the record is inherently biased against finding abrupt faunal turnovers. Conversely, the stratigraphic record of depositional sequences may artificially sharpen boundaries that were in fact more gradual transitions because portions of the record may be lost at sequence-bounding unconformities or interludes of sediment starvation at flooding surfaces (Holland 1995, 1996, 2000; Brett 1995). Given these constraints, it is important to note that some of the faunal turnovers are still observably abrupt (Brett et al. 2009).

Thus, the concept of coordinated stasis encompasses the following aspects of pattern: (a) concurrent taxonomic stasis, (b) ecological stasis, and (c) coordinated biofacies change. It is important to note that each of these is really a separate hypothesis about *pattern*; they may or may not be linked. In retrospect, it may have been excessive to bundle three distinct phenomena under one term; this clearly makes it difficult to verify the concept or, put simply, it makes the notion of coordinated stasis highly vulnerable to falsification, as one might claim that demonstration that any portion of the pattern is false disproves coordinated stasis as a whole. However, like sequence stratigraphy, the concept of coordinated stasis was intended as a heuristic framework for comparing the fossil record, and falsification of a single case study does not automatically undermine the generality of the concept. Moreover, taxonomic and ecological stasis are obviously linked to a degree: ecological stasis, as defined herein to include similar species lists within biofacies, seemingly requires persistence of many of the same or very similar lineages, although mere taxonomic persistence does not necessarily imply that organisms maintain similar ecologies. The importance of the concept is not that it applies everywhere but that it enables a systematic approach to identifying cases that do and do not show similarities of pattern.

What Coordinated Stasis Is Not

The concept of coordinated stasis deals with patterns in the fossil record and, although it may have important implications for ecological-evolutionary processes, it was never intended to present hypotheses about the causes of these patterns. Unfortunately, publication of the coordinated stasis paradigm was preceded by a paper, of which I was a coauthor, which presented a possible cause of long-term ecological-evolutionary stability. That concept was "ecological locking" (Morris et al. 1995). The paper presented the intriguing possibility that interactions in organism communities might in effect "lock up" ecosystems in such a way as to maintain stable ecological structure. Although this is by no means the only possible implication of a pattern of coordinated stasis as subsequently defined by Brett and Baird (1995), it appears that many subsequent researchers conflated the two concepts. This, in fact, seems to have set the stage for much of the debate over coordinated stasis and many supposed falsifications of the latter were actually arguments against ecological locking (see discussions of Buzas and Culver 1994, 1998; Roy and Wagner 1995), but this in no way undermines the concept of coordinated stasis. It is perhaps regrettable that the ecological locking paper preceded that on the definition of coordinated stasis by a few months. Thus, coordinated stasis appeared to be wedded to ecological locking from the outset. There are still important ideas associated with locking that may be viable, incumbency of species being one, but it is simply not the same concept as coordinated stasis.

The original and subsequent formulations of coordinated stasis emphasized that this was an issue of observable patterns and thus testable and separate from hypotheses and speculations about processes. Many of the early responses to coordinated stasis involved the issues of "community unity" (see, for example, Jackson 1994). It was relatively easy to show that biofacies are never exactly the same twice, but this was not the main point.

Does Coordinated Stasis Exist in Its Type Area?

Some of the early contention over coordinated stasis revolved about whether in fact it can be demonstrated at all, and how one might quantify it? Ultimately there was an issue of whether or not this pattern existed even within the original area of the Silurian–Devonian Appalachian Basin. Bonuso et al. (2002a, b) produced two well-documented papers that directly challenged certain aspects of the concept of coordinated stasis in its "type" example: the Middle Devonian Hamilton Group of New York State. They amassed an impressive database of quantitative counts of fossil species in lithologically similar mudrock facies at various levels in the Hamilton Group of central New York State (Syracuse area). On the basis of quantitative analyses of these data, Bonuso et al. concluded that the fossil record of this area displays a pattern that is inconsistent with two of the hypotheses of coordinated stasis. It is important to note that they did not test concurrent rapid local extinction and restructuring. However, they have addressed the other two major aspects of this concept, noted above.

Taxonomic Stasis

The most substantive claim made by Bonuso et al. (2002b) is that many species actually do not persist throughout the Hamilton interval. They supported this with bar graphs showing the frequency with which key taxa were encountered in their samples. This was not simply a case of rare species that might have been accidentally overlooked because Bonuso and colleagues deliberately chose several key taxa that are abundant to dominant in several samples.

There *are* genuine differences between lower and upper Hamilton samples. For example, in over 200 bulk samples that we have processed and nearly a thousand presence-absence samples we have made, we have found, as noted by Bonuso et al., that the gastropod *Bembexia sulcomarginata* indeed shows an abrupt decline from high abundance in many Skaneateles Formation samples to near absence at about the base of the Centerfield Member of the Ludlowville Formation. One might have assumed that this gastropod became locally extinct. Recent work (Bonelli et al. 2006) however showed that it actually occurs extremely rarely in a few middle to upper Hamilton samples and stages a return to moderate abundance in the overlying Tully Formation. The genus and apparently the species thus *do* persist, although the importance of the genus in individual assemblages varies markedly. A genuine example of a dropout involves the bivalve *Ptychopteria (Cornellites) flabellum*, common in the lower to middle Hamilton Group (Oatka Creek, Skaneateles, and Ludlowville formations), but appearing to be completely absent in the upper Hamilton (Moscow Formation).

However, several of the supposed cases of appearance or disappearance of taxa suggested by Bonuso et al. are based on incomplete data, even from the Syracuse area. The majority of taxa cited as occurring abundantly only in the lower or upper Hamilton Group, in fact occur throughout, but only in *appropriate* biofacies. Our data indicate that a majority, but certainly not all, taxa do occur from lowest to highest samples of *appropriate* biofacies (Brett et al. 2007a, b; Ivany et al. 2009).

Ecological Stasis

An important point to draw from the Bonuso et al. results and one we have noted in the past is that many of the Middle Devonian species did show semi-independent ecological behavior, rather than forming tightly integrated communities, although a shared preference for particular paleoenvironments (as per the modifications of Miller's, 1997, "coincident relative stability"; see also Roy and Wagner 1995). For example, it was possible for certain common or even dominant taxa to effectively disappear from biofacies associations without causing fundamental changes in composition of the remaining assemblages. Communities clearly were *not* tightly locked in terms of membership (see also Brett et al. 1996, 2007a, b). In both papers, Bonuso et al. (2002a, b) documented fluctuating relative abundances of species in samples supposedly drawn from an unchanging environment. I do not disagree with their conclusion (or similar inferences of others; see especially Bennington and Bambach 1996) that local communities in similar environments may show substantial variation in specific membership. This conclusion is documented by the careful quantitative studies of Bonuso et al. (2002b) and Bonelli et al. (2006). That the general composition, main cast of characters, species richness, and guild structure remained relatively constant through time (relative with respect to changes seen at boundaries of stasis blocks or ecological-evolutionary subunits) is, in fact, partly supported by the studies by Bonuso et al. as they report general similarity of diversities and membership of biofacies, at least in some cases.

A major assumption made by Bonuso et al. throughout their work is that they were sampling replicates of basically *the same* (or very closely similar) environments. They argue that, in each case, they sampled similar gray, oxic mudstone facies, and that this suggests a control on environment independent of fossil evidence. Against this backdrop of purportedly constant environments, the authors then observe marked fluctuations in the faunal content and argue for major fluctuations in species compositions. But *are* their samples truly representative of a single environment or at least narrow range of similar environments? A great deal of the Hamilton Group in west-central New York could be described as gray, silty mudstone lithofacies. However, as we have previously noted, there are many distinctive biofacies and taphofacies within this broadly similar, "monotonous" lithofacies. Thus, lithology cannot, and should not, be used as a precise proxy of environments; sediment grain size, color, and composition represent only limited aspects of the original environment. Fine-grained mudrocks certainly do suggest dysoxic to oxic, low organic content sediments in generally low energy settings. Not directly recorded, however, are many other variables, such as bathymetry, water temperature, substrate consistency, and oxygen content. Hamilton faunas do not track *lithofacies*; in many instances we have observed nearly identical, contemporaneous faunal associations in samples from somewhat different lithologies. Clearly, the organisms were responding to factors other than those that cause the differences in coloration and other physical properties of the shale.

It is clear that nearly identical lithofacies can yield widely different faunal associations, and different lithologies may contain nearly identical faunal associations. And the most critical conclusion of these observations is that mudrock lithology is *not* a very adequate environmental proxy. Organisms appear to have migrated to follow, or track, their favored parameters as these shifted. If so, they tracked environments, not facies. This has been demonstrated in studies of gradient analysis (e.g., see review in Brett et al. 2007b) in which correlatable trends of fluctuation are based on gradient analyses, for example, DCA (detrended correspondence analysis) scores that reflect substantial and consistent variations in species compositions, despite relatively monotonous lithologies. Thus, we argue that reasonable comparisons of faunas of differing age, required to test for properties of ecological stasis, should be based upon the *most similar biofacies samples*, wherever they occur, and not on samples from a limited geographic region designated as coming from similar environments based on similarities of lithofacies (Brett et al. 2007a).

In recent papers we attempted to address additional issues, raised by Bonuso and others, of documenting concurrent species-level and ecological stasis (Brett et al. 2007a; Ivany et al. 2009). Such stasis can be demonstrated temporally to varying degrees in all Hamilton biofacies associations. These time-averaged associations were, first, compared vertically (temporally) in more than 90 samples through 13 different cycles, ranging from the lowest to highest known occurrences of the Hamilton fauna, including samples of recurrent Hamilton fauna from above the Hamilton Group per se in the upper Tully Formation (Brett et al. 2007a). Biofacies were recognized on the basis of cluster analysis using the Jaccard coefficient of similarity. Significantly, samples of similar composition were clustered together regardless of age with samples from highest to lowest levels grouped together. Detrended correspondence analysis (DCA) showed that samples coded for biofacies grouped closely regardless of age and formed a gradient along DCA axis 1, mirroring closely the vertical stacking order of occurrence of biofacies in single transgressive–regressive cycles. Brett et al. (2007a) also showed that species richness (diversity), species lists, and guild proportions remained similar in all samples of particular biofacies recognized from cluster analysis.

Ivany et al. (2009) further tested for similarity of the diverse coral-brachiopod biofacies in 13 horizons from the lowest to highest examples and spanning perhaps 5 to 5.5 million years. Fossil associations from three major stratigraphic levels representing lower, middle, and upper Hamilton Group were compared. Both taxonomic and ecologic composition were found to be statistically indistinguishable based on several independent tests, including ANOSIM and a maximum likelihood estimation that evaluated stratigraphic turnover using the Bayesian Information Criterion.

Regardless of how they were defined, the associations were found to be similar throughout and simulated datasets yielded patterns consistent with very low rates of extinction (2.6% per million years), much lower than an overall Middle Devonian (Givetian) rate of 11.5% determined from the Paleobiology Database, calculated using the metrics of Foote (2003). The implication of this comparison is that extinction must be concentrated in brief intervals rather than being evenly distributed as a background phenomenon.

The persistence of Hamilton-type "coral bed" biofacies was also examined within a well-documented interval of biotic disruption where such biofacies "survive" the disturbance (Bonelli et al. 2006). The Bellona Bed/West Brook Shale in the upper part of the post-Hamilton Tully Formation preserves a remarkably Hamilton-like fauna occurring above an interval that records a major incursion of non-Hamilton taxa; this suggests the apparent survival of a major Hamilton biofacies as a discrete unit even after significant community restructuring associated with the Taghanic Bioevent (Baird and Brett 2008). Quantitative comparison of the recurrent coral bed association with earlier Hamilton analogs (Bonelli et al. 2006) provided a test for temporal recurrence of this biofacies through both stable and unstable intervals. The results clearly show that whereas a majority of the diverse coral biofacies found in highest Hamilton (Windom Member, Moscow Formation) samples recurs in the upper Tully, following a period of outage of perhaps a few hundred thousand years (lower Tully), the rank order abundance of various taxa was not retained. However, further investigation showed that rank and relative abundance order was not retained even between two adjacent cycles within the Hamilton Group proper. In other words the degree of change in biofacies across this major Taghanic Event disruption was little different from that seen from one cycle to the next in the Hamilton Group itself. Moreover, qualitatively defined "common" taxa remained common and "rare" taxa remained rare.

As a side note, there is no evidence that rare taxa are more prone to evolutionary change than common ones as suggested by McKinney et al. (1996). Indeed, a number of rare species are recorded from lowest to nearly highest samples of particular biofacies, despite the obvious bias against finding them. Though few detailed morphometric studies have been done on these rare taxa, there is limited evidence (for Hamilton trilobite examples, see Lieberman 1994; Lieberman and Kloc 1997) that they are as static as the more abundant taxa.

Thus, although no study shows perfect recurrence of identical communities through time, results of various multivariate statistical studies and modeling, as well as gradient analyses and simple comparisons of taxonomic and paleoecological information, are consistent with predictions of all three aspects of coordinated stasis as originally formulated at least for the Middle Devonian Hamilton Group. These results contrast with those of Bonuso et al. (2002a, b) for the same interval.

The reason for the differing conclusions of various studies is fairly obvious. Bonuso and colleagues sampled very well but only in a relatively confined geographic area in central New York spanning less than one degree of longitude. Conversely, the studies of Brett et al. (2007a) and Ivany et al. (2009) were based on samples for almost the entire outcrop belt from Lake Erie to the Hudson Valley spanning about 8 degrees of longitude and about a degree of latitude (north-south transects also exist in the Finger Lakes region as well as in the Hudson Valley regions of New York). Calcareous mudstone–limestone cycles were sampled wherever they occurred.

This leads to a critical point. Particular environments may or may not recur throughout the stratigraphic column of a given area. Assuming that species and associations of species track their preferred environments (see Brett et al. 2007b), they may therefore not necessarily recur throughout a section in any given area. Thus, for example, in the limited domain of western New York, the lower Hamilton Group is almost completely comprised of dark gray, sublaminated shales with low diversity, dysaerobic faunas, whereas the upper Hamilton comprises bioturbated (oxic), calcareous mudstones and argillaceous limestones. Not surprisingly, any comparison of lower with upper Hamilton faunas would show a dramatic difference. We have come, however, to recognize that many (most) of the faunal elements that we once considered exclusively "upper Hamilton" in the west are present in samples of appropriate facies equivalent to the lower (indeed some of the lowest) Hamilton cycles in eastern New York, central Pennsylvania, and other areas. Fair tests of coordinated stasis must be made on data derived from an entire basin, not a single geographic area within it. Modeling studies, presently in progress, are examining the apparent effect on taxonomic range of only considering data for the Hamilton Group from longitudinal "windows" of varying size along the dominantly east–west outcrop belt (Wall et al. in press). The results are striking: using data derived only from limited geographic areas produces very different patterns of apparent origination and extinction of taxa and show no pattern similar to coordinated stasis, whereas broadening the dataset to a few degrees of longitude yields patterns consistent with coordinated stasis. The implication is clear: patterns of long-term persistence or the lack thereof can only be fairly tested where a relatively large cross-section of a depositional basin can be sampled. Future studies should focus on locating other such examples and testing them carefully.

Coordinated Turnovers

The third aspect of the pattern that requires documentation is that of synchronous abrupt turnovers involving at least local extinction, immigration/emigration, and rapid evolution within endemic lineages. Such turnovers should take place in a fraction of the time of the EE-subunits. Of course it is not always easy to estimate the relative amount of time; Brett et al. (1996) suggested that the best proxy for this issue is to define a series of nested depositional cycles (which may or may not relate to periodic Milankovitch effects) and then to determine the timing of faunal turnover relative to this framework. Also, because of the problem of artificially sharpened faunal boundaries resulting from disconformities (see Walliser 1990; Holland 1996, 2000), the pattern of abrupt turnovers can only be fairly tested in relatively thick, conformable sections with abundant, well-documented faunal occurrences. Thus, for example, we have found highly expanded sections with little or no discontinuity near the base and top of the Hamilton fauna (e.g., Bartholomew and Brett 2007; Baird and Brett 2008). Yet in both cases the faunal change has proven to be abrupt and constrained to single, small-scale cycles.

A major question that has been raised about the EE-subunit boundary events is whether these turnovers are merely local and result from events within, for example, a depositional basin, or result from widespread or even global causes (see, for example,

Miller 1997). Answering this question requires the best possible correlations between biogeographic provinces. Not surprisingly, a definitive answer is difficult. But some progress has been made in the past few years: interregional and international correlations have been improved with refined biostratigraphy, sequence stratigraphy, magnetic susceptibility, and, in some cases, carbon isotopic curves (see, for example, Bartholomew and Brett 2007; DeSantis and Brett 2007; Brett et al. 2011). This has permitted a test of whether or not strong faunal turnovers occur synchronously in different provinces. Thus, Bartholomew and Brett (2007) demonstrated that faunal turnovers of comparable magnitude occurred synchronously in the Middle Devonian of the Appalachian and Michigan Basin provinces.

Recent studies have affirmed that a majority of the Silurian-Devonian EE-subunit boundaries identified initially in the Appalachian Basin (Brett and Baird 1995) are recognizable in other regions (see Brett et al. 2009). For example, the beginning of the Hamilton fauna is coeval with the well-known Kačak Event (DeSantis and Brett 2007) and the end of the fauna corresponds to Taghanic Event (Aboussalam 2003; Baird and Brett 2008). They are synchronous, at least within the constraints of biostratigraphy and sequence stratigraphy; however, the turnovers are not of the same magnitude in all provinces.

One assertion made by Brett and Baird (1995) regarding faunal turnover has not stood up to testing: that is that the addition of large numbers of new taxa was always associated with loss of many incumbents. Although this pattern is common and appears to be the rule in the Appalachian Basin EE-subunits, it is not always true. Holland and Patzkowsky (2007) have shown that the significant change in Late Ordovician benthic faunas termed the "Richmondian Invasion" involves addition of immigrating taxa without local extinction of endemics.

Does Coordinated Stasis Occur in Other Settings?

An agenda for future research should involve comparative consideration of instances in which coordinated stasis does and does not apply. Only in this way is it possible to tease out factors that may promote or militate against stability in ancient biotas. At present, several studies have claimed to test various aspects of coordinated stasis, but these have been conducted at various scales using varied techniques and metrics. Thus, they provide only a preliminary sense of the generality of the patterns. DiMichele et al. (2004) provide an excellent overview of the subject and we intend to make further comparative analyses of various studies, but that topic is beyond the scope of this essay.

Certain authors have observed similar patterns of coordinated taxonomic, ecologic stability, and/or episodic turnover (e.g., DiMichele 1994; DiMichele and Philips 1996; Pandolfi 1996, 2002; Pandolfi and Jackson 2006; Scarponi and Kowalewski 2004; Visaggi 2004; Bonelli and Patzkowsky 2008, Testing for patterns of faunal persistence in the Early Oligocene Byram Formation of Mississippi, Unpublished MS thesis, Syracuse University, 106pp; Bonelli and Patzkowsky 2008). Other researchers have argued that their data show only limited similarities (e.g., Bennington and Bambach 1996; Westrop 1996; Holterhoff 1996; Holland and Patzkowsky 2004; Patzkowsky and Holland 1997, 1999) or that no similarities of pattern exist in their data at all (Stanton and Dodd 1997; Huynh et al. 1999, see review by DiMichele et al. 2004). These studies make clear that coordinated stasis is not a universal pattern but, rather, it seems to apply in some instances in a wide array of environmental settings and not in others.

Abrupt turnovers demarcating blocks of relative stability or at least reduced rates of evolution have also been recognized in numerous other studies, including those of Palmer (1984), Westrop (1996), Patzkowsky and Holland (1997), and Holland and Patzkowsky (2004, 2007). However, Tang and Bottjer (1996) recognized long periods of stasis in several Mesozoic lineages, but without coordinated first or last occurrences. This lack of coordination may reflect an abundance of generalized or eurytopic lineages and/or absence of marked physical events in the Mesozoic of the Western Interior. Conversely, punctuational change in speciation and extinction rates, in line with those observed for the terminations of EE-subunit stable blocks, have also been recognized for broader Phanerozoic datasets by Foote (2003).

Those studies showing patterns most similar to coordinated stasis represent an array of highly diverse environments. Research by Pandolfi (1996) on terraces in Papua New Guinea and Gardiner (2001) in the Bahamas involved recurrent Pleistocene reefal coral and molluscan associations. Samples from different sea-level highstands were very similar, as in the Appalachian Basin Devonian; however, they differed in age by just a few hundred thousand years at most, not the several million years of the Paleozoic examples. It might be argued that coral reef communities are among the most highly structured and show the most tightly integrated networks of interaction of any marine community. They might be anticipated therefore to show strong similarities through time. Indeed, it is among high diversity coral associations that the strongest degrees of similarities were demonstrated in the Devonian examples. However, the work of Scarponi and Kowalewski (2004) in the Po Plain of Italy also showed highly recurrent patterns during a single cycle of late Pleistocene–Holocene age in shallow marine shelf-level bottom assemblages. This study also represented a relatively short period of time. In this case, tracking of shifting shorelines was possible although transgressive and regressive phases showed strong asymmetries as was also demonstrated for parts of the Hamilton Group (Brett et al. 2007a, b). These studies stand in stark contrast to those of many Pleistocene workers, especially in terrestrial environments; these show rather random patterns of species reassortment in, for example, glacial and interglacial times (Gleason 1926; Bennett 1990, 1997; Jackson and Overpeck 2000).

Patterns broadly similar to coordinated stasis have been demonstrated in many other cases. For example, Bennington and Bambach (1996) documented recurrent fossil assemblages through four megacyclothems encompassing several million years in the Pennsylvanian of West Virginia. Although assemblages differed more greatly between samples than within iterative sampling, they still showed very similar species lists in seemingly similar biofacies. Hence, two of the three major aspects of coordinated stasis were observed.

Bonelli and Patzkowsky (2008) documented a high degree of recurrence in Late Missisisippian assemblages from the North American Mid-continent, somewhat consistent with coordinated stasis. DiMichele and Philips (1996) likewise demonstrated recurrent patterns of Carboniferous plant morphotypes that also mimicked many aspects of coordinated stasis in marine communities.

Some aspects of the coordinated stasis pattern are borne out in particular instances and others not. For example, Holland and Patzkowsky (1996) documented abrupt and synchronous turnovers in Late Ordovician faunas of the Cincinnati arch region, but did not find persistence of *species* in intervals between these overturns at a level (60%) consistent with the original concurrent morphological stasis hypothesis. Likewise, Holterhoff (1996) demonstrated substantial recurrence of crinoid species in cyclothems, but also identified certain non-analog assemblages associated with transgressions that had no counterparts at other times.

In other cases, such as Stanton and Dodd's (1997) study of Cenozoic biotas in California, none of the aspects of coordinated stasis were observed; rather, more nearly continuous, asynchronous turnover of species was documented. Just why this and certain other examples show so much higher rates of speciation and extinction remains very unclear.

So, the answer to whether coordinated stasis is demonstrable in other settings is – yes, in some cases. Thus far, relatively few examples have demonstrated the degree of apparent stability in taxonomic composition and ecological properties of the Hamilton Group. It is important to inquire why it appears to apply in this situation and not so well in others. Possibly the Middle Devonian of the Appalachian Basin in New York represents an extreme of faunal stability. At present there are too few parallel studies to be certain. But, if so, it is not clear why this should be the case. The foreland basin was undergoing substantial changes in relation to the continuing Acadian Orogeny. Relative sea level fluctuations of tens of meters displaced facies belts by up to several hundred kilometers over thousands of years. At times the basin was anoxic to dysoxic over a very large area. Increases in sedimentation, associated with regressions, restricted sediment-sensitive organisms to a fraction of their distribution during transgressions. As noted above, profound biogeographic shifts took place during deposition of the lower Tully; much of the Hamilton biota was absent from all sampled portions of the basin. Yet, throughout all of these changes, a majority of lineages persisted and the morphological change was minimal; faunal gradients remained intact and biofacies showed basically similar species lists, diversities, and guild structure.

Alternatively, a lack of persistence of lineages and biofacies in other case studies may reflect incomplete geographic datasets. As noted above, it may only be possible to test for recurrence of faunas if a major portion of the basin is represented in the study area.

To what degree this bias has influenced other studies remains to be evaluated.

Causes and Implications of Coordinated Stasis

Causes of the pattern of coordinated stasis that have been suggested include tracking, incumbency, and ecological locking. Ecological locking (Morris et al. 1995), in which interactions among organisms make communities resistant to change over time, remains one possible mechanism, at least in some types of communities such as coral reefs. However, many arguments have been made against strong interactions in most marine benthic communities.

The issue of incumbency needs to be further considered. It seems likely that once large and/or widespread populations are established they may be difficult to displace by immigrant species of similar ecology. The notion shares some features with ecological locking. This would appear to be a more passive process than most direct competition for resources; the precise dynamics are difficult to square with increasing evidence that competition may not be a strong driving force in evolution. Nonetheless, in many cases incumbents do seem to have an advantage.

Habitat tracking has been emphasized as a mechanism for stasis, especially by Eldredge (2003) who has argued that evolution may be the last option for organisms. The phenomenon of habitat tracking has been considered in some detail by Brett et al. (2007b) and will be only briefly surveyed here. Migration of populations, largely via passive larval drift to follow shifting local environments, may be the primary response of most organisms to environmental change. Even though different species may track individualistically (cf. Gleason 1926; Jackson and Overpeck 2000), many marine organisms have sufficiently similar physical constraints in terms of depth, substrate, and other parameters that they may appear to more or less track as a unit. In other words, species distributions along environmental gradients, especially those related to water depth, may remain relatively stable, but shift spatially as the gradients themselves shift (Brett et al. 2007b). The ability of organisms to track favored environments may be one key to stasis. Tracking may be easier in some environments than others. For example, marine environments with broad shelves or ramps roughly parallel to climatic belts affording enough lateral space to allow migration under various conditions. Most species and associations can persist for prolonged periods provided environmental changes are not too drastic or rapid. Conversely, tracking may be more difficult in other environments that foster more mosaic habitats, or in which environmental gradients run at high angles to climatic belts (see Brett et al. 2007b for further discussion). Under such conditions evolutionary change and local extinction may be more common and coordinated stasis patterns will not be observed.

Perhaps, ironically, it is the near continuous disturbance requiring tracking that produced stasis. Sheldon (1996) has argued in the "plus ça change" model, that directional evolutionary change may be typical of undisturbed environments whereas organisms in disturbed environments may be resilient to morphological change (see also Vrba 1985). Lieberman et al. (2007) and Eldredge et al. (2005) suggest that episodic isolation, followed by renewed hybridization with other populations, disrupts local anagenetic trends, prevents directional change, and maintains effective stasis.

The notion of coordinated stasis has engendered much debate and many excellent studies, both corroborating and falsifying such a pattern in local systems. In this sense the science of evolutionary ecology has been advanced regardless of the relative frequency of "coordinated" patterns in nature. But perhaps the primary importance of the concept of "coordinated stasis", or whatever term is used for the phenomena of concurrent taxonomic and ecological stasis, is in emphasizing the possible stability of ecosystems and the fact that relatively little evolution may occur during vast intervals of geologic time. This, of course, places even more importance on the extinctions and turnovers that mark the ends of relatively stable EE subunits. It will also be important to further investigate the details of faunal turnovers to determine timing and factors that produce breakdown in local ecosystems. The many lesser mass extinctions that punctuate the geologic record may then come into focus as prime movers in the history of life.

Acknowledgments I am indebted to many people who have helped shape my views of evolutionary paleoecology, but I hasten to add that they should not be held accountable for my present ramblings. Gordon Baird has inspired a good deal of my thinking on patterns in the stratigraphic record; likewise, Linda Ivany has guided my thinking on evolutionary paleoecology, as have many former and present students, including Alex

Bartholomew, Heather Baugh, James Bonelli, Mike DeSantis, Forest Gahn, Austin Hendy, Pat McLaughlin, Jocelyn Sessa, and Patrick Wall.

I have benefited greatly from stimulating conversations with colleagues including Tom Baumiller, Niles Eldredge, Mike Foote, the late Stephen Jay Gould, Steve Holland, David Jablonski, Mike Kowalewski, Bruce Lieberman, Dave Meyer, Arnie Miller, Mark Patzkowsky, Dolf Seilacher, Steven M. Stanley, and many others. Arnie Miller reviewed an early draft of a similar paper and suggested many important changes; Linda Ivany reviewed the final draft and made further useful suggestions. I must also acknowledge the supportive atmosphere of the University of Cincinnati for nurturing my interests in paleobiology. My work has also been supported by grants from NSF and the Petroleum Research Fund of the American Chemical Society. I sincerely thank John Talent for encouraging me to finally pull together this retrospective and inviting me to contribute to this volume.

References

Aboussalam ZS (2003) Das «Taghanic-Event» im höheren Mittel-Devon von West-Europa und Marokko. Münstersche Forsch Geol Paläontol 97:1–332

Baird GC, Brett CE (2008) Givetian Taghanic bioevents in New York State: new discoveries and questions. Bull Geosci 83:357–370

Bartholomew AJ, Brett CE (2007) Revised correlations and sequence stratigraphy of the Middle Devonian (Givetian) in Ohio, USA and Ontario, Canada: implications for paleogeography, sedimentology and paleoecology. In: Becker RT, Kirchgasser WT (eds) Devonian events and correlations. Geological society of London special publication, vol 278. Geological Society of London, Bath, pp 105–131

Baumiller TK (1996) Exploring the pattern of coordinated stasis: simulations and extinction scenarios. Paleogeog Palaeoclimatol Palaeoecol 127:135–146

Bennett KD (1990) Milankovitch cycles and their effect on species in ecological and evolutionary time. Paleobiology 16:11–21

Bennett KD (1997) Evolution and ecology, the pace of life, 241 p. Cambridge University Press, Cambridge

Bennington JB, Bambach RK (1996) Statistical testing for paleocommunity recurrence: are similar fossil assemblages ever the same? Palaeogeogr Palaeoclimatol Palaeoecol 127:107–133

Bonelli JR Jr, Patzkowsky ME (2008) How are global patterns of faunal turnover expressed at regional scales? Evidence from the Upper Mississippian (Chesterian Series), Illinois Basin, USA. Palaios 23:760–772

Bonelli JR Jr, Brett CE, Miller AI, Bennington JB (2006) Testing for faunal stability across a regional biotic transition: quantifying stasis and variation among recurring biofacies in the Middle Devonian Appalachian Basin. Paleobiology 32:20–37

Bonuso N, Newton CR, Brower JC, Ivany LC (2002a) Does coordinated stasis yield taxonomic and ecologic stability? Middle Devonian Hamilton Group of central New York. Geology 30:1055–1058

Bonuso N, Newton CR, Brower JC, Ivany LC (2002b) Statistical testing of community patterns: Uppermost Hamilton Group, Middle Devonian (New York State: USA). Palaeogeogr Palaeoclimatol Palaeoecol 185:1–24

Boucot AJ (1975) Evolution and extinction rate controls, 427pp. Elsevier, Amsterdam

Boucot AJ (1990) Silurian and pre-Upper Devonian bio-events. In: Kauffman EG, Walliser OH (eds) Extinction events in earth history. Lecture notes in earth science, vol 3. Springer, pp 125–132

Boucot AJ (1996) Epilogue. Palaeogeogr Palaeoclimatol Palaeoecol 127:339–359

Brett CE (1995) Sequence stratigraphy, biostratigraphy, and taphonomy in shallow marine environments. Palaios 10:597–616

Brett CE, Baird GC (1995) Coordinated stasis and evolutionary ecology of Silurian to Middle Devonian faunas in the Appalachian Basin. In: Erwin DH, Anstey RL (eds) New approaches to speciation in the fossil record. Columbia University, New York, pp 285–315

Brett CE, Miller KB, Baird GC (1990) A temporal hierarchy of paleoecologic processes within a Middle Devonian epeiric sea. In: Miller W III (ed) Paleocommunity temporal dynamics: the long-term development of multispecies assemblages. Paleontological Society Special Publications 5:178–209. The Paleontological Society, Boulder, CO.

Brett CE, Ivany LC, Schopf KM (1996) Coordinated stasis: an overview. Palaeogeogr Palaeoclimatol Palaeoecol 127:1–20

Brett CE, Bartholomew AJ, Baird GC (2007a) Biofacies recurrence in the Middle Devonian of New York State: an example with implications for habitat tracking. Palaios 22:306–324

Brett CE, Hendy A, Bartholomew AJ, Bonelli, JR, McLaughlin PI (2007b) Response of shallow marine biotas to sea level fluctuations: faunal replacement and the process of habitat tracking. Palaios 22:228–244

Brett CE, Ivany LC, Bartholomew AJ, DeSantis MK, Baird GC (2009) Devonian ecological-evolutionary subunits in the Appalachian Basin: a revision and a test of persistence and discreteness. In: Königshof P (ed) Devonian change: case studies in palaeogeography and palaeoecology. Geological Society of London Special Publication, vol 314. Geological Society of London, Bath, pp 7–36

Brett CE, Baird GC, Bartholomew AJ, DeSantis' AJ, Ver Straeten CA (2011) Sequence stratigraphy and a revised sea-level curve for the Middle Devonian of eastern North America. Palaeogeogr Palaeoclimatol Palaeoecol 304:21–53

Bush AM, Markey J, Marshall CR (2004) Removing bias from diversity curves: the effects of spatially organized biodiversity on sampling-standardization. Paleobiology 30:666–686

Buzas MA, Culver SJ (1994) Species pool and dynamics of marine paleocommunities. Science 264:1439–1441

Buzas MA, Culver SJ (1998) Assembly, disassembly, and balance in marine paleocommunities. Palaios 13:263–275

DeSantis MK, Brett CE (2007) Persistent depositional sequences and bioevents in the Eifelian (early Middle Devonian) of eastern Laurentia: Kačak events? In: Becker RT, Kirchgasser WT (eds) Devonian events and correlations. Geological Society of London Special Publication, vol 278. Geological Society of London, Bath, pp 83–104

DiMichele WA (1994) Ecological patterns in time and space. Paleobiology 20:89–92

DiMichele WA, Philips TL (1996) Clades, ecological amplitudes, and ecomorphs: phylogenetic effects and persistence of primitive plant communities in the Pennsylvanian-age tropical wetlands. Palaeogeogr Palaeoclimatol Palaeoecol 127:83–105

DiMichele WA, Behrensmeyer AK, Olszewski TD, Labandeira CC, Pandolfi JM, Bobe R (2004) Long-term stasis in ecological assemblages: evidence from the fossil record. Annu Rev Ecol Evol 35:285–322

Eldredge N (2003) The sloshing bucket: how the physical realm controls evolution. In: Crutchfield JP, Schuster P (eds) Evolutionary dynamics: exploring the interplay of selection, accident, neutrality, and function. Oxford University Press, Oxford, pp 3–32

Eldredge N, Gould SJ (1972) Punctuated equilibria: an alternative to phyletic gradualism. In: Schopf TJM (ed) Models in paleobiology. Freeman, Cooper and Co, San Francisco, CA, pp 82–115

Eldredge N, Thompson JN, Brakefield PM, Gavrilets S, Jablonski D, Jackson JBC, Lenski RE, Lieberman BS, McPeek MA, Miller W III (2005) The dynamics of evolutionary stasis. Paleobiology 31:133–145

Foote M (2003) Origination and extinction through the Phanerozoic: a new approach. J Geol 111:125–148

Gardiner L (2001) Stability of late Pleistocene reef mollusks from San Salvador Island, Bahamas. Palaios 14:372–386

Gleason HA (1926) The individualist concept of the plant association. Bull Torrey Bot Club 53:7–26

Holland SM (1995) The stratigraphic distribution of fossils. Paleobiology 21:92–109

Holland SM (1996) Recognizing artifactually generated coordinated stasis: implications of numerical models and strategies for field tests. Palaeogeogr Palaeoclimatol Palaeoecol 127:147–156

Holland SM (2000) The quality of the fossil record; a sequence stratigraphic perspective. Paleobiology 26:148–168

Holland SM, Patzkowsky ME (2004) Ecosystem structure and stability: middle Upper Ordovician of central Kentucky, USA. Palaios 19:316–331

Holland SM, Patzkowsky ME (2007) Gradient ecology of a biotic invasion: biofacies of the type Cincinnatian Series (Upper Ordovician), Cincinnati, Ohio region, USA. Palaios 22:392–407

Holterhoff PF (1996) Crinoid biofacies in Upper Carboniferous cyclothems, midcontinent North America: faunal tracking and the role of regional processes in biofacies recurrence. Palaeogeogr Palaeoclimatol Palaeoecol 127:47–81

Hunt G (2008) Gradual or pulsed evolution: when should punctuational explanations be preferred? Paleobiology 34:360–377

Huynh T, Sageman B, Kauffman EG (1999) Test of coordinated stasis in the faunal record of Upper Cretaceous strata, Western Interior US. Geol Soc Am Abs Prog 31:A439

Ivany LC (1996) Coordinated stasis or coordinated turnover? Intrinsic versus extrinsic mechanisms for generating pattern. Palaeogeogr Palaeoclimatol Palaeoecol 127:239–256

Ivany LC (1999) So...now what? Thoughts and ruminations about coordinated stasis. Palaios 14:297–300

Ivany L, Brett CE, Baugh HL, Wall P (2009) Coordinated stasis revisited: taxonomic and ecologic stability in the Devonian of New York. Paleobiology 35:499–524

Jackson JBC (1994) Community unity? Science 264:1412–1413

Jackson ST, Overpeck JT (2000) Responses of plant populations and communities to environmental changes of the late Quaternary. In: Erwin DH, Wing SL (eds) Deep time, paleobiology's perspective. Paleobiology 26(4 supp):194–220

Lieberman BS (1994) Evolution of the trilobite subfamily Proetinae and the origin, evolutionary affinity, and extinction of the Middle Devonian proetid fauna of Eastern North America. Bull Am Mus Nat Hist 223:176

Lieberman BS, Brett CE, Eldredge N (1995) A study of stasis and change in two species lineages from the Middle Devonian of New York State. Paleobiology 21:15–27

Lieberman BS, Kloc GM (1997) Evolutionary and biogeographic patterns in the Asteropyginae (Trilobita, Devonian). Bull Am Mus Nat Hist 232:1–127

Lieberman BS, Miller WM III, Eldredge N (2007) Paleontological patterns, macroecological dynamics and the evolutionary process. Evol Biol 34:28–48

Marshall CR (1997) Confidence intervals on stratigraphic ranges with nonrandom distributions of fossil horizons. Paleobiology 23:165–173

McKinney ML, Lockwood JL, Frederick DR (1996) Does ecosystem and evolutionary stability include rare species? Palaeogeogr Palaeoclimatol Palaeoecol 127:191–207

Miller W III (1996) Ecology of coordinated stasis. Palaeogeogr Palaeoclimatol Palaeoecol 127:177–190

Miller AI (1997) Coordinated stasis or coincident relative stability? Paleobiology 23:155–164

Morris PJ, Ivany LC, Schopf KM, Brett CE (1995) The challenge of paleoecological stasis: reassessing sources of evolutionary stability. Proc Natl Acad Sci USA 92:11269–11273

Olszewski TD, Patzkowsky ME (2001) Measuring recurrence of marine biotic gradients: a case study from the Pennsylvanian-Permian mid-continent. Palaios 16:444–460

Palmer AR (1984) The biomere problem: evolution of an idea. J Paleontol 58:599–611

Pandolfi JM (1996) Limited membership in Pleistocene reef coral assemblages from the Huon Peninsula, Papua, New Guinea: constancy during global change. Paleobiology 22:152–176

Pandolfi JM (2002) Coral community dynamics at multiple scales. Coral Reefs 21:13–23

Pandolfi JM, Jackson JBC (2006) Ecological persistence interrupted in Caribbean coral reefs. Ecol Lett 9:818–826

Patzkowsky ME, Holland SM (1997) Diversity dynamics in Middle and Upper Ordovician articulate brachiopods: implications for long-term stability of marine ecosystems. Paleobiology 23:420–443

Patzkowsky ME, Holland SM (1999) Biofacies replacement in a sequence stratigraphic framework: Middle and Upper Ordovician of the Nashville Dome, Tennesee, USA. Palaios 14:301–323

Roy K, Wagner PJ (1995) Communities in the fossil record: coordination or coincidence? Am Paleontol 3:3–4

Scarponi D, Kowalewski M (2004) Stratigraphic paleoecology: bathymetric signatures and sequence overprint of mollusk associations from late Quaternary sequences of the Po Plain, Italy. Geology 32:989–992

Sheldon PR (1996) Plus ça change—a model for stasis and evolution in different environments. Palaeogeogr Palaeoclimatol Palaeoecol 127:209–227

Signor PW III, Lipps JH (1982) Sampling bias, gradual extinction patterns, and catastrophes in the fossil record. In: Silver LT, Schultz PH (eds) Geological implications of impacts of large asteroids and comets on the earth. Geological Society of London Special Publication, vol 190. Geological Society of London, Bath, pp 291–296

Stanton RJ, Dodd JR (1997) Lack of stasis in late Cenozoic marine faunas and communities, central California. Lethaia 30:239–256

Tang CM, Bottjer DJ (1996) Long-term faunal stasis without evolutionary coordination: Jurassic benthic marine paleocommunities, Western Interior, United States. Geology 24:815–818

Vrba ES (1985) Environment and evolution: alternative causes of the temporal distribution of evolutionary events. S Afr J Sci 81:229–236

Wall PM, Ivany LM, Brett CE (in press) The influence of geographic sampling scale and eurytopy on apparent taxonomic turnover rates: a case study in the Hamilton Group (Middle Devonian) of New York. Paleobiology

Walliser OH (1990) How to define "global bioevents". In: Kauffman EG, Walliser OH (eds) Extinction events in earth history. Lecture notes in earth sciences, vol 3. Springer-Verlag, pp 1–3

Westrop SR (1996) Temporal persistence and stability of Cambrian biofacies: Sunwaptan (Upper Cambrian) trilobite faunas of North America. Palaeogeogr Palaeoclimatol Palaeoecol 127:33–46

Whilst This Planet Has Gone Cycling On: What Role for Periodic Astronomical Phenomena in Large-Scale Patterns in the History of Life?

Bruce S. Lieberman and Adrian L. Melott

Abstract

One of the longstanding debates in the history of paleontology focuses on the issue of whether or not there have been long-term cycles (operating over tens of millions of years) in biodiversity and extinction. Here we consider the history of this debate by connecting the skein from Grabau up to 2008. We focus on the evidence for periodicity that has emerged thus far, and conclude that there is indeed some evidence that periodicity may be real, though of course more work is needed. We also comment on possible causal mechanisms, focusing especially on the motion of our solar system in our galaxy. Moreover, we consider the reasons why some scientists have opposed periodicity over the years. Finally, we consider the significance of this for our understanding of evolution and the history of life.

Keywords

Cyclical astronomical phenomena • Solar system • Cosmic ray flux • Biotic versus abiotic forces • Periodicities • Biodiversity • Macroevolutionary patterns • Extinctions • Causal mechanisms

Introduction

Charles Darwin deserves a great deal of the credit for convincing people that there are natural processes continually and uniformly acting to shape biological evolution. One of the eloquent ways he used to try to sway people was by invoking the analogy of how gravity continually acts to maintain the structure of the solar system: hence the genesis of the title of this chapter, which incorporates a fragment from the last sentence of his epochal "On the Origin of Species" (Darwin 1859, p. 490). Of course, scientists have long recognized the clockwork nature of the nearby confines of our solar system; further, since the middle of the twentieth century, the bulk of the biological community has accepted the primary role of natural selection as the shaper of evolution. Thus, Darwin's analogy between astronomy and biology has been universally accepted; however, documenting a causal link between cyclical astronomical phenomena and periodic evolution and extinction has proven far more tenuous and controversial.

B.S. Lieberman (✉)
Department of Geology, and Natural History Museum/Biodiversity Research Center, University of Kansas, Lawrence, KS 66045, USA
e-mail: blieber@ku.edu

J.A. Talent (ed.), *Earth and Life*, International Year of Planet Earth,
DOI 10.1007/978-90-481-3428-1_3, © Springer Science+Business Media B.V. 2012

Here we try to forge this causal link and consider the question of how large-scale, cyclical astrophysical processes may have influenced the history of life at the grand scale, defined here (perhaps arbitrarily) as over the course of tens of millions of years. Such a thesis no longer seems as farfetched as it once did, now that paleontological studies have shown that astronomical objects, such as bolides (possibly at the end of the Cretaceous, Alvarez et al. 1980), and astrophysical phenomena, such as gamma ray bursts (at the end of the Ordovician, Melott et al. 2004), may have precipitated mass extinctions (though not necessarily repeatedly and cyclically). Events such as bolide impacts at the end of the Cretaceous powerfully influence the history of life. Gould (1989) referred to as contingency because they eliminate diverse groups effectively at random and the survivors set the course of subsequent evolution. We concur with Cloud (1948, p. 322) that it is "desirable that workers in several fields of biological science periodically review the principles that they have come to accept and lines of thought that derive from or lead to acceptance of these principles." The issue of periodicity may be just such a case where it would be desirable to review such principles. Maybe it really is time to look to the stars to gain deeper insight into the history of life.

With the role of astronomical phenomena in general, and cyclical astronomical phenomena in particular, is a debate about the role the physical environment plays in influencing the history of life. This is best viewed in the context of a long-standing discourse about the relative roles of biotic versus abiotic forces in evolution: a debate whose focus is beyond the scope of our chapter but that goes back at least to Darwin (1859) and probably to Von Humboldt (1816), de Candolle (1820), and maybe even Buffon (1749–1804). For specific aspects of the discussion from a paleobiological perspective, see Eldredge and Cracraft (1980); Gould and Calloway (1980); Vrba (1980); Ross and Allmon (1990); Vermeij (1993); Benton (1996); Lieberman (2000) and references therein for a more detailed treatment. Although this debate is beyond the scope of this chapter, we would argue that most scientists have moved beyond the view championed by Darwin (1859) and more recently by Dawkins (1976) that much of evolution is caused by competition among organisms. Instead, we would argue that there is ample evidence that, at the grand scale, the history of life and the physical environment, construed broadly to subsume both climate and geology, has powerfully structured large-scale patterns of evolution and extinction. For example, there is Valentine and Moore's (1970) study showing a relationship between the geometry of the Earth's plates, related to plate tectonic changes, and overall diversity in the fossil record; there are more recent studies by Cornette et al. (2002) and Rothman (2001) showing an association between macroevolutionary origination rate and carbon dioxide levels; finally, there is evidence that animal evolution and extinction have been tracking environmental parameters in largely a random walk fashion throughout much of the Phanerozoic (Cornette and Lieberman 2004).

Cyclical Astronomical Phenomena Influencing Biology and Evolution: From Weeks to Months to Millions of Years

In the search for evidence that the physical environment plays an important role in structuring evolution, scientists have long recognized the context of our planet in the broader solar system; that is, although not necessarily explicitly recognized, there has been a nascent appreciation for the role astronomical (or astrophysical) phenomena exert upon biology and evolution. For instance, there are diurnal and seasonal cycles (obviously) controlled by the Earth's rotation and revolution and these cause everything from behavioral changes in organisms to migration to selection and thus evolution (Huntley and Webb 1989 provide an outstanding analysis and documentation of these phenomena at several different hierarchical levels). At even longer, decadal time scales, there are sunspot cycles that affect the Earth's climate and these may drive certain selection mediated shifts in morphology over similar time scales, for example, the classic studies by Grant and Grant (2002) on the Galapagos Finches.

Even as the relative time scale grows, astrophysical phenomena remain important influences on extinction and evolution. For example, the signatures of Milankovitch climatic cycles operating on times scales of tens to hundreds of thousands of years are amply evident in the geological record (e.g., Hays et al. 1976; Anderson et al. 1984; Olsen 1986; etc.). Moreover, these can influence patterns of stability and change within species lineages, causing stasis, speciation, and extinction (Vrba 1985, 1992; Bennett 1990, 1997). This was lucidly developed by Elisabeth Vrba in her Turnover Pulse Hypothesis.

In a fascinating recent result, Van Dam et al. (2006) even uncovered evidence that cycles of evolution and extinction acted on even longer timescales, roughly 2.5 million years, and these were in turn driven by astronomical phenomena. Vrba (1992) had in fact previously argued for the potential existence of such a cycle, but the evidence was somewhat ambiguous. Certainly more study is needed, but this opens a potentially interesting window into astronomically governed cyclical phenomena influencing biology and geology that operate on scales of millions of years. All these different astronomical phenomena operate at several temporal levels; the larger the time scale the larger the hierarchical level of the genealogical and economic biological entities affected (Eldredge 1999; Lieberman 2000). In conclusion, it seems that from the range of days to months to tens of thousands of years, and possibly even to millions of years, there is enough evidence to posit the stamp of cyclical astronomical phenomena on biology. Thus, it seems not wholly beyond the pale to try to uncover evidence for periodic astronomical phenomena acting on even longer time scales that influence biology. One of the best ways to do this is to try to identify the existence of large scale cycles in the history of life: cycles in diversity or extinction or origination.

History of Thought on Long-Term Cycles of Evolution and Extinction

There is a rather long and extensive history of scientists seeking to invoke large-scale (temporal) biological periodicity; aspects of this history are worth exploring here. Moreover, the debate about the existence of periodicity in the history of life is itself periodic; it waxes into and out of favor over a timescale of roughly every 20–30 years (note, not adduced statistically) with "pulses" of watershed publications occurring in the 1930s, the early 1950s, the late 1970s to the mid-1980s, and the 2000s. We provide a review of the literature on large-scale biotic periodicity up through Rohde and Muller's (2005) study, with special emphasis on the earliest studies advocating periodicity; in the next section we then provide a discussion of some of the general criticisms of these studies. Finally, we discuss in greater detail some of the most recent analyses of large-scale biotic periodicity (Lieberman and Melott 2007; Melott 2008) and present possible mechanisms that may explain it.

Although the existence of large-scale biological cycles do not in and of themselves guarantee the existence of periodic, causal astronomical phenomena (as we shall see, scientists have posited a range of other mechanisms from geology to competition to climate), they may help sway the debate one way or the other, especially when the timescale of the astronomical mechanism matches the inferred evolutionary periodicity.

Amadeus Grabau

The paleontologist A. W. Grabau may have been the first scientist to seriously attempt to document the existence of this type of periodicity. This was specifically developed in his "pulsation theory," first presented at the International Geological Congress in 1933 and then developed in detail (Grabau 1934, 1936). Grabau posited that major transgressions and regressions, synchronous across all of the Earth's continents, were connected with swelling and contraction of the seafloor. He further argued that these pulsations would have biological effects, regression narrowing the available area for existence (in the case of marine faunas) and causing a "struggle for existence of ever increasing intensity. This led to the extinction of the weaker and less adaptable, and to modification of the survivors" (Grabau 1936, p. vii). Eventually, when sea-level rises again, "with renewed expansion, the survivors gave birth to an essentially new fauna which increased in number and variety with the expansion of its habitat" (Grabau 1936, p. vii).

The result will be fossil faunas that exist as distinct packages, stacked upon one another, in the geological record: "The succeeding transgressive series... contains a fauna which, though it shows certain relationships to that of the preceding retreatal (regressive) series, is on the whole markedly distinct from it, and still more distantly related to the fauna of the transgressive series of the preceding pulsation" (Grabau 1936, p. vii; also see Grabau 1936, pp. 4–6). Grabau (1934, 1936) went so far as to comment on the duration of these transgressions/regressions, which he called a "pulse-beat": the "duration of such a pulse-beat is between 20–30 million years" (Grabau 1936, p. 1), an interesting number for several reasons that we shall return to. Further, he was prescient enough to recognize that these pulsations would affect both terrestrial and marine faunas, albeit in different ways. In particular, transgressions would benefit marine faunas because they expanded the area available to live, yet be to the detriment of terrestrial faunas because they shrank the area available to live. These ideas were discussed in greater detail, along with additional theories, in Grabau (1940).

Relationship Between Grabau's Pulsation Theory and the Turnover Pulse Hypothesis

In important respects Grabau's pulsation theory matches aspects of Vrba's (1985, 1992) Turnover Pulse hypothesis. In particular, each invokes a set of environmental changes that cause deterioration in the environment as a whole, leading to diminishing geographic-range size for many taxa; as geographic-range sizes decrease, eventually extinction and speciation of the ever more narrowly circumscribed taxa occur; finally, as environmental conditions ameliorate, the taxa expand outwards. There is, however, a difference in the mechanisms precipitating speciation in the pulsation theory and the Turnover Pulse. In the former, it is competition, following the physical environmental changes, that is the driver: the increasing struggle for existence in an ever narrower geographic region. By contrast, in Vrba's Turnover Pulse hypothesis it is allopatric speciation that is the driver. Grabau is arguing, moreover, for significant evolution even as the available area expands again because he claimed that as sea-level rises there is less struggle for existence (in marine faunas), favoring the persistence of less adaptive types (Grabau 1936, pp. 4–5) and thus greater diversity. Again, by contrast, with Turnover Pulse, there should be little if any diversification once climatic conditions improve and species expand their geographic range because the conditions are no longer ripe for allopatric differentiation. Another difference between their ideas is that the transgressions and regressions Grabau was referring to involved significantly more time than the turnovers Vrba was focusing on: these were related to interglacial/glacial climatic cycles and would involve less than one tenth the time of a pulse/beat. Still, there is clearly a connection.

Relationship Between Grabau's Pulsation Theory and Coordinated Stasis

Indeed, because of this connection, Grabau's pulsation theory also shows similarities to Brett and Baird's (1995) coordinated stasis. Interestingly, coordinated stasis was developed in the region where Grabau made his early paleontological discoveries: central and western New York State. Grabau attributed his inferred pattern to competition—the same mechanism Morris et al. (1995) posited to explain the pattern of coordinated stasis: ecological locking.

Newell and Simpson on Biological Periodicity

Norman Newell's ideas on periodicity were first presented at a Geological Society of America meeting (Newell 1949) and elaborated on in print in greater detail subsequently (Newell 1952). He (Newell 1949, 1952) identified peaks and troughs of evolution related to both what he called biological factors (read competition), and also physical factors (Newell 1949, pp. 1911, 1912; Newell 1952, p. 385). The latter would especially involve the expansion and restriction of shallow epicontinental seaways due to the rise and fall of sea-level. Newell's former advisor, R. C. Moore, also sided with him, arguing that the crinoid fossil record suggested that there was "a spasmodic, pulsatory increase and decrease in the census of species (which) reflect(s) real variations in evolution rate between wide extremes" (Moore 1952, p. 352). Further, Moore (1952) also held that these variations in increase and decrease were likely related to the rise and fall, respectively, of sea level.

However, a subtle to significant difference emerges between Newell's thoughts on this and Grabau's, for Grabau was specifically arguing that it was changes in the physical environment that increased competition. Newell was more equivocal on the cause. At first, he posited (Newell 1949, p. 1912) that "several groups show similarity of pattern, suggesting common control," perhaps specifically implicating changes in sea-level, but he also argued that although changes in sea-level might cause extinction (through sea-level fall eliminating habitat of marine organisms) they would not lead to an increase in the rate of evolution (Newell 1949, p. 1912). Moreover, whatever initial support Newell (1949) may have had for the control of the physical environment on biological periodicity was trimmed back in time; he eventually argued that changes in the physical environment will typically not directly affect evolution and, further, throughout most of their history, faunas were at adaptational equilibrium (Newell 1952, pp. 383, 385). Although it cannot be decisively shown, Newell's changed outlook may have been due to the influence of George Gaylord Simpson, one of Newell's colleagues at the American Museum of Natural History (Niles Eldredge, 2008, "personal communication"), one of the founders of the Neo-Darwinian synthesis (Eldredge 1985).

Simpson, certainly by the late 1940s and early 1950s, completely supported the gradualistic, Neo-Darwinian view; this made him ill-disposed to Grabau's ideas on major peaks and troughs of origination and extinction driven by changes in the physical environment. Moreover, his towering intellect and gruff nature made him adept at influencing those surrounding him.

Simpson's viewpoints on this topic were best crystallized in a 1952 publication in the Journal of Paleontology derived from a symposium on the "Distribution of Evolutionary Explosions in Geologic Time". Simpson (1952) did not deny the existence of periodicity. He argued, for instance, that "(t)here is a well-marked periodicity in vertebrate history, especially as regards successive peaks and valleys of high and low proliferation of new groups" (Simpson 1952, p. 370). However, he also suggested that "it seems impossible even to make a start at realistic correlation of so regular and continuous a process" (Simpson 1952, p. 365), i.e., evolution, with abiotic factors, especially geological changes. Further, he noted that although physical processes likely influenced evolution, they were not decisive (Simpson 1952, p. 369). Instead, biotic factors such as competition were likely to be responsible.

Fischer, Raup and Sepkoski, and Rohde and Muller (2005) on Periodicity

Fischer and Arthur's (1977) and Fischer's (1982) studies are significant because they were among the first to revive the debate about biological periodicity and, further, they suggested an environmental causal connection. In particular, Fischer and Arthur (1977) and Fischer (1982) found support for Grabau's ideas and identified a prominent 32-million-year cycle. (Note that this value is remarkably close to the approximate duration of such cycles inferred by Grabau 1936.) Following on the heels of Fischer and Arthur (1977) and Fischer (1982) there are, of course, the studies by Raup and Sepkoski (1984, 1986) and Sepkoski and Raup (1986) (also see Rampino and Stothers 1984) that identified a roughly 26-million-year periodicity: a cyclicity that has been attributed to periodic episodes of bolide or comet impact triggered by the motion of a brown dwarf star in our solar system's vicinity (see Raup 1986). (Stanley [1990] presented the interesting perspective that Raup and Sepkoski's retrieved periodicity may have arisen due to the fact that recovery from major extinction events often requires a significant amount of time, either because the hostile conditions that initially caused the extinction continued for some time, or the surviving fauna tended to diversify at a low rate: for instance, generalists are more apt to survive an extinction event because of their lower rate of extinction, but tend to also have depressed rates of speciation.) Raup and Sepkoski's (1984, 1986) studies were significant for many reasons, but among the most important was that they were the first to use detailed statistical methods, and a comprehensive database, to attempt study of this topic.

The next prominent study supporting periodicity was that of Rohde and Muller (2005). Their study utilized a revised and updated geological time scale not available to Raup and Sepkoski (1984, 1986) and also incorporated some improved taxonomic data in the "Sepkoski dataset" (Sepkoski 2002). They failed to find evidence for a cycle at 26 (or 32) million years but did find strong support for a cycle in biotic diversity operating every 62 million years. Rohde and Muller (2005) left open the question of causal mechanisms for such a long duration cycle, though Rohde (2006) subsequently suggested the causes were connected with geological phenomena.

The Problem of Uniformitarianism and Completeness of the Fossil Record

It is worth noting that just as there has been a long history of identifying evidence for periodicity in the history of life there has been a long history of challenging that very evidence. This typically happens anytime a novel theory or idea is developed and therefore is not that surprising. Some of the motivation for scientists' opposition to periodicity seems to have been, in part, a matter of metaphysics and philosophy. A clear example of this comes from the already-mentioned symposium on the "Distribution of Evolutionary Explosions in Geologic Time" in the Journal of Paleontology already mentioned. Although, in that symposium, Newell (1952) and Simpson (1952) endorsed aspects of periodicity (but not Grabau's causal mechanism), the general tenor of the symposium was largely a rejection of periodicity; Cooper and Williams' (1952) paper is typical in this regard. Some of the authors, however, completely rejected the notion that there might be times of explosive evolution or extinction connected with periodicity (e.g., Camp 1952). This rejection was based on the belief that variations in rates of evolution and extinction, and therefore periodicity, violated the basic tenets of aspects of uniformitarianism (see Gould 1965). This was related to the basic Neo-Darwinian presumption, prevalent at that time, that evolutionary change (and extinction) must always be slow and gradual. Consider that Simpson (1952), one of the architects of the Neo-Darwinian synthesis, and by then a committed gradualist (Eldredge 1985), argued for the continuous rather than pulsed nature of evolutionary origins and also downplayed the existence of mass extinctions in the fossil record.

In some respects, we see parallels between how Grabau's works were treated by many paleontologists in the 1950s and how Goldschmidt's works were treated by specific evolutionary biologists (including the paleontologist George Simpson) in the 1940s. Goldschmidt was the bête noire of the Neo-Darwinian synthesis because his idea of "hopeful monsters" (Goldschmidt 1940) was at odds with certain population-genetic principles. Certainly scientists opposed aspects of Goldschmidt's work for legitimate reasons; there were several problems with his reasoning, but what really got many scientists' goad was that, to Goldschmidt, evolution could not be explained by the simple extrapolation of microevolution to macroevolution: Goldschmidt's view of evolution was not sufficiently "uniformitarian" (Gould *in* Goldschmidt 1982). Similarly, at times Grabau and his "pulsation theory" seems to have become a punching bag because he was espousing ideas that invoked significant variations in rates of evolution and extinction due to variations in the physical environment. Yes, Grabau's pulsation theory may have been outrunning the available evidence (to paraphrase Henbest 1952, p. 318); part of the reason for his theory to be substantially criticized was that some of his ideas were anathema to Neo-Darwinian uniformitarianism.

Other criticisms of periodicity were data and analysis driven, for example, the criticism of Raup and Sepkoski's analyses by Patterson and Smith (1987). Patterson and Smith (1987) legitimately challenged aspects of Raup and Sepkoski's dataset for they discovered that Raup and Sepkoski's extensive (though inadvertent) inclusion of paraphyletic taxa may have clouded their results. Other critiques have surfaced based on the types of methods used to adduce periodicity (see discussion in Sepkoski 1989 and also in Lieberman and Melott 2007, the latter citation discussed more fully below).

Another frequently offered reason for rejecting the existence of periodicity was that there were too many problems with the fossil record (it is too biased or too poor) or that the paleontologic data (i.e., data relating to taxonomic diversity, and biostratigraphic and chronostratigraphic correlations) were so poor that evidence for periodicity based on these data could not be trusted (e.g., Henbest 1952; Camp 1952; Cooper and Williams 1952 in reference to Grabau's 1936 conclusions; and Stanley 1990 in reference to Raup and Sepkoski's 1984, 1986 conclusions). Indeed, from these references, it is clear that this type of criticism of periodicity, that the fossil record (or paleontological data) is (are) compromised, has repeatedly resurfaced. The fact that the fossil record is not perfect has been raised since at least the time of Darwin (1859); it surely represents a legitimate card to play to explain our inability to see something in the fossil record (i.e., perhaps periodicity could arise as an artifact of the geological record). From Raup and Sepkoski's (1984) to Rohde and Muller's (2005), the data utilized were various permutations of the so-called Sepkoski dataset (most recently collected in Sepkoski 2002). Discussion of the validity of the Sepkoski dataset is beyond the scope of this chapter. There have been several studies that have criticized the dataset and questioned aspects of its validity. Further, there have been various studies aimed at addressing whether or not the Sepkoski dataset reflects true variations in diversity through time or rather variations in way the geological record has been compiled (e.g., Adrain and Westrop 2000; Alroy et al. 2001; Peters 2005; Foote 2006; see also discussion in Bambach 2006). Based on these issues, several new taxonomic databases have been developed, the most important being the Paleobiology Database (PBDB); this we will discuss later in this chapter. We argue that uncovering either evidence for periodicity in true diversity or in preserved diversity relating to artifacts of preservation of the geological record would be interesting, albeit for different reasons. The former would imply there was some periodic mechanism controlling the diversity of life on our planet; the latter, that there was some periodic mechanism controlling the geological processes governing the fossil record.

The assertion about the poor quality of paleontologic data in general reduces, in part, to the question of what evolutionary principles the fossil record can be used to study (a topic beyond the scope of this contribution; however, we certainly would argue that the fossil record is a valuable repository, containing much information useful for evolutionary studies and, moreover, is our one true chronicle of the history of life). Raup and Sepkoski (1986, p. 833) argued that if the data were poor the result should degrade toward randomness. In particular, they claimed that uncertainty in databases of paleontologic diversity would be unlikely to lead to periodicity. This seemed reasonable, but the result of Patterson and Smith (1987) provides a cautionary tale.

Re-analyzing Evidence for 62 and 26 Million-Year Periodicity

Summary of Lieberman and Melott's (2007) Analysis

Partly motivated by our desire to assess the resiliency of Rohde and Muller's (2005) evidence for periodicity (and earlier analyses of periodicity), we (Lieberman and Melott 2007) decided to consider their study in greater detail. In particular, we were interested in determining if their results changed when additional and supplementary statistical tests were applied. First, the data from the Sepkoski (2002) dataset (the dataset used by Rohde and Muller 2005) were detrended; this is essential in any analysis looking for periodicity when there is an overlying trend (in this case diversity is increasing through time); details on detrending are provided in Lieberman and Melott (2007). (Note, Rohde and Muller 2005; Cornette 2007 also employed detrending.) Rohde and Muller (2005) used Fourier Spectral Analysis (Brigham 1988) to uncover evidence for periodicity in fossil biodiversity. Lieberman and Melott (2007), by contrast, used the Lomb-Scargle Fourier Transform (Scargle 1982; Laguna et al. 1998); it is more effective at analyzing time-series data where the data points are not evenly spaced. (In the case of fossil diversity data, the samples are not evenly spaced in time.) (More details on the analysis are provided in Lieberman and Melott 2007.) Our results built on the analysis of Cornette (2007) who used the equivalent Gauss-Vanìcek method to re-analyze Rohde and Muller (2005). It is worth mentioning that any (non pathologic) function can be decomposed into a sum of sine waves. The interesting question is whether any stand out above the others, with much higher amplitude. Fourier Analysis is a highly standardized way to approach this, but it requires evenly spaced data. Irregularly spaced data of the fossil record can be used, but interpolation is needed. This can introduce artifacts if not understood by the practitioner. Lomb-Scargle/Gauss-Vanìcek is based on doing a least-squares fit to sinusoids, and the data do not have to be evenly spaced so no interpolation is necessary. The safest thing to do is to use both methods, as did Lieberman and Melott (2007). It turns out that Rohde and Muller (Muller, 2008, "personal communication") also did so but did not report this.

Rohde and Muller (2005) analyzed only biodiversity. Lieberman and Melott (2007) also analyzed Phanerozoic (since 519 Ma) biodiversity but, in addition, considered fractional biodiversity (which considers the relative amount of a diversity change in a given time interval and thus may be more biologically significant). Further, we investigated whether various individual mass extinction events were contributing disproportionately to the evidence for cyclicity. Raup and Sepkoski (1986) and Cornette and Lieberman (2004) indicated this was a possibility. We also investigated whether cyclicity was primarily implicit in the pre-150 Ma or post-150 Ma parts of the data because this appears to mark an important transitional period in the history of life (Bambach 2006). Finally, we investigated origination and extinction (and various metrics of these); Rohde and Muller (2005) also considered these, though in a somewhat different manner (see Lieberman and Melott 2007 for further discussion).

Summary of Lieberman and Melott's (2007) Results

Even using the different analytical method, Lieberman and Melott (2007) continued to find strong evidence for periodicity in biodiversity (significant at the 0.01 level) and also in fractional biodiversity at roughly 62 ± 3 Ma (Fig. 1); the latter is significant at the 0.001 level, matching Rohde and Muller's (2005) identified peak but at a higher level of significance; there is also a peak at roughly 31 ± 1 Ma; it closely matches Fischer and Arthur's (1977) and Fischer's (1982) peak (and thus also near Grabau's (1936) but more informally defined pulse), though only significant at the 0.1 level and thus not treated as statistically significant. This peak may merit further study. Even removing the Ordovician–Silurian, Permian–Triassic, and Cretaceous–Cenozoic mass extinctions from the biodiversity data did not eliminate the strong 62 Ma peak.

Periodicity in biodiversity (and fractional biodiversity) seems to be primarily confined to the time interval 519–150 Ma; post 150 Ma data show no evidence for periodicity (results not shown), but prior to 150 Ma there is a cycle at roughly 61 ± 3 Ma, very close to Rohde and Muller's (2005) 62 Ma cycle, significant at the 0.001 level, and also a cycle at roughly 32 ± 1 Ma, very close to Fischer and Arthur's (1977) and Fischer's (1982) cycle (and thus also in the window of Grabau's 1936 cycle) significant at the 0.05 level (Fig. 2).

Results from analysis of certain metrics of origination reveal significant (at the 0.01 level) cycles at roughly 60 ± 3 Ma and 24 ± 0.5 Ma, the former close to the Rohde and Muller (2005) cycle; the latter close to Raup and Sepkoski's (1984, 1986) cycle (and thus also in the window of Grabau's 1936 cycle) (though in origination, not extinction) (Fig. 3). Analysis of extinction metrics reveal a cycle at roughly 27 ± 1 Ma significant at the 0.02 level (results not shown): effectively indistinguishable from Raup and Sepkoski's (1984, 1986) cycle (and again thus also in the window of Grabau's 1936 cycle) and covering a greater interval of time (Raup and Sepkoski concentrated primarily on the Mesozoic and Cenozoic fossil record). However, relevant for the analysis of extinction, Stigler and Wagner (1987) presented the interesting result that there was a peak in stratigraphic-interval length at roughly 26 Ma; this could have caused Raup and Sepkoski (1984, 1986) to artifactually retrieve a peak in that time interval. Lieberman and Melott's (2007) reanalysis of the data confirms a peak at roughly 27 ± 1 Ma in stratigraphic-interval length, significant at the 0.001 level (Fig. 4). This could imply that the peak is an artifact or that there is a process operating roughly every 27 Ma that causes extinctions to

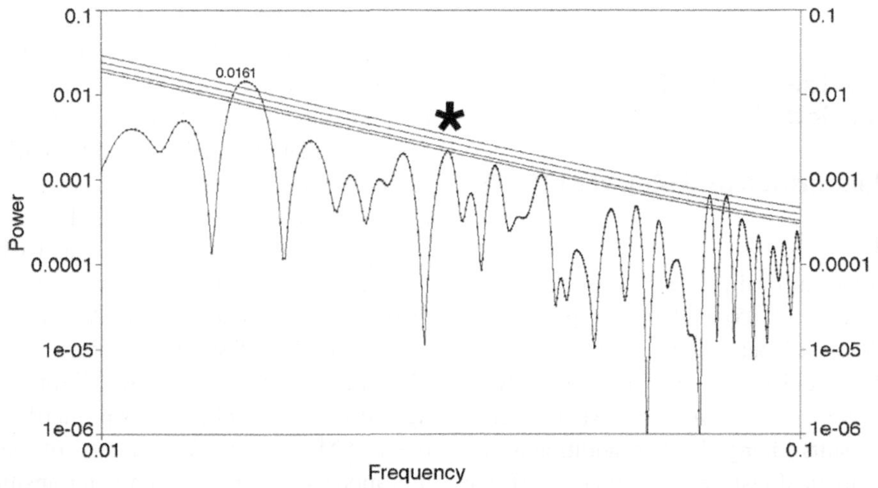

Fig. 1 Analysis of detrended fractional fossil biodiversity fluctuations from Lieberman and Melott (2007). Lines denote 0.1, 0.05, 0.01, and 0.001 levels of significance. Frequencies are given per Ma; there is a peak at a frequency of approximately 0.0161 Ma, equivalent to 62.1 ± 3.1 Ma; this peak is significant at better than the 0.001 level. There is also a peak marked by the "*" at roughly 31 ± 1 Ma, although it is only significant at the 0.1 level. All other significant peaks occur at less than 15 Ma and thus are not relevant because they are below the so called Nyquist frequency (see Lieberman and Melott 2007 for further discussion)

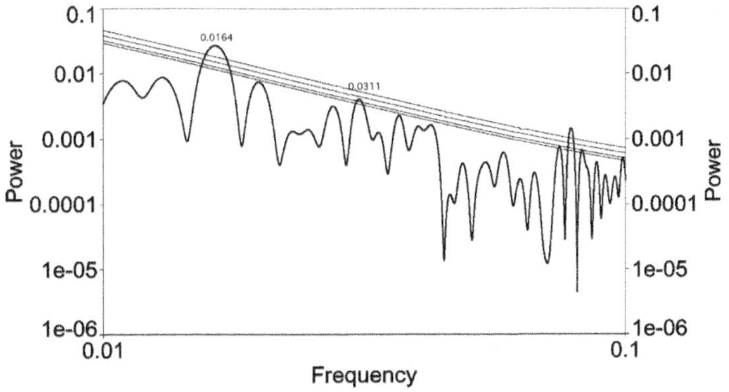

Fig. 2 Analysis of detrended fractional fossil biodiversity fluctuations from 150 to 519 Ma from Lieberman and Melott (2007). Lines denote 0.1, 0.05, 0.01, and 0.001 levels of significance. Frequencies are given per Ma; there is a peak at a frequency of approximately 0.0164 Ma, equivalent to 61.0 ± 3.2 Ma; this peak is significant at better than the 0.001 level. There is also a peak at a frequency of roughly 0.0311 Ma, equivalent to 32.2 ± 1.1 Ma, although it is significant only at the 0.1 level. All other significant peaks occur at less than 15 Ma and thus are not relevant because they are below the so called Nyquist frequency (see Lieberman and Melott 2007 for further discussion)

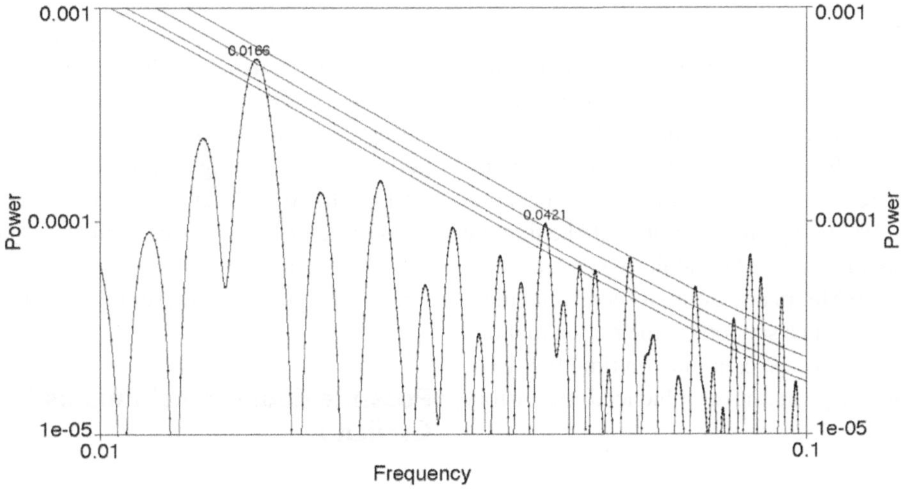

Fig. 3 Analysis of fluctuations in fractional origination intensity from Lieberman and Melott (2007). *Lines* denote 0.1, 0.05, 0.01, and 0.001 levels of significance. Frequencies are given per Ma. There are two peaks significant at the 0.01 level: one at a frequency of approximately 0.0166 Ma, equivalent to 60.1 ± 3.1 Ma; one at a frequency of roughly 0.0421 Ma, equivalent to 23.7 ± 0.5 Ma. All other significant peaks occur at less than 15 Ma and thus are not relevant because they are below the so-called Nyquist frequency (see Lieberman and Melott 2007 for further discussion)

occur and leaves the fossil record broken up into intervals that stratigraphers subsequently identify (see more extensive discussion in Sepkoski 1989 and Lieberman and Melott 2007). Note that there is no peak in stratigraphic interval length at or near 62 Ma suggesting that peak, retrieved in the other analyses, is not an artifact (Fig. 4).

It appears that based on the analysis of the Sepkoski (2002) dataset there is strong evidence for Rohde and Muller's (2005) identified periodicity in biodiversity at 62 Ma; there is more equivocal evidence for periodicity in biodiversity at roughly 32 Ma (vindicating Grabau 1936; Fischer and Arthur 1977; Fischer 1982). There is also evidence for periodicity in origination at roughly 60 Ma (supporting Rohde and Muller's 2005 identified peak) and also roughly 24 Ma (supporting Grabau 1936 and indirectly Raup and Sepkoski 1984, 1986, because they emphasized extinction whereas Grabau emphasized pulses of origination and extinction). Furthermore, there is evidence for periodic extinctions

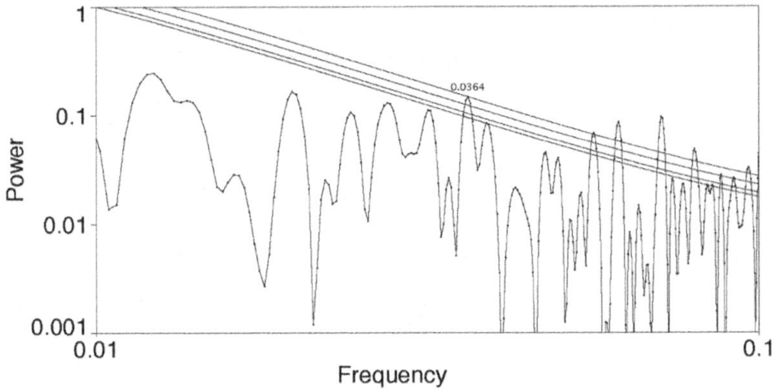

Fig. 4 The power spectrum of stratigraphic interval lengths from Lieberman and Melott (2007). There is a significant peak at the 0.001 level at a frequency of roughly 0.0364 Ma, equivalent to 27.5 ± 0.6 Ma; note there are no peaks at or around 62 Ma

at roughly 26 Ma (again vindicating Grabau 1936, and also Raup and Sepkoski 1984, 1986, but with the important caveat about stratigraphic-interval lengths mentioned). Further, it is worth stating the primary evidence for periodicity comes from the Paleozoic and early to mid-Mesozoic fossil record. This could signify that the result may be artifactual; if these data are of poorer quality than more recent data, it could signify that there are some biological differences between post- and pre-150 Ma organisms, or it could be that the mechanism that once caused periodicity suddenly disappeared post 150 Ma. In any event, it seems that there is at least some evidence supporting the notion that one or more large scale cycles, operating on time frames of tens of millions of years, have influenced the history of life.

Summary of Melott's (2008) Results

Because the validity of the Sepkoski dataset has been perennially criticized, one of us (ALM) performed an additional test to look for evidence of long-term periodicity (Melott 2008); this analysis used the PBDB; it attempts to correct the fossil record of diversity, and specifically improve on the Sepkoski dataset, using various sample standardization techniques in addition to other improvements.

Melott performed an analysis similar to that of Lieberman and Melott (2007) but using the PBDB data provided by John Alroy. He found that both the FFT and Lomb-Scargle analyses showed the existence of a 63-Ma cycle in the PBDB data. The fact that PBDB contains strong corrections for sampling rate suggests that sampling rate is not a causal factor in the existence of this period. Melott (2008) also shows a cross-spectrum, a technique for finding similarities in the periodicities of two different time series; in this case it was the data used by Rohde and Muller (2005) and the PBDB data. The cross-spectrum showed a strong peak at 62 Ma and indicated that not only the period but the timing of the peaks and valleys of the cycle in the two different databases disagreed only by an average of 1.6 Ma. Again, this provides strong support for the notion that there is a strong signature of long-term periodicity present in the history of life.

Possible Causal Mechanisms for 62-Ma Cyclicity

Rohde and Muller (2005) discussed, without endorsing, a wide variety of possible mechanisms that might cause a fairly regular swing in biodiversity over roughly the last 500 Ma. Basically they looked at two types of mechanism: (1) those that can easily impact biodiversity, but for which there is no reason to expect such periodicity; (2) periodic phenomena without any obvious connection to biodiversity.

Astronomical processes are the logical place to look for regular periodic processes over long timescales. Most large-scale motions in the universe are driven by gravity, a weak force acting over large distances, naturally giving rise to long time periods. Bound systems with insignificant friction (space being mostly empty) give processes whose periods are very stable over long times. This leads to the question of what kind of system might be periodic over 62 Ma.

Processes to do with the Earth-Moon system of course take months; inner Solar-system processes take years. There are distant, gravitationally-bound objects called the Oort Cloud, almost certainly the origin of long-period comets. The orbits of these objects may take up to a million years. They could be perturbed, and fall into the inner Solar System, if the Sun had a dim stellar companion with a very long-period orbit (Whitmire and Jackson 1984; Davis et al. 1984). Any object with such a long timescale orbital period would have a larger orbital radius, and be only weakly bound to the Sun by gravity. Passing stars and molecular clouds would likely change the period of such an object, possibly even dislodging it from it orbit (Hills 1984). Therefore, these do not appear to be viable candidates.

Motions of entire galaxies across space take billions of years. For example, our Local Group contains satellite galaxies, but most have quite long orbital periods as well as some that are not particularly regular due to the complicated structure of the group. Hence, these motions are not good candidates as timescales are too long and too irregular.

This leaves the motions of stars within the Galaxy, and here we find a good candidate (Medvedev and Melott 2007). The galaxy is a thin disc, with the Sun and its planets located about two-thirds of the way from the center. The Sun orbits the center of the Galaxy with a time period of about 200 Ma. As it does so, it will move in and out of spiral arms—areas of temporary concentration of stars and gas. Such regions have enhanced star formation, meaning they have many supernovae derived from large, hot, short-lived stars. There is enhanced danger to the biosphere during such crossings. Unfortunately the spiral structure is not very well-known, but not likely to be extremely regular.

Another motion however closely fits a 62 Ma periodicity: as the Sun moves around the Galaxy, it has a wobbling up and down motion, rather like a weight bobbing on a spring (Medvedev and Melott 2007). (In fact, due to a peculiarity of the distribution of mass in the Galaxy, this motion even behaves mathematically in a way not far from that of a weight on a spring.) Gies and Helsel (2005) have provided a solution for the Sun's motion in the past, given our knowledge of its present position, velocity, and the gravitational field of the Galaxy. Interestingly, the period of the vertical motion is about 63 Ma. This suggests a possibly interesting coincidence. But of course, there has to be some kind of effect. The most natural thing is something that happens when we pass through middle of the disc where the mass is more concentrated. There are two problems with this. First, given the approximately 200 light-year amplitude of our "vertical" motion, the mass is not all that much more concentrated at the center than at the extremes. Secondly, something that happens then would happen twice per period, or every 32 Ma. It is possible that the various weaker periodicities in the vicinity of 30 Ma, mentioned earlier, have to do with passage of the Sun through the Galactic midplane on this timescale (e.g., Matese et al. 1996), resulting in an increase in the rate of comet impacts on the Earth. But this could not explain the stronger 62 Ma signal.

In order for this to be responsible for 62 Ma periodicity, something has to happen once per period, which would mean on one side of the galaxy, or when passing through the disc going in one direction. The galaxy itself is reasonably symmetrical, so this doesn't seem to make much sense. However, the external environment isn't symmetrical. The local Universe has a lot more mass on one side of our galactic disc than the other. In particular, the Virgo Cluster of galaxies is the only large mass concentration in our vicinity; its gravitational attraction has given our galaxy a speed of about 200 km/s, falling toward it (Medvedev and Melott 2007). This cluster is located only about 16° off the Galactic north axis. Furthermore, examination of the detrended biodiversity information plotted against the Gies and Helsel (2005) Solar motion (Fig. 5) shows that biodiversity decreases tend to coincide with excursions to galactic north—toward the Virgo Cluster; the cross correlation is significant at $p = 2\times10^{-7}$ (see Medvedev and Melott 2007). That is, not only do the periods of the two cycles agree but the timings of the peaks also agree very strongly as well.

Medvedev and Melott (2007) noticed this coincidence and proposed a mechanism for biodiversity fluctuation. They suggested that the 200 km/s infall toward the Virgo Cluster will produce a shock wave, pushing it toward the galaxy on the north side. This is a shock wave in the plasma (hot ionized gas) which fills space, not too different from the shock that precedes a supersonic jet or the shock wave that the Solar System produces as it moves through the galaxy at a few km/s. A shock in plasma is well-known to produce cosmic rays—protons and other charged elementary

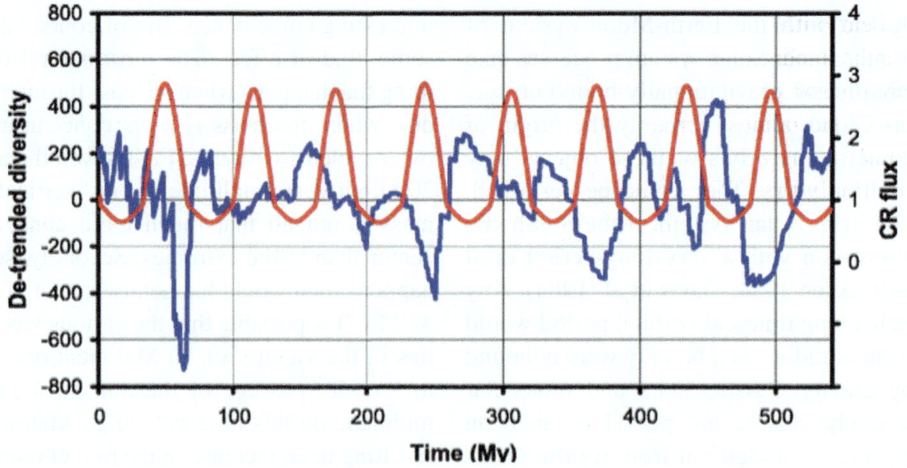

Fig. 5 Detrended diversity variation (blue curve, left scale) as a function of time. Also, normalized cosmic-ray flux calculated from Medvedev and Melott (2007), based on their model of solar system position in galactic arm (*red curve, right scale*). The maxima in the cosmic-ray flux coincide with minima of the diversity cycle. Note also that the onset of diversity decline coincides with rapid increase of the flux. The cross correlation is significant at $p = 2 \times 10^{-7}$ (see Medvedev and Melott 2007)

particles traveling at very high speeds. We are normally exposed to some cosmic rays; these account for a large proportion of the radiation dose we get.

Medvedev and Melott (2007) proposed that the cosmic ray flux at the Earth would increase greatly when we move to galactic north. Thus the times of low biodiversity would correspond to times when the Earth is exposed to a high level of cosmic rays. The increased flux would come because we have moved closer to the source, the shock front, but even more so because there is less magnetic shielding between the Earth and the source.

Detailed computation of the spectrum and intensity of the cosmic rays and their resulting effects has not been done. However, some generalizations are possible. (1) Cosmic rays are a known mutagen and as such can contribute to genetic change, including cancer. A large increase in the rate of such events might lower biodiversity. (2) Cosmic rays ionize the atmosphere. Such ionization can change its chemistry, increasing the concentration of oxides of nitrogen. These in turn catalyze destruction of ozone (O_3). Ozone is the primary shield that keeps destructive Solar UVB radiation away from the Earth's surface. UVB can cause cell damage and break DNA molecules. (3) There is increasing evidence that the same atmospheric ionization can provide nucleation sites for cloud formation. It has been hypothesized that increased cloud formation will change the albedo of the Earth, lowering temperatures. All these possible mechanisms are plausible, but they all need a lot more theoretical work and comparison with data, where possible, before being declared good candidates. Unfortunately, at this time, there are no known long-term isotopic changes that offer hope of a direct measurement of the cosmic-ray variability.

Conclusions

There is tantalizing but diffuse evidence for the possible existence of long-term cycles of biodiversity, possibly related to astronomical phenomena. This evidence has been accumulating over the course of many decades and is part of a larger debate about the role the physical environment plays in causing evolution and also about uniformitarianism. We acknowledge that the issue clearly has not been demonstrably proven by any sense of the imagination; even the most recent round of studies by Rohde and Muller (2005), Cornette (2007), Lieberman and Melott (2007), and Melott (2008) is unlikely to tip the balance, though it may be useful that a possible mechanism to explain 62 Ma periodicity has now been uncovered. The truth is that this issue may never be definitively demonstrated: our focus should, in any case, be directed more to testing hypotheses. However, even though paleontology tends not to be a predictive science we will make one prediction that the future may, or may not, bear out. If the most recent studies on periodicity do not clinch the matter, then we are in for another round of studies focusing on long-term biological periodicity roughly some time in the mid-2020s. In any event, the existence of such cycles would imply that the physical environment, in particular, the environment external to our planetary biosphere, has had a profound and continual influence on macroevolutionary patterns. This may speak to the notion that contingency (*sensu* Gould 1989) rules the history of life, yet there is some repetitive nature to those contingent events.

Acknowledgements We thank NSF DEB-0716162 (to BSL) for support of this research and N. Eldredge and M. Medvedev for discussions.

References

Adrain JM, Westrop SR (2000) An empirical assessment of taxic paleobiology. Science 289:110–112

Alroy J, Marshall CR, Bambach RK, Bezusko K, Foote M et al (2001) Effects of sampling standardization on estimates of Phanerozoic marine diversification. Proc Natl Acad Sci USA 98:6261–6266

Alvarez LW, Alvarez W, Asaro F, Michel HV (1980) Extraterrestrial cause for the Cretaceous–Tertiary extinction. Science 208:1095–1108

Anderson EJ, Goodwin PW, Sobieski TH (1984) Episodic accumulation and the origin of formation boundaries in the Helderberg Group of New York State. Geology 12:120–123

Bambach RK (2006) Phanerozoic biodiversity mass extinctions. Ann Rev Earth Planet Sci 34:127–155

Bennett KD (1990) Milankovitch cycles and their effects on species in ecological and evolutionary time. Paleobiology 16:11–21.

Bennett KD (1997) Evolution and ecology: the pace of life. Cambridge University Press, Cambridge

Benton MJ (1996) On the nonprevalence of competitive replacement in the evolution of tetrapods. In: Jablonski D, Erwin DH, Lipps JH (eds) Evolutionary paleobiology. University of Chicago Press, Chicago, pp 185–210

Brett CE, Baird GC (1995) Coordinated stasis and evolutionary ecology of Silurian to Middle Devonian faunas in the Appalachian Basin. In: Erwin D, Anstey R (eds) Approaches to speciation in the fossil record. Columbia University Press, New York, pp 285–315

Brigham FO (1988) The Fast Fourier transform and its applications. Prentice Hall, Englewood Cliffs

Buffon GL (1749–1804) Histoire Naturelle, Générale et Particulière. Imprimerie Royale, Puis Plassan, Paris

Camp CL (1952) Geological boundaries in relation to faunal changes and diastrophism. J Paleontol 26:353–358

Cloud PE Jr (1948) Some problems and patterns of evolution exemplified by fossil invertebrates. Evolution 2:3223–250

Cooper GA, Williams A (1952) Significance of the stratigraphic distribution of brachiopods. J Paleontol 26:326–337

Cornette JL (2007) Gauss-Vaníсek and Fourier transform spectral analyses of marine diversity. Comput Sci Eng 9:61–63

Cornette JL, Lieberman BS (2004) Random walks in the history of life. Proc Natl Acad Sci USA 101:187–191

Cornette JL, Lieberman BS, Goldstein RH (2002) Documenting a significant relationship between macroevolutionary origination rates and Phanerozoic pCO_2 levels. Proc Natl Acad Sci USA 99:7832–7835

Darwin C (1859) On the origin of species by means of natural selection. Reprint 1st edn. Harvard University Press, Cambridge

Davis M, Hut P, Muller RA (1984) Extinction of species by periodic comet showers. Nature 308:715–717

Dawkins R (1976) The selfish gene. Oxford University Press, New York

de Candolle AP (1820) Géographie botanique. Dict des Sci Nat 18:359–422

Eldredge N (1985) Unfinished synthesis. Oxford University Press, New York

Eldredge N (1999) The pattern of evolution. W. H. Freeman, New York

Eldredge N, Cracraft J (1980) Phylogenetic patterns and the evolutionary process. Columbia University Press, New York

Fischer AG (1982) Long-term climatic oscillations recorded in stratigraphy. In: Berger W (ed) Climate in earth history. National Research Council Studies in Geophysics, National Academy Press, Washington, DC, pp 97–104

Fischer AG, Arthur MA (1977) Secular variations in the pelagic realm. In: Cook HE, Enos P (eds) Deep-water carbonate environments. Soc Econ Paleontols, Minerals Spec Pub 25: 19–50. Tulsa, Oklahoma

Foote M (2006) Substrate affinity and diversity dynamics of Paleozoic marine animals. Paleobiology 32:345–366

Gies DR, Helsel JW (2005) Ice age epochs and the Sun's path through the Galaxy. Astrophys J 626:844–848

Goldschmidt RB (1940) Material basis of evolution. Yale University Press, New Haven

Gould SJ (1965) Is uniformitarianism necessary? Am J Sci 263:223–228

Gould SJ (1982) Introduction. In: Goldschmidt RB (ed) Material basis of evolution. Reprint edn. Yale University Press, New Haven

Gould SJ (1989) Wonderful life. WW Norton, New York

Gould SJ, Calloway CB (1980) Clams and brachiopods; ships that pass in the night. Paleobiology 6:383–396

Grabau AW (1934) Palaeozoic formations in the light of the pulsation theory vol. 1 Taconian and Cambrian pulsation systems, 1st edn. National University of Peking Press, China

Grabau AW (1936) Palaeozoic formations in the light of the pulsation theory vol. 1 Taconian and Cambrian pulsation systems, 2nd edn. National University of Peking Press, China

Grabau AW (1940) The rhythm of the ages. RE Krieger, Huntington, New York

Grant PR, Grant BR (2002) Unpredictable evolution in a 30-year study of Darwin's finches. Science 296:707–711

Hays JD, Imbrie J, Shackleton NJ (1976) Variations in the earth's orbit: pacemaker of the ice ages. Science 194:1121–1132

Henbest LG (1952) Significance of evolutionary explosions for diastrophic division of earth history: introduction to the symposium. J Paleontol 26:299–318
Hills JG (1984) Dynamical constraints on the mass and perihelion distance of Nemesis and the stability of its orbit. Nature 311:636–638
Huntley B, Webb T III (1989) Migration: species' response to climatic variations caused by changes in the earth's orbit. J Biogeog 16:5–19
Laguna P, Moody GB, Mark RG (1998) Power spectral density of unevenly sampled data by least-square analysis: performance and application to heart rate signals. IEEE Trans Biomed Eng 45:698–715
Lieberman BS (2000) Paleobiogeography. Plenum/Kluwer, New York
Lieberman BS, Melott AL (2007) Considering the case for biodiversity cycles: reexamining the evidence for periodicity in the fossil record. PLoS One 2(8):e759
Matese JJ, Whitman PG, Innanen KA, Valtonen MJ (1996) Why we study the geological record for evidence of the solar oscillation about the galactic midplane. Earth Moon Planets 72:7–12
Medvedev MV, Melott AL (2007) Do extragalactic cosmic rays induce cycles in fossil diversity? Astrophys J 664: 879–889
Melott AL (2008) Long-term cycles in the history of life: periodic biodiversity in the Paleobiology Database. PLoS One 3(12):e4044
Melott AL, Lieberman BS, Laird CM, Martin LD, Medvedev MV, Thomas BC, Cannizzo JK, Gehrels N, Jackman CH (2004) Did a gamma-ray burst initiate the late Ordovician mass extinction? Int J Astrobiol 3:55–61
Moore RC (1952) Evolution rates among crinoids. J Paleontol 26:338–352
Morris PJ, Ivany LC, Schopf KM, Brett CE (1995) The challenge of paleoecological stasis: reassessing sources of evolutionary stability. Proc Natl Acad Sci USA 92:11269–11273
Newell ND (1949) Periodicity in invertebrate evolution. Geol Soc Am Bull 60:1911
Newell ND (1952) Periodicity in invertebrate paleontology. J Paleontol 26:371–385
Olsen PE (1986) A 40-million-year lake record of early Mesozoic orbital climatic forcing. Science 234:842–848
Patterson C, Smith AB (1987) Is the periodicity of extinctions a taxonomic artefact? Nature 330:248–251
Peters S (2005) Geologic constraints on the macroevolutionary history of marine animals. Proc Natl Acad Sci USA 102:12326–12331
Rampino M, Stothers RB (1984) Terrestrial mass extinctions, cometary impacts and the Sun's motion perpendicular to the galactic plane. Nature 308:709–712
Raup DM (1986) The nemesis affair. W. W. Norton, New York
Raup DM, Sepkoski JJ Jr (1984) Periodicity of extinctions in the geologic past. Proc Natl Acad Sci USA 81:801–805
Raup DM, Sepkoski JJ Jr (1986) Periodic extinction of families and genera. Science 231:833–836
Rohde RA (2006) 62 million year cycle in biodiversity and associated geological changes. Geol Soc Am Ann Meet Abstr Progr 63–67
Rohde RA, Muller RA (2005) Cycles in fossil diversity. Nature 434:208–210
Ross RM, Allmon WD (1990) Causes of evolution: a paleontological perspective. University of Chicago, Chicago
Rothman DH (2001) Atmospheric carbon dioxide levels for the last 500 million years. Proc Natl Acad Sci USA 98:4305–4310
Scargle JD (1982) Studies in astronomical time series analysis. II. Statistical aspects of spectral analysis of unevenly spaced data. Astrophys J 263:835–853
Sepkoski JJ Jr (1989) Periodicity in extinction and the problem of catastrophism in the history of life. J Geol Soc 146:7–19
Sepkoski JJ Jr (2002) A compendium of fossil marine animal genera. Bull Am Paleontol 363:1–560
Sepkoski JJ Jr, Raup DM (1986) Periodicity in marine extinction events. In: Elliott DK (ed) Dynamics of extinction. Wiley, New York
Simpson GG (1952) Periodicity in vertebrate evolution. J Paleontol 26:359–370
Stanley SM (1990) Delayed recovery and the spacing of major extinctions. Paleobiology 16:401–414
Stigler SM, Wagner MJ (1987) A substantial bias in nonparametric tests for periodicity in geophysical data. Science 238:940–945
Valentine JW, Moores EM (1970) Plate-tectonic regulation of faunal diversity and sea level: a model. Nature 228:659–677
Van Dam JA, Azia HA, Sierra MAA, Hilgen FJ, van den Hoek Ostende LW, Lourens LJ, Mein P, van der Meulen AJ, Pelaez-Campomanes P (2006) Long-period astronomical forcing of mammal turnover. Nature 443:687–691
Vermeij GJ (1993) Evolution and escalation. Princeton University Press, Princeton
Von Humboldt A (1816) On the laws observed in the distribution of vegetable forms. Phil Mag J 47:446
Vrba ES (1980) Evolution, species and fossils: how does life evolve? S Afr J Sci 76:61
Vrba ES (1985) Environment and evolution: alternative causes of the temporal distribution of evolutionary events. S Afr J Sci 81:229–236
Vrba ES (1992) Mammals as a key to evolutionary theory. J Mammal 73:1–28
Whitmire DP, Jackson AA (1984) Are periodic mass extinctions driven by a distant solar companion? Nature 308:713–715

Climate Change Through Time

John Dodson

Abstract

The debate about climate change is fraught with problems; not everyone begins his or her argument from the same base. When pondering the climate of the past, trying to enhance our understanding of past climate variability and how climatic change millions of years ago impacted on the world's biota at that time, and how past climates can help us solve problems based on today's climates, it is essential that all argument begins with an understanding of the Earth's climate system. This brief chapter is aimed at providing such a basic understanding ('level playing field') for such debate.

Keywords

Solar evolution • Earth's systems • Atmosphere evolution • Global heat diffusion • Plate tectonics • Sedimentary record of climate • Greenhouse and glacial intervals • Evaporites • Coals • Earth's climate-change saga • Cultural implications

Introduction

Earth's climate system is powered by the Sun. Incoming solar radiation is absorbed, reflected, transmitted and changed to heat where it is further processed by Earth's physical and biological systems. These occur at different rates of transmission and storage in the atmosphere, water, ice and land cover including vegetation. The long-term averages of these define the climate from place to place. The thumbnail sketch in this chapter gives an overview of selected key events and environments during Earth's history.

Basic physics underpins all climate systems but the number of processes and feedbacks is extremely complex and difficult to model. The atmosphere has evolved chemically and great changes have taken place in the distribution of land, water and land cover including ice. The size and location of mountains and significant plateaus are also important, especially where they intersect major wind patterns.

The heat engine of the climate system has two major elements. One is external to planet Earth and is a function of the evolution of the Sun itself, especially its variability as an energy source, and the patterns in Earth's orbital parameters, especially the shape of the orbit and the tilt and precession of the spinning Earth on its axis. We know about these at the present

J. Dodson (✉)
Institute for Environmental Research, Australian Nuclear Science and Technology Organisation, Sydney, NSW, Australia
e-mail: john.dodson@ansto.gov.au

J.A. Talent (ed.), *Earth and Life*, International Year of Planet Earth,
DOI 10.1007/978-90-481-3428-1_4, © Springer Science+Business Media B.V. 2012

time, but much less clearly for the distant past. We can envisage the start of Earth's climate system from the receipt of solar energy; and the main patterns that follow depend on how this gets redistributed. Figure 1 explains why the most intense energy is received at the Equatorial region, for the simple reason that it is closer to the Sun. It is perhaps surprising that the closeness of the Equator (compared to the Poles) could be that significant compared to the mean distance from the Sun. The other aspect of this is that the curvature of the planet surface means that the incoming radiation is less intense for any given area of planet surface. So it is warmer in the Equatorial region and cooler at the poles. Heat however diffuses from the Equatorial region towards the poles principally through the atmosphere and oceans. Warming from the Equatorial region also causes rising air; this cools at altitude and descends in the mid latitudes to form a series of Sub Tropical Highs (Fig. 2a, b). Finally, heat is transferred much more efficiently through the atmosphere over land than over water, since water is very efficient at absorbing and retaining heat. Today the Southern Hemisphere is water dominated compared to the Northern, and this, amongst other things, ensures that Antarctica is a much colder place than the Arctic. Finally the tilt and precession are important if defining the area of tropical climate and the divergence between seasonal extremes. Earth is a dynamic planet and all of these parameters have changed in time resulting in the patterns we see preserved in the geological record.

The record of past climates is written and recorded in a number of natural archives. Extracting the climate signal from these requires finding archives with recognisable signals and accurately identifying their location

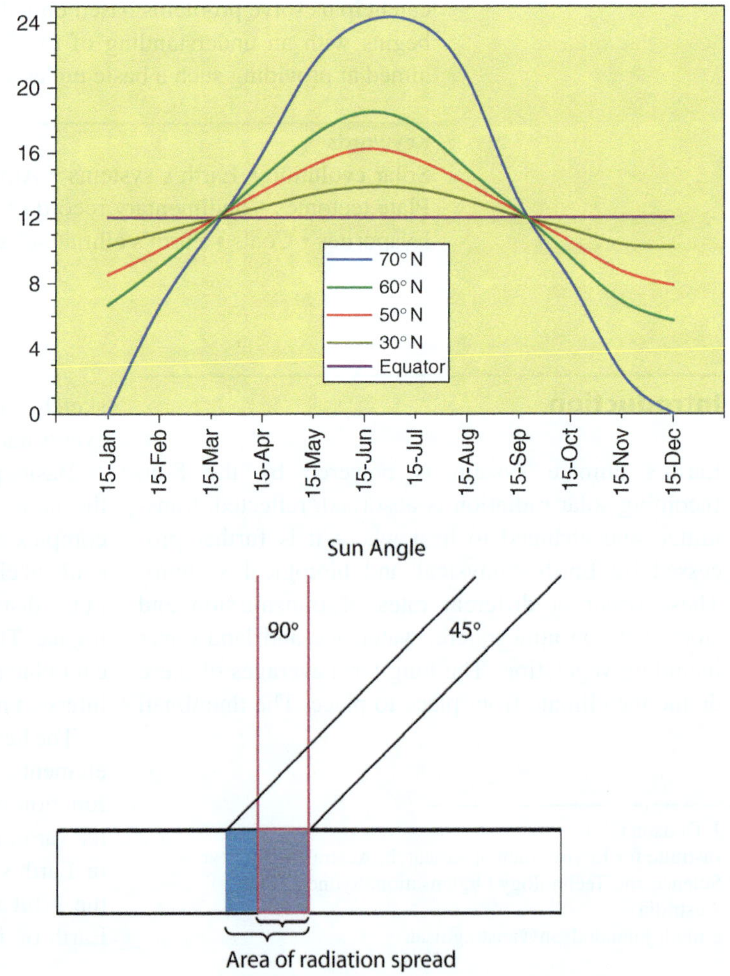

Fig. 1 A simplified view of Earth's climate systems. Note that the most intense energy is received in the equatorial region. The *vertical axis* represents length of day in hours; the *horizontal axis* is time through a year. This assumes both hemispheres have the same surface characteristics (which is not strictly so)

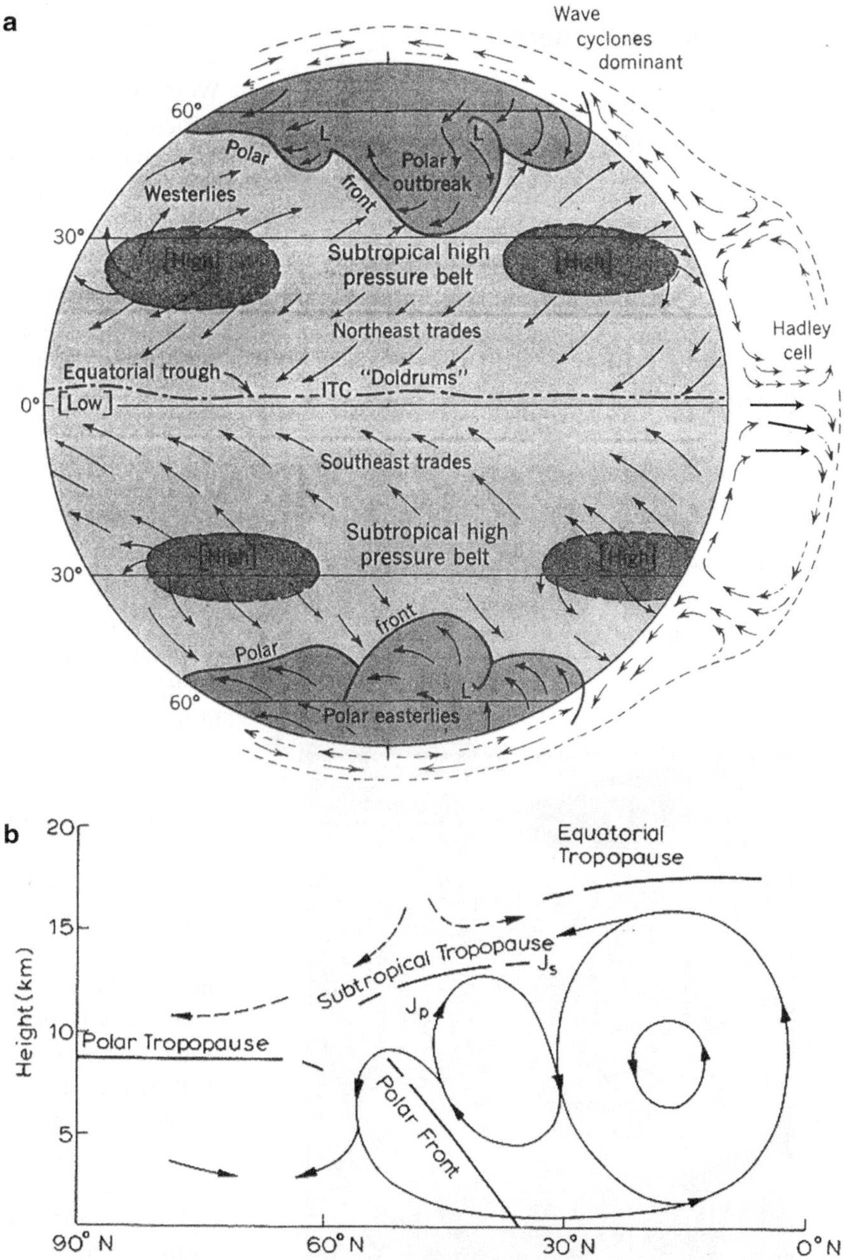

Fig. 2 The main wind and synoptic scale patterns on Earth's surface: (**a**) Subtropical High Pressure Belt; (**b**) The tropopause is the boundary in Earth's atmosphere between the *stratosphere* (*above*) and the *troposphere* (*below*) where air ceases to cool with height and becomes almost completely dry. It is at about 15–17 km over the tropics and 10 km nearer the poles; it varies seasonally and with changes in weather

and age. Some signals are from very localised phenomena, others may be regional, and sets of records are needed to provide some certainty of regional and synoptic scale climate patterns for periods of interest. In overview, these kinds of records reveal that Earth has had a surprising amount of climate variability, indeed much more than we see represented on the planet today. At various times it has been colder, warmer, wetter and drier than we can measure anywhere today. The distribution of macro and micro patterns of climate is absolutely fundamental to the distribution of life forms and ecosystem processes. It follows that since climate types and patterns have changed in the past then they are key elements in helping us understand the evolution of form, function and biogeographic patterns.

Early Earth Atmosphere

We assume that the earliest atmosphere for Earth was like that we see on some other planets in the Solar System. Thus it was initially rich in hydrogen, carbon dioxide, sulphur dioxide, helium and nitrogen; but then the lightest gases like hydrogen and helium were lost due to the relatively low strength of Earth's gravitational field. As Earth cooled, out-gassing products like carbon dioxide and water enriched the atmosphere, and further cooling resulted in water bodies and the first sediments. Life possibly began in these early pools. Many of the early atmospheric gases could dissolve in water and form weak acids. Nitrogen, as N_2, was not soluble and quite quickly became the dominant atmospheric gas.

The oldest known rocks come from Quebec (O'Neill et al. 2008) and are about 4.3 Ga (billion years ago), although zircons even older than this from Western Australia are known. The earliest sediments are perhaps 3.5 Ga (Fig. 3) and are known from Western Australia, Greenland, Canada and parts of Africa. It is thought that the Sun was cooler then than now and since the evidence from the first sediments suggests Earth environments were mainly warm and moist an enhanced greenhouse effect from high quantities of CO_2 may have been important in driving these conditions. Even so there were a few episodes of glaciation, which possibly indicate there were fluctuations in energy output from the Sun. Earliest life forms evolved in very low oxygen and high CO_2 levels, but at some point a form of photosynthesis began to change the nature of the atmosphere from intense greenhouse conditions due to the dominance of carbon dioxide, to one containing ever more oxygen.

Early life was mainly confined to the oceans until photosynthesis had produced enough oxygen for ozone (O_3) to form. Ozone screens out some of the ultraviolet radiation coming from the Sun and this was essential for life to survive on land. It had taken 90% of Earth's history for this to become possible.

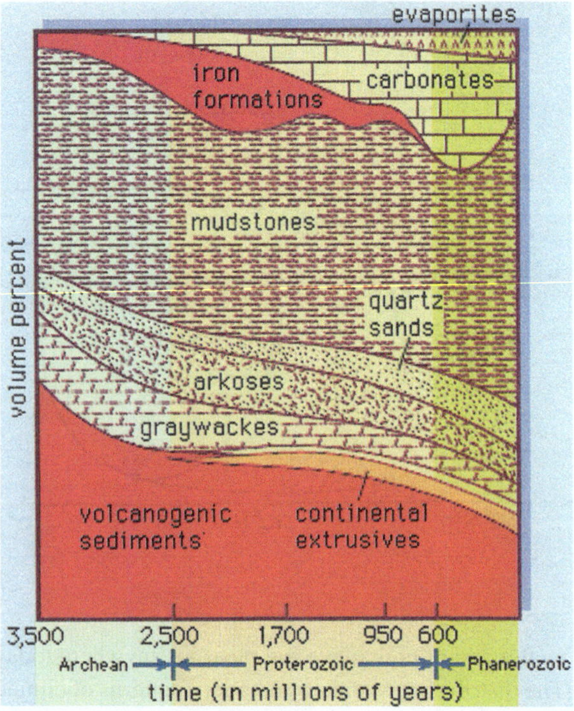

Fig. 3 Relative proportions of various rock types accumulating on Earth's surface between 3500 million years ago (*on the left*) and the present day (*on the right*). Note that, in a general way, carbonates (mainly limestones) and evaporites have been more prominent in the spectrum of sediments in the last one or two billion years

Primitive Oceans

There is some debate about how warm the first oceans were. Were they cool, because the Sun was much less active than present, or were they hot, say warmer than 70°C (Valley et al. 2002; Robert and Chaussiden 2006; Shields and Kasting 2007)? The occurrence of water was very significant for the origin of life and its dependence on it. There are many opinions on how the oceans originated, and these range from impacts of high ice containing comets to water derived from out-gassing from volcanic eruptions. There is a large body of data that suggests the early continents were hostile to life, for example, from UV radiation as described above. The low atmospheric oxygen content is a fair indication that the early oceans were anoxic and acidic. The dissolution of CO_2, SO_2 and other gases would have made the ocean acidic, and chemical and physical weathering of the continents would have made the ocean at least somewhat salty. Carbonates and sulphides were present, and some of the iron oxides formed could not have developed if oxygen was present in the atmosphere (Lane 2002). Work by Canfield on sulphur and oxygen isotopes led to the conclusion that oxygen showed two major build up phases. The first of these was at about 2.7 Ga (to about 1% of present) and a much larger rise from about 2.2 Ga (to about 5–18% of present (Canfield 1998, 1999; Canfield et al. 2000).

First life in the oceans would have been in anaerobic conditions and it was not until near the end of the Archaean that sufficient oxygen was around to exclude pure anaerobes and allow aerobic organisms to prosper.

Iron-Banded Formations: Significance

Banded iron-formations (BIF) are widespread and occur between about 3.8 Ga and 1.8 Ga, then reappear between 0.8 and 0.6 Ga (Klein 2005). Where these are well preserved they are fine to coarse laminations and are thought to be varves or annual layers (Fig. 4). They have a varied mineralogy but they are typically iron-rich and silicon-rich couplets. Their origin is an enigmatic question. They often have pronounced positive anomalies of the element europium (Eu); this is thought to indicate that the source of iron and silicon is from deep ocean hydrothermal activity admixed with seawater and that they were deposited in deep ocean basins (Klein 2005). Others have argued that the Ge/Si ratio indicates the silicon component has a continental origin (Hamade et al. 2003). Attempts to find microbial remains, and hence invoke their role in BIFs, have also proved to be elusive. Lorenz (2002) has argued that a possible explanation might relate to some early electrolyte cycle that was unique to the early oceans.

Despite all the uncertainty of just what BIFs represent, it seems clear they were linked to a stage in Earth's history related to the evolving biosphere and the increasing levels of O_2 in the Precambrian atmosphere. Once the latter had reached a certain threshold the chemistry had changed and BIF ceased.

The Story in Sedimentary Rocks: Limestones, Dolomites, Evaporates, Gypsum, Coal, Reef Limestones and Glacial Deposits

Sediments form under water (or occasionally air) and they form a time sequence. The nature of sediments reveals much about the environment in which they form. Their geochemistry and structure reveals information about the broader and local environments in which they form. For example, poorly sorted sediments consisting of coarse and angular grains indicate short distance high energy transport from mass movement. Sediments which form under water will have a mix of particle sizes which are proportional to the energy of the water flows, and sorting by size will reflect this as well. Many sediments contain fossils which can further inform about the local and sometimes regional environment. Since most plant and animal species are highly tuned to their environment, species assemblages in sediments can be used to describe the regional climate, seasonality, variation in sedimentary processes (for example, freshwater

Fig. 4 Example of a Precambrian banded iron formation from Western Australia. Note the well-expressed seasonal/annual colour banding (photo courtesy of Annette George)

and marine inundations) and whether the conditions have modern analogues. The beauty of sediment facies analysis is that all of these features may be preserved and hence the record of events as well. This is particularly important when we have an accurate chronology to provide an age range for the sediment sequence.

Limestones are dominantly calcium carbonate rocks and they begin to show in abundance on Earth from about 2.7 Ga; they steadily increase in abundance until the Cambrian. These rocks contain precipitates of purely physico-chemical origin as well as carbonates of biological origin. They increase in abundance steadily in time, which probably reflects increasing biological activity, warming from the high partial pressure of CO_2 driven greenhouse effect and increasing concentration of dissolved carbonate in the oceans. Limestones can therefore be interpreted as indications of abundant calcium and carbonates in aquatic environments.

Under warm and hypersaline environments, after primary calcium carbonate has precipitated out of solution, other chemical species can form. There are many types of these but the most abundant are dolomites and gypsum. These form a group called evaporates (Fig. 3). Dolomites are a carbonate of calcium and magnesium. They require warm conditions allowing high rates of evaporation. Gypsum is a hydrated calcium sulphate and forms in warm super-saturated brines. Evaporites begin to become abundant on Earth from about 2.2 Ga, and they increase steadily towards the Silurian. This trend can be interpreted as a general warming and a general increase in oxygen (for sulphate rather than sulphide formations). It is probably related to maturing of the Sun as a source of heat and the expansion of aerobic organisms.

Coals are mostly organic carbon. They form where biological productivity is high and where it exceeds breakdown and decay. The precursor to coals are peats, and with a few exceptions these occur where there is abundant water and water-logging (in which oxygen and therefore aerobic breakdown is limited). They are most abundant in cool climates, at high elevation or latitude, since low temperatures further decrease the rate of decay. The Carboniferous Period (320 to 290 Ma, million years ago) is the major coal forming era of the geologic record.

Reef limestones form in warm water and are dominantly marine (Fig. 5). Since calcium carbonate is less soluble as water temperature rises, plants and animals need methods to dispose of it. Corals, shelled animals and even calcareous plants and plankton have evolved a myriad of ways of doing this. The chemistry and stable isotope ratios of these carbonates reveal much about the ambient temperature conditions under which the reefs formed. The occurrence of reefs is evidence of warm tropical to subtropical seas.

Glacial deposits come in many forms, including ice formed debris structures such as moraines and eskers. Evidence of glaciers also includes erratics, that

Fig. 5 Portion of an ancient reef displayed in Windjana Gorge, Kimberley region of Western Australia. Note well-bedded limestones (*left*) sloping away towards what were deeper waters off massive reef core (or rampart) limestones (*right*) (photo courtesy of Steven Kershaw)

is transported and 'out of place' boulders, ice rafted debris on ocean floors, tills and striations on hard rocks caused by ice dragging debris across a rock surface and also glacial landforms.

Sediments, isotopes, geochemistry and fossils provide a vast array of tools to date and determine the conditions at the various places where they are preserved.

First Glaciations

The Precambrian appears to have long warm periods separated by brief glacial events (Kasting and Howard 2006). The glacial periods were at about 2.4 Ga, and 0.75 and 0.6 Ga. The earliest of these is called the Huronian and was identified from southern Canada. This was centred over Ontario and striations are preserved which show the direction of ice flows from the peak of the ice cap (Pettijohn 1962). The current interpretation of this is that this event coincides with the rise of oxygen. High greenhouse gas concentrations compensated for a weaker Sun and this kept the surface above freezing. Methane was an important component of the greenhouse warming, and the rise in oxygen led to a severe reduction in methane and thus triggered a major glaciation.

The Precambrian glacial evidence from Africa, Australia, Europe and Asia falls between 0.75 and 0.6 Ga, and in some cases these occur as two episodes separated by non-glacial strata. It appears that glaciation at this time was not global and not entirely restricted to highland areas, and this makes it difficult to ascribe an origin. Some scientists have argued that the areas were extensive enough to suggest these were Snowball earth episodes (Pollard and Kasting 2005).

Plate Tectonics

The outer shell of Earth consists of a series of rigid plates (the lithosphere) and these move slowly due to convection currents within the mantle. The interactions of the plates at their boundaries produce huge forces which yield earthquakes and volcanic activity and produce ocean trenches, island chains, mountain ranges and more.

Plate tectonics has played a central role on the nature and distribution of climates. It was noted in the Introduction that water absorbs and retains great amounts of heat. The redistribution of heat to the poles from the Equator is much more efficient over land than oceans, and since Antarctica is located in a water-dominated hemisphere it is much colder than the Arctic. Thus when the plates move the redistribution of land surface alters the heat transfer relationships across the planet. If all of the major land masses were around the Equator, both poles would be very cold. We can track the history of continental movements where we have rocks of sufficient age ranges, and through the use of palaeomagnetism which enables palaeo-latitudes to be estimated. There is much doubt about the positions of the earliest continents due to the fragmentary and discontinuous distribution of earliest rocks. From the Cambrian period forward, there are generally sufficient rocks with palaeomagnetic records to make continental reconstructions relatively uncontroversial. A possible hypothesis for glaciations in the Carboniferous-Permian period, perhaps the longest cold period in Earth's history (a duration of 55 million years), could well be due to the distribution of continents at the time. It was exacerbated by the locking up of large amounts of CO_2 into extensive coal formations by plants. Glaciated areas included parts of Australia, South Africa, Brazil and elsewhere (Fig. 6). These were part of the land-mass Gondwana, and Siberia was probably also affected. Evidence suggests all of these were located in high latitudes (e.g. Eyles et al. 2002; Scheffler et al. 2003). This period had a great impact on the nature of faunas and floras, and, for example, cold water brachiopods characterised areas around Gondwana, and as soon as the glacial period was over there was a great surge in biodiversity as warm water faunas evolved (Chen 2008).

The other direct influence of plate tectonics is its role in consolidation and fragmentation of land masses. This is not strictly a climatic impact but causes advantage and disadvantage in migration and dispersal opportunities for species. The fragmentation of Pangea into Laurasia and Gondwana and then subsequently into the modern day continents were probably big forces in increasing biodiversity and the formation of the biogeographic provinces that now characterise floras and faunas on a global scale.

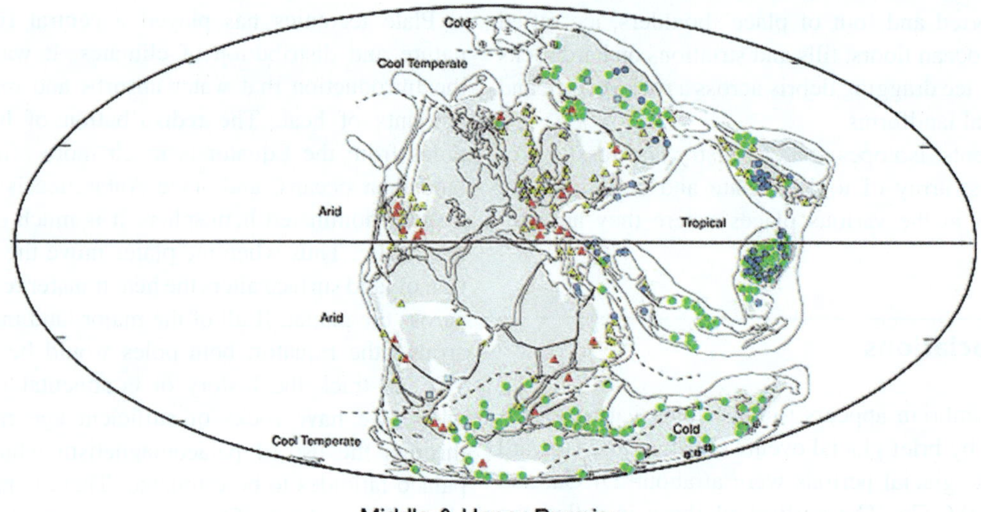

Fig. 6 Continental configuration and evidence of climate in the Carboniferous–Permian period (courtesy of Paleomap Project)

The Pattern of Glaciations and Warm Periods from the Ordovician Until the Late Tertiary

Glaciations of Ordovician—Silurian times are recognised from north and west Africa, Arctic Russia, Canada and France and Spain. The full extent of ice in the Southern Hemisphere is unknown but there are some records from South America. In general all these regions were at high latitudes during this time period.

The following period, the Devonian, has records that mostly indicate warmth. Those few records claiming glaciations at this time are poorly documented.

The Permian was a major period of glaciation—and records are known from Australia, South America, Africa, India and Antarctica. Many of these areas were in high southern latitudes.

Following the Permian climates were apparently quite equable. Coals were abundant in high latitudes and show conditions were cool and wet. Evaporites were common in the mid-latitudes and reefs and bauxites occurred in the Tropics. The Triassic period may have been 10–20°C warmer than present, and certainly much warmer than the Permian. During the Triassic evaporites are less abundant and coals become more abundant. This probably indicates generally cooler and moister conditions. The great abundance of coral suggests the oceans were warm, and the occurrence of ferns to 50–60° latitude suggest warm climates extended well away from the Tropics. This is also the period of large reptiles which also implies warm climates (Fig. 7).

The Cretaceous is a period where geological evidence of climate becomes relatively abundant for the first time. This is because there is a greater degree of preservation and geographical spread of basins which contain key evidence in their sediments. Evaporites were increased in abundance compared to the Jurassic, coals retreated to higher latitudes and corals extended over a greater latitudinal range (Fig. 8). Stable isotopes show oceans were warmer and sea levels were higher. The extension of coals to high latitude is difficult to reconcile with the fact that there must have been long periods of the year with low light conditions. The reasons for the greater warmth may be due to elevated atmospheric CO_2, or possibly solar radiation being higher than present. Attempts have been made to model the climate of the Cretaceous and it is suggested that the palaeogeography may have been the main cause for a weakening of a CO_2 effect and a change in the response of the water cycle (Donnadieu et al. 2006).

By the Tertiary there is abundant geological evidence of climate, and palaeomagnetism allows palaeolatitudes to be established with some certainty. In general the Palaeocene is cooler compared to the Cretaceous and this cooling trend continues into the Eocene and early Oligocene. Palms were present in

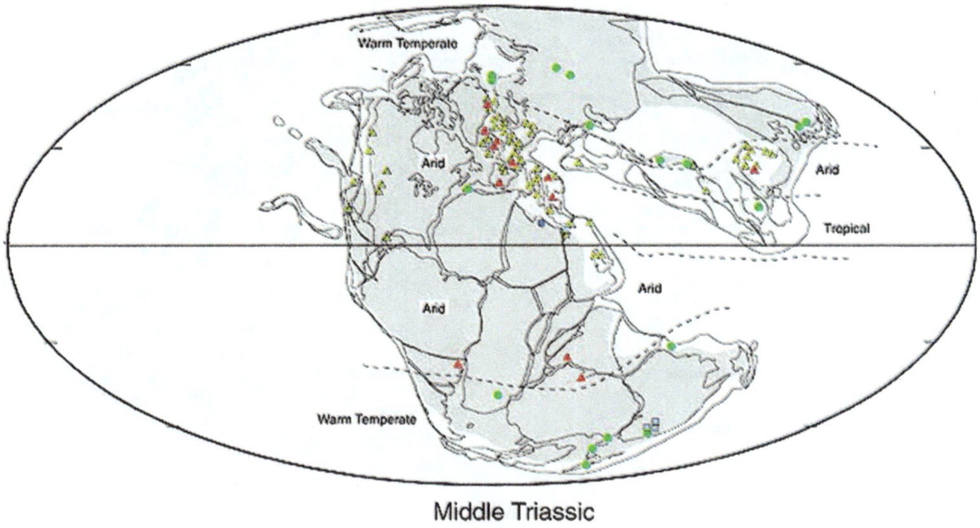

Fig. 7 Continental configuration and evidence of climate in the Triassic period (courtesy of Paleomap Project)

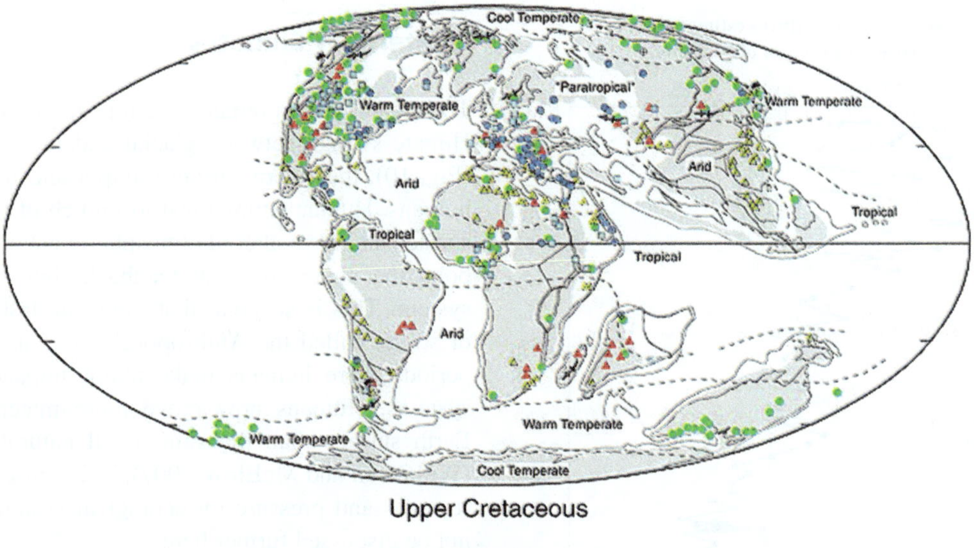

Fig. 8 Continental configuration and evidence of climate in the Cretaceous period (courtesy of Paleomap Project)

relatively high latitudes in Japan, Europe and Canada, suggesting subtropical to tropical floras for those regions. However there are several locations where evaporites and lignites co-occur. Perhaps the best current hypothesis to explain this is that climates generally had greater seasonality, perhaps due to the configuration of Earth's orbit around the Sun.

Around the Oligocene temperatures decreased 4–5°C and sea levels generally fell. The world's temperatures were moving to within glaciation range but glacial ice has not been widely recognised, possibly due to the continental configuration at the time and the fact that there was sufficient ocean through flow through the Indonesian region and between the North and South American continents to keep the oceans mixed and an overall efficient conveyer of heat around the globe.

The late Tertiary, the Miocene (Fig. 9) and Pliocene show marked signatures of cooling. Much of the literature up to a decade ago refers to the late Miocene and Pliocene climates as being relatively benign. However, now we know that these periods show some variability,

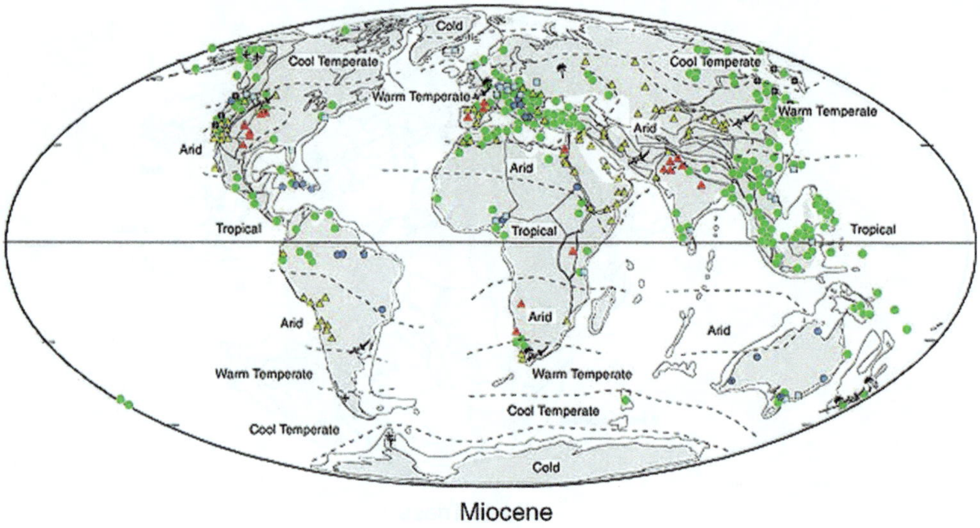

Fig. 9 Continental configuration and evidence of climate in the Miocene period (courtesy of Paleomap Project)

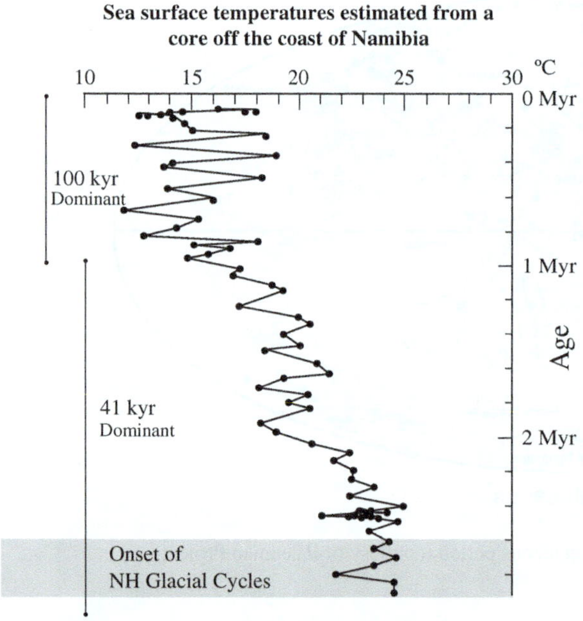

Fig. 10 Sequence of Quaternary climates interpreted from a marine core through sea floor off the coast of Namibia. Note general reduction in temperature from about 2.2 Myr and great variability since that periodit is even more variable from about 0.7 Myr (redrawn from Poore 2007)

so while the overall trends were to cooling, the records are punctuated by warm intervals (e.g. Dodson and Lu 2005; cf. Fig. 10). Antarctica had permanent ice sheets by the mid to late Miocene, and ice sheets in the Arctic were present from the mid to late Pliocene.

Quaternary Climates

The Quaternary is remarkable for the rapid series and climate swings between glacial and interglacial time (Fig. 10); they were certainly important in the rise of humans. Humans now transform much of the planet's surface and have altered atmospheric chemistry to the point that there are great feedbacks into the climate system. This is so great that often the last 100 years or so are called the Anthropocene as a new geologic period where humans make major impact on Earth systems. Humans now move more material on the Earth surface than the sum of all natural processes (Wilkinson and McElroy 2007). This causes huge disruptions and pressure on ecological systems but will not be discussed further here.

The beginning of the Quaternary is now put at about 2.5 mya when a reversal of the Earth's magnetic field occurred makes a useful marker. This is at just after the time that the first permanent ice began to form in the high northern latitudes, and when rather large shifts between glacial maxima and interglacial minima in ice extent began.

It is believed that the shifts are driven primarily by variations in Earth's orbital parameters around the Sun, and these will continue (see Fig. 11). These involved the following:

i. Eccentricity (the amount by which an orbit deviates from a perfect circle) in the case of Earth; this cycle is about 100,000 years.

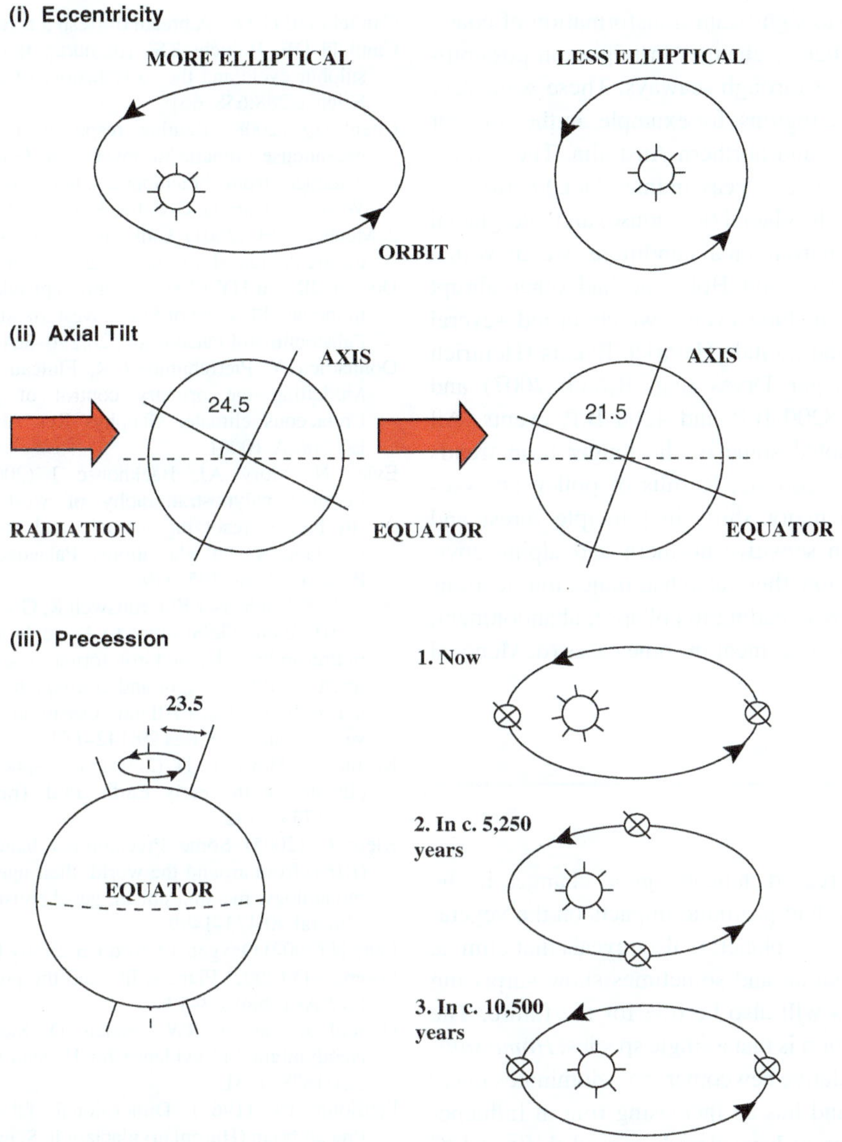

Fig. 11 Variations in Earth's orbital parameters

ii. Axial tilt (the angle between an object's rotational axis and a line perpendicular to the plane of its orbit) in the case of Earth; it has a period of about 41,000 years.
iii. Precession of the equinoxes (slow and continuous change in orientation of an astronomical body's rotational axis, like a toy top spinning) in the case of Earth, a cycle of about 19,000–23,000 years

These perturbations are predicted from Newtons Laws of orbital motion; Milankovitch (in the 1920s) drew attention to their possible role in driving the ice extent (maxima/minima) that is seen in the geological record. This has become well established, initially from records based on stable isotopes of oxygen from ocean sediments, and later from palaeosol-loess sequences in China, and continental lake drilling programs, for example, from Russia (Lake Baikal) and Bolivia. As far as we know these shifts were more rapid and more frequent than for any other geological period. Figure 10 shows one sequence from the Atlantic which represents global climate changes for the whole of Quaternary time.

An important aspect of these changes is that ice sheet extension in glacial periods is fuelled by massive amounts of water being transferred from the oceans, and this has led to significant transformation of coastlines and in particular changes in migration possibilities across land or through seaways. These were very dramatic in some regions, for example, in the corridor between SE Asia and northern Australia. The reverse occurs when ice sheets decay in interglacial periods.

In addition to the glacial (ice-house) and interglacial (warm-house) environmental conditions, we know that the late Pleistocene and Holocene had other abrupt changes. These include events which lasted several hundred years and include Heinrich Events (Heinrich 1988), the Younger Dryas (e.g. Bjőrck 2007) and the short lived 8200 B.P. and 4200 B.P. events. All of these interrupted some of the longer term trends driven by orbital forcing. Results of pollen analyses certainly reveal major shifts in principle forest and herbland taxa in sensitive northern and alpine environments. We know these also had major implications for human cultures, leading to collapse, abandonment, resettlement or replacement in some cases (deMenocal 2001).

The Future

The geological record indicates great changes in the past, which have had profound impacts on the vegetation and fauna of the planet; it also reveals that climate systems are dynamic and sometimes show surprising abruptness. This will also be true for the future. The added complication is that a single species, *Homo sapiens sapiens*, a relative newcomer, now dominates many Earth systems and has an increasing role in influencing climate. Humans have already caused shifts of 1°C or more in temperate regions, and greater changes in high latitudes. These changes are expected to increase in magnitude over the next few decades and the struggle between life forms and climate will continue in its ever fascinating way.

References

Bjőrck S (2007) Younger Dryas Oscillation, global evidence. In: Elias SA (ed) Encyclopaedia Quaternary Science, vol 3. Elsevier, Amsterdam, pp 1985–1993

Canfield DE (1998) A new model of Proterozoic ocean chemistry. Nature 396:450–452

Canfield DE (1999) A breath of fresh air. Nature 400:503–504

Canfield DE, Habicht KS, Thamdrup B (2000) The Archean sulphur cycle and the early history of atmospheric oxygen. Science 288:658–661

Chen ZQ (2008) Benthic response to Permian icehouse-greenhouse climatic turnover from Gondwana and Tethys: evidence from brachiopod faunas in South China and Western Australia. Geophys Res Abs 10:EGU2008-A-07781

deMenocal PB (2001) Cultural responses to climate change during the late Holocene. Science 292:667–673

Dodson JR, Lu HY (2005) Salinity episodes and their reversal in the late Pliocene of South Western Australia. Palaeogeogr Palaeoclimatol Palaeoecol 228:296–304

Donnadieu Y, Pierrehumbert R, Fluteau F, Jacob R (2006) Modelling the primary control of paleogeography on Cretaceous climate. Geophys Res Abs 8:1607-7692/gra/EGU06-A-10061

Eyles N, Mory AJ, Backhouse J (2002) Carboniferous–Permian palynostratigraphy of west Australian marine rift basins: resolving tectonic and eustatic controls during Gondwanan glaciations. Palaeogeogr Palaeoclimatol Palaeoecol 184:305–319

Hamade T, Konhauser KO, Raiswell R, Goldsmith S, Morris RC (2003) Using Ge/Si ratios to decouple iron and silica fluxes in Precambrian banded iron formations. Geology 31:35–38

Heinrich H (1988) Origin and consequences of cyclic ice rafting in the northeast Atlantic Ocean during the past 130,000 years. Quaternary Res 29:142–152

Kasting JF, Howard MT (2006) Atmospheric composition and climate on the early Earth. Phil Trans R Soc Lond B 361:1733–1742

Klein C (2005) Some Precambrian banded iron-formations (BIFs) from around the world: their age, geological setting, mineralogy, metamorphism, geochemistry, and origins. Am Mineral 90:473–1499

Lane N (2002) Oxygen. Oxford University Press, Oxford

Lorenz RD (2002) Planets, life and the production of entropy. Int J Astrobiol 1:3–13

O'Neill J, Carlson RW, Francis D, Stevenson RK (2008) Neodymium-142 evidence for Hadean mafic crust. Science 321:1828–1831

Pettijohn FJ (1962) Dimensional fabric and ice flow, Precambrian (Huronian) glaciation. Science 135:442

Pollard D, Kasting F (2005) Snowball Earth: a thin-ice model with flowing sea glaciers. J Geophys Res 110:C07010

Poore RZ (2007) Pliocene environments. In: Elias SA (ed) Encyclopaedia of Quaternary Science, vol 3. Elsevier, Amsterdam, pp 1949–1958

Robert F, Chaussiden M (2006) A paleotemperature curve for the Precambrian oceans based on the silicon isotopes in chert. Nature 443:969–972

Scheffler K, Hoernes S, Schwark L (2003) Global changes during Carboniferous–Permian glaciation of Gondwana: linking polar and equatorial climate evolution by geochemical proxies. Geology 31:605–608

Shields GA, Kasting JF (2007) Palaeoclimatology: evidence for hot early oceans? Nature 447(7140):E1

Valley JW, Peck WH, King EM, Wilde SA (2002) A cool early Earth. Geology 30:351–354

Wilkinson BH, McElroy BJ (2007) The impact of humans on continental erosion and sedimentation. Geol Soc Am Bull 119:140–156

Development of Intertidal Biotas Through Phanerozoic Time

Markes E. Johnson and B. Gudveig Baarli

Abstract

Changes in the biodiversity and organization of intertidal biotas from rocky, sandy, and muddy shores are summarized on the basis of information in the fossil record for 1622 extinct and extant species through Phanerozoic strata from the Cambrian to the Pleistocene at 361 localities around the world. To enter the database, each fossil species qualified as intertidal in origin based on sedimentological, geological, and other spatial criteria. Among the study sites documented in the scientific literature, 45% are considered former rocky shorelines. Another 31% represent former muddy shores and 24% are indicative of former sandy shores. Rocky-shore biotas demonstrate the greatest change in biodiversity, with species per Cenozoic study site nearly 2.5 times more than found on average at Paleozoic study sites. Key elements of the modern rocky shore biota were in place by Oligocene time and reflect much the same kind of ecological crowding found in that setting today. Coastal mudflat biotas show only a minor increase in biodiversity based on body fossils, although evidence from trace fossils implies an increase in ecological crowding through time. Sandy-shore biotas are the most conservative and least diverse in their development. The influence on intertidal habitat space by global tectonics, sea-level change, and relationship to other ecosystems is considered.

Keywords

Phanerozoic • Recent • Intertidal habitats • Rocky • sandy and muddy-shore biota • Ecological crowding • Sea-level change • Relationship to other ecosystems

When we go down to the low-tide line, we enter a world that is as old as the Earth itself – the primeval meeting place of the elements of earth and water.
Rachel Carson, The Edge of the Sea

Introduction

The transition between land and sea is the most abrupt and coextensive ecological boundary on Earth. Marine invertebrates and algae from the intertidal zone tolerate

M.E. Johnson (✉)
Department of Geosciences, Williams College, Williamstown, MA 01267, USA
e-mail: markes.e.johnson@williams.edu

J.A. Talent (ed.), *Earth and Life*, International Year of Planet Earth,
DOI 10.1007/978-90-481-3428-1_5, © Springer Science+Business Media B.V. 2012

harsh physical conditions over a wide range of latitudes under different climatic settings. Many species from intertidal communities also demonstrate a high degree of biological interaction and interdependence. Marine biologists from North America began a new tradition by making keen observations on species relationships in the context of persistent ecological settings for specific intertidal habitats. On the Pacific seaboard, Ricketts and Calvin (1939) and Ricketts et al. (1985) abandoned the strictures of a formal taxonomic register commonly used in guidebooks. In particular, organisms were described together from shared environmental settings, such as open-coast rocky shores as distinct from protected rocky shores and open-coast sandy beaches as differentiated from sand flats and mudflats in protected embayments. Similarly on the Atlantic seaboard, Carson (1956) struck a bold ecological theme by bringing readers on a grand north-south tour from rocky shores to the sandy beaches on barrier islands to coral shores. Underlying concepts of ecology were emphasized by these pioneers through a novel sense of organization based on environmental parameters. Biological narrative also became more engaging. The dedicated labors of Stephenson and Stephenson (1972) expanded studies on the intertidal life of rocky shores to a global level. Likewise, studies on the dynamics and ecology of sandy beaches were accorded global status through the sustained work of Brown and McLachlan (2002) and McLachlan and Brown (2006). Mud flats have yet to achieve a comparable level of global scrutiny as a biological repository. Case studies for particular geographic areas abound, such as the mud flats of Germany's North Sea coast (Reineck 1975) or California's Bodega Bay (Ricketts et al. 1985).

The natural history of the intertidal zone illustrates countless variations on the theme of biological interactions. Carson (1956), for example, gives us the horse mussel (*Modiolus modiolus*) as a classic umbrella species from the lower intertidal zone on rocky shores. She describes the bivalve's dense tangle of binding byssal threads as providing safe cover for a rich understory of life that includes scale worms, brittle stars, and the young of many echinoderms, crustaceans, and mollusks. Higher in the intertidal zone on rocky shores, Carson (1956) described the predatory efficiency of the dog whelk (*Thais lapillus*) as a critical factor in the shifting competition for living space waged between the blue mussel (*Mytilus edulis*) and the common rock barnacle (*Balanus balanoides*). The barnacle is the preferred prey of this carnivorous gastropod, but when barnacle populations are sufficiently reduced, the blue mussel expands its occupancy to newly opened space and the dog whelk switches to a mussel diet.

Ricketts et al. (1985) tells us that the physical obstacles encountered by sand dwellers on exposed sand beaches are extreme to the limits of insuperability. Yet, we are treated to the wonder of the small bean clam (*Donax gouldii*), which survives the pounding surf to maintain its place in the intertidal zone as a swift and tireless digger. Most improbably, the hiding place of the prolific bean clam is sometimes revealed by the delicate tuffs of a hydroid (*Clytia bakeri*) that protrude above the sand surface but are securely anchored to the clam's umbo region. From the California mudflats also explored by Ricketts et al. (1985), we learn about commensalisms epitomized by the fat innkeeper, the echiuroid worm (*Urechis caupo*). Living in a large U-shaped burrow, this creature maintains such a strong circulatory feeding current that permanent guests representing various species of scale worms, crabs, and small fish enjoy both the dwelling's safety and the host's munificent larder. Whether or not such relationships are typical of the community fabric within the intertidal zone, they are based on observations tested and verified by multiple sources.

More than others from her generation, it was Rachel Carson who sensed the primordial character of the intertidal zone. She understood that the moment Earth achieved a permanent ocean, its margins became a potential habitat for colonization by early life. How those colonizing organisms differed from today's intertidal life or how rapidly the convergence took place, she could only guess. The primary intent of this chapter is to survey the Phanerozoic rock record with respect to fossil biotas from the intertidal zone. In part, this contribution updates the study by Johnson and Baarli (1999) on fossil biotas from former rocky shores. It also represents an expansion to include fossil biotas from former sandy shores, intertidal sand flats, and mud flats. Paleontologists have devoted much effort to tracing the diversity of marine faunas through geological time using familial and generic levels of classification, but the databases so assembled for that purpose (Sepkoski 1981, 1997; Peters and Foote 2001; Stanley 2007) focus almost exclusively on life from subtidal, level-bottom marine habitats. To the extent that intertidal habitats are included in such studies, fossil taxa associated with peritidal settings are sometimes admitted. The rocky-shore setting has a history of neglect by paleontologists because it is widely regarded as an environment of net erosion in a high-energy regime and not one where organisms were likely to be preserved as fossils. The same bias applies to exposed sandy shores as an unlikely setting for fossilization, but sheltered sand flats represent a more stable setting.

The secondary goal of this project is to consider the history of species interactions in the intertidal setting. Some authorities (Bambach and Bennington 1996; Stanley 2008) hold that niche partitioning in marine benthic communities is minimal and that the appearances and disappearances of species in particular microhabitats is a function largely of chance. Others (Vermeij 1987) argue that marine communities show evidence of evolution with the escalation of species interactions through time under the influence of predation and commensalism. To what extent are the claims made by marine biologists such as Ricketts and Calvin (1939), Ricketts et al. (1985), Carson (1956), and Paine (1966, 1984) on the importance of species interactions in the intertidal zone unusual in the context of the broader marine world? How far back in geological time might such relationships be reliably traced?

The world's shorelines encompass about a million kilometers in total length (Bird and Schwartz 1985). With regard to physical geography, the intertidal zone traverses only a few physiographic classes that tally significant distances as relative proportions within this composite coastline. Based on a survey of more than 587,000 km (Johnson 1988a), it is estimated that approximately 33% of the world's present-day coastlines are exposed as rocky shores. A separate canvass of 2200 barrier islands (Pilkey 2003) concluded that barrier islands with outer sandy shores front 12% of all open-ocean shores in the world. Not all sandy shores, however, are related to barrier islands. The 1400-km coastline of Namibia in southwest Africa is 80% dominated by sandy beaches directly attached to the coastal plain, for example, and the 9000-km coastline of India likewise is about 50% dominated by sandy beaches (Bird and Schwartz 1985; Johnson 1988a). Overall, 20% of the world's coastline might be configured as sandy beaches of one grade or another. Considering that the flanks of barrier islands facing coastal plains typically support mud flats, another 12% of the world's coastline probably comes under that designation. Other mud flats commonly are associated with ria coastlines, as found along the shores of the Minas Basin (Nova Scotia), Chesapeake Bay, Delaware Bay, San Francisco Bay, and Cook Inlet (Alaska) in North America. Taken together, 17% of the world's coastline might be regarded as bordered by intertidal mudflats. If such estimates are crudely accurate, rocky shores, sand flats, and mud flats account for perhaps 70% of the world's intertidal zone. These are the principal physiographic categories treated in this study. Other kinds of shores include estuaries, mangrove coasts, and high-latitude ice shelves, none of which come under consideration here. The place of estuaries as a brackish-water ecosystem in geological history has been discussed with respect to trace fossils (Buatois et al. 2005) and microfossils (Schröder-Adams 2006). Traced back to Paleocene times, the comparatively short geological history of the mangrove ecosystem also has received prior attention (Plaziat et al. 2001).

Representative Images of Modern Shorelines

Our experience in today's world shapes how former environments may be perceived with respect to the realities of physical geography. The following images provide iconic views of modern rocky-, sandy-, and muddy-shore environments. A celebrated starting place for the appreciation of rocky shores and rocky-shore biotas is the granite coastline of Monterey Bay near Pacific Grove, California (Fig. 1). It was there Ed Ricketts operated his Pacific Biological Laboratories during the 1930s and 1940s. Granite is widely exposed on the Atlantic coast of Maine and throughout most of the Seychelles Islands in the Indian Ocean. The coast of Sardinia on the Tyrrhenian Sea is notable for its granite exposures. In the western Pacific, parts of Japan's Shikoku Island and some islands off the coast of Queensland, Australia (i.e., Magnetic and Lizard islands) are granitic in origin. Granite shores, however, are surpassed in overall coastal frontage by other igneous rocks such as basalt and andesite. Basalt is the most common rock that forms where continents rift apart, as with Africa and South America astride the South Atlantic Ocean. Andesite shores have a significant distribution along the west coast of South America, where the rock name is derived from the far-reaching Andes Mountains. Resistance to erosion is the key attribute that contributes toward the

Fig. 1 Modern rocky shore from a temperate setting near Pacific Gove on Monteray Bay, northern California, USA. Ed Ricketts (1897–1948) studied the biota from tidal pools formed by coastal exposures of granite in this area

durability of rocky shores. Sedimentary rocks are well represented in modern rocky shores. The strength of sandstone typically depends on the strength of the cement that binds sand grains together. Limestone also makes strong rocky shores, but is vulnerable to karst erosion in a humid climate. Some metamorphic rocks, such as phyllite with high mica content, are prone to easy mechanical breakdown by wave erosion, whereas others such as quartzite are among the hardest, most durable rocks found anywhere.

Sandy shores are familiar, because we enjoy visits to the beach for recreational activities. Beaches near large population centers get heavy use (e.g., Costa del Sol near Barcelona, Spain; Gold Coast near Brisbane, Australia; the sands of Ipanema and Copacabana in Brazil's Rio de Janeiro; extensive beaches on both the Gulf of Mexico and Atlantic coastlines of Florida). The population of La Paz in Baja California Sur, Mexico, frequents playas Pichilingüe and Tecolote but comparable sandy shores occur on nearby uninhabited Isla Cerralvo (Fig. 2). Granite is commonly exposed in the Cape Region of Baja California Sur, where erosion is abetted by thermal springs that follow fractures aligned with many canyons leading to the coast. Numerous beaches dot the circumference of Isla Cerralvo, washed by vigorous long-shore currents that carry an enormous load of silica sand supplied by small fan deltas located at the mouths of island canyons. The sand on such beaches tends to be well sorted for grain size and composition. Not necessarily located near large population centers, some of the world's most popular vacation spots feature "white-sand" beaches derived from the breakage of corals and shells that provide bioclasts of bleached calcium-carbonate material.

Fig. 2 Modern sandy shore from a sub-tropical setting in the Gulf of California west of La Paz on Isla Cerralvo, Baja California Sur, Mexico. Quartz sand is transported by strong long-shore currents that form asymmetrical ripple marks perpendicular to shore. Photo taken after a spring tide with a fall of about 1.5 m

Among some the most frequently visited beaches are those at Trunk Bay on St. John in the U.S. Virgin Islands or nearby at The Baths in the British Virgin Islands; Pink Sand Beach on Harbour Island, The Bahamas; Cancun, on Mexico's Yucatan Peninsula; and numerous beaches throughout the Maldives in the Indian Ocean. Beaches with large volumes of carbonate sand are mostly restricted to the tropics, where the biodiversity of corals and mollusks with hard, calcified parts is high. Abundant calcareous foraminifera contribute small tests to the pink sands found on some beaches in The Bahamas. Golden sands are characteristic of certain beaches on the Malay Peninsula off the Andaman Sea, as found near Krabi, Thailand (Fig. 3). The color comes from a profusion of small bivalve shells that thrive in the intertidal and shallow subtidal zones adjacent to the shore (Fig. 4).

Fig. 3 Modern shelly shore from a tropical setting on the Malay Peninsula off the Andaman Sea near Krabi, Thailand. The beach is composed almost entirely of small mollusk shells living in the intertidal to shallow subtidal zones

Fig. 4 Close-up view of small bivalve shells at low tide on the beach near Krabi, Thailand (pen for scale: 15 cm). Most of these bivalves represent infaunal species

Coastal mud flats are well developed in sheltered areas, such as the leeward side of barrier islands or the inner recesses of enclosed bays. The lower the ratio of sand to mud, the more difficult it is to traverse the intertidal zone on mud flats. Not surprisingly, these coastal areas are less popular with the general public unless some other attraction is available. Birders are attracted to coastal mud flats for the special avifauna it supports. Foraging for clams is another activity that brings visitors to the mud flats. Intertidal sand flats result where the tidal range is high and an adequate source of sand exists. The phenomenal tidal range in the Minas Basin of Nova Scotia draws a regular stream of tourists to witness mud flats and sand flats that stretch far offshore at low water (Fig. 5). It is possible to venture out on the flats during low tides to explore the tracks, trails, and burrows of the many marine invertebrates that thrive there. In the tropics, sheltered lagoons may accumulate white "mud" that is entirely carbonate in origin. Burrow systems, such as those made by callianassid ghost shrimps, are well developed in such settings.

Fig. 5 Modern intertidal mud flat viewed from the harbor near Bass River at low tide, north side of the Minas Basin, Bay of Fundy, Nova Scotia (Canada)

Relationship of Global Tectonics and Coastal Geomorphology

Beyond rudimentary accuracy, it is a challenge to assess the relative proportion of rocky shores to sand flats and mud flats along the worlds's present-day shores. To delimit proportionality among these physiographic classes for the geographies of past worlds is an impossible task. It can be claimed with certainty, however, that the net overall amount of shoreline available during any given interval of geologic time varied according to the general state of global tectonics then in play. Triassic worlds that coalesced around the mega-continent of Pangaea had a composite coastline significantly reduced in total length compared to progressively more recent worlds with smaller, increasingly dispersed continents. In a spatial-metric analysis of 26 paleogeographic maps, adopted from Smith et al. (1994), and compared to the present world, measurable variation was traced with respect to composite shore length back through Cenozoic and Mesozoic times to the Triassic Period (Johnson 2006). Most noteworthy, the assembly of continents during Middle Triassic (Ladinian-Anisian) times shows a net shoreline reduction of 44% compared to today (Fig. 6). With fewer kilometers in circumference available during the assembly of mega-continents, it is evident that a reduction in space for the intertidal zone occurred across all geomorphic classes of coastlines. With the collapse of subduction zones around the periphery of a mega-continent prior to its tectonic disassembly, it is possible that rocky shorelines became especially scarce. Early to Middle Paleozoic geographies with several independent paleocontinents extraneous to Gondwana would have registered longer composite shorelines compared with Triassic Pangaea, but were still significantly less than those found in the present-day world.

The only interval during the last 250 million years during which net shoreline length exceeded the present day occurred near the start of Cenozoic (Thanetian-Danian) time, when an additional 5% of shoreline was added due to a substantial rise in sea level that brought major seaways into the continental interiors of North America, South America, Europe, and Africa (Fig. 7). The invasive seaways of those times promoted the development of ria shores with substantial mud flats. Otherwise, the spatial disposition of rocky shores and sandy shores (with or without barrier islands) was largely controlled by plate tectonics in the context of active as opposed to passive continental margins. These also are known as collisional and trailing margins, respectively. Present-day rocky shorelines typically occur in greater concentrations on tectonically active coasts, compared with passive coasts (Johnson 1988a). For example, the western seaboard of North America from Alaska to Pacific Mexico entails a shoreline of 31,500 km, 50% of which is dominated by rocky shores and only 5% by beaches on barrier islands. The opposite statistics characterize the eastern seaboard of North America. The coast from Maine through the Gulf of Mexico entails a shoreline of 112,500 km, 50% of which is dominated by open sandy shores and only 6% by rocky shores. Estuarine, deltaic, and mangrove shores are particularly well represented along the eastern seaboard to make up much of the difference. Similar statistics are derived for active as opposed to passive shores in South America and

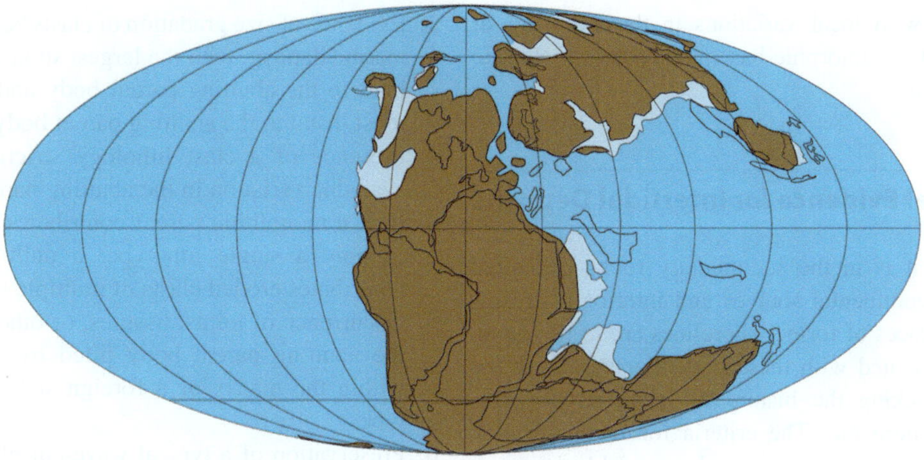

Fig. 6 Pangaean assembly of continents during Middle Triassic (Ladinian-Anisian) times demonstrates a smaller composite shoreline compared to the world today

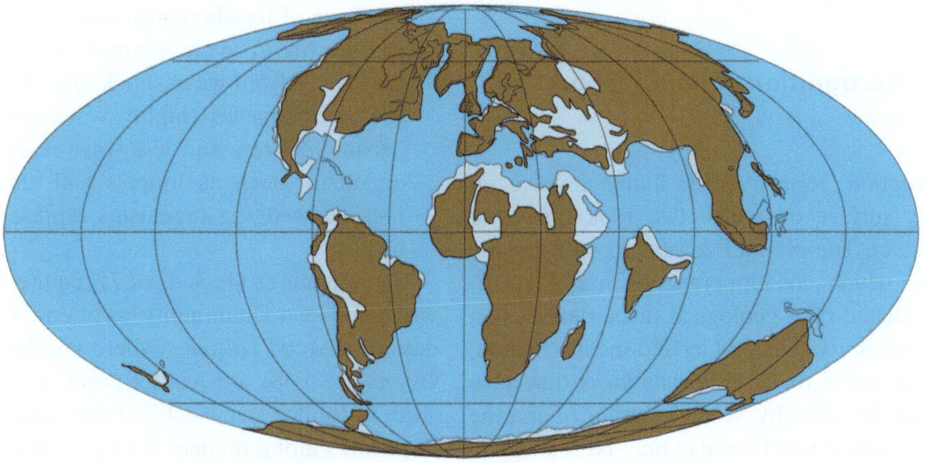

Fig. 7 Early Cenozoic (Thanetian-Danian) distribution of continents during a higher stand in sea level resulted in a modestly larger composite shoreline than found in the world today

Western Europe (Johnson 1988a). In a quantification of the world's 2200 barrier islands in terms of shore length, Pilkey (2003) calculated that open sandy shores occur on trailing continental margins more frequently than on collisional margins by a factor of ten-to-one. In terms of raw numbers of islands, 73% of all barrier islands occur on passive margins. Neo-trailing margins where the tectonic breakup of continents has been geologically recent make an exception to the reciprocal relationship between rocky and sandy shores. The shores of the Red Sea, for example, have a paucity of barrier islands but fringing coral reefs are associated with fault scarps on basement rocks dominated by flood basalt or other igneous rocks.

Over the short term, barrier islands with open-ocean sandy beaches are capable of lateral migration in concert with the rise and fall of sea level across low-relief continental plains on passive margins. The relative longevity of barrier islands due to this kind of rollover activity has been aptly likened to the "tank-tread method" of mobility (Glaeser 1978). In contrast, rocky shores are fixed solidly in place and the role of active tectonic margins is decoupled from the generation of rocky shores during major rises in sea level when continental interiors become extensively flooded. Paleocontinents widely flooded during Early to Middle Paleozoic times, for example, developed rocky shorelines much less dependent on plate tectonics and more in keeping with local variations in the exposure of igneous and metamorphic basement rocks resistant to shore erosion.

Geological Evidence for Intertidal Deposits

Ultimately, it is in the sedimentary rocks deposited mostly on continental shelves and interior platforms where the traces of former shorelines are to be found. Fossils associated with these particular strata are the key to unlocking the history of intertidal life during the Phanerozoic. The criteria for recognition of coastal deposits can be somewhat variable in terms of diagnostic value. The following sections review general guidelines for the identification of strata that accumulated in coastal settings.

Criteria for Recognition of Rocky-Shore Deposits

Without exception, former rocky shores involve an unconformity surface on which two rock types are in direct contact. The original coastal rocks form the basement on which contemporaneous marine sediments accumulated. The geological discordance is an angular unconformity, where older sedimentary strata, layered metamorphic rocks, or basalt flows that have been tilted are overlain by younger strata of marine origin. The shoreline interface also may be found as a nonconformity, in which intrusive igneous rocks were exposed at the Earth's surface and covered by marine strata younger in age. Thirdly, the contact also may be identified as a disconformity between older sedimentary rocks and younger marine strata. All such variations share a history showing evidence of subaerial exposure of a pre-existing rock type and the subsequent burial of an erosion surface by sedimentary rocks deposited in water.

The following features are diagnostic for the recognition of former rocky shorelines preserved in the context of geological unconformities (after Johnson 1988b):

1. Occurrence of a basal conglomerate with a map distribution parallel to the strike of the unconformity.
2. Occurrence of an abutting parent body of rock that serves as the source of eroded clasts.
3. Evidence of size gradation of clasts belonging to the marine deposit, with the largest situated most proximal to the abutting parent body and the smallest most distal to the abutting parent body.
4. Presence of a clast lithology strictly limited by available variation in the abutting parent rock body (i.e., a monolithic parent contributes eroded clasts of only a single lithotype; a multi-lithic parent contributes eroded clasts of multiple lithotypes.
5. Occurrence of joints, fissures, or other openings in the abutting parent body filled by smaller clasts within the matrix of a foreign sedimentary lithotype.
6. Preservation of a typical wave-cut platform or sea cliffs incised in the abutting parent body.
7. Exhumed tide pools, sea stacks, sea caves, or sea arches.
8. Presence of fossils representing organisms specialized for life in an environment with high to moderate energy from wave shock: encrusting organisms such as barnacles (highly wave-resistant) to bryozoans (delicate and less wave-resistant); clinging organisms such as limpets and chitons; encrusting and wedging organisms represented by many bivalves.

Some features (1, 3, 4, 6, 7) require huge outcrops with good lateral continuity, as often found exposed in desert or Arctic latitudes where vegetation is limited. Others (2, 5, 8) are just as prone to occur in exposures of limited size, such as road cuttings or quarries. Exposures along modern rocky shores often preserve a history of events that took place during episodes of higher sea level (i.e., rock terraces and other abutment unconformities). The paleotopography represented by an abutment on an unconformity surface may show dimensions from a few meters in relief to 50 m or more.

An outstanding example of a former rocky shore that embodies many of the criteria enumerated above, is demonstrated by the Upper Ordovician Port Nelson Formation in contact with Proterozoic quartzite near Churchill, Manitoba on the present-day shores of Hudson Bay (Johnson et al. 1988). A large quartzite sea stack surrounded by Upper Ordovician limestone is one of the novelties of this locality (Fig. 8). Due to its position about 45 m seaward from the high tide line, the fossil sea stack is accessible only during low tide.

Fig. 8 An Upper Ordovician rocky shoreline eroded from Proterozoic quartzite near Churchill, Manitoba, Canada, exhumed along the present-day shores of Hudson Bay. The quartzite sea stack in the foreground is 4 m wide and 10 m long

Criteria for Recognition of Tidal Sandy-Shore Deposits

Several kinds of wave-dominated beaches accrue under tidal conditions, subject to variable wave energy by latitude and periodic storm activity. Tropical beaches tend to register lower-energy conditions and to accumulate coarser sand, including bioclastic sand derived from shelly material. Shores in more temperate latitudes generally encounter higher wave energy under stronger tidal cycles with the result that the sand is finer grained and better sorted. Cold water also precludes the prolific accumulation of bioclastic material derived from shelly (carbonate) materials. The two latitudinal extremes are represented by reflective-beaches and dissipative beaches (McLachlan and Brown 2006). Reflective beaches are more typical of tropical latitudes, where the tidal range is small and wave energy is usually low. Under these conditions, virtually all the sediment is maintained on the intertidal beach and backshore. There is no offshore surf zone and wave surge is limited to the beach face. Dissipative beaches are more typical of temperate latitudes, where the tidal range is greater and waves may exceed 2 m in height. The more vigorous the waves, the more a beach is cut back and its sediments are transferred seaward to a surf zone. Under fully dissipative conditions, a beach takes on a flat profile and the sediment load is moved to a succession of sandbars that are oriented parallel to the beach across a broad surf zone. Waves start to break well offshore in the surf zone and may reform and break one or more times before reaching the actual shore.

Storms affect beaches at all latitudes and alter beach morphology on a dynamic basis as a consequence of storm frequency and strength. Thus, an intermediate beach morphology is one in which an ongoing alternation between dissipative and reflective conditions takes place. During a major storm, any given beach (irregardless of latitude) may be planed down to a dissipative state. As calm returns after the storm, sediment starts to shift landward to build a system of long-shore bars and intervening troughs. Initially, a single bar may form in the outer surf zone, but inner bars develop as waves reform and break closer to shore. When conditions become even calmer, beach cusps may start to form, leading to the development of transverse bars (perpendicular to the shore) with intervening rip zones. Flat calm promotes formation of a low-tide terrace, on which ridges and runnels represent the remains of earlier bars and troughs. Given sufficient time in a becalmed state, all or most of the sand on the terrace will shift onto the shore to reconstruct a reflective beach (McLachlan and Brown 2006).

The following features are diagnostic for recognition of strata formed by sediments from tidally influenced beaches somewhat intermediate in state between fully dissipative and fully reflective conditions (after de Vries Klein 1977).

1. Development of cross-stratification parallel to the trend of the sand body and topographic strike of the depositional basin.
2. Bedload transport by tidal currents with bimodal reversals of direction indicated by preservation of herringbone cross-stratification, parallel laminae, and supermature rounding of constituent grains.
3. Asymmetry of bedload transport indicated by preservation of reactivation surfaces, bimodal frequency in the distribution of set thicknesses within cross-stratification, bimodal frequency distribution of dip angle within cross-stratification, and complex internal organization of dunes and sand waves.
4. Late-stage emergence of ebb flow signified by preservation of multi-modal orientations of maximum dip direction within cross-stratification, small current ripples superimposed at 90° or obliquely on larger current ripples, interference ripples, flat-topped current ripples, and washout structures.
5. Slack-water periods signified by preservation of cross-stratification with flasers, wavy bedding, convolute bedding, and current ripples with muddy troughs.
6. Tidal scour indicated by preservation of mud-chip or shell-lag conglomerates at the base of washouts and channels and the presence of flutes and rills.
7. Preservation of trace fossils represented by tracks, trails, and burrows

All features listed above are time-dynamic in the context of rising and falling tides and cannot be found at any specific level within beach sandstone. For the most part, these features are weighted toward higher latitude beach deposits. Tropical beach sandstone includes a more substantial component of shell debris and the overall clast size is larger and more poorly sorted. In areas of high biological productivity and low run-off from land, beach deposits can be entirely bioclastic in origin. Thick stratigraphic sequences are more likely to demonstrate a range of features characteristic of different parts of a tidal cycle or the transition between storm sedimentation and normal, fair-weather sedimentation. Bedding planes provide the best exposure for sedimentary structures representing all variations of ripple marks.

The Silurian Tumblagooda Sandstone is exposed in thick sequences along the Murchison River and the shores of the Indian Ocean in Western Australia's Kalbarri National Park, where cross-bedded sandstone crowded with vertical *Skolithos* burrows are a common feature (Fig. 9). The sands were eroded from the margin of the nearby Yilgarn Block, an Archean terrain dominated by granite and gneiss, and deposited under prograding conditions in shallow, coastal waters.

Criteria for Recognition of Tidal Mud Flat Deposits

Compared with the topic of tidal sand flats, the literature on tidal mud flats is slim. Mud flats most commonly occur today in places lateral to sand flats, where rivers introduce the mud. An extraordinary example involves mud transported from the Amazon River for hundreds of kilometers along the coasts of Surinam and French Guiana on the open northeastern coast of South America (Froidefond et al. 1988). Mud flats

Fig. 9 Outcrop of Silurian Tumblagooda Sandstone (Kalbarri National Park, Western Australia) demonstrating cross stratification with extensive *Skolithos* burrows (average burrow diameter 2 cm)

also extend for considerable distances along the intertidal zone adjacent to chenier plains, over which low, swampy ground prevails. A well-known region where such mud flats have been studied in detail is located on the North Sea coast of Germany, where the flats cover an area 5 to 7 km wide and extend for 450 km (Reineck 1975). This is an open coast subject to frequent storms and high-energy waves. In temperate and high Arctic latitudes, mud flats are composed of dark, organic-rich clay. Shelly material is scarce, but thin sand lenses may be intercalated with the muddy laminae. In the tropics, however, protected bays edged by mangroves accumulate white carbonate mud derived from the fine calcite needles that grow within the tissues of green algae.

The following features are diagnostic for the recognition of strata that originated as tidally influenced mud flats.

1. Stratigraphic context of mudstone and limestone bodies is critical to show fully marine strata below and fully terrestrial deposits above (i.e., thin coal seams)
2. Overall shape of mudstone and limestone bodies tapers outward perpendicular to the strike of the basin margin
3. Internally, the occurrence of rhythmic stratification prevails with laminations that reflect graded deposition from suspension
4. Rhythmites may vary in thickness in relation to duration of the high tide; the succession may follow a pattern of gradual thickening and thinning laminations
5. Preservation of desiccation polygons in both mudstone and limestone
6. Preservation of raindrop impressions in both mudstone and limestone
7. Occurrence of birds-eye structures in limestone
8. Evidence of mud-chip conglomerate intercalated with mud laminae indicates storm activity
9. Common preservation of fecal pellets in both mudstone and limestone and presence of ostracods
10. Trace fossils, such as *Thalassinoides* burrows or *Archaeonassa* trails

Strata representative of mud flats can be extremely thick (i.e., more than 600 m: Ricketts 1994), in which case many of the features listed above may be found in one place. Small, artificial outcrops such as road cuttings and quarries, however, are just as likely to reveal diagnostic features such as desiccation polygons and raindrop imprints.

Interbedded sandstone lenses of Lower Silurian Otsquago Sandstone typically preserve negative prints of tracks and trails from the Sauquoit Shale of eastern New York State, USA (Johnson et al. 1997). Trace fossils include *Archaeonassa fossulata,* the meandering trails of made by a gastropod, and *Diplichnites*, the walking tracks of an arthropod (Fig. 10). The intertidal facies represented by the Sauquoit shale abuts a

Fig. 10 Underside of Lower Silurian Otsquago Sandstone (South Moyer Creek, eastern New York State, USA) with negative imprints of intertidal tracks and trails captured from an original mud surface in the Sauquoit Shale

former shoreline related to the Taconian land area on the margin of Silurian North America.

Method of Study

Separate databases for fossils from rocky-shore, sandy-shore, and muddy-shore deposits were assembled through a review of the published literature using a variety of search engines. A well-established and extensive bibliography on rocky-shore deposits already exists (Johnson 2006), but includes some citations for studies without accompanying paleontological data. The database was purged to retain only those references containing information on fossils, and updated to incorporate reports published since 2006. Core information on fossils affiliated with sandy-shore and muddy-shore deposits was collected using the Paleobiology Database housed at the National Center for Ecological Analysis and Synthesis (NCEAS) in Santa Barbara California. These data were supplemented by information recovered using other specialized search engines. Data on Silurian and Lower Devonian paleocommunities edited by Boucot and Lawson (1999) were particularly useful in the identification of taxa associated with mud flats and sandy shores during those particular intervals.

In collecting data on fossils associated with intertidal zones on rocky, sandy, and muddy shores through Phanerozoic time, we have assumed that the original authors of the studies, from which the data were abstracted, correctly employed the criteria by which those environments can be recognized in the rock record. With respect to rocky-shore deposits, we claim authorship of many contributions on this topic and we have visited many localities described by other authors. We regard the criteria reviewed as diagnostic for former rocky shores (above) as uniformly applied. Thus, the list of fossil rocky-shore species so selected was vetted with a high degree of assurance. Databases assembled here with respect to fossil species from sandy-shore and muddy-shore deposits are the first of their kind. Care was taken to search for explicit descriptors in the published literature (e.g., sand flat, mud flat, foreshore, and littoral zone) pertaining to intertidal deposits, as a precondition before abstracting fossil identifications. In reality, however, these divisions are not always so obvioius. Subtle differences related to wave energy may produce mudstone layers with minor components of interbedded sand, as opposed to dominant sand layers with mud drapes. In all categories, the primary controlling factor has to do with the intensity of wave impact on any given coastline. The size range and the relative sorting of clastic sediments on paleoshores serve as the major indicators of deposition on sheltered as contrasted with exposed environments. Burrowing activity among Recent bivalves is known to be strongly influenced by sediment grains size (Alexander et al. 1993). Likewise, some rocky shores occur in leeward settings protected from direct assault from wind and waves, but the vertical range of the intertidal zone on exposed rocky shores will always be greater due to the heightened reach of wave splash.

All data were entered into a program especially designed to connect with a sorting function through the system Open Office. Subsequent paleontological analysis was organized mainly on the species level for body fossils and trace fossils, although the recognition of temporal trends necessitated grouping species into higher taxa. Basic patterns in the development of rocky-shore, sandy-beach, and mud-flat biotas were outlined using spindle diagrams that track the changing diversity of constituents through time, as for example the history of bivalve richness (Phylum Mollusca, Class Bivalvia) in rocky-shore communities. The higher taxa listed for the various spindle diagrams conform to usages found in the NCEAS Paleobiology Database. Controls on the biodiversity of tidal-zone communities were further examined by comparison with the effects of mass extinctions and the recovery from mass extinctions tabulated more generally for marine benthic communities through Phanerozoic time (Sepkoski 1997). With respect to taxa associated with a sandy substrate, a rating system was employed to differentiate between organisms more common on low-energy sand flats, as opposed to high-energy beach sands.

Results

Table 1 shows distribution across the last 540 million years for data assembled on the number of fossil and extant species associated with former rocky, sandy, and muddy-shore deposits. A total of 330 published studies from the geological and paleontological literature cover a total of 361 study sites spread around

Table 1 Numbers of published studies and study sites for fossil intertidal biotas through Phanerozoic time consulted for this review. Initial number refers to the tally of bibliographic citations; following number (in parentheses) gives number of study sites. Because some studies cover multiple localities of different geological time periods, the totals at the bottom of the table may show a lower sum than otherwise apparent. See Appendices 1–3 for full bibliographic citations

Time interval	Rocky shore	Sandy shore	Muddy shore
Neogene	69 (76)	23 (23)	12 (13)
Paleogene	18 (19)	9 (9)	8 (8)
Cretaceous	24 (25)	9 (9)	13 (15)
Jurassic	9 (11)	6 (6)	7 (7)
Triassic	0	4 (4)	8 (8)
Permian	3 (3)	2 (2)	5 (5)
Carboniferous	3 (3)	1 (1)	10 (10)
Devonian	2 (2)	6 (8)	11 (13)
Silurian	6 (7)	9 (11)	15 (20)
Ordovician	10 (10)	8 (8)	6 (7)
Cambrian	7 (7)	5 (6)	5 (5)
Totals	151 (163)	82 (87)	97 (111)

the world on all continents including Antarctica. From among the 361 study sites, 45% demonstrate former rocky shorelines. Another 31% represent former muddy shores and 24% are indicative of former sandy shores. Separate bibliographies for all publications consulted in this review on former rocky, sandy, and muddy-shore biotas are provided in Appendices 1–3, wherein the citations are arranged by geological period from the Neogene to the Cambrian.

Trends in Rocky-Shore Biotas Through Time

Changes in biodiversity of rocky-shore biotas from the Cambrian forward are registered on the basis of spindle diagrams for 16 components grouped mainly by phylum or class (Fig. 11). The database entails a total of 857 fossil and extant species tallied on a running basis from one time interval to another. This number does not correct for the appearance of any given species in more than one time interval because the primary objective is to track differences in biodiversity among time intervals. Thus, the total number of species recorded in any given cluster of taxa is proportional to the width of the spindle diagram as it crosses from one time interval to another. Based on 163 study sites, there is an average of 5.25 species per site. For the most part, spindle diagrams representing the various clusters of taxa are arranged from left to right to emphasize the staggered order of appearance through time.

Major patterns of rocky-shore biodiversity over time depicted in Fig. 11 are little changed from the earlier analysis of Johnson and Baarli (1999). An important difference results from additional data on the Cambrian Period, documenting first occurrences for sponges, linguliform brachiopods, and gastropods. A modest increase in the number of rocky-shore localities from which the comprehensive taxonomic information was derived also makes some difference. Overall, this gives a more even distribution to the number of localities registered for each time interval. Data of any kind for Triassic rocky shores remain elusive, however, and Paleocene data are still sparse. New data collected since 1999 add more girth to the spindle diagram for Pliocene bivalve faunas. The result is to fill out that particular spindle diagram to support an impression of gradually increasing bivalve diversity on rocky shores through Cenozoic time (Paleocene to Pleistocene epochs). In contrast, the retrenchment in number of gastropod species living on Pliocene rocky shores remains intact. Despite this persistent notch in the gastropod data, the overall increase in combined mollusk biodiversity on rocky shores through the Cenozoic is the most salient trend found in Fig. 11. Second in prominence is the distinct peak in biodiversity that appears during the last half of the Cretaceous Period. This is strongly accentuated among the Upper Cretaceous bivalves (56 species), but also recognized in seven more out of 15 taxonomic clusters including gastropods, rhynchonelliform brachiopods, and scleractinian corals. Pleistocene gastropods register the highest diversity (120 species) of any taxonomic group living on rocky shores during any

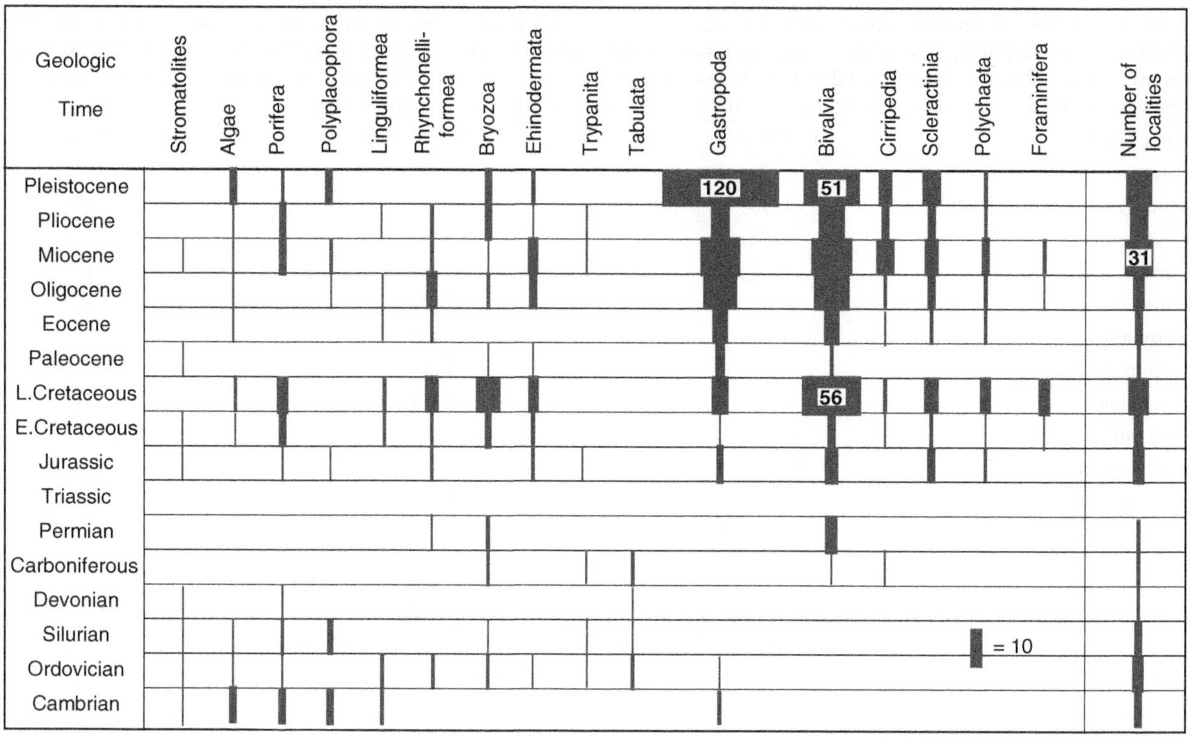

Fig. 11 Spindle diagrams, showing various components of the rocky-shore biota, mostly by phylum or class, track changes in composition and biodiversity through geologic time. Lateral scale for diversity = 10 species

given interval of Phanerozoic time. It is notable, however, that rocky-shore gastropods are out-numbered by bivalves during the Late Cretaceous. In this regard, Stilwell (2003) made the parallel observation that the imbalance between Late Cretaceous bivalves and gastropods from level-bottom communities was completely reversed by a flip-flop in diversification during succeeding Cenozoic times. Counted to eliminate duplications from one time interval to another, the actual number of rocky-shore bivalve species accruing through Phanerozoic time is 198, and the number of gastropods is 231, for a 6 to 7 ratio.

The combined data on rocky-shore biotas through Phanerozoic time support recognition of four major phases in development. These are dominated by Archaic, Paleozoic, Mesozoic, and Modern constituencies. Although elements of these constituencies are sometimes conspicuous for their longevity, younger constituencies add novelty and diversity through time. The Archaic rocky-shore biota is ruled by stromatolites (microbiolites) and encrusting algae. Under unusual conditions typically restricting competition by other organisms, the Archaic biota reappeared on rocky shores during Silurian, Devonian, Paleocene, and Miocene times. A characteristic Paleozoic rocky-shore biota (Table 2) features encrusting linguliform brachiopods, polyplacophorans (chitons), and tabulate corals as body fossils, as well as trace fossils likely to have been made by siphunculid worms (ichnogenus *Trypanites*) and acrothoracian barnacles (ichnogenus *Zapfella*). Among these, encrusting linguliformid brachiopods persist today (craniid brachiopods), but tabulate corals are last known from Carboniferous rocky shores, and disappeared completely from level-bottom communities at the end of the Permian. Boring barnacles survive today, after first appearing on rocky shores during Carboniferous time. *Trypanites* borings have a long history, first appearing during the Ordovician and last known from Pliocene rocky shores.

Rocky-shore biotas were reinvigorated and expanded during Mesozoic times by encrusting bivalves (oysters and rudists), but also boring bivalves (ichnogenus *Gastrochaenolites*). Although brachiopods went into decline as members of level-bottom communities after the Paleozoic mass extinction, thecidean brachiopods (cemented forms) and

Table 2 Paleozoic rocky-shore taxa

Group	Order	Family	Genus/species	Strata
Porifera	Actinostromatda	Actinostromellidae	*Actinostromella*	Ludlow, Silurian
	Clathrodictyida	Actinodictyidae	*Plexodictyon*	Ludlow, Silurian
	Clavulina	Clionaidae	*Cliona*	Lower Cretaceous - Pleistocene
			Clithosa	Miocene
			Entobia	Lower Cretaceous - Pleistocene
	Hexactinellida	Hexactinosa	*Leptophragma*	Upper Cretaceous
	Hexactinosa	Craticulariidae	*Craticularia*	Upper Cretaceous
	Irregulares	Archaeocyathida	*Alconeracyathus*	Terreneuvian, Cambrian
			Archeocyathina	Terreneuvian, Cambrian
			Protopharetra	Terreneuvian, Cambrian
	Lithistida	Not assigned	*Astrobolia*	Upper Cretaceous
			Verruculina	Upper Cretaceous
	Lithonida	Lelapiidae	*Raphidonema*	Lower Cretaceous
	Regulares	Ajacicyathidae	*Dokidocyathus*	Terreneuvian, Cambrian
			Erismacoscinus	Terreneuvian, Cambrian
			Neolculicyathus	Terreneuvian, Cambrian
	Stellispongiida	Endostomatidae	*Corynella*	Upper Cretaceous
		Stellispongiidae	*Oculospongia*	Lower Cretaceous
			Stellispongia	Lower Cretaceous
	Syringostromati-dae	Stachyoditidae	*Stachyodes*	Devonian
Anthozoa	Auloporida	Auloporidae	Auloporid	Ludlow, Silurian
		Syringoporidae	Syringoporid	Pennsylvanian, Carboniferous
	Favositida	Alveolitidae	*Alveolites*	Middle Devonian
		Favositidae	*Michelina*	Carboniferous
	Heliolitida	Not assigned	*Ellisites*	Upper Ordovician
Bryozoa	Cyclostomata	Cerioporina	*Heteropora*	Upper Cretaceous
		Curculionoidea	*Berenicea*	Lower Ordovician – Upper Cretaceous
		Cytididae	*Osculipora*	Upper Cretaceous
		Diastoporidae	*Diastopora*	Upper Cretaceous
		Entalophoridae	*Entalophora*	Upper Cretaceous
		Lichenoporidae	*Multigalea*	Lower Cretaceous
		Multisparsidae	*Reptoclausa*	Upper Cretaceous
		Not assigned	*Homera*	Oligocene
		Oncousoeciida	*Proboscina*	Lower-Upper Cretaceous
	Cyclostomata	Phidoloporidae	*Reteporidea*	Oligocene
		Stomatoporidae	*Voigtopora*	Upper Cretaceous
	Fenestrida	Acanthocladiidae	*Acanthocladia*	Permian
		Fenestellidae	*Fenestella*	Permian
		Septoporidae	*Thamniscus*	Permian
	Trepostomida	Monticuliporidae	*Prasopora*	Upper Ordovician
Linguliformea	Acrotretida	Acrotretidae	*Acrotreta*	Cambrian
			Conotreta	Lower Ordovician
	Craniida	Craniidae	*Crania*	Upper Cretaceous – Pliocene
			Craniscus	Lower Cretaceous – Eocene

Table 2 (continued)

Group	Order	Family	Genus/species	Strata
	Lingulida	Discinidae	*Discinisca*	Upper Cretaceous
		Zhanatellidae	*Hyperobolus*	Lower Ordovician
		Obolidae	*Linguella (Lingulepis)*	Furongian, Cambrian
			Obolus	Cambrian
			Palaeoglossa	Lower Ordovician
Rhynconelliformea	Orthida	Draboviidae	*Nocturnellia*	Lower Ordovician
		Ranorthidae	*Ranorthis*	Lower Ordovician
	Rhynchonellida	Acanthothiridae	*Acanthothiris*	Lower Jurassic
		Notosariidae	*Notosaria*	Oligocene – Miocene
		Rhynchonellidae	*Rynchonella*	Upper Cretaceous
Gastropoda	Euomphalina	Macluritidae	*Scaevogyra*	Furongian, Cambrian
	Eogastropoda	Sinuopeidae	*Sinuopea*	Furongian, Cambrian
Bivalvia	Pholadomyoida	Myopholadidae	*Myopholas*	Lower Jurassic
		Pandoracea	*Entodesma*	Pleistocene
		Periplomatidae	*Offadesma*	Oligocene
		Pleuromyidae	*Pleuromya*	Lower Jurassic
	Pteroidea	Bakevellidae	*Bakevellia*	Permian
Polyplacophora	Neoloricata	Chitonidae	*Chiton*	Oligocene
		Acanthochitonidae	*Craspedochiton*	Miocene
			Cryptoplax	Miocene
			Crytochiton	Pleistocene
		Mopaliidae	*Mopalia*	Pleistocene
		Ischnochitonidae	*Stenoplax*	Pleistocene
		Lepidochitonidae	*Tonicella*	Pleistocene
	Paleoloricata	Mattheviidae	*Chelodes*	Wenlock, Silurian
		Not assiged	*Spicuchelodes*	Wenlock, Silurian
Monoplacophora	Tryblidiida	Tryblidiidae	*Proplina*	Furongian, Cambrian
		Hypseloconidae	*Hypsiloconus*	Furongian, Cambrian
			Proteroconus	Furongian, Cambrian
	Not assigned	Scenellidae	*Kirengella*	Furongian, Cambrian

terebratulid brachiopods (forms characterized by an enlarged foramen to accommodate a stout pedicle) found a new niche on Jurassic and Cretaceous rocky shores. Coralline red algae and scleractinian corals also made appearances on rocky shores during these times (Table 3). Except for rudistid bivalves, all these elements survived the end Cretaceous mass extinction to play ongoing but more limited roles on Modern rocky shores.

Broadly defined, the Modern rocky-shore biota is distinguished by a major expansion of fixed bivalves that adopted byssate and wedging habits to secure a firm foothold in the intertidal zone during Cenozoic times. However, the first big pulse in this development also traces back to Late Cretaceous times (Fig. 12). In this respect, the role of cemented bivalves in contemporary rocky shores has not surpassed that previously attained during the Late Cretaceous (Fig. 12). The Modern rocky-shore expansion realized by bivalves was accompanied by a similar increase in rocky-shore gastropods and balanomorph barnacles. Palmer (1982) recognized that concurrent radiations in generic diversity within three superfamiles of balanomorph barnacles were overshadowed by the radiation of predatory

Table 3 Mesozoic rocky-shore taxa

Group	Order	Family	Genus/species	Strata
Anthozoa	Astrocoeniina	Stylinidae	*Columnastrea*	Upper Cretaceous
	Gorgonida	Not assigned	*Corallium*	Pliocene
			Moltkia	Upper Cretaceous
	Hydrozoa	Stylasterina	*Allopora*	Pleistocene
	Stauriida	Lithostrotionidae	*Lithostrotion*	Lower Jurassic
	Scleractinia	Acroporidae	*Acropora*	Pleistocene
			Montipora	Pleistocene
		Actinacididae	*Actinacis*	Upper Cretaceous
			Allocoeniopsis	Lower Jurassic
		Agariciidae	*Brachyphyllia*	Upper Cretaceous
		Caryophylliidae	*Caryophyllia*	Oligocene
			Cyathoceras	Miocene
			Paracyathus	Upper Cretaceous
			Placoseris	Upper Cretaceous
		Dendrophylliidae	*Dendrophyllia*	Eocene – Miocene
			Tubastreaea	Pleistocene
			Turbinaria	Miocene
		Eupsammidae	*Balanophyllia*	Oligocene – Pleistocene
		Faviidae	*Cladocora*	Pleistocene
			Cyphastrea	Miocene
			Diploria	Pleistocene
			Eohydnophora	Lower Cretaceous
			Goniastrea	Pleistocene
			Leptastrea	Eocene – Pleistocene
			Manicina	Pleistocene
			Montastrea	Pleistocene
			Monticulastrea	Upper Cretaceous
			Solenastrea	Pliocene
			Tarbellastraea	Miocene
		Flabellidae	*Flabellum*	Miocene
			Rhizotrochus	Pleistocene
		Latomeandridae	*Dimorphastrea*	Upper Cretaceous
		Meandrinidae	*Dichocoenia*	Pleistocene
		Montlivaltiidae	*Thecosmilia*	Lower Jurassic
		Not assigned	*Amphastrea*	Lower Cretaceous
		Oculinidae	*Haplohelia*	Oligocene
			Oculina	Miocene
		Pocilloporidae	*Pocillopora*	Pliocene – Pleistocene
		Poritidae	*Alveopora*	Miocene
			Goniopora	Upper Cretaceous
			Porites	Pleistocene
		Siderastraeidae	*Siderastrea*	Pliocene – Pleistocene
		Stylophyllidae	*Heterastraea*	Lower Jurassic
			Phacelostylophyllum	Lower Jurassic
		Synastraeidae	*Leptophyllia*	Upper Cretaceous
		Thamnasteriidae	*Thamnastraea*	Upper Cretaceous

Table 3 (continued)

Group	Order	Family	Genus/species	Strata
Rhynchonelli-formea	Terebratulida	Dielasmatidae	*Dielasma*	Permian
		Loboidothyridae	*Sphaeroidothyris*	Lower Jurassic
		Megathyrididae	*Megathiris*	Oligocene
			Argyrotheca	Oligocene
		Not assigned	*Magas*	Upper Cretaceous
		Sellithyrididae	*Sellithyris*	Lower Cretaceous
		Terebrateliidae	*Austrithyris*	Pliocene
			Neolithyrina	Upper Cretaceous
			Rhynchra	Upper Cretaceous
			Rynchorina	Upper Cretaceous
			Terebratalia	Pliocene
			Terebratella	Oligocene
			Terebratula	Upper Cretaceous
			Terebratulina	Eocene – Pliocene
		Terebratulidae	*Gryphus*	Eocene
			Magasella	Miocene
	Thecideida	Thecidellinidae	*Eothecidellina*	Upper Cretaceous
	Thecideida		*Moorellina*	Lower Cretaceous
	Thecideida		*Praelacazella*	Upper Cretaceous
Bryozoa	?	?	*Inturia*	Upper Cretaceous
	Cheilostomata	Adeonellopsidae	*Adeonellopsis*	Pliocene
		Calloporidae	*Alderina*	Upper Cretaceous
			Biaviculigera	Upper Cretaceous
			Copidozoum	Pleistocene
			Paraellisina	Pleistocene
			Wilbertopora	Upper Cretaceous
		Cellariidae	*Cellaria*	Pliocene
			Melicerites	Upper Cretaceous
		Crepidacanthidae	*Crepidacantha*	Pleistocene
		Cribrilinida	*Leptocheilopora*	Upper Cretaceous
			Cribilaria	Pleistocene
			Pelmatopora	Upper Cretaceous
		Electridae	*Bullaconopeum*	Upper Cretaceous
			Conopeum	Upper Cretaceous
		Hippothoidae	*Hippothoa*	Upper Cretaceous
			Tecatia	Upper Cretaceous
		Lepraliellidae	*Celleporaria*	Pliocene
			Frurionella	Upper Cretaceous
		Lunulitidae	*Indet.*	Paleocene
		Membraniporidae	*Pleuropora*	Upper Cretaceous
		Microporidae	*Micropora*	Upper Cretaceous
		Onychocellidae	*Cheethamia*	Upper Cretaceous
			Onychocella	Upper Cretaceous
		Reteporidae	*Sertella*	Pliocene
		Romancheinidae	*Balantiostoma*	Upper Cretaceous
Bivalvia	Anomalodesmata	Cleidothaerusidae	*Cleidothaerus*	Miocene
		Pholadomyidae	*Goniomya*	Upper Cretaceous

Table 3 (continued)

Group	Order	Family	Genus/species	Strata
			Pholadomya	Lower Jurassic
		Thraciidae	*Thracia*	Pleistocene
	Hippuritoida	Plagioptychidae	*Coralliochama*	Upper Cretaceous
		Caprotinidae	*Praecaprotina*	Lower Cretaceous
		Radiolitidae	*Durania*	Upper Cretaceous
			Radiolites	Upper Cretaceous
	Myoida	Pholadidae	*Jouannetia*	Miocene
			Parapholas	Miocene – Pleistocene
			Penitella	Miocene – Pleistocene
			Aspidopholas	Miocene
			Barnea	Miocene
			Aspidopholas	Miocene
	Mytiloida	Mytilidae	*Lithophaga*	Lower Jurassic – Pleistocene
			Musculus	Oligocene
	Pectinoida	Anomiidae	*Pododesmus*	Pliocene – Pleistocene
			Anomia	Eocene-Miocene
		Pectinidae	*Hinnites*	Pliocene – Pleistocene
		Plicatulidae	*Atreta*	Upper Cretaceous – Miocene
			Plicatula	Upper Cretaceous
		Spondylidae	*Spondylus*	Upper Cretaceous – Pleistocene
	Pterioida	Gryphaeidae	*Amphidonte*	Upper Cretaceous
			Crenostrea	Miocene
			Exogyra	Lower-Upper Cretaceous
			Gryphaea	Lower Jurassic
			Hyotissa	Upper Cretaceous
			Liostrea	Lower Jurassic
			Neopychodonte	Miocene – Pliocene
			Pycnodonta	Upper Cretaceous – Oligocene
		Isognomonidae	*Isognomon*	Upper Cretaceous – Pliocene
		Kolymiidae	*Liebia*	Permian
		Limacea	*Regalilima*	Upper Cretaceous
		Malleidae	*Nayadina*	Eocene
		Ostreidae	*Acutostrea*	Upper Cretaceous – Eocene
			Arctostrea	Upper Cretaceous
			Crassostrea	Oligocene – Miocene
			Liostrea	Lower Jurassic
			Lopha	Upper Cretaceous – Lower Jurassic
			Saccostrea	Pliocene
			Ostrea	Upper Cretaceous – Pleistocene
	Veneroida	Chamidae	*Chama*	Eocene–Pleistocene
		Chamidae	*Pseudochama*	Pleistocene
		Gastrochaenidae	*Gastrochaena*	Upper Cretaceous – Pliocene
		Gastrochaenidae	*Gastrochaenolites*	Pliocene
		Hiatellidae	*Hiatella*	Oligocene – Pleistocene
		Veneridae	*Petricola*	Pliocene – Pleistocene
	Foraminiferida	Allogromiidae	*Placopsilina*	Upper Cretaceous
		Amphisteginidae	*Amphistegina*	Miocene

Table 3 (continued)

Group	Order	Family	Genus/species	Strata
		Lituolidae	*Bdelloidina*	Upper Cretaceous
		Nodosariidae	*Vaginulina*	Miocene
Foraminifera		Not assigned	*Acrulammina*	Upper Cretaceous
		Nubeculariidae	*Nubeculinella*	Lower – Upper Cretaceous
		Nummulitidae	*Heterostegina*	Miocene
		Polymorphinidae	*Bullopora*	Upper Cretaceous
Echinodermata	Arbaciidae	Arbaciidae	*Arbia*	Miocene
	Blastozoa		cystoids	Lower Ordovician
	Cassiduloida	Nucleolitidae	*Nucleolites*	Lower Jurassic
		Pygaulidae	*Trematopygus*	Lower Cretaceous
	Cidaroida	Cidaridae	*Cidaris*	Upper Cretaceous
			Goniocidaris	Miocene
			Phyllacanthus	Miocene
		Histocidaridae	*Histocidaris*	Miocene
	Crinoidea	Isocrinida	*Metacrinus*	Paleocene
	Diadematoida	Diadematidae	*Diadema*	Oligocene
	Echinoida	Echinometridae	*Echinometra*	Pleistocene
		Strongylototidae	*Stronglytocentrus*	Pleistocene
		Toxopneustidae	*Lytechinus*	Oligocene
			Schizechinus	Miocene
	Gorgonidae		*Gorgonid holdfast*	Upper Cretaceous
	Hemicidaroida	Pseudodiadematidae	*Pseudodiadema*	Upper Cretaceous
	Pedinoida	Pedinidae	*Stereopedina*	Miocene
	Phymosomatoida	Not assigned	*Phyosoma*	Upper Cretaceous
		Phymosomatidae	*Trochalosoma*	Upper Cretaceous
	Salenioida	Acrosaleiidae	*Acrosalenia*	Lower Jurassic
		Saleniidae	*Hyposalenia*	Lower Cretaceous
			Novasalenia	Upper Cretaceous
			Salenia	Lower Cretaceous – Oligocene
	Stirodonta	Phymosomatidae	*Glyptocidaris*	Oligocene
		Stomopneustidae	*Phymechinus*	Miocene
	Temnopleuroida	not assigned	*Gagaria*	Oligocene
Polychaeta	Canalipalpata	Sabellidae	*Potamilla*	Lower Jurassic – Miocene
	Not assigned	Not assigned	*Arachnostega*	Upper Cretaceous
	Palpata	Sabellariidae	*Phragmatopoma*	Pleistocene
			Spirobus	Lower Cretaceous – Oligocene
	Sedentarida	Serpulidae	*Serpula*	Upper Cretaceous – Pleistocene
	Serpulimorpha	Not assigned	*Hamulus*	Upper Cretaceous
	Spiomorpha	Not assigned	*Maeandropolydora*	Miocene – Pliocene
	Spionidae	Not assigned	*Polydora*	Oligocene – Pleistocene
	Serpulimorpha	Not assigned	*Pomatoceros*	Upper Cretaceous
			Rotularia	Eocene
			Spiraserpula	Upper Cretaceous

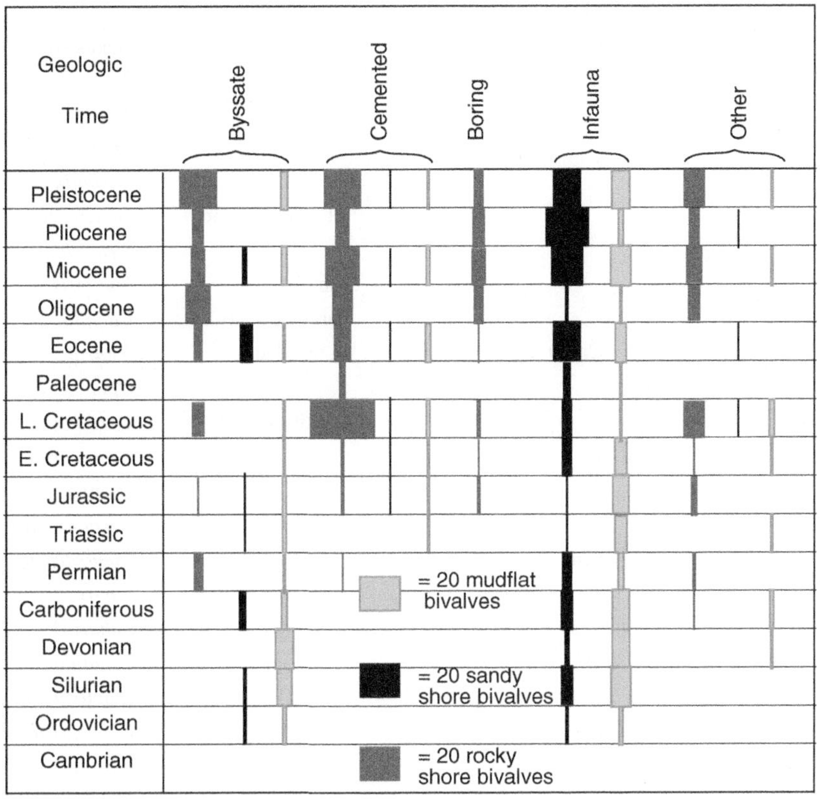

Fig. 12 Spindle diagrams for categories of Class Bivalvia (Phylum Mollusca) showing basic differences in life style on rocky, sandy, and muddy shore with respect to changes in diversity through geologic time. Lateral scale for diversity = 20 species

gastropods, in particular muricacean or drilling gastropods. This predator–prey relationship is exemplified by the present-day dog whelk (*T. lapillus*) and common rock barnacle (*B. balanoides*) that so fascinated Rachel Carson (1956) on the rocky shores of Maine. On the California coast, Ricketts et al. (1985) observed that species of another dog whelk (*Nucella canaliculata*) prefer to feed on the blue mussel (*Mytilus californianus*) over barnacles and higher concentrations of that gastropod occur within the mussel beds. Zweibel and Johnson (1995) reported on the co-occurrence of *Mytilus* and *Nucella* in Pleistocene shell beds from Pacific shores of Baja California. The known fossil representatives of the Modern rocky-shore biota are listed by order and genus in Table 4.

Figure 13 summarizes the combined relationships of Archaic, Paleozoic, Mesozoic, and Modern rocky-shore biotas, the various constituencies for which are distinguished by color and arranged as side-by-side histograms fitted to geological time intervals.

Trends in Sandy-Shore Biota Through Time

Data on changes in sandy-shore biotas through Phanerozoic time are sparse and hard to interpret. The clearest part of the picture (Fig. 14) shows seven out of 13 taxonomic clusters, identified mostly by phylum or class, making an appearance during the Cambrian or Ordovician periods. Three of those are finished before the close of the Carboniferous. For ease of comparison with the development of rocky-shore biotas, the various taxonomic groupings for sandy-shore habitats are shown in the same order from left to right. Unadjusted for rare duplications, the database entails a total of 315 fossil and extant species tallied on a running basis from one time interval to another. Based on 83 study sites, there is an average of 3.6 species per site.

With a record of survival leading through to the Early Cretaceous, linguliform brachiopods show greater staying power than the rhynchonelliform brachiopods that disappeared on sandy shores by the

Table 4 Modern rocky-shore taxa

Group	Order	Family	Genus/species	Strata
Bivalvia	Arcoida	Arcidae	*Acar*	Pleistocene
			Anadara	Miocene – Pliocene
			Arca	Upper Cretaceous, Pliocene
			Barbatia	Upper Cretaceous – Pleistocene
		Glycymerididae	*Glycimeris*	Pliocene
			Striarca	Oligocene – Miocene
		Limopsidae	*Limopsis*	Upper Cretaceous
		Parallelodontidae	*Cucullaria*	Upper Cretaceous
	Carditoida	Astartidae	*Astarte*	Lower Cretaceous
			Opis	Upper Cretaceous
		Carditidae	*Cardita*	Oligocene – Pleistocene
			Cyclocardia	Oligocene
			Glans	Pleistocene
			Purpurocardia	Paleocene
	Limoida	Limidae	*Ctenostreon*	Lower Jurassic
			Lima	Upper Cretaceous
			Limaria	Oligocene
			Limatula	Oligocene
			Plagiostoma	Lower Jurassic
	Mytiloida	Mytilidae	*Choromytilus*	Miocene
			Botula	Miocene
			Brachidontes	Eocene – Pleistocene
			Choromytilus	Miocene
			Modiola	Upper Cretaceous – Oligocene
			Modiolus	Upper Cretaceous – Pleistocene
			Mytilus	Oligocene – Pleistocene
			Perna	Miocene – Pliocene
			Septifer	Oligocene – Pleistocene
			Trichomya	Oligocene
	Pectinoida	Oxytomidae	Indet.	Upper Cretaceous
		Aviculopectinida	*Aviculopecten*	Permian
		Chaenocardiid	*Eurydesma*	Permian
		Deltopectinidae	*Deltopecten*	Permian
		Pectinidae	*Camptonectes*	Upper Cretaceous
			Chlamys	Upper Cretaceous – Pleistocene
			Hinnites	Pliocene
			Lyriochlamys	Upper Cretaceous
			Mesopeplum	Miocene
			Neithea	Upper Cretaceous
			Pecten	Upper Cretaceous
			Serripecten	Miocene
		Pseudomonotidae	*Eumicrotis*	Permian
		Pectidae	*Lentipecten*	Miocene-
	Pterioida	Cercomyidae	*Cercomya*	Upper Cretaceous
		Pinnidae	*Trichites*	Lower Jurassic
		Pteriacea	*Merismopteria*	Permian
		Pteriidae	*Pteria*	Miocene

Table 4 (continued)

Group	Order	Family	Genus/species	Strata
	Sacoglossa	Not assigned	*Berthelinia*	Miocene
	Veneroida	Kellidae	*Kellia*	Pleistocene
		Psammobiidae	Indet.	Upper
		Lucinidae	*Codakia*	Oligocene – Pleistocene
		Trapez	*Coralliophaga*	Pleistocene
		Veneridae	*Chione*	Pleistocene
			Irus	Miocene – Pleistocene
			Periglypta	Pleistocene
Gastropoda	?	?	*Cypressa*	Pleistocene
	Archaeogastropoda	Throchina	*Cookia*	Oligocene – Miocene
		Not assigned	*Erginus*	Eocene
			Corsania	Paleocene
			Diloma	
		Fissurellacea	*Megathura*	Eocene
			Sarmaturbo	Miocene
			Trochacanthus	Upper Cretaceous
	Basommatophora	Siphonariidae	*Siphonaria*	Miocene
	Caeogastropoda	Caprinidae	*Rothpletzia*	Pliocene
	Cephalaspida	Not assigned	*Priscaphander*	Paleocene
		Acteonidae	*Acteon*	Paleocene
		Retusidae	*Retusa*	Oligocene
	Docoglossa	Acmaeidae	*Acmaea*	Oligocene – Pleistocene
			Notoacmea	Pleistocene
			Tectura	Pleistocene
		Lottiidae	*Collisella*	Pleistocene
			Lottia	Pleistocene
		Patellidae	*Ancistromesus*	Pleistocene
			Cellena	Miocene – Pleistocene
			Patella	Lower Jurassic – Pleistocene
			Patelloidea	Eocene – Pleistocene
	Heterobranchia	Pyramidellidae	*Turbonilla*	Oligocene
	Littorinimorpha	Aporrhaidae	*Arrhoges*	Upper Cretaceous
		Calyptraeida	*Calyptraea*	Eocene – Miocene
			Crepidula	Eocene – Pleistocene
			Crepipatella	Pleistocene
			Crucibulum	Miocene – Pleistocene
			Sigapatella	Miocene
		Cypraeidae	*Cypraea*	Miocene – Pleistocene
		Hipponicidae	*Hipponix*	Oligocene – Pleistocene
		Littorinidae	*Lacuna*	Oligocene
			Littorina	Oligocene – Pleistocene
		Naticidae	*Gyrodes*	Upper Cretaceous
			Natica	Upper Cretaceous
		Ranellidae	*Triton*	Miocene
		Rissoidae	*Eatoniella*	Miocene

Table 4 (continued)

Group	Order	Family	Genus/species	Strata
		Strombidae	*Rostellaria*	Miocene
			Strombus	Miocene – Pleistocene
		Triviidae	*Trivia*	Pleistocene
		Velutinidae	*Lamellaria*	Pleistocene
		Vermetidae	*Dendropoma*	Pleistocene
			Petaloconchus	Oligocene – Pleistocene
			Serpulorbis	Pleistocene
			Vermetus	Miocene – Pleistocene
		Xenophoridae	*Xenophora*	Miocene – Upper Cretaceous
	Neogastropoda	Buccinidae	*Burnupena*	Miocene
			Kelletia	Pleistocene
			Pseudofax	Paleocene
		Cancellariidae	*Cancellaria*	Pleistocene
		Conidae	*Conolithes*	Miocene
		Conidae	*Conus*	Miocene – Pleistocene
		Marginellidae	*Volvarina*	Pleistocene
		Mitridae	*Mitra*	Pliocene Pleistocene
		Muricidae	*Plicopupurea*	Pleistocene
			Acanthina	Pleistocene
			Aphorrthais	Upper Cretaceous
			Ceratostoma	Pleistocene
			Morula	Pleistocene
			Hexaplex	Pleistocene
			Lepsiella	Miocene
			Maxwellia	Pleistocene
			Morula	Pleistocene
			Murex	Miocene
			Neorapana	Pleistocene
			Nucella	Miocene – Pleistocene
			Ocenebra	Pleistocene
			Pagodula	Miocene
			Plicopurpura	Pleistocene
			Thais	Pleistocene
		Not assigned	*Roperia*	Pleistocene
			Eoturris	Paleocene
		Pholidotomidae	*Morea*	Upper Cretaceous
		Procheritiidae	Indet.	Lower Jurassic
		Pseudolividae	*Macron*	Pleistocene
			Triumphis	Miocene
		Pseudomelato-midae	*Pseudomelatoma*	Pleistocene
		Turbinellidae	*Tudicla*	Miocene
			Pareora	Miocene
			Turritella	Upper Cretaceous – Oligocene
			Zeacolpus	Paleocene

Table 4 (continued)

Group	Order	Family	Genus/species	Strata
		Turridae	*Clavatula*	Miocene
			Mitromorpha	Pleistocene
		Volutidae	*Volutospina*	Upper Cretaceous
	Neotaenioglossa	Columbellidae	*Amphissa*	Pleistocene
			Columbella	Pleistocene
			Mitrella	Eocene
			Pyrene	Pleistocene
		Fasciolariidae	*Bellifusus*	Upper Cretaceous
			Drilluta	Upper Cretaceous
		Nassariidae	*Nassarius*	Pleistocene
	Neritopsina	Not assigned	*Corsania*	Paleocene
		Neritidae	*Nerita*	Paleocene – Pleistocene
		Neritidae	*Velates*	Paleocene
	Not assigned	Not assigned	*Galeodes*	Miocene
	??	Cyclotrematidae	*Haplochochilas*	Eocene
	Not assigned	Not assigned	*Lemintia*	Oligocene – Miocene
	Prosobranchia	Cerithiops	*Cerithiopsis*	Oligocene
	Ptenoglossa	Epitoniidae	*Amaea*	Oligocene
			Epitonium	Pleistocene
	Pulmonata	Ellobiidae	*Ovatella*	Eocene
	Sorbeoconcha	Cerithiidae	*Bittium*	Pleistocene
			Cerithium	Pliocene – Pleistocene
		Planaxidae	*Hinea*	Oligocene
		Siliquariidae	*Tenagodus*	Miocene
	Vetigastropoda	Fissurellidae	*Diadora*	Eocene – Pleistocene
			Emarginata	Oligocene
			Emarginula	Oligocene – Eocene
			Fissurella	Miocene – Pleistocene
			Lucapinella	Pleistocene
		Haliotidae	*Haliotis*	Oligocene – Pleistocene
		Liotiidae	*Arene*	Eocene
		Pleurotomarae	Pleurotomariids	Lower – Middle Jurassic
		Trochidae	*Calliostoma*	Oligocene – Pleistocene
			Cittarium	Oligocene
			Jujubinus	Oligocene
			Lirularia	Pleistocene
			Monodota	Pleistocene
			Tegula	Miocene – Pleistocene
			trochid	Lower Cretaceous
			Trochus	Lower Jurassic, Pleistocene
		Turbinidae	*Astrea*	Pleistocene
			Austraea	Pleistocene
			Calliomphalus	Upper Cretaceous
			Homalopoma	Eocene – Pleistocene
			Turbo	Pleistocene

Table 4 (continued)

Group	Order	Family	Genus/species	Strata
Cirripeda	Acrothoracica		borings	Carboniferous – Lower Cretaceous
	Campylsolamata	Cornuladae	*Acasta*	Miocene
		Clistadae	*Verruca*	Miocene
	Sessilia	Bathylasmatidae	*Bathylasma*	Upper Cretaceous – Miocene
		Archaeobalanide	*Semibalanus*	Pleistocene
		Tetraclitidae	*Tetraclita*	Pleistocene
	Thoracica	Balanidae	*Armatobalanus*	Miocene – Pleistocene
			Balanus	Pleistocene – Miocene
			Chelonibia	Pleistocene
			Concavus	Pleistocene
			Megabalanus	Miocene – Pleistocene
			Notobalanus	Miocene
			Tasmanobalanus	Miocene
		Brachylepadidae	*Brachylepas*	Upper Cretaceous
		Chthamalidae	*Chthamalus*	Miocene
			Hexelasma	Miocene
		Lepadidae	*Lepas*	Miocene
		Scalpellidae	*Acroscalpellum*	Miocene
			Calantica	Upper Cretaceous – Miocene
			Lithotrya	Pleistocene
			Scalpellum	Miocene
			Virgiscalpellum	Upper Cretaceous

end of the Carboniferous Period. Due to occurrences of *Lingula* and *Glottidia* species on modern sandy shores in many places around the world, it is clear that stock from this ancient family persisted to the present. Early Paleozoic biotas on sandy shores are dominated by rhynchonelliform brachiopods, bryozoans, tabulate corals, echinoderms, and trilobites (Table 5). Among the rhynchonelliform brachiopods, members of the Family Camarotoechiidae show a preference for higher energy environments during the Early Paleozoic. They are replaced by terebratulid brachiopods later on. Some few elements of the Paleozoic biota survive to the present day in this habitat, for example, linguliform brachiopods. Members of the Bivalvia and Gastropoda within Phylum Mollusca demonstrate the most continuous phases of colonization on sandy shores. Species from the bivalve families Modiomorphae and Ambonyciidae, as well as the gastropod orders Bellerophontida and Murchisoniina, are included among the many extinct elements of the Paleozoic biota.

Bivalves from the Family Ostreidae make a first appearance during Mesozoic times on low-energy sandy shores, as represented by species of *Crassostrea*. The number of reports of these is very small. Their status as intertidal dwellers is uncertain, because the shells could have been washed ashore. However, these oysters live on a sandy substrate today and they may represent part of a limited Mesozoic sandy-shore biota (Table 6).

By comparison, it is certain that bivalve taxa within the Order Veneroida became very important during Late Mesozoic times and they continue to fill an innovative place in the Modern sandy-shore biota as infaunal filter feeders, due to the evolution of more efficient siphons with separate inhalent and exhalent tubes. This feature facilitated life in a new niche for deep infauna and provided protection from predators in

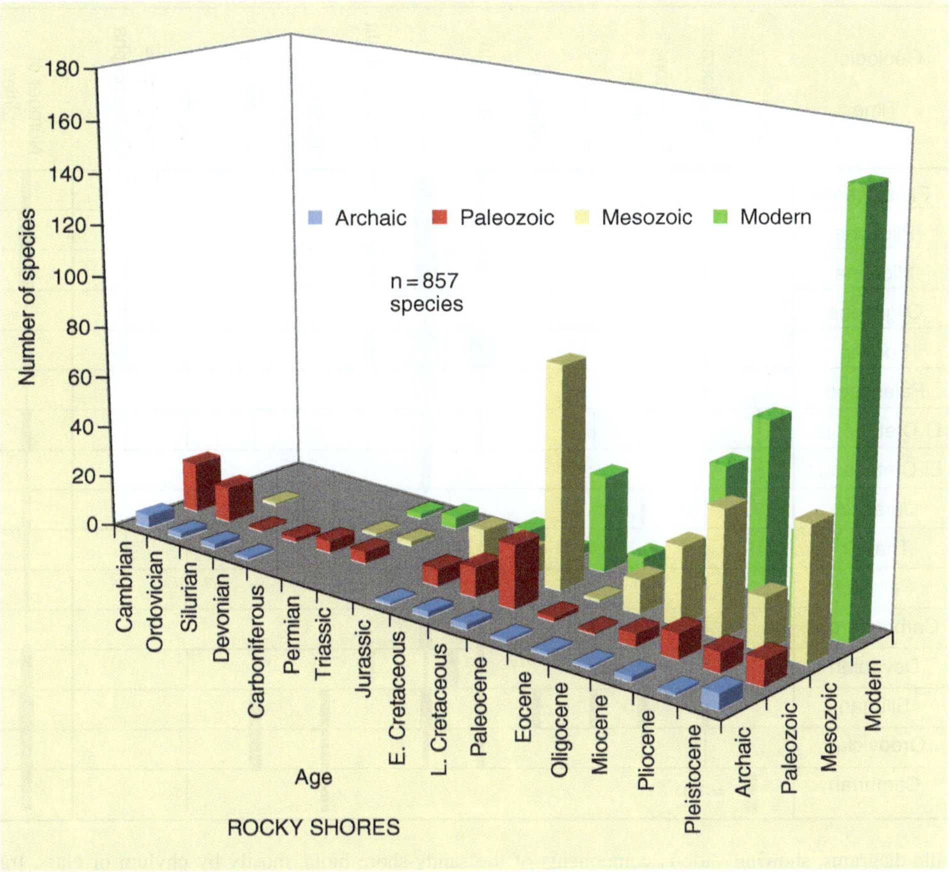

Fig. 13 Changes in the biodiversity of the rocky-shore biota through geologic time arrayed in four overlapping clusters classified as Archaic, Paleozoic, Mesozoic, and Modern representatives

mobile sandy shores. Likewise, undoubted gastropods from the Family Naticidae (moon snails) appear on sandy shores for the first time during the Eocene and make a substantial contribution to the biota as predators. Presence or absence of species living on or within intertidal sand depends on the difference between energy levels on exposed sandy shores compared to sand flats in sheltered bays and consequential differences in clastic grain size (Table 7). Counted to eliminate duplications from one time interval to another, the actual number of sandy-shore bivalve species accruing from Cambrian to Pleistocene time is 113, and the number of gastropods is 50 for an overall 9 to 4 ratio.

Figure 15 summarizes the combined relationships of Archaic, Paleozoic, Mesozoic, and Modern sandy-shore biotas, the various constituencies for which are distinguished by color and arranged as side-by-side histograms fitted to geological time intervals.

Trends in Muddy-Shore Biotas Through Time

The biodiversity of muddy-shore biotas from the Cambrian onwards is recorded on the basis of spindle diagrams for 17 components grouped mostly by phylum or class (Fig. 16). Unadjusted for rare duplications, the database entails a total of 450 fossil and extant species tallied on a running basis from one time interval to another. Based on 111 study sites, there is an average of 4.1 species per site. The width of each spindle is directly proportional to the number of muddy-shore species found for each interval of geologic time. It is notable that most of these taxonomic

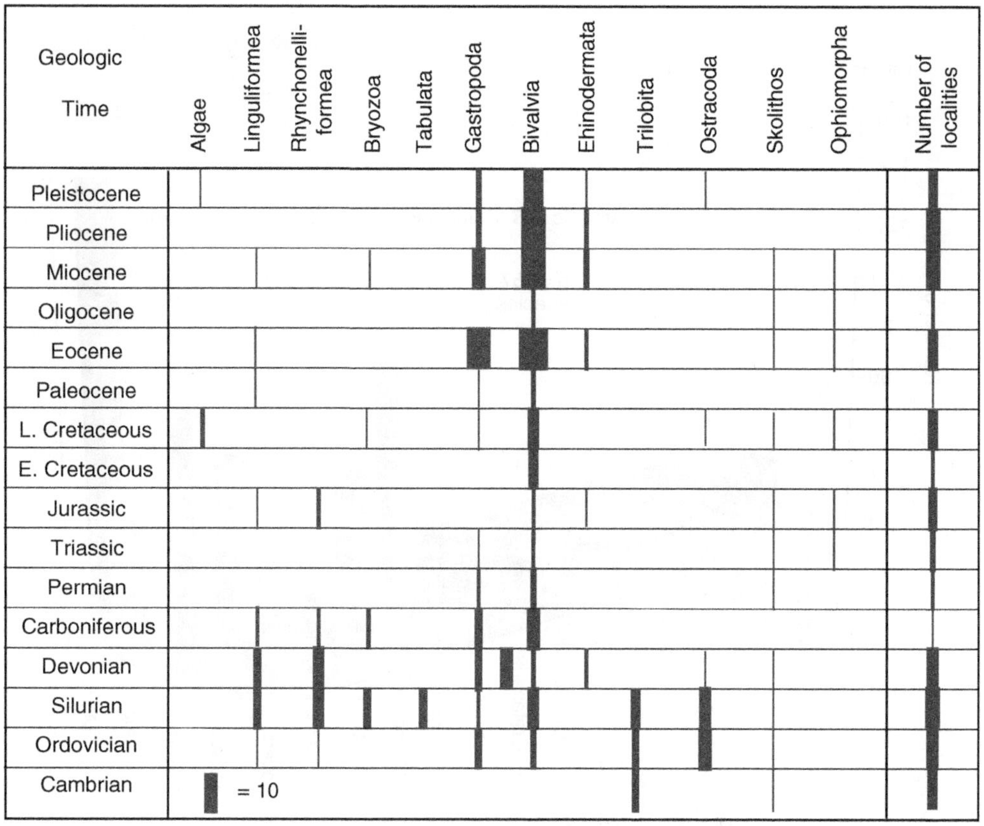

Fig. 14 Spindle diagrams, showing various components of the sandy-shore biota, mostly by phylum or class, track changes in composition and biodiversity through geologic time. Lateral scale for diversity = 10 species

clusters emerged during the Cambrian and Ordovician periods, but nearly half disappeared from coastal mudflats by the end of Carboniferous time. For ease of comparison with the development of rocky-shore biotas, the various taxonomic groupings for muddy-shore habitats are shown in the same order from left to right.

The essential pattern discernable in Fig. 16 is the early experimentation and comparatively short residency of rhychonelliform brachiopods, bryozoans, stromatoporoids, tabulate corals, polyplacophorans, trilobites, and eurypterids on muddy substrates in coastal settings. The highest diversity of any taxonomic group was attained by the eurypterids during the Silurian Period, with minor representation continuing through the Devonian and Carboniferous periods. Certain Bivalvia and Gastropoda within Phylum Mollusca show the most steady and robust presence on muddy shores through nearly all Phanerozoic time. Among infaunal bivalves during the Mesozoic, there is a shift from labial-palp, shallow-infaunal feeders to deeper infaunal taxa due to a radiation within the Order Veneroidea. Gastropods also register a maximum in diversity during the Miocene partly due to an increase in carnivorous and scavenging species in the Neogastropoda. No hint of a Cretaceous expansion is evident in the overall muddy-shore biota, in contrast to what transpired on Cretaceous rocky shores (Figs. 11 and 12). Counted to eliminate duplications from one time interval to another, the actual number of muddy-shore bivalve species accruing from Cambrian through Pleistocene time is 122, and the number of gastropods is 76 for an overall 5 to 3 ratio.

The available data on muddy-shore biotas through Phanerozoic time suggest four phases of development. These are dominated by Archaic, Paleozoic, Mesozoic, and Modern constituencies. Stromatolites (microbiolites) and algae represent the Archaic biota. A distinctive Paleozoic biota (Table 8) is dominated by eurypterids and ostracodes, but also includes a

Table 5 Paleozoic sandy-shore taxa

Group	Order	Family	Genus/species	Strata
Anthozoa	Auloporida	Syringoporidae	*Syringopora*	Wenlock Silurian
	Favositida	Favositidae	*Favosites*	Wenlock Silurian
	Heliolitida	Heliolitidae	*Heliolites*	Wenlock Silurian
	Sarcinulida	Theciidae	*Thecia*	Wenlock Silurian
	Stauriida	Zaphrentidae	*Cyathophyllum*	Wenlock Silurian
Bryozoa	Cyclostomata	Tubuliporidae	*Tobulipora*	Pennsylvanian, Carboniferous
	Cystoporida	Fistuliporidae	*Fistulipora*	Ludlow, Silurian – Pennsylvanian, Carboniferous
	Rhabdomesida	Rhomboporidae	*Rhombopora*	Pennsylvanian, Carboniferous
	Trepostomida	Not assigned	*Amplexopora*	Wenlock, Silurian
		Devonian	*Hallopora*	Ludlow, Silurian
		Not assigned	*Lioclema*	Ludlow, Silurian
Lingulaformea	Lingulida	Discinidae	*Orbiculoidea*	Pridoli (Silurian) – Lower Devonian
		Lingulidae	*Barroisella*	Upper Devonian
			Glottidia	Eocene
			Lingula	Ordovician – Miocene
		Obolidae	*Ectenoglossa*	Upper Ordovician
			Lingulepis	Furongian – Cambrian
			Obolus	Furongian – Cambrian
			Tunisiglossa	Upper Ordovician
		Not assigned	*Dignomia*	Lower Ordovician
		Not assigned	linguloid	Cambrian – Lower Createceous
Rhynchonelliformea	Orthotetida	Derbyiidae	*Derbyia*	Pennsylvanian, Carboniferous
	Productida	Echinoconchidae	*Juresania*	Pennsylvanian, Carboniferous
	Rhynchonellata	Camarotoechiida	*Camarotoechia*	Ludlow, Silurian – Upper Devonian
			Paracamarotoechia	Ludlow, Silurian
		Leptocoeliidae	*Australocoelia*	Lower Devonian
		Orthorhynchulidae	*Orthorhynchula*	Upper Ordovician
		Trigonirhynchiidae	*Microsphaeridiorhynchus*	Pridoli, Silurian
		Wellerellae	*Calcirhynchia*	Lower Jurassic
	Spiriferida	Hedeinopsidae	*Tannuspirifer*	Ludlow, Silurian
		Spiriferinidae	*Spiriferina*	Lower Jurassic
Bivalvia	Carditoida	Carditidae	*Venericardia*[a]	Eocene
	Modiomorphoida	Allodesmatidae	*Ischyrodonta*[b]	Upper Ordovician
		Modiomorphidae	*Modiolepsis*[a]	Upper Ordovician
			Modiolopsis[a]	Ludlow – Pirdoli, Silurian
			Goniophora[a]	Pridoli, Silurian
	Mytiloida	Mytilidae	*Brachiodontes*[a]	Eocene
			Arcuatula[a]	Eocene
			Aulacomya[a]	Eocene
			Botula[a]	Eocene
	Nuculanoida	Malletiidae	*Nuculites*[a]	Pridoli, Silurian
			Palaeoneilo[a]	Lower Devonian
	Nuculoida	Ctenodontidae	*Tancrediopsis*[a]	Upper Ordovician

Table 5 (continued)

Group	Order	Family	Genus/species	Strata
			Ctenodonta[a]	Pridoli, Silurian
		Nuculidae	*Nuculopsis*[a]	Pennsylvanian, Carboniferous
			Phestia[a]	Pennsylvanian, Carboniferous
		Praenucullidae	*Ledopsis*[a]	Pridoli, Silrian
	Pholadomyoida	Grammysiidae	*Fuchsella*[a]	Pridoli, Silurian
			Grammysia[a]	Pridoli, Silurian
		Megadesmidae	*Megadesmus*[c]	Cicuralian, Permian
			Pyramus[c]	Cicuralian, Permian
		Orthodontidae	*Palaeosolen*[a]	Upper Devonian
		Sanguinolitidae	*Australomya*[c]	Cicuralian, Permian
			Wilkingsia[a]	Pennsylvanian, Carboniferous
	Pterioida	Ambonychiidae	*Ambonychia*[a]	Upper Ordovician
		Bakevelliidae	*Phelopteria*[c]	Upper Cretaceous
	Trigonioida	Schizodidae	*Shizodus*[b]	Carboniferous – Permian
		Trigoniida	*Neotrigonia*[c]	Pliocene
			Pterotrigonia[a]	Lower Cretaceous
			Trigonioides[a]	Pennsylvanian, Carboniferous
Gastropoda	Archaeogastropoda	Not assigned	*Burdekina*	Middle Devonian
		Throchacea	*Zethalia*	Pliocene
	Bellerophontida	Bucaniidae	*Bucania*	Upper Ordovician
		Euphemitidae	*Euphemites*	Pennsylvanian – Carboniferous
			Warthia	Cisuralian – Permian
		Plectonotidae	*Plectonotus*	Upper Ordovician – Lower Devonian
		Tropidodiscidae	*Tropidodiscus*	Pridoli (Silurian)
	Euomphalina	Euomphalidae	*Austerum*	Middle Devonian
		Omphalotrochidae	*Labrocuspis*	Middle Devonian
	Murchisoniidae	Eomariidae	*Ptychonema*	Upper Ordovician
		Glossitinidae	*Latitaenia*	Upper Ordovician
			Mourlonia	Cisuralian Permian
			Glabrocingulum	Pennsylvanian – Carboniferous
		Murchosoniacea	*Baylea.*	Pennsylvanian – Carboniferous
			Murchisonia	Pennsylvanian – Carboniferous
Ostracoda	Leperditicopida		*Herrmannina*	Ludlow, Silurian
	Palaeocopida		*Baltonotella*	Middle Ordovician
			Chilobolbina	Middle Ordovician
			Conchoprimitia	Middle Ordovician
			Euprimites	Middle Ordovician
			Kloedenia	Ludlow, Silurian
			Laccochilina	Middle Ordovician
			Piretella	Middle Ordovician
			Piretia	Middle Ordovician
			Sigmobolbina	Middle Ordovician
			Steusloffia	Middle Ordovician
			Tallinella	Middle Ordovician

Table 5 (continued)

Group	Order	Family	Genus/species	Strata
	Platycopida		*Dizygopleura*	Ludlow, Silurian
	Podocopida		*Calcaribeyrichia*	Pridoli, Silurian
			Frostiella	Pridoli, Silurian
			Londinia	Pridoli, Silurian
			Nodibeyrichia	Pridoli, Silurian
Trilobita	Agnostida	Metagnostidae	*Trinodes*	Middle Ordovician
		Remopleurididae	*Remopleurides*	Middle Ordovician
		Saukiidae	*Prosaukia*	Furongian, Cambrian
	Asaphida	Asaphidae	*Asaphus*	Middle Ordovician
			Pseudoasaphus	Middle Ordovician
		Idahoiidae	*Idahoia*	Furongian, Cambrian
		Not assigned	*Briscoia*	Furongian
	Corynexochida	Illaenidae	*Ilaenus*	Middle Ordovician
	Phacopida	Acastidae	*Acaste*	Ludlow, Silurian
			Acastella	Ludlow – Pridoli, Silurian
		Dalmanitidae	*Dalmanites*	Ludlow, Silurian
			Scotiella	Ludlow, Silurian
		Encrinuridae	*Encrinurus*	Ludlow, Silurian
		Homalonotidae	*Homalonotus*	Ludlow, Silurian
			Trimerus	Ludlow, Silurian
	Ptychopariida	Concoryphacea	*Idiomesus*	Furongian, Cambrian
Echinodermata	Arbacioida	Arbaciidae	*Arbacia*	Pliocene
	Cassiduloida	Echinolampadidae	*Echinolampas*	Miocene
	Crinoidea	Pentacrinitidae	*Pentacrinus*	Middle Jurassic
	Diploporita	Macrocystellidae	*Glyptosphaerites*	Middle Ordovician
	Echinoidea	Clypeasteroidae	*Clypeaster*	Miocene
			Dendraster	Pleistocene
		Parechinidae	*Psammechinus*	Pliocene
	Diploporita	Macrocystellidae	*Glytosphaerites*	Middle Ordovician
	Rhombifera	Not assigned	*Echinospjaerites*	Middle Ordovician
	Spatangoida	Echinocardiidae	*Echinocardium*	Pliocene
		Schizasteridae	*Abatus*	Eocene

[a]Low energy (for sand-flat species); [b]Low and high energy (sand-flat and exposed shore species); [c]Exposed sandy-shore species

Table 6 Mesozoic sandy-shore taxa

Group	Order	Family	Genus/species	Strata
Rhynchelliniformea	Terebratulida	Centronellidae	*Proboscidina*	Lower Devonian
		Lobothyridae	*Lobothyris*	Lower Jurassic
		Meganterididae	*Scaphiocoelia*	Lower Devonian
		Meganterididae	*Pleurothyrella*	Lower Devonian
Bivalvia	Pteroida	Ostreidae	*Crassostrea*[a]	Upper Cretaceous – Miocene
			Ostrea[a]	Eocene – Pleistocene

[a]Low energy (for sand-flat species)

Table 7 Modern sandy-shore taxa

Group	Order	Family	Genus/species	Strata
Bivalvia	Arcoida	Arcidae	*Anadare*[a]	Pliocene – Pleistocene
		Cucullaeidae	*Cucullaea*[a]	Eocene
	Cardditoida	Astartidae	*Astartella*[a]	Pennsylvanian – Carboniferous
		Cardiidae	*Cardium*[b]	Miocene – Pliocene
			Cerastoderma[a]	Pliocene
			Protocardia[b]	Upper Triassic
			Venericardia[a]	Eocene
		Crassatellidae	*Cypricardella*[a]	Upper Devonian
	Myoida	Corbulidae	*Corbula*[c]	Upper Cretaceous – Eocene
			Lentidium[b]	Pliocene
			Varicorbula[a]	Eocene
		Myidae	*Mya*[c]	Eocene – Pliocene
	Pectinoida	Anomiidae	*Anomia*[a]	Upper Cretaceous
		Aviculopectini	*Aviculopecten*[a]	Pennsylvanian – Carboniferous
		Pectinidae	*Chlamys*[c]	Eocene – Miocene
			Hilberia[a]	Eocene
			Palliolum[a]	Eocene
	Pterioida	Myalinidae	*Orthomyalina*[a]	Carboniferous – Pennsylvanian
			Septimyalina[a]	Carboniferous – Pennsylvanian
		Pteriidae	*Electroma*[a]	Eocene
			Meleagrinella[b]	Middle Jurassic
			Palliolum[a]	Eocene
			Rhaetavicula[a]	Upper Triassic
	Veneroida	Cardiidae	*Cardium*[c]	Paleocene – Pliocene
			Cerastoderma[a]	Pliocene
			Protocardia[b]	Upper Triassic
			Dinocardium[a]	Miocene
		Corbiculidae	*Tetoria*[a]	
		Donacidae	*Donax*[b]	Miocene – Pleistocene
			Protodonax[b]	Upper Cretaceous
		Hiatellidae	*Cyrtodaria*[b]	Pliocene
			Glycimeris[a]	Eocene
			Panopea[a]	Miocene
		Lucinidae	*Loripes*[c]	Eocene – Pleistocene
		Mactridae	*Mactra*[c]	Eocene – Pleistocene
			Sarmatimactra[a]	Miocene
			Spisula[b]	Eocene – Pliocene
		Mesodesmatidae	*Amesodesma*[b]	Pliocene
			Anapella[a]	Pleistocene
		Semelidae	*Scrobcularia*[b]	Oligocene
		Solecurtidae	*Tagelsu*[a]	Pleistocene
		Solenidae	*Solen*[a]	Miocene
		Tancretiidae	*Tancredia*[b]	Middle Jurassic
		Tellinidae	*Arcopagia*[a]	Pliocene
			Macoma[c]	Pleistocene
			Moerella[a]	Eocene

Table 7 (continued)

Group	Order	Family	Genus/species	Strata
			Tellina[c]	Paleocene – Pleistocene
		Ungulinidae	*Diplodonta*[c]	Eocene
		Veneridae	*Ameghinomya*[a]	Miocene
			Amiantis[c]	Pliocene – Pleistocene
			Anomalocardia[a]	Miocene
			Callista[b]	Eocene
			Chamelea[b]	Pliocene
			Chionopsis[a]	Miocene
			Dosinia[a]	Paleocene – Pliocene
			Dosiniopsis[b]	Late Cretaceous
			Eurhomalea[a]	Eocene
			Katelysia[c]	Pliocene – Pleistocene
			Mercenaria[a]	Miocene
			Pitaria[a]	Miocene
			Protothaca[b]	Pliocene – Pleistocene
			Tawera[b]	Pliocene
			Tivela[b]	Pleistocene
			Venerupis[b]	Pliocene
Gastropoda	Cephalaspidae	Cylichnidae	*Bullinella*	Paleocene
			Scaphander	
	Littorinimorpha	Littorinida	*Littoridina*	Pleistocene
		Naticidae	*Crommium*	Eocene
			Neverita	Pleistocene
			Polinices	Eocene
		Strombidae	*Strombus*	Miocene – Pleistocene
		Struthiolariidae	*Perissodonta*	Eocene
		Eotomariidae	*Ptychonema*	Upper Ordovician
	Neogastropoda	Conidae	*Conus*	Miocene
		Muricidae	*Lenitrophon*	Eocene
			Trophon	Miocene
			Xymene	Eocene
	Neotaenioglossa	Columbellida	*Pyrene*	Miocene
		Melongenidae	*Busycon*	Miocene
		Nassariida	*Nassarius*	Miocene – Pliocene
			Spharonassa	Pleistocene
			Sudonassarius	Eocene
		Turritellidae	*Pareora*	Paleocene
			Turritella	Eocene – Miocene
			Zemacies	Eocene
	Neritopsina	Neritidae	*Agapilia*	Eocene
	Prosobranchia	Goniasmatidae	*Goniasma*	Pennsylvanian – Carboniferous
	Ptenoglossa	Epitoniidae	*Epitonium*	Miocene
	Sorbeoconcha	Cerithiidae	*Cerithium*	Miocene
		Not assigne	*Granulolabium*	Eocene
	Vetigastropoda	Trochidae	*Calliostoma*	Miocene
Cirripeda	Thoracica	Balanidae	*Balanus*	Pliocene

[a]Low energy (for sand-flat species); [b]exposed sandy-shore species; [c]Low and high energy (sand-flat and exposed shore species)

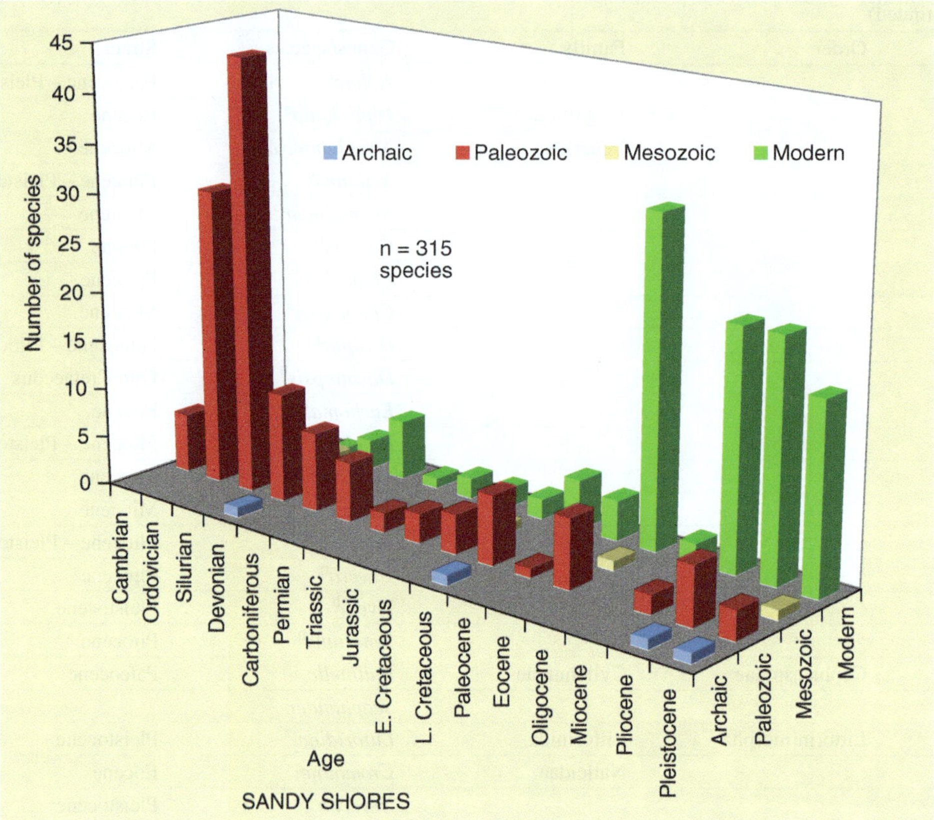

Fig. 15 Changes in the biodiversity of the sandy-shore biota through geologic time arrayed in four overlapping clusters classified as Archaic, Paleozoic, Mesozoic, and Modern representatives

representative showing by bivalves in the Family Nuculacea (labial-palp deposit feeders). The data indicate they ceased to be associated with coastal mud-flat deposits by the end of the Permian. Paleozoic bivalves more conspicuous for their longevity through modern times include the pteroid (winged) bivalves, which register a robust presence in Paleozoic muddy shores and maintain that presence through Mesozoic and Cenozoic times. Ostracodes also qualify as a strongly persistent element of the Paleozoic muddy-shore biota that survived the Paleozoic and Mesozoic extinctions. They are well represented in Silurian muddy shores and made spotty appearances on younger muddy shores through to the Paleocene Epoch. Oysters living on a muddy substrate may represent part of a limited Mesozoic muddy-shore biota, much the same case as for sandy shores (Table 9).

Infaunal siphon feeders dominated by members of the Order Veneracea characterize the modern muddy-shore biota (Table 10). Among the 23 registrations of infaunal bivalves associated with Cenozoic mud-flat deposits in the database, 14 are Venus shells. This agrees with analyses by Stanley (1968), showing that the rapid Cenozoic radiation of infaunal bivalves in level-bottom subtidal habitats far outpaced the coeval generation of new genera in competing families of epifaunal bivalves. Essentially, many species of infaunal bivalves that live on intertidal mudflats also range to deeper, offshore waters. Carnivorous species belonging to the Family Nassariidae are common today as mud snails and are adequately represented in the database for Cenozoic mudflat gastropods. Likewise, detritus feeders belonging to the gastropod Family Turritellidae are found in the fossil database and help signal a modern muddy-shore fauna. Both are representative of the relatively young Neogastropoda (taxonomically regarded as an infraorder).

Figure 17 summarizes the combined relationships of Archaic, Paleozoic, Mesozoic, and Modern muddy-shore biotas, the various constituencies for which are

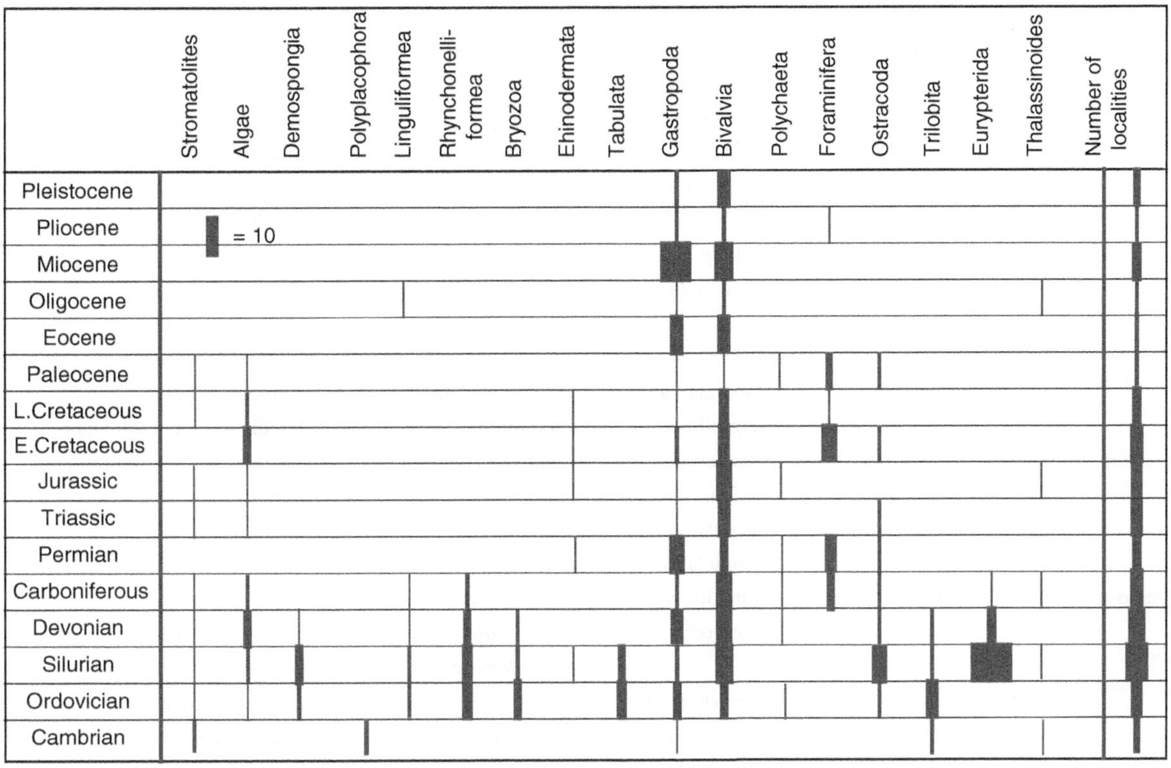

Fig. 16 Spindle diagrams, showing various components of the muddy-shore biota, mostly by phylum or class, track changes in composition and biodiversity through geologic time. Lateral scale for diversity = 10 species

distinguished by color and arranged as side-by-side histograms fitted to geological time intervals.

Fossil Histories of Representative Modern Genera

How far back through geological history is it necessary to search, before life on modern rocky, muddy, and sandy shores would appear quite foreign compared to what we find today? The following sections trace some of the more common and important genera and some species as far back in time as they lead in the strict context of fossils from intertidal deposits.

Tracing Modern Life Strategies for Rocky-Shore Colonization

Modern rocky shores extend to subtropical and temperate latitudes all around the world and typically show well-defined biological zonations. Generally, these are demarcated by littorinid gastropods in the high splash zone, the upper part of the intertidal crowded with barnacles, and the lower parts dominated by mussels. These shores would seem strangely bereft of life without so distinct components. Among herbivorous gastropods, the oldest known fossils attributed to the genera *Littorina* and *Nerita* from an exposed shore occur in deposits of probable Paleogene age in Baja California, Mexico (Woods and Saul 1986). The molecular phylogeny of members belonging to the Family Littorinidae is complex, involving genera such as *Austrolittorina* and *Afrolittorina* with a Southern Hemisphere history (Williams et al. 2003). Based on molecular clocks, the divergence and global spread of this large family can be linked to a much earlier Cretaceous date but the analysis lacks direct support from fossil evidence. The Family Muricidae also has its origins during Mesozoic times and our record of members on rocky shores dates back to the Late Cretaceous with the Genus *Aphorrthais* (Abdel-Gawad 1986). The large family Muricidae specializes in life in the high intertidal zone, where its

Table 8 Paleozoic muddy-shore taxa

Group	Order	Family	Genus/species	Strata
Anthozoa	Auloporida	Bajgoliidae	*Bajgolia*	Upper Ordovician
		Syringoporidae	*Syringopora*	Wenlock, Silurian
	Favositida	Parastriatoporida	*Parastriatopora*	Ludlow, Silurian
	Heliolitida	Coccoserididae	*Coccoceris*	Upper Ordovician
	Lichenariida	Lichenariidae	*Lichenaria*	Middle Ordovician
	Sarcinulida	Billingsariidae	*Billingsaria*	Upper Ordovician
		Syringophyllidae	*Eoflecteria*	Upper Ordovician
			Nyctopora	Upper Ordovician
		Theciidae	*Thecia*	Wenlock, Silurian
	Stauriida	Zaphrentidae	*Cyathophyllum*	Wenlock, Silurian
	Tetradiida	Tetradiidae	*Tetradium*	Middle – Upper Ordovician
Porifera	Chaetetida	Chaetetidae	*Chaetetes*	Mississipian, Carboniferous
	Hexactinellida	Reticulosaidae	*Prismodictya*	Lower Devonian
	Actinostromatida	Densastromidae	*Densastroma*	Ludlow, Silurian
		Pseudolabechidae	*Pseudolabechia*	Ludlow, Silurian
Demospongia	Amphiporida	Amphiporidae	*Amphipora*	Upper Devonian
	Clathrodictyida	Clathrodictyidae	*Clathrodictyon*	Ludlow, Silurian
			Stelodictyon	Ludlow, Silurian
	Labechiida	Stromatoceriidae	*Cystistroma*	Upper Ordovician
	Labechiida	Labechiidae	*Labechiella*	Upper Ordovician
	Labechiida	Labechiidae	*Stratodictyon*	Upper Ordovician
	Not assigned	Not assigned	*Avitastroma*	Ludlow, Silurian
Bryozoa	Cryptostomata	Rhinidictyidae	*Stictopora*	Middle – Upper Ordovician
	Fenestrida-	Phylloporinidae	*Austraphyllo-porina*	Upper Ordovician
	Trepostomida	Not assigned	*Amplexopora*	Wenlock, Silurian
			Batostoma	Upper Ordovician
			Homotrypa	Upper Ordovician
			Monotrypa	Ludlow, Silurian
			Monticulipora	Ludlow, Silurian
			Prasospora	Upper Ordovician
Linguliformea	Lingulida	Lingulidae	*Lingula*	Upper Ordovician – Oligocene
	Trimerellida	Trimerellidae	*Eodinololus*	Upper Ordovician
Rhynchonelliformea	Athyridida	Athyrididae	*Protathyris*	Pridoli, Silurian
	Athyridida	Meristellidae	*Hindella*	Pridoli, Silurian
	Atrypida	Anazygidae	*Zygospira*	Middle Ordovician
		Lissatrypidae	*Atrypoidea*	Ludlow, Silurian
		Protozygidae	*Protozyga*	Upper Ordovician
	Productida	Echinoconchidae	*Juresania*	Pennsylvanian, Carboniferous
		Monticuliferidae	*Linoprotonia*	Mississipian, Carboniferous
	Rhynchonellida	Ancistrorhynchidae	*Ancistrorhyncha*	Middle Ordovician
		Camarotoechiidae	*Camarotoechia*	Pridoli, Silurian – Upper Devonian
			Microsphaeridiorhyncus	Ludlow, Silurian
		Rhynchotrematidae	*Rynchotrema*	Upper Ordovician
		Trigonirhynchiidae	*Rostricellula*	Upper Ordovician
	Spiriferida	Ambocoeliidae	*Crurithyris*	Mississipian, Carboniferous
		Delthyrididae	*Howellella*	Wenlock, Silurian – Lower Devonian

Table 8 (continued)

Group	Order	Family	Genus/species	Strata
		Rhynchospirinidae	*Rhynchospirina*	Pridoli, Silurian
	Strophomenida	Amphistrophiidae	*Mesodouvillina*	Lower Devonian
		Strophomenidae	*Quandogia*	Upper Ordovician
			Strophomena	Middle Ordovician
			Strophomena	Upper Ordovician
	Terebratulida	Meganterididae	*Mutationella*	Lower Devonian
Bivalvia	Modiomorphoida	Allodesmatidae	*Ischyrodonta*	Upper Ordovician
		Cycloconchidae	*Actinodonta*	Ludlow, Silurian
		Modiomorphoidae	*Goniophora*	Wenlock, Silurian – Lower Devonian
			Modiolopsis	Upper Ordovician – Lower Devonian
	Mytiloida	Mytilidae	*Brachiodontes*	Eocene
			Inoperna (Triasoperna)	Upper Triassic
			Modiola	Lower Devonian – Lower Jurassic
			Modioloform	Mississipian, Carboniferous
			Modiolus	Mississipian, Carboniferous – Miocene
			Musculus	Pleistocene
			Mytilus (Mytiloconcha)	Miocene
			Phthonia	Middle Devonian
			Pletomytilus	Lower Devonian
			Promytilus	Pennsylvanian, Carboniferous
			Septifer	Upper Cretaceous
	Nuculanoida	Malletiidae	*Nuculites*	Ludlow, Silurian
			Palaeoneilo	Upper – Lower Devonian
		Nuculanidae	*Polidevcia*	Mississipian, Carboniferous
	Nuculoida	Not assigned	*Caesariella*	Wenlock, Silurian
			Paleostraba	Wenlock, Silurian
		Nuculidae	*Freja*	Wenlock, Silurian
			Nucula	Ludlow, Silurian
			Nuculavus	Cisuralian, Permian
			Nuculoidea	Wenlock, Silurian – Upper Devonian
			Nuculopsis	Mississipian, Carboniferous
		Praenuculidae	*Nuculodonta*	Wenlock, Silurian
	Pholadomyoida	Ceratomyidae	*Ceratomya*	Upper Triassic
		Edmondiidae	*Edmondia*	Pennsylvanian, Carboniferous
		Grammysiidae	*Grammysia*	Ludlow, Silurian
		Permophoridae	*Permophorus*	Pennsylvanian, Carboniferous
		Pleuromyidae	*Pleuromya*	Triassic – Lower Cretaceous
		Sanguinolitidae	*Sanguinolites*	Mississipian, Carboniferous
			Sphenotus	Devonian
			Wilkingia	Permian
	Pterioida	Ambonychiidae	*Ambonychia*	Upper Ordovician
			Mytilarca	Wenlock, Silurian – Lower Devonian

Table 8 (continued)

Group	Order	Family	Genus/species	Strata
		Pterineidae	*Caritodens*	Uooer Ordovician
			Cornellites	Wenlock, Silurian
			Leiopteria	Lower Devonian
			Leptodesma	Lower Devonian
			Limoptera	Lower Devonian
			Molinicola	Wenlock, Silurian
	Solemyoida	Solemyidae	*Janeia*	Wenlock, Silurian
	Trigonioida	Schizodidae	*Schizodus*	Pennsylvanian, Carboniferous – Cisuralian, Permian
		Trigoniidae	*Geratrigonia*	Lower Jurassic
			Steinmannella	Lower Cretaceous
Gastropoda	Bellerophontida	Bellerophontidae	*Belleophontoidea*	Cisuralian, Permian
			Bellerophon	Upper Devonian
			Bucaniopsis	Mississipian, Carboniferous
			Pharkidonotus	Cisuralian, Permian
			Sphenosphaera	Upper Ordovician
		Euphemitidae	*Euphemites*	Cisuralian, Permian
		Plectonotidae	*Plectonotus*	Silurian – Lower Devonian
		Sinuitidae	*Sinuites*	Upper Ordovician
		Tropidodiscidae	*Tropidodiscus*	Upper Ordovician
	Euomphalina	Euomphalidae	*Amphiscapha*	Cisuralian, Permian
			Bicarina	Devonian
			Straparollus	Devonian
		Holopeidae	*Holopea*	Lower Devonian – Cisuralian, Permian
	Mimspirina	Onchochilidae	*Mattherella*	Furongian, Cambrian
	Murchisoniina	Eotomariidae	*Clathrospira*	Upper Ordovician
		Gosseletinidae	*Euconospira*	Cisuralian, Permian
		Lophospiridae	*Lophospira*	Middle Ordovician
			Paupospira	Upper Ordovician
			Worthenia	Cisuralian, Permian
		Loxonematidae	*Loxonema*	Pridoli, Silurian – Devonian
		Murchisoniidae	*Helicospira*	Cisuralian, Permian
			Hormatoma	Pridoli, Silurian
			Murchisonia	Devonian
Trilobita	Agnostida	Remopleurididae	*Remopleurides*	Upper Ordovician
	Asaphida	Asaphidae	*Pseudobasilicus*	Upper Ordovician
	Phacopida	Homalonotidae	*Burmeisteria*	Lower Devonian
		Homalonotidae	*homalonotids*	
		Pliomeridae	*Pliomerina*	Upper Ordovician
		Pterygometopidae	*Calyptaulax*	Middle Ordovician
	Proetida	Bathyuridae	*Bathyurus*	Middle Ordovician
	Ptychopariida	Not assigned	*Eokosovopeltis*	Upper Ordovician
	Redlichiida	Emuellidae	*Balcoracania*	Series 2, Cambrian
Merostomata	Eurypterida	Adelophthalmidae	*Adelophthalmus*	Pennsylvanian, Carboniferous
			Bassipterus	Ludlow, Silurian
		Carcinosomatidae	*Carcinosoma*	Wenlock, Silurian

Table 8 (continued)

Group	Order	Family	Genus/species	Strata
		Dolichopteridae	*Clarkeipterus*	Pridoli, Silurian
			Dolichopterus	Pridoli, Silurian
		Drepanopteridae	*Alkenopterus*	Lower Devonian
			Drepanopterus	Ludlow, Silurian – Lower Devonian
			Moselopterus	Lower Devonian
		Eurypteridae	*Baltoeuryterus*	Wenlock, Silurian
			Eurypterus	Ludlow – Pridoli, Silurian
		Hardieopteridae	*Hardieopterus*	Pridoli, Silurian
		Hughmilleriidae	*Hughmilleria*	Pridoli, Silurian
			Nanahughmilleria	Pridoli, Silurian
			Parahughmileria	Ludlow, Silurian – Middle Devonian
			Waeringopterus	Ludlow, Silurian
		Jaekelopteridae	*Jaekelopterus*	Lower Devonian
		Mixopteridae	*Mixopterus*	Wenlock – Pridoli, Silurian
		Pterygotidae	*Acutiramus*	Pridoli, Silurian
			Erettopterus	Pridoli, Silurian
			Pterygotus	Pridoli Silurian
			Truncatiramus	Ludlow – Pridoli, Silurian
		Rhenopteridae	*Rhenopterus*	Lower Devonian
		Slimoniidae	*Salteropterus*	Pridoli, Silurian
			Slimonia	Pridoli, Silurian
		Stylonuridae	*Tarsopterella*	Pridoli, Silurian
	Xiphosurida		*Indet.*	Silurian, Pridoli
Ostracoda	Leperditicopida		*Hermannina*	Middle Ordovician – Ludlow, Silurian
			Hogmochilina	Lower Devonian
			Kiaeria	Lopingian, Permian
			Leperditia	Middle Ordovician – Lower Devonian
	Not assigned		*Goranella*	Lopingian,
	Not assigned		*Ohesaarina*	Ludlow, Silurian
	Palaeocopida		*Aparchites*	Pridoli, Silurian
			Beyrichia	Wenlock – Ludlow, Silurian
			Judahella	Lopingian, Permian
			Retisacculus	Ludlow, Silurian
			Signetopsis	Ludlow, Silurian
			Welleria	Pridoli, Silurian
	Platycopida		*Dizygopleura*	Pridoli, Silurian
			Frostiella	Pridoli, Silurian
			Indivisia	Lopingian, Permian
			Londinia	Pridoli, Silurian
			Nodibeyrichtia	Pridoli, Silurian
Echinodermata	Holothuroidea	Apodida	*Theelia*	Lopingian, Permian
	Ophiurida	Ophiacanthid	*Ophiolancea*	Lower Cretaceous
	Spatangoida	Hemiasteridae	Indet.	Upper Cretaceous

Table 9 Mesozoic muddy-shore taxa

Group	Order	Family	Genus	Strata
Bivalvia	Hippuritoida	Megalodontidae	*Megalodon*	Triassic
		Radiolitetidae	*Radiolites*	Upper Cretaceous
	Pterioida	Bakevellidae	*Bakevellia*	Middle Triassic – Lower Jurassic
			Gervillea	Lower Jurassic – Lower Cretaceous
		Gryphaeidae	*Ceratostreon*	Upper Cretaceous
			Exogyra	Lower Cretaceous
			gryphaid	Lower Triassic
			Gyrostrea	Upper Cretaceous
			Ilmatogyra	Upper Cretaceous
			Liostrea	Triassic – Lower Jurassic
		Ostreidae	*Ostrea*	Lower Jurassic – Pleistocene
			Sokolowia	Eocene
	Trigonioida	Schizodidae	*Schizodus*	Pennsylvanian, Carboniferous Cisuralian, Permian
		Trigoniidae	*Geratrigonia*	Lower Jurassic
			Steinmannella	Lower Cretaceous

members found refuge from competition in the subtidal (Kirby 2004). Carnivorous species from the genera *Thais* and *Acanthina* (Family Thaididae) trace back no farther than Pleistocene rocky-shore deposits. Other fossil taxa in the same family, such as *Lepisiella* and *Nucella*, occur in rocky-shore deposits of Miocene age in New Zealand (Eagle et al. 1995) and South Africa (Roberts and Brink 2002), respectively.

In colonizing the upper intertidal zone, the acorn barnacles evolved from polliciporid-like barnacles that lost their peduncle and achieved direct attachment of the mantle (capitulum) to a hard substrate. Traced to Upper Cretaceous (Campanian) rocky-shore deposits in Arkansas, USA (Zullo et al. 1987), fossil species belonging to the genera *Brachylepas* and *Virgiscapellus* are among the oldest-known encrusting barnacles. The balanomorph barnacles of today are divided into three major superfamiles recognized as the Balanoidea, Coronumaloidea, and Chthamaloidea. According to Stanley and Newman (1980), living balanoid species outnumber the chthamaloids by a factor of 5:1 and succeeded in displacing the latter group into the upper fringe of the intertidal zone through competitive exclusion only during the last 50 million years since Eocene times. Successful colonization of the upper intertidal zone by the acorn barnacles over this period played a significant role in attracting carnivorous muricacean gastropods to this habitat (Palmer 1982).

Bivalves anchored by byssal attachment in the lower intertidal zone include several gregarious species belonging to the genera *Mytilus* (Fig. 18), *Modiolus*, and *Barbatia*. Many of the intertidal byssate species we recognize today apparently evolved from Devonian modiomorphs, such as *Lyromytilus attenuatus*, living as endobyssates partially submerged in subtidal sediments (Stanley 1972). The oldest known representative of *Mytilus* associated with a former rocky shore comes from the Oligocene of Oregon (Miller and Orr 1988). The oldest representative of *Modiolus* affiliated with a near-shore rocky substrate comes from the Upper Cretaceous of New Zealand (Crampton 1988). Likewise, the oldest representative of *Barbatia* with a near-shore rocky substrate comes from the Upper Cretaceous in Alabama (Bryan 1992).

Overall, it may be argued that the Modern rocky-shore biota shows initial roots in the Cretaceous, but blossomed during the succeeding 66 million years of the Cenozoic Era. This does not invalidate evidence that experimentation along some similar lines occurred

Table 10 Modern muddy-shore taxa

Group	Order	Family	Genus/species	Strata
Bivalvia	Anomalodesmata	Lyonsiidae	*Mytilimeria*	Pleistocene
		Pholadomyidae	*Pholadomya*	Upper Triassic
	Arcoida	Arcidae	*Anadara*	Pleistocene
			Scapharca	Pleistocene
		Cucullaeidae	*Cucullaea*	Lower Cretaceous
		Cyrtodontidae	*Ptychodesma*	Lower Devonian
	Carditoida	Astartidae	*Opisoma*	Lower Jurassic
		Carditidae	*Venericardia*	Paleocene
		Crassatellidae	*Cypricardella*	Upper Devonian
	Hippuritoida	Megalodontidae	*Megalodon*	Upper Triassic – Lower Jurassic
	Myoida	Corbulidae	*Caryocorbula*	Miocene
			Corbula	Eocene
			Ursirivus	Eocene
		Myidae	*Cryptomya*	Pleistocene
	Pectinoida	Aviculopectinoidae	*Aviculopecten*	Pennsylvanian, Carboniferous
		Deltopectinidae	*Streblopteria*	Mississipian, Carboniferous
		Pectinidae	*Chlamys*	Miocene
			Flabellipecten	Miocene
			Volachlamys	Pleistocene
		Posidoniidae	*Daonella*	Middle Triassic
			Posidonia	Middle Triassic
		Pseudomonotidae	*Pseudomonotis*	Pennsylvanian, Carboniferous
			Lyriopecten	Upper Devonian
	Pterioida	Isognomonidae	*Gervilleioperna*	Lower Jurassic
			Isognomon	Miocene
			Mytiloperna	Lower Jurassic
		Lunulacardiidae	*Joachymia*	Ludlow, Silurian
			Mamika	Wenlock, Silurian
		Myalinidae	*Myalina*	Cisuralian, Permian
		Pinnidae	*Pinna*	Lower Cretaceous
	Solecurtidae	Sanguinolitidae	*Sphenotus*	Lower – Upper Devonian
	Veneroida	Arcticidae	*Anisocardia*	Upper Cretaceous
		Cardiida	*Cardium*	Miocene
			Cerastoderma	Pleistocene
			Integricardium	Lower Jurassic
			Obsoletiforma	Miocene
		Corbiculidae	*Corbicula*	Eocene
			Corbula	Eocene
			Filosina	Lower Cretaceous
		Lucinidae	*Loripes*	Pleistocene
			Parvalucina	Miocene
		Mactridae	*Meropesta*	Pleistocene
			Raeta	Pleistocene
		Mesodesmatidae	*Anapella*	Pliocene
		Neomiodontidae	*Eomiodon*	Lower Jurassic
		Pharidae	*Ensis*	Miocene

Table 10 (continued)

Group	Order	Family	Genus/species	Strata
		Semelidae	*Abra*	Miocene
			Ervilia	Miocene
			Theora	Pleistocene
		Solecurtidae	*Solecurtus*	Eocene
		Solenidae	*Leptodomus*	Lower Devonian
		Tellenidea	*Macoma*	Pleistocene
			Tellina	Eocene – Pleistocene
		Ungulinida	*Diplodonta*	Pliocene
		Veneridae	*Dosinia*	Miocene
			Katelysia	Pliocene
			Lirophora	Miocene
			Macrocallista	Miocene
Gastropoda	Not assigned	Not assigned	*Elimia*	Paleocene
	Not assigned	Not assigned	*Natria*	Cisuralian, Permian
	Not assigned	Not assigned	*Peraubium*	Eocene
	Archaeogastopoda	Not assigned	*Indet.*	Lower – Upper Devonian
	Cephalaspidea	Cylichnida	*Acteocina*	Miocene
	Docoglossa	Acmaeidae	*Acmaea*	Miocene
	Heterobranchia	Pyramidellidae	*Pyramidella*	Miocene
	Littorinimorpha	Calyptraeida	*Crucibulum*	Miocene
		Hydrobiidae	*Hydrobia*	Miocene
		Naticidae	*Crommium*	Eocene
			Euspira	Miocene
			Lunatia	Miocene
			Polinices	Miocene
		Rissoidae	*Mohrensternia*	Miocene
		Vermetidae	*Vermetid*	Mississipian, Carboniferous
	Mimspirina	Onychochilidae	*Mattherella*	Furongian, Cambrian
	Neogastropoda	Costellariidae	*Pusia*	Miocene
		Mitridae	*Mitra*	Miocene
			Swainsonia	Miocene
		Muricidae	*Magilus*	Miocene
			Ocenebra	Miocene
		Olividae	*Olivia*	Miocene
	Neogastropoda	Volutidae	*Scaphella*	Miocene
		Eulimidae	*Melanella*	Miocene
		Fasciolariidae	*Cyrtulus*	Miocene
			Latirus	Miocene
		Nassariidae	*Nassarius*	Miocene – Pleistocene
		Not assigned	*Harpagodes*	Upper Cretaceous
			Pirenella	Pleistocene
			Tylochilus	Eocene
		Turritellidae	*Turitella*	Lower Cretaceous – Miocene
	Neritopsina	Naticopsidae	*Naticopsis*	Cisuralian, Permian
		Neritidae	*Nerita*	Miocene
		Neritidae	*Neritina*	Eocene

Table 10 (continued)

Group	Order	Family	Genus/species	Strata
	Prosobranchia	Cerithiopsidae	*Seila*	Miocene
		Meekospiridae	*Meekospira*	Cisuralian, Permian
		Orthonematidae	*Orthonema*	Cisuralian, Permian
		Pseudomelaniidae	*Pseudomelania*	Upper Triassic
	Pulmonata	Amphibolidae	*Salinator*	Pliocene
	Sorbeoconcha	Not assigned	*Granulolabium*	Miocene
		Not assigned	*Vicinocerithium*	Eocene
		Potamididae	*Gantechinobathra*	Eocene
			Potamides	Eocene – Miocene
			Tympanotonos	Eocene
	Vetigastropoda	Trochidae	*Gibbula*	Miocene
			Jujubinus	Miocene

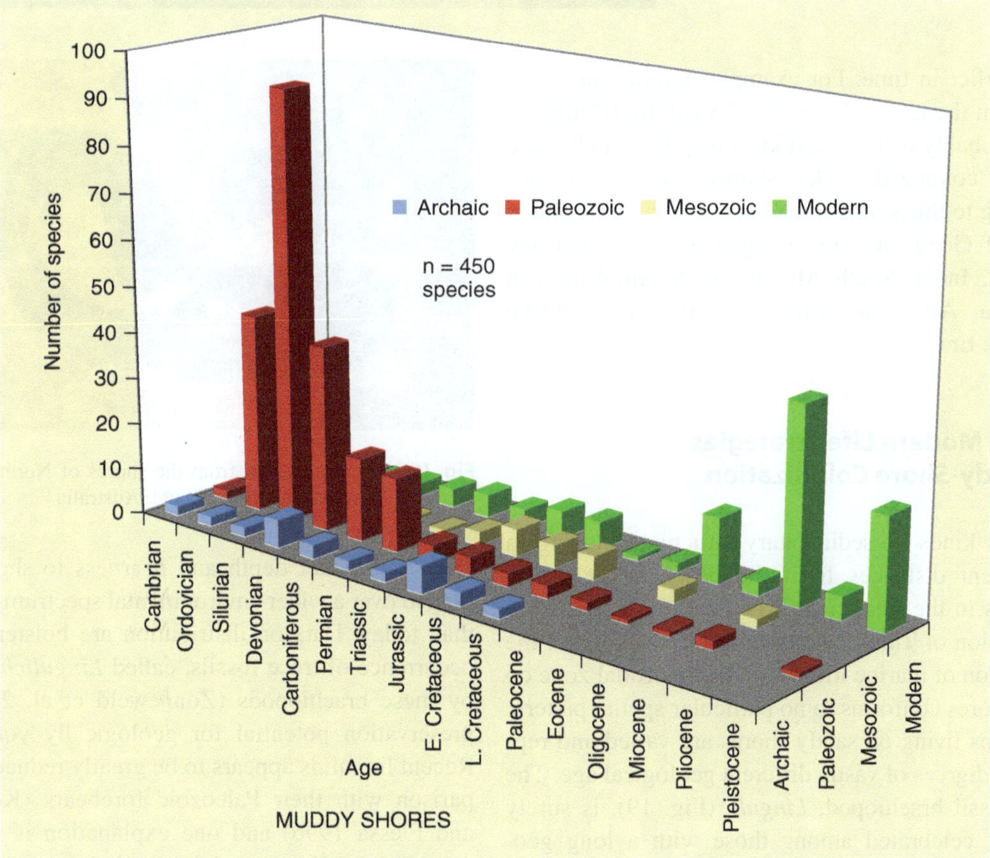

Fig. 17 Changes in the biodiversity of the muddy-shore biota through geologic time arrayed in three overlapping clusters classified as Archaic, Mesozoic, Paleozoic, and Modern representatives

Fig. 18 Rocks in the intertidal zone off the Pacific coast of northern Baja California, Mexico, covered by the byssate bivalve *Mytilus californianus* and under attack by the predator starfish *Pisaster ochraceus*, with surfgrass *Phyllospadix scouleri* in the background

much earlier in time. For example, *Eurydesma playfordi* from the Lower Permian of Australia (Runnegar 1979) probably is the oldest known epibyssate bivalve to have colonized rocky shores. Several species belonging to this genus spread across the high-latitude shores of Gondwana, including parts of present-day Australia, India, South Africa, and South American Argentina. All went extinct near the close of the Paleozoic Era.

Fig. 19 *Lingula anatina* from the shores of North Stradbroke Island in Moreton Bay, Queensland, Australia

Tracing Modern Life Strategies for Sandy-Shore Colonization

Different kinds of sedimentary structures may form at different distances from the beach or at different times in the tidal cycle (see Section "Criteria for Recognition of Tidal Sandy-Shore Deposits"), but the distribution of marine life within the intertidal zone on sandy shores conforms to no particular spatial pattern. Organisms living on sandy shores are varied and represent pedigrees of vastly different geological age. The living fossil brachiopod, *Lingula* (Fig. 19), is surely the most celebrated among those with a long geological record. Recent lingulids are restricted mainly to the intertidal zone, where they exclusively inhabit shallow burrows excavated in sand (Paine 1970). Much the same brachiopods appear in strata back to the Cambrian, although ancient representatives commonly occur in both fine-grained sandstone and shale but rarely limestone. Depositional environments with respect to water depth and nearness to shore clearly ranged over a wider environmental spectrum in the past than today. Data on distribution are bolstered by the occurrence of trace fossils, called *Lingulichnus*, made by these brachiopods (Zonneweld et al. 2007). The preservation potential for geologically younger and Recent lingulids appears to be greatly reduced in comparison with their Paleozoic forebears (Kowalewski and Flessa 1996) and one explanation is that phosphatic mineralization of the shells was stronger in the past than today. Whatever the cause, there is no ambiguity over the fact these primitive brachiopods are far less abundant now than in the past.

Another living fossil is the horseshoe crab, *Limulus*, which remains today a common denizen of sandy shores. The oldest known horseshoe-crab fossil is

Lunatapsis aurora from the Upper Ordovician of northern Manitoba, Canada (Rudkin et al. 2008). This discovery pushed back the record of horseshoe crabs more than 200 million years from the Triassic to the Late Ordovician. Although the text describing this particular record offers no details on the depositional environment in which the Ordovician fossil lived, we know it comes from a silty-lime mud a short distance off a low rocky shoreline. The spotty representations of horseshoe crabs from Ordovician, Triassic, Jurassic, and much younger intervals, belies the fact this tribe of arthropods maintained a presence in nearshore, very shallow-water environments from at least Late Ordovician time onward. The available data on sandy-shore biotas (Appendix 2) simply fail to register what must have existed all along and the same must be said for blank spots in the data on lingulid brachiopods.

Yet a third group of organisms with staying power on sandy shores is denoted by the ichno-fossil genus *Skolithos*. Probably made by many different species of marine polychaets through time, the characteristic vertical burrows for *Skolithos* are recorded in sandy-shore deposits of many ages through the Phanerozoic as far back as the Cambrian. The fact that some time intervals remain blank with respect to the taxa means that appropriate examples remain only to be found. Like *Lingula*, however, it is likely that polychaetes on sandy shores are not as prolific makers of these dwellings today as they once were in Paleozoic times. This impression is based on the shear volume and close packing of fossil *Skolithos* tubes (Fig. 9), the likes of which cannot be seen anywhere today.

Among the Bivalvia, many species in the Order Veneroida are especially well represented on coastal sand flats. These clams are infaunal filter feeders, using siphons to bring food and oxygen-laden water below the sediment surface where they lie hidden from predators. Surf clams in the genus *Donax* have a worldwide distribution and represent a unique innovation filling a niche on high-energy sandy shores with improbably small clams capable of withstanding enormous surf shock. Commonly no more than 1.5 cm in diameter, what these clams lack in size is made-up for by indefatigable digging energy that keeps them from being swept away. Considering how widespread and prolific these bean clams are today, it is surprising that *Donax* has a very poor fossil record. One of the oldest *Donax* fossils in the database traces back to tidal-channel deposits of Miocene age in Delaware, USA (Ward 1998). The very similar *Protodonax*, however, has a much earlier Cretaceous record on sandy shores (Kaufman 1982). *Dosinia* is an example of a much larger venerid clam with an average diameter 10 cm that lives today on sandy shores throughout Europe from the borders of the Norwegian Sea to the Mediterranean. The extant *Dosinia lupinus* has direct ancestors in sandy-shore deposits from the Early Paleocene of New Zealand (Allen 1937), where it co-occurs with the cockle *Cardium*. Cockles are bivalves in the Family Cardiidae with short siphons, but a strong foot that allows them to be more mobile in a sandy substrate. *Cardium* together with the common surf clam, *Spisula*, are examples of substrate specialists capable of burrowing rpidly only within a narrow range of substrate grain sizes (Alexander et al. 1993).

Among the Gastropoda many species in the Family Naticidae also are well adapted for life on coastal sand flats. These snails prey on other shelled mollusks by boring a hole to gain access. Many species in the genus *Polinices* live and hunt on coastal sand flats. They typically possess a large mantle and shuffle through the sand partially buried. Although some *Polinices* species trace back to the Cretaceous, the oldest representative from our database associated with a sandy-shore deposit is the wonderfully named *P. pseudoredemptus* from Upper Eocene to Lower Miocene strata in Austria (Zuschin et al. 2004). *Polinices* species often are crossover snails perfectly capable of living in a sandy-mud substrate.

Tracing Modern Life Strategies for Muddy-Shore Colonization

Modern muddy shores show no obvious biological segregation on a scale related to intertidal zonation, but several examples of organisms from this environment are worthy of attention with regard to their fossil record. In reference to commensals, the most unusual mudflat dwellers are members of the Phylum Echiura, the innkeeper worms. These non-annelid worms have a global distribution in shallow, coastal waters where they construct large U-shaped burrows in mud or fine sand. Lined with mucus, each burrow has two openings commonly 40–100 cm apart, and the burrow may reach a maximum depth of 50 cm. Typically, only three or four species of scale worm, crab, or fish are associated with any one species of echiuran worm. Overall,

35 different living species of marine invertebrates have been identified as echiuran commensals (Anker et al. 2005). The oldest known fossil echiuran, *Priapulites konecniorum*, comes from the Carboniferous Mazon Creek fauna of Illinois, USA (Schram 1973). Living echiurans are variable in size, but may attain a trunk size up to 40 cm in length. By comparison, the Mazon Creek animal was small with a trunk length of only 6 cm. Burrow systems that are U-shaped in morphology are common in the rock record, but there is no example of an echiuran worm preserved within such a burrow. However, the distinctive feeding traces left by the mucus feeding net of modern echiuran worms on the periphery of their burrow are similar to traces preserved in Silurian strata (Risk 1973).

Commensalism also occurs with respect to the burrow systems constructed by ghost shrimps in some coastal mud flats. In Japan, the myid bivalve *Cryptomya truncata* is known to live together with the ghost shrimp in its burrow and the same relationship is based on the co-occurrence of trace fossils and body fossils in a Pleistocene tidal-flat deposit (Nara et al. 2008). As the bivalve genus *Cryptomya* can be traced back to the Miocene, it is inferred that an earlier date for the origin of this behavior may someday be uncovered. The trace fossil *Thallasinoides* is mainly associated with thallasinid shrimp, although other crustaceans are also thought to have made comparable burrows. Examples of *Thallasinoides* trace fossils are interpreted to have been formed in various environments in a full range of water depths from the intertidal to the deep sea, but generally share the same kind of coherent soft sediment. Our database on muddy shores (Appendix 3) records a range from Cambrian to Pliocene ages (Fig. 16). Several instances of the decapod *Mecochirus* occur within *Thallinoides* burrows from the Cretaceous of Portugal (Neto de Carvalho et al. 2007).

Among living solitary bivalves, species assigned to the genera *Macoma* and *Tellina* are examples of infaunal dwellers often found on coastal mud flats. From our database (Appendix 3), the oldest *Macoma* fossil associated with a mudflat deposit is only Pleistocene in age (Yeo and Risk 1981). The oldest *Tellina* fossil connected with mudflat deposits, however, comes from the Upper Cretaceous of Libya (Megerisi and Mamgain 1980). Among living gastropod species from the genus *Nassarius*, the oldest fossil linked to a mudflat deposit comes from the Miocene of Java in Indonesia (Aswan and Ozawa 2006). This very common mud snail is a scavenger with a keen sense of smell that feeds on carrion in the intertidal to shallow subtidal zone. Another gastropod found in large concentrations on mudflats is easily recognized by the high-spired shell characteristic of species in the genus *Turritella*. These are ciliary suspension feeders most often partially buried in soft sediments both in temperate and tropical settings. The oldest *Turritella* fossils associated with coastal mudflat deposits come from the Upper Cretaceous of Poland (Abdel-Gawad 1986).

Discussion

The utility of this overview on intertidal biotas is subject to several factors that shaped the quality of data collected from the fossil record. Certain advantages and disadvantages are inherent in the original depositional settings and the adaptations made by life to those environments as regards the quality of any feasible fossil accumulations through time. Before the relative histories of rocky-shore, sandy-shore, and muddy-shore histories can be compared with one another, the negative and positive influences of these factors must be considered. Before any comparisons can be made with the much more prolific fossil record of normal marine, level-bottom biotas and their march through time, it is necessary to discuss the major similarities and differences between the principal intertidal communities.

Accrual of Fossil Deposits: Weighing the Evidence

The data pool employed for this project amounts to 1622 fossil and related extant species sorted into geological time intervals. In terms of simple ratios, the available data on rocky-shore species outnumber muddy-shore species by nearly 2:1. The imbalance is more striking with sandy-shore biota outnumbered by the rocky-shore biota in a ratio of 2.7 to 1. Each sandy-shore species in the pool stands against roughly 1.4 muddy-shore species. A better picture of intertidal development through time clearly will result from a larger pool in which the numbers of study sites from the different coastal settings are equal.

It is tempting to compare the amount of energy devoted to the study of intertidal biotas as measured by the number of research papers published in each category. In this compilation, there are more than 1.5 papers on rocky-shore biotas for every 1 on muddy-shore biotas. The ratio of rocky-shore studies to sandy-shore studies approaches 2:1. These easy statistics, however, fail to tell the whole story. A more serious detriment is the lack of information from critical time intervals such as the Triassic and Paleocene following the major mass extinctions at the end of the Paleozoic and Mesozoic eras. Investigators doing fieldwork on strata and fossils from all the various intertidal environments have been stymied by this situation.

Some physical conditions may account for why research on these particular intervals is less productive. The assembly of Pangaea as a supercontinent, for example, led to a reduction in the composite coastline during Triassic times. Rocky shorelines also may have been reduced at that time as a result of declining activity in ocean-plate subduction around the margins of the supercontinent as a consequence of tectonic reorganization. Even so, an increase in sandy shores at the expense of rocky shores might have been promoted at the same time under the same circumstances. A record of Triassic sandy and muddy shores does exist, but a dearth of information on fossil content from these settings is hard to overcome. During the Paleogene, however, the global tectonic regime was entirely different from the Triassic. If not a lack of enthusiasm for the subject, other explanations for the poor quality of information on intertidal biodiversity during the Paleogene would be necessary. Environmental degradation linked to the terminal Cretaceous mass extinction may be part of that story.

Factors more inherent in the varying styles of sedimentation on rocky, sandy, and muddy shores, and the different problems met by colonizing organisms in these settings, deserve closer examination with regard to disparity of data. Reliability of rocky-shore data on changing biodiversity is predicated on two conditions that are time independent. One is that rocky-shore deposits can be identified with a high level of confidence as related to basal conglomerate deposits on geological unconformities. The other is that fossils can be associated with a rocky-shore environment with certitude on account of their direct attachment to rock surfaces or their being wedged between boulders and cobbles. Neptunian dikes at various scales and small pocket beaches, for example, represent alternative means for the capture and preservation of rocky-shore invertebrates (Johnson 2006). It can be argued that the high fidelity of the rocky-shore data is mirrored to a certain extent by the in situ nature peculiar to many muddy-shore biota on account of burrowing activities. The echiuran worms, probably many polychaetes, most of the bivalves, and many crabs construct burrows or live permanently below the sediment surface as a means of protection from predators in the muddy intertidal setting. These burrows hold the biota environmentally intact.

Significant problems arise with the sandy-shore biota and its fidelity to an intertidal setting. Anyone who enjoys beach combing knows that the jetsam picked off the beach may have originated far from shore. The various shells collected from a single beach after a storm are sure to represent a mixed assemblage of organisms that occupied different habitats with perhaps subtle but important differences. Due to the highly abrasive environment in which they live Recent bivalves (such as *Donax*, *Spisula*, *Tivela*, *Mesoderma*, *Strigella*, and *Siliqua*) from exposed sandy shores have a poor fossil record.

Even where associated sedimentary structures confirm an intertidal setting for a sand deposit, there is no assurance that the fossils incorporated in those sediments actually lived there. Articulated shells preserved in life position together with sedimentary structures appropriate to an intertidal setting constitute a good test. Bivalve species in the genera *Chione*, *Codakia*, and *Periglypta* commonly are found fully articulated and wedged among cobbles and boulders associated with Pleistocene rocky shores in Baja California (Libbey and Johnson 1997). This attests to the crossover abilities of such bivalves to live both in sand flats and higher-energy settings on rocky shores. With reference to bivalves living in the Gulf of California, Brusca (1980, p. 146) relates that "various species of *Protothaca* and *Chione* are often collected by locals in mud and sand under intertidal rocks, and for this reason are referred to as *almejas piedras*." Thus, we specifically include *Chione, Codakia*, and *Periglypta* in Table 4 as related to the Modern Rocky-shore Biota.

Those who consult the guidebooks on shells also appreciate that many species tolerant of intertidal conditions thrive just as well in offshore, fully subtidal settings. Few living invertebrates are restricted to sandy intertidal deposits, especially in comparison with rocky shores. Those living organisms we know for certain to maintain an affinity with sandy shores pose no problem in making connections with equivalent fossils. Connections become more tenuous with progressively older deposits in which more of the recorded fossils are extinct. The reality may be that open sandy shores with vigorous wave and surf action are especially difficult places in which to live, and have remained so over the long arc of geologic time. It is in this context that the numerically poor database assembled on sandy shores for this study may be reasonably authentic on its own account. Living brachiopods may occur in the intertidal zone, although their optimal environment is in the subtidal. The orders Rhynchonellida and Terrebratulida include the living articulate brachipods. The same orders are predominant in our database. Silurian rhynchonellids show a persistent affinity with well-sorted siliciclastics and grainstones in which lingulid brachiopods commonly are the only other fossils. Fürsich and Hurst (1974) examined Silurian, Jurassic, and Recent rhynchonellids and proposed that the thick, globose shells with strong, angular ribbing, and a stout pedicle represent adaptations to high-energy, turbulent conditions in near-shore settings.

Comparative Histories of the Principal Intertidal Biotas

Comparisons of Figs. 13, 15, and 17 show the major similarities and differences in development of life on rocky, sandy, and muddy shores over Phanerozoic time. All three summaries record persistent but biologically simple forms of life (color-coded blue), such as stromatolites (microbial assemblages) and algal growth, attributed to an Archaic biota little changed from Proterozoic (Precambrian) times. When and where local conditions favored them, Archaic biotas blossomed and left behind spotty traces through the last 540 million years on rocky and sandy shores. On open muddy shores, however, there is no record of an Archaic influence during the Cenozoic.

The relatively large database for rocky-shore life (Fig. 13) presents a history of staggered but persistent biotas identified as Paleozoic, Mesozoic, and Modern in composition. The modern rocky-shore biota (color-coded green) has been numerically dominant at least since the Oligocene, but also achieved a major spike in the Late Cretaceous and shows signs of origins as far back as the Carboniferous. The full details need not be restated (see Section "Tracing Modern Life Strategies for Rocky-Shore Colonization"), but rock-encrusting barnacles, byssally attached bivalves, and a host of herbivorous and carnivorous gastropods make the modern rocky-shore biota what it is, today. Biodiversity is enhanced by the umbrella effect of bivalves with a dense tangle of byssal threads that lends protection to a host of smaller organisms. An ancillary biota with a long history, but one that achieved numerical dominance earlier in time is the Mesozoic biota (color-coded yellow). It is characterized by encrusting invertebrates that include certain brachiopods, oysters, rudists, bryozoans, and scleractinian corals. Most of these elements survive in one guise or another today, but in declining numbers compared to a Late Cretaceous high. An older Paleozoic rocky-shore biota (color-coded red) starts in the Cambrian and continues with a lower level of integration through to the present. It includes other encrusting brachiopods, chitons, tabulate corals, acrothoracian barnacles, and *Trypanites* borings. The portrait of biodiversity in Fig. 13 is one with three main phases of development, all ranked as tiers on rocky shores during the last 65 million years, but with a surviving Mesozoic contribution that overshadows ancient Paleozoic holdovers and a Modern biota that towers above everything else.

The pattern for rocky-shore biotas summarized in Fig. 13 is quite different from those depicted in Figs. 15 and 17 for sandy-shore and muddy-shore biotas. In profile, patterns for the latter two systems are very similar to one another. They share characteristics that present a cradle-shaped profile with a high side of activity in the Paleozoic and a reinvigorated side of activity in the Cenozoic, but not as much in between. Sandy-shore and muddy-shore fossils include older Paleozoic biotas that are strong early on with a level of diversity apparently not surpassed by Modern sandy-shore and muddy-shore biotas. Indeed, sandy-shore and muddy-shore biotas fill significant new niches in Cenozoic times, but the levels of biodiversity shown by green-coded histograms do not exceed earlier phases of experimentation during middle Paleozoic times. To be sure, the high frequencies in the red-coded histograms of Figs. 15 and 17 are indebted to the trove of data on fossils from Benthic Assemblage 1 (shallow end in a range of five shelf-oriented zonations related to water depth) in a global summary of Silurian and Devonian marine paleocommunities (Boucot and Lawson 1999). In the muddy-shore system, the Paleozoic biota includes trilobites, eurypterids, and ostracods from Phylum Arthropoda, most of which became extinct well before the end-Paleozoic mass extinction. The sandy-shore system includes the stalwart lingulid brachiopods and horseshoe crabs, close relatives of which still survive today. If the Paleozoic represents a time of many failed experiments on sandy and muddy shores, the Cenozoic records a period of long-delayed innovations achieved by bivalves that sought refuge below the sediment surface and gastropods that filled unexploited niches in predation.

More subtle in context, the Paleozoic biotas also feature important components represented by trace fossils such as *Skolithos* on sandy shores and *Thalassinoides* on muddy shores. The kind of presence/absence data recorded in spindle diagrams and histograms fails to register the prolific nature of these trace fossils in Paleozoic intertidal deposits compared with younger Cenozoic deposits. Where present, especially in lower Paleozoic strata, these trace fossils are abundant and densely spaced. Unlike body fossils that may be very rare in certain environments, there is no such entity as a single *Skolithos* or *Thalassinoides* burrow. A fundamental change that occurred by later Devonian times was the greening of the planet by plants that enhanced soil development and impeded erosion. Prior to the advent of thick land vegetation, mass wasting of sediments in coastal regions would have occurred at equatorial and temperate latitudes due to sheeted runoff during rainstorms. The implication is that sandy-shore and muddy-shore environments were more widespread in earlier Paleozoic times, before the channelization of water sheds by stream and river systems that resulted in geographically more focused deltas allied with large, individual rivers.

Reflections on Patterns in Level-Bottom and Intertidal Ecosystems

What consensus is there among paleoecologists on the development of marine level-bottom assemblages or communities through Phanerozoic time? How does the organization of rocky-shore, sandy-shore, and muddy-shore communities compare with the major trends observed in the larger record of marine fossils?

The views of paleoecologists on the central questions in the discipline are contentious and sometimes contradictory within camps that see the evolutionary march of major ecosystems from different perspectives. A point of entry to the conversation that lays bare fundamental differences begins with the vocabulary of fossil assemblages versus fossil communities. The difference harkens to a period of intense debate during the early 1900s, when plant ecologists F.O. Clements and H.A. Gleason argued over the degree of biological integration found in plant species that co-occur in living systems with regular frequency (DiMichele 1994). The Clementian camp maintained that communities are highly integrated to the extent that species interactions control the vigor of the community. In opposition, the Gleasonian camp adopted a reductionist view that species coalesce as ephemeral and random collections seeking only the most basic similar needs for resources. Paleoecologists who prefer the term fossil assemblages belong to the Gleasonian tradition. Those who speak of fossil communities claim to see regular and repeating patterns in the kinds of fossils found together in time and space. Often, they also recognize a greater role for species interactions and co-evolution.

Another issue that holds much importance for ecologists and paleoecologists is the question of boundaries. Do collections of species found arrayed with one another in living systems and fossil deposits exhibit distinct borders, or does membership shift on a species-by-species basis in a poorly co-coordinated way through some kind of environmental gradient? The divide is not so strict, but those who see fossils clumped together with vague boundaries often favor the notion of low-grade assemblages as distinguished from high-grade communities composed of inter-dependent species. The question over community evolution is an inquiry about which paleoecologists also hold strong opinions. In one camp, those who find the rock record replete with marine fossil associations dominated by loosely related organisms deny the possibility of community evolution (Bambach and Bennington 1996; Stanley 2008). In the opposing camp, those who find communities duly influenced by predation, competition, and increasing levels of mutualism employ the term community evolution despite the fact that the community is not an entity with genetic heritability in the same sense as an individual or a like population of individuals. Geerat Vermeij is a vocal spokesman for co-evolution in the context of an escalating arms race between prey and predators both in marine and terrestrial systems. His thesis on evolutionary escalation (Vermeij 1987) has stimulated others to re-examine the notion of community evolution and the yet larger question of ecosystem evolution.

Bambach (1993), who later rejected the notion of community evolution, concedes that biomass and energy flow through marine systems increased substantially through Phanerozoic time. He reaches this conclusion by evaluating major shifts through time among invertebrate groups such as functionally modern bivalves versus brachiopods and crabs and lobsters versus trilobites, as well as changes in the volume of bioturbation in marine sediments. He surely qualifies, here, as an advocate of ecosystem evolution (taking the word evolution in a general and non-genetic sense). The insights of Bambach (1993) mesh well with those of Sepkoski (1981, 1997) on the modernization of sequential evolutionary faunas that dominate the marine fossil record. Identification of overlapping Archaic, Paleozoic, Mesozoic, and Modern biotas in our present study on intertidal life owes a debt of inspiration to the research promulgated by Sepkoski (1981, 1997) and those who carried it forward (Peters and Foote 2001; Stanley 2007). By holding to the word "development" over "evolution" in the title of this project, we have tried to remain even-handed regarding some of the central questions frequently debated by paleoecologists who study fossil marine ecosystems. Likewise, we prefer the word "biota" to "community" for much the same reason. This is the place, however, to affirm what we believe the results of our present program signify.

A signal event in our careers as paleoecologists occurred in July 1985 when we stood on the shores of Hudson Bay near the town of Churchill in northern Manitoba, Canada. At low tide, a spectacular unconformity between Proterozoic quartzite and Upper Ordovician dolomite is exposed to reveal a 350-m-long shoreface lined by conglomerate in a maze of quartzite boulders commonly 2 m in diameter. The scene was instantly recognized as a former rocky shore preserved in three-dimensional relief. Only months earlier, we visited The Baths in the British Virgin Islands, where huge granite boulders are bathed in turquoise waters reflected by white carbonate sand and living corals are anchored to the boulders. Fossil corals were predicted to be found attached to the Churchill boulders and the prize was won within minutes of beginning a search for them (Johnson and Baarli 1987). The fossil biota of the Late Ordovician rocky shore at Churchill is fascinating as a harbinger of things to come, but its faunal diversity is low and the level of complexity is entirely Paleozoic (Johnson et al. 1988). The corals adhere like thin skullcaps on quartzite boulders, not as arborescent thickets with nooks and crannies for additional encrusting and nestling species. Adjacent to the boulder-encrusting corals is a facies dominated by monospecific trimerellid brachiopods exposed as a shell pavement. Except in a spatial sense, the brachiopods had no influence on the corals and the corals held no regulation over the brachiopods. The Churchill rocky-shore biota is low grade and represents a largely fortuitous assembly of organisms whose interactions were limited to the meeting of their own environmental needs in the same way that Bambach and Bennington (1996) attribute characteristics to Carboniferous brachiopod associations. The difference from neighboring subtidal occupants, however, was that some tabulate corals took advantage of a high-energy environment on a hard substrate that few other organisms were capable of meeting.

Intertidal shores are different from subtidal, level-bottom systems in several ways. Most important, the reality of gradients in the physical environment is severely truncated. The boundaries between adjacent biotas on modern rocky shores are exceedingly sharp. Among other factors, we are convinced by arguments regarding a refuge in the high intertidal for littorinid gastropods (Kirby 2004), competitive exclusion of one kind of barnacle against another in the upper tidal zone (Stanley and Newman 1980), and the cropping effect by starfish on blue mussels (Paine 1966) in the lower tidal zone, all of which evolved as behaviors to effect a commanding place in the zonation of modern rocky shores. These patterns exhibit remarkable uniformity all around the world, today. The fossil record supports the appearance of a recognizable modern rocky-shore ecosystem well back in the Cenozoic with even earlier organizational signs notable by the Late Cretaceous. There can be no doubt that rocky intertidal surfaces constitute an important resource aggressively partitioned by marine invertebrates with a wide range of functional morphologies. Furthermore, the interwoven fabric of the lower intertidal zone that makes habitat space available for additional species within the interstices of mussel beds gives an important boost to biodiversity.

Sandy-shore biotas are not zoned in the same way as on rocky shores and the fauna often migrates both with the daily tides and seasonally as wave activity changes on the beach. In muddy shores, however, the fauna moves around much less and zonation is dependent on the presence or absence of macrophytes (Raffaelli and Hawkins 1999). The late Cretaceous to Cenozoic invasion of deep-dwelling bivalves in soft coastal substrates was a major innovation that spurred bivalve diversity and resulted in a degree of safety from predators not previously achieved (Stanley 1968). We appreciate the anecdotal experience shared by Stanley (2008) regarding the opportunistic nature of infaunal bivalves in shallow sudtidal settings, where concentrations are known to suddenly appear and then disappear. We are also aware of Pliocene fossil deposits on the Gulf of California in Baja California Sur, Mexico, where diverse bivalve faunas existed in the same places with great fecundity and high lateral continuity over geologically long spans of time (Johnson et al. 2009). Spat falls are controlled by local conditions, and coastal geography probably makes a difference.

The muddy intertidal zone is a rich habitat where digging and burrowing activities provide places of safe refuge and the potential for commensalism. Using resin casts to study burrow systems in mud flats on the Olympic Peninsula of Washington, USA, Gingras et al. (2002) discovered inter-connected structures for a wide range of domiciles that indicate broad commensal activity among bivalves, crabs, and polychaets. This level of complexity in ichnofabric might easily be mistaken as an example of faunal succession if found preserved as interpenetrating trace fossils in the rock record. The intricate ichnofabrics from muddy substrates represent another example of vigorous partitioning of an important resource by marine invertebrates with a wide range of functional morphologies. The same level of activity is not evident in Paleozoic deposits. Echiuran worms have a fossil record that extends to the Carboniferous, but if appreciably smaller than those we are familiar with today, Echiuran ancestors may have lacked the pumping power to be good inn-keepers.

Compared to rocky and muddy shores, it appears that sandy shores have remained poorly colonized and the most Paleozoic in character. Few infaunal bivalves are exclusive to this intertidal setting, but like *Donax*, they thrive and even manage to host commensal hydroids (Ricketts et al. 1985). The problem with high-energy sandy shores is that sedimentologically they fall between a rock and a softer muddy place. In more sheltered beach settings, oysters and even mytilid bivalves may form small clusters attached to the odd cobble or a nest of other shells. On an open sandy shore, however, the constant movement of sand particles impedes colonization both above and below the substrate surface. The general absence of commensal activity on sandy shores compared to rocky and muddy shores indicates that resource partitioning has failed to take hold there on a macro-organic level. Coastal rocks are crowded with life, the mud is full of life, but the sand has remained relatively clean and under utilized as a resource for ecospace.

Conclusions

The effects of physical disturbance and predation can be observed directly as formative influences on present-day rocky, sandy, and muddy shores. Competition and niche partitioning also play significant roles in the exploitation of living space as a prime resource on rocky and muddy shores. Marine biologists Ed Ricketts and Rachael Carson understood and applied complex ecological ideas to their pioneering observations on the intertidal shores of North America dating back to the 1930s, 1940s, and 1950s. While mindful of crucial physical parameters like wave shock and duration of subaerial exposure, these dedicated students of rocky tide pools, sandy shores, and mud flats took delight in describing the intricate biological relationships found among diverse species of marine algae and invertebrates adapted to life in those harsh settings. They never deployed the word "community" to characterize the thriving life of the intertidal zone, but the busy interactions and relationships of the residents was a constant source of fascination that still rings bold in their writings (Ricketts et al. 1985; Carson 1956).

Stanley (2008) has argued that physical disturbance and predation are the overriding factors that regulate the membership of subtidal, level-bottom biotas dominated by marine invertebrates today and in the past. While predation was an increasingly important factor for Paleozoic biotas, there appears to be no evidence for ecological crowding during the Paleozoic Era (Stanley 2007). It generally is acknowledged, however, that a far more sophisticated arms race developed between prey and predators during the Mesozoic marine revolution

(Vermeij 1977; Stanley 2008). Although originally envisioned in terms of shelled gastropods as prey species and various fish and crabs as predators that co-evolved as a result of their interactions, Vermeij (1987) underscored the concept in relation to other organisms in other ecosystems.

The ecosystems of the world's intertidal zone reveal a long history of having been deeply impacted by three grades of physical and biological influence setting them apart from the more extensive subtidal, level-bottom realms of oceanic platforms and shelves.

Plate Tectonics, Global Geography, and Sea Level

On a global scale, all action within the intertidal zone occurs along a vanishingly thin corridor at the juncture between land and sea. The joining together and breakup of megacontinents due to the mechanics of plate tectonics regulates the amount of physical space available for colonization within the intertidal zone. Depending on the degree of coastal smoothness, the contraction of many small continents into one huge continent forces a reduction in available living space by up to 45% or more. During the most recent phases of the Wilson cycle, the break up and dispersal of Pangaea from Late Triassic time forward resulted in a tremendous lengthening of the intertidal zone, although the starting proportions of rocky, sandy, and muddy shores can only be surmised for Triassic geography. Fluctuations in sea level play a lesser role, boosting or diminishing composite shoreline length by a smaller amount. In terms of biological dispersal, a megacontinent with a single conterminous coastline presents far fewer barriers to migration for coastal biotas than multiple, smaller continents separated from one another by long oceanic distances.

Climate and Land Plants

The intertidal zone was different before and after the Devonian Period due to the greening of the planet and the advent of relatively thick plant cover on land. Particularly in climate zones with high rainfall in tropical and temperate latitudes, mass wasting of landscapes inundated pre-Devonian shores with a thick clastic wedge of sediments. Rocky shores wetted by seawater may have been more common in arid landscapes. Post-Devonian shores saw a substantial contraction of sandy and muddy habitats due to the more regular development of major rivers fed by channelized trunk streams. Ria shores with high mud content are more associated with drowned coastal valleys and reflect the overall diminishment of muddy habitat space in post-Devonian worlds.

Ecological Crowding in the Intertidal Zone

Rocky shores have achieved the highest density of ecological crowding by algal and faunal elements of any intertidal setting. Their modernity blossomed over the last 65 million years. Oligocene rocky shores, for example, were similar to today, with a splash zone occupied by littorinid gastropods, a distinct band for barnacles, and another band for byssate bivalves. The diversity of predatory gastropods was on the rise and had a corollary effect on niche partitioning. Older Mesozoic and Paleozoic elements still contribute to modern rocky shores, which adds to the overall biodiversity. During Mesozoic times, however, ecological crowding on rocky shores was nascent. It was not much of a factor during Paleozoic times, when only a few species took advantage of living space, other species were unequipped to handle. Based on fossil counts in our database, the average number of rocky-shore species from the various Cenozoic study sites is nearly 2.5 times greater than found on average at Paleozoic sites.

Coastal mudflats are not so sharply zoned as rocky shores, but effective use of living space by epifaunal and infaunal elements stimulated biodiversity also during Cenozoic time. Scavenging and predatory gastropods living on the surface, or partly submerged in mud, expanded activities during this time. Commensal living among echiuran worms, bivalves, crabs, polychaete worms, and others may be explored through the diversity of burrows and fixed domiciles found in Recent and Cenozoic mudflats. Paleozoic burrow systems from intertidal and shallow subtidal environments collectively reflect much the same range of diversity in morphology, but often occur as monotonous expressions of one or a few morphotypes at any given locality. Based on actual fossil counts, however, the increase in average number of species per Cenozoic study site is marginal over that for Paleozoic sites.

Modern sandy shores reflect the lowest biodiversity of macro-organisms of any of the principal intertidal settings. They retain stellar examples of organisms such as horseshoe crabs and lingulid brachiopods that are successful holdovers from Paleozoic times. Hard substrates on rocky shores present the prospect of a secure anchorage. Mudflats offer the opportunity of refuge within a soft but cohesive substrate. Successful diggers prosper on sand flats, but many live just as comfortably in offshore, subtidal settings. The mobility of sand due to wave surge and coastal currents within the intertidal zone on open sandy shores is particularly difficult for macro-organisms to cope with. Although experimentation appears to have been high in the early Paleozoic, the sandy shore never reached the same potential for ecological crowding as on rocky and muddy shores. Based on the fossil evidence, there is only a slight increase between the average number of species from Cenozoic and Paleozoic study sites.

Acknowledgments We are grateful to Chris Warren, Jonathan Morgan-Leamon, and Sharron Macklin from the Office of Information Technology at Williams College for assistance in creating an electronic filing program suitable to our needs. In particular, Chris Warren formatted the system that allowed us to connect customized files with Open Office. The photograph in Fig. 19 was supplied by Halard L. Lescinsky (Otterbein College); all other photos are by ME Johnson. Steven M. Stanley, Arthur J. Boucot, and Ceith A.W. Crook reviewed an early draft of the manuscript and offered insights and criticisms that led to significant improvements.

Appendix 1: Bibliography on Rocky-Shore Biotas Through Phanerozoic Time

Neogene

Aguirre J, Jiménez AP (1997) Census assemblages in hard-bottom coastal communities: a case study from the Plio-Pleistocene Mediterranean. Palaios 12:598–608

Aigner T (1983) A Pliocene cliff-line around the Giza Pyramids Plateau, Egypt. Palaeogeogr Palaeoclimatol Palaeoecol 42:313–322

Ashby JR, Minch AJ (1987) Stratigraphy and paleoecology of the Mulege embayment, Baja California Sur, Mexico. Ciencias Marinas 13:89–112

Baluk W, Radwanski A (1977) Organic communities and facies development of the Korytnica basin (Middle Miocene; Holy Cross Mountains, Central Poland). Acta Geol Pol 27:85–123

Bosworth TO (1922) Geology of the Tertiary and Quaternary periods in the northwest part of Peru. MacMillan and Co, London

Bromley RG, D'Alessandro A (1987) Bioerosion of the Plio-Pleistocene transgression of southern Italy. Riv Ital Paleontologia e Stratigrafia 93:379–442

Buckeridge JS (1975) The significance of cirripedes to the paleoecology of Motutapu Island. Tane 21:121–129

Burchette TP (1988) Tectonic control on carbonate platform facies distribution and sequence development: Miocene, Gulf of Suez. Sedimentary Geol 59:179–204

Campbell KA, Grant-Mackie JA, Buckeridge JS, Hudson N, Alfaro AC, Hoverd J, Morgan S, Horne N, Banfield A (2004) Paleoecology of an early Miocene, rapidly submerging rocky shore. New Zealand J Geol Geophys 47:731–748

Chappell J, Veeh HH (1978) Late Quaternary tectonic movements and sea-level changes at Timor and Atauro Island. Geol Soc Am Bull 89:356–368

Doménech R, Gibert JMD, Martinell J (2001) Ichnological features of a marine transgression: middle Miocene rocky shores of Tarragona, Spain. Geobios 34:99–107

Donovan SK (1989) Palaeoecology and significance of barnacles in the mid-Pliocene Balanus beds of Tobago, West Indies. Geol J 24:239–250

Donovan SK, Miller DJ (1995) Aspects of ancient rocky shorelines: the Hopegate and Falmouth Formation, Jamaica. Caribbean J Sci 31:174–184

Doyle P, Mather AE, Bennett MR, Bussell MA (1997) Miocene barnacle assemblages from southern Spain and their palaeoenvironmental significance. Lethaia 29:267–274

Eagle MK, Hayward BW (1992) Paleontology and paleoecology of early Miocene sequences in Hays and Tipakuri streams, Northern Hunua Ranges, Auckland. Records Auckland Instit Museum 29:113–133

Eagle MK, Hayward BW, Carter G (1994) Early Miocene rocky shore and coarse sediment fossil communities, Kawau Island, Auckland. Records Auckland Instit Museum 31:187–204

Eagle MK, Hayward BW, Grant-Mackie JA (1995) Early Miocene beach, rocky shore, and enclosed bay fossil communities, Waiheke Island, Auckland. Records Auckland Instit Museum 32:17–44

Fernández J, Rodriguez-Fernández J (1991) Facies evolution of nearshore marine clastic deposits during the Tortonian transgression – Granada Basin, Betic Cordilleras, Spain. Sedimentary Geol 71:5–21

Gibert JM, Martinell J, Doménech R (1996) El Mioceno Marino entre las playas de L'Arrabassada y El Miracle (Tarragona): aspectos paleontológicos e implicaciones sedimentológicas. Acta Geol Hispanica 29:133–148

Gibert JM, Martinell J, Doménech R (1998) Entobia Ichofacies in fossil rocky shores, Lower Pliocene, nortwestern Mediteanean. Palaios 13:476–487

Gutowski J, Machalski M (1984) A new littoral locality within the Middle Miocene (Badenian) Korytnica Basin (Holy Cross Mountains, Central Poland). Acta Geol Polenica 34: 195–201

Hamza FH (1983) Post-Pliocene transgressive phase along the northern part of the Nile Valley, Egypt. Neues Jahrbuch für Geologie und Palaeontologie Monatshefte 6:338–344

Hartley AJ, Jolley EJ (1995) Tectonic implications of Late Cenozoic sedimentation from the Coastal Cordillera of northern Chile (22–24oS). J Geol Soc Lond 152:51–63

Johnson ME, Baarli BG (2005) Erosion and burial of granite rocky shores in the Recent and Late Pleistocene of the Seychelles Islands: physical and biological perspectives. J Coastal Res 21:867–879

Johnson ME, Baarli BG, Scott J (1995) Colonization and reef growth on a Late Pleistocene rocky shore and abrasion platform in Western Australia. Lethaia 28:85–98

Johnson ME, Ledesma-Vázquez J (1999) Biological zonation on a rocky-shore boulder deposit; Upper Pleistocene Bahia San Antonio (Baja California Sur, Mexico). Palaios 14:569–584

Kaye CA (1959) Shoreline features and Quaternary shoreline changes Puerto Rico. U.S. Geol Sur Prof Paper 317-B:1–140

Kröcherta J, Maurera H, Buchner E (2008) Fossil beaches as evidence for significant uplift of Tenerife, Canary Islands. J Afr Earth Sci 51:220–234

Kroh A, Hartzhauser M (1999) An echinoderm fauna from the Lower Miocene of Austria: paleoecology and implications for Central Paratethys paleobiogeography. Annalen des Naturhistorischen Museums in Wien 101 A:145–191

Ledesma-Vázques J, Johnson ME (1994) Late Pliocene abrasion platform from the Cantil Costero Formation of Baja California. Ciencias Marinas 20:139–157

Libby LK, Johson ME (1997) Upper Pleistocene Rocky Shores and intertidal biotas at Playa La Palmita (Baja California Sur, Mexico). J Coastal Res 13:216–225

Macneil FS, Mertie JB, Pilsbry HA (1943) Marine invertebrate faunas of the buried beaches near Nome, Alaska. J Paleontol 17:69–96

Martin JM, Braga JJ, Aguirre J, Betzler C (2004) Contrasting models of temperate carbonate sedimentation in a small Mediterranean embayment: the Pliocene Carboneras Basin, SE Spain. J Geol Soc Lond 161:387–399

Martinell J, Doménech R (1986) Actividad bioerosiva en el Plioceno marino del Emporda (Catalunya). Paleontologia i Evolució 20:247–251

Martinell J, Doménech R (1995) Bioerosive structures on the Pliocene rocky shores of Catalonia (Spain). Rev Esp Paleontol 10:37–44

Masuda K (1968) Sandpipes penetrating igneous rocks in the environs of Sendai, Japan. Trans Proc Palaeontol Soc Jpn 72:351–362

Meco J, Scallet S, Guillou H, Lomoschitz A, Carracedo C, Ballestar J, Betancort J-F, Cilleros A (2007) Evidence for long-term uplift on the Canary Islands from emergent Mio–Pliocene littoral deposits. Global Planetary Change 57: 222–234

Montaggioni LF, Hoang CT (1988) The last interglacial high sea level in the granitic Seychelles, Indian Ocean. Palaeogeogr Palaeoclimatol Palaeoecol 64:79–91

Muhs DR, Szabo BJ (1982) Uranium-series age of the Eel Point terrace, San Clemente Island, California. Geology 10:23–26

Ortlieb L (1991) Quaternary shorelines along the northeastern Gulf of California; Geochronologic data and neotectonic implications. In: Perez-Segura E, Jacaques-Ayala C (eds) Studies of Sonoran geology. Geological Society of America Special Paper, vol 254, pp 95–120

Peck JH (1960) Paleontology and correlation of the Ohlson Ranch Formation. Univ Calif Pub Geol Sci 36:233–242

Pendón JG, Ruiz F, Abad M, González-Regalado ML, Baceta JJ, Tosquella J (2003) The transgressive sequences on foreland margins: a case study pf the Neogene central Guadalqivir basin, southern Spain. Riv Ital Paleontol Stratigrafia 110: 503–515

Pufahl P, Noel PJ, Bone Y, Lukasik JJ (2004) Pliocene sedimentation in a shallow, cool-water, estuarine gulf. Sedimentology 51:997–1027

Radwanski A (1964) Boring animals in Miocene littoral environments of southern Poland. Bulletin Polish Academy of Science, Serie des Sciences Geology et Geogr 12:57–62

Radwanski A (1967) Problems of Miocene littoral structures on the southern slopes of the Holy Cross Mountains, Central Poland. Rocznik Polskiego Towarzystwa Geologicznego 37:169–175

Radwanski A (1968) Tortonian Cliff Deposits at Zahorska Bystrica near Bratislava (Southern Slovakia). Bulletin de L'academie Polonaise des Sciences, Geology et Geogr 16:97–102

Radwanski A (1970) Dependence of rock-borers and burrowers on the environmental conditions within the Tortonian littoral zone of southern Poland. In: Crimes TP, Harper JC (eds) Trace fossils. Letterpress Limited, Liverpool, pp 371–390

Richards GW (1985) Fossil Mediterranean molluscs as sea-level indicators. Geol Mag 122:373–381

Ricketts BD, Ballance PF, Hayward BW, Mayer W (1989) Basal Waitemata Grou lithofacies: rapid subsidence in an early Miocene interarc basin, New Zealand. Sedimentology 36:559–580

Roberts DL, Brink JS (2002) Dating and correlation of Neogene coastal deposits in the Western Cape (South Africa): implications for Neotectonism. S Afr J Geol Geophys 105: 337–352

Russel MP (1991) Modern death assemblages and Pleistocene fossil assemblages in open coast high-energy environments, San Nicolas Island, California. Palaios 6:179–191

Santos A, Mayoral E, Silva CMD, Cachao M, Domenéch R, Martinell J (2008) Trace fossil assemblages on Miocene rocky shores of southern Iberia. In: Wisshak M, Tapanila L (eds) Current developments in bioerosion. Springer, Berlin Heidelberg, New York, pp 431–450

Scott JH, Johnson ME (1993) Lateral variation in the geomorphology of a Pleistocene rocky coastline at Kalbarri, Western Australia. J Coastal Res 9:1013–1025

Silva CM, Cachao M, Martilell J, Doménech R (1999) Bioerosional evidence of rocky palaeoshores in the Neogene of Portugal: environmental and stratigraphic significance. Bull Geol Soc Denmark 45:156–160

Simian ME, Johnson ME (1997) Development and foundering of the Pliocene Santa Ines Archipelago in the Gulf of California: Baja California Sur, Mexico. In: Johnson ME., Ledesma-Vázquez J (eds) Pliocene carbonates and related facies flanking the Gulf of California, Baja California, Mexico. Geological Society of America, Special Paper, vol 318, pp 25–38

Smith JT (1986) Middle Tertiary rocky substrate mollusks from Baja California Sur, Mexico. Am Malacol Bull 4:1–12

Squires DF (1959) Results of the Puritan-American museum of natural history expedition to Western Mexico. Bull Am Museum Nat Hist 118:378–399

Titschack J, Bromley RG, Freiwald A (2005) Plio-Pleistocene cliff-bound, wedge-shaped, warm-temperate carbonate deposits from Rhodes (Greece); sedimentology and facies. Sedimentary Geol 180:29–56

Valentine JW (1980) Camalu: a Pleistocene terrace fauna from Baja California. J Paleontol 54:1310–1318

Valentine JW, Lipps JH (1963) Late Cenozoic rocky terrace faunas from Anacapa Island, California, North America. J Paleontol 37:1292–1302
Walker SE (1995) Taphonomy of modern and fossil intertidal gastropod associations from Isla Santa Cruz and Isla Santa Fe, Galápagos Islands. Lethaia 28:371–382
Watkins R (1990) Paleoecology of a Pliocene rocky shoreline, Salton Trough region, California. Palaios 5:167–175
Wilson MA, Curran HA, White B (1998) Paleontological evidence of a brief global sea-level event during the last interglacial. Lethaia 31:241–250
Ziko A, Hamza F (1987) Bryozoan fauna from a Post-Pliocene outcrop north of the Giza pyramids plateau, Egypt. In: Ross JRP (ed) Bryozoa: present and past, 7th international conference on Bryozoa. Western Washington University, Bellingham, Washington, pp 301–308
Zullo VA (1986) Quaternary barnacles from the Galápagos Islands. Proc Calif Acad Sci 44: 301–308
Zullo VA, Gurswami-Naidu RB (1982) Late Miocene balanid cirripedia from the basal Wilson Ranch Beds ("Merced" Formation), Sonoma County, northern California. Proc Calif Acad Sci 42:525–535
Zwiebel JA, Johnson ME (1995) Late Pleistocene Mytilid and Petricolid bivalves from the open rocky shore of Pacific Baja California (Mexico): unusual preservation of macrofossils. J Coastal Res 11:704–716

Paleogene

Bosel CA, Coombs DS (1984) Foveaux Formation: a warm-water strandline deposit of Landon-Pareora age at Bluff Hill, Southland, New Zealand. New Zealand J Geol Geophys 27:221–223
DeVries TJ (1998) Oligocene deposition and Cenozoic sequence boundaries in the Pisco Basin (Peru). J South Am Earth Sci 11:217–231
Effinger WL (1938) The Gries Ranch Fauna (Oligocene) of western Washington. J Paleontol 12:355–592
Gupta S, Allen PA (1999) Fossil shore, platforms and drowned gravel beaches: evidence for high-frequency sea-level fluctuations in the Distal Alpine Foreland Basin. J Sedimentary Res, Section B: Stratigr Global Stud 69:394–413
Hartkopf C, Stapf KRG (1984) Sedimentologie des Unteren Meeressandes (Rupelium, Tertiär) an Inselstränden im W-Teil des Mainzer Beckens (SW-Deutschland. Mitt Pollichia 7:5–106
Hecker RF (1965) Introduction to paleoecology. Elsevier, New York
Lee DE, Carter RM, King RP, Cooper AF (1983) An Oligocene rocky shore community from Mt. Luxmore, Fiordland. New Zealand J Geol Geophys 26:123–126
Lindberg DR, Squires RL (1990) Patellogastropods (Mollusca) from the Eocene Tejon Formation of southern California. J Paleontol 64:578–587
Linder RA, Durham JW, Orr WN (1988) New late Oligocene echinoids from the central Western Cascades of Oregon. J Paleontol 62:945–958
Lozouet P (2002) First record of the Caribbean genus *Cittarium* (Mollusca, Gastropoda, Trochidae) from the Oligocene of Europe and its paleobiogeographic implications. J Paleontol 76:67–77
Matsubara T (2002) Molluscan fauna of the 'Miocene' Maejima formation in Maejima Island, Okayama Prefecture, Southwest Japan. Paleontol Res 6:127–145
Miller PR, Orr WN (1988) Mid-Tertiary transgressive rocky coast sedimentation: central western Cascade Range, Oregon. J Sedimentary Petrol 58:959–968
Neuffer FO, Rothausen K, Sonne V (1978) Fossilführende Rinnenfüllung im Unteren Meeressand an einer Insel-Steilküste des Mitteloligozan-Meers (Steigerberg bei Eckelsheim, Mainzer Becken). Mainzer Geowissenschaften Mitteilung 6:99–120
Pomoni-Papaioannou F, Solakius N (1991) Phosphatic hardgrounds and stromatolites from the limestone/shale boundary section at Prossilion (Maastrichtian-Paleocene) in Parnassus-Ghiona Zone, Central Greece. Palaeogeogr Palaeoclimatol Palaeoecol 86:243–254
Pyanovskaya IA, Hecker RT (1966) Rocky shores and hardground of the Cretaceous and Paleogene seas in central Kizil-Kum and their inhabitants. In: Organisms and environment in the geological past – a symposium, Moscow, Nauka, pp 222–245
Squires RL, Goedert JL (1994a) New species of Early Eocene small to minute mollusks from the Crecent Formation, Black Hills, southwestern Washington. Veliger 37: 253–266
Squires RL, Goedert JL (1994b) Macropaleontology of the Eocene Crescent Formation in the Little River area, southern Olympic Peninsula, Washington. Nat Hist Museum Los Angeles County Contributions Sci 444:1–32
Stilwell JD, Fordyce EE, Rolfe PJ (1994) Paleocene Isocrinids (Echinodermata: Crinoidea) from the Kauru Formation, South Island, New Zealand. J Paleontol 68: 135–141
Woods AJC, Saul LR (1986) New Neritadae from south-western North America. J Paleontol 60:636–655

Cretaceous

Abdel-Gawad GI (1986) Maastrichtian none-cephalopod mollusks (Scaphopoda, Gastropoda and Bivalvia) of the Middle Vistula Valley, Central Poland. Acta Geol Pol 36: 69–224
Asgaard U, Bromley RG (1991) Population dynamics and autecology of 'Rhynchonella' triangularis, a Late Cretaceous rocky coast brachiopod. In: Mackinnod DI, Lee DE, Campbell JD (eds) Brachiopods through time. A.A. Balkema, Rotterdam, pp 247–252, p 447
Bryan JR (1992) Origin and paleoecology of Maastrichtian rockground and chalk facies in southcentral Alabama. Palaios 7:67–76
Crampton JS (1988) A late Cretaceous near-shore rocky substrate macrofauna from northern Hawkes Bay, New Zealand. N Z Geol Surv Rec 35:21–24
Erlström M, Gabrielson J (1992) Petrology, fossil composition and depositional history of the Ignaberga limestone, Kristianstad Basin, Scania. Sveriges Geologiska Undersökning 80:1–30

Johnson ME, Hayes ML (1993) Dichotomous facies on a Late Cretaceous rocky island as related to wind and wave patterns (Baja California, Mexico). Palaios 8:385–395

Kennedy WJ, Klinger HC (1972) Hiatus concretions and hardground horizons in the Cretaceous of Zululand (South Africa). Paleontology 15:539–549

Krantz R (1972) Die Sponge-Gravels von Faringdon (England). Neues Jahrbuch für Geologie und Palaeontologie 140: 207–231

Lescinsky HL, Ledesma-Vázquez J, Johnson ME (1991) Dynamics of Late Cretaceous Rocky Shores (Rosario Formation) from Baja California, Mexico. Palaios 6:126–141

Lewy Z (1985) Paleoecological significance of Cretaceous bivalve borings from Israel. J Paleontol 59:643–648

Nekvasilová O (1982) Craniidae (Brachiopoda, Inarticulata) from the Lower Cretaceous of Stramberk (Czechoslovakia). Casopis pro mineralogii a geologii 27:127–137

Perkins BF, Stewart CL (1971) Whitestone Quarry. In: Perkin BF (ed) Trace fossils: a guide to selected localities in Pennsylvanian, Permian, Cretaceous and Tertiary rocks of Texas. Louisiana State University School of Geosciences, Miscellaneous Publications, vol 71, pp 17–22

Pietzsch K (1962) Geologie von Sachsen. Veb Deutcher Verlag der Wissenschaften, Berlin, pp 388–389

Pyanovskaya IA, Hecker RT (1966) Rocky shores and hardground of the Cretaceous and Paleogene seas in central Kizil-Kum and their inhabitants. In: Organisms and environment in the geological past – a symposium, Moscow, Nauka, pp 222–245

Rodda PU (1989) A fossiliferous rocky shore environment in the Lower Cretaceous (Albian) of northern California. Geol Soc Am Abstr Programs 21:135–136

Sano S (1991) Discovery of a coral-rudist buildup in the Miyako Group, northeast Japan. Trans Proc Palaeontol Soc Jpn 162:794–800

Skelton PW, Nolan SC, Scott RW (1990) The Maastrichtian transgression onto the northwestern flank of the Proto-Oman Mountains: sequences of rudist-bearing beach to open shelf facies. Geol Soc Spec Publ 49:521–547

Surlyk F, Christensen WK (1974) Epifaunal zonation on an Upper Cretaceous rocky coast. Geology 2:529–534

Wilson MA, Taylor PD (2001) Palaeoecology of hard substrate faunas from the Cretaceous Qahlah Formation of the Oman Mountains. Palaentology 44:21–41

Zitt J (1992) A new occurrence of Upper Cretaceous epibionts cemented to the rocky substrates and bioclasts (locality Radim, Czechoslovakia). Casopis pro mineralogii a geologii 37:145–154

Zitt J, Geys JF (2003) *Novasalnia* gen. nov.: a remarkable Late Cenomanian echinoid from the Bohemian Cretaceous Basin (Czech Republic). Cretaceous Res 24:23–30

Zitt J, Nekvasilová O (1987) Epibionts cemented to rocky bottom and clasts in the Upper Cretaceous of Zelezne Hory Mts. (Czechoslovakia). Casopis Národního Muzea 156:17–35

Zitt J, Nekvasilová O (1996) Epibionts, their hard-rock substrates, and phosphogenesis during the Cenomanian-Turonian boundary interval (Bohemian Cretaceous Basin, Czech Republic). Cretaceous Res 17:715–739

Zullo VA, Russell EE, Mellen FF (1987) Brachylepas Woodward and Virgiscapellus Withers (Cirripedia) from the Upper Cretaceous of Arkansas. J Paleontol 61:101–111

Jurassic

Bucheit AKD RN (2000) 2000 – Initiaion of a carboate Platform: a Comparison Between the Lower Jurassic Broadford Limestone, Isle of Skye, Scotlad, and the Cambrian Honey Creek Limestone, Slick Hills, Oklahoma. Oklahoma Geol Surv Circular 101:57–64

De-La-Beche HT (1846) On the formation of the rocks of south Wales and south western England. Memoires Geol Surv Great Britain 1:1–296

Duff KL, Mckirdy AP, Harley MJ (1985) New sites for old, A student's guide to the geology of the east Mendips. Chapter 10, Fossil rocky sea floors: site descriptions. Nature Conservatory Council 134–143

Dzulynski S (1950) Littoral deposits of the Middle Jurassic south of Krzeszowice. Annales de la Société Géologique de Pologne 19:387–400

Garcia-Hernandez M, Mas JR, Molina JM, Ruiz-Ortiz PA, Vera JA (1988) Episodio de karstification en litorales insulares del Jurasico Superior (Fm. Ammmonitico Rosso, Subbetico Esterno, Provincia de Cordoba). III Coloquio de Estratigrapfia y Paleogeograpfia del Jurasico de España. Logroño 32–35. 10 al 19 de septiembre de 1988

Hoelder H, Hollmann R (1969) Bohrgange mariner Organismen in jurassischen Hart und Felsboden. Neues Jahrbuch für Geologische Palaontologie Abhandlung 133:9–88

Johnson ME, McKerrow WS (1995) The Sutton Stone: an early Jurassic rocky shore deposit in South Wales. Palaeontology 38:529–541

Molina JM, Ruiz-Ortiz PA, Vera JA (1989) Escarpes de Falla Juasicos karstificados: un posible margen bypass de umbral pelagico. Subbetico Externo. XII Congreso Español de Sedimentología. Commuicaciones 149–152

Radwanski A (1959) Littoral structures (cliff, clastic dikes and veins, and borings of Potamilla) in the high-tatric Lias. Acta Geol Pol 9:270–280

Permian

Herrmann A (1956) Der Zechstein am sudwestlichen Harzrand. Geologisches Jahrbuch 72:1–72

Mohr K (1973) Hartz, westlische teil. In: Gwinner MP (ed) Samlung geologischer fürer. Gebrüder Bomtraeger, Berlin, Stuttgart, pp 1–200

Runnegar B (1979) Ecology of Eurydesma and the Eurydesma fauna, Permian of eastern Australia. Alcheringa 3:261–285

Carboniferous

Loope DB (1994) Borings in an oomoldic rockground, Pennsylvanian of southeast Utah. Palaios 9:299–306

Webb GE (1994) Paleokarst, paleosol, and rocky-shore deposits at the Mississippian-Pennsylvanian unconformity, northwestern Arkansas. Geol Soc Am Bull 106:634–648

Wilson MA, Palmer TJ (1998) The earliest Gastrochaenolites (Early Pennsylvanian, Arkansas, USA): an Upper Paleozoic bivalve boring? J Paleontol 72:769–772

Devonian

Cook AG (1995) Sedimentology and depositional environments of the Middle Devonian lower Fanning River Group (Big Bend Arkose and Burdekin Formation), Burdekin Subprovince, North Queensland, Australia. Memoirs Queensland Museum 38:53–91

Fannin NGT (1969) Stromatolites from the middle Old Red Sandstone of western Orkney. Geol Mag 106:77–88

Silurian

Cherns L (1999) Silurian Chitons as indicators of Rocky Shores and lowstand on Gotland, Sweden. Palaios 14:172–179

Johnson ME, Rong JY, Wang CY, Wang P (2001) Continental island from the Upper Silurian (Ludfordian Stage) of Inner Mongolia: implications for eustasy and paleogeography. Geology 29:955–958

Keeling M, Kershaw S (1994) Rocky shore environments in the Upper Silurian of Gotland, Sweden. Geologiska Föreningens i Stockholm Förhandlingar 116:69–74

Kobluk DR (1984) Coastal paleokarst near the Ordovician – Silurian boundary, Manitoulin Island, Ontario. Can Petrol Geol Bull 32:398–407

Pemberton SG, Kobluk DR, Yeo RK, Risk MJ (1980) The boring *Trypanites* at the Silurian Devonian disconformity in southern Ontario. J Paleontol 54:1258–1266

Rong J, Johnson ME (1996) A stepped karst unconformity as an Early Silurian rocky shoreline in Guizhou Province (South China). Palaeogeogr Palaeoclimatol Palaeoecol 121:115–129

Ordovician

Brookfield ME, Brett CE (1988) Paleoenvironments of the Mid-Ordovician (Upper Caradocian) Trenton limestones of southern Ontario, Canada: storm sedimentation on a shoal-basin shelf model. Sedimentary Geol 57:75–105

Desrochers A, James NP (1988) Early Paleozoic surface and subsurface paleokarst: middle Ordovician carbonates, Mingan Islands, Quebec. In: James NP, Choguette PW (eds) Paleokarst. Springer, New York, pp 183–210

Donovan RN, Ragland D, Rafalowski M, McConnell D, Beauchamp W, Marcini WR, Sanderson DJ (1988) Pennsylvanian deformation and Cambro-Ordovician sedimentation in the Blue Creek Canyon, Slick Hills, southwestern Oklahoma. In: Hayward OT (ed) Geological society of America centennial field guide-South-Central Section, vol 4. The Geological Society of America, Inc., Boulder, CO, pp 127–134

Elias RJ, Young GA (2000) Enigmatic fossil encrusting an upper Ordovician rocky shore on Hudson Bay, Canada, is a coral. J Paleontol 74:179–180

Harland TL, Pickerill RK (1984) Ordovician rocky shoreline deposits – the basal Trenton Group around Quebec city, Canada. Geol J 19:271–298

Johnson ME, Baarli BG (1987) Encrusting corals on a late Ordovician to earliest Silurian rocky shore, southwest Hudson Bay, Manitoba, Canada. Geology 15:15–17

Johnson ME, Mu X, Rong J (1998) Enigmatic fossil encrusting an Upper Ordovician rocky shore on Hudson Bay, Canada. J Paleontol 72:927–932

Mergl M (1983) Rocky-bottom fauna of Ordovician age in Bohemia (Arenigian; Prague Basin, Barrandian area). Vestnik Ustredniho ustavu geologickeho 58:333–339

Merlg M (1984) *Marcusodictyon*, an encrusting bryozoan from the lower Ordovician (Tremadocian) of Bohemia. Vestnik Ustresniho ustavu geologickeho 59:171–172

Read JF, Grover GA (1977) Scalloped and planar erosion surfaces, Middle Ordovician limestones, Virginia; Analogues of Holocene exposed karst or tidal rock platforms. J Sedimentary Petrol 47:956–972

Riva J, Pickerill RK (1987) The late mid-Ordovician transgressive sequence and the Montmorency Fault at the Montmorency Falls, Quebec. In: Roy DC (ed) Centennial field guide volume 5: Northeastern section of the geological society of America, vol 5. The Geological Society of America, Inc., Boulder, CO, pp 357–362

Cambrian

Anderson MM (1987) Stratigraphy of Cambrian rocks at Bacon Cove, Duffs, and Manuels River, Conception Bay, Avalon Peninsula, eastern Newfoundland. In: Roy DE (ed) Geological society of America centennial field guide, Northeastern section, vol 5. The Geological Society of America, Inc., Boulder, CO, pp 467–472

Dunbar CO, Rodgers J (1957) Principles of stratigraphy, Wiley, New York, pp 170–171

Edhorn A, Anderson MM (1977) Algal remains in the Lower Cambrian Bonavista Formation, Conception Bay, Southeastern Newfoundland. In: Flügel E (ed) Fossil Algae. Springer, New York, pp 113–123

Martinson A (1956) Kambriska sandstensgångar och Ordoviciska fossilförande block i Nylands skärgård. Nordenskiold-samfundets tidsskrift, pp 89–102

Stinchcomb BL (1975) Paleoecology of two new species of late Cambrian Hypseloconus (Monoplacophora) from Missouri. J Paleontol 49:416–421

Venninn E, Álvaro JJ, Moreno-Eiris E, Perjón A (2003) Early Cambrian coelobiontic communities in tectonically unstable crevices developed in Neoproterozoic andesites, Ossa-Morena, southern Spain. Lethaia 36:53–65

Yochelson EL, Webers GF (2006) A restudy of the Late Cambrian moluscan fauna of Berkey (1898) from Taylors Falls, Minnesota. Minnesota Geol Surv Rep Invest 64:66

Appendix 2: Bibliography on Sandy-Shore Biotas Through Phanerozoic Time

Neogene

Aswan ZY, Ozawa T (2006) Milankovitch 41 000-year cycles in lithofacies and molluscan content in the tropical Middle Miocene Nyalindung Formation, Java, Indonesia. Palaeogeogr Palaeoclimatol Palaeoecol 235:382–405

Belperio AP, Murray-Wallace CV, Cann JH (1995) The last in interglacial shoreline in southern Australia: morphostratigraphic variations in temperate carbonate setting. Quaternary Int 26:7–19

Benvenuti M, Bertini A, Conti C, Dominici S (2007) Integrated analyses of litho- and biofacies in a Pliocene cyclothemic, alluvial to shallow marine succession (Tuscany, Italy). Geobios 40:143–151

Betzler C, Brachert TC, Braga J-C, Marti JM (1997) Nearshore, temperate, carbonate depositional systems (lower Tortonian, Agua Amarga Basin, southern Spain): implications for carbonate sequence stratigraphy. Sedimentary Geol 113:27–53

D'Allesandro A, Massari F, Davaud E, Ghibaudo G (2004) Pliocene – Pleistocene sequences bounded by subaerial unconformities within foramol ramp calcarenites and mixed deposits (Salento, SE Italy). Sedimentary Geol 166:89–144

Del Río CJ, Martínez SA, Scasso RA (2001) Nature and origin of spectacular marine Miocene shell beds of northeastern Patagonia (Argentina): paleoecological and bathymetric significance. Palaios 16:3–25

Dixon RG (2005) On the occurrence of a mollusc fauna in the Red Crag pebble bed from Mason's pit, Great Blakenham, Suffolk. Proc Geol Assoc 116:139–142

Fucks E, Aguirre M, Deschamps CM (2005) Late Quaternary continental and marine sediments of northeastern Buenos Aires province (Argentina): Fossil content and paleoenvironmental interpretation. J South Am Earth Sci 20:45–56

Gammon P (1995) Hautotara Formation, Mangaopari Basin, New Zealand: record of a cyclothemic Pliocene-Pleistocene marine to nonmarine transition. New Zealand J Geol Geophys 38:471–481

James NP, Bone Y (2007) A late Pliocene-early Pleistocene, inner-shelf, subtropical, seagrass-dominated carbonate: Roe Calcarenite, Great Australian Bight, Western Australia. Palaios 22:343–359

Kier PM (1972) Upper Miocene Echinoids from the Yorktown Formation of Virginia and their ecological significance. Smithsonian Contributions Paleobiol 13:1–43

Lee DE, Campbell JD (1987) Ceozoic Records of the genus *Lingula* (Brachiopoda: Inarticuata) in New Zealand. J R Soc N Z 17:17–30

Mandic O, Harzhauser M, Rietzel R, Tibuleac P (2008) Benthic mass-mortality events on a Middle Miocene incised-valley. Facies 54:343–359

Muizon C, DeVries TJ (1985) Geology and paleontology of late Cenozoic marine deposits in the Sacaco area (Peru). Geologische Rundschau 74:547–563

Néraudeau D, Goubert E, Lacoura D, Rouchy JM (2001) Changing biodiversity of Mediterranean irregular echinoids from the Messinian to the present-day. Palaeogeogr Palaeoclimatol Palaeoecol 175:43–60

Norris RD (1986) Taphonomic gradients in shelf fossil assemblages: Pliocene Purisma Formation, California. Palaios 1:256–270

Norton PEP (1975) Paleoecology of the Mollusca of the Tjoernes sequence, Iceland. Boreas 4:97–110

Pendón JG, Ruiz F, Abad M, González-Regalado ML, Baceta JJ, Tosquella J (2003) The transgressive sequences on foreland margins: a case study pf the Neogene central Guadalqivir basin, southern Spain. Riv Ital Paleontologia Stratigrafia 110:503–515

Pufahl P, Noel PJ, Bone Y, Lukasik JJ (2004) Pliocene sedimentation in a shallow, cool-water, estuarine gulf. Sedimentology 51:997–1027

Ramsey KW (1998) Depositional environments and stratigraphy of the Pollack farm site, Delaware In: Benson RN (ed) Geology and paleontology of the lower Miocene Pollack Farm Fossil Site, Delaware. Delaware Geological Survey Special Publication, pp 27–40

Richards GW (1985) Fossil Mediterranean molluscs as sea-level indicators. Geol Mag 122:373–381

Sakinc M, Yaltirac C (2005) Messinian crisis: what happened around the Northeastern Aegean? Mar Geol 221:423–436

Valentine JW (1959) Pleistocene Molluscan Notes. I. The Bay Point Formation at its type locality. J Paleontol 33:685–688

Paleogene

Allan RS (1937) New record of the genus *Lingula* (Brachiopoda) from Tertiary strata in New Zealand. Trans Proc R Soc N Z 65:383–385

Deville J (1996) Palaeontological study of the Neerrepen sands in the Francart Quarry (Tongeren, Belgium). Natuurwetenschappelijk Tijdschrift 75:75–97

Dominici S, Kowalke T (2007) Depositional dynamics and the record of ecosystem stability: early Eocene faunal gradients in the Pyrenean foreland, Spain, Palaios 22:268–284

Elder SR, Hansen TA (1981) Macrofossil Assemblages of the Moodys Branch Formation (Upper Eocene), Louisiana and Mississippi. Gulf Coast Assoc Geol Soc Trans 31:283–287

Oboh-Ikuenobe FE, Obi CG, Jaramilloc CA (2005) Lithofacies, palynofacies, and sequence stratigraphy of Palaeogene strata in Southeastern Nigeria. J Afr Earth Sci 41:79–101

Pufahl P, Noel PJ, Bone Y, Lukasik JJ (2004) Pliocene sedimentation in a shallow, cool-water, estuarine gulf. Sedimentology 51:997–1027

Schäfer A, Hilger D, Gross G, Hocht FVD (1996) Cyclic sedimentation in Tertiary Lower-Rhine Basin (Germany)—the 'Liegendrücken' of the brown-coal open-cast Fortuna mine. Sedimentary Geol 103:229–247

Stilwell JD, Zinsmeister WC (1992) Molluscan systematics and biostratigraphy, Lower Tertiary La Meseta Formation, Seymour Island, Antarctic Peninsula. Antarctic Res Ser 55: 1–192

Sudan CS (2002) Ichnofacies of the Murree Group in Jammu area and their ecological implications during Late Paleogene in the NW Himalaya. J Geol Soc India 60:547–557

Zuschin M, Harzhause M, Mandic O (2004) Spatial variability within a single parautochthonous Paratethyan tidalflat deposit (Karpatian, Lower Miocene – Kleinebersdorf, Lower Austria). Courier Forschungsinstitut Senckenberg 246: 153–168

Cretaceous

Carannante G (2000) Upper Cretaceous ramp limestones from the Sorrento Peninsula (southern Apennines, Italy); micro- and macrofossil associations and their significance in the depositional sequences. Sedimentary Geol 132:89–123

Hernández-Romano U, Aguilera-Franco N, Martínez-Medrano M, Barceló-Duarte J (1997) Guerrero-Morelos Platform drowning at the Cenomanian – Turonian boundary Huitziltepec area, Guerrero State, southern Mexico. Cretaceous Res 18:661–686

Jain SL, Sahni A (1985) Dinosaurian egg-shell fragments from the Lameta Formation at Pisdura, Chandrapur district, Maharashtra. Geosci J 6:211–220

Kamola DL (1984) Trace fossils from marginal-marine facies of the Spring Canyon Member, Blackhawk Formation (Upper Cretaceous), east-central Utah. J Paleontol 58: 529–541

Kaufman EG (1982) Paleobiological facies and coal forming environments. In: Gurgle K (ed) Proceedings, Fifth symposium on the geology of Rocky Mountain coal. Utah Geological and Mineral Survey Bulletin, vol 118, pp 30–31

Komatsu T (1999) Sedimentology and sequence stratigraphy of a tide- and wave-dominated coastal succession: the Cretaceous Goshoura Group, Kyushu, southwest Japan. Cretaceous Res 20:327–342

Rhoads DC, Speden IG, Waage KM (1972) Trophic group analysis of Upper Cretaceous (Maastrichtian) bivalve assemblage from South Dakota. Am Assoc Petrol Geol Bull 56: 1100–1113

Rosetti DF, Netto RG (2005) First evidence of marine influence in the Cretaceous of the Amazonas Basin, Brazil. Cretaceous Res 27:513–528

Schwartz RK, Vuke SM (2007) Tide-dominated facies complex at southern terminus of Sunburst Sea, Cretaceous Kootenai Formation, Great Falls. Montana. Search and Discovery Article #50045:1–3

Jurassic

Fürsich FT, Wilmsen M, Seyed-Emami K (2006) 2006 – Ichnology of Lower Jurassic beach deposits in the Shemshak Formation at Shahmirzad, southeastern Alborz Mountains, Iran. Facies 52:500–610

Schlirf M (2003) Palaeoecologic significance of Late Jurassic trace fossils from the Boulonnais, N France. Acta Geol Pol 53:123–142

Tang CM, Bottjer DJ, Simms MJ (2000) Stalked crinoids from a Jurassic tidal deposit in western North America. Lethaia 33:46–54

Tchoumatchenko P (1972) Thanatocoenoses and biotopes of Lower Jurassic brachiopods in central and western Bulgaria. Palaeogeogr Palaeoclimatol Palaeoecol 12: 27–242

Wignall PB, Sutcliffe OE, Clemsn J, Young E (1996) Unusual shoreface sedimentology in the Upper Jurassic of the Boulonnais, Northern France. J Sedimentary Res, Section A: Sedimentary Petrol Processes, pp 577–586

Wright RP (1974) 1974 Jurassic bivalves from Wyoming and South Dakota: a study of feeding relationships. J Paleontol 48:425–433

Triassic

Fürsich FT, Wilmsen M, Seted-Emami K, Cecca F, Majidifard MR (2005) The upper Shemshak Formation (Toarcian–Aalenian) of the EasternAlborz (Iran): biota and palaeoenvironments duringa transgressive–regressive cycle. Facies 51:365–384

Kelling G, Moshri MA (1977) The orientation of fossil bivalves in a pene-littoral sequence (the Rhaetiian of South Wales). J Sedimentary Petrol 47:1342–1346

Zonneweld J-P, Gingras MK, Pemberton SG (2001) Trace fossil assemblages in a Middle Triassic mixed siliciclastic-carbonate marginal marine depositional system, British Columbia. Palaeogeogr Palaeoclimatol Palaeoecol 166: 249–276

Zonneweld J-P, Beatty TW, Pemberton SG (2007) Lingulide brachiopods and the trace fossil Lingulichnus from the Triassic of Western Canada; implications for faunal recovery after the end-Permian mass extinction. Palaios 22: 74–97

Permian

McCarthy B (1979) Trace fossils from a Permian shoreface-foreshore environment, Eastern Australia. J Paleontol 53:345–356

Minter N, Buatois LA, Lucas SG, Braddy SJ, Smith JA (2006) Spiral-shaped graphoglyptids from an Early Permian intertidal flat. Geology 34:1057–1060

Carboniferous

Mangano MG, Buatois LA, West RR, Maples CG (1998) Contrasting behavioral and feeding strategies recorded by tidal-flat bivalve trace fossils from the Upper Carboniferous of eastern Kansas. Palaios 13:335–351

Devonian

Bjerstedt TW (1987) Latest Devonian earliest Mississippian nearshore Trace-fossil assemblages fro West Virginia, Pennsylvania, and Maryland. J Paleontol 61:865–889

Bradshaw MA, McCartan L (1991) Palaeoecology and systematics of early Devonian bivalves frpm the Horlick formation, Ohio range. Antarctica. Alcheringa 15:1–42

Cook AG (1993) Fletcherviewia septata; a new high-spired, septate gastropod from the Devonian of North Queensland. J Paleontol 67:816–821

Draganits E, Braddy SJ, Briggs DG (2001) A Gondwanan coastal arthropod ichnofauna from the Muth Formation (Lower Devonian, northern India); paleoenvironment and tracemaker behavior. Palaios 16:126–147

Gaillard C, Racheboeuf PR (2006) Trace fossils from nearshore to offshore environments: lower Devonian of Bolivia. J Paleontol 80:1205–1226

Thayer CW (1974) Marine paleoecology in the Upper Devonian of New York. Lethaia 7:121–155, 111–155

Silurian

Baarli BG, Keilen HB, Johnson ME (1999) Silurian communities of the Oslo Region. In: Boucot AJ, Lawson JD (eds) Paleocommunities – a case study from the Silurian and Lower Devonian. World and Regional Geology Series 11. Cambridge University Press, Cambridge, UK, pp 327–349

Boucot AJ (1999) Some Wenlockian-Gedinnian, chiefly brachiopod dominated, communities of North America. In: Boucot AJ, Lawson JD (eds) Paleocommunities – a case study from the Silurian and Lower Devonian. World and Regional Geology Series 11. Cambridge University Press, Cambridge, UK, pp 549–591

Bradsfield KES (1999) Fossil commuities of the late Ludlow and early Prídolí in Wales and the Welsh Borderland. In: Boucot AJ, Lawson JD (eds) Paleocommunities – a case study from the Silurian and Lower Devonian. World Regional Geology Series 11. Cambridge University Press, Cambridge, UK, pp 370–372

Brett CE (1999) Wenlockian Fossil communities in New York State and adjacent areas paleontology and paleoecology. In: Boucot AJ, Lawson JD (eds) Paleocommunities – a case study from the Silurian and Lower Devonian. World Regional Geology Series 11. Cambridge University Press, Cambridge, UK, pp 592–637

Davies NS, Sansom IJ, Turner P (2006) Trace fossils and paleoenvironments of a Late Silurian marginal-marine/alluvial system; the Ringerike Group (lower Old Red Sandstone), Oslo region, Norway. Palaios 21:46–62

Mikulic DG (1999) Silurian trilobite associations in North America. In: Boucot AJ, Lawson JD (eds) Paleocommunities – a case study from the Silurian and Lower Devonian. World Regional Geology Series11. Cambridge University Press, Cambridge, UK, pp 793–798

Racheboeuf PR (1999) Lower Silurian to Lower Devonian communities from the Armorican Massif (western France) and Artois (northern France). In: Boucot AJ, Lawson JD (eds) Paleocommunities – a case study from the Silurian and Lower Devonian. WorldRegional Geology Series 11. Cambridge University Press, Cambridge, UK, pp 260–270

Thomas AT, Lane PD (1999) Trilobite assemblage of the North Atlantic Region. In: Boucot AJ, Lawson JD (eds) Paleocommunities – a case study from the Silurian and Lower Devonian. World Regional Geology Series 11. Cambridge University Press, Cambridge, UK, pp 444–457

Vladimirskaya EV, Kulkov NP (1999) Silurian brachiopod communities of Tuva. In: Boucot AJ, Lawson JD (eds) Paleocommunities – a case study from the Silurian and Lower Devonian. World Regional Geology Series 11. Cambridge University Press, Cambridge, UK, pp 545–548

Ordovician

Aceñolaza GF, Emig EE, Gutiérrez–Marco C (2003) Lingulid shell beds from the Ordovician of Argentina, with notes on other peri–Gondwanan occurrences. Serie de Correlación Geológica 17:237–244

Bretsky P (1969) Central Appalachian Late Ordovician communities. Geol Soc Am Bull 80:193–212

Jaanusson V (1960) The Viruan (Middle Ordovician) of Öland. Bull Geol Instit Univ Uppsala 38:207–287

Kreisa RD (1981) Storm-generated sedimentary structures in subtidal marine facies with examples from the Middle and Upper Ordovician of Southwestern Virginia. J Sedimentary Petrol 51:823–848

Poiré DG, Spalletti LA, Valle AD (2003) The Cambrian-Ordovician siliciclastic platform of the Balcarce Formation (Tandilla System, Aregnetina): facies, trace fossils, palaeoenvironments and sequence stratigraphy. Geol Acta 1:41–60

Popov LE, Ebbestad JOR, Mambetov A, Apayarov FK (2007) A low diversity shallow water lingulid brachiopod/gastropod association from the Upper Ordovician of Kyrgyz Range. Acta Palaeontologica Polonica 52:27–40

Popov LE, Cocks LRM, Nikitin IF (2002) Upper Ordovician brachiopods from the Anderken Formation, Kazakhstan: their ecology and systematics. Bull Br Museum (Nat Hist) Geol Ser 58:13–79

Ramos E, Marzo M, Gibet JM, Khaeri ST, Khoja AA, Bolatti ND (2006) Stratigraphy and sedimentology of the Middle Ordovician Hawaz Formation (Murzuq Basin, Libya). Am Assoc Petrol Geol Bull 90:1309–1336

Cambrian

Driese SG, Byers CW, Dott RH (1981) Tidal deposition in the basal Upper Cambrian Mt. Simon Formation in Wisconsin. J Sedimentary Petrol 51:367–381

Haddox CA, Dott RH (1990) Cambrian shoreline deposits in northern Michigan. J Sedimentary Petrol 60:697–716

Hagadorn JW, Belt TE (2008) Stranded in upstate New York: Cambrian Scyphomedusae from the Ptsdam Sandstone. Palaios 23:424–441

Mangano MG, Buatois LA (2004) Reconstructing Early Phanerozoic intertidal ecosystems: ichnology of the Cambrian Campanario Formation in northwest Argentina. Fossils Strata 51:17–38

Ribecai C, Bagnoli G, Mazzarini F, Musumeci G (2005) Paleontological evidence for Late Cambrian in the Arburese area, SW Sardinia. In: Steemans P, Javaux E (eds) Pre-Cambrian to paleozoic palaeopalynology and palaeobotany. Carnets de Géologie/Notebooks on Geology, Memoir 2005, Brest, pp 45–50

Appendix 3: Bibliography on Muddy-Shore Biotas Through Phanerozoic Time

Neogene

Aswan ZY, Ozawa T (2006) Milankovitch 41 000-year cycles in lithofacies and molluscan content in the tropical Middle Miocene Nyalindung Formation, Java, Indonesia. Palaeogeogr Palaeoclimatol Palaeoecol 235: 382–405

Hetherington R, Reid RG (2003) Malacological insights into the marine ecology and changing climate of the late Pleistocene – early Holocene Queen Charlotte Islands archipelago, western Canada, and implications for early peoples. Can J Zool 81: 626–661

James NP, Bone Y (2007) A late Pliocene-early Pleistocene, inner-shelf, subtropical, seagrass-dominated carbonate: Roe Calcarenite, Great Australian Bight, Western Australia. Palaios 22:343–359

Latal C, Piller WE, Harzhauser M (2004) palaeoenvironmental reconstructions by stable isotopes of Middle Miocene gastropods of the Central Paratethys. Palaeogeogr Palaeoclimatol Palaeoecol 28:371–382

Mandic O, Harzhauser M, Rietzel R, Tibuleac P (2008) Benthic mass-mortality events on a Middle Miocene incised-valley. Facies 54:343–359

Megerisi MF, Mamgain VD (1980) The Upper Cretaceous-Tertiary Formations of northern Libya: a synthesis. Department Geol Res Mining Bull 12:1–85

Nakao K (2006) Fossil molluscan fauna in the Lower Pleistocene Kazusa Formation, Shimabara Peninsula, northwestern Kyushu. Quaternary Res 45:113–121

Nara M, Akiyama H, Itani G (2008) Macrosymbiotic association of the myid bivalve Cryptomya with thalassinidean shrimps: examples from modern and Pleistocene tidal flats of Japan. Palaeogeogr Palaeoclimatol Palaeoecol 261:100–104

Richards GW (1985) Fossil Mediterranean molluscs as sea-level indicators. Geol Mag 122:373–381

Smith JT (1999) Cenozoic marine mollusks and paleogeography of the Gulf of California. In: Dauphi P, Simoneit BRT (eds) The Gulf and Peninsular Province of the Californias. American Association of Petroleum Geologists, Memoir 46, pp 537–666

Ward LW (1998) Molluscs from the Lower Miocene Pollack Farm Site, Delaware. In: Benson RN (ed) Geology and paleontology of the Lower Miocene Pollack Farm fossil site, Delaware. Delaware Geological Survey Special Publication, vol 21, pp 59–133

Yeo RK, Risk MJ (1981) The sedimentology, stratigraphy, and preservation of intertidal deposits in the Minas Basin System, Bay of Fundy. J Sedimentary Petrol 51:252–260

Paleogene

Dominici S, Kowalke T (2007) Depositional dynamics and the record of ecosystem stability: early Eocene faunal gradients in the Pyrenean foreland, Spain. Palaios 22:268–284

Lan X (1997) Paleogene bivalve communities in the western Tarim Basin and their paleoenvironmental implications. Palaeoworld 7:137–157

Lee DE, Campbell JD (1987) Cenozoic records of the genus *Lingula* (Brachiopoda: Inarticuata) in New Zealand. J R Soc N Z 17:17–30

Ricketts BD (1994) Mud-flat cycles, incised channels, and relative sea-level changes on a paleocene Mud-Dominated Coast, Ellesmere Island, Arctic Canada. J Sedimentary Res, Section B: Stratigr Global Stud 64B:211–218

Strougo A, Azab MM (1982) Middle Eocene Mollusca from the basal beds of Gebel Qarara (Upper Egypt), with remarks on the depositional environment of these beds. Neues Jahrbuch fur Geologie und Palaontologie-Monatshefte 1982:667–678

Sudan CS (2002) Ichnofacies of the Murree Group in Jammu area and their ecological implications during Late Palaeogene in the NW Himalaya. J Geol Soc India 60:547–557

Tewaria VC, Stennib B, Puglieseb N, Drobnec K, Riccambonib R, Dolenecd T (2007) Eritidal sedimentary depositional facies and carbon isotope variation across K/T boundary carbonates from NW Adriatic platform. Palaeogeogr Palaeoclimatol Palaeoecol 255:77–86

Vega FJ, Carmen-Perrilliat MD (1995) On some Paleocene invertebrates from the Potrerillos formation (Difunta Group), northeastern Mexico. J Paleontol 69:862–869

Cretaceous

Aguillera-Franco N, Romano UH (2004) Cenomanian–Turonian facies succession in the Guerrero–Morelos Basin, Southern Mexico. Sedimentary Geol 170:135–162

Allmon WD, Coehn PA (2008) Palaeoecological significance of turritelline gastropod-dominated assemblages from the mid-Cretaceous (Albian-Cenomanian) of Texas and Oklahoma, USA. Cretaceous Res 29:65–77

Boscarolli D, Vecchia FMD (1999) The Upper Hauterivian-Lower Barremian dinosaur site of Bale/Valle (SW Istria, Croatia). Natura Nascosta 18:1–5

Callapez PM (2008) Palaeogeographic evolution and marine faunas of the mid-Cretaceous western Portuguese carbonate platform. Thalassas 24:29–52

Dini M, Tunis G, Venturini S (1998) Continental, brackish and marine carbonates from the Lower Cretaceous of Kolone–Barbariga (Istria, Croatia): stratigraphy, sedimentology and geochemistry. Palaeogeogr Palaeoclimatol Palaeoecol 140:245–269

Megerisi MF, Mamgain VD (1980) The Upper Cretaceous-Tertiary formations of northern Libya: a synthesis. Geol Res Mining Bull 12:1–85

Mezga A, Meyer CA, Tesovic BC, Bajraktarevic Z, Gusic I (2006) The first record of dinosaurs in the Dalmatian part (Croatia) of the Adriatic-Dinaric carbonate platform (ADCP). Cretaceous Res 27:735–742

Mitchell-Tapping HJ (1987) Application of the tidal mudflat model to the Sunniland Formation of South Florida. Gulf Coast Assoc Geol Soc Trans 37:415–426

Moro A, Jelaska V (1994) Upper Cretaceous Peritidal deposits of Olib and Ist Islands. Geologia Croatia 47:53–65

Pittman JG (1989) Stratigraphy, lithology, depositional environment, and track type of dinosaur track-bearing beds of the Gulf Coastal Plain. In: Gillette DD, Lockley MG (eds) Dinosaur tracks and traces. Cambridge University Press, Cambridge, pp 135–153

Radley JD, Barker MJ (2000) Lower Cretaceous coastal lagoonal succession (Vectis Formation, Isle of Wight, southern England). Geol Mag 137:193–205

Ruberti D (1997) Facies analysis of an Upper Cretaceous high-energy rudist-dominated carbonate ramp (Matese Mountains, central-southern Italy): subtidal and peritidal cycles. Sedimentary Geol 113:81–110

Scott RW (1979) Depositional model of early Cretaceous Cora-Algal-Rudist reefs. Am Assoc Petrol Geol 63:1108–1127

Shone RWA (1986) New Ophiuroid from the Sundays River Formation (Lower Cretaceous), South Africa. J Paleontol 60:904–910

Jurassic

Azeredo AC, Wright VP, Ramalho MM (2002) The Middle–Late Jurassic forced regression and disconformity in central Portugal: eustatic, tectonic and climatic effects on a carbonate ramp system. Sedimentology 49:1339–1370

Boutakiout E, Hadri M, Milhi A, Nouri J, Díaz-Martínez I, Pérez-Lorente F (2008) Icnitas terópodas y saurópodas del Bajociense en Ifgh (norte de Msemrir, Alto Atlas Central, Marruecos) [Theropod and sauropod footprints from the Bajocian of Ifgh (northern Msemrir, central High Atlas, Morroco)] Geo-Temas 10:1249–1252

Burgess CJ, Lee CW (1978) The development of a Lower Jurassic carbonate tidal flat, Central High Atlas, Morocco. 1: sedimentary history. J Sedimentary Petrol 48:777–793

Fraser N, Bottjer DJ, Fischer AG (2004) Dissecting "Lithiotis" Bivalves: implications for the Early Jurassic reef eclipse. Palaios 10:51–67

Pienkowski G (1991) Liassic sedimentation in Scania, southern Sweden: Hettangian-Sinemurian of the Helsingborg area. Facies 24:39–86

Sugawara K (2004) Brackish and shallow-marine benthic associations of the Lower Jurassic Niranohama Formation in the Shizugawa area, South Kitakami Belt, Northeast Japan. Research Report of the Kochi University. Nat Sci 53:21–40

Vecchia FMD (1995) Jurassic and Cretaceous sauropod evidence in the Mesozoic carbonate platforms of the Southern Alps and Dinarids. GAIA 10:65–73

Triassic

Fürsich FT, Wilmsen M, Seted-Emami K, Cecca F, Majidifard MR (2005) The upper Shemshak Formation (Toarcian–Aalenian) of the Eastern Alborz (Iran): biota and palaeoenvironments duringa transgressive–regressive cycle. Facies 51:365–384

Makhluf IM (2000) Early Triassic intertidal/subtidal patterns of sedimentation along the southern margins of the Tethyan seaway, Jordan. J Asian Earth Sci 18:513–518

Mander L, Twitchett RJ, Benton MJ (2008) Palaeoecology of the Late Triassic extinction event in the SW UK. J Geol Soc Lond 165:319–332

Pérez-Lópex A, López-Garrido AC, Mfirquez-Aliaga A (2003) Ladinian carbonates of the Cabo Cope unit (Betic Cordillera, SE Spain): a tethys-maláguide palaeogeographic gateway. Facies 48:1–8

Preto N, Hinnov LA (2003) Unraveling the origin of carbonate platform cyclothems in the Upper Triassic Dürrenstein Formation (Dolomites, Italy). J Sedimenary Res 73: 774–789

Rodtoguez-Tovar FJ, Péres-Valera F, Pérez-Lopez A (2007) Ichnological analysis in high-resolution sequence stratigraphy: the Glossifungites ichnofacies in Triassic successions from the Betic Cordillera (southern Spain). Sedimentary Geol 198:293–307

Sokac B, Grgasovic T (1995) *Dissocladella bystrickyi* n. sp., a New Calcareous algae (Dasycladacea) from Upper Triassic Doloites of Mt. Medvednica (northern Croatia). Geol Croatia 48:1–17

Zonneveld J-P, Carrelli GG, Riediger C (2004) Sedimentology of the UpperTriassic Charlie Lake, Baldonnel and Pardonet formations from outcrop exposures in the southern Trutch region, northeastern British Columbia. Bull Can Petrol Geol 52:343–375

Permian

Braddy SJ (1995) A new arthrpod trackway and associated invertebrate ichnofauna from the lower Permian Hueco formation of the Robledo mountains, southern New Mexico. In: Lucas SG, Heckert AB (eds) Early Permian footprints and facies. New Mexico Museum of Natural History and Science Bulletin, vol 6. The New Mexico Museum of Natural History and Science, Albuquerque, NM, pp 101–105

Clapham ME, Bottjer DJ (2007) Permian marine paleoecology and its implications for large-scale decoupling of brachiopod and bivalve abundance and diversity during the Lopingian (Late Permian). Palaeogeogr Palaeoclimatol Palaeoecol 49:283–301

Hips K, Pelikan P (2002) Lower Triassic shallow marine succession in the Bükk Mountains, NE Hungary. Geol Carpathica 53:351–367

Maeda H, Mapes RH, Mapes G (2003) Taphonomic features of a lower Permian beached cephalopod assemblage from central Texas. Palaios 18:421–434

Terrell FM (1972) Lateral facies and paleoecology of Permian Elephant Canyon Formation, Grand County, Utah. Brigham Young Univ Geol Stud 19:3–44

Carboniferous

Abegg FE (1988) Carbonate tidal flat in mixed carbonate-siliciclastic Clore Formation (Upper Chesterian in Southern Illinois. Annu Meet Am Assoc Petrol Geol 72:153

Burchette TP, Riding R (1977) Attached vermiform gastropods in Carboniferous marginal marine stromatolites and biostromes. Lethaia 10:17–28

Ferguson L (1962) The paleoecology of a Lower Carboniferous marine transgression. J Paleontol 36:1090–1107
Kues B (1981) A large assemblage of a new eurypterid from the Red Tanks Member, Madera Formation (Late Pennsylvanian-Early Permian) of New Mexico. J Paleontol 55:729
Miller DJ, Eriksson KA (1997) Late Mississippian prodeltaic rhythmites in the Appalachian Basin: a hierarchical record of tidal and climatic periodicities. J Sedimentary Res, Section B: Stratigr Global Stud 67:653–660
Olszewski TD, Patzkowsky ME (2001) Measuring recurrence of marine biotic gradients: a case study from the Pennsylvanian-Permian midcontinent. Palaios 16:444–460
Riding R, Wright VP (1981) Paleosols and tidal-flat/lagoon sequences on a Carboniferous carbonate shelf: sedimentary associations of triple disconformities. J Sedimentary Petrol 51:1323–1339
Sepkoski JJ (1998) Rates of speciation in the fossil record. Philos Trans Roy Soc B: Biol Sci 353:315–326
Somerville D, Strank ARE (1984) Discovery of Arundian and Holkerian faunas from a Dinantian platform succession in north Wales. Geol J 19:85–104
Wescott WA (1982) Depositional setting and history of the Tar Springs Sandstone (Upper Mississippian), southern Illinois. J Sedimentary Petrol 52:353–366

Devonian

Bowen ZP, Rhoads DC, McAlester AL (1974) Marine benthic communities in the Upper Devonian of New York. Lethaia 7:93–120
Cherkesova SV (1999) Lower early Devonian communities from Taimyr. In: Boucot AJ, Lawson JD (eds) Paleocommunities – a case study from the Silurian and Lower Devonian. World Regional Geology Series 11. Cambridge University Press, Cambridge, UK, pp 458–461
Gritsenko VP, Istchenko AA, Konstantinenko LI, Tsegelnjuk PD (1999) Animal and plant communities of Podolia. In: Boucot AJ, Lawson JD (eds) Paleocommunities – a case study from the Silurian and Lower Devonian. Regional Geology Series, vol 11. Cambridge University Press, Cambridge, pp 462–487
Sapelnikov VP, Bogoyavlenskaya OV, Mizens LI, Shuysky VP (1999) Silurian and Early Devonian benthic communities of the Ural-Tien Shan region. In: Boucot AJ, Lawson DJ (eds) Paleocommunities – a case study from the Silurian and Lower Devonian. World Regional Geology Series, vol 11. Cambridge University Press, Cambridge, UK, pp 510–544
Selover RW, Gastaldo RA, Nelson RE (2005) An estuarine assemblage from the Middle Devonian Trout Valley Formation of northern Maine. Palaios 20:192–197
Sepkoski JJ (1998) Rates of speciation in the fossil record. Philos Trans R Soc B: Biol Sci 353:315–326
Thomas AT, Lane PD (1999) Trilobite assemblage of the North Atlantic Region. In: Boucot AJ, Lawson JD (eds) Paleocommunities – a case study from the Silurian and Lower Devonian. World Regional Geology Series, vol 11. Cambridge University Press, Cambridge, UK, pp 444–457
Vai GB (1999) Wenlockian to Emsian communities of the Carnic Alps (Austria and Italy). In: Boucot AJ, Lawson JD (eds) Paleocommunities – a case study from the Silurian and Lower Devonian. World Regional Geology Series, vol 11. Cambridge University Press, Cambridge, UK, pp 282–304
Walker KR, LaPorte LF (1970) Congruent fossil communities from Ordovician and Devonian carbonates of New York. J Paleontol 44:928–944
Wehrmann A, Hertweck G, Brocke R, Jansen U, Konigshof P, Plodowski G, Schindler E, Wilde V, Blieck A, Schultka S (2005) Paleoenvironment of an early Devonian land-sea transition: a case study from the southern margin of the old Red continent (Mosel Valley, Germany). Palaios 20:101–120
Yang W, Harmsen F, Kominz MA (1995) Quantitative analysis of a cyclic peritidal carbonate sequence, the middle and upper Devonian Lost Burro Formation, Death Valley, California – a possible record of Milankovitch climatic cycles. J Sedimentary Res, Section B: Stratigr Global Stud 65B:306–322

Silurian

Baarli BG, Keilen HB, Johnson ME (1999) Silurian communities of the Oslo region. In: Boucot AJ, Lawson JD (eds) Paleocommunities – a case study from the Silurian and Lower Devonian. World and Regional Geology Series, vol 11. Cambridge University Press, Cambridge, UK, pp 327–349
Boucot AJ (1999) Some Wenlockian-Gedinnian, chiefly brachiopod dominated, communities of North America. In: Boucot AJ, Lawson JD (eds) Paleocommunities – a case study from the Silurian and Lower Devonian. World and Regional Geology Series, vol 11. Cambridge University Press, Cambridge, UK, pp 549–591
Brett CE (1999) Wenlockian Fossil communities in New York state and adjacent areas paleontology and paleoecology. In: Boucot AJ, Lawson JD (eds) Paleocommunities – a case study from the Silurian and Lower Devonian. World Regional Geology Series, vol 11. Cambridge University Press, Cambridge, UK, pp 592–637
Calef CE, Hancck J (1974) Wenlock and Ludlow marine comunities in Wales and the Welsh Borderland. Paleontology 17:779–810
Davies NS, Sansom IJ, Turner P (2006) Trace fossils and paleoenvironments of a Late Silurian marginal-marine/alluvial system; the Ringerike Group (lower Old Red Sandstone), Oslo region, Norway. Palaios 21:46–62
Fisher DW (1957) Lithology, paleoecology and paleontology of the Vernon Shale (Late Silurian) in the type area. New York State Museum Bull 364:1–31
Gritsenko VP, Istchenko AA, Konstantinenko LI, Tsegelnjuk PD (1999) Animal and plant communities of Podolia. In: Boucot AJ, Lawson JD (eds) Paleocommunities – a case study from the Silurian and Lower Devonian. World Regional Geology Series, vol 11. Cambridge University Press, Cambridge, UK, pp 462–487
Johnson ME, Tesakov YI, Predtetchensky NN, Baarli BG (1997) Comparison of Lower Silurian shores and shelves in North America and Siberia. In: Klapper G, Murphy MA, Talent JA (eds) Paleozoic sequence stratigraphy, biostratigraphy, and biogeography: studies in honor of J. Granville ("Jess") Johnson. Geological Society of America Special Paper, vol 321, The Geological Society of America, Inc., Boulder, CO, pp 23–46

Liljedahl L (1995) Ecological aspects of a silicified bivalve fauna from the Silurian of Gotland. Lethaia 18:53–66

Mohamad AHB, Tucker EV (1999) Upper Bridgewood Beds: fossil communities. In: Boucot AJ, Lawson JD (eds) Paleocommunities – a case study from the Silurian and Lower Devonian. World Regional Geology Series, vol 11. Cambridge University Press, Cambridge, UK, pp 395–398

Musteikis P, Paskevicius JJ (1999) Brachiopod communities of the Lithuanian Silurian. In: Boucot AJ, Lawson JD (eds) Paleocommunities – a case study from the Silurian and Lower Devonian. World Regional Geology Series, vol 11. Cambridge University Press, Cambridge, UK, pp 305–326

Narbonne GM (1984) Trace fossils in Upper Silurian tidal flat to basin slope carbonates of Arctic Canada. J Paleontol 58: 398–415

Plotnick RE (1999) Habitat of Llandoverian-Lochkovian eurypterids. In: Boucot AJ, Lawson JD (eds) Paleocommunities – a case study from the Silurian and Lower Devonian. World Regional Geology Series, vol 11. Cambridge University Press, Cambridge, UK, pp 106–131

Smosna R, Warshauer SM (1981) Rank exposure index on a Silurian carbonate tidal flat. Sedimentology 28:723–731

Warshauer SM, Smosna R (1977) Paleoecologic controls of the ostracode communities in the Tonoloway Limestone (Silurian; Pridoli) of the Central Appalachians. In: Loffler H, Danielopol D (eds) Aspects of ecology and zoogeography of recent and fossil Ostracoda. W. Junk, The Hague, pp 475–485

Ordovician

Grover G, Read JF (1978) Fenestral and associated vadose diagenetic fabrics of tidal flat carbonates, Middle Ordovician New Market Limestone, southwestern Virginia. J Sedimentary Petrol 48:453–473

Novack-Gottshall PM, Miller AI (2003) Comparative taxonomic richness and abundance of Late Ordovician gastropods and bivalves in mollusc-rich strata of the Cincinnati Arch. Palaios 18:559–571

Patzkowsky ME (1999) Gradient analysis of middle Ordovician brachiopod biofacies: biostratigraphic, biogeographic, and macroevolutionary implications. Palaios 10:154–179

Pratt BR, Noel PJ (1986) The St George Group (Lower Ordovician) of western Newfoundland; tidal flat island model for carbonate sedimentation in shallow epeiric seas. Sedimentology 33:313–343

Walker KR, LaPorte LF (1970) Congruent fossil communities from Ordovician and Devonian carbonates of New York. J Paleontol 44:928–944

Webby BD (1992) Ordovician island biotas: New South Wales record and global implications. J Proc R Soc N S W 125: 51–77

Cambrian

Álvaro JJ, Vennin E, Moreni-Eiris E, Perejón A, Bechtstädt T (2000) Sedimentary patterns across the Lower–Middle Cambrian transition in the Esla nappe (Cantabrian Mountains, northern Spain). Sedimentary Geol 137:43–61

Mangano MG, Buatois LA (2004) Reconstructing Early Phanerozoic intertidal ecosystems: ichnology of the Cambrian Campanario Formation in northwest Argentina. Fossils Strata 51:753–766

Paterson JR, Jaqo JB, Brock GA, Gehling JG (2007) Taphonomy and palaeoecology of the emuellid trilobite Balcoracania dailyi (Early Cambrian, South Australia). Palaeogeogr Palaeoclimatol Palaeoecol 249:302–321

Sprechmann P, Gaucher C, Blanco G, Motana J (2004) Community of the Cerro Victoria formation, Arroyo del Soldado Group (Lowermost Cambrian, Uruguay). Gondwana Res 7:753–766

Vendrasco MJ, Runnegar B (2004) Late Cambrian and early Ordovician stem group chitons (Mollusca: Polyplacophora) from Utah and Missouri. J Paleontol 78:675–689

References

Abdel-Gawad GI (1986) Maastrichtian non-cephalopod mollusks (Scaphopoda, Gastropoda and Bivalvia) of the Middle Vistula Valley, Central Poland. Acta Geol Pol 36: 69–224

Alexander RR, Stanton RJ, Dodd JR (1993) Influence of sediment grain size on the burrowing of bivalves: correlation with distribution and stratigraphic persistence of selected Neogene clams. Palaios 8:289–303

Allen RS (1937) New record of the genus *Lingula* (Brachiopoda) from Tertiary strata in New Zealand. Trans Proc R Soc N Z 65:383–385

Anker A, Marina G-V, Lira C, Carrie J, Palmer AK, Jung M-S (2005) Macro fauna associated with echiuran burrows: a review with new observations of the innkeeper work, *Ochetostoma erythrogrammon* Leuchart and Rüppel, in Venezuela. Zool Stud 44:157–190

Aswan ZY, Ozawa T (2006) Milankovitch 41 000-year cycles in lithofacies and molluscan content in the tropical Middle Miocene Nyalindung Formation, Java, Indonesia. Palaeogeogr Palaeoclimatol Palaeoecol 235:382–405

Bambach RK (1993) Seafood through time: changes in biomass, energetics, and productivity in the marine ecosystem. Paleobiology 19:372–397

Bambach RK, Bennington JB (1996) Do communities evolve? A major question in evolutionary paleoecology. In: Jablonski DJ, Erwin, DH, Lipps JH (eds) Evolutionary paleobiology. University of Chicago Press, Chicago, IL, pp 123–160

Bird CF, Schwartz ML (1985) The world's coastline. Van Nostrand Reinhold, New York

Boucot AJ, Lawson JD (1999) Paleocommunities – a case study from the Silurian and Lower Devonian. World and Regional Geology 11, Cambridge University of Press, Cambridge, UK

Brown AC, McLachlan A (2002) Ecology of sandy shores. Elsevier, Amsterdam

Brusca RC (1980) Common intertidal invertebrates of the Gulf of California. University of Arizona Press, Tucson, AZ

Bryan JR (1992) Origin and paleoecology of Maastrichtian rockground and chalk facies in southcentral Alabama. Palaios 7:67–76

Buatois LA, Gingras MK, Macheachern J, Mángano MG, Zonneveld J-P, Pemberton SG, Netto RG, Martin A (2005)

Colonization of brackish-water systems through time: evidence from the trace-fossil record. Palaois 20:321–347
Carson R (1956) The edge of the sea. Houghton Mifflin Company, Boston
Crampton JS (1988) A late Cretaceous near-shore rocky substrate macrofauna from northern Hawkes Bay, New Zealand. N Z Geol Surv Rec 35:21–24
De Vries Klein G (1977) Clastic tidal facies. Continuing Education Publishing Co., Champaign, IL
DiMichele WA (1994) Ecological patterns in time and space. Paleobiology 20:89–92
Eagle MK, Hayward, BW, Grant-Mackie JA (1995) Early Miocene beach, rocky shore, and enclosed bay fossil communities, Waiheke Island, Auckland. Rec Auckand Instit Museum 32:17–44
Froidefond JM, Pujos M, Andre X (1988) Migration of mud banks and changing coastline in French Guiana. Marine Geol 84:1–21
Fürsich FT, Hurst JM (1974) Envrionmental factors determining the distribution of brachiopods. Palaeontology 17:879–100
Gingras MK, Pickerill R, Pemberton SG (2002) Resin cast of modern burrows provides analogs for composite trace fossils. Palalos 17:206–211
Glaeser JD (1978) Global distribution of barrier islands in terms of tectonic setting. J Geol 86:283–297
Johnson ME (1988a) Why are ancient rocky shores so uncommon? J Geol 96:469–480
Johnson ME (1988b) Hunting for ancient rocky shores. J Geol Educ 36:147–154
Johnson ME (2006) Uniformitarianism as a guide to rocky-shore ecosystems in the geological record. Can J Earth Sci 43:1119–1147
Johnson ME, Baarli BG (1987) Encrusting corals on a latest Ordovician to earliest Silurian rocky shore, southwest Hudson Bay, Manitoba, Canada. Geology 15:15–17
Johnson ME, Baarli BG (1999) Diversification of rocky-shore biotas through geologic time. Geobios 32:257–273
Johnson ME, Skinner DF, Macleod KG (1988) Ecological zonation during the carbonate transgression of a Late Ordovician rocky shore (northeastern Manitoba, Hudson Bay, Canada). Palaeogeogr Palaeoclimatol Palaeoceol 65:93–114
Johnson ME, Ledesma-Vázquez J, Montiel-Boehringer AY (2009) Growth of Pliocene-Pleistocene clam banks (Mollusca, Bivalvia) and related tectonic constraints in the Gulf of California. In: Johnson ME, Ledesma-Vázquez J (eds) Atlas of coastal ecosystems in the western Gulf of California: tracking limestone deposits on the margin of a young sea. University of Arizona Press, Tucson, AZ, pp 104–116
Johnson ME, Tesakov YI, Predtetchensky NN, Baarli BG (1997) Comparison of Lower Silurian shores and shelves in North America and Siberia. In: Klapper G, Murphy MA, Talent JA (eds) Paleozoic sequence stratigraphy, biostratigraphy, and biogeography: studies in honor of J. Granville ("Jess") Johnson. Geological Society of American Special Paper, vol 321, pp 23–46
Kaufman EG (1982) Paleobiological facies and coal forming environments. In: Gurgle K (ed) Proceedings, Fifth symposium on the geology of Rocky Mountain coal. Utah Geological and Mineral Survery Bulletin, vol 118, pp. 30–31
Kirby RR (2004) Genetic variation in a cline in a living intertidal snail arose in the Neogastropoda over 100 million ears ago. J Mol Evol 58:97–105
Kowalewski M, Flessa KW (1996) Improving with age: the fossil record of lingulide brachiopods and the nature of taphonomic megabiases. Geology 11:977–980
Libbey LK, Johnson ME (1997) Upper Pleistocene rocky shores and interidal biotas at Playa La Palmita (Baja California Sur, Mexico). J Coastal Res 13:216–225
McLachlan A, Brown, AC (2006) The ecology of sandy shores. Academic, Amsterdam
Megerisi MF, Mamgain VD (1980) The Upper Cretaceous-Tertiary formations of northern Libya: a synthesis. Geol Res Mining Bull 12:1–85
Miller R, Orr WN (1988) Mid-Tertiary transgressive rocky coast sedimentation: central western Cascade Range, Oregon. J Sediment Petrol 58:959–968
Nara M, Akiyama H, Itani G (2008) Macrosymbiotic association of the myid bivalve *Cryptomya* with thalassinidean shrimps: examples from modern and Pleistocene tidal flats of Japan. Palaeogeogr Palaeoclimatol Palaeoecol 261:100–104
Neto de Carvalho C, Viegas PA, Cachao M (2007) *Thalassinoides* and its producer: populations of *Mecochirus* buried within their burrow systems, Boca du Chapim Formation (Lower Cretaceous), Portugal. Palaios 22:104–109
Paine RT (1966) Food web complexity and species diversity. Am Nat 100:65–75
Paine RT (1970) The sediment occupied by Recent lingulid brachiopods and some paleoecological implications. Palaeogeogr Palaeoclimatol Palaeoecol 7:21–31
Paine RT (1984) Ecological determinism in the competition for space. Ecology 65:1339–1348
Palmer AR (1982) Predation and parallel evolution: recurrent parietal plate reduction in balanomorph barnacles. Paleobiology 8:31–44
Peters SE, Foote M (2001) Biodiversity in the Phanerozoic: a reinterpretation. Paleobiology 27:583–601
Pilkey OH (2003) A celebration of the world's barrier islands. Columbia University Press, New York
Plaziat, J-C, Cavagnetto C, Koeniguer J-C, Baltzer F (2001) History and biogeography of the mangrove ecosystem, based on a critical reassessment of the paleontological record. Wetlands Ecol Manage 9:161–179
Raffaelli D, Hawkins SJ (1999) Intertidal ecology. Springer, Berlin
Reineck, HE (1975) German North Sea tidal flats. In: Ginsburg RN (ed) Tidal deposits – a casebook of recent examples and fossil counterparts. Springer, Berlin, pp 5–12
Ricketts BD (1994) Mud-flat cycles, incised channels, and relative sea-level changes on a Paleocene mud-dominated coast, Ellesmere Island, Arctic Canada. J Sedimentary Resh Sec B 64B:211–218
Ricketts EF, Calvin J (1939) Between Pacific Tides. Stanford University Press, Stanford, CA
Ricketts EF, Calvin, J, Hedgpeth, JW (1985) Between pacific tides. Stanford University Press, Stanford, CA
Risk MJ (1973) Silurian echiuroids: possible feeding traces in the Thorld Sandstone. Science 180:1285–1287

Roberts DL, Brink JS (2002) Dating and correlation of Neogene coastal deposits in the Western Cape (South Africa): implications for Neotectonism. S Afr J Geol Geophys 105:337–352

Rudkin DM, Young GA, Nowland GS (2008) The oldest horseshoe crab: a new xiphosurid from Late Ordovician konservat-Lagerstätten deposits, Manitoba, Canada. Palaeontology 51:1–9

Runnegar B (1979) Ecology of *Eurydesma* and the *Eurydesma* fauna, Permian of eastern Australia. Alcheringa 3:261–285

Schram FR (1973) Pseudocoelomates and a nemertine from the Illinois Pennsylvanian. J Paleontol 47:985–989

Schröder-Adams C (2006) Estuaries of the past and present: a biofacies perspective. Sedimentary Geol 190:289–298

Sepkoski JJ Jr (1981) A factor analytic description of the Phanerozoic marine fossil record. Paleobiology 7:36–53

Sepkoski JJ Jr (1997) Biodiversity: past, present, and future. J Paleontol 7:533–539

Smith AG, Smith, DG, Funnell BM (1994) Atlas of Mesozoic and Cenozoic coastlines. Cambridge University Press, Cambridge, UK

Stanley SM (1968) Post-Paleozoic adaptive radiation of infaunal bivalve mollusks – a consequence of mantle fusion and siphon formation. J Paleontol 42:214–229

Stanley SM (1972) Functional morphology and evolution of byssally attached bivalve mollusks. J Paleont 46:165–212

Stanley SM (2007) An analysis of the history of marine animal diversity. Paleobiology Mem 4:1–55

Stanley SM (2008) Predation defeats competition on the seafloor. Paleobiology 34:1–21

Stanley SM, Newman WA (1980) Competitive exclusion in evolutionary time: the case of the acorn barnacles. Paleobiology 6:173–183

Stephenson TA, Stephenson A (1972) Life between tidemarks on rocky shores. WF Freeman and Company, San Francisco, CA

Stilwell JD (2003) Patterns of biodiversity and faunal rebound following the K-T boundary extinction event in Austral Palaeocene molluscan faunas. Palaeogeogr Palaeoclimatol Palaeoecol 195:319–356

Vermeij GJ (1977) The Mesozoic marine revolution: evidence from snails, predators, and grazers. Paleobiology 3: 245–258

Vermeij GJ (1987) Evolution and escalation. Princeton University of Press, Princeton, NJ

Ward LW (1998) Molluscs from the Lower Miocene Pollack Farm Site, Delaware. In: Benson RN (ed) Geology and paleontology of the Lower Miocene Pollack Farm fossil site, Delaware. Delaware Geological Survey Special Publication, vol 21, pp 59–133

Williams ST, Reid DG, Littlewood DTJ (2003) A molecular phylogeny of the Littorininae (Gastropoda: Littorinidae): unequal evolutionary rates, morphological parallelism, and biogeography of the Southern Ocean. Mol Phylogen Evol 28:60–86

Woods AJ, Saul LR (1986) New Neritidae from southwestern North America. J Paleontol 60:636–655

Yeo RK, Risk MJ (1981) The sedimentology, stratigraphy, and preservation of intertidal deposits in the Minas Basin System, Bay of Fundy. J Sedimentary Petrol 51:252–260

Zonneweld J-P, Beatty TW, Pemberton SG (2007) Lingulide brachiopods and the trace fossil Lingulichnus from the Triassic of Western Canada; implications for faunal recovery after the end-Permian mass extinction. Palaios 22:74–97

Zullo VA, Russell EE, Mellen FF (1987) *Brachylepas* Woodward and *Virgiscapellus* Withers (Cirripedia) from the Upper Cretaceous of Arkansas. J Paleontol 61:101–111

Zuschin M, Harzhauser M, Mandic O (2004) Spatial variability within a single parautochthonous Paratethyan tidal flat deposit (Karpatian, Lower Miocene – Kleinebersdorf, Lower Austria). Cour Forsch –Inst Senckenberg 246:153–168

Zwiebel JA, Johnson ME (1995) Late Pleistocene Mytilid and Petricolid bivalves from the open rocky shore of Pacific Baja California (Mexico): unusual preservation of macrofossils. J Coastal Res 11:704–716

Marine Sclerobiofacies: Encrusting and Endolithic Communities on Shells Through Time and Space

Carlton E. Brett, Trisha Smrecak, Karla Parsons Hubbard, and Sally Walker

Abstract

The concept of sclerobiofacies is defined herein as suites of sclerobiont encrusters and endiont borers (collectively sclerobionts) preserved on skeletons that characterize particular facies/environments. Skeletal components provide biologically standardized substrates; when possible, comparison of encrusting assemblages on fossil shells of the same or closely related eurytopic species provides a degree of substrate control comparable to modern experimentally deployed shells. Taxonomic composition of sclerobiont suites varies rather predictably among marine environments (e.g., based upon depth) but is primarily useful for comparisons of environments within local areas and limited time frames. Parameters that may be used to compare sclerobiofacies across broader spatial and temporal dimensions include: per shell and cumulative species richness (diversity), frequency of encrustation, areal coverage, and guild structure of encrusting taxa. Herein, we summarize characteristic sclerobiofacies in a series of Recent and ancient examples. Modern subtropical marine encrusters, documented on experimentally deployed molluscan shells at sites ranging from 15 to over 200 m, show high biont richness in shallow subtidal areas. Maximal areal coverages in Bahamian samples occur at about 20–30 m, whereas species richness increases downward to the deeper euphotic zone (~75–80 m). Below this level, rapid decline in both richness and percent coverage results in deeper Dysphotic–Aphotic zone samples yielding only a few species with coverage rarely exceeding 5%. Burial is also a key factor such that rapidly buried shells in the Shallow Euphotic zone have species coverages, richnesses, and taxonomic compositions resembling long-exposed shells in deeper areas below the euphotic zone. Shelly substrates from the Cambrian to Early Ordovician exhibit only minor encrustation by solitary attached taxa, especially echinoderms; however, by the Late Ordovician various solitary (e.g., cornulitids, craniid brachiopods) and colonial forms (e.g., trepostome and tubuliporate bryozoans) form distinctive sclerobiofacies. Photic zone-related environments, judged independently on the basis of microendoliths, show overall lower taxonomic richness than modern counterparts. However, they also show common patterns, including a general decrease of richness and percent encrustation from Shallow Euphotic to

C.E. Brett (✉)
Department of Geology, University of Cincinnati, Cincinnati, OH 45221-0013, USA
e-mail: carlton.brett@uc.edu

J.A. Talent (ed.), *Earth and Life*, International Year of Planet Earth,
DOI 10.1007/978-90-481-3428-1_6,

Dysphotic/Aphotic zones. Comparable trends are seen in Middle Devonian exemplars from New York State. Not only were there consistent trends toward lowered diversity/coverage into deep-water settings but also an additional factor related to turbidity and/or sedimentation rate was identified from assemblages at comparable depths arrayed along a distal to proximal gradient with respect to siliciclastic input sources. Carboniferous sclerobiont suites from varied sites in North America show many of the same traits as their Devonian counterparts, although detailed depth zonations are not documented at present. The Permo-Triassic extinctions appear to have had a strong impact on the taxonomic composition of marine sclerobiofacies, although a paucity of studies obscures details of Mesozoic and Cenozoic sclerobiofacies. In general, they appear to have taxonomic compositions and patterns similar to those observed in the Recent. The concept of sclerobiofacies provides another tool for paleoenvironmental analysis. Together with litho-, ichno-, bio-, and taphofacies, the properties of shell encrusting assemblages will yield detailed further insights into ancient environmental gradients.

Keywords

Modern and ancient environments • Ordovician • Devonian • Mississippian • Recent • Experimental • Encrusters • Borers • Microendoliths • Hardgrounds • Sclerobiofacies • Impact of turbidity

Introduction

The ecology and paleoecology of shelly marine benthic communities has generated important discussion of community structure and environmental parameters in both modern and ancient settings (see recent reviews by Bush and Daley 2008; Bambach and Kelley 2008). The epi- and endobionts, or sclerobionts, which encrust or excavate into the shells of living and dead organisms, are an important component of these communities that is preservable precisely in situ (Ager 1963; Richards 1972; Kesling et al. 1980; Liddell and Brett 1982; Vogel 1993; Alexander and Brett 1990; Powers and Ausich 1990; Gibson 1992; Taylor 1990; Lescinsky 1996, 2003; Lescinsky et al. 2002; Rodland et al. 2004; see Taylor and Wilson 2003 for an excellent review and definition of biont terminologies). Marine epi- and endobionts in modern environments have proven to be highly sensitive to environmental parameters, including light levels, sedimentation, substrate type, and productivity (Jackson 1977, 1983; Kay and Keough 1981; Rasmussen and Brett 1985; Martindale 1992; McKinney 1995a, b; Vogel et al. 1999; Lescinsky 1993, 2003) and thus may provide excellent paleoenvironmental proxies. In tandem with other evidence on sedimentology, paleoecology, and taphonomy of their host shells, epi- and endobionts may provide valuable guides to interpretation of ancient environments. Yet, to date, the topic of gradients of sclerobiont biofacies and environmental implications of these micro-communities has received little systematic study.

Herein we propose the concept of "sclerobiofacies", the attributes of preservable suites of skeleton-encrusting and endolithic organisms. Sclerobiofacies may be used to recognize gradients of paleoenvironments. Certain characteristics of sclerobiont assemblages, including diversity, areal coverage, guild structure, and taxonomic composition, may be consistently and uniquely associated with particular marine environments. Patterns of colonization on skeletal substrates may further yield important clues as to exposure time and extents of time-averaging. Moreover, certain of these traits may be compared throughout geologic time, at least since the Late Ordovician when skeletonized encrusters first became abundant.

In this paper we survey the occurrence of distinct sclerobiont assemblages in modern marine tropical environments. We then apply the concept of sclerobiofacies to ancient marine examples from the Paleozoic Era and compare those to other documented

Table 1 Parameters for comparison of sclerobiofacies through time

A. Presence/Absence Metrics
Presence of bionts (overall) per shell:
Number (or %) of shells in a sample that have at least one sclerobiont
Number of biont taxa per shell:
Number of shells with various numbers of taxa (e.g., number of shells with 1, 2, 3. . .taxa)
Number of biont guilds per shell:
Number of shells with various numbers of guilds (e.g., number of shells with 1, 2, 3. . .guilds)
Number of shells per biont taxon:
Number of shells with at least one specimen of a particular biont taxon
Number of shells per guild:
Number of shells with at least one specimen of a particular biont guild
B. Richness/Diversity Metrics
Overall richness:
Total number of taxa integrated over all shells in a particular sample size
Maximum/Minimum Per-Shell Richness:
Lowest and highest number of taxa observed on a single shell in a sample of particular size (assumes approximately equal-sized shell surface areas)
Mean Per-Shell Richness:
Average number of taxa per shell substrate
Taxonomic Diversity:
Number of taxa adjusted for dominance (e.g., Shannon-Weaver index)
C. Areal Coverage Metrics
Overall Mean Coverage:
Average proportion (or percentage) of shell surface area covered by all bionts averaged over number of shells
Mean Per-Shell Coverage of Guilds:
Proportion (percentage) coverage occupied by a given trophic/life habit group averaged over number of shells observed
Mean Per-Shell Coverage by Taxa:
Proportion (percentage) coverage occupied by a given taxon averaged over number of shells observed

examples in communities throughout geologic time. Comparison of analogous sclerobiofacies relative to light/depth zones, through time (Table 1), reveals patterns of temporal change in composition of encrusting communities that occupy similar environments, but also demonstrates some consistencies in pattern that suggest similar controls. Comparison of sclerobiont suites of different geologic ages in the same relative light/depth zone forms the basis for "comparative sclerobiofacies" studies.

Definitions and Parameters of Sclerobiofacies

Sclerobionts are benthic organisms that attach to or encrust outer surfaces of hard substrates, commonly upon another organism. Similarly, endobionts excavate into hard substrates. Taylor and Wilson (2003) proposed the term "sclerobiont" to encompass both epi- and endobionts that encrust and bore into hard substrates, including skeletons of other organisms (sclero- = stony or hard). We chose the term sclerobiont to refer to mixtures of encrusting and endolithic organisms, but use the terms epibiont and endobionts for each of the latter categories in isolation. The majority of sclerobionts discussed herein are epibionts, although etchings of ctenostome bryozoans, *Entobia*, the traces of boring clionid sponges, and worm borings, such as *Vermiformichnus*, are also included in this study. Such trace fossils and skeletons of encrusting sclerobionts are commonly well preserved as fossils, even as internal and external molds where skeletal substrates have been dissolved. These fossils provide a rich and little studied sample of ancient microcommunities that may be highly sensitive to environmental parameters.

What critical parameters of sclerobiofacies can be assessed and compared throughout Phanerozoic communities? Although encrusting communities have undergone substantial changes in their taxonomic composition through time, some aspects of sclerobiont communities can be fairly compared, including species richness and diversity, proportion of surface area covered by skeletonized epibionts and the surface expression of endobiont borings, and guild structure (Table 1).

Richness reflects simply the number of different types of skeletonized and trace fossil taxa (ideally species) in some specified sample size. It can be calculated as average species number per shell or total or average richness integrated over a particular number of shells or a standard surface area of shell substrate (see below).

Aspects of dominance-diversity, however, require additional information on proportional abundance of different encrusting taxa. Because of the clonal/colonial nature of many sclerobionts (e.g., many sponges, corals, bryozoans) surface area of coverage is generally a better metric of relative "importance" of a given taxon than simple counts. Such colonies may vary greatly in size and shape so that the area occupied is a better indicator of their "importance" in encrusting assemblages than the number of colonies. Moreover, sclerobiont coverage on shells or other hard substrates may be dominated by one specific type of epibiont or equally covered by multiple encrusting organisms. This will be referred to herein as community equitability. Likewise, the *proportion of surface area* covered by various sclerobionts is a measurable parameter of encrusting/endolithic communities through time. Available habitat may be measured on a shell or number of shells in a given area, and so the proportion of surface area covered by each sclerobiont type is a more accurate representation of community dominance. Total surface area occupied by all skeletalized epibionts and the excavations of endobionts, expressed as a percentage of total shell area, are another parameter that has been found to vary systematically among environments and are readily compared among analogous environments of different age.

Guild structure of marine organisms has also been compared through time using broadly defined categories in relation to substrate, mode of feeding, and life position (Bush et al. 2007). Sclerobionts are implicitly restricted to fewer guild types because of their inherently limited life habit types.

In the Recent, environmental parameters, such as water depth, light level, nutrient availability, sedimentation, and water energy, may be observed directly along various gradients. To establish sclerobiofacies gradients in ancient environments, however, proxies must be used to establish environmental parameters, especially relative water depth, independently of the fossil assemblages; stratigraphy, facies patterns, and palaeogeography must be adequately known to provide a basis for reconstructing environmental gradients. Suites of light-sensitive microendolithic algae, cyanobacteria, and fungi recognized in modern environments (Vogel 1993; Vogel et al. 1995, 1999, 2000), have been identified in shells as old as the Late Ordovician and Silurian (Glaub and Bundschuh 1997; Bundschuh 2000; Vogel and Brett 2009). Thus, some sclerobiofacies may also be objectively compared between modern and ancient environments based on light zonation—and therefore relative depth.

This paper first summarizes case studies of sclerobiont assemblages along depth gradients in the Recent as exemplars of the sclerobiofacies approach. We then compare the attributes of Middle and Late Paleozoic sclerobiont communities through geologic time highlighting the details that allow the formation of preliminary conclusions on sclerobiofacies through time.

Modern Sclerobiofacies: Case Studies

The following sections summarize three studies that systematically survey sclerobiont assemblages along gradients in modern marine subtropical settings and therefore provide exemplars of sclerobiofacies analysis.

Shelf and Slope Experimental Taphonomy Initiative (SSETI) Studies: Bahamas

The Shelf and Slope Experimental Taphonomy Initiative, (SSETI), was designed to test the efficacy of taphofacies in discriminating depositional environments, i.e., that the taphonomic signature, manifested in skeletal preservation, including suites of shell endo- and sclerobionts, provides unique clues to depositional environments (Parsons-Hubbard et al. 1999, 2000; Callender et al. 2002; Staff et al. 2002; Powell et al. 2008; Brett et al. 2011). A further premise of SSETI is that long-term deployments provide insight into the processes by which skeletal material is preserved or destroyed. Experimental arrays of shells, crabs, sea urchins and wood blocks were deployed in two major suites of depositional environments: subtropical siliciclastic shelf settings in the Gulf of Mexico and subtropical carbonate shelf and submarine escarpment and slope settings in Lee Stocking Island in the Exuma chain of the Bahamas (Fig. 1). SSETI deployed approximately 200 primary experimental arrays in 1993 and 1994. In the case studies below we focus on the Bahamas samples because they are calibrated to light zonation established by microendolith ichnofacies at similar sites along the same transects; Kiene et al. (1995) and Vogel et al. (2000) characterized the distinctive assemblages of particular depths on a transect ranging from near surface to over 800′ of water depth off the "wall", a submarine escarpment at Lee Stocking Island. These researchers characterized distinctive assemblages that appear to correspond to light intensity and classified them as the Shallow Euphotic I (<2 m), Shallow Euphotic II (~2 to 10 m), Shallow Euphotic III (~ 10 to 30 m), Deep Euphotic (~30 to 100 m), Dysphotic (~ 100 to 275 m), and Aphotic (>275 m) zones.

The SSETI experiments were deployed along two preexisting transect lines identified as AA and BA transects, with deployments at depths of approximately 15 and 30 m (50, 100′; Shallow Euphotic zone III), 73–88 m (240–290′, Deep Euphotic zone), 182 m (600′), 213–228 m (700–750′; Dysphotic zone), and 249–264 m (820–870′; Dysphotic–Aphotic zone). Six environments of deposition (EODs) were targeted on the transect. In most cases, multiple sites were chosen for each depth/EOD. Experimental arrays

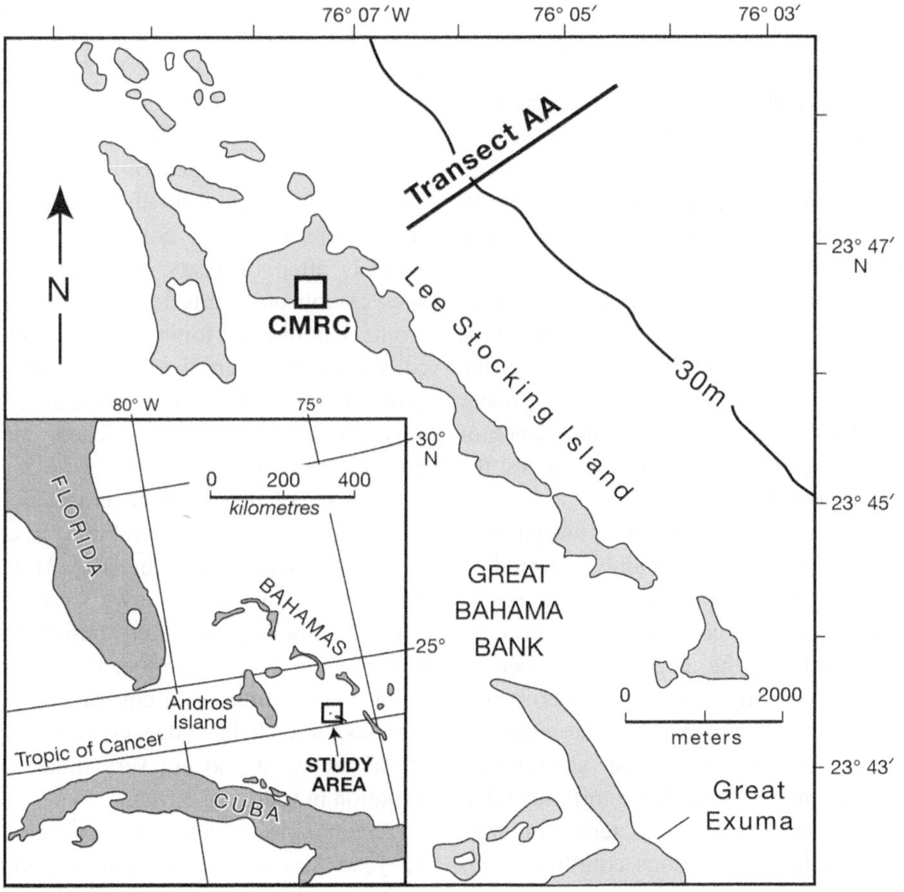

Fig. 1 Location of study area and transect used by the Shelf and Slope Taphonomy Initiative (SSETI) at Lee Stocking Island, Bahamas

contained various skeletal materials including 5–8 species of bivalve and gastropod molluscs, two species of urchins, crabs, and various types of wood.

SSETI recovered experiments at Lee Stocking Island in 1994/5 after 1 year, in 1995/6 after 2 years, and in 1999/2001 after ~8 years. The 12-year collections retrieved in 2006 represent one of the longest experimental deployments of any kind in the marine realm. Analysis of samples conducted on location at Lee Stocking Island included observation of interiors and exteriors of 50 shells per array, encompassing 5–8 species. For the molluscan shells alone, >22,770 sclerobionts (with taxon identifications and estimated percent areal coverage) were recorded. These results will be discussed in more detail in a separate paper (Brett et al. 2011) and will be briefly summarized here (Table 2).

In the shallowest samples, deployed at 18 m (50′) and 32 m (100′), showed the highest overall areal coverage (mean = 18% coverage per shell, with ~24% covered per shell at 18 m and ~10% covered per shell at 32 m). In some cases, especially at the 32 m/100′ locality, shells were deeply covered in sand and had relatively low sclerobiont coverage (Fig. 2). More heavily affected shells, those presumably exposed for long times, had distinctive sclerobiont assemblages dominated by larger foraminifers. Many showed large areas or complete investment by the larger foraminifers *Gypsina*, *Planorbulina*, and *Cornuspiramia*, small sponges, and clionid borings were common (Fig. 2). Serpulids and green algae were less common in the shallowest zones than they were at greater depths. A notable difference between the two localities was that bryozoan diversity was greater in the 30-m sample, along with higher abundance of the encrusting bivalves *Chama* and *Plicatula* and the "worm shell" gastropod *Vermetus.*

Samples from 32 m (100 feet), also assigned to Upper Euphotic III Zone, were mainly covered in sand; their relatively low sclerobiont coverages reflect burial that occurred soon after deployment and continued through much of the experimental period. Thus, sclerobiont compositions of these experimental arrays may not be representative of the potential diversity of these depths. Dominant sclerobionts again include foraminifers (here the ramifying tubular *Cornuspiramia*) are dominant in terms of areal coverage, with abundant filamentous coralline forms (Fig. 2). Clionid sponge borings of two diameters are abundant on many shells; by 12 years these are effective in riddling many shells with galleries that greatly weaken the shells. Bryozoan diversity is higher than in the shallowest samples, although coverage is still low. The encrusting bivalves *Chama* and *Plicatula* are also relatively common, along with the "worm shell" gastropod *Vermetus*. Green algae are also more abundant in this deeper zone, as are serpulid worms.

Samples from 73–88 m (240–290 feet), along the wall, reflect the Deep Euphotic Zone of Kiene et al. (1995) and Vogel et al. (2000). Shells from these sites showed the highest diversity of sclerobionts of any samples, but not the greatest areal coverage. Serpulids are spectacularly common and diverse, showing at least six species (Figs. 2 and 3). Bryozoan diversity is also at its peak, with several taxa present per shell. Foraminifers are less abundant than seen at shallower depths. Sponges of several species are common and clionid borings are still abundant on shells, although total areal cover is substantially diminished from shallower zones. The encrusting bivalves *Chama, Plicatula*, and *Anomia* are present as small individuals on many shells.

Below 280 m (600 feet), conditions are Dysphotic to Aphotic. Bottom waters remain well oxygenated and substrates are generally sandy with scattered coral debris fallen from the escarpment. In general, shells were in pristine condition, differing little from controls after 12 years. Shells sampled from these sites exhibit uniformly low areal coverage by sclerobionts. Clionid and foraminiferal coverage was drastically low, bryozoan diversity was greatly diminished (Fig. 3). Serpulids were minute; many had tubes that appeared to have been breached, indicating taphonomic destruction.

Differences in overall coverage, richness/diversity, and composition of sclerobiont assemblages from different depths are striking (Fig. 3). It is notable that the shells, which show lowest coverage in shallow sites, i.e., those which were buried through some or much of the 12-year period, have diversities, sclerobiont compositions, and coverage similar to those of most exposed shells at deeper sites. Within-depth variations are great and are very much dependent upon variation in sediment cover.

A comparison of samples collected at 1, 2, 8, or 12 years shows that the general trends seen after 1 or 2 years of exposure are still present in later years,

Table 2 Areal coverage of various major groups of encrusting bionts at Lee Stocking Island, Bahamas

	Forams	Algae	Porifera	Clionids	Coral	Serpulids Sabellids	Encrusting Bryozoans	Cemented Mollusks	
15 m Shelf	60.34	31.744	3.44	6.432	0.0964.000	73.260	14.788	4.899	195.005
30 m Shelf	13.28	5.716	1.59	1.279	0.0004	55.601	7.722	1.172	86.363
Wall	22.75	8.107	5.84	1.598	1.926	41.117	16.445	2.821	100.603
Talus	3.16	0.085	0.001	0.231	0.048	9.423	0.234	0.103	13.284
Trough	4.54	0.30	0.067	0.296	0.026	6.929	0.114	0.034	12.3137

Numbers represent grand averages of bivalve shells integrated over ~12 years

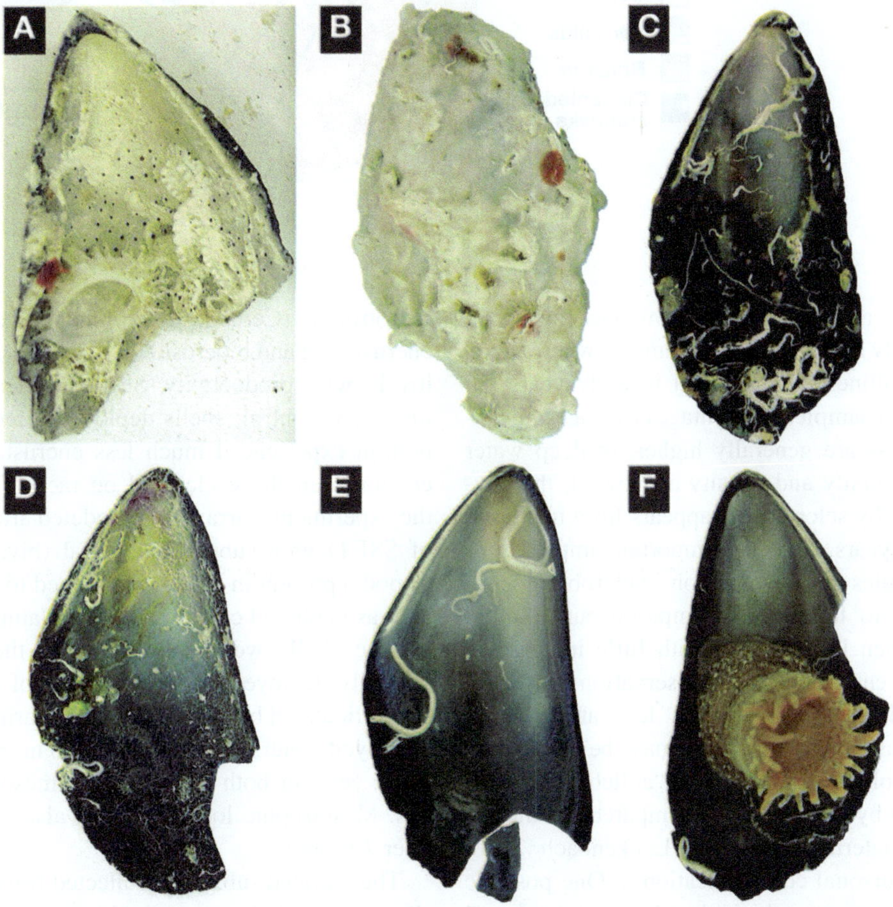

Fig. 2 The concept of sclerobiofacies illustrated by experimentally deployed *Mytilus edulis* valves from various environments of deposition along the AA transect at Lee Stocking Island, Bahamas. All views of valve interiors, except B; all × 1.5. (**a**) Heavily encrusted valve with smooth and knobbed tubes of serpulid annelids, also red foraminifer *Homotrema* and cemented left valve of the bivalve *Chama*. 15 m; loc. AA50RID. (**b**) Exterior of valve almost completely encrusted by the foraminifer *Gypsina*, which has overgrown serpulid tubes and is, in turn, overgrown by other serpulids and the foraminifer *Homotrema*; 15 m; loc. AA50RID. (**c**) Valve heavily encrusted by three species of serpulids as well as small colonies of bryozoans and coated by green algal film; 30 m; loc. AA100SL. (**d**) Valve interior with extensive white branching tubes of the foraminifer *Cornuspiramia*, small bryozoans, and a small *Homotrema*; also note remnant stains of small sabellid agglutinated tubes and thin film of *Ostreobium*; 73 m; loc. AA240LE wall site; (**e**) Nearly pristine valve with a few slender serpulids; 263 m; site: AA865 base of slope. (**f**) Valve with solitary anemone; this biont is not preservable; the valve would be perceived as nearly clean (lacking sclerobionts) in the fossil record; 262 m; AA860BP

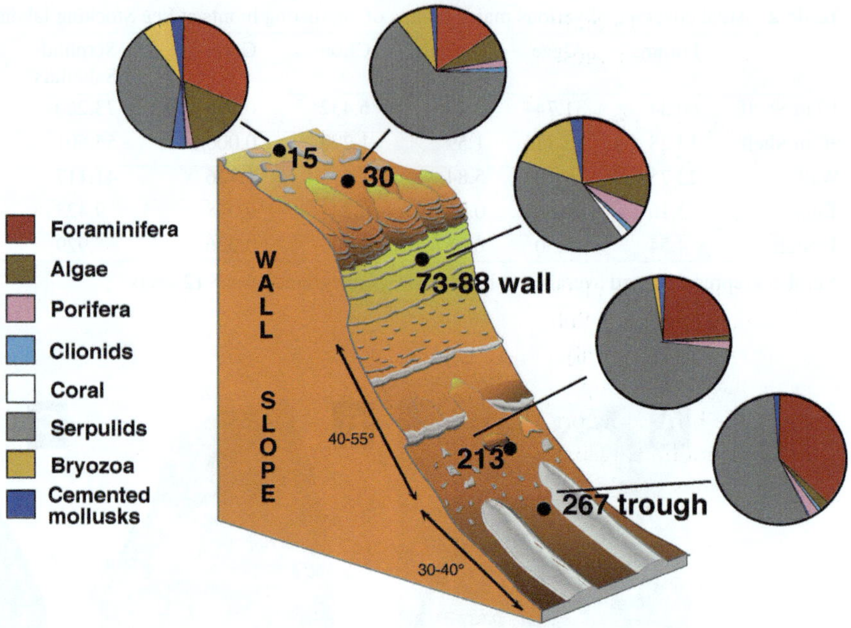

Fig. 3 Schematic diagram of AA transect along narrow shelf, submarine escarpment, base of slope rubble apron, and deep carbonate dunes and troughs. Pie diagrams indicate the integrated relative proportions of bionts on experimentally deployed clean bivalve shells during the 1994–2006 interval

although there are changes in dominance and rank order, especially in shallow water samples where green algae and corallines are dominant in early years but decrease in later samples. Percentage coverage and taxonomic richness are generally higher. In deep water sites where diversity and density are lowest, the coverage of shells by sclerobionts appears little increased over the first years. This has important implications for the dynamics of colonization. Sclerobiont coverage appears to follow an asymptotic curve, rapid initially but then leveling off, with little increase in abundance of encrusters. This observation may suggest that shell material becomes less attractive to colonization as it "ages". This may be analogous to the rapid colonization of fresh "cultch" (air-dried shell material) by oyster spat as compared to "marine aged" shell material (Coen and Luckenbach 2000; EN Powell "personal communication"). One possible explanation is that microbial films may inhibit larval settlement.

Java Sea: Indonesia

Lescinsky et al. (2002) carried out a 2-year taphonomic study using experimentally deployed mollusk shells in the Java Sea, Indonesia (Bromley 1994; Glaub et al. 2001). Shallow eutrophic (Panjang) and mesotrophic (Cemera) environments were gauged for encrustation and bioerosion rates; these correlated positively with productivity. Sedimentation was also an important control; shells deployed directly to the sea bottom experienced much less encrustation and bioerosion than those elevated on racks. The design of the experimental arrays was modeled after the methods of SSETI with substrate material (bivalves and gastropods) present in bags and attached to PVC poles, as well as deployed concrete blocks. Naturally occurring bivalve shells were collected near the two experimentally deployed sites with a mix of recently dead, still-articulated bivalves and older, marine-aged shells. Deployed material was sampled at 3 months, 6 months, and 1 year for both eutrophic and mesotrophic localities. Mesotrophic localities were also collected again after 2 years.

The elevated substrates collected from the eutrophic sites in the study were significantly more encrusted than those from the mesotrophic sites; many elevated substrates were nearly 100% covered by encrusters. The taxa also differed significantly between eutrophic and mesotrophic sites. The eutrophic site was dominated by large bryozoans, barnacles, bivalves, and sponges, some organisms nearing 3.5 cm in size, whereas at the mesotrophic site, bryozoans were the most abundant, bivalves were significantly less common, and barnacles were completely absent. Here serpulid worms and coralline and fleshy algae were more common; the largest encrusting organism was only 0.9 cm.

As a result of the high level of sedimentation in the Java Sea, the substrates that were deployed directly on the seafloor showed a much lower encrustation in the euphotic sites (~2% sclerobiont coverage) than those suspended above the substrate. This is thought to have been a result of cover or complete burial of the substrates by mud. Substrates in the mesotrophic sites were covered by only a small amount of carbonate sand; they showed up to 12% sclerobiont coverage after 2 years. Encrusting organisms were primarily serpulids and coralline algae, similar to what was seen on the elevated mesotrophic substrates.

Brazilian Bight

Rodland et al. (2004) performed an analysis of encrusting organisms on in situ brachiopods across the shelf of the southeastern Brazilian Bight with very shallow localities between 6 and 30 m and deeper localities of 100–500 m. Using in situ brachiopods limits the potential available surface area to a maximum of 100 mm^2; the study did not substantiate a significant correlation between sclerobiont suites and depth. Using Spearman's Rank Correlation Coefficient, Rodland and others did not find statistically significant correlations between encrusting sclerobiont suites and the known depth gradient. However, they do show significant correlations with individual brachiopod size. It is important to note that this study uses a different methodological approach to counting sclerobiont abundance. It does not use areal or surface coverage data in its evaluation of sclerobiont suites. Instead, only occurrence data were used; thus, bryozoans and serpulids were counted equally, although their areal coverage may have been quite different.

Inner shelf localities ranging in depth from 6 to 30 m have a frequency index of 52% on the outer shell surface. The different inner shelf localities display a range of encrusting frequencies between 75 and 100% when both valves are included, with a strong preference for encrustation of shell interiors. The outer shelf localities, with a depth range of 99–500 m, showed a total shell-encrusting frequency index ranging between 0 and 44% at different localities. In both shallow and deep shelf localities, a spatial trend is seen from northern to southern localities; some organisms occurred more frequently in the north than they did in the south. In the southern transect, there is an observable change in encrusting fauna, with serpulid worms and bryozoans decreasing in abundance with depth. This pattern is less predictable in the northern transect—thus trends were not statistically significant.

Ancient Sclerobiont Case Studies

Several past studies have focused on sclerobionts in relation to the life habits of host organisms and sclerobionts (e.g., Ager 1963; Schumann 1967; Rudwick 1970; Richards 1972, 1974; Spjeldnaes 1984; Kesling et al. 1980; Sparks et al. 1980; Bordeaux and Brett 1990; Morris and Felton 1993). Until recently, however, relatively little attention has been paid to the environmental distribution of fossil sclerobionts. Taylor and Wilson (2003) provide an excellent review of studies of epi- and endobionts and the varied applications of sclerobiont data. Here we survey a few key studies pertinent to the issue of sclerobiofacies. This section summarizes case studies of sclerobiont suites of different geologic ages, using comparable relative light/depth zonations.

Late Cambrian Sclerobionts

Cambrian shelly assemblages are distinctive primarily in near absence of encrusting sclerobionts, but may show a distinctive suite of rarely preserved encrusters. In the absence of bryozoans, cornulitids, and spirorbid worm tubes, seen in the later Paleozoic encrusting communities, identifying typical sclerobiofacies is impossible (Table 3). There are also only very rare small boring traces in shells.

Shallowest water settings feature rare examples of trilobite sclerites enclosed within microbial mats as oncolites. "Spongiomorphs" (poriferans?) and holdfasts of eocrinoids have been found in Late Cambrian and Early Ordovician flat pebble hardgrounds from shallow shelf settings (Brett et al. 1983; Guensburg and Sprinkle 2001). These encrusters have not been found on contemporaneous brachiopod shells or other skeletons.

Table 3 Composition of Late Ordovician Sclerobiofacies

Shallow Euphotic—Loc 2—Mt. Auburn Mbr		Deep Euphotic—Loc 13—Fairview Fm		Transitional/Deep Euphotic—Loc 12—Fairview	
trepostome bryozoans (9 species)	14.5	trepostome byozoans (5 species)	6	trepostome Bryozoans (7 species)	1.517857143
ctenostome bryozoans	0	paleotubuliporid bryozoans	0.1667	borings	1.482
paleotubuliporid bryozoans	0	borings	0.43	*Sphenothallus*	0.054
cornulitids	0.58	*Sphenothallus*	0.1		
petrocranids	0.06				
Vermiformichnus borings	0.26			trepostome Bryozoans (10 species)	23.3
Sphenothallus	0.08	Deep Euphotic—Loc 11—Fairview Fm		paleotubuliporid Bryozoans	1.34
crinoid holdfast	0.24	trepostome bryozoans (6 species)	4.625	petrocranids	0.67
algal remnants	0.54	paleotubuliporid bryozoans	0.0625	cornulitids	0.067
Philhedra	0.2	cornulitids	0.0313	*Vermiformichnus* Borings	0.45
		Vermiformichnus borings	0.59	*Sphenothallus*	0.067
Shallow Euphotic—Loc 3—Bellevue		*Sphenothallus*	0.0625		
trepostome bryozoans (10 species)	18.68				
ctenostome bryozoans	0.02				
petrocranids	0.7	Deep Euphotic—Loc 13—Fairview Fm		Dysphotic–Aphotic—Loc 15—Upper Kope Fm	
cornulitids	0.78	trepostome bryozoans (5 species)	6	trepostome bryozoans (4 species)	0.848
Vermiformichnus borings	0.22	paleotubuliporid bryozoans	0.1667	cornulitids	0.176
Sphenothallus	0.12	*Vermiformichnus* borings	0.43	*Vermiformichnus* borings	0.144
crinoid holdfast	0.06	*Sphenothallus*	0.1		
nodular massive bryozoans	4			Dysphotic–Aphotic—Loc 4—lowest Kope Fm	
algal remnant	1.24			trepostome bryozoans (2 species)	0.157894737
		Deep Euphotic—Loc 11—Fairview		*Vermiformichnus* Borings	0.315789474
Shallow Euphotic—Loc 6—Upper Waynesville		trepostome bryozoans (6 species)	4.625		
trepostome bryozoans (8 species)	25.02	paleotubuliporid bryozoans	0.0625		
ctenostome bryozoans	0.76	cornulitids	0.0313		
paleotubuliporid bryozoans	0.68	*Vermiformichnus* borings	0.59		
petrocranid	1.04	*Sphenothallus*	0.0625		
Vermiformichnus borings	1.86				
Sphenothallus	0.08				
crinoid holdfast	0.02				

Evidence for sclerobionts is very scarce in all Cambrian offshore samples examined to date. Although orthid brachiopods are locally fairly common in some shallow shelf facies, we find no evidence of any encrusting epibionts upon those we have examined. There are rare reports of attachment by eocrinoids such as *Gogia*, juvenile helicoplacoids, and edrioasteroids such as *Stromatocystites* to trilobite sclerites, hyoliths, and microbial mats (Lin et al. 2008a, b). Dornbos (2006) provided a summary of those organisms and their evolutionary relationship to the Cambrian Explosion. However, Cambrian and Early Ordovician shelly substrates cannot be assigned to sclerobiofacies at this time.

Late Ordovician Sclerobiofacies, Cincinnati Arch of Ohio, Kentucky, and Indiana

The Late Ordovician rocks of the Cincinnati Arch (Fig. 4) form an excellent source for sclerobiofacies analysis because superbly preserved brachiopods are abundant, and much work has been done independently to establish palaeobathymetry, including sequence-stratigraphy, palaeoecology (Jennette and Pryor 1993; Holland et al. 2001; Miller et al. 2001; Holland and Patzkowsky 2007; Patzkowsky and Holland 2007; Brett et al. 2008; Dattilo et al. 2008;

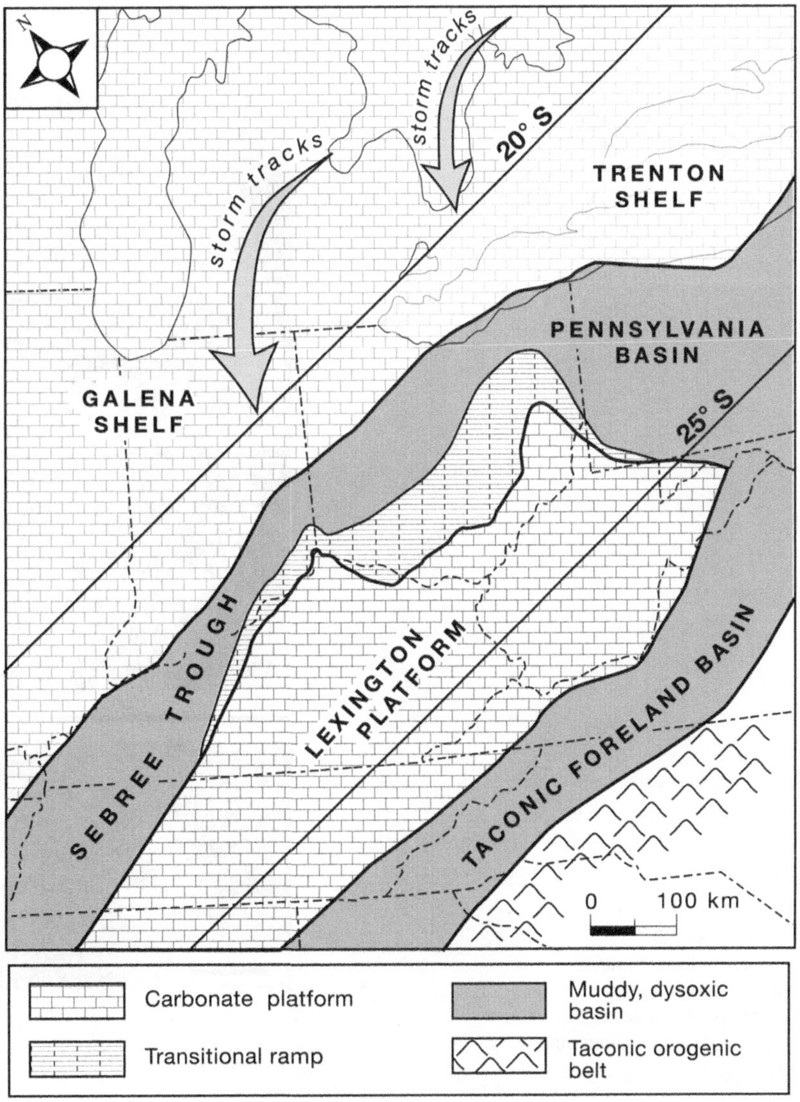

Fig. 4 Study area and stratigraphy for the Upper Ordovician (Cincinnatian) biont assemblages in the Cincinnati Arch area of northern Kentucky, southern Ohio, and southern Indiana

Fig. 4). Moreover, SEM study of Cincinnatian brachiopod shell surfaces by Vogel and Brett (2009) revealed that many microendoliths seen in the modern Bahamas and in other formations throughout much of the Phanerozoic record occurred, perhaps for the first time, in the Late Ordovician of Ohio, Kentucky, and Indiana. Thus, at least some of the many well-preserved sclerobiont assemblages found on brachiopod shells in Late Ordovician rocks can now be assigned tentatively to Shallow Euphotic, Deep Euphotic, and Dysphotic/Aphotic light zonations. With an independent control of palaeobathymetry and the numerous outcrops of fossiliferous rock, the location has the capabilities to display a wide range of facies and relative depth relationships among sclerobiont community suites (Figs. 5 and 6).

Richards (1972) and Alexander and Scharpf (1990) primarily studied Late Ordovician brachiopod palaeoecology; the abundance of sclerobionts found during the study allowed preliminary observations on host preferences; comparisons of taxonomic distribution in different environments were recorded as well as use made of sclerobionts in discerning life habits of brachiopods.

Smrecak and Brett (2008) observed brachiopod sclerobiont characteristics along an Ordovician gradient in the Cincinnati Arch region, characterizing sclerobiont suites based on previous paleoenvironmental studies of the region (cited above) coupled with the light-sensitive microendolith boring suites of Vogel and Brett (2009). The study analyzed percent areal coverage by various sclerobiont taxa, as well as frequency of encrustation, community equitability,

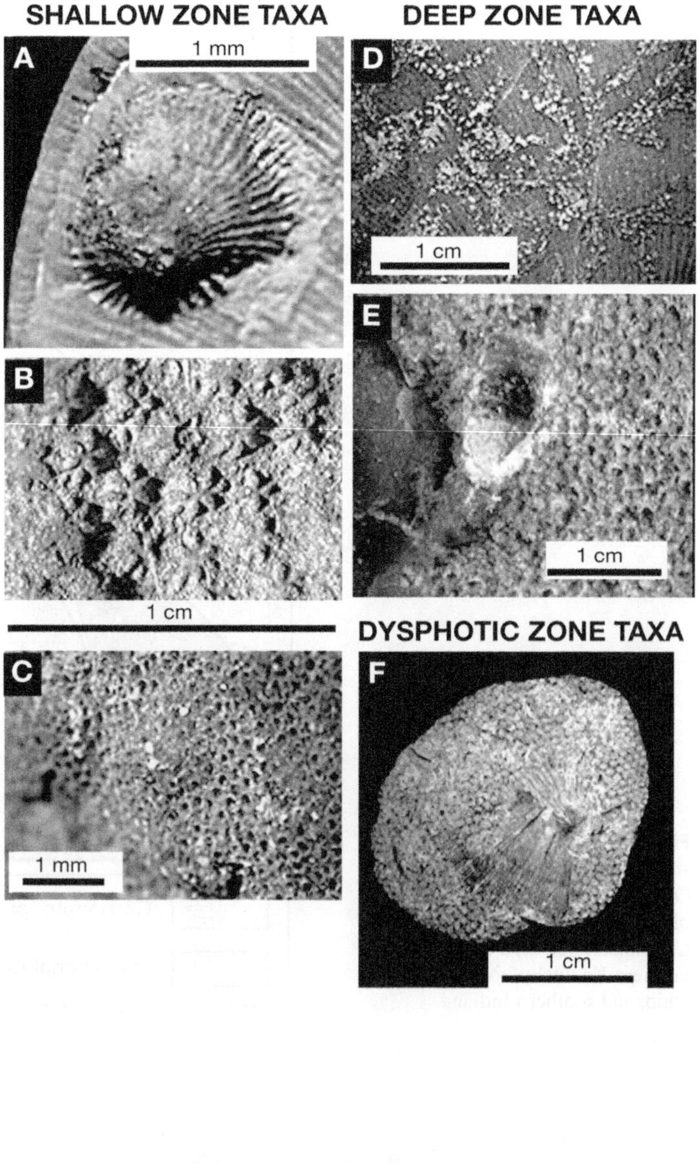

Fig. 5 Common shell-encrusting bionts on brachiopods from three environmental zones in the Upper Ordovician Cincinnati Arch region

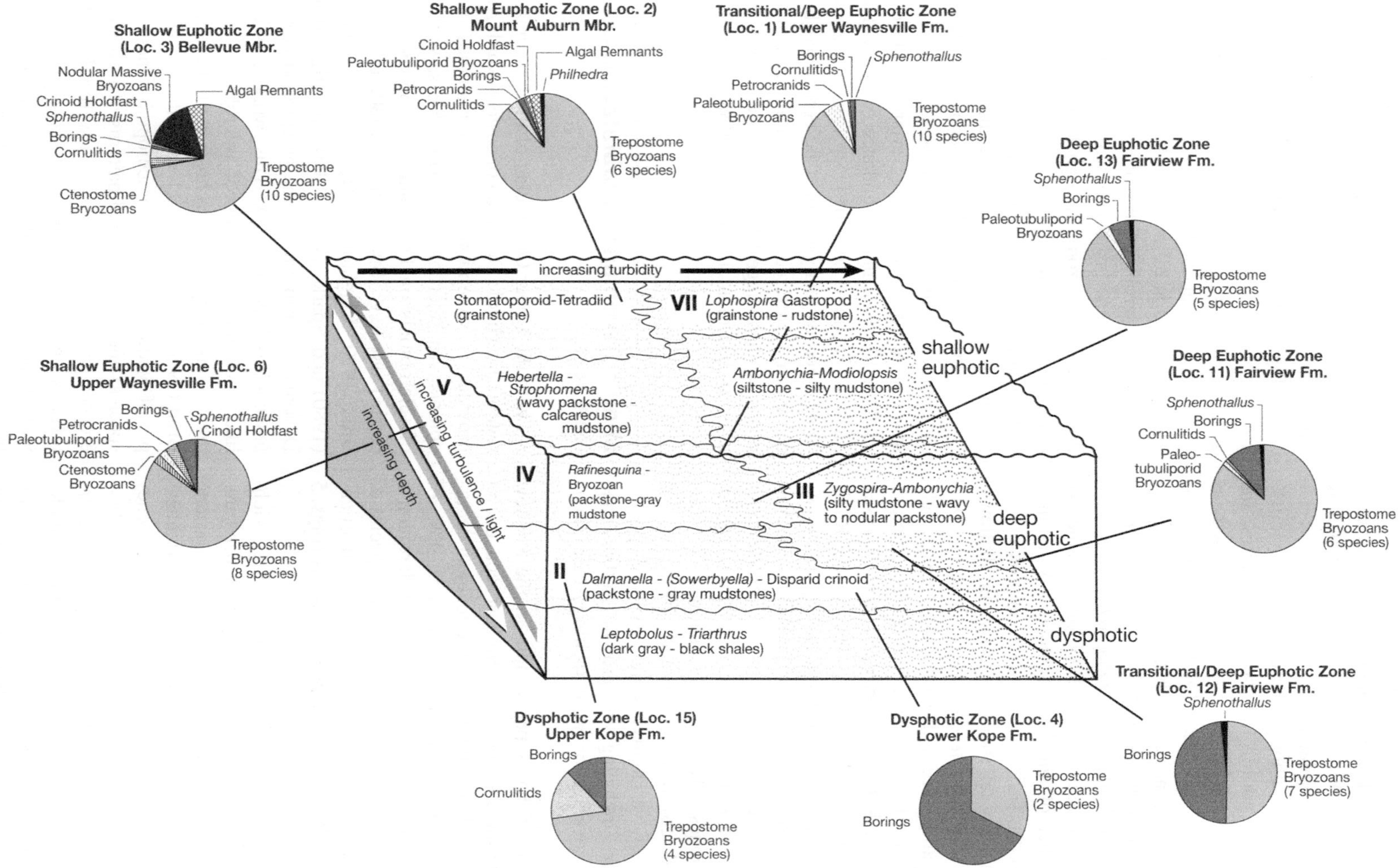

Fig. 6 Biofacies of the Late Ordovician Cincinnatian in relation to relative water depth and turbidity/sedimentation with associated pie diagrams illustrating the proportion of various biont taxa at particular relative depth zones

richness, and number of sclerobiont taxa per shell in each relative depth zone (shallow euphotic, deep euphotic, and Dysphotic/Aphotic) to provide a picture of Ordovician sclerobiofacies.

Localities were chosen based on previously established microendolith light-sensitivity suites; the ubiquitous brachiopod *Rafinesquina* was used as the primary host substrate. To account for possible substrate preferences, and because *Rafinesquina* was not present in the deepest facies, *Platystrophia, Hebertella, Sowerbyella*, and *Dalmanella*, were also observed as host shells for sclerobionts (Table 3). A larger number of smaller brachiopods were analyzed for deeper facies to yield an approximately equal amount of total shell surface as in shallower facies where larger *Rafinesquna* were abundant. Data from each locality were analyzed for dominance, richness, frequency of occurrence, and guild structure. Cluster analyses, de-trended correspondence analyses, and nonparametric analyses, e.g., Spearman's Rank, were used to compare compositions among localities.

Sclerobionts from localities assigned to the Shallow Euphotic zone were censused on the brachiopods *Rafinesquina* and *Platystrophia* derived from argillaceous packstones. These shells showed a total average areal coverage of around 20%, with a high degree of variability; some shells displayed >100% cover (including overgrowth); others had much lower coverage or were lacking sclerobionts altogether.

Total species richness per sample in the Shallow Euphotic zone was 24 species, the highest seen among Ordovician sclerobiont suites. The frequency of occurrence was also very high, near 90% (i.e., 90% of brachiopod shells showed at least one sclerobiont). Dominant taxa in this zone included several species of trepostome bryozoans with small zooecia, the inarticulate brachiopod *Philhedra* and algal remnants (Figs. 5 and 6).

The Deep Euphotic zone showed a high richness, as well, averaging around 19 species per sample, whereas average areal cover dropped to around 8% (Figs. 5 and 6). This somewhat lower richness is associated with a slight reduction in frequency of occurrence—to about 70%. Paleotubuliporoid bryozoans were dominant in this zone, with a different suite of trepostome bryozoans also present in smaller quantities. Cornulitids and the inarticulate brachiopod *Petrocrania* were also numerically abundant but, owing to their small size and solitary nature, were not significant players with respect to areal coverage.

The sclerobiont suite of the Dysphotic/Aphotic Zone was markedly different from those of shallower zones (Figs. 5 and 6). Less than 1% of available shell area was covered by sclerobionts; species richness per sample was comparably low with only nine species present. Less than one third of available host substrates were encrusted at all and then usually by only a single minute sclerobiont. Possible fungal etchings and a few trepostome bryozoans with large zooecia were dominant in this zone, though their occurrence and coverage were minimal.

Cluster and nonparametric analyses also suggest an additional zone between the Shallow Euphotic and Deep Euphotic zones that remains undifferentiated by microendolith analyses. This zone has a slightly lower areal coverage, higher variability of cover per shell, and lacks a high abundance of either dominant Shallow Euphotic Zone or Deep Euphotic Zone taxa. More rigorous analyses must be completed before a cause can be inferred, but sedimentological and taphonomic studies suggest these localities may have been more turbid than the other Shallow Euphotic Zone localities.

Middle Devonian Sclerobiofacies, Central New York

As with the Upper Ordovician of the Cincinnati Arch region, the Middle Devonian of eastern North America provides an exceptional venue for studies of sclerobionts and sclerobiofacies (Fig. 8); several studies of sclerobionts on Devonian brachiopods have been carried out (e.g., Sparks et al. 1980; Bordeaux and Brett 1990; Leighton 2001, 2002). Kesling et al. (1980) and Sparks et al. (1980) investigated Middle Devonian sclerobionts on the brachiopod *Paraspirifer bownockeri* (see Fig. 9). This work dissected the likely settlement patterns of sclerobiont larvae and their relationship to their host shell. Bordeaux and Brett (1990) documented sclerobionts on a large sample of Middle Devonian brachiopods (about a dozen species) and demonstrated patterns of host-shell preferences among sclerobionts that were related to shell textures and the presence or absence of punctae in various species. In all cases, preference of settlement location on the host and physical interaction

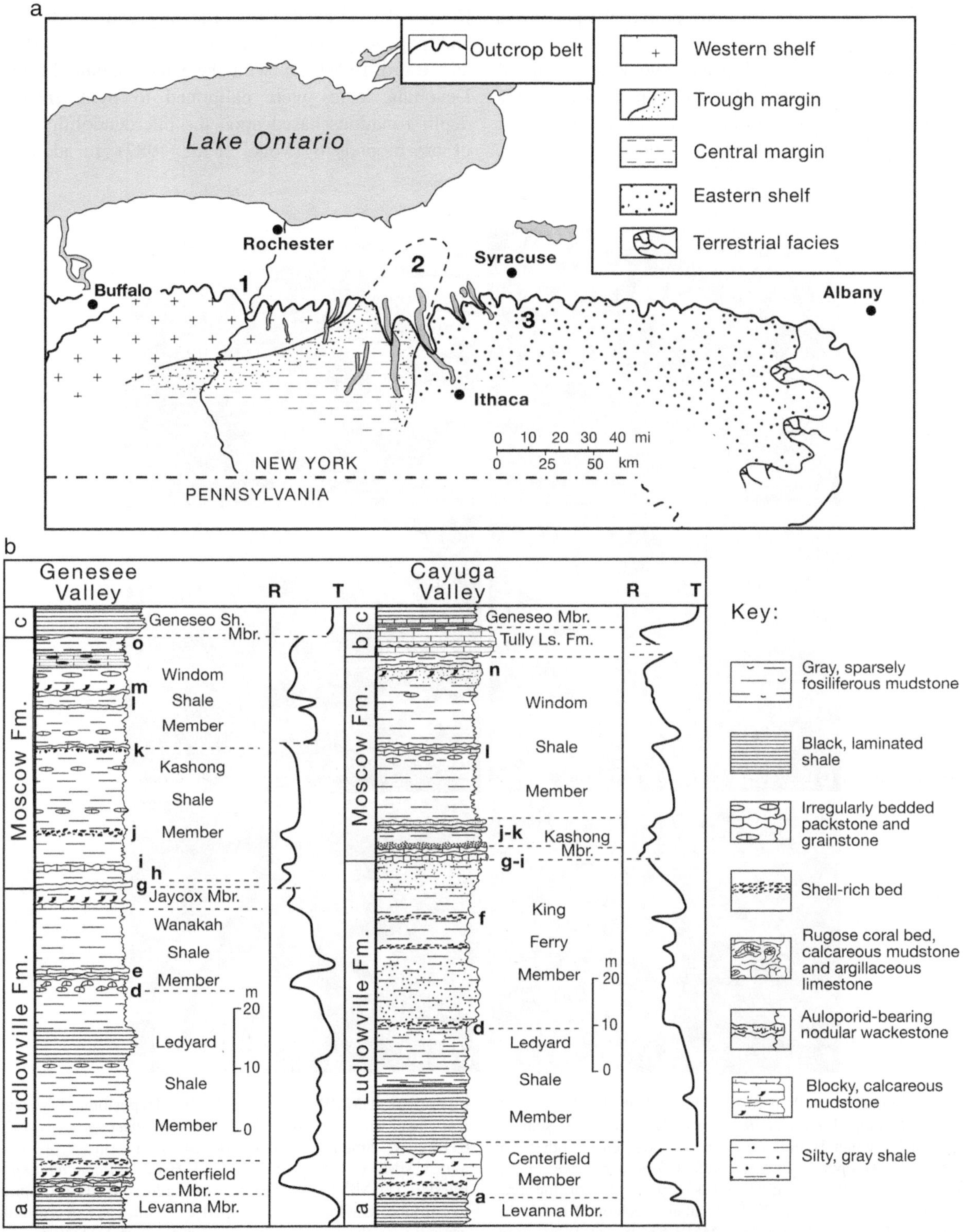

Fig. 7 Study area and stratigraphy for the Middle Devonian. (**a**) Map showing approximate position of foreland basin (2), separating shallow carbonate-rich muddy ramp in western New York (1) from progradational siliclastic shelf (3) in central New York. (**b**) Stratigraphic columns for western and central New York State showing beds from which brachiopods with biont assemblages were obtained

between sclerobionts for that preferred space was documented.

The stratigraphic and paleoecological setting of Middle Devonian Hamilton Group rocks in the Appalachian foreland basin, especially in New York (Fig. 7), have been extensively documented (e.g., Cooper 1957; Brett 1986; Landing and Brett 1991 and papers therein; Brett et al. 1990, 2007; Brett and Baird 1996). As with the Cincinnatian, Middle Devonian facies were calibrated to light-sensitive depth zonations based upon the microendolith study of brachiopods by Vogel et al. (1987). In addition,

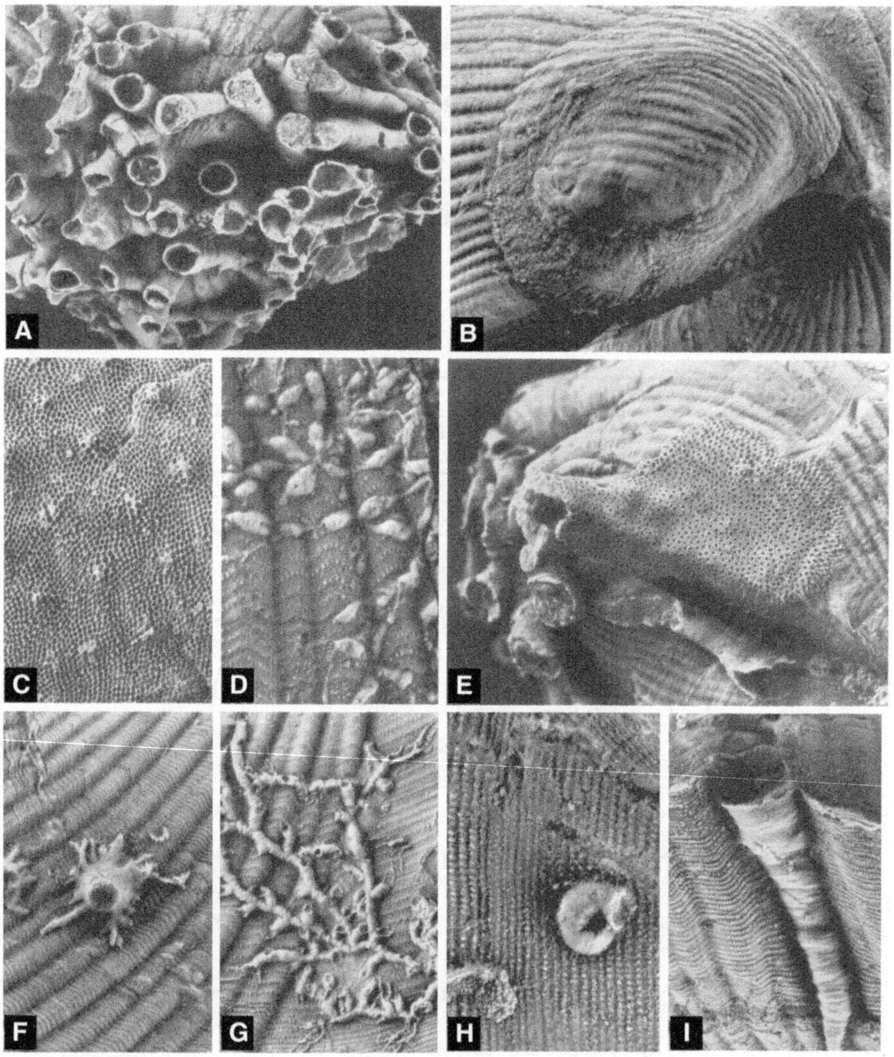

Fig. 8 Common shell-encrusting bionts on the large spiriferid brachiopod *Paraspirifer bownockeri* from the shallow Middle Devonian Hamilton Group. All × 4 unless otherwise noted. (**a**) Auloporid tabulate coral; *Aulopora microbuccinata* Watkins; note that corallites cross commissure of the brachiopod; UMMP 63146. (**b**) Inarticulate brachiopod *Philhedra stewarti* (Fenton and Fenton), UMMP 63298. (**c**) Trepostome bryozoan *Atactotoechus* cf. *A. typicus* Duncan; UMMP 63128. (**d**) ctenostome bryozoan *Ascodictyon fusiforme* (Nicholson and Etheridge), UMMP 63307; × 8. (**e**) Trepostome bryozoan *Leioclema* sp. and coral *Aulopora microbuccinata*, UMMP 63166. (**f**) Small discoidal holdfast of a crinoid or blastoid, UMMP 63175. (**g**) Hederelloid, *Hederella concinnnoides* Bassler; UMMP 63139. (**h**) *Spirorbis planum* Stewart; spiral worm tube; UMMP 63388; X8. (**i**) *Cornulites* sp., worm tube parasitic on the mantle edge of *Paraspirifer*; UMMP 63263. Adapted from Sparks et al. (1980, Plate 10, figs. 1–5; Pl. 11, figs. 4, 3; Pl. 19, fig. 1; Pl. 23, fig. 5)

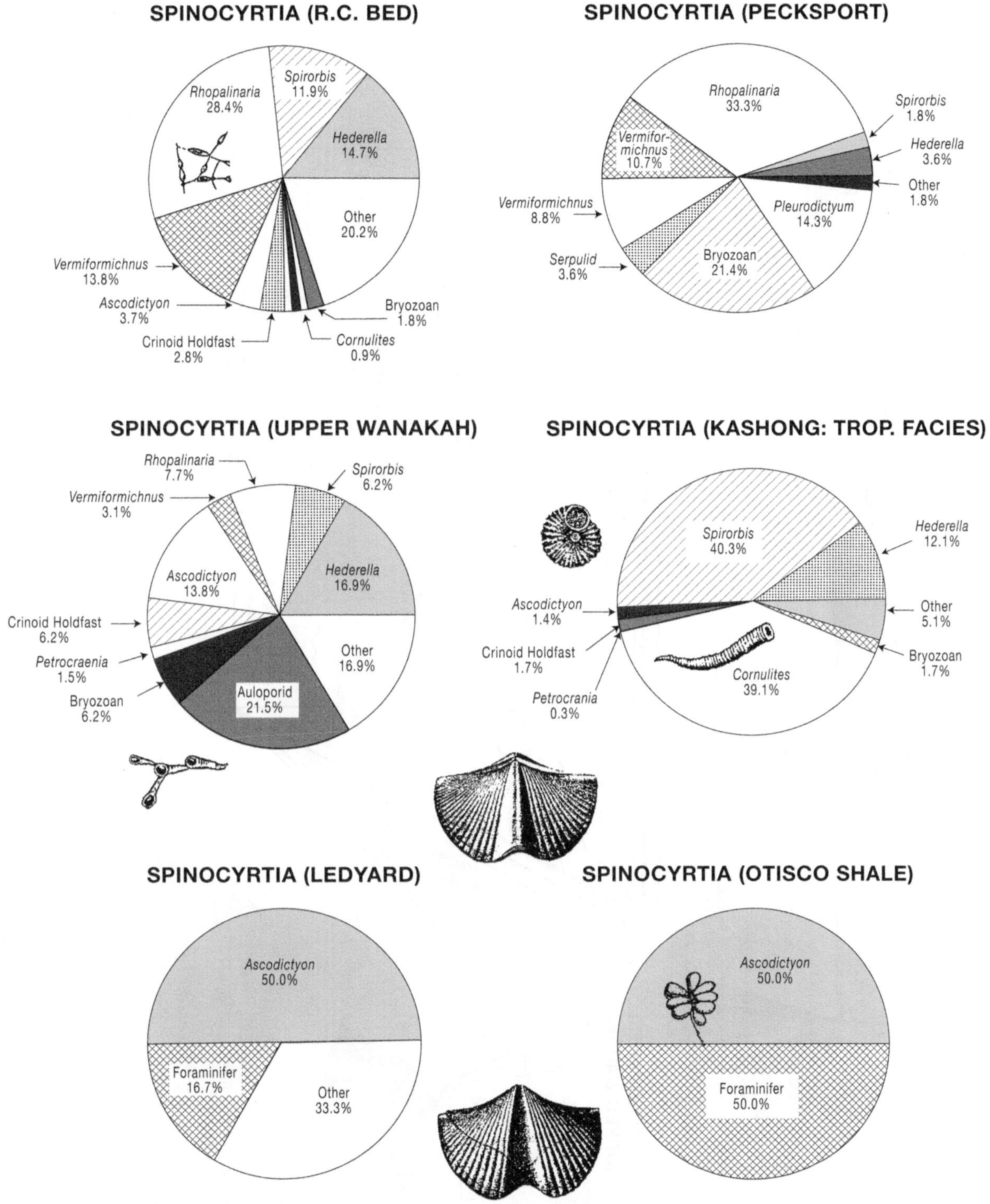

Fig. 9 Sclerobiofacies of the large spiriferid brachiopod *Spinocyrtia granulosa* from various facies of the Middle Devonian Hamilton Group. Pie diagrams show relative proportions of common sclerobionts based on samples of ~20 specimens each. See Fig. 10 for position of these facies

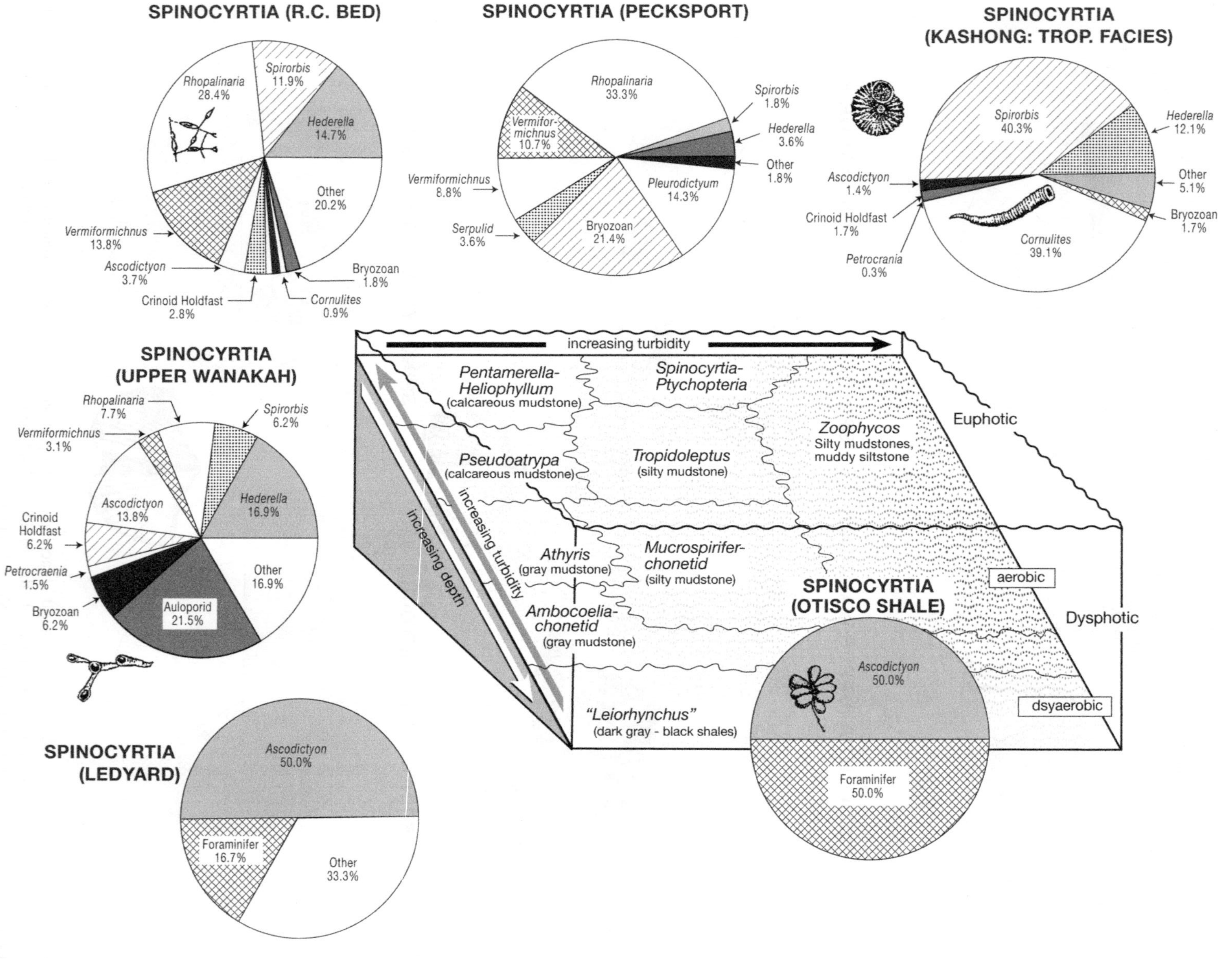

Fig. 10 Biofacies of the Middle Devonian Hamilton Group of western New York in relation to relative water depth and turbidity/sedimentation with associated pie diagrams illustrating the proportion of various biont taxa of *Spinocyrtia* for particular relative depth zones

samples were differentiated according to their estimated position along a gradient of low to high sedimentation, again based on the overall stratigraphic framework and frequency of microendoliths, assuming that at a given water depth the frequency of endolithic borings was related to sedimentation rate (see Vogel et al. 1987 for discussion).

Fortunately, the Hamilton Group also contains several species of brachiopods that occur in multiple differing facies. Preservation of shells is excellent; these shells are commonly heavily encrusted with an array of sclerobionts including auloporid corals, hederelloids, bryozoans, and serpulid and spirorbid worm tubes (Fig. 8). Collections of the large brachiopod *Spinocyrtia* cf. *granulosa* (Conrad) and the closely analogous *Orthospirifer marceyi* (Hall), as well as *Athyris spiriferoides* Conrad and *Ambocoelia umbonata* Hall, spanning an array of facies from their most distal occurrences to the most proximal, were assembled and batches of 20 or more specimens were surveyed for each sample. Counts of sclerobionts and estimates of areal coverage were made as an indication of sclerobiont community structure in each depth zone.

Shallow water brachiopods from the Middle Devonian feature coverages near 20% and frequencies of encrustation between 60 and 100% depending on samples (Brett and Bordeaux "unpublished"; Figs. 9 and 10), at least in higher sedimentation facies (lower in condensed facies). The dominant taxa include many similarities with the Late Ordovician including an abundance of encrusting trepostomes (including monticular forms), cornulitids, *Spirorbis*, *Philhedra*, and other craniids (Figs. 8, 9, and 10). The most abundant encrusters, however, are hederelloids; these do not appear until late in the Silurian. Significantly, Devonian shallow water assemblages from areas of higher sedimentation/turbidity show distinctive sclerobiofacies from those of lower sedimentation.

Specimens of the large spiriferid brachiopod *Spinocyrtia* from relatively condensed, calcareous facies of western New York have a depth-related gradient in sclerobiont coverage and taxonomic richness; this is quite evident (Figs. 9 and 10). Shallowest water specimens from the RC (*Rhipidomella-Centronella*) condensed shell bed of the Kashong Shale (see Bordeaux and Brett 1990 for discussion) actually show lower coverages but have a slightly higher total richness. *Vermiformichnus*, a channel-like boring, is particularly abundant, as well as ctenostome bryozoan etchings—very important in these facies. Corroded surfaces of many valves suggest bioerosion and possible coverage by algae or sponge mats; this may explain the relatively low coverages by bionts in some cases.

For the comparably sized large spiriferid *O. marceyi* (formerly called *Spinocyrtia marceyi*) from similarly shallow water (Shallow Euphotic zone?) but higher sedimentation/more turbid environments, the areal coverage of the host substrate is higher than in the low sedimentation zone (Figs. 9, 10, and 11). The diversity of sclerobionts is somewhat lower in the higher sediment environments, but the types of sclerobionts also change dramatically from ctenostome etchings and *Vermiformichnus* to putatively mud tolerant taxa such as cornulitid and spirorbid worm tubes and hederelloids (Figs. 9 and 10). The lower diversity of species, almost exclusive dominance by filter feeders, near absence of corals except for *Pleurodictyum*, which appears to have been particularly adapted to high sedimentation environments (Brett and Cottrell 1982)—and relatively high abundance of sclerobionts—further suggest eutrophic conditions that favored a thriving, opportunistic assemblage of encrusters (cf. Lescinsky et al. 2002).

Devonian examples of probable Deep Euphotic Zone assemblages show comparable diversity (excluding the strongly split hederelloids from the Silica Shale reported by Sparks et al. 1980). New York examples showed a maximum richness of 14 species, slightly lower than the Ordovician total of 19; this represents a somewhat smaller sample so the difference may not be significant. Coverage averaging around 15–20% is also comparable. Common taxa show some similarities at least at a coarse scale with the early Paleozoic examples: encrusting trepostome bryozoans are moderately common, as are craniid brachiopods, cornulitid tubes, and holdfasts of small crinoids (Figs. 10 and 11). More abundant taxa: auloporid tabulates and hederelloids (possibly phoronids: Taylor and Wilson 2008) are new additions to this assemblage.

Spinocyrtia from the deeper euphotic Wanakah Shale show a maximum of eight taxa per shell and a total generic richness of 14 taxa (Figs. 10 and 11). Ctenostome bryozoans, agglutinated foraminifers, hederelloids, paleotubuliporid bryozoans, auloporid

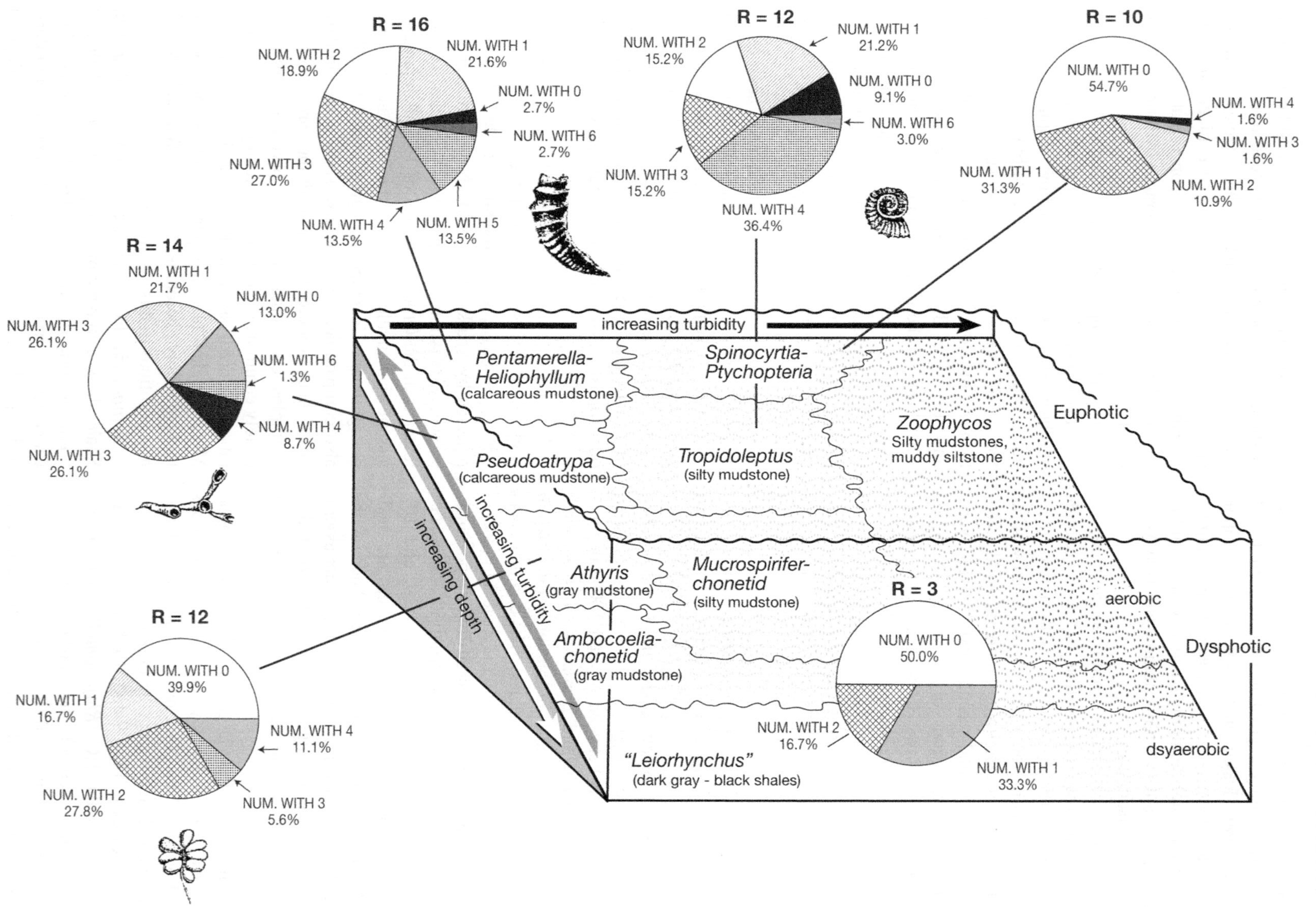

Fig. 11 Biofacies of the Middle Devonian Hamilton Group of western New York in relation to relative water depth and turbidity/sedimentation with associated pie diagrams illustrating the proportions of *Spinocyrtia* shells with various numbers of bionts (e.g., 0 = number of shells with no bionts; 1 = number of shells with one biont, etc.)

tabulate corals, and calcareous worm tubes are the dominant and common sclerobionts. Areal coverage is variable, but can reach nearly 100%, although the average coverage of valves is around 15%.

Relatively small and rare specimens of *Spinocyrtia* obtained from the margin of tolerance in the brachiopod host species in deeper (below storm-wave base), mildly dysoxic, Dysphotic to Aphotic zones in the Ledyard Shale, and Otisco Member were analyzed for the deepest sclerobiofacies (Fig. 10). These show a maximum of three sclerobiont taxa and very low areal coverage. Tubular encrusting foraminifers and ctenostome bryozoan etchings of *Eliasopora* are the only sclerobiont organisms present, with total coverages generally <1%. This is comparable to Late Ordovician Dysphotic sclerobiofacies.

This Devonian example suggests that sclerobiofacies may be useful not only in distinguishing relative light/depth zonations but also more subtle environmental differences related to sedimentation. Although much more controlled data are needed to confirm this, sclerobiofacies may prove useful in distinguishing high sedimentation/eutrophic settings from more starved/oligotrophic settings.

Late Paleozoic Sclerobionts, North American Mid-Continent

Records of encrustation and boring are significantly fewer among late Paleozoic samples (see Brezinski 1984; Sando 1984; Powers and Ausich 1990; Wilson and Palmer 1992; Lescinsky 1994). One notable exception to this is Lescinsky's (1997) study on Carboniferous sclerobiont (mainly epibiont) communities. Mississippian and Pennsylvanian sclerobiont communities from North America represent depositional environments similar (subtropical paleolatitudes, calcareous shale facies, of moderate depth and high faunal diversity) to those examined in the earlier Palaeozoic samples sited above. The common Carboniferous brachiopods included long-hinged spiriferids and *Composita*, comparable respectively to *Spinocyrtia/Orthospirifer/Paraspirifer* and *Athyris* in the Middle Devonian. These brachiopods were studied for diversity, abundance, spatial distribution, and interspecies associations.

Dominant encrusting taxa in examined samples of Carboniferous brachiopods included sheet-like trepostomes and fistuliporoid bryozoans; other epibionts included agglutinated foraminifers, rugose corals, spirorbids, cornulitids, hederelloids, encrusting craniid brachiopods, and some erect bryozoans (Table 2 that shows the areal coverage of a variety of encrusting taxa at Carboniferous localities in the Bahamas).

Mississippian sclerobiont communities are dominated by trepostome, fistuliporoid, and ctenostome bryozoans with trepostomes playing a less dominant role in Lower Mississippian communities and a more dominant role in Upper Mississippian communities (Lescinsky 1997). Pennsylvanian communities are also dominated by fistuliporoid and, to a lesser degree, trepostome bryozoans (Lescinsky 1997). Encrusting foraminifers play a varied role in Pennsylvanian communities, in some instances dominating, in others as a minor component, but present in most (Table 3).

Assuming most samples studied by Lescinsky (1997) were derived from similar offshore shelf environments in the lower euphotic zone, they provide a sense of frequency of encrustation and areal cover in similar habitats late in the Paleozoic. Areal cover in the Lower and Middle Mississippian communities averages around 11% and Upper Mississippian around 18%. Lower Pennsylvanian communities display an average areal coverage near 24%, and Upper Pennsylvanian communities around 16%. To mitigate problems of studied environments not falling appropriately into the Deep Euphotic zone category, combining data to find the average Carboniferous areal coverage seems appropriate; this yields an average areal cover of 17%. Using this technique for Lescinsky's (1997) data, the average encrustation frequency for the Carboniferous was around 47%.

Although it was not Lescinsky's primary objective to examine encrusters across differing environments, there are many aspects of his study which render it comparable in a qualitative sense to those of the earlier Palaeozoic. For example, he controlled for substrate by utilizing a limited number of host brachiopods, primarily larger spiriferids and the smooth athyrid *Composita* as in the Devonian study. The majority of the samples from both the Mississippian and Pennsylvanian appear to be derived from shallow to mid shelf mixed shaley limestones. This is certainly comparable to many Ordovician and Devonian samples. He does not present sufficient data to distinguish carefully among facies, and indeed the similarities of most of the samples in terms of encrusters suggest a limited range of environments comparable to the deeper euphotic zone. Considerable similarities exist with comparable samples from the Hamilton Group and Silica Shale, for example, the high degree of encrustation by sheet-like bryozoans. However, barnacle borings appear abundantly on the Carboniferous brachiopods but are absent on Devonian brachiopods (though present on some platyceratid gastropods; see Baird et al. 1990), whereas hederelloids and auloporid corals are much less common than in the Devonian examples, even though both groups persist.

One set of samples from the Mississippian Pella Shale is distinctive in showing an unusually high relative abundance of the putative annelid tube *Spirorbis* and lesser amounts of encrusting bryozoans. This sample is perhaps comparable to those of the shallow mudstone facies of the Hamilton Group, such as the Kashong Shale, in which spirorbids and cornulitids were particularly abundant while encrusting sheet-like bryozoans were less so.

A second sample among those assemblages analyzed by Lescinsky stands out as distinctly different from the rest. *Composita* specimens from the mid Pennsylvanian Finis Shale showed a low occurrence of bryozoan encrustation and an abundance of agglutinated foraminifera. This sample is comparable to those of assemblages judged to be of the deeper, Dysphotic/Aphotic zone in the Devonian and may well represent an outlier relative to Lescinsky's other samples. In general, Lescinsky did not sample these deeper facies, perhaps because of the paucity of sclerobionts, but this sample suggests persistence of these low diversity foraminifer-dominated samples.

Obviously, more systematic studies of sclerobionts on brachiopods from well-characterized facies are required to better establish sclerobiofacies of the Mississippian and Pennsylvanian. However, these qualitative observations strongly suggest that an array of sclerobiofacies, analogous to those discussed for the Mid Devonian, also persisted into the later Paleozoic.

Mesozoic Sclerobionts

The Permian–Triassic extinction event(s) apparently changed the composition of sclerobiont communities significantly. However, with almost no Permian and Triassic sclerobiont communities reported (Taylor and Wilson 2003), the details of this transition are poorly known. Early Mesozoic sclerobiont assemblages are typified by a specific suite of sclerobiont taxa in level-bottom benthic communities after the Permian–Triassic boundary event(s). Jurassic sclerobiont assemblages vary substantially from Palaeozoic communities. Cementing oyster-like bivalves such as *Plicatula* and *Eopecten* become extremely common, producing thick crusts, along with foraminifers, sponges, and serpulid worms, as well as some ctenostome bryozoans (Kauffman 1978; Harper and Palmer 1993; Fürsich 1980).

Jurassic and Early Cretaceous communities show a dramatic upturn in both epi- and endobiont abundances; their makeup has been shown to be significantly different from earlier Paleozoic communities with inclusion of encrusting bivalves and gastropods (Morton 1965; Fürsich et al. 1994; Harper et al. 1996; Savazzi 1999; Žitt et al. 2003). Organic substrates being used by epi- and endobiont communities at this time change dramatically, incorporating, as well, larger, mobile gastropod and bivalve host substrates.

Cretaceous to modern communities are apparently highly similar, dominated by encrusting foraminifers, serpulid worms, cheilostome bryozoans, and coralline algal encrusters. Cementing bivalves such as oysters remain common. Clionid borings and ctenostome etchings round out the major biont contributors in modern sclerobiofacies; these have not changed drastically since the re-emergence of encrustation in the Jurassic.

Discussion: Sclerobiofacies Through Time

Obviously, the study of sclerobiofacies is in its infancy; accordingly there is as yet only a very limited database upon which to base more general conclusions. We make a very preliminary attempt, however, to compare sclerobiofacies for each of the three major depth/light-related zonations, the Shallow Euphotic, Deep Euphotic, and Dysphotic/Aphotic zones through time using data from the pertinent literature discussed earlier. In the following sections we describe the changing sclerobiont communities through time. These patterns support distinctive groups for the post-Cambrian Palaeozoic Era as opposed to the Mesozoic to Recent (Fig. 12, Table 4).

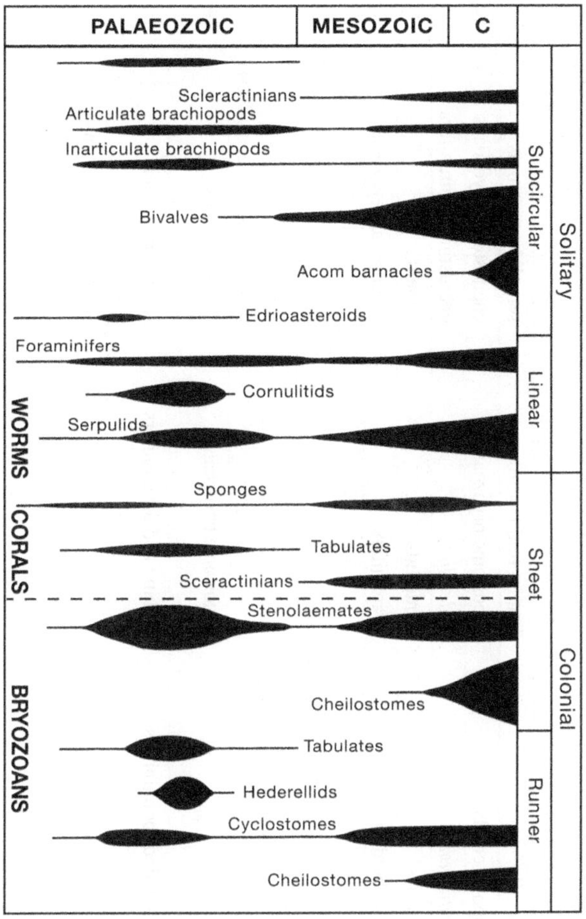

Fig. 12 Temporal ranges and relative abundances of major encrusting biont taxa and growth forms. Modified from Lescinsky (2003)

Cambrian shelly assemblages are distinctive primarily in near absence of any bionts but may show a distinctive suite of rarely preserved encrusters (Fig. 12). The "Cambrian fauna" among sclerobionts is of very low diversity, but includes pelmatozoan echinoderms, microbial mats, and possibly agglutinated foraminifera. Despite the presence of relatively abundant shelly substrates, few skeletonized sclerobionts had yet appeared. There may have been other nonpreservable sclerobionts but, if so, these left no record. The same continued to be true into the Middle Ordovician. Lescinsky (1996) observed no sclerobionts on a large collection of brachiopod shells from the Middle Ordovician (Arenig) Kanosh Shale at Fossil Mountain, Utah.

The Late Ordovician marks the time of major diversification of sclerobiont communities (as with endoliths—see Wilson and Palmer 1992). By the Late Ordovician, diverse preservable encrusting organisms were in existence with greatly increased abundance of bryozoans and calcified tube-dwelling worms as well as encrusting corals, inarticulate brachiopods, holdfast-bearing crinoids, and other pelmatozoans.

Sclerobionts were found to be abundant and diverse on brachiopods and other skeletal remains from the Sandbian (Caradoc; Turinian) Bromide and Benbolt formations of Oklahoma and Virginia, respectively (Lescinsky 1996). Some aspects of modern sclerobiofacies existed for the first time during the Late Ordovician, as did clear-cut variation in microendolith assemblages (Alexander and Scharpf 1990; Lescinsky 1996; Vogel and Brett 2009). For example, we have documented a distinct gradient of sharply decreasing species richness and coverage by sclerobionts in the Dysphotic to Aphotic assemblages. Smrecak and Brett (2008) showed, in Late Ordovician rocks of the Cincinnati arch, areal shell coverage in the Shallow Euphotic zone around 20%, with a frequency of encrustation at 80% (Table 4). On average, every shell possessed at least two sclerobionts, the most common being monticular trepostome bryozoans, trepostome bryozoans with small zooecia, *Constellaria* sp., *Cornulites* sp., and *Philhedra* sp.

The Late Ordovician extinctions appear to have had little effect on sclerobiont assemblages. During the mid-Palaeozoic, as evidenced by Silurian and mid-Devonian assemblages from eastern North America, sclerobiont diversity and coverage levels were similar to those of the Ordovician (Kesling et al. 1980; Liddell and Brett 1982; Alvarez and Taylor 1987; Gibson 1992; Alexander and Brett 1990; Bordeaux and Brett 1990). Many of the same taxa at ordinal or family levels continued to occupy shells that were common in the Late Ordovician, but these were supplemented by newly appearing taxa. New additions included the much more abundant presence of auloporid corals and temporary success of hederelloids (Taylor and Wilson 2008). Cornulitid and spirorbid abundance appears, moreover, to be greater than that of Ordovician assemblages. Partitioning into shallow, intermediate, and deeper assemblages was well developed. Shallow Euphotic zone brachiopods from the Middle Devonian feature coverages similar to those observed in the Late Ordovician (near 20%) and frequencies of encrustation between 60 and 100%, depending on samples (Brett and Bordeaux

Table 4 Comparison of light- (depth-) related sclerobiofacies of different ages

LIGHT/DEPTH ZONE	*ORDOVICIAN* Smrecak and Brett (2008)	*MID PALAEOZOIC* Bordeaux and Brett (ms)	*LATE PALAEOZOIC* Lescinsky (1997)	*MODERN* SSETI and Rodland
	Late Ordovician OH, KY	Mid Devonian Western NY	Miss.-Penn. midcontinent USA	Recent, Lee Stocking Island, Bahamas; Brazilian Bight
UPPER EUPHOTIC				
Generic Richness	max: 24 species	max: 25+ species	max: 20+ species	max: 50
Average % coverage	20%	20–30%	?	18%
Frequency of coverage	80%	n/a		75–100% (Rodland et al. 2004)
Dominant Taxa	*Constellaria*	*Ascodictyon*	(no data)	colonial/larger foraminifers
	monticular trepostomes	*Vermiforichnus*		clionids
	cornuilitids	cornulitids		serpulids
	Vermiformichnus	hederelloids		cheilostome bryozoans
	Philhedra	trepostome bryozoans		encrusting bivalves
LOWER EUPHOTIC				
Generic Richness	max: 19 species	max: 34 species	max: 16 species	max: 70
Average % coverage	12%		17%	14%
Frequency of coverage	66%		47%	0–44% (Rodland et al. 2004)
			1.9	
Dominant Taxa	trepostome bryozoans	*Hederella*	fistuliporoid bryozoans	serpulids/spirorbids
	Cuffeyella	auloporid corals	trepostome bryozoans	cheilostome bryozoans
	paleotubuliporids	*Ascodictyon*	*Hederella*	encrusting foraminifera
	inarticulate brachiopods	spirorbids	cornulitids	clionids
	crinoid holdfasts	crinoid holdfasts	foraminifers	tubuliporid bryozoans
DYSPHOTIC–APHOTIC				
Generic Richness	max: 9 species	max: 3	max:	max: 25
Average % coverage	0.70%	~1.0%		0.90%
Frequency of coverage	6.00%			0–44% (Rodland et al. 2004)
Dominant Taxa	paleotubuliporid bryozoans	*Eliasopora*	ctenostome bryozoans	small serpulids
	large zooecium trepostome bryozoans	agglutinated foraminifers	agglutinated foraminifers	small foraminifers
	ctenostome bryozoans	ctenostome bryozoans		ctenostome bryozoans

"unpublished data"), at least in higher sedimentation facies (lower in condensed facies). The dominant taxa have many similarities with the Late Ordovician including an abundance of encrusting trepostomes, including monticular forms, cornulitids, *Spirorbis*, *Philhedra*, and other craniids. However, the most abundant encrusters are hederelloids; these do not appear until late in the Silurian. Devonian and perhaps Carboniferous shallow water assemblages are further tentatively divisible into "clear" and "turbid" water sclerobiofacies with sheet-like bryozoans dominating space in the cleaner facies and worm tubes dominating in more turbid regions.

Deep Euphotic zones in the Devonian to Carboniferous are characterized by auloporid corals, trepostome bryozoans, and hederelloids. Mississippian and Pennsylvanian sclerobiont communities studied by Lescinsky (1997) show that the frequency of encrustation and areal cover are comparable to similar habitats in the earlier Palaeozoic (Table 4).

Deeper Dysphotic/Aphotic assemblages continue to be dominated by sparse, low-diversity assemblages in the Devonian and apparently also in the later Paleozoic. Areal coverage was generally less than 1% (Table 4).

Details of sclerobiofacies for the Mesozoic remain incompletely known. The loss of many Paleozoic taxa, including trepostome and fistuliporoid bryozoans, tabulate corals, cornulitids, at the Permian–Triassic extinction (Lescinsky 2003) produced major changes in encrusting communities. Among encrusters, only foraminifers, craniid brachiopods, some ctenostome and tubuliporate bryozoans, a few cemented crinoids, and spirorbids (if indeed the later spiral tubes are skeletons of animals comparable to those of the Palaeozoic) persisted across this major boundary (Fig. 12). Conversely, the evolution of new groups of encrusting ostreid and pectenid bivalves early in the Mesozoic permitted population of hard skeletal substrates. Other new taxa had appeared by the late Triassic to Early Jurassic including clionid sponges, scleractinian corals, and serpulid worm tubes. Whereas detailed studies of sclerobiofacies have not yet been carried out, the basic gradients of diversity and coverage were probably maintained, with bivalves, serpulids and tubuliporate bryozoans occupying substantial space in shallow water samples and sparse foraminifer-dominated assemblages persisting in deep water.

The evolution of new groups, especially larger foraminifers and cheilostome bryozoans in the Cretaceous, led to development of sclerobiofacies of modern aspect. Again, detailed studies of Cretaceous and Cenozoic age sclerobiofacies are needed, but some generalities can be made in comparison with modern assemblages.

Modern environments off the coast of between 15 and 88 m are characterized by high generic richness, with over 50 genera of preservable (skeletonized) sclerobionts present. Areal coverage of shells is near 20% on average, despite high levels of energy and sedimentation events throughout the 12 years of study (Tables 2 and 3). This is close to the average coverage noted in studies of shallow assemblages from the Ordovician onward, although the species richness of these assemblages is nearly twice that of early and mid Palaeozoic associations. Dominant encrusting taxa include colonial foraminifers, clionid sponge borings, serpulid worms, cheilostome bryozoans, and encrusting bivalves.

Frequency of encrustation, calculated by Rodland et al. (2004) in the very shallow reaches of the Brazilian Bight between 6 and 30 m, is between 75 and 100%. This is consistent with SSETI shell encrustation frequency, also near 100% in the shallow zones (Brett and Parsons-Hubbard "personal communication"). Encrusting foraminifers and serpulid worms were among the most common sclerobionts seen in Rodland et al. (2004) and by SSETI work, as well.

The Deep Euphotic Zone shows higher generic richness than the shallower with a maximum of 70 different genera present (Table 4). This contrasts with Paleozoic samples in which richness is similar or even declines slightly from shallow to deeper euphotic zones. Most dominant in Bahamian Deep Euphotic zone communities are serpulid and spirorbid worms, followed by cheilostome bryozoans, encrusting foraminifers, some clionid borings, and tubuliporid bryozoans. Areal coverage was calculated to be 14% in the Lower Euphotic Zone; this again is comparable to the few data we have for Paleozoic Lower Euphotic Zone samples. Frequency of encrustation was calculated by Rodland et al. (2004) to range between 0 and 44%, in accord with ancient examples as presently known.

Areal coverage in the Dysphotic Zone off the coast of Lee Stocking Island is 1–5%, with only 25 genera recorded. Dominant sclerobionts include small serpulids, small encrusting foraminifers, and ctenostome bryozoans. Most shells deployed were collected in pristine condition, even after 12 years on the ocean floor (Staff et al. 2002). Rodland et al. (2004) also reported a frequency of coverage between 0 and 44% in deep-water environments. These values are in accord with Early to Middle Palaeozoic samples, though, again, the total richness of skeletonized taxa is greater in the present day.

Conclusions

This review of sclerobiont communities on shells and other skeletal materials in modern and ancient environments permits several very preliminary generalizations. First, skeleton encrusting communities have been relatively well established across a spectrum of environments from the Late Ordovician onward. Prior to that time, shell grounds existed but were not heavily colonized. Indeed, Middle Ordovician hardgrounds support more diverse assemblages than do contemporaneous shell grounds.

Sclerobiofacies, characterized by distinctive suites of sclerobiont taxa and ecological features, can be useful in identifying depositional environments. At least from the Late Ordovician, aspects of species composition, diversity, dominance, areal coverage, and guild structure appear to vary predictably with environment. Parameters related to depth, as well as turbidity and sedimentation rate appear to be particularly important in controlling sclerobiofacies. In particular, light intensity levels, as identified from distinctive endolithic algal, and cyanobacterial and fungal borings appear to be related to differences in sclerobiofacies (Glaub et al. 2001). Some properties of depth-related sclerobiofacies appear persistent over time. The Shallow Euphotic Zone shows relatively high diversities and the highest areal coverage, sometimes up to 100% of preservable organisms; in the Recent, large foraminifers and calcareous algae are dominant organisms in this zone, whereas Early to Middle Paleozoic encrusters on shells from shallow shelf settings are typified by algae, monticular bryozoans, and small corals. The Deeper Euphotic Zone shows the highest diversity although coverage tends to be somewhat lower; it is typified by certain bryozoans, and solitary sclerobionts such as worm tubes. The Dysphotic to Aphotic zone shows marked drop in coverage to very low values; many shells are nearly clean of sclerobionts, others show high dominance (though typically low coverage) by just a few species of small, solitary organisms and small bryozoans. There is some preliminary evidence that at particular depths there may be differentiation of shallow water settings between high and low turbidity settings, the former typified by larger numbers of algae, bryozoans, and corals, the latter by worm tubes and small tubular agglutinated foraminifers.

Many more studies are required to permit detailed comparisons of the evolution of sclerobiont communities through the Phanerozoic. However, we believe that with further refinement, the concept of sclerobiofacies will provide an important tool for palaeoenvironmental reconstruction, more sensitive than the shelly substrates themselves. Such studies will contribute substantially to our understanding of ancient and modern ecosystems and their responses to environmental perturbations of various scales.

Acknowledgments We thank John Talent for encouraging us to submit this paper for the present compendium as this gave us the impetus to collect our ideas about this evolving concept. Anonymous reviewers provided important insights and constructive critiques that helped improve the paper. Our work has benefited greatly from discussions with Mark Wilson and Hal Lescinski. Research on Middle Devonian sclerobionts was aided by Yvette Bordeaux (University of Pennsylvania) when she was an undergraduate at the University of Rochester. Various aspects of this research have been supported by grants from the Donors to the Petroleum Research Fund, American Chemical Society and NSF Grants to C.B. We acknowledge permission to reproduce figures from the University of Michigan Museum of Paleontology.

References

Ager DV (1963) Principles of paleoecology: an Introduction to the study of how and where animals and plants lived in the past. McGraw-Hill, New York

Alexander RR, Brett CE (1990) Symposium on Paleozoic epibionts: an introduction. Hist Biol 4:151–153

Alexander RR, Scharpf CD (1990) Epizoans on Late Ordovician brachiopods from southeastern Indiana. Hist Biol 4:179–202

Alvarez F, Taylor P (1987) Epizoan ecology and interactions in the Devonian of Spain. Palaeogeogr Palaeoclimatol Palaeoecol 61:17–31

Baird GC, Brett CE, Tomlinson JT (1990) Host-specific acrothoracid barnacles on Middle Devonian platyceratid gastropods. In: Alexander R, Brett CE (eds) Paleobiology of epibionts, special issue. Historical biology, vol 4. Taylor and Francis, pp 221–244

Bambach RK, Kelley PH (eds) (2008) From evolution to geobiology: research questions driving paleontology at the start of a new century. Paleontol Soc Spec Publ 14:355

Bordeaux YL, Brett CE (1990) Substrate specific associations of epibionts on Middle Devonian brachiopods—implications for paleoecology. Hist Biol 4:203–220

Brett CE (ed) (1986) Dynamic stratigraphy and depositional environments of the Middle Devonian Hamilton group in New York state part 1. NY State Mus Bull 457:150

Brett CE, Cottrell JF (1982) Substrate selectivity in the Devonian tabulate coral *Pleurodictyum americanum* (Hall). Lethaia 15:248–263

Brett CE, Baird GC (1996) Middle Devonian sedimentary cycles and sequences in the northern Appalachian basin. In: Witzke BJ, Ludvigson GA, Day J (eds) Paleozoic sequence stratigraphy: views from the North American Craton. Geological Society of America Special Paper, vol 306. GSA, Boulder, CO, pp 213–241

Brett CE, Liddell WD, Derstler C (1983) Late Cambrian hard substrate communities from Montana/Wyoming: the oldest known hardground encrusters. Lethaia 16:281–289

Brett CE, Baird GC, Miller KB (1990) A temporal hierarchy of paleoecological processes in a Middle Devonian Epieric sea. In: Miller W III (ed) Paleocommunity temporal dynamics: the long-term development of multispecies assemblages. Paleontological Society Special Publication, vol 5. The Paleontological Society, Boulder, CO, pp 178–209

Brett CE, Bartholomew AJ, Baird GC (2007) Biofacies recurrence in the Middle Devonian of New York state: an example with implications for habitat tracking. Palaios 22: 306–324

Brett CE, Kirchner BT, Tsujita CJ, Dattilo BF (2008) Depositional dynamics recorded in mixed siliciclastic–carbonate marine successions: insights from the Upper Ordovician Kope formation of Ohio and Kentucky, USA. In: Pratt BR, Holmden C (eds) Dynamics of Epeiric Seas. Geological Society of Canada Special Paper, vol 548. The Geological Association of Canada, St Johns, NL, Canada, pp 73–102

Brett CE, Parsons-Hubbard KM, Walker SE, Ferguson C, Powell EN, Staff G, Ashton-Alcox KA, Raymond A (2011) Gradients and patterns of sclerobionts on experimentally deployed bivalve shells: synopsis of bathymetric and temporal trends on a decadal time scale. Palaeogeogr Palaeoclimatol Palaeoecol. Available on-line from 19 May 2011 at doi:10.1016/j.palaeo.2011.05.019

Brezinski D (1984) Upper Misissippian epizoans and hosts from southwestern Pennsylvania. Pa Proc Acad Sci 58: 223–226

Bromley RG (1994) The palaeoecology of bioerosion. In: Donovan SK (ed) The palaeobiology of trace fossils. Wiley, New York, pp 134–154

Bundschuh M (2000) Silurische Mikrobohrspuren. Ihre Beschreibung und Verteilung in verschiedenen Faziesräumen (Schweden, Litauen, Großbritannien und USA). Ph.D. Thesis, FB Geowissenschaften, Fachbereich Geowissenschaften, Johann Wolfgang Goethe-Universität Frankfurt am Main, 129p

Bush A, Bambach RK, Daley GM (2007) Trends in ecospace use in marine fossil assemblages between the mid-Paleozoic and late Cenozoic. Paleobiology 33:76–97

Bush A, Daley GM (2008) Comparative paleoecology of fossils and fossil assemblages. In: Kelley PH, Bambach RK (eds) From evolution to geobiology: research questions driving paleontology at the start of a new century. Paleontological Society Special Publication, vol 14. The Paleontological Society, Boulder, CO, pp 289–318

Callender WR, Staff GM, Powell EN, Parsons-Hubbard KM, Brett CE, Walker SE, Carlson DD, White S, Raymond A, Heise EA (2002) Taphonomic trends along a forereef slope: Lee Stocking Island, Bahamas. I. Location and water depth. Palaios 17:55–65

Coen LD, Luckenbach MW (2000) Developing success criteria and goals for evaluating oyster reef restoration: ecological function or resource exploitation. Ecol Eng 15:323–343

Cooper GA (1957) Paleoecology of the Middle Devonian of the eastern and central United States. In: Ladd HS (ed) Treatise on marine ecology and paleoecology 2, Paleoecology. Geological Society of America Memoirs, vol 67. GSA, Boulder, CO, pp 49–278

Dattilo B, Brett CE, Tsujita CS (2008) Sediment supply vs. storm winnowing in the development of muddy and shelly interbeds from the Upper Ordovician of the Cincinnati region, USA. In: Tapanilla L (ed) Essays in honor of Paul Copper. Canadian Journal of Earth Sciences, vol 45. NRC Publishing, Canada, pp 1–23

Dornbos SQ (2006) Evolutionary paleoecology of early epifaunal echinoderms: response to increasing bioturbation levels during the Cambrian radiation. Palaeogeogr Palaeoclimat Palaeoecol 237:225–239

Fürsich FT (1980) Preserved life positions of some Jurassic bivalves. Paläont Zeit 54:289–300

Fürsich FT, Palmer TJ, Goodyear KL (1994) Growth and disintegration of bivalve-dominated patch reefs in the Upper Jurassic of southern England. Palaeontology 37:131–171

Gibson MA (1992) Some epibiont-host and epibiont-epibiont relationships from the Birdsong Shale Member of the Lower Devonian Ross Formation (west-central Tennesee, USA). Hist Biol 6:112–132

Glaub I, Bundschuh M (1997) Comparative studies on Silurian and Jurassic/Lower Cretaceous microborings. Cour Forsch-Inst Senckenberg 201:123–135

Glaub I, Vogel K, Gektidis M (2001) The role of modern and fossil cyanobacterial borings in bioerosion and bathymetry. Ichnos 8:185–195

Guensburg TE, Sprinkle J (2001) The earliest crinoids: new evidence for the origin of the dominant Paleozoic echinoderms. Geology 29:131–134

Harper EM, Palmer TJ (1993) Middle Jurassic cemented pectenids and the missing right valve of *Eopecten*. J Molluscan Stud 59:63–71

Harper EM, Radley JD, Palmer TJ (1996) Early Cretaceous cementing pectinid bivalves. Cretaceous Res 17:135–150

Holland SM, Miller AI, Meyer DL, Dattilo BF (2001) The detection and importance of subtle biofacies within a single lithofacies: the Upper Ordovician Kope Formation of the Cincinnati, Ohio region. Palaios 16:205–217

Holland SM, Patzkowsky ME (2007) Gradient ecology of a biotic invasion: biofacies of the type Cincinnatian series (Upper Ordovician), Cincinnati, Ohio region, USA. Palaios 22:392–407

Jackson JBC (1977) Competition on marine hard substrata: the adaptive significance of solitary and colonial strategies. Am Nat 111:43–767

Jackson JBC (1983) Biological determinants of present and past sessile animal distributions. In: Tevescz MJS, McCall PL (eds) Biotic interactions in recent and fossil benthic communities. Plenum, New York, pp 29–120

Jennette DC, Pryor WA (1993) Cyclic alteration of proximal and distal storm facies; Kope and Fairview formations (Upper Ordovician), Ohio and Kentucky. J Sedimentary Res 63:183–203

Kauffman E (1978) Evolutionary rates and patterns among Cretaceous Bivalvia. Phil Trans R Soc Lond Ser B 284:277–304

Kay AM, Keough MJ (1981) Occupation of patches in the epifaunal communities on pier pilings and the bivalve *Pinna bicolor* at Edithburgh, South Australia. Oecologia 48:123–130

Kesling RV, Hoare RD, Sparks DK (1980) Epizoans of the Middle Devonian brachiopod *Paraspirifer bownockeri*: their relationships to one another and to their host. J Paleont 34:1141–1154

Kiene WE, Radtke G, Gektidis M, Golubic S, Vogel K (1995) Factors controlling the distribution of microborers in Bahamian reef environments. In: Schuhmacher H, Kiene WE, Dullo W-C (eds) Factors controlling Holocene reef growth: an interdisciplinary approach. Facies, vol 32. Springer, pp 174–188

Landing E, Brett CE (1991) Dynamic stratigraphy and depositional environments of the Hamilton Group (Middle Devonian) Part II. N Y State Mus Bull 469:177

Leighton LR (2001) New example of Devonian predatory drill holes and the influence of brachiopod spines on predator success. Palaeogeogr Palaeoclimatol Palaeoecol 165:53–69

Leighton LR (2002) Inferring predation intensity in the marine fossil record. Paleobiology 28:328–342

Lescinsky HL (1993) Taphonomy and paleoecology of sclerobionts on the scallops *Chlamys hastata* (Sowerby, 1843) and *Chlamys eubida* (Hinds, 1845). Palaios 8:267–277

Lescinsky H (1994) The Late Devonian extinction: effects on shell encrusting communities. Paleobios 16(Suppl 1): 9–10

Lescinsky HL (1996) Early brachiopod associates: epibionts on Middle Ordovician brachiopods. In: Copper PR, Jin J (eds) Brachiopods: Proceedings of the third International Brachiopod Congress, Sudbury, Ontario, Canada (2–5 September 1995). CRC Press, Canada, pp 169–182

Lescinsky HL (1997) Epibiont communities: recruitment and competition on North American Carboniferous brachiopods. J Paleontol 71:34–53

Lescinsky HL (2003) Chapter 4.2.8: Sclerobionts. In: Briggs DEG, Crowther PR (eds) Palaeobiology II. Blackwell, Oxford, pp 460–464

Lescinsky HL, Edinger E, Risk MJ (2002) Mollusc shell encrustation and bioerosion rates in a modern Epeiric sea: taphonomy experiments in the Java Sea, Indonesia. Palaios 17:171–191

Liddell WD, Brett CE (1982) Skeletal overgrowths among epizoans from the Silurian (Wenlockian) Waldron Shale. Paleobiology 8:67–78

Lin J-P, Ausich WI, Zhao Y-L (2008a) Settling strategy of stalked echinoderms from the Kaili Biota (Middle Cambrian), Guizhou Province, South China. Palaeogeogr Palaeoclimatol Palaeoecol 258:213–221

Lin J-P, Ausich WI, Zhao Y-L, Peng J (2008b) Taphonomy, palaeoecological implications, and colouration of Cambrian gogiid echinoderms from Guizhou Province, South China. Geol Mag 145:17–36

Martindale W (1992) Calcified sclerobionts as paleoecological tools: examples from the Recent and Pleistocene reefs of Barbados. Coral Reefs 11:167–177

McKinney F (1995a) Taphonomic effects and preserved overgrowth relationships among encrusting marine organisms. Palaios 10:279–282

McKinney F (1995b) 100 million years of competitive interactions between bryozoan clades: asymmetrical but not escalating. Biol J Linn Soc 56:465–481

Miller AI, Holland SM, Meyer DL, Dattilo BF (2001) The use of faunal gradient analysis for intraregional correlation and assessment of sea-floor topography in the type Cincinnatian. J Geol 109:603–614

Morris RW, Felton SH (1993) Symbiotic association of crinoids, platyceratid gastropods and *Cornulites* in the Upper Ordovician (Cincinnatian) of the Cincinnati, Ohio region. Palaios 8:279–282

Morton JE (1965) Form and function in the evolution of the Vermetidae. Bull Br Mus Nat Hist Zool 11:585–630

Parsons-Hubbard KM, Callender RM, Powell EN, Brett CE, Walker SE, Raymond AL, Staff GM (1999) Rates of burial and disturbance of experimentally-deployed molluscs: implications for preservation potential. Palaios 14: 337–351

Parsons-Hubbard KM, Powell EN, Staff GM, Brett CE, Callender RM, Walker SE (2000) The effect of shell burial on shell preservation and epibiont cover in Gulf of Mexico and Bahamas shelf and slope environments after two years: an experimental approach. In: Aller JY, Woodin SA, Aller RC (eds) Organism–sediment interactions. University of South Carolina, SC, pp 297–314

Patzkowsky ME, Holland SM (2007) Diversity partitioning of a Late Ordovician marine biotic invasion: controls on diversity in regional ecosystems. Paleobiology 29: 573–580

Powell EN, Callendar WR, Staff GM, Parsons-Hubbard KM, Brett CE, Walker SE, Raymond A, Ashton-Alcox K (2008) Molluscan shell condition after eight years on the seafloor—taphonomy in the Gulf of Mexico and Bahamas. J Shellfish Res 27:191–225

Powers RG, Ausich WI (1990) Epizoan associations in a Lower Mississippian paleocommunity (Borden Group, Indiana, USA). Hist Biol 4:245–265

Rasmussen KA, Brett CE (1985) Taphonomy of Holocene cryptic biotas from St. Croix, Virgin Islands: information loss and preservational biases. Geology 13:551–553

Richards RP (1972) Autecology of Richmondian brachiopods (Late Ordovician of Indiana and Ohio). J Paleont 46:386–405

Richards RP (1974) Ecology of the Cornulitidae. J Paleont 48:514–523

Rodland DL, Kowalewski M, Carroll M, Simoes, MG (2004) Colonization of a 'lost world': encrustation patterns in modern subtropical brachiopod assemblages. Palaios 19:381–395

Rudwick MJS (1970) Living and fossil brachiopods. Hutchinson, London, 199p
Sando WJ (1984) Significance of epibionts on horn corals from the Chainman Shale (Upper Mississippian) of Utah. J Paleont 58:185–196
Savazzi E (ed) (1999) Functional morphology of the invertebrate skeleton. Wiley, Chichester, New York, Weinheim, Brisbane, Singapore, Toronto, xi + 706p
Schumann D (1967) Der Lebensweisevon *Mucrospirifer* Grabau 1931 (Brachiopoda). Palaeogeogr Palaeoclimatol Palaeoecol 4:381–392
Smrecak TA, Brett CE (2008) Discerning patterns in epibiont distribution across a Late Ordovician (Cincinnatian) depth gradient. Geological Society of America Abstract with Programs, 42nd Meeting, vol 40, p 424
Sparks DK, Hoare RD, Kesling RV (1980) Epizoans on the brachiopod *Paraspirifer bownockeri* (Stewart) from the Middle Devonian of Ohio. University of Michigan Museum Paleontology, Papers on Paleontology, vol 23, 105p
Spjeldnaes N (1984) Epifaunal as a tool in autecological analysis of Silurian brachiopods. Spec Pap Palaeontol 32:225–235
Staff GM, Callender WR, Powell EN, Parsons-Hubbard KM, Brett CE, Walker SE, Carlson DD, White S, Raymond A, Heise EA (2002) Taphonomic trends along a forereef slope: Lee Stocking Island, Bahamas. II. Time. Palaios 17:50–65
Taylor PD (1990) Encrusters. In: Briggs DEG, Crowther PR (eds) Palaeobiology: a synthesis. Blackwell, Oxford, pp 346–351
Taylor PD, Wilson MA (2003) Palaeoecology and evolution of marine hard substrate communities. Earth-Sci Rev 62:1–103
Taylor PD, Wilson MA (2008) Morphology and affinities of hederelloid "bryozoans". In: Hageman SJ, Key MM Jr, Winston JE (eds) Bryozoan studies 2007: proceedings of the 14th international bryozoology conference, Boone, North Carolina (July 1–8, 2007). Virginia museum of nature history special publication, vol 15, pp 301–309
Vogel K (1993) Bioeroders in fossil reefs. Facies 28:109–114
Vogel K, Brett CE (2009) Record of microendoliths in different facies of the Upper Ordovician in the Cincinnati Arch region USA: the early history of light-related microendolithic zonation. Palaeogeogr Palaeoclimatol Palaeoecol 281:1–24
Vogel K, Golubic S, Brett CE (1987) Endolith associations and their relation to facies distribution in the Middle Devonian of New York state. Lethaia 20:263–290
Vogel K, Bundschuh M, Glaub I, Hofmann K, Radtke G, Schmidt H (1995) Hard substrate ichnocoenoses and their relation to light intensity and marine bathymetry. N Jb Geol Paläont Abh 195:46–91
Vogel K, Balog S-J, Bundschuh M, Gektidis M, Glaub I, Krutschinna J, Radtke G (1999) Bathymetrical studies in fossil reefs, with microendoliths as paleoecological indicators. Profil 16:181–191
Vogel K, Gektidis M, Golubic S, Kiene WE, Radtke G (2000) Experimental studies on microbial bioerosion at Lee Stocking Island, Bahamas and One Tree Island, Great Barrier Reef, Australia: implications for paleoecological reconstructions. Lethaia 33:190–204
Wilson MA, Palmer TJ (1992) Hardgrounds and hardground faunas. Institute Earth science publications, vol 9. University of Wales, Aberstwyth, pp 1–131
Žitt J, Kopáčová, Nekovařík Č (2003) Epibonts of mollusk shells from the Korycany limestones (Upper Cenomanian, Czech Republic). Bull Geosci 78:41–52

Brachiopods and Their Auloporid Epibionts in the Devonian of Boulonnais (France): Comparison with Other Associations Globally

Bruno Mistiaen, Denise Brice, Mikolaj K. Zapalski, and Christian Loones

Abstract

From research undertaken on brachiopods encrusted by auloporids at two intervals in the Frasnian (Beaulieu and Ferques Formations) of Boulonnais (France), we conclude that the most important factors influencing encrustation were usually (in decreasing order): size of the host, morphology, feeding currents, shell orientation in vivo and shell ornament. Shell structure (punctate or impunctate) seems not to have influenced the rate of encrustation. In the cases discussed, most but not all auloporids encrusted their hosts in vivo. The most favourable environments for development of both brachiopods and epibionts seem to have been hard bottoms and thin limestone beds alternating with argillaceous beds. The influence of environmental (sedimentologic) conditions is difficult to precisely specify in terms of the development of encrustation, but seems to have played an important role.

Résumé

Les recherches entreprises sur les brachiopodes encroûtés par des auloporides, dans deux niveaux repérés dans le Frasnien (formations de Beaulieu et de Ferques, Boulonnais, France) nous conduisent aux conclusions suivantes: les facteurs les plus importants influençant les incrustations, sont, en général et par ordre décroissant, la taille de l'hôte, sa morphologie, les courants nourriciers, la position de vie et finalement l'ornementation. La structure de la coquille (ponctuée ou imponctuée) ne semble pas avoir d'influence sur le taux d'incrustation. Dans les exemples analysés, la plupart des auloporides se sont développés du vivant du brachiopode-hôte. L'environnement le plus favorable au développement des brachiopodes et de leurs épibiontes semble correspondre à des fonds marins indurés présentant une alternance de minces dépôts calcaires et argileux. Ce dernier facteur semble jouer un rôle important dans le développement de l'incrustation.

B. Mistiaen (✉)
Laboratoire de Paleontologie stratigraphique FLST and ISA, UMR 8014 CNRS, 59046 Lille cedex, France
e-mail: b.mistiaen@isa-lille.fr

J.A. Talent (ed.), *Earth and Life*, International Year of Planet Earth,
DOI 10.1007/978-90-481-3428-1_7,

Keywords
Devonian • Frasnian • Boulonnais (France) • Iowa • Encrustation • Auloporids • Brachiopods • Ornament • Orientation

Introduction

Brachiopods are common Palaeozoic marine invertebrates, particularly abundant in shallow Devonian seas. They have provided large numbers of fossils and were often associated with other organisms for which they served as hosts. Importantly, the host brachiopods were not merely a substratum for settling, but played an important role in palaeoecological relationships with their epibionts. Symbiosis and commensalism (Ager 1961, 1963; Alvarez and Taylor 1987; Galle and Parsley 2005), competition (Kesling et al. 1980; Le Menn 1989; Zapalski 2005), and parasitism (Biernat 1961; Ager 1963; MacKinnon and Biernat 1970) are most often reported. Various organisms (hard-substrate settlers) showed preference for sculpture or size as important factors (e.g. Brett and Cottrell 1982; Bordeaux and Brett 1990; Zapalski 2005). Such substrate specificity is also known in modern scleractinian corals colonizing gastropod shells (Wilson 1976; Plusquellec 2007).

Two of the present authors (Brice and Mistiaen 1992) studied the epibionts on brachiopod shells from the Upper Devonian (Frasnian) Beaulieu and Ferques formations of the Boulonnais, in northern France (Fig. 1) where encrusted fossils are common compared with fossils from Middle Devonian deposits in the same area.

In that analysis, the authors recognize four lithostratigraphic levels (Fig. 2) with particularly abundant brachiopod epibionts, especially auloporid tabulate corals (*Plexituba briceae*, *P. cucullina*, *Aulocystis* sp. and *Thecostegites bouchardi*). Other kinds of epibionts are also present (in decreasing order of abundance): *Spirorbis*, *Petrocrania*, diverse bryozoans and *Cornulites*.

Since preparation of the paper by Brice and Mistiaen (1992), large numbers of new fossils were collected by one of the authors (CL) from the levels discussed herein (especially from the 2nd level of Brice and Mistiaen 1992). Discussion of part of this material has already been published (Brice and Loomes 2002; Mistiaen 2002; Mistiaen et al. 2002).

The aim of this paper is to focus on two of the four recognized levels (2nd and 4th levels) where encrusted brachiopod hosts and auloporid tabulate epibionts are particularly abundant. Data for these two levels are compared with those for the 1st and 3rd levels, as well as with less extensive data about other levels in the Middle Devonian (Givetian) of Ferques. We also attempt to compare these associations with coeval associations from other parts of the globe.

Analysis of the 2nd and 4th Levels

The two lithostratigraphic levels of the Frasnian (2nd and 4th sensu Brice and Mistiaen 1992) exhibit the highest encrustation rates on brachiopods. They are analysed below with special regard to their
(a) stratigraphic and lithologic setting;
(b) faunal composition (host and epibiont);
(c) rate or frequency of encrustation and the locations of the epibionts;
(d) factors influencing encrustation.

Analysis of the 2nd Level

Stratigraphic and Lithological Settings

The 2nd fossiliferous level (Fig. 2), about 5 m thick, is situated in the middle part of the Pâtures Member (Beaulieu Formation), although it belongs more precisely to the upper part of interval c (= "lithologic set"). The fauna is abundant in several argillaceous beds associated with some thinner calcareous levels. The base of the Pâtures Member can be correlated with the base of the *punctata* Zone, i.e. the former base of the Middle *asymmetricus* Zone according to Bultynck (in Brice et al. 1981; Becker 2002).

Faunal Composition of the 2nd Level

Brachiopods of the 2nd level are abundant and diverse; they represent assemblage 8 of Brice (2003). Of the 28 identified brachiopod species, six (24.4%) are encrusted (Fig. 3): large *Cyrtospirifer*

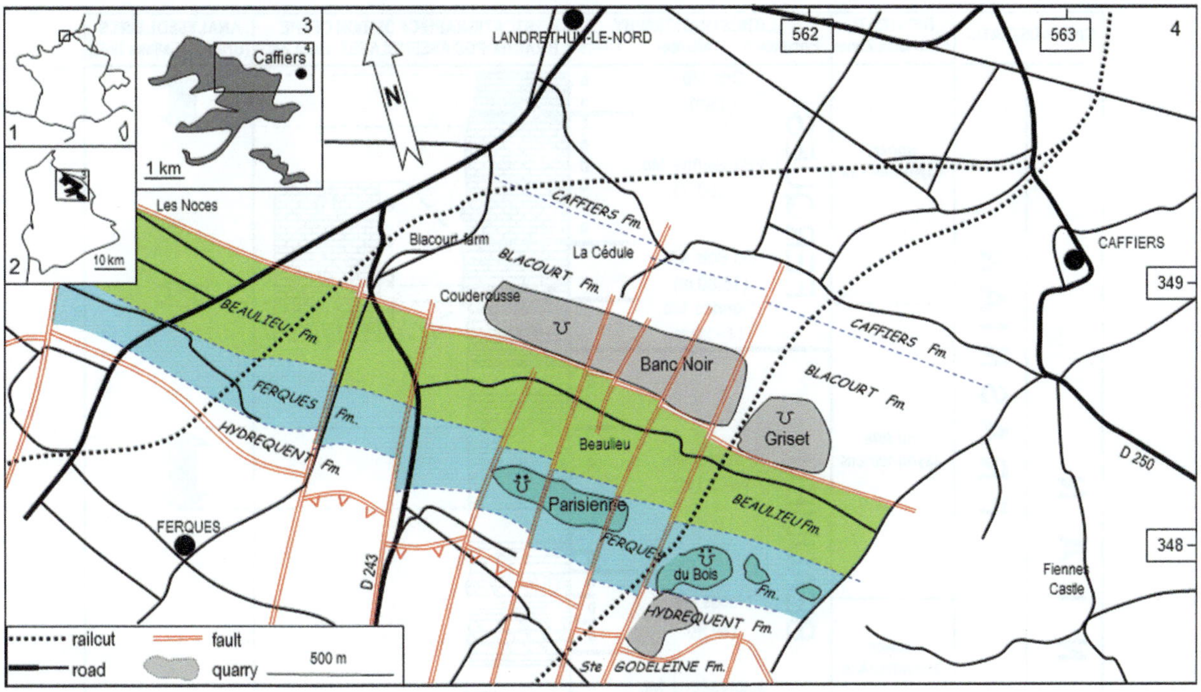

Fig. 1 Geographical and geological setting of the different quarries in Ferques area (Boulonnais, North of France)

syringothyriformis, four medium-sized species *Schizophoria striatula* and three atrypid species, *Desquamatia* (*Seratrypa*) *belliloci*, *Spinatrypina pratorum*, *S.* sp. nov. and the small *Cyrtospirifer archiaci*. Besides common brachiopods, other groups of animals are also present: rare lamellar stromatoporoids, small solitary and phaceloid rugose corals, lamellar and rare massive tabulate corals, crinoids, rare remains of trilobites and ammonoids.

Rate or Frequency of Encrustation and Locations of the Epibionts

P. briceae is the largest and predominant auloporid-encrusting brachiopod in the 2nd level. The coralla of this species are composed of regularly bifurcating, conical corallites 5–7 mm long (Mistiaen 1988). Another auloporid with larger corallites, *Aulocystis* sp., is very rare, but is present (19 corallites) on one shell.

The complete brachiopod shells (327) from this level belong to the six species cited above; 97 of them are encrusted with 2024 auloporid corallites—a mean occurrence of 20.86 corallites per host.

Cyrtospirifer syringothyriformis is the most encrusted brachiopod species—a mean occurrence of 28.91 auloporid corallites per encrusted shell (Fig. 4 and Plate 1, figs. 1–4).

S. striatula is the second most-encrusted species, with a mean occurrence of 17.09 auloporid corallites per encrusted shell (Fig. 5 and Plate 1, figs. 5–8). It is the species with the highest ratio of encrustation (39.8%) of the population.

In contrast, the atrypids (represented here by three species) have a mean occurrence of only 6.88 auloporid corallites per encrusted shell (Fig. 6).

Surprisingly, *Cyrtospirifer archiaci*, a small spiriferid brachiopod, has a mean occurrence of 33.00 auloporid corallites per encrusted shell. However, this species is very rare in the 2nd level; it is represented by only seven specimens, only two of which are encrusted with auloporids. These data should therefore be treated with caution.

Factors Influencing Encrustation

Shell Structure

All encrusted species, except the punctate *S. striatula*, have impunctate shells. Some authors suggest that the punctate structure allowed extrusion of a fouling secretion through the caecae—corresponding, according to

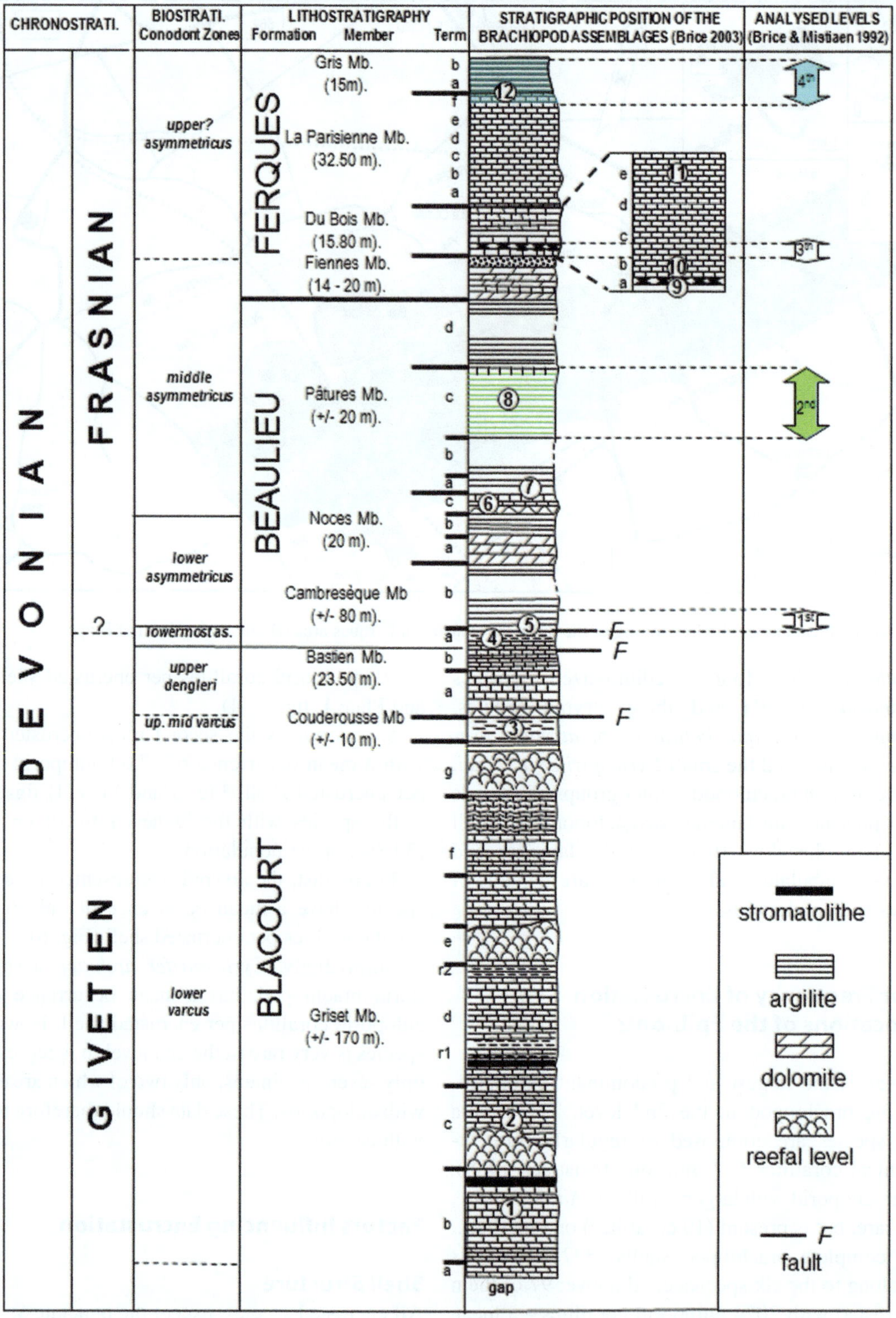

Fig. 2 Lithological subdivisions of the Devonian (Givetian and Frasnian) succession in Ferques area, with location of the four analysed levels with epibionts (Brice and Mistiaen 1992), and position of the 12 brachiopod assemblages (Brice 2003)

2nd level	*Cyrtospirifer syringothyriformis*	*Schizophoria striulata*	atrypid (3 species)	*Cyrtospirifer archiaci*
Systematic group	spiriferid	orthid	atrypid	spiriferid
Shell structure	impunctate	punctate	impunctate	impunctate
Ornament	faint costae	faint ribs	faint ribs growth lines	faint costae
Fixation mode	pedicle	pedicle	pedicle	pedicle
Number of shells with auloporids	**43**	**35**	**17**	**2**
Total number of complete shells	157	88	75	7
% of the population with auloporids	27.4	39.78	22.7	28.6
Total number of auloporid corallites	1224 + 19 * 1243	598	117	66 **
Mean number of auloporid corallites relative to the encrusted shells	28.9	17.1	6.9	33.0

Fig. 3 Brachiopod species of the 2nd level encrusted by auloporids. The six species are cited according to the number of collected encrusted shells. Data of the table are commentated below. *1224 *Plexituba briceae* and only 19 *Aulocystis* sp. **N.B. In Fig. 2 of Brice and Mistiaen (1992), there is an error: the number "2" represents the number of encrusted shells and not the number of all corallites present on the two specimens

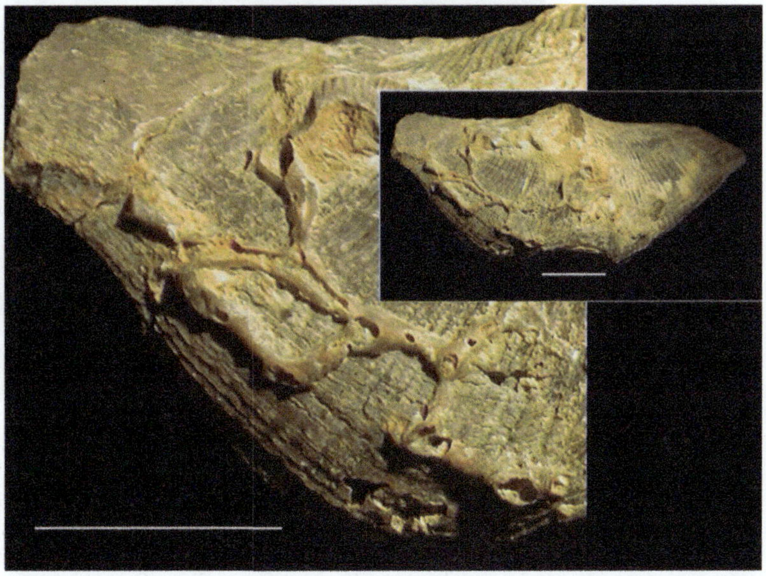

Fig. 4 *Cyrtospirifer syringothyriformis* (Paeckelmann). Specimen MGL 1302, coll. Loones. Beaulieu Fm., des Pâtures Mb., c "term", small temporary trench North of La Parisienne quarry. Dorsal valve. General view and detail showing the development of auloporid epibionts along the growth lines of the brachiopod shell. (Scale bar = 1 cm)

Curry (1983), to a negative factor for encrustation. They state that "terebratulids secrete mucopolysaccharides via their caecae to repel microborers" (cf. Bordeaux and Brett 1990, p. 209). This does not seem to be the case for *S. striatula* which has the largest ratio of encrusted shells (see above).

Size

The six encrusted species are clearly the largest of the 28 species recognized at this level, except *C. archiaci*. The 22 remaining species (all devoid of epibionts), are pentamerids, rhynchonellids, productids of intermediate size, and also a great number of very small

Plate 1 Illustrations of the encrusted brachiopods of the 2nd level. Beaulieu Fm., Pâtures Mb., c "term", Frasnian. [Scale bar = 1 cm]. Figures (1–4) *Cyrtospirifer syringothyriformis* (Paeckelmann) (1) Specimen MGL 1286, coll. Loones. Ventral valve. Encrustation by auloporids, *Petrocrania* sp. and bryozoans. Auloporids are located in the sulcus, preferentially along the anterior commissure but do not cross it; they cover partly *Petrocrania* sp. and are covered by bryozoan (left part of the picture). (2) Specimen MGL 1287, coll. Loones. Dorsal valve. Some auloporids developed along the commissure, none on the ventral valve. (3) Specimen MGL 1288, coll. Loones. Dorsal valve. Encrusted auloporids on a juvenile specimen, the last one corallite along the commissure. (3a) detail. (4) Specimen MGL 1289, coll. Loones. Ventral valve. Encrusted auloporids on the left part of the shell with corallites towards the anterior commissure. Figures (5–8) *Schizophoria striatula* (Schlotheim) (5) Specimen MGL 1290, coll. Loones. Dorsal valve. Encrustation by auloporids, *Petrocrania* sp. and *Spirorbis* sp. Auloporids are preferentially located on the top of the valve and covered *Petrocrania* sp. *Petrocrania* sp., partly covers *Spirorbis* sp. 5a: detail. (6) Specimen MGL 1291, coll. Loones. Dorsal valve. Some auloporids, numerous *Petrocrania* sp. (5), bryozoan and *Spirorbis* sp. several epibionts are again associated on this shell, but relation (anteriority/posteriority) are not clear. (7) Specimen MGL 1292, coll. Loones. Dorsal valve. Auloporids alone, preferentially developed on the top of the valve and growing in different directions. (8) Specimen MGL 1293, coll. Loones. Dorsal valve. Auloporids, bryozoans, and *Spirorbis* sp. Bryozoans are partly developed on *Spirorbis* sp. and auloporids

Fig. 5 *Schizophoria striatula* (Schlotheim). Specimen MGL 1303, coll. Loones. Beaulieu Fm., des Pâtures Mb., c "term", small temporary trench North of La Parisienne quarry. Dorsal valve. Numerous auloporid epibionts. The ventral valve is devoid in encrusted auloporid. (Scale bar = 1 cm)

Fig. 6 *Spinatrypina pratorum* Godefroid. Specimen MGL 1304, coll. Loones. Beaulieu Fm., des Pâtures Mb., c "term", small temporary trench North of La Parisienne quarry. Dorsal valve. Numerous auloporid epibionts in spite of the small size of the shell. Auloporid epibionts are developed on the higher part of the shell, with the corallites growing towards the anterior commissure, the two last corallites crossing the commissure and demonstrating the existence of a post-mortem development. (Scale bar = 1 cm)

Fig. 7 *Cyrtospirifer syringothyriformis* (Paeckelmann). Specimen MGL 1305, coll. Loones. Beaulieu Fm., des Pâtures Mb., c "term", small temporary trench North of La Parisienne quarry. Anterior view. Auloporid epibionts developed on the fold of the dorsal valve, and along the anterior commissure. The ventral valve is devoid in encrusted auloporid. (Scale bar = 1 cm)

species, such as *Eoschuchertella*? *elegans*, *Longispina mailleuxi*, and *Eleutherokomma barroisi*.

It is interesting to compare the two most-represented species at this level, *C. syringothyriformis* and *S. striatula*; they are also the two largest species. *C. syringothyriformis* is the larger species, but is not the host most often chosen (27.4% of the population have auloporids versus 39.8% for *S. striatula*). In contrast, the mean number of corallites per shell is greater for *C. syringothyriformis* (28.91) than for *S. striatula* (17.07). Consequently, encrustation seems to have been easier on *S. striatula* than on *C. syringothyriformis*, although thereafter, the conditions were more favourable for epibionts on the latter. It presents a larger surface for encrustation, but the surfaces are never completely covered. Other factors must therefore have played a role in the process of encrustation.

Auloporids often occur close to the anterior commissure of *C. syringothyriformis* (Fig. 7). In some cases (Fig. 4 and Plate 1, figs. 3, 3a), the development of auloporid corallites seems to follow successive growth lines more or less regularly, a feature already observed by Alvarez and Taylor (1987) and Zapalski (2005). Consequently, the initial encrustation occurred when the shell was smaller. It can be inferred that size is a primary criterion in the choice of a substratum, but not the only one.

Ornament

The two most-encrusted species, *C. syringothyriformis* and *S. striatula*, have subdued well-developed costae and only faint ribs, respectively. Atrypids of a similar size to *S. striatula* are less frequently chosen for encrustation, presumably because their reticulate ornament (faint ribs and growth lines) did not favour settling of coral larvae (Fig. 8a–c).

The ornament of the host shells, in this case, is not an important factor in encrustation.

Shell Morphology and Host Life Position

The six encrusted species were fixed by pedicles, according to their life positions; both valves were therefore available for settling of epibionts. However, usually the dorsal valve—the most convex valve, with a median fold in some species (spiriferids)—is more encrusted (Brice and Mistiaen 1992, figs. 5a, b and 8a). This has already been noted by several authors (Grabau 1899, p. 139; Ager 1961, p. 2).

Feeding Currents

Water currents generated by brachiopods fluctuate according to the position and size of the lophophore. These currents would presumably be greater in spiriferid brachiopods than in the appreciably smaller atrypid or orthid brachiopods. This may explain the distribution of auloporids along the growth lines on the anterior extremity of the fold and along the anterior commissure (Ager 1961, p. 3; Zapalski 2005) where feeding currents occur. It is important to emphasize that very rarely (Fig. 6), a corallite is found crossing the anterior commissure. Some authors (Pitrat and Rogers 1978, p. 1317) stress that the post-mortem development of the auloporid allows it to "spread across the commissure".

This last observation argues in favour of development of epibionts during the life of the host, as suggested previously by several authors (Ager 1961; Schumann 1967; Alexander 1986; Alvarez and Taylor 1987; Cuffey et al. 1995; Fagerstrom 1996). However, this distribution cannot be generalized because auloporid corallites are often variously oriented (Fig. 9).

It should be noted here that the morphology and geometric growth of *P. briceae* (and more generally of all auloporids), with 5–7 mm long corallites, make it impossible for successive corallites to always develop in the best location.

This is confirmed by the observations of several authors: the picture presented by Pitrat and Rogers (1978) undoubtedly shows such positioning of an epibiont. However, Kesling et al. (1980, p. 1145) stated that "in the majority of specimens it is impossible to determine if the epizoan growth is on a live or dead host" (see also Zapalski 2005).

Analysis of the 4th Level

Stratigraphic and Lithologic Settings

The 4th level is situated in the upper part of the Ferques Formation; it corresponds to a specific interval, the upper set of the Parisienne Member (about 6.5 m thick) consisting of nodular limestone, and to the Gris Member (about 15 m thick) consisting of very fossiliferous, argillaceous nodular grey limestone in the lower part; the proportion of limestone and the abundance of fauna decreases towards the top (Brice et al. 1979b). Concerning allocation to conodont

Fig. 8 Relations between encrusted auloporids and different host ornamentations. The three chosen examples show a relative independence in the growth of corallites according to the shell ornamentation. (**a**) *Cyrtospirifer syringothyriformis* (Paeckelmann). Specimen MGL 1306, coll. Loones. Beaulieu Fm., des Pâtures Mb., c "term", small temporary trench North of La Parisienne quarry. Detail of Fig. 9. Auloporid corallites encrusted on lateral part of the dorsal valve. (Scale bar = 1 cm). (**b**) *Schizophoria striatula* (Schlotheim). Specimen MGL 1307, coll. Loones. Beaulieu Fm., des Pâtures Mb., c "term", small temporary trench North of La Parisienne quarry. Ventral valve. Auloporid corallites developed near the sulcus. (Scale bar = 1 cm). (**c**) *Spinatrypina pratorum* Godefroid. Specimen MGL 1304, coll. Loones. Id° to Fig. 6. This species of atrypid is characterized by a poorly reticulate ornamentation. (Scale bar = 1 cm)

zones, the Gris Member has produced *Ancyrognathus triangularis euglypheus* (Coen in Brice et al. 1981). In modern taxonomy (Klapper 1990), this record may refer to either *Ag. tsieni* Mouravieff or *Ag. coeni* Klapper, which range from MN Zone 8 to the middle of MN Zone 11 (Becker 2002).

Faunal Composition of the 4th Level

In the 4th level, the brachiopod fauna (12th assemblage in Brice 2003) is less diverse than in the 2nd level (12 identified species versus 28 in the 2nd level), but the specimens occur in greater abundance. Six species—five presented in 1992 and another, *Spinatrypa longispina*, recently obtained for the first time by one of the authors, CL—are encrusted with auloporids, i.e. 50 versus 24.4% in the 2nd level.

Two particularly large species, *Cyrtospirifer verneuili* (Plate 2, figs. 1–4) and *Athyris murchisoni* (= *Athyris concentrica murchisoni* Brice 1988; Plate 2, figs. 5–8), are common at this level and aggregate 43% of the population (18.8 and 24.5%, respectively; Brice 2003). They are also the most encrusted species.

The remaining encrusted species are *Douvillina dutertrei* (Plate 2, fig. 9), not abundant at this level (2.7% of the assemblage); *Nervostrophia latissima*, a

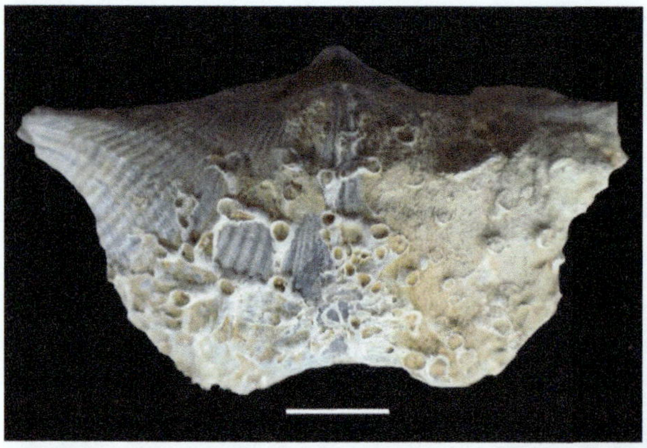

Fig. 9 *Cyrtospirifer syringothyriformis* (Paeckelmann). Specimen MGL 1306, coll. Loones. Beaulieu Fm., des Pâtures Mb., c "term", small temporary trench North of La Parisienne quarry. Dorsal valve. Numerous auloporid corallites developed on the fold and laterally. Some corallites are growing along the commissure and are limited to it, but some other corallites (*left upper part of the photo*) are growing on an opposite direction. (Scale bar = 1 cm)

large brachiopod but not common at this level (1.2% of the assemblage); and *Eoschuchertella ferquensis*—represented by only one encrusted specimen (0.7% of the assemblage) with ten auloporid corallites. The last species was also encrusted with another tabulate auloporid, *T. bouchardi*, a species occurring on eight specimens (16% of the population), covering the entire shell of each specimen.

S. longispina is represented by a single specimen strongly encrusted on the dorsal valve (Fig. 10). In assemblage 12 (Brice 2003), this species is very rare (five specimens), corresponding to 0.3% of the population. This new occurrence is not indicated in Fig. 11.

The associated fauna consists of various tabulate corals (lamellar or encrusting alveolitids, rare thamnoporids), solitary and massive rugose corals, gastropods, and the uncommon large bivalve *Limanomia* (Fig. 12), sometimes encrusted by auloporids, also occurring on the large *S. longispina*.

Rate or Frequency of Encrustation and Locations of the Epibionts

At this level, the encrusting auloporid species also belong to the genus *Plexituba* but to a different species, *"P."? cucullina*, usually with more conical corallites (Mistiaen 1988); a more compact auloporid, *T. bouchardi* (Fig. 13), is associated with *E. ferquensis*.

Of the 706 complete brachiopod shells collected at this level and belonging to these six potentially encrusted species, 84 were encrusted by 1068 auloporid corallites, corresponding to a mean occurrence of 12.71 corallites per shell. Wallace (1969b, p. 111) also reported epifaunal encrustation as being common at this level.

Factors Influencing Encrustation

Shell Structure

All six encrusted species are impunctate or pseudopunctate; none is punctate. Therefore, no distinction can be made based on the significance of this feature at this level.

Size

The six encrusted species are the largest one at this level. Another six species never encrusted; these are very small species such as *Retichonetes armata* (18.9% of the assemblage) or medium-sized species such as the very abundant *Productella subaculeata* (26.9% of the assemblage). *C. verneuili* is one of the largest encrusted species, as well as one of the most encrusted. However, this size-relationship does not apply for other host species. For instance, *D. dutertrei* is usually the smallest of the six encrusted species, but has one of the highest incrustation rates. Among athyrids, the smallest encrusted specimen is *Athyris bayeti* (*sensu* Brice and Mistiaen 1992) with only two auloporid corallites; the small size of the host did not favour *Plexituba*. Similarly, *E. ferquensis* is encrusted, whereas *E.*? *elegans*, common in the 2nd level but very much smaller, is never encrusted. The single encrusted specimen of *S. longispina* (with 21 auloporid corallites) is the largest individual in the population. As in the 2nd level, the size of the host was an important criterion for encrustation, but not the only criterion (cf. *D. dutertrei*).

Ornament

C. verneuili has faint costae, whereas *A. murchisoni* is smooth with faint growth lines. Three other species in the table have similar faint ribs. In contrast, *S. longispina* (one specimen) has faint costae, well-expressed growth lines and well-developed spines along successive growth zones. To date, no epibionts

Plate 2 Illustrations of the encrusted brachiopods of the 4th level. Ferques Fm., La Parisienne quarry, Frasnian. (Scale bar = 1 cm). Figures (1–4) *Cyrtospirifer verneuili* (Murchison) (1–2) Specimen MGL 1294, coll. Loones. (1) Dorsal valve. Auloporids developed on thje left part. Rare *Spirorbis* sp. (2) Ventral valve of the same specimen. Auloporids developed from the middle part towards the anterior commissure. (2a) Majority of corallites are developed in the direction of the anterior commissure (blue arrows) and underline successive growth steps, but some other corallites (*red arrows*) are growing towards posterior part of the shell. (2b) Relation between ornamentation and development of auloporids. Usually the growth correspond to the development of two new corallites in direction 45° relatively to the mother corallite, so each other 90° (blue arrows) but locally, influenced by costae, just one new corallite is developed straight on the precedent (red arrow). (3) Specimen MGL 1295, coll. Loones. Dorsal valve. Detail of the right flank of the Shell. Auloporids exceptionnaly developed on a very small specimen. (4) Specimen MGL 1296, coll. Loones. Dorsal valve. Auloporids, with corallites in one line, developed along the anterior commissure. Figures (5–8) *Athyris murchisoni* Brice. (5) Specimen MGL 1297, coll. Loones. Dorsal valve. Specimen only covered by bryozoans. (6) Specimen MGL 1298, coll. Loones. Ventral valve. Encrustation by auloporids and some *Spirorbis* sp. Auloporids are developed on the two valves, from umbo towards the anterior part. Some corallites cross the anterior commissure, revealing a post mortem development. (7) Specimen MGL 1299, coll. Loones. Dorsal valve. Specimen covered by bryozoans and rare Spirorbis sp. No epibiont on the ventral valve. (7a) detail. (8) Specimen MGL 1300, coll. Loones. Dorsal valve. Auloporid and *Spirorbis* sp., developed on the fold and on the left flank of the shell. (9) *Douvillina dutertrei* (Murchison) Specimen MGL 1301, coll. Loones. Ventral valve. Auloporid developed on the convex part of the shell. (9a) detail

Fig. 10 *Spinatrypa longispina* (Rigaux). Specimen MGL 1308, coll. Loones. Ferques Fm., Gris Mb., La Parisienne quarry. Dorsal valve. Irregularly oriented auloporid corallites, preferentially concentrated on the more convex middle part of the valve. The ventral valve is completely devoid in encrusted auloporid. (Scale bar = 1 cm)

have been observed on this species. Ornament did not influence the rates of encrustation (Plate 2, figs. 9, 9a).

Shell Morphology and the Host Life Position

Two of the six encrusted species, *C. verneuili* and *A. murchisoni*, have biconvex shells and were fixed by pedicles. In the same way, plano-convex *S. longispina* was also fixed by a pedicle. Consequently, these species can carry auloporids on both their dorsal (Plate 2, figs. 1, 3–5, 8) and ventral valves (Plate 2, figs. 2, 6); they are also the species with the highest number of auloporid corallites per encrusted shell.

Two other species, *D. dutertrei* (Plate 2, figs. 9, 9a) and *N. latissima*, have a concave dorsal valve and a convex ventral valve, with no pedicle, just lying free on the sea floor. In most cases, auloporids developed on the convex ventral valve because the host brachiopods were positioned with the convex ventral valve upwards.

Conversely, *E. ferquensis* has a convex dorsal valve oriented upwards and was fixed by a pedicle koskinoïde structure; this agrees with Lescinsky's observations (1995, p. 520) that "strophomenid ... concavo–convex brachiopods lived in the reverse life orientation with the convex valve on top". The mean number of auloporid corallites per shell is, accordingly, lower because one valve was covered.

4thlevel	***Cyrtospirifer verneuili***	***Athyris murchisoni***	***Douvillina dutertrei***	***Nervostrophia latissima***	***Eoschuchertella ferquensis***
Systematic group	spiriferid	athyrid	strophomenid	strophomenid	strophomenid
Shell structure	impunctate	impunctate	pseudopunctate	pseudopunctate	pseudopunctate
Fixation mode	pedicle	pedicle	no pedicle	no pedicle	koskinoïde structure
Number of shells with auloporids	39	28 + 1	10	4	1
Total number of complete shells	244	290+30+7	53	26	51
% of the population with auloporids	15.98	9.06	18.87	15.38	1.96
Total number of auloporid corallites	686	265 + 2	63	21	10
Mean number of auloporid corallites relative to encrusted shells	17.59	9.21	6.30	5.25	10.00

Fig. 11 Table of brachiopod species of the 4th level encrusted by auloporid epibionts. The five species are cited according to the number of collected encrusted shells *Note:* "+" in the *Athyris murchisoni* column corresponds to addition of two other species: *A. bayeti sensu* Brice and Mistiaen (1992) and *A. davidsoni sensu* Brice (2003), presently considered synonyms of *A. murchisoni* (Brice et al. 2009)

Feeding Currents

Feeding currents produced by the brachiopods were probably stronger for the three brachiopod species with pedicles, *C. verneuili, A. murchisoni* and *S. longispina*, all of which had spiral brachidia.

Although the proportion of the population encrusted is not greater for these three species, the mean number of auloporid corallites is clearly higher than for the three other species.

The water current produced by the brachiopod may not have initially favoured settling of small larvae but, after encrustation, this same water current provided more favourable conditions, allowing better encrustation.

Therefore, the encrustation rate clearly differed between the host species. It is interesting to note that the total development of auloporid corallites on a frequently encrusted species (e.g. *N. latissima*) was sometimes not extensive (a mean of 5.25 corallites per encrusted shell), whereas on a species less frequently encrusted (e.g. *A. murchisoni*, 9.7% of the population), the total development of auloporid corallites was more intense (a mean of 9.21 corallites per shell). Consequently, two successive steps can be recognized: facility for first encrustation and capacity for subsequent (secondary) development.

Auloporids on Brachiopods at Other Frasnian Levels

Remarks Concerning the 1st and 3rd Levels (Brice and Mistiaen 1992)

1st Level

The level is located in the lower part of the Beaulieu Formation, interval b of the Cambresèque Member (*falsiovalis* Zone, former lower *asymmetricus* Zone, Bultynck *in* Brice et al. 1979b). At this level, a large spiriferid, *Eodmitria boloniensis*, is very abundant (60.8% of assemblage 5; Brice 2003). Of the 203 brachiopod shells collected, none has auloporids on its surface, whereas they are frequently encrusted with *Petrocrania* sp. (Fig. 14) and rarely with *Spirorbis* sp. At the two previously analysed levels (2nd and 4th), the most encrusted brachiopods were large spiriferids. At this level, the large spiriferids are devoid of auloporids. It seems that large size is an important but not sufficient criterion for settlement of auloporids.

3rd Level

The fauna at this level corresponds to the 9th and 10th assemblages (Brice 2003). The 9th assemblage is located in the basal part; it occurs in a thin (0.6 m) bed, interval a of the du Bois Member—the lower member of the Ferques Formation. The brachiopod fauna is very diverse and abundant. The 10th assemblage is located about 4 m stratigraphically higher; it corresponds to the "brachiopod bed" described by Wallace (1969b). This level is also 0.6 m thick; it is located in interval b of the du Bois Member. According to conodont data, the base of the du Bois Member is approximately *punctata* Zone.

In the 9th assemblage (Brice 2003), where 13 brachiopod species occur, the most abundant species are *Athyris oehlerti* and *C. archiaci*. The small species *C. archiaci* represents 37.3% of the assemblage. Among the 170 specimens collected, only one, the largest, is encrusted. It is particularly important to stress that the specimens of *C. archiaci* collected at the 2nd level are relatively large (Brice and Mistiaen 1992, p. 53). Moreover, the sole specimen encrusted with auloporids in the 3rd level is a specimen significantly larger than average. In addition to the abovementioned species, there are also other very abundant small species: *A. oehlerti* (40.7% of assemblage 9), of which only 2.4% are encrusted. Curiously in this case, *A. oehlerti* is smaller than *C. archiaci* but is four times more encrusted. Wallace (1969a, p. 373) earlier noted that "epifaunal encrustation is moderately common at this horizon".

In the bed containing the 10th assemblage (Brice 2003), the "brachiopod bed", six species of brachiopods occur. *C. verneuili* is a very common species (336 shells collected) and represents 56.3% of the assemblage. Here, all the specimens are smaller (about half or a third of normal size); they form an accumulation of juvenile forms. Brachiopods encrusted with auloporids are less frequent: only 2.7% of the specimens are encrusted; only 12 auloporid corallites were noted (1.3 auloporid corallites per encrusted shell versus 17.6 for the 4th level).

In the same 10th assemblage, five other brachiopod species are devoid of encrusting auloporids, especially *P. subaculeata* (34.3% of the assemblage) and *Spinatrypa silvae* (6.8% of the assemblage).

As claimed before by several authors (Kesling et al. 1980; Sparks et al. 1980; Brice and Mistiaen 1992;

Fig. 12 *Limanomia* sp. Specimen, MGL 1309, coll. Loones. Ferques Fm., Gris Mb., La Parisienne quarry. Dorsal valve. The auloporid epibionts are developed on the two valves of this large shell, independently of the ornamentation (compare also with ornamentation of *Spinatrypa longispina*, Fig. 10). (Scale bar = 1 cm)

Fig. 13 *Eoschuchertella ferquensis* Brice covered by *Thecostegites bouchardi* (Michelin). Specimen MGL 1310, coll. Loones. Ferques Fm., Gris Mb., La Parisienne quarry. *Upper image*, ventral convex valve with outer surface entirely covered by auloporid epibionts (Scale bar = 1 cm). *Lower image*, enlargement of portion of shell edge showing detail and density of auloporid epibionts (Scale bar = 1 cm)

Zapalski 2005) and observed in the two levels analysed above, the dimensions of the host seems to be a positive criterion for substrate selection.

Fig. 14 *Eodmitria boloniensis* Brice. Specimen GFCL 4721. Beaulieu Fm., Cambresèque Mb., b "term", southeast corner of the Griset quarry. Dorsal valve with numerous (13) encrusted *Petrocrania* sp. Note radial ornament of the *Petrocrania* sp. following the ornament of the brachiopod. (Scale bar = 1 cm)

Other Levels in the Devonian (Givetian) of Boulonnais Containing Brachiopods Encrusted with Auloporids

Although encrusted brachiopods are particularly common in the Late Devonian (Frasnian) of Boulonnais, they are also occur but rather rarely in the Middle Devonian (Givetian).

The Couderousse Member (middle member of the Blacourt Formation) dated by conodonts as middle–late *varcus* zones (Brice 2003, p. 408), has numerous brachiopod shells (3rd assemblage; Brice 2003). *Desquamatia* (*Independatrypa*) *sola*, a large atrypid, is the most abundant brachiopod in this level; it occurs with *Cyrtospirifer verneuiliformis*, a very large spiriferid. The two species together represent 71.5% of the brachiopod assemblage (44.8 and 26.7%, respectively). They are both large and impunctate, faintly ribbed brachiopods with a pedicle but, surprisingly, the two species have no auloporids. Another species, *Undispirifer*? *gerolsteinensis grandis*, is also present but not as abundant (5.0% of the assemblage). It also is a species with a large, impunctate shell and a pedicle, but has smooth and large radial undulations and growth lines with small spines. Surprisingly, these three large species have no encrusting auloporids and, generally, no epibionts. One of the authors (CL) recently found three specimens each of the species *C. verneuiliformis* (Fig. 15) and *U.*? *gerolsteinsis grandis* (Fig. 16) encrusted with *Spirorbis* and bryozoans.

In the same level, auloporids are present on another large but infrequent brachiopod, *Stringocephalus* aff. *fontanus* (a terebratulid); it has a smooth and punctate shell. It constitutes 0.6% of the assemblage, with only nine specimens having been collected (Brice et al. 2009). Thus, in this case, the size of the potential support seems to be the relevant criterion for auloporid encrustation. Two other rare stringocephalid species also occur at this level: *Stringocephalus burtini*, with specimens about the same size as *S.* aff. *fontanus*, and *Stringodiscus giganteus*, approximately twice or three times larger than *S.* aff. *fontanus* (Brice et al. 2009). Curiously, these two species have never been found encrusted with auloporids. Consequently, the size of the potential host does not seem to be the main criterion for encrustation in this case. This is interesting because the three concerned species are all stringocephalids, so other factors must have influenced encrustation (ornament, shell morphology, life position and possibly feeding currents).

Fig. 15 *Cyrtospirifer verneuiliformis* (Paeckelmann). Specimen MGL 1311, coll. Loones. Blacourt Fm., Couderousse Mb., Banc Noir quarry. Dorsal valve. Large encrustation by bryozoans. (Scale bar = 1 cm)

Fig. 16 *Undispirifer*? *gerolsteinensis grandis* Brice. Specimen MGL 1312, coll. Loones. Blacourt Fm., Couderousse Mb., Banc Noir quarry. Ventral valve (*bottom left*) encrusted by several *Spirorbis* sp. and small bryozoans. (Scale bar = 1 cm). Larger photos show encrustation by several *Spirorbis* sp. and small bryozoans in detail. (Scale bar = 1 cm)

Auloporids Hosted on Supports Other than Brachiopods in the Devonian of the Ferques Area

Brachiopod shells in the 2nd, 3rd and 4th levels are associated with rugose corals but, curiously, all of them, though they appear to be excellent substrates for encrustation, lack auloporid corallites (Mistiaen et al. 2002). At some other levels in the Boulonnais area, however, auloporids are sometimes fixed to rugose corals or to supports other than brachiopod shells.

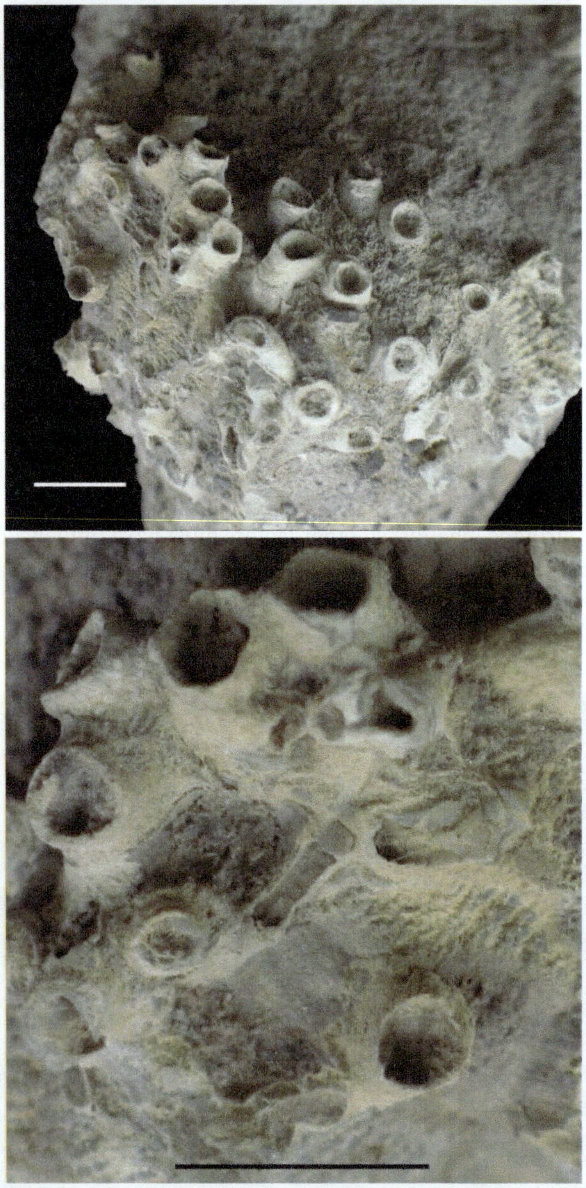

Fig. 17 Solitary rugose coral laterally covered by a colony of *Aulocystis* sp. Specimen MGL 1313, coll. Loones. Blacourt Fm., Griset Mb. C "term", Banc Noir quarry. (Scale bar = 1 cm)

In the Griset Member, interval c, the first reefal level of the Blacourt Formation (Givetian early *varcus* Zone; according to Bultynck *in* Brice et al. 1979a), auloporids are sometimes present on some solitary corals which happen to be very abundant at this level (Fig. 17). Small brachiopods are also present; though not common; they are diverse (12 species in the 2nd assemblage, Brice 2003), but they never host auloporid epibionts.

In interval g, at the top of the Griset Member (third reefal level of the Blacourt Formation), auloporids are developed on some solitary rugose corals (Fig. 18). No brachiopods have yet been found at this level.

In the Couderousse Member, where the brachiopod *S.* aff. *fontanus* is encrusted with auloporids, some rare solitary and colonial rugose corals are also encrusted with auloporids. When settling on a colonial rugose coral (Fig. 19), the auloporids clearly show that development of the epibiont was post-mortem.

In interval b, on top of the Bastien Member (Fig. 2), auloporids are hosted on tabulate corals such as branching *Scoliopora*. In this case, encrustation was also post-mortem.

Comparison with Similar Levels Containing Auloporid-Encrusted Brachiopods Elsewhere in the World During the Devonian

Numerous publications discuss epibionts on brachiopod shells in the Devonian, a period when epibionts—especially auloporids—were particularly numerous and diverse. Extensive but not exhaustive data from the literature are presented in the Appendix, with information on location, level and host.

Records relating to encrusted brachiopods are summarized in Fig. 20.

Figure 20 clearly shows that encrustations by auloporids were

- not frequent during the Early Devonian (about 33% of the data collected in the literature), but there were frequent pleurodictyid tabulates encrusting brachiopod shells at that time (Fuchs and Plusquellec 1982; Plusquellec 2007);
- more frequent during the Middle Devonian (54%);

Fig. 18 Solitary rugose coral, *Acanthophyllum* sp., laterally covered by *Aulocystis* sp. Specimen MGL 1314, coll. Loones. Blacourt Fm., Griset Mb. g "term". (Scale bar = 1 cm). *Top right image*. Detail. Some corallites of the encrusting *Aulocystis* sp. are developed inside the calyx of the rugose coral, demonstrating a post-mortem encrustation. (Scale bar = 1 cm). *Lower right image*. Detail of the lateral encrustation. (Scale bar = 1 cm)

- most frequent during the Late Devonian (close to 64%, but restricted in the Frasnian and apparently absent in the Famennian).

It is interesting to note that most of the Middle Devonian sites (9/15 = 64%) are located in north-eastern USA (Iowa, Ohio, Michigan and New York states), with four other sites in western Europe and two in China.

In terms of the most frequently encrusted kinds of brachiopods, it is clear from the collected data (Fig. 21) that

- spiriferids are predominantly encrusted;
- atrypids, athyrids, strophomenids and orthids are less encrusted;
- terebratulids are rarely encrusted;
- productids, pentamerids and rhynchonellids are never encrusted.

This summary clearly reflects the peculiar abundance of spiriferids during Devonian times. However, other groups, such as rhynchonellids, were also very abundant during the same period and were never encrusted. We now compare the Boulonnais levels analysed above with well-documented horizons elsewhere during the Frasnian, Middle Devonian and Early Devonian: the Frasnian of Iowa (Fenton and Fenton 1924; Wallace 1969a, b, 1978); the Middle Devonian of Iowa (Ager 1961, 1963); the Early Devonian of Spain and the Czech Republic (Alvarez and Taylor 1987).

Frasnian Levels (Iowa, USA) with Brachiopods Encrusted with Auloporids

After studying the Frasnian brachiopods of Ferques, Wallace (1969a, b, 1978) suggested homeomorphy between Devonian brachiopod communities from the upper beds of the "Calcaire de Ferques", Boulonnais, France (corresponding to the 4th level of Brice and Mistiaen 1992), and the brachiopod population in the Lime Creek Formation of Iowa, USA. She emphasized that epibionts are common to both localities. Fenton and Fenton (1924) had previously described several auloporids encrusting brachiopods from the same outcrops of the Hackberry Formation in Iowa (see Appendix). Day (1990, figs. 2, 6), reported *Palmatolepis semichatovae* in Zone 5 from the upper part of the Juniper Hill and lower part of the Cerro Gordo members of the Lime Creek Formation. He wrote (Day op. cit., p. 621) that "*Palmatolepis semichatovae* ranges through the interval that includes brachiopod faunas of most of the *Nervostrophia thomasi* zone and all of the *Douvillina arcuata* and *Cyrtospirifer whitneyi*" zones. These three species can be considered homeomorphs of the species present in Boulonnais, i.e. *N. latissima, D. dutertrei* and *C. verneuili*. It appears therefore that the ages and brachiopod faunas with their epibionts are indeed very similar. Wallace (1978, p. 271) was intrigued by this homeomorphy of the Frasnian brachiopod communities in both areas and concluded that their great similarity may be attributed to "morphologic adaptation to the broadly similar muddy, limey medium shelf environment in which the animals lived".

Middle Devonian Level (North-Eastern America) with Brachiopods Encrusted with Auloporids

The first interesting comparison can be made with the Little Cedar Formation, Givetian, Iowa.

Ager (1961, 1963) published a study on the Cedar Valley Formation spiriferids in which *Spinocyrtia iowensis* encrusted with epifauna were considered to be Frasnian. Cooper (1968), however, regarded the Cedar Valley Formation to be Givetian in age. Subsequently, Witzke et al. (1988, p. 221) and Day (1996, p. 281) confirmed this attribution and noted (Day loc. cit., p. 284) that *Orthospirifer iowensis* in faunal intervals 3–7 of the Little Cedar Formation, the lower part of the Cedar Valley Group, belonging to the Middle *varcus* to lower *subterminus* conodont zones. Ager (1961, p. 2) noted that auloporids were present on 33% of the *S. iowensis*, studied and illustrated by Stainbrook (1943); they are often associated with other epibionts, as occurs in the correlative level in Boulonnais. Ager (1961) also observed that specimens lacking auloporids were obviously smaller and noted the preferred position of the auloporids on the dorsal valve—with a predilection for the central part (fold) of the valve. He observed (op. cit., p. 3) that "*Aulopora* is generally concentrated along the anterior commissure with the apertures of the corallites usually directed anteriorly", and so concluded that the auloporid encrustations developed in this way in the life position. He (Ager 1963) also suggested that the auloporids benefited from the feeding streams of the brachiopods. Consequently, he agreed (op. cit., p. 264) with the suggestion of Yakovlev (1926) that "the relationship may have been one of mutual benefit, with the coral providing protection for the brachiopod by mean of stinging nematocysts". In the population of *S. iowensis*, 89% exhibit encrustation (Ager 1961), whereas only 35% of the population of the atrypid *Atrypa gigantea* that occurs in that horizon show some "symbiosis association" and only one specimen, "an exceptionally large" one, was covered by auloporids along the dorsal anterior margin.

Comparison with other Middle Devonian (Givetian) horizons from north-eastern America can be attempted. Brachiopods are numerous and well studied in the Traverse Group of the Givetian in Michigan (Sanford 1968) and in numerous coeval beds: the Norway Point Formation of Michigan, the Silica Shale and Ten Mile Creek formations of Ohio, the Arkona Shale of Ontario and the Wanakah Shale of New York state (Watkins 1959). Associations with auloporids are frequent in these occurrences. They were first illustrated by Grabau (1899), and subsequently observed by Ehlers and Wright (1955), Watkins (1959), and Pitrat and Rogers (1978).

Important observations concerning these different studies include the following:

- The most frequently occurring brachiopods are several species of the spiriferid genus *Spinocyrtia*: *S. granulosa*, *S. clintoni* and *S. mourante*, encrusted with *Aulocystis*. Kesling et al. (1980) described numerous epizoans (42 species, including *Aulopora microbuccinata*) encrusting *Paraspirifer brownokeri* from the Silica Shale. It is typical that a large proportion of spiriferids is often encrusted. Thus, Watkins (1959, p. 799) noted that "it is difficult to find a specimen of *Spinocyrtia granulosa* that does not have this species [*Aulocystis commensalis*] encrusting it". In the same way, 63% of the population of *S. clintoni* was encrusted by *A. commensalis* (Pitrat and Rogers 1978, p. 1317).

Fig. 19 Colonial rugose coral, *Wapitiphyllum* sp., encrusted by auloporid. Specimen MGL 1315, coll. Loones. Blacourt Fm., Couderousse Mb., Banc Noir quarry. Post-mortem encrustation of the auloporid. (Scale bar = 1 cm)

- The auloporids discussed very commonly encrust the dorsal valve and are oriented towards the anterior commissure, as seen in the illustration by Davidson (1853, Fig. 22) and those of Grabau (1899). Pitrat and Rogers (1978) reported that 88.7% of the studied population of *A. commensalis* encrusting *S. clintoni* were located only on the dorsal valve, 10.73% were located on both valves, and no more than 0.57% (corresponding to only one specimen) was located on the ventral valve. The same authors also called attention to the fact that in 87.57% of cases, the first corallite of *A. commensalis* was systematically positioned in a median ditch on the fold of the shell. They inferred a "commensalism association".

Early Devonian (Spain, Czech Republic) with Brachiopods Encrusted with Auloporids

The Early Devonian is not represented in the Boulonnais, but it is interesting to cite papers on Early Devonian levels with encrustations.

Aulopora sp. encrusting the dorsal valve of *Anathyris phalaena* was described by Alvarez and Taylor (1987) from Ferrones, Asturias, Spain. Other encrusting epibionts were also present (bryozoans, inarticulate brachiopods, *Spirorbis*). As described above, the auloporids were growing along the anterior commissure (op. cit., p. 26).

Galle and Parsley (2005, p. 126), analysing a Pragian shallow-water community from St. Prokop Quarry, Czech Republic, concluded that "none of the 412 brachiopods collected... had visible epibionts", whereas 132 tabulate epibionts (though lacking

Fig. 20 Geological distribution of selected levels with encrusted brachiopods, and ratio with auloporids, according to the annexed table

Series	Stage	Encrusted horizons	Encrusted horizons with auloporids
Late Devonian	Frasnian	14	9
Middle Devonian	Eifelian–Givetian	28	15
Early Devonian	Pragian–Emsian	12	4

GROUP	spiriferids	atrypids	strophomenids	athyrids	orthids	terebratulids
Shell structure	impunctate	impunctate	pseudopunctate	impunctate	punctate	punctate
Ornament	faint costae	faint costae growth lines	faint ribs	growth lines	faint ribs	smooth
No. of genera or species	21	5	5	4	4	1

Fig. 21 Table of relations between the shell structure, the ornamentation and the amount of encrusted auloporid corallites according to the different groups of brachiopods concerned

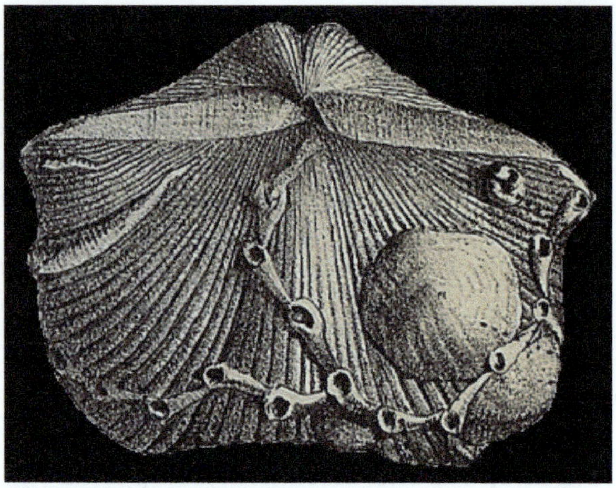

Fig. 22 *Spirifer* sp. from Kwangsi Province (South China) illustrated by Davidson (1853). Dorsal valve encrusted by several epibionts. Auloporid corallites are essentially located along the anterior commissure

auloporids) were present on 509 hyolithids. This occurrence (the Dvorce–Prokop Limestone) corresponds to a carbonate mud mound environment, completely unlike the shaley environments containing numerous encrusted brachiopods discussed earlier.

Conclusions

In the 2nd level, of the various factors considered, the predominant factors influencing encrustation of brachiopods by auloporids seem to be, in decreasing order, the following: size, morphology, feeding currents and shell ornament. In this context, shell structure seems to lack importance. It seems that auloporids always developed on living brachiopods, though sometimes their growth continued post-mortem.

In the 4th level, size again appears the primary factor facilitating encrustation, followed by (in decreasing order) shell morphology, life position of the host and feeding currents. Ornament is a secondary factor. In the 4th level, shell structure had no effect.

The relationship between location of the encrusting auloporids and feeding currents favoured development of epibionts during the host's life. As for the 2nd level, development and distribution of the auloporids along growth lines on the brachiopod shell and near the extremity of the fold (Brice and Mistiaen 1992, figs. 6–7) also indicate development of the auloporids on the host while the latter was alive. Rare occurrences of auloporids crossing the anterior commissure indicate post-mortem growth (Fig. 6).

The lithologic settings are very similar in the two levels analysed—corresponding to analogous environmental conditions. In both levels, brachiopods encrusted with numerous epibionts are present in marls or more-or-less argillaceous limestone and nodular sediments, but become rarer above and below them when the lithology turns to purely argillaceous or shaley sediments indicative of a soft bottom. When large brachiopods are present in such sediments (1st level), auloporid epibionts are absent. This implies these environments correspond to optimal benthic conditions for encrustation (Seilacher 1960), particularly in nutrient-rich water (Wallace 1978, p. 271).

In the Devonian of the Ferques area, epibionts encrusting brachiopods are abundant in some levels during the Frasnian, but rarer during the Givetian, whereas in north-eastern America, epibionts encrusting brachiopods were more abundant during the Givetian. This last conclusion accords with possible migration of faunas between North America and Europe, previously proposed in the context of probable sea-level rises (Brice et al. 1994, p. 303; Rode and Lieberman 2004, p. 355).

It is clear from the data analysed that the encrustation rate of auloporids on Devonian brachiopod shells fluctuated according to several factors.

First, the systematic position of the brachiopod host. For example, similar-sized pentamerids or rhynchonellids are never encrusted, whereas spiriferids are frequently encrusted.

Second, morphologic or biologic features of the shell of the brachiopod host shell. These are (probably in decreasing order of rate of encrustation):

(1) size;
(2) degree of shell convexity, absence or presence of a fold on the valve;
(3) mode of fixation (pedicle, no pedicle, or koskinoïde type);
(4) type of lophophore and position of feeding currents;
(5) type of ornament (strong or faint costae, ribs, growth lines, or smooth shell);
(6) shell structure (impunctate, pseudopunctate or punctate).

The relative importance of some of these features has been suggested previously. For instance, Thayer (1974) discussed size, Ager (1961) morphology and Mistiaen et al. (2006) suggested that large size and rather weaker ornament facilitated encrustation. The factors previously suggested as influencing encrustation are all host features; they do not therefore provide a completely comprehensive explanation. Another important factor, extrinsic to the host, seems to be the lithologic setting, i.e. the surrounding environmental conditions. It is probably the most important factor, but is also the most difficult to specify in a way that would provide a basis for making comparisons. We are thus in agreement with Bose's conclusion (op. cit., p. 70) about determining, if at all possible, the environment in which encrusting of brachiopods occurred.

Appendix

References	Country	Age	Host	Auloporid	EPIBIONTS									
					Porifera Foraminifera	Bryozoan		Crinoid	Brachiopod	*Cornulites*	Worms		Corals	
						Hederella	others				*Spirorbis*	Serpulid	Tabulate corals	Rugose
Davidson (1853)	Chine	Upper	Spirifer	Auloporid					*Crania obsoleta*	*Cornulites*				
	Kwangsi	Devonian												
Gosselet (1894)	Belgique	Frasnian	Spirifer						*Petrocrania*					
Grabau (1899)	USA New York	Givetian	*Spinocyrtia granulosa*	Auloporid										
Fenton and Fenton (1924)	USA Iowa	Frasnian	*Schizophoria iowaensis*	*Aulopora saxivadum*										
			Strophonella reversa[a]	*Aulopora saxivadum*					*Crania*					
			Rugose corals	*Aulopora adnascens*										
			Syringostroma	*Aulopora paucitubulata*										
			Spirifer hungerfordi			*Hederella alternata*			*Crania*					
			Spirifer whitneyi						*Crania*					
			Atrypa devoniana						*Crania*					
Le Maître (1929)	France Avesnois	Eifelien	*Spirifer speciosus*				*Rhopalonaria*							
Grabau (1931)	China Yunnan	Middle Devonian	*Stringocephalus*[a]						*Crania obsoleta*					
			Schizophoria macfarleni[a]						*Crania obsoleta*					
			Atrypa richtofeni						*Crania obsoleta*					
Gekker (1935)	Russia Russian Platform	Upper Devonian	Spiriferid	Auloporid							*Spirorbis*			x?
Ehlers and Wright (1955)	USA Ontario	Middle Devonian	*Spinocyrtia mourante*	Auloporid										
Le Maître (1956)	Afrique du Nord	Siegenien-Eifelien	Crinoid stem										*Hamarilopora minima*	
Gekker (1957)	Russia Russian Platform	Devonian	Spiriferid	*Aulopora*								x		x?
Watkins (1959)	USA Michigan	Givetian	*Spinocyrtia granulosa*	*Aulocystis commensalis*										

(continued)

References	Country	Age	Host	Auloporid	EPIBIONTS									
					Porifera Foraminifera	Bryozoan		Crinoid	Brachiopod	*Cornulites*	Worms		Corals	
						Hederella	others				*Spirorbis*	Serpulid	Tabulate corals	Rugose
Biernat (1959)	Poland	Middle Devonian	*Schizophoria striatula*[a]	*Aulopora* sp.										
	Holy Cross Mts.													
Chang (1959)	Chine	Middle Devonian	*Mucrospirifer*	*Aulopora*										
Ager (1961)	USA Iowa	Givetian	*Spinocyrtia iowensis*	*Aulopora*		*Hederella* cf. *filiformis*	*Paleschara incrustans*				*Spirorbis*			
Ager (1963)	USA Iowa	Givetian	*Spinocyrtia iowensis*	*Aulopora elleri*										
			Eosyringothyris calvini	*Aulopora* cf. *multiramosa*										
Gekker (1966)	Russia Russian Platform	Upper Devonian	Spiriferid	Auloporid								x		x?
Biernat (1966)	Poland Holy Cross Mts.	Middle Devonian	*Schizophoria striatula*[a]	Auloporid			*Ascodictyon*							
			Desquamatia subzonata				*Eliasopora*							
							Ropalonaria							
							Allonema							
Schumann (1967)	Canada	Frasnien	*Mucrospirifer reidfordi*							*Cornulites*				
Bühler (1969)	USA New York	Givetian	*Spinocyrtia granulosa*	*Aulocystis commensalis*										
Wallace (1969a)	France Boulonnais	Frasnian	*Cyrtospirifer verneuili*	*Aulopora*		*Hederella*	*Paleschara*				*Spirorbis*			
			Athyris concentrica											
Hoare and Steller (1969a, b)	USA Ohio Michigan	Eifelian	Articulate brachiopods						*Petrocrania hamiltonensis*					
									Philhedra crenistria					
									Lingulodiscina marginalis					
Rodriguez and Gutschick (1975)	USA Montana Utah	Late Devonian	*Rhipidomella*[a]		Foraminifera		*Fenestella*		*Leptalosia*	*Cornulites*	*Spirorbis*			
			Leptalosia		Foraminifera				*Leptalosia*		*Spirorbis*			
			Schuchertella[a]		Foraminifera										

(continued)

References	Country	Age	Host	Auloporid	EPIBIONTS									
					Porifera Foraminifera	Bryozoan		Crinoid	Brachiopod	*Cornulites*	Worms		Corals	
						Hederella	others				*Spirorbis*	Serpulid	Tabulate corals	Rugose
			Philedra		Foraminifera				*Leptalosia*	*Cornulites*				
			Acanthocrania		Foraminifera					*Cornulites*				
			Horn coral		Foraminifera		*Fenestella*		*Leptalosia*		*Spirorbis*			
			Fenestella		Foraminifera		*Fenestella*							
			Cornulites		Foraminifera		*Fenestella*							
			Platyceras		Foraminifera		*Fenestella*							
			Nautiloid orthoconic		Foraminifera				*Leptalosia*					
			Trilobite		Foraminifera									
			Crinoïd		Foraminifera									
Kesling and Chilman (1975)	USA Michigan	Givetian	Crinoid stems							*Cornulites*				
Marek and Galle (1976)	Czech Republic Bohemia	Pragian	Hyolith *Pterygotheca*										*Hyostragulum mobile*	
Pitrat and Rogers (1978)	USA Michigan	Givetian	*Spinocyrtia clintoni*	*Aulocystis commensalis*		*Hederella*	*Paleschara*			*Cornulites*	*Spirorbis*			
Sparks et al. (1980)	USA Ohio	Givetian	*Paraspirifer bownockeri*	*Aulopora microbuccinata*		*Hederella*	*Eliasopora stellatum*			*Cornulites*				
Kesling et al. (1980)	USA Michigan	Givetian	*Paraspirifer bownockeri*	*Aulopora microbuccinata*	*Clionoides*	*Hederella*	*Ascodictyon*	x	*Petrocrania*	*Cornulites*	*Spirorbis*			
							Eliasopora		*Philedra*					
							Atactoechus		*Lingulodiscina*					
							Leptotrypella		*Orbiculoidea*					
									Cyrtina[a]					
									Mediospirifer					
Brett and Cottrell (1982)	USA New York	Middle Devonian	*Paleozygopleura hamiltoniae*										*Pleurodictyum*	
Fuchs and Plusquellec (1982)	Germany	Siegenian-Emsian	Brachiopods										*Pleurodictyum problematicum*	
Hoare and Walden (1983)	USA Ohio	Middle Devonian	*Paraspirifer bownockeri*								*Vermiforichnus clarki*			
Baird and Brett (1983)	USA New York	Middle Devonian	*Mucrospirifer consobrinus*										*Pleurodictyum* cf. *americanum*	
			Mediospirifer audaculus						*Petrocrania*					

(continued)

References	Country	Age	Host	Auloporid	EPIBIONTS									
					Porifera Foraminifera	Bryozoan		Crinoid	Brachiopod	*Cornulites*	Worms		Corals	
						Hederella	others				*Spirorbis*	Serpulid	Tabulate corals	Rugose
			Spinatrypa spinosa						*Petrocrania*					
			Megastrophia concava					Pelmatozoan						
			Favosites milne-edwardsi				Fenestellid							
			Favosites argus				*x*						Alveolitid	x
			Heliophyllum											*Cystiphylloides*
Alvarez and Taylor (1987)	Spain Asturias Ferrones	Emsian	*Anathyris phalaena*	*Aulopora* sp,		*Hederella*	Trepostome		*Petrocrania*		*Spirorbis*			
							Corynotrypa							
							Fistuliporid							
							Ascodictyon							
Le Menn (1989)	Algeria Ougarta	Upper Emsian	Crinoids	*Aulopora*			Fenestellid							
Baird et al. (1989)	New York Ohio	Eifelian Givetian	Nautiloïd *Michelinoceras*				*Reptaria*						*Alveolites*	
Gourvennec (1989)	France Massif Armoricain	Lower Devonian	*Brachyspirifer rousseaui*	Aulopore					*Cyrtina pauciolicata*[a]		*Spirorbis*			
			Cyrtina intermedia							*Cornulites*				
			Hysterolites venus	Aulopore										
			Brachyspirifer becki	Aulopore						*Cornulites*				
Bordeaux and Brett (1990)	USA New York	Middle Devonian	*Pustulatia pustulosa*											
			Meristella huskinsi											
			Camarotoechia hosfordi											
			Tropidoleptus carinatus[a]			*Hederella*	*Ascodictyon*	x	*Petrocrania*	*Cornulites*	*Spirorbis*		*Pleurodictyum*	
			Camarotoechia sp.											
			Megalowskiella											
			Mediospirifer audaculus											
			Ambocoelia spinosa											
Brice and Mistiaen (1992)	France Boulonnais	Upper Devonian	*Cyrtospirifer verneuili*	*Plexituba cucullina*			x		*Petrocrania*	*Cornulites*	*Spirorbis*		*Thecostegites bouchardi*	
			Cyrtospirifer syringothyriformis,	*Plexituba briceae*			x		*Petrocrania*	*Cornulites*	*Spirorbis*			

(continued)

References	Country	Age	Host	Auloporid	EPIBIONTS									
					Porifera Foraminifera	Bryozoan		Crinoid	Brachiopod	*Cornulites*	Worms		Corals	
						Hederella	others				*Spirorbis*	Serpulid	Tabulate corals	Rugose
		Frasnien	*Cyrtospirifer archiaci*	*Plexituba briceae*			x				*Spirorbis*			
			Eodmitria boloniensis						*Petrocrania*		*Spirorbis*			
			Athyris oehlerti	*Plexituba briceae*							*Spirorbis*			
			Athyris bayeti	*Plexituba cucullina*					*Petrocrania*		*Spirorbis*			
			Athyris murchisoni	*Plexituba cucullina*			x		*Petrocrania*		*Spirorbis*			
			Atrypid (3 species)	*Plexituba briceae*			x		*Petrocrania*		*Spirorbis*			
			Schizophoria striatula[a]	*Plexituba briceae*			x		*Petrocrania*	*Cornulites*	*Spirorbis*			
			Douvillina dutertrei	*Plexituba cucullina*			x				*Spirorbis*			
			Eochuchertella ferquensis	*Plexituba briceae*			x		*Petrocrania*		*Spirorbis*		*Thecostegites bouchardi*	
			Nervostrophia latissima	*Plexituba cucullina*			x		*Petrocrania*		*Spirorbis*			
Gibson (1992)	USA Tennessee	Lower Devonian	*Macropleura macropleura*								*Spirorbis laxus*			
Galle et al. (1994)	Maroc Espagne	Emsien inférieur	Hyolithe *Pterygotheca*										*Hyostragulum ometanum*	
Cuffey et al. (1995)	USA Oklahoma	Lower Devonian	*Meristella atoka*										x	
Lescinsky (1995)	USA Indiana New York Ohio	Middle-Upper Devonian	*Strophodonta demissa*		x	*Hederella*								
			Strophonelloides hybrida											
			Tropidoleptus carinatus[a]											
			Atrypa devoniana											
Galle and Prokop (2000)	Czech Republic	Pragian	Crinoid stem										*Favosites* sp.	
Galle and Ficner (2004)	Czech Republic Moravia	Middle Devonian	*Calceola sandalina*	Auloporid							*Spirorbis*		*Favosites*	
													Alveolitid	

(continued)

References	Country	Age	Host	Auloporid	EPIBIONTS									
					Porifera Foraminifera	Bryozoan		Crinoid	Brachiopod	*Cornulites*	Worms		Corals	
						Hederella	others				*Spirorbis*	Serpulid	Tabulate corals	Rugose
Garcia-Alcalde (2005)	Spain Cantabrian Mts	Upper Emsian	*Tenuicostella comellanensis*	Auloporids			Branched bryozoans					Serpulids	x	
Galle and Parsley (2005)	Czech Republic	Pragian	Echinodermata			x	x							
Gluchowski (2005)	Poland Holy Cross Mts	Upper Eifelian	*Tantalocrinus scutellus*			*Hederella*	*Ascodyction*, cystoporates		*Cyrtina*[a]				*Favosites* sp	*Adradosia* ? sp
							Allonema							
							Eliasopora							
			Schyschcatocrinus creber				"*Ropalonaria*"		Productid					*Adradosia* ? sp
			Gilbertsocrinus vetulus			*Hederella*								*Adradosia* ? sp
			Pentagonistipes petaloides			*Hederella*							*Favosites* sp	
			Cycloocetocrinus			*Hederella*	*Vinella*							
Zapalski (2005)	Poland Holy Cross Mts	Eifelian Givetian	*Kyrtatrypa*	Auloporids										
Mistiaen et al. (2006)	France Boulonnais	Frasnian	*Cyrtospirifer syringothyriformis*	*Plexituba briceae*										
			Desquamatia (Neatrypa) legayi	*Plexituba* sp.										
Bose (2006)	USA Ohio	Eifelian	*Rhipidomella*[a]				x							
Mottequin (2006)	Belgique	Frasnian	*Cyrtospirifer eponae*						*Petrocrania*					
Mottequin (2008)	Belgique	Frasnien	*Schuchertellopsis durbutensis*	Auloporids										
Brice et al. (2009)	France Boulonnais	Givetian	*Stringocephalus aff. fontanus* +	Auloporids							*Spirorbis*			

[a]Punctate brachiopod

x = taxon present; x? = taxon possibly present

References

Ager DV (1961) The epifauna of a Devonian spiriferid. Quart J Geol Soc Lond 117:1–10

Ager DV (1963) Principles of paleoecology. McGraw-Hill, New York, 371p

Alexander R (1986) Resistance to and repair of shell breakage induced by durophages in Late Ordovician brachiopods. J Paleontol 60:273–285

Alvarez F, Taylor PD (1987) Epizoan ecology and interactions in the Devonian in Spain. Palaeogeogr Palaeoclimatol Palaeoecol 61:17–31

Baird GC, Brett CE (1983) Regional variation and paleontology of two coral beds in the Middle Devonian Hamilton Group of Western New York. J Paleontol 57:417–446

Baird GC, Brett CE, Frey RC (1989) "Hitchhiking" epizoans on orthoconic cephalopods: preliminary review of the evidence and its implications. Senckenbergiana Lethaea 69:439–465

Becker RT (2002) Frasnian goniatites from the Boulonnais (France). Ann Soc Géol Nord 9(2ème série):129–137

Biernat G (1959) Middle Devonian Orthoidea of the Holy Cross Mountains and their ontogeny. Palaeontol Pol 10:78

Biernat G (1961) *Diorygma atrypophilia* n. gen. n. sp. A parasitic organism of *Atrypa zonata* Schnur. Acta Palaeontol Polonica 6:17–28

Biernat G (1966) Middle Devonian brachiopods of the Bodzentyn Syncline (Holy Cross Mountains, Poland). Palaeontol Polonica 17:162

Bordeaux YL, Brett CE (1990) Substrate specific associations of epibionts on Middle Devonian brachiopods: implications for Paleoecology. Hist Biol 4:203–220

Bose (2006) Epibionts on brachiopods from the Devonian Dundee Formation of Ohio. Thesis, College of Bowling Green University, 74p

Brett CE, Cottrell JF (1982) Substrate specificity in the Devonian tabulate coral *Pleurodictyum*. Lethaia 15:247–362

Brice D (2003) Brachiopod assemblages in the Devonian of Ferques (Boulonnais, France)—relations to palaeoenvironments and global eustatic curves. Bull Geosci [Czech Geol Surv] 78:405–417

Brice D, Mistiaen B (1992) Epizoaires des Brachiopodes frasniens de Ferques (Boulonnais, Nord de la France). Géobios (Mém Spéc) 14:45–58

Brice D, Loomes CH (2002) Nouvelles données sur des brachiopodes Cyrtospiriferidae, Reticularidae, Uncitidae et Stringocephalidae du Dévonien de Ferques. Boulonnais – France. Ann Soc géol Nord 9(2ème série):91–108

Brice D, Bultynck P, Deunff J, Lobziak S, Streel M (1979a) Données biostratigraphiques nouvelles sur le Givétien et le Frasnien de Ferques (Boulonnais, France. Ann Soc géol Nord 98:325–344

Brice D, Colbeaux JP, Mistiaen B, Rohart JC (1979b) Les Formations dévoniennes de Ferques (Bas Boulonnais, France). Ann Soc géol Nord 98:307–324

Brice D, Coen M, Loboziak S, Streel M (1981) Précisions biostratigraphiques relatives au Dévonien supérieur de Ferques (Boulonnais). Ann Soc géol Nord 100:159–166

Brice D, Milhau B, Mistiaen B (1994) Affinités nord-américaines de taxons dévoniens (Givétien–Frasnien) du Boulonnais, Nord de la France. Migrations et diachronismes. Bull Soc géol France 165:291–306

Brice D, Mottequin B, Loones CH (2009) Découverte de nouveaux brachiopodes dans le Givétien (Dévonien) du Boulonnais (N. France). Ann Soc géol Nord 15(2ème série):1–14

Bühler EJ (1969) Middle Devonian epizoites. Geol Soc Am Abstr 1:4

Chang S-H (1959) Note on the paleoecological relation between *Aulopora* and *Mucrospirifer*. Acta Palaeontol Sin 7:502–504

Cooper GA (1968) Age and correlations of the Tully and Cedar Valley formations in the United States (ed) International symposium on Devonian System, Calgary, Alberta, 1967. Proceedings 2:701–709. Alberta Society of Petroleum Geologists, Calgary, Alberta, Canada

Cuffey CA, Robb AJ, Lembcke JT, Cuffey RJ (1995) Epizoic bryozoans and corals as indicators of life and post-mortem orientations of the Devonian brachiopod *Meristella*. Lethaia 28:139–153

Curry GB (1983) Microborings in Recent brachiopods and the function of caeca. Lethaia 16:119–127, 139–153

Davidson T (1853) On some fossil Brachiopoda of the Devonian Age, from China. Quart J Geol Soc Lond 9:353–359

Day J (1990) The Upper Devonian (Frasnian) conodont sequence in the Lime Creek Formation of north-central Iowa and comparison with Lime Creek ammonoid, brachiopod, foraminifer, and gastropod sequences. J Paleontol 64(4): 614–628

Day J (1996) Faunal signatures of Middle-Upper Devonian depositional sequences and sea level fluctuations in the Iowa Basin. US Midcontinent. Geol Soc Am Spec Pap 306: 277–300

Ehlers GM, Wright JD (1955) The type species of *Spinocyrtia* Frederiks and new species of this brachiopod genus from southwestern Ontario. Contrib Mus Paleontol Univ Michigan 3:1–32

Fagerstrom JA (1996) Paleozoic brachiopod symbioses; testing the limits of modern analogues in palaeoecology. Geol Soc Am Bull 108:1393–1403

Fenton CL, Fenton MA (1924) The stratigraphy and fauna of the Hackberry stage of the Upper Devonian. Univ Michigan Contrib Mus Geol 1:1–260

Fuchs G, Plusquellec Y (1982) *Pleurodictyum problematicum* Goldfuss 1829 (Tabulata, Dévonien). Status, Morphologie, Ontogénie. Geol et Palæontol 15:1–26

Galle A, Ficner F (2004) Middle Devonian *Calceola sandalina* (Linnaeus, 1771) (Anthozoan, Rugosa) from Moravia (Czech Republic): aspects of functional morphology, gerontic growth patterns, and epibionts. *Geodiversitas* 26(1): 17–31

Galle A, Parsley RL (2005) Epibiont relationships on hyolithids demonstrated by Ordovician trepostomes (Bryozoa) and Devonian tabulates (Anthozoa). Bull Geosci Czech Geol Surv 80(2):125–138

Galle A, Prokop RJ (2000) Complex parasitism and symbiosis of crinoid, subepidermal parasite, and tabulate coral. Lower Devonian (Pragian), Barrandian, Czech Republic. Věstník Českėho geologickėho ústavu. 75:441–444

Galle A, Marek L, Vannier J, Racheboeuf PR, Regnault S (1994) Assemblage epibenthique à hyolithe, tabulé épizoaire et ostracode Beyrichiacea du Dévonien inférieur du Maroc et d'Espagne. Rev Paléobiol 13(2):411–425

Garcia-Alcalde JL (2005) Upper Emsian Spinelloidea and Cyrtospiriferoidea (Brachiopoda, Spiriferidina) of the Cantabrian Mountains (N Spain). Géobios 38:69–97

Gekker RF (1935) The phenomena of adherence and attachment among the upper Devonian fauna and flora of the main outcrop. Trans Paleontol Inst Moscow 4:149–280 (in Russian)

Gekker RF (1957) Introduction to Palaeoecology. Gosgeoltekhizdat, Moscow. Bases de la Paléoécologie, French translation by Roger J (1960) Bureau Recherches Géologiques et Minières, Paris, 83 pp (in Russian)

Gekker RF (1966) On living together on organisms in the geological past. In: Gekker RF (ed) Organisms and their environment. Otd Obshch Biol, Akad Nauk, Moscow, pp 14–30 (in Russian)

Gibson MA (1992) Some epibiont–host and epibiont-epibiont relationships from the Birdsong Shale Member of the Lower Devonian Ross Formation (west-central Tennessee, USA). Hist Biol 6(2):73–90

Gluchowski E (2005) Epibionts on upper Eifelian crinoid columnals from the Holy Cross Mountains, Poland. Acta Palaeontol Pol 50(2):315–328

Gosselet J (1894) Etudes sur les variations du *Spirifer verneuili*. Mém Soc géol Nord IV(1):63

Gourvennec R (1989) Brachiopodes Spiriferida du Dévonien inférieur du Massif Armoricain. Biostratigraphie du Paléozoïque 9:281

Grabau AW (1899) The geology of Eighteen-mile Creek. Buffalo Soc Nat Sci 6:97–390

Grabau AW (1931) Devonian Brachiopoda of China: Devonian Brachiopoda from Yunnan and other districts in South China, China Geol Surv, Palaeontologica Sinica ser B 3(3):545

Hoare RD, Steller DL (1969a) A Devonian brachiopod with epifauna. Ohio J Sci 67:291–297

Hoare RD, Steller DL (1969b) Inarticulate brachiopods of the Silica Shale formation (Devonian), Ohio and Michigan. Contrib Mus Paleontol Univ Michigan 22:263–272

Hoare RD, Walden RL (1983) *Vermiforichnus* (Polychaeta) borings in *Paraspirifer bownockeri* (Brachiopoda: Devonian). Ohio J Sci 83(3):114–119

Kesling RV, Chilman RB (1975) Strata and megafossils of the Middle Devonian Silica Formation. Univ Michigan Palaeontol Mus Pap Paleontol 8:1–408

Kesling RV, Hoare RD, Sparks DK (1980) Epizoans of the Middle Devonian brachiopod *Paraspirifer bownockeri*: their relationships to one another and to their host. J Paleontol 54:1141–1154

Klapper G (1990) Frasnian species of the Late Devonian conodont genus *Ancyrognathus*. J Paleontol 64:998–1025

Le Maître D (1929) La faune des couches à *Spirifer cultrijugatus* à Fourmies. Ann Soc géol Nord 54:27–74

Le Maître D (1956) Tabulés des formations dévoniennes du Nord de l'Afrique. CR Acad Sci Paris 243:1339–1342

Le Menn J (1989) Contrôle de la sécrétion stéréomique dans les stolons d'un crinoïde nouveau du Dévonien inférieur d'Algérie. Lethaia 22:395–404

Lescinsky HL (1995) The life orientation of concavo-convex brachiopods: overturning the paradigm. Paleobiology 21:520–551

Mackinnon DI, Biernat G (1970) The probable affinities of the trace fossil *Diorygma atrypophilia*. Lethaia 3:163–172

Marek L, Galle A (1976) The tabulate coral *Hyostraculum*, an epizoan with bearing on hyolithid ecology and systematics. Lethaia 9:51–56

Mistiaen B (1988) Tabulés Auloporida du Givétien et du Frasnien de Ferques (Boulonnais – France). In: Brice D (ed) Le Dévonien de Ferques. Bas-Boulonnais (N. France). Biostratigraphie du Paléozoïque, vol 7. Rennes, France, pp 197–230

Mistiaen B (2002) Stromatopores et coraux tabulés du Membre des Pâtures, Formation de Beaulieu (Frasnien de Ferques, Boulonnais). Ann Soc géol Nord 9(2ème sér): 85–90

Mistiaen B, Becker T, Brice D, Degardin JM, Derycke CL, Loones CH, Rohart JCl (2002) Données nouvelles sur la partie supérieure de la Formation de Beaulieu (Frasnien de Ferques, Boulonnais, France). Ann Soc géol Nord 9(2ème sér):75–84

Mistiaen B, Zapalski M, Brice D (2006) Primary and secondary factors in substrate selection of auloporid tabulates. 2nd International Palaeontological Congress (Beijing July 2006), abstract

Mottequin B (2006) Les Brachiopodes de la Transition Frasnien–Famennien dans le Bassin de Namur-Dinant (Belgique). Thèse de doctorat, Université de Liège, Belgique, p. 427

Mottequin B (2008) Late middle Frasnian to early Famennian (Late Devonian) strophomenid, orthotetid, and atrypid brachiopods from southern Belgium. J Paleontol 82: 1052–1073

Pitrat CW, Rogers FS (1978) *Spinocyrtia* and its epibionts in the Traverse Group (Devonian) of Michigan. J Paleontol 52:1315–1324

Plusquellec Y (2007) Histoire naturelle des Pleurodictyoformes (Cnidaria, Tabulata, Dévonien) du Massif Armoricain et des régions Maghrebo-Européennes principalement. Mém Soc géol minéral Bretagne 32:123

Rode AL, Lieberman BS (2004) Using GIS to unlock the interactions between biogeography, environment, and evolution in Middle and Late Devonian brachiopods and bivalves. Palaeogeogr Palaeoclimatol Palaeoecol 211: 345–349

Rodriguez J, Gutschick RC (1975) Epibiontic relationships on a Late Devonian algal bank. J Paleontol 49:1112–1120

Sanford BV (1968) Devonian of Ontario and Michigan. In: Oswald DH (ed) International symposium on the Devonian System (Calgary 1967) I. Alberta society of petroleum geologists, Alberta, Canada, pp. 973–999

Schumann D (1967) Die Liebensweise von *Mucrospirifer* Grabau, 1931 (Brachiopoda). Palaeogeogr Palaeoclimatol Palaeoecol 3:381–392

Seilacher A (1960) Epizoans as a key to ammonoid ecology. J Paleontol 52:189–193

Sparks DK, Hoare RD, Kesling RV (1980) Epizoans on the brachiopod *Paraspirifer bownockeri* (Stewart) from the Middle Devonian of Ohio. University of Michigan Museum Paleontology, Papers on Paleontol, vol 23, 105p

Stainbrook MA (1943) Spiriferacea of the Cedar Valley Limestone of Iowa. J Paleontol 17:417–450

Thayer CW (1974) Substrate specificity of Devonian epizoa. J Paleontol 8:881–894

Wallace P (1969a) Specific frequency and environmental indicators in two horizons of the Calcaire de Ferques (Upper Devonian), northern France. Palaeontology 12: 366–381

Wallace P (1969b) The sedimentology and palaeoecology of the Devonian of the Ferques inlier northern France. Quart J Geol Soc Lond 125:83–124

Wallace P (1978) Homeomorphy between Devonian brachiopod communities in France and Iowa. Lethaia 11: 259–272

Watkins JL (1959) Middle Devonian auloporid coral from the Traverse Group of Michigan. J Paleontol 33:793–808

Wilson JB (1976) Attachment of the coral *Caryophyllia smithi* S. and B. to tubes of the polychaete *Ditrupa arietina* (Müller) and other substrates. J Mar Biol Assoc UK 56:291–303

Witzke BJ, Bunker BJ, Rogers FS(1988) Eifelian through lower Frasnian stratigraphy and deposition in the Iowa area, central Midcontinent, USA. In: McMillan NJ, Embry AF, Glass DJ (eds) Devonian of the world. Canadian Soc Petrol Geol Mem 14(1):221–250

Yakovlev NK (1926) The phenomena of parasitism, commensalism and symbiosis in the Palaeozoic invertebrates. Ann Soc Paleon Russie 4:113–124 (Russian with English abstract)

Zapalski MK (2005) Paleoecology of Auloporida: an example from the Holy Cross Mountains, Poland. Géobios 38:677–683

Fossil Fish Taphonomy and the Contribution of Microfossils in Documenting Devonian Vertebrate History

Carole J. Burrow and Susan Turner

Abstract

Of late little work has been done on Palaeozoic fossil fish taphonomy especially in the Devonian period, [known also as the Age of Fishes], apart from the best-known Lagerstätten. Konservat deposits in the Silurian are rarer even though they have produced the earliest described complete fossil fish. In recent decades, research including taphonomic studies of fossil assemblages from konzentrat–bonebed deposits, especially in the Devonian in East Gondwana (particularly Australia) have given 'added value' to work on complete remains. Analysis of Australian bonebed and rich fish-bearing deposits has expanded temporal resolution in the mid-Palaeozoic fossil record (sample acuity; stratigraphic completeness) and contributed to solutions of broad-scale problems in evolution, biogeography, and biostratigraphy. Taphonomic studies provide new insights into palaeoenvironmental reconstruction and have palaeobiologic significance including ontogenetic development, population composition, and post-mortem processes.

Keywords

Mass mortalities • Taphonomy • Post-mortem processes • Transport • Degradation • Bonebeds • Biostratigraphy • Palaeobiogeography • Agnatha • Gnathostomata

Introduction

The Devonian Period, the 'Age of Fishes' (416–359 Mya), marked the first major radiation of jawed vertebrates (gnathostomes). Their fossil record is abundant, in large part due to the heavily ossified dermal skeletons of both jawed and jawless fishes of the Devonian. The period saw gradual diversification of the ancestors of extant fishes – actinopterygians (ray-fins), sarcopterygians (lobe-fins), and sharks – but the dominant groups for most of the Devonian were fishes that are now extinct. Jawed placoderms with large plates covering their head and thoracic region were the queens (and kings) of the water worldwide until their extinction at the end of the Devonian, but another group, the acanthodians, were probably distributed even more widely through Early and Middle Devonian waters. Unlike placoderms, the acanthodians had minute scales covering their head, body, and fins;

C.J. Burrow (✉)
Geosciences, Queensland Museum, Hendra, Qld 4011, Australia
e-mail: carole.burrow@gmail.com

J.A. Talent (ed.), *Earth and Life*, International Year of Planet Earth,
DOI 10.1007/978-90-481-3428-1_8, © Springer Science+Business Media B.V. 2012

they are much more commonly found as dissociated elements rather than articulated specimens, often providing palaeontologists with a 'stew' of remains and taxa to sort out.

Fish scales and dental elements from Silurian and Devonian deposits have been studied in Europe for over 160 years; they have been of major importance in determining the composition of vertebrate palaeocommunities as well as contributing to palaeobiogeography and biostratigraphy (Turner and Burrow 1999). Only in the last 40 years have similar studies been undertaken on Australasian material. Rich assemblages of fossil fish microremains of early Lochkovian (earliest Early Devonian) and younger age (e.g. Turner 1993) have been obtained from residues of many marine limestones dissolved in acid in quest of conodonts. These assemblages provide valuable information on the composition of vertebrate communities at both high and low taxonomic levels.

In recent decades, major collaborative projects carried out under the auspices of the UNESCO International Geoscience Program (IGCP) have included research on Palaeozoic vertebrate microremains. The first of such endeavours was the very successful IGCP 328: Palaeozoic Microvertebrate Biochronology and Global Marine–Non-marine Correlation; it ran from 1991 to 1996 (Blieck and Turner 2000) and laid the groundwork for successor projects including IGCP 406: Circum-Arctic Lower–Middle Palaeozoic Vertebrate Palaeontology and Biostratigraphy (1996–2000) and IGCP 491: Middle Palaeozoic Vertebrate Biogeography, Palaeogeography, and Climate (2003–2007), as well as work in related projects such as IGCP 410: The Great Ordovician Biodiversification Event and IGCP 499: Devonian Land–Sea Interaction: Evolution of Ecosystems and Climate (2004–2008).

The relative abundances of microvertebrate allow study of changes in communities from similar environments over time, as well as the differences between coeval communities from, for example, nerashore versus off-shore continental shelf (e.g. Ginter 2001; Ginter et al. 2002). Taxonomic evaluation of the faunas contributes to biostratigraphic and biogeographic investigations, as well as to determining phylogenetic relationships within and between vertebrate groups. Interpretation of the palaeoecology of the vertebrate communities is influenced by knowledge of the invertebrate fauna, as well as the geologic and geographic settings. Most of the vertebrate microfossil assemblages that have been studied are from marine or marginal-marine facies, but someare claimed as freshwater in the absence of such associated invertebrate groups such as conodonts and plants.

Here we review some Palaeozoic occurrences, in particular from the Devonian of Australia, using the analytical approach pioneered by Gross (1950), Denison (1956), and Reif (1976) at a time when Australia (with Antarctica), was the eastern part of the 'southern' to equatorial Gondwana landmass (Young 1981).

Acanthodian scales are nearly universally present in Early Devonian microvertebrate assemblages and are often the most abundant constituents. As shown in northern hemisphere studies (e.g. Gross 1971; Valiukevičius 1994, 1998, 2003), these scales may be assigned to species. In Australia, the Acanthodii are proving the most useful of Early Devonian fishes for biostratigraphy and biogeography (e.g. Burrow 2002; Burrow et al. 2010). Another group with complex micromeric squamation, the agnathan Thelodonti, have also proved valuable stratigraphically through the Silurian–Devonian until their demise in the latest Devonian (e.g. Turner 1997a).

The internal structure of dermal bones, teeth and scales is important in elucidating phylogenetic relationships of early vertebrates, but has been little used in cladistic analyses to date, or else produce conflicting evidence (e.g. Warren and Turner 2006; Friedman 2007). Relationships between different groups of agnathans and early gnathostomes are still unclear, particularly for the Thelodonti, Pituriaspida, Placodermi, Acanthodii, and Chondrichthyes ('sharks'). Thus it is important to use all available data on the development and distribution of these fishes, including microfossil assemblages as well as the much less common articulated fossils, if we are to increase understanding of the phylogeny and evolution of early vertebrates.

Influence of Taphonomic Processes on the Fossil Record

The term 'taphonomy' was coined by I.A. Efremov (1940, p. 93) for the study of the principles governing the transition of organic remains from the biosphere to the lithosphere (Olson 1980; Cadée 1991); it is the study of the way in which organisms die, undergo burial, and become preserved as fossils. Although Efremov was the first to apply the term to these processes, he was not the first taphonomist. In the early twentieth century, German workers such as J. Weigelt (1927, translated 1989) were prominent in the study of decay of organisms and preservation in the fossil record. Weigelt (e.g. 1989) coined the term Fossil Lagerstätten in the 1920s based on his studies of the German Permian Kuferschiefer, one of the oldest-known Palaeozoic bonebeds (Diedrich 2009). Seilacher and his 'school' at Tübigen and others (e.g. Seilacher et al. 1985). The Russian 'school' of taphonomy, led by Efremov (e.g. 1940), was concerned with understanding and attempting to quantify the incompleteness of the fossil record, the main aim of the German 'school' of Aktuopaläontologie was elucidation of palaeoenvironments. Wasmund (1926) introduced the term 'thanatocoenosis' to describe assemblages of organic remains, incorporating allochthonous assemblages (transported from somewhere else), autochthonous assemblages (preserved in life environment), and mixed assemblages. As thanatocoenosis was an ambiguous description (being a burial assemblage, rather than a death assemblage as would be implied by the prefix thanato-), Quenstedt (see Cadée 1991) changed the term to 'taphocoenosis'. The interpretative limits to taphonomy have remained vague – Efremov included fossil diagenesis (i.e. the chemical and mechanical alterations within the sediment until rock formation), but here we will discuss only diagenetic processes occurring soon after burial.

This contribution addresses primarily Australian Devonian fossil fish assemblages with regard to the taphonomic processes that produced them. A general overview of the processes of carcass decay and burial in aquatic environments, with modern analogues of the likely sequence of events during their disintegration, is followed by a description of the basic skeletal structure of the various fish groups represented. Predictions are then checked against fossil remains in various states of preservation.

Whereas fossil assemblages of aquatic invertebrates have been extensively classified into different 'taphofacies' (i.e. groupings of assemblages with similar taphonomic histories (Speyer and Brett 1988), the same cannot be used for fossil fish assemblages. Schäfer (1972) commented on the lack of investigation of processes leading to fossilisation of fishes; such taphonomic studies of specific fossil-fish localities are recent and rare (e.g. McGrew 1975; Antia 1979; Märss 1982; Trewin 1986; Turner 1991; Wilson 1996; Zorn et al. 2005). Despite the relative abundance of fish microfossil assemblages through the vertebrate fossil record, they have received scant attention, even in recent specialist books on vertebrate microfossil assemblages and bonebeds claiming to include detailed databases (Sankey and Baszio 2007; Rogers et al. 2007).

In recent years, there has been a broad division of study into 'microvertebrate' and 'macrovertebrate' assemblages, i.e. assemblages of micro- and macroremains of vertebrates respectively (e.g. Turner and Burrow 1999; Long 1993a).The only known Australian Silurian or Devonian macrovertebrate finds that are indubitably pre-Emsian are the partially articulated teleostome (?acanthodian) *Yealepis douglasi* Burrow and Young 1999, from the Ludlow, and a single undescribed lungfish, *Dipnorhynchus* sp., of latest Pragian age (Campbell and Barwick 2000). Young (1993) attributed the seeming rarity of pre-Emsian vertebrate remains to the global position of Australia at that time: in the subtropical dry belt where, by comparison with modern conditions, one would expect an impoverished fauna. However, this hypothesis runs counter to the abundance of bonebeds, or at least assemblages of vertebrate microremains of pre-Emsian age, and by the wide range of species they contain (e.g. Turner 1993, 1997b). The state of preservation of the known assemblages in Australia seems more likely to have been due to taphonomic processes acting on the original carcasses, rather than reflection of impoverished faunas, though some climatic effect may also be involved. Similar trends occur elsewhere in the Southern Hemisphere where there has been a relative lack of exploration or investigation – the case with some of the bonebeds cited above.

Taphonomic Processes

Primarily physical or chemical taphonomic processes are assumed to have changed little through most of the Phanerozoic, whereas biologically mediated processes have changed considerably (Kidwell and Behrensmeyer 1988). Physical conditions such as anoxia or ice-ages, producing distinctive deposits, have long cycles, e.g. Milankovitch forcing (see Hays et al.

1976, for 'legitimisation' of Milankovitch's theories). Some chemical conditions, particularly in marine environments, have varied since the Palaeozoic. Shallow seas have been saturated with calcium, as evidenced by low-magnesium calcite being the dominant marine cement (Tucker 1991).

Disintegration and Pre-burial Decay

There is no precise published information on the natural patterns of disintegration of Devonian fish. The steps by which natural disintegration of the skeleton of a carcass occurs are determined by the nature of the environment and by mobility between the parts. If, for example, after death in a typical 'normal' setting (i.e. where bacterial breakdown occurs), the lower jaw is attached only by a few ligaments to the upper jaw, it will quickly detach once the softer tissues (muscles, skin) have decomposed. With regard to the external skeleton, fishes with macromeric squamation (typical of the mid-Palaeozoic rather than modern settings) would be expected to maintain carcass integrity better than those with micromeric squamation, partly because large scales tend to overlap and interlock more than do smaller scales. The fossil record is certainly biased in favour of larger animals over small ones – predation, scavenging, abrasion and chemical decomposition would be more likely to destroy all remains of small fish, whereas the more robust parts of larger fish might be preserved (Smith et al. 1988). Some deposits, however, show incredible concentrations of small elements. For example, the Late Silurian 1- to 5-cm-thick Ludlow Bonebed has a concentration of billions of thelodont scales over its c. 100 km^2 extent (as now preserved in the Welsh Borderland). This deposition during storm events suggests staggering numbers of individual fish, even with the conservative estimate of 20,000 scales per 20 cm-long fish (cf. Antia 1979; Turner 1991).

Studies of the disintegration of modern fishes and other vertebrates (e.g. Schäfer 1972; Weigelt 1989) provide illuminating analogues (cf. Figs. 1 and 2 for examples of fresh strandings of modern fish). The most detailed investigations of death, disintegration, and burial of existing marine fishes were carried out by Schäfer (1972). Using North Sea fish species in tanks, Schäfer attempted to maintain constant experimental conditions (temperature, salinity, currents), in order to record the manner of disintegration of carcasses of various species. From Schäfer's and later studies, it is apparent that in aquatic taphonomy, temperature is the most important factor determining the fate of a carcass (Elder and Smith 1988). Low temperatures tend to retard decay of soft tissues and bacterial proliferation. At high temperatures (such as we would expect in the Devonian of East Gondwana) the rate of decomposition of gut contents is accelerated, generating gases faster than they can escape from the carcass (Fig. 1a). Thus, even within a single species, carcass buoyancy depends on the presence or absence of gut contents. Temperature will affect, then, the sequence and rate of disintegration of the carcass – the pattern of disintegration differs depending on the shape of the carcass (e.g. dorsoventrally or laterally flattened or 'round', Fig. 1b–d), and whether the carcass remains lying on the substrate, is neutrally buoyant and hanging above the substrate, or floats at the surface.

High salinity will also affect disintegration by retarding bacterial growth, inhibiting decay of soft tissues, and thus reducing gas production from the gut contents and decreasing the likelihood of the carcass becoming buoyant. It had been assumed for many years that low oxygen concentration in the vicinity of fish remains would inhibit proliferation of bacteria, thus delaying or retarding decay (e.g. Brongersma-Sanders 1957). This might not necessarily be the case. The experimental findings of Allison and Briggs (1991) suggest there is little difference in rate of anaerobic versus aerobic degradation. Anoxia is thought to be more important in impeding scavenging and bioturbation than as an inhibitor of microbial degradation.

Scavenging

The 'natural' schedule of disintegration for an undisturbed carcass can be altered by the effects of scavenging (cf. Behrensmeyer and Hill 1980). If undisturbed, body parts of a carcass often become disarticulated but are still preserved as a loose association (e.g. Barton and Wilson 2005). Scavenging will alter this pattern, and not necessarily in a random manner. For example, the early post-mortem detachment of the lower jaw may be hastened by small scavengers entering the carcass via the mouth (often to eat the soft tissue lining the buccal cavity) in addition to resulting

Fig. 1 Modern fresh fish carcasses showing natural settlement in beach strandings. (**a**) Porcupine fish *Diodon*; (**b**) dasyatid ray; (**c**) shark *Lamna ditropis*; (**d**) triglidid teleost (photos ST)

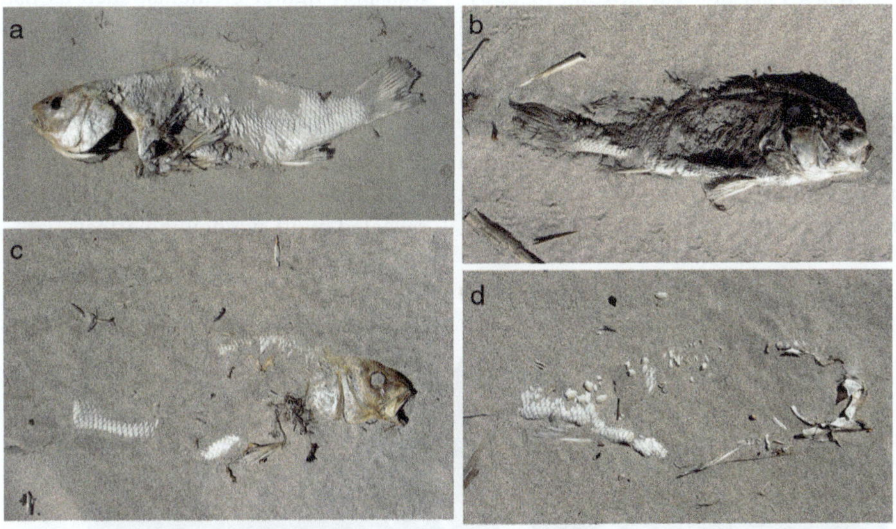

Fig. 2 Stages in disintegration of four stranded freshwater drumfish *Aplodinotus grunniens*, on the shore of Lake Manitoba, Canada, during summer 2003. (**a**) Soft tissues scavenged/disintegrated and washed away; all dermal hard tissues remain in situ; (**b**) scattering of some fin rays, most of the squamation lost/removed on exposed side; (**c**) carcass at a similar stage of disintegration as b, but partially buried; (**d**) carcass after most exposed hard tissues have been scattered/scavenged. (Photographs courtesy of Dr G. Hanke, Royal British Columbia Museum)

from a loose attachment to the upper jaw and/or skull. As well as invertebrate and vertebrate scavengers scattering or devouring fresh remains, bones can also be removed by animals that feed on the bones themselves. Our main challenge is identifying the nature of the scavengers in the Palaeozoic. Only recently, tiny worms have been discovered which devour whole whale bones in deep marine environments (Rouse et al. 2004), and starfish have been found to devour teleost fish skeletons in minutes (Morin 1988). Amongst

Palaeozoic fish, the nipping hand- or spoon-like teeth of *Ageleodus* and *Cynopodius* in the Late Devonian (cf. Garvey and Turner 2006) might have been indicative of a scavenging 'shark'; alternatively, these fish might have been among the first vegetarian fish, grazing on algal mats common in contemporary lagoonal systems.

The structural integrity of bones can be weakened by microbial activity, thus rendering them more susceptible to physical abrasion (Martill 1991). During this time of 'pre-fossilisation', many vertebrate remains, particularly scales, are attacked by sponge, algal or fungal boring (cf. Bystrow 1956; Antia 1979; Goujet and Locquin 1979; Poncet 1993; Fig. 3a–c).

Transport

Most fish carcasses undergo some movement between death and burial, if only by sinking to the substrate. However, carcasses often refloat after fermentation of gut contents. Post-mortem drifting of buoyant carcasses can be prolonged, particularly if the carcasses are large (Schäfer 1972). After the Upper Kellwasser (earliest Famennian) extinction event, fish, especially placoderms, gyracanthid acanthodians, sarcopterygians, and later chondrichthyans, reached a length of up to 3–6 m (e.g. Turner et al. 2005; Ritchie 2007), comparable with that of modern sharks. However, as shown in the debates on the presence or absence of a stomach in thelodonts (cf. van der Brugghen 1994; Märss et al. 2007) as well as of lungs, and a scroll-type spiral valve rather than a stomach in placoderms (Denison 1978), we cannot presume that all Palaeozoic vertebrates had similar gas-producing organs to float a carcass in the same manner as in modern teleosts (cf. Schäfer 1972). Burrow (1996, fig. 9) suggested the process of floating and sinking of dead acanthodians, with loss of their loosely attached heads, as the cause for preservation of all specimens without heads. but having otherwise well-preserved bodies, in the Middle Devonian Bunga Beds fauna of southeastern Australia.

Currents will affect the spatial arrangement of parts of carcasses that disintegrated in situ, or after resettling, if they floated. Whether or not currents have affected the arrangement of the remains will often be revealed by preferred orientation of elongate bones, or by a more pronounced scattering of remains in one direction rather than randomly. Water currents transport lighter elements, such as fish vertebrae and branchial rays, at faster rates than heavier items such as jaws and dermal plates. The shape of the bones will also influence their transport, e.g. cylindrical vertebrae are likely to be rolled away by light currents; rhombic and angular scales are more likely to remain imbricated than rounded ones. Thus current action is often revealed by its winnowing effects.

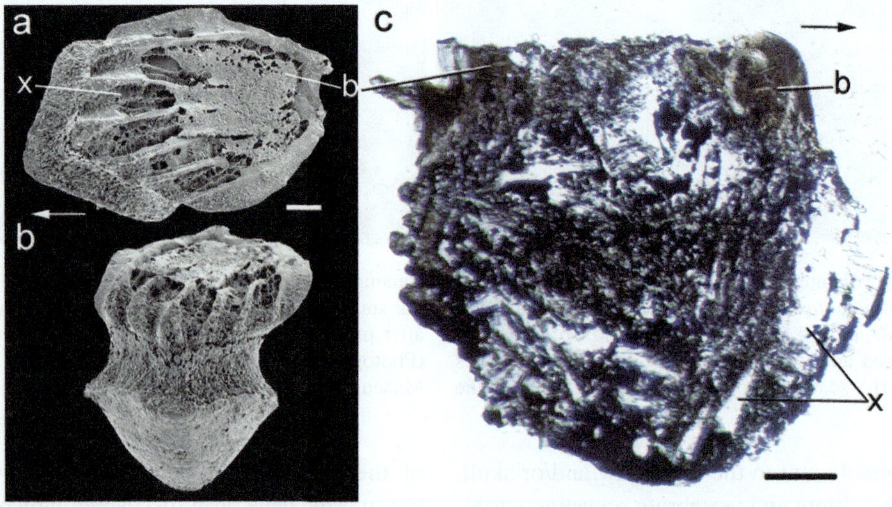

Fig. 3 Diagenetic and scavenging taphonomic processes shown in poorly preserved scales of acanthodian *Nostolepis*? cf. *taimyrica* from the Emsian (Early Devonian) of Wee Jasper, New South Wales. (**a**, **b**) Crown and anterior views of scale WJ2-4; (**c**) vertical longitudinal section ANU V1641.4, showing heavy recrystallisation and borings. Scale bar is 100 μm; *b* boring, *x* recrystallisation; *arrow* points anteriorly. (ANU V = Gavin Young collection, Australian National University)

Burial

Disintegration of a carcass is slowed or even halted by rapid burial. Obrution, or smothering, is more commonly a factor in the preservation of benthic invertebrate faunas than that of fish (Brett and Seilacher 1991). However, early Emsian obrution deposits of the Hunsrück shale of Germany include flattened placoderms, agnathans and one lungfish (Seilacher 1990) killed in a catastrophic event (Brongersma-Sanders 1957), as well as cololites (Seilacher et al. 2001). Such deposits provide a rare insight into entire benthic communities (including fish, sometimes with both macro- and microremains), not usually seen in other fossil occurrences. They can also give insight into the order of the taphonomic processes affecting fish carcasses, as the deposits result from the instantaneous burial of freshly killed carcasses; others were in various states of decomposition. This mixture aids our interpretation of environmental conditions, e.g. the Hunsrück Slate is an instance of low oxygen but highly eutrophic sapropels.

Rapid burial of carcasses affects their disintegration in two main ways: by enhancing sulphate reduction and, correlative with this, by inhibiting or preventing scavenging and bioturbation. In marine sediments, rates of sulphate reduction correlate with rates of deposition, and deposition of degradable organic matter affects sedimentation rate. At higher rates of sediment deposition, less degradable organic matter is decomposed by aerobic respiration, leaving more organic matter available for sulphate reduction (Allison and Briggs 1991).

Degradation

Vertebrate hard tissues include bone (of a number of types), dentine (several types), aspidin, enamel, enameloid or enamel-like tissue, and occasionally mineralised cartilage. In a marine environment, most of the phosphate minerals in skeletons are altered to carbonate fluorapatite, or are dissolved (Lucas and Prévôt 1991). Only in very rare instances are soft tissues preserved; most non-mineralised tissues are readily degraded by microbes. Bones have varying composition, even in the same organism, depending on their function, e.g. whether as protective shells, or as load-bearers. Such differences in mechanical properties influence the potential of a bone to be preserved (Martill 1991). Whereas burial can protect a carcass from scavenging and transport, degradation will continue during and after burial. If bones are buried in sediments depleted of phosphorus, they are quickly destroyed; this also occurs in acidic sediments (Behrensmeyer 1991). In some instances in the fossil record, dissolution did not occur until after infilling of the exoskeleton, forming molds. Even in non-acidic sediments, porous bones are more susceptible to degradation than teeth, because the dentine of the latter is denser and they are often protected by enamel. In calcium carbonate sediments, phosphatic fossils are rare – bone is usually dissolved because phosphorus supply to the sediment is low. The requirements for apatite formation are removal of magnesium, concentration of phosphorus, and non-acidic conditions.

Mass Mortalities and Lagerstätten

Many of the sites that have provided the world's best-preserved fossil fish (and other vertebrates) have resulted from mass mortalities. Studies on the mass mortality of extant marine fishes were summarised and extended by Brongersma-Sanders (1957, see Turner and Cadée 2005). Mass mortality was attributed to various factors including rapid temperature change, extreme changes in salinity, volcanic activity, and planktonic (dinoflagellate or algal) blooms. The latter could be the most common cause, often resulting from upwelling of oceanic waters causing high organic production; the toxins produced by the resultant planktic blooms kill fish and benthic invertebrates (e.g. Thangaraja et al. 2007; Wyatt 2008; Castle and Rodgers 2009). Mass mortalities also occur in freshwater environments, often resulting from deoxygenation events (primarily planktonic blooms), volcanic activity, thermal inversions, variations in salinity (Brongersma-Sanders 1957) and even from reproductive strategies (e.g. Smith et al. 2007). The fish kill events are often related to the El Niño–La Niña climatic cycle (Turner 2002).

When a mass mortality occurs in an enclosed basin (whether freshwater, marginal or marine), sudden influx of organic matter is decomposed by existing aerobic pathways. The resulting depletion of oxygen in the water above the substrate would thus preclude scavenging.

Konservat–Lagerstätten

Assemblages of articulated fossil fish caused by mass mortality occur characteristically in very fine-grained rocks, the consequence of slow sedimentation in calm water or mass death events such as volcanic ash-fall or storm tempestite (e.g. Brongersma-Sanders 1957; Reif 1982). The fossils are usually strongly compressed, giving a more or less two-dimensional preservation of the skeleton and, in rare circumstances, of soft parts as well as partial three-dimensional effects (e.g. Miller et al. 2003). These are Konservat–Lagerstätten, a term first used by Seilacher (1970); in his terminology, bone beds of disarticulated parts (including microfossils) are Konzentrat–Lagerstätten. Few pre-Devonian Konservat–Lagerstätten that preserve fossil fish have been described in detail: the Early Silurian strata of the B–MOTH (Man-on-the-Hill) locality in the Northwest Territories, Canada (Soehn and Wilson 1990), the Early to mid-Silurian Eramosa Lagerstätte in Ontario (von Bitter et al. 2007) and contemporaneous beds in Wisconsin (Mikulic et al. 1985), the Lesmahagow and Hagshaw Inlier assemblages in Scotland (e.g. Ritchie 1985; Märss et al. 2007), and the Late Silurian Gossage Formation and contemporaneous assemblages of the Northwest Territories (e.g. Broad and Lenz 1972). Surprisingly, the Scottish assemblages have mostly been ignored in Lagerstätten studies despite their importance in having produced the first-described (Traquair 1899) complete fossil fish, thelodonts, and anaspids (in large quantity) associated with a diverse arthropod fauna. None of these assemblages include jawed fishes.

The most spectacular of the Devonian assemblages fulfilling the criteria of exceptional preservational quality, specimen abundance, and taxon diversity is in the Early Devonian (Lochkovian) strata of the MOTH locality. Taxa from this site were first described by Dineley and Loeffler (1976) and Bernacsek and Dineley (1977), with many subsequent new descriptions and revisions of taxa (see Wilson et al. 2007). The quality, diversity, and abundance of the fossil vertebrates from this site is highlighted by the so-called Wonder Block (Hanke and Wilson 2006, fig. 2a), which produced eight species representing seven genera. Other Early Devonian Konservat–Lagerstätten that preserve articulated fish are in the Hunsrück Slate in Germany (Bartels et al. 1998), and fine-grained slates of the Arbuthnott Group in the Lower Old Red Sandstone (LORS) of Scotland (Trewin and Davidson 1996). Gnathostomes dominate the fish faunas from these assemblages: placoderms in the Hunsrück Slate, and acanthodians in the LORS. Of younger Devonian occurrences, the freshwater (Givetian) Achanarras and contemporary formations in Scotland (Hamilton and Trewin 1994) and the Lode Quarry Frasnian assemblage in Latvia are notable for preservation of a wide spectrum of taxa, the former including newly described agnathans (e.g. Newman and Trewin 2001; Newman 2002) and the latter with ontogenetic series of placoderm and acanthodian species (Upeniece 2001).

The only marine deposit of this type yet found in the Australian Devonian is the Gogo Formation of Western Australia, with spectacular 3D preservation of fish (discussed in the Section 'Three-Dimensional Preservation'). Several late Middle–Late Devonian freshwater localities are known, including the massive fish kill at Canowindra, New South Wales with many large sarcopterygians and placoderms (Ritchie 2007), the nearby Merriganowry Quarry with thousands of specimens of a single placoderm species (Ritchie 2005), and the Mt Howitt locality in Victoria with its diverse assemblage of placoderm, acanthodian, actinopterygian, and sarcopterygian fishes (e.g. Long 1983, 1985, 1986, 1988a).

Three-Dimensional Preservation

Even more informative than the flattened specimens typically found in mass mortality and/or obrution assemblages are rarer instances of three-dimensional preservation in calcareous nodules. Carbonate concretions of the Late Devonian Cleveland Shale of the USA (Clifton 1957) commonly enclose arthrodire placoderm remains. These pyrite-rimmed concretions are often large (1–3 m diameter) and almost perfectly spherical. The presence of such large dense bodies in what were originally soft sediments has seemed anomalous, but the probable origin is as follows: Criss et al. (1988) considered that these concretions originated in extremely porous sediments near the sediment/water interface as bodies of adipocere ('grave wax'), or low density organic soap, formed from the soft tissues of the fish in rapid early diagenetic growth. During later phases, mineral replacement and recrystallisation transformed the concretions. Their formation also required raid sedimentation (Briggs et al. 1993).

Carbonate concretions of the Late Devonian Gogo Formation of north-western Australia yield specimens preserved in 3D showing exceptionally fine detail of anatomical features which allow new understanding of evolutionary processes and history (Long and Trinajstic 2010). Delicate cartilage with a fine layer of perichondral bone, muscle tissue, nerves, blood vessels (Trinajstic et al. 2007), and even an umbilical cord (Long et al. 2008) have been preserved in situ and revealed by dissolving the concretions in acetic acid. As well as entire skeletons, isolated large bones and semi-articulated skeletons have been preserved in concretions from this site.

Microfossil Bonebeds and 'Bone Sands' (Konzentrat–Lägerstatten)

Compared with other fossil remains from the Palaeozoic, bonebed assemblages of vertebrate microfossils, including scales, teeth, spines, and other bony parts <5 cm diameter, are common worldwide. They provide the oldest record for all vertebrate groups – the only record for some groups. In most cases, these assemblages contain only skeletal remnants robust enough to survive wave and/or current action, transport, scavenging, and even digestion. Recent workers have proposed two terms to describe these deposits: '*microfossil bonebeds*' showing relative concentration of fossils where most ($\geq$75%) elements are $\leq$5 cm in maximum dimension (Eberth et al. 2007) and '*bone sands*' for those comprising fossil elements $\leq$4 mm (Rogers and Kidwell 2007). The first such deposits to be mentioned in the literature were the Late Silurian Ludlow bonebeds (LBB) of west England and Wales (Murchison 1839). The more common environments in which they were deposited include littoral zones, off-shore bars, deltas, estuaries, and inter-reef mini-basins. The resulting accumulations of skeletal debris were often formed during marine transgressions. As sea level rose, sand banks and bars migrated landwards, spreading out to produce the thin layers characteristic of bonebeds. Antia (1979) and Seilacher (1990) hypothesised that many bonebeds were formed during storms that caused mixing and concentration of coarser fragments; thus any specific deposit could consist of fossils from a substantial time-slice. This was undoubtedly the case for those bonebeds with scales and other skeletal elements showing various states of abrasion. It seems, nevertheless, a less likely explanation for assemblages that consist of morphologically and histologically well-preserved microfossils were probably deposited during a short time interval (catastrophic) by a tempestite (cf. Ager 1993). Agassiz (in Murchison 1839) wrote the first taxonomic descriptions of microremains in the LBB assemblages; Pander (1856) subsequently described the fauna from the Late Silurian Ohesaare Formation of Saaremaa Island, Estonia. These beds comprise platy bioclastic limestones in the basal part with a rich shelly fauna and a diverse association of agnathans and fish remains, with each thin layer formed during progressive regressions and transgressions. Many of the vertebrate taxa described and named by Pander were based on isolated scales of thelodont agnathans *Thelodus* spp. and *Phlebolepis* and the gnathostomous acanthodians *Nostolepis striata* and *Gomphonchus sandelensis*, and dermal bone fragments of the putative stem osteichthyan *Lophosteus superbus*; Pander's higher classification of most of these taxa was erroneous. More than a century later, Walter Gross (1969, 1971) described and ascribed isolated teeth, tooth whorls, dermal bone fragments, tesserae, and fin spines to these gnathostomes, none of which are yet known from articulated fish.

Many limestone deposits from the Early Devonian of southeastern Australia yield abundant vertebrate and invertebrate microremains, which appear to have been deposited in relatively brief intervals. The geography of this region at the time – near-shore and off-shore volcanic pedestals and debris flows (e.g. Taube et al. 2005), stromatoporoid-algal reef complexes and other shallow shelf to continental slope environments (e.g. Wolf 1965, environments re-interpreted as based on olistoliths in debris-flow situations by Talent and Mawson 1999) (see also Mawson and Talent 1994, 1997; Talent and Mawson 1994; Sloan et al. 1995). Some of the Devonian fish occurrences in this vast tract may have resulted from mass mortalities, perhaps caused by volcanic effusions and/or algal blooms. The environment would not have favoured good preservation of entire carcasses, but their more resistant parts, such as scales, teeth, and isolated bones including optic capsules were preserved intact (e.g. Burrow 2002, 2006). For micromeric fishes such as thelodonts, acanthodians, and sharks, as noted above, even a small individual could have many thousands of scales covering the body at any one time, and like modern sharks some could have shed both scales and teeth throughout their lifetime. Thus mass-kills of a few (or even one) fish species could have created bonebeds.

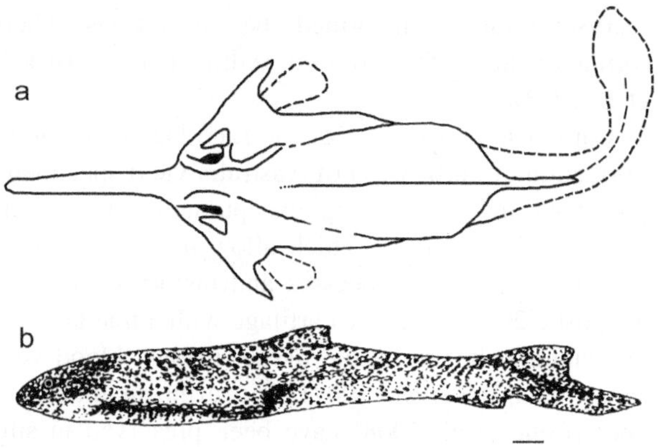

Fig. 4 Agnathan groups found in the Devonian of Australia. (**a**) *Dorsal view* of *Pituriaspis* (after Young et al. 2010, fig. 3B); (**b**) *lateral view* of a turiniid thelodont cf. *Turinia pagei* (sketch by ST)

Structural Diversity, Preservation, and Stratigraphic Record of Australia's Devonian Fishes

Devonian fishes show remarkable diversity in their skeletal structures, both internal and external, hence the common designation of this period as the 'Age of Fishes' (see Long 1995 for a pictorial overview of Devonian fishes). There is no worldwide consensus on the classification of fish-like vertebrates of Devonian age, and so we will examine traditional 'groups' without attempting to arrange them in a formal classification; the terminology follows Janvier (1996, 1997).

Agnatha

Most agnathan groups have not been found in Australia to date, despite their widespread occurrence in the Silurian–Devonan of Europe, Siberia, and North America (the Osteostraci, Pteraspidomorphi, and Anaspida), and in China (the Galeaspida) (e.g. Janvier 1996; Turner 1999). The Arandaspida, the oldest known group of the Pteraspidomorphi, has been found as both microremains and molds of articulated fish in the Ordovician of Gondwana including Australia, but none of these groups has been encountered in the Silurian–Devonian of Australia.

Pituriaspida

This poorly known group of purported agnathans was described by Young (1991) from a few specimens of sandstones of the Cravens Peak Beds (Early to Middle Devonian) in the Georgina Basin of western Queensland. The carapace-like external skeleton is represented only by external molds in fine-grained sandstone; its composition remains unknown (Fig. 4a). Young described this group on the basis of the elongate sutureless armour showing tuberculate ornament, a long narrow rostral process and posterior spine, anterior cornuate processes, oval pectoral fenestrae and an anteroventral branchial opening; the internal skeleton is not known. Current ideas place pituriaspids in close relationship to osteostracans and stem gnathostomes (Janvier 1996), but they have a galeaspid-like median dorsal opening (Young et al. 2010). No scales or other microremains have been attributed to the pituriaspids. If correctly assessed, these are the only post-Ordovician armoured agnathans known from Australia.

Thelodonti

Thelodonts were usually small agnathans (less than 200 mm long), although the largest estimate of length for a partially articulated specimen is about a metre (Turner 1986) (Fig. 4b). Most thelodonts had a large cephalothorax, lateral triangular pectoral flaps, a slim trunk, dorsal and anal fins, and a large hypocercal tail (Turner 1982, 1991). They had an external skeleton of discrete dermal denticles – some (mainly on the head region) non-imbricating; most on the trunk and fins imbricating. Well-preserved specimens are known from the Early Silurian to Early Devonian of Scotland, Estonia, Norway, and North America (Turner 1982, 1986; Märss et al. 2007; Märss and Ritchie 1998), whereas the scales are ubiquitous from Late Ordovician to Late Devonian (e.g. Turner et al. 2004; Märss et al. 2007). For one form, *Lanarkia*, from the Silurian Hagshaw Lagerstätte in Scotland, Turner (e.g. 1992) used the analogy of modern puffer or porcupine fish (Fig. 1a) to hypothesise an erectile, protective function for the spiny scales.

The predominant thelodonts in Gondwana, especially Australia, are the turiniids (Turner 1997a), of which only the type specimen *Turinia pagei* is relatively well preserved (Fig. 3b). Even in articulated fish, little of the endoskeleton has been preserved and thus the internal skeleton is poorly known, but it was presumably mostly unmineralised cartilage. A few of the articulated specimens from Scotland and Norway (Turner and van der Brugghen 1993; Märss and Ritchie 1998) and Canada (Wilson and Caldwell 1998) show evidence of possible cartilaginous endoskeletal supports for the buccopharyngeal and branchial apparatus, caudal fin web with possible cartilaginous dermotrichia, and the cartilaginous notochord extending into the hypochordal lobe of the caudal fin (Märss et al. 2007).

The dermal denticles, or scales, had dentinous, nongrowing crowns and necks, with growing bases of acellular bone-like tissue probably analagous with aspidin. As they grew, the bases were strongly attached to the dermis by partially mineralised collagenous Sharpey's fibres; most taxa are characterised by specialised basal extensions (Turner 1991; Märss et al. 2007). Dentine types exhibited in thelodont scales range from orthodentine to a tissue similar to mesodentine. The scales range in length from 0.1 to 3.5 mm; they show great morphologic variation from taxon to taxon and from region to region across the body of an individual animal, with distinctive head, transitional and body scales, plus special scales in circumorbital, branchial, and pharyngeal areas. Ontogenetic changes in the form of the scales have also been reported (Gross 1967; Turner 1991).

A suitable modern analogue for this group for which the natural order of disintegration of a carcass is known, is a shark such as *Galeorhinus galeus*, as studied by Schäfer (1972). In his study on northern German beaches, the shark's placoid scales came off first, in disorderly heaps. Following this, the cartilaginous skeleton gradually disarticulated. The main similarities between thelodonts and sharks are the general form of their squamation, and their cartilaginous skeleton. The principal difference between their scales is their base structure, being robustly developed in thelodonts with most taxa having a thickened base which could overgrow the pulp cavity and/or have elongate 'roots' resembling those on modern mammalian molar teeth.

No articulated thelodonts have yet been discovered in Australia, but their scale record extends from the early Late Silurian to the Late Devonian in shallow marine to transitional facies (Turner 1997a; Burrow and Turner 2000); some late Early–early Middle Devonian occurrences through central to western Australia are considered to be nonmarine, but some of these may have been reworked from marine incursions (Burrow et al. 2010). All Australian taxa are thelodontid thelodonts; the youngest known record is *Australolepis seddoni* from the late Frasnian of the Virgin Hills and Gneudna formations, north Western Australia (Trinajstic 2001; Trinajstic and George 2005, 2009). Globally, their youngest occurrence is in the Famennian of Iran (Märss et al. 2007).

Gnathostomes

Unlike the Agnatha, nearly all major groups of Devonian gnathostomes are represented in Australia's fossil record. Brazeau (2009) called into question the monophyly of both traditionally recognised extinct classes of gnathostomes, the Placodermi and Acanthodii. However, until Brazeau's analysis can be critically evaluated, we retain these groups.

Chondrichthyes

General studies of modern chondrichthyan hard parts, tooth variation and squamation allow analysis of fossil, mostly disarticulated remains (Fig. 5). Reif (1985) provided a benchmark study of variation in neoselachian scale patterns which serves as a basic analogy for all extinct 'sharks' as well as thelodonts and acanthodians. Other works (e.g. Deynat and Seret 1996; Leidner and Thies 1999) on complete Mesozoic to modern sharks have helped to build knowledge of potential microfossils and patterns of preservation. Few detailed studies have been done on scale and tooth pattern variation in articulated Palaeozoic sharks, although those by Dick (1981) and Maisey (1989) are noteworthy.

Putative chondrichthyan scales are first recorded from the Australian Ordovician, from one of the oldest known bonebeds (e.g. Young 1997; Blieck and Turner 2003), but the first appearance of undisputed

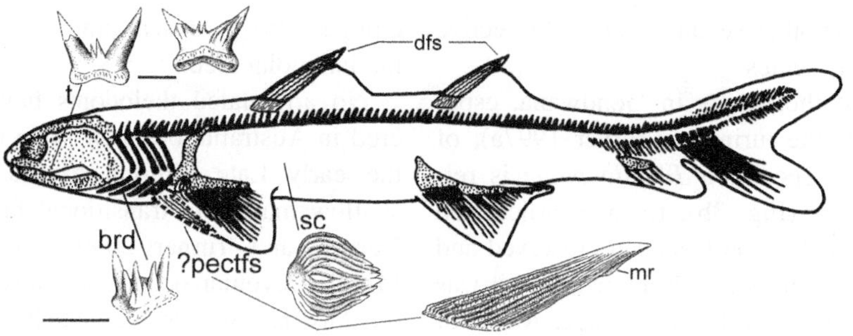

Fig. 5 Hypothetical reconstruction of antarctilamniform shark *Antarctilamna prisca* (after Young 1992, revised to indicate possible pectoral and posterior dorsal fin spines), showing individual elements found as microfossils – brd, branchial denticle; sc, scales; ?pectfs, possible pectoral fin spine; dfs, dorsal fin spines; t, teeth; mr, median ridge. Scale bar is 1 mm for the microfossils

shark teeth is in the earliest Devonian (Lochkovian) of northern Spain (Mader 1986). The only chondrichthyans represented by articulated specimens to the end of the Devonian period are the sharks; crushing teeth of stem holocephalans have been found in the Middle (Darras et al. 2008) and Late Devonian (with the possibly oldest articulated specimens in South Africa). A range of definite holocephalans have been found in the Carboniferous (Mississippian) Bear Gulch Lagerstätten in Montana, USA (e.g. Lund 1986); one putative spine was described from the Silurian of Irian Jaya (Turner et al. 1995). Rays (true batoids) have no fossil record until the Mesozoic but we can look for analogous structures in the 'bradyodonts': eugenodonts, paraselachians, and iniopterygians (e.g. Zangerl 1981).

Well-preserved shark braincases have been found in the Early Devonian of South Africa and eastern Canada (Maisey and Anderson 2001; Miller et al. 2003; Maisey et al. 2009) and the Middle Devonian of Australia and Bolivia (Young 1982; Maisey 2001), but essentially complete specimens do not appear in the fossil record until the Late Devonian (late Famennian) in the famous Cleveland Shale Lagerstätten. Most of the older shark fossils comprise isolated teeth, dermal denticles, fin spines and rare calcified cartilage. Deducing the structure of early sharks is based as much on what is missing from the record as on what is present; e.g. that their internal skeleton was unmineralised cartilage and that the earliest gnathostomes bearing shark-like scales lacked teeth (Hanke and Wilson 1998, 2004, 2010; Wilson and Hanke 1998; Turner and Miller 2005; Wilson et al. 2007).

The shark endoskeleton is composed of cartilage; it usually becomes prismatically calcified in adults in a characteristic tessellate pattern. Isolated braincases from the Devonian show that the neurocranium was not fused to the gill arches or to the upper jaw. The visceral skeleton has weakly calcified branchial arches, plus a mandibular arch cartilage, a palatoquadrate (upper jaw) and Meckel's cartilage (lower jaw). The notochord of Devonian sharks either lacked vertebral centra or these were formed of unmineralised cartilage. Ribs are presumed absent in the earliest sharks (Zangerl 1981), but are present in the putative chondrichthyan *Kathemacanthus* (Hanke and Wilson 2010). Although a single dorsal fin with spine was once considered to have been the primitive elasmobranch condition (Zangerl 1981), more recent discoveries indicate that early sharks retained the general gnathostome condition with two dorsal fins. Partially articulated specimens of the late Middle Devonian shark *Antarctilamna prisca* Young 1982, from southeastern Australia and Antarctica support this interpretation. The discovery of an almost completely articulated skeleton (missing post-anal body with tail) of the Emsian shark *Doliodus* from eastern Canada (Miller et al. 2003), and interpretation of *Antarctilamna from* Antarctica and southeastern Australia (Fig. 5), have shown that some early sharks had paired pectoral fin spines, a feature previously considered to be diagnostic of the Acanthodii. This interpretation will help clarify many pre-Famennian 'ichthyodorulite' (fin spine) taxa, a study that is underway.

Of all fish groups, sharks have changed relatively little since the end of the Devonian: the body shape of the Late Devonian *Cladoselache* closely resembles that of modern nektonic sharks. Using once again as an analogue the disintegration of Schäfer's *G. galeus* carcass, the following events are inferred. The carcass sank to the bottom ventral side down; subsequently the carcass refloated with head and tail hanging down, then sank back to the substrate. Pieces of skin fell off and decomposed, leaving heaps of placoid scales. The abdominal cavity broke open. After 20 days, some remains of the rays of the paired and unpaired fins would remain, along with jaw cartilages and oral teeth, and scattered heaps of placoid scales. The vertebrae would have rolled away.

The skeletal cartilage of sharks is highly distinctive, formed of hexagonal prisms of cartilage linked by short connective fibres. This material is very hard when fresh, but usually these fibres decompose quickly after death, leaving microscopic prisms which may be preserved as a thin film in the sediments or rarely as intact, linked units (e.g. Burrow et al. 2008b, fig. 3). The neurocranium is attached only loosely to other parts of the skeleton; it was the most robust portion of the shark internal skeleton.

Shark teeth consist of a dentine crown usually covered by a hard shiny enameloid substance and a base with vascular canals, attached to the jaw by connective tissue only. The oldest shark teeth found in Australia are those of *Mcmurdodus whitei* from a ?late Emsian/early Eifelian limestone in the Cravens Peak Beds of western Queensland. Surprisingly, the teeth are unlike any other Devonian, or even Palaeozoic, shark teeth, but closely resemble those of the modern deepwater shark *Echinorhinus* (Burrow et al. 2008a). Tessellate prismatic calcified cartilage and bony rings akin to those encasing the lateral line canal in *Echinorhinus* are found in the same microfossil assemblage. Vertebrate remains from this Cravens Peak Beds limestone, though fragmentary, are beautifully preserved, showing no evidence of microscopic or macroscopic scavenging.

A wide spectrum of tooth forms is encountered in younger rocks; most Palaeozoic shark taxa are based on such teeth. In Australia, stratigraphically useful shark teeth are most abundant in the Middle and Late Devonian, with *Phoebodus* and other cosmopolitan taxa found in the Canning and Carnarvon basins of Western Australia (Trinajstic and George 2005, 2009) and eastern Australian basins in northern Queensland (Turner et al. 2000), and in New South Wales (e.g. Young and Turner 2000). The shark teeth record indicates that, if *Mcmurdodus* and *Echinorhinus* are an amazing example of convergence rather than an ancient lineage, then based on teeth, neoselachians appeared first in the Carboniferous (Duffin and Ward 1983) although neoselachian-like placoid scales occur as early as the Silurian (e.g. Turner and Miller 2005). Either date of origin is at odds with the Late Palaeozoic appearance determined by Mesozoic shark workers, who rely only on articulated fossil and Recent sharks to determine lineages (e.g. Maisey et al. 2004).

One cosmopolitan putative chondrichthyan from the Early to Middle Devonian of Australia is *Ohiolepis*, a genus known only by scales; its first identification from Australia was in Schultze (1968). It was probably toothless, without fin spines and with a cartilaginous internal skeleton. Its non-imbricating scales have a convex bone base with a crown of finger-like polyodontodes. Each odontode is formed of subparallel orthodentine tubules leading into a central pulp cavity. Chondrichthyan scales are present in the Early Devonian of Australia (e.g. *Emsolepis* Turner 2004) and much more abundant through the Middle and Late Devonian (Turner et al. 2000), but their taxonomic assignment is more difficult because, based on articulated specimens from other countries (e.g. Williams 1985), the scale types from each area of the fish show little variation within families. Given this proviso, it is still possible to distinguish Cadée antarctilamnid, ctenacanthid, protacrodontid, stethacanthid, phoebodontid, and hybodontid scale forms.

The only near-complete articulated shark specimen yet found in Australian Devonian sediments is from a (Frasnian) Gogo Formation nodule (Long 2007, CJB pers obs). As noted above, specimens of *A. prisca* (Order Antarctilamniformes Ginter et al. 2008), from the ?Eifelian lacustrine flysch deposits of the Bunga Beds, south-eastern New South Wales, are preserved as casts of braincases, casts of fragments of a palatoquadrate (the quadrate part of the cartilage heavily developed), apparent traces of the ceratobranchial on shagreen patches, spines, and teeth. Very few scales are present. The remains are preserved in flat nodules in fine calcareous siltstone, a preservation style consistent with Schäfer's studies: the scales, which have relatively thin bases, probably detached first from the carcass, and were possibly removed by current action before nodule formation began. The braincase and thick parts of the jaw must not have been preserved until after the nodule hardened, but subsequently were dissolved leaving natural molds.

In the light of numerous occurrences of chondrichthyan scales and teeth through the Middle and Late Devonian of Australia, it is surprising that shark fin spines are relatively rare. A few fragments of the spine type with smooth-noded longitudinal ridges – which we consider to be the stem chondrichthyan ornament – have been found at the Cravens Peak Beds *Mcmurdodus* locality (Burrow et al. 2008a, fig. 5A–H). They may be compared with the spines of Lochkovian *Leonodus*, early Emsian *Doliodus* of New Brunswick, and putative stem chondrichthyans from the Lochkovian MOTH site in Canada (e.g. Hanke and Wilson 1998, 2010; pers obs). *Antarctilamna* has robust ctenacanthid-type spines, but their only known occurrence in Australia is at the Bunga Beds locality. The scarcity of fin spine records through Australia's Middle and Late Devonian could indicate that many taxa had lost fin spines ontogenetically. The Gogo shark, for instance, appears to be anacanthous (CBJ pers obs). *Phoebodus*, probably the most useful vertebrate genus for Late Devonian–Early Carboniferous biostratigraphic studies, is known only from isolated teeth and putative scales (Ginter and Turner 1999; Trinajstic and George 2009); it too could have been anacanthous. Many assigned and unassigned 'ichthyodorulites' are, however, relatively common elsewhere in the Devonian world (e.g. Woodward 1891).

Acanthodii

Acanthodians are the first gnathostomes for which evidence of teeth appear in the fossil record (e.g. Hanke et al. 2001a; Burrow 2003a) (Fig. 6). During the Devonian, they were mostly small fusiform fish with a heterocercal tail. The general morphology of the acanthodian endoskeleton is described by Denison (1979) and Janvier (1996). Of the traditionally recognised orders within the group, the Climatiiformes are now considered to be a paraphyletic group; we treat 'climatiiform' families as undetermined at order level.

Regarding the structures most commonly found as fossils, acanthodians had one or two dorsal fins, with fin spines formed of a lamellar inner layer,

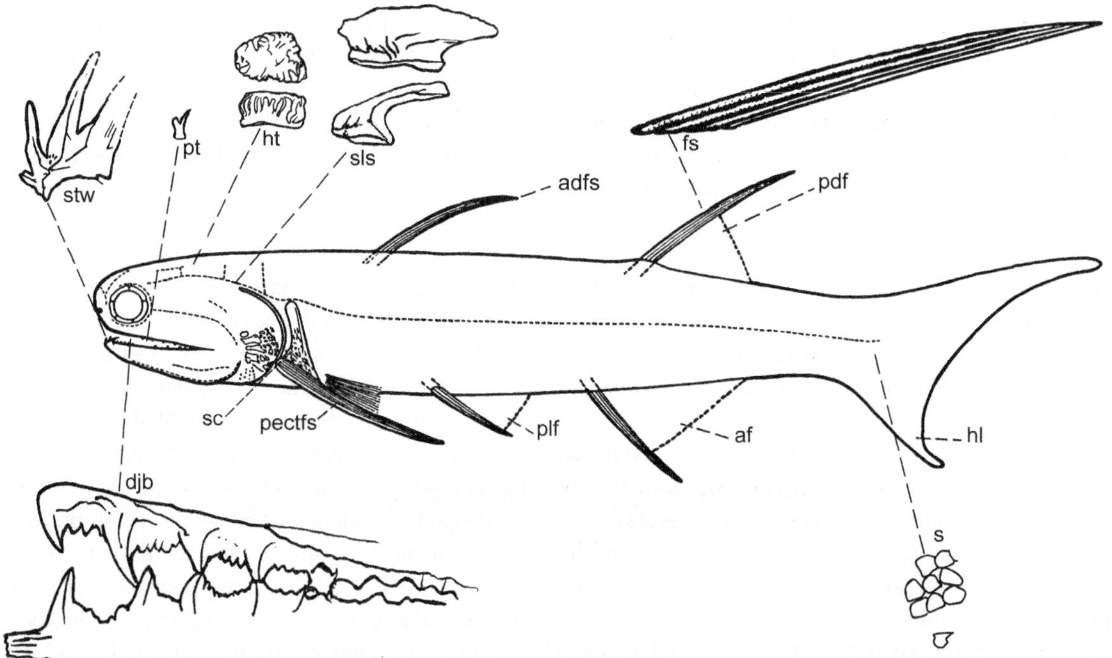

Fig. 6 Individual elements of ischnacanthiform acanthodians found as dissociated microfossils in bonebed assemblages, exemplified by *Ischnacanthus gracilis* (after Watson 1937, Fig. 11) from the Lochkovian slates of Forfarshire, Scotland – adfs, anterior dorsal fin spine; af, anal fin; djb, dentigerous jaw bones (after Ørvig 1973, text-fig. 1c); fs, fin spine; hl, hypocaudal fin lobe; ht, head tessera, lateral and crown views; pectfs, pectoral fin spine; plf, pelvic fin; pdfs, posterior dorsal fin spine; pt, tooth from oral mucosa; s, scales in crown view and one in lateral view; sc, scapulocoracoid; sls, head sensory line scale, lateral and anterior views (after Miles 1966, fig. 11); stw, symphysial tooth whorl (after Ørvig 1967, plate 3.5). Relative magnification of elements: s at 2x ht; ht, pt, sls at 2x djb, fs, stw

osteodentine middle layer, and ortho- or mesodentine outer layer on the leading edges of all fins except the caudal. Some genera had prepelvic and admedian spines located ventrolaterally between the pectoral and pelvic fins, and/or prepectoral spines. Basal fin radials have been noted in the pectoral fins of a few acanthodians, e.g. *Ischnacanthus gracilis* (Watson 1937, fig. 11); basal cartilages have been identified on dorsal fin spines in the ?diplacanthiform *Tetanopsyrus* spp. from the MOTH locality (Hanke et al. 2001b), an ischnacanthid from the Emsian Water Canyon Formation of the central west USA (Burrow 2007a, figs. 6.7–20, 9.3,4), and the Early Devonian Canadian acanthodiform *Promesacanthus eppleri* (Hanke 2008, fig. 6c). Fin spines in climatiids and ischnacanthids were only shallowly inserted in the skin/musculature; spines in all other acanthodians were more deeply inserted. Pectoral and/or prepectoral fin spines are often preserved in articulation with the endoskeletal scapulocoracoid in climatiids and some diplacanthids, even in assemblages of dissociated elements (e.g. Burrow 2007a, figs. 5.10, 11, 6.5), whereas in other articulated examples of Devonian acanthodians, the spine and scapulocoracoid are usually disarticulated during compaction of the fossil. A rare exception to this is found in the microfossil assemblage from the ?late Emsian/early Eifelian limestone outcrop of the Cravens Peak Beds where articulated scapulocoracoids and pectoral fin spines of the mesacanthid acanthodiform *Teneracanthus toombaensis* are preserved with delicate perichondral mineralisation between the basal opening of the spine and the scapulocoracoid (Burrow and Young 2005, figs. 3A–C, 4L,M,R,S). Not surprisingly, fin spines are the most common isolated acanthodian macroremains recorded, given the proviso that at least some Early Devonian fin spines from the Northern Hemisphere still have uncertain affinities (e.g. Hanke and Wilson 1998, 2010).

All acanthodians had a perichondrally ossified scapulocoracoid, a defining feature for the group (Burrow and Turner 2010). The skull had a covering of dermal bones or tesserae and, in some species, the pectoral girdle included dermal elements. The acanthodians are distinguished from the Osteichthyes, or true bony fishes, by their tooth histology (in those species that had teeth). The teeth lack an enamel-like surface and do not show evidence of regular replacement. The dentition of the group varies: climatiid acanthodians had tooth whorls along the upper and lower jaw cartilages; ischnacanthiform acanthodians (Fig. 6) had dentigerous jaw bones, tooth whorls, and isolated teeth. Poracanthodid and acrolepid ischnacanthiforms also have denticulated cones in the mouth or pharynx (Valiukevičius 1992, 2003). Diplacanthiform and acanthodiform acanthodians had no teeth, although some diplacanthiforms appear to have had smooth bony plates on the lower jaw cartilages (Hanke and Davis 2008).

The endocranium was of variable composition, being formed of either unmineralised cartilage (and thus not preserved), or of a granular/globular mineralised cartilage (e.g. ischnacanthiforms, Burrow 2004), or with perichondral bone over mineralised or unmineralised cartilage (advanced acanthodiforms).

The acanthodian body was covered with oblique rows of small, closely packed, rhombic scales; this micromeric scale cover usually extended onto the fins. In most taxa, the scales show an 'onion-skin' (i.e. superpositional) type of growth, and comprise a massive convex base, a neck, and a crown which can be smooth or sculptured. Scales can be divided generally into four histological categories (Valiukevičius 1995), *Nostolepis* type, *Diplacanthus* type, *Acanthodes* type, and *Poracanthodes* type; investigation of the histological structure of isolated scales is usually necessary for determining taxonomic relationships (cf. Valiukevičius and Burrow 2005). There is wide variation in scale morphology on individual fish (particularly in the climatiids) as well as interspecifically. Some poracanthodids have scale crowns with appositional growth zones posteriorly, whereas other acanthodians have crowns with areal growth zones; thus the traditional diagnostic character of 'onion skin' type growth zones in flank scales is not always found.

The classical taxonomic divisions into orders and families within the Acanthodii were based on assemblages in the Early and Middle Devonian Lower Old Red Sandstone quarries of Scotland and *Acanthodes bronni*, the youngest and most derived acanthodian from the Permian of Germany. Most of the LORS acanthodians have been known and studied for more than 150 years (e.g. Woodward 1891). Recently discovered acanthodian, putative chondrichthyan, and stem gnathostome taxa from the Lochkovian of the MOTH locality in Canada have thrown acanthodian and chondrichthyan taxonomy into disorder. Many of these taxa have paired fin spines (previously used as a diagnostic character for the Acanthodii) and scales that would be classified as chondrichthyan if found in microfossil assemblages (e.g. Hanke and Wilson 2004, 2010), compounding the problems posed by the New Brunswick *Doliodus* (see above).

Finding a suitable modern analogue for acanthodians is difficult; the dermal skeleton is perhaps generally comparable to that of sharks, but much of the internal skeleton was mineralised cartilage or perichondrally ossified, rather than purely cartilaginous. From the evidence provided by fossil remains, it appears that the acanthodian skin and scales, with their well-rounded bony bases, maintained their integrity after death better than those of most fossil and modern sharks. Isolated patches of scales are often preserved in their original arrangement (e.g. Woodward 1906).

Myoxocephalus scorpius (sea scorpion, a Recent teleost fish) is an almost fusiform fish with reasonably robust fin rays loosely analogous to acanthodian fin spines. Schäfer's (1972) experiments detailed the disintegration schedule of the fish as follows: the carcass sank to the floor, ventral side uppermost; gas bubbles formed underneath the abdominal skin and eventually broke through. Skin dropped off in pieces from the side of the body. After 4 days, the carcass would have floated to the surface, ventral side up, the abdominal skin would have torn, and the back of the body drooped downwards. The body would have been held at the water surface only by its head and chest section – filled with gas. Unpaired fins would have dropped rays that sank to the bottom. The carcass would then have sunk back to the sea floor and collapsed. In some cases the neurocranium is missing because the skull resurfaces more frequently than other parts and drifts away.

Australia's Devonian acanthodians show a full range of states of preservation: from well-articulated skeletons, through partially articulated skeletons, to assemblages with scales, fin spines, and jaw fragments, and just fin spines or scales. The most common preservation of acanthodian remains is in microvertebrate fossil assemblages. Since Pander's work (1856), it is from scales alone that most acanthodian species have been described.

Some of the best examples of good preservation worldwide are from Lagerstätten in nonmarine or marginal marine deposits, as in the Lower Devonian Arbuthnott Group and Middle Devonian Achanarras Formation of Scotland (Trewin 1986; Hamilton and Trewin 1994; Trewin and Davidson 1999), and from an interval in the Early Carboniferous of the northern Minusinsk Basin of Siberia; it seems to have been caused by a sudden outpouring of volcanic ash (Novitskaya and Obruchev 1964). In Australia, the Frasnian lacustrine Lower Mudstone unit at Mt. Howitt in Victoria has extremely well-preserved acanthodiform (Long 1986) and diplacanthiform acanthodians (Long 1983) in finely varved black shales (over 50 m thick). The interval has been interpreted as mass mortalities caused by volcanic gas emissions, with the fish preserved in the anaerobic bottom of a large lake (Long 1982). Despite the excellent condition of the fossils, the lower jaws of the acanthodiform *Howittacanthus kentoni* are missing. This species had an upper jaw fused to the neurocranium in two places; it would seem that the lower jaw was only loosely attached, separating before any other skeletal parts including the scales – which, as for most acanthodians, were deeply imbedded in the skin. The diplacanthiform *Culmacanthus stewarti* Long 1983, is based on only six specimens, none complete; the holotype is missing the head and pectoral fins, whereas the specimen with the most head plates has just the shaft of a scapulo-coracoid, the anterior dorsal fin spine (labelled as the pectoral fin spine), and patches of squamation (Long 1983, fig. 3).

All but one of the other articulated acanthodians known from the Devonian of Australia are older, from the Eifelian (cf. Young 2007) flysch deposits of the Bunga Beds – which also produced specimens of *Antarctilamna*. The ?ischnacanthid acanthodians show near perfect preservation of the body squamation and deeply imbedded fin spines, but all five known specimens are headless (Burrow 1996). They are prime examples of the disintegration pattern predicted by comparison with Schäfer's (1972) experiments on *Myxocephalus*. Recent collecting from the rich Gogo Formation produced the first and only specimens of three groups: the shark mentioned in Section 'Chondrichthyes', a coelacanth, and an acanthodiform acanthodian (Young et al. 2010). The latter is the only acanthodian known to be preserved in 3D.

Woodward (1906) described acanthodiform acanthodians in a range of preservation states from another nonmarine Australian site, in the ?latest Devonian (Garvey and Turner 2006, see Burrow et al. 2010) Devil's Plain Formation near Mansfield, Victoria. This larger fish were found mostly as scattered remains, whereas smaller fish skeletons were sometimes almost entire, though distorted or disarticulated. If scavenging or current action had caused the scattering, small and large fish would have been equally affected (or perhaps small fish more affected). In the light of Schäfer's (1972) experiments, it is likely that the larger fish were buoyed up by gas, and tended to be more disarticulated before resettling than the smaller fish which may have decomposed in situ. Unlike the acanthodiforms, several specimens of the gyracanthid acanthodian *Gyracanthides murrayi* are almost completely preserved, mostly flattened dorsoventrally though one is laterally compressed, and another is three-dimensional in a nodule. These fish are up to 1 m long, with large, extremely robust pectoral fin spines, plus fin spines on the two dorsal and pelvic fins and scapulocoracoids with thick perichondral bone; the density of these structures might have prevented the carcasses from floating. These specimens of *G. murrayi* are the only fully articulated gyracanthids known. Partially articulated specimens from the Late Devonian of the USA consist mainly of elements of the pectoral girdle complex (e.g. Sullivan et al. 1999).

Bonebed assemblages usually have a regular size; others may be obviously sorted. Martill (1991) recorded five bonebeds in a 20-m section of Permian strata in Texas: one with acanthodian spines, one with acanthodian scales, and three others with assorted assemblages. Assemblages consisting predominantly of spines (sometimes with tooth whorls) have been recorded elsewhere (e.g. from the Devonian of South America by Janvier and Melo 1992). The few such Palaeozoic assemblages found in Australia are Late Silurian (Tharwa Shale, near Canberra; Burrow and Turner 2000) and Early Carboniferous (from the tetrapod-bearing Ducabrook Formation of central Queensland; Turner et al. 2005). One form genus, *Striacanthus,* was proposed (Hills 1931) for dissociated acanthodian fin spines from the presumed Late Devonian to Early Carboniferous Grampians Group of western Victoria. This form is also found as molds in the silicified sandstones of the Early–Middle? Devonian Mulga Downs Group of western New South Wales (e.g. Burrow 2002), and in Frasnian nonmarine deposits of Victoria. The ornament of smooth longitudinal ridges is common to most diplacanthiform acanthodians (Young and Burrow 2004). The Victorian specimens might have derived from *Culmacanthus.* Acanthodian fin spines from the Mulga Downs Group are diplacanthiform and probably are from the same fish as those with scales with *Diplacanthus*-type structure. Most ischnacanthiform taxa are based on isolated dentigerous jaw bones, including the oldest known ischnacanthiform, *Xylacanthus kenstewarti,* from the ?late Wenlock or earliest Ludlow at the B–MOTH locality, Canada (Hanke et al. 2001a). The youngest ischnacanthiform, *Grenfellacanthus zerinae*, is from the late Famennian Hunter Formation near Grenfell, New South Wales (Long et al. 2004). These elements are generally very robust, and are most common worldwide in Late Silurian–Early Devonian marine and marginal marine settings. We presume that ischnacanthiform acanthodians were among the top predators during these times. They persisted through the rise of other predatory fish such as large arthrodire placoderms (up to 6 m long) and onychodontiforms until their demise in the latest Devonian (Burrow 2004). No absolutely complete articulated ischnacanthiforms have yet been found in Australia.

Microvertebrate assemblages with acanthodian scales, and fin and jaw fragments, are common in the Early Devonian (e.g. Turner 1993; Burrow 1997, 2002). The dearth of articulated acanthodians in the Australian fossil record (as noted, at only two late Middle/early Late Devonian localities) highlights the value of their relatively rich record in microfossil bonebeds in both stratigraphic and evolutionary studies. Juozas Valiukevičius of Lithuania has been the world's premier investigator over the last 30 years in taxonomic and biostratigraphic work based mainly on isolated acanthodian scale assemblages from the Baltic and the former USSR (e.g. Valiukevičius 1994, 1998, 2003, 2005). Although the Australian record is sparser, Early Devonian acanthodian scales have proven their stratigraphic worth, particularly in mixed vertebrate and invertebrate assemblages in which the latter are represented only by long-ranging species and in the absence of biostratigraphically useful conodonts (exclusively marine) in shallow water deposits (e.g. Burrow 2002). There are strong cross-Gondwanan faunal connections. Interpreting the phylogenetic relationships and evolutionary development of Australia's acanthodians is also based on these dissociated elements.

Placodermi

Placoderms are now generally considered to be the most basal group of gnathostomes (e.g. Janvier 1996); they are characterised by discrete head and trunk shields that articulate at the neck joint; the plates of the trunk shield completely encircle the thorax. Both shields are composed of overlapping bony plates ornamented with tubercles and/or ridges. All genera had pectoral and pelvic fins, but none is known to have had an anal fin. As basal gnathostomes, they would be expected to have had two dorsal fins as the general condition, but ptyctodontids are the only group known with certainty to have two dorsal fins. Articulated specimens of arthrodires and antiarchs (*contra* Denison 1978, p. 11) having a single dorsal fin. Many placoderms, including taxa without dorsal fins, have median dorsal ridge plates or scales. The spine-like shape of the median dorsal ridge plate in earliest Devonian acanthothoracids (Goujet and Young 2004, fig. 3B, Racheboeuf et al. 2005, fig. 4A–D, Burrow 2006, fig. 10H–L) could indicate homology between these elements and acanthodian anterior dorsal fin spines.

Placoderm dermal bone has a superficial dentine layer in basal forms with a spongiose middle layer and a basal lamellar bone layer. Placoderms have long been considered to lack true teeth; most have occluding denticulated plates or dermal plates. Smith and Johanson (2003) proposed that some advanced arthrodires developed true teeth, but this has not been universally accepted (e.g. Young 2003). Unfortunately the early history of placoderm gnathal bones is not well known (Burrow 2003b). Some acanthothoracids appear to have had convex plates with diagonal rows of tooth-like denticles (Burrow et al. 2006); early arthrodires had long infragnathals (lower jaw bones) with denticle fields (Young et al. 2001). The acanthothoracid form resembles the tooth plates of the stem actinopterygian *Andreolepis* (see Section 'Actinopterygii ("Ray-Finned Fishes")') and lophosteiforms (see Section 'Lophosteiformes').

Seven main sub-groups of placoderms have been found in Australia: arthrodires, antiarchs, phyllolepids, acanthothoracids (Fig. 7), rhenanids, petalichthyids, and ptyctodonts. Early Devonian placoderms found as macrofossils tend to have heavily ossified dermal skeletons, well-ossified braincases, and to have body scales (Burrow and Turner 1999). The most abundant of placoderms, the arthrodires, had axial skeletons with a persistent notochord but with neural and haemal arches ossified (Obruchev 1964). Nearly all placoderms had the anterior elements of the axial skeleton fused into a composite structure, the synarcual, articulating with the occipital condyles. Arthrodires had paired upper jaw plates. Ptyctodonts had reduced bony armour, with body form and dentition similar to that of modern holocephalans; males had bony pelvic claspers. Rhenanids, unlike most other placoderms preserved as articulated specimens, had no large bony plates; instead there was a covering of small tesserae (Burrow and Turner 1999). They had a ray-like appearance with large dorsoventrally flattened pectoral fins; they also had a jaw supported by the hyomandibular (as in modern sharks) and lacked a dermal cervical joint.

It is convenient to subdivide this diverse group into two morphologic subgroups: those with a heavily ossified dermal skeleton of bony plates, and those with a body covering of small bony tesserae and scales.

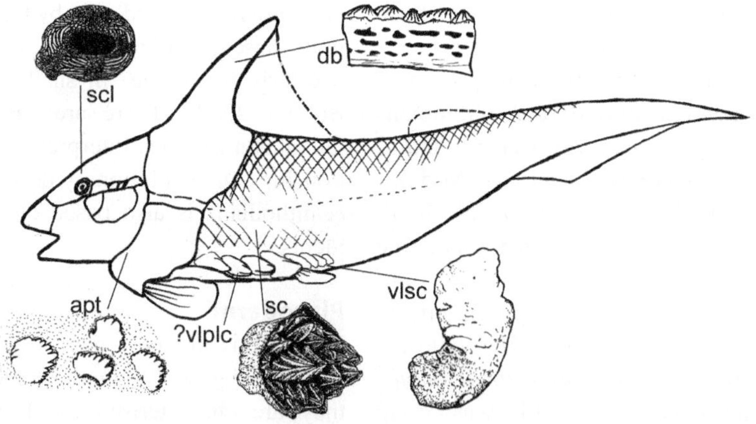

Fig. 7 Hypothetical reconstruction of acanthothoracid placoderm *Murrindalaspis*, showing some of the types of elements found as dissociated elements. To date no gnathal elements have been identified for this genus – apt, postbranchial apronic ornament tubercles; db, vertical section through dermal bone; sc, body scale; scl, sclerotic capsule; vlplc, ventrolateral plate capsule; vlsc, ventrolateral scale (sc, scl, vlsc after Long and Young 1988, figs. 9D, 2B, 9A respectively)

Placoderms with Armour Plates

It is difficult to find a suitable modern analogue for placoderms in this category, though Ivanov et al. (1992) innovatively compared the structure and style of the dermal bone armour with the modern turtle carapace. The sequence of disintegration of the carcass is probably best determined by analysing the states in which the fish have been fossilised. A good range of fully and partially articulated specimens has been described from the Middle Devonian Achanarras fish beds of Scotland (Trewin 1986), preserved in purportedly lacustrine, deep water sediments. The different states of preservation of the heavily armoured placoderms (the arthrodire *Coccosteus* and the antiarch *Pterichthyodes*) include the following: the skeleton wholly intact and articulated with dorsal side uppermost; the skeleton lacking dorsal armour; the head shield fully detached. Trewin hypothesised that the skeletons floated out from shallow waters and sank into deeper waters. It seems unlikely, however, that a carcass would remain fully intact after undergoing floating, transport, then re-sinking. Taking into account the relatively massive exoskeleton, and sedimentary features observed by Trewin – presumed to be drag marks of carcasses – we suggest that the carcasses achieved neutral buoyancy, and hanging just above the substrate, and were then moved downslope by gentle currents. The various states in which the skeletons have been preserved would then reflect the effects of gas buoying up the carcasses. In some instances the head shield held together but separated from the rest of the skeleton (as was observed by Schäfer for disintegrating cetacean carcasses), whereas in other cases ligamentous connections between the dorsal and ventral shield sections must have rotted first.

In Australia, the broadly Emsian Taemas–Wee Jasper fossil sites in southern New South Wales near Canberra have been a rich source of placoderm plates, usually isolated but nevertheless well-preserved. Disarticulation of the plates and encrustation by corals indicate that the encasing sediments were accumulated rather slowly (White 1978) in aerobic conditions. The carcasses would have been attacked by scavengers; the armour plates and scales have extensive microborings. Preservation of plates showing little sign of abrasion accords with them having been deposited in relatively calm waters. It is, however, quite likely they were transported. In Schäfer's (1972) studies, bony plates of similarly encased teleosts *Agonus cataphractus* (pogge) and *Syngnathus acus* (pipe fish), because of their planar shape, were easily carried by currents when shed from carcasses.

Rarely, nodules from the Gogo Formation containing three-dimensional placoderms have scales as well as head shields preserved. These include the ptyctodontid *Campbellodus* (Burrow and Turner 1999, fig. 4H) and the arthrodire *Holonema westolli*, including a specimen prepared with tail squamation in situ (Trinajstic 1999a).

Placoderms with Micromeric Squamation

Placoderms with an external armour of small tesserae and scales would probably have disintegrated in a similar manner to other Devonian fish with micromeric squamation. Scales and bone fragments of these primitive placoderms are common in Australian Early Devonian microvertebrate assemblages (e.g. Turner 1997b). Burrow and Turner (1998, 1999) defined different scale-forms from worldwide assemblages using mainly morphologic and histologic comparison with relatively rare specimens having squamation preserved.

In very rare cases, this type of placoderm has been preserved in an articulated state. The obrution deposits of the Early Devonian Hunsrück Slate, Germany, produce articulated rhenanids (Bartels et al. 1998). The only known preservation of micromerically scaled placoderms in Australia is in microvertebrate assemblages. The thorn ray *Raja clavata* is a suitable modern analogue relative to the disintegration pattern of the dorsoventrally flattened rhenanids. Flattened fish are less likely to generate enough gas to buoy the carcass than deeper-bodied fish, and so the former are more likely to disintegrate in situ. During disintegration of the thorn-ray carcass, dermal denticles are dropped off when the skin is very tough and the bases large, rounded (see Deynat and Seret 1996), and held well in the skin. Once the dermal denticles are free, they separate rapidly according to shape and size, and are sorted into different sediments (Schäfer 1972).

The earliest putative placoderm remains from Australia are scales from a Late Silurian (Ludlow) borehole in Western Australia (Burrow and Turner 2000). Macroremains have not yet been found in Australia in sediments older than Emsian, but small complete plates are known from microfossil assemblages from southeastern Australian Lochkovian limestones (Parkes in Basden et al. 2000; Burrow et al. 2005; Burrow 2006). Placoderm scales and dermal bone fragments are usually the most abundant vertebrate remains in these samples. One such locality in central west New South Wales has yielded plates of small acanthothoracids comparable with the contemporaneous circumArctic *Romundina* (Ørvig 1975), plus *Palaeacanthaspis*, *Kosoraspis*, and *Radotina* from the Czech Republic (Gross 1959), and *Hagiangella* from Vietnam (Racheboeuf et al. 2005), as well as the endemic *Connemarraspis* and the stem buchanosteid *Narrominaspis* (Burrow 2006). The most interesting components of the assemblage are ossified sclerotic, or optic, capsules from at least three of the taxa, *Romundina*, *Narrominaspis,* and *Connemarraspis*. These capsules comprise a ring of dermal bones fused to the ossified sclera which enclosed the rest of the eyeball (e.g. Burrow et al. 2005, figs. 1, 3–6) forming a bone ball that resisted the abrasion and transport which reduced most dermal plates to fragments. Prior to description of these optic capsules, only two other such capsules (both larger) were known, one in an articulated dolichothoracid arthrodire, *Dicksonosteus*, from Spitsbergen (Goujet 1984), and one isolated capsule of acanthothoracid *Murrindalaspis* from Taemas/Wee Jasper (Long and Young 1988). All these specimens are Early Devonian; presumably the sclera was not ossified in more derived taxa, as the ring of dermal bones is occasionally preserved in younger placoderms, but not the sclerotic capsule. Other capsular elements preserved from the same locality are dermal spinal plates with an internal perichondrally ossified layer between the edges of the plates (Burrow 2006, figs. 10H, J–L, 13A–K). If the small fishes from which these elements derived were juveniles, then the dermal median dorsal, pectoral complex, ?pelvic girdle, and possible prepelvic ventrolateral spine capsules could represent the primitive ontogenetic and phylogenetic condition; such has not yet been found in any articulated fossils.

Actinopterygii ('Ray-Finned Fishes')

These fish are characterised by having paired fins supported by a skeleton of ossified finrays without a central axis of bony elements. All articulated actinopterygians known from the Devonian of Australia were fusiform. Usually there is one dorsal fin and, in most early forms, a heterocercal caudal fin (the exception is *Dialipina* from Canada, the oldest stem actinopterygian known from articulated specimens; it which has a diphycercal tail: see Schultze and Cumbaa 2001). The pectoral fins are connected directly to the pectoral girdle.

Devonian actinopterygians had a hyostylic jaw suspension – a palatoquadrate arch movably articulated with the skull by the hyomandibular. The skull comprises many small bones and long, toothed mandibular bones (dentaries) and maxillaries. The teeth are simple and conical in shape with a cap of dense dentine ('acrodin'). The endocranium is often ossified, but not so commonly in early forms as in later ones. The axial skeleton of early actinopterygians is formed by a vertebral column of separate elements partially or wholly surrounding the notochord (Kazantseva 1964). The scales and dermal bones of Devonian actinopterygians are covered with shiny enamel-like tissue ('ganoine'); all except the oldest of known Australian Devonian forms had rhombic scales with pegs and sockets for inter-scale articulation. The scales of basal actinopterygians have three layers: bone overlain by dentine, with an external covering of ganoine pierced by pore canals from the underlying dentine layer (e.g. Schultze 1968).

All Australian Devonian actinopterygians had mesomeric squamation (Figs. 8 and 9); no cheirolepids (with micromeric, deep-based, acanthodian-like scales; see Fig. 7) have been found. Although few modern fish have scales as robust as those of Devonian actinopterygians, the closest analogues are the gar pike *Lepisosteus* and the bichirs, *Polypterus*. Weigelt (1989) observed states of decomposition of gar species at Smither's Lake, Texas, USA, in his pioneering mid-1920s work on organic preservation in modern depositional environments. He undertook his study to understand the preservation of Permian palaeoniscoid fish well known from the copper-rich shales in Germany. There are similar rich fish-bearing economic deposits, this time associated with coals, in eastern Australia; actinopterygians are dominant (Campbell and Phuoc 1983) but there are rare sharks (Leu 1989) and an undescribed coelacanth (e.g. Long 1995).

Although the gar-pike carcasses that Weigelt (1989) observed were out of the water, many had undergone decomposition under water. The lower jaw and opercular region usually detached early, due to decomposition of their connective tissues and scavenging before the tough scaly skin disintegrated. The shape and form of the fish determined the position in which the carcass settled on the substrate, and thus affected the manner

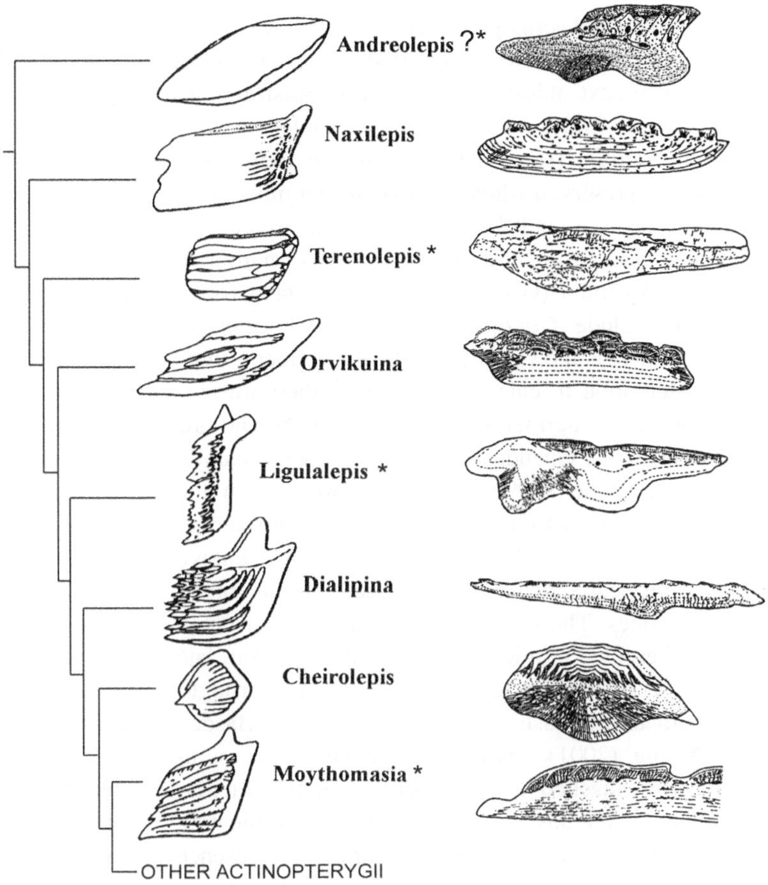

Fig. 8 Phylogram of stem actinopterygian taxa based on scale characteristics (after Schultze 1992, text-fig. 11, with addition of *Terenolepis*) illustrating morphology and histology (*vertical longitudinal section*) of each scale type (*Andreolepis*: Gross 1968; *Naxilepis*: Wang and Dong 1989; *Terenolepis*: Burrow 1995b; *Orvikuina, Ligulalepis, Dialipina*: Schultze 1968; *Cheirolepis*: Gross 1953; *Moythomasia*: Gardiner 1984)

Fig. 9 Lateral view of Late Devonian actinopterygian *Moythomasia* showing scale variants on different areas of the body (after Trinajstic 1999b, Fig. 2; fulcral scale added)

of disintegration of a carcass. The progress of disintegration depended on whether the carcass lay on its side, back, or venter during decomposition because this would influence which area would break open when the carcass distended with gas.

Isolated scales of early actinopterygians are found through the Late Silurian to Late Devonian (and beyond) in Australian marine assemblages; the genera *Andreolepis*?, *Terenolepis*, *Ligulalepis*, *Moythomasia*, and *Mimia* have been identified to date (Fig. 8). The oldest articulated actinopterygians yet to be described, are from the Eifelian Bunga Beds in New South Wales. The next oldest specimens from Australia are *Howqualepis* from Frasnian lacustrine shales near Mt Howitt, Victoria (Long 1988a) – where individuals are mostly preserved whole; they represent most ontogenetic stages – and *Moythomasia durgaringa* and *Mimia toombsi* from the marine Gogo Formation of Western Australia (Gardiner 1984). New taxa found recently include *Gogosardina coatsi* (Choo et al. 2009). Whereas the Gogo Formation specimens show 3D preservation in calcareous nodules, they are also found as scales, dermal bone fragments, teeth and jaw fragments in the Gneudna Formation of the Carnarvon Basin (Trinajstic 1999b; fig. 9), as well as in the Virgin Hills Formation of the Canning Basin (George et al. 2009). All older Australian taxa are known only from scales and isolated, usually fragmentary, dermal bones. The exception is *Ligulalepis*, a taxon well-represented in microfossil assemblages from late Pragian–early Emsian age marine limestones throughout southeastern Australia (e.g. Burrow 1994). Basden and Young (2001) described an isolated neurocranium with attached dermal bones from the Taemas Formation, Wee Jasper. This specimen was the first early osteichthyan found to demonstrate the presence of an eyestalk, previously known only in placoderms and chondrichthyans, but since also identified in a basal sarcopterygian (Zhu et al. 2001). The only Early Devonian acrodin-tipped tooth known is associated with *Ligulalepis* scales from the Troffs Formation (CJB pers obs).

Lophosteiformes

The lophosteiforms are an enigmatic group of fishes represented only by dissociated scales, dermal bone and jaw fragments, and spine-like elements, from disjunct Late Silurian to Early Devonian strata in Australia and circum-Arctic regions. Little evidence has been found of overlap areas or sutures on the dermal bones, so we presume that individual elements were separate. The oldest known occurrence is a single scale from the Coral Gardens Formation (Jeppsson et al. 2007) of the Jack Group (late Ludlow) of northern Queensland. It is identified as belonging to *Lophosteus* Pander 1856, first described from the bountiful latest Silurian (Pridoli) bonebeds of Saaremaa Island, Estonia; its affinities have long been debated (e.g. Burrow 1995a; Gross 1969; Schultze and Märss 2004; Botella et al. 2007). Until now, no definitive osteichthyan characters have been recognised in *Lophosteus*, with most evidence contradicting an osteichthyan affinity. Its dermal bones have open rather than closed sensory line canals. No endochondral bone or dermal lepidotrichia have been found amongst the thousands of *L. superbus* microremains collected over the last 150 years. Rhombic scales with a broad peg-like structure dorsally (characteristic of *Lophosteus*) have also been observed in some placoderm taxa (e.g. Burrow 2006, fig. 8L). A recent identification by Botella et al. (2007, fig. 2a–d) of a maxillary bone with a narrow medial horizontal lamina and a small anterior overlap area from *L. superbus* has been advanced as proving an osteichthyan affinity. Jaws with multiple growth centres and multiple tooth plates are probably a gnathostome plesiomorphy. The more basal placoderms have at least two elements forming each side of the lower and upper jaw biting surfaces; acanthodians have two elements forming each lower jaw cartilage and a single, bipartite, or tripartite upper jaw calcification with separate tooth whorls anterior to the single upper jaw calcification in those taxa that have a dentition (e.g. climatiids, ischnacanthiforms).

Description of the Australian Early Devonian (Lochkovian) species *L. incrementus* Burrow 1995a, was based, as with description of the type species of the genus, on scales, spines, tooth plates, and dermal bone fragments from the Connemarra Formation of central-west New South Wales. Subsequently, additional specimens have been obtained (Burrow 2007b). These and yet more material from recently collected blocks may yield specimens that should help clarify the relationships of this relatively short-lived putative basal osteichthyan group.

Sarcopterygii ('Lobe-Finned Fishes')

Dipnoi (Lungfish)

In dipnoans, the braincase is ossified as a single unit with the palatoquadrate fused to the endocranium (Arratia and Schultze 1991) (Figs. 10 and 11a). Dentition is either a carpet of rasping denticles over the palate and lower jaws or a series of larger tooth-plates. Jarvik (e.g. 1980), however, noted one taphonomic characteristic in Palaeozoic lungfish, that of the 'lost snout' phenomenon (he called it a 'problem'), examples of which have been found in the key Eifelian occurrence at Fish Hill in northern Queensland (Fig. 10a). By the end of the Devonian,

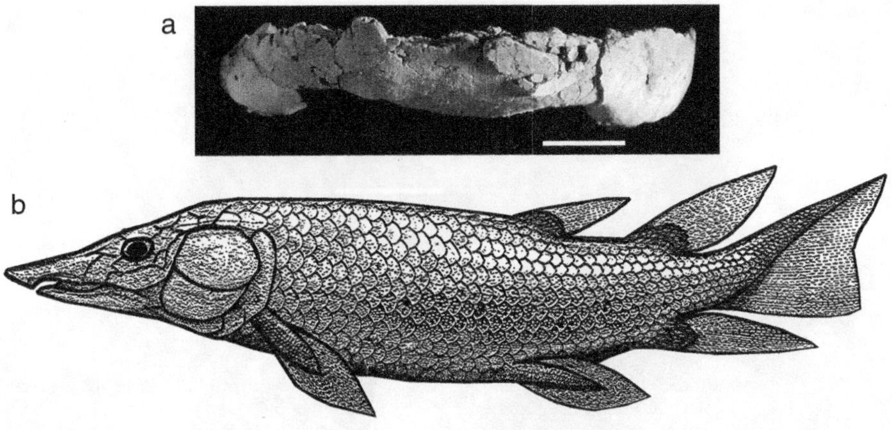

Fig. 10 Dipnoi – (**a**) isolated snout QM F25461 of Dipnoi gen. et sp. nov. from 'Fish Hill' (Eifelian), Broken River Gp (photograph © Dr S. Turner; Queensland Museum Fossil collection); (**b**) *lateral view* of dipnoan *Griphognathus whitei* from the Frasnian Gogo Formation, Western Australia (after Long 1995, p. 165)

different genera filled a wide variety of niches, with a range of trophic strategies from heavy-jawed crushers to suctorial feeders. The skull roof is formed of many small bones. In common with other sarcopterygians, dipnoans had fleshy fins, dermal bone, and scales formed of cosmine, and specialised tissues forming the dentition – in the case of dipnoans, there are mineralised dentines in their toothplates and denticles (Long 1993b).

Unfortunately there have been no studies on the pattern of disintegration of modern lungfish carcasses. In some ways the disintegration of the garfish observed by Weigelt (1989) is also analogous to that of sarcopterygians – the gar carcasses most often lay upside down on the substrate, as is often observed with fossil dipnoans (e.g. *Dipterus* in the Achanarras Formation; Trewin 1986). These skeletons show spinal curvature resulting from gas buoying up the abdominal region, as was also noted by Weigelt (1989) in the carcasses of garfish. The latter often lay with the dorsal surface of the head on the substrate; this is also the aspect in which fossil skulls of sarcopterygians have been preserved. That the head of Devonian sarcopterygians was the most robust part of the body is attested to by their fossil record. Most Devonian dipnoans were known only by cranial anatomy until the discovery of post-cranial skeletal elements in the Gogo Formation, Western Australia (Long 1988b). The most common and best known dipnoan from that formation is *Griphognathus whitei* (Fig. 10b). Isolated lower jaws, tooth plates and skull fragments are the most common fossil remains of sarcopterygians in macrovertebrate assemblages; fragments of scales and dermal bones are found in microvertebrate assemblages. The form of the scales of most sarcopterygians – approximately disc shaped, and non-interlocking, although significantly overlapping – would have favoured their scattering during disintegration of the carcasses. Cosmine-covered scales first appear in the Australian microfossil record in the mid-Pragian; such scales could be from dipnoans, porolepiforms, or osteolepiforms, although only dipnoans are known as macroremains in the Early Devonian. The oldest are ?latest Pragian age from the Taemas/Wee Jasper region (e.g. Campbell and Barwick 2000) and possibly Waratah Bay, Victoria (Long et al. 1994). The oldest dipnoans (in Australia and worldwide) include both tooth-plated taxa and ones with marginal dentition. Unlike thelodonts, acanthodians, placoderms, and Devonian actinopterygians, most dipnoan (and other sarcopterygian) genera (and species) do not have generically distinctive scales (Fig. 11a). Tooth plates are the most distinctive dipnoan fossils, and are often the only taxonomically valuable isolated remains (*contra* Friedman 2007). They have been recorded from several Late Devonian deposits in Australia, including the Gneudna Formation, Western Australia (Basden et al. 2006) and Coffee Hill Member, New South Wales (Jones and Turner 2000).

Onychodontiformes

All predatory lobe-finned fishes (the 'Rhipidistia') had a fully ossified endocranium divided into two parts by a kinetic joint. The upper jaws were attached to the skull by the hyomandibular (amphistylic suspension),

Fig. 11 Sarcopterygian scales in residues from the Emsian?/Eifelian Cravens Peak Beds limestone outcrop – (**a**) dipnoan ANU V3086; (**b**, **c**) onychodontiform scale fragment ANU V3533, with close-up of boundary between tuberculated and border zone; (**d**–**h**) ?rhizodopsid tetrapodomorph scales: (**d**–**g**) thin, broken scale ANU V3534 showing isolated tubercles and fine ridges forming the ornament layer; (**h**) scale ANU V3535 from which outer ornament layer has detached,

and the jaws were well developed with strong teeth. The dermal bones are covered with cosmine, as are the scales.

The onychodonts were large predatory fish, with well developed jaws bearing parasymphysial tooth whorls on the lower jaw, and single teeth set in sockets along both jaws. Scales are rounded in outline, formed of a distinctive cosmine with horseshoe-shaped tubercles on a surface of fine reticulating ridges. The oldest Australian fossils are scales and single smooth teeth of ?early Pragian age (Turner 1997b). More informative material was obtained from microfossil assemblages of the mid-late Pragian Fairy Formation, Victoria, including a disarticulated skull, jaw, palatal, and dental elements assigned by Johanson et al. (2007) to a new taxon *Bukkanodus jesseni*. The teeth of this species have a herring-bone microornament, also typical of isolated teeth found in other late Pragian and younger microfossil assemblages (Turner et al. 2000, plate 2.7). Isolated onychodont jaw bones are known from the Emsian limestones of the Taemas/Wee Jasper region (Lindley 2002), ?Emsian–Eifelian Cravens Peak Beds (Young and Schultze 2005), and the Eifelian of the Broken River region (Turner et al. 2000). Three of the five genera of onychodonts are found in Australia, two being endemic (*Bukkanodus* and *Luckeus* Young and Schultze 2005), and the other two being referred to the type genus *Onychodus*. The most completely known onychodont is *Onychodus jandemarrai* Andrews et al. (2006), based on almost perfectly preserved specimens from nodules in the Frasnian Gogo Formation.

Porolepimorpha

Porolepimorphs were large predators with long feathery pectoral fins, broad heads and small orbits. They had a tooth whorl at the mandibular symphysis. The scales of different species had a range of shapes, from rhombic to round in outline. The group is restricted to the Devonian, and is relatively poorly represented in Australia. Distinguishing between most isolated 'rhipidistian' scales and teeth is often difficult, particularly when they are preserved only as impressions. For example, Young (1987) described such elements from the Late Devonian (Famennian) Knobby Sandstone, east Canning Basin, Western Australia. Young and Goujet (2003) described more substantial possible porolepiform skull and shoulder girdle elements (again, preserved as dissociated impressions) from the ?Emsian/Eifelian Dulcie Sandstone, Northern Territory and Cravens Peak Beds sandstones, west Queensland. Possible cosmine-covered porolepiform scales are found in the Emsian of south-eastern Australia, although some scales identified as such are probably onychodontid (e.g. Basden et al. 2000, fig. 8.10). Ridged scales have been tentatively identified as holoptychiid porolepiform from the north-western Queensland Cravens Peak Beds limestone (Young and Schultze 2005, fig. 9E); this locality is notable for the range of sarcopterygian taxa represented (Fig. 11). Young (2007) identified scales of the cosmopolitan genus *Holoptychius* from the Famennian Worange Point Formation on the south coast of New South Wales.

Actinistia (Coelacanthimorpha)

The first articulated coelacanth remains described from the Devonian of East Gondwana are specimens of *Gavinia* Long 1999, preserved as molds in the Givetian lacustrine shales of Mt Howitt, Victoria, and now also known from scale impressions in the Boyd Volcanic Complex, southern New South Wales (Young 2007). Johanson et al. (2006) reported the oldest known coelacanth *Eoactinistia foreyi*, based on a lower jaw dentary from the Early Devonian (mid-Pragian) Fairy Formation of Victoria. This one specimen has extended the fossil record of coelacanths back 20 My, to the mid-Early Devonian when they are estimated to have diverged from other sarcopterygians. Even older possible coelacanth scales

Fig. 11 (continued) showing underlying network of ?collagenous fibres; (**i–m**) osteolepiform scale varieties (osteolepidid n.g. n.s., Young and Schultze 2005): (**i, j**) fulcral scales ANU V3536 with, and ANU V3537 without, cosmine layer; (**k**) typical rhombic scale ANU V3538 showing groove between cosmine layer and bone base; l, inner surface of scale ANU V3539 showing weak central keel; (**m**) elongate lepidotrichial scale ANU V3540; (**n, o**) ?dipnoan or osteolepiform small bones, ANU V3541 with, and ANU V3542 without, cosmine layer. Scale bar is 1 mm in a, b, d, e, i–n; 100 μm in c, f, g, h, o. (All specimens Australian National University, College of Science Palaeontological Collection; DA Brown Building) have been found in the rich microvertebrate assemblage from the late Lochkovian Connemarra Formation in New South Wales that also yielded placoderm optic and other dermal plate capsules, and *Terenolepis turnerae* and *Lophosteus incrementus* scales, tooth plates and dermal bones mentioned in Sections 'Placodermi', 'Actinopterygii ("Ray-Finned Fishes")', and 'Lophosteiformes' (Burrow et al. 2010). Turner and Young (1987) reported possible coelacanth scales from the limestone outcrop of the Emsian/Eifelian Cravens Peak Beds, western Queensland (Fig. 11d–h). Among new finds by John Long et al. in 2008 is the first coelacanth from the Gogo Formation.

Osteolepimorpha

These are distinguished from other rhipidistians by having an exposed dermal shoulder girdle, and basal scutes on all fins. The scales of basal forms are rhombic with an unornamented dorsal peg and anterior and dorsal edges separated by a groove from the smooth cosmine covering the rest of the scale surface. The scales of different osteolepidid species are mostly similar, although an unnamed taxon from the Cravens Peak Beds limestone is distinguishable from other species by having fine serrations on the posterior edge of its scales (Young and Schultze 2005). Derived osteolepimorphs have cycloid, overlapping scales, with ornament on the exposed section, and a central boss on the base (Long 1985). The oldest Australia specimens represented by more than scales are from the ?nonmarine Eifelian Hatchery Creek Formation, preserved as molds of fronto-ethmoidal shields, jaws, isolated plates, teeth, and scales (Young and Gorter 1981; Hunt 2007). The most completely preserved osteolepimorph is *Gogonasus andrewsi*, with beautiful 3D specimens etched out from nodules of the Gogo Formation, Western Australia (Long et al. 2006). The lacustrine Mt Howitt locality in Victoria has yielded the ?derived osteolepid/basal tristichopterid *Marsdenichthys* (Long 1985; Holland et al. 2010) which has lost the cosmine cover on its dermal bones and scales. Based initially on scale morphology of the two tristichopterids, *Mandageria* and *Cabonnichthys* from the Frasnian Canowindra fossil fish locality, Young (2008) has proposed a Gondwanan clade for this family, contradicting the Laurussian connection previously proposed. This reversal of opinion on relationships based on extensive cladistic analyses of 'whole skeleton' morphology highlights the need to include scale structure (both morphology and histology, when possible) in determining phylogenies. Young's work also emphasises that isolated sarcopterygian scales could be more useful in biostratigraphic and biogeographic studies than some have thought (cf. Turner 1997b). The other osteolepimorph family, the Canowindridae, is an endemic Gondwanan group (e.g. Young 2008) known only from impressions of rare articulated fish from Canowindra (*Canowindra*), Mt Howitt (*Beelarongia*), and Antarctica (*Koharalepis*).

Rhizodontimorpha

The rhizodonts were moderate to very large fish, first appearing in the Late Devonian; their main specialisation was having paddle-like fins with long, stiff, unbranched lepidotrichia (Long 1993b). The only Australian Devonian rhizodontid is *Gooloogongia loomesi* Johanson and Ahlberg (1998), from Canowindra.

Tetrapodomorpha

In recent times more evidence of tetrapod body fossils has been found in the Late Devonian (e.g Long 1995; Blieck et al. 2007). Early forms still possessed a squamation of fish-like, keeled elongate scales, at least ventrally. Recognising such scales and other hard parts poses a challenge, but should be possible in earlier microfaunas. In Australia we have Late Devonian and Early Carboniferous tetrapod-bearing faunas with scales in the Viséan Ducabrook Formation *Ossinodus* Warren and Turner (2004) (Burrow et al. 2010, fig. 6T, U).

Conclusions

Taphonomic processes are often viewed as destructive and capable of causing a major loss of information. Palaeontologists have traditionally favoured study of well-articulated specimens, whereas it is imperfect specimens that provide most information for taphonomic studies, and by the very 'loss' of material can provide information over and above the 'bare bones' data (e.g. Wilson 1989). Because of the bias towards well-preserved fossils, taphonomic data are rarely recorded. Here we have provided a survey of a subject that has received little in the way of systematic research; indeed most data is gleaned from scattered, often incidental, observations. The work of early palaeontologists, with their 'natural history' approach incorporating incidental and anecdotal information, often provides more data for taphonomic studies than work of more modern researchers who have concentrated almost exclusively on phylogenetic analysis.

Comparison of the states in which fossils of Devonian (and other Palaeozoic) fish have been preserved, with that expected from the disintegration of carcasses of modern analogues, appears a worthwhile exercise to rationalise patterns in the fossil record. Definite patterns of preservation are indeed found in the fossil fish record. It seems acceptable then to determine conditions leading to the preservation of fossil assemblages in this way, as the appropriateness of analogies depends on functional similarity and not necessarily on evolutionary closeness. Nevertheless, it is difficult as yet to draw firm conclusions as so little data are available from living organisms. More experimental, process-oriented investigations of necrolysis, examination of bed-by-bed stratification would be helpful, but as most Palaeozoic groups are extinct, comparable study of diagenesis of organic remains in modern environments can/might be a 'red herring'.

Our emphasis in this work has been on thelodont, acanthodian, placoderm, and actinopterygian microremains because they are by far the most useful vertebrate microfossils for biostratigraphic and biogeographical study through the whole Devonian record; chondrichthyan teeth are also important in this regard but relatively sparse in the Australian Devonian record. Burrow et al. (2010) and Young et al. (2010) give overviews of the work that has been completed in the twenty first century on Australasian microfossils and macrofossils and their biogeographical relevance. It is interesting to note that our knowledge of several vertebrate groups (Thelodonti in Australia, Lophosteiformes) relies solely on the microfossil record, where undoubtedly the earliest records of all groups are also to be found.

Taphonomy dictates that most carcasses will be reduced to scattered elements, of which only a small proportion will be preserved in the fossil record. It is hardly surprising that early in the evolution of vertebrate groups, when individuals are few and have a restricted distribution, preservation of body fossils is an extremely rare occurrence. While also reflecting changes in depositional environments and climate, the Australian mid-Palaeozoic record conforms with this view. Of major importance in the Southern Hemisphere, partly perhaps for its 'forcing' of vertebrate evolution (Blieck and Turner 2003; Young 2009) but also for the dearth of record in the Silurian when compared to the Northern Hemisphere, are the effects of the Late Ordovician to Early Silurian ice age and climate extremes.

Despite the Australian mid-Palaeozoic fauna exhibiting endemism at times, there are links across Gondwana to crustal blocks that were probably parts of the Gondwana landmass, as evidenced by similarities between Australian fish microfaunas and those in Spain, Iran, central Russia, and eastern (maritime) North America. Study of the micro- alongside the macrofaunal record maximises our understanding of chronological, geographical, and environmental changes in vertebrate communities.

The broad trends through the Australian Devonian fossil record show placoderms and acanthodians dominant in the mostly marine deposits of the Lower Devonian, with thelodonts in shallow-marginal marine environments; osteichthyans were rare. Sarcopterygians become more common in the Middle Devonian, with an increasing proportion of fossiliferous deposits in marginal marine to freshwater settings. Actinopterygian and chondrichthyan taxa increased in diversity in the late Middle–early Late Devonian, followed by disappearance of all placoderms and thelodonts and virtually all marine acanthodians during the Late Devonian.

Acknowledgements This chapter is dedicated to the life and work of Professor Dr. Wolf-Ernst Reif and Professor Dr. Dolf Seilacher and their students and colleagues at Tübingen who have done so much to illuminate bonebed studies. CJB thanks the Queensland Museum and the Centre for Microscopy and Microanalysis, University of Queensland, for use of facilities. ST thanks the Geosciences, Queensland Museum and Institut für Geowissenschaften, Eberhardt-Karls Universität, Tübingen. We are indebted to Gavin Hanke and Kate Trinasjic for helpful reviews of our manuscript. This is a contribution to IGCP 491: Middle Palaeozoic Vertebrate Biogeography, Palaeogeography and Climate.

References

Ager DV (1993) The new catastrophism: the importance of the rare event in geological history. Cambridge University Press, Cambridge

Allison PA, Briggs, DEG (1991) Taphonomy of nonmineralised tissues. In: Allison PA, Briggs DEG (eds) Taphonomy – releasing the data locked in the fossil record. Plenum Press, New York, pp 455–499

Andrews M, Long J, Ahlberg P, Barwick R, Campbell K (2006) The structure of the sarcopterygian *Onychodus jandemarrai* n. sp. from Gogo, Western Australia: with a functional interpretation of the skeleton. Trans R Soc Edinburgh Earth Sci 96:197–307

Antia DJ (1979) Bone-beds: a review of their classification, occurrence, genesis, diagenesis, geochemistry, palaeoecology, weathering, and microbiotas. Mercian Geol 7:93–174

Arratia G, Schultze H-P (1991) The palatoquadrate and its ossifications: development and homology within osteichthyans. J Morphol 208:1–82

Bartels C, Briggs DEG, Brassel G (1998) The fossils of the Hunsrück Slate. Marine life in the Devonian. Cambridge University Press, Cambridge

Barton DG, Wilson MVH (2005) Taphonomic variations in Eocene fish-bearing varves at Horsefly, British Columbia, reveal 10,000 years of environmental change. Can J Earth Sci 42:137–149

Basden AM, Hocking M, Parkes R, Burrow C, Young GC (2000) Siluro-Devonian microvertebrates from southeastern Australia. In: Blieck A, Turner S (eds) Palaeozoic Vertebrate Biochronology and Global Marine/Non-marine Correlation—Final Report of IGCP 328 (1991–1996). Cour Senckenberg 223:201–222

Basden AM, Trinajstic KM, Merrick JR (2006) Eons of fishy fossils. In: Merrick JR, Archer M, Hickey GM, Lee MSY (eds) Evolution and biogeography of Australasian vertebrates. Australian Scientific Publishing, Sydney, pp 131–157

Basden AM, Young GC (2001) A primitive actinopterygian neurocranium from the Early Devonian of southeastern Australia. J Vert Paleont 21:754–766

Behrensmeyer AK (1991) Terrestrial vertebrate accumulations. In: Allison P, Briggs DEG (eds) Taphonomy: Releasing the data locked in the fossil record. Plenum, New York, pp 291–335

Behrensmeyer AK, Hill AP (eds) (1980) Fossils in the making. Vertebrate taphonomy and paleoecology. Prehistoric Archaeology and Ecology Series, University of Chicago Press, Chicago

Bernacsek GM, Dineley DL (1977) New acanthodians from the Delorme formation (Lower Devonian) of N.W.T., Canada. Palaeontographica A 158:1–25

Blieck A, Turner S (eds) (2000) Palaeozoic vertebrate biochronology and global marine/non marine correlation. Final Report IGCP 328, 1991–1996. Cour Forsch-Inst Senckenberg 223

Blieck A, Turner S (2003) Global Ordovician vertebrate biogeography. Palaeogeogr Palaeoclimatol, Palaeoecol 195: 37–54

Blieck A, Clément G, Blom H, Lelièvre H, Luksevics E, Streel, M, Thorez J, Young GC (2007) The biostratigraphical and palaeogeographical framework of the earliest diversification of tetrapods (Late Devonian). In: Becker RT, Kirchgasser WT (eds) Devonian events and correlations. Subcommission on Devonian Stratigraphy (SDS) Volume in Honour of MR House. Geol Soc London Spec Pap 278, pp 219–235

Botella H, Blom H, Dorka M, Ahlberg PE, Janvier P (2007) Jaws and teeth of the earliest bony fishes. Nature 448:583–586

Brazeau M (2009) The braincase and jaws of a Devonian 'acanthodian' and modern gnathostome origins. Nature 457: 305–308

Brett CE, Seilacher A (1991) Fossil Lagerstätten: a taphonomic consequence of event sedimentation. In: Einsele G, Ricken W, Seilacher A (eds) Cycles and events in stratigraphy. Springer, New York, pp 284–297

Briggs DEG, Kear AJ, Martill DM, Wilby PR (1993) Phosphatization of soft-tissue in experiments and fossils. J Geol Soc 150:1035–1038

Broad DS, Lenz AC (1972) A new Upper Silurian species of *Vernonaspis* (Heterostraci) from Yukon Territory, Canada. J Paleontol 46:415–420

Brongersma-Sanders M (1957) Mass mortality in the sea. Geol Soc Am Mem 67(1):941–1010

Burrow CJ (1994) Form and function in scales of *Ligulalepis toombsi* Schultze, a palaeoniscoid from the Early Devonian of Australia. Rec South Aust Mus 27:175–185

Burrow CJ (1995a) A new lophosteiform (Osteichthyes) from the Lower Devonian of Australia. Geobios MS 19:327–333

Burrow CJ (1995b) A new palaeoniscoid from the Lower Devonian Trundle beds of Australia. Geobios MS 19: 319–325

Burrow CJ (1996) Taphonomic study of acanthodians from the Devonian Bunga Beds (Late Givetian/Early Frasnian) of New South Wales. Hist Biol 11: 213–228

Burrow CJ (1997) Microvertebrate assemblages from the Lower Devonian (pesavis/sulcatus conodont zones) of central New South Wales, Australia. Mod Geol 21:43–77

Burrow CJ (2002) Lower Devonian acanthodian faunas and biostratigraphy of south-eastern Australia. Mem Ass Australas Palaeontols 27:75–137

Burrow CJ (2003a) Redescription of the gnathostome fauna from the mid-Palaeozoic Silverband formation, the Grampians, Victoria. Alcheringa 27:37–49

Burrow CJ (2003b) Comment on "Separate evolutionary origins of teeth from evidence in fossil jawed vertebrates". Science 300:1661b

Burrow CJ (2004) Acanthodians with dentigerous jaws: the Ischnacanthiformes and *Acanthodopsis*. Fossils Strata 50: 8–22

Burrow CJ (2006) Placoderm fauna from the Connemarra Formation (?late Lochkovian, Early Devonian), central New South Wales, Australia. Alcheringa Spec Issue 1:59–88

Burrow CJ (2007a) Early Devonian (Emsian) acanthodian faunas of the western USA. J Paleont 81:824–840

Burrow CJ (2007b) Possible stem osteichthyans from the earliest Devonian of Australia. In: Blom H, Brazeau MD (eds) 40th anniversary symposium on early vertebrates/lower vertebrates (11th international symposium on early vertebrates), Uppsala, Sweden. Ichthyolith Issues SP 10, p22

Burrow CJ, Turner S (1998) Devonian placoderm scales from Australia. J Vert Paleontol 18:677–695

Burrow CJ, Turner S (1999) A review of placoderm scales, and their relevance in placoderm phylogeny. J Vert Paleontol 19:204–219
Burrow CJ, Young GC (1999) An articulated teleostome fish from the Late Silurian (Ludlow) of Victoria, Australia. Rec West Aust Mus Suppl 57:1–14
Burrow CJ, Turner S (2000) Silurian vertebrates from Australia. In: Blieck A, Turner S (eds) Palaeozoic vertebrate biochronology and global marine/non marine correlation. Final Report IGCP 328, 1991–1996. Cour Forsch-Inst Senckenberg, vol 223, pp 169–174
Burrow CJ, Young GC (2005) The acanthodian fauna of the Craven Peaks Beds (Early to Middle Devonian), western Queensland. Mem Qld Mus 51:3–25
Burrow CJ, Turner S (2010) Reassessment of "*Protodus*" *scoticus* from the early Devonian of Scotland. In: Elliott DK, Yu X-B, Maisey J, Maio D (eds) Meeman Chang Festschrift. Friedrich Pfeil, Munich, pp 123–144
Burrow CJ, Jones AS, Young GC (2005) X-ray microtomography of 410 million year old optic capsules from placoderm fishes. Micron 36:551–557
Burrow CJ, Lelièvre H, Janjou D (2006) Gnathostome microremains from the Lower Devonian Jawf Formation, Saudi Arabia. J Paleontol 80:537–560
Burrow CJ, Hovestadt DC, Hovestadt-Euler M, Turner S, Young GC (2008a) New information on the Devonian shark *Mcmurdodus*, based on material from western Queensland, Australia. Acta Geol Pol 58:155–163
Burrow CJ, Turner S, Desbiens S, Miller RF (2008b) Early Devonian putative gyracanthid acanthodians from eastern Canada. Can J Earth Sci 45:897–908
Burrow CJ, Turner S, Young GC (2010) Middle Palaeozoic microvertebrate assemblages and biogeography of East Gondwana (Australasia, Antarctica). Paleoworld 19:37–54
Bystrow AP (1956) On the damage caused by fungi in the skeletal elements of fossil animals. Leningrad Univ Vestnik II(6) Seria Geologo-Geographich (in Russian)
Cadée GC (1991) The history of taphonomy. In: Donovan SK (ed) The processes of fossilization. Columbia, New York, pp 3–21
Campbell KSW, Barwick RE (2000) The braincase, mandible and dental structures of the early Devonian lungfish *Dipnorhynchus kurikae* from Wee Jasper, New South Wales. Rec Aust Mus 52:103–128
Campbell, KSW, Phuoc LD (1983) A Late Permian actinopterygian fish from Australia. Palaeontology 26:33–70
Castle JW, Rodgers JH (2009) Hypothesis for the role of toxin-producing algae in Phanerozoic mass extinctions based on evidence from the geologic record and modern environments. J Environ Geosci 16:1–23
Choo B, Long JA, Trinajstic K (2009) A new genus and species of basal actinopterygian fish from the Upper Devonian Gogo Formation of Western Australia. Acta Zool 9(Suppl 1): 194–210
Clifton HE (1957) The carbonate concretions of the Ohio Shale. Ohio J Sci 57(2):114–124
Criss RE, Cooke GA, Day SD (1988) An organic origin for the carbonate concretions of the Ohio Shale. US Geol Surv Bull 1836:21
Darras L, Derycke C, Blieck A, Vachard D (2008) The oldest holocephalan (Chondrichthyes) from the Middle Devonian of the Boulonnais (Pas-de-Calais, France). C R Palevol 7: 297–304
Denison RH (1956) A review of the habitat of the earliest vertebrates. Fieldiana, Geol 11:359–457
Denison RH (1978) Placodermi. In: Schultze H-P (ed) Handbook of paleoichthology, Part 2. Gustav Fischer Verlag, Stuttgart and New York, pp 1–118
Denison RH (1979) Acanthodii. In: Schultze H-P (ed) Handbook of paleoichthology, Part 5. Gustav Fischer Verlag, Stuttgart and New York, pp 1–62
Deynat P, Seret B (1996) The dermal armature of skates and rays (Chondrichthyes, Elasmobranchii, Batoidea). I: Morphology and arrangement of the dermal denticles. Ann Sci Nat, Zool, Paris 17(2):65–83
Dick JRF (1981) *Diplodoselache woodi* gen. and sp. nov., an early Carboniferous shark from the Midland Valley of Scotland. Trans R Soc Edinburgh, Earth Sci 72:99–113
Diedrich CG (2009) A coelacanthid-rich site at Hasbergen (NW Germany)—taphonomy and palaeoenvironment of a first systematic excavation in the Kupferschiefer (Upper Permian, Lopingian). Palaeobiodiv Palaeoenvir 89: 67–94
Dineley DL, Loeffler EJ (1976) Ostracoderm faunas of the Delorme and associated Siluro-Devonian Formations, North West Territories, Canada. Spec Pap Palaeontol 18: iv+214
Duffin CJ, Ward DJ (1983) Neoselachian sharks' teeth from the Lower Carboniferous of Britain and the Lower Permian of the USA. Palaeontology 26:93–110
Eberth DA, Shannon M, Noland BG (2007) A bonebeds database: classification, biases, and patterns of occurrence. In: Rogers RR, Eberth DA, Fiorillo AR (eds) Bonebeds: genesis, analysis, and paleobiological significance. University of Chicago, Chicago and London, pp 103–219
Efremov IA (1940) Taphonomy: new branch of paleontology. Pan-Am Geol 74:81–93
Elder RL, Smith GR (1988) Fish taphonomy and environmental inference in paleolimnology. Palaeogeogr Palaeoclimatol Palaeoecol 62:577–592
Friedman M (2007) The interrelationships of Devonian lungfishes (Sarcopterygii: Dipnoi) as inferred from neurocranial evidence and new data from the genus *Soederberghia* Lehman, 1959. Zool J Linn Soc 151:115–171
Gardiner BG (1984) The relationships of the palaeoniscoid fishes, a review based on new specimens of *Mimia* and *Moythomasia* from the Upper Devonian of Western Australia. Bull Brit Mus (Nat Hist) Geol 37:1–428
Garvey JM, Turner S (2006) Vertebrate microremains from the presumed earliest Carboniferous of the Mansfield Basin, Australia. Alcheringa 30:43–62
George AD, Trinajstic KM, Chow N (2009) Frasnian reef evolution and palaeogeography, SE Lennard Shelf, Canning Basin, Australia. Geol Soc Lond Spec Pub 314:73–107
Ginter M (2001) Chondrichthyan biofacies in the Late Famennian of Utah and Nevada. J Vert Palaeontol 21: 714–729
Ginter M, Turner S (1999) The early Famennian recovery of phoebodont sharks. Acta Geol Pol 49(2):105–117
Ginter M, Hairapetian V, Klug C (2002) Famennian chondrichthyans from the shelves of North Gondwana. Acta Geol Pol 52:169–215

Ginter M, Liao J-C, Valenzuela-Rios JI (2008) New data on chondrichthyan microremains from the Givetian of the Renanué Section in the Aragonian Pyrenees (Spain). Acta Geol Pol 58:165–172

Goujet D (1984) Les poissons Placodermes du Spitsberg. Arthrodires Dolichothoraci de la Formation de Wood Bay (Dévonien inférieur). Cahiers Paléont, CNRS, Paris

Goujet D, Locquin MV (1979) Découverte de spores fongïques dans les écailles de poissons et d'agnathes paléozoïques: *Mycobystrovia lepidophaga* gen. et sp. nov. 104e Cong nat Soc Savantes Bordeaux, 1979, Sci 1:87–89

Goujet D, Young GC (2004) Placoderm anatomy and phylogeny: new insights. In: Arratia G, Wilson MVH, Cloutier R (eds) Recent advances in the origin and early radiation of vertebrates. Verlag Dr Friedrich Pfeil, Munich, pp 109–126

Gross W (1950) Die paläontologische und stratigraphische Bedeutung der Wirbeltier-Faunen des Old Reds und der marinen altpaläozischen Schichten. Abh dt Akad Wiss Berlin 1949(1):1–130

Gross W (1953) Devonische Palaeonisciden-Reste in Mittel- und Osteuropa. Pal Z 27:85–112

Gross W (1959) Arthrodiren aus dem Obersilur der Prager Mulde. Palaeontogr A 113:1–35

Gross W (1967) Über Thelodontier-Schuppen. Palaeontogr A 127:1–67

Gross W (1968) Fragliche Actinopterygier-Schuppen aus dem Silur Gotlands. Lethaia 1:184–218

Gross W (1969) *Lophosteus superbus* Pander, ein Teleostome aus dem Silur Oesels. Lethaia 2:15–47

Gross W (1971) Downtonische und Dittonische Acanthodier-Reste des Ostseegebietes. Palaeontogr A 136:1–82

Hamilton RFM, Trewin NH (1994) Taphonomy of fish beds from the Upper Flagstone Group of the Middle Old Red Sandstone, Caithness. Scot J Geol 30:175–181

Hanke GF (2008) *Promesacanthus eppleri* n. gen., n. sp., a mesacanthid (Acanthodii, Acanthodiformes) from the Lower Devonian of northern Canada. Geodiversitas 30:287–302

Hanke GF, Davis SP (2008) Redescription of the acanthodian *Gladiobranchus probaton* Bernacsek and Dineley, 1977, and comments on diplacanthid relationships. Geodiversitas 30:303–330

Hanke GF, Wilson MVH (1998) Scale and spine characteristics of Lower Devonian acanthodians and chondrichthyans from northern Canada. J Vert Paleontol 18:48A

Hanke GF, Wilson MVH (2004) New teleostome fishes and acanthodian systematics. In: Arratia G, Wilson MVH, Cloutier R (eds) Recent advances in the origin and early radiation of vertebrates. Verlag Dr. Friedrich Pfeil, Munich, pp 526–537

Hanke GF, Wilson MVH (2006) Anatomy of the early Devonian acanthodian *Brochoadmones milesi* based on nearly complete body fossils, with comments on the evolution and development of paired fins. J Vert Paleontol 26:526–537

Hanke GF, Wilson MVH (2010) The putative chondrichthyans *Kathemacanthus* and *Seretolepis* from the Lower Devonian MOTH locality, Mackenzie Mountains, Canada. In: Elliott DK, Yu J-B, Maisey JG, Maio D (eds) Meeman Chang Festschrift. Verlag Dr Friedrich Pfeil, Munich, pp 159–182

Hanke GF, Wilson MVH, Lindoe LA (2001a) New species of Silurian acanthodians from the Mackenzie Mountains, Canada. Can J Earth Sci 38:1517–1529

Hanke GF, Davis SP, Wilson MVH (2001b) New species of the acanthodian genus *Tetanopsyrus* from northern Canada, and comments on related taxa. J Vert Paleontol 21:740–753

Hays JD, Imbrie J, Shackleton NJ (1976) Variations in the Earth's orbit: pacemaker of the ice ages. Science 194: 1121–1132

Hills ES (1931) The Upper Devonian fishes of Victoria, Australia, and their bearing on the stratigraphy of the state. Geol Mag 68:206–231

Holland T, Long J, Snitting D (2010) New information on the enigmatic tetrapodomorph fish *Marsdenichthys longioccipitus* (Long, 1985). J Vert Palaeontol 30:68–77

Hunt JR (2007) New Middle Devonian osteolepid material (lobe-finned fish) from Hatchery Creek, New South Wales. In: Warren A (ed) CAVEPS 2007 (10–13 April 2007, Museum Victoria, Melbourne). Geol Soc Austr Abstr 85: 18–19

Ivanov A, Pavlov D, Cherepanov G (1992) Some aspects of biomineralization of vertebrate exoskeleton. In: Mark-Kurik E (ed) Fossil fishes as living animals. Academy of Sciences of Estonia, Tallinn

Janvier P (1996) Early vertebrates. Oxford University Press, Oxford

Janvier P (1997) Vertebrata. Animals with Backbones. Version 01 January 1997. http://tolweb.org/Vertebrata/14829/1997.01.01 in The Tree of Life Web Project, http://tolweb.org/

Janvier P, de Melo JHG (1992) New acanthodian and chondrichthyan remains from the Lower and Middle Devonian of Brazil. N Jahrb Geol Pal Monat 4:193–206

Jarvik E (1980) Basic structure and evolution of vertebrates, 2 vols. Academic, London, New York

Jeppsson L, Talent JA, Mawson R, Simpson AJ, Andrew A, Calner M, Whitford D, Trotter JA, Sandström O, Caldon H-J (2007) High resolution Late Silurian correlations between Gotland, Sweden, and the Broken River region, NE Australia: lithologies, conodonts and isotopes. Palaeoclimatol Palaeoecol Palaeogeogr 245:115–137

Johanson Z, Ahlberg PE (1998) A complete primitive rhizodont from Australia. Nature 394:569–573

Johanson Z, Long J, Talent J, Janvier P, Warren J (2006) Oldest coelacanth, from the Early Devonian of Australia. Biol Lett 2:443–446

Johanson Z, Long JA, Talent JA, Janvier P, Warren JW (2007) New onychodontiform (Osteichthyes; Sarcopterygii) from the Lower Devonian of Victoria, Australia. J Paleontol 81:1031–1043

Jones RK, Turner S (2000) Late Devonian fauna from the Columbine Sandstone (Coffee Hill Member), Gap Creek, central New South Wales. Cour Forsch-Inst Senckenberg 223:523–541

Kazantseva AA (1964) Subclass Actinopterygii. In: Obruchev DV (ed) Fundamentals of palaeoichthyology, vol 11 – Agnatha, Pisces. Moscow. (English translation, 1967, Israel program for scientific translations)

Kidwell SK, Behrensmeyer AK (1988) Overview: ecological and evolutionary implications of taphonomic processes. Palaeogeogr Palaeoclimatol Palaeoecol 63:1–13

Leidner A, Thies D (1999) Placoid scales and oral teeth of Late Jurassic elasmobranchs from Europe. In: Arratia G, Schultze H-P (eds) Mesozoic fishes 2. Verlag Dr F. Pfeil, Munich

Leu MR (1989) A Late Permian freshwater shark from eastern Australia. Palaeontology 32:265–286

Lindley ID (2002) Acanthodian, onychodontid and osteolepidid fish from the middle-upper Taemas Limestone (Early Devonian), Lake Burrinjuck, New South Wales. Alcheringa 26:103–126

Long JA (1982) Late Devonian fish taphonomy in Victoria: a cautionary note to biostratigraphers. In: Rich PV, Thompson EM (eds) The fossil vertebrate record of Australasia. Monash Offset Printing, Melbourne, pp 120–127

Long JA (1983) A new diplacanthoid acanthodian from the Late Devonian of Victoria. Mem Ass Australas Palaeontols 1:51–65

Long JA (1985) Osteolepiform fish from the Late Devonian of Victoria, Australia. Alcheringa 9:1–22

Long JA (1986) A new Late Devonian acanthodian fish from Mt. Howitt, Victoria, Australia with remarks on acanthodian biogeography. Proc R Soc Victoria 98:1–17

Long JA (1988a) New palaeoniscoid fishes from the late Devonian and early Carboniferous of Victoria. Mem Ass Australas Palaeontols 7:1–64

Long JA (1988b) Late Devonian fishes from Gogo, Western Australia. Nat Geogr Res 4:436–450

Long JA (1993a) Palaeozoic Vertebrate biostratigraphy and biogeography. Belhaven Press, London

Long JA (1993b) Morphological characteristics of Palaeozoic Vertebrates used in Biostratigraphy. In: Long JA (ed) Palaeozoic Vertebrate Biostratigraphy and Biogeography. Belhaven Press, London, pp 3–24

Long JA (1995) The rise of fishes. University of New South Wales Press, Sydney

Long JA (1999) A new genus of fossil coelacanth (Osteichthyes: Coelacanthiformes) from the Middle Devonian of southeastern Australia. Rec West Aust Mus Suppl 57:37–53

Long JA (2007) Further significant discoveries of fishes from Late Devonian Gogo Formation, Western Australia. In: Warren A (ed) CAVEPS 2007 (10–13 April 2007), Museum Victoria, Melbourne. Geol Soc Aust Abs 85:12–13

Long JA, Trinajstic K (2010) The Late Devonian Gogo formation Lagerstatte of Western Australia: exceptional early vertebrate preservation and diversity. Ann Rev Earth Plan Sci 38:255–279

Long JA, Young GC (1988) Acanthothoracid remains from the Early Devonian of New South Wales, including a complete sclerotic capsule and pelvic girdle. Mem Ass Australas Palaeontols 7:65–80

Long JA, Burrow CJ, Ritchie A (2004) A new ischnacanthid acanthodian from the Upper Devonian Hunter Formation near Grenfell, New South Wales. Alcheringa 28:147–156

Long JA, Campbell KSW, Barwick RE (1994) A phylogenetically significant new dipnoan genus, *Ichnomylax*, from the Early Devonian of Victoria, Australia. J Vert Palaeontol 14:127–131

Long JA, Trinajstic K, Young GC, Senden TJ (2008) Live birth in the Devonian Period. Nature 453:650–652

Long JA, Young GC, Holland T, Senden TJ, Fitzgerald EMG (2006) An exceptional Devonian fish from Australia sheds light on tetrapod origins. Nature 444:199–202

Lucas J, Prévôt LE (1991) Phosphates and fossil preservation. In: Allison PA, Briggs DEG (eds) Taphonomy – releasing the data locked in the fossil record. Plenum, New York

Lund R (1986) The diversity and relationships of the Holocephali. In: Uyeno T, Arai R, Taniuchi T, Matsuura K (eds) Proceedings of the second international conference on Indo-Pacific fishes. Ichthyological Society of Japan, Tokyo, pp 97–106

Mader H (1986) Schuppen und Zähne von Acanthodiern und Elasmobranchiern aus dem Unter-Devon Spaniens (Pisces). Gött Arbeit Geol Paläontol 28:1–59

Maisey JG (1989) *Hamiltonichthys mapesi*, g. and sp. nov. (Chondrichthyes; Elasmobranchii), from the Upper Pennsylvanian of Kansas. Am Mus Novit 2931:42

Maisey JG (2001) A primitive chondrichthyan braincase from the Middle Devonian of Bolivia. In: Ahlberg PE (ed) Major Events in Early Vertebrate Evolution. Taylor and Francis, London and New York, pp 263–288

Maisey JG, Anderson ME (2001) A primitive chondrichthyan braincase from the Early Devonian of South Africa. J Vert Paleontol 21:702–713

Maisey JG, Naylor GJP, Ward DJ (2004) Mesozoic elasmobranchs, neoselachian phylogeny and the rise of modern elasmobranch diversity. In: Arratia G, Tintori A (eds) Mesozoic fishes 3 – systematics, paleoenvironments and biodiversity. Verlag Dr Friedrich Pfeil, Munich

Maisey JG, Turner S, Miller RF (2009) The braincase of the chondrichthyan *Doliodus* from the Lower Devonian Campbellton Formation of New Brunswick, Canada. Acta Zool, Suppl 90:109–122

Märss T (1982) Vanimate selgroogsete leiud Saaremaal (The oldest vertebrate finds in Saaremaa). Eesti Loodus, Oct 1982:646–654 (in Estonian)

Märss T, Ritchie A (1998) Articulated thelodonts (Agnatha) of Scotland. Trans R Soc Edinburgh, Earth Sci 88:143–195

Märss T, Turner S, Karatajute-Talimaa V (2007) Agnatha II – Thelodonti. Volume 1B. In: Schultze H-P (ed) Handbook of palaeoichthyology. Verlag Dr Friedrich Pfeil, Munich, 143pp

Martill DM (1991) Bones as stones: the contribution of vertebrate remains to the lithologic record. In: Donovan SK (ed) The processes of fossilization. Belhaven Press, London, pp 270–295

Mawson R, Talent JA (1994) Age of an early Devonian carbonate fan and isolated limestone clasts and megaclasts, east-central Victoria. Proc R Soc Victoria 106:31–70

Mawson R, Talent JA (1997) Famennian–Tournaisian conodonts and Devonian–early Carboniferous transgressions and regressions in north-eastern Australia. Geol Soc Am Sp Pap 321:189–233

McGrew PO (1975) Taphonomy of Eocene fish from Fossil Basin, Wyoming. Fieldiana Geol 33(14):257–270

Mikulic DG, Briggs DEG, Kluessendorf J (1985) A new exceptionally well preserved biota from the Lower Silurian of Wisconsin, U.S.A. Phil Trans R Soc (Lond) B311:75–85

Miles RS (1966) The acanthodian fishes of the Devonian Plattenkalk of the Paffrath Trough in the Rhineland. Ark Zool 18:147–194

Miller RF, Cloutier R, Turner S (2003) The oldest articulated chondrichthyan from the Early Devonian) period. Nature 425:501–504

Morin JG (1988) Piscivorous behaviour and activity patterns in the tropical ophiuroid *Ophiarachna incrassata* (Ophiuroidea: Ophiodermatidae). In: Burke RD, Mladenov PV, Lambert

P, Parsley RL (eds) Echinoderm biology: sixth international echinoderm conference proceedings. Balkema, Rotterdam, p 832

Murchison RI (1839) The silurian system, 2 vols. John Murray, London, 768p

Newman M (2002) A new naked jawless vertebrate from the Middle Devonian of Scotland. Palaeontol 45:933–941

Newman MJ, Trewin NH (2001) A new jawless vertebrate from the Middle Devonian of Scotland. Palaeontol 44:43–52

Novitskaya LI, Obruchev DV (1964) Class Acanthodei. In: Obruchev DV (ed) Fundamentals of palaeoichthyology vol 11 – Agnatha, Pisces. Moscow (English translation from Russian, 1967, Israel program for scientific translations)

Obruchev DV (1964) Class Placodermi. In: Obruchev DV (ed) Fundamentals of palaeoichthyology vol 11 – Agnatha, Pisces. Moscow (English translation, 1967, Israel program for scientific translations)

Olson EC (1980) Taphonomy: its history and role in community evolution? In: Behrensmeyer AK, Hill AP (eds) Fossils in the making. Vertebrate taphonomy and paleoecology. University of Chicago Press, Chicago, pp 5–19

Ørvig T (1967) Some new acanthodian material from the Lower Devonian of Europe. J Linn Soc (Zool) 47:131–153

Ørvig T (1973) Acanthodian dentition and its bearing on the relationships of the group. Palaeontogr A 143:119–150

Ørvig T (1975) Description, with special reference to the dermal skeleton, of a new radotinid arthrodire from the Gedinnian of arctic Canada. Coll Int Centre Nat Rech Sci 218:41–71

Pander CH (1856) Monographie der fossilen Fische des silurischen Systems der Russisch-Baltischen Gouvernements. Obersilurische Fische. Buchdruckerei der Kaiserlichen Akademie der Wissenschaften, St. Petersburg

Poncet J (1993) Algues perforantes d'une plaque dermique de vertébré Mésodévonien et phénomènes diagénétiques localisés dans le tissu osseux. Rev Micropaléontol 36:355–362

Racheboeuf PR, Janvier P, Phuong TH, Vannier J, Shang-Qi W (2005) Lower Devonian vertebrates, arthropods and brachiopods from northern Vietnam. Geobios 38:533–551

Reif W-E (1976) Sedimentologie und Genese von Bonebeds. Zbl Geol Paläontol Teil II 5/6:252–255

Reif W-E (1982) Muschelkalk/Keuper bone-beds (Middle Triassic, SW-Germany) – Storm Condensation in a Regressive Cycle. In: Einsele G, Seilacher A (eds) Cyclic and event stratification. Springer, New York, pp 299–325

Reif W-E (1985) Squamation and ecology of sharks. Cour Forsch-Inst Senckenb 78:1–255

Ritchie A (1985) *Ainiktozoon loganense* Scourfield, a protochordate? from the Silurian of Scotland. Alcheringa 9:117–142

Ritchie A (2005) *Cowralepis*, a new genus of phyllolepid fish (Pisces, Placodermi) from the Late Middle Devonian of New South Wales, Australia. Proc Linn Soc NSW 125:215–259

Ritchie A (2007) The great Devonian fish kill at Canowindra. In: Merrick JR, Archer M, Hickey G, Lee M (eds), Evolution and zoogeography of Australasian vertebrates. Australian Scientific Publishing, Sydney, pp 159–184

Rogers RR, Eberth DA, Fiorillo AR (eds) (2007) Bonebeds: genesis, analysis, and paleobiological significance. University of Chicago Press, Chicago and London

Rogers RR, Kidwell SM (2007) A conceptual framework for the genesis and analysis of vertebrate skeletal concentrations. In: Rogers RR, Eberth DA, Fiorillo AR (eds) Bonebeds: genesis, analysis, and paleobiological significance. University of Chicago, Chicago and London, pp 1–69

Rouse GW, Goffredi SK, Vrijenhoek RC (2004) *Osedax*: Bone-eating marine worms with dwarf males. Science 305: 668–671

Sankey JT, Baszio S (2007) Vertebrate microfossil assemblages: their role in paleoecology and paleobiogeography. Indiana University, Bloomington

Schäfer W (1972) Ecology and palaeoecology of marine environments. (Original 1962, translated I Oertel). Oliver and Boyd, Edinburgh

Schultze H-P (1968) Palaeoniscoidea-Schuppen aus dem Unterdevon Australiens und Kanadas und aus dem Mitteldevon Spitsbergens. Bull Brit Mus Nat Hist (Geol) 16:343–368

Schultze H-P (1992) Early Devonian actinopterygians (Osteichthyes, Pisces) from Siberia. In: Mark-Kurik E (ed) Fossil fishes as living animals. Tallinn. Academia 1:233–242

Schultze H-P, Cumbaa SL (2001) *Dialipina* and the characters of basal actinopterygians. In: Ahlberg P (ed) Major events in early vertebrae evolution. Palaeontology, phylogeny and development. Taylor and Francis, London, pp 315–332

Schultze H-P, Märss T (2004) Revisiting *Lophosteus*, a primitive osteichthyan. Acta Univ Latv 679:57–78

Seilacher A (1970) Begriff und Bedeuting der Fossil-Lägerstatten (Fossil-Lägerstatten Nr. 1). N Jahrb Geol Pal Monat 1970:34–39

Seilacher A (1990) Taphonomy of fossil Lägerstatten. In: Briggs DEB, Crowther P (eds) Palaeobiology: a synthesis. Blackwell, Oxford, pp 266–270

Seilacher A, Reif W-E, Westphal F (1985) Sedimentological, ecological and temporal patterns of fossil Lagerstätten. Phil Trans R Soc Lond B 311:5–24

Seilacher A, Marshall CH, Skinner CW, Tsuihiji T (2001) A fresh look at sideritic "coprolites". Paleobiology 27(1):7–13

Sloan TR, Talent JA, Mawson R, Simpson AJ, Brock GA, Engelbretsen MJ, Jell JS, Aung AK, Pfaffenritter C, Trotter J, Withnall IW (1995). Conodont data from Silurian–Middle Devonian carbonate fans, debris flows, allochthonous blocks and adjacent autochthonous platform margins: Broken River and Camel Creek areas, north Queensland, Australia. Cour Forsch-Inst Senckenberg 182:1–77

Smith MM, Johanson Z (2003) Separate evolutionary origins of teeth from evidence in fossil jawed vertebrates. Science 299:1235–1236

Smith GR, Stearley RF, Badgley CE (1988) Taphonomic bias in fish diversity from Cenozoic floodplain environments. Palaeogeogr Palaeoclimatol Palaeoecol 63:263–273

Smith GR, Montgomery DR, Peterson NP, Crowley B (2007) Spawning sockeye salmon fossils in Pleistocene lake beds of Skokomish Valley, Washington. Quat Res 68:227–238

Soehn KL, Wilson MVH (1990) A complete, articulated heterostracan from Wenlockian (Silurian) beds of the Delorme Group, Mackenzie Mountains, Northwest Territories, Canada. J Vert Paleontol 10:405–419

Speyer SE, Brett CE (1988) Taphofacies models for epeiric sea environments: Middle Palaeozoic examples. Palaeogeogr Palaeoclimatol Palaeoecol 63:225–262

Sullivan RM, Lucas SG, Randall KA (1999) The scapulocoracoid complex of *Gyracanthus* (Acanthodii:

Climatiiformes) and a reassessment of the pectoral region in the Gyracanthidae. Proc Acad Nat Sci Philadelphia 149:99–108

Talent JA, Mawson R (1994) Condonts in relation to the age and environmental framework of the Burdekin Basin (mid-Devonian), northeastern Australia Cour Forsch-Inst Senckenberg 168:61–81

Talent JA, Mawson R (1999) North-eastern Molong Platform and adjacent Hill End Trough, New South Wales: mid-Palaeozoic conodont data, age-inferences, and implications regarding platform exposure and carbonate sedimentation offshore. Abh Geol Bund (Wien) 54:49–105

Taube A, Mawson R, Talent JA (2005). Repetition of the Mount Morgan stratigraphy and mineralization in the Dee Range, central Queensland, Australia: implications for exploration. Econ Geol 100:375–384

Thangaraja M, Al-Aisry A, Al-Kharusi L (2007) Harmful algal blooms and their impacts in the middle and outer ROPME sea area. Int J Oceans Oceanogr 2:85–98

Traquair RH (1899) Report on fossil fishes collected by the Geological Survey of Scotland in the Silurian rocks of the south of Scotland and supplementary notes. Trans R Soc Edinburgh 39:827–864

Trewin NH (1986) Palaeoecology and sedimentology of the Achanarras fish bed of the Middle Old Red Sandstone, Scotland. Trans R Soc Edinburgh 77:21–46

Trewin NH, Davidson RG (1996) An early Devonian Lake and its associated biota in the Midland Valley of Scotland. Trans R Soc Edinburgh, Earth Sci 86:233–246

Trewin NH, Davidson RG (1999) Lake-level changes, sedimentation and faunas in a Middle Devonian basin-margin fish bed. J Geol Soc London 156:535–548

Trinajstic K (1999a) New anatomical information on *Holonema* (Placodermi) based on material from the Frasnian Gogo formation and the Givetian-Frasnian Gneudna formation, Western Australia. Geodiversitas 21(1):69–84

Trinajstic K (1999b) Scale morphology of the Late Devonian palaeoniscoid *Moythomasia durgaringa* Gardiner and Bartram, 1977. Alcheringa 23:9–19

Trinajstic K (2001) A description of additional variation seen in the scale morphology of the Frasnian thelodont *Australolepis seddoni* Turner and Dring, 1981. Rec West Aust Mus 20:237–246

Trinajstic K, George AD (2005) Devonian (Frasnian) microvertebrate biostratigraphy of north Western Australia. In: Ivanov A, Young G (eds) Middle palaeozoic vertebrates of Laurussia: relationships with Siberia, Kazakhstan, Asia and Gondwana. St Petersburg, Russia, 22–25 August 2005. Ichthyolith Iss, vol 9, pp 34–37

Trinajstic K, George AD (2009) Microvertebrate biostratigraphy of Upper Devonian (Frasnian) carbonate rocks in the Canning and Carnarvon basins of Western Australia. Palaeontology 52:641–659

Trinajstic K, Marshall C, Long J, Bifield K (2007) Exceptional preservation of nerve and muscle tissues in Late Devonian placoderm fish and their evolutionary implications. Biol Lett 3:197–200

Tucker ME (1991) The diagenesis of fossils. In: Donovan SK (ed) The processes of fossilization. Belhaven Press, London, 303pp

Turner S (1982) A new articulated thelodont (Agnatha) from the early Devonian of Britain. Palaeontology 25:879–889

Turner S (1986) *Thelodus macintoshi* Stetson 1928, the largest known thelodont (Agnatha: Thelodonti). Breviora 486:1–17

Turner S (1991) Monophyly and interrelationships of the Thelodonti. In: Chang M-M, Liu Y-H, Zhang G-R (eds) Early vertebrates and related problems of evolutionary biology. Science Press, Beijing, pp 87–119

Turner S (1992) Thelodont lifestyles. In: Mark-Kurik E (ed) Fossil fishes as living animals. Academia 1:20–41

Turner S (1993) Palaeozoic microvertebrate biostratigraphy of Eastern Gondwana. In: Long JA (ed) Palaeozoic vertebrate biostratigraphy and biogeography. Belhaven Press, London, pp 174–207

Turner S (1997a) Sequence of Devonian thelodont scale assemblages in East Gondwana. Geol Soc Am Spec Pap 321: 295–315

Turner S (1997b) Palaeozoic vertebrate microfossils in Australasia. In: Vickers-Rich P, Monagahan JN, Baird RF, Rich TH (eds) Vertebrate palaeontology of Australasia, 2nd edn. Pioneer Design Studios with Monash University of Publications Committee, Melbourne, pp 625–643

Turner S (1999) Jawless fishes. In: Singer R (ed) Encyclopedia of paleontology, 2 vols. FitzR Dearborn Publishers, Chicago, pp 1449–1451

Turner S (2002) Fish kills; a new database for Gondwana. In: Brock GA, Talent JA (eds) 1st international palaeontological congress, IPC-2002, Sydney July 6–10, 2002. Geological Society of Australia, Abstract, vol 68, pp 287–288

Turner S (2004) Early vertebrates: analysis from microfossil evidence. In: Arratia G, Wilson MVH, Cloutier R (eds) Recent advances in the origin and early radiation of vertebrates. Verlag Dr Friedrich Pfeil, Munich

Turner S, Young GC (1987) Shark teeth from the Early Middle Devonian Cravens Peak Beds, Georgina Basin, Queensland. Alcheringa 11:233–244

Turner S, Burrow CJ (1999) Micropaleontology, vertebrate. In: Singer R (ed) Encyclopedia of paleontology. FitzR Dearborn Publishers, Chicago

Turner S, Cadée GC (2005) Dr Margaretha Brongersma-Sanders (1905–1996), Dutch scientist: an annotated bibliography of her work to celebrate 100 years since her birth. Zool Meded Leiden 80:183–204

Turner S, Miller RF (2005) New ideas about old sharks. Am Sci 93:244–252

Turner S, van der Brugghen W (1993) Thelodonti, an important but enigmatic group of fishes. Mod Geol 18:125–140

Turner S, Basden A, Burrow CJ (2000) Devonian vertebrates of Queensland. Cour Forsch-Inst Senckenb 223:487–521

Turner S, Blieck ARM, Nowlan GS (2004) Cambrian-Ordovician vertebrates. In: Webby BD, Droser, M, Feist R, Percival I (eds) IGCP410: the great Ordovician biodiversity event. Columbia University Press, New York

Turner S, Burrow CJ, Warren A (2005) *Gyracanthides hawkinsi* sp. nov. (Acanthodii, Gyracanthidae) from the Lower Carboniferous of Queensland, Australia, with a review of gyracanthid taxa. Palaeontology 48:963–1006

Turner S, Vergoossen JMJ, Young GC (1995) Fish microfossils from Irian Jaya. Mem Austral Assoc Palaeontols 18:165–178

Upeniece I (2001) The unique fossil assemblage from the Lode Quarry (Upper Devonian, Latvia). Mitt Mus Naturk Berlin, Geowiss Reihe 4:101–119

Valiukevičius JJ (1992) First articulated *Poracanthodes* from the Lower Devonian of Severnaya Zemlya. In: Mark-Kurik E

(ed) Fossil fishes as living animals. Academy of Sciences of Estonia, Tallinn, pp 193–214
Valiukevičius JJ (1994) Acanthodians and their stratigraphic significance. In: Cherkesova S, Karatajute-Talimaa V, Matukhin R (eds) Stratigraphy and Fauna of the Lower Devonian of the Tareya Key Section (Taimyr). Nedra, Leningrad (in Russian)
Valiukevičius JJ (1995) Acanthodian histology: some significant aspects in taxonomical and phylogenetical research. Geobios M.S. 19:157–159
Valiukevičius JJ (1998) Acanthodians and zonal stratigraphy of Lower and Middle Devonian in East Baltic and Byelorussia. Palaeontogr A 248:1–53
Valiukevičius JJ (2003) Devonian acanthodians from Severnaya Zemlya Archipelago (Russia). Geodiversitas 25:131–204
Valiukevičius J (2005) Silurian acanthodian biostratigraphy of Lithuania. Geodiversitas 27:349–380
Valiukevičius J, Burrow CJ (2005) Diversity of tissues in acanthodians with "*Nostolepis*"-type histological structure. Acta Palaeontol Pol 50:635–649
van der Brugghen W (1994) Over magen en staarten van thelodonten [On stomachs and tails in thelodonts]. Grondboor en Hamer 4/5:87–94 (In Dutch with English summary)
von Bitter PH, Purnell MA, Tetreault DK, Stott CA (2007) Eramosa Lagerstätte—exceptionally preserved soft-bodied biotas with shallow-marine shelly and bioturbating organisms (Silurian, Ontario, Canada). Geology 35:879–882
Wang N-Z, Dong Z-Z (1989) Discovery of Late Silurian microfossils of Agnatha and fishes from Yunnan, China. Acta Pal Sin 28:192–206
Warren AA, Turner S (2004) A stem tetrapod from the Viséan Ducabrook Formation, Queensland, Australia. Palaeontology 47:1–34
Warren AA, Turner S (2006) Tooth histology patterns in early tetrapods and the significance of dark dentine. Trans R Soc Edinburgh Geol 96:113–130
Wasmund (1926) Biocönose und Thanatocönose, biosoziologische Studie über Lebensgemeinschaften und Totengesellschaften. Arch Hydrobiol 17:1–116
Watson DMS (1937) The acanthodian fishes. Phil Trans R Soc London B 228:49–146
Weigelt J (1989) Recent vertebrate carcasses and their paleobiological implications. (Original 1927; Translated J. Schaefer). University of Chicago Press, Chicago
White EI (1978) The larger arthrodiran fishes from the area of the Burrinjuck Dam New South Wales. Trans Zool Soc London 34:149–262
Williams ME (1985) The "cladodont level" sharks of the Pennsylvanian black shales of central North America. Palaeontographica A 190:83–158
Wilson MVH (1989) Taphonomic processes: information loss and information gain. Geosci Can 15:131–148
Wilson MVH (1996) Taphonomy of a mass-death layer of fishes in the Paleocene Paskapoo formation at Joffre Bridge, Alberta, Canada. Can J Earth Sci 33:1487–1498
Wilson MVH, Hanke GF (1998) Body form and fin spines in species with scales of chondrichthyan growth pattern from the Lower Devonian of the Mackenzie Mountains, Canada. In: Ginter M, Wilson MVH (eds), Circum-Arctic Palaeozoic Faunas and Facies (IGCP 406). Ichthyolith Issues Spec Pub 4, Warsaw, pp 57–58
Wilson MVH, Caldwell MW (1998) The Furcacaudiformes: a new order of jawless vertebrates with thelodont scales, based on articulated Silurian and Devonian fossils from northern Canada. J Vert Paleontol 18:10–29
Wilson MVH, Hanke GF, Märss T (2007) Paired fins of jawless vertebrates and their homologies across the "agnathan"-gnathostome transition. In: Anderson JS, Sues H-D (ed) Major transitions in vertebrate evolution. Indiana University Press, Bloomington, IN, pp 122–149
Wolf KJ (1965) Petrogenesis and palaeoenvironment of Devonian algal limestones of New South Wales. Sedimentology 4:113–178
Woodward AS (1891) Catalogue of the Fossil fishes in the British Museum (Natural History). Part II. Elasmobranchii (Acanthodii), Holocephali, Ichthyodorulites, etc. Brit Mus (Nat Hist), London
Woodward AS (1906) On a Carboniferous fish-fauna from the Mansfield District, Victoria. Aust Nat Mus Melb Mem 1:1–32
Wyatt T (ed) (2008) Harmful Algae news 38:16. http://www.ioc-unesco.org/hab
Young GC (1981) Biogeography of Devonian vertebrates. Alcheringa 5:225–243
Young GC (1982) Devonian sharks from south-eastern Australia and Antarctica. Palaeontology 25(4):817–843
Young GC (1987) Devonian fish remains from Billiluna, eastern Canning Basin, Western Australia. BMR J Aust Geol Geophys 10:179–192
Young GC (1991) The first armoured Agnathan vertebrates from the Devonian of Australia. In: Chang MM, Liu YH, Zhang GR (eds) Early vertebrates and related problems of evolutionary biology. Science Press, Beijing
Young GC (1992) Fossil fishes from Antarctica. In: Tingey RJ (ed) The geology of Antarctica. Oxford University Press, Oxford, pp 538–567
Young GC (1993) Middle Palaeozoic macrovertebrate biostratigraphy of Eastern Gondwana. In: Long JA (ed) Palaeozoic vertebrate biostratigraphy and biogeography. Belhaven Press, London, pp 208–251
Young GC (1997) Ordovician microvertebrate remains from the Amadeus Basin, central Australia. J Vert Paleontol 17:1–25
Young GC (2003) Did placoderm fish have teeth? J Vert Paleontol 23:988–990
Young GC (2007) Devonian formations, vertebrate faunas and age control on the far south coast of New South Wales and adjacent Victoria. Aust J Earth Sci 54:991–1008
Young GC (2008) Relationships of tristichopterids (osteolepiform lobe-finned fishes) from the Middle–Late Devonian of East Gondwana. Alcheringa 32:321–336
Young GC (2009) An Ordovician vertebrate from western New South Wales, with comments on Cambro-Ordovician vertebrate distribution patterns. Alcheringa 33:79–89
Young GC, Burrow CJ (2004) Diplacanthid acanthodians from the Aztec Siltstone (late Middle Devonian) of southern Victoria Land, Antarctica. Fossils Strata 50:23–43
Young GC, Gorter JD (1981) A new fish fauna of Middle Devonian age from the Taemas/Wee Jasper region of New South Wales. Bur Miner Res Bull 209:83–147
Young GC, Turner S (2000) East Gondwana Devonian overview. Cour Forsch-Inst Senckenb 223:453–470

Young GC, Goujet D (2003) Devonian fish remains from the Dulcie Sandstone and Cravens Peak Beds, Georgina Basin, central Australia. Rec West Aust Mus Suppl 53:1–85

Young GC, Schultze H-P (2005) New osteichthyans (bony fishes) from the Devonian of central Australia. Mitt Mus Natur Berlin, Geowiss Reihe 8:13–35

Young GC, Lelièvre H, Goujet D (2001) Primitive jaw structure in an articulated brachythoracid arthrodire (placoderm fish; Early Devonian) from southeastern Australia. J Vert Paleontol 21:670–678

Young GC, Burrow CJ, Long JA, Turner S, Choo B (2010) Devonian macrovertebrate assemblages and biogeography of East Gondwana (Australasia, Antarctica). Palaeoworld 19:55–74

Zangerl R (1981) Chondrichthyes 1. Palaeozoic Elasmobranchii. In: Schultze H-P (ed) Handbook of paleoichthyology, vol 3A. Gustav Fischer Verlag, Stuttgart, pp 1–115

Zhu M, Yu X, Ahlberg PE (2001) A primitive sarcopterygian fish with an eyestalk. Nature 410:81–84

Zorn ME, Caldwell MW, Wilson MVH (2005) Lithological analysis of the vertebrate-bearing beds at the Lower Devonian MOTH locality, N.W.T., Canada: insights to taphonomy and depositional setting. Can J Earth Sci 42: 763–775

Messel Pit Fossil Site – The Legacy of the Environment and Life of the Eocene

Stephan Schaal

Abstract

The Messel Pit, between Frankfurt and Darmstadt, is the richest site in the world for understanding the living environment of the Eocene (about 50 million years ago). It provides unique information about the early stages of the evolution of mammals, and details of exceptionally well-preserved fossils from fully articulated skeletons to hair, feathers, wing membranes, and stomach contents. Many different species of plants and animals are found including crocodiles, bats, birds, reptiles, fish, turtles, insects, and even a herd of primitive horses. This paper traces the story of how the Messel "treasure chest for palaeontology" has been secured as a research resource for scientists, how it has become recognised as a heritage site, and how it is being developed as an educational and recreational facility for the general public.

Keywords

Germany • Eocene • UNESCO World Heritage site • Catastrophic setting (maar) • Lacustrine sedimentation • Oil shale • Sub-tropical/tropical • Exceptional preservation (skeletons to stomach contents) • Mammals to reptiles • Birds • Insects • Plants • Biogeographic linkages

Introduction

Fifty million years ago, during the Eocene, Europe was an archipelago. Although moving northwards with the drifting continental plates, Europe was still located much farther south than it is today. It was even separated from Asia for a long time during the Early Tertiary by oceans: the Tethys Sea in the south, the Atlantic Ocean in the west, and the North Sea. The Turgai Strait north-east of Europe connected the Antarctic Ocean to the Tethys Sea. The palaeoclimate was fairly uniform throughout the world and was conducive to the development of a paratropical rain forest on the European archipelago. In contrast to tropical regions today, this paratropical climate experienced seasons because of its higher latitude.

The Construction of the Chest

During the early Eocene at a latitude of 38°N (the present latitude of southern Spain and Sicily), a natural catastrophe took place when the area around Messel

S. Schaal (✉)
Senckenberg Forschungsinstitut und Naturmuseum Frankfurt, 60325 Frankfurt am Main, Germany
e-mail: stephan.schaal@senckenberg.de

J.A. Talent (ed.), *Earth and Life*, International Year of Planet Earth,
DOI 10.1007/978-90-481-3428-1_9,

experienced extensive tectonic activity. Numerous water-vapour explosions occurred several hundred metres below the surface. When groundwater came into contact with molten rock at about 1200°C, a volcanic crater (known as a "maar" crater) was created on the surface. A mixture of rock debris and volcanic ash formed a ring wall around the crater. The crater was 300 m deep, with a diameter of more than 1 km, and as groundwater and rainwater accumulated in it (Schulz et al. 2002; Felder and Harms 2004), the new freshwater maar lake developed. At least 50 million years ago sedimentation began after this volcanic explosion (Mertz and Reene 2005). During a period of about one million years, lake sediments including breccia, sandstone, siltstone, claystone, and even a laminated algal rock were deposited, together with the fauna and flora of the time; these have been protected from erosion since then (Felder and Harms 2004).

The sediments of the area have been pushed northwards to their present location by tectonic forces as the African plate moved against the European plate. The world climate has cooled considerably and much water is now bound up as ice and snow in the polar regions. As a result of the lowered sea level and continuing orogenesis, the island world of Europe has become the contiguous European continent. The 50-million-year-old maar lake sediments are still moving north with the ongoing movement of Africa against Europe, and are now located in the Federal Republic of Germany, 30 km south of Frankfurt am Main in the state of Hesse.

Man Becomes Interested

From 1884 to the present day, the maar lake sediments of Messel have been mined. The maar lake deposits were discovered in the twentieth century and the sediments, having a high percentage of organic compounds, were identified as oil shale by geologists. The complete sequence of these sediments was called the "Messel Formation" because of their proximity to the village of Messel. The oil shale was open-cast mined from 1884 to 1971; more than one million metric tons of crude oil were recovered during that time. The Messel Pit was formed during mining (Fig. 1), and is 1 km in length, 700 m in width, and currently 60 m in depth. The presence of fossils in this oil shale has been known since 1884. After the cessation of mining in 1971, the pit became available to private fossil collectors, and systematic scientific excavations first became possible in 1975. The Senckenberg Research Institute and the Hessisches Landesmuseum Darmstadt have excavated the Messel Formation since the 1970s and continue to do so today. Fossil species including plants, insects, fish, amphibians, reptiles, and mammals have been found in an extraordinary state of preservation. Complete skeletons, outlines of soft body parts, filaments such as hair, feathers, skin, intestinal contents, and even pregnant females and their foetuses, have been recovered. These offer a unique opportunity to study the adaptive strategies of locomotion, feeding, defence, orientation, and even reproduction. This has not been possible to the same extent at other fossil sites.

Fig. 1 View over the Messel fossil field from the observation platform. (Photo: S. Schaal)

The geological age and its palaeogeographic location of the Messel oil shale play an important role in its unique biota. The Cenozoic Era, to which the Eocene Messel Formation belongs, is known as the Age of Mammals, named after the dominant group of vertebrates to appear after the extinction of the dinosaurs. Because Europe was an archipelago, its mammalian fauna developed in unique ways, supporting relicts from previous epochs and newcomers from modern groups. The land bridges that connected the continents for short periods of time before the sedimentation of the Messel oil shale made the dispersal of various populations possible (Storch 1992). Consequently, the Messel deposits allow us an insight into the history the earth and vertebrate life not previously available.

Unfortunately, with the closure of the Messel mine, it was suggested that the site be used as a waste dump. After a 20-year debate on whether to use the pit as a refuse dump or as a site of palaeontological research, the state of Hesse resolved the issue by purchasing the Messel Pit in 1991, thus guaranteeing its long-term scientific use. The controversy was fuelled over this long period by a court procedure instigated by a civic action group and by the untiring protests and public relations exercises of scientists and scientific institutions all over the world. Recognition of the efforts of the scientists involved in the argument is applauded and one Professor Dr. Willi Ziegler, Director of the Senckenberg Museum at the time, deserves the highest accolade for his tenacity and the energy he put in to the debate. When finally resolved, thanks go to the politicians who, "in the conflict between science and refuse disposal decided in favour of science, as a testament for others" (Schaal and Ziegler 1992). The long-term scientific use of the Messel Pit was finally guaranteed (Schaal and Schneider 1995). The Senckenberg Society of Nature Research has operated the Messel Pit since 1992, and the Messel Research Department of the Senckenberg Museum was established at the same time.

The Messel Pit is now owned by the state of Hesse, including the rights to excavate and mine the oil shale. The Messel mine contains the pit itself, the area of the slag heaps, and an area of sewerage. Its control is incumbent on the Senckenberg Society and the Mining Department of Hesse.

After the Senckenberg Museum became the operator of the pit (1 July, 1992), the Messel Research Department undertook new tasks: planning the operational procedures, safe-guarding the pit, fire prevention, supervision of the stability of the hillside locations, and the regulation of water resources. The costs of the pit are paid by the state of Hesse (50%) and the Federal Republic (50%). To manage these tasks, the Messel Research Department runs a Research Station in Messel. Senckenberg staff in association with a number of geology students excavates during the summer and prepares the fossils for scientific collections in the winter (Fig. 2). A caretaker who is also responsible for the Research Station controls the pit in terms of illegal excavation, fencing, traffic control, and guided tours.

An important milestone was reached in 1995 as *notable increased public awareness* of the area was

Fig. 2 Staff from the Senckenberg Research Institute collecting slabs of Messel slate and inspecting them for fossils. (Photo: S. Tränkner)

apparent. The scientific excavations, with their significant fossil finds, had increased interest in palaeontology; the spectacular preparation of the fossils amazed the community. Research into the oil shale at Messel had been so successful that the Hesse Ministry of the Sciences and Arts applied for the Messel site to be included in the UNESCO World Heritage List. On 12 September 1995, UNESCO accepted the oil-shale pit at Messel onto its World Heritage List, and is now officially known as the "Messel Pit Fossil Site". Messel is the first World Heritage site in Europe recognised for its palaeontological significance. A bibliography containing 1436 publications on Messel research reflects the perspectives and the variety of topics that scientists have addressed with data from the Messel deposits (Schaal et al. 2004). To continue the outreach program of Messel, a Visitors and Information Centre was constructed south of the Messel Pit.

The Contents of the Eocene Trap

For many years, the Messel maar lake acted as a collection and sedimentation pond for mud particles, carcasses, and plant remains. The maar lake sediments contain a mixture of plant and animal remains from the former lake itself and from its surroundings, representing a small part of a large primeval forest. For nearly a million years, the lake supported creatures that were later fossilised. The disintegration of the animals before they were embedded varied, so their remains occur as both complete skeletons and isolated bones. Strong algal growth and poor aeration ensured an extremely low oxygen content in the water and therefore suitable conditions for anaerobic bacteria. This was responsible for the incomplete decay of the organic substances that form kerogen. Because the water was more or less stably stratified, the lake type is described as meromictic.

Generally, only a very small fraction of all the organisms that die become fossils, and even these usually disintegrate. With the exception of species of algae, the density of fossils from the Messel oil shale is relatively low. The rate of fossil finds is as high as it is because during excavation, thousands of years of sedimentation are revealed each week.

The wide spectrum of faunal and floral fossil species found is a feature of the Messel fossils. Skeletons, laterally embedded and unfortunately pressed flat, together with the impressions of biogenic structures,

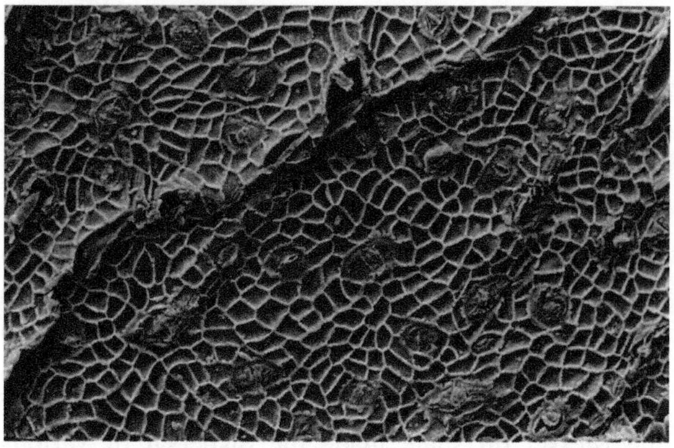

Fig. 3 SEM (Scanning Electric Microscope) micrograph (magnification 210*x*) of food remains from the gut of the primitive Messel horse, *Hallensia matthesi*. Underside of a leaf consumed showing cell walls and stomata. (Photo: R. Richter)

such as skin, hair, and feathers, by have been preserved together by anaerobic bacteria. Paradoxically, the micro-organisms that decompose the soft tissues are also responsible for reproducing the soft-tissue structures (Wuttke 1983). With this information, it is possible to reconstruct the form of the animals and to deduce their living conditions and environments. Even food remnants have been preserved. An analysis of the contents of the fossilised digestive tracts (Fig. 3) and coprolites allow conclusions to be drawn regarding faunal food preferences. These remains provide interesting information about how the organisms fed; conclusions may then be drawn as to their habitats. Food remnants, mainly found with vertebrate fossils, have for the first time been found with leaf- and pollen-eating insects. The insects themselves are pressed completely flat and are therefore deformed.

Plant Fossils

These are the most common macrofossils found in the oil shale. The silicious alga *Melosira* and the green algae *Botryococcus* and *Tetraedron* lived in the open water of the maar lake. *Tetraedron* produced the annual algal blooms responsible for the organic content of the oil shale. Pteridophyta and Gymnospermae are also found, but the angiosperms are dominant. More than 60 plant families are preserved, with their leaves, blossoms, pollen, spores, fruit, and seeds intact (Figs. 4 and 5). Wood fragments, however, are rare. A pollen analysis identified 46 plant families, one-third of which are recorded for the first time in the Middle Eocene.

Fig. 4 A tiny flower (10 mm in diameter), possibly of the Rutaceae (Photo: M. Müller). Note the doughnut-like rows of what might be seeds on the disc of the flower (see Fig. 5, below)

Fig. 5 An arrangement of doughnut-like seeds similar to those seen in the central disc of a rutacean (cf. Fig. 4; photo: S. Tränker). See also Collinson et al. (2010, Fig. 1K) for illustration of a single unit of the suggested wreath of seeds

The diverse flora of the Messel oil shale indicates a humid climate, consistent with its palaeogeographic situation on the European archipelago. The many subtropical plants and tropical plants including palms and lianas, allow us to interpret this Eocene forest as a subtropical to tropical rainforest (Wilde 2006). The presence of crocodiles attests to a climate with temperatures above 10°C (50°F).

Insect Diversity and Other Invertebrates

This is confirmed by the several hundred species of insects that have been discovered, most of which are yet to be described. By the end of 2009, the Senckenberg Museum had recovered 13,659 fossil insects from the Messel oil shale over a period of 35 years. To date, 14 orders have been identified; surprisingly these are mainly land fauna (Wedmann 2005). Some of them, mainly jewel beetles and leaf beetles, are preserved with their structural colours intact giving them an iridescent metallic hue (Fig. 6). Water insects preferring near-shore shallow-water areas with rich vegetation are quite rare because that part of the former lake was not preserved in the oil shale. Only the larval cases of caddis fly larvae are frequently found. Rare records of aquatic insects include dragonfly larvae, adult dragonflies, and damselflies (Odonata), the minute larvae of mayflies (Ephemeroptera), and – most amazingly – over 50 fossils of one species of "water pennies", a peculiarly flattened beetle larvae (Coleoptera; S. Wedmann 2009, "personal communication") have been recovered (Fig. 7). The past presence of a high diversity of aquatic insects can also be inferred indirectly by the analysis of the coprolites of small fish containing many different insect remains (Richter and Wedmann 2005).

Fig. 6 *Psiloptera weigelti*, Family Buprestidae. One of the five species of fossil jewel beetles reported from Messel. Contrary to popular opinion, the colours are not from pigments in the exoskeleton, but are due to physical iridescence whereby a microscopic texture in their cuticle selectively reflects specific wavelengths of light. (Photo: U. Kiel)

Fig. 7 Larvae of the water beetle, *Eubrianax* Kiesenwetter, (a member of the Coleoptera) are known to live today only in fresh, oxygenated water. (Photo: S. Wedmann)

Fig. 8 A species of *Psiloptera*. This family, Buprestidae, rarely found fossilised, is made up of species requiring a warm climate suggesting that the climate at Messel was much warmer in the middle Eocene. (Photo: U. Kiel)

Among the terrestrial groups, compact insects such as beetles (Coleoptera) are frequent, comprising 60% of the insects found, many of which showing the shimmer of their original colours (Fig. 8). Other groups of the insect thanatocoenosis include cockroaches (Blattaria), flies (Diptera), true bugs (Heteroptera), cicadas and hoppers (Homoptera), bees and ants (Hymenoptera), termites (Isoptera), butterflies (Lepidoptera), stick insects (Phasmatodea), grasshoppers (Saltatoria), and even parasites such as Strepsiptera (Lutz 1990; Wedmann 2005). The ant *Formicium giganteum*, the largest ant ever known, is present (Fig. 9); the females have a wing span of up to 15 cm and a body length of about 6 cm. A fossil leaf insect (Phasmatodea) from the Messel deposits represents the oldest fossil record of leaf mimicry in the world (Wedmann et al. 2007). The fossil shows a very specialised morphology and has the same foliaceous appearance as extant male leaf insects. Its morphology is an example of evolutionary stasis. This can probably also be said of its behaviour, because its morphology was presumably perfected for a specific behaviour: the ability to remain perfectly still for long periods or to imitate a leaf swaying in the wind. Another peculiar insect fossil is a bee fly (Diptera), whose extant relatives occur only in Australia (Wedmann and Yeates 2008).

Fig. 9 *Formicium giganteum* from Messel is the largest ant ever found; this female has a wingspan of 13 cm. (Photo: U. Kiel)

Representatives of the weaver ant (Formicidae) are also present in the Messel deposits (Dlussky et al. 2008). Extant weaver ants live in the canopy of tropical forests and are renowned for constructing their nests by weaving together living leaves. However, because no nests have yet been found in the Messel deposits, it is impossible to say whether the fossil weaver ants already displayed this behaviour. However, a phenomenon that indirectly documents this fossil behaviour is the record in the Messel oil shale of cut leaves of several angiosperms. Their similar form suggests that these leaves were cut by Eocene leaf-cutter bees (Hymenoptera) in the manner observed in extant bees (Wedmann et al. 2009).

Vertebrate Fossil Communities

One hundred and thirty-two species of vertebrate fossils are known from the Messel Pit Fossil Site (Morlo et al. 2004). The inhabitants of the lake are the most frequent fossils. Eight species of fishes have been confirmed. Mud fish (genus *Cyclurus*) and bony pike or garfish (genus *Atractosteus*) are commonly found. Perch (genera *Amphiperca*, *Rhenanoperca*, and *Palaeoperca*), and the Teleostei *Thaumaturus* are rarer. Only a single 50-cm specimen of the eel, *Anguilla*, has been found. Non-predatory fish are absent from the fish fauna. In the different oil-shale layers, the composition of the fish fauna varies, indicative of changes in the sediment and other biota of the Messel maar lake (Micklich 2002).

Altogether, 31 species of reptiles have been found, including seven species of crocodiles, most of which are rare as one would expect of such top predators. Among the alligators, only a single species of the fish-eating, lake-dwelling *Diplocynodon* has been identified; it is found frequently and all ages groups are represented. Whereas only single snake vertebrae are found at most fossil sites throughout the world, complete skeletons with their characteristic fragile skulls are recovered at Messel. Five species are members of the family Boidae, like *Palaeopython* (Fig. 10); all grew to 2 m (Schaal 2004). A few specimens of Aniliidae, the pipe snakes, have been found. Lizards, most of them terrestrial, are found less frequently than the aquatic reptiles. Although 3000 species of lizards exist worldwide today, at Messel, only 12 species have been described so far from a few dozen specimens. These belong to the iguana, lacertid, skink, glass lizard, and monitor families. The soft tissue of

Fig. 10 A specimen of *Palaeopython* (subfamily Boinae). This specimen is slightly over 2 m but the skull (*grey* in colour) has been cast from a second specimen. (Photo: S. Tränkner)

Fig. 11 *Messelobatrachus tobieni*, an amphibian famous for its complete soft-tissue preservation. Only a few remnants of bones are preserved, but eyes in the skull can be seen; part of the liver and veins in the thigh are visible. (Photo: E. Haupt)

Fig. 12 Squashed carapaces of two turtles, possibly a species of *Allaeochelys* (subfamily Anisteirnae). (Photo: S. Tränkner)

some amphibians is complete (Fig. 11). Frogs are the most frequent amphibians; three common species are *Eopelobates*, *Messelobatrachus*, and *Lutetiobatrachus*. The salamander *Chelotriton* is a unique find. Turtles (Fig. 12) are other typical reptiles of the former lake; these include still-living aquatic turtles, such as *Neochelys*, *Allaeochelys*, and *Trionyx*.

Vertebrates (Bird Fauna)

The bird fauna, representing nearly 50% of the terrestrial vertebrate taxa, shows an immense diversity of genera and species. It includes more than 30 families, with more than 50 species. They are predominantly small land birds, similar to the ostriches (Struthioniformes), hawks (Falconiformes), crane-like birds (Gruiformes), game birds (Galliformes), nightjars (Caprimulgiformes), swifts (Apodiformes), hoopoes (Upupiformes), and woodpeckers (Piciformes), as well as a number of species yet to be described. Some bird species indicate connections with North and South America. A primitive ostrich of the family Palaeotididae, a large bird, has a present-day relative living in South America. Another example is the flightless bird *Gastornis* of the family Gastornithidae are related to the goose group; it grew to a height of nearly 2 m. Fossils of this bird have been found in North America and, together with the mammals, is a further an indication of a strong dispersal route of the fauna between North America and Europe (Mayr 2000, 2006). The preservation of the fossil birds is extraordinary insofar as many skeletons have feathers (Fig. 13); these are often visible in the form of fossilised layers of bacteria.

Fig. 13 *Scaniacypselus szarskii* Mayr and Peters (1999), a fossil swift thought to have hunted insects over the shallow sea and marshlands above what is now Messel Pit. It measured 9 cm from head to tail and had a wing-span of approximately 24 cm, somewhat smaller than modern swifts. (Photo: S. Tränkner)

Sometimes even patterns, such as banding, have been preserved. One might speculate that the birds have become trapped in the oil and perished.

Mammalian Fossils

Dozens of mammalian finds have made the Messel Pit Fossil Site world famous. Forty-five species are already known and at least 60 species are expected to be described from the Messel Formation laid down during the period when the mammals were developing rapidly, after the extinction of the giant reptiles at the end of the Cretaceous. The number of mammal fossils is large compared with those at other European fossil sites of a similar geological age.

Most of the fossils have been described and assigned to the orders Marsupialia, Edentata, Insectivora, Pholidota, Proteutheria, Lipotyphla, Chiroptera, Primates, Rodentia, Creodonta, Carnivora, Condylarthra, Artiodactyla, and Perissodactyla (Schaal and Ziegler 1992).

Biogeographic Links

Although Europe was an archipelago, separated by oceans for a long time during the Early Tertiary, the geographic and phylogenetic origins of most mammalian species of the Messel Formation can be traced to other continents: Africa, North America,

South America, and Asia. Land bridges connecting the continents for short periods of time made the dispersal of these populations possible. Pangolins show a relationship to Asiatic species, and the Messel pangolins are the oldest fossil record of this group (Storch 1990). Rodents originated in Asia, but the species found in the Messel deposits, *Ailuravus* and *Masillamys* (Fig. 14), show relationships to North American rodents. With the discovery of the oldest and best-preserved edentate, an anteater (*Eurotamandua*), it was possible to reconstruct a connection to South America via Africa, before Lake Messel existed. Carnivores such as Miacidae and a hedgehog-related insectivore, the omnivorous *Macrocranion* and the more robust *Pholidocercus* (Fig. 15), frequent finds in the Messel oil shale, are evidence of the connection to North America (Storch 1990).

Lemur-like adapids, such as *Europolemur* and *Godinotia*, developed in Africa and migrated to Europe. The first discovery of an extraordinary primate

Fig. 14 A rat-like mammal, *Masillamys*, is so well preserved that not only do we see the animal from nose to tail but the outline of the skin also can be seen, and darkening around the gut area are the remains of the animal's last meal. (Photo: S. Tränkner)

Fig. 15 *Pholidocercus hassiacus*, a hedgehog-like creature, with spines on its body but a scale covered head and scales on its long, thick tail. (Photo: E. Haupt)

skeleton in the Messel Formation, the most complete primate from the fossil record, is widely known. It was privately collected and later sold. The primate finds are rare, although the habitat at Messel was well suited to their preservation. The specimen was described as *Darwinius masillae*, belonging to the Cercamoniinae. *Darwinius* (Fig. 16) was an agile, nail-bearing, arboreal quadruped living above the floor of the Messel rainforest. Skull radiography showed a host of teeth developing within the juvenile face. The individual died in her first year of life and probably had a body weight of about 900 g. The find is highly significant. The complete fossil primate includes a soft-body outline and the contents of the digestive tract. Its study allowed the complete reconstruction of its life history, locomotion, and diet. The absence of a toilet claw or a tooth comb demonstrates that *Darwinius* belongs to a larger group of primates, the Adapoidea, representative of the early haplorhine diversification (Franzen et al. 2009).

Two other large groups probably developed in Africa and migrated to Europe: the perissodactyls, with the primeval horses *Eurohippus* and *Hallensia*, the tapir-like *Hyrachyus*, and *Lophiodon*; and the artiodactyls, including *Messelobunodon*, *Aumelasia*, and *Masillabune*. The most famous of all these are the primeval horses (Fig. 17). They differed from modern horses in many features, including their small size, low-crowned cheek teeth, and the numbers of digits (four on the forefoot and three on the hind foot).

A typical endemic mammalian form is the Messel Amphilemuridae, genus *Pholidocercus*, which developed only in Europe. In its biological adaptation,

Fig. 16 *Darwinius massillae* Franzen et al. (2009) is a juvenile, female, early primate. Stained matrix shows the soft-tissue body outline. Abdomen contains organic remains of food in the digestive tract. The specimen was approximately 50 cm long, includingthe tail; at full adult size she would have still weighed less than 2 kg. *Darwinius masillae* is a very significant discovery as it is the most complete primate fossil found to date. (Photo: P Aas)

Fig. 17 A species of *Propalaeotherium? parvulum.* More than 70 specimens of pygmy horses have been recovered from Messel. (Photo: S. Tränkner)

Pholidocercus can be compared with the extant hairy hedgehogs (Fig. 15).

Surprisingly, the most frequent mammals in terms of the number of specimens are the bats, more than 700 finds. Their frequency may be attributable to the carbon dioxide probably associated with the volcanism that caused the maar crater. They contribute to the uniqueness of the Messel fossils because they are rarely found as fossils elsewhere. They are represented by a group of eight species (genera *Hassianycteris*, *Archaeonycteris*, *Palaeochiropteryx*, and *Tachypteron*) and can be allocated to three ecological niches according to their different wing morphologies, gut contents, and inner-ear structures: between the trees, above the tree tops, and close to the ground and foliage. One species of the genus *Tachypteron* can even be assigned to a modern family. A unique direct comparison with extant species is thus possible. Recent progress in three-dimensional microtomography has made possible a detailed examination and analysis of the tiny inner-ear capsule, still intact after 50 million years (Fig. 18). Therefore, extensive inferences about the echolocation of Eocene bats are possible.

Mode of Preservation

Vertebrate fossils of the Messel Formation are normally displayed on yellow artificial resins and not in the dark-coloured oil shale. The oil shale contains approximately 40% water and splits open in the course of drying after it has been mined. Speed is of the essence when an excavation team discovers filigree bones, plants, or insects. All fossils necessarily crumble when they are exposed to drying. Because of undulations or indentations in the host rock, only large fossils (crocodiles, turtles, and mammals, such as the primeval horses and the condylarth *Kopidodon* (Fig. 19)) can be anticipated. When this is the case, wedges and hammers are used to split off large blocks to preserve the finds in these blocks. As a consequence of the high water content of the oil shale, the vertebrate skeletons must be transferred in a time-consuming step from the oil shale to another material. Substantial progress was made in improving this procedure in the 1960s, with the development of synthetic resins. With this transfer method, the fossil is manually freed from one side of the shale. Any hard mineral deposits such as siderite or pyrite are removed with a sand-blasting

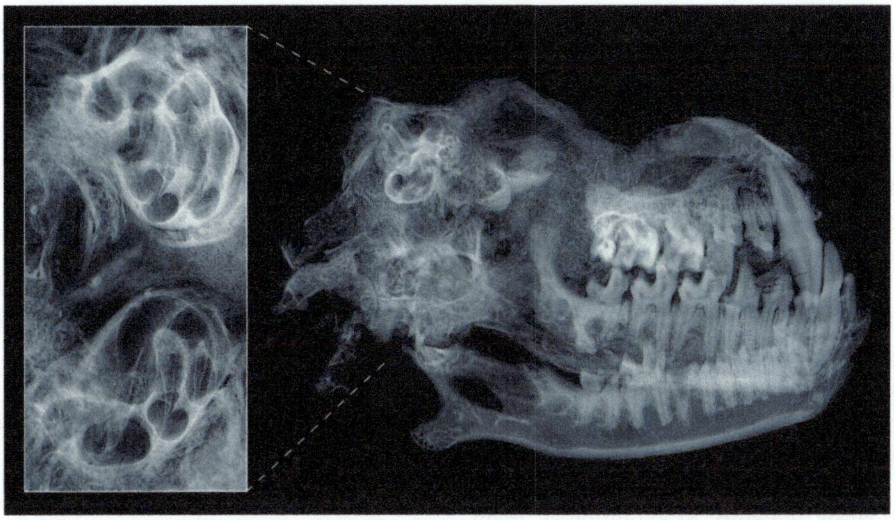

Fig. 18 An X-ray of the skull of a fossil bat, *Hassianycteris messelensis* showing the detailed structure of the cochlea. (Photo: J. Habersetzer)

Fig. 19 *Kopidodon macrognathus*. Compare with Fig. 17. Belongs to a group of basal primates. It is almost complete with bones in place and a dark-stained halo outlining of the animal; this halo is the remains of flesh and fur that has rotted in place. (Photo: S. Tränkner)

device. The shale is fitted with a frame, after which the prepared and dried fossils are coated with several layers of fluid synthetic resin. When the resin, which will become the supporting plate, has hardened, the fossil can be stored in water for many years in this sealed state. If the fossil is to be used for any further scientific research or micro X-ray analysis, the sealing compound is removed from the rest of the shale.

Because plants and insects do not adhere sufficiently to synthetic resin, they can be prepared only on one side. The specimens are placed under a stereomicroscope and are removed from the shale with ultrasonic vibrators or fine needles. When this time-consuming and delicate work has been completed, the fossils must be stored permanently in glycerine.

The Future for the Messel Pit Fossil Site, Hesse, Germany?

In 1971, the Hessian government conceived the idea that the Messel Quarry would be ideal as a garbage pit. Many museums, universities, and private individuals lobbied for this decision to be rescinded and by 1975 the Senckenberg Natural History Museum and Research Institute was granted permission to join the Hessian Regional Museum in excavating and salvaging finds. Since then numerous museums and universities joined the research program to find and conserve fossils from the site. By February 1990, the Hessian Ministers for Environment and Science and Art announced that the Messel quarry would not be used as a dump site. it was designated a cultural monument under the Hessian Heritage Protection Act, thus ensuring long-term availability for scientific research. It was declared a public monument under two agreements: (1) the Conduct of Palaeontological Excavations in Messel Pit with the Senckenberg Society for Nature Research, and: (2) the Scientific and Cultural Use of the Messel Pit Fossil Site with the Society for the Preservation of the Messel Pit Fossil Site. In 1991, the Hessian government purchased the site for scientific use, thus ending the threat of commercial exploitation.

The history of the pit's oil shale as a mineral resource is now considered part of the cultural heritage of Hesse. The fossils gleaned from the Messel pit have opened a window on the story of evolution of mammals and other creatures during the Eocene, providing people, from scientists to a class of 5-year-old children, with inspiration and data to last a lifetime. The future for Messel looks bright.

References

Collinson ME, De Vore ML, Pigg KB (2010) Paleogene floras and global change events. Bull Geosci 85:111–162

Dlussky GM, Wappler T, Wedmann S (2008) New middle Eocene formicid species from Germany and the evolution of weaver ants. Acta Palaeontol Pol 53:615–626

Felder M, Harms FJ (2004) Lithologie und genetische Interpretation der vulkano-sedimentären Ablagerungen aus der Grube Messel anhand der Forschungsbohrung Messel 2001 und weiterer Bohrungen. Cour Forschung Senckenberg 252:151–203

Franzen JL, Gingerich PD, Habersetzer J, Hurum JH, Von Koenigswald W, Smith BH (2009) Complete primate skeleton from the Middle Eocene of Messel in Germany: morphology and paleobiology. Plos One 4(5):e5723, 1–27. doi:10.1371/journal.pone.0005723

Lutz H (1990) Systematische und Paläoökologische Untersuchungen an Insekten aus dem Mittel-Eozän der Grube Messel bei Darmstadt. Courier Forschungsinstitut Senckenberg 124:1–165

Mayr G (2000) Die Vögel der Grube Messel – ein Einblick in die Vogelwelt Mitteleuropas vor 49 Millionen Jahren. Natur Museum 130(11):365–378

Mayr G (2006) Fine feathered fossils of the Eocene – the birdlife of Messel. Vernissage, Series. UNESCO World Heritage Sites 13(21):38–43

Mayr G, Peters DS (1999) On the systematic position of the Middle Eocene swift *Aegialornis szarskii* Peters 1985 with description of a new swift-like bird from Messel (Aves, Apodiformes). N Jb Geol Paläontol Mh 1999:312–320

Mertz DF, Reene PR (2005) A numerical age for the Messel fossil deposit (UNESCO World Heritage Site) derived from $^{40}Ar/^{39}Ar$ dating on a basaltic rock fragment. Cour Forschung Senckenberg 255:67–75

Micklich N (2002) Die Ichthyofauna des eozänen Messelsees – Besonderheiten und paläoökologische Implikationen. Schriftenreihe der Deutschen Geologischen Gesellschaft 21:240–241

Morlo M, Schaal S, Mayr G, Seiffert C (2004) An annotated taxonomic list of the Middle Eocene (MP11) Vertebrata of Messel. Cour Forschung Senckenberg 252:95–108

Richter G, Wedmann S (2005) Ecology of the Eocene Lake Messel revealed by analysis of small fish coprolites and sediments from a drilling core. Palaeogeogr Palaeoclimatol Palaeoecol 223:147–161

Schaal S (2004) *Palaeopython fischeri* n. sp. (Serpentes: Boidae), eine Riesenschlange aus dem Eozän (MP11) von Messel. Courier Forschungsinstitut Senckenberg 252:35–45

Schaal S, Schneider U (1995) Chronik der Grube Messel 1965–1995. In: Schaal S, Schneider U (eds) Chronik der Grube Messel. Kempkes, Gladenbach, pp 195–276

Schaal S, Ziegler W (eds) (1992) Messel – an insight into the history of life and of the Earth. Clarendon Press, Oxford, pp 1–322.

Schulz R, Harms FJ, Felder M (2002) Die Forschungsbohrung Messel 2001: Ein Beitrag zur Entschlüsselung der Genese einer Ölschieferlagerstätte. Zeitschrift für angewandte Geologie 4:9–17

Schaal S, Brahm EE, Habersetzer J, Hebs A, Müller M, Schlosser-Sturm E (2004) Literaturübersicht und Schriftenverzeichnis zur wissenschaftlichen Erforschung der Fossilienfundstätte Messel. Cour Forschung Senckenberg 252:243–245

Storch G (1990) The Eocene mammalian fauna from Messel – a paleobiogeographical jigsaw puzzle. In: Peters G, Hutterer R (eds) Vertebrates in the tropics. Museum Alexander Koenig, Bonn, pp 23–32

Storch G (1992) The mammals of island Europe. Sci Am 226(2):64–69

Wedmann S (2005) Annotated taxon-list of the invertebrate animals from the Eocene fossil site Grube Messel near Darmstadt, Germany. Courier Forschungsinstitut Senckenberg 255:103–110

Wedmann S, Yeates DK (2008) Eocene Records of bee flies (Insecta, Diptera, Bombyliidae, *Comptosia*): their Paleobiogeographic implications and remarks on the evolutionary history of Bombyliids. Palaeontology 51: 231–240

Wedmann S, Bradler S, Rust J (2007) The first fossil leaf insect: 47 million years of specialized cryptic morphology and behavior. Proc Natl Acad Sci USA 4(2):565–569

Wedmann S, Wappler T, Engel M (2009) Direct and indirect fossil records of megachilid bees from the Paleogene of Central Europe (Hymenoptera: Megachilidae). Naturwissenschaften 96:703–712

Wilde V (2006) The Green Eocene—The Diverse Flora of a paratropical climate. Vernissage, Series. UNESCO World Heritage Sites 13(21):14–19

Wuttke M (1983) Weichteil-Erhaltung durch lithifizierte Mikroorganismen bei mittel-eozänen Vertebraten aus den Ölschiefern der Grube Messel bei Darmstadt. Senckenbergiana lethaea 64(5/6):509–527

Part II

Evolution Exemplified by Specific Phyla or Classes

Evolutionary Scenario of the Early History of the Animal Kingdom: Evidence from Precambrian (Ediacaran) Weng'an and Early Cambrian Maotianshan Biotas, China

Jun-Yuan Chen

Abstract

Late Proterozoic (Ediacaran) Weng'an (580 mya) and Early Cambrian Maotianshan (c. 530 mya) faunas of South China, illustrated here, document diverse body plans at phylum and subphylum level and confirm that bilaterians evolved well before the "Cambrian explosion". The Weng'an faunas (from Guizhou), the oldest record of metazoans, consist mainly of embryos with possible affinities to living sponges, cnidarians, and bilaterians, but with adult specimens (though microscopic) of the same groups. The Maotianshan Shale faunas (from Yunnan), remarkably diverse at species level (over 100 species), have great diversity of metazoan body plans, many comparable with those of living groups. Because they occur at or near the evolutionary roots of many animal groups, intermediate forms are present. Evolution of Early Cambrian metazoans was surprisingly rapid. Worm-like ancestral euarthropods elucidate the evolutionary origins of the arthropods. The diverse Maotianshan vertebrates, representing "missing" history between an amphioxus-like ancestor and craniate vertebrates, provide an improved understanding of the early evolution of the vertebrates.

Keywords

China • Guizhou • Yunnan • Chengjiang • Latest Precambrian (Ediacaran)–Early Cambrian • Doushantuo Formation • Weng'an phosphate member • Fossil embryos • Maotianshan Shale (Early Cambrian) • Panarthropod phylogeny • Origin of Chordata and vertebrates (Cristozoa) • Inception of vertebrate brain • Phylogeny of Deuterostomia

Introduction

The sudden appearance of animal fossils at the beginning of the Cambrian was of particular concern to Darwin (*The Origin of Species*, 1859, pp. 313–314); it seemed to be at odds with his theory of evolution through natural selection. He believed that the Cambrian trilobites may have descended from

J.-Y. Chen (✉)
LPS Institute of Geology and Palaeontology of Academia Sinica, Nanjing 210008, People's Republic of China

Institute of Evolution and Developmental Biology, Nanjing University, Nanjing 210093, People's Republic of China
e-mail: chenjunyuan@163.net; chenjy@nju.edu.cn

J.A. Talent (ed.), *Earth and Life*, International Year of Planet Earth,
DOI 10.1007/978-90-481-3428-1_10,

Precambrian ancestors long before the Cambrian. However, no Precambrian ancestor had yet been found at that time. The great evolutionary event called the *Cambrian Explosion* has been a great mystery and a highly controversial issue among researchers.

Like a coin, the Cambrian explosion has two sides: the Cambrian above and the Precambrian below. The magnitude of the event can be understood based on the contrast between the biota and the degree of diversity of the fossils from both sides.

Great advances have been made in Cambrian palaeontology over the past century (e.g., Whittington 1985), especially the discovery of the extraordinarily well-preserved soft-bodied fauna from the Middle Cambrian Burgess Shale (Briggs et al. 1994) and the Lower Cambrian Maotianshan Shale deposits (Chen 2004). The Cambrian side of the "Cambrian explosion" is richly illustrated and contrasts greatly with the Precambrian side, but it should be noted that with new finds, the diversity of both is increasing continuously.

Compared with the Cambrian, our knowledge of Precambrian fossils is very limited. The discovery of the macroscopic Ediacaran biota in Australia in 1946 provided the first evidence of multicellular life before the Cambrian (Glaessner 1984). Members of the macroscopic Ediacaran biota are now found worldwide, and include enigmatic tubular and frond-shaped sessile organisms. They emerged about 570 million years ago, flourished for a period of ca. 30 million years, and suddenly disappeared by the Precambrian/Cambrian transition 542 million years ago. The reasons for the extinction of the Ediacaran biota are not clear and are probably related to the advent of predators. The Ediacaran organisms have been described as metazoans with affinities to the Cnidaria (Glaessner 1959), molluscs (e.g. *Kimberella* Fedonkin and Waggoner 1997) and other bilaterians (Peterson et al. 2008), but most of the fossils are morphologically distinct from modern and even Cambrian life forms. Some palaeontologists do not agree that these creatures can be categorized using the currently accepted biological classification scheme (Seilacher 1989; Seilacher et al. 2003). Because it is difficult to deduce the evolutionary relationships among these Ediacaran organisms, some palaeontologists have suggested that they represent completely extinct lineages and established the kingdom Vendozoa (Seilacher 1984), now renamed the Vendobionta phyllum (Seilacher 1989). Because these enigmatic organisms predominantly left no descendants, the strange forms have been interpreted by some authors as "failed experiments" in multicellular life. The sudden appearance of metazoan diversity in the Cambrian with a great multiplicity of body plans in the animal groups at the phylum and subphylum levels remains a great mystery.

Searching for metazoan fossils from Precambrian times is a challenging task. Once the efficient preservation of soft tissues, as well as bacteria (e.g. Bailey et al. 2007), in phosphate deposits was understood, palaeontologists began to look for metazoan fossils in Precambrian phosphate deposits. In 1998, two parallel teams found fossil embryos and sponges in the Precambrian Duoshantou phosphate deposit in Weng'an, Guizhou, southern China (Li et al. 1998; Xiao et al. 1998; Xiao and Knoll 1999, 2000).

The Weng'an fauna, named after the fossil locality, is the oldest fossil record of metazoans and predominantly includes diverse embryos with affinities to living sponges, cnidarians, and bilaterians. In addition to the fossil embryos, the Weng'an fauna has yielded adult specimens of microscopic metazoans representing sponges, cnidarians, and bilaterians. The fauna lived in the early Ediacaran, immediately after the Snow Ball event, at least 580 million years ago, suggesting that metazoan life, including sponges, cnidarians, and bilaterians, flourished long before the Cambrian explosion and is indeed very deeply rooted.

The great advances in Precambrian palaeontology, especially the intensive study of the Weng'an biota during the past 10 years, have significantly changed our view of the Cambrian explosion. However, the contrast between the Precambrian and Cambrian faunas remains great and the explosive Cambrian evolutionary event remains an enigma.

The recently discovered fossil fauna from the Lower Cambrian Maotianshan Shale deposits in the Chengjiang, Haikou, and Anning areas near Kunming, China, documents the sudden appearance of a diversity of metazoans. These are not only remarkably diverse at the species level, including over 100 species, but illustrate a great morphological diversity of metazoan body plans. Many of these are comparable to those of living groups at the phylum and subphylum levels or were short lived and soon extinct (Gould 1989; Hou et al. 2004; Chen 2004). The close affinities between the 530-million-year-old Maotianshan Shale biota and the 505-million-year-old Burgess Shale biota suggest that these forms were geologically widespread and evolutionarily conserved following the Cambrian explosion. Because the intermediate forms were not well established and few survived, they appear only near the evolutionary roots of the related animal groups, becoming extinct soon after the new, advanced forms emerged. The fauna of the Maotianshan Shale lies close to the roots of many animal groups and intermediate forms living at that time. The discovery of many intermediate forms not only satisfies Darwin's prediction but also casts new light on the evolutionary origins of the animal groups, especially the vertebrates and arthropods.

The body plans (or body forms) of animals are essentially blueprints for the way the body of the organism is laid out (Raff 1996; Hall 1998; Chen 2009). The blueprint is shared by a group of animals ranked together at higher taxonomic levels, including superphyla, phyla, subphyla, classes, and orders. It is characteristic of a unique anatomical body organization that it is evolutionarily conserved and deeply rooted. The sudden appearance of the high animal diversity in the Cambrian corresponds to the sudden appearance of the diverse body plans of the animal groups at the phylum and subphylum levels. There are many living representatives of animals with these body plans, including vertebrates, ascidians, priapulids, sipunculans, arrow worms, possible molluscs, possible annelids, phoronida, and brachiopods (Chen 2004; Hou et al. 2004), many having changed little since the Cambrian (e.g. the ctenophores, priapulids, sipunculans, chaetognaths, tunicates, and inarticulate brachiopods). On the contrary, crest animals (the Cristzoa), including the extinct precraniates and living vertebrates (craniates), evolved rapidly and now include diverse and very distinct nested body plans that emerged later in evolution.

The origin of the vertebrates has long been obscure and controversial, with a lack of fossil evidence for the first vertebrates and the very beginnings of vertebrate evolution. The fossils from the Maotianshan Shale, including *Yunnanozoon*, *Haikouella*, and *Haikouichthys*, are representatives of the missing history of early vertebrate evolution. The study of these fossils contributes to a new understanding of the evolutionary origin of the vertebrates, their heads, brains, head sensorial organs, branchial arches, backbones, and feeding behaviour.

The limb-bearing segmented animals known as panarthropods have been diverse and abundant since the Early Cambrian and include diverse nested body plans representing different stages of evolution. Most of these different body plans appeared in the Early Cambrian, emerging one after another near the root of the evolutionary history of the panarthropods. Their appearance is a remarkable and rare phenomenon, shedding light on the evolutionary origins of various aspects of the arthropods, including the evolution of the head, limbs, and brain.

The affinities of numerous Cambrian animals remain controversial. However, the body plans of these animal groups, especially the stalked sessile and pelagic medusiform lophophore-bearing animals and the vetulicolian organisms, offer clues to their possible affinities with extant animals.

The regulatory network architecture is critical for establishing the body parts and body plan of the animals (Davidson 2006). The evolution of animal body plans provides us with interesting suggestions of how the genomic internal architectures and developmental processes that directed the course of evolution. The sudden birth of the diverse animal body plans suggests that the Cambrian explosion was marked with the diversification event of regulatory network architectures that regulate developmental processes of the animals.

The Oldest Fossil Evidence for the First Occurrence of Animal Life on Earth

Molecular data predict that the divergence of the metazoans (including the calcisponges, cnidarians, and bilaterians) and the separation of the major bilaterian clades (the deuterostomes, ecdysozoans, and lophotrochozoans) occurred deep in time, either within the Ediacaran, following the Cryogenian Marinoan glaciation ending about 635 mya (Hoffman et al. 2004; Condon et al. 2005; Peterson and Butterfield 2005), or 100–200 million years before the Marinoan (Douzery et al. 2004). Recent discovery of possible sponge graded metazoan fossils in pre-Marinoan limestones from South Australia (Maloof et al. 2010) together with evidence from sponge biomarkers indicates that metazoans originated pre-Marinoan time, with a robust presence throughout the Ediacaran (Love et al. 2009).

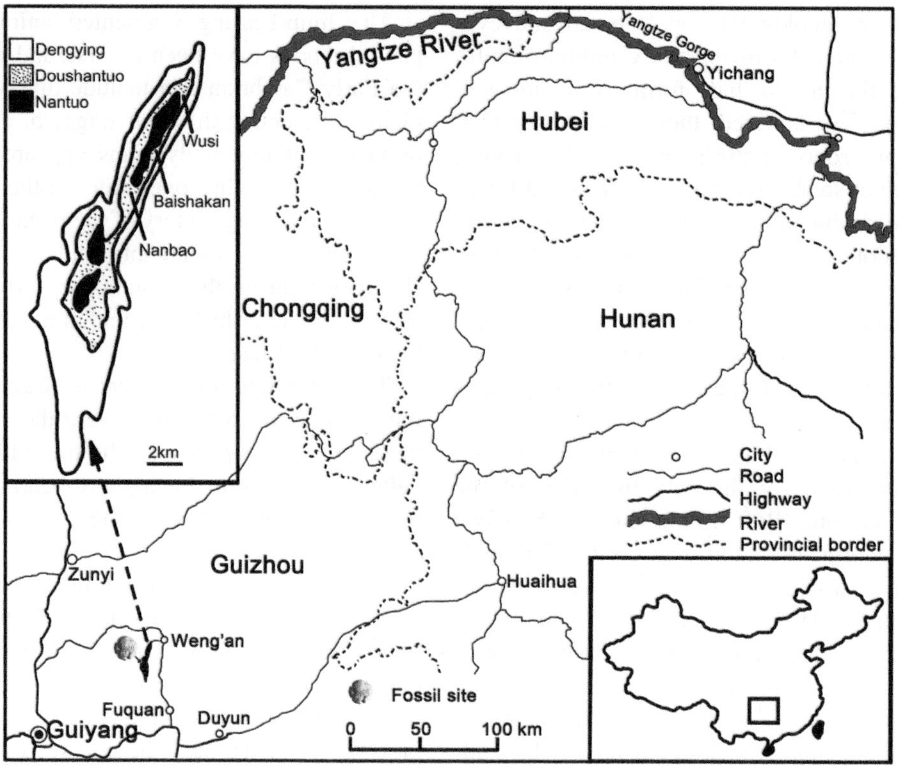

Fig. 1 Main localities producing the oldest known metazoans: the Wusi, Baishakan, and Nantuo quarries (adapted from Chen et al. 2009a)

Chen (2009) in discussing the appearance of diverse animal body plans during the Cambrian explosion states that "The Ediacaran Doushantuo phosphate deposits in Weng'an, Guizhou (Fig. 1) yield diverse metazoans, including sponges (Li et al. 1998), cnidarians (Chen et al. 2000, 2002a; Xiao et al. 2000), and bilateral animals and their related embryos (Chen et al. 2004b; Chen et al. 2006)". This incredible 580-million-year-old Weng'an biota has received great attention because it is the oldest fossil evidence yet found for animal life on Earth.

Geological Setting of the Weng'an Biota

The Doushantuo Formation is widely distributed in southern China and in areas of N. W. China. It is underlain by a glacial deposit known as the Nantuo Formation, laid down during the Marinoan glaciation and thus represents the first deposit after the last extensive glaciation of Snowball Earth. The Doushantuo Formation in Weng'an, Guizhou, crops out along the axis of the Mt Beidou anticline (Fig. 1). It is well exposed in the Wusi quarry (Fig. 1), where the thickness is 49 m, consisting of five different lithological units (Fig. 2). The Weng'an phosphate member of the Doushantuo Formation varies laterally in thickness and lithology on a scale of metres to tens of metres (see Fig. 2).

The Doushantuo Formation has been intensively quarried for phosphates at the Wusi, Baishaikang, and Nanbao quarries. All the specimens examined in the study come from the black and grey facies of the Weng'an phosphate member in these quarries. The quality of preservation varies among the three sites. Wusi (Fig. 3a) and Baishakan (Fig. 3b, c) are the best sites for metazoic embryos (Wusi and Baishakan) and microscopic adults (Baishakan). However, with extensive mining activity, these three quarries have suffered disastrous destruction by extreme sliding activity and have been deeply buried by massive rocks from the overlying strata (Fig. 3d, e). The Weng'an county government had recognized the fossil treasures at this site and erected a monument at the Baishakan quarry (Fig. 3c).

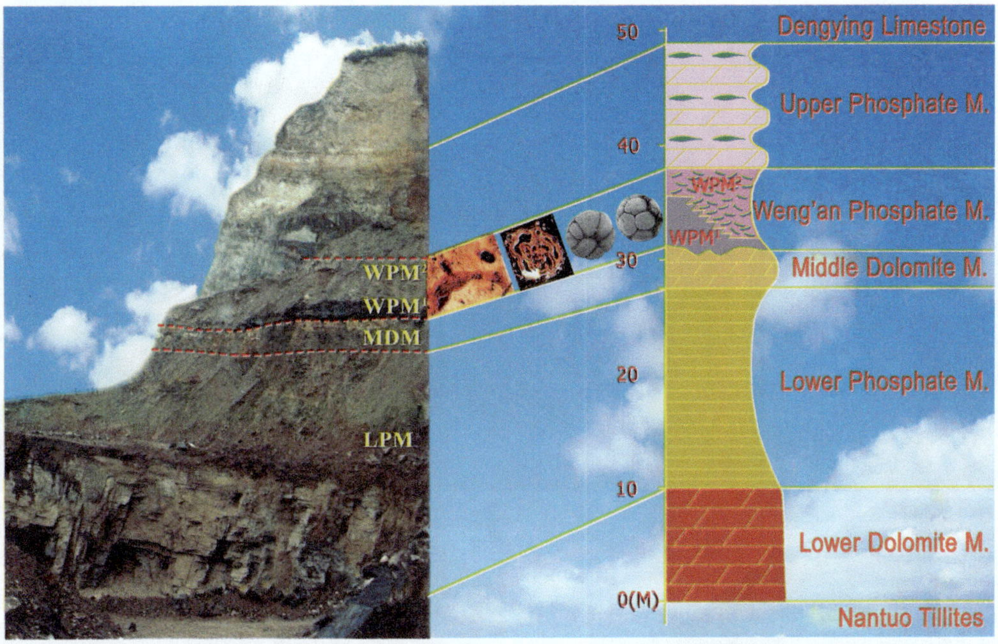

Fig. 2 Natural and diagrammatic profiles of the lithologic sequence of the Ediacaran Doushatuo Formation in the Mt. Beiduoshan area, Wengan, Guizhou. The lithologic profile was photographed from the western part of the Baishakan quarry; the thickness of lithologic units was measured in the Wusi quarry. WPM1 is the *black* facies, WPM2 the *gray* facies of the Weng'an Phosphate Member. MDM = Middle Dolomite Member; LPM = Early Phosphate Member

In the Yangtze Gorges area, about 600 km northeast of Weng'an (Fig. 2), the Lower Dolomite Member of the Doushantuo Formation, interpreted as a cap carbonate, contains an ash bed that has been dated as 635.23 ± 0.57 Ma using U–Pb dating techniques (Condon et al. 2005; Zhu et al. 2007). The top of the Doushantuo Formation in the Yangtze Gorges area contains another ash bed with a U–Pb age of 551.07 ± 0.61 Ma (Condon et al. 2005). Within the Doushantuo Formation at Weng'an, above the basal dolomite, are a lower and an upper phosphorite interval, separated by the Middle Dolomite Member with a minor hiatus above. The upper phosphorite interval, containing the rich Weng'an microfossil fauna, has been interpreted as younger than 580 million years by carbon isotopic and sequence stratigraphic analyses (Condon et al. 2005). However, the upper Doushantuo Formation corresponding to the Weng'an phosphate member has yielded a Pb–Pb age of 599 ± 4 (Barfod et al. 2002). A Pb–Pb age of 576 ± 14 million years has been calculated for the upper phosphate member (Chen et al. 2004a), providing additional evidence that the Doushantuo fauna predates the Ediacaran biota and is thus the oldest record of animal fossils currently known. In summary, the balance of evidence indicates that the Doushantuo microfossils are older than 580 million years and hence are the earliest fossil assemblage displaying affinities to diverse metazoan groups.

Affinities of the Metazoan Embryos and Related Adults

It is not easy to interpret the Doushantuo microfossils using traditional methods, either in petrographic thin sections (one per specimen) or with external scanning electron microscopy (SEM). Many of the microfossils, especially the embryos, are too simple in form to convey significant morphological information insofar as they resemble eggs or very early cleavage stages, although intriguing insights have been made into their early development (e.g. Chen et al. 2006). The diversity represented by these fossils has not yet been fully documented.

A recent technological approach known as propagation phase contrast synchrotron X-ray microtomography (PPC-SR-μCT) provides a unique

Fig. 3 Sites at the Wusi and Baishakan quarries producing the oldest known metazoans were well exposed several years ago but are now deeply buried by massive rock fall from the upper slope. A and B, photographed in 1997 and 2003, respectively, show well-exposed outcrops of the fossil-bearing beds in the Wusi (**a**) and Baishakan quarries (**b**) several years ago; (**c**), shows a stone monument at the Baishakan quarry with illustrations of the Weng'an fauna about the evolutionary dawn of the animal life (photograph taken 2007); **d** and **e** show outcrops at the Wusi (**d**) and Baishakan quarries (**e**), now deeply buried by massive rocks shed from the upper slope (photographs taken by ZJ Yin in 2010)

analytical tool that is increasingly used in studying the Weng'an fossils (Chen et al. 2006, 2009a, b; Donoghue et al. 2006). Microtomography permits the non-destructive computational examination of specimens from any vantage point, the visualization of their internal characters in virtual sections in any plane, and virtual three-dimensional depictions of their internal structures. This technique reveals many structures that are invisible or barely visible with classical absorption contrast-based microtomography (Tafforeau et al. 2006; Chen et al. 2006, 2009a). In the case of the Weng'an fossils, it can show the complex internal features of structures that might previously have been ignored or misinterpreted because of their deceptive exterior forms (Fig. 4). Perhaps most importantly, it facilitates the analysis of the small minority of Doushantuo embryos that retain intrinsically complex structures and large numbers of cells.

The study of the Doushantuo microfossils was initially directed towards algae and acritarchs (Zhang 1989; Zhang and Yuan 1992) and intense interest has been generated in the metazoan embryos and adult forms since 1998 (Xiao et al. 1998; Li et al. 1998). The diversity of the animal forms represented in the microfossil assemblage and their possible phylogenetic affinities are increasing as our knowledge of the faunas continues to expand (Chen et al. 2000).

More recently, great interest has been taken in the discovery of a microscopic putative adult bilaterian, *Vernanimalcula* (Chen et al. 2004a), with a general bilateral antero-posterior organization, a triploblastic structure, and bilateral symmetry. A pair of prominent coelomic structures lies symmetrically on each side of the prominent gut (Fig. 5). An anterior mouth is ventrally directed, followed by a thick pharynx with a multilayered wall and probable muscularity. The pharynx

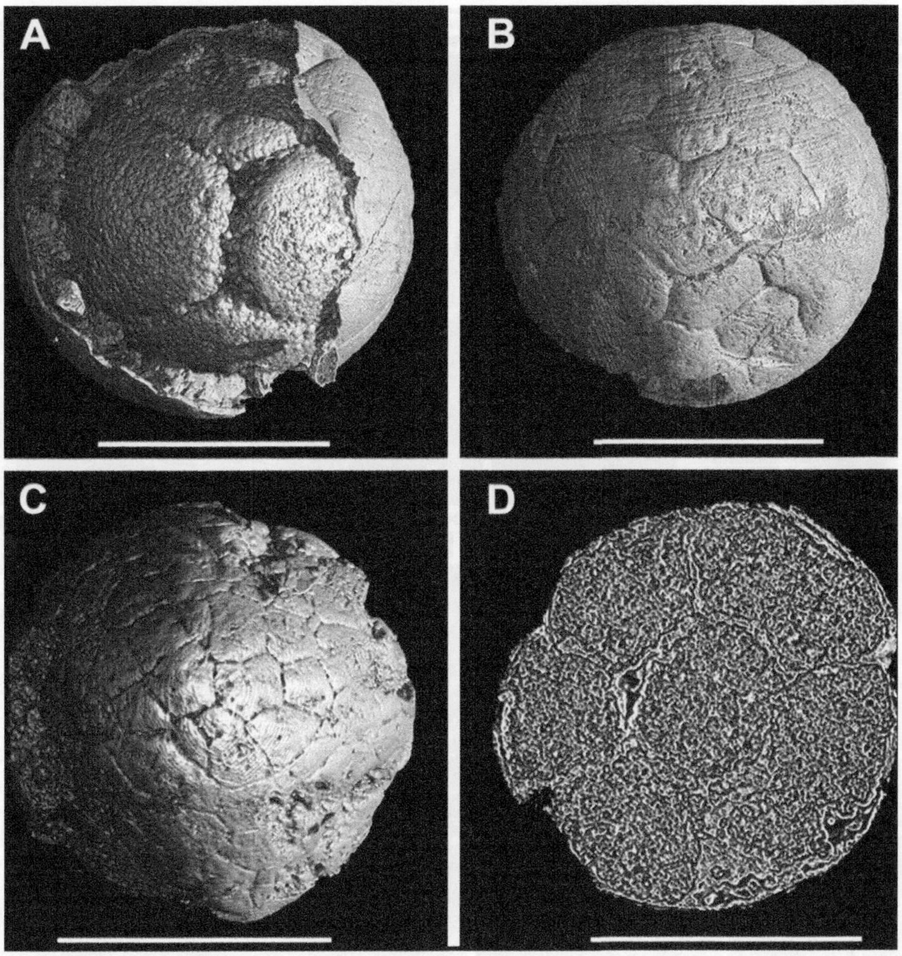

Fig. 4 PPC-SR-μT external view of a "soccer ball"-like 4-cell embryo (**a**) with chorionate cover consisting of large chorion cells (**b**); PPC-SR-μT is an external view of the chorionate cover (**c**) and internal digital section (**d**) of an 8-cell embryo (adapted from Chen et al. 2009a)

opens posteriorly into a large stomach or intestine. The wall of the stomach or intestine is always composed of two tightly apposed, thin, phosphatized layers of tissue. These two layers are separated by a continuous thin partition, interpreted as the remains of a basement membrane. The endodermal layers bounding the stomach (or intestine) are continuous, with a thickened region of the external body wall at the posterior end, i.e. the anus (Fig. 5).

Vernanimalcula is small but structurally complex, representing the earliest adult organism with a bilaterian form so far found in the fossil record. Its simple external morphology and lack of obvious appendages conform to the morphological and developmental predictions of the blueprint of an adult, triploblastic, coelomate animal.

The exquisite preservation of *Vernanimalcula*, however, has led some authors (e.g. Bengtson and Budd 2004) to suspect that these fossils are not real arguing that the authors "fail to take into full account taphonomy and diagenesis". Chen et al. (2004c) believe that Bengston and Budd's argument is not supported by sufficient evidence by counter-arguing that the open spaces in the animal, including its coeloms and the lumina of the mouth and oesophagus, are filled with black amorphous material derived from an organic sedimentary matrix that permeated through the pores of the fossilized wall of the animal while the sediment was still soft and hydrated (Fig. 5a). This indicates that the animal fossilization process began prior to the sediments becoming fully dehydrated. The first-generation cavity fillings in *Vernanimalcula* resulted

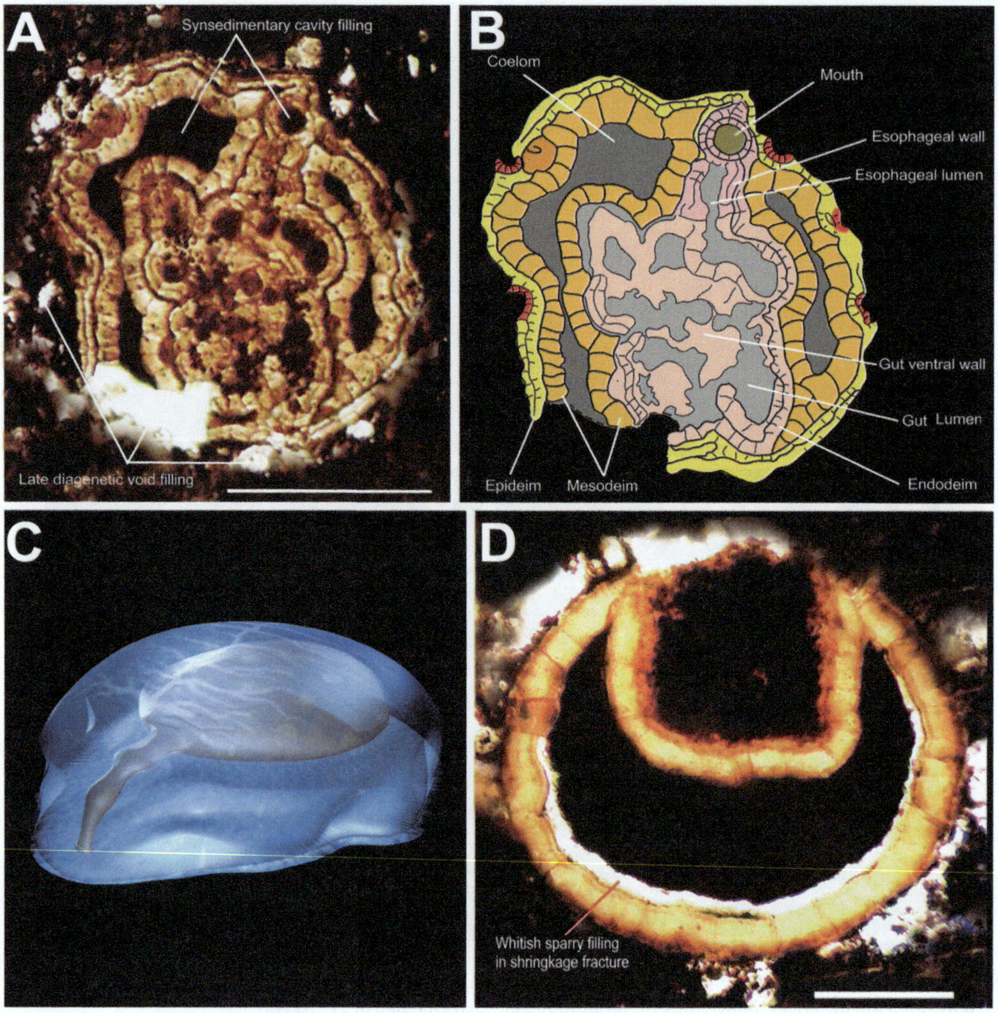

Fig. 5 A–C, image in transmitted light (**a**), illustration (**b**), and restoration (**c**) of a microscopic bilaterian adult *Vernanimalcula guizhouena* from the Precambrian Doushantuo Formation, Weng'an, Guizhou, showing paired coeloms extending the length of the gut, syn-sedimentary organic micritic fillings in the coeloms and late diagenetic vuggy fillings in the rear of the body. (**d**) a coelogastular embryo with possible cnidarian affinities showing organic micrite filling and whitish sparry diagenetic filling in shrinkage fractures (created by dewatering of internal sediments within the spacious blastocoel). Scale bar = 100 μm

from "internal sedimentation" while the matrix was still hydrated. The structures of *Vernanimalcula* were fossilized before internal cementation occurred, arguing against secondary diagenetic noise, and are primarily biological. However, early diagenesis did occur when the filling structure within the interspaces formed as the internal sediments dehydrated, and these usually appear between the fossilized biological wall and the internal sediments (Fig. 5a). A similar diagenetic history can be widely observed in the preservation of the Weng'an-type fossils. The early diagenetic products usually occur as a coating on either the outer or inner surface of the biological wall structure, as diagenetic filling in the narrow interspaces formed by the dehydration of the sediments, or as internal sediments within the biological cavities. A later diagenetic product can also be seen in the specimens examined, as a clean, whitish dolomite filling the voids (Fig. 5a). The voids in the matrix and fossils were probably produced by freshwater dissolution in a late diagenetic stage, destructive in terms of fossil preservation.

In addition to *Vernanimalcula*, the microscopic adult forms in the Weng'an fauna also include sponges, with a tabulate anthozoan form recorded in both thin section (Chen et al. 2002a; Xiao et al. 2000a, b) and with SEM (Xiao et al. 2000a, b) (Fig. 6b–f), and a

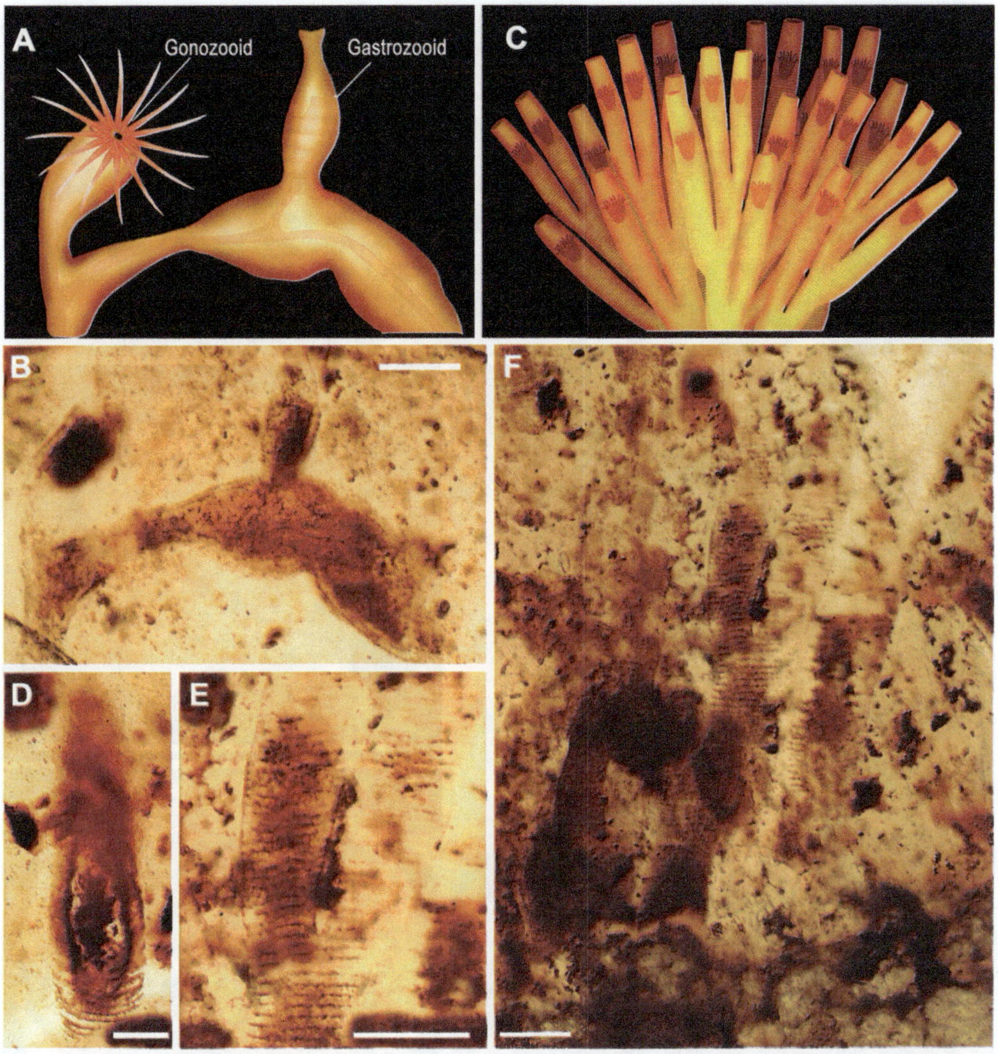

Fig. 6 Microscopic cnidarian adults from the Precambrian Doushantuo Formation, Weng'an, Guizhou. (**a, b**) restoration (**a**) and transmitted-light photograph (**b**) of *Eohydiania wenganense* (holotype); (**c–f**), restoration (**c**) and transmitted-light photograph (**d–f**) of coral-like cnidarian adult *Sinocyclocylicus guizhouensis* (Xue et al. 1992). Scale bar = 100 μm

hydroid-like form (Fig. 6a, b). Cross-sections of possible anthozoan polyps and stalks of similar sie and morphology have also been reported by Chen et al. (2002a) and Chen (2004), and exactly the same forms were discovered earlier by Xiao et al. (2000a, b). Microscopic (<1 mm) putative post-embryonic sponges have also been identified by the dense presence of acid-resistant monaxial spicules, irregular shape and filled with fossilized inclusions (Li et al. 1998; Fig. 7).

In contrast to the rare preservation of adult forms, sub-millimetre embryos are a good size for phosphatization and are common and diverse in the fauna. Many of the embryos retain a metazoan-like embryonic developmental sequence, but show no evidence of patterning organization or cell differentiation, even in the advanced stages of up to 2600 cells (Fig. 8). Their biological affinities remain enigmatic.

Embryos displaying perfect radial symmetry with chorionated covers are common in the Weng'an biota. They are 600–800 μm in diameter. Embryos at the 8- and 16-cell stages have a central blastomere surrounded by peripheral blastomeres extending to the surface, and externally resemble a soccer ball (Fig. 9). These embryos are usually preserved with a

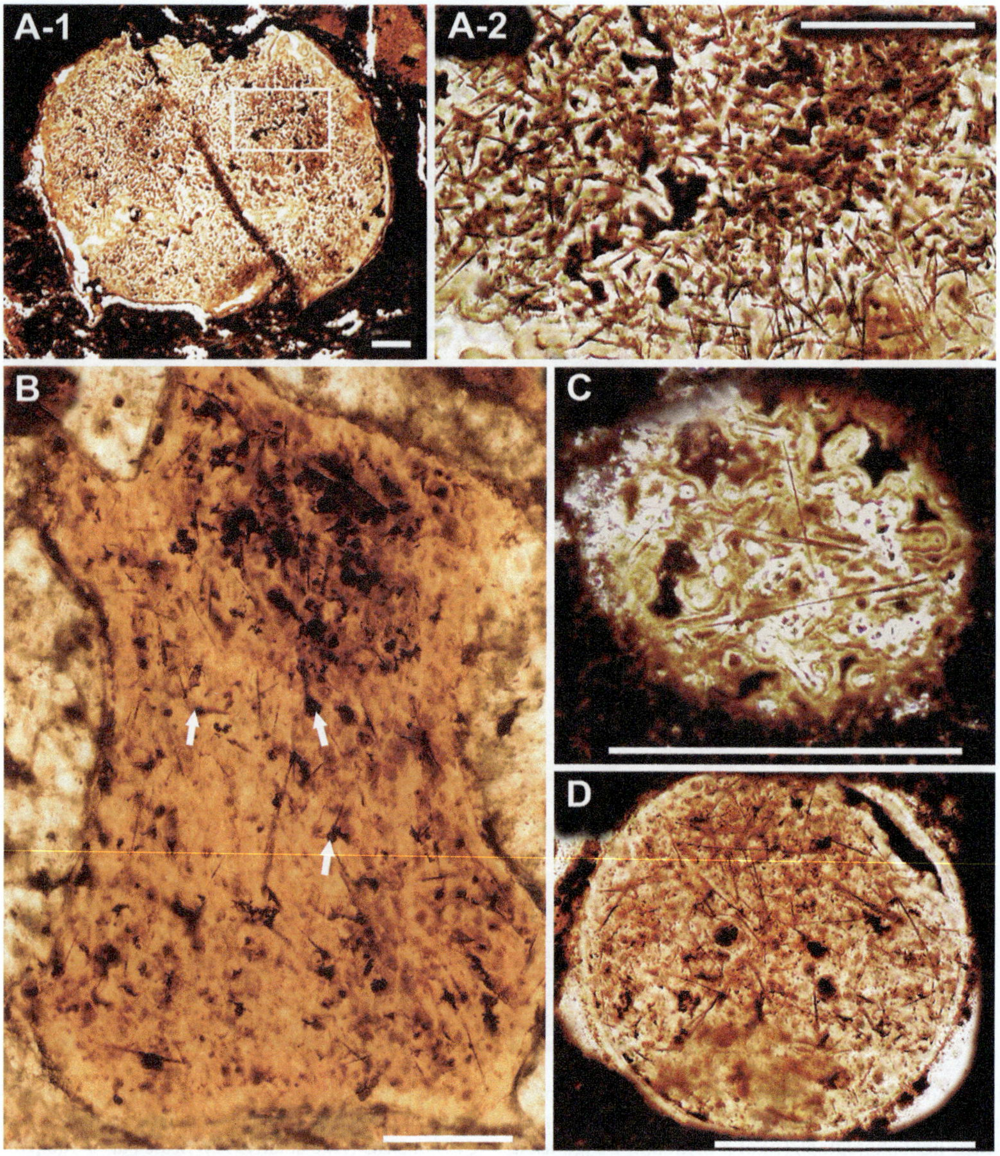

Fig. 7 Microscopic sponges from Precambrian Doushantuo Formation, Weng'an, Guizhou, showing monoaxon spicules densely packed and irregularly arranged in adults (**a**, **b**) and late developing embryos (**c**, **d**); preservation of possible sclerocytes at terminal end in some spicules (**b**, marked with *arrow*). Scale bar = 100 μm

chorion-like cover composed of tightly arranged, flattened cells, forming a regular reticulate pattern when viewed externally. Each cell retains a single depression, probably the original site of a cilium or flagellum. The embryos represent several different developmental stages up to the 32-cell stage. Thin sections show many samples of later developmental stages, with larger numbers of compacted blastomeres appearing to represent more advanced, late-cleavage sponge eggs. Such forms were first recorded in 1998 and interpreted as possible bilaterians by Xiao et al. (1998). They have recently been re-interpreted as probable sponge embryos (Chen et al. 2009b) because similar embryos have been found in modern demosponges, as shown in SEM images (Leys and Ereskovsky 2006).

Embryos with a one-cell-thick hollow gastrula are common in thin sections and an identically organized pattern of embryos has recently been documented

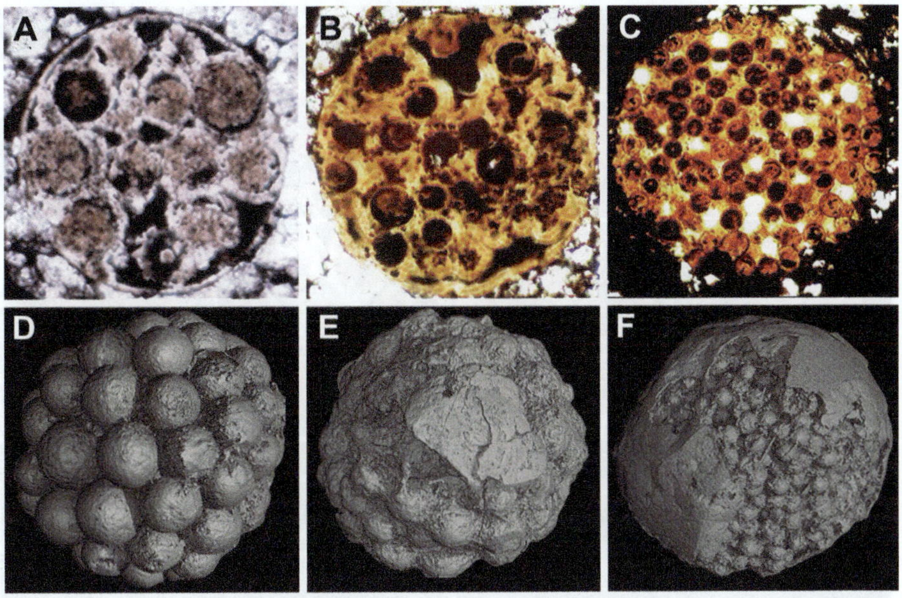

Fig. 8 Embryos in different development stages showing no differentiation of the germ-layer. (**a–c**) transmitted light images at various stages of development; (**d–f**), PPC-SR-μT external view of embryos at various stages of development. Sizes, 380 μm (**a**); 410 μm (**b**); 400 μm (**c**); 500 μm (**d**); 510 μm (**e**); 490 μm (**f**) (adapted from Chen et al. 2009b)

with PPC-SR-μCT technology. The highly organized, one-cell-thick, hollow gastrula displays a strong resemblance to the cnidarian (hydrozoan) gastrula in having a large blastocoel with several delaminated small cells on its floor (Fig. 10).

The bilaterian-like embryos are not common and display diverse embryonic developmental patterns. Some of the embryos resemble bilaterian stereoblastulas undergoing epiboly, as seen in many living bilaterians, with a micromere cap at one end and large endoderm-like blastomeres at the other (Fig. 11). The micromeres in the micromere cap form an ectodermal covering extending down over the endodermal cells. However, a few are remarkable, showing a complex organization with polarities that are comparable to the anterior and posterior poles, dorsal and ventral poles, and left- and right-side symmetry seen in many modern bilaterian embryos. Ectoderm-like cells later appeared around the periphery on all but the ventral and ventrolateral surfaces of the embryo. The cells on both the ventral and ventrolateral surfaces are sub-spherical, suggesting that the external cap did not extend ventrally to cover them.

Two highly organized 32-cell embryos also show a number of bilaterian characters, including the development of anterior–posterior, dorsal–ventral, and right–left polarities (Figs. 12 and 13), and distinct, linearly aligned internal cells, densely packed with inclusions that represent the remains of yolk granules (Fig. 13). The putative yolky internal cells form a contiguous endodermal cord bent into a shallow "S" shape both laterally and ventrodorsally. The cord-like structure could be the precursor of the anterior–posterior gut, a bilaterian character. The organization of the presumptive ectoderm and endoderm in the embryo is not typical of any known animal. The disposition of the putative endodermal cord is unique insofar as it is exposed to the outside at both the anterior and posterior ends. Although the epibolic gastrulation of the large yolky eggs of many animal embryos terminates at a vegetal position, leaving some endodermal cells exposed (i.e. as a blastopore), there are two sites of endodermal exposure in no living animals. However, we should not be surprised that the embryos of the earliest animals are not exactly like those seen after 600 million years of evolution. In addition to the endodermal cord cells, there is a layer of smaller flattened cells that form what may be an ectodermal cap covering the dorsal and lateral surfaces of the embryo. The cells on the ventral surface are larger and subspherical, suggesting that the ectodermal layer may not have extended far enough ventrally to cover them.

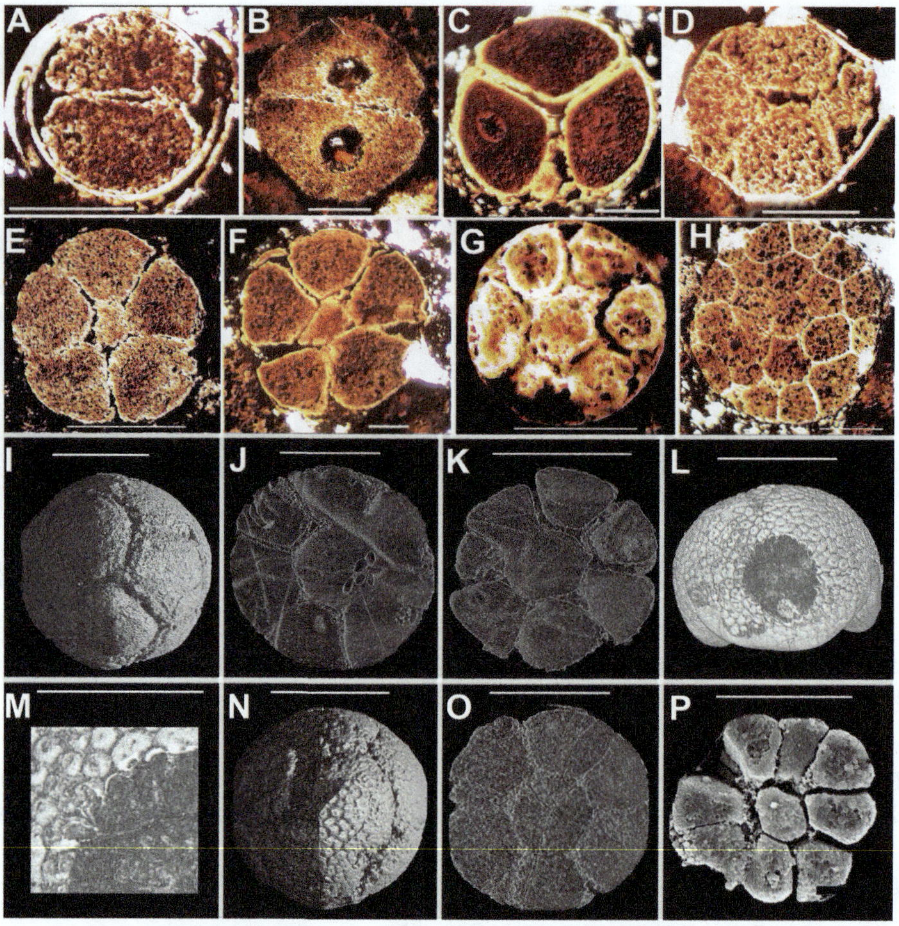

Fig. 9 Possible sponge embryos displaying perfect radial symmetry with chorionated cover (**a–o**) and a blastula in the modern demosponge (**p**). (**a–h**), transmitted light images of 2-cell (**a**, **b**), 4-cell (**c**, **d**), 8-cell (**e**, **f**), 16-cell (**g**), and possibly 64-cell (**h**) embryonic stages showing cellular nuclear (**b**, **c**) and perfect radial symmetry (**e–h**). (**i–o**) PPC-SR-μT images showing external (**i**) and internal views (**j**) of 16-cell embryo; external view (**l**) of a chorionated covered 8-cell embryo and a tangential slice (**m**) through the chorion displaying the 3D nature of the chorion cell and a single depression in each cell (**m**); external view (**n**) and internal section (**o**) mid-way through a 16-cell embryo. P, SEM image of a blastula in the modern demosponge *Halisarca dujardini* (reproduced from fig. 8A in Leys and Ereskovsky 2006). Scale bars = 200 μm (**a–h**), 400 μm (**i–l**, **n**, **o**), 150 μm (**m**), and 40 μm (**p**)

Although composed of a relatively small number of cells, both of these fossil embryos display the complex organizational patterns and differentiated cell types seen in the early gastrulae of some existing invertebrates. The symmetrical and structural characteristics of the embryos suggest that they have anterior–posterior, dorsal–ventral, and right–left axes. The elongated embryo resembles a post-gastrulation bilaterian embryo, still in the last stage of gastrulation. Because of their large yolky eggs, cleavage stages seem similar in both embryos; the yolky endodermal cells are typically internalized by epiboly. During epibolic gastrulation, the presumptive ectodermal cells at the animal pole of the embryo proliferate more rapidly than the yolk-rich endodermal cells and move in a vegetal direction to cover the endoderm.

Despite similarities in their apparent modes of gastrulation, the spatial organizations of the cells of these two fossil embryos are clearly different. These major differences, in internal patterning and in right–left symmetry, suggest that rather than representing different stages of the developmental sequence of the same animal, these two embryos belong to two different animal taxa. This in itself shows that the stage of metazoan evolution represented by these Doushantuo specimens post-dates the last common bilaterian ancestor.

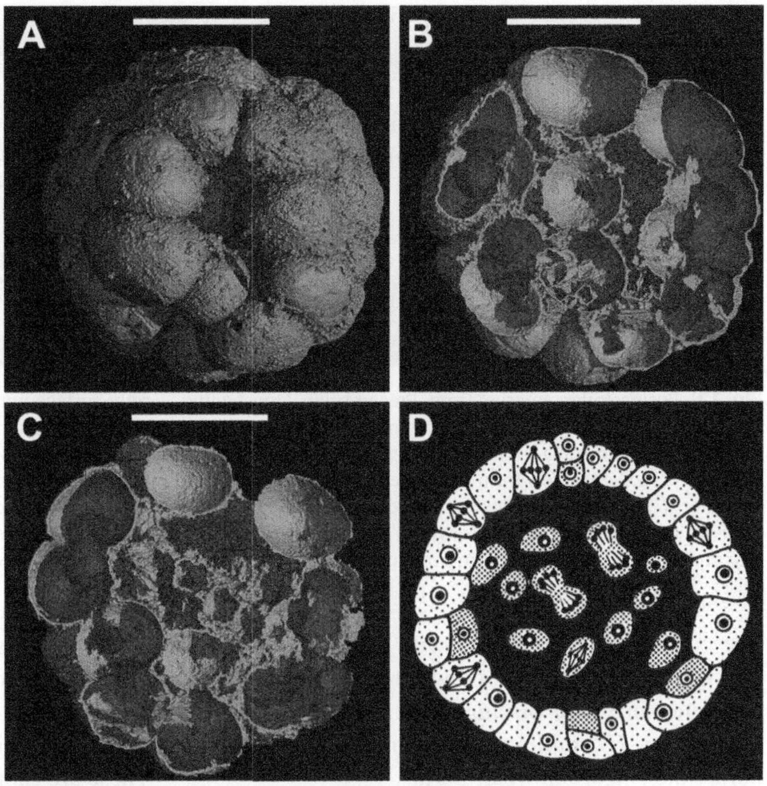

Fig. 10 Fossil gastrula of possible cnidarian from Precambrian Doushantuo Formation, Weng'an, Guizhou. (**a**) PPC-SR-μT external view of blastopore; (**b, c**) successive digital internal sections of embryo in the same orientation—displaying several small cells on the floor of blastocoel (**c**); (**d**) a modern cnidarian gastrula showing small cells on blastocoel floor (reproduced from Pl XII, Fig. 56E of Mergener 1971). Scale bar = 150 μm (adapted from Chen et al. 2009a)

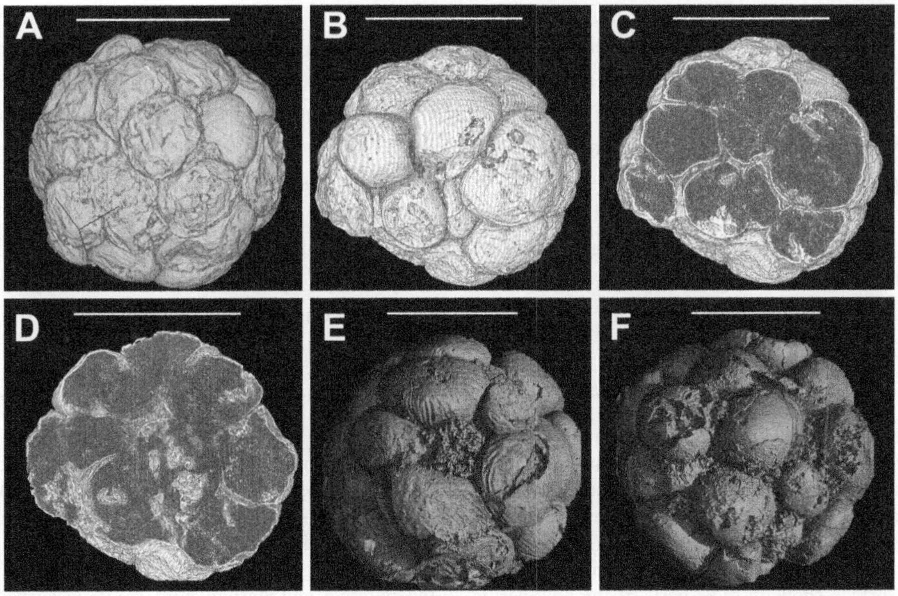

Fig. 11 Bilaterian embryos at late cleavage stage from Precambrian Doushantuo Formation, Weng'an, Guizhou. PPC-SR-μT external view from micromere cap (**a**) and from the opposite ("endoderm") side (**b**) and successive digital sections displaying the larger endodermal blastomere cells (**c**) and micromeres (**d**); (**e, f**) PPC-SR-μT external view from micromere cap (**e**) and from the opposite, endodermal side (**f**). Scale bar = 250 μm (adapted from Chen et al. 2009a)

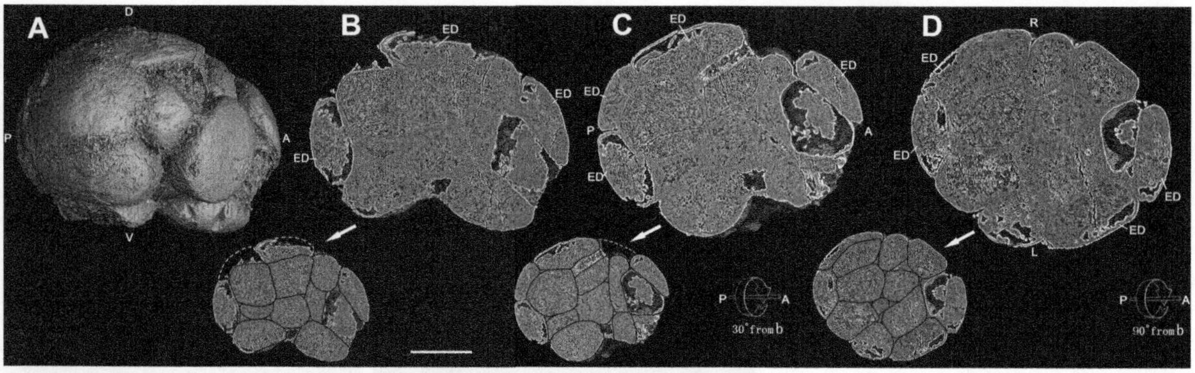

Fig. 12 Postgastrular embryo from Precambrian Doushantuo Formation, Weng'an, Guizhou, displaying bilaterian characters including anterior, dorsal–ventral, and right–left polarities. (**a**) a PPC-SR-μT external view from right side along anterior (*right*)–posterior (*left*) axis, digital internal sections (**b–d**) and the same sections with cell boundaries drawn at reduced size, showing maximal dorsal–ventral asymmetry in section cut half-way from right surface with flat ectodermal cells on dorsal, dorso-anterior and posterior surface (**b**); transitional state at section (**c**) with 30° rotation top-down away from observer from asymmetry (**b**); the plane of left-right bilateral subsymmetry at section (**d**) after 90° rotation from section B. Scale bar = 250 μm. Abbreviations: A, anterior; P, posterior; D, dorsal; V, ventral; R, right side; L, left side; ED, ectodermal cell (modified from Chen et al. 2009b)

The conclusion is consistent with other fossil studies and with the conclusions drawn from molecular phylogenetic analyses, placing the ancestor in the late Cryogenian, 635 million years or more ago, and hence before the end of Snowball Earth.

Both types of embryonic developmental sequences appear to be similar to the duet spiral cleavage found in modern acoel flatworms (Boyer et al. 1996a). Quartet spiral cleavage with a giant polar macromere (Freeman and Lundelius 1992) is seen in the extant prostomian bilaterians, including many lophotrochozoans (Ax 1995) and crustacean ecdysozoans (Anderson 1975).

Duet spiral cleavage patterning is characterized by the unusually early generation of micromere twins, beginning at the four-cell stage (Boyer et al. 1996b; Henry et al. 2000; Baguñà and Riutort 2004). The cleavage division gives rise to the macromere twins forming the first two micromere twins. This is followed by the continuous cleavage of the micromere twins, producing macromere twins at the vegetal pole. These are capped with paired smaller micromere twins in the early-cleavage embryos. Numerous Doushantuo samples are similar to modern duet cleavage embryos, including several at different developmental stages, forming an early developmental sequence similar to that of extant acoel flatworms (Fig. 14a–h; Chen et al. 2009b).

Quartet spiral cleavage establishes a micromere/macromere pole, beginning at the eight-cell stage instead of the four-cell stage as in duet cleavage. D-quadrant specification is an early major developmental event in the establishment of the future axis of

Fig. 13 (continued) section with cell boundaries drawn at reduced size, cut at 29% from the right surface, showing principal endodermal cells E4 and E1 sectioned near the right margin of the endodermal cord; (**c**) 35% from the right surface, showing endodermal cells E5, E4, and E1; (**d**) 43% from right surface. (**e**) 46% from right surface; (**f**) 51% from right surface; (**g**) 53% from right surface showing most parts of the endodermal cord close to the mid-plane of the embryo, and cluster of four cells, apparently daughter cells resulting from mitosis of E3; (**h**) enlarged E3. (**i–j**) digital external (**i**) and transparent view (**j**) from right side along the anterior (*right*: A) and posterior (*left*: P) axis showing anterior (AT) and posterior (PT) terminations (in *red*) of the putative endodermal cord of cells (**i**), and presumed ectodermal cells on dorsal surface, the putative endodermal cord of cells (**j**, in *red*) that has a shallow S-shape dorso-ventrally, and extends through the embryo from anterior to posterior ends. (**k, l**) digital external (**k**) and transparent view (**l**) from dorsal side of the embryo showing cells at the anterior (AT) and posterior (PT) terminations of the putative endodermal cord (in *red*, **k**); the *lines* designated **b–g** represent locations of digital sections; the putative endodermal cord (*red*), is curved into a shallow S-shape laterally. Abbreviations: A, anterior; P, posterior; D, dorsal; V, ventral; L, left side; R, right side; ED, ectodermal cell; AT, anterior termination; and PT, posterior termination. Scale bar on left of H = 250 μm in digital sections (**b–g**), and 181 μm in H, 375 μm in A; sections with drawing of cell boundaries (*at right-bottom*), 250 μm in **i–l** (modified from Chen et al. 2009b)

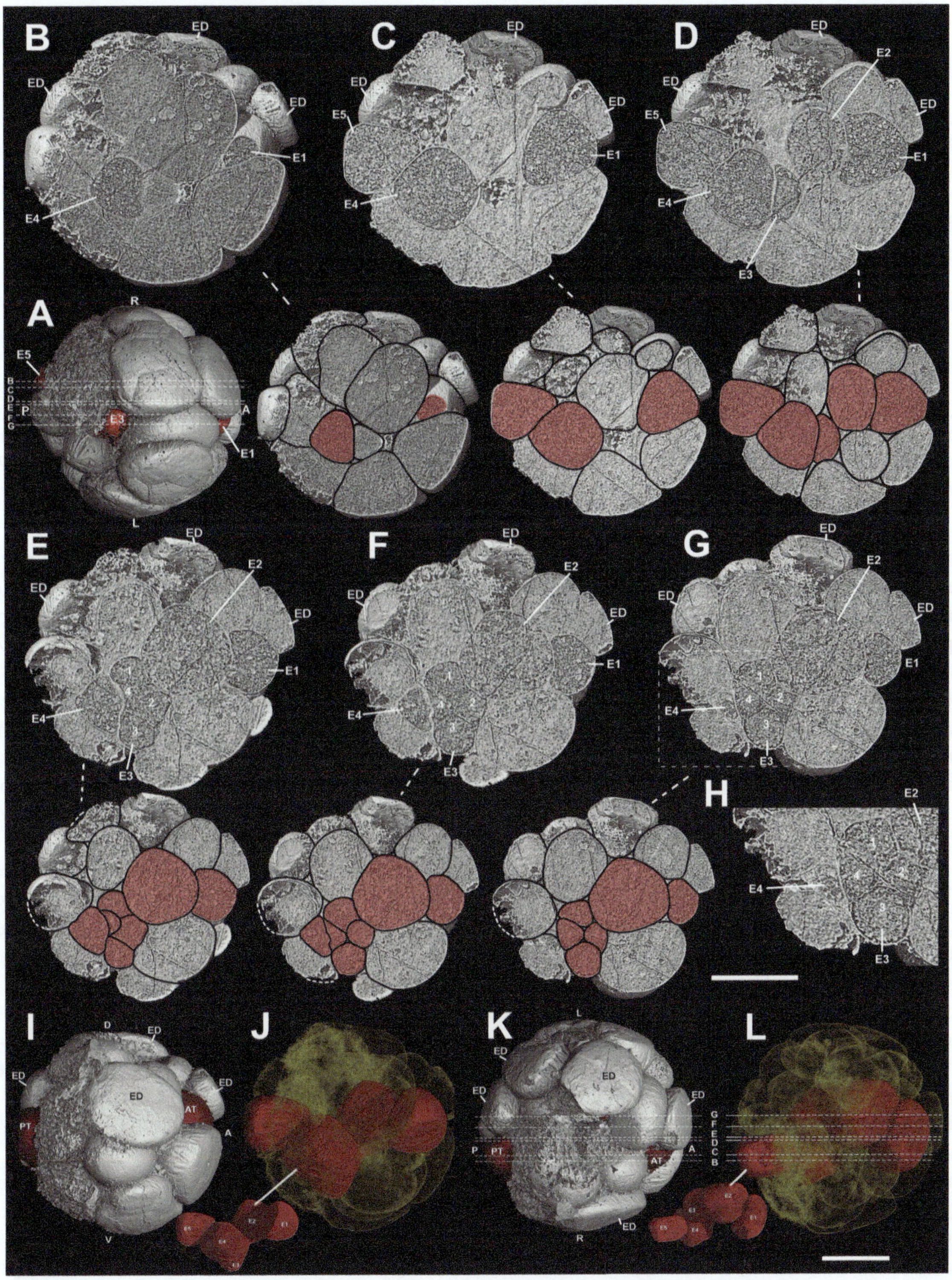

Fig. 13 Postgastrular embryo from Precambrian Doushantuo Formation, Weng'an, Guizhou, displaying bilaterian characters: anterior (*right*)–posterior (*left*), dorso-ventral, and right–left polarities with putative yolky internal cells forming a continuous endodermal cord bent in a wide S-shape both laterally and dorso-ventrally. (**a**) PPC-SR-μT external view from dorsal side with *lines* (**b**–**g**) representing serial digital sections cut at successively deeper layers from the right surface, and the same sections with drawings of cell boundaries (at smaller size) with yolky internal cells (E1–E5) in *red*. (**b**) Internal section, and the same

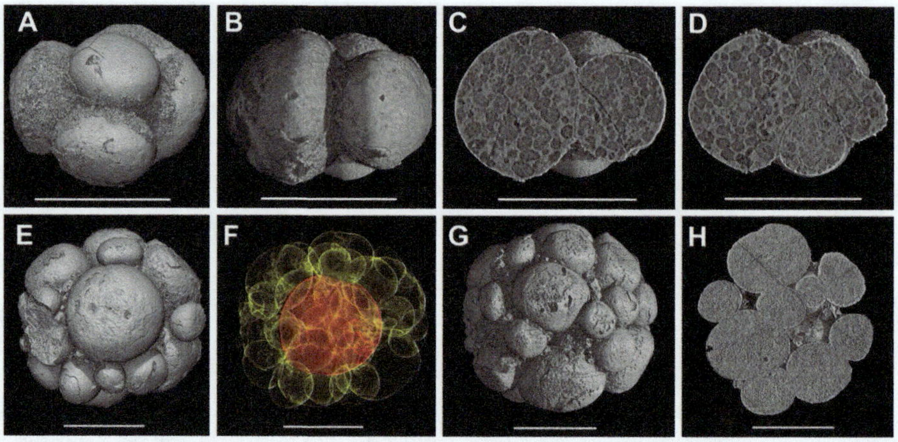

Fig. 14 Spiral cleavage embryos from Precambrian Doushantuo Formation, Weng'an, Guizhou. (**a–d**) represents a 4-cell dual-cleavage embryo; PPC-SR-μT external view from micromere dual (**a**) and from opposite macromere dual (**b**), and successive digital sections deep from macromere dual displaying well-preserved yolk platelets. (**e–h**) a 32-cell embryo with a giant D quadrant showing PPC-SR-μT external (**e**) and transparent view (**f**) of a giant polar macromere, and from opposite micromere cap (**g**), an internal digital section slicing through the micromere cap. Scale bars = 400 μm (**a–d**) and 300 μm (**e–h**)

bilateral symmetry. It is promoted by either an inductive mechanism or cytoplasmic determinants, associated with the equal and unequal cleavage modes, respectively. The inductive mechanism represents the ground state of the quartet cleavage mode, characteristic of morphologically equal cleavage. The cytoplasmic determinants that act in the unequal cleavage mode cause an asymmetric distribution of the cytoplasm, creating one macromere (D cell) larger than the other three by either asymmetric cytokinesis or the formation of an enucleate membrane-bound protrusion (the polar lobe).

The discovery of fossil quartet cleavage embryos with a giant D quadrant reported here (Fig. 14e–h; Chen et al. 2009b) suggests that the asymmetric cytokinesis underlying D-quadrant specification has an ancient Precambrian history. D-quadrant specification is critical in establishing the future axis of bilateral symmetry in the embryo, including many early key bilateral characters, in many lophotrochozoans and crustaceans. In addition to D-quadrant specification, the developmental sequences of early embryos, similar to the polar-lobe-forming embryos of modern spiralians (Chen et al. 2006), have also been recorded in the Weng'an fauna. As seen in some extant spiralian animals, polar lobe formation is one of the symmetry-breaking mechanisms that allow the partitioning of maternal cytoplasmic substances to certain blastomeres and not others. Polar lobe formation leads to unique early cleavage morphologies, including trilobed, J-shaped, and five-lobed structures (Fig. 15). Thus, fossil embryos similar to modern lobe-forming embryos are recognized from the Precambrian Doushantuo Formation phosphates, Weng'an, Guizhou Province, China. These embryos are abundant and form a developmental sequence comparable to the different developmental stages observed in lobe-forming embryos of extant spiralians. These data imply that lobe formation is an evolutionarily ancient process of embryonic specification and evolved before the Weng'an biota event (Chen et al. 2006).

Xiao et al. (2007a, b) have recently reported finding fossil eggs with spirally decorated structures. These are interpreted as post-blastula-stage embryos. However, there is no internal structure and these forms could also be one-cell eggs according to the authors.

The Lower Cambrian Maotianshan Shale: A Unique Window on the Cambrian Explosion

The biota of the Maotianshan Shale at Mt Maotian (Fig. 16) was first recognized for its excellent preservation with the 1984 discovery of naraoiids by X. G. Hou (Zhang and Hou 1985). Fossils from the Qiongzhusi Formation, cropping out widely in eastern Yunnan Province, were known and studied earlier in the twentieth century (e.g. Ho 1942) but the exquisite state of preservation of the fossils from the Maotianshan Shale

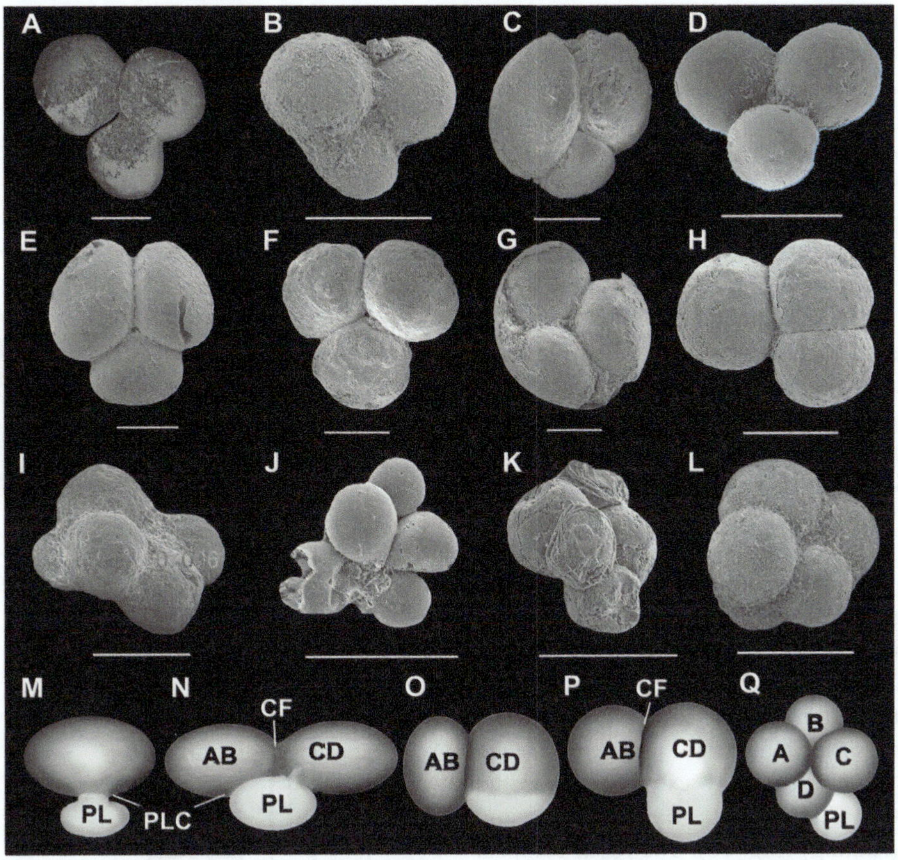

Fig. 15 Polar lobe-forming embryos (**a–l**) from Precambrian Doushantuo Formation, Weng'an, Guizhou, in trefoil stage (**a–g**), J-shaped stage (**h**) and five-lobed embryos (**i–l**). Schematic diagrams (**m–q**) show changes in shape of the PL-forming embryos in some extant molluscs and annelids—egg with the first lobe (**m**); trefoil stage (**n**); two-cell stage (**o**); J-shaped embryo (**p**); and five-lobed embryo (**q**). Scale bar = 250 mm

Fig. 16 Natural profiles of the lithologic sequence of the Early Cambrian at Mt Maotian, Chengjiang, Yunnan. The lithologic profile, photographed from the west, shows location of the classic fossil quarry site and the Museum of Chengjiang Biota (adapted from Chen 2004)

has resulted in it being intensively excavated and studied by many teams of researchers, yielding a constant flow of discoveries that has shed new light on the early evolutionary history of many animal groups and has also triggered extensive scientific debate around their interpretation.

The profile of the Precambrian–Lower Cambrian rocks exposed at Mt Maotian and on its lower slope is shown as a continuous sequence (Fig. 17). It includes the following lithological units from oldest to youngest: Precambrian Baiyanshao Dolomite and Xiaowaitoushan Cherty Dolomite of the Dengyin Formation; Lower Cambrian Zhujiaqing Phosphate Formation and Lower Black Shale and Siltstone units of the Shiyantou Formation of the pre-trilobite deposits, and Upper Black Shale, Maotianshan Shale; and siltstone of the Yu'anshan Formation; and the overlying Canglanpu Formation.

The Maotianshan Shale, 50 m thick at Mt Maotian, crops out broadly in east-central Yunnan (Fig. 18). Several tens of fossil localities have been excavated and are scattered throughout east-central Yunnan. Among these fossil localities, three sites are outstanding and have been extensively excavated and studied: Mt Maotian (Chengjiang), Haikou (Kunming), and Shankou (Anning) (Fig. 18). The fossil compositions at these three sites differ and are known as the Chengjiang, Haikou, and Anning fauna, respectively (Chen 2004).

The Maotianshan Shale consists of fine mud deposits, excellent for the detection of sedimentary event layers (e.g. ash layers or tempestites). Each constituent event layer is 2–3 cm thick, displaying a graded unit at the base and an upper homogeneous fine mud unit. The upper unit is responsible for the exquisite state of fossil preservation.

The taphonomy of these exceptionally well-preserved fossils remains a mystery, although it had been discussed by several authors. They have proposed that storm-generating tempests (Zhang et al. 2009) or mud flows (Chen 2004) were responsible for the extraordinary preservation of the Maotianshan Shale faunas. However, these hypotheses do not accommodate the presence of mobile animals, especially the mobile mud burrowers (such as priapulid worms), active swimmers (early vertebrates, numerous arthropods, and anomalocarinids), and macroscopic pelagic animals (eldoniids, ctenophorans, vetulicolids, and some arthropods), because these animals could readily

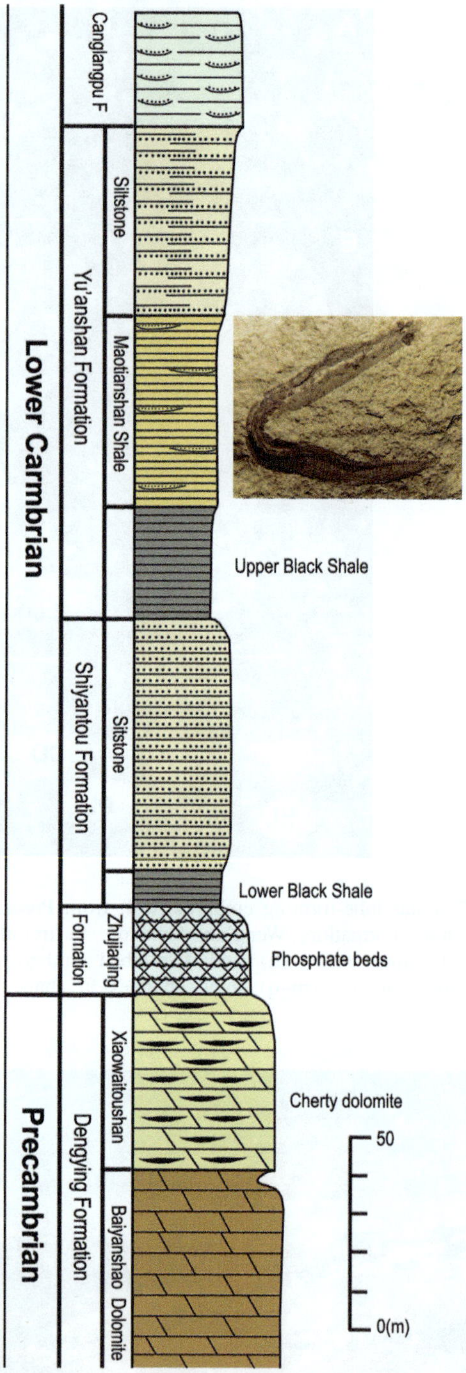

Fig. 17 Diagrammatic profiles of the lithologic sequence of the Early Cambrian in Mt Maotian area, Chengjiang, Yunnan (adapted from Chen 2004)

escape from such thin muddy event deposits. These event mud layers were thin (only 20 cm thick) even before dehydration.

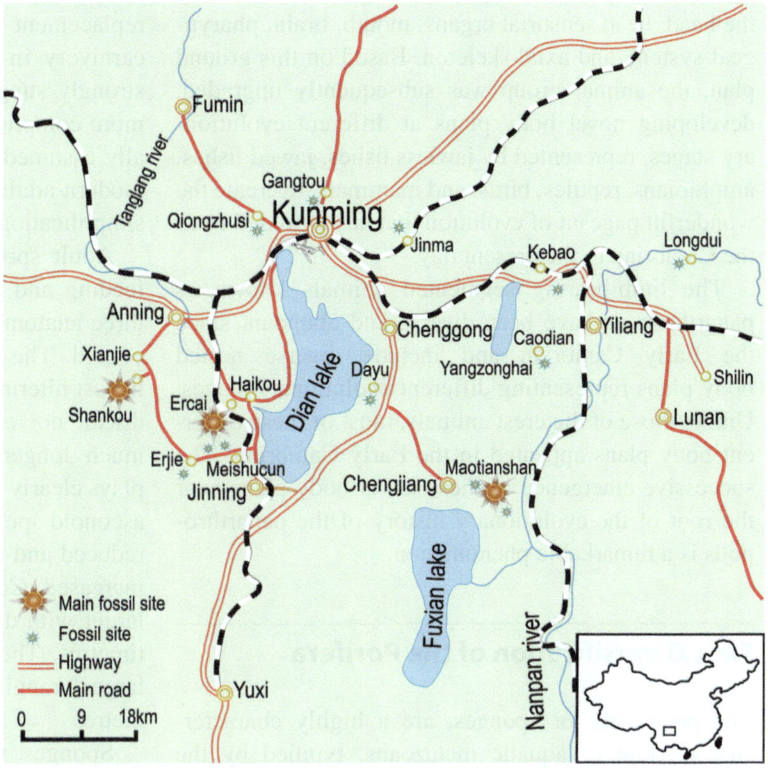

Fig. 18 Location of fossil sites of the Maotianshan Shale fauna

These thin mud layers are apparently insufficient for such excellent preservation. I suggest here that a poisonous event may have played a significant role, together with a sedimentary deposition event, in the exceptional state of the Maotianshan Shale preservation. These organisms may have been poisoned before they were buried by the event deposits.

The Maotianshan Shale biota comprises an extremely diverse faunal assemblage, with more than 100 species described, displaying diverse body plans linked to those of many living phyla and subphyla. In animals, the body plan (or body form) is essentially the blueprint for the way the body of the organism is laid down (Raff 1996; Hall 1998). The blueprint is shared by a group of animals ranked together at a higher taxonomical level (phylum, subphylum, class, or order). It characterizes a unique anatomical body organization, is evolutionarily conserved, and is deeply rooted. The most interesting manifestations of the body plan in animals occur at the taxonomic level of the phylum and sometimes subphylum (especially in the Vertebrata).

The Cambrian explosion was marked by many profound revolutionary changes and innovations, including sudden increases in skeletization, numbers of species, complex ecosystems, intensive arms races, and body plans. Among these, the sudden increase in diverse body plans was the most remarkable, unique change.

The Lower Cambrian Maotianshan Shale lies at the heart of the Cambrian explosion. The shale has preserved body plans of many bilaterian animal groups that are seen in living phyla, including chordates, ascidians, priapulids, sipunculans, arrow worms, possible molluscs, possible annelids, phoronida, and brachiopods (Chen 2004; Hou et al. 2004; Chen 2009) and at the level of the subphyla Vertebrata (Chen 2008) and Tunicata (Chen et al. 2003b). Many of these have changed very little since then (for instance, the ctenophores, priapulids, sipunculans, chaetognaths, tunicates, and inarticulate brachiopods). On the contrary, the animals with backbone are a magical and unique group of animals deeply rooted in the Early Cambrian. This is illustrated well by the outstanding ground body plan of the animals with backbones in the beautifully preserved fossils from the Lower Cambrian Maotianshan Shale. These shed light on the mysterious origins of many aspects of our own group, including the head, head sensorial organs, mouth, brain, pharyngeal system, and axial skeleton. Based on this ground plan, the animal group was subsequently upgraded, developing novel body plans at different evolutionary stages, represented by jawless fishes, jawed fishes, amphibians, reptiles, birds, and mammals, to create the wonderful pageant of evolution that has unfolded from the Cambrian to the present day.

The limb-bearing segmented animals known as panarthropods have been diverse and abundant since the Early Cambrian and include diverse nested body plans representing different evolutionary stages. Unlike those of the crest animals, most of these different body plans appeared in the Early Cambrian. The successive emergence of these novel body plans near the root of the evolutionary history of the panarthropods is a remarkable phenomenon.

First Diversification of the Porifera

The poriferans, or sponges, are a highly characteristic phylum of aquatic metazoans, typified by the presence of choanocytes, a system of branched canals with a collar chamber that transports water throughout the body. The adult has neither a mouth nor an anus. They are widely regarded as an isolated group among the animals. However, the sponges have some characters of intriguing evolutionary significance (Maldonado 2004), including their widely occurring internal fertilization, notable diversity of gastrulation modes, architecturally complex lecithotrophic larvae, their direct development in some groups, and the replacement of choanocyte-based filter feeding by carnivory in some sponges. Together these features strongly suggest that the sponges have a longer and more complicated evolutionary history than traditionally assumed and that the uncomplicated anatomy of modern adult sponges may be the result of secondary simplification.

Adult sponges are uniquely specialized for filter feeding and their body is built according to one of three anatomical designs: asconoid, syconoid, or leuconoid. The asconoid type is the simplest, with the lowest filtering efficiency. The body is small and cylindrical, not exceeding a diameter of 1 mm and not much longer than 10 cm. The syconoid type displays clearly greater filtering efficiency than that of the asconoid sponges because the volume of the atrium is reduced and the area of the flagellated choanoderm is increased. Consequently, these sponges are generally larger with diameters ranging from one to a few centimetres. The body size of leuconoid sponges is the largest, ranging from a few centimetres to more than a metre.

Sponges, widely recorded both in their adult form and as embryos in the Early Ediacaran Weng'an biota, have a deep Precambrian history (Li et al. 1998; Chen et al. 2009b) and diversify greatly until the Cambrian (Fig. 19). The sponges are the major components of the sessile colonies in the Maotianshan Shale fauna. About 30 species belonging to 20 genera have been recorded. Among these, the monaxonid demisponges predominate and are largely represented by complete or nearly complete skeletons. As epifaunal suspension feeders, they are adapted for muddy bottoms, with

Fig. 19 Panorama of diverse sponges in the Early Cambrian

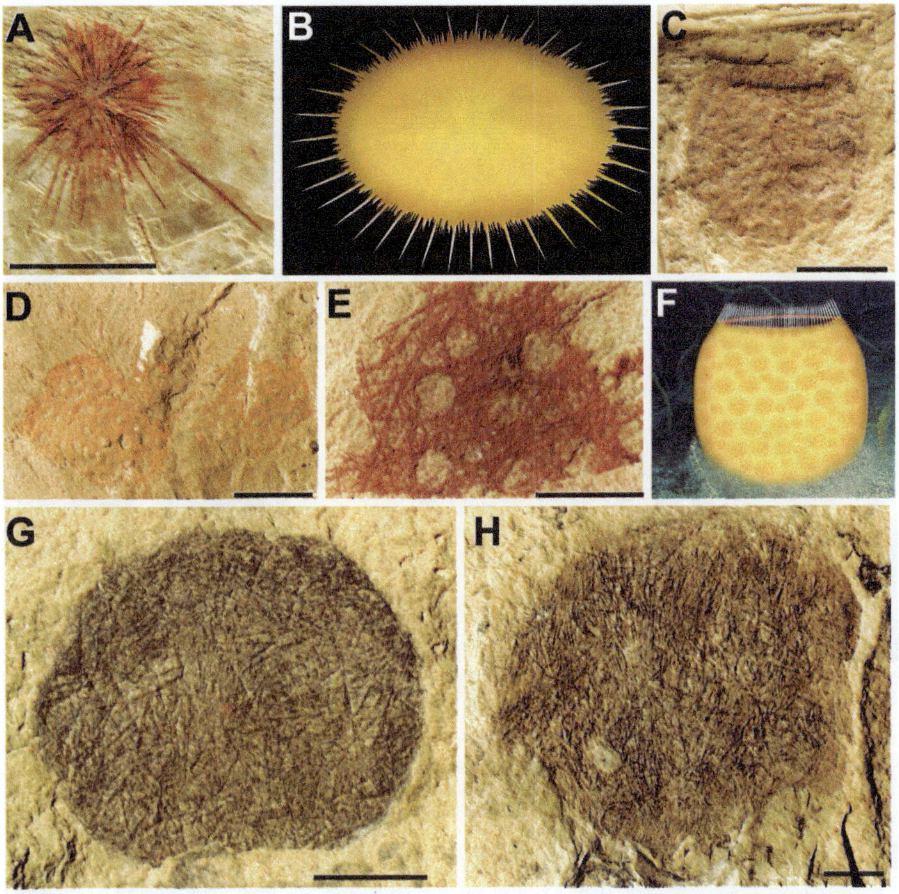

Fig. 20 Small sponges from the Early Cambrian Maotianshan Shale, central Yunnan. (**a**, **b**) photograph and restoration of *Choia xiaolangtianensis*; (**c**–**f**) photographs (**c**–**e**) and restoration (**f**) of *Crumillospongia* sp. nov, and G, H, *Saetaspongia densa*. Scale bars = 0.5 cm (**a**, **c**–**e**), and 0.2 cm (**g**, **h**)

substrates varying from soft to semi-firm, and water energies ranging from extremely low to relatively high. The ecological tiering of the sponges in the Maotianshan Shale varies notably from those of only a few centimetres in some forms, such as *Saetaspongia*, *Choia*, and *Crumillosponge* (Fig. 20), to 30 cm in *Quadrolaminiella* (Fig. 21). The body diameter is only few millimetres in some species, whereas some species are much larger with diameters ranging from one to several centimetres. The largest, *Quadrolaminiella*, has a diameter of more than 10 cm and a length of 30 cm. There is no direct evidence of the design of their filtering chambers, but the size range of these fossil sponges indicates that all three anatomic designs for filter feeding observed in modern sponges had already evolved in Maotianshan times.

The small straw-hat-like sponge known as *Choia* was first recorded in the Middle Cambrian Burgess Shale. In the Maotianshan Shale, it is represented by the species *Choia xiaolantianensis* (Fig. 20a, b). The small disc-like body, about 1–2 cm in diameter, is composed of a single-layered skeleton consisting of a thatch formed by two sets of monaxial spicules, finer and coarser ones, extending radially from the centre of the body, reaching the body margin and far beyond the body margin, respectively. *Choia* is not common in the Maotianshan Shale.

The small sac-like *Crumillosponge* is known from the Middle Cambrian Burgess Shale (Rigby 1986; Walcott 1919). It also appears in the Maotianshan Shale but is uncommon. Figure 20c–f shows new species of the genus from the Maotianshan Shale.

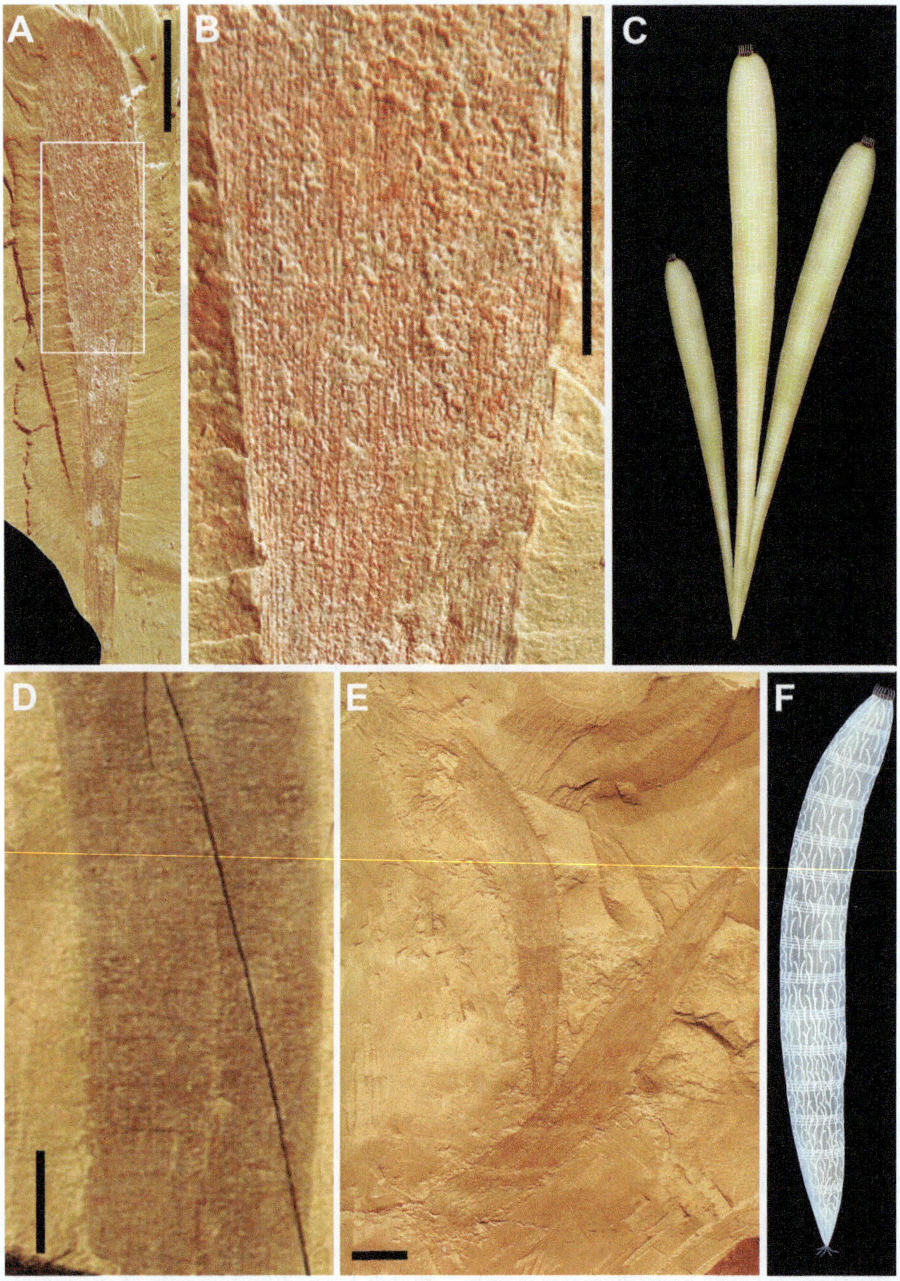

Fig. 21 Elongate sponges from the Early Cambrian Maotianshan Shale, central Yunnan. (**a–c**) photographs (**a**, **b**) and restoration (**c**) of *Leptomites teretiuslulus*, showing double-layered skeleton, with outer layer consisting of long, vertical spicules and inner layer composed of a thatch of tiny horizontally arranged spicules. (**d**) photograph of *Leptomitella conica* showing outer layer of skeleton consisting of a vertical thatch of fine spicules and associated vertical rod and inner layer composed of both bundled and unbundled horizontally arranged spicules. (**e**, **f**) photograph (**e**) and restoration (**f**) of *Paraleptomitella dictyodroma* showing outer layer of skeleton consisting of coarse and curved monaxons and fine vertically arranged spicules, and inner layer formed by bundles of horizontally arranged spicules. Scale bar = 1 cm

The sac-like body is small, 10–15 mm in diameter, topped with a wide opening, the osculum. The osculum is clearly rimmed with small, densely packed spicules along its margin. The wall is perforated with poles, interpreted as ostria, occurring as two differently sized sets: the larger is about 1 mm in diameter and smaller is only 0.3 mm in diameter. The tracts are composed of numerous, parallel, monaxon spicules;

these extend throughout the area between the ostria in various directions. As the base of the body is broadly rounded and there is no evidence of an attachment structure, it seems that the animal was not anchored to the substrate.

The spherical sponge known as *Saetaspongia densa* (Mehl and Reitner in Steiner et al. 1993) is a monospecific genus, recorded only in the Maotianshan Shale (Fig. 20g, h). It is small, up to 3 cm in diameter, and the skeleton is mainly composed of monaxonic diactine spicules. In addition to the monaxonic spicules, there are small spicules in three-, four-, and even six-rayed forms. The flattened outline of the body probably represents an oscular opening. The presence of small polyrayed spicules suggests that the species belongs to the Hexactinellida, the glass sponges. No attachment structure has been found, suggesting that the animal rested on the substrate with the osculum opening upwards.

Sponges with an elongated tube-like body form range in diameter from 0.6 to 2 cm and are about 10 cm in length. They are the dominant forms in the Maotianshan Shale including species of *Leptomitus* Walcott, 1886, *Leptomitella* Rigby, 1986, and *Paraleptomitella* Chen et al. (1989a) (Fig. 21). Their two-layered skeleton is formed by simple monaxial spicules, with a vertical thatch consisting of an outer layer of coarse spicules interspersed with densely packed fine spicules, and a horizontal thatch consisting of an inner layer of fine spicules. These fine spicules are evenly distributed in *Leptomitus* Walcott, 1886 (Fig. 21a–c), but are arranged into bundles or are unbundled in *Leptomitella* Rigby (Fig. 21d–f). Unlike those of *Leptomitus* and *Leptomitella*, the outer layer of the skeleton in *Paraleptomitella* is formed of both curved coarse spicules and densely packed fine spicules.

A vase-shaped sponge known as *Quadrolaminiella* Chen et al. (1990), is the largest of the Maotianshan sponges, with a diameter of 10–20 cm and a length of 30 cm (Fig. 22). It has a thick, four-layered skeleton composed of simple monaxial spicules. The vertical thatch in its outermost layer is composed of coarse and widely spaced spicules that extend vertically through the entire length of the body. The horizontal thatch inside the vertical thatch is formed by fine, dense, and densely packed spicules. The two inner layers are composed of widely spaced coarse spicules interspersed with fine spicules. The thatch of these two layers is composed of spicules that extend diagonally along the body axis but in opposite directions.

A unnamed new species of *Halichondrites* Dawson (1889), from the Maotianshan Shale is similar in size to *Quadrolaminiella* (Fig. 23). Its skeleton is two layered and composed of simple monaxial spicules. The outer layer is a vertical thatch composed of vertically extended and densely packed fine spicules and widely spaced coarser prostalia, extending vertically along the body, although their upper parts extend outward, curving from the skeleton to form a bush-like structure. The inner layer is composed of horizontally extended fine spicules that form bundles.

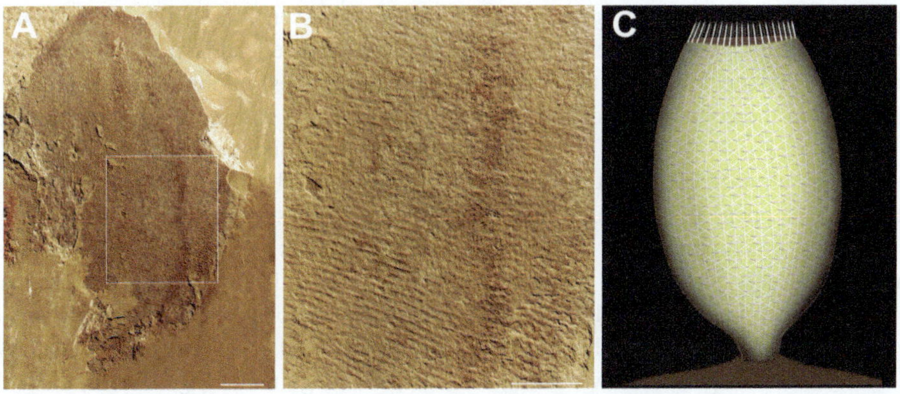

Fig. 22 Photographs (**a**, **b**) and restoration (**c**) of a large ellipsoidal sponge *Quadrolaminiella diagonalis* with 4-layered skeleton from the Early Cambrian Maotianshan Shale, central Yunnan. Scale bars = 2 cm (**a**), 1 cm (**b**)

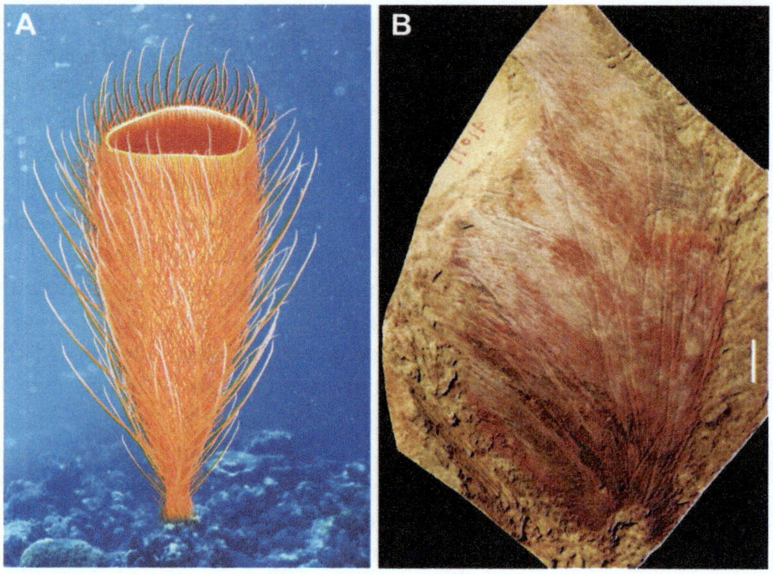

Fig. 23 Restoration and photograph of a large cup-shaped sponge *Halichondrite* sp. nov. from the Early Cambrian Maotianshan Shale, central Yunnan, showing a cup-shaped body with 2-layered skeleton; outer layer consists of axially arranged fine spicules; inner layer formed by bundled horizontally arranged spicules. Coarser spicules (prostalia) are axially extended and protrude radially from the body wall. Scale bar = 1 cm

Fossil Representatives of the Cnidaria

Representatives of the phylum Cnidaria, although not common in the Maotianshan Shale, include a few sea anemones and some other sessile polyps. The sea anemones are represented by *Xianguagia sinica* Chen and Erdtmann (1991); it is entirely soft-bodied and cylindrical in shape. Specimens are commonly preserved in a laterally compacted manner but a few have undergone oral/aboral compaction (Fig. 24). The diameter of the species is up to 3 cm, and the length is up of 6 cm. It consists of a smooth-surfaced pedal disc and a column with external stripe-like vertical structures. The outer surface of the column is ornamented with small knolls arranged linearly and vertically, with two lines in each strip. Sixteen flexible tentacles extend from the oral margin of the column and distally taper to a point.

Archotuba Hou et al. (1999) is a tube dweller (Fig. 25). The tube is open at the wider end and usually attached to a brachiopod shell at the narrow end. The tube usually shows the presence of a funnel-like soft tissue. The closely related species *Cambrorhytium* Conway Morris and Robison (1988) from the Middle Cambrian, is preserved with tentacles on the distal end of the tube and is interpreted as a cnidarian.

Chancelloriids and Possible Affinities

Chancelloriids were a common element in the Cambrian small shelly fossil assemblages and extinct, star-shaped, Cambrian organisms, with hollow, spine-like sclerites. Well-preserved articulated exoskeletons (known as scleritones) are not common and are predominantly restricted to the Lower Maotianshan Shale and the Middle Cambrian Burgess Shale. Chancelloriids have been reconstructed (Janussen et al. 2002; Chen 2004; Randell et al. 2005) as sac-shaped, sessile, benthic animals covered in rows of spiny sclerites. The sclerites protrude from beneath an epidermal layer (Janussen et al. 2002; Chen 2004).

The chancelloriids in the Maotianshan Shale are one of the most important elements among the sessile benthic fauna and are mainly represented by *Allonina phrixothrix* Bengtson and Hou (= *Allonia junyuan* Janussen et al. 2002). The body is sac-like (Fig. 26) and radially symmetrical, narrowing proximally and distally, and is covered with a continuous pliant integument, embedded with regularly disposed hard, mineralized composite spine-like sclerites (Fig. 27). The sclerites usually have three rays, with two rays buried in the integument (Fig. 27c, d), and these two rays form a V shape, with the open side directed distally (Fig. 27a). The other ray protrudes nearly vertically

Fig. 24 Sea anemone *Xianguanqia sinica* from the Early Cambrian Maotianshan Shale, central Yunnan. (**a**, **b**) part and counterpart of a top-laterally compacted specimen showing: fully exposed tenctacles (**a**) and a body column (**b**). (**c**, **d**) laterally compacted specimens showing a smoothly pedal disc, a column with external strip-like longitudinal structures with entire number of 16, each strip-like structure related to a tentacle. (**e**) restoration. Scale bar = 1 cm

from the surface in most parts of the body, except the distal part (Fig. 26). The angle between these external rays and the body axis tends to narrow. At the distal margin of the body, the rays protrude beyond the distal opening, extending either convergently or parallel to the body axis (Fig. 27a).

The inter-sclerite integument has a conspicuous rhombic feature over the entire body (Fig. 27d–f). These rhombic structures are interpreted as the impressions of small, hard, embedded platelets (about 50–60 μm long and 30–40 μm wide; Bengtson and Hou 2001). Bengtson and Hou (2001) observed the platelets to be thicker at the margin and depressed centrally, with a fine rod protruding from the surface.

Affinities

The taxonomic affinities of the chancelloriids are uncertain. They were first described as sponges (Walcott 1920); the sponge hypothesis was challenged with the realization of their complex nature (Bengtson and Missarzhevsky 1981). Further research showed that the "spines" themselves were formed as dermal sclerites, rather than as spicules. In recent studies, the chancelloriids have been compared with either ascidians (Mehl 1996) or cnidarians (Chen 2004; Randell et al. 2005). The ascidian interpretation is based on a comparative study of spicules of the two groups

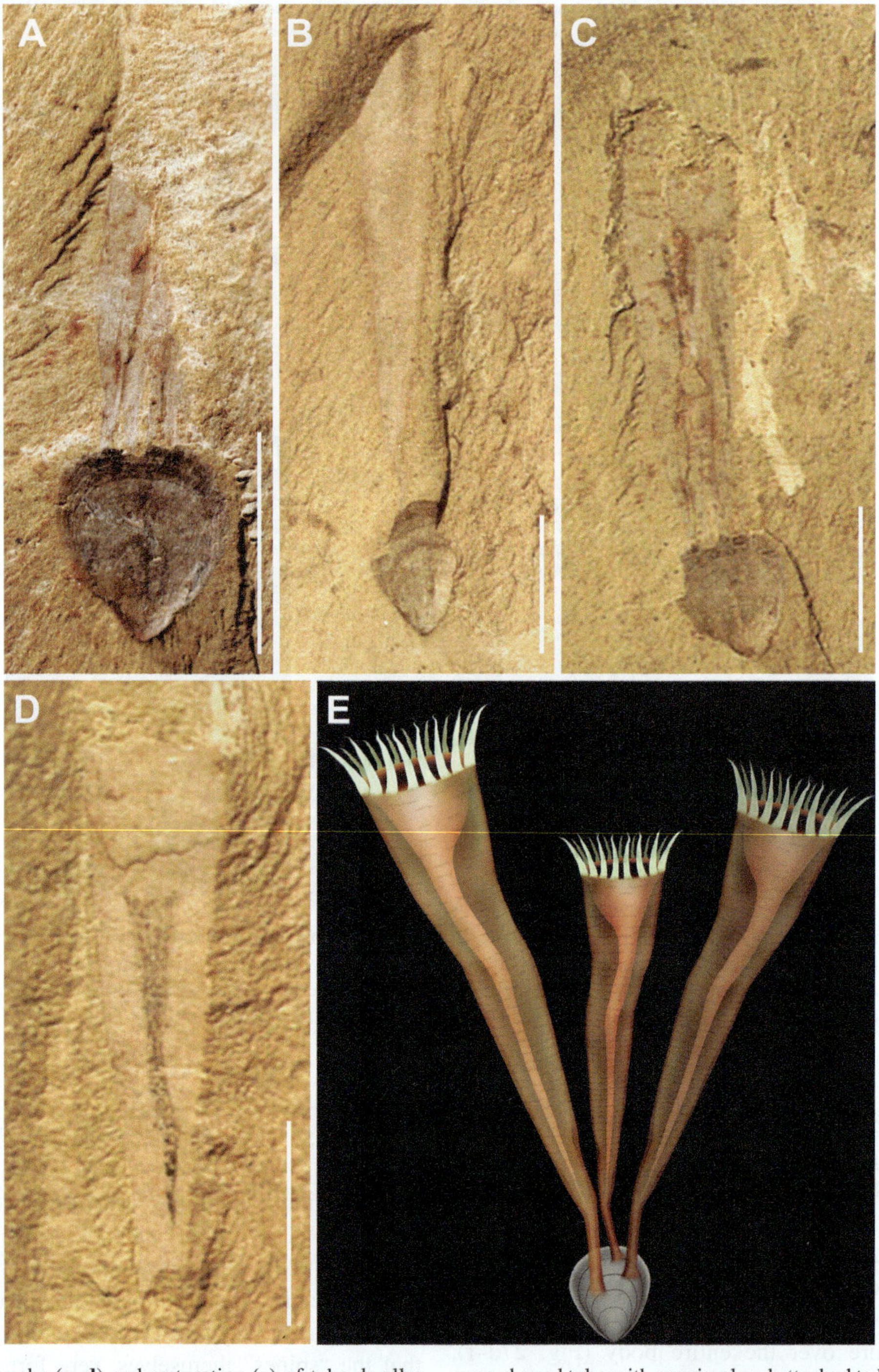

Fig. 25 Photographs (**a–d**) and restoration (**e**) of tube dweller *Archotuba* of possible cnidarian affinities from the Early Cambrian Maotianshan Shale, central Yunnan, showing elongate cone-shaped tube with proximal end attached to brachiopod shell and a distinct elongate cone-shaped organic structure within the tube (**a–d**). Scale bar = 0.5 cm

but as the study was based on evidence from photographs rather than on real specimens and absence of significant ascidian characters, including a branchial cavity or coiled intestine, it challenges this hypothesis. Sclerite-based comparative analyses with ascidians (Mehl 1996) and octocorals (Randell et al. 2005) have provided limited support for the chancelloriid affinities. As observed in *Allonia phrixothrix*, the body

Fig. 26 Laterally compacted *Allonia phrixothrix* from the Early Cambrian Maotianshan Shale, central Yunnan, showing two embedded rays divergin distally and a free ray protruding from body wall nearly vertically. Scale bar = 1 cm

in the chancelloriids is characterized by radial symmetry and a central cavity with a single distal opening, similar to the coelenteron in the cnidarians, and implying cnidarian affinities. The absence of tentacles suggests that chancelloriids were a very unusual animal group but related to the cnidarians. How they fed without tentacles remains unclear although some authors have proposed that the animals were filter feeders, with an inferred apical opening, and that the sclerites acted in a largely defensive role (Janussen et al. 2002; Randell et al. 2005).

Our observations show that the pliant continuous integument was elastic and contractile. The chancelloriids may have fed from circulating water by drawing food-laden water into the coelenteron, with the waste water subsequently expelled from the coelenteron as the body repeatedly expanded and contracted. The external rays of the sclerites at the mouth margin extend either convergently (Fig. 27a, b) or vertically parallel to each other. The divergent angle of the two embedded rays in each sclerite appears to vary. All these characters suggest that the radial rays of the sclerites were moveable, with a role in the contraction or expansion of the body.

Fossil Ctenophora Represents Evolutionary Transition to the Bilateria

The phylum Ctenophora (also referred to as "the comb jellies"), with 80 living species, are predominantly voracious gelatinous, planktonic, spheroidal to oblate predators. Although the phylogeny of the Ctenophora is inconclusive, they have traditionally been interpreted as a sister group of the cnidarians. The presence of mesodermal cells in the mesogloea, muscles in the tentacles, and acrosome-bearing sperm suggest that ctenophores are evolutionarily more advanced than the cnidarians and closely related to the bilaterians (Nielsen 2001).

Since the Cambrian, ctenophorans have been predominantly marine dwellers and are an animal group with a few and morphologically conservative members. Exception to this conservative state are those species in the Middle Cambrian with more than 16 or 24 comb rows (Conway Morris and Collins 1996) and some rather bizarre body forms among modern species (Brusca and Brusca 1990). The body plan, as exemplified by a late-developing embryo (Fig. 28) from a phosphate bed at the base of the Lower Cambrian (Chen et al. 2007b) and adult forms from Maotianshan Shale deposits (Chen and Zhou 1997; Chen 2004; Figs. 29 and 30), is almost identical to the body plan of extant ctenophores, characterized by a spheroidal body with eight comb rows. Like some extant ctenophores, the mouth of the adult is surrounded by a broad skirt-like structure known as the "oral lobe".

The earliest firm evidence of the Ctenophora is an embryo from the Early Cambrian Meishucun phosphate deposit (Fig. 28) (Chen et al. 2007b). The 540-million-old embryo was preserved in a 10-cm-thick bed of black bituminous phosphatic limestone near the base of the Kuanchuanpu Formation, overlying limestones containing the terminal Precambrian diagnostic fossil *Cloudina*. It was pre-mineralized by apatite of the kerogen, is about 190 μm in diameter with an enclosed embryo ($\approx$ 150 μm in diameter) representing a late developmental stage before

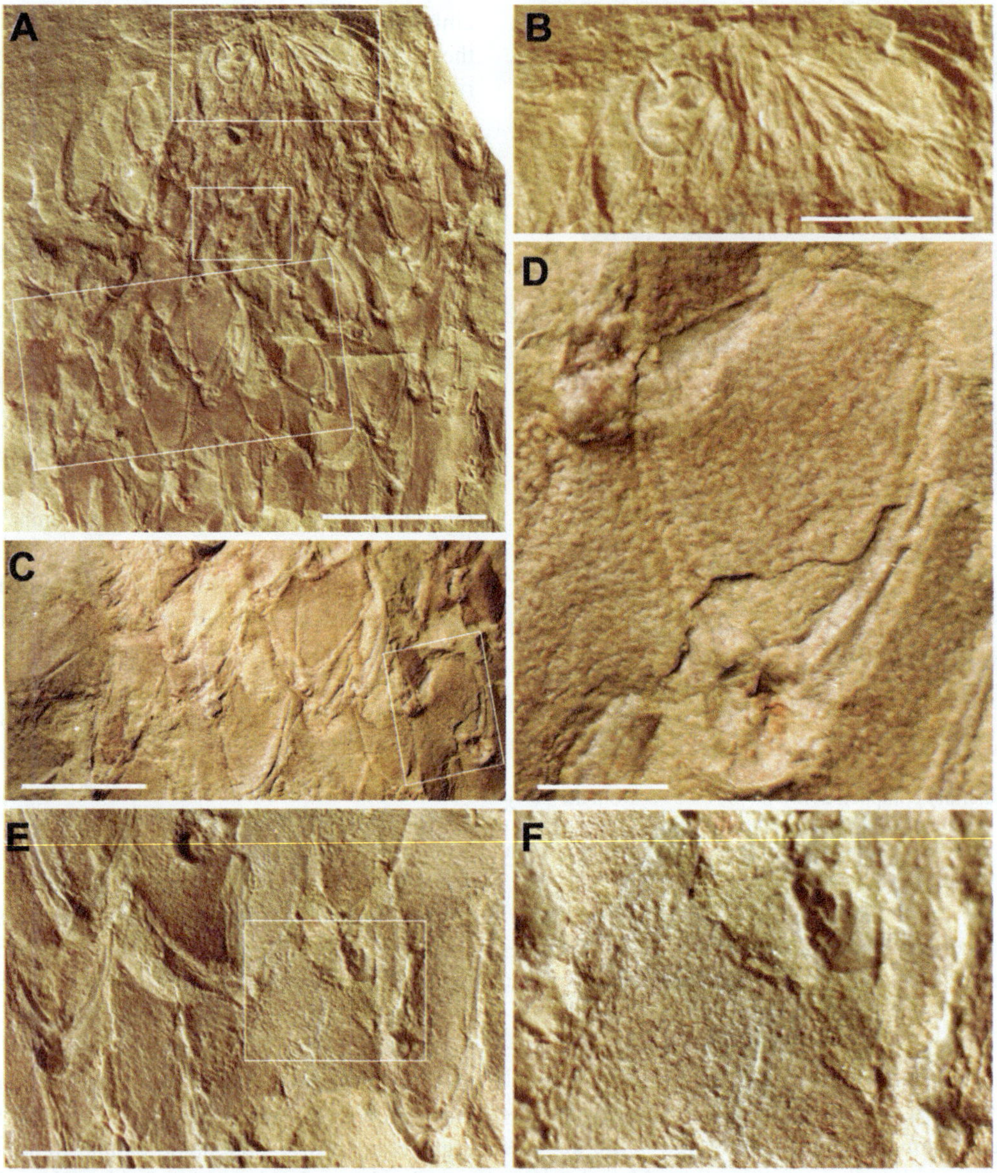

Fig. 27 Distal part of *Allonia phrixothrix* (**a**) and enlargements of the distal margin (**b**) and the integuments (**c–f**) in the distal part of the body, showing the external rays in distal margin extend either convergently or parallel to the body axis and protrude beyond the distal margin (**a**, **b**); two internal rays in each sclerite were embedded in the integument; it displays a conspicuous rhombic texture (**c–f**). Scale bars = 1 cm (**a**), 0.5 cm (**b**, **c**, **e**), 0.1 cm (**d**, **f**)

hatching. The embryo is spheroidal and lacks tentacles, resembling in numerous respects the 10-million-year-younger ctenophore *Maotianoascus octonarius* from the Maotianshan shale. Its aboral pole is represented by an incompletely formed apical (aboral) organ. The comb plates are represented in the embryo by meridional rows, each composed of a narrow strip-like structure that corresponds to the combs of adult ctenophores. Like the comb plates of the larvae and juveniles of modern ctenophores, these meridional rows originate from the apical organ and are attached to the epidermis of the specimen at its aboral end. The fluid-filled, branching, tubular meridional canals in living ctenophores underlying the comb plates are represented in the fossil by branching tubes filled with secondarily emplaced calcite. A complete array of eight comb rows is preserved in the embryo. The total number of comb rows appears to be the same

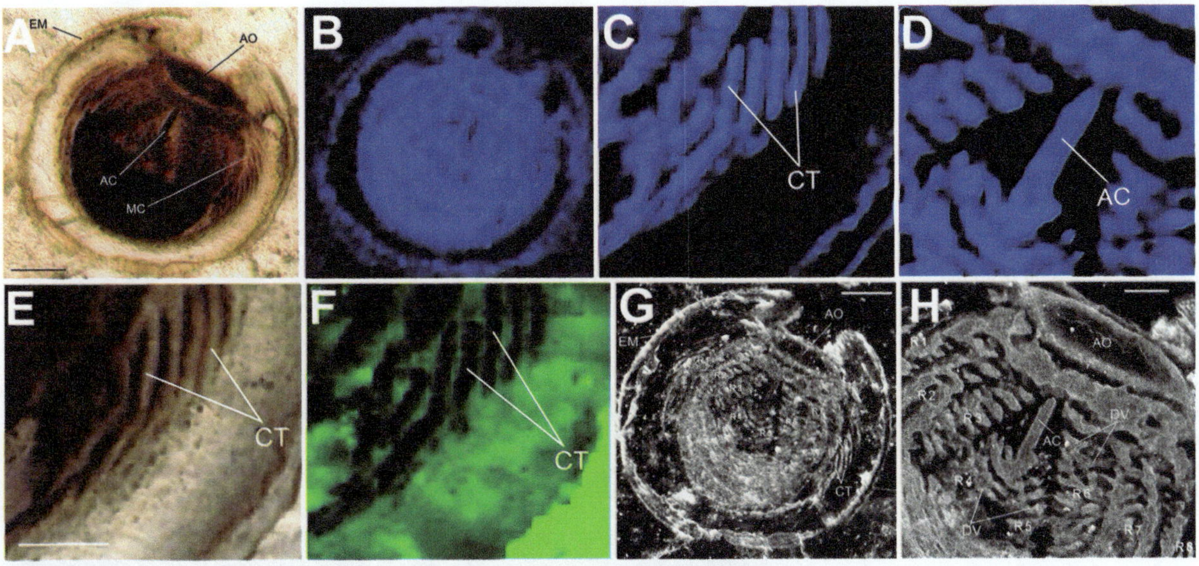

Fig. 28 Optical (**a**), Raman (**b–f**) and Confocal laser scanning microscopy (CLSM) (**g**, **h**) images of a thin section-embedded ctenophore embryo from the Early Cambrian Kuanchuanpu Formation of Ningqiang, Shaanxi. Blue in the Raman images displays the spatial distribution of carbonaceous kerogen; *green* is calcite; and *red* is apatite. Abbreviations: AO, apical (aboral) organ; EM, egg membrane; CT, ctenes; AC, aboral canal; MC, meridional. Scale bar in E = 25 μm (**a**, **b**, **g**); in E it represents 10 μm (**c–f**, **g**) (modified from Chen et al. 2007b)

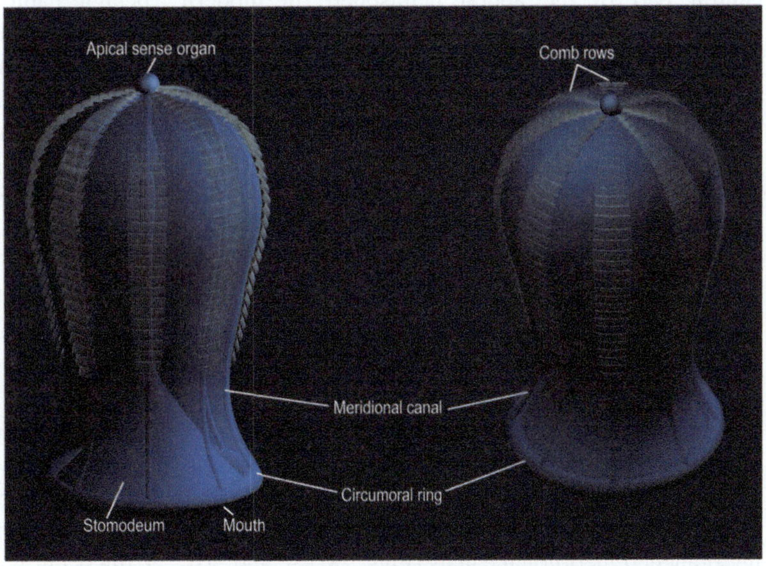

Fig. 29 Restoration of Early Cambrian *Maotianoascus octonarius*

as in the Maotianshan Shale taxon and most modern ctenophores. The individual comb plates (ctenes) are 15–20 μm long and about 1 μm wide. The associated structures, originally hollow and fluid filled (e.g. meridional canals), and the interstices between them, are in-filled with calcite.

Digestion is extracellular in modern ctenophores, occurring in the pharynx, where ingested material passes into a complex system of radiating vascular structures that include the eight meridional canals, one beneath each comb row. The meridional canals in some ctenophores are branched. Composite images

Fig. 30 Photographs of two laterally compacted specimens of *Maotianoascus octonarius* from the Early Cambrian Maotianshan Shale, central Yunnan; it shows a conspicuous aboral organ at the aboral pole; a stomadaeum enlarged in its oral portion with a large mouth; eight rows of ctenes extending from the aboral organ but ending at body constriction before reaching the mouth margin, a meridional canal lies beneath each of the comb rows and extends orally beyond the oral end of the comb row. Scale bar = 1 cm

produced with confocal laser scanning microscopy have documented the presence of eight such branched tube-like structures in the embryo studied here. The central part of the aboral hemisphere in modern ctenophores usually has a tube-like aboral canal (roughly comparable to the distal part of an intestine), a structure represented in the fossil by an unbranched elongate rod, $\approx 6.5\ \mu m$ in diameter that extends orally from the apical organ for a distance of $40\ \mu m$ to where it is buried beneath the branched canals of the comb rows.

Living cydippids are widely accepted as representatives of the ancestral form of the Ctenophora. This implies that the ancestor of the ctenophorans had a spheroidal body form, eight comb rows, and a pair of tentacles. The cydippid hypothesis is challenged by the fossil evidence. Both the ancient 540-million-year-old ctenophore embryo and the 530-million-year-old adult *M. octonarius* from the Maotianshan Shale differ from living cydippids in being atentaculate and having branched meridional canals (Figs. 28, 29, and 30). Instead, these characters are shared by the embryos of extant beroidian ctenophores, to which this fossil seems more closely related. Furthermore, its apical organ is well developed like that of the embryos of most beroidians, although its domed form resembles that of the apical organ of some cydippids. On the basis of the characters exhibited by these embryo and adult fossils, the ancestor of the phylum was an atentaculate, spheroidal organism, most closely allied to extant beroidians; it had a contractile oral lobe and used its comb rows for locomotion. The pelagic habit of the beroidians and most other modern ctenophores suggests that the common ancestor was also pelagic. In contrast with the pelagic interpretation, Shu et al. recently proposed that the early evolution of the ctenophores was marked by a shift from a benthic, sessile existence to a pelagic habit, coupled to a change in the function of their cilia from feeding to locomotion. Such a shift would involve major changes in their basic morphology and ecology and would require many (undocumented) intermediate stages. Based on this hypothesis, Shu et al. (2006) proposed the benthic, frondose *Stromaveris psygmoglena* from the Maotianshan Shale as a stem-group ctenophore. This interpretation is based heavily on the presence of closely spaced branches, inferred to represent precursors of the diagnostic comb rows of ctenophores because they were "probably ciliated". Cilia occur widely in both metazoans and protists, and they have diverse functions, not just in locomotion and feeding. Recourse to the presence of probable cilia as the prime character with which to infer a ctenophore affinity for *S. psygmoglena* is unconvincing. Similarly, the suggestion by Shu et al. that the Ctenophora occupies an intermediate evolutionary position between the sponges and cnidarians is inconsistent with numerous lines of evidence, both anatomical and molecular.

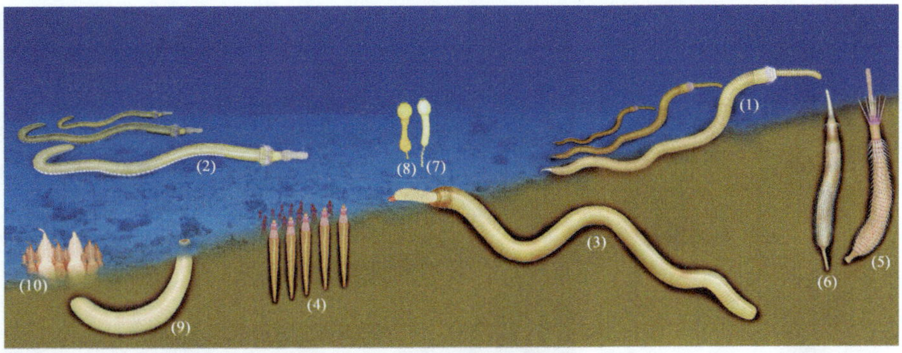

Fig. 31 Panorama of diverse introvert-bearing worms in the Early Cambrian Maotianshan Shale: *Maotianshania* (1), *Circocosmia* (2), *Palaeoscolex* (3), *Paraselkirkia* (4), *Corynetis* (5), *Anningvermis* (6), *Xiaoheiqingella* (7), *Acosmia* (9), and *Sicyophorus* (10)

Three species of ctenophorans belonging to three different genera have been found from the Lower Cambrian Maotianshan Shale. They are *Maotianoascus octonarious* (Chen and Zhou 1997), *Sinoascus palilatus* (Chen and Zhou 1997), and an unnamed new genus and species. *M. octonarious* has a thimble-shaped ovoid body, about 3 cm wide and 4.5 cm long, arranged radially along the oral–aboral axis (Figs. 29 and 30). The oral portion of the stomodeum is enlarged, with a large mouth opening at the oral end. Its aboral side is regularly rounded with a conspicuous aboral organ at the aboral end. Between the two poles are eight well-developed rows of ctenes, originating from the aboral organ but end at the body constriction, long before reaching the mouth margin. Beneath each of the eight comb rows is a meridional canal. The meridional canals extend orally beyond the oral margin of the comb rows, presumably to join a circum-oral canal. Unfortunately, the circum-oral canal is not preserved in our specimen.

S. palilatus is represented by a single specimen characterized by an ovoid body (1.3 cm wide and 2 cm long), possibly radially symmetrical along the oral–aboral axis. Between the two poles are four rows of possible ctenes. The aboral extremity has a depression surrounded by papillae, in which the statocyst was probably located.

Introvert-Bearing Worm-Like Organisms

Introvert-bearing organisms are important members of the benthic communities, especially the endobenthic communities, in the Lower Cambrian Maotianshan Shale fauna (Chen 2004; Huang et al. 2004a, b; Maas et al. 2007a, b; Dornbos and Chen 2008) and the Middle Cambrian Burgess Shale fauna (Conway Morris 1977a, 1993), occurring as at least 11 species in the Maotianshan Shale (Huang 2006) and 3 species in the Burgess Shale (Conway Morris 1979; Fig. 31).

These Cambrian introvert-bearing organisms have been compared with living Priapulida. Their body forms include an elongated sac-like form, a slender vermiform, a loricated form, a dumb-bell-shaped form, and a tubiform, but except for *Acosmia maotiania*, these differ significantly from the body forms of living priapulids. The body of *A. maotiania* (Chen and Zhou 1997; Chen 2004) is elongate and sac like, resembling those of recent priapulids, especially *Priapulus caudatus* (Maas et al. 2007a, b) (Fig. 32). Whether all these introvert-bearing organisms from the Maotianshan Shale are priapulids remains an open question.

Priapulida is a small phylum, with only 18 living species of marine worms. The phylum is a member of the cycloneuralians and is grouped with another four cycloneuralian phyla (Nematoda, Nematomorpha, Kinorhyncha, and Loricifera) to form the monophyletic taxon known as the "Introverta", characterized by introverts and ecdysis. The priapulids are closely related to the kinorhynchs and loriciferans, with which they constitute the Cephalorhyncha, with several apomorphic features, including chitinous scalids on the introvert and two rings of pharyngeal retractor muscles penetrating the brain. Priapulids differ from kinorhynchs and loriciferans in that their introvert has an inversible mouth cone. The terminal mouth in priapulids is located at the anterior end of the resting introvert and opens into a muscular cuticularized pharynx lined with strong cuticle teeth. During feeding, it can

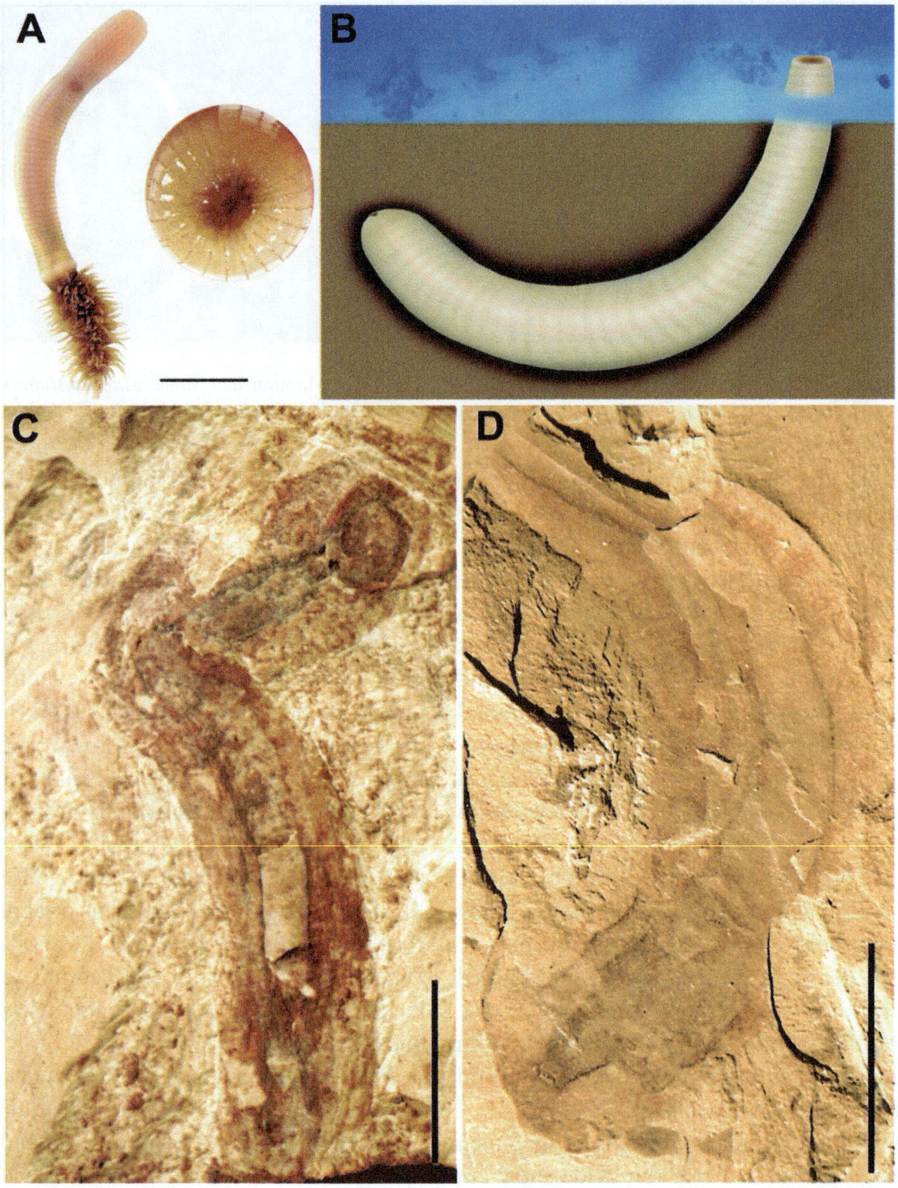

Fig. 32 Extant priapulid: *Priapulus caudatus* (**a**), restoration (**b**) and photographs (**c**, **d**) of Early Cambrian *Acosmia maotiannia*. Scale bar = 1 cm

be everted (turned inside out) through the mouth, forming the mouth cone. The introvert-bearing worms from the Lower Cambrian Maotianshan Shale probably all had the inversible mouth cone typical of priapulids, although they were superficially significantly different from living priapulids. They may represent the stem lineage of the cephalorhynchs.

Loricated forms, including *Palaeopriapulites parvus* (Hou et al. 1999; Chen 2004) and *Sicyophorus rara* (Luo et al. 1999; Hou et al. 1999; Huang et al. 2004c; Chen 2004), from the Lower Cambrian Maotianshan Shale retain a body form resembling that of priapulid larvae, with a thick cuticular lorica (Fig. 33). These loricated forms have been interpreted as either Priapulida by some authors (Hou et al. 1999, 2004; Chen 2004) or Loricifera (Huang 2005). Although these Cambrian forms resemble modern Loricifera in having a life-time cuticular lorica, they

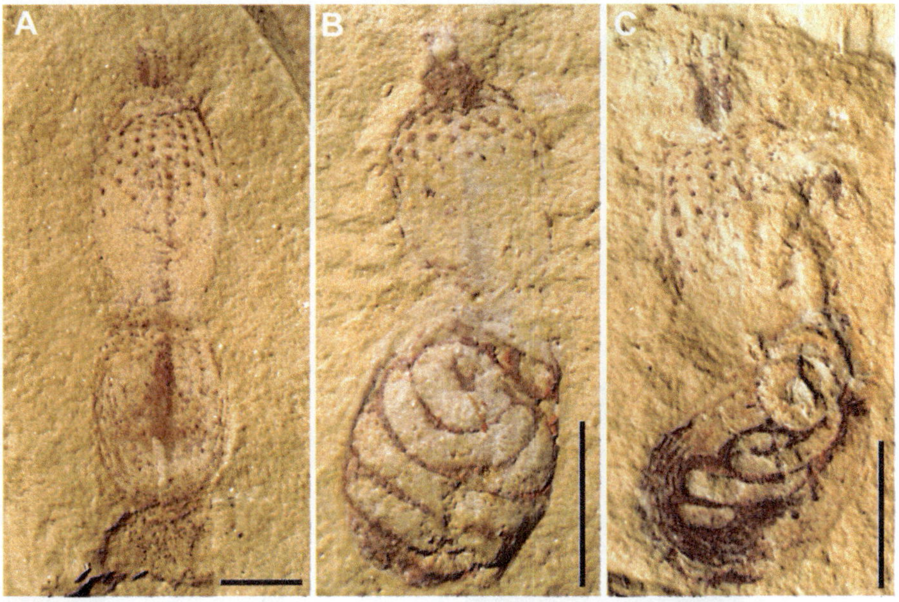

Fig. 33 Loricated introvert-bearing worms from the Early Cambrian Maotianshan Shale. (**a**) a laterally compacted *Palaeopriapulites parvus* showing a bipartite body (introvert and trunk), an introvert bearing 20 rows of spines (larger anteriorly), trunk retaining 10 visible ridges; these suggest the total number of plates was 20 with an identical number of rows of spines which are smaller than those on introvert; intestine simple, straight. (**b**, **c**) laterally compacted *Sicyophorus rara* showing bipartite body with introvert with 20 rows of spines arranged into an alternative order in 14 circles, a conical mouth cone lying front of the introvert; trunk bearing 10 visible plates (total number 20) and a large, tightly coiled gut within the lorica. Scale bar = 0.2 cm

differ from Loricifera in having a different number of plates, a much larger body (10–50 times larger), and in particular an inversible mouth cone. These dumb-bell-shaped species display a striking resemblance to the loricate larvae of extant priapulids, particularly *Tubiluchus corallicola* (Van der Land 1968), with which they share identical numbers of loricate plates. The trunk of the loricate larva of *Tubiluchus* has ten ridges and ten depressions. Living priapulids have a biphasic life cycle and the lorica is only present in the larva, disappearing at metamorphosis. These fossils represent an instance of paedomorphosis, with the larval character retained in the adult.

P. parvus (Hou et al. 1999; Chen 2004) is dumb-bell shaped, with a size range of 3–5 mm (maximally 1 cm). The body is bipartite, separated into the introvert and the trunk. Both are an elongated ovoid shape; the introvert is slightly longer than the trunk bearing about 20 rows of spines increasing in height anteriorly. The trunk cuticle has ten visible longitudinal ridges, suggesting that it had 20 plates in total. The 20 rows of spines on the introvert are much smaller than those on the introvert. The intestine is simple and straight, extending throughout the body (Fig. 33a).

S. rara Luo and Hu (in Luo et al. 1999) (= *Protopriapulites haikouensis* Hou et al. 1999) occurs abundantly in Haikou and Anning, although Luo and Hu considered it rare (Fig. 34). The pharynx is elongated when it is protruded and is armed with tooth- or spine-like outgrowths, arranged in rows and circles, and occurring alternately between adjacent rings. The pharynx is possibly protrusive and retractable, and it bears regularly arranged spines, possibly in rings of five. The introvert has about 20 rows of spines, arranged into an alternative order in 14 circles. The conical structure lying in front of the introvert is the mouth cone, and its surface is smooth. The trunk cuticle is stiff and made up of longitudinal plates, the margins of which are visible as dark ridges. Ten ridges can be seen on one side, so overall there are 20 ridges. It has a large coiled gut within the lorica, making the retraction of the introvert difficult. This is a unique character unknown in other introvert-bearing organisms. *P. parvus* differs from *S. rara* in having a simple

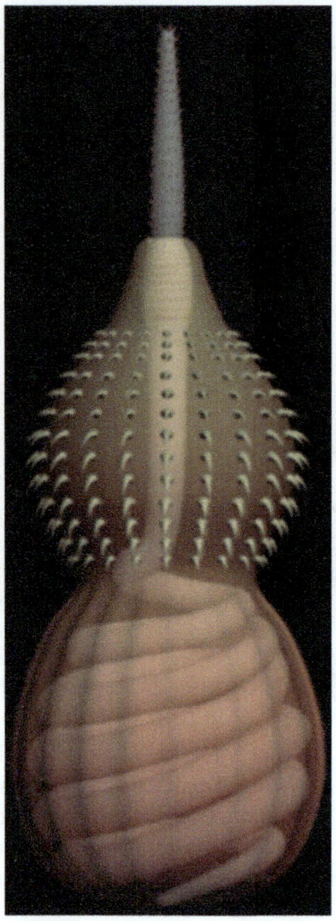

Fig. 34 Restoration of loricated introvert-bearing worm *Sicyophorus rara*

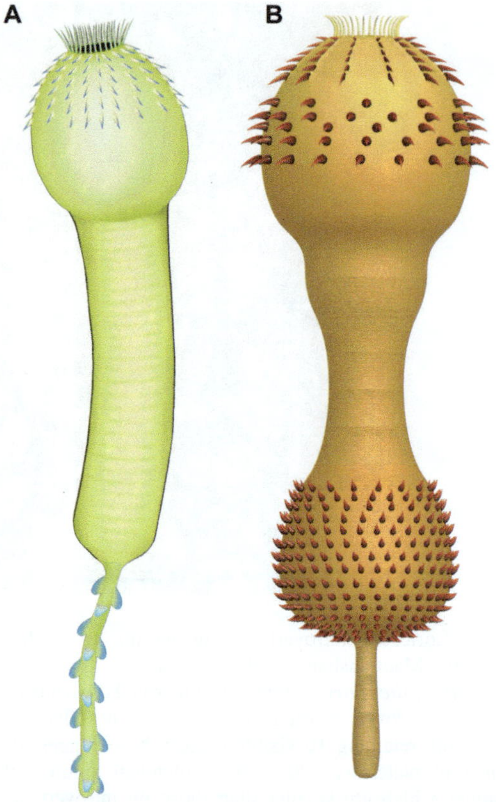

Fig. 35 Restoration of introvert-bearing drum-stick-like worm *Xiaoheiqingella peculiaris* (**a**) and dumb-bell worm *Yunnanpriapulus halteroformis* (**b**)

straight gut and a large introvert, slightly longer than the trunk. The loricate trunk is also present in the early larvae of living priapulids and throughout the lives of living loriciferans. It shares features with the early larvae of living priapulids in having the same number of plates (Lemburg 1999). The loricae of loriciferans all seem to be bilaterally symmetrical and consist of a greater number of plates.

Yunnanpriapulus halteroformis (Huang et al. 2004c) and *Xiaoheiqingella peculiaris* Hu (Chen et al. 2002b) from the Lower Cambrian Maotianshan Shale have drum-stick (Figs. 35a and 36a, b) and dumb-bell forms (Figs. 35b and 36c), respectively. The scalid-bearing introverts of both are sub-spherical, with a corona. The anterior part of the introvert bears 25 rows of spines but the posterior part is smooth in *X. peculiaris*, whereas it has large, irregularly arranged spines in *Y. halteroformis*. The trunk is much longer than the introvert; it is elongated in *X. peculiaris* but has an expanded rear portion in *Y. halteroformis*. It is faintly annulated, and it is adorned with spines in the expanded rear portion in *Y. halteroformis*. The trunk bears a single, slender caudal appendage. The gut is preserved only in *Y. halteroformis*, in which it is a large, locally twisted, tube-like structure.

The slender introvert-bearing vermiform organisms in the Maotianshan Shale include *Maotianshania* (Figs. 37a and 38a, b), *Cricocosmia* (Figs. 37b and 38c, d), and *Palaeoscolex* (Figs. 37c and 38e–g). *Maotianshania* is known in Chengjiang, *Cricocosmia* occurs abundantly in both Haikou and Anning, and *Palaeoscolex* has a wide distribution in the Maotianshan Shale. All three taxa share characters, including an elongated and finely annulated trunk, an introvert, and a protrusile pharynx. They differ from each other by the smooth trunk in *Maotianshania*, the paired sclerites in *Cricocosmia*, and the densely arranged papillae in *Palaeoscolex*.

Because of their superficial resemblance to nemathelminths, they have been interpreted as such by some authors (Hou and Bergström 1994; Hou et al. 1999, 2004; Maas et al. 2007a, b). However, the lack of the characteristic nematoid features in these

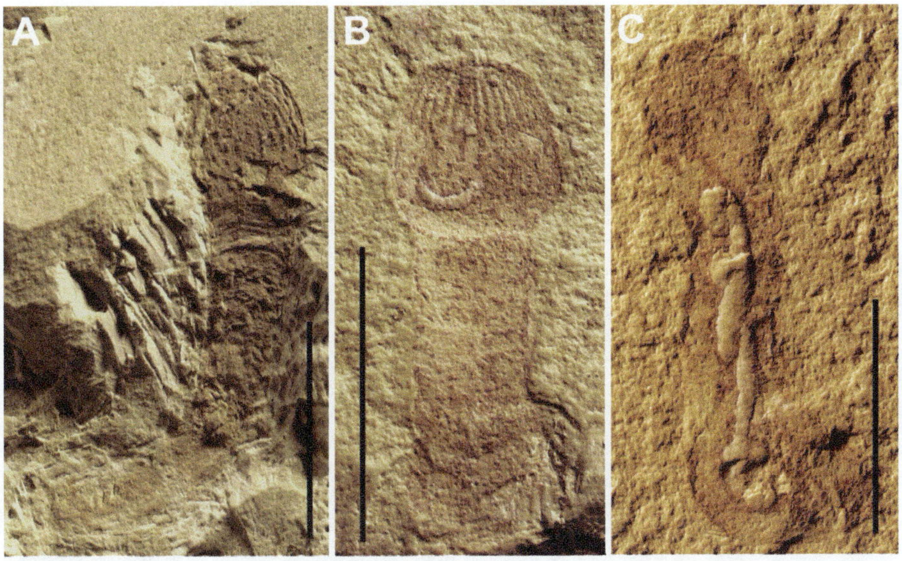

Fig. 36 Introvert-bearing drum-stick-like worm *Xiaoheiqingella peculiaris* (**a**, **b**) and dumb-bell-like worm *Yunnanpriapulus halteroformis* (**c**) from the Early Cambrian Maotianshan Shale. Scale bar = 0.5 cm

Fig. 37 Restoration of slender introvert-bearing worms: *Maotianshani cylindrica* (**a**), *Cricocosmia jinningensis* (**b**) and *Palaeoscolex sinesis* (**c**)

Cambrian vermiform organisms makes the interpretation highly unlikely. The three vermiform organisms from the Maotianshan Shale all have a protrusile pharynx, an exclusively Priapulida character. The nemathelminths include the two living, possibly sister phyla, Nematomorpha and Nematoda. The nematomorphs are a small phylum containing free-living wormiform organisms with larvae that are exclusively parasitic on arthropods (Schmidt-Rhaesa 1996, 1997). They have no introvert and their gut is greatly reduced. However, the larvae bear 2–3 circles of six hooks, possibly homologues of the scalids of Kinorhyncha, Priapulida, and Loricifera (Adrianov and Malakhov 1995; Malakhov and Adrianov 1995; Adrianov et al. 1998). The cuticle in the pharyngeal region is smooth, and chitin is absent in the adult but present in the larva. There are more than 20,000 species of nematodes, constituting more than 90% of recent nemathelminth species. The nematode body is cylindrical and tapered at both ends, and predominantly lacks an introvert. Thus, the considerable differences between them and the nemathelminths refute the interpretation that these Cambrian vermiform worms are nemathelminths.

Fig. 38 Slender introvert-bearing worm *Maotianshani cylindrica* (**a**, **b**), *Cricocosmia jinningensis* (**c**, **d**) and *Palaeoscolex sinesis* (**e**–**g**) from Early Cambrian Maotianshan Shale. Scale bars = 0.5 cm (**a**–**c**, **e**, **g**); 0.2 cm (**d**); 1 cm (**f**)

Maotianshania (Sun and Hou 1987) lived strictly in the Chengjiang area and was apparently adapted as a burrower in soft, wet, muddy bottoms (Figs. 37a and 38a, b). The animals represent a pioneer species and constitute a predominant element in the area, living congregated in the mud layer. The animals are small (3 mm wide and 5 cm long), with a finely annulated trunk (3–4 annulations per millimetre). The gut has a simple tube form with annulations at its rear end and a pair of hooks on the rear surface. The spacing of the gut annulations appears to be identical to that on the body wall, suggesting the trunk annulations are not surface structures.

Cricocosmia jinniense (Hou and Sun 1988; Chen 2004; Maas et al. 2007a, b) occurs abundantly at both Haikou and Shankou, but rarely in Chengjiang (Figs. 37b and 38c, d). It was probably adapted to an epibenthic lifestyle. The animals are elongated (up to 10 cm) and are many times longer than their diameter (average diameter 0.25 mm). They lived in congregations. The inverted introvert is vase-like, with scalids arranged into circles, with the two distal circles larger than the other circles. The pharynx is long, with wider anterior and posterior portions and a constricted mid portion. The trunk is finely annulated, with 120 sclerite pairs. Each trunk annulation carries a pair of sclerites ventrolaterally, except in the anterior part of the trunk. The sclerites are triangular with the narrow part anterior. The ventrolaterally situated sclerites might have facilitated an active type of locomotion on a semi-firm muddy seafloor.

Palaeoscolex (Figs. 37c and 38e–g) is a wide-ranging form from Early Cambrian to Late Silurian strata. In the Maotianshan Shale, it is represented by *Palaeoscolex sinensis* (Hou and Sun 1988); it occurs at numerous localities in the areas of Chengjiang, Jinning, Haikou, and Anning. It has a diameter of 0.5 cm and its length exceeds 10 cm. The trunk is finely annulated, with five annulations per millimetre. The papillae are arranged transversely in lines, with two lines in each annulation. The everted introvert had a conic form. The pharynx has a mid constriction, as seen in *Cricocosmia*. *Palaeoscolex* Whittard (1953) has been assigned to, for example, the Annelida (Whittard 1953), Nematomorpha (Hou and Bergström 1994; Hou et al. 2004), and Priapulida (Chen and Zhou 1997; Conway Morris 1997; Chen 2004; Huang 2005). Isolated palaeoscolecidan annular plates have been widely recorded among small shelly fossils from different areas of the world (e.g. Hinz et al. 1990; Ivantsov and Wrona 2004).

Forms of Pirapulida are characterized by their sac-like body types and are represented by *Corynetis brevis* Luo and Hu (Luo et al. 1999) (Fig. 39a) and *Anningvermis multispinosa* (Huang et al. 2004b, c; Chen 2004) (Fig. 39b). The scalid-bearing introvert in the sac-like form is not well developed. It has a distinct corona, and the protrusile pharynx is highly developed. The trunk is undivided and adorned with papillae/spines that are arranged regularly to irregularly around the trunk. The trunk has an eversible caudal appendage. The gut is a simple tube. *C. brevis* is an uncommon species known only in Haikou and Shakou, and it is about 2 cm long (Fig. 40a–c). Its body is elongated and sac-like, divided into the introvert, trunk, and eversible caudal appendage. The introvert is smooth, except for its anterior termination, which bears a ring of coronal scalids. The protrusile pharynx is long and highly developed. The trunk has an extended rear portion, bearing faint but visible annulations and spines; these are long in the anterior part, short in middle part, and long and slender in rear part. *A. multispinosa* is 6 cm long, with an introvert similar to that of *C. brevis* (Fig. 40d, e). The pharynx is highly developed. The trunk is elongated, with closely arranged

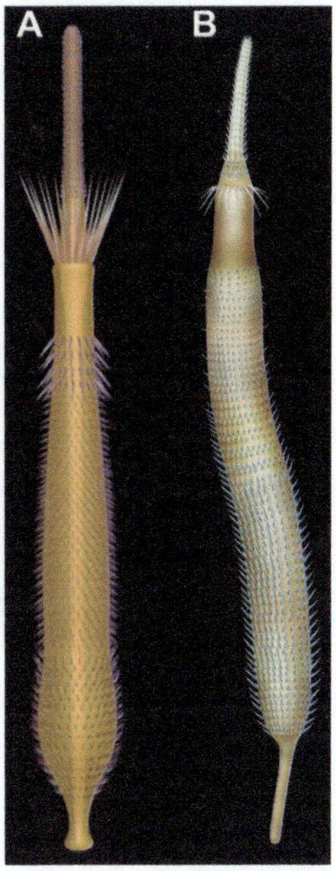

Fig. 39 Restoration of sac-like introvert-bearing worms *Corynetis brevis* (**a**) and *Anningvermis multispinosa* (**b**)

annulations, and is adorned with spines, tilted anteriorly, and becoming larger posteriorly. It has a long and slender caudal appendage and a simple, straight gut.

A sessile priapulid-like animal known as *Selkirkia sinica* Luo et al. (1999) (= *Paraselkirkia jinningensis* Hou et al. 1999) from the Maotianshan Shale has been recorded in abundance in Shankou Haikou (Fig. 41) and was believed to have lived in colonies in firm muddy substrates (Fig. 42). The closely related species known as *Selkirkia columbia* Conway Morris 1977 was first recorded in the Middle Cambrian Burgess Shale. The Chinese specimens are treated as priapulids by several authors (Conway Morris 1977a, c; Hou et al. 1999, 2004; Chen 2004; Huang 2005); they are characterized as having an introvert region and an elongated, conically tapering trunk. Their introvert is distinctly separate from the trunk, has no annulations, and bears scalids in rows and circles, arranged alternately. A corona is present, but it is not as clear as in other forms. The pharynx seems to be protrusive and retractable, and bears spines arranged in circles and rows. The trunk is stiff, and it is presumed to be enclosed in a secreted tube by Hou et al. (1999, 2004). Alternatively, the tube may represent a firm cuticle (Mass et al. 2007).

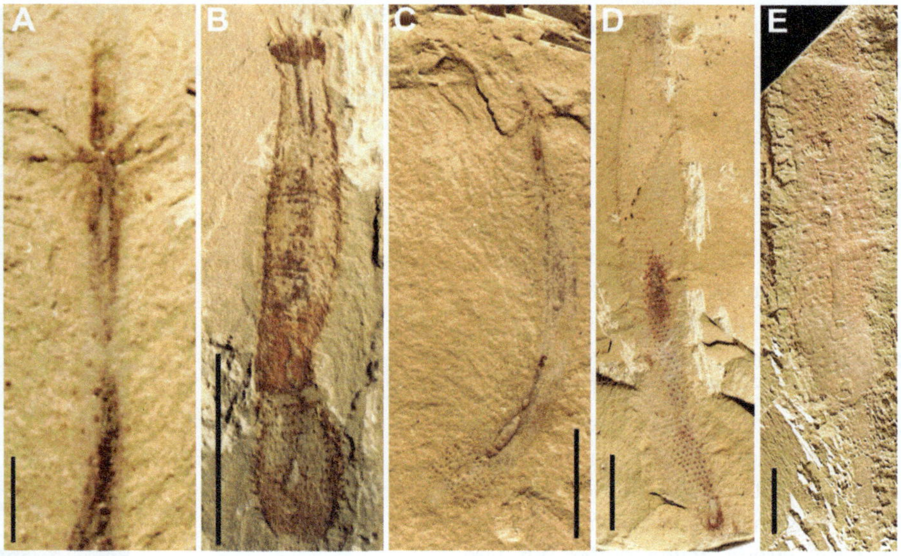

Fig. 40 Sac-like introvert-bearing worms *Corynetis brevis* (**a**–**c**) and *Anningvermis multispinosa* (**d**, **e**) from the Early Cambrian Maotianshan Shale. Scale bars = 0.2 cm (**a**), 0.5 cm (**b**–**e**)

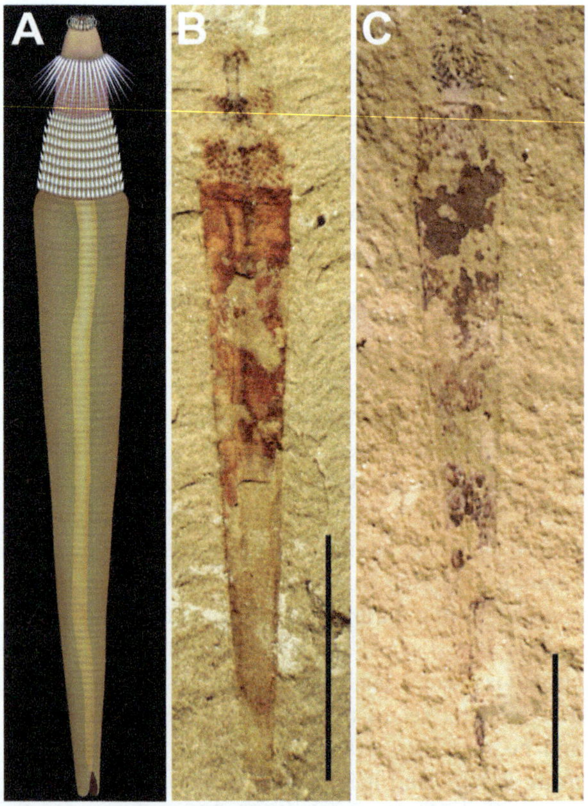

Fig. 41 Restoration (**a**) and photographs (**b**–**c**) of sessile introvert-bearing worm *Selkirkia sinica* from Early Cambrian Maotianshan Shale. Scale bar = 0.5 cm

Sipuncula

Sipunculans (peanut worms) are a small phylum of non-segmented coelomate worms that live exclusively in a variety of marine habitats throughout the world (Cutler 1994).

Although a number of species from different ages have been interpreted as sipunculans, the sipunculan identity of some of these species is highly questionable. These questionable species include *Ottoia prolifica* from the Middle Cambrian Burgess Shale (Walcott 1911a; Howell 1962), *Lechaylus* from the Silurian and Carboniferous (Weller 1925; Muir and Botting 2002), and *Epitrachys rugosus* from the Jurassic (Ehlers 1869). Some authors have proposed sipunculan affinities for hyolithids (Runnegar et al. 1975; Runnegar 1982). Like sipunculans, the extinct mollusc-like hyolithids had a convoluted U-shaped gut, with an anteriorly placed anus. However, hyolithids had a very unusual mineralized exoskeleton, consisting of a pointed shell with an operculum-like anterior lid, and some species had a pair of "helens". These exoskeletal characters make a sipunculan affinity very unlikely.

Three sipunculan worms have recently been recorded from the Lower Cambrian Maotianshan Shale: *Archaeogolfingia caudata*, *Cambrosipunculus tentaculatus*, and *Cambrosipunculus* sp. (Huang et al.

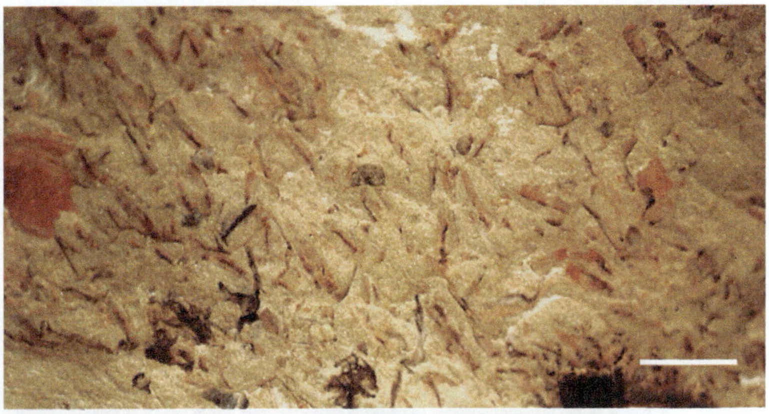

Fig. 42 Sessile introvert-bearing worm *Selkirkia sinica* from Early Cambrian Maotianshan Shale, showing the organisms living colonially in firm mud bottom. Scale bar = 1 cm

Fig. 43 Photographs of extant sipunculan *Phascolosoma granulatum* (**a**) and *Golfigia muricaudata* (**b**) and restoration of Early Cambrian sipunculan *Archaeogolfingia caudata* (**c**) and *Cambrosipunculus tetaculatus* (**d**) (**a** and **b**, courtesy of J Vannier and HY Huang)

2004a; Figs. 43c, d, 44a–c and d–f). Their sipunculan identities are attested by their

- general morphology ("sausage"-shaped body with a slender retractable introvert and a wider trunk);
- external features (e.g. perioral crown of tentacles, and hooks, papillae, and "wrinkled" rings on the body surface); and
- internal features (U-shaped gut and an anus opening near the introvert–trunk junction).

Among the sipunculan worms, *A. caudata* (Huang et al. 2004a), 4 cm long and 0.5 cm wide, is characterized by its relatively elongate introvert, which tapers forward (Figs. 43c and 44a–c). The trunk is as long as the introvert and bears fine ring-like wrinkles along its mid and posterior parts. It terminates posteriorly in a cone-shaped end. The wrinkle pattern on *A. caudata* is similar to that of some living, contracted sipunculans. The trunk is covered with minute spot-like papillae, more densely distributed at the posterior end, as is also typical of many modern sipunculans. The trunk ends in a slender tapered projection, similar in shape to the caudal appendage of modern golfingiid sipunculans (e.g. *Golfinigia muricaudata;* Fig. 43b) (Saiz Salinas 1993; Cutler 1994). The digestive tract is U-shaped, recurving near the posterior end of the trunk.

C. tentaculatus (Huang et al. 2004a) is 4 cm long and 0.4 cm wide and has an elongated body tapering anteriorly, with no clear boundary between the

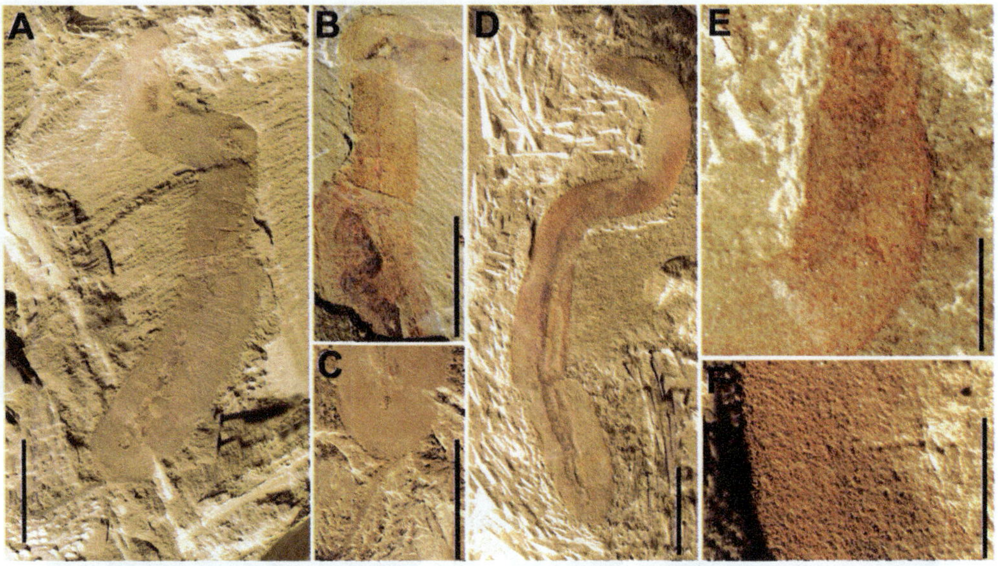

Fig. 44 Photographs of sipunculan *Archaeogolfingia caudata* (**a–c**) and *Cambrosipunculus tetaculatus* (**d–f**) from Early Cambrian Maotianshan Shale. Scale bar = 0.5 cm (**a–d**) and 0.2 cm (**e, f**)

introvert and trunk (Figs. 43d and 44d–f). As in most extant forms, the introvert is armed with anterior hooks that are larger at the anterior ring and smaller posteriorly. The introvert is tipped with a crown of small peripheral tentacles. The crown of tentacles is usually inverted and invisible in most specimens. The trunk is longer than the introvert, with a rounded posterior end, and is decorated with fine papillae. The digestive tract is U-shaped, recurving near the posterior end of the trunk. Both the descending and ascending portions of the tract are quite straight and close to each other. The digestive tract terminates in a circle-like structure representing the anus. *C. tentaculatus* shows a strong similarity to the living sipuncular *Phascolosoma* (Fig. 43a).

These sipuncular fossil forms from the Lower Cambrian Maotianshan Shale display striking similarities to extant sipunculans, especially Golfidiidae. This implies that the sipunculans are not only deeply rooted in the Early Cambrian but have evolved very little since the Early Cambrian.

Worm-Like Organisms with Possible Polychaete Affinities

The evidence of annelids in the Maotianshan Shale is relatively flimsy but the two species collected, *Facivermis yunnanicus* (Hou and Chen 1989) and *Maotianchaeta fuxianella* (Chen 2004), are possibly related to extant polychaetes. *F. yunnanicus* is one of the Cambrian mysteries, with great controversy surrounding its affinities (Figs. 45 and 46). The wormiform organism (6 cm long) has a bipartite body, an anterior appendage-carrying head, and a trunk. The head has a small, short, conic non-segmented part and five segmented parts, each bearing a pair of head appendages protruding ventrolaterally. The anterior non-segmented part bears a terminal mouth. Each appendage has two rows of longitudinally aligned setae, the anterior and posterior rows. They are closely annulated, with 50–60 annulations per centimetre, each bearing a pair of setae, an anterior and a posterior one. The trunk is cylindrical for most of its length, with an expanded rear portion bearing 2–3 circles of small hooks. Parapodium-like structures are preserved, nearly 1 mm long, and extending lateroposteriorly. The gut is a simple tube-like structure with termini at both ends of the body.

M. fuxianella (Chen 2004) is represented by a single, incompletely preserved specimen, about 5.4 cm long (incomplete length) and 0.6 cm wide, with pairs of laterally extended parapodia, which are elongated with distal setae (Fig. 47). Pairs of organic structures are situated at the lateral margin of the trunk, corresponding to the pairs of parapodia. These organic structures may represent the remains of the metanephridia.

Fig. 45 Restoration of *Facivermis yunnanicus*

Although the annelids are broadly accepted as having an ancient origin, reliable polychaetes are first recorded in the Middle Cambrian Burgess Shale. *Facivermis* has a non-segmented acron and possible parapodia, displaying an affinity to the polychaetes. In addition to the acron, the head of *Facivermis* also includes a distinct five-segmented part, each segment bearing a pair of large appendages. A similar segmented head with paired appendage-like structures is also present in some living polychaetes, such as *Odontosyllis enopla*, found in shallow bays around Bermuda (Sterrer 1986). *Maotianchaeta* has been reported from the Maotianshan Shale and is similar to the polychaetes in having parapodia-like structures.

Soft-Bodied Mollusc: *Vetustovermis*

There are 85,000 recognized extant species of molluscs, highly diverse in size, anatomical structure, behaviour, and habitat. There is good evidence for the appearance of shelled molluscs in the Early Cambrian and their diversification into gastropods, cephalopods, and bivalves in the Late Cambrian. However, the evolutionary history of both the molluscs' emergence from the ancestral Lophotrochozoa and their diversification into the well-known living and fossil forms are still vigorously debated.

Ancestral molluscs are widely accepted as having been shell-less, resembling modern aplacophorans, or to have lacked continuous biomineralized exoskeletons, like extant chitons (Brusca and Brusca 2002). Radular grazing trace fossils (*Radulichnus*; Erwin and Davidson 2002; Seilacher 1999, 2003), together with large imprints of possible soft-bodied molluscs (*Kimberella*; Erwin and Davidson 2002; Fedonkin and Waggoner 1997; Seilacher 1999, 2003) from the latest Neoproterozoic Ediacaran indicate that macroscopic soft-bodied molluscs may have a deep Precambrian evolutionary history. The preservation of early shell-less molluscs may have been unlikely, and the recognition of these fossils difficult, and they have been recorded extremely rarely in the literature (Sutton et al. 2001). Radular grazing traces preserved together with circular impressions (about 10 cm) in Lower Cambrian strata suggest the presence of soft-bodied macroscopic molluscs in Cambrian times (Dornbos et al. 2004; Fig. 48).

Although the Lower Cambrian Maotianshan Shale had the potential to preserve soft-bodied fauna, no soft-bodied molluscs have been recorded from it, perhaps because they are difficult to recognize. However, a possible example is *Vetustovermis planus*, first described as an annelid by Glaessner in 1979 and as an arthropod by Luo and Hu in 1999 (Luo et al. 1999). Its molluscan affinities have recently been recognized with the study of new fossil specimens (Chen et al. 2005).

The animal *V. planus* is slug like, flattened dorsoventrally, and 5–6 cm long (maximally 10 cm).

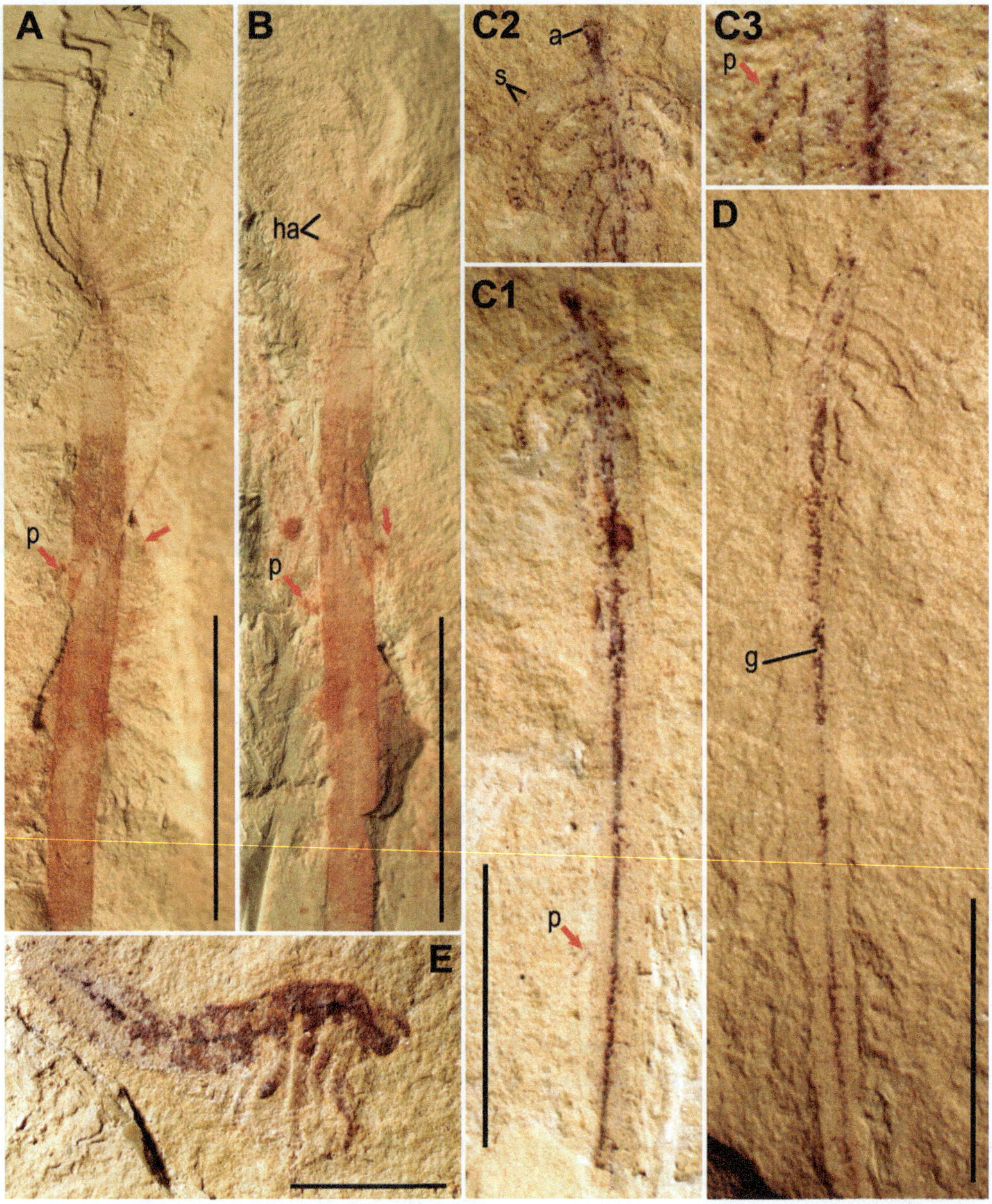

Fig. 46 *Facivermis yunnanicus* from Early Cambrian Maotianshan Shale. (**a**, **b**) part and counterpart of a dorso-ventrally compacted specimen, showing: head tentacles, parapodia (*marked with arrow*) and simply straight gut; (**c1-3**) showing head tentacles (**c1-2**) with setae (**c2**), anterior non-segmented part (acron) (**c2**) and parapodia (**c3**, *marked with arrow*); (**d**) showing head tentacles and gut; (**e**) a laterally compacted specimen, showing head tentacles and gut. Abreviations: g, gut; p, parapodia. Scale bars = 1 cm (**a**, **b**), 0.5 cm (**c1**, **d**), 0.2 cm (**e**)

The rounded head is small but distinct, carrying a pair of large cephalic tentacles projecting from the dorsal surface of its middle part (Figs. 49a, b) and a pair of large, stalked eyes near the base of the tentacles (Fig. 50a, b, d).

The trunk is ovally elongated and broadly triangular anteriorly (Fig. 50a, b, d, g, h), and decorated with hard, short, cone-shaped spicules on the dorsal surface (Fig. 50b, e, h). It is laterally finned and the fins were flexible, extending continuously along most of the trunk, except on its triangular anterior part. About 50 pairs of transversely extending bar-like structures, interpreted as gills, dominate the trunk. The gill bars are usually black, suggesting that they

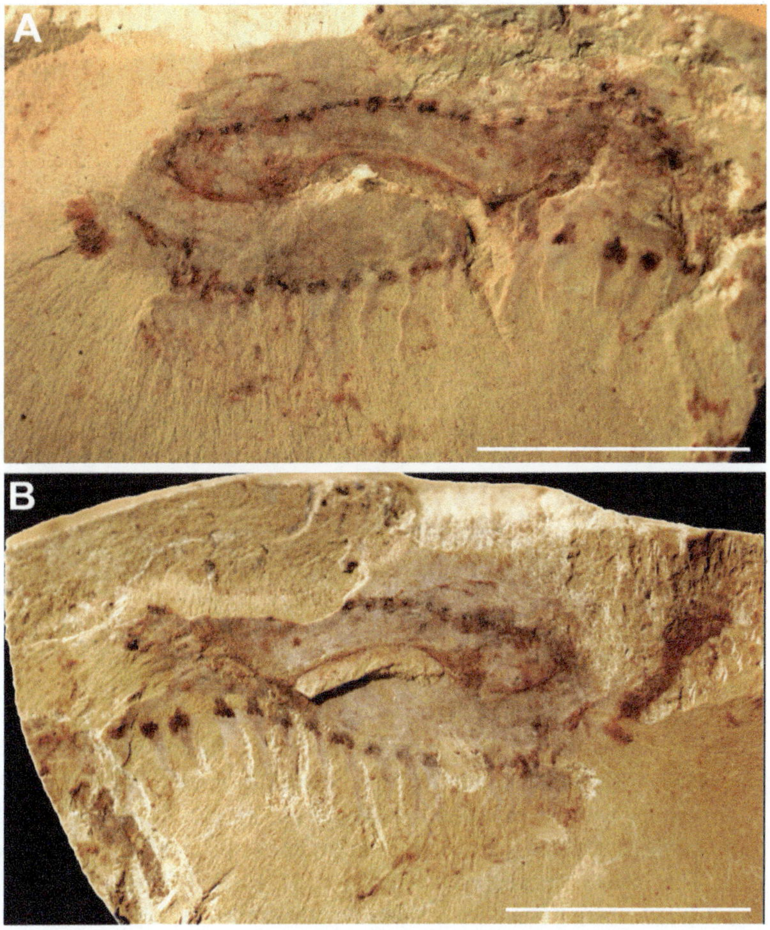

Fig. 47 *Maotianchaeta fuxiannella* from Early Cambrian Maotianshan Shale. (**a**, **b**) part and counterpart of the holotype. Scale bar = 1 cm

are organic. These gills are separated into two lateral sets by a dorsal midline and the lateral margin of the bar series is slightly separated from the trunk margin, as would be expected for gills. These gill bars are more-or-less straight, evenly spaced over the whole trunk (Fig. 50c), tilted anteriorly, and overlap one another as the result of dorsal compaction. The common occurrence of deformation indicates that they were soft and flexible. These soft and flexible transverse gill bars could be gut branches ("gut diverticula"), known in flatworms. However, unlike gut diverticula, these bars are widely separated from the gut (Fig. 50c).

Fig. 48 Rounded impression with radular grazing trace from Early Cambrian phosphate deposits, Jinning, Yunnan. Scale bar = 10 cm

The trunk retains a muscular foot-like structure ventrally. The foot is large, extending to cover most of the ventral surface of the trunk proper except for the most lateral parts of the gill bars (Fig. 50b, g). The lateral margin of the foot is also visible in most specimens split along the dorsal surface, as a series of projections visible beyond the outer margin of the trunk proper. These relationships suggest that the foot was relatively thick.

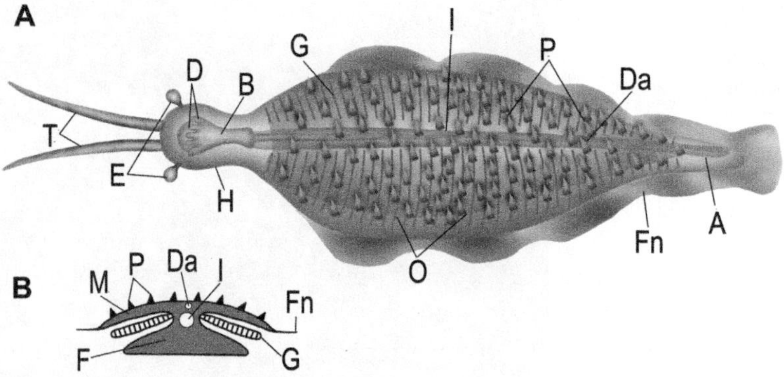

Fig. 49 Anatomical features of the Early Cambrian mollusc-like organism *Vetustovermis planus*. (**a**) dorsal view; (**b**) reconstructed cross-section. Abbreviations: A, anus; B, buccal mass and stomach; D, diverticula; Da, dorsal aorta; E, eye; F, foot; Fn, fin; G, gills; H, head; I, intestine; M, mantle; O, outer margin of foot; P, spine-like structure; T, cephalic tentacles (adapted from Huang et al. 2004a)

The ventral foot shows a striking resemblance to the ventral sole of some modern flatworms. In dorsal view, the gill bars are separated by an axial band-like gut (ca. 3 mm wide) into the right and left gill parts.

V. planus has a complete gut. The head contains a bulb-like structure interpreted as a buccal/stomach mass (Fig. 50a, b). It is commonly preserved in three dimensions, with an expanded, ramified anterior part. The ramified structures are interpreted as gut diverticula. The intestine is represented by a wide, partially mud-filled, three-dimensional tube-like structure (ca. 0.5–1 mm in diameter) that extends axially within the trunk proper with an anal opening just in front of the end of the trunk. The slender dorsally situated, tube-like structure (ca. 0.2 mm in diameter) seen in several specimens is interpreted as a median dorsal aorta, as seen in extant polyplacophorans.

The flattened body and horizontal fins of *V. planus* could be adaptations to support the animal as it moved on the soft seafloor and for gliding or swimming near the seafloor. The large surface area of the gills may have provided effective gas exchange for an active life style. The well-developed sense organs include a pair of large, stalked eyes, and large cephalic tentacles, suggesting that *V. planus* was attuned to the presence of food resources and possible predators.

V. planus (Fig. 49) resembles the molluscs in having a flat foot, a polyplacophoran-like flexible, elongated body, and serial pairs of gills on each side of the trunk (Brusca and Brusca 2002). The dorsal, conic spine-like structures could be homologous to the cuticular spicules of the polyplacophorans and aplacophorans. The variety of characters is not specific to any crown group of molluscs (Brusca and Brusca 2002). For instance, a well-differentiated head is considered to be an advanced character and is absent among the more primitive molluscs. The sluggish head of *V. planus* suggests that the head could have been well developed in the early active pelagic molluscans. The gills in *Vetustovermis* are also simply bar shaped and could represent a primitive form, differing from other molluscan gills. The stalked eyes and tentacles are also present in both arthropods and annelids, although those groups lack the ventral foot, instead using leg-like appendages for locomotion. One might argue that the axial band-like visceral cavity of *V. planus* is a notochord and that the transverse bars are not gills but myomeres, thereby implying that the animal is a chordate. However, no known chordate has stalked eyes or a ventral foot. *V. planus* resembles some planktonic nemertine worms, especially *Nectonemertes* and *Balaenanemertes*, in several respects (Gibson 1972;

Fig. 50 (continued) centre—some of the gills are bent—the imprints of spine-like structures; and a median dorsal aorta. Scale bars are all 5 mm except for E, which is 1 mm. Abbreviations: A, anus; B, buccal mass and stomach; D, diverticula; Da, dorsal aorta; E, eye; F, foot; Fn, fin; G, gills; H, head; I, intestine; M, mantle; O, outer margin of foot; P, spine-like structure; T, cephalic tentacles

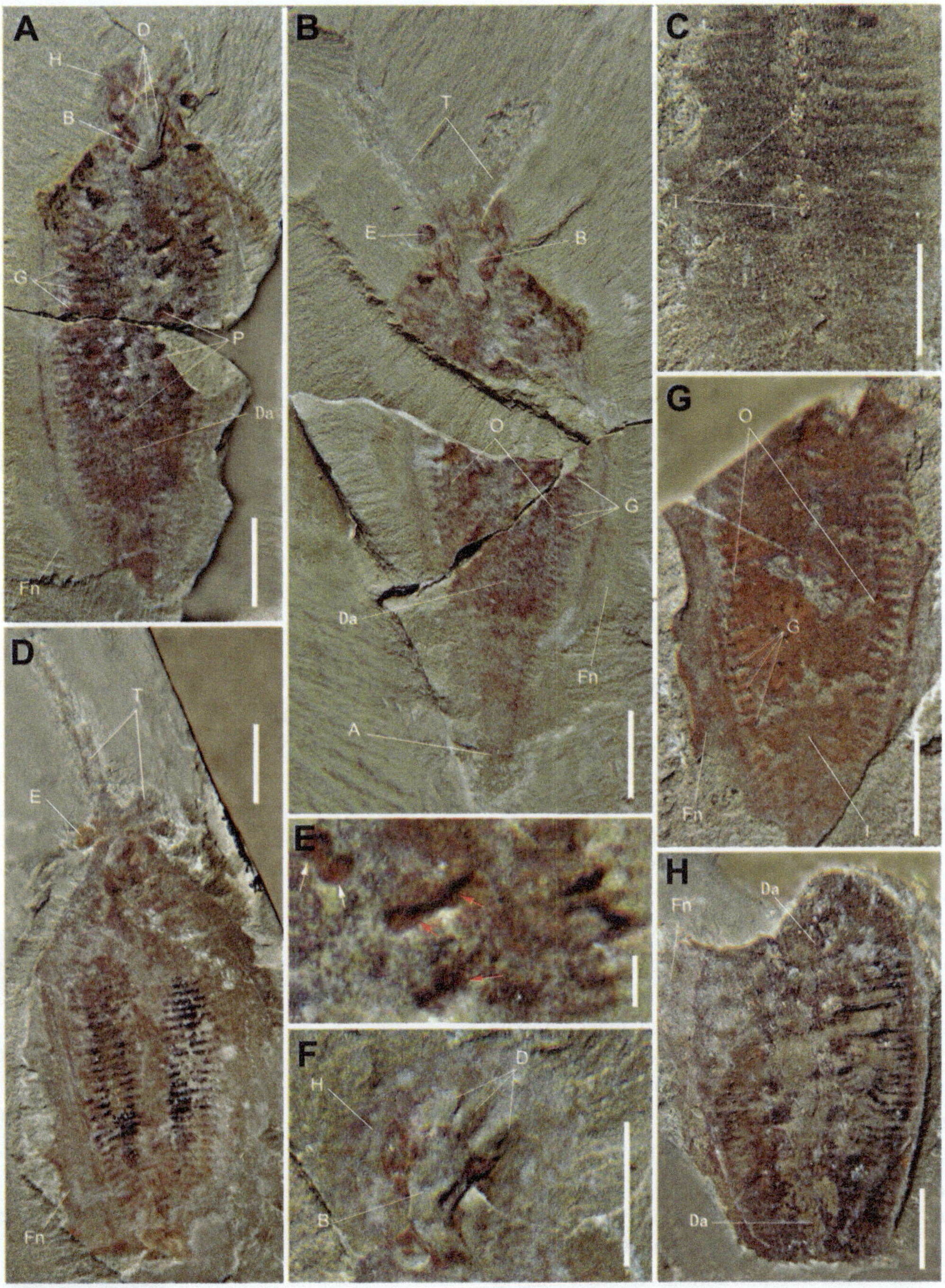

Fig. 50 *Vetustovermis planus* from Early Cambrian Maotianshan Shale. (**a**, **b**) part and counterpart of a nearly complete specimen showing a small round head (**a**) with a pair of cephalic tentacles (**b**) and stalked eyes (**a** and **b**), a buccal mass and stomach with anterior diverticula (**a**), as well as lateral margins of a large ventral foot lying near the outer margins of the trunk (**b**) and imprints of dorsal spine-like structures, dorsal artery, intestine and an anus; (**c**) trunk showing gills and a mud-filled intestine; (**d**) a sagittal plane near the dorsal surface showing cephalic tentacle, eye, lateral fins, and a dorsal artery; (**e**) enlargement of A showing conic (*red arrow*) and rounded (*white arrow*) imprints of spine-like structures. (**f**) head of an animal showing a round head, a three dimensionally preserved buccal mass and stomach with anterior diverticula; (**g**) a sagittal plane near the ventral surface of a trunk showing ventral foot with its outer margins and gills in lateral area of the trunk proper, intestine, and folded lateral fins; (**h**) a sagittal plane near dorsal surface of a trunk showing gills extended axially near the

Hyman 1951). These include an anterior tentacle, lateral fins, ventral foot, and a series of internal gut caecae comparable to the "gills" of *V. planus*. However, nemertines lack a protective cuticle or exoskeleton, as well as stalked eyes, and have a characteristic eversible proboscis. The presence of spine-like protective structures and stalked eyes and the absence of a proboscis do not support a nemertine affinity for the animal. The ventral flat foot also represents a trait that is shared with many turbellarians (free-living flatworms; Hyman 1951; Ruppert et al. 2004). The combined existence of a head, elaborate paired tentacles, and stalked eyes gives no support to an affinity between this animal and the flatworms. Some polychaetes, such as the sea mouse (*Aphrodite aculeate*), also have a ventral muscular creeping sole, and myzostomids have a circular body with fused parapodia; these form a ventral foot-like structure. However, the sea mouse is covered with a thick felt-like or hair layer and myzostomids have numerous extended cirri (Rouse and Pleijel 2001; Ruppert et al. 2004).

Vetustovermis possibly represents an extinct, stem-lineage molluscan, with similarities to the shell-less sea slugs (gastropods that evolved from shelled ancestors) (Yonge 1960). Similar sluggish animals have also been recorded from the Middle Cambrian Burgess Shale. They include *Nectocaris* (Conway Morris 1976), *Amiskwia* (Walcott 1911a), *Odontogriphus* (Conway Morris 1976), and *Pikaia* (Walcott 1911b). *V. planus* also resembles the problematic Carboniferous *T. gregarium* in several important respects (Foester 1979). These include a dorsoventrally flat trunk, lateral fins, distinct visceral mass, stalked eyes, and a series of transverse segments. Foester has tentatively interpreted *T. gregarium* as a prosobranch gastropod (Conway Morris 1976). Among them, *Nectocaris* resembles strikingly to *Vetustovermis* by sharing an open axial cavity with paired gills; wide lateral fins; a single pair of long, prehensile tentacles; and a pair of stalked eyes. Smith and Caron (2010) interprets these naked sluggish animals as stem-lineage cephalopods, although the first shelled cephalopods evolved in late Cambrian (Chen and Teichert 1983a, b).

Chaetognathids (Arrow Worms)

Chaetognatha (arrow worms) is a minor phylum of pelagic marine animals, consisting of about 150 living species. They are found in oceans throughout the world and play an important role in the food web as primary predators. Their body form is reminiscent of an arrow, so they are usually referred to as "arrow worms". However, the fossil record of the Chaetognatha is sparse and the Carboniferous *Paucijaculum samamithion* is the only definitive fossil chaetognath. Some Cambrian protoconodonts could represent the grasping spines of this group. The Middle Cambrian *Amiskwia* has been considered a chaetognath, but with a slug-like head, terminal anus, and cephalic tentacles, it could well be a molluscan.

The body plan of the arrow worms is characterized by a small slender body with strikingly perfect bilateral symmetry. It is divided into a distinct head, finned cylindrical trunk, and a dorsoventrally flattened, finned post-anal tail. The head is ovoid, with a depression that leads to a ventral mouth. The vestibule is flanked laterally with grasping prey-seizing spines. *Eognathacantha ercainella* (Chen and Huang 2002) and *Protosagitta spinosa* Hu 2002 (Chen et al. 2002b) are fossil representatives of the first arrow worms from the Lower Cambrian Maotianshan Shale. *E. ercainella* is represented by a single specimen, about 25 mm long. The animal might have been relatively rigid. It consists of a head, a trunk, and a possible tail (Fig. 51). The rounded head is wider than long, measuring about 2.8 mm in width; it is also wider than

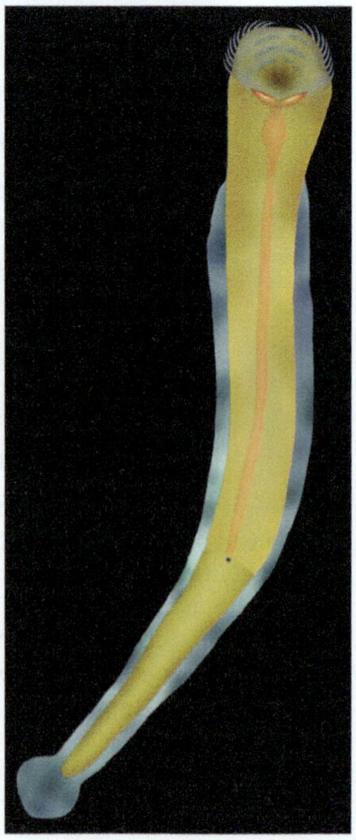

Fig. 51 Restoration of Early Cambrian arrow-worm *Eognathacantha ercainella* from Early Cambrian Maotianshan Shale

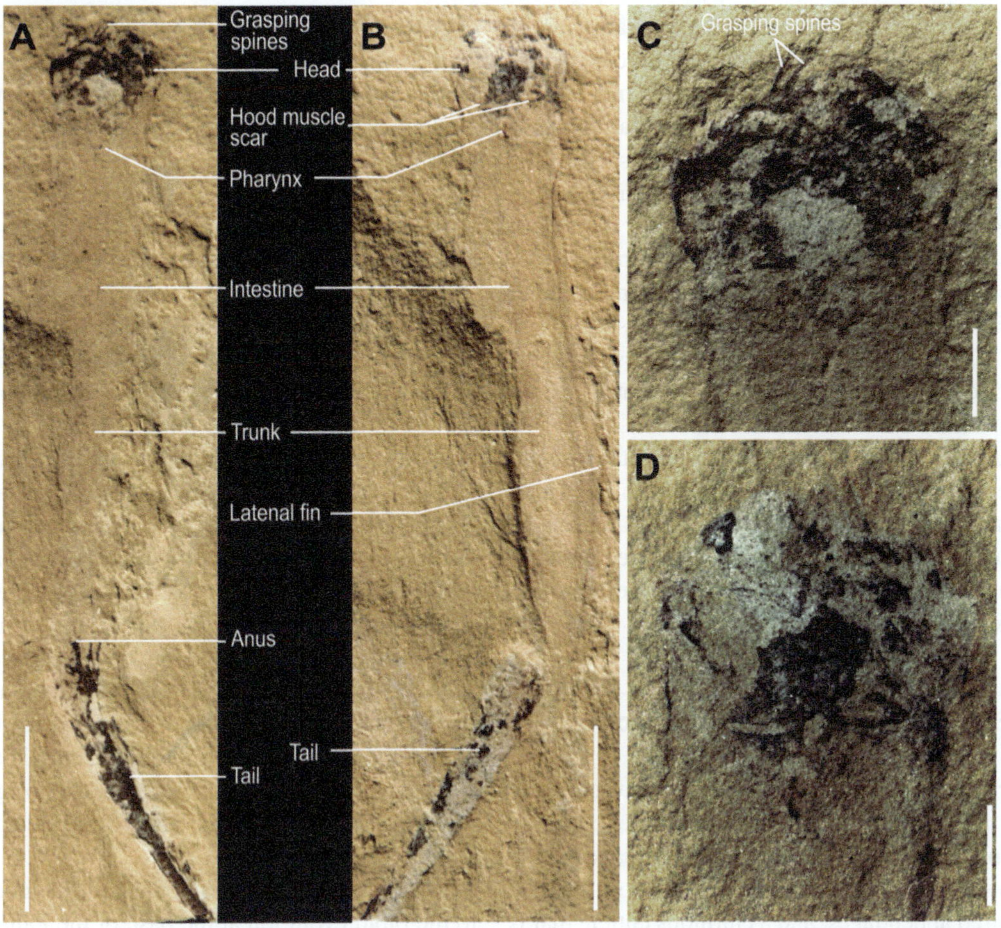

Fig. 52 Arrow worm *Eognathacantha ercainella* from Early Cambrian Maotianshan Shale. (**a**, **b**) part and counterpart of nearly complete specimen. (**c**, **d**) enlargement of anterior portion of part and counterpart. Scale bars = 0.5 cm (**a**, **b**) and 0.1 cm (**c**, **d**)

the rest of the body. Projecting anteriorly and laterally from the head are about 12 slightly recurved grasping spines, each about 900 μm long (Figs. 51 and 52a, b). Like those of modern chaetognaths, the spines have an expanded base and were probably embedded within the tissue. In the anterior part of head, there are numerous smaller spine-like structures (about 200–250 μm long), interpreted as possible teeth. In the middle part of the head, there are numerous blueish papillae-like structures (about 150 μm long) surrounding a small slit-like structure, interpreted as a possible mouth (Fig. 51). The pair of transverse, cup-like structures at the posterior margin of the head is interpreted as the muscle marks of a hood. Such a hood is diagnostic of chaetognaths, and it is an extension of the epidermis, attached ventrally between the head and the trunk. The hood has retractor and extensor muscles. The thick, brown envelope over the head is presumably a cuticle, as in recent chaetognaths. The cuticle disappears at the border between the head and trunk. In contrast to the head, the trunk (2.5 mm wide and 15 mm long) lacks evidence of a cuticle. The trunk epiderm, light yellow in colour presumably has no cuticle. The posterior part of the body is about 8 mm long, with a thick brown envelope; this is considered to be a tapering tail. If the envelope represents a cuticle, it differs from the cuticle-less tails of most living chaetognaths.

Modern chaetognaths usually have paired lateral fins and a wide tail fin, all of which are supported by rays. However, in *E. ercainella*, only a pair of narrow fins without fin rays is preserved. The fins extend continuously from the anterior part of the trunk, throughout the trunk and the tail. It is impossible to determine whether the fins are in the frontal plane of the body, as in modern chaetognaths, or otherwise disposed. The fins have a uniform width of 0.4 mm, except throughout the anterior 2 mm, where they narrow anteriorly. No widening of the tail fin can be seen in the fossil form. A bulbous structure at the anterior end of the trunk is presumably a pharynx (bulbate in modern chaetognaths) narrowing posteriorly to join a slender light-bluish line, interpreted as the intestine running posteriorly, with a possible anus at its posterior end in front of the tail. The position of the possible anus, just anterior to the tail, corresponds to that in extant chaetognaths.

The affinity of *E. ercainella* with chaetognaths is supported by a widened head, distinct from the trunk, its presumed tail (based on the cuticularizaton in the fossil form), grasping spines and possible teeth on the head, and a hood at the border between the head and the trunk.

Limb-Bearing Creatures: Panarthropods

Panarthropods, the first limb-bearing creatures, were diverse in the Early Cambrian. They are probably a monophyletic taxon, sharing a body pattern that includes several important characteristics: a segmented body, paired appendages, and a chitinous cuticle that is moulted during ecdysis (Ruppert et al. 2004).

The presence of pyramid-yolk embryos (Chen et al. 2004d) and late-developing embryos with a segmented germ band (Chen 2004; Steiner et al. 2004) (Fig. 53) in the very basal parts of the Lower Cambrian Meishucun phosphate deposits suggests that the pararthropods are deeply rooted at the very beginning of the Cambrian, about 542 million years ago or even earlier. The exceptionally well-preserved fossil fauna from the 530-million-year-old Maotianshan Shale deposits provides a fossil record of remarkably diverse panarthropods, with a number of body organizations representing different evolutionary stages (Fig. 54). These include the worm-like stem-lineage panarthropods, referred to as Tardiopolypoda (Chen and Zhou 1997), stem-lineage arthropods (proarthropods; Chen 2004; Waloszek et al. 2005), stem-lineage euarthropods, and the possible ancestral form of the two major extant euarthropod groups, the Chelicerata (Chen et al. 2004c) and Mandibulata (Chen et al. 2001). These nested body plans representing different evolutionary stages suggest the presence of step-wise and condensed evolutionary events in the evolutionary pathway from the worm-like ancestor to the euarthropods within a short geological period of about 12 million years in the Early Cambrian.

Worm-Like Stem-Lineage Panarthropods

The Maotianshan Shale has produced Tardipolypoda (Chen and Zhou 1997), worm-like organisms with unjointed limbs. These are the least complex of the Panarthropoda and represent the stem-lineage panarthropods. The group is highly diverse and includes *Luolishania longicruris* (Hou and Chen 1989), *Microdictyon sinicum* (Chen et al. 1989a; Ramsköld and Chen 1998), *Cardiodictyon catenulum* (Hou et al. 1991), *Onychodictyon ferox* (Hou et al. 1991), *Paucipoda inermis* (Chen et al. 1995c), and *Hallucigenia fortis* (Hou and Bergström 1995; Ramsköld and Chen 1998). These taxa share several important characteristics: a soft, segmented body with paired appendages, an undifferentiated head that is either segmented or of non-segmental origin, and a mouth and anus at either end (Ramsköld and Chen 1998; Ramsköld 1992).

An anatomical analysis of the tardipolypods testifies to the rapid evolution at the base of the Panarthropoda as they are separated by morphological gaps into several different groups based on their head plan: a non-segmental head, a one-segmented head, or a two-segmented head (Fig. 54).

The non-segmented-head tardipolypods, including *M. sinicum* (Figs. 55 and 56) and *P. inermis* (Fig. 57), have the least complex head, relatively long in both *M. sinicum* and *P. inermis*, but differing morphologically, being cylindrical in *P. inermis* and narrowing anteriorly in *M. sinicum*. A morphological gap is also great in the trunk anatomy of these two species, in so far as they differ in their numbers of segments and their surface structures; the surface is nearly smooth in *P. inermis* but is adorned with a pair of structurally complex, hard, lateral sclerites on each trunk segment in *M. sinicum*. These sclerites, hard and possibly mineralized, occur widely among Lower and Middle Cambrian rocks. The generic name *Microdictyon* was erected to represent these remarkably complex sclerites by Bengtson, Matthews, and Missarzhevsky in 1981 (see Missarzhevsky and Mambetov 1981;

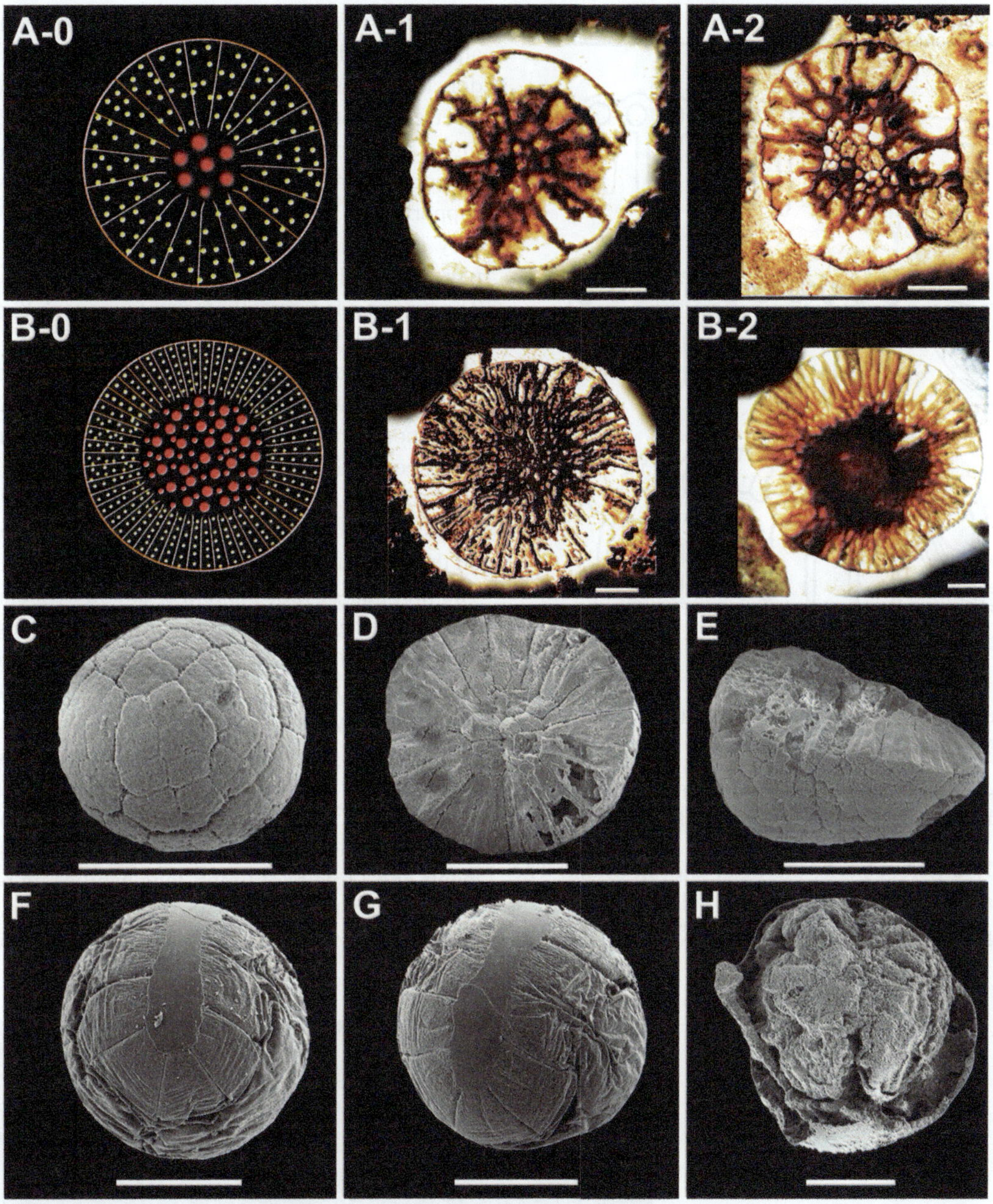

Fig. 53 Yolk-pyramid eggs (**a–e**) and late developing embryos with segmented long germ band (**f–h**) from Early Cambrian Kuanchuanpu formation, Ningqiang, southern Shaanxi. Explanatory (**a-0**) and light transparent images of the eggs (**a-1** and **a-2**) at early yolk-pyramid stage with about 120 yolk-pyramids; Explanation (**b-0**) and light transparent images of eggs (**b-1** and **b-2**) are at late yolk-pyramid stage with about 360 yolk-pyramids. SEM images show the surface (**c**) and internal yolk-pyramid structures of eggs **d–e** and **f–h**; these are SEM images of late-developing eggs showing long segmented germ-band. Scale bar = 200 μm

Bengtson et al. 1986) based on sclerites recovered from rocks in the Atdabanian Stage, Siberia. These remarkable sclerites, with their striking resemblance to complex arthropod eyes, created great confusion before the first complete animal was recovered at 1989 in the Maotianshan Shale and designated *M. sinicum* (Chen et al. 1989a). *M. sinicum* is a worm-like animal, with a maximum length of 9 cm (Figs. 55 and 56). The body is basically cylindrical in shape, except for the tapering head. The body of the animal is differentiated into two

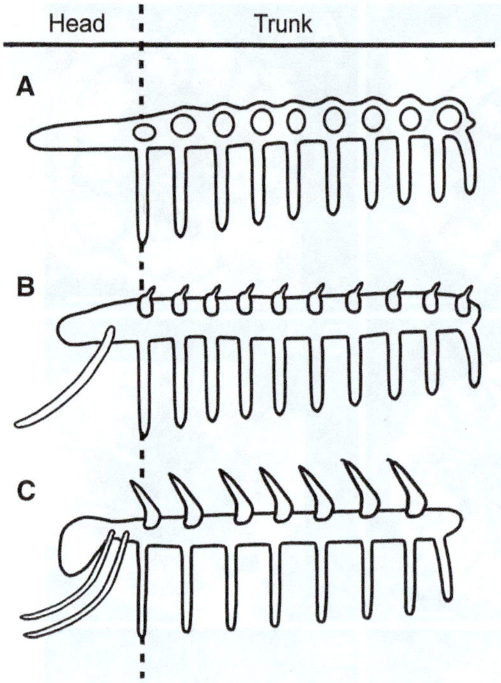

Fig. 54 Body organization in extinct stem lineages in Panarthropoda. (**a**) non-segmented head, exemplified by *Microdictyon sinicum*; (**b**) one-segmented head with a pair of head appendages, exemplified by *Onychodictyon ferox*; (**c**) two-segmented head with two pairs of head appendages, exemplified by *Hallucidegia fortis* (adapted from Chen 2009)

distinctly different parts, the head and the trunk. The region anterior to the first pair of sclerites represents the long head tapering about one-fifth the length of the body to a terminal mouth. The trunk is cylindrical and annulated, with a conspicuously rounded swelling, dorsal to the sclerites. It ends with a small protrusion referred to as the "anal protrusion", with an anus opening at its terminus. The trunk is subdivided into eight subequal segments with nine pairs of sclerites, and is supported by 10 pairs of unjointed limbs. The sclerites vary in both size and morphology, with the anterior pair the smallest, and increasing in size posteriorly. The first pair is vertically elongate, pairs 2–8 are vertically oval with some being tear-drop in shape; the last pair have a sub-quadrant form. The sclerites are situated on the lateral trunk, dorsal to the limbs, and each corresponds to a limb, except for the last pair, which corresponds to two pairs of limbs. Limbs are tubular, marked with faint annulations and have two terminal claws. Bent claws are directed posteriorly, except those on the last pair of limbs, which are directed anteriorly. The first pair of limbs is longer and more slender than the remaining pairs of limbs, sub-equal in length. The centroventrally situated gut is a simple tube opening at both ends of the body. *P. inermis* (Fig. 57) has the least complex body form, shaped as a simple cylinder supported by six pairs of limbs. The claws are oriented as in *M. sinicum*.

O. ferox (Figs. 58 and 59a, b) represents the group having a single, segmented head. The trunk has 10 segments and 11 pairs of limbs, long papillae on body and limbs, and 10 pairs of plates (sclerites) on the dorsal of the body. Anterior to the first pair of light sclerites

Fig. 55 Restoration of Early Cambrian *Microdictyon sinicum*

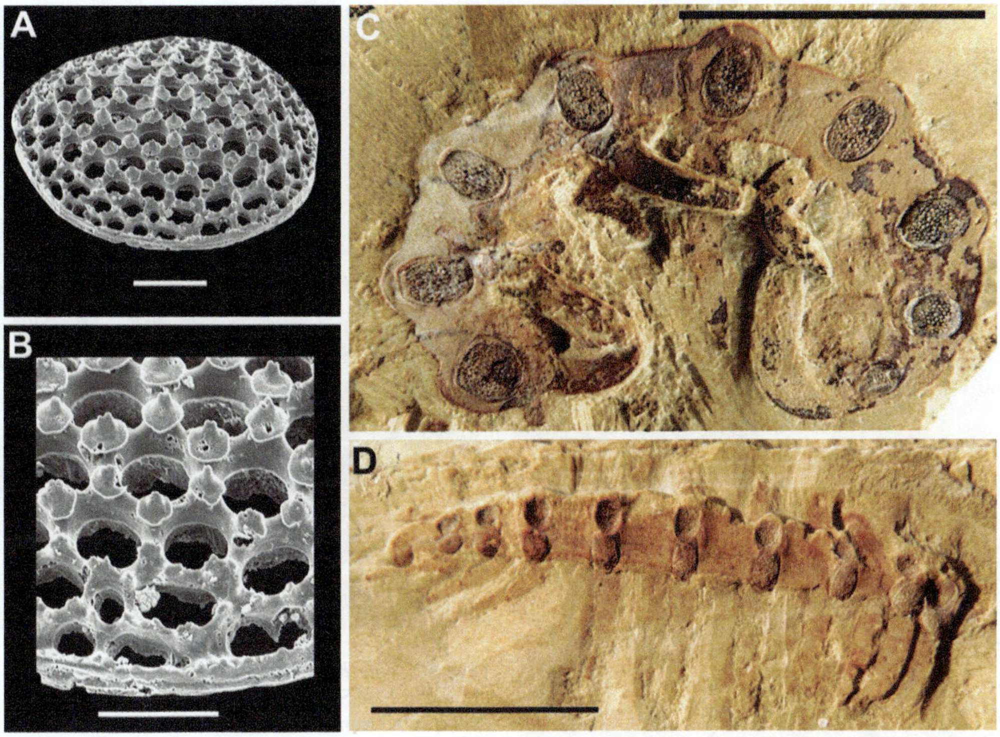

Fig. 56 SEM images of isolated sclerites of *Microdictyon* sp. photographs (**a**, **b**) of laterally (**c**) and dorso-laterally compacted (**d**) complete specimens of *Microdictyon sinicum* from Early Cambrian Moatianshan Shale. Scale bars = 100 μm (**a**, **b**) and 1 cm (**c**, **d**) (**a**, **b**, courtesy of X. G. Zhang)

is a short, smooth, barrel-shaped head, with a mouth opening at the terminus. A pair of head appendages (antennae) extends anteriorly from the lateral rear portion of head. The cylindrical trunk bears ten pairs of unjointed limbs and the same number of dorsolateral sclerites dorsal to the limbs. Each sclerite has an oval base attached to the trunk and a conspicuous spine-like structure protruding dorsally near the dorsal margin of the sclerite. The trunk is annulated, with four annulations per segment, but three in the last segment. Each annulation has approximately ten or more conspicuous cirri aligned transversely in the middle of each annulation. The limbs are relatively stout and annulated, with a pair of terminal claws, directed posteriorly except for the last pair, which is oriented anteriorly. Each limb has 10–12 annulations. Each limb annulation had two cirri, protruding from both the outer and inner sides of the limb. A slender simple gut extends through the entire length of the body to both termini.

L. longicruris is small (c. 2 cm) (Figs. 60 and 61). Its barrel-shaped head has a pair of appendages extending anteriorly. The trunk is cylindrical, supported by 13 pairs of appendages. The segments are wider in the central part, but narrow both anteriorly and posteriorly. There are eight trunk annulations per segment; this decreases to three both anteriorly and posteriorly. The trunk annulations are subequal in width, but are much wider near to the boundary of the two adjacent segemts in order to bear three small dorsal sclerites. The trunk limbs are slender, tapering distally, with 4–5 terminal claws. They are faintly annulated, with a pair of setae on each annulation. The setae are directed both anteriorly and posteriorly.

The two species *C. catenulum* (Figs. 62 and 63a–c) and *H. fortis* (Figs. 64 and 65) represent the two-segmented-head taxon. They display an advanced body plan among the stem-lineage panarthropods. *C. catenulum* is small, at 3 cm in length. The body is cylindrical and bipartite. The head is small, with an elongate oval shape, and a mouth opening anterioventrally. It has two pairs of head appendages, extending anteriorly from the narrow rear portion of the head. The cylindrical trunk is supported by 23–25 pairs of unjointed limbs and is armed with the same number of

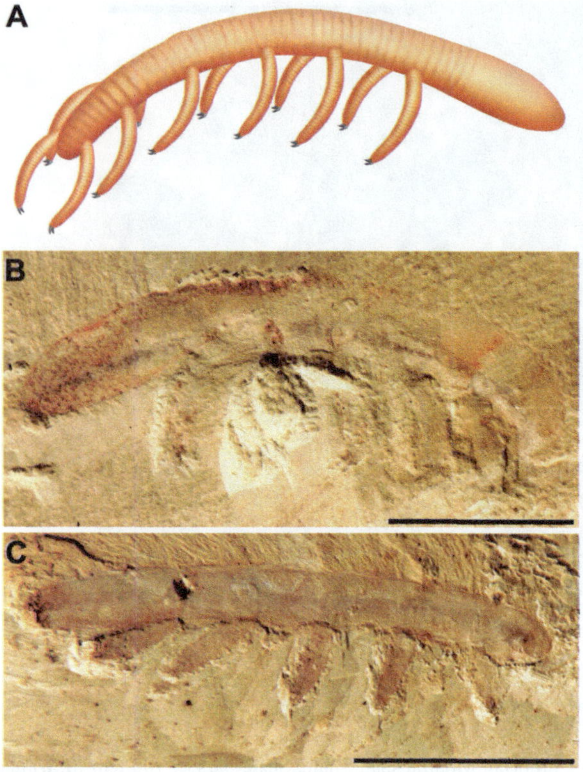

Fig. 57 Restoration (**a**) and photographs of laterally compacted (**b**) and latero-dorsally compacted (**c**) *Paucipodia inermis* from Early Cambrian Maotianshan Shale. Scale bar = 0.5 cm

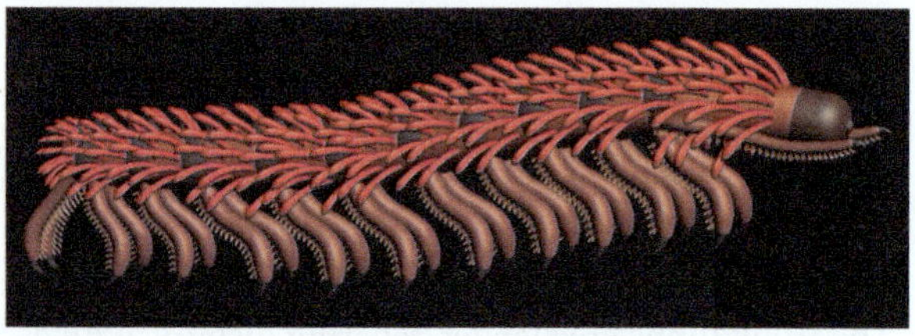

Fig. 58 Restoration of Early Cambrian *Onychodictyon ferox*

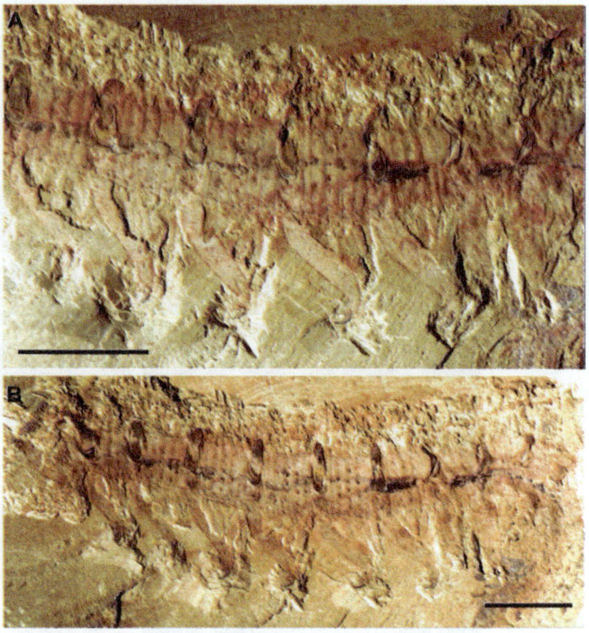

Fig. 59 *Onychodictyon ferox* from Early Cambrian Maotianshan Shale. (**a**, **b**) part and counterpart of a laterally compacted trunk. Scale bar = 1 cm

Fig. 60 Restoration of Early Cambrian *Luolishania longicruris*

sclerite pairs, dorsal to the limbs. The limbs are short with two distal claws directed posteriorly, except for the last pair, which is directed in the opposite direction. The sclerites are in a peltate form in the adult (Fig. 63a) but are V-shaped in the juvenile (Fig. 63c), with the narrow angle directed ventrally. The gut is a simple, straight tube that opens at both ends. *H. fortis* in the Maotianshan Shale resembles the type species *Hallucigenia sparsa* (Walcott 1911a; Conway Morris 1977c) from the Middle Cambrian Burgess Shale, in many of its characters including the shape of its sclerites and the numbers of its head and trunk appendages. However, the oval-shaped head of *H. fortis* differs significantly from that of *H. sparsa* as it resembles the elongated head of *M. sinicum*, narrowing anteriorly. This tail-like "head" has been broadly accepted as a tail. In addition to this anterioposterior reversal, the dorsal spine-like sclerites in *H. sparsa* were thought to be ventral limbs, so the animal was previously interpreted as being in an upside-down position. The animal was popularized with a reconstruction resulting from

Fig. 61 *Luolishania longicruris* from Early Cambrian Maotianshan Shale showing anteriorly extended head appendages (**a**), annulated appendages with a pair of setae in each annulation (**b**), a simple straight gut (**a**, **b**) and each trunk segment bearing three small dorsal sclerites (**a**) (**b**, courtersy of XG Hou). Scale bar = 0.5 cm

misinterpretation (Fig. 66). The complete specimens in the Maotianshan Shale clearly document the body form of *H. fortis* as small (2–3 cm long) with a characteristic two-segmented oval-shaped head, a trunk supported by seven pairs of slender limbs, and was armed with the same number of spine-like sclerites (Figs. 64 and 65). The inner side of the head is kidney shaped, with a narrow neck-like rear portion. Two head appendages extend anteriorly from the rear portion of the head; they are fairly slender and relatively long, extending far beyond the front margin of the head (Fig. 65a). Like *Cardiodictyon*, the mouth of *H. fortis* also opens anterioventrally. The trunk is slender, thick in the middle, and narrowing both anteriorly and posteriorly. Horn-shaped sclerites ornamented with fine striatae, extend dorsolaterally from the trunk, with the concave side directed anteriorly in the anterior sclerites and posteriorly in the posterior ones.

The tardipolypods were interpreted to have been epibenthic animals living on the surface of semi-firm substrates. It is interpreted that most species used their limbs to raise their bodies above the sea floor. *L. longicruris* was the exception by spreading its limbs nearly

Fig. 62 Restoration of *Cardiodictyon catenulum*

horizontally, keeping its body close to the sea floor. The muscular body wall of the animals is like that of living soft-bodied bilateral worm-like organisms, arranged in circular and longitudinal muscle layers. The alternating contraction of the longitudinal and circular muscles around an enclosed fluid-containing space produces locomotion. The rearward-directed terminal claws favour forward locomotion; the claws in the posterior limbs anchor the organism to the surface, while the body is elongated with the contraction of the circular muscles. Alternatively, the claws in the anterior limbs anchor the animal while the body is shortened by the contraction of the longitudinal muscles. The anterior forward-pointing claws on the last pair of limbs suggest that the animal was capable of backward locomotion by anchoring the claws in the last pair of limbs while the longitudinal muscles contracted.

Microdictyon and *Paucipodia* are both found on the dorsal surface of the pelagic medusiform soft-bodied organism known as *Stellostomites*, suggesting that it was a pseudopelagic organism (Fig. 67). Anterior of long, slender appendages are usually directed anterilry, favouring their adhesive attachment to the pelagic host.

The small terminal mouth suggests that tardipolypods fed on microscopic bacteria, algae, and fine organic detritus. Some tardipolypods, such as *Microdictyon* and *Paucipodia*, may have fed on the body fluids of their host, *Stellostomites*, by penetrating the body wall of the host with an oral stylet-like structure, although we have not yet found evidence of stylets. The narrow anterior end of the elongated

Fig. 63 *Cardiodictyon catenulum* from the Early Cambrian Maotianshan Shale. (**a**) lateral compaction of a complete specimen showing an oval head with two head appendages, a simple straight gut and shield-shaped sclerites; (**b**) laterally compacted head and anterior portion of trunk showing two pairs of head appendages; (**c**) laterally compacted juvenile specimen showing triangular sclerites. Scale bars = 1 cm (**a**), 0.5 cm (**b, c**)

Fig. 64 Restoration of Early Cambrian *Hallucigenia fortis*

Fig. 65 *Hallucigenia fortis* from Early Cambrian Maotianshan Shale, showing oval head with two pairs of head appendages (**a, b**), seven pairs of trunk spine-like sclerites and appendages (**b**). Scale bar = 1 cm

head of *M. sinicum* may have allowed the animal to penetrate the soft mud, to feed on buried carcasses.

Sclerites are present in most tardipolypods, except *Paucipodia*. Although the sclerites vary in both morphology and placement, they are, interestingly, all restricted to the trunk, occurring in a metameric pattern. Scleritization is deeply rooted in the stem-lineage panarthropods and is not a novel attribute of arthropod evolution. Sclerites possibly played a protective

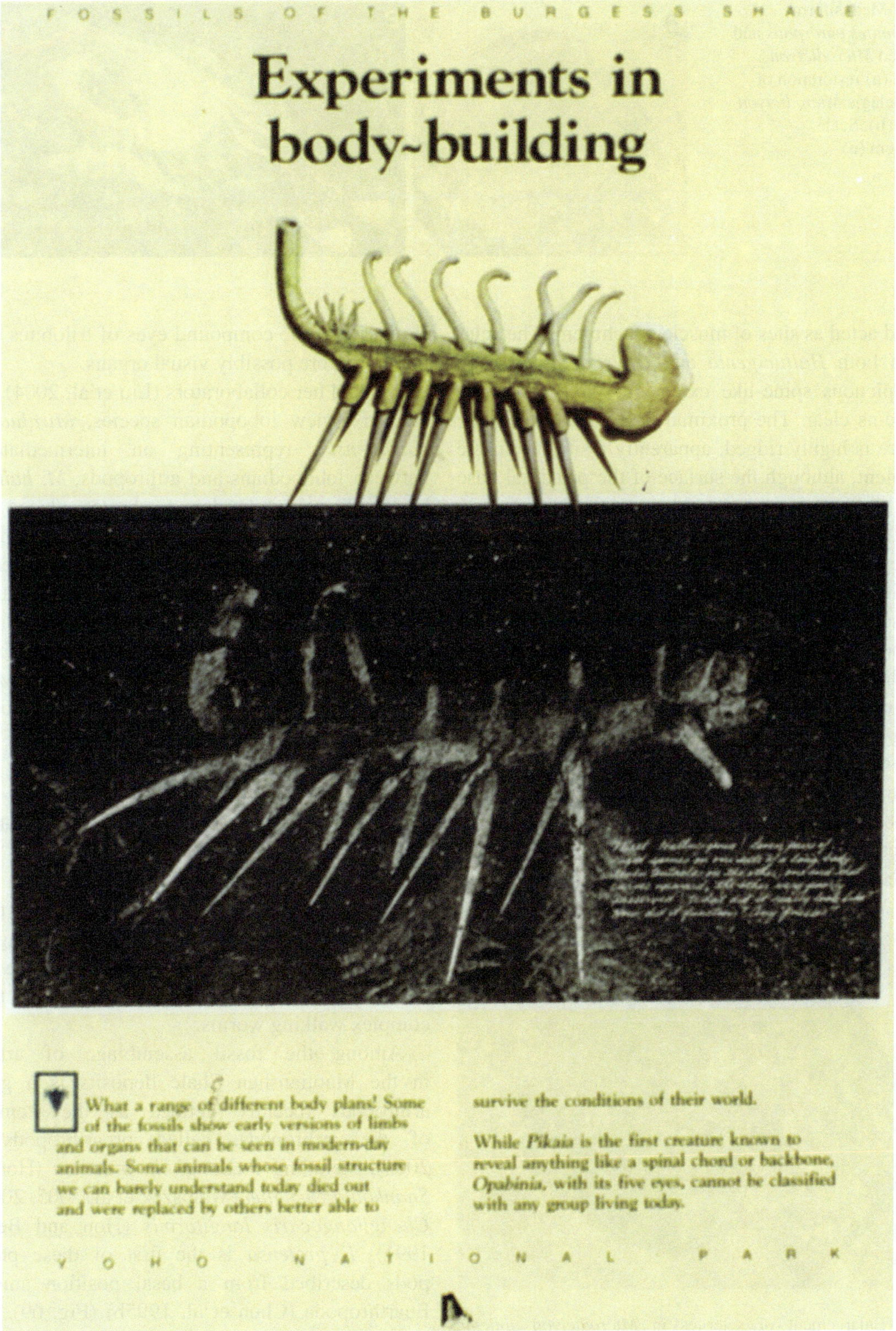

Fig. 66 Fanciful, upside-down reconstructions of the Middle Cambrian *Hallucigenia sparsa* (reproduced from the poster in front of the administrative centre at Park Field, Canada)

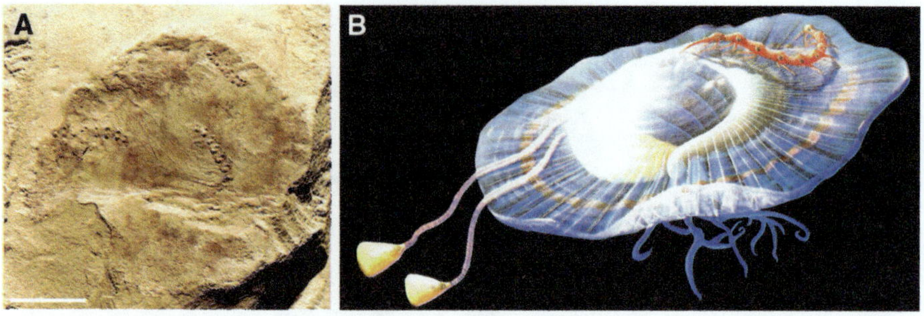

Fig. 67 Medusiform *Stellostomites eurorphus* and associated *Microdictyon sinicum*. (**a**) restoration of pseudopelagic *Microdictyon sinicum* (**b**). Scale bar = 1 cm (**a**)

role and acted as sites of muscle attachment. The sclerites in both *Hallucigenia* and *Onychodictyon* have a conspicuous spine-like extension; their protective role seems clear. The proximal base of the horn-like structure is highly ridged, apparently a site of muscle attachment, although the surface of the proximal base is small.

The sclerites in *Microdictyon* are exceptionally complex, made up of over 100 repeating cylindrical units that extend vertically through the sclerites and resemble the ommatidia of the compound eyes of arthropods (Fig. 56a, b; Fig. 68). The repeating units are hollow; these structures may represent dissolved calcite crystal cones, separated by solid cuticular walls (Zhang and Aldridge 2007). The interspaces between the crystal cones are spiky, mushroom-shaped structures, and the crystal cones have a convex outer surface (Fig. 68). Although the sclerites in *Microdictyon* may have acted as sites of muscle attachment and possibly in protection (Zhang and Aldridge 2007), their striking similarity to the compound eyes of trilobites suggests that they were possibly visual organs.

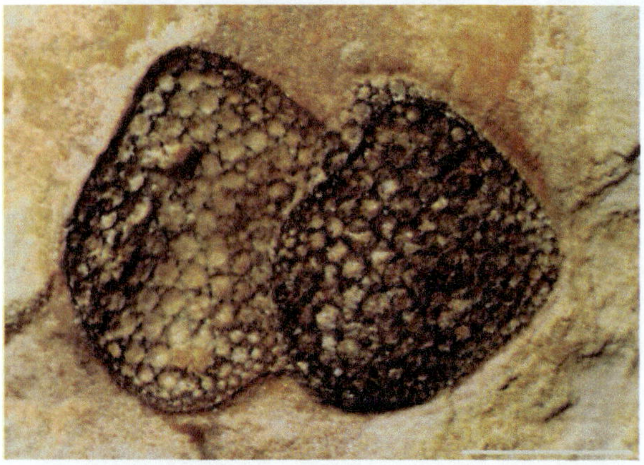

Fig. 68 Enlargement of sclerites in *Microdictyon sinicum* showing convex outer surface of the cylindrical units and nodular-like structure in the interspaces between the cylindrical units. Scale bar = 0.2 cm

Liu and her collaborators (Liu et al. 2004) recently erected a new lobopodian species, *Miraluolishania haikouensis*, representing an intermediate form between lobopodians and arthropods. *M. haikouensis* has small, paired, hardened sclerites, placed dorsally on each pair of appendages, and the "possible dorsal sessile eyes" seem to represent the anteriormost paired sclerites. *M. haikouensis* shares features with *Luolishania*, including its body form, number of appendages, and small, dorsally situated, hardened sclerites. In 2009, Chen reinterpreted *M. haikouensis* as a possible synonym or sister species of *L. longicruris*.

The Evolutionary Transition from Walking Worms to the First Arthropods

The lack of fossil evidence of the first arthropods has resulted in much controversy surrounding studies of arthropod origins. Most researchers agree that the ancestor of the arthropods is found among the least complex walking worms.

Among the fossil assemblage of arthropods in the Maotianshan Shale deposits is a group of arthropods recently defined as the stem group of the arthropods, called proarthropods (Chen 2004), including *Fuxianghuia protensa* (Hou 1987), *Shankouia zhenghei* (Waloszek et al. 2005, 2007), and *Chengjiangocaris longiformis* (Hou and Bergström 1991). *F. protensa* is the first of these proarthropods described from a basal position among the Euarthropoda (Chen et al. 1995b) (Fig. 69). It has a head composed of only two tergite-bearing segments (Chen et al. 1995b; Hou and Bergström 1997), and the second tergite is interpreted as bearing two pairs of

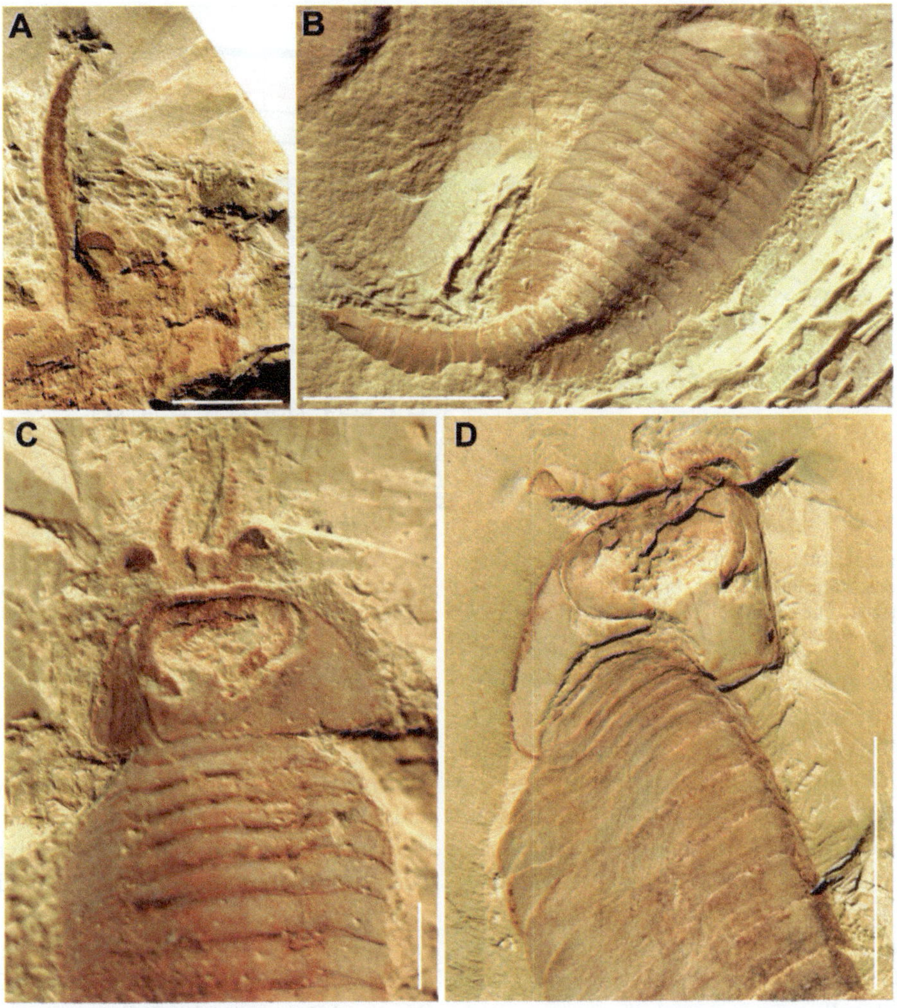

Fig. 69 *Fuxianhuia protensa* from the Early Cambrian Maotianshan Shale showing a small visual segment (**a**, **c**, **d**) with stalked eyes (**a**–**d**); an antennal segment bearing a relatively large head shield (**b**–**d**), a pair of short stout antennae (**a**, **c**, **d**) and a pair of crescent-shaped diverticula lying below the head shield but above the tergite (**c**, **d**). Scale bars = 0.2 cm (**a**, **c**), 1 cm (**b**, **d**)

appendages, representing two segments. However, the identity of the "second pair of limbs" is refuted by evidence of their internal location between the dorsal and ventral cuticles (Waloszek et al. 2005; Fig. 69d). The authors have reinterpreted them as a pair of diverticula in the anterior region of the digestive tract. The head is reinterpreted as two segmented, with an anterior visual segment and a posterior antennal segment. A similar two-segmented head is also recorded in both *C. longiformis* and *S. zhenghei*.

The discovery of a two-segmented head, together with lobopod-like trunk limbs, among the Proarthropods is of great interest, presenting a novel perspective on the evolutionary origin of the arthropods (Fig. 70). The anatomical characters of the proarthropod group establish a bridge linking the tardipolypods and the euarthropods. The two-segmented head of the first arthropods suggests that the arthropod ancestor is among the limb-bearing worm-like organisms resembling the *Hallucigenia* + *Cardiodictyon* clade, and not among the living walking worms. The extant walking worms comprise two phyla (Onychophora and Tardigrada). The phylum Onychophora (or velvet worms) is characterized by a cylindrical body, supported by 13–43 pairs of trunk appendages. The head consists of three head

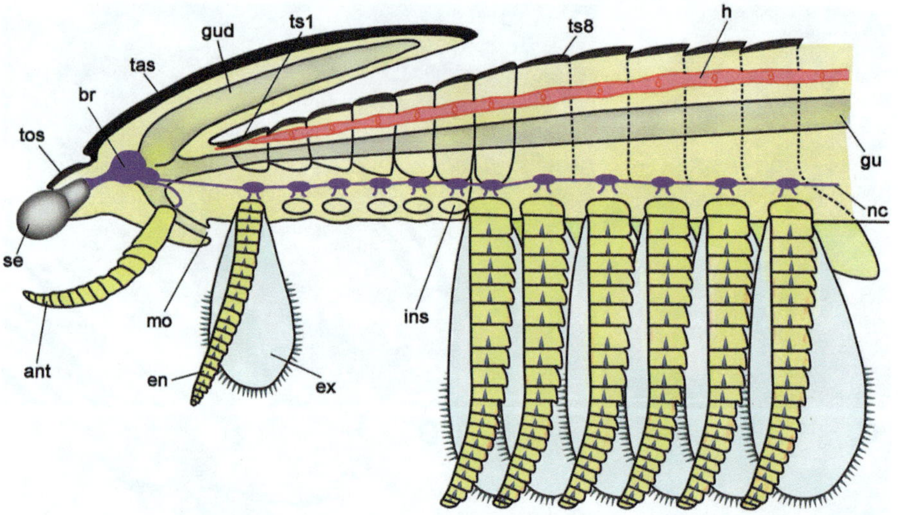

Fig. 70 Schematic restoration of the possible arthropod bauplan. Abbreviations: ant, antennae; br, brain; en, endopod; ex, exopod; gu, gut; gud, gut diverticulum; h, heart; ins, inserter; mo, mouth; nc, neural cord; se, stalked eyes; tas, tergite of antennal segments; tos, tergite of ocular segment; ts, tergite of trunk segments (adapted from Waloszek et al. 2005: Fig. 4)

segments, each carrying a pair of modified appendages corresponding to sensory antennae on the first pair, claw-like mandibles on the second pair, and oral papillae on the third pair. Tardigrades, generally known as "water bears", are microscopic walking worms with a segmented trunk and a short head. The head may have three segments with a terminal mouth and the trunk has 4–5 segments. The trunk legs are short and stubby, with 4–8 retractile claws or an adhesive disc at the tip of each leg. These two groups are highly modified walking worms with three-segmented heads and are unlikely to represent arthropod ancestors.

F. protensa is large, with an approximate length of 10 cm (Figs. 69 and 70). The head consists of a small visual segment, with an oval tergite, jointed posteriorly to the antennal segment (Fig. 70) over which a pair of stalked eyes extends ventrolaterally. The antennal segment is relatively large and covered dorsally with a fairly large tergite extending posteriorly far beyond the head–trunk boundary above the three anterior trunk segments. Extending ventrolaterally from the antennal segment is a pair of short, stout antennae. A pair of crescent-shaped structures extend lateroposteriorly from the second segment on the internal surface of the tergite. Ventral to the structure is a membrane interpreted as a sternite, a structure previously thought to be monoramous limbs, but recently reinterpreted as diverticulae extending from the mid gut in the second segment. The trunk is differentiated into three parts: the thorax, the abdomen, and the telson. The thorax consists of 16–17 segments, each carrying a pair of biramous limbs divided longitudinally into a pair of pleural lobes, and an elongated median axial region that narrows slightly rearward. The pleural lobes are wider in the middle portion but narrow both anteriorly and posteriorly. The cylindrical endopod of the biramous appendage is unjointed and closely annulated, with more than 20 annulations (Hou and Bergström 1997; Bergström and Hou 1998). The exopod is large and flap-like, adorned with short setae. The abdomen is cylindrical, made up of 14 segments lacking appendages. A small caudal tail consists of a pair of flattened paddle-like structures (uropods, similar to those in crustaceans) and a spine-like telson.

Newly discovered *S. zhenghei* is known only from Shankou, Anning (Chen 2004; Waloszek et al. 2005). The species is named after the great navigator and explorer of the Ming Dynasty, Zhenhei, who was born near the fossil site in Shankou village. The animal (Fig. 71) on average is 5–6 cm long, reaching a maximum of 7.5 cm. The head resembles that of *F. protensa*; it has two segments, a small first segment articulated posteriorly with the second segment. The second segment has a large tergite, extending posterior to the six anterior trunk segments. The antennae on the second segment are short and stout, comprising 14–16

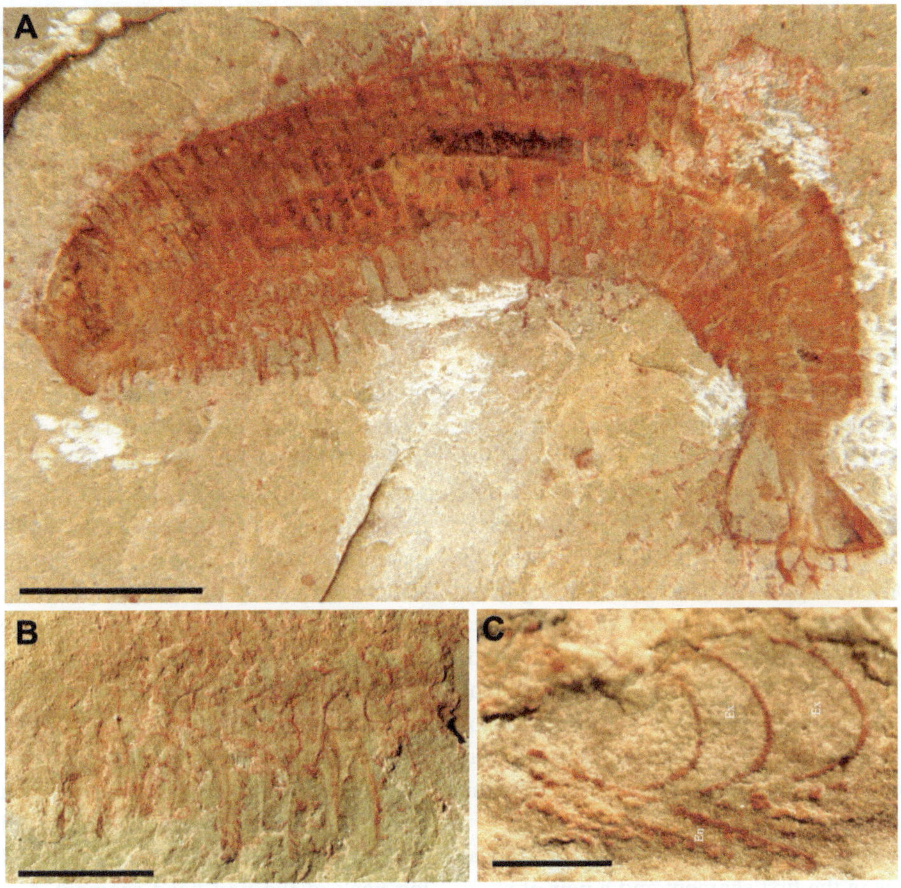

Fig. 71 *Shakouia zhenghei* from the Early Cambrian Maotianshan Shale. (**a**) dorsal view of a complete specimen; (**b**) enlargement showing multiple annulated tubular-shaped limbs; (**c**) enlargement showing flap-like exopods and tubular endopods. Scale bars = 1 cm (**a**), 0.5 cm (**b**), and 0.2 cm (**c**)

segments with no setae. Unlike *F. protensa*, the trunk of this species is not differentiated into a limbless abdomen in its rear portion, but comprises 43 segments, each bearing a pair of biramous appendages. Like *F. protensa*, it is divided longitudinally into pairs of pleural lobes and a median axial region. The pleural lobes narrow anteriorly and posteriorly in its anterior and posterior portions, respectively, but the pleural lobes in its main middle portion are relatively wider. The trunk appendages resemble those of *F. protensa*, comprising large, flap-shaped exopods with short setae at their margins. The endopods are tubular, with at least 17 annulations (Bergström and Hou 1998). The caudal tail is large, comprising a pair of large, flattened paddle-like structures and a spine-like telson. A pair of small, flattened paddle-like structures is also present behind the larger uropods.

Affinities and the Ground State of the Arthropods

Paired, jointed, segmented appendages have previously been accepted as one of the key characters defining the arthropods. The term "Arthropoda" originates from the Greek αρθρον (arthron) meaning "joint", and ποδός (podos) meaning "foot", together meaning "jointed feet". However, the proarthropods have unjointed limbs, inconsistent with the definition of the Arthropoda, although they share the characteristics of a segmented body and a chitinous exoskeleton. The proarthropods have a bi-segmented head, corresponding to the pre-oral part of the arthropod head, the post-oral portion, consisting of at least three segments. Interestingly, the head segmentation shows a close affinity to that of a walking, worm-like stem group of Panarthropods, the *Hallucigenia–Cadiodyctyon* clade, sharing the same number of head segments. The proarthropods most probably represent the missing historical link between the walking worms and the arthropods, thus shedding new light on the evolutionary origins of the arthropods.

Despite sharing similar head segmentation, there is a large morphological gap in the heads of the worm-like ancestors and the first arthropods (exemplified by the proarthropods). The transition from the two-segmented head of the worm-like ancestors to the first arthropod head required quantum leaps through multiple, synchronous events, including the evolution of the first head appendages into stalked eyes; the specialization of the second head appendages into sensorial organs, known as "antennae"; and the displacement of the mouth to a ventral position in the antennal segment.

The head in the basal arthropods, as exemplified by the proarthropods, is small and two segmented, with an anterior visual segment and a posterior antennal segment. It has been proposed that the anterior of ancestral arthropods, known as the "acron", was non-segmented. The acron hypothesis is refuted by the study of proarthropods.

Head sensory organs consist of a pair of stalked compound eyes in the first segment and a pair of short and segmented antennae. Apparently, image-forming eyes had not yet appeared in the tardipolypods; they relied on a simple visual system, known as "eye-spots". Some authors have reported eye-like structures in *Luolishania* (Liu et al. 2004, 2008) but these may, in fact, be dorsal sclerites. The appearance of compound eyes in all three proarthropod species suggests that compound eyes are apomorphic in the arthropods. Whether the original compound eyes were stalked or sessile is an ongoing controversy. The eyes of all proarthropods have a clear, short, tubular stalk. The compound, stalked eye arguably characterizes the ground state of the arthropod eye. Tubular, possibly segmented, eye stalks represent tubular extensions of the body wall, homologous to tubular limbs. This argues for the segmental origin of the anteriormost head region in the arthropods, evolving from the anteriormost, limb-bearing head segment of a worm-like ancestor of the *Hallucigenia–Cadiodyctyon* clade.

The flagelliform antennae have long been accepted as the original arthropod version of the uniramous antennae. Long and multisegmented flagelliform antennae occur widely among the trilobites and extant arthropods; it is assumed that the antennae have a sensorial function. The flagelliform hypothesis, however, comes under scrutiny as the proarthropods with short, stout antennae with a limited number (about 15) of segments (Fig. 69a, c). The short, limb-shaped antennae occurring in all proarthropods and in some stem-lineage euarthropods (*Canadaspis*, *Primicaris*, the bradoriids, and *Ercaia*), represent the ground state of the arthropod antennae (cf. Maas et al. 2004).

In the ground state of the panarthropods, as exemplified by the tardipolypods, the mouth was situated terminally on the anterior segment. The ventral anterior location of the mouth in *Hallucigenia* and *Cadiodyctyon* suggests that the relocation of the mouth has an early history, rooted in the stem-lineage panarthropods. In contrast, the mouth in the ground state of the arthropods, as exemplified by the proarthropods, is located ventrally and opens posteriorly in the rear part of the antennal segment. The displacement of the mouth to a new position represents one of the major events during the evolutionary transition from the worm-like ancestor to the arthropods, although it was initiated in the pre-arthropod evolutionary stage. The looped anterior digestive tract in arthropods argues that the mouth displacement resulted from the curving back of the first head segment. *C. catenulum* and *H. fortis* (Figs. 62, 63, 64, and 65 have an expanded and ventrally curving head and an anterior, ventrally situated mouth, representing an early stage of mouth displacement.

Scleritization occurs broadly among the stem-lineage panarthropods, forming metamerically arranged hard cuticles on the dorsolateral surface (few on the dorsal side) of the trunk, but the main part of the body remains soft. Arthropodization is characterized by the formation of hard, jointed, segmented cuticles that cover the entire dorsal surface, whereas the trunk limbs remain soft. This event, exemplified in the proarthropods, was the evolutionary development allowing for the origination of the arthropods. The formation of jointed sclerites had far-reaching effects on the evolutionary changes in the anatomy, physiology, ecology, and behaviour of the arthropods (Ruppert et al. 2004).

The ground state of the panarthropod appendages was monoramous and unjointed. The trunk appendages in the arthropod ground state are biramous but their endopods remain unjointed (Fig. 71). The head appendages are modified into sensory organs, the stalked eyes and antennae, although they remain in the monoramous state. The exopods are an evolutionary novelty in the arthropods and the first exopods, exemplified in the proarthropods, were flap-like and without gill-like structures, apparently not used for gas exchange but rather in swimming locomotion. The exopods are extended pouches of the body wall adjacent to the base of the limbs. Morphologically, they resembled insect wings, suggesting that the developmental process of the exopods is similar to that of wing development. One might suggest that co-ordinating a genetic cassette for development of the insect wing is very ancient, used by the first arthropods in exopod development and later recruited by insects for wing morphogenesis.

The ground plan of the arthropod trunk appendages is biramous, as exemplified by *S. zhenghei* and possibly *C. longiformis*. There is little differentiation among the different segments, except in *F. protensa*, in which the trunk is differentiated into a limb-bearing thorax and a limbless abdomen. The limbless abdomen is a secondarily derived feature, probably the result of the repressed expression of the *Distalless* gene. The ground plan of the arthropod endopod is a simple tubular form, consisting of multiple segments. It is not very different from the ground plan of the panarthropod limb, except for the pivotal joints between two adjacent segments in the endopods of the arthropod ground plan. The presence of pivotal joints suggests the presence of intrinsic muscles in the first arthropod endopod.

Evolutionary Transition from the First Arthropods to the Euarthropods

Waloszek established the Euarthropoda to include all extant arthropods and a number of extinct arthropod lineages, all characterized by a composite head and bipartite trunk limbs (composed of an expanded basipod and with six or fewer segments in the distal portion) (Waloszek 1999; Maas and Waloszek 2001). A monophyletic status was proposed for the Euarthropoda by the authors.

The significant morphological gap between the proarthropods and euarthropods suggests the presence of an intermediate evolutionary stage. The Lower Cambrian Maotianshan Shale has yielded diverse extinct arthropods that fill the morphological gap between the first arthropods and the euarthropods. These intermediate forms represent different evolutionary stages towards the Euarthropoda in both the head and appendages. For instance, some forms have possibly proarthropod-like two-segmented heads but their trunk appendages are tubular and multisegmented. Some forms have a composite head and tubular, multisegmented limbs. Interestingly, the limbs in some forms have a bipartite design but their basipod is multisegmented. The evolutionary transition towards the euarthropods was much more complex than previously thought and was marked by two parallel events, cephalization and arthropodization. These two events occurred in asynchronous and step-wise ways.

During this transformation, the antennae on the antennal segment either remained unchanged or were specified as a flagelliform sensorial organ or a raptorial limb-like feeding organ. The development of the raptorial limb-like antennae was the key evolutionary novelty that led to the origin of the Chelicerata.

The ground state of the post-oral head of the Euarthropoda, exemplified by the stem-lineage euarthropods in the Maotianshan Shale, is a three-segmented form, and each of the segments carries a pair of indefinitely differentiated biramous appendages, appearing to be homologous to the post-oral head appendages, and trunk appendages. The modification of the post-oral head appendages into feeding organs is the key evolutionary event leading to the Crustacea. This modification occurred in multiple asynchronous steps. As exemplified by *Ercaia minuscule* from the Maotianshan Shale, the modification of the first pair of post-oral biramous appendages into the second antennae was the first event on the path to the Crustacea (Chen et al. 2001), suggesting that the Crustacea is deeply rooted in the evolutionary transition towards the first Euarthropoda.

The design of the post-antennal limbs of the euarthropod ground plan has been defined as bipartite, with an expanded proximal portion (called the "basipod") and a distal portion (consisting of seven or fewer segments). The bipartite post-antennal limb is associated with feeding and locomotion functions (Waloszek and Müller 1990; Waloszek 1993). It has been interpreted as monophyletic and the key novel event in the origin of the Euarthropoda. The Lower Cambrian Maotianshan Shale has yielded many extinct, morphologically disparate arthropod groups, with tubular multisegmented post-oral limbs. They include the first Mandibulata (*E. minuscule*), the stem lineages of the Chelicerata (*Leachoilia*, *Fortiforceps*, and *Tanglangia*), and other extinct euarthropods (*Saperion*, *Canadaspis*, *Isoxys*, bradoriids, and *Primicaris*).

Among the intermediate forms, *Canadaspis* is a macroscopic bivalved arthropod that is well represented in the Middle Cambrian Burgess Shale (Briggs 1992, 1994; Briggs et al. 1994), but is much less common in the Lower Cambrian Maotianshan Shale where it is represented by *Canadaspis laevigata* (Hou and Bergström 1991; Chen 2004) (Fig. 72). The body of the species consists of a small head, thorax, and abdomen, mostly covered by a large, unjointed, bivalved carapace, except for its abdomen, which is not covered by carapace. Its head probably resembles that of the proarthropods, with two segments and a pair of stalked eyes on the first segment and a pair of short antennae on the second segment. The carapace may be comparable to the large head shield of the proathropods, both representing a modified giant second tergite. The thorax comprises about 10 segments, each carrying a pair of biramous appendages. The abdomen is

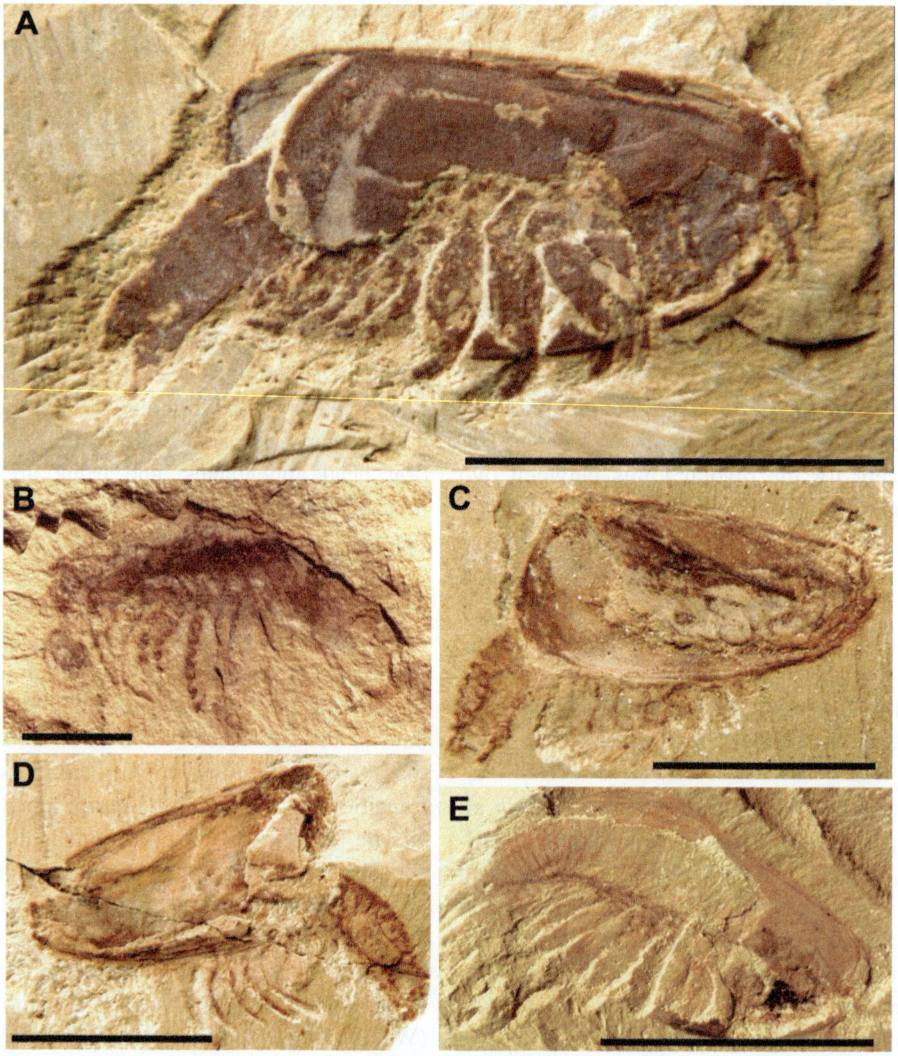

Fig. 72 *Canadaspis laevigata* from the Early Cambrian Maotianshan Shale showing its oval bivalved carapace (**a**, **c–e**) and elongate body (**b**, **e**) attached to the anterior part of the carapace—with short front portion (**a**, **c–e**); the thorax, a region free from the carapace (but beneath it), bears biramous appendages (**a–e**); the abdomen was legless, not covered by the carapace (**a**, **c–e**). Scale bar = 1 cm (**a**, **c–e**), 2 cm (**b**)

rounded and legless, consisting of about ten closely arranged segments, each with a pair of short caudal spines. The exopods are flap like, with short setae at the margins. The endopods are not well preserved in the Lower Cambrian Maotianshan Shale. They probably resemble those in the Middle Cambrian Burgess Shale species *Canadaspis perfecta*, which has 14 limb segments and is bipartite, with an expanded proximal portion (formed by seven proximal limb segments) and a slender distal portion.

Primicaris lavaformis was named by Zhang et al. (2003), and represents larva-like microscopic arthropods of 0.7–5 mm in length (Figs. 73 and 74). The body is entirely covered with a dorsal shield and is differentiated into two parts, the anterior and posterior parts, corresponding to the head and trunk, respectively. The head is broadly rounded with a smooth margin and forms a conspicuous, relatively broad, cephalic doublure fold. It represents the anterior 1/3 to 2/5 of the body, with an elongated hypostome that joins the doublure on its anterior margin and extends across the entire length of the head. A pair of elongated antennae extends anteriorly from the lateral surfaces of the hypostome at 3/5 of its length from the front. The antennae consist of about 15 segments, each carrying conspicuous setae at their joints. The head possibly has two segments. The trunk is characterized by a serrated margin on the dorsal shield, with approximately 15 segments, and each serration of the dorsal shield corresponds to a trunk segment. The trunk appendages are biramous, with elongated and gill-bearing exopods. The endopods are sub-tubular in form and multisegmented.

Cindarella is a mono-specific genus represented by *Cindarella eucalla* (Chen et al. 1996; Ramsköld et al. 1997; Chen 2004), known from a few well-preserved specimens (Figs. 75 and 76). The head shield is hemispherical in outline and extends posterior to the anterior part of the trunk. The head segmentation is not yet clear, but it is considered to have three post-oral segments. The antennae are whip like and extend anteriorly from the sides of the anterior part of the hypostome. The eyes are large and stalked. The trunk is composed of 26 segments and the anterior five segments are "shadowed" by the head shield. The trunk appendages (including the post-oral head appendages) are all biramous. The exopods are gill-bearing and elongated. The endopods are tubular, consisting of 13 segments and a distal claw.

Saperion is a large, rarely occurring arthropod (12 cm long) (Hou et al. 1991; Chen et al. 1996; Hou and Bergström 1997; Chen 2004). The body is totally covered by a single dorsal shield, which is elongated with a broadly rounded margin both anteriorly and posteriorly (Figs. 77 and 78). The large dorsal shield is folded into a wide W-shape transversely and thus strengthened. The first segment of the body has a pair of stalked eyes ventrally and the second segment has a sub-square hypostome and a pair of short, bent antennae, extending laterally and posteriorly from the lateral

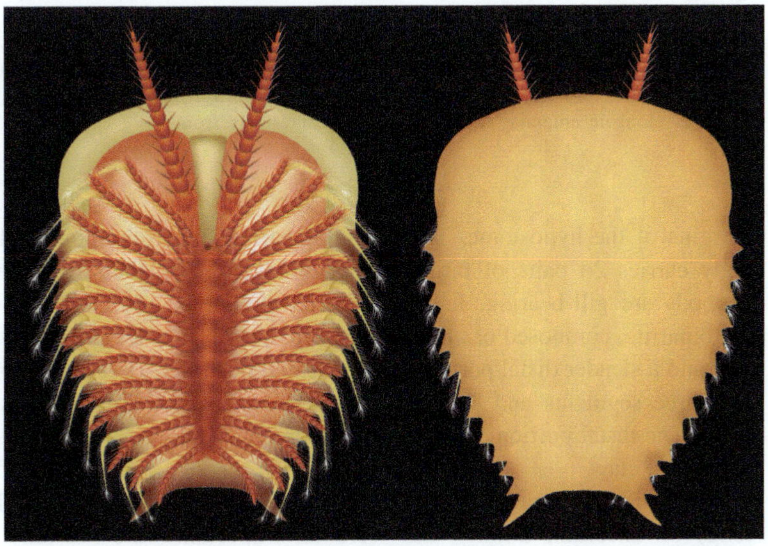

Fig. 73 Restoration of Early Cambrian *Primicaris lavaformis*

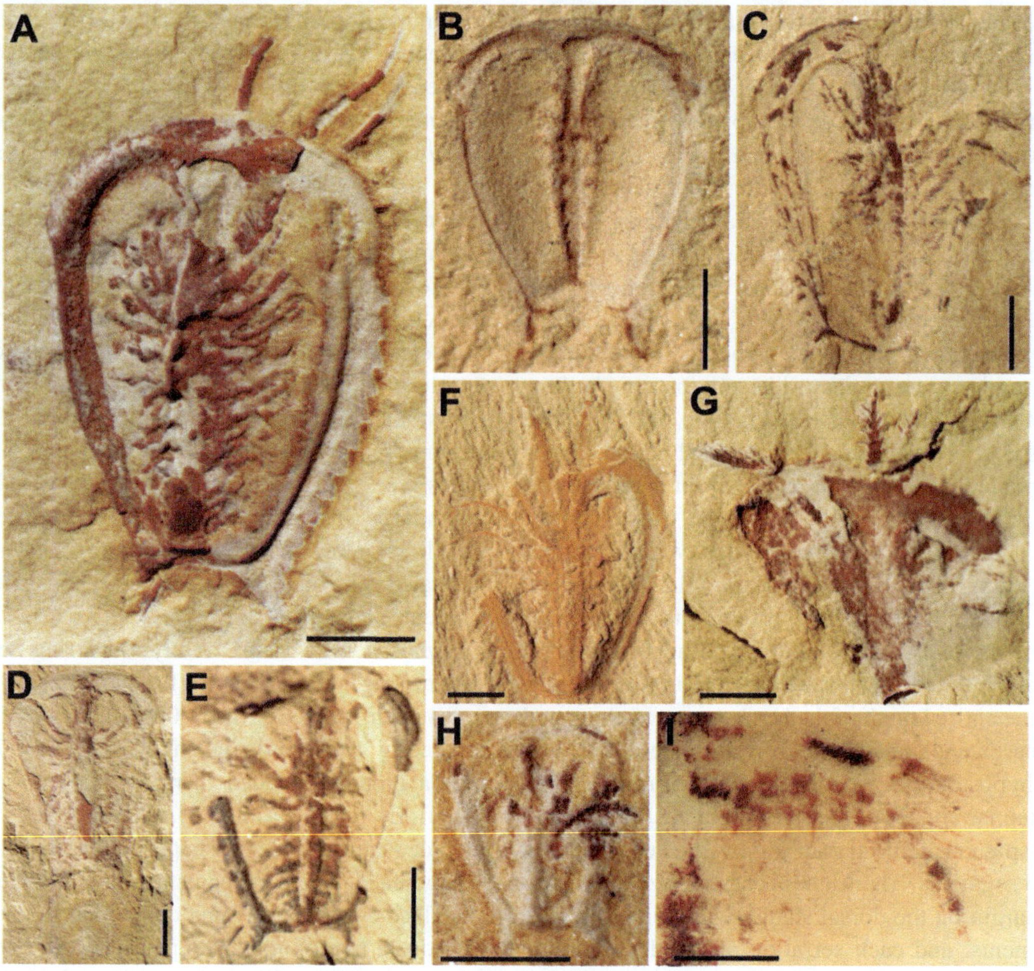

Fig. 74 Photographs of *Primicaris lavaformis* from the Early Cambrian Maotianshan Shale showing its dorsal shield differentiated into anterior and posterior parts (**a–f**, **d–h**); the anterior part was broadly rounded with smooth margin and had a conspicuous relatively broad cephalic doublure (**a–c**, FVH) with an elongate hypostome joined to the anterior margin of doublure and extending across the entire length of the anterior part (**a–b**), a pair of elongate antennae (**a**, **f–g**) extending anteriorly from the lateral sides of the hypostome at about 3/5 of its length from the front (**f**); the posterior part is serrated (**a**, **e**, **f**) and possesses approximately 15 segments; each serration of the dorsal shield corresponds to a trunk segment (**a**, **d**, **e**); the biramous trunk appendages have elongate, gill-bearing exopods and multisegmented sub-tubular endopods (**a**, **c–h**). Scale bars = 1 mm (**a–e**), 0.5 mm (**h**)

margins of the hypostome. The post-oral part of the body carries 26 pairs of biramous appendages. The exopods are gill-bearing. Interestingly, its endopods are bipartite, composed of an expanded proximal portion and a slender distal portion. The proximal portion has three segments and their inner margins are serrated. The distal portion consists of six segments with a terminal claw.

Bradoriids are small, bivalved arthropods (2–4 mm long) of uncertain affinity and were an important component of the Cambrian faunal assemblages before their disappearance in the middle Ordovician (Fig. 79). Bradoriids are currently thought to be a possible sister group of the Crustacea and the phosphatocopids (a sister group to the crown group, Eucrustacea). Their body is entirely covered by a bivalved carapace, which is preserved as either disarticulated or articulated "butterfly"-like valves. Soft-part preservation is uncommon. The bradoriids in the Maotianshan Shale are not only common but also diverse. *Kunmingella*

Fig. 75 Restoration of Early Cambrian *Cindarella eucalla*. (**a**) dorsal view; (**b**) ventral view

Fig. 77 Restoration of Early Cambrian *Saperion glumaceum*

douvillei is the dominant form; its "butterfly"-like articulated carapace is usually seen on the bedding plane in an aggregated state (Fig. 80). The few specimens preserved with soft parts show that the endopods are tubular and multisegmented (Fig. 79c). Some specimens retain clutches of eggs in their appendages (Fig. 79a, b), suggesting that the eggs of the animals were deposited and brooded within egg sacs attached to the appendages.

Bradoriids represent a major element of the primary consumers in the Maotianshan biota and the undigested residues in some giant fossil coprolites predominantly contain small carapaces related to bradoriids

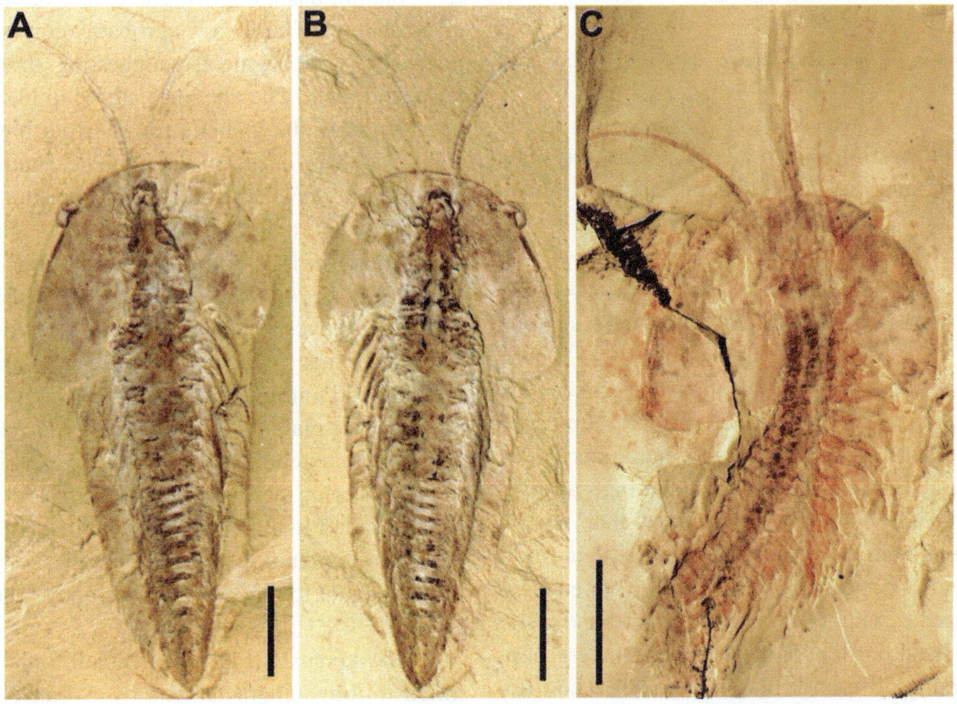

Fig. 76 *Cindarella eucalla* from the Early Cambrian Maotianshan Shale. (**a**, **b**) part and counterpart of holotype; (**c**) dorsal view of a nearly complete specimen. Scale bars = 1 cm (**a**, **b**), 2 cm (**c**)

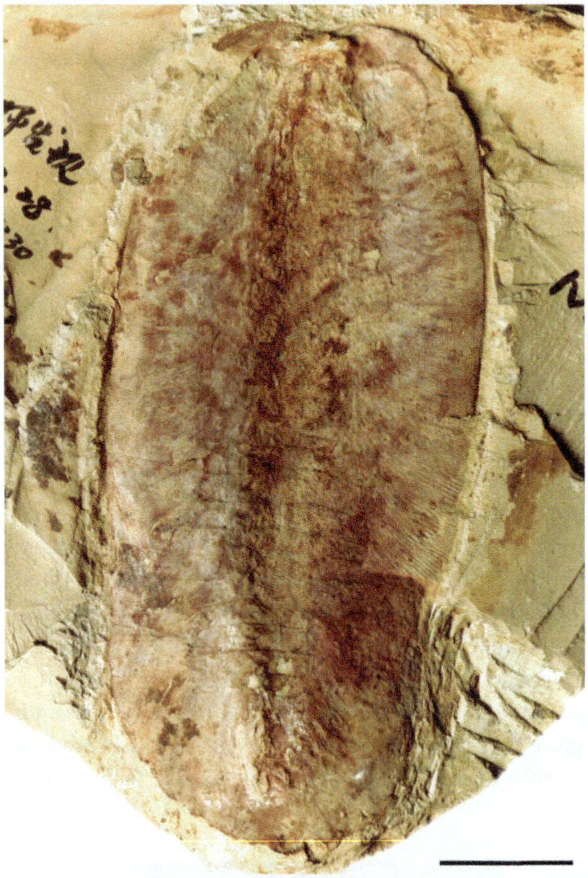

Fig. 78 *Saperion glumaceum* from the Early Cambrian Maotianshan Shale. Scale bar = 2 cm

(Fig. 80b), suggesting that bradoriids were the main diet of many animals, including some unknown giant predators.

Extinct Stem-Lineage Euarthropods

The euarthropods in the Maotianshan Shale are diverse, including several extinct major groups such the trilobites, trilobitomorphs, and naraoiids. They share several key characters, including a five-segmented head; post-oral appendages that are all biramous and homologous other than in their morphology; and endopods in the post-oral appendages that are bipartite, with an expanded basipod and a six-segmented distal portion. However, these three groups are significantly different from one another. The trilobites are characterized by a mineralized exoskeleton and sessile dorsal eyes, three clear axial lobes, and a tripartite body (a head, a jointed thorax, and a fused pygidium). The trilobitomorphs resemble trilobites in sharing a tripartite body, but they differ from trilobites in their non-mineralized exoskeleton and their stalked and ventrally situated eyes. Naraoiids, including *Misszhouia* and *Naraoia*, are characterized by their lack of eyes and a bipartite body composed of a head and trunk.

The trilobites are widely accepted as the dominant components of the macroscopic fauna. Surprisingly, they are minor components of the Maotianshan Shale, accounting for only approximately 5% or less of the number of species and individuals, although their preservation is potentially high. This suggests that the variance in the potential preservation of the organisms plays a critical role in their biased preservation. These trilobites include the following species: *Eoredlichia intermedia*, *Yunnanocephalus yunnanensis*, *Wutinggaspis tingi*, and *Kuanyangia pustulosa*. Their ventral soft parts are presumably often preserved but these soft parts under the hard dorsal exoskeleton are extremely difficult to expose during mechanical preparation.

E. intermedia is a well-known trilobite (Figs. 81, 82, and 83), with a total body length of more than 20 cm. Its cephalon has a conical glabella, incised with three clear furrows. The furrows probably correspond to the post-oral head segments, suggesting three post-oral head segments. The antennae are whip like, extending laterally from the glabella for one-fourth of its length (Fig. 83). The insertion sites of the antennae, corresponding to the antennal segment, are situated slightly anterior to the anterior glabella furrow. The antennae extend anteriorly along the ventral surface and anteriolaterally beyond the front margin of the cephalon. The sessile eyes are large and crescent shaped, with an anterior margin located near the front margin of the glabella and ending posteriorly, at a position equal to the posterior glabella furrow. The free cheek has a long genal spine extending posteriorly to a level equivalent to the posterior margin of the trunk. The thorax has 15 segments and its ninth segment in the axial lobe has a long spine, extending posteriorly, with a length approximately equal to the length of the body. The pygidium is small.

Y. yunnanensis (Fig. 84) is small, only a few centimetres long, characterized by a small free cheek that is narrow and spineless, and a subquadrate glabella with a wide preglabella field, which has three

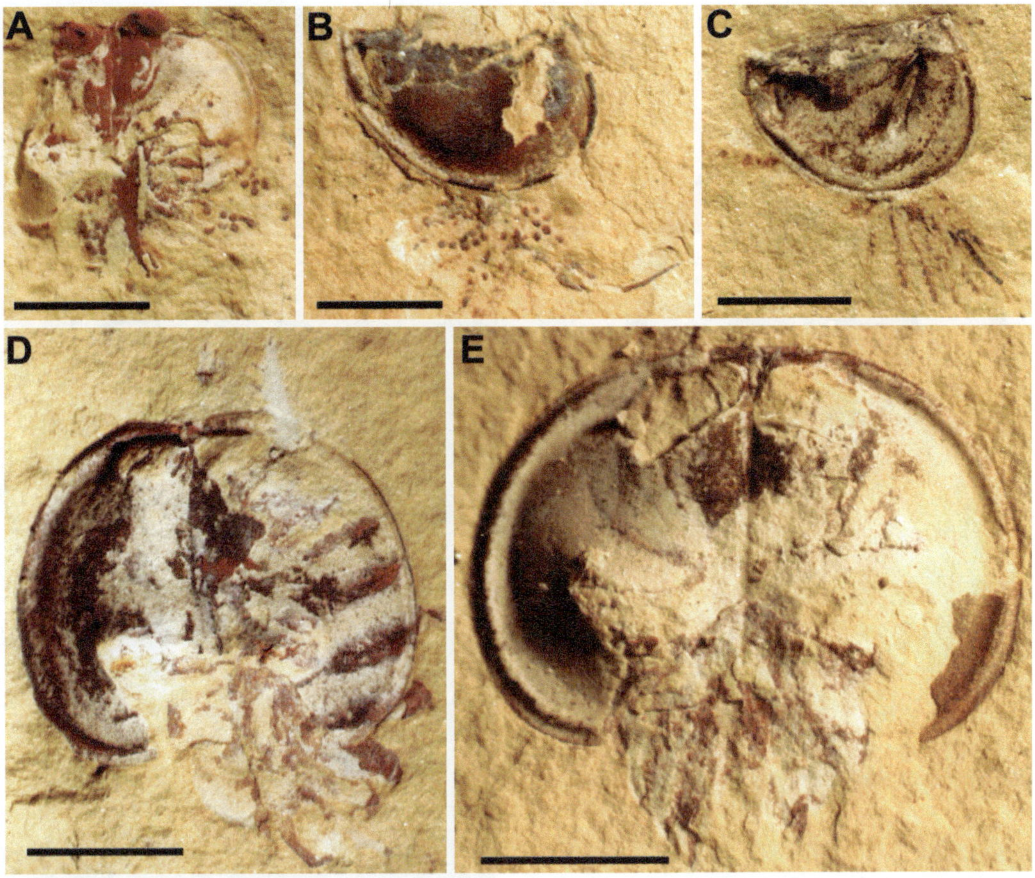

Fig. 79 *Kunmingella douvillei* from the Early Cambrian Maotianshan Shale. (**a**, **b**) dorsal (**a**) and side view (**b**) of specimens showing numerous eggs attached to appendages; (**c**–**e**) side (**c**) and dorsal view (**d**, **e**) showing multisegmented appendages. Scale bar = 2 mm

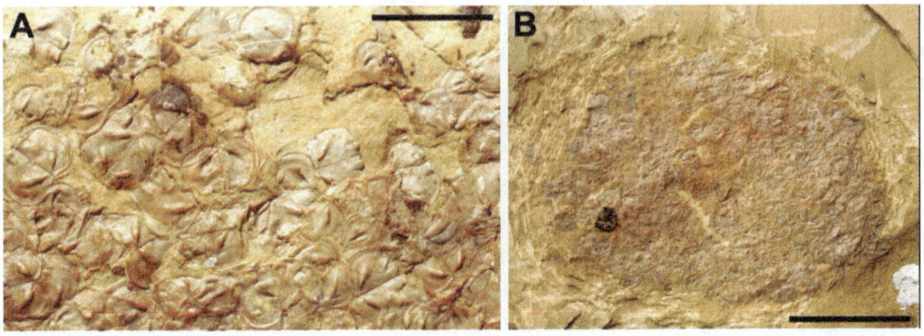

Fig. 80 Aggregated "butterfly" like valves of *Kunmingella douvillei* on a bedding plane (**a**) and a large coprolite containing mainly small carapaces related to bradoriids. Scale bars = 0.5 cm (**a**), 0.2 cm (**b**)

conspicuous furrows. The sessile eyes extend forward posteriolaterally from the lateral surfaces of the anterior glabella. The thorax consists of 14 segments and the pygidium is small.

K. pustulosa (Figs. 85 and 86) is a large trilobite, 10 cm long, characterized by a short cephalon (one-fifth of the body length), which bears a pair of wide free cheeks with short genal spines. The

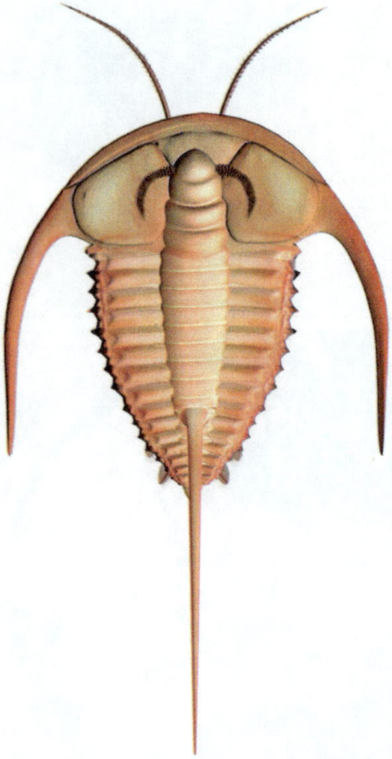

Fig. 81 Restoration of Early Cambrian *Eoredlichia intermedia*

Fig. 82 *Eoredlichia intermedia* from the Early Cambrian Maotianshan Shale displaying a complete caudal spine. Scale bar = 2 cm

hypostome is expanded anteriorly, with a rostral plate directly meeting the doublure. The gut has pairs of segmentally arranged diverticulae. Based on the evidence of the diverticulae, there are three post-oral head segments, coincident with the number of glabella furrows. The thorax consists of 16 segments and the pygidium is small. The dorsal surface bears densely arranged, small but distinct nubbles.

The trilobitomorphs in the Maotianshan Shale include *Kuamaia lata* (Hou 1987; Chen et al. 1996; Hou and Bergström 1997; Chen 2004), *Retifacies abnormalis* (Hou et al. 1989; Hou and Bergström 1997; Chen 2004), and *Acanthomeridian serratum* (Hou et al. 1989), which share several important characters including a tripartite body corresponding to a relatively short five-segmented head with a pair of ventral stalked eyes, the thorax, and the pygidium, and a non-mineralized exoskeleton. *Kuamaia* is a giant, with a maximal length of 40 cm. The head is small, with five segments: a small visual segment carrying a pair of stalked lateral eyes, a relatively larger antennal segment with a ventral hypostome and a pair of whip-like antennae extending from the lateral surfaces of the anterior part of the hypostome, and three post-oral segments, each bearing small biramous appendages. The thorax has seven segments and the pygidium has five pairs of appendages. The visual segment is not fused but jointed to the head shield.

R. abnormalis (Figs. 87 and 88) is relatively large, with a maximum length of 30 cm, and is characterized by a fairly small cephalon (only one-eighth of the body length) with five segments, like that of *Kuamaia*. The thorax has ten segments. The pygidium is relatively large, with up to five indistinct segments, and with multiple spines on the peripheral posterior margin. The almost smooth surface is decorated with a faint polygonal ridged structure. The post-oral appendages are

Fig. 83 *Eoredlichia intermedia* from the Early Cambrian Maotianshan Shale displaying a pair of whip-like antennae. Scale bar = 2 cm

biramous, and with a large plate-like exopod and a stout, seven-segment, bipartite endopod.

Among these stem-lineage euarthropods, the naraoiids are well represented by bulk, and their anatomy is well documented. *Misszhouia longicaudata* (Chen et al. 1997) is an elongated, oval-shaped animal, about 5 cm long (Figs. 89 and 90). The bipartite body consists of a distinct head and a carapace-covered trunk. The head is covered dorsally by a hemispheric head shield, with an oval hypostome complex on the ventral side. A pair of whip-like antennae extend from the anteriolateral surfaces of the hypostome. The animal has no eyes, but has three frontal organs anterior of the hypostome. The three post-oral head segments are defined by the presence of three post-oral biramous appendages. The thorax has 21 pairs of biramous appendages. Each exopod consists of a slender jointed shaft, bearing filaments, with a small distal lobe. The endopods are distinctly bipartite, with an expanded proximal portion (the basipod), with serrations on its inner surface. The distal portion consists of six segments with a terminal claw.

The bipartite *Naraoia* (Figs. 91, 92, and 93) is a close relative of *Misszhouia*. It is common in the Maotianshan Shale, represented by *M.* or *N. spinosa* (Zhang and Hou 1985). It ranges from 1 to 4 cm in length. The head shield is hemispheric; facial sutures appear to be marginal. The head shield extends posteriorly beyond the posterior margin of the head. A pair of antennae extends laterally, rather than anteriorly as they do in *Misszhouia*. A pair of long, ramifying, distensible diverticulae occurs along the midgut, possibly functioning in both food storage and digestion, as suggested by recent analogues. The trunk is covered dorsally by a carapace and consists of approximately 14 segments. The pygidium may bear a pair of distinct rear lateral spines and three small lateral spines.

Naraoia is thought to have lived in water-saturated soft mud and was an opportunistic intermittent feeder. The animals are usually preserved in an enrolled state, with the head parallel to the bedding plane and the thorax extending vertically deep into the surface. The enrolling behaviour may indicate that they could dig into the soft substrate to exploit a potential food source or to protect the organs on the underside of the cephalon from the fine mud and from predation.

Fossils Related to the Evolutionary Transition Towards the Mandibulata

The extant arthropods, including the Crustacea, Hexapoda, Myriapoda, and Chelicerata, can be grouped into two deeply rooted clades, the Mandibulata and the Chelicerata. The extant clade Mandibulata is composed of the Hexapoda, Myriapoda, and Crustacea. The head segmentation and related appendages of the Hexapoda and Myriaopoda are homologous to those of the Crustacea, except for the first post-oral segment, which is known as an "intercalary" segment rather than the second antennal segment. Until recently, the intercalary segment has long been regarded as a structure without an appendage, but the labrum has been reinterpreted as a fused pair of appendages of the intercalary segment (Finkelstein and Perrimon 1991; Popadic et al. 1998; Boyan et al. 2002) and the head has been reinterpreted

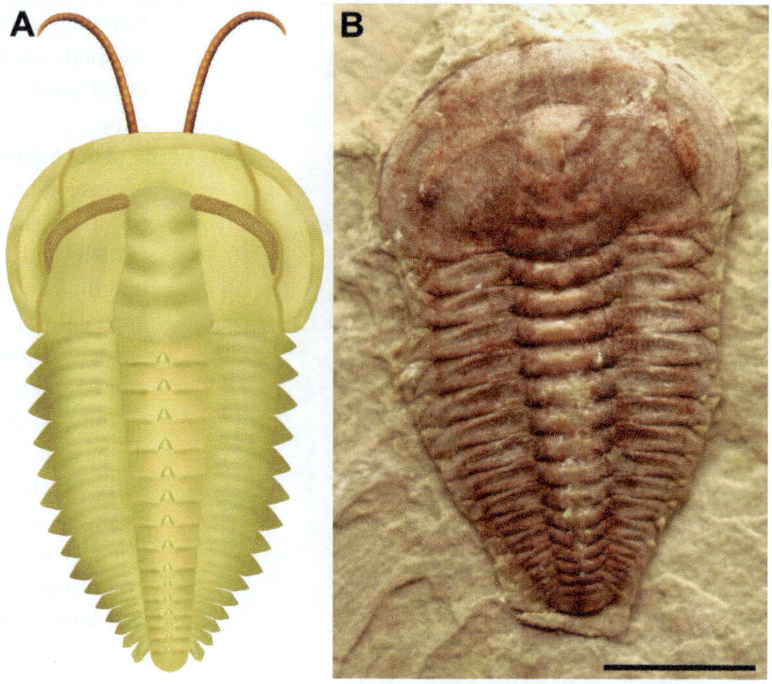

Fig. 84 Restoration (**a**) and photograph (**b**) of *Yunnanocephalus yunnanensis* from the Early Cambrian Maotianshan Shale. Scale bar = 5 cm

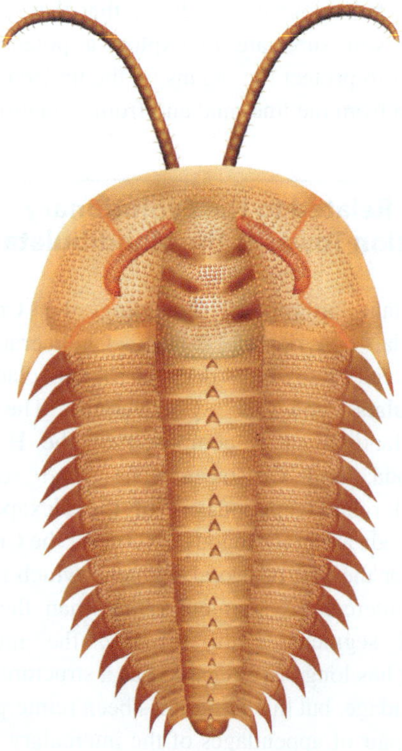

Fig. 85 Restoration of Early Cambrian *Kuanyangia pustulosa*

Fig. 86 *Kuanyangia pustulosa* from the Early Cambrian Maotianshan Shale. Scale bar = 2 cm

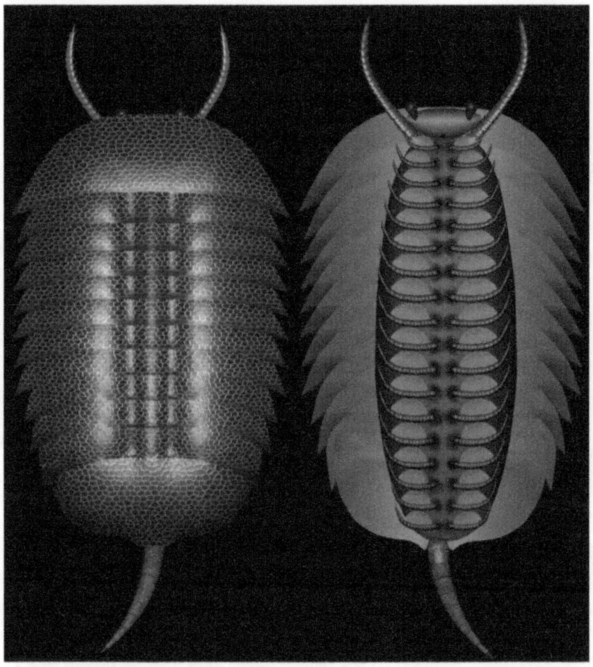

Fig. 87 Restoration of Early Cambrian *Retifacies abnormalis*. (**a**) dorsal view; (**b**) ventral view

as having six segments (Zacharias et al. 1993; Minelli 2001). The six segments of the insect head are (from anterior to posterior) the ocular, antennal, intercalary, mandibular, maxillary, and labial segments. The insect head segmentation and the related structures show striking homologies to those of Crustacea, suggesting that the insects (possibly including the Myriapoda) originated among the Crustacea or from a crustacean-like ancestor. The evolution of the second antennae into a feeding organ, known as the "labrum", with anterior displacement and the fusion of the paired second antennae, was one of the major transitions in the evolution of the insects.

E. minuscule (Figs. 94, 95, and 96), known from the Maotianshan Shale, is small (2–4 mm) and represents an ancestral form of Mandibulata (Chen et al. 2001). It has a six-segment head and an elongated, untagmatized trunk bearing serially repeated, biramous appendages (long and segmented endopods and flap-like exopods). The visual segment of the head is small and is covered by a rounded tergite jointed posteriorly to the head shield. A pair of stalked eyes extends ventrolaterally from the visual segment (Fig. 96a, b). The head shield is an elongated oval shape, overlying a pair of antennae and four pairs of post-oral biramous appendages. The antennae are elongated, with 10–11 segments and setae in the joints. The first post-oral appendages are highly modified to form large setose fans, which possibly function in feeding and are comparable to the second antennae in living crustaceans (Figs. 95 and 96d). The trunk consists of 13 segments. The appendages, including the trunk and post-oral head appendages, are also biramous and apparently homologous to another, except for the last pair, which is segmented and fringed with setae. The last paired appendages are homologous to the other appendages but are comparable to the uropods seen in various living crustaceans. The endopods of both the trunk and post-oral head are sub-tubiform and consist of 14 segments.

The early member of Mandibulata has an isolated visual segment resembling that of the proarthropods and some intermediate forms between the proarthropods and euarthropods. The antennae resemble those in the proarthropods in their general morphology but bear setae, suggesting that they are more advanced than the setae-lacking antennae of the proarthropods. The second antennae are a newly derived character, composed of a relatively small 4- to 5-segment endopod and a large crescent-shaped exopod with prominent setae on its margin. Posterior to the A2, there are usually three biramous head appendages, resembling trunk appendages. The endopods of the post-A2 appendages are tubiform, consisting of 14 segments. The primitive nature of the post-A2 head appendages in *E. minuscule* suggests that the clade Mandibulata is deeply rooted in the evolutionary transition from the proarthropods to the euarthropods.

The recent discovery of an epipodite-bearing eucrustacean *Yicaris* from Yonshan, Yunnan (not far from the fossil site of *E. minuscule*), within a horizon almost contemporaneous with the Lower Cambrian Maotianshan Shale (Zhang et al. 2007), indicates that the first appearance of the Eucrustacea was much earlier than we previously thought.

The Stem-Lineage Chelicerata

The chelicerates are widely accepted as the unquestioned, basic monophyletic group of the Arthropoda. They share a distinct and conservative body plan, including features such as the prosomal–opisthosomal subdivision and the cheliceral segment, with a pair of morphologically unique chelicerae. The Chelicerata

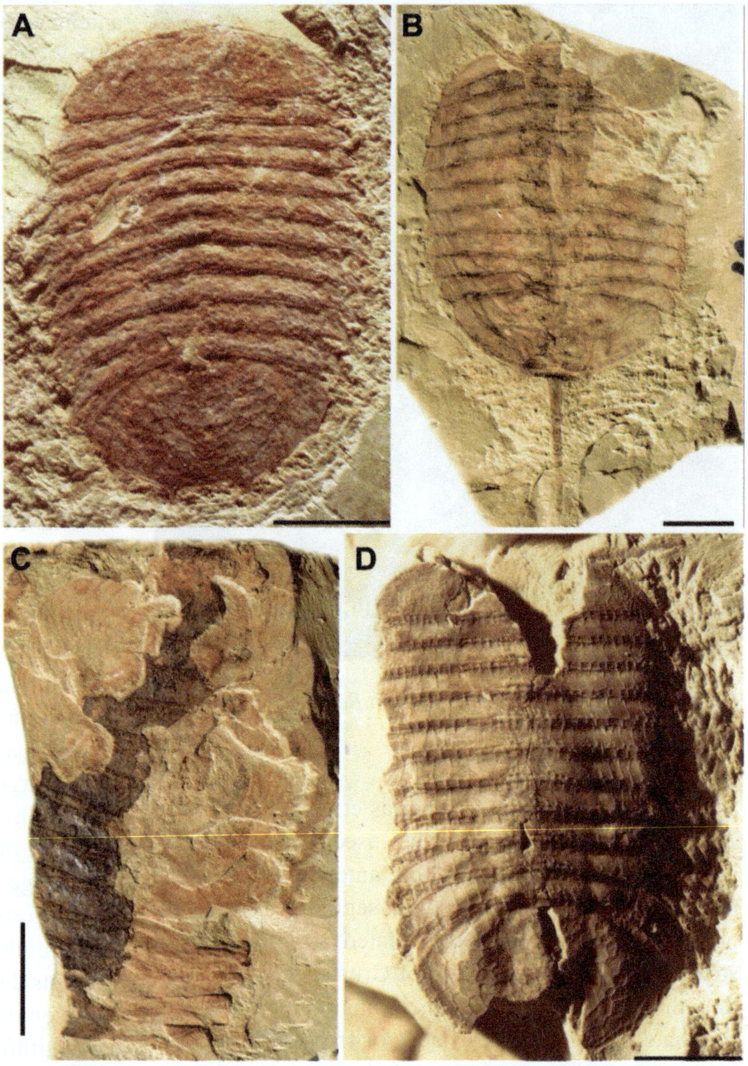

Fig. 88 *Retifacies abnormalis* from the Early Cambrian Maotianshan Shale. (**a**, **b**) dorsal view of two nearly complete specimens with preservation of complete segmented caudal spine (**b**); (**c**) partially preserved trunk showing trunk appendages; (**d**) dorsal view of a nearly complete specimen showing polygonal external structure. Scale bar = 1 cm

is named after the pair of modified, uniramous head appendages called the "chelicerae". The chelicerae consist of only three components: the proximal peduncle and two distal peduncles, forming the chela. The chelicerae have long been regarded as modified second antennae, but on the basis of molecular developmental and neurobiological studies, they have now been reinterpreted as modified first antennae. The ground plan of the chelicerate head consists of a two-segmented anterior head and five-segmented posterior head (Dunlop 1999). On the anterior head, the ocular segment bears dorsal, sessile, compound eyes and the antennal segment bears a pair of anteriorly pointing, raptorial, uniramous appendages, the chelicerae. The posterior head bear five pairs of biramous appendages, as exemplified by the Silurian *Offacolus kingi* (Sutton et al. 2002).

The large modified antennae known as the "great appendages" have long been recorded from the Early to Middle Cambrian. These "great appendages", each comprising a proximal arm and a distal claw, are similar to the chelicerae in their general morphology. Despite this similarity, the possible evolutionary relationship between the peculiar Cambrian arthropods with "great appendages" and the Chelicerata was not recognized until the fossil *Haikoucaris ercaiensis* was found in 2004 (Chen et al. 2004e) (Figs. 97 and 98). The animal, recorded from the Lower Cambrian Maotianshan Shale, has a pair of chelicera-like structures that fill the morphological gap between the Cambrian "great appendages" and the chelicerae of the crown-lineage Chelicerata. The chelicera-like structure consists of a proximal two-segmented arm and a distal

Fig. 89 Ventral and dorsal views (**a**, **b**) of the Early Cambrian *Misszhouia longicaudata*

three-fingered claw, resembling the chelicera but has one more segment in both the arm and claw; thus, it is similar to the "great appendage" except that it has one fewer segment than the claws of most Cambrian "great appendage" creatures. Because of their homology to the chelicerae, Chen (2004) referred to these "great appendages" as prochelicerae, interpreting them as the precursors of the chelicerae.

Further, a group of giant Cambrian predators, known as "anomalocaridids", also has a pair of large pre-oral "great appendages" (Hou et al. 1999). The "great appendages" of the anomalocaridids are not only much larger than any others but have a multisegmented, spine-bearing arm. However, the small distal claw (usually with four segments) of the anomalocaridids displays a striking resemblance to that in the non-anomalocaridid "great appendage" arthropods. Chen (2004) suggested that the "great appendages" of the anomalocaridids and non-anomalocaridids are closely related and are both representatives of extinct stem-lineage Chelicerata (also referred to as "Prochelicerata"). In addition to their larger "great appendages", the anomalocaridids have a unique mouth, which is large, ventrally facing, and armed with a ring or several rings of hard, sharp teeth.

The Prochelicerata is an extinct animal group, broadly recorded in the Lower to Middle Cambrian from China, Canada, USA, Poland, and Australia. They died out after the Middle Cambrian.

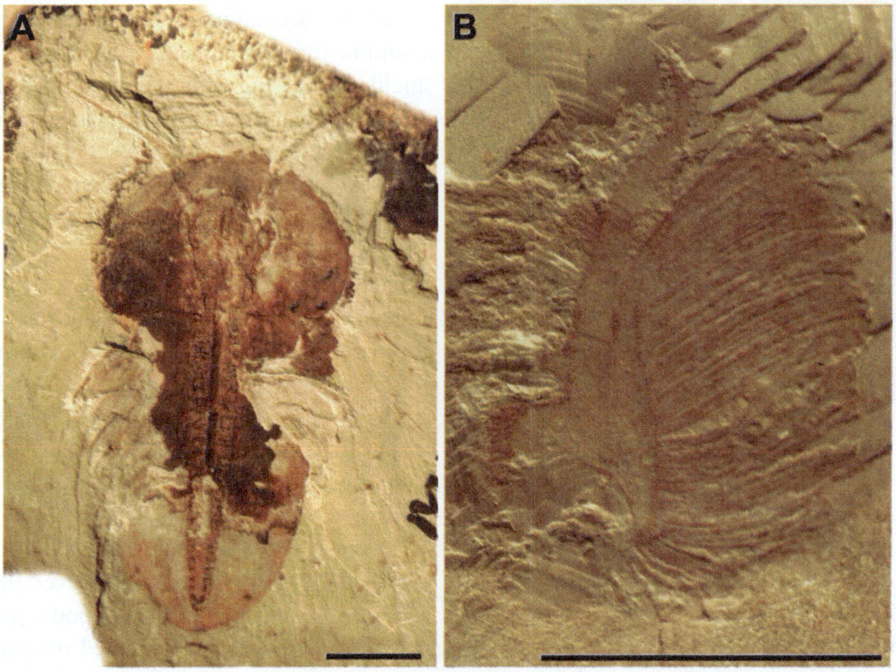

Fig. 90 *Misszhouia longicaudata* from the Early Cambrian Maotianshan Shale. (**a**) dorsal view of a nearly complete specimen; (**b**) bipartite trunk appendage. Scale bar = 1 cm

Fig. 91 Restoration of Early Cambrian *Naraoia spinosa*

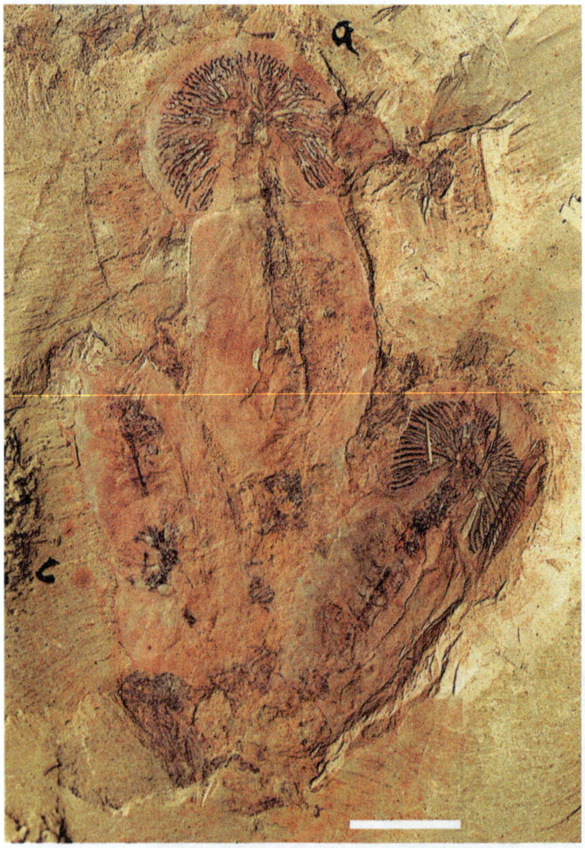

Fig. 92 *Naraoia spinosa* from the Early Cambrian Maotianshan Shale. Scale bar = 1 cm

The head ground plan in Chelicerata, as exemplified by the Prochelicerata, is a five-segmented model. The five segments are arranged, from anterior to posterior, as the ocular segment, the procheliceral segment, and the three post-antennal segments. Each of the three post-antennal segments has a pair of biramous appendages identical to those on the trunk. Their endopod is usually simple and multisegmented, with least differentiation along the main axis (e.g. *Fortiforceps foliosa*, *Leachoilia illecebrosa*, and *Tanglangia longicaudata*), but *H. ercaiensis* has only seven segments, probably grouped into two functional portions. The primitive design of the limb in most Prochelicerata suggests that the Chelicerata is deeply rooted among the stem-lineage arthropods. The transformation of the short, stout antennae into bipartite prochelicerae was the first key event leading to the evolution of Chelicerata lineage. The transformation of the Prochelicerata head into the Euchelicerata head was marked by two probably synchronous events, the reduction of the procheliceral segments and the recruitment of two more segments from the trunk.

Non-anomalocaridid Procheliceratids

The diverse non-anomalocaridid procheliceratids are small, only a few centimetres long. In the Maotianshan Shale, they include *H. ercaiensis* (Chen et al. 2004e), *Leanchoilia illecebrosa* (Hou 1987), *Jianfengia multisegmentalis* (Hou 1987), *T. longicaudata* (Luo et al. 1999), and *F. foliosa* (Hou and Bergström 1997).

All the non-anomalocaridid procheliceratids have a similar five-segmented head, covered by a cephalic shield and bearing ventral compound eyes in the visual segment, a pair of anteroventral and anteriorly pointing prochelicerae in the second segment, and three post-oral segments with biramous appendages, which are similar to one another and are very similar to the trunk appendages.

The trunks of all non-anomalocaridid procheliceratids are generally similar, having relatively short tergopleurae with various number of segments: 13 in *H. ercaiensis* and *T. longicaudata*, 12 in *L. illecebrosa*, 20 in *Fortiforceps*, and 22 in *J. multisegmentalis*. The shape of the tail end varies, being lanceolate and ringed with short spines in *L. illecebrosa* but fan-like in *J. multisegmentalis*.

The post-oral appendages are all biramous and morphologically homologous. However, their endopods vary in design from the euarthropod type in *H. ercaiensis* to a tubular, multisegmented design, with various numbers of segments (about 9 in *L. illecebrosa*, more than 10 in *T. longicaudata*, and 15 in *F. foliosa*).

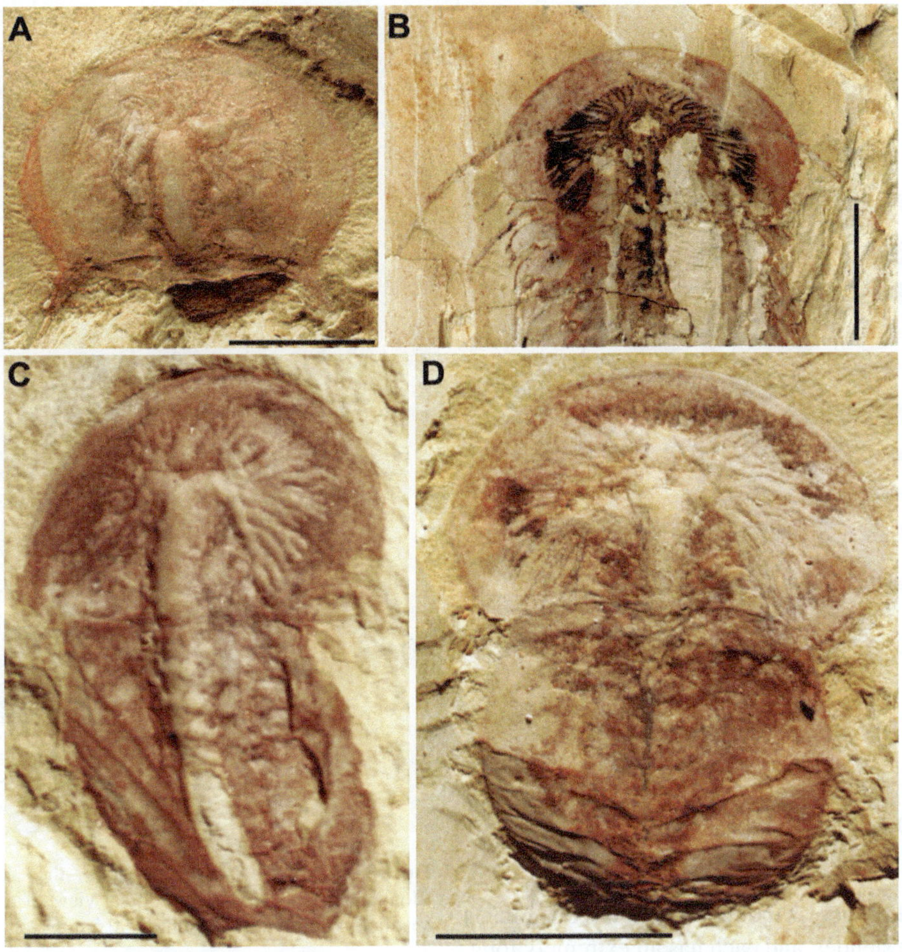

Fig. 93 *Naraoia spinosa* from the Early Cambrian Maotianshan Shale. (**a**) enrolled preservation with head parallel to bedding plane; (**b**) anterior portion of body, showing ramifying diverticulae and a laterally extended left antenna; (**c**) dorsal view of a nearly complete specimen showing 3D diverticulae and gut; (**d**) dorsal view of a bent body. Scale bars = 0.5 cm (**a**, **d**), 1 cm (**b**), and 0.2 cm (**c**)

Among the prochelicerates, *H. ercaiensis* is extremely interesting in terms of the morphological gap in the chelicerae of the Cambrian "great appendage" arthropods and the living chelicerates. The animal, about 38 mm long, has a pair of prominent, stout prochelicerae accounting for slightly more than one-fourth of its body length (Figs. 97 and 98). The raptorial prochelicera is a proximal arm composed of two segments. The arm is oriented anteroventrally and then bends dorsally at the joint between the segments. The distal claw has three segments and its very prominent proximal segment is drawn out dorsodistally into a stout, slightly curved spine (fixed spine). The next segment is smaller and rounded, and articulates against the proximal claw segment; a dorsodistally oriented, fixed spine, because of its articulation against the basal segment, can swing against the proximal spine. The last segment, forming the widened base of its spine, is slightly curved against the second spine, acting as a movable finger, swinging against the fixed spine below. The spines in the claw can move apart or close against each other (Chen et al. 2004e).

L. illecebrosa (Figs. 99 and 100) is a small (1–3 cm long), commonly occurring arthropod in the Maotianshan Shale. The head has a narrow front margin, with two pairs of small stalked eyes, a short subquadrate hypostome, and a pair of prochelicerae on the lateral surfaces of the hypostome, extending anteriorly. Each prochelicera comprises a two-segmented arm and a four-segmented claw. The segments in the claw are

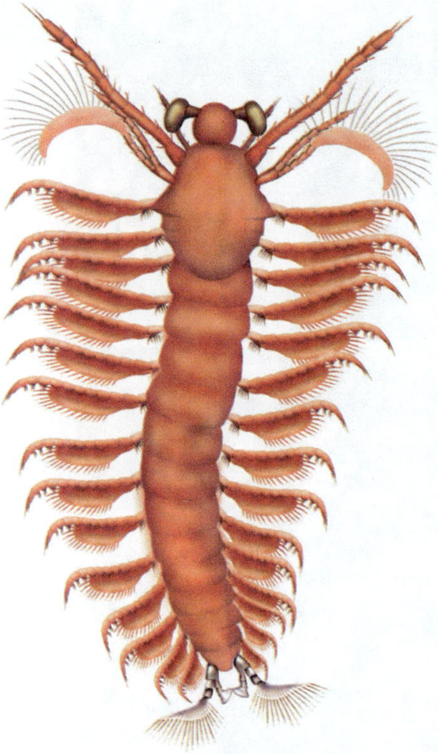

Fig. 94 Restoration of Early Cambrian *Earcaia minuscule* (from Chen 2004)

Fig. 95 *Earcaia minuscule* from the Early Cambrian Moatianshan Shale showing dorsal view of a nearly complete specimen (holotype). Scale bar = 0.5 mm

relatively wide, whereas the subsequent three segments become more slender, and each has an extremely long, filamentous spine, almost the length of the body or longer. These spines can be positioned close together or wider apart (Hou and Bergström 1997).

J. multisegmentalis (Figs. 101 and 102) is a rare arthropod, 2 cm long with an elongated body. The head is small and sub-oval, with a pair of stalked eyes and a pair of prochelicerae, with a similar number of segments to those in the head of *L. illecebrosa*; the arm is longer. The claw has four, slightly more regular, annular segments and straight spines. The trunk consists of 22 segments; the tail is long with a terminal spine.

T. longicaudata (Figs. 103 and 104) occurs only in Haikou. It is 2–4 cm long, with an elongated body, small head, slender trunk, and flat tail spine, drawn out from the posterior end of the trunk. The head shield is sub-quadrate in outline, with a separate visual segment jointed to its posterior head shield. It has a pair of lateral eyes situated on the visual segment. The prochelicerae are relatively large, comprising a two-segmented arm and a four-segmented claw. Each of the claw segments is drawn out to form a short spine. There are possibly three pairs of post-oral biramous appendages. The trunk has 13 segments, each with a pair of biramous appendages.

F. foliosa (Figs. 105 and 106) is a rarely occurring arthropod. It is about 4 cm long, with a slender body comprising a small head, an elongated segmented trunk, and a telson. The head is covered by an oval head shield, and has a pair of stalked eyes and a pair of prochelicerae. The prochelicerae are relatively stout, and consist of a two-segmented arm and a four-segmented claw. The trunk has 20 segments, each with a pair of biramous appendages, each with a flap-shaped exopod and a tubiform endopod, consisting of 15 segments. The conspicuous telson consists of a pair of tubular uropods and a tail spine.

Isoxys is a bivalved arthropod distributed worldwide in the Cambrian and occurring commonly in the Lower Cambrian Maotianshan Shale (Fig. 107a–e). It is predominantly represented by empty bivalved carapaces, with two prominent spines of various lengths

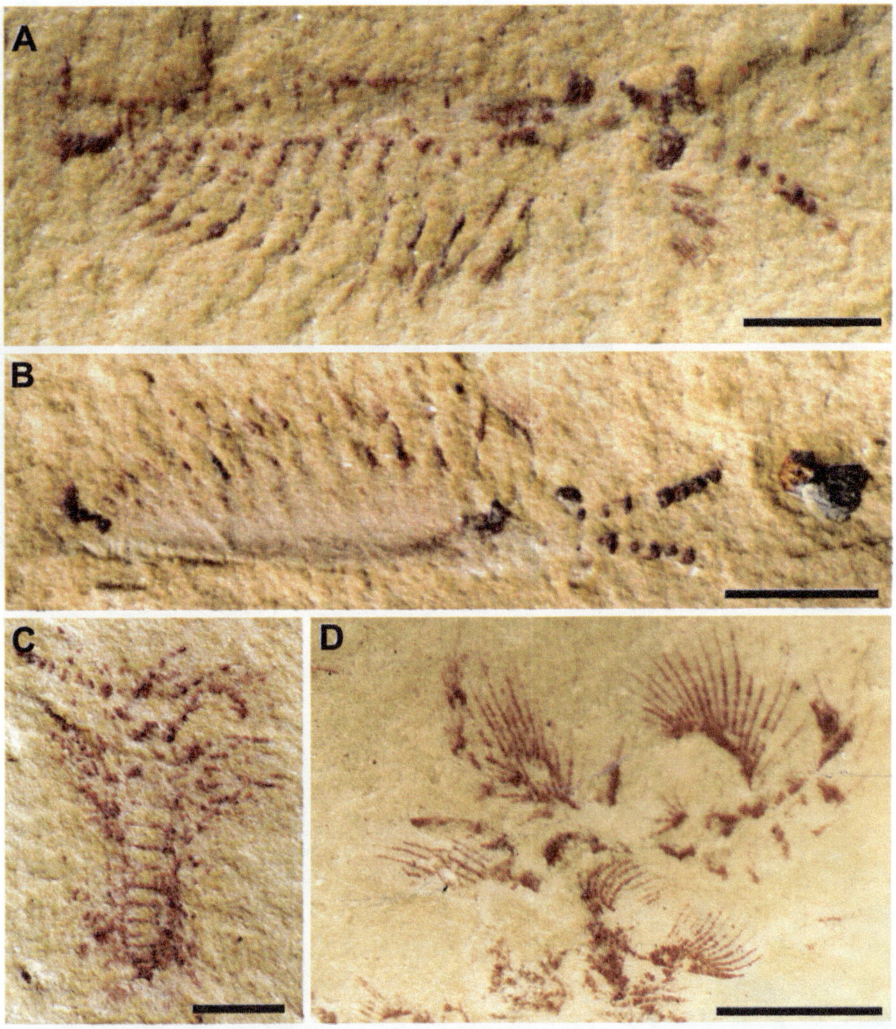

Fig. 96 *Earcaia minuscule* from the Early Cambrian Moatianshan Shale showing eyes (**a**, **b**), 1st pair (**a**–**c**) and 2nd pair (**c**, **d**) of antennae, multisegmented endopods (**a**), and uropods (**a**, **b**). Scale bar = 500 μm

protruding both anteriorly and posteriorly. The dorsal outline is usually straight, but a few are weakly convex. The carapace has no articulated hinge, and possibly represents a giant head shield derived from the tergite in the second head segment. The soft parts of *Isoxys* are rarely preserved and are known only from the Lower Cambrian Maotianshan Shale (Chen 2004) and Middle Cambrian Burgess Shale (Garcia-Bellido et al. 2009). They have an elongated body, covered largely by carapace. The head is presumably short and two segmented, and the first segment carries a pair of large stalked eyes. The appendages on the second head segment were thought to have a regular antennal design, but surprisingly, the Middle Cambrian species, *Isoxys acutangulus*, documents the presence of "great appendage"-like structures suggesting that *Isoxys* is a member of the Prochelicerata. Based on the specimens from the Lower Cambrian Maotianshan Shale, the trunk of *Isoxys* has numerous (at least 14) segments. Each segment carries a pair of biramous appendages. The endopods are tubular, with numerous segments (approximately 14).

Affinity

The affinity of the prochelicerates and the Euchelicerata is confirmed by the common chelicera-like structures in the antennal segment (Fig. 108). The chelicerae are a "chelicerate-only" character,

Fig. 97 Restoration of *Haikoucaris ercaiensis*

Fig. 98 *Haikoucaris ercaiensis* from the Early Cambrian Maotianshan Shale. (**a** and **b**) counterparts of holotype

homologous to the antennae in non-chelicerate arthropods. They are characterized by a type of bipartite design, with a proximal arm and a spine-bearing distal claw. In extant Euchelicerata, the chelicerae tend to be reduced in size or totally lost. The four-segmented chelicera is interpreted as the ground state in the eucheliceratid chelicerae. Both the arm and claw have two segments, but the arm of most euchelicerates has

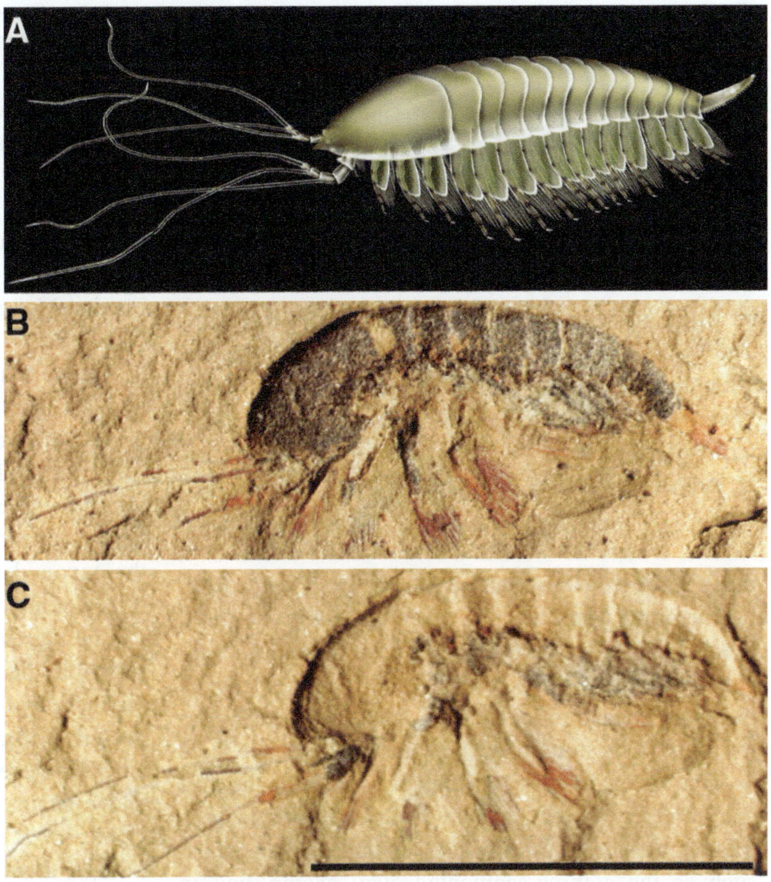

Fig. 99 *Leanchoilia superlata* from the Early Cambrian Maotianshan Shale. (**a**) restoration; (**b**, **c**) side view of nearly complete specimens showing 4-segmented claw with three filamentous spines. Scale bar = 0.5 cm

Fig. 100 *Leanchoilia superlata* from the Early Cambrian Maotianshan Shale. (**a**) side view of a specimen; (**b**) ventral view of a specimen, showing four attachment marks of stalked eyes. Scale bar = 1 cm

Fig. 101 Restoration of *Jianfengia multisegmentalis*

only a single segment through secondary degeneration. The two-segmented claw is a pincer-like structure, with the spine from the first segment acting as the "fixed finger" and the curved conical spine of the terminal segment as the "movable finger", which can be moved against the "fixed finger" to grasp and crush food.

The chelicerae of *Haikoucaris* differ from both those of the extant euchelicerates and from the prochelicerae of the Cambrian and other non-anomalocaridid prochelicerates in having one more and one less "finger" in the claw, respectively. The difference between the chelicera and prochelicera is phylogenetically significant, suggesting that the euchelicerate chelicera evolved through stepwise reductions in both the size and number of the segments in the claw, from a four-segmented claw through an intermediate state, such as the a three-segmented claw exemplified in *Haikoucaris*.

The Anomalocaridids: Cambrian Monsters

The anomalocaridids are a group of early giant predators that shared a single pair of pre-oral "great appendages" (Hou et al. 1999). Their fossils are recorded in the Lower to Middle Cambrian from China, Canada, USA, Poland, and Australia. They were the largest of the Cambrian animals, reaching a maximal length of 2 m, and died out after the Middle Cambrian.

The animals are characterized by both a single pair of large, pre-oral raptorial "great appendages" and a large rounded mouth, armed with a ring or several rings of hard, sharp teeth. These pre-oral limb-like appendages are bipartite, resembling those in the other "great appendage" animals. Both are included within the taxon designated Prochelicerata by Chen et al. (Chen et al. 2004e; Chen 2004, 2009).

The "great appendages" of the anomalocaridids resemble the prochelicerae of the non-anomalocaridid procheliceratids, with which they share a bipartite design (a distal claw and an arm) and a four-segmented claw. However, the arm in the anomalocaridids is larger and contains more segments, except in *Parapeytoia yunnanensis* (Hou et al. 1995; Chen 2004), which has a single arm. It was widely believed that the anomalocaridids used these giant "great appendages" to grasp

Fig. 102 *Jianfengia multisegmentalis* from the Early Cambrian Maotianshan Shale. A and B, counterparts of holotype. Scale bar = 1 cm

Fig. 103 Restoration of Early Cambrian *Tanglangia longicaudata*

their prey and bring it to their mouths. Our observation casts doubt on the argument because the "great appendages" cannot either grasp prey or bend to bring the captured prey to the animal's mouth.

The large rounded mouth armed with numerous teeth is an "anomalocaridid-only" character. The mouth, unlike those of other arthropods, faces ventrally. The teeth are arranged in a circle and those in the outer ring are attached to the margin of the mouth and those in the inner rings to the wall of the oesophagus. The teeth are flat, with a wider proximal end and a serrated, free distal end. The up-and-down motion of the distal free ends of the teeth changes the size of the mouth aperture from maximal to minimal, to capture, crush, and release the prey into the oesophagus.

The mouth cannot be closed entirely but presents a relatively large, rectangular aperture, even in its maximally closed state, an indication that the animal could not feed on small prey. The ringed teeth, like cutting nippers, could easily crack relatively large, exoskeleton-protected prey by moving the free distal end from a lower to a horizontal position. In addition to the outer ring, some anomalocaridids have several rings of teeth in their oesophagi. The oesophageal teeth are arranged in the same way as in the outer ring but they become smaller with increasing depth within the oesophageal cavity. The motion of the teeth not only cracks and processes the prey into pieces, but pulls the prey deeper into the oesophageal cavity.

Anomalocaridids have stalked compound eyes. Their tails have two long, caudal bars and several fin-like structures on the sides of the caudal bars, to maintain their balance when swimming at high speed. They represent the top predators of Cambrian times. It was thought that the anomalocaridids hunted trilobites. However, the undigested residues among the large coprite fossils rarely contain trilobite exoskeletal fragments but predominantly consist of the carapaces of bivalved arthropods related to *Chuandianella* (a sister group *Waptia* from the Burgess Shale).

The anomalocaridids are truly gigantic creatures for the Cambrian. Based the size of the great appendage and the mouth, the maximal size of the complete animal reaches lengths of 60 cm in *Anomalocaris* to a staggering 2 m in *Parapeytoia*. The anomalocaridids are also extremely agile compared with many other sea-dwelling creatures of the time. The animal is segmented and the body has 11 pairs of laterally extended flap-like structures, possibly homologous to exopods. The body probably moved in a wave-like manner, similar to that of present-day rays, allowing it to move at great speed or to "hover" silently (Chen et al. 2007c).

Three different species of anomalocaridids are known from the Lower Cambrian Maotianshan Shale: *Anomalocaris saron* (Chen et al. 1994; Briggs 1994; Hou et al. 1995), *P. yunnanensis* (Hou et al. 1995), and *Amplectobelua symbrachiata* (Chen et al. 1994; Hou et al. 1995). *A. saron* (Fig. 109) is the smallest anomalocaridid, with a maximum length of 60 cm. A completely preserved specimen shows a streamlined

Fig. 104 *Tanglangia longicaudata* from the Early Cambrian Maotianshan shale showing elongate spine (**a**, **c**), multisegmented endopods (**b**) and prochelicerae (**d**). Scale bars = 1 cm (**a–c**) and 0.5 cm (**d**)

body with its maximum width one-third from the anterior and a telson consisting of paired, fin-like tail fans and two fine, elongated caudal furca, providing efficient balance while swimming (Chen et al. 1994; Chen 2004). The mouth is ringed with 32 rounded, evenly arranged teeth, but probably no pharyngeal teeth. The trunk has 11 pairs of laterally extended flaps, but no limbs. The great appendages are composed of a large 11-segmented arm and a small 4-segmented claw. *A. symbrachiata* (Fig. 110) is larger than *A. saron*, with a maximum length of about 1 m (Chen 2004). Its "great appendage" is similar to that of *A. saron* but the spines in the proximal fourth segment are much larger. In contrast to *A. saron*, the front of the body in *A. symbrachiata* (Fig. 111) is much wider and the eyes lie laterally on the head, rather than in a frontal position. *P. yunnanensis* has a maximum length of 2 m and is the largest of the anomalocaridids. In contrast to the two other anomalocaridids, the formidable mouth of *P. yunnanensis* is not only gigantic (25 cm in diameter) but is armed with a ring of teeth on the mouth and numerous rings of teeth within the pharynx (Fig. 112). The great appendages are similar to those of other non-anomalocaridid prochelicerids but relatively shorter, composed of a large one-segmented arm and a four-segmented claw (Fig. 113a, b). Each segment of the claw carries blade-shaped spine. The trunk limbs in *P. yunnanensis* resemble those in the

Fig. 105 Restoration of *Fortiforceps foliosa*

euarthropods in being bipartite, with a huge basipod and a seven-segmented distal limb (Fig. 113c, Hou et al. 1999, 2004).

The anomalocaridids were extremely agile. Like present-day rays, the flexible lateral flaps along their bodies could probably be moved in a wave-like formation, allowing the animals to move at great speeds or glide smoothly through the water to approach their prey silently (Chen et al. 2007c).

A long-standing view holds that the anomalocaridids fed on hard-bodied animals, including trilobites (Nedin 1999). Some Cambrian trilobites have been found with round or W-shaped "bite" marks, matching the shape of the mouthparts of *Anomalocaris*. Stronger evidence that the anomalocaridids consumed trilobites is provided by fossilized faecal pellets that contain trilobite parts. However, studies of large faecal pellets from the Maotianshan Shale (Vannier and Chen 2005) have shown that the undigested particles within the faecal pellets (Fig. 114b, c) predominantly contain

Fig. 106 *Fortiforceps foliosa* from the Early Cambrian Maotianshan Shale. (**a** and **b**) counterparts of a nearly complete specimen. Scale bar = 1 cm

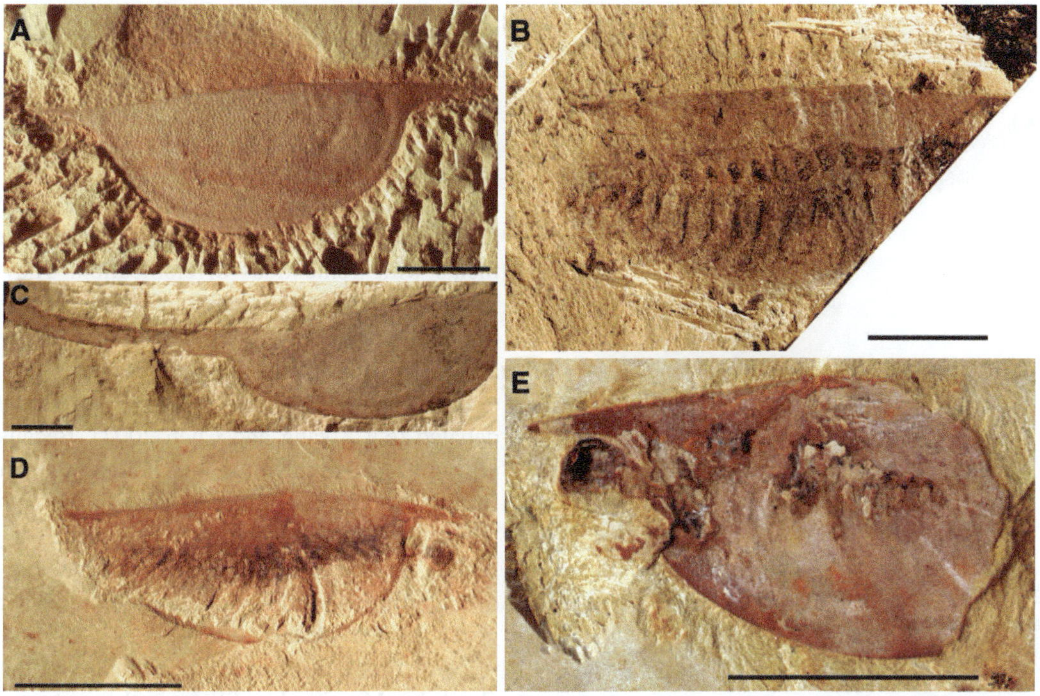

Fig. 107 *Isoxys* from the Early Cambrian Maotianshan Shale. (**a**) *Isoxys auritus* showing polygonal structure on surface of carapace; (**b**) a newly molted *Isoxys* showing multisegmented limbs; (**c**) *Isoxys paradoxus* showing an elongated rear projection of the carapace; (**d–e**) *Isoxys curvirostratus* showing stalked eye. Scale bar = 1 cm (**a–d**) and 0.2 cm (**e**)

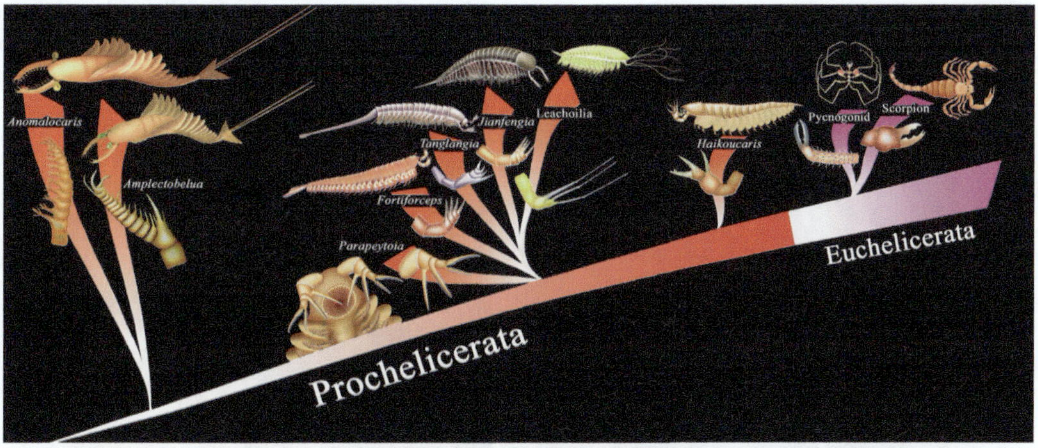

Fig. 108 Scenario for the evolutionary origin of the Chelicerata

the carapaces of either *Chuandienella* (a close relative of *Waptia* of the Burgess Shale; Fig. 114a) or microscopic bradoriids (Fig. 80b). In addition to arthropod carapaces, some faecal pellets predominantly contain exoskeletons related to hyoliths (Chen 2004). Because the mouth in the anomalocaridid cannot close fully and a large central aperture remains in the maximally closed state, we infer that the anomalocaridids could not fed on microscopic arthropods such as bradoriids. Furthermore, hyolithids were epibenthic organisms, so it is unlikely that these hyolithid faecal pellets were produced by anomalocaridids. The producers of these giant faecal pellets containing bradoriids and hyolithids remain a mystery. Thus, *Chuandienella* and some other arthropods were the main diet of the anomalocaridids during the Early Cambrian.

Fig. 109 A nearly complete *Anomalocaris saron* from the Early Cambrian Maotianshan Shale. Scale bar = 2 cm

The Evolutionary Scenario of the Panarthropods

The Arthropoda represents a vast assemblage of segmented animals with exoskeletons and jointed appendages. Their tremendous adaptive diversity has allowed them to survive in many and various environments with tremendous diversification. They are the most successful animal phylum on our planet in terms of population size and species diversity, accounting for about 80% of all known animal species.

The evolutionary origins of this group are hotly debated (e.g., Wang and Chen 2004). Because living arthropods are highly modified, belonging to the crown lineages of the euarthropods, and because the fossil record is incomplete, the large gap in the early evolutionary history of the arthropods has made discussion of the arthropod origins highly speculative and poorly grounded in evidence.

New fossil discoveries, especially in the Lower Cambrian Maotianshan Shale, in the past two decades have profoundly changed our understanding of the origins and evolution of the arthropods. The fossils, including the worm-like ancestors, the first arthropods (or "proarthropods") with unjointed limbs, the intermediate forms between the proarthropods and

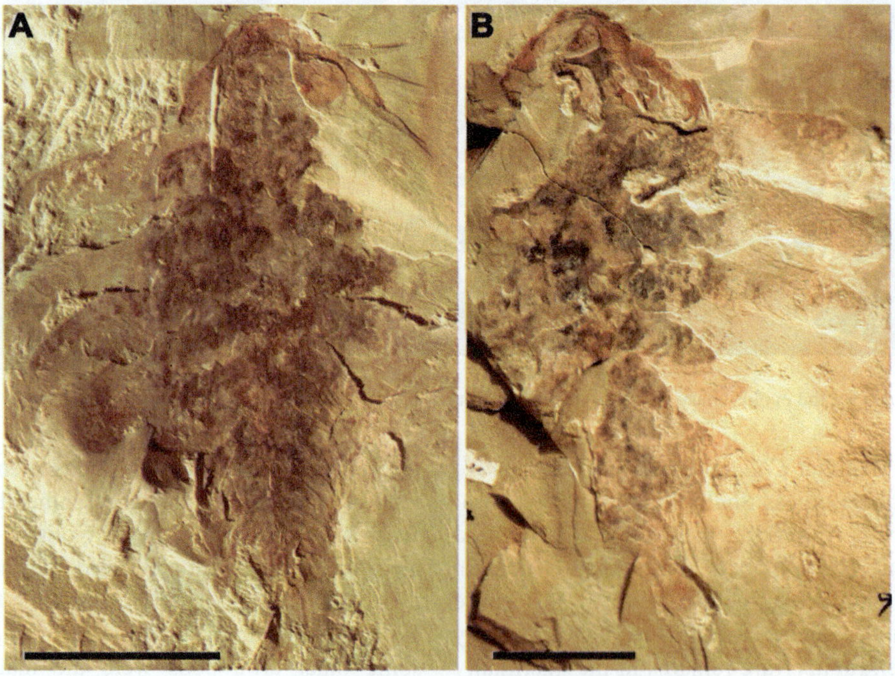

Fig. 110 Two nearly complete *Amplectobelua symbrachiata* from the Early Cambrian Maotianshan shale. Scale bars = 2 cm (**a**) and 3 cm (**b**)

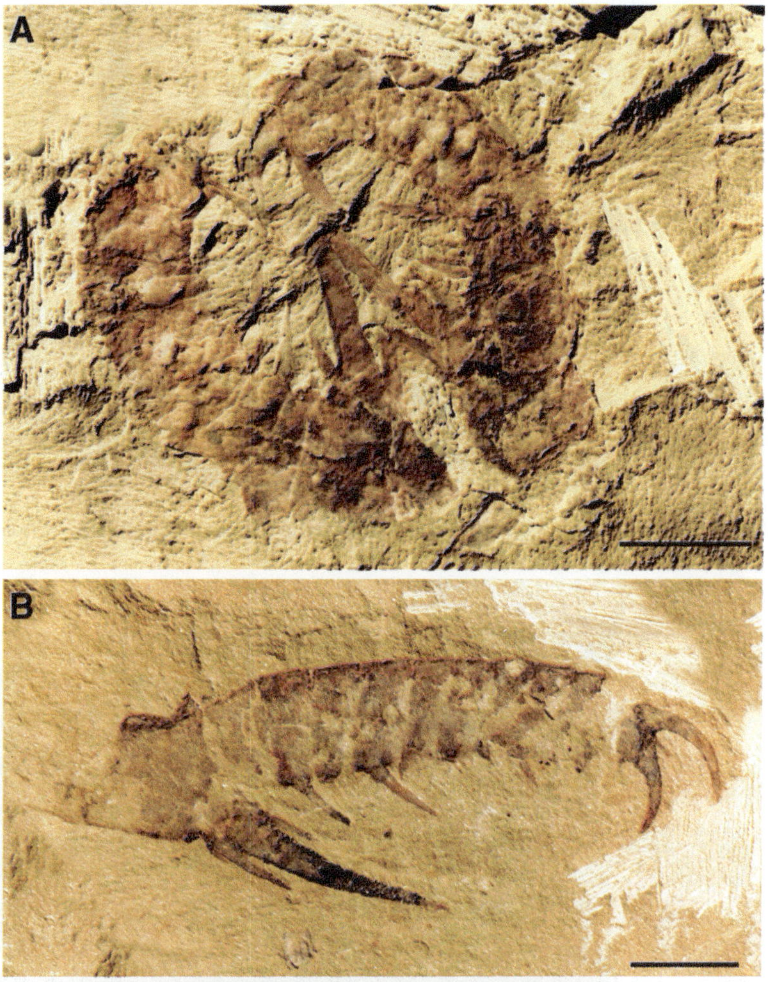

Fig. 111 Formidable "great appendages" of *Amplectobelua symbrachiata* from the Early Cambrian Maotianshan shale. Scale bar = 1 cm

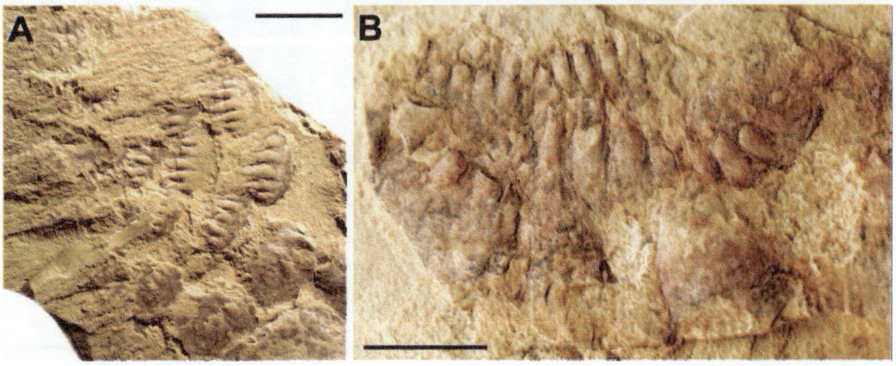

Fig. 112 Formidable mouth apparatus of *Parapeytoia yunnanensis* from the Early Cambrian Maotianshan Shale. Scale bar = 1 cm

euarthropods, the stem-lineage euarthropods ancestral to the Mandibulata, and the stem-lineage Chelicerata, shed new light on the mysterious history of the origin and early evolution of the arthropods. They provide evidence of a grand scenario, with a complex texture of the step-wise evolution that progressed in multiple parallel lines, involving cephalization, sclerotization, arthropozation, arthropodization, segmentation, and tagmatization.

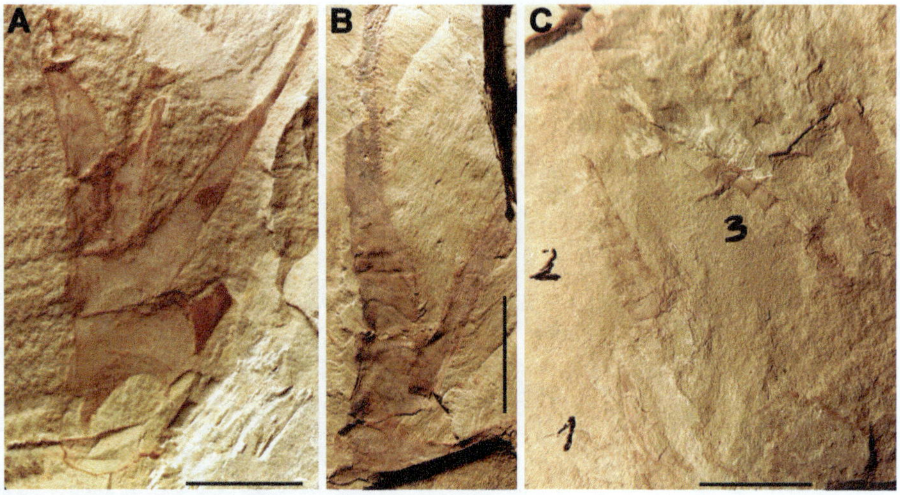

Fig. 113 "Great appendages" (**a**, **b**) and limbs (**c**) in *Parapeytoia yunnanensis* from the Early Cambrian Maotianshan Shale. Scale bar = 1 cm

Fig. 114 *Chuandianella ovata* and coprolites from the Early Cambrian Maotianshan Shale. (**a**) *Chuandianella ovata*; (**b**, **c**) giant coprolites with undigested carapaces similar to those of *Chuandianella ovata*. Scale bar = 1 cm

Cephalization: Origin of the Arthropod Head and the Evolution of Arthropod Head Segmentation

The head accommodates the sensorial and feeding organs externally and the brain, oesophagus, and midgut internally. Cephalization lies at the core of arthropod evolution. Fossils delineate the step-wise evolution of the arthropods in multiple parallel lines, including head segmentation, morphological and functional specialization, and the modification of the head appendages for sensory and feeding roles.

The head is segmented externally, the segmentation providing evidence with which to trace the pathway of its evolutionary changes. However, the segmental homology and segmental identity with the extant arthropod head have been obscured by the fusion, modification, and relocation of these segments in some taxa, especially in the insects. Consequently, the head segmentation of the arthropods has been an enduring puzzle. Comparative studies of representative fossils from the Lower Cambrian Maotianshan Shale at different evolutionary stages have contributed to a new understanding of the evolutionary origin of the arthropod head.

Cephalization is rooted in the early history of the panarthropods. The worm-like animals (tardipolypods) in the Maotianshan Shale document the development of the panarthropod head and its differentiation into three different patterns, the non-segmented head, the one-segmented head, and the two-segmented head, although these early heads are scarcely distinguishable from the trunk (Fig. 115-1–3). It is unclear which of the three heads represents the ground plan of the panarthropod head.

The head is the result of an early tagmosis event, a simple process that organized the segments into an anterior head and a posterior trunk. Tagmosis coevolved with sclerotization. Interestingly, the first sclerotization occurred in the domain related only to the trunk appendages. The head represents the front part, anterior to the anterior sclerites. The head appendages are also usually more laterally situated, more slender, and longer than those on the trunk.

The least complicated proarthropod head represents the ground plan of the arthropod head. It is short, consisting of two segments corresponding to the visual and antennal segments of the anterior head region

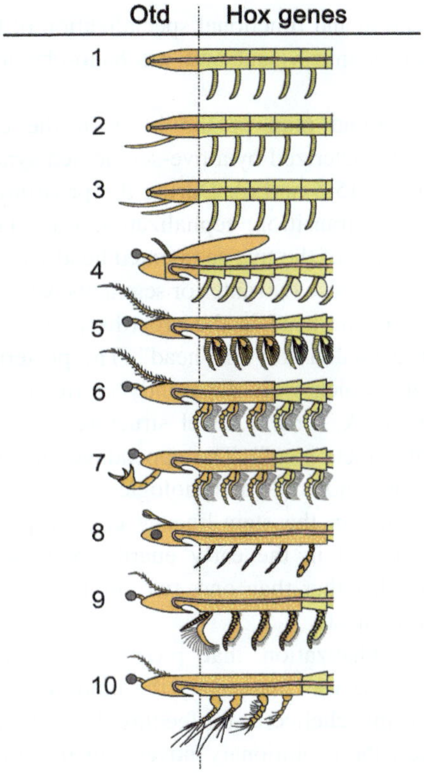

Fig. 115 States of the head in the stem lineage of Panarthropoda with non-segmented head (1), one-segmented head (2) and two-segmented head (3); Proarthropoda (4) with two-segmented head; transitional forms (5) from Proarthropoda and Euarthropoda; stem-lineage euarthropods (6); stem lineage of the Chelicerata (7); Euchelicerata (8); stem lineage of the Mandibulata (9) and Eucrustacea (10)

in the euarthropod syncephalon (Fig. 115-4). The two-segmented primitive head is known as the "procephalon". The visual segment has a pair of stalked eyes and the antennal segment has a pair of short antennae.

Although the head in the first arthropods had the same number of segments as observed in the worm-like ancestors (*Cardiodictyon+Hallucigenia* clade), the morphological gap in the heads of the worm-like ancestors and the first arthropods is very great. The cephalization that occurred in the transition from the worm-like ancestor to the arthropods was a great evolutionary leap, marked by multiple parallel changes that included the relocation of the terminal mouth to a new position (below the second segment) with the backward folding of the visual segment, and the modification and functional specialization of the head appendages into visual and mechano-chemosensory organs.

The ground plan of the head in the euarthropods is characterized by a five-segmented syncephalic head (Fig. 115-5 and 6). During the proarthropod-to-euarthropod transition, cephalization was marked by the formation of the five-segmented head with a bipartite design, whose two anterior segments were virtually identical to the head in the proarthropods, known as the "procephalon" or "pro-head". The posterior three segments, which were apparently recruited from the anterior trunk, form a novel structure, here referred to as the "meta-head". The appendages in the meta-head are biramous and morphologically resemble those in the trunk in the stem-lineage euarthropods. Why the meta-head in the early euarthropods was three segmented, rather than one, two, or four segmented, remains a puzzle.

The cephalization that progressed towards the Cheliceratida was marked by the modification of the antennae into chelicerae for feeding (Fig. 115-7 and 8). However, the evolutionary novelty in the origination of the Mandibulata was the modification of the meta-head appendages into feeding organs. In addition to the morphological and functional modifications to the head appendages, cephalization was also marked by a second round of rearward expansion to form the four-segmented meta-head in the Mandibulata (Fig. 115–9 and 10) and the five-segmented meta-head in the Euchelicerata.

The fossil evidence shows that the evolutionary origin of the Mandibulata head progressed in two separate events. *Ercaia* represents the first event, which was marked by the modification of the first pair of appendages in the meta-head into antennae-like structures. The second event occurred shortly after the first event and is represented in the first eucrustaceans (Zhang et al. 2007; Siveter et al. 2001). This event involved the modification of the second to fourth pairs of appendages into feeding organs (designated the mandible and maxilla 1 and 2, respectively). The Crustacea is the stem group of the Mandibulata, and the insects and myriapods evolved from the crustaceans.

Cephalization that occurred during the evolutionary transition from the crustacean-like ancestor to the insects and myriapods was marked by the modification of the second antenna-bearing segment into the intercalary segment, a segment that has long been considered to be limbless. Because the head of the Insecta has been highly modified and some of the head structures have been displaced, the segmentation of the head and the identity of the head structures have been controversial. The labrum (or upper lip), a plate-like process anterior to the mouth field, is the most unexpected structure. It led some authors to speculate that the anteriormost part of the head is of non-segmental origin. The term "acron" was introduced to represent the non-segmental region. The acron includes the labrum only, according to some authors (Snodgrass 1935; Rempel 1975; Scholtz 2001), but it has been interpreted as representing a fused anterior part of the head that includes the labrum, eye, and protocerebrum by others (Brusca and Brusca 2002). Based on the later interpretation, the pre-oral part of the head is composed either of a non-segmental part and two segments or of a non-segmental part and one segment.

The appendicular origin of the labrum is well supported by both embryological (Eastham 1930; Butt 1960; Rempel 1975) and molecular developmental studies (Schmidt-Ott and Technau 1992). Whether the labrum represents the anteriormost segment or alternatively the intercalary segment and whether the insect head has six or seven segments are ongoing debates. Molecular and neural studies (Boyan et al. 2002; Popadic et al. 1998) have proposed that the labrum is the fused paired appendages of the intercalary segment, so that the modification of the second antennae into a feeding organ was one of the key evolutionary novelties in the origin of the Tracheata (Hexapoda+Myriapoda).

The cephalization in the Chelicerata lineage is marked by the modification of the antennal appendages into feeding organs, known as the "chelicerae". The evolution of the chelicerae progressed in multiple steps. The anomalocaridid type of "great appendages" probably represents the ground plan of the chelicerae, characterized by large, multisegmented arms and four-segmented claws. The extant chelicerae evolved from the ground plan through two transitional steps, a design with a four-segmented claw and a two-segmented arm (in most stem-lineage chelicerates) and a design with a three-segmented claw and a two-segmented arm (in *Haikoucaris*).

The meta-head in the living Chelicerata has five segments. In addition to the modification of the antennae into feeding organs, the cephalization in the

Chelicerata had evolved with a second round of rearward expansion in the meta-head. A seven-segmented mode characterizes the ground plan of the euchelicerates and suggests that the second expansion event occurred in the Silurian, when the first-known seven-segmented chelicerate head is reported in the fossil record (Sutton et al. 2002).

Tagmatization and the Origin of the Molecular Axial Bipartite Domains

Head and trunk bipartization characterizes the tagmatization that occurred during the transition from the worm-like ancestor to the arthropods. However, the head/trunk bipartite design of the body was already clearly recognizable in the worm-like ancestors (as exemplified by the *Cardiodictyon+Hallucigenia* clade). The trunk in the proarthropod *Fuxianfuia* displays a clear bipartite pattern, with a limb-bearing thorax and limbless abdomen, suggesting that the reorganization of the trunk segments into bipartite regions occurred in the proarthropod evolutionary stage. The fusion of some or all of the thoracic segments with the head to form a cephalothorax occurred in many crown-lineage chelicerates.

Molecular and genetic research has confirmed that all bilaterian animals, including the arthropods and humans, have a common genetic mechanism, with the *Hox* cluster genes patterning the anterioposterior axis along the main axis of the body, except in its anteriormost part, which is outside the *Hox* gene expression domain. Recently, it has become clear that the anterior margin of the *Hox* gene domain is conserved among different extant arthropod groups and lies at a position close to the border between the anterior and posterior developing head of the extant arthropods, but is not tied to the morphological boundary between the head and the trunk (Fig. 115-5–10). The anterior head, outside the *Hox* gene expression domain, is comprised of two segments, the anterior ocular and the posterior antennal segments, innervated by the protocerebrum and the deutocerebrum, respectively. The *otd*/*otx* genes are both involved in the patterning of the anterior head and the rostral brain. The boundary of the two developmental subdivisions of the body (the anterior *otd*/*otx*-specified and posterior *Hox*-specified region) is conserved not only across the different extant arthropod groups but also throughout the different bilaterian animals. It is difficult to understand why the molecular boundary does not match the morphological head–trunk boundary. Fossils may provide the answer.

The morphological boundary of the head and trunk is not conservative. The trunk in the proarthropods represents the post-antennal part of the body and the boundary between the head and trunk well matches the molecular boundary between the *Hox* gene expression region anteriorly and the *ems*/*otd* gene expression region posteriorly. The arthropod head has a tendency to expand with the posterior displacement of the morphological head–trunk boundary, so the molecular boundary has a more anterior position than the morphological boundary. Expansion characterizes the evolution of the arthropod head with the gain of a head segment when the head shifted posteriorly beyond the border. The original position of the head–trunk boundary in the arthropods, as exemplified by the proarthropods, was situated at the rear margin of the antennal segment, at a much more anterior position than that in the extant arthropods. The implication is that the molecular boundary between the anterior *otd*/*otx* domain and the posterior *Hox* domain is highly conservative and ancient and that its relationship to head segmentation has not changed since its first establishment in the earliest arthropods.

The ancient head–trunk boundary has vanished morphologically and has virtually left no distinct morphological mark, except in a few extant arthropod species. For instance, in developing *Drosophila*, a transient structure known as the "cephalic furrow" may represent a relic of the ancient head–trunk boundary. The cephalic furrow lies at the junction of the patterning systems that define the procephalic and segmental germs (Costa et al. 1993; Vincent et al. 1997) and the cells in the transient structure all unfold back onto the surface of the embryo, contributing to the ectoderm in later stages (Costa et al. 1993). The cephalic furrow is interpreted as an early organizer region analogous to the midbrain–hindbrain border (MHB) in the vertebrates (Ball-Cuif and Wassef 1995; Joyner 1996). The antennal segment forms a posterior part of the early arthropod head and bears a very large and posteriorly extended cephalic shield. The large and posteriorly extended tergite of the antennal segment is characteristic of the first arthropod head, and a similar structure also reappears in some extant crustaceans (Casanova et al. 2002).

Sclerotization and Arthropozation

The ground state in the exoskeleton of the panarthropods is characterized by a soft and flexible cuticle. The hardening of the cuticle to form hard sclerites is accomplished by sclerotization, which is deeply rooted at the base of panarthropod evolution. As exemplified by the tardipolypodians in the Maotianshan Shale, the early sclerotization event was restricted to the trunk, forming paired, morphologically diverse sclerites dorsal to the trunk appendages.

Arthropozation characterizes the arthropods and represents a significant event during the transition from the worm-like ancestors to the arthropods. Arthropozation is characterized by the hardening of the cuticle into separated plates, which are both metamerically arranged and transversely arranged into four parts (a dorsal tergite, a pair of pleurates, and a ventral sternite). These sclerites are connected by soft, flexible cuticle, known as "articular membrane". Arthropozation is accomplished by the formation of the articular membrane (otherwise known as the "joint"). Early arthropozation was restricted to the head and trunk, as exemplified in the proarthropods of the Maotianshan Shale, and excluded the trunk limbs, which remained soft bodied.

Arthropodization and the Evolution of the Trunk Limb

"Arthropoda", from the Greek word for "jointed feet", refers to those organisms with jointed limbs. Arthropodization is the development of segmented and jointed trunk limbs and is thought to have been a novelty of the arthropods. The first arthropods, exemplified by the proarthropods, had a segmented and jointed body, but soft-bodied and unjointed trunk limbs. This indicates that arthropozation and arthropodization were not synchronous and that arthropozation predated arthropodization.

The diverse arthropod fauna, together with their worm-like ancestors, provides a unique window through which to establish the evolutionary development of the arthropod appendages. It is suggested that the arthropods derived from a worm-like ancestor that closely resembled the "*Hallucigenia*+*Cardiodictyon*" clade. The presumed worm-like ancestor had a simple, elongated body divided into serial segments. Each segment had a pair of soft, unjointed, uniramous appendages, with a single-tube form. These appendages had multiple annulations and lacked an intrinsic musculature. The preservation of a central cavity in some appendages suggests that these limbs were hydraulic. The appendages were paired and undifferentiated, except for the more dorsal, slender, head appendages.

The evolutionary origin of the Arthropoda was marked by the transformation of the head appendages into sensorial organs, and by the transformation of the uniramous trunk appendages (Fig. 116-1) into biramous appendages (Fig. 116-2). The jointed limbs arose during the transition from the proarthropods to the euarthropods. The first jointed limbs (Fig. 116-3) were tubular, tapered slowly, and were composed of 17–20 regular annulus, connected to one another by pivot joints (Hou et al. 1999, 2004). The presence of a pivot joint suggests the presence of

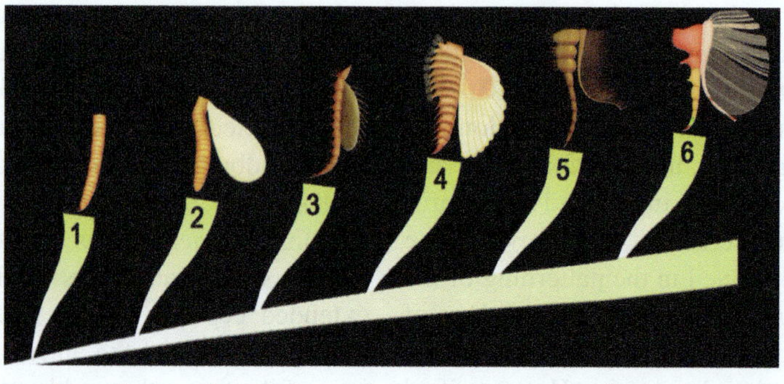

Fig. 116 Evolutionary scenarios in trunk limbs of Panarthopoda. (1) unjointed monoramous limb; (2) biramous appendages with unjointed exopod; (3) biramous appendages with multisegmented exopod; (4) biramous appendages with multisegmented, bipartite exopod; (5) biramous appendages with multisegmented, bipartite exopod but fewer segments in the proximal portion; (6) biramous appendages with bipartite exopod

intrinsic muscles and that these annuli were true segments.

The biramous trunk is a novelty of the arthropods, and the first biramous trunk appendages were characterized by an unjointed endopod and gill-less exopod. The ground plan of the jointed endopod was tubiform and multisegmented. The euarthropod limb has fewer segments (seven or fewer) with strong distal–proximal differentiation, forming a bipartite design: a distal slender part and an expanded basipod. The euarthropod limb developed from the multisegmented tubiform, through intermediate stages. The limb in the intermediate stage is characterized by a bipartite design, with a multisegmented and expanded proximal part (Fig. 116-4, 5). It is proposed that the evolution of the euarthropod limb progressed in two separate steps: distal–proximal axial differentiation and the fusion of the multisegmented, expanded part into a single basipod.

Arthropodization occurred during the evolutionary transition from the proarthropods to the euarthropods and involved the formation of segmented and jointed trunk limbs. Euarthropoda has been defined as monophyletic, both in terms of its five-segmented head and its post-antennal limb design (Waloszek 1999; Maas and Waloszek 2001). The authors propose that the post-antennal limb in Euarthropoda consists of a rigid, anterioposterior-compressed proximal part (the basipod) and a seven-segmented endopod (Fig. 116-6). However, the bipartite design of this limb along its main axis does not always appear in the stem groups of the Euarthropoda. In numerous species of the distally related stem groups of the Euarthropoda (*Canadapsis, Isoxus, Ercaia, Primicaris, Tanglangia, Fortiforceps, Leachoilia, Saperion*, and the bradoriids), the post-antennal limbs do not display the expected bipartite design, but a simple, tubular form with slight differentiation along the main axis and 9–15 segments. This implies that the bipartite limb design is not monophyletic. The anterioposterior-compressed basipod with its expanded and serrated inner edge was probably associated with feeding activity, breaking large food items into pieces, and with the anterior transportation of food to the mouth. The origin of the expanded basipod with its serrated inner edge is a pivotal event in the change from microphagous to macrophagous feeding behaviour.

Proximodistal specification is the basis of the evolutionary change in limb design. This change drove the limb towards a progressive increase in its differentiation along the main axis, in a stepwise manner from the simple, multisegmented, tubular limb to the bipartite design, via intermediate steps. The limb design during these intermediate steps is exemplified by *Canadaspis* (Fig. 116-4) and *Saperion* (Fig. 116-5), which display clear proximodistal domain differentiation, with an expanded, multisegmented, proximal domain. The transformation of the intermediate form into a limb with two functional domains (Fig. 116-6) probably resulted from the fusion of the multisegmented, proximal domain into a single-segmented proximal domain, known as the "basipod".

Both the limbs and antennae of extant arthropods express the same proximodistal axis patterning genes during their development, including *Distal-less* (*Dll*), *dachshund* (*dac*), and *homothorax* (*hth*) (Dong et al. 2001). The products of these patterning genes either act together or act in mutually exclusive ways. The mutually exclusive regulatory gene products allow the development of sharp boundaries between domains. Both the limb and antennae, as exemplified in the proarthropods, have a simple tubular shape, with the least proximodistal differentiation in the early arthropods. This mutually exclusive genetic regulatory interaction underlying limb or antennae patterning had not yet appeared in the early arthropods. The presence of bipartite antennae, known as "prochelicerae", in the stem groups of the Chelicerata and bipartite limbs among the stem groups of the Euarthropoda implies that this mutually exclusive regulatory interaction of the patterning gene products evolved to define the sharp boundary between these domains in both the antennae and limbs.

The differentiation of the head and trunk appendages suggests that two axial molecular domains, such as the *otd*/*ems* and *Hox* gene domains, had been established at the very beginning of arthropod evolutionary history. The *otd*/*ems* genes in the head region were selectively expressed to modify the simple, tubular head appendages into stalked compound eyes and antennae, respectively. The *Hox* genes were selectively expressed in the trunk region and were responsible for biramous appendage development. The exopod is a novel structure in the arthropods and its flap-like form represents an extended pouch of the body wall, which resembles the insect wing in its general morphology. Its evolution was presumably regulated by a gene cassette similar to that involved in the developing wings of extant insects.

The speciation of the appendages in arthropods evolved roughly in an anterioposterior direction along the body axis during arthropod evolution. The first event of appendage speciation was the transformation of the first appendage into a visual organ, the eye stalk. The following event was the independent modification of the short limb-like structure in the second segment into a whip-like sensorial or grasping organ, at a transitional stage from the proarthropods to the euarthropods.

The speciation of the post-antennal appendages in the head of the Mandibulata involved three separate events. The earliest event, exemplified by *Ercaia*, was the transformation of the first post-oral appendage into the second antenna. The recent discovery of Early Cambrian eucrustaceans from both north-eastern Yunnan, China (Zhang et al. 2007), and Shropshire, England (Siveter et al. 2001) suggests that the origin of the gnathal appendages followed soon after the origin of the second antenna. The gnathal appendages (called the mandible, maxilla 1, and maxilla 2) are feeding organs, comprising structures modified from the second to fourth post-antennal appendages.

The appendages also differentiated in a caudal-to-anterior direction. A loss of appendages characterizes this anterior-directed differentiation, linked to the process of abdominalization (Boxshell 2004). Abdominalization may have occurred repeatedly in the evolutionary history of the arthropods. *Fuxianhuia* is a fossil representative of the earliest abdominalization event at the very beginning of arthropod evolutionary history.

Centralization of the Nervous System

The developmental regulatory genes of the orthodenticle (*otd/otx*) gene family and the *Hox* gene family not only act in the specification of the embryonic, anterioposterior body axis but are also involved in brain development in both insects and mammals. The *otd/otx* genes are involved in the development of the rostral part of the brain and the *Hox* genes in its caudal development. The molecular and genetic similarities in the brain development of insects and mammals have led some researchers to draw the conclusion that a tripartite organization of the embryonic brain was already established in the last common urbilaterian ancestor of the protostomes and deuterostomes (Hirth et al. 2003). Other studies refute the early origin of the tripartite brain organization (Chen 2008, 2009). The arthropods have recognizable head segments, and these and the corresponding appendages appear closely related to brain regionalization. The relationship between the head segment and brain regionalization is consistent across distally related groups of extant arthropods implying that the relationship is very highly conservative and ancient, and established at the first appearance of the arthropods. The visual system in the ocular segment is an important component of the protocerebrum, and the olfactory centres of the antennal segment are the major components of the deutocerebrum. All these systems and the related protocerebrum and deutocerebrum were already present in the first arthropods, as exemplified by the Lower Cambrian proarthropods.

The caudal brain comprises the posterior suboesophageal ganglion and three sub-oesophageal neuromeres; these are connected to the second antennal segment and the three gnathal segments of the posterior head, respectively. The caudal brain, connected to the post-antennal head segments, apparently had not developed in the first arthropods, suggesting the caudal brain is a secondarily derived structure probably originating from the modification of the trunk ganglia. In extant insects, the caudal brain (tritocerebrum) is composed of the mandibular, maxillary, and labial neuromeres; these are connected to the feeding organs, including the labrum, mandibular, maxillary, and labial organs. The stem groups of the Euarthropoda had only three post-oral segments and related unspecialized appendages. These post-oral head segments and their related appendages were strikingly homologous to those in the trunk. It is unlikely that the caudal brain had evolved in the early euarthropods.

Grand Evolutionary Scenario

Panarthropoda represents a monophyletic taxon that presumably derived from a segmented appendageless worm-like ancestor, as shown in Fig. 117 (Fig. 117-1). The monoramous unjointed trunk limbs were the key novelty in the evolutionary origin of this group. Cephalization and sclerotization occurred within the trunk during the early evolution of the panarthropods and led to the formation of three principal

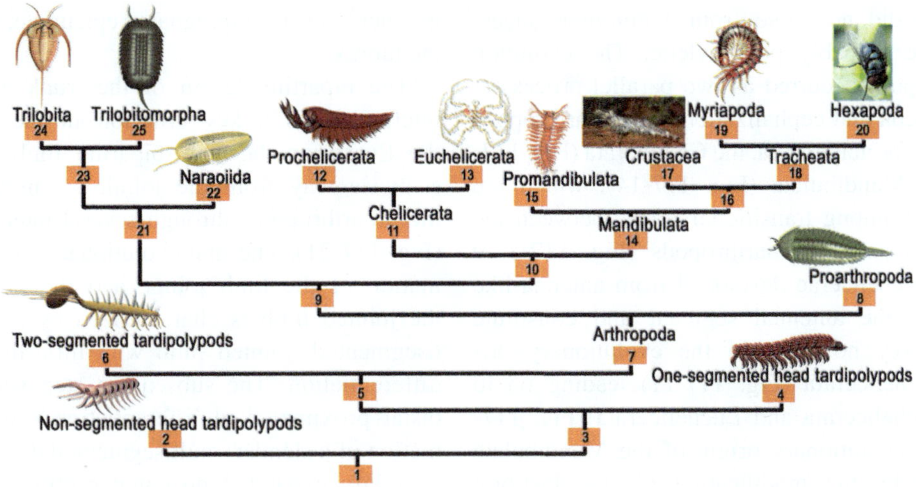

Fig. 117 Diagram of phylogenetic kinship of the high ranking clades among the Panarthropoda. (1) common ancestor of Panarthropoda characterized by paired trunk limbs which were monoramous and unjointed; (2) stem-lineage Panarthropoda with non-segmented head; (3) origination of segmented head; (4) stem-lineage Panarthropoda with one-segmented head; (5) origination of two-segmented head; (6) stem-lineage Panarthropoda with two-segmented head; (7) evolutionary origin of the arthropods (with key novelties: relocation of mouth, transformation of 1st pair of head appendages into stalked eyes, 2nd pair of head appendages with antennae, head and trunk arthropodization, and transformation of trunk limbs from monoramous to biramous; (8) the first arthropods (proarthropods); (9) evolution towards euarthropods; (10) evolution towards crown lineage arthropods; (11) evolutionary origin of Chelicerata; (12) stem-lineage Chelicerata; (13) Euchelicerata; (14) evolutionary origin of Mandibulata—marked by transformation of the 1st pair of post-oral appendages into 2nd antennae; (15) the *Ercaia*-like first Manbibulata; (16) evolutionary origin of the Eumandibulata (marked by transformation of the 2nd to 4th post-oral head appendages into feeding organs); (17) Crustacea representing the stem group of the Eumandibulata; (18) evolutionary origin of Tracheata—marked by transformation of the 2nd antennae into a feeding organ (the labrum), formation of a novel tracheal system and degeneration of the exopods; (19) Myriapoda representing the stem-lineage of the Tracheata; (20) Hexapoda arising from a myriapod-like ancestor, presumed to be through parallel lines of evolutionary change including tagmatization of the trunk leading to formation of a three-segmented thorax and limbless abdomen, origin of water-conservation structures (the spiracle with closing mechanism, excretion of uric acid, and the waxy-impregnated epicuticle), and origin of wings on the thorax; (21) extinct euarthropods characterized by having bipartite post-oral limbs; (22) Naraoiids characterized by having a bipartite body, non-mineralized exoskeleton and absence of eyes; (23) origin of tripartite bodied euarthropods; (24) Trilobita characterized by having a mineralized exoskeleton, sessile dorsal eyes, presence of three axial lobes and tripartite body; (25) Trilobitomorpha by non-mineralized exoskeleton, stalked and ventrally situated eyes

stem lineages characterized by different types of head design: the non-segmented-head group (Fig. 117-2), the one-segmented-head group (Fig. 117-4), and the two-segmented-head group (Fig. 117-6). The phylogenetic relationships among the three different groups are unclear. The non-segmented-head group is presumed to be the primitive one, whereas the two-segmented-head clade represents an advanced group. The arthropods presumably derived from a worm-like ancestor with a two-segmented head (Fig. 117-6) resembling *Hallucigenia+Cardiodictyon*. The transition from the worm-like ancestor to the arthropods (Fig. 117-7) was an evolutionary leap achieved via numerous synchronous events. These included arthropozation, cephalization, and the transformation of monoramous limbs to biramous limbs. Arthropozation represents an advanced evolutionary stage of sclerotization, characterized by the presence of a hardened cuticle on both the trunk and head, but not on the limbs. The hard cuticle is divided into plates (sclerites) separated segmentally and transversely, and connected by flexible cuticle known as the articular membrane. Cephalization occurring at the origin of the arthropods was marked by evolutionary novelties such as the relocation of the mouth opening from the anterior terminus to a ventral position at the level of the second segment, and the modification of the head appendages to visual organs, including stalked eyes and antennae. The trunk region was modified by the transformation of the monoramous limbs into biramous limbs.

The fossils in the Maotianshan Shale show that the arthropods are a monophyletic taxon, but that the euarthropods did not arise from a common ancestor and were probably polyphyletic. The evolution of the arthropods occurred as two parallel processes: arthropodization and cephalization. Living arthropods consist of two principal taxa, the Chelicerata (Fig. 117-11) and the Mandibulata (Fig. 117-14), which are deeply rooted among transitional stages between the proarthropods and the euarthropods (Fig. 117-8, 9 and 10). The chelicerae developed from antenna-like structures on the antennal segment and constitute one of the key novelties in the evolutionary origin of the Chelicerata (Fig. 117-11) leading on to the taxa Prochelicerata and Euchelicerata (Fig. 117-12, 13). The evolutionary origin of the Mandibulata was marked by the modification of the post-oral head appendages into antenna-like structures, known as the second antennae, as exemplified by *Ercaia* from the Maotianshan Shale. This animal has a pair of second antennae, but unlike extant Crustacea, its other post-oral head appendages remain similar to the trunk appendages suggesting that the evolution of the Mandibulata proceeded in a step-wise manner towards Promandibulata (e.g. Fig. 117-15). The transformation of the first post-oral head appendage into an antenna-like structure occurred earlier than the evolutionary transformation of the post-oral second to fourth head appendages into feeding organs, leading to the origin of the Eucrustacea and Crustacea (Fig. 117-16 and 17). The transformation of the second antennae into a feeding organ, known as the "labrum", represents one of the important evolutionary novelties leading to the origin of the Tracheata (Fig. 117-18). The origin of the Tracheata was also marked by the innovation of the tracheal system and the degeneration of the exopods. The Tracheata comprises two possibly monophyletic taxa, the Myriapoda (Fig. 117-19) and the Hexapoda (Fig. 117-20). The Myriapoda probably represents a stem group of the Tracheata (Fig. 117-18), with a bipartite body, a head, and an elongated untagmatized trunk containing numerous segments, each bearing a pair of limbs. The tracheal system is at a primitive stage, with spiracles that usually cannot be closed and are not well protected from water loss. The Hexapoda (Fig. 117-20) probably arose from a myriapod-like ancestor, along parallel lines of change, including the following: tagmatization of the trunk leading to the formation of the three-segmented thorax and limbless abdomen; water conservation structures including spiracles with a closure mechanism; the excretion of uric acid; a wax-impregnated epicuticle; and wings on the thorax.

The bipartite design of the trunk appendages is interpreted as a key criterion in the definition of the Euarthropoda. This bipartite limb evolved in a stepwise way from the jointless annulated limb of the proarthropods, through several intermediate steps (Fig. 117-21). The first evolutionary step was the formation of the limb joints, and the ground plan of the jointed limb is characterized by a slender, multisegmented, jointed limb with little distal–proximal differentiation. The subsequent step was marked by distal–proximal axial differentiation leading to the formation of a slender, multisegmented distal portion and a greatly expanded proximal portion. The bipartite limb with a single, expanded basipod and a multisegmented, slender distal portion, described as the "euarthropod type of limb", probably arose from a bipartite limb with a multisegmented, expanded proximal portion through the fusion of the segments in the proximal portion. The fossils in the Maotianshan Shale show that the ancestral forms of both the Chelicerata and Mandibulata had slender and multisegmented trunk limbs. The euarthropod type of limb apparently evolved independently in the two taxa. Interestingly, the Trilobita (Fig. 117-24), Trilobitomorpha (Fig. 117-25), and Naraoiida (Fig. 117-22), the three distinct groups considered the stem-lineage euarthropods, all have trunk limbs with the typical euarthropod design (Fig. 117-23).

Vetulicolians and Their Affinities

Vetulicolians are large Cambrian animals with a tripartite body composed of a simple cone-shaped head, a larger expanded, segmented thorax, and a dorsoventrally flattened abdomen. The mouth and anus open at opposite ends of the body. The thorax is enclosed by an exoskeleton composed of a pair of flexible pleural membranes and dorsal and ventral plates. The exoskeleton is modified to form lateral segmentally arranged gill chambers. These shared features suggest that the vetulicolids are a monophyletic group. They are an extinct group, known in both the Lower Cambrian Maotianshan Shale and the Middle Cambrian Burgess Shale. In the Burgess Shale, they are represented by the highly modified species *Banffia constricta* (Walcott 1911b; Caron 2006).

The vetulicolids include *Vetulicola cuneata* (Hou 1987), *Didazoon haoae* (Shu et al. 2001b), *Yuyuanozoon magnificissimi* (Chen et al. 2003a), *Heteromorpha confusa* (Chen and Zhou 1997; Luo et al. 1999), and *Pomatrum ventralis* Lu and Hu 1999 (= *Xidazoon* Shu et al. 1999a) from the Lower Cambrian Maotianshan Shale and *B. constricta* (Walcott 1911b; Caron 2006) from the Middle Cambrian Burgess Shale. Of these, *V. cuneata* occurs commonly in the Maotianshan Shale and is characterized by a large, laterally compressed thorax and a dorsoventrally flattened abdomen. The thorax is sub-rectangular in lateral view, with a compressed cross section, and a height twice as great as its width. The thorax has five fused segments, each with a pair of gill-bearing chambers (Figs. 118 and 119). The exoskeleton comprises four plates transversely comparable to the dorsal plate, the pair of lateral flexible pleural membranes, and the ventral plate in Arthropoda. The dorsal and ventral plates are both finned in their median planes and are thick, consisting of a thin, dark outermost layer (Fig. 119d) and a thick, pitted inner layer (Fig. 119a), comparable to the epicuticle and procuticle in the arthropods, respectively. The pitted structure of the inner layer in *V. cuneata* indicates the presence of microscopic canals. Identical microscopic canals are also present in the inner layer of the arthropod sclerites (Brusca and Brusca 2002). In addition to fine pits, the inner layer also displays a distinct transverse linear structure (Fig. 119a); this may represent the segment boundary.

The lateral pleural membranes are thin and flexible, extending across the middle of the thorax for its entire length. They are contractive and usually preserved in the contracted state (Fig. 119a, b, d). Figure 119c shows a fully exposed pleural membrane.

Each of the thoracic segments has a pair of pear-shaped, gilled chambers with their narrower ends pointing anteriorly (Fig. 120). As in many living crustaceans, the gill chambers are formed by invaginations of the exoskeleton and have a wall with gill filaments lodged on its inner side; probably protecting the fragile gill filaments from damage. Similar gill-bearing branchial chambers occur widely in extant decapod crustaceans. Anterior to the thorax are two pairs of sclerites, possibly protecting the soft-bodied head.

Fig. 118 Reconstruction of *Vetulicola cuneata*

The abdomen is limbless, oar-like, and compacted dorsoventrally. It consists of seven jointed segments, each boxed by a dorsal plate, with two narrow lateral pleural joints and a ventral plate, exactly as in the arthropods (Brusca and Brusca 2002). The abdomen, rounded anteriorly and increasingly flattened posteriorly, inserts dorsally into the rear end of the thorax. The segments are equal in length, apart from the terminal one which is semi-circular in shape and much longer than the others. The gut is regularly bead like (Fig. 121), with an anus opening ventrally near the rear end of the last segment (Fig. 119d).

P. ventralis is usually 7–10 cm long (20 cm maximum length), with an elongated thorax, consisting of five segments, each rounded in cross section (Fig. 122). Each segment has a pair of gill chambers. The abdomen resembles that of *Vetulicola* and is subdivided into seven equal segments, except for the last, much longer one. The head is exposed externally, with no protection from anterior sclerites as seen in *Vetulicola*, and it is small, short, and conic, only 3–4 mm long and 2 cm in diameter. Its gut is identical to that of *Vetulicola*, with a bead-like shape.

Heteromorphus confusa is soft bodied, usually 3.5–5 cm long, maximally 6 cm (Fig. 123). It has a small, rounded head protruding anteriorly from the thorax. The thorax is rectangular in side view, laterally flattened, and consists of four segments, each bearing a pair of lateral gill chambers. Like the Middle Cambrian *B. constricta,* the abdomen has 20 closely arranged segments. The segments are of equal length, except for the last, much longer one; this is semi-circular in outline when viewed from the dorsal surface. The gut is simple and straight, ending ventrally at an anus near the terminal end of the last segment.

Y. magnicissimi is a large vetulicolid, about 20 cm long (Fig. 124). A short, rounded head projects anteriorly from the thorax. The thorax is an elongated oval in outline, composed of five segments, each bearing a pair of gill chambers. The abdomen consists of seven segments, probably sub-rounded in cross section.

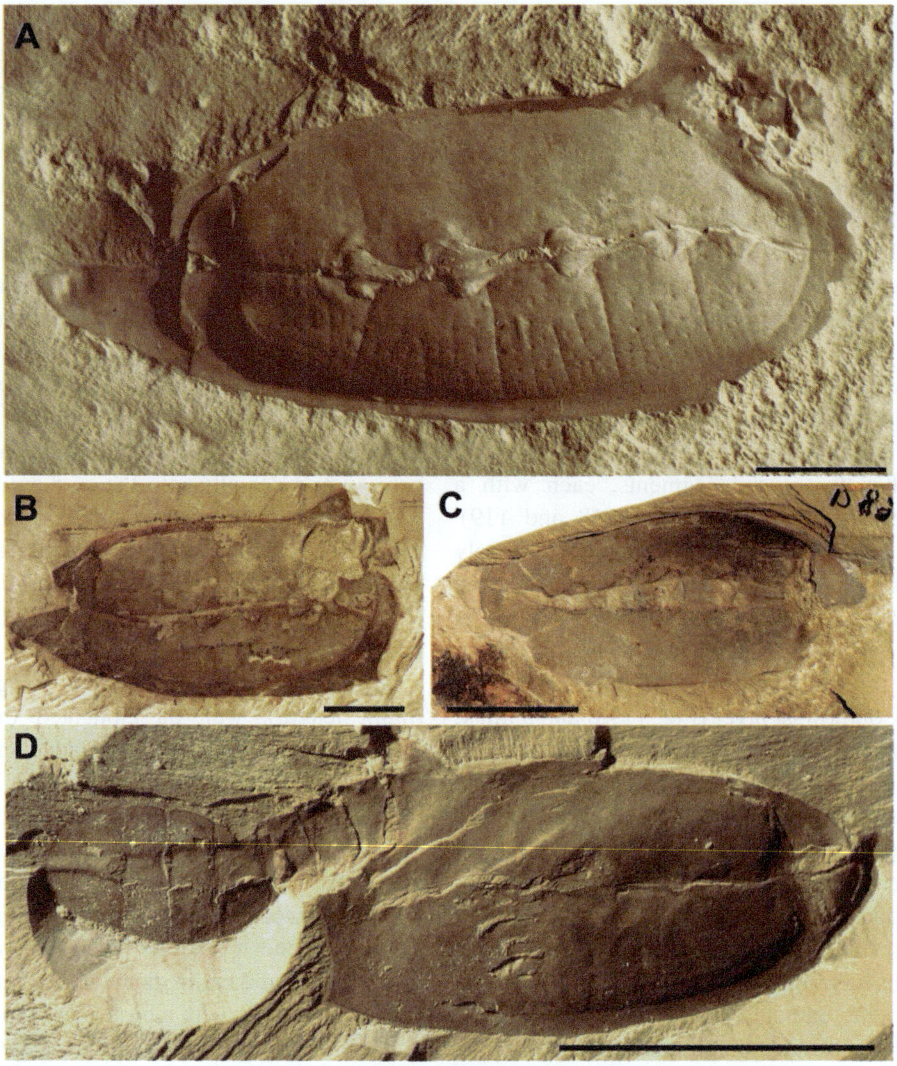

Fig. 119 *Vetulicola cuneata* from the Early Cambrian Maotianshan Shale showing fused segmented structure (**a**), pitted exoskeleton (**a**), contracted (**a**, **b**, **d**) and exposed lateral pleural membrane (**c**), and gilled chambers (**a**–**d**) of thorax, and limbless oar-like abdomen with jointed segments. Scale bars = 1 cm (**a**, **b**), 2 cm (**c**) and 3 cm (**d**)

Fig. 120 Pear-shaped gilled chamber and gills within the chamber in *Vetulicola cuneata* from the Early Cambrian Maotianshan Shale. Scale bar = 1 cm

Fig. 121 A laterally compacted abdomen of *Vetulicola cuneata* from the Early Cambrian Maotianshan Shale showing bead-like gut. Scale bar = 1 cm

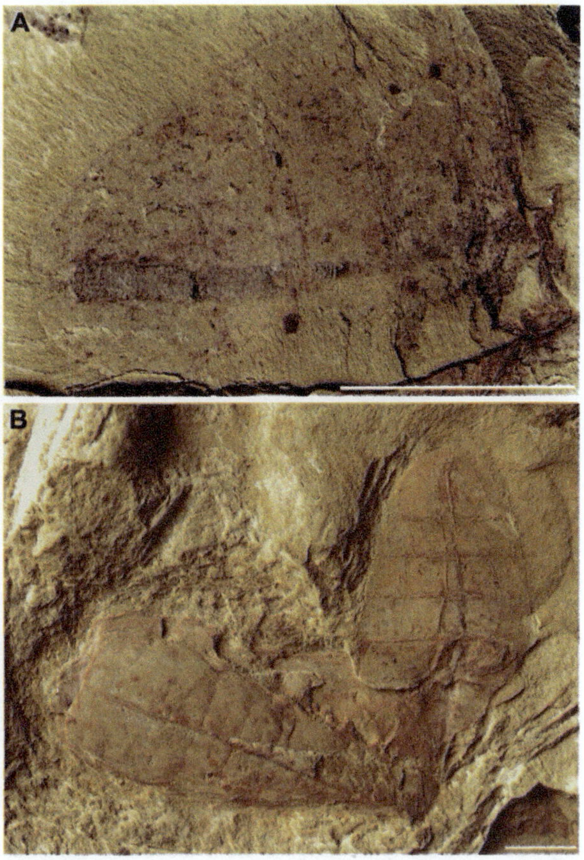

Fig. 122 *Pomatrum ventralis* from the Early Cambrian Maotianshan Shale. (**a**) dorso-ventrally compacted abdomen showing a simple straight gut; (**b**) dorso-ventrally compacted, nearly complete specimen showing a short, conical mouth, segmented thorax, a jointed abdomen, and fine straight gut with terminal anus. Scale bar = 1 cm

Affinities

The vetulicolids are a monophyletic taxon, sharing a tripartite body comprising a small non-segmented head, a fused segmented thorax with paired gill-bearing branchial chambers, and a segmented limbless abdomen. However, the affinities of the vetulicolids are controversial. They are interpreted as arthropods or an arthropod-related animal group (Chen and Zhou 1997; Bergström and Hou 2003; Chen 2004; Briggs et al. 2005; Caron 2006), or as either basal (Shu et al. 2001b) or more derived deuterostomes, probably close to the tunicates (Gee 2001; Lacalli 2002; Aldridge et al. 2007), or even as vertebrates (Shu et al. 1999a).

Their arthropod affinities are based on the segmented and jointed exoskeleton, and the transverse division of four skeletal plates: two lateral pleural plates and a dorsal and a ventral hard plate. The gill chambers are segmentally arranged in the pleural region, walled with membrane, and embedded within the exoskeleton, resembling those of many living decapod crustaceans. The gill chambers are apparently modified exoskeleton, representing pharynx-unrelated structures, in contrast to the pharynx-related gill pouches in the vertebrates. The exoskeleton of *Vetulicola* is not only thick, but is formed of a thin, dark outermost layer and a thick, pitted inner layer, comparable to the exoskeleton in the arthropod epicuticle (Brusca and Brusca 2002). However, the differences between the vetulicolids and the arthropods are also great. For instance, the vetulicolid head is short, simple (lacking eyes and other structures), and non-segmented, with a mouth situated at the terminal end. The thorax and appendages are fully enveloped by the exoskeleton (the gilled chambers are interpreted

Fig. 123 *Heteromorphus confuse* from the Early Cambrian Maotianshan Shale. (**a**) restoration; (**b**) a nearly complete specimen, showing gilled chambers in thorax and segmented abdomen; (**c**) a nearly complete specimen showing gilled chamber in the thorax and a flattened abdomen; (**d**) a dorsoventrally compacted abdomen, showing a large terminal segment, hemicircular in outline. Scale bar = 1 cm

Fig. 124 *Yuyuanozoon magnicissimi* from the Early Cambrian Maotianshan Shale. Scale bar = 6 cm

as modified exopods). We here propose that the vetulicolids are an extinct panarthropod group with a close affinity to the arthropods.

The loss of the endopods in the thorax and the development of modified exopods concealed within the exoskeleton suggest that the group may have adapted for pelagic life.

A Stalked Animal: *Dinomischus* and Its Possible Affinities

Dinomischus is a Cambrian mystery, known from both Middle Cambrian Burgess Shale, Canada (Conway Morris 1977b), the Kali biota, Guizhou, China (Peng et al. 2006), and the Lower Cambrian

Maotianshan Shale, Yunnan, China. *Dinomischus* in the Maotianshan Shale is represented by *D. venustus* (Chen et al. 1989b; Chen and Erdtmann 1991; Chen 2004). It is a solitary sessile organism, ranging from a few centimetres to 10 cm in height, and attaches to the sea floor by a stalk (Fig. 125). The stalked calyx-like body looks loosely like a flower and is enclosed by a whorl of 18 short petal-like plates. The petals are solid, plate-like structures (Fig. 126), smooth externally with elongated folds to provide strength (Fig. 127c). The inner surfaces of the petals are transversely wrinkled (Fig. 126), presumably representing the attachment sites for ciliary bands, playing a role in creating currents that facilitate food collection and its transport to the mouth. These petals are preserved in either an upright state (Fig. 126) or extended horizontally (Fig. 127), suggesting that they are jointed to the calyx at their proximal ends. These movable petal plates form a feeding bowl of adjustable size.

In addition to its petal plates, *D. venustus* had an elongated, hard sclerite plate (Fig. 126), extending distally from the lateral margin of the calyx and presumably jointed to the lateral margin of the calyx at its proximal end. This plate probably acted as a backstop against the current, suggesting that the animal favoured environments in which the current was oriented in a single direction. The flower-shaped feeding bowl plays a role in down-current feeding. The mouth and anus occur next to one another within the feeding bowl. The flow bringing food enters the feeding bowl from the rear, passing through the spaces between the petal plates and flowing away with the waste water. The ciliary bands on the inner surface of the petal plates act in the food collecting by wafting food particles and prey downward to the mouth.

Fig. 125 Restoration of *Dinomichius venustus* (from Chen 2004)

Affinities

Dinomischus shows close affinity to the extant Kamptozoa (= Entoprocta), sharing a number of key characters, including a stalked calyx-like body, a downstream feeding mechanism, and a U-shaped gut with the mouth and anus positioned within the feeding bowl. Kamptozoa (= Entoprocta) is a small phylum, represented by 150 living species, all less than 1 mm in size, much smaller than *Dinomischus*. Most living kamptozoans are colonial in habit and their bodies are very soft.

Brachiopods

The Brachiopoda constitute a phylum of solitary marine animals; they are also known as "lamp shells". Although there are only 300 living species today, they have rich evolutionary history as they were the most diverse and prolific group of invertebrates in the Palaeozoic seas. The soft body is protected by a pair of mineralized dorsoventrally oriented valves referred to as the brachial or dorsal valve and the pedicle or ventral valve. The former houses the brachidium (lophophore) and gut while in the latter is the stalk or pedicle for attachment to the substrate. The pedicle valve has the opening from which the predicle itself protrudes. In the few forms without a pedicle, the pedicle (or ventral valve) is cemented directly to a hard substrate.

The phylogenetic position of the Brachiopoda remains controversial. Although molecular studies have excluded the brachiopods from the deuterostomes, they do share a number of characters that are

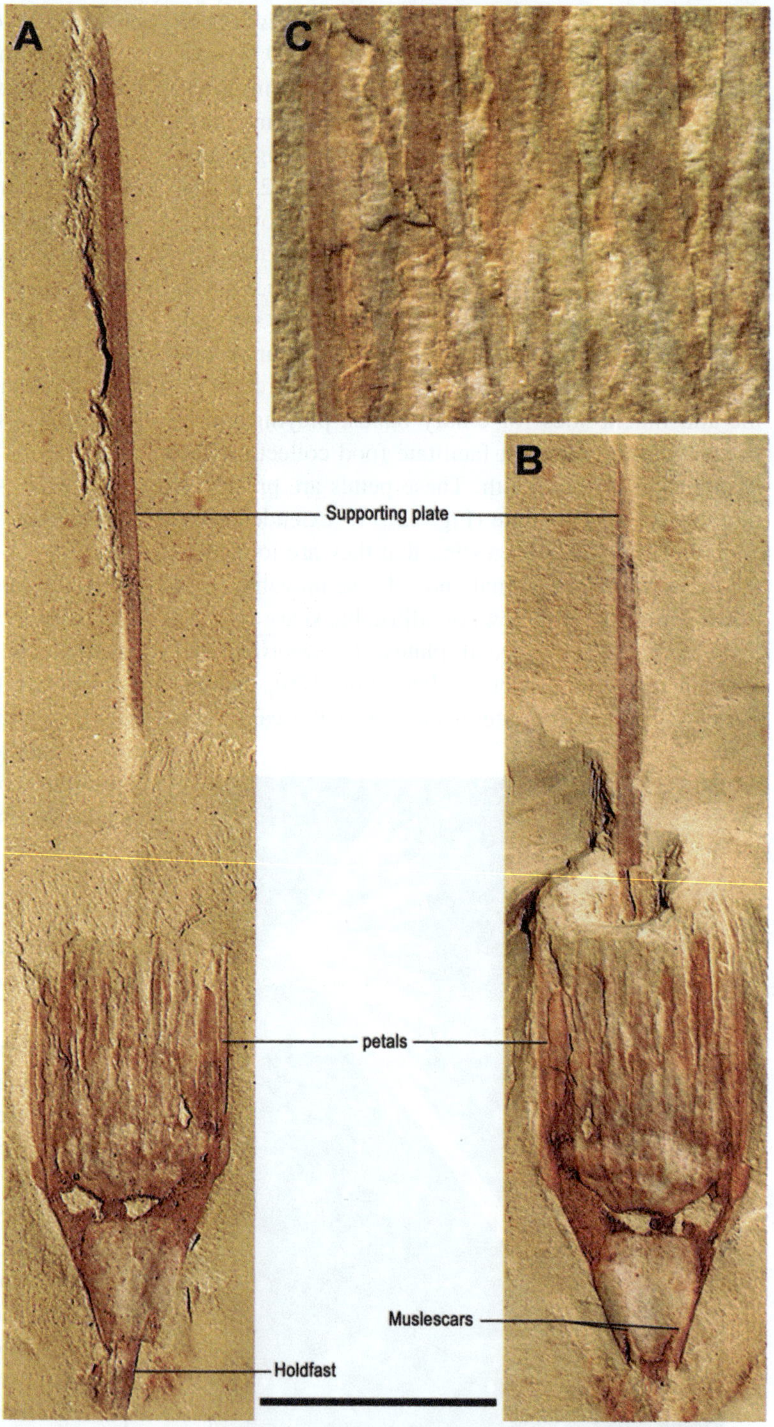

Fig. 126 *Dinomichius venustus* from the Early Cambrian Maotianshan Shale. Part (**a**) and counterpart (**b**) of holotype, and enlargement of petals, showing wrinkled inner surface (**c**). Scale bar = 1 cm

unique to the deuterostomes (Nielsen 2001). These characters include an apical organ containing several to many serotonergic cells, in contrast to the protostomes, which have only a few (2–3) serotonergic cells; an apical organ either absent or not incorporated into the adult nervous system; the archimeric body typical of the deuterostomes, with a well-developed protocoel in the lava of *Crania*, a mesosome that sends canals to

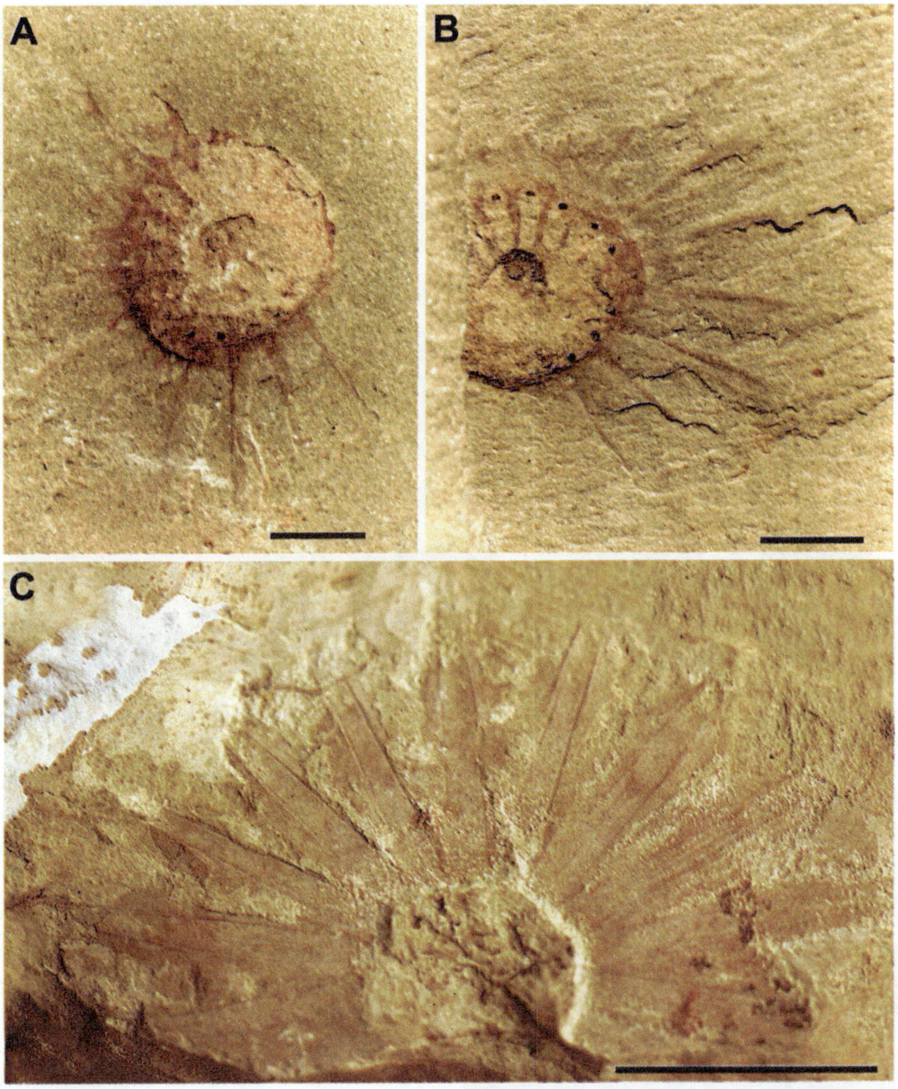

Fig. 127 *Dinomichius venustus* from the Early Cambrian Maotianshan Shale showing the feeding bowl in widely extended state. Scale bars = 0.2 cm (**a**, **b**) and 1 cm (**c**)

the tentacles, and a blastopore situated at the posterior end and closed in later larva, with the mouth opening at the opposite end in living *Crania*.

Brachiopods in the Maotianshan Shale occur as two different forms: infaunal or endobenthic lingulaceans and epibenthic discinaceans. The shells in both forms are chitinophosphatic and have a pedicle. The endobenthic lingulaceans anchor to the bottom of their burrows by relatively long, elastic pedicles, whereas the pedicle in the epibenthic discinacean *Heliomedusa* is short, protruding through a slit in the pedicle valve and attaching to a hard substrate.

The lingulaceans in the Maotianshan Shale are represented by *Lingulella chengjiangensis* (Jin et al. 1993), *Lingulellotreta malongensis* (Rong 1974; Holmer et al. 1997; Zhang et al. 2004, 2006), and *Longtancunella chengjiangensis* (Hou et al. 1999). *L. chengjiangensis* (Fig. 128) is common in the Maotianshan Shale and its shell is small (8–12 mm long), elongated, triangular, and gently convex. The embo is rounded, about 1 mm in diameter. The pedicle valve has a narrow but relatively tall pseudointerarea, with a pedicle foramen at its peak. The pedicle is slender, fairly long. Internal muscles form a shield-shaped

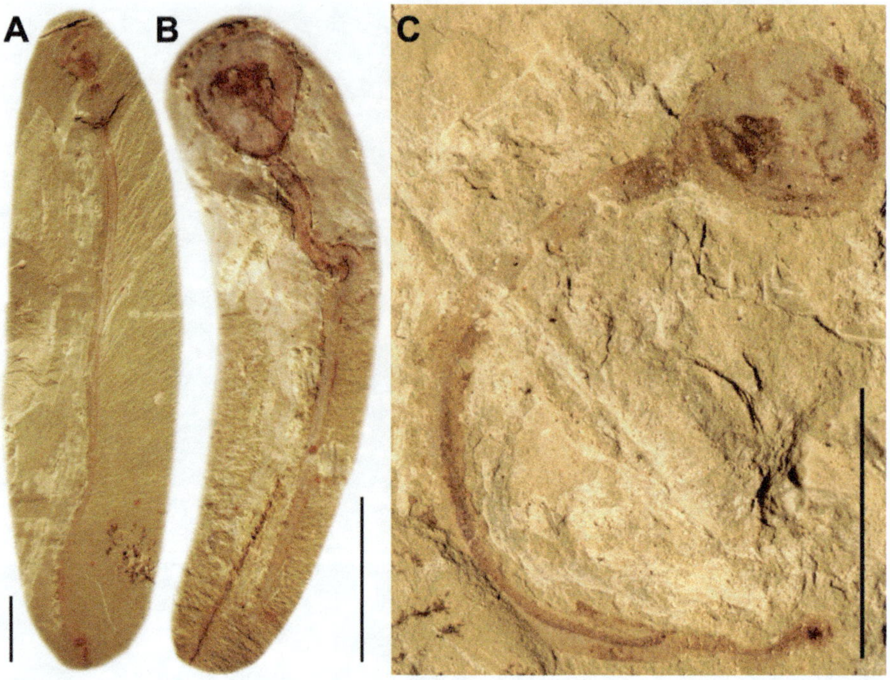

Fig. 128 *Ligulella chengjiangensis* from the Early Cambrian Maotianshan Shale showing the elongate shell with a slender, fairly long pedicle. Scale bar = 1 cm

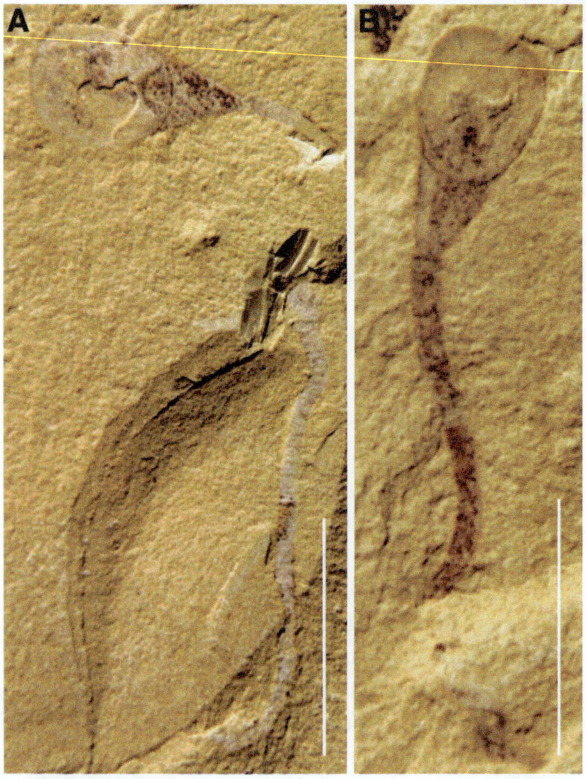

Fig. 129 *Lingulellotreta malongensis* from the Early Cambrian Maotianshan Shale. Scale bar = 0.5 cm

muscular area, a heart-shaped umbonal scar, and a small, triangular, central muscle scar.

L. malongensis (Fig. 129) has a small (7 mm) oval-shaped shell that is long and distinctly convex, adorned with discontinuous costae. The pseudointerarea in the pedicle valve is distinctly concave, with a pedicle groove in the apical part of the homeodelthyrium. The pseudointerarea is short, with a medial slit in the brachial valve. The interior of the brachial valve has a fan-shaped muscle scar divided into two parts by a narrow groove and enclosed by two lateral ridges posterolaterally. *L. chengjiangensis* has a large oval shell, 2 cm in diameter, adorned with fine, radially extended ribs, with a short, stout pedicle (Fig. 130).

Heliomedusa orienta from the Lower Cambrian Maotianshan Shale (Figs. 131 and 132) is here regarded as a discinoid brachiopod and is the dominant form in the epibenthic community of the Maotianshan Shale fauna. It was first described as a medusa by Sun and Hou (1987) as is reflected in the genus name; in 1992 it was reinterpreted as a craniopsoid brachiopod by Jin and Wang (1992). Based on its many discinoid characters, it was redescribed as a discinoid brachiopod by Chen et al. (2007a). These characters include a longitudinally oval pedicle foramen; a short but stout

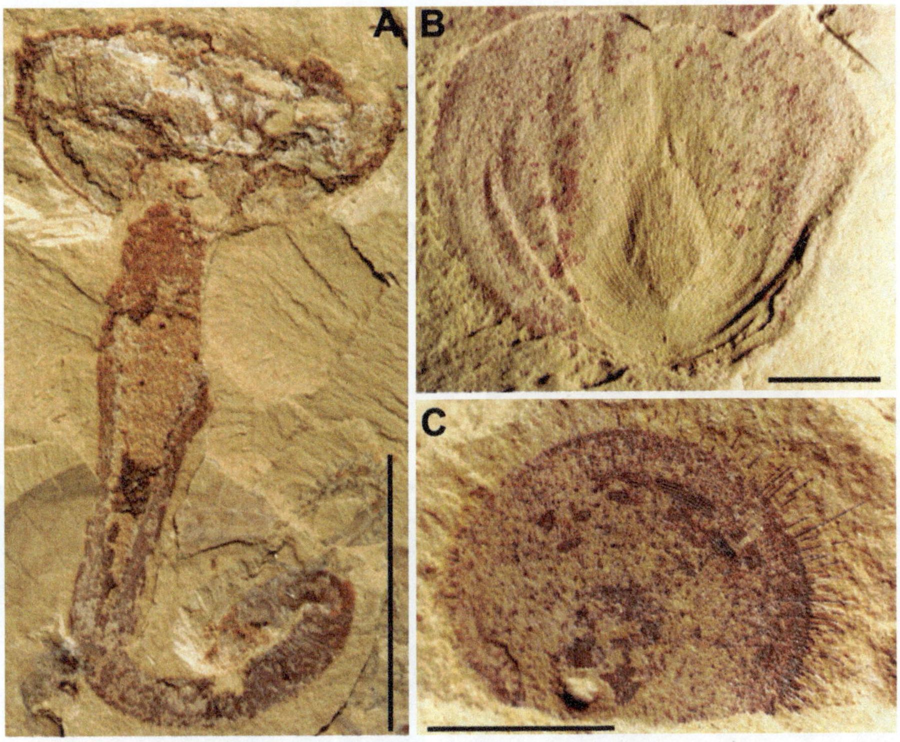

Fig. 130 *Longtancunella chengjiangensis* from the Early Cambrian Maotianshan Shale. (**a**) showing compressed shell and stout pedicle; (**b**, **c**) showing radially extended ribs

pedicle descending ventrally from the pedicle valve; a paired rectus in the large pedicle canal; elongated scars of paired anterior adductor muscles; and a median septum connected to a U-shaped ridge on the ventral interior. The shell is thin and sub-circular in outline. The convex dorsal valve has a sub-marginal umbo with mixo-peripheral growth, and a unique pseudointerarea protruding from the posterior end. The ventral valve is flat, with a slight outward bulge in the anterior and lateral regions around the foramen, and a sub-central umbo with holo-peripheral growth anterior to the anterior rim of the elongate, oval-shaped pedicle foramen. The entire mantle borders of both valves are aligned with the setae. The lophophore occupies a large portion of the mantle cavity and comprises two brachia, in a simple design referred to as "trochophore", with fewer than 20 hollow, ciliary tentacles in each brachium when young, become increasingly numerous in the adult (Zhang et al. 2003). The muscle scars comprise a pair of elongated anterior adductors and a pair of small posterior adductors. The pedicle is short, extending ventrally from the foramen in the anterior region of the posterior sector of the ventral valve.

Non-chordate Deuterostomes

The deuterostomes are morphologically well-defined groups of bilaterian animals. The criteria that define the non-chordate deuterostomes include the following: the blastopore becomes the anus in the adult; the apical organ has long cilia in its sensorial cells and contains many serotonergic cells that do not become integrated into the adult nervous system or contribute to the development of the adult nervous system; and the body is tripartite, including a prosome, mesosome, and metasome. The mesosome is a well-defined region, bearing tentacles. It is paired, surrounding the pharynx/oesophagus, and sends canals into the tentacles. In echinoderms, the left pouch is specialized as the hydrocoel, with radial canals extending into the podia. They probably represent a monophyletic group that

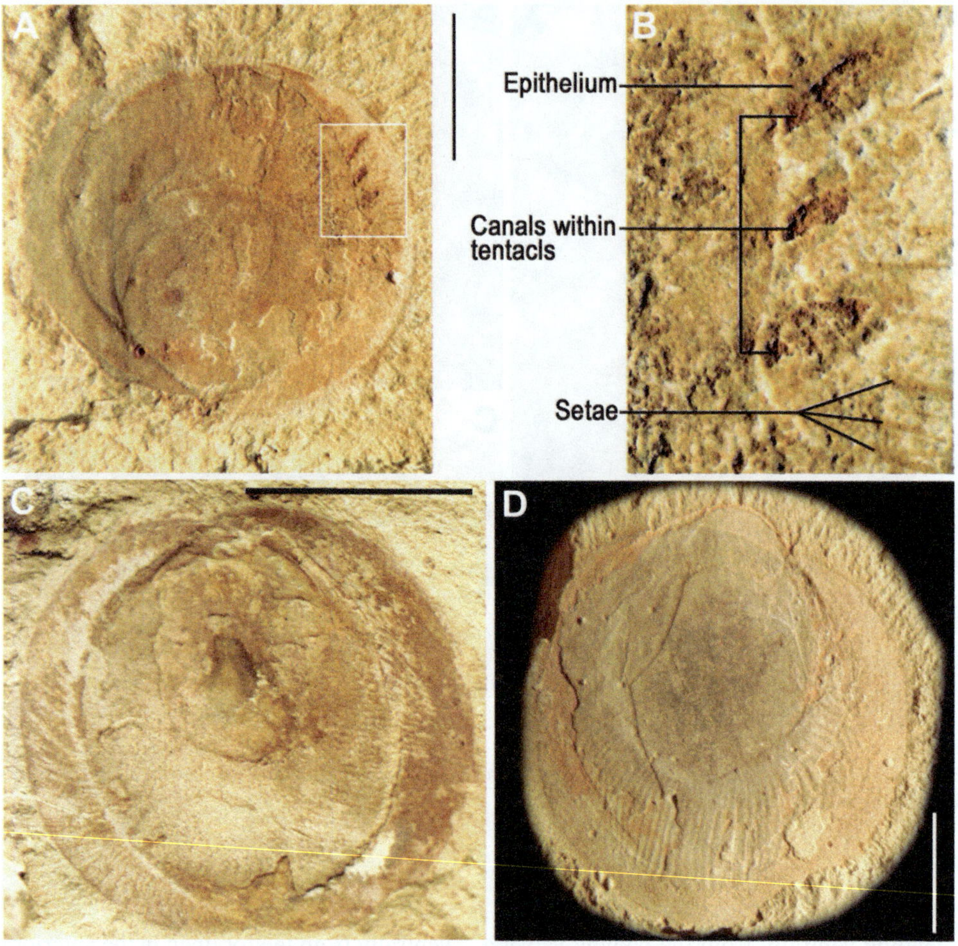

Fig. 131 *Heliomedusa orienta* from the Early Cambrian Maotianshan Shale. (**a**, **b**) showing hollow tentacles and base of setae; (**c**, **d**) showing closely arranged tentacles. Scale bars = 0, 5 cm (**a**, **c**) and 1 cm (**d**)

arose from a common ancestor with bilateral symmetry, from which they evolved with diverse body plans as adaptations to different lifestyles.

The non-chordate lower deuterostomes evolved in the Early Cambrian or even earlier, and diversified in the Early Cambrian. The Lower Cambrian Maotianshan Shale lies close to the root of deuterostome evolution, and preserves many distinct stem-lineage deuterostomes. However, their relationships to modern phyla are difficult to ascertain and have been a controversial issue.

In addition to brachiopods, two major groups of enigmatic Early Cambrian animals, a sessile (*Phlogites*) and a pelagic group (eldoniids), are possible deuterostomes. They have a crown of hollow ciliated tentacles (also called a "lophophore") encircling the mouth. The anus lies outside the tentacle crown but near it.

Sessile Lower Deuterostome *Phlogites* (=*Cheungkongella*)

Phlogites Luo and Huo 1999 (=*Cheungkongella* Shu et al. 2001a) is a tentaculate bulb-shaped animal on a stalk (Fig. 133) with a spacious body cavity containing a coiled gut (Fig. 133a, b). The mouth opens within the feeding area and the anus at the end of the intestine is situated on the side of the calyx body (Fig. 133a, b); the latter is referred to as a "cloacal siphon", although

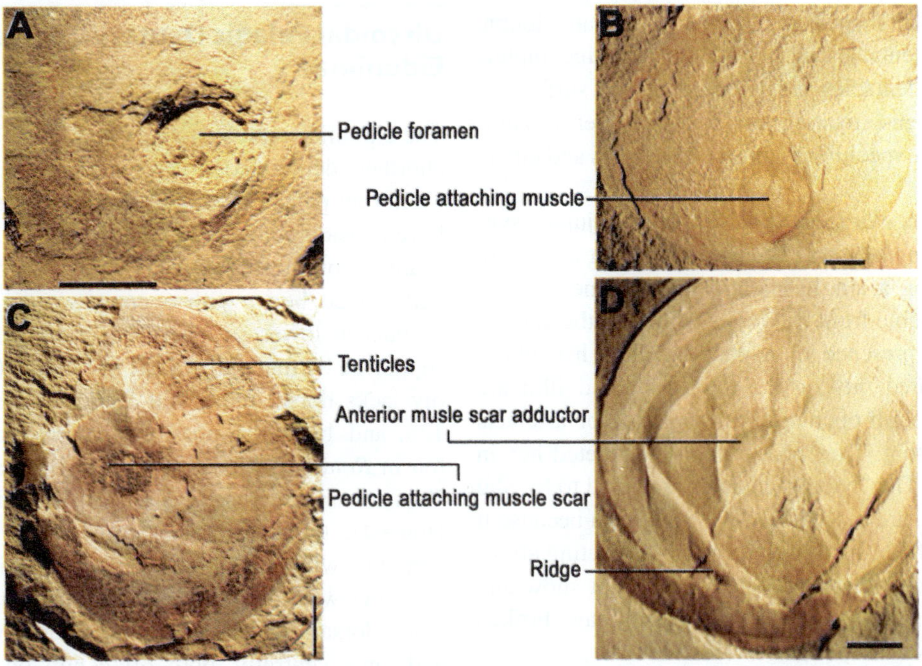

Fig. 132 *Heliomedusa orienta* from the Early Cambrian Maotianshan Shale. (**a**) enlargement of ventral valve showing pedicle foramen; (**b**) dorsal valve showing pedicle attachment scar; (**c**) dorsal valve showing tentacle and pedicle attachment muscle scar; (**d**) dorsal valve showing anterior adductor muscle scar and ridge. Scale bar = 0.2 cm

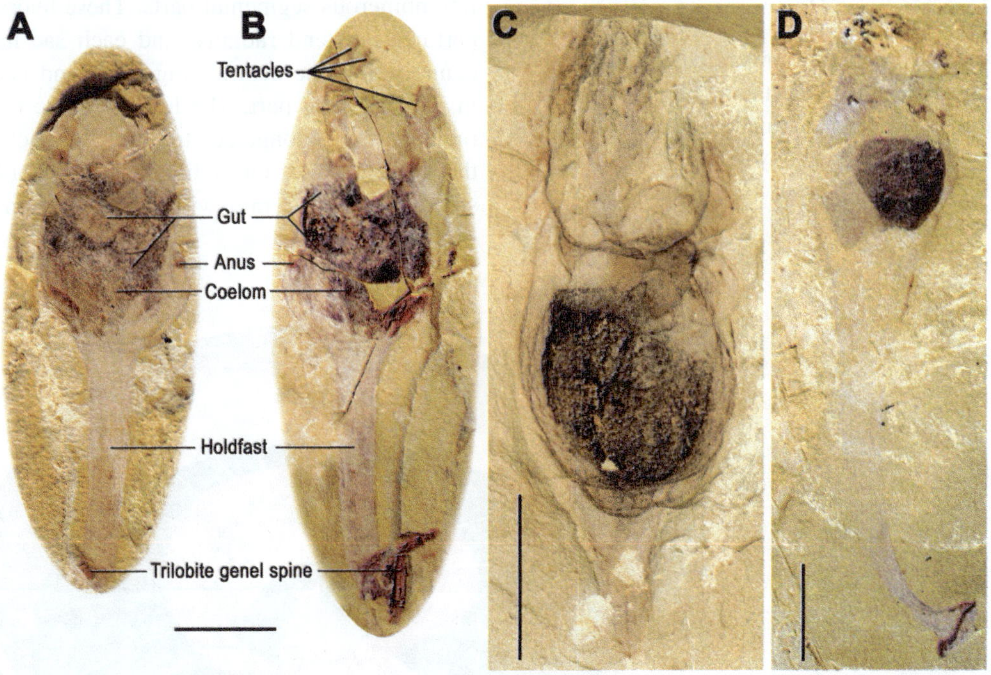

Fig. 133 *Phlogites longus* from the Early Cambrian Maotianshan Shale. (**a**, **b**) counterparts of a nearly complete specimen showing branched tentacles, coiled gut within the coelomic cavitiy, anus on side of the body and a holdfast attached to a trilobite genal spine; (**c**, **d**) showing branched tentacle (**c**) and large coelomic cavity (**c**, **d**). Scale bars = 1 cm (**a**–**c**) and 0.5 cm (**d**)

it is not funnel shaped as that of the tunicate siphon. The hollow tentacles are branched; only incomplete parts of the tentacles are exposed at the surface of most specimens, making the total number remains unclear. Some specimens suggest three arms and others suggest five.

Cheungkongella (Shu et al. 2001a) is a junior synonym of *Phlogites* (Luo and Hu, in Luo et al. 1999). It is based on a single incomplete specimen that is indistinguishable from *Phlogites*, although the authors claimed the presence of a pharynx. The irregularly arranged dotted pits on the darkly stained film are claimed to be gill openings, but appear to be artefacts of rock splitting. The "pharynx" is interpreted herein as a coelom. The "buccal tentacles" referred to by Shu et al. (2001a) are more likely to be a mouth because it opens at the body surface, rather than projecting above it. The *Cheungkongella* specimen does not show any obvious arms, perhaps because its arms have broken off or are covered by rock.

Phylogenetic Affinity

Phlogites is interpreted as a lower deuterostome, sharing an archimeric body plan. Its mesocoel is externally recognizable by its large, branched, hollow tentacles; the metacoel is a spacious body cavity with an encased coiled gut.

Discoidal Pelagic Deuterostomes: Eldoniidae

The eldoniids represent an extinct, monophyletic non-chordate deuterostome group, dominant among the Cambrian pelagic community, although some authors have considered them benthic (Dzik et al. 1997). They share a common body plan, characterized by a low bell-shaped body, with a convex upper surface and a concave lower surface. The upper surface is soft, supported by fine dorsal radii in *Eldonia* (which possibly lacks dorsal radii) and *Stellostomites* (Figs. 134, 135, and 136) or a possibly mineralized hard skeleton in *Rotadiscus* (Figs. 137, 138, 139, and 140) and *Pararotadiscus* (Zhao and Zhu 1994; Zhu et al. 2002). However, their internal anatomy is asymmetric and complex, with a sub-horizontally clockwise-coiled sac (when viewed from above). The alimentary canal is situated dorsally in the coiled sac, ending with a mouth and anus ventrally, both off centre on one side of the disc and close to each other. A crown of hollow, branched tentacles encircles the mouth. A hydraulic sac lies ventrally in the alimentary canal and is divided into numerous segmental parts. These hydraulic compartments extend radially, and each sac has a canal connected to the central ring canal and is bifurcated in its peripheral part. The lower surface is soft, with radial rows of elongated structures that correspond to the distribution of the radial canals (Fig. 140). These structures are 0.2 mm wide and 0.35 mm long, but

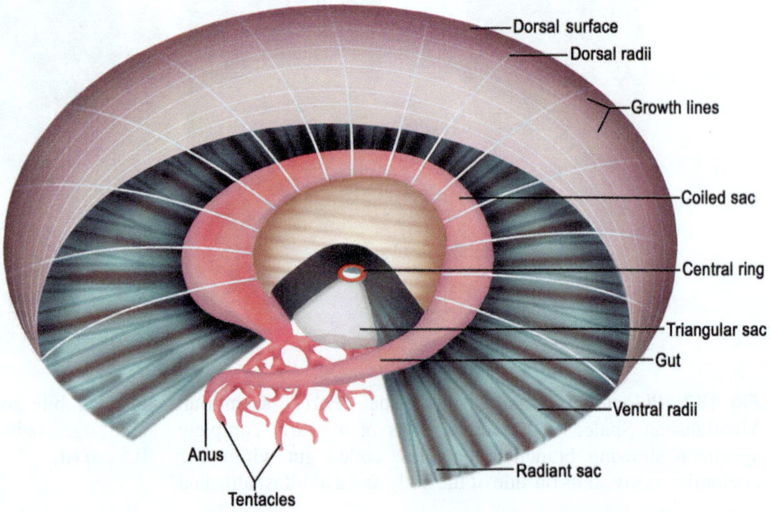

Fig. 134 Restoration of Early Cambrian medusiform organism *Stellostomites eurorphus*

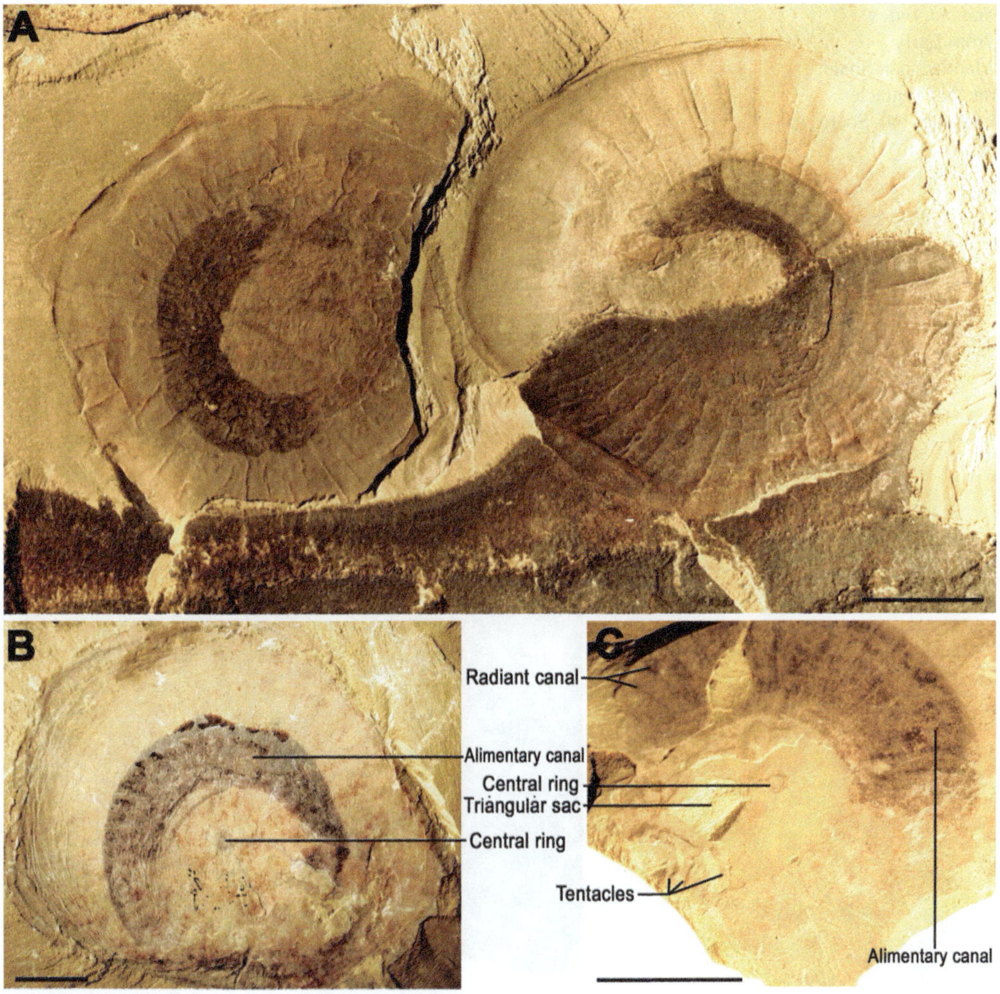

Fig. 135 *Stellostomites eurorphus* from the Early Cambrian Maotianshan Shale showing: imbricate preservation (**a**), clockwise coiled sac with alimentary canal lying dorsally in the sac (**a** "right one", **b**, **c**), tentacles (**c**), central ring (**b**, **c**), radial canals (**c**), and triangular sac (**c**). Scale bars = 2 cm (**a**) and 1 cm (**b**, **c**)

the innermost one is much larger. They are comparable to the tube feet in echinoderms. The eldoniids represent an extinct basal deuterostome group with a complex water-vascular system. The system consists of tube feet, a radiating arrangement of internal canals and sacs, and a circum-oral ring canal at the centre of the discoidal body. Radial canals, bifurcating midway, extend to the peripheral margin of the discoidal body. Each of the radially extended vascular sacs has a bifurcated canal; its ventral surface has organic structures that are tubular outgrowths of the body wall, comparable to the tube feet of living echinoderms. These structures are arranged in a radial row, with a bifurcated shape on the surface ventral to the radial canal. The bodies of the eldoniids are radially symmetrical externally, coiled, and sac like, encasing a complete gut. The gut is highly differentiated into a mouth, oesophagus, stomach, intestine, and terminal anus. The mouth is surrounded by a pair of branched, hollow, and possibly ciliated tentacles, forming a horse-shoe-like feeding bowl with the opening facing forward. The animal is an up-current feeder. The current enters the feeding bowl from the open side, carrying food items, but leaves through the open spaces between the tentacles on the closed side of the tentacle crown.

Eldoniids are recorded in both the Middle Cambrian Burgess Shale in Canada (*Eldonia ludwigi*) and recently in the Lower Cambrian Maotianshan

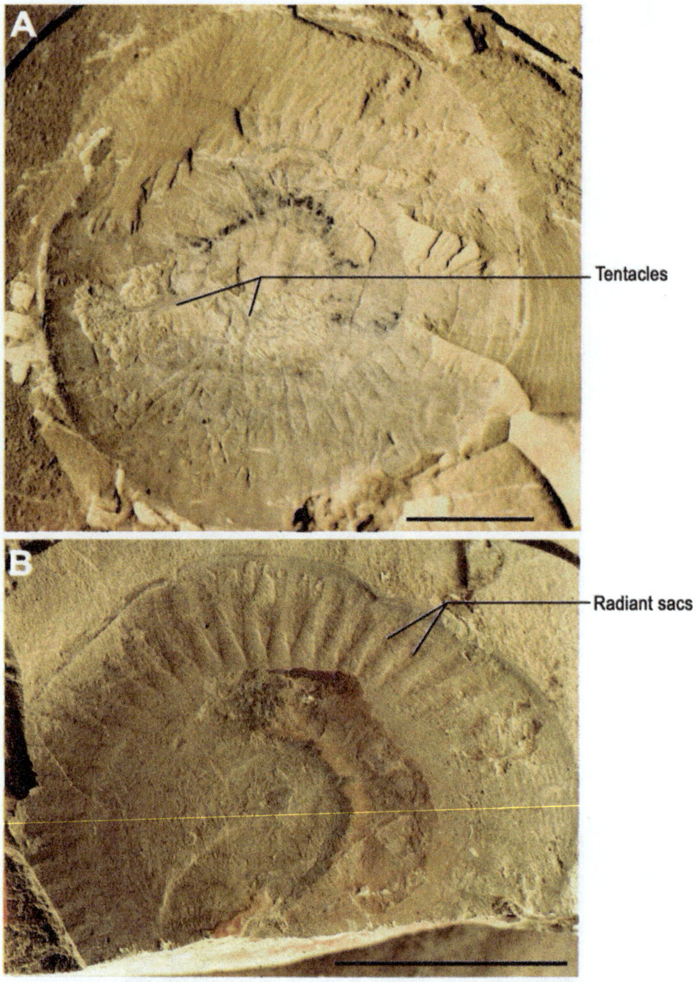

Fig. 136 *Stellostomites eurorphus* from Early Cambrian Maotianshan Shale, showing branched tentacles (**a**) and radiant sacs (**b**). Scale bar = 2 cm

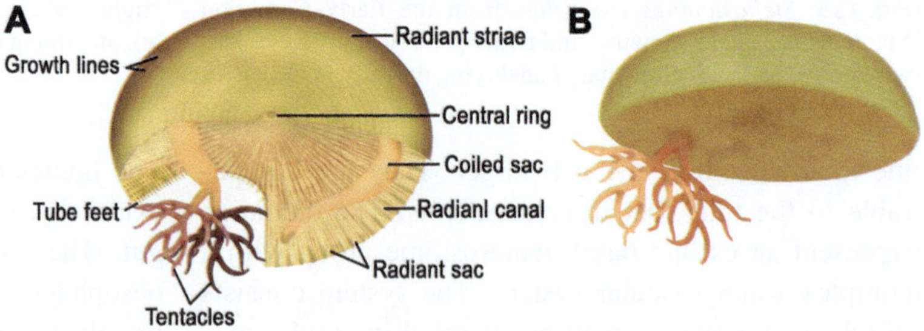

Fig. 137 Restoration of *Rotadiscus grandis*

Shale in Yunnan in southern China (*Stellostomites eurorphus*, *Rotadiscus grandis*; Chen et al. 1995d, 1996; Chen 2004) and the Middle Cambrian strata at Guoizhou (*Pararotadiscus guizhouensis*; Zhu et al. 2002).

S. eurorphus (=*Yunnanomedusa eleganta*) (Figs. 134, 135, and 136) is the predominant element in the pelagic community of Chengjiang, but it is much less common in Anning and is absent in Haikou. The animal has a low bell shape, weakly sclerotized, with a convex upper surface and a concave lower surface. A major part of the body is a sub-horizontally clockwise-coiled (top view) sac (Figs. 134 and 135), which encases a gut that extends parallel to and along the top surface of the sac and opens externally at both ends onto the lower surface of the discoidal body. The gut

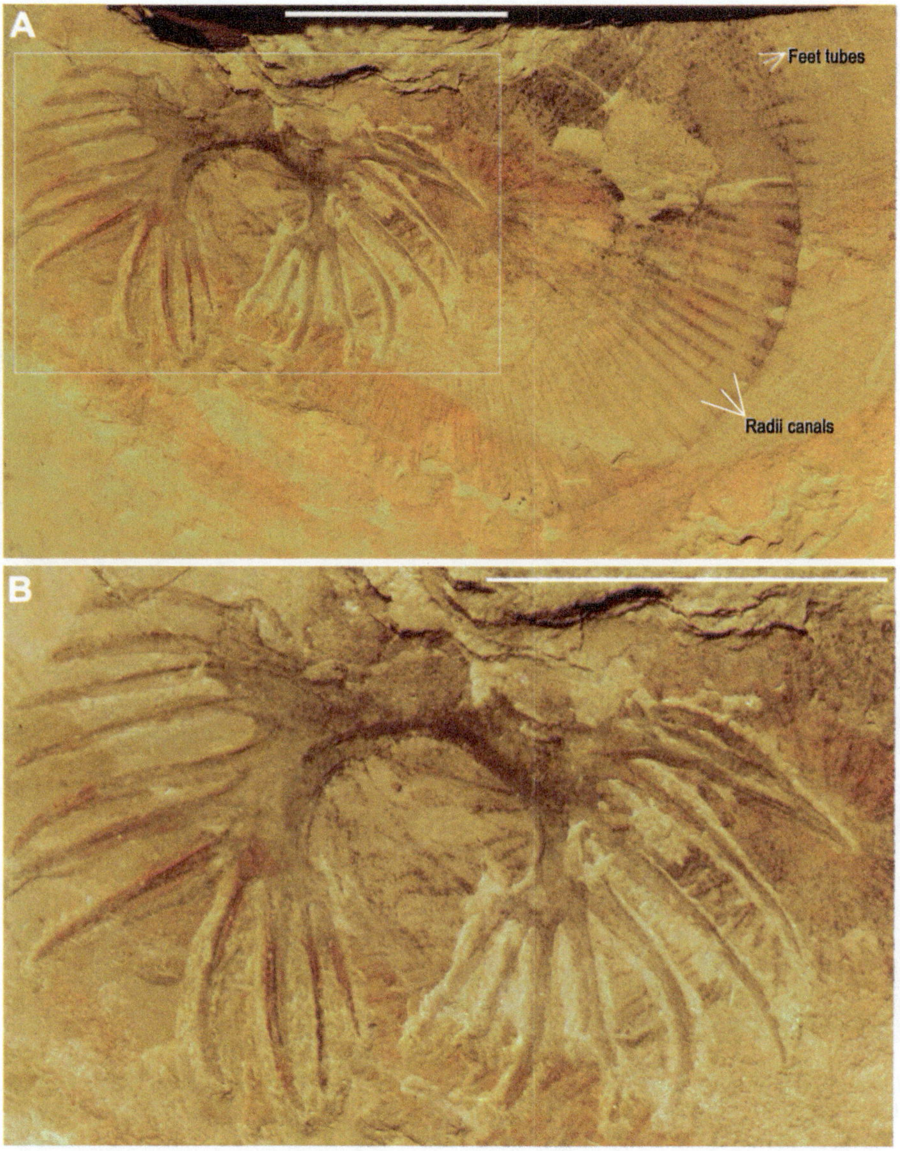

Fig. 138 *Rotadiscus grandis* from the Early Cambrian Maotianshan Shale showing: horseshoe-shaped tentacle crown and branched hollow tentacles (**a**, **b**), radial canals (**a**) and tube feet (**a**). Scale bar = 2 cm

is differentiated into the mouth, oesophagus, expanded stomach, and intestine. The mouth is surrounded by a pair of branched tentacles (Figs. 135c and 136a); these are hollow, possibly ciliated, and form a U-shaped feeding structure with a wide anterior opening, suggesting that the animal was an up-current feeder. The current not only brings food particles to the open side but also pushes the tentacles from the open side, creating an anticlockwise circulation in the discoidal body.

The spacious coiled sac is a well-defined structure located about midway between the centre and the margins of the disc and encasing the alimentary canal. The enclosed alimentary canal lies dorsally within the sac and is differentiated into the oesophagus, stomach, and intestine (Fig. 135b, c). It is spiral, generally in the same horizontal plane except for its anterior mouth and posterior anus, which open ventrally. The mouth is surrounded by a horse-shoe-shaped tentacle crown (Figs. 134, 135c, and 136a).

Fig. 139 Dorsal view of a nearly complete *Rotadiscus grandis* from the Early Cambrian Maotianshan Shale showing growth lines and radially extended fine ribs. Scale bar = 2 cm

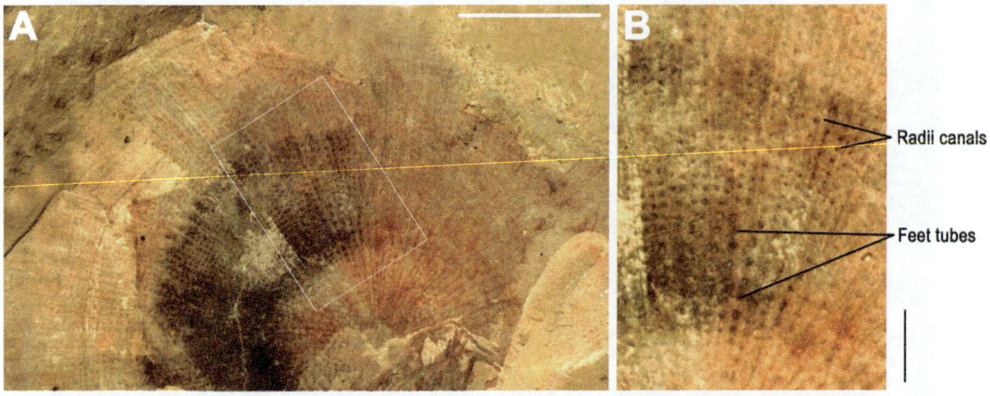

Fig. 140 *Rotadiscus grandis* from the Early Cambrian Maotianshan Shale showing tube feet and radial canals. Scale bars = 2 cm (**a**) and 0.5 cm (**b**)

The proposed anus opens at a point about midway between the centre and outer margin, with the anterior oral end located more centrally. The up-current feeding activity, with anticlockwise circulation, could effectively avoid the mixing of waste-laden water from the anus with food-laden water within the feeding bowl. The oesophagus narrows rapidly anteriorly and ends with a terminal constriction. The stomach is the most prominent part of the alimentary canal; it is strongly curved, occupying the broadest part of the alimentary canal diminishing in width both posteriorly and anteriorly. A conspicuous feature of the structure is its black stain, usually absent from the adjacent parts of the intestine and from the underlying sac. The posterior end shows an abrupt contraction and the black stain usually terminates at this point. The posterior part of the digestive tract is elongated and is roughly as long as the stomach, with width equal to about one-half the maximum diameter of the stomach. The lower compartment of the sac, referred to here as the "coiled sac proper", seems to be a vascular organ enclosed ventrally by radial sacs. The coiled sac is preserved as a yellowish structure, with no indication of black stain.

The alimentary canal and the coiled sac proper appear to be of similar sizes. They overlap completely for their entire lengths as the result of vertical compaction, so only one of them is exposed at the surface (Fig. 135a). However, in a number of cases, a split reveals both structures. When compaction is in an oblique direction, the two structures only partially overlap, producing an elongated zone of overlap within the outline of the sac (Fig. 135a on the right, and b, c). The borders are well defined and probably represent the mesenteria of the alimentary canal.

The water-vascular system is highly developed, consisting of a central ring and radially extended sacs and canals. Each of the radially extended vascular sacs has a bifurcated canal and rarely preserved black dotted structures representing tubular outgrowths of the body wall, comparable to the tube feet of living echinoderms. The sclerites are apparently external structures, attached to the outer surface of the ventral cuticle. These structures are arranged in radial rows, with a bifurcated shape on the surface ventral to the radial canal. They are usually detached after the death of the animal. In most cases, they are either entirely absent or only partly preserved. They appear to have a sub-circular base and are elongated distally, 0.2 mm wide and 0.35 mm long, except for the innermost one, which is considerably larger.

The animal has two sets of radially extended structures, a dorsal and a ventral set, which function in support and as water-vascular structures, respectively. There are 44 ventral radii radiating from the central ring canal to the periphery on the same plane throughout the ventral part of the coiled sac. Their course, which is otherwise strictly radial, tends to diverge from radial in the area adjacent to the ventral depression. A pair of auxiliary canals is added to each of the primary canals at its mid length, approximately at the outer margin of the coiled sac.

The ventral radial canals are encased by radially extended sacs. Each of the sacs encases a branched ventral canal. In most cases, the dorsal surface is supported from below by the ventral canals and tends to maintain a positive relief, whereas the original shape of their ventral surfaces is completely obliterated.

The central part of the discoidal body has a cavity that is enclosed by a rigid, ridged membrane dorsally and by a soft membrane ventrally. The ventral membrane tends to be lower ventrally, forming a triangular depression (Fig. 135c). The central cavity is probably a component of the water-vascular system.

The tentacle crown consists of two arm-like ridges, tentacle-carrying organs that curve into a horse-shoe-like form in outline. The two tentacles growing out from the organs are branched dichotomously many times, forming fairly fine branches distally.

The wall of the convex, inferred dorsal surface of the animal is distinctly defined, displaying fine concentric lines in some specimens. These concentric lines are interpreted as growth lines and are evenly spaced, at about 0.1 mm, throughout the entire dorsal surface. Identical growth lines are also observed in *Rotadiscus*. The wall of the concave ventral surface of the body is thinner but also bears fine concentric lines, interpreted as growth lines, densely spaced at the same distance as seen on the dorsal epidermis.

An annular band, interpreted here as possible concentric muscle markers, is formed by a group of concentric raised riblets, about 2–4 mm wide, located at one-fifth of the disc radius from its outer margin. It lies ventral to the dorsal set of radial canals but dorsal to the radial sacs. The riblets of the annular band appear to be rigid and rough.

The dorsal epidermis is soft and supported by a set of 88 fine radii, which are stiff and extend radially to the periphery of the discoidal body. They are about 0.1–0.2 mm in diameter.

R. grandis (Sun and Hou 1987) is a large discoidal eldoniid (15 cm in diameter), having a hard, possibly mineralized dorsal cover (Fig. 137). The dorsal cover has both fine, radially extending lines and concentric growth lines (Fig. 139). The animal has only one set of radii, corresponding to the ventral radii in *S. eurorphus*. There are 88 radial canals, extending from the central canals and ending in a pair of auxilliary canals midway along the radius (Fig. 138a). Ventral to the radii are hollow, linearly arranged tube feet, 0.5 mm in diameter (Fig. 140).

Like *S. eurorphus*, the major body region of *R. grandis* is a horizontally extended, clockwise-coiled sac, encasing the alimentary canal, which is differentiated into a mouth, an oesophagus, a stomach, an intestine, and an anus. The mouth is surrounded by a horse-shoe-shaped tentacle crown composed of two tentacle-bearing organs, each with 13 tentacles (Fig. 138b).

Affinities

The phylogenetic affinity of the eldoniids, especially *Eldonia*, has evoked much controversy since 1911, when it was originally described as a holothurian (Walcott 1911a) or considered an echinoderm (Clark 1912; Croneis and McCormack 1932). The interpretation of *Eldonia* as a holothurian, introduced by Walcott (1911a), remained popular for a long time (Durham 1974). Duncan Friend evoked it again, pointing to its similarities to a group of extant deep-sea pelagic holothurians (see Gee 1992), although the echinoderm affinity of *Eldonia* has been queried by other authors (Paul and Smith 1984; Chen et al. 1995d; Dzik 1989). The marginally accreted external skeleton and no sign of either pentamerous symmetry or an endoskeleton in the eldoniids suggest that they cannot be attributed to the Echinodermata, but rather to the pre-echinoderm deuterostomes (Conway Morris 1998).

The eldoniids have a well-developed water-vascular system, including a central ring canal, radial sacs and canals, and tube feet. A clockwise-coiled sac encases the digestive tract. The group is not consistent with any recognized deuterostome. The eldoniids show affinity to *Phlogites* in having a large main body cavity, housing a coiled gut, but they differ from *Phlogites* in having a highly developed water-vascular system, including a central ring canal, radial sacs and canals, and tube feet. They show close affinities to the echinoderms with which they share several key characters, such as external radial symmetry, a coiled alimentary canal, and a highly developed water-vascular system that includes a central ring canal, radial canals, and tube feet. However, the differences between them are great. For instance, eldoniids are perfectly radially symmetrical; they lack an endoskeleton; and they have marginal excrementary growth, suggesting that the eldoniids represent a pre-echinoderm stage in deuterostome evolution (Conway Morris 1998).

The eldoniids also share characters with the brachiopods, such as the marginal excrementary growth of the exoskeleton, but their differences from them are great. For instance, the body of the brachiopod is well protected both dorsally and ventrally by mineralized valves, and its main body cavity (metacoel) sends extensions into the mantle folds and stalk, and the water-vascular system is not developed. For these reasons, we reinterpret the eldoniids as a stem-lineage deuterostome group that evolved with the echinoderms from a common ancestor, but adapted to a pelagic lifestyle rather than to a sessile lifestyle.

Non-vertebrate Chordates: *Shankouclava*

Tunicates (sea squirts) and their planktonic relatives form a subphylum (Urochordata) of the Chordata, but they resemble other chordates only very slightly. They are usually sessile (e.g. Fig. 141a), and with the exception of the planktonic form (Appendicularia), their chordate characters, including the notochord, hollow dorsal nerve cord, and post-anal tail, only occur in the transitory larva, called the "tadpole".

The subphylum Urochordata was previously interpreted as the most basal branch of the known

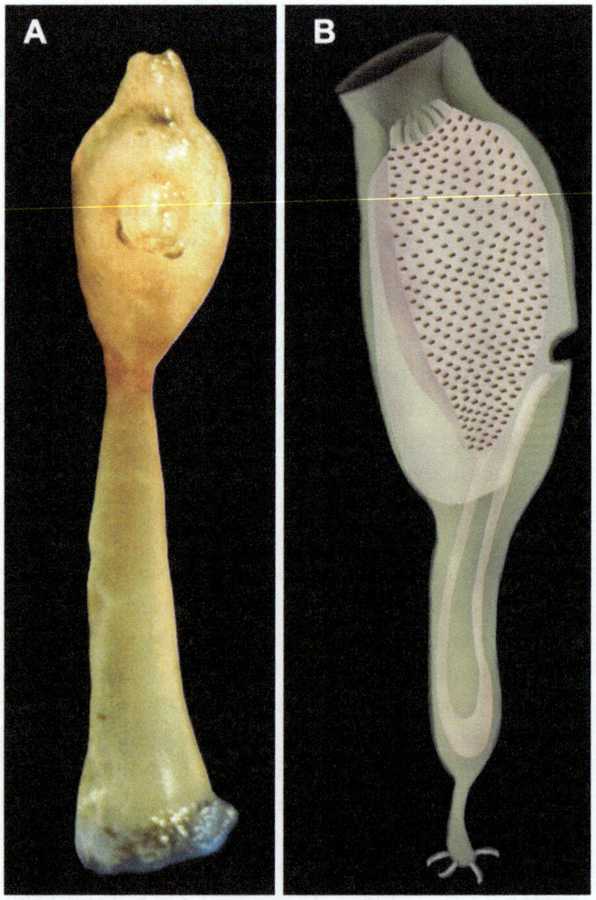

Fig. 141 Extant urochordate (**a**) and restoration of Early Cambrian urochordate *Shankouclava anningense*. (**a**, courtesy of D. Weber)

chordates. However, recent molecular data have shown that this subphylum is a sister group to the Vertebrata (Delsuc et al. 2006; Schubert et al. 2006) and that both evolved from an amphioxus-like ancestor but adapted to different lifestyles, a sessile inactive lifestyle in the tunicates and a more active lifestyle in the vertebrates.

The discovery of both fossil vertebrates and tunicates in the Lower Cambrian Maotianshan Shale implies that chordates had already differentiated into three major groups during the Early Cambrian or even earlier, comparable to the extant cephalochordates, tunicates, and vertebrates probably explaining why the tunicates have diverged morphologically furthest from the chordate ground plan.

The fossil record of the tunicates is also fairly limited. The Lower Cambrian tunicate *Shankouclava anningense* (Chen et al. 2003b) is the only known reliable fossil representative of the tunicates. The Lower Cambrian *Phlogites* (= *Cheungkongella*) is said to be a tunicate (Shu et al. 2001a), but it differs from the tunicates in lacking a pharynx and by the presence of long branched tentacles (a distinctly non-tunicate feature) (Shu et al. 2001a), and probably represents a basal branch of the lower deuterostomes (Caron et al. 2010). Another so-called tunicate fossil, *Jaekelocarpus*, from the Carboniferous is covered with large calcite plates instead of a tunic, and its supposed branchial bars have an unusual coronal orientation, so a tunicate identity is unlikely.

Shankouclava shankouense from the Lower Cambrian Maotianshan Shale (Shankou, Anning, near Kunming) is a solitary sessile tunicate. It is a soft-bodied, club-shaped animal, elongated (2–4 cm long), and pointed proximally, with attachment stolons indicating it was sessile and rooted to the ocean floor by these structures (Fig. 141b). The body is bipartite, with a barrel-shaped anterior part and a triangularly elongated posterior part, called the "abdomen". The anterior part, which occupies more than half the length of the body, is dominated by a large pharynx, with up to 30 branchial bars (Fig. 142). The branchial bars branch to form a rectangular network with stigmata (branchial openings). An oral siphon opens subventrally, containing oral tentacles at its base. An endostyle band extends along the mid-ventral part of the pharynx, and a sac-like atrium surrounds the pharyngeal basket (Fig. 143a). The alimentary canal is U-shaped, with the anterior gut (oesophagus) running posteriorly from the pharynx and forming a loop, called the "stomach", near the base of the abdomen. From there, the posterior gut (intestine) extends anteriorly, ending near an atrial pore lying two-thirds of the way back from the anterior pharyngeal margin.

This Cambrian animal resembles the tunicates to some extent, in its tunic and mantle, the large perforated pharyngeal region with numerous transversely oriented structures, branchial bars, the sac-like peripharyngeal atrium, the oral siphon with what appear to be oral tentacles at its base, the dorsal atrial pore, and the elongated endostyle band on the mid-ventral part of the pharynx. Its resemblance to the single zooids of modern colonial aplousobranch tunicates, such as *Clavelina* and *Amaroucium*, is striking. Like *Shankouclava*, the bodies of *Clavelina* and *Amaroucium* are slender and differentiated longitudinally into two parts of sub-equal length, with a notched constriction between them. *Shankouclava* differs from the aplousobranchs in that its atrial pore opens further posteriorly on the pharyngeal roof.

However, the discovery of the Lower Cambrian *Shankouclava* (Chen et al. 2003b) indicates that the tunicates had already diverged morphologically from the main evolutionary course of the vertebrates by the Early Cambrian. Despite its departure from the main vertebrate evolutionary line for more than half a billion years, and the lack of any crucial features in either its body plan (Lacalli 2005) or genome (Holland 2007) that link it directly to the vertebrates, the tunicates still offer many interesting clues to the evolutionary origin of the vertebrates, with respect to the neural crest cells (Jeffery et al. 2007), brain (Dufour et al. 2006), heart (Simões-Costa et al. 2005; Davidson and Levine 2003), and eye lens (Shimeld et al. 2005).

The First Vertebrate Fossils Fill a Gap in the History of Evolution of the Vertebrates

The Cristozoa (also known as "crest animals") represents those animals containing a neural crest and its derivatives and includes all known craniates and their immediate pre-craniate ancestors. The pre-craniate crest animals are all extinct and are preserved only in the Early Cambrian strata of Yunnan (south-western China). The fine anatomical details of *Haikouella*

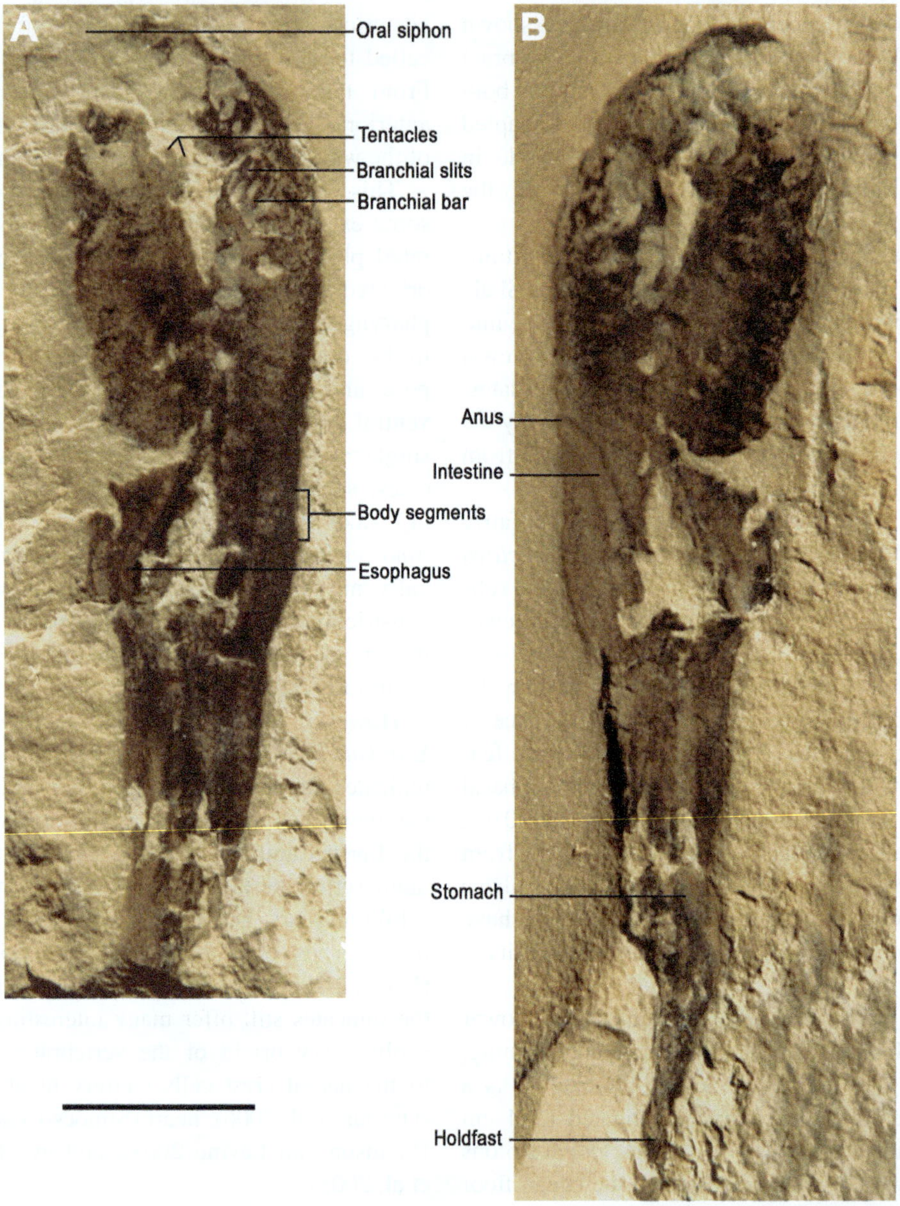

Fig. 142 Holotype of *Shankouclava anningense* from Early Cambrian Maotianshan Shale. (**a**) and (**b**) are counterparts. Scale bar = 0.5 cm

and the slightly more advanced species *Yunnanozoon* show many characters unique to the vertebrates, but lack a skull and most other elaborated placodal and neural-crest derivations, throwing new light onto the previously missing history at the very beginning of cristozoan evolution. A comparative study of the cephalochordate amphioxus suggests that pre-craniate evolution was marked by a series of innovations, including muscular ventilation with gill-bearing and jointed branchial arches; paired head sensorial organs, including paired eyes and nostrils; a relatively large, slightly differentiated brain; protovertebrae; and some derivatives of the neural crest cells. However, the pre-craniate crested animals lack ears, a clear telencephalon, and a skull. Furthermore, a comparison of the brains of amphioxus and the craniates suggests that the tripartite brain with a telencephalon is not the older structure, but a novelty of the craniates.

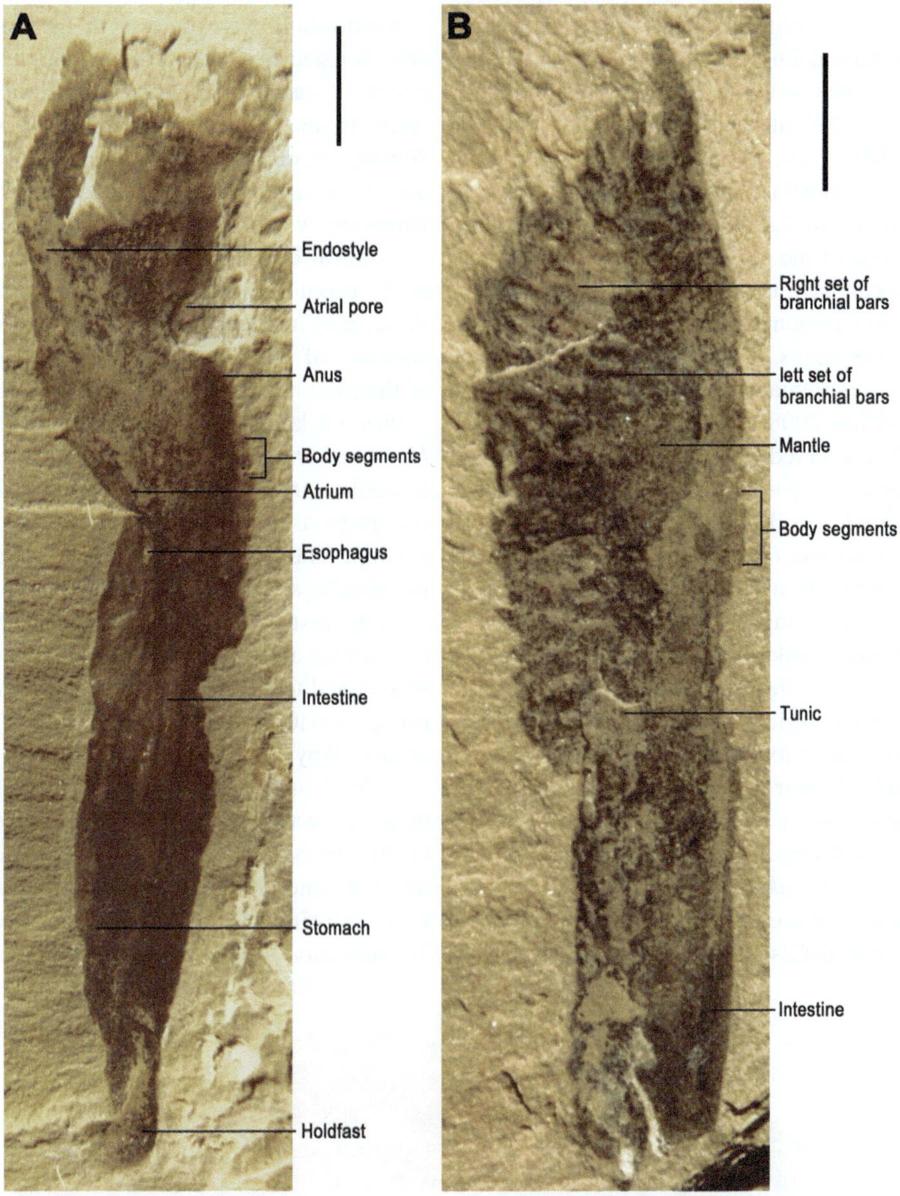

Fig. 143 *Shankouclava anningense* from the Early Cambrian Maotianshan Shale. (**a**) laterally compacted specimen showing endostyle and coiled gut with anal opening on opposite side to endostyle; (**b**) side view of a partially preserved specimen showing tunic, mantle, and branchial bars (both *left* and *right* sets). Scale bar = 0.2 cm

The vertebrae and skull are widely accepted as the two major features defining the vertebrates. However, these two criteria did not evolve synchronously. The skull was previously considered an evolutionarily older structure than the vertebrae (Janvier 1978). In contrast, the Lower Cambrian fossils *Haikouella lanceolata*, *Yunnanozoon lividum*, and *Haikouichthys ercaicunensis* all have protovertebrae but no clear skull, suggesting that the vertebrae are evolutionarily deeply rooted, representing a structure older than the skull. The lack of vertebrae in the living hagfishes is probably a derived feature.

To avoid the ambiguity entailed when either the vertebrae or skull are used as characters defining the vertebrates, Holland and Chen (2001) proposed the term "Cristozoa", or crest animals, to represent those animals with a neural crest and its derivatives. These Cambrian animals had jointed cartilagenous branchial arches, pharyngeal teeth, and evidence of muscular ventilation, implying that they had definitive neural crest cells and were crest animals (Holland and Chen 2001; Mallatt and Chen 2003; Holland and Holland 2001). The Cristozoa includes the craniates and their immediate ancestors, the pre-craniate Cristozoa. Among the Early Cambrian crest animals, *Haikouella* is one of the most remarkable. Its soft tissues and organs are not only preserved in extraordinary detail, but it also represents a crucial evolutionary position close to the nexus between an amphioxus-like ancestor and the vertebrates, at the very beginning of the Cristozoa (Chen 2008, 2009).

These early preserved cristozoans have numerous vertebrate characters: paired eyes, nostrils (preserved only in *Haikouella*), a distinct head with upper and lower lips (only seen in *Haikouella*), jointed branchial arches with paired gill rays, pharyngeal denticles, and protovertebrae, supporting a close affinity with the vertebrates. However, unlike vertebrates, these animals lack a skull, ears, a clear telencephalon, and most other elaborated placodal and neural-crest derivations. They are also similar to the living cephalochordate, amphioxus, in both their general external aspects and in some specific characters, including their oral tentacles, endostyle, and large notochord. All these factors strongly suggest that *Haikouella* and *Yunnanozoon* represent a missing evolutionary link at the dawn of the vertebrates, before the rise of the craniate vertebrates.

Cristozoans were diverse in the Early Cambrian, and included *H. lanceolata*, *Y. lividum*, and *H. ercaicunensis*, each representing a different evolutionary stage in the evolution of the craniate vertebrates. Among them, *H. lanceolata* is the most primitive and its occurrence is restricted to Haikou (Fig. 144), Kunming, where it is fabulously preserved in great anatomical detail. It also represents a basal position in the evolution of the crest animals. The anatomy of these early cristozoans is crucial to our new understanding of the ground plans of the vertebrates and of their brains, heads, head sensorial organs, branchial arches, backbones, and feeding behaviour (Fig. 144). *H. lanceolata* is 2.5–3 cm long (maximally 4 cm) and lanceolate in outline, flattened ventrally in its anterior portion but compressed laterally in its posterior portion. The pharyngeal system has only six post-mandibular arches and the three mandibular arches are missing. Instead of mandibular arches, *Haikouella* has mandibular aortae. It has only four pairs of gonads. The pharyngeal teeth are situated ventrally in the mid pharyngeal cavity. The myotome consists of 25 myomeres, separated by simple straight myosepta.

Y. lividum (Fig. 145) is restricted to the Chengjiang area. It was previously thought by Hou et al. (1991) to be problematic, with unknown affinities, and was interpreted as a possible chordate by Chen et al. (1995a). However, Chen recently reinterpreted *Yunnanozoon* as an early crest animal, with close

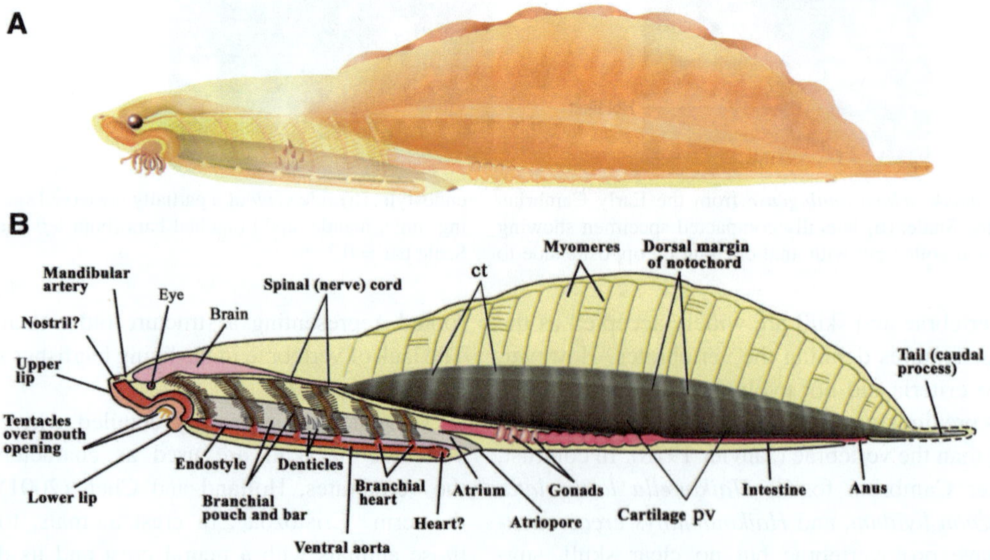

Fig. 144 Restoration (**a**) and illustration of anatomical characters (**b**) in *Haikouella lanceolata*

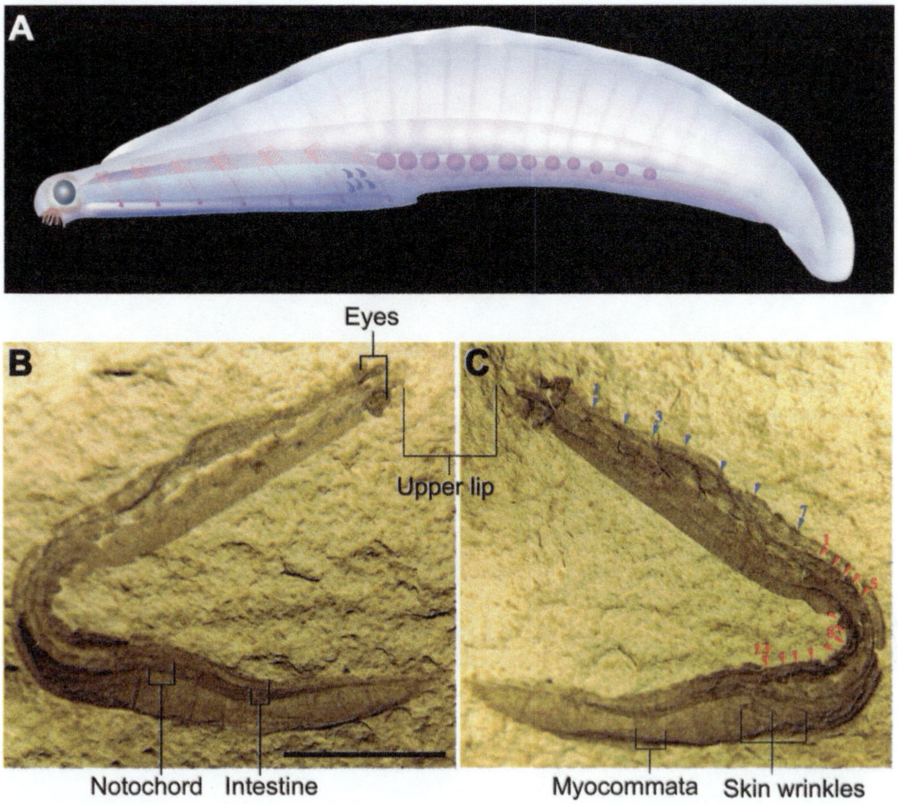

Fig. 145 *Yunnanozoon lividum* from the Early Cambrian Maotianshan Shale. (**a**) restoration; (**b**, **c**) counterparts of a nearly complete specimen. Scale bar = 0.5 cm

affinity to *H. lanceolata*. The animal has paired mandibular branchial arches (Fig. 145), larger lateral eyes (Fig. 150) than *H. lanceolata*, and much larger pharyngeal denticles (Fig. 153C–D), suggesting that *Yunnanozoon* is evolutionarily more advanced than *H. lanceolata* (Chen 2004, 2008, 2009). The animal is 3–4 cm long, lanceolate in form, and compressed laterally. The pharyngeal system consists of seven pairs of segmented branchial arches, thicker than those of *H. lanceolata*, separated by wide branchial slits. Each segment of the branchial arches bears a pair of gill rays. Simple, cone-shaped denticles, about 1 mm long, occur on the pharyngeal floor at the base of the pharynx in *Yunnanozoon*. The myotomes are divided by simple straight myosepta into 23 repeated segmental structures, known as muscle segments ("myomeres"). The 13 pairs of gonads occur on the lateral surface of the foregut (Fig. 145b, c).

H. ercaicunensis (Shu et al. 1999b; Fig. 146) co-occurs with *Haikouella*, restricted to the Haikou area. It is lanceolate and only 2.5 cm long. The head is small, with a pair of lateral eyes, a possible nasal capsule, and an auditory capsule. The pharynx consists of 6–7 pairs of branchial arches. The myosepta are W-shaped. *H. ercaicunensis* has a prominent dorsal fin with fin radials, similar but not the same as those of the hagfish and lamprey. The 13 circular structures along the base may be gonads. The animal is interpreted as a primitive craniate, but it has no clear skull and its craniate identity is questionable. However, the animal has W-shaped myosepta, large lateral eyes, and a prominent dorsal fin with fin radials (Fig. 146b), suggesting that it is the most advanced form of the Lower Cambrian crest animals, representing a turning point in the transition from procraniates to craniates.

Although some researchers have considered *Yunnanozoon* and *Haikouella* hemichordates or lower deuterostomes, or problematic organisms with unknown affinities (Shu et al. 1996, 2003b; Conway Morris 2000; Smith et al. 2001), their conflicting views have been clarified by a number of studies (Mallatt and Chen 2003; Chen 2008, 2009).

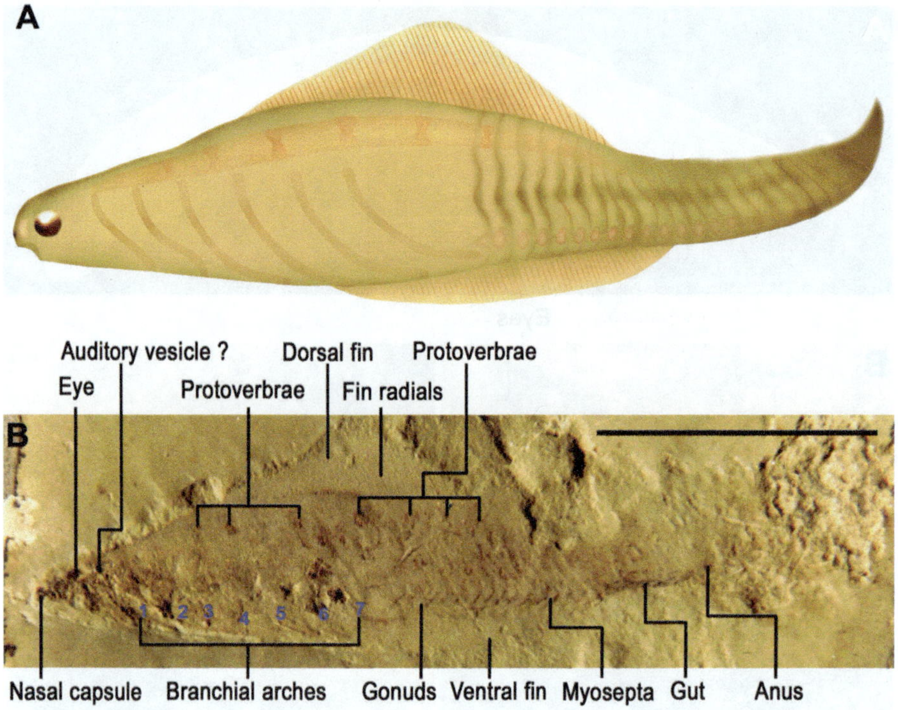

Fig. 146 *Haikouichtys ercaicunensis* from the Early Cambrian Maotianshan Shale. (**a**), restoration; (**b**) holotype. Scale bar = 0.5 cm

Haikouella Sheds Light on the Ground-Plan of the Head of Crested Animals

The head accommodates the sensorial and feeding organs externally and the brain internally. It is fundamental to the evolutionary changes from the amphioxus-like ancestor to the craniate. The head in the ancestral condition, as exemplified by the living amphioxus, is simple and characterized by a short, blunt, anterior projection, known as the "rostrum" (Fig. 147), with which the animal pushes aside the sand while burrowing (Fig. 148). The pre-mandibular oral cavity has a ventrally located mouth, surrounded by tentacles. The mouth is protected dorsally by the rostrum but there is no lower lip (Fig. 147c). None of the elaborated paired sensorial organs of the craniates existed in the amphioxus-like ancestor. In contrast to the simple ancestral head, the craniate head is a complex and unique structure, including a cartilagenous or bony skull that encases the brain, paired sensory organs, and an oral cavity, with a premandibular mouth surrounded with upper and lower lips. The great contrast between the simple amphioxus-like head and the complex craniate head implies the existence of unknown intermediate forms.

The head in the early crest animals, as exemplified by *Haikouella* (Fig. 149) and *Yunnanozoon*, probably fills the gap between the simple amphioxus-like head and the complex craniate head. It has paired eyes, possible olfactory organs, a large brain, and a mouth surrounded by upper and lower lips (Fig. 149), and is thus much more complex than the head of amphioxus (Fig. 147c). However, the head in the early crest animals lacked a clear skull and ears, so it was simpler than that of the craniates.

The mouth is surrounded by both an upper dorsal lip and a lower ventral lip in the early crest animals (*Haikouella*): these might have functioned to prevent the escape of the semi-sessile soft-bodied prey upon which the animals presumably fed. The lower lip is a novelty in the vertebrates, but the upper lip is deeply rooted in an amphioxus-like ancestor. The anteriorly extended rostrum in amphioxus might be the precursor of the upper lip. It is a robust structure, strengthened by the anteriorly extended notochord (Fig. 147c), and presumably has multiple functions, including in the protection of the mouth during burrowing with its

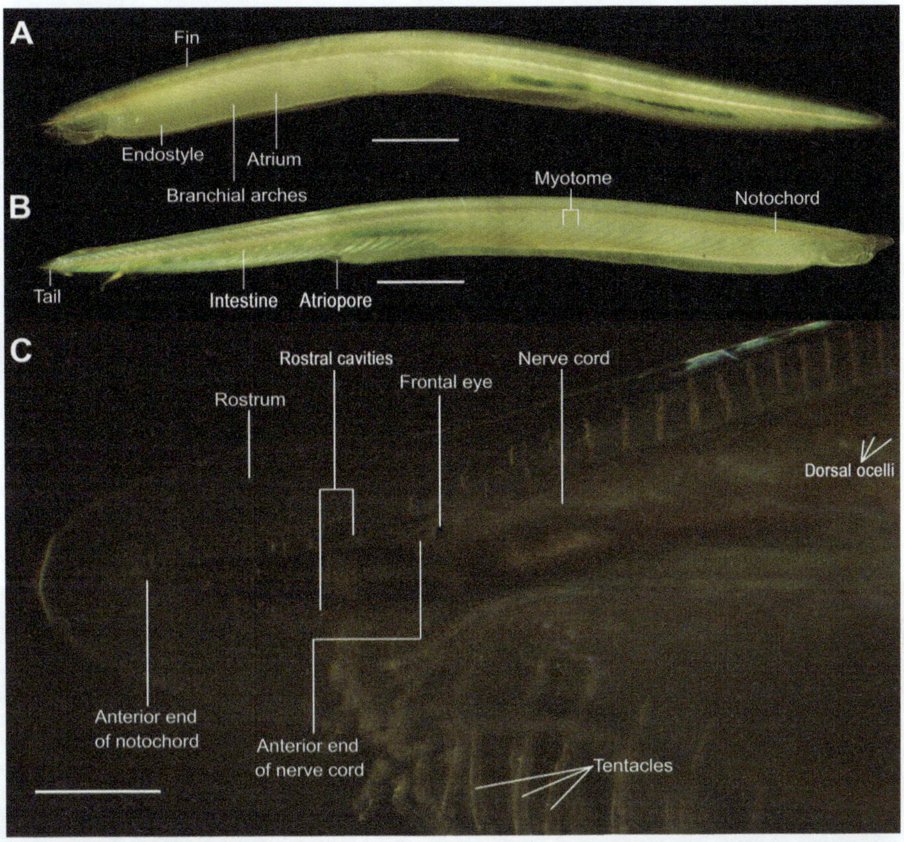

Fig. 147 Extant cephalochordate *Branchiostoma belcheri* Gray from the South China Sea. (**a**, **b**) side views (*left* and *right*) of an individual; (**c**) view from the *left side* of anterior portion of the body. Scale bars = 2 mm (**a**, **b**) and 200 μm (**c**)

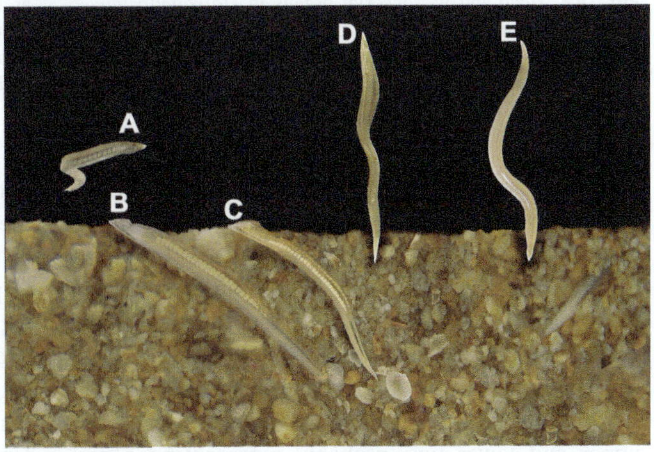

Fig. 148 Lifestyle of the extant cephalochordate *Branchiostoma belcheri*. (**a**) swimming; (**b**, **c**) burrowing in sand with head exposed; with tail (**c**, **d**) and head (**e**) first, respectively, while burrowing. (**d**, **c**) with the tail (**d**) and head (**e**), respectively, first while burrowing

pointed end (Fig. 148e) and possibly in the collection of olfactory information.

The oral cavity in the early Cristozoa, as exemplified by *Haikouella*, is situated in a premandibular position, with a circular mouth opening surrounded by an oral skeletal ring and tentacles, exactly as in amphioxus (Fig. 147c). The premandibular oral cavity and mouth (known as the "old mouth") represent the primitive condition in the chordates. The "old mouth" occurs not only in the cephalochordates and tunicates

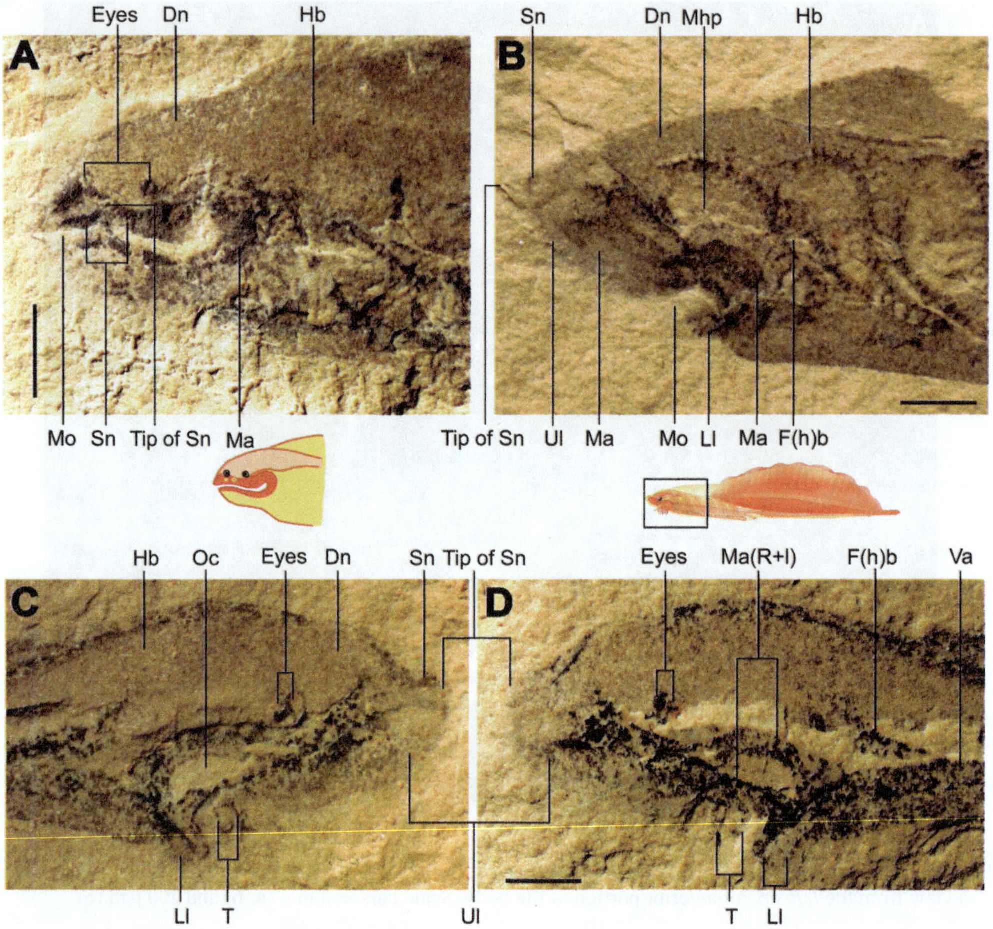

Fig. 149 Head and its sensorial organs in *Haikouella lanceolata* from the Early Cambrian Maotianshan Shale showing paired eyes (**a**, **c**, **d**), paired nostrils (**a**–**d**), diencephalon and hindbrain (**a**, **b**), mouth (**a**–**d**), tentacles (**c**, **d**), upper and lower lips (**b**–**d**), oral cavity (**c**, **d**), and gill rays internal to the branchial arches (**b**). Abbreviations: Dn, diencephalon; Hb, hindbrain; Mo, mouth; Nt, nostrils; Sn, snout; Mhp, first (mandibulo-hyoid) gill pouch; Ul, upper lip; Ll, lower lip; F(h)b, first (hyoid) branchial arches; Ma, mandibular aorta *Note:* **d** is counterpart of **c**. Scale bar = 1 mm (adapted from Chen 2008)

but has also continued into more advanced forms, including the pre-craniate crest animals, agnathans, and some lower gnathans (Mallatt 1996). The oral cavity in *Haikouella* has an external mouth opening at in front, with a diameter of about 1 mm, identical to that in amphioxus. The mouth is surrounded by a skeletal ring that supports the oral tentacles (Fig. 149c, d).

The oral-labial region in *Haikouella* has distinct upper and lower lips, probably representing a novelty in the crest animals. The oral-labial region in the cephalochordate amphioxus lacks a homologue of the vertebrate lower lip but has a large rostrum projecting dorsoanteriorly. The rostrum in amphioxus is not strengthened by a muscular structure, as seen in the crest animals, but is supported by the notochord, extending anteriorly to the pointed end of the rostrum (Fig. 147c). The rostrum is probably used by amphioxus to burrow with its pointed end and is presumably the precursor of the upper lip in the early Cristozoa. The upper lips in *Haikouella* are large and paired, probably representing muscular structures (Fig. 149c, d). An identical structure also appears in the lamprey larva, hagfishes, and some soft-boned fishes. Unlike the rostrum with its pointed end in amphioxus, the upper lips in *Haikouella* are broadly rounded and do not function for burrowing, but are probably used to probe for prey with the help of its olfactory senses. These prey might include algae, embryos, larvae, and small animals living near the surface or shallowly within the sediment. The lower lip in *Haikouella* is a conspicuous structure and has none of the homologous structures seen in the protochordates, but appears in the lower craniates, including the jawless fishes (larval lamprey and hagfishes) and some soft-boned fishes (chimaeroids and some sharks; Mallatt 1996). It is a novelty in the crest animals (Mallatt and Chen 2003).

Oral tentacles occur in the cephalochordate amphioxus and some living ascidians (Brusca and Brusca 2002) and also in the Early Cambrian tunicate *S. anningense* (Chen et al. 2003b). They are used to prevent larger food and sand particles from entering the mouth. The tentacles in amphioxus are numerous (up to 40), very fine (40 μm in diameter), and narrowly spaced (about 40 μm; Fig. 147c), suggesting that amphioxus feeds on food grains smaller than 40 μm, mainly single-cell algae. Oral tentacles also occur in *Haikouella*, presumably resembling those in amphioxus and forming a screen to prevent larger food and sand particles entering the mouth. However, they are relatively more robust (250 μm in diameter) and fewer in number (about 12 in total), with relatively wider spacing, with about 300 μm between two adjacent tentacles (Fig. 149c, d). The tendencies of the oral tentacles to become increasingly robust, to become more widely spaced, and to decrease in number presumably correspond to a shift to increasingly larger food grains during the transition from an amphioxus-like pre-cristozoan ancestor to the cristozoan animals. The wider spacing in *Haikouella* indicates that the animals could ingest coarser food grains, up to 300 μm in size, including sea weeds, embryos, larvae, and microscopic animals.

Head Sensory Organs

The ground plan for the head sensory organs in the crest animals has been a controversial issue. The rise of uniquely vertebrate sensory organs has been postulated in a serial transformation model by Butler (2000, 2006) and other authors (Northcutt and Gans 1983; Gans 1993), in opposition to the notion that all the craniate sensory organs elaborated simultaneously when the pre-craniates became predators.

Amphioxus lacks a vertebrate-type head sensory organs but has a simple non-image-forming front eye (Fig. 147c), possibly the precursor of paired eyes. It also lacks an identifiable olfactory organ and has no hearing organ, but some molecular data suggest that the rostrum in amphioxus has an olfactory sense (Glardon et al. 1998).

Haikouella has both paired eyes (Fig. 149a–c) and possible paired nostrils (Fig. 149a), and it conforms well to Butler's model, predicting that the eyes are older structures and all the craniate sensory organs were not elaborated simultaneously (Butler 2000, 2006). The eyes in the early crest animals, as exemplified by *Haikouella*, *Yunnanozoon*, and *Haikouichthys*, are regularly rounded and relatively large, at about 0.2–0.4 mm in *Haikouella* (Mallatt and Chen 2003; Chen 2004), 0.6 mm in *Yunnanozoon* (Chen 2004), and 0.4 mm in *Haikouichthys* (Shu et al. 1999b, 2003a).

The visual organs in amphioxus are represented by the front eye (Fig. 147c), located at the rostral end of the nerve cord and consisting of a pigment cup, two rows of uniciliate sensory cells (presumably photoreceptors) behind it, and two rows of neurons further behind them (resembling retinal amacrine and bipolar cells, respectively) (Lacalli et al. 1994; Lacalli 1996; Wicht and Lacalli 2005). The front eye is presumably the homologue of the vertebrate paired eyes based on the general similarity of their cellular organization and the similarity of the front eye to the developing vertebrate eye, represented by a single medial primordium. The front eye is very small and simple (only 10 μm in diameter in the larva and about 20 μm in the adult), with its neurons arranged transversely in single rows. As the eye has only a one-dimensional receptor field, it is not an image-forming organ (Lacalli et al. 1994). In contrast to amphioxus, the large eyes in these Early Cambrian cristozoans might resemble those of extant vertebrates, as they have many more neurons packed in layers, probably forming two-dimensional receptor fields allowing proper image development (Lacalli et al. 1994).

The origin of the eyes, especially the complex, image-forming eyes, has received special attention. The tool-kit of genes that participate in building the eyes are shared among the bilateral animals, placing the origin of the eyes in the common ancestor of the bilateral animals. In contrast to this prediction, image-forming paired eyes are not present in most animal groups, but arose independently among only a few, distantly related animal groups, including the molluscs, arthropods, chordates, and annelids. The sudden emergence of image-forming eyes among these distantly related animal groups is intriguing and it has recently been interpreted as a response to increased predation and visual hunting (Fortey et al. 1997; Parker 1998). However, the first arthropods, as exemplified by the Lower Cambrian *Fuxianhuia* and *Shankouia*, were definitely not active predators but inactive suspension and microphagous feeders, although these animals had large, paired, image-forming eyes. The paired eyes are a novelty of the crest animals and the first crest animals, exemplified by *Haikouella* and *Yunnanozoon*, were definitely not active predators. The development of elaborated, image-forming eyes in both the arthropods and crest animals in Cambrian times was probably triggered by an intensified "arms race" in predators and prey, with the emergence of visual hunting in the predators and of visual feeding and escape in the prey.

Chemosensory Faculty

This is important for synchronizing sexual maturation, for reproductive behaviour, and for feeding. The system for collecting olfactory information has an ancient origin. However, amphioxus lacks an identifiable olfactory organ and its sensitivity to chemical stimuli remains unknown (Lacalli 2004). Pax6, a marker of the olfactory epithelium, is expressed only in the developing rostrum ectoderm (Glardon et al. 1998) and the dopaminergic neurons present in the vertebrate olfactory bulb occur in the anterior nerve cord of amphioxus (Moret et al. 2004). The rostral pathway in amphioxus may be homologous to the olfactory nerve (Lacalli 2004). The preservation of possible paired nostrils (Fig. 149a–d) and related olfactory bulbs (Figs. 150 and 151b–c) in *Haikouella* argues for the origination of an elaborate olfactory system, including paired nostrils and paired enlarged olfactory bulbs, which appeared simultaneously with elaborate paired eyes in the early evolutionary stages of Cristozoa. The primitive nostrils, exemplified in *Haikouella*, are paired (Fig. 149a), with a dorsal and anterior location at the anterior margin of the head (Figs. 149b–d and 151c). Although the nostrils in the adult lamprey are unpaired, they do develop from paired olfactory pits (Johnels 1948; Mallatt 1996).

The Brain in the First Cristozoa: Evolutionary Origin of the Vertebrate Brain

The centralization of the central neural system is characterized by the expansion of the anterior part of the neural cord, leading to the formation of the brain. The tripartite brain shares similar gene expression patterns across the major animal groups, suggesting that the tripartite gene expression ground plan arose very early.

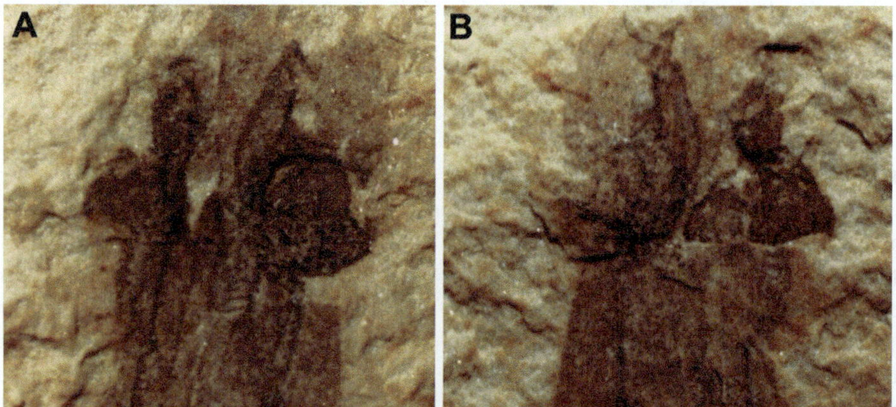

Fig. 150 Pair of large eyes in *Yunnanozoon lividum*. (**a**) and (**b**) are counterparts of the nearly complete specimen, cf. Fig. 142A, B in its anterior part. Scale bar = 1 mm

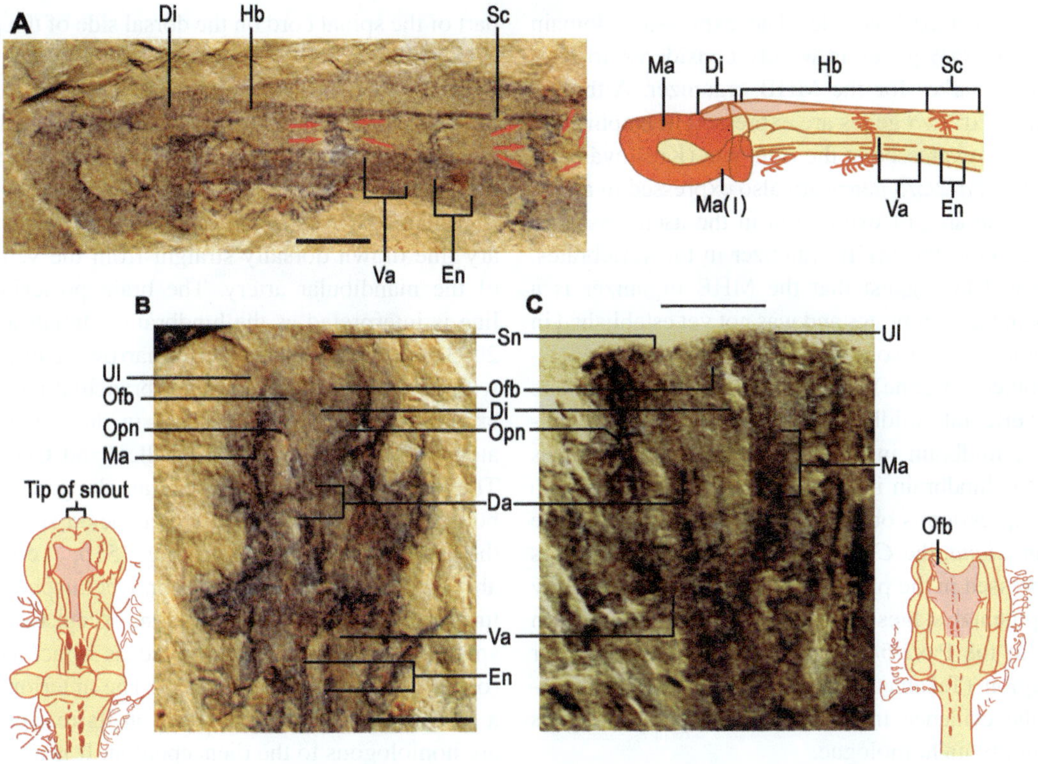

Fig. 151 Brain in *Haikouella lanceolata* from the Early Cambrian Maotianshan Shale. (**a**) a laterally compacted specimen showing side view of brain; (**b**, **c**) two dorso-ventrally compacted heads, showing ventrodorsal view of the brain. Abbreviations: Dn, diencephalon; Hb, hindbrain; Sc, spiral cord; Va, ventral aorta; Da, dorsal aorta; En, endostyle; Ofb, olfactory bulb; Sn, snout; Ul, upper lip; Ma, mandibular aorta; Opn, optic nerves. (modified from Mallatt and Chen 2003). Scale = 1 mm

In contrast to the ancient origin of the tripartite gene expression ground plan, the nervous system might have centralized later (Lowe et al. 2006; Chen 2008, 2009), independently among the Deuterostomia, Ecdysozoa, and Lophotrochozoa. A recent study of fossil arthropods has prompted the surprising suggestion that the tripartite brain is not ancient at all (Chen 2009). The brain in the first arthropods (known as "proarthropods") has only two divisions (Waloszek et al. 2005), corresponding to the protocerebrum and deuterocerebrum (Chen 2009). The tritocerebrum brain is a derived character and did not arise until the evolution of the euarthropods.

The tripartite brain is also not an old structure in the deuterostomes. The deuterostomes, such as the echinoderms and hemichordates, have no brain at all. The nervous system might not have become centralized until the emergence of the chordates. In the tunicates and cephalochordates, the brain is not only very small but also incomplete. In the living cephalochordate amphioxus, the anterior part of the nerve cord is weakly expanded; the minimally enlarged part has been described as a "cerebral vesicle" and shown to be morphologically and micro-anatomically different from the rest of the nerve cord (Fig. 147c). The "cerebral vesicle" can be divided into an anterior and a posterior part, with infundibular cells lying between the two subdivisions. The posterior termination is defined by the posterior end of the lamellar body. The cerebral vesicle embeds structures such as the front eye, lamellar cells, and infundibular cells; these may have homologues in the vertebrates occurring within the diencephalon (Lacalli et al. 1994; Lacalli 1996). The expression patterns of *AmphiOtx* (Williams and Holland 1998) and *AmphiDll* (Holland et al. 1996) agree well with the conclusions drawn from micro-anatomical studies (Lacalli et al. 1994). It is possible that the cerebral vesicle may also include a homologue of the vertebrate midbrain (Holland and Holland 1998), but there is no micro-anatomical support for this hypothesis. If a midbrain homologue exists at all in amphioxus, Lacalli et al. (1994) prefer to locate it behind the cerebral vesicle. The expression domain of the *Pax 2/5/8* genes is widely considered to be a molecular marker for the MHB organizer. Although the *AmphiPax2/5/8* genes are expressed in amphioxus, they are not expressed at the right site (Kreslová et al. 2002). The *Pax2/5/8* genes are also expressed in ascidians, but the area of expression in the ascidians does not function as the MHB organizer in the vertebrates. All these data suggest that the MHB organizer is a novelty in the vertebrates and was not yet established in the protochordate evolutionary stage. *Dmbx1* (a PRD-class homeobox gene) is one of the molecular markers of the vertebrate midbrain; it is expressed in the presumptive midbrain in the early developmental stages and in the hindbrain in later stages. In amphioxus, no *Dmbx* expression is observed in the neural tube. In the ascidian *Ciona*, the *CiDmbx* gene is detectable, but is not expressed in the presumptive midbrain in the early developmental stages, expressed instead in a region caudal to the Pax2/5/8-positive cells, in the anterior Hox region (Takahashi and Holland 2004). In summary, the evidence indicates that the protochordates lack a midbrain homologue.

Based on gene expression patterns, the posterior limit of the region homologous to the forebrain+midbrain (Holland and Holland 1998; Shimeld and Holland 2000, 2005) lies at the level of the boundary between somites 2 and 3 in amphioxus, and the homologue of the posterior hindbrain lies at the boundary between somites 4 and 5 (Holland et al. 1992); the latter interval yields very little information about the anatomical counterpart within the vertebrate hindbrain. The transcription factor *Krox-20* is an important molecular marker of the vertebrate hindbrain. However, a *Krox-20*-related gene, *AmphiKrox*, in amphioxus is expressed in different developing regions, including the anterior cerebral vesicle and a club-shaped gland, but its expression does not overlap with the expression domains of *AmphiHox-1*, *-2*, *-3*, and *-4*. The roles of Krox-20 in hindbrain segmentation and hindbrain development have not yet been established in amphioxus (Knight et al. 2000). Considered together, the evidence suggests that the hindbrain is absent in amphioxus and that its brain contains only a homologue of the vertebrate diencephalon.

The fabulously preserved outlines of the brain in *Haikouella* provide novel evidence of the ground plan of the cristozoan brain (Figs. 149a–b and 151). The brain occurs as a conspicuous enlargement of the front part of the spinal cord on the dorsal side of the pharynx (Figs. 149a–b and 151a). It is about 4 mm long, 0.4 mm thick, and 0.5 mm wide and is much larger (relative to its body size) than that in amphioxus (Fig. 147c), but equal in size to those of other vertebrates, including the jawless and bony fishes, amphibians, and reptiles. It contains two morphological parts, with the boundary line drawn dorsally straight from the ventral part of the mandibular artery. The brain posterior to the line is interpreted as the hindbrain (Mallatt and Chen 2003) and occurs as a posterior narrowing of the brain, with its end at the third (vagal) branchial bar, situated exactly at the same position as in the jawless craniates (Figs. 149b and 151a; Mallatt and Chen 2003). The brain anterior to the line is an elongated and bulbous feature. Dorsoventrally compacted specimens of the anterior part of the brain (Fig. 151b, c) confirm the absence of large, paired, cerebral hemispheric structures related to the telencephalon, except for a pair of small bulges that are interpreted as olfactory nerves (or bulbs; about 0.1 mm thick). The anterior part has a pair of lateral eyes embedded at its rear end; these are homologous to the diencephalon. It is about 1 mm long, 0.3–0.4 mm thick (Fig. 151a), and 0.5 mm wide. The front margin of the brain in *Haikouella* resembles that of amphioxus and the brains in both do not reach the front margin of the head, but end at a position about 1.5 mm from the front margin in *Haikouella* (Figs. 149b and 151a) and at a rear part of the rostrum, about 0.5 mm from the anterior margin, in amphioxus (Fig. 147c).

Deuterostomes, such as the echinoderms and hemichordates, have no brain. The centralization of the deuterostome nervous system most likely evolved in the common ancestor of the chordates, with the occurrence of the first visible but miniature simple brain in the protochordates, as seen in extant cephalochordates and tunicates. The brain ground plan in the chordates, exemplified in the extant protochordate tunicates and cephalochordates, is not only small but also compact and incomplete, and its possible vertebrate homology lies basically within the diencephalon, in both tunicates (Lacalli and Holland 1998) and cephalochordates (Lacalli et al. 1994; Lacalli 1996). However, there is little or no morphological evidence for homologues of the vertebrate telencephalon, midbrain, or hindbrain.

Brain evolution during the pre-cristozoan-to-cristozoan transition might have been triggered by an increasingly active lifestyle and was marked by the expansion and rearward extension that led to the formation of a morphologically recognizable hindbrain. The brain ground plan in Cristozoa, as exemplified by *Haikouella*, is a relatively large and elongated structure consisting of two morphologically distinct parts, homologous to the craniate diencephalon and hindbrain. A possible midbrain might lie between the diencephalon and the hindbrain. Compared with the size of the brain in the cephalochordate amphioxus, the brain in the first cristozoans, as exemplified by *Haikouella*, is extremely large, although amphioxus and *Haikouella* are the same size. There is no morphologically recognizable telencephalon homologue in the first Cristozoa. Whitlock and Westerfield (2000) demonstrated that the olfactory nerves are closely related to the telencephalon during embryonic development. The presence of possible paired olfactory nerves (or bulbs) in *Haikouella* (Fig. 151b, c) implies that the olfactory nerves (or bulbs) are the precursors of the telencephalon. The emergence of a morphologically clear telencephalon is a novelty in the craniates.

The anterioposterior gene expression patterns along the dorsal nerve cord in ascidians and vertebrates share a similar basic tripartite ground plan, comprising an anterior *Otx* expression region (the forebrain and midbrain in the vertebrates), a *Pax2/5/8*-expressing central region (MHB), and a *Hox*-expressing region (hindbrain and spinal cord in the vertebrates; Takahashi and Holland 2004). A similar tripartite gene expression pattern appears in the non-chordate deuterostomes, the hemichordates (Lowe et al. 2006), and the arthropods (Hirth et al. 2003) suggesting that the tripartite gene expression ground plan arose very early, even before the evolution of the chordates, from the common ancestor of the protostomes and deuterostomes. In contrast to the ancient origin of the tripartite gene expression ground plan, the nervous system might have centralized later (Lowe et al. 2006), independently among the Deuterostomia, Ecdysozoa, and Lophotrochozoa.

Branchial System in the Cristozoan Ground Plan

Water is 800 times denser than air and the dissolved oxygen in water is 30 times more dilute than in air. The cost of respiration is rather high, accounting for up to 30% of the total metabolic energy (Young 1981). The adaptations for a more active lifestyle that evolved during the transition from an amphioxus-like ancestor towards the crest animals very probably demanded an increased metabolic rate. The replacement of the gill-less pharyngeal system with a gill-based pharyngeal system and the change in the ventilation mechanism from ciliary to muscular are two of the major landmarks in the transition from an amphioxus-like ancestor to the crest animals, as predicted by Northcutt and Gans (1983).The change allowed a more active lifestyle in terms of feeding, oxygen intake, and locomotion, with an increasing metabolic rate. In the ancestral gill-less pharyngeal system, as exemplified by amphioxus, water is pumped through the pharynx by cilia on the branchial arches and these cilia-bearing arches must be arranged close to each other to form narrow, regularly spaced, branchial slits (Fig. 152a). The secondary growth of new additional branchial arches in later developmental stages, with the larger numbers of closely spaced branchial arches in the adult, characterizes an ancestral gill-less pharyngeal system. The gill-less branchial arches do not function in oxygen uptake (Schmitz et al. 2000), but are used for food intake. The gill pharyngeal system is a novelty of the crest animals. When the gill-less ancestor became a more active swimmer, the increased metabolic rate required must have triggered the selection of animals with a more efficient respiratory system, and thus the development of gills and gill rays (gill-supporting structures). The development of gills with increasing resistance to water flow might have prompted an evolutionary change from ciliary ventilation to the more powerful, muscular ventilation pump and the replacement of the thin, unjointed branchial arches with thicker, segmented, cartilagenous branchial arches. These are stronger and more capable of the elastic recoil that expands the pharynx to draw in water during inhalation. Fewer, more widely spaced branchial arches, as exemplified in *Haikouella* (Fig. 152e–g) and *Yunnanozoon* (Fig. 152h), would reduce the resistance to the water flow passing under the pressure of muscular ventilation, characterizing the transition from the amphioxus-like ancestor to the crest animals.

It was argued by Mallatt in 1996 that the common ancestor of the craniates had both internal and external branchial arches, as observed in the anterior pharynx of the living hagfish *Myxine glutinosa* (Holmgren 1946).

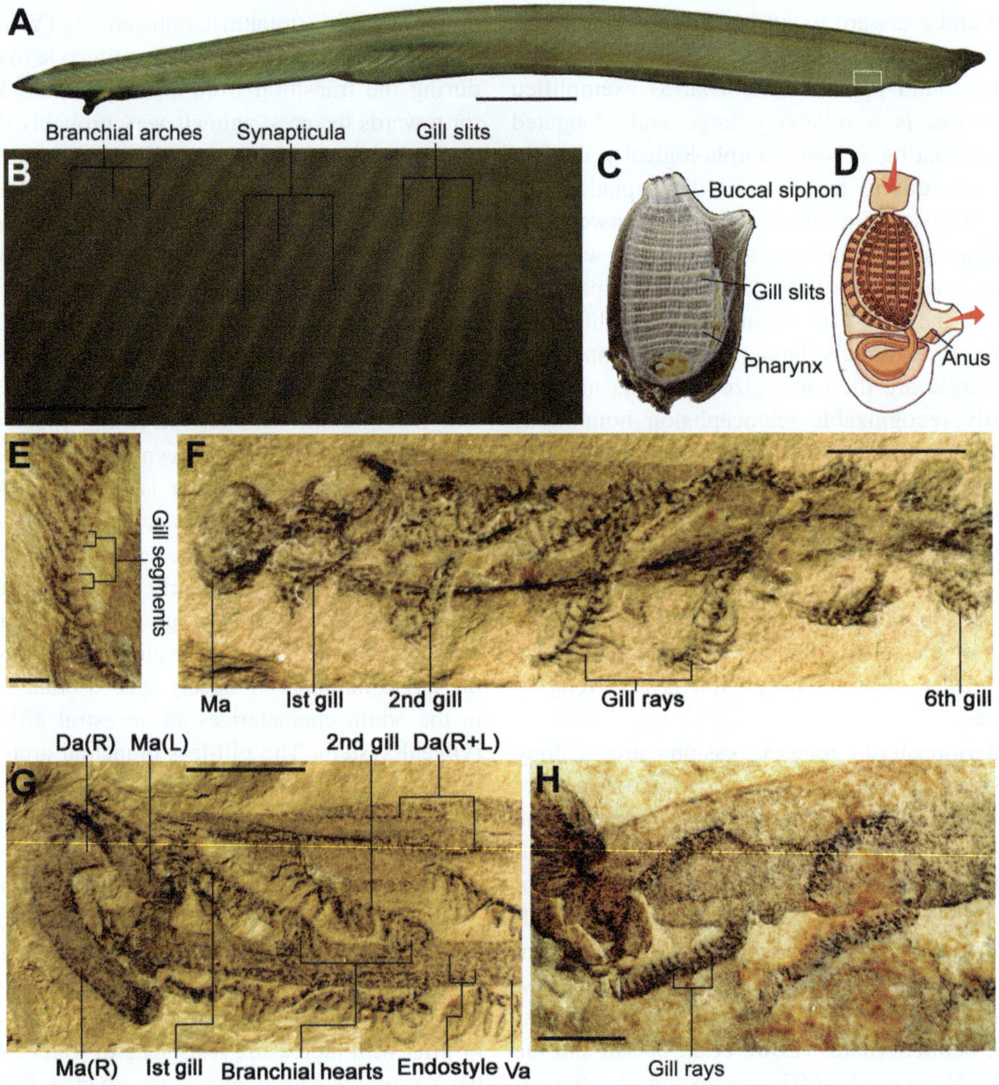

Fig. 152 Branchial system in living Cephalochordata (**a**, **b**), Urochordata (**c**, **d**) and Early Cambrian Cristozoa (**e–h**). A, view from right side of *Branchiostoma belcheri* showing closely arranged branchial arches; (**b**) enlargement showing closely arranged branchial arches and synapticula between branchial arches; (**c**, **d**) living urochordates showing closely arranged branchial arches connected by transverse bars with gill-slits consistent in size; (**e**) showing segmented branchial arch with a pair of gill rays in each segment in *Haikouella*; (**e–g**) Branchial system in *Haikouella* showing brachial manidibular aorta and branchial arches with paired gill rays; (**h**) branchial arches with paired gill rays in *Yunnanozoon lividum*. Abbreviations: Ma, mandibular aorta; Da, dorsal aorta; Va, ventral aorta. (**h**, reproduced by permission from D. Y. Huang). Scale bars = 2 mm (**a**, **f–h**), 200 μm (**b**), and 100 μm (**e**)

However, the adult lamprey has only unjointed, external branchial bars. The internal branchial arches are proposed to have been secondarily lost in the lamprey, whereas those in some other pre-gnathostomes were enlarged and subdivided into associated ventilation muscles (Mallatt 1996).

The early Cristozoa, including *Haikouella* and *Yunnanozoon*, have only one set of branchial arches with paired gill rays, interpreted by Mallatt and Chen (2003) to project externally from the branchial arches and thus to be homologous to the external branchial arches of the craniates. Further observation has shown that most of the intact gill rays in *Haikouella* project internally rather than externally from the branchial arches (Fig. 152f, g), and thus play a role similar to those of the internal branchial arches. The branchial arches are segmented with joints, which might provide both the branchial arches and their gill-rays with the flexibility to allow the reorientation of the gill rays, to adapt to different levels of feeding activity. The paired gill rays can form a narrow angle and project posteriorly, with one placed within the pharyngeal lumen and the other outside the lumen but within the atrium (Fig. 154a). This suggests that the branchial arches in the early Cristozoa are homologous to the internal branchial arches in function, although they may have played roles in both internal and external functions in some cases.

The branchial arches in both the cephalochordates (Fig. 152a, b) and tunicates (Fig. 152c, d) lie close together, maintaining very narrow branchial slits of a consistent size. As the animals become larger, they must add new branchial arches as they grow to meet the increasing ventilatory demands of filter feeding. The branchial arches are fairly thin, non-cellular structures usually connected to each other by fine transverse bars (called "synapticula" in amphioxus) that limit the size of the gill slits (Fig. 152b); the upper size limit of food particles is fixed by the size of the gill slits and includes microbes and phytoplankton.

The transformation of an amphioxus-like invertebrate ancestor into the evolutionary line of vertebrate animals was probably driven by an increasingly active lifestyle and an increasingly higher metabolic rate, triggering a series of consequent changes in the branchial system. A central event in the changes was the shift from ciliary ventilation to respiratory muscular ventilation with gills (Northcutt and Gans 1983). The branchial arches of *Haikouella* and *Yunnanozoon* are not only broad but also segmented with joints, and each segment has a pair of thin, blade-shaped, skeletal gill rays (Fig. 152e–h; Chen et al. 1999; Mallatt and Chen 2003; Chen 2004; Chen and Huang 2006). The same structures occur in the bony fishes, in which they support the respiratory gill filaments and gill septa (Hughes 1984; Laurent 1984) and act as insertion sites for the branchiomeric muscles (Dunnel-Erb et al. 1993). The occurrence of gill rays, possibly with gills, in *Haikouella* and *Yunnanozoon* (Mallatt and Chen 2003) indicates that muscular ventilation was an early structure, representing a novelty in the crest animals.

Paired gill rays are known only in the bony fishes, but are absent in the agnathans and primitive jawed fishes such as Chondrichthys. Chondrichthys has gill rays but they are larger and unpaired. The paired gill rays are not considered to be a primitive feature in the crest animals. However, the occurrence of paired gill rays in both *Haikouella* and *Yunnanozoon* suggests that paired gill rays are a primitive character, occurring at the very beginning of the crest animals, and that their absence in the agnathans is ascribable to secondary degeneration. Alternatively, these early gill rays may not be homologous to those of the bony fishes (Mallatt and Chen 2003).

The respiratory muscular ventilation in *Haikouella* and *Yunnanozoon* is confirmed not only by gill bars but also by the branchial arches, which are robust, jointed, bendable, and presumably capable of elastic recoil. They are fewer in number, with wider slits (or pouches) to lower the resistance to the water flow passing under the pressure of muscular ventilation (Mallatt and Chen 2003). The branchial arches in *Haikouella* are about 120 μm in diameter, slightly thicker than those in the living larval sea lamprey, but much thicker than those of amphioxus, although they have similar body sizes. The branchial arches measure about 80 μm in diameter in the larval sea lamprey, but only 30 μm in amphioxus. They are abundantly jointed and consist of numerous stacked discs, resembling those of the branchial arches made by cartilage cells in the larval lamprey and embryonic fishes (Pardos and Benito 1988; Kimmel et al. 1998; Holland and Chen 2001; Mallatt and Chen 2003), suggesting the branchial arches to be cellular, composed of cartilage. Cartilage and multijointed branchial arches are novelties in the Cristozoa (Meulemans and Bronner-Fraser 2007).

There are few branchial arches in the early developmental stages of protochordates (including the cephalochordates and tunicates) and hemichordates in their early developmental stages, but they increase in number as the animal grows. In contrast, the vertebrates not only have much fewer branchial arches (usually no more than seven) but they also always lack secondarily derived branchial arches, appearing only in later development (Jefferies 1986; Mallatt and Chen 2003; Chen 2008). A small but constant number of branchial arches is a novelty of the Cristozoa and presumably represents a larval version of the pre-Cristozoa ancestor, resulting from a paedomorphic process (Chen 2008).

There are fewer than ten branchial arches in all the crest animals, with the exception of the agnathans, with usually up to 15 branchial arches (or pouches) in the living hagfishes and 45 in the fossil hagfishes (Janvier 2004). However, the early crest animals, as exemplified by *Haikouella* and *Yunnanozoon*, have only six post-mandibular arches, plus the mandibular arch. Interestingly, the mandibular arch is absent in *Haikouella* but is represented by the mandibular artery (Fig. 152g; Mallatt and Chen 2003), suggesting that the vertebrate branchial arches did not arise synchronously and that the mandibular arch is a structure slightly younger than the other branchial arches. There are seven pairs of branchial arches in *Yunnanozoon* (Fig. 145b, c); this is probably the prototype of the crest animals. More than seven branchial arches, as seen in the agnathans, is a secondarily derived character.

Atria

The pharynx of the early crest animals, as represented by *Haikouella* and *Yunnanozoon*, is protected by a pair of water chambers known as "atria", indicated by the presence of an atrial pore and its related mud-filled cavity (Fig. 153a) and by the mud-filled atrial chamber in a dorsoventrally compacted specimen. These atria are a plesiomorphic character and also occur in the protochordates, including the ascidians and cephalochordates, but disappear in the craniates. They represent water chambers on the sides of the pharynx, receiving the exhalent ventilation water leaving the branchial slits of the pharyngeal lumen and convey it to the outside through the atriopore in the cephalochordates (Fig. 147b) or through an exhalent siphon in the tunicates.

The atria might function to protect the delicate branchial arches from muddy and sandy substrata during swimming and burrowing (Holland and Chen 2001) and to protect the delicate gill rays from damage by water currents during swimming. The atria are present in amphioxus but absent in the craniates, although they are still present in the early crest animals, as exemplified by *Haikouella* (Fig. 153a) and *Yunnanozoon*, suggesting that the early crest animals were capable of burrowing temporarily into the soft mud when the animal encountered a predator.

Endostyle

The thyroid gland is a craniate-only character and one of the largest endocrine glands in the group. It produces thyroid hormones, principally thyroxine (T4) and triiodothyronine (T3), to regulate the rate of metabolism and affect the growth and rate of function of many other body systems. The endostyle is the precursor of the craniate thyroid gland, producing both thyroid and digestive glands, and adding mucus and digestive enzymes to trapped food grains in the pharynx during suspension feeding. The mucoproteins contain monoiodothyronine (T1), diiodothyronine (T2), T3, and T4, perhaps regulating the metabolic rate and

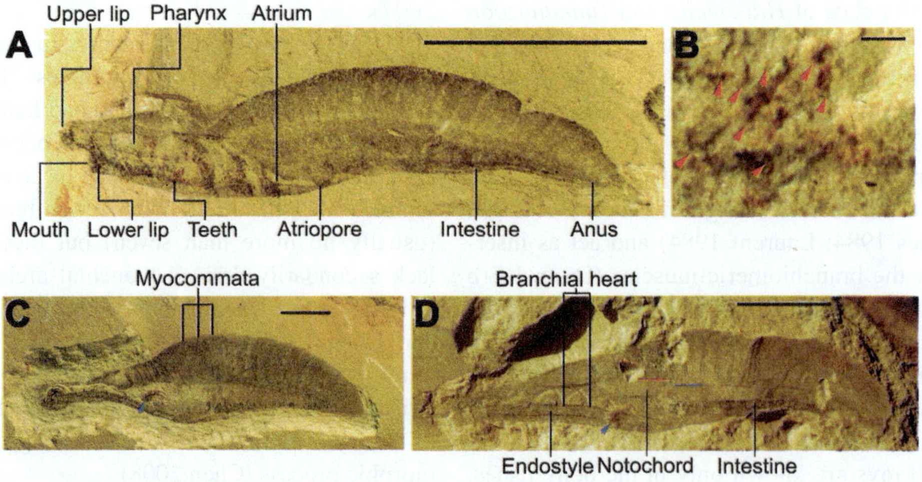

Fig. 153 Pharyngeal teeth (*marked with arrows*) in *Haikouella lanceolata* (**a**, **b**) and *Yunnanozoon lividum* (**c**, **d**) from the Early Cambrian Maotianshan Shale. Scale bars = 5 mm (**a**, **c**, **d**) and100 μm (**b**)

affecting the growth and functions of body systems, as in vertebrates. However, T3 is relatively more abundant in amphioxus.

The endostyle is an evolutionary precursor of the thyroid gland and occurs as a longitudinal ciliated groove on the ventral wall of the pharynx, producing mucus to gather food particles and to assist in transporting food to the oesophagus. The endostyle is a chordate-only structure; it is present in the protochordates (including the cephalochordates and urochordates) and existed continuously in the early crest animals, as exemplified by *Yunnanozoon* (Fig. 153d) and *Haikouella* (Figs. 151a, b and 152g). In some living primitive craniates, such as the lamprey, the endostyle remains in the larval stage only (Barrington and Sage 1972), suggesting the thyroid gland to be an evolutionary novelty in the craniates.

Pharyngeal Teeth

Simple cone-shaped denticles occur on the pharyngeal floor in *Yunnanozoon* and *Haikouella* (Chen et al. 1999; Holland and Chen 2001) making it a possibility for the teeth to be evolutionarily deeply rooted, representing an apomorphic character in the Cristozoa that emerged before the onset of the mineralized dermal skeleton. The teeth in the early cristozoans, as exemplified by *Haikouella* and *Yunnanozoon*, are simply cone-shaped, situated on the pharyngeal floor (Fig. 153). The teeth in *Haikouella* are small (only 0.1 mm), located medially on the floor of the pharyngeal cavity (Fig. 153a, b), but they are much larger (about 1 mm each) when situated at the rear of the pharynx in *Yunnanozoon* (Fig. 153c, d). The pharyngeal teeth are probably a primitive feature in the cristozoans and are also present in conodont animals. Pharyngeal teeth that predate the origination of skin scales challenge the scale-origination hypothesis of teeth evolution (Romer 1962).

When larger food grains entered the pharynx, animals used the intra-pharyngeal denticles to pierce and break them down before they entered the gut (Chen et al. 1999; Holland and Chen 2001; Mallatt and Chen 2003). The small (only 0.1 mm) intra-pharyngeal denticles in *Haikouella* are probably allowed food particles of up to 0.3 mm to be processed. Food may have included embryos and larvae. The much larger (1 mm) intra-pharyngeal denticles of *Yunnanozoon* suggest that the animal was able to feed on food items larger than 1 mm, up to 3–4 mm, including embryos and small animals such as bradoriids, *Ercaia*, and *Primicaris* (Chen 2004). These food animals were not only small (in the range of 2–4 mm for bradoriids and *Ercaia* and 1–7 mm for *Primicaris*) but were also extremely abundant in the Maotianshan Shale fauna.

Evolution of the Axial Skeleton

The notochord is an axial, non-bony, hydrostatic skeleton and a chordate-only character. The notochord appears only in the tail of the larval stages of the tunicates but remains well developed throughout life in the cephalochordates. In the cephalochordates, the notochord is a large, distinct structure, extending throughout the entire length of the body (Fig. 147a, b). The notochord exists in vertebrates, but it tends to degenerate, and is reduced or almost entirely lost in the adult form.

The protochordates burrowed in clean sand, as demonstrated by the living amphioxus. Its body tapers to a point at both the anterior and posterior ends, allowing the animal to push aside sand with either the head or tail end first while burrowing. These burrowing activities may have triggered the emergence of the axial skeleton, known as the "notochord", allowing the animal to resist deformation when burrowing in the sand, either head or tail first (Fig. 148). The notochord of amphioxus is the living representative of the first axial skeleton, extending to both body termini. The function of the early axial skeleton was not only to resist axial deformation while burrowing but also for the attachment of the muscles used in burrowing or while swimming with a burst of rapid lateral undulations of the body.

The notochord exists in the vertebrates, but they have also developed a novel axial skeleton known as the "vertebrae", bony elaborate joints, encasing both the notochord and spinal cord. The bony vertebrae have made the soft notochord superfluous, degenerating in the adult until it was almost entirely lost. The notochord in the Cristozoa ground plan, as exemplified by *Haikouella* and *Yunnanozoon*, resembles the large and distinct structure of amphioxus. Unlike that in amphioxus, the notochord of *Haikouella* disappears anteriorly at about the dorsal location of branchial

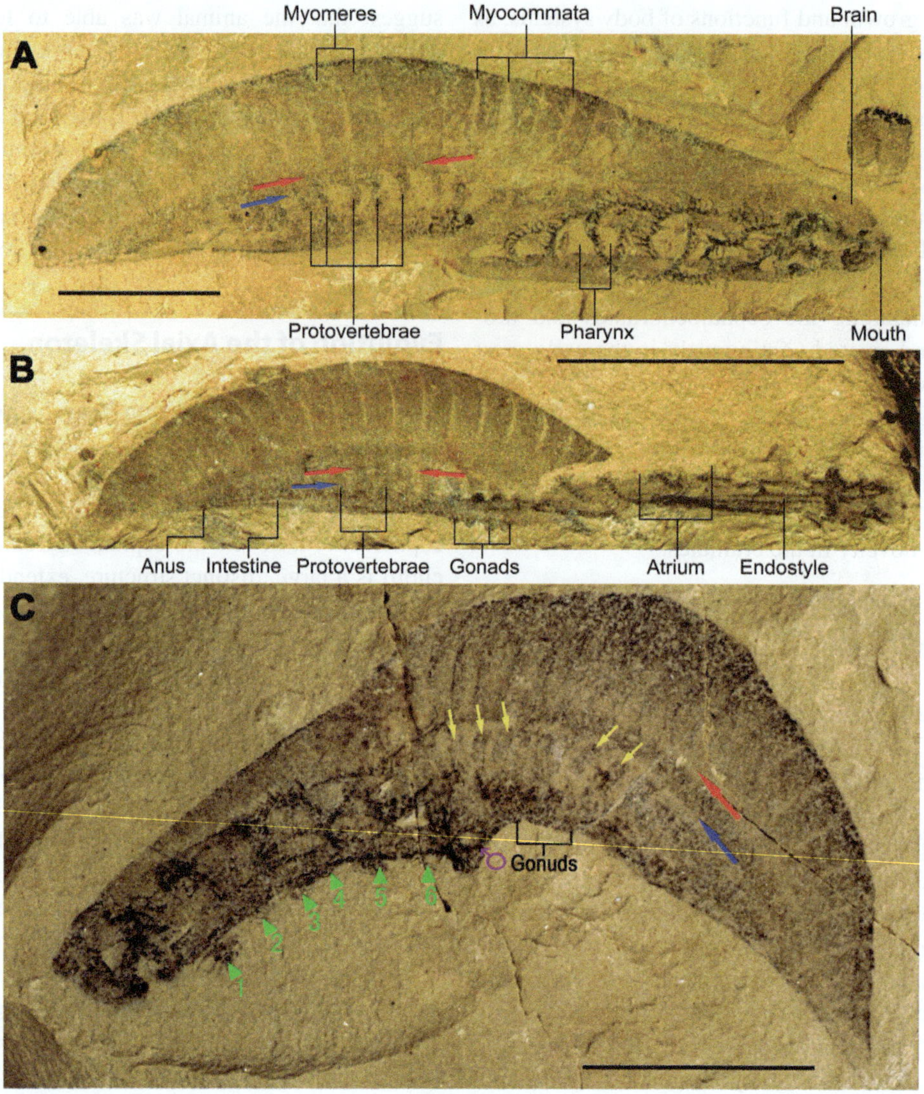

Fig. 154 Laterally compressed *Haikouella lanceolata* from the Early Cambrian Maotianshan Shale showing protovertebrae (**a**–**c**), notochord (**a**, **b**), brain (**a**), posteriorly directed gill rays (**a**), four pairs of gonads (**b**, **c**), atrium (**b**, **c**), intestine (**b**, **c**), anus (**b**), tail (**b**), mouth (**a**). *Red arrow* represents the dorsal margin of protovertebrae; *blue arrow* indicates dorsal margin of notochord. Scale bar = 5 mm

arch 3, whereas in amphioxus, it extends continuously to the anteriormost end of the body.

The vertebral column is a vertebrate-only character. Its origin remains controversial, and it was thought to be a younger structure than the skull. *Haikouella* and *Yunnanozoon* have segmented cartilagenous structures (Fig. 154) that represent the precursor of the vertebral column of advanced vertebrates and are comparable to the protovertebrae in the living lamprey (Chen and Li 2000; Holland and Chen 2001; Young 1981), suggesting that the vertebrae evolved before the origination of the elaborated skull.

Musculoskeleton

Myotomes are divided by myosepta (or myocommata) into repeated, segmented structures known as "myomeres" (muscle segments). These are a chordate-only character, present only in the extant

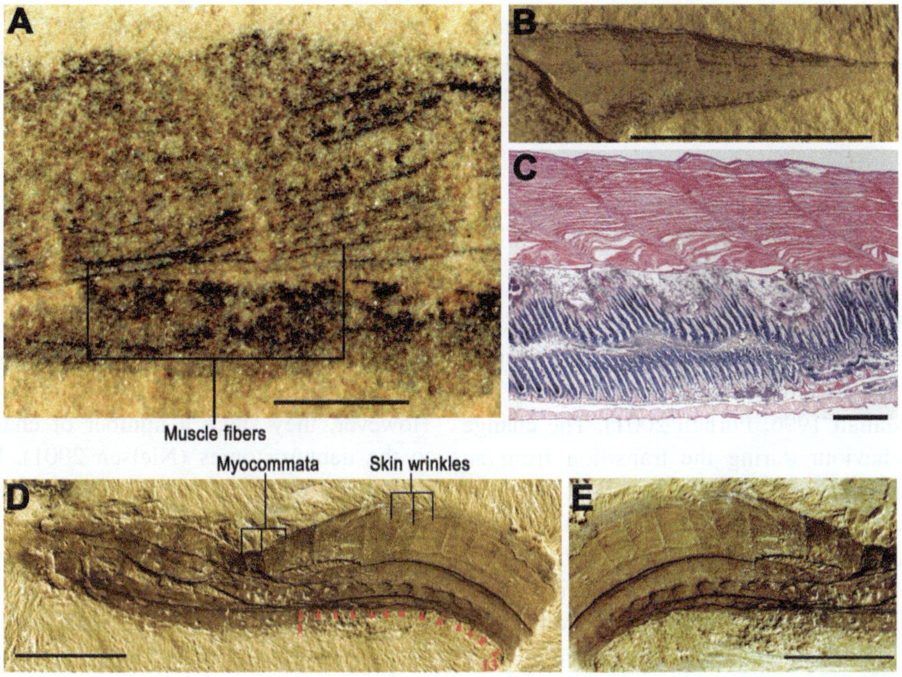

Fig. 155 Musculature in the first crest animals *Haikouella lanceolata* (**a**, **b**) and *Yunnanozoon lividum* (**d**, **e**) and in the extant cephalochordate *Branchiostoma belcheri* (**c**) showing myocommata (**a**–**e**), myomeres with muscle fibres (**a**–**e**), and skin on top of both the myomeres and myocommata (**d**, **e**). Scale bar = 0.5 mm (**a**, **c**) and 5 mm (**b**, **d**–**e**)

cephalochordates, craniates, and larval tunicates. In these taxa, the myosepta have a complex V- or W-shape. In *Haikouella* and *Yunnanozoon*, there are about two dozen myomeres, separated by myosepta in a simple straight form (Fig. 155), as in the young larvae of the living cephalochordates and tunicates. Because of their unusual simplicity and dorsal location, they were regarded as a large, segmented fin by Shu et al. (1996) and as cuticular rings by Conway Morris (2000). The preservation of the muscle fibres in some *Haikouella* specimens (Fig. 155b, c) and the muscle fibres and wrinkled skin overlying the muscles in some *Yunnanozoon* specimens (Fig. 155c, d) confirm that these segmental structures are truly myomeres (Chen and Li 1997; Mallatt and Chen 2003). The young larvae of lancelets (Fig. 155c) and tunicates have the same simple, straight, segmental muscles and swim well (Holland and Chen 2001). Simple myosepta may represent the ground plan of the crest animals. The presence of W-shaped myosepta in the Lower Cambrian *Haikouichthys* (Shu et al. 1999b) indicates that the W-shaped myosepta had a very early origin, evolving independently and directly from the simple straight form at an advanced evolutionary stage of the precraniates or at the beginning of the evolutionary history of the craniates. The complex V-shaped myosepta in the cephalochordates and the W-shaped myosepta in the crest animals presumably represent parallel evolution.

Positional relationships of the musculoskeletal structure, the axial skeleton, the neural cord, and the branchial structure among the chordates are confused. The "dual animal" hypothesis of Romer (1972) proposed that the vertebrates evolved from a tadpole larva-like ancestor, in which the body was segregated into a visceral head and a somatic trunk. Romer's hypothesis pushes tunicates to the central stage of vertebrate origins. However, the ascidian body plan is so derived that in some respects it is not very useful; for discussion of the origin of the vertebrate body plan, see Lacalli (2005).

Amphioxus and *Haikouella* display an anterior–posterior body organization that is very different from that of the "dual animal" of Romer. In amphioxus, the musculoskeletal structure extends anteriorly at a level corresponding to the rostral margin of the cerebral vesicle (a homologue of the vertebrate diencephalon), whereas the rostral margin in *Haikouella* lies at a level corresponding to the possible MHB in the vertebrates. That the visceral head and somatic trunk in both the amphioxus-like pre-cristozoan ancestor and the early crest animals were not segregated but integrated is a possibility.

Evolution of Feeding Behaviours

As predicted by many authors, changes in feeding behaviour evolved with the changes in ventilation, before any true predator arose (Northcutt and Gans 1983; Mallatt 1996; Purnell 2001). The change in feeding behaviour during the transition from an amphioxus-like ancestor towards the crest animals, as exemplified by *Haikouella* and *Yunnanozoon*, is marked by the ability to handle increasingly larger food items before a true predator arose. The maximal food size of these early crest animals ranged from about 300 μm to a few millimetres. The emergence of elaborate olfactory organs, as exemplified in *Haikouella,* allowed these early crest animals to become selective feeders on food with higher nutritional value, probably including the embryos, immobile larvae, and small invertebrates available on the seafloor. The large muscular upper lips of the early crest animals could form an oral hood to drape over a cluster of interesting food items on the sea bottom, and their muscular ventilation allowed the animals to then suck the prey items through the mouth into the pharynx.

Evolutionary Tree of the Deuterosomes and the Approach Towards a Backbone

Although the deuterostomes are a morphologically well-defined group and probably monophyletic, there is no agreement about the shape of their evolutionary tree. Figure 156 suggests that the deuterostomes derived from a tentaculate worm-like ancestor (1) in two basic branches: the tentaculate deuterostomes (11) and the living pharyngeal deuterostomes (2). Tentaculate feeding was presumably the plesiomorphic character of the deuterostomes. *Phlogites* (12) from the Lower Cambrian Maotianshan Shale, with least complexity, probably represents the ancestral form of the tentaculate sessile deuterostomes. *Phlogites* exemplifies the ground plan of the tentaculate deuterostomes insofar as it is sessile and characterized by a spacious cavity that encloses a coiled gut (12). A set of large hollow and branched tentacles surrounding the mouth forms a feeding bowl. The anus is situated close to the mouth but outside the feeding bowl.

From a *Phlogites*-like ancestor (13), the two major tentaculate deuterostome branches arose: the phoronids+brachiopods (14, 16) and the eldoniids+echinoderms (19). The phylogenetic positions of both the Brachiopoda and Phoronida remain controversial. Molecular studies exclude the brachiopods from the deuterostomes and both Brachiopoda and Phoronida are usually interpreted as protostomes. However, they share a number of characters unique to the deuterostomes (Nielsen 2001). Phoronida has a secreted chitinous tube (15) and Brachiopoda has mineralized dorsal and ventral valves (16). The eldoniid+echinoderm clade (17) is characterized by a highly developed water-vascular system that includes radial canals, a central ring, and tubular outgrowths of the body wall (tube feet). This clade is adapted for two different lifestyles, a pelagic lifestyle in the eldoniids (18) and a sessile lifestyle in the echinoderms (19).

Pharyngeal feeding, with the related branchial slits, is an apomorphic character in the deuterostomes. The hemichordates (3) lie at the evolutionary base of the pharyngeal deuterostome clade, representing different evolutionary stages in the transition from tentaculate feeding to pharyngeal feeding. Among the extant hemichordates, the pterobranchs are predominantly tentaculate feeders, except for *Cephalodiscus*. With a single pair of gill slits, these represent the early evolutionary history of pharyngeal feeding. Enteropneusts (= acorn worms) are all pharyngeal feeders, representing an advanced evolutionary stage of the hemichordates. The enteropneust body is characterized by an anterior proboscis, a small middle collar, and a long posterior trunk. The animal uses its muscular proboscis for burrowing with retrograde peristaltic waves that transport the sand rearward.

The Chordata (4) is the sister group of the Hemichordata (3); both groups are adapted for burrowing, but in different ways. Unlike the hemichordate enteropneusts, which use a muscular proboscis for burrowing, the first chordates (5), exemplified by the extant amphioxus, developed a lancelet-like body that narrows to a point at both ends, burrows with the help of an axial skeleton (notochord). The ground plan of the notochord in chordates, as exemplified by

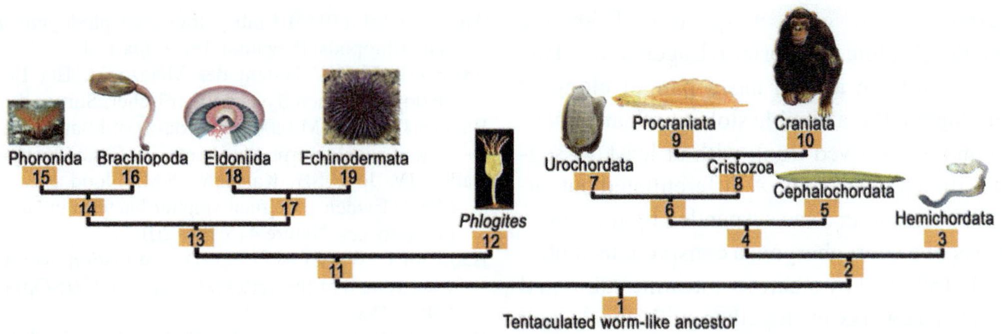

Fig. 156 Phylogenetic kinship of the high ranking clades among the Deuterostomata. (1) tentaculated worm-like ancestor; (2) evolutionary origin of the pharyngeal deuterostomes; (3) Hemichordata representing the transitional stage from tentaculate feeding to pharyngeal feeding; (4) Chordata with key novelties including axial skeleton (notochord), musculature system and centralization of central nerve system; (5) cephalochordates, the stem group of the Chordata; (6) evolutionary origin of the Urochordata+Cristozoa clade; (7) Urochordata (adapted for an inactive lifestyle); (8) Cristozoa (adapted for a more active lifestyle) marked by key novelties including whole genome duplication, neural crest cells and related producers, protovertebrae, gill-pharyngeal system, muscular ventilation system, brain size expansion, image-forming lateral eyes, and possible nostrils; (9) Procraniata representing the stem lineage of Cristozoa; (10) craniate cristozoans marked by presence both of a complex skull and telecephalon. (11) origin of tentaculate deuterostomes; (12) *Phlogites*-like animals (characterized by a spacious cavity enclosing a coiled gut; a set of large hollow and branched tentacles surrounding the mouth forms a feeding bowl (anus is situated close to the mouth but outside of the feeding bowl); it represents the stem lineage of the tentaculate deuterostomes; (13) origin of advanced tentaculate deuterostomes; (14) origin of common ancestor of the phononids+brachiopod clade; (15) Phoronida (having a secreted chitinous tube); (16) Brachiopoda characterized by mineralized dorsal and ventral valves; (17) origin of common ancestor of eldoniid+echinoderm clade—characterized by a highly developed water-vascular system including radial canals, central ring, and tubular outgrowths of the body wall (the tube feet); (18) eldoniids (adapted for a pelagic lifestyle); (19) echinoderms (adapted for a sessile lifestyle)

amphioxus, is large, extending the entire length of the body. The notochord plays a role not only in resisting axial deformation under stress while the animal burrows through sand, either head or tail first, but also in muscle attachment while the animals burrow or swim with bursts of rapid lateral undulations of the body. The early Cristozoa (9) are soft bodied; predatory avoidance behaviour was critical to their survival—burrowing is one of the behaviours used to avoid predation. Development of the fast-start response is another major predation-avoidance behaviour in which the body accelerates rapidly from a perceived threat. The complex musculature plays a critical role in the fast-start response, as seen in all fishes today (Westneat et al. 1998).

The Cristozoa+Urochordata clade (6) arose from an amphioxus-like ancestor and branched dichotomously into two groups, the urochordates (7) and the crest animals (8) at a time before the first round of whole-genome duplication in the vertebrates. The Urochordata is probably the sister group of the Cristozoa and diverged from the evolutionary lineage leading to the latter before the whole-genome duplication event. The origin of the Urochordata (7) was presumably driven by selection pressure that adapted them to an inactive sessile lifestyle. They not only diverged morphologically from the main evolutionary pathway to the Cristozoa, but their genome displays substantial reduction, contrary to the expansion observed in Cristozoa.

The transition from an amphioxus-like ancestor to the Cristozoa (8) was probably triggered by selection pressure for a more active lifestyle. The evolutionary key novelties that led to the transformation of the amphioxus-like ancestor to the Cristozoa included whole-genome duplication, neural crest cells and related producers, protovertebrae, the gill–pharyngeal system, the muscular ventilation system, brain expansion, image-forming lateral eyes, and possible nostrils. The transition from the procraniate cristozoans (9) to the craniate cristozoans (10) was marked by the clearly complex skull and telencephalon.

Molecular and morphological evidence unite the hemichordates and echinoderms as the Ambulacraria, but their earliest history remains almost entirely conjectural. This is on account of the morphological disparity of the ambulacrarians and a paucity of obvious stem-groups. We describe here a new taxon

Herpetogaster collinsi gen. et sp. nov. from the Burgess Shale (Middle Cambrian) Lagerstätte. This soft-bodied vermiform animal has a pair of elongate dendritic oral tentacles, a flexible stolon with an attachment disc, and a re-curved trunk with at least 13 segments that is directed dextrally. A differentiated but unlooped gut is enclosed in a sac suspended by mesenteries. It consists of a short pharynx, a conspicuous lenticular stomach, followed by a narrow intestine sub-equal in length. This new taxon, together with the Lower Cambrian *Phlogites* and more intriguingly the hitherto enigmatic discoidal eldoniids (Cambrian-Devonian), form a distinctive clade (herein the cambroernids). Although one hypothesis of their relationships would look to the lophotrochozoans (specifically the entoprocts), we suggest that the evidence is more consistent with their being primitive deuterostomes, with specific comparisons being made to the pterobranch hemichordates and pre-radial echinoderms. On this basis some of the earliest ambulacrarians are interpreted as soft-bodied animals with a muscular stalk, and possessing prominent tentacles.

Acknowledgments This work was supported by the National Basic Research Program of China (grants 2007CB815800 and 2006CB806400) and the National Science Foundation of China (grant 41023008). I thank John Talent, editor of this volume, for helpful comments and polishing the English of the manuscript; he and several assessors, including SQ Dornbos, Ruth Mawson, Peter Cockle, Janine Miller, and Karen Novotny, also contributed importantly to editing of the manuscript. Technical assistance was given by HZ Wu and XZ Li. This paper presents a wide coverage of research advanced by stimulating collaboration and discussion with many colleagues and former students, especially E Davidson, D Bottjer, MG Hadfield, L Gang, F Gang, and P Tafforeau on the Weng'an biota; J Mallat, D Waloszek, A Maas, MY Zhu, DY Huang, SQ Dornbos, J Vannier, N Holland, and SH Chuang on the Maotianshan Shale biota.

References

Adrianov AV, Malakhov VV (1995) Comparative-morphological analysis of the organization of cephalorhynch worms, the phylogeny and the system of the phylum Cephalorhyncha. 5. Phylogeny and system. Zool Zh 74:19–27

Adrianov AV, Malakhov VV, Spiridonov SE (1998) Fine morphology of the larvae of hairworm *Gordius* sp (Nematomorpha). Dokl Akad Nauk 361:558–561

Aldridge RJ, Hou XG, Siveter DJ, Gabbott SH (2007) The systematics and phylogentic relationships of vetulicolians. Palaeontology 50:131–168

Anderson DT (1975) Embryology and phylogeny in annelids and arthropods. Pergamon Press, Oxford

Ax P (1995) Das System der Metazoa I. Ein Lehrbuch der phylogenetischen Systematik. Fischer, Stuttgart

Baguñà J, Riutort M (2004) The dawn of bilaterian animals: the case of acoelomorph flatworms. BioEssay 26:1046–1057

Bailey JV, Joye SB, Kalanetra KM, Flood BE, Corsetti FA (2007) Evidence of giant sulphur bacteria in Neoproterozoic phosphorites. Nature 445:198–201

Ball-Cuif L, Wassef M (1995) Determination events in the nervous system of the vertebrate embryo. Curr Opin Genet Dev 5:450–458

Barfod GH, Albarede F, Knoll AH, Xiao S, Telouk P, Frei R, Baker J (2002) New Lu-Hf and Pb-Pb constraints on the earliest animal fossils. Earth Plan Sci Lett 201: 203–212

Barrington EJW, Sage M (1972) The endostyle and thyroid gland. In: Hardistym MW, Potter IC (eds) The biology of lampreys, vol 2. Academic, London, pp 105–134

Bengtson S, Budd G (2004) Comment on "Small bilaterian fossils from 40 to 55 million years before the Cambrian". Science 306:1291a

Bengtson S, Hou XG (2001) The integument of Cambrian chancelloriids. Acta Palaeontol Pol 46:1–22

Bengtson S, Matthews SC, Missarzhevsky VV (1986) The Cambrian netlike fossil *Microdictyon*. In: Hoffman A, Nitecki MH (eds) Problematic Fossil Taxa vol 5. Oxford, NY and Clarendon, Oxford, pp. 97–115

Bengtson S, Missarzhevsky V (1981) Coeloscleritophora—a major group of enigmatic Cambrian metazoans. US Geol Surv Open-file Rep 81–743:19–21

Bergström J, Hou XG (1998) Chengjiang arthropods and their bearing on early arthropod evolution. In: Edgecombe GD (ed) Arthropod fossils and phylogeny. Columbia University, New York, pp. , 151–184

Bergström J, Hou XG (2003) Cambrian arthropods: a lesson in convergent evolution. In: Legakis A, Sfenthourakis S, Polymeni R, Thessalou-Legaki M (eds) The new panorama of animal evolution. Proceedings of the XVIII International Congress of Zoology Pensoft, Sofia, Moscow, pp. 89–96, xvi + 738 pp

Boxshell GA (2004) The evolution of arthropod limbs. Biol Rev 79:253–300

Boyan GS, Williams JLD, Posser S, Bräunig P (2002) Morphological and molecular data argue for the labrum being non-apical, articulated, and the appendage of the intercalary segment in the locust. Arthrop Struct Dev 31:65–76

Boyer BC, Henry JQ, Martindale MQ (1996a) Dual origins of mesoderm in a basal member of the spiralian clade: cell lineage studies in the polyclad turbellarian *Hoploplana inquilina*. Dev Biol 179:329–338

Boyer BC, Henry JQ, Martindale MQ (1996b) Modified spiral cleavage: the duet cleavage pattern and early blastomere fates in the acoel turbellarian *Neochildiafusca*. Biol Bull 191:285–286

Briggs DEG (1992) Phylogenetic significance of the Burgess Shale crustacean *Canadaspis*. Acta Zool (Stockholm) 73:293–300

Briggs DEG (1994) Giant predators from the Cambrian of China. Science 264:1283–1284

Briggs DED, Erwin DH, Collier FJ (1994) The fossils of the Burgess Shale. Smithsonian Inst Press, 238p
Briggs DEG, Lieberman BS, Halgedahl SL, Jarrard RD (2005) A new metazoan from the Middle Cambrian of Utah and the nature of the Vetulicolia. Palaeontology 48:681–686
Brusca RC, Brusca GJ (2002) Invertebrates, 2nd edn. Sinauer Associates, Massachusetts
Butler AB (2000) Chordate evolution and the origin of craniates: an old brain in a new head. Anat Rec 261:111–125
Butler AB (2006) The serial transformation hypothesis of vertebrate origins: comment on "the new head hypothesis revised". J Exper Zool (Mol Dev Evol) 306B:419–424
Butt FH (1960) Head development in the arthropods. Biol Rev (Cambridge Phil Soc) 35:43–91
Caron JB (2006) *Banffia constricta*, a putative vetulicolid from the Middle Cambrian Burgess Shale. Trans R Soc Edinburgh Earth Sci 96:95–111
Caron JB, Conway Morris S, Shu DG (2010) Tentaculate fossils from the Cambrian of Canada (British Columbia) and China (Yunnan) interpreted as primitive deuterostomes. Plos One 5(3):e9586
Casanova B, Jong L, Moreau X (2002) Carapace and mandibles ontogeny in the Dendrobranchiata (Decapoda), Euphausiacea, and Mysidacea (Crustacea): a phylogenetic interest. Can J Zool 80:296–306
Chen JY (2004) The Dawn of animal world. Jiangsu Publishing House of Science and Technology, Nanjing, 366p
Chen JY (2008) Early crest animals and the insight they provide into the origin of craniates. Genesis 6:623–639
Chen JY (2009) The sudden appearance of diverse animal body plans during the Cambrian explosion. Int J Dev Biol 53:733–751
Chen JY, Erdtmann BD (1991) Lower Cambrian fossil Lagerstätte from Chengjiang, Yunnan, China: insights for reconstructing early metazoan life. In: Simonetta AM, Conway Morris S (eds) The early evolution of metazoa and the significance of problematic taxa. Cambridge, pp. 57–75
Chen JY, Huang DY (2002) A possible Lower Cambrian chaetognath (arrow worm). Science 298:197
Chen AL, Huang DY (2006) Gill rays found on the Early Cambrian primitive vertebrate Yunnanozoon. Acta Palaeontol Sin 45(3):345–350
Chen JY, Li CW (2000) Distant ancestor of mankind unearthed: 520 million year-old fish-like fossils reveal early history of vertebrates. Science Progress 83:123–133
Chen JY, Teichert C (1983a) Cambrian cephalopods. Geology 11:648–650
Chen JY, Teichert C (1983b) Cambrian cephalopoda of China. Palaeontographica, Abt A, Bd 181
Chen JY, Zhou GQ (1997) Biology of the Chengjiang Fauna. Bull Nat Mus Nat Sci 10:11–106
Chen JY, Hou XG, Lu HZ (1989a) Early Cambtian netted scale-bearing worm-like sea animal. Acta Palaeont Sinica 28:1–26
Chen JY, Hou XG, Lu HZ (1989b) Lower Cambrian leptomitids (Demospongea), Chengjiang, Yunnan. Acta Palaeontol Sin 28:17–31
Chen JY, Hou XG, Li GX (1990) New Lower Cambrian demosponges, *Quadrolaminiella* gen. nov. from Chengjiang, Yunnan. Acta Palaeontol Sin 29:402–413
Chen JY, Ramsköld L, Zhou GQ (1994) Evidence for monophyly and arthropod affinity of Cambrian giant predators. Science 264:1304–1308
Chen JY, Dzick J, Edgecombe GD, Ramsköld L, Zhou GQ (1995a) The earliest chordate from Early Cambrian, Yunnan, China. Nature 377:720–722
Chen JY, Edgecombe GD, Ramsköld L, Zhou GQ (1995b) Head segmentation in Early Cambrian *Fuxianhuia*: implications for arthropod evolution. Science 268:1339–1343
Chen JY, Zhou GQ, Ramsköld L (1995c) A new Early Cambrian onychophoran-like animal, *Paucipodia* gen. nov., from the Chengjiang fauna, China. Trans R Soc Edinburgh Earth Sci 85:275–282
Chen JY, Zhu MY, Zhou GQ (1995d) The early Cambrian medusiform metazoan *Eldonia* from Chengjiang Lagerstätte. Acta Palaeontol Pol 40:213–244
Chen JY, Zhou GQ, Zhu MY, Yeh KY (1996) The Chengjiang Biota—a Unique Window of the Cambrian explosion. National Museum of Natural Science, Taichung, Taiwan, 222p
Chen JY, Edgecombe GD, Ramsköld L (1997) Morphological and ecological disparity in naraoiids (Arthropoda) from the Early Cambrian Chengjiang fauna, China. Rec Australian Mus 47:1–24
Chen JY, Huang DY, Li CW (1999) An Early Cambrian craniate-like chordate. Nature 402:518–522
Chen JY, Oliveri P, Li CW, Zhou GQ, Gao F, Hagadom JW, Peterson KJ, Davidson EH (2000) Precambrian animal diversity: putative phosphatized embryos from the Doushantuo Formation of China. Proc Natl Acad Sci USA 97: 4457–4462
Chen JY, Vannier J, Huang DY (2001) The origin of crustaceans: new evidence from the Early Cambrian of China. Proc R Soc Lond B 268:1–7
Chen JY, Oliveri P, Gao F, Dornbos SQ, Li CW, Bottjer DJ, Davidson EH (2002a) Precambrian animal life: probable developmental and adult cnidarian forms from southwest China. Develop Biol 248:182–196
Chen LZ, Luo HL, Hu SX, Yin GY, Jiang ZW, Wu ZL, Li F, Chen AL (2002b) Early Cambrian Chengjiang Fauna in Eastern Yunnan, China. Yunnan Science and Technology Press, Kunming, 199p, 28 pls [Chinese, with English summary]
Chen AL, Feng HZ, Zhu MY, Ma DS, Li M (2003a) A new vetulicolian from the Early Cambrian Chengjiang fauna in Yunnan of China. Acta Geol Sinica 77:281–287
Chen JY, Huang DY, Peng QQ, Chi HM, Wang XQ, Feng M (2003b) The first tunicate from Early Cambrian of south China. Proc Nat Acad Sci 100(14):8314–8318
Chen DF, Dong WQ, Zhu FQ, Chen XP (2004a) Pb-Pb ages of Neoproterozooic Doushantuo phosphorites in South China; constraints on early metazoan evolution and glaciation events. Precambrian Res 132:123–132
Chen JY, Bottjer DJ, Oliveri P, Dornbos SQ, Gao F, Ruffins S, Chi H-M, Li C-W, Davidson EH (2004b) Small bilaterian fossils from 40 to 55 Million Years before the Cambrian. Science 305:218–222
Chen JY, Oliveri P, Davidson E, Bottjer DJ (2004c) Response to comment on "Small bilaterian fossils from 40 to 55 million years before the Cambrian". Science 306:1291b
Chen JY, Braun A, Waloszek D, Peng QQ, Maas A (2004d) Lower Cambrian yolk-pyramiod embryos from southern Shaanxi, China. Progr Nat Sci 14(2):167–172
Chen JY, Waloszek D, Maas A (2004e) A new 'great-appendage' arthropod from the Lower Cambrian of China

and homology of chelicerate chelicerae and raptorial antero-ventral appendages. Lethaia 37:3–20

Chen JY, Huang DY, Bottjer DJ (2005) *Vetustovermis* and its possible affinities. Proc R Soc B 272:2003–2007

Chen JY, Bottjer DJ, Davidson EH, Dornbos SQ, Gao X, Yang YH, Li CW, Li G, Wang XQ, Xian DC, Wu HJ, Hwu YK, Tafforeau P (2006) Phosphatized polar lobe-forming embryos from the Precambrian of Southwest China. Science 312:1644–1646

Chen JY, Huang DY, Chuang SH (2007a) Reinterpretation of the Lower Cambrian brachiopod *Heliomedusa orienta* Sun and Hou 1987, as a discinid. J Paleont 81:38–47

Chen JY, Schopf JW, Bottjer DJ, Zhang CY, Kudryavtsev AB, Tripathi AB, Wang XQ, Yang YH, Gao X, Yang Y (2007b) Raman spectra of a Lower Cambrian ctenophore embryo from SW Shaanxi, China. Proc Nat Acad Sci USA 106:6289–6292

Chen JY, Waloszek D, Maas A, Braun A, Huang DY, Wang XQ, Martin Stein M (2007c) Early Cambrian Yangtze Plate Maotianshan-shale macrofauna biodiversity and the evolution of predation. Palaios 254:250–272

Chen JY, Bottjer DJ, Davidson EH, Li G, Gao F, Cameron AR, Hadfield MG, Xian DC, Tafforeauf P, Jia QJ, Sugiyamag H, Tang R (2009a) Phase contrast synchrotron X-ray microtomography of Ediacaran (Doushantuo) metazoan microfossils: phylogenetic diversity and evolutionary implications. Precambrian Res 173:191–200

Chen JY, Bottjer DJ, Li G, Hadfield MG, Gao F, Cameron AR, Zhang CY, Xian DC, Tafforeauf P, Liao X, Yin ZJ (2009b) Complex embryos displaying bilaterian characters from Precambrian Doushantuo phosphate deposits, Weng'an, Guizhou, China. Proc Natl Acad Sci USA 106: 19056–19060

Clark HL (1912) Fossil holothurian. Science 35:274–278

Condon D, Zhu M, Bowring S, Wang W, Yang A, Jin Y (2005) U-Pb ages from the Neoproterozoic Doushantuo Formation, China. Science 308:95–98

Conway Morris S (1976) A new Cambrian lophophorate from the Burgess Shale of British Columbia. Palaeontology 19:199–222

Conway Morris S (1977a) Fossil priapulid worms. Spec Pap Palaeont 20:95p

Conway Morris S (1977b) A new entoproct-like organism from the Burgess Shale of British Columbia. Palaeontology 20:833–845

Conway Morris S (1977c) A new metazoan from the Burgess Shale of British Columbia. Palaeontology 20:623–640

Conway Morris S (1977d) A redescription of the Middle Cambrian worm *Amiskwia sagittiformis* Walcott from the Burgess Shale of British Columbia. Paläont Zeit 51: 271–287

Conway Morris S (1993) Ediacaran-like fossils in Cambrian Burgess Shale-type faunas of North America. Palaeontology 36:593–635

Conway Morris S (1997) The cuticular structure of the 495-Myeold type species of the fossil worm *Palaeoscolex piscatorum* (?Priapulida). Zool J Linn Soc 119:69–82

Conway Morris S (1998) Crucible of creation: the Burgess Shale and the rise of animals. Oxford University Press, Oxford

Conway Morris S (2000) The Cambrian "explosion": slow-fuse or megatonnage? Proc Natl Acad Sci USA 97: 4426–4429

Conway Morris S, Collins DH (1996) Middle Cambrian ctenophores from the Stephen Formation, British Columbia. Phil Trans R Soc Lond B 351:279–308

Conway Morris S, Robison RA (1988) More soft-bodied animals and algae from the Middle Cambrian of Utah and British Columbia. Univ Kansas Paleont Contrib Pap 122:1–48

Costa M, Sweeton D, Wieschaus E (1993) Gastrulation in Drosophila: cellular mechanisms of morphogenetic movements. In: Bate M, Hartenstein V (eds) The development of *Drosophila melanogaster*. Cold Spring Harbor Laboratories, Long Island, NY, pp 425–465

Cutler EB (1994) The Sipuncula—their systematics, biology, and evolution. Cornell University, Ithaca, NY

Darwin C (1859) On the origin of species. John Murray, London

Davidson B, Levine M (2003) Evolutionary origins of the vertebrate heart: Specification of the cardiac lineage in *Ciona intestinalis*. Proc Natl Acad Sci USA 100: 11469–11473

Davidson EH (2006) The regulatory genome: gene regulatory networks in development and evolution. Academic, San Diego, CA

Dawson JW (1889) New species of fossil sponges from the Siluro-Cambrian at Little Metis on the lower St. Lawrence. Trans R Soc Canada 7(Sect 4):31–55

Delsuc F, Brinkmann H, Chourrout D, Philippe H (2006) Tunicates and not cephalochordates are the closest living relatives of vertebrates. Nature 439:965–968

Dong PD, Chu J, Panganiban G (2001) Proximodistal domain specification and interactions in developing *Drosophila* appendages. Development 128:2365–2372

Donoghue PCJ, Bengtson S, Dong XP, Gostling N, Huldtgren T, Cuningham JA, Yin C, Yue Z, Peng F, Stampanoni J (2006) Synchrotron X-ray tomographic microscopy of fossil embryos. Nature 442:680–683

Dornbos SQ, Chen JY (2008) Community palaeoecology of the Early Cambrian Maotianshan Shale biota: ecological dominance of priapulid worms. Palaios 258:200–212

Dornbos SQ, Bottjer DJ, Chen JY (2004) Evidence for seafloor microbial mats and associated metazoan lifestyles in Lower Cambrian phosphorites of Southwest China. Lethaia 37: 127–137

Douzery EJP, Snell EA, Bapteste E, Delsuc F, Philippe H (2004) The timing of eukaryotic evolution: Does a relaxed molecular clock reconcile proteins and fossils? Proc Natl Acad Sci USA 101:15386–15391

Dufour HD, Chettouh Z, Deyts C, de Rosa R, Goridis C, Joly JS, Brunet JF (2006) Precraniate origin of cranial motoneurons. Proc Natl Acad Sci USA 103:8727–8732

Dunlop JA (1999) Pasando revista a la evolución de los quelicerados. In: Melic A, De Haro JJ, Mendez M, Ribera I (eds) Evolución y filogenia de Arthropoda. Bol Soc Entomol Aragonesa 26:255–272

Dunnel-Erb S, Bailly Y, Laurent P (1993) Pattern of gill innervation in two teleosts, the perch and the trout. Can J Zool 71:18–25

Durham JW (1974) Systematic position of *Eldonia ludwigi* Walcott. J Paleontol 48:750–755

Dzik J (1989) Is fossil evidence consistent with traditional views of the early metazoan phylogeny? In: Simonetta AM, Conway Morris S (eds) The early evolution of metazoa and the significance of problematic taxa. Cambridge University Press, Cambridge, pp 47–56

Dzik J, Zhao YL, Zhu MY (1997) Mode of life of the Middle Cambrian eldonioid lophophorate *Rotadiscus*. Palaeontology 40:385–396

Eastham LES (1930) The embryology of *Pieris rapae*—organogeny. Phil Trans R Soc Lond Biol Sci 219:2–50

Ehlers E (1869) Ueber fossile Würmer aus dem lithographischen Schiefer in Bayern. Palaeontographia 17:145–175

Erwin DH, Davidson EH (2002) The last common bilaterian ancestor. Development 129:3021–3032

Fedonkin MA, Waggoner BM (1997) The Late Precambrian fossil *Kimberella* is a mollusc-like bilaterian organism. Nature 388:8–871

Finkelstein R, Perrimon N (1991) The molecular genetics of head development in *Drosophila melanogaster*. Development 112:899–912

Foester MW (1979) A reappraisal of *Tullimonstrum*. In: Nitecki MH (ed) Mazon creek fossils. Academia, New York, pp 269–302

Fortey RA, Briggs DEG, Wills MA (1997) The Cambrian "explosion" recalibrated. BioEssays 19:429–434

Freeman G, Lundelius JW (1992) Evolutionary implications of the mode of D quadrant specification in coelomates with spiral cleavage. J Evol Biol 5:205–247

Gans C (1993) Evolutionary origin of the vertebrate skull. In: Hanken J, Hall BK (eds) The Skull, vol 2. University of Chicago, Chicago, pp 1–35

Garcia-Bellido DC, Vannier J, Collins D (2009) Soft-part preservation in two species of the arthropod *Isoxys* from the Middle Cambrian Burgess Shale of British Columbia, Canada. Acta Palaeontol Pol 54:699–712

Gee H (1992) Something completely different. Nature 358: 456–457

Gee H (2001) On being vetulicolian. Nature 414:407–409

Gibson R (1972) Nemerteans. London, Hutchinson

Glaessner MF (1959) Precambrian Coelenterata from Australia, Africa and England. Nature 183:1472–1473

Glaessner MF (1979) Lower Cambrian Crustacea and annelid worms from Kangaroo Island, South Australia. Alcheringa 3:21–31

Glaessner MF (1984) The Dawn of animal life: a biohistorical study. Cambridge University Press, Cambridge, 244p

Glardon S, Holland LZ, Gehring WJ, Holland ND (1998) Isolation and developmental expression of the amphioxus Pax-6 gene (AmphiPax-6): insights into eye and photoreceptor evolution. Development 125:2701–2710

Gould SJ (1989) Wonderful life: the Burgess Shale and the nature of history. Norton, New York

Hall BK (1998) Evolutionary developmental biology. Chapman and Hall, London, New York

Henry JQ, Martindale MQ, Boyer BC (2000) The unique developmental program of the acoel flatworm, *Neochildiafusca*. Dev Biol 220:285–295

Hinz I, Kraft P, Mergl M, Muller KJ (1990) The problematic *Hadimopanella, Kaimenella, Milaculum* and *Utahphospha* identified as sclerites of Palaeoscolecida. Lethaia 23: 217–221

Hirth F, Kammermeier L, Frei E, Walldorf U, Noll M, Reichert H (2003) An urbilaterian origin of the tripartite brain: developmental genetic insights from *Drosophila*. Development 130:2365–2373

Ho CC (1942) Phosphate deposits of Tungshan, Chengjiang, Yunnan. Bull Geol Surv China 35:97–106

Hoffman KH, Condon DJ, Bowring SA, Crowley JL (2004) U-Pb zircon date from the Neoproterozooic Ghaub Formation, Namibi: constraints on Marinoan glaciation. Geology 32:817–820

Holland LZ (2007) A chordate with a difference. Nature 447:1153–1155

Holland LZ, Holland ND (1998) Developmental gene expression in amphioxus: new insights into the evolutionary origin of vertebrate brain regions, neural crest, and rostrocaudal segmentation. Am Zool 38:647–658

Holland LZ, Holland ND (2001) Evolution of neural crest and placodes: amphioxus as a model for the ancestral vertebrate. J Anat 199(1, 2):85–98

Holland ND, Chen JY (2001) Origin and early evolution of the vertebrates: new insights from advances in molecular biology, anatomy, and palaeontology. Bioessays 23:142–151

Holland PWH, Holland LZ, Williams NA, Holland ND (1992) An amphioxus homebox gene: Sequence conservation, spatial expression during development and insights into vertebrate evolution. Development 116:653–661

Holland ND, Panganiban G, Henyey EL, Holland LZ (1996) Sequence and developmental expression of AmphiDII, an amphioxus Distalless gene transcribed in the ectoderm, epidermis and nervous system. Development 122:2911–2920

Holmer LE, Popov LE, Konova SP, Rong JY (1997) Early Cambrian *Lingulellotreta* (Lingulata, Brachiopoda) from south Kazakhstan (Malyi Karatau Range) and south China (eastern Yunnan). J Paleont 71:577–584

Holmgren N (1946) On two embryos of *Myxine glutinosa*. Acta Zool Stockholm 27:1–90

Hou XG (1987) Three new large arthropods from Lower Cambrian, Chengjiang, eastern Yunnan. Acta Paleontol Sin 26:272–285

Hou XG, Bergström J (1991) The arthropods of the Lower Cambrian Chengjiang fauna, with relationships and evolutionary significance. In: Simonetta AM, Conway Morris S (eds) The early evolution of metazoa and the significance of problematic taxa. Cambridge University, Cambridge, pp. 179–187

Hou XG, Bergström J (1994) Palaeoscolecid worms may be nematomorphs rather than annelids. Lethaia 27:11–17

Hou XG, Bergström J (1995) Cambrian lobopodians—ancestors of extant onychophorans? Zool J Linn Soc 114:3–19

Hou XG, Bergström J (1997) Arthropods of the Lower Cambrian Chengjiang fauna, southwest China. Fossils Strata 45:1–116

Hou XG, Chen JY (1989) Early Cambrian arthropod-annelid intermediate sea animal, *Luolishania* gen. nov. from Chengjiang, Yunnan. Acta Palaeont Sinica 28: 207–213

Hou XG, Sun WG (1988) Discovery of Chengjiang fauna at Meishucun, Jinning, Yunnan. Acta Palaeontol Sin 27:1–12

Hou XG, Chen JY, Lu HZ (1989) Early Cambrian new arthropods from Chengjiang, Yunnan. Acta Palaeontol Sin 28: 42–57

Hou XG, Ramsköld L, Bergström J (1991) Composition and preservation of the Chengjiang fauna: a Lower Cambrian soft-bodied biota. Zool Scripta 20:395–411

Hou XG, Bergstöm J, Wang HF, Feng XH, Chen AL (1999) The Chengjiang Fauna: exceptionally well preserved animals from 530 million years ago. Yunnan Science and Technology Press, 170p

Hou XG, Aldridge RJ, Bergström J, Siveter DJ, Siveter DJ, Feng XH (2004) The Cambrian Fossils of Chengjiang, China: the flowering of early animal life. Blackwell, Malden, MA, and Oxford

Howell FB (1962) Worms. In: Moore RC (ed) Treatise on invertebrate paleontology, Pt W, Miscellanea. New York, Geological Society of America, New York, and University of Kansas, Lawrence, pp 144–177

Huang DY (2005) Early Cambrian worms from SW China, morphology, systematics, lifestyles and evolutionary significance. PhD thesis, University of Lyon, 247p

Huang DY (2006) The early body plan, origin and evolutionary radiation of Priapulida. In: Rong JY, Fang ZJ, Zhou ZH, Zhan RB, Wang XD, Yuan XL (eds) Originations, radiations and biodiversity changes—evidence from the Chinese fossil record. Science Press, Beijing, pp 125–137, 845–846

Huang DY, Chen JY, Vannier J, Saiz Salinas JI (2004a) Early Cambrian sipunculan worms from southwest China. Proc R Soc Lond B 271:1671–1676

Huang DY, Vannier J, Chen JY (2004b) Anatomy and lifestyles of Early Cambrian priapulid worms exemplified by *Corynetis* and *Anningvermis* from the Maotianshan Shale (SW China). Lethaia 37:21–33

Huang DY, Vannier J, Chen JY (2004c) Recent Priapulidae and their Early Cambrian ancestors: comparisons and evolutionary significance. Geobios 37:217–228

Hughes GM (1984) General anatomy of the gills. In: Hoar WS, Randall DJ (eds) Fish physiology X, Gills, Pt A, anatomy, gas transfer, and acid-base regulation. Academic, New York, 72p

Hyman LH (1951) The invertebrates, vol 2. McGraw-Hill, New York

Ivantsov AY, Wrona R (2004) Articulated palaeoscolecid sclerite arrays from the Lower Cambrian of eastern Siberia. Acta Geol Pol 54:1–22

Janussen D, Steiner M, Zhu MY (2002) New well-preserved scleritomes of Chancelloridae from the Early Cambrian Yuanshan Formation (Chengjiang, China) and the Middle Cambrian Wheeler Shale (Utah, USA) and paleobiological implications. J Paleont 76:596–606

Janvier P (1978) Les nageoires paires des ostéostracés et la position systématique des céphalaspidomorphes. Ann Paléont (Vertébrés) 64(2):113–142

Janvier P (2004) Early specializations in the branchial apparatus of jawless vertebrates: a consideration of gill number and size. In: Arratia G, Wilson MVH, Cloutier R (eds) Recent advances in the origin and early radiation of vertebrates. Verlag Dr. Friedrich Pfeil, München, pp 25–52

Jefferies RPS (1986) The ancestry of the vertebrates. British Museum, London (Nat Hist)

Jeffery WR, Strickler AG, Yamamota Y (2007) Migratory neural crest-like cells from body pigmentation in a urochordate embryo. Nature 431:696–699

Jin YG, Wang HY (1992) Revision of the Lower Cambrian Brachiopod *Heliomedusa* Sun and Hou 1987. Lethiaia 25:35–49

Jin YG, Hou XG, Wang HY (1993) Lower Cambrian pediculate lingulids from Yunnan, China. J Paleont 67:788–798

Johnels A (1948) On the development and morphology of the skeleton of the head of *Petromyzon*. Acta Zool Stockholm 29:139–278

Joyner AL (1996) Engrailed, Wnt and Pax genes regulate midbrain–hindbrain development. Trends Genet 12:15–20

Kimmel CB, Miller CT, Kruze G., Ullmann B, Bremiller EA, Larison KD, Snyder HC (1998) The shaping of pharyngeal cartilages during early development of the zebrafish. Dev Biol 203:245–263

Knight RD, Panopoulou GD, Holland PWH, Shimeld M (2000) Amphioxus Krox gene: insights into vertebrate hindbrain evolution. Dev Genes Evol 210(1):517–521

Krešlová J, Holland LZ, Schubert M, Burgtorf C, Beneš V, Kozmik Z (2002) Functional equivalency of amphioxus and vertebrate Pax258 transcription factors suggests that the activation of mid-hindbrain specific genes in vertebrates occurs via the recruitment of Pax regulatory elements. Gene 282:143–150

Lacalli TC (1996) Frontal eye circuitry, rostral sensory pathways and brain organization in amphioxus larvae: evidence from 3D reconstructions. Phil Trans R Soc Lond B 351:243–263

Lacalli TC (2002) Vetulicolians—are they deuterostomes? Chordates? Bioessays 24:208–211

Lacalli TC (2004) Sensory systems in amphioxus: a window on the ancestral chordate condition. Brain Behav Evol 64:148–162

Lacalli TC (2005) Protochordate body plan and the evolutionary role of larvae. Can J Zool 83:216–224

Lacalli TC, Holland LZ (1998) The developing dorsal ganglion of the salp *Thalia democratica*, and the nature of the ancestral chordate brain. Phil Trans R Soc Lond B 53:1943–1969

Lacalli TC, Holland ND, West JE (1994) Landmarks in the anterior central nervous system of amphioxus larvae. Phil Trans R Soc B 344:165–185

Laurent P (1984) Gill internal morphology. In: Hoar WS, Randall DJ (eds) Fish physiology X. Gills, Pt A, anatomy, gas transfer, and acid-base regulation. Academic, New York, pp 73–183

Lemburg C (1999) Hypothesen zur Phylogenie der priapulida underen Bedeutung fur der Nemathelminthes. Cuvillier Verlag, Göttingen, 393p

Leys SP, Ereskovsky AV (2006) Embryogenesis and larval differentiation in sponges. Can J Zool 84:262–287

Li CW, Chen JY, Hua TE (1998) Precambrian sponges with cellular structures. Science 279:879–882

Liu JN, Shu DG, Han J, Zhang ZF (2004) A rare lobopod with well-preserved eyes from Chengjiang Lagerstätte and its implications for origin of arthropods. Chinese Sci Bull 49:1063–1071

Liu JN, Shu DG, Han J, Zhang ZF (2008) Comparative study of Cambrian lobopods *Miraluolishania* and *Luolishania*. Chinese Sci Bull 53:87–93

Love D, Grosjean E, Stalvies C, Fike DA, Grotzinger JP, Bradley AS, Kelly AE, Bhatia M, Meredith W, Snape CE, Bowring SA, Condon DJ, Summons RE (2009) Fossil steroids record the appearance of Demospongia during the Cryogenian period. Nature 457:718–721

Lowe CJ, Terasaki M, Wu M, Freeman RM, Runft L, Kwan K, Haup S, Aronowicz J, Lauder E, Gruber C, Smith M, Kirschner M, Gerhart J (2006) Dorsoventral patterning in hemichordates: insights into early chordate evolution. PLoS Biol 4:1603–1619

Luo HL, Hu SX, Cheng LZ, Zhang SS, Tao YH (1999) Early Cambrian Chengjiang Fauna from Kunming Region, China. Yunnan Science and Technology Press, Kunming, 129p, 32pls

Maas A, Waloszek D (2001) Cambrian derivatives of the early arthropod stem lineage, pentostomids, tardigrades and lobopodians—an "Orsten" perspective. Zool Anzeiger 240:451–459

Maas A, Huang DY, Chen JY, Waloszek D, Braun A (2007a) Maotianshan Shale nemathelminths—morphology, biology, and the phylogeny of Nemathelminthes. Palaios 22: 288–306

Maas A, Huang DY, Chen JY, Waloszek D, Braun A (2007b) Maotianshan Shale nemathelminths-new information about their morphology and biology, and phylogeny of Nemathelminthes. Palaeogeogr, Palaeoclimatol, Palaeoecol 254:285–303

Maas A, Waloszek D, Chen JY, Braun A, Wang XQ, Huang DY (2004) Phylogeny and life habits of early arthropods—predation in the early Cambrian sea. Progr Natl Sci 14:158–166

Malakhov VV, Adrianov AV (1995) Cephalorhyncha—a new phylum of the animal kingdom. KMK Scientific Press, Moskva, 199p

Maldonado M (2004) Choanoflagellates, choanocytes, and animal multicellularity. Invert Biol 123:1–22

Mallatt J (1996) Ventilation and the origin of jawed vertebrates: a new mouth. Zool J Linn Soc 117:329–404

Mallatt J, Chen JY (2003) Fossil sister group of craniates: predicted and found. J Morph 258:1–31

Maloof AC, Rose CV, Beach R, Samuels BM, Calmet CC, Erwin DH, Poirier GR, Yao N, Simons FJ (2010) Possible animal-body fossils in pre-Marinoan limestones from South Australia. Nat Geosci 3:653–659

McCormack CC (1932) Fossil Holothuroidea. J Paleont 6:111–148

Mehl D (1996) Organization and microstructure of the chancelloriid skeleton: implications for the biomineralization of the Chancelloridae. Bull Inst Océan Monaco no sp 14:377–385

Mergener H (1971) Cnidaria. In: Reverberi G (ed) Experimental embryology of marine and fresh-water invertebrates. North-Holland, Amsterdam, pp 1–84

Meulemans D, Bronner-Fraser M (2007) Insights from amphioxus into the evolution of vertebrate cartilage. Plos One 8:e787 (electronic)

Minelli A (2001) A three-phase model of arthropod segmentation. Dev Genes Evol 211:509–521

Missarzhevsky VV, Mambetov AM (1981) Stratigraphy and fauna of the Precambrian–Cambrian boundary beds of Malyj Karatau. Trudy Geol Inst AN SSSR 326:1–90

Moret F, Guilland JC, Coudouel S, Rochette L, Vernier P (2004) Distribution of tyrosine hydroxylase, dopamine and serotonin in the central nervous system of amphioxus: implications for the evolution of catecholamine systems in vertebrates. J Comp Neurol 468:135–150

Muir L, Botting P (2002) A Lower Carboniferous sipunculan from the Granton Shrimp Bed, Edinburgh. 46th Palaeontol Assoc Ann Mtng: 62 (poster). Department of Earth Sciences, University of Cambridge

Nedin C (1999) *Anomalocaris* predation on nonmineralized and mineralized trilobites. Geology 27:987–990

Nielsen C (2001) Animal evolution. Oxford University Press, Oxford

Northcutt RG, Gans C (1983) The genesis of neural crest and epidermal placodes: a reinterpretation of vertebrate origins. Q Rev Biol 58:2–28

Pardos F, Benito J (1988) Blood vessels and related structure in the gill bars of *Glossobalanus minutus* (Enteropneusta). Acta Zool (Stockh) 69:87–94

Parker AR (1998) Color in Burgess Shale animals and the effect of light on evolution in the Cambrian. Proc R Soc Lond B 265:967–972

Paul CRC, Smith AB (1984) The early radiation and phylogeny of echinoderms. Biol Rev 59:443–481

Peng J, Zhao YL, Lin JP (2006) *Dinomischus* from the Middle Cambrian Kaili Biota, Guizhou, China. Acta Geol Sinica 80:498–501

Peterson KJ, Butterfield NJ (2005) Origin of the Eumetazoa: testing ecological predictions of molecular clocks against the Proterozoic fossil record. Proc Natl Acad Sci USA 102:9547–9552

Peterson KJ, Cotton JA, Gehling JG, Pisani D (2008) The Ediacaran emergence of bilaterians: congruence between the genetic and the geological fossil records. Phil Trans R Soc B 363:1435–1443

Popadic A, Panganiban G, Rusch D, Shear WA, Kaufman TC (1998) Molecular evidence for the appendicular origin of the labrum and other structures. Dev Genes Evol 208: 142–150

Purnell MA (2001) Scenarios, selection, and the ecology of early vertebrates. In: Ahlberg PE (ed) Major events in early vertebrate evolution: palaeontology, phylogeny, genetics and development. Taylor and Francis, London, pp 187–208

Raff RA (1996) The shape of life: genes, development, and the evolution of animal form. University of Chicago, Chicago

Ramsköld L (1992) Homologies in Cambrian Onychophora. Lethaia 25:443–460

Ramsköld L, Chen JY (1998) Cambrian lobopodians: morphology and phylogeny. In: Edgecombe G (ed) Arthropod fossils and phylogeny. Columbia, pp 107–150

Ramsköld L, Chen JY, Edgecombe GD, Zhou GQ (1997) *Cindarella* and the arachnate clade Xandarellida (Arthropoda, Early Cambrian) from China. Trans R Soc Edinburgh, Earth Sci 88:19–38

Randell RD, Lieberman BS, Hasiotis ST, Pope M (2005) New chancelloriids from the Early Cambrian Sekwi Formation with a comment on chancelloriid affinities. J Paleont 79:987–996

Rempel JG (1975) The evolution of the insect head: the endless dispute. Quaest Entomol 11:7–25

Rigby JK (1986) Sponges of the Burgess shale (Middle Cambrian), British Columbia. Palaeontographica Canadiana 2:1–105

Romer AS (1962) The vertebrate body, 3rd edn. Saunders, Philadelphia

Romer AS (1972) The vertebrate as dual animal—somatic and visceral. Evol Biol 6:121–156

Rong JY (1974) Cambrian brachiopods. In: Nanjing Inst Geol Palaeontol Acad Sinica (ed) Handbook of stratigraphy and palaeontology in Southwest China. Science Press, Beijing, pp 113–114

Rouse GW, Pleijel F (2001) Polychaetes. Oxford University Press, Oxford, 354pp

Runnegar B (1982) A molecular-clock date for the origin of the animal phyla. Lethaia 15:199–205

Runnegar B, Pojeta J, Morris NJ, Taylor JD, Taylor ME, McClung G (1975) Biology of the Hyolitha. Lethaia 8:181–191

Ruppert EE, Fox RS, Barnes RD (2004) Invertebrate zoology, 7th edn. Brooks/Cole-Thomson Learning, Belmont, CA

Saiz Salinas JI (1993) Sipuncula. In: Ramos MA (ed) Fauna Iberica, vol 4. Museo Nacional de Sciencias Naturales, Madrid, pp 1–200

Schmidt-Ott U, Technau G (1992) Expression of en and wg in the embryonic head and brain of *Drosophila* indicates a refolded band of seven segment remnants. Development 116:111–125

Schmidt-Rhaesa A (1996) Zur Morphologie, Biologie und Phylogengie der Nematomorpha, pp 1–276

Schmidt-Rhaesa A (1997) Nematomorpha. Gustav Fischer, Stuttgart, 128p

Schmitz A, Gemmel M, Perry S (2000) Morphometric partitioning of respiratory surfaces in amphioxus (*Branchiostoma lanceolatum pallas*). J Exp Biol 203:3381–3390

Scholtz G (2001) Evolution of developmental patterns in arthropods—the analysis of gene expression and its bearing on morphology and phylogenetics. Zoology 103:99–111

Schubert M, Escriva H, Xavier-Netro J, Laudet V (2006) *Amphioxus* and tunicates as evolutionary model systems. Trends Ecol Evol 21:269–277

Seilacher A (1989) Vendozoa: organismic construction in the Proterozoic biosphere. Lethaia 22:229–239

Seilacher A, Grazhdankin D, Legouta A (2003) Ediacaran biota: the dawn of animal life in the shadow of great protists. Paleont Res 7:43–54

Shimeld SM, Holland PWH (2000) Vertebrate innovations. Proc Natl Acad Sci USA 97:4449–4452

Shimeld SM, Holland ND (2005) Amphioxus molecular biology: insight into vertebrate evolution and developmental mechanisms. Can J Zool 83:100

Shimeld SM, Purkiss AG, Dirks RPH, Rateman OA, Slingsby C, Lunsen NH (2005) Uborchordate bg-crystalin and the evolutionary origin of the vertebrate eye lens. Curr Biol 15:1684–1689

Shu DG, Zhang XL, Chen L (1996) Reinterpretation of *Yunnanozoon* as the earliest known hemichordate. Nature 380:428–430

Shu DG, Conway Morris S, Zhang XL, Chen L, Li Y, Han J (1999a) A pipscid-like fossil from the Lower Cambrian of southern China. Nature 400:746–749

Shu DG, Luo HL, Conway Morris S, Zhang XL, Hu SX, Chen L, Han J, Zhu M, Li Y, Chen LZ (1999b) Lower Cambrian vertebrates from south China. Nature 402:42–46

Shu DG, Chen L, Han J, Zhang XL (2001a) An early Cambrian tunicate from China. Nature 411:472–473

Shu DG, Conway Morris S, Han J, Chen L, Zhang XL, Zhang ZF (2001b) Primitive deuterostomes from the Chengjiang Lagerstätte (Lower Cambrian, China). Nature 414: 419–424

Shu DG, Conway Morris S, Han J, Zhang ZF, Yasui K, Janvier P, Chen L, Zhang XL, Liu JN, Li Y, Liu HQ (2003a) Head and backbone of the Early Cambrian vertebrate *Haikouichthus*. Nature 421:526–529

Shu DG, Conway Morris S, Zhang ZF, Liu JN, Han J, Chen L, Zhang XL, Yasui K, Li Y (2003b) A new species of *Yunnanozoon* with implications for deuterostome evolution. Science 299:1380–1384

Shu DG, Conway Morris S, Han J, Li Y, Zhang XL, Hua H, Zhang ZF, Liu JN, Guo J-F, Yao Y et al (2006) Lower Cambrian vendobionts from China and early diploblast evolution. Science 312:731–734

Simões-Costa MS, Vasconcelos M, Sampaio AC, Cravo R., Linhares VL, Hochgreb T, Yan CYI, Davidson B, Xavier-Neto J (2005) The evolutionary origin of cardiac chambers. Dev Biol 277:1–15

Siveter DJ, Williams M, Waloszek D (2001) A phosphatocopid crustacean with appendages from the Lower Cambrian. Science 293:479–481

Smith MR, Caron JB (2010) Primitive soft-bodied cephalopods from the Cambrian. Nature 465:469–472

Smith MP, Sanson IJ, Cochrane KD (2001) The Cambrian origin of vertebrates. In: Ahlberg PE (ed) Major events in early vertebrate evolution: palaeontology, phylogeny, genetics and development. Taylor and Francis, London, pp 76–84

Snodgrass RE (1935) Principles of insect morphology. McGraw-Hill, New York

Steiner M, Zhu MY, Li GX, Qian Y, Erdtmann BD (2004) New early Cambrian bilaterian embryos and larvae from China. Geology 32:833–836

Steiner M, Mehl D, Reitner J, Erdtmann BD (1993) Oldest entirely preserved sponges and other fossils from the lowermost Cambrian and a new facies reconstruction of the Yangtze Platform (China). Berliner Geowiss Abh 9: 293–329

Sterrer W (1986) Marine Fauna and Flora of Bermuda: a systematic guide to the identification of marine organisms. Wiley, New York

Sun WG, Hou XG (1987) Early Cambrian medusae from Chengjiang, Yunnan, China. Acta Palaeontol Sin 26: 257–271

Sutton MD, Briggs DEG, Siveter DJ, Siveter DJ (2001) Invertebrate evolution (Communications arising): *Acaenoplax*—polychaete or mollusc? Nature 410:461–463

Sutton MD, Briggs DEG, Siveter DJ, Siveter DJ, Orr PJ (2002) The arthropod *Offaculus kingi* (Chelicerata) from the Silurian of Herefordshire, England: computer based morphological reconstructions and phylogenetic affinities. Proc R Soc Lond B 269:1195–1203

Takahashi T, Holland PWH (2004) Amphioxus and ascidian Dmbx homeobox genes give clues to the vertebrate origins of midbrain development. Development 131:3285–3294

Van der Land J (1968) A new aschelminth, probably related to the Priapulida. Zool Meded 42(22):237–250

Vannier J, Chen JY (2005) Early Cambrian food chain; new evidence from fossil aggregates in the Maotianshan Shale biota, SW China. Palaios 20:3–26

Vincent A, Blankenship JT, Wieschaus E (1997) Integration of the head and trunk segmentation systems controls cephalic furrow formation in *Drosophila*. Development 124:3747–3754

Walcott CD (1911a) Cambrian Geology and Paleontology. Middle Cambrian holothurians and medusae. Smithsonian Misc Coll 57(3):41–68

Walcott CD (1911b) Cambrian geology and paleontology: middle Cambrian annelids. Smithsonian Misc Coll 57(5): 109–142

Walcott CD (1920) Cambrian geology and paleontology IV (6). Middle Cambrian Spongiae. Smithsonian Misc Coll 67(6):261–364

Waloszek D (1993) The upper Cambrian *Rehbachiella* and the phylogeny of Branchiopoda and Crustacea. Fossils Strata 32:1–202

Waloszek D (1999) On the Cambrian diversity of Crustacea. In: Schram FR, Von Vaupel Klein JC (eds) Crustaceans and the biodiversity crisis. Proceedings of the Fourth International Crustacean Congress (Amsterdam, The Netherlands, July 20–24, 1998), vol 1. Brill, Leiden, pp 3–27

Waloszek D, Müller KJ (1990) Upper Cambrian stem-lineage crustaceans and their bearing upon the monophyletic origin of Crustacea and the position of *Agnostus*. Lethaia 23:409–427

Waloszek D, Chen JY, Maas A, Wang XQ (2005) Early Cambrian arthropods—new insights into arthropod head and structural evolution. Arthropod Syst Dev 34:189–205

Waloszek D, Maas A, Chen JY, Stein M (2007) Evolution of cephalic feeding structures and the phylogeny of Arthropoda. Palaios 254:273–287

Wang XQ, Chen JY (2004) Possible developmental mechanisms underlying the origin of the crown lineages of arthropods. Chinese Sci Bull 49(1):49–53

Weller S (1925) A new type of Silurian worm. J Geol 33: 540–544

Westneat MW, Hale ME, McHenry MJ, Long JH (1998) Mechanics of the fast-start: muscle function and the role of intramuscular pressure in the escape behavior of *Amia calva* and *Polypterus palmas*. J Exper Bio 201:3041–3055

Whittard WF (1953) *Palaeoscolex piscatorum* gen. et sp. nov., a worm from the Tremadocian of Shropshire. Quart J Geol Soc Lond 109:125–133

Whittington HB (1985) The Burgess Shale. Yale University, New Haven

Wicht H, Lacalli TC (2005) The nervous system of amphioxus: structure, development, and evolutionary significance. Can J Zool 83:122–150

Williams NA, Holland PWH (1998) Molecular evolution of the brain. Brain Behav Evol 52:177–185

Xiao SH, Knoll AH (2000) Phosphatized animal embryos from the Neoproterozoic Doushantuo Formation at Weng'an, Guizhou, South China. J Paleont 74:767–788

Xiao SH, Zhang Y, Knoll AH (1998) Three-dimensional preservation of algae and animal embryos in a Neoproterozoic phosphorite. Nature 391:553–558

Xiao SH, Yuan XL, Knoll AH (2000a) Eumetazoan fossils in terminal Proterozoic phosphorites? Proc Natl Acad Sci USA 97:13684–13689

Xiao SH, Yuan XL, Knoll AH (2000b) Eumetazoan fossils in terminal Proterozic phosphorites, South China. Lethaia 32:219–240

Xiao SH, Hagadorn JW, Zhou C, Yuan X (2007a) Rare helical spheroidal fossils from the Doushantuo Lagerstätte: Ediacaran animal embryos come of age? Geology 35:115–118

Xiao SH, Zhou CM, Yuan XL (2007b) Undressing and redressing Ediacaran embryos. Nature 446:198–201

Xue TS, Tang TF, Yu CL (1992) Discovery of oldest skeleton fossils from Upper Sinian Doushantuo Formation in Weng'an Guizhou, and its significance. Acta Palaeontol Sin 31:530–539

Yonge CM (1960) General characters of mollusca. In: Moore RC (ed) Treatise on invertebrate paleontology: part I, Mollusca, vol. 1. Geological Society of America and Kansas University, Lawrence, KS, pp 13–136

Young JZ (1981) The life of vertebrates, 3rd edn. Clarendon Press, Oxford

Zacharias D, Williams JLD, Meier T, Reichert H (1993) Neurogenesis in the insect brain: cellular identification and molecular characterization of brain neuroblasts in the grasshopper embryo. Development 118:941–955

Zhang Y (1989) Multicellular thallophytes with differentiated tissues from Late Proterozoic phosphate rocks of South China. Lethaia 22:113–132

Zhang WT, Hou XG (1985) Preliminary notes on the occurrence of the unusual trilobite *Naraoia* in Asia. Acta Palaeontol Sin 24:591–595

Zhang Y, Yuan XL (1992) New data on multicellular thallophytes and fragments of cellular tissues from Late Proterozoic phosphate rocks, South China. Lethaia 25: 1–18

Zhang XG, Aldridge RJ (2007) Development and diversification of trunk plates of the Lower Cambrian lobopodians. Palaeontology 50:401–415

Zhang XG, Hou XG, Emig CC (2003) Evidence of lophophore diversity in Early Cambrian Brachiopoda. Proc R Soc Lond B (Suppl) 270:65–68

Zhang XG, Siveter D, Waloszek D, Maas A (2007) An epipodite-bearing unusual crown-group crustacean from Lower Cambrian. Nature 448:595–598

Zhang ZF, Han J, Zhang XL, Liu JN, Guo JF, Shu DG (2006) Note on the gut preserved in the Lower Cambrian *Lingulellotreta* (Lingulata, Brachiopoda) from southern China. Acta Zool 88:65–70

Zhang ZF, Han J, Zhang XL, Liu JN, Shu DG (2004) Soft tissue preservation in the Lower Cambrian linguloid brachiopod from South China. Acta Palaeontol Pol 49:259–266

Zhao YL, Zhu MY (1994) Medusiform fossils of Kaili fanna from Taijiang, Guizhou. Acta Palaeontol Sin 33: 272–280

Zhu MY, Zhao YL, Chen JY (2002) Revision of the Cambrian discoidal animals *Stellostomites eumorphus* and *Pararotadiscus guizhouensis* from South China. Geobios 35:165–185

Zhu MY, Zhang JM, Yang AH (2007) Integrated Ediacaran (Sinian) chronostratigraphy of South China. Palaeogeogr Palaeoclimatol Palaeoecol 254:7–61

The Ordovician Radiation: Macroevolutionary Crossroads of the Phanerozoic

Arnold I. Miller

Abstract

The Ordovician Radiation although noted for being one of the most important intervals of global diversification of life is less well known for the part it plays in paleoecological behavioral and molecular aspects of the development of life. This chapter seeks to document the features and uniqueness of the Ordovician Radiation with respect to its effects on diversity at all levels and to review the gamut of mechanisms suggested as causes of the Ordovician Radiation.

Keywords

Prominence of epicontinental seas • Increased terranes/microcontinents • Increased diversity of biota • Increasing sea-floor abundances • Increasing brachiopod dominance • Whiterock Fauna • Ashgill mass extinction • Biogeographic differentiation

Introduction

If anyone with an interest in the evolution and diversification of life were asked to draw up a list of the five most significant biological events or intervals in Earth history, it is almost a given that the Ordovician Radiation would not be on it. Even if we restricted ourselves to the Phanerozoic eon chances are that the list would include many of the other events covered in this volume such as the Cambrian Explosion, which was associated with the first major diversification of multicellular life; and the "big five" Phanerozoic mass extinctions most notably the End Permian and End-Cretaceous events, which are thought to have had overarching roles in reshaping the composition of Earth's biota. But coming as it did in the aftermath of its Cambrian big sibling, the Ordovician Radiation typically escapes the attention of many observers.

And yet the Ordovician Radiation was a profound biological transition in its own right, the full nature of which is still being unmasked. While its most obvious feature is a massive increase in the number of marine animal orders, families, genera, and species, which in itself marks it as one of the two most unfettered intervals of global diversification at lower taxonomic levels in the history of life, lurking beneath this veneer is an array of paleoecological behavioral and perhaps molecular transitions that hint at an event of significant breadth and complexity. For this reason the Ordovician Radiation has become a focal point of research worldwide has garnered the attention in the past decade of two international working groups that

A.I. Miller (✉)
Department of Geology, University of Cincinnati, Cincinnati, OH 45221-0013, USA
e-mail: arnold.miller@uc.edu

J.A. Talent (ed.), *Earth and Life*, International Year of Planet Earth,
DOI 10.1007/978-90-481-3428-1_11,

have actively pumped out research papers and special volumes and has even earned an alternative name in the literature: the Great Ordovician Biodiversification Event (GOBE; Webby 2004).

In many respects the Ordovician has become something of a natural laboratory for the study of biotic transitions in the context of paleoecology and paleogeography and for analyses of the complex relationship between physical and biological transitions on regional through global scales. With this in mind the purpose of this chapter is twofold: (a) to convey the major features and uniqueness of the Ordovician Radiation not only with respect to its effects on diversity at the global level but also its ecological manifestations down to the community level and (b) to review the spectrum of potential mechanisms that have been proffered some quite recently as causes of the Ordovician Radiation. As we will see, however, just as the Ordovician Radiation was itself multifaceted, so too might be its causes.

Anatomy of the Ordovician Radiation from Top to Bottom

Global-Scale Taxonomic Richness

Over the past 15 years a near-total overhaul of the global Ordovician timescale (see Bergström et al. 2009 for a useful overview) coupled with new approaches to the statistical correction of global diversity curves for sampling inhomogeneities (e.g., Miller and Foote 1996; Alroy et al. 2001, 2008) suggests that iconic depictions of the Ordovician Radiation on a global scale (e.g., Sepkoski 1981, 1997) are in need of some tweaking. Nevertheless, while aspects of the timing, duration, and extent of global diversification are still being investigated the broad scale pattern is clear (Fig. 1; Webby 2004; Harper 2006): at family and genus levels the global taxonomic richness of marine animals increased three- or fourfold relative to the Cambrian. While a synoptic compilation is not yet available at species level because of uncertainties in species-level assignments within many higher taxa, it seems likely that the increase at the species level was even more impressive given the hierarchical nature of classification (Valentine 1970): for every genus that was added at least one species was also added and for many genera certainly more.

Interestingly a roughly threefold Ordovician increase in richness can also be observed at the order level (Sepkoski 1978) but additions at class and especially phylum level were far more muted. The origin and addition of marine animal phyla and classes was strongly front-loaded in the Phanerozoic concentrated mostly in the Cambrian period (Valentine 1969; Erwin et al. 1987) and pointing to the uniqueness of the biotic transitions that marked the Ediacaran and Cambrian periods. This suggests that something rather different was afoot during the Ordovician Radiation than the undeniably profound events that marked the earlier origination and initial diversification of the large majority of phyla. But numbers of taxa aside the Ordovician Radiation was transformative in its own right as evidenced even in global-scale compilations such as those in Fig. 1. Beneath this synoptic level a profound taxonomic transition among *Evolutionary Faunas* (Sepkoski 1981) was talking place: taxa comprising the *Cambrian Evolutionary Fauna* most notably trilobites began to decline from their peak levels of the Cambrian. By contrast elements of the *Paleozoic Evolutionary Fauna* including rhynchonelliform brachiopods, stenolaemate bryozoans, rugose and tabulate corals, nautiloid cephalopods, graptolites, and a variety of echinoderms were the main contributors to the Ordovician Radiation, not only providing the large majority of Ordovician biodiversity but also dominating sea floors for much of the balance of the Paleozoic Era. The *Modern Evolutionary Fauna* most notably bivalves and gastropods also exhibited a significant Ordovician diversity increase although the zeniths of these and several other taxa occurred after the Paleozoic.

Observations about evolutionary faunas admittedly mask a more nuanced set of transitions lurking beneath these broad taxonomic designations. Among trilobites, for example, while it may be true that their highest aggregate diversity occurred in the Cambrian, a subset of trilobite taxa dubbed the *Whiterock Fauna* was clearly diversifying in the mid to Late Ordovician in apparent concert with elements of the Paleozoic Evolutionary Fauna (Adrain et al. 1998, 2004). Nevertheless, the transitions among the three evolutionary faunas effectively summarize the most important biotic transitions of the Ordovician and, as we will see, reflect a variety of major changes to benthic marine communities and ecosystems.

Global Taxon–Age Distributions

While it is obviously important to investigate changes in the taxonomic make-up of the biota in space and through time, it is also essential to come to grips with changes in the nature of the taxa themselves. In this respect an important effect of the Ordovician Radiation was its impact on the global age-distribution of the extant pool of taxa (Fig. 2). At the genus (and presumably species) level there was very little carryover of taxa from the Cambrian to the Ordovician. With Ordovician diversification rising, the initial set of genera was by definition geologically *young*; earliest Ordovician genera had not yet been extant for lengthy intervals of geological time. As the Ordovician Radiation progressed, not only did new genera continue to originate but also other genera persisted so that by the end of the Ordovician the plurality of extant

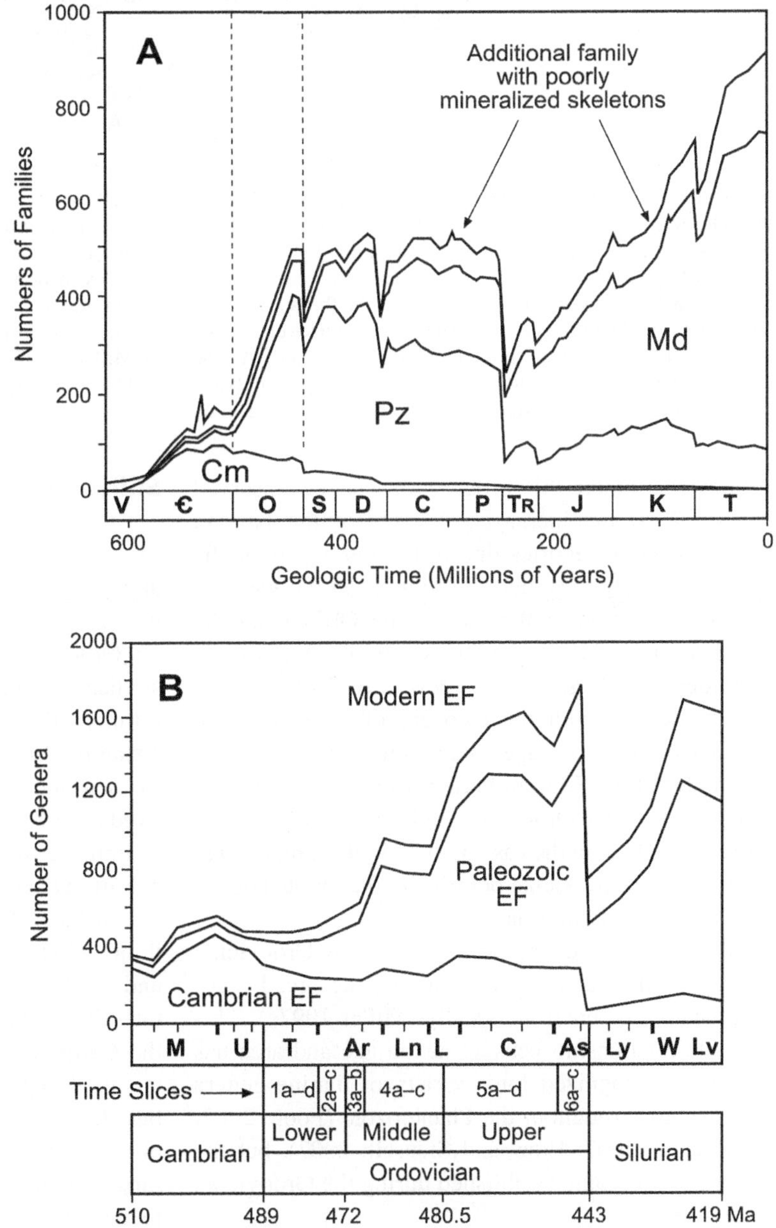

Fig. 1 Major transitions in the global richness of taxa through geologic time as depicted originally by J.J. Sepkoski Jr. (**a**) Family richness through the Phanerozoic Eon with the Ordovician demarcated with *vertical dashed lines*. (**b**) Genus richness through the Early Paleozoic. In both cases a three- to fourfold increase is indicated during the Ordovician. Note also the major transition among evolutionary faunas as described in the text (after Webby 2004)

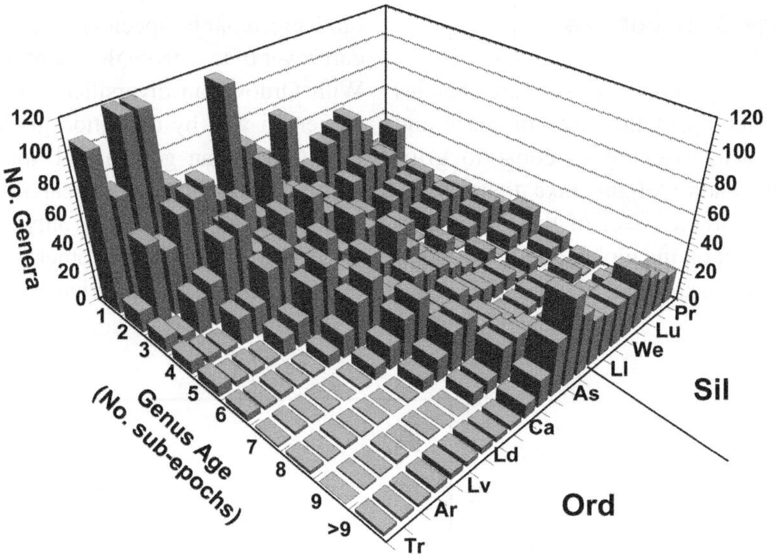

Fig. 2 Change in the age-distribution of marine genera through the Ordovician and Silurian (determined with data from Sepkoski's global compendium of marine animal genera; only genera whose first and last global appearances were resolved to sub-epoch were used in this analysis). At the start of the Ordovician the large majority of genera were relatively young with very little carryover from the Cambrian Period. As the Ordovician progressed additional genera first appeared but a large number of already-extant genera persisted resulting in an overall aging of the biota as evidenced by the increasing heights of bars in older age categories through the Ordovician. This distribution was "reset" to a limited degree by the Late Ordovician mass extinction but a large number of older genera persisted across the boundary. The Ordovician Radiation was clearly unique in its filling of "age space" (figure based on the design of Miller (1997c) but is more highly resolved stratigraphically and has been expanded to include the Silurian)

genera was older; this progression can be observed in Fig. 2 by the increasing heights of the bars in the older age categories during the latter part of the Ordovician. Crossing into the Silurian in the wake of a mass extinction at the end of the Ordovician the age-distribution was "reset" to some extent and there is evidence in Fig. 2 of a "new wave" of younger taxa associated with the recovery of diversity from the extinction whose ages then increased through the remainder of the Silurian Period. But many older taxa remained from the Ordovician and the overall pattern is more subtle than in the case of the initial filling of "age space" in the Ordovician pointing to the uniqueness of the Ordovician Radiation.

Moreover as genera aged during the Ordovician Radiation, they also became more widespread, geographically and environmentally (Miller 1997c). This intimate relationship between taxon age and area has now been recognized for a variety of marine genera and species throughout the Phanerozoic (Foote 2007; Foote et al. 2007; Liow and Stenseth 2007), but the pattern first became established during the Ordovician Radiation.

Shifting Life Habits

The global taxonomic transitions among evolutionary faunas discussed earlier were manifestations of major changes to the ecology of communities living on Ordovician sea floors. In a series of pioneering studies, Bambach (1983, 1985) assessed the "ecospace utilization" of the three evolutionary faunas by categorizing the life habits of major higher taxa with respect to their modes of feeding and where and how they lived in reference to the sediment–water interface (e.g., swimming in the water column, floating in the water column attached to a hard substrate, and reclining on the surface of the sea floor burrowing within the sea floor). Collectively the taxa comprising the Paleozoic and Modern Fauna utilized a wider range of ecospace than their Cambrian predecessors; benthic elements of the Cambrian Fauna were limited primarily to habitats in close proximity to the sediment–water interface but the Paleozoic Fauna—the most prolific contributors to the Ordovician Radiation—included a wide range of attached and reclining sessile suspension-feeders that exploited habitats that were sometimes

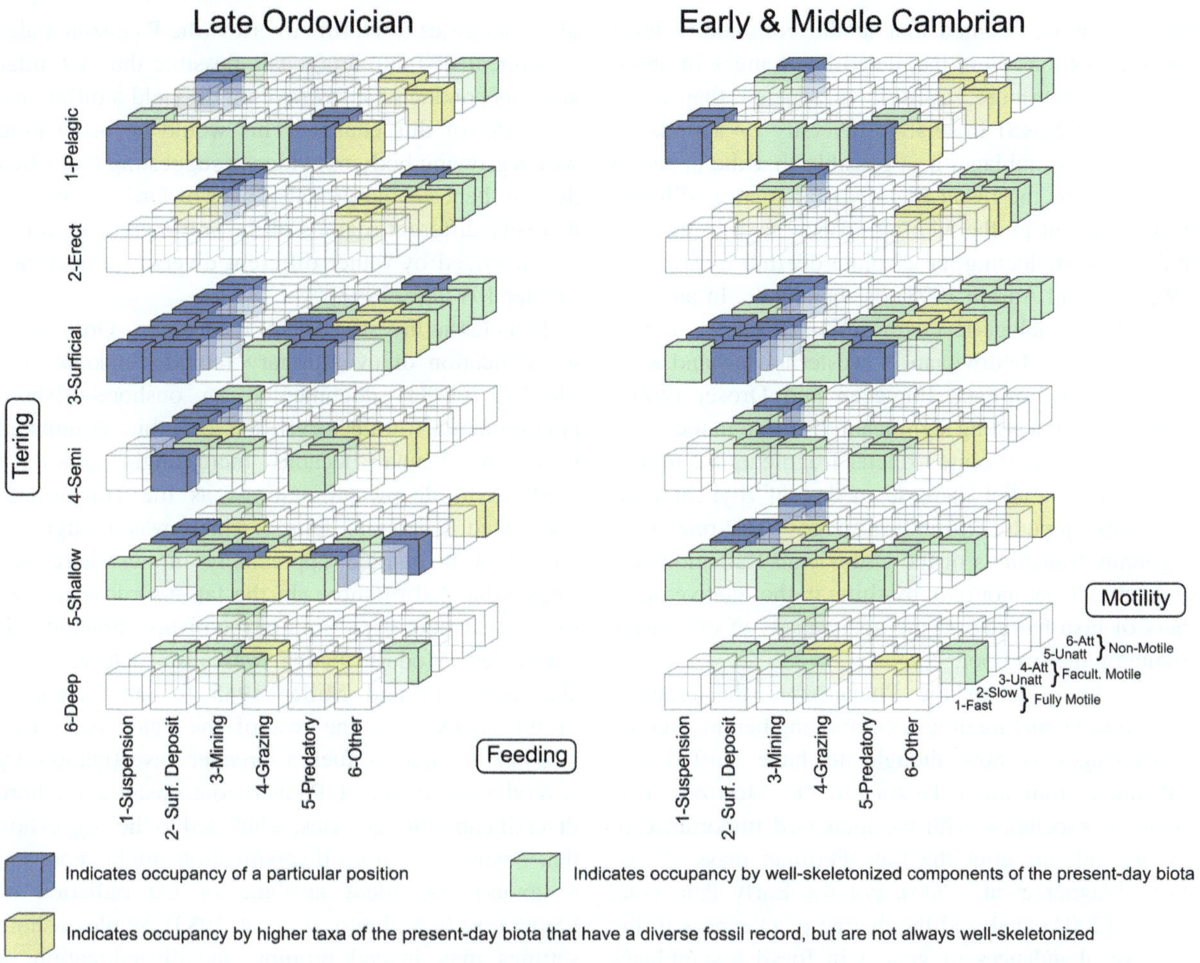

Fig. 3 Change in ecospace utilization from the Cambrian (*right*) to the Late Ordovician (*left*) for the *skeletonized subset* of the biota based on the conceptualization of Bambach et al. (2007). Feeding strategies are depicted along the "front" axis motility along the axis that goes "back" into the page and tiering above (1) at (3–4) and below (5–6) the sediment–water interface along the vertical axis. Blue boxes indicate occupancy of a particular position; green boxes indicate occupancy by *well-skeletonized* components of the *present-day* biota; yellow boxes indicate occupancy by higher taxa of the present-day biota that have a diverse fossil record but are not always well-skeletonized. Note the greater number of blue boxes for the Late Ordovician particularly at tiers 3 and 4 (after Bambach et al. 2007, figures 5B and 7A)

well above the sediment–water interface. Elements of the Modern Fauna in turn included a number of burrowers that even in the Ordovician began to further exploit habitats below the sediment–water interface. From an ecological perspective, an important outcome of the Ordovician Radiation therefore was the increased occupation and utilization of ecospace at above and below the sediment–water interface (Bottjer and Ausich 1986), a finding that Bambach has recently formalized more rigorously with the use of a conceptual cube for documenting ecospace-utilization changes through time, the three dimensions of which capture changes in feeding motility and height above and below the sea floor (Fig. 3; Bambach et al. 2007; Bush et al. 2007).

A Revolution in Relative Abundance

Continuing on the theme of community-level manifestations of the Ordovician Radiation, analyses of individual assemblages have permitted unprecedented glimpses into changes in the relative abundances of major taxonomic contributors to Ordovician biotas. Regardless of the changes in diversity that were hallmarks of the Ordovician Radiation we should not take it for granted that at the community level these transitions were paralleled by changes in abundance. Furthermore, even if it appears that a set of taxa decreased in relative diversity *or* abundance within an assemblage, it is possible that the apparent decrease was the result literally of statistical dilution brought about by the addition of new taxa to the mix rather than reflecting an absolute decline in diversity (Westrop and Adrain 1998) or abundance. In an analysis of assemblage-level transitions from the Lower to the Middle Ordovician in western Utah and eastern Nevada, however, Finnegan and Droser (2005) demonstrated a strong significant decrease in the absolute density of trilobite skeletal fragments >2 mm in size and a parallel increase in that of rhynchonelliform brachiopods. This in turn suggests that one of the signature transitions of the Ordovician Radiation was indeed reflected not only by shifts in the relative numbers of taxa but also in the abundances of organisms living on the sea floor.

On the scale of the Phanerozoic the relative-abundance structure of taxa comprising benthic marine assemblages is now thought to have shifted significantly from the Paleozoic to the Mesozoic perhaps in association with the increased importance of mobile infauna after the Late Permian mass extinction (Wagner et al. 2006). For the Early Paleozoic, Peters (2004) analyzed the degree of evenness in the relative abundances of genera in fossil assemblages from North America representing offshore biotas and found that evenness was significantly elevated (i.e., relative abundances of constituent genera were more equitable) in the Ordovician compared to the Cambrian (but see Adrain and Westrop 2005). Interestingly in contrast to the more continuous nature of other transitions through the Ordovician, Peters' analyses suggest that the overall elevation in evenness was concentrated in the Late Cambrian with very little change thereafter. There is some evidence, however, of an evenness increase *within* for taxa of the Paleozoic fauna.

Geographic Dissection

Although Peters' finding (Peters 2004) may, indeed, suggest some sort of decoupling in the timing of changes in evenness relative to other transitions that were hallmarks of the Ordovician Radiation it may also be that this reflects a limit in the geographic scope of the analyses to North America (the Paleozoic paleocontinent of Laurentia); it is possible that a similar analysis conducted elsewhere might yield a difference in timing of this change. This would be in keeping with a growing body of evidence suggesting that when delving beneath the global veneer of Ordovician biodiversity different regions of the world are sometimes characterized by rather different diversity trajectories (Miller 1997a, b, 1998).

In a classic investigation of the Cambro-Ordovician diversification of evolutionary faunas, Sepkoski and Sheehan (1983) documented an onshore–offshore environmental trajectory: the trilobite-dominated Cambrian Fauna became increasingly restricted offshore to deepwater settings as the brachiopod-dominated Paleozoic fauna, which was thought to have first diversified appreciably in nearshore settings expanded offshore. In the latest Ordovician, an initial nearshore expansion of the bivalve/brachiopod-dominated Modern Fauna was also detected (see also Sepkoski and Miller 1985 for an expansion of this analysis to the rest of the Paleozoic). This analysis helped to fuel a broader investigation by several researchers of Phanerozoic onshore–offshore diversification trajectories, which led to the suggestion that onshore–offshore diversification might represent a taxon-transcendent attribute of the radiation of higher taxa (e.g., Jablonski et al. 1983). While onshore settings may indeed promote the diversification of higher taxa with major order-defining evolutionary novelties (Jablonski and Bottjer 1991; Jablonski 2005), it has also become clear that the initial depictions for the Ordovician were oversimplifications with respect to both the environmental variables that were at play and the global veracity—or not—of the basic onshore–offshore pattern. Importantly the datasets used for these pioneering analyses were limited almost entirely to Laurentia. By dissecting these data into subsets from siliciclastic and carbonate environmental settings, Miller (1988) showed that for bivalve molluscs the important environmental variable governing their distributions was not position in an onshore–offshore gradient but rather the amount of siliciclastic influx into the environment (see below). Perhaps more importantly in subsequent investigations of Ordovician paleoenvironmental diversification trajectories in other parts of the world such as the Paleozoic paleocontinent of South China, onshore–offshore patterns have emerged that are rather different than those documented for Laurentia (Zhan and Harper 2006; Zhan et al. 2006; Zhan and Jin 2008).

In the light of these differences, it should probably not be a surprise that as more and more analyses of Ordovician diversification are conducted at regional scales the world over, virtually every region appears to carry its own unique signatures of diversification governed in large measure by physical characteristics of the region in question. In a series of studies conducted throughout eastern-midcontinent North America, Holland and Patzkowsky have demonstrated that Middle and Upper Ordovician biotas were highly responsive to regional paleoceanographic (Patzkowsky and Holland 1993) and paleoclimatic (Holland and Patzkowsky 2007) transitions that induced rapid transitions in faunal composition and the nature of faunal gradients. In western Argentina, Waisfeld and colleagues (Waisfeld and Sanchez 1996; Waisfeld et al. 1999, 2003) have documented a remarkable set of differences in the diversity of brachiopods, trilobites, bivalves, and sponges among basins that are now in close proximity to one another but that in the Ordovician were widely separated and characterized by dramatically different sedimentary and oceanographic regimes.

When tracing the Ordovician diversification trajectories of major higher taxa around the world in a paleogeographic and paleoenvironmental context, much the same picture emerges. In fact, as researchers have begun to dissect in detail the paleogeography of Ordovician diversification in major taxa, it can be said confidently that rare is the higher taxon whose diversification unfolded uniformly around the Ordovician world. Following on the work of Babin (1993) and Cope and Babin (1999) on bivalves, Novack-Gottshall and Miller conducted a comparative global paleogeographic analysis of the Ordovician diversity trajectories of bivalves and gastropods (Novack-Gottshall and Miller 2003a) followed by a regional comparison of their diversity and abundance with respect to lithological patterns in the type Cincinnatian (Novack-Gottshall and Miller 2003b). At all scales analyzed, bivalves were more diverse and abundant in siliciclastic settings whereas gastropods were more highly concentrated in carbonate environments (Fig. 4). These differences can be observed in the relative distributions and diversity trajectories of bivalves and gastropods among paleocontinents comprising the Ordovician world and it is just as clear in their distributions and abundances on a regional scale among the strata of the type Cincinnatian (see also Fang 2006). For bivalves the importance of siliciclastics can be observed directly in the paleogeographic trajectory of their Ordovician diversification: their initial radiation took place in siliciclastic settings at high southern latitudes marginal to Gondwana whereas they did not diversify appreciably in lower-latitude settings of Laurentia until the Middle and Late Ordovician as sediments in eastern Laurentia became dominated increasingly by siliciclastics supplied by the Taconic Orogeny.

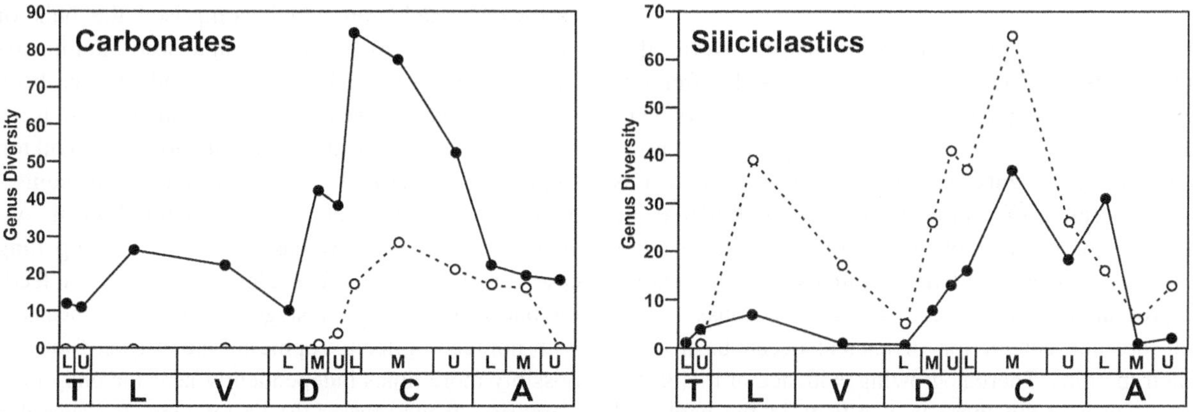

Fig. 4 Comparative global diversity trajectories for bivalves (*dotted lines*) and gastropods (*solid lines*) in siliciclastic and carbonate settings through the Ordovician. Note the greater preponderance of gastropods in carbonates and bivalves in siliciclastics (after Novack-Gottshall and Miller 2003a, figure 7)

A perusal of diversification trajectories for other higher taxa suggests similarly dynamic paleogeographic and paleoenvironmental trajectories during the Ordovician. An obvious challenge is to compare and contrast the trajectories of different taxa. In some cases it may be discovered that phylogenetically disparate taxa had similar paleogeographic or paleoenvironmental trajectories, suggesting that despite their evolutionary differences they shared similar physical imperatives during their radiations. Popov et al. (in Harper et al. 2004), for example, recognized that like bivalves, lingulide brachiopod assemblages initially occupied settings marginal to Gondwana and then proliferated elsewhere. Chitinozoans also exhibited Ordovician paleogeographic affinities reminiscent of bivalves including a major pulse of diversification in Laurentia during the Late Ordovician and an earlier diversity peak in Gondwana; however, chitinozoans were not as limited as bivalves were in Laurentia prior to the Late Ordovician (Paris et al. 2004; Achab and Paris 2007 provide an extended discussion of the relationship between chitinozoan diversity in relation to the dynamic geochemistry and paleogeography of Ordovician seas). Just as intriguing are taxa exhibiting trajectories that might be viewed as quite the opposite: initial diversification at low latitudes followed by radiation in high-latitude settings. Plectambonitoid brachiopods appear to exhibit such a pattern (Cocks in Harper et al. 2004).

Ordovician Terrane Faunas

One aspect of the Phanerozoic marine world that often gets overlooked is the fundamental difference between two major milieus: *epicontinental seas*, very broad shallow seas that covered large portions of now-emergent continents, and *open-ocean settings* such as coastal regions of continents and islands. Whereas the former have declined precipitously in importance since the Mesozoic era, the latter are almost certainly underrepresented in the Paleozoic record because of the closure and subduction of ancient ocean basins and their margins. There is growing evidence of major differences between these two milieus in the nature of circulation, regionalization of water masses propensity to become stratified, and the breadth of environmental belts (Slingerland 1986; Wells et al. 2005a, b; Allison and Wells 2006; Panchuk et al. 2006; Peters 2007). Taken together, these differences have bolstered suggestions that they exhibited noteworthy differences in the nature of taxonomy and extinction during the Ordovician and at other times (e.g., Sheehan 1975, 2001, 2008; Miller et al. 2009).

Among the growing sources of open-ocean data are the many terranes/microcontinents that are now known to have dotted Ordovician seas. As the Paleozoic progressed, a large number of these terranes accreted onto other terranes and paleocontinents (see, for example, Cocks and Torsvik 2007 for a detailed description of the complex history of accretion onto Siberia) and the Ordovician may have represented something of a zenith in the existence of terranes as independent entities with unique biotas (e.g., Harper 1992; Owen et al. 1992; Harper et al. 1996). Some of these terranes such as the Argentinean Precordillera (see Finney 2007) have proven critical to understanding fundamental aspects of Ordovician paleogeography and tectonics. An important focus of future investigation, however, will be to not only continue documenting the biogeographic affinities and/or uniqueness of these terranes but also determine whether the kinetics of origination and extinction of terrane faunas were different collectively from those on the inboard epicontinental-sea portions of Ordovician paleocontinents.

Molecular Insights into the Ecological Revolution

As a group the major taxonomic contributors to the Ordovician Radiation which comprised the bulk of the Paleozoic Evolutionary Fauna were dominated by epifaunal suspension feeders and it has been suggested that the proliferation of suspension feeders was associated with a diversification of plankton (Signor and Vermeij 1994). Based on a combination of molecular-clock and paleontological evidence, Peterson (2005) has presented a compelling case that taxa with life cycles that include a feeding planktotrophic-larval stage evolved from ancestors with nonfeeding lecitotrophic larvae at least four and possibly more times independently near the end of the Cambrian or the start of the Ordovician. Because this coincides closely in time with the onset of the proliferation of epifaunal suspension feeders, it suggests a major role for escalation (*sensu* Vermeij 1987) in this important transition: larvae that were positioned higher in the water column would effectively escape the reaches of benthic predators including epifaunal suspension feeders living on or near the sediment–water interface.

Peterson's findings dovetail well with those of Nützel and colleagues (Nützel and Fryda 2003; Nützel et al. 2006) who provided compelling evidence of significant decreases in the sizes of the protoconchs of benthic molluscs after the Cambrian, implying a transition in these taxa to feeding planktotrophic larvae. That said, Nützel and colleagues were quick to point out that this does not necessarily explain *why* there was an explosive diversification of both planktotrophic larvae *and* suspension feeding at that time and they proposed a fundamental physical trigger linked to an increased nutrient supply brought about by increased tectonic activity in the Ordovician. Taken as a whole this transition serves as an instructive example of what is almost certainly the complex interrelationship between physical and biological transitions associated with the Ordovician Radiation and as such provides an especially appropriate lead-in to a discussion of its causes.

Possible Causes of the Ordovician Radiation

It should already be evident from the foregoing discussion that a lot happened during the Ordovician Radiation. Nevertheless, perhaps because synoptic global diversity curves like those in Fig. 1 portray a fairly continuous transition through the interval, there has been a tendency to search for a singular cause for the Ordovician Radiation much as one might search for the cause of a global mass extinction event. We would do well, however, to appreciate that the Ordovician Radiation need not have had a singular "cause" (Botting and Muir 2008) and that it might have been shaped quite literally by a combination of global-scale and regional agents, some perhaps related to conditions established in the Cambrian and others that were distinctly Ordovician in aspect.

As suggested by Miller (2004) it might be appropriate to view the Radiation as having two different *facets* that require unique explanations: the overall *shape* of the Ordovician diversity trajectory may have been caused and governed by one set of factors whereas the underlying biological transitions—the *biological selectivity* of the Radiation—may relate to an entirely different set of mechanisms. With respect to the shape of the global trajectory the dramatic diversity increase in the early to middle portion of the period should perhaps rightfully be viewed as *the* signature characteristic of the Radiation. From a macroevolutionary standpoint, however, the rapid decrease in the overall rate of diversification *before* the onset of the Late Ordovician mass extinction (perhaps well before—see Miller and Foote 1996) is just as compelling: while many higher taxa undoubtedly diversified throughout the Ordovician this was counterbalanced by the decline of other higher taxa so that total trajectory began to flatten noticeably by the Late Ordovician (Fig. 1). Furthermore some higher taxa that radiated initially in the Ordovician and then continued to diversify thereafter nevertheless exhibited significant declines in their total and per-taxon rates of diversification after the Ordovician (e.g., bivalves; Miller and Sepkoski 1988). To some researchers this suggests that global marine diversity was density-dependent: as the world's oceans began to fill with taxa, diversification was impeded as the carrying capacity for the available finite ecospace was approached akin to the limitations exhibited among terrestrial biotas on islands (Sepkoski 1978, 1979, 1984; Miller and Sepkoski 1988). Although other researchers have questioned this view (e.g., Benton 1995, 1997; Stanley 2007), the important point is that the nature of the diversity trajectory during the Ordovician and at other times might well have been subject at least in part to global imperatives that had little to do with the unique conditions that fomented the radiation of particular higher taxa.

Cambrian Roots Climate Oxygenation Nutrients and Substrates: A Tangled Web?

It almost goes without saying that the Ordovician Radiation was dependent in some way on the Cambrian Explosion: most taxonomic players in the Radiation were already on the scene as the Ordovician unfolded. Among the major contributors to the Ordovician Radiation, bryozoans were the only phylum that did not first appear earlier. In itself, however, this tells us little about the relationship between the two intervals. There is probably no way to know, for example, whether the Cambrian Explosion resulted in a genetic setup in and of itself for a subsequent diversification at lower taxonomic levels making the Ordovician Radiation all but inevitable with the passage of additional time. As an alternative, however, we *can* ask whether physical or biological conditions during the Late Cambrian and Ordovician were in some way uniquely conducive to the Radiation. Interestingly a major positive global excursion in $\delta^{13}C$ (Saltzman et al. 2000) is now known to have occurred during the Late Cambrian, the causes and consequences of which are subjects of considerable current discussion. There appears to have been a global extinction of trilobites around the time of the initiation of the excursion suggesting that both had a common cause. Subsequently, however, the excursion in $\delta^{13}C$ following the initial onset of the SPICE event may have actually coincided with a diversity *increase* in marked contrast to the more typical Phanerozoic pattern of diversity decrease observed at or near the zeniths of positive $\delta^{13}C$ excursions (Saltzman et al. 2000). Saltzman et al. (2000, 2004) demonstrated that the SPICE event was associated with a eustatic sea-level decline with an increase in the weathering rates of newly exposed cratonic regions further amplified by an increase in tectonic activity causing an increase in the burial and sequestration of isotopically light organic carbon in marine settings marginal to the exposed cratons. Recently Saltzman et al. (2007) have posited a link between these transitions and a parallel positive shift observed in some localities in sulfur isotopic ratios (see Gill et al. 2007), implying increased pyrite burial and with it increasing oxygenation of the atmosphere, which may be attributable to an increase in primary productivity. If there was indeed an increase in the availability of oxygen, this might have played some—as yet unspecified—role in the Radiation; with the rise of multicellular animals already accomplished, however, this would seem less critical than it might have been during the Cambrian Explosion.

Harking back to the earlier discussion of the Late Cambrian–Early Ordovician multi-lineage advent of planktotrophic larvae, however, a concomitant increase in primary productivity fueled by a diversification at that time of phytoplankton would be especially intriguing because of its potential effects up the food chain for suspension feeders (Signor and Vermeij 1994) and other taxa. Recently Servais et al. (2008) have described the onset of a dramatic increase during the Late Cambrian/Early Ordovician in the morphology and diversity of acritarchs major contributors to the Early Paleozoic phytoplankton.

This in turn inevitably begs the question of what it was that might have fueled an increase in primary productivity in the first place and ultimately a diversification of phytoplankton. Given that there was an increase in tectonic activity that continued into the Ordovician with the closing of the ancient Iapetus Ocean among other events it has been suggested that the erosion of newly emergent mountains and volcanoes may have supplied nutrients to marine settings (e.g., Miller and Mao 1995). It was also proposed that submarine volcanoes (Vermeij 1995) or perhaps a superplume (Barnes 2004) may have been sources of nutrients. In a broad assessment of the potential role of tectonic activity in fomenting diversification during the radiation, Connolly and Miller (2002) did not find strong support for this possibility. This does not preclude the possibility that either nutrient or siliciclastic input could have acted locally or regionally to promote diversification especially of particular taxa. Certainly the regional studies highlighted earlier point to the importance of substrate and paleoceanographic differences in this respect. The role of substrate transitions in promoting diversification may be subtle requiring careful assessments of the *changing* substrate affinities of taxa in a changing world. In an Ordovician world that was becoming increasingly siliciclastic rich, for example, the affinities of rhynchonelliform brachiopods for siliciclastics appear to have been *increasing* even though the group as a whole exhibited a greater affinity for carbonates relative to trilobites and perhaps other major Ordovician groups (Miller and Connolly 2001).

Climatic Cooling or Warming as Agents of Radiation?

Needless to say given present-day concerns about the possible link between global warming and climate, the association of climate change with major changes in diversity has long been studied with strong links now demonstrated for a number of biodiversity transitions among terrestrial and marine biotas throughout the Phanerozoic. It should come as no surprise therefore that there is considerable interest in the possible relationship between climate and diversification during the Ordovician. Fortey and Cocks (2005), for example, have used evidence of faunal migration and changing degrees of endemism to diagnose a warming event in the Late Ordovician prior to the onset of the end-Ordovician mass extinction; researchers are now investigating the importance of this event for Ordovician diversity trends. More provocatively, a team of researchers (Trotter et al. 2008) has recently presented a new analysis of Ordovician oxygen isotopic records preserved in the apatite of conodonts to argue that there was a decline of some 10°C in sea-surface temperatures from the Early to Middle Ordovician reaching levels comparable to those observed in tropical settings today (~30°C). Undoubtedly this new analysis will be thoroughly vetted over the next several years but it should come as no surprise that the authors argue that this amelioration in temperature played a central role in the Ordovician Radiation notwithstanding the fact that the diversification and its signature biotic transitions appear to have been well underway prior to the Middle Ordovician.

Impact-Induced Diversification?

Equally provocative is the suggestion of a link between the Ordovician Radiation and a series of impacts (Schmitz et al. 2008). There is now strong evidence from geochemical and petrographic analyses that an excessive concentration of chromite grains preserved in Middle Ordovician limestones of Sweden and China are of extraterrestrial origin and relate to a previously documented break-up of a Low-iron chondrite in the asteroid belt. In Baltoscandia this concentration closely coincides with an upturn in the diversification trajectory of brachiopod species and there is also regional evidence in the surrounding interval of several Middle to Late Ordovician impact craters with other Ordovician craters known from around the world. Furthermore several Ordovician sections that are rich in chromite are also known to contain unusual lithologies and fossil biotas, perhaps indicative of significant disruptive events. Schmitz et al. (2008) suggested that an increase in the number of impacts during the Ordovician may have promoted diversification by increasing the heterogeneity of habitats on the sea floor. While this is somewhat speculative and the stratigraphic and potential geographic relationships between impacts and the timing of biotic turnover still need to be more firmly established—as is also the case for the cooling hypothesis described in the previous section—there will clearly be opportunities in the coming years to investigate these patterns in much more detail.

Concluding Remarks

Once overlooked by many researchers the Ordovician Period has now become an interdisciplinary proving ground not only for the investigation of the kinetics of global diversification but also for the dissection of the relationships between the history of life and physical transitions to the Earth at multiple scales. Although there remains much to learn about the intimacy of these relationships, the progress of the past two decades has been nothing short of spectacular. While it might never be possible to *definitively* determine its cause or causes, research on the Ordovician Radiation has nevertheless enriched our fundamental understanding of the kinetics of macroevolutionary and paleoecological transitions.

Acknowledgments My research on the Ordovician Radiation and broader themes in the history of Phanerozoic diversity has been generously funded over the years by a series of research grants from the National Aeronautics and Space Administration Program in Exobiology and Evolutionary Biology and a grant from the National Science Foundation Program in Biocomplexity. I am very grateful to John Talent for the invitation to contribute a chapter to this volume.

References

Achab A, Paris F (2007) The Ordovician chitinozoan biodiversification and its leading factors. Palaeogeogr Paleoclimatol Palaeoecol 245:5–19

Adrain JM, Westrop SR (2005) Evenness and diversity of lower Paleozoic trilobite faunas of Laurentian North America. Geol Soc Am (Abstr Prog) 37(7):461

Adrain JA, Fortey RA, Westrop SR (1998) Post-Cambrian trilobite diversity and evolutionary faunas. Science 280:1922–1925

Adrain JM, Edgecombe GD, Fortey RA, Hammer O, Laurie JR, McCormick T, Owen AW, Waisfeld BG, Webby BD, Westrop SR, Zhen Y (2004) Trilobites. In: Webby BD, Paris F, Droser ML, Percival IG (eds) The great Ordovician biodiversification event. Columbia University Press, New York, pp 231–254

Allison PA, Wells MR (2006) Circulation in large ancient epicontinental seas: what was different and why? PALAIOS 2:513–515

Alroy J, Marshall CR, Bambach RK, Bezusko K, Foote M, Fürsich FT, Hansen TA, Holland SM, Ivany LC, Jablonski D,

Jacobs DK, Jones DC, Kosnik MA, Lidgard S, Low S, Miller AI, Novack-Gottshall PM, Olszewski TD, Patzkowsky ME, Raup DM, Roy K, Sepkoski JJ, Sommers MG, Wagner PJ, Webber A (2001) Effects of sampling standardization on estimates of Phanerozoic marine diversification. Proc Natl Acad Sci USA 98(11):6261–6266

Alroy J, Aberhan M, Bottjer DJ, Foote M, Fürsich FT, Harries PJ, Hendy AJW, Holland SM, Ivany LC, Kiessling W, Kosnik MA, Marshall CR, McGowan AJ, Miller AI, Olszewski TD, Patzkowsky ME, Peters SE, Villier L, Wagner PJ, Bonuso N, Borkow PS, Brenneis B, Clapham ME, Fall LM, Ferguson CA, Hanson VL, Krug AZ, Layou KM, Leckey EH, Nurnberg S, Powers CM, Sessa JA, Simpson C, Tomasovych A, Visaggi CC (2008) Phanerozoic trends in the global diversity of marine invertebrates. Science 321(5885):97–100

Babin C (1993) Rôle des plates-formes gondwaniennes dans les diversifications des mollusques bivalves durant l'Ordovicien. Bull Soc Géol France 164(2):141–153

Bambach RK (1983) Ecospace utilization and guilds in marine communities through the Phanerozoic. In: Tevesz MJS, McCall PL (eds) Biotic interactions in recent and fossil benthic communities. Plenum, New York, pp 719–746

Bambach RK (1985) Classes and adaptive variety: the ecology of diversification in marine faunas through the Phanerozoic. In: Valentine JW (ed) Phanerozoic diversity patterns: profiles in macroevolution. Princeton University Press, Princeton, NJ, pp. 191–253

Bambach RK, Bush AM, Erwin DH (2007) Autecology and the filling of ecospace: key metazoan radiations. Palaeontology 50:1–22

Barnes CR (2004) Was there an Ordovician superplume event? In: Webby BD, Paris F, Droser ML, Percival IG (eds) The great Ordovician biodiversification event. Columbia University Press, New York, pp 77–83

Benton MJ (1995) Diversification and extinction in the history of life. Science 268(5207):52–58

Benton MJ (1997) Models for the diversification of life. Trends Ecol Evol 12(12):490–495

Bergström SM, Chen X, Gutiérrez-Marco JC, Dronov A (2009) The new chronostratigraphic classification of the Ordovician system and its relations to major regional series and stages and to δ13C chemostratigraphy. Lethaia 42:97–107

Botting JR, Muir LA (2008) Unravelling causal components of the Ordovician radiation: the Builth Inlier (central Wales) as a case study. Lethaia 41:111–125

Bottjer DJ, Ausich WI (1986) Phanerozoic development of tiering in soft substrata suspension-feeding communities. Paleobiology 12:400–420

Bush AM, Bambach RK, Daley GM (2007) Changes in theoretical ecospace utilization in marine fossil assemblages between the mid-Paleozoic and late Cenozoic. Paleobiology 33:76–97

Cocks LRM, Torsvik TH (2007) Siberia the wandering northern terrane and its changing geography through the Palaeozoic. Earth Sci Rev 82:29–74

Connolly SR, Miller AI (2002) Global Ordovician faunal transitions in the marine benthos: ultimate causes. Paleobiology 28:26–40

Cope JCW, Babin C (1999) Diversification of bivalves in the Ordovician. Geobios 32:175–185

Erwin DH, Valentine JW, Sepkoski JJ Jr (1987) A comparative study of diversification events: the early Paleozoic versus the Mesozoic. Evolution 41:1177–1186

Fang ZJ (2006) An introduction to Ordovician bivalves of southern China with a discussion of the early evolution of the Bivalvia. Geol J 41:303–328

Finnegan S, Droser ML (2005) Relative and absolute abundance of trilobites and rhynchonelliform brachiopods across the Lower/Middle Ordovician boundary eastern Basin and Range. Paleobiology 31:480–502

Finney SC (2007) The parautochthonous Gondwanan origin of the Cuyania (greater Precordillera) terrane of Argentina: a re-evaluation of evidence used to support an allochthonous Laurentian origin. Geol Acta 5(2):127–158

Foote M (2007) Symmetric waxing and waning of marine invertebrate genera. Paleobiology 33(4):517–529

Foote M, Crampton JS, Beu AG, Marshall BA, Cooper RA, Maxwell PA, Matcham I (2007) Rise and fall of species occupancy in Cenozoic fossil mollusks. Science 318(5853): 1131–1134

Fortey RA, Cocks LRM (2005) Late Ordovician global warming; the Boda event. Geology 33:405–408

Gill BC, Lyons TW, Saltzman MR (2007) Parallel high-resolution carbon and sulfur isotope records of the evolving Paleozoic marine sulfur reservoir. Palaeogeogr Palaeoclimatol Palaeoecol 256:156–173

Harper DAT (1992) Ordovician provincial signals from Appalachian-Caledonian terranes. Terra Nova 4:204–209

Harper DAT (2006) The Ordovician biodiversification: setting an agenda for marine life. Palaeogeogr Palaeoclimatol Palaeoecol 232:148–166

Harper DAT, MacNiocaill C, Williams SH (1996) The palaeogeography of early Ordovician Iapetus terranes: an integration of faunal and palaeomagnetic constraints. Palaeogeogr Palaeoclimatol Palaeoecol 121:297–312

Harper DAT, Cocks LRM, Popov LE, Sheehan PM, Bassett MG, Copper P, Holmer LE, Jin J, Rong JY (2004) Brachiopods. In: Webby BD, Paris F, Droser ML, Percival IG (eds) The great Ordovician biodiversification event. Columbia University Press, New York, pp 157–178

Holland SM, Patzkowsky ME (2007) Gradient ecology of a biotic invasion: Biofacies of the type Cincinnatian series (Upper Ordovician) Cincinnati Ohio region USA. PALAIOS 22:392–407

Jablonski D (2005) Evolutionary innovations in the fossil record: the intersection of ecology development and macroevolution. J Exp Zool B Mol Dev Evol 304B(6): 504–519

Jablonski D, Bottjer DJ (1991) Environmental patterns in the origins of higher taxa: the post-Paleozoic fossil record. Science 252(5014):1831–1833

Jablonski D, Bottjer DJ, Sepkoski JJ Jr, Sheehan PM (1983) Onshore–offshore patterns in the evolution of Phanerozoic shelf communities. Science 222(4628):1123–1125

Liow LH, Stenseth NC (2007) The rise and fall of species: implications for macroevolutionary and macroecological studies. Proc R Soc B Biol Sci 274(1626): 2745–2752

Miller AI (1988) Spatio-temporal transitions in Paleozoic Bivalvia: an analysis of North American fossil assemblages. Hist Biol 1:251–273

Miller AI (1997a) Comparative diversification dynamics among palaeocontinents during the Ordovician Radiation. Geobios Mémoire Spécial 20:397–406

Miller AI (1997b) Dissecting global diversity trends: examples from the Ordovician radiation. Annu Rev Ecol Syst 28: 85–104

Miller AI (1997c) A new look at age and area: the geographic and environmental expansion of genera during the Ordovician radiation. Paleobiology 23:410–419

Miller AI (1998) Biotic transitions in global marine diversity. Science 281:1157–1160

Miller AI (2004) The Ordovician radiation: towards a new global synthesis. In: Webby BD, Paris F, Droser ML, Percival IG (eds) The great Ordovician biodiversification event. Columbia University, New York, pp 380–388

Miller AI, Connolly SR (2001) Substrate affinities of higher taxa and the Ordovician radiation. Paleobiology 27:768–778

Miller AI, Foote M (1996) Calibrating the Ordovician radiation of marine life: implications for Phanerozoic diversity trends. Paleobiology 22:304–309

Miller AI, Mao S (1995) Association of orogenic activity with the Ordovician radiation of marine life. Geology 23(4): 305–308

Miller AI, Sepkoski JJ Jr (1988) Modeling bivalve diversification: the effect of interaction on a macroevolutionary system. Paleobiology 14:364–369

Miller AI, Aberhan M, Buick DP, Bulinski KV, Ferguson CA, Hendy AJW, Kiessling W (2009) Phanerozoic trends in the global geographic disparity of marine biotas. Paleobiology 35:612–630

Novack-Gottshall PM, Miller AI (2003a) Comparative geographic and environmental diversity dynamics of gastropods and bivalves during the Ordovician radiation. Paleobiology 29:576–604

Novack-Gottshall PM, Miller AI (2003b) Comparative taxonomic richness and abundance of late Ordovician gastropods and bivalves in mollusc-rich strata of the Cincinnati Arch. PALAIOS 18(6):559–571

Nützel A, Fryda J (2003) Paleozoic plankton revolution: evidence from early gastropod ontogeny. Geology 31:829–831

Nützel A, Lehnert O, Fryda J (2006) Origin of planktotrophy – evidence from early molluscs. Evol Dev 8(4):325–330

Owen AW, Harper DAT, Romano M (1992) The Ordovician biogeography of the Grangegeeth terrane and the Iapetus suture zone in eastern Ireland. J Geol Soc 149:3–6

Panchuk KM, Holmden CE, Leslie SA (2006) Local controls on carbon cycling in the Ordovician midcontinent region of North America with implications for carbon isotope secular curves. J Sediment Res 76:200–211

Paris F, Achab A, Asselin A, Chen XH, Grahn Y, Nõlvak J, Obut O, Samuelsson J, Sennikov N, Vecoli M, Verniers J, Wang XF, Winchester-Seeto T (2004) Chitinozoans. In: Webby BD, Paris F, Droser ML, Percival IG (eds) The great Ordovician biodiversification event. Columbia University Press, New York, pp 294–311

Patzkowsky ME, Holland SM (1993) Biotic response to a middle Ordovician paleoceanographic event in eastern North America. Geology 21:619–622

Peters SE (2004) Evenness of Cambrian–Ordovician benthic marine communities in North America. Paleobiology 30:325–346

Peters SE (2007) The problem with the Paleozoic. Paleobiology 33:165–181

Peterson KJ (2005) Macroevolutionary interplay between planktic larvae and benthic predators. Geology 33:929–932

Saltzman MR, Ripperdan RL, Brasier MD, Lohmann KC, Robison RA, Chang WT, Peng SC, Ergaliev EK, Runnegar B (2000) A global carbon isotope excursion (SPICE) during the Late Cambrian: relation to trilobite extinctions organic-matter burial and sea level. Palaeogeogr Palaeoclimatol Palaeoecol 162:211–223

Saltzman MR, Cowan CA, Runkel AC, Runnegar B, Stewart MC, Palmer AR (2004) The late Cambrian spice δ13C event and the Sauk II–SAUK III regression: new evidence from Laurentian basins in Utah Iowa and Newfoundland. J Sediment Res 74:366–377

Saltzman MR, Young S, Gill B, Lyons T, Kump L, Runnegar B (2007) Increased carbon isotope fractionation during the Late Cambrian SPICE event and a pulse of atmospheric oxygen. Geol Soc Am (Abstr Prog) 39(6):24

Schmitz BD, Harper AT, Peucker-Ehrenbrink B, Stouge S, Alwmark C, Cronholm A, Bergstrom SM, Tassinari M, Xiaofeng WF (2008) Asteroid breakup linked to the great Ordovician biodiversification event. Nat Geosci 1(1): 49–53

Sepkoski JJ Jr (1978) A kinetic mode of Phanerozoic taxonomic diversity I. Analysis of marine orders. Paleobiology 4:223–251

Sepkoski JJ Jr (1979) A kinetic model of Phanerozoic taxonomic diversity II. Early Phanerozoic families and multiple equilibria. Paleobiology 5:222–251

Sepkoski JJ Jr (1981) A factor analytic description of the Phanerozoic marine fossil record. Paleobiology 7:36–53

Sepkoski JJ Jr (1984) A kinetic model of Phanerozoic taxonomic diversity III. Post-Paleozoic families and multiple equilibria. Paleobiology 10:246–267

Sepkoski JJ Jr (1997) Biodiversity: past present and future. J Paleontol 71:533–539

Sepkoski JJ Jr, Miller AI (1985) Evolutionary faunas and the distribution of Paleozoic benthic communities in space and time. In: Valentine JW (ed) Phanerozoic diversity patterns: profiles in macroevolution. Princeton University Press, Princeton NJ, pp 153–190

Sepkoski JJ Jr, Sheehan PM (1983) Diversification faunal change and community replacement during the Ordovician radiations. In: Tevesz MJS, McCall PL (eds) Biotic interactions in recent and fossil benthic marine communities. Plenum Press, New York, pp 673–717

Servais T, Lehnert O, Li J, Mullins GL, Munnecke A, Nützel A, Vecoli M (2008) The Ordovician biodiversification: revolution in the oceanic trophic chain. Lethaia 41(2):99–109

Sheehan PM (1975) Brachiopod synecology in a time of crisis (late Ordovician–early Silurian). Paleobiology 1: 205–212

Sheehan PM (2001) The Late Ordovician mass extinction. Annu Rev Earth Planet Sci 29:331–364

Sheehan PM (2008) Did incumbency play a role in maintaining boundaries between late Ordovician brachiopod realms? Lethaia 41:147–153

Signor PW, Vermeij G (1994) The plankton and the benthos: origins and early history of an evolving relationship. Paleobiology 20:297–319

Slingerland R (1986) Numerical computation of co-oscillating paleotides in the Catskill Epeiric sea of eastern North-America. Sedimentology 33:487–497

Stanley SM (2007) An analysis of the history of marine animal diversity. Paleobiology 33:1–55

Trotter JA, Williams IS, Barnes CR, Lecuyer C, Nicoll RS (2008) Did cooling oceans trigger Ordovician biodiversification? Evidence from conodont thermometry. Science 321(5888):550–554

Valentine JW (1969) Patterns of taxonomic and ecological structure of the shelf benthos during Phanerozoic time. Palaeontology 112:684–709

Valentine JW (1970) How many marine invertebrate fossil species? A new approximation. J Paleontol 44: 410–415

Vermeij GJ (1987) Evolution and escalation: an ecological history of life. Princeton University Press, Princeton, NJ

Vermeij GJ (1995) Economics volcanoes and Phanerozoic revolutions. Paleobiology 21:125–152

Wagner PJ, Kosnik MA, Lidgard S (2006) Abundance distributions imply elevated complexity of post-Paleozoic marine ecosystems. Science 314(5803):1289–1292

Waisfeld BG, Sanchez TM (1996) "Cambrian Fauna" versus "Palaeozoic Fauna" in the lower Ordovician of Western Argentina: provincialism and environment interactions. Geobios 29:401–416

Waisfeld BG, Sanchez TM, Carrera MG (1999) Biodiversification patterns in the early Ordovician of Argentina. PALAIOS 14:198–214

Waisfeld BG, Sanchez TM, Benedetto JL, Carrera MG (2003) Early Ordovician (Arenig) faunal assemblages from western Argentina: biodiversification trends in different geodynamic and palaeogeographic settings. Palaeogeogr Palaeoclimatol Palaeoecol 196:343–373

Webby BD (2004) Introduction. In: Webby BD, Paris F, Droser ML, Percival IG (eds) The great Ordovician biodiversification event. Columbia University Press, New York, pp 1–37

Wells MR, Allison PA, Hampson GJ, Piggott MD, Pain CC (2005a) Modelling ancient tides: the upper Carboniferous epi-continental seaway of Northwest Europe. Sedimentology 52:715–735

Wells MR, Allison PA, Piggott MD, Pain CC, Hampson GJ, De Oliveira CRE (2005b) Large sea small tides: the LATE Carboniferous seaway of NW Europe. J Geol Soc 162: 417–420

Westrop SR, Adrain JM (1998) Trilobite alpha diversity and the reorganization of Ordovician benthic marine communities. Paleobiology 24:1–16

Zhan RB, Harper DAT (2006) Biotic diachroneity during the Ordovician radiation: evidence from South China. Lethaia 39:211–226

Zhan RB, Jin JS (2008) Onshore migration of a deep-water brachiopod fauna from the lower Ordovician Tonggao formation Jiangnan slope Southeastern Guizhou province South China. Can J Earth Sci 45:141–157

Zhan RB, Jin J, Rong JY (2006) Beta-diversity fluctuations in early-Mid Ordovician brachiopod communities of South China. Geol J 4:271–288

Phylogeny of Palaeozoic Gastropods Inferred from Their Ontogeny

Jiří Frýda

Abstract

Gastropods are not only one of the most diverse groups of living animals (the second after Insecta) occuring in marine, freshwater as well as terrestrial environments, but also have a rich fossil record extending back to the Cambrian. Because a third of all gastropod families are extinct, understanding of gastropod evolution and phylogeny necessarily involves study of both fossil and living species. Knowledge of the latter can be obtained from anatomical, morphologic, and molecular data, but for extinct forms virtually the only data source is the shell; typically it presents us with few characteristics. For both living and fossil gastropods, elucidation of ontogenic strategies is of prime importance in understanding the high-level phylogeny of this enormously diverse group. The analysis of Palaeozoic gastropods presented here relies heavily on ontogenies based on shell characteristics. It is argued that these results coordinate well with phylogenies of living gastropods inferred from the wider aggregate of anatomical, morphological, and molecular data. On the other hand, the analysis has highlighted problems with published phylogenies of living gastropods and, moreover, has produced evidence for the existence of several order-rank and long-lasting gastropod lineages forming an important part of Palaeozoic gastropod faunas but which failed to cross the Permian–Triassic boundary. Clearly, for understanding the phylogeny of the Gastropoda, it is imperative that the history of fossil gastropod clades be included.

Keywords

Ontogenetic characters • Shell microstructure • Polyphyletic groups • Heterobranchia • Caenogastropoda • Bellerophontoidea • Macluritoidea • Mimospirina • Archaeogasropoda • Euomphalomorpha • High-level phylogenies

Introduction

Efforts to understand the origin and evolution of living organisms have been an important part of the scientific activities of many thousands of biologists during the 150 years since publication of the famous

J. Frýda (✉)
Faculty of Environmental Sciences, Czech University of Life Sciences Prague, 165 21 Suchdol, Czech Republic

Czech Geological Survey, 118 21 Prague 1, Czech Republic
e-mail: bellerophon@seznam.cz

J.A. Talent (ed.), *Earth and Life*, International Year of Planet Earth,
DOI 10.1007/978-90-481-3428-1_12,

work of Charles Darwin on the origin of species. However, even before then, biologists had followed the taxonomic principles of Carl Linnaeus and united living organisms in a hierarchic system, thus suggesting that some organisms are more closely related than others. Intensive research for more than 250 years resulted in the collection of enormous sets of morphological, anatomical, molecular, and ecological data on living organisms, which have frequently been used for the interpretation of their phylogeny using different methodological approaches. Like living organisms, extinct taxa have also been studied intensively by zoologists and botanists, but relatively quickly, a new group of researchers, geologists, and palaeontologists took a leading role, with the study of fossil organisms. The intensity of the cooperation and the exchange of scientific knowledge among palaeontologists on one side and neontologists (zoologists and botanists) on the other side varied considerably over time and also depended on the groups of organisms studied. Palaeobotanists and palaeontologists examining extinct vertebrates have generally cooperated more intensively with neontologists than have palaeontologists studying the fossil records of invertebrates. The astonishing biodiversity and morphological disparity of extinct invertebrates and the very frequent lack of knowledge about their anatomy have limited the opportunities to place many fossil groups in the phylogenetic schemas proposed by zoologists. The main consequence of this unhappy situation has been the poor communication between invertebrate zoologists and palaeontologists. Therefore, both groups of scientists have often analysed evolution and phylogeny based on incomplete information. One of the most diversified phyla, Mollusca, represents a typical example of such a group. Recent decades have been characterized by increasing efforts to unite both the neontological and palaeontological concepts of the taxon and to better our understanding of molluscan evolution and phylogeny (e.g. Ponder and Lindberg 1997, 2008).

In this chapter, the Palaeozoic evolution and phylogeny of the class Gastropoda is discussed, with special focus on the pattern of their early shell ontogeny. This feature seems to be a fruitful source of phylogenetic information for the analysis of the high-level phylogeny of the class and can be studied in both living and extinct species. The chapter starts with a review of current opinions about the phylogenetic position of the class Gastropoda and its relationships to other molluscan classes. In the following section, an overview of all the available characters of fossil gastropods is given, with a discussion of their significance in the analysis of their higher phylogenetic relationships. The distributions of these characters in the Palaeozoic gastropods are summarized and the phylogeny inferred from these characters is compared with phylogenetic models based on the characters of living gastropods.

Position of the Gastropods Within the Molluscan Clade

To determine the correct position of the class Gastropoda within the phylum Mollusca, we must first determine the animal taxon that is the sister group of the phylum. This knowledge should help us to understand the polarity of the evolutionary changes in the molluscan groups and to recognize their plesiomorphic features. Recent molecular data have suggested that the Mollusca, together with the Nemertea, Entoprocta, Sipuncula, and Annelida, form a clade (designated 'Trochozoa') belonging to a larger clade called the 'Lophotrochozoa'. However, the position of the Trochozoa within this clade is equivocal, and so the identity of the molluscan sister taxon cannot be easily inferred from the available molecular studies.

For more than a century, authors have also debated whether the Mollusca had a segmented last common ancestor (e.g. Scheltema 1996; Nielsen 2001). The organization of the polyplacophoran body, with eight shell plates, is interpreted frequently as evidence supporting this hypothesis. However, the majority of recent studies (Haszprunar 2000; Haszprunar and Wanninger 2000; Okusu 2002; Nielsen et al. 2007) have shown that the most primitive extant molluscan groups, Solenogastres and Caudofoveata, show no signs of segmented organization. Moreover, the most recent data clearly support the hypothesis that the Mollusca and Entoprocta are sister groups and suggest that the basal Mollusca were primarily non-segmented animals (Haszprunar and Wanninger 2008; Haszprunar and Schaefer 1997; Giribet et al. 2006; Nielsen et al. 2007; Todt et al. 2008; Haszprunar et al. 2008; Simison and Boore 2008).

Like all trochophoran groups, these primitive molluscs developed a trochophore larva in their biphasic life cycle. The lecithotropic development of these larvae (i.e. the larva is not dependent on an external food source, but on the yolk of its egg) also seems to represent the primitive condition in the Mollusca. These primitive molluscs were probably small, worm-like animals, dorsally protected by a chitinous cuticle with embedded aragonitic spicules or scales (Salvini-Plawen 1985, 1990, 2003; Chaffee and Lindberg 1986; Salvini-Plawen and Steiner 1996; Haszprunar 1992, 2000).

Living molluscs are divided into eight classes (Fig. 1), which are generally considered to be monophyletic. The simplest body morphology is found in two molluscan groups, Solenogastres and Caudofoveata, which are most frequently interpreted

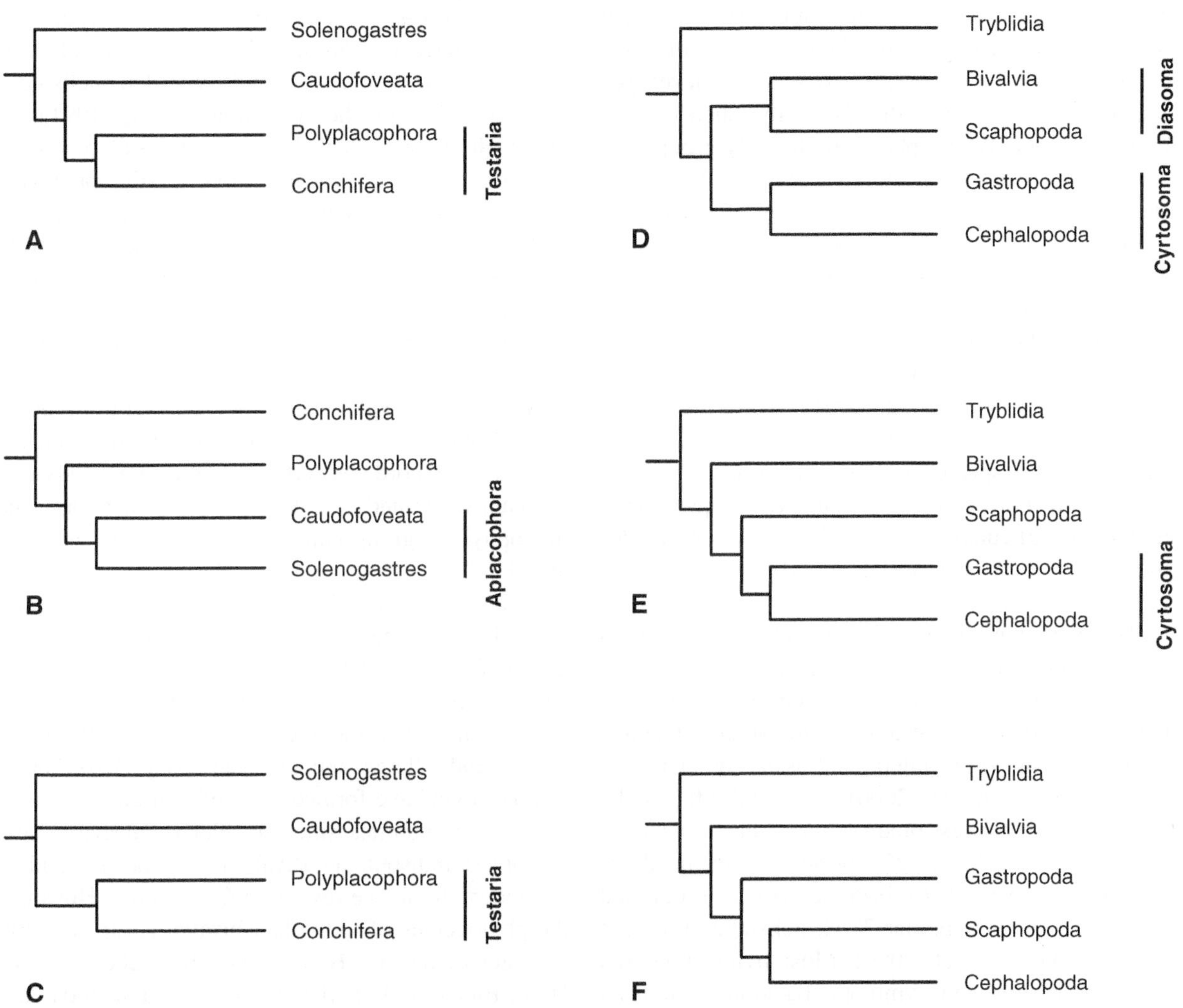

Fig. 1 Living molluscs are divided into eight classes, generally considered to be monophyletic. Differing views are held regarding their phylogenetic relationships. The following clades are defined within the phylum Mollusca: Conchifera (Tryblidia + Bivalvia + Scaphopoda + Cephalopoda + Gastropoda), Testaria (Conchifera + Polyplacophora), Aplacophora (Solenogastres + Caudofoveata), Diasoma (Bivalvia + Scaphopoda), and Cyrtosoma (Gastropoda + Cephalopoda). (**a–c**) Alternative phylogenic placements of Solenogastres, Caudofoveata, and Polyplacophora relative to Conchifera. (**a**) Solenogastres and Caudofoveata are basal groups, with the Solenogastres the first molluscan offshoot (Salvini-Plawen and Steiner 1996); (**b**) Aplacophora is a derived, monophyletic clade, the sister group to Polyplacophora (Scheltema 1996); (**c**) Solenogastres and Caudofoveata are independent clades with unresolved relationships to Testaria (Salvini-Plawen 2003). (**d–f**) Alternative phylogenies for the Conchifera. (**d**) Diasoma and Cyrtosoma are sister groups (Runnegar 1996); (**e**) Cyrtosoma is a clade and Diasoma is a paraphyletic group (Haszprunar 2000; Wanninger and Haszprunar 2001, 2002). (**f**) Scaphopoda and Cephalopoda are sister groups (Waller 1998; Steiner and Dreyer 2003)

as basal molluscan offshoots. Solenogastres contains about 250 species, which feed nearly exclusively on Cnidaria. The Caudofoveata contains about 150 species living in soft sediment. These two exclusively marine molluscan groups have been interpreted as sister taxa (Scheltema and Schander 2006, but see Haszprunar 2008) and united in the clade Aplacophora. In contrast to all other molluscan classes, the bodies of the aplacophorans are not primarily covered by shell or shell plates (the shell-less state of some higher molluscs, like slugs and squids, has developed secondarily). Amongst the primarily shell-bearing molluscs, the Polyplacophora are the only group with several shell plates (living polyplacophorans have eight shell plates). This exclusively marine group is adapted to the hard substrates of shallow or intertidal habitats, although a few species have been reported from deep-water environments. In contrast to the aplacophoran taxa, the Polyplacophora have a relatively rich fossil record extending back to the Early Palaeozoic (see summary in Cherns et al. 2004). About 1000 living species have been described and the number of described fossil species is of the same order. The fossil record of the Polyplacophora is large but frequent post-mortem disarticulation of the shell plates precludes the placement of some fossils into distinct species or genera.

The remaining five molluscan classes, Tryblidia (= Monoplacophora), Bivalvia, Scaphopoda, Cephalopoda, and Gastropoda, are often united in the group Conchifera (Fig. 1) because some shared features (synapomorphies) are interpreted as implying monophyly (see Haszprunar 2000). The Tryblidia is the group with the lowest biodiversity of all living molluscan classes. About 25 species living in deep, cold waters have hitherto been described (Warén and Bouchet 2001; Haszprunar 2008; Schwabe 2008; but see also Wilson et al. 2009). Most living Tryblidia feed on detritus and symbiotic bacteria have been reported in the mantle epidermis in a few of them (Haszprunar et al. 1995; Haszprunar 2008). The number of fossil tryblidian species is much higher than the number of living tryblidians and the oldest presumed members of this molluscan group have been found in Cambrian strata (see Knight et al. 1960; Horný 1963a; Wahlman 1992; Cherns et al. 2004). However, interpreting the tryblidian fossil record is difficult because the number of available shell features is limited. Cap-shaped shells similar to those of Tryblidia are also well known in another molluscan class, Gastropoda. The placement of certain fossil taxa into either the Tryblidia or the Gastropoda is decided from their shell features, which are believed to be diagnostic for both classes. The geometry of the muscle scars in fossil shells has been the most frequently used 'diagnostic' character (e.g. Peel and Horný 1999). However, this feature has a limited phylogenetic signal (see below). Based on the 'muscle scars' hypothesis, several Palaeozoic groups (like the bellerophontiform molluscs) with coiled, multi-whorled shells have been included in the Tryblidia (Horný 1991a, b, 1993; Horný and Peel 1996). The fossil record of the Tryblidia also shows a large stratigraphic gap, which includes the entire Mesozoic. Highly diversified Palaeozoic tryblidians have been identified from shallow-water environments, and this habitat differentiates those taxa from all living Tryblidia, which are only known from deeper waters (Schwabe 2008). The lack of a tryblidian fossil record in the Mesozoic strata has been explained by their adaptation to deep-water environments, which generally produce only a poor fossil record. Nevertheless, the hypothesis that the Palaeozoic and extant Tryblidia do not form one monophyletic group cannot be rejected and must be tested.

In contrast to the Tryblidia, members of the class Bivalvia represent a commonly recognized animal group of about 20,000 species living in marine, brackish, and freshwater environments. The Bivalvia have a long and rich fossil record (Cox et al. 1969) that also includes large extinct groups (e.g. Kříž 2007). The Bivalvia have formed a significant component of marine communities since the Early Palaeozoic and the oldest indisputable Bivalvia have been found in Cambrian strata (see review in Cope 2004). However, the phylogenies of individual bivalve groups are still a matter of debate (Bieler 2006; Kříž 2007; Giribet 2008; Bieler et al. 2010). The class Scaphopoda contains about 600 living species of exclusively marine animals living in soft sediment. Their fossil record is long but relatively poor (Knight et al. 1960; Reynolds and Steiner 2008). About 1000 living species belong to the class Cephalopoda, but the diversity of the cephalopods was much higher in the past (Teichert et al. 1964). The oldest cephalopods come from Cambrian strata (Frey et al. 2004; Nishiguchi and Mapes 2008) and for about 400 Ma (from the Ordovician to Cretaceous) they formed a significant part of marine faunas. In contrast to the exclusively marine cephalopods, the Gastropoda are the only molluscan class to have colonized most marine, freshwater, and terrestrial environments. The gastropods are also the most diversified molluscan class, with more than 100,000 extant species. Their fossil record is also very rich since the Ordovician (Wenz 1938–1944; Knight et al. 1960; Wagner 1999; Frýda and Rohr 2004). The oldest recorded gastropods are from Cambrian strata, but the determination of gastropod affinities among the Cambrian taxa is complicated (Runnegar 1983, 1996; Runnegar and Pojeta 1974; Parkhaev 2008).

As discussed above, the Mollusca are a highly diversified animal phylum, the species of which can be found in almost all environments, and its fossil record has been very rich since the beginning of the Palaeozoic. These facts have stimulated efforts to understand their phylogeny since the nineteenth century. However, the classical approach, using morphological and anatomical data (including the results of modern microanatomical methods), has resulted in different phylogenetic models, which challenge us to resolve the molluscan phylogeny with molecular biological methods. For this reason, a brief description of those models is given separately here.

Published analyses of the phylogenetic relationships of the molluscan classes based on molecular data are quite contradictory (e.g. Passamaneck et al. 2004; Giribet et al. 2006). Most published molecular studies have focused on information from the molluscan mitochondrial genome. Unfortunately, some recent papers (Dreyer and Steiner 2004; Boore et al. 2004) have suggested that mitochondrial data are very limited in their ability to resolve the relationships between the main molluscan lineages. Hitherto, the first multi-gene analysis that included all the molluscan classes (Giribet et al. 2006) resulted in a phylogenetic topology refuting the monophyly of the classes Gastropoda and Bivalvia. The Tryblidia were also placed within the Polyplacophora. This phylogenetic model is difficult to accept because it conflicts with many morphological, anatomical, and palaeontological data.

Phylogenetic analyses based on morphological data have produced a better consensus. The monophyly of the Conchifera (Tryblidia + Bivalvia + Scaphopoda + Cephalopoda + Gastropoda) and the basal position of the Tryblidia are well supported (Fig. 1). The cladistic analyses of morphological data published by Salvini-Plawen and Steiner (1996) and Haszprunar (2000) have shown that the Solenogastres and Caudofoveata are basal groups, with Solenogastres the first molluscan offshoot and that the taxon Aplacophora must be considered paraphyletic (Fig. 1a). Conversely, the Aplacophora is considered to be a sound monophyletic clade that originated by progenetic derivation from a polyplacophoran-like ancestor (Fig. 1b) in the model published by Scheltema and Schander (2006). It is noteworthy that even after the intensive and protracted (more than a century) systematic collection of morphological and anatomical data, we are still unable to resolve the phylogenetic relationships of the living molluscan classes. The monophyly of the Conchifera is generally accepted and this clade is characterized by two synapomorphies – a single shell and paired statocysts (Haszprunar et al. 2008). Figure 1d–f illustrates the three principal models of the conchiferan taxa. The Tryblidia are considered to represent the basal conchiferan group in all models, even though it is difficult to identify synapomorphies among the remaining conchiferan taxa (Haszprunar et al. 2008). In some phylogenetic models, the Bivalvia are united with the Scaphopoda in the clade Diasoma, and the extinct Rostroconchia are considered a common stem group (e.g. Runnegar 1996, but see Peel 1991a, b). Similar to the Bivalvia and Scaphopoda, the Cephalopoda and Gastropoda are also presumed to be sister groups, united in the taxon Cyrtosoma (Salvini-Plawen 1980). The clade Cyrtosoma seems to be supported by several synapomorphies (Haszprunar et al. 2008). An on-going study of the shell microstructures and textures of living and fossil molluscan shells has given additional support to the monophyly of the Cyrtosoma (Frýda et al. 2009). The analysis of the nacre in many tens of fossil and living molluscs has clearly shown that all nacre-bearing gastropods and many fossil cephalopod lineages share quite unique nacre (Fig. 2), which is unknown in other molluscan groups. The present morphological, anatomical, biomineralogical, and palaeontological data thus strongly suggest that the cephalopods are a sister group of the gastropods.

Molecular data have hitherto given only equivocal results and it is questionable how much molecular studies will contribute in the future to the resolution of the class-level molluscan phylogeny because of the difficulties caused by the rapid evolution and radiation of the conchiferan groups in the Cambrian (see discussion in Simison and Boore 2008). The clade Cyrtosoma (Cephalopoda + Gastropoda) seems to be

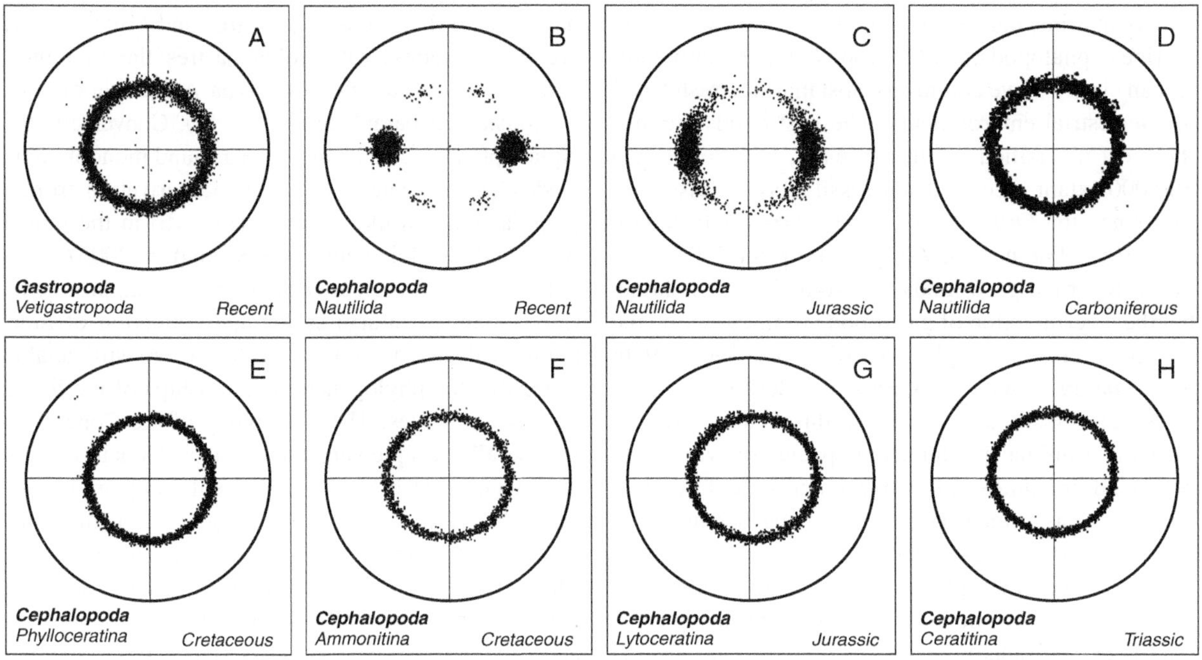

Fig. 2 Results of crystallographic textural analysis of the nacre (mother-of-pearl) of living and fossil cephalopods and gastropods (presented as pole figures of crystallographic direction [101] in stereographic projection) supporting monophyly of the clade Cyrtosoma (Gastropoda + Cephalopoda). Nacre is known only in molluscs; it has been recorded in four classes: Gastropoda, Cephalopoda, Bivalvia, and Tryblidia. Gastropods and cephalopods share the same microstructural type of nacre (columnar nacre), which is unknown in other molluscan groups. However, gastropod and cephalopod nacres are considered to have different crystallographic textures (i.e. the orientation of the individual nacreous platelets), as shown at figures (**a**) (a recent vetigastropod) and (**b**) (*Nautilus*). A continuing study has shown that the nacre of living and fossil gastropods is crystallographically uniform (preferred orientation of aragonite with [001] as axis of the texture, and perpendicular directions [100] and [010] random). The crystallographic direction [101] of the individual nacreous platelets in this type of texture forms a characteristic circle in stereographic projection (**a**). This textural pattern of gastropod nacre has remained unchanged for at least the last 300 Ma (since the Carboniferous). However, previous studies of nacre in living cephalopods (**b** – species of *Nautilus*) have revealed a different crystallographic texture: perfect preferred orientation of all axes, providing a single crystal-like X-ray pattern. This has been interpreted as evidence of the unrelated origins of the nacre of the classes Gastropoda and Cephalopoda. However, this continuing study has shown that the perfect preferred orientation of *Nautilus* nacre is the result of a long macroevolutionary process during which the original random directions [100] and [010] of the individual nacreous platelets in fossil nautilids assumed more preferred orientations (compare figures **b**, **c**, and **d**). Carboniferous nautilids (**d**) had the same nacreous crystallographic texture as that in living and fossil gastropods (**a**). The same type of nacre is also documented in Mesozoic ammonoids belonging to the suborders Phylloceratina (**e**), Ammonitina (**f**), Lytoceratina (**g**), and Ceratitina (**h**). Presence of the same type of nacre in members of the Cephalopoda and Gastropoda is interpreted as new evidence of the close evolutionary relationship of these classes and additional support for monophyly of the clade Cyrtosoma. (**a**) Recent vetigastropod *Haliotis asinina*; (**b**) Recent *Nautilus* sp.; (**c**) Late Jurassic nautiloid *Paracenoceras* sp.; (**d**) Carboniferous coiled nautiloid *Metacoceras* sp.; (**e**) Cretaceous ammonite *Phylloceras velledae* (Phylloceratina); (**f**) Cretaceous ammonite *Puzosia* sp. (Ammonitina); (**g**) Jurassic ammonite *Hemilytoceras fraasi* (Lytoceratina); (**h**) Triassic ammonite *Anaxenaspis oleneken-sis* (Ceratitina). (Based on Frýda et al. 2004, 2006b, 2007 and unpublished data)

well supported. An analysis of the early ontogenetic patterns across the phylum Mollusca has shown that the oldest gastropods probably underwent lecitotrophic development, so they were not dependent on an external food source during their larval ontogeny.

Definition of the Class Gastropoda

Gastropoda forms the second largest animal class and the largest molluscan group, containing about 80% of all living molluscs (Haszprunar et al. 2008). These show great diversity in their external forms, anatomy, behaviour, and physiology, and therefore they can be characterized by only a few uniting characters (apomorphies). Nevertheless, published phylogenetic analyses based on morphological and anatomical data clearly support the monophyly of the Gastropoda. However, it is noteworthy that the monophyly of the Gastropoda is not supported by the most complete analysis of the available molecular data (Giribet et al. 2006). The diagnostic features of the Gastropoda include the torsion of their soft bodies during the early larval stage and probably also the formation of a single shell with an operculum, at least in the larval stage. The former character is considered to be the most characteristic of the class and has been frequently used for its definition. The torsion of the gastropod soft body during the early larval stage produces a crossing of their nerve connections, a bending of the intestine, and a twisting of the mantle cavity (together with the associated structures, including the ctenidia, anus, kidney openings, etc.) anteriorly over the gastropod head. During this torsion, the visceral mass rotates up to 180° with respect to the head and foot as the result of a combination of muscular action and differential growth. Some gastropod groups (opisthobranchs) show a rotation of only about 90°, but this is the probable consequence of a later secondary untwist during post-larval ontogeny (Page 1995). Ponder and Lindberg (1997) have discussed all the apomorphies of the Gastropoda in detail and, as well as those mentioned above, they listed additional features like the presence of a tubular protoconch, a cone crossed-lamellar shell microstructure, a single tentacle nerve, and cephalic tentacles.

The definition of the class Gastropoda by several anatomical and morphological characters is useful for their recognition among living taxa, but unfortunately provides very limited recognition in the fossil record. Evidence of torsion of the soft body, inferred from the geometry of their muscle scars, has been the character most commonly used to determine the gastropod affinities in fossil molluscs (e.g. Horný and Peel 1996; Peel and Horný 1999). Apart from this torsion, the presence of a tubular protoconch, a cone crossed-lamellar shell microstructure, and a spiral operculum are the only characters potentially useful for the recognition of fossil gastropods. However, as shown below, the utility of any of the above-mentioned characters in the analysis of the fossil record is limited. Generally, in most cases, it is impossible to find direct evidence (i.e. the presence of any of synapomorphic characters of the class Gastropoda) in fossil gastropods. However, apomorphies of the smaller groups of living gastropods can be used for their recognition in the fossil record and so also for the indirect determination of their gastropod nature.

The discovery of fossilized soft body parts in fossil gastropods (e.g. Sutton et al. 2006) is so rare that they have almost no influence on the formulation of models of their phylogeny. Therefore, the only structures upon which to establish an understanding of the evolution and phylogeny of extinct gastropods are their shells and rarely their opercula.

Gastropod Shell

Gastropods include about 100,000 living species but the number of extinct species must have been much higher, and estimates of gastropod diversity suggest about 13,000 extant and fossil genera (Bieler 1992). In contrast to living gastropods, our knowledge of fossil gastropods comes from only the limited number of characters observable in their fossilized shells. In the following section, these characters are discussed briefly, with a focus on their significance in the analysis of their high-level phylogeny.

Gastropod Teleoconch

In the shell-bearing gastropods, the shell grows during almost the whole of their ontogeny. The part of the shell formed during the embryonic and larval stages is called the 'protoconch' and the part that grows after metamorphosis is called the 'teleoconch'. During more than 500 million years of evolution, the gastropods developed shells with various shapes and ornamentation, ranging in size from about 1 mm up to more than 1 m (Fig. 3; see also Wenz 1938–1944; Knight et al. 1960). The shell and its ornamentation may be broadly linked to the mode of gastropod life (e.g. origin of limpet-shaped shells in unrelated gastropod groups living on intertidal rocks). Several studies have shown that the range of the morphological diversity in gastropod shells has changed during their long evolution (Wagner 1995, 1999, 2001). An interesting observation was made by Vermeij (1977, 1987), who showed that since the Jurassic–Cretaceous, gastropods

Fig. 3 During more than 500 million years of evolution, the gastropods have developed shells with various shapes and ornament. Variations in this extraordinary morphological diversity of shell forms in living gastropods are shown: (**a**) high-spired, *Mitra mitra* (Caenogastropoda, Neogastropoda, Muricoidea); (**b**) turbiniform, *Liguus vittatus* (Heterobranchia, Orthalicoidea); (**c**) spinose fusiform, *Chicoreus ramosus* (Caenogastropoda, Neogastropoda, Muricoidea); (**d**) strombiform, *Lambis chiragra* (Caenogastropoda, Littorinimorpha, Stromboidea); (**e**) ovoid, *Olivancillaria gibbosa* (Caenogastropoda, Neogastropoda, Olivoidea); (**f**) conoidal, *Conus litteratus* (Caenogastropoda, Neogastropoda, Conoidea); (**g**) involute, *Cypraecassis rufa* (Caenogastropoda, Littorinimorpha, Tonnoidea); (**h**) fusiform, *Pleuroploca trapezium* (Caenogastropoda, Neogastropoda, Muricoidea); (**i**) convoluted, *Cyprea tigris* (Caenogastropoda, Littorinimorpha, Cypraeoidea); (**j**) turriculate, *Terebra* sp. (Caenogastropoda, Neogastropoda, Conoidea); (**k**) discoidal, *Architectonica perspectiva* (Heterobranchia, Architectonicoidea); (**l**) irregularly coiled, *Siliquaria ponderosa* (Caenogastropoda, Sorbeoconcha, Cerithioidea); (**m**) limpet, *Megathura crenulata* (Vetigastropoda, Fissurelloidea)

belonging to a wide variety of clades reacted to the increasing selective pressure exerted by durophagous (shell-crushing) predators by building stronger external structures and narrower apertures with dentition. A similar morphological trend was observed in gastropod shells with openly coiled whorls or widely umbilicate shells. The latter morphologies were relatively common in the Palaeozoic gastropods (see Wenz 1938–1944; Knight et al. 1960), but their frequency gradually decreased during the Mesozoic and Cenozoic. Vermeij (1987) pointed out that such shells are mechanically weaker and more susceptible to successful attack by durophagous predators than are the non-umbilicate morphologies that dominate the benthic gastropod faunas today.

An almost identical shell morphology can be found in unrelated gastropod lineages (e.g. limpet-shaped shells). Therefore, similarities in teleoconch characters often result from shell-shape convergence rather than as a manifestation of a close phylogenetic relationship. This fact demonstrates the low relevance of teleoconch morphological characters for the higher classification of the gastropods. Wagner (2001) and Schander and Sundberg (2001) showed that teleoconch characters are more homoplastic than are soft-tissue characters. Conversely, Wagner (2000) pointed out that a hierarchy of shell characters is not significantly less useful than a hierarchy of the skeletal characters used in low-level analyses of vertebrates and echinoderms, and significantly more useful than the characters used in low-level trilobite studies. Regardless these limitations, the phylogenetic analyses based on the large dataset of teleoconch morphological characters allow notable conclusions to be drawn and new hypotheses to be made about higher level phylogeny (e.g. Wagner 1999).

The teleoconchs (i.e. post-metamorphic shells) thus retain a strong phylogenetic signal among closely related species or genera, but because of their severe architectural restrictions, they generally provide no useful characters for the definition of the large gastropod clades.

Gastropod Protoconch

A characteristic feature of gastropod ontogeny is the formation of a tubular protoconch before metamorphosis. This feature is also one of the few apomorphies of the class and can be used for the determination of gastropod affinities in fossil molluscs. The main groups of living gastropods differ in their early development, which is reflected in their protoconch morphologies (Bandel 1982, 1997). The more ancient gastropod groups (Patellogastropoda and Archaeogastropoda) have the simplest shell ontogeny and their protoconch has only an embryonic shell (protoconch I), which is followed by the teleoconch (Bandel 1982; Sasaki 1998). Bandel (1982, 1997) noted that the biomineralization of the so-called archaeogastropod-type protoconch (Fig. 4a, b, g) is different from the protoconchs of other gastropods. He argued that the apparent coiling of those protoconchs is the result of the physical deformation of the non-mineralized shell. Subsequent calcification stabilizes this form and sometimes produces characteristic lateral pouches on the sides of the protoconch (but also see Page 1997). Lindberg (1981, 1988) noted that the protoconchs of vetigastropods (archaeogastropods) and patellogastropods are not identical because the latter show a

Fig. 4 The main gastropod groups differ in their early development; this is reflected in their protoconch morphology. The Archaeogastropoda have the simplest shell ontogeny. Their protoconchs have only an embryonic shell (protoconch I), followed by a teleoconch (**a**, **b**, and **g**). The protoconchs of more advanced gastropods, such as the Neritimorpha (**c** and **j**), Caenogastropoda (**e**, **f**, **h**, and **i**), and Heterobranchia (**d**), consist of an embryonic shell (protoconch I) and a subsequent larval shell (protoconch II). In most caenogastropods, the larval shells have different ornament from that of the teleoconchs; both shells are coiled in the same direction. In contrast, the protoconchs of the Heterobranchia, which display planktotrophic development, are coiled in the opposite direction to the coiling of the teleoconchs; the embryonic and subsequent larval shells are often smooth (**d**). (**a**) Late Triassic *Wortheniella coralliophila* (Vetigastropoda); (**b**) Recent *Anatoma proxima* (Vetigastropoda); (**c**) Recent *Smaragdia* sp. (Neritoidea); (**d**) heterostrophic larval shell of Recent *Mathilda* sp. (Architectonicoidea, Heterobranchia); (**e**, **f**) Recent caenogastropod protoconch (Coralliophilidae, Muricoidea); (**g**) Devonian *Diplozone innocens* (Murchisonoidea, Archaeogastropoda); (**h**) Early Devonian *Balbiniconcha cerinka* (Subulitoidea, Caenogastropoda); (**i**) Recent *Hipponix* sp. (Vanikoroidea, Caenogastropoda); (**j**) Late Triassic *Pseudorthonychia alata* (Pseudorthonychiidae, Neritimorpha)

left offset. However, this feature is also documented in living vetigastropods (Hadfield and Strathmann 1990), fossil archaeogastropods (Bandel 1993; Frýda 1997; Frýda and Blodgett 1998, 2001), and macluritoideans (Frýda and Rohr 2006). The protoconchs of more advanced gastropods (Neritimorpha, Caenogastropoda, and Heterobranchia) consist of an embryonic shell (protoconch I) and a subsequent larval shell (protoconch II). In most caenogastropods, the larval shells have different ornamentation from that of the teleoconchs (Fig. 4e, f, h, i) and both shells are coiled in the same direction (this condition is designated 'homeostrophic'). In contrast, in the Heterobranchia, which display planktotrophic development, the protoconch is coiled in the opposite direction to that of the teleoconch (Fig. 4d). Such shells are termed 'heterostrophic'. Frýda and Blodgett (2001) pointed out that the shell heterostrophy of the Heterobranchia and Archaeogastropoda is not homologous. Page (1995) has suggested that the entire larval shell of the heterobranchs is homologous to the larval shell of the 'archaeogastropods' and to protoconch I of the caenogastropods. However, the latter opinion is not generally accepted and must be reinvestigated. Both the embryonic and larval shells of the Heterobranchia are typically smooth, so distinguishing them is difficult. The Neritimorpha form typical, strongly convoluted protoconchs during planktotrophic development, which are homeostrophic (Fig. 4c, j). Higher gastropods with non-planktotrophic development (some marine, freshwater, and terrestrial gastropods) have simplified their early ontogeny and thus also the morphology of their protoconchs. The latter strategy has been documented from the Devonian (400 Ma).

Operculum

The larvae of all marine gastropods develop a lid-like structure (operculum) that closes the aperture of the larval shell. The operculum is also present in the majority of gastropods during their post-metamorphic stages, but some gastropods have lost it as part of their different ecological strategies (e.g. marine limpets and the majority of terrestrial gastropods). The post-metamorphic operculum evolved very early in gastropod evolution and it has been documented in Early Ordovician Macluritoidea (Knight et al. 1960) and Middle Ordovician archaeogastropods (Rohr and Frýda 2001). The operculum of living gastropods is usually horny (corneous) and may be tightly (multi-spiral) or loosely (pauci-spiral) coiled or concentric. Some gastropod groups have calcareous operculae (Fig. 12d); only this opercular type is found in the fossil record.

Shell Structure

Most gastropod shells are composed of an outer organic layer (periostracum) and an inner, usually much thicker, calcified layer. The colour pattern typical of many gastropod shells is formed by different organic pigments, which are limited to the periostracum and the uppermost calcified layer. Traces of the colour pattern have also been found in fossil gastropods (Fig. 5) since the Palaeozoic. The inner layers

Fig. 5 The colour pattern typical of many gastropod shells is formed by different organic pigments, restricted to the periostracum and uppermost calcified layer. Traces of the colour pattern may also be found at the latter site on fossil gastropod shells, as seen on the apical (**a**) and lateral (**b**) views of the shell of the German Middle Devonian (about 400 Ma) neritimorph *Paffrathopsis subcostata*

of gastropod shells consist of minute calcium carbonate crystals (aragonite or calcite, or very rarely vaterite) in an organic matrix. There are many structural types of gastropod shells (Carter and Clark 1985; Carter and Hall 1990) and, in general, the more ancient groups exhibit more diverse shell structures (Fig. 6). The Patellogastropoda have the most complex shell microstructures (Fig. 6e, f). Data on the shell microstructures of patellogastropod shells are useful for establishing their taxonomy (MacClintock 1968; Fuchigami and Sasaki 2005). In the fossil record, these data also represent the only available character for the identification of patellogastropods because their simple limpet shells lack any diagnostic characters (except their protoconchs). In contrast, the majority of the higher gastropods have developed simple aragonitic shells with a crossed-lamellar structure (Bandel 1990). Some shell microstructures are restricted to certain groups (e.g. nacre to Vetigastropoda) and this can be used for their identification in the fossil record (Fig. 6c, d). Nacreous and crossed-lamellar microstructures (Fig. 6a–d) have been known since the Palaeozoic.

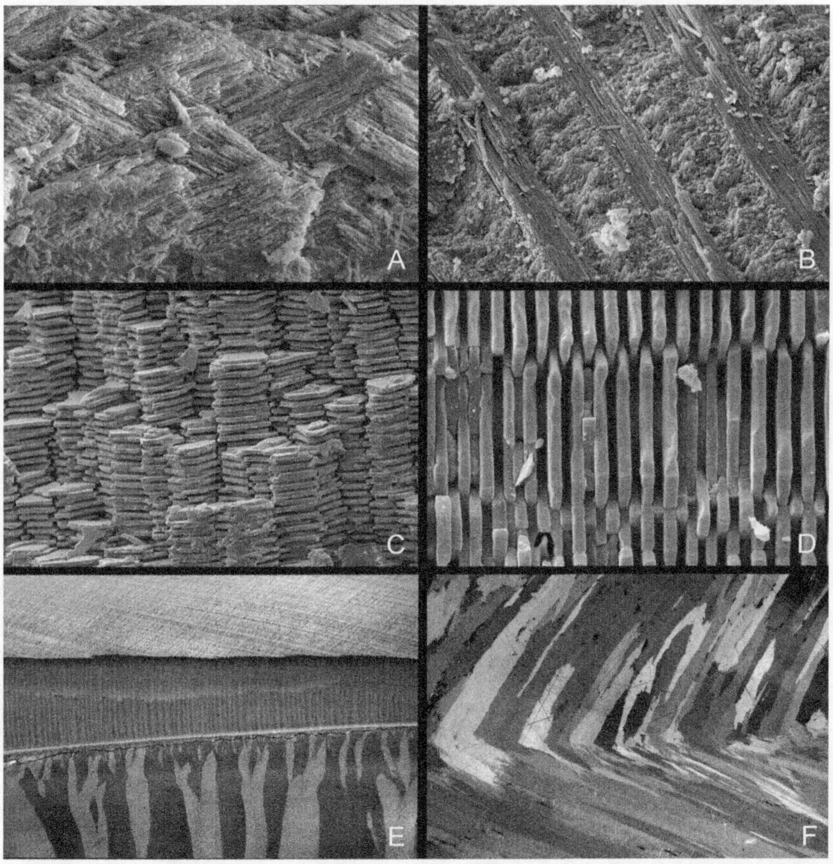

Fig. 6 Inner layers of gastropod shells consist of minute calcium carbonate crystals in an organic matrix. Of the many structural types of gastropod shells, the crossed-lamellar (**a**, **b**) and nacreous (**c**, **d**) microstructures are considered to be the most primitive; both are known since the Palaeozoic. The Patellogastropoda have the most complex shell microstructure (**e**, **f**). Some shell microstructures are restricted to certain groups (e.g. nacre to Vetigastropoda); this can be used in their identification in the fossil record. (**a**, **b**) Aragonitic crossed-lamellar microstructure in (**a**) the Carboniferous *Amphiscapha catilloides* (Euomphaloidea) and (**b**) the Late Triassic *Serpulospira pustulosa* (Archaeogastropoda?); (**c**, **d**) aragonitic nacreous microstructure (columnar nacre) in (**c**) Late Cretaceous *Sensuitrochus ferreri* (Porcellioidea, Archaeogastropoda) and (**d**) Late Triassic *Wortheniella coralliophila*; (**e**, **f**) radial sections of patellogastropod shells showing complex aragonitic and calcitic microstructures, (**e**) *Nipponacmea concinna* (Lottiidae) and (**f**) *Cellana nigrolineata* (Nacellidae)

Shell Coiling

The majority of gastropod shells are asymmetrically (anisostrophically) coiled and most of them are right-handed (dextral). Although left-handed (sinistral) shells seem to be less frequent, they are known to have occurred in several unrelated gastropod groups during the more than 500-million-year evolution of the gastropods (Wenz 1938–1944; Knight et al. 1960; Pchelintsev and Korobkov 1960; Vermeij 1975). Two of the oldest presumed gastropod clades, Mimospirina and Macluritoidea (see below), contain only species with sinistrally coiled shells. Symmetrical shells have only developed in some gastropod groups and they are either uncoiled (limpet-like) or symmetrically coiled (isostrophically) like those in the Palaeozoic Bellerophontoidea and some Porcellioidea (Fig. 7). The symmetrically coiled shells of the first group have

Fig. 7 Independent development of similar shell shapes in unrelated gastropod groups complicates the use of these characters for phylogenetic analyses. Example of the extraordinary morphological diversity of fossil shells within a small gastropod clade (Devonian and Carboniferous Porcellioidea). All teleoconchs of the porcellioideans are sinistrally coiled (**a–e**, **i**), except in those that secondarily developed bilaterally symmetrical shells (**f–h**, **j**, **k**). Some shells have an apertural slit, producing a selenizone (**a**, **b**, **f–h**, **j**, **k**) or tremata (**i**), whereas some have no slit (**c–e**). The latter feature has been used as a high-rank character in classical gastropod classifications (e.g. Knight et al. 1960). (**a**, **b**) Shell of the Emsian *Alaskiella medfraensis* showing inclined heterostrophy; (**c**, **d**) high-spired shell of the Emsian *Alaskacirrus bandeli*; (**e**) low-spired shell of the Lochkovian *Anoriostoma sinistra*; (**f**, **j**, **k**) bilaterally symmetrical teleoconchs of the Early Carboniferous *Martinidiscus woodwardii*. Note the dextrally coiled protoconch and the beginning of the teleoconch (**f**), which is the characteristic feature of all porcellioidean gastropods; (**g**, **h**) bilaterally symmetrical teleoconchs of the Early Carboniferous *Porcellia puzo*, ornamented on both sides of the whorl with a series of prominent nodes; (**i**) shell of the Pragian *Perryconcha pulchra* bearing apertural tremata

been considered to represent a primitive stage (e.g. Knight 1952). In contrast, the bilaterally symmetrical coiled shell in the Porcellioidea (Fig. 7h–k) is a derived shell character (Frýda 1997). Similarly, the development of uncoiled patelliform shells in different post-Palaeozoic gastropod groups is also a derived shell character. This is probably also true for the Patellogastropoda (= Docoglossa), which are considered to be the oldest gastropod offshoot (see the discussion in Ponder and Lindberg 1997).

Regardless of the manner of shell coiling, the gastropod viscera are torted relative to the head–foot axis. This torsion of the soft body is one of the most typical and thus most diagnostic characters of the class Gastropoda. The soft-body torsion is independent of shell coiling. The dextrality and sinistrality of the body torsion are also determined very early in embryonic development, and once determined, they are never reversed (Robertson 1993). Therefore, shell-less gastropods or those with bilaterally symmetrical shells (Fig. 7h–k) also have torted viscera. Considering these facts, there are four possible permutations: anatomically dextral and sinistral animals may also have dextrally or sinistrally coiled shells. These four types of geometric arrangement in the shell-bearing gastropods (Fig. 8) were recognized more than 100 years ago (Pelseneer 1893) and may be described by the terms discussed in detail by Knight (1952). The arrangement in an anatomically dextral animal with a dextrally coiled shell is designated 'dextral orthostrophic' (Fig. 8a). The mirror image of this (i.e. anatomically sinistral animal with a sinistrally coiled shell) is designated 'sinistral orthostrophic' (Fig. 8b). Conversely, the arrangement in an anatomically dextral animal with a sinistrally coiled shell is called 'dextral hyperstrophic' and its mirror image is called 'sinistral hyperstrophic' (Fig. 8c, d). These terms are, of course, applicable only to the shell-bearing gastropods. The vast majority of gastropods are dextrally orthostrophic, followed by sinistrally orthostrophic ones. Nevertheless, as summarized by Robertson (1993), dextrally hyperstrophic and sinistrally hyperstrophic gastropods are also known among modern gastropods. The identification the above-mentioned arrangements among the fossil gastropods is much more difficult because the manner of coiling of their soft bodies can be inferred only indirectly from the coiling of the operculum if it is present or in some case, from the apertural shape (Knight 1952; Robertson 1963, 1993; Peel and Horný 1996). The dextral orthostrophic gastropods have sinistrally (counter-clockwise) coiled opercula and vice versa (Figs. 8a, b and 12d). Similarly, the opercula of dextral hyperstrophic gastropods are coiled sinistrally (counter-clockwise) and vice versa in sinistral hyperstrophic gastropods (Fig. 8c, d). However, in some gastropods, shell coiling may be even more complex. In several gastropod groups, the shell whorls

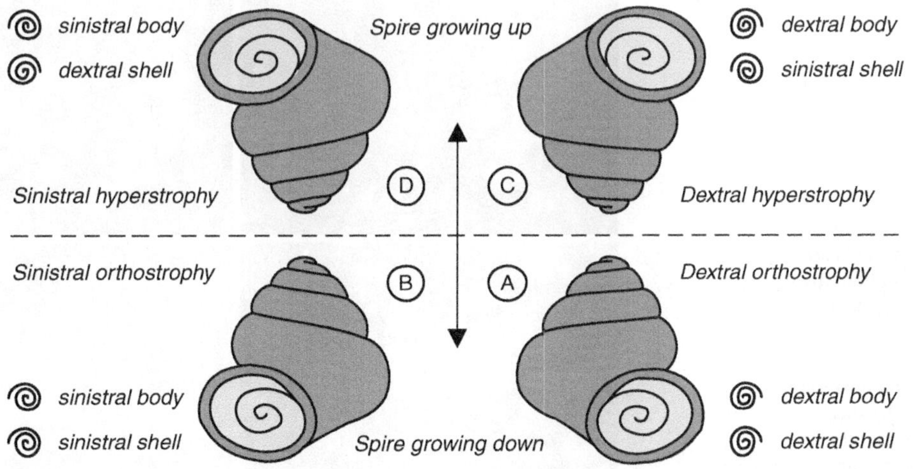

Fig. 8 Schematic diagram showing the four possible relationships between shell coiling and body asymmetry in the shell-bearing gastropods. 'Orthostrophy' (**a** and **b**) means that anatomically dextral (or sinistral) animals occupy dextrally (or sinistrally) coiled shells. Conversely, the term 'hyperstrophy' (**c** and **d**) is used when the handedness of the shell and that of the soft body are different (*see text for explanation*)

coil in one direction for part of their life and in the other direction for the other part. The term 'heterostrophy' (Figs. 4d and 7a, b) has been used for this condition (e.g. Knight 1941, 1952). If all the shell whorls are coiled in the same direction (i.e. dextrally or sinistrally), we call the shells 'homeostrophic' (Fig. 4h). Some authors use the term 'heterostrophy' in a narrower sense, for a condition of the protoconch wherein the whorls appear to be coiled in the opposite direction to those of the teleoconch (Knight et al. 1960). This condition is also called 'larval hyperstrophy' (Robertson 1993). A more detailed description and the different uses of the term 'heterostrophy' can be found in Dzik (1983) and Hadfield and Strathmann (1990). Shell heterostrophy occurs in several Palaeozoic gastropod lineages and it has hitherto been documented in the Heterobranchia (since the Devonian; Frýda and Blodgett 2001; Bandel 2002a), Porcellioidea (Fig. 7; since the Silurian; Frýda 1997), and Macluritoidea (occurring only in the Ordocivian; Frýda and Rohr 2006). The Ordovician Macluritoidea have been interpreted as dextrally hyperstrophic, from their sinistrally coiled teleoconchs and opercula (e.g. Knight 1952; Knight et al. 1960). Frýda and Rohr (2006) identified shell heterostrophy in the Early Ordovician macluritoidean gastropods and pointed out that it represents the oldest occurrence of this shell feature in the class Gastropoda.

Muscle Scars

The gastropod shell is attached to the soft body by muscles, which may leave distinct scars on the inner shell surface (Fig. 9). The geometry of the muscle scars has frequently been used as a diagnostic feature for the distinction of the torted (i.e. gastropods) and untorted states of Palaeozoic molluscs (Runnegar 1981; Horný 1963b, 1991a; Horný and Peel 1996; Harper and Rollins 2000). However, recent anatomical studies of living gastropods have shown that the larval muscles that participate in torsion and the post-larval

Fig. 9 Muscle scars in the Early Devonian *Orthonychia protei*. Gastropod shells are attached to the soft body by muscles, which may leave distinct scars on the inner shell surface. (**a**) Inner cast of a shell showing the position of the muscle scar; (**b**) specimen with partly preserved thick shell with characteristic growth lines; (**c**, **d**) detailed views of two specimens of the same species showing intraspecific variability in the muscle scars

muscles develop quite independently (Wanninger et al. 2000). Therefore, the muscle scar patterns sometimes observable in fossil molluscan shells may be a good ecological indicator, but have limited significance for phylogenetic studies.

Major Palaeozoic Gastropod Groups

Our present knowledge of the early shell ontogeny, stratigraphic ranges, and classification of the major Palaeozoic gastropod clades is summarized in the following paragraphs.

First Gastropods

The age of the first gastropods and the age of the first molluscs have been matters of intensive debate in the last 40 years, during which many new mollusc-like fossils from the Cambrian have been discovered (e.g. Halkieria, Merismoconcha, etc.). It is generally believed that some fossils that occurred just below the Precambrian–Cambrian boundary belong to the Mollusca (Yochelson 1978; Runnegar 1983; Bengtson and Conway Morris 1992; Fortey et al. 1996). However, support for their molluscan affinities is relatively weak, so the age of the first gastropods is unknown. Among the many Cambrian mollusc-like organisms, some have been interpreted as gastropods (see detailed discussion in Parkhaev 2008). The morphologies of the shells of the Cambrian 'molluscs' are very different from those of the post-Cambrian molluscs, which complicates their recognition and interpretation. The effort to interpret these molluscs has given rise to different models of the evolutionary relationships within the Mollusca (see Parkhaev 2008 for a review). Even though these models are controversial, the class Gastropoda has generally been accepted as the sister group of the classes Cephalopoda and Tryblidiida ('Monoplacophora'). As discussed above, the torsion of the soft body is considered one of the main diagnostic characters of the Gastropoda. However, this anatomical feature cannot be used as diagnostic in fossil gastropods and only indirect characters, such as anisostrophic shell coiling and muscle scar positions, have been used. Different groups of the Cambrian 'molluscs' have been variously interpreted as untorted or torted molluscs and thus as belonging or not belonging to the class Gastropoda, respectively (see overview in Parkhaev 2008). Generally, it is accepted that the first indisputable gastropods appeared in the Late Cambrian and had lecithotrophic development (Nützel et al. 2006, 2007a).

Bellerophontiform Molluscs

Characteristic features: The typical feature of the bellerophontiform molluscs is the bilateral symmetry of their shells (Fig. 10), which may range in size from a few millimetres to more than 10 cm (Knight et al. 1960; Kues and Duchene 1990). Their bilaterally symmetrical shells led some researchers to interpret the bellerophontiform molluscs as untorted molluscs, and thus not members of the class Gastropoda. Coiled and bilaterally symmetrical shells also occur in different living gastropod groups, but they evolved those shells secondarily from anisostrophically coiled shells. The same process has also been reported in extinct Palaeozoic gastropod groups (e.g. Porcellioidea, Euomphaloidea). Another typical feature of the bellerophontiform molluscs is the development of a sinus, slit, or tremata in the middle of their shell dorsum (Fig. 10). This feature has been used in their family-level taxonomy and it is interpreted as evidence of the paired and bilaterally symmetrical organization of their soft bodies.

Bellerophontid adult shells are formed of aragonite and have a characteristic complex crossed-lamellar microstructure (MacClintock 1968). The absence of nacre is an additional characteristic feature of the bellerophontiform molluscs, distinguishing them from the majority of archaeogastropods (see below). However, it is noteworthy that data on the shell microstructure come from just one (Bellerophontoidea) of several lineages of the bellerophontiform molluscs. Like the shell microstructure data, there is little knowledge of their early shell ontogeny (Fig. 10h, i). Nevertheless, these data and the results of a phylogenetic analysis based on teleoconch data suggest that the bellerophontiform molluscs are a polyphyletic group (Frýda 1999a; Wagner 2002).

Stratigraphic and geographic distributions: Bellerophontiform molluscs formed a significant part of many Palaeozoic faunal communities but only a few genera survived the Permian–Triassic crisis. The youngest bellerophontiform molluscs are reported

Fig. 10 Palaeozoic bellerophontiform molluscs. (**a**, **b**) Shell of the Givetian *Bellerophon vasulites*, the type species of *Bellerophon*; (**c**, **d**) discoidal shell of the Pragian *Tropidodiscus delanouei* with a narrow selenizone; (**e**) shell of the Lochkovian cyrtonellid *Kolihadiscus tureki*, with a prominent apertural sinus; (**f**, **g**) Early Carboniferous *Bellerophon umbilicatus* with a very narrow selenizone; (**h**, **i**) larval shell of Silurian *Bellerophon* sp.; the diameter of the first whorl is about 0.05 mm

from the Upper Triassic strata, when they probably became extinct (Yochelson 1967). The identification of the oldest bellerophontiform molluscs in the Cambrian faunas is difficult (see Parkhaev 2008). Bellerophontiform molluscs have been reported from the Palaeozoic strata of all palaeocontinents.

Classification: The class-level assignment of the bellerophontoidean molluscs has been discussed frequently during the past 50 years. The unknown nature of the soft parts of these molluscs, with their coiled, bilaterally symmetrical shells, has divided palaeontologists, who have variously argued that the bellerophontiform molluscs were untorted, exogastrically oriented tryblidians (= monoplacophorans), torted, endogastrically oriented gastropods, or a polyphyletic combination of both (see Wahlman 1992; Horný 1991a; Frýda 1999b for a review). There is no consensus about the classification of the bellerophontiform molluscs, even though they have been the most frequently discussed group of Palaeozoic molluscs. A list of the families included in the bellerophontiform molluscs and one of their most recently proposed classifications can be found in Bouchet et al. (2005). A more detailed discussion of the classification of the bellerophontiform molluscs and the often controversial interpretation of their nature have been summarized by Horný (1963b, 1991a), Wahlman (1992), Peel (1991a, b), Frýda (1999b), and Frýda et al. (2008a).

Macluritoidea

Characteristic features: The Macluritoidea were the largest Palaeozoic gastropods, and in some Ordovician communities, even the most common gastropods (Rohr et al. 1992). The typical feature of the macluritoideans is their sinistrally coiled shell (Fig. 11), with a thick calcified operculum. The spiral operculum of *Maclurites* is sinistrally (counter-clockwise) coiled. As noted by Knight (1952), this fact had already been interpreted by Woodward (1854) as evidence of its dextral hyperstrophic coiling. Detailed discussions of the dextral hyperstrophy of the Macluritoidea can be found in several papers (Knight 1952; Knight et al. 1960; Yochelson 1984, 1990; Linsley and Kier 1984; Peel and Horný 1996). According to Knight et al. (1960), the thick shell of the macluritoideans is formed of outer calcitic and thick inner aragonitic layers. The microstructure of the macluritoidean shells is poorly known but the latter authors point out that nacre, the

Fig. 11 Palaeozoic Macluritoidea (**a–d**, **g**, **h**) and Onychochiloidea (**e**, **f**). (**a–d**) Different views of the sinistrally coiled teleoconch of the Early Ordovician *Macluritella stantoni*; (**g**, **h**) detailed views of the early whorls in the same shell, showing distinct dextral coiling. The latter species is one of the oldest macluritoidean gastropods. (**e**) Sinistrally coiled shell of the Emsian onychochilid *Voskopiella barborae*; (**f**) sinistrally coiled shell of the Emsian clisospirid *Trochoclisa bouceki*

characteristic microstructure in the majority of living and fossil archaeogastropods, is absent in the macluritoideans. Frýda and Rohr (2006) reported the first data on the early shell ontogeny in Macluritoidea. Their study of the oldest macluritid gastropod, *Macluritella stantoni* Kirk, 1927 (Fig. 11), revealed that its early whorls are openly and dextrally coiled, in contrast to those in the later teleoconch, which are sinistrally coiled, as in all macluritoideans. They pointed out that the Macluritoidea evolved from the dextrally orthostrophic gastropods and their dextral hyperstrophy is a derived rather than a primary feature.

Stratigraphic and geographic distributions: Members of the Macluritoidea are known only from the Ordovician strata and they belong to the well-known group of Ordovician gastropods (Knight 1952; Rohr 1979, 1994; Rohr and Gubanov 1997; Frýda and Rohr 2004). The oldest Macluritoidea (in the sense of Gubanov and Rohr 1995) are known from the Floian and reached their highest diversity in the early Middle Ordovician. In the middle Darriwilian, their generic diversity dropped rapidly and remained low until the late Katian, when the Macluritoidea became extinct. They lived in warm, shallow marine waters and are found in both carbonate and siliciclastic facies on all Ordovician palaeocontinents situated in tropical regions (Rohr 1979; Rohr et al. 1992; Gubanov and Rohr 1995; Gubanov and Tait 1998).

Classification: The Macluritoidea include only one family, Macluritidae, with several genera. The taxonomic position of the macluritoideans has often changed. Wenz (1938–1944) placed the family Macluritidae together with Euomphalidae, Omphalocirridae, Platyacridae, Cirridae, Oriostomatidae, and Poleumitidae within the superfamily Euomphalacea (= Euomphaloidea). Later, Cox and Knight (1960) placed Macluritoidea, together with Euomphaloidea, in the suborder Macluritina of the Archaeogastropoda (Knight et al. 1960). The concept of Macluritina was later criticized by several authors (Morris and Cleevely 1981; Dzik 1983; Linsley and Kier 1984; Yochelson 1984). Similarly, opinions about the generic composition of the Macluritoidea have changed several times (Linsley and Kier 1984; Yochelson 1984; Rohr et al. 1992; Rohr 1994; Wagner 2002; Frýda and Rohr 2004).

Mimospirina

Characteristic features: The group Mimospirina contains species of the families Clisospiridae and Onychochilidae (Fig. 11e, f). A typical feature of those molluscs is their small, trochiform, sinistrally coiled shell. Some species (Clisospiridae) have a distinct lamella-forming shell periphery at the base of their whorls. The compositions and microstructures of their shells are still unknown. Members of Mimospirina have multi-whorled protoconchs with large embryonic shells (Dzik 1983; Frýda 1999b), which are smooth and also sinistrally coiled.

Stratigraphic and geographic distributions: Species of the Mimospirina are relatively rare components of the Cambrian–Devonian marine communities. Since the Ordovician, they lived like the macluritoidean in warm, shallow marine waters and are found in both carbonate and siliciclastic facies. The palaeogeographic distribution of the Ordovician Mimospirina was summarized by Frýda and Rohr (1999), who showed that the highest biodiversity of the Ordovician clisospirids and onychochilids is found in the Baltic faunas. The generic diversity of the Mimospirina was low until the early Middle Ordovician. From the beginning of the Darriwilian, their diversity started to increase strongly until the middle of the Late Ordovician, when their diversity reached its peak (Frýda and Rohr 2004). The latest Ordovician was a time of a sharp reduction in their diversity. The Mimospirina were affected by the crisis at the Ordovician–Silurian boundary, but in contrast to some gastropod groups (e.g. Macluritoidea), they survived. The diversity of the Mimospirina increased during the Silurian and reached another peak during the Early Devonian. They were later affected by the Basal Choteč Bioevent (close to the Emsian–Eifelian boundary) and became extinct. The Mimospirina have been reported from the Cambrian to the Early Devonian strata of many areas, but they reached their highest diversity in Baltica and Perigondwana (Barrandian area).

Classification: The taxon Mimospirina was established by Dzik (1983) as a new suborder of the Archaeogastropoda containing members of the families Clisospiridae and Onychochilidae. Wenz (1938–1944) regarded both families as unrelated taxa within the Archaeogastropoda. This opinion was later followed by Knight et al. (1960), but in contrast to Wenz (1938–1944), they considered onychochilids to be dextral hyperstrophic gastropods like macluritids. For this reason, Knight et al. (1960) transferred the family Onychochilidae into the superfamily Macluritacea (= Macluritoidea), which together with the superfamily Euomphalacea (= Euomphaloidea) was placed in the suborder Macluritina. Clisospirids were interpreted as sinistral orthostrophic gastropods with an uncertain higher systematic position. Discoveries of many new clisospirid and onychochilid taxa have revealed that the two families (Clisospiridae and Onychochilidae) are closely related (Horný 1964; Golikov and Starobogatov 1975; Peel 1986; Wängberg-Eriksson 1979).

On the basis of a functional analysis, Linsley and Kier (1984) proposed uniting the Clisospiroidea, Macluritoidea, and possibly the Euomphaloidea into a new order Hyperstrophina in a new class Paragastropoda. The class Paragastropoda was considered to represent untorted molluscs. Similarly, based on a phylogenetic analysis of their teleoconch characters, Wagner (1999) linked the Cambrian 'mimospirines' to pelagielloids and found that this clade was only distantly related to the gastropods. However, the quite different early shell ontogenies of the members of Clisospiroidea (Dzik 1983; Frýda and Rohr 1999) and Euomphaloidea (Yoo 1994; Bandel and Frýda 1998; Nützel 2002) suggest that the class Paragastropoda is an artificial group with no zoological validity (Frýda 1999b). Conversely, these facts support Dzik's (1983) concept of the Mimospirina.

The multi-whorled protoconch of *Mimospira* represents a true larval shell (protoconch II) and its large size and shape suggest the presence of a non-planktotrophic larval stage, during which the larval shell was formed. This fact argues against the placement of Mimospirina in the Archaeogastropoda (Frýda 1999b, 2001).

Archaeogastropoda

Characteristic features: The use of the name 'Archaeogastropoda' (Fig. 12) has changed several times (see Hickman 1988), reflecting its different definitions. 'Archaeogastropoda' is here used in the sense of Bandel (1997), thus including the clade Vetigastropoda Salvini-Plawen, 1980 (for superfamily content, see Bouchet et al. (2005) and Geiger et al. (2008)), some hot-vent taxa, and also many extinct taxa. The taxon Vetigastropoda (often used by zoologists as an equivalent name for the Archaeogastropoda) was defined by its anatomical characters, and its author also included in it extinct taxa, like the Macluritoidea (see Salvini-Plawen 1980,

Fig. 12 Palaeozoic archaeogastropods. (**a**) Juvenile shell of the Early Devonian *Diplozone innocens* (Murchisonoidea); (**b**) high-spired shell of the Early Devonian *Murchisonia pragensis* (Murchisonoidea); (**c**) trochiform, slit-bearing shell of the Givetian *Devonorhineoderma orbignyana* (Eotomarioidea); (**d**) shell of the Early Devonian *Australonema* sp. with spiral operculum *in situ* (Gyronematidae); (**e**) highly ornamented shell of the Givetian *Murchisonia coronata* (Murchisonioidea); (**f**) discoidal shell of the Eifelian *Stusakia pulchra*; (**g**) sinistrally coiled shell of the Early Carboniferous *Agnesia acuta* (Porcellioidea); (**h**) shell of the Emsian *Quadricarina noklebergi* with a complex selenizone (Eotomarioidea); (**i**) the Emsian *Balbinipleura krawczynskii* (Eotomarioidea); (**j**) shell of the Emsian *Decorospira lepaini* lacking a selenizone (Microdomatoidea); (**k**) turbiniform shell of the Givetian *Kitikamispira armata* (Gyronematidae); (**l**) shell of the Givetian *Euryzone delphinuloides* with a wide selenizone (Eotomarioidea); (**m**) juvenile shell of the Emsian *Zlichomphalina* sp. showing the boundary between the embryonic shell and the teleoconch – indicated by the *white arrow* (Eotomarioidea)

for details). The name 'Vetigastropoda' is not used here because only the membership of living taxa can be tested. Bandel (1982, 1997) used some protoconch features for his redefinition of the Archaeogastropoda, and these characters can be used to define not only living but also fossil and extinct archaeogastropods. Of course, this definition is not without its own problems, but it is used here for its applicability to both living and fossil gastropods.

The archaeogastropod protoconch (Fig. 4a, b, g) is formed by nearly one whorl (Bandel 1982; Sasaki 1998; Hickman 1988, 1992). The protoconch is initially organic and egg-shaped, then it is spirally deformed and finally mineralized (Bandel 1982; Hickman 1992, but see also Page 1997). Studies of the early shell ontogeny of the vetigastropods (the largest archaeogastropod group) show that their larvae do not feed on plankton (e.g. Bandel 1982). This protoconch type is also present in the Patellogastropoda (Docoglossa) and in the 'cocculiniform' gastropods (Bandel 1982; Haszprunar 1988b, 1998; Sasaki 1998). The first group differs from the Archaeogastropoda in its complex shell microstructure (Fig. 6e, f; MacClintock 1968) and the second group may be a sister group of the Vetigastropoda or even derived from them (see Haszprunar 1988b; McArthur and Harasewych 2003, but see also Aktipis et al. 2008). Another characteristic feature of the majority of the Archaeogastropoda is the presence of nacre in their shells (Fig. 6c, d). Among living gastropods, only Vetigastropoda have a nacreous shell layer, whereas it is absent in some other groups including Lepetodrilidae, Fissurellidae, Scissurellidae Phasianellidae, Halistylinae, and Skeneidae. However, some extinct gastropod groups had nacre (e.g. Porcellioidea; Kiel and Frýda 2004). Whether those extinct archaeogastropods (Fig. 12) also belong to the Vetigastropoda cannot be tested because of the lack of any anatomical features. Taking these data together, the definition of the Archaeogastropoda by their protoconch characters and/or by the presence of nacre means that this taxon includes many more gastropod groups than the vetigastropod crown group.

Stratigraphic and geographic distributions: The oldest presumed Archaeogastropoda identified by their teleoconch morphology are known from the Late Cambrian (Knight et al. 1960; Tracey et al. 1993; Wagner 2002). The oldest evidence based on the occurrence of their typical protoconch type comes from the Early Devonian (Fig. 12m; Frýda and Bandel 1997; Frýda and Manda 1997; Frýda 1999b). Frýda and Rohr (2004) noted that the slit- or sinus-bearing archaeogastropods ('Murchisonoidea' and 'Pleurotomarioidea') form the oldest gastropod groups and are known from the latest Cambrian (Tracey et al. 1993). The generic diversity of the first archaeogastropods increased moderately from the beginning of the Ordovician until the Middle/Late Ordovician, when their generic diversity increased at a much greater rate. The number of genera doubled from the early Late Ordovician to the later Late Ordovician (Frýda and Rohr 2004). This observation fits well with the fact that the slit-bearing Archaeogastropoda are a dominant gastropod group in Silurian strata. In contrast to the slit- or sinus-bearing Archaeogastropoda, there is no reliable evidence of slit-lacking Archaeogastropoda in Lower Ordovician strata (Frýda and Rohr 2004). However, since the Middle Ordovician, the diversity of the archaeogastropod taxa that lacked apertural slits or sinuses increased, and since that time to the Recent, they have shown similar changes in their diversity to those of the slit- or sinus-bearing archaeogastropods (Wenz 1938–1944; Knight et al. 1960; Tracey et al. 1993). As a dominant group of the Palaeozoic gastropods (Fig. 12), the Archaeogastropoda had a cosmopolitan distribution from the Ordovician to the Recent.

Classification: The concept of the Archaeogastropoda has changed many times and its different usages may be found in the most recent scientific studies. Thiele (1929–1931) placed the members of the Docoglossa, Zeugobranchia, Trochoidea, Neritoidea, and Cocculinoidea in his Archaeogastropoda. This concept of the Archaeogastropoda was later slightly modified by Wenz (1938–1944), who also added two groups of Palaeozoic gastropods, Tryblidioidea and Bellerophontoidea, to his Archaeogastropoda. Knight et al. (1960) considered the majority of Palaeozoic gastropods to belong to the Archaeogastropoda, which were divided into seven suborders: Bellerophontina, Macluritina, Pleurotomariina, Patellina, Trochina, Neritopsina, and Murchisoniina. However, a quite different concept of the Archaeogastropoda was published by Golikov and Starobogatov (1975).

Fretter Graham, and McLean (1981) noted that the anatomy of a newly discovered limpet *Neomphalus fretterae* from hot springs was an extraordinary combination of archaeogastropod and 'mesogastropod' characters, combined with some unique features. For this reason, McLean (1981) placed his genus *Neomphalus* in a new superfamily Neomphaloidea of a new suborder Euomphalina, which also contains the Palaeozoic members of the Euomphaloidea. The suborder Euomphalina was regarded as an independent group derived from the early pleurotomaroideans and belonging to the Archaeogastropoda (McLean 1981; Yochelson 1984). Bandel (1982, 1992) redefined the subclass Archaeogastropoda using protoconch characters. He also showed that the Neritimorpha have a quite different type of early shell formation (Fig. 4c) and that they cannot be regarded as Archaeogastropoda. Conversely, Haszprunar (1985) placed the Vetigastropoda, Docoglossa, and Neritopsina (= Neritimorpha) in the Archaeogastropoda. Later, Haszprunar (1988a), in accordance with Golikov and Starobogatov (1975) and Lindberg (1988), considered the Docoglossa to be the first (archaeo-)gastropod offshoot. Hickman (1988) discussed the structure of the taxon Archaeogastropoda and suggested a narrower use of the name 'Archaeogastropoda' for the members of the Pleurotomarioidea, Fissurelloidea, and Trochoidea. Later, Haszprunar (1993) again discussed the inner structure and taxonomic position of the Archaeogastropoda and suggested that the Archaeogastropoda represents a paraphyletic taxon.

As mentioned by many authors (see Tracey et al. 1993; Bandel and Frýda 1996; Bandel and Geldmacher 1996 for a review and discussion), the family-level classification of the fossil and living Archaeogastropoda requires a complex modern revision. The most recent classification and its discussion can be found in Bouchet et al. (2005) and Geiger et al. (2008).

Euomphalomorpha

Characteristic features: The taxon Euomphalomorpha (Fig. 13) was established by Bandel and Frýda (1998) for Palaeozoic gastropods united in Euomphaloidea because of the unusual morphology of their protoconchs, which differ from those of all living gastropod groups. The protoconch is egg-shaped, about 0.1–0.2 mm in width, and is succeeded by an openly coiled whorl (Fig. 13f), which usually leaves a central gap before the next whorls touch it, in both taxa with trochospirally coiled teleoconchs and those with planispirally coiled teleoconchs (Yoo 1994; Bandel and Frýda 1998; Frýda et al. 2006a). The Euomphalomorph protoconch consists of only embryonic shell, like those of the archaeogastropods and patellogastropods, but in contrast to the latter, it is not spirally deformed (Fig. 13f). In addition to this feature, the euomphalomorph gastropods lack a nacreous shell layer. Their shell consists of an outer calcitic layer and an inner aragonitic layer composed of needle-like biocrystallites oriented in spherulite sectors, resembling an aragonitic crossed-lamellar microstructure (Batten 1984; Bandel and Frýda 1998).

The teleoconchs of the euomphalomorph gastropods are rather variable. The juvenile shells are typically planispirally coiled and later their teleoconch whorls remain so or become trochospirally coiled (Fig. 13a–d). The teleoconch whorls commonly do not overlap much and in some euomphalomorph taxa, they are openly coiled (Bandel and Frýda 1998). The euomphalomorph teleoconch generally has a wide-opened umbilicus and the margin of the aperture is straight or there is a sinus in the upper part of the outer lip. Recently, Frýda et al. (2006a) showed that some euomphalomorph taxa developed an apertural slit (Fig. 13e).

Stratigraphic and geographic distributions: The Euomphalomorpha are known only from Palaeozoic strata. The oldest species (judging from their teleoconch features) come from the Upper Cambrian (Wagner 2002). However, until the beginning of the Middle Ordovician, the number of genera was

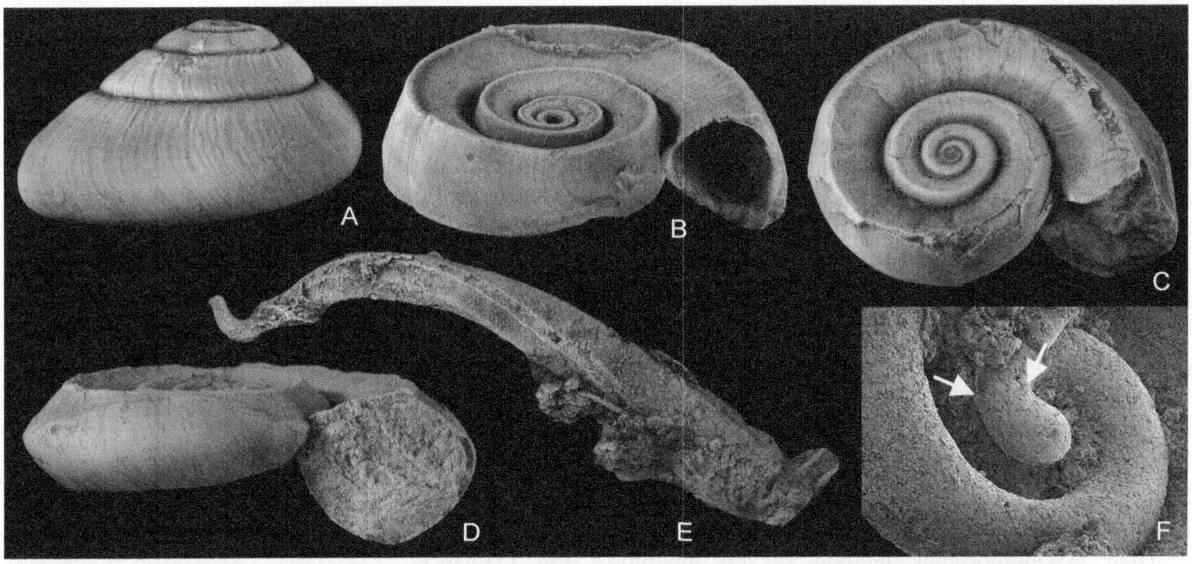

Fig. 13 Euomphaloidea. (**a**) Low-spired shell of the Early Carboniferous *Straparollus fallax*; (**b–d**) discoidal shell of the Early Carboniferous *Euomphalus pentangulatus*, type species of *Euomphalus*; (**e**) openly coiled shell of the Eifelian *Odontomaria cindiprellerae* with a wide selenizone; (**f**) detailed view of the first whorl of the Givetian *Odontomaria gracilis* showing the embryonic shell (protoconch I) formed by a large, short, slightly curved tubular shell with a diameter about 0.12 mm and a length about 0.2 mm. The boundary between the embryonic shell and the teleoconch is indicated by the *white arrows*

relatively low. Later, their diversity increased and reached its maximum in the early Middle Ordovician (Frýda and Rohr 2004). Their biodiversity subsequently dropped and remained approximately constant until the end of the Ordovician, when they were affected by the Ordovician–Silurian boundary crisis. The euomphalomorph gastropods later underwent several radiation events from the Middle Devonian to the end of the Palaeozoic, when they became extinct. The Euomphalomorpha belong to one of the dominant groups of the Late Palaeozoic gastropods. They are known from almost all regions, but their diversity was the highest in palaeotropical regions with carbonate-dominated facies.

Classification: Yochelson (1956) interpreted the Euomphaloidea to have derived from the Macluritoidea in Early Ordovician times and he placed three families (Euomphalidae, Helicotomidae, and Omphalotrochidae) into this superfamily. This concept was followed by Knight et al. (1960) and later slightly modified by the addition of two families, Omphalocirridae and Lesueurillidae, in the classification of Bouchet et al. (2005). A different classification was published by Wagner (2002) based on a phylogenetic analysis of the teleoconch characters. He proposed that his euomphalomorph clade contained some Euomphallidae, Euomphalopteridae, some Helicotomidae, Pseudophoridae, and Planitrochidae.

Patellogastropoda

Characteristic features: All living patellogastropods are limpets with cap-shaped shells ranging in size from a few millimetres to about 20 cm. Cap-shaped shells have been independently developed in almost all major gastropod lineages. Therefore, this shell type cannot be used as diagnostic for the Patellogastropoda. The patellogastropods are almost exclusively marine gastropods, with one species occurring in brackish and freshwaters (Lindberg 1990). Some patellogastropods have been found at both cold seeps and hydrothermal vent sites. However, the vast majority of patellogastropods typically occurred on intertidal rock substrates.

Their synapomorphies include several anatomical features (see Ponder and Lindberg 1997; Sasaki 1998; Lindberg 2008). Some of their shell microstructures (Fig. 6e, f) also represent synapomorphies (MacClintock 1967; Fuchigami and Sasaki 2005), which could be used for their identification in the fossil record.

Stratigraphic and geographic distributions: The Patellogastropoda have been considered to represent a very ancient gastropod group, but there is no indisputable evidence for their occurrence in the Palaeozoic. Some presumed Palaeozoic patellogastropods, like the Ordovician *Floripatella rousseaui* described by Yochelson (1988), probably do not belong to the class Gastropoda. Similarly, the patellogastropod affinity of members of the family Damilinidae, known from Devonian marine strata, was inferred only from their muscle scar geometry (Peel and Horný 1999). The earliest patellogastropod limpet verified by its characteristic shell microstructure is known from the Upper Triassic (Hedegaard et al. 1997). There is a relatively good fossil record of the patellogastropods from the Cretaceous to the Recent (Akpan et al. 1982; Kollman and Peel 1983; Lindberg 1988; Kase and Shigeta 1996). Living patellogastropods occur on intertidal rock substrates on all continents.

Classification: Since the first analyses of gastropod relationships, the Patellogastropoda have been considered quite different from the other gastropods. For this reason, they were placed in their own group, apart from the rest of the gastropods (Cyclobranchia by Cuvier 1817; Docoglossa by Troschel 1865). The monophyly of the Patellogastropoda seems to be well supported by the analysis of morphological data, but it becomes uncertain when only molecular data are analysed (see the discussion in Lindberg 2008). The internal classification of the patellogastropods is also unstable (see Lindberg 1988, 2008; Ponder and Lindberg 1997; Sasaki 1998; McArthur and Harasewych 2003; Nakano and Ozawa 2004). The Patellogastropoda is divided into three typical groups, Patelloidea, Nacelloidea, and Lottioidea, and one possibly related, deep-sea group, Neolepetopsoidea McLean, 1990 (e.g. Bouchet et al. 2005, but see also Lindberg 2008 and references therein).

Neritimorpha

Characteristic features: The clade Neritimorpha (Fig. 14) can be characterized by several synapomorphies, including its anatomical features (Ponder and Lindberg 1997; Sasaki 1998; Lindberg 2008) and its unique protoconch morphology (Fig. 4c). The latter character is the only feature that can be used to identify

Fig. 14 Neritimorpha (**a–d**, **f**, **m**) and Cyrtoneritimorpha (**e**, **g–l**, **n**, **o**). (**a**) Middle Devonian *Paffrathia lotzi*; (**b**, **c**) highly ornamented shell of the Late Triassic *Bandelopsis oregonensis*; (**d**) Late Triassic *Weitschatopsis pulchra*; (**e**) Carboniferous *Orthonychia parva* with preserved cyrtoneritimorph protoconch; (**f**) Early Devonian *Oriostoma* sp.; (**g**) Middle Devonian *Eifelcyrtus blodgetti* with preserved cyrtoneritimorph protoconch; (**h**) juvenile shell of the Early Devonian *Vltaviella reticulata*; (**i**) irregular shell of the Early Devonian *Orthonychia protei*; (**j**) Early Devonian *Platyostoma janthinoides*; (**k**) Silurian cyrtoneritimorph protoconch; (**l**) Late Silurian cyrtoneritimorph *Krameriella hornyi*; (**m**) Late Triassic *Pseudorthonychia alata*; (**n**) larval shell of the Carboniferous *Orthonychia parva*, the boundary between the embryonic shell and protoconch II is indicated by the *white arrow*; (**o**) the Early Devonian cyrtoneritimorph *Praenatica cheloti*

neritimorph gastropods in the fossil record. The neritimorph protoconch consists of embryonic and strongly convoluted larval shell (Bandel 1982, 1992, 2007). Another typical feature of the neritimorph gastropods is their eccentric operculum and the internal structure of their shells. However, the shapes of the neritimorph opercula show greater morphological diversity among the fossil taxa (Kaim and Sztajner 2005; Bandel and Heidelberger 2001; Kano 2006; Bandel 2008). The typical shell of living neritimorphs differs from those of the other coiled gastropods in the internal whorls of the shell, because they are absorbed during later ontogeny. However, this feature is only typical of the majority of living neritimorphs (all except species of *Neritopsis* [Bandel 2007]). However, the vast majority of fossil neritimorphs (identified by their typical protoconch shape) did not absorb the internal whorls of their shells (see Bandel 2007). The shell microstructure of the Neritimorpha has hitherto been studied in only a few species. Generally, the shell microstructure varies little and there are two basic shell layers, the outer homogeneous calcitic layer and the inner aragonitic layer, with a co-marginal crossed-lamellar or cone complex crossed-lamellar microstructure (Hedegaard 1990; Bandel and Frýda 1999; Sasaki 2001).

Stratigraphic and geographic distributions: The stratigraphic distribution of the Neritimorpha is uncertain (Frýda et al. 2008a and discussion and references therein). If their protoconch is used as a diagnostic character, then they can be traced from the Recent back to the Triassic (Bandel and Frýda 1999; Bandel 2007; Frýda et al. 2008b). If their teleoconch shapes (Knight et al. 1960; Frýda 2001; Nützel et al. 2007b) or their opercula (Kaim and Sztajner 2005) are used, then they can be traced from the Recent back to the Ordovician (Frýda and Rohr 2004; Frýda et al. 2009). Protoconch data on the presumed Palaeozoic neritimorphs suggest that the typical neritimorph protoconch of the living marine taxa with planktotrophic development (Fig. 4c) originated during the Triassic (Nützel et al. 2007b; Bandel 2007; Frýda et al. 2009). Living marine neritimorphs are recorded from continental shelves and their habitats range from tropical intertidal beach rocks to the deep sea, including cold seeps and hydrothermal vent environments (Lindberg 2008). They are most diverse in tropical and warm temperate environments and the marine and freshwater neritimorphs are typically associated with hard substrates. Members of the Neritimorpha also colonize terrestrial environments (see Kano et al. 2002 and references therein).

Classification: The monophyly of the Neritimorpha has become widely recognized (Salvini-Plawen and Haszprunar 1987; Haszprunar 1988a; Ponder and Lindberg 1997; Sasaki 1998; McArthur and Harasewych 2003; Lindberg 2008). The living Neritimorpha include the Neritopsidae, Neritidae, Neritiliidae, Helicinidae, Hydrocenidae, Phenacolepadidae, and Titiscaniidae (Kano et al. 2002; Bouchet et al. 2005). The classification of fossil neritimorphs is more complex and the number of taxa involved varies in different models (Bandel 1992, 2007, 2008; Nützel et al. 2007b). Bandel (2007) showed that the species diversity of the neritimorphs was much higher in Triassic reef environments than in any modern reefs (Fig. 14a–d, m). He noted that some Triassic species can be placed in the superfamilies Neritopsoidea and Neritioidea, but not in any of the modern families. In contrast, some Triassic neritimorphs (subfamilies Ampezzonaticopsinae Bandel, 2007 and Hologyrinae Bandel, 2007) may belong to the family Naticopsidae. Bandel (2007) noted that the naticopsid lineage is extinct, so it has no connection to the modern Neritimorpha, like the vast majority of Late Triassic neritimorphs. The principal problem in the classification of the fossil neritimorphs is whether to place in the Neritimorpha only those taxa that share the strongly convoluted neritimorph protoconch (i.e. the group called 'Cycloneritimorpha' by Frýda 1998a, b) or also their stem groups (see the discussion in Frýda et al. 2009).

Cyrtoneritimorpha

Characteristic features: The taxon Cyrtoneritimorpha (Fig. 14) was established by Frýda (1998a, b, 1999b) for presumed neritimorph gastropods bearing unusual protoconchs with an openly coiled, fish-hook-like morphology. The diagnostic feature of the clade Cyrtoneritimorpha is their protoconch, which consists of embryonic and larval shells, with a long straight beginning formed by uncurved, tubular embryonic shell and the first part of the larval shell (Fig. 14e, g, h, k, n). This straight part of the cyrtoneritimorph protoconch attains about half the total protoconch width (Frýda et al. 2009).

The teleoconchs of the cyrtoneritimorph gastropods vary (Fig. 14) and include limpet-like shells (*Orthonychia*), naticiform shells (*Praenatica*), and turbiniform shells (*Krameriella*). The total morphological diversity of the cyrtoneritimorph teleoconchs cannot yet be analysed because only a few cyrtoneritimorph shells with well-preserved protoconchs have been found (Frýda et al. 2009). No data exist on their shell microstructures.

Stratigraphic and geographic distributions: Based on their protoconchs, the Cyrtoneritimorpha can be traced from the Early Ordovician to the Late Permian, and thus lasted for about 250 million years (Frýda 1998a, b, 1999c). These protoconchs sometimes occur together with other types of gastropod protoconchs in dissolved rock residues (typically from the separation of conodonts). The very rich fossil material containing isolated cyrtoneritimorph protoconchs from different stratigraphic levels of Ordovician to Lower Devonian strata show that the cyrtoneritimorphs were common in the Early Palaeozoic gastropod faunas.

Since the Middle Devonian, the relative numbers of cyrtoneritimorph protoconchs in these residues decline rapidly and there is no record of their occurrence after the Permian–Triassic boundary crisis. The documented trend towards decreasing proportions of all types of open-coiled protoconchs throughout the Palaeozoic may explain this reduction in the frequency of cyrtoneritimorph protoconchs (Nützel and Frýda 2003; Frýda et al. 2008a). The limited record of adult cyrtoneritimorph shells with well-preserved protoconchs precludes a more detailed analysis of their geographic distribution, but all the available data, which are drawn from carbonate facies, indicate warm, shallow marine waters. However, this may be a collection artefact (attributable to the preferred selection of carbonates for the isolation of conodont samples).

Classification: Although cyrtoneritimorph protoconchs are common in the Early Palaeozoic, there are only several gastropod genera in which the protoconchs found in the adult teleoconch. Families that belong to or may belong to the Cyrtoneritimorpha are discussed by Frýda et al. (2009).

Perunelomorpha

Characteristic features: The taxon Perunelomorpha was established for members of the superfamily Peruneloidea Frýda and Bandel, 1997, to emphasize their phylogenetic position (Frýda 1998a, b, 1999b, d). The diagnostic feature (apomorphy) of the perunelomorph gastropods is their protoconch shape. The homeostrophic perunelomorph protoconch consists of embryonic and larval shells and the first protoconch whorl is openly coiled. In this feature, the perunelomorph protoconchs resemble cyrtoneritimorph protoconchs, but the protoconch whorls of the perunelomorph gastropods are curved to spiral from their beginning (Fig. 15a, b, i, j). In contrast to those of the Perunelomorpha, the cyrtoneritimorph protoconchs have a long straight beginning formed by uncurved, tubular embryonic shell and the first part of the larval shell (Fig. 14). The Euomphalomorpha also have an openly coiled first whorl, but they differ from both the Perunelomorpha and Cyrtoneritimorpha by their lack of larval shell (their protoconch consists of only embryonic shell; Fig. 13f). The perunelomorph protoconchs are smooth or bear fish-bone-like ornamentation (Fig. 15; Nützel and Cook 2002; Bandel 2002b). The teleoconch shapes of the perunelomorph gastropods vary from high-spired shells subulitimorph shells (Chuchlinidae Frýda and Bandel 1997; Imoglobidae Nützel et al. 2000) to globular and to low-spired shells (Perunelidae Frýda and Bandel 1997; Sphaerodomidae Bandel 2002b). The perunelomorph teleoconchs are smooth or ornamented with distinct colabral ribs (Fig. 15). There are no data on the shell microstructures of the perunelomorph gastropods.

Stratigraphic and geographic distributions: The relatively high diversity of the perunelomorph gastropods can be inferred from the common occurrences of their protoconchs in residues after the dissolution of Lower Palaeozoic rocks. They can be traced from the Early Ordovician to the Carboniferous (Frýda 1999b; Frýda and Rohr 2004; Nützel et al. 2006, 2007a). Even though the Perunelomorpha were common in Palaeozoic faunas, only a few taxa have been hitherto described based on their adult shells. They have been reported from Europe, North America, and Australia (Frýda and Bandel 1997; Frýda and Manda 1997; Frýda 1999b; Nützel et al. 2000; Nützel and Cook 2002; Bandel 2002b).

Classification: The Perunelomorpha consist of four families, Chuchlinidae, Perunelidae, Imoglobidae, and Sphaerodomidae, united in the superfamily Peruneloidea Frýda and Bandel, 1997 (Bouchet et al. 2005).

Caenogastropoda

Characteristic features: The Caenogastropoda constitutes the most diverse gastropod group living today, and includes about 60% of living gastropod species. The morphological diversity of their bodies and shells is extraordinary (Figs. 3 and 15). Caenogastropods can be found in almost every marine, many freshwater, and some terrestrial environments. Because of their exceptional morphological, ecological, physiological, and behavioural diversity, there are only a limited number of synapomorphies in this clade. They mainly include anatomical features like the presence of a single monopectinate gill, with skeletal rods or 1–2 marginal radular teeth (for a complete list, see Ponder and Lindberg 1997; Ponder et al. 2008). Apart from these features, the protoconch morphology is a good synapomorphy of the living and fossil caenogastropods. The caenogastropod protoconch consists of embryonic and larval shells and typically has several whorls (Fig. 4e, f). This feature

Fig. 15 Palaeozoic Perunelomorpha (**a–d**, **f**, **g**, **i–j**) and Caenogastropoda (**e**, **h**, **k–m**). (**a**, **b**) Silurian peruneloid larval shells; (**c**) Early Devonian *Havlicekiela parva*; (**d**) Early Devonian *Zenospira pragensis*; (**e**) Early Devonian *Prokopiconcha bisinuata*; (**f**) Early Devonian *Chuchlina minuta*; (**g**) apical view of *Zenospira pragensis* showing the perunelomorph protoconch; (**h**) Early Devonian *Pragoscutula wareni* with preserved larval shell; (**i**) apical view of *Havlicekiela parva* showing the perunelomorph protoconch; (**j**) apical view of Carboniferous *Imogloba* sp.; (**k**, **l**) two species of the Middle Devonian genus *Macrochilina*; (**m**) larval shell of Early Devonian *Balbiniconcha cerinka*

distinguishes the caenogastropods from the patellogastropods, archaeogastropods, and loxonematoidean gastropods (Fig. 16c, d, f), which never developed larval shells. Members of the loxonematoidean superfamily are considered to form the ancestral group of the caenogastropods and/or the heterobranch gastropods (e.g. Knight et al. 1960).

The caenogastropod protoconchs are dextrally or sinistrally coiled, but their coiling is always in the same direction as that of their teleoconchs (they are orthostrophic). This feature distinguishes the caenogastropods from the heterobranch gastropods (Fig. 4d). The protoconch of the Neritimorpha is also orthostrophic but it is strongly convoluted (Fig. 4c). The larval shell of the caenogastropods is often highly ornamented (Fig. 4e, f). Many studies (e.g. Bandel 2002b; Nützel 1998; Riedel 2000) have shown that the caenogastropod protoconchs display less evolutionary change than their teleoconchs and therefore they are often a rich source of characters for phylogenetic analysis and provide apomorphies for families or even higher taxa.

In contrast to those of the Patellogastropoda and Archaeogastropoda, the composition and microstructure of the caenogastropod shells are simple (Bandel 1990). The caenogastropod shells are mainly aragonitic with a crossed-lamellar shell microstructure. The occurrence of calcite in their shells

Fig. 16 Palaeozoic Heterobranchia (**a**, **b**, **e**, **g**) and Loxonematoidea (**c**–**d**, **f**). Members of the latter superfamily are considered to form the ancestral group of caenogastropods and/or heterobranch gastropods in some phylogenetic models. (**a**, **b**) Early Devonian *Kuskokwimia moorei*; (**c**) Early Devonian palaeozygopleurid *Palaeozygopleura bohemica*; (**d**) Early Devonian loxonematid *Pragozyga costata*; (**e**) heterostrophic larval shell of Carboniferous streptacidoidean; (**f**) protoconch of Early Devonian loxonematoidean *Bohemozyga kettneri*; (**g**) Middle Devonian *Heteroloxonema moniliforme*

is relatively rare. Caenogastropods share their simple composition and crossed-lamellar microstructure with the heterobranch gastropods.

Stratigraphic and geographic distributions: The oldest caenogastropods diagnosed by their protoconch morphologies come from Lower Devonian strata (Frýda 1999b, 2001; Cook et al. 2008). However, their fossil record may be much longer if it is inferred from their teleoconchs (Frýda and Rohr 2004; Frýda et al. 2008a and references therein). Since the Late Palaeozoic, the caenogastropods have formed an important or dominant part of marine gastropod faunas. Their rapid radiation (mainly Neogastropoda and other predatory gastropods) after the Cretaceous–Palaeocene boundary crisis led to their strong dominance and high diversity in almost all gastropod communities (see Sepkoski and Hulver 1985; Taylor et al. 1980; Vermeij 1977, 1987, for details).

Living marine caenogastropods are most diverse in tropical and warm temperate environments, but their habitats range from intertidal to deep-sea (including cold seeps and hydrothermal vents), freshwater, and some terrestrial environments (see details in Ponder et al. 2008).

Classification: Living caenogastropods include many highly diversified groups with long fossil records. However, about one third of the approximately 190 families are extinct (Bouchet et al. 2005). Their phylogenetic relationships have often remained unresolved. The most recent classification of the Caenogastropoda can be found in Bouchet et al. (2005; but see also Ponder and Warén 1988; Ponder and Lindberg 1997; Riedel 2000; Colgan et al. 2007; Ponder et al. 2008).

Heterobranchia

Characteristic features: The morphological diversity of the heterobranch bodies and shells is as large as that of the caenogastropods. They live in almost all marine and terrestrial environments (the great majority of land gastropods are heterobranchs). The total number of living heterobranchs is estimated to be over 30,000, including mainly terrestrial species. The monophyly of the Heterobranchia is well supported by many anatomical characters (see the list in Ponder and Lindberg 1997) and by their protoconch morphology. The protoconchs of marine heterobranchs are coiled in a different direction from that of their teleoconchs (i.e. they are heterostrophic; Fig. 4d) and consist of embryonic and larval shells (but also see Page 1995). The latter feature allows their recognition in the fossil record. The heterobranch shell is mainly aragonitic with a crossed-lamellar microstructure. Several large groups of marine and terrestrial heterobranch gastropods have lost their shells.

Stratigraphic and geographic distributions: The protoconch morphology seems to be the only good feature with which to diagnose the heterobranch affinities to extinct heterobranchs (Bandel 2002a). The oldest record, based on their protoconch morphology, comes from Lower Devonian strata (Fig. 16a, b; Frýda and Blodgett 2001, 2004). There are only two other records from the Middle Devonian (Fig. 16g; Frýda 2000; Bandel and Heidelberger 2002). During the Late Palaeozoic, the diversity of the heterobranchs increased and since the Mesozoic, they have been present in almost all environments, as are living heterobranchs (Bandel 2002a). Their geographic ranges include almost all oceanic and terrestrial environments on all continents.

Classification: The individual groups of heterobranch gastropods are well characterized by unique features, which make their diagnosis relatively simple. However, the relationships between those groups are often uncertain because of the very high diversity of their bodies. This fact has caused discrepancies in their internal classification. Details of their classification can be found in Bouchet et al. (2005), Wägele et al. (2008), and Mordan and Wade (2008).

Comparison of Phylogenies Inferred from Palaeontological and Neontological Data

During the last two decades, opinions on the phylogenetic relationships of the living gastropod clades have converged onto the following topology: (Patellogastropoda + (Archaeogastropoda + (Neritimorpha + (Caenogastropoda + Heterobranchia)))). This phylogenetic model is generally accepted and the main recent efforts of zoologists (see also Lindberg and Guralnick 2003; Page 2006; Ponder and Lindberg 2008) have focused on the classification of species to superfamily-level taxa. However, for a correct understanding of the gastropod phylogeny, we must include in our model not only living taxa but also extinct taxa, which are commonly omitted from our consideration (mainly by molecular biologists). It is noteworthy that the vast majority of gastropods are extinct, so the absence of those taxa from the phylogenetic model is unacceptable. In the following section, the gastropod phylogeny inferred from the pattern of ontogeny of the Palaeozoic gastropods is compared with the phylogenies based on data from living gastropods.

Mimospirina – Gastropods or Something Else?

The Cambrian–Devonian Mimospirina have been interpreted as Archaeogastropoda (Dzik 1983), a possible ancestral group of the Heterobranchia (Ponder and Lindberg 1997), or as a molluscan clade that does not belong to the class Gastropoda (Linsley and Kier 1984; Wagner 1999). It is easier to define the Mimospirina negatively (what they are not) than positively (what they are). The morphology of the multi-whorled protoconch of Mimospirina, which consists of both embryonic and larval shell, testifies against the interpretation of this clade as belonging to the Archaeogastropoda (Frýda 1999b). Ponder and Lindberg (1997, p. 203) speculated that the heterobranch ancestors may have arisen from a lineage with dextrally hyperstrophic (i.e. sinistrally coiled) shells that were significantly spired. Those authors suggested that the macluritoideans or sinistrally hyperstrophic onychochilids (like *Mimospira*) are examples of a possible ancestor. However, the multi-whorled protoconchs of the Mimospirina are sinistrally coiled, as are their teleoconchs, and no operculum has been found in these molluscs. Therefore, there is no evidence that the Mimospirina were 'sinistrally hyperstrophic' (Frýda 1999b; Frýda and Rohr 1999, 2004; Frýda and Blodgett 2001). On the basis of a functional analysis, Linsley and Kier (1984) proposed uniting the Onychochiloidea (= Mimospirina), Macluritoidea, and possibly the Euomphaloidea in a new order Hyperstrophina of the new class Paragastropoda. The class Paragastropoda is considered to represent untorted molluscs. However, the very different protoconch morphologies of Mimospirina (Dzik 1983; Frýda 1999b), Euomphaloidea (Yoo 1994; Bandel and Frýda 1998; Nützel 2002; Frýda et al. 2006a), and Macluritoidea (Frýda and Rohr 2006) indicate that the class Paragastropoda is a paraphyletic group (Frýda 1999b).

However, it is much more difficult to solve the question: are the Mimospirina gastropods or not? As pointed out by Linsley and Kier (1984), the teleoconch morphology of the Mimospirina differs in some characteristics from that of most gastropods. This has also been demonstrated by Wagner (1999), who linked Cambrian mimospirines to pelagielloids based on a phylogenetic analysis of their teleoconch characters, and found that this clade was only distantly related to gastropods.

The age of the oldest member of the Mimospirina also seems to support a phylogenetic position outside the class Gastropoda. However, the tubular embryonic shell, which is considered one of the few apomorphies of the class Gastropoda (Ponder and Lindberg 1997), is well developed in Mimospirina. It is not possible to test the torsion of the soft bodies of these molluscans. Our present level of knowledge allows us to interpret the Mimospirina as a basal gastropod group or as a sister group to this class (Fig. 17).

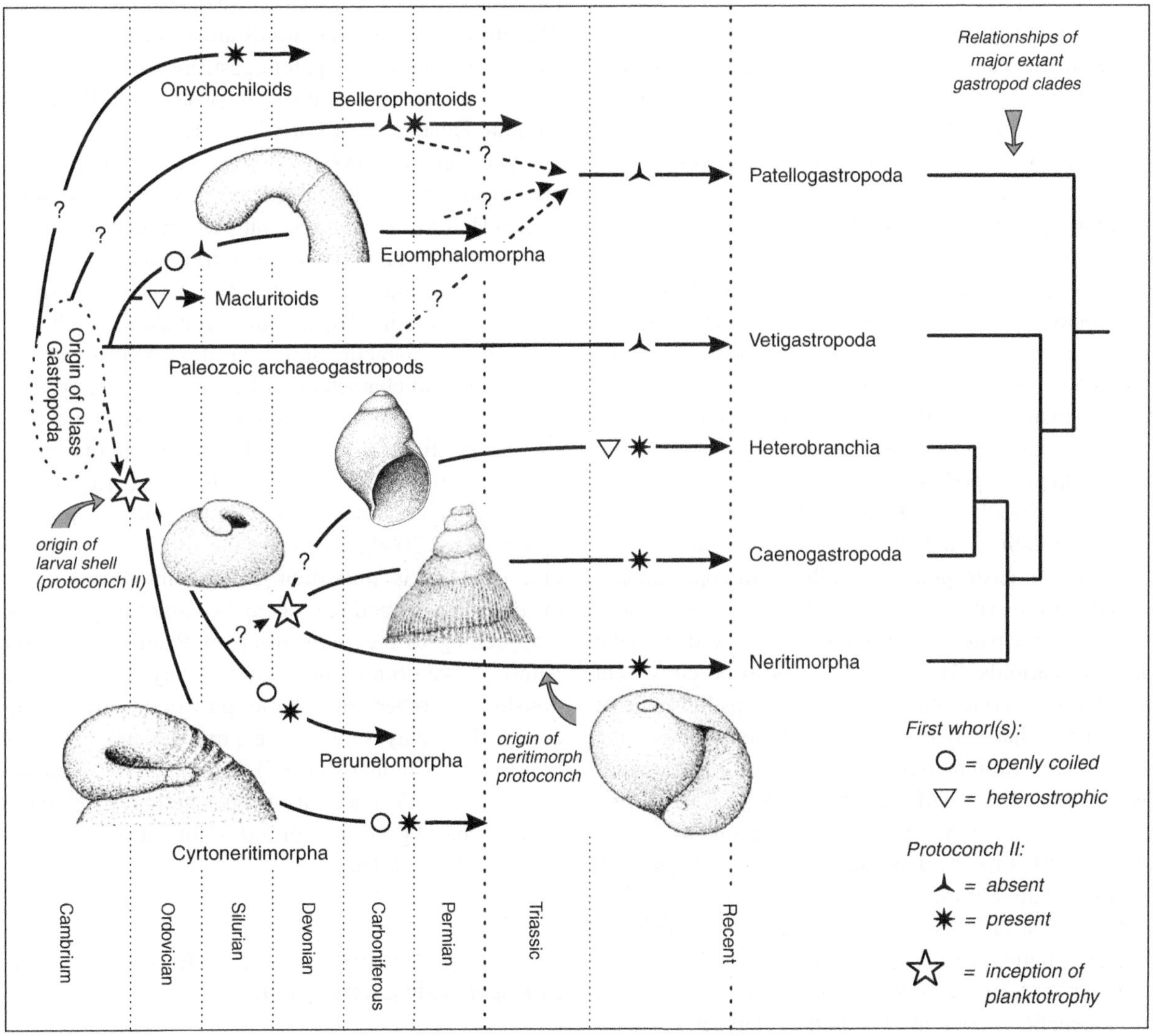

Fig. 17 Comparison of gastropod phylogenies inferred from palaeontologic and neontologic data. *Left side* of the diagram shows hypothetical phylogenies of Palaeozoic gastropods derived mainly from patterns of their early shell ontogenies (see text for details). Relationships of the major extant gastropod clades inferred from anatomical, morphologic, and molecular data are shown on the *right*

Patellogastropoda – First Gastropod Offshoot or Derived Group?

The Patellogastropoda have been considered to represent a sister group to all other living gastropods (Golikov and Starobogatov 1975; Ponder and Lindberg 1997; Aktipis et al. 2008 and references therein). The only exceptions are several molecular studies that place them in a derived clade with some Vetigastropoda (e.g. Colgan et al. 2003). However, the prevailing opinion that the patellogastropods form the first offshoot of the gastropods is not supported by palaeontological data. The oldest indisputable evidence of the Patellogastropoda is from Late Triassic strata (Fig. 17; Hedegaard et al. 1997). Therefore, there is no reliable fossil record for the first half of their presumed evolution as an independent gastropod clade (i.e. data are missing for about 250 Ma). This controversy has been explained by either their adaptation to intertidal environments, which produce a relatively poor fossil record, or by the opinion that the Palaeozoic patellogastropods had coiled shells and have not yet been recognized among the Palaeozoic gastropods. The nature of the shells of the early patellogastropods is still a matter of debate. Haszprunar (1988a) hypothesized that the limpet shell is a plesiomorphy retained by the patellogastropods. However, Ponder and Lindberg (1997; also Lindberg and Ponder 2001) argued that this was a parallelism and that the last common ancestor of the extant gastropods was coiled. This opinion had already been expressed by Knight (1952), who suggested that the patellogastropods were derived from the pleurotomarioids. This model seems to be consistent with the protoconch nature of both groups (absence of true larval shell). If this model is accepted, then the very complex microstructure (Fig. 6e, f) of the patellogastropod shells (MacClintock 1968; Bandel and Geldmacher 1996; Sasaki 1998; Fuchigami and Sasaki 2005) must have evolved secondarily, perhaps with their adaptation to intertidal environments.

Ponder and Lindberg (1997) expressed the opinion that the patellogastropods evolved from the Palaeozoic platyceratoidean gastropods (Figs. 9 and 14i–k). The phylogenetic model in which the platyceratoideans represent the possible coiled ancestors of the patellogastropods is supported only by their common regularly foliated shell microstructure. However, there is some doubt that this microstructure is really identical in both groups (see Bandel and Geldmacher 1996). Recent descriptions of the protoconch morphologies of the Paleozoic platyceratids (Frýda et al. 2009) also argue against this phylogenetic model. The patellogastropods share the same pattern of early shell ontogeny with archaeogastropods like Vetigastropoda (Bandel 1982; Sasaki 1998) insofar as their protoconchs develop without a true larval shell (protoconch II). However, Frýda et al. (2009) documented that the platyceratids form a diphyletic group, uniting taxa with openly coiled cyrtoneritimorph protoconchs and taxa with tightly coiled protoconchs (Fig. 14), but both groups developed true larval shells (protoconch II). Therefore, if the platyceratoids are a stem group of the patellogastropods, the patellogastropods lost the ability to develop true larval shells (protoconch II) during the Late Palaeozoic or Early Triassic.

Sutton et al. (2006) reported an unusually well-preserved gastropod from the lower Silurian (Wenlock series) Herefordshire Lagerstätte of England, with fossilized soft tissues. This gastropod was assigned to the Platyceratidae (*Platyceras*? sp.) on the basis of its simple sub-circular aperture, trochiform shell, and absence of ornament. Sutton et al. (2006) noted that the anatomical characters of their *Platyceras*? sp. suggest a patellogastropod affinity rather than a neritopsine [= neritimorph] affinity. However, the placement of the gastropod from the Herefordshire Lagerstätte in the Platyceratidea is not sufficiently supported (Frýda et al. 2009). However, the discovery made by Sutton et al. is important for the interpretation of the patellogastropod phylogeny because this gastropod may belong to the stem group of Patellogastropoda. Although we do not understand its phylogenetic relationship to other Palaeozoic gastropod groups, its coiled shell may support the phylogenetic model of Ponder and Lindberg (1997) (Fig. 17). In summary, the phylogenetic position of the Patellogastropoda as the most basal gastropod group is not yet supported by palaeontological data.

Archaeogastropoda – The Oldest Indisputable Gastropods

As mentioned above, members of the taxon Vetigastropoda cannot be recognized in the fossil record because they are defined by their soft body characters. Only two characters could potentially be used as diagnostic for fossil Vetigastropoda: the presence of nacre in their shells and their protoconch morphology. However, the presence of a nacreous layer cannot be used as a diagnostic character for all fossil vetigastropods because it is absent in some living vetigastropods (see above). Furthermore, the vetigastropod protoconch morphology (see above) occurs in many extinct Palaeozoic gastropod groups that definitely do not constitute the vetigastropod crown group (e.g. extinct Porcellioidea; Fig. 7). If we define the taxon Archaeogastropoda by the archaeogastropod-type protoconch (as suggested by Bandel 1982, 1997), we can separate this larger clade (including all Vetigastropoda) from the rest of the Palaeozoic gastropods.

The Archaeogastropoda can be traced from the beginning of the Ordovician to the Recent and unites the oldest indisputable gastropods (Frýda and Rohr 2004; Wagner 1999; Frýda et al. 2008a). The morphological variability and biodiversity of these gastropods reached their maxima in the Palaeozoic (Wenz 1938–1944; Knight et al. 1960). The relationships between them and many living archaeogastropod groups (e.g. Kunze et al. 2008) is unfortunately only poorly known.

Living vetigastropods mainly represent groups originating after the Permian–Triassic crisis (Geiger et al. 2008). However, some of them may represent relicts of Palaeozoic groups. Efforts to understand the phylogeny of living vetigastropods without rich fossil material necessarily produce phylogenetic trees with unstable topologies (compare Geiger and Thacker 2005; Colgan et al. 2003; Lindberg et al. 2004). Future efforts to clarify the archaeogastropod phylogeny should include not only living taxa but also fossil taxa.

Euomphalomorpha – Extinct Basal Group or Sister Group of Archaeogastropoda

Simple protoconchs formed only of embryonic shell, as in the archaeogastropods, are also documented in the euomphalomorph gastropods (Figs. 13 and 17; Bandel and Frýda 1998; Nützel 2002; Frýda et al. 2006a). The euomphalomorph protoconch is not deformed during late embryonic stages to a spiral coil, as is typical in fossil and living archaeogastropods (see Bandel 1982 for details). Frýda (1999c, 2001) demonstrated that the morphology of the archaeogastropod protoconch has not changed since the Devonian. The absence of protoconch II (true larval shell) in both the archaeogastropods and euomphalomorph gastropods suggests that the extinct Palaeozoic Euomphalomorpha may represent a sister group of the Archaeogastropoda (Fig. 17; see also the model of Wagner 1999).

Palaeozoic or Mesozoic Origin of the Neritimorph Crown Group?

Recent phylogenetic analyses based on morphological, anatomical, and molecular data have placed the Neritimorpha as the sister group of the Apogastropoda (Fig. 17; clade including Caenogastropoda and Heterobranchia). Living marine neritimorph gastropods, with a planktotrophic developmental stage, form a unique protoconch that is strongly convoluted (Fig. 4c; Bandel 1982). Bandel (1982) studied the ontogeny and protoconch morphology of extant neritimorphs in detail and concluded that they represent a separate evolutionary lineage, distinct from those of other archaeogastropods. This conclusion was later supported by anatomical and molecular data (e.g. Haszprunar 1988a; Healy 1988; Sasaki 1998; Ponder and Lindberg 1997; Kano et al. 2002). Ponder and Lindberg (1997), like Haszprunar (1988a), considered the Neritimorpha to represent an 'early archaeogastropod offshoot', so according to their interpretation, the neritimorphs form a sister group to all gastropods except the Patellogastropoda. This model implies that planktotrophy originated twice in the evolution of the Gastropoda (Haszprunar 1995): in the Neritimorpha and in the Apogastropoda (uniting the Caenogastropoda and Heterobranchia). However, new molecular studies support the earlier published position of the Vetigastropoda as a sister group to the Neritimorpha, Caenogastropoda, and Heterobranchia, and the monophyly of the Neritimorpha (Colgan et al. 2003; McArthur and Harasewych 2003; Kano et al. 2002; Aktipis et al. 2008). The typical neritimorph protoconch has been known since the Triassic (Fig. 17; Bandel and Frýda 1999; Bandel 2007) and is a good diagnostic character for the neritimorph crown group (however, the protoconch of the terrestrial, freshwater, and marine neritimorphs with lecitotrophic development is secondarily simplified). Bandel (2007) showed that several modern neritimorph groups can be traced from the Late Triassic (Fig. 14) to the Recent.

However, the Palaeozoic history of the neritimorph gastropods is puzzling. There is not a single piece of evidence for the occurrence of typical neritimorph protoconchs in Palaeozoic strata. Hypotheses on the possible ancestral group of the post-Palaeozoic neritimorphs have been inferred only from the similarities between the teleoconchs of some Palaeozoic gastropods and those of modern neritimorphs. Palaeozoic naticopsid and platyceratid gastropods have been most frequently suggested as the ancestral groups of all modern neritimorphs (Wenz 1938–1944, Knight et al. 1960; Bandel 1992). However, recent studies of the protoconch morphologies of the latter groups have revealed a much more complex evolutionary pattern. Naticopsid protoconchs are simple, smooth, and orthostrophic (Yoo 1994; Nützel and Mapes 2001; Bandel 2002b; Nützel et al. 2007b). The same protoconch type has also been documented in some Silurian and Devonian platyceratids (Frýda et al. 2009). The Palaeozoic naticopsid and platyceratid gastropods may represent the ancestral groups of the neritimorphs, as has already been suggested, but this model presumes that the typical strongly convoluted neritimorph protoconch evolved from a simple orthostrophic protoconch. Palaeontological data thus suggest that the typical neritimorph protoconch probably originated after the Permian–Triassic mass extinction event, but before the Late Triassic (Fig. 17). If the latter model is correct, the oldest neritimorphs appeared much earlier (Frýda and Rohr 2004), but later than the archaeogastropods, which is consistent with neontological models. Nevertheless, the oldest common ancestor of the living neritimorphs (the neritimorph crown group) was probably of post-Palaeozoic (Triassic) age.

Caenogastropoda – A Sister Group of Heterobranchia or Neritimorpha

Neontological models of gastropod phylogeny place the Caenogastropoda and Heterobranchia in the clade Apogastropoda (Ponder and Lindberg 1997). Both groups are well documented since the Early Devonian based on their protoconch morphology (Frýda 2001; Frýda and Blodgett 2001, 2004; Cook et al. 2008). Since the Late Palaeozoic, the caenogastropods have included several clades and the protoconch morphology of many tens of species is well documented (see references in Nützel 1998; Bandel 2002b; Frýda et al. 2008a; Ponder et al. 2008). Members of the Zygopleuroidea, Cerithioidea, Soleniscidae, and Meekospiridae form the most diverse groups of Late Palaeozoic caenogastropods. The Late Palaeozoic caenogastropods also include the non-marine families Anthracopupidae, Dendropupidae, and Palaeocyclophoridae (Bandel 2002b). However, the caenogastropod-type protoconch has been documented in only two Devonian gastropod groups, in the subulitiform gastropods and in the genus *Pragoscutula*, which has a limpet-like shell (Fig. 15). The phylogenetic position of the latter genus is still poorly understood (Frýda 2001; Cook et al. 2008). In contrast, the Palaeozoic subulitiform gastropods are traditionally placed in the superfamily Subulitoidea, and represent a highly variable group known from the Ordovician to the Permian (Knight et al. 1960). However, Late Palaeozoic and Devonian subulitiform gastropods share only superficial similarities to the Ordovician *Subulites* and are probably not directly related (Wagner 1999; Nützel et al. 2000; Frýda 2001). The oldest caenogastropod protoconchs are found in the Devonian subulitiform gastropods (e.g. *Prokopiconcha* and *Balbiniconcha*; Fig. 15), which are closely related to the late Palaeozoic 'caenogastropod-like subulitids' (Frýda 2001; Nützel et al. 2000). As mentioned above, the simple, smooth, orthostrophic, multi-whorled protoconchs are found not only in the Devonian subulitiform gastropods but also in presumed Palaeozoic neritimorphs (Nützel et al. 2007b; Frýda 2001; Frýda et al. 2009). The teleoconchs of some Devonian 'subulitiform gastropods' (e.g. *Decorochilina* and *Macrochilus*; Fig. 15) resemble the teleoconchs of the presumed Late Palaeozoic neritimorphs. Thus, there is no difference between the protoconch morphologies of the oldest caenogastropods and those of the presumed neritimorphs. The palaeontological data suggest that caenogastropods are a sister group of the neritimorphs rather than of the heterobranch gastropods (Fig. 17). In this context, it should be interesting to test the hypothesis of Page (1995) that the entire larval shell of the heterobranchs is homologous to the larval shell of the 'archaeogastropods' and protoconch I of the caenogastropods. If her hypothesis is correct, then this fact could support a caenogastropod + neritimorph clade.

Perunelomorpha – Extinct Basal or Sister Group of Caenogastropoda

Among the Devonian gastropods with subulitiform teleoconchs are two groups with different protoconch morphologies. The first unites taxa with simple, orthostrophic, multi-whorled, tightly coiled protoconchs, as is typical for post-Palaeozoic caenogastropods, and these gastropods have been interpreted as the oldest caenogastropods (Frýda 2001; Frýda and Blodgett 2004). However, the second group of Devonian subulitoidean gastropods developed true larval shells, with an openly coiled first whorl (Fig. 15; Frýda and Bandel 1997; Frýda and Manda 1997; Frýda 1999b). Gastropods with this type of protoconch morphology are united in the Peruneloidea and have been documented from the Ordovician to the Carboniferous (Nützel et al. 2000; Frýda and Rohr 2004; Nützel and Cook 2002; Frýda et al. 2008a; Bouchet et al. 2005). Frýda (1998a, 1999b, 2001) suggested that the Ordovician–Carboniferous Peruneloidea (Perunelomorpha) may represent the basal caenogastropods or their sister group. If the model in which the caenogastropods and neritimorphs are sister groups is correct, then the perunelomorph gastropods may also be the ancestral or stem group of the neritimorphs. However, the oldest members of the caenogastropod crown group are probably not older than the Early Carboniferous or Devonian (Fig. 17).

Heterobranchia – A Group of Unknown Origin

The only character that can be used to recognize the Heterobranchia in the fossil record is the morphology of their protoconch, which coils in a different direction from that of the teleoconch (Fig. 4d). Like the Caenogastropoda, the Heterobranchia can also be traced by their typical protoconch morphology to the Early Devonian (Fig. 16; Frýda and Blodgett 2001, 2004; Bandel and Heidelberger 2002). The heterobranch gastropods have been fairly diverse and abundant since the Early Carboniferous. Palaeozoic heterobranchs include the Streptacidoidea (Allogastropoda), Stuoraxidae (Architectonicoidea), and possible Devonian stem taxa, such as *Palaeocarboninia* and *Kuskokwimia* (Frýda and Blodgett 2001; Bandel 2002a; Bandel and Heidelberger 2002; Nützel 2002). The relationship between the Palaeozoic heterobranchs (Fig. 17) and the other Palaeozoic gastropods is unclear. Similarly, the Palaeozoic heterobranch lineages (if any) that belong to the heterobranch crown group are also unknown. The oldest indisputable members of the heterobranch crown group are documented from Triassic strata (Bandel 2002a).

Cyrtoneritimorpha – Extinct Group of Palaeozoic Gastropods

The Cyrtoneritimorpha comprises gastropods with openly coiled, fish-hook-like protoconchs consisting of embryonic and larval shells, with a long straight beginning formed by uncurved, tubular, embryonic shell and the first part of the larval shell (Fig. 14). This group can be traced from the Ordovician to the Permian (about 250 Ma), with no distinctive changes in the protoconch morphology (Fig. 17). The stratigraphic distribution of the cyrtoneritimorph gastropods is known mainly from the occurrence of their characteristic protoconchs and only a few adult shells have hitherto been found with preserved cyrtoneritimorph protoconchs (see Frýda et al. 2009 for review). Therefore, the morphological diversity of the cyrtoneritimorphs is still poorly understood. The first cyrtoneritimorph protoconchs found were associated with neritimorph-like shells (Knight 1934; Frýda and Manda 1997; Bandel and Frýda 1999). For this reason, the Cyrtoneritimorpha were interpreted as the possible ancestral group of the living neritimorph gastropods (Frýda 1998a, b, 1999b; Bandel and Frýda 1999). Additional discoveries of cyrtoneritimorph teleoconchs and well-preserved protoconchs of Palaeozoic platyceratid and naticopsid gastropods (Frýda and Heidelberger 2003; Nützel et al. 2007b; Frýda et al. 2009) led to another interpretation of the phylogenetic position of the Cyrtoneritimorpha as an independent extinct clade of Palaeozoic gastropods (Frýda and Heidelberger 2003; Frýda et al. 2008b).

Comparison of the Phylogenies Inferred from Neontological and Palaeontological Data

Palaeontological data support a basal position for the Vetigastropoda (Archaeogastropoda), even though the oldest members of the vetigastropod crown group probably come from Mesozoic strata. The archaeogastropods were the dominant group of the Palaeozoic gastropods and unite many extinct taxa and taxa with hitherto unknown relationships to the living archaeogastropods (Fig. 17). The ancestors of the clade Patellogastropoda should be sought amongst these gastropods. The majority of phylogenetic analyses based on living taxa have produced topologies that suggest that the Patellogastropoda represents the most basal gastropod clade. This conclusion is not yet supported by palaeontological data. The oldest indisputable patellogastropods are known from the Mesozoic (Late Triassic). The lack of patellogastropods with limpet-like shells in Palaeozoic strata (therefore, for more than 200 Ma) supports the hypothesis that they had Palaeozoic ancestors with coiled shells or suggests that they were not the first gastropod offshoot (Fig. 17). The type of nacre common to the Cephalopoda and the majority of vetigastropods (see Fig. 2) also supports the basal position of the Archaeogastropoda and a more derived position for the patellogastropods.

The present data do not refute the hypothesis that some living hot-vent taxa and some vetigastropod groups may be the descendants of some Palaeozoic groups. If this is so, then the oldest members of the vetigastropod crown group may come from the Palaeozoic. Extinct Ordovician Macluritoidea and Ordovician–Permian Euomphalomorpha probably represent sister or basal clades of the Archaeogastropoda (Fig. 17).

The palaeontological data support a close relationship between the Caenogastropoda and Neritimorpha and do not refute their close relationship to the Heterobranchia (Fig. 17), as has been suggested by recent neontologist models. The protoconch morphology of the Palaeozoic caenogastropods and presumed neritimorphs supports their interpretation as sister groups. However, the origin of the Heterobranchia, the oldest members of which come from Early Devonian strata like those of the caenogastropods, is not clear. It is equally probable that they originated from some group of the caenogastropod–neritimorph clade or from another Early Palaeozoic gastropod group. Extinct Perunelomorpha, ranging from the Ordovician to the Carboniferous, probably represent a basal or sister group of the caenogastropod–neritimorph clade.

The phylogenetic positions of several extinct Palaeozoic groups, including the bellerophontiform molluscs, mimospirids, and cyrtoneritimorphs, are uncertain. The bellerophontiform molluscs probably constitute polyphyletic groups, as has been suggested by a phylogenetic analysis of their teleoconch morphology and the differences in their early shell ontogeny. The Cambrian–Devonian Mimospirina may not even belong to the class Gastropoda.

The development of protoconch II (true larval shell), which is known in living Neritimorpha, Caenogastropoda, and Heterobranchia, has been interpreted by zoologists as a feature associated with the origin of planktotrophy. Analyses of the protoconch morphologies of the Palaeozoic gastropods suggest a more complex evolutionary pattern. The protoconchs formed by both embryonic and true larval shells have been documented in the Palaeozoic Mimospirina, Bellerophontoidea, Cyrtoneritimorpha, Perunelomorpha, Caenogastropoda, Neritimorpha, and Heterobranchia (Fig. 17). The development of the larval shell in these groups is evidence for a prolongation of the pre-metamorphic ontogeny, which was probably associated with an ability to use external food sources. However, this strategy was probably not linked to the use of plankton as food in all these groups. The strategy of using an external food source before metamorphosis probably originated close to the Cambrian–Ordovician boundary. The origin of planktotrophy (i.e. the use of plankton as a food source) is probably much younger (Fig. 17) and associated with the Silurian–Devonian macroevolutionary trend in protoconch morphology (continuous disappearance of protoconchs with a first openly coiled whorl and the appearance of modern protoconch morphologies in different lineages), which occurred during the so-called Palaeozoic plankton revolution.

Conclusions

The analysis of data on the ontogeny of the Palaeozoic gastropods has given some support to the phylogenies inferred from the anatomical, morphological, and molecular data of living gastropods. However, this analysis has also raised several new questions (see above). Therefore, the collection of new data on the early shell ontogenies of the Palaeozoic gastropods has considerably increased our understanding of their phylogenetic positions and has contributed much new information to the phylogeny of the living gastropod clades. The first results of the analysis of the microstructures and crystallographic textures of fossil molluscan shells have clearly shown the great potential of these data for the phylogenetic analysis of the major clades.

In summary, studies of the Palaeozoic fossil record of the class Gastropoda, based on new approaches (studies of the early ontogeny, microstructures, and crystallographic textures of gastropod shells, etc.) not only will resolve long-standing questions regarding gastropod evolution but may also produce new data with which to test the phylogenetic models of zoologists. They may also reveal the existence of large, highly diversified, and long-lasting gastropod clades that are now extinct. An understanding of the phylogeny of the clade Gastropoda is impossible without the inclusion of those gastropod clades.

Acknowledgements This work was supported by grants from the Grant Agency of the Czech Republic (210/12/2018) and the Czech–American Cooperation Programme (Kontakt ME08011). The paper benefited greatly from the constructive review and linguistic help of Prof. John Talent (Australia) and from technical help with illustrations by Sangeetha Sathiamurthy (India). I thank both of them very much.

References

Akpan EB, Farrow GE, Morris N (1982) Limpet grazing on Cretaceous algal-bored ammonites. Palaeontology 25: 361–367

Aktipis SW, Giribet G, Lindberg DR, Ponder WF (2008) Gastropoda – an overview and analysis. In: Ponder W, Lindberg DL (eds) Phylogeny and evolution of the Mollusca. University of California Berkeley, Los Angeles, CA, pp 201–237

Bandel K (1982) Morphologie und Bildung der frühontogenetischen Gehäuse bei conchiferen Mollusken. Facies 7:1–198

Bandel K (1990) Shell structure of the Gastropoda excluding Archaeogastropoda. In: Carter JG (ed) Skeletal biomineralization: patterns, processes and evolutionary trends, vol 1. Van Nostrand Reinhold, New York, pp 117–134

Bandel K (1992) Platyceratidae from the Triassic St. Cassian Formation and the evolutionary history of the Neritomorpha (Gastropoda). Palaeont Zeit 66(3–4):231–240

Bandel K (1993) Evolutionary history of sinistral archaeogastropods with and without slit (Cirroidea, Vetigastropoda). Freiberger Forschungshefte C450:41–81

Bandel K (1997) Higher classification and pattern of evolution of the Gastropoda. Cour Forsch Inst Senckenberg 201: 57–81

Bandel K (2002a) About the Heterostropha (Gastropoda) from the Carboniferous and Permian. Mitt Geol-Paläontol Institut Univ Hamburg 86:45–80

Bandel K (2002b) Reevaluation and classification of Carboniferous and Permian Gastropoda belonging to the Caenogastropoda and their relation. Mitt Geol-Paläont Inst Univ Hamburg 86:81–188

Bandel K (2007) Description and classification of late Triassic Neritimorpha (Gastropoda, Mollusca) from St. Cassian Formation, Italian Alps. Bull Geosci 82(3):215–274. doi:10.3140/bull.geosci.2007.03.215

Bandel K (2008) Operculum shape and construction of some fossil Neritimorpha (Gastropoda) compared to those of modern speces of the subclass. Vita Malacol 7:19–36

Bandel K, Frýda J (1996) *Balbinipleura*, a new slit bearing archaeogastropod (Vetigastropoda) from the early Devonian of Bohemia and the early Carboniferous of Belgium. N Jb Geol Paläont Monatshefte 6:325–344

Bandel K, Frýda J (1998) Position of Euomphalidae in the system of the Gastropoda. Senckenberg Lethaea 78:103–131

Bandel K, Frýda J (1999) Notes on the evolution and higher classification of the subclass Neritimorpha (Gastropoda) with the description of some new taxa. Geol Palaeont 33:219–235

Bandel K, Geldmacher W (1996) The structure of the shell of *Patella crenata* connected with suggestions to the classification and evolution of the Archaeogastropoda. Freiberg Forsch C 464:1–71

Bandel K, Heidelberger D (2001) The new family Nerrhenidae (Neritimorpha, Gastropoda) from the Givetian of Germany. N Jb Geol Paläont Mh 12:705–718

Bandel K, Heidelberger D (2002) A Devonian member of the subclass Heterostropha (Gastropoda) with valvatoid shell shape. N Jb Geol Paläont, Mh 2002:533–550

Batten RL (1984) The calcitic wall in the Paleozoic families Euomphalidae and Platyceratidae (Archeogastropoda). J Paleontol 58(5):1186–1192

Bengtson S, Conway Morris S (1992) Early radiation of biomineralizing phyla. In: Lipps JH, Signor PW (eds) Origin and early evolution of the metazoa. Plenum Press, New York, pp 447–481

Bieler R (1992) Gastropod phylogeny and systematics. Annu Rev Ecol Syst 23:311–338

Bieler R (2006) Bivalvia – a look at the branches. Zool J Linn Soc 148:221–552

Bieler R, Carter JG, Coan EV (2010) Classification of bivalve families, 113–133. In: Bouchet P, Rocroi J-P (eds) Nomenclator of Bivalve Families. Malacologia 52(2):1–184

Boore JL, Medina M, Rosenberg LA (2004) Complete sequences of the highly rearranged molluscan mitochondrial genomes of the scaphopod *Graptacme eborea* and the bivalve *Mytilus edulis*. Mol Biol Evol 21:1492–1503

Bouchet P, Rocroi JP, Frýda J, Hausdorf B, Ponder W, Valdes A, Warén A (2005) Classification and nomenclator of gastropod families. Malacologia 47:1–368

Carter JG, Clark GR (1985) Classification and phylogenetic significance of Molluscan shell microstructure. In: Broadbent TW (ed) Mollusks: notes for a short course. University of Tennessee, Department of Geological Sciences, Studies in Geology, vol 13, pp 50–71

Carter JG, Hall RM (1990) Part 3. Polyplacophora, Scaphopoda, Aracheogastropoda and Paragastropoda (Mollusca). In:

Carter JG (ed) Skeletal biomineralisation: patterns, processes and evolutionary trends. II Atlas and Index. Van Nostrand Reinhold, New York, pp 29–51, pls 122–134

Chaffee C, Lindberg DR (1986) Larval biology of early Cambrian molluscs: the implications of small body size. Bull Mar Sci 39:536–549

Cherns L, Rohr DM, Frýda J (2004). Polyplacophoran and symmetrical univalve mollusks. In: Webby BD, Paris F, Droser ML, Percival IG (eds) The great Ordovician biodiversification event. Columbia University Press, New York, pp 179–183

Colgan DJ, Ponder WF, Beacham E, Macarana JM (2003) Gastropod phylogeny based on six segments from four genes representing coding or non-coding and mitochondrial or nuclear DNA. Molluscan Res 23(2):123–148

Colgan DJ, Ponder WF, Beacham E, Macaranas J (2007) Molecular phylogenetics of Caenogastropoda (Gastropoda: Mollusca). Mol Phylogenet Evol 42:717–737

Cook A, Nützel A, Frýda J (2008) Two Mississippian caenogastropod limpets from Australia and their meaning for the ancestry of the Caenogastropoda. J Paleontol 82(1): 183–187

Cope JCW (2004) Bivalve and rostroconch mollusks. In: Webby BD, Paris F, Droser ML, Percival IG (eds) The great Ordovician biodiversification event. Columbia University, pp 196–208

Cox LR, Knight JB (1960) Suborders of Archaeogastropoda. Proc Malacol Soc London 33:262–264

Cox LR, Newell ND, Boyd DM, Branson CC, Casey R, Chavan A, Coogan AH, Dechaseaux C, Fleming CA, Haas F, Hertlein LG, Kauffman EG, Keen AM, LaRocque A, McAlester AL, Moore RC, Nuttall CP, Perkins BF, Puri HS, Smith LA, Soot-Ryen T, Stenzel HB, Trueman ER, Turner RD, Weir J (1969) Treatise on invertebrate paleontology. Mollusca 6, vols 1, 2, Bivalvia. Geological Society of America and University of Kansas, Boulder and Lawrence, KS, 952pp

Cuvier G (1817) Le règne animal distribué d'après son organisation, tome 2 contenant les reptiles, les poissons, les mollusques, les annélides., Deterville, Paris, xviii + 532 pp

Dreyer H, Steiner G (2004) The complete sequence and gene organization of the mitochondrial genome of the gadilid scaphopod *Siphonondentalium lobatum* (Mollusca). Mol Phylogenet Evol 31:605–617

Dzik J (1983) Larval development and relationships of *Mimospira*; a presumably hyperstrophic Ordovician gastropod. Geol Foereningen i Stockholm Foerhandlingar 104:231–239

Fortey RA, Briggs DEG, Wills MA (1996) The Cambrian evolutionary 'explosion': decoupling cladogenesis from morphological disparity. Biol J Linn Soc 57:13–33

Fretter V, Graham A, McLean JH (1981) The anatomy of the Galapagos rift limpet. *Neomphalus fretterae.* Malacologia 21:337–361

Frey RC, Beresi MS, Evans DH, King AH, Percival IG (2004) Nautiloid Cephalopods. In: Webby BD, Paris F, Droser ML, Percival IG (eds) The great Ordovician biodiversification event. Columbia University Press, New York, pp 209–213

Frýda J (1997) Oldest representatives of the superfamily Cirroidea (Vetigastropoda) with notes on their early phylogeny. J Paleontol 71:839–847

Frýda J (1998a) Did the ancestors of higher gastropods (Neritimorpha, Caenogastropoda, and Heterostropha) have an uncoiled shell? In: Bieler R, Mikkelson PM (eds) World congress of malacology (Chicago), Abstracts. UNITAS Malacologica, p 107

Frýda J (1998b) Higher classification of the Paleozoic gastropods inferred from the early shell ontogeny. In: Bieler R, Mikkelson PM (eds) World congress of malacology (Chicago), Abstracts. UNITAS Malacologica, p 108

Frýda J (1999a) Suggestions for polyphyletism of Paleozoic bellerophontiform molluscs inferred from their protoconch morphology. 65th Annual Meeting, American Malacological Society (Pittsburgh), Abstracts

Frýda J (1999b) Higher classification of Paleozoic gastropods inferred from their early shell ontogeny. J Czech Geol Soc 44:137–152

Frýda J (1999c) Shape convergence in gastropod shells: an example from the early Devonian *Plectonotus (Boucotonotus)–Palaeozygopleura* community of the Prague Basin (Bohemia). Mitt Geol–Paläontol Inst Univ Hamburg 83:179–190

Frýda J (2000) Some new Givetian (late Middle Devonian) gastropods from the Paffrath area (Bergisches Land, Germany). Mem Queensland Mus 45:359–374

Frýda J (2001) Discovery of a larval shell in Middle Paleozoic subulitoidean gastropods with description of two new species from the early Devonian of Bohemia. Bull Czech Geol Surv 76:29–37

Frýda J, Bandel K (1997) New Early Devonian gastropods from the *Plectonotus (Boucotonotus)–Palaeozygopleura* community in the Prague Basin (Bohemia). Mitt Geol Paläont Inst Univ Hamburg 80:1–57

Frýda J, Blodgett RB (1998) Two new cirroidean genera (Vetigastropoda, Archaeogastropoda) from the Emsian (late Early Devonian) of Alaska with notes on the early phylogeny of Cirroidea. J Paleontol 72:265–273

Frýda J, Blodgett RB (2001) The oldest known heterobranch gastropod, *Kuskokwimia* gen. nov., from the Early Devonian of west-central Alaska, with notes on the early phylogeny of higher gastropods. Bull Czech Geol Surv 76:39–53

Frýda J, Blodgett RB (2004) New Emsian (Late Early Devonian) gastropods from Limestone Mountain, Medfra B–4 Quadrangle, west-central Alaska (Farewell Terrane), and their paleobiogeographic affinities and evolutionary significance. J Paleontol 78:111–132

Frýda J, Heidelberger D (2003) Systematic position of Cyrtoneritimorpha within the class Gastropoda with description of two new genera from Siluro-Devonian strata of Central Europe. Bull Czech Geol Surv 78:35–39

Frýda J, Manda S (1997) A gastropod faunule from the *Monograptus uniformis* graptolite Biozone (early Lochkovian, Early Devonian) in Bohemia. Mitt Geol–Paläontol Inst Univ Hamburg 80:59–121

Frýda J, Rohr DM (1999) Taxonomy and paleobiogeography of the Ordovician Clisospiridae and Onychochilidae. Acta Univ Carol Geol 43:405–408

Frýda J, Rohr DM (2004) Gastropods. In: Webby BD, Paris F, Droser ML, Percival IG (eds) The great Ordovician biodiversification event. Columbia University, New York, pp 184–195

Frýda J, Rohr DM (2006) Shell heterostrophy in early Ordovician Macluritella Kirk, 1927, and its implications for phylogeny and classification of Macluritoidea (Gastropoda). J Paleontol 80:264–271

Frýda J, Rieder M, Klementová M, Weitschat W, Bandel K (2004) Discovery of gastropod-type nacre in fossil cephalopods: a tale of two crystallographic textures. In: Wells FE (ed) World congress of malacology (Perth 2004), Western Australia, p 43

Frýda J, Heidelberger D, Blodgett RB (2006a) Odontomariinae, a new Middle Paleozoic subfamily of slit-bearing euomphaloidean gastropods (Euomphalomorpha, Gastropoda). N Jb Geol Paläont Abh 4:225–248

Frýda J, Rieder M, Weitschat W, Týcová P, Haloda J (2006b) Evolution of nacre: a new evidence for close phylogenetic relationship of classes Cephalopoda and Gastropoda. In: Qun Y, Yongdong W, Weldon EA (eds) Ancient life and modern approaches. Second International Palaeontological Congress Abstracts, Geological Publishing House, Beijing, pp 51–52

Frýda J, Weitschat W, Týcová P, Haloda J, Mapes RH (2007) Crystallographic textures of cephalopod nacre: its evolution, time stability, and phylogenetic significance. – Seventh international symposium cephalopods present and past (Sapporo, Japan, 2007) Abstracts. Hokkaido University, Hokkaido, pp 56–57

Frýda J, Nützel A, Wagner PJ (2008a) Paleozoic gastropods. In: Ponder W, Lindberg DL (eds) Phylogeny and evolution of the Mollusca. University of California, Berkeley, Los Angeles, CA, pp 239–270

Frýda J, Racheboeuf PR, Frýdová B (2008b) Mode of life of early Devonian *Orthonychia protei* (Neritimorpha, Gastropoda) inferred from its post-larval shell ontogeny and muscle scars. Bull Geosci 83:491–502

Frýda J, Racheboeuf PR, Frýdová B, Ferrová L, Mergl M, Berkyová S (2009) Platyceratid gastropods – stem group of patellogastropods, neritimorphs or something else? Bull Geosci 84:107–120

Frýda J, Bandel K, Frýdová B (2009) Crystallographic texture of Late Triassic gastropod nacre: evidence of long-term stability of the mechanism controlling its formation. Bull Geosci 84:747–756

Fuchigami T, Sasaki T (2005) The shell structure of the Recent Patellogastropoda (Mollusca: Gastropoda). Palaeontol Res 9:143–168

Geiger DL, Thacker CE (2005) Molecular phylogeny of Vetigastropoda reveals non- monophyletic Scissurellidae, Trochoidea, and Fissurelloidea. Molluscan Res 25:47–55

Geiger DL, Nützel A, Sasaki T (2008) Vetigastropoda. In: Ponder W, Lindberg DL (eds) Phylogeny and evolution of the Mollusca. University of California, Berkeley and Los Angeles, CA, pp 297–330

Giribet G (2008) Bivalvia. In: Ponder W, Lindberg DL (eds) Phylogeny and evolution of the Mollusca. University of California, Berkeley, Los Angeles, CA, pp 105–141

Giribet G, Okusu A, Lindgren AR, Huff SW, Schrödl M, Nishiguchi ML (2006) Evidence for a clade composed of molluscs with serially repeated structures: monoplacophorans are related to chitons. Proc Natl Acad Sci USA 103:7723–7728

Golikov AN, Starobogatov YI (1975) Systematics of prosobranch gastropods. Malacologia 15:185–232

Gubanov AP, Rohr DM (1995) Paleogeography of the Macluritidae (Ordovician–Gastropoda). Ordovician Odyssey. Short papers for the 7th International Symposium on the Ordovician System, pp 461–464

Gubanov AP, Tait J (1998) *Maclurites* (Mollusca) and Ordovician palaeogeography. Schriften Staat Mus Mineral Geol Dresden 9:140–142

Hadfield MG, Strathmann MF (1990) Heterostrophic shells and pelagic development in trochoideans: implications for classification, phylogeny and palaeoecology. J Molluscan Stud 56:239–256

Harper JS, Rollins HB (2000) The bellerophont controversy revisited. Am Malacol Bull 15:147–156

Haszprunar G.(1985) The fine morphology of the osphradial sense organs of the Mollusca. Part 1: Gastropoda–Prosobranchia. Philos Trans R Soc Lond B 307: 457–496

Haszprunar G (1988a) On the origin and evolution of major gastropod groups, with specialreference to the Streptoneura. J Moll Stud 54:367–441

Haszprunar G (1988b) Anatomy and relationships of the bone-feeding limpets Cocculinella minutissima (Smith) and Osteopelta mirabilis Marshall (Archaeogastropoda). J Moll Stud 54:1–20

Haszprunar G (1992) The first molluscs – small animals. Boll Zool 59:1–16

Haszprunar G (1993) The Archaeogastropoda: a clade, a grade or what else? Am Malac Bull 10:165–177

Haszprunar G (1995) On the evolution of larval development in the Gastropoda, with special reference to larval planktotrophy. Notiz CISMA XVI:5–13

Haszprunar G (1998) Superorder Cocculiniformia. In: Beesley PL, Ross GJB, Wells A (eds) Mollusca: the southern synthesis. Fauna of Australia, vol 5B. CSIRO, Melbourne, pp 653–664

Haszprunar G (2000) Is the Aplacophora monophyletic? A cladistic point of view. Am Malac Bull 15(2):115–130

Haszprunar G (2008) Monoplacophora (Tryblidia). In: Ponder W, Lindberg DL (eds) Phylogeny and evolution of the Mollusca. University of California, Berkeley, Los Angeles, CA, pp 97–104

Haszprunar G, Schaefer K (1997) Anatomy and phylogenetic signifi cance of *Micropilina arntzi* (Mollusca, Monoplacophora, Micropilinidae fam. nov.). Acta Zool (Stock.) 77:315–334

Haszprunar G, Wanninger A (2000) Molluscan muscle systems in development and evolution. J Zool Syst Evol Res 38:157–163

Haszprunar G, Wanninger A (2008) Fine structure of the creeping larva of *Loxosomella murmanica*: additional evidence for a clade of Kamptozoa (Entoprocta) and Mollusca. Acta Zool (Stockh) 89:137–148

Haszprunar G, Schaefer K, Warén A, Hain S (1995) Bacterial symbionts in the epidermis of an Antarctic neopilinid limpet (Mollusca, Monoplacophora). Philos Trans R Soc Lond B 347:181–185

Haszprunar G, Schander C, Halanych KM (2008) Relationships of the higher molluscan taxa. In: Ponder W, Lindberg DL (eds) Phylogeny and evolution of the Mollusca. University of California, Berkeley, Los Angeles, CA, pp 19–32

Healy JM (1988) Sperm morphology and its systematic importance in the Gastropoda. In: Ponder WF (ed) Prosobranch phylogeny. Malacol Rev Suppl 4:251–266

Hedegaard C (1990) Shell structures of the recent Archaeogastropoda. Thesis, Department of Ecology and Genetics, University of Aarhus, Denmark, 1, 154p, 2 Atlas, 78 pls

Hedegaard C, Lindberg DR, Bandel K (1997) Shell microstructure of a Triassic patellogastropod limpet. Lethaia 30: 331–335

Hickman CS (1988) Archaeogastropod evolution, phylogeny and systematics: a re-evaluation. In: Ponder WF (ed) Prosobranch phylogeny. Malacol Rev 4 Suppl:17–34

Hickman CS (1992) Reproduction and development of trochacean gastropods. Veliger 35:245–272

Horný R (1963a) Lower Paleozoic Monoplacophora and Patellid Gastropoda (Mollusca) of Bohemia. Sborník Ústř Úst Geol Oddíl Paleontol 28:7–84

Horný R (1963b) Lower Paleozoic Bellerophontina (Gastropoda) of Bohemia. Sbornik ÚÚG 2:58–164

Horný R (1964) New lower Paleozoic gastropod genera of Bohemia (Mollusca). Čas Nár Muz, Ř přírod 133:211–216

Horný R (1991a) Vývoj názorů na klasifikaci cyklomyárních měkkýšů (Mollusca, Monoplacophora). [Problems of classification of the cyclomyan molluscs (Mollusca, Monoplacophora) – a historical review.] Čas Nár Muz Ř přírod 157[1988](1–4):13–32. Praha

Horný RJ (1991b) Shell morphology and muscle scars of *Sinuitopsis neglecta* Perner (Mollusca, Monoplacophora). Čas Nár Muz Ř prírod 157:81–105

Horný RJ (1993) Shell morphology and mode of life of the Lower Devonian cyclomyan *Neocyrtolites* (Mollusca, Tergomya). Čas Nár Muz Ř prírod 162:57–66

Horný RJ, Peel JS (1996) *Carcassonnella*, a new Lower Ordovician bellerophontiform mollusc with dorsally located retractor muscle attachments (Class Tergomya). Bull Czech Geol Surv 71:305–331

Kaim A, Sztajner P (2005) The opercula of neritopsid gastropods and their phylogenetic importance. J Molluscan Stud 71:1–9

Kano Y (2006) Usefulness of the opercular nucleus for inferring early development in neritimorph gastropods. J Morphol 267:1120–1136

Kano Y, Chiba S, Kase T (2002) Major adaptive radiation in neritopsine gastropods estimated from 28S rRNA sequences and fossil records. Proc R Soc Lond B Biol Sci 269: 2457–2465

Kase T, Shigeta Y (1996) New Species of Patellogastropoda (Mollusca) from the Cretaceous of Hokkaido, Japan and Sakhalin, Russia. J Paleontol 70:762–771

Kiel S, Frýda J (2004) Shell structure of Late Cretaceous *Sensuitrochus ferreri* (Cirridae, Gastropoda). J Paleontol 78:795–797

Kirk E (1927) *Pycnodesma*, a new molluscan genus from the Silurian of Alaska. US Nat Mus Proc 71(20):1–9

Knight JB (1934) The gastropods of the St. Louis, Missouri, Pennsylvanian outlier: VII. The Euomphalidae and Platyceratidae. J Paleontol 8:139–166, pls. 20–26

Knight JB (1941) Paleozoic gastropod genotypes. Geol Soc Am (Spec Pap) 32:510pp

Knight JB (1952) Primitive fossil gastropods and their bearing on gastropod classification. Smithsonian Misc Coll 117:1–56

Knight JB, Cox LR, Batten RL, Yochelson EL (1960) Systematic descriptions. In: Moore RC (ed) Treatise on invertebrate paleontology. Part I. Mollusca 1. University of Kansas, Lawrence, pp 169–309

Kollman HA, Peel JS (1983) Paleocene gastropodsfrom Nűgssuaq, West Greenland. Grønlands Geol Undersøg Bull 146:1–115

Kříž J (2007) Origin, evolution and classification of the new superorder Nepiomorphia (Mollusca, Bivalvia, Lower Paleozoic). Palaeontology 50:1341–1365

Kues BS, Duchene HR (1990) The largest late Paleozoic bellerophontid gastropod. J Paleontol 64(3):478–479

Kunze T, Brückner M, Beck F, Heß M, Haszprunar G (2008) Skeneimorph gastropods in Neomphalida and Vetigastropoda – a preliminary report. In: Geiger D, Ruthensteiner B (eds) Micromolluscs: methodological challenges – exiting results. Zoosymposia 1:119–131

Lindberg DR (1981) Is there a coiled ancestor in the docoglossan phylogeny? Ann Rept Western Soc Malacol 12:15

Lindberg DR (1988) The Patellogastropoda. In: Ponder WF, Warén A (eds) Prosobranch Phylogeny: proceedings of the 9th international malacological congress. Malacol Rev 4

Lindberg DR (1990) The systematics of *Potamacmaea fluvatilis* Blandford: a brackishwater patellogastropod (Patelloidinae: Lottiidae). J Molluscan Stud 56:309–316

Lindberg DR (2008) Patellogastropoda, Neritimorpha, and Cocculinoidea. In: Ponder W, Lindberg DL (eds) Phylogeny and evolution of the Mollusca. University of California, Berkeley, Los Angeles, CA, pp 271–296

Lindberg DR, Guralnick RP (2003) Phyletic patterns of early development in gastropod molluscs. Evol Dev 5:494–507

Lindberg DR, Ponder WF (2001) The influence of classification of the evolutionary interpretation of structure – a reevaluation of the evolution of the pallial cavity of gastropod molluscs. Org Divers Evol 1:273–299

Lindberg DR, Ponder WF, Haszprunar G (2004) The Mollusca: relationships and patterns from their first half-billion years. In: Cracraft J, Donoghue MJ (eds) Assembling the tree of life. Oxford University Press, Oxford, pp 252–278

Linsley RM, Kier WM (1984) The Paragastropoda: a proposal for a new class of Paleozoic Mollusca. Malacologia 25: 241–254

MacClintock C (1967) Shell structure of patelloid and bellerophontoid gastropods (Mollusca). Peabody Mus Nat Hist Bull 22:140pp

MacClintock C (1968). Shell structure of patelloid and bellerophontoid gastropods (Mollusca). Peabody Mus Nat Hist Bull 22:1–140

McArthur AG, Harasewych MG (2003) Molecular systematics of the major lineages of the Gastropoda. In: Lydeard C, Lindberg DR (eds) Molecular systematics and phylogeography of mollusks. Smithsonian Books, Washington, DC, pp 140–160

McLean JH (1981) The Galapagos rift limpet *Neomphalus*: relevance to understanding the evolution of a major Paleozoic–Mesozoic radiation. Malacologia 21:291–336

McLean JH (1990) Neolepetopsidae, a new docoglossate limpet family from hydrothermal vents and its relevance to patellogastropod evolution. J Zool 222:485–528

Mordan P, Wade CH (2008) Heterobranchia II – Pulmonata. In: Ponder W, Lindberg DL (eds) Phylogeny and evolution of the Mollusca. University of California, Berkeley, Los Angeles, CA, pp 409–426

Morris NJ, Cleevely RJ (1981) *Phanerotinus cristatus* (Phillips) and the nature of euomphalacean gastropods, Molluscans. Bull Brit Mus Nat Hist (Geol) 35:195–212

Nakano T, Ozawa T (2004) Phylogeny and historical biogeography of limpets of the order Patellogastropoda based on mitochondrial DNA Sequences. J Mollusc Stud 70:31–41
Nielsen C (2001) Animal evolution, interrelationships of the living phyla, 2nd ed. Oxford University Press, Oxford, 563pp
Nielsen C, Haszprunar G, Ruthensteiner B, Wanninger A (2007) Early development of the aplacophoran mollusc *Chaetoderma*. Acta Zool 88:231–247
Nishiguchi MK, Mapes RH (2008) Cephalopoda. In: Ponder W, Lindberg DL (eds) Phylogeny and evolution of the Mollusca. University of California, Berkeley, Los Angeles, CA, pp 163–199
Nützel A (1998) Über die Stammesgeschichte der Ptenoglossa (Gastropoda). Berliner Geowissenschaftliche Abhandlungen. Reihe E. Paläobiologie 26(1):1–225
Nützel A (2002) An evaluation of the recently proposed Palaeozoic gastropod subclass Euomphalomorpha. Palaeontology 45:259–266
Nützel A, Cook AG (2002) *Chlorozyga*, a new caenogastropod genus from the early Carboniferous of Australia. Alcheringa 26:151–157
Nützel A, Frýda J (2003) Paleozoic plankton revolution: evidence from early gastropod ontogeny. Geology 31: 829–831
Nützel A, Mapes RH (2001) Larval and juvenile gastropods from a Carboniferous black shale: palaeoecology and implications for the evolution of the Gastropoda. Lethaia 34:143–162
Nützel A, Erwin DH, Mapes RH (2000) Identity and phylogeny of the late Paleozoic Subulitoidea (Gastropoda). J Paleontol 74:575–598
Nützel A, Lehnert O, Frýda J (2006) Origin of planktotrophy – evidence from early molluscs. Evol Dev 8:325–330
Nützel A, Lehnert O, Frýda J (2007a) Origin of planktotrophy – evidence from early molluscs: a response to Freeman and Lundelius. Evol Dev 9:312–317
Nützel A, Frýda J, Yancey TE, Anderson JR (2007b) Larval shells of late Palaeozoic naticopsid gastropods (Neritopsoidea: Neritimorpha) with a discussion of the early neritimorph evolution. Paläont Zeit 81:213–228
Okusu A (2002) Embryogenesis and development of *Epimenia babai* (Mollusca Neomeniomorpha). Biol Bull 203:87–103
Page LR (1995) The ancestral gastropod larval form is best approximated by hatching stage opisthobranch larvae: evidence from comparative developmental studies. In: Wilson HW, Stricker SA, Shinn GL (eds) Reproduction and development of marine invertebrates. Johns Hopkins University, Baltimore, MD, pp 206–223
Page LR (1997) Ontogenetic torsion and protoconch form in the archaeogastropod *Haliotis kamtschatkana*: evolutionary implications. Acta Zool 78:227–245
Page LR (2006) Modern insights on gastropod development: reevaluation of the evolution of a novel body plan. Integr Comp Biol 46:134–143
Parkhaev PY (2008) The early Cambrian radiation of Mollusca. In: Ponder W, Lindberg DL (eds) Phylogeny and evolution of the Mollusca. University of California, Berkeley, Los Angeles, CA, pp 33–69
Passamaneck YJ, Schander C, Halanych KM (2004) Investigation of molluscan phylogeny using large-subunit and small-subunit nuclear rRNA sequences. Mol Phylogenet Evol 32:25–38
Pchelintsev VF, Korobkov LA (1960) Mollusca – Gastropoda. In: Orlov YA (ed) Osnovy paleontologii. Akademia Nauk, Moscow, pp 1–360
Peel JS (1986) Systematics and mode of life of a new Silurian *Clisospira* (Mollusca) from North Greenland. Rapp Grønlands Geol Undersøg 128:65–74
Peel JS (1991a) Functional morphology of the Class Helcionelloida *nov*., and the early evolution of the Mollusca. In: Simonetta A, Conway Morris S (eds) The early evolution of the metazoa and the significance of problematic taxa. Cambridge University Press, Cambridge, UK, pp 157–177
Peel JS (1991b) The classes Tergomya and Helcionelloida, and early molluscan evolution. Grønlands Geol Undersøg Bull 161:11–65
Peel JS, Horný R (1996) Sinistral hyperstrophic coiling in a Devonian gastropod from Bohemia with an *in situ* operculum. Palaeontology 39:709–718
Peel JS, Horný R (1999) Muscle scars and systematic position of the Lower Palaeozoic limpets *Archinacella* and *Barrndicella* gen. new. (Mollusca). J Czech Geol Soc 44:97–115a
Pelseneer P (1893) Introduction à l'étude des Mollusques. Ann Soc R Malacol Belg 27:31–243
Ponder WF, Lindberg DR (1997) Towards a phylogeny of gastropod molluscs: an analysis using morphological characters. Zool J Linn Soc 119:83–265
Ponder WF, Lindberg DR (2008) Molluscan evolution and phylogeny – an introduction. In: Ponder W, Lindberg DL (eds) Phylogeny and evolution of the Mollusca, University of California, Berkeley, Los Angeles, CA, pp 1–17
Ponder WF, Warén A (1988) Classification of the Caenogastropoda and Heterostropha – a list of the family-group names and higher taxa. In: Ponder WF (ed) Prosobranch phylogeny. Malacol Rev Suppl 4:288–328
Ponder WF, Colgan DJ, Healy JM, Nützel A, Simone LRL, Strong EE (2008) Caenogastropoda. In: Ponder W, Lindberg DL (eds) Phylogeny and evolution of the Mollusca. University of California, Berkeley, Los Angeles, CA, pp 331–383
Reynolds PD, Steiner G (2008) Chapter 7. Scaphopoda. In: Ponder W, Lindberg DL (eds) Phylogeny and evolution of the Mollusca. University of California, Berkeley Los Angeles, CA, pp 143–161
Riedel F (2000) Ursprung und Evolution der "höheren" Caenogastropoda. Eine paläobiologische Konzeption. Berliner Geowiss Abh, Reihe E. Paläobiologie 32:1–240
Robertson R (1963) Abnormal dextral hyperstrophy of postlarval *Heliacus* (Gastropoda: Architectonicidae). Veliger 6:76–79
Robertson R (1993) Snail handedness. National Geog Res Explorn 9:104–119
Rohr DM (1979) Geographic distribution of the Ordovician gastropod *Maclurites*. In: Gray J, Boucot AJ (eds) Historical biogeography, plate tectonics, and the changing environment. Oregon State University, Corvallis, pp 45–52
Rohr DM (1994) Ordovician (Whiterockian) gastropods of Nevada – Bellerophontoidea, Macluritoidea, and Euomphaloidea. J Paleontol 68:473–486
Rohr DM, Frýda J (2001) A new Ordovician gastropod and operculum from the Czech Republic. J Paleontol 75: 461–462

Rohr DM, Gubanov AP (1997) Macluritid opercula (Gastropoda) from the Middle Ordovician of Siberia and Alaska. J Paleontol 71:394–400

Rohr DM, Blodgett RBB, Furnish WM (1992) *Maclurina manitobensis* (Whiteaves) (Ordovician Gastropoda); the largest known Paleozoic gastropod. J Paleontol 66:880–884

Runnegar B (1981) Muscle scars, shell form and torsion in Cambrian and Ordovician univalved molluscs. Lethaia 14:311–322

Runnegar B (1983) Molluscan phylogeny revisited. Mem Assoc Austral Paleontol 1:121–144

Runnegar B (1996) Early evolution of the Mollusca: the fossil record. In: Taylor JD (ed) Origin and evolutionary radiation of the Mollusca. Oxford University Press, Oxford, pp 77–87

Runnegar B, Pojeta J (1974) Molluscan phylogeny: the paleontological viewpoint. Science 186:311–317

Salvini-Plawen LV (1980) A reconsideration of systematics in the Mollusca (phylogeny and higher classification). Malacologia 19:249–278

Salvini-Plawen LV (1985) Early evolution and the primitive groups. In: Wilbur K (ed) The Mollusca 10, Evolution. Academic Press, New York, pp 59–150

Salvini-Plawen, LV (1990) Origin, phylogeny and classification of the phylum Mollusca. Iberus 9:1–33

Salvini-Plawen LV (2003) On the phylogenetic significance of the aplacophoran Mollusca. Iberus 21:67–97

Salvini-Plawen LV, Haszprunar G (1987) The Vetigastropoda and the systematics of streptoneurous gastropods (Mollusca). J Zool A 211:747–770

Salvini-Plawen LV, Steiner G (1996) Synapomorphies and plesiomorphies in higher classification of Mollusca. In: Taylor JD (ed) Origin and evolutionary radiation of the Mollusca. Oxford University Press, Oxford, pp 29–51

Sasaki T (1998) Comparative anatomy and phylogeny of the recent Archaeogastropoda (Mollusca: Gastropoda). University Museum, University of Tokyo, Tokyo, Bulletin No. 38, pp 1–223

Sasaki T (2001) Macro- and microstructure of shell and operculum in two recent gastropod species, *Nerita (Theliostyla) albicilla* and *Cinnalepeta pulchella* (Neritopsina Neritoidea). Palaeontol Res 5:21–31

Schander C, Sundberg P (2001) Useful characters in gastropod phylogeny: soft information or hard facts? Syst Biol 50: 136–141

Scheltema AH (1996) Phylogenetic position of Sipuncula, Mollusca and the progenetic Aplacophora. In: Taylor JD (ed) Origin and evolutionary radiation of the Mollusca. Oxford University Press, Oxford, pp 53–58

Scheltema AH, Schander C (2006) Exoskeletons: tracing molluscan evolution. Venus 65:19–26

Schwabe E (2008) A summary of reports of abyssal and hadal Monoplacophora and Polyplacophora (Mollusca). Zootaxa 1866:205–222

Sepkoski JJ Jr, Hulver ML (1985) An atlas of Phanerozoic clade diversity diagrams. In: Valentine JW (ed) Phanerozoic diversity patterns. Princeton University Press, Princeton, NJ, pp 11–39

Simison WB, Boore JL (2008) Molluscan evolutionary genomics. In: Ponder W, Lindberg DL (eds) Phylogeny and evolution of the Mollusca. University of California, Berkeley and Los Angeles, CA, pp 447–461

Steiner G, Dreyer H (2003) Molecular phylogeny of Scaphopoda (Mollusca) inferred from 18S rDNA sequences: support for a Scaphopoda–Cephalopoda clade. Zool Scripta 32: 343–356

Sutton MD, Briggs DEG, Siveter DJ, Siveter DJ (2006) Fossilized soft tissues in a Silurian platyceratid gastropod. Proc R Soc B Biol Sci 273:1039–1044

Taylor JD, Morris NJ, Taylor CN (1980) Food specialization and the evolution of predatory prosobranch gastropods. Palaeontology 23:375–409

Teichert C, Kummel B, Sweet WC, Stenzel HB, Furnish WM, Glenister BF, Erben HK, Moore RC, Nodine Zeller DE (1964) Treatise on invertebrate paleontology. Part K. Mollusca 3. Cephalopoda – general features, Endoceratoidea–Actinoceratoidea, Nautiloidea, Bactritoidea. Geological Society of America and University of Kansas Press, Boulder, CO and Lawrence, KS, 519pp

Thiele J (1929–1931) Handbuch der Systematischen Weichtierkunde, vol 1. Gustav Fischer, Jena

Todt C, Okusu A, Schander C, Schwabe E (2008) Solenogastres, Caudofoveata and Polyplacophora. In: Ponder W, Lindberg DL (eds) Phylogeny and evolution of the Mollusca. University of California Berkeley, Los Angeles, CA, pp 71–96

Tracey S, Todd JA, Erwin DH (1993) Mollusca: Gastropoda. In: Benton MJ (ed) The fossil record. Chapman & Hall, London, pp 131–167

Troschel FH (1865) In Troschel FH (1856–1879) Das Gebiß der Schnecken zur Begründung einer natürlichen Klassifikation, vol 2. Nicolai, Berlin, pp 1–48

Vermeij GJ (1975) Evolution and distribution of left-handed and planispiral coiling in snails. Nature 254:419–420

Vermeij GJ (1977) The Mesozoic marine revolution: evidence from snails, predators, and grazers. Paleobiology 3: 245–258

Vermeij GJ (1987) Evolution and escalation: an ecological history of life. Princeton University Press, Princeton, NJ, 527pp

Wägele H, Klussmann-Kolb A, Vonnemann V, Medina M (2008) Heterobranchia I. The Opisthobranchia. In: Ponder W, Lindberg DL (eds) Phylogeny and evolution of the Mollusca. University of California Berkeley, Los Angeles, CA, pp 385–408

Wagner PJ (1995) Diversity patterns among early gastropods – contrasting taxonomic and phylogenetic descriptions. Paleobiology 21:410–439

Wagner PJ (1999) Phylogenetics of the earliest anisostrophically coiled gastropods. Smithsonian Contrib Paleobiol 88:1–132

Wagner PJ (2000) Exhaustion of cladistic character states among fossil taxa. Evolution 54:365–386

Wagner PJ (2001) Gastropod phylogenetics: progress, problems and implications. J Paleontol 75:1128–1140

Wagner PJ (2002) Likelihood tests of general phylogenetic hypotheses: how many times did bellerophont molluscs evolve? Geol Soc Am (Abstr Prog) 34:261

Wahlman GP (1992) Middle and Upper Ordovician symmetrical univalved molluscs (Monoplacophora and Bellerophontina) of the Cincinnati Arch region. US Geol Surv Prof Pap 1066–O:O1–O213

Waller TR (1998) Origin of the molluscan class Bivalvia and a phylogeny of major groups. In: Johnston PA, Haggard JW (eds) Bivalves: an eon of evolution. University of Calgary, Calgary, AB, pp 1–45

Wängberg-Eriksson K (1979) Macluritacean gastropods from the Ordovician and Silurian of Sweden. Sver Geol Unders Ser C 72:1–33

Wanninger A, Haszprunar G (2001) The expression of an engrailed protein during embryonic shell formation of the tusk-shell, *Antalis entalis* (Mollusca, Scaphopoda). Evol Dev 3:312–321

Wanninger A, Haszprunar G (2002) Muscle development in *Antalis entalis* (Mollusca, Scaphopoda) and its significance for scaphopod relationships. J Morphol 254: 53–64

Wanninger A, Ruthensteiner B, Haszprunar G (2000) Torsion in *Patella caerulea* (Mollusca, Patellogastropoda): ontogenetic process, timing, and mechanisms. Invert Biol 119:177–187

Warén A, Bouchet P (2001) Gastropoda and Monoplacophora from hydrothermal vents and seeps; new taxa and records. Veliger 44:116–231

Wenz W (1938–1944) Gastropoda. In Schindewolf OH (ed) Handbuch der Paläzoologie Bd 6. Borntraeger, Berlin, pp 1–7, 1639pp

Wilson NG, Huang D, Goldstein MC, Cha H, Giribet G, Rouse GW (2009) Field collection of *Laevipilina hyalina* McLean, 1979 from southern California, the most accessible living monoplacophoran. J Molluscan Stud 75:195–197

Woodward SP (1854) A manual of the Mollusca; or, rudimentary treatise of recent and fossil shells. London, Weale, 486pp

Yochelson EL (1956) Euomphalacea, Trochonematacea, Pseudophoracea, Anomphalacea, Craspedostomatacea, and Platyceratacea, Permian Gastropoda of the southwestern United States. Bull Am Mus Nat Hist 110:173–275

Yochelson EL (1967) Quo vadis, Bellerophon? In: Teichert C, Yochelson EL (eds) Essays in paleontology and stratigraphy. Kansas University, Lawrence, KS, pp 141–161

Yochelson EL (1978) An alternative approach to the interpretation of the phylogeny of the ancient mollusks. Malacologia 17:165–191

Yochelson EL (1984) Historic and current considerations for revision of Paleozoic gastropod classification. J Paleontol 58:259–269

Yochelson EL (1988) A new genus of Patellacea (Gastropoda) from the Middle Ordovician of Utah: the oldest known example of the superfamily. New Mexico Bur Mines Min Res Mem 44:195–200

Yochelson EL (1990) Billings' second operculum; a late Early Ordovician *Maclurites* (Gastropoda) from western Newfoundland and the Canadian Arctic. Can J Earth Sci 27:669–676

Yoo EK (1994) Carboniferous Mollusca from the Tamworth Belt, New South Wales, Australia. Rec Aust Mus 46: 63–120

Palaeozoic Innovations in the Micro- and Megafossil Plant Record: From the Earliest Plant Spores to the Earliest Seeds

Philippe Steemans, Elodie Petus, Pierre Breuer, Paula Mauller-Mendlowicz, and Philippe Gerrienne

Abstract

Recently, major advances have been made in understanding terrestrialization processes and the development of early vegetation. This chapter reviews the major steps in the evolution of early land plants, with focii on cryptospores, trilete spores and on the meso- and megafossil remains of Silurian and Devonian plants. The major morphological innovations of plants and their spores are described. Cryptospores are the earliest fossil record of a terrestrial vegetation cover; the oldest indisputable specimens are observed from Darriwilian (mid-Ordovician) strata in Saudi Arabia and the Czech Republic. The biological affinities of cryptospores are discussed. It is generally agreed that cryptospores were derived mainly from 'bryophytes', but some cryptospores may have arisen from early tracheophytes or their immediate ancestors. The earliest trilete spores are Ordovician in age. Most trilete spores are considered to have tracheophyte affinities; we discuss possible relationships between trilete spores and several cryptospores in dyads or monads. The earliest record of plant mesofossils comes from Middle Silurian strata. The evolution and affinities of the major groups of Late Silurian and Devonian land plants are presented within a phylogenetic and a stratigraphic framework.

Keywords

Cryptospores • Trilete spores • Meso-megafossil plants • Propagule dispersion • Tripapillate forms as biostratigraphic markers • Embrophytes • Spermatophyte adaptive radiation

For me, the most exciting events in land plant evolution were completed by end of the Devonian Period. The origin of the embryophytes (bryophytes + vascular plants) is shrouded in controversy, and represents one of the greatest palaeontological mysteries.
J.B. Richardson (1996a)

Introduction

The colonization of land by plants, fungi and animals – also known as the terrestrialization – was undoubtedly one of the most crucial advances in the history of life

P. Steemans (✉)
University of Liège, Paléobotanique, Paléopalynologie et Micropaléontologie, B-4000 Liège 1, Belgium
e-mail: p.steemans@ulg.ac.be

J.A. Talent (ed.), *Earth and Life*, International Year of Planet Earth,
DOI 10.1007/978-90-481-3428-1_13,

on Earth. The exact nature and timing of the early steps remain conjectural, but evidence has accumulated of the existence of complex terrestrial ecosystems in mid-Palaeozoic times. The oldest evidences of terrestrial plants are in the form of dispersed propagules (cryptospores and trilete spores) and phytodebris (tubular structures and fragmentary cuticles) in the sediments. These are at least Ordovician in age; plant megafossils are not known before the Middle Silurian. In this chapter, we review the major steps in the evolution of the early terrestrial vegetation.

The Miospores

Cambrian Palynomorphs

Ambiguous Cambrian palynomorphs from the Bright Angel Shale, Arizona, USA, have been described (Taylor and Strother 2008) as '... the desiccation-resistant spores of cryptogams belonging to the charophyte–embryophyte lineage'. They are characterized by multilaminate spore walls, characteristic of some extant liverworts and Palaeozoic cryptospores. The authors suggested such these spores could have evolved via the fusion of separate, multiple laminae, possibly of the primitive plant sporoderm type. The Cambrian palynomorphs include dyads and tetrads similar to younger cryptospores, but most differ morphologically by their much smaller size, by the presence of polyads (with a variable number of spores) and by the presence of envelopes enclosing individual spores of the polyads rather than the envelope of younger cryptospores enclosing the whole dyad or tetrad. As stated by Taylor and Strother (2008), 'Their overall morphology cannot prove their embryophytes affinity, but laminate wall ultrastructure stands out as a single character they appear to have in common with the liverworts ... For now, they might be considered as "protoembryophytes" ...'.

Cryptospores

The oldest evidence for the first land plants are propagules, termed cryptospores, because of their primitive aspect, the lack of a trilete mark, and because their parent plants were previously unknown (Richardson et al. 1984). In 2000, Steemans proposed that cryptospores should be included in miospores as they most probably are produced by embryophytes and more precisely by 'bryophytes' (Wellman et al. 2003). Cryptospores show unusual features when compared with the younger spores of tracheophytes. They are preserved either as monads or as obligate dyads or as tetrads, indicating that they remain permanently fused (Fig. 1). At the border of their contact area, each spore of a dyad or tetrad shows an annular thickening. Cryptospores are either naked or enclosed within a thin envelope, or membrane. The envelope and/or the wall of the cryptospores may be ornamented by various elements. The ornament is usually inconspicuous in the oldest species, but in Late Silurian and Early Devonian specimens, it becomes more and more similar to that of trilete spores. Dyads and tetrads are termed fused or unfused according to the respective absence or presence of a superficial area of attachment between the spores; the exact nature of the association of the spores is not clearly known (Wellman 1996). Taphonomy may also play a role in the aspect of the attachment area, presumably depending on how the specimens have been compressed during the fossilization processes. For example, the tetrads of the text fig. 5B and F in Wellman (1996) show at the same time typical sutures of *Cheilotetras caledonica* (Fig. 2.1) and of *Tetrahedraletes medinensis* (Fig. 2.2-3). *Note: All miospore and plant taxa and their authors are listed in Tables 1 and 2.*

Evolution of the Cryptospore Assemblages Through Time

The biological affinities of the Cambrian palynomorphs remain controversial, but the embryophytic origin of the palynomorphs observed in the Darriwilian (Llanvirn) assemblages from Saudi Arabia (Strother et al. 1996) and from the Czech Republic (Vavrdová 1990a, b) is widely accepted. Llanvirnian cryptospores are very different from and much more diverse than the Cambrian palynomorphs, including tetrads, dyads and monads, naked or enclosed in an envelope, laevigate or ornamented. Assemblages from the Sandbian and Katian (Caradoc in Wellman 1996), the Katian or Hirnantian (e.g. Richardson 1988; Steemans 2001; Vavrdová 1984, 1988; Wang et al. 1997) and the early Llandovery (e.g. Burgess 1991; Mizusaki et al. 2002; Richardson 1988; Steemans et al. 2000; Steemans and Pereira 2002; Wellman et al. 2000) exhibit strong similarities. Biostratigraphic criteria based on cryptospores from the Llanvirn up to the mid-Llandovery are rare.

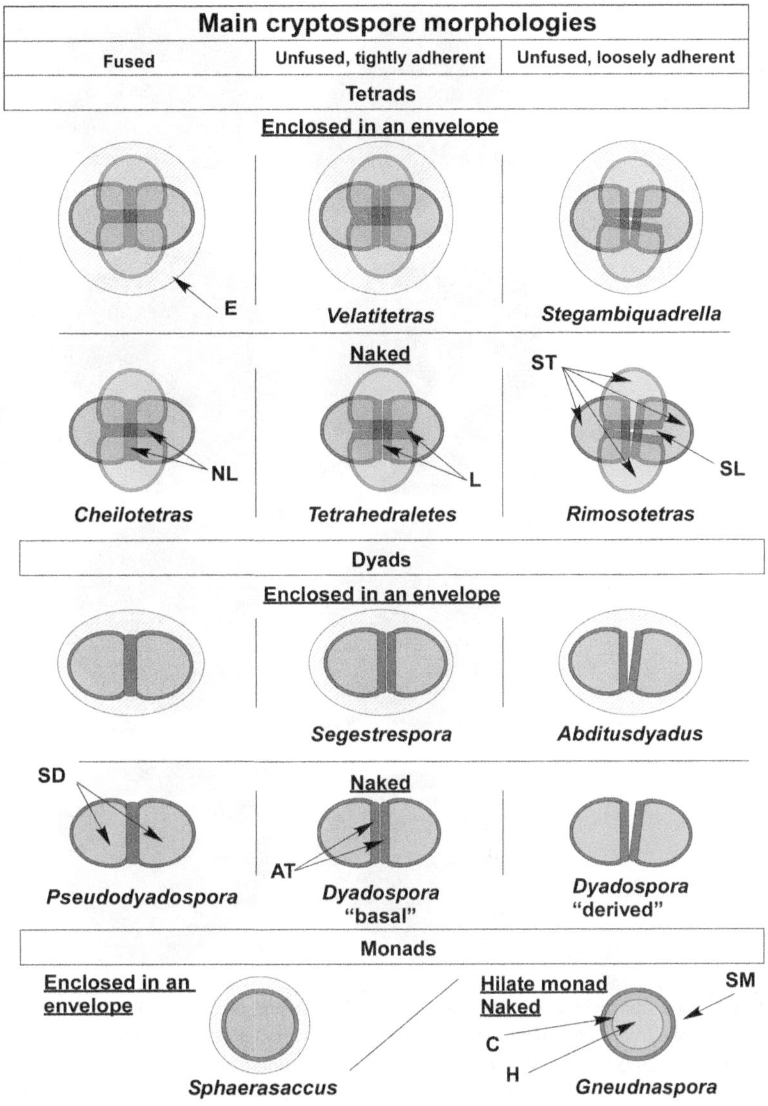

Fig. 1 Simplified graphic representation of the main cryptospore morphological features (modified from Richardson 1996b). *AT* annular thickening (equatorial ring bordering the area of attachment of tetrads and dyads); *C* curvaturae (line bordering the hilum); *E* envelope (inclosing tetrads, dyads or monads); *H* hilum (contact area of spores stem from dissociated dyads); *L* line of attachment (two or four spores are attached together. L is the line of attachment between the spores. Each AT of the spores are separated by the L); *NL* no line of attachment (no surface of attachment between the spores. Only one AT is visible); *SL* clear separation between the spores of tetrads or of dyads along the L; *SD* the two spores of the dyad; *SM* the monad spore (hilate and alete spore); *ST* the four spores of the tetrads

Strangely, the Hirnantian glaciation, causing massive extinctions in many fossil groups, did not affect the cryptospore biodiversity. The wide climatic tolerance of the parent plants demonstrated by the almost ubiquitous cryptospore distribution (from high, even from glacial sediments, to low latitudes) could explain why climatic changes did not affect their biodiversity. Lands exposed after the melting of the ice sheet could immediately be colonized by the cryptospore producing plants that had survived in refugia Steemans and Wellman (2004). From the Wenlock to the Přídolí, cryptospores assemblages diversified once more (Steemans 2000; Steemans and Wellman 2004). During the Devonian, cryptospore populations decreased at a continuous rate. An abundant but low diversity assemblage of Middle Devonian cryptospores has recently been recorded (Breuer 2007). Contrariwise, the cryptospore biodiversity decreased dramatically below the Aeronian/Telychian boundary (Steemans 2000).

Main Morphological Events Within Cryptospores

(1) Naked dyads and tetrads are usually unornamented and their spores are tightly attached to each other.
(2) Morphology of the cryptospores changes greatly after the decline in biodiversity observed below the Aeronian/Telychian boundary (bioevent in Fig. 3).

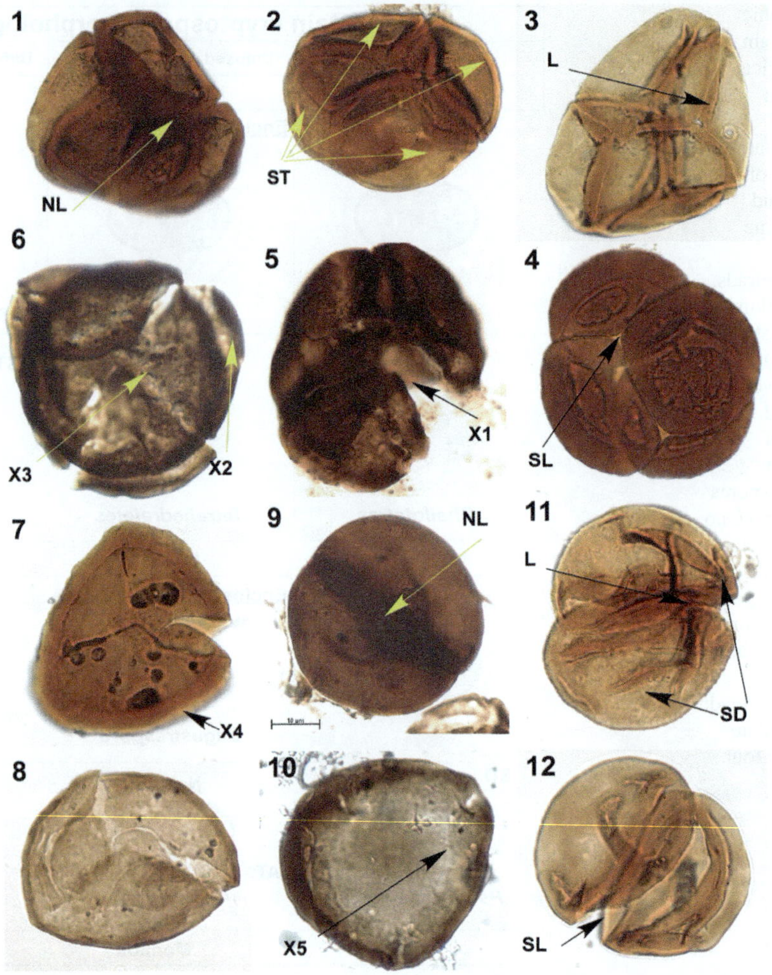

Fig. 2 For abbreviations, see Fig. 1. (1) *Cheilotetras caledonica* Wellman and Richardson 1993. Llandovery, Paraná Basin, Paraguay; diameter size: 53 μm. (2–3) *Tetrahedraletes medinensis* (Strother and Traverse) Wellman and Richardson 1993. Llandovery, Paraná Basin, Paraguay. Specimen in (2) shows a thick exine; diameter size: 58 μm. Specimen in (3) has a very thin exine; diameter size: 64 μm. (4) *Rimosotetras problematica* Burgess 1991 with each spores separated from the others. The distal face is abnormally thick. Llandovery, Paraná Basin, Paraguay; diameter size: 49 μm. (5) Dissociated tetrad. Silurian of Argentina. One of the four spores is absent (X1); diameter size: 57 μm. (6–7) *Imperfectotriletes patinatus* Steemans et al. 2000. Specimen in (6), Silurian of Argentina, shows remains of the tetrad (X2). The pseudo – trilete mark is formed by tears (X3); diameter size: 30 μm. Specimen in (7), Llandovery, Paraná Basin, Paraguay, shows a thick distal face or patina (X4). (8) *Imperfectotriletes varvdovae* (Richardson) Steemans et al. 2000. Llandovery, Saudi Arabia. The exine is thinner than that of *I. patinatus*; diameter size: 29 μm. (9) *Pseudodyadospora laevigata* Johnson 1985. Llandovery, Paraná Basin, Brasil. There is no line of separation between the two spores; diameter size: 35 μm. (10) *Gneudnaspora* (*Laevolancis*) *chibrikovae* Breuer et al. 2007. Llandovery, Paraná Basin, Paraguay. The border of the pseudohilum is torn (X5); diameter size: 38 μm. (11) *Dyadospora murusattenuata* Strother and Traverse 1979 emend. Burgess and Richardson (1991). The exine is thin and strongly folded. The line of attachment shows tendency to be separated into two monads; diameter size: 45 μm. (12) *Dyadospora murusdensa* Strother and Traverse 1979 emend. Burgess and Richardson (1991). The exine is thick. The two spores of the dyads are clearly separated and could be confused with *Gneudnaspora divellomedia* (see Fig. 7); diameter size: 40 μm

The oldest Ordovician assemblages are composed mainly of tetrads and dyads. Monads are rare (Figs. 3, 4, 5, and 6). Some monads display tears with a pseudo 'Y' shape that looks like an imperfect trilete mark so have been interpreted as spores resulting from the disaggregation of tetrads (Steemans et al. 2000). They are considered to be cryptospores and are named *Imperfectotriletes* (Fig. 2.6-8). Numerous dyads (e.g.

Table 1 Miospore taxa and their authors

Taxa	Authors
Abditusdyadus	Wellman and Richardson (1996)
Acinosporites lindlarensis morphon	Richardson et al. (1993)
Acontotetras inconspicuis	Richardson (1996c)
Ambitisporites	Hoffmeister (1959)
Ambitisporites avitus-dilutus morphon	Steemans et al. (1996)
Ancyrospora	Richardson emend Richardson (1996b)
Apiculiretusispora	(Streel) Streel (1967)
Archaeozonotriletes	Naumova emend. Allen (1965)
Archaeozonotriletes chulusnanus morphon	Steemans et al. (1996)
Artemopyra	Burgess and Richardson (1991)
Artemopyra brevicosta	Burgess and Richardson (1991)
Breconisporites	Richardson et al. (1982)
Cheilotetras caledonica	Wellman and Richardson (1993)
Chelinohilates	Richardson (1996c)
Chelinohilates erraticus	Richardson (1996c)
Chelinospora	Allen (1965)
Chelinospora cantabrica	Richardson et al. (2001)
Chelinospora cassicula	Richardson and Lister (1969)
Chelinospora poecilomorpha	(Richardson and Ioannides) Richardson et al. (2001)
Clivosispora verrucata convoluta	McGregor and Camfield (1976)
Clivosispora verrucata verrucata	McGregor and Camfield (1976)
Coronaspora	Rodríguez (1978)
Cymbohilates	Richardson (1996c)
Cymbohilates baqaensis	Breuer et al. (2007)
Cymbohilates cymosus	Richardson (1996c)
Cymbohilates disponerus	Richardson (1996c)
Cymbosporites	Allen (1965)
Cymbosporites dammamensis	Steemans (1995)
Cymbosporites proteus	McGregor and Camfield (1976)
Cymbosporites stellospinosus	Steemans (1989)
Dibolisporites	Richardson (1964)
Dictyotriletes	Naumova (1939) ex Ishchenko (1952)
Dyadospora	Strother and Traverse (1979)
Dyadospora murusattenuata	Strother and Traverse (1979) emend. Burgess and Richardson (1991)
Dyadospora murusattenuata-murusdensa morphon	Steemans et al. (1996)
Dyadospora murusdensa	Strother and Traverse (1979) emend. Burgess and Richardson (1991)
Emphanisporites	McGregor (1961)
Emphanisporites protophanus	Richardson and Ioannides (1973)
Gneudnaspora divellomedia	(Chibrikova) Balme (1988)
Gneudnaspora divellomedia var minor	Breuer et al. (2007)
Grandispora	Hoffmeister et al. emend. McGregor (1973)
Hispanaediscus	Cramer (1966)
Hispanaediscus verrucatus	(Cramer) Burgess and Richardson (1991)
Hystricosporites	McGregor (1960)
Imperfectotriletes	Steemans et al. (2000)
Imperfectotriletes patinatus	Steemans et al. (2000)
Imperfectotriletes varvdovae	(Richardson) Steemans et al. (2000)

Table 1 (continued)

Taxa	Authors
Laevolancis chibrikovae	Steemans et al. (2000)
Perotrilites	Couper emend. Evans (1970)
Retusotriletes	Naumova emend. Streel (1964)
Rhabdosporites	Richardson emend. Marshall and Allen (1982)
Rimosotetras problematica	Burgess (1991)
Rugosisporites	Dufka (1995)
Scylaspora	Burgess and Richardson (1995)
Segestrespora	Burgess (1991)
Segestrespora membranifera	(Johnson) Burgess (1991)
Stegambiquadrella contenta	Johnson (1985)
Streelispora	Richardson and Lister (1969)
Streelispora newportensis	(Chaloner and Streel) Richardson and Lister (1969)
Synorisporites	Richardson and Lister (1969)
Synorisporites cf. *S.? libycus*	in Burgess and Richardson (1991)
Synorisporites maculosus	Dufka (1995)
Synorisporites verrucatus	Richardson and Lister (1969)
Tetrahedraletes	Strother and Traverse (1979)
Tetrahedraletes medinensis	(Strother and Traverse) Wellman and Richardson (1993)
Velatitetras	Burgess (1991)
Velatitetras retimembrana	(Miller and Eames 1982) Wellman and Richardson (1996)
Zonotriletes	Luber and Walts (1938)

Segestrespora, Fig. 7.5) and tetrads (e.g. *Velatitetras*, Fig. 7.7) are enclosed by an outer laevigate or ornamented envelope. Ornament on the outer envelope is usually made up of small grana, a low reticulum or low rugulae. Dyads and tetrads may have exactly the same ornamentation; those from the early Llandovery are almost absent above this limit. The cryptospores enclosed in an outer envelope become very rare. The spores in dyads become progressively less attached, sometimes displaying a clear separation between them. Monads are more and more abundant.

The first ornamented naked dyads and monads evolved during the Llandovery becoming very abundant during the Wenlock (e.g. *Hispanaediscus verrucatus*, Fig. 7.11) evolving simultaneously with the first cryptospores with a hilum ridged by radial muri (e.g. *Artemopyra brevicosta*, Fig. 7.9). The first distal ornament consisted of verrucae, regular, irregular or merging into convoluted muri. During the Pridoli and the Lochkovian, ornamented patinate cryptospores (*Chelinohilates*, Fig. 8.1-3; *Cymbohilates*, Fig. 8.7-8, 9?) diversified and became abundant.

Trilete Spores

The Oldest Trilete Spores

The oldest known trilete spores are Llandoverian specimens of Hoffmeister (1959) from Libyan sediments. Much later, rare trilete spores were isolated from Hirnantian samples from Turkey (Steemans et al. 1996). In both cases, the trilete spores belong to the simple unornamented *Ambitisporites avitus-dilutus* morph (Fig. 9.1-3).

Very recently, a diverse assemblage of trilete spores (Steemans et al. 2008a, 2009; Wellman et al. 2008) has been described from Saudi Arabian strata. These have been independently dated as being Ordovician in age (late Katian/Hirnantian boundary) by chitinozoans (Paris et al. 2008) and by acritarchs (Le Hérissé et al. 2008). The assemblage includes laevigate trilete spores (*Ambitisporites*) as well as ornamented trilete spores (mainly *Synorisporites*) and forms illustrated in Fig. 9.1-3, 6. Previously non-ornamented trilete spores had been recorded prior to the Homerian Stage (Burgess and Richardson 1991; Dufka 1995). The observation of a diverse assemblage of Ordovician

Table 2 Plant taxa and their authors

Taxa	Authors
Aglaophyton	D.S. Edwards (1986)
Archaeopteris	Dawson (1871)
Asteroxylon	Kidston and Lang (1920)
Baragwanathia	Lang and Cookson (1935)
Bathurstia	Hueber (1972)
Calamophyton	Leclercq and Andrews (1960)
Chaleuria	Andrews et al. (1974)
Cooksonia banksii	Habgood et al. (2002)
Cooksonia caledonica	Edwards (1970)
Cooksonia paranensis	Gerrienne et al. (2001)
Cooksonia pertoni	Lang (1937)
Crossia	Beck and Stein (1993)
Drepanophycus	Göppert (1852)
Eospermatopteris	Goldring (1924)
Equisetum	Linnaeus (1753)
Estinnophyton	Fairon-Demaret (1978)
Foozia	Gerrienne (1992)
Hollandophyton	Rogerson et al. (2002)
Huvenia	Hass and Remy (1991)
Ibyka	Skog and Banks (1973)
Junggaria	Dou and Sun (1983)
Leclercqia	Banks et al. (1972)
Lyonophyton	Remy and Remy (1980)
Metacladophyton	Wang and Geng (1997)
Moresnetia	Stockmans (1948)
Oocampsa	Andrews et al. (1975)
Pertica	Kasper and Andrews (1972)
Pseudosporochnus	Leclercq and Banks (1962)
Psilophyton	Dawson (1859)
Rellimia	Leclercq and Bonamo (1973)
Remyophyton	Kerp et al. (2004)
Rhacophyton	Crépin (1875)
Rhynia	Kidston and Lang (1917)
Runcaria	Stockmans (1968)
Salopella	Edwards and Richardson (1974)
Sciadophyton	Steinmann (1929)
Sennicaulis	Kenrick et al. (1991)
Sphenophyllum	Brongniart (1822)
Stockmansella	Fairon-Demaret (1986)
Taeniocrada dubia	In Hueber (1982)
Tortilicaulis	Edwards (1979)
Zosterophyllum	Penhallow (1892)

trilete spores raises the question of their biological affinities: Were they produced by 'bryophytes' or by tracheophytes (refer to material online in Steemans et al. 2009)?

There is a clear relationship between tracheophytes and trilete spores, but not with cryptospores. The diversity of trilete spores is very low up to the Aeronian/Telychian transition, coincincidental approximately with the radical modification in the cryptospore assemblages. After this bioevent (Fig. 3), the diversity trilete of spores increases strongly. The earliest plant megafossils are of Homerian age: they are *Cooksonia*-type 'rhyniophytoids' (Edwards and Feehan 1980). The earliest demonstrated tracheophytes are of Přídolí and Lochkovian ages (Edwards 2003), a period of time during which the trilete spore biodiversity explodes. During Devonian times, trilete spores and tracheophytes diversify simultaneously, whereas there is a strong decrease in the diversity of cryptospores.

Trilete spores have been observed in situ from 'rhyniophytoids', i.e. plants resembling the simplest vascular plants but lack evidence of vascular tissue (Edwards et al. 1994), and in tracheophytes. Rare trilete marks have been observed in extant 'bryophyte' spores (Gray et al. 1985). In many cases, the trilete mark is merely a scar formed during ontogeny by contact with the other members of the tetrad and is not a functional germination suture: such spores are termed cryptotrilete. Gray et al. (1985) estimates that only some 2.5% of extant 'bryophyte' genera produce trilete spores. Many of these are cryptotrilete forms. True trilete marks are characterized by a suture or a suture with associated lips; the latter are usually discernible in fossil trilete spores. Many of those observed in the Ordovician from Saudi Arabia display such a true trilete mark, and therefore suggest tracheophyte affinities for their parent plants.

Ambitisporites and *Synorisporites* with irregular 'thickenings' similar to those seen on Ordovician specimens have also been observed in sporangia from several Late Silurian and Early Devonian specimens of *Cooksonia pertoni*. The following laevigate and ornamented spores have been identified: *Aneurospora* (Fig. 9.4-5), *Retusotriletes* (Fig. 9.7-8), *Streelispora newportensis* (Fig. 10.5) and *Synorisporites verrucatus* (Edwards and Richardson 1996; Edwards et al. 1999; Fanning et al. 1988; Habgood et al. 2002; Rogerson et al. 1993). *Synorisporites* with irregular 'thickening' is very close to *Synorisporites maculosus* in Dufka (1995) and to *Synorisporites* cf. *S.*? *libycus* in Burgess and Richardson (1991). Interestingly, it seems that a progressive

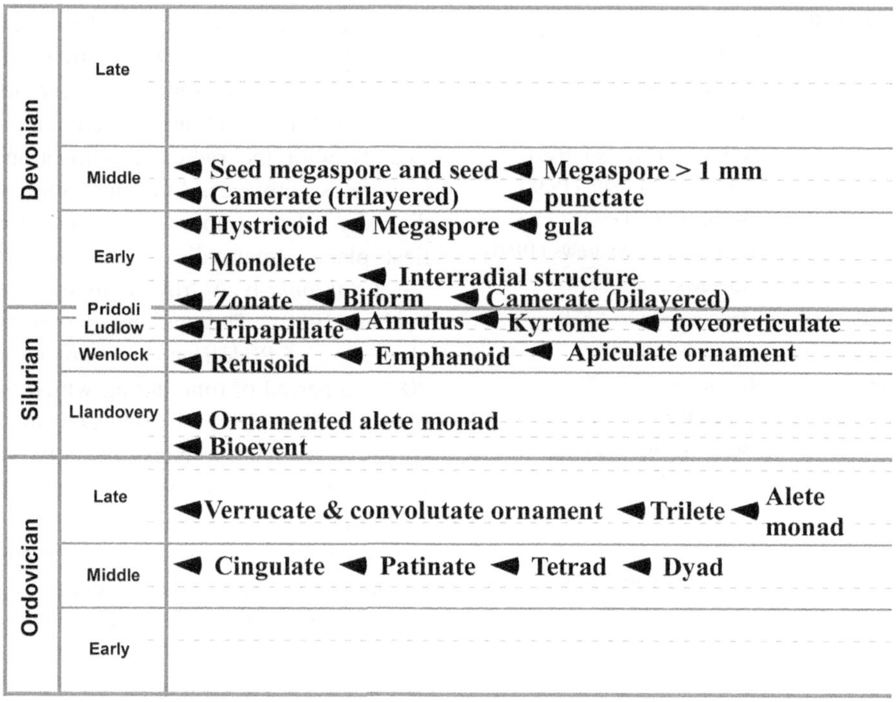

Fig. 3 Approximate time of appearance of the main morphological features and sculpturing types among cryptospores, monolete and trilete spores. For bioevent (see Section 'Main Morphological Events Within Cryptospores' and Steemans 2000)

morphological transition exists from the laevigate *Ambitisporites* to the ornamented *Synorisporites* in the Ordovician material (Steemans et al. 2009). This suggests that they may have been produced by the same plant and reinforces the possibility that the oldest trilete spores of the Ordovician were produced by tracheophytes similar to *Cooksonia* or by their ancestors.

The 'bryophyte' or tracheophyte affinities of the Ordovician trilete spores reported from Saudi Arabia will probably remain conjectural for a long time. Alternatively, the Saudi Arabian spores may have been produced by plants with a grade of organization transitional between 'bryophytes' and tracheophytes.

Main Morphological Events Within Trilete Spores

The earliest trilete spores are simple (Figs. 3, 4, 5, and 6), cingulate (with an equatorial thickening), laevigate (*Ambitisporites*) or ornamented with verrucae (*Synorisporites*). As stated earlier, these types of ornamented Ordovician trilete spores have recently been discovered from a single locality at the late Katian/Hirnantian boundary in Saudi Arabia (Steemans et al. 2008a; Wellman et al. 2008). *Ambitisporites* has also been observed in Hirnantian strata from Turkey (Steemans et al. 1996), and from a Rhuddanian (Silurian) Saudi Arabian locality (Steemans et al. 2000). The earliest occurrences of the genus are diachronous from east to west as shown in Steemans (2000).

The first patinate (i.e. with a thin proximal face and a thick distal face) forms are laevigate and called *Archaeozonotriletes* (Fig. 8.5-6). They have been observed from several localities of Telychian age, for example, in Saudi Arabia (Wellman et al. 2000), Libya (Richardson 1988), Paraguay (Steemans and Pereira 2002) and the United Kingdom (Burgess 1991).

The first large-scale diversification of trilete spores occurs during the Homerian. It is interesting to note that the earliest *Cooksonia*-like mesofossils (tracheophytes?) are contemporaneous. Laevigate (*Retusotriletes*, Fig. 9.7-8) or ornamented (*Apiculiretusispora*, Fig. 9.9-11) retusoid forms appear. Radial muri (*Emphanisporites*, Figs. 7.10 and 10.4, 6) or irregular rugulate muri *Rugosisporites* (in Dufka 1995) or *Scylaspora* (in Burgess and Richardson 1995) on the proximal face are observed for the first

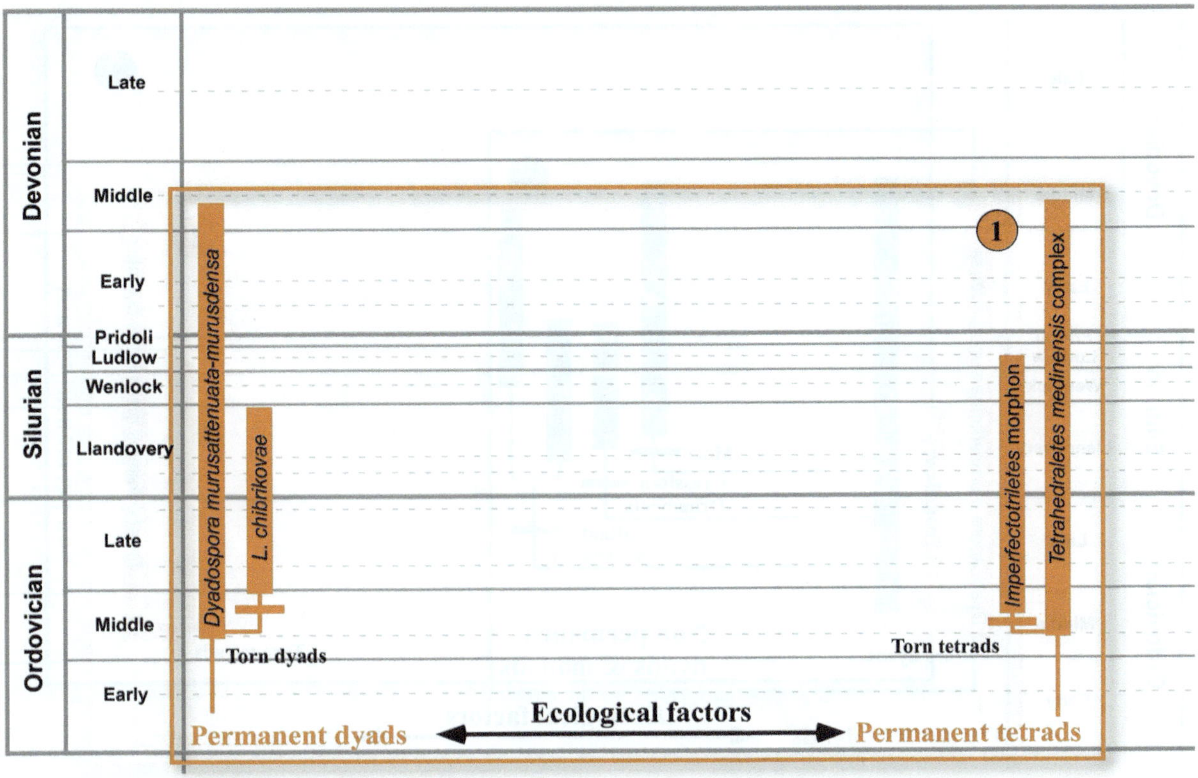

Fig. 4 Figures 4, 5 and 6 show the four successive layers illustrating approximate stratigraphic range (*the vertical rectangles*) of the main taxa of cryptospores and trilete spores. *Lines* linking rectangles symbolize possible morphological relationships between taxa. *Horizontal small rectangles* intersecting the lines explain the event responsible of the morphological modification between the taxa. The four layers symbolize steps of the morphological evolutions through time for same group of morphologies and their direct descendant. Links exist between the layers. Layer 1 shows the approximate stratigraphic range of the two main simplest taxa of permanent tetrad and dyad cryptospores. They can be torn to release artificial monads with a pseudotrilete mark (Imperfectotriletes)

time. Very rare foveolate Homerian specimens have also been observed (Richardson and Ioannides 1973).

Tripapillate forms, such as in *Synorisporites papillensis* (Fig. 9.6), have long been an important biostratigraphic marker of the Ludfordian/Přídolí boundary. It has now been demonstrated that this proximal structure (see above paragraph) is present during the Ludlow (Rubinstein and Steemans 2002). That kyrtomes on the proximal surface, as in *Coronaspora cromatica* (Fig. 10.2), seem to have appeared during the Přídolí (Richardson et al. 2001) needs further consideration. The distal ring (annulus), as in *Amicosporites* (Fig. 10.3), is another important structure first appearing during the Přídolí (Richardson and Ioannides 1973). Although a possible *Aneurospora*, cingulate and ornamented (Fig. 9.4-5), has been observed in Ordovician sediments from Saudi Arabia, it suggests that the actual diversification of the genus must have occurred during the late Ludlow. The patinate spores ornamented with coni and spines, as in *Cymbosporites* (Figs. 8.9, 10 and 10.1), or with verrucae, as in *Chelinospora* (Fig. 8.4), first appeared in the Ludlow. The late Ludlow and Přídolí interval is characterized by very similar species ornamented with coarse verrucae or convolute coarse muri, as in *Chelinospora poecilomorpha*, *Clivosispora verrucata* var. *verrucata* or *C. verrucata* var. *convoluta*. It is difficult to separate the latter species as they most probably all belong to a single morphon. The reticulate structure on the distal patina, as in *Chelinospora cantabrica*, first occurred during the Přídolí.

The Lochkovian age is marked by two important innovations: the zonate and the camerate trilete spores. The zonate spores as in *Zonotriletes* spp. (Fig. 10.10)

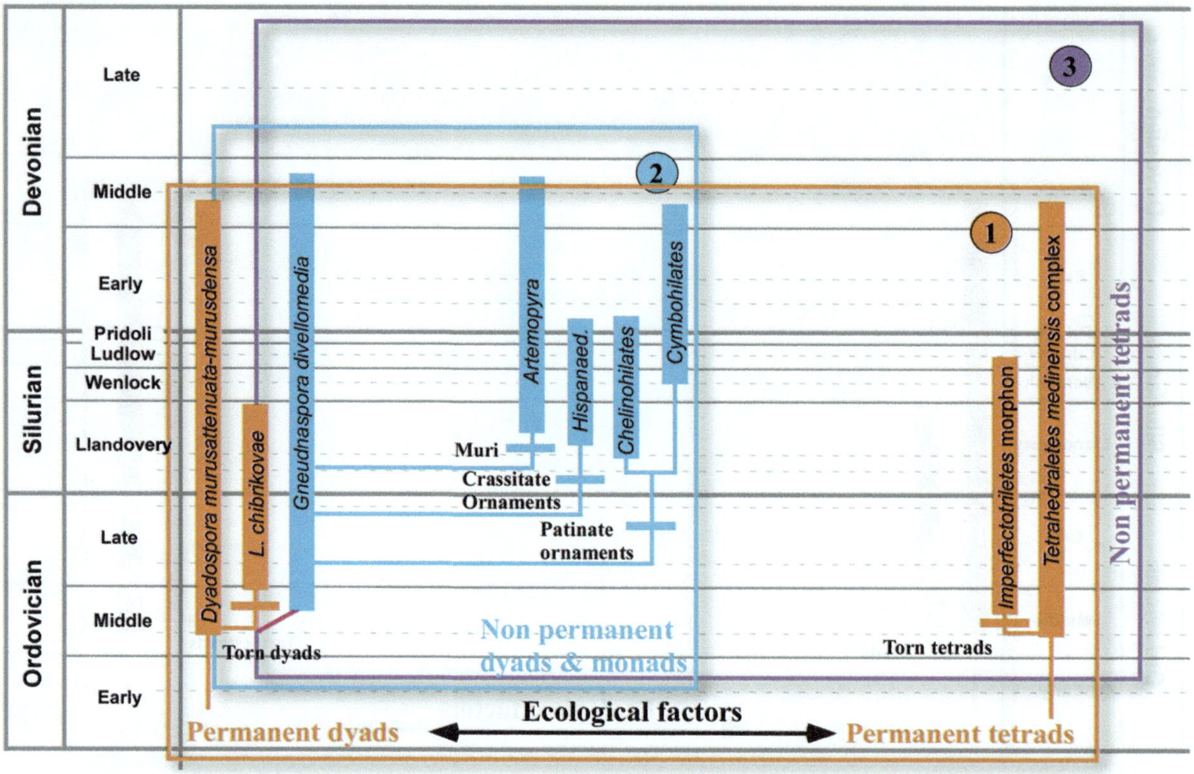

Fig. 5 (See also Fig. 4 caption) Layer 2 shows the approximate stratigraphic range of the taxa of non-permanent dyads and the monads stem from them. The definition of the different taxa relates to presence of patina or equatorial crassitude or ornament. Layer 3 contains no taxa. It symbolizes a putative stage in which all plant producing cryptospores illustrated on the layers 1 and 2 shift from permanent to non-permanent tetrads. Non-permanent tetrads are preserved in the sporangia. When they are at maturity, tetrads split and isolated trilete spores are dispersed

seem to appear first in Gondwanan localities during the early Lochkovian (Jardiné and Yapaudjian 1968) and later in the Lochkovian on Euramerica as exemplified by *Cirratriradites diaphanus* (Fig. 10.9). Camerate spores appeared very early in the Lochkovian with *Perotrilites* (Fig. 10.11) and with the good biostratigraphic marker *S. newportensis* (Fig. 10.5) on the Old Red Sandstone area (Richardson and Lister 1969; Streel et al. 1987). On the Gondwanan continent, rare camerate specimens of *Grandispora* (Fig. 10.12) are known from the early Lochkovian of Brasil (Steemans et al. 2008b) and Saudi Arabia (Steemans et al. 2007). *Dictyotriletes* (Fig. 10.8) became very abundant from the Lochkovian onwards. No retusoid spores ornamented with biform ornament such as *Dibolisporites* (Fig. 9.12) have been observed prior to mid-Lochkovian times (Steemans 1989).

The earliest monolete spores, for example, *Latosporites* (Fig. 11.1), are known from more or less coeval localities near the Pragian/Emsian transition from Saudi Arabia (Breuer et al. 2007), the United Kingdom (Wellman 2006) and Brazil (Mendlowicz Mauller et al. 2007).

Well-marked differentiation of the spore into sexine and nexine was common during mid-Emsian times, initially with *Rhabdosporites* and, shortly after, by large forms such as *Grandispora* (Fig. 11.8).

The late Emsian is characterized by the incoming of bifurcate ornament in species of the genera *Ancyrospora* (Fig. 11.4-5) and *Hystricosporites*.

From the Silurian to the Emsian, all spores increase in size. The Emsian is marked by a very important step in the evolution of the vegetation: the advent of heterospory (Andrews et al. 1974; Gerrienne 1996; Meyer-Berthaud and Gerrienne 2001). Apart from the megaspores mentioned by Richardson and McGregor (1986) from sediments now regarded as being of early Pragian to earliest Emsian in age, the earliest

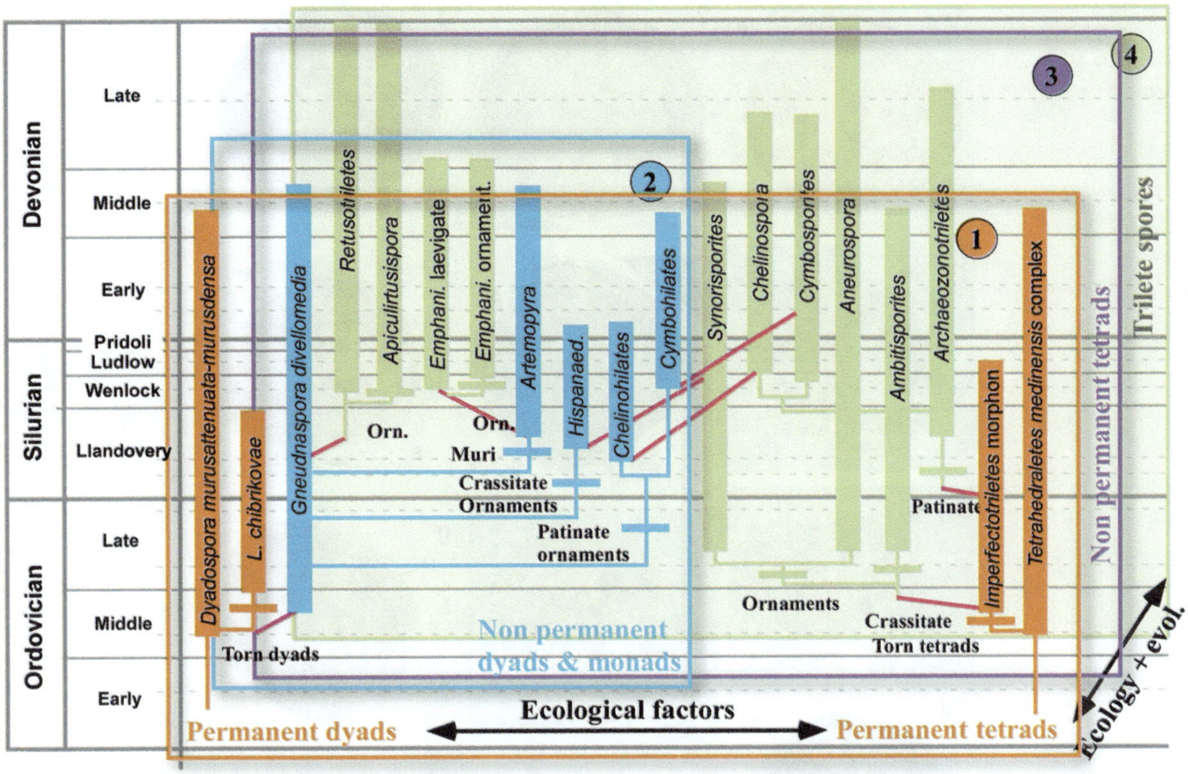

Fig. 6 (See also Fig. 4 caption). The layer 4 contains the trilete spores resulting from the dissociation of non-permanent tetrads

unequivocal incoming of dispersed megaspores (>200 μm) occurred at the end of the Emsian (Traverse 2007). Whilst the functional definition of a megaspore is clear, recognition on morphological basis only of a dispersed fossil spore as a true megaspore is difficult. The diameter size of 200 μm defining a megaspore is arbitrary, and it is not a definitive proof of heterospory. It is also widely accepted that large dispersed spores smaller than 200 μm could be small megaspores. It has been suggested (Hemsley et al. 1999) that 115 μm in diameter size could be a more realistic limit. In some cases, spores of very similar morphologies show a bimodal diameter size curve as shown in Poland (Turnau and Jakubowska 1989). The 1 mm diameter size boundary (Fig. 11.11) is crossed during the Givetian (Ville de Goyet et al. 2007).

Several Middle Devonian camerate species clearly show differentiation of the nexine, into two outer layers (Breuer 2007, Fig. 11.10). During the Middle Devonian, different patinate/crassitate taxa show for the first time a clear punctuation of their exine (Breuer 2007).

A further step in the heterospory evolution is the seed-megaspore, where three aborted spores stay attached to the megaspore (Marshall and Hemsley 2003).

From Cryptospores to Trilete Spores: A Possible Scenario

Dyads or Tetrads: Could They Be Produced by the Same Plant?

The oldest assemblages of cryptospores are composed mainly of tetrads and dyads, naked or enclosed in an envelope. The most common and morphologically simple cryptospores belong to *T. medinensis* and the *Dyadospora murusattenuata-murusdensa* morphon. They are composed of four and two cells, respectively, both laevigate and naked, and display wide morphological variation as expressed by their large range of both of their diameter and wall thickness. The lack of distinctive morphological characters and as all intermediate forms exist between the different

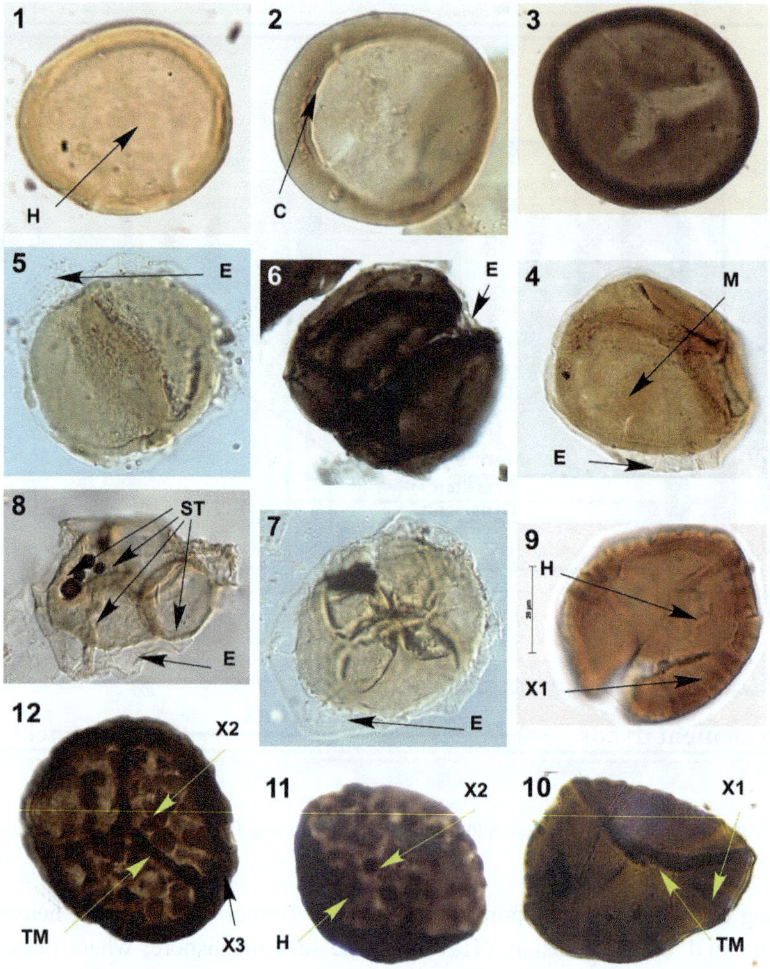

Fig. 7 For abbreviations, see Fig. 1. (1–3) *Gneudnaspora* (*Laevolancis*) *divellomedia* var. minor Breuer et al. 2007 showing different exine thickness. (1, 2) Emsian, Saudia Arabia; diameter sizes: 27 and 29 μm. (3) Lochkovian, Brittany; diameter size: 35 μm. (4) *Sphaerasaccus glabellus* Steemans et al. 2000. Llandovery, Paraná Basin, Paraguay; diameter size: 47 μm. (5) *Segestrespora laevigata* Burgess 1991 Hirnantian, Chad; diameter size: 40 μm. (6) *Abditusdyadus laevigatus* Wellman and Richardson 1996. Lochkovian, Saudi Arabia. The two cells of the dyad are separated; diameter size: 40 μm. (7) *Velatitetras laevigata* Burgess 1991. Hirnantian, Chad; diameter size: 48 μm. (8) *Stegambiquadrella contenta* Johnson 1985. Hirnantian, Chad. The four spores of the terad are clearly dissociated inside of the envelope; diameter size: 52 μm. (9) *Artemopyra brevicosta* Burgess and Richardson 1991. Ludlow/Přídolí, Saudi Arabia. Short radial muri are visible at the internal limit of the hilum (X1); diameter size: 22 μm. (10) *Emphanisporites protophanus* Richardson and Ioannides 1973. Late Silurian/Early Devonian, Argentina. The trilete mark is visible (TM), the contact area is ornamented by radial muri similar to those of *A. brevicosta* in (9); diameter size: 34 μm. (11) *Hispanaediscus verrucatus* Cramer 1966. Lochkovian, Brittany. The distal face is ornamented by rounded verrucae (X2); diameter size: 18 μm. (12) *Synorisporites verrucatus* Richardson and Ioannides 1973. Lochkovian, Saudi Arabia. The trilete mark is well visible (TM), the verrucae on the distal face (X2) are similar to those of *H. verrucatus*, the equatorial crassitude here is thicker (X3); diameter size: 30 μm

morphologies, the only distinguishing character is the tetrad or dyad configuration. Considering the long stratigraphic range of the two taxa from the Middle Ordovician to the Middle Devonian, and their worldwide palaeogeographic distribution, it seems clear they have been produced by different plants. This has been confirmed by the high variability of their ultrastructure wall (Edwards et al. 1996; Taylor 1995, 1996, 1997, 2002, 2003; Wellman et al. 1998b).

Fig. 8 For abbreviations, see Fig. 1. (1–3) *Chelinohilates erraticus* Richardson 1996c. Lochkovian, Saudi Arabia. Specimen in (1) is in tetrad configuration (image in reflected light); diameter size: 53 μm. Specimen in (2) is in dyad configuration, the envelope is ornamented by convolute muri which become parallel over the area of contact between the two spores (X1); diameter size: 58 μm. Specimen in (3) is in monad configuration; diameter size: 40 μm. (4) *Chelinospora cassicula* Richardson and Lister 1969, the specimen shows the same convolute muri on the distal face than those of *C. erraticus*. The trilete mark (TM) is clearly visible on the proximal face; diameter size: 30 μm. (5–6) *Archaeozonotriletes chulusnanus* morphon Steemans et al. 1996. Emsian, Saudi Arabia. Specimens showing a strong patina (X2) and a thin proximal face; diameter size: 52 μm (5) and 35 μm (6). (7–8) *Cymbohilates baqaensis* Breuer et al. 2007. Pragian/Emsian, Saudi Arabia. Specimens showing tears on the hilum; diameter size: 38 μm (7) and 31 μm (8). (9) *Cymbohilates baqaensis* Breuer et al. 2007 or *Cymbosporites dammamensis* Steemans 1995. Pragian/Emsian, Saudi Arabia. The tears on the proximal face looks like a trilete mark, but the curvaturae are perfectly circular and do not show the typical invagination at the end of the rays of the trilete mark (X3); diameter size: 35 μm. (10) *Cymbosporites dammamensis* Steemans 1995. Pragian/Emsian, Saudi Arabia. The trilete mark and the curvaturae are well visible. Invaginations at the end of the rays of the trilete mark are present; diameter size: 36 μm. (11–12) *Cymbohilates disponerus* Richardson 1996c. Lochkovian, Brittany, France. Specimen in (11) is in dyad configuration; diameter size: 55 μm. Specimen in (12) is a monad closely similar to *Cymbosporites proteus* McGregor and Camfield 1976 (Fig. 10.1); diameter size: 27 μm

Hemsley (1994) suggested ontogenetic pathways to explain how a given sporangium might produce a great variety of cryptospores, depending on the timing of sporopollen in deposition during meiosis. It can even be hypothesized that monads, dyads or tetrads, naked or enclosed in an envelope may be formed in the same sporangium. Ultrastructural studies, however, give conflicting results. In some cases, dyads and tetrads with identical morphologies exhibit ultrastructural differences suggesting that they were produced

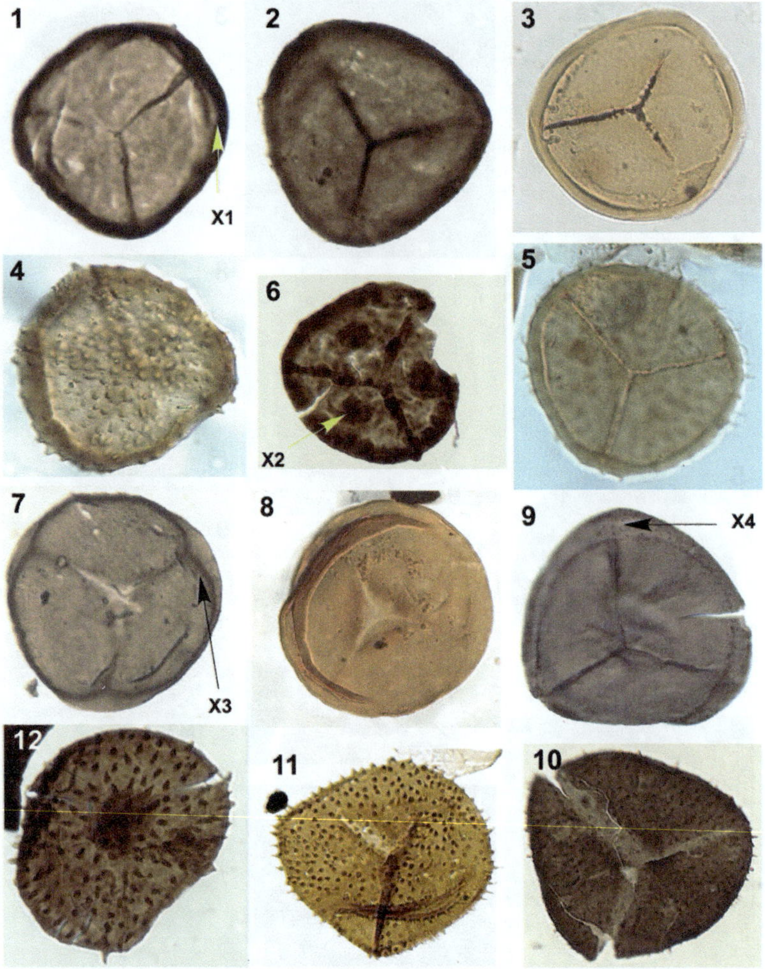

Fig. 9 For abbreviations, see Fig. 1. (1–3) Various aspects of *Ambitisporites avitusdilutus* morphon Steemans et al. 1996. Lochkovian, Brittany, France (1–2); diameter size: 28 and 33 μm. Pragian, Saudi Arabia (3); diameter size: 54 μm. The equatorial crassitude displays variable thicknesses. The lightest specimens are close to the morphological aspect of Retusotriletes. The darker specimens may more easily be confused with Archaeozonotriletes. (4) *Aneurospora* cf *A. bollandensis* in Steemans et al. 2008a. Lochkovian, Amazon Basin, Brasil; diameter size: 35 μm. (5) *Aneurospora richardsonii* Richardson et al. 2001. Lochkovian, Amazon Basin, Brasil; diameter size: 31 μm. (6) *Synorisporites papillensis* McGregor 1973. Lochkovian/Pragian, Saudi Arabia. A papilla is present on each contact area; diameter size: 37 μm. (7–8) Various aspects of *Retusotriletes* spp. Lochkovian, Brittany, France (7); diameter size: 38 μm. Pragian/Emsian, Saudi Arabia (8); diameter size: 40 μm. The darkest specimens (7) are close to *Ambitisporites* as their curvaturae appear thicker, appearing as an equatorial crassitude. (9–11) Various aspects of *Apiculiretusispora* spp. Lochkovian, Brittany, France (9, 10). Pragian/Emsian, Saudi Arabia (11).The ornaments (X4) are clearly situated outside the contact areas (photographed in reflected light in 9); diameter size: 47 μm. Specimens in (10) have longer spines; diameter size: 35 μm. Specimen in (11) is ornamented by spines with a slightly bulbous base close to that of some species of *Dibolisporites*; diameter size: 40 μm. (12) *Dibolisporites eifeliensis* McGregor 1973. Pragian, Saudi Arabia; diameter size: 50 μm

by distinct mother plants (Taylor 2002). Other tetrads and dyads with similar morphologies, for example, the dyad *Segestrespora membranifera* and the tetrad *Velatitetras retimembrana*, show striking ultrastructural similarities (Taylor 1996), strongly suggesting that a same plant or group of plants, possibly because of variable ecological factors, could produce either mainly tetrads or mainly dyads.

The Tetrahedraletes Complex

The tetrad is the most common spore association produced in embryophyte sporangia. Except for obligate

Fig. 10 For abbreviations, see Fig. 1. (1) *Cymbosporites proteus* McGregor and Camfield 1976. Lochkovian, Brittany, France. Specimen displaying a aspect similar to *Cymbohilates disponerus* Richardson (1996) (Fig. 8.12); diameter size: 32 μm. (2) *Coronaspora cromatica* Jansonius and Hills 1979. Lochkovian, Saudi Arabia. Three kyrtomes are visible on the proximal face (X1) and a foveolate structure on the proximal and distal areas. Similar specimens that may belong to the same species have been erroneously published in Steemans 1995 as *Chelinospora arabiensis*; diameter size: 37 μm. (3) *Amicosporites streelii* Steemans 1989. Lochkovian, Ardenne, Belgium. Presence of a distal annulus (X2) and verrucae inside it; diameter size: 42 μm. (4) *Emphanisporites annulatus* McGregor 1961. Emsian, Saudi Arabia. The annulus (X2) is on the distal face and the radial muri (X3) on the proximal face; diameter size: 46 μm. (5) *Streelispora newportensis* Richardson and Lister 1969. Lochkovian, Ardenne, Belgium. Three papillae (X4) are visible on the proximal face. The outer exine of this camerate species is folded. The folds (X5) link the papillae and the trilete mark. Ornamentation is not clearly visible on this specimen; diameter size: 26 μm. (6) *Emphanisporites zavallatus* Richardson et al. 1982 var *gedinniensis* Steemans and Gerrienne 1984. Lochkovian, Ardenne Belgium; diameter size: 37 μm. (7) *Brochotriletes foveolatus* Naumova 1953. Pragian/Emsian, Saudi Arabia. The foveolate structure (X6) is on the distal face; diameter size: 54 μm. (8) *Dictyotriletes subgranifer* McGregor 1973. Pragian/Emsian, Saudi Arabia. The reticulum is on the distal face (X7); diameter size: 62 μm. (9) *Cirratriradites diaphanus* Steemans 1989. Pragian/Emsian, Saudi Arabia. There are three papillae on the contact surfaces and an equatorial zona (X8); diameter size: 43 μm. (10) *Zonotriletes* sp. Emsian, Brasil. This specimen is characterized by a large equatorial zona; diameter size: 72 μm. (11) *Perotrilites laevigatus* Steemans 1989. Lochkovian, Amazon Basin, Brasil. The outer sexine (X9) is visible at the equator; diameter size: 40 μm. (12) *Grandispora* sp. A in Steemans et al. 2008a. Lochkovian, Saudi Arabia. The outer sexine is folded (X9) and ornamented by coni; diameter size: 29 μm

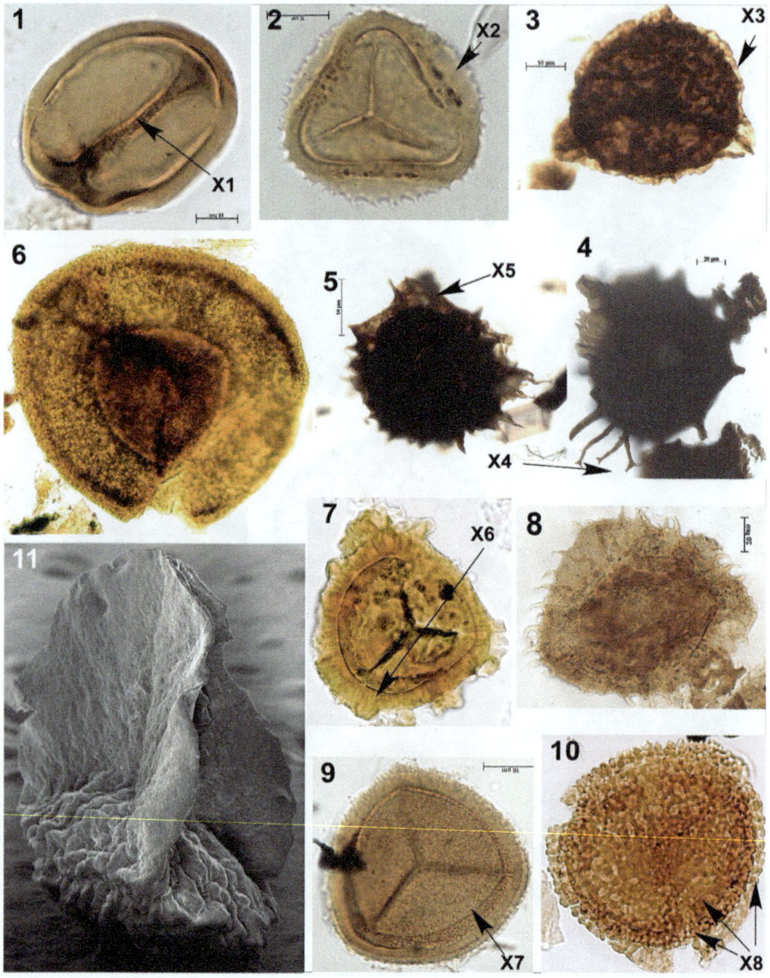

Fig. 11 For abbreviations, see Fig. 1. (1) *Latosporites ovalis* Breuer et al. 2007. Emsian, Paraná Basin, Brasil. The proximal face is characterized by a monolete mark (X1); diameter size: 56 μm. (2) *Camarozonotriletes sextantii* McGregor and Camfield 1976. Eifelian, Namur Basin, Belgium. The equatorial interradial areas (X2) are ornamented by long spines; diameter size: 33 μm. (3) *Samarisporites triangulatus* Allen 1965. Givetian, Namur Basin, Belgium. The equatorial zona is narrower in the interradial equatorial areas (X3) in the interradial area; diameter size: 52 μm. (4–5) *Ancyrospora ancyrea* Richardson 1964. Givetian, Namur Basin, Belgium. The top of the elements is clearly bifurcate (X4) in (4); diameter size: 47 μm. The gula (X5) is developed above the trilete mark; diameter size: 45 μm. (6) *Contagisporites optivus* Owens 1971. Givetian, Saudi Arabia. Megaspore; diameter size: 220 μm. (7) *Lophozonotriletes media* Taugourdeau-Lantz 1967. Givetian, Ghadames Basin, Tunisia. The punctuatae (thin canals) are clearly visible in the thick equatorial crassitude (X6); diameter size: 80 μm. (8) *Grandispora douglastownense* McGregor 1973. Middle Devonian, Bolivia; diameter size: 92 μm. (9) *Geminospora lemurata* Balme emend. Playford 1983. Givetian, Ronquières, Belgium. The inner exine (X7) is thinner than the outer exine; diameter size: 40 μm. (10) *Acinosporites* sp. 1 in Breuer et al. 2007. Eifelian, Saudi Arabia. The specimen is camerate and is characterized by three exine layers (X8); diameter size: 84 μm. (11) *Lagenicula* sp. Givetian, Ghadames Basin, Libya. Megaspore larger than 1 mm in equatorial diameter with a very high gula; diameter size: 1861 μm

cryptospore tetrads, the spores at maturity are usually liberated from the sporangium as trilete or monolete isolated spores. Permanent tetrads are produced by a limited number of extant 'bryophytes'.

Even if it cannot be demonstrated, there might be a link between obligate tetrads and true trilete spores. The spores from Tetrahedraletes sometimes show a tendency to become detached. Another example is *Rimosotetras problematica* (Fig. 2.4), an obligate tetrad with loosely attached spores. Sometimes, the tetrad is physically destroyed, releasing individual monads with a torn proximal face. The tears may mimic a trilete mark: the resemblance is at times so close that it is impossible to distinguish a pseudo trilete mark from a true functional trilete mark. Those spores were first attributed to ?*Ambitisporites* (Richardson 1988). Later, the new genus *Imperfectotriletes* was established (Steemans et al. 2000) to accommodate such spores.

Tetrads are heavier and larger than monads and are statistically dispersed closer to the parent plant than isolated spores. Gray et al. (1985) suggested that the reproductive advantage of cryptospore obligate tetrads was the juxtaposition of male and female gametophytes in the propagules, conferring advantage for founder populations in stressed environments (see also Section 'From Alete Monads Totrilete Spores'). The colonization of new areas, in more favourable conditions, could presumably be improved by the production of individual spores, which disperse further because they are lighter and smaller than tetrads. This reproductive strategy allows genetic exchanges with individuals dispersed from other areas.

Imperfectotriletes patinatus (Fig. 2.6-7) is patinate (having a thick distal face), and *Imperfectotriletes varvdovae* (Fig. 2.8) is crassitate (with an equatorial crassitude). There is no clear separation between both species as all intermediate wall thickenings exist. Optically, there are obvious similarities between the cryptospore *I. patinatus* and the trilete spore *Archaeozonotriletes chulusnanus* morphon (patinate trilete spores Fig. 8.5-6) and between the cryptospore *I. varvdovae* and trilete spore *A. avitusdilutus* morphon (equatorially crassitate trilete spores Fig. 9.1-3).

For all taxa, the laevigate trilete spore appears before its ornamented morphologic equivalent. For example, the laevigate *Archaeozonotriletes* first occurs during the Telychian and when it becomes ornamented with muri or similar structures, it is named *Chelinospora*; this event first occurs during the Ludlow. If the ornament is spines or coni, the specimens are called *Cymbosporites*. Ornamented *Ambitisporites* are either *Synorisporites* or *Aneurospora* according to the type of ornament present on the distal face.

The *Dyadospora* Complex

It was shown by Gray (1993) that no living plant produces dyads, except as a result of meiotic irregularities. According to Hemsley (1994), theoretically it is possible for plants to produce dyads if they favoured meiotic dispersion strategy of their propagules. The two more abundant dyads are *D. murusattenuata* (Fig. 2.11) and *D. murusdensa* (Fig. 2.12), characterized by a thin and a thick wall respectively. Between both, all variation of thickness exists and a morphon has been created to accommodate the two species (Steemans et al. 1996).

As mentioned above for tetrads, the production of dyads as propagules might be disadvantageous for dispersion and reproduction, and the production of monads possesses important advantages. This could explain why spores in dyads show a progressive tendency to be less and less attached, to be finally completely separated in two alete monads. In Ordovician samples, some monads with a torn proximal face have clearly produced by the physical separation of dyads. It can be hypothesized that tetrads were modified in isolated *Imperfectotriletes* monads by the same process. The monads physically isolated from tightly attached dyads are called *Gneudnaspora* (*Laevolancis*) *chibrikovae* (Fig. 2.10). In younger sediments, spores from dyads are clearly separated and, when they are found as monads, they are called *Gneudnaspora* (*Laevolancis*) *divellomedia* or *plicata* (Fig. 7.1-3), according to their wall thickness (Breuer et al. 2007). Once more, there is a continuous, progressive, morphological variation of their wall thickness.

From Alete Monads to Trilete Spores

Monads are rare during the Llandovery: some were ornamented (Wellman et al. 2000). Monads become abundant during the Wenlock and later. Four genera have been created according to the type of sculpture and the wall thickness. *Artemopyra* has radial muri on the proximal face, *Hispanaediscus* is equatorially crassitate and distally ornamented by verrucae, *Chelinohilates* and *Cymbohilates* are patinate: the former ornamented by muri or similar elements and the latter by spines and cones.

There are clear morphological relationships between several hilate monads and trilete spores: some ornamented monads are very similar to trilete spore species, for example, *Artemopyra brevicosta* (Fig. 7.9) is the cryptospore equivalent to the trilete spore *Emphanisporites protophanus* (Fig. 7.10), while the cryptospore *H. verrucatus* (Fig. 7.11) corresponds to the trilete spore *S. verrucatus* (Fig. 7.12). Sometimes, a progressive transition is observable between the hilate monad and the trilete spore, for example, the Lochkovian and Pragian *Cymbohilates baqaensis* (Fig. 8.7-8, 9?) and *Cymbosporites dammamensis* (Fig. 8.9?, 10). Some specimens of *Cymbohilates* display a torn hilum mimicking a pseudotrilete mark, thus making identification of the specimens difficult. In all cases, the cryptospore appears first, rapidly followed by the corresponding trilete spore. However, the presence of a trilete mark on spores having a cryptospore equivalent organized in a dyad is puzzling. As the necessary condition for the presence of a trilete mark is the tetrad configuration, it seems impossible that a monad isolated from a dyad could display a trilete mark. Therefore, it is inferred that plants that produced dyads dispersed as isolated alete monads (e.g. *H. verrucatus*) were also able to produce tetrads. The reason as to how the shift from dyads to tetrads occurred remains conjectural, but, tentatively, it again appears to relate to variation of ecological conditions. As the released tetrads have not been observed in the palynological record, they were presumably not obligate, and the spores were dispersed as monads. Although obligate tetrads become rare in the Early Devonian, an unusual abundance of obligate tetrads was noticed in Lochkovian samples from the Arbuthnott Group in Scotland (Lavender and Wellman 2002). The authors noticed that '... perhaps land plants of Lochkovian age were rather flexible in terms of the state in which their spores were released'. We suggest that, presumably under harsh conditions, the basal obligate tetrad configuration could be superior to the derived monad configuration. Indeed, the obligate tetrad configuration of mature spores has a genetic basis. Two genes are required for the spore separation during the normal development (Rhee and Sommerville 1998). The occurrence of abnormal permanent tetrad configuration in living plants is the most common mutation due to modification by atmospheric air composition (e.g. presence of volcanic ash and gases) or increasing radiation (UV) caused by less effective protection by the ozone layer. Several observations support this suggestion. Specimens similar to Cymbohilates cymosus are known under three different configurations (Richardson 1996b): tetrad, alete monad and a trilete monad (then called *Cymbosporites stellospinosus*). *Chelinohilates erraticus* (Fig. 8.1-3) is known as alete monad, dyad, tetrad and a trilete monad (then called *Chelinospora cassicula*, Fig. 8.4). *Cymbohilates disponerus* (Fig. 8.11-12) is known as an alete monad and dyad; it is also very similar to the tetrad *Acontotetras inconspicuis* and to the trilete spore *Cymbosporites proteus* (Fig. 10.1).

Synthetic Diagram

The scenario leading from cryptospores to trilete spores is summarized in Figs. 4, 5, and 6. We emphasize that this figure has no strict phylogenetic implications. The palynological work is based on morphotaxa and not on Linnean species. The diagram shows the evolution of the spore morphologies through time; it does not imply any ancestor/descendant relationship. The figure is comprised of four superposed and related transparent layers. The first layer contains the permanent tetrads and dyads and is the result of their physical separation. The second layer includes the loose dyads transformed into monads and their various morphologies. The third layer symbolizes the theoretical phase when all sporangia, probably because of ecological factors, produce mainly non-permanent tetrads (and no more non-permanent dyads) however; as the tetrads are not permanent, they cannot be observed, and accordingly, layer 3 contains no taxa. The fourth layer contains only trilete spores.

The diagram shows that several morphotaxa of trilete spores (*Cymbosporites*, *Chelinospora* and *Synorisporites*) may be derived in two ways, either directly from tetrad-producing plants or indirectly from dyad-producing plants. Ecological factors most probably determine the type of propagules produced by the sporangia. In a given genus, the species may be different according to the strategy followed by the mother plant. In contrast, retusoid and emphanoid spores seem to be produced only by a plant lineage that originally produced only permanent dyads. Similarly, *Archaeozonotriletes*, *Ambitisporites* and *Aneurospora* seem to be produced only by a plant lineage that originally released only permanent tetrads.

It is sometimes very difficult, especially with coalified palynomorphs, to estimate the nature and the thickness of the spore wall. Retusoid forms may be confused with equatorially crassitate forms, which in turn may be confused with patinate forms. Therefore, the Synthetic diagram (Figs. 4, 5, and 6) tries to reflect all possibilities.

From Permanent to Non-permanent Dyads and Tetrads

Layers 2 and 3 in Figs. 5 and 6 raise the question of how permanent dyads or tetrads might be transformed into non-permanent ones. The oldest dyads and tetrads seem to be tightly attached over the whole contact area between the spores, making it difficult to isolate one of them without tearing the contact area(s) of the isolated spore or of but the other spores in the tetrads or the dyads. On some isolated *Imperfectotriletes*, the remains the tetrad or of dyad wall have been observed. The development of haptotypic elements on the proximal surface has possibly been made easier by the division of the spores from tetrads and dyads by decreasing the area of the surfaces in contact (Fig. 12). Indeed, labra and curvaturae are raised in relief above the proximal face: this presumably decreases the surface of contact and therefore decreases also the strength of attraction between spores in the dyads and tetrads.

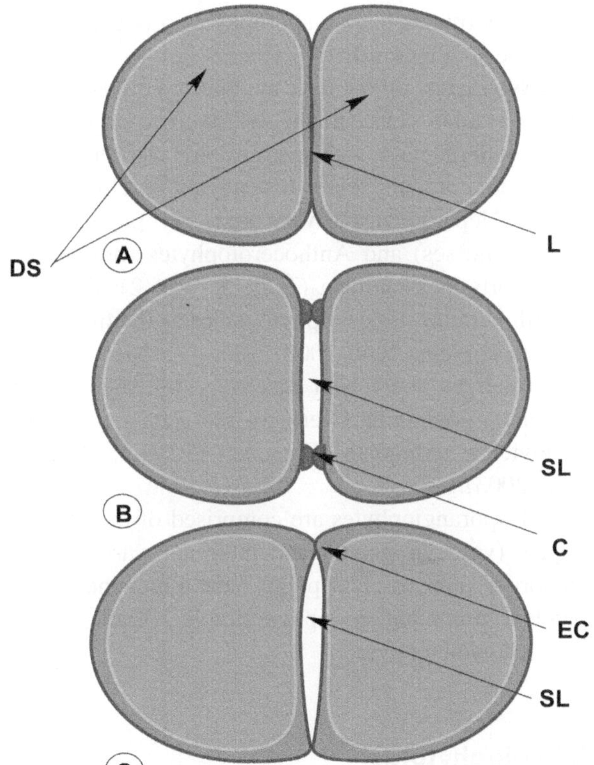

Fig. 12 (See Fig. 1 for abbreviations) **a** A dyad with a line of attachment; **b** The surface of attachment between the spores could be reduced by the presence of, for example, curvaturae slightly raised in relief on the surface of attachment; **c** The presence of a proximo-equatorial crassitude (EC) could also result in a separation between the spores

What Happens with the Cryptospores Enclosed in a Membrane?

There appears to be no place in Synopsis diagram (Figs. 4, 5, and 6) for cryptospores enclosed in an envelope. According to the hypothesis expressed above, all the morphologies displayed by the most basal trilete spores could be obtained from the sole naked *Dyadospora* and *Tetrahedraletes*. Nevertheless, some taxa enclosed in an envelope seem to have followed the same evolutionary strategy, the dyads and tetrads progressively separating as monads. Indeed, *Abditusdyadus laevigata* (Fig. 7.6) and *Stegambiquadrella contenta* (Fig. 7.8) are respectively dyads and tetrads formed by spores slightly separated and enclosed in an envelope. The presence of the envelope may be interpreted as having prevented the development of a trilete mark. The abundance of cryptospores enclosed in envelopes declined drastically near the Aeronian/Telychian boundary, when trilete spores were rapidly diversifying (Steemans 2000). On the contrary, the naked *Tetrahedraletes* and *Dyadospora* remained common at least up to the Lochkovian, suggesting that the presence of an envelope was disadvantageous, maybe in preventing a rapid germination of the spore. Plants producing cryptospores enclosed in an envelope were thus not able to compete with plant producing alete and trilete spores, and finally became extinct. Some could have found temporary refuge around lakes or swamps as in the Lochkovian Lorne area, Scotland, where primitive cryptospores have been observed (Wellman and Richardson 1996).

Considerations of In Situ Cryptospores and Trilete Spores

The cryptospore *H. verrucatus* (Fig. 7.11) is very similar to the trilete spore *S. verrucatus* (Fig. 7.12). As *S. verrucatus* has been observed in situ from *C. pertoni* sporangia (Fanning et al. 1988), it is clear that this trilete spore is produced by tracheophytes. What is then the real biological relationship between *H. verrucatus* and *S. verrucatus*? Are they the result of simple morphological convergence between cryptospores and trilete spores? Are some Ordovician and Silurian alete spores produced by tracheophytes or their immediate ancestors rather than by 'bryophytes'? Wellman (1996) suggested that cryptospores could derive from two sources: bryophytes and 'rhyniophytoid' plants. It has been demonstrated that the alete spore *Gneudnaspora divellomedia* var minor (Fig. 7.1-3) is produced by Lochkovian 'rhyniophytoids' *Laevolancis divellomedia* (Wellman et al. 1998a). As *G. divellomedia* var minor has recently been observed in the Ordovician (Steemans et al. 2008a; Wellman et al. 2008), does this mean that 'rhyniophytoids' evolved during the Ordovician, while plant meso- or mega-fossil remains are unknown below the Homerian? In fact, *G. divellomedia* var minor is a very simple laevigate monad suggesting that different taxa could possibly have been given the same name even though they derive from plants that are not necessarily closely related. Wellman et al. (1998b) have shown that detailed morphological, anatomical and ultrastructural analysis reveals subtle differences between the Lochkovian specimens of *G. divellomedia* allow them to be grouped into five distinct types. This could reflect their intraspecific variability. *G. divellomedia* var. minor are monads that are the separated products of dyads (*D. murusdensa-murusattenuata* morphon) and are occasionally seen as loosely attached dyads. The dyads *D. murusdensa* (Fig. 2.12) have also been observed in sporangia from 'rhyniophytoids' (Wellman et al. 1998a). One of the plant specimens has a bifurcating sporophytic axis, evidence against an affinity with extant 'bryophytes'.

This conflicting information supports Wellman's (1996) earlier suggestion that cryptospores are most probably produced by a polyphyletic group of embryophytes, and that they are not only restricted to 'bryophytes'. It seems that, at least during the Late Ordovician, the vegetation had reached a level of evolution allowing for wider diversification than previously believed. The kind of propagules produced by the basal embryophytes are controlled by ecological factors and probably do not reflect their biological affinities alone.

Concluding their study of Early Devonian spore assemblages from the Scottish Midlands, Lavender and Wellman (2002) said 'perhaps land plants of Lochkovian age were rather flexible in terms of the state in which their spores were released ...'. This seems equally applicable to Ordovician and Silurian land plants; perhaps views of the development of earlier vegetation need further consideration.

Early Evolution of Land Plants

Land plants (embryophytes; Fig. 13) are distinguished from their green algal ancestors by several characters, most of them being related to terrestrial habitats. These characters are a multicellular sporophyte (diploid generation), pluricellular sexual organs, a pluricellular sporangium, the cuticle and the sporopollenin (Kenrick and Crane 1997a and references therein). The name of the clade derives from 'embryo' (Table 3, trait 1), the name given to the young sporophyte growing within the archegonium (female sexual organ). Land plants are also characterized by diplobiontic life cycles, i.e. an alternation of multicellular gametophytic and sporophytic generations. Extant land plants include the paraphyletic 'bryophytes' and monophyletic polysporangiophytes (plants having sporophytes producing more than one sporangium).

'Bryophytes' do not have annular or helically thickened vascular cells (tracheids); their sporophyte is small, unbranched and matrotrophic on the dominant gametophyte. The 'bryophytes' include three clades: Hepatophytes (liverworts), bryophytes sensu stricto (mosses) and Anthocerotophytes (hornworts). Liverworts lack stomata (Table 3, trait 2); these are identified in most recent phylogenies as the first divergent land plants (Qiu 2008). This has been recently confirmed by the recent discovery of fragments of sporangia containing liverwort-like spores from Late Ordovician sediments (Caradoc, 450 Ma; Wellman et al. 2003).

Polysporangiophytes are comprised of the tracheophytes (vascular plants) and their non-vascular precursors. Extant vascular plants have a branched, independent, dominant sporophyte (Table 3, trait 3) and a reduced gametophyte.

'Rhyniophytoids'

The oldest fertile axial land plant fossils are reported from the Middle Silurian (Edwards and Feehan 1980). They are extremely small consisting of isotomously branched axes terminating in cup-shaped *Cooksonia*-type sporangia. Abundant plant remains are known from younger horizons (Late Silurian–Early Devonian). All those earliest land plants are informally called 'rhyniophytoids' (plants that resemble

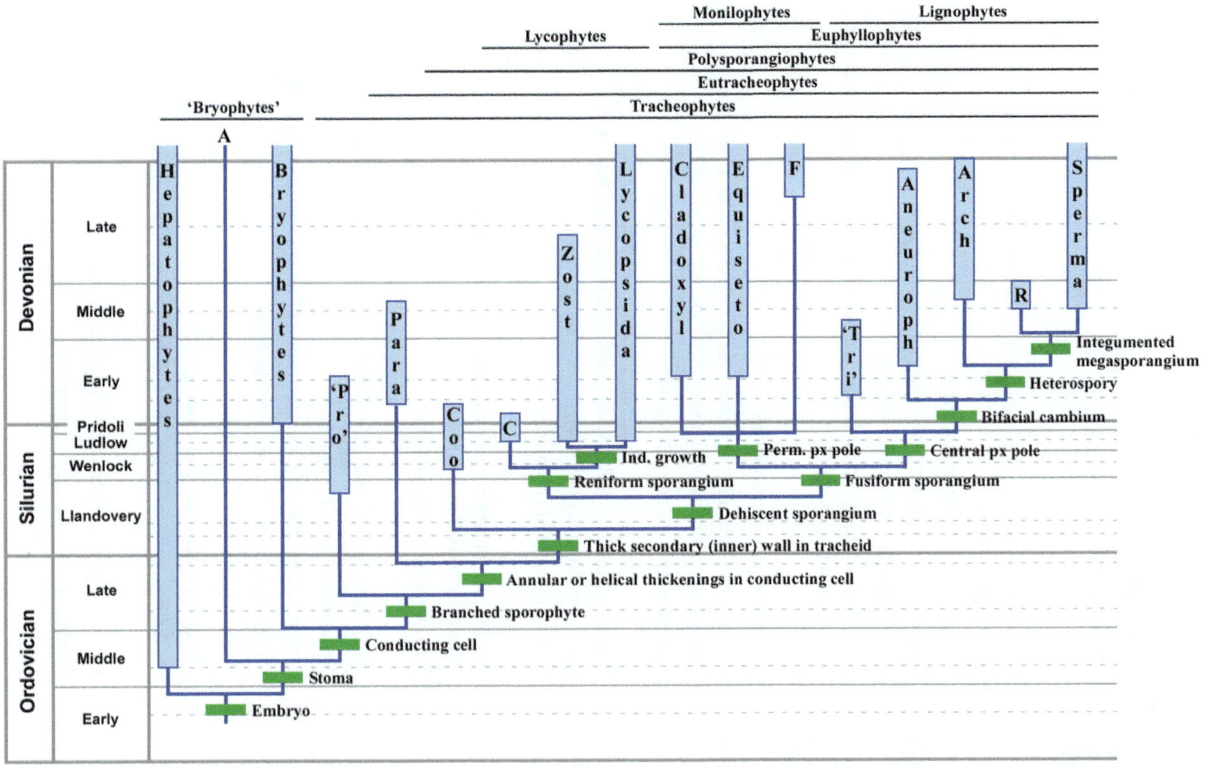

Fig. 13 (Based principally on Kenrick and Crane 1997a). Suggested phylogenetic relationships and stratigraphic distribution of early embryophytes. Taxa in quotes are para- or polyphyletic. *A* Anthocerotophytes; *Aneuroph* Aneurophytales; *Arch* Archaeopteridales; *R Runcaria*; *C Cooksonia caledonica* (new genus name required); *Cladoxyl* Cladoxylopsida; *Coo Cooksonia pertoni* and close species; *Equiseto* Equisetophytes; *F* ferns; *Para* Paratracheophytes; *'Pro'* 'Protracheophytes'; *Sperma* Spermatophytes (seed plants); *'Tri'* Trimerophytina; *Zost* Zosterophyllophytes. *Ind.* indeterminate; *Perm.* permanent; *px* protoxylem

the simplest vascular plants but lack evidence of vascular tissue). As an example, the Early Devonian (early Lochkovian; 415 Ma) Brown Clee Hill locality (Shropshire, England; see Glasspool et al. 2006 and references therein) has yielded a large number of 'rhyniophytoids' exquisitely preserved as charcoal. 'Rhyniophytoids' are difficult to describe accurately, as they are often very fragmentary and because the number of useful characters is low. The two following genera are 'rhyniophytoids'. *Tortilicaulis* Edwards is a plant characterized by branched axes terminating in spiralling, branched or unbranched, elongated sporangia. *Salopella* Edwards and Richardson has dichotomous axes bearing terminal fusiform, branched or unbranched sporangia opening into two equal valves. *Sporangia* of *Salopella* may be borne by pairs. 'Rhyniophytoids' most probably comprise the precursors of many younger lineages.

As most 'rhyniophytoids' consist of an unbranched axis with a terminal sporangium, it is impossible to tell if they are 'bryophytes' or polysporangiophytes. The liverwort affinities of the in situ cryptospores similar to *Tetrahedraletes* described by Wellman et al. (2003) have been inferred from their ultrastructure, but, generally, the presence of in situ cryptospores has long been considered as indicating affinities with 'bryophytes'. Cryptospores have also been described from sporangia of 'rhyniophytoids' with bifurcating axes (Edwards et al. 1999; Wellman et al. 1998b). Sporophytes of 'bryophytes' are typically unbranched, indicating that some non-'bryophyte' plants were also cryptospore producers.

Aglaophyton

Aglaophyton major (Kidston and Lang) described by DS Edwards (1986 and Fig. 14a) is a polysporangiophyte devoid of annular or helically thickened conducting cells; hence it is not yet a tracheophyte. It has been

Table 3 First occurrence of key traits of the embryophytes described in this chapter

Trait no.	Trait	Earliest occurrence	Comments
1	Embryo	Ordovician (Wellman et al. 2003), possibly Cambrian (Taylor and Strother 2008)	Synapomorphy of embryophytes.
			First ontogenetic stages of the sporophyte, occurring within the gametophytic tissues.
2	Stoma	Late Silurian (Pridoli)	Synapomorphy of stomatophytes.
		In *Hollandophyton* (Rogerson et al. 2002)	Epidermal pore associated with a pair of guard cells
3	Branched sporophyte	Middle Silurian (Wenlock)	Synapomorphy of polysporangiophytes.
			Branching of the sporophyte axis implies the presence of more than one sporangium per sporophyte
		In *Cooksonia* (Edwards and Feehan 1980)	
4	S-type conducting cell	Early Devonian (Pragian)	Typical of paratracheophytes.
		In *Sennicaulis* (Kenrick and Crane 1997a)	Annular and helically thickened conducting cells with a very thin decay-resistant inner wall and a thick 'spongy' outer wall
5	Thick lignified inner layer in conducting cell	Early Devonian (Lochkovian)	Synapomorphy of eutracheophytes.
		In *Zosterophylls* (Edwards 2003)	Various types exist (all have in common a thick inner wall, not demonstrated in *Cooksonia*)
6	Exarch primary xylem maturation in stem	Early Devonian (Lochkovian).	Synapomorphy of lycophytes.
		In *Zosterophylls* (Edwards 2003)	Occurs also in lignophyte roots, and in *Sphenophyllum* (Carboniferous-Permian)
7	Two-valved ovoid-reniform sporangium	Late Silurian (late Ludlow)	Synapomorphy of lycophytes.
		In cf. *Bathurstia* (Kotyk et al. 2002)	Also characterized by a specialized dehiscence mechanism
8	Microphyll	Late Silurian (Ludlow)	Synapomorphy of *Lycopsida*.
		In *Baragwanathia* (Rickards 2000)	Leaf resulting from the vascularization of an epidermal outgrowth
9	Anisotomous branching	Late Silurian (Pridoli)	Occurs in lycophytes and euphyllophytes
		In *Junggaria* (Cai et al. 1993)	Allows indeterminate growth
10	Unifacial vascular cambium	Middle Devonian (Givetian)	Occurs in lycophytes and monilophytes.
		In *Metacladophyton* (Wang and Geng 1997)	Produces secondary xylem inwards but no secondary phloem
11	Heterospory	Late Early Devonian (Late Emsian)	Occurs in almost all lineages of eutracheophytes.
		In *Chaleuria* (Andrews et al. 1975)	Production of micro and megaspores

Table 3 (continued)

Trait no.	Trait	Earliest occurrence	Comments
12	Dichotomous lateral appendage with determined growth	Early Devonian (Pragian)	Probable synapomorphy of basal euphyllophytes.
		In *Psilophyton* (Edwards 2003)	Precursor of megaphyll
13	Megaphyll	Late Devonian (Frasnian)	Probable synapomorphy of euphyllophytes.
		In *Archaeopteris* (Gensel and Andrews 1984)	Leaf resulting from planation and webbing of lateral dichotomous branches
14	Elongate sporangia borne in pairs	Early Devonian (Pragian)	Characteristic of basal euphyllophytes.
		In *Psilophyton* (Edwards 2003)	Sporangium with dehiscence slit occurring along one side only.
			Sporangium C-shaped in transverse section.
			Sporangia often borne in pairs
15	Differenciation of exine in sexine and nexine	Late Early Devonian (Late Emsian)	Probable synapomorphy of lignophytes
		In *Oocampsa* (Wellman and Gensel 2004)	
16	Mesarch protoxylem confined to the lobes of the xylem strand	Middle Devonian (Givetian)	Synapomorphy of monilophytes
		In *Ibyka* (Gensel and Andrews 1984)	
17	Bifacial vascular cambium	Middle Devonian (Eifelian)	Synapomorphy of lignophytes.
		Probable in *Crossia* (Beck and Stein 1993)	Produces secondary xylem inwards but secondary phloem outwards
18	Central protoxylem (other strands along the midplanes of xylem ribs)	Early Devonian (Emsian)	Synapomorphy of basal lignophytes
		In the unnamed plant of Gensel (1984)	
19	Monomegaspory	Middle Devonian (Givetian)	Occurs in lycophytes; in equisetophytes and in spermatophytes.
		In *Runcaria*? (Gerrienne et al. 2004)	Meiosis results in four cells, but only one is functional
20	Indehiscent mega-sporangium	Middle Devonian (Givetian)	Occurs in spermatophytes.
		In *Runcaria* (Gerrienne et al. 2004)	The megasporangium of heterosporous lycophytes is dehiscent
21	Integumented meg-asporangium	Middle Devonian (Givetian)	Occurs in lycophytes and in spermatophytes.
		In *Runcaria* (Gerrienne et al. 2004)	The 'integument' of the lycophyte megasporangium is a lateral expansion of the megasporophyll.
			The origin of the spermatophyte integument is controversial

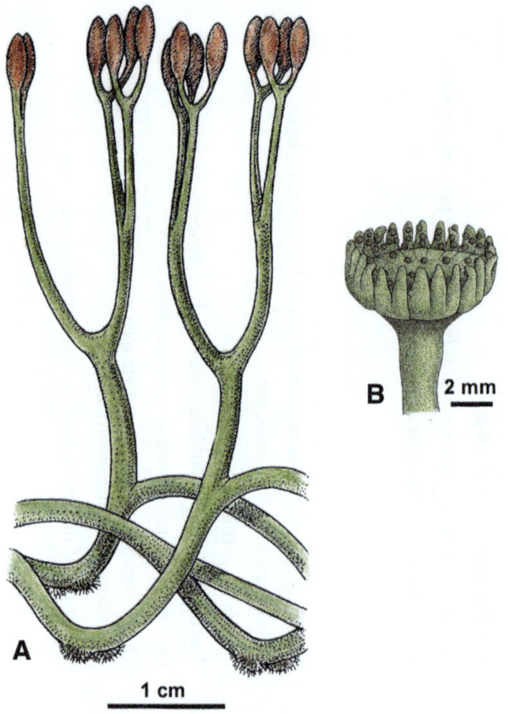

Fig. 14 **a** (modified from DS Edwards 1986). Reconstruction of *Aglaophyton* (sporophyte); **b** (modified from Remy et al. 1993). Reconstruction of *Lyonophyon* (gametophyte)

informally placed in a group called 'protracheophytes' (Kenrick and Crane 1997a, b; Kenrick and Crane 1991). *Aglaophyton* is a monospecific genus described on the basis of exceptionally well-preserved permineralized specimens from the Rhynie Chert (Edwards et al. 1994; Kenrick and Crane 1997a; Taylor et al. 2005). The age of the Rhynie outlier is latest Pragian or earliest Emsian (Wellman 2006). The complex history of this species, previously referred to the genus *Rhynia* is summarized by Kenrick and Crane (1997a).

Simple early land plants such as *Aglaophyton*, mostly characterized by terminal sporangia were previously grouped in the *Rhyniophyt(in)a* (Banks 1975) and presented as the simplest vascular plants. We now know that *Rhyniophyt(in)a* in its various definitions is an unnatural assemblage (Kenrick and Crane 1997a) that includes precursors of vascular plants (protracheophytes) and vascular plants. We believe that the terms *Rhyniophyta*, *Rhyniophytina* or *Rhyniophytes* are ambiguous and should be abandoned.

The sporophyte of *A. major* is comprised of naked isotomously branched axes. A prostrate rhizome with non-septate rhizoids has been described (Edwards 1986). A terminal sporangium is present on all distal axis segments. The sporangium is fusiform and radially symmetrical. No obvious dehiscence line is visible. Exceptionally well-preserved spores have been found in the sporangium of *Aglaophyton*. They are trilete, laevigate and show a distinctive thinning associated with the trilete mark; they can be accommodated in the dispersed spore genus *Retusotriletes* (Wellman et al. 2006). Interestingly, the genus *Retusotriletes* is morphologically close to the monad *Laevolancis* and the dyad *Dyadospora*, both found in situ in 'rhyniophytoid' sporangia, including a specimen with a branched axis (Wellman et al. 1998a) suggestive of polysporangiophyte affinities.

A. major plays an important role in our understanding of the evolution of land plants for two reasons:

(1) It has features unique to vascular plants (branched, independent sporophyte) and characters that recall 'bryophytes' (terminal sporangium, absence of annular or helically thickened conducting cells).
(2) The sporophyte and the gametophyte of the plant are known. The gametophyte has been called *Lyonophyton rhyniensis* Remy and Remy 1980 (Fig. 14b). Male and female gametophytes are distinct (Taylor et al. 2005). Male gametophytes (antheridiophores) are upright, unbranched and approximately 20 mm high. Female gametophytes (archegoniophores) are of the same size but they are dichotomously branched. The corresponding *Aglaophyton/Lyonophyton* is based on similarities in epidermal cells, stomatal structure, internal anatomy and wall structure of conducting cells.

Paratracheophytes

Paratracheophytes have been recently defined by Gerrienne et al. (2006). The division includes the plants that were previously called Rhyniaceae (Kenrick and Crane 1991) or Rhyniopsida (Kenrick and Crane 1997a). Paratracheophytes are a small distinctive group of vascular plants; they are a sister group of the eutracheophytes. Paratracheophytes include sporophytes (*Rhynia* [Fig. 15], *Stockmansella*, *Huvenia*), gametophytes (*Remyophyton* – gametophyte of *Rhynia*, *Sciadophyton*) and morphotypes (*Sennicaulis hippocrepiformis*, *Taeniocrada dubia*) (Gerrienne et al. 2006). Based on current evidence, paratracheophytes evolved during the Early Devonian

Fig. 15 (Modified from Edwards 1980) Reconstruction of *Rhynia*

and became extinct during the Late Devonian. Their synapomorphies include the following characters: (1) distinctive adventitious branching (*Rhynia*-type) (Kenrick and Crane 1997a); (2) abscission or isolation layer at base of sporangium (Kenrick and Crane 1997a); (3) sporangium attached to a 'pad' of tissue (Kenrick and Crane 1997a); (4) alternation of more or less isomorphic, independent sporophytic and gametophytic generations (Gerrienne et al. 2006). S-type conducting cell (Kenrick and Crane 1991) are obviously also an important character of the clade. S-type conducting cells (Table 3, trait 4) are annular or helically thickened cells characterized by a two-layered cell wall comprising a very thin decay-resistant inner layer and a thick, spongy outer layer. They differ from the tracheids in eutracheophytes characterized by a thick decay-resistant (lignified) inner layer (Table 3, trait 5). Whether the conducting cells in paratracheophytes and eutracheophytes are homologous or analogous characters is not known.

The gametophyte of *Rhynia* was recently described from the Rhynie Chert (Kerp et al. 2004) as 'remarkably small in comparison to those of other Rhynie chert plants'. The gametophyte consists of a dense stand of erect or prostrate antheridia- or archegonia-bearing axes.

On the basis of axial gametophyte/sporophyte combinations in *Aglaophyton* and in paratracheophytes, it is currently accepted that the alternation of isomorphic generations is the plesiomorphic condition among all early land plants, including basal eutracheophytes (Kenrick 2000). The reduction of the gametophyte would have occurred later, in more derived eutracheophytes (Kenrick 2000; Kenrick and Crane 1997a).

Cooksonia

The genus *Cooksonia* Lang (1937) (Fig. 16) includes leafless plants, a few centimetres high, with isotomously branched axes, and terminal sporangia. In the type species, *C. pertoni*, a gradual transition between the subtending axis/sporangium is visible;

Fig. 16 (Modified from Gerrienne et al. 2001) Reconstruction of *Cooksonia paranensis*. The sporophyte is presented attached to a thalloid gametophyte

the sporangium is terminal and trumpet-shaped. Some in situ conducting cells (C-type tracheids) have been described from three specimens (Edwards et al. 1992; Edwards 2003). The stratigraphic range of the genus is Middle Silurian–Early Devonian. *Cooksonia* is currently considered as the earliest known simplest eutracheophyte.

According to the type of in situ spores, four subspecies of *C. pertoni* have been created (Habgood et al. 2002):

C. pertoni ssp. *pertoni* (in situ spores: *Ambitisporites* sp.),

C. pertoni ssp. *synorispora* (in situ spores: *S. verrucatus*),

C. pertoni ssp. *apiculispora* (in situ spores: *S. newportensis* and *Aneurospora* sp.) (Fanning et al. 1988) and

C. pertoni ssp. *reticulispora* (in situ spores:? *Synorisporites* sp.).

The gametophyte of *Cooksonia* is unknown, but Gerrienne et al. (2006) have recently illustrated an exceptionally large specimen of *Cooksonia paranensis* consisting of five dichotomous axes attached at their base to a small thalloid(?) structure. A possible interpretation of this fossil would be a cluster of five individual sporophytes still attached to the remains of a small female or bisexual, prothallium-like, gametophyte. The presence of such a thalloid gametophyte is illustrated in Fig. 16.

Lycophytes

The name lycophytes is here used in a broad sense, as defined by Gensel and Berry (2001). It includes *Zosterophyllopsida*, prelycophytes such as *Baragwanathia* and *Drepanophycus*, and *Lycopsida* (*Lycopodiales*, *Selaginellales* and *Isoetales*). The lycophytes have a long geologic history extending back to the Late Silurian (Kotyk et al. 2002). Some lycophytes were arborescent and extremely abundant during most of the Carboniferous, when they were the dominant group in the extensive equatorial swamps covering Euramerica and North China. Today lycophytes are cosmopolitan, divided into seven herbaceous genera and basal within vascular plants. Their synapomorphies are the ex-arch primary xylem maturation in the stem (Table 3, trait 6), an ovoid-reniform sporangium opening into two valves (Table 3, trait 7) and, for the most advanced members of the group, the microphyll (Table 3, trait 8). Microphylls are leaves with a single vein that do not alter the shape of the stele when it is produced. Sporangia of typical lycophytes are borne singly on adaxial surface of fertile microphylls.

A Possible Precursor: *C. caledonica* Edwards

C. caledonica (Fig. 17) is a small plant with dichotomous axes terminated in a reniform sporangium opening into two valves. It is known from the Lochkovian of Britain (Edwards 1970) and Brasil (Gerrienne et al. 2001).

A range of plants with terminal sporangia have been named *Cooksonia*. Some of them such as *C. paranensis* Gerrienne et al. and *C. banksii* Habgood et al. show the same sporangial construction as the type species *C. pertoni*, i.e. a trumpet-like sporangium resulting from the subtending axis widening. On the contrary, the sporangium of *C. caledonica* Edwards is strikingly different, being reniform and showing a distal dehiscence line that opens the sporangium into two equal valves; *C. caledonica*, therefore, should be attributed to another genus. On the basis of the presence of this reniform sporangium, it is here tentatively presented as a precursor of the lycophytes.

Zosterophyllopsida

Zosterophylls (Zosterophyllopsida) are a well-known group of early vascular plants, with a subcosmopolitan distribution. They range in age from the late Ludlow (Ludfordian; Kotyk et al. 2002) to the Late Devonian (Frasnian; Hueber and Grierson 1961). Zosterophylls were an important component of the Early Devonian vegetation. More than 15 Zosterophyll genera are

Fig. 17 *Cooksonia caledonica* (Early Devonian of Brasil)

Fig. 18 (Modified from Kenrick and Crane 1997a) Reconstruction of *Zosterophyllum*

known. The most typical representative genus is *Zosterophyllum* Penhallow (Fig. 18). Zosterophylls were medium-sized herbaceous plants 10–30 cm high (generally 20 cm). They are characterized by predominantly isotomous axis branching, although some genera exhibit anisotomous branching (Table 3, trait 9). Other characteristics are exarch maturation of xylem, and reniform lateral sporangia. Axes are smooth or covered with enations of various shapes; distal segments often exhibit circinate vernation. Sporangia are distributed along the distal parts of the stem or more or less densely aggregated in terminal spikes. All zosterophylls are homosporous. Simple in situ spores have been recovered several times and are generally attributed to the *Retusotriletes*/*Calamospora* genera of sporae dispersae. Nothing is known about the gametophytes of these plants.

Lycopsida

The oldest members of this class appeared from latest Silurian to early Devonian (Gensel and Berry 2001; Kotyk et al. 2002). They are not 'true' lycopsids as they lack some typical lycopsid characters. *Asteroxylon* Kidston and Lang (1920) (Fig. 19), from the Rhynie Chert, has an exarch xylem, but its microphyll-like enations are not vascularized, as the vascular trace extends only to their base. *Drepanophycus* and *Baragwanathia* have an exarch xylem and microphylls, but their sporangia are borne directly on the stem and

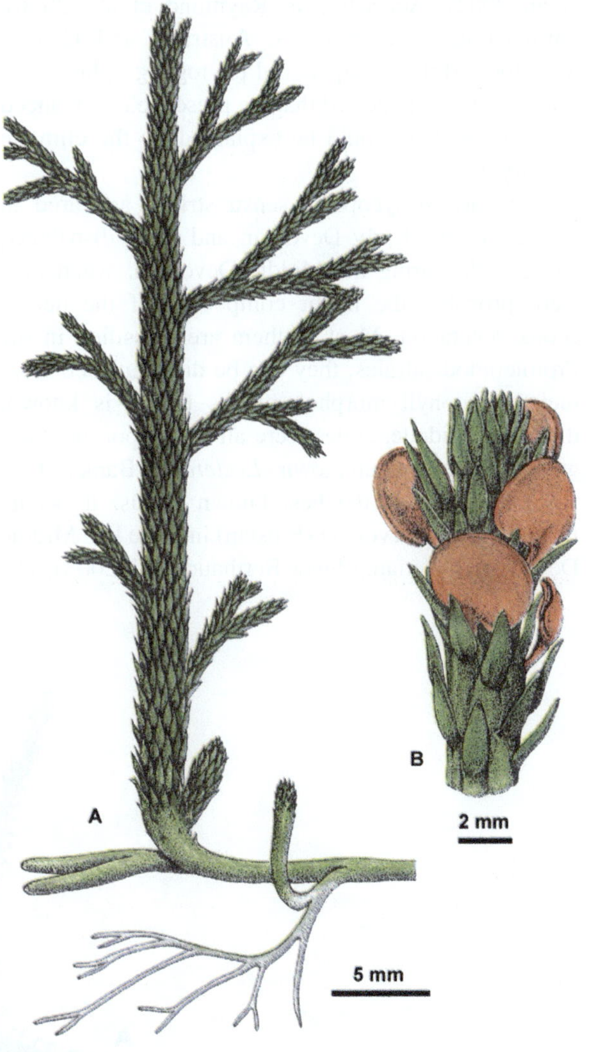

Fig. 19 (Modified from Kidston and Lang 1920) Reconstruction of *Asteroxylon*. (**a**) Gross view; (**b**) detail of a fertile axis

are not associated with a leaf. Those basal lycopsids are sometimes included in the Drepanophycales (Taylor et al. 2009). All are homosporous; their gametophytes are unknown.

Baragwanathia is known from several localities from Australia and Canada (Hueber 1983). Its stratigraphic distribution has long been controversial (Rickards 2000). It extends from the Gorstian (early Ludlow, Silurian) to the Pragian (Early Devonian). The Silurian age has been indirectly confirmed by the recent discovery of a rich and diversified Late Silurian flora at Bathurst Island (Arctic Canada; Kotyk et al. 2002). According to Raymond et al. (2006), during Late Silurian times, Australia and Canada were located in two equatorial phytogeographic units. These authors suggested that the presence of advanced early floras there could be explained by the climatic conditions.

The earliest lycopsids sensu stricto appeared at the end of the Early Devonian and were distributed world-wide during the Middle Devonian, when they were probably the major component of the herbaceous vegetation. Most of them are classified in the Protolepidodendrales; they can be distinguished from their microphyll morphology. As far as is known, the Protolepidodendrales were all homosporous; their gametophytes are unknown. *Leclercqia* Banks et al. 1972 (Fig. 20) is the best known genus. It occurs from the Early Devonian (Emsian) into the late Middle Devonian (Givetian; Meyer-Berthaud et al. 2003). The microphylls of *Leclercqia* are easily identifiable as they have five forks: the long, central segment bears two lateral dichotomizing projections. The central segment is vascularized and abaxially recurved. The in situ spores are referable to the *Acinosporites lindlarensis* morphon sensu (Richardson et al. 1993).

During late Devonian times, several arborescent lycopsid genera evolved. Arborescent lycopsids were by far the most conspicuous elements in the Carboniferous swampy landscapes. They are responsible for the production of the extensive amount of biomass that was trapped in coal. Most of those genera belong to the Isoetales, an order that is still represented today. Among other features, those plants are characterized by (1) the presence of a unifacial vascular cambium (Table 3, trait 10), responsible for the production of some secondary xylem, (2) a pseudobipolar growth involving rhizomorphic root system (Kenrick and Crane 1997a). The arborescent lycopsids illustrate one of the many ways to make a tree: the production of large amount of periderm (external tissues of the stem, including cork cambium and its internal and external derivatives). During the Carboniferous, in parallel to the development of the tree habit, the number of heterosporous lycopsid taxa increased. All arborescent lycopsids are characteristically heterosporous (Table 3, trait 11); the number of functional megaspores is reduced to a single one in several genera. The development of their gametophytes was also endosporic (Kenrick and Crane 1997a; Taylor et al. 2009).

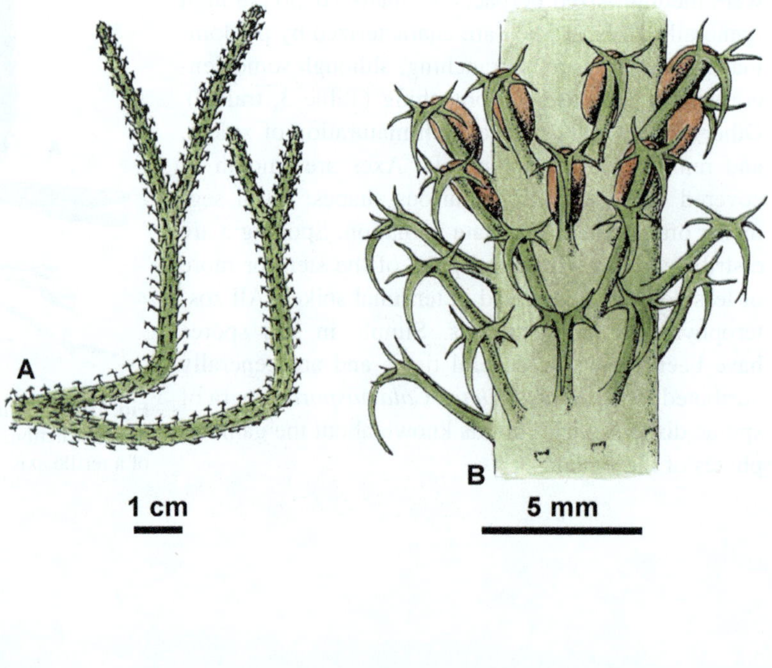

Fig. 20 (Modified from Gensel 1992) Reconstruction of *Leclercqia*. (**a**) Gross view; (**b**) detail of a fertile axis

Euphyllophytes

Within the tracheophytes, euphyllophytes are the sister group to the lycophytes. It is an extremely important group of vascular plants, representing more than 270,000 living species. Euphyllophytes are characterized by a large number of apomorphies (Kenrick and Crane 1997a), including the presence of megaphylls (leaf with multiple or branched veins; Table 3, trait 13), and pairs of elongate sporangia, each sporangium having a dehiscence slit along one side only (Table 3, trait 14). Euphyllophytes include two clades, the monilophytes (ferns sensu lato) and the lignophytes (seed plants and precursors). Both groups are discussed below. The earliest euphyllophytes evolved during Early Devonian times and belong to the 'Trimerophytina'. Those plants did not possess megaphylls, but vegetative or fertile iso- and anisotomous lateral appendages spirally inserted on their stems. Those appendages had a determinate growth (Table 3, trait 12) and are here considered as homologous to megaphylls.

'Trimerophytina' and Derivatives

Plants included in the 'Trimerophytina' are known worldwide from a large number of Early Devonian localities. Although the subdivision has been demonstrated to be paraphyletic (Kenrick and Crane 1997a), all the 'Trimerophytina' share many features. Most are medium-sized plants, ranging from several tens of centimeters to more than 1 m high. Their axes branch isotomously and anisotomously and bear lateral isotomous and anisotomous lateral appendages with determined growth. Fertile lateral branching systems terminate in clusters of fusiform sporangia borne in pairs. *Psilophyton* (Fig. 21) is the best known genus. Its stem comprises a large centrach protostele with P-type tracheids (tracheids with scalariform pitting, the pit aperture being covered with a perforate sheet; Kenrick and Crane 1997a). The main axis of *Pertica* also bears dichotomously branched laterals: some are vegetative; others are fertile and terminate in ellipsoidal sporangia borne in pairs. The anatomy of the genus is unknown. *Psilophyton* and *Pertica* were homosporous and produced spores of the *Apiculiretusispora* type.

Fig. 21 (Modified from Banks 1975) Reconstruction of *Psilophyton*. (**a**) Gross view; (**b**) detail of a fertile axis

The late Emsian (late Early Devonian) plant *Oocampsa* Andrews et al. 1975 is morphologically close to *Pertica* and generally considered as an advanced 'trimerophyte'. The plant is of great interest because its spores, attributed to the genus *Grandispora*, are bilayered (differentiation between an external sexine and in internal body covered by exine) (Table 3, trait 15), a condition comparable to what is found in younger plants of the lignophyte lineage (Andrews et al. 1975; Wellman and Gensel 2004).

Chaleuria Andrews et al. is another plant that might be a 'trimerophyte' derivative. It was described from the late Early Devonian strata. Its main axis bears at least two orders of branches, the second order bearing dichotomous sterile or fertile ultimate appendages. *Chaleuria* produced spores of variable size that fall into two intergrading size classes, considered by Andrews et al. (1974) to represent 'incipient heterospory'. The large spores of *Chaleuria* have been compared to *Apiculiretusispora* or *Cyclogranisporites*, the small ones to *Streelispora* (Andrews et al. 1974).

An unnamed, well-preserved plant axis was described by Gensel in 1984 from a late Early Devonian outcrop in Gaspé, Quebec, Canada. Although poorly known, the plant is noted for its well-preserved anatomy indicating 'a level of stem-leaf differentiation more advanced than any known for trimerophytes or other taxa of this age' (Gensel 1984, Nature 309, 785). It has a triradiate stele with four protoxylem strands, including a central one. Traces supplying the lateral appendages show some bilateral symmetry and were interpreted as precursors of megaphylls (Gensel 1984).

Traditionally, plants grouped within the paraphyletic 'Trimerophytina' are considered the earliest euphyllophytes, and hence to have given rise to both monilophytes and lignophytes. On the basis of the presence of a central protoxylem pole (see Section 'Lignophytes'), we have chosen another evolutionary history, with the 'Trimerophytina' complex basal to the lignophytes only (Fig. 13). This implies that the last common ancestor of the monilophytes and the lignophytes has still to be identified.

Monilophytes

Monilophytes are a major clade that unites most extant vascular plants without seeds, apart from the lycophytes. The clade has one clear synapomorphy, mesarch protoxylem confined to the lobes of the xylem strand (Table 3, trait 16). It was first recognized in a cladistic analysis of extinct and living taxa (Kenrick and Crane 1997a). Strong support for the results of the analysis came from other phylogenetic studies (Pryer et al. 2001; Qiu et al. 2007). Extant monilophytes includes equisetophytes (horsetails), psilotophytes (whisk ferns) and all eusporangiate and leptosporangiate ferns (Pryer et al. 2001). Two major monilophyte groups existed during the Devonian, the Cladoxylopsida and the Iridopteridales. Both are extinct.

Cladoxylopsida

Cladoxylopsida range from Early Devonian (Emsian; Gerrienne 1992) to the Mississippian (Meyer-Berthaud et al. 2007). They are characterized by a dissected vascular system consisting of a variable number of xylem plates. Those xylem segments are disposed in a variety of ways but are generally radially elongate. Generally, permanent protoxylem strands are positioned at the end of the xylem segments. The presence of secondary xylem has been proposed by Hilton et al. (2003), but most genera produce only primary vascular tissues. The last-order axes of the cladoxylopsids bear variously branched lateral appendages that can be vegetative and/or fertile; in this case, the ultimate segments of the appendages terminate in pairs of ellipsoidal sporangia. In most genera, lateral appendages are distributed in irregular helices. They are whorled in the proximal orders of axes in two genera: *Pietzschia* and *Polyxylon* (Meyer-Berthaud et al. 2007). Pseudosporochnalean cladoxylopsids such as *Calamophyton* (Fig. 22) or *Pseudosporochnus* (Fig. 23) share a similar body plan consisting of an upright trunk bearing densely inserted branches (Meyer-Berthaud et al. 2007). These branches divide characteristically in a digitate manner, a pattern resulting from very close dichotomies.

Cladoxylopsids were major contributors to terrestrial floras worldwide during the second half of the Devonian Period. *Eospermatopteris*, the Gilboa tree from the Middle Devonian of New York State (USA), has been recently demonstrated to belong to pseudosporochnalean cladoxylopsids (Stein et al. 2007). The plant was at least 8 m tall, with a long trunk that produced a crown of short, erect branches. It is considered as the earliest tree (Meyer-Berthaud and Decombeix 2007). *Pseudosporochnus* Leclerq and Banks (Fig. 23), although smaller, had a similar arborescent habit (Berry and Fairon-Demaret 2002).

Fig. 22 A branch of *Calamophyton* (Middle Devonian of Belgium)

Fig. 23 (Modified from Berry and Fairon-Demaret 2002) Reconstruction of *Pseudosporochnus*

The internal anatomy of the trunk of *Eospermatopteris* is unknown, but on the basis of the absence of secondary vascular tissue in *Pseudosporochnus* (Stein and Hueber 1989), it is suggested that both plants illustrate another way to make a tree: the multiplication of primary vascular strands combined with the presence of abundant sclerenchyma.

In situ spores attributed to the genus *Dibolisporites* have been isolated from *Foozia* (Gerrienne 1992 and *Calamophyton* (Leclercq and Andrews 1960) sporangia. All cladoxylopsids are thought to have been homosporous; their gametophytes are unknown.

Iridopteridales

Iridopteridales range from the Middle to Late Devonian, possibly Mississippian (Meyer-Berthaud et al. 2007). They share a number of characters with the cladoxylopsids, including the possession of whorled lateral appendages and of a highly dissected, ribbed protostele containing peripheral permanent protoxylem strands. They were widespread medium-sized plants. When known, the fertile appendages were shown to terminate in pairs of ellipsoidal sporangia (Berry and Stein 2000). Another important feature of the order is the iterative branching (Cordi and Stein 2005). Iridopteridales are apparently close to the Cladoxylopsida, but both groups can be distinguished from each other by a number of characters including the iterative versus hierarchical architecture (Meyer-Berthaud et al. 2007).

The phylogenetic position of Cladoxylopsida and Iridopteridales remains questionable; their precise placement is hampered by the difficulty of identifying phylogenetically significant characters and by the fragmentary nature of fossil material (Taylor et al. 2009). It has been suggested that Iridopteridales may be closely related to horsetails (equisetophytes) (Stein et al. 1984).

Equisetophytes (Sphenophytes; Horsetails)

Equisetophytes are characterized by a number of synapomorphies, including the presence of whorled appendages (branches and leaves) and of a sporangiophore with distinctive morphology. They first appeared in the Devonian, a possible precursor being *Estinnophyton* Fairon-Demaret, from the Early Devonian of Belgium. Equisetophytes were rather abundant during the Carboniferous and have since then experienced a gradual decline. *Equisetum* is the only extant genus. The record of Devonian equisetophytes is scarce in comparison with that of cladoxylopsids and Iridopteridales.

Other Devonian Monilophytes (= Early Fern-Like Plants Sensu Taylor et al. 2009)

As for equisetophytes, the Devonian record of other monilophytes is generally scarce. *Rhacophyton* Crépin 1875 (Fig. 24) is a notable exception: it has been described from world-wide numerous Late Devonian localities, where it is sometimes the dominant element. Its taxonomic position is still under debate, but most authors consider it as a basal fern. It was a bushy plant, 1–2 m tall, with stems bearing large fronds and numerous pinnule-like structures. The main axis of the frond comprises a central, clepsydroid-shaped strand of primary xylem that was sometimes surrounded by secondary xylem (Taylor et al. 2009). The *Rhacophyton* isospores are attributed to the *Diducites plicabilis–Auroraspora varia sporae dispersae* complex, characteristic of downstream or upstream 'coal' swamps (Maziane et al. 2002). The gametophyte of *Rhacophyton* is unknown.

Fig. 24 (Modified from Mauseth 2003) Reconstruction of *Rhacophyton*

Lignophytes

The lignophytes are the plants that possess a bifacial vascular cambium (Table 3, trait 17), producing secondary phloem (inner bark) towards the outside and secondary xylem (wood) towards the inside (Crane 1985). Another important character is the presence of a central protoxylem strand and of other strands along the midplanes of xylem ribs, derived by branching from the central one (Beck and Stein 1993) (Table 3, trait 18). Important also is the one-to-one correspondence of the peripheral protoxylem strands and the lateral appendages borne by the shoot (Beck and Stein 1993). The clade includes all seeds plants and some of their free-sporing precursors. Today, with more than 260,000 living species (and probably much more), the spermatophytes are the most diverse and the most important group of living plants. Extant seed plants (spermatophytes) include five lineages: coniferophytes, cycadophytes, ginkgophytes, gnetophytes and angiosperms (flowering plants). All evolved later than the Devonian and will not be considered here.

The earliest lignophytes evolved during the Middle and Late Devonian. They had a gymnospermous-type of secondary xylem, but they did not produce seeds. They are collectively known as 'progymnosperms', an informal taxon that gathers several orders, among which are the Aneurophytales and the Archaeopteridales that are discussed here.

Aneurophytales

Aneurophytales range in age from the late Emsian to the Frasnian (Gerrienne et al. 2010). Members of the order are characterized by three-dimensional branching systems with lateral dichotomous appendages in helical or decussate arrangement (Fig. 25; the illustrated specimen belongs to the genus *Rellimia*). The primary vascular system consists of a deeply lobed mesarch protostele, with a central protoxylem strand and others occurring along the mid-planes of the

Fig. 25 Fertile (*upper left*) and vegetative (*lower right*) specimens of *Rellimia* (Middle Devonian of Belgium)

lobes and near their tip. The sporangia are located on ultimate divisions of the much-branched, highly distinctive, lateral fertile appendages. Aneurophytales are homosporous, and produce bilayered spores attributed to *Rhabdosporites*; their gametophytes are unknown. Their phylogenetic position remains controversial, but on the basis of their homosporous status, they are often considered as sister to a group including the heterosporous Archaeopteridales and the spermatophytes.

The Stenokoleales are anatomically close to the Aneurophytales known exclusively from permineralizations and range in age from Middle Devonian to Early Carboniferous (Beck and Stein 1993). They are characterized by a three-ribbed protostele, numerous protoxylem strands including a central one and others arranged along the mid-planes of the arms of the protostele, pairs of traces to lateral appendages, and a helical phyllotaxis of those pairs of appendages. Secondary xylem is sometimes present in small amounts (Matten 1992; Beck and Stein 1993), but has not been so far described in detail. In the absence of morphological data, especially on reproductive parts, the Stenokoleales are currently left as incertae sedis. They are not represented in Fig. 13.

Archaeopteridales

The earliest arborescent representatives of the lignophytes belong to the Archaeopteridales, an order reported from the late Middle Devonian to the earliest Carboniferous. The advent of *Archaeopteris* (Fig. 26) was an important breakthrough in the evolution of plants. *Archaeopteris* experienced an extraordinary success during the Late Devonian, dominating the forest ecosystems and having a cosmopolitan distribution (Edwards et al. 2000). *Archaeopteris* was a large tree. Its trunk included a significant amount of secondary xylem that was in many respects comparable to that of extant gymnosperms. *Archaeopteris* illustrate

Fig. 26 (Modified from Stewart and Rothwell 1993) Reconstruction of *Archaeopteris*

a third way to make a tree: the production of secondary vascular tissue, mainly xylem. The tree habit was accompanied by the acquisition of long-lived roots (Meyer-Berthaud and Decombeix 2007), which had major implications on the elaboration of early soils and complex microbial communities (Algeo et al. 2001). Vegetative leaves of *Archaeopteris* exhibit extensive interspecific variation, their margins ranging from highly dissected to nearly entire. *Archaeopteris* was heterosporous and free-sporing. Fertile leaves were generally non-laminate and bear up to 40 fusiform micro- and/or megasporangia. The microspores belong to the genera *Cyclogranisporites* or *Geminospora*; the megaspores are 16–32 per sporangium and conform to those of the genera *Contagisporites* or *Biharisporites* (Taylor et al. 2009). The presence of endosporic megagametophytes has been described but not illustrated (Scheckler et al. 1997).

Spermatophytes (Seed Plants)

There are obviously strong selecting forces in favour of heterospory, as it evolved several times, in fact in almost all tracheophyte lineages (Bateman and DiMichele 1994). In the spermatophyte lineage, heterospory is accompanied by a series of modifications involving the megasporangium and the part of the sporophyte immediately surrounding it. The number of functional megaspores drops to one (Table 3, trait 19), and the indehiscent megasporangium (Table 3, trait 20) is enclosed in sporophytic tissues called integument (Table 3, trait 21). The resulting structure is called an ovule (or a seed, after fertilization) or preovule if the integument is dissected. The ovule is the characteristic element of the spermatophyte lineage. In the earliest representatives of the group, the ovule or preovule is borne in a further external envelope called the cupule, and the megasporangium exhibits an apical modification probably related to the capture of microspores.

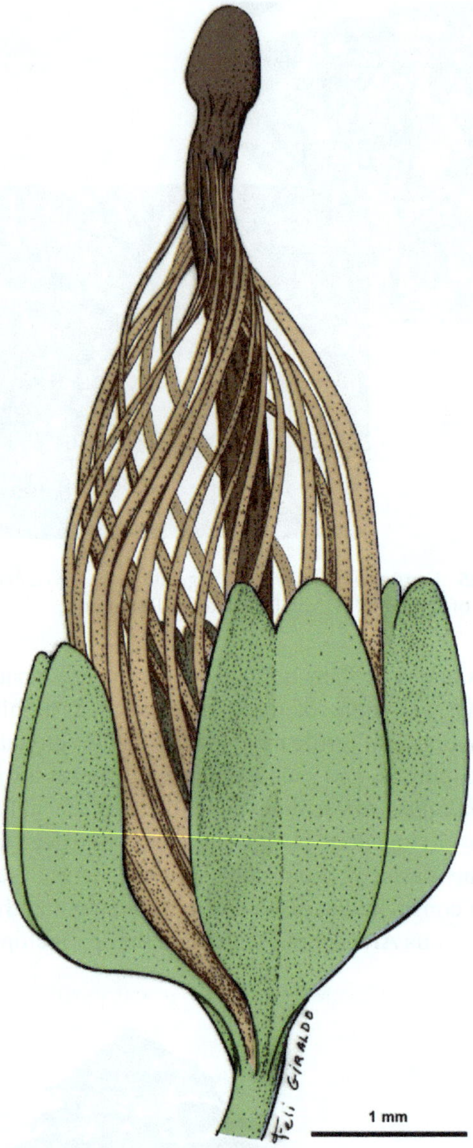

Fig. 27 (Modified from Gerrienne et al. 2004) Reconstruction of *Runcaria*

Runcaria, a Precursor of Seed Plants

The Middle Devonian (mid-Givetian, 390 Ma) *Runcaria* (Fig. 27) is a small, radially symmetrical megasporangium. The integument is dissected into 16–20 spiralling lobes. The structure is surrounded by a segmented cupule. The megasporangium bears an unopened distal extension protruding above the integument (Gerrienne et al. 2004). This suggests that the early evolution of the integument was related to megasporangium protection rather than to pollination. *Runcaria* has been termed a 'proto-ovule' because it does not show the typical apical modification of the megasporangium shown by most of the younger pre-ovules (see below). *Runcaria* evolved during the Middle Devonian and may be considered as a precursor of those younger plants with pre-ovules. Its already complex structure also suggests that the spermatophyte lineage originated earlier than mid-Givetian times.

Late Devonian Pre-spermatophytes

A number of pre-ovules are known from Late Devonian strata distributed worldwide. They show a variety of morphologies, but they all share the hydrasperman modification (Rothwell 1986) of the megasporangium, also called a nucellus. The apex of the nucellus is transformed into a lagenostome, consisting of a pollen chamber surmounted by a cylindrical structure called the salpinx. The pollen chamber contains a central parenchymatous dome (Prestianni 2005). Late Devonian pre-ovules show various degrees of fusion of the integumentary lobes and/or various types of cupules, and have been classified into five distinctive morphological types (Prestianni 2005). *Moresnetia* Stockmans (Fig. 28) has highly dissected cupules borne at the tip of large branching systems; each cupule contains up to four pre-ovules. The diversity of the Late Devonian pre-ovules most probably illustrates the earliest adaptive radiation of spermatophytes.

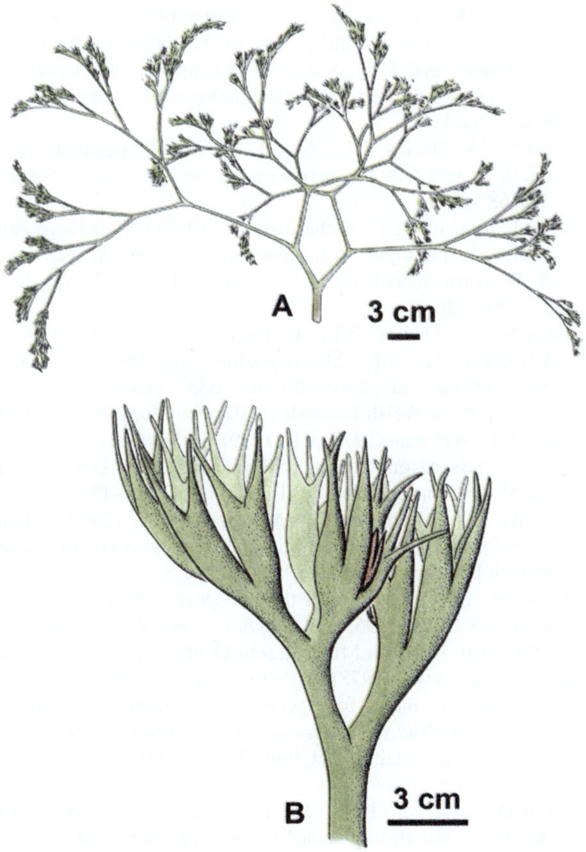

Fig. 28 (Modified from Fairon-Demaret and Scheckler 1987) Reconstruction of *Moresnetia*

Brief Summation of Early Evolution of Land Plants

From the Late Silurian (Prídolí, 418 Ma) onwards, land plant populations evolved from patchy stands of *Cooksonia*-type plants (a few centimeters high, isotomously branched, with terminal sporangia) to worldwide distributed forests dominated by *Archaeopteris* trees at the end of the Devonian (Famennian, 360 Ma) or gigantic lycophytes and equisetophytes during the Carboniferous (359–299 Ma). On their way to the greening of the continents, plants evolved indeterminate growth, secondary tissues including wood and bark, long-lived roots, leaves, heterospory and seeds. All those innovations allowed them to increase size, to enhance propagule dispersion and efficiency and to colonize almost all terrestrial habitats. This had a global-scale impact on geophysical and geochemical cycles that peaked during the late Devonian, when Earth experienced one of the five major extinction events.

Acknowledgements One of us (P.S.) thanks Saudi Aramco for permission to publish pictures of spores from Saudi Arabian samples. The editor of this volume, John Talent, is thanked for inviting us to contribute to this volume and for his careful editing of the chapter. We thank Maurice Streel (University of Liège, Belgium) and Merrell Miller (Aramco, Saudi Arabia) for their constructive remarks. The manuscript has been critically and very constructively reviewed by C. Wellman (Sheffield, UK), J. Marshall (Southampton) and J. Galtier (Montpellier, France). Ruth Mawson, Peter Cockle, Janine Miller and Karen Novotny have also contributed importantly to checking and polishing the English of the manuscript. P. Steemans and P. Gerrienne are FRS-FNRS Research Associates.

References

(Including taxa references)

Algeo TJ, Scheckler SE, Maynard JB (2001) Effects of the Middle to Late Devonian spread of vascular land plants on weathering regimes, marine biotas and global climate. In: Gensel PG, Edwards D (eds) Plants invade the land. Evolutionary and Environmental Perspectives, Columbia University Press, New York, pp 13–236

Allen KC (1965) Lower to Middle Devonian spores of North and Central Vestspitsbergen. Palaeontology 8:687–748

Andrews HN, Gensel PG, Forbes WH (1974) An apparently heterosporous plant from the Middle Devonian of New Brunswick. Palaeontology 17:387–408

Andrews HN, Gensel PG, Kasper A (1975) A new fossil plant of probable intermediate affinities (trimerophyte-progymnosperm) Can J Bot 53:1719–1728

Banks HP (1975) Reclassification of Psilophyta. Taxon 24:401–413

Banks HP, Bonamo PM, Grierson JD (1972) *Leclercqia complexa*, gen. et sp. nov., a new lycopod from the late Middle Devonian of eastern New York. Rev Palaeobot Palynol 14:19–40

Bateman RM, Dimichele WA (1994) Heterospory – the most iterative key innovation in the evolutionary history of the Plant kingdom. Biol Rev 69:345–417

Beck CB, Stein WE (1993) *Crossia virginiana* gen. et sp. nov., a new member of the Stenokoleales from the Middle Devonian of southwestern Virginia. Palaeontographica 229B:115–134

Berry CM, Fairon-Demaret M (2002) The architecture of *Pseudosporochnus nodosus* Leclercq et Banks: a Middle Devonian cladoxylopsid from Belgium. Int J Plant Sci 163:699–713

Berry CM, Stein WE (2000) New Iridopteridalean from the Devonian of Venezuela. Int J Plant Sci 161:807–827

Breuer P (2007) Devonian Miospore Palynology in Western Gondwana: an application to oil exploration. PhD Thesis, University of Liège

Breuer P, Al-Ghazi A, Al-Ruwaili M et al. (2007) Early to Middle Devonian miospores from northern Saudi Arabia. Rev Micropalaeontol 50:27–57

Brongniart A (1822) Sur la classification et la distribution des végétaux fossiles. Mém Mus Hist Nat 8:203–204

Burgess ND (1991) Silurian cryptospores and miospores from the type Llandovery area, south-west Wales. Palaeontology 34:575–599

Burgess ND, Richardson JB (1991) Silurian cryptospores and miospores from the type Wenlock area, Shropshire, England. Palaeontology 34:601–628

Burgess ND, Richardson JB (1995) Late Wenlock to early Pridoli cryptospores and miospores from south and south-west Wales, Great Britain. Palaeontographica 236:1–44

Cai CS, Dou YW, Edwards D (1993) New observations on a Pridoli plant assemblage from Xinjiang, Northwest China, with comments on its evolutionary and paleogeographic significance. Geol Mag 130:155–170

Cordi J, Stein WE (2005) The anatomy of *Rotoxylon dawsonii* comb. nov (*Cladoxylon dawsonii*) from the upper Devonian of New York State. Int J Plant Sci 166:1029–1045

Cramer FH (1966) Palynomorphs from the Siluro-Devonian boundary in N. W. Spain. Notas y Comunicaciones del Instituto Geologico y Minero de Espana 85:71–82

Crane PR (1985) Phylogenetic analysis of seed plants and the origin of angiosperms. Annals of the Missouri Botanical Garden72:716–793

Crépin F (1875) Description de quelques plantes fossiles de l'étage des Psammites du Condroz (Dévonien supérieur). B Acad R Sci Let Beaux-Arts Belg 38:356–366

Dawson JW (1859) On fossil plants from the Devonian rocks of Canada. Q J Geol Soc London 15:477–488

Dawson JW (1871) The fossil plants of the Devonian and upper Silurian Formations of Canada. Geological Survey of Canada, Montréal

Dou YW, Sun ZH (1983) Devonian plants. In: House GP (ed) Palaeontological Atlas II. Late Palaeozoic Section. Geological Publishing House (In Chinese), Beijing, pp 561–562

Dufka P (1995) Upper Wenlock miospores and cryptospores derived from a Silurian volcanic island in the Prague Basin (Barrandian area, Bohemia). J Micropalaeontol 14:67–79

Edwards D (1970) Fertile Rhyniophytina from the Lower Devonian of Britain. Palaeontology 13:451–461

Edwards D (1979) A late Silurian flora from the Lower Old Red Sandstone of south-west Dyfed. Palaeontology 22:23–52

Edwards DS (1980) Evidence for the sporophytic status of the Lower Devonian plant *Rhynia gwynne-vaughanii* Kidston and Lang. Rev Palaeobot Palynol 29:177–188

Edwards DS (1986) Aglaophyton major, a non-vascular land-plant from the Devonian Rhynie Chert. Bot J Linn Soc 93:173–204

Edwards D (2003) Xylem in early tracheophytes. Plant Cell Environ 26:57–72

Edwards D, Feehan J (1980) Records of *Cooksonia*-type sporangia from late Wenlock strata in Ireland. Nature 287:41–42

Edwards D, Richardson JB (1974) Lower Devonian (Dittonian) plants from the Welsh Borderland. Palaeontology 17: 311–324

Edwards D, Richardson JB (1996) Chapter 14A – Review of in situ spores in early land plants. In: Jansonius J, McGregor DC (eds) Palynology: principles and applications. American Association of Stratigraphic Palynologists Foundation, Salt Lake City, UT

Edwards D, Davies KL, Axe L (1992) A vascular conducting strand in the early land plant *Cooksonia*. Nature 357:683–685

Edwards D, Fanning U, Richardson JB (1994) Lower Devonian coalified sporangia from Shropshire: *Salopella* Edwards & Richardson and *Tortilicaulis* Edwards. Bot J Linn Soc 116:89–110

Edwards D, Davies KL, Richardson JB et al. (1996) Ultrastructure of *Synorisporites downtonensis* and *Retusotriletes* cf *coronadus* in spore masses from the Pridoli of the Welsh Borderland. Palaeontology 39:783–800

Edwards D, Wellman CH, Axe L (1999) Tetrads in sporangia and spore masses from the upper Silurian and Lower Devonian of the Welsh Borderland. Bot J Linn Soc 130:111–156

Edwards D, Fairon-Demaret M, Berry CM (2000) Plant megafossils in Devonian stratigraphy: a progress report. Cour Forsch Inst Senckenberg 220:25–38

Evans PR (1970) Revision of the miospore genera *Perotrilites* Erdtm ex Couper 1953 and *Diaphanospora* Balme & Hassell 1962. Bull Aust Bur Miner Resour Geol Geophys 116:65–82

Fairon-Demaret M (1978) *Estinnophyton gracile* gen. et sp. nov., a new name for specimens previously determined *Protolepidodendron wahnbachense* Kräusel & Weyland, from the Siegenian of Belgium. Bull Acad R Belg Class Sci 64:597–609

Fairon-Demaret M (1986) *Stockmansella*, a new name for *Stockmansia* Fairon-Demaret (fossil). Taxon 35:334

Fairon-Demaret M, Scheckler SE (1987) Typification and redescription of *Moresnetia zalesskyi* Stockmans, 1948, an early seed plant from the upper Famennian of Belgium. Bull Inst R Sci Nat Belg 57:183–199

Fanning U, Richardson JB, Edwards D (1988) Cryptic evolution in early land plant. Evol Trends Plants 2:13–24

Gensel PG (1984) A new Lower Devonian plant and the early evolution of leaves. Nature 309:785–787

Gensel PG (1992) Phylogenetic relationships of the zosterophylls and lycopsids – evidence from morphology, paleoecology, and cladistic methods of inference. Ann Miss Bot Gard 79:450–473

Gensel PG, Andrews HN (1984) Plant life in the Devonian. Praeger, New York

Gensel PG, Berry CM (2001) Early lycophyte evolution. Am Fern J 91:74–98

Gerrienne P (1992) The Emsian plants from Fooz-Wepion (Belgium). 3. *Foozia minuta* gen et spec nov, a new taxon with probable cladoxylalean affinities. Rev Palaeobot Palynol 74:139–157

Gerrienne P (1996) Contribution à l'étude paléobotanique du Dévonien inférieur de Belgique: les genres nouveaux Ensivalia et Faironella. Acad R Belg, Publ Clas Sci, Collect in 4°, 3è sér.:1–94

Gerrienne P, Bergamaschi S, Pereira E et al. (2001) An Early Devonian flora, including *Cooksonia* from the Paraná Basin (Brasil). Rev Palaeobot Palynol 116:19–38

Gerrienne P, Meyer-Berthaud B, Fairon-Demaret M et al. (2004) *Runcaria*, a Middle Devonian seed plant precursor. Science 306:856–858

Gerrienne P, Dilcher DL, Bergamaschi S et al. (2006) An exceptional specimen of the early land plant *Cooksonia paranensis*, and a hypothesis on the life cycle of the earliest eutracheophytes. Rev Palaeobot Palynol 142:123–130

Gerrienne P, Meyer-Berthaud B, Lardeux H et al. (2010) First record of *Rellimia* Leclercq & Bonamo (Aneurophytales) from Gondwana, with comments on the earliest lignophytes. Geol Soc London, Spec Publ 339:81–92

Glasspool IJ, Edwards D, Axe L (2006) Charcoal in the early Devonian: a wildfire-derived Konservat-Lagerstatte. Rev Palaeobot Palynol 142:131–136

Goldring W (1924) The upper Devonian forest of seed ferns in eastern New York. N Y State Mus Bull 251:50–72

Göppert HR (1852) Fossile flora des Ubergangsgebirges. Verhandlungen Kaiserlichen Leopoldinisch Carolinischen Akad Naturforsch Suppl Bandes 14:1–299

Gray J, Chaloner WG, Westoll TS (1985) The microfossil record of early land plants; advances in understanding of early terrestrialization, 1970–1984. In: Chaloner WG, Lawson JD (eds) Evolution and environment in the late Silurian and early Devonian. Philos Trans R Soc Lond B 309:167–198

Gray J (1993) Major Paleozoic land plant evolutionary bioevents. Palaeogeogr Palaeoclimatol Palaeoecol 104: 153–169

Habgood KS, Edwards D, Axe L (2002) New perspectives on *Cooksonia* from the Lower Devonian of the Welsh Borderland. Bot J Linn Soc 139:339–359

Hass H, Remy W (1991) *Huvenia kleui*, nov. gen., nov. spec. – ein Vertreter der Rhyniaceae aus dem höheren Siegen des Rheinischen Schieferbegirges Argu Palaeobot 8:141–168

Hemsley AR (1994) The origin of the land plant sporophyte: an interpolational scenario. Biol Rev 69:263–273

Hemsley AR, Scott AC, Collinson ME (1999) The architecture and functional biology of freely dispersed megaspores. In: Kurmann MH, Hemsley AR (eds) The evolution of plant architecture. Royal Botanic Gardens, Kew, London, pp 253–277

Hilton J, Geng BY, Kenrick P (2003) A novel late Devonian (Frasnian) woody Cladoxylopsid from China. Int J Plant Sci 164:793–805

Hoffmeister WS (1959) Lower Silurian plant spores from Libya. Micropalaeontology 5:331–334

Hueber FM (1972) Early Devonian land plants from Bathurst Island, District of Franklin. Geol Surv Can Pap 71:1–17

Hueber FM (1982) *Taeniocrada dubia* Kr. and W. its conducting strand of helically strenghtened tubes. Bot Soc Am Misc Ser Publ 162:58–59

Hueber FM (1983) A new species of *Baragwanathia* from the Sextant formation (Emsian), northern Ontario. Bot J Linn Soc 86:57–79

Hueber FM, Grierson JD (1961) On the occurrence of *Psilophyton princeps* in the early upper Devonian of New York State. Am J Bot 48:473–479

Jansonius JJ, Hills LV (1979) Genera file of fossil spores, supplement. Special publication, Department of Geology, University of Calgary, Canada

Jardiné S, Yapaudjian L (1968) Lithostratigraphie et palynologie du Dévonien-Gothlandien gréseux du Bassin de Polignac (Sahara). Rev Inst Fr Pet 23:439–469

Johnson NG (1985) Early Silurian palynomorphs from the Tuscarora Formation in central Pennsylvania and their paleobotanical and geological significance. Rev Palaeobot Palynol 45:307–360

Kasper AE, Andrews HN (1972) *Pertica*, a new genus of Devonian plants from Northern Maine. Am J Bot 59: 897–911

Kenrick P (2000) The relationships of vascular plants. Philos Trans R Soc Lond B Biol Sci 355:847–855

Kenrick P, Crane PR (1991) Water-conducting cells in early fossil land plants: implications for the early evolution of tracheophytes. Bot Gaz 152:335–356

Kenrick P, Crane PR (1997a) The origin and early diversification of land plants. A cladistic study. Smithsonian Inst press, Washington and London, 441pp

Kenrick P, Crane PR (1997b) The origin and early evolution of plants on land. Nature 389:33–39

Kenrick P, Edwards D, Dales RC (1991) Novel ultrastructurein water-conducting cells of the Lower Devonian plant *Sennicaulis hippocrepiformis*. Palaeontology 34:751–766

Kerp H, Trewin NH, Hass H (2004) New gametophytes from the early Devonian Rhynie chert. Trans Roy Soc Edinburgh 94:411–428

Kidston R, Lang WH (1917) On Old Red Sandstone plants showing structure, from the Rhynie Chert Bed, Aberdeenshire. Part I. *Rhynia gwynne-vaughanii*, Kidston and Lang. Trans Roy Soc Edinburgh 51:761–784

Kidston R, Lang WH (1920) On Old Red Sandstone Plants showing structure, from the Rhynie Chert bed, Aberdeenshire. Part III. *Asteroxylon mackiei* Kidston & Lang. Trans Roy Soc Edinburgh 52:643–680

Kotyk ME, Basinger JF, Gensel PG et al. (2002) Morphologically complex plant macrofossils from the late Silurian of Arctic Canada. Am J Bot 89:1004–101

Lang WH (1937) On the plant-remains from the Downtonian of England and Wales. Philos Trans R Soc Lond B Biol Sci 227:245–291

Lang WH, Cookson IC (1935) On a flora, including vascular land plants, associated with *Monograptus*, in rocks of

Silurian age, from Victoria, Australia. Philos Trans R Soc Lond B Biol Sci 224:421–449

Lavender K, Wellman CH (2002) Lower Devonian spore assemblages from the Arbuthnott Group at Canterland Den in the Midland Valley of Scotland. In: Steemans P, Servais T, Streel M (eds) Palaeozoic palynology: a special issue in honour of Dr. Stanislas Loboziak. Rev Palaeobot Palynol 118: 157–180

Leclercq S, Andrews HN (1960) *Calamophyton bicephalum*, a new species from the Middle Devonian of Belgium. Ann Mo Bot Gard 47:1–23

Leclercq S, Banks HP (1962) *Pseudosporochnus nodosus* sp. nov., a Middle Devonian plant with cladoxylalean affinities. Palaeontographica B 110:1–34

Leclercq S, Bonamo PM (1973) *Rellimia thomsonii*, a new name for *Milleria* (*Protopteridium*) *thomsonii* Lang 1926 Emend. Leclercq and Bonamo 1971. Taxon 22:436–437

Le Hérissé A, Al-Ruwaili M, Miller MA et al. (2008) Late Ordovician organic-walled protists from the QUSAIBA 1 shallow core (NW Saudi Arabia). 12th International Palynological Congress, Bonn, August 30–September 5, 2008

Linnaeus C (1753) Species plantarum, exhibentes plantas rite cognitas, ad genera relatas, cum differentiis specificis, nominibus trivialibus, synonymis selectis, locis natalibus, secundum systema sexuale digestas. Holmiae, Impensis Laurentii Salvii. [L. Salvius, Stockholm.]. [1 May 1753] [Starting point for Spermatophyta, Pteridophyta, Sphagnaceae, Hepaticae, Fungi (incl. slime moulds and lichen-forming fungi) and Algae (pro parte)]

Luber AA, Walts IE (1938) Classification and stratigraphical value of spores of some Carboniferous coal deposits in the USSR. Trudy Tsentral'nogo Naucho-Issledovatd' Skogo Geologo-Razvedochnogo Instituta, Leningrad 50:186–195 [in Russian]

Marshall JEA, Allen KC (1982) Devonian miospore assemblages from Fair Isle, Shetland. Palaeontolology 25:277–312

Marshall JEA, Hemsley AR (2003) A Mid Devonian seed-megaspores from East Greenland and the origin of the seed plants. Palaeontolology 46:647–670

Matten LC (1992) Studies on Devonian plants from New York State: *Stenokoleos holmesii* n. sp. from the Cairo flora (Givetian) with an alternative model for lyginopterid seed fern evolution. Cour Forsch Inst Senck 147:77–85

Mauseth JD (2003) Botany: an introduction to plant biology, 3rd edn. Jones and Bartlett Publishers, Sudbury, MA

Maziane N, Higgs KT, Streel M (2002) Biometry and paleoenvironment of *Retispora lepidophyta* (Kedo) Playford 1976 and associated miospores in the latest Famennian nearshore marine facies, eastern Ardenne (Belgium). Rev Palaeobot Palynol 118:211–226

McGregor DC (1960) Devonian spores from Melville Island, Canadian Arctic Archipelago. Palaeontology 3(1):26444

McGregor DC (1961) Spores with proximal radial pattern from the Devonian of Canada. Geol Surv Can Bull 76(i–ix):1–11

McGregor DC (1973) Lower and Middle Devonian spores of eastern Gaspé, Canada. I. Systematics. Palaeontographica 142B:177

McGregor DC, Camfield M (1976) Upper Silurian? To Middle Devonian spores of the Moose River Basin, Ontario. Geol Surv Can Bull 263:1–63

Mendlowicz Mauller P, Machado Cardoso TR, Pereira E et al. (2007) Resultados Palinoestratigráficos do Devoniano da Sub-Bacia de Alto Garças (Bacia do Paraná – Brasil). In: Carvalho IS, Cassab RCT, Schwanke C, Carvalho MA et al. (eds) Paleontologia: cenários de Vida, Interciência, Rio de Janeiro

Meyer-Berthaud B, Decombeix AL (2007) Palaeobotany – a tree without leaves. Nature 446:861–862

Meyer-Berthaud B, Gerrienne P (2001) *Aarabia*, a new early Devonian vascular plant from Africa (Morocco). Rev Palaeobot Palynol 116:39–53

Meyer-Berthaud B, Fairon-Demaret M, Steemans P et al. (2003) The plant *Leclercqia* (Lycopsida) in Gondwana: implications for reconstructing Middle Devonian paleogeography. Geol Mag 140:119–130

Meyer-Berthaud B, Soria A, Young GC (2007) Reconsidering differences between Cladoxylopsida and Iridopteridales: evidence from *Polyxylon australe* (Upper Devonian, New South Wales, Australia). Int J Plant Sci 168:1085–1097

Mizusaki AM, Melo JHG, Lelarge ML et al. (2002) Vila Maria Formation, Paraná Basin, Brasil – an example of integrated geochronological and palynological dating. Geol Mag 139:453–463

Naumova S (1953) Spore-pollen assemblages of the upper Devonian of the Russian Platform and their stratigraphic significance. Trans Inst Geol Sci, Acad Sci USSR 143:204

Owens B (1971) Miospores from the Middle and Early Upper Devonian rocks of the Western Queen Elisabeth Island, Artic Archipelago. Geol Surv Can Bull 70:1–157

Paris F, Verniers J, Miller MA et al. (2008) Biostratigraphy of Late Ordovician chitinozoans from the QUSAIBA shallow core (NW Saudi Arabia). 12th International Palynological Congress, Bonn, August 30–September 5, 2008

Penhallow DP (1892) Additional notes on Devonian plants from Scotland. Can Rec Sci 5:1–13

Playford G (1983) The Devonian miospore genus *Geminospora* Balme 1962: a reappraisal based upon topotypic *G. lemurata* (type species). Mem Ass Austral Palaeontol 1:311–325

Prestianni C (2005) Early diversification of seeds and seed-like structures. Note Geol Mem 2005/02:Abstract 06

Pryer KM, Schneider H, Smith AR et al. (2001) Horsetails and ferns are a monophyletic group and the closest living relatives to seed plants. Nature 409:618–622

Qiu YL (2008) Phylogeny and evolution of charophytic algae and land plants. J Syst Evol 46:287–306

Qiu YL, Li LB, Wang B et al. (2007) A nonflowering land plant phylogeny inferred from nucleotide sequences of seven chloroplast, mitochondrial, and nuclear genes. Int J Plant Sci 168:691–708

Raymond A, Gensel PG, Stein WE (2006) Late Silurian phytogeography. Rev Palaeobot Palynol 142:165–192

Remy W, Remy R (1980) *Lyonophyton rhyniensis* nov. gen. et nov. sp., ein Gametophyt aus dem Chert von Rhynie (Unterdevon, Schottland). Arg Palaeobot 6:27–72

Remy W, Gensel PG, Hass H (1993) The gametophyte generation of some early Devonian land plants. Int J Plant Sci 154:35–58

Rhee SY, Sommerville CR (1998) Tetrad pollen formation in quartet mutants of *Arabidopsis thaliana* is associated with persistence of pectic polysaccharides of the pollen mother cell wall. Plant J 15:79–88

Richardson JB (1964) Middle Old Red Sandstone spore assemblages from the Orcadian basin north-east Scotland. Palaeontolology 7:559–605

Richardson JB (1988) Late Ordovician and Early Silurian cryptospores and miospores from northeast Libya. In: El-Arnauti A, Owens B, Thusu B (eds) Subsurface palynostratigraphy of northeast Libya. Garyounis University Publications, Benghazi, Libya, pp 89–110

Richardson JB (1996a) Chapter 18A. Lower and Middle Palaeozoic records of terrestrial palynomorphs. In: Jansonius J, McGregor DC (eds) Palynology: principles and applications. American Association of Stratigraphic Palynologists Foundation, Salt Lake City, pp 555–574

Richardson JB (1996b) Abnormal spores and possible interspecific hybridization as a factor in the evolution of Early Devonian land plants. Rev Palaeobot Palynol 93:333–340

Richardson JB (1996c) Taxonomy and classification of some new Early Devonian cryptospores from England. In: Cleal CJ (ed) Studies on early land plant spores from Britain. Special Papers in Palaeontology, Great Britain

Richardson JB, Ioannides NS (1973) Silurian palynomorphs from the Tanezzuft and Acacus Formations, Tripolitania, North Africa. Micropalaeontology 19:257–307

Richardson JB, Lister TR (1969) Upper Silurian and Lower Devonian spore assemblages from the Welsh Borderland and South Wales. Palaeontolology 12:201–252

Richardson JB, McGregor DC (1986) Silurian and Devonian spore zones of the Old Red Sandstone continent and adjacent regions. Geol Surv Can Bull 364:1–79

Richardson JB, Streel M, Hassan A et al. (1982) A new spore assemblage to correlate between the Breconian (British Isles) and the Gedinnian (Belgium). Ann Soc Géol Belg 105: 135–143

Richardson JB, Ford JH, Parker F (1984) Miospores, correlation and age of some Scottish Lower Old Red Sandstone sediments from the Strathmore region (Fife and Angus). J Micropal 3:109–124

Richardson JB, Bonamo PM, McGregor DC (1993) The spores of *Leclercqia* and the dispersed morphon *Acinosporites lindlarensis* Riegel: a case of gradualistic evolution. B Nat Hist Mus London (Geol) 49:121–155

Richardson JB, Rodríguez RM, Sutherland SJE (2001) Palynological zonation of Mid- Palaeozoic sequences from the Cantabrian Mountains, NW Spain: implications for interregional and interfacies correlation of the Ludford/Přídolí and Silurian/Devonian boundaries, and plant dispersal patterns. B Nat Hist Mus London (Geol) 57:115–162

Rickards RB (2000) The age of the earliest club mosses: the Silurian Baragwanathia flora in Victoria, Australia. Geol Mag 137:207–209

Rodríguez RM (1978) Mioesporas de la Formación San Pedro/Furada (Silúrico Superior-Devónico Inferior), Cordillera Cantabrica, NO de España. Palinol Núm Extra 1:407–433

Rogerson ECW, Edwards D, Davies KL et al. (1993) Identification of in situ spores in a Silurian *Cooksonia* from the Welsh Borderland. In: Collinson ME, Scott AC (eds) Studies in palaeobotany and palynology in honour of Professor WG Chaloner, FRS. Special Papers in Palaeontology. Palaeontological Association, London, 49:17–30

Rogerson C, Edwards D, Axe L, Davies KL (2002) A new embryophyte from the Upper Silurian of Shropshire, England. In: Wyse Jackson PN, Parkes MA, Wood R (eds), Studies in palaeozoic palaeontology and biostratigraphy in honour of Charles Hepworth Holland, Sp Pap Palaeontology. Palaeontological Association, London, 67:233–249

Rothwell GW (1986) Classifying the earliest Gymnosperms. In: Spicer RA, Thomas BA (eds) Systematic and taxonomic approaches in palaeobotany, Systematics Association Special Volume, Clarendon Press, Oxford, pp 137–162

Rubinstein CV, Steemans P (2002) Miospore assemblages from the Silurian-Devonian boundary, in borehole A1–61, Ghadamis Basin, Libya. In: Steemans P, Servais T, Streel M (eds) Paleozoïc Palynology: a special issue in honour of Dr. Stanislas Loboziak. Rev Palaeobot Palynol 118: 397–421

Scheckler SE, Snigirevskaya NS, Hill SA (1997) Endosporic gametophytes of the heterosporous Late Devonian (Famennian) progymnosperm *Archaeopteris archetypus* (= *A. obtusa*). Am J Bot 84:141

Skog JE, Banks HP (1973) *Ibyka amphikoma*, gen. et sp. nov., a new protoarticulate precursor from the late Middle Devonian of New York state. Am J Bot 60:366–380

Steemans P (1989) Palynostratigraphie de l'Eodévonien dans l'ouest de l'Europe. Serv Géol Bel, Bruxelles

Steemans P (1995) Silurian and Lower Emsian spores in Saudi Arabia. Rev Palaeobot Palynol 89:91–104

Steemans P (2000) Miospore evolution from the Ordovician to the Silurian. Rev Palaeobot Palynol 113:189–196

Steemans P (2001) Ordovician cryptospores from the Oostduinkerke borehole, Brabant Massif, Belgium. Géobios 34:3–12

Steemans P, Gerrienne P (1984) La micro- et macroflore du Gedinnien de la Gileppe, Synclinorium de la Vesdre, Belgique. Ann Soc Géol Belg 107:51–71

Steemans P, Pereira E (2002) Llandovery miospore biostratigraphy and stratigraphic evolution of the Paraná Basin, Paraguay – palaeogeographic implications. Bull Soc Géol France 173:407–414

Steemans P, Wellman CH (2004) Miospores and the emergence of land plants. In: Webby B, Paris F, Droser ML, Percival IG (eds) The Great Ordovician biodiversification event. Columbia University Press, New York, pp 361–366

Steemans P, Le Hérissé A, Bozdogan N (1996) Ordovician and Silurian cryptospores and miospores from Southeastern Turkey. Rev Palaeobot Palynol 93:35–76

Steemans P, Higgs KT, Wellman CH (2000) Cryptospores and trilete spores from the Llandovery, Nuayyim-2 Borehole, Saudi Arabia. In: Al-Hajri S, Owens B (eds) Stratigraphic palynology of the Palaeozoic of Saudi Arabia. GeoArabia, Bahrain, pp 92–115

Steemans P, Wellman CH, Filatoff J (2007) Palaeophytogeographical and palaeoecological implications of a spore assemblage of earliest Devonian (Lochkovian) age from Saudi Arabia. Palaeogeogr Palaeocl Palaeoecol 250:237–254

Steemans P, Wellman CH, Miller MA et al. (2008a) An Ordovician cryptospore and trilete spore assemblage from Saudi Arabia. 12th International Palynological Congress. GeoUnion Alfred-Wegner-Stiftung, Bonn, August 30–September 5, pp 266

Steemans P, Rubinstein C, Melo JHG (2008b) Siluro-Devonian miospore biostratigraphy of the Urubu River area, western Amazon Basin, northern Brasil. Géobios 41:263–282

Steemans P, Le Hérissé A, Melvin J et al. (2009) Origin and radiation of the earliest vascular land plants. Science 324:353

Stein WE, Hueber FM (1989) The anatomy of *Pseudosporochnus – Pseudosporochnus hueberi* from the Devonian of New York. Rev Palaeobot Palynol 60:311–359

Stein WE, Wight DC, Beck CB (1984) Possible alternatives for the origin of Sphenopsida. Syst Bot 9:102–118

Stein WE, Mannolini F, Hernick LV et al. (2007) Giant cladoxylopsid trees resolve the enigma of the Earth's earliest forest stumps at Gilboa. Nature 446:904–907

Steinmann G (1929) Die organischen Einschlüsse. In: Steinmann G, Elberskirsch W (eds) Neue bemerkenswerte Funde im älternen Unterdevon des Wahnbachtales bei Siegburg. Sitzungbericht naturhistorischer, Verein Preussen Reinland and Westfalen Germany, 21/22:1–74

Stewart WN, Rothwell GW (1993) Palaeobotany and the evolution of plants. Cambridge University Press, Cambridge

Stockmans F (1948) Végétaux du Dévonien Supérieur de la Belgique. Mém Mus R Hist Nat Bel 110:1–85

Stockmans F (1968) Végétaux mésodévoniens récoltés aux confins du Massif du Brabant (Belgique). Mém Inst R Sci Nat Belg 159:1–49

Streel M (1964) Une association de spores du Givétien inférieur de la Vesdre, à Goé (Belgique). Ann Soc Géol Belg 87: 1–30

Streel M (1967) Associations de spores du Dévonien inférieur belge et leur signification stratigraphique. Ann Soc Géol Belg 90:11–53

Streel M, Higgs K, Loboziak S et al. (1987) Spore stratigraphy and correlation with faunas and floras in the type marine Devonian of the Ardenno-Rhenish regions. Rev Palaeobot Palynol 50:211–229

Strother PK, Traverse A (1979) Plant microfossils from the Llandoverian and Wenlockian rocks of Pennsylvania. Palynology 3:1–21

Strother PK, Al-Hajri S, Traverse A (1996) New evidence for land plants from the lower Middle Ordovician of Saudi Arabia. Geology 24:55–59

Taugourdeau-Lantz J (1967) Spores nouvelles du Frasnien du Bas Boulonnais (France). Rev Micropal 10:48–60

Taylor WA (1995) Ultrastructure of *Tetrahedraletes medinensis* (Strother and Traverse) Wellman and Richardson, from the Upper Ordovician of Southern Ohio. Rev Palaeobot Palynol 85:183–187

Taylor WA (1996) Ultrastructure of Lower Paleozoic dyads from Southern Ohio. Rev Palaeobot Palynol 92:269–279

Taylor WA (1997) Ultrastructure of Lower Paleozoic dyads from southern Ohio II: *Dyadospora murusattenuata*, functional and evolutionary considerations. Rev Palaeobot Palynol 97:1–8

Taylor WA (2002) Studies in cryptospore ultrastructure: variability in the tetrad genus *Tetrahedraletes* and type material of the dyad *Dyadospora murusattenuata*. Rev Palaeobot Palynol 119:325–334

Taylor WA (2003) Ultrastructure of selected Silurian trilete spores and the putative Ordovician trilete spore *Virgatasporites*. Rev Palaeobot Palynol 126:211–223

Taylor WA, Strother PK (2008) Ultrastructure of some Cambrian palynomorphs from the Bright Angel Shale, Arizona, USA. Rev Palaeobot Palynol 151:41–50

Taylor TN, Kerp H, Hass H (2005) Life history biology of early land plants: deciphering the gametophyte phase. Proc Natl Acad Sci USA 102:5892–5897

Taylor TN, Taylor EL, Krings M (2009) Paleobotany – the biology and evolution of fossil plants. Elsevier Academic Press, Burlington

Traverse A (2007) Paleopalynology. Springer, New York

Turnau E, Jakubowska L (1989) Early Devonian miospores and age of the Zwolen Formation (Old Red Sandstone facies) from Ciepielów IG-1 borehole. Ann Soc Geol Poloniae 59:391–416

Vavrdová M (1984) Some plant microfossils of the possible terrestrial origin from the Ordovician of central Bohemia. Vest Ceskeho Geol Ust 59:165–170

Vavrdová M (1988) Further acritarchs and terrestrial plant remains from the Late Ordovician at Hlasna Treban (Czechoslovakia). Cas Mineral Geol 33:1–10

Vavrdová M (1990a) Coenobial acritarchs and other palynomorphs from the Arenig/Llanvirn boundary, Prague basin. Vest Ceskeho Geol Ust 65:237–242

Vavrdová M (1990b) Early Ordovician acritarchs from the locality Myto near Rokycany (late Arenig, Czechoslovakia). Cas Mineral Geol 35:239–250

Ville de Goyet FD, Breuer P, Gerrienne P et al. (2007) Middle Devonian (Givetian) megaspores from Belgium (Ronquières) and Libya (A1–69 borehole). In: Steemans P, Javaux EJ (eds) Recent advances in Palynology. Carnets Géol/Notebooks Geol, Brest

Wang Y, Geng BY (1997) A new Middle Devonian plant: *Metacladophyton tetraxylum* gen. et sp. nov. Palaeontographica B 245:85–102

Wang Y, Li J, Wang R (1997) Latest Ordovician cryptospores from southern Xinjiang, China. Rev Palaeobot Palynol 99:61–74

Wellman CH (1996) Cryptospores from the type area for the Caradoc Series (Ordovician) in southern Britain. Palaeontolology 55:103–136

Wellman CH (2006) Spore assemblages from the Lower Devonian 'Lower Old Red Sandstone' deposits of the Rhynie outlier, Scotland. Trans R Soc Edinburgh 97: 167–211

Wellman CH, Gensel PG (2004) Morphology and wall ultrastructure of the spores of the Lower Devonian plant *Oocampsa catheta* Andrews et al., 1975 Rev Palaeobotany and Palynology 130:269–295

Wellman CH, Richardson JB (1993) Terrestrial plant microfossils from Silurian inliers of the Midland Valley of Scotland. Palaeontolology 36:155–193

Wellman CH, Richardson JB (1996) Sporomorph assemblages from the 'Lower Old Red Sandstone' of Lorne Scotland. In: Cleal CJ (ed) Studies on early land plant spores from Britain. Special Papers of Palaeontology, Great Britain. Palaeontological Association, London, 55:41–101

Wellman CH, Edwards D, Axe L (1998a) Permanent dyads in sporangia and spore masses from the Lower Devonian of the Welsh Borderland. Bot J Linn Soc 127:117–147

Wellman CH, Edwards D, Axe L (1998b) Ultrastructure of laevigate hilate cryptospores in sporangia and spore masses from the Upper Silurian and Lower Devonian of the Welsh Borderland. Philos Trans R Soc Lond B Biol Sci 353:1983–2004

Wellman CH, Higgs KT, Steemans P (2000) Spore assemblages from a Silurian sequence in Borehole Hawiyah-151 from Saudi Arabia. In: Al-Hajri S, Owens B (eds) Stratigraphic Palynology of the Palaeozoic of Saudi Arabia. GeoArabia, Bahrain, pp 116–133

Wellman CH, Osterloff PL, Mohiuddin U (2003) Fragments of the earliest land plants. Nature 425:282–285

Wellman CH, Kerp H, Hass H (2006) Spores of the Rhynie chert plant *Aglaophyton* (*Rhynia*) *major* (Kidston and Lang) D.S. Edwards, 1986. Rev Palaeobot Palynol 142: 229–250

Wellman CH, Steemans P, Miller MA (2008) Trilete spores from the Ordovician of Saudi Arabia: earliest evidence for vascular plants and their immediate predecessors ("pro-tracheophytes"). 12th International Palynological Congress. GeoUnion Alfred-Wegner-Stiftung, Bonn, August 30–September 5, 2008, p 304

Tentaculitoids – An Enigmatic Group of Palaeozoic Fossils

Eberhard Schindler

Abstract

A class without a firmly agreed name, without agreement on the life habits of some of its orders, and with some imprecise information on stratigraphic ranges of salient groups makes the Tentaculitoida a group that is not easy to handle. The roots of the class extend back into the Ordovician; the tentaculitoids flourished in the Silurian and even more so in the Devonian. Whereas thick-shelled Tentaculitida were widespread and have been well documented from both the Silurian and Devonian, Dacryoconarida became exceptionally diverse in the Early and Middle Devonian and the Homoctenida having their acme in the Frasnian. Whether or not isolated populations of tentaculitoids ranged into the mid-Famennian or even into the Carboniferous, the main collapse in their diversity was connected with the Kellwasser Crisis, a profound extinction event near the Frasnian–Famennian boundary in the Late Devonian. This review is not exhaustive; it is limited to a selection of specific questions. Because of contradictory results from study of the composition and structure of tentaculitoid shells, the potentially compelling source of classificatory information cannot yet be used for systematics. Uncertainties regarding the life habit of the order Tentaculitida are briefly probed. Examples of rare taxa and unanticipated associations are considered.

Keywords

Tentaculitoida • Tentaculitida • Dacryoconarida • Homoctenida • Styliolinidae • Nowakiidae • Problematic affinities • Life habits • Facies distribution • Biostratigraphic use

Too big for the ones, too small for the others, they [the tentaculitoids – E.S.] have neither received attention by the micro- nor by the macro-palaeontologists.
(translated from H. Lardeux 1969, p. 3)

Introduction

Despite being abundant and widespread during the mid-Palaeozoic, study of the extinct class Tentaculitoidea is bedevilled by uncertainties and still

E. Schindler (✉)
Senckenberg Forschungsinstitut und Naturmuseum Frankfurt, 60325 Frankfurt am Main, Germany
e-mail: eberhard.schindler@senckenberg.de

J.A. Talent (ed.), *Earth and Life*, International Year of Planet Earth,
DOI 10.1007/978-90-481-3428-1_14, © Springer Science+Business Media B.V. 2012

unsolved questions. Its position in the animal kingdom has yet to be uneqivocally fixed; it is not clear which systematic names should be used for the class and its orders. There is no consensus as to whether or not the class, as currently understood, constitutes a monophyletic taxon. Moreover, despite imprecision regarding the inception and demise of many taxa, numerous tentaculitoids have proved useful in biostratigraphy as well as for analysis of facies and palaeoenvironmental patterns.

Tentaculitoids are small conical calcareous shells that are circular in transverse section. Most were straight, but a few were curved or even coiled. In cases when the calcareous shell has been dissolved, external imprints (moulds) may reproduce the sculpture of the outer shell, particularly in pelitic claystones and shales, though also in limestones and marls. In some representatives, the overall sculpture of the outer shell is repeated on the inner side – reproduced on steinkerns – but in others the inner wall is smooth and does not mirror the outer shell ornament. Figure 1 shows the general appearance of tentaculitoids.

It is not possible to exhaustively survey the characteristics and what these imply for these curious fossils. What is presented are some general remarks and a few specific observations. It is also not possible in this overview to deal with all tentaculitoid families. And there are no general explanations – schemes of bauplans – extended charts of systematics and stratigraphic range charts. Such information must be sought in the cited literature, specifically the major monographs of specialists on the group.

One of the most prominent features of any group of calcareous fossils is their shell mineralogy and structure. General accounts on tentaculitoid shells should be sought in monographic works by tentaculitoid workers, e.g. Fisher (1962), Blind (1969), Lardeux (1968, 1969), Larsson (1979), Lindemann and Yochelson (1984, 1992, 1994), Yochelson and Lindemann (1986), Alberti (1993 and subsequent papers) and Farsan (1994). My own investigations, based on details of the conical shells and of their external and internal moulds, are presented below. To date there is no accepted record of preservation of any soft tissue – despite the name 'tentaculitoids' having been derived (erroneously) from either the suggestion of 'crinoid tentacles' (von Schlotheim 1820) or of tentacles protruding from the youngest (apertural) parts of the tentaculitoid shell (for details see Fisher 1962). Reports about

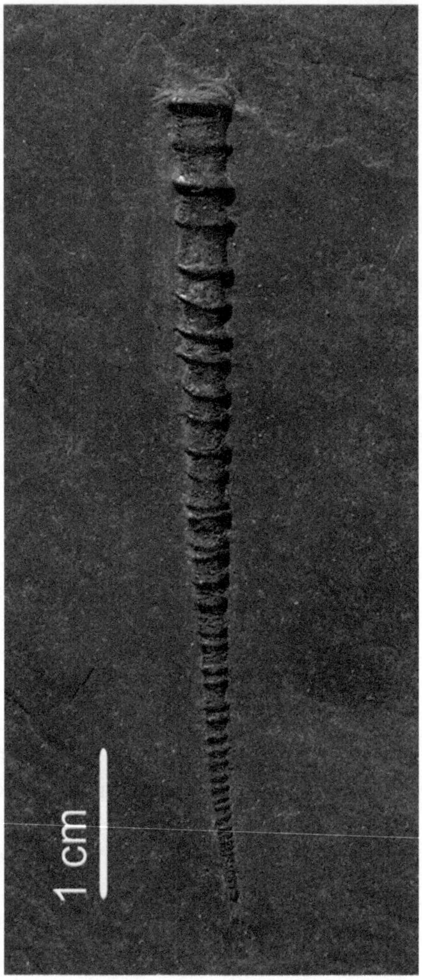

Fig. 1 *Tentaculites* sp. – a representative of the Tentaculitida from the Lower Devonian of Kaub (Hunsrück, Rheinisches Schiefergebirge, Germany) – refigured from Schindler (2005)

the presence of preserved soft tissue in tentaculitoids (Blind and Stürmer 1977) were rejected by Yochelson (1989) – see discussion of this topic and literature therein. Recent authors, nevertheless, have argued for organic material having been preserved in the residues from palynological sample processing (Wood et al. 2004; Filipiak and Jarzynka 2009).

In an old and remarkable microscope (!) study on material from erratic boulders from Nemmersdorf in East Prussia, Dewitz (1883) noted the presence of a "dépot organique", but this report was overlooked by subsequent workers. It was not until 86 years later that Blind (1969) suggested organic sheets on the outer and inner surfaces of the shells of Tentaculitida. He also suggested they had an operculum, but this was ignored by subsequent workers – 40 more years of research has not produced an undisputed find. Concerning von Schlotheim's assignment to the crinoids, it should be mentioned that during the early days of systematic studies in palaeontology, anything 'spiny' was thought to be crinoidal – not just tentaculitoids.

Systematics

The class Tentaculitoidea is one of the rare groups of organisms not to have a generally accepted common name! Though Walch (1775) had already mentioned and figured them, it was not until von Schlotheim (1820) introduced the name 'tentaculi(toi)ds' that the group received a name, even though he regarded them as a kind of crinoid. No name-change was proposed until the mid-twentieth century when detailed research got underway. Lyashenko (1957) created the name 'Tentaculitoidea', regarded by her as a superorder within the class 'Coniconchia' – proposed by her in 1955 and viewed as including the Hyolithoidea, a suggestion that did not gain acceptance. Fisher (1962) in the 'Treatise on Invertebrate Palaeontology' proposed the 'Cricoconarida' with two orders: Tentaculitida (including the families Tentaculitidae, Homoctenidae, and Uniconidae) and the Dacryoconaridae (including, among others, the Nowakiidae and the Styliolinidae). Bouček (1964) introduced the order Homoctenida – but also the Coleolida and the Cornulitida. In his profound monograph, Lardeux (1969) created the widely accepted classification; he regarded Lyashenko's superorder Tentaculitoidea as a class embracing three orders, the Tentaculitida, the Dacryoconarida, and the Homoctenida. The classification of Lardeux (1969) is used in this chapter. Nevertheless, in 1993, Alberti was still able to point out, with good reason, that 'generation of the class itself lies [not only] in the dark, but also the phylogenetic relations between the orders are still widely unsolved'.

The introduction of new subclasses and superorders (especially in the Tentaculitida and Homoctenida) by Farsan (1994) has not gained wide acceptance, mainly because taxonomic attribution is based on embryonic parts of the shells (their morphology, shell generation, structure, and ornament of early ontogenetic stages in thin sections of limestones!). As it is possible only in very favourable cases – rarely fulfilled – to have limestone preservation (and in addition to have large samples of specimens showing a sufficient number of embryonic parts of juvenile shells cut in perfectly-oriented thin sections), most tentaculitoid workers have sidestepped Farsan's approach and therefore have not been in a position to follow his systematics.

It is not the goal of this chapter to probe all possible approaches that may help resolve tentaculitoid systematics. Comparing and contrasting the development of shells in the various tentaculitoid orders – generally regard as a high-ranked character for systematics – will, however, be broached herein through examples. Contradictory results (possibly present knowledge engenders more confusion than assumptions of the past!) do not imply that future studies, especially if more effort is devoted to investigating shell materials, might not lead to helpful criteria for high-level systematics. Examples from all three orders orders (the first two examples representing families of the Dacryoconarida) are as follows:

1. *Styliolinidae*. Based on the observation that some styliolinid shells – in contrast to other dacryoconarid taxa at family and order level – consist of a single layer of primary calcite, Lindemann and Yochelson (1984, 1992, 1994) and Yochelson and Lindemann (1986) suggested separation of the styliolinids from other 'tentaculitoids'; they regarded, as did Bouček (1964), the difference in shell material to be of high systematic value. Consequently, they placed the styliolinids in a separate class as 'incertae sedis'. Our first attempts to gain more knowledge about tentaculitoid shells focused on a combination of 'regular' light microscopy, scanning electron microscopy (SEM), and cathodeluminescence (CL) applied to a selection of specimens having what seemed to be especially elegant preservation (Schindler and Neuser 1993b). However, prior information on styliolinid shell structure had been contradictory: Yochelson and Lindemann (1986) had demonstrated that only a single shell layer of 10 μm thickness occurred in specimens from the Middle and Upper Devonian of New York, as well as in topotypes of *Styliolina nucleata*, the type species of *Styliolina* from the eastern Urals of Russia (see Karpinsky 1884), whereas Alberti (1975) reported a tripartite styliolinid shell from the Early Devonian of the Frankenwald in Germany. The same discrepancy arose with our previously unpublished data: Whereas Fig. 2a (styliolinid shell viewed by

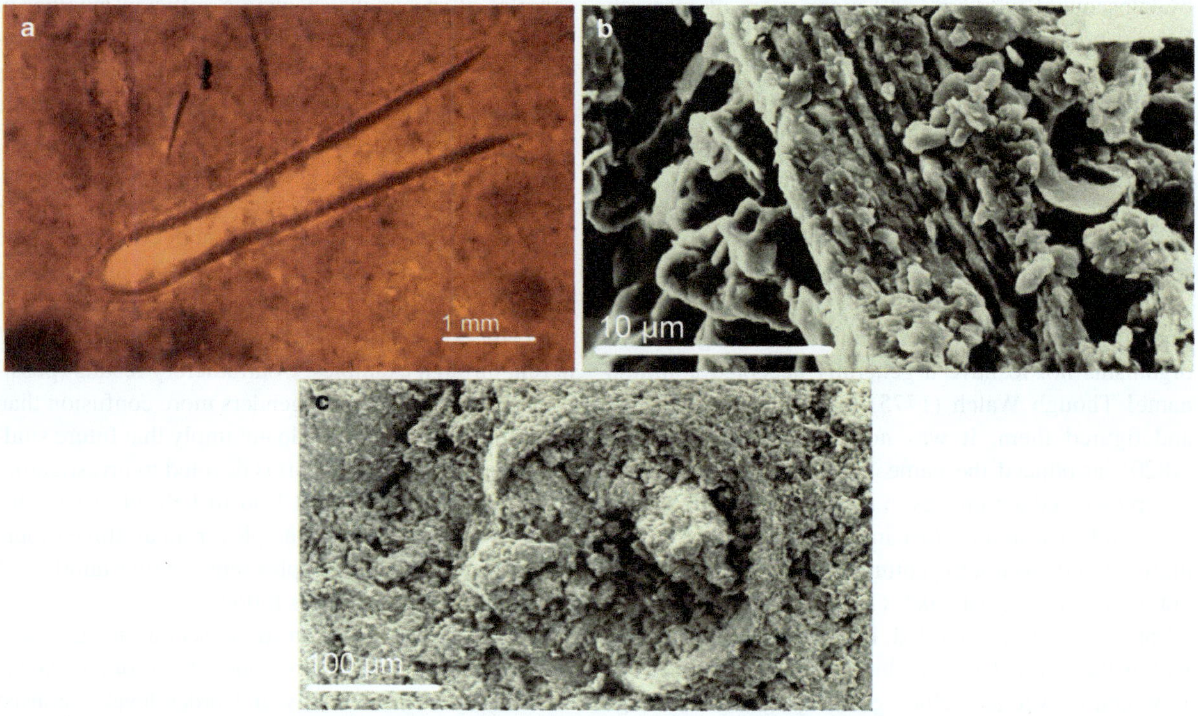

Fig. 2 (**a**) Longitudinal section of a styliolinid specimen (under cathodeluminescence) from the Frasnian of Bou Tchrafine (Tafilalt, Morocco). The unique dark (*brownish*) colour of the shell suggests presence of a single shell layer; the bright (*yellow*) luminescence of the thin cover seemingly reflects a 'micritic envelope' due to former organic tissue. (**b**) SEM photo of a layered styliolinid shell from a specimen of the same section as in (**a**). (**c**) SEM photo of a styliolinid from the same section as in (**a**) showing growth of crystals from the wall into the styliolinid conch

CL from the Frasnian of Bou Tchrafine, Tafilalt, Morocco) seemed consistent with a single shell layer, a specimen from the same section studied by SEM indicated a layered shell (Fig. 2b) – an observation now duplicated by Lindemann (personal communication 2010) from SEM studies on material from New York. In both cases the CL luminescence colours differ from the matrix; both occurrences have crystals growing into the styliolinid conch from the inner wall (Fig. 2c) – these suggest primary calcitic shell material.

2. *Nowakiidae*. Parallel with the previous example, the CL photo of a nowakiid shell from the type locality of the Zlichov Limestone (Lower Emsian) of the Barrandian area, Czech Republic (Fig. 3a), accords with a single shell layer, but SEM photos of another specimen from the same site (Fig. 3b, c) indicate a multi-layered shell, consistent with other reports (e.g. Yochelson and Lindemann 1986). Another feature seen in these figures is a 'micritic envelope' on the inner surface of the shell; this might have originated from previous organic tissue, as observed for homoctenid shells as well (see below and Schindler and Neuser 1993a). The bright yellow luminescent lines in the CL photo (Fig. 3a) are thought likewise to represent micritic layers.
3. *Order Tentaculitida*. The existence of primary calcitic shell material can be shown in specimens from the Frasnian of Hofermühle (Velbert Anticline, Eastern Rheinisches Schiefergebirge, Germany) display, using 'regular' light microscopy (Fig. 4a), preservation of internal features, notably pores (or more correctly pseudopores) in the shell – this would have not been the case if primary shell material (e.g. aragonite) had transformed into calcite. A CL photo (Fig. 4b) produced the same result.
4. *Order Homoctenida*. CL photos of shells of homoctenids from the Frasnian Lower Kellwasser Horizon of the Brausnberg section (Berga Anticline, Thüringisches Schiefergebirge, Germany) could lead to a different conclusion (Schindler and Neuser 1993a). The fact that both

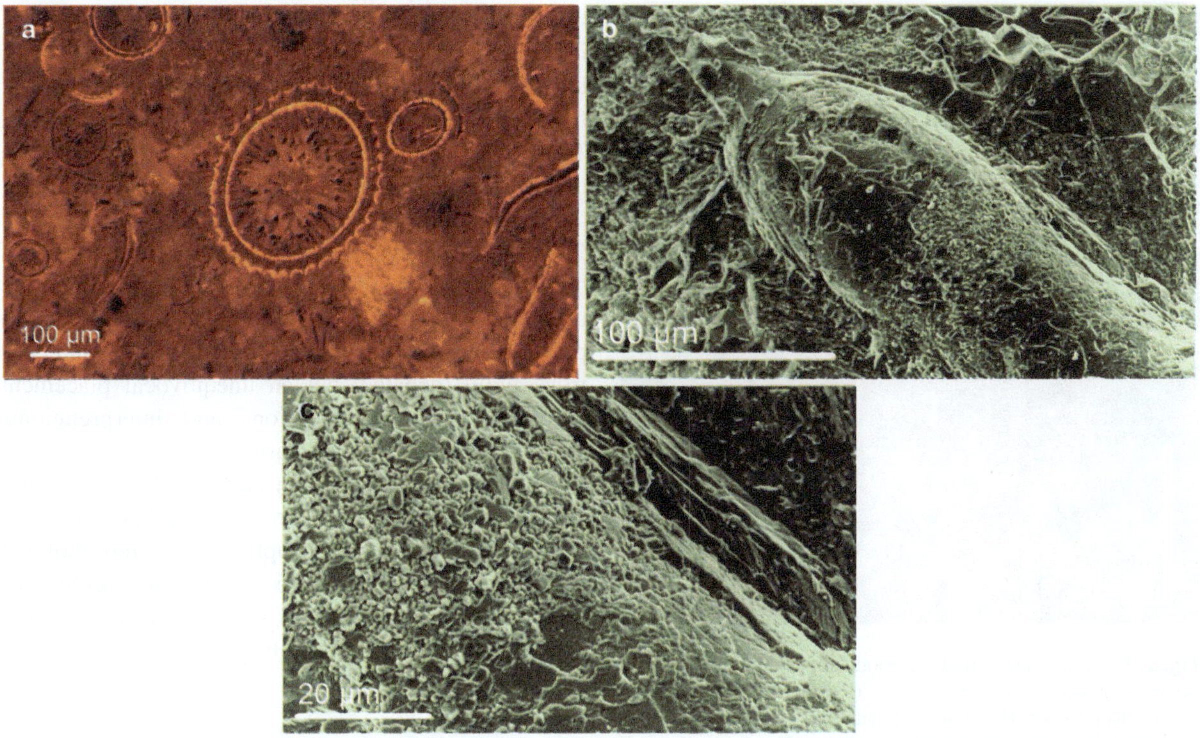

Fig. 3 (**a**) Transverse section of nowakiid specimens (under cathodeluminescence) from the type locality of the Lower Emsian Zlichov Limestone (Barrandian area, Czech Republic). The dark (*brownish*) colour of the shell suggests presence of a single shell layer; the bright (*yellow*) luminescence of the thin coverings of the inner and outer surface of the shell apparently reflects 'micritic envelopes' corresponding to former organic tissue; note also the clearly visible growth of crystals from the inner wall into the conch. Note also differences in thickness of the transverse section of the more adult specimen (wider transverse section) and the thinner shells of the more juvenile specimens. (**b**) Juvenile part of a nowakiid specimen from the same locality as (**a**) showing layered structure of the shell; note also the micritic layer forming the inner surface of the shell visible on the steinkern – it seemingly corresponds to the bright luminescent 'envelopes' in (**a**). (**c**) Enlargement of (**b**)

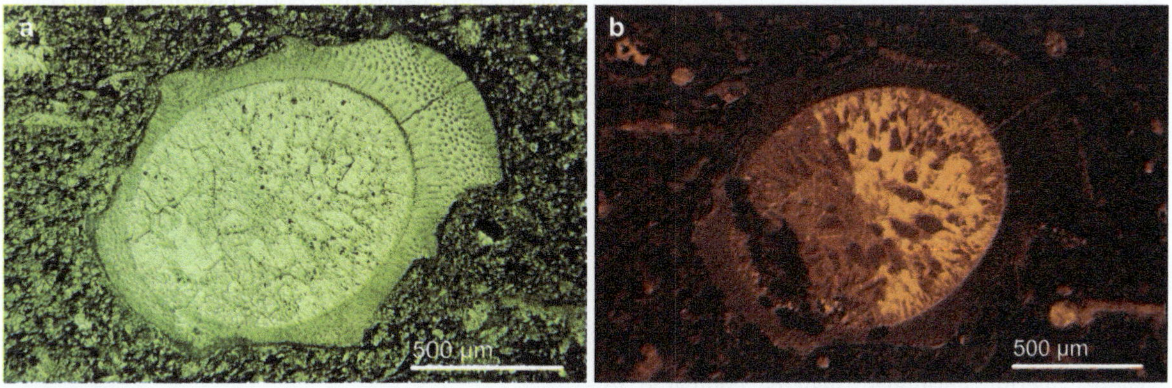

Fig. 4 (**a**) Transverse section of a specimen (ordinary light microscopy) of *Dicricoconus* sp. from the Frasnian of the Hofermühle Section (Velbert Anticline, Rheinisches Schiefergebirge, Germany). The shell comprises pseudopores indicating presence of primary calcitic material; faint structures would have been obliterated if the original shell material had been aragonitic. (**b**) Same picture as in (**a**), but under cathodeluminescence; it supports the observations in (**a**)

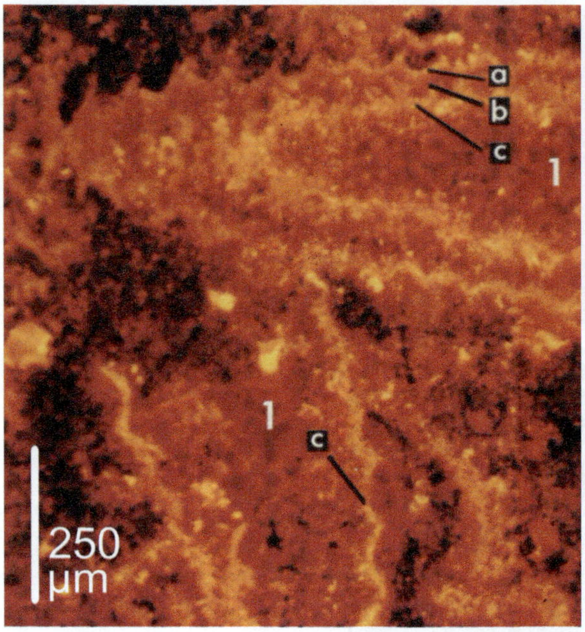

Fig. 5 Oblique section under cathodeluminescence of a specimen of *Homoctenus tenuicinctus* (Reomer 1950) from the Lower Kellwasser Horizon at Braunsberg (Berga Anticline, Thüringisches Schiefergebirge, Germany). Note that the calcitic material of the shell (b), the cement (blocky) filling the conch (1), and the matrix show the same luminescence, possibly indicating that the shell has been transformed from primary aragonite to calcite. The two bright (*yellow*) luminescing seams (a and c) apparently represent 'micritic envelopes' like those seen in other tentaculitoid orders; they may relate to former organic tissue. Photo refigured from Schindler and Neuser (1993a)

the calcitic shell material and the calcitic matrix have the same luminescence (Fig. 5) at the present day may be construed as indicating the primary shell material could have been aragonitic. If consideration is limited to shell material, the homoctenids could be argued as far removed from other major tentaculitoid groups. Interpretation of one of the CL features, i.e. the bright yellow reflecting outer and inner margins of the homoctenid shell (Fig. 5) to represent the previously mentioned 'micritic envelopes'; these may indicate former organic tissue (as discussed above for the nowakiids – for further discussion see Schindler and Neuser 1993a). These 'micritic envelopes' may be compared to comparable observations made on thick-shelled Tentaculitida by Blind (1969); he interpreted thin layers in thin sections as former organic tissue.

The 'shell problem' aside, the suggested classificatory position of tentaculitoids in the animal kingdom has varied widely. As mentioned earlier, they were initially regarded as crinoids (von Schlotheim 1820), but were associated with other high-rank taxa, e.g. foraminifers, cephalopods, annelids and 'worms'. They were considered to be spines of brachiopods or trilobites, as molluscs, treated as a discrete major taxon, or taxonomically discarded as 'incertae sedis'. Based on the suggested planktonic mode of life of dacryoconarids – similar to that of extant planktonic molluscs, they were long thought to be related to pteropods. Mollusc affinities are still highly (and controversially) debated, but an unequivocal placement requires further investigations and interpretations, surely including shell structure.

Throughout the twentieth century, the class experienced a progressive deletion of taxa that did not fit the emerging model or bauplan. Thus, morphologic attributes of the hyolithids, cornulitids, coleolids, and others are no longer relevant in considering classification of the tentaculitoids.

Stratigraphic Range, Appearance, and Demise

As with systematics, the stratigraphic range of the tentaculitoids is still not precisely known, especially regarding the when-and-where of their demise. Their origin dates back to the Ordovician. Specimens of the thick-shelled order Tentaculitida, the oldest group of tentaculitoids, were reported from the Early Ordovician Chepultepec Formation of Virginia (Lower Canadian = Tremadocian; Fisher and Young 1955; Fisher 1962) and have been found in strata of the same age in other regions (Fisher 1962). This age assignment is also mentioned in Corbett and Banks (1974), Yochelson and Lindemann (1986), and Wittmer (2009b). The Middle Ordovician information-gap suggests the group was neither widespread nor abundant until the Late Ordovician when an obvious blooming occurred, exemplified by the Late Ordovician tentaculitoids of the Cincinnati Arch (Waynsville Formation, type Cincinnatian) documented by Wittmer (2009a, b). The thin-shelled order Homoctenida contributed to marine faunas from the Early Devonian onwards, especially when, if we follow Alberti (1993, 1997a), the Paranowakiidae are included in the Homoctenida. He suggested the order is also present in the Silurian, as was suggested earlier by Fisher (1962). According to Fisher, the other major thin-shelled order, the Dacryoconarida, is known from the Silurian onwards though with a significant expansion in diversity during the Early Devonian (Alberti 1993). For detailed information the reader is referred especially to publications by Alberti (inter alia 1993, 1997a, b, 1998, 2000). The report of a styliolinid in Ordovician rocks from Arctic North America (Miller et al. 1954 and cited in Wittmer 2009b) has been queried by the authors themselves. They believe it possible that imprecise sampling may have occurred when the material was collected in the mid-1820s (Miller et al. 1954, p. 25). The dacryoconarids certainly achieved maximum diversity during the Middle Devonian whereas the homoctenids reached their acme in the Frasnian.

Even more puzzling than the origin of the major tentaculitoid groups is their demise. Most of the entire class Tentaculitoidea vanished from the geological record in the Late Devonian, but the demise of most taxa at family- or order-level was not isochronous (see, for example, Yochelson and Kirchgasser 1986; Schindler 1990a, b; Walliser 1996; Bond 2006). Most tentaculitoids nevertheless became extinct (or at least severely affected) during the Kellwasser (KW) Event at the end of the Kellwasser Crisis, near the Frasnian–Famennian (F–F) boundary. A few papers from the last decades of the twentieth century and from the first decade of the twenty-first century dealing with this faunal turnover are relevant, among them those of McLaren (1970), Walliser (1984, 1996, 2003), Sandberg et al. (1988), Schindler (1990a, b, 1993), Buggisch (1991), Becker and House (1994), McGhee (1996), House (1997, 2002), Morrow (2000), Over (2002), Joachimski and Buggisch (2002), Crick et al. (2002), Bond et al. (2004), Joachimski et al. (2002, 2009), Racki (2005), and Gereke (2007). Apart from some disputed finds reported in older literature, it transpires that the homoctenids indeed crossed the F–F boundary, but ranged only into the basal bed of the Famennian – linkage of their demise to environmental deterioration at the end of the KW Event has been discussed elsewhere (Schindler 1990a, b).

There are reports of even younger tentaculitoids. Occurrences of styliolinids and homoctenids in mid-Famennian strata (homoctenids up to the *rhomboidea* conodont-zone; styliolinids up to the *marginifera* Zone) were reported by Li (2000) from South China where they occur with undisputed conodonts of those time intervals (possible reworking was denied by the author). Shul'ga et al. (1994) reported tentaculitoids with co-occurring Early Carboniferous goniatites in the L'vov–Volhynian Basin (Ukraine). It is surprising that these exclusively marine groups would have been deposited in 'clay and carbonate facies of a shallow coastal sea' (authors' words) interfingering with coal beds! Subsequent reworking, perhaps even more than once, cannot be ruled out until sedimentologic evidence demonstrates autochthoneity of the two groups of fossils. Even more puzzling is Niko's (2000) report of possible tentaculitoids of Carboniferous/Permian age from Central Japan. Though the beds in which the tentaculitoids have been found are bounded top and bottom by faults (Niko 2000, p. 381), co-occurring Sakmarian corals provide an age for the association. Again, because a detailed sedimentologic study has not been presented (to minimise or exclude the possibility of reworking), the age of the tentaculitoids remains questionable.

This notwithstanding, numerous dacryoconarids and homoctenids evolved rapidly, were near cosmopolitan in occurrence, and therefore have great value for long-distance stratigraphic alignments. This is especially true for the Devonian. For example, Alberti indicated their biostratigraphic value in many papers not only from the 1960s to 1980s, but also in his later major monographs (Alberti 1993, 1997a, b, 1998, 2000 q.v. for earlier references). A tentaculitoid-based zonation was developed in his earlier papers and was subsequently updated. Many other authors have used tentaculitoids for biostratigraphy but, because of the vastness of the published literature, I refer only to salient items: Lyashenko (1959, 1967), Zagora (1962a, b, 1964), Bouček (1964), Lardeux (1968, 1969), Lütke (1974, 1979, 1985), Larsson (1979), Sauerland (1983), Lukeš (1989), Truyóls-Massoni and Garciá-Alcalde (1994), Bultynck et al. (2000), Jansen et al. (2007), Kim (2008), and Wittmer (2009b). Recently, the significance of curved and coiled tentaculitoids has been applied in the discussion of the Lochkovian–Pragian boundary (Koren' et al. 2007).

Facies and Environmental Application – Life Habit

The different orders of the tentaculitoids had – at least in part – very different modes of life and therefore occur in different lithofacies. Whereas thick-shelled

Tentaculitida were clearly benthic and are found mainly in siliciclastic sediments such as sandstones and siltstones as well as in shallow-water limestones and marls, the dacryoconarids and homoctenids tend to be found predominantly in sedimentary rocks of deeper, in other words pelagic, environments. Most probably, they were plankton with a preference for deep-water limestones and pelitic rocks (a widespread argument for the deep-water assignment is the often irregular distribution of these conical shells in fine-grained shales and limestones). There even exist lithostratigraphic terms referring to the occurrence of tentaculitoids in the lithostratigraphic succession of the German Early Devonian: the 'Tentaculitoid Nodular Limestone' (Tentakuliten-Knollenkalk in German) of the late Lochkovian to basal Emsian and the 'Tentaculitoid Shale' (Tentakuliten-Schiefer in German) found in most Emsian pelagic situations (i.e. mainly in the Frankenwald area and in the Thüringisches Schiefergebirge). Exemplarily for many more stratigraphers dealing with the term, a paper by Zagora (1962b) subdivided the Tentakuliten-Schiefer. The paper cites a wealth of older references from the days when the term was introduced (Richter 1851; Gümbel 1863). For a more recent summary demonstrating continuing use see Zagora et al. (2008).

Widely differing views have been presented regarding the life style of the tentaculitoids. In an appealing sketch, Fisher (1962, Fig. 51 on p. W105) figured more or less every possible mode of life. Whereas for dacryoconarids and homoctenids, a planktonic life habit within the pelagic facies realm has generally gone undisputed – though with exceptions, notably of Yochelson and Lindemann (1986) who claimed that styliolinids may also have lived in very shallow water – the mode of life of the Tentaculitida has always been much debated. The principal strategies suggested have been nektobenthonic (by several authors, including Fisher 1962), sessile (e.g. Blind 1969, 1970; Hurst and Hewitt 1977), and even endozoan (Blind 1969; Farsan 1994). Farsan (1994, Fig. 8 on p. 22) assumed a larval development favouring the curved shells of his Trompetoconarida. Schubert (1996) rejected the view that tentaculitids stood upright in sediment, and also and especially the endozoan hypothesis. Interestingly, however, some specimens within a 'pavement' of tentaculitids preserved in an upright position were reported recently (Cornell et al. 2003) from Early Devonian (Lochkovian) limestones of New York.

An interesting aspect concerning the mode of life of the tentaculitoids is the documentation (Berkyová et al. 2007) of injury and repair of dacryoconarids and homoctenids connected with unsuccessful predation in the Emsian of the Barrandian area.

Rare Tentaculitoids

As mentioned in Section 'Systematics', systematic assignment is not always easy. Sometimes it is difficult to place a species in an appropriate genus – or even in what could be the right order. An example of this is *Tentaculites deubeli* erected by Zagora (1964). Dimensions as well as the general appearance are consistent with referring this form to *Homoctenus*, in other words, transferring it from the order Tentaculitida to the order Homoctenida. Figure 6 shows a specimen from the Sessacker Trench (Frasnian) of the Dill Syncline in the Rheinisches Schiefergebirge (Germany). It is residual uncertainty regarding its taxonomic location, but '*T.*' *deubeli* probably belongs in *Homoctenus*.

Problems often occur with very rare taxa. For example, for some of the many taxa erected by GP Lyashenko (e.g. 1955, 1959, 1985) from the Russian Platform, the number of specimens mentioned in her papers is limited to only one or two per taxon (often from core material). However, even such rare taxa as her genera *Polycylindrites* (erected in 1955) or *Dneprovskites* (erected in 1985) and, until now, known exclusively from the type area, do in fact

Fig. 6 '*Tentaculites*' *deubeli* Zagora 1964. The size and 'appearance' seem to be like homoctenids. Specimen from the Sessacker Trench (Frasnian) of the Dill Syncline (Rheinisches Schiefergebirge, Germany)

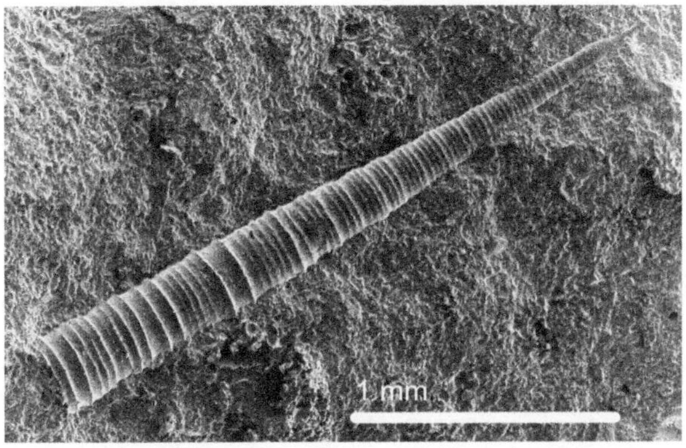

Fig. 7 *Polycylindrites* sp. (Polycylindritidae, Homoctenida) from the Sessacker Trench (Frasnian) of the Dill Syncline (Rheinisches Schiefergebirge, Germany)

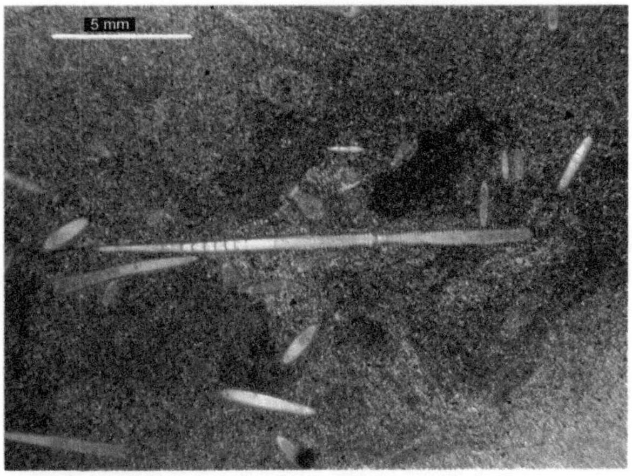

Fig. 9 Exceptionally well-preserved *Tentaculites* sp. from Frasnian strata in drill hole Trent 1/68 (3083.2 m). Note the numerous septa in the juvenile part of the shell; not all of these are visible because of the slight curvature of the specimen

occur in the Rheinisches Schiefergebirge of Germany (Figs. 7 and 8); the latter occurrences have yet to be documented.

Material belonging to the Tentaculitida have been obtained from deep boreholes on the islands of Rügen, Hiddensee, and Usedom, and even from an offshore location (Vorpommern, NE Germany). Some of the wells reached depths of more than 5500 m (the one on Usedom Island close to 7500 m) where calcareous siltstones, marls, and limestones of Frasnian age were penetrated – for general description see Zagora (2008); for detailed description of the cores see Zagora and Aehnelt (2009). Preservation of the rich tentaculitid fauna from this unusual source is spectacular (Fig. 9). The material has yet to be documented.

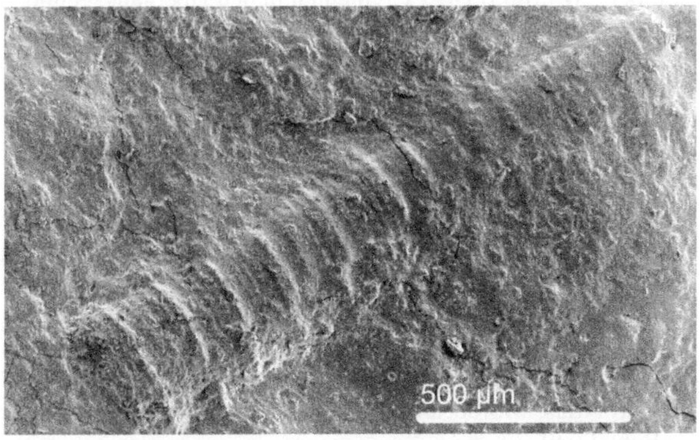

Fig. 8 *Dneprovskites* sp. (Polycylindritidae, Homoctenida) from the Grube Christiane Section (Frasnian) near Diemelsee-Adorf (Rheinisches Schiefergebirge, Germany)

General Problems and Conclusions

As indicated in the 'Introduction' section, a complete overview about the tentaculitoids was not intended in this contribution. Besides pointing out a number of studies and subsequent references, it was also thought to discuss specific items of interest, especially where they shed light on current problems and how these might be tackled. For example, many tentaculitoid workers hope that knowledge about the composition and structure of the conical shells (even if perhaps methodologically complicated) must be undertaken seriously in quest of solutions to higher-rank systematic assignments. As pointed out, different shells are present in what were thought to be closely related groups. There is potential for making use of shell material and shell structure by a combination of methods, specifically SEM, CL, and 'regular' light microscopy.

Life habit, especially of the thick-shelled Tentaculitida, still offers a wide field for discussions. Although the conchs are almost always found lying with their long axes parallel to the sediment surface, some interpretations – and even recent field observations – require careful checking of the taphonomic situation.

And as shown above, exciting observations are still to be made and documented; these may come from rare taxa or when rare tentaculitoid-bearing environments are discovered.

At present, it is questionable if investigation of shell material and structure may provide a tool for systematic assignment either within the class or where to locate the class within the animal kingdom. Eventually, the general position of the class Tentaculitoidea – or parts of it – has to be reorganised within the tree of life. But for this, surely more – and clearly focused – research is needed.

Acknowledgements I am especially grateful to four colleagues: the editor of this volume, John Talent, for his patience while I wrote this chapter and for checking and improving the final version; Dick Lindemann (Saratoga Springs) for constructive comments on a draft manuscript; Rolf Neuser (Bochum) for taking the cathodeluminescence photos; and Michael Ricker (Senckenberg, Frankfurt) for help with photography and electronic handling of the images.

References

Alberti GKB (1975) Zur Struktur der Gehäusewand von *Styliolina* (Dacryoconarida) aus dem Unter-Devon von Oberfranken. Senck leth 55:505–511

Alberti GKB (1993) Dacryoconaride und homoctenide Tentaculiten des Unter- und Mittel-Devons I. Cour Forsch-Inst Senckenberg 158:229pp

Alberti GKB (1997a) Planktonische Tentakuliten des Devon. I. Homoctenida BOUČEK 1964 und Dacryoconarida FISHER 1962 aus dem Unter- bis Ober-Devon. Palaeontographica Abt A 244:85–142

Alberti GKB (1997b) Planktonische Tentakuliten des Devon. II. Dacryoconarida FISHER 1962 aus dem Unter- und Mittel-Devon. Palaeontographica Abt A 246:1–32

Alberti GKB (1998) Planktonische Tentakuliten des Devon. III. Dacryoconarida FISHER 1962 aus dem Unter-Devon und oberen Mitteldevon. Palaeontographica Abt A 250:1–46

Alberti GKB (2000) Planktonische Tentakuliten des Devon. IV. Dacryoconarida FISHER 1962 aus dem Unter-Devon. Palaeontographica Abt A 256:1–23

Becker RT, House MR (1994) Kellwasser events and goniatite successions in the Montagne Noire with comments on possible causations. Cour Forsch-Inst Senckenberg 169:45–77

Berkyová S, Frýda J, Lukeš P (2007) Unsuccessful predation on Middle Paleozoic plankton: shell injury and anomalies in Devonian dacryoconarid tentaculites. Acta Palaeont Polonica 52:407–412

Blind W (1969) Die systematische Stellung der Tentakuliten. Palaeontographica Abt A 133:101–145

Blind W (1970) Epizoen als Anzeiger der Lebendstellung von *Tentaculites* sp. N Jb Geol Paläont Abh 136:243–261

Blind W, Stürmer W (1977) *Viriatellina fuchsi* KUTSCHER (Tentaculoidea) mit Sipho und Fangarmen. N Jb Geol Paläont Mh 1977:513–522

Bond D (2006) The fate of the homoctenids (Tentaculitoidea) during the Frasnian-Famennian mass extinction (Late Devonian). Geobiology 4:167–177

Bond D, Wignall PB, Racki G (2004) Extent and duration of marine anoxia during the Frasnian-Famennian (Late Devonian) mass extinction in Poland, Germany, Austria and France. Geol Mag 141:173–193

Bouček B (1964) The tentaculites of Bohemia. Czechoslovak Academy of Sciences, Praha

Buggisch W (1991) The global Frasnian-Famennian "Kellwasser Event". Geol Rdsch 80:49–72

Bultynck P, Lardeux H, Walliser OH (2000) On the correlation of middle-Emsian. Subcomm Devonian Stratigr Newsl 17:10–11

Corbett KD, Banks MR (1974) Ordovician stratigraphy of the Florentine Synclinorium, southwest Tasmania. Pap Proc Roy Soc Tasmania 107:207–238

Cornell SR, Brett CE, Sumrall CD (2003) Paleoecology and taphonomy of an edrioasteroid-dominated hardground association from tentaculitid limestones in the Early Devonian of New York: a Paleozoic rocky peritidal community. Palaios 18:212–224

Crick RE, Ellwood BB, Feist R, El Hassani A, Schindler E, Dreesen R, Over DJ, Girard C (2002) Magnetostratigraphy susceptability of the Frasnian/Famennian boundary. Palaeogeogr Palaeoclimatol Palaeoecol 181:67–90

Dewitz J (1883) Bemerkungen über Tentaculiten. Z. Naturwiss 56: 80–87

Farsan NM (1994) Tentaculiten: ontogenese, systematik, phylogenese, biostratonomie und morphologie. Abh senckenberg naturforsch Ges 547:125pp

Filipiak P, Jarzynka A (2009) Organic remains of tentaculitids: new evidence from Upper Devonian of Poland. Acta Palaeontol Polonica 54:111–116

Fisher DW (1962) Small conoidal shells of uncertain affinities. In: Moore RC (ed) Treatise on invertebrate paleontology, part W. Miscellanea, pp 98–143 Geol Soc America and Kansas University, Lawrence, 260pp

Fisher DW, Young RS (1955) The oldest known tentaculitid – from the Chepultapec Limestone (Canadian) of Virginia. J Paleontol 29:871–875

Gereke M (2007) Die oberdevonische Kellwasser-Krise in der Beckenfazies von Rhenoherzynikum und Saxothuringikum (spätes Frasnium / frühestes Famennium, Deutschland). Kölner Forum Geol Paläont 17:228pp

Gümbel CW (1863) Über Clymenien in den Übergangsgebilden des Fichtelgebirges. Palaeontographica 11:85–165

House MR (1997) The late Devonian mass extinction: the Frasnian/Famennian crisis. Hist Biol 12:291–301

House MR (2002) Strength, timing, setting and cause of Mid-Palaeozoic extinctions. Palaeogeogr Palaeoclimatol Palaeoecol 181:5–25

Hurst JM, Hewitt RA (1977) On tubular problematica from the type Caradoc (Ordovician) of England. N Jb Geol Paläont Abh 153:147–169

Jansen U, Lazreq N, Plodowski G, Schemm-Gregory M, Schindler E, Weddige K (2007) Neritic–pelagic correlation in the Lower and basal Middle Devonian of the Dra valley (S Anti-Atlas, Moroccan Pre-Sahara). Geol Soc London Spec Publ 278:9–37

Joachimski MM, Buggisch W (2002) Conodont apatite $\delta^{18}O$ signatures indicate climatic cooling as a trigger for the Late Devonian (F-F) mass extinction. Geology 30:711–714

Joachimski MM, van Geldern R, Breisig S, Day J, Buggisch W (2002) Oxygen isotope evolution of biogenic calcite and apatite during the Middle and Upper Devonian. Int J Earth Sci (Geol Rdsch) 93:542–553

Joachimski MM, Breisig S, Buggisch W, Talent JA, Mawson, R, Gereke M, Morrow JR, Day J, Weddige K (2009) Devonian climate and reef evolution: insights from oxygen isotopes in apatite. Earth Planet Sci Lett 284:599–609

Karpinsky A (1884) Die fossilen Pteropoden am Ostabhange des Ural. Mém Acad Sci 7th Ser 32:1–21

Kim AI (2008) Dacryoconaride biostratigraphy of the Devonian deposits of South Tien Shan. In: Kim AI, Salimova, FA, Meshchankina, NA (eds) Global alignments of Lower Devonian carbonate and clastic sequences. Contr Internat Conf, SDS/IGCP Project 499 Joint Field Meeting, pp 51–55

Koren' TN, Kim AI, Walliser OH (2007) Contribution to the biostratigraphy around the Lochkovian-Pragian boundary in Central Asia (graptolites, tentaculites, conodonts). Senck leth 87(2):187–219

Lardeux H (1968) Les Tentaculites d'Europe occidentale et d'Afrique du Nord: Quelques résultats. Bull Soc Etud Sci Anjou VI:59–78

Lardeux H (1969) Les Tentaculites d'Europe occidentale et d'Afrique du Nord. Cahiers Paléont, Centre Nat Rech Sci 238pp

Larsson K (1979) Silurian tentaculitids from Gotland and Scania. Fossils Strata 11:180

Li YX (2000) Famennian tentaculitids of China. J Paleontol 74:969–975

Lindemann RH, Yochelson EL (1984) Styliolines from the Onondaga Limestone (Middle Devonian) of New York. J Paleontol 58:1251–1259

Lindemann RH, Yochelson EL (1992) *Viriatellina* (Dacryoconarida) from the Middle Devonian Ludlowville Formation at Alden, New York. J Paleontol 66:193–199

Lindemann RH Yochelson EL (1994) Redescription of *Styliolina* (*incertae sedis*); *Styliolina fissurella* (Hall) and the type species *S. nucleata* (Karpinsky). In: Landing E (ed) Studies in Stratigraphy and Paleontology in Honor of Donald W. Fisher. New York State Museum Bull 481:149–160, New York State Educational Department, Albany, NY

Lütke F (1974) Die Verbreitung von Tentaculiten (Dacryoconarida) aus dem Unterdevon des SW-Harzes und ihre Korrelation mit der Conodonten-Parachronologie. N Jb Geol Paläont Abh 145:183–206

Lütke F (1979) Biostratigraphical significance of the Devonian Dacryoconarida. Spec Pap Palaeont 23:281–289

Lütke F (1985) Devonian tentaculites from Nevada (USA). Cour Forsch-Inst Senckenberg 75:197–226

Lukeš P (1989) Tentaculitids from the Lower/Middle Devonian section in Praha-Barrandov. Věstn ústr Ústav geol 64:193–206

Lyashenko GP (1955) Novyje danyje o systematike tentakulitov, novakii i styliolin. Autoreferat doklada, B MOIP, Otdel Geol 29:94–95

Lyashenko GP (1957) Novyje roda devonskich tentaculitov. Doklad Akad nauk SSSR 116:141–144

Lyashenko GP (1959) Konikonchi Devon zentralnych i vostotschnych oblasteiy Russkoi Platformy. Gostoptechizdat, VNIGNI, Leningrad, 220pp

Lyashenko GP (1967) Conichonchia (Tentaculitida, Nowakiida, Styliolinida) and their importance in Devonian biostratigraphy. In: Oswald DH (ed) International symposium on the Devonian System 2, Alberta Soc Petrol Geol, pp 897–903

Lyashenko GP (1985) Novyje devonskije konikonchi Vostotschno-Evropeijskoj Platformy. In: Stratigrafitscheskije isledovanija prirodnych rezervuarov nefti i gaza. VNIGNI, pp 19–27

McGhee GR (1996) The Late Devonian mass extinction. The Frasnian/Famennian Crisis. Columbia University Press, New York

McLaren DJ (1970) Presidential address: time, life, and boundaries. J Paleontol 44:801–815

Miller AK, Youngquist W, Collinson C (1954) Ordovician cephalopod fauna of Baffin Island. Geol Soc Am Mem 62:1–234

Morrow JR (2000) Shelf-to-basin lithofacies and conodont paleoecology across Frasnian-Famennian (F-F, mid-Late Devonian) boundary, Central Great Basin (Western U.S.A.). Cour Forsch-Inst Senckenberg 219:1–57

Niko S (2000) Youngest record of tentaculitioids: *Hidagaienites* new genus from the Carboniferous-Permian boundary in central Japan. J Paleontol 74:381–385

Over DJ (2002) The Frasnian-Famennian Boundary in the Appalachian Basin, Michigan Basin, Illinois Basin, and southern continental margin, central and eastern United States. Palaeogeogr Palaeoclimatol Palaeoecol 181:153–170

Racki G (2005) Toward understanding Late Devonian global events: few answers, many questions. In: Over DJ, Morrow JR, Wignall PB (eds) Understanding Late Devonian and Permian-Triassic biotic and climatic events: towards an integrated approach. Elsevier, Amsterdam, pp 5–36

Richter R (1851) Erläuterungen zu der geognostischen Übersichtskarte des ostthüringischen Grauwackengebirges. Z dt geol Ges III:536–553

Sandberg CA, Ziegler W, Dreesen R, Butler JL (1988) Late Frasnian mass extinction: conodont event stratigraphy, global changes, and possible causes. Cour Forsch-Inst Senckenberg 102:263–307

Sauerland U (1983) Dacryoconariden und Homocteniden der Givet- und Adorf-Stufe aus dem Rheinischen Schiefergebirge (Tentaculitoidea, Devon). Göttinger Arb Geol Paläont 25:86

Schindler E (1990a) The Late Frasnian (Upper Devonian) Kellwasser Crisis. Lect Notes Earth Sci 30:151–159

Schindler E (1990b) Die Kellwasser-Krise (hohe Frasne-Stufe, Ober-Devon). Göttinger Arb Geol Paläont 46:115

Schindler E (1993) Event-stratigraphical markers within the Kellwasser Crisis near the Frasnian/Famennian boundary (Upper Devonian) in Germany. Palaeogeogr Palaeocliamtol Palaeoecol 104:115–125

Schindler E (2005) Tentakuliten – Fossilien paläozoischer Meere. Natur und Museum 135:182–183

Schindler E, Neuser RD (1993a) Apical part and shell structure of *Homoctenus tenuicinctus* (Cricoconarida; Devon). Göttinger Arb Geol Paläont 58:107–114

Schindler E, Neuser RD (1993b) Schalenstruktur bei Homocteniden und verwandten devonischen Cricoconariden. Paläont Ges, 63. Jahrestag, 21. 26. September, Abstrakte, Paläozoikum, 19

Schubert M (1996) Zur Lebensweise dickschaliger Tentakuliten. N Jb Geol Paläont Abh 201:1–16
Shul'ga VF, Shimanskiy VN, Dorodnova IV (1994) Lower Carboniferous tentaculites-like organisms from the L'vov–Volhynian Basin. Paleontol J 28 (1A):97–106
Truyóls-Massoni M, Garciá-Alcalde JL (1994) Faune rhéno-bohémienne (dacryoconarides, brachiopodes) à la limite Emsien inférieur/supérieur au Cabo la Vela (Asturies, Espagne). Geobios 27:221–241
von Schlotheim E (1820) Die Petrefactenkunde auf ihrem jetzigen Standpunkte durch die Beschreibung seiner Sammlung versteinerter und fossiler Überreste des Thier- und Pflanzenreichs der Vorwelt erläutert. Gotha
Walch J (1775) Von einer noch unbekannten Tubulitenart. Der Naturforscher 7:211–213
Walliser OH (1984) Geologic processes and global events. Terra Cogn 4:17–20
Walliser OH (1996) Global events in the Devonian and Carboniferous. In: Walliser OH (ed) Global events and event stratigraphy. Springer, Berlin, pp 225–250
Walliser OH (2003) Die biotische Frasne-Krise und der Kellwasser-Event im Oberdevon. In: Hansch W (ed) Katastrophen in der Erdgeschichte – Wendezeiten des Lebens. museo, vol 19. Städtische Museen, Heilbronn, Germany, pp 100–117
Wittmer JM (2009a) Spacial and temporal patchiness of tentaculitoid-rich fossil assemblages in the type Cincinnatian. Cincinnati Mus Center Sci Contr 3:354–355
Wittmer JM (2009b) The regional and global diversity of tentaculitoids: a new look at the spatio-temporal and paleogeographic history of an enigmatic group. Master Thesis, University of Cincinnati, 194pp
Wood GD, Miller MA, Bergström SM (2004) Late Devonian (Frasnian) tentaculite organic remains in palynological preparations, Radom-Lublin region, Poland. Mem Assoc Australasian Palaeontologists 29:253–258
Yochelson EL (1989) Reconsideration of possible soft parts in dacryoconarids (*incertae sedis*) from the Hunsrück-Schiefer in Western Germany. Senck leth 69:381–390
Yochelson EL, Kirchgasser WT (1986) The youngest styliolines and nowakiids (Late Devonian) currently known from New York. J Paleontol 60:689–700
Yochelson EL, Lindemann RH (1986) Considerations on systematic placement of the styliolines (*incertae sedis*: Devonian). In: Hoffmann A, Nitecki MH (eds) Problematic fossil taxa. Oxford University Press, New York, pp 45–58
Zagora K (1962a) Zur biostratigraphischen Bedeutung der Tentaculiten im thüringischen Unter- und Mitteldevon. Geologie 11:548–556
Zagora K (1962b) Zur Gliederung des Mitteldevons an der SE-Flanke des Schwarzburger Sattels (Thüringisches Schiefergebirge). Wiss Z Friedrich-Schiller-Univ Jena 11:359–365
Zagora K (1964) Tentaculiten aus dem thüringischen Devon. Geologie 13:1235–1273
Zagora K (2008) Rügen/Hiddensee (und Usedom). Schriftenr dt Ges Geowiss 52:574–577
Zagora K, Aehnelt, M (2009) Oberdevon in NE-Deutschland – Ergebnisse von 15 Tiefbohrungen im Küstengebiet Vorpommerns. Z geol Wiss 37:153–205
Zagora K, Weyer D, Zagora I (2008) Unter- und Mitteldevon im Thüringischen Schiefergebirge. Schriftenr dt Ges Geowiss 52:51–58

Palaeozoic Ammonoids – Diversity and Development of Conch Morphology

Dieter Korn and Christian Klug

Abstract

Palaeozoic ammonoids are not only valuable index fossils, but are also valuable for discriminating global crises in biodiversity such as the end-Eifelian Kačák Event, the Frasnian Kellwasser Events, the late Famennian Hangenberg Event, and the Permian-Triassic Event. They are also useful for solving various questions in palaeobiology; their suture lines and conch geometry permit morphometric studies using multivariate and shape analyses. Results from such analyses accord with results derived from taxa-based diversity curves (Hangenberg Event), but may contrast with them in other cases (the Kellwasser and Permian-Triassic Events). The disparity curve of conch shapes in Palaeozoic ammonoids is markedly discontinuous. Character unfolding at the beginning of their evolutionary history in the Early Devonian was rapid; the subsequent evolutionary development exhibits phases of wide morphologic extent (i.e. high disparity: middle to late Famennian, late Viséan to Bashkirian, Artinskian to Wordian) alternating with intervals of less morphologic extent (i.e. low disparity: earliest Famennian, earliest Tournaisian). Palaegeographic distribution of morphospace occupation is discussed in the context of the early rediversification and the end-Permian endemism and diversity-decline.

Keywords

Global biodiversity crises • Kačák Event (end-Eifelian) • Kellwasser events • Hangenberg Event • End-Permian Event • Morphometric and morphospace studies (from multivariate and shape analyses) • Diversity dynamics • Conch geometries • Preservation of diversity

Introduction

Preservation of biological diversity, globally, is arguably the most important problem facing mankind. This requires greatly improved understanding of diversity changes such as life crises, their causes, and their

D. Korn (✉)
Museum für Naturkunde, Leibniz-Institut für Evolutions und Biodiversitätsforschung an der Humboldt-Universität zu Berlin, D-10115 Berlin, Germany
e-mail: dieter.korn@museum.hu-berlin.de

J.A. Talent (ed.), *Earth and Life*, International Year of Planet Earth,
DOI 10.1007/978-90-481-3428-1_15,

spatial and temporal extent. These are not only problems for biologists, but also the pre-eminent task of palaeontologists who are concerned with deep time – improving understanding of the patterns and processes of global change through the deep (e.g. the Palaeozoic) past. This is the backdrop, better knowledge of which will provide better appreciation of present day global biodynamics. Ammonoids, because of species richness, rapid evolutionary unfolding of conch characters, wide geographic distribution and long history, are particularly suited for elucidation of these diversity patterns and fluctuations. The history of the ammonoid group, taken as a whole, may be regarded as having been like a natural laboratory – having an importance to palaeontology analogous to the way research with vinegar flies (*Drosophila*) in laboratories all over the world has been important in the history of modern genetics. As Seilacher (1989) has said, the ammonoids, by presenting us with so much knowledge about evolutionary patterns during the Palaeozoic (Fig. 1) became the '*Drosophila* of Palaeontology'.

The mathematically calculable geometry of spirally coiled ammonoid conchs has long attracted the attention of scientists. Robert Hooke (1705, published posthumously), one of the most ingenious seventeenth century philosophers of science, was fascinated by the logarithmic coil of ammonite shells and their regularly arranged septa. He stated that 'the Shell is of a true Conical Figure from the *Base* to the *Apex* ... That this Cone is turned into a *Voluta* or Spiral Cone ... That this Spiral being a true proportional Spiral, is continually at certain distances intercepted by Diaphragmes; so that those Diaphragmes being taken as Bases of several Cones, the Cones shall be found to diminish in a series Geometrically Proportional'.

The first mathematical analyses of the conchs of Recent *Nautilus* and of ammonites were during the mid-nineteenth century (Moseley 1838; Naumann 1840a, b, 1845, 1849; Müller 1850, 1853; Sandberger and Sandberger 1850–1856; Sandberger 1855). Surprisingly, these pioneering studies were not followed up until more than 100 years later when David Raup, in three innovative articles (Raup and Michelson 1965; Raup 1966, 1967), outlined the theoretical framework for further investigations. He analysed various types of mollusc conchs in terms of their geometrical properties and defined principal morphological conch parameters such as the whorl expansion rate, i.e. the opening rate of the whorl spiral.

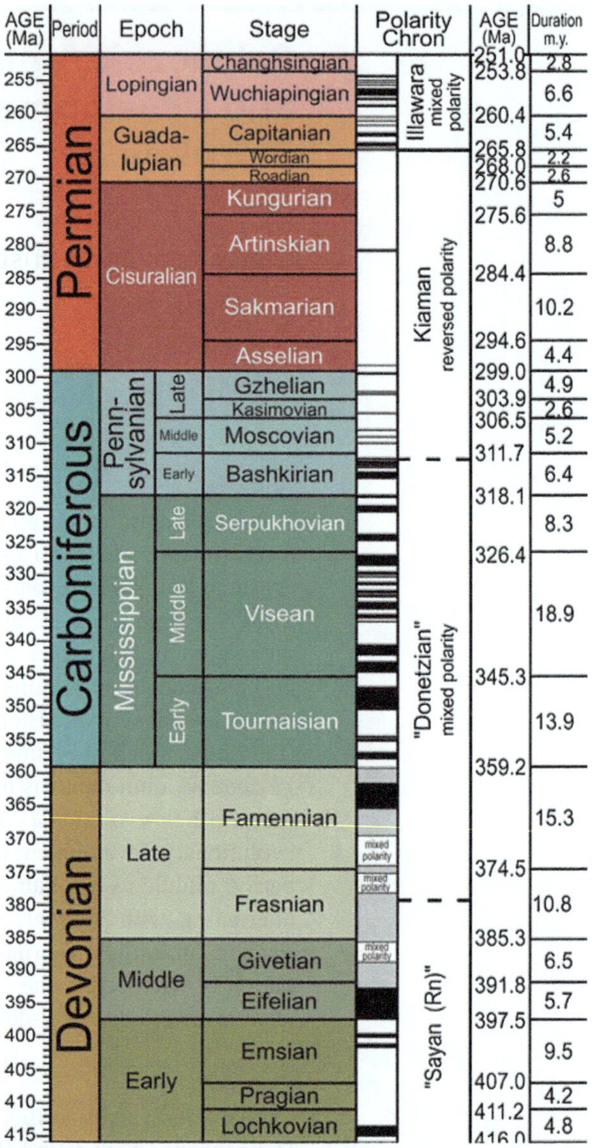

Fig. 1 Geochronology and chronostratigraphy of the Devonian to Permian (after Gradstein et al. 2004)

Mathematical investigations were presented by Kullmann and Scheuch (1970, 1972) as well as Kant. In a series of articles, the latter (e.g. Kant 1973a, b; Kant and Kullmann 1988) focused his attention on the basic morphometric parameters: conch radius, whorl width, whorl height, and umbilical width, and their development during ontogeny. According to his studies, allometric growth was particularly common during the Carboniferous and Permian.

Application of these essentially theoretical studies to palaeontology was initiated, using Carboniferous ammonoids, by Saunders and Swan (1984) and Saunders and Work (1996). Within ammonoids of the Namurian stage, they discriminated eight morphotypes characterized by the ratio of whorl expansion rate to umbilical width.

Morphological transformations and morphospace occupation of successive ammonoid faunas about the Devonian-Carboniferous Boundary were documented by Korn (2000). Detailed investigation of the whorl expansion rate and umbilical width falsified the common assumption, accepted for instance by Nikolaeva and Barskov (1994), that the earliest Carboniferous goniatites rapidly occupied the morphospace vacated by the demise of the clymeniid ammonoids.

Recent studies of morphometrics and morphospace-occupation of Palaeozoic ammonoids were undertaken in describing and discussing morphological trends in Early and Middle Devonian ammonoids (Korn and Klug 2001, 2003). Korn and Vöhringer (2004) used as an example the earliest Carboniferous species *Gattendorfia crassa* Schmidt (1924) for outlining ontogeny and intraspecific variability. Villier and Korn (2004) analysed the Permian demise of the ammonoids; Korn (2004) evaluated the putative *Annulata* Event (Late Devonian) by means of ammonoid morphospace; and Korn et al. (2004a) analysed the relationship between 'cuboid' and 'normal' Carboniferous ammonoids. Morphometrics have figured importantly in investigations of various ammonoid groups (Korn and Tilsley 2002; Korn and Weyer 2003; Ebbighausen et al. 2004; McGowan 2004a, b, 2005; McGowan and Smith 2007; Gottobrio and Saunders 2005; Klug et al. 2006; Korn and Klug 2007; Barskov et al. 2008; Saunders et al. 2008; Ubukata et al. 2008).

Diversity studies of Palaeozoic ammonoids have been undertaken several times; the general ammonoid diversity pattern was outlined repeatedly by House (1981, 1985; see Fig. 2). Such spindle diagrams show that their evolutionary history was strikingly

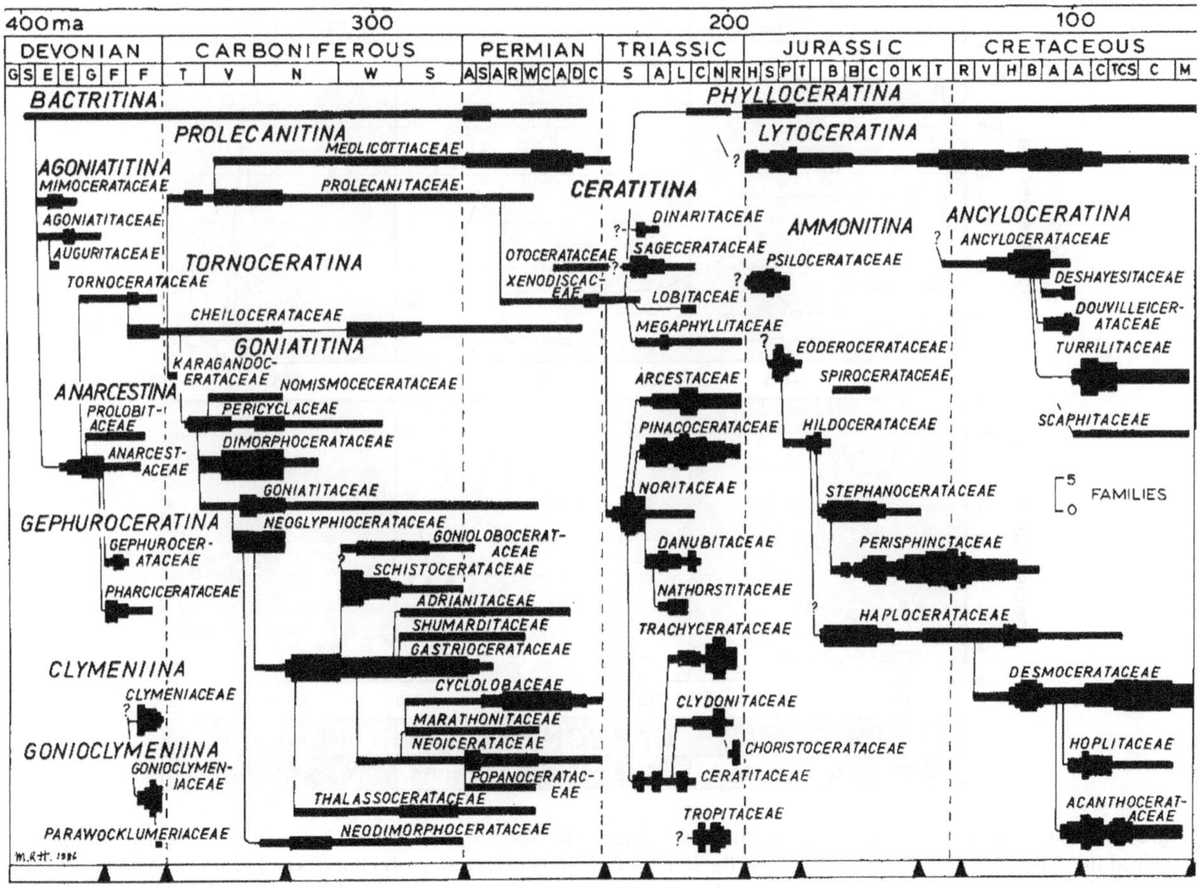

Fig. 2 The system of the Palaeozoic ammonoids and the stratigraphic distribution of the major taxonomic units (from House 1981)

discontinuous, punctuated by several extinction events with subsequent recoveries. Kullmann (1983), however, concluded that the main changes in diversity of Late Devonian and Carboniferous ammonoids were associated with diversification events, extinction events playing only a subordinate role. Studies of Palaeozoic ammonoid diversity were summarized by Becker and Kullmann (1996).

Simple counts of species, genera, and families reflect diversity fluctuations during a specific interval (Fig. 3). Several such investigations have been published, especially for the early Late Devonian (Becker et al. 1989, 1993; Becker and House 1994) and about the Devonian-Carboniferous Boundary (Korn 1986, 1993, 2000; Korn and Kullmann 1988; Kullmann 2000).

Because of the scarcity of soft-tissue preservation in ammonoids, knowledge of their anatomy is poor, usually limited to traces of the digestive tract, muscle remains, and, exceptionally, also their gills (Lehmann 1981, 1985). The external morphology is completely unknown and thus, only inferences can be made. Applying the phylogenetic bracket on the number of arms, Recent embryonic nautilids and belemnites

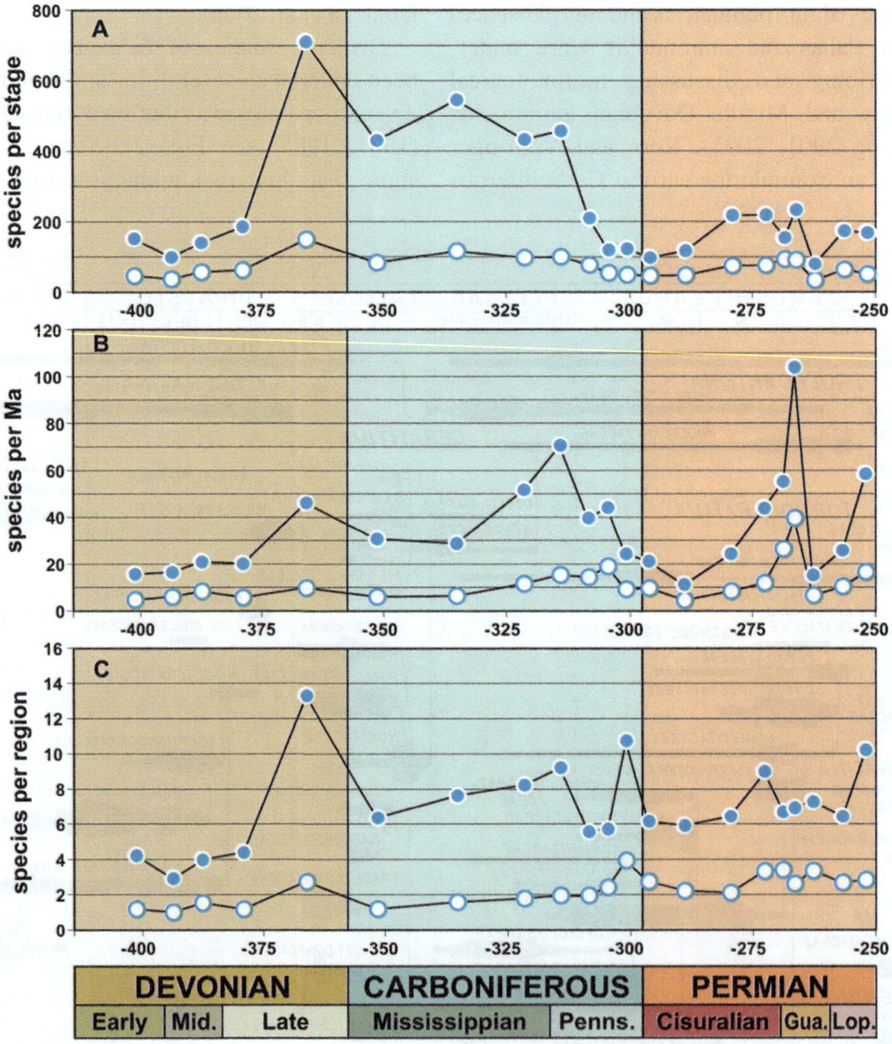

Fig. 3 Diversity patterns of Palaeozoic ammonoids; data obtained from the AMMON database (Korn and Ilg 2008). (**a**) Species (*blue dots*) and genus (*white dots*) richness of each stage. (**b**) Species and genus richness (mean value) per millions of years in each stage. (**c**) Species and genus richness (mean value) per region (usually defined as a sedimentary basin) in each stage

may be used. Belemnites and some other coleoids are known to have had ten arms (Johnson and Richardson 1968; Reitner and Urlichs 1983; Doguzhaeva et al. 2007; Doguzhaeva et al. 2010); nautilid embryos have five pairs of arm-buds, each with a pair of arms at an age of 3 months (Shigeno et al. 2008); thus, ten arms seem likely. Another unresolved problem is the presence or absence of a hood. In nautilid embryology, the hood develops from a clearly discernible bud (Shigeno et al. 2008); a non-mineralized hood may have formed similarly in ammonoids.

Other indications of the presence of a hood in ammonoids were discussed by Klug et al. (2004); they argued that most ammonoids probably had a black layer like that of Recent nautilids, indirectly attached via the dorsal mantle.

Regarding the mode of life of ammonoids, we follow Klug and Korn (2004) and Klug et al. (2008); they suggested that morphological transformations around the transition from orthocones to fully coiled ammonoids reflects adaptation to a rather nektonic mode of life, limited to the middle and late growth stages. Mapes and Nützel (2008) argued convincingly that ammonoids had floating egg masses and planktonic juveniles.

The Conch of Palaeozoic Ammonoids and Their Ontogeny

Palaeozoic ammonoids display a wide range of conch geometries (Fig. 4). It is now generally accepted that the subclass Ammonoidea was derived from bactritoid cephalopods and that the transformation from uncoiled to coiled conchs happened rather quickly in the lineage leading to the oldest ammonoids: from the base of the late Early Devonian Emsian stage (Fig. 1) (Schindewolf 1933; Erben 1964a, b; Bogoslovsky

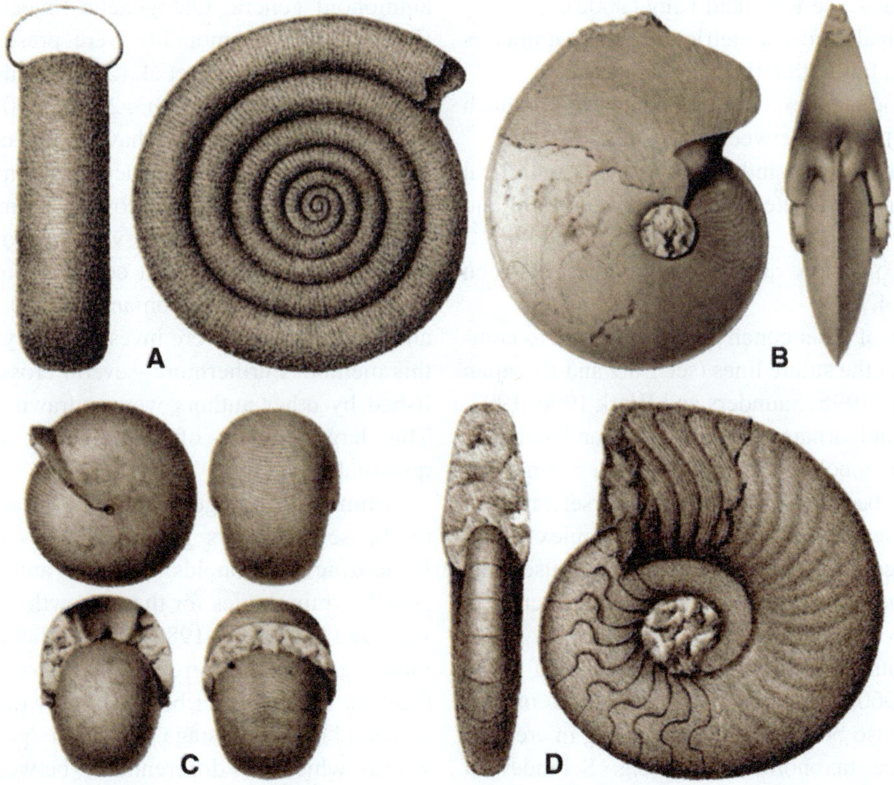

Fig. 4 Examples for the range of Palaeozoic ammonoid morphology. (**a**) Morph A: lenticular, widely umbilicate conch with low aperture (*Phenacoceras*); (**b**) Morph B: lenticular, narrowly umbilicate conch with high aperture (*Carinoceras*); (**c**) Morph C: globular, narrowly umbilicate conch with low aperture (*Prolobites*); (**d**) Morph D: lenticular to globular, moderately umbilicate with moderately high aperture (*Pseudoclymenia*) (after Sandberger and Sandberger 1850–1856 and Gümbel 1863)

1969; Chlupàč and Turek 1983; Klug 2001a; Klug and Korn 2004). This means that all the earliest ammonoids had incomplete coiling, i.e. there was an umbilical window in the openly coiled conchs. By the late Emsian, however, a spectrum of conch morphologies had been elaborated comparable to or even exceeding those, which evolved during any subsequent Palaeozoic time-interval.

Palaeozoic ammonoid conchs share several common characters:

- All possess a planispiral conch; trochospiral forms are not known.
- All conchs, except for the earliest ammonoids, are convolute, meaning that the whorls embrace preceding whorls to some degree.
- The protoconch usually has a diameter of approximately 0.5–0.7 mm; only the earliest ammonoids had a protoconch diameter reaching 1.2 mm.
- A large number of different conch shapes can be discriminated from extreme serpenticonic to globular and cadyconic conchs; for a summary of the Devonian taxa, see Korn and Klug (2002).
- The principal morphometric conch parameters investigated fall within the following ranges:
  - Conch width index (CWI = conch width/conch diameter) ranges between 0.08 and 1.40.
  - Umbilical width index (UWI = umbilical width/conch diameter) ranges between 0.00 and 0.85.
  - Whorl expansion rate (WER) ranges between 1.30 and 4.75.

Distribution of other conch parameters such as complexity states of the suture lines (see Lutz and Boyajian 1995; Saunders 1995; Saunders and Work 1996, 1997; Allen 2007) and ornament is irregular and not predictable from conch geometry. Evolute forms, for instance, may be smooth-shelled or coarsely ribbed; they may possess a very simple or a complex suture line. It is particularly the suture line that, because of its long-term trends in morphological unfolding, enables classification of Palaeozoic Ammonoidea into major taxonomic units (Wedekind 1918; Schindewolf 1929; Ruzhencev 1960). However, overestimation of this character has also been practiced, resulting in creation of polyphyletic taxonomic units (e.g. Schindewolf 1950; discussion by Korn 2003).

In addition to the possibility to address phylogenetic questions, mollusc shells also provide great opportunities for studying their entire ontogeny; this is true also for Palaeozoic ammonoids. Nevertheless, study of the role of development in ammonoid evolution has been studied less frequently than one might expect. Aspects of development were studied by earlier authors such as Clarke (1899a), who studied ammonitella of Devonian ammonoids, and later Erben (e.g. 1962, 1964b, 1965), who often used ontogenetic changes of characters in his discussions. Later studies, focused on ontogenetic development of Palaeozoic ammonoid conchs, were undertaken by Kullmann and Scheuch (1970, 1972), Kant (1973a, b), as well as Kant and Kullmann (1988). These investigators also used cross-sections from which the main conch parameters were obtained for detailed analysis.

They used the Raupian conch parameters: conch radius, umbilical radius, and whorl width to calculate the whorl expansion rate (W), umbilical width index (U), and whorl width index (S). More recently, Klug (2001b), Korn and Klug (2001, 2007), Korn and Vöhringer (2004), and Ebbighausen and Korn (2007) portrayed ontogenetic trajectories of some Devonian ammonoid genera. Ontogenetic trajectories of some Carboniferous ammonoids were presented in recent years by Ebbighausen et al. (2004), Klug et al. (2006), and Korn et al. (2002, 2003, 2004a, 2007).

In the last 25 years, we have produced a large number of cross-sections of Palaeozoic ammonoid conchs. Almost every description of an ammonoid assemblage has profited from the visualization and analysis of ontogenetic changes in conch geometry. In total, approximately 200 Devonian and 300 Carboniferous ammonoid species were investigated by application of this method. Furthermore, several cross-sections published by other authors were redrawn and measured. This large amount of data enables comprehensive quantitative studies.

Ammonoid ontogenies can be grouped according to the several ways growth changes occur in most Palaeozoic ammonoids. Various authors have proposed terminologies for these growth stages. Whereas Kant and Kullmann (1988) differentiated only between mono- and diphasic growth in Devonian ammonoids, Clausen (1969) described four growth stages: protoconch (1), 'larval stage' (2), and 'postlarval stage' within which he differentiates between 'inner' (3) and the 'adult whorls' (4) based on Late Devonian ammonoids. Klug (2001b) suggested the following five growth stages, separated by four changes in growth.

Early Embryonic Period (Before the 1st Change of Growth)

Embryonic development of all Palaeozoic ammonoids commences by formation of an ovoid to globular and straight to curved protoconch (Branco 1880; Clarke 1896; Schindewolf 1933; Erben 1964a, b; Landman et al. 1996; Klofak et al. 1999). Its size varies between 0.5 and 1.1 mm in length and between 0.6 and 1.5 mm in height (Chlupàč and Turek 1983). This ovoid to globular shape inherited from orthocerid to bactritoid ancestors (Kröger 2007) is important from both ecologic and phylogenetic points of views (see also the next section). The earliest ammonoids of Early Emsian age have a large, more or less straight, slightly ovoid protoconch with a more or less straight initial shaft originating phylogenetically from their cyrtoconic to orthoconic ancestors. Coiling increased rapidly during early ammonoid phylogeny, first in the middle whorls, and then extended to the first and last whorls (Klug and Korn 2004). As shown by Klofak et al. (2007), the protoconch of some early ammonoids (e.g. *Mimagoniatites*) is covered by subparallel lines probably formed during shell growth. There are diverging opinions on the formation of these growth line-like structures (for reviews of current hypotheses see Landman et al. 1996, and Klofak et al. 1999, 2007). We have difficulties in seeing other ways to form such structures other than by accretionary growth. Most Middle Devonian and younger ammonoids have smooth protoconchs; various modes of formation have been suggested for these.

Late Embryonic Period (Between the 1st and 2nd Change of Growth)

The boundary between the two embryonic growth stages is clearly discernible: Transition from the protoconch to the first whorl or the initial shaft in the early ammonoid ammonitellae and bactritellae is usually characterized by a ring of concave shell between the protoconch-sphere and the more or less curved cone of the initial whorl or shaft. The end of the late embryonic period can be identified by the nepionic constriction (Landman et al. 1996 or Klug 2001b and references therein). The shell comprising those parts formed in the first two growth stages is termed ammonitella or bactritella, respectively, and measures roughly 1.5–3 mm in bactritoids (Erben 1964a; Mapes 1979; Kröger 2008) up to 2.6 mm in diameter in the earliest ammonoids and as little as 0.52 mm in Cretaceous scaphitids (Landman et al. 1996). In loosely coiled ammonoids with nearly straight initial whorls, the not-yet-coiled ammonitella measured probably more than 2.5 mm, with values of over 5 mm conceivable (compare, e.g. Bogoslovsky 1969, pl. 1). This reflects an evolutionary trend towards a decrease in ammonitella size (especially when extended to Early Palaeozoic cephalopods).

Another important evolutionary trend is the transition from straight initial shell tubes via curved initial whorls to fully coiled initial whorls. This represents one of the most important left wall-effect in ammonoid evolution (compare Gould 1997 and Korn and Klug 2003). This transition was finished by late Eifelian time, when all ammonoids had closed umbilici; by contrast most Emsian taxa have umbilical windows of varying sizes.

The small protoconchs and ammonitellas, bactritellas and orthoceridellas, respectively, reflect high reproductive rates compared to the embryonic shells of many pre-Devonian cephalopods (e.g. endocerids). Occurrences of abundant embryonic shells of these cephalopods in the Carboniferous led Mapes and Nützel (2008) to conclude that the egg-masses probably floated and the hatchlings lived planktonically. Taken together, this would explain phenomena such as the high value of ammonoids (and several other cephalopods) as index fossils – due to their relatively rapid rate of geographic dispersion. This is attested to by the correspondence of shell volume, gas volume of the first chambers present at hatching, and the volume of the initial body chamber with neutral buoyancy (Landman et al. 1996).

Juvenile or Neanic Period (Between the 2nd and 3rd Change of Growth)

Around hatching, the nepionic constriction is formed. This represents a growth pause (like *alte Mundränder* or megastriae in later ontogeny). Many Palaeozoic ammonoid taxa had rather evolute whorls with low ww/dm ratios and low whorl expansion rates, usually between 1.4 and 1.9, e.g. for the Devonian *Rherisites* (Fig. 5) and *Manticoceras* (Fig. 6). This might usually include one, two or three whorls (compare Klug

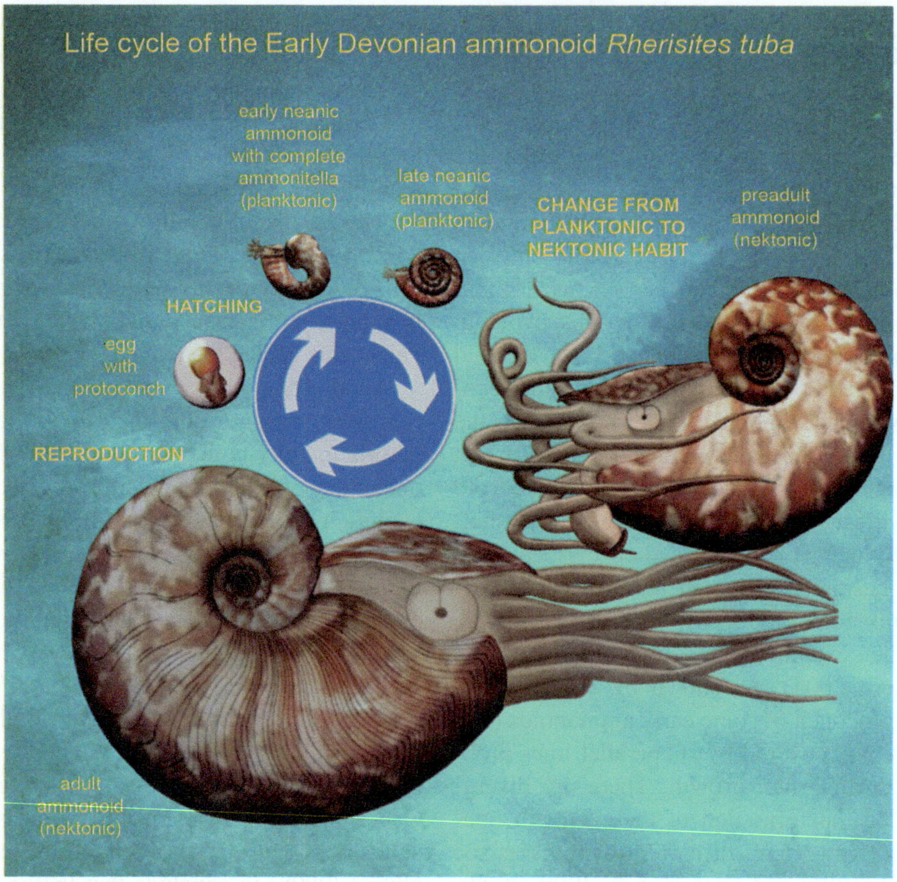

Fig. 5 Life reconstruction showing the ontogenetic stages of *Rherisites tuba* Klug, 2001 (after Klug 2001b)

2001b). Subsequently, the whorl cross-section changed more or less gradually, hampering precise identification of the end of the juvenile period. Most ammonoids, however, display a more or less distinct, although sometimes subtle, increase in WER, usually at diameters around 10 mm or less. Speculatively, this third growth change might coincide with the end of the fully planktonic phase and the beginning of nektoplanktonic movements. This is also likely because at this size, according to Jacobs and Chamberlain (1996), swimming movements began to be effective. In some ammonoid taxa, this growth change is virtually undetectable. This growth stage corresponds to the neanoconch postulated by Westermann (1996).

An interesting special example of a morphologically discrete neanoconch occurs in the Carboniferous genus *Entogonites* (Korn et al. 2005); its neanoconch displays approximately quadrate whorls, whereas the preadult and adult whorls have a normal round form (Fig. 7).

Preadult Period (Between the 3rd and 4th Change of Growth)

In many ammonoids, the WER rises after formation of the neanoconch; in *Rherisites* it reaches record values as high as 4.73 (Klug 2001b). Additionally, the ww/dm- or the wh/dm-ratio increases markedly. This is usually the growth stage where, because typical characters of taxa become more distinct, ammonoids become 'determinable'. The term 'preadult' is reasonable since this was the phase wherein the animals accumulated mass in preparation for reproduction – for which increased nutrition was required. The limit between the preadult and the adult or mature period

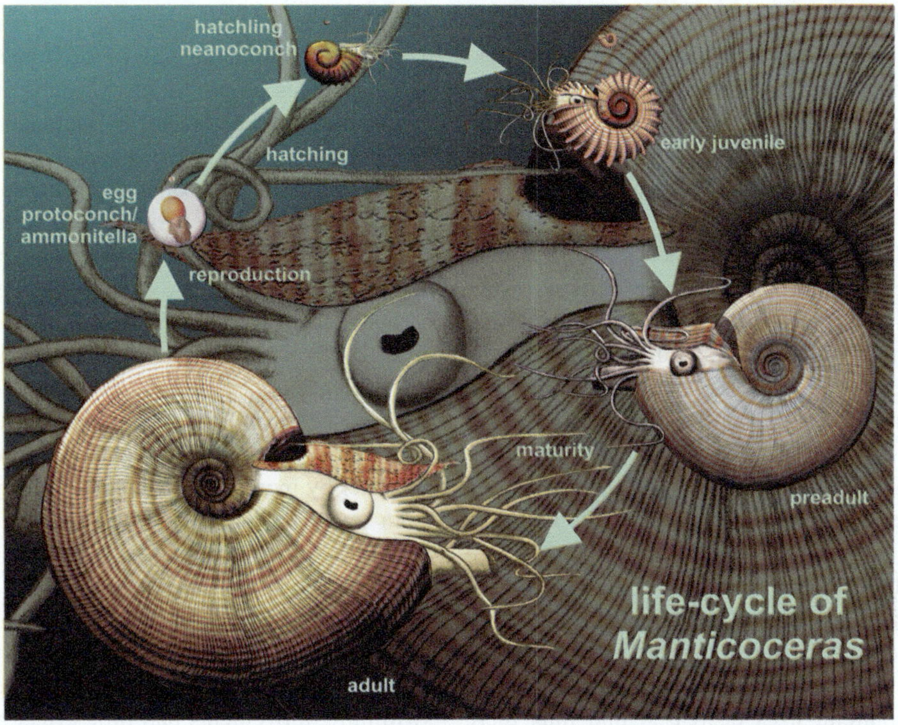

Fig. 6 Life reconstruction showing the ontogenetic stages of *Manticoceras* sp. (after Korn and Klug 2007)

Fig. 7 Life reconstruction of *Entogonites saharensis* Korn, Klug and Mapes, 2005. The high number of arms is pure speculation

is again difficult to locate on an exact spot on a whorl. Usually the adult whorls can be identified by a significant change in WER and, often, in the uw/dm ratio (usually an increase in the latter case). The whorl expansion rate tends to increase before maturity in ammonoid taxa with generally low WER such as in anarcestids and tends to decrease in forms with high WER such as in agoniatitids (Klug 2001b).

Adult or Mature Period (After 4th Change of Growth)

This is presumably the most conspicuous growth period, probably because of increased interest in sexual dimorphism, terminal countdown (see Seilacher and Gunji 1993), mature modifications, etc. As mentioned earlier, several geometric changes occur at the beginning of this growth period. Body chamber lengths probably changed in most ammonoids, as reflected by a decrease in WER in groups with generally high WER and an increase in WER in groups with generally low WER. Another common effect is the umbilical egression, i.e. a widening of the umbilicus, visible, e.g. in many wocklumeriids (see, e.g. Ebbighausen and Korn 2007). Further growth changes commonly related to maturity were listed for Recent nautilids by Ward (1987); most of these occur very similarly in ammonoids (Davis et al. 1996; Klug 2004; Klug et al. 2007).

This was also the growth stage when sexual dimorphism, indicated by shell shape, developed (Makowski 1962; Callomon 1963, 1981); this character may have been more frequent than reported due to subtle differences of shape between the sexes of various species. Additionally, there may have been differences in soft-part morphology not reflected by the shells. Sexual dimorphism has been suggested only rarely in Palaeozoic ammonoids (e.g. Makowski 1962; Davis et al. 1996); the oldest ammonoids suggested by these authors to be sexually dimorphic are the Frasnian genera *Manticoceras* and *Tornoceras*. We are of the opinion that sexual dimorphism may have occurred also in pre-Frasnian ammonoids, but clear evidence is lacking so far. Mapes (1979) reported an example of sexual dimorphism among Carboniferous ammonoids.

Another interesting but largely speculative aspect is the amount of eggs produced by a female ammonoid. This must also have changed significantly during the course of evolution of Devonian ammonoids, simply because ammonitella-sizes decreased and thus there may have been more and smaller eggs requiring less ovary volume. Korn and Klug (2007) published estimates of egg numbers for Frasnian *Manticoceras*: up to roughly 35,000 per adult female (at a conch diameter of 400 mm) and up to about 5000 in the Eifelian *Exopinacites* (at a conch diameter of 300 mm). Early, loosely coiled ammonoids, such as the Emsian *Erbenoceras*, had much lower body-chamber volumes and larger eggs; the maximum number of eggs did not exceed a couple of hundreds at an adult shell diameter of 150 mm (Korn and Klug 2007).

Conch Geometry, Orientation of the Aperture and Swimming Throughout Ontogeny

The above described ontogenetic changes in shell geometry coincided with more or less profound changes in orientation of the aperture (Klug 2001b). Additionally, there was a general evolutionary trend in orientation of the aperture from orthoconic orthocerids and bactritoids (e.g. *Devonobactrites*) via cyrtoconic bactritoids (e.g. *Cyrtobactrites*) to crioconic (e.g. *Erbenoceras*) and fully coiled ammonoids (e.g. *Fidelites*) from downward to obliquely downward, and from obliquely upward to upward (Klug and Korn 2004). Because of increased distance between the centre of mass and the opening of the funnel (hyponome), these changes imply a phylogenetically improving manoeuvrability and also an increase in swimming velocity – because of horizontal alignment of the centre of mass and the opening of the funnel.

The change in shell orientation had another side effect: the amount of decoupled chamber liquid increased greatly in the course of the evolution of early coiled ammonoids from straight bactritoids (Klug et al. 2008). This required some essential innovations to enable chamber emptying, probably facilitated by organic sheets lining the walls of the youngest chamber. This is also reflected by changes in septa formation and later by increasing septal complexity.

In many Devonian ammonoids, the orientation of the aperture changed ontogenetically in the same way, usually from an oblique upward to an upward position where, aligned horizontally with the centre of mass, their horizontal swimming velocity was optimal. This makes sense because, as small neanic animals, swimming movements had little effect and it is likely, as suggested above, that the young ammonoids were planktonic organisms while as adults, swimming was important for reproduction.

Diversity Dynamics of Palaeozoic Ammonoids

The following summary of Palaeozoic ammonoid history is based on the AMMON database (Korn and Ilg 2008). The large-scale diversity curve with stages as units suffers from the units being of unequal lengths (Fig. 3a); the Wordian is the shortest interval (2.2 Ma) and Viséan the longest (18.9 Ma). Another drawback of the diversity curve is the differing number of occurrences that have produced ammonoids from the various stages. Accordingly, the diversity per millions of years and per region is also plotted (Fig. 3b, c).

The stages between the Famennian and Bashkirian are the most diverse when genus and species richness per stage are plotted, with the Famennian represented by 700 species and about 150 genera (Fig. 3a). The plot illustrating species per millions of years and stage shows that the Famennian is still the richest of the Devonian stages, but the number of species is higher in the Serpukhovian, Bashkirian, Roadian, and Wordian (Fig. 3b). When the number of occurrences (regions and areas) is subdivided by the number of species, the Famennian far outranks other Devonian stages (Fig. 3c).

Devonian

The Devonian is noteworthy for being an interval during the evolutionary history of organisms that was punctuated by several variously severe global events (House 1985); it includes one of the so-called Big Five mass extinction events, the Upper Kellwasser Event at the Frasnian-Famennian boundary, but also other significant global biotic crises such as the late Givetian Taghanic Event and the end-Devonian Hangenberg Event. The class Ammonoidea was notably impacted by some of these crises; they experienced several minor extinctions and an almost complete extinction near the close of the Devonian, each followed by more or less rapid recoveries. They display various diversity patterns in their morphological and taxonomic development through that era. A general overview of the Devonian ammonoid species and their stratigraphic position has been provided by Korn and Klug (2002).

The ammonoids emerged in the Early Devonian Emsian stage from uncoiled or cyrtoconic bactritoids and were soon present in rather high diversity and abundance. Already from the early Emsian (Zlíchovian) substage, more than 60 species have been described, with the richest occurrences in Germany (Schindewolf 1933; Erben 1964a), the Czech Republic (Chlupàč and Turek 1983), Morocco (Klug 2001a, 2002), Uzbekistan (Bogoslovsky 1969) and South China (Ruan 1981a). Many of the species closely resemble each other, and therefore this species number may be an artefact of taxonomic oversplitting (De Baets et al. 2009). This has its origin in the often poor preservation, the low specimen numbers, and, partially, the lack of modern revisions.

Only 35 species are known from the late Emsian (Dalejan), with the reduction putatively caused by the preceding 'Daleje Event'. Diversity in the following Eifelian stage is higher with about 75 known species, followed by 55 species in the early Givetian (probably reduced by the end-Eifelian Kačák Event), and 40 in the late Givetian after the Taghanic Event. It has to be kept in mind, however, that some of these time intervals have not been revised in terms of ammonoid taxonomy, and that the numbers presented here are subject to correction in the future. This is particularly true for the late Givetian interval, from which still-to-be-revised assemblages have been documented from Morocco (cf. Petter 1959; Bensaïd 1974).

The Frasnian is characterized by a rapid increase in ammonoid diversity; 200 ammonoid species have been reported from this stage. Again, there is a serious problem with taxonomic practice; in this case there is oversplitting of some of the taxa reported from this interval. One such example can be seen with the manticoceratid ammonoids (suborder Gephuroceratina; see Fig. 2) for which at least 85 species names are available (Korn and Klug 2002, 2007). They belong to *Manticoceras* and closely related genera or subgenera, all embraced in the family Gephuroceratidae. Although intraspecific variability in coiling among these ammonoids has been known for a long time (Clarke 1899b), subsequent authors have paid relatively little attention to this phenomenon when describing assemblages from various regions using what are best described as parataxonomies. This is particularly true for the immense number of species attributed to *Manticoceras* (Korn and Klug 2007) described in numerous monographs by Bogoslovsky (1958, 1969), Clarke (1899b), Clausen (1969), Glenister (1958), Miller (1938), Petter (1959), and Wedekind (1913, 1918). These monographs provide an overview of the putative diversity of the Gephuroceratidae but, if one reduces the morphological range of the genus *Manticoceras* s. str., for instance by separation of the oxyconic as *Carinoceras*, the more globular as *Sphaeromanticoceras*, etc. (as done by Becker and House 1993), more than 40 nominal species remain. Considering that the ornament of all these nominal species is very similar and that all have similar suture lines, such elaborate discrimination of so many 'species' is a meaningless exercise (Korn and Klug 2007).

The end-Frasnian was marked by a drastic reduction in ammonoid diversity connected with the global Kellwasser Event (e.g. House 1981, 1985; Becker and House 1994). This event must be understood as a two-phased event, of which the first (lower Kellwasser Event) reduced the diversity to some degree, and the second (upper Kellwasser Event) caused near extinction of most groups (see, e.g. McGhee 1988; Buggisch 1991). Among the organisms of the shelf habitats, it was particularly the Ammonoidea that suffered to a high degree. Only a few lineages survived the crisis, e.g. the archoceratids, the last representatives of the suborder Gephuroceratina. These are still present in the earliest Famennian, when they became extinct without descendants. Additionally, some lineages of the suborder Tornoceratina survived; one of these, surviving into the Late Devonian, became the root stock from which subsequent ammonoids evolved (Becker 1995).

In the early Famennian, the Tornoceratina (order Goniatitida) diversified and were widely distributed along the equatorial and subequatorial shelf regions. Many of the species are morphologically similar; two major groups can be discriminated: (1) the 'conservative' tornoceratids with high aperture and (2) the cheiloceratids with modification of the conch towards more globular forms with lower apertures. A total of 220 ammonoid species have been described from this interval, but many are apparently synonyms (see Bogoslovsky 1971, and Becker 1993a for overviews). The high diversity might be spurious due to extreme splitting of tornoceratids and cheiloceratids by Sobolew (1914), who alone erected about 80 'species', many of which would have to be placed in synonymy.

The mid-Famennian witnessed tornoceratids exploring a wider and greater variability of conch shape and giving rise to the clymeniid ammonoids (House 1970; Korn 1992; Gottobrio and Saunders 2005) by shifting the siphuncle from the ventral to a dorsal position. Diversity increased remarkably with this innovation; 260 species have been reported from this interval. There is again a clear bias from taxonomic practice (e.g. Sobolew 1914), but possibly less than for the early Famennian. Ammonoids of mid-Famennian age have been reported from many regions, with the assemblages generally dominated by clymeniids (see Wedekind 1914; Bogoslovsky 1971, and Czarnocki 1989 for an overview). The *Prolobites delphinus* and the *Platyclymenia annulata* zones in the mid-Famennian mark the climax in the diversification of Devonian ammonoids; about 100 species are known from each of these zones.

There are probably about 265 valid species known from the late Famennian of Central Europe, the South Urals, North Africa, etc. The order Clymeniida still predominated in communities having a wide variety of conch geometries, ornament, and suture lines (for an overview cf. Wedekind 1914; Schindewolf 1937; Petter 1960; Bogoslovsky 1981; Czarnocki 1989; Nikolaeva and Bogoslovsky 2005).

The end-Devonian Hangenberg Event brought the class Ammonoidea close to total extinction. Only two or three of the various ammonoid lineages survived; one clymeniid lineage survived the event but declined immediately thereafter without descendants; one goniatite lineage was the basis for all Carboniferous and subsequent ammonoid evolution. Immediately after the event, faunas are extremely low in diversity (approximately ten species are known so far), composed only of *Acutimitoceras* and closely related forms as well as a few clymeniids (Schmidt 1924; Korn 1984, 1993; Korn et al. 2004b).

Carboniferous

After the almost complete extinction of the Ammonoidea during the Hangenberg Event near the Devonian-Carboniferous boundary, a rather sudden increase in diversity occurred. About 100 species are known from the oldest genus-zone of the Carboniferous (*Gattendorfia-Eocanites* Genus Zone; early Tournaisian–Vöhringer 1960; Ruan 1981a; Korn 1994; Bockwinkel and Ebbighausen 2006; Ebbighausen and Bockwinkel 2007). The most diverse assemblages have been found in the Rhenish and Thuringian Mountains, north-western Africa and South China. According to current knowledge, only two ammonoid lineages survived the event: (1) the family Cymaclymeniidae with two species, neither diversifying up into the Carboniferous, and (2) the family Prionoceratidae, with two branches surviving the event, the involute Prionoceratinae, and the Acutimitoceratinae with evolute inner whorls (e.g. Korn et al. 2004b). Only the latter became extremely successful; more than 90% of the ammonoid species of the *Gattendorfia-Eocanites* Genus Zone belong to this subfamily; its descendants probably gave rise to all subsequent Carboniferous and younger ammonoids.

The middle Tournaisian is the least documented interval of Early Carboniferous ammonoid history; the only records are scattered and rather incomplete (e.g. Miller and Collinson 1951; Gordon 1986; Korn et al. 2002). In this time interval, some early Tournaisian ammonoid groups are no longer present and new morphologies arose rapidly (e.g. Gordon 1986). Among innovations was subdivision of the external lobe; this occurred in several independent lineages and was perpetuated by almost the entire Ammonoidea. Konovalova (2005) reviewed the ammonoid diversity dynamics of the Tournaisian; she reported approximately 40 species from the 'Middle' Tournaisian (*Goniocyclus-Protocanites* Genus Zone) and 50 from the early Late Tournaisian (*Goniocyclus-Protocanites* Genus Zone).

The subsequent history of the Early Carboniferous ammonoids is markedly discontinuous; high diversity and low-diversity intervals alternate. The first of these periods is the Late Tournaisian (see for instance Kusina 1980 for an overview), an interval from which approximately 140 ammonoid species have been described from various regions including Ireland, Central Europe, the North Urals, the Tyan' Shan', and North Africa (e.g. Foord 1901, 1903; Delépine 1940, 1941; Holzapfel 1889; Schindewolf 1951; Kusina 1980; Librovitch 1927; Popov 1968; Pareyn 1961). After a rather rapid decrease in the Early and Middle Viséan (about 30–40 species; see Riley 1996), diversity increased almost continuously towards the latest Viséan to early Serpukhovian. From this interval (Brigantian to Arnsbergian of the British chronostratigraphic scheme), a high number of species has been described: Brigantian: 280 species, Pendleian: 290 species, Arnsbergian: 205 species (see Gordon 1965; Ruzhencev and Bogoslovskaya 1971; Ruan 1981b, and Korn 1988 for an overview). It is the most diverse interval within the history of Palaeozoic ammonoids, thought to have been caused by increasing provincialism connected with the Variscan Orogeny (Korn 1997). By far the richest occurrences are from the South Urals (Russia and Kazakhstan); similar but impoverished assemblages have been described from the Cantabrian Mountains of Spain (Kullmann 1961, 1962, 1963), Algeria (Pareyn 1961), and the Anti-Atlas of Morocco (Korn et al. 1999; Klug et al. 2006). Faunas with a different spectrum of genera are known from other areas, for instance the Subvariscan region (e.g. Korn 1988, 1997; Korn and Ebbighausen 2008) and the American Midcontinent (e.g. Gordon 1965).

A combined total of 150 species have been reported from the Chokieran and Alportian (see Ruzhencev and Bogoslovskaya 1978) after an extinction event near the Early-Late Carboniferous (Mississippian-Pennsylvanian) boundary. The richest assemblages are again from the South Urals, followed by Central Asia. These two regions retained prime importance through subsequent chronostratigraphic units, i.e. Kinderscoutian, Marsdenian, and Yeodonian. From the last, 210 ammonoid species are known, suggesting diversity had recovered to about that of the Serpukhovian. Or, to put it another way, the ammonoid provincialism reported for the Serpukhovian persisted. Provincialism may be at the root of this high diversity.

Ammonoid faunas from the succeeding stratigraphic units of the Carboniferous are scattered; thus, it is difficult to estimate their diversity. From the Desmoinesian stage, for instance, ammonoids have been recorded from only a few places in the world, e.g. the American Midcontinent (Oklahoma, Texas, Canada, Japan, Verkhoyan); only about 30 species are known from the Desmoinesian (see Boardman et al. 1994 for an overview).

The late Carboniferous is characterized by establishment of two regions in which ammonoids are diverse. The first is again the South Urals area (Ruzhencev 1950); the second is the American Midcontinent (Boardman et al. 1994). It is mainly from these two regions that 100 species are known from the Kasimovian and 120 from the Gzhelian. This automatically leads to the question, whether the high diversity of these assemblages has its origin in oversplitting of taxa, which is a common phenomenon, not just among Palaeozoic ammonoid taxa.

Permian

In contrast to the Devonian and Carboniferous, evolution of the Permian Ammonoidea (for a summary see Leonova 2002) was not affected by many crises during the Permian, but the end-Guadalupian and end-Permian extinction events were severe, causing almost-extinction of the entire clade (for a comparison of diversity and disparity, see Villier and Korn 2004).

Transition from the Carboniferous to Permian was not accompanied by a profound faunal change; the evolutionary lineages from the last two Carboniferous stages remained present in the Early Permian. The South Urals and American Midcontinent remain the most important regions from whence Early Permian (Cisuralian) ammonoid faunas have been recorded, supplemented subsequently by Central Asia and Timor (see Ruzhencev 1951, 1952, 1956; Leonova et al. 1989; Leonova in Leven et al. 1992; Miller and Furnish 1940; Haniel 1915, and Smith 1927 for an overview). The approximate numbers of species per Early Permian stage are Asselian 90, Sakmarian 110, Artinskian 210, and Kungurian 215. The Early Permian increasing diversity may be explained by the onset of provincialism with vicariance evolution in the various regions (American Midcontinent, Urals, Central Asia, South China) and also by the more widespread occurrence of ammonoid-bearing localities (species-area effect).

With destruction of shelf areas of the South and Middle Urals by the Uralian Orogeny, some ammonoid lineages characteristic for that region became extinct, for instance the paragastrioceratid and the metalegoceratid ammonoids (order Goniatitida). This was probably the reason for the decline in diversity between the Early and Middle Permian (Guadalupian). Only about 140 ammonoid species are known from the Roadian, the earlier high diversity was re-established with a total of 230 species in the ensuing Wordian. Wordian ammonoids were probably the geographically most widespread during the Permian; rich and similarly composed assemblages are known from Sicily (Gemmellaro 1887), Timor (Haniel 1915; Smith 1927), Texas (Miller and Furnish 1940), Arctic Canada (Nassichuk et al. 1965), and elsewhere. The Capitanian then saw again a reduction to less than 100 species.

The Middle-Late Permian (Guadalupian to Lopingian) transition is marked by a selective extinction event (Villier and Korn 2004) that particularly reduced the orders Goniatitida and Prolecanitida, but the Ceratitida were much less impacted; the Ceratitida even profited from the event. The Lopingian is dominated by ceratitid ammonoids representing more than 90% of the species diversity (e.g. Zhao et al. 1978; Leonova 2002). Two independent lineages were present, the araxoceratids and the xenodiscids. The first of these was particularly common in the Wuchiapingian, the second in the Changhsingian. Both lineages survived into the Triassic, but only the xenodiscids proved successful, forming the rootstock for the Mesozoic ammonoids.

Methods of Morphometrics and Morphospace Investigation in Ammonoids

It is well known that the development of taxonomic diversity (with all its possible biases) and morphologic disparity may be unconnected (Foote 1993, 1996, 1997). Rapid generation of morphologic disparity is a common feature in the early evolution of lineages (Erwin 1994; see Korn and Klug 2003 for Early Devonian ammonoids), though morphologic deceleration can be observed in long-living lineages.

Analysis of morphological disparity can overcome some of the sampling bias of the fossil record (Wills 2001); it also compliments patterns of taxonomic diversity. By focusing on shape differences, morphologic disparity is relatively independent of taxonomy; it allows comparison of samples in which a variable proportion of taxa are preserved or sampled. Estimates of disparity consider the distribution of taxa and proportions of morphologic space occupied during successive time intervals.

Quantification of morphospace occupation of ammonoids has been achieved in particular for mid-Carboniferous assemblages (Saunders and Swan 1984), for the Permian-Triassic transition (McGowan 2004a, b; Villier and Korn 2004), for the Middle Triassic ammonoids of the German Muschelkalk (Klug et al. 2005), and for Early Jurassic ammonoids (Neige et al. 2001; Dommergues et al. 2002; Navarro et al. 2005), but also for longer time intervals such as the entire Palaeozoic (Saunders et al. 2004, 2008).

Preparation and Digitization Techniques

Cross-sections have the highest potential for providing a large data-set for morphometric analysis of ammonoid conchs. With a reasonable number of cross-sections, ontogenetic development as well as intraspecific variability can be analysed for a number of species or complete assemblages. Three steps are necessary to establish a morphometric database:

1. Production of a series of high-precision cross-sections of ammonoids, i.e. oriented sections that meet the protoconch at its largest diameter in the coiling centre. Any deviation from the coiling centre will lead to imprecise data particularly of the inner whorls.
2. Photographic or scanning reproduction of the cross-sections or, if necessary, of acetate peels, and subsequent digitization of the sections by using dedicated computer software.
3. Recording of conch dimensions and calculation of ratios for compilation of a database of morphometric data. From computer drawings of the

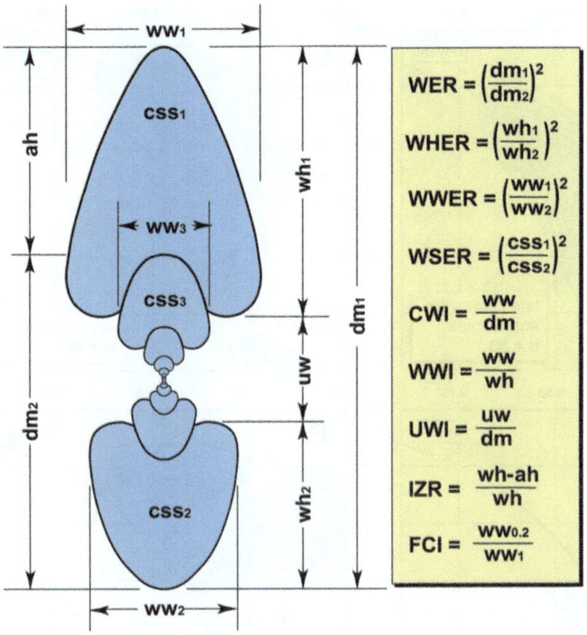

Fig. 8 Parameters obtained from cross-sections of ammonoids (exemplified by a Frasnian gephuroceratid ammonoids) as well as conch ratios and expansion rates calculated from the parameters (from Korn and Klug 2007)

cross-sections, three of the basic conch parameters can immediately be obtained for each half volution (Fig. 8):

- conch diameter (dm)
- whorl width (ww)
- whorl height (wh)

Using these basic parameters, secondary parameters can be computed as follows:

- umbilical width (uw) $= dm_1 - wh_1 - wh_2$
- apertural height (ah) $= dm_1 - dm_2$
- imprint zone width (iz) $= wh_1 - ah$ or $wh_1 - (dm_1 - dm_2)$

Expansion rates (growth rates) and conch proportions were calculated in the following way by using the three basic conch parameters:

- whorl expansion rate (WER) $= (dm_1/dm_2)^2$ or $[dm_1/(dm_1 - ah)]^2$
- umbilical width index (UWI) $= uw/dm_1$ or $(dm_1 - wh_1 - wh_2)/dm_1$
- imprint zone rate (IZR) $= wh_1 - ah/wh_1$ or $(wh_1 - (dm_1 - dm_2))/wh_1$
- conch width index (CWI) $= ww_1/dm_1$
- whorl width index (WWI) $= ww_1/wh_1$

Morphospace Analysis Using the Cardinal Conch Parameters

In the first step of morphospace analysis, the three cardinal conch parameters – presumed to be most important for the life of the ammonoid animals – were analysed.

- The expansion rate of the conch diameter (whorl expansion rate; WER) – being a measure of the opening rate of the whorl spiral. Since the WER is a proxy for the length of the body chamber and, thus, the orientation of the aperture of the conch, it is of significance for the living animal and may allow ecological interpretations (Saunders and Work 1996; Klug 2001b). Ontogenetic changes in the WER reflect a change in organization of the animal during lifetime.
- The conch width index (CWI). It determines the general shape of the ammonoid conch and is important for its hydrodynamic properties.
- The umbilical width index (UWI). It determines, in addition to the whorl expansion rate, the shape of the whorl cross-section and thus the geometry of the space in which the living animal was accommodated.

The data obtained by these calculations can be illustrated in bivariate plots, of which the WER/UWI and WER/CWI plots are most informative (Fig. 9). Distribution contour lines help to visualize differences between compared assemblages (Saunders and Swan 1984; Nikolaeva and Barskov 1994; Korn 2000; Korn and Klug 2003; Barskov et al. 2008).

Fourier Shape Analysis

Cross-section shapes of ammonoids are a valuable data source for morphologic studies of this group. They provide an immense amount of biometric data (Reyment and Kennedy 1991; Korn and Klug 2003); their preparation is regarded as a standard method for studying ammonoids. One of the major characters for such studies is the shape of the whorl outline defining the space that accommodated the living animal. Standard techniques, such as operating with so-called Raupian parameters (whorl expansion rate, whorl width, umbilical width) cover only part of the morphological range. For instance, the degree of whorl overlap, the curvature of flanks, and the presence of grooves and keels are not

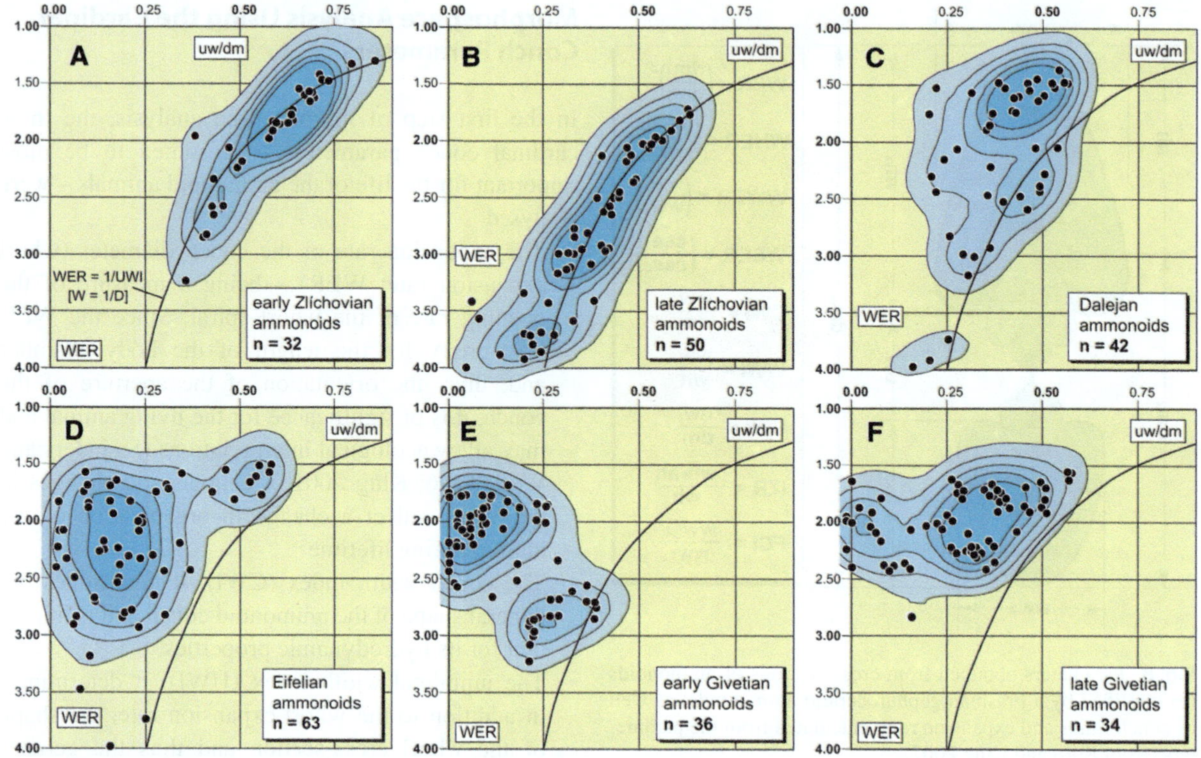

Fig. 9 Bivariate plots showing the morphospace occupation of the cardinal conch parameters (WER and UWI) are pair-wise illustrated for six successive Early and Middle Devonian time, intervals shown sequentially A→F (from Korn and Klug 2002). See text for results of this analysis

taken into account. For this reason we were seeking an alternative approach towards a more comprehensive analysis of morphologic features. One such alternative method is outline analysis; beside landmark-based analysis, it provides the opportunity to describe the complete spectrum of morphometric characters in geometric objects (Rohlf and Marcus 1993; Bookstein 1998; Adams et al. 2004). Plane closed curves, as given in whorl outlines of ammonoids, are well suited for this type of morphometric study. Therefore, we created a morphospace in combination with perimeter-based Fourier Analyses (e.g. Zahn and Roskies 1972; Ferson et al. 1985; Foote 1989). This approach allowed for detailed evaluation of disparity in ammonoids.

We used whorl cross-sections of approximately 700 Devonian and earliest Carboniferous ammonoid species for Fourier Analysis. We used only those species where a complete whorl outline was available; several species are thus missing in the analysis. Furthermore, we used only the shape of the most characteristic (usually the final) whorl cross-section for analysis.

The scanned and digitized images of the whorl cross-sections were, for precise outline tracing, aligned and filled with black. Outline tracing was conducted with the aid of the tpsdig program (Rohlf 2004) by transforming each outline shape into 500 *x*/*y*-coordinates. Transforming these whorl outline data into periodical functions was achieved with the hangle and hmatch programs introduced by Crampton and Haines (1996), creating 22 Fourier coefficients that represented 11 of the 12 created harmonics. The first harmonic merely represents a circle among all cases of closed outlines and can thus be neglected.

A principal component analysis (PCA) was performed as a dimension-reducing tool for processing the Fourier coefficients (see Joliffe 2002). This was done using the program PAST (PAlaeontological STatistics; Hammer et al. 2008) by applying the co-variance algorithm for data analysis. Of the 22 principal components, the first three already explain 79% of the total variance (PC1: 33.8%; PC2: 25.4%; PC3: 19.7%). The distribution of data points within the first two most informative axes shows the 'average-most' whorl cross-sections, i.e. almost circular shapes with a small indentation zone in the centre of the occupied area.

Principal Component Analysis (PCA)

Principal component analysis can be achieved by using very different characters of the conch geometry (Joliffe 2002). This data analysis allows the two-dimensional illustration of morphospace occupation based on more than two characters; temporally or spatially separated assemblages can be compared. Therefore, a number of investigations can be carried out by using such multivariate analyses, i.e. stratophenetic or palaeobiogeographic studies.

An empirical morphospace for the Palaeozoic ammonoids can be calculated by use of the three conch parameters WER, CWI, and UWI listed above (see Figs. 10 and 11 for the morphospace occupation of the Devonian ammonoids). The use of these parameters is profitable because these characters are independent to a large degree, but not completely.

When the distribution of the two most informative principal components 1 and 2 are illustrated, all species plot within a triangular field. The most extreme morphologies have a position in its corners, and the most average morphology in the centre. Four main morphs can be separated in the scatter plot:

- Morph A – extremely platyconic forms with very low WER and widely opened umbilicus (represented by *Phenacoceras*).
- Morph B – extremely lentiform conchs with high WER and very narrow umbilicus (represented by the extreme form *Carinoceras*).
- Morph C – globular forms with very low WER and closed umbilicus (represented by *Prolobites*).
- Morph D – the intermediate area contains the remaining taxa with moderate values in all parameters.

This triangular field delimits the morphological range that is possible for gyroconic, advolute, and convolute ammonoids; its margins are largely defined by geometric constraints:

- The WER of the analysed ammonoids ranges between 1.30 and 4.75 – a smaller value is practically not possible because of the lack of space for the living animal (WER = 1.00 is a circle, not a spiral). A WER higher than 5.00 is theoretically possible, but not realized in Palaeozoic ammonoids.
- The CWI ranges between 0.08 and 1.40 – a smaller CWI is hardly possible because of difficulties in accommodation of the animal; otherwise, ammonoid conchs cannot be infinitively wide for the same reasons.

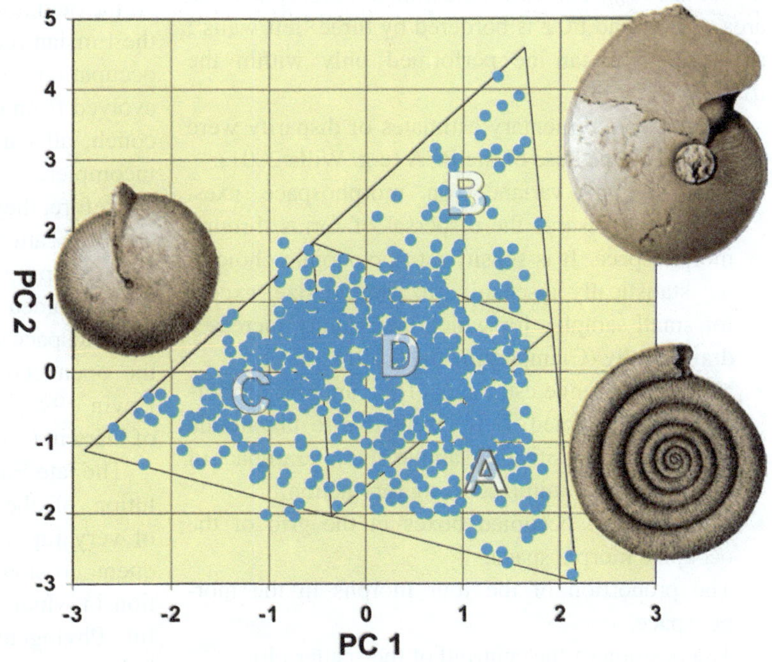

Fig. 10 Empirical morphospace of the Devonian ammonoids calculated using the three conch parameters WER, CWI, and UWI, together with representative examples of ammonoids and their position within the morphospace segment. When PC1 and PC2 are illustrated, all species are plot in a triangular field from which four main morphs can be discerned. See text for discussion of the morphs

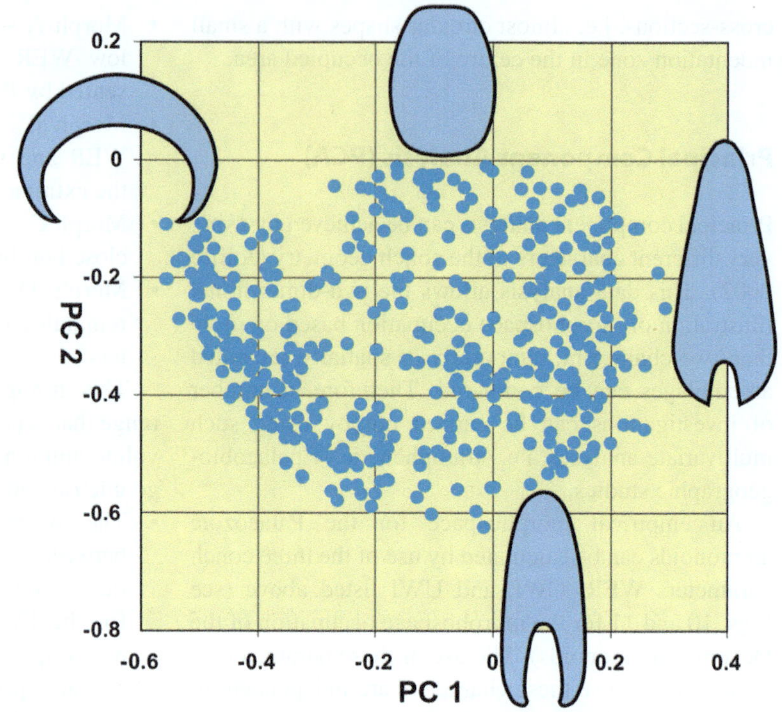

Fig. 11 Empirical morphospace of the Devonian ammonoids calculated using whorl cross-section outlines by performing a Fourier Analysis, together with representative examples of ammonoids and their position within the morphospace segment

- The UWI ranges between 0.00 (closed umbilicus) and approximately 0.75 – a wider opened umbilicus is practically impossible in convolute ammonoids.

The list expresses that each character has a geometric limitation on one side ('left wall' effect according to Gould 1997). Therefore, the triangular morphospace area of PC1 and PC2 is bordered by three 'left walls'; an acquisition can be performed only within the triangle.

Several complimentary estimates of disparity were calculated on the base of the PCA (e.g. Wills 2001):

- The sum of variance on morphospace axes. Variance measures the dispersal of forms through morphospace. It is sensitive to taxonomic choices, but statistically insensitive to sample size, except for small samples for which uncertainty increases dramatically (Ciampaglio et al. 2001).
- The sum of range estimates of the amount of morphospace occupied. It is insensitive to taxonomic choices, but sensitive to sample size, the impact of which can be minimized using rarefaction.
- The count of occupied boxes in the grid of the occupied morphospace.
- The proportion of the four morphs in the morphospace.
- The position of the centroid of the scatter plot.

Morphometric Evolution of Palaeozoic Ammonoids

Origin of the Ammonoidea and Early Diversification

At the beginning of ammonoid evolution at the base of the Emsian (early Zlíchovian substage), morphospace occupation was rather limited. Since the ammonoids evolved from orthocone cephalopods by coiling of the conch, all early forms were widely umbilicate with incomplete coiling (Erben 1964b; Klug 2001a, b). Therefore, they plot in one distinct corner of the WER-uw/dm scatter plot (Fig. 9a; Korn and Klug 2003). Due to rapid subsequent morphological evolution, the Ammonoidea show accelerated diffusion into other morphospace areas (Fig. 9b–f). This pattern supports the open ecospace hypothesis (e.g. Valentine 1980; Erwin 1994), which roots in the evolutionary theory of Darwin (1859).

The late Early and Middle Devonian character evolution of the Ammonoidea is remarkable because of very rapid development of characters and subsequent erosion of morphospace leading to canalization in which only a selected area remained successful. Phylogeny of the early ammonoids shows two independent lineages beginning in the late Emsian and continuing into the Middle and Late Devonian, of which the agoniatitid branch is characterized by a rather slow morphological diversification of conch geometry. At the same time, the anarcestid branch evolved rapidly and became the rootstock of most of the subsequent Middle and Late Devonian ammonoid clades.

It is noteworthy that even within the morphologically rather plastic anarcestid lineage, character variability ranges within distinct limits, e.g. the whorl expansion rate only rarely exceeds a value of 2.25 and thus only rarely touches the agoniatitid morphospace (Korn and Klug 2003). Furthermore, all Early Devonian anarcestids as well as their Middle Devonian descendants possess strongly embracing whorls. Morphologic plasticity includes, in addition to the rapid increase in the number of sutural elements, mainly the width of the umbilicus and the shape of the whorl cross-section (Korn and Klug 2003).

Three major evolutionary patterns concerning shell geometry can be discriminated in the evolutionary history of Early and Middle Devonian ammonoids (Fig. 12):

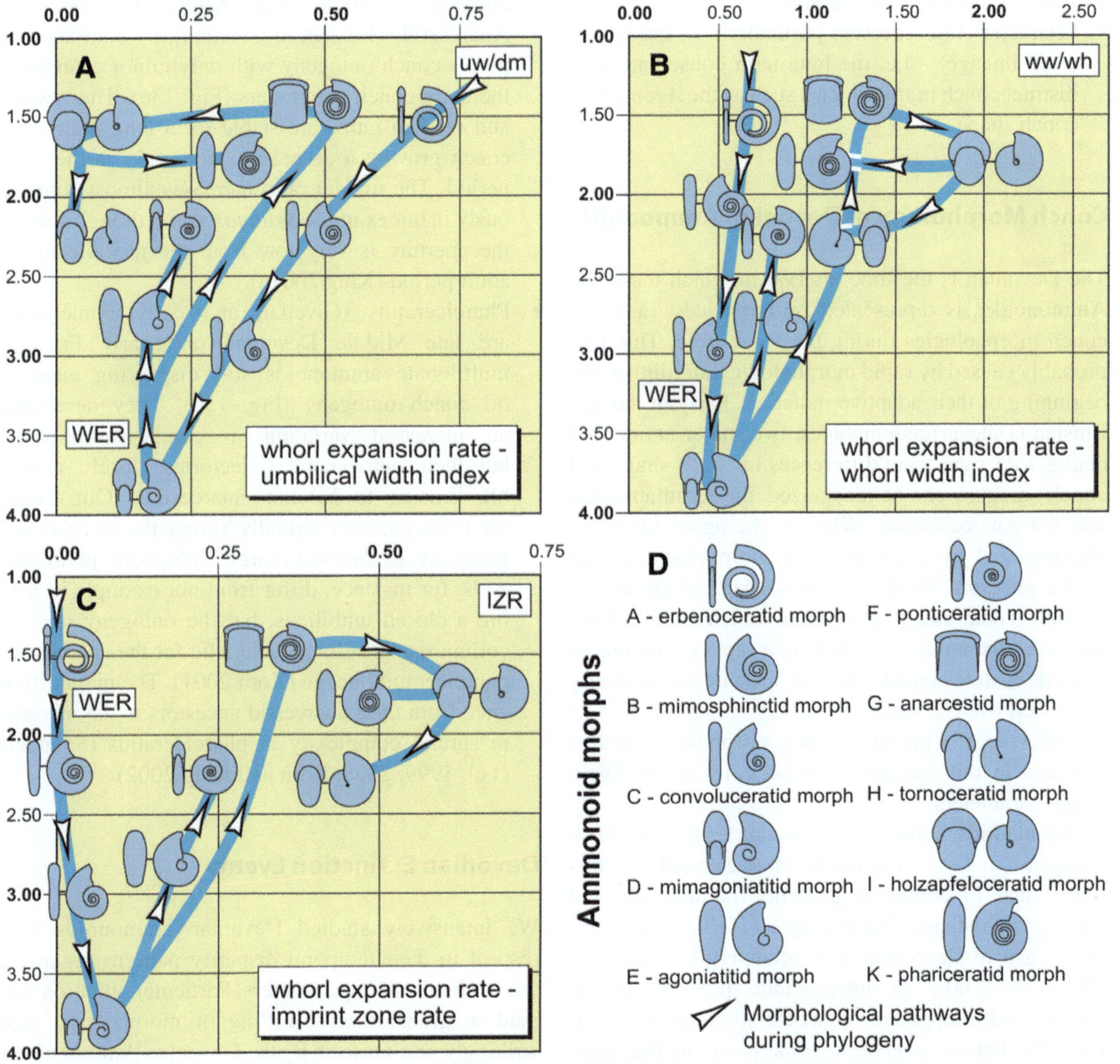

Fig. 12 (**a–c**) Early and Middle Devonian morphologic evolution and morphologic pathways based on comparison of whorl expansion rate (WER) with umbilical width index (**a**), with whorl width index (**b**), and with whorl width index (**c**). Morphological pathways during phylogeny are shown in (**d**). (after Korn and Klug 2003) See text for results of this analysis

- Long-term morphologic trends (mainly present in the most ancestral ammonoids, i.e. the Zlíchovian anetoceratids and mimagoniatitids) – i.e. expressed in the gradually increasing degree of coiling and the subsequent size reduction and the eventual disappearance of the umbilical window by the formation of a tightly coiled ammonitella without perforation.
- Short-term morphologic trends and character reversals (preferably present in the anarcestid lineage) – i.e. the opening of the umbilicus in the anarcestids and its closure in descendant lineages such as the sobolewiids and pharciceratids.
- Character stasis (present particularly in the agoniatitid lineage) – i.e. the long-term conservation of distinct conch morphologies such as the *Agoniatites* conch shape.

Conch Morphology of Devonian Ammonoids

The Devonian is the time interval in which the clade Ammonoidea is represented by the widest range of conch morphologies during the Palaeozoic. This was probably caused by rapid morphologic unfolding at the beginning of their adaptive radiation. Early in the late Emsian (Dalejan) assemblages, two major ammonoid clades with significant differences in conch shape and conch ontogeny can be recognized, the Agoniatitaceae and the Anarcestaceae. Whereas the agoniatitids are characterized by a conservative morphology similar to the ancestral mimagoniatitids (particularly in the laterally compressed subinvolute conchs with a high aperture), the anarcestids display a rapid development towards evolute conchs with low aperture and ventrally depressed whorl cross-section.

Differences in the adult conch morphology are also manifest in the ontogenetic trajectories of the main conch parameters.

Agoniatitids (Emsian to Givetian) show a conch ontogeny that is particularly characterized by different rates of coiling in juvenile, intermediate, and adult growth stages. Our example *Fidelites* (Fig. 13a) shows a juvenile period with relatively low aperture (WER = 2.00); an intermediate preadult period follows, with appreciable acceleration of the coiling rate, before stagnation took place in the adult growth period. In this pattern, they closely resemble the mimagoniatitids, of which the genus *Rherisites* has been analysed in great detail (Klug 2001a).

- Gephuroceratids (latest Middle Devonian and Frasnian) show a very similar conch ontogeny, even when they possess different conch shapes. Our example (*Manticoceras*; Fig. 13b) not only possesses a similar conch shape to the agoniatitid *Fidelites*, but also the tripartite postembryonic ontogeny of the coiling rate. Close resemblance of conch ontogeny including similar ammonitella sizes was the reason for suggesting close phylogenetic relationships between agoniatitids and gephuroceratids (Korn 2001; Korn and Klug 2001a, 2002).
- Anarcestids (Emsian to Givetian), by contrast, display a conch ontogeny with only minor changes in the main conch parameters (Fig. 13c). The ww/dm and ww/wh ratios are stable for a long interval of conch growth; a decrease occurs only in the adult period. The uw/dm ratio increases almost continuously in our example *Sellanarcestes*, though slowly; the aperture is very low from the juvenile to the adult period (Klug 2001a).
- Pharciceratids (Givetian and early Famennian) are late Middle Devonian and early Frasnian multilobate ammonoids still displaying anarcestid conch ontogeny (Fig. 13d). They developed an increased variation in conch morphology, but their ontogenetic trajectories closely resemble Emsian to Eifelian anarcestids. Our example (*Pharciceras*) virtually shows the same conch ontogeny as *Sellanarcestes*. Advanced pharciceratids, for instance, differ from our example in having a closed umbilicus, but the ontogeny of their coiling rate is still characteristic for the anarcestid–pharciceratid lineage (Korn 2001). The main difference from their anarcestid ancestors is the increase in sutural complexity in pharciceratids (Saunders et al. 1999, 2008; Korn and Klug 2002).

Devonian Extinction Events

We intensively studied Devonian ammonoids with respect to their temporal disparity patterns regarding the cardinal conch parameters. Particular attention was paid to global events. Filling of morphospace was strikingly non-uniform by the Devonian Ammonoidea. Not only extinction events but also rapid adaptive

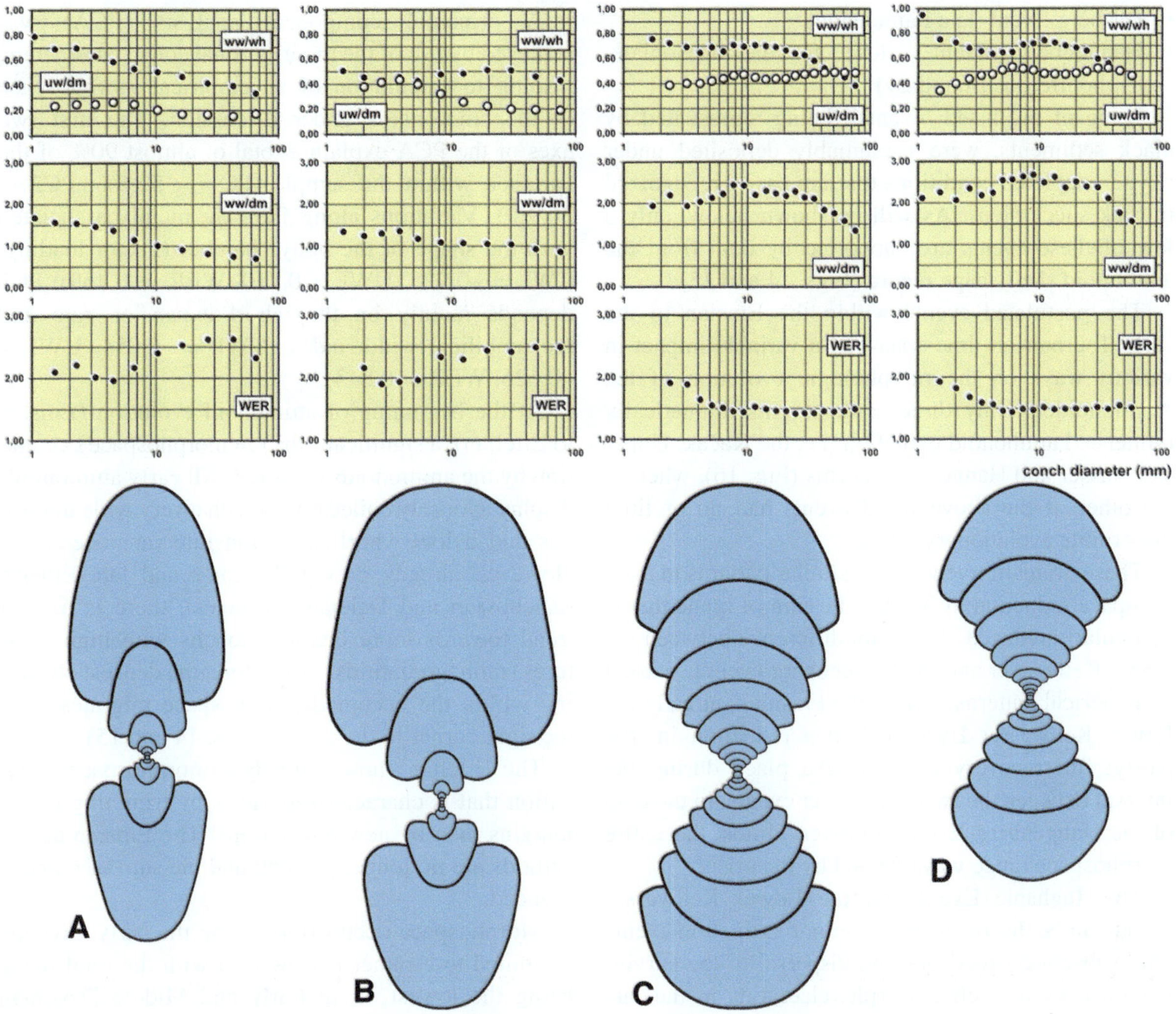

Fig. 13 Ontogenetic development of conch geometry in some Devonian ammonoids (after Korn and Klug 2001, 2007; Klug 2002). The diagrams represent (from *top* to *bottom*) development of the conch width index (ww/dm), umbilical width index (uw/dm), whorl width index (ww/wh), and whorl expansion rate (WER). (**a**) *Fidelites clariondi* (Petter 1959); (**b**) *Manticoceras* sp.; (**c**) *Sellanarcestes wenkenbachi* (Kayser 1884); (**d**) *Pharciceras* sp.

radiations were responsible for discontinuities in morphologic disparity through time. Various modes of erosion of the morphospace during extinction events and subsequent diffusion of the surviving lineages are recognizable. It is likely that most of these patterns can be explained by a branching random walk with logistic diversification (see for a comparison Pie and Weitz 2005).

The time interval from the Emsian (Early Devonian) to the early Tournaisian (Early Carboniferous) includes a number of geological events, defined primarily by punctuations in the diversity history of the Ammonoidea (e.g. House 1985, 2002; Walliser 1985; Kauffman and Walliser 1988). The most important of these are as follows:

- Choteč Event (close to the Emsian–Eifelian boundary)
- Kačák Event (close to the Eifelian–Givetian boundary)
- Taghanic Event (mid-Givetian)
- Frasnian Event (Middle–Late Devonian boundary)
- Lower Kellwasser Event (late Frasnian)
- Upper Kellwasser Event (Frasnian–Famennian Boundary)
- Enkeberg Event (early Famennian)
- *Annulata* Event (middle Famennian)

- Dasberg Event (late Famennian)
- Hangenberg Event (close to the Devonian–Carboniferous Boundary)

Some of the listed 'events', being represented by black sediments, were presumably deposited under oxygen depleted conditions that are easily recognized in rock successions. As will be shown below, only a few of these events are supported by data from the ammonoid shell-shape record (Figs. 14 and 15).

The postulated events within the Emsian to the early Tournaisian time-span had a variable impact in various ways on the morphologic evolution of the Ammonoidea. Only three of the events dramatically influenced ammonoid evolution, i.e. the Kačak, Upper Kellwasser and Hangenberg events (Fig. 16), whereas the other of the above listed events had no or little discernible evolutionary impact.

These dramatic events show similar patterns in morphospace reduction; it is often the sum of range that is particularly reduced. The main difference between the Upper Kellwasser and the Hangenberg event is caused by historical patterns: within the Frasnian, already the Lower Kellwasser Event led to a reduction in disparity; full recovery did not take place during the interval between the two Kellwasser events. In the case of the Hangenberg Event, however, almost the entire morphospace range was affected by the crisis.

The Taghanic Event and the Lower Kellwasser Event can be discriminated as minor extinction events with subsequent recovery and dispersal of the surviving lineages in each case (pharciceratids in the late Givetian, archoceratids in the late Frasnian).

The putative Frasnes, Enkeberg, *Annulata*, and Dasberg Events cannot be detected in the morphospace evolution of the ammonoid shell shape. There is no clear signal that would allow a connection with any physical perturbation. These results correspond well with a similar conclusion based on examination of Cretaceous ammonoids by Monnet et al. (2003).

We measured a total of approximately 700 Early Devonian to early Tournaisian ammonoid species to gain morphometric data on a large scale. Particular interest was paid to three cardinal conch parameters: (1) Conch width index (CWI = relative thickness of the conch), (2) Umbilical width index (UWI = relative width of the umbilicus), and (3) Whorl expansion rate (WER = coiling rate of the whorl spiral). We used these parameters, computed according to the study by Korn (2000; compare also Raup 1967).

A principal component analysis (PCA) was achieved by use of the three cardinal conch parameters, leading to calculation of an empirical morphospace of the ammonoids under discussion. The first two axes of the PCA explain a total of almost 90% of the variance within the sample (PC1 = 52.6%; PC2 = 36.9%). Variations along PC1 are mainly associated with the shape of the body chamber (Factor loading: CWI = –0.925, UWI = 0.710, WER = 0.469); PC2 depends mainly on the whorl expansion rate and the umbilical width index (CWI = –0.054, UWI = –0.626, WER = 0.843).

At the beginning of ammonoid evolution (Emsian to Eifelian), a significant shift in morphospace occupation by the ammonoids occurred. All early ammonoids display a loosely coiled conch with a very wide umbilicus and a low whorl-expansion rate (anetoceratids). However, already during the early and late Emsian (Zlíchovian and Dalejan substages), there is a clear trend towards more involute conchs with high apertures (mimagoniatitids). Therefore, the densest clustering within the occupied morphospace migrates from the right corner to the centre (Figs. 14 and 15).

The Eifelian shows already a morphospace occupation that is characterized rather by trimming of the margins than by new innovations. The Emsian anetoceratids are no longer present, and the sum of range is reduced.

Morphospace occupation during the early Givetian is limited to a rather narrow area with the total range being the lowest of all Early and Middle Devonian intervals (Fig. 16); the interval can be regarded as a low-disparity period. The transition from the early into the late Givetian (marked by the Taghanic Event) is not influenced by a major extinction event; most of the morphologic disparity survived with their evolutionary lineages. During the late Givetian, diffusion took place in the direction of more widely umbilicate forms (morph A).

Kellwasser Crisis and Recovery

There is almost no fluctuation in morphologic disparity from late Givetian to mid-Frasnian; no 'Frasnian Event' is apparent (Fig. 16). The sum of range increases slightly during this interval; there is a minor shift towards forms with high aperture and narrow umbilicus (morph B). At the Lower Kellwasser Event,

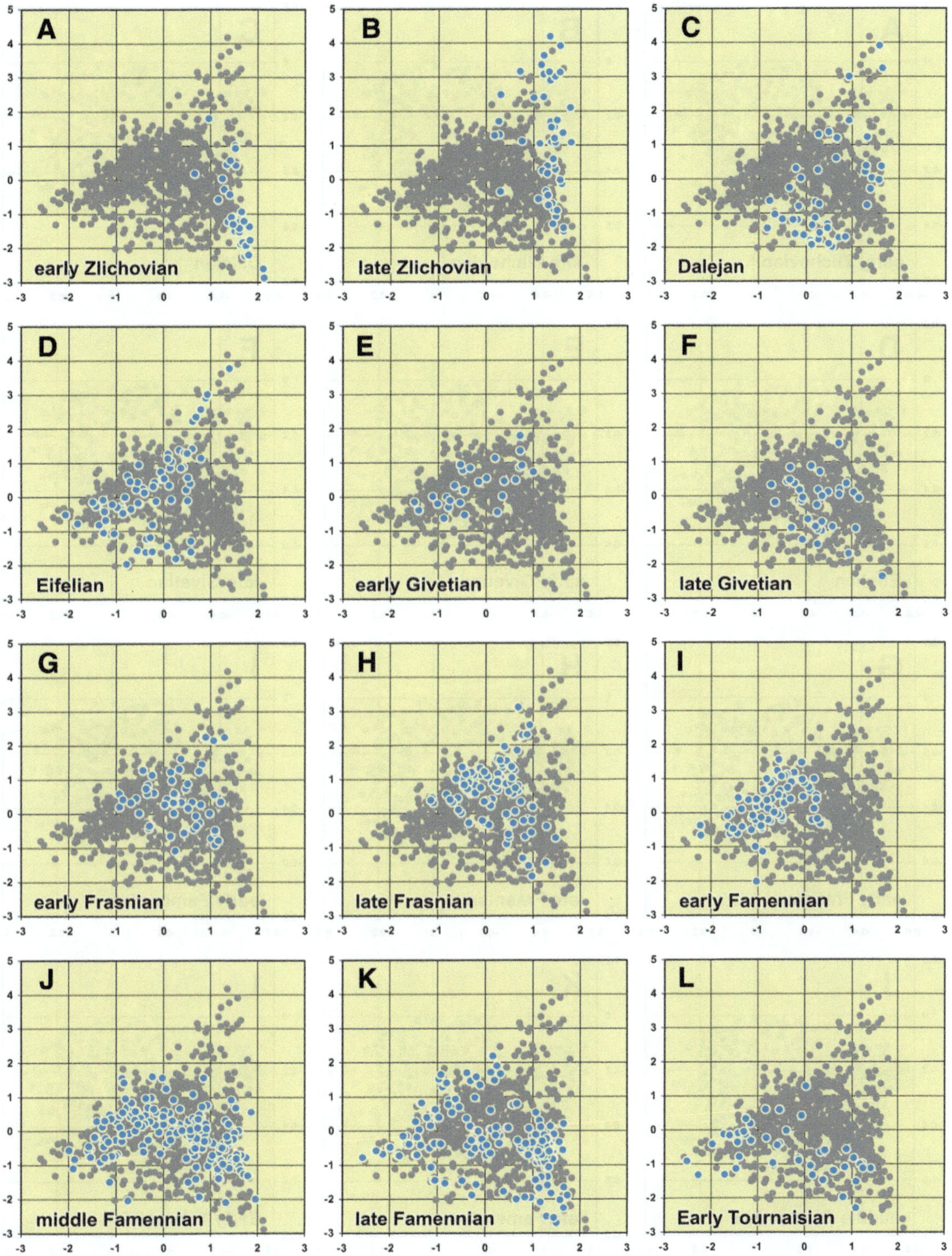

Fig. 14 Eleven Early to Late Devonian and one Early Tournasian intervals arranged sequentially A→L, showing morphospace occupation of ammonoids and the transition into the earliest Carboniferous by application of the method of a principal component analysis using conch parameters whorl expansion rate, conch width index, and umbilical width index. See text for results of these analyses

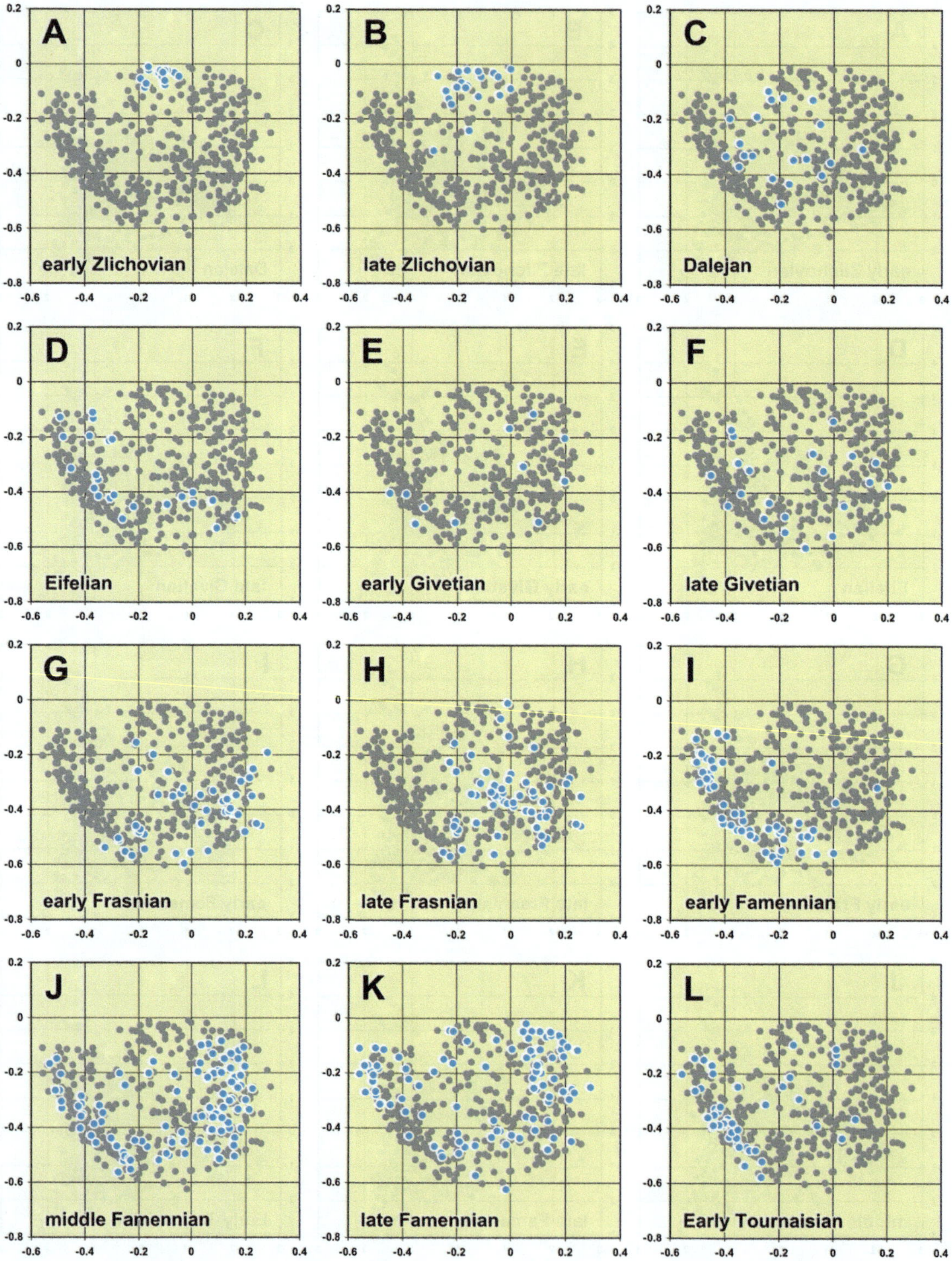

Fig. 15 Morphospace occupation of Early to Late Devonian ammonoids and the transition into the earliest Carboniferous by application of the method of principal component analysis using the whorl cross-section outlines calculated by Fourier Analysis. See text for results of this analysis

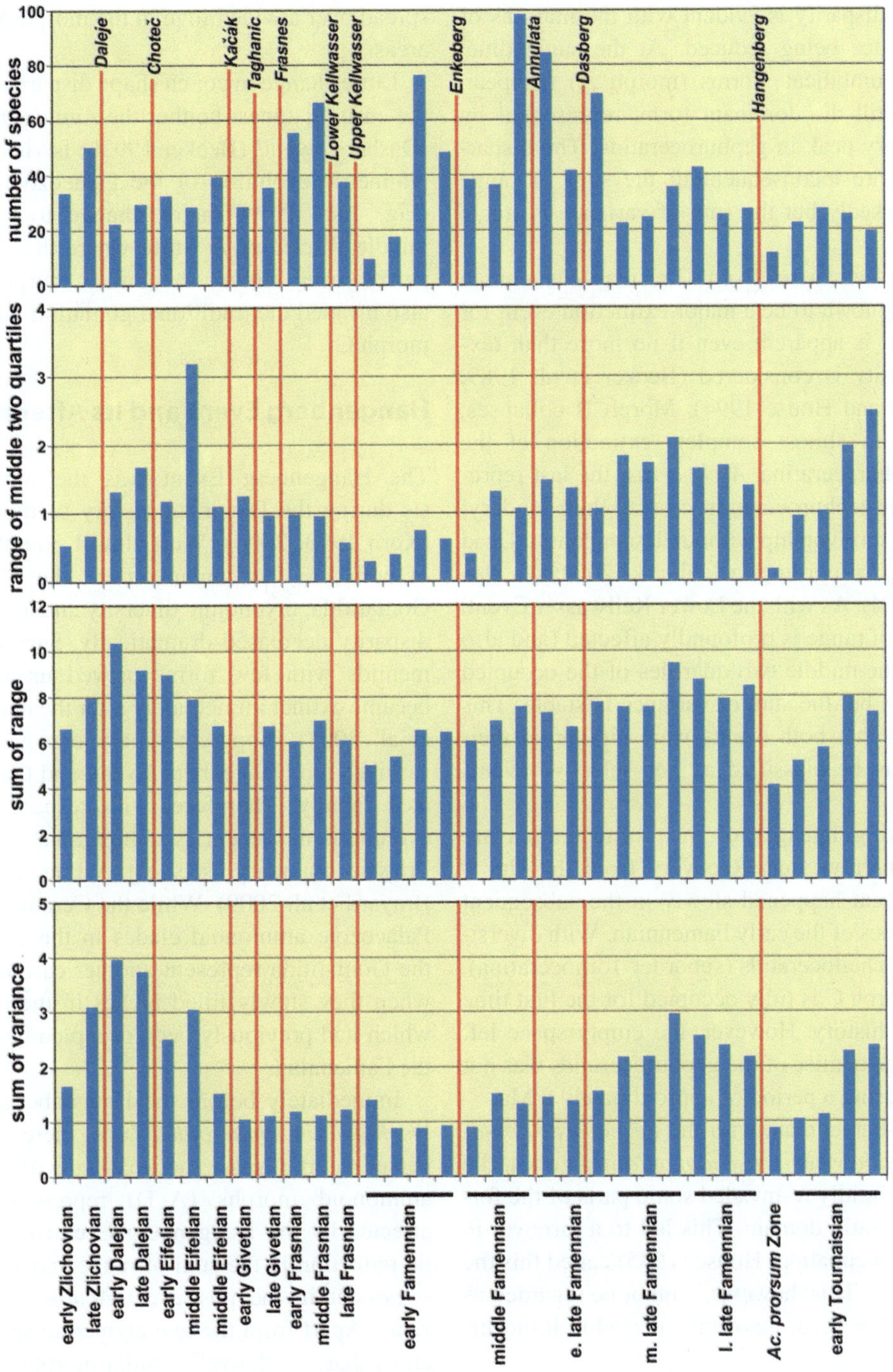

Fig. 16 Morphospace occupation of Early to Late Devonian ammonoids and transition into the earliest Carboniferous: (**a**) diversity fluctuations are illustrated as counts of occupied boxes of principal components 1 and 2 calculated by the cardinal conch parameters; (**b**) range of the two middle quartiles of principal components 1 and 2 calculated by the cardinal conch parameters; (**c**) and range of the two middle quartiles of principal components 1 and 2 calculated by Fourier Analysis. Note: The column to the *left* of the table shows the comparative time range of important stages from early Zlichovian to early Tournasian. The principal extinction events are shown by red horizontal lines across the columns

a decrease in disparity is evident with the margins of the morphospace being reduced. At the same time, some widely umbilicate forms (morph A) reappear. Morph B is still the dominant form, represented by the evolutionary peak in gephuroceratids. The disparity measures are inconsequential; the sum of range decreases markedly but the sum of variance is stable (Fig. 16c, d).

The Upper Kellwasser Event (Frasnian–Famennian Boundary) is known to be a major extinction event for ammonoids; it is apparent even if no more than taxonomic diversity is considered (Becker et al. 1989, 1993; Becker and House 1994). Morph B collapses, concurrent with almost complete extinction of the suborder Gephuroceratina. *Archoceras*, the last representative of the gephuroceratids, crossed the boundary; it exemplifies survivorship without descendants ('Dead Clade Walking' according to Jablonski 2002; see also Korn et al. 2004). As with the Lower Kellwasser Event, only the sum of range is profoundly affected (and also the range of the middle two quartiles of the occupied morphospace), but the sum of variance is stable. This pattern shows that both events were similar in their nature and can be classified as non-selective events (Foote 1993).

The surviving lineages of the Tornoceratina are limited to a narrow area. Recovery from the Upper Kellwasser Event happened slowly in the subsequent ammonoid zones of the early Famennian. With diversification of the cheiloceratids (suborder Tornoceratina), the area of morph C is fully occupied for the first time in ammonoid history. However, the empty space left behind after the demise of the gephuroceratids was not replenished during a period of approximately 8 Ma.

In the mid-Famennian, with the *Pernoceras dorsatum* and *Pseudoclymenia pseudogoniatites* zones, the tornoceratids timidly re-invaded some parts of the former gephuroceratid domain. This led to a turnover in morphospace occupation; House (1985) called this the 'Nehden Event'. This, however, cannot be regarded as a crisis but rather as an innovative period in tornoceratid evolution.

With the entry of the rapidly diversifying clymeniids in the *Prolobites delphinus* zone, ammonoids can be viewed as having fully recovered from the Upper Kellwasser Crisis; a peak of morphologic range can be recognized. Tornoceratids and cheiloceratids maintain their morphospace (mainly morph C); the clymeniids spread over a wide range in the morph A and morph D areas.

Little change in conch shape disparity can be seen in the ensuing zones; neither the *Annulata* Event nor the 'Dasberg Event' (Becker 1993b) is visible in the morphometric evolution of the Palaeozoic Ammonoidea (Fig. 16c, d). A major change took place in the middle *Wocklumeria* Stufe where the clymeniids dispersed over almost the entire morph A field and also invaded the traditional goniatitid morphology, i.e. morph C.

Hangenberg Event and Its Aftermath

The Hangenberg Event was the most severe crisis during the Devonian history of the Ammonoidea (Korn 1993, 2000). With almost complete extinction of both late Famennian orders, the Clymeniida and Goniatitida, taxonomic diversity and also morphologic disparity decreased dramatically. Survival of the clymeniids with few forms proved unsuccessful; they became extinct immediately after the main event (Korn et al. 2004). Apparently only one goniatite lineage survived the Hangenberg Event, and that lineage radiated rapidly. Therefore, with respect to ammonoid evolution, this event is comparable to the Permian-Triassic boundary (see, e.g. Villier and Korn 2004; Brayard et al. 2009). While the Ceratitida replaced the Palaeozoic ammonoid clades in the earliest Triassic, the Goniatitida represent another case of incumbency when they slowly filled niches in the Carboniferous, which had previously been occupied by clymeniids in the Famennian.

Immediately before the Hangenberg Event, in the *Wocklumeria sphaeroides* Zone, there was rather dispersed morphospace occupation with all four main ammonoid morphs (A–D) represented. All were affected by the Hangenberg Event; it caused a drastic reduction in the sum of range and sum of variance – evident in the post-event *Acutimitoceras prorsum* Zone. Apart from the two clymeniid species, all goniatites share a relatively similar morphology and hence plot in a narrow area (Figs. 14 and 16).

Within the first four Carboniferous ammonoid zones (*Acutimitoceras acutum* to *Paragattendorfia patens* Zone), the gap in conch shape morphospace, caused by the demise of the clymeniids and tornoceratids, was only partially refilled. The range of morphospace occupation of preceding Late Devonian zones, however, was not attained during the early Tournaisian.

The Basal Carboniferous Recovery

Morphometric analyses of Carboniferous ammonoids have been undertaken several times (Saunders and Swan 1984; Nikolaeva and Barskov 1994; Saunders and Work 1996; Saunders et al. 1999).

In a case study, a stratophenetic analysis has been made of the ammonoid faunas subsequent to the transition from the latest Devonian into earliest Carboniferous cephalopod limestones of the Rhenish Mountains (Korn 1986, 2000). The investigation concentrated on the whorl expansion rate (WER) and the umbilical width index (UWI), two characters very important in ammonoids since they indicate body-chamber length (and hence orientation in the water column) and mobility as well as shape of the space occupied by the animal. The study led to the conclusion that the Hangenberg Event caused an almost complete reworking of the morphological spectrum adopted by ammonoids. All clymeniids as well as tornoceratids became extinct at or immediately after the Hangenberg Event; the morphospace vacated by these groups was replenished, though incompletely, by the surviving prionoceratid ammonoids (Fig. 17).

Recovery after the Hangenberg Event was rapid; eight species of prionoceratids with open umbilicus in juvenile stages are known already in the latest Devonian *Acutimitoceras prorsum* Zone after the event. *Acutimitoceras* and *Nicimitoceras* occur seemingly instantaniously with a new conch geometry that had probably developed earlier. Lack of *Mimimitoceras*, the only genus so far known to survive the Hangenberg Event, in the *Ac. prorsum* Zone, and its reappearance in the *Ac. acutum* Zone, is interpreted as a reflection of migration.

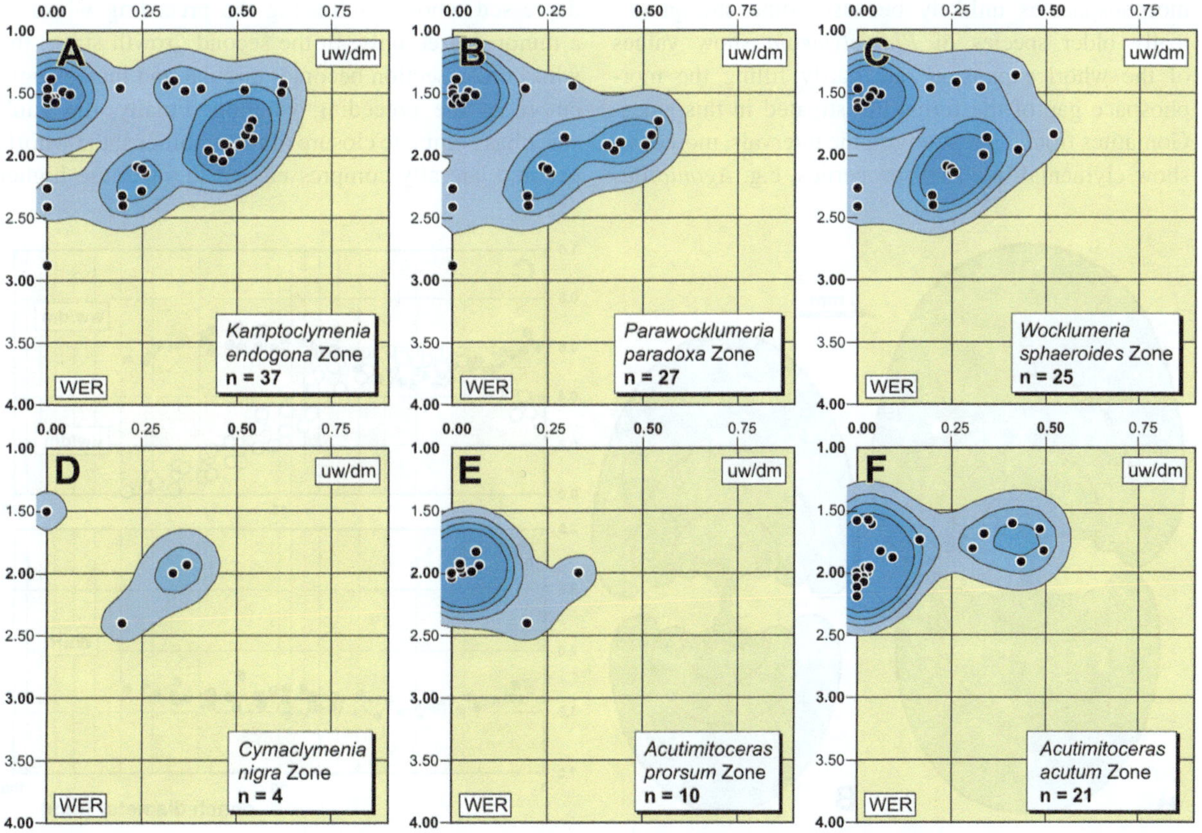

Fig. 17 Morphospace occupation using the example of the WER–UWI couplet across the Devonian–Carboniferous Boundary sequentially from the *Kamptoclymenia endogona* Zone to the *Acutimitoceras acutum Zone* (after Korn 2000). Note the sharp decrease in the *Cymaclymenia nigra* Zone (Hangenberg Event) and subsequent recovery

During the four earliest Carboniferous zones, from the *Acutimitoceras acutum* Zone to the *Paragattendorfia patens* Zone, conch diversification did not increase remarkably, as can be seen from the very similar contour diagrams for these zones (Fig. 17). It can be seen that a group of widely umbilicate goniatites became well established and that intermediates became unimportant.

It is remarkable that the goniatites of the *Gattendorfia* Stufe differ so conspicuously in their WER/UWI proportions from the clymeniids of the Late Devonian *Wocklumeria* Stufe. The bivariate plot of the WER/UWI values of all the clymeniids from the *Wocklumeria* Stufe and goniatites from the *Gattendorfia* Stufe demonstrates the significant differences in the position of the morphospace occupied by these groups (Fig. 17).

There appears to be no simple explanation for this striking discordance. A possible reason that clymeniids could only, because of their dorsal siphuncle (Gottobrio and Saunders 2005), occupy distinct morphospace is unlikely because some stratigraphically older species of *Platyclymenia* show values of the whorl expansion rate partly filling the morphospace gap of the forms investigated in this study. Goniatites from other stratigraphic intervals, moreover, show clymeniid-like shell properties, e.g. *Agoniatites* and *Manticoceras* resemble *Cymaclymenia*, and *Anarcestes* resembles *Kamptoclymenia*.

Conch Morphology of Carboniferous Ammonoids

Almost all Carboniferous ammonoids belong to a single monophyletic clade originating in the earliest Tournaisian prionoceratids. One genus that is probably very close to the initial radiation is *Acutimitoceras*; its conch ontogeny has been studied several times (Korn 1984; Korn et al. 2007) on the basis of material from the Rhenish Mountains and the Anti-Atlas.

Ac. hilarum is a late Devonian species with strikingly allometric conch growth compared with many Carboniferous ammonoids (Korn et al. 2004, 2007). Its ontogeny passes through three stages discriminated by proportions of the whorl width, umbilical width, and whorl expansion rate (Fig. 18). The first growth stage is characterized by slowly expanding, ventrally depressed whorls embracing the preceding whorl to a minor degree only. In the second growth stage, the whorl cross-section becomes circular and increasingly embraces the preceding volution. Finally, the adult stage has complete closure of the umbilicus; the whorls become laterally compressed and the aperture higher

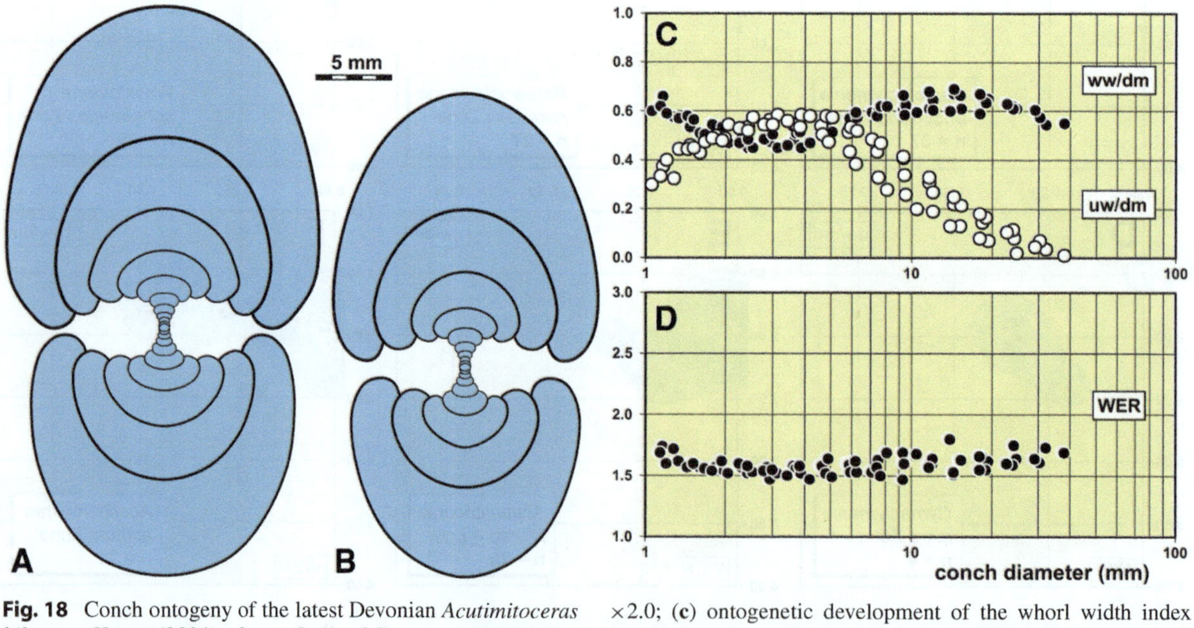

Fig. 18 Conch ontogeny of the latest Devonian *Acutimitoceras hilarum* Korn (2004) from Lalla Mimouna (Anti-Atlas, Morocco) on the basis of cross-sections. (**a**, **b**) Cross-sections, ×2.0; (**c**) ontogenetic development of the whorl width index (ww/dm) and umbilical width index (uw/dm); (**d**) ontogenetic development of the whorl expansion rate (WER)

(terminal whorl expansion rate is 1.75). Variability in the species occurs mainly in the timing of umbilical closure; this appears to be its most plastic character (Fig. 18c).

In a case study, Korn and Vöhringer (2004) analysed the ontogeny of *G. crassa* from the basal-most Early Carboniferous (Mississippian) Hangenberg Limestone of the Rhenish Mountains. It is located phylogenetically near the origin of most of the Carboniferous Ammonoidea and can thus be treated as representative for many lineages. In general, the ontogeny of *Gattendorfia* resembles that of *Acutimitoceras*, the main difference being that the umbilicus is not completely closed in adult *Gattendorfia*. Analysis of 15 high-precision cross-sections of this species documented the small intraspecific variability of some of the conch shape characters and the remarkably high intraspecific variability in other characters (Fig. 19). Whereas whorl expansion rate, umbilical width, and conch thickness vary within narrow limits, expansion rates of the whorl height and whorl width are quite plastic. Intraspecific variability is particularly high in growth stages where ontogenetic changes took place rapidly, meaning that intraspecific variability is mainly a matter of differences in timing of ontogenetic changes.

Cravenoceras leion (Fig. 20) from basalmost Serpukhovian strata of Derbyshire has an adult morphology superficially similar to *Gattendorfia* as well as similar ontogenetic trajectories (Korn and Tilsley 2002). Ontogenetic changes in outline of the whorl cross-section are conspicuous, reflecting major

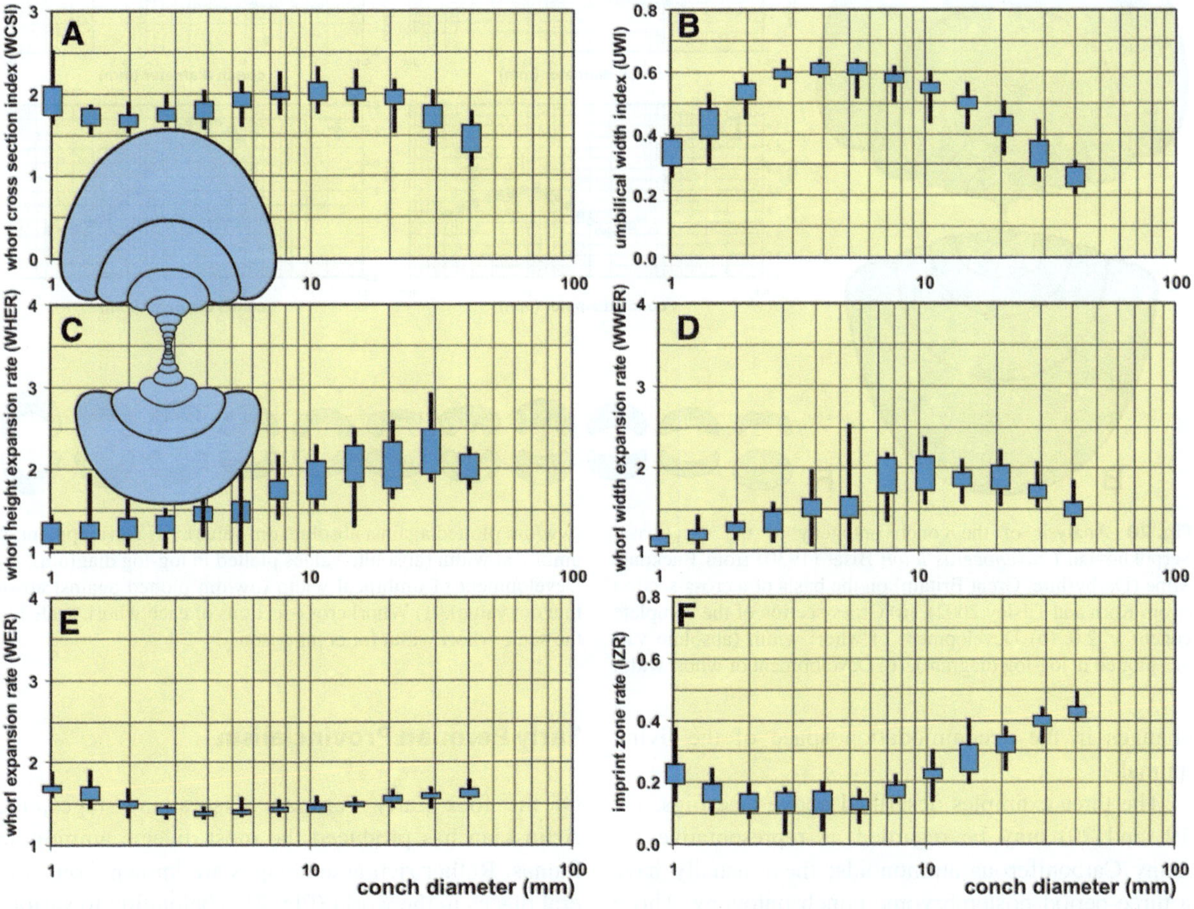

Fig. 19 Conch ontogeny and intraspecific variability of a sample of *Gattendorfia crassa* (Schmidt 1924) from the basal Early Carboniferous Hangenberg Limestone of the Rhenish Mountains (after Korn and Vöhringer 2004). Box-and-whiskers diagrams (*bold black line* refers to the total range; *blue boxes* refer to extension of the middle two quartiles) show ontogenetic variations of intraspecific variability of the major conch ratios and expansion rates (as shown in *black* along the side of each of the six graphs (**a**–**e**). For example, the whorl expansion rate (**e**) and umbilical width (**b**) vary within narrow limits, whereas the expansion rates of whorl height (**c**) and whorl width (**d**) are decidedly variable. See text for further discussion

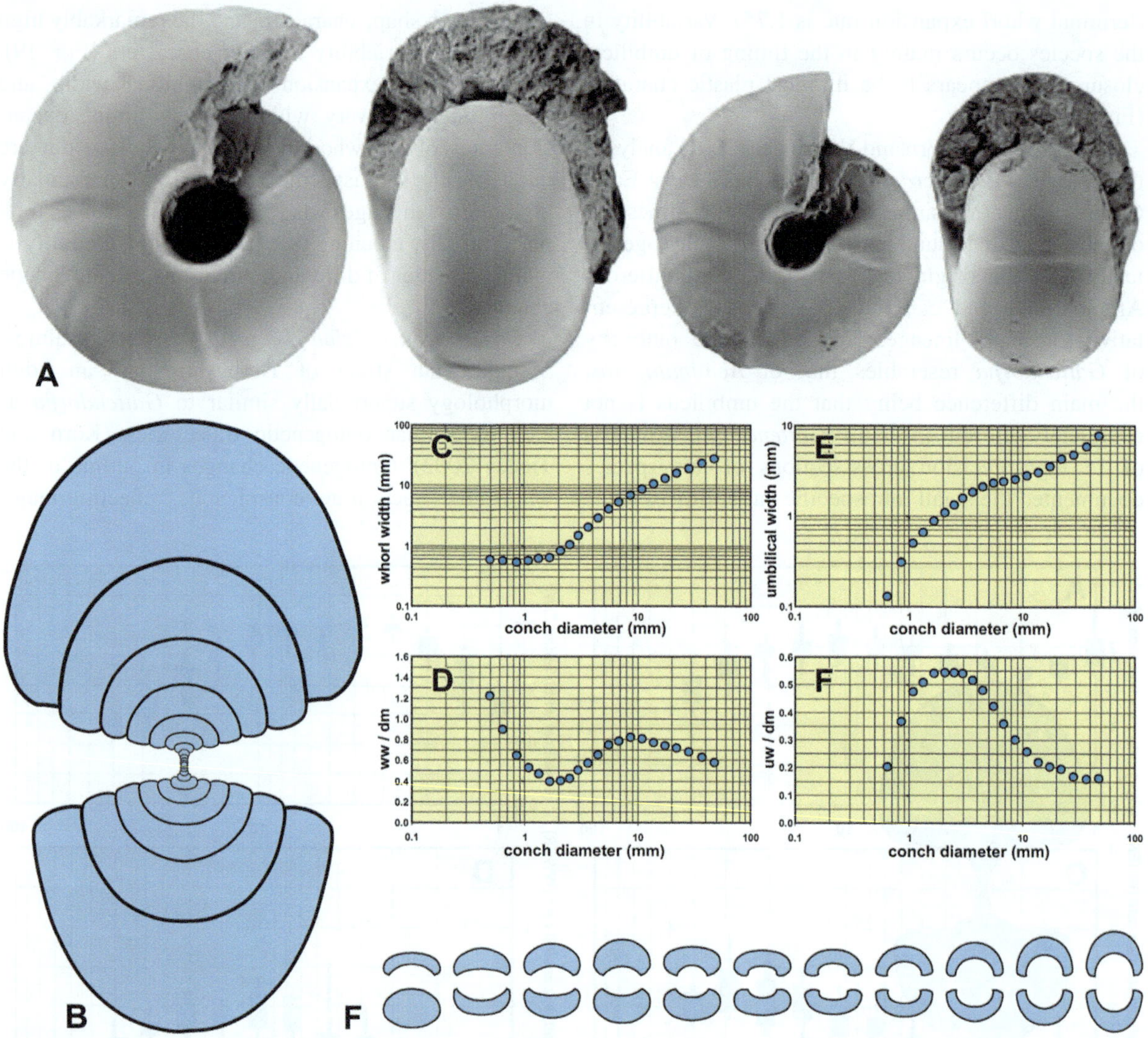

Fig. 20 Analysis of the conch ontogeny of the very basal Serpukhovian *Cravenoceras leion* Bisat (1930) from Backdale Mine (Derbyshire, Great Britain) on the basis of a cross-section (from Korn and Tilsley 2002). (**a**) Cross-section of the complete conch, × 2.0. (**b**) Development of whorl width (absolute values plotted in log-log diagram). (**c**) Development of whorl width (ww/dm plotted against absolute dm value). (**d**) Development of umbilical width (absolute values plotted in log-log diagram). (**e**) Development of umbilical width (uw/dm plotted against absolute dm value). (**f**) Whorl cross-sections of each whorl, scaled to the same whorl width for comparison

changes in the accommodation space of the living animal.

The three examples described above (see Figs. 18, 19, and 20) may be regarded as representative for many Carboniferous ammonoids; these usually have a three-period postembryonic conch ontogeny. Those three periods are expressed to varying degrees and also differ in their length in Carboniferous ammonoid taxa.

Early Permian Provincialism

Of the four Early Permian (Cisuralian) stages, the Artinskian has produced the most diverse ammonoid faunas. Rather rich assemblages are known from several places in the world (Fig. 21a) belonging to various climatic zones and palaeogeographic realms:

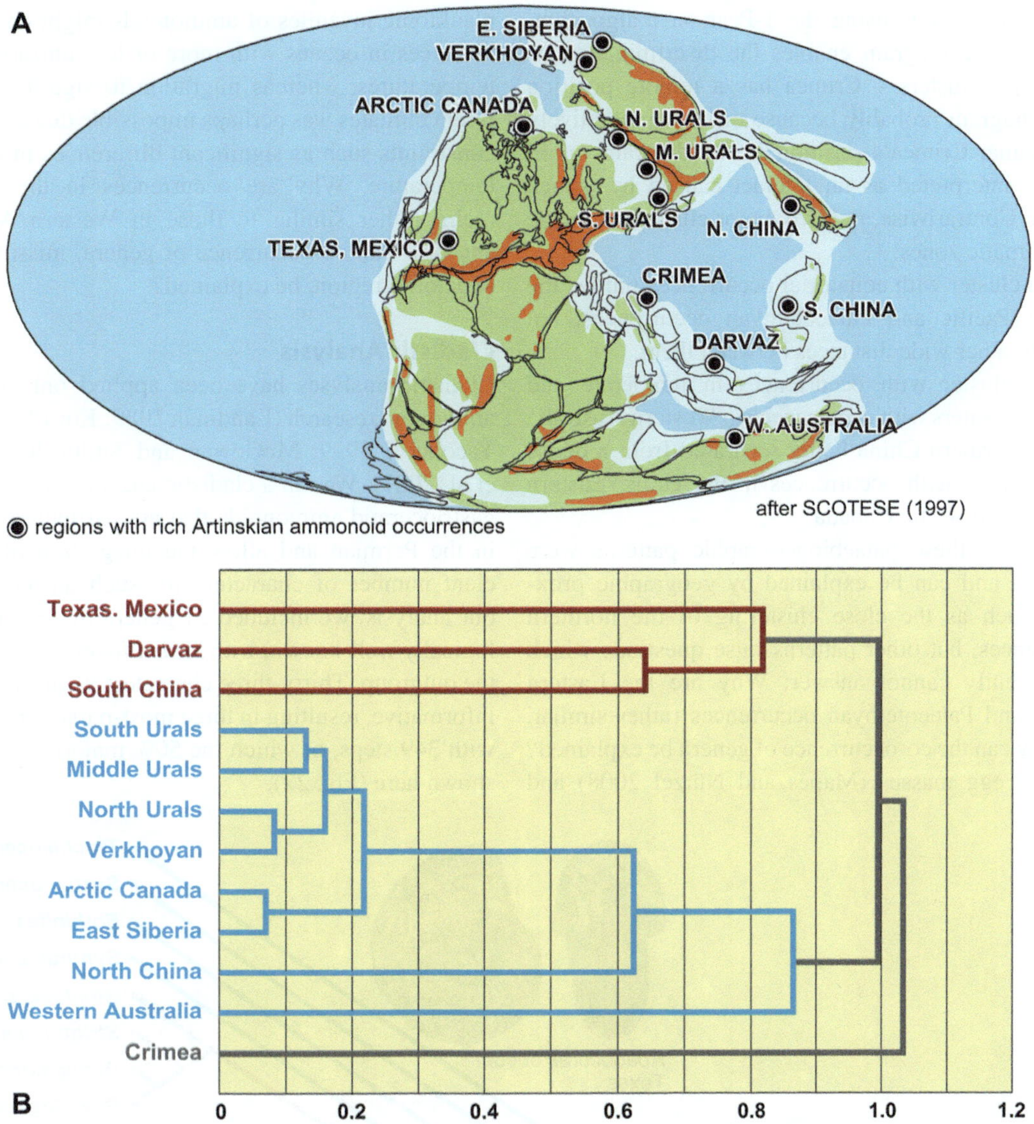

Fig. 21 (**a**) Early Permian palaeogeography (after Scotese 1997) and the most important regions with Artinskian ammonoids. (**b**) Cluster analysis of the main regions with Artinskian ammonoids based on the occurrence and species diversity of genera, calculated with 1-Pearson-r algorithm. *Brown colour* refers to warm water faunas' *blue colour* refers to temperate and cold waters

- Palaeotethyan Realm (with Crimea, Central Asia, and South China);
- East Pacific Realm (with Texas and adjacent areas);
- Northern Boreal Realm (with occurrences in the Urals, East Asia, and Arctic Canada);
- Southern Hemisphere (with Western Australia).

The rich faunas from the listed regions enable application of several quantitative methods of biogeography, of which some will be used here for testing some palaeogeographic hypotheses.

Cluster Analysis

We analysed the number of species per genus and region of the 112 most important regions (Fig. 21a) by cluster analysis using the 1-Pearson r algorithm. The resulting diagram enables the discrimination of two major clusters – Crimea has a remote position in the diagram probably because of its comparatively poor fauna; Crimea's position in the diagram might thus be interpreted as an artefact caused by a lack of data. Contrariwise, the two major clusters coincide with climatic zones:

1. One cluster with equatorial occurrences unifies the East Pacific and Palaeotethyan occurrences, but with rather wide distances between them.
2. One cluster with occurrences in temperate and colder waters with occurrences in Western Australia and Northern China being separated from a dense subcluster with occurrences in the Urals, Eastern Asia, and Arctic Canada.

Some of these palaebiogeographic patterns were expected and can be explained by geographic proximity, such as the close clustering of the northern occurrences, but other patterns raise questions which we currently cannot answer: Why are the Eastern Pacific and Palaeotethyan occurrences rather similar, and how can the co-occurrence of genera be explained? Floating egg masses (Mapes and Nützel 2008) and planktonic juveniles of ammonoids might cover large distances in oceans with more or less uniformly high temperatures, whereas migration to significantly different climates was perhaps impossible due to ecologic constraints such as significant differences in seawater temperature. Why are occurrences in the northern realm rather similar to those in Western Australia, and how can co-occurrence of genera, missing in the equatorial region, be explained?

Cladistic Analysis

Cladistic analyses have been applied only rarely in ammonoid research (Landman 1989; Korn 1997, 2001; Yacobucci 1999; McGowan and Smith 2007; Neige et al. 2007). We ran a cladistic analysis with the paragastrioceratid ammonoids that are particularly diverse in the Permian and allow the integration of a sufficient number of characters for such an analysis. In our analysis, we included 21 genera that are morphologically well-known, with *Trochilioceras* representing the outgroup. Thirty-three characters were shown to be informative, resulting in three most-parsimonious trees with 349 steps, of which the 50% majority rule tree is shown here (Fig. 22).

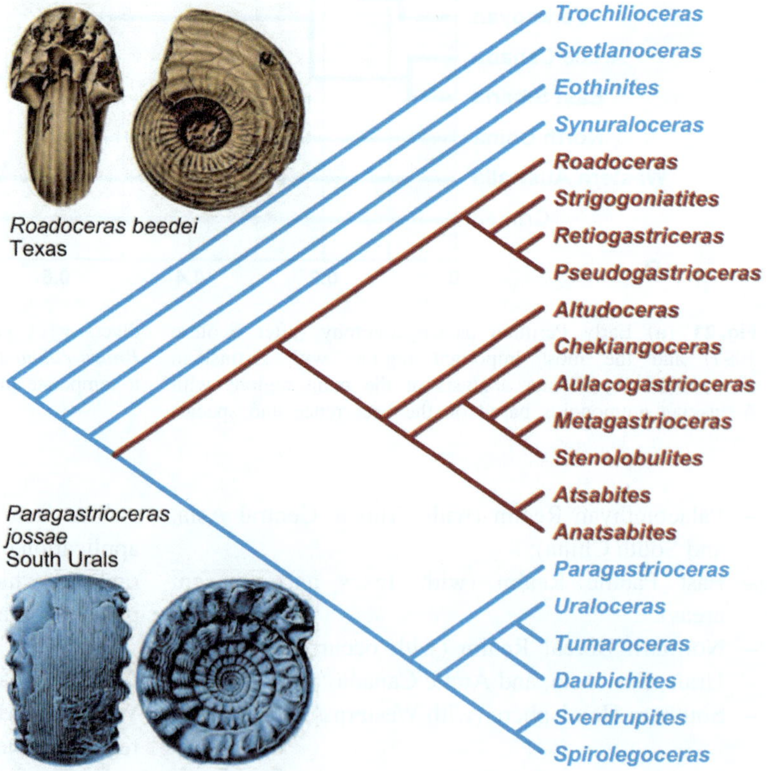

Fig. 22 Cladistic analysis of the Permian ammonoid family Paragastrioceratidae using PAUP 50% majority rule tree of three most parsimonious trees; 33 informative characters; tree length 349 steps; consistency index 0.39. *Brown colour* refers to genera occurring in warm water faunas; *blue colour* refers to temperate and cold waters

The cladogram shows that two major monophyletic units can be separated, one consisting of the subfamily Paragastrioceratidae and the descendant family Spirolegoceratidae, and the other one consisting of the two subfamilies Pseudogastrioceratinae and Aulacogastrioceratinae. The palaeogeographic distribution of these two clades is separated: the paragastrioceratids occur in the boreal province and the pseudogastrioceratids are characteristic for the equatorial belt. The results gained from phylogenetic analysis from one of the major Permian ammonoid groups corroborate the results from the cluster analysis outlined above.

Morphospace Analysis

In a third attempt, we analysed the conch morphology of all Permian ammonoids with respect to spatial and temporal patterns (mainly based on the monographic descriptions by Miller and Furnish 1940; Ruzhencev 1956; Bogoslovskaya 1962; Andrianov 1985; Leven et al. 1992, etc.). Of the main conch parameters, it is especially the whorl expansion rate (WER) and the umbilical width index (uw/dm), which, when plotted in bivariate morphospace diagrams, produced the most informative results (Fig. 23). All 180 investigated species display a distribution with three more densely occupied clusters, representing the three major ammonoid morphs of the Early Permian:

- *Paragastrioceras* Morph: Subevolute to evolute conchs (uw/dm = 0.30–0.65) with low aperture (WER = 1.35–1.90);
- *Almites* Morph: Involute to subinvolute conchs (uw/dm = 0.00–0.20) with low aperture (WER = 1.40–1.90);
- *Artinskia* Morph: Involute to subinvolute conchs (uw/dm = 0.00–0.20) with high aperture (WER = 2.20–3.30).

Separation of the morphs by their conch shape is supported by patterns in the suture line and, to a lesser degree, by shell ornament (Fig. 24). The *Paragastrioceras* Morph has a simple goniatitic suture and often possesses coarse radial or spiral ornament. The *Almites* Morph often shows increasing sutural complexity with addition of elements and serration of

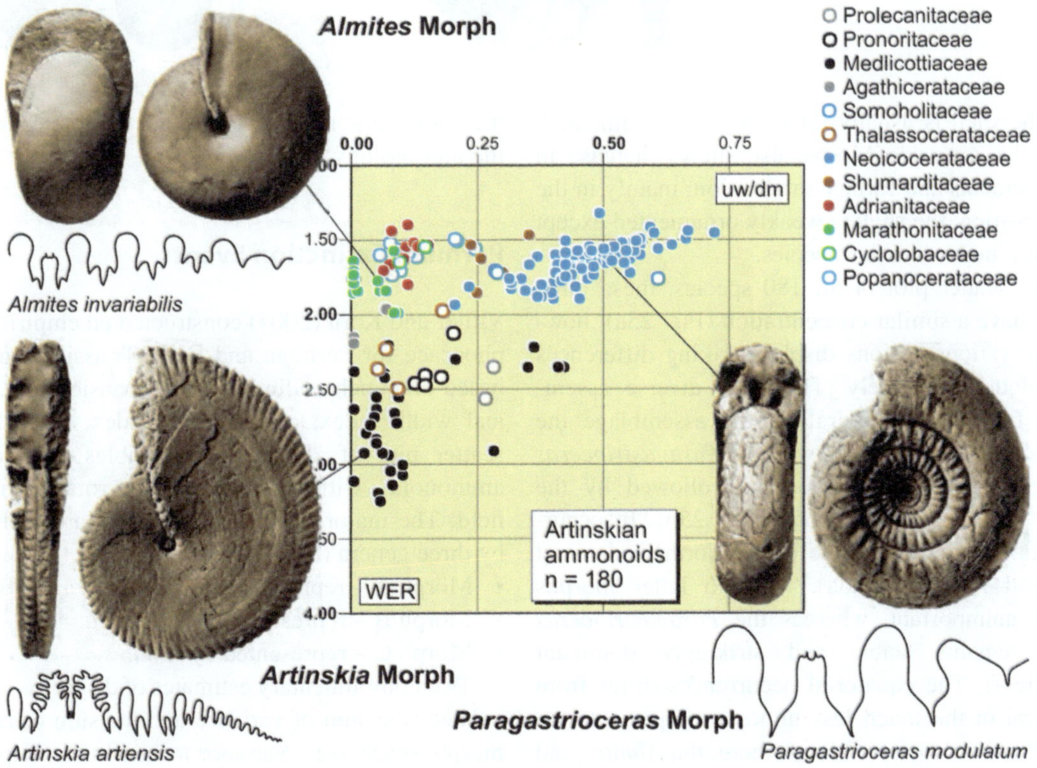

Fig. 23 Bivariate plots showing the WER and uw/dm values of Artinskian ammonoids with respect to their attribution to superfamilies

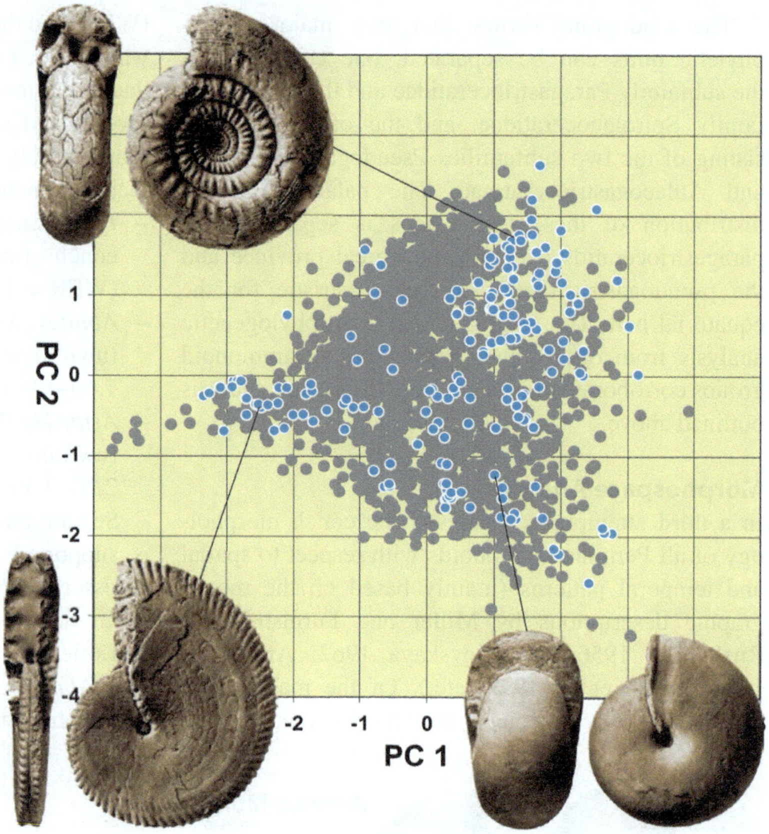

Fig. 24 Bivariate plots showing the WER and uw/dm values of Artinskian ammonoids in various regions

lobes; the shell is usually rather weakly ornamented. Finally the *Artinskia* Morph also shows increase in suture elements and further subdivision, mainly in the ventral portion; the shell is weakly ornamented except for ventral nodes in a few species.

In the scatter plot of all 180 species, these three clusters have a similar concentration (Fig. 25a); however, the various regions display striking differences when plotted separately. The most diverse assemblage is from the South Urals. In this assemblage, the three morphs are separated, with the *Paragastrioceras* Morph being the most important followed by the *Almites* and *Artinskia* Morphs (Fig. 25b). In occurrences farther north (Middle Urals, North Urals, East Asia, and Arctic Canada), the two latter morphs become unimportant, whereas the *Paragastrioceras* Morph remains stable and strikingly dominant (Fig. 25c–f). The equatorial occurrences differ from the boreal in the much less important representation of the *Paragastrioceras* Morph; here, the *Almites* and *Artinskia* Morphs are predominant (Fig. 25g–i). The results confirm the results of cluster diversity analysis; the main ammonoid provinces are clearly separated by the morphology pattern.

Permian Extinction Events

Villier and Korn (2004) constructed an empirical morphospace for Permian and Early Triassic ammonoids based on conch coiling (whorl expansion rate, umbilical width index, conch width index; Fig. 25). The scatter plot of all species resembles the Devonian ammonoids, with an anti-clockwise rotated triangular field. The major conch morphologies are represented by three genera (highlighted in Fig. 25):

- Morph A – represented by *Paragastrioceras*.
- Morph B – represented by *Artinskia*.
- Morph C – represented by *Almites*.

Two complimentary estimates of disparity were calculated: the sum of variance and the sum of range on morphospace axes. Variance measures the dispersal of forms through morphospace. It is sensitive to taxonomic choices, but statistically insensitive to sample

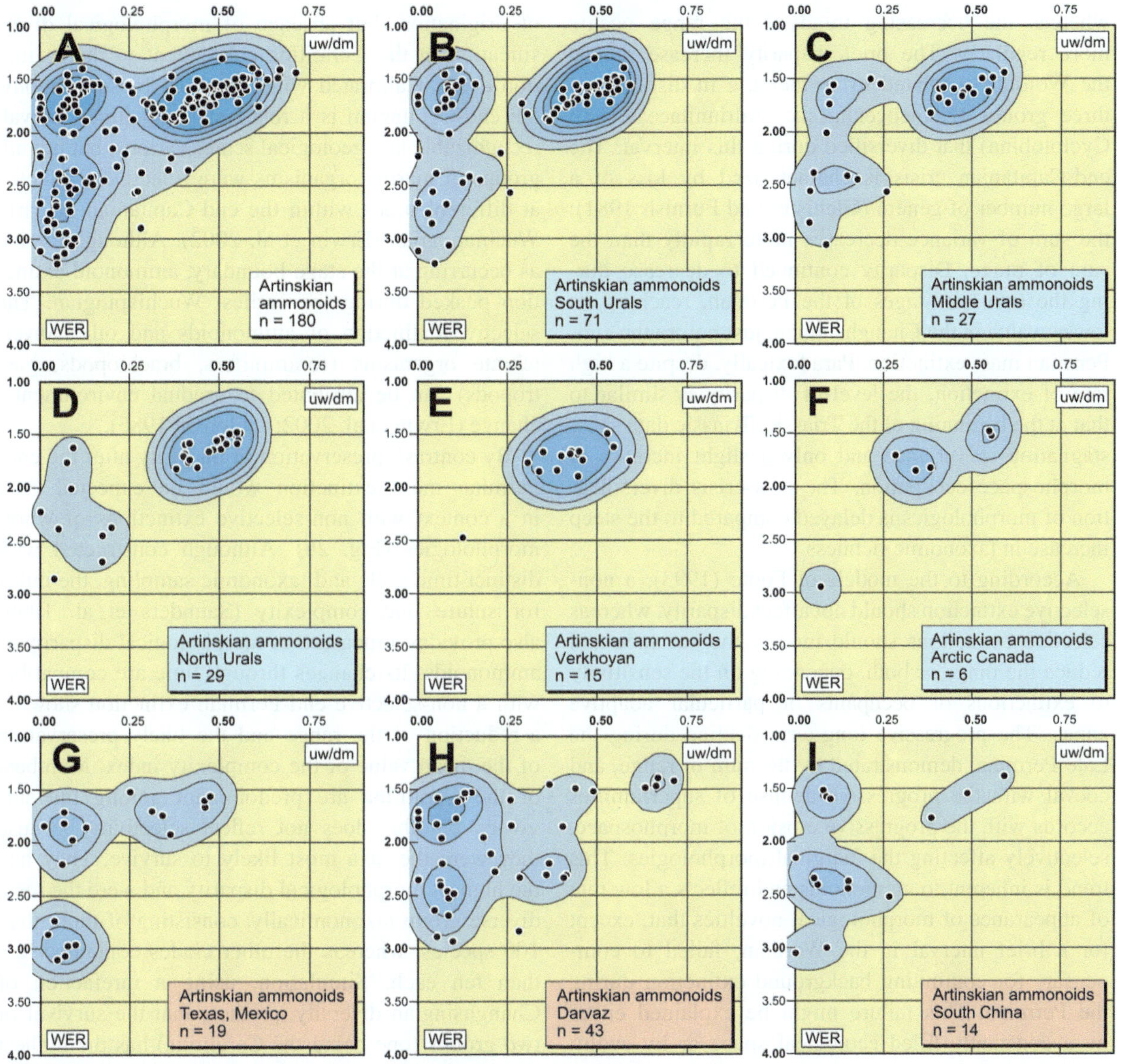

Fig. 25 Morphospace occupation using conch parameters: whorl expansion rate, umbilical width index, whorl width index, and imprint zone rate of ammonoids from the latest Carboniferous to Early Triassic (after Villier and Korn 2004). See text for further discussion

size, except in small samples for which uncertainty increases dramatically (Ciampaglio et al. 2001). The sum of range parallels the amount of morphospace occupied. It is insensitive to taxonomic choices, but sensitive to sample size, the impact of which can be minimized using rarefaction.

The temporal patterns of disparity that we found are broadly similar to the diversity curve for superfamilies (Spearman rank correlation test $p = 0.011$). However, at low taxonomic levels, diversity and morphological disparity are independent (no statistical support for correlation, $p = 0.718$ for genera, $p = 0.740$ for species) and track different information. Disparity increased in the Late Carboniferous to Early Permian and remained stable during the first three Permian stages. During the period from Artinskian to the end of the Permian (approximately 30 Ma), disparity decreased, interrupted only by a brief increase during the Wordian. However, patterns of the two disparity estimates differ substantially. Fluctuations in variance are more pronounced whereas the decreasing trend of the range occurs more regularly. The brief disparity increase during the Wordian is related to an increase in disparity in three groups (Neoicocerataceae, Adrianitaceae, and Cyclolobina) that diversified during this interval. The end-Capitanian crisis is characterized by loss of a large number of genera (Glenister and Furnish 1981); the sum of variance decreases more rapidly than the sum of range. Disparity continued to decrease during the two last stages of the Permian, reaching its lowest value in the Changhsingian, just before the end-Permian mass extinction. Paradoxically, despite a high rate of extinction, the level of disparity is similar to that at the beginning of the Triassic. Triassic data show stagnation in variance and only a slight increase in morphospace occupation. The post-crisis diversification of morphologies is delayed compared to the steep increase in taxonomic richness.

According to the models of Foote (1993), a non-selective extinction should not affect disparity, whereas selective extinctions should modify the variance and reduce the range or both, depending on the sensitivity to extinctions of occupants of particular adaptive zones. The progressive long-term decline during the Late Permian, demonstrated by the sum of range, and coeval with the progressive demise of superfamilies accords with the progressive erosion of morphospace, selectively affecting the marginal morphologies. This trend is inherent to ammonoids and reflects a low rate of appearance of morphological novelties that, except for a brief interval in the Wordian, failed to compensate for continuing background extinction during the Permian. This failure might be explained either by a constantly filled ecological space or by evolutionary and developmental properties of the organisms (Ciampaglio 2004).

The decrease in morphological variance during the Capitanian crisis reflects reworking of the morphospace whose margins were trimmed back, as expressed by decrease in the sum of range, leaving surviving forms mainly clustered in its central part (Glenister and Furnish 1981). Thus, the end-Capitanian crisis was selective with most of the cases affecting the goniatite morphologies in three groups (Adrianitaceae, Cyclolobina and Thalassocerataceae) and laterally compressed forms (Medlicottiaceae). All these clades suffered during the Capitanian event or became extinct. The Wuchiapingian is distinguished from the Capitanian by its high rate of origination, but absence of morphological diversification in the Ceratitina subsequent to the extinction event. Calibrated with conodont biostratigraphy, the end-Capitanian is a relatively long time interval, recognizable at a geological scale. Various, but not all, groups of marine organisms were selectively affected at different times within the end-Capitanian or early Wuchiapingian (Erwin et al. 2002). Although treated as occurring at the stage boundary, ammonoid extinction peaked during the earliest Wuchiapingian. The selective extinction of ammonoids and other invertebrate organisms (foraminifers, brachiopods, gastropods) can be attributed to gradual environmental change (Erwin et al. 2002; Jablonski 1986).

By contrast, preservation of disparity after the end-Permian mass extinction would be expected only in a context with non-selective extinctions of whorl morphologies (Fig. 26). Although constructed with distinct time scale and taxonomic sampling, the index for suture line complexity (Saunders et al. 1999) also provides a measure of morphological disparity of ammonoids. Its changes through time are compatible with a non-selective end-Permian extinction showing a reduction of the range and the likely preservation of the mean value of the complexity index. Members of the Ceratitina are predominant among the survivors, but this does not reflect selectivity because they were the taxa most likely to survive. They had the highest morphological disparity and were the most diverse group taxonomically, consisting of more than 100 species, whereas the other clades contained less than ten each. Simulation, using a rarefaction of Changhsingian diversity, predicts that the survival of two groups (one being the Ceratitina) has the highest likelihood when 87% of species are randomly killed; this matches previously estimated extinction rates for the end Permian crisis (Erwin et al. 2002).

The end-Capitanian and the end-Changhsingian have been recognized as two distinct mass extinction events, the latter being marked by its intensity, but this model cannot explain all aspects of the end-Permian extinctions. Variation through time of morphological disparity suggests three independent patterns for Permian ammonoids: a long-term reduction in disparity, a high level of selective extinction at the end of the Capitanian, and a non-selective extinction at the end of the Permian. The pattern at the end of the Capitanian corresponds to a model of background extinction despite the high level of extinction. Only

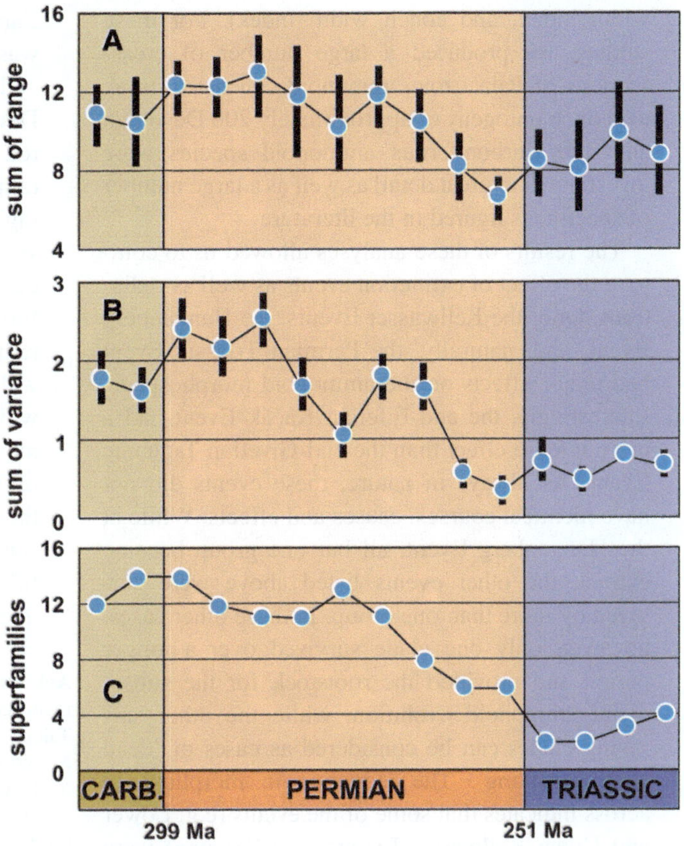

Fig. 26 Morphospace occupation of latest Carboniferous to Early Triassic ammonoids demonstrating maintenance of morphological disparity within the latest Carboniferous and Early Permian, the selective decrease caused by the end-Guadalupian event, the non-selective decrease by the end-Permian event, and Early Triassic recovery (after Villier and Korn 2004). See text for further discussion

the end-Permian event matches the model of a 'mass extinction regime' (Jablonski 1986), arguing for a catastrophic cause consisting of a brief but major event, independent of earlier variations in diversity, with a worldwide effect and, in the main, non-selective demise of taxa.

Conclusions

The taxonomic diversity of ammonoids (number of taxa preserved) provides an incomplete picture of the extinction pattern (a strongly biased fossil record) during the Palaeozoic. The analysis of morphological disparity (the variety of shell-shape) is a powerful complementary tool for testing hypotheses about the selectivity of extinction and permits recognition of three distinct patterns.

Taxonomic diversity and morphological disparity are often strikingly uncoupled, but analyses of morphological disparity can overcome part of the sampling bias of the fossil record. By focusing on shape differences, morphological disparity is relatively independent of taxonomy and allows comparison of samples in which a variable proportion of taxa are preserved or sampled. Estimates of disparity consider the distribution of taxa and proportions of morphologic space occupied during successive time intervals.

Ammonoids are suitable subjects for analyses of extinction dynamics as they show high fluctuations in taxonomic diversity during their history and record numerous extinction events and subsequent recoveries. We analysed diversity dynamics of ammonoids from a distribution of approximately 4000 species ranging from the Emsian stage (Early Devonian) to the Early Triassic. The data are derived from the database AMMON, a compendium of Paleozoic species of ammonoids encompassing taxonomy, morphologic data, and geographic and stratigraphic occurrences.

We constructed an empirical morphospace for Devonian to Early Triassic ammonoids based on conch morphology (whorl expansion rate, umbilical width index, and conch width index). For these studies, we produced a large number of cross-sections of Palaeozoic ammonoid conchs. In total, the conch ontogeny of approximately 200 Devonian and 300 Carboniferous ammonoid species were investigated in great detail as well as a large number of specimens figured in the literature.

The results of these analyses allowed us to compare the effect of extinction events as well as radiations. Only the Kellwasser Events, the Hangenberg Event, and, naturally, the Permian Triassic Event had major effects on the ammonoid morphospace. Interestingly, the end-Eifelian Kačák Event had a more intense effect than the end-Givetian Taghanic Event. As always in nature, these events did not have identical courses, causes and effects. While at the Hangenberg Event, all but one group became extinct, the other events listed above were survived by more than one group. In these other cases, however, only one clade survived over a longer period and provided the rootstock for the subsequent ammonoid evolution, while the other surviving clades can be considered as cases of 'dead clades walking'. The reduction in morphospace across indicates that some of the events (e.g. Lower and Upper Kellwasser Events, end-Permian) were non-selective relative to particular morphotypes. In contrast, the end-Capitanian extinction was morphologically selective, with laterally compressed and 'goniatite' forms preferentially eliminated.

The extinction-processes apparently also differed. This aspect is, however, difficult to assess due to, e.g. sampling and preservational biases ('Signor-Lipps effect': Signor and Lipps 1982). As far as the available data are concerned, it appears as if the Kačák and Hangenberg Events were rather sudden while the end-Permian and the end-Frasnian events had rather stepwise courses.

During the recovery phase, the morphospace-re-occupation also displays different courses. While the Early Triassic rediversification was rather fast, the Early Famennian and Early Carboniferous diversifications appear to have been stepwise. In all cases, it has to be taken into account that forms that lived exclusively in deeper waters of the oceans are hardly known because of the non-extant record of unaltered pre-Triassic oceanic sediments. In the Early Carboniferous, a slowly increasing endemism was documented.

The initial radiation of ammonoids in the Early Devonian differs significantly from the Palaeozoic rediversification events. This initial radiation was rather rapid, and within roughly 5 Ma, forms including loosely coiled taxa with strong ornamentation, smooth oxycones with more or less closed umbilicus, various platycones but also rather broad forms evolved. This radiation is unrivalled by later radiations which may be explained by the left-wall effect, caused by the ammonoid ancestors which had orthoconic shells. This ancestral shell morphology is still reflected in the shell shape of the first ammonoids sensu strictu. Hopefully, the results of our analyses verify Seilacher's statement, that ammonoids are nearly as useful as Recent *Drosophila* to address evolutionary questions.

Acknowledgements We are grateful for revision of the manuscript by Sonny Walton, Berlin, and, in particular, to John Talent, Sydney. We greatly appreciate the thorough reviews by Margret Yaccobucci, Bowling Green, Ohio, Neil Landman, New York, and Royal Mapes Athens, Ohio.

DK is indebted to the Deutsche Forschungsgemeinschaft (DFG) for financial support of this and several other research projects (Ko 1829/1-1, Ko 1829/2-1, Ko 1829/3-1, Ko 1829/6-1, Ko 1829/8-1).

CK thanks the Swiss National Science Foundation for the financial support of the research project No. 200021-113956/1 that helped provide data for the analyses published here.

References

Adams DC, Rohlf FJ, Slice DE (2004) Geometric morphometrics: ten years of progress following the 'revolution'. Ital J Zool 71:5–16

Allen EG (2007) Understanding ammonoid sutures: new insight into the dynamic evolution of Paleozoic suture morphology. In: Landman NH, Davis RA, Manger WM, Mapes RH (eds) Cephalopods – present and past. Springer, New York, pp 159–180

Andrianov VN (1985) Permskie i nekotorye kamennougolnye ammonoidei Severo-Vostoka Azii. Trudy Geologicheskogo Instituta (Novosibirsk), Izd Nauka Sibirsk Otdel 1985:1–180

Barskov IS, Boiko MS, Konovalova VA, Leonova TB, Nikolaeva SV (2008) Cephalopods in the marine ecosystems of the Paleozoic. Paleontol J 42:1167–1284

Becker RT (1993a) Stratigraphische Gliederung und Ammonoideen-Faunen im Nehdenium (Oberdevon II) von Europa und Nord-Afrika. Cour Forschungsinst Senckenb 155:1–405

Becker RT (1993b) Anoxia, eustatic changes, and Upper Devonian to lowermost Carboniferous global ammonoid diversity. In: House MR (ed) The ammonoidea: environment, ecology, and evolutionary change. Systematics Association Special, vol 47. Clarendon, Oxford, pp 115–163

Becker RT (1995) Taxonomy and evolution of Late Famennian Tornocerataceae (Ammonoidea). Berl geowiss Abh E16: 607–643

Becker RT, House MR (1993) New early Upper Devonian (Frasnian) goniatite genera and the evolution of the "Gephurocerataceae". Berl geowiss Abh E9:111–133

Becker RT, House MR (1994) Kellwasser Events and goniatite successions in the Devonian of the Montagne Noire with comments on possible causations. Cour Forschungsinst Senckenb 169:45–77

Becker RT, Kullmann J (1996) Paleozoic ammonoids in space and time. In: Landman NH, Tanabe K, Davis RA (eds) Ammonoid paleobiology. Topics in Paleobiology, vol 13. Plenum, New York, pp 711–753

Becker RT, Feist R, Flajs G, House MR, Klapper G (1989) Frasnian-Famennian extinction events in the Devonian at Coumiac, southern France. C R Acad Sci 309:259–266

Becker RT, House MR, Kirchgasser WT (1993) Devonian goniatite biostratigraphy and timing of facies movements in the Frasnian of the Canning Basin, Western Australia. In: Hailwood EA, Kidd RB (eds) High resolution stratigraphy. Geol Soc London Spec Pub 70:293–321

Bensaïd M (1974) Etude sur des Goniatites à la limite du Dévonien moyen et supérieur, du Sud marocain. Notes Serv Géol Maroc 36:82–137

Boardman DR, Work DM, Mapes RH, Barrick JE (1994) Biostratigraphy of Middle and Late Pennsylvanian (Desmoinesian–Virgilian) ammonoids. Kansas Geol Surv Bull 232:1–121

Bockwinkel J, Ebbighausen V. (2006) A new ammonoid fauna from the *Gattendorfia-Eocanites* Genozone of the Anti-Atlas (Early Carboniferous; Morocco). Fossil Record 9(1):87–129

Bogoslovskaya MF (1962) Artinskie ammonoidei Srednego Urala. Trudy Paleontol Inst Akad Nauk SSSR 103:1–103

Bogoslovsky BI (1958) Devonskie ammonoidei Rudnogo Altaya. Trudy Paleontol Inst Akad Nauk SSSR 64:1–155

Bogoslovsky BI (1969) Devonskie ammonoidei. I. Agoniatity. Trudy Paleontol Inst Akad Nauk SSSR 124:1–341

Bogoslovsky BI (1971) Devonskie ammonoidei. II. Goniatity. Trudy Paleontol Inst Akad Nauk SSSR 127:1–228

Bogoslovsky BI (1981) Devonski ammonoidei, III. Klimenii (Podotryad Gonioclymeniina). Trudy Paleontol Inst Akad Nauk SSSR 191:1–122

Bookstein FL (1998) A hundred years of morphometrics. Acta Zool Acad Sci Hung 44:7–59

Branco W (1880) Beiträge zur Entwicklungsgeschichte der fossilen Cephalopoden. Palaeontographica 27:17–81

Brayard A, Gilles E, Bucher H, Monnet C, Brühwiler T, Goudemand N, Galfetti T, Guex J (2009) Good genes and good luck: ammonoid diversity and the end-Permian mass extinction. Science 325:118–1121

Buggisch W (1991) The global Frasnian-Famennian "Kellwasser Event". Geol Rundsch 80:49–72

Callomon JH (1963) Sexual dimorphism in Jurassic ammonites. Trans Leic Lit Phil Soc 57:21–56

Callomon JH (1981) Dimorphism in ammonoids. In: House MR, Senior JR (eds) The ammonoidea. Systematics Association Special, vol 18. Academic, London, pp 257–273

Chlupàč I, Turek V (1983) Devonian goniatites from the Barrandian area, Czechoslovakia. Rozpr Ústr úst Geol 46:1–159

Ciampaglio CN (2004) Measuring changes in articulate brachiopod morphology before and after the Permian mass extinction event: do developmental constraints limit morphological innovation? Evol Develop 6(4): 260–274

Ciampaglio CN, Kemp M, McShea DW (2001) Detecting changes in morphospace occupation patterns in the fossil record: characterization and analysis of measures of disparity. Paleobiology 27:695–715

Clarke JM (1896) Notes on the early stages of certain goniatites. Univ State New York, State Mus Rep 50(2):163–170

Clarke JM (1899a) Notes on the early stages of certain goniatites. New York State Mus Ann Rep 50 (for 1896): 165–169

Clarke JM (1899b) The Naples fauna (fauna with *Manticoceras intumescens*) in Western New York. New York State Mus Ann Rep 50 (for 1896):31–161

Clausen C-D (1969) Oberdevonische Cephalopoden aus dem Rheinischen Schiefergebirge. II. Gephuroceratidae, Beloceratidae. Palaeontogr A 132:95–178

Crampton JS, Haines AJ (1996) Users' manual for programs HANGLE, HMATCH, and HCURVE for the Fourier shape analysis of two-dimensional outlines. Inst Geol Nucl Sci Rept 96/37. Lower Hutt, New Zealand

Czarnocki J (1989) Klimenie Gór Świętokrzyskich. Prace Panst Inst Geol 127:1–91

Darwin C (1859) On the origin of species by means of natural selection, or the preservation of favoured races in the struggle for life. J. Murray, London

Davis RA, Landman NH, Dommergues J-L, Marchand D, Bucher H (1996) Mature modifications and dimorphism in ammonoid cephalopods. In: Landman NH, Tanabe K, Davis RA (eds) Ammonoid paleobiology. Topics in Paleobiology, vol 13. Plenum, New York, pp 463–539

De Baets K, Klug C, Korn D (2009) Anetoceratinae (Ammonoidea, Early Devonian) from the Eifel and Harz Mountains (Germany) with a revision of their genera. N Jb Geol Paläontol Abh 252:361–376

Delépine G (1940) Les goniatites du Dinantien de la Belgique. Mém Mus Royal Hist Nat Belgique 91:1–91

Delépine G (1941) Les goniatites du Carbonifère du Maroc et des confins Algéro-Marocains du sud (Dinantien-Westphalien). Not mém Serv Géol Protect Rép Franç Maroc, Dir Prod Industr Min, Div Min Géol 56:1–111

Doguzhaeva LA, Mapes RH, Mutvei H (2010) Evolutionary patterns of Carboniferous coleoid cephalopods based on their diversity and the morphological plasticity. In: Tanabe K, Shigeta Y, Sasaki T, Hirano H (eds), Cephalopods Present and Past. Tokai University, Hadano, Japan, pp 171–180

Doguzhaeva LA, Mapes RH, Mutvei H (2007) A Late Carboniferous coleoid cephalopod from the Mazon Creek Lagerstätte (USA), with a radula, arm hooks, mantle tissues, and ink. In: Landman NH, Davis RA, Mapes RH (eds) Sixth international symposium cephalopods present and past. New

insights and fresh perspectives. Springer, The Netherlands, pp 121–143
Dommergues JL, Montuire S, Neige P (2002) Size patterns through time: the case of the Early Jurassic ammonite radiation. Paleobiology 28:423–434
Ebbighausen V, Bockwinkel J (2007) Tournaisian (Early Carboniferous/Mississippian) ammonoids from the Ma'der Basin (Anti-Atlas, Morocco). Fossil Record 10:125–163
Ebbighausen V, Korn D (2007) Conch geometry and ontogenetic trajectories in the triangularly coiled Late Devonian ammonoid *Wocklumeria* and related genera. N Jb Geol Paläontol Abh 244:9–41
Ebbighausen V, Bockwinkel J, Korn D, Weyer D (2004) Early Tournaisian ammonoids from Timimoun (Gourara, Algeria). Mitt Mus Nat-kde Berl Geowiss R 7:133–152
Erben HK (1962) Über die „forme elliptique" der primitiven Ammonoidea. Paläont Zeit 36, Suppl 1, H. Schmidt-Festband: 38–44, Springer, Stuttgart
Erben HK (1964a) Bactritoidea. In: Moore RC (ed) Treatise on invertebrate paleontology, part K, Mollusca, vol 3. Geological Society of America, New York, University of Kansas Press, Lawrence/Kansa, pp K491–K507
Erben HK (1964b) Die Evolution der ältesten Ammonoidea (Lieferung I). N Jb Geol Paläontol Abh 120(2): 107–212
Erben HK (1965) Die Evolution der ältesten Ammonoidea. II, N Jb Geol Paläontol Abh 122(3):275–312
Erwin DH (1994) Early introduction of major morphological innovations. Acta Palaeontol Pol 38:281–294
Erwin DH, Bowring SA, Yin Y (2002) End-Permian mass extinctions: a review. Geol Soc Amer Spec Pap 356:363–383
Ferson S, Rohlf FJ, Koehn RK (1985) Measuring shape variation of two-dimensional outlines. Syst Zool 34:59–68
Foord AH (1901) Monograph of the Carboniferous Cephalopoda of Ireland, Part IV, Containing the families Solenocheilidae (concluded) and Glyphioceratidae. Palaeontogr Soc 55:127–146
Foord AH (1903) Monograph of the Carboniferous Cephalopoda of Ireland, Part V, Containing the families Glyphioceratidae (concluded) and Prolecanitidae, with title-page and index. Palaeontogr Soc 57:147–234
Foote M (1989) Perimeter-based Fourier analysis: a new morphometric method applied to the Trilobite cranidium. J Paleontol 63:880–885
Foote M (1993) Discordance and concordance between morphological and taxonomic diversity. Paleobiology 19: 185–204
Foote M (1996) Models of morphological diversification. In: Jablonski D, Erwin DH, Lipps JH (eds) Evolutionary paleobiology. University of Chicago Press, Chicago, pp 62–86
Foote M (1997) The evolution of morphological diversity. Ann Rev Ecol Syst 28:129–152
Gemmellaro GG (1887) La Fauna dei Calcari con Fusulina della valle del Fiume Sosio (nella Provincia di Palermo). Giorn sci nat econom 19:1–106
Glenister BF (1958) Upper Devonian ammonoids from the *Manticoceras* Zone, Fitzroy Basin, Western Australia. J Paleontol 32:50–96
Glenister BF, Furnish WM (1981) Permian ammonoids. In: House MR, Senior JR (eds) The ammonoidea. Systematics Association Special, vol 18. Academic, London, pp 49–64
Gordon M Jr (1965) Carboniferous cephalopods of Arkansas. Geol Surv Prof Pap 460:1–322
Gordon M (1986) Late Kinderhookian (Early Mississippian) Ammonoids of the Western United States. J Paleontol Mem 19:1–36
Gottobrio WE, Saunders WB (2005) The clymeniid dilemma: functional implications of the dorsal siphuncle in clymeniid ammonoids. Paleobiology 31:233–252
Gould SJ (1997) Full house: the spread of excellence from Plato to Darwin. Harmony Books, Nevada City, CA
Gradstein FM, Ogg JG, Smith AG, Agterberg FP, Bleeker W, Cooper RA, Davydov V, Gibbard P, Hinnov LA, House MR, Lourens L, Luterbacher HP, McArthur J, Melchin MJ, Robb LJ, Shergold J, Villeneuve M, Wardlaw BR, Ali J, Brinkhuis H, Hilgen FJ, Hooker J, Howarth RJ, Knoll AH, Laskar J, Monechi S, Plumb KA, Powell J, Raffi I, Röhl U, Sadler P, Sanfilippo A, Schmitz B, Shackleton NJ, Shields GA, Strauss H, Van Dam J, van Kolfschoten T, Veizer J, Wilson D (2004) A geologic time scale 2004. Cambridge University Press, Cambridge
Gümbel CW (1863) Über Clymenien in den Übergangsgebilden des Fichtelgebirges. Palaeontogr 11:85–165
Hammer O, Harper DAT, Ryan PD (2008) PAST – PAlaeontological STatistics, ver. 1.77. http://folk.uio.no/ohammer/past/past.pdf. Accessed 11 Feb 2008
Haniel CA (1915) Die Cephalopoden der Dyas von Timor. In: Wanner J (ed) Paläontology of Timor, vol 36, Suppl 1, E. Schweizerbart'sche, Stuttgart
Holzapfel E (1889) Die Cephalopoden-führenden Kalke des unteren Carbon von Erdbach-Breitscheid bei Herborn. Palaeontol Abh NF 5:1–74
Hooke R (1705) Lectures and discourses of earthquakes, and subterraneous eruptions. Explicating the causes of the rugged and uneven face of the earth; and what reasons may be given for the frequent finding of Shells and other Sea and Land Petrified Substances, scattered over the whole Terrestrial Superficies. In: The Posthumous Works of Robert Hooke. Containing his Cutlerian Lectures, and other discourses, read at the Meetings of the Illustrious Royal Society. Richard Waller, London
House MR (1970) On the origin of the clymenid ammonoids. Palaeontology 13:664–676
House MR (1981) On the origin, classification and evolution of the early Ammonoidea. In: House MR, Senior JR (eds) The ammonoidea. Systematics Association Special, vol 18. Academic, London, pp 3–36
House MR (1985) Correlation of mid-Palaeozoic ammonoid evolutionary events with global sedimentary perturbations. Nature 213:17–22
House MR (2002) Strength, timing, setting and cause of mid-Palaeozoic extinctions. Palaeogeogr Palaeoclimatol Palaeoecol 181:5–25
Jablonski D (1986) Background and mass extinctions – the alternation of macroevolutionary regimes. Science 231(4734):129–133
Jablonski D (2002) Survival without recovery after mass extinctions. Proc Nat Acad Sci USA 99:8139–8144
Jacobs DK, Chamberlain JA (1996) Buoyancy and hydrodynamics in ammonoids. In: Landman NH, Tanabe K, Davis RA (eds) Ammonoid paleobiology. Topics in Paleobiology, vol 13. Plenum, New York, pp 43–63

Johnson RG, Richardson ES Jr (1968) Ten-armed fossil cephalopod from the Pennsylvanian in Illinois. Science 159:526–528

Joliffe IT (2002) Principal component analysis, 2nd edn. Springer, New York

Kant R (1973a) Allometrisches Wachstum paläozoischer Ammonoideen: Variabilität und Korrelation einiger Merkmale. N Jb Geol Paläontol Abh 143(2):153–192

Kant R (1973b) Untersuchungen des allometrischen Gehäusewachstums paläozoischer Ammonoideen unter besonderer Berücksichtigung einzelner "Populationen". N Jb Geol Paläontol Abh 144:206–251

Kant R, Kullmann J (1988) Changes in the conch form in the paleozoic ammonoids. In: Wiedmann J, Kullmann J (eds) Cephalopods – present and past. Schweizerbart, Stuttgart, pp 43–49

Kauffman EG, Walliser OH (1988) Global bioevents – abrupt changes in the global biota. Episodes 11:289–292

Klofak SM, Landman NH, Mapes RH (1999) Embryonic Development of primitive ammonoids and the monophyly of the ammonoidea. In: Olóriz F, Rodríguez-Tovar FJ (eds) Advancing research on living and fossil cephalopods. Springer, New York, pp 23–45

Klofak SM Landman NH, Mapes RH (2007) Patterns of embryonic development in early to middle Devonian ammonoids. In: Landman NH, Davis RA, Manger WM, Mapes RH (eds) Cephalopods – present and past. Springer, New York, pp 15–26

Klug C (2001a) Early Emsian ammonoids from the eastern Anti-Atlas (Morocco) and their succession. Paläontol Z 74:479–515

Klug C (2001b) Life cycles of some Devonian ammonoids. Lethaia 34:215–233

Klug C (2002) Quantitative stratigraphy and taxonomy of late Emsian and Eifelian ammonoids of the eastern Anti-Atlas (Morocco). Cour Forschungsinst Senckenb 238:1–109

Klug C (2004) Mature modifications, the black band, the black aperture, the black stripe, and the periostracum in cephalopods from the Upper Muschelkalk (Middle Triassic, Germany). Mitteilungen aus dem Geologisch-Paläontologischen Institut der Universität Hamburg 88:63–78

Klug C, Korn D (2004) The origin of ammonoid locomotion. Acta Palaeontol Pol 49(2):235–242

Klug C, Korn D, Richter U, and Urlichs M (2004) The black layer in cephalopods from the German Muschelkalk (Middle Triassic). Palaeontology 47:1407–1425

Klug C, Schatz W, Korn D, Reisdorf A (2005) Morphological fluctuations of ammonoid assemblages from the Muschelkalk (Middle Triassic) of the Germanic Basin – indicators of their ecology, extinctions, and immigrations. Palaeogeogr Palaeoclimatol Palaeoecol 221:7–34

Klug C, Döring S, Korn D, Ebbighausen E (2006) The Viséan sedimentary succession at the Gara el Itima (Anti-Atlas, Morocco) and its ammonoid faunas. Fossil Record 9:3–60

Klug C, Brühwiler T, Korn D, Schweigert G, Brayard A, Tilsley J. (2007) Ammonoid shell structures of primary organic composition. Palaeontology 50:1463–1478

Klug C, Meyer E, Richter U, Korn D (2008) Soft-tissue imprints in fossil and Recent cephalopod septa and septum formation. Lethaia 41:477–492

Konovalova VA (2005) Ammonoid evolution at the Tournaisian-Visean boundary. Paleontol J 39(Suppl 5):S538–S557

Korn D (1984) Die Goniatiten der Stockumer *Imitoceras*-Kalklinsen (Ammonoidea; Devon/Karbon-Grenze). Cour Forschungsinst Senckenb 67:71–89

Korn D (1986) Ammonoid evolution in late Famennian and early Tournaisian. Ann Soc Géol Belgique 109:49–54

Korn D (1988) Die Goniatiten des Kulmplattenkalkes (Cephalopoda, Ammonoidea; Unterkarbon; Rheinisches Schiefergebirge). Geol Paläontol Westf 11:1–293

Korn D (1992) Relationship between shell form, septal construction and suture line in clymeniid cephalopods (Ammonoidea; Upper Devonian). N Jb Geol Paläontol Abh 185: 115–130

Korn D (1993) The ammonoid faunal change near the Devonian-Carboniferous boundary. Ann Soc Géol Belgique 115:581–593

Korn D (1994) Devonische und karbonische Prionoceraten (Cephalopoda, Ammonoidea) aus dem Rheinischen Schiefergebirge. Geol Paläontol Westf 30:1–85

Korn D (1997) Evolution of the Goniatitaceae and Viséan-Namurian biogeography. Acta Palaeontol Pol 42:177–199

Korn D (2000) Morphospace occupation of ammonoids over the Devonian-Carboniferous boundary. Paläontol Z 74:247–257

Korn D (2001) Morphometric evolution and phylogeny of Palaeozoic ammonoids. Early and Middle Devonian. Acta Geol Pol 51:193–215

Korn D (2003) Typostrophism in Palaeozoic ammonoids. Paläontol Z 77:445–470

Korn D (2004) The mid-Famennian ammonoid succession in the Rhenish Mountains: the "*annulata* Event" reconsidered. Geol Quart 48:245–25

Korn D, Klug C, Mapes R (2005) The Lazarus ammonoid genus *Goniatites*, the tetrangularly coiled *Entogonites*, and Mississippian biogeography. J Paleontol 79:356–365

Korn D, Ebbighausen V (2008) The Early Carboniferous (Mississippian) ammonoids from the Chebket el Hamra (Jerada Basin, Morocco). Fossil Record 11:83–156

Korn D, Ilg A (2008) AMMON. http://www.wahre-staerke.com/ammon/

Korn D, Klug C (2001) Biometric analyses of some Palaeozoic ammonoid conchs. Berliner geowisschenschaftliche Abhandlungen E36:173–187

Korn D, Klug C (2002) Ammoneae Devonicae. Fossilium Catalogus I: Animalia, vol 138. Backhuys, Leiden, pp 1–375

Korn D, Klug C (2003) Morphological pathways in the evolution of Early and Middle Devonian ammonoids. Paleobiology 29:329–348

Korn D, Klug C (2007) Conch form analysis, variability, morphological disparity, and mode of life of the Frasnian (Late Devonian) ammonoid *Manticoceras* from Coumiac (Montagne Noire, France). In: Landman NH, Davis RA, Mapes RH (eds) Cephalopods – present and past. Springer, New York, pp 57–85

Korn D, Kullmann J (1988) Changes in Clymeniid diversity. In: Wiedmann J, Kullmann J (eds) Cephalopods – present and past. Schweizerbart, Stuttgart, pp 25–28

Korn D, Tilsley J (2002) Stratigraphically important ammonoids from the Visean-Namurian Boundary at Backdale Mine, Derbyshire. Proc Yorksh Geol Soc 54:111–119

Korn D, Vöhringer E (2004) Allometric growth and intraspecific variability in the basal Carboniferous ammonoid *Gattendorfia crassa* Schmidt, 1924. Paläontol Z 78: 425–432

Korn D, Weyer D (2003) High resolution stratigraphy of the Devonian-Carboniferous transitional beds in the Rhenish Mountains. Mitt Mus Nat-kde Berl Geowiss R 4:79–124

Korn D, Klug C, Mapes RH (1999) Viséan and Early Namurian Ammonoids from the Tafilalt (Eastern Anti-Atlas, Morocco). Abh Geol B-Anst 54:345–375

Korn D, Klug C, Ebbighausen V, Bockwinkel J (2002) Palaeogeographic meaning of a Middle Tournaisian ammonoid fauna from Morocco. Geol Palaeontol 36:79–86

Korn D, Bockwinkel J, Ebbighausen V, Klug C (2003) Palaeobiogeographic and evolutionary meaning of an early Late Tournaisian ammonoid fauna from the Tafilalt of Morocco. Acta Palaeontol Pol 48:71–92

Korn D, Klug C, Mapes RH (2004a) Cuboid Carboniferous ammonoids. Mitt Geol Paläontol Inst Univ Hamb 88:79–97

Korn D, Belka Z, Fröhlich S, Rücklin M, Wendt J (2004b) The youngest African clymeniids (Ammonoidea, Late Devonian) – failed survivors of the Hangenberg Event. Lethaia 37:307–315

Korn D, Bockwinkel J, Ebbighausen V (2007) The Tournaisian and Viséan ammonoid stratigraphy in North Africa. N Jb Geol Paläontol Abh 243(2):127–148

Kröger B (2007) On the origin of bactritoids. Paläontol Z 81:316–327

Kröger B (2008) Nautiloids before and during the origin of ammonoids in a Siluro-Devonian section in the Tafilalt, Anti-Atlas, Morocco. Spec Pap Palaeontol 79:5–110

Kullmann J (1961) Die Goniatiten des Unterkarbons im Kantabrischen Gebirge (Nordspanien). I. Stratigraphie. Paläontologie der U.O. Goniatitina Hyatt. N Jb Geol Paläontol Abh 113(3):219–326

Kullmann J (1962) Die Goniatiten der Namur-Stufe (Oberkarbon) im Kantabrischen Gebirge, Nordspanien. Abh Akad Wiss Lit Mainz, math-naturwiss Kl 1962:259–377

Kullmann J (1963) Die Goniatiten des Unterkarbons im Kantabrischen Gebirge (Nordspanien). II. Paläontologie der U.O. Prolecanitina Miller and Furnish. Die Altersstellung der Faunen. N Jb Geol Paläontol Abh 116:269–324

Kullmann J (1983) Maxima im Tempo der Evolution karbonischer Ammonoideen. Paläontol Z 57:231–240

Kullmann J (2000) Ammonoid turnover at the Devonian-Carboniferous boundary. Rev Paléobiol, Spec Vol 8:169–180

Kullmann J, Scheuch J (1970) Wachstums-Änderungen in der Ontogenese paläozoischer Ammonoideen. Lethaia 3:397–412

Kullmann J, Scheuch J (1972) Absolutes und relatives Wachstum bei Ammonoideen. Lethaia 5:129–146

Kusina LF (1980) Saurskie ammonoidei. Trudy Paleontol Inst Akad Nauk SSSR 181:1–108

Landman NH (1989) Iterative progenesis in Upper Cretaceous ammonites. Paleobiology 15:95–117

Landman NH, Tanabe K, Shigeta Y (eds) (1996) Ammonoid paleobiology. Topics in Paleobiology 13. Plenum, New York, 857pp

Lehmann U (1981) Ammonite jaw apparatus and soft parts. In: House MR, Senior JR (eds) The ammonoidea. Systematics Association Special, vol 18. Academic, London, pp 275–287

Lehmann U (1985) Zur Anatomie der Ammoniten: Tintenbeutel, Kiemen, Augen. Paläontol Z 59:99–108

Leonova TB (2002) Permian ammonoids: classification and phylogeny. Paleontol J Suppl 36:1–114

Leonova TB, Dmitriev VYu, Shimanskiĭ VN (1989) Rannepermskie ammonoidei Yugo-Vostochnogo Pamira. Trudy Paleontol Inst Akad Nauk SSSR 235:1–198

Leven EYa, Leonova TB, Dmitriev VYu (1992) Perm' Darvaz-Zaalaiskoi zony Pamira: fuzulinidy, ammonoidei, stratigrafiya. Trudy Paleontol Inst Akad Nauk SSSR 253:1–203

Librovitch LS (1927) Nizhnekamennougolnye golovonogie iz rayona ozera Son-kul (Tyan Shan). Mat obsch prikl geol, Geol Kom 74:1–55

Lutz TM, Boyajian GE (1995) Fractal geometry of ammonoid sutures. Paleobiology 21:329–342

Makowski H (1962) Problem ofsexual dimorphism in ammonites. Paleont Pol 12:1–92

Mapes RH (1979) Carboniferous and Permian Bactritoidea in North America. Univ Kansas Paleontol Contributions 64:1–75

Mapes RH, Nützel A (2008) Late Palaeozoic mollusc reproduction: cephalopod egg-laying behavior and gastropod larval palaeobiology. Lethaia 18. doi:10.1111/j.1502-3931.2008.00141

McGhee GR (1988) The Late Devonian extinction event – evidente for abrupt ecosystem collapse. Paleobiology 14(3):250–257

McGowan AJ (2004a) The effect of the Permo-Triassic bottleneck on Triassic ammonoid morphological evolution. Paleobiology 30:369–395

McGowan AJ (2004b) Ammonoid taxonomic and morphologic recovery patterns after the Permian-Triassic. Geology 32:665–668

McGowan AJ (2005) Ammonoid recovery from the Late Permian mass extinction event. C R Palevol 4:517–530

McGowan AJ, Smith AB (2007) Ammonoids across the Permian/Triassic boundary: a cladistic perspective. Palaeontology 50(3):573–590

Miller AK (1938) Devonian ammonoids of America. Geol Soc Am Spec Pap 14:1–262

Miller AK, Collinson C (1951) Lower Mississippian ammonoids of Missouri. J Paleontol 25:454–487

Miller AK, Furnish WM (1940) Permian ammonoids of the Guadalupe Mountain Region and adjacent areas. Geol Soc Amer Spec Pap 26:1–242

Monnet C, Bucher H, Escarguel G, Guex J (2003) Cenomanian (early Late Cretaceous) ammonoid faunas of western Europe. Part 2: the timing and patterns of faunal changes and the end-Cenomanian anoxic event. Ecl Geol Helv 96: 381–398

Moseley H (1838) On the geometrical forms of turbinated and discoid shells. Philos Trans Roy Soc London 1838:351–370

Müller JHT (1850) Beiträge zur Conchyliometrie. Ann Phys Chem 81:533–544

Müller JHT (1853) Zweiter Beitrag zur Conchyliometrie. Ann Phys Chem 90:323–327

Nassichuk WW, Furnish WM, Glenister BF (1965) The Permian ammonoids from Arctic Canada. Bull Geol Surv Canada 131:1–56

Naumann CF (1840a) Beitrag zur Conchyliometrie. Ann Phys Chem 50(2):223–236

Naumann CF (1840b) Ueber die Spiralen der Ammoniten. Ann Phys Chem 51(2):245–259

Naumann CF (1845) Ueber die wahre Spirale der Ammoniten. Ann Phys Chem 64(4):538–543

Naumann CF (1849) Über die logarithmische Spirale von *Nautilus Pompilius* und *Ammonites galeatus*. Ber Verh Kgl Sächs Ges Wiss Leipz 2:26–34

Navarro N, Neige P, Marchand D (2005) Faunal invasions as a source of morphological constraints and innovations? The diversification of the early Cardioceratidae (Ammonoidea; Middle Jurassic). Paleobiology 31:98–116

Neige P, Elmi S, Rulleau L (2001) Crisis – what crisis? Quantification of the morphological disparity of ammonites at the Lias-Dogger boundary. Bull Soc Géol France 172:257–264

Neige P, Rouget I, Moyne S (2007) Phylogenetic practices among scholars of fossil cephalopods, with special reference to cladistics. In: Landman NH, Davis RA, Manger WM, Mapes RH (eds) Cephalopods – present and past. Springer, New York, pp 3–14

Nikolaeva SV, Barskov IS (1994) Morphogenetic trends in the evolution of Carboniferous ammonoids). N Jb Geol Paläontol Abh 193:401–418

Nikolaeva SV, Bogoslovsky BI (2005) Devonskie ammonoidei: IV. Klimenii (podotryad Clymeniina). Trudy Paleontol Instituta Ross Akad Nauk 287:1–214

Pareyn C (1961) Les Massifs Carbonifères du Sahara Sud-Oranais. Tome II. Paléontologie stratigraphique. Publ Centre Rech Sahar, Sér Géol 1:1–244

Petter G (1959) Goniatites Dévoniennes du Sahara. Publ Serv Carte Géol de Alg, N S. Paléontol 2:1–313

Petter G (1960) Clymènies du Sahara. Publ Serv Carte Géol de Alg, N S, Paléontol 6:1–58

Pie MR, Weitz JS (2005) A Null Model of Morphospace Occupation. Am Nat 166(1):E1–E13

Popov AV (1968) Viseiskie ammonoidei severnogo Tian′-Shania i ikh stratigraficheskoe znachenie. Akademia Nauk Kirgizogo SSR, Trudy Institut geologii. 1–116

Raup DM (1966) Geometric analysis of shell coiling: general problems. J Paleontol 40:1178–1190

Raup DM (1967) Geometric analysis of shell coiling: coiling in ammonoids. J Paleontol 41:43–65

Raup DM, Michelson A (1965) Theoretical morphology of the coiled shell. Science 147:1294–1295

Reitner J, Urlichs M (1983) Echte Weichteilbelemniten aus dem Untertoarcium (Posidonienschiefer) Südwestdeutschlands. N Jb Geol Paläontol Abh 165:450–465

Reyment RA, Kennedy WJ (1991) Phenotypic plasticity in a Cretaceous ammonite analyzed by multivariate statistical methods – a methodological study. Evol Biol 25:411–426

Riley NJ (1996) Mid-Dinantian ammonoids from the Craven Basin, northwest England. Spec Pap Palaeontol 53:1–87

Rohlf FJ (2004) tpsDig. http://life.bio.sunysb.edu/morph/index.html. Accessed 21 Nov 2007

Rohlf FJ, Marcus LF (1993) A revolution in morphometrics. Trends Ecol Evol 8:129–132

Ruan Y (1981a) Devonian and earliest Carboniferous Ammonoids from Guangxi and Guizhou. Mem Nanj Inst Geol Paleontol Acad Sin 15:1–152

Ruan Y (1981b) Carboniferous Ammonoid faunas from Qixu in Nandan of Guangxi. Mem Nanj Inst Geol Paleontol Acad Sin 15:153–232

Ruzhencev VE (1950) Verkhnekamennougol'nye ammonity Urala. Trudy Paleontol Inst Akad Nauk SSSR 29:1–220

Ruzhencev VE (1951) Nizhnepermskie ammonity yuzhnogo Urala. 1. Ammonity Sakmarskogo Yarusa. Trudy Paleontol Inst Akad Nauk SSSR 33:1–186

Ruzhencev VE (1952) Biostratigrafiya Sakmarskogo Yarusa v Aktyubinskoy Oblasti Kazakhskoy SSR. Trudy Paleontol Inst Akad Nauk SSSR 42:1–87

Ruzhencev VE (1956) Nizhnepermskiye ammonity Yuzhnogo Urala. II. Ammonity Artinskogo yarusa. Trudy Paleontol Inst Akad Nauk SSSR 60:1–275

Ruzhencev VE (1960) Printsipy sistematiki, sistema i filogeniya paleozoiskikh ammonoidei. Trudy Paleontol Inst Akad Nauk SSSR 83:1–331

Ruzhencev VE, Bogoslovskaya MF (1971) Namyurskiy etap v evolyutsii ammonoidey. Rannenamyurskie ammonoidei. Trudy Paleontol Inst Akad Nauk SSSR 133:1–382

Ruzhencev VE, Bogoslovskaya MF (1978) Namyurskiy etap v evolyutsii ammonoidey. Pozdnenamyurskie ammonoidei. Trudy Paleontol Inst Akad Nauk SSSR 167:1–336

Sandberger G (1855) *Clymenia subnautilina* (nova species), die erste und bis jetzt einzige Art aus Nassau. Jb Ver Naturkde Herz Nassau 10:127–136

Sandberger G, Sandberger F (1850–1856) Die Versteinerungen des rheinischen Schichtensystems in Nassau. Mit einer kurzgefassten Geognosie dieses Gebietes und mit steter Berücksichtigung analoger Schichten anderer Länder. Wiesbaden I–XIV:1–564

Saunders WB (1995) The ammonoid suture problem – relationships between shell and septum thickness and suture complexity in Paleozoic ammonoids. Paleobiology 21(3):343–355

Saunders WB, Swan ARH (1984) Morphology and morphologic diversity of mid-Carboniferous (Namurian) ammonoids in time and space. Paleobiology 10:195–228

Saunders WB, Work DM (1996) Shell morphology and suture complexity in Upper Carboniferous ammonoids. Paleobiology 22:189–218

Saunders WB, Work DM (1997) Evolution of shell morphology and suture complexity in Paleozoic prolecanitids, the root-stock of Mesozoic ammonoids. Paleobiology 23:301–325

Saunders WB, Work DM, Nikolaeva SV (1999) Evolution of complexity in ammonoid sutures. Science 286(5440):760–763

Saunders WB, Work DM, Nikolaeva SV (2004) The evolutionary history of shell geometry in Paleozoic ammonoids. Paleobiology 30:19–43

Saunders WB, Greenfest-Allen E, Work DM, Nikolaeva SV (2008) Morphologic and taxonomic history of Paleozoic ammonoids in time and morphospace. Paleobiology 34:128–154

Schindewolf OH (1929) Vergleichende Studien zur Phylogenie, Morphologie und Terminologie der Ammoneen-Lobenlinie. Abh Preuss geol L.-Anst N F 115:1–102

Schindewolf OH (1933) Vergleichende Morphologie und Phylogenie der Anfangskammern tetrabranchiater Cephalopoden. Eine Studie über Herkunft, Stammesentwicklung und System der niederen Ammoneen. Abh Preuss geol L.-Anst N F 148:1–115

Schindewolf OH (1937) Zur Stratigraphie und Paläontologie der Wocklumer Schichten (Oberdevon). Abh Preuss geol L.-Anst N F 178:1–132

Schindewolf OH (1950) Grundfragen der Paläontologie. Geologische Zeitmessung. Organische Stammesentwicklung. Biologische Systematik. Schweizerbart, Stuttgart, 506pp

Schindewolf OH (1951) Über ein neues Vorkommen unterkarbonischer Pericyclus-Schichten im Oberharz. N Jb Geol Paläontol Abh 93:23–116

Schmidt H (1924) Zwei Cephalopodenfaunen an der Devon-Carbongrenze im Sauerland. Jb Preuss geol L-Anst 44(for 1923):98–171

Scotese CR (1997) Paleogeographic Atlas, PALEOMAP Progress Report 90-0497. Department of Geology, University of Texas, Arlington, Texas, 45pp

Seilacher A (1989) Why are nautiloid and ammonoid sutures so different? N Jb Geol Paläontol Abh 177:41–69

Seilacher A, Gunji YP (1993) Morphogenetic countdown: another view on heteromorph shells in gastropods and ammonites. N Jb Geol Paläontol Abh 190:73–101

Shigeno S, Sasaki T, Moritaki T, Kasugai T, Vecchione M, Agatal K (2008) Evolution of the cephalopod head complex by assembly of multiple molluscan body parts: evidence from *Nautilus* embryonic development. J Morphol 269: 1–17

Signor PW III, Lipps JH (1982) Sampling bias, gradual extinction patterns, and catastrophes in the fossil record. In: Silver LT, Schultze PH (eds) Geological implications of impacts of large asteroids and comets on the Earth. Geological Society of American Special Publications, vol 190. Boulder, CO, pp 291–296

Smith JP (1927) Permian ammonoids of Timor. Jb Mijnw Nederl-Indie 55(for 1926):1–58

Sobolewski D (1914) Nabroski po filogenii goniatitov [Synopsis of the goniatite phylogeny]. Izvestia Varshavskogo politekhnicheskogo instituta, 1–191. Warsaw

Ubukata T, Tanabe K, Shigeta Y, Maeda H, Mapes RH (2008) Piggyback whorls: a new theoretical morphologic model reveals constructional linkages among morphological characters in ammonoids. Acta Palaeontol Pol 53:113–128

Valentine JW (1980) Determinants of diversity in higher taxonomic categories. Paleobiology 6:444–450

Villier L, Korn D (2004) Morphological disparity of ammonoids and the mark of Permian mass extinctions. Science 306(5694):264–266.

Vöhringer E (1960) Die Goniatiten der unterkarbonischen *Gattendorfia*-Stufe im Hönnetal (Sauerland). Fortschr Geol Rheinld Westf 3:107–196

Walliser OH (1985) Natural boundaries and commission boundaries in the Devonian. Cour Forschungsinst Senckenb 75:401–408

Ward PD (1987) The natural history of *Nautilus*. Allen & Unwin, Boston, MA

Wedekind R (1913) Die Goniatitenkalke des unteren Oberdevon von Martenberg bei Adorf. Sb Gest Naturforsch Fr Berlin 1913:23–77

Wedekind R (1914) Monographie der Clymenien des Rheinischen Gebirges. Abh kgl Ges Wiss Gött, Math-Phys Kl, N F 10(1):1–73

Wedekind R (1918) Die Genera der Palaeoammonoidea Goniatiten). Mit Ausschluß der Mimoceratidae, Glyphioceratidae und Prolecanitidae. Paläontogr 62: 85–184

Westermann GEG (1996) Ammonoid life and habitat. In: Landman NH, Tanabe K, Davis RA (eds) Ammonoid paleobiology. Topics in Paleobiology, vol 13. Plenum, New York, pp 607–707

Wills MA (2001) Morphological disparity: a primer. In: Adrain JM, Edgecombe GD, Lieberman BS (eds) Fossils, phylogeny, and form – an analytical approach. Kluwer Academic/Plenum Press, New York, pp 55–144

Yacobucci MM (1999) Plasticity of developmental timing as the underlying cause of high speciation rates in ammonoids – an example from the cenomanian Western Interior seaway of north America. In: Oloriz F, Rodriguez-Tovar FJ (eds) Advancing research on living and fossil cephalopods. Kluwer Academic/Plenum, New York, pp 59–76

Zahn CT, Roskies RZ (1972) Fourier descriptors for plane closed curves. IEEE Trans Computers 21:269–281

Zhao J, Liang X, Zheng Z (1978) Late Permian cephalopods from South China. Palaeontol Sin Ser B 12:1–194

Quantitative Approach to Diversity and Decline in Late Palaeozoic Trilobites

Rudy Lerosey-Aubril and Raimund Feist

Abstract

Quantitative data reveal complex evolution in late trilobite diversity. In the Middle to early Late Devonian, a series of extinction events led to dramatic taxonomic impoverishment of the trilobites. In the Famennian, when only two orders remained, originations began to compensate for the still high extinction rates, marking the start of a remarkable diversification. Although interrupted by the major Hangenberg turnover, the general diversification trend accelerated in the Tournaisian, whereas extinctions became modest. Originations diminished markedly during the Viséan and Serphukovian, causing this diversity to decrease to the level observed in the Frasnian. It was never much higher thereafter, despite massive restructuring of the trilobite communities in the early Pennsylvannian; this allowed for progressive domination of the ditomopygines. After another decline in the Kasimovian, a period of stasis occurred with very low diversity levels and almost no renewal. The last burst of diversification occurred in the Wordian, but ceased rapidly in the Capitanian when degradation of environmental conditions began to inhibit originations. Thus, extinction of the Trilobita at the end of the Permian resulted from disappearance of merely a handful of genera.

Keywords

Trilobites • Devonian-Permian extinctions • Quantitative analysis • Early Devonian diversity peak • Mid-Devonian–Frasnian decrease • Famennian–Tournaisian diversity acceleration • Viséan–Serpukhovian marked diversity decrease • Last diversification Wordian • Decreasing diversity Capitanian to end-Permian total extinction

Introduction

The extraordinary evolutionary success of the arthropods is incontestable. Present in virtually all habitats, frequently in great abundance, arthropods represent by far the most-diversified group of metazoans on Earth. Their abundance, taxonomic diversity, and omnipresence all demonstrate that these organisms

R. Lerosey-Aubril (✉)
Senckenberg Forschungsinstitut und Naturmuseum Frankfurt, 60325 Frankfurt am Main, Germany
e-mail: rudy.leroseyaubril@senckenberg.de

J.A. Talent (ed.), *Earth and Life*, International Year of Planet Earth,
DOI 10.1007/978-90-481-3428-1_16, © Springer Science+Business Media B.V. 2012

have a unique capacity for adaptation. However, about 250 million years ago, a particularly diversified and emblematic group of Palaeozoic arthropods became extinct, the trilobites.

In contrast to most representatives of the phylum that appeared during the Cambrian explosion, trilobites diversified rapidly to become a dominant element in marine benthic communities. Throughout the Early to early Middle Palaeozoic, their diversity fluctuated with an apogee recorded in the Middle Cambrian and minimum values during the Silurian (Adrain 2008). Another peak of trilobite diversity occurred during the Early Devonian with values similar to those of the Late Cambrian. Probably because the diversity and abundance of these organisms decreased dramatically thereafter, much less attention has been accorded to their late evolutionary history, although this history offers a unique opportunity to study an exceptional event in the history of life: the complete extermination of a major group of arthropods. It has long been known that a series of biological events during the Middle to Late Devonian caused a dramatic reduction in trilobite diversity at high taxonomic levels so that the class came to be represented by the single order, Proetida, throughout the Late Palaeozoic. However, to date, neither the various steps that led to the almost inevitable extinction of the Trilobita during the Permo-Triassic mass extinction nor the impacts of the major climatic, eustatic, and tectonic events that characterized the late Middle to Late Palaeozoic have been described in detail.

In this chapter, we take advantage of the recent revision of the stratigraphic framework of the Devonian (Bultynck 2000), Carboniferous (Heckel and Clayton 2006), and Permian (Owens 2003) to develop a precise quantitative description of the evolution of trilobite diversity from the Middle Devonian to the terminal Permian, based on a recently compiled taxonomic database. Because we consider the extinction of the trilobites to be a historical event that must be restored to its historical context, we then attempt to interpret the observed diversity patterns with special reference to global biotic, climatic, eustatic, and tectonic events.

Methodology

To allow a quantitative approach to the evolution of trilobite diversity from the Middle Devonian to the terminal Permian, a taxonomic database has been created that re-groups all the genera and subgenera known from this period. The data for the Devonian were initially extracted from earlier works (essentially Briggs et al. 1988; Feist 1991; and Chlupáč 1994) and then updated and completed with more recent data. Additions to the database have allowed us to consider almost twice the number of taxa considered by Chlupáč (1994). The database for the Carboniferous and Permian is entirely new and was compiled from primary sources. Throughout the compilation, no distinction was made between genera and subgenera; all are cited as 'genera' below. Occurrences for which the assignment and/or the age of the taxa were questionable were not used to describe the diversity patterns presented here, although curves representing the evolution of raw generic diversity seem to be comparable whether or not these occurrences are considered. The time intervals considered are the stages throughout the study. All in all, our database re-groups 425 taxa in total: Mid–Late Devonian 226; Carboniferous 177; Permian 35), this stratigraphic distribution has been reviewed and updated with reference to the most recent stratigraphic framework proposed by the International Commission of Stratigraphy (Ogg et al. 2008).

To propose an accurate description and analysis of the diversity patterns of the Middle Devonian to Permian trilobites, we have used various metrics and evaluated the concordance of the results achieved. Four series of measures were used: raw taxonomic richness, standing diversity, normalized diversity (i.e. raw diversity divided by the duration of time interval), and origination/extinction/turnover rates. For standing diversity (SDv1, SDv2), origination rate (OR1, OR2, SOR), extinction rate (ER1, ER2, SER), and turnover rate (TuR1, TuR2, STR), the metrics defined by Harper (1996; see his Table 1 for definitions) have been used.

Description

Diversity

In the later part of the Palaeozoic, trilobite diversity reached its maximum in the Early Devonian when trilobites were nearly as globally diverse as during the Late Cambrian (Adrain 2008). Following this period of abundance, culminating in the terminal Emsian, the Middle Devonian begins with still high raw ordinal and familial diversities (Fig. 1a). However, thereafter

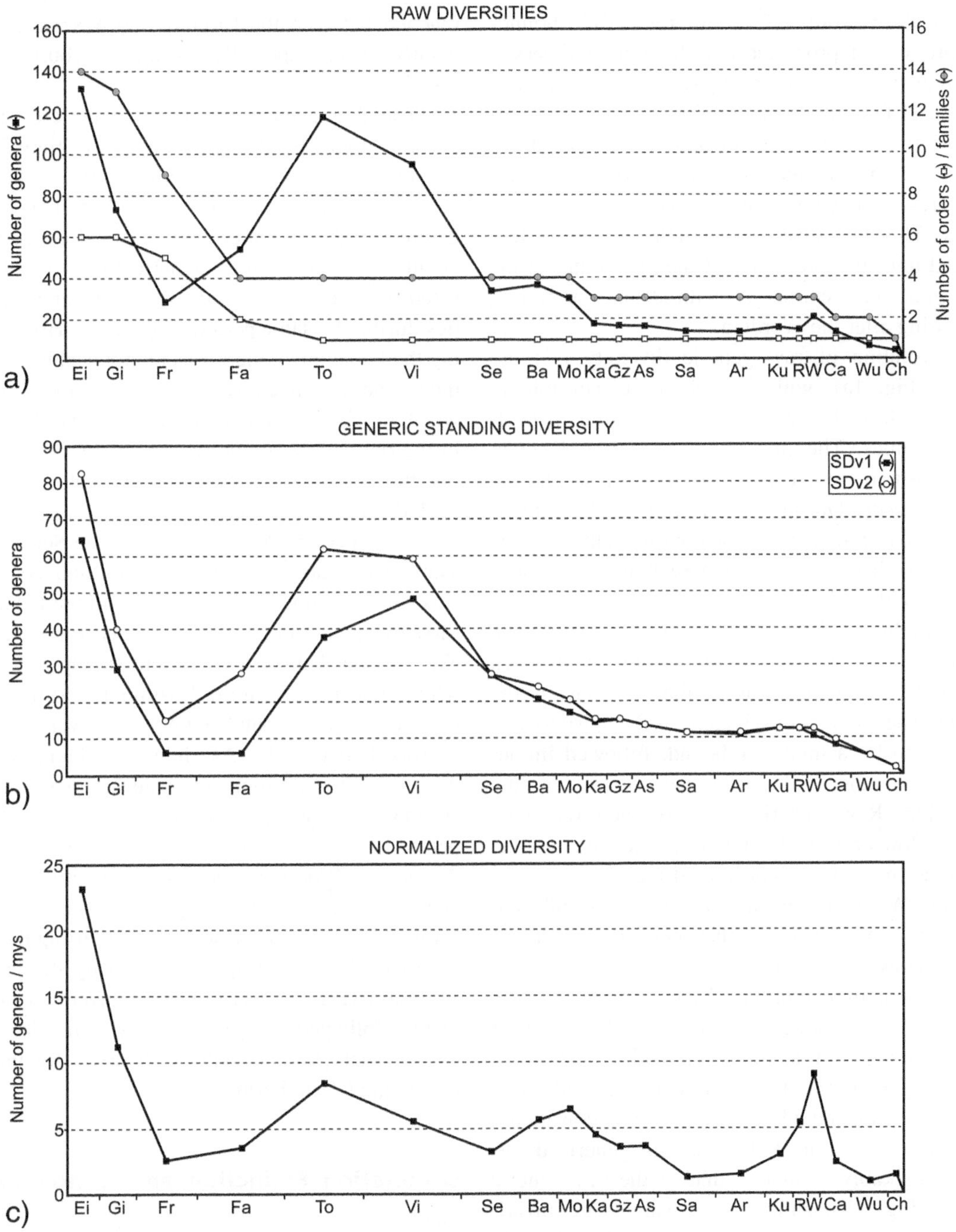

Fig. 1 Evolution of trilobite diversity from the Eifelian (Middle Devonian) to the Changhsingian (Lopingian). (**a**) Ordinal, familial, and generic raw diversities. (**b**) Generic standing diversity. SDv1 and SDv2 are the two different metrics defined by Harper (1996). (**c**) Raw generic diversity normalized to the duration of the time interval in millions of years. Except for the Roadian (R) and the Wordian (W), the names of the stratigraphic stages are abbreviated to their first two letters: Eifelian (Ei), Givetian (Gi), Frasnian (Fr), Famennian (Fa), Tournaisian (To), Viséan (Vi), Serpukhovian (Se), Bashkirian (Ba), Moscovian (Mo), Kasimovian (Ka), Gzhelian (Gz), Asselian (As), Sakmarian (Sa), Artinskian (Ar), Kungurian (Ku), Capitanian (Ca), Wuchiapingian (Wu), Changhsingian (Ch)

and up to the Devonian–Carboniferous (D/C) boundary, there was a pronounced reduction in diversity of the higher taxonomic categories when extinctions (5 orders, 11 families, 30 subfamilies) greatly outnumbered originations (0 orders, 1 family, 9 subfamilies). Thus, at the onset of the Upper Palaeozoic, trilobites were represented by only one order (Proetida) and four families. This low diversity at high taxonomic levels lasted until the extinction of the group during the end-Permian mass extinction, with all but one family disappearing after the Wordian.

The evolution of raw generic diversity is more complex (Fig. 1a), with two phases discernible in the Middle through Late Devonian and four in post-Devonian times. The first phase, from the Eifelian to the Frasnian, was characterized by a substantial loss of diversity. During the Famennian, their diversity increased again, a trend that continued through the Tournaisian when values reach more than 85% of those observed in the Eifelian. This period was followed by an initially moderate decline during the Viséan, becoming more severe during the Serpukhovian when the number of genera was similar to that in the Frasnian. The early Pennsylvannian (Bashkirian) was characterized by a modest rebound, followed immediately by another fall during the Moscovian and Kasimovian. Raw generic diversity then remained invariably low until the beginning of the Guadalupian ('middle Permian'). A final limited increase occurred during the Wordian, but once again, it was followed immediately by a decline that persisted until the Permian–Triassic boundary.

The pattern of evolution of the standing generic diversity of the trilobites from the Middle Devonian to the end of the Permian differs markedly when stratigraphic singletons (i.e. taxa confined to a single time interval) are considered. When they are considered (SDv2), the evolution of the standing generic diversity is remarkably similar to that of the raw generic diversity (Fig. 1b), although the values are invariably lower. The main differences include relatively higher levels in the Viséan and Serpukhovian and a certain flattening of the curve thereafter; it lacks the clear diversity rebounds that occurred in the Bashkirian and Wordian. When the stratigraphic singletons are ignored (SDv1), the standing and raw diversities differ to a greater extent (Fig. 1b). The post-Viséan part of the curve is even more flattened; the values reached between the Eifelian and Viséan are significantly lower, especially during the Tournaisian and Famennian.

Normalizing measures of raw generic diversity by dividing them by the duration of the time intervals (in millions of years) markedly modifies the pattern of evolution of generic diversity (Fig. 1c). From the Eifelian to the Frasnian there was a sharp continuous decline in the normalized generic diversity; this is mirrored in the evolution of the raw and standing diversities during the same period (Fig. 1a, b). However, during the Famennian, there was no noticeable increase in the normalized diversity or even during the Tournaisian when its rise remained modest compared with rises in the raw and standing diversities. Its peak was only one-third the value of the Eifelian peak. The normalized diversity then decreased evenly to a minimum of 3.14 genera per million years in the Serpukhovian. The Bashkirian rebound in the normalized diversity was more pronounced; it even continued during the Moscovian; the decline commenced only during the Kasimovian. Unlike the raw and standing diversities, which plateaued from the Kasimovian to the Roadian (Fig. 1a, b), the normalized generic diversity dropped sharply between the Asselian and Sakmarian. This decline led to extremely low values; this situation persisted until the Kungurian and the beginning of the last period of increasing diversity. This period culminated during the Wordian when the diversity per million years was the highest in all post-Devonian times, confirming that diversification did occur during the Middle Permian; this is perhaps suggested by the pattern of evolution (Fig. 1a). However, a severe drop immediately followed this Wordian diversification lead to an extremely low normalized diversity for the whole Lopingian ('Late Permian').

Origination, Extinction, and Turnover Rates

Once again, considering (OR1, ER1, TuR1) or ignoring (OR2, ER2, TuR2, SOR, SER, STR) stratigraphic singletons substantially modifies the evolutionary patterns of origination, extinction, and turnover rates. When singletons are considered, the origination rate (OR1; Fig. 2a) had a moderately low value (0.4) in the Middle Devonian, increasing strongly throughout the Late Devonian to reach 0.96 during the Famennian.

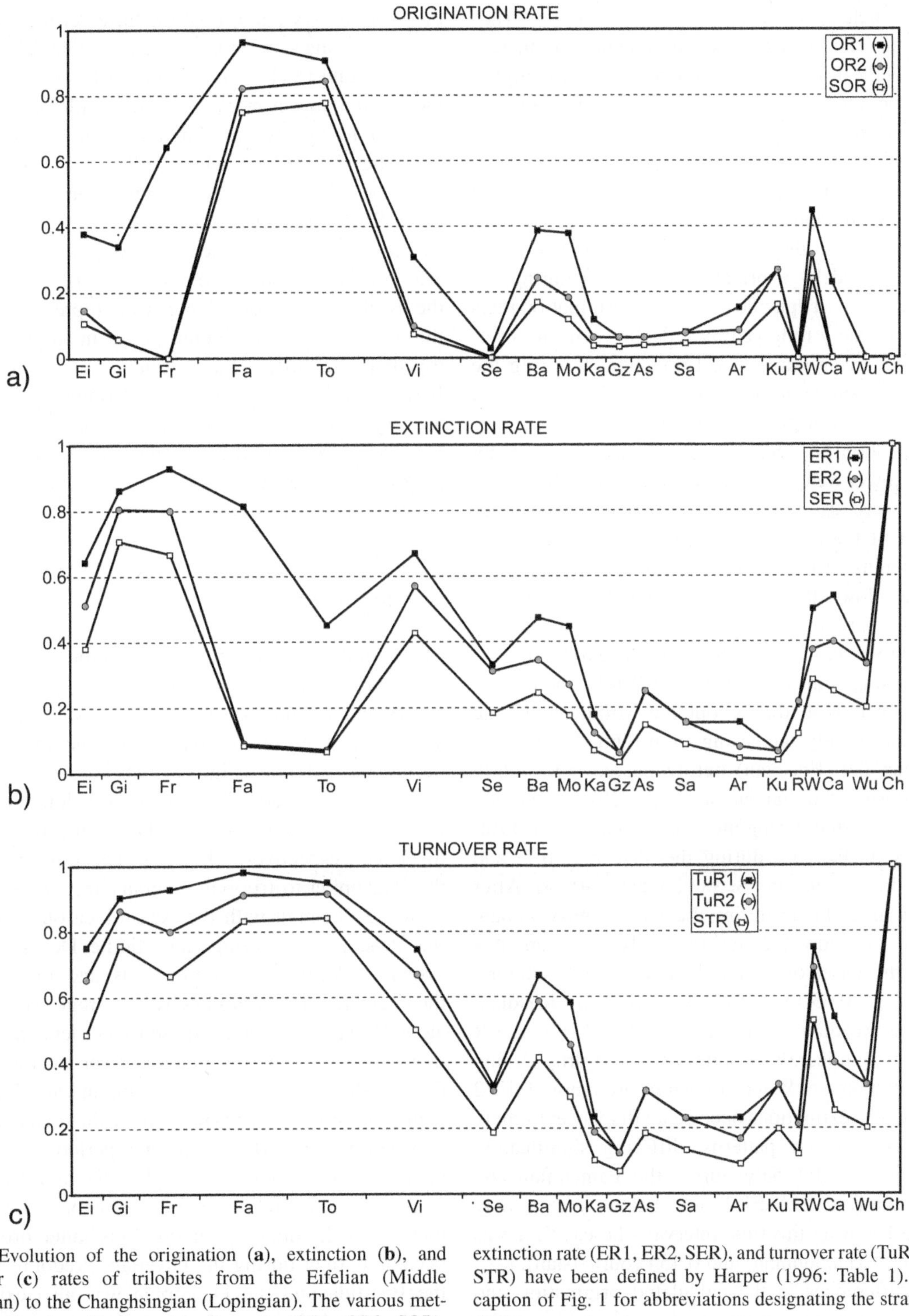

Fig. 2 Evolution of the origination (**a**), extinction (**b**), and turnover (**c**) rates of trilobites from the Eifelian (Middle Devonian) to the Changhsingian (Lopingian). The various metrics used to describe the origination rate (OR1, OR2, SOR), extinction rate (ER1, ER2, SER), and turnover rate (TuR1, TuR2, STR) have been defined by Harper (1996: Table 1). See the caption of Fig. 1 for abbreviations designating the stratigraphic stages

Roughly unchanged during the Tournaisian, the origination rate fell drastically thereafter, approaching zero by the end of the Mississippian. Markedly higher origination rates (0.4) occurred during the Bashkirian–Moscovian interval, but only temporarily because a strong fall occurred during the Kasimovian marking the onset of a period of low OR1 (~0.1) that lasted until the Sakmarian. An increase then occurred from the Artinskian to the Wordian (OR1 = 0.45), but it was interrupted by a brutal drop to zero during the Roadian. This last episode of elevated origination rates was immediately followed by a decline with no origination occurring during the Lopingian. Ignoring the stratigraphic singletons leads to very similar patterns of evolution of the origination rate (OR2) and the standing origination rate (SOR), the main difference being slightly lower values for SOR (Fig. 2a). These two patterns were also similar to that observed with OR1 for most of the post-Devonian period, despite a certain flattening of the curves. However, OR2 and SOR were much lower than OR1 during the Middle and Late Devonian, particularly during the Frasnian when they declined to zero, whereas OR1 was close to 0.65.

When singletons are considered (ER1), the extinction rate was high throughout the Middle and Late Devonian, initially increasing from 0.65 (Eifelian) to more than 0.9 (Frasnian) before decreasing slightly to 0.8 during the Famennian (Fig. 2b). A marked decrease occurred during the Tournaisian (~0.45) followed by an increase during the Viséan (~0.7) and another drop during the Serpukhovian (~0.35). After a plateau around 0.45 during the early Pennsylvanian, ER1 fell to below 0.2 during the Kasimovian; this marked the onset of a period of low but fluctuating values (from 0.05 to 0.25) lasting until the Roadian. From the Roadian to the end of the Permian, ER increased significantly despite a marked drop during the Wuchiapingian. When singletons are ignored (ER2 and SER), the extinctions once again display extremely similar evolutionary patterns, differing significantly from those of ER1 only during the Famennian and Tournaisian (Fig. 2b). ER2 and SER were extremely low (~0.1) during this time interval whereas ER1 was between 0.8 (Famennian) and 0.45 (Tournaisian).

The differences in the calculated origination and extinction rates that occur when stratigraphic singletons are or are not considered are somewhat attenuated when the turnover rates are calculated. Indeed, whether singletons are considered (TuR1) or not (TuR2, STR), the evolutionary patterns of the turnover rates are fundamentally the same (Fig. 2c), except that the standing turnover rate (STR) has constantly lower values than the simple turnover rates (TuR1, TuR2). From the Eifelian to the Tournaisian the turnover rates were all rather high, even when TuR2 and STR showed a period of lower values centred on the Frasnian when TuR1 remained above 0.9. The turnover rates decreased during the Viséan and Serpukhovian, reaching low values during the latter stage. In contrast, the Bashkirian–Moscovian period represents one of the three post-Mississippian peaks in turnover rates, together with those in the Wordian and Changhsingian. It was followed by an interval of rather low (~0.35) to very low (~0.1) turnover rates lasting until the Roadian. Similarly, the Wordian and Changhsingian peaks in turnover rates flank a period of lower but still significant turnover rates.

Composition

The evolution of trilobite composition from Mid to Late Palaeozoic times is shown in Fig. 3. With 6 orders and 14 families, the trilobites still represented a diversified group of marine benthic invertebrates during the Eifelian (Fig. 1a). The orders Phacopida and Proetida already dominated the trilobite communities, representing 34 and 44% of the raw generic diversity, respectively (Fig. 3). The contribution of the Phacopida to trilobite diversity increased slightly until the Frasnian with the Acastoidea outnumbering the Phacopoidea during the Middle Devonian. The Proetida, dominated by proetoids and in particular the dechenellines until the Givetian, regressed moderately. However, all in all, these two orders constituted more than 75% of the generic diversity throughout the Middle and early Late Devonian. Similarly, the contribution of the Corynexochida fluctuated slightly between 5.5 and 10% during this period. In contrast, contributions of the Odontopleurida and Harpetida increased moderately during the same period, whereas that of the Lichida decreased. This latter order also became extinct during the Givetian. Predominance of the Proetida and the Phacopida was also expressed at the familial level (Fig. 4). Whereas the other orders were represented by a single family from the Eifelian to the Frasnian, four proetid families and two to six phacopid families occurred during this period

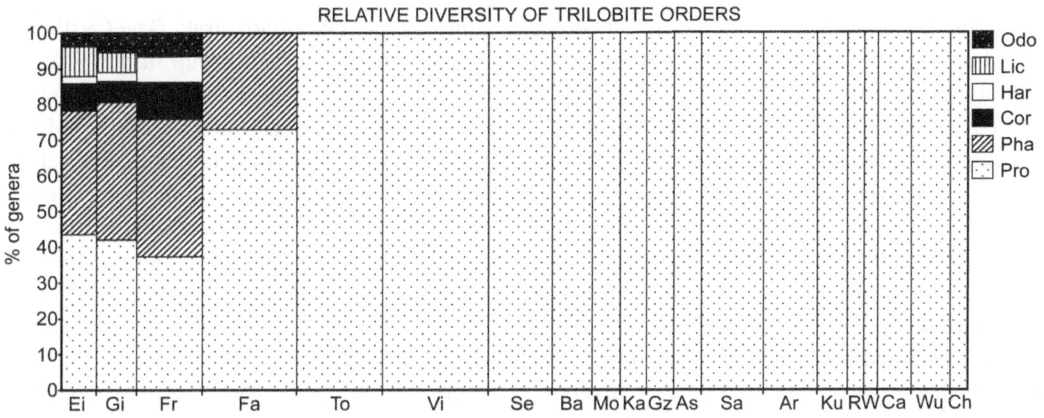

Fig. 3 Relative diversities of the trilobite orders from the Eifelian (Middle Devonian) to the Changhsingian (Lopingian). Abbreviations of orders: Corynexochida (Cor), Harpetida (Har), Lichida (Lic), Odontopleurida (Odo), Phacopida (Pha), Proetida (Pro). See the caption of Fig. 1 for the abbreviations designating the stratigraphic stages

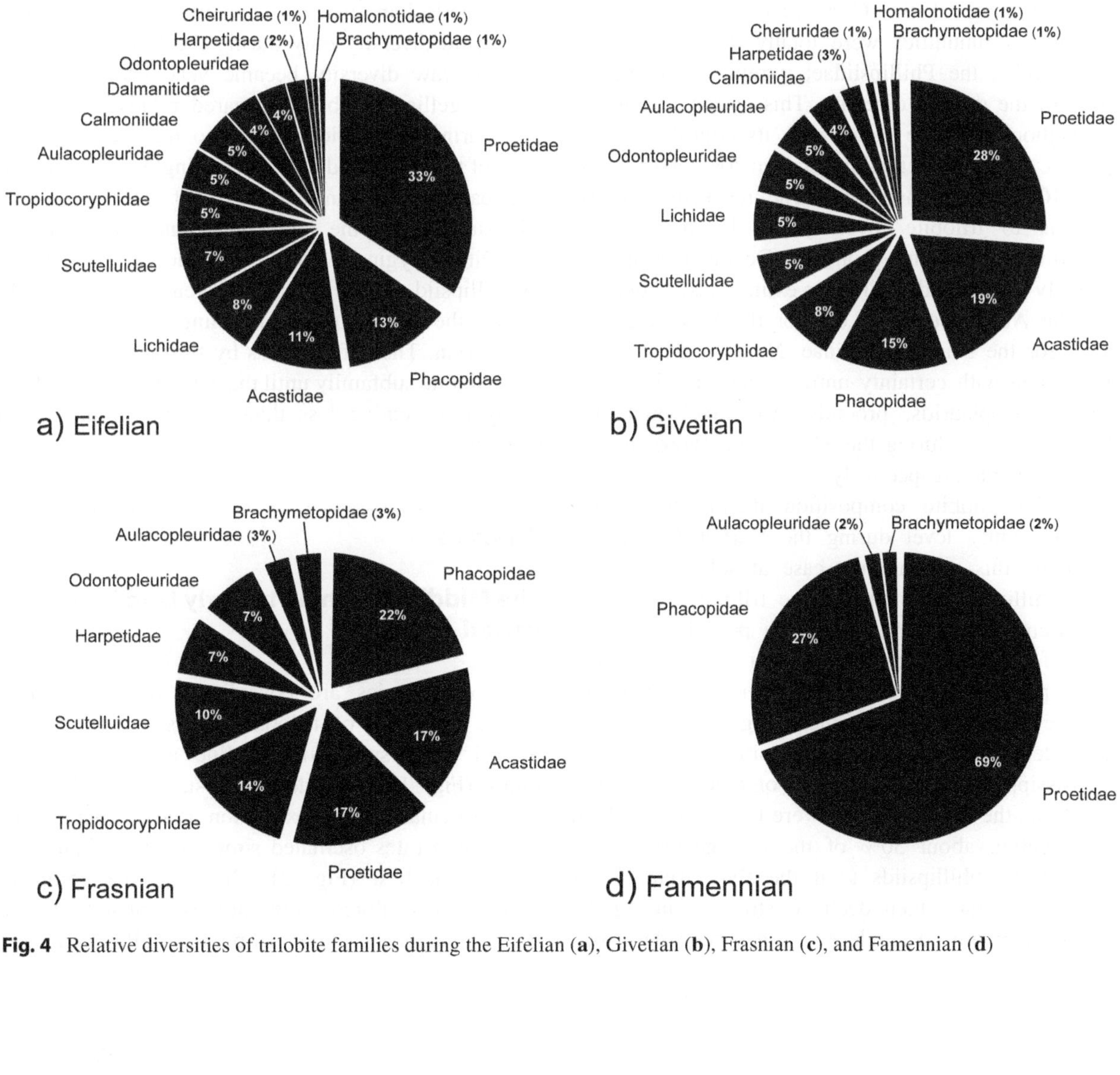

Fig. 4 Relative diversities of trilobite families during the Eifelian (**a**), Givetian (**b**), Frasnian (**c**), and Famennian (**d**)

(Fig. 4a–c). In this regard, the Frasnian is noted for phacopids because it was associated with a significant fall in raw familial diversity. The Phacopidae and Proetidae were the only families that invariably displayed high generic diversity until the Frasnian (Fig. 4a–c).

The Famennian was characterized by a severe decline in trilobite diversity at high taxonomic levels; only two orders and four families were present during this period (Figs. 1a, 4d). End-Frasnian extinctions greatly benefited the Proetida; they then contributed more than 70% of the total generic diversity (Figs. 3, 4d). The Proetidae diversified strongly to constitute nearly 85% of the raw generic diversity during the Eifelian.

The Tournaisian was another period when the trilobite composition was profoundly modified. Not only was the Proetida the sole order present at the beginning of the Carboniferous (Fig. 3), but trilobite communities were totally dominated by a new family, the Phillipsiidae; it represented almost 90% of the generic diversity. This domination lasted throughout the Late Palaeozoic, its contributions to generic diversity fluctuating between 70% (Artinskian) and 100% (Changhsingian). The respective contributions to trilobite diversity of the three families directly inherited from the Devonian remained roughly unchanged until their extinctions: about 2% for the Aulacopleuridae, 6% for the Proetidae, and 11% for the Brachymetopidae. None of these families lasted with certainty until the terminal Permian. The aulacopleurids, proetids, and brachymetopids became extinct during the Moscovian, Wordian, and Wuchiapingian, respectively.

If the trilobite composition fluctuated negligibly at family level during the Carboniferous and Permian, this was not the case at subfamily level. The Phillipsiidae dominated the trilobite communities tremendously throughout this period, so evolution of trilobite composition at that time can largely be explained by changes in the relative diversity of the phillipsiid subfamilies (Fig. 5). Phillipsiid trilobites were particularly diversified in the early Mississippian with a maximum of nine subfamilies. Of these, the Archegoniinae were the most abundant, representing about 30% of the raw generic diversity of the phillipsiids until the Bashkirian. Their relative diversity then declined strongly during the Moscovian; only a single genus has been described from younger (Wordian) strata. The Cystispininae also were a rather important group in the Tournaisian, but its relative diversity decreased progressively thereafter until its extinction at the end of the Mississippian. Weaniine trilobites were rather diverse at the beginning of the Carboniferous; they remained so until their extinction at the end of the Guadalupian, with a peak of relative diversity during the Artinskian. However, the Linguaphillipsiinae and Phillipsiinae were minor contributors to the Tournaisian diversity of the trilobites; both groups became extinct before the end of the Mississippian. The stratigraphic range of the Griffithidae extended into the Moscovian, but their relative diversity remained rather modest throughout this period, exceeding 10% only during the Serpukhovian. Lastly, the Ditomopyginae, Bollandiinae, and Cummingellinae were the only subfamilies to survive until the terminal Permian. Among them, the Bollandiinae was a significant contributor to phillipsiid diversity only during the Lopingian when trilobite raw diversity became very low. Likewise, cummingelline trilobites appeared relatively diversified during this period; they also represented about 20% of the phillipsiid diversity during the Kasimovian and early Cisuralian. However, of all the phillipsiid subfamilies, the most important was definitely the Ditomopyginae. The contribution of this group to phillipsiid generic diversity increased progressively during the Mississippian, reaching 25% during the Bashkirian. Thereafter it was by far the most diversified trilobite subfamily until the extinction of the class, rarely representing less than 50% of all phillipsiid diversity.

Analysis

The Middle Devonian to Early Late Devonian Period

Whatever metrics are considered, this first phase of the evolution of trilobite diversity was characterized by a substantial decline at all taxonomic ranks (Fig. 1). This decline resulted mainly from the particularly high extinction rates, whereas the origination rates oscillated simultaneously from 0.35 to 0.65 at best (Fig. 2). However, the compositions of the trilobite communities changed little at the ordinal level; the Phacopida and Proetida were

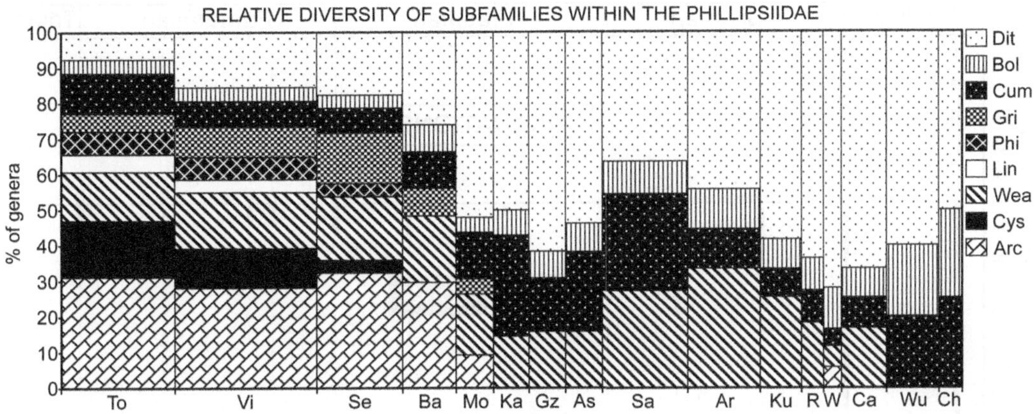

Fig. 5 Relative diversities of subfamilies within the trilobite family Phillipsiidae from the Tournaisian (Early Carboniferous) to the Changhsingian (Lopingian). Abbreviations of subfamilies: Archegoninae (Arc), Bollandiinae (Bol), Cummingellinae (Cum), Cystispininae (Cys), Ditomopyginae (Dit), Griffithidinae (Gri), Linguaphillipsiinae (Lin), Phillipsiinae (Phi), Weaniinae (Wea). See caption of Fig. 1 for abbreviations designating stratigraphic stages

enormously dominant throughout this time interval (Fig. 3).

Figure 1 shows the evolution of trilobite diversity as a continuous decline from the Eifelian to the Frasnian but, in fact, this evolution was stepwise with more or less drastic diversity changes alternating with intervals of stasis. Severe environmental perturbations repeatedly affected the marine environments during this period, generating severe restructuring within marine ecosystems on a global scale (Walliser 1996). Because most of these bioevents did not strictly coincide with stratigraphic stage boundaries and because several of these events occurred within a single stage, a correct assessment of the evolutionary dynamics of trilobite diversity during this first phase requires examination of shorter time intervals. Eight biotic events, with a mean extinction rate of 30 genera per event, can be recognized from the Eifelian to the end-Frasnian (Fig. 6). The terminal Emsian and basalmost Eifelian was a time of apogee for taxa such as the Asteropyginae, Calmoniidae, Odontopleuridae, Phacopinae, Synphoriinae, and Scutelluidae (Chlupáč 1994). The series of environmental perturbations that affected their habitats during the Middle and early Late Devonian reduced their diversity to a such extent that three of the five orders present in the Frasnian were represented by a single genus immediately before the Upper Kellwasser Event.

Both Eifelian events, the Choteč Event and the Kačák Event, might have resulted from global eustatic rises (Johnson et al. 1985) coupled to establishment of oxygen-depleted conditions in benthic habitats. As a consequence, the degree of diversification remained roughly stable compared with that in Early Devonian times, though it was rather low in groups such as the Homalonotidae, Cheiruridae, and Brachymetopidae. However, adaptive innovations occurred in the Proetidae, Acastidae, and Lichidae; the last included double the number of genera in the Eifelian compared to what it had at the end of the Early Devonian.

With the terminal mid-Givetian Taghanic Onlap, the mean sea-level reached the highest stands of all Mid to Late Palaeozoic times (Haq and Schutter 2008), leading to a worldwide expansion of the pelagic environment and to the demise of a great variety of near-shore neritic and peri-reefal habitats. Trilobites diversified particularly on the inner platform, especially in near-shore to coastal environments; these exhibited abrupt lateral variations in their physico-chemical conditions. The abolition of these habitats might largely explain the severe extinctions associated with the Taghanic Event; this was connected with a considerable reduction in generic and specific trilobite abundances (Fig. 6) and also to the disappearance of several emblematic families such as the Calmoniidae, Cheiruridae, and Lichidae, together with major proetoid (Eremiproetinae, Dechenellinae) and aulacopleurid (Aulacopleurinae, Cyphaspidinae) subfamilies. In North America, the Homalonotidae, Dechenellinae, and certain Asteropyginae (*Greenops*) also become extinct, but

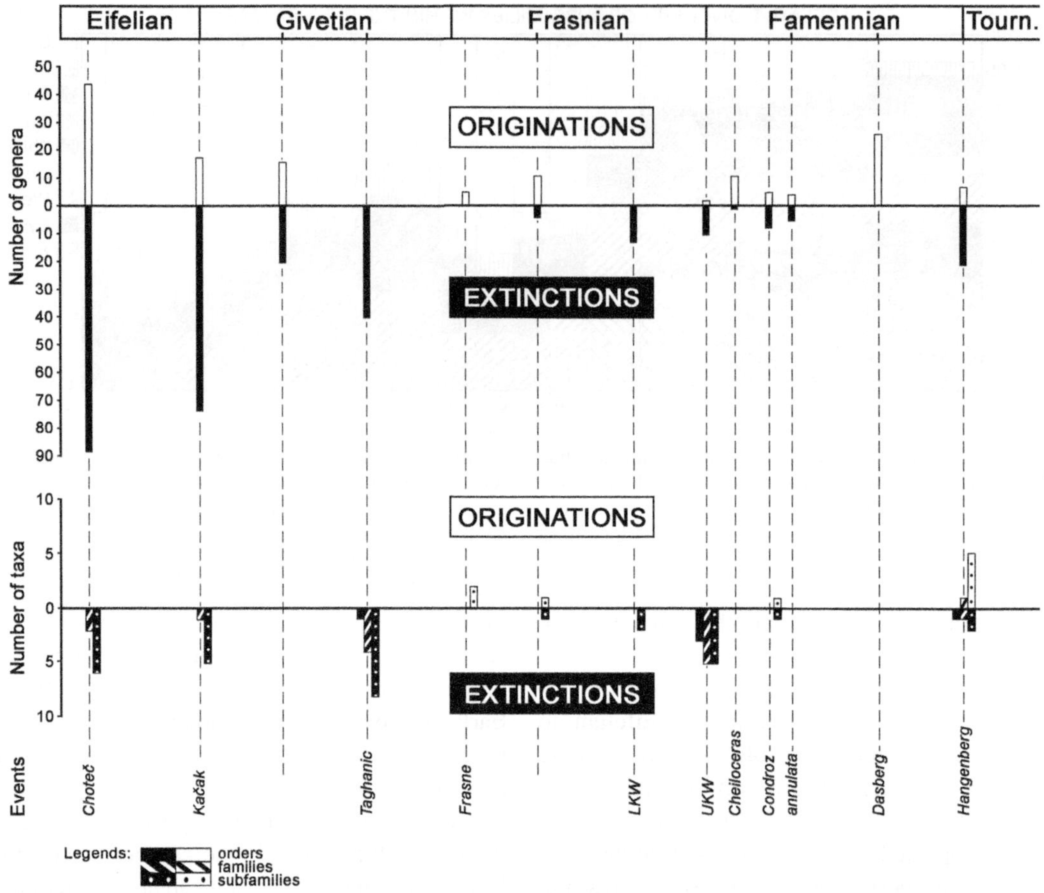

Fig. 6 Number of originations and extinctions of genera (*upper part*) and subfamilies, families, and orders (*lower part*) during the Middle and Late Devonian biotic events. Tournaisian is abbreviated to Tourn

probably a little later than the middle *varcus* Zone, although still within the Givetian. For unknown reasons, the Taghanic Event apparently affected trilobite communities, in particular those of the European/North African realm, earlier than other benthic biotas such as corals/stromatoporoids (within the Late *varcus* Zone) and brachiopods (between the Early and Late *hermanni–cristatus* zones) (Walliser 1996).

A few taxa survived this world-wide ocean deepening and successfully adapted to the deeper and more homogeneous environmental conditions in the stylioline–cephalopod realm of the late Givetian. In contrast to the pre-Taghanic Early and Mid Devonian trilobite communities – which were rather diverse, locally extremely abundant, and typically distributed in geographically restricted biofacies – the surviving faunas exhibited a pronounced reduction in generic and specific diversity concomitant with an increase in cosmopolitanism. Therefore, all the families/subfamilies that occurred in the stylioline–cephalopod realm were composed of single genera and radiated into the Late Devonian. In summary, with extremely low generic origination (zero) and extinction rates, the post-Taghanic Givetian times marked the onset of a period of stability with regard to trilobite diversity; this persisted during most of the Frasnian (Fig. 6).

At the beginning of the Late Devonian, the Frasne or *Manticoceras* Event was not marked by notable extinctions of trilobites (Fig. 6), whereas it had an important impact on organisms such as ammonoids (extinction of most pharciceratids, appearance of *Manticoceras*), brachiopods, corals, and stromatoporoids (Walliser 1996). Among the 27 known trilobite genera in the Frasnian, only 5 originated in the early Frasnian; 11 others appeared during the mid-Frasnian. Both generic and specific diversities remained relatively low during the 10 Ma or so of the global sea-level high stand that preceded the terminal Frasnian extinction events (Haq and Schutter 2008). This stability in trilobite diversity may have mirrored the homogenization of the environmental conditions on the expanding outer platforms, resulting from the loss of oceanic barriers (e.g. the remnants of the Rheïc ocean) caused by rapprochement of the epicontinental margins. In these outer shelf environments, the biotic evolutionary responses to extrinsic constraints generated a paedomorphic trend towards eye reduction in many lineages, as in the proetoid *Pteroparia* and *Palpebralia* (Feist 1991, 1995). Although no longer present on the African and American continents, large-eyed asteropygines experienced a period of pronounced specific diversification in the early and mid-Frasnian neritic brachiopod-rich organo-clastic environments extending from Europe to Southern Asia (Morzadec 1992). Interestingly, *Bradycryphaeus*, the only asteropygine to reach the base of the Upper Kellwasser Extinction level, lived within the off-shore cephalopod milieu.

The terminal Frasnian Kellwasser Extinction Event particularly affected trilobites adapted to life in deep epibenthic habitats. Of the five orders present during the Frasnian (Corynexochida, Harpetida, Odontopleurida, Phacopida, and Proetida), only the Phacopida and Proetida persisted into the Famennian. In European and North African localities, the disappearance of the trilobite communities coincided with a period of eustatic instability; it triggered a sudden spread of oxygen-depleted depositional conditions in the deeper outer-shelf areas. Such hypoxic deposits, the so-called Lower and Upper Kellwasser horizons, occurred at the end of conodont Zone 12 and during Zone 13c, respectively (Girard et al. 2005). Precise correlations of the last occurrences, which reached the base of the Upper Kellwasser extinction level, have been obtained for several sites in Europe (Feist and Schindler 1994) and North Africa (Feist 2002); they are uniformly fine-grained, mostly calcilutitic, cephalopod- and dacryoconarid-rich deposits representative of off-shore depositional environments at or beyond the limit of light penetration. By comparison, late Frasnian trilobites from near-shore deposits before and contemporaneous with the Kellwasser Extinction levels remain poorly known. At present, no data exist for the Americas or Asia, for example. In this context, the faunas from the Canning Basin (NW Australia) are essential because they occur in precisely dated sections where well-oxygenated conditions prevailed throughout a continuous Frasnian–Famennian transition (George and Chow 2002). These comprise carbonate ramp and basinal deposits with biogenic debris flows; these originated from immediately adjacent shallow reefal and peri-reefal areas. In contrast to the European and North African sections, the continuous Frasnian–Famennian transitional beds are devoid of any Kellwasser-type hypoxic or anoxic intercalations. The terminal Frasnian extinction event can nevertheless be recognized by a biotic loss contemporaneous with and of similar intensity to those observed elsewhere (McNamara and Feist 2008). The late Frasnian Kellwasser biocrises differed markedly from each other in their intensities, as shown by both the number of extinctions and the taxonomic ranks concerned (Fig. 6). In the Lower Kellwasser Event, 13 of the 22 genera present in the mid-Frasnian strata vanished, whereas all eight families crossed into the late Frasnian. Of these, five families and all the remaining ten genera became extinct at the base of the Upper Kellwasser Event. These included *Eskoharpes*, *Gondwanaspis*, and *Telopeltis*, the single last representatives of the orders Harpetida, Odontopleurida, and Corynexochida, respectively. This demonstrates that the Upper Kellwasser Event dramatically affected trilobite biodiversity in higher rank categories but was in fact rather insignificant in the lower ranks, with even fewer genera (10) disappearing at that time than during the period immediately preceding the Lower Kellwasser Event.

Eye reduction occurred in a great number of unrelated lineages that evolved contemporaneously to form the atheloptic assemblages before the terminal Frasnian extinction: *Acuticryphops*, *Trimerocephaloides* (Phacopidae), *Telopeltis* (Scutelluidae), *Chlupaciparia*, *Pteroparia* (Tropidocoryphidae), *Canningbole*, *Palpebralia* (Proetidae), and *Cyphaspis* (Aulacopleuridae). This phenomenon coincided with and may be a reflection of the global rise in sea level culminating in an appreciable increase in water depth immediately before the Upper Kellwasser Event (e.g. Girard and Feist 1997). At this time, 67% of the known lower rank taxa had reduced eyes or were blind. All these genera and species disappeared in Zone 13b when both the outer platform and peri-reefal environments were perturbed by the major regressive pulse of the Kellwasser Event (Girard et al. 2005; McNamara and Feist 2008).

The Famennian

The second phase of the evolution of trilobite diversity in the Devonian considered in this study differed markedly from the preceding phase. Indeed, generic trilobite diversity stopped declining and even increased, depending on the metrics considered (Fig. 1) – due to a spectacular increase in origination rate to such a level that the still high extinction rate was compensated (Fig. 2a, b). The presence of only two orders (Fig. 3), survivors of the Kellwasser mass extinction, and the complete domination of the trilobite communities by the Proetidae (Fig. 4d) were also characteristic of this phase.

Like the preceding period, the Famennian was punctuated by a series of events whose impacts on trilobite diversity differed in intensity and nature (Walliser 1996; Fig. 6 herein). These events are, in chronological order, as follows: the *Cheiloceras* and Condroz–Enkeberg Events (early Famennian), the *annulata* Event (middle Famennian), and the Dasberg and Hangenberg Events (late Famennian). With the start of the Famennian, the post-event trilobite recovery was slow; no forms have been found in the first conodont zone. Hitherto 11 genera (seven phacopids and four proetids) have been reported from the early Famennian of Europe, North Africa, Southern Urals, Iran, and NW Australia, although no trilobites of that age are yet known from the Americas or eastern Asia. The earliest Famennian trilobite was the blind phacopid *Nephranops;* it occurred in European off-shore sites as early as the Middle *triangularis* Zone. Although it appears in slightly younger strata in the Canning Basin (Upper *triangularis* Zone), the trilobite *Houseops* is evidence that oculated forms were present not long after the Upper Kellwasser Event (Feist et al. 2009). This genus may have derived from an oculated lineage that survived the Kellwasser Events by living in shallow-water reefal to peri-reefal refugia unaffected by hypoxia.

Strikingly, the proetids reappeared later than the phacopids in all outcrops studied. The earliest Famennian proetids appeared within the Middle *crepida* Zone, three conodont zones later than the appearance of the phacopids. The late–early Famennian saw the true onset of the post-Kellwasser rediversification of the trilobites, as well as that of other marine organisms, such as the ammonoids ('*Cheiloceras* Event', Walliser 1985). Two types of communities were present in photic off-shore environments: blind proetids of cryptogenic origin such as *Formonia*, *Frithjofia*, and *Calybole*, found in basinal facies, and the first cyrtosymbolines, which had large eyes. This realm was extremely common at that time, expanding from Europe/North Africa to the Southern Urals and Eastern Asia on the one hand, and to Iran and NW Australia on the other. Moreover, it can be speculated that the radiation and final predominance of the Cyrtosymbolinae during the late early and middle Famennian times somewhat mirrored the development of uniform biofacies on shallow outer platforms, accompanying convergence of continental margins all around the proto-Tethyan 'gulf'. During this period, several phacopid taxa, both blind and oculated, appeared contemporaneously with the proetids.

After a mid-Famennian interval characterized by lower rates of generic origination, both the phacopids and proetids experienced a second major diversification period in the late Famennian (Dasberg Event; Fig. 6). At that time, about 70% of the genera known from the Famennian were still present. As was the case just before the Kellwasser Event, the overwhelming majority of these trilobites were blind or had reduced eyes and some, such as *Chaunoproetus* and *Helioproetus*, constituted the final stage in the evolutionary trend towards eye reduction (Yuan 1988). Whereas these forms originated and diversified in the European, Kazakhstanian, and Eastern Asian deeper off-shore clymenid realm, normally sighted cyrtosymbolines, such as *Pseudowaribole*, the problematic proetine *Perliproetus*, and the phacopids *Omegops*, *Rabienops*, and *Phacops*, were widely distributed in contemporaneous shallow shoal areas in Europe, Southern Urals, and Iran/Afghanistan or on the expanded shallow carbonate platforms in southern Morocco and Central and Eastern Asia. In the latter realm, the reappearance of brachymetopids and proetines after some 30 Ma of pseudo-extinction since mid-Givetian times constitutes a good example of the 'Lazarus' effect. Strikingly, both communities, with the exception of the Americas and Australia, exhibited a high degree of cosmopolitanism.

The terminal Devonian Hangenberg Extinction Event had an extremely severe effect on trilobite evolution, as demonstrated by extermination of the order Phacopida and the high level of generic extinctions, comparable only to those observed during the Middle Devonian events (Fig. 6). The reduced-eyed and blind taxa among the Proetida became extinct without exception, most likely trapped by the sudden rise in sea level and the spread of oceanic anoxia. The survivors probably originated from conservative, normally sighted communities in shallow shoal refugia that were beyond the devastating effects of the oxygen-depleted conditions at the onset of the Hangenberg black shale. The Hangenberg Event recalls the terminal Frasnian mass extinction in terms of both survivorship and post-event recovery. However, unlike the situation after the Upper Kellwasser Event, post-event recovery was immediate and worldwide after the Hangenberg biocrisis, even in North America where *Pudoproetus* originated in the Late *praesulcata* Zone before the defined D/C boundary (Feist and Petersen 1995). The total worldwide absence of reduced-eyed trilobites in this terminal Devonian period may indicate a considerable eustatic shallowing. The first phillipsiids, representatives of the subfamilies Archegoninae and Weaniinae, developed contemporaneously at that time and crossed the D/C boundary without any loss before their main radiations in the Early Carboniferous (Brauckmann et al. 1993). A major biotic turnover in trilobite diversity was associated with the Hangenberg Event, i.e. shortly before the internationally defined D/C boundary. The subsequent D/C lineages have yet to be defined within the continuously radiating trilobite lineages of 'Carboniferous aspect' (Chlupáč et al. 2000).

The Early Carboniferous

The Carboniferous began with high levels of raw, standing, and to a lesser extent normalized diversities (Fig. 1) resulting from a high origination rate associated with a moderately low extinction rate (Fig. 2a, b). This period was characterized by a great diversity of available habitats, an element that probably played a major role in the emergence and radiation of the Phillipsiidae after the late Famennian Hangenberg Extinction Event. Indeed, facies constraints on the composition of the trilobite communities at that time are evident (Hahn 1990; Owens 1990). For example, the 'Culm facies', representing the bathyal zone, has yielded rich and diversified faunas composed mainly of reduced-eyed or blind representatives of the two most-diversified subfamilies of the Tournaisian: the Archegoniinae and the Cystispininae (Hahn 1990; Yuan and Xiang 1998). The other phillipsiid subfamilies present during this stage occupied shallower environments represented by limestone facies (e.g. the 'Carboniferous limestone' of Central Europe). Though the 'Culm facies' developed in Asia (Hahn et al. 1989; Yuan and Xiang 1998; Hahn and Hahn 2005) and especially in Europe, where it has been intensively studied (e.g. Hahn and Hahn 1981), limestone facies dominated the Mississippian of North America (Brezinski 1998) and Australia (Engel and Morris 1997). In these shelf habitats, the palaeogeographic context might have favoured diversification of the trilobites. Indeed, the trilobite faunas of the limestone facies in Europe are mainly composed of phillipsiines, griffithidines, cummingellines, and bollandiines (Phillipsiidae; Hahn 1990). In contrast, the Australian faunas are particularly rich in linguaphillipsiines and weaniines (Phillipsiidae), as well as crassiproetines (*Conophillipsia*, Proetidae; Engel and Morris 1997), and although present worldwide, ditomopygines dominated the trilobite communities in North America.

Trilobite diversity fell during the Viséan (Fig. 1) with a significant reduction in origination rate and a simultaneous increase in extinction rate (Fig. 2a, b). However, the raw and standing diversities were still rather important; the distributions and compositions of trilobite communities remained grossly unchanged. The origination rate approached zero during the Serpukhovian, whereas the extinction rate, although diminishing, remained at around 0.35 (Fig. 2a, b). Accordingly, trilobite diversity decreased strongly to a value similar to those observed during the Late Devonian (Fig. 1a). The continuous fall in trilobite diversity during the Middle/Late Mississippian was accompanied by a continuous reduction in turnover rate (Fig. 2c). In Eastern and Central Europe, as well as in the British Isles, extinction of the diversified 'Culm' faunas resulting from abolition of deeper water basinal environments during the Sudetian orogenic phase might have accounted for a large part of the fall in trilobite diversity during the late Early Carboniferous (Hahn 1990; Owens 1990). Brezinski (1999) also noted a major fall in trilobite diversity in the USA during the middle Viséan (mid-Meramecian). Following a period of exceptional diversity, his stage 2, which lasted until the end-Serpukhovian, was characterized by an extremely limited number of genera. He attributed this significant drop in trilobite diversity to a transition from a transgressive period to a regressive period (Brezinski 1999). In eastern Australia, local volcanic activity led to the first episode of reduced diversity at the beginning of the Viséan. After moderate diversification, a second episode of diversity decline at the end of the Viséan might have resulted from the cumulative effects of a major local volcanic episode and rapid cooling marking the onset of a period of glaciation in Australia (Dickins 1996; Engel and Morris 1997).

As with the trilobites, the generic diversity of the marine invertebrates as a whole was rather high in the Tournaisian, declining throughout the Early Carboniferous (Alroy et al. 2008: Fig. 1). A long-term drop in sea level apparently paralleled this reduction in diversity of marine invertebrates (Haq and Schutter 2008). The late Mississippian–early Pennsylvanian has long been recognized as a period of significant changes in marine ecosystems. Although varying in intensity and duration, mid-Carboniferous events have been described in ammonoids (Saunders and Ramsbottom 1986), brachiopods (Raymond et al. 1990; Shen et al. 2006), rugose corals (Kossovaya 1996; Wang et al. 2006), reefs (Flügel and Kiessling 2002), and conodonts (Ziegler and Lane 1987; Nemirovskaya and Nigmadganov 1994). These perturbations of the biosphere were concomitant with an increase in the magnitude of glacio-eustatic fluctuations; the late Viséan represented the initial phase of ice accumulation; it was followed by a period of widespread glaciation that lasted until the beginning of the Bashkirian (Dickins 1996; Rygel et al. 2008). During this glacial episode, glacio-eustatic fluctuations as great as 40–100 m might have occurred (Rygel et al. 2008). Important eustatic changes recorded during the late Mississippian were already regarded as a major eustatic event by Saunders and Ramsbottom (1986), with significant repercussions for faunal diversity. However, the assumption of these and other authors (e.g. Ziegler and Lane 1987; Weems 1992) that a major extinction event occurred during the Serpukhovian is not really supported by our data for trilobites. Indeed, even if a marked drop in trilobite generic diversity could be observed during this stage, it was already preceded by a diversity decline during the Viséan (Fig. 1a, b). This is particularly obvious when the durations of the time intervals are considered (Fig. 1c). Moreover, the origination and turnover rates had already declined during the Viséan (Fig. 2a, c). A similar pattern of diversity has been described for the Carboniferous brachiopods of South China by Shen et al. (2006). These authors concluded that the 'mid-Carboniferous extinction' might have been a gradual process, experienced over a long time, and was mainly associated with a restructuring of brachiopod communities. We regard the decline in trilobite generic diversity and extinction of three subfamilies (Cystispininae, Linguaphillipsiinae, and Phillipsiinae) during the late Mississippian (late Viséan–Serpukhovian) to have been similarly gradual. Global climatic and eustatic changes (i.e. the glacial episode initiated during the Viséan) are thus regarded as the main factors responsible for evolution of trilobite diversity at that time. However, tectonic evolution (e.g. in Europe) or volcanic activity (e.g. in Australia) might also have had significant effects on trilobite diversity locally.

The Bashkirian–Moscovian Period

The pattern of the evolution of trilobite generic diversity during the Bashkirian and Moscovian is not clear. Whereas the raw diversity remained roughly unchanged, the standing diversity declined moderately and the normalized diversity increased (Fig. 1a–c). However, the origination, turnover, and to a lesser extent the extinction rates increased significantly during the Bashkirian and remained roughly unchanged during the Moscovian; this points to a certain renewal within the trilobite communities. Thus, about 70% of the trilobite genera that occurred in the Kasimovian originated after the Serpukhovian. Considering that the origination rate in the Kasimovian was extremely low, this suggests that most of this renewal occurred during the 11 million years or so of the Bashkirian–Moscovian interval, a period only slightly exceeding the Serpukhovian stage (Ogg et al. 2008). This renewal of trilobite communities was accompanied by continued restructuring initiated during the late Early Carboniferous. Ditomopygines become particularly common, whereas archegoniines became rare (Fig. 5) until their simultaneous extinction with another phillipsiid subfamily, the Griffithidinae, and the family Aulacopleuridae during the Moscovian. This second phase in evolution of the diversity and composition of post-Devonian trilobites ended with a reduction in generic diversity and falls in the origination, extinction, and turnover rates during the Kasimovian (Figs. 1 and 2).

The diversified faunas of Spain (Gandl 1987, 2011) and China (Hahn et al. 1989; Yuan and Li 2000) illustrate particularly well the transitional aspects of the early Late Carboniferous. These faunas were dominated by cummingelline and ditomopygine phillipsiids and brachymetopids, but still included a notable proportion of archegoniine trilobites. However, in the USA, the early Late Carboniferous was associated with a renewal within the trilobite communities, rather than a modification of their subfamilial compositions. The prevalence of ditomopygines was already apparent in the late Mississippian trilobite communities composed of species of *Kaskia* and *Paladin* (Brezinski 2003, 2008). These trilobites were replaced during the late Bashkirian by another moderately diversified fauna composed exclusively of ditomopygines (*Sevillia*, *Ditomopyge*, and *Ameura*; Brezinski 1999). Although poorly known (Hahn and Hahn 1991, 1992), the late Early Carboniferous trilobites of Alaska were more diversified than those of the USA, comprising representatives of four phillipsiid subfamilies. They were replaced in the early Late Carboniferous by a fauna composed of brachymetopids and once again, various phillipsiids (Hahn and Hahn 1991, 1992, 1993). The Ditomopyginae only slightly dominated this trilobite association in terms of generic diversity, but the genus *Ditomopyge* was by far its most abundant component (Hahn and Hahn 1991). Therefore, the Bashkirian–Moscovian period in Alaska was associated with both a renewal of trilobite communities and the onset of the ditomopygine predominance, at least in terms of abundance. In eastern Australia, the Late Carboniferous trilobites were only represented by a single species of the brachymetopid *Australosutura* (Engel and Morris 1997). This species belonged to the presumably cold-water *Levipustula* fauna (Campbell 1961; see Dickins 1996 for a different interpretation). A similar fauna has been described in the Bashkirian of Argentina; it also included a representative of *Australosutura* (Hahn et al. 2001). However, diverse species of the ditomopygine *Ameura* have also been found in the Late Carboniferous of South America. These trilobites, known from different regions of Brazil, are apparently of Moscovian age (Hahn et al. 2001); this suggests that trilobites were also mainly represented by ditomopygines in this part of South America at that time. Lastly, two early Late Carboniferous trilobite faunas showed unusual compositions with regard to the Ditomopyginae. Weber (1937) has described Bashkirian and Moscovian trilobite faunas from western Russia (Urals, Moscow Basin) and the Ukraine (Donetz Basin) composed exclusively of ditomopygines. This unique situation might be due to these faunas having come from shale facies (Hahn et al. 1989). In contrast, contemporaneous faunas in Japan were completely devoid of ditomopygine trilobites (Kobayashi and Hamada 1980) but were especially rich in cummingellines. They therefore anticipate the importance this group would have in later periods.

A rather modest increase in diversity of marine invertebrates apparently occurred in the early Late Carboniferous (Alroy et al. 2008: Fig. 1). This increase was particularly well exemplified by brachiopods from South China (Shen et al. 2006); they displayed a notable increase in origination and turnover rates indicative of a radiation. Rugose corals from the same region also diversified during this period and as with the trilobites, they illustrate a profound restructuring of benthic communities with replacement of large dissepimented solitary corals by compound corals (Wang et al. 2006). According to Wang et al. (2006), these compositional changes in marine benthic communities were associated with a transgression that led to formation of a uniform carbonate platform in most areas of South China. Major transgressive events were also associated with the Bashkirian and Moscovian in the Tethys Himalaya (India, Nepal, and Tibet) and in the peri-Gondwanan belt (Garzanti et al. 1998). On the global scale, the Bashkirian and most of the Moscovian have been considered an interval of low-magnitude eustatic sea-level changes indicating a reduction in the worldwide volume of glacial ice (Rygel et al. 2008; see also Isbell et al. 2003). This period also marked the onset of a long-term rise in sea level lasting until the beginning of the Permian (Haq and Schutter 2008). In summary, restructuring of the trilobite communities and, more generally, the benthic marine communities that occurred during the early Late Carboniferous might have resulted from favourable eustatic and climatic conditions.

The Kasimovian–Roadian Period

This long interval (about 40% of the Late Palaeozoic), which encompassed the late Late Carboniferous, the Cisuralian, and the earliest Guadalupian, was characterized by invariably low raw and standing generic diversities (Fig. 1) and low origination, extinction, and turnover rates (Fig. 2). The only notable changes during this period were two modest peaks in the turnover rate corresponding to an increase in the extinction rates of the Asselian and an increase in the origination rates of the Kungurian. In terms of composition, cummingellines and weaniines seem to have been relatively more diversified during the Kasimovian–Sakmarian and Sakmarian–Roadian intervals, respectively (Fig. 5), but the raw diversity was too low to allow any reliable estimation of the relative diversities of the phillipsiid subfamilies. Basically, the trilobite communities remained roughly unchanged during most of this period. For example, more than 80% of the genera present in the Asselian originated or already existed in the Moscovian. The extinction of four of those genera by the end of the Asselian was not accompanied by an equivalent origination, and therefore led to a slight drop in diversity. It was only during the Kungurian that definite renewal began, but it remained modest until the Wordian.

This third phase in the evolution of trilobite diversity in the Late Palaeozoic, characterized by low diversity, few changes, and a predominance of ditomopygines, is particularly well illustrated in the USA. After the extinction of *Sevillia* during the Moscovian, only two ditomopygine genera (*Ditomopyge* and *Ameura*) occurred during most of the Late Carboniferous (Brezinski 1999). More diversified trilobite faunas were present during the terminal Carboniferous and at the beginning of the Permian. However, these faunas did not result from local diversification but testify to an invasion of the American biota by pandemic trilobites (Brezinski 1999). It is also noteworthy that the last diversification phase in the history of the trilobites had begun by the end of the Cisuralian in the USA (see stage 4 of Brezinski 1999), whereas it became obvious no sooner than the Wordian on the global scale. Recently, Owens (2003) noted that all genera that occurred in the Tethyan Realm during the early Cisuralian were inherited from either the Early Carboniferous or the Late Carboniferous. Our data suggest that the genera that originated during the Bashkirian or later actually represented the overwhelming majority of trilobites on the global scale. Interestingly, some trilobites occurred at high latitudes after the Asselian. In particular, the reappearance of trilobites in south-eastern Australia (Tasmania) in the late Sakmarian (Wass and Banks 1971) might indicate a certain amelioration of the climatic conditions locally. However, their persistence during the Artinskian, together with the appearance of *Anujaspis* in high northern latitudes (Balashova 1960) might indicate somewhat warmer climatic conditions on a more global scale.

The end-Carboniferous and early Cisuralian times (Gzhelian to early Sakmarian) are recognized as the largest and most widespread period of Late Palaeozoic continental glaciation (Isbell et al. 2003; Montañez et al. 2007; Rygel et al. 2008; Tabor and Poulsen 2008). According to Alroy et al. (2008), this period was associated with a particularly low diversity of marine invertebrates; its end might have allowed a considerable increase in their diversity, culminating during the Guadalupian. The late Sakmarian to Artinskian stages thus represent a transitional interval associated with compositional changes within marine benthic communities (e.g. fusulinids: Leven et al. 1996; reefs: Weidlich 2002; rugose corals: Kossovaya et al. 2001; also see Clapham and James 2008). Whereas a slight fall in the normalized trilobite diversity (Fig. 1c) and a modest increase in the extinction and turnover rates (Fig. 2b, c) can be tentatively associated with the climax of the Asselian glaciation, no significant changes in the trilobite diversity parameters really indicate the onset of an icehouse/greenhouse climatic transition in the mid-Cisuralian. As mentioned above, this latter transition is nevertheless suggested by the geographical distribution of the trilobites, especially the reappearance of taxa at high latitudes. The modest increases in the origination and turnover rates (Fig. 2a, c) observed in the Kungurian might indicate the onset of the last diversification phase of trilobite history. However, the raw diversity had reached such low values after the Kasimovian that any interpretation of the evolution of trilobite diversity during this period must be made with great caution.

The Wordian–Changhsingian Period

In contrast to the preceding phase, the last stage in the evolution of trilobite diversity was associated with notable changes. As indicated by the rise in the raw and normalized generic diversities (Fig. 1a, c) and the simultaneous increase in origination rates (Fig. 2a), the Wordian represented a period of trilobite diversification. The renewal within trilobite communities is exemplified by the fact that 45% of the genera present during this stage were new. However, as mentioned above, this renewal occurred earlier in the USA (i.e. during the Kungurian; Brezinski 1992, 1999) with the appearance of an endemic fauna exclusively composed of ditomopygines (Lerosey-Aubril 2008). It is also noteworthy that most of the new genera belonged to the Ditomopyginae. Although modest, this diversification of trilobites in the Wordian mirrored a general increasing trend in the diversity of marine invertebrates, initiated in the mid-Cisuralian (Alroy et al. 2008).

More than 50% of the genera that originated during the Wordian were restricted to this time interval. This observation stresses another characteristic of this stage: the rather high extinction rates (Fig. 2b). These remained high in the Capitanian and the Wuchiapingian, whereas origination rates decreased strongly during these two stages (Fig. 2a). The end-Guadalupian was thus characterized by an important reduction in the generic diversity of trilobites (Fig. 1; Owens 2003) and therefore deserves particular attention. A major drop in diversity has also been recorded in brachiopods (Shen and Shi 2002), corals (Wang and Sugiyama 2000), gastropods (Erwin and Pan 1996; Pan 2004), fusulinids (Stanley and Yang 1994; Ota and Isozaki 2006), reefs (Weidlich 2002), and other taxonomic groups (e.g. Yin et al. 2007) which has led several authors to propose a mass extinction in the late Guadalupian (e.g. Stanley and Yang 1994; Shen and Shi 2002; Ota and Isozaki 2006; Retallack et al. 2006; Ali and Wignall 2007; Weidlich and Bernecker 2007). Whether it should be recognized as an independent extinction event is still contentious (Clapham et al. 2009), but most authors now admit that the end of the Guadalupian marks the onset of a period of major environmental perturbation. Global cooling ('Kamura Event'; Isozaki et al. 2007a, b; Clapham et al. 2009), a major regression (Erwin 1993; Shi and Shen 2000; Shen and Shi 2002; Tong 2004; Pan 2004; Yin et al. 2007), a massive spread of deep-water anoxia (Bottjer et al. 2008; Clapham et al. 2009), a catastrophic methane release (Retallack et al. 2006), and an eruption of a large volume of basalts ('Emeishan Large Igneous Province'; Ali et al. 2002; Ali and Wignall 2007) are all environmental changes identified/suspected in the late Guadalupian that might have ultimately resulted in a reduction in biodiversity. Considering the low diversity of trilobites in the Permian, it might be unwise to use their diversity patterns to favour one or other of these hypotheses. However, our data show that the origination rate decreased strongly during the Capitanian and Wuchiapingian, whereas there was no increase in the extinction rate (Fig. 2a, b). Clapham et al. (2009) have recently described a similar pattern in several other groups of marine invertebrates; they interpret this as evidence that end-Guadalupian diversity losses resulted from a lack of originations rather than from increased extinctions. According to them, this low level of origination in the Capitanian–Wuchiapingian can only be explained by (1) cooling, (2) the spread of anoxic deep-water masses, or (3) a reduction in area and provinciality of the marine habitat – attributable to tectonic activity.

The Changhsingian was also marked by an absence of originations; the trilobites were then represented by only four genera – *Acropyge* (Ditomopyginae), *Carniphillipsia* (Ditomopyginae), *Kathwaia* (Bollandiinae), and *Paraphillipsia* (Cummingellinae) – to which might be added *Pseudophillipsia* (Ditomopyginae) and *Cheiropyge* (Brachymetopidae), known as several species of 'Lopingian' age. Although the trilobites persisted in several peri-Tethyan regions (Lerosey-Aubril 2008), they were extremely diminished in term of their diversity and were apparently unable to accommodate the dramatic environmental upheavals of the end-Permian. Therefore, they became extinct after 270 million years of a remarkable evolutionary history.

Conclusion

The time period considered in this study encompasses more than 145 My (Ogg et al. 2008) and is therefore particularly rich in biotic, climatic, eustatic, geochemical, and tectonic events. For example, the dynamic convergence of Lower Palaeozoic continental blocks, ending with formation of the mega-continent Pangaea during the Permian, certainly constituted a singular tectonic context (Scotese and McKerrow 1990). Dramatic climatic changes also occurred in this period, including the transition from a greenhouse to an icehouse climatic system around the Early–Late Carboniferous boundary and vice versa during the Cisuralian. This multiplicity of major environmental changes makes it difficult to assess the precise impact of each of these events on the biosphere or to discern how they interacted. Consequently, it is not possible to determine with certainty which of these changes ultimately influenced trilobite diversity.

However, in the Middle and Late Devonian, the stratigraphic precision of the quantitative data allows us to recognize that eustatic variations had a major influence on trilobite diversity, even if the origins of these eustatic changes are not always clear. The expansion of outer-shelf domains during transgressions or the increases in near-shore facies belts during sea-level low stands apparently had important repercussions on biotic diversity, determining the available benthic habitats successfully invaded by biotas during adaptive radiations. The trilobites were particularly sensitive to bathymetric changes, as demonstrated by the generalized trend to eye reduction within trilobite communities during major transgressive phases (e.g. Frasnian, Famennian) or the very many extinctions associated with abrupt regressive events (e.g. Upper Kellwasser Event, Hangenberg Event). Therefore, it seems reasonable to infer that the intensity and dynamics of eustatic oscillations had direct consequences for trilobite evolutionary history at that time. Deepening trends led to important morphologic and ecologic adaptations of different taxa to life in deep-benthic habitats; brutal regressive pulses triggered massive extinctions within these specialized communities. From the Eifelian to the Frasnian, a series of extinction events led to drastic impoverishment of trilobite diversity, involving in the first place, and nearly exclusively, the lower taxonomic categories (genus and species). It was only at the very end of this period, when trilobite generic diversity had dropped to the lowest level of Early to Mid Palaeozoic times, that higher ranks were severely affected – at the end-Frasnian biocrisis. In the Famennian, when the group was represented by only two orders, originations started to compensate for the still high extinction rate, marking the onset of remarkable lower rank diversification. Although this second phase ended with the major turnover caused by the terminal Famennian Hangenberg Event, the general diversification trend continued and even accelerated in the Tournaisian when extinction rates became more moderate.

The Late-Palaeozoic diversity patterns described here suggest there may have been a relationship between the evolution of trilobite diversity and the main climatic changes in post-Devonian times. When glacial conditions occurred, whether they were associated with individual events (Viséan and Serphukovian glaciations, Capitanian Kamura Event) or with an icehouse period (Kasimovian–Sakmarian), the origination rate fell dramatically (Fig. 2a), leading to a drop in diversity or its maintenance at extremely low levels (Fig. 1). However, the stratigraphic resolution of our recently compiled database is not sufficient to explore this relationship further, because it does not allow identification of rapid fluctuations in diversity or their correlation with abrupt climatic changes. In this regard, it would be interesting to consider time intervals shorter than stages in the Early Carboniferous because these are notably longer than those of the Late Carboniferous and Permian. This should permit better appreciation of the dynamics of the diversity fluctuations during this crucial period in the late evolutionary history of the trilobites. Another potentially important improvement to the database would be incorporation of palaeobiogeographic data; this would allow us to assess the completeness of the fossil record for each interval and to search for palaeobiogeographic signals of major events, especially climatic events. It would also be worth including data on ecologic diversity from both morphologic and taphonomic studies because they might allow us to describe the evolution of the roles of trilobites within Late Palaeozoic ecosystems.

Finally, this quantitative study has pointed out a particularly interesting aspect of the late evolutionary history of the trilobites. After the terminal Frasnian Kellwasser Event, there was an obvious decoupling of trilobite diversity at the generic level and trilobite diversity at the familial and ordinal levels. The spectacular increase in generic diversity observed in the Tournaisian, with values not attained since the Eifelian, demonstrates that trilobites still had an intrinsic and significant capacity to adapt at that time. This increased diversity involved only the Proetida, a conservative group that had hitherto maintained its very simple morphological bauplan since Cambrian times. It then underwent a burst of homeomorphic variation, adapting to various ecological behaviours and habitats formerly occupied by morphologically different groups, such as the Phacopida. Therefore, extinction of this class of emblematic arthropods at the end of the Permian was not, as frequently proposed, an almost inevitable fate after restriction of the group to one single order after the Late Devonian biocrises. Instead, it more probably represents the ultimate result of major environmental changes that affected the biosphere during the Late Palaeozoic. Much more attention must be paid to Permo-Carboniferous trilobites if we are to fully understand this extinction.

Acknowledgements Many people have generously helped us gather the literature required to create the database used in this study; we are very grateful to all of them. We especially thank Martin Basse (Forschungsinstitute Senckenberg), Carsten Brauckmann (Technischen Universität Clausthal), James Cook, Gerhard Hahn, and Thomas A. Hegna (Yale University) for kind assistance in this regard. We are also indebted to Euan N.K. Clarkson and an anonymous referee for helpful comments and to the editor, John Talent, for major linguistic improvements in our text. This is a contribution of the Senckenberg Forschungsinstitut und Naturmuseum (Frankfurt am Main) and UMR 5554, CNRS (Montpellier) to IGCP Project 596 "Climate change and biodiversity patterns in the Mid-Palaeozoic (Early Devonian to Late Carboniferous)".

References

Adrain JM (2008) A global species database of Trilobita: progress, results, and revision of the Treatise. In: Rabano I, Gozalo R, Garcia-Bellido D (eds) Advances in Trilobite research. Cuadernos del Museo Geominero 9:27–28

Ali JR, Wignall PB (2007) Comment on 'Fusiline biotic turnover across the Guadalupian–Lopingian (middle–upper Permian) boundary in mid-oceanic carbonate build-ups: biostratigraphy of accreted limestone in Japan' by Ayano Ota and Yukio Isozaki. J Asian Earth Sci 30:199–200

Ali JR, Thompson GM, Song X et al (2002) Emeishan basalts (SW China) and the 'end-Guadalupian' crisis: magnetobiostratigraphic constraints. J Geol Soc London 159:21–29

Alroy J, Aberhan M, Bottjer DJ et al (2008) Phanerozoic trends in the global diversity of marine invertebrates. Science 321:97–100

Balashova R (1960) Some early Permian trilobites from the north-east of the USSR. Materialy po Geologii i Polezn'm Iskopaem'p Severo-Vostoka SSSR. Magadan 14:74–82

Bottjer DJ, Clapham ME, Fraiser ML et al (2008) Understanding mechanisms for the end-Permian mass extinction and the protracted Early Triassic aftermath and recovery. Geol Soc Am Today 18:4–10

Brauckmann C, Chlupáč I, Feist R (1993) Trilobites at the Devonian–Carboniferous boundary. In: Streel M, Sevastopoulo MG, Paproth E (eds) Devonian–Carboniferous boundary. Ann Soc Géol Belgique 115:507–518

Brezinski DK (1992) Permian trilobites from West Texas. J Paleontol 66:924–943

Brezinski DK (1998) Trilobites from Lower Mississippian starved basin facies of the southern United States. J Paleontol 72:718–725

Brezinski DK (1999) The rise and fall of late Paleozoic trilobites of the United States. J Paleontol 73:164–175

Brezinski DK (2003) Evolutionary and biogeographic implications of phylogenetic analysis of the late Paleozoic trilobite genus *Paladin*. In: Lane PD, Siviter D, Fortey RA (eds) Trilobites and their relatives. Spec Pap Palaeontol 70:363–375

Brezinski DK (2008) Phylogenetics, systematics, paleoecology, and evolution of the trilobite genera *Paladin* and *Kaskia* from the United States. J Paleontol 82:511–527

Briggs DEG, Fortey RA, Clarkson ENK (1988) Extinction and the fossil record of the arthropods. In: Larwood GP (ed) Extinction and survival in the fossil record. Clarenton Press, Oxford

Bultynck P (2000) Subcommission on Devonian stratigraphy. Recognition of Devonian series and stage boundaries in geological areas. Cour Forsch-Inst Senckenberg 225:1–347

Campbell KSW (1961) Carboniferous fossils from the Kuttung rocks of New South Wales. Palaeontology 4:428–474

Chlupáč I (1994) Devonian trilobites – evolution and events. Geobios 27:487–505

Chlupáč I, Feist R, Morzadec P (2000) Trilobites and standard Devonian stage boundaries. In: Bultynck P (ed) Subcommission on Devonian stratigraphy. Fossil groups important for boundary definition. Cour Forsch-Inst Senckenberg 220:87–98

Clapham ME, James NP (2008) Paleoecology of Early–Middle Permian marine communities in eastern Australia: response to global climate change in the aftermath of the Late Paleozoic ice age. Palaios 23:738–750

Clapham ME, Shen S, Bottjer DJ (2009) The double mass extinction revisited: reassessing the severity, selectivity, and causes of the end-Guadalupian biotic crisis (Late Permian). Paleobiology 35:32–50

Dickins JM (1996) Problems of a Late Palaeozoic glaciation in Australia and subsequent climate in the Permian. Palaeogeogr Palaeoclimatol 125:185–197

Engel BA, Morris LN (1997) Biostratigraphy of eastern Australian Carboniferous trilobites. Geologica Palaeontol 31:67–96

Erwin DH (1993) The Great Paleozoic crisis, life and death in the Permian. Columbia University, New York

Erwin DH, Pan HZ (1996) Recoveries and radiations: gastropods after the Permo–Triassic mass extinction. In: Hart MB (ed) Biotic recovery from mass extinction events. Geol Soc Spec Pub 102:223–229

Feist R (1991) The Late Devonian trilobite crises. In: Buffetaut E, Flessa K, Hallam A (eds) Innovations and revolutions in the biosphere. Hist Biol 5:197–214

Feist R (1995) Effect of paedomorphosis in eye reduction on patterns of evolution and extinction in trilobites. In:

McNamara KJ (ed) Evolutionary change and heterochrony. Wiley, New York

Feist R (2002) Trilobites from the latest Famennian Kellwasser crisis in North Africa (Mrirt, central Moroccan Meseta). Acta Palaeontol Pol 47:19–26

Feist R, Petersen MS (1995) Origin and spread of *Pudoproetus*, a survivor of the Late Devonian trilobite crisis. J Paleontol 69:99–109

Feist R, Schindler E (1994) Trilobites during the Frasnian Kellwasser crisis in European Late Devonian cephalopod limestones. In: Königshof P, Werner R (eds) Willi Ziegler-Festschrift II. Cour Forsch-Inst Senckenberg 169:195–223

Feist R, McNamara KJ, Crônier C, Lerosey-Aubril R (2009) Patterns of extinction and recovery of phacopid trilobites during the Frasnian–Famennian (Late Devonian). Geol Mag 146:12–33

Flügel E, Kiessling W (2002) Patterns of Phanerozoic reef crises. In: Kiessling W, Flügel E, Golonka J (eds) Phanerozoic reef patterns. SEPM Spec Pub 72:691–733

Gandl J (1987) Die Karbon-Trilobiten des Kantabrischen Gebirges (NW Spanien), 4: Trilobiten aus dem höheren Namur und tieferen Wesfal. Abh Senckenberg Naturforsch Ges 543:1–79

Gandl J (2011) Los trilobites Carboníferos de la Cordillera Cantábrica (España noroccidental), 5: Trilobites [Carboniferous trilobites of the Cantabrian Mountains (NW Spain), 5: upper Westphalian] Abhandlungen der Senckenberg Gesellschaft für Naturforschung, Band 569:143pp

Garzanti E, Angiolini L, Brunton H et al (1998) The Bashkirian 'Fenestella shales' and the Moscovian 'Chaetetid shales' of the Tethys Himalaya (South Tibet, Nepal and India). J Asian Earth Sci 16:119–141

George AD, Chow N (2002) The depositional record of the Frasnian/Famennian boundary interval in a forereef succession, Canning Basin, Western Australia. Palaeogeogr Palaeoclimatol 181:347–374

Girard C, Feist R (1997) Eustatic trends in conodont diversity across the Frasnian–Famennian boundary in the stratotype area, Montagne Noire, Southern France. Lethaia 29: 329–337

Girard C, Klapper G, Feist R (2005) Subdivision of the terminal Frasnian *linguiformis* Conodont Zone, revision of the correlative interval of Montagne Noire Zone 13, and discussion of stratigraphically significant associated trilobites. In: Over JR, Morrow JR, Wignall PB (eds) Understanding Late Devonian and Permian–Triassic biotic and climatic events: towards an integrated approach. Elsevier, Amsterdam, pp 181–198

Hahn G (1990) Palaeobiogeographic distribution and biostratigraphic significance of Lower Carboniferous trilobites: a review. Cour Forsch-Inst Senckenberg 130:199–205

Hahn G, Hahn R (1981) Kulm-Trilobiten und ihr Lebensraum. Natur Museum 111:355–361

Hahn G, Hahn R (1991) Trilobiten aus dem Karbon von SE-Alaska, Teil 1. Geologica Palaeontol 25:147–191

Hahn G, Hahn R (1992) Trilobiten aus dem Karbon von SE-Alaska, Teil 2. Geologica Palaeontol 26:99–133

Hahn G, Hahn R (1993) Neue Trilobiten-Funde aus dem Karbon und Perm Alaskas. Geologica Palaeontol 27:141–163

Hahn G, Hahn R (2005) Kulm trilobites (Lower Carboniferous) from S-China. Palaeontol ZH 79:371–375

Hahn G, Hahn R, Yuan J-L (1989) Trilobites from the Upper Carboniferous (Westphalian A) of S-China (N-Guangxi). Geologica Palaeontol 23:113–203

Hahn G, Brauckmann C, Gröning E (2001) Carboniferous and Permian trilobites in South America. Acta Geol Leopold 24:259–270

Haq BU, Schutter SR (2008) A chronology of Paleozoic sea-level changes. Science 322:64–68

Harper CW Jr (1996) Patterns of diversity, extinction and origination in the Ordovician–Devonian Stropheodontacea. Hist Biol 11:267–288

Heckel PH, Clayton G (2006) The Carboniferous system. Use of the new official names for the subsystems, series, and stages. Geol Acta 4:403–407

Isbell JL, Miller MF, Wolfe KL, Lenaker PA (2003) Timing of late Paleozoic glaciation in Gondwana: was glaciation responsible for the development of northern hemisphere cyclothems? In: Chan MA, Archer AA (eds) Sedimentary giants – extreme depositional environments. Geol Soc Am Spec Pap 370:5–24

Isozaki Y, Kawahata H, Minoshima K (2007a) The Capitanian (Permian) Kamura cooling event: the beginning of the Paleozoic–Mesozoic transition. Palaeoworld 16:16–30

Isozaki Y, Kawahata H, Ota A (2007b) A unique carbon isotope record across the Guadalupian–Lopingian (Middle–Upper Permian) boundary in mid-oceanic paleo-atoll carbonates: the high-productivity 'Kamura event' and its collapse in Panthalassa. Global Planet Change 55:21–38

Johnson JG, Klapper G, Sandberg CA (1985) Devonian eustatic fluctuations in Euramerica. Geol Soc Am Bull 96:57–587

Kobayashi T, Hamada T (1980) Carboniferous trilobites of Japan in comparison with Asian, Pacific and other faunas. Palaeontol Soc Japan Spec Pap 23:1–132

Kossovaya OL (1996) The Mid-Carboniferous rugose coral recovery. Geol Soc London Spec Pub 102:187–199

Kossovaya OL, Guseva EA, Lukin AE et al (2001) Middle Artinskian (Early Permian) ecological events: a case study of the Urals and northern Timan. Proc Estonian Acad Sci Geol 50:95–113

Lerosey-Aubril R (2008) Trilobite biogeography and Permian biochores. In: Rabano I, Gozalo R, Garcia-Bellido D (eds) Advances in trilobite research. Cuadernos del Museo Geominero 9:225–228

Leven EY, Bogoslovskaya MF, Ganelin VG et al (1996) Reorganization of marine biota during the mid-Early Permian epoch. Stratigr Geo Correl 4:57–66

McNamara KJ, Feist R (2008) Patterns of trilobite evolution and extinction during the Frasnian/Famennian mass extinction, Canning Basin, Western Australia. In: Rabano I, Gozalo R, Garcia-Bellido D (eds) Advances in Trilobite research. Cuadernos del Museo Geominero 9:269–274

Montañez IP, Tabor NJ, Niemeier D et al (2007) CO_2-forced climate and vegetation instability during Late Paleozoic deglaciation. Science 315:87–91

Morzadec P (1992) Evolution des Asteropyinae (Trilobita) et variations eustatiques au Dévonien. Lethaia 25:85–96

Nemirovskaya T, Nigmadganov I (1994) The mid-Carboniferous conodont events. Cour Forsch-Inst Senckenberg 168:319–333

Ogg JG, Ogg G, Gradstein FM (2008) The concise geologic time scale. Cambridge University, Cambridge

Ota A, Isozaki Y (2006) Fusiline biotic turnover across the Guadalupian–Lopingian (Middle–Upper Permian) boundary in mid-oceanic carbonate build-ups: biostratigraphy of accreted limestone in Japan. J Asian Earth Sci 26:353–368

Owens RM (1990) Carboniferous trilobites: the beginning of the end. Geol Today 6:96–100

Owens RM (2003) The stratigraphical distribution and extinctions of Permian trilobites. Spec Pap Palaeontol 70:377–397

Pan H (2004) Remarks on Permian extinction and Triassic recovery of gastropods. In: Rong JY, Fang ZJ (eds) Mass extinction and recovery, evidence from the Palaeozoic and Triassic of South China. University of Science and Technology, China Press, Beijing [in Chinese with English summary]

Raymond AC, Kelley PH, Lutken CB (1990) Dead by degrees: articulate brachiopods, paleoclimate and the Mid-Carboniferous extinction event. Palaios 5:111–123

Retallack GJ, Metzger CA, Greaver T et al (2006) Middle–Late Permian mass extinction on land. Geol Soc Am Bull 118:1398–1411

Rygel MC, Fielding CR, Frank TD et al (2008) The magnitude of Late Paleozoic glacioeustatic fluctuations: a synthesis. J Sediment Res 78:500–511

Saunders WB, Ramsbottom WHC (1986) The mid-Carboniferous eustatic event. Geology 14:208–212

Scotese CR, McKerrow WS (1990) Revised world maps and introduction. In: McKerrow, WS, Scotese CR (eds) Palaeozoic palaeogeography and biogeography. Geol Soc Mem 12:1–21

Shen SZ, Shi GR (2002) Paleobiogeographical extinction patterns of Permian brachiopods in the Asian–western Pacific region. Paleobiology 28:449–463

Shen S-Z, Zhang H, Li W-Z et al (2006) Brachiopod diversity patterns from Carboniferous to Triassic in South China. Geol J 41:345–361

Shi GR, Shen SZ (2000) Asian–western Pacific Permian Brachiopoda in space and time: biogeography and extinction patterns. In: Yin HF, Dickins JM, Shi GR, Tong JN (eds) Permian–Triassic evolution of Tethys and western circum-Pacific. Elsevier, Amsterdam, pp 327–352

Stanley SM, Yang X (1994) A double mass extinction at the end of the Paleozoic era. Science 266:1340–1344

Tabor NJ, Poulsen CJ (2008) Paleoclimate across the Late Pennsylvanian–Early Permian tropical paleolatitudes: a review of climate indicators, their distribution, and relation to paleophysiographic climate factors. Palaeogeogr, Palaeoclimat, Palaeoecol 268:293–310

Tong JN (2004) Evolution of foraminiferid groups through the Paleozoic–Mesozoic transition in South China. In: Rong JY, Fang ZJ (eds) Mass extinction and recovery, evidence from the Palaeozoic and Triassic of South China. University of Science and Technology of China Press, Beijing [in Chinese with English summary], pp 702–719

Walliser OH (1985) Natural boundaries and commission boundaries in the Devonian. Cour Forsch-Inst Senckenberg 7:401–408

Walliser OH (1996) Global events in the Devonian and Carboniferous. In: Walliser OH (ed) Global events and event stratigraphy in the Phanerozoic. Springer, Berlin, pp 225–250

Wang X-D, Sugiyama T (2000) Diversity and extinction patterns of Permian coral faunas of China. Lethaia 33:285–294

Wang X-D, Wang X-J, Zhang F et al (2006) Diversity patterns of Carboniferous and Permian rugose corals in South China. Geol J 41:329–343

Wass RE, Banks MR (1971) Some Permian trilobites from eastern Australia. Palaeontology 14:222–241

Weber VN (1937) Trilobity kamennougolnych i permskych otloshenij SSSR. 1. Kamennougolnye trilobity. [Trilobites of the Carboniferous and Permian system of the USSR. 1. Carboniferous trilobites]. Monogr Paleontol SSSR 71:1–160 [in Russian with abridged English version].

Weems RE (1992) The 'terminal' Triassic catastrophic event in perspective: a review of Carboniferous through Early Jurassic vertebrate extinction patterns. Palaeogeogr Palaeoclimatol Palaeoecol 94:1–29

Weidlich O (2002) Permian reefs re-examined: extrinsic control mechanisms of gradual and abrupt changes during 40 My of reef evolution. Geobios MS 24:287–294

Weidlich O, Bernecker M (2007) Differential severity of Permian–Triassic environmental changes on Tethyan shallow-water carbonate platforms. Global Planet Change 55:209–235

Yin H, Feng Q, Lai X et al (2007) The protracted Permo–Triassic crisis and multi-episode extinction around the Permian–Triassic boundary. Global Planet Change 55:1–20

Yuan JL (1988) Proetiden aus dem jüngeren Oberdevon von Süd-China. Palaeontogr A 201:1–102

Yuan J-L, Li Y (2000) Biostratigraphy of Carboniferous trilobites in China. J Stratigr 24:283–289 [in Chinese with English summary]

Yuan J-L, Xiang L-W (1998) Trilobite fauna at the Devonian–Carboniferous boundary in South China (S-Guizhou and N-Guangxi). Nat Mus Natur Sci, Spec Publ 8:1–281

Ziegler W, Lane HR (1987) Cycles in conodont evolution from Devonian to mid-Carboniferous. In: Aldridge RJ (ed) Palaeobiology of conodonts. Horwood Press, Chichester, pp 148–163

Devonian Cladid Crinoid Evolution, Diversity, and First and Last Occurrences: Summary Observations

Gary D. Webster

Abstract

New discoveries and systematic revisions since the 1978 publication of the crinoid volumes of the Treatise on Invertebrate Paleontology provide new information about Devonian cladid crinoid evolution, diversity, and first and last occurrences. Summarizing these previous studies resulted in numerous conclusions, including the following. The overall diversity of the Devonian cladids does not reflect the diversity acmes and extinction patterns of the individual superfamilies within the Cyathocrinida and Dendrocrinida. Cyathocrinoidea and Codiacrinoida genera were minor elements in the diversity of Devonian Cyathocrinida that were dominated by Gasterocomoidea (acme in the Eifelian) genera. The Gasterocomidae and Cupressocrinitidae are a clade restricted to the Devonian and Early Mississippian. Primitive Dendrocrinids were replaced in the Devonian by the Glossocrinoidea (referred to as Transitional Dendrocrinids); the Advanced Dendrocrinids and articulates first occurred in the Devonian. Morphologic and stratigraphic differences justify systematic differentiation between these groups. The last occurrences of cladid genera were greater in the Emsian and Givetian than they were in the Frasnian with the exception of the gasterocomids. Advanced cladids with plesiomorphic conical cups are thought to be derived from the Transitional Dendrocrinids, whereas the evolution of those with bowl-shaped cups is uncertain, but may be from the cyathocrinids. Time gaps in the ranges of carry-through genera and the sparse record in the Lochkovian and Pragian suggest stratigraphic intervals for future research to help resolve evolutionary and taxonomic questions in the cladids.

Keywords

Cyathocrinida • Reassignment of genera • Lochkovian–Pragian data deficiency • Emsian–Eifelian diversity maximum • End-Givetian extinction • Famennian rebound • Predicted latest Devonian bothocrinoids

G.D. Webster (✉)
School of Earth and Environmental Sciences, Washington State University, Pullman, WA 99164-2812, USA
e-mail: webster@wsu.edu

J.A. Talent (ed.), *Earth and Life*, International Year of Planet Earth,
DOI 10.1007/978-90-481-3428-1_17, © Springer Science+Business Media B.V. 2012

Introduction

Evolutionary Trends

General evolutionary trends in the cladid crinoids were summarized for the cyathocrinids and dendrocrinids by Lane (1978). He listed the major differences between the Ordovician cyathocrinids and dendrocrinids, noted the difficulty in proposing a common ancestor for both groups, and suggested that the discovery of older echinoderms would hopefully provide an answer to the origin of each. In addition, he considered the dendrocrinids to be a reasonably homogeneous group that gave rise to the poteriocrinids in the Devonian. He considered the key morphologic characters for phyletic derivation of the poteriocrinids to be the appearance of muscular articulation between the radials and first primibrachs and development of pinnules from non-pinnulate arms with closely spaced ramules; secondary value was placed on the relative width of the radial articular facets (considered of perhaps dubious value) and the number of anal plates in the cup. Furthermore, he considered the major difference between the dendrocrinids and poteriocrinids to be the lack of muscular articulation and pinnules in the dendrocrinids while recognizing that not all crinoids assigned to the Poteriocrinina in the Treatise of Invertebate Paleontoloogy (Moore and Teichert 1978) had these features. For example, *Poteriocrinites* and *Springericrinus* lack muscular articulation and *Rhenocrinus* and *Charientocrinus* have ramules, not pinnules. Webster and Maples (2008) considered *Poteriocrinites* and *Springericrinus* to belong to the Primitive Dendrocrinids, *Charientocrinus* to be a Transitional Dendrocrinid, and *Rhenocrinus* an Advanced Dendrocrinid with pinnules as they redefined them.

Strimple (1978) summarized the evolutionary trends for the poteriocrinids. He noted the change in cup shape from conical with upflaring infrabasals to bowl to discoid with or without a deep basal invagination. He recognized the development of the bowl-shaped cup in the decadocrinids in the Devonian. In addition Strimple (1978) described the evolutionary changes in the anals starting with a reduction in the number in the cup with the ultimate loss of the anals in the most advanced forms. He also recognized evolutionary trends in the arms and columnal shapes in the poteriocrinids. Most of the evolution in the poteriocrinids occurred in post-Devonian time.

Webster and Maples (2006) reviewed the anal condition in the cladids (excluding codiacrinids) and described a number of subconditions in the individual arrangement of the anals of various numbers while agreeing with the previously recognized trend toward reduction in number of anals in the cup and ultimate elimination of anals from the cup. A number of these subconditions occur in Devonian cladids in evolutionary stages between more primitive and more advanced subconditions. Webster and Maples (2008) reviewed the evolution of radial facet width and articulation, brachial type, and arm appendages in the cladids (excluding codiacrinids). They referred to the Glossocrinoidea as a clade and considered it to be transitional between the Primitive Dendrocrinids and the Advanced Dendrocrinids most of which were previously included in the poteriocrinids (see discussion of classification below). Webster and Maples (2008) agreed with Lane (1978) that most Primitive Dendrocrinids lacked muscular articulation and pinnules. Webster and Maples (2008) considered that these conditions evolved in the Transitional Dendrocrinids and are present in all Advanced Dendrocrinids; they agreed with Lane (1978) that pinnules evolved repeatedly polyphyletically in the dendrocrinids and that most cladids previously assigned to the poteriocrinids (Advanced Dendrocrinids herein) were derived within the dendrocrinids. The Devonian was an important time for this evolution, as recognized by Moore and Laudon (1943), Kammer and Ausich (2006), among others.

Extinctions

A mass extinction in the Late Devonian at the Frasnian–Famennian boundary has been considered one of the five major extinction events in the Paleozoic as described by McGhee (1996). Although several phyla were deeply affected by that event, the crinoids, blastoids, and bryozoans did not suffer a major extinction (Webster et al. 1998). New discoveries or revision of earlier faunas of Devonian echinoderms in China (Chen and Yao 1993; Lane et al. 1996; Waters et al. 2003; Webster and Waters 2009), Australia (Jell 1999; Jell and Jell 1999; Jell et al. 1988; among others), Argentina (Haude 2004), South Africa (Jell and Theron 1999), England (Lane et al. 2001a). Germany (Lane et al. 2001b; Hauser 2003, 2007;

Bohatý 2005, 2006a, b, 2009), Iran (Webster et al. 2007), Morocco (Webster et al. 2005; Webster and Becker 2009), and North America (Webster and Hafley in Webster et al. 1999; McIntosh 2001; Webster et al. 2003) provided new information on the cladid genera as discussed below. Waters and Webster (2009) noted that the Famennian crinoid and blastoid diversity was five times greater than that recognized in the Sepkoski (2002) compilation based on the recent discoveries and revision of older specimens, which changed the apparent extinction patterns and recovery of the Late Devonian echinoderms.

Classification

Since publication of the Treatise of Invertebrate Paleontology (Moore and Teichert 1978) studies of the cladid crinoids included the discovery of many new taxa and additional specimens of previously described taxa and undesignated, mentioned, or listed specimens. These discoveries have added to the cladid database by providing significant morphologic details and when combined with various other investigations of the cladids have altered many of the earlier concepts of the cladids. These include significant modifications within the classification at various levels, better definition of previously generalized evolutionary trends, and paleogeographical distribution.

With minor exceptions (i.e., recognition of new families or reassigning a genus to a different family) crinoid specialists generally followed the higher level crinoid classification given in the Treatise of Invertebrate Paleontology (Moore and Teichert 1978) until Simms and Sevastopulo (1993) proposed a revised classification when investigating the origin of the articulate crinoids. Their revisions did not recognize the Subclass Inadunata by elevating the Orders Disparida and Cladida to subclass level, recognizing the Cyathocrinina, Flexibilia, and Articulata as infraclasses within the Cladida and including the Hybocrinida as an Incertae Sedis ("Subclass") of the crinoids, also removing them from the Inadunata.

Simms and Sevastopulo (1993) recognized that the Paleozoic included "stem articulates" that they left in the cladids, restricting the Articulata to the post-Palaeozoic, which was also followed by Hess et al. (1999) and Kammer and Ausich (2006). Webster and Jell (1999) described the first isocrinid articulate from the Early Permian (Artinskian) of Western Australia and they also recognized the order Ampelocrinida, including the Ampelocrinidae and Corythocrinidae, among some other Late Paleozoic cladid families, as the stem articulates based on the presence of muscular articulation in the arms, among other features, and assigned them to the Articulata. This removed them from the "stem articulates" status and elevated them to a recognized taxonomic group.

The Suborder Poteriocrinina, which included a large percentage of the post-Devonian cladids, has also been questioned as a suborder by numerous investigators (e.g., Kelly 1982; Webster and Lane 1987; McIntosh 2001) because they recognized that these crinoids were polyphyletically derived from the dendrocrinid or cyathocrinid cladids. Although most crinoids considered to be primitive poteriocrinids in the earlier classification evolved in the Devonian Webster et al. (2003) proposed that most of them evolved from the Glossocrinoidea (Middle Silurian to Early Mississippian) and referred to the glossocrinids as "Transitional Dendrocrinids".

Several systematic reallocations have been made within the cladids during the past 20 years. These include reassigning genera from the Cyathocrinida to the Dendrocrinida (McIntosh and Brett 1988), transferring families from the Poteriocrinida into the Gasterocomoidea (Webster and Hafley in Webster et al. 1999), shifting genera from one family to another within the Gasterocomoidea (Bohatý 2006a; Webster and Maples 2008), among others. With the incorporation of these revisions, the classifications used in Tables 1, 2, 3, and 4 differ from the classification in the Treatise (Moore and Teichert 1978) because they incorporate most, but not all, modifications by later investigations. A complete revision of the classification of the cladids is beyond the scope of this review.

Data Base

The Devonian cladids (excluding the flexibles) discussed below are divided into five groups:

1. Cyathocrinida
2. Primitive Dendrocrinida
3. Transitional Dendrocrinida (Glossocrinoidea)
4. Advanced Dendrocrinida
5. Articulata

Table 1 Devonian Cyathocrinida. First column lists genera and author(s) for Devonian Cyathocrinida families and superfamilies. Next four columns give the number of species for each genus recognized in pre-Devonian, Devonian, post-Devonian, and total number of species. Arrows in pre- and post-Devonian columns indicate occurrence of genera in the family or superfamily in pre- and post-Devonian times. Presence (X) or non-recognition of the genus (void) are indicated for each of the seven Devonian epochs. A question mark behind the X indicates the generic identification is questioned. A question mark without the X indicates the age is uncertain

Age	CYATHOCRINIDA	No pre-Dev species	No Devonian species	No post-Dev species	Total no species	Pre-Devonian	Lochkovian	Pragian	Emsian	Eifelian	Givetian	Frasnian	Famennian	Post-Devonian
	Cyathocrinitoidea					<--								-->
	Cyathocrinitidae					<--								-->
M Sil – E Miss	*Cyathocrinites* Miller (1821)	4	12	42	58	<--				X	X	X	X	-->
M Sil – E Perm	*Gissocrinus* Angelin (1878)	24	3	1	27	<--		X	X	?				-->?
	Euspirocrinidae					<--								-->
L Sil – E Dev	*Kooptoonocrinus* Jell and Holloway (1983)	1	2		2	<--	X							
E Dev	*Manocrinus* Haude (2004)		1		1		?	?	?					
E Dev	*Monaldicrinus* Jell and Theron (1999)		1		1				X					
E Dev	*Othozecrinus* Jell and Theron (1999)		1		1				X					
M Dev – E Perm	*Parisocrinus* Wachsmuth and Springer (1880)		4	5	9					X	X		X	-->
E Dev – M Dev	*Vasocrinus* Lyon (1857)		7		7				X	X	X			
	Gasterocomoidea					<--								
	Gasterocomidae													
E Dev – M Dev	*Gasterocoma* Goldfuss (1839)		17		17				X	X	X			
E Dev – M Dev	*Ancyrocrinus* Hall (1862)		3		3			X		X	X			
M Dev	*Arachnocrinus* Meek and Worthen (1866)		5		5					X	X			
E Dev	*Corynecrinus* Kirk (1934)		1		1				X					
M Dev	*Lecythocrinus* Müller (1859)		1		1					X				
E Dev	*Mictocrinus* Goldring (1923)		1		1				X					
E Dev – M Dev	*Myrtillocrinus* Sandberger and Sandberger (1850–1856)		5		5				X	X				
M Dev	*Nanocrinus* Müller (1856)		1		1					X				
E Dev	*Schultzicrinus* Springer (1911)		2		2				X					
M Dev	*Scoliocrinus* Jaekel (1895)		1		1					X	X			
M Dev	*Tetrapleurocrinus* Wanner (1942)		1		1					X				
M Dev	*Trapezocrinus* Haude (2007)		4		4						X			

Table 1 (continued)

Age	CYATHOCRINIDA	No pre-Dev species	No Devonian species	No post-Dev species	Total no species	Pre-Devonian	Lochkovian	Pragian	Emsian	Eifelian	Givetian	Frasnian	Famennian	Post-Devonian
	Cupressocrinitidae													
E Dev – L Dev	*Cupressocrinites* Goldfuss Steininger (1831)		19		19			X	X	X	X	X	X	
M Dev	*Abbreviatocrinites* Bohatý (2005)		11		11					X	X			
M Dev	*Procupressocrinus* Jaekel (1918)		2		2						X	X		
M Dev	*Rhopalocrinus* Wachsmuth and Springer (1880)		1		1					X				
M Dev	*Robustocrinites* Bohatý (2005)		3		3					X	X			
	Crotalocrinitidae					<--								
M Sil- E Dev	*Crotalocrinites* Austin and Austin (1843)	6	1?		6	<--	X?							
E Dev	*Parapernerocrinus* Yakovlev (1949)		1		1				X					

Age	CYATHOCRINIDA	No pre-Dev species	No Devonian species	No post-Dev species	Total no species	Pre-Devonian	Lochkovian	Pragian	Emsian	Eifelian	Givetian	Frasnian	Famennian	Post-Devonian
E Dev	*Pernerocrinus* Bouska (1946)		4		4			X	X					
	Petalocrinidae					<--								
E Dev	*Vadarocrinus* Prokop (1983)		1		1			X						
	Codiacrinoidea					<--								-->
	Codiacrinidae					<--								-->
M – L Dev	*Codiacrinus* Schultze (1867)		5		5				X	X		X		
E Dev	*Gemmacrinus* Prokop and Petr (1989)		1		1			X						
L Dev	*Ovalocrinus* Chen and Yao (1993)		1										X	
L Dev	*Quasicydonocrinus* Chen and Yao (1993)		2		2								X	
E Dev – L Dev	*Elicrinus* Prokop (1973)		2		3			X	X				X?	
	Sphaerocrinidae													
M Dev	*Sphaerocrinus* Roemer (1851)		3		3					X	X			
	Streblocrinidae													-->
M Dev	*Streblocrinus* Koenig and Meyer (1965)		1		1						X			
L Dev – E Miss	*Pentececrinus* Koenig and Niewoehner (1959)		1	1	1								X	-->
	Family uncertain													
E Dev	*Briseocrinus* Prokop and Petr (1991)		1		1				X					
E Dev	*Treocrinus* Prokop and Petr (1991)		1		1				X					

Table 2 Devonian Primitive Dendrocrinida. First column lists genera and author(s) for Devonian primitive dendocrinid families and superfamilies. Next four columns give the number of species for each genus recognized in pre-Devonian, Devonian, post-Devonian, and total number of species. Arrows in pre- and post-Devonian columns indicate occurrence of the Devonian genus or non-Devonian genera in the family or superfamily in pre- and post-Devonian times. Presence (X) or non-recognition of the genus (void) are indicated for each of the seven Devonian epochs. A question mark behind the X indicates the generic identification is questioned

Age	PRIMITIVE DENDROCRINIDA	No pre-Dev. Species	No Devonian Species	No post-Dev. Species	Total no species	Pre-Devonian	Lochkovian	Pragian	Emsian	Eifelian	Givetian	Frasnian	Famennian	Post-Devonian
	Dendrocrinoidea					<--								-->
	Dendrocrinidae					<--								-->
E Dev	*Dracrinus* Webster and Becker in press		1		1				X	X				
M Dev	*Eckidocrinus* Jell and Theron (1999)		1		1					X				
E Dev	*Parisangulocrinus* Schmidt (1934)		6		6		X		X					
	Botryocrinidae					<--								-->
E Dev – M Miss	*Costalocrinus* Jaekel (1918)		6	1	7				X	X	X			-->
E Dev	*Kopficrinus* Goldring (1954)		2		2				X					
E Dev	*Situlacrinus* Breimer (1962)		1		1				X					
	"Botryocrinidae" "orphans"													
E Dev	*Gastrocrinus* Jaekel (1895)		9		9				X					
E Dev	*Imitatocrinus* Schmidt (1934)		1		1				X					
M Dev	*Jahnocrinus* Jaekel (1918)		1		1						X			
L Dev	*Parabotryocrinus* Yakovlev (1941)		1		1							X		
L Dev	*Pskovicrinus* Rozhnov and Arendt (1984)		1		1							X		
E – M Dev	*Rhadinocrinus* Jaekel (1895)		4		4				X		X			
	Plicodendrocrinidae					<--								
L Sil – E Dev	*Holmesocrinus* Jell (1999)	1	1		2	<--	X							
	Poteriocrinitidae					<--								-->
M Sil – E Perm	Poteriocrinites Miller (1821)	1	20	32	53	<--	X?		X	X	X	X	X	-->
E Dev	*Denariocrinus* Schmidt (1942)		1		1				X					
M Dev	*Ornatuscrinites* Hauser (2007)		1							X				

Table 2 (continued)

Age	PRIMITIVE DENDROCRINIDA	No pre-Dev. Species	No Devonian Species	No post-Dev. Species	Total no species	Pre-Devonian	Lochkovian	Pragian	Emsian	Eifelian	Givetian	Frasnian	Famennian	Post-Devonian
	Thalamocrinidae					<--								
M Sil – E Dev	*Thalamocrinus* Miller and Gurley (1895)	6	3		7	<--	X		X					
M Sil – M Dev	*BactrocrinitesSchnur* Steininger (1849)	2	20		22	<--	X		X	X	X			
L Dev	*Belanskicrinus* Strimple and Levorson (1969)		1		1							X		
L Dev	*Cradeocrinus* Goldring (1923)		5		5							X		
E Dev	*Eifelocrinus* Wanner (1916)		6		6				X					
E Dev	*Follicrinus* Schmidt (1934)		3		3			X	X					
L Dev	*Linobrachiocrinus* Goldring (1939)		1		1							X		
E – M Dev	*Nuxocrinus* McIntosh (1983)		3		3				X		X			
E Dev – L Dev	*Pagecrinus gracilis* Kirk (1929)		2		2				X			X		
E – M Dev	*Pyrenocrinus* McIntosh (1983)		4		4				X		X			
E Dev	*Sacrinus* Jell and Theron (1999)		1		1				X					
E Dev	*Sigambrocrinus* Schmidt (1942)		2		2				X					
	Mastigocrinoidea					<--								-->
	Mastigocrinidae					<--								-->
M Sil – E Dev	*Antihomocrinus* Schmidt (1934)	2	6		7	<--	X		X					
M Sil – E Dev	*Dictenocrinus* Jaekel (1918)	2	12		14	<--	X		X					
E Dev – E Miss	*Lasiocrinus* Kirk (1914)		5	1	6		X		X	X		X		-->
	Merocrinoidea					<								
	Cupulocrinidae					<								
E Dev	*Stewbrecrinus* Jell (1999)		1		1		X							
	Superfamily and Family Uncertain													
E Dev	*Kophinocrinus* Goldring (1954)		2		2				X					

Table 3 Devonian Transitional Dendrocrinida. First column lists genera and author(s) for Devonian transitional dendocrinid families and superfamilies. Next three columns give the number of species for each genus recognized in pre-Devonian, Devonian, and total number of species. Arrows in pre- and post-Devonian columns indicate the occurrence of non-Devonian genera in the family or superfamily in pre- and post-Devonian times. Presence (X) or non-recognition of the genus (void) are indicated for each of the seven Devonian epochs

Age	TRANSITIONAL DENDROCRINIDA	No pre-Devonian species	No Devonian species	Total no species	Pre-Devonian	Lochkovian	Pragian	Emsian	Eifelian	Givetian	Frasnian	Famennian	Post-Devonian
	Glossocrinoidea				<--								-->
	Glossocrinidae												-->
M – L Dev	*Glossocrinus* Goldring (1923)		6	6							X	X	
L Dev	*Catactocrinus* Goldring (1923)		2	2							X	X	
M Dev	*Charientocrinus* Goldring (1923)		3	3						X	X		
L Dev	*Liparocrinus* Goldring (1923)		1	1							X		
	Amabilicrinidae												-->
L Dev	*Bufalocrinus* Webster et al. (2003)		1	1								X	
L Dev	*Coquinacrinus* Webster and Becker in press		1	1							X		
L Dev	*Embolocrinus* Webster and Becker in press		1	1							X		
M – L Dev	*Hallocrinus* Goldring (1923)		1	1							X		
L Dev	"*Mrakibocrinus* Webster, Becker and Maples (2005)"		1	1								X	
L Dev	"*Moroccocrinus* Webster, Becker and Maples (2005)"		1	1								X	
	Corematocrinidae												-->
L Dev	*Corematocrinus* Goldring (1923)		1	1							X		
L Dev	*Maragnicrinus* Whitfield (1905)		1	1							X		
	Rutkowskicrinidae				<--								
M Dev	*Rutkowskicrinus* McIntosh (2001)		2	2						X			
M Dev	*Decorocrinus* McIntosh (2001)		2	2						X			
E Dev	*Iteacrinus* Goldring (1923)		3	3				X			X		
L Sil.- E Dev	*Nassoviocrinus* Jaekel (1918)	2	7	9	<--		X	X		X			
E Dev	*Propoteriocrinus* Schmidt (1934)		6	6				X					
M – L Dev	*Quantoxocrinus* Webby (1965)		3	3						X		X	
E Dev	?*Schmidtocrinus* Haarmann (1920)		1	1						X			

Table 4 Devonian Advanced Dendrocrinida. First column lists genera and author(s) for Devonian advanced dendocrinid families and superfamilies. Next four columns give the number of species for each genus recognized in the Devonian, post- Devonian and total number of species. Arrows in post-Devonian column indicate occurrence of the Devonian genus or non-Devonian genera in the family or superfamily in pre- and post-Devonian times. Presence (X) or non-recognition of the genus (void) are indicated for each of the seven Devonian epochs. A question mark behind the X indicates the generic identification is questioned

Age	ADVANCED DENDROCRINIDA	No Devonian species	No post-Dev. Species	Total no species	Lochkovian	Pragian	Emsian	Eifelian	Givetian	Frasnian	Famennian	Post-Devonian
	Rhenocrinoidea											
	Rhenocrinidae											
E Dev	*Rhenocrinus* Jaekel in Schmidt (1906)	2		2			X					
	Scytalocrinoidea											-->
	Scytalocrinidae											-->
L Dev – M Miss	*Scytalocrinus* Wachsmuth and Springer (1880)	2	22	24							X	-->
L Dev – M Miss	*Histocrinus* Kirk (1940)	1	6	7							X	-->
"L Dev – M Penn"	*Hydriocrinus* Trautschold (1867)	6	5	11							X	-->
L Dev – M Miss	*Hypselocrinus* Kirk (1940)	1	17	18							X?	-->
L Dev	*Julieticrinus* Waters et al. (2003)	1		1							X	
L Dev – E Miss	*Prininocrinus* Goldring (1938)	1	1	2						X		-->
	Aphelecrinidae											-->
L Dev – L Miss	*Cosmetocrinus*	1	4	5							X	-->
	Bridgerocrinidae											-->
L Dev – E Miss	*Bridgerocrinus* Laudon and Severson (1953)	5	4	9						X	X	-->
L Dev – M Miss	*Holcocrinus* Kirk (1945)	1	6	7							X	-->
	Cercidocrinidae											-->
L Dev	*Emhacrinus* Webster and Waters (2009)	1		1							X	
	Sostronocrinidae											-->
L Dev – E Miss	*Sostronocrinus* Strimple and McGinnis (1969)	6	2	8							X	-->
L Dev	*Amadeusicrinus* Waters et al. (2003)	1		1							X	
	Decadocrinoidea											-->
	Decadocrinidae											-->
M Dev.- M Miss	*Decadocrinus* Wachsmuth and Springer (1880)	26	10	36					X	X	X	-->
L Dev	*Graubauicrinus* Waters et al. (2003)	4		4							X	
M Dev	*Zostocrinus* Kirk (1948)	1		1					X			
	Lophocrinoidea											-->
	Pachylocrinidae											-->
L Dev – M Miss	*Pachylocrinus aequalis* (Hall 1862)	1	24	25							X	-->
	Uncertain family											
M Dev	*Struveicrinites* Hauser (1998)	1		1				X				
L Dev	*Tarassocrinus* Webster et al. (1999)	1		1							X	

This subdivision is used for practical purposes and only the Transitional Dendrocrinids are considered a clade. Stratigraphic ranges given for each genus are based on the ages given for species assigned to that genus as compiled by Webster (2003; unpublished compilation 1999–2009). The number of species of each genus is given for pre-Devonian, Devonian, post-Devonian, and total number where appropriate in Tables 1, 2, 3, and 4. Subspecies are not included in the compilation because they usually have the same stratigraphic range as the nominal species. Devonian epoch names follow the geologic time scale of Gradstein and Ogg (2004). Although some species used for compiling these epochs occur in a small part of the epoch, the range is given as the complete epoch. Rarely a species is recognized in more than one epoch and this is usually at the boundary interval of two epochs or at a period boundary. Thus, a listing including one pre-Devonian and four Devonian species could give a total number of four if one species is recognized in the latest Silurian and earliest Devonian. A grand total of Devonian cladid genera (111) and species (355) are divided among the five groups as follows: Cyathocrinida 39, Primitive Dendrocrinids 33, 134; Transitional Dendrocrinids 19, 44; 134; Advanced Dendrocrinids 19, 37; and Articulata 1, 6. The evolutionary trends discussed for each group are based largely on the type species for each Devonian genus because there may be a few species that should not be recognized within a particular genus and there may be some species that should be placed in synonomy with other species within a particular genus or species of another genus. If the type species is of pre- or post-Devonian age the Devonian species is the basis of the discussed evolutionary trend. Species revisions await later investigation, but are not considered to significantly alter the observations recognized herein—except they could increase the apparent diversity of the Devonian cladid genera.

Cyathocrinida

Devonian cyathocrinid representatives are assigned to three superfamilies, Cyathocrinoidea, Gasterocomoidea, and Codiacrinoidea (Table 1). Each superfamily is discussed independently.

Cyathocrinoidea

The Cyathocrinoidea has two families, Cyathocrinidae and Euspirocrinidae, with Devonian genera and both families have pre- and post-Devonian genera (Table 1). Two genera of the Cyathocrinidae with Devonian species are *Cyathocrinites* and *Gissocrinus.*

Frest (1977) assigned most of the Silurian species of *Cyathocrinites* to subgenera resolving some of the systematic problems within the genus. Webster and Maples (2006, 2008) elevated the subgenera to the genus level because each was easily distinguishable from *Cyathocrinites.* At present 58 species are assigned to *Cyathocrinites*, most of which are of Mississippian age; all 58 are in need of systematic revision. The 12 Devonian species of *Cyathocrinites* range from the Eifelian into the Famennian. The lack of species recognized in the Lochkovian to Emsian represents a carry-through of the genus from the Silurian, although, all pre-Devonian species may belong to one or more new genera. *Cyathocrinites* was not affected by the Fransian–Famennian extinction and radiated in the post-Devonian.

Most (24) of the 27 species assigned to *Gissocrinus* are of pre-Devonian age. The Eifelian age of one of the three Devonian species is questioned, and the post-Devonian species is a questionable identification. The third species of *Gissocrinus*, of Emsian age, may represent the last occurrence of the genus, but I question the generic assignment of all of the Devonian species (see below).

Cyathocrinites retains most of the plesiomorphic characters of earlier cyathocrinids, which include a bowl-shaped cup, angustary unifascial radial facets, rectilinear brachials, and no arm appendages; however, bifascial radial facets had evolved on the Mississippian type species of the genus. Until the Devonian species assigned to *Cyathocrinites* are revised no evolutionary trends are recognized herein. Devonian species assigned to *Gissocrinus* have the first branching on the second or third primibrachial. The first branching on the Silurian type species of *Gissocrinus* is on the single primibrachial. In the cladids the general evolutionary trend is for the first branching of the arms to become lower not higher. Therefore I question the generic assignment of the Devonian species of *Gissocrinus* and believe that a systematic review of all of the species of *Gissocrinus* is needed.

Silurian euspirocrinids generally have a medium cone- or bowl-shaped cup, unifascial angustary radial facets, rectilinear brachials, arms branching multiple times, no arm appendages, three or four anals in the cup, an elongate anal tube, and round pentameric or holomeric stems. *Kooptoonocrinus* differs by developing apical pits with single stellate interplate ridges on the conical cup and has a single large anal. *Manocrinus, Monaldicrinus*, and *Vasocrinus* have low conicial cups, arms extended laterally, and round holomeric stems. *Manocrinus and Monaldicrinus* lack the development of an arm in the A-ray and have a large laterally projecting anal tube. *Manocrinus* has armlets on every third or fifth brachial. *Vasocrinus* has a weakly developed transverse ridge on either side of a central canal in the bifascial radial facet. The distal arms and tegmen are unknown on *Vasocrinus*, but the laterally projecting proximal parts of the arms suggest that it may have had a similar development as *Manocrinus* and *Monaldicrinus*. *Othozecrinus* has a bowl-shaped cup with bulbous plates, more than three anals, strongly rounded rectilinear brachials, and isotomously branching arms. *Parisocrinus* retained the basic plesiomorphic characters of the euspirocrinids, including a pentameric round stem, but differs by development of the central pore in the bifascial radial facet.

The stratigraphic distribution of the eight Devonian genera of Cyathocrinitoidea shown in Table 1 is summarized to show the diversity and first and last occurrences of the numbers of genera in the data-table and line-graph of Fig. 1. As may be noted therein, the Devonian cyathocrinitids are poorly represented in the Lochkovian and Pragian radiating to their greatest diversity (four genera) as well as greatest number of first and last occurrences in the Emsian. This was followed by an end-Emsian extinction event that reached a zero low in the last occurrences in the Eifelian, rebounded slightly in the Givetian and dropped to zero in the Frasnian and Famennian. First occurrences attained a zero low in the Givetian and Frasnian to begin a rebound in the Famennian. The overall diversity declined to the Eifelian and Givetian level (three genera) before an extinction event at the end of the Givetian resulted in a low in the Frasnian (one genus). They rebounded to an intermediate level (two genera) in the Famennian. Overall the cyathocrinitoids were never a highly diverse group in the Devonian and were not affected by the Frasnian–Famennian extinction event.

Gasterocomoidea

Literature in which gasterocomoid genera were first recognized reflects their geographic distribution, dominantly Europe. Only four of the 21 gasterocomoid genera are reported from North America, 17 of the 21 genera were described prior to 1950, and four genera have been described in the past 10 years (Table 1). The more recent literature (Hauser 1997, 2001, 2007, 2008; Bohatý 2001, 2005, 2006a, b, 2009; Webster et al., 2008) reemphasizes the common occurrence of this superfamily in Europe.

Several taxonomic revisions have recently been proposed for the Gasterocomoidea. The Cupressocrinitidae were assigned to the Poteriocrinina (Moore and Teichert 1978) and transferred to the Gasterocomoidea by Webster and Hafley (in Webster et al. 1999). Bohatý (2006a) revised the Gasterocomoidea transferring *Lecythocrinus* and *Tetrapleurocrinus* to the Gasterocomidae and considered *Cestocrinus, Corynecrinus*, and *Othozecrinus* to belong to the Euspirocrinidae. Although I tentatively agree with assigning *Cestocrinus* and *Othozecrinus* to the Euspirocrinidae, I include *Corynecrinus* in the Gasterocomidae because it has the accessory canals in the column and other cup and arms characters similar to *Lecythocrinus* and *Ancyrocrinus*. *Ancyrocrinus* was assigned to the Botryocrinidae (McIntosh and Schreiber 1971) and orphaned when McIntosh (1984) excluded it when he revised the Botryocrinidae. Webster and Maples (2008) considered *Ancyrocrinus* a lecythocrinid based on the similarity of the crown to other lecythocrinids and the presence of the four supplementary axial canals in the column. Webster et al. (2008) included *Ancyrocrinus* in the Gasterocomidae recognizing that the lecythocrinids were included therein by Bohatý (2006a). *Arachnocrinus* was transferred to the Gasterocomidae by Webster et al. (2008), based on the cup, arm, and column characters. *Parabarycrinus*, a Tournaisian genus, was assigned to the Barycrinidae by Chen and Yao (1993). Restudy of the specimen revealed that it has an anal opening within the radial circlet surrounded by numerous small plates and axial canal with three accessory canals. It is reassigned to the Gasterocomidae by Webster et al. (2009).

The Gasterocomoidea range from the Ordovician into the earliest Mississippian with most genera restricted within the Devonian. Devonian taxa (21 genera and 85 species) are assigned to four families,

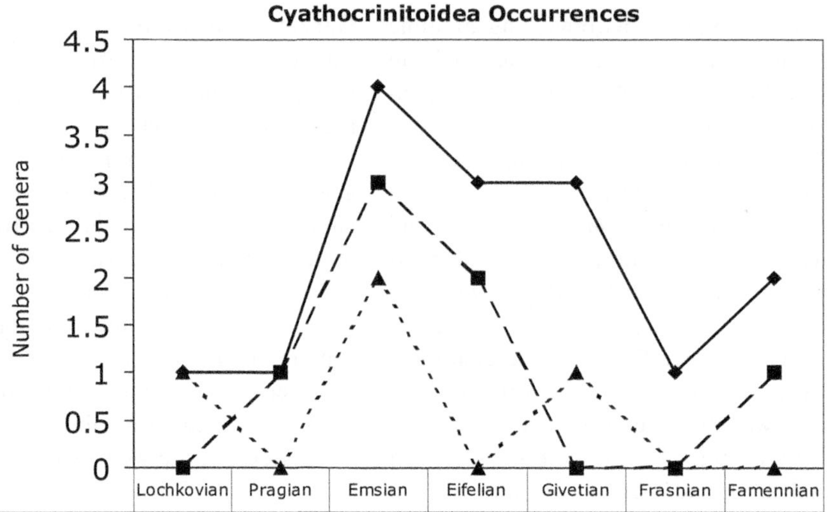

	Lochkovian	Pragian	Emsian	Eifelian	Givetian	Frasnian	Famennian
Diversity	1	1	4	3	3	1	2
First Occurrence	0	1	3	2	0	0	1
Last Occurrence	1	0	2	0	1	0	0

Fig. 1 Cyathocrinitoidea Occurrences. Shows total diversity and first and last occurrences of the eight Devonian cyathocrinitoid genera. Data bank summarized from Table 1. Questionable age occurrences not included in diversity

with all genera except *Crotalocrinites* restricted to the Devonian (Table 1). The Gasterocomidae is the most diverse with 12 genera. The Cupressocrinitidae has five genera. The Crotalocrinitidae (three genera) and Petalocrinidae (one genus) are the least diverse gasterocomid families at both genus and species level in the Devonian.

The oldest gasterocomids are the crotalocrinitids. The Lochkovian crotalocrinitid, *Crotalocrinites bashkiricus* Yakovlev (1949), is based on a partial cup and stem fragment that may represent a new genus transitional between *Crotalocrinites* and *Pernerocrinus*. The fragmentary cup has multiple arm facets all the way across the radial, whereas they are grouped centrally, not extending laterally to the extremities of the facet in *Crotalocrinites*. *Pernerocrinus* evolved from a crotalocrinitid ancestor by massive modification of the cup plates, arms adjoined laterally all the way around the crown, and attachment to the substrate with an extensive root system. *Parapernerocrinus* has the multiple arms in lateral contact, a massive cup with very wide columnals with a large lumen and may have evolved from the Lochkovian crotalocrinitd rather than *Pernerocrinus*.

As discussed above, the Gasterocomidae have had several genera included in the family by Bohatý (2006a) and Webster et al. (2008). Genera of the Gasterocomidae show the greatest diversity of evolution of any family in the Gasterocomoidea. This is in part the result of recognition and inclusion of 12 Devonian genera in the family. *Ancyrocrinus, Corynecrinus, Lecythocrinus, Mictocrinus*, *Tetrapleurocrinus*, and *Trapezocrinus* retain the most plesiomorphic characters in the cylindrical crown, conical cup, angustary radial facets, transversely rounded rectilinear brachials, and isotomous arm branching, whereas *Corynecrinus, Lecythocrinus*, and *Trapezocrinus* also retain the most primitive character of more than three anals below the radial summit. *Lecythocrinus* has a more advanced medium low conical cup, whereas *Corythocrinus* has a high conical cup with very low infrabasals, and *Trapezocrinus* a medium bowl-shaped cup. These three genera probably evolved from a euspirocrinid ancestor in Late Silurian or Early Devonian time and should perhaps be separated and returned to the Lecythocrinidae.

Ancyrocrinus has an intermediate anal arrangement with two anals below the radial summit and evolved the anchor stem attachment. *Mictocrinus* has the most apomorphic anal arrangement of a single anal in the medium-low bowl-shaped cup.

Gasterocoma, Arachnocrinus, Nanocrinus, Myrtillocrinus, and *Schultzicrinus* have bowl-shaped cups, an entoneural canal through the radial, and all except *Myrtillocrinus* have the anal opening through the cup within the radial circlet. *Gasterocoma, Arachnocrinus*, and *Schultzicrinus* have similar cups with differing arm structure. *Gasterocoma* has pinnulate arms with multiple branches. *Arachnocrinus* has non-pinnulate multiple arms, with large bulbous axillary brachials. *Schultzicrinus* has apomorphic atomous arms. *Nanocrinus* and *Myrtillocrinus* have widely flaring arms with those of *Myrtillocrinus* flaring horizontally from the cup whereas those of *Nanocrinus* are flaring slightly upward. The holotype of *Nanocrinus* lacks an arm in the A-ray and may be an abnormal specimen. The holotype of *Scoliocrinus* is a poorly preserved theca with the anal opening through the cup below the radial summit. Bohatý (2005) thought it might be a synonym of *Tetrapleurocrinus*, whereas Haude (2007) reported crowns with two fan-like arms and a horizontal anal tube. Recently discovered specimens from the Eifel type region affirm the diagnoses of Jaekel (1895) and Haude (2007) (Bohatý personal communication 2009). These six genera show evolutionary modification of the arms in the gasterocomoids, with the two fan-like arms of *Scoliocrinus* representing the most extreme evolution.

The Cupressocrinitidae (five genera) are the most advanced gasterocomids with most genera recognized by a low to medium bowl cup (exception *Procupressocrinus*), trifascial radial facets, no anals in the cup (except *Rhopalocrinus*, a single anal), atomous arms, and where known hyperpinnulate. Bohatý (2006b) subdivided the Cupressocrinitidae into the Cupressocrininae and Rhopalocrininae, the latter a monotypic subfamily. *Cupressocrinites* ranges from the Pragian into the Famennian and is the longest ranging gasterocomid genus. *Halocrinites* and *Robustocrinites* are probably derived from *Cupressocrinites* in the late Pragian or early Eifelian. *Halocrinites* developed a mutilamellar exoplacoid layer (Bohatý 2005, 2006b, 2009 [as *Abbreviatocrinites*]); the brachials of *Robustocrinites* extended into a U-shaped cross section transversely. Both are not known after the Givetian. *Procupressocrinus* ranges from the Eifelian into the Givetian and has a longer cup than other cupressocrinitids. *Rhopalocrinus* is restricted to the Eifelian; it retains the plesiomorphic character of a single anal in the cup. The evolution of *Procupressocrinus* and *Rhopalocrinus* is uncertain; they probably evolved from the same lineage as *Cupressocrinites.* Except for *Rhopalocrinus* the cupressocrinitds were not affected by the Frasnian–Famennian extinction. Geographically the cupressocrinitids are known only from Europe and Asia with the exception of a poorly preserved crown of *Cupressocrinites* sp. reported from Colorado (Webster and Hafley in Webster et al. 1999).

Vadarocrinus, the only Devonian (Pragian) petalocrinid, is based on an arm plate that is narrower than those of the Silurian progenitor *Petalocrinus.* It has not been reported known after the Pragian. The petalocrinids probably evolved from the crotalocrinitids by fusing of the brachials into a solid plate in each ray.

The stratigraphic distribution of the 21 genera of gasterocomoids given in Table 1 is summarized to show the diversity and first and last occurrences of the numbers of genera in Fig. 2. That figure shows that the gasterocomoids radiated after their first Devonian occurrence in the Lochkovian to attain their greatest diversity (12 genera) and several first occurrences (eight) in the Eifelian. They were nearly decimated by the end-Givetian extinction (last occurrences seven), barely survived the Frasnian–Famennian extinction (one additional extinction), and occurred last in the Famennian (one genus) and Tournaisian (one genus). The genera of Gasterocomidae and Cupressocrinidae represent a clade restricted to the Devonian and earliest Mississipian with an acme in the Eifelian, whereas the Crotalocrinidae and Petalocrinidae originated in the Silurian and became extinct in the Emsian.

Codiacrinoidea

Codiacrinids range from the Middle Silurian into the Permian where they attained their greatest diversity and geographic dispersal. Devonian codiacrinids include eight genera assigned to three families (Codiacrinidae five genera; Sphaerocrinidae one genus; Streblocrinidae two genera) and two genera of uncertain family (Table 1); they include 18 species. Species of *Codiacrinus* reported from the Pridoli (Jell and Holloway 1983) and Lochkovian (Jell 1999) from Australia are not considered to belong to the genus.

No evolutionary trends are recognized within the Devonian codiacrinids because the genera are diverse and of possible polyphyletic origin or are microcrinoids that may be juvenile forms of uncertain megacrinoids. Major evolution of the codiacrinids

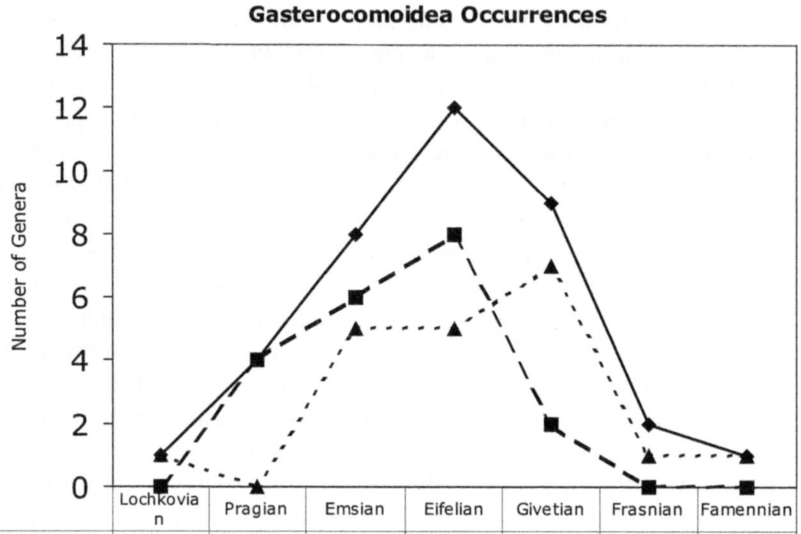

	Lochkovian	Pragian	Emsian	Eifelian	Givetian	Frasnian	Famennian
Diversity	1	4	8	12	9	2	1
First occurrence	0	4	6	8	2	0	0
Last Occurrence	1	0	5	5	7	1	1

Fig. 2 Gasterocomoidea Occurrences. Shows total diversity and first and last occurrences of the 21 Devonian gasterocomoid genera. Data bank summarized from Table 1

occurred in post-Devonian time. Most Devonian codiacrinids are small, of variable cup shape, have three infrabasals, an anal opening below or at the radial summit, and a round stem. The genera need a major study and are discussed individually.

Codiacrinus has a globose cup with interplate ridges and angustary unifascial radial facets. The interplate ridges suggest it was not derived from the globose, non-ridged Silurian *Thyridocrinus*. The origin of *Codiacrinus* is uncertain.

Gemmacrinus was left in uncertain classification by Prokop and Petr (1989) and assigned to the Codiacrinidae by Hauser (2003). The cup shape of *Gemmacrinus* is variable including an elongate high cone, an inverted pear, and a globose bowl. The plenary radial facets have a transverse ridge and are deeply set below the distal ends of the radials. This is an apomorphic character for this Pragian form.

Ovalocrinus and *Quasicydonocrinus* were reported from strata assigned an unspecified Middle Devonian age by Chen and Yao (1993). Conodonts from these beds are of Famennian age (Wang et al. 2007). *Ovalocrinus* has a pyriform cup, fused infrabasals (incorrectly reported as five infrabasals by Chen and Yao 1993), lacks ornament, and has trifascial peneplenary radial facets. *Quasicydonocrinus* has a medium bowl-shaped cup, three infrabasals, an enlarged posterior basal, and trifascial peneplenary radial facets.

Elicrinus has a medium- to high-conical cup, interplate ridges, and trifascial angustary radial facets. The trifascial radial facets in *Ovalocrinus, Quasicydonocrinus*, and *Elicrinus* are apomorphic evolutionary characters and not common in the codiacrinids. These three genera may not belong in the codiacrinids or may belong in a new family.

The Sphaerocrinidae is a monotypic family with the removal from the family of *Thalamocrinus* by McIntosh and Brett (1988) and *Palaeocrinus* by Ausich (1998). *Sphaerocrinus* has a small to large globose cup, three or five infrabasals, angustary radial facets, an anal opening below or at the radial summit, and multiple small plates within the tegmen including ambulacrals and orals. These characters have affinity with some gasterocomids as well as some codiacrinids. It is here considered to belong to the Codiacrinoidea with the development of three infrabasals while retaining the plesiomorphic characters of the multiple small plates within the tegmen and five infrabasals in some species. *Sphaerocrinus* probably evolved from an unknown Silurian or an Early Devonian codiacrinid or gasterocomid. Additional study of *Sphaerocrinus* is needed to resolve its origin. *Sphaerocrinus* became extinct in the Givetian.

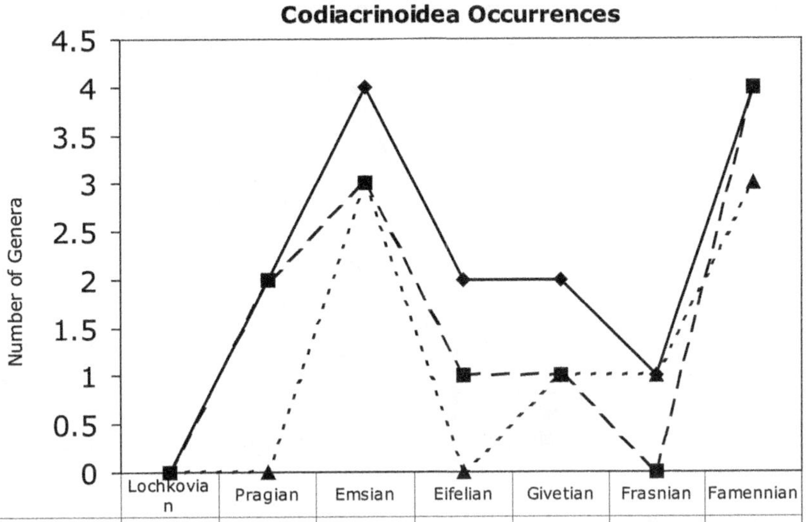

	Lochkovian	Pragian	Emsian	Eifelian	Givetian	Frasnian	Famennian
Diversity	0	2	4	2	2	1	4
First occurrence	0	2	3	1	1	0	4
Last Occurrence	0	0	3	0	1	1	3

Fig. 3 Codiacrinoidea Occurrences. Shows total diversity and first and last occurrences of the ten Devonian codiacrinoid genera. Data bank summarized from Table 1

Streblocrinus and *Pentececrinus* are very small conical-shaped microcrinoids with five infrabasals; they are based on single specimens. The holotype of *Streblocrinus* is damaged in the anal area (anals unknown), but has three radials and five large orals. *Pentececrinus* lack radials and has one anal and five elongate orals. Sevastopulo (2008) described the problem of recognition of juvenile stages of megacrinoids and microcriniods. Lacking growth stages it is unknown if *Streblocrinus* and *Pentececrinus* are based on juveniles of uncertain megacrinoids or growth stages of microcrinoids. Thus their origin is uncertain. *Streblocrinus* is known only from the Givetian and *Pentececrinus* occurs in the Devonian–Mississippian boundary beds.

Briseocrinus and *Treocrinus* are microcrinoids reported from the upper Emsian (Prokop and Petr 1991). *Briseocrinus* has a high-cone shape, fine irregular surface sculpture, and distally adjoined radials bearing one or rarely two arm facets. *Treocrinus* has a medium-cone shape, granular ornament, three radials with arm facets, and five large orals. The specimens of both of these genera may be juveniles of uncertain megacrinoids; their origin and evolution are uncertain.

The codiacrinids first Devonian occurrence (two genera) is in the Pragian (Fig. 3), although the superfamily originated in the Silurian. They radiated in the Pragian to attain a diversity acme (four genera) in the Emsian, were affected by an extinction event at the end of the Emsian to a diversity of two genera in the Eifelian and Givetian, affected by an extinction event at the end of the Givetian, declined to a low in the Frasnian (one genus), radiated in the Famennian matching the Emsian acme, and continued into the Carboniferous and Permian. Only one (*Pentececrinus*) of the ten Devonian genera continued into the Carboniferous.

Cyathocrinida Summary

Comparing the diversities of the three superfamilies of Devonian cyathocrinids shown in Fig. 4, it is obvious that the greater number of genera of the gasterocomoids controls the diversity acme in the Eifelian (17 genera), only slightly higher than that of the Emsian (16 genera) in which both the cyathocrinitoids and codiacrinoids reached their acme. Both the cyathocrinitoids and codiacrinoids declined slightly in the Eifelian while the gasterocomoids continued to diversify. All three superfamilies were affected by an extinction event at the end of the Givetian declining to a low diversity level in the Frasnian. Although the overall diversity of the cyathocrinids increased after the Frasnian–Famennian extinction event, the codiacrinoids had double (four genera) the diversity

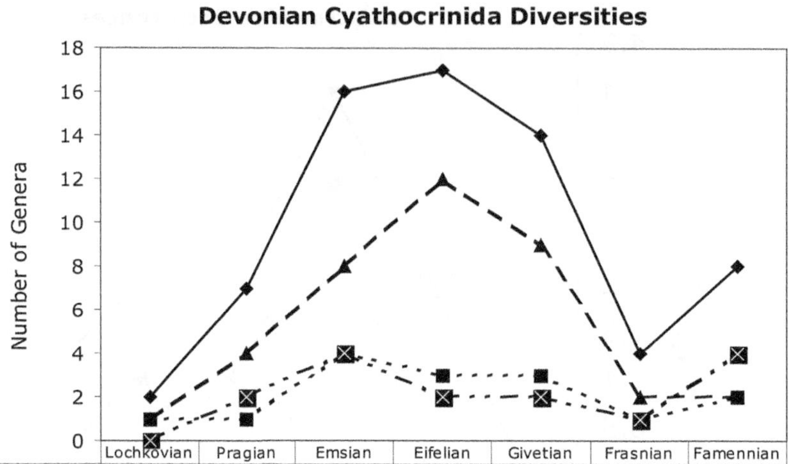

	Lochkovian	Pragian	Emsian	Eifelian	Givetian	Frasnian	Famennian
Total	2	7	16	17	14	4	8
Cyathocrinitoidea	1	1	4	3	3	1	2
Gasterocomoidea	1	4	8	12	9	2	2
Codiacrinoidea	0	2	4	2	2	1	4

Fig. 4 Devonian Cyathocrinida Diversities. Compares diversity of the Devonian genera of each of the three Cyathocrinida superfamilies to total diversity. Diversity data from data-tables Figs. 1, 2, and 3

of both the cyathocrinitoids and gasterocomoids (two genera each). The Devonian could be referred to as the age of the gasterocomoids among the cyathocrinids, but it must be remembered that their palaeogeographic distribution is dominantly Europe and Asia, whereas the cyathocrinitoids and codiacrinoids are also well represented in North America.

Primitive Dendrocrinida

Devonian Primitive Dendrocrinids (Table 2) include 33 genera, 28 of which are in five families and one "Botryocrinidae orphans" in the Dendrocrinoidea, three genera in the Mastigocrinoidea, and one genus each in the Merocrinoidea and an uncertain superfamily. These families represent the waning Primitive Dendrocrinids most of which became extinct before the end of the Devonian. Six (one questionably) of the nine genera of Primitive Dendrocrinids recognized in the Lochkovian also occur in the Middle or Late Silurian. Only three genera (*Costalocrinus, Poteriocrinites*, and *Lasiocrinus*) of the 33 Devonian genera are recognized in the Mississippian, although other genera of the Poteriocrinitidae and Mastigocrinidae that do not occur in the Devonian have been identified in the Mississippian.

The Poteriocrinitidae are included in the Dendrocrinoidea, following Webster and Maples (2008), because the type species has bifascial radial facets and the arm appendages are not known. Some species assigned to *Poteriocrinites* should be assigned to other genera in the Transitional or Advanced Dendrocrinids because they have muscular articulation radial facets and pinnules. Webster and Maples (2008) recognized only two Mississippian Primitive Dendrocrinid genera (*Hebohenocrinus* and *Springericrinus*) with muscular articulating radial facets. Because distal arms are not known on *Hebohenocrinus*, they questioned the assignment to the Primitive Dendrocrinids. *Springericrinus* has pinnules and should probably be assigned to the Amabilicrinidae in the Transitional Dendrocrinids.

Most Devonian Primitive Dendrocrinids retained the conservative plesiomorphic characters of a medium- to high-conical cup, three or four anals in the cup (exceptions in Botryocrinidae and Thalamocrinidae with two anals), angustary radial facets, unifascial or bifascial radial facets, and rectilinear brachials lacking appendages. Slightly cuneate brachials developed polyphyleticly with *Denariocrinus* (Poteriocrinitidae) in the Emsian and *Pskovicrinus* (Botryocrinidae "orphans") in the Frasnian. Ramules evolved polyphyleticly with *Pagecrinus* (Thalamocrinidae) in the Emsian,

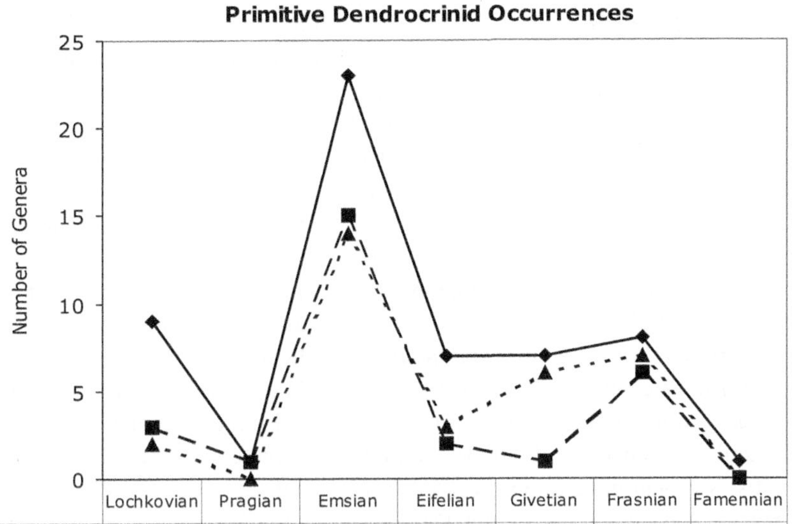

	Lochkovian	Pragian	Emsian	Eifelian	Givetian	Frasnian	Famennian
Diversity	9	1	23	7	7	8	1
First occurrence	3	1	15	2	1	6	0
Last Occurrence	2	0	14	3	6	7	0

Fig. 5 Primitive Dendrocrinid Occurrences. Shows total diversity and first and last occurrences of the 33 Devonian Primitive Dendrocrinid genera. Data bank summarized from Table 2

Eckidocrinus (Dendrocrinidae) in the Eifelian, and *Cradeocrinus* (Thalamocrinidae) in the Frasnian, though *Pagecrinus* and *Cradeocrinus* may have evolved from a common ancestor.

The most dramatic evolution in the Primitive Dendrocrinids occurred in the radial and brachial facets. Most Devonian Primitive Dendrocrinids have angustary or peneplenary radial facets. *Dracrinus*, a recently discovered Emsian dendrocrinid (Webster and Becker 2009), has articulation facets on both sides of each short rectilinear brachial. These facets have a transverse ridge across the entire width of the plenary facet, a broad shallow apical pit merging aborally with radiating crenulae extending to the outer margin of the facet. Two large muscle fields border the narrow intermuscular furrow adorally of the transverse ridge. Externally the brachials appear to have symplexy articulation. A second specimen of *Dracrinus*, recently discovered in the Eifelian in Morocco, has the secundibrachials preserved and shows that they also have articular facets on both sides of each brachial. The more distal parts of the arms are not known; the proximal arms through the sixth secundibrachial lack appendages suggesting that *Dracrinus* lacked arm appendages.

The diversity and first and last occurrences of the 33 genera of Devonian primitive dendrocrinids are summarized from Table 2 and given in Fig. 5. The line-graphs show that the first occurrences (three genera), last occurrences (two genera), and diversity (nine genera) in the Lochkovian reflect a general carry-over of Silurian genera into the Devonian. Primitive Dendrocrinids are virtually unknown in the Pragian with one first occurrence and a diversity of one. Yet seven of the Primitive Dendrocrinid genera known in the Lochkovian are carry-throughs, also known in the Emsian when the Primitive Dendrocrinids attained their Devonian acme in diversity (23 genera), first occurrences (15 genera) and last occurrences (14 genera). An extinction event at the end of the Emsian resulted in a diversity of seven genera in the Eifelian and Givetian, with a slight increase (eight genera) in the Frasnian, whereas the Frasnian–Famennian extinction event nearly eliminated the Primitive Dendrocrinids (Famennian diversity of one genus). Three genera of the Primitive Dendrocrinids carried through into the Carboniferous, but Primitive Dendrocrinids are minor elements in the Carboniferous and Permian faunas when compared to the Advanced Dendrocrinids.

Transitional Dendrocrinida

Webster et al. (2003) considered the Glossocrinoidea a clade while Webster and Maples (2008) referred to them as the Transitional Dendrocrinids. The Glossocrinoidea range from the Silurian (one genus) into the Mississippian (seven genera) with 19 Devonian genera and 44 species (Table 3). All genera, except *Nassoviocrinus*, are restricted to one period; most are restricted to a single epoch.

The Rutkowskicrinidae were recognized as a family by McIntosh (2001) and are considered to have evolved from *Nassoviocrinus* in the Silurian and given rise to the other three families (Webster et al. 2003). All three families have Mississippian genera that evolved from Devonian representatives, but none of the Devonian genera are recognized in the Mississippian whereas the Mississippian genera became extinct therein.

Rutkowskicrinids retained the plesiomorphic characters of a small medium- to high-conical cup, three or four anals in the cup, angustary radial facets, and rectilinear brachials. The type of radial facet is unknown for most Transitional Dendrocrinids, but is bifascial where known in the Glossocrinidae and Rutkowskicrinidae, muscular trifascial in the Amabilicrinidae, and questionably muscular trifascial in the Corematocrinidae. Slightly cuneate brachials bearing ramules developed on *Schmidtocrinus* and pinnules developed in the Givetian on four other rutkowskicrinids. Peneplenary radial facets are common in all glossocrinoids except the rutkowskicrinids with angustary radial facets. Glossocrinidae genera retained rectilinear brachials; all except *Catactocrinus* (ramule bearing) have pinnules. Slightly to moderately cuneate brachials developed in the corematocrinids and amabilicrinids in the Late Devonian and became strongly cuneate in the Mississippian genera. Most Transitional Dendrocrinids have an elongate anal sac formed of columns of ribbed, plicate, hexagonal plates interlocking laterally. Many of the anal tubes are recurved. McIntosh (2001) considered the anal plates and sac to be one of the unifying characters of the rutkowskicrinids. The same type of anal sac occurs in all other glossocrinoid families, but in the Mississippian a few genera (e.g., *Cydrocrinus*) evolved with balloon-shaped anal sacs or elongate tubes of pentagonal plates.

Figure 6 shows the diversity and first and last occurrences of the Devonian Transitional Dendrocrinids as summarized from Table 3. Although the first occurrences and diversity of the Transitional Dendrocrinids remained relatively low in the Pragian (one genus each) and Emsian (first occurrences two, diversity three genera) they are not known from the Eifelian; they radiated through the Givetian to a high in the Frasnian (diversity ten genera, first occurrences nine genera, last occurrences eight genera). The diversity

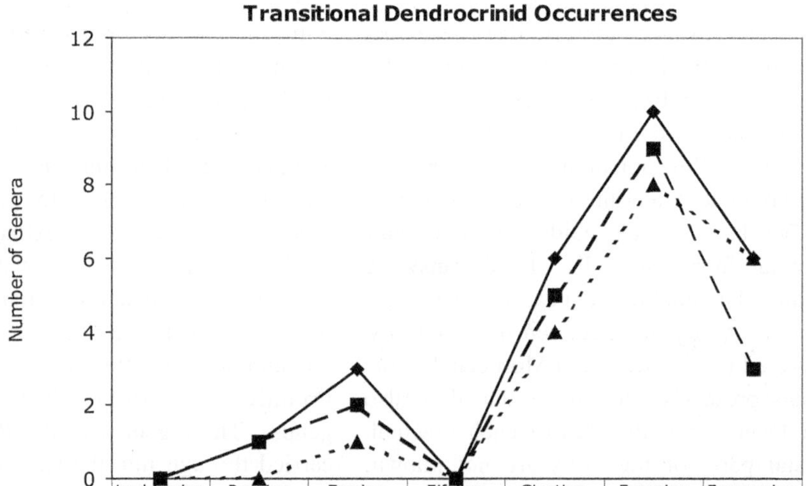

	Lochkovian	Pragian	Emsian	Eifelian	Givetian	Frasnian	Famennian
Diversity	0	1	3	0	6	10	6
First occurrence	0	1	2	0	5	9	3
Last Occurrence	0	0	1	0	4	8	6

Fig. 6 Transitional Dendrocrinid Occurrences. Shows total diversity and first and last occurrences of the 19 Devonian Transitional Dendrocrinid genera. Data bank summarized from Table 2

and last occurrences dropped to six genera whereas the first occurrences dropped to three genera in the Famennian. Thus the Transitional Dendrocrinids were strongly affected by Emsian–Eifelian and Frasnian–Famennian extinction events.

Advanced Dendrocrinida

As noted above the Advanced Dendrocrinids were previously assigned to the Poteriocrinida but are now included in the Cladida under the dendrocrinids. No suborder has been proposed for these crinoids because they are under study and their evolutionary lineage is not established. The Devonian Advanced Dendrocrinids are assigned to four superfamilies, nine families (one uncertain), 19 genera, and 37 species (Table 4). None of the superfamilies have pre-Devonian genera but all except the Rhenocrinoidea have post-Devonian representatives, with ten of the Devonian genera continuing into the post-Devonian.

Rhenocrinus is the only Advanced Dendrocrinid recognized in the Early Devonian. It retains the plesiomorphic conical cup, four anals, an elongate anal tube, rectilinear basals, and arms branching isotomously well above the cup. The type of radial facet is unknown. The four anals are a primitive feature, whereas the presence of pinnules is an advanced feature. *Rhenocrinus* was included in the Advanced Dendrocrinids by Webster and Maples (2008) because it has pinnules although they were considered ramules on every second brachial by Moore et al. (in Moore and Teichert 1978). *Rhenocrinus* probably evolved from a euspirocrinid with evolution of the pinnules.

The second oldest (Eifelian) Advanced Dendrocrinid (*Struveicrinites*) is not assigned to a family because it is based on a partial crown, has both primitive and advanced characters, and may not have pinnules. Angustary radial facets and uniserial rectilinear brachials are plesiomorphic characters, whereas it is uncertain if the bowl-shaped cup with a deep basal invagination is a plesiomorphic or apomorphic character; the one anal, and arms branching on the second primibrachial, are apomorphic characters. If it lacks pinnules, it would be considered an evolved Primitive Dendrocrinid with invaginated bowl-shaped cup. The derivation of *Struveicrinites* is uncertain, but it may be from a cyathocrinid.

Givetian Advanced Dendrocrinids are the decadocrinids *Decadocrinus* and *Zostrocrinus*. *Decadocrinus* retains three anals in the cup and an elongate anal tube; it has a medium bowl-shaped cup, apomorphic characters of plenary radial facets, first arms branching on the second primibrachial, and slightly to moderately cuneate brachials with pinnules. The distal brachials of *Zostocrinus* are unknown, but otherwise the cup structure is similar to *Decadocrinus*. *Graubauicrinus* and *Holcocrinus* evolved in the Late Devonian. *Graubauicrinus* has a cup structure similar to *Decadocrinus*, but the arms have very short first primibrachials; distal brachials are only slightly cuneate. The derivation of the decadocrinids is uncertain but the bowl-shaped cup accords with a possible cyathocrinid ancestor.

Two Frasnian Advanced Dendrocrinids are *Bridgerocrinus* (Bridgerocrinidae) and *Prininocrinus* (Scytalocrinidae). *Bridgerocrinus* retains the plesiomorphic characters of a small conical cup, three anals in the cup, and an elongate anal tube. It has the apomorphic characters of plenary radial facets, arm branching on the second primibrachial, and pinnule-bearing slightly cuneate brachials. *Bridgerocrinus* is thought to have evolved from one of the Transitional Dendrocrinids such as *Quantoxocrinus* by lowering of the first branching of the arms and development of the slightly cuneate brachials. *Bridgerocrinus* probably gave rise to *Holcocrinus*, which has a more bowl-shaped cup, a single anal in the cup, and a single primibrachial. *Holcocrinus* was transferred to the Decadocrinidae by Webster and Kues (2006) because it has cuneate brachials. It differs from other decadocrinids by the presence of a single elongate axillary primbrachial and presence of a single anal. Webster and Lane (2007) recognized the Bridgerocrinidae based on several genera with elongate primibrachials. In addition, they considered primitive forms to have a conical cup, three anals, and two primibrachials whereas advanced forms have bowl-shaped cups, some with a single anal and a single primibrachial. *Holcocrinus* is here placed in the Bridgerocrinidae and *Eireocrinus* is accepted as a junior synonym of *Holcocrinus*, as recognized by Ausich and Sevastopulo (2001). Both *Bridgerocrinus* and *Holcocrinus* continue into the Mississippian.

Prininocrinus is the earliest scytalocrinid (Frasnian) recognized in the Devonian; all other

Devonian scytalocrinid genera are Famennian in age. *Prininocrinus*, *Histocrinus*, *Julieticrinus*, and *Scytalocrinus* have bowl-shaped cups; all except *Scytalocrinus* have two primibrachials. *Prininocrinus* has rectilinear brachials; the others have slightly to moderately cuneate brachials except *Julieticrinus*, which has laterally expanded flanges on alternate sides of the brachials. *Prininocrinus* was probably derived from an uncertain Transitional Dendrocrinid by modification of the cup to a medium bowl, reduction of the arms to two per ray, with the first branching on the second primibrachial. *Scytalocrinus* may have evolved from *Prininocrinus* by loss of one primibrachial and development of slightly cuneate brachials. *Julieticrinus* could have evolved from *Prininocrinus* by development of the flanges on the brachials while *Histocrinus* evolved cuneate brachials. The other two Famennian scytalocrinids, *Hypselocrinus* and *Hydriocrinus* have plesiomorphic small high cone-shaped cups and apomorphic conditions of a single primibrachial and cuneate brachials. They were probably derived from one of the Transitional Dendrocrinids and may be of different lineage than *Prininocrinus*. Only one of the Devonian scytalocrinid genera (*Julieticrinus*) does not continue into the Mississippian where the scytalocrinids reached their acme. Problems with classification of the genera assigned to the Scytalocrinidae have been discussed by Webster (1997), Ausich and Sevastopulo (2001) and Webster et al. (2003). These include the Devonian as well as Mississippian genera and are beyond the scope of this review.

Cosmetocrinus, the earliest aphelecrinid, retains the plesiomorphic characters of a conical cup with three anals and multiple isotomously branched arms with transversely rounded brachials; it has apomorphic characters in the plenary radial facets, first branching of the arms on the first or second primibrachial and moderately cuneate brachials. It is thought to have evolved from a Transitional Dendrocrinid by development of lower first branching of the arms and cuneate brachials. *Cosmetocrinus* continues into the Mississippian.

Emhacrinus, the earliest cercidocrinid, retains the conical cup with three anals but has plenary radial facets, endotomously branching arms with the first branching on the third primibrachial, and moderately cuneate brachials. It is also thought to have evolved from a Transitional Dendrocrinid, but is restricted to the Famennian.

Devonian species assigned to *Sostronocrinus* (Sostronocrinidae) may be incorrectly assigned to the genus. The type species of *Sostronocrinus* is of Mississippian age and has a conical cup, deep apical pits, stellate ornament, peneplenary radial facets, three anals, recurved anal tube, slightly cuneate brachials, first branching on the second or third primibrachial, four arms per ray, and bears pinnules. Famennian species questionably assigned to *Sostronocrinus* from Germany (Haude and Thomas 1989) have a similar cup but the arms first branch on the first or second (most commonly) primibrachial and have medium cuneate brachials. These differences are a more advanced character than on the Mississippian type species of the genus, which branch on the second or third primibrachial and have slightly cuneate brachials. Specimens from England originally described by Whidborne (1898) and reassigned to *Sostronocrinus* by Lane et al. (2001a) have a straight-sided cup and two primibrachials. The straight-sided cup differs from the type species, which has a cup with deep apical pits and stellate ornament. One of the northwestern China species (*S. minutus*) assigned to *Sostronocrinus* by Waters et al. (2003) was reassigned to *Bridgerocrinus* by Webster and Lane (2007); the other species (*S. quadrabrachiatus*) has a bowl-shaped cup, the first arm branching on the first (rarely) or fourth primibrachial, and a pentagonal stem. All these characters differ from the type species. I consider the German and English species assigned to *Sostronocrinus* to be derived from a Transitional Dendrocrinid. Derivation of the northwestern China species, *S. quadrabrachiatus*, is uncertain, but the bowl-shaped cup may suggest an unknown cyathocrinid. All these species are considered not to belong to the genus and may not be the progenitor of *Sostronocrinus*.

Amadeusicrinus first appears in the Famennian of northwestern China and has a bowl-shaped cup, peneplenary trifascial radial facets, three anals, first branching on the fourth primibrachial, slightly cuneate brachials, 20 arms, pinnules, and a pentagonal stem facet. The derivation of *Amadeusicrinus* is uncertain, but the bowl-shaped cup may suggest an unknown cyathocrinid.

Famennian specimens from England originally described by Whidborne (1898) and questionably reassigned to *Pachylocrinus* by Lane et al. (2001a) differ from the type species by having moderately deep apical pits on the cup and the first branching of the arms on the first to several primibrachials instead of the third primibrachial. Lane et al. (2001a) thought this might be a progenitor within the *Pachylocrinus* lineage. Although I agree that it might be a progenitor, I consider it to represent a new genus and that it was probably derived from a Transitional Dendrocrinid.

Specimens from the late Famennian of Colorado (U.S.A.) assigned to *Tarassocrinus* and an indeterminate family by Webster and Hafley (in Webster et al. 1999) have a medium bowl-shaped cup, plenary radial facets, three anals, an elongate tegmen, arms branching on primibrachials one to nine with one additional branching in some rays, slightly cuneate brachials which may become biserial distally, and a circular stem. The origin of this taxon is uncertain, but the biserial nature of the brachials is an apomorphic character, as are the plenary radial facets.

The first occurrence of the Advanced Dendrocrinids is in the Emsian as shown in Fig. 7, summarized from Table 4. Advanced Dendrocrinids had a low diversity and first occurrence until the Famennian when they radiated and attained their Devonian acme of diversity (15 genera), first occurrences (13 genera), and last occurrences (5 genera). Only one genus had a last occurrence in the Givetian. The Late Devonian was clearly the time of evolution, not extinction, for the Advanced Dendrocrinids because they were virtually not affected by the Frasnian–Famennian extinction event. This is in agreement with the stratigraphic distribution of the Dendrocrinida (i.e., "Dendrocrinoidea" of More and Laudon 1941, Fig. 2; among others) that the Late Devonian evolution of the Advanced Dendrocrinids (poteriocrinids of most later authors, e.g., Lane 1978) was the beginning of the group that were to dominate Late Carboniferous and Permian crinoid faunas.

Articulata

Webster and Jell (1999) recognized *Logocrinus* as the earliest articulate crinoid, based on the paired muscular articulation in the brachials among other characters. *Logocrinus* retains the plesiomorphic characters of a small high-conical cup, three anals, slender externally rounded brachials, and slightly pentagonal stem. It has apomorphic characters in the plenary radials, muscular trifacial radial facets, paired muscular articulating cuneate brachials, and bears pinnules. It was derived from one of the earlier Devonian glossocrinoids.

Evolution of the plenary radial facets and paired muscular articulating brachials was critical for use of the arms in walking on the arms after undoing the holdfast—as described in modern articulate crinoids by Baumiller and Messing (2005). This critical Middle Devonian evolutionary development was the start of the Articulata and separates them from the advanced cladids. *Logocrinus* occurs in the Givetian and Frasnian and was an apparent victim of the Frasnian–Famennian extinction event.

Dendrocrinida Summary

Figure 8 is a line-graph comparison of diversities of the Primitive, Transitional, and Advanced Dendrocrinids (data-bank from Figs. 5, 6, and 7 with addition of the one articulate genus) with the total dendrocrinid diversity for each Devonian epoch. The total diversity suggests that the Lochkovian dendrocrinds (nine genera) were decimated by an extinction event at the end of the Lochkovian with only two genera in the Pragian, radiated to an acme of 27 genera in the Emsian, crashed at the end Emsian extinction event to another low in the Eifelian (nine genera), and increased through the Givetian and Frasnian to a second high in the Famennian (22 genera). The diversity of the Primitive Dendrocrinids closely parallels that of the total diversity until the Givetian when it decreases to slightly less than half (seven genera) of the total (16 genera) and is barely represented in the Famennian by one genus. The Transitional Dendrocrinid genera are a small part of the total diversity until they make up significant parts of the total diversity in the Givetian (six genera), Frasnian (ten genera), and Famennian (six genera). Genera of the Advanced Dendrocrinids are a minor element until the Famennian where they made up approximately two-thirds (15 of 22 genera) of the total diversity. Advanced Dendrocrinids replaced the Primitive Dendrocrinids through the Transitional Dendrocrinids in the Devonian. With only one genus present in the Eifelian and Givetian, Devonian articulates were never a significant part of the total diversity; however, this marks their first occurrence.

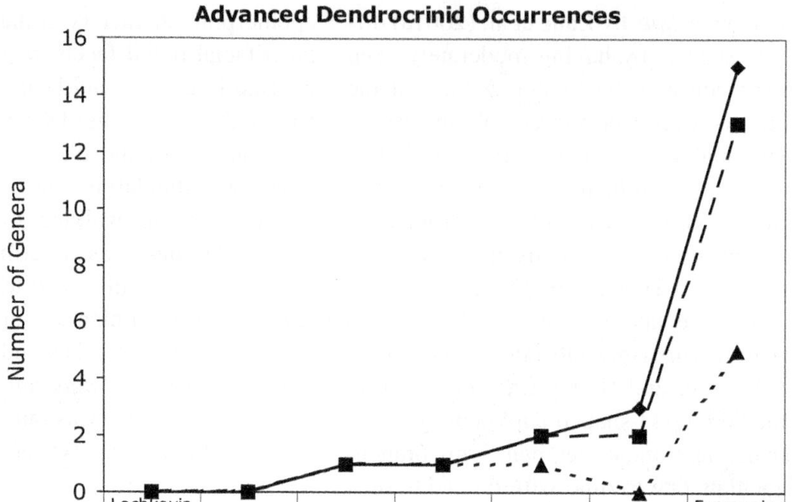

	Lochkovian	Pragian	Emsian	Eifelian	Givetian	Frasnian	Famennian
Diversity	0	0	1	1	2	3	15
First occurrence	0	0	1	1	2	2	13
Last Occurrence	0	0	1	1	1	0	5

Fig. 7 Advanced Dendrocrinid Occurrences. Shows total diversity and first and last occurrences of the 19 Devonian Advanced Dendrocrinid genera. Data bank summarized from Table 2

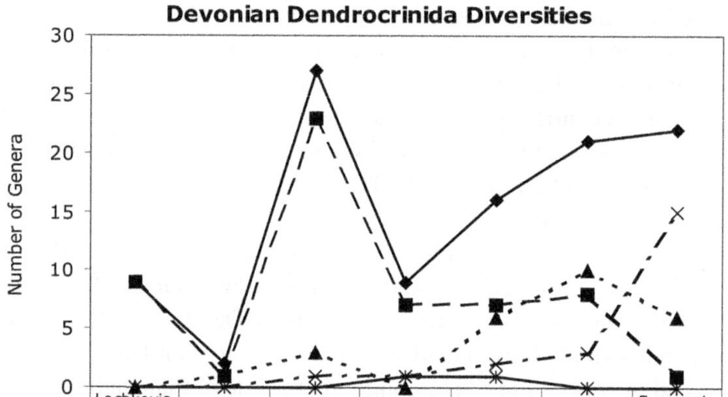

	Lochkovian	Pragian	Emsian	Eifelian	Givetian	Frasnian	Famennian
Total	9	2	27	9	16	21	22
Primitive Dendrocrinids	9	1	23	7	7	8	1
Transitional Dendrocrinids	0	1	3	0	6	10	6
Advanced Dendrocrinids	0	0	1	1	2	3	15
Articulata	0	0	0	1	1	0	0

Fig. 8 Devonian Dendrocrinida Diversities. Compares diversity of the Devonian genera of each of the three Dendrocrinida subgroups to that of the total diversity. Diversity data from data-tables Figs. 5, 6, and 7 with addition of one articulate

A comparison of all dendrocrinid first and last occurrences with the diversity as summarized from Tables 2, 3, and 4 and Figs. 5, 6, and 7 shows that the first and last occurrences essentially parallel the total diversity throughout the Devonian with some diversion in the Famennian (Fig. 9). That diversion represents the decline of the Primitive and Transitional Dendrocrinids.

Cladid First and Last Occurrences

Figure 10 shows that the diversity of the Devonian genera of the Cladida was relatively low in the Lochkovian (10 genera) and Pragian (nine genera), increased dramatically to an acme in the Emsian (43 genera), was nearly cut in half by an end-Emsian extinction event,

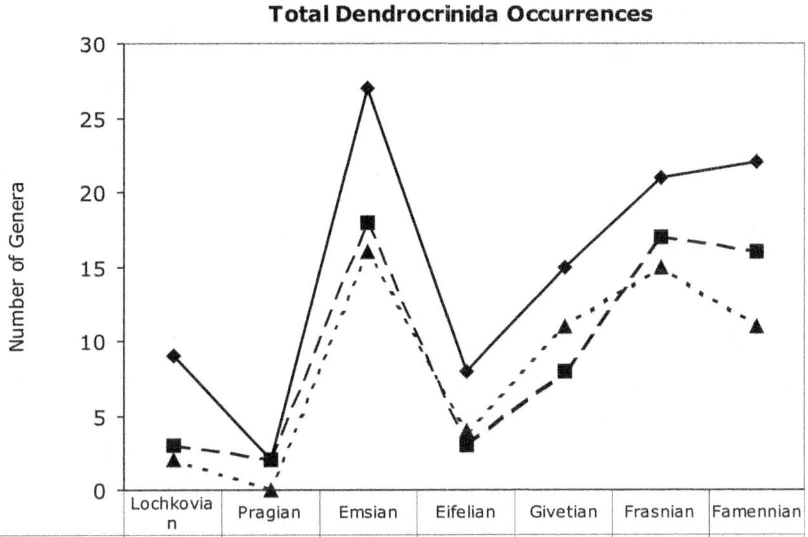

	Lochkovian	Pragian	Emsian	Eifelian	Givetian	Frasnian	Famennian
Diversity	9	2	27	8	15	21	22
First occurrence	3	2	18	3	8	17	16
Last Occurrence	2	0	16	4	11	15	11

Fig. 9 Total Dendrocrinida Occurrences. Shows total diversity and first and last occurrences of all Devonian dendrocrinids. First and last ccurrences compiled from Figs. 5, 6, and 7 with addition of one articulate

and maintained a moderate number of 25–30 genera throughout the Eifelian to Famennian. This compiled figure does not show the differences in the first and last occurrences and standing diversities of each of the Devonian superfamilies of the Cyathocrinida and Dendrocrinida as described above. It does not show that some cladid superfamilies had their Devonian acme in the Eifelian or Famennian instead of the Emsian, some superfamilies are restricted to the Devonian, or that some superfamilies became extinct

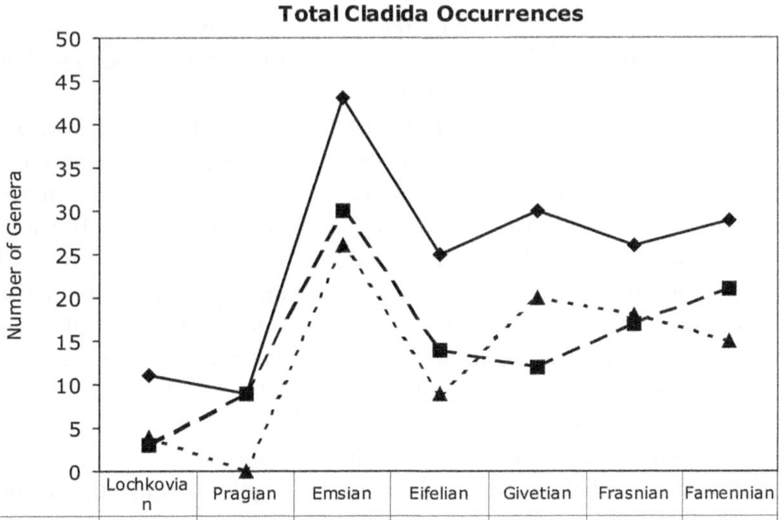

	Lochkovian	Pragian	Emsian	Eifelian	Givetian	Frasnian	Famennian
Diversity	11	9	43	25	30	26	29
First occurrence	3	9	30	14	12	17	21
Last Occurrence	4	0	26	9	20	18	15

Fig. 10 Total Cladida Occurrences. Shows total diversity and first and last occurrences of Devonian cladids. Data compiled from Figs. 1–9

in the Devonian. Figure 10 should be used with caution and in conjunction with Tables 1, 2, 3, and 4 and Figs. 1–9.

Conclusions

Overall, the diversity of the Devonian Cyathocrinida peaked in Emsian and Eifelian, declined slightly in the Givetian, was drastically affected by a major extinction event at the end of the Givetian, and began a rebound in the Famennian. Subdivision of the Cyathocrinida into the three superfamilies is clearly justified on the basis of morphology, but the evolutionary lineages of and within these superfamilies needs critical review.

Cyathocrinoidea genera were a minor part of the total Cyathocrinida diversity throughout the Devonian.

The Gasterocomoidae are divisible into two subgroups based on morphologic differences in the cup and arms. Group one includes the Crotalocrinidae and Petalocrinidae; they are more common in the Silurian. Group two includes the Gasterocomidae and Cupressocrinitidae, a Late Devonian/Early Mississippian clade. They differ extensively from the Crotalocrinidae and Petalocrinidae in the structure of the cup and arms. These two groups should probably be separated into different superfamilies.

The Codiacrinoidea needs serious review to determine the origin of the various types of radial facets present on Devonian genera currently assigned to the superfamily.

The overall diversity of the Devonian genera of Dendrocrinida had two lows (Pragian and Eifelian) following extinction events at the end of the Lochkovian and Emsian, respectively. There were also two peaks, highest in the Emsian and second highest in the Famennian. Although included in the data bank, the overall diversity summary does not show that the Primitive Dendrocrinids were nearly wiped out in the Devonian and replaced by the Advanced Dendrocrinids which first occurred in the Devonian. Neither does it show the numerous first occurrences of the Transitional Dendrocrinids with development of trifascial radial facets and pinnules nor does it show development of paired muscular articulating cuneate brachials in the articulates.

These morphological differences justify and allow subdividing the Dendrocrinida into four groups with the following major morphologic differences:

(1) Primitive Dendrocrinids lacking trifascial radial facets and pinnules;
(2) Transitional Dendrocrinids with plesiomorphic conical cups, trifacial radial facets, and pinnules (this would include part of the former Poteriocrinida genera included in the Advanced Dendrocrinida above;
(3) Advanced Dendrocrinida with plesiomorphic bowl-shaped cups, trifascial radial facets, and pinnules (this would include part of the former Poteriocrinida genera included in the Advanced Dendrocrinida above); and
(4) Articulata with paired muscular articulating cuneate brachials.

In the course of compiling the data bank for this chapter, there are several other observations that resulted in several other conclusions or areas for future study of the Cladida.

In the Primitive Dendrocrinids, an analysis of the "Botryocrinidae orphans" is needed to determine both their lineage and their relationship to one another and other Primitive Dendrocrinids.

In the Scytalocrinoidea, the Scytalocrinidae and Sostronocrinidae may be derived from a different lineage than the Aphelecrinidae, Bridgerocrinidae, and Cercidocrinidae.

The Decadocrinidae may be in a lineage akin to that of the Scytalocrinidae and Sostronocrinidae, whereas the Pachylocrinidae may be in a lineage related to that of the Aphelecrinidae, Bridgerocrinidae, and Cercidocrinidae. These two lineages are suggested on differences in cup shape.

Devonian cladids retaining plesiomorphic characters of the conical cup and apomorphic characters in the arms are morphologically most like the Glossocrinoidea (Transitional Dendrocrinids) from which they were most likely derived. Several of these families continued to radiate in post-Devonian time to be part of the Late Paleozoic Crinoid Macroevolutionary fauna of Ausich and Kammer (2006) in the Pennsylvanian and Permian.

Derivation of Devonian cladids with bowl-shaped cups is uncertain. These forms would require modification of the conical cup to a bowl-shape or need an ancestor with a bowl-shaped cup. They may be derived from an unknown cyathocrinid akin to *Gissocrinus* or *Parisocrinus*, genera with bowl-shaped cups and three or more anals in the cup. Such an ancestral lineage would require slight modification of the cup and polyphyletic evolution of the muscular radial facets and pinnules. If such a lineage were established, this would support the previously suggested polyphyletic evolution for the post-Devonian cladids. It would also require recognition of two major lineages within post-Devonian cladids.

Reviewing the Devonian cladid genera and species suggests that several of the species assigned to various genera are not properly identified and need reassignment or the genus needs revision with an expanded interpretation to include characters present on one of more of the species assigned to that genus. These revisions should not include species with apomorphic characters not present on the type species, which is stratigraphically younger than the Devonian species.

Genera with numerous species (some or many of which may be pre- or post-Devonian) typically are in need of systematic revision, including cladistic analyses. This includes *Bactrocrinites, Crotalocrinites, Cyathocrinites, Gasterocoma*, and *Poteriocrinites*. When these revisions are done they will probably result in greater generic diversity in the Devonian cladids and should clarify some evolutionary lineages.

The time gaps for carry-through genera show the incompleteness of the fossil record and suggest stratigraphic intervals that should be investigated to help complete the record. Until recently the Famennian was considered to have an extremely meager crinoid record. New discoveries in the past 15 years have expanded that record extensively, as noted by Waters and Webster (2009).

Generic occurrences shown on Tables 1, 2, 3, and 4 indicate that the records for the Lochkovian and Pragian are quite meager. This results in the question: Was there an extinction event at the end of the Silurian and Lochkovian or is the cladid crinoid record of these epochs poor because of environmental controls or lack of discovery of fossiliferous localities? Discovery of faunas of these ages will improve that record in the future.

Another question that arose in the course of this study is why there are no blothrocrinids in the Famennian. The blothrocrinids (Early Mississippian–Early Permian) have a conservative plesiomorphic conical cup, three anals, and the pinnulate arms first branch well above the cup. When looking at the Devonian occurrence and morphology of related scytalocrinids, aphelecrinids, and cercidocrinids, it suggests there should be blothrocrinids in the latest Devonian because they may be the progenitors of some of these related taxa.

The Devonian was a critical time in the evolution and extinction of the cladids. This review expands on the summaries of Lane (1978) and Strimple (1978) and suggests additional areas of research to resolve some of the remaining problems with cladid evolution and first and last occurrences.

Acknowledgments Jan Bohatý provided new insights into the gasterocomids and Tom Kammer suggested addition of the graphs. Helpful reviews by both are gratefully acknowledged. Paul Olin kindly gave advice on construction of the graphics.

References

Angelin NP (1878) Iconographia Crinoideorum: in stratis Sueciae Siluricis fossilium. Samson and Wallin, Holmiae, pp 1–62

Ausich WI (1998) Phylogeny of Arenig to Caradoc crinoids (Phylum Echinodermata) and suprageneric classification of the Crinoidea. Univ Kansas Paleontol Contrib (n. s.) 9:1–36

Ausich WI, Kammer TW (2006) Stratigraphical and geographical distribution of the Mississippian (Lower Carboniferous) Crinoidea from England and Wales. Proc Yorkshire Geol Soc 56:91–109

Ausich WI, Sevastopulo GD (2001) The Lower Carboniferous (Tournaisian) crinoids from Hook Head, County Wexford, Ireland. Palaeontologr Soc Mon 216:1–136, Suppl 1:1–34; Suppl 2:1–11

Austin TS, Austin TJ (1843) Description of several new genera and species of Crinoidea. Ann Mag Nat Hist ser 1 11:195–207

Baumiller TK, Messing CG (2005) Crawling in stalked crinoids: *in situ* observations, functional morphology, and implications for Paleozoic taxa. Geol Soc America Abstr Prog 37(5):65

Bohatý J (2001) Wachstumsanomalien mitteldevonischer Crinoidenkelche aus den Kalkmulden der Eifel (Rheinisches Schiefergebirge). Greifswalder Geowissenschaftliche Beiträge 9:79

Bohatý J (2005) Doppellagige Kronenplaten: Ein neues anatomisches Merkmal paläozoischer Crinoiden und Revision der Familie Cupressocrinitidae (Devon). Paläont Zeit 79(2):201–225

Bohatý J (2006a) Seltene Gasterocomidae (Crinoidea) aus dem Mittel-Devon der Eifel (linksrheinisches Schiefergebirge, Deutschland). N Jahrb Geol Paläontol Abh 239:399–443

Bohatý J (2006b) Neue Cupressocrinitidae (Crinoidea) aus den mitteldevonischen Kalkmulden der Eifel (linksrheinisches Schiefergebirge, Deutschland). Senckenbergiana Lethaea 86:151–189

Bohatý J (2009) Pre- and postmortem skeletal modifications of the Cupressocrinitidae (Crinoidea, Cladida). J Paleontol 83:45–62

Bouska J (1946) Celed Crotalocrinidae (Angelin) v ceském siluru a devonu [On Crotalocrinidae (Angelin) from the Silurian and Devonian of Bohemia]. Rozpravy II. Trídy Ceské Akademie 56(4):1–24

Breimer A (1962) A monograph on Spanish Paleozoic crinoidea. Leidse Geol Med 27:1–190

Chen Z-T, Yao J-H (1993) Palaeozoic echinoderm fossils of western Yunnan, China, vol 1. Geological Publishing House, Beijing, 102p

Frest TJ (1977) *Cyathocrinites* from the Silurian (Wenlock) strata of southeastern Indiana. Fieldiana Geol 35(7):109–136

Goldfuss GA (1826–33) Petrefacta Germaniae, tam ea, Quae in Museo Universitatis Regiae Borussicae Fridericiae Wilhelmiae Rhenanea, serventur, quam alia quaecunque in Museis Hoeninghusiano Muensteriano aliisque, extant, iconibus et descriptions illustrata. Abbildungen und Beschreibungen der Petrefacten Deutschlands und der Angränzende Länder, unter Mitwirkung des Herrn Grafen Georg zu Münster, herausgegeben von August Goldfuss. 1:1–242 (1826–1833), Divisio prima. Zoophytorum reliquiae, 1–114; Divisio secunda. Radiariorum reliquiae, 115–221 [Echinodermata]; Divisio tertia. Annulatorium reliquiae. Atlas of plates, 1–199. Arnz & Co., Düsseldorf

Goldfuss GA (1839) Beitäge zur Petrefactenkunde, Acta Natur. Curiosorum. Nova Acta Physico-medica Academie Caesar Leopoldino-Carolinae Naturae Curiosorum 19: 329–364, Breslau & Bonn

Goldring W (1923) The Devonian crinoids of the state of New York. New York State Mus Mem 16:1–670

Goldring W (1938) Devonian crinoids from the Mackenzie River Basin (N.W.T.) Canada. Bull Am Paleontol 24(81):1–23

Goldring W (1939) *Linobrachiocrinus*, new name for *Linocrinus* Goldring, not Kirk, 1938. J Paleontol 13:354

Goldring W (1954) Devonian crinoids: new and old, II. New York State Mus Circ 37:1–51

Gradstein FM, Ogg JG (2004) Geologic time scale 2004 – why, how, and where next! Lethaia 37:175–181

Haarmann E (1920) Die Botryocriniden und Lophocriniden des rheinischen Devons. Jahrb Preussischen Geol Landesanstalt 41(1, hf 1):1–87

Hall J (1862) Preliminary notice of some of the species of Crinoidea known in the Upper Helderberg and Hamilton groups of New York. New York State Cabinet of Natural History 15th Ann Rep, pp 87–125

Haude R (2004) Morphology and palaeobiology of echinoderms in the Lower Devonian of the Argentine Precordillera. In: Heinzeller T, Nebelsick JH (eds) Echinoderms. Müchen, Taylor and Francis Group, London, pp 417–419

Haude R (2007) Mode of life of fan-hand-like *Scoliocrinus* according to functional form and syntopy with two other irregular crinoids from the Middle Devonian of the Rhenish Massif (Germany). Ann Paléont 93:291–316

Haude R, Thomas E (1989) Ein Oberdevon–Unterkarbon-Profil im Velberter Sattel (nördliches Rheinisches Schiefergebirge) mit neuen Arten von (?) *Sostronocrinus* (Echinodermata). Bull Belg Vereniging Geol 98: 373–383

Hauser J (1997) Die Crinoiden des Mittel-Devon der Eifler Kalkmulden. Privately pub, Bonn 1, 273p

Hauser J (1998) Neue Crinoidenfunde aus dem Mitteldevon der Eifel. Fossilien 98(4):245–249

Hauser J (2001) Neubeschreibung mitteldevonischer Eifelcrinoiden aus der Sammlung Schultze (Museum of Comparative Zoology, The Agassiz Museum, Harvard University, Massachusetts, USA) nebst einer Zusammenstellung der Eifelcrinoiden (Holotypen) der Goldfuss-Sammlung. Privately pub, Bonn, 86p

Hauser J (2003) Über *Jaekelicrinus* und andere Crinoiden aus dem Frasnium (Oberdevon) vom Südrand der Dinant Mulde (Ardennen, Belgien). Privately pub, Bonn, 59p

Hauser J (2007) Die Crinoidenwelt der Eifel vor 350.000.000 Jahren. Privately pub, Bonn, 83p

Hauser J (2008) Crinoiden und Begleitfauna des Ahbachiums der Rommersheimer Trasse. Privately pub, Bonn, 80p

Hess H, Ausich WI, Brett CE, Simms MJ (1999) Fossil crinoids. Cambridge University, 275p

Jaekel O (1895) Beiträge zur Kenntniss der palaeozoischen Crinoiden Deutschlands. Paläont Abh 7:116p

Jaekel O (1918) Phylogenie und System der Pelmatozoen. Paläeont Zeit 3(1):128p

Jell PA (1999) Silurian and Devonian crinoids from central Victoria. Mem Queensland Muse 43:1–114

Jell PA, Holloway DJ (1983) Devonian and ?Late Silurian palaeontology of the Winneke Reservoir site, Christman Hills, Victoria. Proc Roy Soc Victoria 95:1–21

Jell PA, Jell JS (1999) Crinoids, a blastoid and a cyclocystoid from the Upper Devonian reef complex of the Canning Basin, Western Australia. Mem Queensland Mus 43:201–236

Jell PA, Theron JN (1999) Early Devonian echinoderms from South Africa. Mem Queensland Mus 43:115–200

Jell PA, Jell JS, Johnson BD, Mawson R, Talent JA (1988) Crinoids from Devonian limestones of eastern Australia. Mem Queensland Mus 25:355–402

Kammer TW, Ausich WI (2006) The "age of crinoids": a Mississippian biodiversity spike coincident with widespread carbonate ramps. Palaios 21:238–248

Kelly SM (1982) Origin of the crinoid orders Disparida and Cladida: possible inadunate cup plate homologies. Third North American Paleontological Convention Proc 1: 285–290

Kirk E (1914) Notes on the fossil crinoid genus *Homocrinus* Hall. Proc US Nat Mus 46:473–483

Kirk E (1929) *Pagecrinus*, a new crinoid genus from the American Devonian. Proc US Nat Mus 75(22):1–4

Kirk E (1934) *Corynecrinus*, a new Devonian crinoid genus. Proc US Nat Mus 83(2972):1–7

Kirk E (1940) Seven new genera of Carboniferous Crinoidea Inadunata. J Washington Acad Sci 30:321–334

Kirk E (1945) *Holcocrinus*, a new inadunate crinoid genus from the Lower Mississippian. Am J Sci 243:517–521

Kirk E (1948) Two new inadunate crinoid genera from the Middle Devonian. Am J Sci 246:701–710

Koenig JW, Meyer DL (1965) Two new crinoids from the Devonian of New York. J Paleontol 39:391–397

Koenig JW, Niewoehner W (1959) *Pentececrinus*, a new microcrinoid from the Louisiana Formation of Missouri. J Paleontol 33:462–470

Lane NG (1978) Cladida, Cyathocrinina and Dendrocrinina. In: Moore RC, Teichert C (eds) Treatise on invertebrate paleontology, Part T, Echinodermata 2, Crinoidea, 3 vols.

Geological Society of America and University of Kansas, pp 294–298, 1027
Lane NG, Waters JA, Maples CG, Marcus SA, Liao Z-T (1996) A camerate-rich Late Carboniferous (Moscovian) crinoid fauna from volcanic conglomerate, Xinjiang, People's Republic of China. J Paleont 70:117–128
Lane NG, Maples CG, Waters JA (2001a) Revision of Late Devonian (Famennian) and some Early Carboniferous (Tournaisian) crinoids and blastoids from the type Devonian area of North Devon. Palaeontology 44:1043–1080
Lane NG, Maples CG, Waters JA (2001b) Revision of Strunian crinoids and blastoids from Germany. Paläont Zeit 75:233–252
Laudon LR, Severson JL (1953) New crinoid fauna, Mississippian, Lodgepole Formation, Montana. J Paleont 27:505–536
Lyon SS (1857) In: Lyon SS, Cox ET, Lesquereux L (eds) Palaeontological report. Geol Rept Kentucky 3:467–497
McGhee GR Jr (1996) The Late Devonian Mass Extinction. Columbia University, New York, 303p
McIntosh GC (1983) *Nuxocrinus* and *Pyrenocrinus*, two new Devonian cladid inadunate crinoid genera. J Paleont 57:495–513
McIntosh GC (1984) Devonian cladid inadunate crinoids: family Botryocrinidae Bather, 1899. J Paleont 58:1260–1281
McIntosh GC (2001) Devonian cladid crinoids: families Glossocrinidae Goldring, 1923, and Rutkowskicrinidae new family. J Paleont 75:783–807
McIntosh GC, Brett CE (1988) Occurrence of the cladid inadunate crinoid *Thalamocrinus* in the Silurian (Wenlockian) of New York and Ontario. Roy Ontario Mus, Life Sci Contrib 149:1–17
McIntosh GC, Schreiber RL (1971) Morphology and taxonomy of the Middle Devonian crinoid *Ancyrocrinus bulbosus* Hall, 1862. Univ Michigan Mus Paleont, Contrib 23: 381–403
Meek FB, Worthen AH (1866) Descriptions of invertebrates from the Carboniferous system. Illinois Geol Surv 2(2):143–411
Miller JS (1821) A natural history of the Crinoidea, or lily-shaped animals; with observations on the genera, *Asteria, Euryale, Comatula* and *Marsupites*. Bryan & Co, Bristol, England, 150p, numerous unnumbered plates
Miller SA, Gurley WFE (1895) New and interesting species of Palaeozoic fossils. Illinois State Mus Bull 7:1–150, numerous unnumbered plates 189
More RC, Laudon LR (1941) Symbols for crinoid parts. J Paleont 15:412–423
Moore RC, Laudon LR (1943) Evolution and classification of Paleozoic crinoids. Geol Soc Am Spec Pap 46:1–151
Moore RC, Teichert C (eds) (1978) Treatise on invertebrate paleontology, Part T, Echinodermata 2, Crinoidea. 3 vols, Geological Society of America, University Kansas, 1027p
Müller J (1856) Über neue Crinoiden aus dem Eifeler Kalk. K, Akad Wiss Berlin Monatsber 353–356
Müller J (1859) Bericht über die zur Bekanntmachung geeigneten Verhandlungen der Königlich Preussischen Akademie der Wissenschaften zu Berlin. K Preussischen Akad Wiss Berlin Monatsber (1858), pp 185–198
Prokop RJ (1973) *Elicrinus* n. gen from the Lower Devonian of Bohemia. Ústřed Ústavu Geol Vstnik 48:221–223
Prokop RJ (1983) *Vadarocrinus vassa* gen. et sp. n. (Crinoidea, Petalcrinidae) from the Lower Devonian of Bohemia. Casopis Národního Muzea, Rada Priródovedná 152(4):187–188
Prokop RJ, Petr V (1989) New Silurian and Devonian crinoid taxa from Bohemia. Casopis pro Mineralogii a Geologii 34(1):45–57
Prokop RJ, Petr V (1991) New microcriniods from theTrebotov Limestone of the Bohemian Lower Devonian (Dalejan = upper Emsian). Casopis pro Mineralogii a Geologii 36(2–3):131–136
Roemer CF (1851) Beiträge zur Kenntniss der fossilen fauna des Devonischen gebirges am Rhein. Naturhist Verein Preuss Rheinl u. Westfalens, Verhandl 8:357–376
Rozhnov SV, Arendt YuA (1984) Novyi rod krinoidei iz verkhnego devona Glavnogo devonskogo polya [A new genus of Upper Devonian crinoid from the main Devonian field]. Paleont Zhur 4:118–121
Sandberger G, Sandberger F (1850–1856) Die Versteinerungen des rheinischen Schichtensystems in Nassau. Wiesbaden, Kreidel and Niedner, 564 p, 41pls
Schmidt WE (1906) Der oberste Lenneschiefer zwischen Letmathe und Iserlohn. Deutsche Geol Gesell Verhandl 1905, 57:498–566
Schmidt WE (1934) Die Crinoideen des rheinischen Devons Teil I: Die Crinoideen des Hunsrückschiefers. Preuss Geol Landesanst Abh 163:1–149
Schmidt WE (1942) Die Crinoideen des Rheinischen Devons Teil II. A. Nachtrag zu: .Die Crinoideen des Hunsrückschiefers. B. Die Crinoideen des Unterdevon bis zur *Cultrijugatus* -Zone (mit Ausschluss des Hunsrückschiefers). Reichstelle Bodenforscch Abh, n. s., 182 (1941):1–253
Schultze L (1867) Monographie der Echinodermen des Eifler Kalkes. K Akad Wiss Berlin Mathematisch-Naturwiss Kl 26:113–230
Sepkoski JJ (2002) A compendium of fossil marine animal genera. Bull Am Paleont 363:1–560
Sevastopulo GD (2008) Chapter 3. Paleobiology of Carboniferous microcrinoids. In: Ausich WI, Webster GD (eds) Echinoderm paleobiology. Indiana University, Bloomington, pp 55–69, 1–456
Simms MJ, Sevastopulo GD (1993) The origin of articulate crinoids. Palaeontology 36: 91–109
Springer F (1911) Some new American fossil crinoids. Mem Museum Comparative Zoology, Harvard University 25(3):117–161
Steininger J (1831) Bermerkungen über die Versteinerungen, welche in dem Übergangskalkgebirge der Eifel gefunden warden, 44p, Trier
Steininger J (1849) Die Versteinerungen des Uebergangs-Gebirges der Eifel. Jahresbericht über den Schul-Cursus 1848–49 an dem Gymnasium zu Trier, 1–50 Trier
Strimple HL (1978) Evolutionary trends among Poteriocrinina. In: Moore RC, Teichert C (eds) Treatise on invertebrate paleontology, Part T, Echinodermata 2, Crinoidea, 3 vols, Geological Society of America, University of Kansas, pp 298–301, 1027
Strimple HL, Levorson CO (1969) Two Upper Devonian crinoids. In: Fossil crinoid studies. University of Kansas Paleontological Contribuations, Paper, vol 42(4), pp 17–20

Strimple HL, McGinnis MR (1969) New crinoid from the Gilmore City Formation, Lower Mississippian of Iowa. In: Fossil crinoid studies. University of Kansas Paleontological Contribuations, Paper, vol 42(5), pp 21–22

Trautschold H (1867) Einige crinoideen und andere Thierreste des jungeren Bergkalks im Gouvernement Moskau. Soc Impér Nat Moscou Bull 40(2, 3):49p

Wachsmuth C, Springer F (1880) Revision of the Palaeocrinoidea pt 1, The families Ichthyocrinidae and Cyathocrinidae Pt. I. The families Ichthyocrinidae and Cyathocrinidae. Acad Nat Sci Philadelphia Proc 1879:226–378

Wang XD, Sugiyama T, Cao C, Li Y (2007) Peri-Gondwanan Carboniferous to Permian sequences in the Baoshan Block, west Yunnan—faunal, climatic, and geographic changes. 16th International Congress on the Carboniferous and Permian, Guidebook Excursion C1, pp 1–45

Wanner J (1916) *Eifelocrinus* und *Peripterocrinus*, nom. nov. Wanner, J. Zeitschrift Deutsche Geolog Gesell, Band 68:200–203

Wanner J (1942) Einige neue Krinoiden aus dem Mittel-Devon der Eifel. Decheniana, A B, Festschrift, 101 AB:25–38

Waters J, Webster GD (2009) A reevaluation of Famennian echinoderm diversity: implications for patterns of extinction and rebound in the Late Devonian. In: Königshof P (ed) Devonian change: case studies in palaeogeography and palaeoecology. Geological Society of London Special Publications 314, pp 149–161

Waters J, Maples CG, Lane NG, Marcus S, Liao Z-T, Liu L, Hou H-F, Wang J-X (2003) A quadrupling of Famennian pelmatozoan diversity: new Late Devonian blastoids and crinoids from northwest China. J Paleont 77:922–948

Webby BD (1965) *Quantoxocrinus*, a new Devonian inadunate crinoid from west Somerset. Palaeontology 8:11–15

Webster GD (1997) Lower Carboniferous echinoderms from northern Utah and western Wyoming. Utah Geol Surv Bull 128, Paleontology Series 1:1–65

Webster GD (2003) Bibliography and index of Paleozoic crinoids, coronates, and hemistreptocrinoids, 1758–1999. Geol Soc America, Spec Pap 363:2335 pp GSA website: http://crinoid.gsajournals.org/crinoidmod/

Webster GD, Becker RT (2009) Devonian (Emsian to Frasnian) crinoids of the Dra Valley, western Anti-Atlas Mountains, Morocco. In: Königshof P (ed) Devonian change: case studies in palaeogeography and palaeoecology. Geological Society of London, London, Special Publications 314, pp 131–148

Webster GD, Jell PA (1999) New Carboniferous crinoids from eastern Australia. Mem Queensland Mus 43:237–278

Webster GD, Kues BS (2006) Pennsylvanian crinoids of New Mexico. New Mexico Geol 28(1):3–36

Webster GD, Lane NG (1987) Crinoids from the Anchor Limestone (Lower Mississippian) of the Monte Cristo Group southern Nevada. Univ Kansas Paleont Contrib Paper 119:1–55

Webster GD, Lane NG (2007) New Permian crinoids from the Battleship Wash patch reef in southern Nevada. J Paleont 81:951–965

Webster GD, Maples CG (2006) Cladid crinoid (Echinodermata) anal conditions. a terminology problem and proposed solution. Palaeontology 49:187–212

Webster GD, Maples CG (2008) Cladid crinoid radial facets, brachials, and arm appendages: a terminology solution for studies of lineage, classification, and paleoenvironment. In: Ausich WI, Webster GD (eds) Echinoderm paleobiology. Indiana University, Bloomington, pp 196–226

Webster GD, Lane NG, Maples CG, Waters JA, Horowitz AS (1998) Frasnian–Famennian extinction was a non-event for crinoids, blastoids and bryozoans. Geol Soc Am Abstr Progr 30(7):30–31

Webster GD, Hafley DJ, Blake DB, Glass A (1999) Crinoids and stelleroids (Echinodermata) from the Broken Rib Member, Dyer Formation (Late Devonian, Famennian) of the White River Plateau, Colorado. J Paleont 73:461–486

Webster GD, Maples CG, Mawson, R, Dastanpour M (2003) A cladid-dominated Early Mississippian crinoid and conodont fauna from Kerman Province, Iran and revision of the glossocrinids and rhenocrinids. J Paleont Mem 60, 77(supp. to part 3):1–35

Webster GD, Becker RT, Maples CG (2005) Biostratigraphy, paleoecology, and taxonomny of Devonian (Emsian and Famennian) crinoids from southeastern Morocco. J Paleont 79:1052–1071

Webster GD, Maples CG, Yazdi M (2007) Late Devonian and Early Mississippian echinoderms from central and northern Iran. J Paleont 81:1105–1118

Webster GD, Yilmaz I, Kozlu H (2008) A new Middle Devonian gasterocomid crinoid from central Turkey and revision of the Gasterocomidae. Palaeoworld 17:12–20

Webster GD, Waters J, Chen X (2009) Revision of the Chen and Yao Devonian to Permian crinoids from western Yunnan. Palaeobiodiversity Palaeoenviron 89:119–160

Webster GD, Waters J (2009) Late Devonian echinoderms from the Hongguleleng Formation of northwestern China. In: Königshof P Devonian change: case studies in palaeogeography and palaeoecology. Geological Society of London, London, Special Publication 314, pp 263–287

Whidborne GF (1898) A monograph of the Devonian fauna of the south of England. V3: the fauna of the Marwood and Pilton Beds. Palaeontogr Soc Monog 52, 3(3):236p

Whitfield RP (1905) Notice of a new crinoid and a new mollusk from the Portage rocks of New York. Am Mus Nat Hist Bull 21(2):17–20

Yakovlev NN (1941) Morskie lilii Glavnogo Devonskogo Polya I. v: Fauna Glavnogo Devonskogo Polya [Crinoids of the main Devonian Field 1] In: Fauna of the Main Devonian Field I]. Akad Nauk SSSR Izvestii Paleontologicheskii Institut, pp 323–331

Yakovlev NN (1949) O sushchestvovanii v verkhnem silure i nizhnem devone SSSR morskikh lilii sem Crotalocrinitidae [On the existence of crinoids of the family Crotalocrinitidae in the Upper Silurian and Lower Devonian of Russia]. Vserossiiskoe Paleontologicheskogo Obshchestva Ezhegodnik 13:14–29

Palaeoecology, Aerodynamics, and the Origin of Avian Flight

Sankar Chatterjee and R. Jack Templin

Abstract

Data from hundreds of small, exquisitely preserved feathered coelurosaurs and early birds from the Early Cretaceous Jehol Group, Liaoning Province, China, suggest that avian flight probably began in the trees (arboreal or trees-down theory) rather than on the ground (cursorial or ground-up theory). The quality of preservation of the Jehol biota and the sediment in which they were preserved accords with recurrent mass mortality events, the animals being asphyxiated, buried by ash falls and swiftly preserved. The inferred palaeoecology of the Jehol biota indicates they lived in a forest environment bordering a large lake. New information on the origin of flight by the Jehol fossils is presented; it includes various transitional stages of flight—from wingless, tree climbing coelurosaurs to parachuting, to gliding, to fully winged, active flying birds. The fossils show development of adaptations necessary to enable wing-assisted climbing: highly recurved claws on hands and feet; long fingers, wrist joints that swivelled, and stiffened tails. Several feathered paravian coelurosaurs, exemplified by *Epidendrosaurus*, *Epidexipteryx*, and *Scansoriopteryx* with wings suitable for climbing, became arboreal, yet were flightless. Their arboreal lifestyles contradict arguments previously advanced supporting the cursorial (ground-up) theory of the origin avian flight. The cursorial model fails to explain climbing adaptations of protobirds, different stages of flight, and development of neurosensory specializations in early birds. Rather, the climbing adaptations of protobirds support an arboreal setting for the evolution of flight. These adaptations include, for example, different stages of flight from a perch, gradual brain enlargement for three-dimensional orientation, acquisition of vision and acute sight, and neurosensory specialization for hearing and balance. We have identified six evolutionary stages of avian flight represented by phylogeny and transitional fossils—arboreal leaping, parachuting, biplane gliding, monoplane gliding, undulating flight, and manoeuvring flapping flight. Arboreal life is suggested to have promoted enlargement of the brain, increased visual acuity,

S. Chatterjee (✉)
Museum of Texas Tech University, Lubbock, TX 79409, USA
e-mail: SANKAR.CHATTERJEE@ttu.edu

J.A. Talent (ed.), *Earth and Life*, International Year of Planet Earth,
DOI 10.1007/978-90-481-3428-1_18,

and development of more sophisticated vision. A computer model to simulate the flight performance of protobirds and early birds has been developed that corroborates the argument for the above evolutionary pathway.

Keywords
Early Cretaceous • Liaoning (China) • Jehol Group • Feathered coelurosaurs • Early birds • Palaeoecology • Mass mortality events • Hypothesized asphyxiation • Arboreal origin of avian flight • Enlargement of the brain • Computer modelling

Introduction

Evolution of powered flight in birds is generally regarded as a key adaptive breakthrough that took them into another direction from the rest of the dinosaurs of the Mesozoic world. The low expenditure of energy in flying relative to that of walking or running made early birds more versatile than earth-bound dinosaurs. Flight allowed birds to escape from predators, to exploit new resources, to nest in trees, to search for mates, to migrate rapidly to more favourable habitats, and to exploit the best seasons for reproduction and survival.

The ancestry of birds and the origin of avian flight have aroused passionate debate. Part of that debate stems from lack of transitional fossils and lack of unequivocal information, vital for the documentation of their arboreal or terrestrial origins. The origin of flight in extinct organisms must necessarily be approached from interpretation of fossil remains. The controversy remained at a stalemate for several decades. Two factors rekindled the debate: the theropod hypothesis regarding the origin of birds and the newly discovered transitional fossils documenting different stages in early experiments in flight. The discovery of extraordinary feathered coelurosaurs and early birds in China, if interpreted correctly, may settle the century-old argument whether avian flight began in the trees ("trees-down" theory) or on the ground ("ground-up" theory).

The debate on the origin of birds began soon after discovery of a well-preserved skeleton of *Archaeopteryx* in 1861, the most primitive and ancient bird with intact wings (the "Urvogel") in the Late Jurassic Solnhofen Limestone in Germany. This crow-sized, toothed bird with powerful forelimbs with claws, perching feet, and a long bony tail was purchased by the British Museum of Natural History and described by Richard Owen (1863). Huxley (1868), Darwin's ardent ally, emphasized the importance of this fossil as a missing link between reptiles and birds and recognized its dinosaurian heritage. Darwin (1866, 4th edition, p. 367) referred briefly to *Archaeopteryx* in *The Origin of Species* as "that strange bird, *Archaeopteryx*". Perhaps Darwin knew that more than a single specimen would be needed to bolster his evolutionary theory, or perhaps he was trying to avoid confrontation with Owen. When a small coelurosaurian theropod, *Compsognathus,* was found in the Solnhofen Limestone with other specimens of *Archaeopteryx*, Huxley recognized the uncanny resemblance between the two skeletons and noted that were it not for those remarkable feather imprints surrounding the skeleton of *Archaeopteryx,* it would be identified as a coelurosaur. Huxley pondered whether the skeletal similarities between the two resulted from close relationship or from convergent evolution. The most crucial evidence for their phylogenetic relationship came from their bipedal posture and the striking skeletal similarities in the pelvic girdle, hind limb, and ankle structures. Using this compelling anatomical evidence, Huxley (1870) presented a radical proposal that birds are glorified reptiles and suggested an evolutionary link between avian evolution and the evolution of flight.

Although Huxley's "theropod theory" was initially accepted, most biologists discredited his idea soon after the publication of Heilmann (1926), in which he suggested that birds evolved not from dinosaurs but from a primitive branch of archosaurs that included the pseudosuchians. The central argument against a theropod origin was that no theropods had been found with clavicles fused to form the furcula of modern birds. Thus, the presence of the furcula in birds would be anomalous, had they evolved from theropods. Heilmann's theory was widely accepted.

In the past three decades, the debate concerning theropod–bird relationship was renewed as a consequence of discoveries of fossil material and new analyses. For example, Ostrom (1976) showed extensive osteological similarities between the skeletons of the dromaeosaurid theropods, such as *Deinonychus,* and *Archaeopteryx*; he pointed out that some theropods indeed possessed clavicles, their absence in most theropod specimens reflecting the vagaries of preservation of the extremely delicate clavicles. Gauthier (1986), with detailed cladistic analysis, showed that dromaeosaurs and *Archaeopteryx* are more closely related to each other than to any archosaur and put *Archaeopterix* and later birds in a new clade, Avialae. Largely as a consequence two publications (Ostrom 1976; Gauthier 1986), most researchers have accepted the theropod origin of birds (Chiappe and Witmer 2002). Most recent phylogenetic analyses accord with the Avialae being a monophyletic clade, the sister group to the maniraptoran theropods such as the deinonychosaurs. Birds are now considered to be flying dinosaurs (Bakker 1986; Paul 2002; Chiappe 2007).

Once the theropod ancestry of *Archaeopteryx* was accepted, the logical next step was to address the question of the origin of avian flight within this new phylogenetic framework, but the questions remained because of lack of fossils showing the transition from flightless ancestors to flying species of birds, specifically, the development of its feathers and elongation of the forelimb to increase its span. These adaptations for flight did not occur simultaneously but required intermediate stages.

Until recently, *Archaeopteryx* provided the earliest unambiguous example of feathers similar to those of modern birds. For this reason, the origin of powered flight was often equated with the origin of *Archaeopteryx* (Ostrom 1976), with its fully developed wings with asymmetric flight feathers. Understanding the steps in the origin of avian flight, required pre-*Archaeopteryx* stages of the flight apparatus, namely wings beginning to develop by elongation of the forelimbs of the theropods, and development of feathers from body integuments.

Recent discovery of a series of hundreds of small, exquisitely preserved feathered coelurosaurs and early birds in the Early Cretaceous Jehol Group of the western part of Liaoning Province in northeast China (Fig. 1a) provided the crucial evidence bridging the morphological and evolutionary gaps between non-flying theropods and flying birds. All died suddenly in catastrophic events ~125 Ma (million years ago) (Chang 2008). They show various transitional stages—from wingless, tree-dwelling theropods to fully winged, active flyers.

The Jehol Coelurosaurs

The Jehol Group, dated by Chang et al. (2009), consists of two units: the Yixian Formation [129.7 ± 0.5 Ma for the base of the unit] and Jiufotang Formation [122.1 ± 0.3 Ma for the lowermost part of the unit]. These new dates show that the entire Yixian Formation was deposited within the Early Cretaceous over an interval of ~7 Ma. The biota includes vertebrates, including fish, amphibians, turtles, choristoderes, squamates, pterosaurs, and ornithischians, representing one of the most diverse assemblages of terrestrial Cretaceous vertebrates any site in the world; they provide many missing links in evolutionary history (Chang 2008).

The taphonomy of the Liaoning fossil site has not had intense scrutiny, but associated plant fossils indicate a forest environment along the shore of an ancient lake. Because the skeletons were preserved in ash beds, undisturbed and fully articulated (Fig. 1b), it is suggested the animals were engulfed in clouds of volcanic ash from a nearby volcano, asphyxiated by poisonous gas, and buried. Their remains are so numerous and so well preserved that the Liaoning site has been dubbed a "Cretaceous Pompeii" (Chatterjee and Templin 2005). Because so many animals appear to have been preserved in sleeping or resting positions, death by asphyxiation seems likely.

It is remarkable that the skeletons of terrestrial animals—thousands of them—entered the lake intact or as imprints of delicate soft tissues including external integuments (skin, scales, feathers, fur), viscera, and stomach contents in amazing detail. It appears that the habitat of the tetrapods around the lake shore and the burial ground on the lake bottom were different, involving recurrent mud and/or ash [lahars] slides, presumably triggered by torrential rain, carrying entire communities, to become a death assemblages on the lake bottom, where rapid burial promoted anoxic conditions, and halted bacterial decay of the soft tissues (Zhou et al. 2003) (Fig. 1c, d).

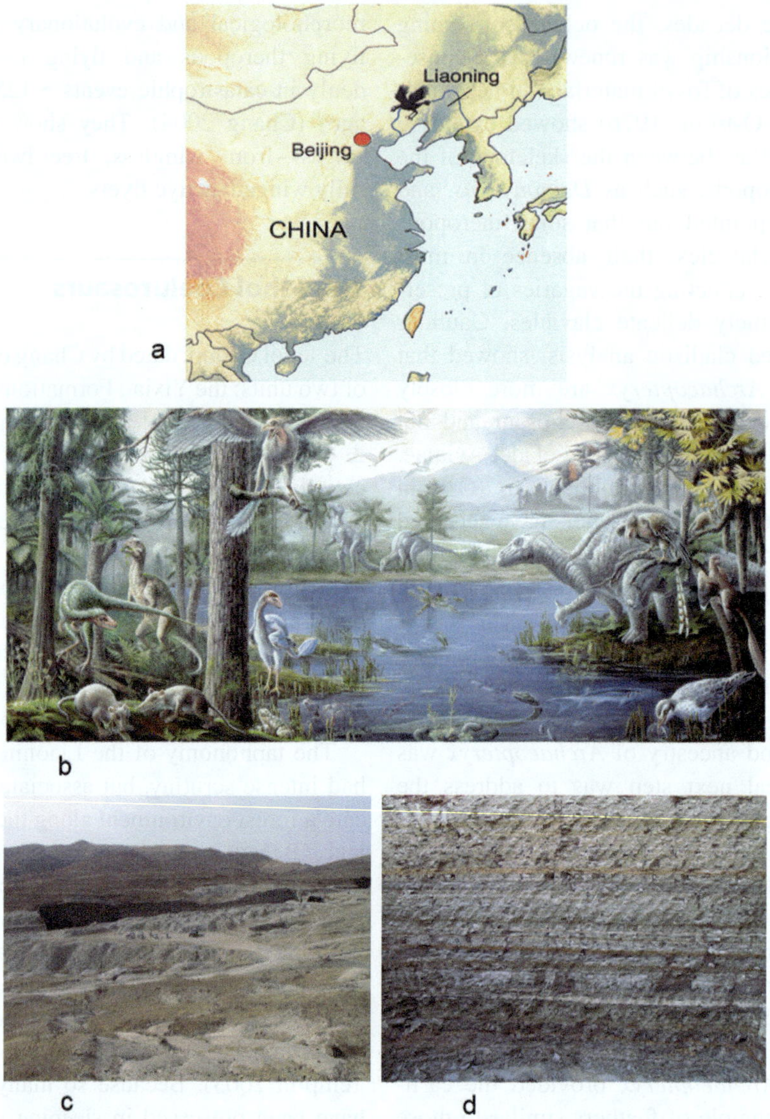

Fig. 1 (a) Location of fossil locality in northwestern China. (**a**, **b**) Composite palaeoecological reconstruction of the Jehol biota showing dinosaurs, feathered coelurosaurs, early birds, and mammals that lived in and near ancient lakes; the killer volcano erupted at a distance that would recurrently kill and bury the community with ash beds. (**c**) View of Sihetan locality (*lower part* of the Yixian Formation) in Liaoning Province, showing tuffaceous sandstones and mudstones from which excellent feathered coelurosaurs and early birds were obtained. (**d**) Section of the Yixian Formation showing lacustrine deposits (*gray*) with intercalated tuffs or ash beds (*yellow*) (courtesy of Zhonghe Zhou)

The most celebrated fossils from the Jehol biota are undoubtedly the non-avian theropods morphologically spanning the transition from basal coelurosaurians to ornithurine birds. Many of the small, feathered coelurosaurs acquired avian characters and are successively closer to early birds. One of the most compelling lines of evolutionary evidence from the Jehol coelurosaurs is the unique preservation of a range of primitive feathers and the evolution of avian flight. These feathered coelurosaurs, "living fossils" in the Early Cretaceous ecosystem, evolved in isolation as long-lived relicts, in an isolated refugium when China was an island continent—a sort of ancient Madagascar populated by primitive relict species (Luo 1999). This hypothesis explains the presence of so many endemic and primitive animals that are otherwise known only from the Jurassic of Europe.

The Beginnings of Avian Flight

Three theories have been suggested regarding the origin of avian flight within the theropod hypothesis of the origin of birds:

(1) In the cursorial (ground-up) theory of Ostrom (1986) (Fig. 2a, b), flight is viewed as developing in running theropods through a series of short jumps; as these jumps became greater, the wings were used for balance and propulsion, with the animals beginning to fly without an intermediate gliding stage (Nopsca 1923; Ostrom 1979; Gauthier and Padian 1985; Balda et al. 1985; Padian and Chiappe 1998; Burgers and Chiappe 1999).

(2) In the wing-assisted incline running (WAIR) theory (Fig. 2c), protobirds are viewed as beginning to climb overhanging slopes by flapping their wings and eventually taking off (Dial 2003).

(3) In the arboreal (trees-down) theory (Fig. 2d), is viewed as originating in small arboreal theropods leaping from branch to branch or tree to tree, steadying themselves with outstretched wings, and eventually parachuting from heights, gliding, and finally acquiring powered flight (Bock 1985; Chatterjee 1997; Elzanowski 2002; Paul 2002; Chatterjee and Templin 2003).

Each theory relies on speculation regarding the palaeoecology and functional adaptation of *Archaeopteryx* and its progenitors.

Cursorial Theory

This theory centres on the premise that the immediate theropod ancestors of birds were strictly terrestrial bipeds using their wings as bilateral stabilizers during jumps into the air in pursuit of prey. As the jumps lengthened, the wings became increasingly used for balance and propulsion—and the animals began to fly actively without passing through an intermediate gliding stage (Fig. 2a). The cursorial theory is untenable because the skeletal anatomy of the progenitors of *Archaeopteryx* show no adaptations necessary for taking off from the ground and remaining airborne, such as large wings or a capacity for vigorous flapping. The strongest criticisms of the cursorial theory are based on ecologic, adaptive, and mechanical grounds (Bock 1985, 1986; Norberg 1990; Rayner 1991; Chatterjee 1997). The theory is marred by functional and adaptive gaps from pre-flight to actively flapping flight stages—requiring flight to have evolved almost by saltation without gliding stages. It does not explain the origin of the various pre-flight adaptations, such as small size, the origin of feathers, brain enlargement, large eyes, vision and hearing acuity, neuromuscular coordination, elongation of the forelimb, and dorsoventral flight strokes (Bock 1985, 1986; Norberg 1990; Chatterjee 1997; Goldsmith 2006).

Another version of the cursorial hypothesis is that the birds, such as the maniraptoran theropods, were fully terrestrial, their forelimbs being used to subdue and kill prey—these predatory motions are viewed as giving rise to the flight stroke (Ostrom 1979; Gauthier and Padian 1985; Padian and Chiappe 1998). There are several inconsistencies in the prey-catching scenario of the maniraptorans, with their specialized swivelling wrist joint. Development of the semilunate carpal, an avian type of linkage system between the elbow and wrist joints, came about in the maniraptorans. As the elbow folded, the hand of the forelimb would have moved farther away from the mouth so, like modern birds, the hands of dromaeosaurs could not be used for predation and feeding. The predatory movements of the forelimbs would be directed forward to grasp the prey, whereas flight movement is essentially dorsoventral; it is inconceivable that this propalinal motion of the hand gave rise to dorsoventral flight strokes. The hands of dromaeosaurs were doing something other than catching prey. As in extant birds, they could be folded at the sides of the body during terrestrial locomotion. The elongated forelimbs and swivelling wrist joints in the maniraptorans are interpreted as features evolved for climbing (Chatterjee and Templin 2004a). A long-standing criticism of the cursorial theory is that large, complicated structures such as wings must have evolved all at once from running to flapping, in one great evolutionary lunge without any intermediate stages.

The latest version of the cursorial theory views *Archaeopteryx* as being capable of take off from a running start, using its wings as a primary thrust generator (Burgers and Chiappe 1999)—on beginning to run, starting to flap vigorously, gaining considerable velocity from its wing thrust and allowing it to take off. It was proposed that generation of thrust, but not lift, was crucial during ground takeoff (Fig. 2b). However, there are inconsistencies in the

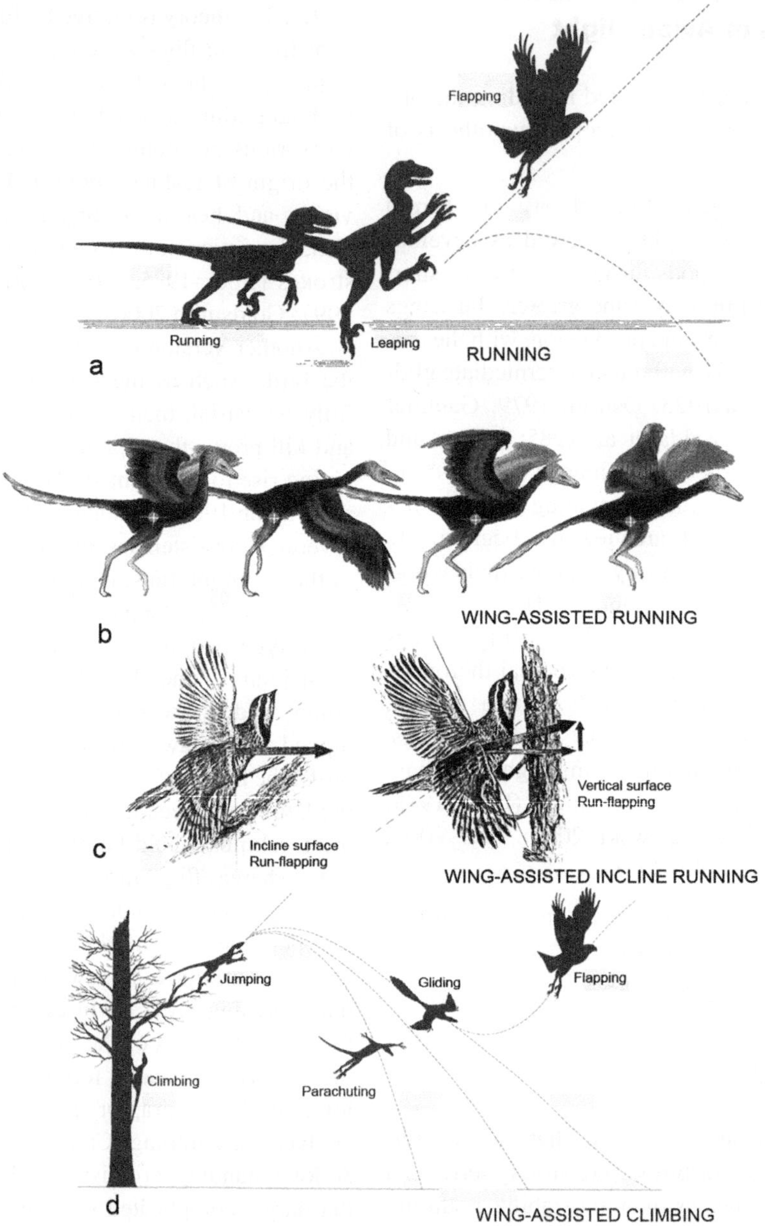

Fig. 2 Four models for the origin of avian flight. (**a**) Running: for the cursorial model, a running theropod leaping into the air to become an active flier without an intervening gliding stage (after Chatterjee 1997). (**b**) Wing-assisted running: a modified version of the cursorial model, showing the takeoff sequence of *Archaeopteryx*, where running speed increases with thrust generated by the wings so the bird begins to take off (simplified from Burgers and Chiappe 1999). (**c**) Wing-assisted incline running: a partridge climbs a steep slope by flapping its wings to aid traction, generating a force perpendicular to the plane of wing movement (simplified from Dial 2003). (**d**) Wing-assisted climbing: a progenitor of *Archaeopteryx* climbing assisted by wings, jumping, then parachuting from trees using gravity as the source of power, then beginning to glide. Flapping begins to prolong powered flight (after Chatterjee 1997)

wing-assisted running (WAR) model. *Archaeopteryx* required not only aerodynamic feathers for take off, but also a skeletal architecture allowing its muscles to produce the powerful flapping for it to remain in the air.

There are biological limitations to generation of the vigorous flapping required by *Archaeopteryx* to take off, for example, lack a modern avian supracoracoideus pulley, the primary elevator of the wing—necessary for rapid wing upstroke (Poore et al. 1997). With a limited upstroke, *Archaeopteryx* could not execute the powerful downstroke necessary to generate sufficient thrust for a ground takeoff. Its flight apparatus was weakly developed compared with that of modern flying birds. For example, the coracoid was too small to produce a wide range of flapping motion and its sternum, the site of attachment of the flight muscles, was not ossified. Its pelvis, synsacrum, and hind limbs were primitively developed and were not sufficiently strong enough to generate the powerful leg thrust required for a ground takeoff (Chatterjee and Templin 2003). Moreover, its narrow and solid primary rachises suggest poor flapping ability (Nudds and Dyke 2010).

Using a computer simulation model, we showed that for *Archaeopteryx*, takeoff would have been more efficient and cost effective from a perch than from the ground (Chatterjee and Templin 2003). If *Archaeopteryx*, with its fully developed wings and airworthy feathers, found a ground takeoff difficult how would they generate sufficient power to become airborne from the ground without the supracoracoideus pulley?

The cursorial theory is based on the assumption that protobirds were obligatorily terrestrial until after the evolution of active flight. This view became no longer tenable as evidence for the arboreal lifestyles of the feathered coelurosaurs from China began to emerge (Zhang et al. 2002; Xu et al. 2003). The cursorial theory does not adequately explain the origin of encephalization and the neurosensory specializations associated with three-dimensional perceptual control or the transition between rudimentary and fully developed flight (Bock 1985). In summary, biomechanical, neurological, and palaeoecological studies fail to support the cursorial theory. In contrast, fossil evidence supporting the arboreal theory is overwhelming, as discussed below.

Wing-Assisted Incline Running Model

In an interesting experiment, Dial (2003) showed that many living ground-dwelling birds, such as the chukar partridge (*Alectoris chukar*), tend to seek refuge at height when threatened (Fig. 2c). They vigorously beat their wings to run up steep, even vertical substrates, increasing the load on their hind limbs to reach these heights. As they run uphill, the chukars flap their wings strongly up and down at a different angle from that used when flying, thus improving the traction of their legs on the vertical surface. After trimming the remiges of chukars of various ages, Dial demonstrated that without feathers, the birds could not run up slopes steeper than 60°. In contrast fully feathered chukars could climb vertical substrates. Dial thus concluded that flapping does not lift the bird, but rather the opposite: it presses the bird into the inclined surface to increase its traction, to keep its feet from slipping, "acting like the spoiler on a racing car" (Dial 2003, p. 403). Later, Bundle and Dial (2003) performed experiments on a number of chukars using two accelerometers to measure the acceleration in the forward and vertical directions. They concluded that vertical ascent is possible for a chukar when it simultaneously flaps its wings and runs on its legs.

Although Dial has discovered a novel method by which poorly flying birds can get off the ground, its relevance to the origin of avian flight is unclear at the moment. Bundle and Dial (2003) acknowledge that the behaviour they observed in young partridges is not necessarily the basal condition of birds or their ancestors. They believe that this climbing behaviour of the partridge is an escape strategy. Lacking the supracoracoideus pulley and the keeled sternum, the protobirds, such as the feathered dromaeosaurs and *Archaeopteryx*, could not perform these complex wing movements to generate the thrust required for a vertical ascent, as proposed by Dial. It appears that the WAIR motion involves fully developed wing movement and is only possible when birds have learnt how to perform complex wing movements, after acquiring the supracoracoideus pulley, which was absent in *Archaeopteryx* and its immediate ancestors (Poore et al. 1997). It is unlikely that protobirds could generate enough thrust against gravity to prevent slipping from vertical substrates; they must have used their claws to cling to tree trunks. In fact, Dial (2003) has shown that young partridges move their wings differently for climbing steep slopes from whence they fly.

The ability of a young partridge to climb vertical surfaces is a twist on the arboreal theory, in which the initial height must be gained by climbing before the bird launches from a perch. The WAIR model does not specify how early birds took to the air. It fails to explain why the protobirds and *Archaeopteryx* retained a grasping hand equipped with sharp and recurved claws, if the wings were not used for climbing. To understand the origin of avian flight, a scansorial baby hoatzin is a better analogue than a chukar because its hands are equipped with sharp claws supported by elongated penultimate phalanges, like those of the protobirds and *Archaeopteryx*, for grasping cylindrical vertical trunks. The wing-assisted climbing (WAC) adaptation is prevalent among hoatzin chicks, which crawl and clamber through tree branches with their wing claws, just as protobirds might have done millions of years ago (Fig. 3a). Their grasping hands are exceedingly similar to those of the maniraptoran coelurosaurs with swivelling wrist joints and terminal claws. These claws, which persist in the inner fingers (I and II) for the first 3 months of life (Fig. 3b), are absent in the adults but provide the youngster with just the extra grip it needs to climb back up to the nest area (Chatterjee 1997). Much anatomical evidence supports the WAC adaptations of the protobirds, such as the laterally facing glenoid, elongated forelimbs, semilunate carpal with a swivelling wrist joint, elongated penultimate phalanx, and sharp recurved claws, as in the young hoatzin, which are typical of trunk-climbing birds (Chatterjee and Templin 2003, 2004a, b). Feduccia (1993) suggests the proportions and curvatures of the manual phalanges of *Archaeopteryx* attest to their climbing ability. The lengths of the proximal and penultimate phalanges differ among terrestrial and scansorial birds. The penultimate phalanx, which supports the claw, is longer than the proximal phalanx in scansorial animals, which increases the diameter of the grasping hand. If the wing were used solely for WAIR flapping motion by the early birds, elongated penultimate phalanges and sharp claws would be superfluous in those animals. Their presence attests to their adaptation to WAC, not WAIR, during the evolution of true flight. Moreover, the lack of fusion or rigidity in the critical regions of the pelvis, synsacrum, tibiotarsus, and tarsometatarsus indicate that feathered coelurosaurs similar to those from China (the presumed ancestors of birds) were not specialized for running uphill like a chukar (Chatterjee and Templin 2003). Most of these coelurosaurs had arboreal habitats, as discussed below. This new anatomical and ecological evidence is in serious conflict with the WAIR model of flight.

Arboreal Origin of Avian Flight

The arboreal theory was first proposed by Marsh (1880), who observed that a large number of Holocene birds lived in trees or bushes and that there are many modern parachuting and gliding tetrapods, all of which are arboreal. Marsh believed that flight evolved from arboreal protobirds, whose scales developed into rudimentary feathers. These feathers were used as parachutes, slowing the descent of the animals as they leaped from branches and controlling their jumps and falls. Other workers, such as Heilmann (1926), Bock (1985, 1986), and Feduccia (1996), have presented more detailed versions of the arboreal theory using pseudosuchians as the putative ancestors of birds.

The arboreal theory assumes that protobirds lived in trees before the evolution of active flight. The scenario begins with small, wing-assisted climbing protobirds that began to leap from branch to branch, then to parachute by spreading their forelimbs to increase their resistance and to slow their descent, then to glide, and finally to fly with wing strokes. The central theme of the arboreal theory is the use of gravity as a source of power by the protobirds to convert potential energy to kinetic energy.

Chatterjee (1997) and Chatterjee and Templin (2003, 2004a, 2005, 2007a) have revived the arboreal theory in a theropod phylogenetic framework using biomechanical and palaeoecological interpretations of the Chinese feathered coelurosaurs. Most recent researchers of avian flight now support this new version of the arboreal theory, which is further elaborated here, supplemented with new evidence (Elzanowski 2002; Paul 2002; Xu et al. 2003; Zhang et al. 2002, 2008). Even Chiappe (2007), a staunch supporter of the cursorial theory, has endorsed our model as a plausible sequence by which flight in birds might have arisen.

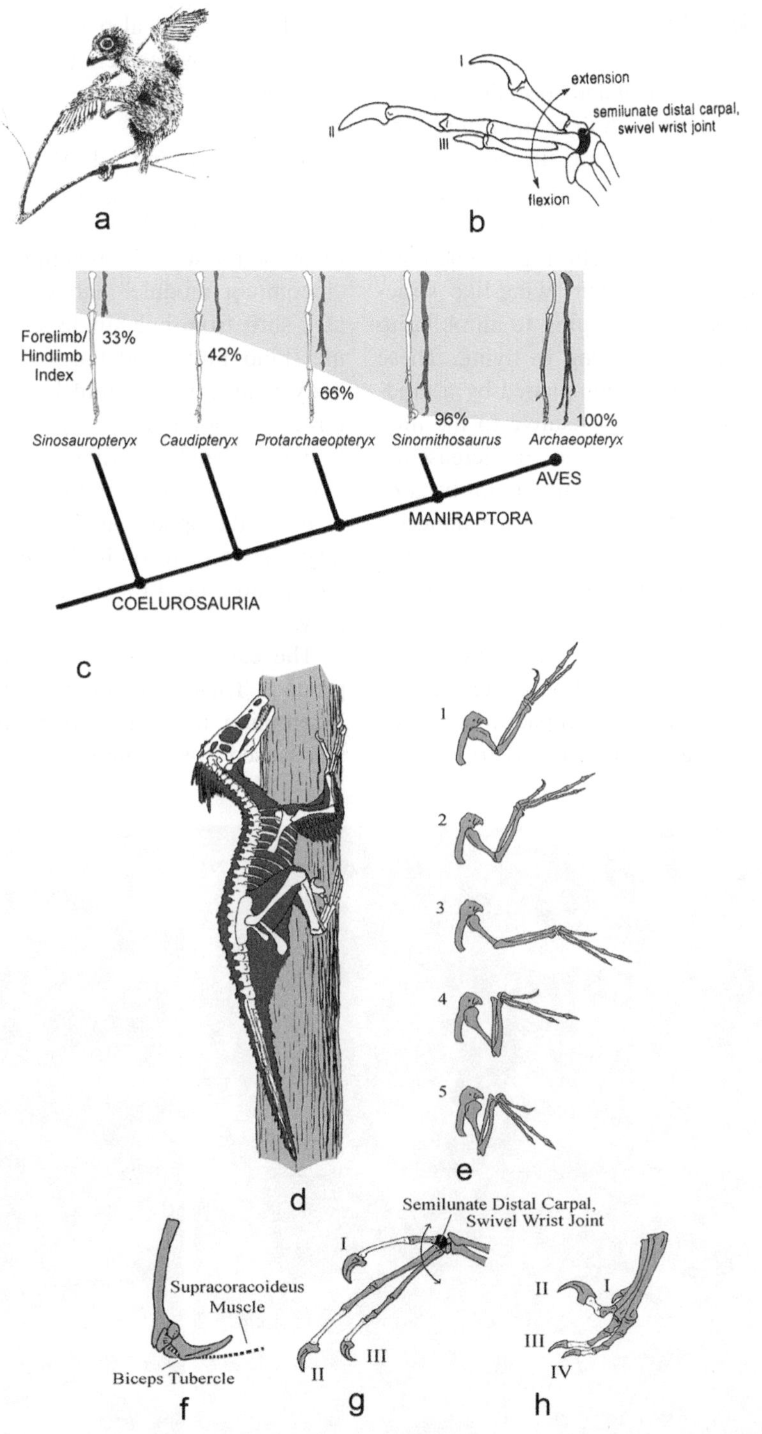

Fig. 3 Wing-assisted climbing adaptations. (**a**) Illustration of a hoatzin (*Opisthocomus*) chick climbing trees with its wing claws. (**b**) Hand of a young hoatzin showing the swivelling wrist joint and terminal claws used for climbing trees. (**c**) Diagram showing gradual lengthening of the forelimbs in relation to hindlimbs in arboreal coelurosaurs, initially for climbing and later for flying. (**d**) Climbing adaptations of maniraptorans such as *Sinornithosaurus* employing climbing adaptations including elongated forelimbs, swivel wrist joint, and recurved claws for grasping trunks, a caudally directed pubis, and a stiffened tail as a supporting prop. (**e**) Sequence of a synchronized forelimb cycle (1–5) employed during climbing—precursor to flight strokes: up-and-forward and down-and-backward. (**f**) Shoulder girdle showing the laterally facing glenoid for dorsoventral wing excursion, and a biceps tubercle for the inception of the acrocoracoid process. (**g**) Grasping hand with a longer penultimate phalanx and recurved claws for hooking and clinging to trunks. (**h**) Pes, showing sharp claws for anchoring to tree trunks

Wing-Assisted Climbing

Birds acquired two independent and specialized methods of locomotion: walking with their hind limbs and flying with their forelimbs. During the evolution of flight in the transition from theropods to birds, the hind limbs did not change significantly. The forelimbs underwent progressive structural and functional modifications as they evolved into wing-like structures, in the transitions from grasping to climbing to balancing to parachuting to gliding to flying. These adaptive modifications were accompanied by a gradual lengthening of the forelimbs relative to the hind limbs in the same way as coelurosaurs, to increase the wingspan, so that the forelimbs and hind limbs became equal in length (Fig. 3c) (Chatterjee 1997; Chatterjee and Templin 2004a). Simultaneously, the feathered coelurosaurs became relatively small (Fig. 4a); life in the trees would favour the smaller size of animals. Small size has several advantages, including reducing the energy expended for vertical climbing, support by smaller branches, and reduced impact force when hitting the ground during accidental falls from trees. Small size could also be associated with insulation of the body by feathers (Fig. 4b) and may constitute a prerequisite for the early evolution of flight (Bock 1985).

SM Gatsey and KP Dial (1996) interpreted the evolution of avian flight in terms of changes in "the sequence of modifications of the primitive tetrapod locomotor system through time". They coined the term "locomotor module" and viewed the origin of flight as a shift from one single terrestrial locomotor module (hind limbs and tail), to the ancestral theropod body plan plus an aerial locomotor component, to a three locomotor module plan as seen in extant birds (wings, tail and hind limbs). The arboreal theory helps to explain the evolution of locomotor module of birds.

Basal theropods, such as ceratosaurs, were obligatory bipeds with forelimbs decoupled from terrestrial locomotion, used for the capture and manipulation of prey.

The earliest birds spent time in trees and on the ground. Their hind limbs were used for terrestrial locomotion but both sets of limbs and a rigid tail were used for climbing trees. Maniraptorans such

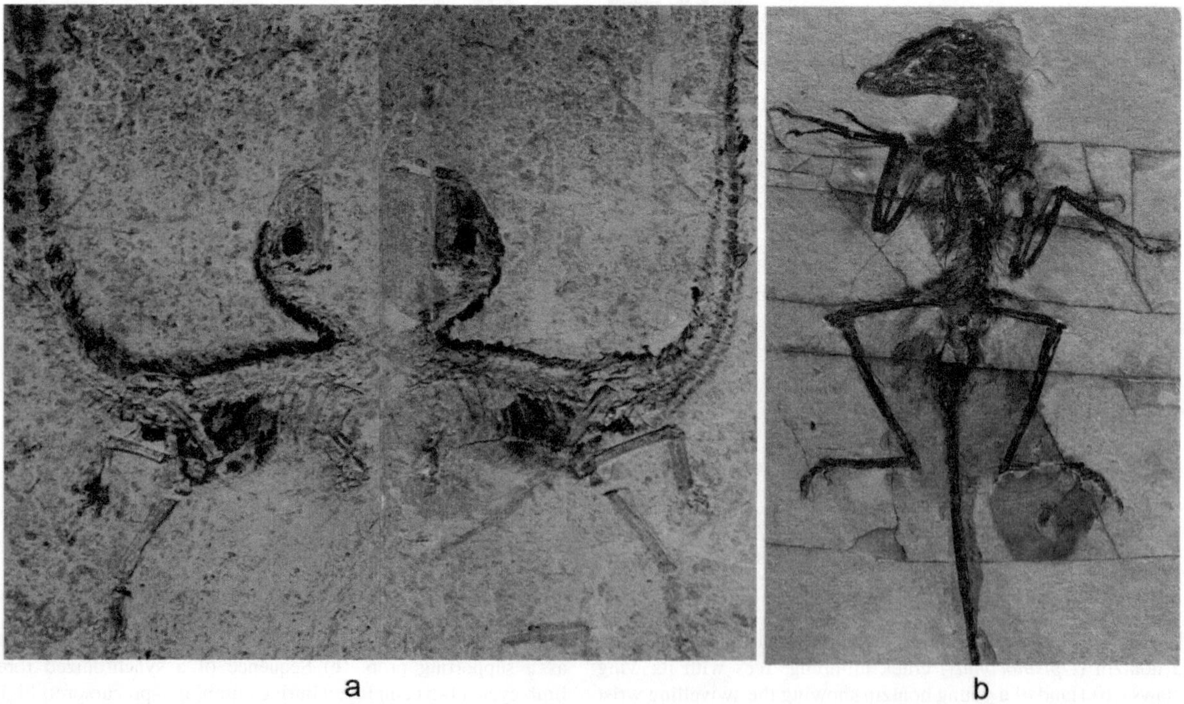

Fig. 4 (**a**) *Sinosauropteryx*, a chicken-sized coelurosaur showing filamentous body covering like downy feathers; slab and counter slab. (**b**) Juvenile *Sinornithosaurus*, nicknamed "Dave", showing enlargement of contour feathers on its wings (courtesy of Xing Xu)

as *Sinornithosaurus* (Fig. 4b) display characters for climbing vertical trunks, such as the laterally facing glenoid for dorsoventral humeral excursion, a biceps tubercle for inception of the supracoracoideus tendon, elongated forelimbs for grasping trunks, ossification of the sternum for attachment of the pectoralis and supracoracoideus muscles, swivelling wrist joints to allow movement of the hands in the plane of the forearm, longer penultimate phalanges and recurved claws for grasping and hooking onto trunks, a caudally directed pubis to flatten the body, and a stiffened tail as a supporting prop (Chatterjee 1997; Chatterjee and Templin 2004a). Ancestral birds may have climbed the vertical trunks of trees in a similar way to modern flying squirrels—using their limbs alternately in unison; first the paired forelimbs and then the paired hind limbs, using their clawed hands and feet as crampons and their rigid tail for additional support. While climbing, the outstretched forelimbs were pressed against the sides of the trunk, recurved claws providing a hooking grip; the forelimbs and hind limbs moved in a reciprocal manner. In the recovery phase, the forelimbs were outstretched and swung upward, and the manus was extended to grasp the trunk. The weight of the body was carried by highly flexed hind limbs. In the next propulsive phase, the hands and forelimbs were gradually flexed to propel the animal, whereas the hind limbs were extended to move the animal upwards. The highly flexed hind limbs provided much of the propulsive thrust, the swivelling wrist joints helping the hands to flex to accommodate this upward movement. The sequence of such a strong, synchronized forelimb cycle during climbing was a precursor to a flight stroke: up-and-forward and down-and-backward (Fig. 3d–h). This climbing transition allowed the forelimb to eventually gain a novel function within the coelurosaurs, namely flight. Such an increase in modularity from one to three modules allowed birds to exploit ecospace adapting to a wide range of life styles.

Several of the new Chinese coelurosaurs are closer to *Archaeopteryx* phylogenetically than are the earlier coelurosaurs; these exhibit obligatory arboreal habits, suggesting that trees became "safe havens" for these dinosaurs. For example, *Epidendrosaurus* (Zhang et al. 2002) and *Epidexipteryx* (Zhang et al. 2008) from the Middle to Late Jurassic of Inner Mongolia had developed WAC hands with curved claws and an extremely elongated third finger, possibly for probing for insects in trees (Fig. 5a). In these animals, the forelimb is longer than the hind limb, and the toes show long penultimate phalanges and perching feet for grasping branches, features associated with arboreal adaptation. Another closely related form exhibiting similar obligatory arboreal adaptations is *Scansoriopteryx* (Czerkas and Yuan 2002). Although belonging to a sister group of *Archaeopteryx*, *Scansoriopteryx* lacked any flight feathers, became secondarily flightless, but remained arboreal for safety, probably because of its diminutive size. It developed fluffy down feathers as it reverted to a flightless form, just as has occurred with modern flightless birds.

In contrast, other arboreal forms, such as *Anchiornis* from the earliest Late Jurassic of Liaoning Province (Hu et al. 2009), *Microraptor* (Xu et al. 2003) from the Early Cretaceous of Liaoning Province (Fig. 5b). *Pedopenna* (Xu and Zhang 2005) from the Middle to Late Jurassic of Inner Mongolia developed long contour feathers along the entire length of the metatarsus; these feathers extended below the level of the feet, thus hindering terrestrial locomotion (Fig. 6). The presence of metatarsus feathers in unequivocal maniraptorans such as *Epidendrosaurus*, *Epidexipteryx*, *Scansoriopteryx*, *Anchisaurus*, *Pedopenna*, and *Microraptor* indicates their obligatory arboreal habitat. It is likely that *Archaeopteryx* and its immediate ancestors were arboreal and began flying from trees.

Neurosensory Evidence

The abundance of plant remains found in association with the coelurosaur fossils (see Fig. 1b) is consistent with their being dwellers in the forest environments of Liaoning. Because impressions found in endocasts can reveal relative brain sizes and surface features, such as the boundaries between the brain components (cerebral hemispheres, optic lobes, and cerebellum), gross aspects of the theropod brain architecture can be compared with those of living reptiles and birds. The architecture of the brains of coelurosaurs, their relatively large brains, and their visual acuity are good indicators of arboreal habitats in the vertebrates (Jerison 1973; Chatterjee 1997). Compared with all the other dinosaurs, the coelurosaurs have larger brains (Hopson 1980) and larger eye sockets, possibly in response to a new set of selection pressures in arboreal niches (see Chatterjee 1997, fig. 10.6). Spending much of their lives in trees, these would have required large

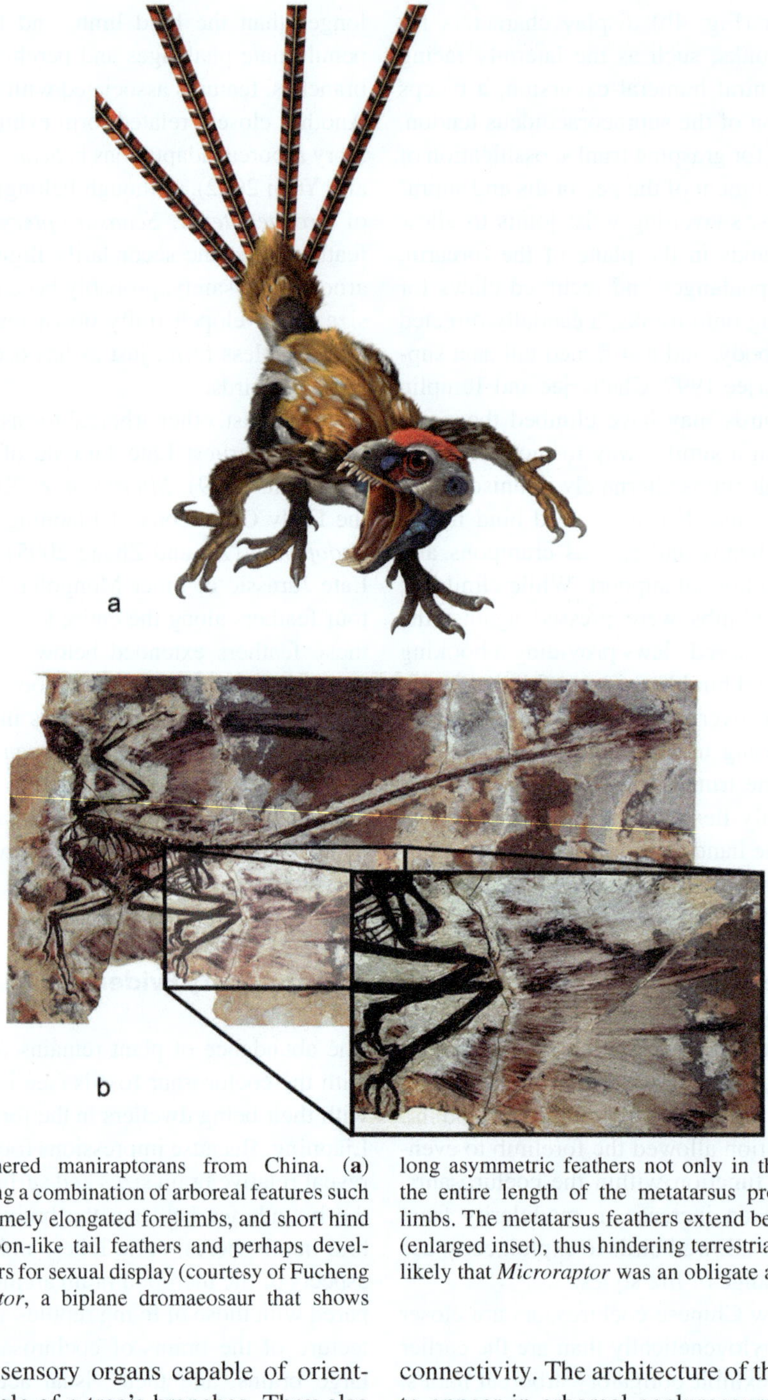

Fig. 5 Arboreal feathered maniraptorans from China. (**a**) *Epidexipteryx* displaying a combination of arboreal features such as grasping hand, extremely elongated forelimbs, and short hind limbs; it had four-ribbon-like tail feathers and perhaps developed extravagant colours for sexual display (courtesy of Fucheng Zhang). (**b**) *Microraptor*, a biplane dromaeosaur that shows long asymmetric feathers not only in the hand, but also along the entire length of the metatarsus projecting from the hind limbs. The metatarsus feathers extend below the level of the feet (enlarged inset), thus hindering terrestrial locomotion. It is most likely that *Microraptor* was an obligate arboreal form

brains and refined sensory organs capable of orienting them in the tangle of a tree's branches. They also developed the typical avian architecture of the brain with the cerebral hemispheres expanded for enhanced senses and intelligence, and in contact the cerebellum dorsally—thus displacing the optic lobes ventrally and laterally. Across their phylogeny, avian brains show broad similarities in architecture and internal connectivity. The architecture of the avian brain begins to appear in arboreal coelurosaurs; it is a good index with which to distinguish between terrestrial and arboreal forms of theropods. In typical terrestrial theropods such as allosaurs and tyrannosaurs, the brains were small and the brain architecture had the typical reptilian pattern, with the cerebral hemispheres, optic lobes and cerebrum arranged serially on the dorsal

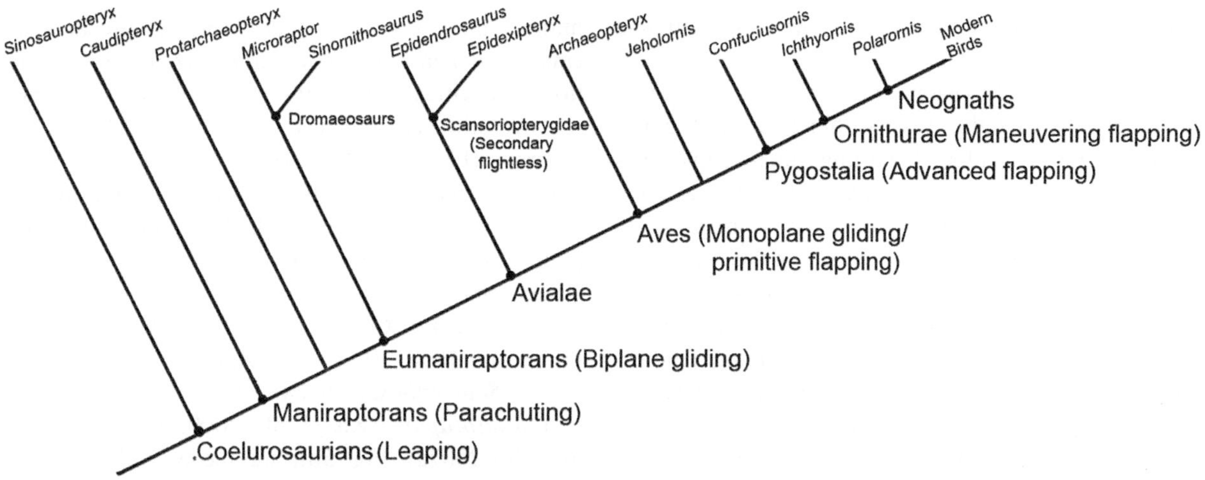

Fig. 6 The "palaeoecologic" evolution of birds and their flight. The sequence of events (from arboreal leaping, to parachuting, to gliding, to primitive flying, to advanced manoeuvring flying) are plotted against salient genera involved in the origin of avian flight (modified from Xu et al. 2003; Chiappe 2007; Chatterjee and Templin 2004a; Chatterjee 2002)

surface; the intervening optic lobes preclude any contact between the cerebrum and cerebellum (Jerison 1973; Hopson 1980; Chatterjee 1991, 1997).

Brain architecture and vision provide the strongest evidence for the arboreal lifestyle of the coelurosaurs. In the 1970s, Jerison (1973) developed the concept of the encephalization quotient (EQ) as a simple way of measuring an animal's intelligence working on the assumption that smarter animals have larger brain to body ratios than less intelligent ones. The EQ measures of the size of a creature's brain relative to the size of the rest of its body, and compares this ratio against other species of roughly the same size. From 0.25 in the early coelurosaurs, the EQ began to increase considerably to 0.54 in modern birds (Chatterjee 1991, 1997; Chatterjee and Templin 2004a). In contrast, typical terrestrial theropods, such as ceratosaurs and allosaurs, had relatively small brains, with EQ around 0.15. It is assumed that this was because they lived in a neurologically less-demanding world (Chatterjee 1997). The arboreal ancestral birds required larger brains than the brains of other terrestrial dinosaurs, just as the earliest primates required larger brains than those of other mammals of their time (Jerison 1973). The neurological distinction between arboreal and terrestrial theropods is analogous to that of the arboreal cats and terrestrial dogs, the former having a good three-dimensional sense and relatively larger and more complex brains (Bock 1985). Progressive enlargement of the brain and enhanced vision in the coelurosaurs are attributed to invasion of a new adaptive zone, the complex arboreal niches such as those of living primates (Jerison 1973; Chatterjee 1997). The confusingly mottled background of leaves, branches, and other foliage at different levels of trees might have provided a strong selection pressure for enlargement of the brain in order to process complex three-dimensional visual information.

Flight in birds is primarily guided by vision that provides spatial information at sufficient speed and resolution to guide flight. Together with encephalization, the visual resolving power of ancestral birds was greatly improved in aerial niches as is evident by the enlargement of their optic lobes—the major site for visual processing in the brain. The orbits were enlarged and, positioned frontally, resulted in broad, overlapping stereoscopic vision. Vision played a key role in early avian evolution.

Colour vision among vertebrates is a result of having specialized light receptor structures known as rods and cones at the back of the eye in the retina (Allman 1999). Rods are extremely sensitive to even dim light but provide relatively coarse, colourless images (similar to night vision goggles).

Birds have, perhaps, the most advanced visual system of any vertebrate. Modern diurnal birds have a wide visual spectral range and are tetrachromatic, whereas catarrhine primates (including us) are trichromatic; birds can see red, blue, green, and ultraviolet light and their combined hues, whereas we are insensitive to ultraviolet wavelengths (Hill and McGraw 2006).

The origin of vision in birds remains enigmatic but the Liaoning fossils offer interesting clues. Identification of colour-imparting microscopic sacs inside the feathers of Liaoning coelurosaurs, called melanosomes, correspond precisely in shape to structures associated with specific colours in the feathers of living birds indicating that these theropods had developed vibrant colour plumages (Zhang et al. 2010; Li et al. 2010). Some reconstructed Liaoning feathers appear to have developed a plethora of extravagant colour; they retain vivid, banded patterns and elongate, ribbon-like tails in many forms, indicating that vision had already developed in the ancestral birds for appreciation of ornament, species recognition and sexual display (Fig. 5a). Many of the patterns may have been invisible to terrestrial predators. The development of vision in birds might have originated from the arboreal protobirds, conferring several advantages: camouflage, foraging for ripe fruit or insects, providing food for chicks, and finding a suitable mate.

Arboreal coelurosaurs not only developed visual acuity but also acquired an enhanced auditory system for balance and hearing to supplement their sense of smell—as is evident from the endocasts of the inner ear region containing an elaborate vestibular system and elongated cochlea (Hopson 1980; Chatterjee 1991, 1997). The large vestibular system, the built-in gyroscope required for three-dimensional orientation, is a prerequisite for active flight. Similarly, a well-developed inner ear with an elongated lagena, as seen in the coelurosaurs and modern birds, is associated with hearing acuity and vocal behaviour. Refined hearing might have played an important role in such archaic bird activities as communication, alerting others of impending danger, mating, the care of young ones, and social behaviour. A capacity for three-dimensional orientation and hearing acuity would be more likely to have evolved in arboreal animals than in ground-dwellers. The earliest birds depended increasingly on sight and sound, and less on smell—as is apparent from their reduced olfactory lobes in endocasts. A sense of smell is less important in arboreal animals.

Materials and Methods

In the Jehol biota, non-avian coelurosaurs are presently represented by ten genera; with new discoveries, the number increases every year. Most coelurosaurs are small: chicken-sized to a turkey-sized. Size reduction is a critical step in the origin of flight. To trace the evolutionary sequence of flight, we follow the diagram (Fig. 6a). The six taxa listed show transitional stages in the flight apparatus and feather development leading to an *Archaeopteryx*-like wing planform. Phylogenetic analysis (Fig. 6) has demonstrated that the taxa, *Sinosauropteryx* (Chen et al. 1998) (Fig. 4a), *Caudipteryx* and *Protarchaeopteryx* (Ji et al. 1998) (Fig. 7b), *Anchiornis* (Hu et al. 2009), *Sinornithosaurus* (Xu et al. 1999a) (Fig. 4b), and *Microraptor* (Xu et al. 2000; Chatterjee and Templin 2007a) (Fig. 5b), are successively closer to *Archaeopteryx*.

As in birds, recurrent flightlessness was common among archaic birds when they began learning to fly (Paul 2002). We exclude *Beipiaosaurus* (Xu et al. 1999b), a basal therizinosaur, from the flight analysis because it is more than 2 m long, the largest Liaoning theropod, and may have become secondarily flightless (this is suggested by its filamentous feathers, large size, bulky body, and reduced wings). Secondary flightlessness raises a complex but interesting issue in terms of the origin of flight. How do we differentiate between primary terrestrial coelurosaurs and secondarily flightless species that reverted to a terrestrial lifestyle from their arboreal habitats? The answer may be found in the study of modern flightless birds such as ostriches, rheas, and emus. It is generally believed that ratites are secondarily flightless descendants of their flying ancestors and after losing their power of flight, they evolved into medium-sized grazing animals (Feduccia 1996). The flight apparatus of ratites is a response to their paedomorphic terrestrial adaptation with development of reduced wings, fluffy feathers, a flat sternum, and lack of ossification around the ilioischiadic fenestra. However, their brain architecture has retained clues to their arboreal and flying heritage: encephalized brain and acute eyesight (Cornfield et al. 2008). Although ratites have relatively smaller brains and cerebral hemispheres than neognaths; their brain architecture is very similar to the brains of extant flying birds. Therefore, brain architecture may provide an important clue with which to discriminate between primary terrestrial forms and secondarily flightless groups.

Secondarily flightless coelurosaurs, such as *Beipiaosaurus*, *Deinonychus*, and *Velociraptor*, became large and efficient hunters on the ground

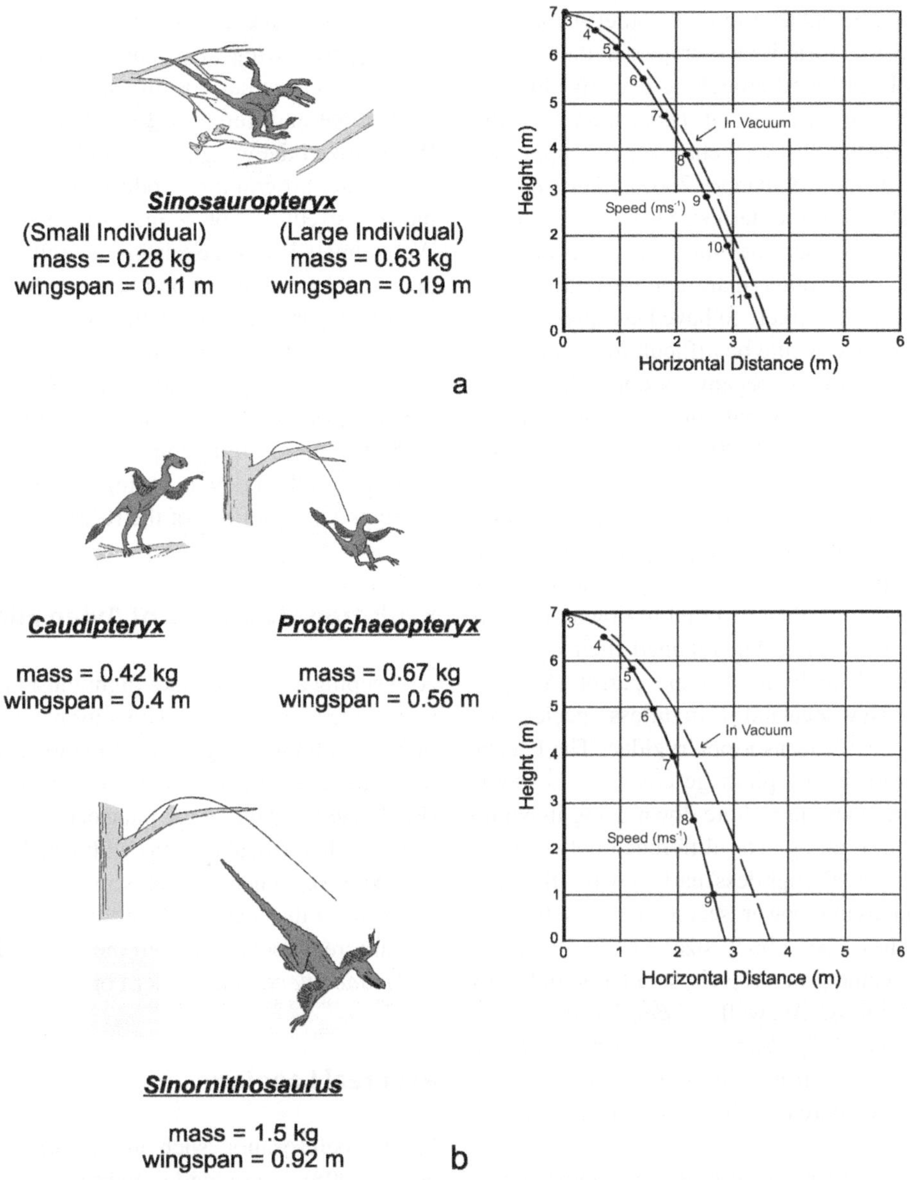

Fig. 7 (**a**) The leaping trajectory of *Sinosauropteryx*—without proper wing aerodynamics, forces had little effect on its jumping performance. The animal is viewed as leaping from branch to branch to achieve balance while steadying its wings. It could jump from a height of 2 m onto hard ground without injury (after Chatterjee and Templin 2004a). (**b**) The parachuting trajectory of *Caudipteryx*, *Protarchaeopteryx*, and *Sinosauropteryx* from a perch (a single curve is shown because their trajectories are essentially the same) With development of narrow incipient wings, these animals could have parachuted and retarded their fall from height (modified from Chatterjee and Templin 2004a)

but their arboreal heritage is evident in retention of large brains, large eyes, downy feathers, elongate forelimbs with sharp claws, grasping hands, and stiff tails (Chatterjee and Templin 2004a). In contrast, other large theropods, such as the allosaurs and tyrannosaurs, display no evidence of having passed through an arboreal stage in their early evolutionary history: they appear to have been always terrestrial—this is reflected in their small brains and large bodies (Chatterjee 1997). Because all maniraptorans developed relatively large brains with an avian architecture, they were either arboreal or became secondarily cursorial from arboreal ancestors. Another secondarily flightless terrestrial form is an oviraptorid, the skeleton of which has been found crouching over its clutch of eggs. Advocates of the cursorial theory have cited this evidence of ground-nesting behaviour as supporting the cursorial model (Chiappe 2007). This nesting behaviour points to the terrestrial lifestyle of the oviraptorids; these became large and secondarily flightless like some other maniraptorans. Moreover, arboreal nests are less likely to have been preserved in the fossil record and would be difficult to identify.

Some maniraptorans became secondarily flightless but retained an arboreal lifestyle, for example, *Epidexipteryx* (Zhang et al. 2008) (Fig. 4a) from the Late Jurassic deposits of China, and *Scansoriopteryx* (Czerkas and Yuan 2002) from the Jehol Group, included in the family Scansoriopterygidae, close relatives of *Archaeopteryx* (Fig. 6). They became secondarily flightless when they reverted to downy feathers to become fully arboreal but retained their climbing wing (Fig. 5a). The living kakapo parrot (*Strigops habrotilus*) of New Zealand, a flightless species, is a modern analogue of Scansoriopterygidae. The kakapo has unusual soft downy plumage covering the body; it can climb trees and parachute down using its wings for balance and braking. The difference between terrestrial and arboreal flightless early birds—that lost their incipient flying power secondarily—is that the former group has a large body size, whereas the arboreal species became progressively smaller, with highly elongated forelimbs. As well as *Beipiaosaurus*, we have also excluded *Epidendrosaurus*, *Epidexipteryx*, and *Scansoriopteryx* from our performance analysis because these feathered coelurosaurs became secondarily flightless.

Aerodynamic theory is useful for understanding the basic physics of avian flight. The continuously morphing wing planform, and the twist, rotation, and elastic deformation of the feathers during a wing beat cycle make calculation of instantaneous forces virtually impossible. Unlike aeroplanes, flying vertebrates obtain both lift and thrust by beating their wings. In this respect, they are like helicopters, which depend on their rotors for both lift and thrust. In this study, we used helicopter streamtube momentum theory to calculate the flight performances of the protobirds and early birds. A full account of the method, as applied to animals, is given elsewhere (Templin 2000; Chatterjee and Templin 2004a). Various flight parameters of the coelurosaurs and early birds, such as wingspan, wing area, and body length, were calculated from actual specimens or high-fidelity casts (Chatterjee and Templin 2003, 2004a, 2006, 2007a). We digitized the wing margin in its dorsal aspect with a computer programme produced by Rohlf <http://life.bio.sunysb.edu/morph/> and estimated the wing area. We estimated the mass using a multivariate analysis proposed by Atanassov and Strauss (2002). To analyse the flight performance of the coelurosaurs and early birds, we used two computer algorithms, ANFLTPWR (animal flight power) and ANFLTSIM (animal flight simulation), based on the streamtube model (Templin 2000; Chatterjee and Templin 2003, 2004a, 2007a, b). In all cases, the body and wing drag coefficients were computed as functions of the Reynolds number.

Evolutionary Stages of Avian Flight

A sweeping adaptation like flight must have evolved in many stages, over many generations, with each stage of the evolving wing being functional and of adaptive value, setting the stage for the next adaptation (Bock 1985). Phylogeny, palaeoecology, and aerodynamics offer critical insights with which we can retrace the likely sequence of events in evolution of flapping wings from the scansorial forelimbs of small, bipedal, and arboreal coelurosaurs to the powered flight stages of *Archaeopteryx* and modern birds (Fig. 6a).

Arboreal Leaping

Small coelurosaurs, such as *Sinosauropteryx* (Chen et al. 1998), were agile, lightly built animals that probably leaped from branch to branch as an escape strategy or to reach adjacent trees. Such behaviour established the animal's practice of taking to the air temporarily. When nearing a landing place, the forelimbs were outstretched for balance or to grab a branch. At this time, the outstretched forelimbs would have conferred enhanced stability and body control.

Simple, hair-like filaments covering the body of *Sinosauropteryx* may have been precursors to feathers providing insulation in the cooler microclimates of trees (Fig. 4a). These insulating proto-feathers may have reduced loss of body heat and had initial aerodynamic effects. The coat of proto-feathers conferred smooth and streamlined body contours on *Sinosauropteryx*, reduced frictional drag on the animal as it leapt among the tree branches, and provided additional protection during any accidental fall.

The encephalization quotient of *Sinosauropteryx* at this stage as deduced from basal coelurosaurs was 0.25 (Chatterjee 1997). We calculated the leaping trajectory of *Sinosauropteryx* from a perch at 3 m/s from an initial height of 7 m; also shown (Fig. 7a) for comparison is the theoretical trajectory in a vacuum, indicating that for an animal of this size with no wings, aerodynamic forces have little effect on its jumping performance. *Sinosauropteryx* could not slow during its fall. Speed rapidly increased to 6 m/s after falling from a height of 2 m, and to 10 m/s after about a 5 m drop—probably too high for safe landing unless the landing surface were compliant, such as a branch or soft, plant-covered ground (Templin 2000). The animal could probably jump from a height of 2 m onto a hard substrate without injury (Chatterjee and Templin 2004a, b).

Parachuting

When parachuting, leaping archaic birds might have used outstretched forelimbs and tails to maximize drag and slow descent, softening the impact of landing in much the same way as extant sugar gliders and some squirrels. Some recent tree frogs have been shown to be parachuters. In a fall with the descendant angle >45°, they launch themselves with a jump, hold the limbs out to the side, and control their orientation so that their flat ventral surface acts as an aerobrake for slowing the fall. Additionally, some parachuting frogs spread their webbed feet to act as a flight surface. The ability of archaic birds to leap from a great height without injury would have been an important adaptation in the early evolution of flight. Eventually, stiff and collapsible flight surfaces may have evolved in the forelimbs and tails in archaic birds associated with the development of contour feathers. In *Caudipteryx*, *Protarchaeopteryx*, and *Sinornithosaurus* (Fig. 6b, c), there is evidence of contour feathers on the remiges and retrices, and down-like feathers on the body (Ji et al. 1998; Xu et al. 2001). Although these contour feathers are long, vaned, and barbed, they are symmetric, indicating these animals probably could not fly or glide effectively (Feduccia and Tordroff 1979), their proto-wings and feathers being insufficiently large to generate enough lift for flight. Our study suggests *Caudipteryx*, *Protarchaeopteryx*, and *Sinornithosaurus* may have had a high wing loading and low aspect ratio. Their small proto-wings most probably evolved as a safety device for arboreal manoeuvring to increase drag and reduce the impact of their fall.

We have computed the parachuting trajectories of *Caudipteryx*, *Protarchaeopteryx*, and *Sinornithosaurus* with narrow wings (Fig. 7b), using the ANFLTSIM programme (Templin 2000). Because the separate trajectories of all three genera were essentially the same a single curve is presented in Fig. 7b; it displays speed versus height-loss, with the theoretical curve shown for a trajectory in a vacuum (under the same initial conditions). The difference between it and the actual curve reflects aerodynamic forces. Comparison of the two curves shows that even small wings may have a significant effect in "saving" falling animals of this size. The terminal velocity in this case is approximately 11 m/s, and is 2 m/s less after a fall of 7 m from a perch (Chatterjee and Templin 2004b). In our previous study (Chatterjee and Templin 2004a), we calculated the gliding performance of *Sinornithosaurus* (Xu et al. 1999b) assuming that it had a "more modern" wing size, and found that it was capable of phugoid gliding. Downy feathers covered the body of *Sinornithosaurus*, but contour feathers in the hands and tail were not preserved in the fossil. Because *Sinornithosaurus* is more derived than *Caudipteryx* or *Protarchaeopteryx* and has a relatively long wingspan, we assume that it had acquired a gliding surface. The recent discovery of additional material from a juvenile specimen suggests that *Sinornithosaurus* developed two types of branching structures: filaments joined in a basal tuft, and a feather with a central rachis and serially fused barbs forming a pair of symmetrical vanes (Xu et al. 2001). However, it lacked the sturdy contour feathers with sufficient structural integrity to form an effective gliding surface. We recalculated the flight performance of *Sinosauropteryx* (Fig. 6b) with its narrow wings and found that it could parachute but not glide. Possibly *Sinornithosaurus* (Fig. 6c) was reverting to flightlessness as had some other feathered maniraptorans; this is reflected in its fluffy wing feathers.

Biplane Gliding

The next stage was biplane gliding, wherein the pull of gravity provides the force required for forward airborne movement that preceded powered flight. Gliding is a simpler and cheaper way of flying than flapping the wings—most seabirds, with long and narrow wings, are primarily gliders. Gliding evolved independently in many groups of vertebrates, including lizards, colugo, phalangers, anomalurid rodents, pterosaurs, and birds (descendant angle <45°). Gliders stretch their motionless wings to form one lifting surface and passively descend through the air in response to gravity. An arboreal theropod launched from on high, was deflected from the line of fall, its horizontal travel increased.

The biplane wing planform has been identified in several Chinese paravian coelurosaurs including *Anchiornis* (Hu et al. 2009) that had long contour feathers on their lower legs and feet. *Anchiornis* was a small, crow-shaped troodontid known from the Late Jurassic of China and is older than *Archaeopteryx*; it shows two sets of wings, one on the forelimb, the other on the hindlimb. Its primary feathers in the forewing were as long as the secondaries with curved and symmetrical vanes indicating it poorer gliding ability. Large contour feathers also cover the whole length of the lower leg of *Anchiornis*; its feathers on the metatarsus were short, symmetric and oriented perpendicular to the toes forming an additional ventral wing. Thus the two sets of wings in *Anchiornis* suggests its biplane wing design, but the symmetric nature of the primary feathers suggests that *Anchiornis* was an unpowered glider where two sets of wings would generate lift and retard the terminal velocity when deployed from a perch by creating drag, but could not produce thrust.

The next stage was powered biplane gliding where the pull of gravity and asymmetric primary wings in both hands and feet would generate thrust and extend airborne time. Of all the feathered coelurosaurs from China, *Microraptor* is probably the most unusual in development of a biplane wing planform, possibly a precursor to the monoplane design of *Archaeopteryx* and later birds (Xu et al. 2003, Chatterjee and Templin 2005, 2007a). Refinements in its flight apparatus seem to reflect the gliding capacity of *Microraptor* (Xu et al. 2003). Its shoulder girdle is similar to that of *Archaeopteryx* where the glenoid, facing laterally, allowed gentle extension of the wing; a scapulacoracoid with ends oriented at an acute angle to each other; a single, enlarged sternum for attachment of flight muscles, ossified sternal ribs and well developed uncinate processes for resisting compressive forces on the thoracic cavity imposed during downstroke; development of the swivel wrist joint for an automatic wing-folding mechanism; strongly bowed outer metacarpal; a flattened digit for attachment of the primaries; and elongated asymmetric vanes in the flight feathers.

The two sets of wings, one above the other, may have functioned much like the fixed wings of a biplane (Fig. 8a). The distribution of feathers across the dorsal wings of the "biplane" resemble those on the wings of *Archaeopteryx*, but the presence of flight feathers on the feet, forming ventral wings, is the most unusual feature. A set of long asymmetric feathers with hooked barbs on the forelimbs and hind limbs (Fig. 8b, c_1, c_2) of *Microraptor* was an innovation. The leading edge of each long primary feather was narrower than the trailing edge, helping to streamline the body in flight. The hooked, interlocking barbs gave strength and flexibility to the feather and prevented air from passing through it during flight. Asymmetric feathers are essential for flight (Feduccia and Tordroff 1979); their presence on the metatarsals would have made walking or running on the ground nearly impossible.

Other than *Microraptor*, the biplane wing planform has been identified in other paravians including *Pedopenna* that had long contour feathers on their lower legs and feet (Xu and Zhang 2005). *Pedopenna,* a small maniraptoran dinosaur, from the Middle Jurasssic of China, from a close sister group to *Archaeopteryx,* is identified by their hind legs possessing shorter, but symmetric, contour feathers on the metatarsus. Since these feathers show fewer aerodynamic adaptations than the similar hindwings of *Microraptor*, and appear to be less stiff, they may be vestigial, representing an early stage of losing hindwings before evolving into monoplane configuration, as seen in *Archaeopteryx*. The long metatarsal feathers of paravian theropods such as *Anchiornis, Microraptor, and Pedopenna* are out of kilter with the cursorial theory of the origin of flight (Xu et al. 2003).

Gliding performance can be measured by a polar curve constructed by plotting horizontal speed against sinking speed—determined primarily by wing loading so that lighter, larger gliders descend more slowly than do smaller, heavier ones. We used the ANFLTSIM programme to calculate the possible gliding performance

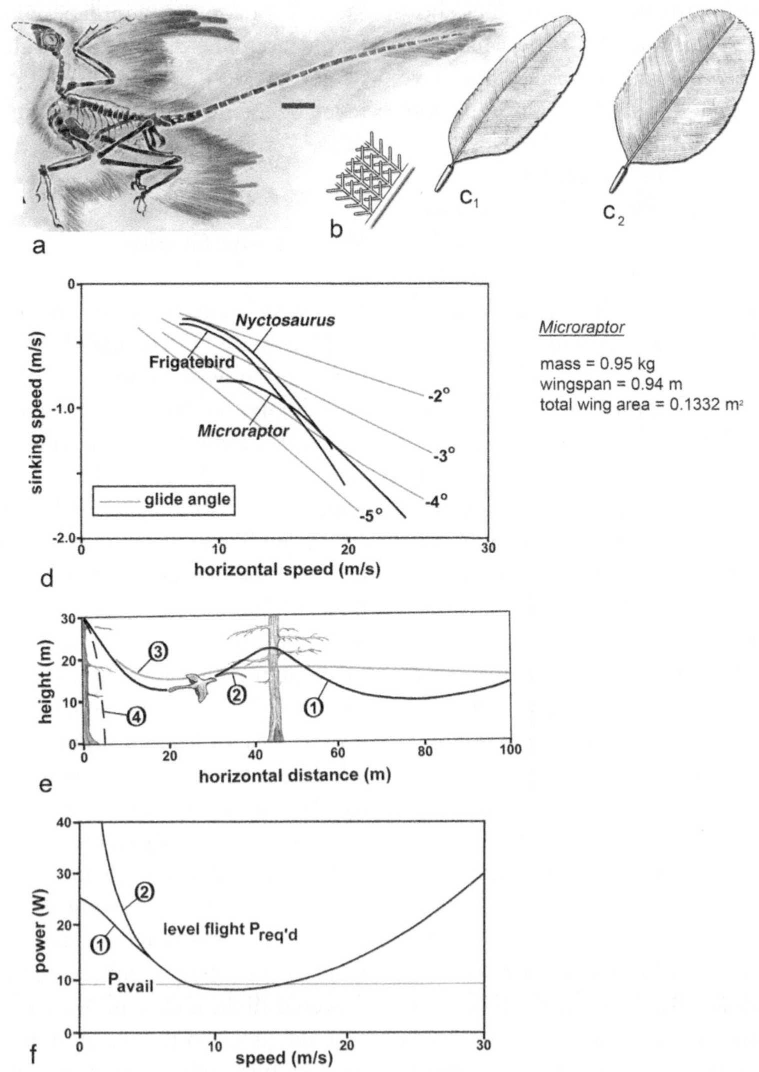

Fig. 8 (**a**) Holotype of *Microraptor gui* showing two sets of wings. (**b** and **c**) Interlocking barbules of the long, asymmetric feathers of the hand and foot *Microraptor*. (**d**) Symmetric feathers of wing and hindlimbs of other coelurosaurs. Graphs of flight potential: Glide polars of *Microraptor* compared with a seabird (frigate bird, $M = 1.5$ kg) and a pterosaur (*Nyctosaurus*, $M = 1.86$ kg); the polar curve accords with that *Microraptor* having been an efficient glider. (**e**) Glide paths of *Microraptor* from a perch; curve 1 shows phugoid gliding, curve 2 shows a final rapid pitchup with high drag, curve 3 gliding path with the pitch damper on, curve 4 a parachuting trajectory. (**f**) Power curves (steady level flight) for *Microraptor*; the horizontal line represents the estimated maximum continuous power available; two curves are shown for level flight: the U-shaped curve is the power required for level flight, whereas the horizontal curve represents the power available. Because they converge at speeds greater than 6 m/s, an incipient level flight was possible for *Microraptor* (after Chatterjee and Templin 2007a, b)

of *Microraptor* compared with that of a seabird (frigate bird) and a pterosaur of similar size (*Nyctosaurus*) (Fig. 8d). Results show that *Microraptor* was probably a moderately efficient glider compared with the long-wing-span frigate bird or the Cretaceous pterodactyloid, *Nyctosaurus*. Predicted glide paths of *Microraptor* from a perch suggest that its biplane wings were adapted for undulatory phugoid gliding between trees (Fig. 8e). Evolution as observed in *Microraptor*, may be an adaptation for increasing gliding efficiency. The aerodynamic efficiency of modern biplane aircraft is inferior to that of monoplanes because of the drag of the supporting struts and wires that required to stiffen the lightweight biplane wing assemblies. Structural scaling rules show that stresses decrease steadily as the size is reduced, so that the wings of *Microraptor* would not have required external bracing. It was demonstrated more than 70 years ago (e.g. Prandtl and Tietjens 1934) that the total induced drag of a cantilevered plane is theoretically smaller than that of a monoplane of the same span and with the same total lift. For the *Microraptor* biplane configuration, the improvement is a maximum lift/drag ratio, or glide angle, which would have been small (perhaps about 5%) but not negligible.

With the ANFLTPWR programme, we calculated the possible steady level flight capacity of *Microraptor* (Fig. 8f). Because the two curves, *power required* and *power available*, nearly coincide at flight speeds of 9–15 m/s, limited steady flight may have been theoretically possible. Therefore, *Microraptor* may have been primarily a glider that had begun to acquire a limited amount of flapping flight.

Phylogenetically, *Microraptor* is currently considered to belong to the Paraves (Sereno 1997), a group that also includes troodontids, dromaeosaurs, and *Avialae* (Figs. 6 and 9); basal members of all three groups had long contour feathers on their lower legs and feet. Surprisingly, *Archaeopteryx* had long contour feathers on the hindlimbs, especially in the tibial region, as seen in the Berlin specimen. It apparently lacked the metatarsal feathers, or the ventral wings; this was compensated by development of large forewings and long frond-like asymmetric retrices to provide additional lift surfaces. We were intrigued by the biomechanical functions of the leg feathers in several Chinese maniraptorans, *Archaeopteryx*, and modern raptors (Fig. 9a) and concluded that these feathers conferred streamlining and stretching effects on the cylindrical tibia to reduce drag during flight (Chatterjee and Templin 2007a). Like modern raptors, these maniraptorans probably kept their hind legs in a z-configuration during their preparation for aerial attack and, while carrying prey, dangling their tibiae in a vertical plane. The leg feathers kept their prey-catching structures streamlined during flight.

The unusual biplane wing configuration of *Microraptor* raises an interesting evolutionary possibility: did avian flight go through a *Microraptor*-like biplane stage in development of a monoplane configuration, like that of *Archaeopteryx*? The phylogeny of the *Paraves* indicates a gradual shift in the evolution of flight from combined biplane wings to a progressively larger monoplane wing, as paralleled in the evolution of aircraft design (Fig. 9b, c). [Possibly *Microraptor* invented the biplane 125 million years before invention of the Wright 1903 Flyer (Fig. 10)!]

Monoplane Gliding

The next stage was the beginning of monoplane gliding flight, documented by *Archaeopteryx* from the Late Jurassic Solnhofen Limestone of Germany (Wellnhofer 1990). This magpie-sized basal bird still retained a long bony tail with frond-like retrices for stability and lift, as well as developing fully formed wings with a long span, and asymmetric flight feathers on its hands and tail to provide greater lift and a larger flight surface. Its forewings became larger than those of *Microraptor*, but the feathers on the foot were lost on the hindlimbs forming, mimicking, in essence, the structure of a gliding monoplane. The long, asymmetic flight feathers on hands and long bony tail provided large lift and flight surfaces, formed a novel allegiance of locomotor module, and offered stability. *Archaeopteryx* shows long contour feathers on the hind limb, especially in the tibial region; it apparently lacked the metatarsus feathers, which were compensated for by larger forewings and long asymmetric tail feathers for additional lifting surfaces. The arrangement of the flight feathers suggests some degree of powered flight ability in *Archaeopteryx*, but the wings lack the alula on the freely moving first digit used by modern birds when flying at low speed or when landing. The lack of an alula may have been traded for climbing claws, another piece of evidence consistent with the arboreal theory (Fig. 5b). During an early stage of active flight, the capacity of these birds to become fully airborne increased. It is easy to envisage, from the example of *Archaeopteryx*, how flapping flight could have developed from a gliding animal that was already adapted for controlling wing movement during takeoff from a perch and landing (Norberg 1990). Flapping flight entails a downstroke, when the wing is brought downward and forward to provide thrust, and an upstroke that prepares the wing for the subsequent downstroke. Powered flight requires much more energy than does gliding. Energy comes from the flight muscles moving the wings up and down to keep the animal airborne against the pull of gravity.

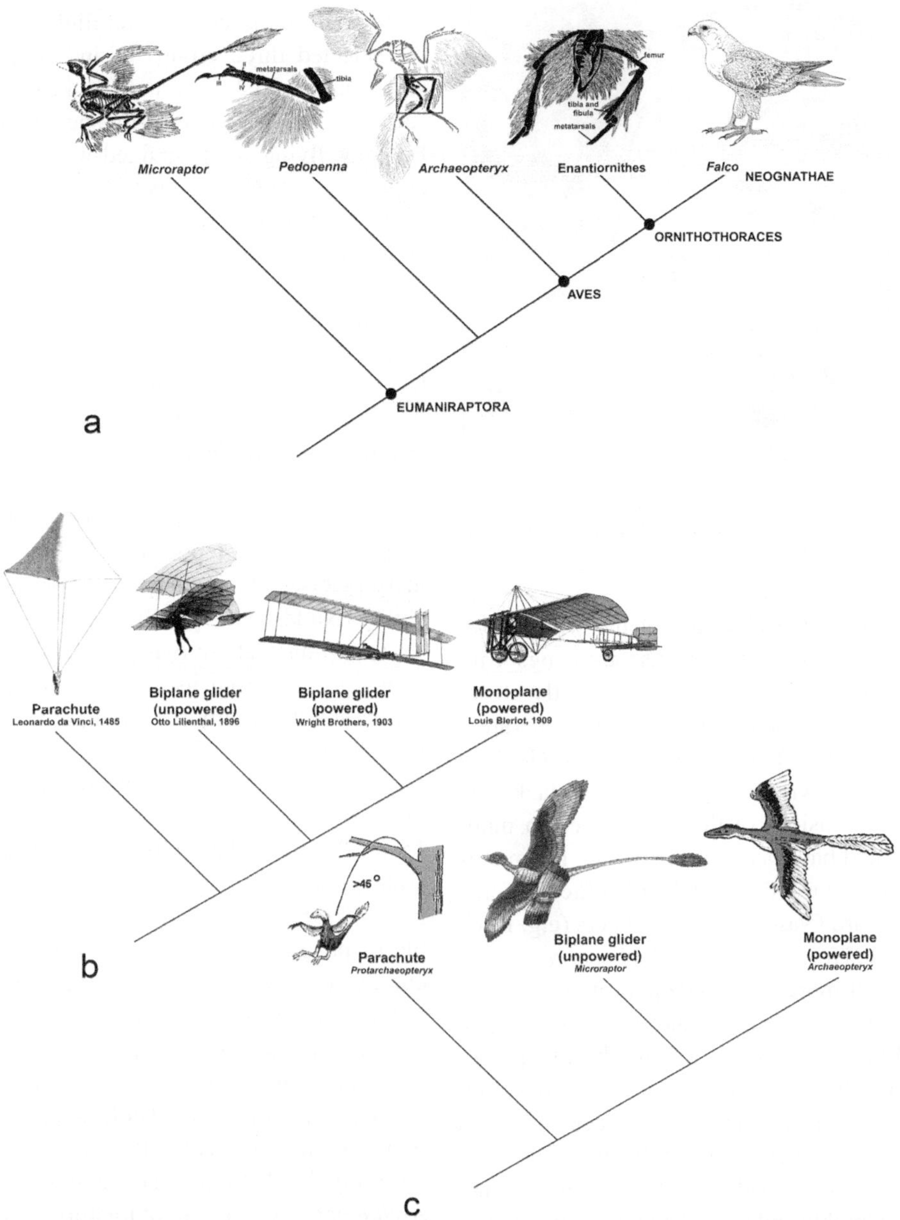

Fig. 9 (**a**) Paravian theropods showing distribution of the leg feathers in selected taxa (modified from Chatterjee and Templin 2007a). (**b–c**) The early evolution of aircraft, from parachute to biplane glider to monoplane, shows uncanny convergence with the early evolution of avian flight as documented by fossil evidence. Man learned to fly independently by progressing through stages analogous to those of the protobirds millions of years earlier, with no prior knowledge. He started with the parachute, then moved onto short glides, which gradually led to longer glides with powered biplane wings, and then to the monoplane, refined over the years. From this early monoplane stage, aeronautical technology diverged very rapidly from the avian model

The flight apparatus of *Archaeopteryx* remained virtually unchanged from that of its maniraptoran predecessors (Fig. 5c). For example, it lacked the bony sternum that houses the flight muscles. The pectoralis and supracoracoideus muscles of birds are the principal depressor and elevator of the wing, respectively. However, *Archaeopteryx* lacked the supracoracoideus pulley, the primary elevator of the wing required to execute the upstroke, a condition necessary for ground takeoff and active-flapping flight (Poore et al. 1997). Its glenoid faces laterally instead of dorsolaterally, the position seen in modern

Fig. 10 Reconstruction of *Microraptor* showing its biplane wing planform—compare with the Wright 1903 Flyer in the background (Credit: Jeff Martz)

birds, indicating limited elevation of the wing above the horizontal position. It could raise its wings with its deltoideus muscles to some extent, slightly above the horizontal position, but not enough to generate a powerful thrust in the following downstroke (Jenkins 1993). Because the wrist of *Archaeopteryx* lacks the interlockings and fusions of the carpus seen in modern birds, it would have been unable to stop the manus hyperpronating during a downstroke of sufficient force to propel it forward (Vasquez 1992). Others (e.g. Bock 1986; Ostrom 1976) consider that *Archaeopteryx* was at best a glider, or more probably a poor flyer.

Several features in the brain and inner ear indicate clearly that *Archaeopteryx* was a flying bird (Fig. 5a). It shows the bird-like brain organization with development of large, separated optic lobes for keen sight, and enlargement of the cerebellum for balance and control. Its brain had developed typical avian architecture and was considerably enlarged from non-flying precursors where the estimated EQ was 0.34 (Chatterjee 1997). Recent computed tomographic X-ray scanning images suggest that *Archaeopteryx* had developed the movement and balance control required for powered flight and not just gliding; this is evident from the topography of the inner ear with large semicircular canals with vestibular apparatus (Alonso et al. 2004). The brain is relatively bigger than those of non-flying coelurosaurs. Our flight analysis reflects the neural specialization of *Archaeopteryx* and suggests that it acquired an initial stage of powered flight (Chatterjee and Templin 2003, 2004a).

To analyse the steady-level flight of *Archaeopteryx*, we selected three groups of individuals of different size (mass = 0.2, 0.4, and 0.6 kg), representing the Eichstätt, London, and Berlin specimens, respectively. For any flying animal, fixed-wing aircraft, or helicopter, the plot of power required for level flight against airspeed is a U-shaped curve, higher at both ends than at the middle. Using the ANFLTPWR programme, we generated three power curves for *Archaeopteryx*, representing the three size groups—each U-shaped curve is the power required for steady-level flight and the corresponding horizontal line on the right side of the curve is the estimated maximum steady power available (Fig. 11a). Because the predicted available power exceeds the minimum power at each size, *Archaeopteryx* appears to have been capable of initial stages of powered flight. The narrow primary rachises of *Archaeopteryx* contradict its flapping ability (Nudds and Dyke 2010).

Assuming that *Archaeopteryx* was a tree dweller, as is apparent from its climbing wings and reversed hallux and the highly recurved pedal claws for perching on branches (Feduccia 1993), we calculated the gliding performance of the above three specimens of *Archaeopteryx* launching from a height of 15 m. We used the ANFLTSIM programme to generate glide polar curves (Fig. 11b). It appears that, like *Microraptor*, *Archaeopteryx* was probably capable of phugoid gliding between trees, possibly as an energy-conservation strategy. We also calculated the relative energy costs between a ground takeoff and an arboreal takeoff for *Archaeopteryx*, using these three individuals. In all cases, considering the energy required, an arboreal takeoff appears to have been more efficient and cost effective than a takeoff from the ground (Chatterjee and Templin 2004b). Possibly *Archaeopteryx* was an obligatory monoplane glider and had developed the ability to take short flights between trees, utilizing a novel method phugoid gliding without expending a great deal of energy. It would launch from a perch to initiate gliding losing height at first, and then sweep up with a large amplitude undulation to initiate phugoid gliding for swinging from tree to tree. Many modern birds such as crows use a similar strategy.

Another long-tailed primitive bird in the Jehol biota, *Jeholornis*, is similar to *Archaeopteryx* in its gross anatomy, but its tail is much longer than that of *Archaeopteryx* and its tail feathers are shaped

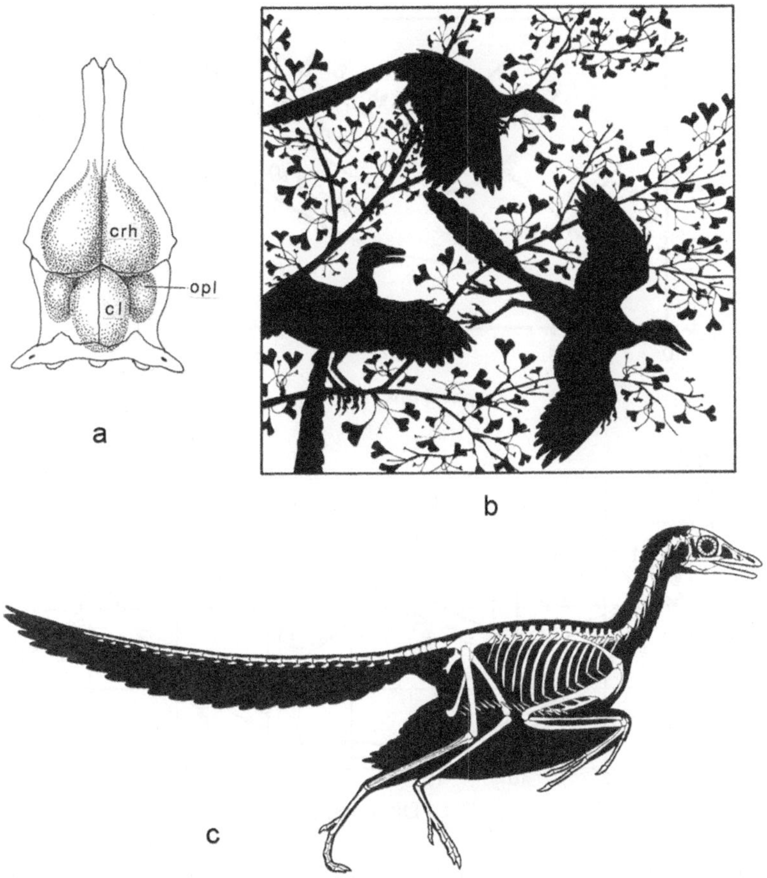

Fig. 11 (**a**) Morphology of brain of *Archaeopteryx* in dorsal view showing the three important components arranged in avian fashion: large cerebral hemispheres (crh), cerebellum (cl), and optic lobes (opl). (**b**) Suggested palaeoecology of *Archaeopteryx*; its claw geometry and swivelling wrist joints suggest that it was primarily an arboreal bird. (**c**) Skeletal restoration of *Archaeopteryx*, based mainly on the London specimen (after Chatterjee 1997)

like distal fans, as seen in the dromaeosaurs (Zhou and Zhang 2003). The hallux is reversed and the claws, large and curved, indicate arboreal adaptation. Compared with *Archaeopteryx*, *Jeholornis* shows some advancement in the flight apparatus: the glenoid faces dorsolaterally, the coracoid is strut-like, the sternum is ossified with lateral trabeculae, there is development of the interclavicular sac, the carpometacarpus is fused with a bowed metacarpal III (presumably one on each side); it has a shortened and more robust digit II, more suitable for attachment of flight feathers. *Jeholornis*, like *Archaeopteryx*, was a monoplane glider but with a long bony tail, more efficient.

Undulating Flight

The fourth stage in evolution of avian flight is the beginning of powered flight as documented by *Confuciusornis*, a beaked bird from the Jehol Group (Hou et al. 1995). As seen in *Archaeopteryx*, a well-designed tail with long, narrow, frond-like retrices is one of the most effective stabilizers; it is an inbuilt and essentially permanent feature. Long bony tails give a flyer passive stability, providing steady flight and avoiding unintentional changes (Alexander 2002). Stability was reduced by transforming the long, bony tail into a short tail with few vertebrae and a short pygostyle with caudal vertebrae fused into a single bone supporting the fan of tail feathers and the musculature—as seen in the clade Pygostylia. The pygostyle has an important function in controlling flight, steering, and braking. With loss of the tail, the Pygostylia became unstable but more manoeuvrable.

The basal radiation of the Pygostylia is represented by *Confuciusornis* (Fig. 12), one of the most abundant bird fossils in the Jehol biota—more than a thousand skeletons have been collected. With loss

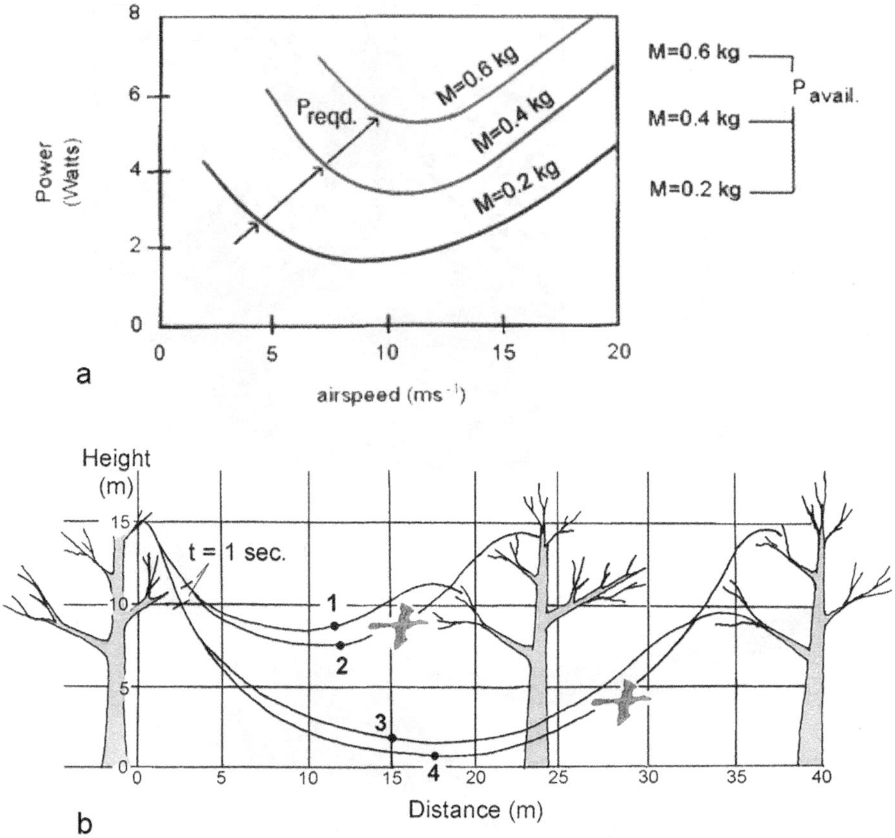

Fig. 12 Flight performances of three individuals of *Archaeopteryx* with different masses and body sizes: Eichstätt specimen ($M = 200$ g), London ($M = 400$ g), and Solnhofen ($M = 600$ g). (**a**) Power curves (steady level flight) for the three specimens of *Archaeopteryx*. In all three cases, the *power-available* horizontal curves (*right*) intersect the U-shaped *power-required* curves, indicating that *Archaeopteryx* achieved powered flight. (**b**) Phugoid gliding paths of *Archaeopteryx* from a height of 15 m with a launch speed of 2 m/s (*horizontal*). Phugoid gliding allowed *Archaeopteryx* to make short flights between trees without muscular effort

of its tail, *Confuciusornis* would be less stable during level flight but more manoeuvrable. A notable feature of *Confuciusornis* was that some individuals, possible males, possessed a pair of long, streamer-like tail feathers (Fig. 12), similar to those of birds of paradise. These may have been for sexual display. Other individuals lacking these long feathers may have been females—consistent with sexual dimorphism in some species of the genus.

Several refinements in the flight apparatus of *Confuciusornis* from an *Archaeopteryx*-like body plan are known. The relatively large wings, the arrangement of flight feathers, robust furcula, large deltoid crest on the humerus, large ossified sternum with a subdued keel, and partially fused wrist bones, suggest that downstroke had already developed in *Confuciusornis* to generate both lift and thrust, thereby enabling powered flight. Its hand had more flexing ability that helped during upstroke to reduce the wingspan. However, orientation of the glenoid in lateral position and lack of the supracoarcoideus pulley indicate that the upstroke was still not perfected in *Confuciusornis*. Takeoff from level ground would be difficult at this stage because of the absences of the supracoarcoideus pulley and narrow primary rachises, which were not hollow as in living birds (Nudds and Dyke 2010). Most likely *Confuciusornis* was a feeble flapper toward the gliding end, but the vigorous flapping flight of modern birds was unlikely. It was clearly an arboreal bird, as evidenced by long, curved and laterally compresses pedal claws and the proportions of its pedal phlanges. It would take off from a perch and remain airborne by an alternating sequence of flapping and gliding. The bird must glide to achieve the upstroke position by vertical air resistance. The height lost during gliding was regained in the next downstroke. Many living birds, such as starlings, bee-eaters, swifts, swallows, and fulmars employ this kind of undulating flight path to save muscular energy (Burton 1990). The partially fused synsacrum, tibiotarsus, and tarsometotarsus in *Confuciusornis* had strengthened the landing gear. The truncated bony tail and a long pygostyle made *Confuciusornis* less stable during level flight but more agile and manoeuvrable. It is easy to envisage from the example of *Confuciusornis* how undulating flight could have given rise to true powered flight as seen among enantiornithines and *Ichthyornis.*

Manoeuvring Flapping Flight

The final stage in evolution of avian flight would be to achieve the ability for manoeuvring comparable to that seen in modern birds and bats. There is a distinct dichotomy among the Pygostylia, one branch leading to enantiornithines ("opposite" birds), the other towards the ornithurines (modern birds) in Cretaceous ecosystems. The former were distributed globally until their extinction at the end of the Cretaceous (Fig. 13). Their flight ability was enhanced relative to the *Confuciusornis* stage. They occupied various niches: the early forms were small sparrow-sized perching birds; some became aquatic, and others adapted to the lifestyle of shore birds. Relative to ornithurines, they had a reversed articular arrangement between the scapula and coracoid that formed the glenoid cavity and developed a primitive triosseal canal entirely confined to the coracoid, allowing a wide range of upstroke movements. The coracoid being strut-like, amplified the range of the wingstroke. They developed an alula; it delayed stalling during slow flight during taking off and landing. Although the sternum was relatively short compared with those of ornithurine birds, it had a keel for housing the flight muscles. Therefore enantiornithines possessed a more advanced flight apparatus than *Confuciusornis*, enabling them to flap their wings vigorously. However, they retained many primitive features in the pelvis and hind limbs, indicating their "landing gear" was not perfected at this stage.

The ornithurine birds, such as *Ichthyornis* from the Upper Cretaceous of Kansas, were terrestrial and possessed sophisticated flight skills, nearly identical to modern birds (Fig. 14). In this group, although the skull is primitive with retention of teeth on the maxilla and dentary, the flight apparatus is modern in appearance. The sternum was enlarged with development of a strong keel for attachment of the flight muscles. The wings were very long and powerful, the glenoid faced dorsolaterally allowing large excursions of the dorsal and ventral movements of the wings; development of an alula prevented stalling during takeoff and landing. The ribcage was strengthened by

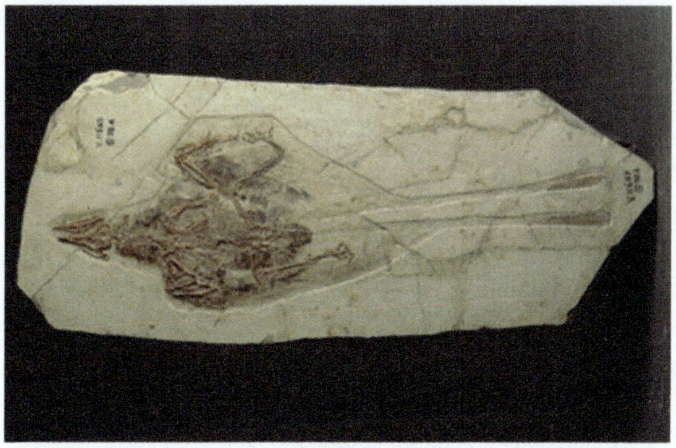

Fig. 13 A crow-sized basal bird, *Confuciusornis sanctus,* with a pygostyle and two streamer-like tail feathers, housed in the Palaeozoological Museum of China in Beijing (image provided by Zhonghe Zhou, one of the palaeontologists who discovered and described this basal bird)

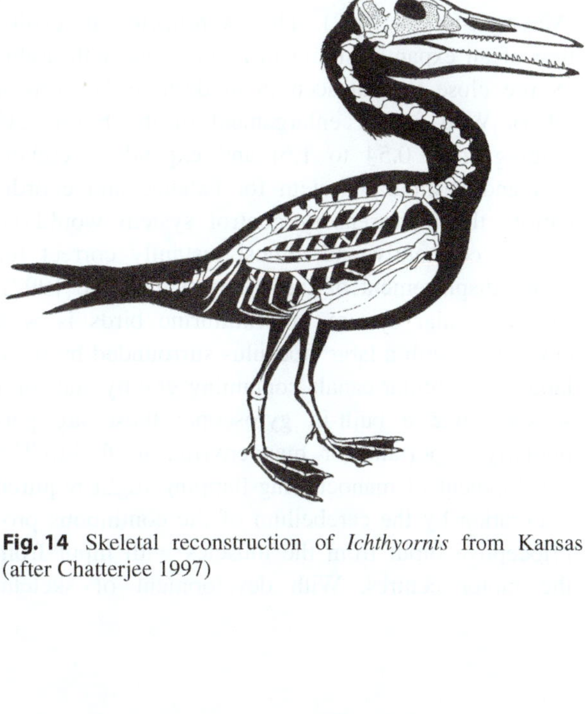

Fig. 14 Skeletal reconstruction of *Ichthyornis* from Kansas (after Chatterjee 1997)

ossification of the uncinate process. The pelvis and hind limbs were highly modified like those of modern birds by a large degree of bone fusion at the joints. Ground landing was also perfected at this stage with development of the synsacrum, the fused tibiotarsus, and the fused tarsometatarsus (Chatterjee 1997). The wrist was modified in such a fashion that the manus could be locked in place during the downstroke (Vasquez 1992). As a result, various subtle and complex flight-related movements of the wings could be performed. That they had achieved long-distance manoeuvring flight may have contributed to the wide distribution of *Ichthyornis*.

That evolution has favoured manoeuvrability at the expense of passive stability during flight is a recurrent theme in the evolution of both birds and pterosaurs. Bird size may be the most important indicator of manoeuvrability because the gap between aerobic power available and the minimum power required for steady flight increases (percentage-wise) as size decreases; the gap allows for increased manoeuvrability. Small birds are more agile because of the large margin of available power compared with power required for steady flight (Templin 2000).

Brain evolution and neural control are also important for manoeuvrability. In contrast to passive stability, manoeuvrability allows for change in direction (Alexander 2002). Ornithurine birds developed a specialized nervous system required for active stabilization and control of manoeuvring during flight (Maynard-Smith 1952). This is reflected in evolution of an expanded brain in *Ichthyornis* with architecture close to that seen in modern birds (Jerison 1973). With further enlargement of the brain (EQ ranging from 0.54 to 1.6) and expanded cerebellum and vestibular system for balance and coordination, the neurosensory control system would be able to constantly and almost instantly correct for minor displacement of the wings (Chatterjee 1997). The vestibular system in ornithurine birds is well developed, with a large flocculus surrounded by three large semicircular canals containing gravity and rotor sensors—like a built-in gyroscope; these are particularly important for manoeuvring in flight. The development of manoeuvring-flapping flight required integration by the cerebellum of the continuous proprioceptive input from the muscles with input from the motor centres. With development of skeletal and neurological innovations, ornithurine birds, as exhibited by *Ichthyornis*, were able to make long-distance manoeuvring flights.

Conclusions

The evidence presented here best supports the arboreal theory of evolution of avian flight. The anatomy of small, feathered coelurosaurs from China suggests both scansorial and arboreal habitats. Climbing adaptations of the archaic birds led to gradual elongation of forelimbs, increasing the wingspan and development of rudimentary flight strokes. Similarly, the arboreal life of archaic birds promoted encephalization, visual acuity, and neurosensory specializations necessary for flight. The hypothetical flight stages in the arboreal theory—climbing to leaping to parachuting to gliding to early flapping to manoeuvring flight—are corroborated by a series of transitional fossils and their flight performance. The arboreal model is a straightforward explanation for the evolution of avian flight.

Acknowledgments We thank John Talent for inviting us to contribute to this volume and many Chinese colleagues, especially Zhonghe Zhou, Xing Xu, and Fucheng Zhang, for demonstrating the Liaoning fossil locality to us, allowing us to examine some of the spectacular feathered coelurosaurs, for sharing their ideas with us, and providing us with excellent photos and drawings. Phil Currie, Bill Mueller and other colleagues reviewed earlier versions of the manuscript. Ruth Mawson, Peter Cockle, Robyn Izberg, Colin McKenzie, Janine Miller, Karen Monk, and Trevor Worthy kindly improved the manuscript. Kyle McQuilkin, Mike Nickell, and Jeff Martz produced the illustrations. Our research was supported by Texas Tech University.

References

Alexander DE (2002) Nature's flyers. Johns Hopkins University Press, Baltimore

Allman JM (1999) Evolving brains. Scientific American Library, New York

Alonso PD, Milner AC, Ketcham RA, Cookson MJ, Rowe TB (2004) The avian nature of the brain and inner ear of *Archaeopteryx*. Nature 430:666–669

Atanassov MN, Strauss RE (2002) How much did *Archaeopteryx* and *Quetzalcoatlus* weigh? Mass estimation by multivariate analysis of bone dimensions. J Vert Paleont 22(Supp. to No 3):33A

Bakker RT (1986) The dinosaur heresies. William Morrow, New York

Balda RP, Caple G, Wills RR (1985) Comparisons of the gliding and flapping sequence with the flapping to gliding sequence. In: Hecht MK, Ostrom JH, Viohl G, Wellnhofer P (eds) The beginnings of birds. Freunde des Jura Museum, Eischstätt, pp 267–277

Bock WJ (1985) The arboreal theory for the origin of birds. In: Hecht MK, Ostrom JH, Viohl G, Wellnhofer P (eds) The beginnings of birds. Freunde des Jura Museum, Eischstätt, pp 267–277

Bock WJ (1986) The arboreal origin of avian flight. In: Padian K (ed) Origin of birds and evolution. California Academy of Sciences, San Francisco, pp 57–73

Bundle MW, Dial KP (2003) Mechanics of wing-assisted incline running (WAIR). J Exp Biol 206:4553–4564

Burgers P, Chiappe LM (1999) The wing of *Archaeopteryx* as a primary thrust generator. Nature 399:60–62

Burton R (1990) Bird flight. Facts on File, New York

Chang M (ed) (2008) The Jehol fossils. Academic, Amsterdam

Chang S, Zhang H, Renne PR, Fang Y (2009) High-precision $^{40}Ar/^{39}Ar$ age for the Jehol biota. Palaeogeogr Palaeoclimatol Palaeoecol 280:94–104

Chatterjee S (1991) Cranial anatomy of new Triassic bird from Texas. Phil Trans Roy Soc Lond B 332:277–346

Chatterjee S (1997) The rise of birds. Johns Hopkins University Press, Baltimore, MD

Chatterjee S (2002) The morphology and systematics of *Polarornis*, a Cretaceous loon (Aves: Gaviidae) from Antarctica. In: Zhou Z, Zhang F (eds) Proceedings of the 5th international meeting of the society of Avian paleontology and evolution. Science Press, Beijing, pp 125–155

Chatterjee S, Templin RJ (2003) The flight of *Archaeopteryx*. Naturwiss 90:27–32

Chatterjee S, Templin RJ (2004a) Feathered coelurosaurs from China: new light on the arboreal origin of avian flight. In: Currie PJ, Koppelhaus EB, Shugar MA, Wright JL (eds) Feathered dragons: studies on the transition from dinosaurs to birds. Indiana University Press, Bloomington, IN, pp 1–64

Chatterjee S, Templin RJ (2004b) Posture, locomotion, and paleoecology of pterosaurs. Geol Soc Am Spec Paper 376:1–64

Chatterjee S, Templin RJ (2005) How did dinosaurs begin to fly? Nat Hist 114:54–55

Chatterjee S, Templin RJ (2006) A new fossil loon from the Late Cretaceous of Antarctica and early radiation of foot-propelled diving birds. J Vert Paleont Abs 26:49A

Chatterjee S, Templin RJ (2007a) Biplane wing planform and flight performance of the feathered dinosaur *Microraptor gui*. Proc Natl Acad Sci USA 104:1576–1580

Chatterjee S, Templin RJ (2007b) The aerodynamics of *Argentavis*, the world's largest flying bird from the Miocene of Argentina. Proc Natl Acad Sci USA 104:12398–12403

Chen PJ, Dong Z, Zhen S (1998) An exceptionally well-preserved theropod dinosaur from the Yixian Formation of China. Nature 391:147–152

Chiappe LM (2007) Glorified dinosaurs: the origin and early evolution of birds. Wiley, Hoboken, NJ, 192pp

Chiappe LM, Witmer LM (eds) (2002) Mesozoic birds. University of California Press, Berkeley

Cornfield JR, Wild JM, Hauber ME, Parsons S, Kubke MF (2008) Evolution of brain size in the palaeognath lineage, with an emphasis on New Zealand ratites. Brain Behav Evol 71:87–99

Czerkas SA, Yuan C (2002) An arboreal maniraptoran from northwestern China. In: Czerkas S (ed) Feathered dinosaurs and the origin of flight. The Dinosaur Museum, Blanding, Utah, pp 63–95

Darwin CH (1866) On the origin of species, 4th edn. John Murray, London

Dial KP (2003) Wing-assisted incline running and the evolution of flight. Science 299:402–404

Elzanowski A (2002) Biology of the basal birds and the origin of avian flight. In: Zhou Z, Zhang F (eds) Proceedings of the 5th international meeting of the society of Avian paleontology and evolution. Science Press, Beijing, pp 211–226

Feduccia A (1993) Evidence from claw geometry indicating arboreal habits of *Archaeopteryx*. Science 259: 790–793

Feduccia A (1996) The origin and evolution of birds. Yale University Press, New Haven, CT

Feduccia A, Tordroff HB (1979) Feathers of *Archaeopteryx*: asymmetric vanes indicate aerodynamic function. Science 203:1021–1022

Gatsey SM, Dial KP (1996) Locomotor modules and the evolution of avian flight. Evolution 50:331–340

Gauthier J (1986) Saurischian monophyly and the origin of birds. In: Padian K (ed) Origin of birds and evolution. California Academy of Sciences, San Francisco, CA, Mem 8:1–55

Gauthier J, Padian K (1985) Phylogenetic, functional and aerodynamic analysis of the origin of birds and their flight. In: Hecht MK, Ostrom JH, Viohl G, Wellnhofer P (eds) The beginnings of birds. Freunde des Jura Museum, Eischstätt

Goldsmith TH (2006) What birds see. Sci Am 298:68–75

Heilmann G (1926) The origin of birds. Witherby, London

Hill GE, McGraw KJ (eds) (2006) Bird colouration: function and evolution. Harvard University Press, Cambridge

Hopson JA (1980) Relative brain size in dinosaurs: implications for dinosaurian endothermy. In: Thomas RDK, Olson EC (eds) A cold look at the warm-blooded dinosaurs. West View Press, Boulder, CO, pp 287–310

Hou L, Zhou Z, Martin LD, Feduccia A (1995) A beaked bird from the Jurassic of China. Nature 377:616–618

Hu D, Hou L, Zhang L, Xu X (2009) A pre- *Archaeopteryx* troodontid theropod from China with long feathers on the metatarsus. Nature 461:640–643

Huxley TH (1868) Remarks upon *Archaeopteryx lithographica*. Proc R Soc Lond 16:243–248

Huxley TH (1870) Further evidence of the affinity between dinosaurian reptiles and birds. Proc Geol Soc Lond 26: 12–31

Jenkins FA (1993) The evidence of avian shoulder joint. Am J Sci 293A:253–267

Jerison HJ (1973) Evolution of the brain and intelligence. Academic, New York

Ji Q, Currie PJ, Norell MA, Ji SA (1998) Two feathered dinosaurs from northeastern China. Nature 393:753–761

Li Q, Gao KA, Vinther J, Shawkey MD, Clarke JA, D'Alba L, Meng O, Briggs DEG, Prum RO (2010) Plumage colour patterns of an extinct dinosaur. Science 327:1369–1372

Luo Z (1999) A refugium for relicts. Nature 400:23–25

Marsh (1880) Odontornithes: a monograph on extinct toothed birds of North America. Rep Geol Expl Fortieth Parallels 7:1–201

Maynard-Smith J (1952) The importance of nervous system in the evolution of animal flight. Evolution 6:127–129
Nopsca F (1923) On the origin of flight of birds. Proc Zool Soc Lond 1907:223–236
Norberg UM (1990) Vertebrate flight. Springer, Berlin
Nudds RL, Dyke GJ (2010) Narrow primary feather rachises in *Confuciusornis* and *Archaeopteryx* suggest poor flight ability. Science 328:887–889
Ostrom JH (1976) *Archaeopteryx*, and the origin of birds. Biol J Linn Soc 8:91–182
Ostrom JH (1979) Birds flight: how did it begin? Am Sci 67:46–56
Ostrom JH (1986) The cursorial origin of avian flight. In: Padian K (ed) Origin of birds and the evolution if flight. California Academy of Sciences, San Francisco, CA, pp 73–81
Owen R (1863) On the *Archaeopteryx* of von Meyer, with a description of the fossil remains of a long-tailed species from the lithographic stone of Solnhofen. Phil Trans R Soc Lond 153:33–47
Padian K, Chiappe LM (1998) The origin of birds and their flight. Sci Am 278:38–47
Paul GS (2002) Dragons of the air. Johns Hopkins University Press, Baltimore, MD
Poore SO, Sanchez-Harman A, Goslow GE Jr (1997) Wing upstroke and the evolution of flapping flight. Nature 387:799–802
Prandtl I, Tietjens OG (1934) Applied hydro- and aeromechanics. Dover Publications, New York
Rayner JMV (1991) Avian flight and the problem of *Archaeopteryx*. In: Rayner JMV, Wooton RJ (eds) Biomechanics and evolution. Cambridge University Press, New York, pp 183–212
Sereno PC (1997) The origin and evolution of dinosaurs. Ann Rev Earth Planet Sci 25:435–489
Templin RJ (2000) The spectrum of animal flight: insects to pterosaurs. Prog Aero Sci 36:393–436
Vasquez RJ (1992) Functional osteology of the avian wrist and the evolution of flapping flight. J Morph 211:159–268
Wellnhofer P (1990) *Archaeopteryx*. Sci Am 262:70–77
Xu X, Zhang F (2005) A new maniraptoran dinosaur from China with long feathers on the metatarsus. Naturwiss 92:173–177
Xu X,Wang XI, Wu XC (1999a) A dromaeosaurid dinosaur with a filamentous integument from the Yixian Formation of China. Nature 401:262–266
Xu X, Tang ZL, Wang X (1999b) A therizinosaurid dinosaur with integumentary structure from China. Nature 399:350–354
Xu X, Zhou Z, Wang X (2000) The smallest known non-avian theropod dinosaur. Nature 408:705–708
Xu X, Zhou Z, Prum RO (2001) Branched integumental structures in *Sinornithosaurus* and the origin of feathers. Nature 410:200–204
Xu X, Zhou Z, Wang X, Kuang X, Zhang F, Du X (2003) Four-winged dinosaurs from China. Nature 421:335–340
Zhang F, Zhou Z, Xu X, Wang X (2002) A juvenile coelurosaurian theropod from China indicates arboreal habits. Naturwiss 89:394–398
Zhang F, Zhou Z, Xu X, Wang X, Sullivan C (2008) A bizarre Jurassic maniraptoran from China with elongated ribbon-like feathers. Nature 455:1105–1108
Zhang F, Kearns SL, Orr PJ, Benton MJ, Zhou Z, Johnson D, Xu X, Wang X (2010) Fossilised melanosomes and the color of Cretaceous dinosaur and birds. Nature 463:1075–1078
Zhou Z, Zhang F (2003) *Jeholornis* compared to *Archaeopteryx*, with a new understanding of the earliest avian evolution. Naturwiss 90:220–225
Zhou Z, Barrett PM, Hilton J (2003) An exceptionally preserved Lower Cretaceous ecosystem. Nature 421:807–814

Part III

Global Extinction Events and Biocrises

An Analysis of the Ireviken Event in the Boree Creek Formation, New South Wales, Australia

Peter D. Molloy† and Andrew J. Simpson

Abstract

Conodont data from a section through the Boree Creek Formation in New South Wales, the best sequence in Australia extending through the Llandovery–Wenlock boundary, represent strata of the Ireviken Event. A significant number of conodont datum planes within the sequence indicating step-wise extinctions can be recognised and correlated with the sequence through the Visby Beds, in Gotland, Sweden, the most comprehensively analysed Ireviken Event sequence globally. This intercontinental correlation, involving both north and south hemispheres, adds to the data set concerning cyclicity in Silurian marine strata and postulated models concerning causality.

Keywords

Silurian • Llandovery • Wenlock • Australia • Boree Creek Formation • Conodonts • Ireviken Event • Stepped extinctions • Stratigraphic alignments • Cyclicity

Introduction: The Late Silurian Events

The abiotic cause of major or 'the big five'-type extinctions has been the subject of much scientific literature (e.g. Alvarez et al. 1980; Veron 2008). Over the last two decades, however, Silurian conodonts have been the focus of lesser known, but nonetheless, global extinction studies. While the extinction of conodont faunas had been recognised for some time (e.g. Aldridge 1976; Aldridge and Jeppsson 1984), it was Jeppsson (1987, 1990, 1997a, b) who first commented on the relationship between patterns of marine sedimentation and the chronology of Silurian conodont faunas. He also noted that major extinctions of Silurian taxa were preceded by anomalous sedimentation at the close of both the *amorphognathoides* and the *siluricus* zones in a number of geographically widespread localities. He named the late Llandovery–early Wenlock extinction event the Ireviken Event and the Late Ludfordian extinction event the Lau Event, respectively. Both events are made up of a series of intervals of time during which several datum points can be established allowing comparison of the events from one locality to another, for example, from Australia's Broken River sequences to those found on Gotland, Sweden. A comparative analysis of conodont, isotopic

A.J. Simpson (✉)
Department of Earth and Planetary Sciences, Museum Studies and Centre for Ecostratigraphy and Palaeobiology (MUCEP), Macquarie University 2109, Australia
e-mail: andrew.simpson@mq.edu.au

Peter D. Molloy: Deceased

J.A. Talent (ed.), *Earth and Life*, International Year of Planet Earth,
DOI 10.1007/978-90-481-3428-1_19, © Springer Science+Business Media B.V. 2012

and lithological data for these events from localities from separate palaeogeographic continents and different hemispheres of the globe (Jeppsson et al. 2007) has demonstrated significant correlation supporting the interpretation of such events as of global significance. For a detailed discussion of the Lau Event, see Jeppsson et al. (2007) and 'Late Ludfordian Correlations and the Lau Event', this volume).

Jeppsson's Model

Jeppsson (1990) noted cyclicity in the Silurian sequences and published a model to explain the observed changes in these marine sequences. Previous models explained the observed faunal and lithological record as derived from either lateral facies displacement from transgression/regression patterns resulting from sea level change and/or tectonic factors, shifting climate belts or a combination of both. These assume relatively homogenous oceanic conditions through time. Jeppsson (1990) argued that these, in isolation, are inadequate to explain the observed record. The model implies oceanic heterogeneity through intervals of instability punctuating periods of two distinctively different forms of oceanic quiescence.

Jeppsson's model considered alternating marine conditions separated by boundary strata characterised by increased faunal turnover (Jeppsson 1990). The two marine regimes were named Primo and Secundo episodes (Jeppsson *in* Aldridge et al. 1993). These episodes were believed to represent cool and warm high latitude climates respectively, with climatic conditions affecting ocean chemistry, particularly the ability of deep waters to store carbon dioxide, thus affecting the nutrient supply supporting planktic communities with implications for the diversity of the fauna. Jeppsson noted that major evolutionary events in the history of the oceans were linked with changes in the ocean at global scale. Jeppsson's Primo episodes were characterised by a more abundant and diverse conodont fauna and restricted carbonate sedimentation; Secundo episodes were characterised by impoverished conodont faunas and more extensive carbonate sedimentation. Recently other researchers have focussed on similar scenarios taking their examples from more recent times. Regarding the difficulties in explaining how major benthic foraminiferal extinctions, for example, at the Paleocene–Eocene boundary, may have occurred, Thomas (2007) notes:

> possible causes include changing oceanic productivity, lowered oxygenation, and carbonate corrosivity, but none of these factors occurred globally, and survival in refugia followed by repopulation would have prevented extinction of cosmopolitan species. Global warming might have been the most important cause of extinction, but mechanisms are not understood.

At times of global warming, it is assumed that primary productivity increases; transfer of food to the ocean floor would be more prevalent during warm periods. Benthic faunal turnover during episodes of global cooling may reflect the increased seasonality of primary productivity and increased delivery of organic matter to the seafloor giving rise to an extinction event.

Jeppson's model was devised largely as a means of explaining the ranges of Silurian conodont taxa from the intensively sampled Gotland sequence. The best documented and most profound event is the late Llandovery–early Wenlock Ireviken Event. The Gotland Ireviken Event represents the transition from the Snipklint Primo Episode to the Vattenfallet Secundo Episode (Aldridge et al. 1993) as seen in the Visby Beds. Between the Primo and Secundo pillars of the Ireviken Event, a series of smaller events, boundaries and datum points, give the Event an internal structure that has enabled significantly enhanced, finer biostratigraphic resolution (Jeppsson 1990, 1997a; Aldridge et al. 1993). The identification of eight datum points within the Ireviken Event based on the appearance, disappearance – temporary retreat to refugia – or extinction of key taxa was detailed by Jeppsson (*in* Aldridge et al. 1993). Furthermore, Aldridge et al. (1993) and Jeppsson (1993) identified a number of datum planes within a single event representing a stepwise extinction pattern considered to be caused by fluctuating conditions before the stable conditions of one episode or another prevailed. Such fluctuating conditions were linked to Milankovich cycles (Jeppsson 1993) and discussed in some detail according to the nature of both the preceding and subsequent episode (Jeppsson 1998).

The other well-known event from the Silurian is the Late Silurian Lau Event. A comparative analysis of conodont, isotopic and lithological data for this event from localities from separate palaeogeographic continents and different hemispheres of the globe (Jeppsson et al. 2007) has demonstrated significant correlation supporting the interpretation of the Lau Event as of global significance. For a detailed discussion of the Lau Event, see Jeppsson et al. ('Late Ludfordian Correlations and the Lau Event', this volume).

The Ireviken Event in Australia

Detailed summaries of Silurian strata of Australia appear in Talent et al. (1975, 2003). Simpson (1995) discussed the most prospective carbonate sequences in Australia spanning the late Llandovery to early Wenlock. The most complete conodont documentation spanning this interval is from the Boree Creek Formation; conodonts from other localities (Quarry Creek Limestone, the Liscombe Pools Limestone and the Glendalough Formation from mid-western New South Wales) are more sparse

Conodont faunas of late Llandovery–early Wenlock age similar to those documented from the Visby Beds of Gotland were described by Bischoff (1986) from the Boree Creek Formation. Platform equipped and ramiform conodont taxa from several localities in central-western New South Wales including the Boree Creek Formation and the overlying Borenore Limestone were illustrated by Bischoff (1986). His distribution tables showed high yields of a broad spectrum of conodont taxa as well as a major depletion at about the late Llandovery–early Wenlock boundary. The original documentation of conodonts from Boree Creek by Bischoff did not include coniform taxa, although he published two additional papers each documenting the presence of an unusual form (Bischoff 1997, 1998). Some of the datum points, however, comprising event episodes recorded in Gotland and Estonia are recognised by their impact on coniform taxa (Jeppsson and Mannik 1993). There was a need to reinterpret some of the conodont taxonomy of Bischoff (1986).

From an area about 1 km west of the Borenore Caves, near Orange, west-central New South Wales, Talent and Mawson sampled a short section [BOC] through the Boree Creek Formation, for limestones suitable for ^{18}O and ^{13}C analyses in order to establish the occurrence of the Ireviken Event by a considerable shift in the δ13C values in whole-rock limestone samples. Their results set out in Andrew et al. (1994) showed a 'distinct and sustained' increase in ^{18}O and ^{13}C values between 9.6 and 10 m above the base of their BOC section (Talent et al. 1993: 149, fig. 7). This equates with sample B39 of Bischoff [Bischoff's B Section] from (1986, tables 7, 8) from an interval in which 83% of conodont species disappear near the end of the *amorphognathoides* Zone, a level corresponding to datum level 2 in the type section of the Ireviken section on the island of Gotland. These results were confirmed by Molloy's (2006) comprehensive re-examination of the area and the conodont faunas through the Boree Creek Formation.

From his B section, Bischoff proposed the *Pterospathodus amorphognathoides–Pterospathodus latus* Assemblage Zone to cover the time interval prior to the major extinctions of the Ireviken Event based on these New South Wales faunas. He defined the upper boundary of this zone with the extinction of a number of taxa such as *Apsidognathus* and *Aulacognathus* (Bischoff 1986, p. 54). Their disappearance is equated with datum 2 of the Ireviken Event (Jeppsson 1997b) and is considered close to the Llandovery–Wenlock boundary. Interestingly, Bischoff (1986) did not include the near-disappearance of *Pterospathodus* taxa as part of his definition of the upper boundary of this assemblage zone, known to extend through into the Wenlock to the sixth datum of the Ireviken Event (Jeppsson 1997a, b). Simpson (1995) argued that this assemblage zone was based on a misinterpretation of the phylogeny of *Pterospathodus* and suggested the retention of the original biostratigraphic nomenclature and concept of Walliser (1964).

As previously mentioned, the aim of this study was to test the feasibility of the Jeppsson (1990) model of an extinction event from investigating the spectrum of conodont faunas from the Boree Creek Formation and to identify how many datum points for the Ireviken Extinction Event could be discriminated in conodont faunas from a locality in the Southern Hemisphere. The Boree Creek Formation was recollected for both conodonts (Peter Molloy) and brachiopods (James Valentine) – the last to test the impact on benthic communities.

Previous Investigations of the Boree Creek Formation

The stratigraphy of the Boree Creek Formation has been considered in several papers, principally by Walker (1959), Sherwin (1971), Talent et al.

(1975, 1993, 2003), Pickett (1982), Bischoff (1986), Holloway and Lane (1998), Jell and Talent (1989), Cockle (1999) and Valentine et al. (2003). Broader relationships of the formation to other stratigraphic units were dealt with by Talent et al. (1975) and Pickett (1982), respectively, in Australia and New South Wales. Bischoff (1986) documented platform equipped and ramiform conodonts from both the Boree Creek Formation (including its basal interval) and the overlying Borenore Limestone. Sherwin (1971) had

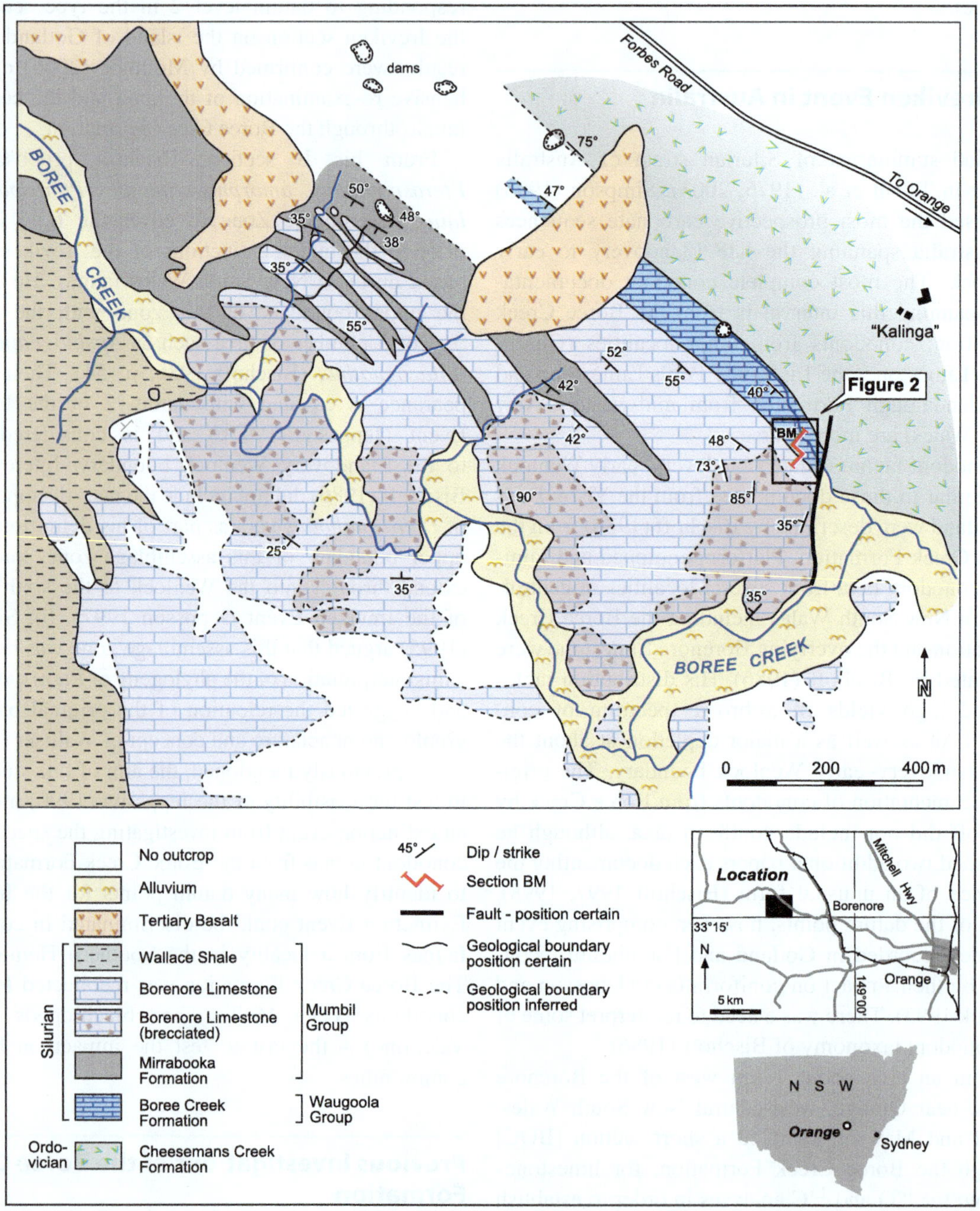

Fig. 1 Regional geological map of the Boree Creek area (based on Cockle 1999, text-fig. 1), location of Fig. 2 with stratigraphic section is indicated in the boxed area

divided the Boree Creek Formation into three informal units: Limestone A (oldest), tuffaceous trilobite beds, and Limestone B (youngest); this threefold subdivision was followed by later authors. Six units can be discriminated in the vicinity of the sequence that formed the target for the present investigation; how far these may be traced laterally is problematic, largely because of the generally substantial soil cover throughout the Boree Creek–Cheeseman's Creek area.

The geology of the area and the location of the BOC section sampled are shown in Figs. 1 and 2. The sequence of lithologies through the section is shown in Fig. 3. Bischoff (1986) sampled a section across the easternmost exposure of Boree Creek Formation on the property of 'Kalinga', 1 km west of the Borenore Caves. He concluded that his stratigraphically lowest 31 samples (Bischoff's B00 to B29) and 10 samples from the overlying red limestone (his samples (B29 to B38) spanned the *amorphognaoides–latus* Zone. Bischoff considered the red and light grey units of the present study to belong to a single unit, though noting that stratigraphically younger units (his samples B39–B45/46) had different conodont faunas from lower in the sequence. These he referred to as the 'lowermost *ranuliformis* Zone'. He failed to obtain conodonts from his samples B42 to B62 from the tuffaceous trilobite interval so accepted Sherwin's (1971) 'Limestone B' for the carbonates immediately above the tuffaceous trilobite beds, extending up to the disconformity with the Borenore Limestone. He dated this interval of partly dolomitic limestones (Bischoff's samples B63 to B89) as well as the lower beds of the Borenore Limestone (B118) to be within the *ranuliformis* Zone.

As referred to earlier, as part of a study of mid-Palaeozoic extinction events, Talent et al. (1993) undertook whole-rock analyses for carbon and oxygen isotopes for samples from the BOC section, essentially portion the B section studied by Bischoff (1986). They

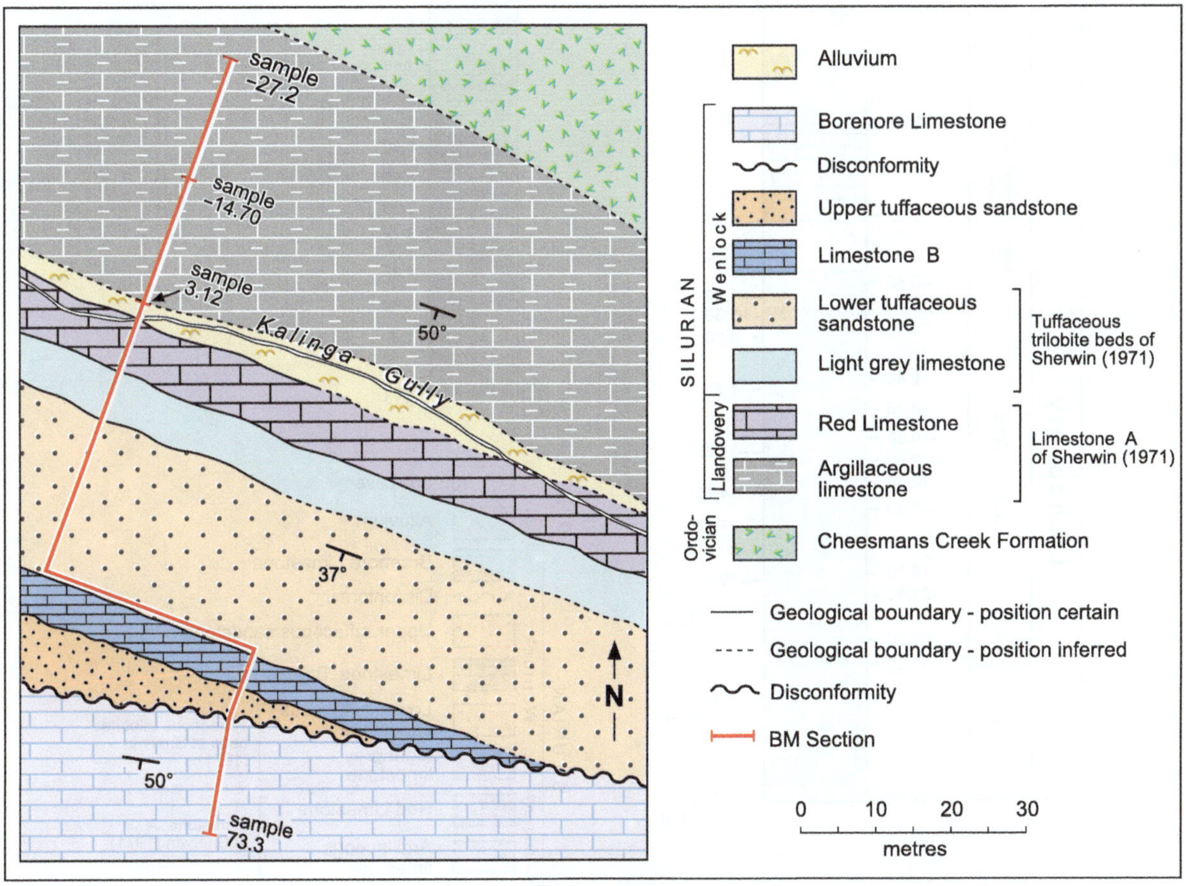

Fig. 2 Detailed geological map of the study area showing the location of section BM

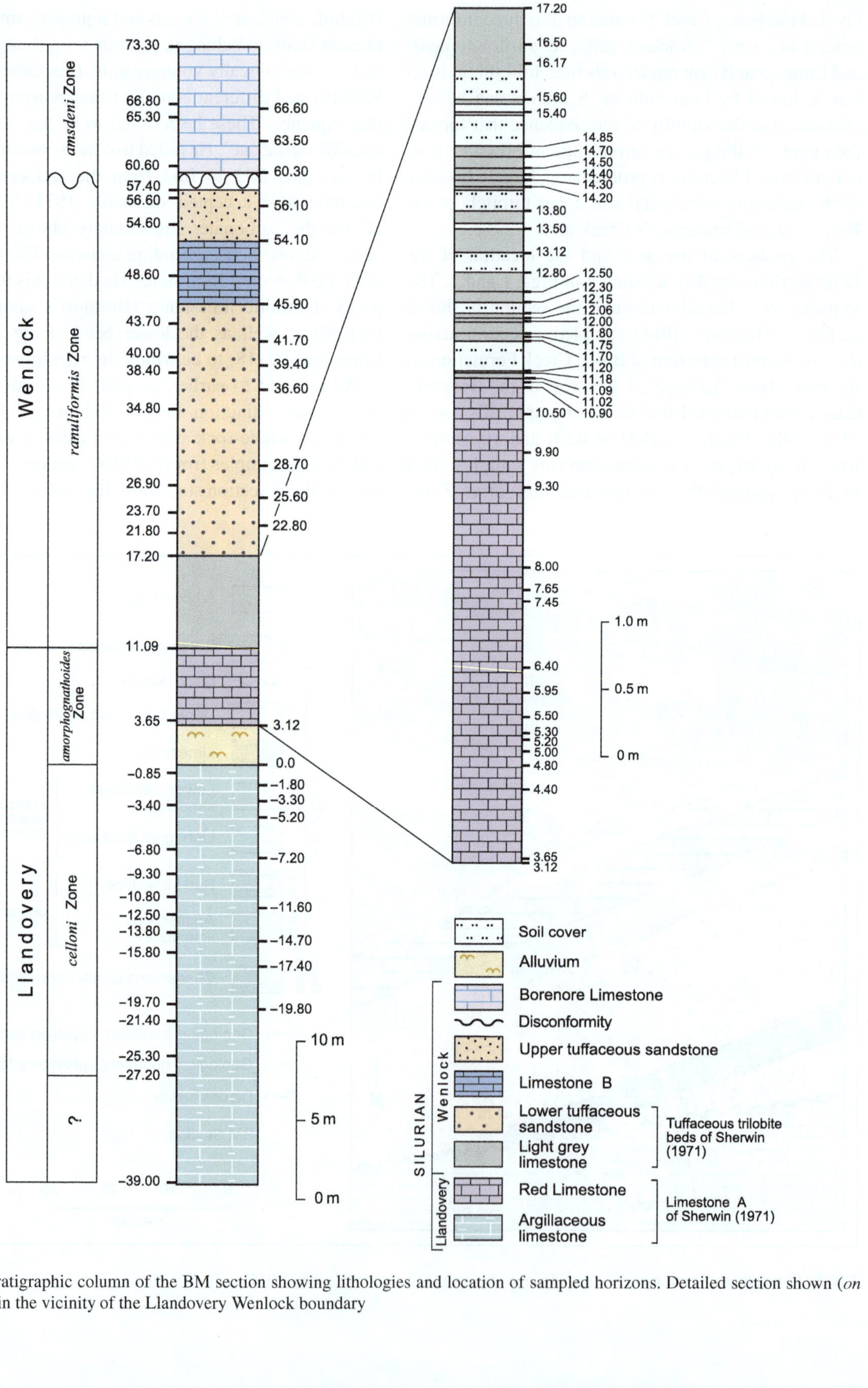

Fig. 3 Stratigraphic column of the BM section showing lithologies and location of sampled horizons. Detailed section shown (*on the right*) in the vicinity of the Llandovery Wenlock boundary

established a major isotopic excursion corresponding to Jeppsson's (1990, 1997b, 1998) datum 2 of the Ireviken Event on Gotland, Sweden. Molloy (2006) resampled the area for conodont data from his BM section and analysis of the fauna in association with Simpson, for example.

In his study of the Boree Creek area, Cockle (1999) remapped the area (Fig. 1) and acid-leached nine samples from the basal 49 m of dark grey limestone immediately below the section sampled for isotopes by Talent et al. (1993). These samples produced a small fauna including *Apsidognathus tuberculatus tuberculatus* Walliser 1964, *Distomodus staurognathoides* (Walliser 1964), *Panderodus greenlandensis* Armstrong, 1990, *Panderodus recurvatus* (Rhodes 1953) and *Panderodus unicostatus* Branson and Mehl 1933. Cockle (1999, p. 114) believed the assemblage to represent the *amorphognathoides* Zone (late Llandovery–earliest Wenlock) age. Two conodonts from this basal unit recovered in this study and identified tentatively as juvenile *Pterospathodus celloni* (Walliser) suggest a slightly older age, *celloni* Zone for this underlying limestone.

Valentine (in Valentine, Brock and Molloy 2003) reported the results of the investigation into brachiopod faunas from the Boree Creek Formation sampled herein for conodonts (Molloy 2006). While the main purpose of this work was the documentation of linguiliformean brachiopods, Molloy (in Valentine et al. 2003) noted that a number of datum points could either be tentatively or definitely identified from the associated conodont data. Datum point 1 was placed at BM 8.0, datum points 2 and 3 were placed between 11.09 and 11.15, datum point 4 was tentatively identified at BM 13.8 (essentially on the basis of a lithological change) and datum point 6 was placed between BM 14.85 and BM 15.4. More significantly they reported changes in the diversity and abundance of brachiopod faunas coincident with their interpretation of Ireviken datum points in the sequence. Molloy (2006) confirmed these datum points and compared them to Jeppsson's earlier work on the Ireviken Event in Sweden (see Table 1). Further work may necessitate a revision of these conclusions (Valentine, personal communication 2009). In this study we recognise a new datum point below datum point 1; we believe datum point 1 to be at a slightly higher level; we discriminate between datum points 2 and 3; we do not recognise datum point 4; and we recognise datum point 6 at a higher level (see 'Results: Conodonts and Datum Point Data' section below). This highlights the requirement for adequate conodont collections before interpreting the architecture of the event.

Geology of the Boree Creek Formation

The Boree Creek Formation overlies Ordovician volcanics (Cheesemans Creek Formation). Cockle (1999, p. 112) reported the contact as an unconformity but, due to soil cover, the nature of the contact is problematic in the immediate vicinity of the section sampled for this study. Six informal lithologic units are discriminated in the sampled section BM – cf. Fig. 3 (modified from Valentine et al. 2003, fig. 3). The 6th unit is a second tuffaceous sandstone discovered during resampling. The units, from oldest to youngest, are as follows:

1. a basal dark grey limestone interval,
2. a red limestone interval,
3. a light grey limestone interval,
4. a tuffaceous sandstone interval,
5. 'limestone B', and
6. a second tuffaceous sandstone interval unconformably overlain by the Borenore Limestone.

The lowest interval consists of about 40 m of dark grey, highly fossiliferous limestone on the north-eastern flank of Kalinga Gully. The lowest sample (BM – 39) has an admixture of Silurian and Ordovician conodonts – the latter consistent with reworking from nearby Ordovician limestones – but, from BM 27.20 to the top of the section, only Silurian conodonts have been obtained. Macro-fossils from this interval include tabulate corals, articulate brachiopods, and stromatoporoids; these accord with deposition on a low energy shallow shelf. Low conodont yields and the general lack of index species causes uncertainty as to whether this interval is *celloni* or *amorphognathoides* Zone, but the possible juveniles of *P. celloni* (see above) suggest the former rather than the latter.

The basal limestone interval is overlain by about 5 m of sparry, red limestone forming a prominent outcrop low on the south side of Kalinga Gully, but alluvium in Kalinga Gully obscures the contact between these two intervals. Lack of fine-grained material in the sparry, red limestone accords in general with a higher energy environment. Conodont yields from this unit are high. Association

Table 1: Tabulation of data used by Jeppsson (1998, table 3) data for Datum points for the Ireviken Event in Sweden compared with data derived for the same event in the Boree Creek Formation, Australia (Molloy 2006)

Datum points: Jeppsson and Männik (1993), Jeppsson (1998)		Boree creek FM: Molloy (2006)	Ireviken event, Sweden: Jeppsson (1998)	Additional comments
Not identified	8	Incoming: *Kock. walliseri*	Extinction: *Dist. staurognathoides*	
Possibly present	7	Not evident in Australia	Not evident in Sweden	Present in Estonia Jeppsson and Männik (1993) *Pa. unicostatus* (a Lazarus species) returns between Datum 7 and 8 in Sweden
Not identified	6.2	Not evident in Australia	Drop off: *Pa. unicostatus* Incoming: *Daps. sparsus*	
Definite	6	Extinction: *Pa. greenlandensis; Ps. triconis; Dist. staurognathoides* Drop off: Incoming:	Drop off: *Pa. recurvatus*	*Dist. staurognathoides* extends to Datum 8 in Sweden
Definite	5	Extinction: *Ps. boreensis* Drop off: *Oul. rect. rectangularis*	Extinction: *Ps. boreensis; Ps. tricornis*	In Australia, *Ps. tricornis* extends to Datum 6
Not identified	4	No evident in Australia	Extinction: *Oul. petilus; Oz. s. sp. A*	
Possibly present	3.3	Extinction: *Pa. langkawiensis*	Extinction: *Pa. langkawiensis*	
Tentatively identified	3	Extinction: ?*Apsid.* cf. *A. ruginosus* Drop off: *Pa. langkawiensis*	Extinction: *Ic. inconstans; Pt. amorphognathodus*	No specimens of *Pt. amorphognathodus* found in Australia
Definite	2	Extinction: *Asp. borenorensis; Ans. misha; Pa. deuteroconus; Wa.* cf. *Wa. santiclari; Wa.* n. sp. A; *Asp.* n. sp. A.; *Aula. chapini,* Coniform groups A to E; *Oul. r. eorectangularis; Pyr. latus; Pyr obliquus.* Drop off: *Pa. greenlandensis*	Extinction: *Pa. greenlandensis Apsid. ruginosus*	In Australia, *Pa. greenlandensis* (Lazarus species) returns between Datum 6 and 8
Highly possible	1	Extinction: *Pseudobelodella silurica*	Extinction: *Nudibelodina sensitiva*	*Nudibelodina sensitiva* is not found in Australia
New Datum Point		Incoming: *Aula. chapini; Kock. ranuliformis*	Incoming: *Kock. ranuliformis*	

Note: Column 1 shows Datum point numbers used by Jeppsson (1998) and brief comment as to its presence or otherwise in the Boree Creek Formation
Abbreviations used: Ans. Ansella, *Apsid.* Apsidognathus, *Asp.* Aspelundia, *Aula.* Aulacognathus, *Daps.* Dapsilodus, *Dist.* Distomodus, *Ic.* Icriodella, *Kock.* Kockelella, *Oul.* Oulodus, *Oz.* Ozarkodina, *Pa.* Panderodus, *Pt.* Pterospathodus, *Ps.* Pseudooneotodus, *Pyr.* Pyrsognathus, *Wa.* Walliserodus

of *Pterospathodus rhodesi* Savage, *Pterospathodus pennatus procerus* (Walliser), *Apsidognathus tuberculatus* Walliser and *Aulacognathus chapini* Savage accords with a late Llandovery–early Wenlock *amorphognathus* Zone age.

Three subtle changes in lithology of the sparry, red limestone are visible on freshly broken surfaces. The base of the unit (samples BM 3.12 to BM 6.40) is dark red and sparry with prominent darker red stylolites. Just above sample BM 6.40 is a contact (Fig. 3) with less compact red limestone; the latter continues through to sample BM 9.25. The limestone is again compact with a greenish hue (due to change in stylolite material) from there to the top of the interval at sample BM 11.09. These subtle lithological changes are also reflected in conodont abundance and diversity, both of which decline from the middle part of the unit (samples BM 6.40 to BM 9.25). A similar pattern is evident in the distribution lists presented by Bischoff (1986, tables 7 and 8). He obtained a large number of specimens from his samples B31 to B 33, followed by a decline in numbers from samples B34 and B35, followed by high yields and increased diversity in his samples B36 to B38. None of this affects attribution to the *amorophognathoides* Zone, but is significant when considering the location of Jeppsson's (1997b) datum points within the Boree Creek sequence.

Overlying the red limestone unit is a 5-m sequence of clean, sparry, light-grey limestones (samples BM 11.15 to BM 17.20). Presence of trilobite remains and lack of fine-grained, argillaceous material suggests a shallower, higher energy environment. As most of this unit is soil-covered, a trench was excavated revealing conformity between the red limestone and overlying grey limestone units (between samples BM 11.09 and 11.15). The uppermost beds of this light grey limestone interval (BM 16.80 and 17.20) are distinctively finer grained and very dark red in colour and, incidentally, rich in trilobite remains. Deep soil cover prevented excavation of the boundary between these beds and the overlying tuffaceous sandstone.

A major change in conodont diversity occurs in the light grey limestone between samples BM 11.09 and 11.15, the suggested location of Datum 2 of the Ireviken Event; see below). The former is the last occurrence of approximately 40% of species occurring in the red limestone unit. From BM 11.18 to 17.20, the conodont fauna is dominated by *P. pennatus procerus* (Walliser), *Kockelella ranuliformus* (Walliser), *D. staurognathoides* (Walliser) and species of *Oulodus*, the assemblage indicating the lower *ranuliformis* Zone (early Wenlock).

The overlying 17-m interval of tuffaceous sandstone produced trilobite, brachiopod and coral remains but failed to produce conodonts. The boundary between the lower and upper *ranuliformis* zones may occur within this interval.

Limestone B, a ridge-forming sequence of dolomitic limestones overlain by the upper interval of tuffaceous sandstones, was estimated by Bischoff (1986, p. 36) to be 25 m thick, but is certainly much thinner along the section-line (Fig. 3). Conceivably, the two tuffaceous sandstone intervals may be construed as a single interval enclosing a lenticular limestone body (Limestone B) thickening westwards. Conodonts from Limestone B are low in diversity and abundance. The presence of *K. ranuliformis* (Walliser), *K. walliseri* (Helfrich), *K. latidentata* Bischoff and *Pseudooneotodus bicornis* Drygant accord with allocation to the upper *ranuliformis* Zone and therefore above the eighth and final datum point of the Ireviken Event (Jeppsson 1997b).

Methodology: Plan of Attack

A stratigraphic sequence through the Boree Creek Formation was sampled on 'Kalinga' property, 1.5 km NNW of Borenore Caves and 22 km WNW of Orange in central-western New South Wales commencing at grid reference 7975 2005 on Molong 1:50,000 topographic sheet. The BM section (Fig. 2) of the current study is located on the easternmost exposure of the formation embracing and extending the sections of Talent et al. (1993) and Cockle (1999). It is very close to and parallel with (if not in fact identical in part to) Bischoff's (1986) section allowing alignment of much conodont and brachiopod data obtained from the present study with data from previous studies.

A total of 93 bulk samples were collected, 83 from the Boree Creek Formation and 9 from the basal beds of the disconformably overlying Borenore Limestone, to establish which species ranged from the Boree Creek Formation into younger strata. The following techniques were used by Molloy (2006). Note: Valentine joined forces with Molloy in the fieldwork as he required for his studies, the lingulacean brachiopods that could be found in the conodont residues.

Bed-by-bed sampling of the BM section was carried out. Because a largish proportion of the BM section is covered by soil, trenching was undertaken, focused on the top of the red limestone unit (sample BM 10.90) through to the first occurrence of the tuffaceous sandstone unit (BM 17.20). The trench was excavated in such a way that every bed in the grey, lensoidal limestone unit overlying the red limestone unit could be sampled. Samples averaging 5 kg were collected from each bed. Horizons in key areas with low conodont yields were more voluminously sampled – sample sizes were shown on the relevant distribution tables. Initially, because emphasis had focused on discrimination of the datum points immediately prior to the isotopic excursion (cf. Talent et al. 1993), it was not thought necessary to sample the basal limestone unit.

Subsequently, it seemed that establishing the age of the basal limestone unit would assist understanding the broader biostratigraphic context; accordingly, the sampled section was increased in length. This resulted in use of negative sample numbers.

All samples were dissolved in a 10% acetic acid. To obtain something like maximum conodont recovery, the method of buffered acetic acid leaching, as outlined by Jeppsson et al. (1999), was followed as closely as laboratory facilities would allow. Dissolved solutions were washed through a 2.5-mm sieve to collect larger limestone fragments for retreatment. All water and fine sediment discarded at each washing stage was passed through a 63-μm sieve to prevent loss of juvenile specimens. Conodonts were recovered from residues by standard techniques (Anderson et al. 1995).

Results: Conodonts and Datum Point Data

Discussion of the conodont fauna in each of the six informal lithological units identified in the Boree Creek Formation was examined and compared with the datum points identified by Jeppsson (1997b) for the Ireviken Event in Sweden. While not all species of conodonts found in Gotland are found in Australia, there is a markedly high degree of correlation. The tabulation of the datum point data for the Ireviken Extinction Event from Sweden compared to data from the Boree Creek Formation shown in Table 1 is from Molloy (2006).

The abundance of conodont elements found in the basal dark grey, argillaceous limestone unit is relatively low suggesting that our knowledge of the fauna is not yet complete. Elements of 7 coniform taxa and 11 non-coniform taxa have been found, three of which, *Panderodus* n. sp. A, *Apsidognathus* n. ssp. and *Pterospathodus celloni* (Walliser), are unique to this unit. The fauna is of Llandovery *celloni* Zone age though the boundary between this fauna and the overlying *amorphognathoides* Zone fauna in the overlying red limestone unit has not been determined due to the low abundance and lack of exposure of beds between the two units. This limestone represents a pre-Ireviken Event fauna.

The overlying red limestone unit has both a high diversity and abundance of conodont elements. Elements of 11 coniform taxa and 18 non-coniform taxa occur in it with 5 of them unique to the unit: *Ansella mischa* Bischoff, *Panderodus deuteroconus* Bischoff, *A. tuberculatus lobatus* Bischoff, *A. chapini* Savage and *Aspelundia* n. sp. A.

?New Datum 0. Prior to Datum 1 in the BM section there is the possibility of another Datum Point not recognised by Jeppsson (1997b) in the collections from Gotland. At sample BM 7.65 there is a noticeable drop in diversity and introduction of two new species, *A. chapini* Savage and *K. ranuliformis* (Walliser). A visible change in the lithology within the unit was noted in the field at this point (see above). The drop in diversity is only temporary and primarily affects the coniform taxa *P. deuteroconus*, *P. langkawiensis*, *P. recurvatus*, *Pseudobelodella silurica* and *Walliserodus* cf *W. sancticlairi*.

The introduction of *A. chapini* also occurs at this point in both the current study and in Bischoff's (1986) section. The first occurrence of *K. ranuliformis* is found at this point in Bischoff's section but not until BM 11.02 in the current study. This difference may be purely an artifice of sampling. *K. ranuliformis* is known to be a minor constituent of pre-Irevekin Event faunas. Aldridge (1985) reported the species in small numbers from Llandovery strata in Great Britain, Armstrong (1990) reported the closely related form *Kockelella* cf. *K. ranuliformis* from the Llandovery of Greenland.

Datum 1: The location for Datum 1 occurs between samples BM 9.90 and BM 10.50. In sample BM 9.90 conodonts are high in both abundance and diversity and the sample marks the last occurrence of *Pseudobelodina silurica* Armstrong. Sample BM 10.50 immediately following shows a sudden decrease in both abundance and diversity. Jeppsson (1997b, fig. 17.2) uses the loss of *P. silurica* and *Nudibelodina sensitiva* (the latter not recovered in this study) as his markers for Datum 1. Following sample BM 10.50 the conodont faunas increase in abundance up to sample BM 11.09, marking the transition between the red limestone unit and conformably overlying light grey limestone unit. A number of species in samples from the topmost beds of the red limestone, such as *Walliserodus curvatus*, show a sudden increase in size.

Jeppsson (1997b) places the first datum as between Bischoff (1986) samples B35 and B36 primarily on the appearance of *K. ranuliformis*. He (Jeppsson 1997b) also notes that the first datum is often marked by the replacement of populations of *Pseudooneontodus tricornis* with *P. bicornis*. In both Bischoff's (1986) study and this one, *P. tricornis* occurs in very small numbers below and above this datum, but *P. bicornis* does not appear until much higher in the sequence (BM 16.7 – this study; B52 – Bischoff 1986).

Datum 2: In the Boree Creek Formation, this datum point is identified as occurring between samples BM 11.09 and BM 11.18. It marks the most severe loss of species during the Ireviken Extinction Event in the Boree Creek Formation with the coniform taxa *A. mischa*, *Walliserodus* cf *W. saincticlairi* and *Walliserodus* n. sp. A becoming extinct. *Decoriconis fragilis* also retreats at this level and reappears higher in the sequence. A large number of non-coniform taxa become extinct at the same level, including *A. chapini, A. tuberculatus, Aspelundia borenorensis,* the '*Carniodus carnulus*' group, *Oulodus rectangularis* n. ssp. and *Pterospathodus* n. sp. A.

The overlying light grey limestone unit has a different lithology to the red limestone unit. As noted earlier, the limestones from this unit are light grey, sparry and contain little argillaceous material. The limestones occurs as blocks up to a metre long, with a thickness of up to 20 cm. Residues from this unit contain a high percentage of trilobite remains and an increase in crystals of pyrite.

The conodont fauna from the light grey limestone unit contains relatively few species but there is a large increase in abundance and most elements also increase in size. The fauna is dominated by *O. rectangularis rectangularis* Bischoff and *P. pennatus procerus* (Walliser). Only seven species of coniform conodonts and nine species of non-coniform conodonts continue through into the grey limestone following the loss of taxa at Datum 2. Two non-coniform species, namely; *Apsidognathus* cf. *A. ruginosus* Mallibard and Aldridge, *O. rectangularis rectangularis* Bischoff and two coniform taxa, *P. bicornis* Drygant and *Dapsilodus obliquicostatus* (Branson and Mehl), appear in the grey limestone unit, the coniforms are found only in the upper beds of the grey limestone unit (sample BM 16.17).

This datum at Boree Creek represents the most dramatic reduction in conodont diversity. Jeppsson (1997b) also recorded significant loss of platform equipped taxa from many contemporaneous sequences. He (Jeppsson 1997b) also reported some differences in the nature of *Panderodus panderi* specimens before and after this datum on Gotland. A marked drop in abundance at datum 2 of this taxon is noted for the Boree Creek section. Jeppsson (1997b) identified this datum between Bischoff's (1986) samplers B38 and B39 based on the disappearance of the same platform equipped taxa noted above. Bischoff (1986) recorded the first appearance of *Oulodus sinuosis* and *Ozarkodina excavata excavata* in sample BO 39.

Datum 3: This is tentatively established between samples BM 12.00 and 12.06 where there is a decrease in abundance of most species and the last occurrence of *Apsidognathus* cf. *A. ruginosus*. This datum is only established tentatively due to the relatively few specimens of *Apsidognathus* cf. *A. ruginosus* found, however, the marked decrease in abundance is down to approximately 17% yields in comparison with the underlying sample for both coniform and non-coniform taxa. Jeppsson (1997b) placed this datum between Bischoff's (1986) B39 and B40 samples; a similar but less marked reduction in abundance is recorded. Datum 3 is also equated with the extinction of *P. amorphognathoides* (Jeppsson 1997b). Bischoff (1986) recorded the highest Pa element attributed to this taxon in his sample B39, although his distribution charts record other (non-Pa) elements sporadically in some higher samples. We believe some taxonomic revision is required for specimens of this genus from the Boree Creek Formation.

Datum 3.3: This datum was only recognised through comparative analysis of the sequences of Gotland and Estonia (Jeppsson and Mannik 1993) and is additional to the original eight datum points identified. It is marked by the disappearance of *Panderodus langkawiensis* in both sequences. This datum can possibly be established between samples BM 12.90 and BM 13.12 which marks another drop in overall abundance of many taxa. Sample BM 12.90 marks the last occurrence of *P. langkawiensis*. *W. curvatus* whilst only a very minor constituent of the fauna in this study disappears after sample BM 13.12.

Datum 4: This datum point was established (Jeppsson 1997b) based on the disappearance of certain panderodontid taxa, some of which reappear higher in the sequence. It is also marked by the extinction of *Oulodus petilus petilus* (Jeppsson 1997b). This datum point could not be identified within the Boree Creek sequence from either new collections in this study or by Jeppsson (1997b) in examining the data from Bischoff (1986).

Datum 5: This datum is established between samples BM 15.40 and BM 15.60, as BM 15.40 marks the last occurrence of *Pseudooneotodus boreensis* Bischoff and *Panderodus* n. sp. Jeppsson and Männik. There is also an overall decline in abundance of other taxa. Sample BM 16.17 shows a slight resurgence in abundance of most species and *P. pennatus procerus* particularly. It also marks the introduction of two new coniform taxa, *D. obliquicostatus* and *P. bicornis*. Samples immediately after BM 16.17 show a decline in abundance. Jeppsson (1997b) was unable to identify this datum in the collections of Bischoff (1986) as sampling was not close enough for discrimination of this data point.

Datum 6: The final occurrence of *P. pennatus procerus, D. staurognathoides, Dapsilodus praecipuus, Pseudooneotodus panuarensis* Drygant and *Pseudooneotodus tricornis* Drygant in sample BM 17.20 marks Datum 6. Sample BM 17.20 is just prior to the first occurrence of the overlying trilobite rich sandstone unit. No conodonts were found in this unit. *O. excavata excavata, P. bicornis* and *P. panuarensis* continue in limestone above the sandstone unit. Jeppsson (1997b) recognised this datum in the collections of Bischoff (1986) between Bischoff's samples B45 and B46 with the extinction of *P. procerus*.

Datum 6.2 and Datum 7: Neither of these datum points could be determined with certainty in the Boree Creek Formation. However, overlying the tuffaceous sandstone is a 5-m-thick dark grey, dolomitic limestone unit. The conodont fauna from this unit has limited abundance and diversity, and most of the specimens are broken. As noted above it consists of *K. ranuliformis, O. excavata excavata, P. bicornis* and *Decoriconus fragilis* that survived the biocrisis at the end of the light grey limestone unit. *Kockelella walliseri* (Helfrich) makes a first appearance in this limestone; this taxon is only recorded from younger post-Ireviken Event strata (Jeppsson 1997b). *Panderodus greenlandensis* Armstrong, also appears at this level, having been absent from the sequence since prior to Datum point 1. Close comparison of these faunas is required to ascertain any morphological differences; they have only been recovered in small numbers to date.

Jeppsson and Männick (1993) indicated that Datum 6.2 was based on the fall-off or disappearance of *P. unicostatus*. This taxon was recovered in the penultimate sample (BM 17.20) below the tuffaceous sandstone, but was not recovered in the overlying limestone. Jeppsson (1997b) noted that datum 7 is generally equated with the extinction of *D. stauragnathoides* and further commented that the identification of these younger datum points can be difficult as would be expected when working with a fauna that has already been substantially impoverished. *D. staurognathoides* also occurs in sample BM 17.20, but not in younger strata. Jeppsson (1997b) indicated that datum 7 could be identified in the collections of Bischoff (1986) between samples B46 and B47, the former recording the final occurrence of *D. staurognathoides*. Bischoff's (1986) samples B47 to B51 were barren, however, and can be undoubtedly equated with the interval of tuffaceous sandstone. A more realistic conclusion therefore is that these datum points (i.e. 6.2 and 7) were masked by the interval of siliciclastics in the Boree Creek Formation.

Datum 8: This datum, considered the final marker of the Ireviken Event, is recognised on Gotland with the disappearance of *P. unicostatus* (Jeppsson 1997b) and the appearance of a number of post-event taxa. This datum could not be determined in the Boree Creek Formation in this study and neither did Jeppsson (1997b) recognise it in the collections of Bischoff (1986). As noted above, *K. walliseri* was recovered in small numbers from the limestone overlying the tuffaceous sandstone interval. While this taxon is generally associated with post-event strata, it has been recorded within the range of *Pterospathodus* (Helfrich 1980; Over and Chatterton 1987; Uyeno 1990) but is clearly only recovered in younger strata to date in the Boree Creek Formation.

Overlying the Limestone beds is a second tuffaceous unit that extends up to the base of the overlying Borenore Limestone. The ranges of *K. ranuliformis, O. excavata excavata, Pseodooneontodus bicornis*, and *P. greenlandensis* extend from the uppermost carbonates of the Boree Creek Formation into the lower carbonates of the Boreenore Limestone. *W. curvatus* reappears in the Borenore Limestone having been absent (only ever recovered in this study in very small numbers) since around Datum 3.3. *D. obliquicostatus* makes a first appearance in the basal samples of the Borenore Limestone. None of this data, however, assists with the discrimination of the final two datum points of the Ireviken Event.

Outcomes and Discussion

1. This study has provided an improved framework for 'micro-correlation' through the Ireviken Extinction Event (late Llandovery–Wenlock) as identified in the Boree Creek Formation, Borenore, Australia with that identified by Jeppsson (1990, 1997a, b, 1998) in the Visby Beds, on the island of Gotland, Sweden. The most highly developed study of the internal architecture of the step-wise extinction patterns of the Ireviken Event are based on extensive collections of conodonts from the Visby beds at Ireviken 3 on Gotland. Jeppsson and Männik (1993) have undertaken detailed micro-correlation of Ireviken Event strata between Gotland and Estonia.
2. Silurian studies that were subsequent to the proposal of the oceanic model (Jeppsson 1990), have included notes on the recognition of the events around the globe (Aldridge et al. 1993; Jeppsson et al. 1995; Jeppsson 1997b; Jeppsson and Aldridge 2000). For the Ireviken Event, Jeppsson (1997b) gave brief notes on 12 sequences world-wide, in most of these only a couple of datum points were identified, usually those associated with the most severe impact on conodont faunas. Jeppsson (1997b) recognised five datum points in the Boree Creek material, even though as noted, most of these sequences other than Gotland were not collected with a view to elaborate or even test for the existence of Ireviken Event architecture. This study was aimed at testing for this evidence and we have confirmed the presence five of these datum points (1, 2, 3, 5 and 6), identified another two (0 and 3.3), and demonstrated that the previous identification of datum point 7 is in fact masked by the trilobite rich tuffaceous sandstone interval.
3. This micro-correlation between Gotland, Sweden, and Boree Creek, Australia, is intercontinental. As already widely agreed, the global nature of these events is now essentially irrefutable dispelling the notion of the Silurian period being one of stable greenhouse conditions (Calner and Eriksson 2006). The step-wise extinction patterns associated with the intermittent collapse of food chains is only one characteristic of these events; their global nature according to Jeppsson's (1990) model implies that intensive and detailed sampling of contemporaneous carbonate sequences should reveal close to the full set of datum points comprising an event, barring local tectonic anomalies and eustatic changes preventing preservation.
4. As noted by Jeppsson (1997b) Gotland is the only area to be studied intensively with conodont numbers a quantum level higher than in all other areas. Some of the datum points involve the extinction or the disappearance/reappearance of extremely rare taxa and it may require years of collection building before such datum points can be identified in other areas. It is also worth remembering that subtle palaeogeographic differences might also mean the involvement of different taxa in datum point changes in different parts of the world. There are a number of taxa in the Boree Creek section that have not been obtained elsewhere (e.g. *Ansella mischa*); such endemism may make intercontinental micro-correlation more challenging. Table 1 summarises the nature individual datum points for the Ireviken Event in Sweden and in the Boree Creek Formation, Australia.
5. The impact on other fossil groups near the close of the Llandovery has been noted by others, e.g. polycheates (Bergman 1989; Eriksson 1997; Eriksson 2006) conularids (Jerre 1993), foraminifera (Mabillard and Aldridge 1985; Watkins et al. 1999), graptolites (Melchin 1994) trilobites (Chatterton et al. 1990; Mikulic and Kluessendorf 1999) corals (Kaljo et al. 1996), brachiopods (Erlfeldt 2005; Valentine et al. 2003) and microplankton (Gelsthorpe 2004).

Many extinction events are also associated with distinctive shifts in carbon isotope signatures and changes to patterns of sedimentation. Talent et al. (1993) recognised the potential for correlations with this data (including a signature for the Ireviken Event in Boree Creek). Signatures for the Ireviken and other events all indicating sharp positive excursions have subsequently been published (e.g. Saltzman 2001; Munnecke et al. 2003; Martma et al. 2005; Kaljo and Martma 2006; Jeppsson et al. 2007). Some have argued that glacio–eustacy provides an adequate mechanism to explain both isotopic signatures and extinction patterns (e.g. Kaljo et al. 2003; Johnson 2006).

While the role of carbon dioxide as a driver of climate change is broadly accepted, whether these significant excursions represent major exchange of carbon between ocean and atmosphere accompanying oceanic overturn as predicted in Jeppsson's original model (Jeppsson 1990) is likely. While this paper provides additional information on the correlation of Ireviken datum points between continents, only data on the precise timing of the relationship between isotopic excursion; conodont extinctions and contemporaneous sedimentary changes can help identify whether the forcing mechanism is involved in the cyclical replacement of deep ocean waters. Nevertheless, given the detailed nature of these developing datasets, it is perhaps simpler to assume some form of mixing and overturn rather than an essentially vertically static ocean for much of Silurian time. Jeppsson's model (1990) provides ambitious and worthwhile challenges to the interdisciplinary nature of Palaeozoic studies.

The global current system is driven by vertical mixing of ocean water masses at high latitudes in both the northern and southern hemispheres. The melting of sea ice appears to be the most important factor controlling the salinity and density of ocean surface waters near the poles (Rothrock et al. 2008; Serreze et al. 2003; Alley et al. 1995). As the water freezes, salt is expelled, freezes and forms sea ice. Dense brine sinks and flows down the continental shelf of Antarctica to form Antarctic Bottom Water – the densest water in open oceans. This water flows outward from the Southern Ocean and through other ocean basins as part of the global ocean circulation 'conveyor belt' that distributes heat, nutrients and gases around the world. Jeppsson's model just might be relevant. As sea ice at the poles is diminishing rapidly, the overturning circulation may slow, decreasing the ocean's ability to deal with atmospheric carbon dioxide thus affecting the distribution of heat around the globe.

Acknowledgements Keith Goodridge cheerfully gave free access to limestone occurrences in his property, *Kalinga,* for sampling the Boree Creek Formation and provided much good cheer as we attacked our field work. James Valentine shared much of the fieldwork with Peter. Our program has benefited from association with Prof. Lennart Jeppsson who showed us his remarkable Gotland sequences, generously provided access to his extensive conodont collections, and was always quick to provide insightful advice on all aspects of conodont taxonomy, especially with regard to Silurian taxa. He and Ann-Sofi Jeppsson provided extraordinarily generous hospitality in Lund. Prof John Talent and Ruth Mawson have provided us with excellent supervision for both our doctoral studies and have worked with us as friends and colleagues, showing us what real research is all about.

We are especially grateful to our friends from the Macquarie University Centre for Ecostratigraphy and Palaeobiology who have been involved with this and other conodont endeavours: David Mathieson, George Wilson, Margaret Harvey, Robyn Isberg, Karen Novotny, and Col MacKenzie. Peter Cockle deserves special mention and was especially generous with advice on computing matters. The camaraderie, humour and assistance provided by a gang of MUCEP friends all helped generate a cheerful environment in which to work. The elegant drafting of Figs. 1, 2, and 3 is by our talented friend Dean Oliver.

References

Aldridge RJ (1976) Comparison of macrofossil communities and conodont distribution in the British Silurian. In: Barnes CR (ed) Conodont paleoecology, Geol Assoc Canada, Spec Paper 15:91–104

Aldridge RJ (1985) Conodonts from the Silurian System of the Btitish Isles. In: Higgins AC, Austin RL (eds) A stratigraphical index of Conodonts. Ellis Horwood, Chichester, pp 68–92

Aldridge RJ, Jeppsson L (1984) Ecological specialists among Silurian conodonts. Spec Pap Palaeontol Lond 32:141–149

Aldridge RJ, Jeppsson L, Dorning KJ (1993) Early Silurian oceanic episodes and events. J Geol Soc 150:501–513

Alley RB, Finkel RC, Nishiizumi K, Anandakrishnan S, Shuman CA, Mershon GR, Zielinski GA, Mayewski PA (1995) Changes in continental and sea-salt atmospheric loadings in central Greenland during the most recent deglaciation. J Glaciol 41:503–514

Alvarez L, Alvarez W, Asaro F, Michel HV (1980) Extraterrestrial cause for the Cretaceous–Tertiary extinction. Science 208:1095–1108

Anderson MA, Dargan G, Brock GA, Talent JA, Mawson R (1995) Maximising efficiency of conodont separations using sodium polytungstate solution. Cour Forsch-inst Senckenb 182:515–521

Andrew AS, Talent JA, Mawson R, Whitford DJ, Hamilton, PJ (1994) Comparison of carbon isotopic responses associated with Silurian-Devonian mass extinction events. Erlanger geol Abh 122:4

Armstrong HA (1990) Conodonts from the Upper Ordovician–Lower Silurian carbonate platform of north Greenland. Bull Grφnlands Geologiske Undersøgelese 159:1–151

Bergman CF (1989) Silurian paulinitid polychaetes from Gotland. Fossils Strata 25:1–128

Bischoff GCO (1986) Early and Middle Silurian conodonts from midwestern New South Wales. Cour Forsch-inst Senckenb 89:1–337

Bischoff GCO (1997) Ansella mischa n. sp. (Conodonta) from the late Llandoverian and early Wenlockian strata of midwestern New South Wales. N Jb Geol Paläont Mh H8: 477–488

Bischoff GCO (1998) New species of Panderodus (Conodonta) from late Llandoverian and early Wenlockian strata of midwestern New South Wales. N Jb Geol Paläont Abh 210: 267–288

Calner M, Eriksson ME (2006) Silurian research at the crossroads. GFF (Geologiska Föreningens i Stockholm Förhandlingar) 128:73–74

Chatterton BDE, Edgecombe GD, Tuffnell PA (1990) Extinction and migration in Silurian trilobites and conodonts of northwestern Canada. J Geol Soc Lond 147:703–715

Cockle P (1999) Conodont data in relation to time, space and environmental relationships in the Silurian (late Llandovery–Ludlow) succession at Boree Creek (New South Wales, Australia). Abhandlungen der Geologischen Bundesanstalt 54:107–133

Eriksson ME (1997) Lower Silurian polychaetaspid polychaetes from Gotland, Sweden. GFF 119(3):213–230

Eriksson ME (2006) The Silurian Ireviken Event and vagile benthic faunal turnovers (Polychaeta; Eunicida) on Gotland, Swden. GFF 128(2):91–96

Erlfeldt A (2005) Brachiopod faunal dynamics during the Silurian Ireviken Event, Gotland, Sweden. Examensarbete i geologi vid Lunds universitet, Nr 199, 22pp

Gelsthorpe DN (2004) Microplankton changes through the early Silurian Ireviken extinction event on Gotland, Sweden. Rev Paleobot Palynol 130:89–113

Helfrich CT (1980) Late Llandovery–Early Wenlock conodonts from the upper part of the Rose Hill and the basal part of the Mifflinton Formations, Virginia, West Virginia, and Maryland. J Paleontol 54:557–569

Holloway DJ, Lane PD (1998) Effaced styginid trilobites from the Silurian of New South Wales. Palaeontology 4:853–896

Jell JS, Talent JA (1989) Australia: the most instructive sections. In: Holland CH, Bassett MG (eds) A global standard for the Silurian system. Nat Mus of Wales, Geol Ser 9: 183–200

Jeppsson L (1987) Lithological and conodont distributional evidence for episodes of anomalous oceanic conditions during the Silurian. In: Aldridge RJ (ed) Palaeobiology of conodonts. Ellis Horwood, Chichester, pp 129–145

Jeppsson L (1990) An oceanic model for lithological and faunal changes tested on the Silurian record. J Geol Soc Lond 147:633–674

Jeppsson L (1993) Silurian events: the theory and the conoidonts. Proc Estonian Acad Sci USA 42:23–27

Jeppsson L (1997a) A new latest Telychian, Sheinwoodian and Early Homerian (Early Silurian) Standard Conodont Zonation. Trans R Soc Edinburgh: Earth Sci 88: 91–114

Jeppsson L (1997b) The anatomy of the mid-Early Silurian Ireviken event and a senario for P–S events. In: Brett CE, Baird GC (eds) Palaeontological events–stratigraphic, ecological and evolutionary implications. Columbia University Press, New York, pp 451–492

Jeppsson L (1998) Silurian Oceanic events: summary of general characteristics. In: Landing E, Johnson ME (eds) Silurian cycles. New York State Mus Bull 491:239–257

Jeppsson L, Aldridge RJ (2000) Ludlow (late Silurian) oceanic episodes and events. J Geol Soc, London 157:1137–1148

Jeppsson L, Männik P (1993) High-resolution correlations between Gotland and Estonia near the base of the Wenlock. Terra Nova 5:348–358

Jeppsson L, Aldridge RJ, Dorning KJ (1995) Wenlock (Silurian) oceanic episodes and events. J Geol Soc, London 152: 487–498

Jeppsson L, Anehus R, Fredholm D (1999) The optimal acetate buffered acetic acid technique for extracting phosphatic fossils. J Paleontol 73:964–972

Jeppsson L, Talent JA, Mawson R, Simpson AJ, Andrew A, Calner M, Whitford DJ, Trotter JA, Sandstrom O, Caldon HJ (2007) High Resolution Late Silurian Correlations between Gotland, Sweden and the Broken Rive Region, NE Australia: lithologies, conodonts and isotopes. Palaeogeogr Palaeoclimatol Palaeoecol 245: 115–137

Jerre F (1993) Conulariid microfossils from the Lower Visby Beds of Gotland, Sweden. Palaeontology 36:403–424

Johnson ME (2006) Relationship of Silurian sea level fluctuations to oceanic episodes and events. GFF 128:115–121

Kaljo D, Martma T (2006) Application of carbon isotope stratigraphy to dating the Baltic Silurian rocks. GFF 128: 123–129

Kaljo D, Boucot AJ, Corfield RM, Herisse AL, Koren TN, Kriz J, Mannik P, Mars T, Nestor V, Shaver RH, Siveter DJ, Viira V (1996) Silurian bio-events. In: Walliser OH (ed) Global events and event stratigraphy in the Phanerozoic. Springer, Berlin, pp 174–224

Kaljo D, Martma T, Mannik P, Viira V (2003) Implications of Gondwana glaciations in the Baltic late Ordovician and Silurian and a carbon isotopic test of environmental cyclicity. Bull de la Societe Geologique de France 174: 59–66

Mabillard JE, Aldridge RJ (1985) Microfossil distribution across the base of the Wenlock Series in the type area. Palaeontology 28:89–100

Martma T, Brazauskas A, Kaljo D, Kaminskas D, Musteikis P (2005) The Wenlock–Ludlow carbon isotope trend in the Vidukle core, Lithuania, and its relations with oceanic events. Geol Q 49:223–234

Melchin MJ (1994) Graptolite extinction across the Llandovery–Wenlock boundary. Lethaia 27:285–290

Mikulic DG, Kluessendorf J (1999) Stasis and extinctions of Silurian (Llandovery – Wenlock) trilobite associations related to oceanic cyclicity. J Paleontol 73:320–325

Molloy PD (2006) Analysis of the Ireviken extinction event in the Boree Creek Formation, New South Wales, Australia. Unpublished PhD thesis, Macquarie University, Australia, 239pp

Munnecke A, Samtleben C, Bickert T (2003) The Ireviken Event in the lower Silurian of Gotland, Sweden – relation to similar Palaeozoic and Proterozoic events. Palaeogeogr Palaeoclimatol Palaeoecol 195:99–124

Over DJ, Chatterton DE (1987) Silurian conodonts from the southern Mackenzie Mountains, northwest Territories, Canada. Geolog et Palaeontol 21:1–49

Pickett JW (1982) The Silurian system in New South Wales. Bull Geol Surv NSW 29:1–264

Rothrock DA Percival DB, Wensnahan M (2008) The decline in arctic sea-ice thickness: separating the spatial, annual, and interannual variability in a quarter century of submarine data. J Geophys Res 113:C05003. doi:10.1029/2007JC004252

Saltzman MR (2001) Silurian $\delta^{13}C$ stratigraphy: a view from North America. Geology 29:671–674

Serreze MC, Maslanik JA, Scambos TA, Fetterer F, Stroeve J, Knowles K, Fowler C, Drobot S, Barry RG, Haran TM (2003) A record minimum arctic sea ice extent and area in 2002. Geophys Res Lett 30(3):10

Sherwin L (1971) Stratigraphy of the Cheesemans Creek district, New South Wales. Rec Geol Surv, NSW 13:199–237

Simpson AJ (1995) Silurian conodont biostratigraphy in Australia: a review and critique. Cour Forsch-inst Senckenb 182:325–346

Talent JA, Berry WBN, Boucot AJ (1975) Correlation of the Silurian rocks of Australia, New Guinea and New Zealand. Geol Soc Am Spec Pap 150:108pp

Talent JA, Mawson R, Andrew AS, Hamilton PJ, Whitford DJ (1993) Middle Palaeozoic extinction events: faunal and isotopic data. Palaeogeogr Palaeoclimatol Palaeoecol 104: 139–152

Talent JA, Mawson R, Simpson AJ (2003) Silurian of Australia and New Guinea. In: Landing E, Johnson ME (eds) Silurian lands and seas–palaeogeography outside of Laurentia. New York State Museum Bulletin, vol 493. New York State Museum, Albany, NY, pp 181–200

Thomas E (2007) Cenozoic mass extinctions in the deep sea: what perturbs the largest habitat on Earth? Geol Soc Am Spec Pap 424:1–23

Uyeno TT (1990) Biostratigraphy and Conodont Faunas of Upper Ordovician through Middle Devonian Rocks, Eastern Arctic Archipelago. Geol Surv Canada Bull 401:2–210

Valentine JL, Brock GA, Molloy PD (2003) Linguliformean brachiopod faunal turnover across the Ireviken Event (Silurian) at Boree Creek, central-western New South Wales, Australia. Cour Forsch-inst Senckenb 242:301–327

Veron JEN (2008) Mass extinctions and ocean acidification: biological constraints on geological dilemmas. Coral Reefs 27:459–472

Walker DB (1959) Palaeozoic stratigraphy of the area to the west of Borenore, N.S.W. J Proc R Soc NSW 93:39–46

Walliser OH (1964) Conodonten des Silurs. Abhandlungen des Hessischen Landesamtes fur Bodenforschung 41:1–106

Watkins R, Walsh KA, Kuglitsch JJ (1999) Silurian (Llandovery/Wenlock) foraminiferida in carbonate environments of southwestern Wisconsin. J Paleontol 73:540–546

Isotope Geochemistry and Plankton Response to the Ireviken (Earliest Wenlock) and *Cyrtograptus lundgreni* Extinction Events, Cape Phillips Formation, Arctic Canada

Paula J. Noble, Alfred C. Lenz, Chris Holmden, Monika Masiak, Matthew K. Zimmerman, Simon R. Poulson, and Anna Kozłowska

Abstract

Several Canadian Arctic Silurian stratigraphic sections from the basinal facies of Cape Phillips Formation have been sampled across the Llandovery–Wenlock and early–late Homerian (late Wenlock) boundary intervals for integration of biotal (graptolite, radiolarian, palynomorph) and geochemical (^{13}C) data for two well-known extinction events, the Ireviken and *lundgreni* Extinction (LEE) events. Graptolites, abundant and well preserved, provide a refined biostratigraphic base for other paleontologic and geochemical data. They were globally affected by both extinction events: about 64% reduction for the Ireviken and 90–95% for the LEE. Recovery from the LEE event was slow and diversity low through the late Homerian. Radiolarians—diverse (28 species), abundant, and beautifully preserved through the early Homerianare sharply reduced slightly below the LEE boundary. Data for the late Homerian are more scattered, but it appears that diversity was low; few early Homerian taxa crossed the extinction boundary and new taxa appeared. Palynological studies around the LEE interval are at a preliminary level, but it appears that chitinozoans and microflora (acanthomorph acritarchs, prasinophytes, sphaeromorphs) were impacted by the extinction event. Chitinozoans, though seldom abundant, appear to disappear briefly across the LEE boundary, as do palynomorphs. Amorphous organic matter is abundant in the upper part of the *lundgreni* Zone; it is much less common in the early and middle–late Homerian and common in the latest part. Stable isotope geochemistry shows well-marked, positive excursions in the $\delta^{13}C_{organic}$ fraction associated with the Ireviken event and LEE. The Ireviken excursion (C1) curve has a sharp base, reaches a peak in the early Wenlock, and then tapers more slowly. The LEE excursion (C4) peaks at, or slightly below, the early–late Homerian boundary. Both are positive excursions. Considering the limits of biostratigraphic placement of the boundaries, they were close to or coincident with regressions, particularly across

P.J. Noble (✉)
Department of Geological Sciences and Engineering,
University of Nevada, Reno, NV 89557-0138, USA
e-mail: noblepj@unr.edu

J.A. Talent (ed.), *Earth and Life*, International Year of Planet Earth,
DOI 10.1007/978-90-481-3428-1_20, © Springer Science+Business Media B.V. 2012

the LEE interval. The $\delta^{13}C_{organic}$ excursions are greater for inshore sections compared with the offshore section. The most parsimonious explanation for increased carbon content is accelerated weathering of carbonates exposed during a lowstand.

Keywords
Silurian • Llandovery • Wenlock • Arctic Canada • Cape Phillips Formation • Radiolarians • Graptolites • Isotope data • Ireviken Event • *lundgreni* Extinction

Introduction

The Silurian strata of the autochthonous provinces in the central and southeastern Canadian Arctic Islands are characterized by extensive platform carbonates, and their more oceanward graptolitic facies equivalents. The best known and most extensive of the basinal units is the Cape Phillips Formation, a graptolite-bearing limestone and calcareous mudstone unit that ranges from the late Ordovician into the early Devonian and is distributed throughout the Cape Phillips embayment on numerous islands (Thorsteinsson 1959; de Freitas et al. 1999). Herein, we summarize stable isotopic and paleontologic data that we have amassed from the Cape Phillips Formation in sections that span the latest Llandovery through early Gorstian. These data show a number of positive isotope excursions, two of which are linked with global extinction events and with eustatic sea-level drops. For a comprehensive description of the Silurian System in the Arctic Islands, particularly the development and correlation of reef track systems, see de Freitas et al. (1999) and de Freitas and Nowlan (1998).

The first of the two global events recognized in the Wenlock of the Cape Phillips Formation is the Ireviken Event, a major conodont extinction event that occurs in the early Sheinwoodian *murchisoni* through *riccartonensis* graptolite zones (Loydell 2007a, b). The Ireviken Event was first recognized in Lower Visby Formation of Gotland, Sweden (Jeppsson 1987, 1990, 1997). Trilobites are also impacted (Chatterton et al. 1990; Mikulic and Kluessendorf 1999; Calner et al. 2004), as are chitinozoans (Nestor 1997; Nestor et al. 2002; Gelsthorpe 2004), while brachiopods appear affected to a lesser extent (Talent et al. 1993). There has been some confusion as to the degree of impact on graptolites during the Ireviken event, in part owing to the difficulties of correlating shallow shelly facies to the deeper graptolitic facies; however, there is a modest extinction, the *murchisoni* event, documented in the Barrandian sequences (Štorch 1995); this slightly post-dates the severest part of conodont event 2-datum in Gotland occurring near the base of the *murchisoni* Zone (Loydell 2007a, b). A large positive isotopic excursion has been found in association with the Ireviken Event in many places, including Baltica (e.g., Kaljo et al. 1997, 2003; Kaljo and Martma 2006; Wenzel and Joachimski 1996; Bickert et al. 1997; Loydell 1998; Munnecke et al. 2003), Avalonia (Corfield et al. 1992; Loydell and Frýda 2007), North America (Saltzman 2001; Cramer and Saltzman 2005), and in New South Wales (Talent et al. 1993). Correlation of conodont, chitinozoan, and graptolite data from the Aizpute-41 Core, Latvia, indicates that the positive excursion peaks in the *murchisoni* Zone, toward the end of the Ireviken conodont event (Loydell et al. 2003; Štorch 1995; Loydell 2007a, b). The event has been associated with either a widespread regression (e.g., Wenzel and Joachimski 1996; Kaljo et al. 2003) or, alternatively, a deglaciation phase following a regression (Cramer and Saltzman 2005). Loydell (2007a, b) associates the Ireviken event with a lowstand, followed by a transgression in the *murchisoni* Zone.

The second of these events, the *Cyrtograptus lundgreni* Extinction Event (LEE = *lundgreni* Extinction Event), coincides with the boundary between the early and late Homerian. It is a global extinction event recognized first and principally in the graptolite community (Jaeger 1991; Koren 1991). The LEE has been considered highly selective by most workers, largely affecting the plankton community but having relatively little effect on shallow benthic organisms (Kaljo et al.

1995). Recent revisions in correlation of type Wenlock material suggest that shelly benthos are also affected by this event, although the extent is not fully studied (Calner et al. 2004). The effects on graptolites were profound, resulting in a reduction from about 50 species globally in the early Homerian to two or three in the directly overlying late Homerian *P. dubius–G. nassa* Zone (Lenz and Kozłowska-Dawidziuk 2001, 2002; Lenz and Kozłowska 2006). Palynomorphs are also affected, showing a drop in both abundance and diversity across the LEE throughout Baltica (Nestor 1997; Porębska et al. 2004). Trilobites were also affected during the LEE, with six genera becoming extinct globally and six major clades being eliminated from northern Laurentia (Adrain and Edgecombe 1997; Adrain 2000).

The graptolites of the late Wenlock and early Ludlow of the Cape Phillips Formation are well preserved, thus providing a stable biostratigraphic foundation for associated organisms, geochemical data, and for documenting extinction pattern across the LEE (e.g., Lenz 1993a; Lenz and Kozłowska-Dawidziuk 2001, 2002, 2004; Lenz and Kozłowska 2006). Abundant calcareous concretions occur throughout the Cape Phillips, yielding three-dimensional graptolites of unprecedented preservation and also beautifully preserved radiolarian and sponge spicule faunas. Radiolarian faunas of the Cape Phillips Formation are very well preserved, and while older and younger faunas have been examined (e.g., Holdsworth 1977; Goodbody 1982, 1986; MacDonald 1998, 2003, 2006; Renz 1988), the only discussion of the radiolarian response during the LEE appears in Lenz et al. (2006). For this study, several well-exposed Wenlock and early Ludlow sections in Arctic Canada (Fig. 1) have been simultaneously sampled for graptolites, radiolarians, palynomorphs, and for geochemical analyses, particularly variations in $\delta^{13}C$. Three of these sections are close to the carbonate shelf, while the fourth is more than100 km ocean ward.

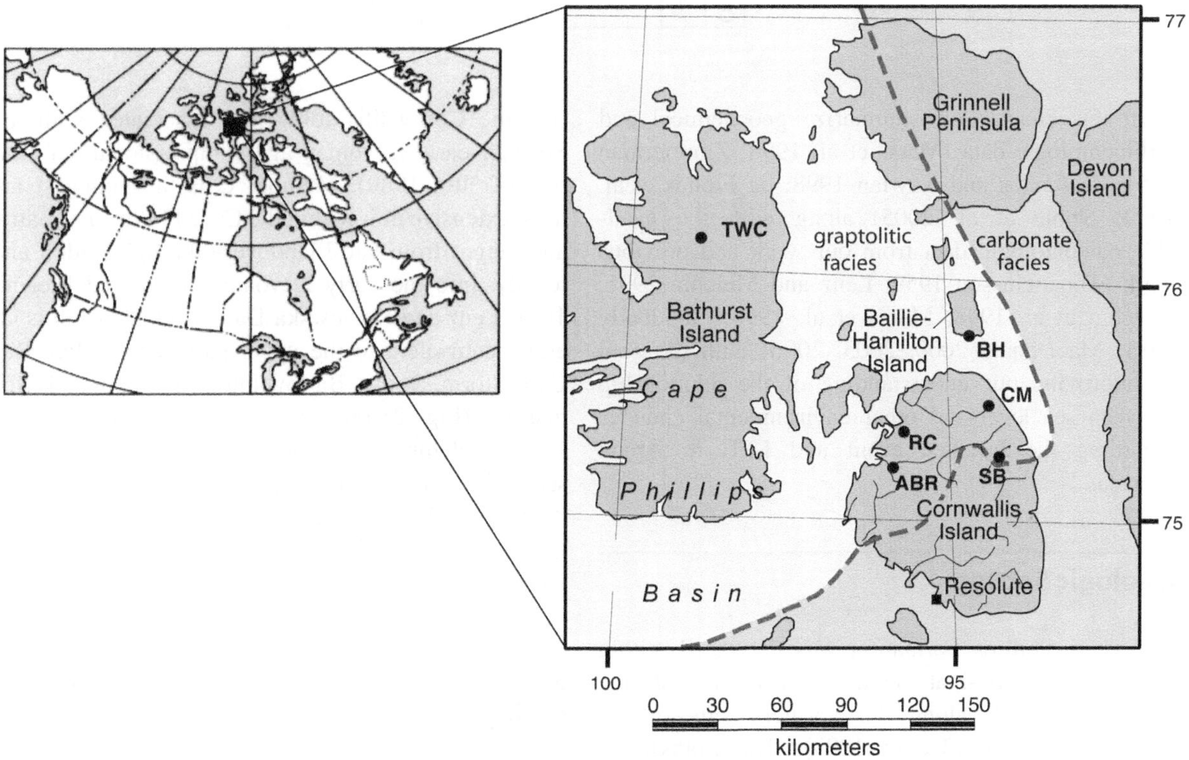

Fig. 1 The central Arctic Islands (after Noble and Lenz 2007) showing localities discussed. *ABR* Abbott River, *SB* Snowblind Creek, *RC* Rookery Creek, *CM* Cape Manning, *BH* Baillie Hamilton, *TWC* Twilight Creek. *Dashed line* delineates Wenlockian Shelf Break between platform carbonate facies to the south and east and basinal graptolitic facies to the west (position taken from Melchin 1989)

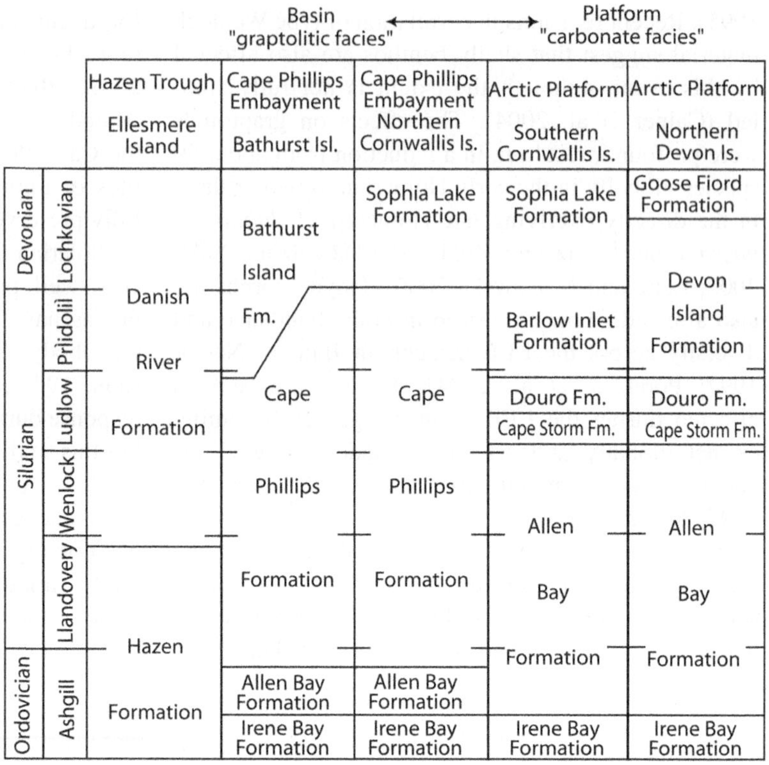

Fig. 2 Chart correlating late Ordovician–early Devonian lithostratigraphy in the central Arctic islands. After Adrain and Edgecombe (1997). *Fm* Formation

In this chapter we summarize geochemical and sedimentologic data (Märss et al. 1998; Zimmerman 2001; de Freitas and Nowlan 1998; de Freitas et al. 1999; Noble et al. 2005) along with the available paleontologic data from our work and previous work (Thorsteinsson 1959; Lenz and Melchin 1991; Melchin et al. 1998; Märss et al. 1998; Goodbody 1981; MacDonald 2000, 2003, 2006) with the goal of improving our understanding of the relationship between sea level and other environmental changes regarding the Ireviken event and LEE in Arctic Canada.

Geologic Setting

The Cape Phillips Formation was deposited within a deep-water basinal setting oceanward of the Franklinian shelf along the northern side of the Laurentian craton (Trettin 1989). The Franklinian shelf accumulated carbonate and clastic sediments from late Cambrian through Devonian time in a largely passive margin setting. These strata are mildly deformed in a series of broad N–S trending folds (i.e. the Cornwallis fold Belt) associated with late Silurian–early Devonian age compressional deformation (Trettin 1989). Member C is the youngest and most extensive unit within the Cape Phillips Formation and ranges from late Llandovery through Pridoli and, locally, into the early Devonian (Lenz and Melchin 1991; Lenz and Kozłowska-Dawidziuk 2004). It is laterally equivalent to the upper part of the Allen Bay, Cape Storm, Douro, Barlow Inlet, and Sophia lake formations (Fig. 2) on the Arctic shelf (Thorsteinsson 1959; Melchin 1989) and is exposed along uplifted sea cliffs and incised stream valleys on several of the Arctic Islands including Cornwallis, Little Cornwallis, Bathurst, Baillie Hamilton, Ellesmere, and Devon Islands.

Stratigraphy and Lithology of the Cape Phillips Formation

The Cape Phillips Formation consists of three members (Fig. 3). Member A (Ashgilian–base of Telychian) consists of fine-grained gray dolomite and alternating argillaceous limestone locally with cherty limestone.

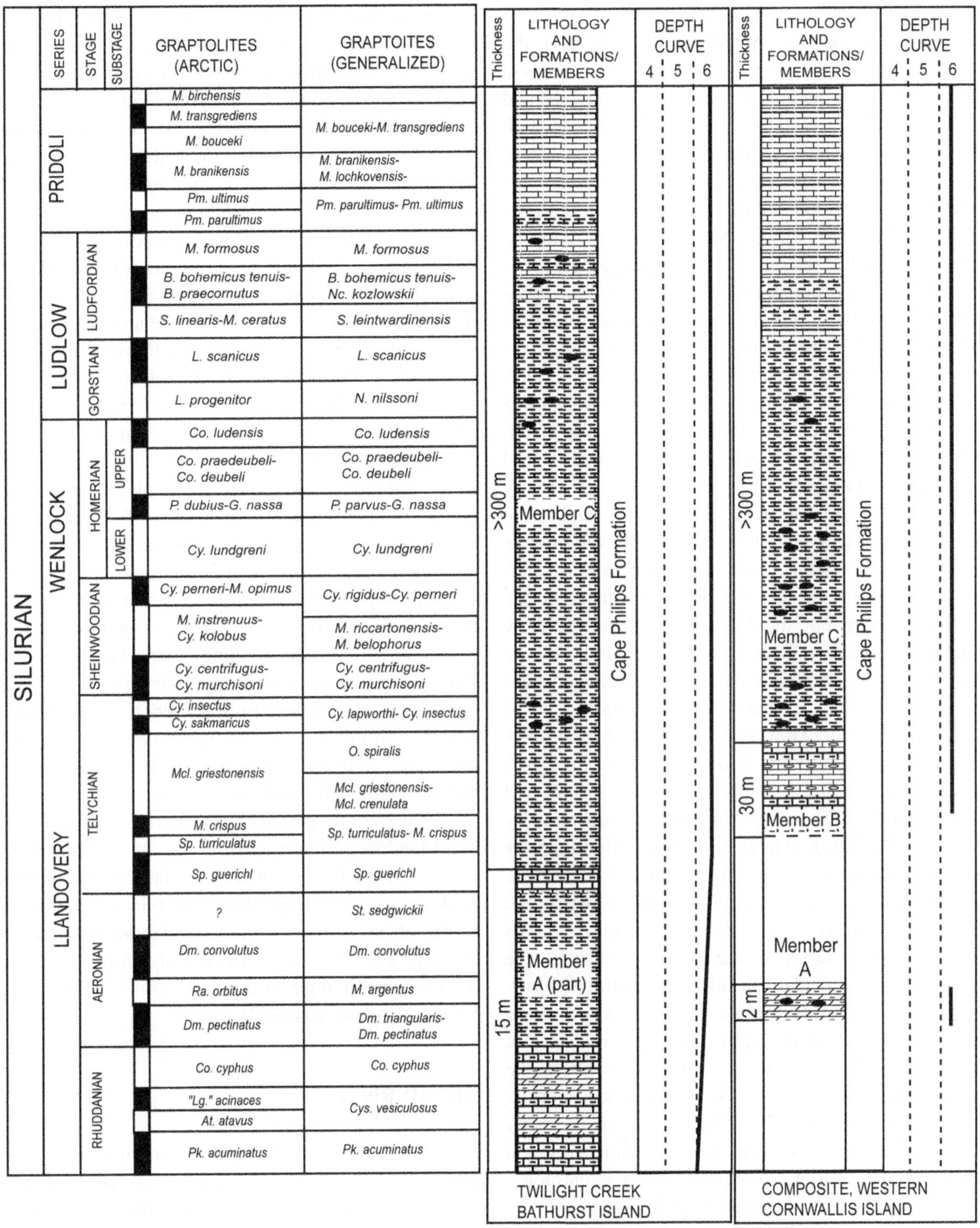

Fig. 3 Chart showing correlation between Wenlock, Ludlow, and Pridoli generalized graptolite zonation of Koren et al. (1996) and Arctic graptolite zonation adopted from Melchin (1989), Melchin et al. (1998), Lenz and Melchin (1991), Lenz (1995), and Lenz and Kozłowska-Dawidziuk (2004). Correlation of graptolite zonations is based on Melchin et al. (1998) for the Llandoverian and Wenlock and Lenz and Kozłowska-Dawidziuk (2004) for the Ludlow. Composite graphic log of the Cape Phillips Formation from northern Bathurst Island (principally from Twilight Creek) and western Cornwallis Island (composite Abbott River and Rookery Creek) shows age of members and depth curve. Thicknesses of members A and B are from data in Thorsteinsson (1959) and Melchin (1989). See Fig. 4 for lithologic key

Member A is fairly thin (<50 m) and distributed widely throughout the Arctic islands wherever the Cape Phillips Formation is present (Thorsteinsson 1959). Overlying member A locally on Cornwallis and southern Bathurst Islands is a cherty limestone with abundant concretions, member B, which appears to be confined to a short interval in the Telychian and is restricted geographically (Melchin 1989). In places where member B is absent, Melchin (1989) reports a thin limestone bed restricted in age to latest Aeronian–early Telychian, occurring between the *Dm. convolutus* and *Sp. turriculatus minor* graptolite zones (= part of the *Sp. guerichi* Zone in later publications).

Member C ranges from Telychian through Pridoli and locally may be as young as early Devonian (Lenz and Melchin 1991; Lenz and Kozłowska-Dawidziuk 2004). It is the thickest (>300 m) and most widespread member, consisting of thinly bedded, laminated silty calcareous mudstone and calcareous or dolomitic shale, and is locally interbedded with minor amounts of bioclastic limestone. Calcareous concretions ranging from several cm to 10's of cm in diameter are common throughout the Llandovery–lowest Ludlow interval of member C and provide the bulk of the radiolarian and graptolite data. The Llandovery and Wenlock parts of member C are essentially free of a clay component. However, beginning in the early Ludfordian (mid-Ludlow; *S. linearis–M. ceratus* Zone) an increased siliclastic component is recognized, and sedimentation rates appear to increase substantially (Lenz and Kozłowska-Dawidziuk 2004). Rhythmically bedded argillaceous shale and laminated calcarenite and calcilutites are common in the Ludlow. At Baillie Hamilton Island (Fig. 1), 585 m of member C was measured from the top of the Telychian through the lower part of the Pridoli. In this section, the Wenlock is approximately 160 m thick and the Ludlow about 400 m thick. The stratigraphy and sedimentology are discussed in detail from three localities where the Wenlock part of member C is exposed nearly in its entirety: Abbott River, Twilight Creek, and Rookery Creek. These sections were sampled extensively and simultaneously for geochemistry and paleontology; their ages are well constrained by graptolites.

Twilight Creek

Twilight Creek on Bathurst Island (Fig. 1) is located more than 100 km northwest of the shelf break and represents a deeper basinal facies than the other two sections discussed later. Two hundred fifty meters of section was measured in a series of stream-cuts, of which the Wenlock is represented by 102 m of near-continuous exposure (Fig. 3). Stage boundaries were determined by author ACL through field determinations of flattened graptolites found in place. The lithology is fairly uniform and comprises organic-rich thin-bedded platy calcareous mudstone and fissile calcareous shale (Fig. 4). Calcareous concretions averaging 8 cm in diameter are found in abundance near the base of the section in the Llandovery but are largely absent from the Wenlock. Phosphatic concretions occur within a meter interval at the Wenlock–Ludlow boundary and average 2 cm in diameter. Flattened graptolites occur on bedding planes and are the source of age control for this section. The Llandovery–Wenlock boundary is placed at ~22 m, above the last observed occurrence of *C. sakmaricus* at 21 m, and below the first in-place occurrence of Sheinwoodian graptolites at 26 m.

Petrographic examination shows millimeter-scale laminations of silty quartz-rich layers interspersed with finer-grained cryptocrystalline calcite layers (Fig. 5). Pyrite is common, occurring as framboids or small cubes of 1 mm or less in diameter. X-ray diffraction reveals that the Homerian is mineralogically homogenous and composed almost entirely of calcite with minor quartz silt, and virtually no clay component (Fig. 6). Radiolarians were recovered from unit A in the Llandovery interval, from the *M. spiralis* Zone, but not from the Sheinwoodian part of the section. Radiolarian data from the Homerian are scant due to the absence of concretions and bioclastic carbonate beds and are limited to several samples clustered within a meter interval directly below the Wenlock–Ludlow Boundary (Fig. 3).

Abbott River

Exposed in a series of sequential stream-cuts along a tributary of Abbott River is a succession of over 350 m of member C that spans the top of the Sheinwoodian through the Ludlow. Herein, we present the stratigraphy and geochemistry from the Homerian through the base of the Gorstian portion only (Fig. 4). Abbott River is located on the west side of Cornwallis Island (Fig. 1) and represents a more proximal facies than Twilight Creek, being situated near the basin margin,

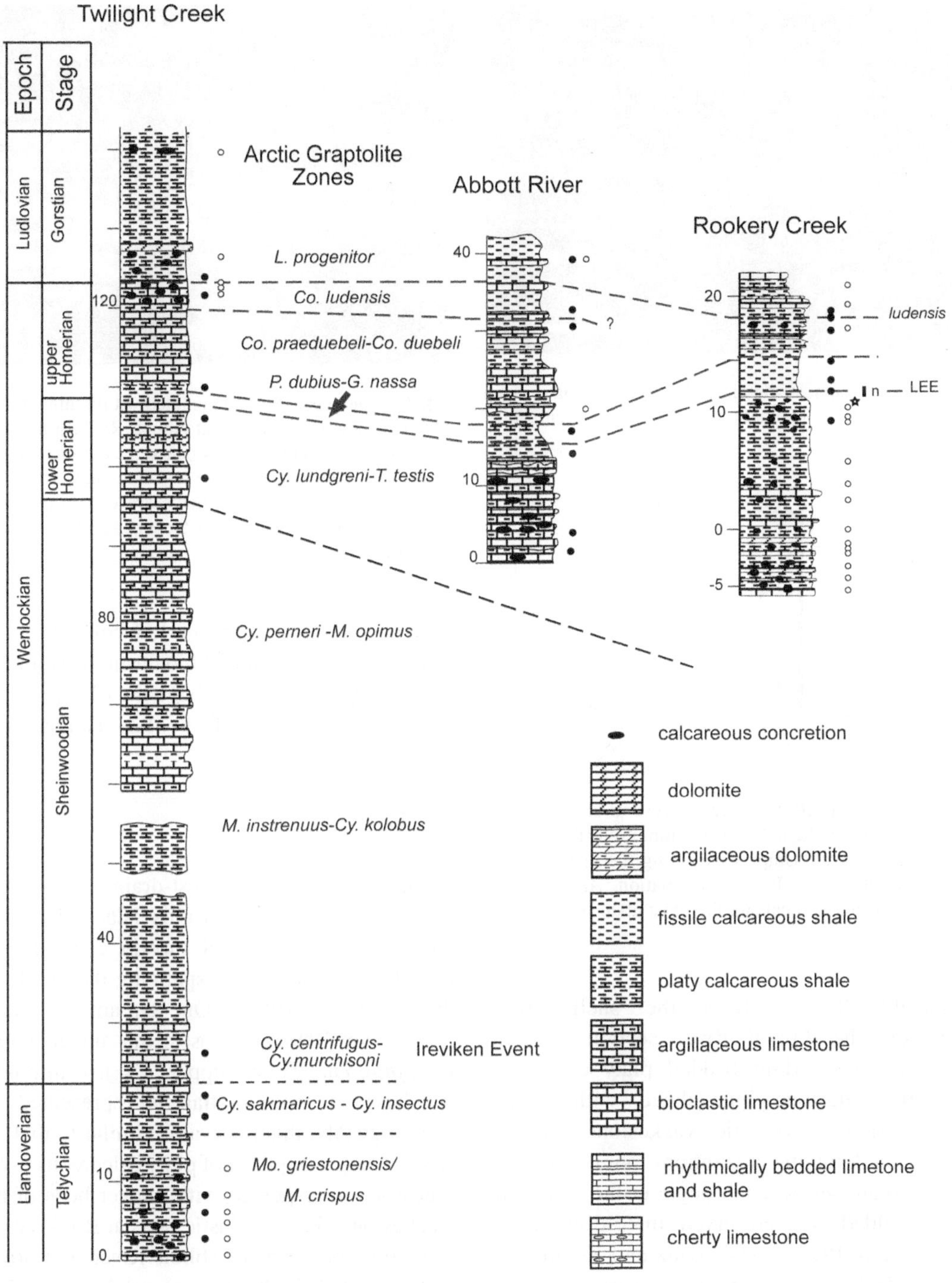

Fig. 4 Biostratigraphic correlation of measured sections from Twilight Creek, Abbott River, and Rookery Creek. *Dashed lines* mark the start of the Ireviken, LEE, and *ludensis* extinction events. See Noble et al. (2005) for discussion of the *ludensis* event. Position of radiolarian samples is shown by open circles and graptolite samples by solid circles positioned on right side of each stratigraphic column. Star positioned at 11.5 m in the Rookery Creek section marks the one barren radiolarian sample discussed in the text, and "n" at 12 m marks the nautiloid mass mortality interval

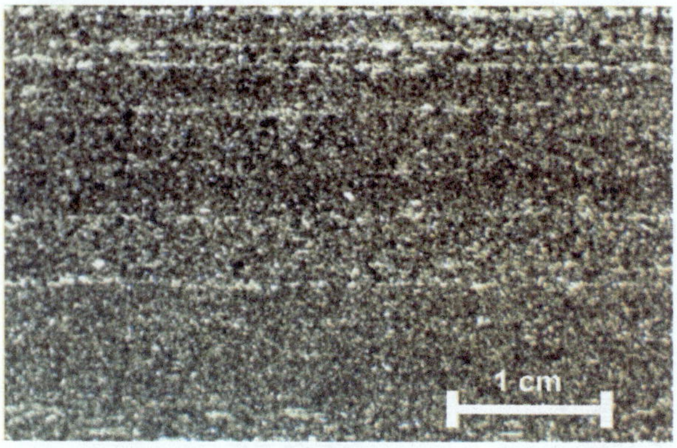

Fig. 5 Photomicrograph of laminated silty carbonate from Twilight Creek showing laminations varying in percentage of quartz silt

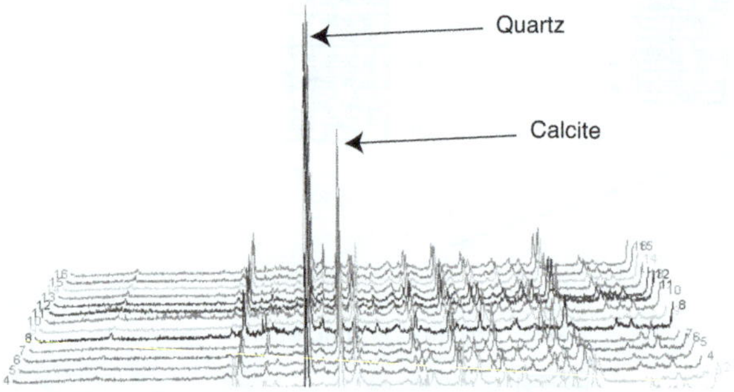

Fig. 6 Compilation of XRD spectra for several samples from the Wenlock of the Twilight Creek section. Note the similarity between peaks for all samples indicating the mineralogical homogeneity of the Cape Phillips Formation. *Arrows* show strong peaks of quartz and calcite, the most abundant minerals

Fig. 7 A bioclastic bed intercalated within thinner bedded laminated calcareous mudstone at Abbott River. It has an undulose base, is graded, and contains coarse (2 cm or less) pentamerid brachiopod debris at the base and fines upward. Pelmatozoan, sponge, and graptolite fragments are common in these beds; radiolarians are occasional in the finer-grained tops. Lens cap for scale

approximately 25 km from the shelf break. Lithologically, the Abbott River section is composed of thin- to medium-bedded platy calcareous to dolomitic mudstone intercalated with thicker and coarser-grained bioclastic wackestone (Fig. 7). Calcareous and dolomitic concretions ranging from 10 to 50 cm in diameter occur throughout the Homerian and have yielded well-preserved three-dimensional graptolites (Lenz 1993b, 1994; Lenz and Kozłowska-Dawidziuk 2001). Graptolites also occur commonly in flattened form on bedding plane surfaces and collectively provide good age control for this section. Although concretions are abundant, radiolarian recovery from Abbott River is low because many of the partially dolomitic concretions show calcite replacement of radiolarian tests, making it impossible for them to be extracted and examined. Radiolarians have been recovered from a few concretions and from some of the finer-grained bioclastic beds. In particular, radiolarians from 0.5 m above the LEE provide critical data from the base of the late Homerian and are discussed later.

Rookery Creek

Rookery Creek is a west-draining stream on the northwest side of Cornwallis Island. Exposed along the creek is a series of mostly continuous stream-cuts through member C spanning the late Llandovery through the Gorstian. One stream-cut (section 01-RC-1) completely exposes the upper part of the *C. lundgreni* Zone (upper 18 m) through which the base of the Gorstian is represented (Fig. 4). Lithologically, this interval is similar to that at Abbott River, being composed of platy calcareous to dolomitic mudstone interspersed with thicker beds of bioclastic wackestone. The bioclastic beds at Rookery Creek are less common than at Abbott River and are medium to thick-bedded (20–80 cm) calcirudites rich in pelmatozoan and sponge debris. Calcareous concretions of variable size ranging from 5 to 30 cm in diameter are common throughout the early Homerian (Fig. 8), but are much less common in the late Homerian and Ludlow at Rookery Creek. Well-preserved radiolarians and graptolites have been recovered from these concretions and provide excellent paleontologic control.

Fig. 8 Interval of laminated platy calcareous mudstone at Rookery Creek containing abundant small (4–8 cm diameter) calcareous concretions (*arrows*). Hammer (42 cm long) placed on a bedding plane surface

The late Homerian at Rookery Creek is fairly thin (6 m) and largely devoid of calcareous concretions. Only one radiolarian sample was recovered from near the top of the late Homerian, and an additional sample was recovered from a bioclastic bed several meters above the base of the Gorstian. Graptolite data from Rookery Creek are also extensive, with specimens occurring in concretions and also flattened on bedding surfaces.

Paleontology

The paleontologic data discussed herein have been amassed largely from the early and late Homerian relating to the LEE. Our paleontologic data relating to the Ireviken event are restricted to field identifications of flattened graptolites from the Twilight Creek section (Fig. 3) and Llandoverian radiolarians and are discussed later in this chapter along with published data by other workers (Melchin 1994; MacDonald 2000, 2003, 2006).

Graptolites

The graptolites around the Llandovery–Wenlock boundary are well developed, and boundary sequences are documented in several sections on Cornwallis Island, including the type section of the Cape Phillips Formation, as well as on Baillie Hamilton and Bathurst Islands (Lenz and Melchin 1991), although to date, only the flattened graptolites have been studied. Published and unpublished data (Lenz and Melchin 1991; Melchin et al. 2000; Senior 2005) show that the *Cyrtograptus sakmaricus* and the succeeding *C. insectus* zones of the latest Llandovery are overlain by the *C. centrifugus C. murchisoni* zones going in the lowest Wenlock. The *C. centrifugus* interval is very condensed and in some sections is relegated to only a few beds. Interestingly, rich fish faunas occur within the *C. insectus* Zone of the Cape Phillips-type section and on Baillie Hamilton Island (Märss et al. 1998, personal observations) and coincide with the start of the isotopic excursion.

Graptolite collections across the LEE are extensive and come from five main sections on Cornwallis, Bathurst, Baillie Hamilton, and Ellesmerre Islands (Fig. 9). Stratigraphic occurrences from two of these richly fossiliferous localities appear herein and are representative of the regional occurrences. The *C. lundgreni* Zone was an interval in which a diversity of greater than thirty-five species was more or less maintained to the top of the zone (Lenz and Kozłowska 2006), although there is some evidence that three retiolitid genera disappeared about midway through the zonal interval (see the following section). The faunas include the monograptids, retiolitids, and cyrtograptids (Lenz and Melchin 1991; Lenz 1993a; Lenz 2001; Lenz and Kozłowska-Dawidziuk 2001, 2004). The monograptid genera comprise *Monograptus*,

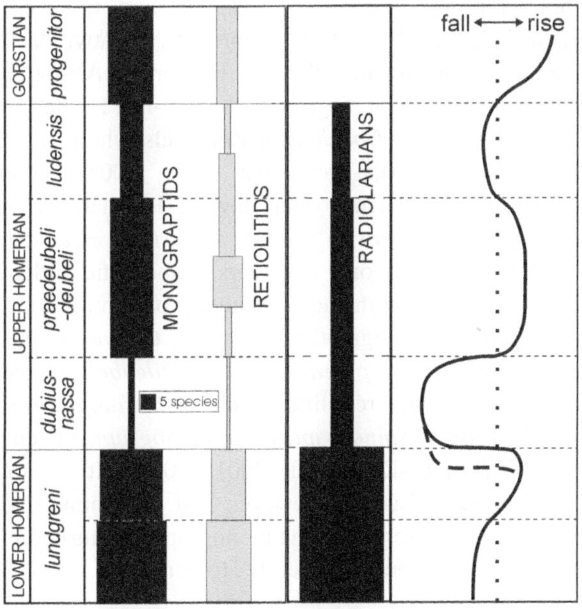

Fig. 9 Relative abundance of graptolite and radiolarian taxa

Streptograptus, *Pristiograptus*, *Monoclimacis*, and *Testograptus*, the last-named confined to the early Homerian. The retiolitids are relatively diverse and include *Gothograptus*, *Cometograptus*, *Sagenograptus*, and *Eisenackograptus*, with all but the last named confined to the *lundgreni* Biozone. Also present are the long-ranging retiolitid genera *Sokolovograptus*, *Pseudoplectograptus*, *Retiolites*, *Stomatograptus*, and *Paraplectograptus*, the last three being long-ranging but apparently extinguished in the early half of the *lundgreni* Biozone. The last graptolite group is that of the cyrtograptids consisting of the genera *Diversograptus* and *Cyrtograptus*. The most prominent cyrtograptid species are *C. lundgreni* and *C. radians* that range through the *C. lundgreni* Zone (Lenz and Melchin 1991; Lenz 1993a, 2001; Lenz and Kozłowska-Dawidziuk 2001, 2004).

Unlike other places in the world where a distinct step-wise extinction is noted across the *C. lundgreni* Zone, the LEE in the Canadian Arctic shows only a weak two-stepped event that is coincident with the end of the *C. lundgreni* Zone. Present data suggest, however, that the retiolitids *Retiolites*, *Stomatograptus*, and *Paraplectograptus* disappeared prior to the end of the biozone (Lenz and Kozłowska 2006).

The late Homerian sequence is, by contrast, characterized by a low-diversity fauna, especially in the immediate post-extinction strata, where only the retiolitid *Gothograptus nassa* and the monograptids *Pristiograptus dubius* and the very closely related *P. jaegeri* are present, the two last-named continuing into the late Homerian. Although *Gothograptus nassa* ranges into the mid-late Homerian (*Co. praedeubeli–Co. deubeli* Zone) elsewhere (e.g., Baltica: Kozłowska-Dawidziuk et al. 2001), it is apparently restricted to the lowest late Homerian (*P. dubius–G. nassa* Zone) in Arctic Canada. Moderate diversity only reappears in about the middle of the substage with the appearance of morphologically novel monograptids such as *Colonograptus* (including *Co. praedeubeli–Co. deubeli*) and *Lobograptus*, and retiolitids such as *Baculograptus*, *Papiliograptus*, *Spinograptus*, and *Doliograptus* (Lenz and Kozłowska-Dawidziuk 2001, 2002). It is not until the early Ludlow, however, that taxonomic and morphologic diversity of the monograptids returned to about the level seen in the early Homerian.

Radiolarians

Radiolarian data pertaining to the LEE are largely from section 01–RC1 along Rookery Creek, where the upper 18 m of the early Homerian through the base of the Gorstian are fully exposed (Figs. 4 and 9). At Rookery Creek, the radiolarian data from the *C. lundgreni* Zone are extensive; samples taken on average every 1 m yielded well-preserved radiolarians. Late Homerian radiolarian data are more sparse. At Rookery Creek the late Homerian is represented by a fairly thin interval (6 m), and only one sample, approximately 0.5 m below the top of the Homerian, yielded radiolarians. Abbott River, Cape Manning, and Snowblind Creek have yielded additional late Homerian data, collectively providing a glimpse of the faunas from this interval.

Radiolarians from the *C. lundgreni* Zone are abundant and diverse, consisting of twenty-eight species. At Rookery Creek, they are found ubiquitously in every concretion examined throughout the *C. lundgreni* Zone, with the exception of concretions at the 11 m horizon in section 01–RC1, which is approximately 1 m below the top of the early Homerian. The radiolarian fauna from the *C. lundgreni* Zone is comprised of abundant inaniguttids, including three or more species of *Zadrappolis*, *Inanihella tarangulica* group taxa, and one species of *Plussatispila*. Ceratoikiscids (five species) and palaeoscenidiids (eight species) also make up an important component. Entactiniids (three to four species), secuicollactines (one to two species), and haplotaeniatumids (two species) are also present and make up a smaller component of the fauna. *Praespongocoelia parva* makes its first appearance in the upper part of the *C. lundgreni* Zone, 8 m below the LEE, and persists as an abundant component of the fauna until the highest sample just below the LEE at 10.5 m in section 01–RC1 (Fig. 4). As stated earlier, the 11-m horizon at Rookery Creek is devoid of radiolarians, and the LEE is placed at 12 m, directly above the highest occurrence of an apparently early Homerian monograptid.

Radiolarians recovered from the late Homerian are more scattered, and come from a sample in the *P. dubius–G. nassa* Zone at Abbott River, another in the *Co. praedeubeli–Co. deubeli* Zone at Cape Manning, a third from within a meter of the top of the late Homerian at Rookery Creek, and a fourth from within 2 m of the top of the late Homerian at Snowblind Creek (u. Homerian thickness at least 30 m). The Abbott River sample contains a low-diversity fauna consisting of three inaniguttid species, including *Inanihella tarangulica*, and one species each of *Ceratoikiscum* and *Palaeoscenidium*, and an undescribed species of spongy spumellarian. Only two of the inaniguttid species found in the *C. lundgreni* Zone persist into the late Homerian. The fauna from this sample is well preserved, yet sparse, and true diversity estimates may be hampered by the low numbers of specimens recovered. The Cape Manning sample contains *Ceratoikiscum*, *Zadrappolus*, *I. tarangulica*, and abundant *Futobari*, a genus recognized from the Silurian of Japan. At Rookery Creek, a sample from a concretion at 17.5 m, 0.5 m below the Wenlock–Ludlow boundary (Fig. 4), contains abundant radiolarians, most of which are assigned to the *I. tarangulica* group. Also present are *Zadrappolus*, two undescribed species of spongy spumellarians, and two species of ceratiokiscids. A sample from a bioclastic bed at Snowblind Creek approximately 4 m below the Wenlock–Ludlow boundary also yields abundant radiolarians, most of which are one species of undescribed spongy spumellarian, along with rare occurrences of *I. tarangulica*. The difference in faunal composition between the samples at Rookery Creek and Snowblind Creek may be a function of facies. Snowblind Creek is more proximal to the basin margin and the radiolarian-bearing sample is from a bioclastic limestone rich in sponge and pelmatozoan debris. The predominance of spongy spumellarians in this bioclastic bed may reflect gravity transport of a shallower water assemblage. Additional analysis of faunas from coeval bioclastic and concretion samples will need to be conducted in order to confirm this hypothesis.

Palynomorphs

Palynomorph data included herein are from preliminary analyses on the Homerian through early Gorstian interval (*C. lundgreni* through *L. progenitor*) by A. Soufiani (principally chitinozoans) from Abbott River and Twilight Creek provided in a service report to author PJN in 2000. Additional analyses and interpretations of phytoplankton and organic matter are made by author MM on material collected later from Rookery Creek and Abbott River. The results are too preliminary and sketchy to document the extent of the palynomorph changes in association with the LEE, due largely to sparse recovery and variable preservation, although the chitinozoans and possibly the entire palynological assemblage appear to be impacted. Twilight Creek samples contain abundant amorphous organic matter, but have produced poorly preserved and rare palynomorphs and are not discussed further, whereas Abbott River and Rookery Creek have produced more promising results.

In both Abbott River and Rookery Creek data sets, a poor palynological assemblage is observed containing amorphous organic matter, rare organic-walled microfossils ranging from highly degraded to well-preserved sphaeromorphs, probably *Leiosphaeridia* spp., some questionable cryospores, prasinophytes, and rare chitinozoans. One sample (RC01–2, 2.5 m, *C. lundgreni* Zone) contains a few so-called giant acritarchs, such as *Polygonium* sp. Poor sphaeromorphs (mainly leiosphaeres) occur throughout but range in preservation from almost completely degraded to well preserved. Preservation in some samples is too poor to distinguish the acanthomorphic acritarchs with short process from prasinophyte algae. In other samples, however, the sphaeromorphs are sufficiently well preserved to indicate that the lack of acritarchs is probably primary and not the result of preservation.

Chitinozoans appear to be affected by the LEE, although data are sparse and preliminary. In the Abbott River section (Fig. 4), they are well preserved but sparse from 0 to 13 m (upper part of *C. lundgreni* Zone), absent from 13.5 to 15.5 m, then reappear at 16 m and persist throughout the late Homerian (A. Soufiane written communication 2000). Chitinozoan disappearance predates the highest occurrence of *C. lundgreni* Zone graptolite taxa, which occurs at 14.5 m. The 15.5 m interval, in particular, is barren of all organic-walled microfossils. No chitinozoans, scolecodonts, acritarchs, nor graptolite fragments were observed in the 15.5 m sample (A. Soufiane written communication 2000). In the Abbott River section, the base of the late Homerian is placed at 16 m, coinciding with the first appearance of Homerian chitinozoan species *Linochitina odiosa* Laufeld (A. Soufiane written communication 2000). The lowest late Homerian graptolites in this section occur at 17.5 m (*Co. deubeli*), although additional collections of graptolites from an auxillary section along the opposite side of the creek (99–AB 1a section, not illustrated herein) provide increased graptolite resolution across the LEE and show that the lowest 1 m of the late Homerian represents the *P. dubius–G. nassa* Zone, which has then been projected into the section illustrated herein at 16 to 17 m (Fig. 4).

Attempts were made to quantify the relative abundance of organic matter observed in palynological samples and may serve as a rough index of plankton abundance. At Abbott River, amorphous organic matter is abundant in the upper part of the *C. lundgreni* Zone but then is uncommon in the *P. dubius–G. nassa* and *Co. praedeubeli–Co. deubeli* zones. Organic matter becomes common to abundant in the *Co. ludensis* Zone, the top biozone of the late Homerian. Likewise, at Rookery Creek, organic matter is uncommon in the *Co. praeduebeli–Co. deubeli* Zone. The observed drop in abundance in the early part of the late Homerian may be related to the recovery phase following the LEE, but data are too preliminary to provide more than speculation at this point.

Geochemistry

Stable carbon isotope geochemistry, and limited elemental geochemistry (X-ray fluorescence), was conducted on the Homerian at three localities to determine if there is any geochemical signature associated with the biotic events, and to help identify correlative or causative factors (Zimmerman 2001; Noble et al. 2005; Lenz et al. 2006). Elemental geochemical analyses were restricted to Twilight Creek and reveal no change across the LEE (Noble et al. 2005; Zimmerman 2001). Stable isotope geochemistry shows a more interesting story in that a positive excursion in the $\delta^{13}C_{organic}$ fraction is closely associated with the LEE in all three sections (Noble et al. 2005; Lenz et al. 2006) and the Ireviken Event in the section at Twilight Creek that spans the late Llandovery–early Wenlock interval.

A total of six positive $\delta^{13}C$ excursions occur in the latest Llandovery through early Ludlow part of the Cape Phillips Formation and are denoted C1–C6 (Noble et al. 2005; Figs. 10, 11, and 12). Excursions are sharpest and best developed in the $\delta^{13}C_{organic}$ fraction, but at Twilight Creek and Abbott River, there is sufficient general agreement between the $\delta^{13}C_{organic}$ and $\delta^{13}C_{carb}$ curves to conclude that the values indicate the samples have not been altered substantially by diagenesis. In the study sections analyzed, the $\delta^{13}C_{organic}$ fraction is believed to be more reflective of changes in $\delta^{13}C_{DIC}$(dissolved organic carbon) of contemporaneous seawater, as recorded by the fixation of organic carbon by local primary producers, whereas the bulk of the sedimentary carbonate fraction appears to be detrital and was most likely sourced from the shelf platform margin. The greater noise in the $\delta^{13}C_{carb}$ signature probably reflects transport and mixing processes involving carbonate sediment of different ages from the platform and differences in local C-cycling processes (Panchuk et al. 2005a, b) between the platform and deeper basin (Noble et al. 2005; Melchin and Holmden 2006). In addition, the occasionally very low $\delta^{13}C_{carb}$ values at Twilight Creek and Abbott River are attributed to the production of diagenetic cements driven by organic matter oxidation under sulphate-reducing conditions (Coniglio and Melchin 1995; Melchin and Holmden 2006). At Rookery Creek, there is poor agreement between the $\delta^{13}C_{carb}$ and $\delta^{13}C_{organic}$ curves, and mean values of $\delta^{13}C_{carb}$ are much lower than corresponding values at Abbott River and Twilight Creek (Figs. 10, 11, and 12), indicating that Rookery Creek may have been more extensively affected by diagenesis. The $\delta^{13}C_{organic}$ values at Rookery Creek closely correspond with those from Abbott River and Twilight Creek and appear to be unaffected by diagenesis. Herein, we focus on excursions C1, in the early Sheinwoodian, and C4 and C5, which are associated with the LEE and the post-LEE interval.

Excursion C1 occurs at Twilight Creek, where positive shifts of 2.6 and 2.2‰ occur, respectively, in the $\delta^{13}C_{organic}$ and $\delta^{13}C_{carb}$ at the base of the Wenlock (Fig. 10). The C1 excursion begins in the latest Telychian *C. sakmaricus* Zone at 20 m and reaches a maximum at 28 m in the earliest Sheinwoodian. Conodonts were not collected, and so the placement of the Ireviken conodont extinction cannot be confirmed locally; however, its coincidence with isotope stratigraphy in other parts of the world suggest that it should occur within this 8 m interval. The C1 has an asymmetrical shape characterized by a sharp rise in $\delta^{13}C$ to the peak of the excursion, followed by a gradual decline in values that continues throughout most of the earlier half of the Wenlock, and eventually stabilizing at a baseline $\delta^{13}C$ value that is 2‰ lower than the pre-excursion one (Fig. 10). Both the magnitude and the shape of the excursion are similar to

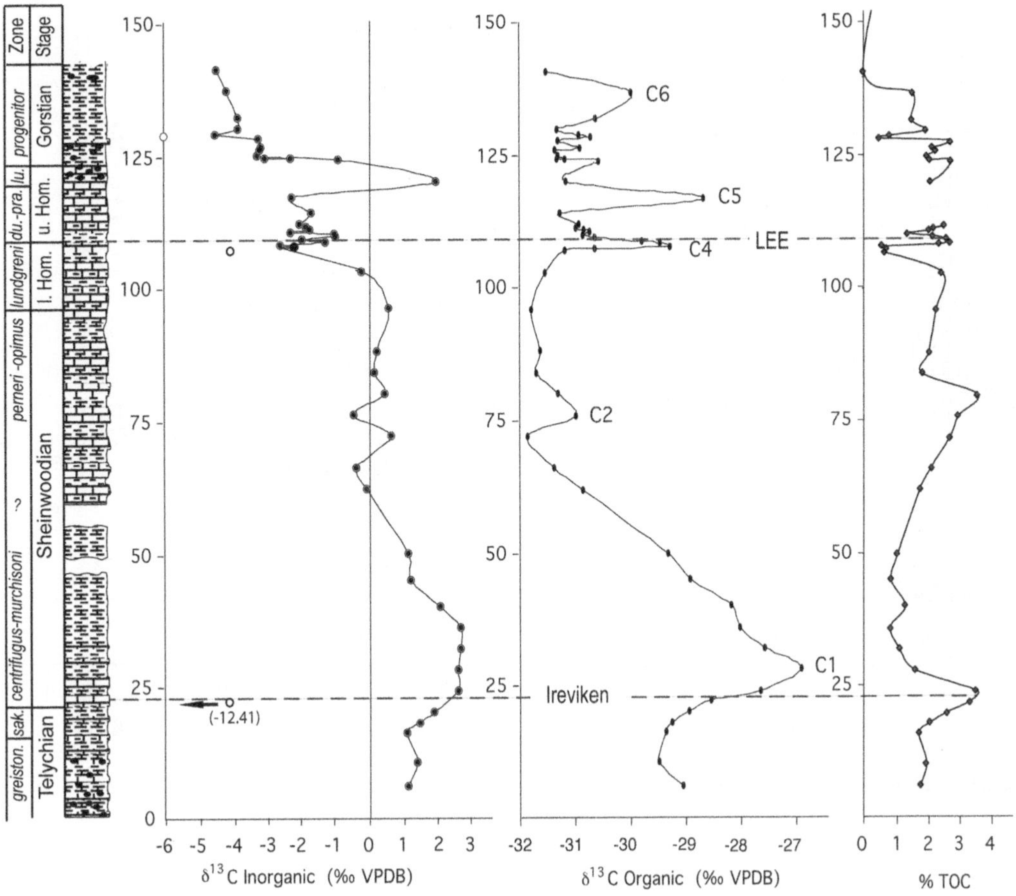

Fig. 10 Geochemical profiles for Twilight Creek

that of the excursions reported from Baillie Hamilton Island (Märss et al. 1998), and from other regions, including Gotland (Bickert et al. 1997; Samtleben et al. 2000), Latvia and Estonia (Kaljo et al. 1997, 1998), and Nevada (Saltzman 2001).

Similarly, a positive $\delta^{13}C$ excursion associated with the LEE is observed in all three sections (Figs. 10, 11, and 12). The onset of the excursion begins just prior to our placement of the early–late Homerian boundary in all three sections, and the maximum peak value occurs within 1 m of the boundary. Our precision in the placement of the boundary is 1 m at Rookery Creek, 3 m at Abbott River, and is within 4 m at Twilight Creek. The excursion is bifurcated with the lower peak labeled C4 and the upper peak C5. The magnitude of C4 ranges from 2.3 to 3.4‰ (Figs. 10, 11, and 12) and shows greater magnitude in the two sections more proximal to the basin margin (Abbott River = 3.4‰, Rookery Creek = 3.35‰) than for the more oceanward Twilight Creek section. The upper peak, C5, occurs within the *praedeubeli–deubeli* Zone. The shape of the excursion is best seen in the Rookery Creek profile (Fig. 12), most likely due to higher sampling resolution, at 0.5 m intervals throughout the late Homerian, as opposed to 1 m intervals throughout most of the late Homerian at Twilight Creek and Abbott River. The Rookery Creek excursion exhibits an asymmetrical curve with a rapid onset and gradual decline in $\delta^{13}C$ values throughout much of the late Homerian. At Rookery Creek, C5 is bifurcated, and a positive baseline shift of approximately 1‰ from pre-excursion values is seen in the late Homerian that persists into the Gorstian.

In contrast to the LEE, no distinctive change in carbon isotope signature is observed at the Wenlock–Ludlow boundary(*ludensis* Event) in any of the sections. Fluctuations in $\delta^{13}C$ are seen at all three sections, but neither excursions nor baseline shifts are noted (Figs. 10, 11, and 12). These data are in agreement

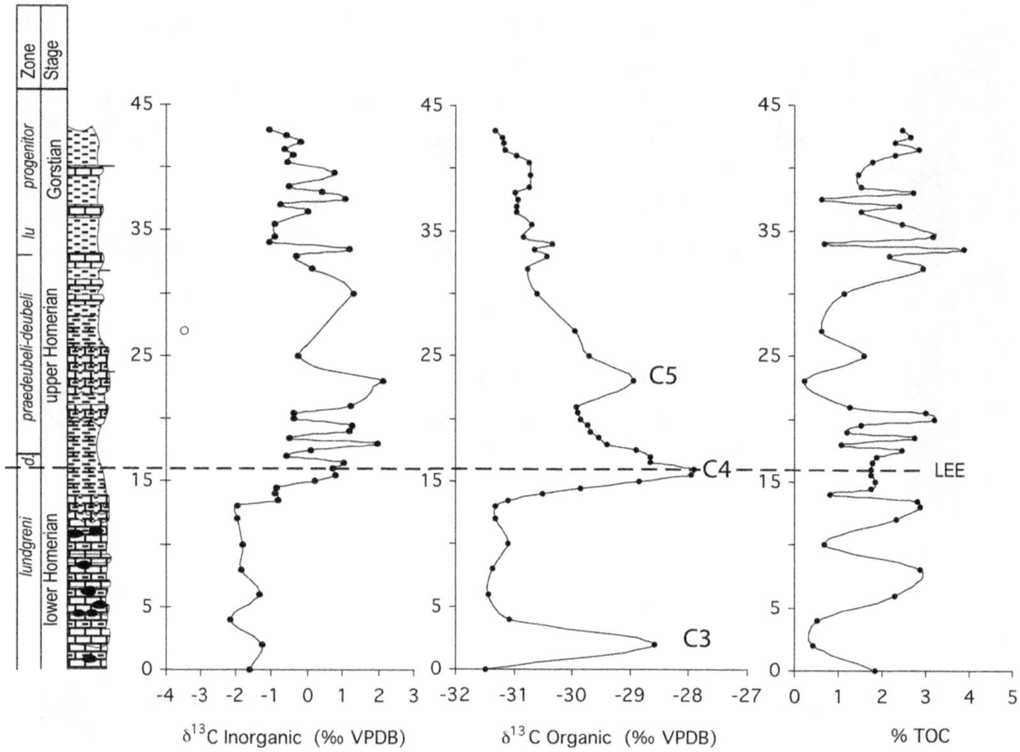

Fig. 11 Geochemical profiles for Abbott River

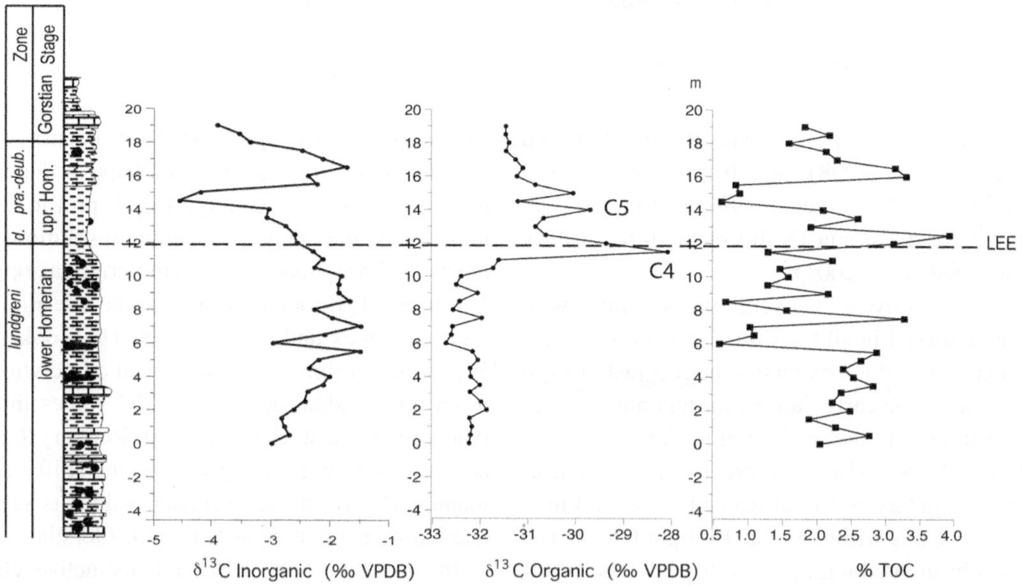

Fig. 12 Geochemical profiles for Rookery Creek

with isotopic data presented by Märss et al. (1998) from Baillie Hamilton and Cornwallis Islands. In addition, two other smaller positive excursions are noted at Twilight Creek (Fig. 10), one in the C. *perneri–M. opimus* Zone (C2), and another in the Gorstian (C6). For a more detailed discussion of the geochemical profiles at Twilight Creek and Abbott River, including methodology, their significance, and their correlation to other geographic areas, see Noble et al. (2005).

Timing and Impact of Ireviken Event and Lee in the Arctic

The graptolite extinction toward the end of the Ireviken event, the *C. murchisoni* event, is recognized globally and results in a 64% extinction at the species level (Melchin et al. 1998). The effects of the Ireviken event on radiolarians have not been discussed directly, but data from MacDonald (2003, 2006) show a major faunal turnover from the late Llandovery to the Wenlock, although sample control is not sufficiently high resolution to determine if the radiolarian response coincides with the conodont extinction, or follows with the graptolites. Our radiolarian data from Twilight Creek are restricted to the pre-Ireviken interval, and consequently we cannot add further to the radiolarian story other than to corroborate the composition of a diverse pre-Ireviken radiolarian fauna documented by MacDonald (2003, 2006). One significant contribution pertains to the presence of the C1 excursion and its position relative to the Llandovery–Wenlock boundary. The onset of C1 begins just prior to the top of the Telychian, with the excursion maximum occurring approximately 6 m above our biostratigraphically placed Llandovery–Wenlock boundary. The start of the excursion appears to be older in the Arctic than indicated in Baltica (Kaljo and Martma 2006; Loydell 2007a, b) and warrants recollection of this section to confirm the position of the excursion relative to conodonts, graptolites, and potentially overlooked nondepositional surfaces.

The relative sequence of biotic, lithologic, and isotopic events relating to the LEE is also observed. At Rookery Creek, radiolarians temporarily disappear between 10.5 and 11 m, above which is a lithologic change beginning at 11.5 m from platy calcareous shale to fissile shale. Coincident with this subtle lithologic change is a nautiloid mass mortality in which the surfaces of several beds from 11.5 m to just above 12 m are littered with orthocone nautiloid shells. The last appearance of *lundgreni* Zone graptolites occurs at 12 m and marks the graptolite extinction, which is a single-stepped or, at best, a weakly two-stepped event, as opposed to the distinct two-stepped event seen in Gotland (Calner et al. 2004) and the three-stepped event seen in Poland (Porębska et al. 2004). The peak of the isotopic excursion in all three sections appears to slightly pre-date the LEE, beginning within 1 or 2 m of the top of the *lundgreni* Zone and reaching its maximum just at or slightly below the LEE.

Radiolarians appear to be strongly affected by the LEE, although our ability to determine the extent of the effect is somewhat thwarted by the paucity of late Homerian radiolarian data. We currently have no data from the first meter above the LEE from the *dubius–nassa* Zone. The lowest late Homerian sample from Abbott River, taken at 0.5 m above the LEE, provides the best insight. Although radiolarians are sparse, they are well preserved and show a dramatically different fauna as compared with that from the *lundgreni* Zone.

Sea-Level Fluctuations, C-Cycling, and Climate Change

Marine extinction events throughout Earth's history have been consistently associated with either eustatic sea level rise or fall, presumably because resulting perturbations in global climate and oceanic and atmospheric patterns contribute to ecologic stress (Hallam and Wignall 1999). Both the Ireviken event and LEE appear to be consistent with this observation, although the precise timing of eustatic change relative to the event and the proposed mechanism responsible for the event are still open to debate. Each of these events is discussed in more detail in the following sections.

Ireviken Event

A positive isotopic excursion has been recognized in association with the Ireviken event in many parts of the world aside from the Arctic, including parts of Baltica (e.g., Bickert et al. 1997; Kaljo et al. 1997; Calner and Säll 1999), New South Wales (Andrew et al. 1994; Talent et al. 1993), the United States (Saltzman 2001; Cramer and Saltzman 2005; Kleffner et al. 2005), and several other localities worldwide (Azmy et al. 1998; Heath et al. 1998), yet the driving mechanism behind the excursion and the extinction event is still under debate. Some workers link the event to a global regression (e.g., Wenzel and Joachimski 1996; Kaljo et al. 1997, 2003; Melchin 1994; Noble et al. 2005), based on sedimentologic evidence at or around the Llandovery–Wenlock boundary, including global changes in sedimentation patterns, unconformities, and Gondwanan glacial deposits (Grahn and Caputo 1992; Caputo 1998). Similarly a major regression is indicated on global and regional sea-level curves (Johnson and McKerrow 1991; Johnson 1996). In contrast, a transgression is also widely recognized at the base of the Wenlock in some regions (Barrick 1997; Baarli et al. 2003; de Freitas et al. 1999) and recently has been interpreted to coincide with the positive excursion, rather than post-date it (Cramer and Saltzman 2005). The Ireviken event appears to post-date the youngest glacial deposits in Bolivia, Peru, and Argentina, which are confined to the latest Llandovery (Diaz-Martinez 1997, 1998; Diaz-Martinez and Grahn 2007). Moreover, the youngest tillites from Brazil, which have previously been linked to the Ireviken event (e,g., Kaljo et al. 2003) and considered to range from late Llandovery to earliest Wenlock (Grahn and Caputo 1992) may well be confined to the Llandovery (Cramer and Saltzman 2005). Loydell (2007a, b), citing a higher resolution of correlation between graptolites, conodonts, and their lithofacies, indicates that the Ireviken conodont extinction occurs at a highstand and is then followed shortly thereafter by a regression associated with both the graptolite extinction and the positive excursion.

In the Arctic, a sedimentologic shift is recorded in the platform facies from stromatoporoid-coral reefs to coral-microbial reefs in the latest Llandovery, but age control in the early Sheinwoodian platform is not sufficient to identify specific unconformity surfaces that might equate to shelf exposure (de Freitas and Nowlan1998; de Freitas et al. 1999). As an added complication, the onset of Caledonian tectonism is coincident with the sedimentologic shift (de Freitas et al. 1999), thus strongly overprinting any potentially eustatic signature. A cross section from shelf to basin on east-central Ellesmere Island (de Freitas and Nowlan 1998, p. 335, Fig. 4) shows a more basinal tongue transgressing onto the shelfal strata in the early Wenlock. In a regional synthesis, the major transgressive event is included within the latest Telychian–earliest Sheinwoodian *amorphognathoides* conodont zone (de Freitas et al. 1999), but placed in the middle, not at the end of this zone, a position that would correspond to the beginning of the Ireviken event. The exact timing of this transgressive event, its potential diachroneity, and relationship to the Ireviken event thus need to be constrained better in the Arctic platform facies. The Ireviken event at Twilight Creek is recognized only isotopically and based on our resolution appears to commence in the latest Llandovery but reaches its maximum only in the earliest Wenlock. No sedimentologic change is observed at Twilight Creek, but none can be expected in basinal facies found 100 km outboard of the basin margin. The exact timing of sea-level change relative to the Ireviken event, therefore, remains elusive.

C. lundgreni Extinction Event

The LEE in the Canadian Arctic coincides closely with a subtle change in lithofacies that has been previously tied to either progradation and/or regression (Noble et al. 2005). The facies change consists of an increase in the frequency and relative proportion of bioclastic beds from the early to the late Homerian. Bioclastic beds, ranging from 6 cm to 1 m, commonly have irregular (scoured?) bases, locally display grading and/or poor sorting, and contain an abundance of coarse shelf-derived debris (Fig. 7). The bioclastic beds are intercalated within an otherwise uniformly bedded sequence of basin or basin-margin limestones and shales and are interpreted as single-event debris flows. The positioning and numbers of these flows appear to be related to the proximity of the section to the shelf break. Snowblind Creek is only a few kilometers west of the shelf rim, and most or all of the late Homerian is composed of debris flows, whereas the underlying early Homerian *lundgreni* Biozone is apparently undisturbed by debris flows, although the contact between the early and late Homerian is not completely exposed. At Abbott River, a section slightly more outboard (Fig. 1), the uppermost few meters of the *C. lundgreni* Zone (total thickness of biozone at least 50 m), and lower one-quarter of the late Homerian (total thickness about 25 m) are dominated by 6 to 20 cm thick debris flows that display graded bedding. In the Rookery Creek section (Fig. 1) debris flows are minor, and the *P. dubius–G. nassa* Zone interval is encompassed within about a 1.5 m interval of dolomitic shale (Fig. 3) populated by myriads of *P. dubius*. The most distal section—that at Twilight Creek (Fig. 1)—shows little variation, the dominant lithology being evenly bedded shale or dolomitic shale containing only flattened graptolites; the extinction boundary here is recognized only by the abrupt change in graptolite faunas.

Direct physical evidence of a shallowing event is somewhat equivocal, because there are no subaerial exposure surfaces in the basinal facies. Age control on shelfal strata for the Wenlock–Ludlow is very broad and insufficient to either date or trace any one of a number of discontinuity surfaces (de Freitas et al. 1999). In addition, increased "carbonate factory" production and subsequent progradation have also been linked to gradual rise in the sea level, as carbonate systems respond very differently to eustatic change relative to clastic systems. The interpretation of the bioclastic beds as a manifestation of a eustatic sea-level drop is best supported by observations made elsewhere. In Gotland, the top of the *lundgreni* Zone is marked by a subaerial exposure surface and local erosion of the carbonate platform, indicating a minimum sea-level drop of 16 m (Calner and Säll 1999). Similarly, there is evidence for a lowstand in many other places throughout Baltica, and in Avalonia, Bohemia, and New South Wales (Loydell 1998; Johnson 1996; Baarli et al. 2003; Kříž et al. 2003). In summary then, sedimentologic evidence for shallowing in the Cape Phillips basin is based on the increased frequency of debris flows, which are most detectable near the carbonate shelf–basin transition sections, and appear to coincide with, or begin only very slightly before, the LEE.

Discussion

Aside from the exact timing of the onset of the Ireviken event and positive $\delta^{13}C$ excursion relative to glaciation/deglaciation, questions remain concerning the specific C-cycling processes that were perturbed by the inferred environmental changes, and whether epeiric sea and ocean carbon reservoirs were equally affected. A common and widely held view is that positive excursions are the result of the increased burial of organic carbon. Such a view is compatible with times of glaciation, where an invigorated meridonal overturning circulation may have increased nutrient upwelling, thus stimulating increased primary productivity (e.g., Kaljo et al. 1998). This in turn may allow more organic carbon to be buried in marine sediments, which will increase the $\delta^{13}C_{DIC}$ value of seawater. One major problem with relying on increased organic carbon burial as a means of accomplishing the large magnitude Silurian $\delta^{13}C_{DIC}$ shifts is that burial rates must be increased measurably, especially in cases where the large ocean carbon reservoir is assumed to have shifted in $\delta^{13}C_{DIC}$ value by the same amount indicated by the excursions recorded in epeiric sea deposits. In modeling the positive excursion associated with the end of Ordovician extinction and Hirnantian glaciation, Kump et al. (1999) calculated that organic carbon burial rates would have to increase by 50–75% in order to drive the associated 6 per mil excursion. Such a dramatic increase in organic carbon burial rate should show up quite clearly in the sedimentologic record as an increase in the total organic carbon content of the sediments hosting the excursion. The converse is true, however, with respect to the Ireviken event. Carbonate strata that record the Ireviken event are organically lean, causing some authors to suggest that the carbon was sequestered in the deep marine environment (Cramer and Saltzman 2005, 2006). Likewise, there is no obvious increase in organic carbon contents of the strata recording the LEE, again relegating carbon sequestration to the deep ocean (e.g., Wenzel and Joachimski 1996) and the primary producers to far offshore environments. Twilight Creek is sufficiently distal and deep to record this predicted increase in total organic carbon (TOC) in the deep ocean basin, and yet there is no correlation between positive excursions and TOC. In some instances, there is a corresponding increase, whereas in other cases the correlation is negative, leading us to conclude that carbon burial is not the driving mechanism behind the excursions. Melchin and Holmden (2006) reached a similar conclusion for the Hirnantian carbon isotope event citing the lack of evidence for increased organic carbon content at Dobb's Linn, where the sediments are deep water and considered to have been deposited on oceanic crust.

An alternative mechanism for explaining positive isotopic shifts relates to changes in the carbonate weathering flux (Kump et al. 1999). Changes in carbonate weathering can be accomplished by the emergence of carbonate shelves during lowstand events. Melchin and Holmden (2006) support this model for the large isotopic excursion recognized in the late Ordovician of the Canadian Arctic, and this model is also favored by the authors herein for the Ireviken event and LEE (Noble et al. 2005) because of the lack of correlation between TOC and $\delta^{13}C$ values. While productivity may increase, it is not required to explain the large positive shifts. The carbonate weathering model is further supported by a difference in magnitude of the shifts recorded between distal and proximal basin margin sections. Studies of local C cycling in epeiric basins show that local influences play a significant role in governing seawater $\delta^{13}C$ values (Holmden et al. 1998; Immenhauser et al. 2002, 2003; Noble et al. 2005; Panchuk et al. 2005a, b; Melchin and Holmden 2006), where shallower regions show a greater sensitivity to sea level–forced changes in local C-cycling processes. In particular, sea-level drop causes carbonate sediments to be locally exposed or submerged, which in turn changes the local-scale isotope value of the carbon weathering flux because carbonate sediment is ^{13}C-enriched compared to other sources of carbon contributing to continentally derived DIC (atmospheric CO_2 and oxidized organic matter). Furthermore, restricted seawater circulation between epeiric seas and contemporaneous oceans, and the short decadal scale residence time of carbon in epeiric seas (Panchuk et al. 2005b), means that an epeiric sea may respond more quickly and dramatically to local-scale C-cycle forcings than the contemporaneous ocean where the impact of C-flux changes on the carbon isotope balance of the ocean is damped by the large ocean carbon reservoir.

A model that predicts extinctions to be linked with climate destabilization from one of two stable states, a Primo and a Secundo state, was proposed by Jeppsson (1990, 1998), where the Primo state is characterized by humid conditions in low latitudes, but fairly cold conditions in high latitudes, and high productivity and diversity in plankton/nekton. The Secundo state is characterized by arid conditions at low latitudes, reef expansion, and oligotrophic conditions that reflect lower plankton diversity. Bickert et al. (1997) expanded on Jeppsson's model to include isotopic response at low latitudes. Rather than evoking carbon burial as a means of changing seawater $\delta^{13}C$ values, they attribute positive excursions associated with Secundo events (i.e., arid state) to changes in shelf circulation patterns wherein the high evaporation rate of proximal shelf waters with originally low $\delta^{13}C$ values causes surface ocean waters with higher $\delta^{13}C$ values to be drawn into the proximal shelf to replace them (Munnecke et al. 2003). The shift in shelf sea circulation pattern and associated perturbation in local-scale C-cycling would be recorded as a positive shift in $\delta^{13}C$ value of shelf proximal carbonates. A flaw in this scenario that was pointed out by Melchin and Holmden (2006) is that during the Secundo state, positive excursions are only predicted for the shelf, not the offshore basin, where only the surface mixed layer of the ocean remains high in $\delta^{13}C$ value. Our data from Twilight Creek show that positive $\delta^{13}C$ excursions occur in both basin margin and more offshore basinal study sections, albeit at different magnitudes, which is inconsistent with the predicted pattern of circulation changes.

Conclusions

Isotopic and paleontologic changes are recorded in association with the Ireviken event and LEE in the Cape Phillips Formation in Arctic Canada. A large positive $\delta^{13}C$ excursion associated with the Ireviken event begins at the base of the Wenlock boundary at Twilight Creek at a position that appears slightly older than the start of this excursion in Baltica (Kaljo and Martma 2006). Radiolarians also appear to be impacted during or immediately following the Ireviken event, based on an evaluation of published accounts (MacDonald 2003, 2006) of pre- and post-Ireviken faunas. The LEE in the Arctic is marked by a single-stepped to weakly two-stepped extinction in the graptolite community at the top of the *C. lundgreni* Zone (top of the early Homerian). Radiolarians and palynomorphs are also affected by the LEE, showing both strong faunal turnover and a diversity drop. Both radiolarians and chitinozoans show an earlier response to the LEE than the graptolites, disappearing $\sim$1 m below the youngest *C. lundgreni* Zone graptolites and 0.5 to 1 m below the positive isotope excursion. The magnitude of the impact for radiolarians appears to be a significant diversity drop from 28 species in the *C. lundgreni* Zone to 6 species in the *P. dubius–G. nassa* Zone, with only two species persisting from the early to the late Homerian. This result is somewhat qualified by the scant radiolarian data from the base of the late Homerian. The extent of the impact on palynomorphs is not completely understood due to sparse occurrence and the preliminary nature of the palynomorph data. A positive $\delta^{13}C$

excursion is recorded in association with the LEE at three localities, with the magnitude being greater by approximately 1‰ in the sections more proximal to the basin margin. An increase in carbonate weathering flux is proposed as the most likely scenario for the positive excursions because no obvious correlation exists to support increased carbon burial, either in lithofacies or in TOC values.

Acknowledgments This project was funded by National Science Foundation grants EAR 9870431, 9972845, and 0107139; logistical support was provided by the Polar Continental Shelf Project. M Desilets and B Peccoraro assisted with X-ray diffraction analysis; A Soufiane provided preliminary palynomorph data for the 1998 field collections at Abbott River and Twilight Creek; M. Jones and C Stott assisted in the field; funding for ACL was provided through a Natural Sciences and Engineering Research Council research grant (Canada); and partial funding for MKZ came from Geological Society of America Grants-in-aid. We sincerely thank John Talent for inviting us to contribute to this volume and are grateful to reviewers Michael Melchin, Art Boucot, and Lennart Jeppsson for constructive criticism of our manuscript.

References

Adrain JM (2000) Regional Silurian trilobite turnover coincident with graptolite mass extinction. Geological Society of America, 2000 annual meeting, Abstracts With Programs 32:367

Adrain JM, Edgecombe GD (1997) Silurian (Wenlock) calymenid trilobites from the Cape Phillips Formation, central Canadian Arctic. J Paleontol 71:237–261

Andrew AS, Hamilton PJ, Mawson R, Talent JA, Whitford DJ (1994) Isotopic correlation tools in the mid-Palaelozoic and their relation to extinction events. Aust Pet Explor Assoc J 34:268–277

Azmy K, Veizer J, Basset MG, Copper P (1998) Oxygen and carbon isotopic composition of Silurian brachiopods: implications for coeval seawater and glaciations. Geol Soc Am Bull 110:1499–1512

Baarli BG, Johnson ME, Antoshkina A (2003) Silurian stratigraphy and palaelogeography of Baltica. In: Landing E, Johnson ME (eds) Silurian lands and seas: palaelogeography outside of Laurentia. New York State Museum Bull 493:3–34

Barrick JE (1997) Wenlock (Silurian) depositional sequences, eustatic events, and biotic change on the southern shelf of North America. In: Klapper G, Murphy MA, Talent JA (eds) Palaelozoic sequence stratigraphy, biostratigraphy, and biogeography: studies in honor of J. Granville ("Jess") Johnson. Geol Soc Amer Spec Pap 321:47–65

Bickert T, Pätzold J, Samtleben C, Munnecke A (1997) Palaeoenvironmental changes in the Silurian indicated by stable isotopes in brachiopod shells from Gotland, Sweden. Geochimica et Cosmochimica Acta 61:2717–2730

Calner M, Säll E (1999) Transgressive oolites onlapping a Silurian rocky shoreline unconformity. Geologiska Föreningens I Stockholm Förhandlinger 121:91–100

Calner M, Jeppsson L, Munnecke A (2004) The Silurian of Gotland – Part I: review of the stratigraphic framework, event stratigraphy, and stable carbon and oxygen isotope development. Erlanger geologische Abhandlungen – Sonderland 5:113–131

Caputo MV (1998) Ordovician–Silurian glaciations and global sea-level changes. In: Landing E, ME Johnson (eds) Silurian cycles: linkages of dynamic stratigraphy with atmospheric, oceanic and tectonic changes. New York State Mus Bull 491:15–25

Chatterton BDE, Edgecombe GD, Tuffnell PA (1990) Extinction and migration in Silurian trilobites and conodonts of northwestern Canada. J Geol Soc (London) 147:703–715

Coniglio M, Melchin MJ (1995) Petrography and isotope geochemistry of diagenetic carbonates in the lower Cape Phillips Formation, Cornwallis Island, Arctic Archipelago, Canada. Bull Canadian Petrol Geol 43:251–266

Corfield RM, Siveter DJ, Cartlidge JE, McKerrow WS (1992) Carbon isotope excursion near the Wenlock–Ludlow (Silurian) boundary in the Anglo–Welsh area. Geology 20:371–374

Cramer BD, Saltzman MR (2005) Sequestration of ^{12}C in the deep ocean during the early Wenlock (Silurian) positive isotope excursion. Palaeogeogr Palaeoclimatol Palaeoecol 219:333–349

Cramer BD, Saltzman MR (2006) Fluctuations in epeiric sea carbonate production during Silurian positive isotope excursions: a review of proposed palaeoceanographic models. Palaeogeogr Palaeoclimatol Palaeoecol 245:37–46

De Freitas TA, Nowlan GA (1998) A new, major Silurian reef tract and overview of regional Silurian reef development, Canadian Arctic and north Greenland. Bull Can Petrol Geol 48:327–249

De Freitas TA, Trettin HP, Dixon OA, Mallamo M (1999) Silurian system of the Canadian Arctic Archipelago. Bull Can Petrol Geol 47:136–193

Diaz-Martinez E (1997) Latest Ordovician–Early Silurian glaciation and carbonate deposition in the Bolivian Central Andes. V Reunión International del Proyecto 351 del PICG, La Coruña, Libro de Resúmenes y Excursiones, pp 51–53

Diaz-Martinez E (1998) Silurian of Peru and Bolivia: recent advances and future research. Temas Geologico–Mineros ITGE 23:69–75

Diaz-Martinez E, Grahn Y (2007) Early Silurian glaciation along the western margin of Gondwana (Peru, Bolivia, and northern Argentina): Palaeoecologic and geodynamic setting. Palaeogeogr Palaeoclimatol Palaeoecol 245: 62–81

Gelsthorpe DN (2004) Microplankton changes through the early Silurian Ireviken extinction event on Gotland, Sweden. Rev Palaeobot Palynol 130:89–203

Goodbody QH (1981) Silurian Radiolaria from the Cape Phillips formation, Canadian Arctic Archipelago. Unpublished MSc thesis, University of Alberta, Edmonton, Alberta, Canada, 388 pp

Goodbody QH (1982) Silurian Radiolaria from the Cape Phillips formation, Canadian Arctic Archipelago. Proceedings of the

Third North American Palaelontological Convention, vol 1, pp 211–216

Goodbody QH (1986) Wenlock Palaeoscenidiidae and Entactiniidae (Radiolaria) from the Cape Phillips formation of the Canadian Arctic Archipelago. Micropalaelontology 32:129–157

Grahn Y, Caputo MV (1992) Early Silurian glaciations in Brazil. Palaeogeogr Palaeoclimatol Palaeoecol 99:9–15

Hallam A, Wignall PB (1999) Mass extinctions and sea-level changes. Earth Sci Rev 48:217–250

Heath RJ, Brenchley PJ, Marshall JD (1998) Early Silurian carbon and oxygen stable-isotope stratigraphy of Estonia: implications for climate change. In: Landing E, Johnson ME (eds) Silurian cycles: linkages of dynamic stratigraphy with atmospheric, oceanic and tectonic changes. New York State Mus Bull 491:313–323

Holdsworth BK (1977) Palaeozoic Radiolaria: stratigraphic distribution in Atlantic borderlands. In: Swain FM (ed) Stratigraphic micropalaelontology of Atlantic Basin and Borderlands. Elsevier, Amsterdam, pp 167–184

Holmden C, Creaser RA, Muehlenbachs K, Leslie SA, Bergström SM (1998) Isotopic evidence for geochemical decoupling between ancient Epeiric seas and bordering oceans: implications for secular curves. Geology 26:567–570

Immenhauser A, Kenter JAM, Gansse G, Bahamonde JR, Vliet VA, Saher MH (2002) Origin and significance of isotope shifts in Pennsylvanian carbonates (Asturias, NW Spain). J Sed Res 72:82–94

Immenhauser A, Porta DG, Kenter JAM, Bahamonde JR (2003) An alternative model for positive shifts in shallow-marine carbonate $\delta^{13}C$ and $\delta^{18}O$. Sedimentology 50:1–7

Jaeger H (1991) Neue Standard-Graptolithenzonenfolge nach der "Grossen Krise" an der Wenlock/Ludlow–Grenze (Silur). Neues Jahrbuch für Geologie und Paläontologie, Abhandlungen 182:303–354

Jeppsson L (1987) Lithological and conodont distributional evidence for episodes of anomalous oceanic conditions during the Silurian. In: Aldridge RJ (ed) Palaeobiology of conodonts. Ellis Horwood, Chichester, pp 129–145

Jeppsson L (1990) An oceanic model for lithological and faunal changes tested on the Silurian record. J Geol Soc London 147:663–374

Jeppsson L (1997) The anatomy of the mid–early Silurian Ireviken event and a scenario for P–S events. In: Brett CE, Baird GC (eds) Palaelontological events: stratigraphic, ecological, and evolutionary implications. Columbia University Press, New York, pp 451–492

Jeppsson L (1998) Silurian oceanic events: summary of general characteristics. In: Landing E, Johnson ME (eds) Silurian cycles: linkages of dynamic stratigraphy with atmospheric, oceanic and tectonic changes. New York State Mus Bull 491:239–257

Johnson ME (1996) Stable cratonic sequences and a standard for Silurian eustacy. In: Witzke BJ, Ludvigson GA, Day JE (eds) Palaelozoic sequence stratigraphy: views from the North American craton. Geol Soc Am Spec Pap 306:203–311

Johnson ME, McKerrow WS (1991) Sea level and faunal changes during the latest Llandovery and earliest Ludlow (Silurian). Hist Biol 5:153–169

Kaljo D, Martma T (2006) Application of carbon isotope stratigraphy to dating the Baltic Silurian rocks. GFF Geol Soc Sweden 128:123–129

Kaljo D, Boucot AJ, Corfield RM, Leherissé A, Koren TN, Kříž, J, Männik T, Nestor V, Shaver RH, Siveter DJ, Viira V (1995) Silurian bio-events. In: Walliser OH (ed) Global events and event stratigraphy in the phanerozoic. Springer, Berlin, pp 173–224

Kaljo D, Kiipli T, Martma T (1997) Carbon Isotope event markers through the Wenlock–Pridoli sequence at Ohesaare (Estonia) and Priekule (Latvia). Palaeogeogr, Palaeoclimatol, Palaeoecol 132:211–223

Kaljo D, Kiipli T, Martma T (1998) Correlation of carbon isotope events and environmental cyclicity in the east Baltic Silurian. In: Landing E, Johnson ME (eds) Silurian cycles: linkages of dynamic stratigraphy with atmospheric, oceanic, and tectonic changes. New York State Mus Bull 491: 297–312

Kaljo D, Martma T, Mannik P, Viira V (2003) Implications of Gondwana glaciations in the Baltic late Ordovician and Silurian and a carbon isotopic test of environmental cyclicity. Bulletin de la Societé géologique de France 174:59–66

Kleffner MA, Cramer BD, Saltzman MM (2005) First documentation of early Wenlock Ireviken positive delta (super 13) C (sub CARB) excursion in the type area of the Niagaran provincial series; Irondequoit Limestone, Rochester Shale, Decew Dolomite, Gasport Dolomite, and Goat Island Dolomite are all Sheinwoodian (early Wenlock) in age. Abstr Prog Geol Soc Am 37:79

Koren TN (1991) The *lundgreni* extinction event in central Asia and its bearing on graptolite biochronology within the Homerian. Proc Estonian Acad Sci Geology 40:74–78

Koren TN, Lenz AC, Loydell DK, Melchin MJ, Štorch P, Teller L (1996) Generalized graptolite zonal sequence defining Silurian time intervals for global palaelogeographic studies. Lethaia 29:59–60

Kozłowska-Dawidziuk A, Lenz AC, Štorch P (2001) Upper Wenlock and Lower Ludlow (Silurian), post-extinction graptolites, Všeradice section, Barrandian area, Czech Republic. J Paleontol 75:147–164

Kříž J, Degardin JM, Ferretti A, Hansch W, Gutiérrez Marco JC, Paris F, Piçarra JM, Robardet M, Schönlaub H-P, Serpagli E (2003) Silurian stratigraphy and palaelogeography of Gondwanan and Perunican Europe. In: Landing E, Johnson ME (eds) Silurian lands and seas: palaelogeography outside of Laurentia. New York State Mus Bull 493:105–179

Kump LR, Arthur MA, Patzkowsky ME, Gibbs MT, Pinkus DS, Sheehan PM (1999) A weathering hypothesis for glaciation at high atmospheric pCO_2 during the late Ordovician. Palaeogeogr Palaeoclimatol Palaeoecol 152:173–187

Lenz AC (1993a) Late Wenlock–Ludlow (Silurian) graptolite extinction, evolution, and biostratigraphy: perspectives from Arctic Canada. Can J Earth Sci 30:491–498

Lenz AC (1993b) Upper Wenlock and Ludlow (Silurian) Plectograptinae (retiolitid graptolites), Cape Phillips formation, Arctic Canada. Bull Am Paleont 104:1–54

Lenz AC (1994) New upper Homerian (uppermost Wenlock, Silurian) monograptids from Arctic Canada. Can J Earth Sci 31:1779–1784

Lenz AC (1995) Upper Homerian (Wenlock, Silurian) graptolites and graptolite biostratigraphy, Arctic Archipelago, Canada. Can J Earth Sci 32:1378–1392

Lenz AC (2001) Upper Wenlock (Silurian) graptolites of Arctic Canada: pre-extinction, *lundgreni* Biozone fauna. Palaeontographica Canadiana 20:61 pp

Lenz AC (2002) Late Wenlock and early Ludlow graptolite extinction, evolution, and diversification: a reassessment. Palaeont Assoc Spec Papers 67:171–183

Lenz AC (2004) Ludlow and Pridoli (Upper Silurian) Graptolites from the Arctic Islands, Canada. NRC Research Press Ottawa, Ontario, Canada, 141 pp

Lenz AC, Kozłowska A (2006) Graptolites from the *lundgreni* Biozone (lower Homerian, Silurian), Arctic Islands, Canada: new species and supplementary material. J Paleontol 80: 616–637

Lenz AC, Kozłowska-Dawidziuk A (2001) Evolutionary developments in the Silurian Retiolitidae (Graptolites). J Czech Geol Surv 46:227–238

Lenz AC, Kozłowska-Dawidziuk A (2004) Ludlow and Pridoli (Upper Silurian) Graptolites from the Arctic Islands, Canada. NRC Research Press, Ottawa, Canada, 141 pp

Lenz AC, Melchin MJ (1991) Wenlock (Silurian) graptolites, Cape Phillips formation, Canadian Arctic islands. Trans Roy Soc Edinburgh Earth Sci 82:211–237

Lenz AC, Noble PJ, Masiak M, Poulson SR, Kozłowska A (2006) The *lundgreni* extinction event: integration of palaelontological and geochemical data from Arctic Canada. GFF Geol Soc Sweden 128:153–158

Loydell DK (1998) Early Silurian sea-level changes. Geol Mag 135:447–471

Loydell DK (2007a) Early Silurian positive $\delta^{13}C$ excursions and their relationship to glaciations, sea-level changes, and extinction events. Geol J 42:531–546

Loydell AJ (2007b) Carbon isotope stratigraphy of the upper Telychian and lower Sheinwoodian (Llandovery–Wenlock, Silurian) of the Banwy River section, Wales. Geol Mag 144:1015–1019

Loydell DK, Mannik P, Nestor P (2003) Integrated biostratigraphy of the Lower Silurian of the Aizpute-41 Core, Latvia. Geol Mag 140:205–229

Macdonald EW (1998) Llandovery Secuicollactinae and Rotasphaeridae (Radiolaria) from the Cape Phillips formation, Cornwallis Island, Arctic Canada. J Paleontol 72: 585–604

Macdonald EW (2000) Radiolaria from the lower Silurian of the Cape Phillips Formation, Arctic Canada. Program with Abstracts Ninth Meeting International Association of Radiolarian Palaelontologists, Blairsden, CA, pp 1–48

Macdonald EW (2003) Radiolaria from the lower Silurian of the Cape Phillips formation, Cornwallis Island, Nunavut, Canada. PhD Dissertation, Dalhousie University, Halifax, Nova Scotia, 370 pp

Macdonald EW (2006) A preliminary radiolarian biozonation for the lower Silurian of the Cape Phillips formation, Nunavut, Canada. Can J Earth Sci 43:205–211

Märss T, Caldwell M, Gagnier P, Goujet D, Männik P, Martma T, Wilson M (1998) Distribution of Silurian and Lower Devonian vertebrate microremains and conodonts in the Baillie–Hamilton and Cornwallis Island sections, Canadian Arctic. Proc Estonian Acad Sci Geol 47:51–76

Melchin MJ (1989) Llandovery graptolite biostratigraphy and palaelobiogeography, Cape Phillips formation, Canadian Arctic Islands. Can J Earth Sci 26:1726–1746

Melchin MJ (1994) Graptolite extinction at the Llandovery–Wenlock boundary. Lethaia 27:285–290

Melchin MJ, Holmden C (2006) Carbon isotope chemostratigraphy in Arctic Canada: sea level forcing of carbonate platform weathering and implications for Hirnantian global correlation. Palaeogeogr Palaeoclimatol Palaeoecol 234: 186–200

Melchin MJ, Koren TN, Štorch P (1998) Global diversity and survivorship patterns of Silurian graptoloids. In: Landing E, Johnson ME (eds) Silurian cycles. New York State Mus Bull 493:165–182

Melchin MJ, Heath RJ, Jowett DMS, Senior SJH, Barnes CR, Marshall JD (2000) Correlation of the graptolite and conodonts zonations and carbon isotope signal through the late Llandovery and early Wenlock, Cornwallis Island, Arctic Canada. Palaeontology Down Under 2000, Geol Soc Australia Abs 61:64–65

Mikulic DG, Kluessendorf J (1999) Stasis and extinction of Silurian (LLandovery–Wenlock) trilobite associations related to oceanic cyclicity. J Paleontol 73: 320–325

Munnecke A, Samtleben C, Bickert T (2003) The Ireviken event in the lower Silurian of Gotland, Sweden – relation to similar Palaeozoic and Proterozoic events. Palaeogeogr Palaeoclimatol Palaeoecol 195:99–124

Nestor V (1997) Reflection of Wenlock oceanic episodes and events on the chitinozoan succession of Estonia. Eesti Teaduste Akadeemia Toimetised, Geoloogia 46(3): 119–126

Nestor V, Einasto R, Loydell DK (2002) Chitinozoan biostratigraphy and lithological characteristics of the Lower and Upper Visby boundary beds in the Ireviken 3 section, northwest Gotland. Eesti Teaduste Akadeemia Toimetised, Geoloogia 51(4):215–226

Noble PJ, Lenz AC (2007) Upper Wenlock Ceratoikiscidae (Radiolaria) from the Cape Phillips Formation, Arctic Canada. J Paleontol 81:1044–1052

Noble PJ, Zimmerman MK, Holmden C, Lenz AC (2005) Early Silurian (Wenlockian) $\delta^{13}C$ profiles from the Cape Phillips formation, Arctic Canada and its relation to biotic events. Can J Earth Sci 42:1419–1430

Panchuk KM, Holmden C, Leslie S (2005a) Local controls on carbon cycling in the Midcontinent region of North America with implications for carbon isotope secular curves. J Sed Res 76:200–211

Panchuk KM, Holmden C, Kump LR (2005b) Sensitivity of the epeiric sea carbon isotope record to local-scale carbon cycle processes: tales from the Mohawkian Sea. Palaeogeogr, Palaeoclimatol, Palaeoecol 228:320–337

Porębska A, Kozlowska-Dawidzuik A, Masiak M (2004) The *lundgreni* event in the Silurian East European platform, Poland. Palaeogeogr Palaeoclimatol Palaeoecol 213: 271–294

Renz GW (1988) Silurian Radiolaria of the genus *Ceratoikiscum* from the Canadian Arctic. Micropalaelontology 34:260–267

Saltzman MR (2001) Silurian $\delta^{13}C$ stratigraphy: a view from North America. Geology 29:671–674

Samtleben C, Munnecke A, Bickert T (2000) Development of facies and C/O-isotopes in transects through the Ludlow of Gotland: evidence for global and local influences on a shallow–marine environment. Facies 43:1–38

Senior SJH (2005) Silurian graptolites of the genus *Cyrtograptus* from the Cape Phillips formation, Canadian Arctic

Archipelago, Nunavut. University of Western Ontario, London, Canada, 307 pp, unpublished PhD thesis

Štorch P (1995) Biotic crises and post-crisis recoveries recorded by Silurian planktonic graptolite faunas of the Barrandian area (Czech Republic). GeoLines (Praha) 3:59–70

Talent JA, Mawson R, Andrew AS, Hamilton PJ, Whitford DJ (1993) Middle Palaelozoic extinction events: faunal and isotopic data. Palaeogeogr Palaeoclimatol Palaeoecol 104:139–152

Thorsteinsson R (1959) Cornwallis and Little Cornwallis Islands, district of Franklin, Northwest Territories. Geol Surv Canada, Mem 294:133 pp

Trettin HP (1989) The Arctic Islands. In: Bally AW, Palmer AR (eds) The geology of North America. Decade of North American Geology A, pp 349–370

Wenzel B, Joachimski MM (1996) Carbon and oxygen isotopic composition of Silurian brachiopods (Gotland/Sweden): palaeoceanographic implications. Palaeogeogr Palaeoclimatol Palaeoecol 122: 143–166

Zimmerman MK (2001) Stable isotope and elemental geochemistry of early Silurian (Wenlockian) bio-events: Cape Phillips formation, Arctic Canada. University of Nevada Reno, Geological Sciences, unpublished MSc thesis

Late Ludfordian Correlations and the Lau Event

Lennart Jeppsson, John A. Talent, Ruth Mawson, Anita Andrew, Carlo Corradini, Andrew J. Simpson, Jane Wigforss-Lange, and Hans Peter Schönlaub

Abstract

Changes in whole conodont faunas and $\delta^{13}C$ values are combined to achieve high-resolution correlations of Upper Silurian successions in many areas (primarily Gotland, Skåne, Lithuania, Bohemia, Austria, Sardinia and Queensland); other areas are correlated with lower precision. Four of the widely recognised subzones average considerably less than 0.1 Ma and a fifth interval less than 0.01 Ma. The main constraints on resolution and precision now achievable are the amount of, and the precision in, new or previously published data from each local section; centimetric scale collecting would be worthwhile in many sections. Some stratal characters are more widespread during certain intervals than might have been expected, for example, the presence of oncolites and algal coatings in the Icriodontid Zone and the lower part of the *O. snajdri* Zone. Similarly, the *Dayia navicula* bloom in the Upper *P. siluricus* Subzone was widespread. So also were muddy-sandy sediments followed by oolite low in the *O. snajdri* Zone; crinoids flourished widely when the lower part of the Lower Icriodontid Subzone was formed. Closely spaced samples show that, in some intervals, the now well-known $\delta^{13}C$ spike was modified considerably by fluctuations; that is, it is not a smooth plateau. The best-documented fluctuations (19 analyses) are 2.5 smooth cycles with up to 2.7‰ in amplitude during ca 12,000 years or less in the base of the *O. snajdri* Zone. An enigma is the depletion of the spike in some sections, especially in the best Bohemian section.

Keywords

Silurian • Mid-Ludfordian • Gotland • Skåne • Bohemia • Austria • Sardinia • Australia • Conodonts • Lithostratigraphy • Chemostratigraphy • $\delta^{13}C$

Introduction

Major late Ludfordian changes in faunas and stable carbon isotope ratios are here used for high-resolution correlations that show lithologic changes in some areas to have been coeval. High-resolution correlations are

L. Jeppsson (✉)
Department of Geology, University of Lund,
SE 223 62 Lund, Sweden
e-mail: Lennart.Jeppsson@geol.lu.se

J.A. Talent (ed.), *Earth and Life*, International Year of Planet Earth,
DOI 10.1007/978-90-481-3428-1_21, © Springer Science+Business Media B.V. 2012

of special interest since once the chronology of the local changes are known, it is possible, for example, to more rigorously test competing models regarding the cause of the mass extinction, sedimentologic and isotopic changes during part of this interval, the Lau Event (Jeppsson 1990, 1998; Jeppsson and Aldridge 2000; Loydell 1998). High-resolution correlations of this interval on Gotland (Sweden; Jeppsson 2006) with the Broken River region of northeast Australia (Jeppsson et al. 2007) are here extended to other well-known areas.

Viewed globally, graptolites are less commonly encountered in the late Ludfordian sequences than, for example, from the late Wenlock; this is due, at least in part, to a greater abundance of shallow-marine carbonates in the Ludlow, though conodont-bearing carbonates are especially prominent, for example, in the Early Silurian of Burma, Spiti (Himalayan India) and Sardinia (see later: Additional Areas). Because six of the authors of this chapter have more experience with conodonts than graptolites, we use a detailed conodont zonation in addition to $\delta^{13}C$ changes for correlations.

Specialists on various major clades have drawn attention to mass extinctions during the Ludfordian and named them, for example, the Cardiola event, podoliensis event, bioevent 3, kozlowski event, C_3, *A. hedei*-event, pentamerid event (formally suppressed in favour of the name 'Lau Event' by Talent in Jeppsson et al. 2007). Each of these names has been used for a specific clade being studied. We prefer to use a single name and to use standard stratigraphic terminology, that is, names based on place names. We therefore refer to it as the Lau Event (Jeppsson 1993), independent of what clade or phenomenon we discuss—mass extinction is only one of the many possible effects of a global event. Events were intervals of time with several identifiable datum points. The most pronounced effects in two clades may be chronologically offset relative to one another due to different datum points. In other words, high-resolution correlations may reveal slight differences in timing, despite having been connected with the Lau Event. The 'Cardiola event' and the '*A. hedei*-event' differ, according to definition. They equate with a longer interval of time, approximately the *P. siluricus* Zone; only portions of them were connected with the Lau Event.

Some of the taxonomic names used here are in fact groups of closely related subspecies or species, for example, *Ozarkodina confluens*, *O. excavata* and species of *Panderodus*. Biological characters such as co-occurrence in collections without any intermediates have been used in older intervals to conclude that several forms should be construed as discrete species (e.g. Jeppsson 2006; unpublished data). However, because there is a large number of available names, selection of species should be undertaken with care following attentive reexamination of type material, coupled with application of the rules of priority. And this should be undertaken before new names are introduced. The isotope-analytical methods used for this investigation are the same as we have used previously (Jeppsson et al. 2007).

Each of us is a specialist on different regions and/or methods. CC, AS and LJ collected and analysed their local areas for conodonts and drafted that part of the text. Most of the collections referred to have been compared with those from Gotland. AA measured all the new isotope data; CC collected all Sardinian samples, reidentified the conodonts there as well as *Kockelella* species from Cellon; HPS collected conodonts from Mušlovka; AS identified those from Queensland and JAT and RM collected from most other areas and thoroughly revised the manuscript including the linguistics. JW-L collected most of the samples used for isotope analysis from Skåne, the rest were from reference pieces from LJ's samples. LJ identified some of the conodonts from Cellon and the Czech Republic and, with the other authors, edited the local descriptions and correlations and wrote much of the other text—for further details see the various areas mentioned later.

Zones and Other Intervals Used as Proxies for Time Units

Biozones are chiefly of interest as proxies for units of time (though by modern definition, zones are units of rocks). The recognised local base of a zone will depend on the distance between samples and the size of the critical collections, if formal definitions are applied (e.g., between the stratigraphically lowest local finds of successive index taxa). Closer-spaced samples and/or larger collections may change the position of the recognised zonal base significantly. Hence, the position of the nearest reliable absence must be taken into account in order to calculate uncertainties with respect to the recognised base; see also Marshall (e.g. 1990) regarding confidence intervals.

Though Walliser's (1964) conodont zonation has been widely used, 42 years of accumulated knowledge of ranges and distribution of stratigraphically important taxa has necessitated refinement of zonal boundaries (Jeppsson 2006) in order to remove salient ambiguities. New boundary definitions and new subzones were based on large collections from the relatively expanded sequence on Gotland, Sweden (see that section, mentioned later, as well as Jeppsson 2006 and references therein). Many of the conodont species are rare or decrease in frequency up-sequence. For instance, the frequency of *P. siluricus* in the Upper *P. siluricus* Subzone on Gotland is below 0.01 %; hence range ends based on smaller collections elsewhere are likely to be truncated. Similarly, first/last finds of *Pa. latialata* cannot be used to identify zonal boundaries in the absence of supporting data from other taxa. Even collections with a few thousands of conodonts each may not be large enough; most or all published collections from a particular interval have relatively few specimens. Moreover, the first appearance of *O. snajdri* is uncertain; the base of that zone must be defined on other changes. In order to overcome such problems we have used all kinds of data available to identify as precisely as possible where a boundary would likely be found. For example, the zonal position has been deduced from the whole fauna, not from the more or less rare finds of a single taxon. Re-examination of published collections shows that most of these units can be discriminated, in some cases resulting in substantial changes. The following numbers are also used in our reviews of local successions to facilitate recognition of coeval units (Fig. 1).

1. *The main part of the Po. siluricus Zone* – This has a very diverse fauna; the yield/kg is high in most studied areas.
2. *The Upper Po. siluricus Subzone* – Stepwise extinctions and disappearances of rarer taxa decreased the diversity. Thickness <10% of the zone. The uppermost part is unit 3 (immediately below).
3. *The O. excavata faunal interval* – *Ozarkodina excavata* has a markedly higher relative frequency than in 1 and 2. Thickness <1% of the *Po. siluricus* Zone. *The 3/4 discontinuity.*
4. *The Lower Icriodontid Subzone* – Collections are dominated by a few surviving taxa.
5. *The Middle Icriodontid Subzone* – At the base, the slender form of *Panderodus equicostatus* reappeared.
6. *The Upper Icriodontid Subzone* – The faunas were strikingly unbalanced, with a single species sometimes forming >90% of the collections. Dominance of a single species, in many areas *Pa. equicostatus*, makes for ready recognition. This is possible even in small collections if frequencies or numbers of specimens are given, species of *Panderodus* are identified and collections are closely spaced—for another kind of fauna see Skåne, mentioned later.
7. *The O. snajdri Zone* – Since *O. snajdri* has been reported from older strata as well, the base is defined by the abrupt change to a more balanced fauna; the change, even in relatively large faunas, may appear as an increase in the number of species.

Studied Areas

Gotland, Sweden

The interval of interest here has been revised recently (Jeppsson 2006) and new $\delta^{13}C$ data published (Jeppsson et al. 2007). Facies change from shallow water limestone in the northeast to distal marls in the southwest. About 200 collections from the interval analysed herein include about 100,000 teeth.

1. *Main part of the P. siluricus Zone* – The När Fm consists of at least 57.8 m of interbedded marls and argillaceous limestones. Rare graptolites occur, chiefly on western Gotland (Hede 1942). Crinoidal limestones, representing only the lower part of the formation, cap some of the hills in the far NE.

At least 23 conodont species have been found on Gotland (in c. 100 collections aggregating several 10,000 s of specimens; for localities see Jeppsson 2006). The list starts with the more frequent taxa; facies preferences are marked with (m) and (l) for taxa with pronouncedly higher frequency in the marl and limestone areas, respectively; (e) indicates taxa found in the older part of this interval but not in the younger: *Oz. confluens*, *Pa. serratus*, a new ssp. of *Pa. greenlandensis* (l) *Oz. excavata*, *Coryssognathus dubius*, *Polygnathoides siluricus*, *Oulodus* cf. *excavatus* (l), *Pa. panderi* (m), *Silurognathus maximus* (Jeppsson 2006) (e), *Oul. s. siluricus* (e), *K. ortus* (e), *Pa. gracilis* (frequent only in the most distal marls), *Decoriconus*,

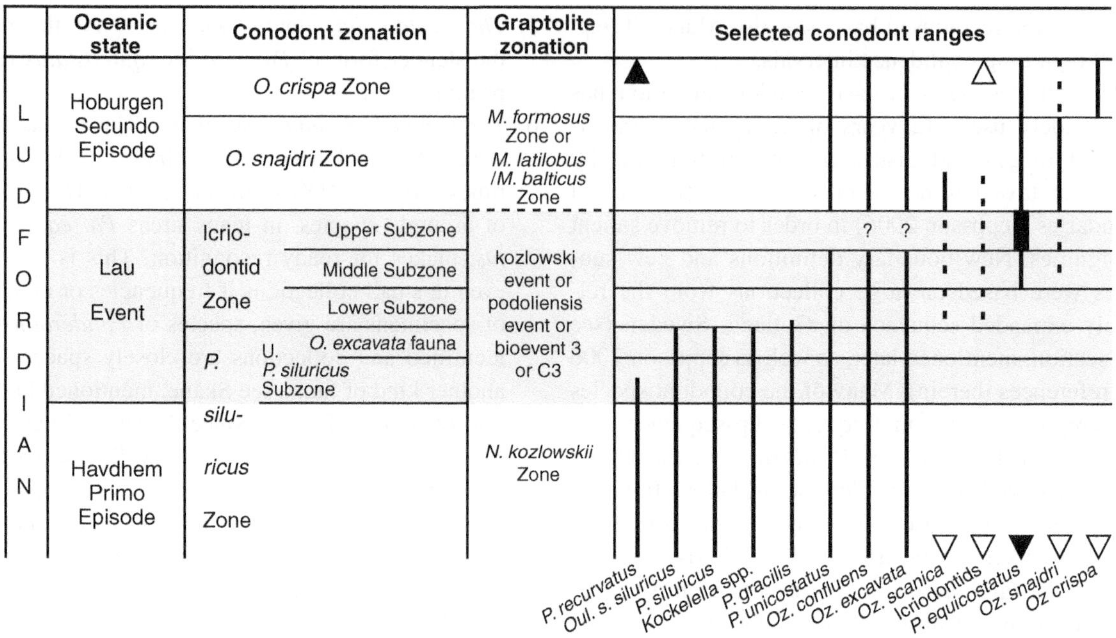

Fig. 1 Conodont zonation and selected conodont ranges. *Dashed line* marks more local ranges (or extremely low frequency in some areas); *arrows* indicate ranges outside of the discussed interval; *open arrow* records of a similar, perhaps ancestral/descendent taxon. A *thick line* indicates the species is strongly dominant in that interval. Correlation with the graptolite zonation is also indicated (*dashed line* indicates data with low correlation precision)

Ps. beckmanni, *Belodella* sp., *B. mira*, *Aldridgeodus minimus* (Jeppsson in Calner 2008). The δ^{13}C values are mostly between 0.00 and 0.80 (to 1.68) ‰, increasing slightly upwards.

2. *Upper P. siluricus Subzone* – The Botvide Mbr is the distinct upper part (2.15 m at the type locality) of the När Fm. Both fauna and lithology change markedly through the member. Its upper, more weathering-resistant part is rich in coquinas of the brachiopod *Dayia*, and markedly dolomitic (Munthe 1902). In the most distal outcrop, Bodudd 1, argillaceous mudstones with scattered thin limestone interbeds include at least one *Dayia* coquina and a distinct 26 mm bed of pale argillaceous dolomite.

Extinctions and disappearances through this interval reduced the previous diverse fauna markedly (details in Jeppsson 2006). *Pa. gracilis* spread to the proximal marls, *Pa. unicostatus* and *Oulodus s. siluricus* returned, the latter only to become extinct. Similarly, a new subspecies of *Kockellella* returned and became extinct, seemingly leaving ecospace for *K.* aff. *sardoa*. *Po. siluricus* was probably present only intermittently, extremely briefly. Through this subzone at Botvide 1, δ^{13}C values increase from 1.5 to 2.4‰ VPDB (c. 0.7‰/m).

3. *O. excavata faunal interval* – *Shaleria* aff. *ornatella* covers bedding planes of the uppermost decimetre(s) of the Botvide Mbr (Munthe 1902); at Nyan 2, the easternmost locality, these 0.21 m are thin-bedded, dolomitic and hard.

Through most of the *P siluricus* Zone, *Oz. c. confluens* is usually more abundant than *Oz. excavata*. However, conodonts are rarer, small-teeth dominant, and *Oz. excavata* is more frequent than *Oz. confluens* in most samples from this interval. Extinctions continued; both *Polygnathoides* and *Kockelella* became extinct.

At Nyan 2, δ^{13}C increased abruptly from c. 2–2.5 to c. 3.5‰ VPDB, that is, slightly more than 1‰ at a markedly rusty surface c. 0.12 m above the faunal change. The increase is c. 25–27 cm below the top of the När Fm at Bodudd 1. Hence, erosion (see the following section) in most localities farther east is regarded as having been slight.

The 3/4 discontinuity – The basal contact of the Eke Fm on eastern Gotland has been described as a submarine discontinuity-level with phosphorite and glauconite coatings (Spjeldnaes 1950) overlain by rounded pebbles of Hemse Marl (see next paragraph for more details).

4. *Lower Icriodontid Subzone* – In more proximal sections, the basal metres of the Eke Fm consist of oncoidal crinoidal limestone, passing upwards into oncoidal marls. Palaeokarst has affected the lower Eke Fm (Cherns 1982). An as yet unlocated discontinuity therefore exists higher up, probably only a few metres higher (Eriksson and Calner 2008). Bodudd 1, the most distal locality, differs by lacking an erosion surface, basal conglomerate and crinoidal limestone; that is, the succession seems to be complete. Instead, the lower Eke consists of 2.55 m of mostly rubbly weathering mudstones.

A few hardy survivors characterise this low-diversity conodont fauna: *Oz. confluens*, *Oz. excavata* and *Pa. unicostatus* with rarer *Ps. beckmanni* and *Decoriconus*. Near the base, *C. dubius*, *Pa. gracilis* and *Pa. serratus* have been found, but a redeposited origin (e.g., in a pebble of conglomerate origin) cannot be excluded. *Pa. serratus* returned later. $\delta^{13}C$ rose through this interval from c. 3.5 to c. 6.55‰.

5. *Middle Icriodontid Subzone* – Initially, oncoidal marls spread to Bodudd 1, and *Pa. equicostatus* reappeared (absent since the very earliest Ludfordian). Other taxa include *Oz. confluens*, *Oz. excavata*, *Pa. serratus*, *Ps. beckmanni*, *Decoriconus* and rare *Belodella* sp. At Bodudd 1, *Pa. equicostatus* reappeared at +0.65/0.70 m where ^{13}C is 7.05‰, but 5.57‰ at +1.16/1.20 m.

6. *Upper Icriodontid Subzone* – The upper Eke Fm consists of calcareous limestone and muddy limestone rich in thin, large shells in many outcrops. *Pa. equicostatus* is strikingly dominant (typically >90%). *Decoriconus*, *Ps. beckmanni* and *Oulodus* sp. are regular and frequent. *Oz.* cf. *scanica*, *Dapsilodus obliquicostatus* and *Icriodus* sp. are rare and/or intermittent. $\delta^{13}C$ ranges from 7.66 to 9.17‰ with $\delta^{13}C$ apparently increasing through most of this interval.

7. *O. snajdri Zone* – The Burgsvik Fm consists of 47 m of sandstone, with intercalated sandy claystone; in the top (to 7.18 m) intercalated with or overlain by oolite (Hede 1919, 1921; Jeppsson 2006). Bivalves are a salient component of the fauna (Hede 1921, p. 73; Liljedahl 1994). In the shallowest areas, only some ten metres of oolite represent the Burgsvik (Hede 1925, pp. 30–33). The lower part of the Hamra Fm consists of oncolitic marls and limestones, at least part of which belongs in this zone.

The first distinct *O. snajdri* Zone fauna includes *Oz. snajdri*, *Oz. wimani*, *Oz. confluens*, *Oul. novoexcavatus*, *Pa. equicostatus*, *Ps. beckmanni*, *Decoriconus*, *Oz. scanica*, *Icriodus* sp. and *Dapsilodus obliquicostatus*. The former five are all well represented. *Oz. excavata* did not return until in the upper (?uppermost) Burgsvik Fm and *Pa. unicostatus* in the Hamra Fm. New taxa appear higher up, for example, *Oz. remscheidensis* and, in the *O. crispa* Zone, *Oul. elegans* and *Ctenognathodus confluens*.

In the Burgsvik Fm $\delta^{13}C$ values from 5.28 to 10.54‰ have been reported. A few analyses indicate a drop in the average upwards through this zone and beyond: 4.62‰ to c. 7.60‰ in the lower Hamra, 4.06‰ 5.87‰ in the upper Hamra and as low as 0.52‰ close to the end of the Ludlow.

Skåne, Southernmost Sweden

There are two areas with exposures in Skåne, c. 340 km southwest of Gotland. Despite their proximity to Gotland, high-resolution correlations have not been achieved previously. The following data regarding localities and lithostratigraphy are from Jeppsson and Laufeld (1987 [1986]), the conodonts and biostratigraphy mainly from Jeppsson (1975 [1974]). Subsequent collecting has probably tripled the number of teeth and allowed better comparison with the richer faunas from Gotland. Improved identifications of rare species have enabled using the new standard zonation in preference to the local faunal sequence previously used. The base of the 'younger *Oz. scanica* fauna' has previously been identified in both the uppermost part of Klinta 2 and in Klinta 4 (Jeppsson 1975 [1974]) indicating an overlap, but there are too few conodonts in the two collections from Klinta 2 for reliable conclusions.

JW-L re-excavated the Klinta sections of Jeppsson and Laufeld (1987 [1986]) and described the lithologies (Wigforss-Lange 1999, 2007). She selected reference levels (Jeppsson and Laufeld had measured from the bottom of each trench). Reference levels tied to the stratigraphy are better; we therefore use her reference levels here. Based on sedimentologic characteristics, she suggested a correlation with considerable overlap. As an overlap was expected, appropriate distances were chosen so that this could be carefully probed.

Isotope data herein provide additional evidence (for measured levels and isotope data, see Fig. 2). From the following section, $\delta^{13}C$ data suggest that Klinta 5 and 3 overlap as shown (Fig. 2), but that the intervals sampled for isotope analysis in Klinta 3, 2, 4 and 1 do not overlap and that Klinta 4 does not overlap with Bjärsjölagård 2. The distances between each of the sections are similar (c. 30 m, except 4 and 1; 20 m). Hence, it is likely that the unsampled gaps are minor; perhaps no more than what is found in Fig. 2. Thus, the isotope data indicate that most sections do not overlap and that there must be some kind of disturbance between some of the sections (JW-L prefers her previous correlation). The correlation in Fig. 2 results in repetition of, for example, the oolitic strata—if this indicates cyclic sedimentation, we may ask if it were due to Milankovitch cyclicity.

In the quarry Bjärsjölagård 2, Fredrik Jerre and two students sampled the section accessible behind the right side of the water (locality Bjärsjölagård 2b, herein, Fig. 3) for scolecodonts in 1987. Here we are using conodont data exclusively. The upper part of their measured section, in the inner corner, has partly collapsed. Their reference level is the base of the 0.16 m limestone-bench forming the floor of the most deeply eroded part of the section, 1.69 m above the water table. In 2000, LJ and Peep Männik collected supplementary microfossil samples below that level as well as 60 isotope samples from there and from the 3.16 m of strata above it, some metres to the right.

In the Klinta area, the Klinta Fm includes the Lunnarna, Bjär and Bjärsjö members. Except for the Bjärsjölagård Mbr in the Bjärsjölagård-Öved area, the sequence is markedly less calcareous than that on Gotland.

1. *Po. siluricus Zone* – The Lunnarna Mbr (Klinta Fm) is at least 65 m thick (Wigforss-Lange 1999). Most of its $\delta^{13}C$ data are between 0.33 and 0.56‰. At Lunnarna 1, *Po. siluricus* is found together with, in order of frequency, *Oz. excavata, Oz. confluens, C. dubius, Panderodus serratus, Oulodus* sp., *Ozarkodina* sp. and *Ps. beckmanni*. Most of it, or perhaps the entire member, correlates with the När Fm on Gotland.

In the Bjär Mbr at Klinta 6, $\delta^{13}C$ is 1.53‰ in Kl69-7LJ (Wigforss-Lange 1999, Table 1; JW-L herein: misprinted as KL67-9). We do not have enough data to tell if this short section belongs here or in the next interval or spans their boundary.

2. *Upper Po. siluricus Subzone* – There is a very argillaceous dolomite at c. –3.45 m in Klinta 5. On Gotland at least the upper part of this subzone is dolomitic (Munthe 1902: 272–3); see preceding section regarding a dolomite at Bodudd 1. Considered alone, this similarity, or any of those mentioned later, is not enough for a reliable correlation, but our correlation is strengthened by their occurrence in the same order. It places the base of the Icriodontid Zone c. 4.7 metres above the dolomite; thus, strata coeval with the Botvide Member are more than twice as thick as at Botvide. This agrees well with the lower rate of sedimentation at Botvide 1 and Nyan 2, indicated by the several rust layers in those sections. Collection Kl67-40LJ&SL includes one fragment of *Ctenognathodus*. The fauna includes *Oz. excavata, Oz. confluens, C. dubius* and *Panderodus serratus*.

$\delta^{13}C$ increases from c. 1.6‰ at the base of Klinta 5 to ca 4‰ at the top of the subzone. At first the rise is stable but, in the upper c. 2 m, recorded values fluctuate up to more than 2‰ around an increasing trend. For example, it increases from 2.16 to 5.00‰ and back to 3.04‰ through 0.55 m (4 samples) and drops from 4.00 to 2.77‰ and rises back to 4.87‰ just above the top of this interval.

3. *O. excavata faunal interval* – *O. confluens* is more frequent than *O. excavata* in all samples from Klinta 5

Fig. 2 Conodont zones and isotope data from Klinta, Skåne, southernmost Sweden. Sampling for conodonts and for isotope studies was undertaken at various times. Some of the saved parts of the 1967 and 1969 samples were also isotopically analysed. These serve as tests of possible errors in correlation between data from sampling at different times—these proved to be very small. The strata are very uneven in thickness over the 0.1–0.5 m (laterally) scale; hence, correlations between the sections excavated by LJ and SL (Jeppsson and Laufeld 1986—data in *grey* with a *ring* as symbol) and those of JW-L probably change between slightly expanded and slightly condensed over a 0.1 m scale. Minor differences in the measured carbon isotope data may be due to this but some may be genuine, reflecting that denser sampling may have indicated additional minor oscillations in the curve. The only exception is that samples from Klinta 1 form two curves; it was not possible to relocate this locality; isotope data indicate that the 1967 samples represent a slightly younger interval than that of JW-L. Klinta 5 and Klinta 3 overlap; to facilitate discriminating these data, Klinta 5 text is in italics and the metre-lines are dashed. The metric level (measured by JW-L) of each isotope sample is given between the two curves; corresponding isotope values are at the same vertical level—provided there was space to do so

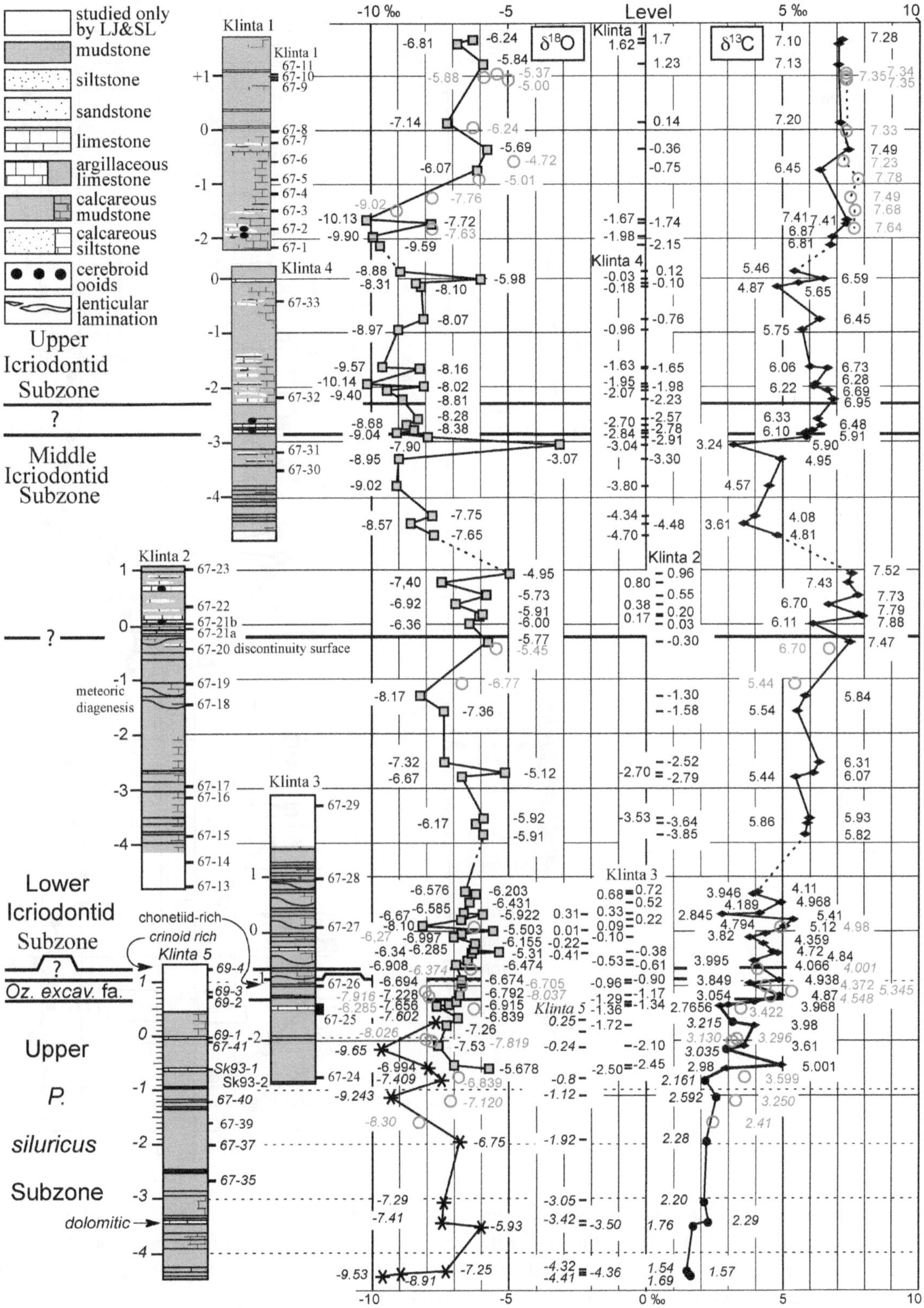
studied only by LJ&SL
mudstone
siltstone
sandstone
limestone
argillaceous limestone
calcareous mudstone
calcareous siltstone
cerebroid ooids
lenticular lamination
Upper Icriodontid Subzone
?
Middle Icriodontid Subzone
Lower Icriodontid Subzone
?
Oz. excav. fa.
Upper P. siluricus Subzone
dolomitic
meteoric diagenesis
chonetiid-rich
crinoid rich
Klinta 1
Klinta 4
Klinta 2
Klinta 3
Klinta 5
67-20 discontinuity surface
-10 ‰
-5
Level
5 ‰
10
δ18O
δ13C
0 ‰

Fig. 2

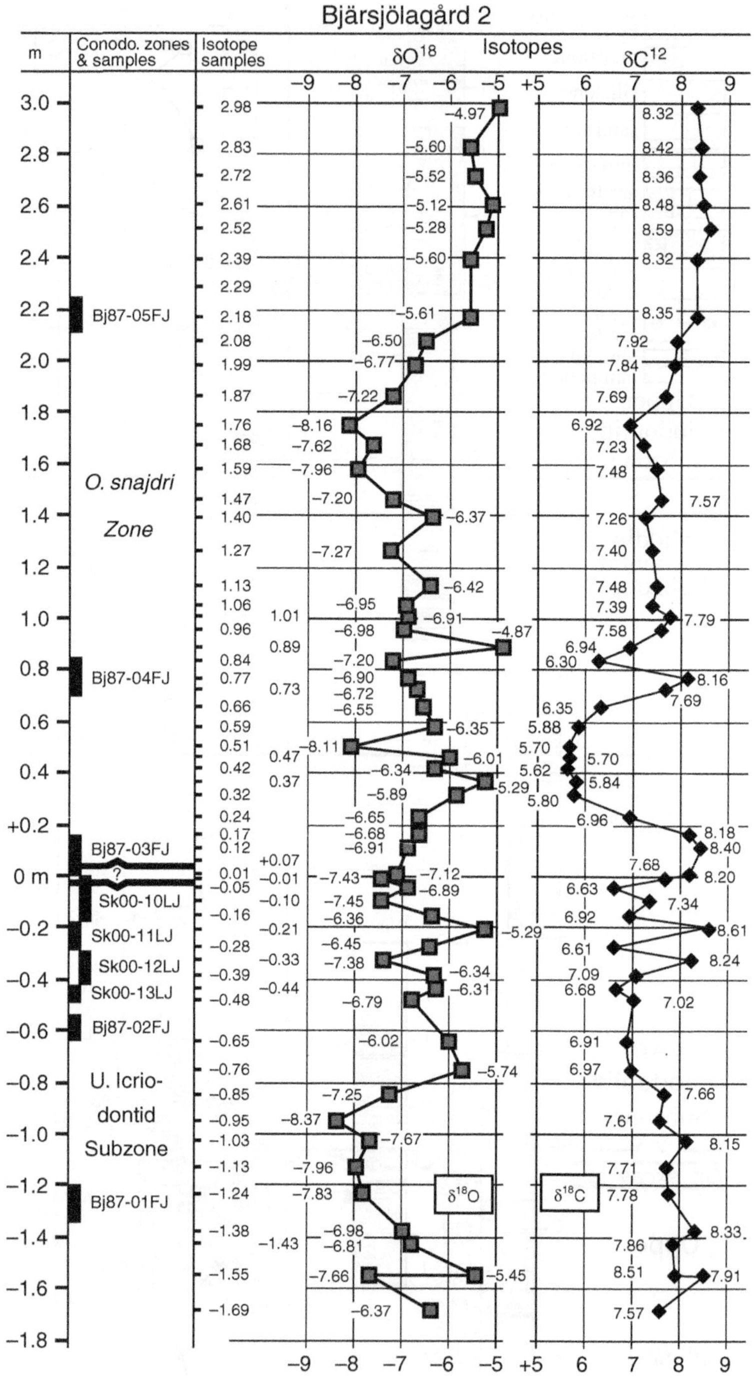

Fig. 3 Conodont zonation and isotope data from Bjärsjölagård 2 in Skåne, southernmost Sweden. Samples +0.07 and 2.29 could not be analysed

except in samples Kl69-2LJ and Kl69-3LJ (Jeppsson 1975 [1974]), from ca 4.1 m above the dolomite. This interval has thus the same conodont faunal characteristics as in the COG (Coral Gardens section, Broken River, Queensland; Jeppsson et al. 2007) and Gotland sections. Sample Kl67-26LJ&SL from Klinta 3, according to $\delta^{13}C$ data, correlates with this interval (Fig. 2) and is rich in chonetid brachiopods (Jeppsson and Laufeld 1987 [1986])—another similarity with Gotland.

4–6. *Icriodontid Zone* – In Bjärsjölagårdsborrningen 2, $\delta^{13}C$ rises from 0.38‰ at 68.15 m to 6.01‰ in the next sample (at 58.40 m) (Wigforss-Lange 1999); hence, the base of this zone intervenes. The Icriodontid Zone is somewhere between 25 and 35 m in thickness in that core, since the base of the *O. snajdri* Zone most probably correlates to c. 33 m.

At c. 1.5 m below the tentative top of the Lower Icriodontid Zone, $\delta^{13}C$ increases c. 2‰; the following c. 1.5 m are more calcareous, initially oolitic, followed by a decrease in the curve. Two more such intervals are found higher up; these subcycles form a major part of the deviations from a smoothly increasing $\delta^{13}C$ curve.

4. *Lower Icriodontid Subzone* – At Klinta 5, *O. scanica* appeared in Kl69-4 from a crinoid-rich lens or layer. Crinoids are found commonly at the base of this subzone. The fauna also includes *O. confluens*, *O. scanica* and *D. dubius*. The increasing $\delta^{13}C$ trend continued through this subzone but with weaker fluctuations, reaching 7.47‰ near the top. Some fluctuations in the lower part are well supported, and so is the rise just below the top; we consider these changes genuine.

5. *Middle Icriodontid Subzone* – In addition to the taxa in the subjacent subzone, in the upper part of Klinta 2, *Pa.* cf. *equicostatus* and *Ps. beckmanni* occur in Kl67-21aLJ&SL, and *Pa. equicostatus* in Kl67-23LJ&SL. Rare oolites appeared at the assumed base of this subzone. Low in it, $\delta^{13}C$ reached 7.88‰ (three samples above 7.7‰) but then fell to around 4‰ and remained there to the presumed top.

6. *Upper Icriodontid Subzone* – This subzone (earlier the 'younger *Oz. scanica* fauna') yields only *Oz. scanica* and rare *Panderodus equicostatus*. Klinta 4 is the best section for studying its lower boundary. Further collecting, including recollecting Kl 67-32 LJ&LS (Sk93-4LJ, 18.0 kg; ca 1200 conodont teeth picked hitherto) and Kl67-33LJ&LS (Sk93-3LJ, 37.2 kg, c. 4 000 teeth, hitherto) showed that the observed low diversity was not an artefact of too small collections during the 1970s. The same species are found on Gotland in the upper Eke Fm and, further, a single species is strongly dominant, although not the same species. From a minimum of 3.24‰ just below the base, $\delta^{13}C$ rose through the following 0.8 m, first rapidly and then gradually to 6.95‰. Higher up in Klinta 4, $\delta^{13}C$ reached 7.49‰.

In the core Bjärsjölagårdsborrningen 2, c. 175 m S of the quarry Bjärsjölagård 2. $\delta^{13}C$ is 8.32‰ in the basal part of the Bjärsjölagård Member at 52.05 m. The $\delta^{13}C$ excursion culminated with 8.91‰ in an argillaceous mudstone from 43.75 m, and 9.16‰ in a limestone/oncoid sample from 42.08 m (Wigforss-Lange 1999). The former lithology is closest to the whole-rock samples analysed for our study; hence that ratio should be comparable with our data.

Slightly younger strata at Bjärsjölagård 2b include the topmost 1.55 m of this subzone. Sample Bj87-01FJ from the limestone at –1.55/–0.93 m and in the mudstone at –0.18/0.00 m (Sk00-10LJ) yielded the typical fauna. Through these 1.55 m, the trend in $\delta^{13}C$ decreases from c. 8.2‰ at –1.55 m to an average of 6.96‰ in the 3 samples from –16/15, –10/9 and –5/4 cm. Both the low diversity conodont fauna and $\delta^{13}C$ data fit well with the upper Eke Fm on Gotland.

7. *O. snajdri Zone* – Bjärsjölagård 2b is the only available section where both the base of the *O. snajdri* Zone is known with a precision of a few centimetres and the stable carbon isotope changes are known from more than one sample/decimetre. Directly above the reference level, at 0.00/+0.16 m (Bj87-03FJ), a diverse fauna (previously the '*Oz. wimani* fauna') included, as on Gotland, *Oz. scanica, Oz wimani, Oz. confluens, Oul. novoexcavatus* and *Panderodus* sp. Oncolitic limestone returned at 2.11/2.15 m (Bj87-05FJ) and continued to the top of this section.

Two and a half smooth $\delta^{13}C$ cycles with amplitudes of up to 2.78‰ occur in the basal metre of this zone (19 analyses; see Fig. 3). Considering the frequency of each species in the samples adjacent to the reference level, the number of specimens extracted, thickness of the two samples, and so on, the zonal boundary is somewhere between –2 and +5 cm from the reference level. Thus, the start of these cycles may not have preceded (by a couple of centimetres) but coincided with the zonal boundary. The minimum at the top of the Icriodontid Zone was followed by the first rise (1.77‰) in 0.15 m or less of strata. It started somewhere between –4 cm (with 6.63‰) and –1 cm. It rose to 8.40‰ at +11/12 cm. The next 0.65 m recorded a full cycle: The first drop continued down to 5.62‰ at 0.41/0.42 m; the second rise culminated with 8.16‰ at 0.76/0.77 m. The second drop reached 6.30‰ in the next sample, from 0.83/0.84 m. The third smooth rise continued to 7.79‰ at 0.99/1.02 m. The succeeding 1.98 m recorded fluctuations of decreasing amplitude around a rising trend of c. 0.28‰/m to c. 8.36‰.

Three samples from a section to the right of the water (W.-L.; Bjärsjölagård 2r herein) yielded, from below, 11.21‰ in a stromatolite, 6.63‰ in a brachiopod and 9.65‰ in a crinoid ossicle (Wigforss-Lange

1999). Our whole rock analyses agree well with the brachiopod value – generally considered to best reflect isotopic changes in seawater.

Above the Bjärsjölagård Lst Mbr, in the core Bjärsjölagård BH 2, Wigforss-Lange (1999) recorded a drop. It probably started only a metre or a few metres above the strata exposed at Bjärsjölagård 2b. It continued through the succeeding 20 m to –0.61‰ at 11.47 m.

The Broken River Region, Queensland, NE Australia

During the late Silurian, the Broken River region formed part of the East Gondwana margin. The COG section through the Coral Gardens Formation commences at the base of the formation (dip 77°), continuing for 84 m, reaching into the basal Jack Hills Limestone (dip c. 85°). Seventy-four samples for conodonts and 229 for isotope analysis resulted in high-resolution correlations with Gotland (cf. Jeppsson et al. 2007 for discussion of data and their sources). Many age-diagnostic late Silurian conodont taxa occur in low frequencies.

1. *Main part of the P. siluricus Zone* – Members 1 and 2 of the Coral Gardens Fm consist of thin, interbedded, nodular marly limestones, siltstones and fine-grained micaceous arenites with abundant in situ coral heads and stromatoporoids, brachiopods, medium- and high-spired gastropods, bryozoans, ostracods and disarticulated crinoid remain. Faunas throughout these members consist mainly of *Oz. e. excavata* and *Panderodus* ssp. Coniform taxa predominate from COG 16 through to COG 43.1 but, particularly above COG 23.2 m, *Oulodus s. siluricus*, *Coryssognathus dubius*, *Pa. panderi* and *Pa. serratus* occur in small numbers. Presumably, equivalent strata have yielded rare teeth of *Kockelella ortus*, *K. variabilis*, *Oulodus* sp., *Polygnathoides siluricus*, *K. maenniki* and *Ancoradella ploeckensis*. δ^{13}C increases c. +0.03‰/m, reaching 1.65‰ at 43.4 m.

2. *Upper Po. siluricus Subzone* – Lower Member 3, COG 43.5–44.7 m, consists of thinly bedded (c. 5 to 10 cm) dark bioclastic flaggy micrites outcropping more prominently than members 1 and 2 and markedly more calcareous. Panderodontids dominate the conodont faunas, which include *Pa. serratus*, *Pa panderi*, *Ozarkodina e. excavata*, *Oulodus s. siluricus*, *Polygnathoides siluricus*, *Decoriconus fragilis*, *Belodella resima* and *B. anomalis*. The last record of *Po. siluricus* is from COG 44.70 m, that is, just below the isotope excursion (see the following section); this is c. 0.1–0.2 m higher than on Gotland. Such range discrepancies are probably due to competitive exclusion with *Kockelella* (see Jeppsson 2006).

3. *O. excavata faunal interval* – It started somewhere between COG 44.70 and 44.74 m. In 44.74–44.8 and 44.8–44.85, *Oz. excavata* has a frequency of c.96 %; its teeth are smaller than those from underlying strata and have more closely packed denticulation. *Panderodus* sp. is present, but *Oz. confluens* has not been found. δ^{13}C shifts from 2.06 to 3.7‰ between 44.70 and 44.74/44.80 m, 16–20 cm below the top of this interval at 44.90 m.

4. *Lower Icriodontid Subzone* – A part of middle Member 3, COG 44.9–47.3 m, differs from older strata in being purer, oncolitic, containing crinoid remains and having a markedly poorer conodont yield. The lowest sample yielded a population of more normal *Oz. excavata*. The *Oz. excavata*/panderdontid ratio decreased from 4:1 to well below 1:1. ?*Oz. confluens* and *Pseudooneotodus beckmanni* appeared. The rise in δ^{13}C continued through this interval from 3.93 to 6.02‰.

5. *Middle Icriodontid Subzone* – Part of Mbr 3, to COG 50.6 m, is oncolitic like the strata below. *Pa. equicostatus* appeared in COG 47.3–47.4 m, occurring with *Ozarkodina excavata*. A similar conodont fauna was obtained from COG 50.15–50.2, but the *Panderodus* teeth are not identifiable at species level. The ratios of *Panderodus* to *Ozarkodina* teeth are closely similar in both samples, but the numbers obtained to date are very small. Another sample reported by Simpson (2000) yielded a single fragmentary Pa element of ?*Pedavis latialata*. This was from earlier reconaissance sampling at the Broken River Crossing (JS Jell, University of Queensland); the precise stratigraphic position can not be determined with any precision. It was obtained with a fauna dominated by *Panderodus* (*P. equicostatus* and *P. unicostatus*) with subordinate elements of *Ozarkodina excavata*. The rise in δ^{13}C continued through this interval, reaching 6.88‰ at its top.

6. *Upper Icriodontid Subzone* – Uppermost Mbr 3, COG 50.6–51.4 m, consists of oncoidal shales. Member 4, 51.4–68.15 m, is characterised by micaceous siltstone to fine sandstone with trilobites,

brachiopods and bivalves, for example, at 52.05 m. The zonal boundary may be at either boundary of the latter interval or within it (Jeppsson et al. 2007). No conodonts have been found hitherto in this COG interval, but other sections have produced impoverished panderodontid faunas. $\delta^{13}C$ varies between 5.19 and 7.03‰.

7. *O. snajdri Zone* – Member 4 consists of sandy limestone; Mbr 5 (68.15–81.2 m) is partly oolitic with some partings. The lowermost Jack Hills Limestone, to 84.1 m, consists of well-bedded calcarenite and succeeding parts of massive, poorly bedded, dark blue-grey, gorge-forming limestones. At the Broken River Crossing, poor faunas include *Oz. excavata, Pa. equicostatus* and ?*Decoriconus fragilis*. At 68.15–68.25 m, $\delta^{13}C$ jumped to 8.75‰ and then remained above 8‰ to the top, reaching 8.98‰ at 73.35 m. The base of this zone is probably somewhere near 55 m (Jeppsson et al. 2007). The lowermost part of that zone in Skåne is characterised by large $\delta^{13}C$ fluctuations; COG $\delta^{13}C$ data between 55.3 and 60 m are too scattered (more than 4‰) to form a curve (Jeppsson et al. 2007); some of this scatter may indicate that such disturbances were widespread.

Sardinia, Italy

Sardinia was part of the North Gondwana margin during the Silurian (Ferretti and Serpagli 1996). The early Ludlow–latest Pridoli (*Oz. ex. hamata* to *Oul. e. detortus* zones) Ockerkalk on SE Sardinia consists of about 25 m of monotonous blue-grey nodular argillaceous limestone, weathering to ochre due to its high pyrite content (Corradini et al. 1998a, 2002a). Neither macroscopic nor thin-section analyses have revealed any variations through the interval of interest, even though the rate of sedimentation increased upwards (c. 60% of the Ockerkalk is Pridoli in age). Macrofossils are limited to very rare cephalopods and, in the upper part, *Scyphocrinites* (stems appearing within the *snajdri* Zone and loboliths occurring at the base of the *detortus* Zone). Microbiofacies analysis has revealed ostracods, brachiopods, thin-shelled bivalves, trilobites, gastropods and sponge spicules (Barca et al. 1995; Ferretti and Serpagli 1996). Acid leaching produced conodonts, rare phyllocarids (mainly mandibles) and micro-brachiopods. Barca et al. (1995) interpreted it as deposited in a quiet pelagic deep-water environment below normal wave-base with bioclastic input increasing upwards.

More than 130 samples, each 2–10 kg, from several sections produced on average c. 25 (up to 170) conodont teeth per kg with unbuffered technique and a 100 μm screen (Barca et al. 1995; Corradini et al. 1998b, 2009a; Serpagli et al. 1998). The teeth are black with CAI of 5–5.5; many are broken or texturally altered by regional metamorphism or diagenesis. In the following section we use data about conodonts (identifications reconsidered) and sediments from the Silius I section (SIL; Fig. 4 herein) supplemented with data from other sections, mainly Genna Ciuerciu section (GCIU; Corradini et al. 1998b, 2002b, 2009b) located a few hundred metres SSW of Silius I (the thickness of biozones, the conodont ranges and their abundance are as previously presented). The oxygen isotope data show the effect of metamorphism and re-equilibration, so caution is needed in interpreting the C isotope data.

1. *P. siluricus Zone* – The base is below SIL 7 (at 3.2 m) and the top above SIL 12. Conodonts from SIL 7 to SIL 12 inclusive (1639 teeth), in approximate order of frequency, are as follows: *Oz. excavata, Oz. confluens, Oulodus* sp., *P. siluricus, K. ortus sardoa, K maenniki, K. variabilis ichnusae* (only in SIL 7 and 8), *Ps. beckmanni* and, relatively rarely, *Ps. bicornis bicornis*. Yield is 20–50 teeth/kg; *Kockelella* and *Po. siluricus* are frequent, but *Panderodus* is absent. *Ps. bicornis* seems to be a Lazarus taxon, re-appearing in SIL 9. Some 'anomalous' teeth have been reported from this interval (Corradini et al. 1995). From the top of SIL 7 to SIL 12, $\delta^{13}C$ fluctuates between 0.82 and +0.55‰ around a slightly increasing trend.

In contrast, SIL 13–SIL 15 yielded 6–12 conodont teeth/kg, *Panderodus* is regular, *Kockelella* absent, *Po. siluricus* rare and the relative frequency of the two *Pseudooneotodus* taxa reversed. Identification of *Po. siluricus* is based on ramiform teeth only but is supported by a single fragmentary Pa element from the nearby Genna Ciuerciu section, 0.60 m below the first recorded *Pd. latialata*. Above SIL 12 to SIL 16 the $\delta^{13}C$ trend decreases slightly from c. +0.37 to c. –0.50‰.

2. *Upper P. siluricus Subzone* – The base is somewhere between SIL 12 and SIL 16. If the base is directly below SIL 13 and the top directly above SIL 15, then the subzone constitutes c. 13 % of the *Po. siluricus* Zone. If the base or top are farther apart, the

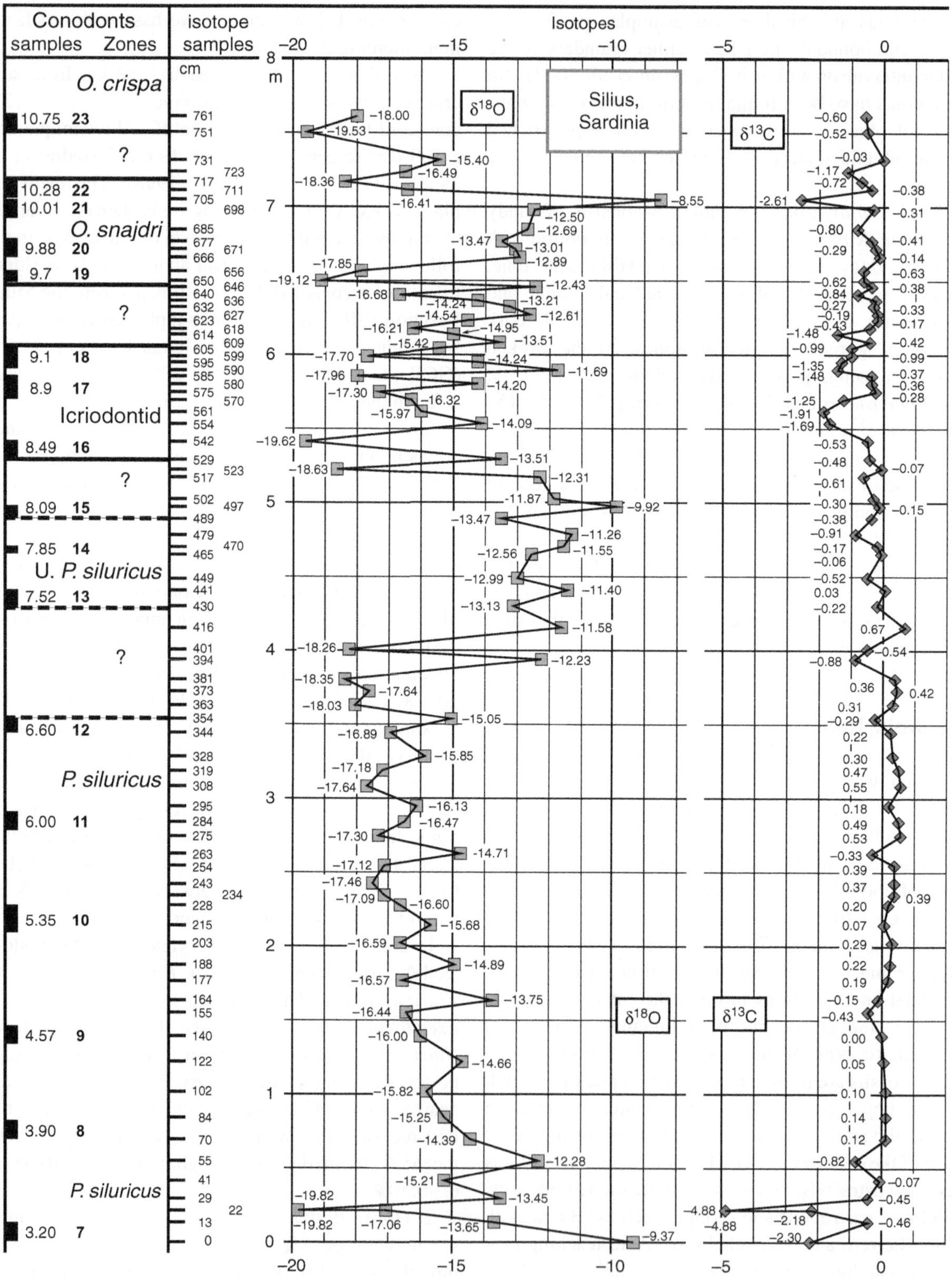
Conodonts
samples Zones
isotope
samples
Isotopes
cm
m
Silius,
Sardinia
δ18O
δ13C
O. crispa
O. snajdri
Icriodontid
U. P. siluricus
P. siluricus
P. siluricus
10.75 23
10.28 22
10.01 21
9.88 20
9.7 19
9.1 18
8.9 17
8.49 16
8.09 15
7.85 14
7.52 13
6.60 12
6.00 11
5.35 10
4.57 9
3.90 8
3.20 7
–20
–15
–10
–5
0

Fig. 4

ratio increases; it may exceed 25 %. On Gotland, that subzone makes up c. 3.4 % of the zone. That ratio corresponds only to the unsampled interval between SIL 15 and SIL 16. Hence, identifying the base of this subzone with the major faunal change between SIL 12 and SIL 13 is rather uncertain.

3. *O. excavata interval* – The expected thickness is so small that it is unlikely to have been sampled.

4–6. *Icriodontid Zone* – *Oz.* cf. *scanica* has now been identified in SIL 16 (+8.49 m). It appears at, or near, the base of this zone in Skåne. *Oz. confluens* is present, but *Pd. latialata* appeared in SIL 18 (+9.1 m). On Gotland another icriodontid appeared, and *Oz. confluens* is absent in the Upper Icriodontid Subzone. The yield was only 4–5 teeth/kg in SIL 16–17. From slightly above SIL 16 into the lowest part of SIL 17, there is a $\delta^{13}C$ minimum reaching down to −1.91‰ at +8.82 m and then rising to −0.28‰ at +8.95 m. It changes little within the interval of SIL 17.

7. *The O. snajdri Zone* – *Oz. snajdri* was found in SIL 21 (+10.1 m) in a frequency of c. 2 %. Its absence from SIL 20 (36 specimens) and SIL 22 (23 specimens) is thus not significant. The yield in these four samples is 15–47 teeth/kg. *Oz. crispa* and *Oz. remscheidensis* in SIL 23 (+10.75 m) indicate the next zone. $\delta^{13}C$ dropped abruptly from −0.37‰ at 9.05 m to −1.48‰ at 9.10 m (basally in SIL 18), and increased at a decreasing rate to c. −0.27‰ at 9.56 m, then rose, rapidly at first but levelled off to c. −0.18‰ at 9.50 m. This rise resembles the coeval rise at Bjärsjölagård 2 in Skåne, except for the numerical difference.

Cellon, Austria

Our isotope data (Fig. 5) agree well with those of Wenzel (1997, fig. 8). The minor difference between the data from the same bed may be due to each bed representing a rather long interval of time. Only 25 $\delta^{13}C_{carb.}$ and five $\delta^{13}C_{org.}$ analyses are available; hence, all comments rely on every single sample being undisturbed, and that the time of any change has a too-wide uncertainty for high-resolution correlations; future sampling needs to be at centimetric scale. The data is, however, enough for concluding that there is no $\delta^{13}C_{carb.}$ spike resembling those in less strongly heated areas.

1. *P. siluricus Zone* extends through 2.87 m, from Bed 21 through Bed 24A (Walliser 1964). The strata consist of cycles of grey micrite, laminated strata and black ferruginous shale (Histon and Schönlaub 1999). Conodonts were abundant and diverse. CC has restudied Walliser's *Kockelella* specimens and identified *K. maenniki* in beds 21, 21A, 22, 22B, 22C and 23 and *K. variabilis ichnusae* from 17E to 20A and in 21A and 22A. *K. o. absidata* has its last occurrence in bed 21, questionably in 23A. *K. o. sardoa* occurs in 18D, 21 and 22C only. $\delta^{13}C_{carb.}$ decreased about 1‰ at or near above the base of this zone and rose again higher up. $\delta^{13}C_{org.}$ increased slightly from below this zone to Bed 24.

2. *Upper P. siluricus Subzone* – Bed 24 (with *K.* cf. *variabilis* and *Po. siluricus*, but without *K. ortus*; Walliser 1964, Table 2, II) and Bed 24A (with *K. ortus*) might represent the Upper *P. siluricus* Subzone. If so, then their range order is the same as on Gotland in this subzone. However, the records are uncertain, since reworking occurred in the *Po. siluricus* Zone (Histon and Schönlaub 1999: 266), and moreover, CC did not note any *Kockelella* specimens in these samples (and no specimens have been illustrated). Alternatively, erosion below Bed 25 (see 4–6 mentioned later) may have destroyed any strata representing this interval. Regardless of which alternative is correct, beds 24 and 24A may represent this subzone.

3. *O. excavata interval* – Considering relative thicknesses elsewhere, this interval would be expected to be a few centimetres at the most. Considering the erosive base of the succeeding bed, this interval may have been removed.

The 3/4 discontinuity – Bed 25 has a sharp erosive basal contact (Kreutzer and Schönlaub 1997, p. 103).

Fig. 4 Conodont and isotope data of whole-rock samples from Silius 1 section, Sardinia, Italy. Conodont zonation, oxygen (*squares*) and carbon (*diamonds*) isotope data. Minor uncertainty as regards zonal intervals is indicated—for further comments, see the text. The metric level and conodont sample numbers are given (for their contents, see our text and quoted references). The isotope-sample levels are given with both a short dash and the metric level (in centimetres). Similarly the recorded isotope data are given near the symbol; repeated analyses of a sample (e.g. that at 22 cm) are indicated by repetition of the recorded value, if identical. Conodont samples were measured from the base of the section, isotope samples from the base of conodont sample 7, the oldest one from which *P. siluricus* has been reported

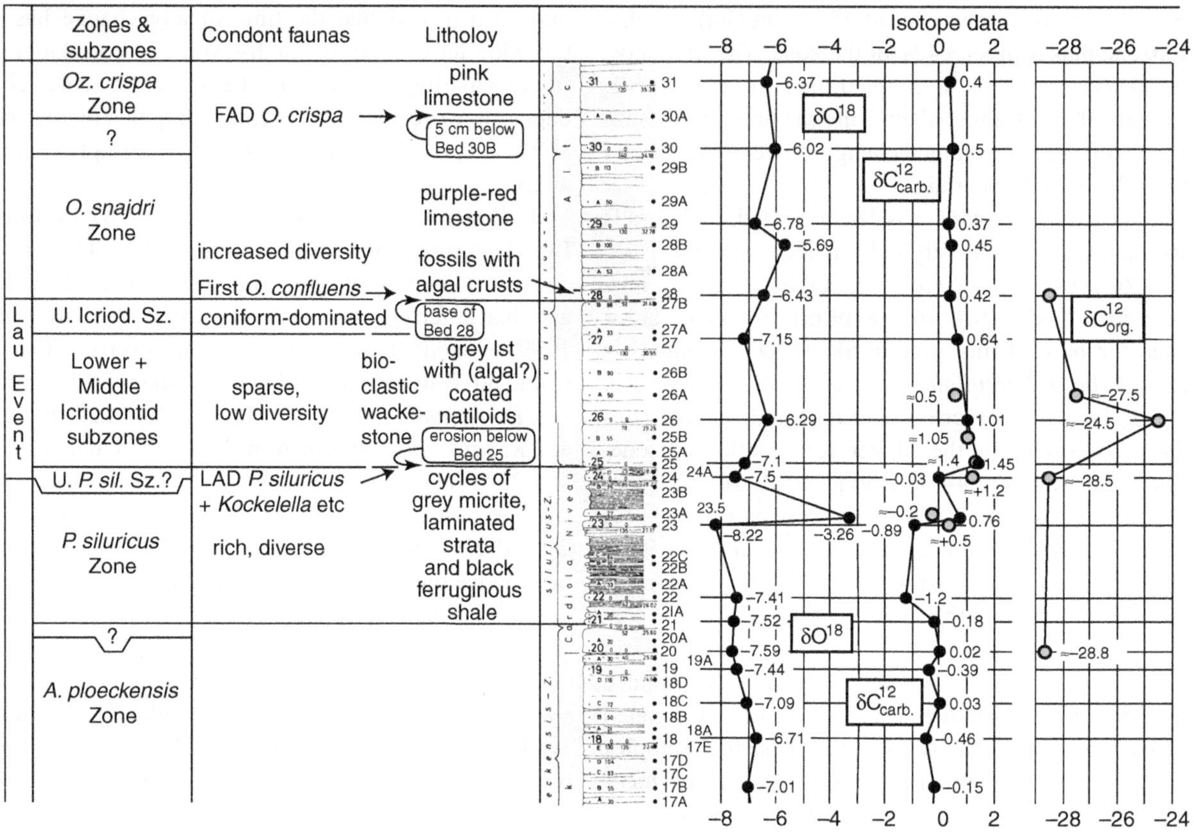

Fig. 5 Conodont and isotope data from the Cellon section, Austria. Oldest and youngest finds refer to Walliser's collections; for rare taxa, real ends of ranges ends may be markedly different. Conodont data and the section are from Walliser 1964; lithological descriptions from Kreutzer and Schönlaub (1997) and Histon and Schönlaub (1999): isotope data marked with a ring with grey centre were remeasured from Fig. 8 in Wenzel (1997). Dots = our data

4–6. *Icriodontid Zone* – Beds 25–27 consist of bioclastic wackestone (Kreutzer and Schönlaub 1997) including nautiloids with a coating that may be algal in origin (Histon and Schönlaub 1999). *Pd. latialata* appeared in Bed 25 (Walliser 1964); the last specimen was reported from Bed 28A, but neither LJ nor CC saw any in that collection. Thus, the proven range is Beds 25–27 (2.4 m), although considering collection sizes and its low frequency, a higher range end is not ruled out. The proven range top differs from the top of the *Pd. latialata* Zone as drawn previously: below Bed 30, 5.71 m thick (Walliser 1964, Plate 1; Histon and Schönlaub 1999), and below Bed 30A, 6.36 m (Walliser 1964, Table 2II).

A $\delta^{13}C_{org.}$ spike reaches ≈–24.5‰ in Bed 26 and then decreases. One sample from the *Pd. latialata* Zone in the Rauchkofel section also shows this spike (Wenzel 1997). Our samples show a positive $\delta^{13}C_{carb.}$ spike of about 1.5‰ beginning in Bed 25 and decreasing to about 0.6‰ within this zone but not returning to subzero values until in the Pridoli. Both spikes declined quickly, although they differ in the $\delta^{13}C_{carb.}$ spike culminating well before the $\delta^{13}C_{org.}$ spike. The $\delta^{13}C_{org.}$ increase is coeval with the $\delta^{13}C_{carb.}$ rise on Gotland and in COG.

6. *Upper Icriodontid Subzone* – *Panderodus* and *Belodella* dominate in Walliser's collection from Bed 27A and *Panderodus* alone in Bed 27B (compare Gotland, COG and Mušlovka).

7. *O. snajdri Zone* – The lithology changes from grey to purple-red limestone at the base of Bed 28 (Histon and Schönlaub 1999). In Bed 28 fossils have algal crusts (Kreutzer and Schönlaub 1997). *Oz. confluens* first appeared in that bed and remained regular at least into Bed 29A (Walliser 1964; compare its return on, e.g. Gotland at the base of the *O. snajdri* Zone).

The strata change from purple-red to pink 5 cm below the base of Bed 30B (Histon and Schönlaub 1999). *Oz. crispa* appears in Bed 30A, ranging into Bed 32 – c. 2.1 m (Walliser 1964, Table 2.II). $\delta^{13}C_{carb.}$ seems to fluctuate through the rest of the Ludlow.

Mušlovka, Czech Republic

The following summary is based on bed numbers, lithologies, macrofaunal lists and selected conodont records of Kříž and Schönlaub (1980), Kříž (1992), personal comments from Kříž to LJ, LJ's notes when Schönlaub showed his conodont collections in 1994 and isotope data in Lehnert et al. (2003; re-digitised from their Fig. 2). Further, Hermann Jaeger (in mail 1985-12-05) noted that the youngest *M. fritschi linearis* occurs 0.5 m below Schönlaub's first sample (in the figure of Kříž and Schönlaub 1980; Jaeger's find was, erroneously, marked as between samples 3 and 4). Graptolites are found only at a few levels, according to Jaeger (as quoted by Kříž and Schönlaub 1980). Hence, we assume that the graptolite zonation given by Lehnert et al. (2003) was extrapolated from other localities. We lack data needed to include uncertainty intervals for these zones, but we have included them in our Fig. 6 with the correction mentioned previously.

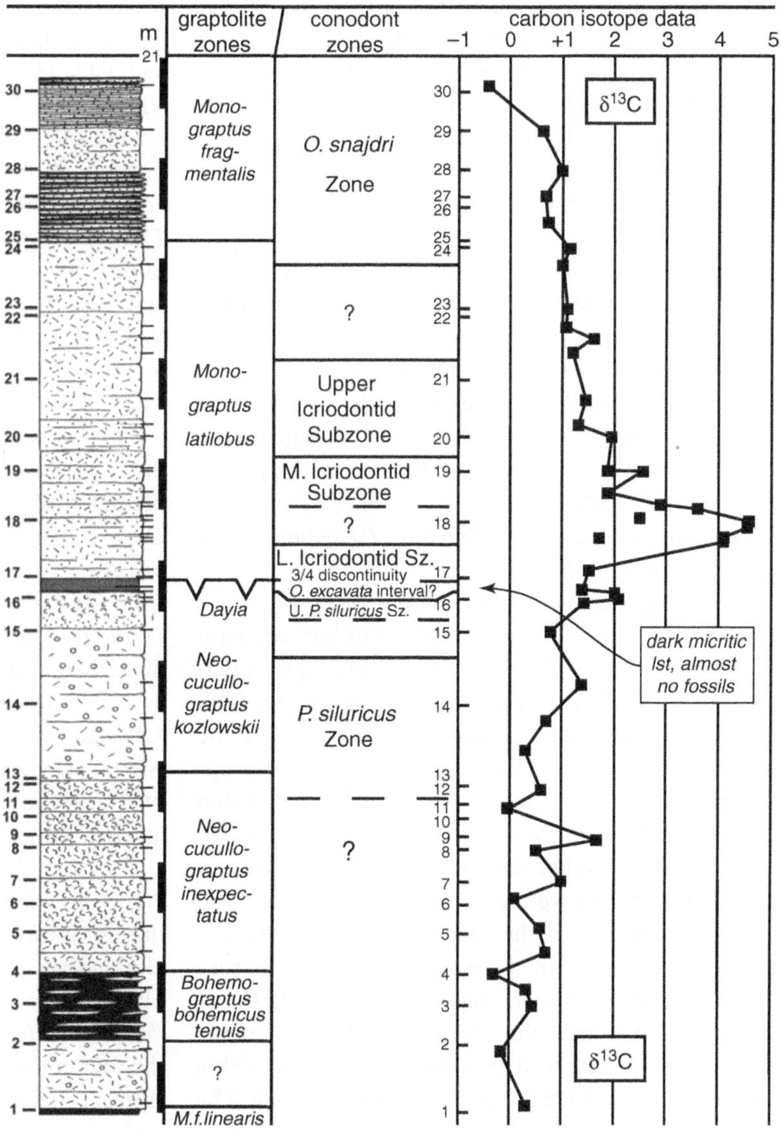

Fig. 6 Conodont and isotope data from Mušlovka section, Bohemia, based in part on figures and data in Kříž and Schönlaub (1980), Kříž (1992), Lehnert et al. (2003) and personal comments from H Jaeger and J Kříž to LJ. *Dashes* between the column and the metre scale indicate isotope-sampling levels; bed-numbers are repeated in the conodont zonation field

1. *P. siluricus Zone* – A diverse conodont fauna with unquestionable *Po. siluricus* occurs in Beds 12, 13 and 14. Bed 12 is a grey biosparite with brachiopods, while beds 13–14 are grey nodular biosparite with cephalopods. Beds 13–15 are 2.70 m. $\delta^{13}C$ increased from c.+0.4 to c.1.4‰.

2. *Upper P. siluricus Subzone* – Bed 16 and perhaps Bed 15. These yielded a less diverse conodont fauna. Bed 15 is a grey nodular cephalopod limestone. Bed 16 is 0.70 m of grey biomicrite to biosparite with very abundant *Dayia* (Kříž 1992; cf. coquinas of *Dayia* in this subzone on Gotland). $\delta^{13}C$ decreased to c. 0.8‰ in Bed 15 and rose through Bed 16 to c. 2.1 and 2.0‰.

3. *O. excavata interval* – Probably the 0.25 m of dark muddy micrite above Bed 16 is this interval. It has sparse unidentified macrofossils (Bouček 1937; Kříž 1992, pers. comm.). $\delta^{13}C$ dropped to c. 1.4‰.

The 3/4 discontinuity – A discontinuity surface occurs above the dark micrite (e.g. Lehnert et al. 2003). Discontinuity surfaces have been reported from other levels, for example, below Bed 16 (Manda and Kříž 2006).

4. *Lower Icriodontid Subzone* – Bed 17, grey to pink crinoidal biosparite with diverse brachiopods and the *Ananaspis fecunda–Coniproetus* trilobite assemblage (also in beds 18–24). Beds 17–24 aggregate 6.60 m.

5. *Middle Icriodontid Subzone – Pa. equicostatus* appears in Bed 19 (cf. in Bed 18). $\delta^{13}C$ culminated in Bed 18 at 4.6‰ (Lehnert et al. 2003) and dropped to below 2‰ in Bed 19.

6. *Upper Icriodontid Subzone – Pa. equicostatus* dominates in beds 20 and 21.

7. *O. snajdri Zone* – At least from Bed 24 (i.e. not from Bed 19; see the introduction about how the base of this zone is now identified); same lithology as Bed 17. Beds 25–27 consist of crinoidal biosparite, with calcareous shale intercalations. The *Oz. crispa* Zone was reported in Bed 33 f. $\delta^{13}C$ drops gradually up-section to c. +1‰ in Bed 24 and to c. –0.4‰ in Bed 30. The early culmination of the $\delta^{13}C$ is unexpected (conodont and isotope samples were not collected at the same time, but we are not aware of any reason to doubt identification of the strata).

In deeper water, in the Kosov section, $\delta^{13}C$ rose from ca –0.1‰ below the reported range of *P. siluricus* to c. 2.5‰ near the recorded top of that range (Lehnert et al. 2007). It then rose to c. 6.4‰ within less than 1 m reaching 8‰ some metres higher up. The curve is closely similar to those from Skåne, Gotland and Queensland. Similarities with the former may also include the brief drops just below the tops of the *P. siluricus* Zone and the Middle Icriodontid Subzone, if there is no random noise in the data (20 sampled levels), that is, if every one-unit-deviation from a smooth curve is reliable. Unfortunately, conodont data to confirm the former correlation are poor and, for the latter, absent.

Lithuania

The following identifications are based on direct comparison between Antanas Brazauskas's conodont collections (with an eye to lithology data, but taking conodont data as compelling) from Lithuanian cores and LJ's from Gotland, and isotope data from the Viduklé core by Kaljo and Martma (2006; though the ranges seem to have been based on the same core, there are differences). The Viduklé-61 lithologies and conodont faunas are very similar to those on Gotland. Thus, limestones, the middle part of the Nova Beds of the Etelhem Secundo Episode, occur between 1199 and 1181.5 m (the conodont fauna below them, in the lower Nova Beds, equates well with the equivalent interval on Gotland).

1. *P. siluricus Zone – P. siluricus* was found in the five collections between 1356.4 and 1345.9 m in the Tauragé core from western Lithuania. The conodont faunas are very similar to those on Gotland, e.g. *O. confluens*, *O. excavata*, *P. gracilis*, *P.* cf. *recurvatus*. A very similar fauna occurs in the Viduklé-61 core from 1160.6 to 1158.6 m (*P. siluricus* in both samples); *D. dubius* was found at 1161.6 and 1158.6 m. In the upper part of the Nova Beds, in a Viduklé core, $\delta^{13}C$ is c. –1.5‰. In the uppermost Nova Beds, $\delta^{13}C$ rises to –1.5‰ and to 1.9‰ at the boundary with the Mituva Fm where Kaljo and Martma (2006) marked the last record of *P. siluricus* (in Viduklé-61, that boundary was drawn at 1167.8 m).

2. *Upper P. siluricus Subzone* – In the Viduklé-61 core, lithologies and changes therein are very similar to those in the equivalent interval on eastern Gotland, including presence of the *Dayia navicula* bloom. Only a few specimens of *O. excavata*, *O. confluens*, *Panderodus* and *Decoriconus* were found in the four collections (together more than 50 specimens) from 1156.6 to 1150.6 m.

The 3/4 discontinuity – East of Viduklé, in the Geluva, Sutkai and Ledai cores, there is a major gap. However, it also includes parts of the Icriodontid Zone, so erosion during the karst-forming sea-level drop seems to have cut down at least to this discontinuity.

4–6. *The Icriodontid Zone* – Oncolites appeared at 1152.6 m.

6. *Upper Icriodontid Subzone* – In the Viduklé-61 core, *Pa. equicostatus* dominates strongly in 5 samples of oncoid-rich marls from 1148.6 to 1143.4 m; only one fragment each of *Dapsilodus* and a ramiform, perhaps *O. scanica*, were seen. Kaljo and Martma indicated that as with Gotland, the first record of *Pa. equicostatus* preceded the $\delta^{13}C$ maximum, c. 8.1‰. In the Tauragé core *Pa. equicostatus* dominates at 1337.8 m.

7. *O. snajdri Zone* – In the Tauragé core at 1324.0 m, *O confluens*, a form of *O. scanica*, *Ps. beckmanni*, *P. recurvatus* and so on were found. At 1321.0 m, *O. remscheidensis* (with alternating denticulation on the S teeth) appeared, and *Pa. gracilis* returned. Both panderodontids continued into succeeding samples, confirming their presence. In the Viduklé-61 core, as on Gotland, neither of these panderodontids returned after the Lau Event but, as with Gotland, the appreciable dominance of *Pa. equicostatus* ended, for example, *O. confluens*, *O. excavata* and *Pa. equicostatus* were among c. 20 specimens recovered from the sample at 1140.3 m. Rare oolites occur between 1146.5 and 1139.5 m. $\delta^{13}C$ had decreased to c. 4.5‰ when *O. scanica* appeared; *O. wimani* appeared only slightly higher up (Kaljo and Martma 2006).

Latvia

Dimitriy Kaljo kindly provided us with the measured ratios used in the 1997 figures quoted hereafter. In the Priekule core (Kaljo et al. 1997) $\delta^{13}C$ is 1.3‰ at 1038.4 m; it rose via 4.2‰ at 1035.3 m to +5.9‰ at 1028.5 m, dropping to 2.2‰ at 1023.9 m and then stabilised close to 1‰ or less. The entire spike is in the Nova Beds (1025–1038 m), between *Bohemograptus* ex. gr. *bohemicus* at 1040.6 m, and *Ludensograptus latilobus* at 975.8 m. On Gotland, *B. bohemicus* has been reported (e.g. Hede 1942) in the När Fm (see the following 'Further areas' regarding *L. latilobus*).

Samtleben et al. (1996, p. 290) found 'exactly the same' $\delta^{13}C$ ratios across different facies on Gotland. If that conclusion can be extended to the facies in this core, then the ratios in the Nova Beds correlate them with the lower Eke Fm; the unsampled uppermost Nova might perhaps yield $\delta^{13}C$ ratios closer to those in the upper Eke (see the following section for an alternative interpretation). Either way, the Priekule Nova Beds correlate with the gap above the Paadla, as known in outcrops and in the Ohesaare core (Jeppsson et al. 1994), not with that mentioned later.

In the Ventspils core, *Neobeyrichia scissa* co-occurs with *N. lauensis* at c. 475 m (Sarv 1982); on Gotland this co-occurrence is characteristic for the När Fm (data in Martinsson 1967). $\delta^{13}C$ rose from 0.8‰ at 470.2 m to 2.5‰ at 468.7 m and reached 5.0‰ at 465.4 m (in uppermost Mituva Fm) (Kaljo et al. 1998: 305). *Oz.* aff. *snajdri* appeared at 464.2 m (Viira 1982), and $\delta^{13}C$ decreased through 4.0‰ at 458.7 m and 2.1‰ at 456.4 to become mainly below 1‰ during the latest Ludlow. In the Pavilosta core, *P. siluricus* occurs at 756.4/756.0 m (Viira in Jeppsson et al. 1994), a $\delta^{13}C$ maximum of 4.2‰ was found at 739.0 m (Kaljo et al. 1998), and *O.* aff. *snajdri* appeared at 622.7 m (Viira 1982).

Kaljo et al. (1997 p. 217), noting that the amplitude of other $\delta^{13}C$ changes was less in the Priekule core than in the Estonian Ohesaare core, accepted this as an authentic difference and ascribed it to a difference in water depth. Absence of $\delta^{13}C$ differences due to facies on Gotland (see previous section) excludes a gradual decrease with depth. A depth-difference requires a boundary layer between shallow and deeper water in the Baltic Basin. Before this can be taken as proven, other potential causes need to be excluded. For example, there may be gaps in the sequence, or the maximum may not have been sampled (there is 4.6 m gap between the observed Priekule maximum and the next sample). Four studies of this $\delta^{13}C$ maximum on Gotland illustrate the sampling-interval effect. Wenzel and Joachimski (1996) with 4 Eke localities recorded 8.05‰, Samtleben et al. (1996) with five Eke localities c. 8‰, but Samtleben et al. (2000) with 19 localities, raised the maximum to 8.80‰. Using conodont biostratigraphy to select localities, Jeppsson et al. (2007) found a brief interval with up to 9.17‰ in the Eke and 10.58‰ slightly higher up.

Additional Areas

The conodont *Po. siluricus* has been identified widely in more or less calcareous facies. Its extinction early in the Lau Event can be identified globally. For example, Walliser (1964) found it in *Spain* and in *Algeria* (with '*M. nilssoni*' and '*M. colonus*', so this is either an earliest Ludlow find, far older than the *P. siluricus* Zone, or the graptolites need to be restudied). Rexroad and Craig (1971) indicated that the top of the *Po. siluricus* Zone and a succeeding interval occur in the Bainbridge section in *Missouri* (restudied and confirmed by LJ). Barrick et al. (2009), incidentally, have identified the Lau Event in southern Oklahoma, eastern Missouri and western Tennessee; lithologies appear to display some similarity to those in Gotland. Männik and Małkowski (1998) described a similar interval in the Gołdap core from *NE Poland*.

In *Britain*, the Lau Event began close to the base of the Lower Whitcliffe Fm, that is, it was middle-late Ludfordian in age (Jeppsson and Aldridge 2000). Lawson (1960) noted that *Dayia navicula* is 'present only in the basal beds' of the Whitcliffe fauna, and that it 'occurs quite commonly in the lower part of the Whitcliffe Beds', 'but has never been found in the Upper Whitcliffe Beds'.

In *Poland* the c. 25 m with impoverished graptolite faunas between a *Neocucullograptus kozlowskii* fauna and one with *Pseudomonoclimacis latilobus* was included in the *N. kozlowskii* Zone by Urbanek (1997). The impoverished conodont faunas in the Icriodontid Zone is approximately coeval with those impoverished graptolite faunas, the *P. siluricus* Zone with at least part of the *N. kozlowskii* fauna, and the *O. snajdri* Zone with the *M. balticus*/*P. latilobus* Zone (in the lower *M. formosus* Zone) (Jeppsson and Aldridge 2000). These changes can be used to identify this interval in graptolitic facies. Kozłowski (2003) described a succession in the Holy Cross Mountains changing from shale to sandstones, oolites and oncolites, named it the Jadowniki Sandstones and Oolite Formation and, based on trilobites, concluded that it is of late Ludfordian age. This interval has also been identified in the Tien-Shan of *Kazakhstan* (Koren' 1993), and in the *Czech Republic* (Štorch 1995).

In *Estonian* outcrops there is a gap at this level (Jeppsson et al. 1994) as there probably is in other shallow areas. This accords with a low-stand and glaciation connected with the Lau Event. The older (missing) strata may have been lost by erosion during the low-stand absence of the younger ones being due to non-deposition.

Salzman (2001) found the risen $\delta^{13}C$ signature in 3 samples from *Oklahoma*. In *Nevada* the peak (2 samples) was within the range of the graptolite *S. chimaera* and the conodont *K. variabilis* and at the base of the *P. siluricus* Zone (biostratigraphic data from Klapper and Murphy 1975 [1974], and Berry and Murphy 1975). A similar severe mismatch between isotope data and biostratigraphic data was found lower down (several Ludlow graptolites reported from an interval of the middle Homerian Mulde Event Excursion; see also Cramer et al. 2006). Moreover, the conodont and graptolite data disagree (Uyeno 1981 [1980]; Jeppsson 2006). Cramer et al. (2006) discussed four possible explanations for the isotope and graptolite mismatches but found none of them convincing; they concluded that recollecting is needed. Pending such work (or some kind of corrections from one or more of the three authors of the biostratigraphic data), those biostratigraphic data should not be trusted. The maximum $\delta^{13}C$ recorded by Saltzman was smaller than that recorded from Gotland; however, the 5 samples with enriched $\delta^{13}C$ values have markedly different values indicating a considerable probability that the true maximum may not have been sampled. That is, the sampling-interval effect (see Latvia, as mentioned previously) might have influenced the maximum recorded.

A discontinuity was reported between the Duoro Fm and the Drake Bay Fm in the *Canadian Arctic Islands* (Uyeno 1981 [1980]). *Po. siluricus* occurs in the Duoro Fm and *Pedavis latialatus* in the basal beds of the Drake Bay Fm; hence, the discontinuity is one of the two formed during the Lau Event, or a combination of both discontinuities.

The *Yass Basin* sequence in *Australia* was the site of the first late Silurian conodont investigation in Australia (Link and Druce 1972). The transition from the Hume Limestone Member of the Silverdale Formation to the overlying Black Bog Shale, particularly the Yarwood Siltstone Member, clearly represents the chronologic interval in question despite a pioneering sampling regime that lacks the refinement required for precise correlations. The Hume Limestone contains abundant macrofauna and is strongly crinoidal in some areas: It has produced a rich, high diversity conodont fauna typical of the *Po. siluricus* Zone. Taxa include *Po siluricus*, *Oz. excavate*, *Oz. confluens* and

C. dubius. Many other species typical for this interval could be determined by a modern taxonomic treatment of the faunas obtained previously. In contrast, the significantly impoverished conodont fauna from the overlying strata of the Yarwood Siltstone was described as 'dominated by long ranging panderodids' (Link and Druce 1972: 13). They (Link and Druce 1972) equated the Yarwood Siltstone with Walliser's (1964) *I. latialata* Zone despite absence of the index form but based essentially on the extinction of many taxa from the pre-existing interval. It is anticipated that with detailed sampling through this sequence and modern taxonomic analysis, much of the internal architecture of the Lau Event may be recognised.

Commentary

The Ludlow lasted 4.2 Ma (Ogg et al. 2008; Cramer et al. 2010); it is 380 m thick on Gotland (Jeppsson 2006). Units 2 to 6 in the preceding lists (= the Lau Event) are c. 15 m thick; that is, all five units were deposited during only 0.177 ± 0.209 Ma, if the sedimentation rate remained consistent through this time interval. That is, biostratigraphic correlations with a precision in the order of 50 ka or better are now possible during that event. Even a large part of previously published conodont successions permits identification of these subzones (in some cases after updating of the species lists). We are confident that future studies based on adequately large samples processed scientifically (see Jeppsson 2005) will provide such precision, or better, through this event, wherever conodonts can be found.

Conodont ranges and frequencies in different areas are now known well enough to conclude that using whole faunas provides high-resolution correlations (we assume that other major clades show the same pattern and can provide improved resolution if handled similarly). In contrast, relying only on oldest and youngest finds is problematic for several reasons, especially if rare taxa in small collections are accorded significance.

The trend in conodont faunal changes was widespread: high diversity and high abundance faunas in the main part of the *P. siluricus* Zone followed by low diversity and lower abundance faunas in the Icriodontid Zone, followed in turn during the *O. snajdri* Zone by increased diversity, including new taxa. The dominance of coniform conodonts, mostly the *P. equicostatus* group, in the Upper Icriodontid Zone is also widespread, although in Skåne, *Oz. scanica* dominates strongly, and *P. equicostatus* is subordinate. An unexpected detail is that the *Dayia* bloom on Gotland in the Upper *P. siluricus* Zone was also found in Bohemia, Lithuania and probably in Britain.

Some lithologic successions are closely similar on different palaeo-continents, for example, in COG in Queensland on Gondwana, on eastern Gotland and in Lithuania on Baltica. Even the lithologic and conodont faunal changes in them were synchronised. These successions are even more similar than those of western and eastern Gotland. This similarity indicates that, in similar settings, oceanic changes due to the Lau Event were strong enough to override local influences.

Deposition of oncoids and coated grains was widespread, starting at the base of the Lower Icriodontid Zone on eastern Gotland, in the COG section, and in Lithuania. On western Gotland it started at or close to the base of the Middle Icriodontid Subzone, continuing into the *O. snajdri* Zone. The exact starting time in Skåne is not known, but here too it continued into the same zone. Oncolites are found in Poland too. Algal crusts have been reported from the strongly condensed section at Cellon, low in the *O. snajdri* Zone. Sandstones, or silty-sandy sediments, seem to be coeval on Gotland, in COG, Poland, and in Lithuania. Oolites formed briefly during an early part of the *O. snajdri* Zone on Gotland, in COG, Poland, and in Lithuania. These and several other characteristics are evidence of widespread anachronistic facies due to the Lau Event (cf. Calner 2005).

The 3/4 discontinuity is found on eastern Gotland, in Bohemia and in Cellon; to what extent there may have been similarities of water depth has not been spelled out—one of us (CC) believes there were differences in lithologies that may imply differing water depths. The report of younger palaeokarst on eastern Gotland indicates an as yet unlocated discontinuity higher up; it means that there may be undiscovered gaps in shallow settings there and elsewhere where erosion was strong enough, the strata between these two or more discontinuity surfaces can of course be absent, so that only a single discontinuity surface would be found. Data from different sections indicate that the interval near this discontinuity was very complicated, perhaps with repeated minor? sea-level fluctuations producing discontinuity surfaces.

The $\delta^{13}C$ spike was found in most areas. The most detailed isotope curve hitherto from that interval is from Klinta in Skåne. On Gotland, in Skåne, COG and Kosov, the maximum is between 8 and 10.5‰ – one of the largest during the Phanerozoic. Based on good biostratigraphic data, the spike can be shown to have culminated and began to decline in the *O. snajdri* Zone on Gotland. Similarly, in the COG section, it continued rising to the last sample. Contrastingly, in Mušlovka, Bohemia, it rose only to 4.6‰ (from 1.4‰), culminating already in the Middle Icriodontid Subzone. The Bohemian correlations add another enigmatic detail to the $\delta^{13}C$ spike connected with the Lau Event—a very large difference between two sections only a few kilometres apart. Some of the other lower maxima may also be real, but others may be due to gaps in the succession, or result from too few samples, or be due to recrystallisation. In Cellon, there is a pronounced difference between the $\delta^{13}C_{carb.}$ and the $\delta^{13}C_{org.}$ spikes. The $\delta^{13}C_{org.}$ spike resembles the $\delta^{13}C_{carb.}$ spike from other areas more than the $\delta^{13}C_{carb.}$ and seems to be more stable when the temperature has been high enough to turn the conodonts black.

More detailed curves show rapid large fluctuations in some intervals. The best-documented ones are in the basal part of the *O. snajdri* Zone (Fig. 3) with amplitudes of up to 2.7‰ lasting about 6000 years or less per cycle. Thus, something unusual took place early in the *O. snajdri* Zone, that is, just after the Lau Event.

Conclusions

Sequences discussed earlier represent different palaeocontinents and very different facies. Faunal differences were therefore expected and were found, although they are not large enough to impede correlations. To a large extent correlations herein rely on collections extracted for projects with other requirements (this includes most of our own samples), for example, many collections are too small to reliably record taxa with a frequency of only a few percent or less. Hence, only by using all taxa is it possible to reliably identify the biostratigraphic position. Strong faunal evidence thereby identifies most intervals in most areas, even if boundary positions can be imprecise.

Correspondence between the timing, and in some cases even the amplitude, of the $\delta^{13}C$ excursion occurs widely. Comparison of the maximum $\delta^{13}C$ recorded on Gotland by different authors shows that it increases with increasing number of analyses (as would be assumed intuitively). This effect must be considered when there are no biostratigraphic data to tell that the correct interval has been sampled. Close similarities were also found in some faunal and lithologic characters. The main limit on how widely some of the characters can be identified is lack of enough data. Even in the best-studied, most expanded sequences, eastern Gotland and COG, strata from the smallest recognised intervals are only c. 0.2 m thick.

The interval following the end of the Lau Event includes two and a half smooth $\delta^{13}C$ cycles with amplitudes of up to 2.78‰ in the basal metre of the *O. snajdri* Zone in Skåne (Fig. 3). In COG, the increasing trend in $\delta^{13}C$ reached 6.97‰ at 54.10 m where it was interrupted by a decrease through 5.96‰ at 55.97 m to 5.59‰ at 56.02 m and then rose again to 6.03‰ at 56.16 m. The interval of 1.87 m between the first two samples did not yield any useful carbonates; hence the strata with the drop/rise cycle are somewhere between 0.3 and 2.06 m thick. Changes were rapid in the 0.2 m with three samples, as occcurred at Bjärsjölagård. All considered, it seems probable that the base of the *O. snajdri* Zone is close to 55 m and that the rapid $\delta^{13}C$ changes reflect the same unstable conditions as were recorded in Skåne. Other curves are based on much fewer data points, and the biostratigraphic data are more imprecise. Thus it is uncertain if any large difference between consecutive samples represents the same brief unstable interval. However, since the rapid large fluctuations found on Baltica were most probably recorded on Gondwana too, these fluctuations seem not to be local but widespread. These fluctuations were dated at Bjärsjölagård as starting at the end of the Lau Event. Therefore they indicate that the change in the oceanic regime that ended the event was not smooth but represented about 10,000 years. Many Silurian oceanic changes were connected with the events (Jeppsson and Aldridge 2000, and references therein); a search for such brief but large isotopic fluctuations is needed to tell if the end of the Lau Event was special in this respect, or if such perturbations were normal. Continuation of the anachronistic facies through the lower part of the *O. snajdri* Zone indicates that recovery took some further time. Scattering of the measured $\delta^{13}C$ values near the top of the Burgsvik Fm on Gotland (Jeppsson et al. 2007) seems to indicate that the return to normal conditions also included a brief interval of sub-Milankovitch oceanic fluctuations. A general conclusion is that sampling intervals representing about 1000 years can result in much interesting data, at least if collected from sections with a good biostratigraphic control and lacking evidence for substantial gaps.

The difference between the Mušlovka and Kosov (in Bohemia) $\delta^{13}C$ curves was interpreted by Lehnert et al. (2007) as due to a major hiatus at Mušlovka at the (herein: 3/4) discontinuity. The previously unpublished coniform conodont data (mentioned previously) indicate a much less extensive hiatus. Recent carbonate platforms can record a $\delta^{13}C$ depletion of 'as much as 4‰ relative to open-ocean water' (Patterson and Walter 1994). Kosov may be described as more open ocean and Mušlovka perhaps as part of a carbonate bank, but the Bohemian $\delta^{13}C$ difference reached at least 7‰, and the question remains: Why only Mušlovka records such a depletion and not, for example, eastern Gotland? On the other hand, Bohemia and Gotland were at different palaeolatitudes. Patterson and Walter's data were latitudinally limited and may not have included the maximal effect. The $\delta^{13}C$ disturbance due to the Lau Event was much larger than what has been recorded in Cainozoic strata.

Based on their fraction of the total Ludlow thickness on Gotland, the time represented by each subzone during the Lau Event and correlated widely herein, averages much less than 0.1 Ma. Similarly, the *O. excavata* faunal interval is probably less than 0.01 Ma. Some other changes, potentially useful for correlation, have a similar time scale, for example, the isotopic cycles in the *O. snajdri* Zone in Skåne. However, the limit for a global correlation relying on a waterborne signal is the oceanic mixing time – today c. 1000 years. The scale of signals spread by the atmosphere is less than a year. A resolution of 1000 years requires sampling of the order of a few hundreds of samples/section from just before the Lau Event to just after it. We have achieved that degree of resolution in the isotope curves for parts of the studied interval in one or more sections, if all measured concentrations are the original ones (see Jeppsson et al. 2007 for further data). With centimetre-scale collecting, this can be achieved even where the event interval may be condensed to only 2 m. Similarly, resolution in the conodont data can be increased considerably, though this is expensive and time consuming. Thus, although correlations herein are a major step forward regarding resolution and precision, future improvements are possible and expected to yield interesting results.

Acknowledgements Otto Walliser permitted us to use all data from his Cellon collections but declined co-authorship. Antanas Brazauskas very kindly gave access to and generously provided much information on Lithuanian bore cores. Sven Laufeld, Rickard Anehus and Peep Männik assisted LJ during parts of the fieldwork. Git Klintvik Ahlberg processed and picked many of the conodont samples. Fredrik Jerre measured and sampled the section at Bjärsjölagård 2b for scolecodonts and gave the conodonts to LJ. Correspondence with Jiri Kříž and Dim Kaljo helped interpret the Bohemian and Latvian sequences. Nadia Talent assisted in sampling the Botvide section for isotopes; Terry Sloan, the late Ross Talent and Glenn Brock helped sample various Broken River sections. The Swedish Natural Science Council and the Swedish Research Council funded much of LJ's field and laboratory work, especially conodont extraction, administration, rent for working space and similar university fees. Research grants from the Australian Research Council and Macquarie University to RM and JAT enabled sampling for isotopes at Botvide, and sampling various Broken River sections for conodont and isotope data, as well as funding the isotope analyses and conodont extraction from Australian samples. To all our sincere thanks.

References

Barca S, Corradini C, Ferretti A, Olivieri R, Serpagli E (1995) Conodont biostratigraphy of the "Ockerkalk" (Silurian) from southestern Sardinia. Rivista Italiana Paleont Stratigr 100:459–476

Barrick JE, Kleffner MA, Gibson MA, Peavey FN, Karlsson HR (2009) The mid-Ludfordian Lau Event and carbon isotope excursion in southern Laurentia – preliminary results. Soc Paleontol Italiana Bolli 49:13–34

Berry WBN, Murphy MA (1975) Silurian and Devonian graptolites of central Nevada. Univ California Pub Geol Sci 110:109 pp

Bouček (1937) Stratigrafie siluru v dalejském údolí u Prahy a v jeho nejblizsím okolí (also in French). Rozpr Ceské Akad Ved a Umeni Praha Tr II 46(27):1–20

Calner M (2005) A Late Silurian extinction event and anachronistic period. Geology 33:305–308

Calner M (2008) Silurian global events – at the tipping point of climate change. In: Elewa AMT (ed) Mass extinctions. Springer, Berlin, Heidelberg, pp 21–58

Cherns L (1982) Paleokarst, tidal erosion surfaces and stromatolites in the Silurian Eke Formation of Gotland, Sweden. Sedimentology 29:819–833

Corradini C, Olivieri R, Serpagli E (1995) Possible relationships between anomalous conodonts and Silurian oceanic episodes. N Jahrb Geol Paläontol Monats 12:737–746

Corradini C, Ferretti A, Serpagli E (1998a) The Silurian and Devonian sequence in SE Sardinia. In: Serpagli E (ed) Sardinia field-trip guide-book, ECOS VII. Giorn Geol 60(Spec Issue):71–74

Corradini C, Ferretti A, Serpagli E, Barca S (1998b) The Ludlow-Pridoli section "Genna Ciuerciu" west of Silius. In: Serpagli E (ed) Sardinia field-trip guide-book, ECOS VII. Giorn Geol 60:112–118

Corradini C, Ferretti A, Serpagli E (2002a) The Gerrei Tectonic Unit (SE Sardinia, Italy). Soc Paleontol Italiana Rendiconti 1:69–76

Corradini C, Ferretti A, Serpagli E (2002b) The "Ockerkalk" limestone in the Genna Ciuerciu Section (SE Sardinia, Italy). Soc Paleontol Italiana Rendiconti 1:255–260

Corradini C, Ferretti A, Corriga MG, Serpagli E (2009a) The reference section of the Sardinian Ockerkalk: the Silius section. Soc Paleontol Italiana Rendiconti 3:209–216

Corradini C, Ferretti A, Corriga MG, Serpagli E (2009b) Loboliths (crinoids) and conodont biostratigraphy of the Genna Ciuerciu section (SE Sardinia). Soc Paleontol Italiana Rendiconti 3:217–224

Cramer BD, Kleffner MA, Saltzman MR (2006) The late Wenlock 'Mulde' positive carbon isotope ($\delta^{13}C_{carb}$) excursion in North America. Geologiska Föreningens i Stockholm Förhandlingar 128:85–90

Cramer BD, Brett CE, Melchin MJ, Männik P, Kleffner MA, McLaughlin PI, Lloydell DK, Munnecke A, Jeppsson L, Corradini C, Brunton FR, Saltzman MR (2010) Revised correlation of Silurian provincial series of North America with global and regional chronostratigraphic and $\delta^{13}C_{carb}$ chemostratigraphy. Lethaia. doi:10.1111/j.1052-3931.2010.00234.x

Eriksson MJ, Calner C (2008) A sequence stratigraphical model for the late Ludfordian (Silurian) of Gotland, Sweden: implications for timing between changes in sea level, palaeoecology, and the global carbon cycle. Facies 54:253–276

Ferretti A, Serpagli E (1996) Geological outline, community sequence and paleoecology of the Silurian of Sardinia. Rivista Italiana di Paleontologia e Stratigrafia 102:353–362

Hede JE (1919) Djupborrningen vid Burgsvik på Gotland 1915. Paleontologisk-stratigrafiska resultat. Sveriges Geologiska Undersökning C298:1–59

Hede JE (1921) Gotlands Silurstratigrafi. Sveriges Geologiska Undersökning C305:100p

Hede JE (1925) Berggrunden (Silursystemet). In: Munthe H, Hede JE, von Post L (eds) Beskrivning till kartbladet Ronehamn. Sveriges Geologiska Undersökning Aa156: 14–51

Hede JE (1942) On the correlation of the Silurian of Gotland. In: Lunds Geologiska Fältklubb, 1892–1942, pp 205–229 [also in Meddelanden från Lunds Geologisk-Mineralogiska Institution 101]

Histon K, Schönlaub HP (1999) Taphonomy, palaeoecology and bathymetric implications of the nautiloid fauna from the Silurian of the Cellon section (Carnic Alps, Austria). Abh Geol Bundesanstalt (Wien) 54:259–274

Jeppsson L (1975 [1974]) Aspects of Late Silurian conodonts. Fossils Strata 6:79p

Jeppsson L (1990) An oceanic model for lithological and faunal changes tested on the Silurian record. J Geol Soc Lond 147:663–674

Jeppsson L (1993) Silurian events: the theory and the conodonts. Proc Estonian Acad Sci 42:23–27

Jeppsson L (1998) Silurian oceanic event. A summary of general characteristics. In: Landing E, Johnson ME (eds) Silurian cycles: linkages of dynamic stratigraphy with atmospheric, oceanic and tectonic changes, James hall centennial volume. N Y State Mus Bull 491:239–257

Jeppsson L (2005) Biases in the recovery and interpretation of micropalaeontological data. In: Purnell MA, Donoghue PCJ (eds) Conodont biology and phylogeny – interpreting the fossil record. Spec Pap Palaeontol 73:57–71

Jeppsson L (2006) Conodont-based revisions of the Late Ludfordian on Gotland, Sweden. Geologiska Föreningens i Stockholm Förhandlingar 127:273–282

Jeppsson L, Aldridge RJ (2000) Ludlow (late Silurian) oceanic episodes and events. J Geol Soc London 157:1137–1148

Jeppsson L, Laufeld S (1987 [1986]) The Late Silurian Öved-Ramsåsa Group in Skåne, southern Sweden. Sveriges Geologiska Undersökning, Ser Ca 58:1–45 [also as Lund Publications in Geology 55]

Jeppsson L, Viira V, Männik P (1994) Silurian conodont-based correlations between Gotland (Sweden) and Saaremaa (Estonia). Geol Mag 131:201–218

Jeppsson L, Talent JA, Mawson R, Simpson AJ, Andrew AS, Calner M, Whitford DJ, Trotter JA, Sandström O, Caldon H-J (2007) High-resolution Late Silurian correlations between Gotland, Sweden, and the Broken River region, NE Australia: lithologies, conodonts and isotopes. Palaeogeogr Palaeoclimatol Palaeoecol 245:115–137

Kaljo D, Martma T (2006) Application of carbon isotope stratigraphy to dating the Baltic Silurian rocks. Geologiska Föreningens i Stockholm Förhandlingar 128:123–129

Kaljo D, Kiipli T, Martma T (1997) Correlation of carbon isotope event markers through the Wenlock–Pridoli sequence at Ohesaare (Estonia) and Priekule (Latvia). Palaeogeogr Palaeoclimatol Palaeoecol 132:211–223

Kaljo D, Kiipli T, Martma T (1998) Carbon isotope events and environmental cyclicity in the east Baltic Silurian. In: Landing E, Johnson ME (eds) Silurian cycles: linkages of dynamic stratigraphy with atmospheric, oceanic and tectonic changes. James Hall Centennial Volume. N Y State Mus Bull 491:197–312

Klapper G, Murphy MA (1975 [1974]) Silurian–Lower Devonian conodont sequence in the Roberts Mountains Formation of central Nevada. Univ California Publ Geol Sci 111

Koren' T (1993) Main event levels in the evolution of the Ludlow graptolites. Stratigr Geol Correl 1:44–52 (in Russian)

Kozłowski W (2003) Age, sedimentary environment and palaeogeographical position of the Late Silurian oolitic beds in the Holy Cross Mountains (Central Poland). Acta Geol Polonica 53:341–357

Kreutzer LH, Schönlaub HP (1997) The depositional environment. In: Schönlaub HP (ed) IGCP project 421 North Gondwanan Mid Palaeozoic Biodynamics inaugural meeting Vienna (Sept 17–21, 1997). Berichte Geol Bundesanstalt (Wien) 40:99–105

Kříž J (1992) Silurian field excursions Prague Basin (Barrandian), Bohemia. Natl Mus Wales, Geol Ser 13:111pp

Kříž J, Schönlaub HP (1980) Stop 1. Daleje Valley, Muslowka Quarry section Ludlovian-Pridolian. In: Schönlaub HP (ed) Second European Conodont symposium (Ecos II) guide-book, abstracts. Abh Geol Bund (Wien) 35:153–157

Lawson JD (1960) The succession of shelly faunas in the British Ludlovian. Int Geol Congr, Rept Twenty-First Session, 1960, Part VII, Proc Sec 7 Ordovician and Silurian Stratigraphy and Correlations, Copenhagen, pp 114–125

Lehnert O, Fryda J, Buggisch W, Manda S (2003) A first report of the Ludlow Lau event from the Prague Basin (Barrandian, Czech Republic). In: Ortega G, Aceñolaza GF (eds) Proc 7th international Graptolite conference and field meeting of the international subcommission on Silurian Stratigraphy. Ser Correl Geol 18:139–144

Lehnert O, Fryda J, Buggisch W, Munnecke A, Nützel A, Kříž J, Manda S (2007) $\delta^{13}C$ records across the late Silurian Lau event: new data from middle palaeo-latitudes of northern peri-Gondwana (Prague Basin, Czech Republic). Palaeogeogr Palaeoclimol Palaeoecol 245:227–244

Liljedahl L (1994) Silurian nucoloid and modiomorfid bivalves from Sweden. Fossils Strata 33:1–89

Link AG, Druce EC (1972) Ludlovian and Gedinnian conodont stratigraphy of the Yass Basin, New South Wales. Bur Min Res Geol Geophys Bull 134:136pp

Loydell DK (1998) Early Silurian sea-level changes. Geol Mag 135:447–471

Manda S, Kříž J (2006) Environmental and biotic changes in subtropical isolated carbonate platforms during the Late Silurian kozlowskii Event, Prague Basin. Geologiska Föreningens i Stockholm Förhandlingar 128:161–168

Männik P, Małkowski KC (1998) Silurian conodonts from the Goldap core, Poland. In: Szaniawski H (ed) Proc Sixth European Conodont symposium (ECOS VI). Paleontologia Polonica 58:141–151

Marshall CR (1990) Confidence intervals on stratigraphic ranges. Paleobiology 16:1–10

Martinsson A (1967) The succession and correlation of ostracode faunas in the Silurian of Gotland. Geologiska Föreningens i Stockholm Förhandlingar 89:350–386

Munthe H (1902) Stratigrafiska studier öfver Gotlands silurlager. Geologiska Föreningens i Stockholm Förhandlingar 24: 272–273

Ogg JG, Ogg G, Gradstein FM (2008) The concise geological time scale. Cambridge University, Cambridge

Patterson WP, Walter LM (1994) Depletion of ^{13}C in seawater $\sum CO_2$ on modern carbonate platforms: significance for the carbon isotopic record of carbonates. Geology 22:865–888

Rexroad CB, Craig WW (1971) Restudy of the conodonts from the Bainbridge formation (Silurian) at Lithium Missouri. J Paleont 45:468–703

Salzman MR (2001) Silurian $\delta^{13}C$ stratigraphy: a view from North America. Geology 2001 August:671–674

Samtleben C, Munnecke A, Bickert T, Pätzold J (1996) The Silurian of Gotland (Sweden). facies interpretation based on stable isotopes in brachiopod shells. Geol Rundschau 85:278–292

Samtleben C, Munnecke A, Bickert T (2000) Development of facies and C/O-isotopes in transects through the Ludlow of Gotland: evidence for global and local influences on a shallow-marine environment. Facies 43:1–38

Sarv L (1982) On ostracode zonation of the East Baltic Upper Silurian. In: Kaljo D, Klaamann E (eds) Ecostratigraphy of the East Baltic Silurian. Academy of Sciences of the Estonian SSR, pp 71–77

Serpagli E, Corradini C, Ferretti A (1998) Conodonts from a Ludlow–Pridoli section near the Silius Village. In: Serpagli E (ed) Sardinia field-trip guide-book, ECOS VII; Giornale di Geologia 60(spec issue):104–111

Simpson A (2000) Silurian to basal Devonian conodonts from the Broken River Crossing, northern Australia. Rec Western Australian Mus Suppl 58:145–162

Spjeldnaes N (1950) On some vertebrate fossils from Gotland with some comments on the stratigraphy. Arkiv Mineral Geol 1(8):211–218

Štorch P (1995) Upper Silurian (upper Ludlow) graptolites of the *N. inexpectatus* and *N. kozlowski* zones from Kosov Quarry near Beroun (Barrandian area, Bohemia). Vestnik Ceskeho Geologickeho Ustavu 70:65–89

Urbanek A (1997) Late Ludfordian and early Prídolí mono-graptids from the Polish Lowland. In: Urbanek A, Teller L (eds) Silurian graptolite faunas in the East European Platform: stratigraphy and evolution. Palaeontol Polonica 56: 23–57

Uyeno TT (1981 [1980]) Stratigraphy and conodonts of Upper Silurian and Lower Devonian rocks in the environs of the Boothia Uplift, Canadian Arctic Archipelago. Part II, Systematic study of conodonts. Geol Surv Canada Bull 292:39–75

Viira V (1982) Late Silurian shallow and deep water con-odonts of the East Baltic. In: Kaljo D, Klaamann E (eds) Ecostratigraphy of the East Baltic Silurian. Estonian SSR, Academy of Sciences, pp 79–87

Walliser OH (1964) Conodonten des Silurs. Abh Hessischen Landesamtes Bodenforsch Wiesbaden 41:106pp

Wenzel B (1997) Isotopenstratigraphische Untersuchungen an silurischen Abfolgen und deren paläozeanographische Interpretation. Erlanger Geol Abh 129:117pp

Wenzel B, Joachimski MM (1996) Carbon and oxygen isotopic composition of Silurian brachiopods (Gotland/ Sweden): palaeoceanographic implications. Palaeogeogr Palaeoclimatol Palaeoecol 122:143–166

Wigforss-Lange J (1999) Carbon isotope $\delta^{13}C$ enrichment in Upper Silurian (Witcliffian) marine calcareous rocks in Scania, Sweden. Geologiska Föreningens i Stockholm Förhandlingar 121:273–279

Wigforss-Lange J (2007) Tidal facies in the Upper Silurian Öved-Ramsåsa Group of Scania, Sweden: linkages of radial and cerebroid ooids and evaporite tracers to subtidal, lagoonal environment. Geologiska Föreningens i Stockholm Förhandlingar 129:7–16

The Late Middle Devonian (Givetian) Global Taghanic Biocrisis in Its Type Area (Northern Appalachian Basin): Geologically Rapid Faunal Transitions Driven by Global and Local Environmental Changes

James J. Zambito IV, Carlton E. Brett, and Gordon C. Baird

Abstract

The late Middle Devonian 'Taghanic (Pharciceras) Event' was originally named by Michael House for goniatite turnovers in the New York Appalachian Basin during deposition of the Tully Limestone; subsequently, it has been associated with the extinction of most of the long-lasting 'Hamilton Fauna' in this region. Stratigraphic and paleoecologic research in the type area has revealed at least three main faunal transitions, recognized as discrete bioevents: (1) replacement of much of the endemic 'Hamilton Fauna' (a subset of the Eastern Americas Realm) with the previously equatorial 'Tully Fauna' (a subset of the Old World Realm); (2) subsequent extermination of most of the 'Tully Fauna' and recurrence of the 'Hamilton Fauna', coincident with the eustatic sea-level rise termed the 'Taghanic Onlap'; and (3) extinction of much of the 'Hamilton Fauna' and return of some 'Tully' taxa along with a further incursion of Old World Realm taxa during continued rise in global sea level. The Taghanic Biocrisis is currently recognized globally as a series of pulsed biotic transitions and extinctions, ultimately resulting in an end to previous faunal provinciality and appearance of a global cosmopolitan fauna. We review the current knowledge of these faunal transitions in the type area with respect to geologically rapid global and local environmental changes observed using a high-resolution stratigraphic framework across the entire onshore-offshore environmental gradient. Globally recognized environmental changes, specifically temperature increases, changes between arid and humid intervals, rapid sea-level fluctuations, and widespread black shale deposition account for the faunal transitions discriminated in the type area, but only in the context of regional basin dynamics associated with basin morphology and the degree to which estuarine-type watermass circulation patterns were operating, resulting in salinity variation as a dominant control on faunal distribution. Herein, we outline the interplay between global and local environmental changes that served as driving forces behind the local incursions and extinctions, including the demise of the long-stable 'Hamilton Fauna'.

J.J. Zambito IV (✉)
Department of Earth and Atmospheric Sciences, Central Michigan University, Mt. Pleasant, MI 48859, USA
e-mail: zambi1jj@cmich.edu

J.A. Talent (ed.), *Earth and Life*, International Year of Planet Earth,
DOI 10.1007/978-90-481-3428-1_22,

Keywords
New York • Hamilton group • Tully formation • Genesee group • Geneseo formation • Taghanic Biocrisis • Subevents • Recurrent faunas • Extinctions • Faunal transitions • Eustasy

Introduction

During the *Global Taghanic Biocrisis* (*GTB*), Middle Devonian faunas worldwide underwent a major extinction in apparent conjunction with eustatic sea-level rise, intensified oceanic circulation, increased warming, carbon-cycle changes, reduced oxygenation in epicontinental seas, and increased aridity (Johnson 1970; Murphy et al. 2000; House 2002; Aboussalam 2003; Joachimski et al. 2004, 2009; Hüneke 2006, 2007; van Geldern et al. 2006; Marshall et al. 2011; Aboussalam and Becker 2011). House (2002) estimated that the impact of this biocrisis on biodiversity may have been as great, or greater, than the well-known Frasnian–Fammenian crisis. Johnson (1970) and Boucot (1988) describe this time as the end of an established Devonian faunal provinciality, resulting in cosmopolitan faunas worldwide. Moreover, McGhee (1982, 1989, 1996) and Sandberg et al. (2002) describe this interval as a time of faunal crisis, lasting approximately 7 million years and culminating in the Frasnian–Fammenian extinction. Studies of marine invertebrate global biodiversity trends at various taxonomic scales through the Phanerozoic recognize the *GTB* interval as the beginning of a period of protracted extinction with no clear end until the beginning of the Permian, the duration of which has likely prevented this biocrisis from being recognized as a mass extinction (Raup and Sepkoski 1982; Alroy et al. 2008). Nevertheless, the more frequent events that still result in faunal restructuring and replacement at regional to global scales may have a greater aggregate effect on the evolution of life than the five or six most prominent first-order life crises (Walliser 1990; Brett and Baird 1995; Brett et al. 1996; Miller 1998).

The type area of the *GTB* is the Appalachian Basin deposits of New York State (Fig. 1). The *GTB* (House 1985) is named after the Taghanic Onlap of Johnson (1970), which in turn is named for the exposures along Taughannock Creek, near the Village of Trumansburg in central New York State (Fig. 2). Taughannock and Taghanic, along with several other variants, have all been accepted names and spellings for this creek, the waterfalls along it, and the sections exposed therein (Hall 1843; Kurtz 1883; Cooper and Williams 1935). Interestingly, the *GTB* and the Taghanic Onlap were both named for the paleontologic and lithologic changes seen at the base of the Tully Formation. However, as pointed out in Baird and Brett (2003), the greatest transgression of this interval actually commences with deposition of the upper part of the Tully Formation and continues into the overlying Geneseo Formation (Fig. 2).

In describing the biotic changes observed during the *GTB*, we prefer the terminology outlined in Walliser (1996) and refined in Aboussalam and Becker (2011): a biocrisis is a perturbation interval spanning more than one biozone and comprised of a series of bioevents that each generally occur within a single biozone, black shale interval, and/or one transgressive-regressive couplet. In the type area of this extinction, the Taghanic Biocrisis has been extensively documented at high-stratigraphic resolution as a series of at least three bioevents involving faunal incursions, replacements, recurrences, and extinctions, resulting ultimately in the disappearance of numerous ammonoid, coral, and brachiopod taxa through an interval of approximately 500,000 years (House 2002, and references therein; Baird and Brett 2003, 2008; Kaufmann 2006; Zambito et al. 2007, 2009). These transitions include, in upward succession, (1) replacement of much of the endemic 'Hamilton Fauna' (a subset of the Eastern Americas Realm) with the previously equatorial 'Tully Fauna' (a subset of the Old World Realm) during the *Lower Tully Bioevent*; (2) subsequent extermination of most of the 'Tully Fauna' and recurrence of the 'Hamilton Fauna' during the *Upper Tully Bioevent*, coincident with the eustatic sea-level rise referred to as the 'Taghanic Onlap'; and (3) extinction of much of the 'Hamilton Fauna' and return of some 'Tully' taxa along with further incursion of Old World Realm taxa during continued rise in global sea level during the *Geneseo Bioevent*.

The Taghanic Biocrisis is now recognized globally as a series of pulsed biotic transitions and extinctions, ultimately resulting in an end to previous

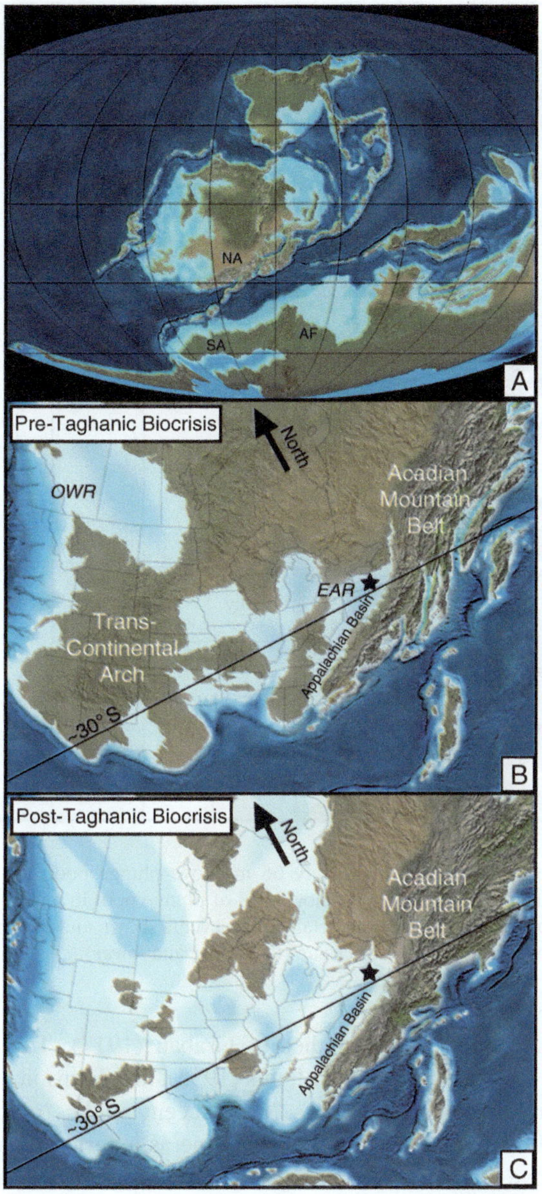

Fig. 1 Paleogeographic reconstruction of the globe (**a**) and study area, respectively, before (**b**) and after (**c**) the Taghanic Biocrisis. The Appalachian Basin strata of New York State (type area, represented by '*star*') that record this biocrisis were deposited approximately 30° south of the paleo-equator. OWR and EAR denote the positions of 'Old World Realm' and 'Eastern Americas Realm' faunas, respectively; note the position of the Trans-Continental Arch ('Continental Backbone' of Johnson 1970) relative to these faunas. Abbreviations NA, SA, and AF stand for North America, South America, and Africa, respectively. Paleomaps adapted from Blakey (2010)

faunal provinciality and the appearance of a global cosmopolitan fauna (Johnson 1970; Boucot 1988; Oliver 1990; Feist 1991; Day 1996; Day et al. 1996; Brice et al. 2000; Aboussalam and Becker 2001, 2011; House 2002; Aboussalam 2003, and references therein). As a globally recognized biocrisis, it is therefore imperative to fully understand the interplay between global and local environmental changes in the type area where this biocrisis is recorded in a relatively complete stratigraphic succession along an entire onshore–offshore gradient. Only with complete understanding of the faunal patterns and environmental changes in the type area can we accurately understand this biocrisis at the global scale. In the following section, we review the faunal transitions observed during the *GTB* in the type area by reconstructing biofacies spectra before, during, and after the biocrisis; outline the interplay between global and local environmental changes driving these transitions; and discuss the implications of these observations for evolutionary paleoecology.

Geologic Setting

The type area of the late Middle Devonian Taghanic Biocrisis is located in the northern Appalachian Basin deposits of New York and nearby states, herein referred to as the NAB (northern Appalachian Basin) (Figs. 1 and 2; see also House 1985; House 2002; Baird and Brett 2003). At this time, the NAB was located approximately 30° south latitude. Regional strata were deposited in a foreland basin that formed during the Acadian Orogeny as the Laurentian and Avalonian terranes converged obliquely (Fig. 1; see also Ettensohn 1985; Ettensohn et al. 1988; ver Straeten and Brett 1995, 1997). Erosion of the collisional highlands produced the classic progradational complex known widely as the 'Catskill Delta'; it advanced in a generally westward direction and largely filled the foreland basin by the early Mississippian. The *GTB* occurs during the transition between the second and third collisional tectophases of the Acadian Orogeny (Ettensohn 1985; Ettensohn et al. 1988; ver Straeten and Brett 1995, 1997). During each tectophase, orogenic activity occurred along the eastern seaboard of

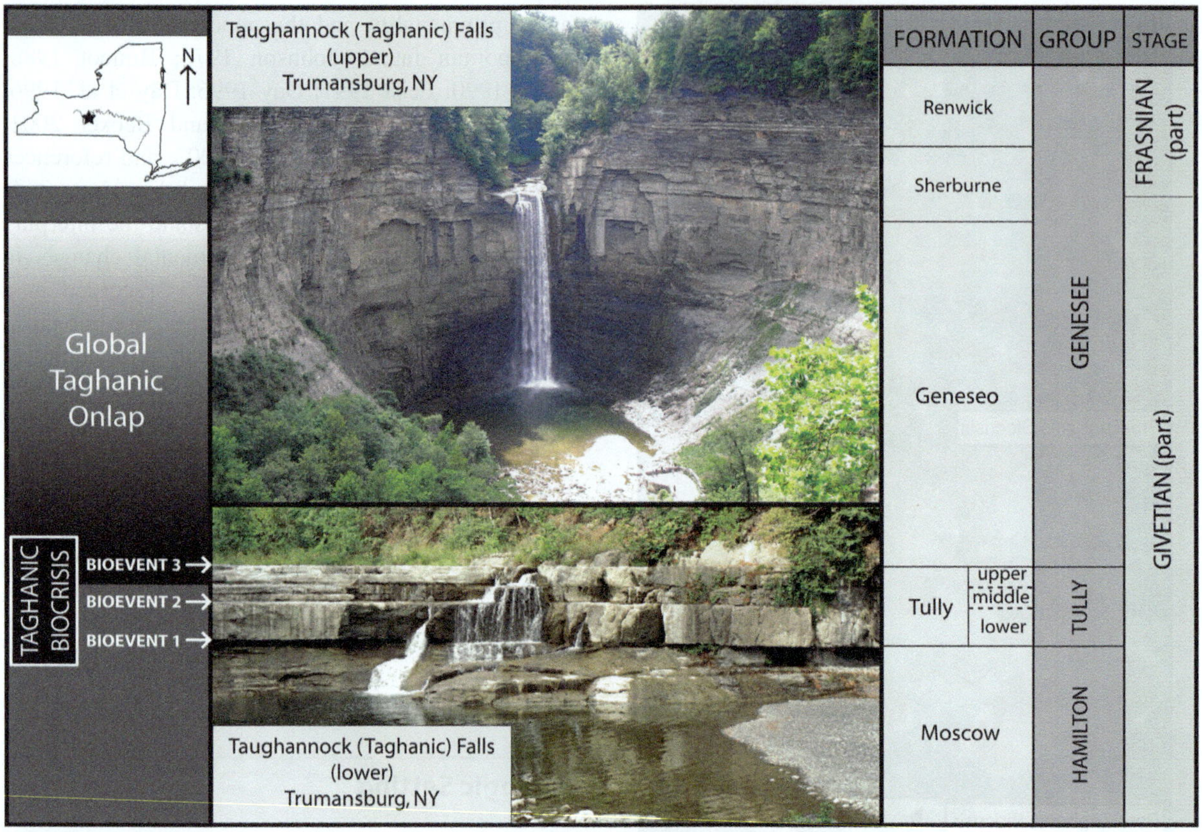

Fig. 2 Type section of the Global Taghanic Biocrisis, Taughannock (Taghanic) Falls, Trumansburg, Tompkins County, New York (represented by '*star*' on inset figure of Tully Group outcrop belt). Lower falls is approximately 6 m high. Lip of upper falls is approximately 66 m above plunge pool; top of Tully Group is slightly below level of plunge pool in picture of upper falls. See text for details of bioevents shown and informal divisions within Tully Group (i.e., *upper* and *lower*)

North America (Laurentia) causing subsidence of the Appalachian Basin in response to tectonic loading on the crust and a resultant influx of sediment into the basin from erosion of the complementing mountains.

In the NAB, the *GTB* is recorded primarily in the sediments of the Tully Group, but immigrations and extinctions related to the *GTB* can be observed first within the upper Hamilton Group and then continue into the lower Genesee Group (Sessa 2003; Baird et al. 2003; Baird and Brett 2003, 2008; Zambito et al. 2007, 2009). These deposits can be observed in an east-west outcrop belt across New York State that is relatively complete stratigraphically due to near-continuous subsidence during this time and, furthermore, represents a complete onshore–offshore gradient as the outcrop belt is normal to depositional strike (Fig. 3). The Tully Formation has received detailed investigation through the years; the reader is referred to the studies of Grabau (1917), Trainer (1932), Cooper and Williams (1935), and Johnson and Friedman (1968, 1969). More recent studies of the Tully Formation interpret this time as a period of relative tectonic quiescence when the second tectophase was waning and sediment input to the basin was relatively minimal (Heckel 1973; Baird et al. 2003; Baird and Brett 2003, 2008).

Sediment supply to the basin was further controlled by syn-depositional formation of a fault-controlled 'clastic trap,' resulting in three distinct yet concurrent depositional facies for the Tully Formation (Fig. 4; Heckel 1973; Baird et al. 2003; Baird and Brett 2003), including (1) a broad epicontinental shelf across western and central New York and northwest Pennsylvania where the anomalous Tully limestone deposit accumulated under clastic sediment-starved conditions; (2) in east-central New York and in central Pennsylvania, a

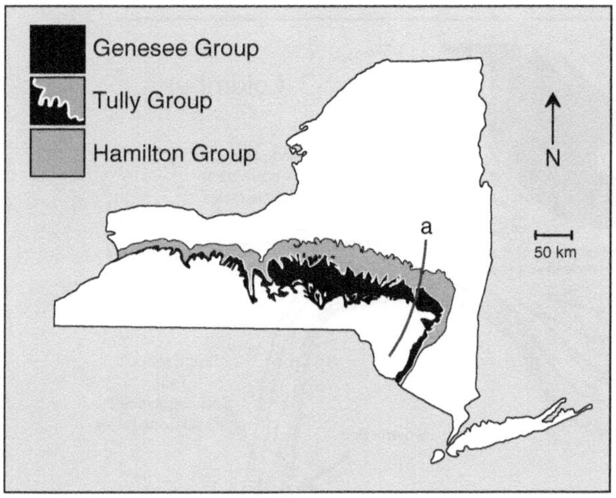

Fig. 3 Outcrop belt for the Hamilton, Tully, and Genesee groups in the Appalachian Basin deposits of New York State. *Line* labeled 'a' represents the approximate position of the shoreline during deposition of the Tully Group. *Star* denotes type locality (see Fig. 2)

structural trough was developed separating the platform from areas of clastic sediment supply from the east and southeast characterized by sparsely fossiliferous dysoxic facies; and (3) an eastern shelf area connected to paralic habitats along the paleocoastline that was variably supplied by terrigenous sediments from that coast, which, together with the sediments deposited in the trough, has been referred to as the TFCC (Tully Formation Clastic Correlative) (Figs. 4 and 5; Heckel 1973; Baird et al. 2003; see also Heckel 1997, and ver Straeten and Brett 1997 for alternative views on the 'clastic trap' formation).

In comparison, depositional lithofacies during the preceding upper Hamilton Group were similar to but much more gradational than Tully deposits. For example, the upper Hamilton Group of New York also displays an eastern shelf, centrally located dysoxic basin, and western carbonate-rich facies. However, during upper Hamilton times the basin exhibited a gently dipping ramp profile, resulting in an overall siliciclastic-dominated basin in which marker beds representing discrete time intervals can be traced along the entire onshore–offshore gradient (Fig. 5; Brett and Baird 1985, 1994). With onset of the third tectophase of the Acadian Orogeny, coincident with a global sea-level rise (Taghanic Onlap), the basin once again changed drastically as it subsided, and migrated westward. Input of siliciclastic sediment into the western NAB resumed and greatly intensified, manifested in the deep-water, anoxic deposits of the Geneseo black shale and subsequent formations and progradation of a distinct slope setting (Fig. 5; Johnson 1970; Zambito et al. 2007, 2009).

Taghanic Biocrisis in the Type Area: Sequence of Faunal Transitions

Pre-biocrisis Interval

Boucot et al. (1969), Johnson (1970), Boucot (1975, 1988), and Koch and Boucot (1982) suggested that brachiopod-dominated faunas of the Devonian could be divided globally into biogeographic 'realms,' inferring that the distribution of realms was controlled by climatic (temperature) gradients between the equator and poles. At the onset of the *GTB*, these realms in North America consisted of the equatorial Old World Realm (OWR), and the higher latitude, slightly cooler water, Eastern Americas Realm (EAR) (Fig. 1b). Prior to the Taghanic Biocrisis, genera that comprised the 'Tully Fauna' (a subset of the OWR Cordilleran province) occurred in what is now the western United States and western Canada in settings more equatorial than the Appalachian Basin. The 'Tully Fauna' that eventually came to occupy the NAB was a low-diversity, primarily benthic fauna, dominated by brachiopods and almost entirely devoid of mollusks (see faunal lists of Cooper and Williams 1935; Heckel 1973). Prior to this time, the NAB was occupied by the highly diverse 'Hamilton Fauna' (a subset of the EAR Appohimchi Province) comprised of various functional forms, conspicuous encrusting biotas and large trace fossils, for a period of approximately 4–5 million years (Brett and Baird 1995; Brett et al. 1996). During this time, the Hamilton Fauna was characterized by a pattern of long-term stability of biofacies associations among fossil communities (coordinated stasis), comprised of species lineages that appear to track their preferred habitat rather than adapt to locally changing conditions (Brett and Baird 1995; Brett et al. 1996, 2007a, b; Brett et al. 2009; Ivany et al. 2009). Additionally, detailed morphological studies of various organisms indicate morphological stasis through the Hamilton interval as well (Sorauf and Oliver 1976; Lieberman 1994; Lieberman et al. 1995).

Toward the end of this stable interval, pre-biocrisis incursions of OWR taxa are observed within the

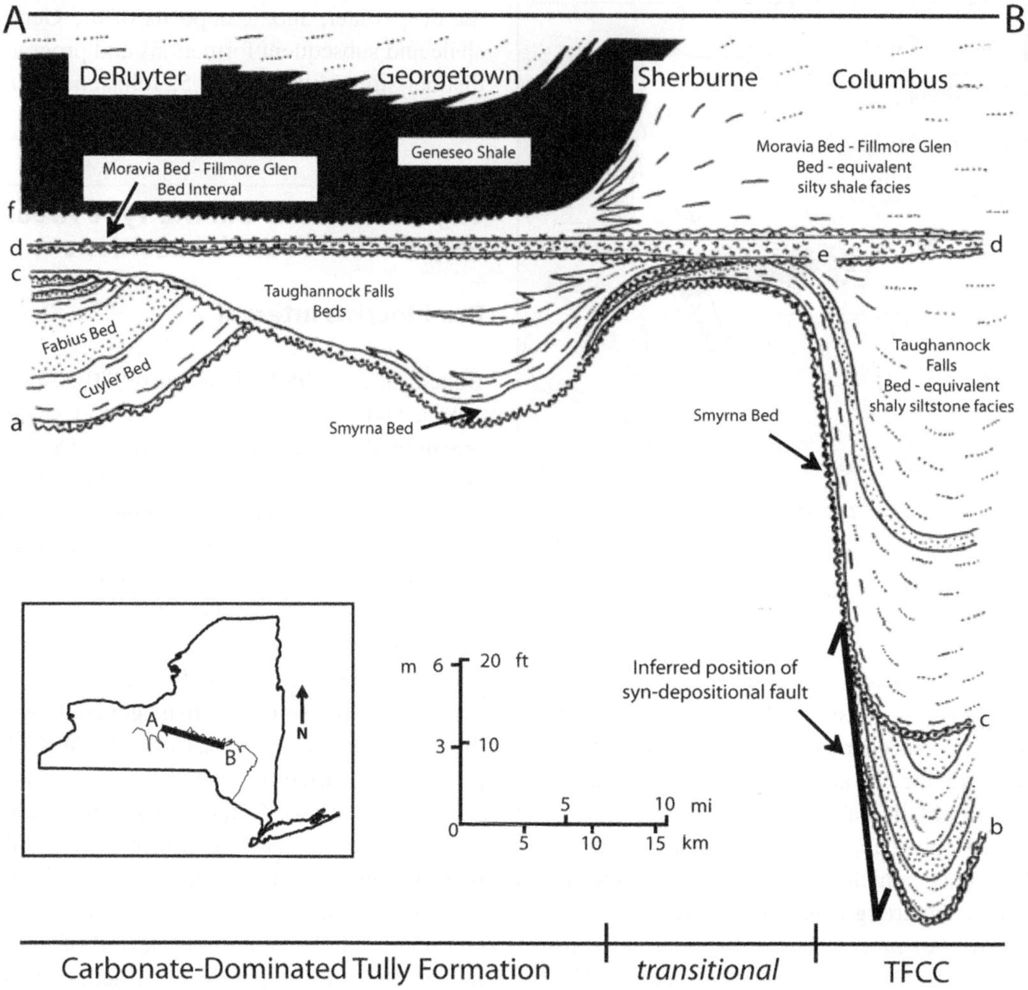

Fig. 4 Regional cross-section of the Tully and basal Genesee Group across east-central New York State showing inferred correlations of Tully Group divisions with equivalents in the Tully Formation Clastic Correlative succession (TFCC). Location of section along outcrop belt is shown in inset figure. The underlying Hamilton Group is not shown for ease of Tully unit visualization. The transect shows the rapid eastward thickening of the Tully clastic correlative succession across a possible growth fault east of Sherburne, New York; the sediment-starved, western (structurally upthrown) side of this structure ('Sherburne High') bounded a local, structurally downthrown, basinal trough which served as a clastic trap. Note the prominent disconformable nature of the mid-Tully sequence unconformity (**c**) and the upper Tully sequence unconformity (**d**). Note also the striking thinness and uniformity of thickness of the Bellona Bed/West Brook Shale (**e**) relative to underlying and overlying units (see text). The Tully-Genesee contact (**f**) is unconformable across part of the region as shown. Lettered features include the following: (**a**) DeRuyter Bed; (**b**) chamositic-sideritic bed believed to be equivalent to DeRuyter Bed; (**c**) Smyrna Bed floored by mid-Tully sequence disconformity; (**d**) upper Tully sequence disconformity; (**e**) Bellona Bed/West Brook Shale; and (**f**) black shale-roofed corrosional discontinuity with associated detrital pyrite marking the Tully-Geneseo contact. Figure adapted from Baird and Brett (2003)

upper Hamilton Group (Fig. 6). Beginning during the deposition of the Amsdell Beds and correlative Fisher Gully submember, *Emanuella praeumbona* (an OWR genus) becomes common in upper Hamilton Group dysoxic dark gray shales and siltstones. Notably, the upward change from oxic shelf to dysoxic shale facies within the Sheds through Highland Forest submembers suggests that diverse brachiopod facies typical of the Hamilton Group occupied shallower, better oxygenated settings, whereas OWR taxa replaced typical dysaerobic Hamilton biofacies (Baird and Brett 2003; Sessa 2003).

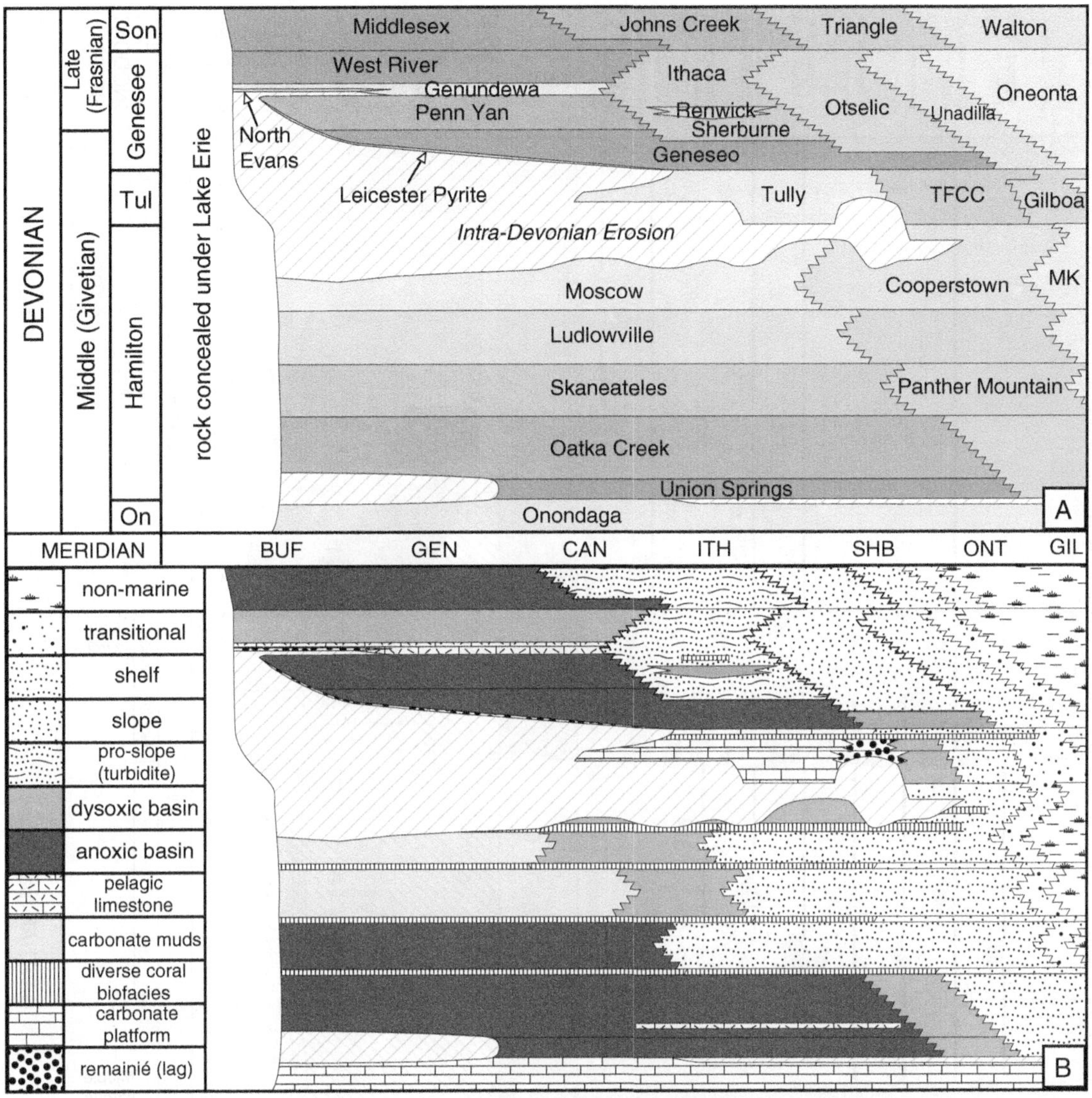

Fig. 5 (**a**) Time-stratigraphic chart for the strata of New York that record times prior to, during, and following the Taghanic Biocrisis, including strata of the Onondaga (On), Hamilton, Tully (Tul), Genesee, and Sonyea (Son) groups. For the most part, stratigraphic divisions within groups are at the level of formation. Meridians are Buffalo (BUF), Geneseo (GEN), Canandaigua (CAN), Ithaca (ITH), Sherburne (SHB), Oneonta (ONT), and Gilboa (GIL). (**b**) Environments of deposition overlain on the chronostratigraphic framework of part a of this figure. Note that slope and pro-delta slope environments did not exist prior to the Genesee Group, and diverse coral biofacies (for the most part) do not occur following the Taghanic Biocrisis. Figure is adapted from Rogers et al. (1990) using data and information from the following references: Cooper and Williams (1935), Rickard (1975, 1981), Brett et al. (1986), Grasso (1986), Brett and Baird (1985, 1994), Bridge and Willis (1991, 1994), House (2002), Baird and Brett (2003, 2008), Baird et al. (2006), Desantis et al. (2007), Zambito et al. (2007, 2009) and J. Zambito (unpublished data). For a complete list of diverse coral and brachiopod horizons of the Hamilton Group, see Fig. 4 of Brett et al. (2007a)

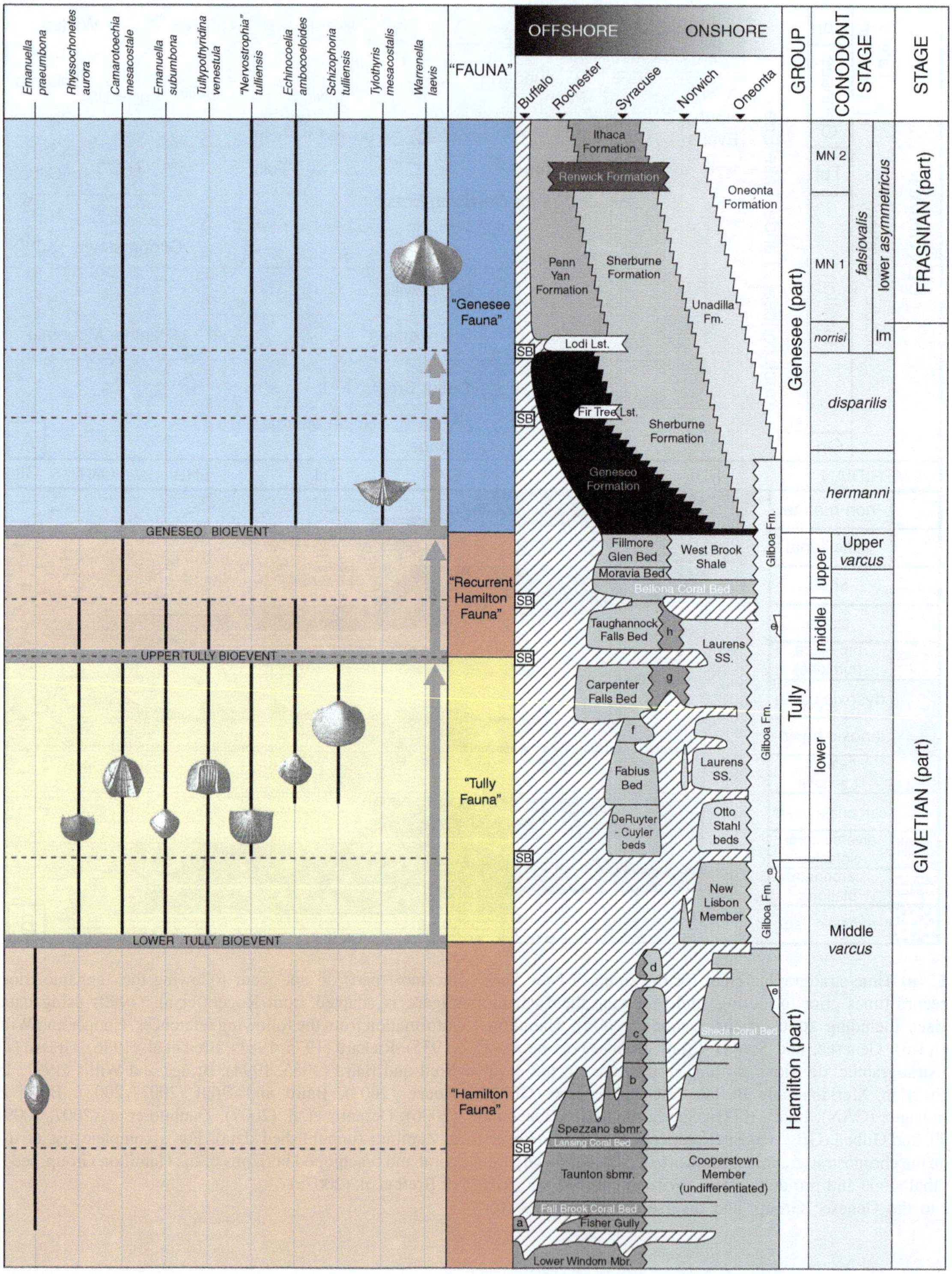

Fig. 6 Time-stratigraphic chart showing the ranges of 'Old World Realm' taxa in the New York State Appalachian Basin during the Global Taghanic Biocrisis. Bioevents (i.e., Lower Tully, Upper Tully, and Geneseo) are marked at the respective horizons for which each bioevent commences; however, the effects of each bioevent (migrations) are seen over a given interval. SB stands for sequence boundary. Note that, for a number of taxa, ranges cross sequence boundaries (dashed lines), suggesting that the faunal transitions observed are real, and not an artifact of unconformities. Lettered units in '*white*' represent

Lower Tully Bioevent

The onset of Taghanic biotic instability (*Lower Tully Bioevent*) is marked by step-wise incursion of 'Tully Fauna' taxa into the NAB as observed in the New Lisbon Member (uppermost Hamilton Group) and the lower Tully Formation (Figs. 6 and 7). The New Lisbon Member succession is characterized by the first occurrence of the OWR taxa *Camarotoechia mesacostalis*, *Rhyssochonetes aurora*, and *Tullypothyridina venustula*. Although *C. mesacostalis* is abundant within the dysoxic, flaggy siltstone facies of the New Lisbon succession, *R. aurora* and *T. venustula* occur only rarely and, furthermore, these OWR associations are gradational through the upward-shallowing New Lisbon succession to generalistic taxa of the 'Hamilton Fauna' in more oxic facies (Fig. 7; Baird et al. 2003; Sessa 2003; Baird and Brett 2008).

The incursion of the 'Tully Fauna' becomes further pronounced in the overlying lower Tully Formation where *R. aurora* and *T. venustula* become more common, and within the succeeding TFCC in which the first appearance of OWR taxa *'Nervostrophia' tulliensis, Schizophoria tulliensis*, and *Echinocoelia ambocoeloides* is observed. By this time, the 'Tully Fauna' is found in oxic, outer shelf facies, and the 'Tully Fauna' prominently occupies both the carbonate and siliciclastic (TFCC) portions of the onshore–offshore gradient (Figs. 6 and 7) (Cooper and Williams 1935; Heckel 1973; Baird et al. 2003; Baird and Brett 2008).

Upper Tully Bioevent

The second phase of this biocrisis, termed the *Upper Tully Bioevent*, represents the loss of the majority of the Tully Fauna from the Appalachian Basin and recurrence of the 'Hamilton Fauna' in the NAB (Figs. 6 and 7). To date, there has been little progress in identifying the refugia of the Hamilton during lower Tully times; possibilities include subsequently eroded areas of the Canadian Shield and South America (see the following discussion, and Boucot et al. 1986; Brett et al. 2007a). The Upper Tully Event commences with deposition of the Taughannock Falls Bed and correlative TFCC in which a return of dysoxic to mid-shelf 'Hamilton Fauna' is observed (Baird et al. 2003; Baird and Brett 2003). Facies occupied by the 'Hamilton Fauna' during this time are observed to grade laterally into low-diversity communities of the 'Tully Fauna' in more basinal TFCC facies in east-central New York and Pennsylvania (Baird and Brett 2008). Therefore, the recurrence of the 'Hamilton Fauna,' like the incursion of 'Tully Fauna,' was step-wise and, furthermore, rather symmetrical; 'Tully taxa' first appear in more basinal facies relative to 'Hamilton taxa,' and the 'Hamilton' taxa first recur in more proximal facies relative to 'Tully' taxa. Additionally, it is observed that the generalistic taxa of each fauna are the first to appear with each faunal transition, followed later by more specialized forms.

The most diverse representation of the Hamilton faunal elements that recur in the NAB is best observed in the Bellona-West Brook Beds; and consists of over 100 species of corals, bryozoans, brachiopods, mollusks, trilobites, and echinoderms (see Cooper and Williams 1935; Heckel 1973 for extensive faunal lists). Quantitative paleoecological analysis across various spatial scales showed taxonomic stability between the recurrent 'Hamilton Fauna' of the upper Tully Formation and older occurrences of diverse coral-brachiopod facies within the Hamilton Group (Bonelli et al. 2006). This recurrence, while extensive in that almost all 'Hamilton Fauna' biofacies are represented, is geologically short-lived because the Taghanic Onlap continued rendering the basin inhospitable to most benthos, due to overspread of

Fig. 6 (continued) diverse coral-brachiopod communities of Hamilton taxa. Lettered units include: (a) Amsdell Bed, (b) Gage Gully submember, (c) Sheds submember, (d) Highland Forest submember, (e) non-marine (red bed) tongues extending westward into nearshore marine deposits (Gilboa Fossil Forest), (f) Vesper Bed, (g) Smyrna Bed (chamosite), and (h) the 'Taughannock Falls Chamosite Bed' of Heckel (1973). Diagonal lines represent areas of syn-depositional (intra-Devonian) erosion. Stratigraphy based on Baird and Brett (2008), Brett and Baird (1994), and Zambito et al. (2007, 2009). Conodont zonation is based on summary in Klapper (1981) and House (2002). Faunal information compiled from Williams (1884, 1890), Kindle (1896, 1906), Prosser (1899), Cooper and Williams (1935), Heckel (1973), Baird et al. (2003), Baird and Brett (2003, 2008), Sessa (2003), Bonelli et al. (2006), Zambito et al. (2007, 2009), and J. Zambito (unpublished work). Stratigraphic succession adapted from Baird and Brett (2008). Fossil pictures adapted from Linsley (1994) and Baranov and Alkhovik (2006)

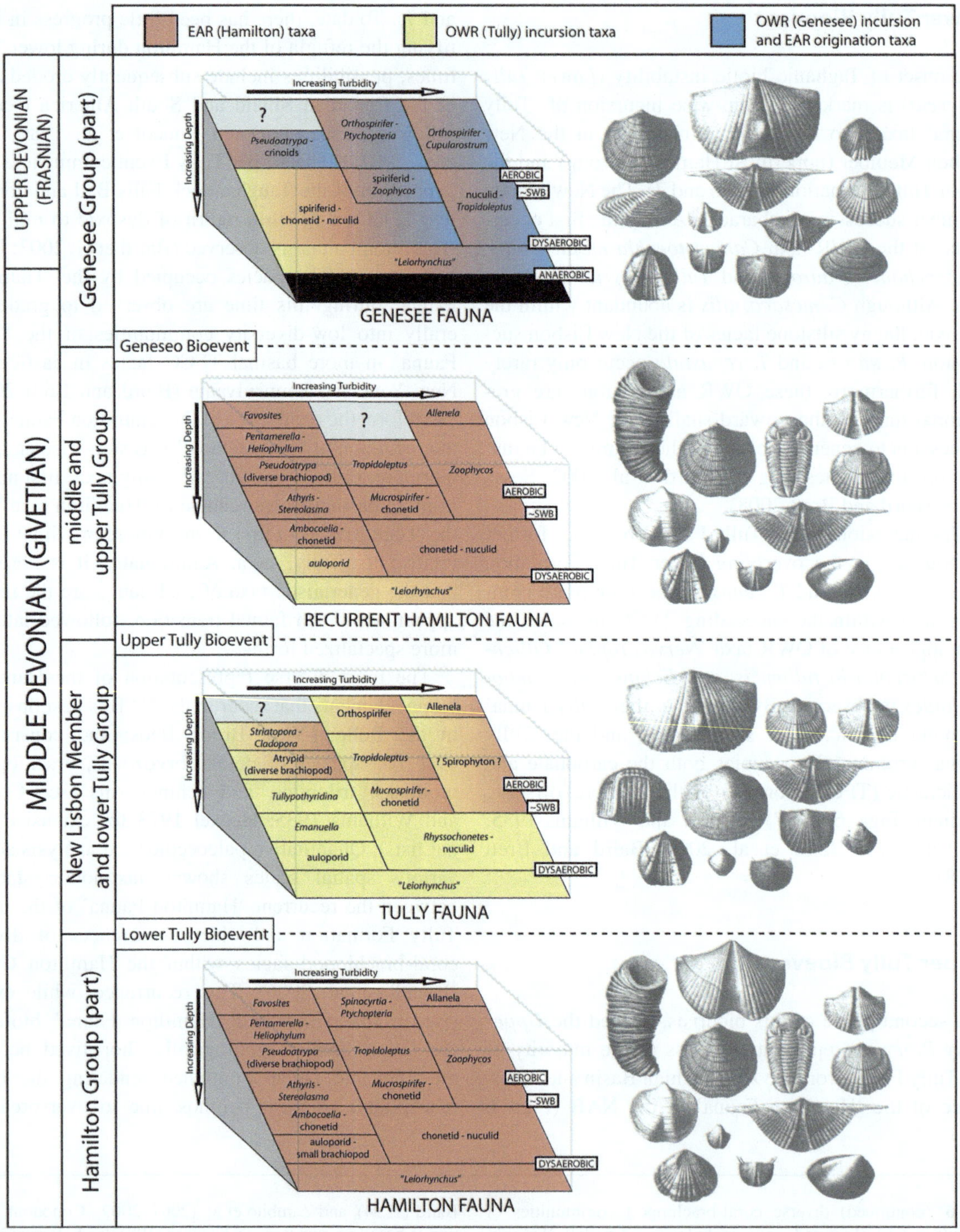

Fig. 7 New York Appalachian Basin biofacies spectrum through the Taghanic Biocrisis. Abbreviations include EAR, 'Eastern Americas Realm' and OWR, 'Old World Realm.' Portions of the spectrum denoted by '?' indicate unknown biofacies resulting from non-deposition, erosion, or lack of outcrop for that facies. Note selective, depth/salinity-controlled diachroneity, respectively, for OWR biota, recurrent 'Hamilton Fauna,' and key Genesee Fauna elements indicated by spectral color shading. Fossil plates are used to illustrate representative taxa of the faunas and are adapted from Linsley (1994). The biofacies spectrum of the 'Hamilton Fauna' was first proposed by Brett et al. (1990) and demonstrated quantitatively by

near-basin-wide, post-Tully dysoxia and local anoxia during Geneseo black shale deposition (Figs. 5 and 7). It is important to note that the most dysoxic portions of the 'Tully Fauna' continue to persist through the interval of the recurrent 'Hamilton Fauna,' most notably *Camarotoechia mesacostalis* in place of the typical *Eumetabolotoechia multicostatum* community of the Hamilton Group. Additionally, and in a general sense similar to when the 'Tully Fauna' occupies the NAB, the recurrent 'Hamilton Fauna' also occupies both carbonate and siliciclastic (TFCC) portions of the onshore–offshore gradient. For example, the Moravia Beds are dominantly carbonate, but the West Brook Shale (with peak development of the recurrent 'Hamilton Fauna') is anomalously shaly. Even the Bellona Bed with its coral fauna is very impure carbonate at best (Figs. 5 and 6; Heckel 1973).

Geneseo Bioevent

The third phase of the biocrisis, the *Geneseo Bioevent*, occurred during the subsequent, post-Tully, highstand part of the Taghanic Onlap. This resulted in extinction of most Hamilton taxa observed in the recurrent 'Hamilton Fauna' of the upper Tully Formation (at least locally), as generalistic 'Hamilton' taxa replace the diverse biofacies in the transitional Moravia Beds occuring between the Bellona-West Brook interval and the Geneseo black shale. This was a time characterized by speciation within 'Hamilton' genera (see Linsley 1994, for more information), and additional immigration of taxa from the western US and Canada (OWR) such as *Tylothyris mesacostalis*, including the reappearance of a few 'Tully' taxa (*'Nervostrophia' tulliensis* and *E. ambocoeloides*); these taxa are observed only in shallow settings at this time due to the basin being inhospitable to benthos. This was part of a global trend toward faunal cosmopolitanism observed until the Frasnian–Fammenian Extinction approximately 7 million years later (Fig. 7) (Johnson 1970; McGhee 1982, 1989, 1996, 1997; Boucot 1988; Sandberg et al. 2002). Shown in Fig. 7, the Genesee Group biofacies spectrum is composed of a cosmopolitan fauna of generalistic taxa with poorly defined biofacies; for example, there were no Hamilton-like analogs of the diverse coral and brachiopod assemblages in the Genesee Group (Zambito et al. 2007, 2009; J. Zambito 'unpublished data').

Post-biocrisis Interval

The cosmopolitan fauna that resulted from the aftermath of the Taghanic Biocrisis occupied the NAB until the Frasnian–Fammenian extinction, with only minor faunal changes through that time (see Zambito et al. 2007 and 2009 for some of these changes). Earlier workers, most notably H.S. Williams and his students, described a recurrent 'Hamilton Fauna' in pro-deltaic deposits of the Frasnian Ithaca Formation near Ithaca, New York, long after the 'Hamilton Fauna' was thought to have gone extinct with deposition of the Geneseo black shale (Williams 1884, 1906, 1913; Kindle 1896; Williams et al. 1909). Recognition of these anachronous taxa is essentially the observation of generalist 'Hamilton' taxa within the cosmopolitan fauna of the Upper Devonian; we observe these forms to persist in nearshore settings during the deposition of the Geneseo black shale in offshore portions of the basin. The 'recurrence' of Hamilton forms in the Ithaca Formation is therefore the result of survival of these taxa in nearshore settings while the offshore portions of the basin were inhospitable; hence, these biotas would later expand back into offshore settings when hospitable conditions returned (Cooper and Williams 1935; Zambito et al. 2007, 2009; J. Zambito 'unpublished data').

Taghanic Biocrisis in the Type Area: Global and Local Environmental Changes

Whereas, at first glance, a change from predominantly siliciclastic to carbonate and then returning to siliciclastic depositional environments (Hamilton, Tully, and Genesee groups, respectively) seems a reasonable explanation for the earlier noted faunal changes, reconstruction of a detailed stratigraphic framework

Fig. 7 (continued) Brett et al. (2007a). The biofacies spectrum for the lower and upper Tully Group is based on observations from Baird et al. (2003), Baird and Brett (2003, 2008) and Baird (unpublished data). The lower Genesee Group biofacies spectrum is based on observations from Boyer and Droser (2007), Zambito et al. (2007, 2009), and J. Zambito (unpublished data) along a complete onshore–offshore gradient shows that both the 'Hamilton' and 'Tully' faunas occupied comparable onshore–offshore environmental gradients, including both siliciclastic and carbonate depositional settings (Figs. 5 and 7; Heckel 1973; Baird and Brett 2003, 2008). Additionally, stratigraphic ranges of taxa cross large-scale unconformities (sequence boundaries), in particular sequence boundaries that correspond with the onset of bioevents and onshore–offshore faunal gradients. These are mirrored in shallowing and deepening depositional sequences in a given outcrop; they are consistent with the observed transitions being not artifacts of a biased preservation of certain environments (*sensu* Holland 2000).

Widely recognized global environmental changes during this interval that could be responsible for the faunal transitions in the type area include the following: (1) sea-level rise and intensified oceanic circulation (Johnson 1970; Hüneke 2006, 2007); (2) reduced oxygenation in epicontinental seas (Murphy et al. 2000; Aboussalam 2003); (3) carbon-cycle changes and increased nutrient influx (Aboussalam and Becker 2011), (4) increased warming (Joachimski et al. 2004, 2009; van Geldern et al. 2006), and (5) increased aridity and rapid climate shifts (Marshall et al. 2011). In the following section, we place the faunal transitions and extinctions outlined in the reconstructed biofacies spectra (Fig. 7) in the context of these global environmental changes.

Eustatic Sea-Level Rise and Faunal Incursion

The first appearance (incursion) of 'Old World Realm' genera into the NAB occurs prior to the GTB and concurrently with 'Hamilton Fauna' during deposition of dysoxic dark shales, during the uppermost part of the Hamilton Group succession when the generalistic genus *Emanuella* is first observed in the NAB (Fig. 6). The uppermost Hamilton Group strata probably record the second-highest sea-level conditions for the biocrisis interval, followed only by the later post-Tully (Genesee) part of the Taghanic Onlap (Johnson et al. 1985; Baird and Brett 2003). The rise in sea level during the latest Hamilton Group deposits, and continuing throughout the GTB, likely resulted in increased linkage between the NAB and western North America, thereby facilitating immigration of the OWR taxa. The scenario of transgression-induced immigration has been invoked for numerous Devonian Laurassian benthic immigrations (Johnson 1970; Racki 1993; Brice et al. 1994; May 1995; Day 1996, 1998). During the Lower Tully Bioevent, further incursion of OWR taxa occurs; these incursions are also likely due to increased basin linkage (Figs. 1 and 6). Whereas increased sea level explains the migration pathways of the faunal incursions, it cannot alone explain replacement of the 'Hamilton Fauna' by the 'Tully Fauna' in the NAB; this is consistent with a major paleoenvironmental control driving this change.

Basin Eutrophication (Reduced Oxygenation and Enhanced Nutrient Flux)

Globally, it has been observed that with many Paleozoic transgressions in epeiric seas, bottom-water stagnation occurs and low-oxygen conditions prevail in offshore settings, often associated with biotic crises (Walliser 1996; Hallam and Wignall 1999). In the NAB, deoxygenation of bottom waters could be further enhanced by increased input of organic matter to the basin as a result of terrestrial colonization of plants seen in the first appearance of the Gilboa fossil forest during the latest part of the Hamilton Group (Fig. 6; Bridge and Willis 1994; Bartholomew 2002 and references therein). In fact, it is during this time that plant material first becomes common in basinal settings, continuing for the remainder of the Devonian. Furthermore, it is associated with relatively organic-rich, dysaerobic, and relatively nearshore facies (Prosser 1899; Cooper and Williams 1935; J. Zambito 'unpublished data'). Therefore, the increased nutrient flux to marine settings as a result of enhanced physical and chemical soil weathering by vegetation may have resulted in enhanced burial of organic carbon and resultant increases in bottom water oxygen depletion observed at this time in the NAB (Algeo et al. 1995; Algeo and Scheckler 1998).

During the *Lower Tully Bioevent*, we observe a step-wise replacement of offshore portions of the Hamilton fauna leading to the near-exclusion of all but some generalistic 'Hamilton' taxa in nearshore settings (Fig. 7). This pattern of replacement suggests there was some environmental change occurring in deeper water that was precluding occupation by Hamilton forms and promoting incursion of the 'Tully Fauna,' while simultaneously not affecting shallow-water habitats to the same degree. That these immigrants were generalistic OWR taxa would be expected, as generalistic taxa are typical of dysoxic settings and would have an easier time tolerating any differences between the NAB and their endemic area, as has been shown in studies of migrations through the late Devonian (Rode and Lieberman 2004).

This pattern of replacement, though, suggests that low-oxygen conditions alone could not explain these incursions. This is because any explanation for this initial incursion and subsequent establishment of OWR taxa must account for the predominance of similar dysoxic shale facies in the Hamilton Group that were occupied by dysaerobic biofacies of the 'Hamilton Fauna' prior to this incursion for a period of 4–5 million years (Figs. 5 and 7). Whereas eustatic sea-level rise and increased nutrient influx likely led to bottom water oxygen deficiency in the NAB, some other aspect of these settings must have allowed for 'Tully Fauna' incursion, while simultaneously resulting in 'Hamilton Fauna' exclusion. Furthermore, at the peak of *Lower Tully Bioevent*, 'Tully' taxa are observed in oxic settings and, during the onset of the *Upper Tully Bioevent*, dysaerobic 'Hamilton' biofacies are more proximal than dysaerobic 'Tully' biofacies; both of these observations suggest an additional environmental control on faunal distribution, not oxygen deficiency alone. Additionally, carbon-cycle changes during the *GTB* are seen as a positive carbon isotope excursion beginning with the *Upper Tully Bioevent* in sections from Morocco representing ocean-facing settings (Aboussalam 2003; Aboussalam and Becker 2011): Assuming these changes are global, this would suggest that increased carbon burial and therefore possibly lower oxygen levels were more prevalent—beginning with the recurrence of the 'Hamilton Fauna' and after the occupation of the 'Tully Fauna' in the NAB. Application of these ocean-facing environmental patterns to the NAB epeiric setting should be done cautiously; they suggest that reconstruction of a carbon isotope curve for the type area is necessary.

It therefore appears that some other environmental change, possibly in conjunction with reduced oxygenation of the NAB, is responsible for the observed faunal transitions during the *Lower* and *Upper Tully Bioevents.* However, as suggested by previous studies and discussed in detail in the following section, the pervasive dysoxic/anoxic settings associated with the *Geneseo Bioevent* did have an effect on the faunal patterns observed during the *GTB*.

Global Warming

Recent reconstructions of temperature changes through the Taghanic Biocrisis have depicted this event as a period of global warming, but no study to date has produced a high-resolution temperature curve during the Taghanic Biocrisis interval that can be directly compared to the faunal changes seen in the Tully Formation type area, given that current temperature reconstructions are based on oxygen isotopic changes within conodont apatite from areas other than the NAB that are much less stratigraphically complete (Joachimski et al. 2004, 2009). The position of the OWR in more equatorial settings prior to and during the biocrisis, and the subsequent immigration of these taxa into the NAB immediately before and during the Lower Tully Biocrisis, supports the hypothesis that changes in climatic gradients, specifically warming of the NAB, could have driven these incursions (Johnson 1970; Boucot 1988). In particular, an episode of global warming, coupled with transgressive connecting of basins, would have expanded the ranges of equatorial taxa into higher latitudes thereby increasing cosmopolitanism.

As outlined in Fig. 7, during the *Lower Tully Bioevent* the offshore portions of the 'Hamilton Fauna' were replaced in the NAB by the 'Tully fauna' and, additionally, the diverse coral and brachiopod assemblages of the 'Hamilton Fauna' are absent from the NAB altogether. If these portions of the 'Hamilton Fauna' were forced to emigrate from the NAB due to warming, then one might expect that higher-latitude (cooler) locations would be an ideal location for the Hamilton Fauna to find refuge. As shown in Fig. 1, possible locations of higher-latitude refugia include South America and Africa. Unfortunately, correlative rocks in South America have either been eroded away or represent non-marine environments, so that there is no record of marine faunas at high latitudes in the Americas during the *GTB.* Faunal lists from the same-age rocks in Africa do not contain the emigrated Hamilton taxa (Barrett 1988; Barrett and Isaacson

1988; de Melo 1988; Isaacson and Sablock 1988, Brice and Latréche 1998; P.E. Isaacson 2008, 'personal communication'; Racheboeuf et al. 2001, 2004).

Although the predicted high-latitude refugia have yet to be identified for the 'Hamilton Fauna,' the hypothesis of shifting latitudinal climatic gradients is nonetheless supported by the temperature reconstructions of a period of warming as well as latitudinal affinities of taxa (Boucot 1988; Joachimski et al. 2004, 2009; but see Brand et al. 2008); a higher-resolution temperature reconstruction, preferably from the type area, is needed to demonstrate the corresponding hypothesis that a brief return to cooler temperatures allowed recurrence of Hamilton taxa during the *Upper Tully Bioevent.* It should be noted that the most recent temperature reconstruction (Joachimski et al. 2009) indicates that the entire Hamilton Group-through-Frasnian succession, inclusive of the Taghanic interval, was a period of almost steady temperature increase, and that the 'Hamilton Fauna' occurred during some of the coolest times in all of the Devonian. This suggests that if a reduction in climatic gradient associated with warming did indeed drive the faunal transitions of the *GTB*, some temperature threshold must have been reached during the latest Hamilton Group that allowed for faunal incursions and replacements. If global warming was driving the OWR incursions, there would also need to be an explanation of why 'warm' water occurs more in deeper-water settings, as suggested by the distribution of 'Tully' taxa in offshore settings, down-ramp of generalistic 'Hamilton' taxa, during the *Lower* and *Upper Tully Bioevents* (Fig. 7).

Increased Aridity and Rapid Climate Shifts

Previous studies have suggested that the *GTB* interval was a period of generally increased aridity (Day et al. 1996; Witzke and Bunker 1997; Marshall et al. 2011). Marshall and others (2011), in their work on the Old Red Sandstone, also suggested that during this overall arid interval, the *GTB* was a period of rapidly fluctuating cool arid and hot pluvial events, corresponding with times of intra-Tully erosion (unconformities) and carbonate deposition, respectively. Based on the correlation of these climate changes with the depositional sequences of the Tully Group and faunal transitions, they speculated that OWR incursions were driven by these hot pluvial events. This agrees well with the hypothesis that OWR incursions are driven by shifts in climatic gradients associated with warming and basin linkage resulting from the rise in sea level; however, under this model, the recurrent 'Hamilton Fauna' would occur during an inferred hot pluvial event.

Aridity, associated with increased marine salinity, may have some control on faunal distribution and migration. The work of Boucot (1988) suggests that OWR taxa were adapted to higher-salinity environments; he placed these taxa within an arid belt containing evaporite deposits (cf. discussion in Boucot 1975: 269–282, 346). Additionally, within the EAR, the 'Hamilton Fauna has been a problem ever since it was first described' (Boucot 1975: 320–321) because it contains an admixture of earlier faunas (see also Brett et al. 2009). Presumably this accords with the Appalachian Basin 'Hamilton Fauna' having required a unique environmental setting; we suggest that relatively low salinity, as a function of proximity to freshwater sources (Acadian Hinterlands) and restricted nature of the Northern Appalachian Basin was a likely characteristic of 'Hamilton' times (Boucot 1975; Algeo et al. 2007).

Watermass Circulation Patterns

If salinity were a controlling factor, it would be useful to model the influence of freshwater influx on the relatively restricted NAB. The most appropriate model for the NAB is that of estuarine-type circulation (*sensu* Algeo et al. 2007). Due to the restricted nature of the NAB marine embayment, oceanic water influence was rather minimal (Fig. 1); therefore, we have adapted the super-estuarine model of marine anoxia presented by Algeo et al. (2008; see also Witzke 1987). Our model (Fig. 8) differs because the super-estuarine model invokes a connection to ocean settings, whereas the NAB did not have, so the NAB was not influenced by the oceanic thermocline. Applied to the NAB, this model of estuarine-type circulation would have operated as a result of influx of higher-salinity water from the lower-latitude portions of the North American craton where evaporitic settings were intermittently developed throughout the *GTB* interval, such as within the Iowa, Elk Point, and Great Basin successions (Day et al. 1996; Witzke and Bunker 1997). The higher-salinity watermass would have been

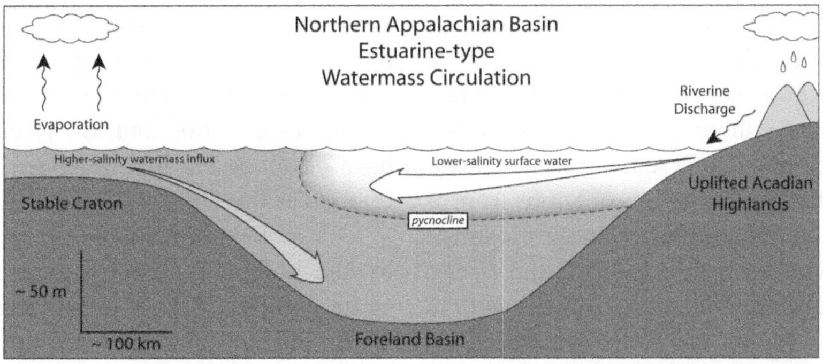

Fig. 8 Generalized model of estuarine-type watermass circulation pattern for the Appalachian Basin, after Witzke (1987) and Algeo et al. (2007, 2008). Riverine discharge (*white arrow*) from the Acadian Hinterlands provides freshwater sources to the foreland basin, forming a low-salinity surface watermass. Shallow water evaporitic settings on the North American craton (see Fig. 1) would provide a denser, more saline watermass (*gray arrow*) to the foreland basin. Together, these watermasses would be expected to form a pycno(halo)cline which, upon interacting with the oxycline of the epeiric sea, might result in a 'boundary layer' of fluctuating oxygen and salinity

denser than the overlying lower-salinity watermass resulting from freshwater runoff from the Acadian hinterlands. This layered structure would have supported a halocline which, in turn, would have enhanced the oxycline, already strengthened by enhanced nutrient influx. Upwelling, commonly associated with estuarine circulation, would also have been possible. The resulting pycnocline would have acted more like a boundary layer where salinity and oxygen levels were constantly fluctuating (Fig. 8).

Under this model, the 'Hamilton Fauna' would have been relegated to shallower, less saline waters, and the 'Tully Fauna' to deeper, more saline, and possibly warmer waters; this explains the facies distribution and step-wise replacements observed during the *GTB*. Furthermore, the distribution of these faunas would be expected to overlap in biofacies composed of generalistic taxa; this is what is observed during the *Upper Tully* Bioevent. Additionally, this model is compatible with the observed global warming, sea-level rise, and aridity observed at this time; all these changes would result in decreased barriers to migration between basins as well as influx of a cratonic watermass into the NAB. Whereas epeiric seas, and in particular the NAB, were subject to large deviations from normal marine conditions because of their restricted nature (Witzke 1987; Algeo et al. 2007, 2008), it should be noted that both the 'Hamilton' and 'Tully' faunas contain what are considered to be fully marine taxa, such as trilobites, echinoderms, brachiopods, and corals. This is consistent with each fauna being comprised of a spectrum of taxa adapted to different salinity levels; alternatively, this might suggest that absolute deviation in salinity from normal marine conditions was rather small. This model may also be applicable to times of OWR incursion into the EAR areas during the late Eifelian and prior to appearance of the 'Hamilton Fauna' during which basinal facies underwent OWR incursion and turnover (Appalachian and Michigan basins), yet shallow-water settings did not experience the same level of incursion and turnover, such as the platform settings in Ohio and Ontario described by Desantis et al. (2007). During Hamilton time prior to the *GTB*, rare incursions of OWR ambocoeliid brachiopods also occurred in biofacies composed of generalistic taxa occupying basinal settings (Goldman and Mitchell 1990).

Environmental Changes Superimposed on Local Basin Dynamics

As noted earlier, we suggest that the presence of OWR and EAR taxa in the NAB was controlled mainly by salinity and nutrient flux, and that the distribution of these taxa along an onshore–offshore (shallow–deep) environmental gradient would suggest that some form of watermass stratification was in place, such as an estuarine-type circulation model. Application of this model to the NAB can be tested by placing it in the context of the geologically rapid changes in Appalachian Basin morphology that occurred during the *GTB*. In Fig. 9 (see also Figs. 4 and 5), we outline these major changes in basin dynamics: (A) During

Hamilton times, the basin displayed a gently sloping, ramp-style morphology; (B) coincident with the onset of the *Lower Tully Bioevent,* a fault-bounded 'clastic trap' formed in nearshore portions of the basin with associated development of a sediment-starved and nutrient-poor carbonate-bank setting offshore from it to the west; (C) by the time of maximum development of the recurrent 'Hamilton Fauna,' during the *Upper Tully Bioevent*, the clastic trap had been filled and the basin once again resembled the ramp-style morphology typical of the Hamilton Group; and (D) with the onset of the third tectophase of the Acadian Orogeny, the basin rapidly subsided and was filled with detrital sediments derived from the rejuvenated Acadian hinterlands. For the first time since the initial appearance of the 'Hamilton Fauna,' the NAB took on a shelf–slope profile due to these increased sedimentation and progradation rates (see Fig. 5; Rogers et al. 1990).

In the following section, we outline how the changes in NAB morphology might amplify the effects of the estuarine-type watermass-circulation model, the supporting lithologic observations for this amplification, and correspondence of hypothesized watermass circulation patterns to the faunal transitions. Application of the estuarine-type circulation model with regard to changes in basin morphology is shown in Fig. 10. Inferred as a generally arid time undergoing increasing global warming and sea-level rise, the lower Tully Group would be deposited under relatively reduced riverine influx, and relatively high input of watermass influx from cratonic settings. A pycnocline would develop between these two watermasses and would be further enhanced by the 'silled' basin formed as a result of the position of the 'Sherburne High' (Figs. 4 and 10a; see for reference Witzke 1987; Algeo et al. 2008). The position of the pycnocline in this model provides two differing depositional settings that correspond to the onshore–offshore (Hamilton–Tully) biofacies distributions (Fig. 7). It is further possible that anti-estuarine circulation, due to increased aridity and evaporation and low riverine influx, may also have developed at this time (*sensu* Witzke 1987, and references therein).

The model shown in Fig. 10a is further supported by the unique lithologic characteristics observed in strata containing the 'Tully Fauna' (Baird and Brett 2003, 2008). Chamosite, rare in the Hamilton Group, is observed at numerous horizons in the New Lisbon Member and lower Tully Formation. It characteristically occurs as mud-supported, sand-size, discoidal, black grains within intensely bioturbated thin layers in association with diagenetic siderite. It is most concentrated in the Smyrna Bed both along and east of the 'Sherburne High' (see Figs. 4, 5 and 6) (Heckel 1973; Baird et al. 2003; Baird and Brett 2008). Tully chamosite beds generally contain very few fossils; these when present are often highly corroded. The Smyrna Bed appears to record only moderate- to low-energy conditions, despite its designation as the 'Tully oolite' in literature (Cooper and Williams 1935; Heckel 1973; Baird and Brett 2003). Whereas the origin of chamosite-bearing beds is poorly understood, some generalizations about its depositional requirements can be made. In particular, sediment-starvation, reducing conditions, warm humid climates and high sea-level events have all been implicated as conditions favoring this type of deposit (Huber and Garrels 1953; Maynard 1986; Kim and Lee 2000; Baird et al. 2003). These reconstructions accord well with the model presented in Fig. 10a, especially when considering that the Tully Group is inferred to have been deposited during warm humid periods during an overall aridity event. Synchronous reduced sediment input, high sea-level conditions, and development of

Fig. 9 Basin dynamics through the Taghanic Biocrisis. Basin profile models represent the extent of the basin observed along the Tully Group outcrop belt from the paleoshoreline to the westernmost extent of outcrop (see Fig. 3); vertical dimension of model is approximately 60 m for a, b, and c and 125 m for d. 'TO' represent the initiation of the Taghanic Onlap, and 'TP3' signifies the onset of the third tectophase of the Acadian Orogeny (see text for references). Relative sea-level curve is adapted from Baird and Brett (2003, 2008), and Zambito et al. (2009). Stratigraphic labels are outlined in Fig. 5. (**a**) During the majority of the Hamilton Group, the New York Appalachian Foreland Basin was a gently sloping ramp. (**b**) With the onset of syn-depositional faulting during the lower Tully Group, a 'clastic trap' was formed as a result of generation of the 'Sherburne High'; a carbonate platform-type setting formed to the west (Heckel 1973; Baird and Brett 2003). (**c**) By the time of initial onset of the Taghanic Onlap (Bellona-West Brook beds interval), the 'clastic trap' has become filled with sediment, and the basin once again exhibits a ramp-profile similar to that in Hamilton Group times (see Fig. 7). (**d**) With a drastic increase in subsidence and progradation due to onset of the third tectophase of the Acadian Orogeny, a well-defined shelf and slope is formed

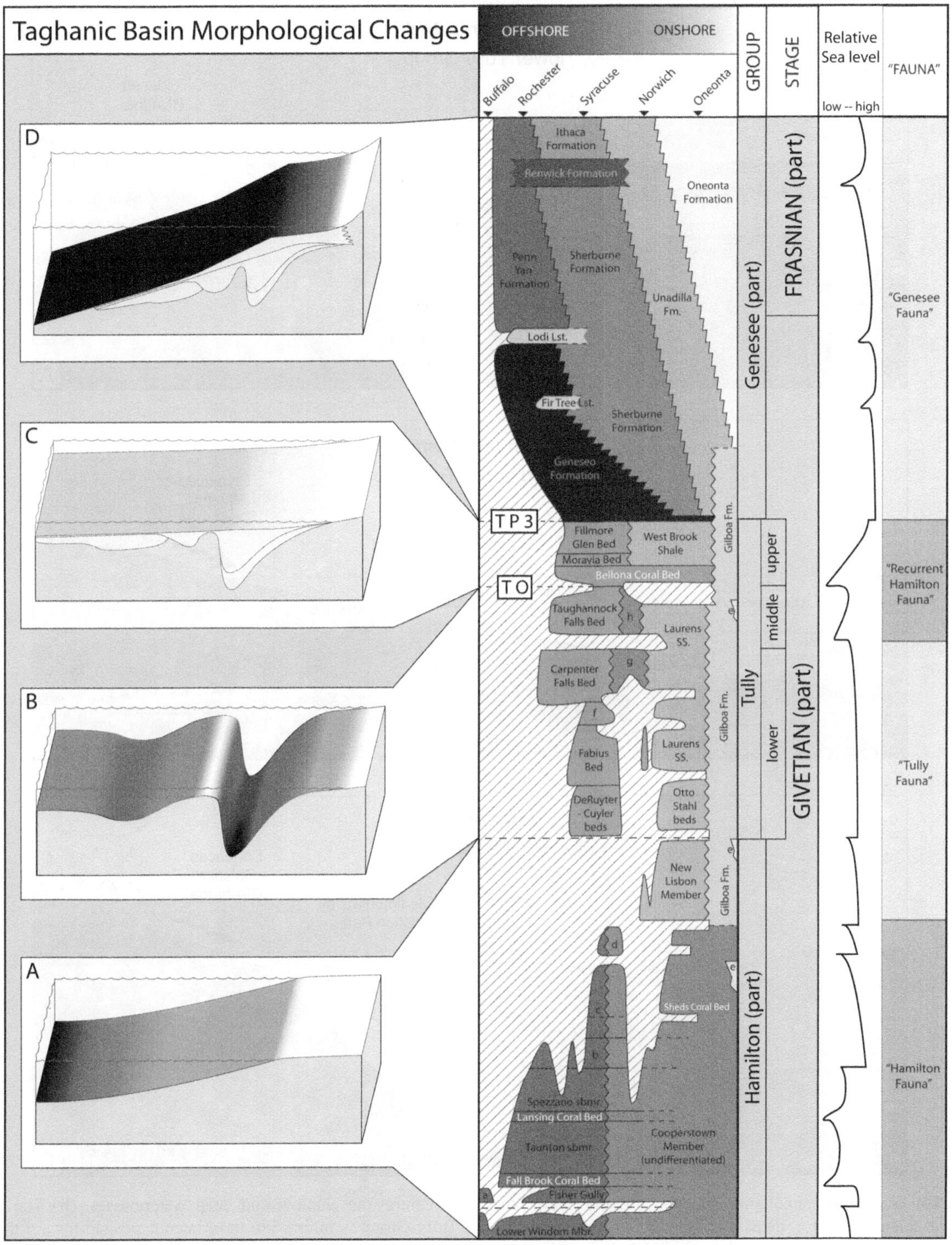

Taghanic Basin Morphological Changes
OFFSHORE
ONSHORE
Buffalo
Rochester
Syracuse
Norwich
Oneonta
GROUP
STAGE
Relative Sea level
low -- high
"FAUNA"
D
C
B
A
Ithaca Formation
Renwick Formation
Oneonta Formation
Penn Yan Formation
Sherburne Formation
Unadilla Fm.
Lodi Lst.
Fir Tree Lst.
Sherburne Formation
Geneseo Formation
TP3
TO
Fillmore Glen Bed
Moravia Bed
West Brook Shale
Bellona Coral Bed
Gilboa Fm.
Taughannock Falls Bed
Laurens SS.
Carpenter Falls Bed
Fabius Bed
Laurens SS.
DeRuyter - Cuyler beds
Otto Stahl beds
New Lisbon Member
Sheds Coral Bed
Spezzano sbmr.
Lansing Coral Bed
Taunton sbmr.
Cooperstown Member (undifferentiated)
Fall Brook Coral Bed
Fisher Gully
Lower Windom Mbr.
Genesee (part)
Tully
upper
middle
lower
Hamilton (part)
FRASNIAN (part)
GIVETIAN (part)
"Genesee Fauna"
"Recurrent Hamilton Fauna"
"Tully Fauna"
"Hamilton Fauna"

Fig. 9

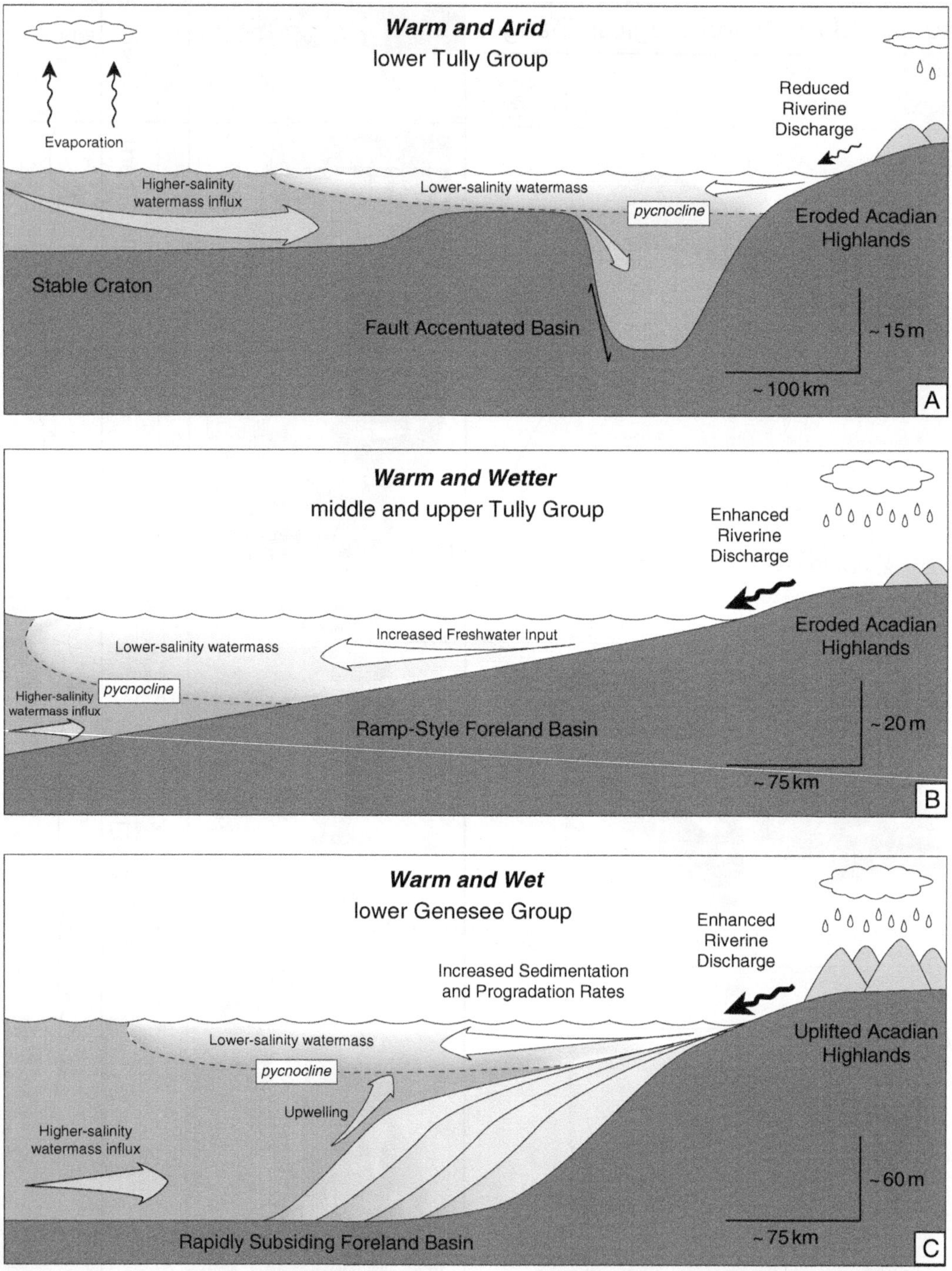

Fig. 10 Watermass circulation models for the New York Appalachian Basin for times during and after the Taghanic Biocrisis. (**a**) Watermass circulation model during the lower Tully Group, inferred to have been a period of increased aridity (Marshall et al. 2011). Increased aridity would have led to reduced runoff, resulting in a rather shallow, lower-salinity surface layer and an elevated pycnocline. Basin morphology, in the form of a fault-controlled sill, further accentuates development of contrasting shallow and deep watermasses. (**b**) The upper Tully Group is inferred to have been deposited toward the end of the arid interval (Marshall et al. 2011); this would have resulted in an increased influx of freshwater runoff into the basin forcing the pycnocline into deeper settings, accentuated by a basin topography similar to that of previous Hamilton times (**c**) By the time of deposition of the Genesee Group, the aridity interval had ended (Marshall et al. 2011) and freshwater

a 'mixing front' between the two watermasses at or near the 'Sherburne High' would explain the unique chamositic character of the Smyrna Bed (Baird et al. 2003; Baird and Brett 2008; Marshall et al. 2011). Originally, Baird and Brett (2003) modeled the development of numerous, repeated, eutrophication-driven oxygen minimum zone events impinging on the top and east flank of the 'Sherburne High' in a sediment-starved, warm water regime to explain the Smyrna Bed 'oolite.' However, the stratified watermass model, advanced herein, is similarly suited as an explanation for the chamosite texture and distribution.

Additionally, the Smyrna Bed also contains stromatolites and syneresis cracks; these have not been found in the intervals before or after the *GTB* (Baird et al. 2003). The most commonly held view is that syneresis cracks are the result of contraction of the swelling clay lattices due to salinity changes in pore waters, possibly associated with mat structures (Jüngst 1934; White 1961; Burst 1965; Plummer and Gostin 1981; Pflueger 1999, however see Pratt 1998 for an alternate interpretation). Modern stromatolites form in physically stressed environments, such as hypersaline and high energy environments with shifting oolitic sand substrates and can be found in both shallow- and deep-water settings (Monty 1971; Hofmann 1973; Logan et al. 1974; Playford 1980; Dravis 1983; Böhm and Brachert 1993; George 1999). Again, the position of a mixing front close to the 'Sherburne High' would provide an environment conducive to stromatolites and syneresis crack formation.

Finally, the Tully limestone is itself a dramatic, lithologic anomaly within the overall Catskill Delta siliciclastic succession. Tully Group carbonates differ drastically from Hamilton and Genesee Group carbonates in that the former are dominantly calcilutites and micrites with dilute concentrations of well-preserved fossils; the latter consist of local shell accumulations during siliciclastic sediment starvation (Heckel 1973; Brett and Baird 1994; Zambito et al. 2007, 2009). Heckel (1973) also observed that carbonate grains within the Tully Group are polycrystalline with various shapes and sizes; he suggested that these grains are possibly pieces of skeletal grains that have undergone extreme comminution. However, no algal structures were observed within the Tully carbonates and, furthermore, skeletons are typically non-abraded. Due to these observations, Heckel (1973) could not rule out physiochemical precipitation of at least some of the Tully carbonate grains, *sensu* whitening events (Cloud 1962; Wells and Illing 1964; Shinn et al. 1989). We suggest that physiochemical precipitation of at least some of 'Tully' carbonate occurred and was facilitated by the watermass mixing front at the interface of the higher-salinity water-mass from cratonic areas and the lower-salinity freshwater runoff from the Acadian hinterlands (Fig. 10a). Under extreme aridity, the possible formation of anti-estuarine circulation during Tully deposition may also have occurred. Anti-estuarine circulation would have resulted in carbonate-rich, nutrient-poor conditions, such as are typical of the Tully (Witzke 1987; Heckel 1973; Baird et al. 2003).

Coincident with the onset of the *Upper Tully Bioevent*, the 'clastic trap' had almost been completely filled with sediment; dysaerobic biofacies of 'Hamilton Fauna' are observed up-ramp from the 'Tully Fauna.' This suggests that environmental conditions in shallow settings began to change to more Hamilton-like conditions before those in deeper settings (Figs. 4 and 7) (Baird and Brett 2008). We attribute this step-wise recurrence of 'Hamilton' biofacies, ultimately culminating in the diverse coral and brachiopod biofacies of the Bellona-West Brook Beds, to return of a ramp-style basin morphology and the end of the aridity event (Figs. 9c and 10b; Marshall et al. 2011). Though estuarine-type circulation would still have been in effect, an increase in freshwater runoff and the ramp-style basin would have decreased the elevation of the pycnocline, resulting in a larger habitable area for the 'Hamilton Fauna,' while maintaining a mixing front. This would have resulted in micrite precipitation continuing until transgressive overspread of regional anoxia associated with post-Tully flexural basin collapse, eustatic highstand, and interbeds

Fig. 10 (continued) runoff was further accentuated by the topography of the uplifted Acadian Highlands, also resulting in increased sediment delivery to the basin. The eustatic sea-level rise (Taghanic Onlap of Johnson 1970) was also in full effect, leading to an influx of higher-salinity bottom waters into the basin and resulting in a highly elevated pycnocline. Furthermore, basin topography would have facilitated upwelling with the black shales of the Geneseo Shale as the Taghanic Onlap progressed. The continued maintenance of the mixing front (pycnocline) and associated watermasses of differing characteristics (especially salinity) is suggested by persistence of 'Tully' taxa in the deepest, most dysaerobic settings well into the Genesee Group; in fact, the *Eumetabolatoechia* biofacies of the 'Hamilton Fauna' never recurs and was permanently replaced by the 'Tully Fauna' equivalent *Camarotoechia* biofacies (Fig. 7). The sequence boundary at the base of the Bellona-West Brook beds, preceding the Taghanic Onlap, probably also played a role in the return of diverse 'Hamilton' assemblages, because the sequence boundary unconformity essentially 'leveled the sea floor' removing any basin topographic remnant of the 'Sherburne High,' thereby allowing the low-salinity watermass to reach further into the basin. This relief profile began to return during deposition of the Moravia and Fillmore Glen portions of the uppermost Tully (Fig. 4).

With the continuation of the Taghanic Onlap and onset of the third tectophase of the Acadian Orogeny, the sea level continued to rise, the Appalachian Basin subsided drastically, sedimentation and progradation were greatly increased, and the basin morphology permanently changed to a shelf-slope-basin profile that was present until the Mississippian when the basin became completely filled (Figs. 5, 9 and 10c).

The *Geneseo Bioevent* is recorded in the transition between the uppermost Tully Group (Moravia and Fillmore Glen Beds) and the lowest Geneseo Formation (Fig. 5) when black shale deposition took the place of carbonates in the west and coarser siliciclastics in the east. Based on the estimates of Ettensohn (1985), the Geneseo Shale is interpreted as recording a maximum highstand event for the NAB. Boyer and Droser (2007) report minimal bioturbation and low-diversity fossil assemblages for the Geneseo Shale, indicative of depleted bottom-water oxygen levels. Geochemical analysis of the Geneseo Shale by Murphy et al. (2000) and Formolo and Lyons (2007) suggests that anoxic, and possibly even euxinic, conditions would have been favored by relatively short-term fluctuations in the position and strength of the pycnocline and upwelling conditions; this may be expected if freshwater runoff and sediment supply were influenced by seasonal or some other climatic oscillation (Figs. 8 and 10c). Witzke (1987) suggests that upwelling may be enhanced under estuarine-type circulation, especially when the basin morphology includes a shelf-slope break, thereby increasing productivity and black shale deposition. The transitional nature of interbedded micritic limestones and black shale of the Moravia Beds into the micrite-free, organic-rich Geneseo black shale likely represents the onset of upwelling conditions and increased nutrient influx to bottom waters.

Reconstruction of a high-resolution stratigraphic framework along an onshore–offshore gradient during this time suggests that dysoxic conditions extended to at least storm-wave base (shelf settings) during the initial phases of the Taghanic Onlap—as is evidenced in nearshore Geneseo Formation equivalents by rusty-weathering fossiliferous gutter casts in otherwise fossil-poor dark gray to black silty shales and, also, dark-gray silty shales containing dysaerobic communities interbedded with swaley cross-bedded sandstones interpreted as shore-zone storm-wave current deposits with slightly more diverse and aerobic communities (Fig. 5) (Johnson and Friedman 1969; Bridge and Willis 1994; Zambito 'unpublished data'). Additionally, the Taghanic Onlap also increased linkages between depositional basins, resulting in incursions of OWR taxa such as *Tylothyris mesacostalis* and the recurrence of *'Nervostrophia' tulliensis* and *E. ambocoeloides* (Fig. 6). This continued linkage of the NAB with the cratonic watermass following the Taghanic Onlap (Fig. 1c), the rapid subsidence and removal of much of the basin from the mixing effects found above storm-wave base, and basin morphology induced upwelling and isolation from oceanic water input. These collectively ensured that estuarine-type circulation driven by cratonic watermass influx persisted in the Appalachian Basin throughout the remainder of its duration (see Algeo et al. 2007).

Sediment input into the basin greatly increased, concurrently with this exceptionally elevated oxycline, precluding low-turbidity communities in the NAB such as diverse coral assemblages, even when oxygen was available; indeed the extinction rate for rugose corals was higher during the Taghanic than at the Frasnian–Fammenian extinction (Oliver 1990). With the exception of dysaerobic auloporid communities, the only coral bed known from the Genesee Group is in the Firestone beds' interval of the Ithaca Formation; by this time, additional OWR incursions had occurred. This relatively low-diversity coral assemblage had practically no resemblance to the 'Hamilton Fauna' (Zambito et al. 2007, 2009). Additionally, only a few coral beds are known from post-Geneseo strata; all are low diversity (McLean and Sorauf 1988). Sediment-input-related faunal changes are also seen in the increase in the relative proportion of infaunal bivalves relative to sessile brachiopods at that time (Thayer 1972, 1974).

These observations suggest that during the *Geneseo Bioevent*, the amount of available ecospace (both in terms of habitable area and total biofacies spectrum occupancy) in the NAB was greatly reduced by the concurrent elevation of the oxycline and increased sedimentation input into the basin (Figs. 5 and 7). Furthermore, those biofacies that persisted in the NAB underwent significant restructuring as a result of the decrease in habitable area and incursions of OWR taxa; a surprising complement of 'Hamilton' taxa survived in nearshore settings only to later reoccupy the basin when hospitable conditions returned to offshore settings (Zambito et al. 2007, 2009). With continued progradation, the combination of an elevated pycnocline and high-sediment influx left a majority of 'Hamilton' taxa with nowhere to go but become extinct.

Conclusions and Implications

The application of watermass circulation models to global environmental changes and local basin dynamics to understanding the Global Taghanic Biocrisis in the New York Appalachian Basin provides a relatively simple, and as we argue, plausible, and well-supported explanation for the faunal incursions, replacements, recurrences, and extinctions observed. Under this model, summarized in Fig. 11, the incursion and subsequent replacement of the 'Hamilton Fauna' by the 'Tully Fauna' was driven by the influx of higher-salinity waters into the NAB during a time of overall increased aridity and was facilitated by basin linkage and the breakdown of latitudinal climatic gradients associated with global sea-level rise and warming as well as local basin topography associated with a fault-induced scarp (*Lower Tully Bioevent*). Subsequently, the end of the aridity event and return to the typical ramp-style basin profile of Hamilton times resulted in recurrence of the 'Hamilton Fauna' in the NAB by increasing the amount of the NAB affected by a lower-salinity watermass of riverine runoff (*Upper Tully Bioevent*). Finally, increased subsidence and sediment input into the NAB in conjunction with a highly stratified water column and periodic upwelling resulted in permanent changes to the basin, resulting in absence of Hamilton-type environments. The 'Hamilton Fauna,' with the exception of generalist forms, underwent major extinction (*Geneseo Bioevent*).

After occupying the Appalachian Basin for approximately 4–5 million years during which little ecological change or speciation occurred, the 'Hamilton Fauna' extinguished as a result of combined global (sea-level rise and warming) and local (basin dynamics and watermass circulation) environmental changes. Faunal transitions recognized by reconstructing biofacies spectra and corresponding changes in basin morphology during the *GTB* suggest that there was a distinct environmental suite (other than lithofacies) that the 'Hamilton' and 'Tully' faunas each preferred and tracked temporally. 'Habitat tracking' has been suggested as a mechanism by which communities comprised of species with similar preferred habitats (biofacies) migrate (through generations) along with a preferred environment (facies belts) rather than adapt to locally changing conditions (Brett et al. 2007a, b). This may be the primary explanation for the pattern of 'coordinated stasis' in the fossil record (Brett and Baird 1995; Brett et al. 1996). The pattern presented here of 'habitat tracking' at the faunal level suggests there is a hierarchical scale at which tracking and the resultant ecological stasis occur.

A continuing discussion in the paleoecologic literature is whether faunas represent cohesive communities that are adapted to distinctive environmental conditions or whether faunas are comprised of species that are tracking their preferred habitats individually (Brett et al. 1996; Patzkowsky and Holland 1999; DiMichele et al. 2004; Brett et al. 2007b). The patterns observed in the biofacies spectrum during the faunal transitions studied herein suggest that, indeed, faunas do represent cohesive communities but, at the same time, individual species still track habitats individually. For example, we have suggested from the patterns observed that both the 'Hamilton' and 'Tully' faunas each consist of cohesive groups of species; these had preferred salinity ranges. The boundary between these faunas is blurred by the generalistic taxa of each fauna commonly overlapping in space and time, that is, during the initial

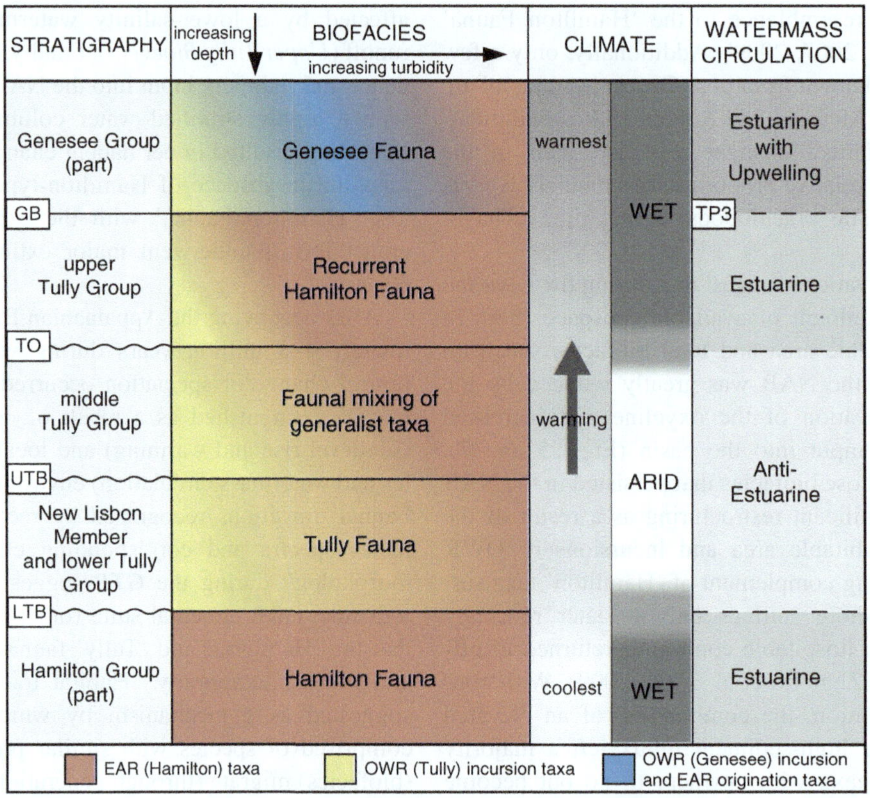

Fig. 11 Summary of the faunal, climate, and watermass changes discussed in this chapter. Aridity changes from Marshall et al. (2011). Generalized temperature change is from Joachimski et al. (2009). See Fig. 7 for more detailed description of the faunas. EAR, OWR, LTB, UTB, TO, GB, and TP3 = Eastern Americas Realm, Old World Realm, Lower Tully Bioevent, Upper Tully Bioevent, Taghanic Onlap, Geneseo Bioevent, and Tectophase 3 of the Acadian Orogeny, respectively

'Tully Fauna' incursions in the latest Hamilton Group, and the initial 'Hamilton Fauna' recursions during the middle Tully Group (Figs. 6 and 7).

Following the *Geneseo Bioevent*, the resultant 'Genesee Fauna,' considered as a subset of the global cosmopolitan fauna, consists, for the most part, of generalist 'Hamilton' and 'Tully' taxa, species originating from 'Hamilton' genera, and a few, new OWR incursions. At this point, it is evident that the elevated extinction and origination rates observed at this time are related to habitat contraction associated with the pycnoclinal rise and increased sediment input followed by habitat expansion as progradation allowed progressive portions of the basin to be colonized. One of our current goals is to quantify these evolutionary rates, both temporally and spatially, through the extinction and subsequent recovery.

The patterns described suggest that the demise of the recurrent 'Hamilton Fauna' during the *Geneseo Bioevent* was a combination of global and local environmental changes. One thing that remains to be understood is whether, or to what degree, the local effects of the Acadian Orogeny were possibly the driving mechanism for the globally observed sea-level rise and warming. As a next step, we must compare our observed faunal and environmental patterns to those in Taghanic successions described elsewhere, particularly where contextual paleolatitude, tectonic, and water circulation parameters are clearly different. This approach should be the best way for us to assess the importance of the Taghanic Biocrisis as a global event.

Acknowledgments We thank the editor, John Talent, for his patience with us in getting this manuscript completed. Comments by reviewers R. Feist and A. Simpson greatly enhanced the readability of the manuscript and conveyance of ideas therein. Various discussions with RT Becker and JEA Marshall greatly improved our understanding of the Taghanic

Biocrisis at a global scale in both marine and terrestrial settings. We are also indebted to TJ Algeo for greatly enhancing our understanding of carbonate depositional environments and watermass circulation models. Our research into the type area of the Taghanic Biocrisis was supported at various times by student grants to JJZ from the American Association of Petroleum Geologists, the American Museum of Natural History, the Evolving Earth Foundation, the Geological Society of America, the Mid America Paleontological Society, the Paleontological Society, the Schuchert and Dunbar Grants In Aid Program at the Yale Peabody Museum, Sigma Xi, the Society for Sedimentary Geology (SEPM), and the University of Cincinnati, Department of Geology. CEB has been supported by NSF EAR 9219807, EAR 9996178, and the USGS STATEMAP program. GCB has been supported by the Pennsylvania Department of Natural Resources (Geological Survey).

References

Aboussalam SZ (2003) Das “Taghanic-Event” im höheren Mittel-Devon von West-Europa und Marokko. Münst Forsch Geol Paläontol 97:1–332

Aboussalam SZ, Becker RT (2001) Prospects for an upper Givetian substage: fossil record. Mitteilungen Museum Naturkunde Berlin 4:83–99

Aboussalam SZ, Becker RT (2011) The global Taghanic Biocrisis (Givetian) in the eastern Anti-Atlas, Morocco. Palaeogeogr Palaeoclimatol Palaeoecol 304(1–2):136–164

Algeo TJ, Scheckler SE (1998) Terrestrial-marine teleconnections in the Devonian; links between the evolution of land plants, weathering processes, and marine anoxic events. Phil Trans R Soc London Biol Sci 353:113–130

Algeo TJ, Berner RA, Maynard JB, Scheckler SE (1995) Late Devonian oceanic anoxic events and biotic crises; “rooted” in the evolution of vascular land plants? GSA Today 5:45, 64–66

Algeo TJ, Lyons TW, Blakey RC, Over DJ (2007) Hydrographic conditions of the Devono-Carboniferous North American Seaway inferred from sedimentary Mo-TOC relationships. Palaeogeogr Palaeoclimatol Palaeoecol 256:204–230

Algeo TJ, Heckel PH, Maynard JB, Blakey R, Rowe H (2008) Modern and ancient Epeiric seas and the super-estuarine circulation model of marine anoxia. In: Holmden C, Pratt BR (eds) Dynamics of Epeiric seas: sedimentological, paleontological and geochemical perspectives. Geological Association Canada Special Publication, 7–38

Alroy J, Aberhan M, Bottjer DJ, Foote M, Füsich FT, Harries PJ, Hendy AJW, Holland SM, Ivany LC, Kiessling W, Kosnik MA, Marshall CR, McGowan AJ, Miller AI, Olszewski TD, Patzkowsky ME, Peters SE, Villier L, Wagner PJ, Bonuso N, Borkow PS, Brenneis B, Clapham ME, Fall LM, Ferguson CA, Hanson VL, Krug AZ, Layou KM, Leckey EH, Nürnberg S, Powers CM, Sessa JA, Simpson C, Tomašoných, Visaggi CC (2008) Phanerozoic trends in the global diversity of marine invertebrates. Science 321:97–100

Baird GC, Brett CE (2003) Shelf and off-shelf deposits of the Tully Formation in New York and Pennsylvania: faunal incursions, eustasy and tectonics. Cour Forsch-Inst Senckenberg 242:141–156

Baird GC, Brett CE (2008) Late Givetian Taghanic bioevents in New York State: new discoveries and questions. Bull Geosci 83:357–370. doi:10.3140/bull.geosci.2008.04.357

Baird GC, Bartholomew AJ, Brett CE (2003) Late Middle Devonian biotic and sedimentologic events in east-central New York; Tully Formation clastic correlative succession in the Sherburne-Oneonta area. New York State Geol Assoc Guidebook 75:1–54

Baird GC, Brett CE, Kirchgasser WT, Over DJ (2006) An early late Devonian bone bed-pelagic limestone succession; the North Evans-Genundewa limestone story. New York State Geol Assoc Guidebook 78:354–395

Baranov VV, Alkhovik TS (2006) Brachiopods of the family Ambocoeliidae (Spiriferida) from the Givetian of southern Verkhoyansk region (northeastern Russia). Paleontol J 40:162–167

Barrett SF (1988) The Devonian system in Colombia. In: McMillan NJ, Embry AF, Glass DJ (eds) Devonian of the world. Canadian Soc Petrol Geols Mem 14 (1): 705–717

Barrett SF, Isaacson PE (1988) Devonian paleogeography of South America. In: McMillan NJ, Embry AF, Glass DJ (eds) Devonian of the World. Canadian Soc Petrol Geols Mem 14 (1): 655–668

Bartholomew AJ (2002) Correlation of high order cycles in the Marine–Paralic transition of the upper middle Devonian (Givetian) Moscow formation, Eastern New York State MS thesis University of Cincinnati, 114p

Blakey R (2010) NAU geology, paleogeographic reconstructions. http://jan.ucc.nau.edu/~rcb7/RCB.html

Böhm F, Brachert TC (1993) Deep-water stromatolites and *Frutexites* Maslov from the early and middle Jurassic of S-Germany and Austria. Facies 28:145–168

Bonelli JR Jr, Bennington JB, Brett CE, Miller AI (2006) Testing for faunal stability across a regional biotic transition; quantifying stasis and variation among recurring coral-rich biofacies in the Middle Devonian Appalachian Basin. Paleobiology 32:20–37

Boucot AJ (1975) Evolution and extinction rate controls. Elsevier, New York, NY, 427 p

Boucot AJ (1988) Devonian biogeography; an update. In: McMillan NJ, Embry AF, Glass DJ (eds) Devonian of the world. Canadian Soc Petrol Geols Mem 14:211–227

Boucot AJ, Brett CE, Oliver WA Jr, Blodgett RB (1986) Devonian faunas of the Sainte-Helene Island breccia, Montreal, Quebec, Canada. Can J Earth Sci 23:2047–2056

Boucot AJ, Johnson JG, Talent JA (1969) Early Devonian brachiopod zoogeography. Geol Soc Am Spec Pap 119:107

Boyer DL, Droser ML (2007) Devonian monospecific assemblages: new insights into the ecology of reduced-oxygen depositional settings. Lethaia 40:321–333

Brand U, Azmy K, Jiang G, Lee X (2008) Global Taghanic and Givetian seawater records: an amelioration of faunal realms, climatic conditions and high levels of atmospheric carbon dioxide. Geological Society of America Abstracts with Programs—Annual Meeting, Houston, TX

Brett CE, Baird GC (1985) Carbonate-shale cycles in the Middle Devonian of New York; an evaluation of models for the origin of limestones in terrigenous shelf sequences. Geology (Boulder) 13:324–327

Brett CE, Baird GC (1994) Depositional sequences, cycles, and foreland basin dynamics in the late Middle Devonian (Givetian) of the Genesee Valley and western Finger Lakes region. In: Brett CE, Scatterday J (eds). New York State Geol Assoc Guidebook 66:505–586

Brett CE, Baird GC (1995) Coordinated stasis and evolutionary ecology of Silurian to Middle Devonian faunas in the Appalachian Basin. In: Erwin DH, Anstey RL (eds) New approaches to speciation in the fossil record. Columbia University, pp 285–315

Brett CE, Baird GC, Miller KB (1986) Sedimentary cycles and lateral facies gradients across a Middle Devonian shelf-to-basin ramp, Ludlowville Formation, Cayuga Basin. New York State Geol Assoc Guidebook 58:81–127

Brett CE, Baird GC, Miller KB (1990) A temporal hierarchy of paleoecologic processes within a Middle Devonian epeiric sea. Spec Publ (Paleont Soc) 5:178–209

Brett CE, Baird GC, Bartholomew AJ (2007a) Biofacies recurrence in the Middle Devonian of New York State; an example with implications for evolutionary paleoecology. Palaios 22:306–324

Brett CE, Bartholomew AJ, Bonelli JR Jr, Hendy AJW, McLaughlin PI (2007b) Response of shallow marine biotas to sea-level fluctuations; a review of faunal replacement and the process of habitat tracking. Palaios 22:228–244

Brett CE, Ivany LC, Bartholomew AJ, DeSantis MK, Baird GC (2009) Devonian ecological-evolutionary subunits in the Appalachian Basin; a revision and a test of persistence and discreteness. Geol Soc Spec Publ 314:7–36

Brett CE, Ivany LC, Schopf KM (1996) Coordinated stasis; an overview. Palaeogeogr Palaeoclimatol Palaeoecol 127:1–20

Brice D, Latréche S (1998) Brachiopodes du Bassin d'Illizi (Sahara algerien oriental) près de la limite Givétien-Frasnien. Geobios 31:437–454

Brice D, Milhau B, Mistiaen B (1994) Affinités nord-americaines de taxons Dévoniens (Givétien–Frasnien) du Boulonnais, nord de la France; migrations et diachronismes. Bull Soc Géol France 165:291–306

Brice D, Carls P, Cocks LRM, Copper P, Garcia-Alcalde JL, Godefroid J, Racheboeuf PR (2000) Brachiopoda: subcommission on Devonian stratigraphy; fossil groups important for boundary definition. Cour Forsch-Inst Senckenberg 220:65–86

Bridge JS, Willis BJ (1991) Middle Devonian near-shore marine, coastal and alluvial deposits, Schoharie Valley, central New York State. New York State Geol Assoc Guidebook 63:131–160

Bridge JS, Willis BJ (1994) Marine transgressions and regressions recorded in Middle Devonian shore-zone deposits of the Catskill clastic wedge. Geol Soc Am Bull 106:1440–1458

Burst JF (1965) Subaqueously formed shrinkage cracks in clay. J Sed Petrol 35:348–353

Cloud PE Jr (1962) Environment of calcium carbonate deposition west of Andros Island, Bahamas. U.S. Geol Surv Prof Pap 350:138p

Cooper GA, Williams JS (1935) Tully formation of New York. Geol Soc Am Bull 46:781–868

Day J (1996) Faunal signatures of Middle–Upper Devonian depositional sequences and sea-level fluctuations in the Iowa Basin: U.S. Midcontinent. In: Witzke BJ, Ludvigson GA, Day J (eds) Paleozoic sequence stratigraphy: views from the North American Craton. Geol Soc Am Spec Pap 306:277–300

Day J (1998) Distribution of latest Givetian–Frasnian Atrypida (Brachiopoda) in central and western North America. Acta Palaeontol Pol 43:205–240

Day J, Bunker BJ, Norris W, Uyeno T, Witzke BJ (1996) Middle-Upper Devonian relative sea-level histories of central and western North American interior basins. In: Witzke BJ, Ludvigson GA, Day J (eds) Paleozoic sequence stratigraphy: views from the North American Craton. Geol Soc Am Spec Pap 306:259–275

de Melo JHG (1988) The Malvinokaffric realm in the Devonian of Brazil. In: McMillan NJ, Embry AF, Glass DJ (eds) Devonian of the world. Canadian Soc Petrol Geol Mem 14 (1): 669–703

Desantis MK, Brett CE, ver Straeten CA (2007) Persistent depositional sequences and bioevents in the Eifelian (early Middle Devonian) of eastern Laurentia; North American evidence of the Kačak events? In: Becker RT, Kirchgasser WT (eds) Devonian events and correlations. Geol Soc London Spec Pub 278:83–104

DiMichele WA, Behrensmeyer AK, Bobe R, Labandeira CC, Olszewski TD, Pandolfi JM, Wing SL (2004) Long-term stasis in ecological assemblages; evidence from the fossil record. Ann Rev Ecol Evol Syst 35:285–322

Dravis JJ (1983) Hardened subtidal stromatolites, Bahamas. Science 219:385–386

Ettensohn FR (1985) The Catskill Delta complex and the Acadian Orogeny; a model. In: Woodrow DL, Sevon WD (eds) The Catskill Delata. Geol Soc Amer Spec Pub 201:39–49

Ettensohn FR, Miller ML, Dillman SB, Elam TD, Geller KL Swager DR, Markowitz G, Woock RD, Barron LS (1988) Characterization and implications of the Devonian-Mississippian black-shale sequence, eastern and central Kentucky, USA; pycnoclines, transgression, regression and tectonism. In: McMillan NJ, Embry AF, Glass DJ (eds) Devonian of the World. Canadian Soc Petrol Geol Mem 14 pt II: 323–345

Feist R (1991) The Late Devonian trilobite crises. Hist Biol 5:197–214

Formolo MJ, Lyons TW (2007) Accumulation and preservation of reworked marine pyrite beneath an oxygen-rich Devonian atmosphere; constraints from sulfur isotopes and framboid textures. J Sed Res 77:623–633

George AD (1999) Deep-water stromatolites, Canning Basin, northwestern Australia. Palaios 14:493–505

Goldman D, Mitchell CE (1990) Morphology, systematics, and evolution of Middle Devonian Ambocoeliidae (Brachiopoda), western New York. J Paleontol 64:79–99

Grabau AW (1917) Stratigraphic relationships of the Tully limestone and the Genesee shale in eastern North America. Geol Soc Am Bull 28:945–958

Grasso TX (1986) Redefinition, stratigraphy and depositional environments of the Mottville Member (Hamilton Group) in central and eastern New York. Bull New York State Mus 457:5–31

Hall J (1843) Geology of New York. Part IV, Comprising the survey of the fourth geological district. Carrol & Cook, Albany, NY, 683p

Hallam A, Wignall PB (1999) Mass extinctions and sea-level changes. Earth-Sci Rev 48:217–250

Heckel PH (1973) Nature, origin, and significance of the Tully Limestone; an anomalous unit in the Catskill Delta, Devonian of New York. Geol Soc Am Spec Pap 138:244

Heckel PH (1997) Overview of the Tully Limestone. In: Brett CE, ver Straeten CA (eds) Devonian cyclicity and sequence stratigraphy in New York State. Fieldtrip Guidebook University of Rochester, Rochester, NY, pp 79–85

Hofmann HJ (1973) Stromatolites; characteristics and utility. Earth-Sci Rev 9:339–373

Holland SM (2000) The quality of the fossil record; a sequence stratigraphic perspective. Paleobiology 26:148–168

House MR (1985) Correlation of mid-Palaeozoic ammonoid evolutionary events with global sedimentary perturbations. Nature (London) 313:17–22

House MR (2002) Strength, timing, setting and cause of mid-Palaeozoic extinctions. Palaeogeogr Palaeoclimatol Palaeoecol 181:5–25

Huber NK, Garrels RM (1953) Relation of pH and oxidation potential to sedimentary iron mineral formation. Econ Geol 48:337–357

Hüneke H (2006) Erosion and deposition from bottom currents during the Givetian and Frasnian: response to intensified oceanic circulation between Gondwana and Laurussia. Palaeogeogr Palaeoclimatol Palaeoecol 234: 146–167

Hüneke H (2007) Pelagic carbonate ooze reworked by bottom currents during Devonian approach of the continents Gondwana and Laurussia. In: Viana AR, Rebesco M (eds) Economic and palaeoceanographic significance of contourite deposits. Geol Soc London Spec Pub 276:299–328

Isaacson PE, Sablock PE (1988) Devonian system in Bolivia, Peru and northern Chile. In: McMillan NJ, Embry AF, Glass DJ (eds) Devonian of the world. Canadian Soc Petrol Geol 14 (1):719–728

Ivany LC, Brett CE, Wall HLB, Wall PD, Handley JC (2009) Relative taxonomic and ecologic stability in Devonian marine faunas of New York State; a test of coordinated stasis. Paleobiology 35:499–524

Joachimski MM, Breisig S, Buggisch W, Day J, van Geldern R (2004) Oxygen isotope evolution of biogenic calcite and apatite during the Middle and Late Devonian. Int J Earth Sci 93:542–553

Joachimski MM, Breisig S, Buggisch W, Talent JA, Mawson R, Gereke M, Morrow JR, Day J, Weddige K (2009) Devonian climate and reef evolution: insights from oxygen isotopes in apatite. Earth Plan Sci Lett 284:599–609

Johnson JG (1970) Taghanic onlap and the end of North America Devonian provinciality. Geol Soc Am Bull 81:2077–2105

Johnson KG, Friedman GM (1968) Depositional environments of Tully Limestone and clastic equivalents (Upper Devonian), east-central New York State. Geol Soc Am Spec Pap 101(Abstracts for 1966):263–264

Johnson KG, Friedman GM (1969) The Tully clastic correlatives (Upper Devonian) of New York State; a model for recognition of alluvial, dune(?), tidal, nearshore (bar and lagoon), and offshore sedimentary environments in a tectonic delta complex. J Sed Petrol 39:451–485

Johnson JG, Klapper G, Sandberg CA (1985) Devonian eustatic fluctuations in Euramerica. Geol Soc Am Bull 96:567–587

Jüngst H (1934) Zur geologischen Bedeutung der Synärese. Ein Beitrag zur Entwässerung der Kolloide im werdenden Genstein. Geol Rundschau 25:312–325

Kaufmann B (2006) Calibrating the Devonian time scale; a synthesis of U-Pb ID-TIMS ages and conodont stratigraphy. Earth-Sci Rev 76:175–190

Kim Y, Lee YI (2000) Ironstones and green marine clays in the Dongjeom Formation (Early Ordovician) of Korea. Sed Geol 130:65–80

Kindle EM (1896) The relation of the fauna of the Ithaca group to the faunas of the Portage and Chemung. Bull Am Paleontol 2:56

Kindle EM (1906) Notes on the range and distribution of *Reticularia laevis*. J Geol 14(3):188–193

Klapper G (1981) Review of New York Devonian conodont biostratigraphy. In: Oliver WA Jr, Klapper G (eds) Devonian Biostratigraphy of New York Pt 1, Text. International Union of Geological Sciences, Subcommission on Devonian Stratigraphy, Washington DC, pp 57–66

Koch WF II, Boucot AJ (1982) Temperature fluctuations in the Devonian Eastern Americas Realm. J Paleontol 56:240–243

Kurtz DM (1883) Ithaca and its resources: being an historical and descriptive sketch of the "forest city" and its magnificent scenery, Glens, Falls, Ravines, Cornell University, and the Principal Manufacturing and Commercial Interests. Journal Association Book and Job Print, Ithaca, NY, 121p

Lieberman BS (1994) Evolution of the trilobite subfamily Proetinae Salter, 1864, and the origin, diversification, evolutionary affinity, and extinction of the Middle Devonian proetid fauna of eastern North America. Bull Am Mus Nat Hist 223:176p

Lieberman BS, Brett CE, Eldredge N (1995) A study of stasis and change in two species lineages from the Middle Devonian of New York State. Paleobiology 21:15–27

Linsley DM (1994) Devonian paleontology of New York; containing the brachiopods, bivalves, rostroconchs, gastropods, tergomyans, ammonoids, trilobites, eurypterids and phyllocarids, based on the lithographs of James Hall and John Clarke. Palaeontol Res Inst Spec Pub 21:472

Logan BW, Hoffman P, Gebelein CD (1974) Algal mats, cryptalgal fabrics, and structures, Hamelin Pool, Western Australia. In: Logan BW, Read JF, Hagan GM (eds) Evolution and diagenesis of quaternary carbonate sequences, Western Australia. Am Assoc Petrol Geol Mem 22:140–194

Marshall JEA, Brown JF, Astin TR (2011) Recognising the Taghanic Event in the Devonian terrestrial environment and its implications for understanding land-sea interactions. Palaeogeogr Palaeoclimatol Palaeoecol 304(1–2):165–183

May A (1995) Relationship among sea-level fluctuation, biogeography and bioevents of the Devonian: an attempt to approach a powerful, but simple model for complex long-range control of biotic crises. Geolines (Praha) 3:38–49

Maynard JB (1986) Geochemistry of oolitic iron ores, an electron microprobe study. Econ Geol 81:1473–1483

McGhee GR Jr (1982) The Frasnian–Famennian extinction event: a preliminary analysis of Appalachian marine ecosystems. In: Silver LT, Schultz PH (eds) Geological implications of impacts of large asteroids and comets on Earth. Geol Soc Am Sp Pap 190:491–500

McGhee GR Jr (1989) The Frasnian–Famennian extinction event. In: Donovan SK (ed) Mass extinctions: processes and

evidence. Columbia University Press, New York, NY, pp 133–151
McGhee GR Jr (1996) The Late Devonian mass extinction; the Frasnian/Famennian crisis. Columbia University Press, New York, NY, 303p
McGhee GR Jr (1997) Late Devonian bioevents in the Appalachian Sea; immigration, extinction, and species replacements. In: Brett CE, Baird GC (eds) Paleontological events; stratigraphic, ecological, and evolutionary implications. Columbia University, New York, NY, pp 493–508
McLean RA, Sorauf JE (1988) The distribution of rugose corals in Frasnian outcrop sequences of North America. In: McMillan NJ, Embry AF, Glass DJ (eds) Devonian of the world. Canadian Soc Petrol Geol 14 (1):379–396
Miller AI (1998) Biotic transitions in global marine diversity. Science 281:1157–1160
Monty CLV (1971) An autoecological approach of intertidal and deep water stromatolites. Ann Soc Géol Belgique 94:265–276
Murphy AE, Brett CE, Hollander DJ, Lyons TW, Sageman BB (2000) Black shale deposition and faunal overturn in the Devonian Appalachian Basin; clastic starvation, seasonal water-column mixing, and efficient biolimiting nutrient recycling. Paleoceanography 15:280–291
Oliver WA Jr (1990) Extinctions and migrations of Devonian rugose corals in the eastern Americas realm. Lethaia 23:167–178
Patzkowsky ME, Holland SM (1999) Biofacies replacement in a sequence stratigraphic framework; Middle and Upper Ordovician of the Nashville Dome, Tennessee, USA. Palaios 14:301–317
Pflueger F (1999) Matground structures and redox facies. Palaios 14:25–39
Playford PE (1980) Environmental controls on the morphology of modern stromatolites at Hamelin Pool, Western Australia. Ann Rep West Aust Geol Surv 1979:73–77
Plummer PS, Gostin VA (1981) Shrinkage cracks; desiccation or synaeresis? J Sed Petrol 51:1147–1156
Pratt BR (1998) Syneresis cracks; subaqueous shrinkage in argillaceous sediments caused by earthquake-induced dewatering. Sed Geol 117:1–10
Prosser CS (1899) Classification and distribution of the Hamilton and Chemung series of central and eastern New York, Part 2. Ann Rep Regents New York State Mus 2, Rep 51:65–315
Racheboeuf PR, Girard C, Lethiers F, Derycke C, Herrera ZA, Trompette R (2001) Evidence for Givetian Stage in the Mauritanian Adrar (West Africa); biostratigraphical data and palaeogeographic implications. Newslett Stratigr 38:141–162
Racheboeuf PR, Gourvennec R, Deynoux M, Brice D (2004) The Devonian of the Hodh area (Islamic Republic of Mauritania); paleontology and stratigraphy. J Paleont 78:98–110
Racki G (1993) Brachiopod assemblages in the Devonian Kowala Formation of the Holy Cross Mountains. Acta Palaeontol Pol 37:297–357
Raup DM Sepkoski JJ Jr (1982) Mass extinctions in the marine fossil record. Science 215:1501–1503
Rickard LV (1975) Correlation of Silurian and Devonian Rocks in New York State. NY State Mus Sci Service Map Chart Ser 24:16p
Rickard LV (1981) The Devonian system of New York State. In: Oliver WA Jr, Klapper G (eds) Devonian biostratigraphy of New York Pt 1 (Text). International Union of Geological Sciences, Subcommission on Devonian Stratigraphy, Washington DC, pp 5–21
Rode AL, Lieberman BS (2004) Using GIS to unlock the interactions between biogeography, environment, and evolution in Middle and Late Devonian brachiopods and bivalves. Palaeogeogr Palaeoclimatol Palaeoecol 211:345–359
Rogers WB, Isachsen YW, Mock TD, Nyahay RE (1990) New York State geological highway map. New York State Museum and Science Service Educational Leaflet, No. 33, 1 sheet
Sandberg CA, Morrow JR, Ziegler W (2002) Late Devonian sea-level changes, catastrophic events, and mass extinctions. In: Koeberl C, MacLeod KG (eds) Catastrophic events and mass extinctions: impacts and beyond. Geol Soc Am Spec Pap 356:473–487
Sessa J (2003) The dynamics of rapid, asynchronous biotic turnover in the middle Devonian Appalachian Basin of New York. MS thesis. University of Cincinnati, 80p
Shinn EA, Steinen RP, Lidz BH, Swart PK (1989) Whitings, a sedimentologic dilemma. J Sed Petrol 59:147–161
Sorauf JE, Oliver WA Jr (1976) Septal carinae and microstructure in middle Devonian Heliophyllum (Rugosa) from New York State. J Paleontol 50:331–343
Thayer CW (1972) Marine paleoecology of the Upper Devonian Genesee group of New York. PhD thesis. Yale University, 240p
Thayer CW (1974) Marine paleoecology in the Upper Devonian of New York. Lethaia 7:121–155
Trainer DW Jr (1932) The Tully limestone of central New York. New York State Mus Bull 291:43p
van Geldern R, Alvarez F, Day J., Jansen U, Joachimski MM, Ma XP, Yolkin EA (2006) Carbon, oxygen and strontium isotope records of Devonian brachiopod shell calcite. Palaeogeogr Palaeoclimatol Palaeoecol 240:47–67
Ver Straeten CA, Brett CE (1995) Lower and Middle Devonian foreland basin fill in the Catskill Front; stratigraphic synthesis, sequence stratigraphy, and Acadian Orogeny. New York State Geol Assoc Guidebook 67:313–356
Ver Straeten CA, Brett CE (1997) Lower and Middle Devonian foreland basin fill in the Catskill Front: stratigraphic synthesis, sequence stratigraphy, and the Acadian Orogeny. In: Brett CE, ver Straeten CA (eds) Devonian cyclicity and sequence stratigraphy in New York State, Fieldtrip Guidebook, International Union of Geological Sciences, Subcommission on Devonian Stratigraphy. University of Rochester, Rochester, NY, pp 23–54
Walliser OH (1990) How to define "global bio-events". Lect Notes Earth Sci 30:1–4
Walliser OH (1996) Global Events in the Devonian and Carboniferous. In: Walliser OH (ed) Global events and event stratigraphy in the phanerozoic. Springer, Heidelberg, pp 225–250
Wells AJ, Illing LV (1964) Present-day precipitation of calcium carbonate in the Persian Gulf. In: Wolff KW (ed) Developments in sedimentology, vol 1. Elsevier, Amsterdam, pp 429–435
White WA (1961) Colloid phenomena in sedimentation of argillaceous rocks. J Sed Petrol 31:560–570

Williams HS (1884) On the fossil faunas of the upper Devonian along the meridian of 76 degrees 30′, from Tompkins County, New York, to Bradford County, Pennsylvania. Geol Soc Am Bull 3:51–86

Williams HS (1890) The Cuboides zone and its fauna; a discussion of methods of correlation (with discussion by C. D. Walcott). Geol Soc Am Bull 1:481–500

Williams HS (1906) The Devonian section of Ithaca, New York, Part I. J Geol 14:579–598

Williams HS (1913) Recurrent *Tropidoleptus* zones of the Upper Devonian in New York. US Geol Surv Prof Pap 79:103

Williams HS, Tarr RS, Kindle EM (1909) Description of the Watkins Glen–Catatonk District, New York. US Geol Surv Geol Atlas US, Folio 69:242p

Witzke BJ (1987) Models for circulation patterns in epicontinental seas applied to Paleozoic facies of North America Craton. Paleoceanography 2:229–248

Witzke BJ, Bunker BJ (1997) Sedimentation and stratigraphic architecture of a Middle Devonian (late Givetian) transgressive–regressive carbonate–evaporite cycle, Coralville Formation, Iowa area. In: Klapper G, Murphy MA, Talent JA (eds) Paleozoic sequence stratigraphy, biostratigraphy, and biogeography: studies in Honor of J. Granville ("Jess") Johnson. Geol Soc Am Sp Pap 321: 67–88

Zambito JJ IV, Baird GC, Bartholomew AJ, Brett CE (2007) Reexamination of the type Ithaca Formation; correlations with sections in western New York. New York State Geol Assoc Guidebook 79:83–105

Zambito JJ IV, Baird GC, Brett CE, Bartholomew AJ (2009) Depositional sequences and paleontology of the Middle–Upper Devonian transition (Genesee Group) at Ithaca, New York: a revised lithostratigraphy for the northern Appalachian Basin. In: Over DJ (ed) Studies in Devonian Stratigraphy: proceedings of the 2007 international meeting of the subcommission on Devonian Stratigraphy and IGCP 499. Paleontogr Am 63:49–69

The Permian—A Time of Major Evolutions and Revolutions in the History of Life

Monica J. Campi

Abstract

The Permian was a critical time during the evolution of life on Earth. It was a time of significant global changes to climatic systems, continental arrangement, palaeobiogeography and ecosystem function. In the marine realm it saw a major transition from the brachiopod-dominated Palaeozoic invertebrate fauna to one dominated by bivalves. This transition began during the Permian and was finalized by the major ecosystem reorganization that was brought about by the end-Permian mass extinction.

Keywords

Permian • Tethys • Pangaea • Gondwana • Rift-drift • Continental accretion • Micro-continents • Basaltic vulcanism • Palaeoclimate • Anoxia • Extinctions • provinciality

Events in the Permian

Continental movements initiated the formation of the supercontinent Pangea during the Carboniferous, a process that continued into the Permian. Several micro-continents (e.g., Lhasa Block, Sibumasu terrane and Qiangtang terrane) rifted off the northern margin of Gondwana (modern-day Western Australia, India and Arabia) during the Permian and drifted northwards through the Tethys Ocean towards the equator (Fig. 1).

M.J. Campi (✉)
Department of Primary Industries, GeoScience Victoria, Melbourne, Vic 3001, Australia

School of Life and Environmental Sciences, Deakin University, Burwood, Vic 3125, Australia
e-mail: monica.campi@dpi.vic.gov.au

This rift-drift sequence had important implications for palaeobiogeography as discussed further later.

There were major episodes of volcanic activity during the Permian, with the production of two large igneous provinces, the Siberian and Emeishan Flood Basalts, occurring at the end-Permian and end-Guadalupian boundaries, respectively. The Siberian Traps may have erupted more than 14×106 cubic km of both ash and lava (Dobretsov 2005) and has been linked to the major end-Permian biotic crisis (Kamo et al. 2003). The Emeishan Basalts erupted across the southwest area of the South China Block (covering large parts of modern-day Sichuan, Guizhou and Yunnan provinces of the People's Republic of China and extending into northern Vietnam (Hoa et al. 2008)). These are not as extensive and less well known but have recently been suggested to have been erupted close to the Guadalupian/Lopingian boundary (Zhou

J.A. Talent (ed.), *Earth and Life*, International Year of Planet Earth,
DOI 10.1007/978-90-481-3428-1_23, © Springer Science+Business Media B.V. 2012

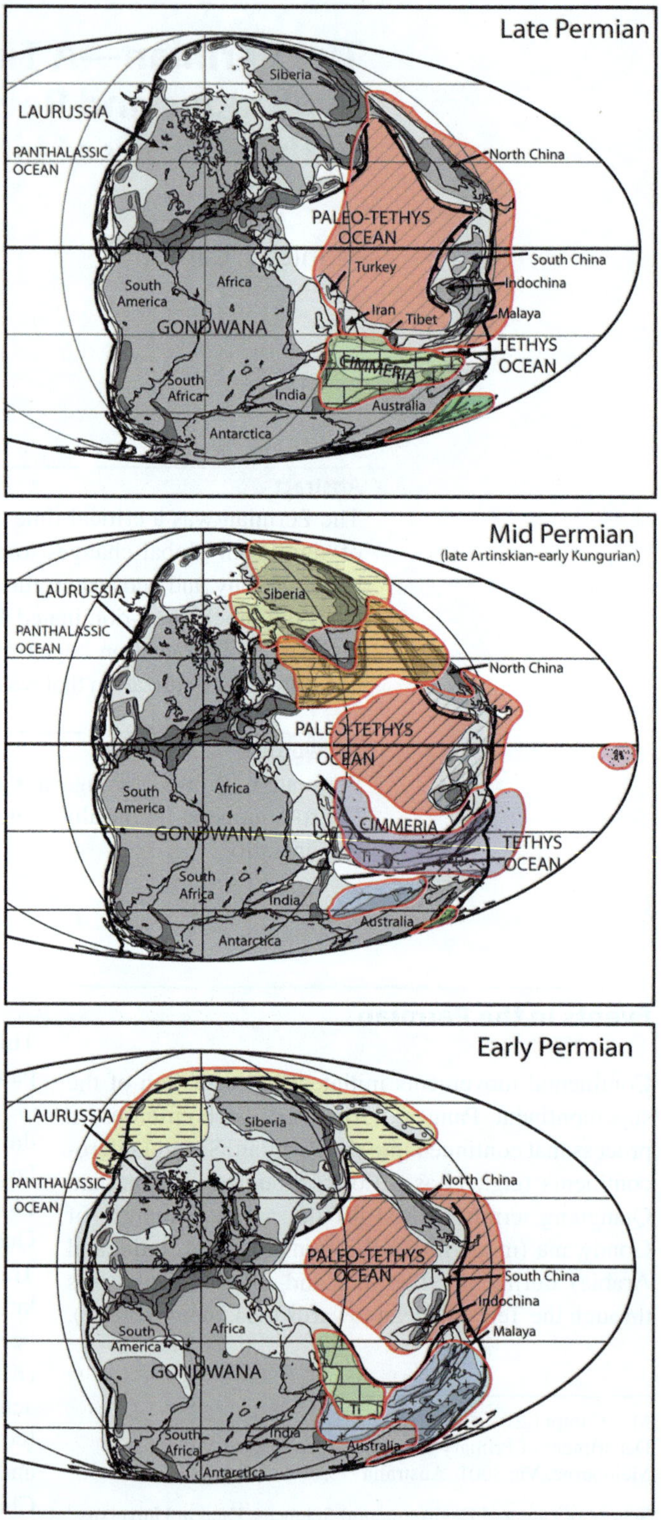

Fig. 1 Marine invertebrate provinciality through the Permian. Modified from Shi (1998); base map after Scotese (2002)

et al. 2002; Ali et al. 2005). It has been suggested that the eruption of these basalts took place closer to the end of the Permian (Chung et al. 1998), for at least the last phase of the eruptions. This, however, has largely been discounted. Stratigraphical placement of the basalts between the Guadalupian Maokou Formation and the overlying Wuchiaping Limestone of Lopingian age support an eruption time close to the G/L boundary, at least for the main eruptive sequence (Zhou et al. 2002). The Emeishan eruption has also been linked to the Guadalupian mass extinction (Wignall et al. 2009).

Other volcanism occurred in the Sydney Basin on the eastern margin of Australia (then the eastern margin of Gondwana). This suite of up to seven major volcanic episodes is intercalated with the shallow marine Broughton Formation and is thought to be Wordian/Roadian in age (Tye et al. 1996). This sequence was erupted through the Clyde Coal Measures (Sakmarian?), which is significant when the volcanogenic heating of the carbon-rich coals is considered as a mechanism for the liberation of methane and carbon dioxide (Retallack and Jahren 2008), and associated flow-on effects influencing the global carbon cycle, and links to global warming and other environmental effects.

Another substantial influence on the global carbon cycle through the late Palaeozoic was the development of vast forests in the Carboniferous and Permian coal swamps, which sequestered huge volumes of CO_2. Burial rates of organic carbon rose dramatically during the Carboniferous and stayed high through the majority of the Permian (Berner 2002) and can be linked to climatic cooling and development of global ice-house conditions of the Late Palaeozoic ice age. Ice sheets were well developed in parts of Gondwana by the Bashkirian (Late Carboniferous; Cocks 2007), and the major phase of glacial retreat occurred during the Sakmarian (Korte et al. 2005). An additional trigger for the Late Palaeozoic ice age was the closure of the equatorial connection between the Panthalassic and Palaeo-Tethys Oceans in the mid-Carboniferous (Cocks 2007). This affected oceanic circulation and global heat transfer, and along with the large expanse of land at high latitudes, it was conducive to the development and longevity of ice sheets.

There was a significant increase in O_2 levels in the Late Carboniferous and into the Permian when compared to the Early Palaeozoic and Meso/Cenozoic (see O_2 and CO_2 curves in Graham et al. (1995)). The development of extensive forests of lycophytes (dominant until the Late Carboniferous) and pteridosperms (seedferns) also contributed to increased O_2. These earliest seed-bearing plants became dominant in the Southern Hemisphere (e.g., *Glossopteris*). Species reliant on stable, humid conditions characterised the Late Carboniferous, followed by a transition to more xeromorphic species adapted to seasonally dry, semi-arid environment by the Middle Permian (Montañez et al. 2007).

The development of the ability to digest plant material through acquisition of microbial gut flora by both vertebrates in the Permian and invertebrates (established in the Devonian, Labandeira 2005) signalled a large change in the carbon cycling on land. Herbivory allows the carbon present in the leaves of plants to be incorporated into other living organisms, rather than partially decomposing before being consumed by detritivores, or being sequestered via the transition from peat to coal.

Climate

There was a major change in climate from icehouse conditions of the Carboniferous-Permian glaciation to greenhouse conditions at the Permian-Triassic boundary. Glaciation probably also helped to dry out the climate, as indicated by the deposition of extensive red-bed formations in Gondwana (in the Karoo Basin, South Africa, and Paraña Basin in Brazil and Argentina). New research indicates that the Late Palaeozoic ice age was not a simple event, with evidence now suggesting that the world probably went through a sequence of glacial/interglacial cycles much like those that occurred during the Pleistocene glaciation (Fig. 2). The main phase of glacial retreat and collapse of the continental ice sheets occurred in the Sakmarian, although there was a brief return to colder conditions during the Kungurian and early Roadian (evidence from glendonites from the southern Sydney Basin, Frank et al. 2008; ikaites in the Chihsia Formation in South China, Shi and Grunt 2000; and palaeotemperature data from oxygen isotopes in brachiopod shells). Further evidence for an increasingly arid climate in the Early and Middle Permian comes from oxygen isotopes from low-latitude brachiopod shells (Grossman et al. 2008).

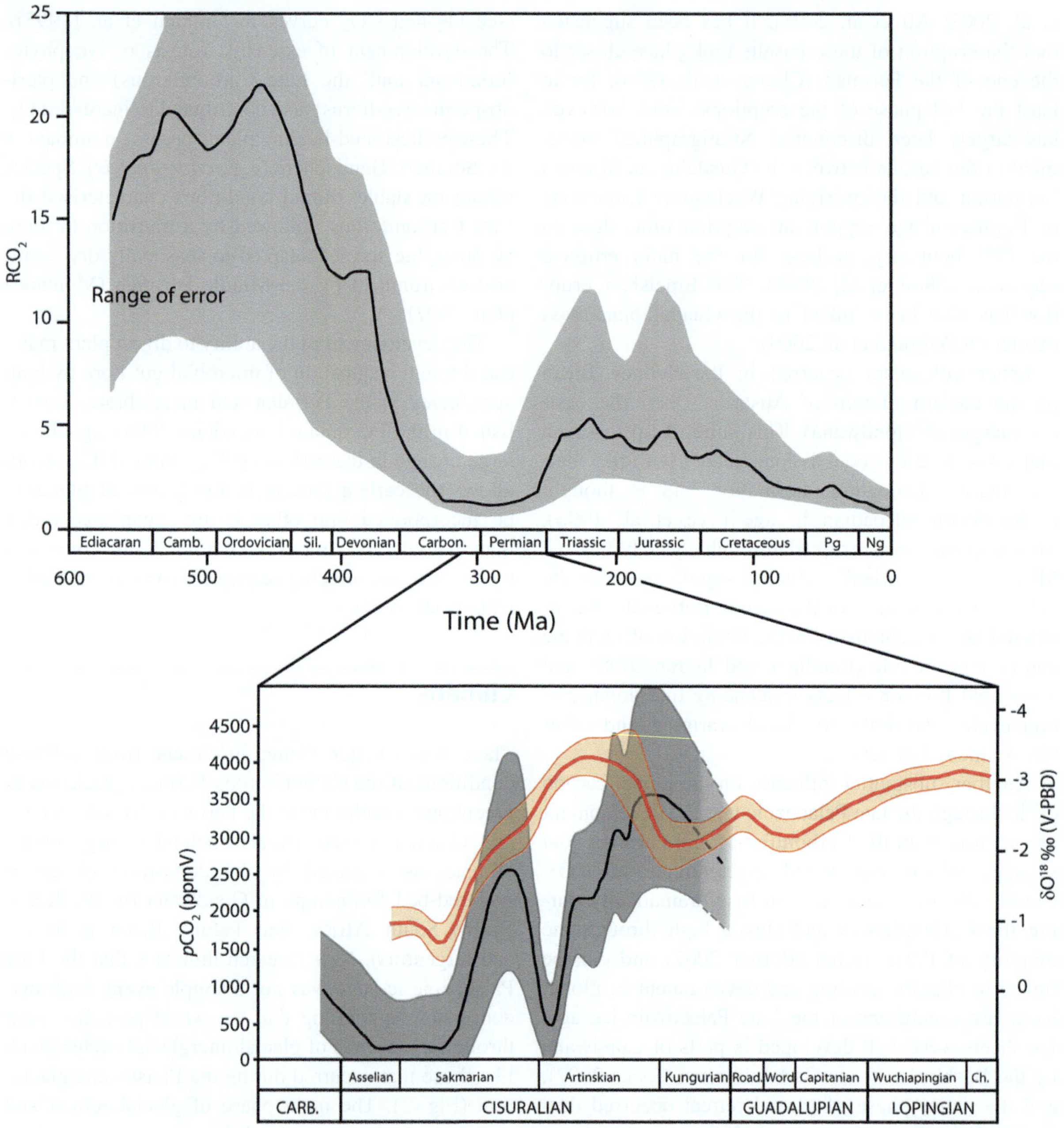

Fig. 2 CO_2 and δO^{18} curves for the Phanerozoic (*top*, based on Berner 2002) and Permian (*bottom*, CO_2 curve based on Montañez et al. 2007, δO^{18} from Korte et al. 2005)

There was also a change from a high polar-equatorial temperature gradient in the Early Permian to a much lower gradient in the Late Permian (GR Shi 'personal communication'), which has significant implications for global circulation of the ocean-atmosphere system. A high polar-equator gradient is the main driving force behind vigorous oceanic circulation and associated movement of oxygenated water from the surface to deep water and of the upwelling of nutrient-rich bottom water to the ocean surface. Evidence for cool to cold-water upwelling systems is found in the Kungurian Chihsia Formation of South China (Shi and Grunt 2000), where ikaites are found. A lower gradient in the Late Permian is supported by the extensive evidence of a shutdown of the ocean circulation system in the later part of the Permian (Isozaki 1994, 1997), and ocean stagnation and anoxia, even 'superanoxia' or a 'Strangelove ocean' (Kump 1991). These changes to global ocean circulation have been suggested in the past as causes of the end-Permian biotic crisis and the extended 'bleak time' before recovery occurred in the Anisian. Further evidence comes from black shales sequences in otherwise radiolarian rich-red cherts from deep-sea sequences preserved in Japan (Isozaki 1994), abundant pyrite in many Changhsingian deposits, the sulfur isotopic record, and the loss of benthic fauna.

The Carboniferous and Early Permian was a significant minima in the long-term (Phanerozoic) trend in atmospheric pCO_2 (Berner 1998), with levels similar to those today. Levels rose steeply during the Permian, reaching concentrations $\sim$4 times that of the present-day atmosphere by the Permian-Triassic transition. This Permo-Carboniferous minima also coincided with a period of significant glaciation. One of the major drivers behind the low concentrations of atmospheric CO_2 was the sequestration of organic matter by extensive swamp forests that developed during the Carboniferous.

A higher-resolution picture of the interaction between pCO_2 and Permo-Carboniferous climate from palaeosols and shallow-water brachiopods (Montañez et al. 2007) reveals a more-complicated story, showing a climate undergoing a sequence of glacial/interglacial cycles, much like those that have occurred during the Pleistocene glaciation. CO_2 values were similar to modern pCO_2 in the Asselian and Sakmarian (the early part of the Cisuralian), with a rapid, but short-lived, peak of 2500 ppmV in the late Sakmarian (probably corresponding to the collapse of the major Gondwanan ice sheets), a brief return to $\sim$500 ppmV in the earliest Artinskian, and levels between 2000 and 3500 ppmV for the remainder of the Artinskian and Kungurian.

Tectonic Changes

Major tectonic changes included the final stages of the amalgamation of Pangea with associated uplift, climatic change (drying) and volcanism (e.g., eastern Australia and the Large Igneous Provinces of the Emeishan Basalts and Siberian Traps). There was also a series of micro-continents or 'terranes' that rifted off the northern margin of Gondwana during the Permian and drifted northwards across the Palaeo-Tethys (e.g., Sibumasu, Lhasa, Qiangtang, see Fig. 1). Detailed discussion of these tectonic changes have been covered elsewhere (e.g., Scotese 2002; Metcalfe 1998, 2002; Ziegler et al. 1997), and their impact on the global palaeobiogeography is discussed in the following section.

Major Evolutionary Landmarks

Terrestrial Environments

In the terrestrial realm one of the major evolutionary events that occurred during the Permian was the evolution of the therapsids from their pelycosaur forebears. Larger vertebrate herbivores first appeared close to the Permo-Carboniferous boundary and became common during the Permian. This was possible through the development of complex symbioses to digest plant material. As discussed earlier, this development had a significant impact on the global carbon cycle. Gymnosperms (naked-seed plants) became widespread also and formed thick sequences of coal across Gondwana.

Terrestrial ecosystems dried out towards the end of the Permian (Retallack et al. 2006), which lead to changing floral communities and in places an increased influx of terrigenous sediment into the oceans (Sephton et al. 2005). This influx of sediment has been associated with changes to marine organisms and communities; for example, it has been linked to the miniaturization of marine invertebrates (He et al. 2007).

There were two terrestrial extinctions in the Late Permian, based on evidence from South Africa and Antarctica (Retallack et al. 2006). Changes in the hydrological regime, from high- to low-sinuosity river systems, were interpreted as a result of the extinction of terrestrial plants at both the end of the Guadalupian and the end of the Permian. Marker beds of massive sandstones occur at both times within the coal seam sequences (back-arc marsh environments), while the loss of plants led to a destabilization of the soil, which was subsequently washed into river systems in flash-flooding events, possibly contributing to the increased sediment discharge into the oceans mentioned earlier (Sephton et al. 2005).

The terrestrial record of southwest China documents a large shift from spore-dominated (both in abundance and diversity) to pollen-dominated palynological assemblages across the Permian-Triassic boundary (PTB) (Peng and Shi 2009). The pattern of this shift mirrors changes seen in the marine environment, with an overall gradual loss of diversity from the Wuchiapingian through to the end-Changhsingian, an acceleration of loss of diversity across the PTB into the earliest Triassic, with the final disappearance of Permian relicts in the Early Triassic. The pattern of species disappearance is not synchronous across the area, and the decline through the Changhsingian does not support the argument for a globally synchronous collapse of terrestrial ecosystems at the end-Permian as suggested by Twitchett (2001). A change similar to that found in Gondwana is also found in the hydrological regime (Retallack et al. 2006) in southwest China (Peng and Shi 2009), as only root fossils, and not leaves or other plant parts, are preserved in that regime in the PTB interval in southwest China. This suggests a change in taphonomy of this interval, possibly indicating higher-flow regimes removing leaves and other plant re-mains from surface and a more arid climate facilitating the decay of leaf material before it can become incorporated into either the soil or fluvial sediments.

Marine Environments

A number of significant changes to marine life occurred during the Permian, for example, the change from the brachiopod-dominated Palaeozoic Invertebrate Fauna to bivalve-dominated Modern Fauna. This change began during the Permian (Clapham and Bottjer 2007; Clapham et al. 2006) and was likely to have been finalized by the end-Permian extinction. According to Knoll et al. (2007), it preferentially affected organisms with low metabolic rates, little or no carbonate buffering ability and reliance on calcium carbonate skeletal framework.

The tectonic and climatic changes outlined earlier also affected marine organisms, as exemplified by changes in marine biogeographic provinciality:

1. changes from high provinciality in the Early Permian,
2. development of transitional zones in the Kungurian,
3. low provinciality in the Late Permian, to ultimately,
4. a cosmopolitan 'disaster' fauna in the immediate aftermath of the end-Permian biotic crisis.

There were at least two major biotic transitions during the Permian. The most notable of these was the catastrophic end-Permian mass extinction, where up to 95% of marine species disappeared (see Wignall 2007 for summary). The second event occurred at the Guadalupian-Lopingian boundary, and evidence is still equivocal about the severity and extent of this event, and it is further discussed later. There has been some suggestion of a third major faunal turnover during the late Cisuralian, reported as affecting foraminifera and several other marine groups (Leven et al. 1996; Yang et al. 2005), although this 'event' has not been extensively studied in depth. It is also unclear if it represents a true extinction event caused by environmental perturbations or faunal turnover.

Unique reef styles developed during the Permian, with reef mound, constructed by calcareous algae, sponges, bryozoa, brachiopods, rugose corals and crinoids. This is in contrast to earlier reefs (i.e., those of the Devonian) that were constructed by stromatoporoids and tabulate corals, and later reefs built by rudist bivalves (Cretaceous) and scleractinian corals (Modern). Algal mounds and patch reefs were common in the tropical Palaeo-Tethys (e.g., South China, Tong et al. 1998, Greece, Flügel and Reinhart 1989) and extensive, well-developed reef systems formed on the continental shelf of western Pangea (Guadalupe Mountains, Texas).

Two examples of these changes are examined in more detail: (1) changing marine provinciality and (2) changes in brachiopod diversity through the Permian.

Marine Provinciality

Marine invertebrates have been well studied during the Permian and have been used as tools to explore the changing global palaeobiogeography through this period. In particular, the provinciality of the Brachiopoda is now well established, through the work of Shi and colleagues (Shen and Shi 2000; Shen et al. 2000; Shi 2000, 2006; Shi and Archbold 1993, 1995; Shi et al. 1995). As mentioned earlier, the Permian was a time of significant changes in climate: from icehouse conditions with extensive polar ice caps in the Cisuralian to a warmer greenhouse climate at the Permian/Triassic transition. These climatic changes had major effects on the distribution of organisms, as illustrated by the changing provinciality seen in the Asia-Western Pacific region for example (Shi 1998).

Three palaeobiogeographic realms have been recognised in the Permian (Fig. 1, Table 1):

Table 1 Permian marine biotic provinces of the Asia-Western Pacific region in space and time

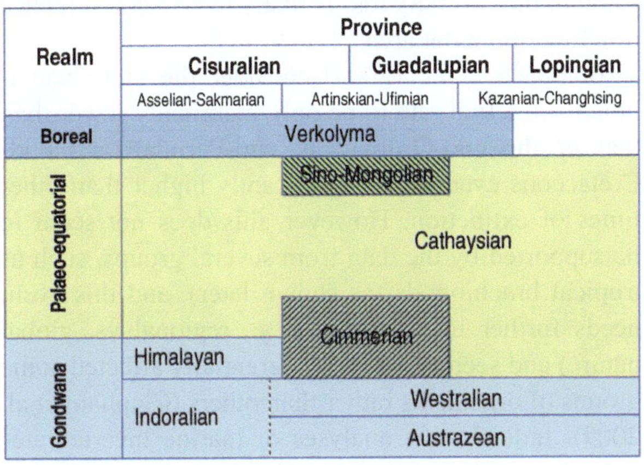

Diagonal stripes indicate transitional provinces and blue fill indicates cool water provinces. Modified from Shi (1998)

1. the Gondwanan realm, encompassing most of the southern part of Pangea (Australia, India, Antarctica, southern Africa and southern South America);
2. the Boreal realm, extending across the Arctic region; and
3. the Palaeo-equatorial realm divided into two regions, separated by the Panthalassan Ocean. The Asian Region incorporated South China and Indochina, the Tarim Basin, North China at times, parts of Japan, central Asia and the Mediterranean region, while the American Region included Texas, Mexico, central America and northern South America. The separation of the Palaeo-equatorial realm into two discrete regions possibly occurred because of the presence of biogeographic barriers: the Pangean supercontinent and the vast Panthalassa Ocean.

Numerous provinces have been recognised within these realms (e.g., see Shi 1998), with eight provinces in the Asian-Western Pacific region at various times during the Permian (Table 1).

The palaeobiogeographic provinces were not stable through the Permian because of climatic fluctuations and tectonic activity moving micro-continents through the Palaeo-Tethys Ocean (Fig. 1). In the early Cisuralian (Asselian-Sakmarian), four marine provinces can be recognised (Fig. 1; Shi and Archbold 1993), with the Indoralian and Himalayan provinces in the Gondwanan realm, the Cathaysian province (present throughout the Permian) in the Palaeo-equatorial realm and the Verkolyma province in the Boreal realm.

During the Asselian, palaeotemperature data from brachiopods indicate that the tropics were still as warm as today (24–35°C for Early Asselian or 24–33°C for the Middle Asselian), the range in temperature varied according to the extent of the Gondwanan ice caps (Angiolini et al. 2009). This supports the evidence from palaeobiogeographic studies (e.g., Shi 1998) indicating that the warm palaeo-tropical belt became reduced in extent, rather than temperature, during the Cisuralian, and that the polar-equatorial temperature gradient increased, driving vigorous ocean circulation.

Climatic amelioration after the retreat of the Late Palaeozoic ice age, plus the northwards drift of continental fragments that had rifted off central northern Gondwana, led to changing provinciality into the Artinksian and Kungurian. Two transitional provinces with unique character developed in the temperate zones: the Cimmerian province can be recognised in the Southern Hemisphere and the Sino-Mongolian Province in the Northern Hemisphere (Fig. 1). These provinces were characterised by a mixture of cold-cool water and palaeo-equatorial genera, in addition to endemic genera.

Recently, Roadian-Wordian palaeobiogeography has been updated to recognise the complexity of the tropical area (Shen et al. 2009), with the establishment of a new 'Panthalassan realm' proposed for the Mino terrane (part of modern-day Japan). This terrane, composed of exotic limestone-greenstone blocks now located in Jurassic mélange, is thought to have been an oceanic reef-seamount complex in the Middle Permian (Tazawa and Shen 1997). Shen et al. (2009) place this station to the north of the equator directly between the Cathaysian and Grandian provinces of the Palaeoequatorial realm. It has the highest similarity to the West Texas fauna (Grandian Province of Shen et al. 2009), sharing six genera (*Ametoria*, *Cartorhium*, *Cenorhynchia*, *Coscinophoria*, *Eolyttonia* and *Lepidospirifer*), and it also shares four genera with the Cimmerian Province (southern transitional zone—*Anomaloria*, *Calliprotina*, *Cartorhium* and *Eolyttonia*). The only links to the Cathaysian and Sino-Mongolian-Japanese (northern transitional zone) provinces are through cosmopolitan genera (*Martinia*, *Compressoproductus* and *Enteletes*). Based on the stronger links to the southern transitional zone rather than the northern transitional zone, as suggested by Tazawa (2007), I propose that it should be placed to the south of the equator (Fig. 1).

Continued climatic warming into the Lopingian lead to the disappearance of the transitional zones and with a poleward migration of the Palaeoequatorial realm (Fig. 1; Shi 1998). The Boreal realm disappeared at the end of the Guadalupian (Table 1, Fig. 3) during the closure of the Ural Seaway and the exposure of a large part of this region. The Austrazean province became restricted to isolated terranes that have now been incorporated into New Zealand.

Palaeoequatorial taxa tend to be smaller, while coolwater faunas from the Gond-wanan and Boreal realms tend to be larger, and often have lower diversity. The Boreal and Gondwanan realms also share many taxa, which indicates a bipolar distribution. This pattern of distribution suggests migration along deep-water coolwater currents (e.g., Shi and Grunt 2000).

In general, the cool- and cold-water Gondwanan and Boreal realms had highest diversity in the Asselian-Sakmarian (Fig. 3), with the Verkolyma province (Boreal realm) disappearing entirely by the Wuchipingian, probably due to regression and exposure of large areas of Siberia at the end of the Guadalupian (Fig. 3c). The Palaeo-equatorial realm increased in diversity through the Permian, with highest diversity reached in the Changhsingian.

At the very end of the Permian, a low-diversity 'disaster' fauna of cosmopolitan distribution was present in the immediate aftermath of the end-Permian biotic crisis (Chen et al. 2005a), which was probably related to the end-Permian extinction rather than climatic factors.

Brachiopod Extinctions

The end-Permian biotic crisis was devastating for the Brachiopoda, with the demise of several of the major orders (the Productida, Orthotethida, Orthida and Spiriferida), and there has been suggestion of an end-Guadalupian extinction as well, although the significance of this event is yet to be clarified (Clapham et al. 2009).

Evidence for an end-Guadalupian mass extinction is equivocal; early evidence from South China brachiopods indicated that there was a significant extinction at the family level (Shen and Shi 1996), but this is not supported by a more recent compilation of generic data (Sun and Shen 2004, Fig. 3). Outside the Tethys, this was a time of change from marine to terrestrial conditions around the Pangean margin, and therefore some of the diversity loss is likely to be a consequence of a loss of habitat area.

Bambach et al. (2004) consider the end-Guadalupian extinction rate to be only marginally lower than that of the end-Ordovician, end-Permian and end-Cretaceous events and significantly higher than other times of extinction. However, this does not seem to be supported by the data from several groups, such as tropical brachiopods (as shown later), and this event needs further investigation (e.g., regional vs. global nature) and seems to have preferentially affected some groups of organisms rather than others (Clapham et al. 2009). Indeed, new analyses of marine invertebrates indicate that there was no 'extinction event' at the end of the Guadalupian (Clapham et al. 2009) but rather a reduction in diversity due to low origination rates and a reduction of provinciality.

Brachiopod diversity through the Permian has been extensively analysed (e.g., Rong and Shen 2002; Shen and Shi 1996, 2002; Shen et al. 2006; Sun and Shen 2004). Data presented for both family- and generic-level diversity patterns of brachiopods in South China through the Permian have previously indicated the presence of at least two major diversity declines during this time (Shen and Shi 1996).

In the Asia-Pacific region, five stages of brachiopod evolution can be recognised through the Permian and the Triassic (Shen and Shi 2002; Sun and Shen 2004). A stable stage stretched from the Carboniferous into the Middle Permian (Kufengian), followed by an extinction stage from the Lengwuan to the Griesbachian (encompassing the pre-Lopingian and end-Permian biotic crises). A 'bleak stage' of low diversity and delayed recovery in the Early Triassic followed the end Permian mass extinction. Recovery and radiation from the mass extinction occurred over several stages of the Triassic, with a further mass extinction stage at the Norian-Rhaetian boundary. However, further analysis of Sun and Shen's (2004) data does not support an extinction stage of the length suggested (end of the Guadalupian to the earliest Triassic), as there is no change in family-level diversity in the later half of the Permian (Fig. 4) and a significant rise in generic diversity in the Lopingian (Fig. 5). In South China, brachiopod diversity reaches its highest level in the Changingian (Fig. 5). The 'pre-Lopingian biotic crisis' does not appear to be significant at the generic or

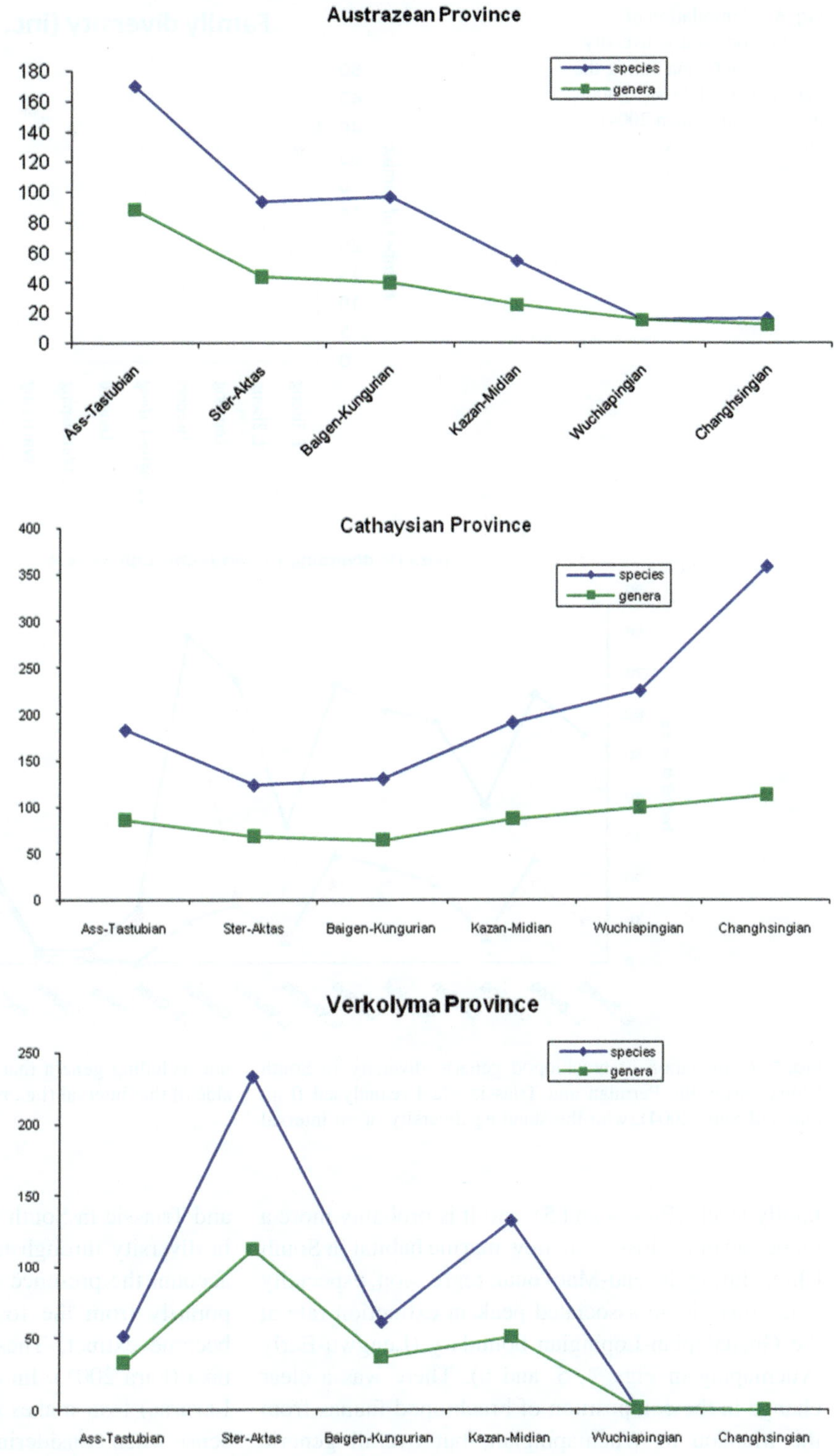

Fig. 3 Comparison of Permian brachiopod diversity across the Asia-Western Pacific region (data compiled from Shi and Shen 1999)

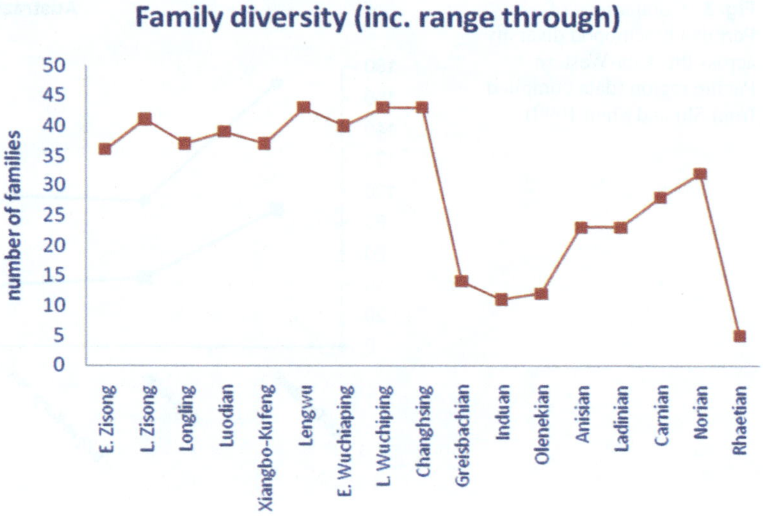

Fig. 4 Compilation of brachiopod family diversity from South China during the Permian and Triassic (data from Sun and Shen 2004). Diversity includes 'range-through' taxa

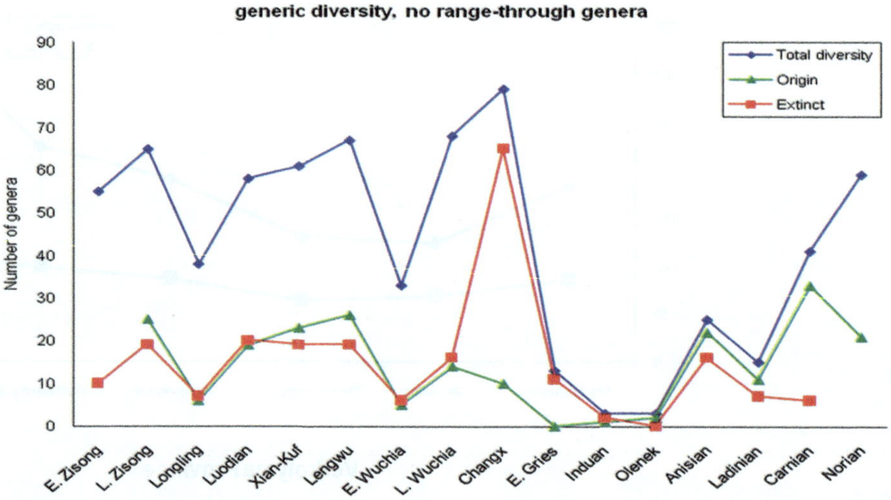

Fig. 5 Compilation of brachiopod generic diversity in South China during the Permian and Triassic (data reanalysed from Sun and Shen 2004), with the standing diversity of an interval not including genera that were present in the time slice either side of that interval (i.e., no 'range-through' taxa)

family levels (Figs. 4 and 5), and it is probably more a reflection of the loss of shallow marine habitat in South China during the end-Maokouan regression, especially when there is no associated peak in extinction rate at the Guadalupian-Lopingian boundary (Lengwu-Early Wuchiaping in Figs. 4, 5, and 6). There was a clear change in the composition of brachiopod faunas from the Maokou to Wuchiapingian, but overall generic diversity appears to have remained relatively stable.

Data from South China (Sun and Shen 2004) have been reanalysed herein (Figs. 4, 5, and 6) to show changes in brachiopod diversity through the Permian and Triassic in South China. When analysing changes in diversity through time, it is important to take into account the presence of taxa that may disappear temporarily from the fossil record but haven't actually become extinct. These have been termed 'Lazarus' taxa (Fara 2001). Including these 'range-through' (or Lazarus) taxa makes environmental and evolutionary sense when considering changes in diversity and looking for extinction events. A temporary absence from the fossil record can be explained by the population falling in number below a point where there is little possibility for individuals to be fossilised. With

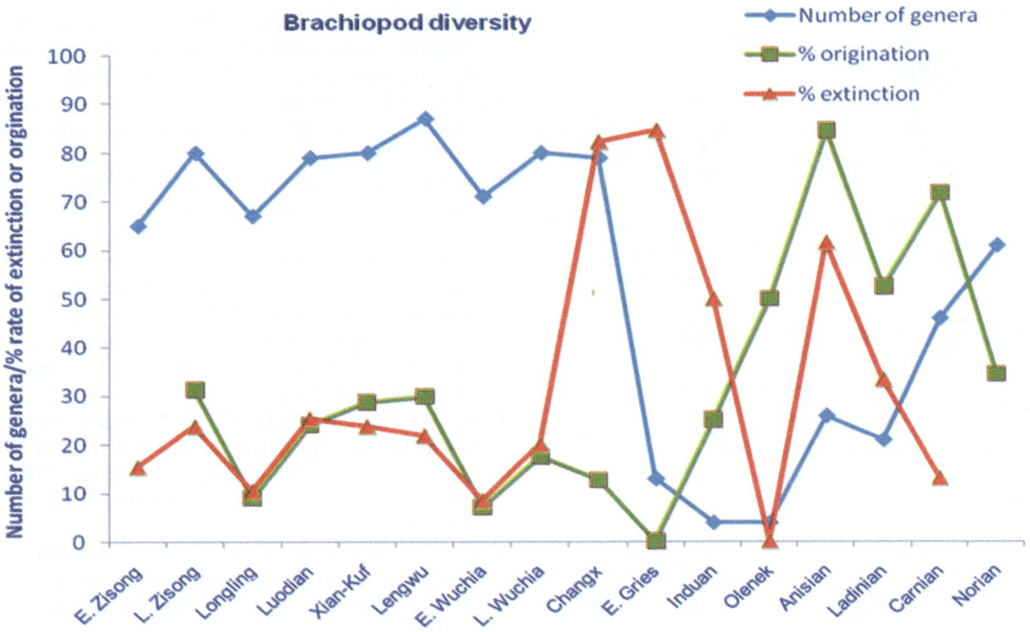

Fig. 6 Brachiopod diversity pattern for the Permian and Triassic of South China when 'range-through' taxa are included. Data reanalysed from Sun and Shen (2004)

changed conditions, the population might regenerate or the taxa involved might persist in a suitable environment elsewhere.

Once 'range-through' genera are included in the analysis of brachiopod diversity through the Permian and Triassic of South China (Fig. 6), it is clear that there is only one extinction event in this time in South China, occurring at the end of the Changhingian and corresponding to the end-Permian biotic crisis. This is substantiated both by the dramatic drop in generic diversity from the Changhsingian to the Greisbachian (with a loss of 82% of genera) and by the spike in the percentage of genera disappearing at the end of the Changhsingian. The analysis of extinction and origination rates in the Permian and Triassic indicates that the end-Permian biotic crisis had a major impact on the evolution of brachiopods in South China, as both were fairly stable through the Permian, whilst origination and extinction rates become uncoupled and fluctuate considerably during the Triassic indicating continued stresses and environmental instability into the Triassic; this contributed to the difficulty of recovery of ecosystem function after the end-Permian crisis.

There is no denying that the end-Permian extinction event caused major upheaval and reorganisation of ecosystems. A further illustration of these changes is a comparison of the diverse Late Permian benthic community from South China (Fig. 7A) with that present in the immediate aftermath of the extinction, the 'holdover' fauna that then disappeared by the earliest Triassic (Fig. 7B). For example, the pre-extinction fauna at Chuanmu, Sichuan (Campi and Shi 2005), contained more than 70 identifiable species of brachiopod, in addition to ammonoids, crinoids, trilobites, bivalves and ostracods, whereas the 'holdover' fauna contained only seven quite-small brachiopod species and one bivalve.

The causes of the end-Permian extinction have been extensively examined elsewhere (see Wignall 2007 for a good summary). The current consensus is that the proximal cause was the Siberian Traps volcanism leading to a cascade of disastrous environmental changes: global warming, elevated pCO_2 and associated hypercapnia and calcification crisis, oceanic anoxia and acid rain.

This extinction event set the scene for life in the Triassic, as the evolutionary slate was almost wiped clean. The extinction caused major reorganisation of marine and terrestrial ecosystems, and environmental conditions remained poor for several million years into the Triassic. The timing of recovery after the end-Permian biotic crisis is still unclear, as there

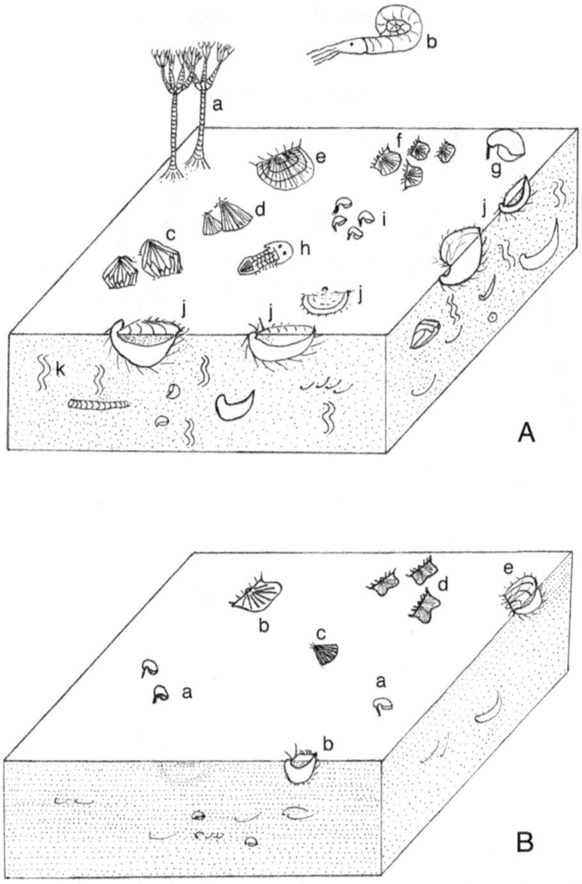

Fig. 7 Comparison of pre-extinction (late Changhsingian) and post-extinction marine communities in South China. (**A**) Pre-extinction—(**a**) crinoids; (**b**) ammonoid; (**c**) *Prelissorhynchia*; (**d**) *Acosarina*; e, *Schuchertella*; (**f**) chonetids; (**g**) *Spirigerella*; (**h**) trilobite; (**i**) *Crurithyris*; (**j**) productids (e.g., *Spinomarginifera* and *Haydenella*); (**k**) bioturbation. (**B**) Post-extinction holdover taxa—(**a**) *Crurithyris*; (**b**) *Cathaysia*; (**c**) *Acosarina*; (**d**) chonetids; (**e**) *Spinomarginifera*. From Campi and Shi (2005)

are isolated appearances of new taxa in the late Griesbachian (within c. 1 million years) (Chen et al. 2002, 2005b). The reappearance of complex and stable ecosystems took much longer, with some estimates of up to 10 million years.

References

Ali JR, Thompson GM, Song X-Y, Wang Y-L (2005) Emeishan large igneous province, SW China. Lithos 79: 475–489

Angiolini L, Jadoul F, Leng MJ, Stephenson MH, Rushton J, Chenery S, Crippa G (2009) How cold were the Early Permian glacial tropics? Testing sea-surface temperature using the oxygen isotope composition of rigorously screened brachiopod shells. J Geol Soc 166:933–945

Bambach RK, Knoll AH, Wanf SC (2004) Origination, extinction, and mass depletions of marine diversity. Paleobiology 30:522–542

Berner RA (1998) The carbon cycle and CO_2 over Phanerozoic time: the role of land plants. Philos Trans R Soc London, Ser B 353(1365):75–82

Berner RA (2002) Examination of hypotheses for the Permo-Triassic boundary extinction by carbon cycle modelling. Proc Natl Acad Sci USA 99:4172–4177

Campi MJ, Shi GR (2005) New Lopingian (Late Permian) rugosochonetid species from Sichuan, South China. Alcheringa 29:275–285

Chen ZQ, Shi GR, Kaiho K (2002) A new genus of rhynchonellid brachiopod from the Lower Triassic of South China and implications for timing the recovery of Brachiopoda after the end-Permian mass extinction. Palaeontology 45: 149–164

Chen ZQ, Kaiho K, George AD (2005a) Early Triassic recovery of the brachiopod faunas from the end-Permian mass extinction: a global review. Palaeogeogr Palaeoclimol Palaeoecol 224:270–290

Chen ZQ, Kaiho K, George AD (2005b) Survival strategies of brachiopod faunas from the end-Permian mass extinction. Palaeogeogr Palaeoclimatol Palaeoecol 224: 232–269

Chung SL, Jahn BM, Wu GY, Lo CH, Cong BL (1998) The Emeishan Flood Basalt in SW China: a mantle plume initiation model and its connection with continental breakup and mass extinction at the Permian-Triassic Boundary. In: FlowerMFJ, Chung SL, Lo CH, Lee TY (eds) Mantle dynamics and plate interactions in East Asia, vol 27. American Geophysical Union, Washington, DC, pp 47–58

Clapham ME, Bottjer DM (2007) Permian marine paleoecology and its implications for large-scale decoupling of brachiopod and bivalve abundance and diversity during the Lopingian (Late Permian). Palaeogeogr Palaeoclimatol Palaeoecol 249:283–301

Clapham MED, Bottjer J, Powers CM, Bonuso N, Fraiser ML, Marenco PJ, Dornbos SQ, Pruss SB (2006) Assessing the ecological dominance of Phanerozoic marine invertebrates. Palaios 2:431–441

Clapham ME, Shen SZ, Bottjer DJ (2009) The double mass extinction revisited: reassessing the severity, selectivity, and causes of the end-Guadalupian biotic crisis (Late Permian). Paleobiology 35:32–50

Cocks LRM (2007) Blowing hot and cold in the Palaeozoic. Proc Geol Assoc 118:225–237

Dobretsov NL (2005) 250 Ma large igneous provinces of Asia: Siberian and Emeishan traps (plateau basalts) and associated granitoids. Russian Geol Geophys 46:847–868

Fara E (2001) What are Lazarus taxa? Geol J 36:291–303

Flügel E, Reinhart J (1989) Uppermost Permian reefs in Skyros (Greece) and Sichuan (China): implications for the Late Permian extinction event. Palaios 4:502–518

Frank TD, Thomas SG, Fielding CR (2008) On using carbon and oxygen isotope data from glendonites as paleoenvironmental proxies: a case study from the Permian system of Eastern Australia. J Sed Res 78:713–723

Graham JB, Dudley R, Aguilar NM, Gans C (1995) Implications of the late Palaeozoic oxygen pulse for physiology and evolution. Nature 375:117–120

Grossman EL, Yancey TE, Jones TE, Bruckschen P, Chuvashov B, Mazzullo SJ, Mii HS (2008) Glaciation, aridification, and carbon sequestration in the Permo-Carboniferous: the isotopic record from low latitudes. Palaeogeogr Palaeoclimatol Palaeoecol 268:222–233

He W, Shi GR, Feng Q, Campi MJ, Gu S, Bu J, Peng Y, Meng Y (2007) Brachiopod miniaturization and its possible causes during the Permian-Triassic crisis in deep water environments, South China. Palaeogeogr Palaeoclimatol Palaeoecol 252:145–163

Hoa TT, Izokh AE, Polyakov GV, Borisenko AS, Anh TT, Balykin BA, Phuong NT, Rudnev SN, Van VV, Nien BA (2008) Permo-Triassic magmatism and metallogeny of Northern Vietnam in relation to the Emeishan plume. Russian Geol Geophys 49:480–491

Isozaki Y (1994) Superanoxia across the Permo-Triassic boundary: record in accreted deep-sea pelagic chert in Japan. Pangea: Global Environments and Resources. Canadian Soc Petrol Geol Mem 17:805–812

Isozaki Y (1997) Permo-Triassic boundary superanoxia and stratified superocean: records from lost deep sea. Science 276(5310):235–238

Kamo SL, Czamanske GK, Amelin Y, Fedorenko VA, Davis DW, Trofimov VR (2003) Rapid eruption of the Siberian flood-volcanic rocks and evidence for coincidence with the Permian-Triassic boundary and mass extinction at 251 Ma. Earth Planet Sci Lett 214:75–91

Knoll AH, Bambach RK, Payne JL, Pruss S, Fischer WW (2007) Paleophysiology and end-Permian mass extinction. Earth Planet Sci Lett 256:295–313

Korte C, Jasper T, Kozur HW, Veizer J (2005) δO^{18} and δC^{13} of Permian brachiopods: a record of seawater evolution and continental glaciation. Palaeogeogr Palaeoclimatol Palaeoecol 224:333–351

Kump LR (1991) Interpreting carbon-isotope excursions: Strangelove oceans. Geology 19:299–302

Labandeira CC (2005) Invasion of the continents: cyanobacterial crusts to tree-inhabiting arthropods. Trends Ecol Evol 20:253–262

Leven EY, Bogoslovskaya MF, Ganelin VG, Grunt TA, Leonova TB, Reimers AN (1996) Reorganisation of marine biota during the mid-Early Permian Epoch. Stratigr Global Correl 4:57–66

Metcalfe I (1998) Palaeozoic and Mesozoic geological evolution of the SE Asian region: multidisciplinary constraints and implications for biogeography. In: Hall R, Holloway JD (eds) Biogeography and geological evolution of SE Asia. Backhuys Publishers, Leiden, pp 25–42

Metcalfe I (2002) Permian tectonic framework and palaeogeography of SE Asia. J Asian Earth Sci 20:551–566

Montañez IP, Tabor NJ, Niemeier D, DiMichele WA, Frank TD, Fielding CR, Isbell JL, Birgenheier LP, Rygel MC (2007) CO_2-forced climate and vegetation instability during Late Paleozoic deglaciation. Science 315(5808): 87–91

Peng YQ, Shi GR (2009) Life crises on land across the Permian-Triassic boundary in South China. Global Planetary Change 65:155–165

Retallack GJ, Jahren AH (2008) Methane release from igneous intrusion of coal during Late Permian extinction events. J Geol 116:1–20

Retallack GJ, Metzger CA et al. (2006) Middle-Late Permian mass extinction on land. Geol Soc Am Bull 118:1398–1411

Rong JY, Shen SZ (2002) Comparative analysis of the end-Permian and end-Ordovician brachiopod mass extinctions and survivals in South China. Palaeogeogr Palaeoclimatol Palaeoecol 188:25–38

Scotese CR (2002) PALEOMAP website. http://www.scotese.com., Accessed 12 May 2008

Sephton MA, Looy CV et al. (2005) Catastrophic soil erosion during the end-Permian biotic crisis. Geology 33:941–944

Shen SZ, Shi GR (1996) Diversity and extinction patterns of Permian Brachiopoda of South China. Hist Biol 12:93–110

Shen SZ, Shi GR (2000) Wuchiapingian (early Lopingian, Permian) global brachiopod palaeobiogeography: a quantitative approach. Palaeogeogr, Palaeoclimatol, Palaeoecol 162:299–318

Shen SZ, Shi GR (2002) Paleobiogeographical extinction patterns of brachiopods in the Asian-western Pacific region. Paleobiology 28:449–463

Shen SZ, Archbold NW et al. (2000) Changhsingian (Late Permian) brachiopod palaeobiogeography. Hist Biol 15: 121–134

Shen SZ, Zhang H et al. (2006) Brachiopod diversity patterns from Carboniferous toTriassic in South China. Geol J 41:345–361

Shen SZ, Xie JF et al. (2009) Roadian-Wordian (Guadalupian, Middle Permian) global palaeobiogeography of brachiopods. Global Planetary Change 65:166–181

Shi GR (1998) Aspects of Permian marine biogeography: a review of nomenclature and evolutionary patterns, with particular reference to the Asia-Western Pacific region. Palaeoworld 9:97–112

Shi GR (2000) Terrane rafting enhanced by contemporaneous climatic amelioration as a mechanism of vicariance: Permian marine biogeography of the Shan-Thai terrane in southeast Asia. Hist Biol 15:135–144

Shi GR (2006) The marine Permian of East and Northeast Asia: an overview of biostratigraphy, palaeobiogeography and palaeogeographical implications. J Asian Earth Sci 26:175–206

Shi GR, Archbold NW (1993) Distribution of Asselian-Tastubian (Early Permian) circum-Pacific brachiopod faunas. Assoc Australas Palaeontol Mem 15:343–351

Shi GR, Archbold NW (1995) Palaeobiogeography of Kazanian – Midian (Late Permian) Western Pacific brachiopod faunas. J Southeast Asian Earth Sci 12: 129–141

Shi GR, Grunt TA (2000) Permian Gondwana-Boreal antitropicality with special reference to brachiopod faunas. Palaeogeogr Palaeoclimatol Palaeoecol 155:239–263

Shi GR, Shen SZ (1999) A compendium of Permian brachiopod faunas of the Western Pacific region 7. A statistical summary and report. School of Aquatic Science and Natural Resources Management, Deakin University Technical Paper, pp. 1–92

Shi GR, Archbold NW et al (1995) Distribution and characteristics of mixed (transitional) mid-Permian (Late Artinksian –

Ufimian) marine faunas in Asia and their palaeobiogeographical implications. Palaeogeogr Palaeoclimatol Palaeoecol 114:241–271

Sun DL, Shen SZ (2004) Permian-Triassic brachiopod diversity patterns in South China. In: Rong JY, Fang ZJ (eds) Mass extinctions and recovery. Evidences from the Palaeozoic and Triassic of South China, vol 2. University of Science and Technology Press, Hefei, pp 543–570

Tazawa JI (2007) Middle Permian brachiopod faunas of Japan and their significance for understanding the Paleozoic-Mesozoic tectonics of the Japanese Islands. In: Wong TE (ed) Proceedings of XVth international congress on carboniferous and Permian stratigraphy. Royal Netherlands Academy of Arts and Sciences, Amsterdam, pp 565–573

Tazawa JI, Shen SZ (1997) Middle Permian brachiopods from Hiyomo, Mino Belt, Central Japan: their provincial relationships with North America. Science Reports of Niigata University, Series E (Geol and Mineral), vol 12, pp 1–17

Tong JN, Shi GR, Lin DX (1998) Evolution of the Permian and Triassic reef ecosystems in South China. Proc R Soc Vict 110:385–399

Twitchett RJ (2001) Incompleteness of the Permian-Triassic fossil record: a consequence of productivity decline. Geol J 36:341–353

Tye SC, Fielding CR et al. (1996) Stratigraphy and sedimentology of the Permian Talaterang and Shoalhaven Groups in the southernmost Sydney Basin, New South Wales. Aust J Earth Sci 43:57–69

Wignall PB (2007) The end-Permian mass extinction – how bad did it get? Geobiology 5:303–309

Wignall PB, Sun YD, Bond DPG, Izon GT, Newton RJ, Védrine S, Widdowson M, Ali JR, Lai X, Jiang H, Cope H, Bottrell SH (2009) Volcanism, mass extinction, and carbon isotope fluctuations in the Middle Permian of China. Science 324(5931):1179–1182

Yang XN, Liu JR et al. (2005) Early Permian bioevent in the fusulinacean fauna of South China. Lethaia 39:1–16

Zhou M-F, Malpas J, Sone X-Y, Robinson PT, Sun M, Kennedy AK, Lesher CM, Keays RR (2002) A temporal link between the Emeishan large igneous province (SW China) and the end-Guadalupian mass extinction. Earth Planet Sci Lett 196:113–122

Ziegler AM, Hulver ML, Rowley DB (1997) Permian world topography and climate. In: Martini IP (ed) Late glacial and post-glacial environmental changes: quaternary, carboniferous-Permian and Proterozoic. Oxford University Press, New York, NY, pp 111–146

Millennial Physical Events and the End-Permian Mass Mortality in the Western Palaeotethys: Timing and Primary Causes

Enzo Farabegoli and Maria Cristina Perri

Abstract

This chapter focuses on the nature and pattern of four transgressive–regressive depositional cycles (C1–C4) across the Permian–Triassic Boundary (PTB) in the Dolomites, on their timing and on the possible causal relationships with four mass-mortality events (E0–E3), which, considered together, constitute the end-Permian extinction event in the western Palaeotethys. The duration of the investigated interval is ca. 200 ky; the duration of each cycle ranged from less than 20 ky to ca. 100 ky; and the magnitude of the sea level changes ranged from 5 to 15 m. Each mass-mortality event affecting the shallow marine environments of the western Palaeotethys corresponds with a regressive phase lasting a few millennia. The oldest mortality event (E0) at the top of Cycle 1 (i.e., the top of the Ostracod Unit) in the Southern Alps is aligned with the regressive Bed 24e of the Meishan D section in the eastern Palaeotethys; it is usually considered the actual end-Permian extinction event. The same cooling/fall-stand has been identified in various sites along the shallow-marine Gondwana margin. In the Southern Alps, E0 is mostly masked by stressed conditions typical of the regional carbonate tidal flat. During the following transgression and high-stand periods of Cycle 2 (i.e., Bulla Member), the shallow marine environment became re-populated by ca. 200 species referred to ca. 30 genera. At the top of Cycle 2, the sea level fell 10 m or less in a few millennia; it started the most devastating mass-mortality event (E1) in the Southern Alps. This mortality event lasted less than 20 millennia; it continued briefly during the trangressive phase of the following Cycle 3—which brackets the Bellerophon-Werfen formational boundary (BWB). This interval, aligned with Beds 26–27a at Meishan in the eastern Palaeotethys, was deposited in a deeper and distal environment. About 90% of the marine skeletal biomass disappeared at the end of E1. The acme of mortality event, E1, corresponded

E. Farabegoli (✉)
Dipartimento di Scienze della Terra e Geologico-Ambientali, Alma Mater Studiorum – Università di Bologna, 40126 Bologna, Italy
e-mail: enzo.farabegoli@unibo.it

J.A. Talent (ed.), *Earth and Life*, International Year of Planet Earth,
DOI 10.1007/978-90-481-3428-1_24,

with a submarine chemical-corrosion event, followed locally by subaerial exposure and pedogenesis. The mass-mortality event on land slightly predates—or is nearly coeval—with the mass-mortality event in shallow marine environments. The intensity of submarine corrosion became almost imperceptible at the foreshore–offshore boundary. The sea level rose ca. 15 m during Cycle 3 when the shallow marine environment, mostly over-saturated in carbonate but punctuated by short periods of vadose or submarine dissolution, transgressed rapidly more than 40km inland over the corroded bedrock, depositing oolite shoals and microbialite. The subsequent mortality events E2 and E3 are obviously of less intensity. E2, ca. 20 ky after E1, corresponds to a regressive interval associated with the first appearance of *Hindeodus parvus* (i.e., the Permian–Triassic Boundary). It seems to be the acme of colonisation of the shallow sea floor by cyanobacteria (stromatolites). E3, ca. 10 ky after the Permian–Triassic Boundary, corresponds to the last occurrence of Permian-type red algae in the Dolomites area. Whereas the end of E3 is gradual in the shoreface, it appears to have been abrupt in the lower foreshore, probably because of general conditions of less-ventilated and suboxic conditions. We hypothesise that a few local palaeo-environmental factors (e.g., distance of stressing factors from the source area, the pattern of atmospheric and marine palaeocurrents, and reduction of the shallow coastal area due to retreat of the coastline) concurred to modulate the intensity and duration of mortality events in space and time. Data suggest that increased warming was of primary importance in controlling the mortality tail but doesn't allow us to confirm or deny other local or general concurrent causes, such as up-welling of anoxic oceanic waters from the Palaeotethys. We interpret the cause of the mass-mortality events in the Dolomites area as having been a composite "top-down" mechanism with acid-rain events devastating the Permian-type life on continental and, subsequently, in shallow marine environments during millennial periods of cooling and regression of the Bellerophon sea. The ultimate causal factor was, very probably, large atmospheric perturbations connected with volcanism. Most of the sparse surviving biota disappeared immediately after the beginning of the following transgression—because of rapid global warming produced by greenhouse conditions, with only minor, repeated, episodes of acid rains. These stressed conditions contributed to inhibiting recovery of the long and efficient shallow-marine food chain. Because the magnitude of mass-mortality event E0 in the Dolomites and in much of the Gondwana margin is appreciably lower than the coeval one in Meishan, the first may have acted as refugia. Mass-mortality event E1 affected the shallow-marine western Palaeotethys for only a few millennia after E0. In the eastern Palaeotethys, coeval Beds 26–27a of Meishan were deposited from lower foreshore to marine shelf, lacking any clear record of anoxic conditions. It is the same for the coeval short-term parasequences in many sites along the Gondwana margin. We interpret the different magnitude of extinction on the shelves as due to different levels of temperature and excessive carbon dioxide (pCO_2) in the seawater. The rapid demise of taxa (occurring concordantly with the diachronous major mortality events) caused local severing of food chains, mostly of small suspension feeders, resulting in the "Lilliput" faunas (*sensu* Twitchett 2005) of event E2 in the Dolomites. This aligns with Beds 27c–d at Meishan—these beds were deposited in the lower foreshore and marine shelf environments under suboxic to dysoxic bottom conditions. It seems unlikely that the disappearance of red algae in the western Palaeotethys was connected with dysoxic conditions; increased temperature seems a more likely factor. Doubtless a medley of different mechanisms, including rapid fluctuations in marine salinity, operated variously as regards time and space and produced the end-Permian extinction—occurring over a time span of less than 100 ky.

Keywords
Permian • Triassic • Palaeotethys • Gondwana • Dolomites • Shelf environments • Highstands • Cyclicity • Mass mortality

Introduction

The quasi-centennial debate on the causes of the end-Permian extinction is shifting from a simple uniparameter to a multi-parameter cause in which short-term and long-term changes of each factor concurred to disrupt the global equilibrium of Late Permian life (cf. Racki and Wignall 2005 with references). The debate has given rise to a thousand papers or more, often based on several datasets of differing nature (litho-, bio-, chrono-, chemo- and magnetostratigraphy; sedimentology; facies analysis; and palaeogeography), leading to contrasting or partly overlapping solutions for this enigma. Moreover, it is not unusual to find a set of data collected in one stratigraphic section contrasting with data of the same nature collected from the same section by different researchers. This puts geologists at risk of shifting towards a "nihilist" scientific position, well exemplified by the doubts expressed by Hermann (2004). Contrasting sea-level-change curves at the PTB have been proposed: Holser and Magaritz's (1987) curve shows a regression, whereas Golonka and Ford (2000) show a transgression. Hermann concluded with a few questions: Which aspects prove one of these viewpoints is correct? Are there regions with regression, and others with transgression? Is there a global eustatic signal or a peak in tectonic activity? Is there a consistent sedimentological pattern from place to place? Or is there confusion in every field of geology, including geochemistry.

The majority of recent papers have focused on the protracted recovery (4–5 My) in the oceans (Botthjer et al. 2008, *cum bibl.*), but the timing and the ultimate cause of the end-Permian extinction remains an unresolved enigma. All authors agree that the extinction process started several million years before the end-Permian extinction; nearly all insist that it was not a chance process. If so, two fundamental questions must be broached: (1) What phenomena, acting over a geologically very short period, produced the global palaeobiologic catastrophe?; (2) Did a single-phase, global catastrophe really exist that rapidly eliminated, step by step, the clades in territory after territory. If so, what was the duration and timing of each mass-mortality episode, and what was the duration of the global extinction event?

Because shallow marine areas are highly sensitive to environmental changes and underwent profound mass mortality, detailed and precise data about biostratigraphic alignments, palaeogeographic distributions and patterns of palaeobiologic variations are fundamental if one is to propose well-constrained evolutionary steps in extinction-and-recovery scenarios.

The biostratigraphy across the P–T boundary is presently based primarily on conodonts (hindeodids and isarcicellids) that lived in coastal and offshore marine environments. This enables correlating the main physical variations (e.g., sea-level changes, types of marine bottoms, energy levels and nutrient availability) with coeval patterns of biotas and palaeophysiology of fossils, and to infer, with a reasonable confidence, the complex and synergic relationships connected with mortality or survival.

The Dolomites, a ca. 15,000 km^2 area (Fig. 1), is a significant part of the Southern Alps formerly lying in the north-western part of the Palaeotethys. In the last 150 years, more than 100 well-exposed shallow marine successions in the Dolomites, bracketing the Permian–Triassic Boundary (PTB), have been studied. Many data of good-quality are now available regarding the stratigraphic position and architecture of the strata containing the PTB, the position of the major mass-mortality events and the composition and palaeophysiology of biotas pre-and post-extincton(s) and primary survival strategies (Knoll et al. 2007). Most of the authors agree with Venzo (1955) and Bosellini (1964) that the major mass-mortality event is located at or immediately above the Bellerophon–Werfen formational boundary (BWB). We find it astonishing that after such a huge number of papers, the most debated issue is—excluding the ultimate cause(s) of the end-Permian extinction—whether the depositional succession is continuous or discontinuous (i.e., interrupted by disconformities or unconformities) and in the latter case, the duration of the hiatuses.

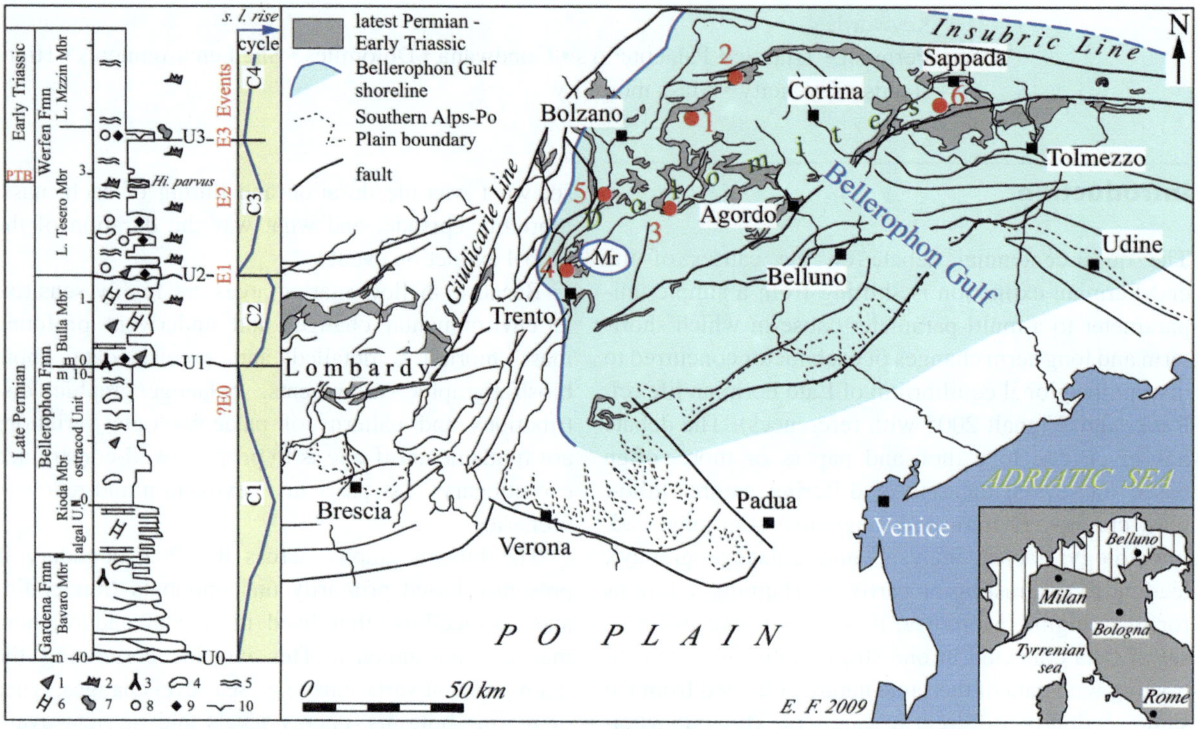

Fig. 1 Distribution of latest Permian–Early Triassic terrains in the Southern Alps and location of examined sections: 1. Bulla, 2. Sass de Putia W, 3. Tesero, 4. Masi Saracini, 5. Montagna, 6. Casera Federata (*right*)—schematic stratigraphic column about the PTB in the Dolomites (*left*): 1. gastropods, 2. conodonts, 3. centimetre-sized roots in living position, 4. ostracods, 5. microbialites and stromatolites, 6. green algae, 7. red algae, 8. ooids, 9. lithoclasts and reworked fossils and 10. a crack

Most authors agree with Venzo's (1955) hypothesis that the BWB is transitional and conformable (e.g., Mostler 1982; Broglio Loriga et al. 1983; Brandner et al. 1986; Broglio Loriga et al. 1986a, b; Holser and Magaritz 1987; Noé 1987; Brandner 1988; Buggisch and Noè 1988; Magaritz et al. 1988; Posenato 1988, 1989, 1990, 1991, 1998, 2001; Baud et al. 1989; Hallam 1989; Broglio Loriga and Cassinis 1992; Wignall and Hallam 1992, 1996; Wignall and Twitchett 1996; Hallam and Wignall 1997, 1999; Kozur 1998; Beretta and Cimmino 1998; Beretta et al. 1999; Neri and Posenato 1999; Scholger et al. 2000; Beretta and Radrizzani 2001; Rampino et al. 2002; Newton et al. 2004; Korte and Kozur 2005; Chen et al. 2006, 2009; Gorjan et al. 2007; Groves et al. 2007; Horaceck et al. 2007a, b; Horacek 2010).

Only a few authors agree with Bosellini's interpretation (1964) that the BWB is a disconformity with an erosional microkarstified surface separating the Bellerophon Formation from the overlying oolites of the Werfen Formation. Assereto et al. (1973) followed Bosellini's interpretation and believed that the basal oolitic sheet of the Werfen Fm. transgressed northwards over the Bellerophon Fm in the entire Dolomites. A hiatus, spanning two biozones, divides the two units. The latter was a rather rash conclusion, because there was no detailed and shared shallow marine-based biostratigraphy available at that time. Farabegoli and Perri (1998) identified and depicted the unconformable BWB in the Bulla section and recognised the PTB by the First Appearence Datum (FAD) of *Hi. parvus* ca. 1.35 cm above the BWB. They (Perri and Farabegoli 2003) and Farabegoli et al. (2007) suggested that the duration of the disconformity-related hiatus was not geologically determinable; it was referred to the Lower *praeparvus* Zone and, they believed, probably represented only a few millennia. They defined the Bulla Mbr of the Bellerophon Fm as the 1–1.5 m of dark packstone underlying the BWB disconformity, interpreting the base of the Bulla Mbr as also disconformable. That the latter contact was disconformable was previously pointed out by

Farabegoli and Viel (1982a, b) in the Masi Saracini section; Hallam and Wignall (1999) noted it in the Siusi section but erroneously attributed to the BWB.

If the deposition of the very shallow marine succession across the end-Permian extinction event is really punctuated by disconformable surfaces, their origin (tectonic, eustatic, or other), extent (regional vs. global) and links with mass mortality events have to be explained. The depositional pattern and disconformities can have serious implications for the duration of the extinction event(s); in fact, if "the time involved was geologically insignificant the extent of this time span is critical in testing any hypothesis of the cause or causes of the PT extinction" (Olson 1989).

This chapter focuses on the nature and pattern of a few depositional cycles across the PTB, each spanning 10–100 ky. The major mass-mortality event in the paleo-equatorial Dolomites seems to have been controlled by an intense climatic cooling and by acid rains during a widespread episode of emersion and pedogenesis, then by warming and the consequent transgressive phase. Olson (1989, p. 29) stated that if the "cooling cycle hypothesis is validated and detectable over the world, then a world-wide basis of correlation would exist, applicable to both marine and non-marine circumstances". So, the final purpose of our analysis is to have more precise timing of the mass-mortality events in the Southern Alps, to discriminate between their natures (i.e., top-down negative effects vs. base-up effects) and to have a detailed and well-constrained regional network of physical and palaeobiologic events that can be validated or refuted at world-scale by using a complete database of coeval shallow-marine successions.

Methods

About 50 stratigraphic sections across the Bellerophon–Werfen formations (BWB) in the Dolomites (Fig. 1), 20–200 m thick, were measured; their sedimentology described in detail; their biofacies and lithofacies (macro- and microfacies) sampled; and their stable isotopes determined. A few sections were studied in very great detail with continuous samples being collected up to 1 m in length, obtained with a rock saw with four-stroke engine. The set of data available in the literature is considered and discussed. We particularly considered six stratigraphic sections (Bulla, Sass de Putia W, Tesero, Masi Saracini, Montagna and Casera Federata), which we refer to three special shallow-marine subenvironments. Correlations between the end-Permian to Early Triassic events in the Dolomites and coeval events in the palaeoworld are mainly taken from the literature.

Data

Six stratigraphic columns (Bulla, Fig. 2; Sass de Putia W, Fig. 3; Tesero, Fig. 4; Masi Saracini, Fig. 5; Montagna, Fig. 6; and Casera Federata, Fig. 7) are described below. The temporal correlations of the columns are based on biostratigraphy and lithostratigraphy and particularly on inferred physical events.

Bulla Section

The Bulla section (Fig. 2) is probably the most studied section, but it is also probably the most contentious. A homonymous section, located ca. 1 km to the north, was first studied for its conodont stratigraphy (Huchkriede 1958; Staesche 1964). The section presented here, located along an abandoned tract of the Ortisei–Bulla road, has been analysed several times for its litho–biostratigraphy (Mostler 1982; Perri and Andraghetti 1987; Perri 1991; Farabegoli and Perri 1998; Perri and Farabegoli 2003; Farabegoli et al. 2007), palaeomagnetism (Newton et al. 2004, as far as the neighbouring Siusi section), stable isotope stratigraphy (Korte and Kozur 2005; Korte et al. 2005; Gorjan et al. 2007; Horaceck et al. 2007a, b), foraminifers, micropalaeontology (Groves et al. 2007), brachiopods (Posenato 2008), ostracods Crasquin et al. 2008) and pollen. Farabegoli et al. (2007) proposed the Bulla section as the parastratotype for the Permian–Triassic Boundary (PTB) in the shallow-marine western Palaeotethys.

The lower unit in the section (Ostracod Unit, Changhsingian) is represented by ca. 15 m of grey silty dolomitic mudstone–wackestone (Fig. 2a), alternating with subordinate grey shale and rare slightly dolomitic limestone beds. It corresponds to the Rioda Member p.p. (Farabegoli and Viel 1982b, Farabegoli and Viel in Broglio Loriga et al. 1986a) and to the Ostracod Unit (Neri et al. 1986) of the Bellerophon Formation (Fig. 1). The fossil assemblage is poor, dominated

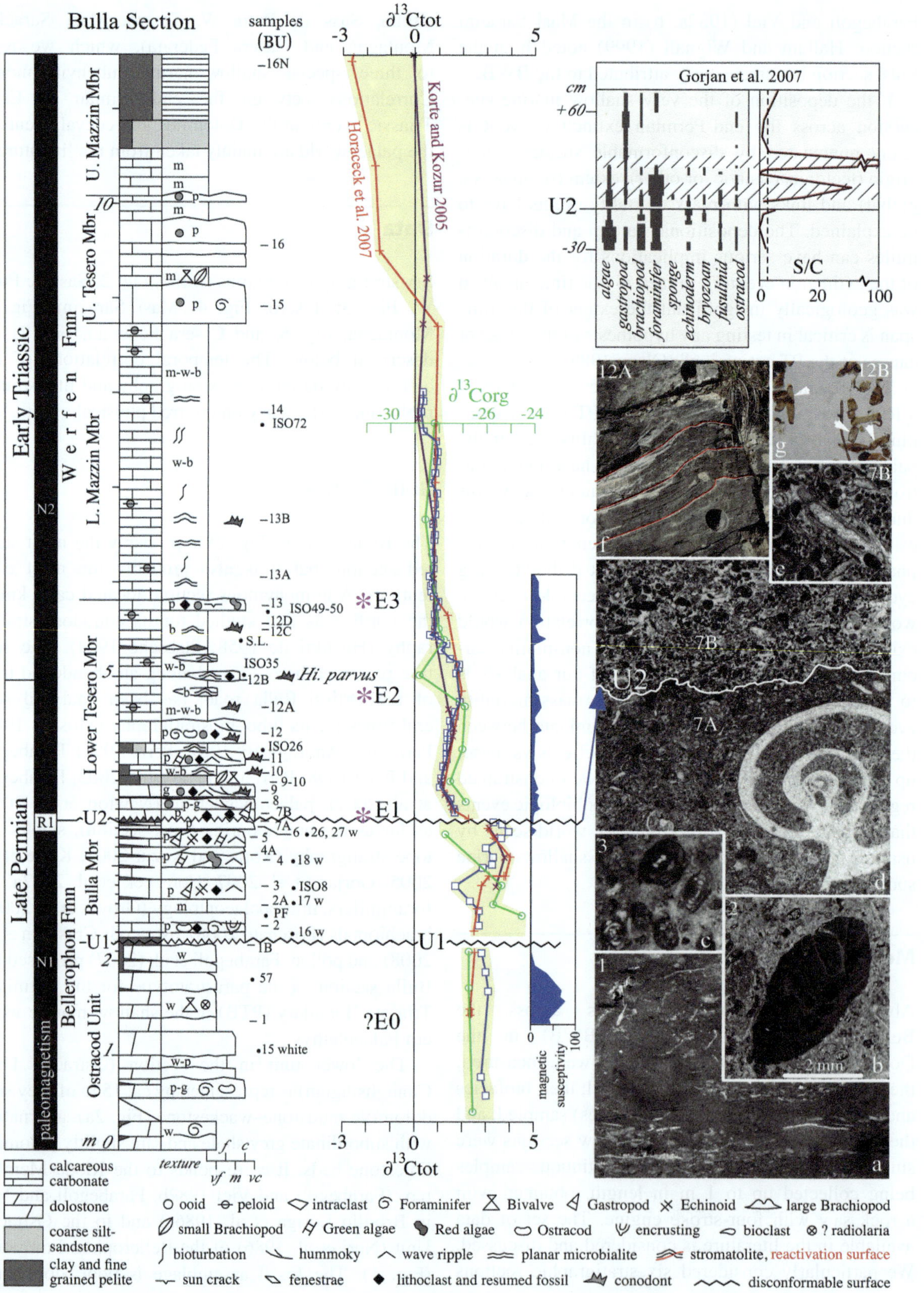
Bulla Section
samples (BU)
∂^{13}Ctot
Gorjan et al. 2007
Korte and Kozur 2005
Horaceck et al. 2007
∂^{13}Corg
Early Triassic
Late Permian
Werfen Fmn
Bellerophon Fmn
U. Mazzin Mbr
U. Tesero Mbr
L. Mazzin Mbr
Lower Tesero Mbr
Bulla Mbr
Ostracod Unit
paleomagnetism
N2
R1
N1
U2
U1
E1
E2
E3
?E0
Hi. parvus
magnetic susceptivity
S/C
algae
gastropod
brachiopod
foraminifer
sponge
echinoderm
bryozoan
fusulinid
ostracod
texture
calcareous carbonate
dolostone
coarse silt-sandstone
clay and fine grained pelite
ooid
peloid
intraclast
Foraminifer
Bivalve
Gastropod
Echinoid
large Brachiopod
small Brachiopod
Green algae
Red algae
Ostracod
encrusting
bioclast
? radiolaria
bioturbation
hummoky
wave ripple
planar microbialite
stromatolite, reactivation surface
sun crack
fenestrae
lithoclast and resumed fossil
conodont
disconformable surface
2 mm

Fig. 2

by ostracods, with rare bivalves (pectinoids and *Permophorus*-like shells), millimetre-size gastropods and five species of Lagenidae (*Protonodosaria*? sp., *Geinitzina taurica*, *Nodosinelloides* sp., *Nodosaria elabugae* and *Frondina* cf. *permica*) (Groves et al. 2007). The ostracod fauna is characteristically low diversity, ranging from 2 to 5 species, and high number of specimens (63 to 950). Up-section the assemblages pass from normal marine shallow water to stressed conditions and are dominated for about 92% by *Cavellina bellerophonella* and *Sargentina postacuta* (Crasquin et al. 2008). If the *Paratirolites* sp. found by Posenato and Prinoth (2004) in a slab of an ostracod-rich facies 6 km north of Bulla correlates with this unit, then it would have been deposited during the penultimate Permian ammonoid zone. No conodonts have been found. Scholger et al. (2000) showed that the uppermost part of the Ostracod Unit (ca. 1 m, described as "dolomite with root traces") has reversed polarity.

The top of the unit is truncated by a sharp surface (U1), paraconformably overlain by the Bulla Member of the Bellerophon Formation, here about 130 cm thick (Fig. 2). U1 is evident in the ostracod assemblage mainly by a change in composition rather than by significant extinctions, even though a lot of biodiversity and abundance is documented.

The lowest beds (BU2–BU4, ca. 40 cm) usually contain sub-angular litho- and bioclasts with reddish, oxidised margins, interpreted as reworked from a subaerial bedrock. The rounded 4 mm micritic pebble in sample BU2A (Fig. 2b) is a good sample of a reworked grain. The bioclasts consist predominantly of miliolids (Fig. 2c) and coiled tubes attributed to the polychaete annelid *Spirorbis*; it is now regarded as a possible lophophorate of the genus *Microconchus* (Posenato 2008 cum ref.); fusulinids and calcareous algae are subordinate. Dark bioclasts with calcareous algae (*Gymnocodium bellerophontis* Rothpletz, *Mizzia velebitana* Schubert, *Macroporella preromangica* Praturlon) are associated with autochthonous bivalves, gastropods, ostracods, a few calcispheres, crinoid ossicles, echinoid spines, fish teeth and foraminifers. The number of washed but not reworked fossils (echinoids, crinoids and calcareous algae) increases upwards. The overlying 85 cm is characterised by a predominance of fusulinids such as *Nankinella* cf. *quasihunanensis* (Sheng), a few *Nankinella* cf. *discoides* (Lee) and *Sichotenella* sp. (some abraded and filled with dark micrite); calcareous algae are represented by *Gymnocodium bellerophontis*, *Macroporella preromangica* and *Mizzia velebitana*. There are associated bivalves, gastropods—mainly represented by large shells of *Bellerophon* sp. and *Catinella depressa* (Stache)—ostracods, rare calcispheres and scolecodonts (BU4 and BU4A). Groves et al. (2007) identified 21 species of lagenid foraminifers referred to 12 genera: *Calvenzina*, *Cryptoseptida*, *Frondina*, *Geinitzina*, *Langella*, "*Nodosaria*", *Nodosinelloides*, *Pachyploia*, *Pseudolangella*, *Protonodosaria?*, *Rectostipulina* and *Robuloides*. Unfortunately, it is nearly impossible to stratigraphically locate their samples precisely in the parastratotype Bulla section (Farabegoli et al. 2007) because, although they used the sample numbers painted on the beds by Farabegoli and Perri (1998), their relative vertical distances do not correspond to ours. The problem is particularly important across the BWB because they did not recognise the well-known disconformable surface. Some ostracods (*Acratia* and *Bairdia*) have been found in residues from acid leaching during a search for conodonts (BU6–BU6A). The highest diversity in ostracod faunas, occurring in the Bulla Member, is linked with the transgressive interval; 48 species were identified from BU2A to BU6. This is followed by

Fig. 2 (continued) Bulla section. Palaeomagnetism and magnetic susceptivity after Scholger et al. (2000). Isotopic profiles after Korte and Kozur (2005) (*magenta*), Horaceck et al. (2007a, b) (*red*) and unpublished data after Qi Yu-ping: C_{tot} (*blue*), C_{org} (*green*). Gorjan et al. (2007) interpreted the positive peak in the S/C curve (high-right) as two poor oxygenation-redox episodes. (**a**) Bioturbated, *grey* silty dolomitic wackestone with ostracods and rare bivalves (BU1). (**b**) Sub-angular, small litho- and bioclasts with oxidised margins and a rounded 4 mm micritic pebble (BU2). (**c**) Miliolids are fragmented and show intense post-mortem chemical corrosion (BU3). (**d**) Disconformable erosional regional surface (U2) divides a packstone with foraminifers, echinoids and *Bellerophon* (BU7A) from the overlying packstone-grainstone rich in autochthonous and washed fossils, small oolites and bio- and lithoclasts (BU 7B). (**e**) Detail of BU7B with dark, rounded, small litho- and bioclasts (fusulinids) and washed, geopetally filled, calcareous algae. (**f**) Planar to domal stromatolite bed-set separated by paraconformable–disconformable surfaces (*red*), interpreted as a phase of interruption and reactivation of growth (BU12A). (**g**) *Tympanocysta* sp. flora at the PTB (BU12B) decrease in number of specimens in BU7A (Crasquin et al. 2008). With the exception of one species, all these species are absent in the underlying Ostracod Unit. The dominance of Bairdioidea confirms the previous interpretation that the depositional environment of the Bulla Member represented more open sea. Neither large unfragmented brachiopods nor conodonts have been found in the Bulla Member of the Bulla section.

The BWB is a disconformable erosional surface (U2 in Fig. 2d) deepening a few centimetres over a lateral distance of ca. 5 m. Farabegoli et al. (2007) argued that a rapid sea-level fall and a very short period of subaerial exposure produced this erosional surface, as well as subaerial cementation of the old sea bottom.

The overlying basal Lower Tesero Member (ca. 20 cm, BU7B) is composed of autochthonous and washed fossils mixed with bio- and lithoclasts, certainly partly reworked from the Bulla Member (Fig. 2d, e). These pass gradually upwards to small oolites (BU8) and then (BU9) to slightly fossiliferous, larger and well-formed oolites (figured by Farabegoli et al. 2007). Between BU7A and 7B, a drastic drop in the ostracod assemblages occurs; 12 species become extinct. All the Lower Tesero Member is characterised by very low diversity and abundance of ostracods (Crasquin et al. 2008). Conodonts occur only in bed BU9, ca. 30 cm from the base of the Werfen Formation, extending up to BU13B (3.3 m from the base).

The bioturbated silty interval BU9–BU10 contains a fossil assemblage, superficially appearing to be "Lower Triassic"; it is very much dominated by *Eumorphotis*, an epifaunal byssate cosmopolitan aviculopectinid, and subordinately by the millimetre-size gastropod sp. *Bellerophon* (Posenato 2009 cum ref.). Representatives of the non-dysaerobic *Eumorphotis* are here noticeably small. The assemblage contains very rare byssate bivalves, namely *Bakevellia, Towapteria* and *Promyalina;* the burrowing *Unionites*; the small gastropod *Coelostylina* and two byssate brachiopods *Ombonia* and *Orbicoelia* (Posenato 2009). BU9 and BU10 have a few ostracod species present in the Ostracod Unit and absent in the Bulla Member.

The conodont faunules are composed mainly of hindeodids and by forms with morphologic peculiarities (from which the isarcicellids developed) as well as appearance of new species such as *Hindeodus pisai, Isarcicella peculiaris* and *I. prisca*—selected as marker for the base of the Upper *praeparvus* Zone (Perri and Farabegoli 2003; Farabegoli et al. 2007). *Hi. typicalis* and *Hi. praeparvus* co-occur in these assemblages.

Resampling of a few beds produced Pa elements of *Hi. eurypyge* in BU12 not previously found in the section.

The succession continues with slightly fossiliferous oolitic beds decreasing upwards in number and thickness; these alternate with microbialitic beds with, upwards, a change from lenticular to columnar stromatolites generated on a tidal flat (Fig. 2f; and fig. 6 in Farabegoli et al. 2007). Shallow lenticular microbialites are exclusive to the interval BU13–BU15 (lower Mazzin Member), followed by oolitic beds alternating with microbialitic mudstone (Upper Tesero Member). The overlying siltstone is referred to as the Upper Mazzin Member.

The first appearance datum of *Hindeodus parvus*, that is, the PTB, in sample BU12B co-occurs with a flora of abundant *Tympanocysta* sp. (Fig. 2g), a green alga (often reported as a fungus) living in salty lagoons and estuaries. Perri and Farabegoli (2003) divided the interval immediately above the BWB, shown in Fig. 2, into two conodont zones, the Upper *praeparvus* and *parvus* zones. In the Upper Mazzin Member after an interval of about 25 m barren of conodonts, recurring conodont faunules allow discrimination of three conodont biozones defined by the entry of the markers *Isarcicella lobata, I. staeschei* and *I. isarcica.* Those biozones are followed by the *aequabilis* Zone, defined by the shallow-water multielement species *Hadrodontina aequabilis* (Plate 1).

Farabegoli et al. (2007) distinguished three rapid extinction phases (E1, E2 and E3), each spanning 10–30 ky. E1 brackets the disconformity U2 and corresponds to a 65% drop in abundance and diversity of fusulinids and algae, and to extinction of large brachiopods and molluscs. This extinction phase corresponds roughly to the end-Permian extinction event detected by Gorjan et al. (2007) in bed 8; it is interpreted as having been produced by two pulses of dysoxic–euxinic marine waters (the two etched bands in Fig. 2) formed elsewhere. However, Gorjan et al. (2007) did not detect the U2 disconformity and therefore overestimated the percentage of surviving individuals, particularly foraminifers (cf. diagram in Fig. 2). Event E2, which nearly corresponds to the PTB, reflects a strong reduction in foraminifers and small carbonate-shelled brachiopods and the extinction of green calcareous algae. Event E3, which nearly corresponds to bed sample BU15, reflects the extinction of red calcareous algae.

The four available $\delta^{13}C_{tot}$ and $\delta^{13}C_{org}$ curves are illustrated in Fig. 2. The curves have been interrupted to correspond to the two paraconformable–disconformable surfaces U1 and U2. All the curves differ by only 1 PDB, with two exceptions: the upper part of the column, probably because of the scarcity of data available to both Korte and Kozur (2005) and Horaceck et al. (2007a, b), and the Bulla Member, represented by many data but from greatly differing sources (foraminifers, brachiopods, reworked bio-lithoclasts, terrestrial plants, etc.). Moreover, the $\delta^{13}C_{org}$ values in the Ostracod Unit are systematically lower than the $\delta^{13}C_{tot}$ values, whereas this occurs only near the top of the Bulla Member, corresponding to the spike of *Tympanocysta* sp. (Fig. 2g) at the PTB. This may be explained by local influxes of estuarine waters and/or terrestrial plants rich in ^{12}C.

The negligible values for magnetic susceptibility detected by Scholger et al. (2000) in the Bulla Member (Fig. 2) may be explained, at least in part, by the presence of reworked clasts, but other factors cannot be excluded.

Sass de Putia W Section

The Sass de Putia W section (Fig. 3) is located near the Passo delle Erbe, ca. 200 m west of the Sass de Putia section (PK), described by Broglio Loriga et al. (1986b, 1988) and Posenato (1998). We recently found *Hi. parvus* very low in the section for the first time. Therefore, this section is at present the most representative of the end-Permian events in the central sector of the Bellerophon Gulf.

The lower part of the column in Fig. 3 is an interval rich in dark grey algal packstone. It corresponds to the Algal Assemblage Unit (Neri et al. 1986) or to the Rioda Member p.p. (Farabegoli and Viel in Broglio Loriga et al. 1986a) of the Bellerophon Formation. It is followed upwards by the Ostracod Unit, a transgressive–regressive shallowing unit. The transgressive tract is a bivalve-rich packstone with subordinate grainstone containing transported reefoidal ?bryozoans (Fig. 3b). The middle part of the unit (ca. 12 m) is a regressive tract, mostly covered by talus composed of silty carbonate. The top part is composed of silty carbonate alternating with wave-rippled siltstone rich in ostracods (beds 1 and 2 in Fig. 3c–e).

Cirilli et al. (1998) have already shown the upper Ostracod Unit in the nearby, more-distal Seres section (interval 2, 12 m thick), containing palynomaceral material and fungi increasing upwards, whereas vitrinite frequently shows corroded edges. The amorphous organic matter is drastically reduced until it disappears near the top; the palynomorphs and vitrinite are commonly bleached.

The paraconformable–disconformable surface U1 divides the Ostracod Unit from the overlying Bulla Member, here 140 cm thick. The very base (bed 3) is composed of a framework of centimetre-sized branches as well as trees, mostly in living positions, anchored into formerly loose sediment (Fig. 3c–e). The spaces between the carbonaceous matter (replaced by carbonate during diagenesis) are filled with a fine-grained packstone rich in ostracods, foraminifers and reworked lithoclasts. The overlying beds, WPK4–WPK10 (Fig. 3e), are composed of bioclastic wackestone and packstone alternating with thin layers of dark siltstone rich in millimetre–centimetre-size plant debris. The carbonate biofacies is dominated by fusulinids and calcareous algae and by the large articulated brachiopod *Comelicania*. Bed WPK10 records a well-diversified brachiopod assemblage (Broglio Loriga et al. 1988; Posenato 1998, 2001). Bed WPK10B, the topmost 5 cm of bed WPK10, is a grainstone with the most diversified brachiopod fauna (six genera: *Comelicania*, *Comelicothyris*, *Janiceps*, *Ombonia*, *Orthothetina*, ?*Septospirigerella*; Posenato, 2009), but of reduced size, plus undetermined nautiloids, foraminifers and broken and reworked bio–lithoclasts. Large disarticulated *Comelicania* are very frequent in the top of bed WPK10B (Fig. 3f). Little holes (see arrow on an individual) are referable to incipient bioerosion during life. The fragments of *Comelicania* valves are sub-rounded and show intense post-mortem chemical corrosion and bio-erosion (Fig. 3g). The vertical succession of the sedimentary facies suggests a very rapid sea-level fall from foreshore (bed WPK9) to a regime of discontinuous but high hydraulic energy in the upper shoreface (bed WPK10). The augmented biodiversity may accord with a change from a warm-humid to a temperate-humid climate, but other interpretations are possible.

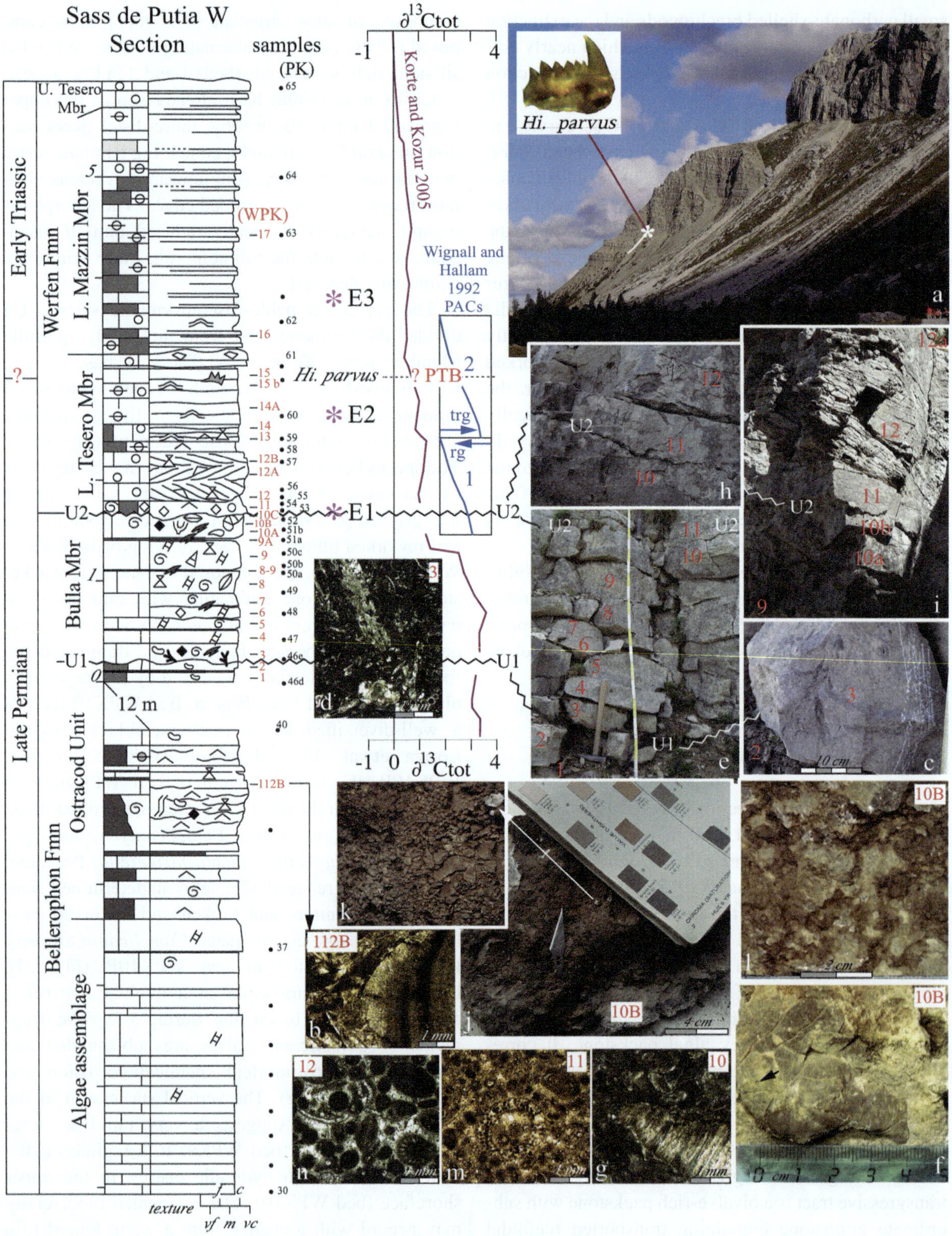
Sass de Putia W
Section
samples
(PK)
Early Triassic
Late Permian
Werfen Fmn
Bellerophon Fmn
U. Tesero Mbr
L. Mazzin Mbr
L. Tesero Mbr
Bulla Mbr
Ostracod Unit
Algae assemblage
(WPK)
Hi. parvus
? PTB
E1
E2
E3
U1
U2
12 m
texture
vf m vc
f c
Korte and Kozur 2005
Wignall and Hallam 1992 PACs
trg
rg
-1
0
4
∂13Ctot
112B
10B
12a
4 cm
3 cm
10 cm
1 mm
0 cm 1 2 3 4

Fig. 3

The BWB in this section has always been considered transitional (cf. Broglio Loriga et al. 1986a, b, 1988) or of doubtful origin (Farabegoli et al. 2007). Recently, Farabegoli found clear evidence that the boundary surface is not a normal diastem between layers, but is rather a paraconformable–unconformable contact. In fact, the uppermost surface of the Bulla Member (U2 in Fig. 3e, h, i) is irregularly shaped at centimetric scale, because it is incised by small channels at decimetre–metre spacing attributed to a subaerial drainage network. The U2 surface is also covered by rare, scattered lenses of a brownish breccia, less than 1 m in diameter and less than 1 cm thick (Fig. 3j). The breccia is nearly monomictic (Fig. 3k, l) in that it is composed of 99% brownish sub-angular claystone clasts and 1% sub-rounded grey carbonate clasts. The claystone clasts are heterometric (1–100 mm) and well compacted (3l). This bed is the remnant of a clayey colluvium, probably transported by a debris flow over a short distance from the parent palaeosoil. The colour of the clayey clasts ranges from moderate brown to light brown (5YR 4/4 and 5YR 5/6, respectively, in the Rock-Color Chart; Fig. 3k). It is usually associated with pedogenetic processes, generated in a moderately humid, temperate climate. The uppermost and thinner clasts (Fig. 3k) are here interpreted as clayey colluvium, probably transported over a short distance from the parent palaeosoil. The clasts of the lowermost breccia are larger, and the margins fit perfectly (Fig. 3l). We interpret this as the product of an in situ fragmentation process referable, for example, to a very short-term event of moderate cryoclastism.

The ravinement surface that marks the subsequent transgression of the oolitic Werfen Formation contributed to the nearly complete erosion of the parent palaeosoil and preserved the clayey breccia in some slightly depressed land areas (Fig. 3j, k). In other words, the latter erosional surface did not have sufficient energy and/or time to deepen significantly into the underlying, sub-aerially eroded and weathered carbonate bedrock (i.e., U2). Where the palaeosoil is lacking, the two surfaces cannot be distinguished.

The thickness of the Tesero Member occurring between the BWB and PTB is the same as in the Bulla section. The basal layer WPK11 (0–12 cm thick) of the Tesero Member is composed of an autochthonous fauna with small brachiopods (an *Ombonia-Orthothetina* assemblage, Posenato 2001), molluscs, foraminifers and calcareous algae mixed with broken litho- and bioclasts, passing gradually upwards to small oolites (Fig. 3m). A cross-bedded oolitic co-set (WPK12) of ca. 50 cm, with well-formed oolites, herringbone cross-bedding (Fig. 3i) and wave ripples, has locally completely eroded the underlying bed WPK11, which is reflected in the large enclosed lithoclasts (Fig. 3n). It corresponds to filling of a shallow, approximately S–N trending, palaeo-tidal channel with ebb floods from S to N. It is worth noting that this channel is traceable for more than 3 km to the south and is clearly visible in the Misci section where it was completely described by Bosellini (1964). Since then, the same section has been discussed many times, but under two different names: the Seres section (Cirilli et al. 1998; Beretta and Cimmino 1998; Beretta and Radrizzani 2001) and the Val Badia section (Sephton et al. 2002; Koeberl et al. 2004). None of these authors recognised the disconformities; they located the BWB at the base of the tidal channel (i.e., the

Fig. 3 (continued) Sass de Putia W section. (**a**) Location of section on the northern side of Sass de Putia and an image of *Hi. parvus* we found in sample WPK15. (**b**) A lenticular grainstone intercalated in the Ostracod Unit contains transported, microbored reefoidal ?bryozoans (WPK112B). (**c**) The centimetre-sized trunks in living position was anchored into loose sediment by rootlets (WPK112B) at the very base of the Bulla Mbr. (**d**) Enlargment of the structure of the calcified trunk at the very base of the Bulla Mbr, just above surface U1. (**e**) The Bulla Mbr is bounded by surfaces U1 and U2 and here consists of beds 3–11. (**f**) *Top view* of a large disarticulated *Comelicania* at the very top of the Bulla Mbr; *arrow* indicates a small hole interpreted as caused by unsuccessful bioerosion. (**g**) A large sub-rounded fragment of a *Comelicania* valve showing intense post-mortem chemical corrosion and bio-erosion. (**h**) Small scours spaced at decimetre–metre scale give an irregular shape to the U2 surface. (**i**) The image shows clearly the irregular shape of the U2 surface, the filling of a thin channel (bed 11) and filling of a tidal channel with an inclined bed-set 12. (**j**) A mm-thick red-brown clayey siltstone locally covering the U2 surface. (**k**) The enlargement shows the brecciated structure of the red-brown soil which we interpret as transported over a very short distance and/or the in situ product of a moderate cryoclastism. (**l**) *Top view* of bed 10B showing how some red-brown clayey breccia and basal hematite films are preserved in some slightly depressed areas. (**m**) Microfacies of the basal layer of the Tesero Member (WPK11). The small calcareous alga is autochthonous because it has an unrotated geopetal filling of micrite and small oolites. The rounded dark nuclei of the oolites and the fusulinid test on the right are presumed to be derived from the Bulla Mbr. (**n**) The microfacies from layer 12 of the Tesero Member is composed of lithoclasts, reworked bioclasts, micritised oolites and an unreworked bivalve with a geopetal filling co-set WPK12). The WPK13A assemblage consists of bivalves (*Eumorphotis, Bakevellia, Unionites*); it is dominated by *Eumorphotis* (Posenato 2009).

Less conodont data are available from the Sass de Putia W section compared with the Bulla section. Despite having acid-leached several samples from the Bulla and Lower Tesero members for conodonts, only two were productive. Conodonts occur first in level WPK14 of this section at ca. 90 cm from the base of the Werfen Formation, with one very small Sc element of *Hindeodus* (Fig. 3a).

Bed WPK15, at ca. 140 cm above the base of the Werfen Formation, produced an interesting faunule with Pa elements of *Hindeodus parvus, Hi. eurypyge, Hi. inflatus* and Sc elements of *Hindeodus* (Plate 1). Korte and Kozur (2005) report, as unpublished data of Kozur, its discovery throughout the Tesero Member, with the exception of the basal 50 cm, of *Hi. praeparvus* and *Stepanovites dobruskinae* elements. Considering that in their stratigraphic column the Tesero Member is ca. 200 cm thick, they found *Hi. praeparvus* in the highest 150 cm where, at the top, *Hi. parvus* has now been found. *Hi. praeparvus* co-occurring with *Stepanovites dobruskinae* have been reported also in the Lower Tesero Member in the Bulla section, just below and in the level of the first occurrence of *Hi. parvus* (Perri and Farabegoli 2003).

The occurrence of *Hi. parvus* cannot be yet considered the first indubitable entry of the species in the section, even though lithostratigraphic and conodont alignments with the Bulla section allow the assertion that if the level of WPK15 does not coincide with the PTB, it must be very close.

The vertical distribution of skeletal biomass allowed us to recognise three extinction phases E1, E2 and E3, clearly comparable to the ones identified by Farabegoli et al. (2007) in the Bulla section. Phase E1 brackets the unconformity U2, including beds WPK10 and WPK11; it corresponds to the disappearance of large brachiopods (Broglio Loriga et al. 1988; Rampino et al. 2002) and a reduction in faunal size (Fig. 3). Phase E2 is just below the first occurrence of *Hi. parvus* and marks the disappearance of the bivalves, *Bellerophon waceki*, myalinids and *Orbicoelia* and a reduction in foraminifers. Overall, the diversity of this bivalve-dominated assemblage is very similar to that of the Bulla section, although the occurrence of rare articulates in the latter could reflect some slight environmental differences (Posenato 2009). The bioevent E3, 70–80 cm above the ?PTB, corresponds to the last appearance of calcareous algae and foraminifers with Permian affinities.

In the light of these observations, the boundaries of the two punctuated aggradational cycles (PACs; Fig. 3) detected by Wignall and Hallam (1992) in the Sass de Putia section and used by them to explain the regressive–transgressive phases that caused the end-Permian extinction must be revised.

The $\delta^{13}C_{tot}$ curve measured by Korte and Kozur (2005) is redrawn in Fig. 3. On the whole, it does not differ significantly from the curves measured in the Bulla section (2), except for a rapid reduction in the correspondence of bed WPK10, that is, at about the beginning of the fall-stand. This curve also declines immediately above WPK15 (?PTB), confirming the decline recorded by the $\delta^{13}C_{org}$ curve of the Bulla section. This minimum has also been detected in the Val Badia (Misci-Seres) section (Sephton et al. 1999; Koeberl et al. 2004).

Tesero Section

The Tesero section (Fig. 4) is located along the road joining the Tesero village with the national road no. 67 Val di Fiemme (Fig. 4a, c). It is one of the most-studied sections since Leonardi's (1935) pioneering study (Bosellini 1964; Assereto et al. 1973; Pasini 1985; Neri

Fig. 4 (continued) Tesero section. (**a**) Rocky wall along the road joining Tesero village with road no. 67 Val di Fiemme. (**b**) Enlargment of 4a showing thin-bedded whitish dolostone of the Ostracod Unit with thin rootlets in life position and sun cracks. (**c**) Oolitic microbialite of the Tesero Mbr pinching out to the northeast. (**d**) Location of some samples on a detailed section of the Bulla (*below*) and Lower Tesero (*above*) members. White arrows highlight some pockets in correspondance with the U2 paraconformable surface (details in **f** and **g**). (**e**) Alga- and echinoderm-dominated packstone in the lower part of the Bulla Mbr. (**f**) The erosional paraconformable surface U2 locally displays centimetre-sized pockets. (**g**) Enlargment of a pocket filling: the brownish material contains "residual" calcareous fragments and is permeated with hematite. (**h**) A millimetre-thick breccia with plate-shaped centimetric clasts cemented by brown-reddish hematite, locally overlying the very top of the Bulla Mbr (T7B, TS6) and disconformity U2. (**i**) Microfacies of Lower Tesero Member (TS8). Packstone–grainstone with small oolites and calcareous algae. Some oolites have a dark micritic sub-rounded nucleus or a local internal half-moon-type dissolution structure with geopetal filling (*red arrow*), often rotated

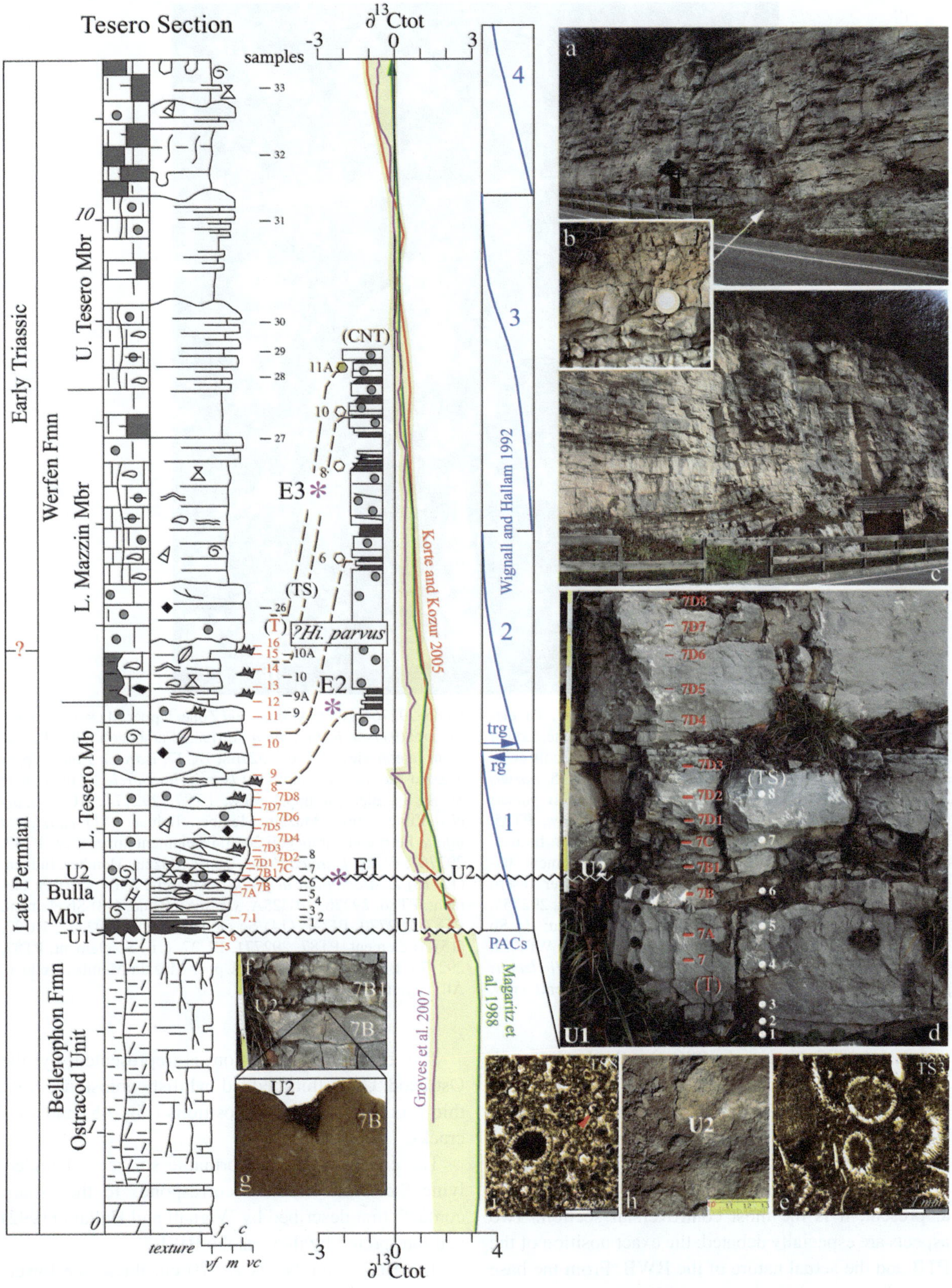

Fig. 4

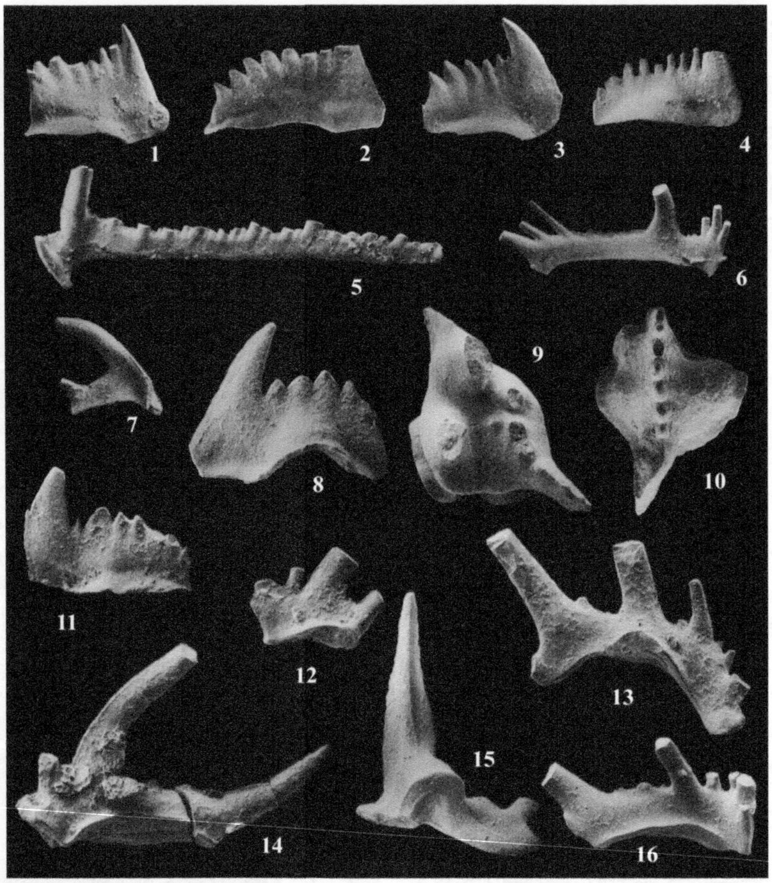

Plate 1 (1) *Hindeodus eurypyge* Nicoll et al. 2002, Pa element lateral view, PT85, 292751, WPK15, *parvus* Zone, Lower Tesero Member, Induan; (2) *Hindeodus eurypyge* Nicoll et al. 2002, Pa element lateral view, PT85, 292753, WPK15, *parvus* Zone, Lower Tesero Member, Induan; (3) *Hindeodus parvus* (Kozur and Pjatakova 1976), Pa element lateral view, PT85, 292752, WPK15, *parvus* Zone, Lower Tesero Member, Induan; (4) *Hindeodus inflatus* Nicoll et al. 2002, Pa element lateral view, PT85, 292754, WPK15, *parvus* Zone, Lower Tesero Member, Induan; (5) *Hindeodus* sp., Sc element, PT85, 292755, WPK15, *parvus* Zone, Lower Tesero Member, Induan; (6) Sb element, PT85, 292757, (7) Sa element, PT85, 292756, WPK15, *parvus* Zone, Lower Tesero Member, Induan; (8) *Hindeodus parvus* (Kozur and Pjatakova 1976), Pa element lateral view, PT86, 292764, BU25A, *staeschei* Zone, Upper Mazzin Member, Induan; (9) *Isarcicella lobata* Perri and Farabegoli 2003, Pa element lateral view, PT86, 292768, BU27, *isarcica* Zone, Upper Mazzin Member, Induan; (10) *Isarcicella isarcica* (Huchkriede 1958), Pa element upper view, PT87, 292774, BU27, *isarcica* Zone, Upper Mazzin Member, Induan; (11) *Isarcicella inflata* Perri and Farabegoli 2003, Pa element upper view, PT87, 292767, BU27, *isarcica* Zone, Upper Mazzin Member, Induan; (12–16) *Hadrodontina aequabilis* Staesche 1964, (12) Pa element, PT86, 292763, BU25A, *staeschei* Zone; (13) Sb element, PT87, 292770, BU27, (14) Sc element, PT87, 292772, BU27, (15) M element, PT87, 292771, BU27, (16) Sc element, PT87, 292773, BU27, *isarcica* Zone, Upper Mazzin Member, Induan. All magnifications are × 82

and Pasini 1985; Neri et al. 1986; Broglio Loriga et al. 1988; Perri and Andraghetti 1987; Noé 1987; Perri 1991; Magaritz et al. 1988; Wignall and Hallam 1992; Wignall et al. 1996; Neri and Posenato 1999; Nicora and Perri 1999; Perri and Farabegoli 2003; Chen et al. 2005, 2006; Groves et al. 2007; Posenato 2009), but at present, it is the most controversial section. Two aspects are especially debated: the exact position of the PTB and the actual nature of the BWB. From the base upwards, the section consists as follows.

The base of the section is represented by the Ostracod Unit: thin-bedded whitish dolostone, with thin root apparati in life positions (Fig. 4b) and suncracks.

The erosional paraconformable surface U1 underlying the Bulla Member corresponds to the "sharp contact" first described by Wignall and Hallam (1992) and interpreted by them as the BWB.

The Bulla Member is ca. 70 cm thick; the lowermost part of the member is dark grey claystone, 15 cm thick (Fig. 4d), with rare ostracods; it is overlain by mudstone with bivalves, followed rapidly by alga- and echinoderm-dominated packstone (Fig. 4e) arranged as continuous lenticular beds, 5–10 cm thick, stacked in silty millimetric layers highlighting well-preserved current and symmetric wave ripples (e.g., bed 3 and the next bed upwards).

The upper part of the unit (bed T7B or TS6) is dark grey fossiliferous packstone, rich in lagenids and miliolids. Bed T7B (Fig. 4f) is a calcareous algae-dominated grainstone, passing upwards to echinoid-dominated packstone; the very top of the unit (Fig. 4f, g) is truncated by the erosional paraconformable surface U2, which locally displays centimetre-sized pockets (not previously described), filled with brownish material (Fig. 4f, g). In some areas, the crests of the ripples at the top of bed 7B have escaped the erosional processes. Broglio Loriga et al. (1988) and Groves et al. (2007) showed that the top bed contains varied faunas and floras, decreasing dramatically in bed T7B1.

The pockets are filled with carbonate sandstone cemented by brown–reddish hematite (Fig. 4g). Elsewhere, the disconformity U2 is overlain by a millimetre-thick breccia cemented by brown–reddish hematite (Fig. 4h); the centimetre clasts are plate-shaped, angular, heterometric and polymictic and are overlain (or laterally substituted) by a coarse arenite. The breccia is here interpreted as the "basal conglomerate" of the transgressive Tesero Member.

The lowermost Tesero Member (beds T7B1, 7C, 7D1 and 7D2-TS89) is packstone–grainstone, ca. 25 cm thick, containing small oolites, mostly with a dark micritic sub-rounded nucleus (Fig. 4i); the oolites have a local internal half-moon-type dissolution structure; the geopetals are often rotated (red arrows show fillings in Fig. 4i). Small calcareous algae, filled with dark micrite (Fig. 4i), are present at various levels, as are small *Bellerophon waceki* with geopetal filling.

The oolitic co-set of the basal Tesero Member onlaps the Bellerophon Formation bedrock; its thickness decreases from 170 to 90 cm over a distance of less than 100 m (Fig. 4c). Major changes in facies and thickness of the Lower Tesero Member along the road have been described by Neri (1999, Fig. 4); he illustrated a composite Tesero section composed of interfingering oolitic calcarenite, marls, mudstone and microbialite lenses. The correlation scheme we present in Fig. 4 is a simplified version of that of Broglio Loriga et al. (1988), modified in the interval across the BWB, integrating subsequent logs of Neri et al. (1986), whose samples are indicated by the letter T of Perri (1991), and by TS of Neri and Posenato in Beretta et al. (1999), and by CNT and TES for conodont samples collected by Nicora and Perri (1999), following the numbering of Beretta et al. (1999) (see Fig. 8 and Section VI of Fig. 4 in Neri and Posenato 1999), which contains well-studied faunas (Broglio Loriga et al. 1988; Chen et al. 2006; Posenato 2009). Because the areal distribution of the biota depends strictly on shallow-marine microenvironments, it is of fundamental importance to understand whether their vertical ranges depend on local changes in facies, or in fact to a mass-extinction event.

Two groups of brachiopod-dominated faunas still showing Palaeozoic affinities have been identified (Posenato 2009). The older (CNT6, CNT8 and CNT10) occurs above the top of the lowermost oolite co-set and is dominated by *Orbicoelia*. The latter is dominated by *Streptorhynchus* (CNT11A); it records the last occurrence of articulates in the Tesero section (Posenato 2009).

The sedimentary structures corresponding to the *Orbicoelia* and *Streptorhynchus* assemblage indicate deposition at fairweather wave base; individuals predominantly attached to the tops of the stromatolite domes are observed in some "reefal" beds of the Bulla section (Farabegoli et al. 2007).

Conodont resampling confirmed the association described in Perri and Farabegoli (2003) but failed to produce *Hindeodus parvus* below the stratigraphic interval barren in conodonts (2.30 to 11 m from the base of the Werfen Formation). A new *datum* is the occurrence (TS10) of *Hi. eurypyge*, not previously found in the section. At 1.3 m, a Pa element of *Hi. changxingensis* has been reported by Nicora and Perri (1999). Perri and Farabegoli (2003) have included the species in the genus *Isarcicella* because of its swollen aspect; they have continued to cite it as *Is. changxingensis*. The lowest first occurrences of *Is. changxingensis* are located slightly above the main extinction event and below the PTB at 12 cm in the GSSP Meishan D section (Wang 1995; Nicoll et al. 2002; Jiang et al. 2007), at 18 cm in the Zhongzhai section in South China (Metcalfe and Nicoll 2007) and 15 cm in the Zal section at Abadeh in Iran (Korte et al. 2004). In other sections, it occurs first in association with *Hindeodus parvus* or slightly above. The entry of

Is. changxingensis can be utilised for identifying the interval between the main extinction event and the PTB and, mainly, for being near the PTB.

At Tesero, *Hindeodus parvus* enters as soon as conodonts recur at the top of the conodont-barren interval and by the presence of *Is. changxingensis* at 1.3 m from the base of the Werfen Formation. We believe that the boundary should be located closer to the entry of *Is. changxingensis,* at the top of the Lower Tesero Member, as it is at Bulla, where the PTB has been identified at 1.3 m from the base of the Werfen Formation. Available conodont data, including the presence of *Is. changsingensis*, suggest, in fact, that the more probable first occurrence of *Hi. parvus* slightly pre-dates sample CNT11A or TS 10A, and is not located in the Mazzin Member ca. 10 m above the BWB—as proposed by Kozur (1988).

The two $\delta^{13}C$ curves, proposed recently by Korte and Kozur (2005) and Groves et al. (2007), are very similar to the original curve proposed by Magaritz et al. (1988). Major differences are apparent in the Ostracod Unit curves, which cannot be easily or unequivocally explained. However, the different curves for the lower Tesero Member can be explained in terms, *inter alia*, of differential sampling, rapid lateral changes in facies and different compositions of the transgressive tract.

It is noteworthy that previous authors have never discriminated either the discontinuity U2 or the different mortality phases forming the end-Permian extinction.

Masi Saracini Section

The Masi Saracini section (Farabegoli and Viel 1982a) is located ca. 300 m north of Masi Saracini (see Google Earth Map) and ca. 2 km south of Vigo Meano. This section represents a late-Permian shallow-marine marginal area located a few kilometres east of the NW–SE-trending shoreline of the upper Bellerophon Gulf (Assereto et al. 1973). In this marginal area, the continental to deltaic sediments of the Gardena Sandstone lie only a few dozen metres below the Bellerophon Formation, whereas the thickness of the latter is ca. 10% of what is found in basinal areas (e.g., Cadore and Carnia: Bosellini and Hardie 1973; Farabegoli and Viel 1982b). Because of rapid lateral facies transition, poorly constrained by biostratigraphic data, several litho- and biostratigraphic, stable-isotopic stratigraphic and depositional sequences have been proposed (Farabegoli and Viel 1982a, b; Neri 1999; Visscher et al. 1996, 2004; Massari and Neri 1997; Looy et al. 1999; Sephton et al. 2002; Watson et al. 2005). Moreover, early diagenetic phases during the Early Triassic produced intense dolomitisation in marginal areas as well as, locally, sulphide-bearing (including sphalerite) strata-bound ores—which extensively obliterated depositional structures and textures.

The log shown in Fig. 5 can be summarised as follows, from the base upwards.

The Bavaro Member of the Gardena Formation (see Fig. 1) is a terrigenous unit composed of alternating grey sandstone and pelites (Fig. 5a), rich in both thin trees in the living position and transported centimetre-sized fragments of trunks (Fig. 5b). The transition to the overlying unit is an interval of ca. 12 m, mostly covered by slope talus.

The Ostracod Unit of the Bellerophon Formation is ca. 1.9 m (Fig. 5c) of whitish, bioturbated silty–sandy microcrystalline dolomite (Fig. 5d) with centimetre-size fenestrae, alternating with lenticular-bedded greenish pelites with small wave

Fig. 5 (continued) Masi Saracini Section. (**a**) Alternating grey sandstone and pelites of the Bavaro Mbr of the Gardena Fmn. (**b**) Transported centimetre-size fragments of trunks in the bioturbated grey sandstone of the Bavaro Mbr. (**c**) The uppermost part of the Ostracod Unit (scale bar = 1m), the Bulla Mbr (between U1 and U2) and the Lower Tesero Mbr. (**d**) An example of bioturbated silty–sandy microcrystalline dolomite of the Ostracod Unit of the Bellerophon Fmn; the filling is enriched in pyrite crystals. (**e**) The sharp, paraconformable surface U1 dividing the Ostracod Unit (*greenish pelites*) from the overlying Bulla Mbr (*dark pelites*). (**f**) The packstone–grainstone at very top of the Bulla Mbr (MS5.3) contains echinoids, brachiopods, bivalves and well-preserved foraminifers. (**g**) The sharp, paraconformable, erosional surface U2 divides locally the Bulla Mbr (*below*) from the overlying Tesero Mbr of the Werfen Fmn. (**h**) Less than 20 m north of site **g**, the same erosional surface U2 appears disconformable to unconformable. (**i**) At microscopic scale, surface U2 looks sharp, irregular and erosional. (**j**) Domal, centimetric-scale columnar stromatolites alternating with mm-thick lenticular siltstone with wave ripples (Lower Tesero Mbr). (**k**) The domal to centimetre-columnar stromatolites contain rare *Bellerophon waceki.* (**l**) Lenticular siltstone beds contain frequent, sparse valves of suspension feeders; feeble marine currents oriented the concavity of the valves predominantly downwards

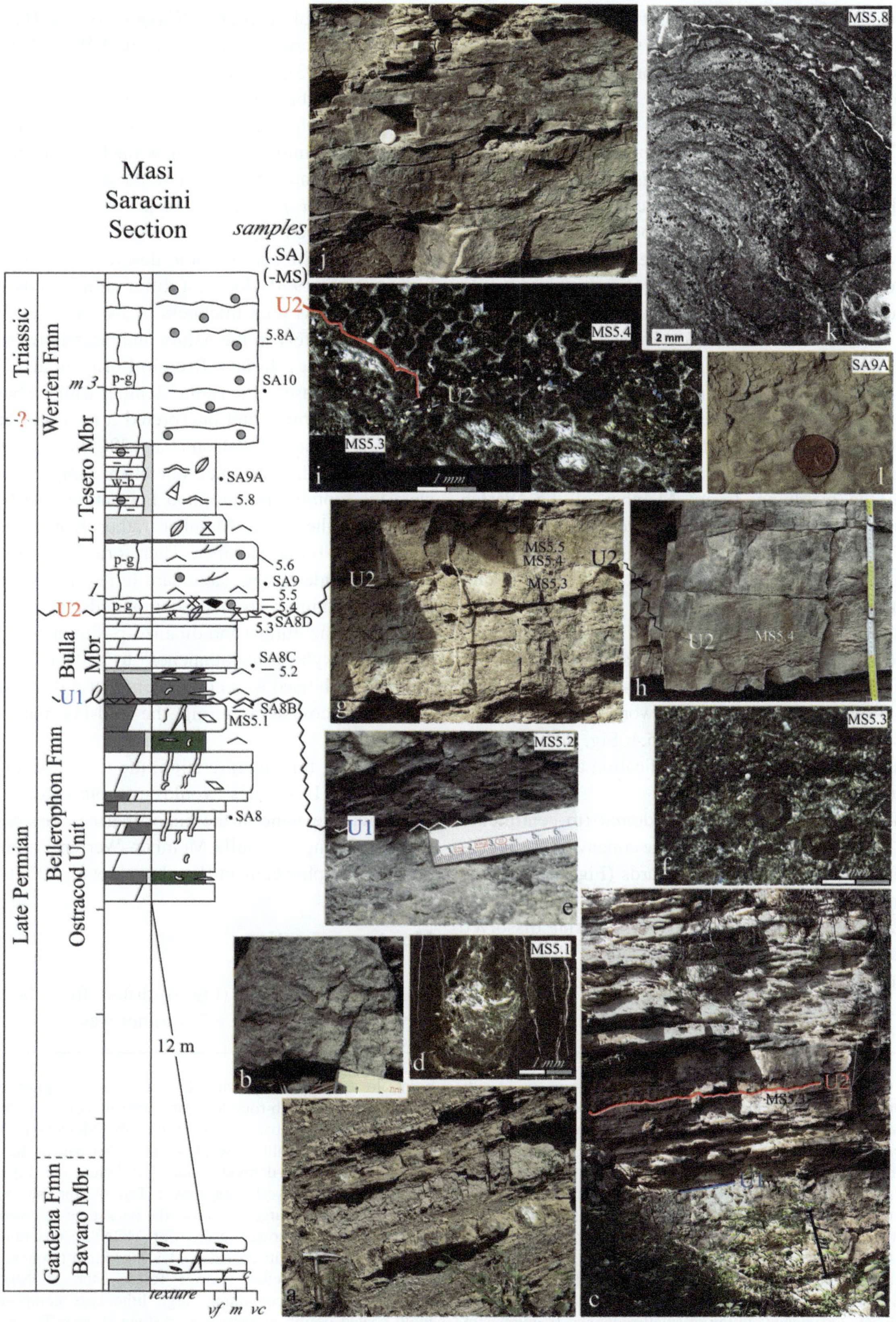
Masi
Saracini
Section
samples
(.SA)
(-MS)
Triassic
Late Permian
Werfen Fmn
L. Tesero Mbr
Bulla Mbr
Bellerophon Fmn
Ostracod Unit
Gardena Fmn
Bavaro Mbr
U2
U1
SA10
SA9A
SA9
SA8D
SA8C
SA8B
SA8
MS5.1
12 m
texture
vf m vc
MS5.8
MS5.4
MS5.3
SA9A
MS5.5
MS5.2
MS5.1
1 mm
2 mm

Fig. 5

ripples; the scarce fossil content is represented by a few ostracods, rare fish teeth and vertical roots.

A sharp, erosional, paraconformable surface (U1) divides the Ostracod Unit from the overlying Bulla Member (Fig. 5e).

The very base of the Bulla Member is an interval ca. 25 cm thick of dark grey (brownish where weathered) lenticular-bedded pelites, very rich in organic matter, including a few millimetric lithoclasts at the base. It is overlain by very fine sandstone rich in plant debris; it passes gradually upwards to centimetre-size, evenly lenticular-bedded dolomitised packstone–wackestone rich in echinoids and undetermined ghosts of fossils. The very top of the member is the packstone–grainstone MS5.3, containing echinoids, brachiopods, bivalves and well-preserved foraminifers (Fig. 5f).

A sharp erosional surface (U2), paraconformable (Fig. 5g, i) to unconformable (Fig. 5f), divides it from the overlying Tesero Member of the Werfen Formation. This surface looks macroscopically like U2 detected in the Sass de Putia section, but there is no overlying soil. Where the surface has not been substituted with an irregularly shaped stylolite or with a millimetre-thick layer of brownish insoluble oxides, it looks sharp and irregular at microscopic scale (Fig. 5i).

The overlying layer is a grainstone composed of multi-coated oolites with rare reworked lithoclasts and fossils from the very base (MS 5.4; Fig. 5i); wave ripples and hummocky cross stratification are intrastratal structures.

The overlying siltstone has domal (to centimetric scale) columnar stromatolites; the concavity of shells is oriented predominantly downwards (Fig. 5i). The log terminates with oolite grainstone.

Despite the large quantity of material dissolved in acid, no conodonts were found in this section. In the Strigno section, deposited in a comparable palaeogeographic context, *Hi. typicalis* and *Hi. pisai* have been reported 2.3 m above the BWB (Perri and Farabegoli 2003).

From available data, *Hi. pisai*, first described from the Southern Alps, seems to be a short-ranging species, occurring first immediately below the PTB and becoming extinct immediately above it; the occurrence of *Hi. parvus* with *Hi. pisai* could be utilised for indicating proximity to the PTB.

The Vigo Meano section described by Broglio Loriga et al. (1990) does not differ significantly from this in either its total thickness or its lithology, but it differs completely in the nature and boundaries of the minor lithozones. In fact, Broglio Loriga et al. (1990) confused the base of the Bulla Member with the base of the Tesero Member and interpreted it as a continuous lithologic boundary. Watson et al. (2005) assumed that the organic-rich marls (i.e., our black sample MS5.2) were the topmost Bellerophon Formation, just below the base of the Tesero Member, rather than the actual base of the Bulla Member. Therefore, Watson et al. (2005) considered the "abundant terrestrially-sourced oxygen containing aromatic compounds such as xanthones, dibenzofurans and dibenzo-p-dioxin as well as abundant alkylated naphthalenes" to be "land-derived material mobilised as a consequence of a destabilised terrestrial ecosystem during the end-Permian mass extinction".

Following this interpretation, the terrestrial extinction in the Dolomites would just pre-date (a few millennia) the acme of the major marine mass extinction bracketing the Bulla Member–Werfen Formation lithostratigraphic contact (Farabegoli et al. 2007).

Montagna Section

The Montagna section (Fig. 6) differs from the Masi Saracini section in only a few minor ways.

Fig. 6 (continued) Montagna section. (**a**) Vuggy, sandy–silty dolomicrite in the upper part of the Ostracod Unit of the Bellerophon Fmn. (**b**) Under the microscopic, mm dome-shaped microbialitic laminae appear (MON2). (**c**) Dolomitic wackestone rich in ostracods (MON3). (**d**) Uppermost part of the Ostracod Unit has grey lenticular beds of silty marls alternating with whitish, bioturbated, silty–sandy microcrystalline marly dolomite and dolomite. (**e**) Sharp contact U1 between the Ostracod Unit and the Bulla Mbr. Top surface of the Ostracod Unit is cut by small roots. (**f**) Undulating micritic marly dolomite laminae locally topped by rare ostracods. (**g**) Carbonate-terrigineous filling of small roots and fractures simulates neptunian dykes in calcareous layers (MON12A). Arrow points to a small red-brown sphalerite grain of epidiagenetic origin. (**h**) Small, sub-rounded, carbonate lithoclasts in brownish silty marl (very base of the Bulla Mbr, MON4). (**i**) Small foraminifer in dolomitised wackestone (upper part of the Bulla Mbr, MON7). (**j**) Ill-defined surface U2 divides the Bulla Mbr (*below*) from the overlaying Lower Tesero Mbr (see **e**). (**k**) Ghosts of medium-large oolites in the packstone–grainstone at the very base of the Lower Tesero Mbr (MON8A). (**l**) Grainstone containing large, multi-coated oolites showing a frequent inner half-moon-type structure (Lower Tesero Mbr, MON9). (**m**) Oolitic grainstone containing large lithoclasts composed of multi-coated oolitic grainstone-packstone (Lower Tesero Mbr, MON10)

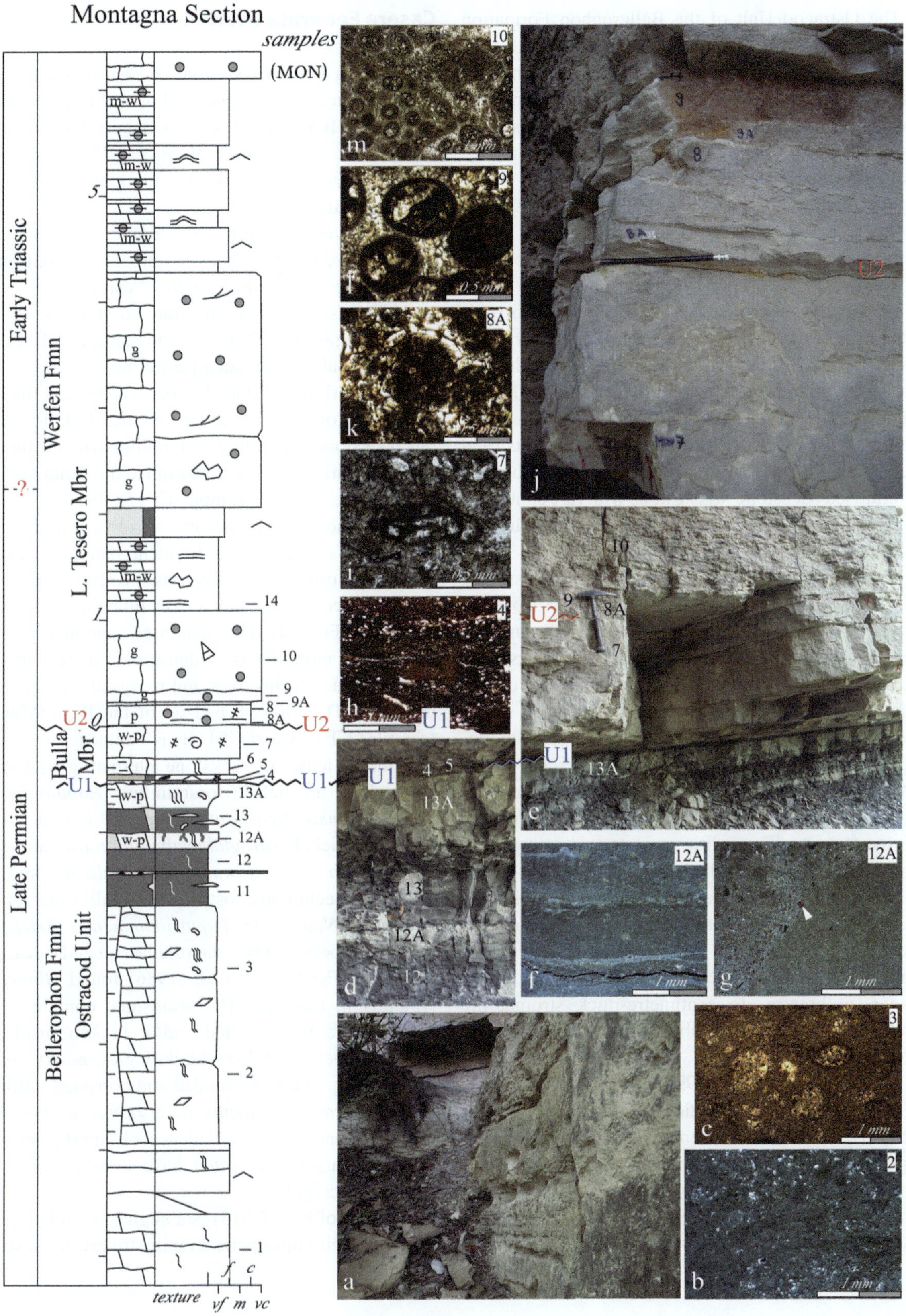
Montagna Section
samples
(MON)
Early Triassic
Late Permian
Werfen Fmn
L. Tesero Mbr
Bulla Mbr
Bellerophon Fmn
Ostracod Unit
U1
U2
texture
vf
m
vc
f
c

Fig. 6

The Ostracod Unit of the Bellerophon Formation is a vuggy sandy-silty dolomicrite (Fig. 6a) with dome-shaped microbialitic laminae (Fig. 6b) and levels rich in ostracods (Fig. 6c), passing upwards (Fig. 6d, e) to grey lenticular beds of silty marls alternating with whitish, bioturbated, silty–sandy microcrystalline marly dolomite and dolomite with undulating micritic laminae (Fig. 6f). The top surface of these more calcareous layers is cut by small roots and fractures (Fig. 6g) simulating neptunian dykes.

The upper calcareous layer (bed MON13A) is truncated by a sharp surface (U1) and is paraconformably overlain by the Bulla Member. A thin, brownish, silty marl bed with millimetric, sub-rounded, carbonate lithoclasts (Fig. 6h) passes upwards to a very fine sandstone (sample MON5), rich in plant fragments, and then gradually to centimetric-size, lenticular-bedded dolomitised packstone-wackestone with rare small fossils (Fig. 6i). The total thickness of the Bulla Member is 55 cm.

A slightly undulating, paraconformable surface (U2; Fig. 6e, j) divides the Bulla Member from the overlying Tesero Member of the Werfen Formation. The very base of the Tesero Member is a packstone–grainstone with ghosts of medium to large oolites (Fig. 6k). It continues upwards with large, multicoated oolites showing a frequent inner half-moon-type structure (Fig. 6l). These were leached by vadose waters during a very early diagenetic phase but were later rotated in situ and/or transported by waves and marine currents. The oolitic grainstone of sample 10, which contains large lithoclasts composed of multicoated oolitic grainstone (Fig. 6m), also points to shallow-marine corrosional, erosional and transport processes attributable to low- and high-energy storm waves. The succession continues upwards with subplanar, millimetre- to centimetre-thick stromatolite strata, predominantly referable to a sub-tidal carbonate environment.

Broglio Loriga et al. (1990) and Massari and Neri (1997) described this section in detail but did not recognise the two sharp erosional contacts bracketing the Bulla Member. Moreover, they attributed the pelitic interval that we here refer to the top of the Ostracod Unit and the overlying Bulla Member as Werfen Formation.

Casera Federata Section

The Casera Federata section, ca. 7 m thick (Fig. 7), crops out in a small abandoned quarry (Fig. 7a, detail of the BWB in Fig. 7b) located near the border of the Cadore and Carnia regions, ca. 75 km northeast of Bulla (Fig. 1). The succession may be described schematically as follows.

The lower part, referable to the Ostracod Unit, is composed of silty, bioturbated fossiliferous wackestone–packstone with ostracods, small brachiopods, bivalves and rare fragments of calcareous algae (Fig. 7c). The top sample (CF12) is a calcareous-alga-dominated packstone.

The overlying Bulla Member, ca. 66 cm thick, is composed of centimetric–decimetric dark fossiliferous packstone alternating with millimetric–centimetric layers of dark marl. The transition to the Bulla Member corresponds here to sample CF13, that is, a packstone characterised by occurrence of fusulinids, small foraminifers (miliolids are frequent, lagenids are rare), encrusting organisms, ostracods, bivalves, small brachiopods, echinoids and a few lithoclasts, probably reworked (Fig. 7d). The fusulinids have nearly disappeared in sample CF14, a dense packstone with miliolids, ostracods, bivalves and recrystallised bioclasts (Fig. 7e). The topmost bed-set of the Bulla Member (samples CF15A and CF15B) is composed mainly of centimetric lenses of packstone alternating with grainstone. Both are rich in foraminifers (*Robuloides* sp., *R. gourisiensis*, *Neoendothyra* sp., *Globulina vorderschmitti* Reichel, *Paraglobuvalvulina mira* Reitlinger, *Paraglobulina* cf. *septulifera*, *Geinitzina* sp. and miliolids), echinoids, ostracods, small brachiopods, fusulinids (*Nankinella* cf. *quasihunanensis*, *Reichelina* sp.), gastropods and rare dasycladacean and calcareous red algae. Farabegoli et al. (1986) found *Hindeodus typicalis* in this bed. The grainstone contains lithoclasts and a few abraded and/or broken fusulinids; their chambers are filled with dark grey, nearly opaque micrite (Fig. 7f). Sub-rounded and corroded millimetric *Comelicania* fragments are also present. Fossils in the packstone show no evidence of significant transport; the larger ones often have unrotated geopetal fillings (Fig. 7g, h).

The top of bed (CF15) is a slightly undulating surface (U2) with millimetre–centimetre erosional scours

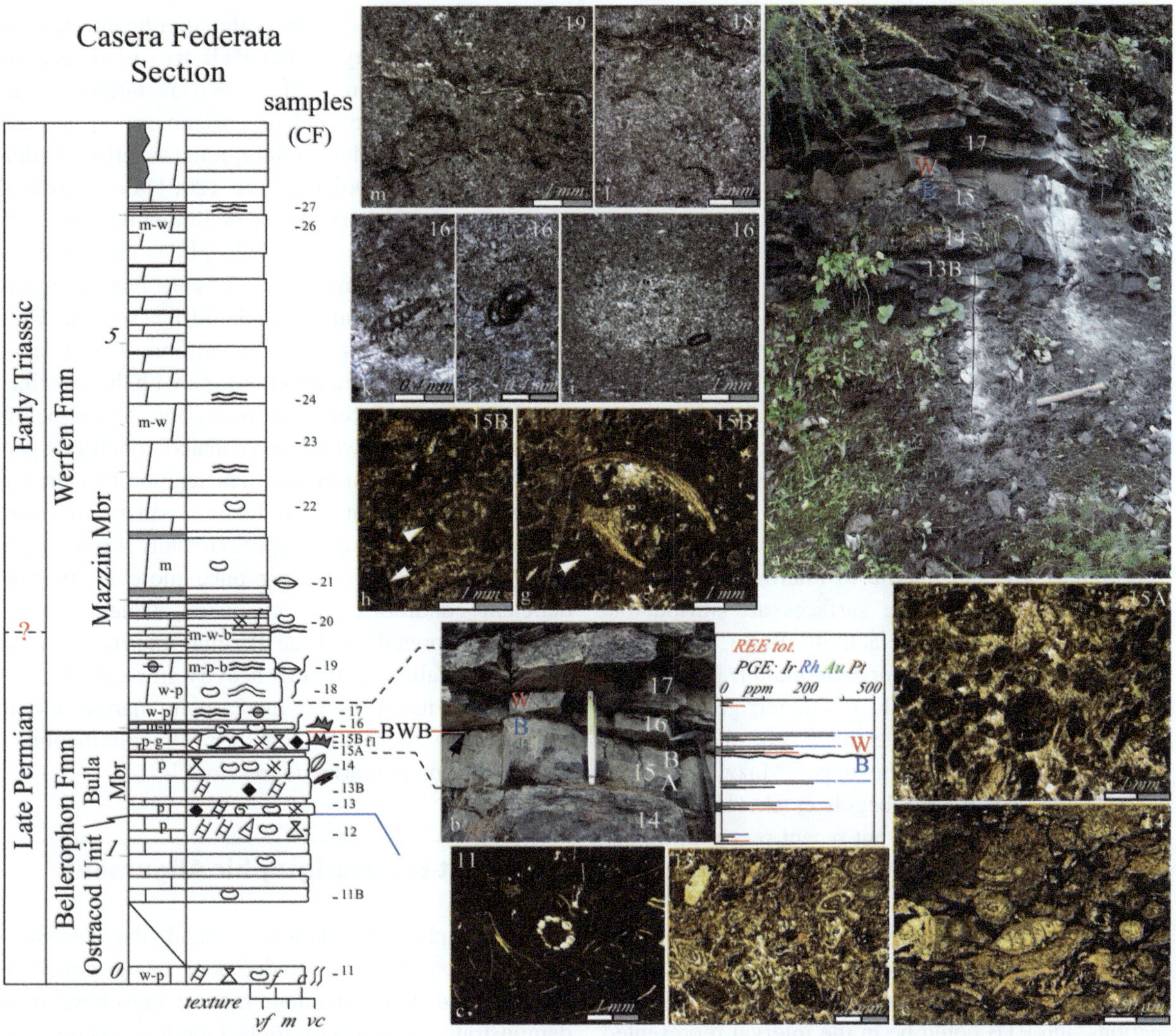

Fig. 7 Casera Federata section. (**a**) Small abandoned quarry; the location of some samples is shown. (**b**) Location of the samples around the Bellerophon–Werfen formational boundary. On the *right* are shown the average content in REE (Rare Earth) and PGE (Palladium group elements) (after Oddone and Vannucci 1988). (**c**) Silty, bioturbated fossiliferous packstone with ostracods, bivalves and calcareous algae (Ostracod Unit, CF11). (**d**) Packstone containing fusulinids, small foraminifers, encrusting organisms, ostracods, bivalves, small brachiopods, echinoids and lithoclasts, probably reworked (lowermost Bulla Mbr, CF13). (**e**) Dense packstone with miliolids, ostracods, bivalves and recrystallised bioclasts; fusulinids have nearly disappeared (middle-upper Bulla Mbr, CF14). (**f**) Grainstone is rich in microfossils and lithoclasts; it contains abraded and/or broken fusulinids—their chambers are filled with *dark grey*, nearly opaque micrite (upper Bulla Mbr, CF15A). (**g**) The brachiopod has an unrotated geopetal filling; the white arrow points to a corroded, transported fusulinid (?reworked lithoclast) (topmost Bulla Mbr, CF15B). (**h**) The fusulinid and micro-foraminifer look poorly reworked in the packstone (topmost Bulla Mbr, CF15B). (**i**) Mudstone with less than 1% skeletal fossils and rare bioturbation structures (lowermost Mazzin Mbr, CF16). (**j**) One of the rare, small miliolids in mudstone (lowermost Mazzin Mbr, CF16). (**k**) A rare, small lagenid in mudstone (lowermost Mazzin Mbr, CF16). (**l**) Recrystallised millimetre-high mounds showing discontinuous microbialitic laminae and beds (Lower Mazzin Mbr, CF18). (**m**) Recrystallised sub-planar microbialitic laminae containing a disarticulated valve of a small brachiopod (Lower Mazzin Mbr, CF19)

(black arrow in Fig. 7b). The overlying 1–2 cm of lenticular silty marl is grey in colour and is referred to the Mazzin Member of the Werfen Formation, so the BWB corresponds to the very top of bed CF15, as in Farabegoli et al. (1986, fig. 12). The overlying Mazzin Member succession is a centimetre–decimetre, even or slightly undulating set of mudstone–wackestone beds alternating with rare millimetric intervals of grey marlstone. Skeletal fossils (less than 1% of the rock) are represented nearly exclusively by small ostracods; bioturbation is less than 5% (Fig. 7k); very rare small foraminifers are found in bed CF 16 (Fig. 7j, i). The microbialitic laminae and beds are usually planar, but bed CF18 shows millimetre-high mounds (Fig. 7l); sample CF19 also has a disarticulated valve of a small brachiopod (Fig. 7m). All the beds contain small, scattered, pyrite crystals, the sizes of which seem to correlate with the micrite to microsparite carbonate. The bed and laminae surfaces are attributable to changes in microbialitic activity. Bed CF20 shows silty laminae, rich in dark organic matter. Framboidal pyrite is frequent in bed CF21. The terrigenous input increases to ca. 6 m above the BWB; the resulting facies is comparable to the upper Mazzin Member in the Bulla section. Farabegoli et al. (1986) found *Hi. typicalis* in bed CF16, but recent re-sampling has yielded no further conodonts. Therefore, at present, we do not know the exact position of the PTB. If we cautiously assign a sedimentary rate to the microbialites of 50% of the rate in the nearly coeval oolite–terrigenous–stromatolite interval in the Bulla or Sass de Putia sections, the PTB could be in the interval represented by samples CF19–20, that is, 60–70 cm above the well-defined BWB.

The stable carbon isotopes of the Casera Federata samples have never been analysed, so it is at present impossible to compare the trend in this deeper, nearly continuous, marine succession with either that of the coeval shallow marine succession in the Dolomites or with the equivalent, nearly continuous deposits in the Idrijca Valley (cfr. Ramovs 1986; Dolenec et al. 2001; Dolenec and Vokal 2003) and the Karavanke Mountains (Dolenec et al. 1999).

Oddone and Vannucci (1988) analysed the major, minor and trace elements across the BWB in the Casera Federata section (see element distribution on the right of Fig. 7b) and concluded that "the average content of PGE (Palladium Group) and REE (Rare Earth) elements reflect a primitive elemental incorporation in marine sediments, even if the significant increase of PGE in proximity of the PTB was of diagenetic origin". The iridium in the BWB is interpreted as of terrigenous origin, and its enrichment in the beds bracketing the boundary is interpreted as reflecting diagenetic alteration. Dolenec et al. (2001) demonstrated a similar trend in enrichment of lithophile rare earth elements, siderophiles and chalcophiles in an interval ca. 60 cm thick (from –50 to +10 cm), bracketing the lithostratigraphic boundary in the Idrijca section. They suggested more or less anoxic conditions during deposition of the Late Permian limestone with the culmination of these anoxic conditions at about 50 cm below the boundary, after which oxygenated conditions persisted into the earliest Induan. The temporal pattern for cerium indicates that the reduction/oxidation (redox) conditions changed again approximately 4.6 m above the P–T boundary, resulting once more in oxygen-deficient conditions. In conclusion, Dolenec et al. (2001) interpreted the Late Permian marine fauna in the Idrijca Valley section as heavily stressed and dramatically reduced by an increase in marine anoxia (50 cm below the BWB), together with culmination of the Late-Permian marine regression.

Conodont Biostratigraphic Alignment

Biostratigraphic correlations around the Permian–Triassic Boundary are undertaken using gondolellid and hindeodid conodont faunas; they have been demonstrated to be useful tools. Gondolellids are considered to have been inhabitants of offshore, oxygen-rich, basinal environments, whereas hindeodids are considered to have inhabited warm to cold pelagic environments including shallow water habitats (Lai et al. 2001, Jiang et al. 2007). The two lineages provide a double-biozonation, each aligned with the other. Because hindeodids are virtually ubiquitous, even though occasionally rare, a species of the genus *Hindeodus* has been chosen for identifying the base of the Triassic. The hindeodid (or shallow water) biozonation has accordingly been regarded as having special value.

In the Southern Alps, the end Permian–Triassic successions consist of shallow-water depositional environments with unique conodont faunas represented by scarce assemblages of hindeodids. Despite the scarcity it was possible to align levels marked by entries of guide species with equivalent beds in the GSSP Meishan D section and to propose six conodont biozones: Lower *praeparvus*, Upper *praeparvus, parvus, lobata, staeschei, isarcica* and *aequabilis* zones (Perri and Farabegoli 2003) (Fig. 8). In the Bellerophon Formation only the uppermost part produced a few conodont faunules. Pa elements of *Hindeodus typicalis,* the stratigraphically lowest species encountered in the Alpine Late Permian sequences, have been found about 45 m below the BWB in the Ansiei section, 60 km E of Bulla. That particular level belongs to the Casera Razzo Member of the Bellerophon Formation, the member below the Ostracod Unit and the Bulla Member; it was referred to the Late Changhsingian by Farabegoli et al. (2007) because of the presence of *Hi. typicalis* whose FAD falls in the Changhsingian *Cl. changxingensis* Zone (*sensu* Ding et al. 1996). In the same Ansiei section, deeper water environments compared with those of the Bulla section—40 m above the level with *Hi. typicalis* in the Bulla Member—a few Pa elements of *Hi. praeparvus* defines the Lower *praeparvus* Zone of Perri and Farabegoli (2003), aligned with Beds 25–26 of the Meishan D section (Farabegoli et al. 2007). The biozone corresponds to the lower portion of the *Cl. meishanensis–Hi. praeparvus* Zone of Orchard and Krystyn (1998) and of the *Cl. meishanensis–Hi. latidentatus* Zone (*Cl. meishanensis* Fauna) of Yin et al. (2001) and to the upper part of the *Hi. praeparvus* Zone of Jiang et al. (2007).

In the Southern Alps, conodonts become quite abundant in the Lower Tesero Member, 30–40 cm above the base of the Werfen Formation, that is, of the main extinction event E1 embracing the top of the Bulla Member, unconformity U2 and the first few cms of the subsequent Tesero Member. Why should it be there that conodonts increase in abundance and diversity? Wu (2005) was the first to report replacement of the *Hindeodus*-dominated conodont fauna by the *Clarkina*-dominated conodont fauna in Bed 27a in the deep-water Meishan D section where gondolellids underwent a drastic drop in diversity and abundance in favour of hindeodids. A more detailed study of an abundant conodont fauna collected from the Meishan A section (Jiang et al. 2007) confirmed the bioevent. In the Southern Alps, the scarcity of conodonts can be attributed to stressed environmental conditions during deposition of the Bellerophon Formation, but not more stressed than those of the Lower Tesero Member. In the transgressive phase of the Bulla Member (deposition in a more open sea), we would have expected to find a rich conodont fauna. But in that interval, hindeodids constituted only a very low percentage of the conodonts, viewed globally, at that time. The Bulla Member has been aligned with Beds 25–26 of Meishan D section (Farabegoli et al. 2007).

Jiang et al. (2007) identified three subsequent conodont assemblages. The first, derived from Beds 24a–26, is characterised by high diversity and abundance and is dominated by 90% gondolellids; hindeodids were clearly subordinate and had little chance of living in the upper Bellerophon Sea. In Assemblage 2 of Jiang et al. (2007), namely, in Beds 27–28, hindeodids replaced gondolellids and became dominant. Permian forms belonging to *Neogondolella* co-occur with Triassic ones represented by *Hindeodus* and *Isarcicella.* Farabegoli et al. (2007) aligned Beds 27a–27b, the base of Assemblage 2 (with more than 90% hindeodids) with interval BU9–BU12A of the Bulla section. Above the BWB, the transgressive basal 30–40 cm of the Werfen Formation produced no conodonts in the Bulla, Tesero and Sass de Putia W sections; this may reflect either locally stressed environmental conditions or global scarcity of hindeodids. In the Bulla section, Gorjan et al. (2007) detected three troughs in $\delta^{34}S$ sulfate, the last two coinciding with the dysoxic-euxinic intervals immediately below bed BU9 (Fig. 2), the first sample with conodonts. Bed BU9 has been correlated with Bed 27a, exactly where at Meishan (and on a global level) hindeodids became highly dominant in conodont assemblages—with more possibility of them being found in shallow-water environments, such as the sea in which the Lower Tesero Member accumulated. The contemporaneous increase in diversity among hindeodids could reflect recovery after the main extinction (E1) affected the gondolellids. The "aftermath recovery" was virtually instantaneous for conodonts.

Availability of vacated niches allowed diversification in a conodont fauna with morphologic peculiarities (which developed into isarcicellids) as well as the appearance of new species such as *Hi. pisai, Hi. eurypyge, Hi. inflatus, Is. prisca, Is. peculiaris* and *Is. changsingensis,* most of them having short ranges. *Is. prisca* was chosen by Perri and Farabegoli (2003) as the marker for defining the base of the Upper *praeparvus* Zone. *Hi. pisai,* originally described from the Southern Alps (Perri and Farabegoli 2003), helps

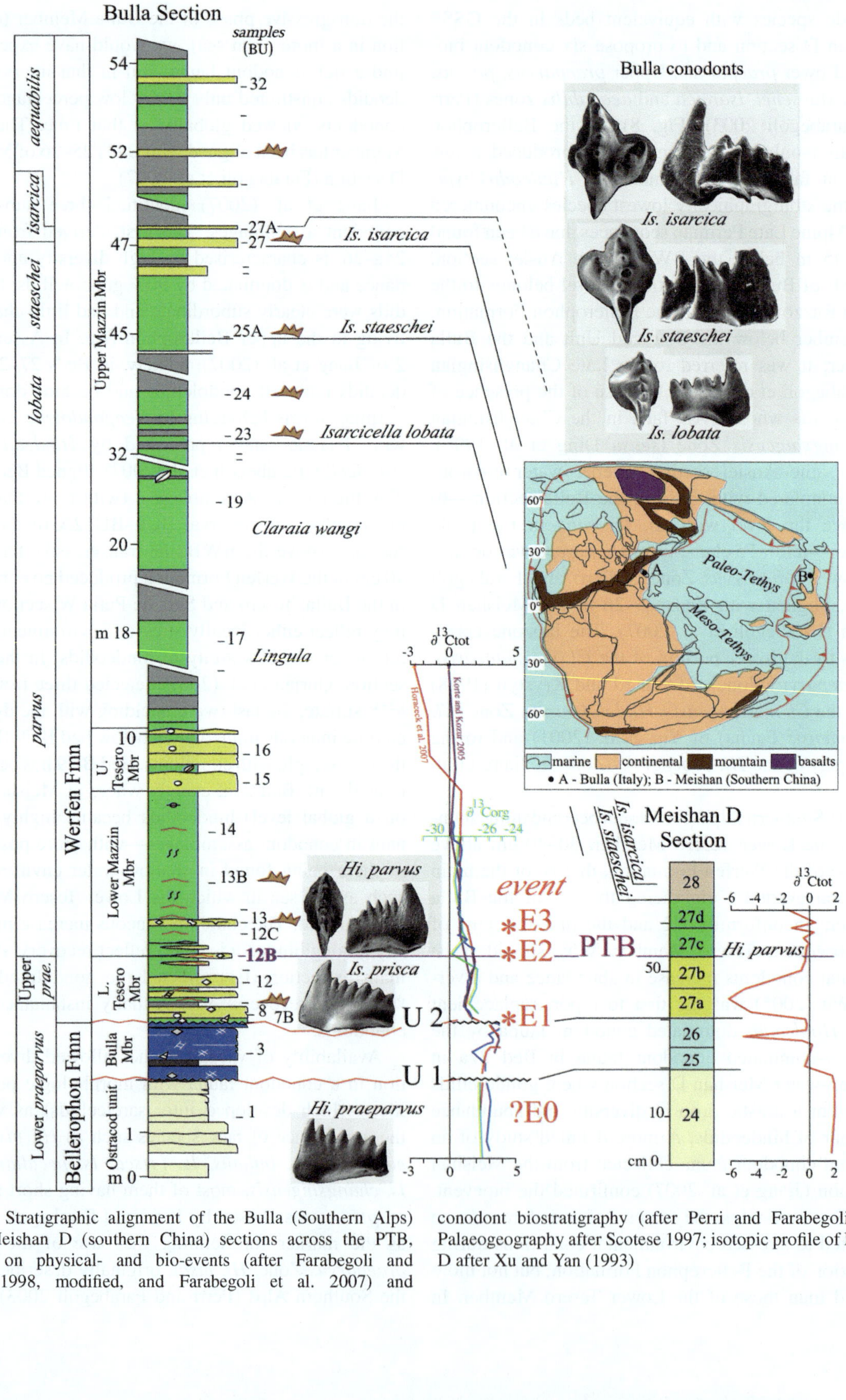

Fig. 8 Stratigraphic alignment of the Bulla (Southern Alps) and Meishan D (southern China) sections across the PTB, based on physical- and bio-events (after Farabegoli and Perri 1998, modified, and Farabegoli et al. 2007) and conodont biostratigraphy (after Perri and Farabegoli 2003). Palaeogeography after Scotese 1997; isotopic profile of Meishan D after Xu and Yan (1993)

in defining the biozone and proximity to the PTB, as does *Is. changsingensis.* Jiang et al. (2007) have referred some elements, found in the Meishan A section, to *Hi. pisai*. We agree with their identifications except for the element obtained from Bed 24d (Plate 4, Fig. 25) having teeth too wide and a different general shape. That element aside, the species in China displays identical ranges to those in the Southern Alps, entering after the main extinction event in Bed 27b in the interval Bed 27a–27b aligned by Farabegoli et al. (2007) with the sequence BU9–BU12A, with *Hi. pisai* having its first occurrence in BU12A. The range of *Hi. pisai* overlaps the PTB, becoming extinct soon afterwards. *Is. changsingensis* presents a very similar distribution with entry in the same correlated intervals below the PTB, identifying the Upper *praeparvus* Zone in the Southern Alps, and the *Hi. changsingensis* Zone (Jiang et al. 2007) at Meishan. Wu (2005) and Jiang et al. (2007) utilised *Hi. changsingensis* as a biozonal marker species; it occurs in the Southern Alps in the Tesero section 1.3 m above the BWB in the Upper *praeparvus* Zone and was assigned by Perri and Farabegoli (2003) to the genus *Isarcicella* because of morphologic peculiarities for the genus such as the thickness and swelling of the basal cavity and of the attachment area of the denticles giving the conodont a swollen appearance. The Upper *praeparvus* Zone corresponds to the upper part of the *Cl. meishanensis–Hi. praeparvus* Zone of Orchard and Krystyn (1998), to the upper part of the *Hi. latidentatus–Cl. maishanensis* Zone (*H. typicalis*) of Yin et al. (2001) and to the *Hi. changxingensis* Zone of Wu (2005) and of Jiang et al. (2007).

In the Southern Alps, the occurrence of *Hi. parvus* identifies the PTB and the base of the *parvus* Zone in the Bulla section (BU12B) at 1.30 m from the BWB base and possibly in the Sass de Putia W section (WPK15) at 1.40 m. In Tesero, the occurrence of *Is. changxingensis* at 1.30 m from the BWB allows the supposition that the PTB should be very close. BU12b and possibly WPK15 have been aligned with Bed 27c at Meishan.

In the Southern Alps, *Is. lobata* defines the *lobata* Zone located below the *staeschei* and *isarcica* zones (Perri and Farabegoli 2003). The species has been found also in Slovenia (Kolar-Jurkovsek and Jurkovsek 2007) and in Turkey (Nicora pers. com.). In Meishan, some elements have been assigned to *Is. lobata* in Beds 28 and 29, but they seem not to present the typical shape of the species. Before publication of the large collection of Jiang et al. (2007), it was easy to align layers BU25A (Bulla) and TS23 (Tesero), the base of the *staeschei* Zone, to Bed 28 in Meishan D. The element figured by Yin et al. (2001) from Bed 28 of Meishan as *Is. staeschei* falls within the intraspecific range of variability of the species. Elements referred to *Is. staeschei* by Jiang et al. (2007, Plate 5, Fig. 8, Bed 27d; Figs. 9 and 11, Bed 28; Figs. 12 and 13, Bed 29) are very dissimilar from the average morphology of the species; they could be referred possibly to a new species or could represent transitional forms to *Is. staeschei*. In our opinion, the photos of the elements assigned to *Is. staeschei* found in Beds 27d, 28 and 29 are very distinctive, showing peculiarities not shared with any other published conodont fauna. We prefer to align the base of the *staeschei* Zone with Bed 28. An element of *Is. staeschei* from the Selon section (bed 19) at 10–15 cm above the PTB falls within the intraspecific variability of the species (Shen et al. 2006). The base of the *isarcica* Zone identified in the Southern Alps in samples BU27 (Bulla) and TS26 (Tesero) should be aligned with Bed 29 at Meishan where the nominal species has been found even though no adequate photos have yet been published.

Discussion

The gross lithostratigraphy presented in this study of the shallow-marine succession, 20–30 m thick, bracketing the BWB and/or the PTB in the Dolomites is similar to that described by the majority of authors in the last 30 years. The upper part of the Bellerophon Formation, that is, the Ostracod Unit, is overlain by the Bulla Member, a very thin unit rich in shallow-marine Permian taxa. This is in turn overlain by the lower part of the Werfen Formation (Lower Tesero and Mazzin members), which contains only a few taxa with mixed Permian–Triassic affinities. Here, we highlight three major differences between our analysis and the data in the literature: (1) the lower and upper boundaries of the Bulla Member are disconformable–unconformable in the shallow-marine environment, whereas they are conformable in the deeper, foreshore–upper shelf successions, as at Casera Federata; (2) the positions of the PTB and the mass-extinction phases are more precisely defined at 0.5–2.5 m above the BWB and (3) the timing of the major physical events in the region provides clues to the possible causes of the mass mortality.

The Ostracod Unit is a lenticular, but informally defined, lithostratigraphic unit, up to few dozen metres thick. The upper part was predominantly deposited in marginal marine environments as carbonate–terrigenous marsh–lagoon and carbonate–terrigenous tidal-flat deposits, whereas locally (e.g., the Sass de Putia W section), the lower part of this unit encloses a fully marine high-energy facies. The Ostracod Unit constitutes a shallowing-upwards sequence, culminating with peritidal carbonate-dominated deposits, locally enclosing nodules of sulfates. Wignall and Hallam (1992) interpreted the upper part of this unit as a PAC (Fig. 4), that is, an aggrading depositional sequence formed during a longer, transgressive period. Our regional analysis shows that the depositional wedge of the upper Ostracod Unit prograded tens of kilometres southwards and eastwards in a time interval of 100–200 thousand years, but did not completely fill the shallow-marine Bellerophon Gulf (Fig. 1). Because both the marginal marine areas west of the Adige Valley and a few isolated palaeotopographic areas (e.g., Mount Rosà) emerged (Farabegoli and Viel 1982a, b; Neri 1982; Massari and Neri 1997; Perri and Farabegoli 2003), and considering the presence of coarse terrigenous material in the carbonate deposits, a more convincing interpretation is that this shallowing sequence was controlled by a moderate, rather long, fall-stand plus a low-stand period. Assuming a near-constant subsidence rate of a few centimetres per year, and given the difficulty in invoking global tectonics as the primary cause of the falling sea level because the time interval is too short, it is reasonable to attribute this limited fall (probably less than a dozen metres) to the thermal contraction consequent on global cooling. Therefore, a major question arises: Was the cooling period caused by a "normal" orbital change, or were different and more complex processes active? We also note that the marginal marine areas, such as displayed by the Montagna and Masi Saracini sections, seem to record limited but significant changes in palaeoenvironmental conditions, as indicated by the marls and carbonate layers that alternate several times, and by the stronger input of terrigenous sediment. Therefore, it seems that the climatic changes on the land were the primary factors influencing the marginal Bellerophon Sea.

That these stressful environmental conditions were responsible for destabilisation of this equatorial terrestrial ecosystem is clearly demonstrated by the land-derived material from lower vascular plants, fungi and lichens, that is, xanthones and dibenzofurans, together with numerous alkylated naphthalenes (Watson et al. 2005) in the greenish marls of the Vigo Meano (Masi Saracini) section. However, Cirilli et al. (1998) have already shown the environmental deterioration both on land and in the nearshore zone during deposition of the Ostracod Unit in the more distal Seres area. It is therefore possible to formulate the previous question more precisely: Which factor(s), other than cooling, could have destabilised both the land and the marginal marine belt? We will return to this question after a discussion of the BWB.

The lower boundary of the Bulla Member (surface U1) can be considered in four different ways: (1) a clearly disconformable–unconformable contact in the very marginal areas (Masi Saracini; Montagna); (2) a sharp contact, dividing dark pelites referable to a salty marsh from the underlying lithified peritidal carbonate bedrock (Tesero); (3) a decimetre-thick level of roots and plants in living positions, overlain by washover-derived calcarenite composed of coeval small foraminifers mixed with reworked clasts (Sass de Putia W); (4) a fossiliferous packstone, also containing reworked clasts, paraconformably and conformably covering lagoon deposits (e.g., Bulla, Casera Federata). This is consistent with a short low-stand period (millennial), during which the negative vertical excursion of the sea level was limited to a few metres during which the shallow-marine Bellerophon Gulf was not completely filled.

The Bulla Member is a very thin, 50–150 cm tabular unit deposited in a coastal environment ranging from shoreface to foreshore. The unit is arranged as an asymmetric transgressive–regressive depositional cycle, widespread on a regional scale. The transgressive tract occupies 90% of the sequence. Four faunal and floral assemblages gradually succeed one another upwards with deepening of the sea: (1) marginal marine ostracod wackestone; (2) echinoderms and small foraminifers (*Hemigordius*) in packstone; (3) fusulinid packstone; and (4) packstone with (red) calcareous algae. The large brachiopod *Comelicania* lived on deep, high-energy bottoms, often rich in land-derived silt and organic matter. The uppermost bed-set is alternating grainstone and packstone dominated by calcareous algae, foraminifers, echinoderms and plant debris deposited in a shoreface environment, locally visited by nautiloids (e.g., Sass de Putia). The presence of transported and reworked clasts, the increased size of the land-derived material and having the topmost facies dominated by small foraminifers and/or echinoids (a few centimetres where not eroded by the following U2 surface) are clear evidence for more normal fair-weather wave base. Intermittent-energy conditions on the marine bottom were present only in slightly deeper successions (e.g., Casera Federata). The duration of this thin fall tract, 5–15 cm thick, probably did not exceed three thousand years. Moreover, the frequent carbonate clasts, corroded during sedimentation and/or the early diagenetic phase, suggest that the pCO_2 of the marine water increased markedly during this period. This was a selective process of corrosion, nearly completely destroying some clasts (e.g., small foraminifers) and some moderately (fusulinids), but preserving clasts less-sensitive to corrosion, such as echinoderm plates.

This fall tract culminated with a well-documented low-stand. The sea fell a few dozen metres (or less) and exposed the nearly flat belt (up to 50 km) of the previous shallow-marine Bellerophon Gulf to sub-aerial agents; only the deeper foreshore remained submerged, near-normal fair-weather wave base (Casera Federata). The land was covered by a sheet of residual material, mostly of sand size, produced by sub-aerial disaggregation in situ of the bedrock, whereas the slopes of the small inherited islands (e.g., M. Rosà, Fig. 1) produced blocks. The fine–medium-sized sediment was then carried to the sea through the centimetre–decimetre drainage network and was distributed predominantly on the bottom of the coastal shallow-marine area. The emergent areas, with a scarcely developed drainage network, developed a brown soil (e.g., Sass de Putia area); fresh, well-oxygenated waters are essential for this pedogenetic process. It is noteworthy that this soil is comparable to the Pleistocene palaeosoils developed in the mid-latitude Mediterranean areas during the moderate–warm interglacial stages. Its reduced thickness could have resulted from the shorter timespan (millennial vs ca. 20–50 ky). The isolated, small (centimetre-sized) holes filled by pockets of residual litho- and bioclasts (mostly echinoids), embedded in a hematitic matrix, found in the Tesero area are here interpreted as a different shape of structure produced by the same pedogenetic process. In this case, relatively acid superficial waters flowed centripetally for a few decimetres and corroded the central part of the contributing area, creating small blind holes. The carbonate grains disaggregated from the walls of the holes and/or were transported here from the vicinity after having fallen into this acid bath and then contributed to its rapid buffering. This process can explain satisfactorily why the holes usually have a cylindrical shape, whereas a cruet shape is very rare. Echinoid crystals tended to selectively escape this repeated but short (days) corrosion process.

During the first phase of the subsequent rapid transgression over the land, waves tended to erode the palaeosoil and the basal sand cover and to smooth the topmost carbonate bedrock locally. Only rare, small lenses of the pedogenetic cover escaped erosion, mostly in shallow hollows and pockets. Locally, the ravinement surface was covered by storm-derived washover deposits. This sheet, composed of residual debris mixed with newly formed small oolites, had a higher potential for preservation, probably because of the high rate of retreat of the shoreline (m/dozen years) over the flat palaeotopography. This latter proposition is consistent with the shallow tidal channel (sample WPK12), more than 3 km long, along the alignment of the Seres and Sass de Putia W sections. The topographic highs were transgressed later. In this case, the more prolonged subaerial erosion and elevation of the bedrock above the sea favoured excavation of the small karst network described by Bosellini (1964) and Assereto et al. (1973) in the Ostracod Unit bedrock (cf. Farabegoli et al. 2007). The large, well-formed oolites filling the small karst holes are good evidence of this delayed transgression. However, the duration of this "prolonged" transgression, accompanied by increasing carbonate precipitation in the shallow-marine environment, was in fact very short (centennial–millennial) on the geological scale.

By contrast, the Casera Federata area represents a slightly deep and distal marine never-emergent area, far from the oolite-production sites. The thin residual calcareous–terrigenous sediment overlying the minutely irregular, submarine paraconformity U2 was very probably produced in situ by a centennial–millennial event of elevated pCO_2 or underwent very limited transport by weak marine currents. The submarine leaching is attributable to substitution of the normal marine water by slightly acidic freshwater draining from the land surrounding the Bellerophon

Gulf, plus the direct contribution of acid rains over the sea surface. During the following transgression, the bottom was below the normal fair-weather wave base and was covered by thin, soft microbialitic layers, binding carbonate mud, peloids and rare microfossils. The growth of domal and columnar stromatolites was probably nearly completely inhibited by the low intensity of visible-spectrum light. The *Hi. typicalis* conodont fauna, found both below and above the BWB (Farabegoli et al. 1986; Perri and Farabegoli 2003), accords with nearly continuous local deposition.

The palaeoenvironmental interpretation of areas such as Bulla is problematic because the relatively deeper (cm) holes with irregular walls characterising the surface of U2 do not contain clasts derived unequivocally from the brownish palaeosoil, but only residual coarse sand mixed with rare, newly-formed, small oolites. This can be explained in three not completely conflicting ways: (a) the soil never developed because of the peculiar palaeogeographic location, similar to some present-day upper intertidal and splash belts of poorly sloping rocky shores; (b) the soil developed over the carbonate bedrock, but was later completely washed out by intense and/or prolonged wave action during the transgression and (c) the marine bottom was cemented and then corroded by fluxes of acid water in the submarine environment, caused by drainage from the neighbouring land, after which the bottom was covered with coarse residual sand containing small oolites increasing in size upwards. It is realistic to suppose that the duration of exposure differed in different sectors because, given the same absolute rise in sea level and ignoring the different subsidence rates, the exposure would depend predominantly on the initial relative elevation of the bottom, the slope and so on.

The low-stand created a very shallow Bellerophon Sea over a very flat bottom, thus causing important (but unknown) changes in the pattern of marine currents and/or a reduction in their velocity. The expected final result was dramatic diminution of the buffering capacity of the marine water plus enhanced spatial variability in geochemistry of the marine water. Unfortunately, the widespread emersion at regional scale limited detailed geochemical sampling of the low-stand deposits. Geochemical analysis of the major, minor and trace elements of the low-stand tract highlights a generic environmental change towards more stressed conditions. In contrast, the large amount of geochemical isotopic data measured by various authors in the top centimetres of the Bulla Member, that is, the fall-stand tract, show significant variability ranging from 1‰ to 3‰ PDB (Bulla section: 2‰–3‰; Tesero: 1‰–2‰ and Sass de Putia W: 1‰; compare Figs. 2, 3 and 4). Because $\delta^{13}C_{tot}$ values depend most strongly on the relative proportions of different terrestrial and marine organisms that lived around and in the Bellerophon Gulf, the high variability detected shows irrefutably that environmental changes at regional scale played a role as important as that of global changes. Conversely, the strict time coincidence (millennial) between the fall-stand/low-stand period, deterioration of the continental environment with concomitant drastic impoverishment of the flora and mass mortality in the shallow seas requires a unique ultimate cause which we propose to be as follows. Millennial global cooling caused a significant drop in the atmospheric temperature and the temperature of shallow-marine waters in the northern tropical belt of Paleo-Tethys. We suggest that the combined effects of low temperature and acid rain episodes on land caused impoverishment of the plant cover, increasing weathering and soil erosion. At that time, the reduced temperature of the coastal waters in the smaller and shallower Bellerophon Gulf slowed the marine circulation, so waves became the primary factor in oxygenation of the coastal marine waters. Under these stressed conditions, the repeated flows from the land of slightly acidic freshwater significantly increased the pCO_2 of the marine water. Corrosion of the carbonate marine bottom decreased seawards from the shore, becoming imperceptible near the lower foreshore-offshore transition. However, the lower pH and lower temperature had a devastating effect on organisms with carbonate skeletons living at a depth of less than 30–50 m in a marine belt 10–50 km wide. In fact, the estimated mass mortality (E1) of genera across the BWB (cf. Farabegoli et al. 2007), based on a database compiled from various authors, ranged from 60 to 70%. However, if the estimate is based on the disappearance of carbonate-skeleton biomass, the mass mortality exceeded 90%.

In conclusion, it seems reasonable to propose the hypothesis that the ultimate cause of the major devastation was very probably a millennial perturbation of the atmosphere, mediated by local terrestrial and marine physiographic changes. The major mass mortality was isochronous at regional scale but was probably slightly diachronous (millennial) at the global scale. In fact, the fall and rise in sea level can be reasonably considered isochronous at global scale, although the intensity of destabilisation of the marine ecosystems depended on many other factors including temperature change, the range of the local vertical excursion of sea level with respect to physiography of the marine bottom, the intensity and duration of acid rains, the size of the continental area, the pattern of the continental hydrographic network, the distance of biota from the shoreline, the depth of the sea and so on.

The 50–300 cm succession that overlies the BWB and comprises the PTB was deposited during a period of global warming. Based on the parallel conformation of the two Triassic conodont biozones (*parvus* and *staeschei* zones)—geochronologically well-constrained in Meishan (Mundil et al. 2001)—with the same two biozones in Bulla (Perri and Farabegoli 2003; Farabegoli et al. 2007), the duration of this short interval ranges from 15 to 30 ky. This interval was at least 2–3 times longer than the previous fall-/low-stand, and the warming was more intense than the cooling, so the maximum retreat of the shoreline exceeded 10–20 km landwards at the top of the Bulla Member transgression. The rapid increase in marine carbonate precipitation (oolites) is clear evidence of carbonate-oversaturated shallow-marine water. The coeval disappearance of Permian taxa was nearly complete (the last were red algae) and was probably the result of extremely rapid climatic changes towards abnormally high temperatures. Therefore, the rapid cyanobacterial conquest of the marine bottom is, ex post, an expected result. In sub- to intertidal areas around the oolite-production sites, chemical and biochemical sedimentation overcame the subsidence, and the vertical stromatolites grew upwards in domal to columnar shapes. Oolite storm layers, covering the cyanobacterial films and permanently excluding them from the light, were primary limiting factors in distribution of microbialites. Instead, in the relatively deeper (5–10 m) marine environments, far from the oolite-production sites, the cyanobacterial soft films aggraded slowly. Rare attempts at domal and columnar growth were rapidly aborted, probably because of scarcity of light on the bottom. In this way, the sedimentation did not compensate for subsidence, and the sea depth gradually increased. Only a few small foraminifers that had previously adapted to these deeper conditions survived, but they disappeared immediately afterwards. Because the lower foreshore bottom was predominantly oxic to slightly dysoxic (a few short anoxic events probably post-dated this interval), the ultimate cause of their disappearance would have been the concurrent increase in sea depth, reduced light on the bottom and warmer water. The very short food chain was a direct consequence of food scarcity connected with the abrupt mass mortality. It comprised cyanobacteria, rare red algae, rare small benthic cone-shaped foraminifers, calcispheres of unknown origin, small suspension feeders (brachiopods and bivalves, most of which were attached to the hard microbialitic substrate, although some had an infaunal life), fish and conodonts and small grazing–browsing gastropods such as *Holopella* sp. and *Bellerophon waceki*, which were probably at the top of this food chain. Changes in the brachiopod and bivalve assemblages did not follow a clear stratigraphic order, but rather alternated with each other. This suggests relatively steady environmental conditions modulated by moderate environmental changes, different substrates, marine currents of differing strengths, warming and/or pH oscillations and so on. The sum of all these factors can satisfactorily explain the second (E2) and third (E3) phases of mortality, E3 occurring just above the PTB.

Finally, the lack of rapid recovery during the short BWB–PTB interval or immediately after it could simply be due to persistence of stressed environmental conditions along the northern margin of Palaeotethys after the E1 mass mortality event. A second plausible explanation is that the western Palaeotethys acted as the last refuge for shallow marine biota in the Northern Hemisphere, from the base of the Bulla Member (Farabegoli et al. 2007), without organisms migrating from the east.

Local and/or Global Causes of the End-Permian Extinction

Every hypothesis about the causes of the global environmental change leading to the end-Permian extinction will be credible if each accord with the acquired data and, furthermore, give satisfactory answers to three fundamental questions: (1) Was the end-Permian extinction a single worldwide isochronous event or, on the contrary, was the end of the Permian palaeo-world caused by the sum of a few regional, more or less diachronous short-term (centennial–millennial) mass-mortality phases?; (2) If the latter is true, how could a single global process have been split in time and space and caused local continental and marine mass-mortality events, and where were the refugia for future Lazarus taxa located? Compelling answers to these questions are dependent on the third question, which is at present much debated: (3) Why the 5 My delay in recovery of the biosphere?

The first question concerns our capacity for discriminating a fine-tuned time-scale in which the duration of each subdivision is essentially comparable with the physical, chemical or biological short-term events recognised in the *Hi. praeparvus–Hi. parvus* interval, that is, a time-interval of 200–300 ky. Since the ca. 70–90 m depositional succession, we analysed in the Southern Alps highlights (incompletely) four depositional cycles, three to four mass extinctions and one to two chemical events (partly coincident), the necessary time unit does not exceed 10 ky. This result cannot be discriminated using present-day technologies. In fact: (1) The duration of the shortest time interval detected by geochronometric methods is as long as the entire succession here analysed; (2) the palaeomagnetic stratigraphy in the Southern Alps has poor time resolution and, furthermore, its validity at global scale is debated and (3) the stable isotope stratigraphy in the Southern Alps seems to be derived from local factors superimposed on global factors, so it is not possible to detect a common signal of less than ca. 100 ky.

Thus, *in primis* we can reconstruct a few local stratigraphies based on well-known physical and biological events, and then we can try to integrate these different *tesserae* in a global framework of relative-time-scale events.

The first important *tessera* in the framework is certainly the FAD of *Hi. parvus,* perfectly defined in the Bulla section, and roughly identified in other sections of the Southern Alps. It ranges from 130 cm (in the distal and in the more proximal marine sections) to ca. 250 cm (in intermediate marine sections) above the BWB. Thus, in order to reduce errors, it is better to compute the short-term events, starting from the unequivocal identification of the PTB, and then proceed downwards and upwards.

As concerns shallow marine environments in the western Palaeotethys, we have shown that the atmospheric changes produced acid rains that flowed seawards from land, thus changing profoundly the pCO_2 of marine waters; the intensity of corrosion on the sea bottom seems to decrease from the shoreface seawards and became imperceptible at the upper off-shore boundary. So, it can be expected that the signature of the marine geochemistry and consequent bioevents (comprising the mass mortality) must be different in number and magnitude depending on, among other environmental factors, the distance from the shoreline, the marine topography and the marine currents. In other words, it is extremely probable that the end-Permian mass extinction in the shallow marine realm was the final result of a multi-phased phenomenon. The duration of each regional mortality event was from a few millennia to a few dozen millennia in length; the strongest mass-mortality event in a region (i.e., the bio-event locally described as the end-Permian extinction) shows, at global scale, a diachroneity of a few dozen millennia.

So, the only way to obtain worldwide alignment of the short-term bioevents discriminated in very distant regions is to delineate a framework of short-term physical events—specifically the local shallow-marine sedimentary cycles or sequences—and to fix them with scattered good-quality bio-, chrono- and magnetostratigraphic data. Farabegoli et al. (2007) applied this method and, based on the FAD of *Hi. parvus*, correlated the peritidal sample BU12B in the Bulla section with the lower shoreface-offshore Bed 27c in the Meishan D section. Moreover they pointed out that, below this *datum*, the two sections contain the top surfaces of two coeval very short-term depositional cycles: (1) The regressive top of the Bulla Member and the U2 erosional surface in the Bulla section were aligned with the regressive silty Bed 26 in Meishan D section; (2) The regressive top of the Ostracod Unit and the U1 surface in the Bulla section were aligned with the regressive interval Beds 24e in the Meishan D section. Tong et al. (2007) referred climate changes in southern China to a global cause, that is, to Sun–Earth orbital forcing. This alignment implies that the major mass mortality (i.e., *mass-extinction* A.A.) detected at the top (or 1.5 cm below) offshore bed 24e in the Meishan D section predates by one sedimentary cycle the major mass mortality (mass-extinction phase E1 of Farabegoli et al. 2007) in the shallow marine western Tethys. Farabegoli et al. (2007) came to the conclusion that the western Palaeotethys acted like a refugium. Following the acid-rain-based mechanism of mass mortality we proposed previously, and assuming that the Meishan D section was very probably in a more distal position with respect to the South China palaeoshoreline than Bulla to the Southern-Alps Permian land, the opposite result would be expected. So, others factors were involved in the process. We suggest the major and most probable one was relative proximity of the Eimeshan Trap causing immensely intense acid rains (cf. Liang 2002), increasing the pCO_2 in a wider and deeper belt around the land and culminating with the volcano-derived ash-bed 25. Moreover, the sharp contact ash-bed 25 (silt bed 26) doesn't offer a direct and unequivocal confirmation of corrosion processes of the marine bottom. Instead, clear evidence of intense and repeated corrosion events on the marine bottom are recorded by the overlying bed-set 27. In fact, the top surfaces of beds 27a and 27b in the Meishan D section (the latter bed immediately predates the PTB) show clear evidence of corrosion and submarine microbialitic aggradation, plus bioturbation. Similar intrastratal structures characterise the upper surface of beds 27c and 27d (the first corresponding to the FAD of *Hi. parvus*). The presence of the same intrastratal structures in the North Pingdingeshan succession ca. 200 km south-west of Meishan is clear evidence of the regional extent of increased pCO_2, up to the *isarcica* Zone.

In the Southern Alps, individual submarine erosion surfaces are lacking across the PTB. A plausible explanation is that, at that time, the temperature of the shallow western Palaeotethys was exceptionally high, favouring deposition of (a) biotic carbonate marine precipitates, usually as oolites. Rapid aggradation of the oolitic banks located many km away from land ("oolite factories") elevated their tops to or above the sea surface. The oolites underwent repeated leaching in the intertidal zone. This was caused by seasonal episodes of acid rain on the sea, each episode being rapidly buffered by marine water oversaturated in calcium carbonate, locally alternating with selective processes of dolomitisation during early burial. Just after the PTB, and well before the end of the Early Induan transgression, the E3 phase of mass mortality corresponds with the complete disappearance (?extinction) of red algae, that is, the last Permian-type taxa. Increased flows of acid freshwater and fine-sized terrigenous input probably contributed to a schizohaline coastal-marine environment, inhibiting the premature, but very improbable, recovery of western Palaeotethys from the east.

Bioturbation, present during the *parvus* Zone in South China and along the north-west margin of Palaeotethys (Bukk Mountains, Slovenia and Southern Alps), indicates that the marine waters, including the deeper offshore bottom, were at least partly oxygenated. Therefore, it seems very unlikely that bottom-up transport of anoxic–euxinic deep marine waters could have been the primary cause for the end-Permian and Early Triassic mass-mortality events along this coastal margin. Moreover, the hypothesis that repeated releases of methane hydrate were the ultimate cause for the repeated mass mortality events in Palaeotethys cannot be confirmed by identification of multiple negative carbon-isotope excursions at millennial scale. It is difficult to imagine how such a mechanism could have been activated.

Further evidence for possible time correlation of short-term climatic events, such as sea-level changes and associated chemical or biotic events, comes from the comparative analysis of peri-Gondwana sections located in the Southern Hemisphere along the margin of the Neo- (or Meso-) Tethys, coeval Western Tethys and the equatorial south China and Iran sections (Kozur 2007). Farabegoli et al. (2007) suggested that a more humid period produced the terrigenous input of the Bavaro Member in the Bellerophon Gulf, culminating with the U1 surface; they aligned this unit with the terrigenous, brachiopod-rich upper Chhidru Formation in the Salt Range. The upper Chhidru Formation has a *Colaniella*-dominated fauna, whereas the upper part of the formation is characterised by some typical cool-water conodonts and some bellerophontid archaeogastropods. The uppermost unit of the Chhidru Formation, that is, the White Sandstone Unit, was deposited in a coastal terrigenous tidal flat (Mertmann 2003). Its brachiopod fauna consists of palaeoequatorial warm-water elements with an admixture of temperate genera. Moreover, fusulinids and small species of foraminifers disappeared in the uppermost part of the White Sandstone Unit; all these data can be interpreted in terms of a falling sea followed by a low-stand during a cool period. This period can be aligned with the shallowing upper part of the Ostracod Unit in the Southern Alps, culminating with the disconformable surface U1.

Progressive harshening of environmental conditions in the Southern Alps may explain the upward increase in higher-plant debris and fungi, and increasing corrosion of palynomorphs and vitrinite fragments in proximal marine sediments. This can be interpreted as due to the tree canopy on land undergoing rapid impoverishment. This interpretation is supported by many data of different kinds. The ghosts of gypsum crystals dispersed in the carbonate-terrigenous sediment of the upper Ostracod Unit may be interpreted as the early diagenetic product of two non-contrasting processes: (1) Circulation of hyperhaline waters connected with evaporation of marine waters on tidal flats; (2) circulation of marine waters enriched in SO_4 by oxygenation of acid-rain-derived H_2S. Liang (2002) gives mineralogical evidence that the gypsum at the base of Bed 25 in Meishan was derived from oxygenation of hydrated sulphuric acid caused by heavy acid rain and marine acidification—produced by volcanic eruption or, alternatively, a bolide impact. Newton et al. (2004) argued for a rapid positive shift of sulphur isotope values in the upper 6 m of the Ostracod Unit in the Siusi Section (a few km north of the Bulla section). Kaiho et al. (2002) detected this positive shift of the sulphate/sulphur isotope ratio a few cm below Bed 25 at Meishan and in Balvany, Hungary; Kaiho et al. (2006) explained this as a very complicated, polyphase phenomenon: sulphur injection(s) from the mantle causing enrichment of sulphate in the (oxygenated) ocean; later, the sulphate was reduced to sulphide in a largely anoxic stratified ocean. This is clearly a very improbable mechanism for the peritidal signature of the upper part of the Ostracod Unit in the Southern Alps. Therefore, in our opinion, oxygenation of hydrated sulphuric acid caused by repeated, strong volcano-derived acid rain is a candidate for having had a fundamental role at global scale.

In the Southern Alps, the total duration of this uncondensed depositional cycle is estimated as ca. 100 ky whereas the duration of the fall-/low-stand tracts was between 30 and 50 ky, that is, a time span significantly long to greatly change the chemistry of the atmosphere and ocean. Zhang et al. (1996) suggested less than 25 m of shallowing at Meishan. Based on architecture of the depositional prism in the Southern Alps and the average depositional rate of the uppermost Bellerophon Formation, we estimated a ca. 10 m fall in sea level. It is notable that the peculiar harsh environmental conditions in the Southern Alps masked the rate of mortality of marine taxa (many of them reappeared in the Bulla Member). It was about 50–60% in equatorial Meishan (Kaiho et al. 2006), but was moderate in the Southern Hemisphere Salt Range sections. These are clear evidence that the end-Permian extinction was not a single event, but rather the final result of a few regional mortality events, each diachronous at centennial–millennial scale, rapidly succeeding in various sectors of the Palaeotethys.

Farabegoli et al. (2007) aligned the fall-stand/low-stand tracts truncated by the sharp surface U2 in the Southern Alps with Bed 26 in Meishan D, a bed rich in fine terrigenous components and shallow marine fossils (brachiopods, bryozoans, ostracods and foraminifers; cf. Kaiho et al. 2006); they aligned it also with the erosional surface at the base of the lower Kathwai Member in the Salt Range (Pakistani-Japanese Group 1985). Shen et al. (2006) argued for a slightly earlier beginning of this physical event at the base of the ash-derived clay bed 25 in Meishan, but maintained alignment of the sequence with the base of the lower Kathwai Member. One of the authors (EF), using a revised version of the physical-based 3D software SIMSAFADIM (Bitzer and Salas 2001), and hypothesising a constant sedimentation rate of 0.5 mky^{-1} and ca. 20 ky for deposition of the Bulla Member, came up with a value of ca. 5 m sea-level rise followed by a 12 m fall-stand over 2–3 ky and, then, a long low-stand tract for 5 ky. By using the same software, the duration of the rising-/high-stand tract bracketing the U2–FAD of *Hi. parvus* in the Southern Alps was about 12 ky long and its magnitude was ca. 15 m. These values were probably very similar in both hemispheres of the Palaeotethys realm. By comparing the Southern Alps and coeval Salt Range successions, a few similarities and local differences emerge. For example, the brachiopods in the basal Kathwai Member are typical of warm seawater (cf. Shen et al. 2006 with references), as it was for the oolitic, shallow marine, lowermost Tesero sea; both environments experienced mass mortality. In particular, three fusulinid species (*Nankinella* sp., *Codonofusiella* sp. and *Reichelina* sp.) disappeared at that time in the Salt Range (Mertmann 2000) but, by using the cycle boundary (fall-stand tract) as a time marker, it is possible to affirm that the disappearance of *Reichelina* sp. predates their disappearance in the Southern Alps by a few millenia, whereas the range of *Nankinella* sp. is the same; *Codonofusiella* was not present in the Southern Alps (cf. Farabegoli et al. 2007). So, diachroneity of disappearances occurred at millennial scale, probably closely tracking regional environmental and facial changes.

As regards the exact location and timing of paraconformable–disconformable boundaries of the shallow-marine, short-term sequences bracketing the end-Permian extinction interval, Shen et al. (2006) detected two SB (separating three marine sequences) predating the FAD or FO of *Hi. parvus* in the peri-Gondwana Kashmir, Tulong and Qubu sections. They aligned these SB with the two SB detected at Meishan and the Salt Range—these we discussed and slightly modified above. Unfortunately, the sequence stratigraphy they used in describing the environmental evolution is clearly an oversimplification (see Fig. 17B in Shen et al. 2006), and thus cannot be used for time correlation and understanding the dynamics of rapid worldwide events. Likewise for the condensed shallow-marine section at Selong where the erosional surface at the base of a caliche bed with some "marine influence" seems to be the younger SB masking the first SB, but different stratigraphic interpretations are possible. Thus, in our opinion, these peri-Gondwana areas need more intense study.

Lehrmann et al. (2005) described a few shallow-marine sequences (Dijang, Rungbao and Heping sections) from the carbonate Great Bank of Guizhou (South China). A sharp contact, with no evidence of sub-aerial erosion or diagenesis, separates the Late Permian Wujiaping Formation from overlying calcimicrobial framestone of open-marine origin lacking recognizable Permian macrofossils. At Heping, *Hi. parvus* was found in a calcimicrobial framestone 65 cm above the erosional surface. They interpreted these as continuous sections, in spite of the erosional surface, comparable with the U2 surface in the Southern Alps. Whereas the echinoderms and articulate brachiopods in the calcimicrobial framestone recall those of the Lower Tesero around the PTB, in our opinion, the upper metre of the Wujiaping Formation in the Dajiang and Heping sections, composed of foraminifers and fragmented shell material, was a response to the shallowing tract, as was the peritidal facies underlying the U2 surface in the Southern Alps. The sedimentation rate of the interval between the erosional surface and the entry of *Hi. parvus* and of the *parvus* Zone is more or less 50% of the one at Bulla. Payne et al. (2007) confirmed that the previously described erosional surface occurred in a submarine setting and aligned it from Palaeotethys (South China) to shallow marine Neo- (Meso-) Tethys (Taskent section, Turkey) and the Panthalassa superocean (shallow marine, Takachiho section, Japan). But, in our opinion, the primary correlative criteria they utilised (i.e., "the truncation surface at the top of Upper Permian skeletal carbonates in each section, corresponding to the last occurrence of Upper Permian macrofossils") isn't "per se" a real time marker of the claimed chemical-biotic event. On the contrary, at the Taskent section, the thin oolite interval at the top of the Çekiç Dagi Formation is clear evidence that the pulse of high carbonate saturation predates (or is coeval with) the mass-mortality erosional surface. In other words, this was a diachronous "event", or the area suffered from more than one event, as happened in the Southern Alps.

In any case, given the successions are roughly constrained by the FAD of *Hi. parvus*, these Neotethys and Panthalassa superocean sections need more intense study. Payne et al. (2007) hypothesised that the heating of coals during emplacement of the Siberian Traps produced rapid release of thermogenic methane (cf. Vermeij and Dorritie 1996)—this unbuffered carbon release from sedimentary reservoirs of the deep ocean was viewed as responsible for global erosion followed immediately by a pulse of high carbonate saturation. It seems more probable to us that the precipitation of stromatolites, as in the Southern Alps, or of microbial limestone containing upward-growing carbonate crystal fans extending over a time span of up to 100 ky was caused by more than one event. For example, a possible alternative or concurrent cause for coeval decrease in ∂^{13}C would be isotopically light carbon from release of gas hydrates and oxidation of organic carbon associated with falling sea level and exposure of shallow shelves (Holser et al. 1989; Erwin 1993; Grossman 1994; Martin and Macdougall 1995). The association of fall-stand-∂^{13}C decrease with the erosion surface, and raise-stand-∂^{13}C decrease, accord with a model having concurrent mechanisms.

The detailed stratigraphy available today for equatorial regions indicates that the Spitsbergen sections need restudy. Wignall et al. (1996), incidentally, interpreted the Spitsbergen sections as "a complete record of the environmental and faunal changes during the crisis interval". They suggested "the mass extinction crisis in Spitsbergen is coincident with the extensive development of oxygen-poor conditions in the water column" and that there was alignment of the "timing and nature of the events with the crisis seen in lower latitude Tethyan settings". Unfortunately, at least the latter affirmation cannot be confirmed; the detail they used for their stratigraphic analysis was insufficient to discriminate millennial depositional intervals and to recognise sea-level changes of dozens of metres.

We hope to have given a first, partial answer to Hermann's (2004) doubts. The end-Permian extinction remains a profound enigma. The best procedure for finding a solution is to avoid any hypothetical or ideological approach at global scale, but to focus on short-term (millennial) regional problems, then to verify the proposed local solutions (*tessera*) through independent methods, to implement the stratigraphic timescale and, finally, to build, piecemeal, a global solution rigorously respecting time in addition to the relative timing of the events. This last condition is essential for avoiding confusion as to what were causative processes (i.e., kill mechanism of Knoll et al. 2007) and what were the results of the extinction event(s) (Krull et al. 2004), and, moreover, to explain why recovery was delayed for an additional 4 to 6 My (cf. Corsetti et al. 2005).

Conclusions

(1) A few sections, 25 to 40 m thick, bracketing the PTB boundary in the Southern Alps have been revised. The duration of the investigated interval is 100–200 ky. The depositional environments range from continental through terrigenous-carbonate peritidal to shallow-marine shoreface and foreshore, from metazoan-rich to metazoan-poor.

(2) Four short time-transgressive-regressive sedimentary cycles and four rapid mass-extinction phases (E) have been described. The first three cycles occurred in the Late Permian Lower and Upper *praeparvus* zones, the younger cycle during the Early Triassic *parvus* Zone.

(3) Sedimentary Cycle 1 includes two members (Bavaro Member and Ostracod Unit) of the Gardena and Bellerophon Formations; its duration was ca. 100 ky. The transgressive tract is locally represented by continental sandstone whereas the high-stand tract is a fully marine, metazoan-rich, high-energy upper foreshore facies or open to restricted fossil-poor lagoonal facies; the fall- and low-stand tracts, a few metres thick, include very shallow marine to peritidal microbialites alternating with fine-grained terrigenous sediments with gypsum ghosts. Along the margin of the Bellerophon Sea, Cycle 1 ends with the erosional subaerial unconformity U1; it changes distally to a paraconformable surface. The estimated sea-level drop is in the order of 10 m. The cycle is interpreted as a humid warm period changing to a relatively cold and semiarid period, 10–20 ky interval; the latter is interpreted as a period of acid rains (probably because of volcano-derived H_2S outpourings into the atmosphere) discharged into the Bellerophon Gulf from the neighbouring land. We therefore suggest the ultimate cause of E0 in Southern Alps was cold, semiarid, acid-rain conditions concurrent with cooling and rapid reduction of the shallow coastal area.

3.1 At global scale, this relatively cold, low-stand phase aligns with the low-stand and "extinction-event" of Bed 24e in the eastern Tethys (GSSP Meishan D section). In the Southern Alps (western Palaeotethys), the facies changes upwards to stressed peritidal conditions masking E0. This cooling fall-stand is displayed at various sites along the shallow-marine Gondwana margin, but the magnitude of the mass extinction is markedly reduced compared with Meishan. So, E0 may in fact be a diachronous, multi-phase mass-mortality event. It, at first, profoundly affected the eastern Palaeotethys, whereas the western Palaeotethys and the Gondwana margin continued for a few millennia as refugia. Palaeogeographic and palaeoenvironmental factors, comprising distance from source area and nature of the stressing factors, combined to modulate the intensity and duration of this event in space and time, so that in some areas it appears to have been the "extinction event", whereas in other areas many Permian-type taxa reappeared immediately after the supposed extinction event.

(4) Sedimentary Cycle 2 is 50–150 cm thick; it corresponds to the Bulla Member of the Bellerophon Formation; and its duration was less than 20 ky. The high-stand tract is shoreface and foreshore facies rich in Permian-type metazoans. It was followed by a millennial sea-level drop whose magnitude is estimated at a dozen metres. Over most of the Bellerophon Sea, the fall-stand tract was truncated by the sub-aerial erosional surface U2, locally overlain by a thin brownish-red palaeosoil. The U2-palaeosoil corresponds to the boundary between the Bellerophon and Werfen formations; it can be traced down to lower foreshore environments as a submarine dissolution surface. The mass-mortality event E1 is mostly coeval with and brackets the U2 surface. The E1 event was

responsible for disappearance of about 90% of the metazoan biomass in the western Palaeotethys. The destruction (mass mortality) of the forest canopy on land is coeval with the sea-level drop and the marine mass-mortality event. The cycle is interpreted as a humid warm 10–20 ky period passing up to a temperate interval lasting a few ky. We suggest the ultimate causes of the E1 in the Southern Alps were the cooling climate coinciding with repeated episodes of intense acid rains that devastated most of the trees on land and, by increasing the pCO_2 along the margins of the Bellerophon Gulf, most of the coastal benthic fauna. In other words, the acme of the mortality event in the Southern Alps was mostly of top-down type. Unfortunately, given the unknown palaeobathymetry of the gulf, it is impossible to estimate the reduction of the shallow coastal area due to retreat of the coastline and the role of this in the marine mortality event.

4.1 At global scale, this relatively cold, low-stand phase and the "extinction-event" E1 are aligned with the low-stand of Bed 26 in the Eastern Tethys (Meishan D section), where it caused a relatively low intensity "mortality event". This cooling-fall-stand has been discriminated in various sites along the shallow-marine Gondwana margin, characterised everywhere by mass-extinction events of magnitude lower than that in the Southern Alps. Given that the magnitude of this mortality phase differs greatly in different marine areas, it is reasonable to conclude that local environmental factors were fundamental in modulating the biotic response to the overall global cause.

(5) The subsequent 100–260 cm Cycle 3 consists of the Lower Tesero Member. It extends from the very base of the Werfen Formation just above the PTB; its duration was less than 20 ky. The sea level rose about 15 m; the fall was less than 10 m. This implies that at the end of the cycle only very marginal areas of the Southern Alps returned to the initial peritidal conditions. On land, the tree cover was probably almost totally depleted; such conditions persisted many tens of thousands of years. A minor percentage of Permian-type marine taxa survived the acme of mortality, reappearing in the ventilated shoreface at the beginning of the transgression. Nevertheless, most of these disappeared in a few millennia, associated with marine waters oversaturated with $CaCO_3$ (which determined oolite growth) alternating with short acid fluxes. The "tail" of mass-mortality E2 is lacking in the less-ventilated lower foreshore because of persisting suboxic to dysoxic conditions and rapid colonisation connected with the microbialites. The duration of mass-extinction event E2 bracketing the BWB was ten thousand years or less. Available data are consistent with increased warming, having been an important factor controlling the "mortality tail" but do not let us validate or rule out other local or general concurrent causes (from slow down of the local marine currents to up-welling of anoxic oceanic waters—from the Palaeo- or Neo-Tethys). The very shallow marine oolite bank and stromatolites in the Southern Alps record the last signal of previous refugial areas and, at the same time, the first weak recovery signal. The rapid demise of taxa in the Southern Alps truncated the food chain at the level of small suspension feeders ("Lilliput" brachiopods and/or bivalves). It is not clear if the small suspension feeders disappeared during the shortfall tract or during the transgressive tract of the following cycle.

5.1 The coeval Beds 27a-c of Meishan were deposited in a deeper and distal environment referable to the lower foreshore and marine shelf, lacking both a clear record of anoxic conditions and a signal of recovery of magnitude comparable with the one in the Southern Alps; the latter signal was already present in many sites along the Gondwana margin.

(6) The 100–300 cm Cycle 4 consists of the middle-upper part of the Lower Mazzin Mamber and, locally, the lower part of the Upper Tesero Member; its duration is not well constrained but lasted more than 30 ky. Extinction event E3 is recorded in the shallow marine, transgressive lower part of the cycle; it affected mostly solenoporacean algae. We therefore suggest that increased warming was the factor controlling this last mortality event in the Southern Alps but cannot validate or deny other local or general concurrent causes.

6.1 The nearly coeval Bed 27d of Meishan D section was deposited in the lower foreshore and marine shelf environment with suboxic to dysoxic conditions at the bottom.

A more humid period started in the Southern Alps during the upper *parvus* Zone, continuing upwards through the *lobata, staeschei, isarcica* and *aequabilis* zones with deposition of carbonate mud and fine terrigenous sediments in schizohaline foreshore to shelf marine environments. Gradual transition to the Andraz Member of the Werfen Formation indicates a fall- and low-stand period with semiarid conditions prevailing in peritidal environments.

(7) The examined Southern Alps succession is reasonably constrained by palaeomagnetic data, by stable isotope stratigraphy and geochronology, and through correlation with Meishan, but in many cases available data are insufficient for global-scale precise alignments of individual depositional cycles whose duration was a few tens of thousands of years. Future high-precision sedimentologic and biostratigraphic studies may enable confident discrimination of the cycle and sequence stratigraphy at global scale.

(8) All the data show that the end-Permian extinction was in fact the sum of individual mortality events affecting specific palaeogeographic regions and palaeoenvironments and that these were of different magnitude and timing and were at scales of a few millennia or tens of thousands of years. In places, the mortality events on land were coeval with those in the surrounding shallow seas but, at global scale, the end-Permian extinction events on land differ significantly in magnitude and timing from corresponding events in the sea. Therefore, even though one may consider that volcanism was probably the primary, long-term factor, very different kill mechanisms were responsible for the "diachronous" mortality events in the various regions. We suggest that the main cause of the mortality events in the benthic faunas and floras in near-shore marine environments was conjunction of a cool climate (temperate at the equator) with repeated episodes of acid rains (that markedly increased the sea-water pCO_2) turning rapidly to very warm climates with acid rains and marine oversaturation in $CaCO_3$.

Acknowledgements We are deeply indebted to John A. Talent for careful and critical revision of the manuscript and for fruitful suggestions. Yu-ping Qi (Chinese Academy of Sciences, Nanjing) helped in collecting and made available chemical data on samples from the Bulla section. The services of three University of Bologna people were indispensable for completing this manuscript: Fabio Gamberini for producing numerous large-format thin sections, Claudio Dalmonte for photos of the thin sections and Paolo Ferrieri for photos of the conodonts. Our research was funded by RFO (Ricerca Fondamentale Orientata) 2007–2009 grants to E Farabegoli and MC Perri.

References

Assereto R, Bosellini A, Fantini Sestini N, Sweet WC (1973) The Permian–Triassic boundary in the Southern Alps (Italy). In: Logan A, Hills LV (eds) The Permian and Triassic systems and their mutual boundary. Canadian Soc Petrol Geol Mem 2:176–199

Baud A, Holser WT, Magaritz M (1989) Permian-Triassic of the Tethys: carbon isotope studies. Geol Rundschau 78: 649–677

Beretta C, Cimmino F (1998) La serie di Seres. (Val Badia); studio petrografico e geochimico del limite Permo-Trias. Mem Soc Geol Italy 53:359–385

Beretta C, Radrizzani S (2001) The Bellerophon-Werfen boundary in the western Dolomites (Italy) – petrographical studies and a new interpretation. Ann Mus Civ Nat Brescia 25:125–131

Beretta C, Cimmino F, Cirilli S, Neri C, Nicora A, Perri MC, Pirini Radrizzani C, Posenato R, Radrizzani S (1999) The P–T boundary in the Tesero section, western Dolomites (Trento). In: Cassinis G et al (eds) Stratigraphy and facies of the Permian deposits between Eastern Lombardy and the Western Dolomites. International Field Conference on "The Continental Permian of the Southern Alps and Sardinia (Italy). Regional Reports and General Correlations" (Brescia, 15–27 Sept 1999), Field Trip Guidebook, pp 90–109

Bitzer K, Salas R (2001) Simulating carbonate and mixed carbonate-clastic sedimentation using predator-prey models. In: Merriam D, Davis JC (eds) Geologic modeling and simulation: sedimentary systems. Kluwer Academic/Plenum, New York, pp 169–204

Bosellini A (1964) Stratigrafia, petrografia e sedimentologia delle Facies carbonatiche al limite permiano-trias nelle Dolomiti occidentali. Mem Mus di Storia Naturale della Venezia Tridentina 15(2):106

Bosellini A, Hardie LA (1973) Depositional theme of a marginal marine evaporite. Sedimentology 20:5–27

Botthjer DJ, Clapham ME, Fraiser ML, Powers CM (2008) Understanding mechanisms for the end-Permian mass extinction and the protracted Early Triassic aftermath and recovery. GSA Today 18(9):4–10

Brandner R (1988) The Permian–Triassic boundary section in the Dolomites (Southern Alps, Italy), San Antonio section. Bet Geol Bundesanst (Wien) 15:49–56

Brandner R, Donofrio DA, Krainer K, Mostler H, Nazarow MA, Resch W, Stingl V, Weissert H (1986) Events at the Permian–Triassic boundary in the southern and northern Alps. Italian IGCP project 203, Field Conference on Permian and Permian–Triassic Boundary in the South-Alpine Segment of the Western Tethys, and Additional Regional Reports (Brescia, 4–12 July 1986), Abstract Book, p. 15

Broglio Loriga C, Cassinis G (1992) The Permo-Triassic boundary in the southern Alps and in adjacent Periadriatic regions. In: Sweet CS, Yang Z, Dickins JM, Yin H (eds)

Permo-Triassic events in the Eastern Tethys. World and Regional Geology, Cambridge, pp 78–97

Broglio Loriga C, Masetti D, Neri C (1983) La Formazione di Werfen (Scitico) delle Dolomiti occidentali: sedimentologia e biostratigrafia. Riv Ital Paleont Strat 58:501–598

Broglio Loriga C, Conti MA, Farabegoli E, Fontana D, Mariotti N, Massari F, Neri C, Nicosia U, Pasini M, Perri MC, Pittau P, Posenato R, Venturini C, Viel G (1986a) Upper Permian sequence and P/T boundary in the area between Carnia and Adige valley. Italian IGCP Project 203, Field Conference on Permian and Permian–Triassic Boundary in the South-Alpine Segment of the Western Tethys, and Additional Regional Reports (Brescia, 4–12 July 1986), Field Guidebook, pp 23–28

Broglio Loriga C, Neri C, Pasini M, Posenato R (1986b) Stop 2—The upper Bellerophon Fm. and the P/T boundary, northern slope of the Sass de Putia Mt. Italian IGCP Project 203 (ed) Field Conference on Permian and Permian–Triassic Boundary in the South-Alpine Segment of the Western Tethys, and Additional Regional Reports (Brescia, 4–12 July 1986), Field Guidebook, pp 88–94

Broglio Loriga C, Neri C, Pasini M, Posenato R (1988) Marine fossil assemblages from Upper Permian to lowermost Triassic in the western Dolomites (Italy). In: Cassinis G (ed) Permian and Permian–Triassic boundary in the South-Alpine Segment of the Western Tethys, and additional regional reports. Mem Soc Geol Italy, vol 34, pp 5–44 [Imprint 1986]

Broglio Loriga C, Góczán F, Haas J, Lenner K, Neri C, Oravecz Scheffer A, Posenato R, Szabó I, Tóth Makk A (1990) The Lower Triassic sequences of the Dolomites (Italy) and Transdanubian mid-mountains (Hungary) and their correlation. Mem Soc Geol Padova 42:41–103

Buggisch W, Noè S (1988) Upper Permian and Permian–Triassic boundary of the Carnia (Bellerophon Formation, Tesero Horizon, northern Italy). In: Cassinis G (ed) Permian and Permian–Triassic boundary in the South-Alpine segment of the Western Tethys, and additional regional reports. Mem Soc Geol Italy, vol 34, pp 91–106 [Imprint 1986]

Chen ZQ, Kaiho K, George AD (2005) Survival strategies of brachiopod faunas from the end-Permian mass extinction. Palaeogeogr Palaeoclimatol Palaeoecol 224:232–269

Chen ZQ, Kaiho K, George AD, Tong J (2006) Survival brachiopod faunas of the end-Permian mass extinction from the Southern Alps (Italy) and South China. Geol Mag 143:301–327

Chen J, Beatty TW, Henderson CM, Rowe H (2009) Conodont biostratigraphy across the Permian–Triassic boundary at the Dawen section, Great Bank of Guizhou, Guizhou Province, South China: implications for the Late Permian extinction and correlation with Meishan. J Asian Earth Sci 36:442–458

Cirilli S, Pirini Radrizzani C, Ponton M, Radrizzani S (1998) Stratigraphical and paleoenvironmental analysis of the Permian–Triassic transition in the Badia valley. Palaeogeogr Palaeoclimatol Palaeoecol 138:85–113

Corsetti FA, Baud A, Marenco PJ, Richoz S (2005) Summary of Early Triassic carbon isotope records. Compte Rendue Palevol 4(Académie des Sciences 2005):473–486

Crasquin S, Perri MC, Nicora A, Wever P (2008) Ostracods across the Permian–Triassic boundary in the western Tethys: the Bulla parastratotype (Southern Alps, Italy). Riv Ital Paleont Stratigr 114:233–262

Ding M, Zhang K, Lai X (1996) Evolution of *Clarkina* lineage and *Hindeodus–Isarcicella* lineage at Meishan section, South China. In: Yin H (ed) The Palaeozoic–Mesozoic boundary candidates of the global stratotype section and point of the Permian–Triassic boundary. China University of Geosciences, Wuhan, pp 65–71

Dolenec M, Vokal B (2003) Carbon and sulfur isotope anomalies across the Permian-Triassic boundary (PTB) in W Slovenia. Goldschmidt Conference Abstracts: A81, Kurashiki, Japan

Dolenec T, Lojen S, Buser S, Dolenec M (1999) Stable isotope event markers near the Permo-Triassic boundary in the Karavanke Mountains (Slovenia). Geol Croat 52:77–81

Dolenec T, Lojen S, Ramovs A (2001) The Permian–Triassic boundary in western Slovenia/Idrijca Valley section: magnetostratigraphy, stable isotopes, and elemental variations. Chem Geol 175:175–190

Erwin DH (1993) The great Paleozoic crisis: life and death in the Permian. Columbia University, New York, NY, 327pp

Farabegoli E, Viel G (1982a) Litostratigrafia della Formazione di Werfen (Trias inferiore) delle Dolomiti Occidentali. L'Industria Mineraria 6:3–14

Farabegoli E, Viel G (1982b) Il Permo-Scitico delle Alpi Meridionali. Relazione finale Contratto C.R.E.S.T. 092.79.7 MPPI. Samim-Pertusola, 3.1–3.48

Farabegoli E, Perri MC (1998) Permian/Triassic boundary and Early Triassic in the Bulla section (Southern Alps, Italy): lithostratigraphy, facies and conodont biostratigraphy. Giorn Geol 3, 60, Spec Pub ECOS VII Southern Alps Field Trip, pp 292–311

Farabegoli E, Levanti D, Perri MC (1986) The Bellerophon Formation in the southwestern Carnia. Lavardet-Rioda and Casera Federata Section. In: Italian IGCP project 203 (ed) Field conference on Permian and Permian–Triassic boundary in the South-Alpine segment of the Western Tethys, and additional regional reports. 4–12 July 1986, Field guide-book, Brescia, pp 69–75

Farabegoli E, Perri MC, Posenato R (2007) Environmental and biotic changes across the Permian–Triassic boundary in western Tethys: the Bulla parastratotype, Italy. In: Yin H, Warrington G, Xie S (eds) Environmental and biotic changes during the Paleozoic–Mesozoic transition. Global Planetary Change 55:1–3, 109–135

Golonka J, Ford D (2000) Pangean (Late Carboniferous–Middle Jurassic) paleoenvironment and lithofacies. Palaeogeogr Palaeoclimatol Palaeoecol 161:1–34

Gorjan P, Kaiho K, Kakegawa T, Niitsuma S, Chen ZQ, Kajiwara Y, Nicora A (2007) Paleoredox, biotic and sulfur-isotopic changes associated with the end-Permian mass extinction in the western Tethys. Chem Geol 244:483–492

Grossman EL (1994) The carbon and oxygen isotope record during the evolution of Pangea: Carboniferous to Triassic. Geol Soc Am Spec Pap 288:207–228

Groves JR, Rettori R., Payne JL, Boyce MD, Altiner D (2007) End-Permian mass extinction of Lagenide foraminifers in the Southern Alps (Northern Italy). J Paleont 81:415–434

Hallam A (1989) The case for sea-level changes as a dominant causal factor in mass extinction of marine invertebrates. Phil Trans R Soc London, B 325:437–455

Hallam A, Wignall PB (1997) Mass extinctions and their aftermath. Oxford University Press, 309pp

Hallam A, Wignall PB (1999) Mass extinctions and sea-level changes. Earth-Sci Rev 48:217–250
Hermann S (2004) Sea level changes around the P/T boundary – the story behind, 10p. www.geo.tu-freiberg.de/oberseminar/os03_04/herrmann.pdf
Holser WT, Magaritz M (1987) Events near the Permian-Triassic boundary. Modern Geol 11:155–180
Holser WT, Schönlaub HP, Attrep M, Boeckelmann K, Klein P, Magaritz M, Orth CJ, Fenninger A, Jenny C, Kralik M (1989) A unique geochemical record at the Permian/Triassic boundary. Nature 337:39–44
Horaceck M, Brandner R, Abart R (2007a) Carbon isotope record of the P/T boundary and the Lower Triassic in the Southern Alps: evidence for rapid changes in storage of organic carbon. Palaeogeogr Palaeoclimatol Palaeoecol 252:347–354
Horaceck M, Richoz S, Brandner R, Krystyn L, Spötl C (2007b) Evidence for recurrent changes in Lower Triassic oceanic circulation of the Tethys: the $\delta^{13}C$ record from marine sections in Iran. Palaeogeogr Palaeoclimatol Palaeoecol 252:355–369
Horacek M, Povoden E, Richoz S, Brandner R (2010) High-resolution carbon isotope changes, litho- and magnetostratigraphy across Permian–Triassic Boundary sections in the Dolomites, N-Italy. New constraints for global correlation. Palaeogeogr Palaeoclimatol Palaeoecol 290: 58–64
Huchkriede R (1958) Die conodonten der mediterranen Trias und ihr stratigraphischer Wert. Paläontol Z 32:141–175
Jiang H, Lai X, Luo G, Aldridge R, Zhang K, Wignall P (2007) Restudy of conodont zonation and evolution across the P/T boundary at Meishan secion, Changxing, Zhejiang, China. Global Planetary Change 55:1–3, 39–55
Kaiho K, Kajiwara Y, Miura Y (2002) End-Permian catastrophe by a bolide impact: evidence of a gigantic release of sulfur from the mantle: comment and reply. Geology 30:856
Kaiho K, Chen ZQ, Kawahata H, Kajiwara Y, Sato H (2006) Close-up of the end-Permian mass extinction horizon recorded in the Meishan section, South China: sedimentary, elemental, and biotic characterization and a negative shift of sulfate sulfur isotope ratio. Palaeogeogr Palaeoclimatol Palaeoecol 239:396–405
Knoll AH, Bambach RK, Payne JL, Pruss S, Fischer WW (2007) Paleophysiology and end-Permian mass extinction. Earth Plan Sci Lett 256:295–313
Koeberl C, Farley KA, Pueckner-Ehrenbrink B, Sephton MA (2004) Geochemistry of the end-Permian extinction event in Austria and Italy: no evidence for an extraterrestrial component. Geology 32:1053–1056
Kolar-Jurkovsek T, Jurkovsek B (2007) First record of *Hindeodus–Isarcicella* population in Lower Triassic of Slovenia. Palaeogeogr Palaeoclimatol Palaeoecol 252:72–81
Korte C, Kozur H (2005) Carbon isotope stratigraphy across the Permian/Triassic boundary at Jolfa (NW-Iran), Peitlerkofel (Sas de Putia, Sass de Putia), Pufels (Bula, Bulla), Tesero (all three Southern Alps, Italy) and Gerennavár (Bükk Mts, Hungary). J Alp Geol 47:119–135
Korte C, Kozur HW, Joachimski MM, Strauss H, Veizer J, Schwark L (2004) Carbon, sulfur, oxygen and strontium isotope records, organic geochemistry and biostratigraphy across the Permian/Triassic boundary in Abadeh, Iran. Int J Earth Sci (Geol Rundschau) 93:565–581
Korte C, Torsten J, Kozur HW, Veizer J (2005) $\delta^{18}O$ and $\delta^{13}C$ of Permian brachiopods: a record of seawater evolution and continental glaciation. Palaeogeogr, Palaeoclimatol Palaeoecol 224:333–351
Kozur HW (1988) The Permian of Hungary. Z Geol Wiss 16:1107–1115
Kozur HW (1998) Some aspects of the Permian–Triassic boundary (PTB) and of the possible causes for the biotic crisis around this boundary. Palaeogeogr Palaeoclimatol Palaeoecol 143:227–272
Kozur HW (2007) Biostratigraphy and event stratigraphy in Iran around the Permian–Triassic Boundary (PTB): implications for the causes of the PTB biotic crisis. Global Planetary Change 55:155–176
Kozur HW, Pjatakova M (1976) Die Conodontenart *Anchignathodus parvus* n.sp., eine wichtige Leitform der basalen Trias.–Koninkl. Nederland Akademie van Wetenschappen-Amsterdam Proc, Ser B 79(2):123–128
Krull ES, Druke D, Kessel B, Yu YY, Li R (2004) Stable carbon isotope stratigraphy across the Permian-Triassic boundary in shallow marine carbonate platforms, Nanpanjiang Basin, south China. Palaeogeogr Palaeoclimatol Palaeoecol 204:297–315
Lai X, Wignall P, Zhang K (2001) Palaecology of the conodonts *Hindeodus* and *Clarkina* during the Permian–Triassic transitional period. Palaeogeogr Palaeoclimatol Palaeoecol 171:63–72
Lehrmann DJ, Payne JL, Enos P, Wei J, Yu YY, Bowring SA, Ramezani J, Orchard MJ, Montgomery P, Schrag DP, Knoll AH (2005) End-Permian extinction and biotic recovery: a complete record from an isolated carbonate platform, the Great Bank of Guizhou, south China. Albertiana 33:54–56
Leonardi P (1935) Il Trias Inferiore delle Venezie. Istituto di Geologia della Regia. Università di Padova Mem 11:11–36
Liang H (2002) End-Permian catastrophic event of marine acidification by hydrated sulfuric acid: mineralogical evidence from Meishan Section of South China. Chinese Sci Bull 47(16):1393–1397
Looy CV, Brugman WA, Dilcher DL, Visscher H (1999) The delayed resurgence of forests after the Permian–Triassic ecologic crisis. Proc Natl Acad Sci USA 96:13857–13862
Magaritz M, Bär R, Baud A, Holser WT (1988) The carbon–isotope shift at the Permian/Triassic boundary in the Southern Alps is gradual. Nature 331:337–339
Martin EE, Macdougall JD (1995) Sr and Nd isotopes at the Permian/Triassic boundary: a record of climate change. Chem Geol 125:73–99
Massari F, Neri C (1997) The infill of a supradetachment (?) basin: the continental to shallow-marine Upper Permian succession in the Dolomites and Carnia (Italy). Sed Geol 110:181–221
Mertmann D (2000) Foraminiferal assemblages in Permian carbonates of the Zaluch Group (Salt Range and Trans Indus Ranges, Pakistan). N Jb Geol Paläontol Mh 3:129–146
Mertmann D (2003) Evolution of the marine Permian carbonate platform in the Salt Range (Pakistan). Palaeogeogr Palaeoclimatol Palaeoecol 191:373–384
Metcalfe I, Nicoll RS (2007) Conodont biostratigraphic control on transitional marine to non-marine Permian–Triassic boundary sequences in Yunnan–Guizhou, China. Palaeogeogr Palaeoclimatol Palaeoecol 252:56–65

Mostler H (1982) Bozener Quarzporphyr und Werfener Schichten. In: Mostler H (ed) Excursionsführer zur 4. Jahrestagung der Österreichischen Geologischen Gesellschaft, Seis am Schlern Südtirol. H. Kowatsch, Innsbruck, pp 43–79

Mundil R, Metcalfe I, Ludwig KR, Renne PR, Oberli F, Nicoll RS (2001) Timing of the Permian–Triassic biotic crisis: implications from new zircon U/Pb age data (and their limitations). Earth Plan Sci Lett 187:131–145

Neri C (1982) Il paleoalto di Lavis e i suoi rapporti con la serie Permo-Scitica. Ann Univ Ferrara 8:21–27

Neri C (1999) The Permian–Triassic Boundary in the central Southern Alps: an overview. In: Cassinis G et al. (eds) Stratigraphy and facies of the Permian deposits between eastern Lombardy and the western Dolomites. Field Trip Guidebook, International field conference on "The Continental Permian of the Southern Alps and Sardinia (Italy), Regional Reports and General Correlations", 15–25 September 1999, Brescia, Italy, Pavia, pp 17–18

Neri C, Pasini M (1985) A mixed fauna at the Permian–Triassic boundary, Tesero section, western Dolomites (Italy). Boll Soc Paleontol Italy 23:113–117

Neri C, Posenato R (1999) Sedimentary and bio-chronostratigraphical aspects of the P/T boundary in the eastern Southern Alps. In: Cassinis G (ed) Stratigraphy and facies of the Permian deposits between Eastern Lombardy and the Western Dolomites; International congress on the continental Permian of the Southern Alps and Sardinia (Italy), Brescia, 15–25 Sept 1999, Field Trip Guidebook, pp 35–42

Neri C, Pasini M, Posenato R (1986) The Permian–Triassic boundary and the early Scythian sequence – Tesero section, Dolomites. In: Italian IGCP project 203 (ed) Field conference on Permian and Permian–Triassic boundary in the South-Alpine segment of the Western Tethys, and additional regional reports, Brescia 4–12 July 1986, Field guidebook, pp 123–128

Newton RJ, Pevitt E., Wignall PB, Bottrell SH (2004) Large shifts in the isotopic composition of seawater sulphate across the Permo-Triassic boundary in northern Italy. Earth Planet Sci Lett 218:331–345

Nicoll RS, Metcalfe I, Wang GY (2002) New species of the conodont Genus *Hindeodus* and the conodont biostratigraphy of the Permian–Triassic boundary interval. J Asian Earth Sci 20:609–631

Nicora A, Perri MC (1999) The P/T Boundary in the Tesero section, western Dolomites (Trento). 3.3 Bio-and chronostratigraphy: conodonts. In: Cassinis G et al. (eds) Stratigraphy and facies of the Permian deposits between eastern Lombardy and the western Dolomites. Field Trip Guidebook – 23–25 September, 1999. International field conference on "The continental Permian of the Southern Alps and Sardinia (Italy), Regional reports and general correlations" 15–25 September 1999, Brescia, Italy, Pavia, pp 97–100

Noé SU (1987) Facies and paleogeography of the marine Upper Permian and of the Permian–Triassic boundary in the Southern Alps (Bellerophon Formation, Tesero Horizon). Facies 16:89–142

Oddone M, Vannucci R (1988) PGE and REE geochemistry at the B-W boundary in the Carnic and Dolomite Alps (Italy). Mem Soc Geol Italy 34:129–139

Olson EC (1989) Problems of Permo-Triassic terrestrial vertebrate extinctions. Hist Biol 2:17–35

Orchard MJ, Krystyn L (1998) Conodonts from the lowermost Triassic of Spiti, and new zonation based on Neogondolella succession. Riv Italy Paleontol Stratigr 104:341–382

Pakistani-Japanese Group (1985) Permian and Triassic Systems in the Salt Range and Surghar Range, Pakistan. In: Nakazawa K, Dickins JM (eds) "The Tethys" her paleogeography and paleobiogeography from Paleozoic to Mesozoic. Tokai University, pp 221–312

Pasini M (1985) Biostratigrafia con i foraminiferi del limite Formazione a Bellerophon-Formazione di Werfen fra Recoaro e la Val Badia (Alpi Meridionali). Riv Ital Paleontol Stratigr 90(4):481–510

Payne JL, Lehrmann DJ, Follett D, Seibel M, Kump LR, Riccardi A, Altiner D, Sano H, Wei J (2007) Erosional truncation of uppermost Permian shallow-marine carbonates and implications for Permian-Triassic boundary events. Geol Soc Am Bull 121:957–959

Perri MC (1991) Conodont biostratigraphy of the Werfen Formation (Lower Triassic) Southern Alps, Italy. Boll Soc Paleontol Italy 30:23–46

Perri MC, Andraghetti M (1987) Permian-Triassic boundary and Early Triassic conodonts from the Southern Alps, Italy. Riv Ital Paleont e Stratigr 93:291–328

Perri MC, Farabegoli E (2003) Conodonts across the Permian–Triassic boundary in the Southern Alps. Cour Forsch-Inst Senckenberg 245:281–313

Posenato R (1988) The Permian/Triassic boundary in the western Dolomites, Italy. Review and proposal. Ann Univ Ferrara (N. S.), Sez Scienze della Terra 1:31–45

Posenato R (1989) Chronological and geographical distribution of the Fam. Comelicaniidae Merla, 1930 (Brachiopoda). Riv Ital Paleontol Stratigr 94:383–400

Posenato R (1990) La crisi biologica permo-triassica nelle Dolomiti occidentali. In: Robba E (ed) Proceedings of the fourth symposium on the ecology and palaeoecology of Benthic communities. Museo regionale di Scienze Naturali Torino, pp 69–93

Posenato R (1991) Endemic to cosmopolitan brachiopods across the P/Tr boundary in the Southern Alps (Italy). Proceedings of Shallow Tethys (3–23 Sept. 1990). Saito Ho-on Kai Spec Pub, vol 3. Toko Printing Co, Sendai, pp 125–139

Posenato R (1998) The gen. *Comelicania* Frech, 1901 (Brachiopoda) from the Southern Alps: morphology and classification. Riv Italy Paleontol Stratigr 104:43–68

Posenato R (2001) The athyridoids of the transitional beds between Bellerophon and Werfen formations (uppermost Permian, Southern Alps, Italy). Riv Italy Paleontol Stratigr 107:197–226

Posenato R (2008) Global correlations of mid Early Triassic events: the Induan/Olenekian boundary in the Dolomites (Italy). Earth-Sci Rev 91:93–105

Posenato R (2009) Survival patterns of macrobenthic marine assemblages during the end-Permian mass extinction in the western Tethys (Dolomites, Italy). Palaeogeogr Palaeoclimatol Palaeoecol 280:150–167

Posenato R, Prinoth H (2004) Orizzonti a nautiloidi e a brachiopodi della Formazione a Bellerophon (Permiano Superiore) in Val Gardena. Geo Alp Innsbruck-Bolzano 1:71–85

Racki G, Wignall PB (2005) Late Permian double-phased mass extinction and volcanism: an oceanographic perspective. In: Over DJ, Morrow JR, Wignall PB (eds) Understanding Late Devonian and Permian–Triassic biotic and climatic events: towards an integrated approach. Elsevier, Oxford, pp 263–297

Ramovs A (1986) Marine development of the Zazar Beds at the lowermost Scythian beds. In: Italian IGCP project 203 (ed) Field conference on Permian and Permian-Triassic boundary in the South-Alpine segment of the Western Tethys, and additional regional reports. 4–12 July 1986, Brescia, Field guide-book, pp 39–42

Rampino MR, Prokoph A, Adler AC, Schwindt DM (2002) Abruptness of the end-Permian mass extinction as determined from biostratigraphic and cyclostratigraphic analyses of European western Tethyan sections. Geol Soc Am Spec Pap 356:415–427

Scholger R, Mauritsch HJ, Brandner R (2000) Permian–Triassic boundary magnetostratigraphy from the Southern Alps (Italy). Earth Planet Sci Lett 176:495–508

Scotese CR (1997) PALEOMAP Paleogeographic Atlas, PALEOMAP Progress Report 90:22. Department of Geology, University of Texas

Sephton MA, Looy CV, Veefkind RJ, Visscher H, Brinkhuis H, de Leeuw JW (1999) Cyclic diaryl ethers in a Late Permian sediment. Org Geoch 30:267–273

Sephton MA, Looy CV, Veefkind RJ, Brinkhuis H, Leeuw JW, Visscher H, (2002) Synchronous record of δ13C shifts in the oceans and atmosphere at the end of the Permian. In: Koeberl C, MacLeod KG (eds) Catastrophic events and mass extinctions: impacts and beyond. Geol Soc Am Spec Pap 356:455–462

Shen SZ, Cao CQ, Henderson CM, Wang XD, Shi GR, Wang Y, Wang W (2006) End-Permian mass extinction pattern in the northern peri-Gondwanan region. Palaeoworld 15:3–30

Staesche U (1964) Conodonten aus dem Skyth von Südtirol. N Jb Geol Palaont Abh 119:247–306

Tong J, Zhang S, Zuo J, Xiong X (2007) Events during Early Triassic recovery from the end-Permian extinction. Global Planetary Change 55:66–80

Twitchett RJ (2005) The Lilliput Effect in the aftermath of the end- Permian extinction event. Albertiana 33:79–81

Venzo GA (1955) Il Permiano superiore nei dintorni di Egna (Alto Adige) e le sue dolomie calcaree a tessitura oolitica. Mem Mus Stor Nat Venezia Tridentina 10:1–14

Vermeij GJ, Dorritie D (1996) Late Permian extinctions. Science 274:1549–1552

Visscher H, Brinkhuis H, Dilcher DL, Elsik WC, Eshet Y, Looy CV, Rampino MR, Traverse A (1996) The terminal Paleozoic fungal event: evidence of terrestrial ecosystem destabilization and collapse. Proc Natl Acad Sci USA 93: 2155–2158

Visscher H, Looy CV, Collinson ME, Brinkhuis H, van Konijnenburg-van Cittert JHA, Kurschner WM, Sephton MA (2004) Environmental mutagenesis during the end-Permian ecological crisis. Proc Nat Acad Sci USA Early Edition: 1–5. www.pnas.org/cgi/10.1073/pnas.0404472101

Wang CY (1995) Conodonts of the Permian–Triassic boundary beds and biostratigraphic boundary (In Chinese with English abstract). Acta Palaeontol Sinica 34:129–151

Watson JS, Sephton MA, Looy VC, Gilmour I (2005) Oxygen-containing aromatic compounds in a Late Permian sediment. Org Geochem 36:371–384

Wignall PB, Hallam A (1992) Anoxia as a cause of the Permian–Triassic mass extinction: facies evidence from northern Italy and the western United States. Palaeogeogr Palaeoclimatol Palaeoecol 93:21–46

Wignall PB, Hallam A (1996) Facies change and the end-Permian mass extinction in SE Sichuan, China. Palaios 11:587–596

Wignall PB, Twitchett RJ (1996) Oceanic anoxia and the end Permian mass extinction. Science 272:1155–1158

Wignall PB, Kozur H, Hallam A (1996) On the timing of palaeoenvironmental changes at the Permo-Triassic (P/Tr) boundary using conodont biostratigraphy. Hist Biol 12: 39–62

Wu YS (2005) Conodont evolution during Permian–Triassic transition in mid-low latitudes: a close-up view. Albertiana 33:93–94

Xu D, Yan Z (1993) Carbon isotope and iridium event markers near the Permian/Triassic boundary in the Meishan section, Zhejiang Province, China. Palaeogeogr Palaeoclimatol Palaeoecol 1104:171–176

Yin H, Zhang K, Tong J, Yang Z, Wu S (2001) The global stratotype section and point (GSSP) of the Permian–Triassic boundary. Episodes 24:102–114

Zhang K, Tong J, Yin H, Wu S (1996) Sequence stratigraphy near the Permian–Triassic Boundary at Meishan section, south China. In: Yin H (ed) The Palaeozoic-Mesozoic boundary candidates of the global stratotype section and point of the Permian–Triassic Boundary. China University of Geosciences, Wuhan, pp 72–83

The Cretaceous–Tertiary Mass Extinction, Chicxulub Impact, and Deccan Volcanism

Gerta Keller

Abstract

After three decades of nearly unchallenged wisdom that a large impact (Chicxulub) on Yucatan caused the end-Cretaceous mass extinction, this theory is facing its most serious challenge from the Chicxulub impact itself, as based on evidence in Texas and Mexico and from Deccan volcanism in India. Data generated from over 150 Cretaceous–Tertiary (KT) boundary sequences to date make it clear that the long-held belief in the Chicxulub impact as the sole or even major contributor to the KT mass extinction is not supported by evidence. The stratigraphic position of the Chicxulub impact ejecta spherules in NE Mexico and Texas and the impact breccia within the crater on Yucatan demonstrate that this impact predates the KTB by about 300,000 years. Planktic foraminiferal and stable isotope analyses across the primary impact ejecta layer reveal that not a single species went extinct as a result of this impact and no significant environmental changes could be determined. The catastrophic effects of this impact have been vastly overestimated. In contrast, recent advances in Deccan volcanic studies indicate three volcanic phases with the smallest at 67.5 Ma, the main phase at the end of the Maastrichtian (C29r), and the third phase in the early Danian C29r/C29n transition (Chenet et al. 2007). The main phase of eruptions occurred rapidly, was marked by the longest lava flows spanning 1500 km across India, and ended coincident with the KT boundary. The KT mass extinction may have been caused by these rapid and massive Deccan lava and gas eruptions that account for ~80% of the entire 3500 m thick Deccan lava pile.

Keywords

Cretaceous • Maastrichtian • Paleogene • Danian • Mass extinctions • Foraminifers • Stable isotopes • Chicxulub impact • Deccan volcanism

Introduction

One of the main conclusions of the year 2000 assessment was that most studies on the Cretaceous–Tertiary (KT) mass extinction in the marine realm concentrated almost exclusively on the narrow interval surrounding

G. Keller (✉)
Department of Geosciences, Princeton University, Princeton, NJ 08544, USA
e-mail: gkeller@princeton.edu

J.A. Talent (ed.), *Earth and Life*, International Year of Planet Earth,
DOI 10.1007/978-90-481-3428-1_25, © Springer Science+Business Media B.V. 2012

the iridium anomaly globally or the impact spherule layers in Mexico and the Caribbean (Keller 2001). This fixation on the narrow Cretaceous–Tertiary boundary (KTB) interval can be traced to the publications of the bolide impact theory in 1980 (Alvarez et al. 1980), the discovery of the Chicxulub crater on Yucatan in 1991 (Hildebrand et al. 1991), and the subsequent media and public fascination with it. But it was fueled mainly by the large number of scientists that were convinced that a meteorite caused the global extinction of the dinosaurs and many other groups at the end of the Cretaceous. Keller wrote (2001, p. 817):

> This theory unquestionably has great sex appeal. The largest and most fascinating creatures that ever roamed the Earth were wiped out in a single day in a ball of fire caused by a meteorite impact that leaves behind the crater of doom. No wonder the hypothesis captured the minds and hearts of the public. But apart from the appeal to the imagination, the theory is also supported by undeniable geochemical and geophysical evidence of a bolide impact on the Yucatan Peninsula about 65 million years ago. The existence of an impact crater alone, however, neither proves nor explains the demise of the dinosaurs, or the mass extinction of any other groups. . . .Ultimately, the validity of any mass extinction hypothesis depends on how well it explains the paleontological record.

This statement is as true today as it was in 2000—only more so. During the past decades, the bells and whistles that warned of a more complex extinction scenario inherent in the fossils and sedimentary records were largely ignored. In the 1980s and 1990s scientists who discovered evidence inconsistent with the impact extinction scenario or disagreed with the impact theory were generally derided as old-fashioned gradualists, incompetent, and much worse. In Luis Alvarez' famous interview with *New York Times*' Malcolm Browne, he said, "I don't like to say bad things about paleontologists, but they're really not very good scientists. They're more like stamp collectors" (New York Times, January 19, 1988). If accumulating a large database to test the validity of a theory is equated with stamp collecting, then many scientists fall in this category. Having analyzed over 150 KT sequences worldwide (Fig. 1) certainly counts the author and collaborators among them. A large database is necessary to find out what really happened globally and to test the impact theory.

The Alvarez et al. (1980) impact theory gained its strongest support from the Iridium anomaly in a thin clay layer that separates Cretaceous and Tertiary sediments at the Gubbio section in Italy and marks the mass extinction in planktic foraminifera. This iridium anomaly was attributed to a large meteorite impact, a link that was established in the 1970s with the Apollo lunar regolith samples (Morgan et al. 1975). Soon the link between impacts (e.g., meteorites) and extinctions implied by the Alvarez theory became the frequently proposed scenario for all mass extinction events, even though no evidence could be found that directly linked impacts and mass extinctions in any of the other extinction events (e.g., Courtillot 1999; Wignall 2001; Courtillot and Renne 2003; Keller 2005). By the early 1990s, the twin discoveries of the Chicxulub crater on Yucatan (Hildebrand et al. 1991) and impact spherule ejecta in Haiti and NE Mexico (Izett et al. 1991; Smit et al. 1992, 1996) were widely believed to prove the KT impact theory—the smoking gun had been found. Or was it not?

Despite the irrefutable evidence of a large meteorite impact on Yucatan, nagging doubts persisted fueled by a plethora of evidence that could not be reconciled with the Chicxulub impact as the sole cause for the mass extinction. A review of the evidence that led to this premature conclusion and the evidence that contradicts it help us understand this nearly three-decades-old controversy. In the past few years, new evidence increasingly points away from Chicxulub and toward the other KT catastrophe—Deccan volcanism—as the likely smoking gun.

Database and Methods

Over 150 KT sequences have been analyzed to date by the author, collaborators, and many students over the past 20 years (Fig. 1). The largest concentration of KT sequences was analyzed throughout Central America (Fig. 2) in order to evaluate the age and biotic consequences of the Chicxulub impact. Over 45 sections were analyzed in NE Mexico by a dozen students and their advisors (Thierry Adatte, Gerta Keller, and Wolfgang Stinnesbeck) who systematically mapped, sampled, and studied outcrops containing the sandstone complex with Chicxulub impact spherules over 50 km^2 (Lindenmaier 1999; Schilli 2000; Affolter 2000; Schulte 1999; Schulte et al. 2003, 2006; Lopez-Oliva 1996; Lopez-Oliva, Keller 1996; Harting 2004). Another 18 sequences were analyzed in southern Mexico, Guatemala, Belize, Haiti, and Cuba. These

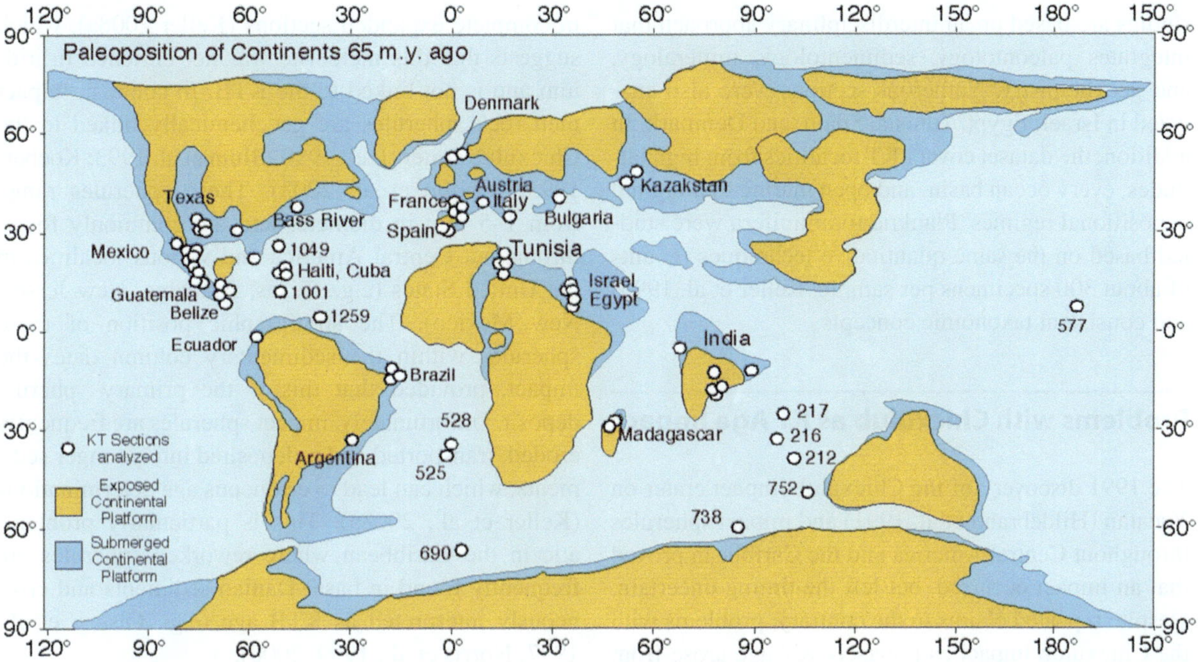

Fig. 1 Localities showing Cretaceous–Tertiary boundary sequences analyzed worldwide by Keller and collaborators (modified after Keller and Pardo 2004)

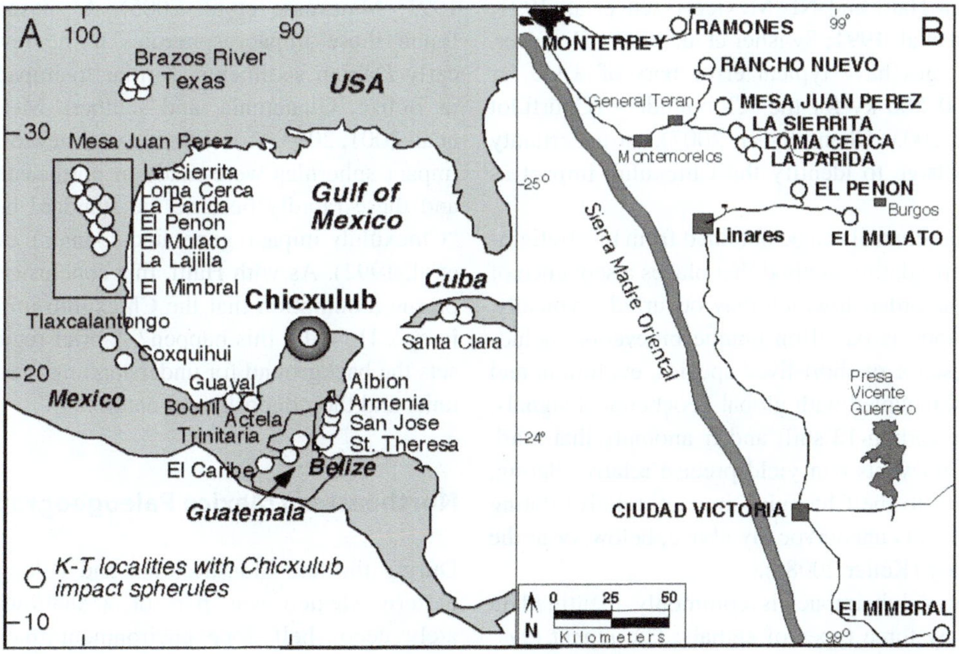

Fig. 2 Localities showing Cretaceous–Tertiary boundary sequences analyzed by Keller and collaborators in the Caribbean, Central America, and southern USA that contain Chicxulub impact glass spherule ejecta (modified after Keller et al. 2003a)

studies are based on an interdisciplinary approach that integrates paleontology, sedimentology, mineralogy, and geochemistry. Numerous sections were also analyzed in Israel, Egypt, Tunisia, Spain, and Denmark. In addition, the dataset covers KT localities from high latitudes, every ocean basin, and open marine to marginal depositional regimes. Planktic foraminifera were studied based on the same quantitative techniques (counts of about 300 specimens per sample, Keller et al. 1995) and consistent taxonomic concepts.

Problems with Chicxulub as KT Age Impact

The 1991 discovery of the Chicxulub impact crater on Yucatan (Hildebrand et al. 1991) and impact spherules throughout Central America and the Caribbean proved that an impact occurred, but left the timing uncertain. Despite repeated claims to the contrary, problems with the Chicxulub impact as precisely KT age arose from the very beginning. Impact craters are very difficult to date. The first Chicxulub age dates were based on $^{39}Ar/^{40}Ar$ ages from melt rock within the impact breccia (suevite) of the Yucatan crater and from melt rock spherules in Haiti and NE Mexico (Izett et al. 1991; Sigurdsson et al. 1991; Swisher et al. 1992). However, $^{39}Ar/^{40}Ar$ ages have typical error bars of $\pm 1\%$, or about $\pm$300–500 ky for the KT boundary (Courtillot and Renne 2003; Chenet et al. 2007), an uncertainty that is too large to identify the Chicxulub impact as KT age.

Better age control can be obtained from biostratigraphy, a relative dating method that places a sequence of events in the order in which they occurred. Typically, biostratigraphy is based on unique bioevents, including the presence of short-lived species, evolution, and extinctions. Coupled with global geochemical signals, such as the carbon-13 shift and Ir anomaly that mark the KTB, bioevents can yield precise relative dating. With respect to the Chicxulub impact, relative dating can place events unequivocally above, below, or at the KT boundary (Keller 2008a).

The Chicxulub impact is commonly identified in sediments by three types of signals: (1) *impact (suevite) breccia*, which is restricted to proximal areas surrounding the crater, (2) *impact spherules*, and (3) the *Ir anomaly*. However, no Ir anomaly has ever been documented in association with other Chicxulub ejecta (e.g., suevite or impact spherules), and no Chicxulub ejecta has ever been found at the KTB in complete expanded sections (Keller 2008a), which suggests that this meteorite was not enriched in iridium and is not linked to the KTB. In contrast, impact melt rock spherules are geochemically linked to the Chicxulub crater (Izett 1990; Blum et al. 1993; Koeberl 1993; Schulte et al. 2003). These spherules range from 1–5 mm in diameter and are commonly found throughout Central America and various localities in the United States (e.g., Texas, Alabama, New Jersey, New Mexico). The stratigraphic position of these spherules within the sedimentary column dates the impact, provided that this is the primary spherule deposit. Unfortunately, impact spherules are frequently eroded, transported and redeposited into younger sediments, which can lead to erroneous age determinations (Keller et al., 2008a). This is particularly problematic in the Caribbean where reworked spherules are frequently found in basal Danian sediments and erroneously interpreted as KTB age (e.g., Olsson et al., 1997; Norris et al., 1999, 2000).

For example, the first discovery of Chicxulub impact spherules was in Haiti at or near the KT boundary and was considered proof that this impact is KT in age (Izett et al. 1991; Maurasse and Sen 1991; Maurrasse et al. 2005). A subsequent study found these impact spherules to be reworked into early Danian sediments, similar to impact spherules in Belize, Guatemala, and southern Mexico (Keller et al. 2001, 2003a, b). Within a year new discoveries of impact spherules were made in northeastern Mexico, and these rapidly became the standard bearer of the "Chicxulub impact generated tsunami event" (Smit et al. 1992). As with Haiti, this conclusion was based on the assumption that the Chicxulub impact is KTB in age. How did this happen? A brief regional history sets the background for understanding these sandstone units and their depositional nature.

Northeastern Mexico Paleogeography

During the late Maastrichtian and Paleogene, northeastern Mexico was part of a shallow to moderately deep shelf-slope environment that received a steady influx of terrigenous clastic sediments from the Laramide Orogeny and uplift of the Sierra Madre Oriental to the west (Sohl et al. 1991; Galloway et al. 1991). Along the western Gulf of Mexico, a series of basins formed in the present states of Veracruz and Tamaulipas and in the southern and eastern parts of

Nuevo Leon. Within these basins rhythmically bedded marls, shales, and thin layers of volcanic ash were deposited in outer neritic to upper bathyal pelagic environments. These sedimentary rocks belong to the Campanian to Maastrichtian Mendez Formation and the Paleocene Velasco Formation.

During this time, the western Gulf of Mexico was characterized by several large submarine canyon systems that formed along the prograding unstable continental margin. The best known of these are the Lavaca and Yoakum paleocanyons in Texas and the Chicontepec paleocanyon system in central east Mexico. It is in the submarine channels of these canyon systems that the thick sandstone complexes of northeastern Mexico and Texas were deposited during the late Maastrichtian. There are several such sandstone units interbedded in the Maastrichtian Mendez Formation in NE Mexico (e.g., a prominent late Maastrichtian sandstone unit is exposed outside Linares, Nuevo Leon). Each is correlative with a major sea-level fall during the Maastrichtian that resulted in erosion and increased detrital input from the rising Sierra Madre Oriental. But only the sandstone complex closest to the KT boundary has Chicxulub impact spherules at the base. Today these channel-fill deposits, including the one near Linares, form the top of low-lying hills throughout northeastern Mexico.

Fig. 3 The Cretaceous–Tertiary sequence at El Mulato, NE Mexico. The KT boundary in NE Mexico is above a thick sandstone complex with Chicxulub impact spherules at the base. This sandstone complex infills submarine channels. Based on the assumption that the Chicxulub impact is KT in age, the sandstone complex is commonly interpreted as impact-generated tsunami deposit. Deposition during a sea-level fall is more consistent with the evidence

NE Mexico Sandstone Complex—Tsunami or Sea-Level Change?

After the discovery of impact spherules in Haiti an intensive search was underway to discover similar deposits in Mexico. The first discovery was at El Mimbral in NE Mexico, where a 1 m thick spherule layer is present below a massive 2 m thick sandstone that is overlain by alternating sand, silt, and shale layers. The KTB and Ir anomaly are above these sediments. Similar deposits were observed at El Mulato and El Peñon (Figs. 3 and 4). The spatial separation between impact spherules and the KTB in these outcrops presented a major problem. If we assume that the impact struck Yucatan at KTB time and caused the mass extinction, then the impact spherules must be KT in age, and the sandstone complex in between must have been deposited within days. The simple solution would be an impact-generated mega-tsunami (Smit et al. 1992). This interpretation had already been proposed for similar sandstone units in KT sequences along the Brazos River in Texas (Bourgeois et al. 1988). In this scenario the glass spherules rained from the sky within hours of the impact and settled on the ocean floor, the impact-generated tsunami waves arrived soon after and deposited the massive sandstone, which was followed by the waning tsunami waves depositing the alternating sand, silt, and shale layers, with the iridium settling during the subsequent weeks (Fig. 4, Smit et al. 1992, 1996; Smit 1999; Kring 2007). This scenario provided complete support for the KT impact theory and was an instant hit in the early 1990s and persists to this day (see Schulte et al. 2010). But what is the evidence?

Scientists searching for verification of the impact theory soon discovered dozens of outcrops with sandstone complexes similar to El Mimbral throughout northeastern Mexico with the best localities at El Peñon, El Mulato, La Lajilla, Loma Cerca, La Sierrita, and Mesa Juan Perez (Figs. 2 and 3). At each locality a thick (1 m to 8 m) sandstone complex infills submarine channels with a lateral continuity of about 100–250 m. In most areas, the sandstone complexes form the flat tops of low-lying hills with the overlying sediments eroded. But in a few localities the overlying sediments are preserved (e.g., El Mimbral, La Lajilla,

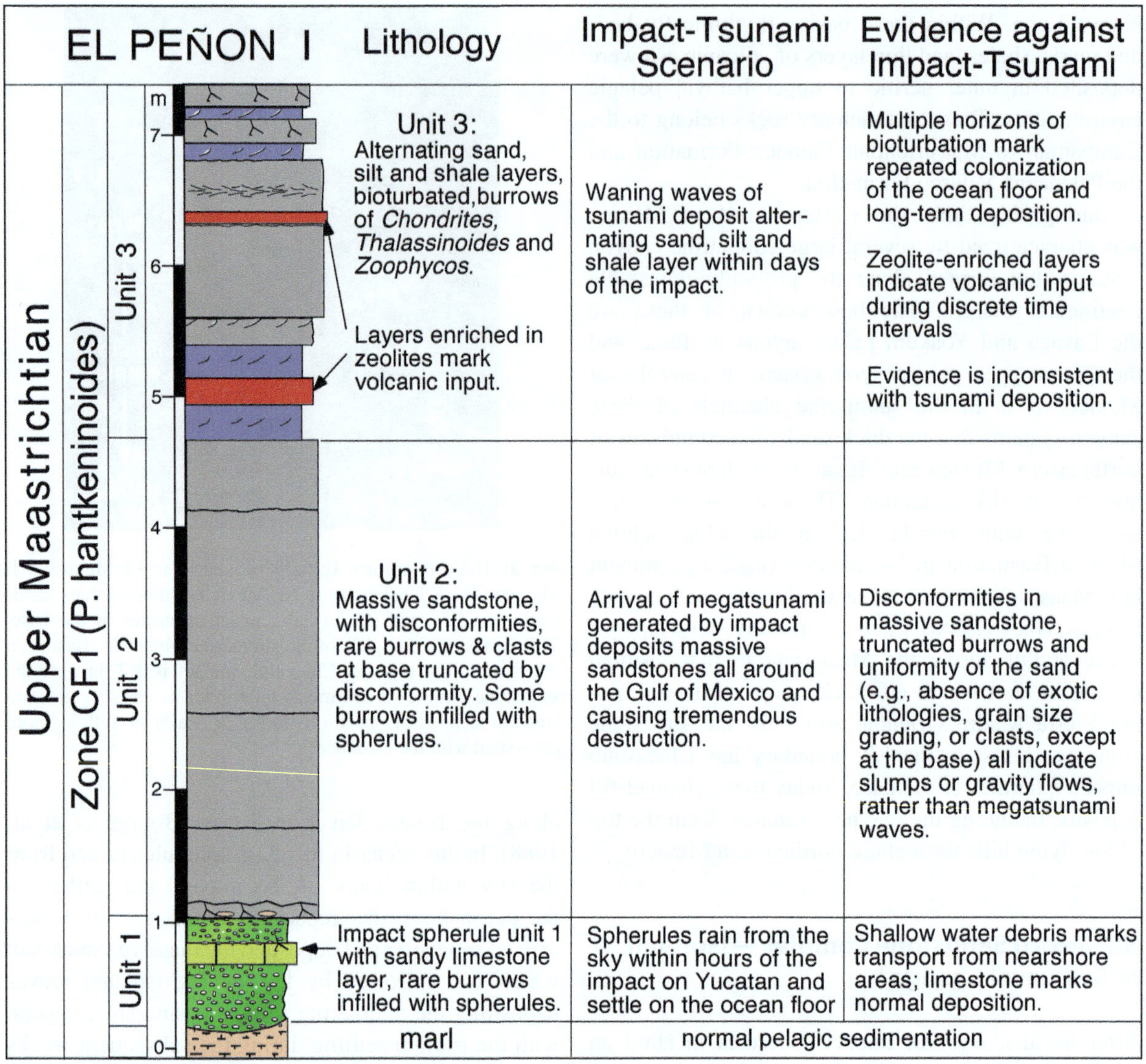

Fig. 4 Lithology and description of the sandstone complex at El Peñon with the impact-tsunami interpretation and the facts that contradict this interpretation

El Mulato, Fig. 3). The sandstone complex consists of three distinctive lithological units, although the thickness is highly variable within an outcrop and from one area to another and some units may be missing.

Unit 1: The thickest sandstone complex with all three lithological units present is known from the classic El Peñon locality (Fig. 4). Unit 1 is the spherule-rich layer at the base of the sandstone complex and is about 1 m thick. Sediments consist of green altered impact glass spherules, 1 to 5 mm in size, and characterized by abundant vesicles (Fig. 5a–e). These spherules have been identified as microtektites produced by melting and quenching of terrestrial rocks during a hypervelocity impact (Izett 1990; Izett et al. 1991; Sigurdsson et al. 1991; Blum et al. 1993; Koeberl 1993). The impact spherules are mixed with abundant quartz sand, shallow water foraminifera, and plant debris, which indicate erosion and transport from shallow nearshore areas. Deposition occurred in an upper bathyal environment at depths >500 m (Keller et al. 2003a, 2004a, b; Alegret et al. 2001).

In many outcrops of NE Mexico, including El Mimbral, La Lajilla, El Peñon, and Rancho Grande, the spherule-rich unit 1 is separated into two spherule

Fig. 5 Characteristics of the spherule unit 1 at the base of the sandstone complex. (**a**) Rock of the spherule-rich layer. (**b**) Remnant spherule glass preserved after acid treatment. (**c**) Close-up of spherule-rich layer showing altered impact glass. (**d**) Thin-section micrograph of spherule in debris matrix. (**e**) Thin-section micrograph of spherules in calcite cement. (**f**) J-shaped burrow from sandy limestone within spherule unit 1. (**g**) Sandy limestone layer in spherule-rich unit 1 at El Peñon with location of burrow truncated by erosion (close-up in **f**). (**h**) Sandy limestone between spherule layers at El Mimbral with sculpted upper surface (adapted from Keller 2008a)

layers by a 15–25 cm thick sandy limestone layer (SLL) (Figs. 4 and 5, Keller et al. 1997). Rare J-shaped burrows infilled with spherules and truncated by erosion are present (Fig. 5f–h, Keller et al. 1997; Ekdale and Stinnesbeck 1998). This limestone layer indicates a long period of normal sediment deposition. The second spherule layer disconformably overlies the sandy limestone and indicates a renewed influx of shallow-water debris. These lithological characteristics are inconsistent with deposition within hours as proposed by the tsunami scenario (Smit et al. 1992, 1996; Smit 1999). Smit argued that the limestone layer was tectonically injected into the spherule layer by the impact disturbance, a feat that is difficult to contemplate over the vast distance in which these layers are found in the same stratigraphic unit 1.

Unit 2: Unit 2 is a massive sandstone that disconformably overlies unit 1 (Fig. 4). There is no grading, no significant change in grain size, no exotic debris, and near absence of clasts, except for some marl clasts at the base. At El Peñon, occasional J-shaped burrows infilled with spherules are present near the base. At El Mimbral a large deposit of plants, leaves, and wood marks the base of unit 2. Disconformities within the sandstone indicate several depositional episodes. These lithologic characteristics indicate slumps or gravity flows. Unit 2 lacks the chaotic sediments that are associated with tsunami deposition.

Fig. 6 (**A**) Alternating sand, silt, and shale layers of unit 3 disconformably overlying the sperule unit 1 at La Lajilla. (**B**) Burrows of unit 3 at El Peñon. (**a**) *Thalassinoides*, (**b**) *Chondrites*, (**c**) *Zoophycos*, (**d**) fine grained, laminated layers of unit 3 with common burrows (**e**) of *Chondrites* (*vertical view*), (**f**) bird's-eye view. Multiple burrowed horizons within unit 3 indicate that the ocean floor was colonized by invertebrates during sedimentation, which is inconsistent with tsunami deposition over hours to days (modified after Keller et al. 1997)

Unit 3: Unit 3 consists of alternating layers of sand, silt, and laminated shale (Fig. 6a, b), which are topped by a rippled sandy limestone (RPL). The characteristics of the alternating layers are clearly visible in the resistant sand layers jutting out from the outcrops as shown for La Lajilla, where unit 3 overlies unit 1 (Fig. 6a). Sand layers show variable features, including rippled, wavy, and convolute bedding. Trace fossils are abundant, particularly in the fine silt and shale layers where *Chondrites* burrows are abundant (Fig. 6b, e–f). At El Peñon, trace fossils are particularly abundant in the sandy top layers, including *Thalassinoides, Zoophycos Chondrites* (Fig. 6). Ekdale and Stinnesbeck (1998) observed at least three successive burrowing episodes that indicate unit 3 was deposited over a long period of time with repeated colonization of the ocean floor. This is inconsistent with tsunami deposition over a period of hours to days.

Two distinct layers enriched in zeolites are present near the base and top of unit 3 at El Peñon, La Sierrita, El Mulato, La Lajilla, and El Mimbral, as well as in the spherule unit 1 and early Danian sediments at El Mimbral (Figs. 4 and 7, Adatte et al. 1996). These layers indicate significant volcaniclastic influx. Because these layers are correlatable over long distances, they represent discrete volcanic input during normal sedimentation, which is inconsistent with tsunami deposition over a period of hours to days.

Evidence of unit 3 thus indicates long-term deposition inconsistent with the tsunami scenario. Smit et al. (1994, p. 100) argued that the "correct interpretation of the many burrowing structures (*Chondrites*,

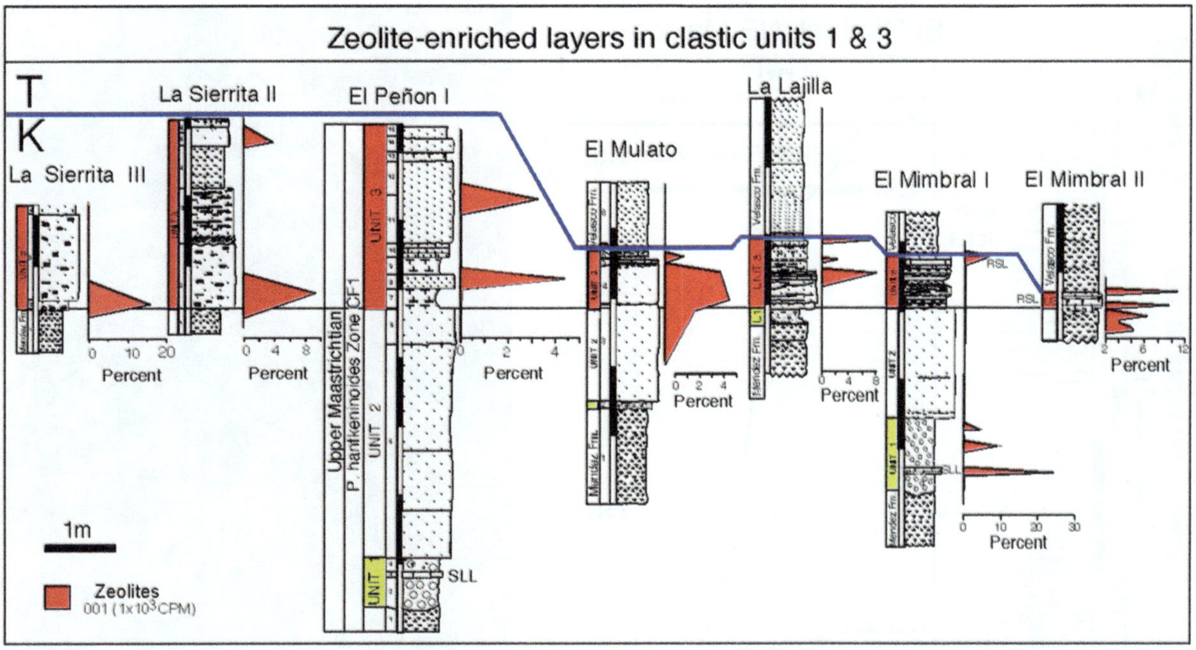

Fig. 7 Correlation of the sandstone complex in NE Mexico showing the continuity of two zeolite-enriched layers within unit 3. These zeolite-enriched layers mark discrete episodes of volcanoclastic influx that are inconsistent with tsunami deposition over hours to days (modified from Adatte et al. 1996)

Zoophycos) that occur on and in the top of the KT clastic complex .... (is that they) clearly penetrate from above *after* deposition of the clastic complex." They further suggest that vertical tubes of unknown origin, up to 3 m long and 1 cm diameter, reach down into unit 3 and branch out horizontally in the fine-grained layers. Though vertical tubes representing water escape structures are occasionally observed, these do not explain the multiple-burrowing horizons truncated by erosion, or the small *Chondrites* traces (Fig. 6b). Water-escape structures are not inconsistent with the tsunami interpretation, but repeated colonization of the ocean floor is, because it requires long-term deposition. In addition, volcanic influx (e.g., zeolite-enriched layers) requires a discrete source and undisturbed deposition. Thus, both burrows and discrete volcanic influx negate the tsunami scenario.

What is a plausible alternative interpretation of this sandstone complex? Keller et al. (1994b) and Adatte et al. (1996) proposed a major sea-level fall with a concomitant subaerial unconformity (sequence boundary (SB) at the base of the sandstone complex) and low sea level (LST) during which units 1 and 2 were deposited via erosion and transport of nearshore sediments (unit 1) and gravity flows and slumps (unit 2, Fig. 8). Unit 3 was deposited during the subsequent sea-level rise (TST), which reached the maximum flooding surface (MFS) at the KTB. This scenario is in agreement with the well-known Late Maastrichtian sea-level changes (e.g., Haq et al. 1987, 1988; Baum and Vail 1988; Donovan et al. 1988; Loutit et al. 1988).

Underlying the controversy over impact-tsunami or sea-level fall interpretations is not the nature of sediment deposition, but whether the Chicxulub impact is KT age. If deposition of the sandstone complex occurred over a long time period, then the Chicxulub impact predates the KTB and cannot be the cause for the KT mass extinction. The fact that the KT mass extinction and Ir anomaly are always above the sandstone complex compounds the problem (Smit et al. 1992; Stinnesbeck et al. 1993; Keller et al. 1994a, b; Rocchia et al. 1996). In addition, there is frequently a thin layer of Maastrichtian marl between the sandstone complex and the KTB (e.g., La Lajilla, El Mulato, La Parida), which further jeopardizes the tsunami interpretation (Lopez-Oliva and Keller 1996). A simple solution was to redefine the KTB as the Chicxulub impact, which automatically places the KTB at the spherule layer at the base of the sandstone complex

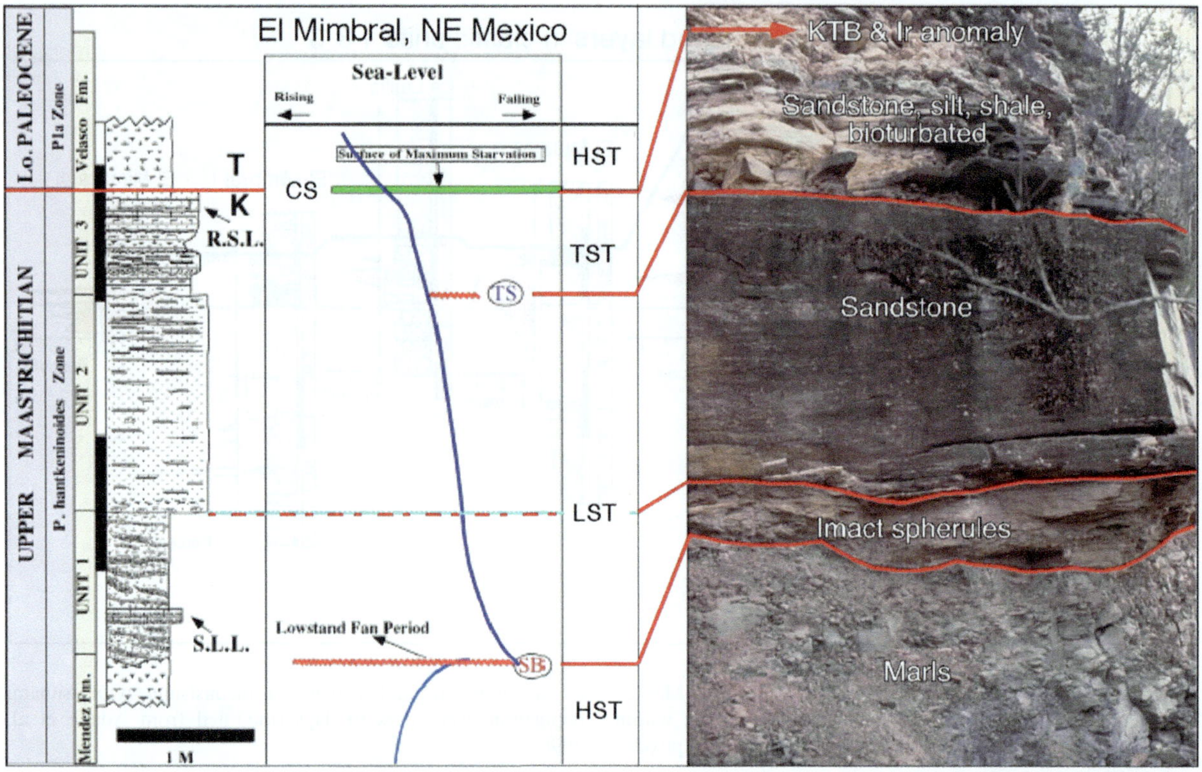

Fig. 8 Sea-level changes inferred from the lithology of the sandstone complex at El Mimbral, NE Mexico. The erosional base of the channel marks a sequence boundary (SB) and the lowest sea level. The sandstone complex was deposited during the following slowly rising sea level (modified from Adatte et al. 1996)

(Smit et al. 1992, 1994, 1996, 2004; Smit 1999; Molina et al. 2006; Arenillas et al. 2006; Kring 2007; Schulte et al., 2010). This circular reasoning has not solved the problem of the age of the Chicxulub impact.

Texas: Sandstone Complex—Tsunami or Sea-Level Change?

The impact-tsunami hypothesis was first proposed by Bourgeois et al. (1988) for a sandstone complex that infills inner shelf channels near the KT boundary along the Brazos River in Texas (Fig. 9). The sandstone complex in this area is very similar to the sandstone complex in NE Mexico, including reworked impact spherules at the base, multiple burrowed horizons within the sandstone, and the KT boundary and Ir anomaly above it (Hansen et al. 1987, 1993; Keller 1989; Yancey 1996; Schulte et al. 2006; Keller et al. 2007a). Impact-tsunami proponents placed the KT boundary at the base of the sandstone complex based on the assumption that the impact is KT in age (Bourgeois et al. 1988; Smit et al. 1996; Hansen et al. 1993; Schulte et al. 2006, 2008), a practice that was followed in NE Mexico sections.

However, one cannot test the hypothesis that the Chicxulub impact caused the KT mass extinction by defining the impact as the K–T boundary, as Schulte et al. (2008, 2010) proposed (see Keller et al. 2008a). In our global analyses of over 150 KT sequences, the most consistent markers are the mass extinction in planktic foraminifera, the first appearance of Danian species, and the $\delta^{13}C$ shift. The $\delta^{13}C$ shift is a global oceanographic signal and therefore provides an independent check on paleontological and impact criteria, which is critical to avoid circular reasoning. In the Brazos sections these KT defining criteria are well documented above the sandstone complex (Jiang and Gartner 1986; Keller 1989; Barrera and Keller 1990; Keller et al. 2007a, 2008a, 2009a).

Detailed sedimentologic analysis of the sandstone complex in Brazos sections revealed long-term deposition, including multiple storm-related erosion surfaces followed by re-colonization of the sea floor

Fig. 9 The Cretaceous–Tertiary sequence of the Darting Minnow Creek (DMC), a tributary of the Brazos River, Falls County in Texas. The KT boundary is well above the sandstone complex with Chicxulub impact spherules at the base which infills submarine channels. This deposit was originally interpreted as impact-generated tsunami (Bourgeois et al. 1988) but is now recognized as sea-level lowstand. (**a**) clasts at base of sandstone complex. (**b**) Thin section of clasts show impact spherules. (**c**) Dessication racks in clasts are infilled with impact spherules and reveal a history of Chicxulub ejecta well prior to erosion and redeposition at the base of the sandstone complex (modified from Keller et al. 2007a)

during a low sea level (Fig. 10) (Yancey 1996; Gale 2006; Keller et al. 2007a). Two to three upward-fining spherule-rich layers with abundant shell hash and glauconite are common at the base of the sandstone complex along with clasts of the underlying sediments. In some outcrops, large clasts (5–10 cm) were found at the base of the spherule deposits (Fig. 9a). Some clasts contain common impact spherules (B) and cracks infilled with spherules and sparry calcite (Fig. 9c). These spherule-rich clasts demonstrate the presence of an older spherule deposition event with sediments lithified and subsequently exposed, at times subaerially, to erosion followed by redeposition in submarine channels (Keller et al. 2007a).

Above the sandstone complex deposition of dark grey mudstones and claystones with few macrofossils and burrows infilled by pyrite mark dysoxic conditions and a rising sea level (Fig. 10). The KT boundary is identified in the lower part of this dark claystone. Texas sections thus demonstrate deposition of the sandstone complex during a low sea level prior to the KTB, similar to NE Mexico (Fig. 4). The reworked spherules in multiple layers and clasts indicate that the Chicxulub impact must predate the KT boundary (discussed in Section "Texas—Primary Spherule Layer").

Deep-Sea-Spherule Deposits

Advocates for a KT age of the Chicxulub impact frequently point to the stratigraphic proximity of spherules and an Ir anomaly in some deep-sea marine sequences (e.g., Blake Nose, Bass River, Demerara Rise) as unequivocal evidence that Chicxulub caused the KT mass extinction (Olsson et al. 1997; Norris et al. 1999, 2000; Klaus et al. 2000; MacLeod et al. 2007). Differential settling of iridium is often used to explain the proximity of the Ir anomaly at or a few cm above the spherule layer. This argument is plausible only if there is continuous sedimentation and no inverse grading or other disturbances in the spherule layer, and no disconformities due to erosion or non-deposition—all are features evident at Bass River, Blake Nose, and Demerara Rise (Keller 2008a).

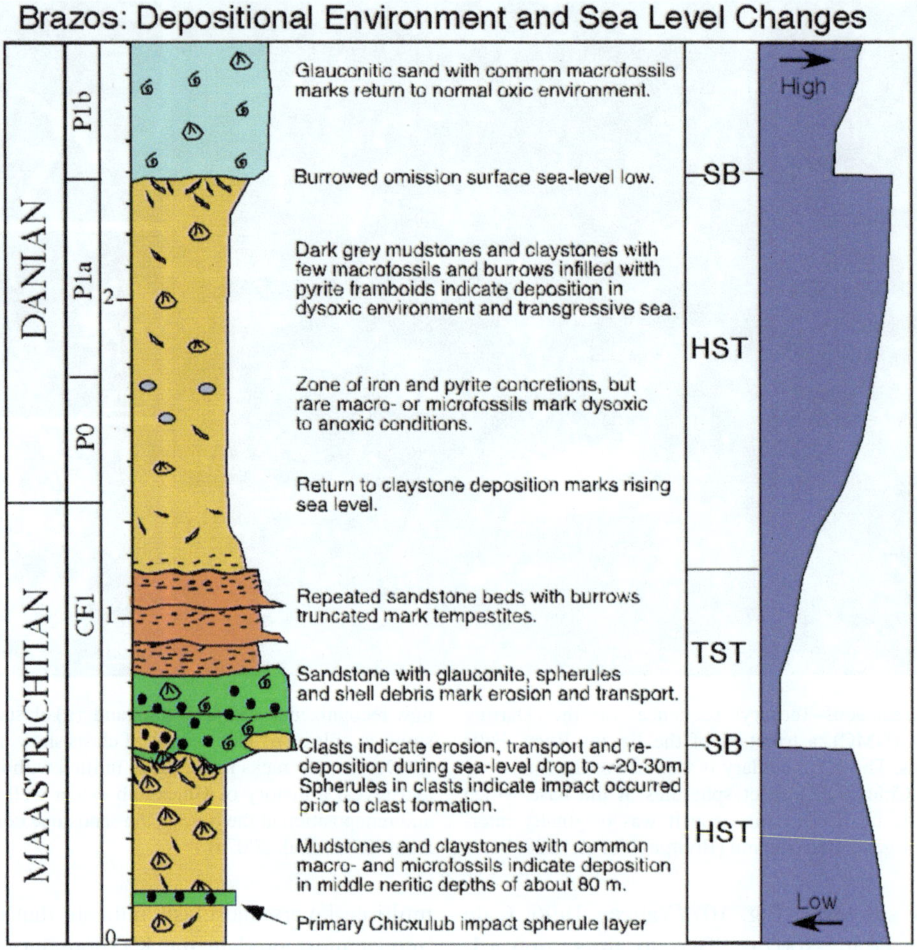

Fig. 10 Lithologic description and sea-level interpretation of Brazos sections (modified from Gale 2006)

The Bass River core of New Jersey was described as a complete KT boundary sequence that provides "unequivocal evidence" that the Chicxulub impact is KT in age (Olsson et al. 1997). However, this shallow water (~100 m) sequence is demonstrably incomplete. The Maastrichtian consists of highly bioturbated glauconitic clay with only an 8 cm interval representing the *Micula prinssi* (CC26) zone, which spans the last 500 ky of the Maastrichtian (Fig. 11). This indicates that most of the uppermost Maastrichtian is missing. The spherule layer is poorly sorted with clasts containing Maastrichtian microfossils that indicate erosion, transport, and redeposition. Above the spherule layer the first 280 ky of the Danian (zones P0 and P1a, *Parvularugoglobigerina eugubina*) is represented by just 8 cm with an assemblage indicative of the upper part of this interval (subzone P1a(2), Fig. 11), which indicates that most of the early Danian is missing. The juxtaposition of the spherule layer, the clasts, and early Danian sediments is thus due to a major hiatus, which allows no conclusion as to the timing of the Chicxulub event.

At Blake Nose ODP Site 1049 the close proximity of the spherule layer, Ir anomaly, and early Danian planktic foraminifera have also been claimed as proof that the Chicxulub impact is KT age (Norris et al. 1999, 2000; Klaus et al. 2000; Martínez-Ruiz et al. 2002). The spherule layer is 15 cm thick and disconformably overlies mottled, mixed carbonate ooze of the basal *M. prinsii* zone and planktic foraminiferal zone CF3 (Fig. 11). The spherule layer contains clasts of limestone, dolomite, chert, and schist, which indicate eroded and transported sediments. A rich lower Danian planktic foraminiferal assemblage of subzone

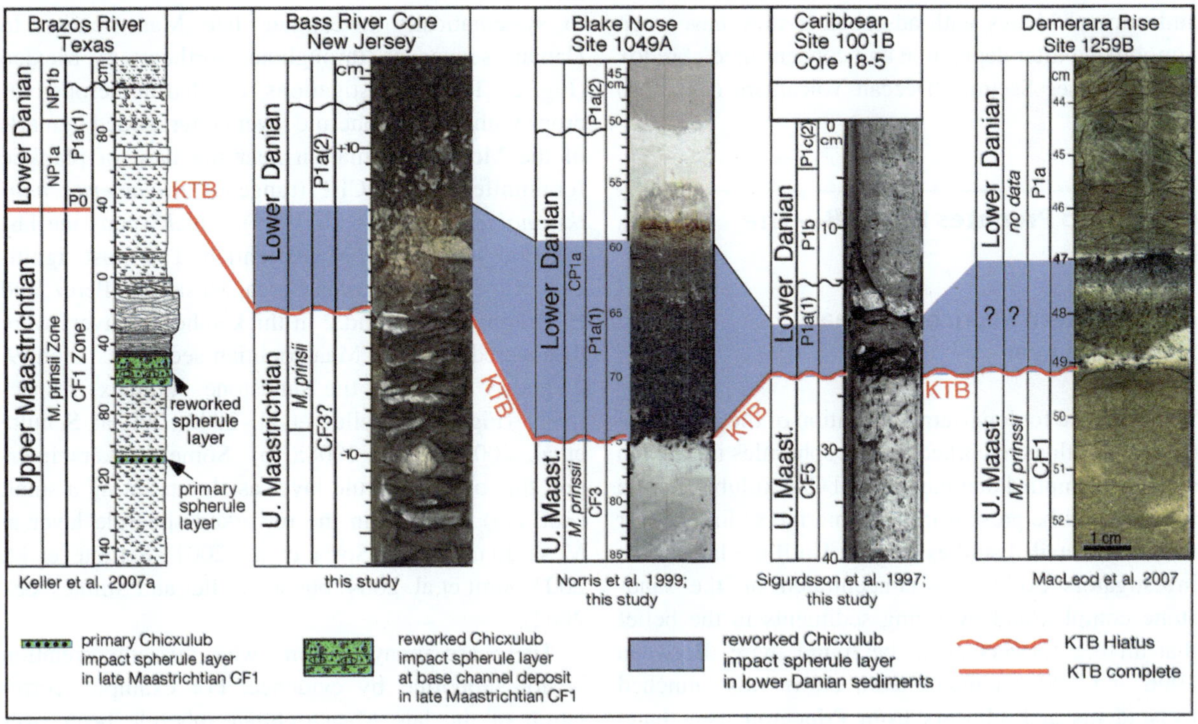

Fig. 11 Deep-sea Cretaceous–Tertiary sequences are often claimed to be complete and proof that the Chicxulub impact is KT in age. However, they are generally condensed and incomplete. Comparison with the more complete and expanded sequences along the Brazos River in Texas reveals the juxtaposition of Danian sediments with impact spherules as an artifact of an incomplete record in the Northwest Atlantic and Caribbean

P1a(1) is present in the spherule layer along with reworked Cretaceous species. This indicates that the spherules were eroded from older rocks and redeposited in the lower Danian. At the top is a 3 mm thick limonitic layer followed by 3–4 cm thick dark grey mottled and bioturbated ooze with abundant quartz and limestone chips that also indicate erosion. A mix of Cretaceous foraminifera and abundant *Parvularugoglobigerina eugubina* (48% in 58–61 cm interval) are also present above the reworked spherule layer. The Ir anomaly (1.3 ppb) within these sediments therefore does not represent primary deposition. Thus at Blake Nose, the stratigraphic record of Site 1049 is incomplete and the juxtaposition of the spherule layer and Ir anomaly therefore provides no proof that Chicxulub is KT age.

Demerara Rise ODP Site 1259 recovered a similar KT transition as Site 1049 on Blake Nose. A 2 cm thick spherule layer separates chalk of the late Maastrichtian *M. prinsii* and CF1 zones from overlying early Danian claystone with an Ir anomaly (1.5 ppb, Fig. 11) (MacLeod et al. 2007). The first Danian species appear about 3 cm above the spherule layer and mark the *Parvularugoglobigerina eugubina* P1a zone. The characteristic KT boundary clay layer is missing. To date no quantitative faunal data are available to assess the continuity of this stratigraphic record. Given the discontinuous records at Blake Nose and Bass River to the north, the Demerara Rise sedimentation may have undergone similar erosion by currents and the juxtaposition of spherules, Ir anomaly, and the KTB may be an artifact of a condensed record.

These deep-sea sequences cannot be considered proof that the Chicxulub is KT age or has the same origin as the Ir anomaly. Stratigraphic comparison with complete and expanded sequences along the Brazos River reveals that they are not only highly condensed but also have significant hiatuses at the KT boundary (Fig. 11). The spherule-rich condensed layers span the equivalent of the primary and reworked spherule layers in Brazos sections, plus the sandstone complex. In addition, the basal Danian appears missing in all three deep-sea sections. The presence of an Ir anomaly is no guarantee that the KTB is present because Ir and mineral phases with adsorbed Ir may have been remobilized after deposition and concentrated at redox boundary (see Section "Deccan Volcanism Linked to KTB").

Chicxulub Predates the KTB—The Evidence

Northeastern Mexico—Primary Spherule Layer

The evidence for long-term deposition of the sandstone complex with its reworked impact spherules in Mexico and Texas implied that the original Chicxulub spherule layer should be present in older marine sediments for which no detailed studies existed. Until the late 1990s investigators exclusively concentrated on the sandstone complex and overlying sediments in the belief that all the answers could be found there. Between 1996 and 2002, a major team effort was launched by students and advisors from Princeton, Neuchatel (Switzerland), and Karlsruhe (Germany) Universities to systematically investigate late Maastrichtian to Danian sequences throughout northeastern Mexico (Fig. 2). These investigations revealed numerous outcrops with impact spherule layers interbedded in marls of the Mendez Formation near the base of planktic foraminiferal zone CF1 (range of *Plummerita hantkeninoides*, Pardo et al. 1996), which spans the last 300,000 years of the Maastrichtian. The most significant of these are at Loma Cerca, Mesa Juan Perez, and El Peñon where 1 and 2 m thick spherule layers were discovered in upper Maastrichtian sediments about 9, 2, and 4 m below the sandstone complex, respectively (Fig. 12) (Keller et al. 2002, 2003a; Schulte et al. 2003; Keller 2008a, b). Some workers interpret this older spherule layer as slump, citing a small (<60 cm) fold within the reworked spherule layer at Mesa Juan Perez (Soria et al. 2001; Schulte et al. 2003; Smit et al. 2004, but see Keller and Stinnesbeck 2002).

There are many reasons why this interpretation is not supported by evidence. For example, correlation of the late Maastrichtian spherule layer over 35 km from El Peñon to Loma Cerca and Mesa Juan

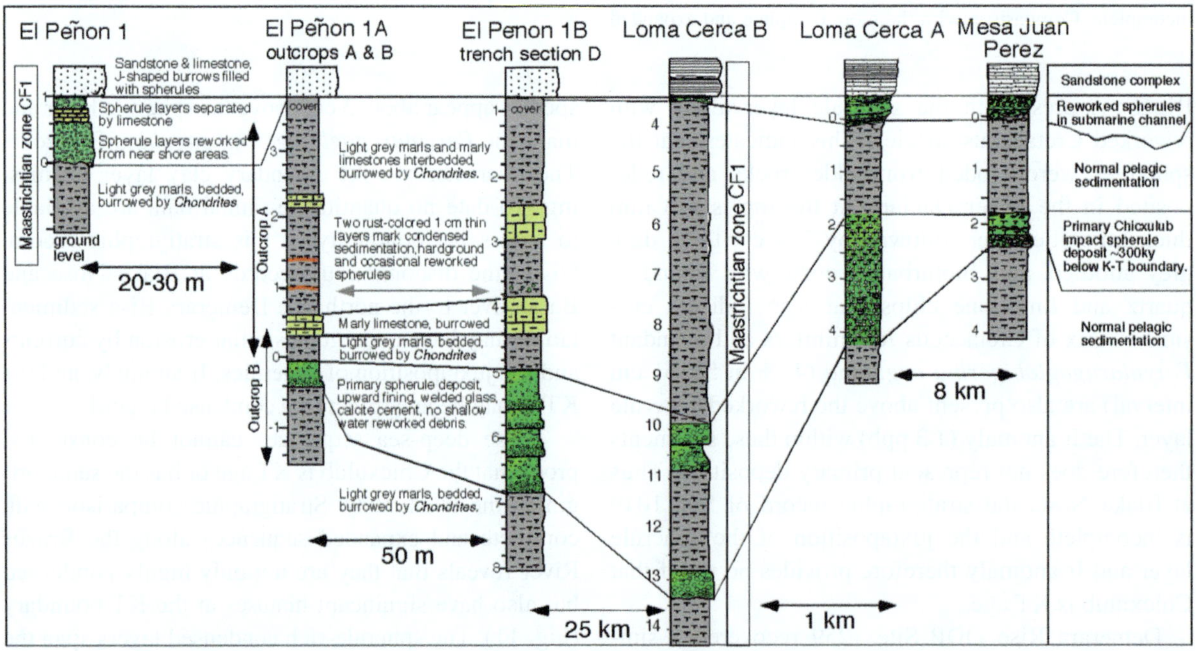

Fig. 12 Correlation of El Peñon 1 outcrops showing similar horizontal deposition of marls, marly limestone, and primary impact spherule layer, which were deposited in a submarine channel centered at El Peñon 1B. El Peñon can be correlated to Loma Cerca and Mesa Juan Perez sections at 25 and 35 km to the north, respectively. Variable erosion in submarine channels accounts for the reduced marl layer at Loma Cerca A and Mesa Juan Perez (modified from Keller et al. 2009c)

Perez reveals that slumps or tsunami deposition cannot account for such continuous emplacements (Fig. 12). Spherule layers at these sections are interbedded in pelagic marls of the Mendez Formation that show no evidence of disturbance. They are normally bedded, bioturbated pelagic marls that at El Peñon contain marly limestone layers as well as a condensed red clay that truncates burrows (e.g., hardground, hiatus) and a bentonite layer (Keller et al. 2009c). Such bedding and distinct lithologies are not characteristics of slump deposits. The lateral continuity of the Maastrichtian spherule layer at El Peñon can be traced for at least 80 m parallel to the sandstone complex above, which indicates normal deposition (Fig. 13). Over this distance the spherule layer remains a coherent unit, but thins from 2 m to 1 m and then to 25 cm, which suggests deposition in a depression or submarine channel. In the Mesa Juan Perez and Loma Cerca area, the variable vertical distance between spherules at the base of the sandstone complex and the late Maastrichtian spherule layer interbedded in marls below is largely due to the variable erosion in the submarine channels at the base of the sandstone complex.

With the slump argument Soria et al. (2001) and Schulte et al. (2003, 2010) contend that the late Maastrichtian spherule layer is the same as the reworked spherule layer(s) at the base of the sandstone complex. For the El Peñon section it was even suggested that seismic shaking from the impact caused marls to liquefy permitting the spherules to sink 4–5 m

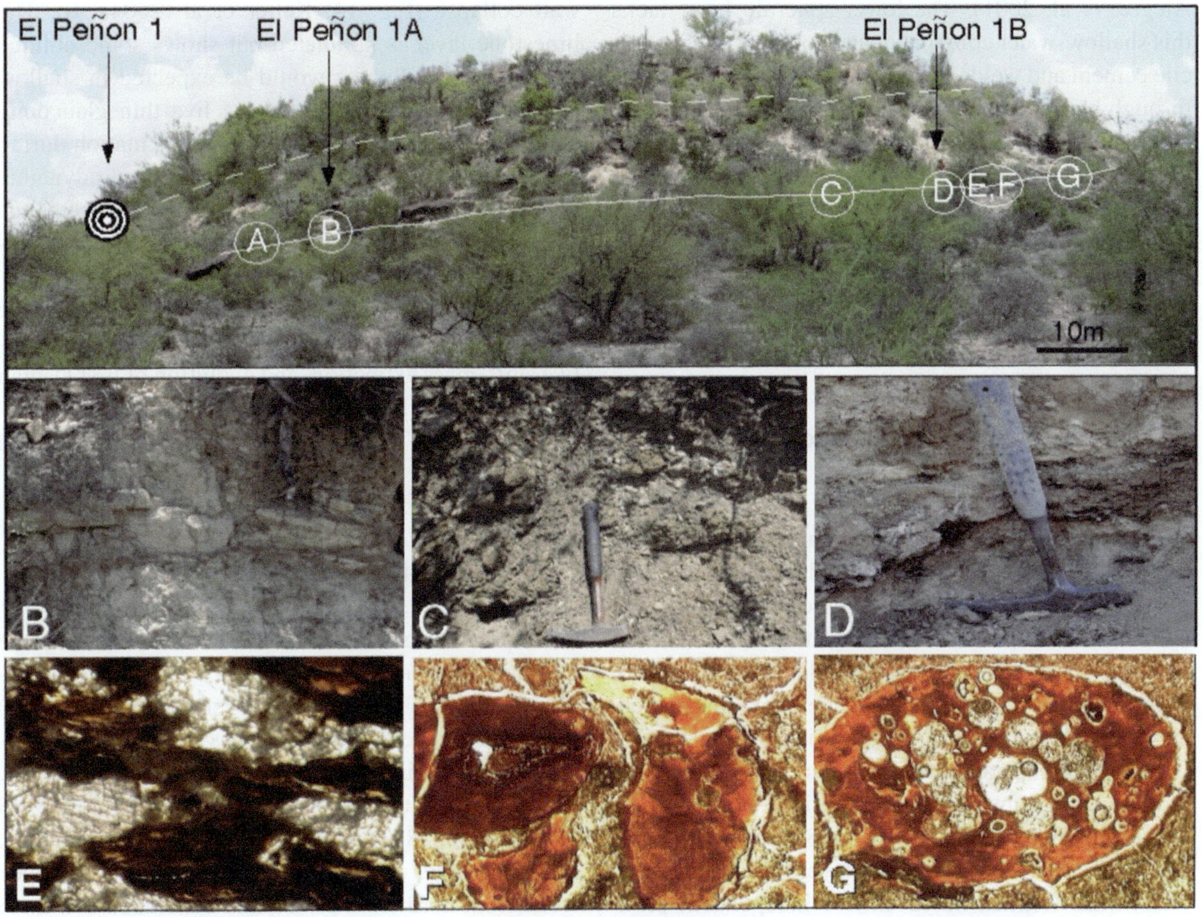

Fig. 13 El Peñon hill showing the base of the sandstone complex with reworked spherules (*dashed line*) dipping to the northeast. Parallel to the sandstone complex and 4–5 m below is the primary Chicxulub impact spherule ejecta layer (*solid line*) interbedded in late Maastrichtian marls. Locations **b**, **c**, and **d** illustrate exposures of the spherule layer. Welded impact spherules (**e**) and concave–convex contacts (**f**) indicate deposition while still hot. Oval spherule (**g**) with characteristic air bubbles. *Circle* marks classic El Peñon 1 outcrop (modified from Keller et al. 2009c)

and accumulate in the dense 2 m thick layer in which they are now found. Unexplained in this scenario is how this liquefaction/sinking would happen. How resettling could occur in a dense spherule layer with only calcite cement, with distinct sublayers of welded amalgamated glass (Fig. 13e–g). And how resettling could exclude the clastic shallow water debris, wood, plants or shallow-water foraminifera that characterize the reworked spherule layers from which they supposedly originated.

The two spherule layers, in fact, reflect very different depositional histories. The upper spherule layers at the base of the sandstone complex are eroded from shallow nearshore areas and transported across the shelf and slope to rest in submarine canyons at >500 m depth (Smit et al. 1992; Keller et al. 1994a, b; Alegret et al. 2001). The lower spherule layer lacks this shallow-water component and contains mainly calcite cement and welded glass that suggest rapid settling through the water column. In the upper part of the spherule unit spherules are dispersed in a marl matrix suggesting some current reworking. We consider this the primary impact spherule ejecta layer (Keller et al. in press). The stratigraphic position of this spherule layer near the base of zone CF1 is the basis for estimating the age of the Chicxulub impact as predating the KT boundary by about 300 ky.

Chicxulub Crater Well Yaxcopoil-1—Impact Breccia and KTB

Yaxcopoil 1 (Yax-1), drilled within the Chicxulub crater in 2001–2002, was expected to yield the final proof that this impact occurred precisely 65 m.y. ago and caused the mass extinction at the Cretaceous–Tertiary (KT) boundary (Dressler et al. 2003, 2004). Instead, contrary evidence was discovered. Yaxcopoil-1 recovered a 100 m thick impact breccia (suevite) with upward fining clasts and current bedding in the upper 15 m. Above the breccia is a 50 cm thick laminated limestone layer with the KT boundary at the top. The age of this limestone layer turned out to be latest Maastrichtian based on five independent proxies (sedimentologic, biostratigraphic, magnetostratigraphic, stable isotopic, and iridium) that revealed the Chicxulub impact predates the KT boundary by about 300,000 years and could not have caused the mass extinction (Keller et al. 2004a, b). This is demonstrated by the presence of five bioturbated glauconite layers, planktic foraminiferal assemblages of the latest Maastrichtian zone CF1, magnetostratigraphic chron 29R, and characteristic late Maastrichtian stable isotope signals (Fig. 14). Critics proposed that this limestone interval represents chaotic backwash and crater infill from the Chicxulub impact (Arz et al. 2004; Dressler et al. 2003, 2004; Smit et al. 2004). Jan Smit maintains to this day that there are no foraminifera in these sediments and the ones identified as such are just dolomite crystals, despite the independent confirmation by his own consultants (M. Caron) and Arz et al. (2004).

The 50-cm thick limestone layer shows no evidence of chaotic deposition from backwash and crater wall collapse (Fig. 14, Keller et al. 2004a, b). The limestone layer is laminated but shows some oblique 1 cm thick bedding as would be expected in shallow platform sediments. Each of the five thin glauconite clay layers represents condensed sedimentation during tens of thousands of years under low hydrodynamic conditions. Bioturbation is common and indicates the seafloor was colonized during deposition. The pre-KT age was based on late Maastrichtian biostratigraphy (zone CF1), magnetostratigraphy, and stable isotopes, all followed by the KT characteristic negative shift, mass extinction, and evolution of first Danian species. Chicxulub crater core Yaxcopoil-1 thus supports the pre-KT age findings from NE Mexico and Texas.

Correlation of Yucatan Wells

Yaxcopoil-1 was not the first crater evidence that the Chicxulub impact predates the KT boundary. Lopez-Ramos (1973, 1975) and Meyerhoff et al. (1994) described Cretaceous assemblages in limestones overlying breccias in Yucatan Pemex wells Y6 and C1, which are close to Yaxcopoil-1 (Yax-1) (Fig. 15). Ward et al. (1995, p. 873) re-examined the geophysical logs, sedimentology, and biostratigraphy of seven Pemex wells from northern Yucatan and concluded that there is evidence that "the impact breccia unit is overlain by about 18 m of uppermost Maastrichtian marls," suggesting an impact before the Cretaceous–Tertiary boundary.

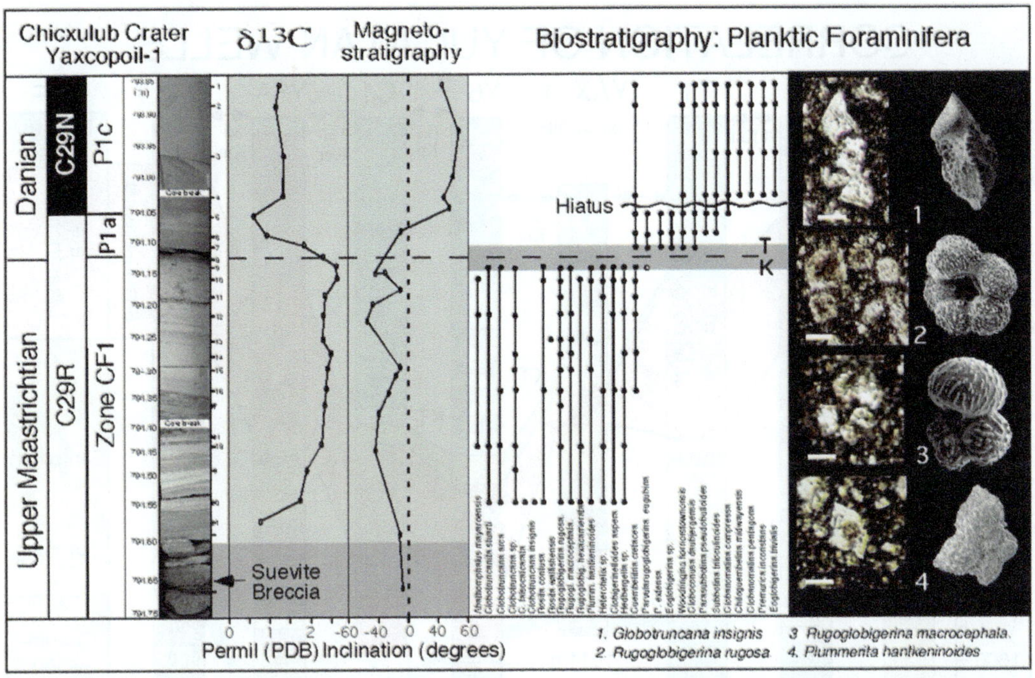

Fig. 14 Chicxulub impact crater core Yaxcopoil-1: Stratigraphy of the KT transition from the impact (suevite) breccia to the KT boundary shows a 50-cm thick limestone layer with five thin glauconite clay layers deposited over tens of thousands of years. Magnetostratigraphy and planktic foraminiferal assemblages indicate sediment deposition above the impact breccia occurred during C29R and zone CF1, which spans the last 300 ky of the Maastrichtian. Foraminifera are illustrated from thin sections of the dolomitic limestone (*left column*). For comparison, SEM illustrations are shown of the same species (*right column*). Modified from Keller et al. (2004a)

Stratigraphic correlation of Yucatan Wells shows a predominance of limestones, dolomites, and evaporites (anhydrites) below the impact breccia unit, which marks very shallow carbonate platform environments (e.g., sabkha, lagoons). An organic-rich dark shale layer was deposited during the late Cenomanian marine transgression and ocean anoxic event OAE2 (*Rotalipora cushmani* zone–*Whiteinella archeocretacea* zone) and can be correlated from wells Y2 to T1, Yax-1, Y5A, and Y4 (Fig. 15, Ward et al. 1995; Stinnesbeck et al. 2004). The stratified nature of these late Cretaceous sediments with correlatable biohorizons over several hundred kilometers within and outside the impact crater indicates normal sedimentation below the impact breccia, rather than chaotic distribution of megablocks. Moreover, Ward et al. (1995) and Stinnesbeck et al. (2004) concluded that the sediments are well preserved without the strong thermal alteration one would expect from a large impact.

The thickness of the impact breccia in the Yucatan wells varies from 100 m to 500 m, and some cores have a significant melt rock component (Y5A, C1, Y6, T1, Fig. 15). The petrography and geochemistry of the breccia unit in Yax-1 is detailed in Dressler et al. (2004). Ward et al. (1995) observed a layer of evaporite (anhydrite) in the breccia of well Y2 and thick (50–100 m) layers of dolomite in the breccia and melt rock of wells Y6, Y4 and C1 (Fig. 15). They could not determine whether these sediments are normally stratified or represent chaotic blocks. Limestone and dolomite layers were also observed within the breccia unit of the UNAM wells (Stinnesbeck "unpublished data").

Sediments above the impact breccia are predominantly limestones and marls, which indicates a deeper carbonate platform environment. Dolomites prevail only in wells Y5A and Y4 (Fig. 15). The contact between the breccia and overlying sediments is often not available for study in PEMEX or UNAM wells. As noted earlier, Cretaceous planktic foraminiferal assemblages are documented from wells Y6 and C1 by Lopez-Ramos (1973, 1975), and Meyerhoff et al. (1994), and from Yax-1 by Keller et al. (2004a, b)

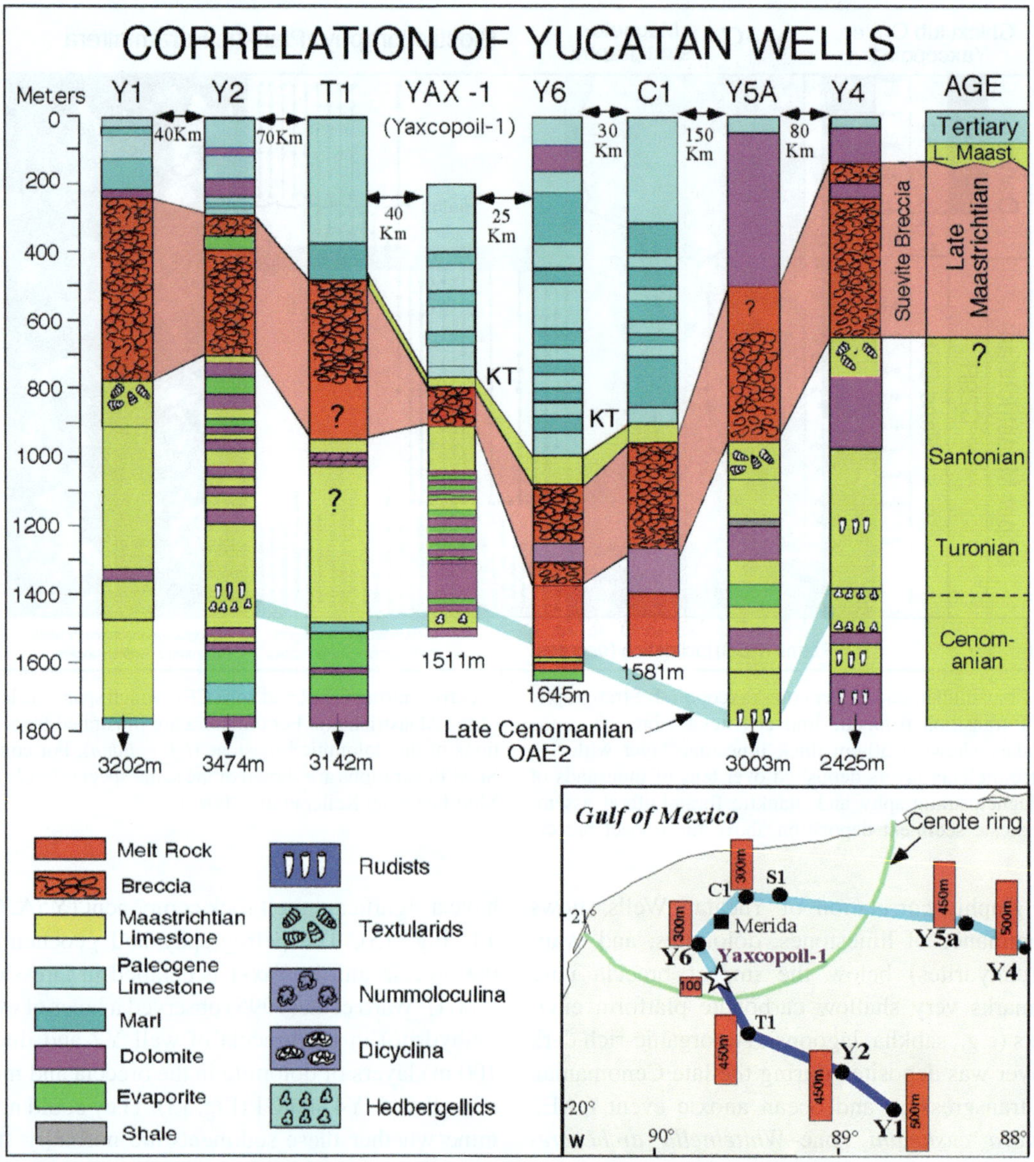

Fig. 15 Stratigraphic correlation of Yaxcopoil-1 (Yax-1) and PEMEX wells across northern Yucatan. Correlation is based on lithology, biostratigraphy, and electric logs (modified from Ward et al. 1995). Note the Maastrichtian-age limestone layer overlying the impact breccia in Yax-1, Y6, and C1 (modified from Ward et al. 1995)

(Fig. 14). These Cretaceous sediments indicate a pre-KT age for the Chicxulub impact, which correlates with observations in NE Mexico and Texas.

Texas—Primary Spherule Layer

The pre-KT age determined from the crater core Yaxcopoil-1 for the Chicxulub impact caused renewed controversy. After more than a dozen years of hailing Mexico's KT localities as providing unambiguous evidence in support of a KT age for the Chicxulub impact (e.g., Smit et al. 1996; Smit 1999), Mexico was now claimed to be too close to the impact crater to provide good age control (Smit et al. 2004; Kring 2007). This critique could be answered by KT localities along the Brazos River of Falls County, Texas, where sediment deposition occurred in a shallow middle to inner neritic environment without any trace of tectonic disturbance. The Texas KT localities were restudied, and two new wells drilled by DOSECC (Drilling, Observation and Sampling of Earths Continental Crust). The results provide confirmation not only of the pre-KT age of the Chicxulub impact but also of the sandstone complex and associated sea-level fall (Fig. 10) (Keller et al. 2007a).

In well Mullinax-1 (Mull-1), the KT boundary is 80 cm above the sandstone complex, including upward fining sediments (Fig. 16). The sediments consist of dark organic-rich bedded mudstones and claystones with burrows frequently infilled by pyrite, small shells, and small Cretaceous planktic foraminifera (mostly low-oxygen-tolerant heterohelicids and *Guembelitria cretacea*, Keller et al. 2009a). The KT boundary is marked by the mass extinction, by the $\delta^{13}C$ shift, and immediately above by the evolution of Danian species. These results were duplicated in outcrops (Keller et al. 2008a). The only significant difference between six sequences analyzed is the interval between the sandstone complex and the KTB, which varies from 20 to 80 cm due to erosion and paleotopography.

Fig. 16 New core Mullinax-1: GK and Jerry Baum examine the spherule-rich sandstone complex, upward fining, and 80 cm of dark organic-rich burrowed and bedded mudstone with late Maastrichtian microfossils (planktic foraminifera, nannofossils, palynomorphs), followed by the KT boundary mass extinction and characteristic carbon isotope shift (modified from Keller 2008b)

The spherule-rich clasts at the base of the sandstone complex described in Section "Texas: Sandstone Complex—Tsunami or Sea-Level Change?" (Fig. 9) indicate the existence of an older primary spherule layer. This primary Chicxulub impact spherule layer was discovered 45–65 cm below the sandstone complex in the Cottonmouth Creek section as a 2–4 cm thick yellow clay layer consisting of impact glass now altered to cheto smectite (Fig. 17, Keller et al. 2007a). Two reworked spherule layers at the base of the sandstone complex above show the same alteration to cheto smectite, which consists of nearly 100% Mg smectite. The same alteration is also observed in the three upward fining spherule layers of Mull-1, as well as in spherule-rich deposits throughout Central America (Keller et al. 2003b). Planktic foraminifera indicate deposition of the Brazos primary impact spherule layer near the base of zone CF1 predating the KT boundary by about 300,000 years similar to NE Mexico and Yucatan crater well Yaxcopoil-1 (Keller et al. 2007a, 2008a, 2009a). The pre-KT age of the Chicxulub impact was thus confirmed in three different and widely separated localities with the primary impact spherule layer discovered in NE Mexico and Texas.

In the Brazos sections, the primary Chicxulub impact spherule layer was deposited in a middle neritic environment at about 80 m depth and prior to the sea-level fall to inner neritic depth of less than 30 m, as indicated by the diversity and increased abundance of benthic foraminifera and low diversity and abundance of planktic foraminifera, beginning with the sandstone complex (Fig. 10). This sea-level fall exposed nearshore areas to erosion, including the primary spherule deposit, and scoured submarine channels into which the eroded and reworked sediments were transported and redeposited. These reworked sediments include glauconite, shell hash, impact spherules, and lithified clasts with spherules from the primary spherule deposit (Fig. 9). In deeper areas or depressions where submarine scour was more limited, the primary Chicxulub impact spherule layer was preserved (Fig. 17).

Fig. 17 The Cottonmouth Creek (a tributary of the Brazos River) shows the primary Chicxulub impact spherule layer now altered to Cheto smectite (yellow layer) between 45 and 60 cm below the sandstone complex with its two reworked spherule layers at the base. Note that the glass spherule alteration to Cheto smectite is the same in the all three layers. The KT boundary at this locality is about 40 cm above the top of the sandstone complex (modified from Keller et al. 2009a)

Kill Effect of the Chicxulub Impact

Planktic Foraminifera as Environmental Proxies

The Brazos and NE Mexico sections provide an unprecedented opportunity to evaluate the biotic effects of the Chicxulub impact in comparison with the KT mass extinction in both upper slope and marginal shelf environments within 600–1000 km of the Chicxulub impact crater. Planktic foraminifera are the best microfossil group to evaluate biotic and environmental changes across the KT transition for several reasons. These single-celled marine microorganisms form shells of calcium carbonate in isotopic equilibrium with the seawater in which they lived, and their shells thus record climate and productivity conditions during their lifetimes. Today, as in Cretaceous times, they live in the upper water mass of oceans from high to low latitudes and are very sensitive to variations in temperature, salinity, oxygen, and nutrients, which are reflected in species abundance variations, extinctions, and evolutionary diversification. Their high-evolutionary turnover rates at times of biotic crises have made them ideal index fossils for high-resolution biostratigraphic dating and correlations.

At the KTB planktic foraminifera suffered the most devastating mass extinction with 2/3 of the species extinct (first documented by Keller 1988) and 1/3 surviving into the early Danian, with only one long-term survivor (e.g., Keller 1996, 2001; Molina et al. 2006). Evolution of the new early Danian species began almost immediately after the mass extinction and was followed by rapid evolutionary diversification. No other microfossil group records such immediate and devastating biotic effects. For all these reasons, planktic foraminifera are the ideal fossil group for evaluating the biotic effects of catastrophes, such as impacts and volcanism, as well as long-term climate and sea-level changes.

Biotic Effects of the Chicxulub Impact

In the Brazos sections, sea level fell from about 80 m to 20–30 m depth and even lower in some areas depending on the paleotopography (Fig. 10). The primary spherule layer was deposited in deeper waters, the reworked spherule layers during the low sea level

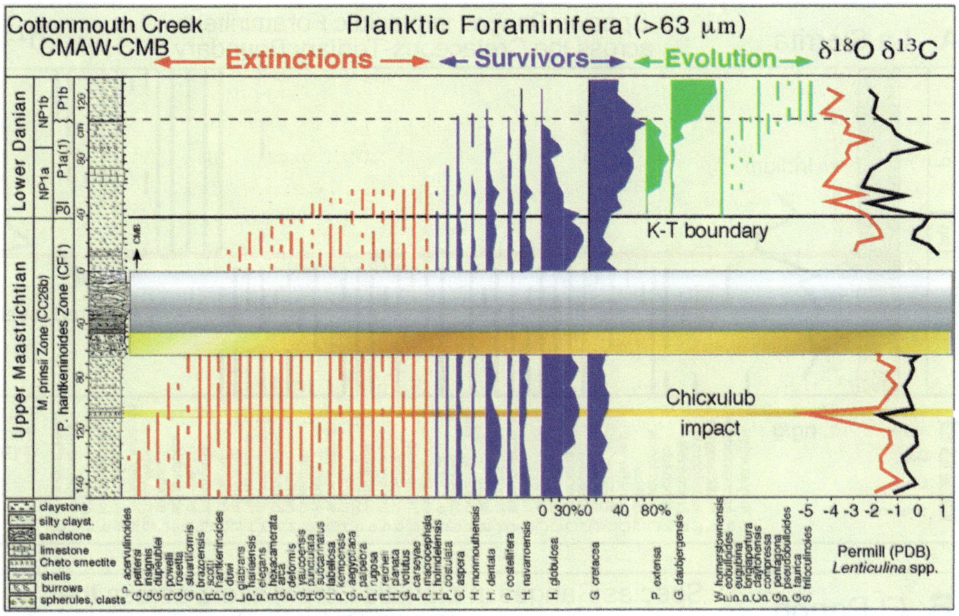

Fig. 18 Cottonmouth Creek KT transition with planktic foraminiferal ranges and species abundances across the primary Chicxulub impact layer (Cheto smectite), sandstone complex with reworked spherules, and KTB above. Most Cretaceous species persist to the KTB marked by the $\delta^{13}C$ shift and first Danian species. Gradual decrease in species richness is due to sea-level fall and exclusion of deeper dwelling larger species. Negative $\delta^{13}C$ and $\delta^{18}O$ excursions in primary Chicxulub impact layer is not associated with significant changes in diversity or abundance of species and may be diagenetic in origin (modified from Keller et al. 2009a)

(sandstone complex) and the KTB during the subsequent sea-level rise. In this shallow, but variable environment, it is now possible to evaluate the biotic effects of the Chicxulub impact, the sea-level fall, and the KTB event based on diversity and abundances of planktic foraminifera (Fig. 18).

Across the primary impact spherule layer planktic foraminifera show no significant changes. No species went extinct and no significant environmental changes occurred as a result of the Chicxulub impact (Keller et al. 2009a). The $\delta^{13}C$ shift at the primary impact spherule layer may be due to diagenetic alteration, although short-term environmental effects due to the impact cannot be ruled out (Fig. 18). The four species that disappeared at or below the primary spherule layer, as well as rare and sporadically present species, are deeper dwellers and known to become extinct at the KTB. Their early disappearance or sporadic occurrence is due to the shallowing sea.

The sea-level drop to <30 m marked by the sandstone complex resulted in high stress conditions for all species, as indicated by the rare and sporadic occurrences of species known to go extinct by the KTB. In this shallow environment all larger, deeper dwelling species had disappeared, though some dwarfed forms survived (Fig. 18) (Keller et al. 2009a). Most of the surviving species also became dwarfed as a result of the high stress conditions (MacLeod et al. 2000; Keller and Abramovich 2009).

The KT boundary is marked by the mass extinction, characteristic $\delta^{13}C$ shift, and evolution of first Danian species (Fig. 18). A number of small species (e.g., heterohelicids, hedbergellids) survived into the early Danian as indicated by their Danian stable isotope signals (Barrera and Keller 1990). This gradual species extinction pattern across the impact spherule layer, sandstone complex, and the KT boundary appears unequivocally one of a gradually changing environment related to the fall in the sea level from middle to inner neritic depths. The observed mass extinction pattern is consistent with shallow sequences worldwide (e.g., Denmark, Tunisia, Argentina, Keller et al. 1993, 1998, 2007b).

In NE Mexico, the biotic effects of planktic foraminifea were analyzed across the primary Chicxulub impact spherule layer in the upper-slope environment of El Peñon and Loma Cerca A and B (Keller et al. 2002; Keller et al. 2009c). No biotic effects are apparent across the primary impact spherule layer (Fig. 19). No species went extinct even within

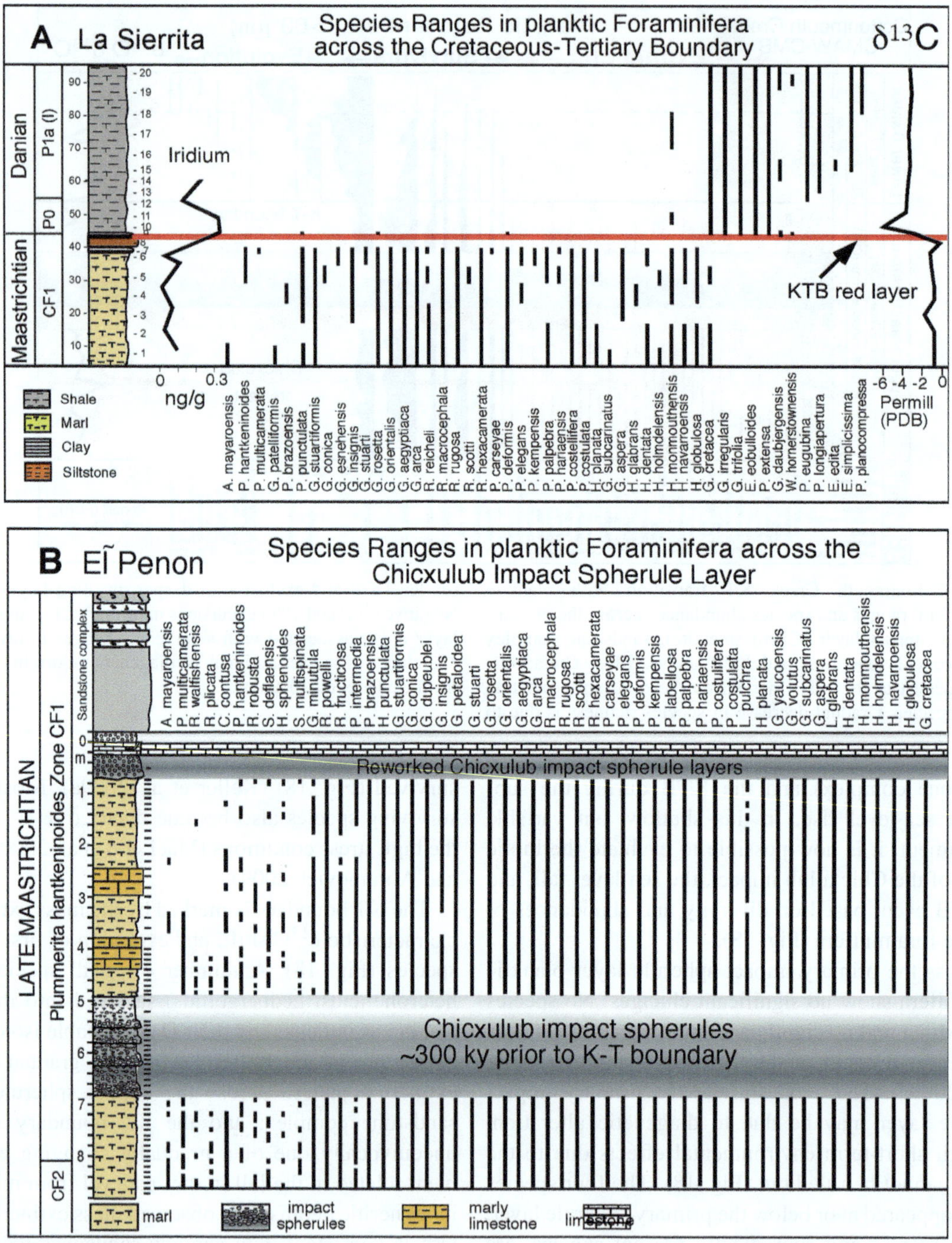

Fig. 19 (**a**) The KT transition at La Sierrita in NE Mexico shows the characteristic mass extinction pattern including a red layer with a small Ir anomaly and carbon isotope shift, but no impact evidence. (**b**) A 2 m thick primary impact spherule layer is present in zone CF1, 4–5 m below the reworked impact spherules at the base of the sandstone complex. This impact spherule layer predates the KTB by ~300 ky. No species extinctions occurred at this time (modified from Stinnesbeck et al. 1996; Keller et al. 2009c)

600 km of the Chicxulub impact crater. In contrast, the mass extinction is characterized by the extinction of all large, complex tropical and subtropical species at or shortly before the KT boundary, as shown for the La Sierrita section about 25 km north of El Peñon (Fig. 19). This mass extinction pattern is consistent with deeper marine sequences worldwide. The absence of species extinctions or any significant environmental effects associated with the primary impact spherule layer in Mexico and Texas indicates that the destructive effects of the Chicxulub impact have been vastly overestimated.

A survey of impact craters and mass extinctions over the past 500 years reveals that apart from the K–T boundary, none of the five major mass extinctions are associated with an impact (Courtillot 1999; Wignall 2001; Keller 2005). The Chicxulub crater with a maximum diameter of 180 km is the largest known impact. Other well-studied impacts that show no extinctions or significant other biotic effects include the 90–100 km in diameter late Eocene Chesapeake Bay and Popigai craters and the 100–120 km in diameter, the late Triassic Manicouagan and the late Devonian Alamo and Woodleigh craters (Montanari and Koeberl 2000; Poag et al. 2002; Keller 2005).

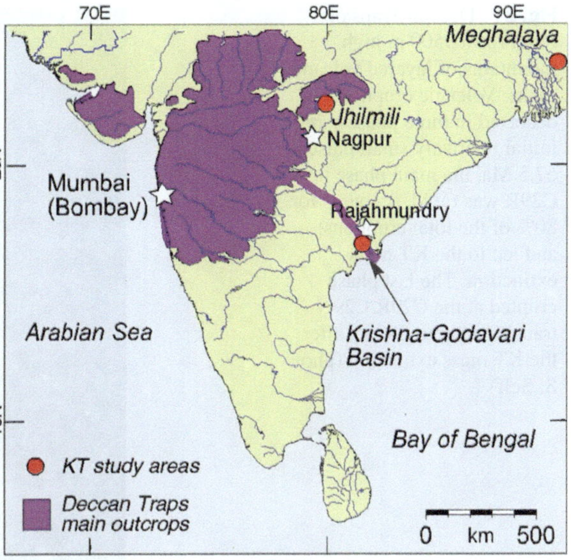

Fig. 20 Map of India with main Deccan Volcanic Province. The longest lava flows ranged 800–1000 km across India and out to the Gulf of Bengal. The KT boundary was first identified in Rajahmundry quarries in intertrappean sediments between these lava flows, which mark the main phase of Deccan volcanism. The KTB has also been identified in intertrappean sediments at Jhilmili and in sections of Meghalaya (modified from Keller et al. 2009d)

Deccan Volcanism and the KT Mass Extinction

Deccan volcanic eruptions once covered most of India and even today with much of it eroded or buried, exposures still cover an area the size of France (Fig. 20). Lava eruptions piled up flow after flow to several thousand meters thick, which today still form 3500 m high mountains (Fig. 21). Some massive eruptions reached up to 1500 km across India and out to the Gulf of Bengal, forming the longest lava flows known on Earth (Self et al. 2008a). It is therefore no surprise that scientists advocated Deccan volcanism as the major contributor or cause of the KT mass extinction even before the impact theory was proposed (McLean 1978) and through the 1980–1990s (McLean 1985; Courtillot et al. 1986, 1988; Courtillot 1999). More recently continental flood basalts (CFB) have been correlated with most major mass extinctions (Wignall 2001), leading Courtillot and Renne (2003) to suggest that this may be the general cause of mass extinctions. But acceptance of CFB volcanism as the catastrophe that led to the extinction of the dinosaurs and many other groups has lagged for two main reasons: (1) the popularity of the Chicxulub impact theory as the cause for the mass extinction and (2) the absence of a direct link between the mass extinction and Deccan volcanism. Indeed, until very recently it was not possible to pinpoint the KTB within the Deccan lava pile (Keller et al. 2008b).

Age of Deccan Volcanism

The Deccan Traps (named for the stepwise sequence of successive lava flows) and its correlation with the KTB mass extinction is the most extensively studied CFB volcanism. However, a direct correlation has been difficult to establish because the estimated duration of Deccan volcanism varied from less than one to several million years based on paleomagnetic studies and $^{40}Ar/^{39}Ar$ dating (Courtillot et al. 1986, 1988, 2000; Duncan and Pyle 1988; Vandamme et al. 1991; Vandamme and Courtillot 1992; Venkatesan et al. 1993, 1996; Baksi et al.1994; Raju et al. 1995; Hofmann et al. 2000; Widdowson et al. 2000; Sheth

Fig. 21 Deccan Traps of India form 3500 m high mountains of layered volcanic rocks. Volcanic eruptions occurred in three phases: an initial relatively small pulse at 67.5 Ma, the main phase in C29R was rapid, accounts for 80% of the total eruptions, and led to the KT mass extinction. The last phase erupted at the C29R/C29N transition about 280 ky after the KT mass extinction (photo S. Self)

et al. 2001). In the past few years, significant advancements have been made with respect to the age of volcanic eruptions based on comprehensive $^{40}K/^{40}Ar$ and $^{40}Ar/^{39}Ar$ dating (e.g., Knight et al. 2003, 2005; Pande et al. 2004; Baksi 2005; Chenet et al. 2007, 2008) and geochemical characterization of different eruptions (Jerram and Widdowson 2005; Jay and Widdowson 2008; Jay et al. 2009).

Based on these studies, Chenet et al. (2007, 2008) proposed that Deccan volcanism occurred in three short mega-pulses with the initial smaller pulse around 67.5 Ma near the C30R/C30N transition, the largest pulse in C29R and the last pulse in the early Danian spanning the C29R/C29N transition. Nearly 80% of the total Deccan eruptions are estimated to have occurred during the main phase in C29R, possibly over a very short time period (e.g., on the order of thousands of years). This mega-pulse consists of several major eruptive events with volumes ranging from 20,000 km^3 to 120,000 km^3, attaining a thickness up to 200 m and emplaced over hundreds of km (Chenet et al. 2008). The longest lava flows reached 1500 km across India to Rajahmundry and out to the Bay of Bengal (Fig. 20).

Deccan Volcanism Linked to KTB

Until recently, the best link to the KTB was based on lava flows that straddle thick intertrappean sediments exposed in quarries of Rajahmundry (Fig. 22). These lava flows, known as lower and upper Rajahmundry traps, are reversed and normal polarity zones in C29R and C29N, respectively (Subbarao and Pathak 1993). The best age determinations to date are based on $^{40}K/^{40}Ar$ (absolute) and $^{40}Ar/^{39}Ar$ (relative) dates of plagioclase separates (Fig. 23) (Knight et al. 2003, 2005; Baksi 2005; Chenet et al. 2007). Error bars for these radiometric ages are large (1% or 0.6 m.y.), which permit no determination of the KTB position. Nevertheless, radiometric ages for the upper trap are within the C29R/C29N transition. But the age for the lower trap is not well constrained, though it still overlaps with C29R of Cande and Kent (1991). Based on paleomagnetic, radiometric, and geochemical data, the upper and lower Rajahmundry traps can be correlated to the Mahalabeshwar and Ambenali Formations, respectively, in the main Deccan volcanic province (Chenet et al. 2007; Jay and Widdowson 2008; Jay et al. 2009).

Despite the improved dating of upper and lower traps in the Rajahmundry quarries, a direct link between volcanism and the mass extinction remained elusive without marine intertrappean sediments with age diagnostic microfossils. A search for marine sediments in the main Deccan province remained futile because intertrappean beds were largely deposited in terrestrial to fluvio-lacustrine environments (Cripps et al. 2005). Previous studies reported marine microfossils spanning the KTB in cores from the Krishna-Godavari Basin (Fig. 20), though the precise position

Fig. 22 Intertrappean sediments between lower (C29R) and upper (C29R/C29N) lava flows in the Gauriputnam Quarry of Rajahmundry. These intertrappean sediments were deposited in shallow estuarine to marine sediments that contain planktic foraminifera of the earliest Danian (zones P0–P1a). Four of the longest lava flows mark the end of the main phase of the Deccan eruptions in C29R and the KT mass extinction in the Rajahmundry area and Krishna–Godavari Basin

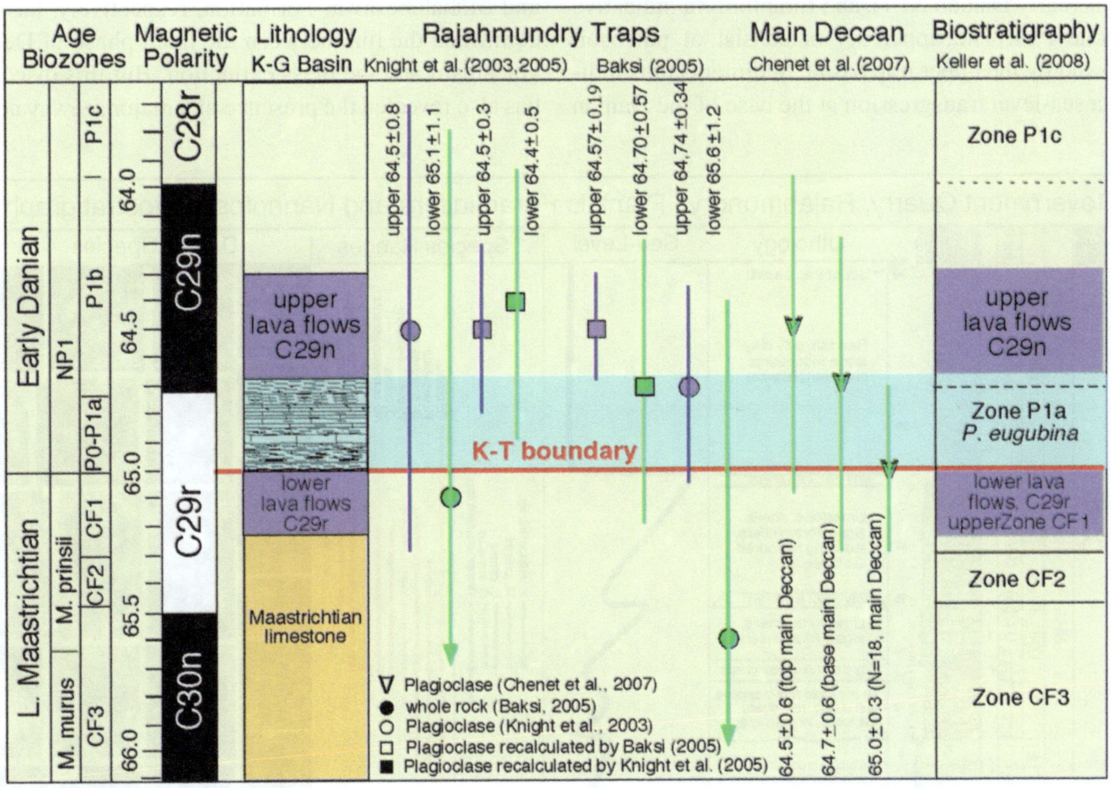

Fig. 23 $^{40}K/^{40}Ar$ and $^{40}Ar/^{39}Ar$ ages of the lower and upper Rajahmundry Deccan traps yield ages with an accuracy of 1% or ± 0.3 m.y., which is too imprecise to locate the KT boundary. Planktic foraminiferal biostratigraphy places the lower trap in the latest Maastrichtian with the uppermost of four lava flows at the KT boundary and mass extinction (modified from Keller et al. 2008b)

of the KTB could not be determined (Raju et al. 1994, 1995; Bhandari 1995; Jaiprakash et al. 1993; Saxena and Misra 1994; Prasad and Pundeer 2002; Khosla and Nagori 2002).

Keller et al. (2008b) concentrated on the thick intertrappean sediments of the Rajahmundry quarries (Fig. 22), which revealed the first direct link between the KT mass extinction and Deccan volcanism based on planktic foraminifera. In the Rajahmundry area intertrappean sediments in quarries show similar biostratigraphy and depositional environments as shown for the Government Quarry (Fig. 24). Sediment deposition above the lower trap basalt begins with dolomitic mudstone with clasts containing the small earliest Danian planktic foraminifera indicative of Zone P0 or base of zone P1a. In the area 4 m above, sediments vary from silty claystone with rare shells and foraminifera to limestones with common shells, calcareous nannofossils (early Danian NP1), and foraminifera indicative of zone P1a. The upper 3.4 m consist of paleosoil followed by the upper trap basalt. Sediments thus indicate a sea-level transgression at the base of the Danian followed by fluctuating estuarine to inner neritic environments and a return to terrestrial deposition well prior to the arrival of the upper trap basalt (Fig. 24).

Biostratigraphic data thus indicate that the lower Rajahmundry trap directly underlies earliest Danian sediments. This indicates that the mass extinction coincides with the end of the main phase of Deccan volcanism as represented by the last of at least four of the longest (800–1000 km) lava flows that are observed in the Krishna-Godavari Basin (Jaiprakash et al. 1993; Self et al. 2008a, b).

A surprising discovery of early Danian zone P1a planktic foraminifera was made in intertrappean sediments at Jhilmili, in Central India (Fig. 20, Keller et al. 2009b, d). These intertrappeans were considered to be terrestrial and of late Maastrichtian age. The presence of early Danian zone P1a planktic foraminifera between lower and upper trap basalts of the Ambenali and Mahalabeshwar Formation, respectively, has thus confirmed the link between the main phase of Deccan volcanism and the mass extinction. But this discovery has also revealed the presence of a major seaway across

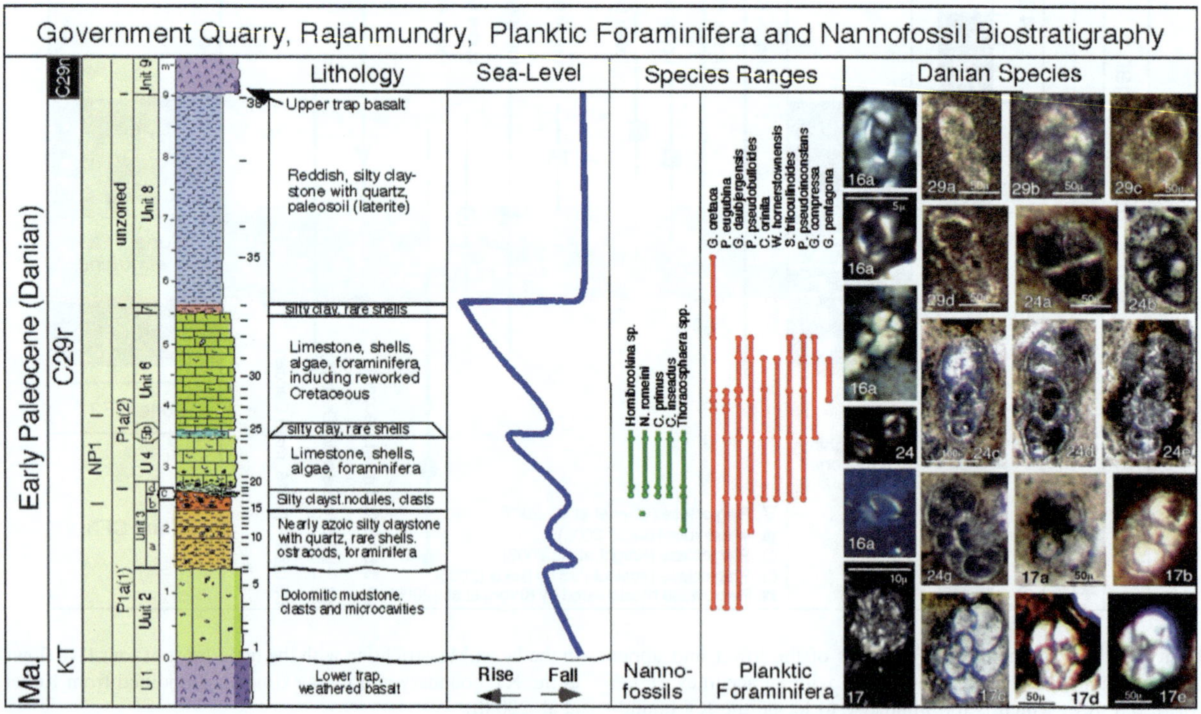

Fig. 24 Lithology, biostratigraphy, and sea-level changes of intertrappean sediments between lower and upper Traps at the Government Quarry, Rajahmundry. Sediment deposition after the main phase of volcanic eruptions occurred in a shallow estuarine to marine environments with repeated emersions. Planktic foraminifera and calcareous nannofossils mark this time as earliest Danian zones P0–P1a, immediately following the mass extinction (modified from Keller et al. 2008b)

India during the KT transition. The seaway likely followed the Narmada-Tapti rift extending from the west 800 km across India and existed already during the late Cenomanian marine transgression (Chiplonkar and Badve 1968; Badve and Ghare 1977; Sahni 1983).

Discussion

Chicxulub Controversy

Over the past 15 years numerous discoveries revealed that the Chicxulub impact is not KT age and therefore could not have caused the mass extinction. The evidence is multi-disciplinary, including paleontology, biostratigraphy, magnetostratigraphy, stable isotope geochemistry, trace element geochemistry, and mineralogy. Moreover, the findings have been duplicated in numerous KT sequences in NE Mexico (e.g., Adatte et al. 1996; Keller et al. 1997, 2003a, b; Stüben et al. 2005), confirmed in the Chicxulub crater well Yaxcopoil-1, and Pemex wells (Ward et al. 1995; Keller et al. 2004a, b), and once again duplicated and confirmed in outcrops and cores along the Brazos River in Texas (Gale 2006; Keller et al. 2007a, 2008a, 2009a). The interdisciplinary nature of the studies, the duplication and confirmation of results in three widely separated regions (Yucatan, NE Mexico, Texas), and very different environments (e.g., impact crater, slope environment, and shallow middle to inner shelf) provide the best evidence that these findings are not artifacts of the rock record (see Keller 2008a) but represent the age and environmental conditions during the KT transition.

Ordinarily, just a fraction of the evidence accumulated would suffice to make a convincing argument. But this is the KTB interval, and even an overkill of accumulated evidence showing that Chicxulub is not the KT killer will leave many scientists firmly believing in the impact theory as the sole cause for the mass extinction. A commonly expressed sentiment to this mountain of evidence is this: *I can't believe it. How could so many be so wrong for so long?* Counterarguments have been weak. They range from outright denial that the evidence exists (e.g., foraminifera and glauconite layers don't exist in crater core Yaxcopoil-1, Smit et al. 2004) to arguing for regional tectonic and seismic disturbance based on no more than occasional small folds (<60 cm) (Soria et al. 2001; Schulte et al., 2003; Smit et al. 2004) and to redefining the KTB based on impact evidence (Smit et al. 1996; Smit 1999; Schulte et al. 2006, 2008; Arenillas et al. 2006; Molina et al. 2006). The latter invariably leads to circular reasoning: Chicxulub is KT age, therefore the presence of Chicxulub impact spherules marks the KTB (see Schulte et al. 2008 and reply by Keller et al. 2008a). A group of 41 authors recently published an article in Science concluding that Chicxulub is precisely KTB in age and the cause of the KTB mass extinction as originally proposed 30 years ago (Schulte et al. 2010). Data to the contrary, as presented in this paper were largely ignored or misrepresented in reaching this conclusion.

The controversy over the cause of the end-Cretaceous mass extinction, which began 33 years ago, thus rages on supported by the popular consensus that the Chicxulub impact caused the mass extinction. Any evidence to the contrary is generally greeted with disbelief, citing the lack of consensus. But any decades-old controversy will never achieve consensus nor is consensus a precondition to advance science and unravel truth. What is necessary is careful documentation of results that are reproducible and verifiable. But convincing scientists that a long-held belief in the impact theory is wrong demands extraordinary documentation of verifiable evidence. This review shows that this step has been taken and the extraordinary documentation of verifiable evidence has been amassed.

Impact Spherules and KTB

In principle, the stratigraphic position of Chicxulub impact glass spherules in any sedimentary sequence should mark the time of this impact. In reality, impact spherules are frequently eroded, transported, and redeposited after their primary deposition. As a result, multiple impact spherule layers are common in late Maastrichtian sediments of north and central Mexico and Texas, early Danian sediments of southern Mexico, Guatemala, Belize, and Haiti (Keller et al. 2003a, b, 2009c), and deep-sea sections of Bass River, New Jersey, and Blake Nose (Fig. 11). Therefore, only the stratigraphically oldest spherule layer represents the time of the impact. This means that a careful search for the oldest and primary spherule layer must be conducted and its stratigraphic position determined relative to the KTB. But the primary spherule layer may only be rarely preserved due to erosion. To date the oldest and primary Chicxulub impact spherule layer has been discovered in late Maastrichtian zone CF1 sediments and predating the KTB by about 300,000 ± 30,000 years in Texas and El Peñon and Loma Cerca (Fig. 12) in NE Mexico (Keller et al. 2007a, 2009a, c).

Why are Chicxulub impact spherules so commonly reworked? Their stratigraphic positions yield clues. The most prominent and best-known spherule deposits infill the base of submarine channels in the deep-water (>500 m) continental slope sections of NE Mexico (see Section "NE Mexico Sandstone Complex—Tsunami or Sea-Level Change?") and shallow shelf sections of the Western Interior Seaway that is now Texas (Section "Texas: Sandstone Complex—Tsunami or Sea-Level Change?"). In both areas scoured channels are infilled with a spherule-rich unit at the base with abundant glauconite, shallow water debris (plants, wood fragments, shallow water benthic foraminifera, and shell fragments), and clasts from the underlying sediments (Fig. 4). Above it are sandstones that represent debris flows in the slope environment (Keller et al. 1997) and hummocky cross-bedded current deposition in the shallow Texas sections (Gale 2006; Keller et al. 2007a). Deposition of these spherule-rich sandstone complexes in NE Mexico and Texas occurred during the latest Maastrichtian sea-level fall and subsequent early sea-level rise (Fig. 8).

Multiple spherule-rich layers are common in early Danian sediments of Guatemala, Belize, and Haiti within zone P1a that marks the first 180 ky of the Paleocene (Keller et al. 2001, 2003b). These deposits are associated with the sea-level falls of zone P1a, at the P0/P1a, subzone boundary P1a(1)–P1a(2), and at the top of zone P1a (Keller and Stinnesbeck 1996; Keller et al. 2007a). With each drop in sea-level erosion, transport and redeposition of sediments, including impact spherules, occurred in the shallow platform environments of Central America. On a global scale, these sea-level falls are associated with hiatuses (MacLeod and Keller 1991; Keller and Stinnesbeck 1996).

In deep-sea sections, sea-level falls are generally associated with condensed sedimentation and hiatuses because of intensified bottom current circulation as first documented on a global basis by Keller and Barron (1983). This also appears to be the case in the deep-sea sections of Demerara Rise, Blake Nose, and Bass River, where the Danian sediments and impact spherules are juxtaposed (Fig. 11). The multiple Chicxulub impact spherule layers in late Maastrichtian and early Danian sediments thus can be easily explained by sea-level changes and associated erosion, transport, and redeposition of sediments. The stratigraphic position of impact spherule layers therefore does not a priori indicate the time of the Chicxulub impact, but most often reflects erosion and redeposition during subsequent sea-level changes. Only in NE Mexico and Texas has the primary impact spherule deposit been discovered to date.

Iridium Anomaly and the KTB

Anomalous concentrations of iridium at or near the KTB are frequently used as KT defining criteria and proof that an impact caused the mass extinction. The Chicxulub impact is generally assumed to have been this impact, although no Ir anomaly or other PGE concentrations have ever been observed in Chicxulub impact spherule layers (review in Keller 2008a). The widespread presence of an Ir anomaly at the KTB has led Keller et al. (2002, 2003a) and Stüben et al. (2005) to suggest that a second large impact at KT time may have been the cause. But the absence of any biotic effects attributable to the Chicxulub impact in Mexico and Texas suggests that even a larger impact alone would not likely have been sufficient to cause the KT mass extinction. In addition, there is currently no credible evidence of a second larger impact at KT time. Another problem is that the KT Ir anomaly is frequently not just a single anomaly as commonly reported, but multiple anomalies of diverse origins (e.g., impact, volcanic, redox conditions, Graup et al. 1989; Stüben et al. 2005; Grachev et al. 2005; Keller 2008a) that have yet to be fully understood. The Ir anomaly can thus no longer be considered sufficient credible evidence for a large impact at the KT boundary.

Ir anomalies are generally interpreted as indictors of extraterrestrial sources, such as asteroids, comets, cosmic dust, or impact ejecta. This has been justified on the basis that iridium is highly depleted in Earth's crust but enriched in some asteroids and comets and in Earth's interior, from where it is brought to the surface by volcanic activity. Because volcanic iridium is expected to have extruded over a long time period, a sharp peak in iridium concentrations is generally interpreted as a cosmic marker. However, sharp peaks most often result from condensed sedimentation. In addition, Ir anomalies typically occur in organic-rich shales or clays, which serve as low-permeability redox boundaries. In these sediments Ir can move both upward and downward from redox boundaries (Gilmore et al. 1984, Tredoux et al. 1988, Sawlowicz 1993, Wang et al. 1993).

The complex behavior of iridium and other PGEs in sedimentary environments is still poorly understood, but the variability in distribution and multiple anomalies caution against considering it as solely a cosmic interpretation (Gertsch et al. 2011). Now that Deccan volcanism is understood to have occurred in massive rapid eruptions over very short time periods with the main eruptions centered at KT time, Ir anomalies may well contain a major component of volcanic origin.

Deccan Volcanism and KTB

The critical link between Deccan volcanism and the KT mass extinction has now been established in intertrappean sediments in Rajahmundry quarries of southeastern India and subsequently confirmed in intertrappean sediments at Jhilmili in central India (Keller et al. 2008b, 2009b). Paleontologic, paleomagnetic, radiometric, and geochemical studies have linked the KT mass extinction to the longest lava flows on Earth, spanning over 1500 km across India (e.g., Knight et al. 2003, 2005; Chenet et al. 2007, 2008; Jay and Widdowson 2008; Keller et al. 2008b, 2009b; Self et al. 2008a, b). The main phase of Deccan volcanism, which encompasses 80% of the Deccan lava pile, can now be positively linked to the KT mass extinction. In the Rajahmundry area at least four closely spaced lava flows reached across India and out to the Gulf of Bengal.

Environmental consequences of these massive eruptions were likely devastating. Based on rare gas bubbles preserved in Deccan volcanic rocks, Self et al. (2008a) estimate annual gas rates released by Deccan lavas at many times the rate of anthropogenic emissions of SO_2 and more than an order of magnitude greater than the current global background volcanic emission rate. Chenet et al. (2007, 2008) estimate gas emissions based on volume of Deccan lavas. Concentrating on the largest 30 eruption pulses, they estimated that each pulse injected up to 150 GT of SO_2 gas over a very short time (decades). Thus each one of 30 Deccan eruption pulses injected SO_2 in quantities at least the equivalent of the Chicxulub impact (e.g., 50–500 GT). The total Deccan volcanic eruptions are estimated to have released 30 to 100 times the amount of SO_2 released by the Chicxulub impact.

It is not just the sheer volume of SO_2 injection, but also the rapid succession of volcanic eruptions that would have compounded the adverse effects of SO_2 leading to severe environmental consequences (e.g., cooling, acid rain, extinctions), preventing recovery and likely causing a run-away effect. By comparison, CO_2 emissions (greenhouse gases) would have been small compared with the mass already in the atmosphere during the Cretaceous and would have had more limited effects. It is granted these estimates have large error margins for both Deccan volcanic emissions and the Chicxulub impact, but they serve as basis for comparison.

How Deccan volcanism affected the environment and how it may have led to the mass extinction of dinosaurs and other organisms in India and globally are still unknown. Documentation of the biotic and environmental effects of Deccan volcanism is still in the early stages. Age correlation between terrestrial sequences with dinosaur fossils and the KTB remains problematic without marine microfossils. Age control for the onset of the main phase of Deccan volcanism that ended at the KTB still remains to be determined along with the immediate biotic and environmental effects leading up to the KT mass extinction. Despite the many unanswered questions, Deccan volcanism is closely linked to the KT mass extinction, whereas the Chicxulub impact is not.

Conclusions

Evidence from Texas, NE Mexico, and the Chicxulub impact crater on Yucatan demonstrates that the Chicxulub impact and KT mass extinction are two separate and unrelated events with the Chicxulub impact predating the KTB by about 300,000 years. Planktic foraminifera, which suffered the most severe mass extinction at the KT boundary, show that not a single species went extinct as a result of the Chicxulub impact in either Mexico or Texas. The biotic and environmental effects of this large impact have been vastly overestimated. Chicxulub thus joins all other known large impacts, such as Alamo, Woodleigh, Manicouagan,

Popigai, and Chesapeake, that caused no mass extinctions (review in Keller 2005).

The cause for the KT mass extinction must lie with the other end-Cretaceous catastrophe—the massive continental flood basalt eruptions of India as originally suggested by McLean (1978, 1985) and Courtillot et al. (1986, 1988). Recent advances in Deccan volcanic studies revealed three volcanic eruption phases with the smallest phase at 67.5 Ma, the main phase at the end of the Maastrichtian (C29R), and the third phase in the early Danian C29R/C29N transition (Chenet et al. 2007, 2008). The main phase of eruptions occurred rapidly, was marked by the longest lava flows (Self et al. 2008b), and ended coincident with the KT mass extinction (Keller et al. 2008b, 2009b). All indications thus point to the main phase of Deccan volcanism as the cause for the KT mass extinction. How and why this volcanic phase caused such devastation globally remains to be explored.

Acknowledgments The KT studies in Texas, Mexico, and throughout Central America that have led to the discovery of the complex history of the Chicxulub impact and the KT mass extinction are the result of long-term collaborations particularly with Thierry Adatte, Zsolt Berner, Doris Stueben, and Wolfgang Stinnesbeck along with many students. Thierry Adatte was also instrumental in Deccan studies along with Silvia Gardin, Annachiara Bartolini, and Sunil Bajpai. I'm grateful to all of them. I also thank Vincent Courtillot, Yves Gallet, and Frédéric Fluteau for many discussions during a sabbatical visit at the Institute de Physique du Globe de Paris. I'm grateful also to reviewers Karl Föllmi and William D. MacDonald for their comments to improve the manuscript. The material of this study is based upon work supported by the US National Science Foundation through the Continental Dynamics Program, Sedimentary Geology and Paleobiology Program, and Office of International Science and Engineering's India Program under NSF Grants EAR-0207407 and EAR-0447171.

References

Adatte T, Stinnesbeck W, Keller G (1996) Lithostratigraphic and mineralogical correlations of near-K/T boundary clastic sediments in northeastern Mexico: implications for megatsunami or sea level changes? Geol Soc Am Spec Paper 307:197–210

Affolter M (2000) Etude des depots clastiques de la límite Cretace–Tertiaire dans la region de la Sierrita, Nuevo Leon, Mexique, MS thesis. Geological Institute University of Neuchatel, Neuchatel, Switzerland, 133p

Alegret L, Molina E, Thomas E (2001) Benthic foraminifera at the Cretaceous–Tertiary boundary around the Gulf of Mexico. Geology 29:891–894

Alvarez LW, Alvarez W, Asaro F, Michel HV (1980) Extraterrestrial cause for the Cretaceous–Tertiary extinction: experimental results and theoretical interpretation. Science 208:1095–1108

Arenillas I, Arz JA, Grajales-Nishimura JM, Murillo-Muneton G, Alvarez W, Camargo-Zanguera A, Molina E, Rosales-Dominguez C (2006) Chicxulub impact event is Cretaceous/Paleogene boundary in age: new micropaleontological evidence. Earth Planet Sci Lett 249:241–257

Arz JA, Alegret L, Arenillas I (2004) Foraminiferal biostratigraphy and environmental reconstruction at the Yaxcopoil-1 drillcore, Chicxulub crater, Yucatań Peninsula. Meteoritics Planet Sci 39:1099–1111

Badve RM, Ghare MA (1977) Palaeoecological aspects of the Bagh Beds, India. Recent Res Geol 4:388–402

Baksi AK (2005) Comment on "$^{40}Ar/^{39}Ar$ dating of the Rajahmundry Traps, easternIndia and their relations to the Deccan Traps" by Knight et al. [Earth Planet Sci Lett 208(2003):85–99]. Earth Planet Sci Lett 239:368–373

Baksi AK, Byerly GR, Chan LH, Farrar E (1994) Intracanyon flows in the Deccan province, India: case history of the Rajahmundry Traps. Geology 22:605–608

Barrera E, Keller G (1990) Stable isotope evidence for gradual environmental changes and species survivorship across the Cretaceous/Tertiary boundary. Paleoceanography 5:867–890

Baum GR, Vail PR (1988) Sequence stratigraphic concepts applied to Paleogene outcrops, Gulf and Atlantic basins. In: Wilgus CK, Hastings BS, Kendall CG, Posamentier HW, Ross CA, Van Wagoner JC (eds) Sea-level: an integrated approach. Soc Econ Paleontol Mineral, Spec Pub, vol 42, pp 309–327

Bhandari A (1995) Ostracodes from the inter-trappean beds near Duddukuru, Andhra Pradesh and a note on their age and paleoecological significance. Indian J Petrol Geol 4:89–107

Blum JD, Chamberlain CP, Hingston MP, Koeberl C, Marin LE, Schuraytz BC, Sharpton VL (1993) Isotopic comparison of K–T boundary impact glass with melt rock from the Chicxulub and Manson impact structures. Nature 364: 325–327

Bourgeois I, Hansen TA, Wiberg PL, Kauffman EG (1988) A tsunami deposit at the Cretaceous–Tertiary boundary in Texas. Science 241:567–570

Cande S, Kent DV (1991) Revised calibration of the geomagnetic polarity timescalefor the Late Cretaceous and Cenozoic. J Geophys Res 100:159–190

Chenet A-L, Quidelleur X, Fluteau F, Courtillot V (2007) $^{40}K/^{40}Ar$ dating of the main Deccan large igneous province: further evidence of KTB age and short duration. Earth Planet Sci Lett 263:1–15

Chenet A-L, Fluteau F, Courtillot V, Gerard M, Subbarao KV (2008) Determination of rapid Deccan eruptions across the KTB using paleomagnetic secular variation: (I) Results from 1200 m thick section in the Mahabaleshwar escarpment. J Geophys Res 113:B04101

Chiplonkar GW, Badve RM (1968) Age and affinities of the Bagh fauna – a Reassessment. Symposium Deccan Trap Country Special Issue. Bull Indian Nat Sci Acad (issued 1973) 45:19–29

Courtillot V (1999) Evolutionary catastrophes: the science of mass extinctions. Cambridge University Press, Cambridge, 171p
Courtillot VE, Renne PR (2003) On the ages of flood basalt events. Curr Res Geosci 335:113–140
Courtillot V, Besse J, Vandamme D, Montigny R, Jaeger JJ, Capetta H (1986) Deccan flood basalts at the Cretaceous/Tertiary boundary. Earth Planet, Sci Lett 80:361–374
Courtillot V, Feraud G, Maluski H, Vandamme D, Moreau MG, Besse J (1988) Deccanflood basalts and the Cretaceous/Tertiary boundary. Nature 333:843–846
Courtillot V, Gallet Y, Rocchia R, Féraud G, Robin E, Hofmann C, Bhandari N, Ghevariya ZG (2000) Cosmic markers, ^{40}Ar/^{39}Ar dating and Paleomagnetism of the KT sections in the Anjar area of the Deccan large igneous province. Earth Planet, Sci Lett 182:137–156
Cripps JA, Widdowson M, Spicer RA, Jolley DW (2005) Coastal ecosystem response to late stage Deccan Trap volcanism: the post-K–T boundary (Danian) palynofacies of Mumbai (Bombay), west India. Paleogeogr Paleoclimatol Palaeoecol 216:303–332
Donovan AD, Baum GR, Blechschmidt GL, Loutit TS, Pflum CE, Vail PR (1988) Sequence stratigraphic setting of the Cretaceous–Tertiary boundary in central Alabama. In: Wilgus CK, Hastings BS, Kendall CG, Posamentier HW, Ross CA, Van Wagoner JC (eds) Sea-level: an integrated approach. Soc Econ Paleont Mineral, Spec Pub, vol 42, pp 299–307
Dressler BO, Sharpton VL, Morgan J, Buffler R, Smit J, Stoeffler D, Urrutia J (2003) Investigating a 65-Ma-old smoking gun: deep drilling of the Chicxulub impact structure. EOS Trans, Am Geophys Union 84(14):125–130
Dressler BO, Sharpton VL, Schwandt CS, Ames D (2004) Impactites of the Yaxcopoil-1 drilling site, Chicxulub impact structure: petrography, geochemistry and depositional environment. Meteoritics Planet Sci 39:857–878
Duncan RA, Pyle DG (1988) Rapid eruption of the Deccan flood basalt at the Cretaceous/Tertiary boundary. Nature 333: 841–8443
Ekdale AA, Stinnesbeck W (1998) Ichnology of Cretaceous–Tertiary (K/T) boundary beds in northeastern Mexico. Palaios 13:593–602
Gale A S (2006) The Cretaceous–Palaeogene boundary on the Brazos River, Falls County, Texas: is there evidence for impact-induced tsunami sedimentation? Proc Geol Assoc 117:173–185
Galloway WE, Bebout DG, Fisher WL, Dunlap JB Jr, Cabrera-Castro R, Lugo-Rivera JE, Scott TM (1991) Cenozoic. In: Salvador A (ed) The Gulf of Mexico basin: the geology of North America, vol J. Geol Soc Am, Boulder, CO, pp 245–324
Gertsch B, Keller G, Adatte T, Berner Z (2011, in press) Platinum Group Elements (PGE) geochemistry of Brazos sections, Texas, USA. In: Keller G, Adatte T (eds) The End-Cretaceous Mass Extinction and the Chicxulub Impact in Texas, SEPM Special Publication No. 100
Gilmore JS, Knight JD, Orth CL, Pillmore CL, Tschudy RH (1984) Trace element patterns at a non-marine Cretaceous–Tertiary boundary. Nature 307:224–228
Grachev AF, Korchagin OA, Kollmann HA, Pechersky DM, Tsel'movich VA (2005) Anew look at the nature of the transitional layer at the K/T boundary near Gams, Eastern Alps, Austria, and the problem of the mass extinction of the biota. Russian J Earth Sci 7(6)ES600:1–45
Graup G, Spettel B, Herm D, Weidich KF (1989) Mineralogy and phase-chemistry of an ir-enriched pre-K/T layer from the Lattengebirge, Bavarian Alps, and significance for the KTB problem. Earth Planet Sci Lett 95:271–290
Hansen T, Farrand RB, Montgomery HA, Billman HG, Blechschmidt G (1987) Sedimentology and extinction patterns across the Cretaceous–Tertiary boundary interval in east Texas (USA). Cretaceous Res 8:229–252
Hansen TA, Upshaw B III, Kauffman EG, Gose W (1993) Patterns of molluscan extinction and recovery across the Cretaceous–Tertiary boundary in east Texas; report on new outcrops. Cretaceous Res 14:685–706
Haq BU, Hardenbol J, Vail PR (1987) The chronology of fluctuating sea level since the Triassic. Science 235:1157–1167
Haq BU, Hardenbol J, Vail PR (1988) Mesozoic and Cenozoic chronostratigraphy and cycles of sea-level change. In: Wilgus CK, Hastings BS, Kendall CG, Posamentier HW, Ross CA, Van Wagoner JC (eds) Sea-level: an integrated approach. Soc Econ Paleontol Mineral Spec Pub, vol 42, pp 71–108
Harting M (2004) Zum Kreide/Tertiar-Ubergang in NE Mexico: geochemische characterisiering der Chicxulub Impactejekta. Karlsruhe University, 131p
Hildebrand AR, Penfield GT, Kring DA, Pilkington M, Jacobsen S, Boynton WV (1991) Chicxulub crater: a possible Cretaceous/Tertiary boundary impact crater on the Yucatan Peninsula, Mexico. Geology 19:867–871
Hofmann C, Féraud G, Courtillot V (2000) ^{40}Ar/^{39}Ar dating of mineral separates and whole rocks from the Western Ghats lava pile: further constraints on duration and age of the Deccan traps. Earth Planet Sci Lett 180:13–27
Izett GA (1990) The Cretaceous/Tertiary boundary interval, Raton Basin, Colorado and New Mexico, and its content of shock-metamorphosed minerals: evidence relevant to the K/T boundary impact-extinction hypothesis. Geol Soc Am, Spec Pap, vol 249, 100 pp
Izett G, Dalrymple GB, Snee LW (1991) ^{40}Ar/^{39}Ar age of K–T boundary tektites from Haiti. Science 252:159–1543
Jaiprakash BC, Singh J, Raju DSN (1993) Foraminiferal events across the K/T boundary and age of Deccan volcanism in Palakollu area, Krishna–Godavari Basin, India. J Geol Soc India 41:105–117
Jay AE, Widdowson M (2008) Stratigraphy, structure and volcanology of the south-east Deccan continental flood basalt province: implications for eruptive extent and volumes. J Geol Soc 165:177–188
Jay AE, MacNiocaill C, Widdowson M, Self S, Turner W (2009) NewPalaeomagnetic data from the Mahabaleshwar Plateau, Deccan Flood Basalt Province, India: implications for the volcanostratigraphic architecture of Continental Flood Basalt Provinces. J Geophys Res 166:1–12. doi:10.1144/0016-76492007-150
Jerram DA, Widdowson M (2005) The anatomy of Continental Flood Basalt Provinces: geological constraints on the processes and products of flood volcanism. Lithos 79:385–405
Jiang MJ, Gartner S (1986) Calcareous nannofossil succession across the Cretaceous–Tertiary boundary in east-central Texas. Micropaleontology 32:232–255

Keller G (1988) Extinctions, survivorship and evolution across the Cretaceous/Tertiary boundary at El Kef, Tunisia. Marine Micropal 13:239–263

Keller G (1989) Extended K/T boundary extinctions and delayed populational change in planktic Foraminiferal faunas from Brazos River Texas. Paleoceanography 4:287–332

Keller G (1996) The K/T mass extinction in planktic foraminifera biotic constraints for catastrophe theories. In: MacLeod N, Keller G (eds) The Cretaceous–Tertiary mass extinction: biotic and environmental events. Norton Press, New York, pp 63–100

Keller G (2001) The end-Cretaceous mass extinction: year 2000 assessment. J Planet Space Sci 49:817–830

Keller G (2005) Impacts, volcanism and mass extinctions: random coincidence or cause and effect? Australian J Earth Sci 52:725–757

Keller G (2008a) Impact stratigraphy: old principle, new reality. Geol Soc Am, Spec Pap 437:147–178

Keller G (2008b) Cretaceous climate, volcanism, impacts and biotic effects. Cretaceous Res 29:754–771

Keller G, Abramovich S (2009) Lilliput Effect in late Maastrichtian planktic foraminifera: response to environmental stress. Palaeogeogr Palaeoclimatol Palaeoecol 284:47–62

Keller G, Barron JA (1983) Paleoceanographic implications of Miocene deep-sea hiatuses, Geol Soc Am Bull 94:1–30

Keller G, Pardo A (2004) Disaster opportunists Guembelitridae: index for environmental catastrophes. Marine Micropal 53:83–116

Keller G, Stinnesbeck W (1996) Sea-level changes, clastic deposits and megatsunamis across the Cretaceous–Tertiary boundary. In: MacLeod N, Keller G (eds) Cretaceous–Tertiary mass extinctions. WW Norton & Co, New York, pp 415–450

Keller G, Stinnesbeck W (2002) Slumping and a sandbar deposit at the Cretaceous–Tertiary boundary in the El Tecolote section (Northeastern Mexico): an impact-induced sediment gravity flow – comment. Geology 30: 382–383

Keller G, Barrera E, Schmitz B, Mattson E (1993) Gradual mass extinction, species survivorship and long-term environmental changes across the Cretaceous–Tertiary Boundary in high latitudes. Geol Soc Am Bull 105:979–997

Keller G, Stinnesbeck W, Lopez-Oliva JG (1994a) Age, deposition and biotic effects of the Cretaceous/Tertiary boundary event at Mimbral, NE Mexico. Palaios 9:144–157

Keller G, Stinnesbeck W, Adatte T, MacLeod N, Lowe DR (1994b) Field guide to Creataceous–Tertiary boundary sections in northeastern Mexico, vol 827. Lunar and Planetary Institute, LPI Contribution, Houston,110p

Keller G, Li L, MacLeod N (1995) The Cretaceous/Tertiary boundary stratotype section at El Kef, Tunisia: how catastrophic was the mass extinction? Palaeogeogr Palaeoclimatol Palaeoecol 119:221–254

Keller G, Lopez-Oliva JG, Stinnesbeck W, Adatte T (1997) Age, stratigraphy and deposition of near K/T siliciclastic deposits in Mexico: relation to bolide impact? Geol Soc Am Bull 109:410–428

Keller G, Adatte T, Stinnesbeck W, Stuben D, Kramar U, Berner Z, Li L, von Salis Perch-Nielsen K (1998) The Cretaceous–Tertiary transition on the shallow Saharan Platform in southern Tunisia. GeoBios 30:951–975

Keller G, Adatte T, Stinnesbeck W, Stueben D, Berner Z (2001) Age, Chemo- and Biostratigraphy of Haiti spherule-rich deposits: a Multi-event K–T scenario. Canadian J Earth Sci 38:197–227

Keller G, Adatte T, Stinnesbeck W, Affolter M, Schilli L, Lopez-Oliva JG (2002) Multiple spherule layers in the late Maastrichtian of northeastern Mexico. Geol Soc Am Spec Pub 356:145–161

Keller G, Stinnesbeck W, Adatte T, Stüben D (2003a) Multiple impacts across the Cretaceous–Tertiary boundary. Earth-Sci Rev 62:327–363

Keller G, Stinnesbeck W, Adatte T, Holland B, Stueben D, Harting M, De Leon C, de la Cruz J (2003b) Spherule deposits in Cretaceous–Tertiary boundary sediments in Belize and Guatemala. J Geol Soc London 160:783–795

Keller G, Adatte T, Stinnesbeck W, Rebolledo-Vieyra M, Urrutia Fucugauchi J, Kramar U, Stueben D (2004a) Chicxulub crater predates K–T mass extinction. Proc Natl Acad Sci USA 101:3753–3758

Keller G, Adatte T, Stinnesbeck W, Stüben D, Berner Z, Harting M (2004b) More evidence that the Chicxulub impact predates the K/T mass extinction. Meteoritics Planet Sci 39:1127–1144

Keller G, Adatte T, Harting M, Berner Z, Baum G, Prauss M, Tantawy AA, Stueben D (2007a) Chicxulub impact predates the K–T boundary: new evidence from Texas. Earth Planet Sci Lett 255:339–356

Keller G, Adatte T, Tantawy AA, Berner Z, Stueben D (2007b) High Stress Late Cretaceous to early Danian paleoenvironment in the Neuquen Basin, Argentina. Cretaceous Res 28:939–960

Keller G, Adatte T, Baum G, Berner Z (2008a) Reply to 'Chicxulub impact predates K–T boundary: new evidence from Brazos, Texas' comment by Schulte et al. Earth Planet Sci Lett 269:620–628. doi:10.1016/j.epsl.2007.12.025

Keller G, Adatte T, Gardin S, Bartolini A, Bajpai S (2008b) Main Deccan volcanism phase ends near the K–T boundary: evidence from the Krishna-Godavari Basin, SE India. Earth Planet Sci Lett 268:293–311. doi:101016/j.epsl.2008.01.015

Keller G, Abramovich S, Berner Z, Adatte T (2009a) Biotic effects of the Chicxulub impact, K–T catastrophe and sea-level change in Texas. Palaeogeogr Palaeoclimatol Palaeoecol 271:52–68. doi:10.1016/j.palaeo.2008.09.007

Keller G, Sharma R, Khosla A, Khosla SC, Bajpai S (2009b) Early Danian planktic foraminifera from Intertrappean beds at Jhilmili, Chhindwara district, Madhya Pradesh, India. J Foram Res 39:40–55

Keller G, Adatte T, Pardo Juez A, Lopez Oliva JG (2009c) New evidence concerning the age and biotic effects of the Chicxulub impact in NE Mexico. J Geol Soc London 166:393–411

Keller G, Adatte T, Bajpai S, Mohabey DM, Widdowson M, Khosla A, Sharma R, Khosla SC, Gertsch B, Fleitmann D, Sahni A (2009d) K–T transition in Deccan Traps of central India marks major marine seaway across India. EPSL 282:10–23

Khosla SC, Nagori ML (2002) Ostracodes from the Intertrappean beds (Early Paleocene) of the east coast of India. Paleontol Res 6:191–210

Klaus A, Norris RD, Kroon D, Smit J (2000) Impact-induced mass wasting at the K–T Boundary, Blake Nose, western North Atlantic. Geology 28:319–322

Knight KB, Renne PR, Halkett A, White N (2003) $^{40}Ar/^{39}Ar$ dating of the Rajahmundry traps, Eastern India and their relationship to the Deccan Traps. Earth Planet Sci Lett 208:85–99

Knight KB, Renne PR, Baker J, Waight T, White N (2005) Reply to $^{40}Ar/^{39}Ar$ dating of the Rajahmundry Traps, Eastern India and their relationship to theDeccan Traps: discussion by AK Baksi. Earth Planet Sci Lett 239: 374–382

Koeberl C (1993) Chicxulub crater, Yucatan: Tektites, impact glasses, and the geochemistry of target rocks and breccias. Geology 21:211–214

Kring DA (2007) The Chicxulub impact event and its environmental consequences at the Cretaceous–Tertiary boundary. Palaeogeogr Palaeoclimatol, Palaeoecol 255:4–21

Lindenmaier F (1999) Geologie und Geochemie an der Kreide/Tertiär-Grenze im Nordosten von Mexiko. Diplomarbeit, Institut für Regionale Geologie, Karlsruhe, 90p

Lopez-Oliva JG (1996) Stratigraphy of the Cretaceous–Tertiary (K–T) boundary transition in northeastern and east-central Mexico. Princeton University, Princeton, 246p

Lopez-Oliva JG, Keller G (1996) Age and stratigraphy of near-K/T boundary clastic deposits in NE Mexico. Geol Soc Am, Spec Pap 307:227–242

Lopez-Ramos E (1973) Estudio geológico de la Peninsula de Yucatan. Boletin Associacion Mexicana de Geologos Petroleros 25:23–76

Lopez-Ramos E (1975) Geological summary of the Yucatan Peninsula. In: Nairn AEM, Stehli FG (eds) The ocean basins and margins. The Gulf of Mexico and the Caribbean, vol 3. Plenum Press, New York, pp. 257–282

Loutit TS, Hardenbol J, Vail PR, Baum GR (1988) Condensed sections: the key to age dating and correlation of continental margin sequences. In: Wilgus CK, Hastings BS, Kendall CG, Posamentier HW, Ross CA, Van Wagoner JC (eds) Sea-level changes: an integrated approach. Soc Econ Paleontol Mineral Spec Pub, vol 42, pp. 183–213

MacLeod N, Keller G (1991) Hiatus distributions and mass extinctions at the Cretaceous/Tertiary boundary. Geology 19:497–501

MacLeod N, Ortiz N, Fefferman N, Clyde W, Schulter C, MacLean J (2000) Phenotypic response of foraminifera to episodes of global environmental change. In: Culver SJ, Rawson P (eds) Biotic response to global environmental change: the last 145 million years. Cambridge University Press, Cambridge, pp. 51–78

MacLeod KG, Whitney DL, Huber BT, Koeberl C (2007) Impact and extinction in remarkably complete Cretaceous–Tertiary boundary sections from Demerara Rise, tropical western North Atlantic. Geol Soc Am Bull 119:101–115. doi:10.1130/B25955.1

Martínez-Ruiz F, Ortega-Huertas M, Palomo-Delgado I, Smit J (2002) Cretaceous–Tertiary boundary at Blake Nose (Ocean drilling Program Leg 171B): a record of the Chicxulub impact ejecta. Geol Soc Am, Spec Pap 356:189–200

Maurasse FJMR, Sen G (1991) Impacts, tsunamis and the Haitian Cretaceous–Tertiary boundary layer. Science 252:1690–1693

Maurrasse FJMR, Lamolda MA, Aguado R, Peryt D, Sen G (2005) Spatial and temporal variations of the Haitian K/T boundary record: implications concerning the event or events. J Iberian Geol 31:113–133

McLean DM (1978) A terminal Mesozoic "Greenhouse": lessons from the past. Science 201:401–406

McLean DM (1985) Deccan Traps mantle degassing in the terminal Cretaceous marine extinctions. Cretaceous Res 6: 235–259

Meyerhoff AA, Lyons JB, Officer CB (1994) Chicxulub structure: a volcanic sequence of Late Cretaceous age. Geology 22:3–4

Molina E, Alegret L, Arenillas I, Arz J A, Gallala N, Hardenbol J, von Salis K, Steurbraut E, Vandenberghe N, Zaghbib-Turki D (2006) The global boundary stratotype section and point for the base of the Danian stage (Paleocene, Paleogene, "Tertiary", Cenozoic) at El Kef, Tunisia – original definition and revision. Episodes 29:263–273

Montanari A, Koeberl C (2000) Impact stratigraphy. Lecture Notes in Earth Sciences, no. 93. Springer, Heidelberg, Germany, 364p

Morgan JW, Ganapathy R, Krähenbühl U (1975) Meteoritic trace elements in lunar rock 14321, 184. Geoch Cosmoch Acta 39:261–264

Norris RD, Huber BT, Self-Trail J (1999) Synchroneity of the K–T oceanic mass extinction and meteorite impact: Blake Nose, western North Atlantic. Geology 27:419–422

Norris RD, Firth J, Blusztajn Ravizza G (2000) Mass failure of the North Atlantic margin triggered by the Cretaceous–Paleogene bolide impact. Geology 28:1119–1122

Olsson RK, Miller KG, Browning JV, Habib D, Sugarmann PJ (1997) Ejecta layer at the Cretaceous–Tertiary boundary, Bass River, New Jersey (Ocean Drilling Program Leg 174AX). Geology 25:759–762

Pande K, Pattanayak SK, Subbarao KV, Navaneethakrishna P, Venkatesan TR (2004) $^{40}Ar/^{39}Ar$ age of a lava flow from the Bhimashankar Formation, Giravali Ghat, Deccan Traps. Proc Indian Acad Sci 113:755–758

Pardo A, Ortiz N, Keller G (1996) Latest Maastrichtian and K/T boundary foraminiferal turnover and environmental changes at Agost, Spain. In: MacLeod N, Keller G (eds) The Cretaceous–Tertiary mass extinction: biotic and environmental effects. Norton Press, New York, pp. 157–191

Poag CW, Plescia JB, Molzer PC (2002) Ancient impact structures on modern continental shelves: the Chesapeake Bay, Montagnais, and Toms Canyon craters, Atlantic margin of North America. Deep Sea Res Pt II(49):1081–1102

Prasad B, Pundeer BS (2002) Palynological events and zones in Cretaceous–Tertiary boundary sediments of Krishna-Godavari and Cauvery basins, India. Palaeontographica Abt B 262:39–70

Raju DSN, Jaiprakash BC, Kumar A, Saxena RK, Dave A, Chatterjee TK, Mishra CM (1994) The magnitude of hiatus and sea-level changes across K/T boundary in Cauvery and Krishna-Godavari basins, India. J Geol Soc India 44: 301–315

Raju DSN, Jaiprakash BC, Kumar A, Saxena RK, Dave A, Chatterjee TK, Mishra CM (1995) Age of Deccan volcanism across KTB in Krishna-Godavari Basin: new evidences. J Geol Soc India 45:229–233

Rocchia R, Robin E, Froget L, Gayraud J (1996) Stratigraphic distribution of extraterrestrial markers at the Cretaceous–Tertiary boundary in the Gulf of Mexico area: implications for the temporal complexity of the event. In: Ryder G, Fastovsky D, Gartner S (eds) The Cretaceous–Tertiary boundary event and other catastrophes in Earth history. Geol Soc Am, Boulder, Colorado, Spec Pap, vol 307, pp. 279–286

Sahni A (1983) Upper Cretaceous Palaeobiogeography of Peninsular India and the Cretaceous–Paleocene transition: the vertebrate evidence. In: Maheshwari HK (ed) Cretaceous of India. Indian Association. Palynostratigraphers, Lucknow, pp. 128–140

Sawlowicz Z (1993) Iridium and other platinum-group elements as geochemical markers in sedimentary environments. Paleogeogr Paleoclimatol Paleoecol 104:253–270

Saxena RK, Misra CN (1994) Time and duration of Deccan volcanism in the Rezole area, Krishna-Godavari Basin, India. Curr Sci 66:74–76

Schilli L (2000) Etude de la limite K–T dans la région de la Sierrita, Nuevo Leon, Mexique, MS thesis, Geological Institute, University of Neuchatel, Neuchatel, Switzerland, 138p

Schulte P (1999) Geologisch-sedimentologische Untersuchungen des Kreide/Tertiär (K/T)-Übergangs im Gebiet zwischen La Sierrita und El Toro, Nuevo Leon, Mexiko: Diplomarbeit, Universität Karlsruhe, Institut für Regionale Geologie, Karlsruhe, Germany, 134p

Schulte P, Stinnesbeck W, Steuben D, Kramar U, Berner Z, Keller G, Adatte T (2003) Multiple slumped? Chicxulub ejecta deposits with iron-rich spherules and quenched carbonates from the K/T transition, La Sierrita, NE Mexico. J Int Earth Sci 92:114–142

Schulte P, Speijer RP, Mai H, Kontny A (2006) The Cretaceous–Paleogene (K–P) boundary at Brazos, Texas: sequence stratigraphy, depositional events and the Chicxulub impact. Sed Geol 184 77–109

Schulte P, Speijer RP, Brinkhuis H, Kontny A, Caleys P, Galeotti S, Smit J (2008) Comment on the paper, 'Chicxulub impact predates K–T boundary: new evidence from Brazos, Texas' by Keller et al. (2007). Earth Planet Sci Lett 269:613–619. doi:10.1016/j.epsl.2007.11.066

Schulte P, and 40 others (2010) The Chicxulub asteroid impact and mass extinction at the Cretaceous–Paleogene boundary. Science 327:1214–1218

Self S, Jay AE, Widdowson M, Keszthelyi LP (2008a) Correlation of the Deccan and Rajahmundry Trap lavas: are these the longest and largest lava flows on Earth? J Vocanol Geothermal Res 172:3–19

Self S, Blake S, Sharma K, Widdowson M, Sephton S (2008b) Sulfur and Chlorine in Late Cretaceous Deccan Magmas and Eruptive gas release. Science 319:1654–1657

Sheth HC, Pande K, Bhutani R (2001) $^{40}Ar/^{39}Ar$ age of a national geological monument: the Gilbert Hill basalt, Deccan Traps, Bombay. Curr Sci 80:1437–1440

Sigurdsson H, Bonté P, Turpin L, Chaussidon M, Metrich N, Steinberg M, Pradel P, D'Hondt S (1991) Geochemical constraints on source region of Cretaceous–Tertiary impact glasses. Nature 353:839–842

Smit J (1999) The global stratigraphy of the Cretaceous–Tertiary boundary impact ejecta. Ann Rev Earth Planet Sci 27:75–113

Smit J, Montanari A, Swinburne NHM, Alvarez W, Hildebrand A, Margolis SV, Claeys P, Lowerie W, Asaro F (1992) Tektite-bearing deep-water clastic unit at the Cretaceous–Tertiary boundary in northeastern Mexico. Geology 20: 99–104

Smit J, Montanari A, Alvarez W (1994) Tsunami-generated beds at the KT boundary in northeastern Mexico. In: Keller G, Stinnesbeck W, Adatte T, MacLeod N, Lowe DR (eds) Field guide to Cretaceous–Tertiary boundary sections northeaster Mexico, vol 827. Lunar and Planetary Institute, Houston, pp. 95–101

Smit J, Roep TB, Alvarez W, Montanari A, Claeys P, Grajales-Nishimura JM, Bermudez J (1996) Coarse-grained clastic sandstone complex at the K/T boundary around the Gulf of Mexico: deposition by tsunami waves induced by the Chicxulub impact. Geol Soc Am Spec Pap 307:151–182

Smit J, van der Gaast S, Lustenhouwer W (2004) Is the transition to post-impact rock complete? Some remarks based on XRF scanning electron microprobe and thin section analyses of the Yaxcopoil-1 core. Meteoritics Planetary Sci 39:1113–1126

Sohl NF, Martinez RE, Salmerón-Ureña P, Soto-Jaramillo F (1991) Upper Cretaceous. In: Salvador A (ed) The Gulf of Mexico basin: the geology of North America. Journal of Geological Society of America, Vol J. Boulder, CO, pp. 205–244

Soria AR, Llesa CL, Mata MP, Arz JA, Alegret L, Arenillas I, Melendez A (2001) Slumping and a sandbar deposit at the Cretaceous–Tertiary boundary in the El Tecolote section (northeastern Mexico): an impact-induced sediment gravity flow. Geology 29:231–234

Stinnesbeck W, Barbarin JM, Keller G, Lopez-Oliva JG, Pivnik D, Lyons J, Officer C, Adatte T, Graup G, Rocchia R, Robin E (1993) Deposition of channel deposits near the Cretaceous–Tertiary boundary in northeastern Mexico: Catastrophic or "normal" sedimentary deposits? Geology 21:797–800

Stinnesbeck W, Keller G, Adatte T, Lopez-Oliva JG, MacLeod N (1996) Cretaceous–Tertiary boundary clastic deposits in northeastern Mexico: impact tsunami or sea level lowstand? In: MacLeod N, Keller G (eds) Cretaceous–Tertiary mass extinctions: biotic and environmental changes. WW Norton & Company, New York, pp. 471–518

Stinnesbeck W, Keller G, Adatte T, Harting M, Stueben D, Istrate G (2004) Yaxcopoil-1 and the Chicxulub impact. Int J Earth Sci (Geol Rundsch) 93:1042–1065

Stüben D, Kramar U, Harting M, Stinnesbeck W, Keller G (2005) High-resolution geochemical record of Cretaceous–Tertiary boundary sections in Mexico: new constraints on the K/T and Chicxulub events. Geochemica et Cosmochimica Acta 69(10):2559–2579

Subbarao KV, Pathak S (1993) Reversely magnetized flows, Rajahmundry, Andhra Pradesh. J Geol Soc India 41:71–72

Swisher CC, Grajales-Nishimura JM, Montanari A, Margolis SV, Claeys P, Alvarez W, Renne P, Cedillo-Pardoa E, Maurrasse FJ-MR, Curtis GH, Smit J, McWilliams MO (1992) Coeval $^{40}Ar/^{39}Ar$ ages of 65 million years ago from Chicxulub crater melt rock and Cretaceous–Tertiary boundary tektites. Science 257:954–958

Tredoux M, de Wit MJ, Hart RJ, Linsay NM, Verhagen B, Sellschop JPF (1988) Chemostratigraphy across the Cretaceous–Teriary boundary and a critical assessment of the iridium anomaly. J Geol 97:585–605

Vandamme D, Courtillot V (1992) Paleomagnetic constraints on the structure of the Deccan Traps. Phys Earth Planet Int 74:241–261

Vandamme D, Courtillot V, Besse J, Montigny R (1991) Paleomagnetism and age determinations of the Deccan traps (India): results of a Nagpur–Bombay traverse and review of earlier work. Rev Geophys Space Phys 29: 159–190

Venkatesan TR, Pande K, Gopalan K (1993) Did Deccan volcanism pre-date the Cretaceous/Tertiary transition? Earth Planet Sci Lett 119:181–189

Venkatesan TR, Pande K, Ghevariya ZG (1996) ^{40}Ar–^{39}Ar ages of lava flows from Anjar, Western Deccan Province, India and its relation to the Cretaceous–Tertiary boundary events. Curr Sci 70:990–996

Yancey TE (1996) Stratigraphy and depositional environments of the Cretaceous–Tertiary boundary complex and basal section, Brazos River, Texas. Gulf Coast Assoc Geol Soc Trans 46:433–442

Wang K, Attrep M Jr, Orth CJ (1993) Global iridium anomaly, mass extinction and redox change at the Devonian–Carboniferous boundary. Geology 21:1071–1074

Ward W, Keller G, Stinnesbeck W, Adatte T (1995) Yucatan subsurface stratigraphy: implications and constraints for the Chicxulub impact. Geology 23:873–876

Widdowson M, Pringle MS, Fernandez OA (2000) A post K–T boundary (Early Paleocene) age for Deccan-type feeder dykes, Goa, India. J Petrol 41:1177–1194

Wignall PB (2001) Large igneous provinces and mass extinctions. Earth Sci Rev 53:1–33

Survival, but. . .! New Tales of 'Dead Clade Walking' from Austral and Boreal Post-K–T Assemblages

Jeffrey D. Stilwell and Eckart Håkansson

Abstract

Our knowledge of postmass extinction biotic trajectories is at the mercy of available data and detailed research on the governing factors of differential extinction/survivorship patterns of fossil biotas. Some taxa managed—barely—to survive major extinction events, but only for the short-term, becoming extinct at variable times in the following geologic stage, having succumbed to myriad natural forces generated by severe paleoenvironmental perturbations. These 'Dead Clade Walking' (DCW) organisms should be included in investigations on the resultant effects of the extinction bottleneck and subsequent rebound phase(s). Significantly, even though the 'big five' mass extinctions of the Phanerozoic are distinguished primarily by their overwhelming intensities, their magnitudes—and thus importance for shaping the present-day biosphere—have been systematically underestimated, when time frames including immediate, post-apocalyptic DCW taxa are included in survivorship/extinction analyses. Our research from recent studies of Austral and Boreal invertebrates and vertebrates in relation to the Cretaceous–Tertiary (K–T) boundary provides alluring new evidence of the DCW phenomenon, including the short-term survivorship of ammonoid cephalopods and possible non-avian dinosaurs into the dawn of the Cenozoic.

Keywords

Cretaceous • Maastrichtian • Paleogene • Danian • Cheilostome bryozoans • Ammonoids • Gastropods • Postmass extinction • Survival patterns

Introduction

An intriguing, yet confounding, outcome of mass extinction events is the post-apocalypse differential survivorship patterns of fossil fauna and flora. It is well known from the fossil record that some groups managed to persist for many millions of years into subsequent geologic stages (a few even becoming

J.D. Stilwell (✉)
Applied Palaeontology and Basin Studies Group, School of Geosciences, Monash University, Clayton, Vic 3800, Australia

Centre for Evolutionary Research, Australian Museum, Sydney, NSW, Australia
e-mail: Jeffrey.Stilwell@sci.monash.edu.au

J.A. Talent (ed.), *Earth and Life*, International Year of Planet Earth,
DOI 10.1007/978-90-481-3428-1_26,

present-day 'living fossils'), whereas others only just survived the extinction into the dawn of the following stage or died out at some point during the few million years thereafter. Previous work has focused primarily on the complex biodiversity patterns through 'deep time' of pre-extinction taxa, boundary events, and the diversification of animal and plant groups in the ecological vacuum of the catastrophe aftermath. Since the new millennium, there has been growing interest in our understanding of the enhanced, long-term picture of mass extinctions and evolutionary dynamics that extend well beyond the extinction interval of the 'Big Five' Phanerozoic calamities, or perhaps more correctly 'Big Six,' if we include the present, ominously accelerating extinction intensities comparable to the last major episode at the Cretaceous–Tertiary (K–T) boundary at 65 Ma.

A significant aspect of these crises in Earth history is that some of the extinction 'winners' battle on through the major pulse(s) of calamity—documented both in large or small individual numbers at species, genus, and higher taxonomic levels—only to fall prey to extermination a short interval after the evolutionary bottleneck. This has been appropriately termed the 'Dead Clade Walking' (DCW) phenomenon by Jablonski (2002), after the best-selling 1994 book *Dead Man Walking* by Sister Helen Prejean (a Roman Catholic nun). Dead Clade Walking refers to the elevated extinction rates of surviving taxa of major extinction crises directly following the extirpation, with particular monophyletic lineages succumbing to natural forces not long thereafter in the post-extinction interval. Similar titles of short-term, mass extinction survivors include 'failed crisis progenitors' of Kauffmann and Erwin (1995), 'hold-over taxa' of Hallam and Wignall (1997), and 'failed survivors' of Korn et al. (2004), but we prefer the use herein of DCW. Importantly, this phenomenon means that many extinction intervals are in fact much higher in terms of intensity. Magnitudes in the literature are probably underestimates, especially if longer intervals, including the post-extinction stage, are included in the analyses. As stated by Jablonski (2002) on his study of the pre-human 'Big Five' mass extinctions, the coupled, resultant effects of the extinction bottleneck and survival/recovery intervals are probably equally significant in the observed diversity patterns; thus, the governing factors of the fates of DCW taxa reflect our need to better assess post-extinction biotic trajectories and to test hypotheses that provide plausible reasons for their demise.

This chapter details supporting evidence of DCWs from Danian invertebrate and vertebrate assemblages from Boreal and Austral regions, specifically from Denmark, the Netherlands, New Zealand–Chatham Islands, and Antarctica. Specifically, it is our aim to reveal the diversity of higher organisms that demonstrate the DCW phenomenon, not all discussed previously in the literature, including intriguing (and tantalizing) new evidence of groups that have long thought to have been exterminated right at the Cretaceous–Tertiary boundary 65 Ma.

Selected Case Studies of 'Dead Clade Walking' from High-Latitude Danian Assemblages

Evidence for Ammonoid Survival into the Early Paleocene (Danian)

Ammonoids—cosmopolitan, nektic, and cephalopod mollusk predators of the late Paleozoic and Mesozoic—are believed to have met their demise at the close of the Cretaceous, in some regions extending, albeit in low diversity, right to the dire end of the latest Maastrichtian, such as in Denmark (Birkelund 1979, 1993) and Antarctica (Zinsmeister 2001; Stilwell 2003). Ammonites have long been held as a textbook example of a major victim of the K–T mass extinction, together with the non-avian dinosaurs. Despite the recorded occurrence of ammonites in the lower Danian Cerithium Limestone ('Cerithiumkalk', now Cerithium Limestone Member of the Rødvig Formation (Surlyk et al. 2006)) in Stevns Klint, Denmark, more than 80 years ago by Rosenkrantz (1924), most subsequent publications up to now have cited their occurrence in the basal part of the type Danian as no more than reworked late Maastrichtian taxa from the underlying chalk (e.g., Rasmussen 1971; Birkelund 1979, 1993; Machalski 2002) (Fig. 1). An exception to the prevailing view of these ammonites being no more than re-deposited Maastrichtian forms is a paper by Surlyk and Nielsen (1999), who in their work, 'The last ammonite?' bravely challenged the deeply seated paradigm of terminal Cretaceous extinction of this major invertebrate clade. Although there was no detailed evidence for their survival into

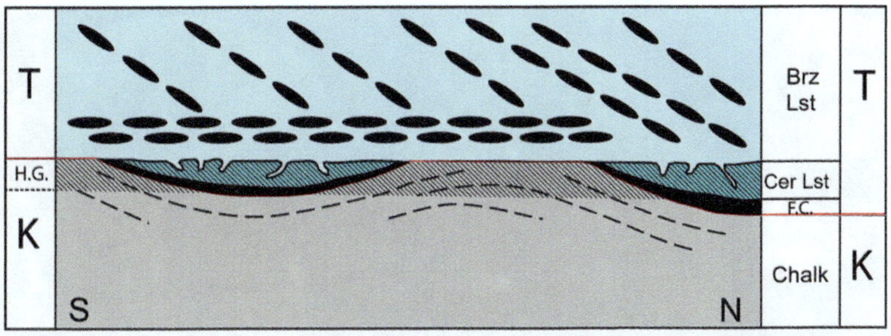

Fig. 1 Diagrammatic sketch of the K–T boundary stratigraphy of Stevns Klint, Danish Basin. Sequence characterized by uppermost Maastrichtian chalk, *solid red line* represents K–T boundary itself with iridium anomaly. [F.C. = boundary Fish Clay; Cer Lst = lower Danian Cerithium Limestone; Brz Lst = Danian Bryozoan Limestone; H.G.= hardground.] (figure adapted from Machalski and Heinberg 2005)

the Danian in Denmark presented by Surlyk and Nielsen (1999)—their hypothesis subsequently having been challenged by Machalski (2002)—the seed was planted to conduct further detailed studies of the reputed baculitid and scaphitid ammonites recovered from the Cerithium Limestone. The tables have now turned, with all but unequivocal isotopic, sedimentologic, and compositional data, revealing that specific individuals of ammonites are indigenous survivors of the K–T extinction, albeit in much lower numbers compared with late Maastrichtian assemblages. Extremely detailed studies by Machalski and Heinberg (2005) and Håkansson et al. (2008) have shed new light on the reputed ammonoid survivors of the K–T event in the Cerithium Limestone (Fig. 2).

Resting directly on top of iridium-bearing Fish Clay ('Fiskeler', now Fiskeler Member of the Rødvig Formation (Surlyk et al. 2006)) at the K–T boundary in the Stevns Klint succession south of Copenhagen, the lowermost Danian carbonate unit Cerithium Limestone Member of the Rødvig Formation (Surlyk et al. 2006) is exposed over almost 10 km. It was deposited in minor basins as infill in the <10 m to tens of meters-wide depressions between the emergent crests of the uppermost Cretaceous (uppermost Maastrichtian) bryozoan mounds or bioherms (Fig. 1). The top of the Cerithium Limestone is consistently truncated by well-developed hard-ground preserving aragonitic skeletons as moulds, and the observed maximum thickness of the Cerithium Limestone rarely exceeds 0.7 m. A rich diversity of foraminifera dates the Cerithium Limestone to the earliest part of the Danian, specifically to the P0 to P1C foraminiferal zones (*sensu* Berggren and Miller 1988; Keller 1988) with sections in the south a little older, reflecting the slightly diachronous nature of the unit (Heinberg et al. 2001; Rasmussen et al. 2005). A total of 11 heteromorph specimens of *Baculites vertebralis* Lamarck (1801), and *Hoploscaphites constrictus* (Sowerby 1817) were collected from some 64 bulk samples totaling more than 250 kg, indicating in all likelihood that ammonites are not particularly rare in the Cerithium Limestone (Machalski and Heinberg 2005). The best preserved specimen from the Cerithium Limestone, an adult macroconch (female) of *H. constrictus* (Fig. 2, MGUH27366), is comparable morphologically to the specimens recorded from the equally cemented, truncated tops of the underlying Maastrichtian bioherms. It reveals no evidence of reworking, and it has an infilling that is '...indistinguishable from the surrounding Danian matrix, yields a "Danian" isotopic signal, and contains a calcareous dinocyst, typical of Danian strata in the area' (Machalski and Heinberg 2005, p. 107).

The association of other invertebrate taxa in the Cerithium Limestone provides additional clues and corroboration for the ammonites being Danian in age. Bryozoans, which constitute more than 10% by weight of the uppermost Maastrichtian mounds, are exceedingly rare in the Cerithium Limestone, while for instance micromorphic brachiopods are totally absent from this unit. Bivalves and gastropods, on the other hand, are represented by rich, yet very different faunas in the two parts of the cemented hard-ground encompassing the K–T boundary (Heinberg 1999, in prep.; T. Hansen 2009, pers. comm.). If a significant reworking component was invoked for the ammonites in the basal Danian limestone, other taxa should be

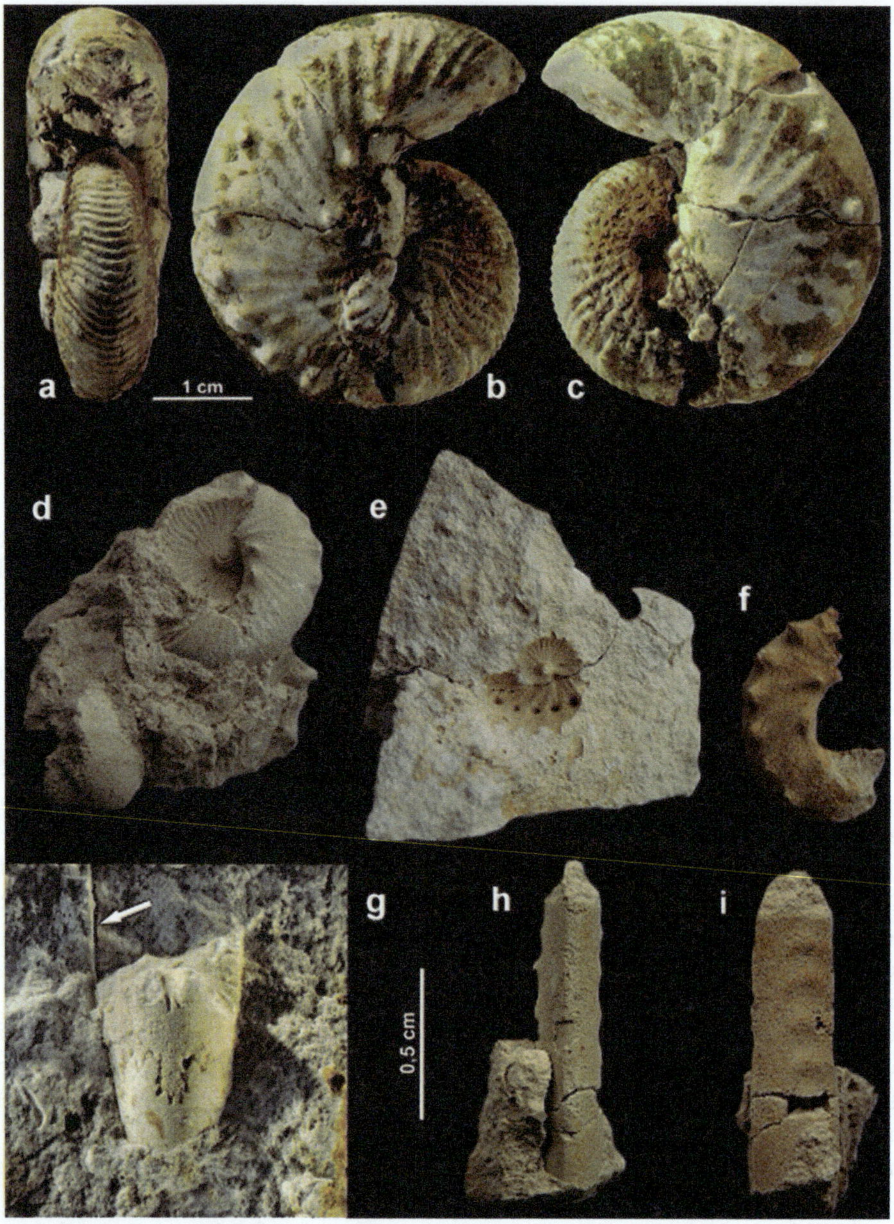

Fig. 2 (**a–e**) Ammonites *Hoploscaphites constrictus* (Sowerby 1817) from the Cerithium Limestone, lowermost type Danian, Stevns Klint region, Denmark, Danish Basin; (**f–h**) *Baculites vertebralis* de B. Lamarck (1801), from the Cerithium Limestone, same localities; **i** is side view of **h**

reworked as well from underlying units into this limestone. On the contrary, the abundance of evidence points to minimal reworking and thus to ammonites surviving into the Danian of Denmark for up to perhaps 200 k.y., as indicated by the estimated age of the hardground truncating the basins of the Cerithium Limestone (Machalski and Heinberg 2005). This well-corroborated occurrence of ammonites in the type Danian of Denmark is mirrored by similar occurrences in the Netherlands (Fig. 3), and the Atlantic Coastal Plains of the United States indicating a pattern that requires further detailed scrutiny (Håkansson et al. 2008; Machalski et al. 2009).

Abundant baculitids together with rare scaphitids, both typically with intact apertures, are recorded from the Meerssen Member (presumed lowermost Danian)

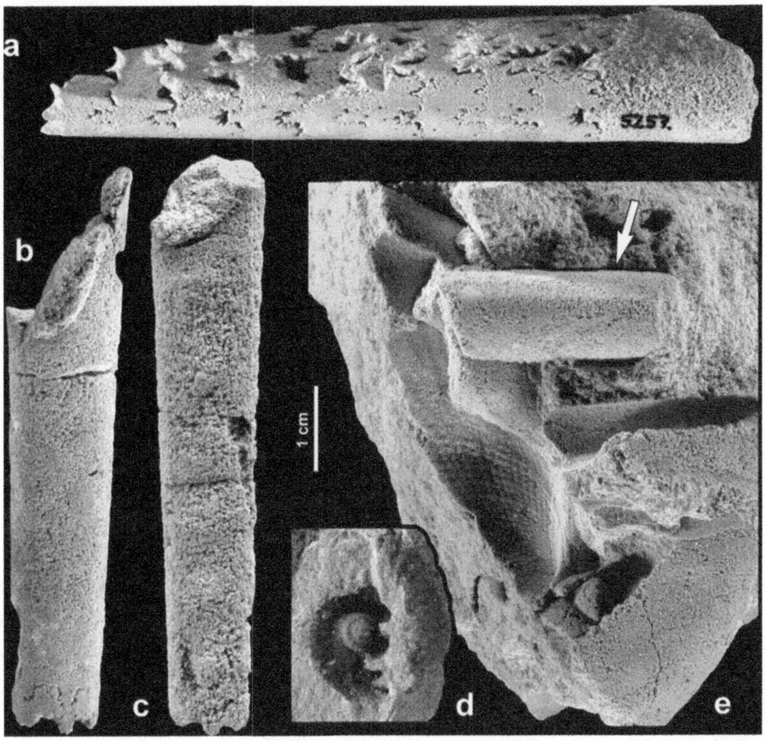

Fig. 3 Representative early Danian ammonites coated with ammonium chloride from the Meersen Member, Maastricht Formation, Netherlands. All specimens are *Baculites cf. vertebralis*, except for (**d**), which is *Hoploscaphites cf. constrictus*

of the Maastricht Formation in the Netherlands (Fig. 3; see Jagt et al. 2003, and references therein). Interestingly, the ammonite taxa recorded from the Meersen Member deviates from those of the Cerithium Limestone, conceivably reflecting the fact that the two localities represent fairly different near-coastal conditions, even though they are not far apart. From further afar, in New Jersey, on the eastern seaboard of the North American continent, Niel Landman and co-workers described an even richer ammonite fauna, in part conspecific with the fauna from the Netherlands, from strata *above* a distinct Ir-anomaly (Landman et al. 2007). Thus, a total of 12 ammonite taxa have now been recorded with a confirmed or suspect Danian origin, comprising 7 scaphitids, 3 baculitids, one *Pachydiscus*, and one *Spenodiscus*. In all three cases do the ammonites disappear for good during the very basal Danian?

Disappearance of a Bizarre, New Zealand Gastropod *Conchothyra* by Mid-Late Danian

The solely Austral gastropod family Struthiolariidae ('ostrich foot shells') is represented by diverse taxa ranging from late Cretaceous to recent from New Zealand, Australia, Antarctica (including sub-Antarctic islands), and southern South America. The earliest recorded occurrence of the group is in the Campanian–Maastrichtian of South Island, New Zealand, where it comprises two very distinct species of *Conchothyra* Hutton, 1877 (*C. parasitica* Hutton 1877, and *C. marshalli* Trechmann 1917), in interpreted shallow shelf deposits (Stilwell 1994). Some horizons abound with these shells in nearly monotypic concentrations of thousands of these massive, dense individuals, such as in the Waimakariri River sequence, where intervals are about 1 m thick of little other than these greatly thickened, globose shells (see Figs. 4 and 5). No *Conchothyra*-like gastropod(s) has been reported in rocks older than Campanian in New Zealand or elsewhere, and the relationship of this bizarre gastropod to other members of the Struthiolariidae and Stromboidea remains poorly understood. The abrupt appearance of *Conchothyra* in the fossil record limits the detail to which the phylogeny and early origins of the Struthiolaridae can be worked out, as no putative ancestors are present—is the genus an early, short-lived offshoot or progenitor of Cenozoic struthiolariid groups? The morphological similarity of *Conchothyra* with Cretaceous members

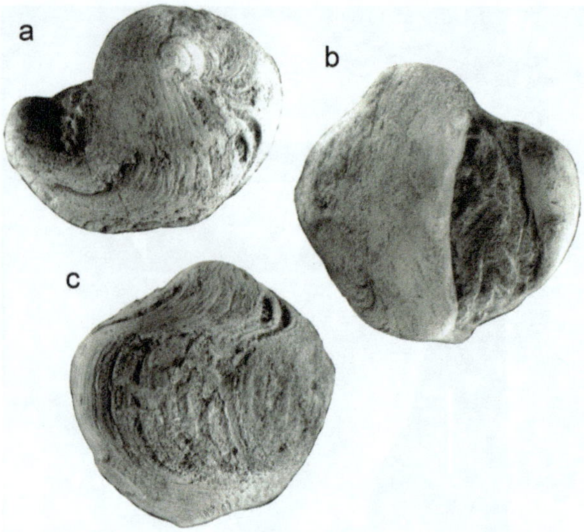

Fig. 4 (**a**–**c**) Late Cretaceous extremely thick, heavily callused, struthiolariid gastropod, *Conchothyra parasitica* Hutton (1877), from the Conway Formation, Waimakiriri River Gorge, South Canterbury, South Island, New Zealand; (**a**) apical, (**b**) apertural, and (**c**) abapertural views; height of specimen 48 mm

of the Aporrhaidae (Stromboidea) lies in all likelihood with an early or early late Cretaceous aporrhaid ancestor structurally and phylogenetically intermediate between *Conchothyra* and *Tephlon* Popenoe (1983) (Stilwell 1994).

The only other record of *Conchothyra* is in Danian deposits of the Wangaloa Formation of South Island near Dunedin and also Wangaloa; these are dated as around ca. 63–62 Ma (Stilwell 1993; Buchanan et al. 2007). *Conchothyra australis* (Marshall 1916) (Fig. 6) is characterized by pronounced reduction in calcification of the shell; it is not rare in the Wangaloa Formation, but not particularly abundant either, where it is associated with relatively high-diversity molluscan and vertebrate assemblages of >100 taxa, representing the stabilization phases following the K–T mass extinction event a few millions years after the calamity. *Conchothyra* is not recorded in any rocks younger than mid- to late Danian, and the ideas of why there was such a pronounced morphological change in the group across the K–T are explored in the following section.

Conchothyra is interpreted to have been an epifaunal or semi-infaunal deposit feeder or browser on fine, nutritive deposits (e.g., diatoms, dinoflagellates, and other detritus, similar to extant forms) that would have moved in/on the benthos with ease, given its powerful and muscular foot. We know of no other comparable gastropod with such a thick shell, so the snail must have ploughed steadily through shallow marine habitats of the 'Zealandian' shelf. The extremely thick shell may have served as a predatory defense against Cretaceous durophagous predators (e.g., ammonites, belemnites, teleost fish, rays, and marine reptiles). It could be argued that the increased shell thickness in *Conchothyra* evolved so that an epifaunal lifestyle, unlike extant members of the family, was possible; hence, protection from predators was perhaps at a maximum in earlier forms of Struthiolariidae, especially given the known diversity of apex predators in the New Zealand–Chatham Islands regions during the late Cretaceous due to interpreted nutrient-rich upwelling centres (Stilwell 1994; Consoli and Stilwell 2009). To

Fig. 5 JDS collecting the *Conchothyra* bed as a PhD student; extraordinary monotypic concentration of thousands of individuals of *C. parasitica*, representing a parauthochthonous assemblage of shallow marine macroinvertebrates from the Conway Formation cropping out along the Waimakiriri River. Photo courtesy Ewan Fordyce, University of Otago

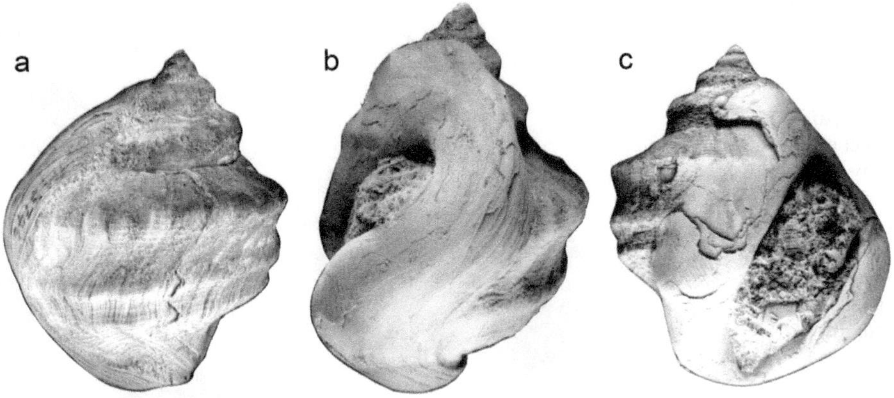

Fig. 6 (**a–c**) Well-preserved individual of the mid-Danian struthiolariid survivor, *Conchothyra australis* (Marshall 1916), from the Wangaloa Formation, south-eastern Otago, South Island, New Zealand, inferred to be a lineal descendant of *Conchothyra parasitica* from the late Cretaceous, abapertural, lateral, and apertural views, respectively of specimen, height 42 mm

invoke a semi-infaunal or even more unlikely infaunal mode of life seems unnecessary; the massive shell of *Conchothyra* would probably have been cumbersome beneath the sea bottom surface and passage through the sediment difficult depending on the mass of soft tissue. Also, a very thick shell would have been unnecessary if *Conchothyra* were infaunal, especially against vagile, nektic predators, even skates and rays.

Observed marked reduction in callus extent, shell thickness, and hence, overall shell volume in *Conchothyra* over the Cretaceous–Paleogene interval may reflect a decrease in predation pressure across the extinction boundary, especially in light of the extinction of many cephalopods and marine reptiles, all of which may have preyed on this snail. If rays (species recorded in both late Cretaceous and Danian in New Zealand) were a principal predator on *Conchothyra*, one could argue that the shell thickness and volume should have stayed relatively comparable, but this is not observed. The cause(s) of the apparent extinction of *Conchothyra* in the Danian of New Zealand is explored herein.

Loss of or decrease in suitable shallow shelf habitat and/or temperature change may been factors influencing the eventual demise of this group which presumably had poor larval dispersal capabilities given its probable non-planktotrophic mode of development. As no members of the family have been recognized in the Chatham Islands region, including taxa of *Conchothyra*, even though similar habitats are found, dispersal of *Conchothyra* larvae would seemingly have been restricted to within a relatively narrow geographic zone. Indeed, the group is known only from southern Marlborough to southern Canterbury on the South Island in the late Cretaceous, while in the Danian, *C. australis* is known from Otago and much further south in the late Danian, probably reflecting New Zealand's northward movement into lower latitudes and hence, the southward migration of the group along the 'Zealandian' paleocoastline. Thus, as mentioned earlier, the very restricted geographic zone of *Conchothyra* may be related to poor dispersal capabilities, but also to differential tolerances toward sea-surface temperatures, food availability after the mass extinction, competition, substrate preferences, predation, or a combination of these factors and possible others. However, the suggested decrease in predation on the snail across the K–T apparently did not ensure its long-term survival after the Danian.

Long-Ranging, Barremian to Earliest Danian Gastropod *Vanikoropsis*

One of the cosmopolitan and most long-ranging gastropods in Cretaceous, Greenhouse Earth, shallow marine environments is the Barremian to Danian, vanikorid gastropod *Vanikoropsis* Meek (1876), with oldest records dating back to Barremian–Aptian of Japan, and subsequent, widespread Cretaceous records from Europe, West India, North America, Australia, New Zealand, and Antarctica (see reviews by Stilwell et al. 2004, 2008, and references therein). This once geographically widespread genus was reduced

Fig. 7 (**a–c**, **e**) Antarctic earliest Danian gastropod survivor, *Vanikoropsis arktowskiana* (Wilckens 1910), and d, *Amauropsis notoleptos* Stilwell et al. (2004), one of many opportunistic bloom taxa of the repopulation phase following the mass extinction. Both species from Seymour Island. (**a–b**) height 44.5 mm; (**c**, **e**) height 34 mm; (**d**) height 8 mm

at the K–T to a single species, *Vanikoropsis arktowskiana* (Wilckens 1910), that crossed the boundary in Antarctica (Fig. 7), surviving for perhaps 300,000 years, and then disappearing by the onset of the recovery phase and subsequent diversification of many molluscan groups, the origins of which are the subject of major reviews by Stilwell (2003 and in review, *Bulletins of American Paleontology*). *Vanikoropsis* is a prime example of a DCW taxon. In the López de Bertodano and overlying Sobral formations on Seymour Island, *V. arktowskiana* ranges from 900 to 1201 m from the bottom of the section, where some 100 specimens have been collected. A radiometric date of 64.701 Ma by McArthur et al. (1998) at c. 1130 m indicates that *V. arktowskiana* survived into the earliest part of the Danian. The only other species that disappears in the same interval is the widespread and long-ranging Antarctic bivalve, *Lahillia larseni* (Sharman and Newton 1897), the only mollusk to form monotypic floods of individuals immediately after the extinction, where it represents an opportunistic bivalve, infilling vacant ecospace along the Antarctic Peninsula shallow shelf environments. However, the genus did not disappear entirely, as is also recorded in the Danian of New Zealand, Selandian–Thanetian of Australia, Eocene of Antarctica, and into the Miocene of South America (reviewed by Stilwell 1994). In the lowermost range of the species at the end of the Cretaceous in Antarctica, *L. larseni* is usually found in relatively sparse numbers at many localities in the López de Bertodano Formation, but this species, together with *V. arktowskiana*, disappeared, coincident with the diversification and appearance of many new benthic taxa above 1200 m (Stilwell et al. 2004). Thus, it is surmised that these once-important elements of the shallow marine Antarctic Cretaceous assemblages constituted a local holdover community which was, in turn, outcompeted by the influx of migrant taxa in the early Danian.

New Evidence of Danian(?) Non-Avian Dinosaurs from Chatham Island

Controversy rages in scientific circles regarding the presence of post-K–T occurrences of Paleocene (Danian) non-avian dinosaurs from various widespread localities around the globe, including especially the Hell Creek Formation of western USA and China (see e.g., Fassett et al. 2002; Fassett 2007). The most convincing evidence for these non-avian dinosaur DCWs is derived from the Hell Creek Formation, where a well-preserved hadrosaur bone was discovered in sediments deposited about 1 m.y. after the K–T, in association with Danian-aged microfossils (ibid.). Robertson et al. (2004) inferred from their detailed study of differential survivorship of fauna and flora in the 'early hours' of the Cenozoic that few large animals would have been able to survive the intense, global heat pulse following the Chicxulub bolide impact, especially given the postulated ballistically re-entering ejecta that would have covered the planet directly after the blast. According to these researchers, '. . .lack of shelter would have been lethal.' (ibid., p. 760), precluding the survival of larger non-avian dinosaurs. But, what about smaller theropod dinosaurs that could have sheltered in natural cavities (caves, large tree trunks, underwater/underground hiding spaces along rivers, streams, etc.)? New evidence of burrowing non-avian dinosaurs makes it likely that at least some of these reptiles were able to survive the effects of the calamity for an interval of time (Varricchio et al. 2007). Was there an intensity gradient to the heat pulse from the impact in low latitudes along the Yukatan Peninsula, meaning that the higher-latitude regions could have been buffered from the after-effects to a degree? Evidence in the marine realm of higher survivorship rates in mollusk taxa along Antarctic Peninsula at the K–T boundary (Stilwell 2003; Stilwell et al. 2004) is suggestive of lessened 'impact' of the event in the high southern latitudes, but what about on land? We present here some new data from the Chatham Islands, SW Pacific, which are equally compelling regarding the possible survival of small 1–2 m long theropod dinosaurs for a short interval into the Danian of Zealandia.

In 2003 the senior author (JDS) and his team, through a National Geographic Society–sponsored expedition, discovered the first record of non-avian dinosaurs on the remote Chatham Islands about 865 km east of Christchurch, New Zealand, which represents emergent parts of the mostly submerged Chatham Rise (Stilwell et al. 2006). These remains in the Takatika Grit, plus tectonic evidence, corroborate the hypothesis that a ca. 900–1000 km land bridge extended from what is now New Zealand to the easternmost part of the Chatham Rise. This low-lying, narrow, finger-like tract of emergent land, a crustal extension of the New Zealand subcontinent (part of 'Zealandia'), hosted coastal, temperate environments teeming with diverse life, including many plants and animals. The Chatham Rise in the late Cretaceous was blanketed in a forest, dominated by clubmoss and conifers, and small- to medium-sized theropod dinosaurs, ranging from 1 m to 8–10 m long; other groups were also represented (JDS 'unpublished data').

The larger theropod remains include a proximal pedal phalanx II–1 (or III–1) that is 18.5 cm long and 10.2 cm wide and probably belonged to a large neotheropodan theropod, such as a basal allosaurid, with a relatively slender pes comparable to *Carnotaurus* (Stilwell et al. 2006). Other bones (centra, partial tibia) represent medium-sized individual theropods up to 4 m long. All of the bones have been recovered from the richly fossiliferous Takatika Grit, which is a thin unit (<10 m thick) comprising very hard, glauconitic quartzofeldspathic grit made up of sandstones and siltstones with a minor volcanogenic component, deposited in shallow shelf settings, sitting noncomformably on basement schist. Overlying the main dinosaur-bearing beds in stratigraphically younger sediments about 1 m above, called informally the Greensand Package II, a theropod manual phalanx and manual ungual have been collected in beds that have been dated by microfossils to be early Danian in age (Wilson et al. 2005). As stated by Stilwell et al. (2006, p. 245), these bones exhibit no evidence at all of reworking and are very well preserved (Fig. 8). The ungual (claw) is not strongly recurved like birds and is more characteristic of some ornithomimosaurs (Stilwell et al. 2006; Julia Clarke 2009, 'personal communication'), and the phalanx is similar to several non-specific theropod groups (e.g., troodontids, oviraptorosaurids, ornithomimids). Given that these small non-avian theropod remains are from deposits dated as belonging to the dawn of the Danian, the possibility remains that smaller theropods survived the K–T extinction event for a short time slice on a remote strip of forested land in the SW Pacific. The well-preserved

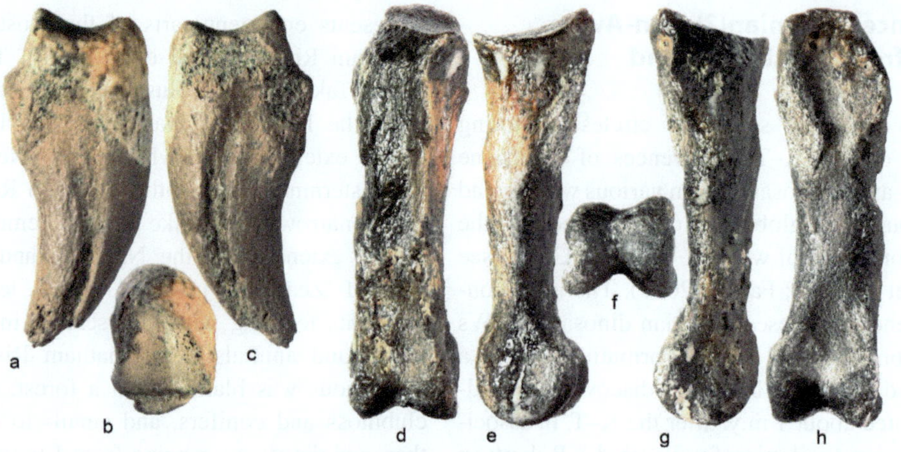

Fig. 8 (**a–h**) Non-avian theropod dinosaur elements from the early Danian from the Takatika Grit, Chatham Islands, Southwest Pacific. (**a–c**) small theropod manual ungula length 15.25 mm, possibly an ornithomimosaur. (**d–h**) theropod manual phalanx of medium-sized animal length 33.6 mm

nature of the material—although admittedly meager at this time—warrants further investigation; this is currently well underway.

Cheilostome Bryozoans—An Example of Wholesale Clade Replacement?

Cheilostome bryozoans are arguably the last of the major skeletized groups in the modern marine fauna to enter the global stage. From a slow beginning in the later part of the Jurassic the cheilostomes went through a dramatic burst of evolution in the late Cretaceous with an increase in number of families from less than five at the beginning of the Cenomanian to almost 40 in the Maastrichtian. This evolutionary success is by and large confined to the shallow shelf seas of Europe, most notably to the north European Chalk Sea estimated to house more than 2500 cheilostome species toward the end of the Cretaceous. Despite the usual distortion of perspective surrounding anything European, the highly localized nature of this burst does seem quite real. Following this period of multitudes, subsequent Paleocene faunas are scarce and far between, even in Europe; only in the narrow seaway crosscutting Europe from Denmark to the Crimean peninsula through the early Paleocene, Danian Stage, do bryozoans seem to thrive (Voigt 1962). From the beginning of the Eocene a second burst of evolution commenced amongst the cheilostomes, leading up to the present situation with about 100 well-skeletized families. So, a classic case of recovery it would seem. But, in view of the scarcity of Paleocene bryozoan faunas, the actual patterns of recovery for the cheilostome bryozoans are not known in detail. Only in Denmark and Crimea are rich early Danian faunas known, and only in a single section—Nye Kløv, Denmark—has the faunal development through the K–T biotic turnover been investigated at the species level (Håkansson and Thomsen 1999). Sufficient evidence exists, however, to maintain that the pattern established here is indeed valid for the Danish Basin as a whole, a basin sporting at least 10 K–T boundary successions without biostratigraphical hiati at the boundary, including world-famous Stevns Klint. Perhaps as many as 1500 species of cheilostome bryozoans are present in the Upper Maastrichtian–Danian strata of this small basin, and based on the figures established from Nye Kløv, their survival rate is about 20% at the species level, around 60% at the genus level, and almost 90% at the family level (Fig. 9). In the Danish Basin, as well as in Crimea and Kazakhstan, cheilostome bryozoans recover within a very short period of time, reaching a maximum diversity already in early Danian (NP 2) time, with a complete dominance of surviving Mesozoic lineages (Håkansson and Thomsen 1999). The abrupt end to this remarkable recovery coincided with the terminal Danian collapse of the carbonate platform, which had dominated northern and eastern Europe throughout the late Cretaceous (Surlyk 1997), with all but a handful of lineages becoming extinct (Fig. 9; Håkansson and Thomsen 1999).

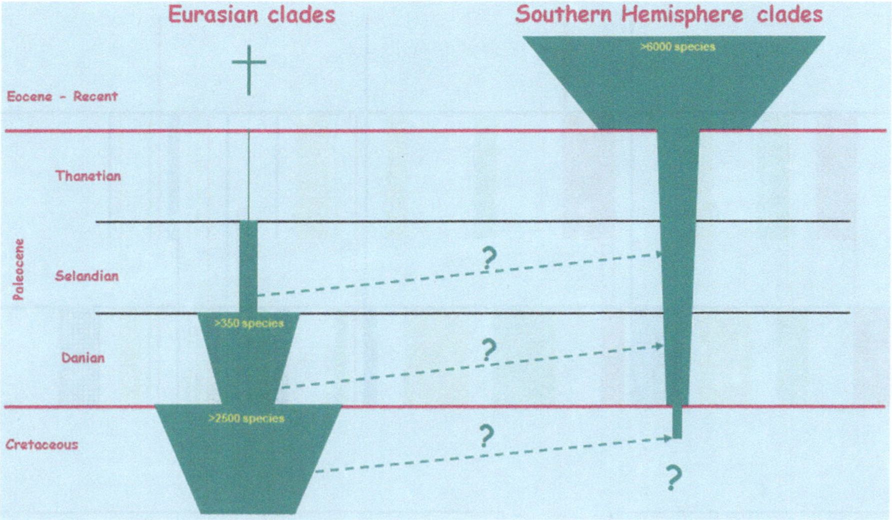

Fig. 9 Diagrammatic illustration of the distribution of cheilostome bryozoans across the Maastrichtian–Danian (i.e., K–T) and Danian–Seladian boundaries in the Danish Basin (here illustrated with superfamily-level clades). A fairly uniform extinction and recovery pattern within each clade is obvious, with a brief refugial interval followed by a rapid recovery, levelling out at a somewhat lower diversity level maintained throughout the Danian. With the collapse of the carbonate regime characterizing this basin (as well as large parts of Northern and Eastern Europe) from the early part of the late Cretaceous, only a handful of lineages persist in low numbers into the middle Paleocene to disappear before the Eocene. Note the comparatively low number of extinctions at the K–T boundary (cross-hatched bar) at this taxonomic level, as well as the very limited amount of new elements in the Danian fauna (data in part from Håkansson and Thomsen 1999)

Recent, renewed investigations of the very limited late Cretaceous faunas outside Europe as well as the scattered Paleocene faunas on a global scale have revealed a distribution pattern which may put the fragmenting Gondwana super-continent into a more central position. Thus, results from ongoing studies on the detailed evolution within a number of selected families (Thomsen and Håkansson 1995; EH in prep.; EH et al. in prep.) have opened the way for an entirely novel model for the evolution of the cheilostome bryozoans across the K–T boundary, involving iterative evolution of fundamental specializations in cheilostomes, such as frontal wall architecture and secondary calcification (Fig. 10; Håkansson 2007 and in prep.). In this model most or all of such specializations already present in the Upper Cretaceous faunas of the European Chalk Sea were lost in a series of extinction events at the Maastrichtian–Danian (K–T), the Danian–Selandien, and the Paleocene–Eocene boundaries (Håkansson E and Thomsen E in prep.). In conjunction with this evolutionary drama in Europe less advanced forms migrated out into other parts of the globe, where they founded the new lineages which in turn—perhaps related to the end-Paleocene temperature spike—spread out to form the basis for the second burst in cheilostome evolution, replacing their Mesozoic 'look-alikes,' also in Europe (Fig. 10; Håkansson 2007 and in prep.).

Discussion

It has long been a popular pastime to speculate about what distinguished groups succumbing to mass extinction events from those who survived—although it has been an area with more suggestions than arguments. In our view it would seem more productive to differentiate such speculations, distinguishing the discussions in such a way as to consider groups that experience a long-term survival separately from those that survive only for a short period—the 'Dead Clade Walking' concept of Jablonski (2002). In other words, it may be more productive to modify and supplement the 'classic' question *What is the common denominator for groups surviving mass extinctions?* toward a pair of complementary questions: *Is there a the common denominator for groups surviving mass extinctions?* and *Is there a common denominator for groups surviving mass extinctions but failing to recover?* But with

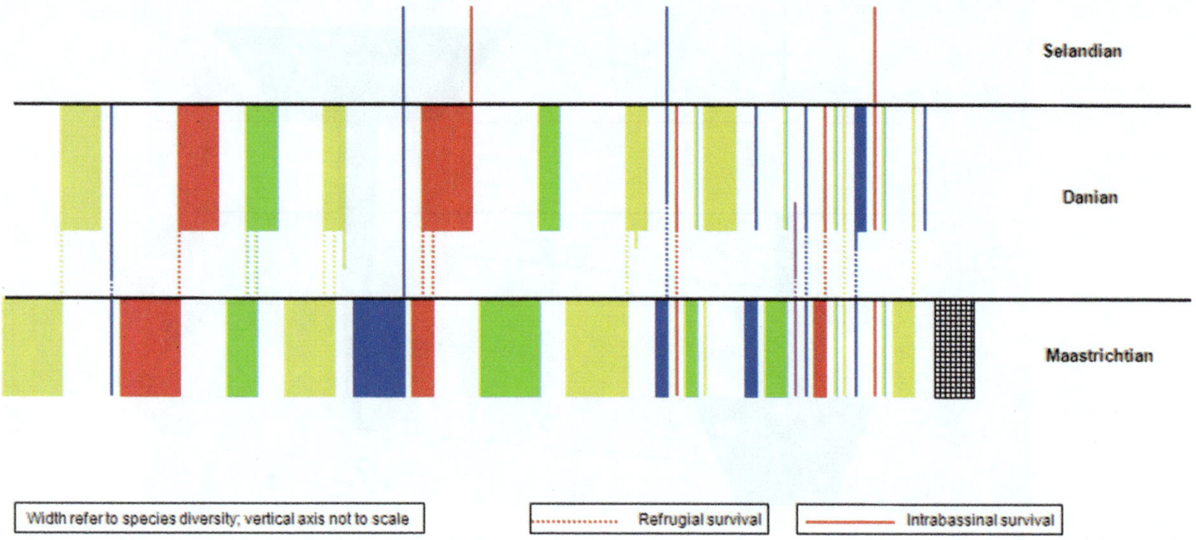

Fig. 10 Tentative model for disjunct evolution of the gymnolaemate bryozoan order Cheilostomata through the Mesozoic–Cenozoic biotic transition. From a slow start in the late Jurassic all but a single of the fundamental frontal wall designs known in modern cheilostomes had evolved by the latest Cretaceous, associated with a very high diversity (conservative diversity estimates in *yellow*), and with a near-exclusive European focus. This primary group of cheilostomes experienced a stepwise extinction through the Paleocene. The second burst of cheilostome radiation, suggested to originate in the Southern Hemisphere, was characterized by repetitive re-invention of the fundamental types of frontal wall architecture already present in the late Cretaceous. This second group of cheilostomes spread around the World from the Eocene onward to dominate the present-day faunas. (Based on work presented by Håkansson 2007; E Håkansson and E Thomsen in prep.)

anything other than fairly broad-scaled investigations presently being limited, well-founded answers to either question do not appear imminent.

Much, of course, is known about the reaction of various taxonomic groups to the global ecological collapses associated with the major extinction events, including the K–T boundary. But this accumulated knowledge vary immensely in level of detail, as demonstrated in the numerous summaries of the fate of biota at the K–T boundary (e.g., MacLeod et al. 1997) sparked by the discussion of the impact hypothesis published by the Alvarez group (Alvarez et al. 1980). So even at the present level of knowledge the emerging picture comes out as somewhat ambiguous, with some of the biotic response commencing before the actual boundary, and with some of the 'classic' extinction victims lingering on for shorter periods of time (e.g., Birkelund and Håkansson 1982; Keller 1988)—the Dead Clades Walking.

The series of cases presented here serve to illustrate just how varied responses to a particular 'global event' (*in casu* the K–T boundary) might be in different taxonomic groups when considered in detail. The disparity illuminated by these examples, sharing only their fairly high paleolatitude amongst ambient ecological parameters, may suggest that a more structured approach will be necessary in order to address the type of questions posted earlier. Previous considerations seem to have been concerned primarily with terminology and categorization (Kauffmann and Erwin 1995; Hallam and Wignall 1997; Korn et al. 2004); hence we shall here direct attention toward the more qualitative aspects of such categories.

Rapid radiation as well as very limited radiation may be found within both groups experiencing long-range survivorship and groups surviving only for a short period—'Dead Clades Walking' (see Fig. 11). The evolutionary history of long-ranging, and therefore by most popular definitions 'successful,' survivors may be categorized as experiencing either significant radiation (e.g., the spectacular radiation of dinosaurs and ammonites following the Triassic–Jurassic mass extinction, commonly associated with

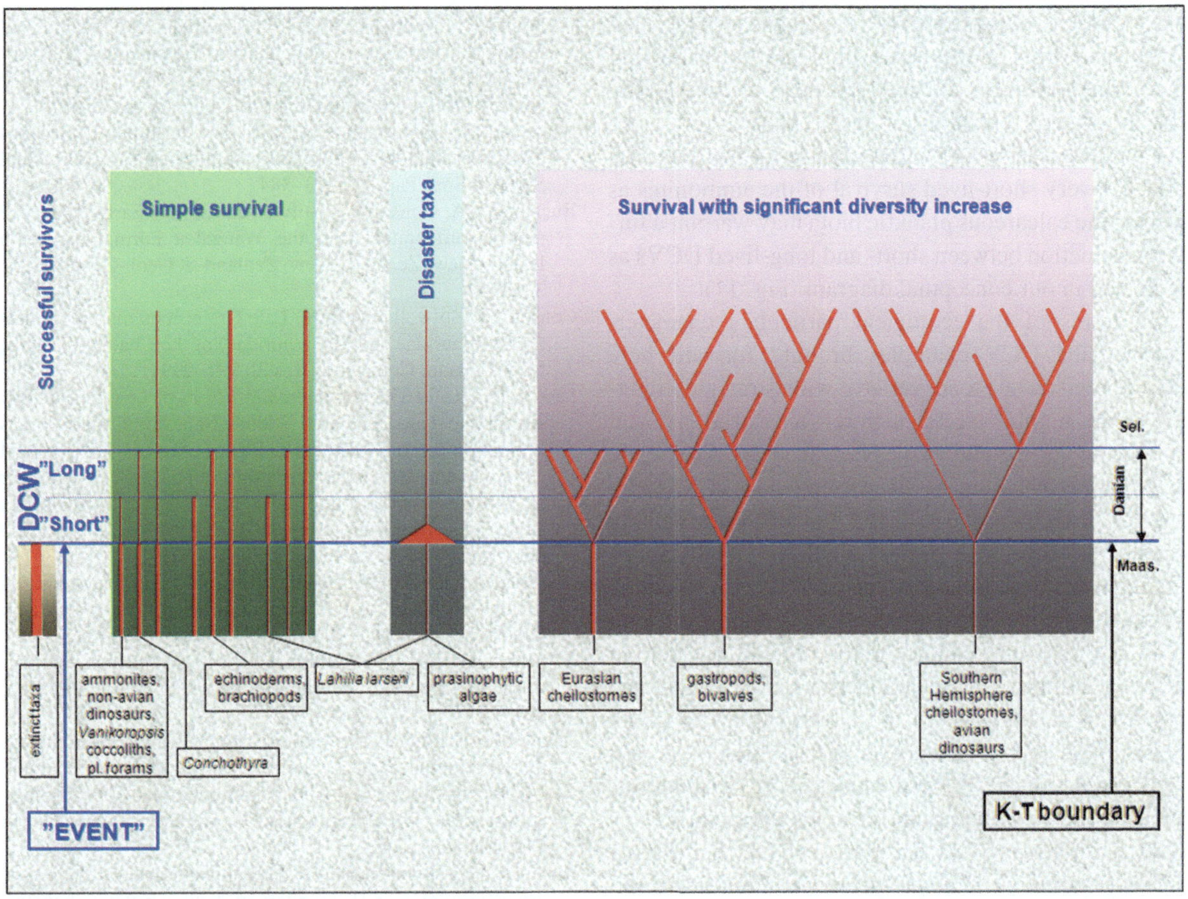

Fig. 11 Conceptual diagram showing clade response to a significant extinction event. Primary-level distinction is between extinction and successful survival. A secondary level of distinction between successful survivors may be a continuation with limited change, restricted continuation, or radiation. 'Dead Clade Walking' conditions if extended to the first post-extinction stage may be found within all three categories; however, we know of no examples of 'short term' DCVs experiencing radiation. The position of cases discussed in the text and other examples related to the K–T boundary are indicated in *black boxes*

terms like 'rulers of a new world' or similar) or very limited radiation (e.g., the inarticulate brachiopod *Lingula* and its relatives, surviving each and every Phanerozoic mass extinction but never experiencing any noteworthy radiation, usually characterized by the term 'living fossils').

Parallel to this, also short-ranging survivors (DCWs) may either go through a rapid, albeit brief, phase of active evolutionary radiation before their disappearance or simply vanish after a short interval of time. Both types of survival of DCWs in the Danian are illustrated by the examples discussed here: cheilostome bryozoans experiencing rapid evolution at a low taxonomic level involving hundreds of new species through the Danian, and ammonites as well as non-avian dinosaurs fading away after a very brief presence in the basal Danian. Further examples may be the equally rapid fading away of the Maastrichtian plankton (coccoliths as well as forams) in the basal Danian or the impressive, low-level taxonomic diversification of ceratitid ammonites continuing throughout the Triassic.

The relative longevity of the single ammonite clades to diversify after the near-extinction of the ammonites at both the Permo-Triassic and Triassic–Jurassic boundaries is in strong contrast to the very short-lived 'hanging in there' after their equally severe fate at the K–T boundary, illuminating the need for a concise time frame to the 'Dead Clade Walking' concept. The time frame defined by David Jablonski in his quantitative approach is simple and operational: Only taxa lost in the stage immediately following the extinction event fall into the dead clade walking (DCW) category (Jablonski 2002). And we fully concur with that. However, in relation to the K–T extinction, the very short-lived survival of the ammonites as well as the calcareous planktic biota may warrant a further distinction between short- and long-lived DCVs as indicated in our conceptual diagram (Fig. 11).

In terms of the ecological particulars characteristic of the DCV examples brought forward here do we register a broad variety, with strategies ranging from highly vagile hunters—terrestrial as well as marine—to sessile suspension feeders. And while *Lahillia larseni* may be characterized as a 'disaster species' (and thus an ecological generalist presumably capable to cope with almost anything) totally dominating vacant ecospace in the area of the Antarctic Peninsula in the wake of the K–T boundary event, it is also a prime candidate for the DCV label. In marked contrast, the echinoderms and cheilostome bryozoans taking over in the North European Chalk Sea immediately after this event are all highly specialized types, adapted toward muddy, soft sea bottoms, such as the bourgueticrinid crinoids *Bourgueticrinus danicus* and *Democrinus maximus* and the cheilostome species *Lekythoglena ampullacea*, *Pavolunulites n. sp.*, and *Danocella haakanssoni* (Håkansson and Thomsen 1999; Kjaer and Thomsen 1999).

Acknowledgments We thank the following for their support and assistance throughout this investigation: Draga Gelt and Ben Doherty (Monash University); Bill Zinsmeister (Purdue University); Ewan Fordyce (University of Otago, Dunedin, New Zealand); Chris Consoli (Geoscience Australia); Claus Heinberg (Roskilde, Denmark); John Jagt (Natuurhistorisch Museum Maastricht, the Netherlands). This research was funded by a Research Initiatives grant from Monash University to JDS, and support from the Geological Institute, Copenhagen, Denmark, and University of Western Australia, Perth, to EH.

References

Alvarez LW, Alvarez W, Asaro F, Michel HV (1980) Extraterrestrial cause for the Cretaceous/Tertiary extinction. Science 208:1095–1108

Berggren WA, Miller KG (1988) Paleogene tropical planktonic foraminiferal biostratigraphy and magnetobiochronology. Micropaleontology 34:362–380

Birkelund T (1979) The last Maastrichtian ammonites. In: Birkelund T, Bromley RG (eds) Cretaceous–Tertiary boundary events. I. The Maastricthian and Danian of Denmark. University of Copenhagen, Copenhagen, pp 51–57

Birkelund T (1993) Ammonites from the Maastrichtian White Chalk of Denmark. Bull Geol Soc Denmark 40:33–81

Birkelund T, Håkansson E (1982) The terminal Cretaceous extinction in Boreal shelf seas – a multicausal event. Geol Soc Am Spec Pap 190:373–384

Buchanan LA, Consoli CP, Stilwell JD (2007) Early Paleocene marine vertebrates from the Wangaloa Formation, South Island, New Zealand. New Zealand J Geol Geophys 50: 33–37

Consoli CP, Stilwell JD (2009) Late Cretaceous marine reptiles (Elasmosauridae and Mosasauridae) of the Chatham Islands, New Zealand. Cretaceous Res 30:991–999

Fassett JE (2007) The documentation of in-place dinosaur fossils in the Paleocene Ojo Alamo Sandstone and Animas Formation in the San Juan Basin of New Mexico and Colorado mandates a paradigm shift: dinosaurs can no longer be thought of as absolute index fossils for end-Cretaceous strata in the Western Interior of North America. New Mexico Geol 29:56

Fassett JE, Zielinski RA, Budahn JR (2002) Dinosaurs that did not die: evidence for Paleocene dinosaurs in the Ojo Alamo Sandstone, San Juan Basin, New Mexico. In: Koeberl, C, MacLeod KG (eds) Catastrophic events and mass extinctions: impacts and beyond. Geol Soc Am Spec Pap 356: 307–336

Håkansson E (2007) The role of frontal wall architecture in the development of a cheilostome phylogeny – and the Cretaceous Tertiary boundary. International Bryozoology Association, 14th International Conference, Boone, North Carolina. Abstracts

Håkansson E, Thomsen E (1999) Benthic extinctions and recovery patterns at the K/T boundary in shallow water carbonates, Denmark. Palaeogeogr Palaeoclimatol Palaeoecol 154:67–85

Håkansson E, Machalski M, Heinberg C, Jagt J (2008) Ammonites from the type Danian of Denmark. Abstract, 33rd International Geological Congress, Norway, 6th–14th August 2008

Hallam A, Wignall PB (1997) Mass extinctions and their aftermath. Oxford University Press, Oxford

Heinberg C (1999) Lower Danian bivalves, Stevns Klint, Denmark: continuity across the K/T boundary. Palaeogeogr Palaeoclimatol Palaeoecol 154:87–106

Heinberg C, Rasmussen JA, Håkansson E (2001) Planktic foraminifera from the lowermost type Danian of Stevns Klint, Denmark. Palaeontol Assoc, 45th Annual Meeting, Abstracts, 15, Geological Museum, Copenhagen

Hutton FW (1877) Report on the geology of the north-east portion of the South Island, from Cook Straits to the Rakaia. Reports of the Geological Explorations during 1873–4, Wellington, New Zealand, pp 27–58

Jablonski D (2002) Survival without recovery after mass extinctions. Proc Natl Acad Sci USA 99(12):8139–8144

Jagt JWM, Smit J, Schulp A (2003) ?Early Paleocene ammonites and other molluscan taxa from the Ankerpoort–Curfs quarry (Geulhem, southern Limburg, the Netherlands). In: Lamolda MA (ed) Bioevents: their stratigraphic records, patterns and causes, Caravaca, 3rd–8th June 2003:113. Ayuntamiento de Caravaca de la Cruz

Kauffmann E, Erwin D (1995) Surviving mass extinctions. Geotimes 40.3:14–17

Keller G (1988) Extinction, survivorship and evolution of planctic foraminifera across the Cretaceous/Tertiary boundary at El Kef, Tunesia. Mirine Micropaleontol 13: 239–263
Kjaer CR, Thomsen E (1999) Heterochrony in bourgueticrinid sea-liliesrelated to the K/T boundary event. Paleobiology 25:29–40
Korn D, Belka Z, Fröhlich S, Rücklin M, Wendt J (2004) The youngest African clymeniids (Ammonoidea: Late Devonian) – failed survivors of the Hangenberg Event. Lethaia 37:307–315
Lamarck J-BPA de M de (1801) Système des Animaux sans vertèbres. The Author, Dèterville, Paris, vii+432 pp
Landman NH, Johnson RO, Garb MP, Edwards LE, Kyte FT (2007) Cephalopods from the Cretaceous/Tertiary boundary interval on the Atlantic Coastal Plain, with a description of the highest ammonite zones in North America. Part III. Manasquan River Basin, Monmouth County, New Jersey. Bull Am Mus Nat Hist 303:1–122
Machalski M (2002) Danian ammonites: a discussion. Bull Geol Soc Denmark 49:49–52
Machalski M, Heinberg C (2005) Evidence for ammonite survival into the Danian (Paleogene) from the Cerithium Limestone at Stevns Klint, Denmark. Bull Geol Soc Denmark 52:97–111
Machalski M, Jagt JWM, Heinberg C, Landman NH, Håkansson E (2009) Dańskie amonity – obecny stan wiedzy I perspektywy badań (Danian ammonites – the present state of knowledge and perspectives for future research [in Polish]. Przegląd Geol 57:486–493
MacLeod N, Rawson PF, Forey PL, Banner FT, Boudagher-Fadel MK, Bown PR, Burnett JA, Chambers P, Culver S, Evans SE, Jeffrey C, Kaminski MA, Lord AR, Milner AC, Milner AR, Morris N, Owen E, Rosen BR, Smith AB, Taylor PD, Urquhart E, Young JR (1997) The Cretaceous–Tertiary biotic transition. J Geol Soc 154:265–292
Marshall P (1916) Some new fossil gastropods. Trans New Zealand Inst 48:100–119
McArthur JM, Thirwall MF, Engkilde M, Zinsmeister WJ, Howarth RJ (1998) Strontium isotope profiles across the K/T boundary sequences in Denmark and Antarctica. Earth Planet Sci Lett 160:179–192
Meek FB (1876) A report on the invertebrate Cretaceous and Tertiary fossils of the Upper Missouri Country. U.S. Geol Surv Territories, Reports, 9, 629pp
Popenoe (1983) Cretaceous Aporrhaidae from California: Aporrhainae and Arrhoginae. J Paleontol 57: 742–765
Prejean H (1994) Dead man walking. Random House, New York
Rasmussen HW (1971) Echinoid and crustacean burrows and their diagenetic significance in the Maastricthian—Danian of Stevns Klint, Denmark. Lethaia 4:191–216
Rasmussen JA, Heinberg C, Håkansson E (2005) Planktonic foraminiferid biostratigraphy of the lowermost Danian strata at Stevns Klint, Denmark. Bull Geol Soc Denmark 52: 113–131
Robertson DS, McKenna MC, Toon OB, Hope S, Lillegraven JA (2004). Survival in the first hours of the Cenozoic. GSA Bull 116:760–768
Rosenkrantz A (1924) Nye iagttagelser over Certhiumkalken I Stevns Klint med bemækninger om Grænsen mellem Kridt og Tertiær. Meddelelser fra Dansk Geologisk Forening 6: 28–31
Sharman G, Newton ET (1897) Notes on some additional fossils collected at Seymour Island, Graham's Land, by Dr. Donald and Captain Larsen. Proc R Soc Edinburgh 22:58–61
Sowerby J (1817) The mineral conchology of Great Britain, vol. 2, pls. 151–184, A, 185–186. The Author, London
Stilwell JD (1993) New Early Paleocene Mollusca from the Wangaloa formation of South Island, New Zealand. J Paleontol 67(3):360–369
Stilwell JD (1994) Latest Cretaceous to earliest Paleogene molluscan faunas of New Zealand: changes in composition as a consequence of the break-up of Gondwana and extinction. Unpublished PhD dissertation, The University of Otago, Dunedin, New Zealand, 1630 pp, 84 plates
Stilwell JD (2003) Patterns of biodiversity and faunal rebound following the K–T boundary extinction event in Austral Palaeocene molluscan faunas. Palaeogeogr Palaeoclimatol Palaeoecol 195:319–356
Stilwell JD, Consoli CP, Sutherland R, Salisbury D, Rich TH, Vickers-Rich PA, Currie PJ, Wilson GJ (2006) Dinosaur sanctuary on the Chatham Islands, Southwest Pacific: first record of theropods from the K–T boundary Takatika Grit. Palaeogeogr Palaeoclimatol Paleoecol 230:243–250
Stilwell JD, Hope S, Consoli CP (2008) Origins of post–K–Pg Danian macroinvertebrates and vertebrates from the southern high latitudes: new data from Antarctica and New Zealand–Chatham Islands. Abstract, 33rd International Geological Congress, Norway, 6th–14th August 2008
Stilwell JD, Zinsmeister WJ, Oleinik AE (2004) Early Paleocene Mollusks of Antarctica: systematics, paleoecology and paleobiogeographic significance. Bull Am Paleontol 367: 89 pp
Surlyk F (1997) A cool-water carbonate ramp with bryozoan mounds: Late Cretaceous Danian of the Danish Basin. In: James NP, Clarke JAD (eds) Cool-water carbonates. Special Publications – SEPM (Society for Sedimentary Geology), vol 56, pp. 293–307
Surlyk F, Nielsen JM (1999) The last ammonite? Bull Geol Soc Denmark 46:115–119
Surlyk F, Damholt T, Bjerager M (2006) Stevns Klint, Denmark: uppermost Maastrichtian chalk, Cretaceous–Tertiary boundary, and lower Danian bryozoan mound complex. Bull Geol Soc Denmark 54:48pp
Thomsen E, Håkansson E (1995) Sexual versus asexual dispersal in clonal animals – examples from cheilostome bryozoans. Paleobiology 21:496–508
Trechmann CT (1917) Cretaceous Mollusca of New Zealand. Geological Magazine n.s. Decade vi, 4:294–305, 337–342
Varricchio DJ, Martin AJ, Katsura Y (2007) First trace and body fossil evidence of a burrowing, denning dinosaur. Proc R Soc B, Biol Sci 274:1361–1368
Voigt E (1962) Werchnemelowyje Mshanki europeiskoi tschasti SSSR i nekotorych sopredelnych oblastai. Moskowo Universiteta, 124 pp (in Russian)
Wilckens O (1910) Die Anneliden, Bivalven und Gastropoden der Antarktischen Kreideformations In: Wissenschaftliche

Ergebnisse der Schwedischen Südolar-Expedition 1901–1903 unter leitung von Dr. Otto Nordenskjöld. Band III. Geologie und Paläontologie, 132pp, 4 pls. Lithographisches Institut des Generalstabs, Stockholm
Wilson GJ, Schiøler P, Hiller N, Jones CM (2005) Age and provenance of Cretaceous marine reptiles from the South Island and Chatham Islands, New Zealand. New Zealand J Geol Geophys 48:377–387
Zinsmeister WJ (2001) Late Maastricthian short-term biotic events on Seymour Island, Antarctic Peninsula. Geology 109:213–229

Fungi, a Driving Force in Normalization of the Terrestrial Carbon Cycle Following the End-Cretaceous Extinction

Vivi Vajda

Abstract

Geologists have long recognized the magnitude, abruptness, and the global pattern of the major biotic turnover across the Cretaceous–Paleogene (K–Pg) boundary approximately 65.5 million years ago. It was associated with one of the most catastrophic events in the history of life, involving mass mortality in both terrestrial and marine ecosystems globally; vast amounts of dead biomass covered the Earth's surface. Of prime importance are data from distal boundary sites. These reveal much about the pattern of extinction, whether overprinted or not by local effects such as cratering and post-impact tsunamis. There are few instructive Cretaceous–Paleogene boundary sequences in terrestrial settings. Such sites include the western interior of the USA, Canada, and New Zealand. The pattern of vegetation turnover in these areas is characterized by a so-called fern-spike: abrupt and marked increase in relative abundance of fern spores in palynological assemblages. The latest Cretaceous palynofloras were a rich and diverse mix of angiosperms and gymnosperms with fern, lycophyte, and moss spores. In New Zealand, the boundary is characterized by total dominance of fern spores, notably in undisturbed, fine-grained sediments at the Moody Creek Mine where turnover from the latest Maastrichtian flora consists of a thin fungal layer followed by a step-wise recovery succession of several ground- and tree-fern taxa. Identification of this 'fungal-spike' accords with fungi having played an important role in the devastated terrestrial ecosystems. Fungi may even have facilitated normalization of the global carbon cycle.

Keywords

Cretaceous • Paleogene • Maastrichtian • New Zealand • Moody Creek • Mass mortality • Terrestrial • Palynology • Fungae • Fern-spike • Carbon-cycle normalization

V. Vajda (✉)
Department of Earth and Ecosystem Sciences, Lund University, 223 62 Lund, Sweden
e-mail: vivi.vajda@geol.lu.se

Fungi in the Geological Record

Fossilized fungal components such as spores, fruiting bodies, and hyphae are often obtained from palynological preparations of sedimentary rocks.

J.A. Talent (ed.), *Earth and Life*, International Year of Planet Earth,
DOI 10.1007/978-90-481-3428-1_27, © Springer Science+Business Media B.V. 2012

Fungi have a long geological record; protein sequence analyses indicate that major lineages of fungi were already present during the Precambrian, 1000 million years ago (Heckman et al. 2001). The earliest well-documented fossil fungi, however, are the Glomales-like forms found in mid-Ordovician rocks (age: 460 million years) in Wisconsin, USA (Redecker et al. 2000), whereas the earliest well-documented ascomycotan fossils have been identified in the Scottish Rhynie Chert (Devonian: 400 Ma; Taylor et al. 1999). Fungi, coevolving with plants, assisted colonization of barren terrestrial ecosystems by higher plants (Pirozynski and Malloch 1975; Bonfante and Genre 2008). An example from the Rhynie chert is the early land plant *Aglaophyton major* hosting fungal structures within the plant's cells (Taylor et al. 1995).

Contemporary fungi are mainly decomposers; this was arguably their main function in early ecosystems. They also take part in other processes, for example, serving as geochemical agents contributing to rock weathering by mobilizing mineral constituents with inorganic or organic acids, or by redox attack on Fe and Mn minerals (Ehrlich 1998). These processes contributed to early soil formation and possibly facilitated colonization of land by plants.

Fungi did not become abundant and diverse in the palynological record until the Cretaceous when, interestingly, their diversification coincided with the origin and diversification of angiosperms. Studies of fossil fungi contribute to the knowledge of plant-fungi-fauna interactions in ancient habitats. They shed light on the role of fungi as an important part of the terrestrial carbon cycle in these ancient terrestrial ecosystems.

The Cretaceous–Paleogene Mass-Extinction Event

One of the three or four largest Phanerozoic mass-extinctions occurred at the end of the Cretaceous, 65.5 million years ago (Sepkoski 1990; Schulte et al. 2010); this event has been linked to the possible impact of an asteroid in what is today the Yucatan Peninsula of Mexico (Alvarez et al. 1980). It has been argued that this event resulted in the extinction of perhaps 70% of all species including non-avian dinosaurs, marine reptiles such as the plesiosaurs, and the air-borne reptiles or pterosaurs. Apart from leaving a crater 180 km wide and 2 km deep in the Earth's surface, the impact is argued to have caused a series of environmental disasters, some dramatic and instantaneous, such as tsunamis and emplacement of a thick blanket of ejected rubble and ash around the Caribbean region (Ocampo et al. 2006; Schulte et al. 2010). Other more subtle and long-term effects have been argued to include increased acid rain from sulphur aerosols ejected into the atmosphere (Pope et al. 1994). These, together with the dust and soot from wild fires, would have reduced insolation, resulting in major traumas for ecosystems globally. The sedimentary signature of this event includes enrichment of sediments with cobalt, nickel, zinc, and especially the rare earth element, iridium. This package of chemical data enables high-precision identification of the boundary between the Cretaceous and the Paleogene systems as well as the recovery succession in marine and terrestrial sediments globally.

The purpose of this contribution is documentation of a remarkable Cretaceous–Paleogene palynological sequence in southwestern New Zealand that provides what I believe to be a compelling insight into what occurred to floras through the K–T boundary event, tens of thousands of kilometres away from the classic K–T boundary sites in western Europe and the Caribbean area of North America.

The region in question consisted of an extensive peat mire, hosting a lush vegetation incorporating a canopy stratum of podocarps, Huon pines, *Araucaria*, and several angiosperm genera such as *Nothofagus* (southern beech) with an understorey of ground ferns and tree ferns (Vajda et al. 2001; Pole and Vajda 2009)—perhaps not very different to the present New Zealand wet temperate forests (Fig. 1).

Studies of the sediments spanning the K–T boundary at the Moody Creek Mine in the Greymouth Coalfield, New Zealand (Figs. 2 and 3) produced interesting results previously outlined by Vajda and McLoughlin (2004). The rocks sampled at that site were sliced into millimetre-thick layers to obtain the highest-possible time resolution. Subsequently, 10–20 g of sediment from each sample were crushed and subjected to palynological processing; the final residue was mounted on glass slides for transmitted light microscopy.

The samples pre-dating the K–T boundary include a diverse pollen and spore flora with representatives of at

Fig. 1 Typical temperate rainforest of west-coast New Zealand, an analogue for the Late Cretaceous vegetation of the region

least 80 species of plants: mainly ferns, conifers, and a few angiosperms. The boundary layer, identified by the rare-element anomaly, shows a complete absence of spores and pollen from photosynthetic plants. Instead, the thin layer contains only fungal spores and filaments (Fig. 4) together with dispersed wood tracheids. It has not yet been possible to relate these fungal spores to living taxa because fossil spores are often assigned to artificial form-genera based on the number of cells and pores (Elsik et al. 1983).

The 'fungal-spike' is interpreted to represent a short period of a few months to a few years when sunlight was blocked by atmospheric dust and aerosols (Vajda and McLoughlin 2004, 2007). Reduced sunlight would have caused mass mortality of photosynthetic plants, paving the way for proliferation of saproptrophs such as fungi (Idnurm and Heitman 2005). A likely scenario is that of devastated forests with saproptrophs such as bracket fungi (Polyporaceae) thriving on fallen logs and other woody debris (Fig. 5).

This razor-sharp change in the palyno-assemblages is further exemplified by an extinction of key pollen species including taxa such as *Proteacidites palisadus*, *Tricolpites brevicolpus*, *T. lilliei*, and *Nothofagidites kaitangata* (Fig. 6; Vajda and Raine 2003, 2010). Zonate lycophyte spores have been shown to be important stratigraphic markers in New Zealand Cretaceous–Paleocene assemblages (Raine 2008) and tentatively so in other geographic regions.

The stratigraphic extent of the fern-spike in the New Zealand sections ranges between 25 cm (mid-Waipara) and 40 cm (Moody Creek Mine and Compressor Creek), reflecting different sedimentation rates at each site. The fern-dominated interval was followed by re-establishment of angiosperm and gymnosperm dominance, mainly represented by the pre-impact taxa.

In the sediments directly overlying the fungal layer, a typical recovery succession is evident: Ground ferns were the first to reappear, followed by tree fern species, then a gradual return to more complex vegetation dominated by conifers and flowering plants.

Through photosynthesis, plants convert carbon dioxide to biomass and vice versa. Carbon stored in trees and soils is released to the atmosphere by combustion and through biological decomposition. When vegetation recovers, plants withdraw carbon from the atmosphere and store it, again temporarily, in trees and soils. The overall process, the carbon cycle, here spelled out in simplicity, keeps ecosystems balanced. The scenario following a mass extinction event is, however, far from balanced. As a consequence of the mass mortality of plants and animals following the end-Cretaceous event, ecosystems were greatly changed with, globally, a huge amount of dead biomass—in terrestrial contexts, predominantly plant matter. A large component of this plant matter consisted of cellulose lignin, a compound that resists degradation. Fungi, particularly white-rot and brown-rot basidiomycetes, have extraordinary enzymatic capacities and can degrade both cellulose and highly resistant lignin-rich materials, such as tracheid secondary walls (Deacon 1997). Saprotrophic decomposition, the complementary process in the carbon cycle, is a key process in all ecosystems—next to primary production, it is the second largest flux in the global organic matter cycle (Sterner and Elser 2002). During decomposition, the organic matter would not only release CO_2 into the atmosphere, but important nutrients, such as nitrogen and phosphorus, would

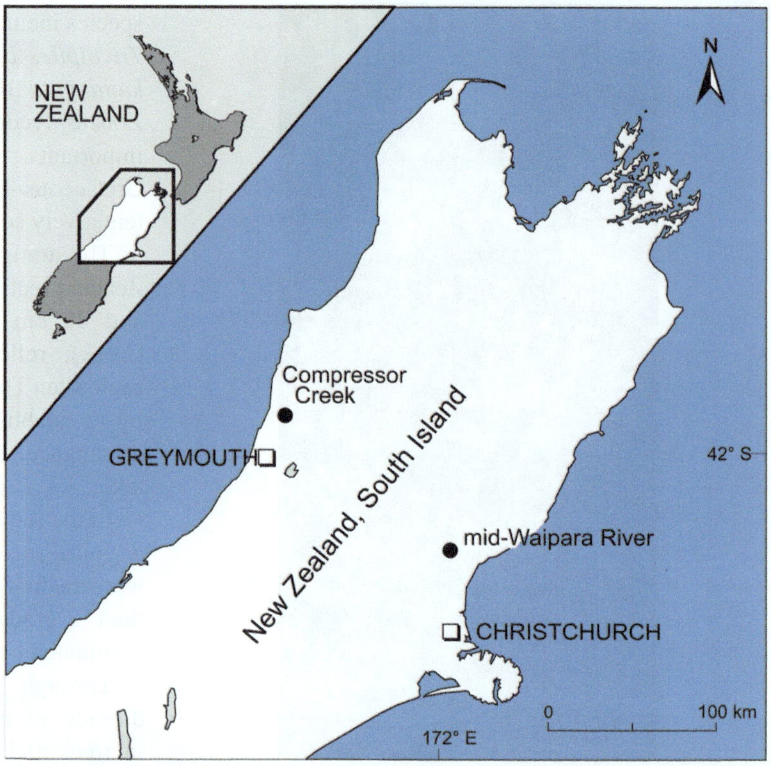

Fig. 2 Location of the Cretaceous–Paleogene (K–Pg) exposures at Moody Creek Mine (Greymouth Coalfield), Compressor Creek and mid-Waipara River, New Zealand

Fig. 3 Cretaceous–Paleogene boundary site at Moody Creek Mine, New Zealand (*arrow* indicates the boundary)

be released into the soil, ready to be used by the recovery succession of plants. Without doubt, fungi played an extremely important role in balancing the carbon cycle, and perhaps ameliorating the climate, in the aftermath of the end-Cretaceous global extinction event.

Dinosaur Extinction

Possible causes for the non-avian dinosaur extinction have been intensely debated for many years. Vajda et al. (2001) showed that vegetation at the end of the

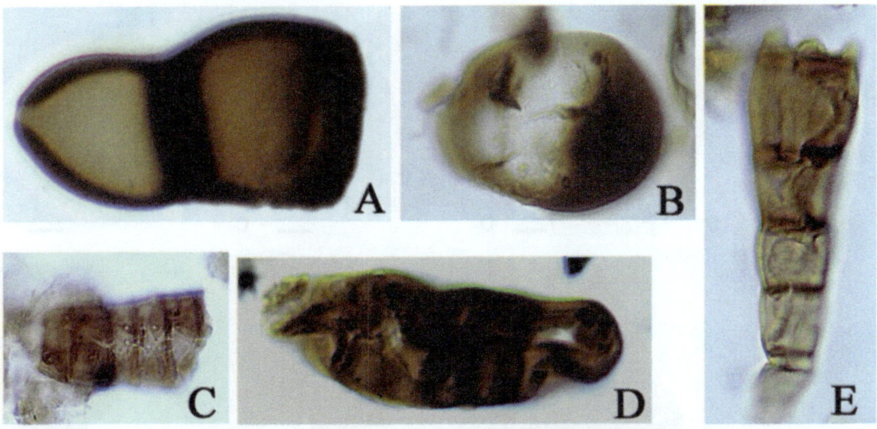

Fig. 4 Light micrographs showing fungal spores from the Cretaceous–Paleogene boundary, New Zealand. Magnification 1000×. (**a**) Monoporate bicellate spore (aff. *Didymoporisporonites*); (**b**) *Monosporites* sp.; (**c**–**e**) Various forms of *Pluricellaesporites*

Fig. 5 A possible modern analogue of the fungi that proliferated after the extinction event 65 million years ago

Cretaceous suffered a global mass-mortality event (not to be confused with mass-extinction—where species disappear permanently). Mass dieback of plants may also have been connected with the extinction of the polar dinosaurs. These, though demonstrably able to have withstood relatively cool temperatures and periods of low light intensity, were obviously unable to survive the abnormally prolonged post-impact winter conditions. The demise of carnivorous dinosaurs may have been due to lack of food rather than to prolonged darkness and cooling after obliteration of the herbivores. Following discovery of the fungal layer in New Zealand (Vajda and McLoughlin 2004), other theories have been proposed.

Uncontrolled proliferation of fungi can lead to devastating diseases affecting plants and animals (including humans). Casadevall (2005) argued that the dramatic increase in fungal activity following the end-Cretaceous extinction event could have contributed to extinction of the non-avian dinosaurs—due to increased levels of airborne spores leading to high levels of infection. In a study on dinosaur eggshells from the Late Cretaceous Hugang Formation in central China, Gong et al. (2008) suggested that endolithic fungi could have contributed to the extinction of non-avian dinosaurs. The endolithic fungi, identified in fossil eggshells, could have inhibited normal development of the dinosaur foetus. These alternative theories are intriguing, but global environmental calamities such as reduced insolation, deforestation, extensive wildfires, and changes in atmospheric chemistry could have been sufficient to exterminate 70% of the Earth's species.

Whereas the K–Pg extinction rates were catastrophic among large faunal groups, the extinction of plants was moderate, not reaching above 15% in the New Zealand record (Vajda and Raine 2003). Extinction was greatest among angiosperms; their species disappeared abruptly at the boundary and are therefore excellent biostratigraphic markers. This may further be related to the negative effect on insect

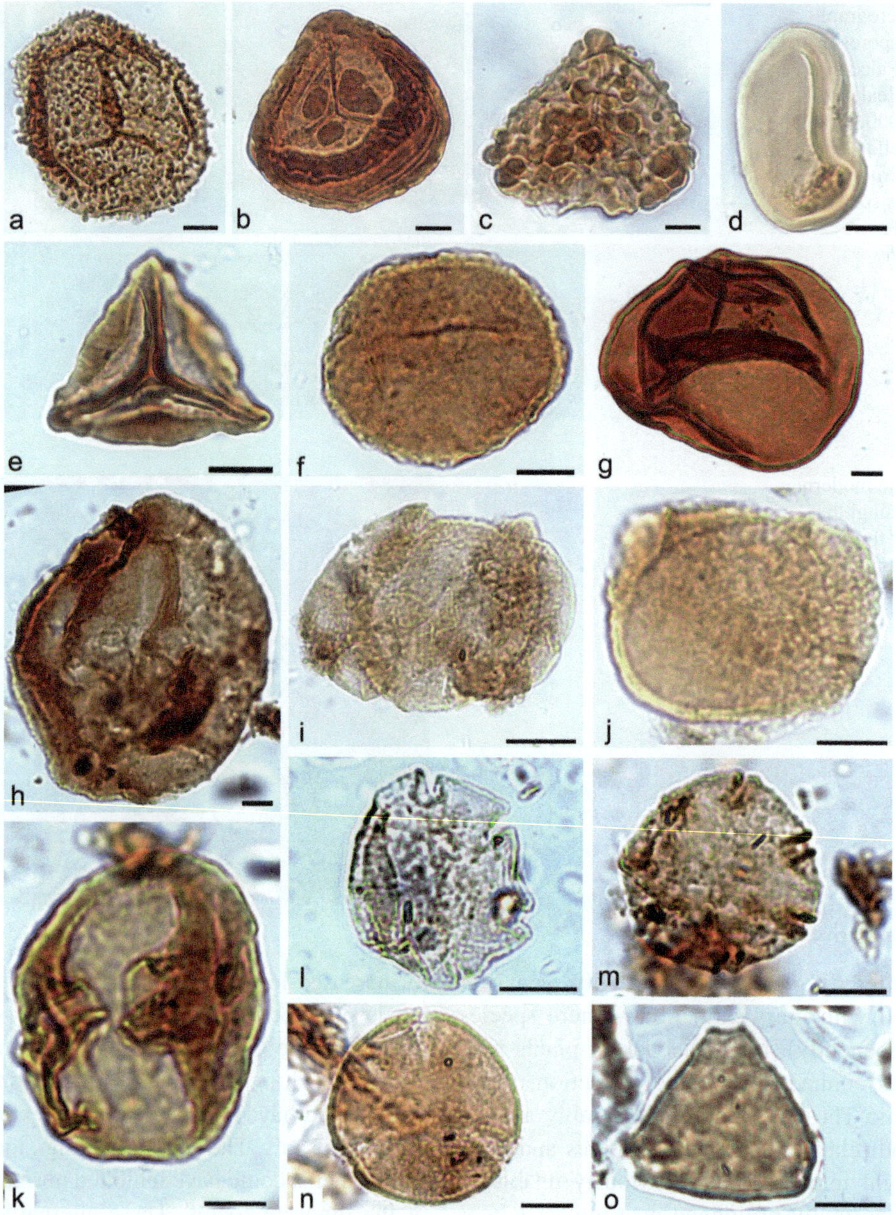

Fig. 6 Light micrographs of representative spores and pollen from the K–Pg boundary interval from the Compressor Creek section, New Zealand. Scale bars = 10 μm. (**a**) *Baculatisporites comaumensis*, (**b**) *Cyatheacidites annulatus*, (**c**) *Cibotiidites tuberculiformis*, (**d**) *Laevigatosporites ovatus*, (**e**) *Gleicheniidites senonicus*, (**f**) *Peromonolites bowenii*, (**g**) *Cyathidites australis*, (**h**) *Araucariacites australis*, (**i**) *Podocarpidites ellipticus*, (**j**) *Dacrydiumites praecupressinoides*, (**k**) *Phyllocladidites mawsonii*, (**l**) *Nothofagidites kaitangata*, (**m**) *Nothofagidites waipawaensis*, (**n**) *Tricolpites lilliei*, (**o**) *Proteacidites scaboratus*

populations. The relatively low extinction rate among plants related to the K–Pg event indicates that the time of upheaval was short (a few years at most), with plants recovering from seed banks and rhizomes as soon as light levels permitted photosynthesis (Vajda and McLoughlin 2007). These few years, with primary production shut down, would have been enough to exterminate large percentages of the terrestrial and marine biota, providing a base for proliferation of fungi.

Acknowledgements I am grateful to Solid Energy New Zealand for access to the Greymouth coalfield sample locality, and especially Rob Boyd for his support. Ian Raine and Christopher Hollis (GNS) gave field assistance in field and shared their knowledge of the geology of New Zealand. Comments by Erland Bååt (Department of Microbial Ecology, Lund University) improved the manuscript. This research was supported by the Swedish Research Council (VR) and the Royal Swedish Academy of Sciences through the Alice and Knut Wallenbergs Foundation.

References

Alvarez LW, Alvarez W, Asaro F, Michel HV (1980) Extraterrestrial cause for the Cretaceous–Tertiary extinction. Science 208:1095–1108

Bonfante P, Genre A (2008) Plants and arbuscular mycorrhizal fungi: an evolutionary developmental perspective. Trends Plant Sci 13:492–498

Casadevall A (2005) Fungal virulence, vertebrate endothermy, and dinosaur extinction: is there a connection? Fungal Genet Biol 42:98–106

Deacon JW (1997) Modern mycology. Blackwell, Oxford

Ehrlich HL (1998) Geomicrobiology: its significance for geology. Earth Sci Rev 45:45–50

Elsik W, Baesemann CB, Graham AK, Hopkins WS, Jansonius J, Jarzen DM, Norris G, Peterson ET, Srivastava SK, Sweet AR, Traverse A (1983) Annotated glossary of fungal palynomorphs. Am Assoc Stratigr Palynologists Contrib 11:1–35

Gong Y, Xu R, Hu B (2008) Endolithic fungi: a possible killer for the mass extinction of Cretaceous dinosaurs. Sci China, Ser D: Earth Sci 51:801–807

Heckman DS, Geiser DM, Eidell BR, Stauffer RL, Kardos NL, Hedges S (2001) Molecular evidence for the early colonization of land by fungi and plants. Science 293:1129–1133

Idnurm A, Heitman J (2005) Light controls growth and development via a conserved pathway in the fungal kingdom. PLOs Biol 3:615–626

Ocampo A, Vajda V, Buffetaut E (2006) Unravelling the Cretaceous–Paleogene (KT) catastrophe: evidence from flora fauna and geology. In: Cockell C, Koeberl C, Gilmour I (eds) Biological processes associated with impact events. Springer Science, Dordrecht. pp 203–227

Pirozynski KA, Malloch DW (1975) The origin of land plants: a matter of mycotropism. Biosystems 6: 153–164

Pole M, Vajda V (2009) A new terrestrial Cretaceous–Paleogene site in New Zealand – turnover in macroflora confirmed by palynology. Cretaceous Res 30:917–938

Pope K, Baines K, Ocampo A, Ivanov B (1994) Impact winter and the Cretaceous/Tertiary extinctions: results of a Chicxulub asteroid impact model. Earth Planet Sci Lett 128:719–725

Raine JI (2008) Zonate lycophyte spores from New Zealand Cretaceous to Paleogene strata. Alcheringa 32:99–127

Redecker D, Kodner R, Graham LE (2000) Glomalean fungi from the Ordovician. Science 289:1920–1921

Schulte P, Alegret L, Arenillas I, Arz JA, Barton PJ (2010) The Chicxulub Impact and the mass extinction at the Cretaceous–Paleogene boundary. Science 327:1214–1218

Sepkoski J (1990) The taxonomic structure of periodic extinction. In: Sharpton V, Warda P (eds) Global catastrophes in earth history. Geol Soc Am Spec Pap 247: 33–44

Sterner RW, Elser JJ (2002) Ecological stoichiometry. Princeton University Press, Princeton

Taylor TN, Hass H, Kerp H (1999) The oldest fossil ascomycetes. Nature 399:648

Taylor TN, Remy W, Hass H, Kerp H (1995) Fossil arbuscular mycorrhizae from the Early Devonian. Mycologia 87: 560–573

Vajda V, McLoughlin S (2004) Fungal proliferation at the Cretaceous–Tertiary boundary. Science 303:1489

Vajda V, McLoughlin S (2007) Extinction and recovery patterns of the vegetation across Cretaceous–Palaeogene boundary—a tool for unravelling the causes of the end-Permian mass-extinction. Rev Palaeobot Palynol 144:99–112

Vajda V, Raine JI (2003) Pollen and spores in marine Cretaceous/Tertiary boundary sediments at mid-Waipara River, north Canterbury, New Zealand. New Zealand J Geol Geophys 46:255–273

Vajda V, Raine JI (2010) A palynological investigation of plesiosaur-bearing rocks from the Upper Cretaceous Tahora Formation, Mangahouanga, New Zealand. Alcheringa 34:359–374

Vajda V, Raine JI, Hollis CJ (2001) Indication of global deforestation at the Cretaceous–Tertiary boundary by New Zealand fern spike. Science 294:1700–1702

Part IV
Palaeobiogeography

Changes in the Pattern of Brachiopod Biogeography in Northern Asia Through Early and Middle Devonian Times

John A. Talent, Yevgeniy A. Yolkin†, Valentina N. Yolkina†, Tat'yana P. Kipriyanova, and Rimma T. Gratsianova†

Abstract

Changes in the biogeographic pattern obtained by mathematical investigation of the brachiopod database from five of the seven Devonian stages, Lochkovian–Givetian, are presented as evidence for substantial relative movements of several of the crustal blocks during Devonian times in what now constitutes northern Asia. Such movements—identified by the end results of evolution and extinction in isolation—would, surely, have substantially pre-dated development of the biogeographic contrasts revealed by mathematical analysis. How much time would have been required to develop such contrast remains problematic. Strong biotic linkage existed between the eastern and western flanks of the Urals through the Lochkovian and Pragian with significant linkage to the Salair and Altai. Linkage between faunas from the two flanks of the Urals weakened for the early and late Emsian but was re-established for the Eifelian and early Givetian; this may be interpreted as perhaps connected with a prior interlude of substantial (though not overwhelmingly wide) separation of the East and West Urals terranes along the Urals Suture. Biotic linkage of the Urals terranes with the Salair and Altai weakened in the Emsian, increased in the Eifelian, then declined and, curiously, contrasting with what happened over most of the globe, disappeared in the Givetian—consistent with major separation of the crustal blocks in question. Strong linkage between the Taymyr and Settedaban regions and significant linkage between them and the Salair and Altai were maintained through the Lochkovian and Pragian. Faunal interaction with the Salair and Altai became inconsequential from the Pragian onwards; this is interpreted as possibly reflecting substantial separation of these terranes from the old Siberian continent, or as a reflection of increased marine-biogeographic differentiation around the old continental margin.

J.A. Talent (✉)
Earth and Planetary Sciences, Macquarie University 2109, Australia
e-mail: jatalent32@gmail.com

Yevgeniy A. Yolkin, Valentina N. Yolkina, Rimma T. Gratsianova: Deceased

J.A. Talent (ed.), *Earth and Life*, International Year of Planet Earth,
DOI 10.1007/978-90-481-3428-1_28, © Springer Science+Business Media B.V. 2012

Weakening of linkage through the Lochkovian and Pragian for the Novaya Zemlya region accords with increasing and, finally, near-total biologic isolation. The Kazakhstan blocks displayed weak linkage to the western Urals in the early Lochkovian, then developed strong linkage to Mongolia (late Lochkovian) and the Far East (Pragian and early Emsian) but then (late Emsian to early Givetian) became completely isolated biogeographically. Additional information, including data from other major groups of Devonian organisms (rugosans, molluscs and trilobites) from the various crustal blocks, is needed for increased precision.

Keywords
Northern Asia • Crustal blocks • Urals • Siberian palaeocontinent • Settedaban • Mongolia • Relative movements • Devonian • Brachiopods • Biogeography

Introduction

Our aim in the present exercise has been to mathematically analyse part of our large brachiopod data set (Talent et al. 2001) through about 35 million years (beginning Early Devonian to end-Middle Devonian), specifically for crustal blocks of northern Eurasia, in quest of the main trends in biogeography through that time interval.

We focused on brachiopods as the basis for our initiative because of our experience with the phylum but most importantly because they far exceeded any other phylum in numbers and in taxonomic diversity during the time interval that concerns us. Our biogeographic approach was summarised (Yolkin et al. 1985) and subsequently presented at length with our published database (Talent et al. 2001).

It has been accepted that pieces of a biogeographic jigsaw arranged logically will form a coherent picture, but the picture is not necessarily final; there may be other ways of putting the same pieces together to produce other coherent pictures. In science, whatever the coherent picture, such exercises lead to formulation of further questions and, hopefully, only fine-tuning of the picture.

We have presented progress reports at several conferences (e.g., Talent et al. 1988; Yolkina et al. 1999; Kipriyanova et al. 2005; Yolkin et al. 2005a, b, 2007) and have spelled out our mathematical procedure in some detail when dealing with the Pragian (mid-Early Devonian) brachiopod faunas for the entire hemisphere (Yolkina et al. 2000; Yolkin et al. 2011). This was done in relation to the pattern of crustal blocks hypothesised for the time interval of our study (Fig. 1). In the present exercise we consider the results of this process in relation to eight 'time-slices'—early and late Lochkovian, Pragian, early and late Emsian, Eifelian and early and late Givetian—focusing on the crustal blocks of northern Eurasia.

Methods

As noted earlier (Yolkina et al. 2000; Talent et al. 2001), the project required major exercises in establishing the best-possible stratigraphic framework for the Asia–Australia hemisphere under consideration (cf. Talent and Yolkin 1987; Yolkin et al. 1987; see Fig. 2 for stratigraphic alignments for the region considered here), and establishing consistency of taxonomy for all described and figured brachiopods (latest Silurian through to end-Middle Devonian), throughout the hemisphere. These exercises had to be carried through to completion before advanced numerical methods of analysis could be meaningfully broached. The brachiopod taxa for the latest Silurian to end-Givetian of the Asia–Australia hemisphere were allocated to four categories: (1) Species identifications accepted; (2) Additional generic identifications (but not species) accepted; (3) 'Species' requiring more information before acceptance; and (4) Identifications rejected (with reasons given for all except our very first listings).

In earlier reports (Talent et al. 1987a, b; Yolkina et al. 2000) we considered our data in relationship to the pattern of crustal blocks hypothesised

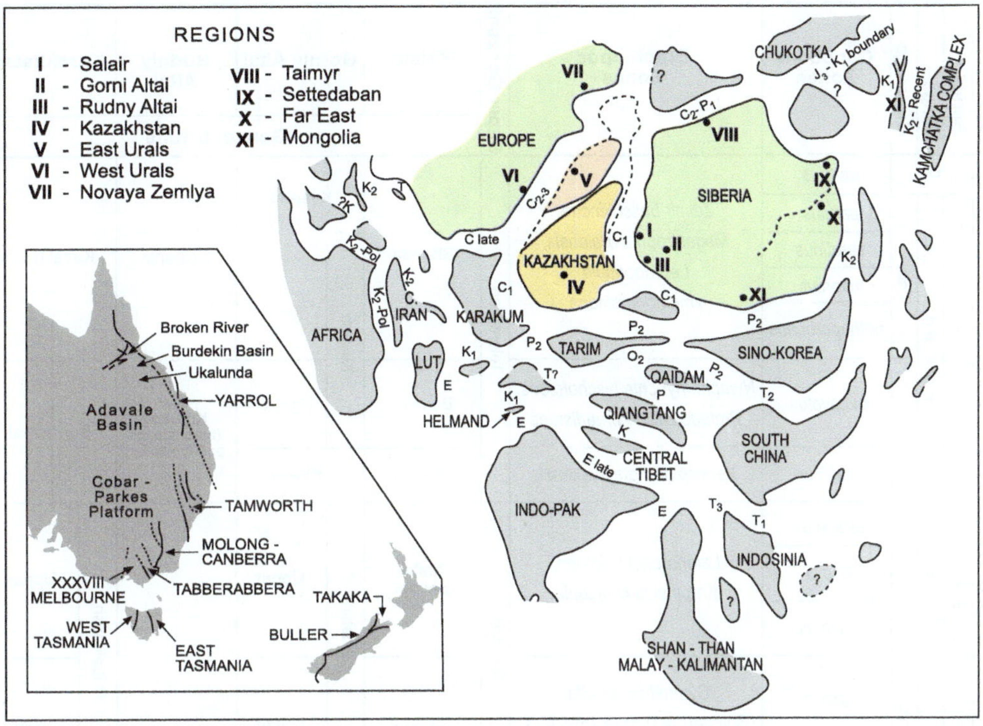

Fig. 1 Pattern of crustal blocks of northern Asia (from Talent et al. (1987b), Fig. 5, and Talent et al. (2001), Fig. 2—modified to suggest composite nature of 'Kazakhstan/Kazakhstania' crustal blocks: precise number and configurations not known

to have existed during Devonian times for the Asia–Australia hemisphere. We suggested that the eastern slope of the Urals could be part of our Block II, Kazakhstania/Kazakhstan, though acknowledging that the latter could be composite, as suggested by Krasnov et al. (1993). Subsurface data (unpublished) obtained in recent years has increased certainty that our Block II is indeed composite.

A cartoon we published earlier (Talent et al. 1987b, 2001) is a pull-apart diagram displaying, in a simplified way, major terranes and mobile belts of the Asia–Australia hemisphere (Fig. 1). More detail is now available from detailed syntheses for western Siberia (e.g., Sennikov et al. 2001 for the Altai–Sayan region) and grand-scale syntheses for the Russian Far-East driven, inter alia, by the late Leonid Parfenov (Nokelberg et al. 1994; Fujita et al. 2009 for extensive bibliography), but we prefer to use our cartoon, a Devonian view, modified only with respect to terranes revealed by drilling of the West Siberian Plain—in order to retain continuity with earlier stages of our investigation.

As with our analysis of Pragian brachiopods for the Asia–Australia Hemisphere (Yolkina et al. 2000), the data were analysed using classification algorithms developed in the Siberian Branch of the Academy of Sciences, Novosibirsk: the package of programs OTEKS (Zagoruiko et al. 1986) for data analysis—included in software of the expert system EXNA (Yolkina and Zagoruiko 1991, 1992). The theory behind use of these programs leading to generation of a minimum spanning tree (MST), tested by program KRAB, is elaborated in our Pragian analysis (Yolkina et al. 2000).

It is possible from the taxonomy algorithm to obtain taxa of indeterminate shape using the shortest unclosed path, joining all points of the initial set into a minimum spanning tree (MST, Fig. 3, Pragian) and having a special function F of taxonomy quality (Yolkina and Zagoruiko 1969). This idea was accomplished in the algorithm KRAB (Yolkina and Zagoruiko 1969, 1978). The program KRAB divides the set of initial objects (points) into K taxa of indeterminate shape. At first, all points are joined by a minimum spanning tree by

System	Series	Stage	Substage	Conodont Zones	Brachiopod Zones	Super Gorizont	Salair	Gorniy Altai	Rudniy Altai	Kazakhstan
							Gorizont/Suite			
DEVONIAN	D2			*partitus*		Telengitian	Shanda	Kirsanov	Losishin	Kazakhstan
	Early	Emsian	Late	*patulus*	*Zdimir baschkincus, Megastrophia uralensis, Leptodontella zmeinogorskiana*					
				serotinus				Kuvash		
				inversus						
				nothoperbonus						
			Early	*excavatus*	*Nymphorhynchia bischofoides, Protodouvillina praedistans*		Belova	Tarbuzin	Kryokov (brachiopods poorly known)	Sardazhal
				kitabicus	*Xenospirifer gurjevskensis*		Salairka	Kireev		
		Pragian		*pireneae*	*Latonotoechia latona, Karpinskia conjugula*	Beltirian	Malyi Bachat	Yakushin: Upper	Korbalikhin (without fauna)	Pribalkhash: Upper
				kindlei						
				sulcatus						
		Lochkovian	Late	*pesavis*	*Cymostrophia alfa. Spirigerina supramargineali sibirica*		Krekov	Lower		Lower
				delta						
			Early	*eurekaensis*	*Tenuiatrypa subsalairica, Uncinulus krekovensi*	Kaybalian	Peetz	Remnyov		Kara-Espe: Kokbaytal
				postwoschmidti woschmidti	*Protathyris sibirica, Howellella laeviplicata*		Tom-Chumysh			Aynasuy

Fig. 2 Correlation chart for the principal brachiopod-bearing sequences of the 11 regions considered. Conodont zones, 'super gorizonts' and 'gorizonts' are according to EA Yolkin, brachiopod zones according to RT Gratsianova

using the special program GRAPH which, being part of KRAB, can also be used separately. It is possible to build the MST using Euclidean or Jaccard distances. The program KRAB, by an exclusive search, checks all variants and calculates the quality estimation F for each variant.

If the number of objects and requested taxa are relatively large, the exclusive search becomes cumbersome. To reduce the search, program KRAB selects the 'pretenders': the arcs most likely to be among the boundary ones.

Salient Features of Mathematically Generated Biogeographic Patterns for Northern Eurasia

Early Lochkovian

There was a high level of similarity between the northern and southern regions of the west flank of the Urals, with weaker linkages to Novaya Zemlya, to the Salair and Gorniy Altai, and to the Kazakhstan block(s).

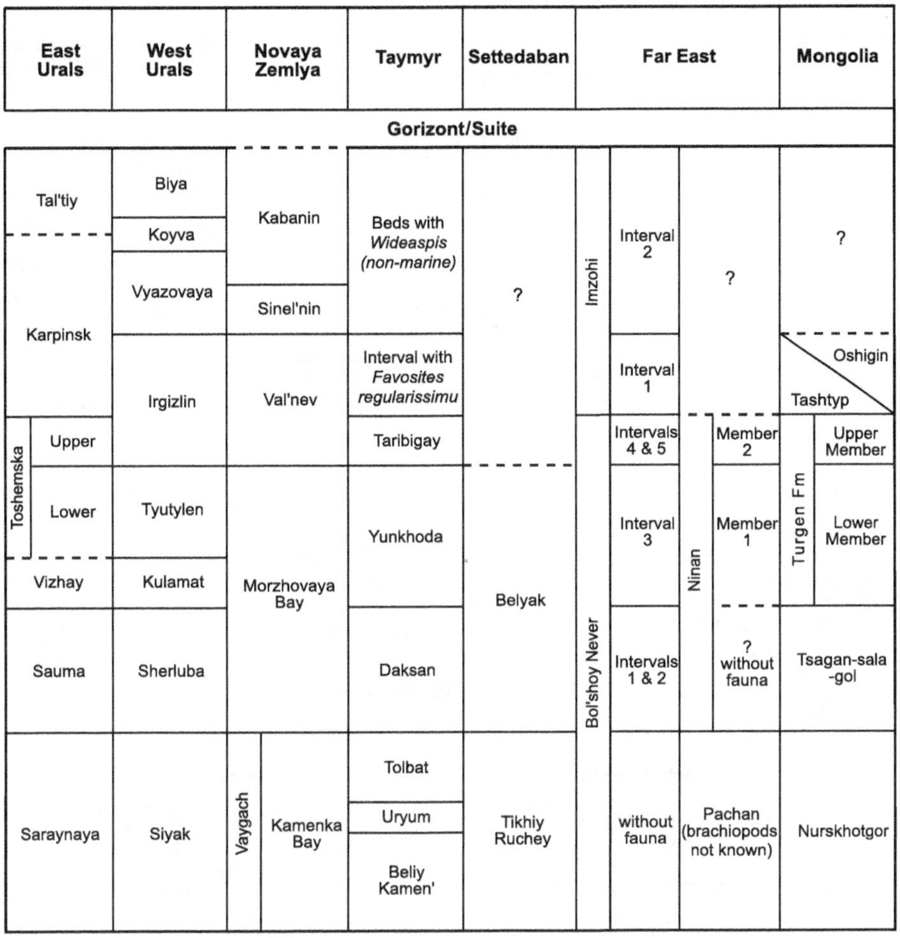

Fig. 2 (continued)

There is presently a lack of unequivocal brachiopod data for Mongolia for this interval (Fig. 4).

Late Lochkovian

The Mongolia and Kazakhstan blocks have moderately strong linkage to each other (presumed to have been similar for early Lochkovian), with modest linkage to the Salair. Linkage between the Taymyr and Settedaban regions is obvious, as is significant linkage from them to the Salair. Linkage between the northern and southern regions of the west flank of the Urals remains pronounced, with weaker linkage to Novaya Zemlya. There are no data for the Far East but, in view of the subsequent strong faunal as well as stratigraphic linkage between the Far East, Mongolia and Kazakhstan during the Pragian, a similar strong linkage between the three regions may well have existed during the Late Lochkovian.

Pragian

The Altai and Salair regions were strongly linked to both the East and West Urals—consistent with greatly increased proximity—but the Novaya Zemlya region displays weak/weakening linkage. The Kazakhstan, Mongolia and Far East regions are strongly linked, as was previously suggested non-mathematically (Talent et al. 1987a, b), but these, as a unit, have only weak (arguably weakening) linkage to the Altai–Salair–Urals regions. Linkage between the Taymyr and Settedaban regions remained strong, but, curiously, linkage with other regions almost faded out.

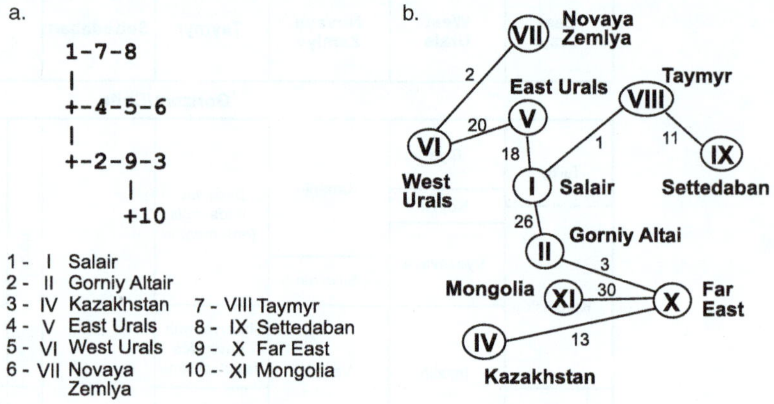

Fig. 3 Minimum spanning tree (MST) (Program KRAB) (from Yolkina et al. 2000)

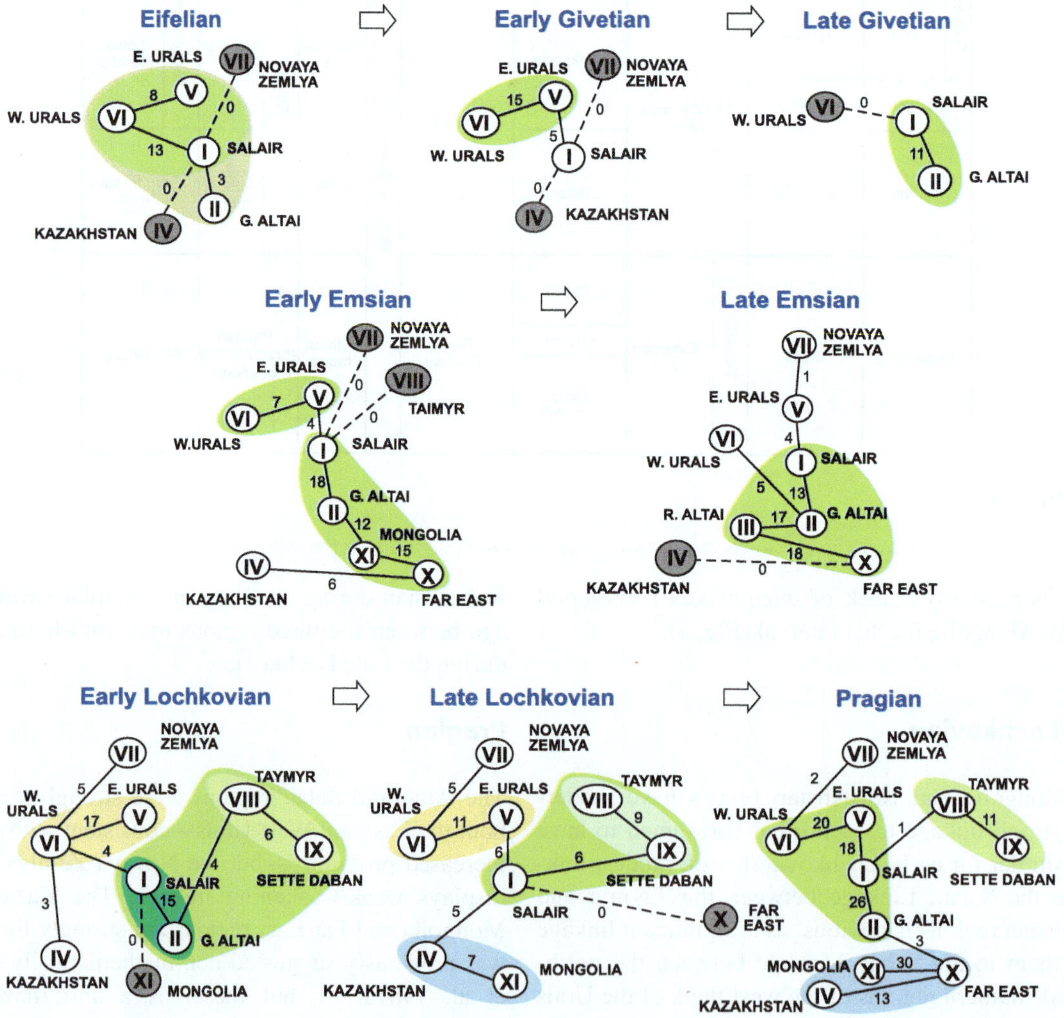

Fig. 4 Minimum spanning trees (MSTs) based on Jaccard coefficients for brachiopod faunas of the regions of northern Asia generated by the KRAB program for eight intervals: early Lochkovian, late Lochkovian, Pragian, early Emsian, late Emsian, Eifelian, early Givetian and late Givetian. Roman numerals are the same as those shown for the north Asian crustal blocks, Fig. 1. G Altai = Gorniy Altai; E and W Urals = eastern and western flanks of the Urals. *Circles with red* infilling represent crustal blocks having zero brachiopod taxa in common with other crustal bocks in the same minimum spanning tree

Early Emsian

Strikingly, linkage between the Altai and Salair on one hand and the Urals regions (both the East and West flanks of the Urals) plummeted by the beginning of the early Emsian, remaining low through the early and late Emsian, and recovered (as far as the East Urals is concerned) during the Eifelian.

Striking also, and coinciding in time, is the strong linkage implying proximity and ease of biotic interaction that developed between the Altai and Salair regions on the one hand and the Mongolia and Far East regions on the other, with significant, though reduced, linkage to the Kazakhstan block(s).

This pattern was maintained through the early and late Emsian. It may be viewed as according with a major interval of crustal dilation, the Gorniy AltaiSalair regions becoming located more than 1000 km (cf. Talent 1985) away from the East Urals region with it, in turn, increasingly separated from the West Urals regions.

Late Emsian

Though brachiopod data are available from several horizons in the Rudniy Altai, high-quality data are presently available only for the late Emsian; they demonstrate close faunistic linkage at that time to the Far East and to the Gorniy Altai. Kazakhstan lost faunal connection with the Gorniy Altai–Salair regions at about the middle of the Emsian; it retained isolation through the Eifelian and early Givetian. This would accord with profound tectonic displacement (a major episode, perhaps, of tectonic 'extrusion') from between the Urals and Altai regions/crustal blocks.

Eifelian

Faunal interaction between the Salair and East Urals region was re-established during the Eifelian times but decreased during the early Givetian.

Givetian

The continued high level of taxonomic linkage of the Salair and Gorniy Altai during the early and late Givetian contrasted with the profound isolation of these from the south-eastern Urals; this may have set in during early Givetian times.

One may be excused for assuming that the Novaya Zemlya region may have formed part of the faunal fabric along the west flank of the Urals throughout the mid-Palaeozoic. Our mathematical analysis is consistent with significant isolation during the Pragian and with profound isolation from at least early Emsian through to about late Eifelian and perhaps early Givetian, but caution is required. Though a cursory glance at the Emsian–early Givetian brachiopod faunas may suggest zero linkage, there has in fact been little documentation of faunas younger than late Emsian from Novaya Zemlya. How far may the Novaya Zemlya region have been separated from other crustal blocks under discussion during the Pragian–Givetian? Because of the total contrast for Emsian onwards, one might hazard a guess of more than 1000 km (cf. Talent 1985; see also 'Cautionary Note below') and for the onset of this to have taken place appreciably prior to the Pragian–Emsian boundary.

Interaction between the Taymyr and Settedaban faunas persisted through the Lochkovian and Pragian. By the early Emsian, profound isolation seems to have brought about cessation of interaction with the Salair faunas.

Onset of the profound isolation of the faunas of the Kazakhstan Block(s) ('Kazakhstania') from the faunas of other crustal blocks under consideration might appear to have commenced no later than during the early Emsian, but generation of the faunal contrasts implies a substantial interval of evolution and extinction in isolation before the faunal contrasts would have become evident (see the previous section).

There are many areas where present-day biogeographic contrasts and similarities were obviously connected with major tectonic separations or conjunctions—for example, New Zealand from Australia, New Caledonia from Australia and Norway from Greenland—or crustal blocks being brought closer, for example, Sulawesi to Kalimantan, Chatham Island region to NewZealand.

How long does it take (approximately) for disappearance of all or most biotic linkages between marine faunas when they become too far apart for larval stages to cross deep oceanic waters (of the order of 1000 km)? Or how long does it take when a land barrier has developed, as in the case of Panama. How rapid has development of biotic contrast been since closure of the seaway between the Pacific and Atlantic oceans though Panama?

Such questions need approximate answers before questions about more accurate timing of separation or juxtaposition of major crustal blocks may be broached from palaeobiogeographic data. One problem is predicting (fairly compellingly) how many marine provinces there may have been along latitudinally oriented coastlines in the deep past—in the present case, for the margins of the old Siberian continent. This might be broached in a very general way by assuming the latitudinal extent of present-day provinces (despite uncertainties of boundaries and definitions) may provide a rough approximation to the average latitudinal extent of marine provinces in the past. Consider the latitudinal extent of provinces along the east and west coasts of the Americas, northeast Asia and eastern Australia. These, of course, are much influenced by gross patterns of oceanic circulation, but present-day shallow marine provinces, apart from the extensive tropical provinces (Indo Pacific, Guinean, Caribbean and Peruvian provinces), range from c. 600 km latitudinally (Carolinian and Virginian provinces), to a general figure of 1300 to 1800 km (e.g., Oregonian, Peronian, Lusitanian, Nova Scotian and Japonic provinces) or even more (e.g., Labradorian, Beringian).

Cautionary Note

If the displacement of crustal blocks exceeds the width of open ocean that organisms are capable of crossing during their larval stages, biogeographic contrast should increase through time—as a consequence of evolutionary divergence and extinctions—until, in the limiting case of total isolation, 100% faunal contrast is eventually achieved. Inevitably, there will be a lag in time (on a scale that may reflect several millions of years of isolation) between the tectonic-displacement event(s) and the slow achievement of substantial or total biogeographic contrast. Interpretation of the apparent chronology of the mathematically generated patterns presented here for northern Eurasia (Fig. 4) should therefore attempt to take this 'time-lag effect' into account.

Acknowledgements Three of our friends (EAY, VNY, RTG), co-authors of this protracted biogeographic initiative, passed away in rapid succession while this manuscript was being processed for publication. We (JAT, TPK) dedicate this paper to them.

References

Fujita K, Prokopiev AV, Stone DB (2009) Biography: Leonid M. Parfenov (1937–2002) Stephan Mueller Spec Pub Ser 4:5–9. Copernicus Publications for European Geoscience Union

Kipriyanova T, Kipriyanov AA Jr, Yolkina VN (2005) The computer system 'Palaeobiogeography' for processing paleontological data. Devonian Terrestrial Marine Environments: from Continent to Shelf (IGCP Project 499/SDS Joint Field Meeting Contributions to International Conference, Novosibirsk, July 25–August 9, 2005), 79–81. 'Geo' Branch of Novosibirsk Publishing House, Siberian Branch, Russian Academy of Science

Krasnov VI, Isaev GD, Astashkina VF, Bochkarev VS, Dubatolov VN, Kul'kov NP, Makarenko SN, Miretskaya NM, Nefedov VA, Peregoedov LG, Savina NI, Saev VI, Serdyuk ZYa (1993) Regional'naya stratigraficheskaya schema paleozoiskikh obrazovaniy neftegazonosnykh rayonov Zapadno–Sibirskoy ravniny. Stratigrafiya paleogeografiya fanerozoya Sibiri. Novosibirsk, SNIIGGiMS, pp 47–78

Nokelberg WJ, Parfenov LM, Monger JWH, Baranov BV, Byalobzhesky SG, Bundtzen TK, Feeney TD, Fujita K et al (1994) Circum-North Pacific Tecton-Stratigraphic Terrane Map. US Geol Surv Open-File Rept 94–714, 221p, 4 map sheets

Sennikov NV, Yolkin EA, Iwata K, Izokh NG, Khlevnikova TV (2001) The Gondwana Ordovician–Silurian terranes in the structure of the Altai–Sayan folded area and the timing of their accretion towards the margin of the Siberia continent. Gondwana Res 4:779–780

Talent JA (1985) Provintsialism i raskhozhgenie kontinental'nykh blokov v proshlom [Provincialism and separation of continental blocks in the past]. In: Simakov KV (ed) Ekostratigrafiya, paleobiogeografiya i stratigraficheskie granitsy (XIV Tikookeanskiy nauchnyy kongress, Khabarovsk, 1979) Akademiya Nauk SSSR, Magadan, pp 59–90 [in Russian]

Talent JA, Yolkin EA (1987) Transgression–regression patterns for the Devonian of Australia and southern West Siberia. Cour Forsch-inst Senckenberg 92:235–249

Talent JA, Gratsianova RT, Shishkina GR, Yolkin EA (1987a) Devonian faunas in relation to crustal blocks: Kazakhstan, Mongolia, northern China. Cour Forsch-Inst Senckenberg 92:225–233

Talent JA, Gratsianova RT, Yolkin EA (1987b) Prototethys: fact or phantom? Palaeobiogeography in relation to the crustal mosaic for the Asia–Australia hemisphere in Devonian–Early Carboniferous times. In: McKenzie KG (ed) Shallow Tethys, vol 2. AA Balkema, Rotterdam, pp 87–111

Talent JA, Gratsianova RT, Yolkin EA (1988) Global 'events', life-crises and provinciality in the Asia–Australia–Pacific hemisphere: a mid Palaeozoic perspective. Mezhdunarodnyy simpozium "Glubinnoe stroenie Tikhogo Okeana i ego kontinental'nogo obramleniya" (Blagoveshchensk Aug 1988), Tezisy, pp 88–90

Talent JA, Gratsianova RT, Yolkin EA (2001) Latest Silurian (Pridoli) to Middle Devonian of the Asia–Australia hemisphere: rationalization of brachiopod taxa and faunal lists; correlation charts. Cour Forsch-Inst Senckenberg 236:221p

Yolkin EA, Talent JA, Gratsianova RT (1985) Avstralo–Aziatskie pozdnesiluriyskie–rannekamennougol'nye brakhiopody i trilobity (osnovny problemy taksonomii i biostratigrafii b svyazi paleobiogeograficheskim rayonirovaniem) Australo–Asiatic Late Silurian–Early Carboniferous brachiopods and trilobites (principal problems of taxonomy and biostratigraphy in defining of palaeobiogeographic regions). Geologiya i geofizica 12:3–8 [in Russian]

Yolkin EA, Talent JA, Thanh T-D, Erina MV, Gratsianova RT, Kim AI, Mawson R (1987) Alignment of reference stratigraphic scales for the Devonian of Siberia, Central Asia, Vietnam and Australia. Proceedings First Conference on Geology of Indochina (Ho-Chi-Minh, Vietnam, 1986), vol 1, pp 59–77

Yolkina VN, Yolkin EA, Talent JA, Gratsianova RT, Kipriyanova TP, Kipriyanov AA Jr (1999) Application of data analysis methods (FOREL and KRAB) in palaeogeography: progress report. Abstracts, Peshawar meeting IGCP 421 (5–20 December 1999), pp 32–34

Yolkin EA, Talent JA, Kipriyanova TP, Gratsianova RT, Yolkina VN (2005a) Quantitative estimation of palaeobiogeographic relations between regions of northern Eurasia during the Early Devonian. Devonian Terrestrial and Marine Environments from Continent to Shelf (IGCP 499/Subcommission on Devonian Stratigraphy; Novosibirsk, 25 July–9 Aug 2005), Abstracts, pp 143–146

Yolkin EA, Talent JA, Kipriyanova TP, Gratsianova RT, Yolkina VN (2005b) Deciphering distances and orientation of blocks of the Earth's crust of northern Eurasia in the Early Devonian using fundamental quantitative estimates of palaeobiogeographic connections between regions. Geodynamic evolution of the lithosphere of the Central-Asiatic mobile belt (from ocean to continent). Materials for a meeting of the Irkutsk Institute of the Earth's Crust, Siberskie otdelenie, Russkie Akademiya Nauk 1:89–92 [in Russian]

Yolkin EA, Talent JA, Izokh NG, Kipriyanova TP, Kipriyanov AA, Gratsianova RT, Yolkina VN, Shcherbanenko TA (2007) Rayonirovanie ranne-srednedevonskikh morey Evrazii na osnove kolichestvennykh otsenok paleobiogeograficheskikh svyazey mezhdu regional'nymi vidovymi assotsiatsyami brakhiopod. Verkhniy paleozoy Rossii: stratigrafiya i paleogeografiya. Materialy Bserossiyskoy konferentsii (25–27 sentyabrya 2007 g.). Kazan', Kazanskiy Gosudarstvennyy Universitet, pp 102–107

Yolkin EA, Talent JA, Kipriyanova TP, Yolkina VN, Gratsianova RT, Shcherbanenko TA, Kipriyanov AA Jr (2011, in press) Biogeography of Early Devonian brachiopods from North Eurasia: (computer modeling and interpretation problems) [in Russian]. Novosti paleontologii i stratigrafii (Prilozhenie k zhurnalu 'Geologiya i geofizika' 52) 15

Yolkina VN, Yolkin EA, Talent JA, Gratsianova RT, Kipriyanova TP, Kipriyanov AA Jr (2000) Brachiopod biogeography of the Asia–Australia hemisphere during Pragian (Early Devonian) times. In: Mawson R, Talent JA, Long JA (eds) Rec West Aust Mus Supp 58:349–384

Yolkina VN, Zagoruiko NG (1969) Kolichestvenniye kriterii kachestva taksonomii I ich ispol'zovanie v prozesse prinyatiya reshenij. Vychislitel'nye sistemy 36:29–46

Yolkina VN, Zagoruiko NG (1978) Some classification algorithms developed at Novosibirsk. RAIRO Informatique/Comput Sci 12:37–46

Yolkina VN, Zagoruiko NG (1991) Block analysis of data using the expert system EKSNA. Vychislitel'nye sistemy 144:54–175. Novosibirsk [in Russian]

Yolkina VN, Zagoruiko NG (1992) Block analysis of data using the expert system EKSNA (Conclusion). Vychislitel'nye sistemy 145:3–28. Novosibirsk [in Russian]

Zagoruiko NG, Yolkina VN, Lbov GS, Emel'yanov SV (1986) Package for using the program OTEKS, 160 pp. Moskva, Finansy i statistika [in Russian]

The Paleobiogeography of Pennsylvanian Crinoids and Blastoids

Johnny Waters and Gary D. Webster

Abstract

Historically, Pennsylvanian crinoid and blastoid paleogeography has been poorly constrained. Recent discoveries of Pennsylvanian faunas from Queensland (eastern Australia), Algeria, Western China, and Iran contribute to a worldwide data set that permits a more complete analysis of Pennsylvanian echinoderm paleogeography. Cluster analysis of generic occurrence groups from North America, China, the UK/Ireland, and Russia define a paleogeographic group based on cladid domination of the echinoderm communities. In contrast, many echinoderm communities in the Paleo-Tethyan realm show increased contributions from camerates and blastoids. Crinoids are noted for being endemic, rarely occurring on more than one tectonic block. Approximately 60% of Pennsylvanian crinoid genera occur on a single block. Crinoid clades differ in their patterns of endemism. Cladids are the most endemic (60%), followed by the flexibles, the disparids, and the camerates, each of which have approximately 50% endemic genera. Among genera found in a single geographic area, North America contains the most endemic faunas (64%), followed by Russia (40%) and Australia (30%). Other geographic regions show significantly lower rates of endemism ranging from 0 to 16%. North American endemics are predominantly cladids, a pattern not seen in the other areas. Although poorly known until recently, these Pennsylvanian Paleo-Tethyan echinoderm communities were the precursors to the well-studied, but still enigmatic, Permian echinoderm communities from Southeast Asia, particularly Timor.

Keywords

Pennsylvanian • Crinoids • Blastoids • Cluster analysis • Paleobiogeography • Endemism

Introduction

Ausich and Kammer (2001) reviewed the study of crinoids in the twentieth century and outlined challenges for the twenty-first century, among which was

J. Waters (✉)
Department of Geology, Appalachian State University, Boone, NC 28608, USA
e-mail: watersja@appstate.edu

J.A. Talent (ed.), *Earth and Life*, International Year of Planet Earth,
DOI 10.1007/978-90-481-3428-1_29, © Springer Science+Business Media B.V. 2012

a better understanding of Paleozoic crinoid paleobiogeography. The authors noted that the discovery of new faunas outside North America and Europe was key to realizing this goal. Camerate crinoids dominated the Middle Paleozoic Crinoid Macroevolutionary Fauna from the Early Silurian to the Middle Mississippian (Ausich et al. 1994; Baumiller 1994). The rise of the Late Paleozoic Crinoid Macroevolutionary Fauna signaled a rise in cladid crinoids and decrease in dominance in camerates that began in the Late Mississippian (Kammer and Ausich 2006) and continued into the Pennsylvanian at least in North America and Europe. Kammer and Ausich (2006) summarized diversity trends among clades of Mississippian crinoids. They concluded that the Age of Crinoids ended with the Serpukhovian extinction event. In general most camerate crinoids and blastoids were rare in Pennsylvanian faunas, but this interval is sandwiched between Mississippian and Permian faunas in which both groups are very abundant, very diverse, or both. Kammer and Ausich (2006) discussed the biodiversity spike of crinoids in the Mississippian and concluded that changing patterns of dominance between camerate crinoid clades and cladid crinoid clades were the result of changing sedimentation patterns that occurred during the Mississippian, particularly in North America and Europe. Although data presented herein carried into the Pennsylvanian, detailed discussion of patterns of clade richness ended at the Serpukhovian extinction event.

In this chapter, we examine temporal and spatial patterns of Pennsylvanian crinoid clade diversity to illustrate differing patterns of crinoid distribution between the Mississippian and Pennsylvanian. Our goal for this chapter is to examine patterns of Carboniferous crinoid diversity and paleobiography and ask the following question: Was the Tethyan realm a refugium for Pennsylvanian camerate crinoids and blastoids? In addition, this chapter serves to fill the gap in our understanding of crinoid paleobiogeography between the studies of Lane and Sevastopulo (1990) for Lower Carboniferous crinoids and Webster (1998) and Webster and Lane (2007) for Permian crinoid faunas.

Recent discoveries of Pennsylvanian faunas from Queensland (northeastern Australia), Algeria, Western China, and Iran suggest that this pattern may be muted in the Tethyan realm.

Crinoid Data Set

We have developed a crinoid data set independent of that used by Kammer and Ausich (2006). Their data set updated the information in Sepkoski (2002) using the crinoid compilation of Webster (2003) and appropriate newer literature. Because we sought paleobiogeographic information lacking in Sepkoski (2002) and Kammer and Ausich (2006), we compiled our data set directly from Webster (2003) augmented by additional unpublished information through 2008 also compiled by Webster. Our data set records the temporal and geographic distribution of each Pennsylvanian crinoid genus. Because we are interested in the origins of the Pennsylvanian crinoid faunas, we recorded each occurrence of each genus whether in pre-Pennsylvanian time, Pennsylvanian time, or post-Pennsylvanian time. We recognize 204 genera of crinoids that lived in some interval of the Pennsylvanian and 334 discrete spatial-temporal occurrences. Pennsylvanian crinoids include 141 genera of cladids, 34 genera of camerates, 10 genera of articulates, 10 genera of disparids, and 9 genera of exibles (Fig. 1).

We "lumped" geographic occurrences into the following regions: North America (primarily the United States), the UK/Ireland (England, Scotland, Ireland), Europe, Russia, North Africa (Morocco, Algeria, Tunisia), China, Australia, Japan, South America, and the Middle East (Pakistan, Iran). Figure 2 shows a Pennsylvanian paleogeographic reconstruction of these areas. Temporal comparisons among the geographic areas are complicated by local chronostratigraphic nomenclature and nonstandard correlations. For example, the North American subdivisions of the Pennsylvanian (Morrowan, Atokan, Desmoinesian, Missourian, and Virgilian) do not match up particularly well with the Standard Pennsylvanian Stages (Bashkirian, Moscovian, Kasimovian, and Gzhelian). We encoded the ranges of each crinoid genus in the local chronostratigraphic scheme recorded in Webster (2003) and used the GeoWhen database (http://www.stratigraphy.org/geowhen/index.html) in the www.stratigraphy.org website for correlations. With the exception of the large, diverse Pennsylvanian faunas from North America, the faunas from other paleographic regions were not sufficiently diverse to subdivide temporally within the Pennsylvanian.

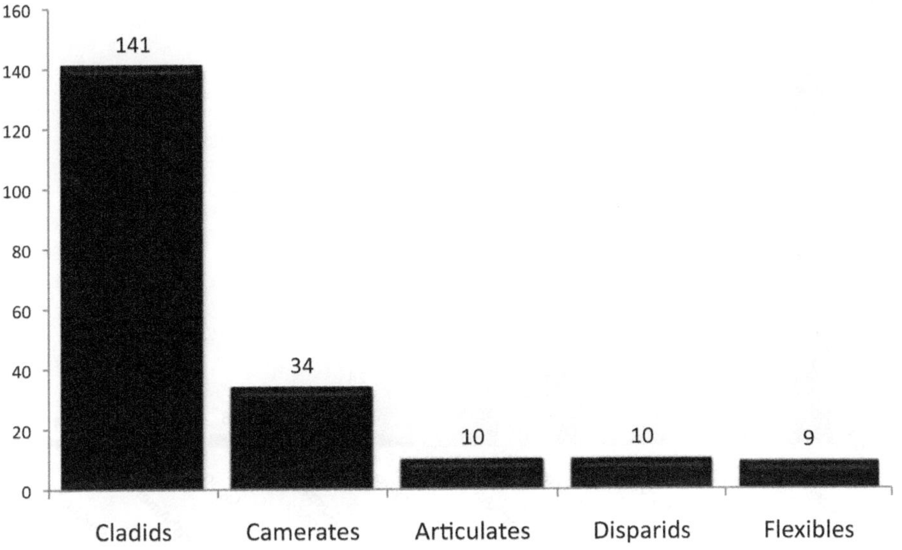

Fig. 1 Bar chart showing the diversity of major crinoid clades in the Pennsylvanian

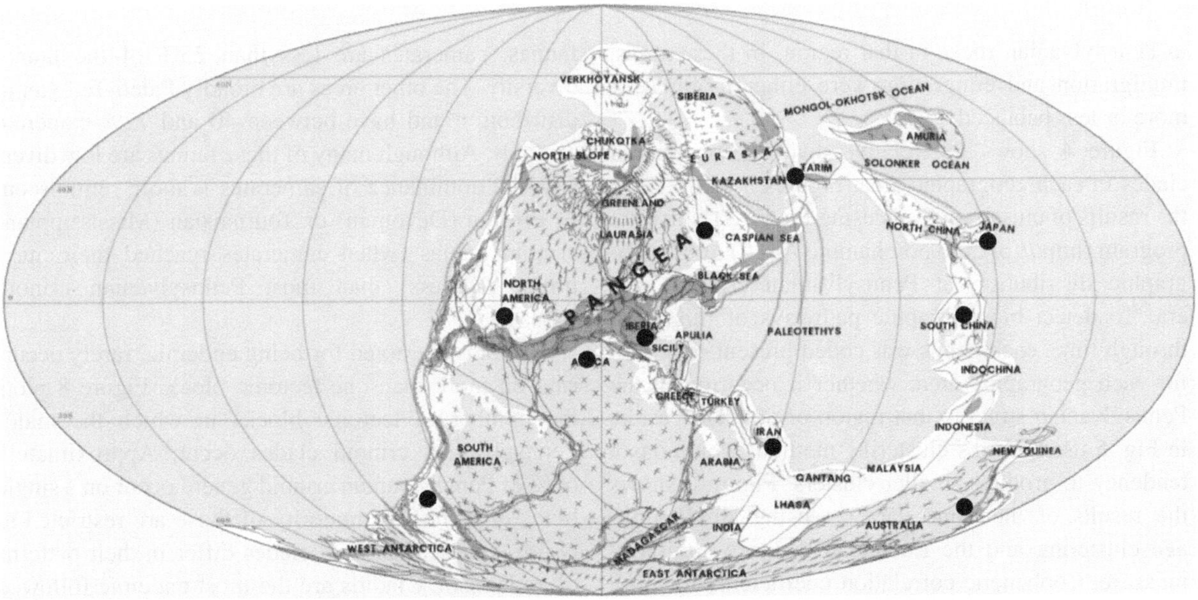

Fig. 2 Late Carboniferous paleogeographic map showing the distribution of Paleo-Tethyan echinoderm faunas containing camerate crinoids and blastoids. Base map from Golonka (2002)

Figure 3 shows the geographic distribution of Pennsylvanian crinoid genera. The first column in each area is the total number of Pennsylvanian genera found in that region, although not necessarily in Pennsylvanian-aged rocks. The second column shows the number of genera found in Pennsylvanian strata in that region. The UK/Ireland and Europe stand out as areas in which significant numbers of Mississippian crinoids emigrated to some other geographic region in the Pennsylvanian, but do not occur in Pennsylvanian faunas in these areas themselves. In contrast, the other geographic regions show reasonable consistency between the number of Pennsylvanian crinoids found in a region and the number of genera found

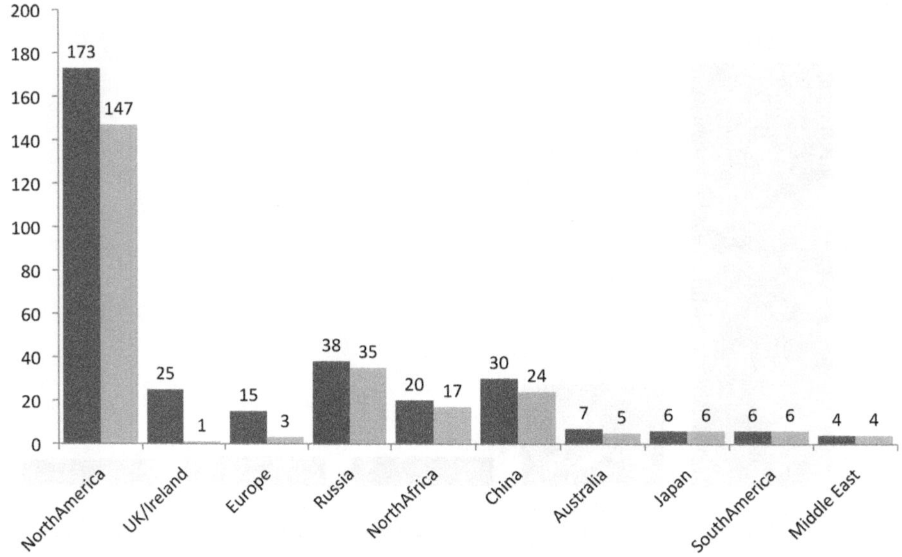

Fig. 3 Bar chart of total number of crinoid genera separated by paleobiogeographic regions. Column one in each region is the total number of Pennsylvanian crinoid genera found in that region, although not necessarily in the Pennsylvanian. Column two for each region is the number of crinoid genera found in Pennsylvanian strata in that region

in Pennsylvanian rocks in that region. In these areas, immigration and emigration were either minimal or more or less balanced.

Figure 4 shows the relative diversity of crinoid clades in each geographic area. Figures 5 and 6 show the results of cluster analysis using the PAST software program (http://folk.uio.no/ohammer/past/) of the geographic distribution of Pennsylvanian crinoid genera. To detect biogeographic pathways of migration through time, each genus was coded present or absent for each geographic area, whether it occurred in the Pennsylvanian strata in that region or not. The results in Fig. 5 used Ward's clustering method, which has a tendency to produce smaller clusters. Figure 6 shows the results of the same data set using single linkage clustering and the Euclidean distance similarity measure. Cophenetic correlation coefficients for both analyses were greater than 0.9. Figure 7 shows the same data set analyzed using the Principle Coordinates Analysis algorithm in PAST. Both multivariate techniques show similar results.

By virtue of its significantly higher diversity, North America separates from the other regions in the cluster analysis (Figs. 4 and 5). The other two large groups separate based on the relative dominance of cladid crinoids in the Pennsylvanian faunas. China, the UK/Ireland, and Russia are all cladid dominated with 60, 60, and 58% cladids, respectively. In these faunas, camerates are less than 25% of the faunal diversity. The other areas are broadly Paleo-Tethyan in distribution and have between 40 and 72% camerate crinoids. Although many of these faunas are low diversity, the dominance of camerates is more reminiscent of Emsian (Devonian) or Tournaisian (Mississippian) crinoid faunas (when camerates reached their maximum richness) than most Pennsylvanian crinoid faunas.

Crinoids are noted for being endemic, rarely occurring on more than one tectonic block. Figure 8 plots the number of tectonic blocks in which the major Pennsylvanian crinoid clades occur. Approximately 60% of Pennsylvanian crinoid genera occur on a single block, and the vast majority of those are restricted to North America. Crinoid clades differ in their patterns of endemism. Cladids are the most endemic followed by the exibles, the disparids, and the camerates, each of which has approximately 50% endemic genera. The camerates and the exibles are the most widely distributed Pennsylvanian crinoid group with more than one-third of the genera found in three or more geographic areas. In contrast the cladids have less than 25% of their diversity similarly distributed. Because the Paleo-Tethyan terranes have additional camerate genera known from material too poorly preserved to allow identification to genus (e.g., Hashimoto 2001, 2005), the difference in patterns of endemism likely

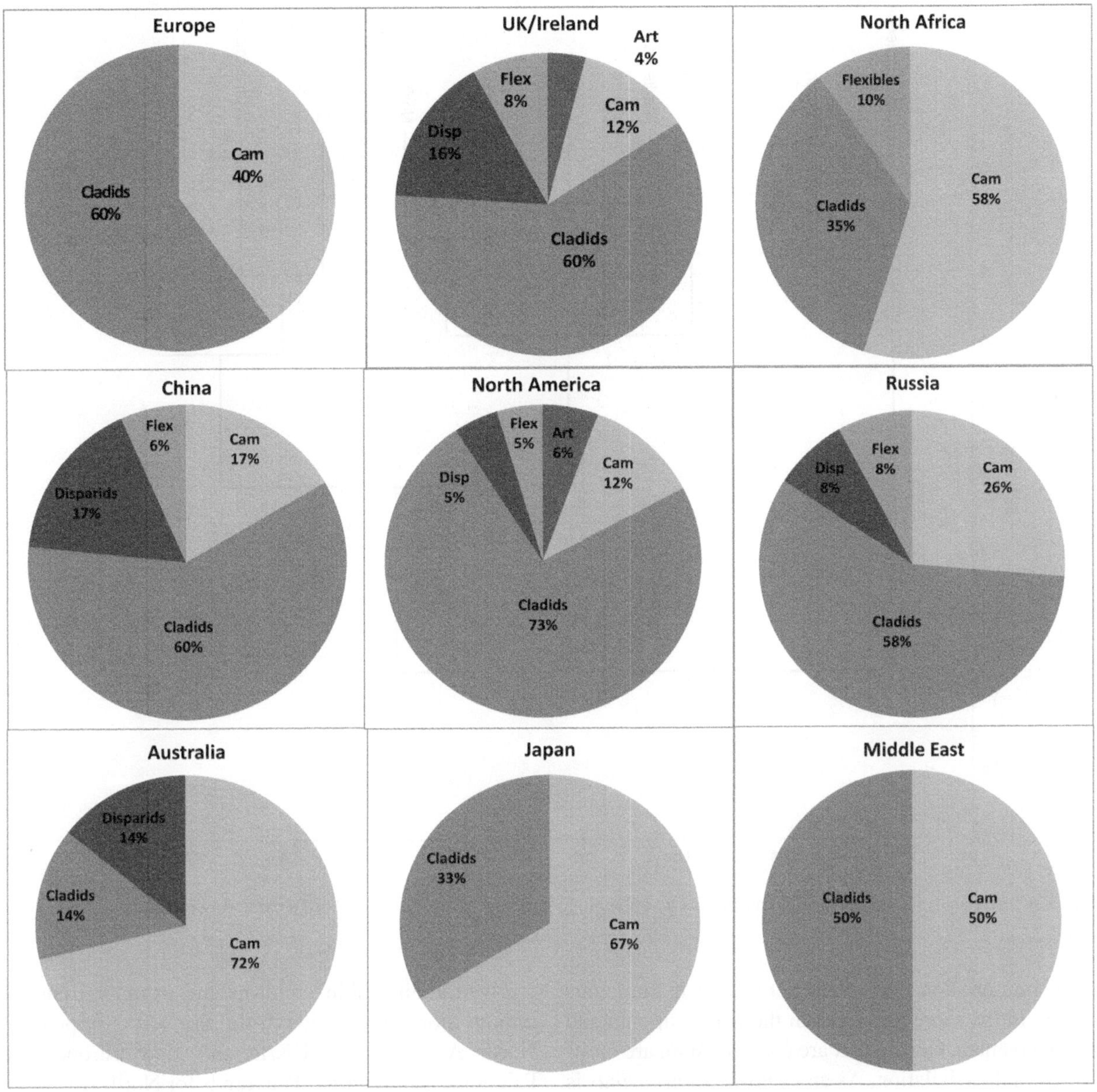

Fig. 4 Pie diagrams showing diversity of crinoid clades in each geographic area

is more pronounced than what the current data set indicates.

Among genera found in a single geographic area, North America contains the most endemic faunas (64%) followed by Russia (40%) and Australia (30%). Other geographic regions show significantly lower rates of endemism ranging from 0 to 16%. North American endemics are predominantly cladids, a pattern not seen in the other areas.

Table 1 shows emigration and immigration patterns among crinoid genera between the Mississippian (the "From" column) and Pennsylvanian (the "To" columns). Data were tabulated by counting instances in which Mississippian genera from one area are found in the Pennsylvanian of another area. For genera found in multiple Mississippian areas and a single Pennsylvanian area, each pair is tallied as an immigration event in the absence of more refined

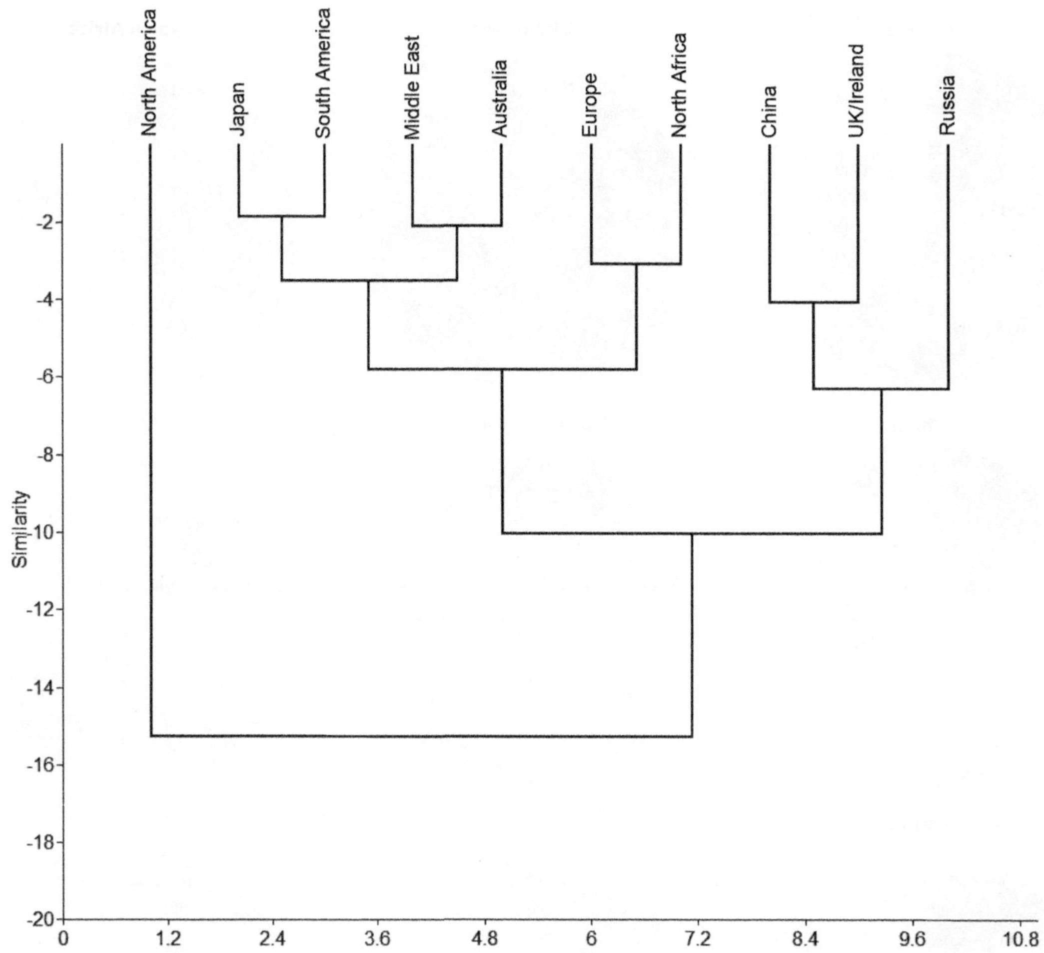

Fig. 5 Cluster analysis of Pennsylvanian crinoids by paleobiogeographic locations using Ward's clustering method

data. Immigration events are not tallied if the genus occurs in the same area in both the Mississippian and Pennsylvanian. Genera that are cosmopolitan are omitted from the tabulation. As an example *Blairocrinus* is known from the Mississippian of North America and the Pennsylvanian of North Africa. This is recorded as one of the five "From" North America "To" North Africa immigration events. *Sampsonocrinus* is recorded from the Mississippian of Australia, England, and the United States and the Pennsylvanian of North Africa. Because we do not know the actual biogeographic pathway, each of these is tallied separately. Although this analysis overstates the absolute numbers of taxa migrating through time, we are more concerned with the patterns, which are not impacted by this data tabulation.

Overall, the table confirms the primary trend of immigration of crinoids from the Mississippian of North America, the UK/Ireland, and Europe into Paleo-Tethyan regions. The trends for North American genera suggest comparable levels of immigration and emigration. Interestingly, emigration of genera is to Russia and the regions of Paleo-Tethys, while immigration of taxa is primarily from the UK/Ireland and Europe. In contrast UK/Ireland and Europe show little immigration, but significant emigration of taxa broadly throughout the Pennsylvanian world.

In summary, Pennsylvanian crinoid diversity was significantly highest in North America, but the fauna was dominated by cladid endemics. We suggest that the evolving Appalachian orogen and the onset of

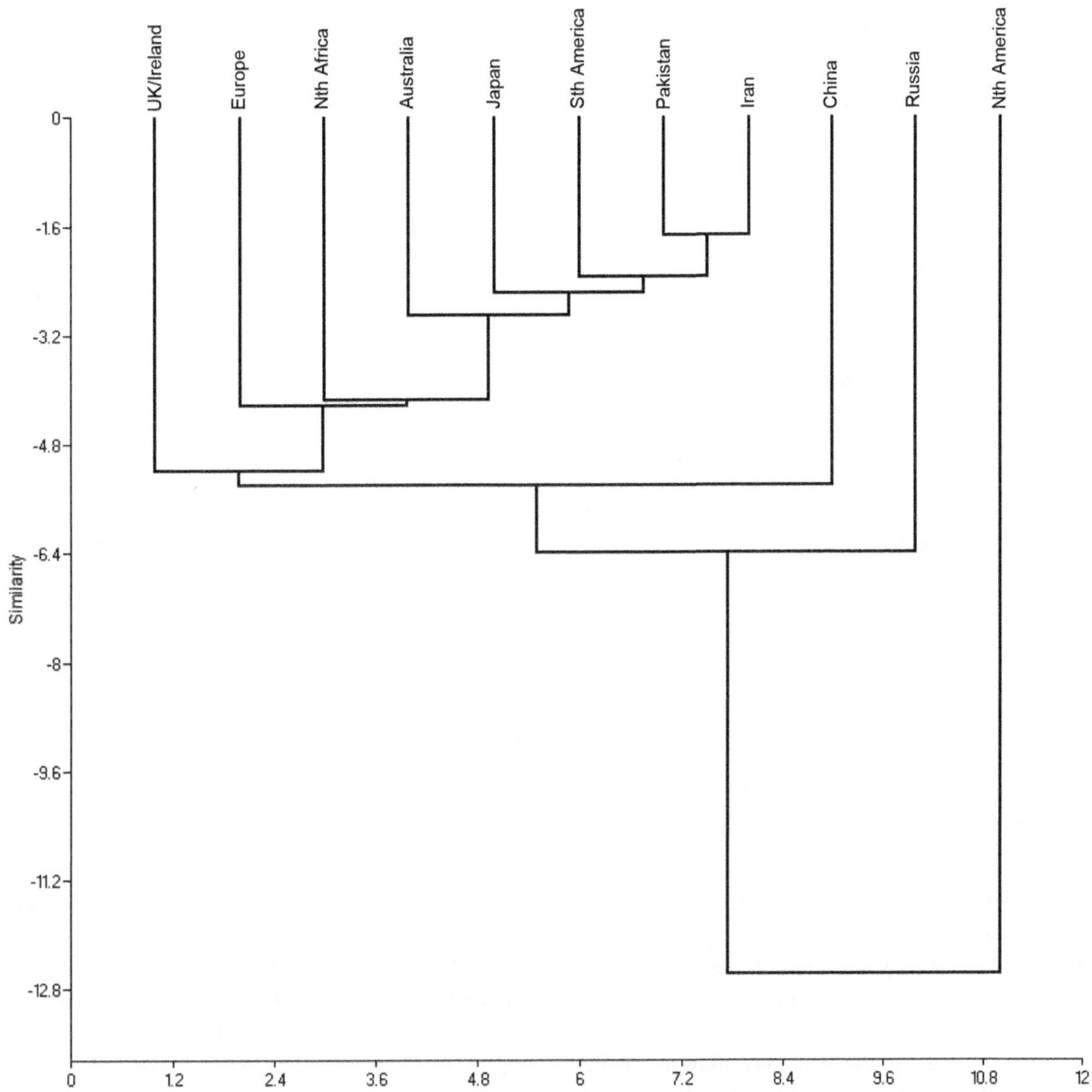

Fig. 6 Cluster analysis of Pennsylvanian crinoids by paleobiogeographic locations using single linkage clustering and Euclidean Distance similarity measure

cyclic sedimentation caused by Gondwanan glaciation produced environments favored by cladids but prevented their large-scale emigration to other parts of the world during Pennsylvanian time. The Mississippian of the UK/Ireland and Europe was a time of origin for Pennsylvanian crinoids, but the majority of the genera found in Pennsylvanian-aged rocks immigrated to the Paleo-Tethyan realm. Russia exhibited the second highest level of Pennsylvanian endemism, but unlike North America, almost half the endemic genera were camerates.

Pennsylvanian Cladid Distribution

As major elements of the Late Paleozoic Macroevolutionary Fauna (Baumiller 1994), cladid crinoids diversified significantly from the Mississippian through the Pennsylvanian and Permian (Lane and Sevastopulo 1990). The cladids can be divided into two distinct groups: the primitive cladids (Cyathocrinina and the Dendrocrinina), which lack pinnules, and the pinnulate advanced cladids, formerly included in

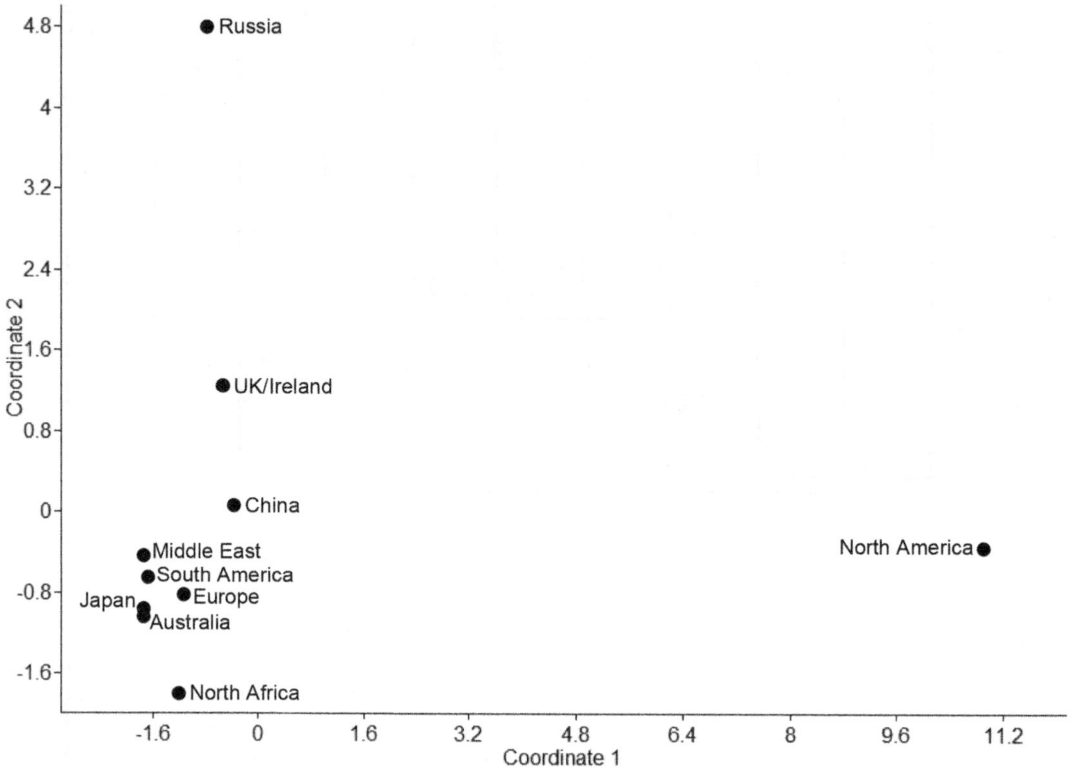

Fig. 7 Principle Coordinates Analysis of the Pennsylvanian crinoid dataset. The segregation of North America, Russia, China, and the UK/Ireland results from cladid crinoid domination of these faunas

the Poteriocrinina (Moore and Teichert 1978) and referred to as advanced dendrocriids by Webster and Maples (2008). Two cyathocrinid genera are found in the Pennsylvanian, *Cyathocrinites* and *Protencrinus*. *Cyathocrinites* is a long-ranging (Silurian to Permian) cosmopolitan genus. In the Pennsylvanian, it has been collected in Russia, the UK and Europe. *Protencrinus* is known from the Pennsylvanian of Russia and the United States. The other large group of cyathocrinine crinoids are the codiacrinids, which typically are small, have unbranched non-pinnulate arms (or were armless), and are thought to have attached to firm substrates with a short, rigid stem (Sevastopulo 2005). Nine genera of these codiacrinids (*Cydonocrinus*, *Lecythiocrinus*, *Coenocystis*, *Dichostreblocrinus*, *Lageniocrinus*, *Lampadosocrinus*, *Neolageniocrinus*, *Abrachiocrinus*, and *Clistocrinus*) have been found in Pennsylvanian faunas, all occur in multiple geographic areas, and all survive into the Permian of Paleo-Tethys.

As stated earlier, most Pennsylvanian cladid crinoids are endemic. The vast majority of these genera are advanced dendrocrinids occurring in North America, the center of the cladid explosive radiation in the Carboniferous. Although perhaps an extreme example, Pennsylvanian crinoid faunas from New Mexico (Webster and Kues 2006) are illustrative of North America as a whole. The Pennsylvanian stratigraphy of New Mexico was developed in a complex paleogeography with sporadic uplift and erosion and shows complex patterns of facies development related to climate change and eustatic sea level uctuations controlled by Gondwana glaciation. At the species level, each depositional basin has an endemic fauna. Migration from and into New Mexico to other depocenters of Pennsylvanian sedimentation was sporadic. In general, crinoid faunas from North America are dominated by genera from the cladid superfamilies Cromyocrinoidea, Erisocrinoidea, and Pirasocrinoidea.

With the exception of *Phanocrinus*, a long-lived phanocrinid found in North America, Scotland, and Morocco, all Pennsylvanian members of the Cromyocrinoidea belonged to the Cromyocrinidae. The Pennsylvanian was a time of explosive growth

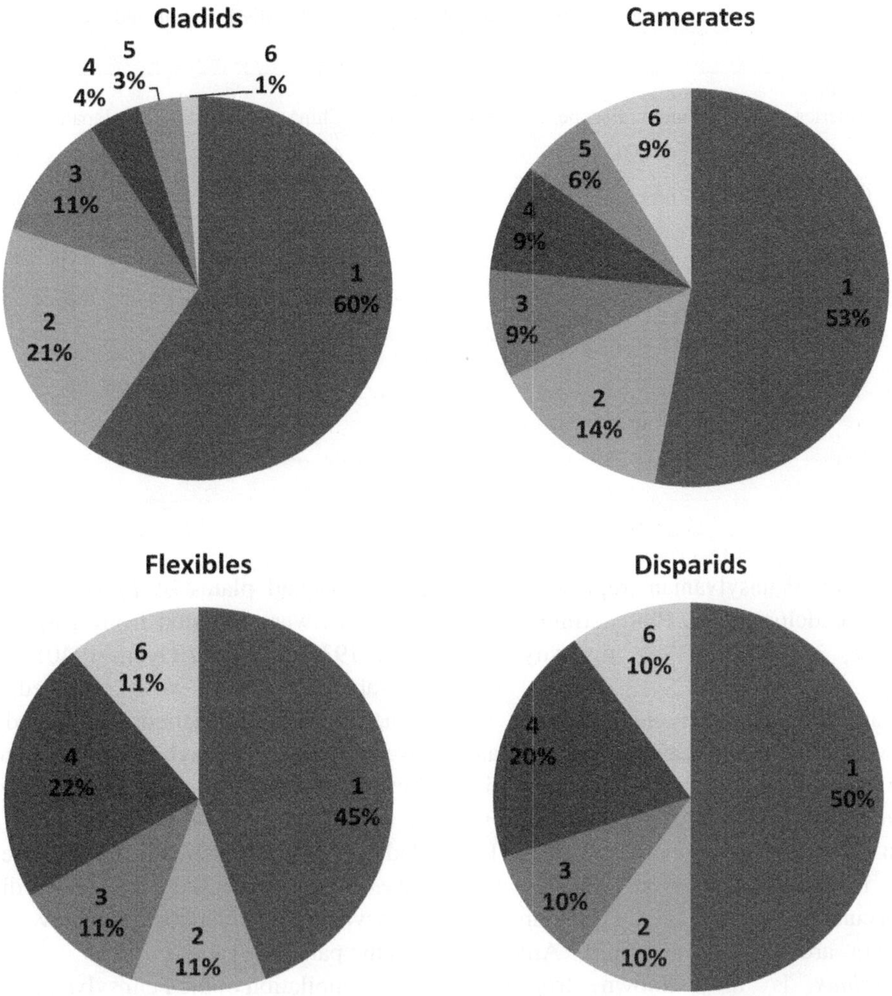

Fig. 8 Pie diagrams of Pennsylvanian crinoid clades showing the number of tectonic blocks on which they occur. Most genera are endemic, occurring in a single area. Genera assigned to the Articulates are all endemic and all occur in North America

in this family with 3 genera in the Mississippian, 18 genera in the Pennsylvanian, but only 4 genera surviving into the Permian (Webster 1981). Many of these genera were endemic to North America, but *Cromyocrinus*, *Dicromyocrinus*, *Mathericrinus*, *Metacromyocrinus*, *Moapacrinus*, *Mooreocrinus*, and *Parulocrinus* are more geographically widespread. Of this group, only *Moapacrinus* and *Parulocrinus* survived into the Permian. Recently two new genera of cromyocrinids, *Binariacrinus* and *Bassocrinus*, have been described from the Moscovian of the Tien Shan by Lane et al. (1996) and Webster et al. (2009). Pennyslvanian-aged cladids in the superfamily Erisocrinoidea are divided into 7 families. The Catacrinidae contains 8 genera, all endemic to North America with the exception of *Delocrinus*, which is also found in the Pennsylvanian of China. The Erisocrinidae has 6 genera of Pennsylvanian crinoids. Erisocrinus is very widely distributed and survived into the Permian. *Sinocrinus* has been found in the USA, Europe, and China. Three genera endemic to the USA are *Eperisocrinus*, *Euerisocrinus*, and *Exaetocrinus*. The genus *Akiyoshicrinus* is endemic to Japan. The Graphiocrinidae is represented by two Pennsylvanian genera. *Contocrinus* is endemic to North America, but *Graphiocrinus*, is long lived (Tournaisian–Artinskian) and shows a clear pattern of migration from the Mississippian of the United States and the UK/Ireland to the Pennsylvanian of Pakistan and Russia. The other families of the

Table 1 Showing immigration and emigration among tectonic blocks between the Mississippian and the Pennsylvanian

	To									
From	North America	UK/Ireland	Europe	Russia	North Africa	China	Australia	Japan	South America	Middle East
North America			1	6	5	4		2	1	2
UK/Ireland	6			6	2	5	1	1	1	3
Europe	2			2	2	1		1		
Russia	2				1	1				
North Africa								1		
China	1				1			1		
Australia					2					
Japan										
South America										
Middle East										

Erisocrinoidea with Pennsylvanian representatives, Diphuicrinidae, Paradelocrinidae, Protencrinidae, and the Arkacrinidae, are mostly low diversity and endemic.

Pennsylvanian-aged pirasocrinoid crinoids are confined to two families, the Laudonocrinidae and the Pirasocrinidae. Three genera of laudonocrinids are endemic to North America, but *Bathronocrinus* is also known from the Moscovian of China. The Pirasocrinidae is significantly more diverse with 21 genera occurring in the Pennsylvanian. Of these, 19 genera are endemic to North America. *Metaperimestocrinus* is also known from the Moscovian of China and *Plaxocrinus* is known from Brazil.

Pennsylvanian Camerate Distribution

Diplobathrid camerates were most diverse and abundant in the Ordovician to Devonian. Few genera survived into the Mississippian (Webster 2003, unpublished tabulation) and diplobathrids are unknown thereafter. Monobathrid camerate crinoids were not as diverse as the diplobathrids in the Ordovician, but were approximately equally diverse in the Silurian and Devonian. The monobathrids had their greatest diversity in the Mississippian, but few genera survived into the Pennsylvanian (Bassler and Moodey 1943; Ausich and Kammer 2001). For years, camerates were virtually unrecognized in the Pennsylvanian except for columnals and plates of *Platycrinites*. Yet Permian camerates were reported from Timor (Wanner 1916, 1924, 1937; de Marez Oyens 1940) and showed considerable diversity. As work continued, the recognition of acrocrinids and dichocrinids added to the known diversity of the Pennsylvanian camerates and showed they were present in a few faunas from various parts of the world, but were greatly outnumbered by the cladid crinoids. Discoveries during the past 20 years have greatly increased the known diversity of the Pennsylvanian camerates and show strong paleogeographic patterns (Table 1).

A compilation of the Pennsylvanian camerate distribution based on cup plates, calices, and crowns is given in Table 2. It should be noted that some genera are questionably identified and other taxa are only referred to in more general terms. These taxa may be incorrectly identified, but are thought to generally belong to the assigned family or superfamily and provide a better estimate of the diversity of the Pennsylvanian camerates.

Comparison of the faunas from the various tectonic blocks listed in Table 1 shows that *Platycrinites* and genera of the Acrocrinidae have a wide paleogeographic distribution. *Platycrinites* was essentially cosmopolitan in the equatorial belt during the Early and Middle Pennsylvanian, but has not been reported from the Late Pennsylvanian except for columnals from North America (Pabian and Strimple 1980). The acrocrinids evolved rapidly in the Early Pennsylvanian of North America, spread to Australia, China, and

Table 2 Compilation of Pennsylvanian camerate crinoid and blastoid (in brackets) distribution by tectonic plates. No relative age relationships are indicated within the listings except as divided into Early, Middle, and Late Pennsylvanian. Taxa are arranged systematically, and numbers in parentheses behind a taxon indicate the number of different genera recognized by the authors for that column

	Japan	China	Russia	Australia	Algeria	Iran	North America
Late							*Discocrinus*
	Periechocrinacea						*Planacrocrinus*
							Caucacrocrinus
							Dinacrocrinus
							[*Pentremites*]
							[*Angioblastus*]
Middle	*Megistocrinus*?			*Denarioacrocrinus*	*Dichocrinus*		*Eretmocrinus*
	Paragaricocrinid	Paragaricocrinid	*Paracroscrinus*				
	Amphoracrinus	Hexacrinid					*Dichocrinus*
	Actinocrinites	*Actinocrinites*	*Platycrinites*				*Planacrocrinus*
	Diatorcrinus?						*Dinacrocrinus*
	Blairocrinus?	Acrocrinid			*Platycrinites*		*Globacrocrinus*
		Platycrinites					*Metacrocrinus*
	Platyacrocrinus	[unident blastoid]					*Platycrinites*
				[*Malchiblastus*]			*Ensulacrinus*
							Amphipsalidocrinus
	Paragaricocrinid						
	Actinocrinites				Actinocrinitids (3)	Actinocrinid	*Platyacrocrinus*
					Aacocrinus		*Planacrocrinus*
					Sampsonocrinus		*Globacrocrinus*
					Blairocrinus	*Camptocrinus*	
Early					*Pimlicocrinus*	*Platycrinites*	*Dichocrinus*
	Platycrinites						
			Amphipsalidocrinus			Hapalocrinid	*Platycrinites*
		Amphipsalidocrinus			*Platycrinites*	*Amphipsalidocrinus*	*Amphipsalidocrinus*
		[new genus; Xinxu]			*Eucladocrinus*		[*Pentremites*]
						[*Iranoblastus*]	[*Angioblastus*]

Literature sources are Japan (Yabe and Sugiyama 1934; Hashimoto 2001, 2005; Hayashaka 1924); China (Lane et al. 1996); Russia (Arendt 2002; Arendt and Stupachenko 1983; Holtedahl 1911; Yakovlev and Ivanov 1956); Australia (Webster and Jell 1999); Algeria (Termier and Termier 1950); Iran (Webster et al. 2004); and North America (Bowsher and Strimple 1986; Itano et al. 2003; Lane and Sevastopulo 1982; Moore and Strimple 1969, 1973; Strimple 1975; Strimple and Heckel 1978; Strimple and Moore 1971; Strimple and Watkins 1969; Webster and Lane 1970)

Russia in the Middle Pennsylvanian, and continued their diversification in North America throughout the Pennsylvanian.

Dichocrinus was present throughout the Pennsylvanian of North America and has been reported from Algeria in the Middle Pennsylvanian (Termier and Termier 1950; Webster et al. 2004). *Amphipsalidocrinus* has also been reported from the Early and Middle Pennsylvanian of North America (Weller 1930) and Early Pennsylvanian of Russia (Arendt 2002) and Iran (Webster et al. 2001). *Amphipsalidocrinus* is a microcrinoid with a calyx slightly over 2 mm in length in the adult stages (Sevastopulo 2008). It is often overlooked in many crinoid faunas because the microcrinoids were not studied. We would expect the paleogeographic distribution of the genus to increase with future studies to show that it is cosmopolitan.

Except for the occurrence of *Eretmocrinus* from the Middle Pennsylvanian of North America (Itano et al. 2003), no other batocrinids are reported from North America or elsewhere after the Mississippian. The Pennsylvanian occurrence of *Eretmocrinus* is interpreted as a Lazarus taxon.

The Pennsylvanian reports of paragaricocrinids, amphoracrinids, actinocrinitids, hexacrinitids, and camptocrinids are all restricted to the Paleo-Tethys. The Early Pennsylvanian fauna from Algeria (Webster et al. 2004) contains four actinocrinitids identified to genus and three unnamed actinocrinitid genera from three horizons. An unnamed actinocrinitid was also reported from Iran by Webster et al. (2001). New collections of the Iranian material show that it contains columnals and cup plates of a camptocrinid. Hashimoto (2001) recognized 3 taxa (*Actinocrinites*, an actinocrinitid, and a paragarcocrinid) in Morrowan strata of Japan.

Middle Pennsylvanian camerate faunas described from northwestern China (Lane et al. 1996; Webster et al. 2009) and Japan (Yabe and Sugiyama 1934; Hashimoto 2001) have *Actinocrinites* and unidentified paragarocrinids in common. The Chinese fauna also contains an indeterminate hexacrinitid, whereas the Japanese fauna contains *Amphoracrinus*, ?*Megistocrinus*, ?*Diatorocrinus*, and ?*Blairocrinus*. The only Late Pennsylvanian camerate from the Paleo-Tethyan realm is a periechocrinoid from Japan (Hashimoto 2005).

All of these recognized and unidentified camerate genera from the Paleo-Tethys belong to families that were diverse and abundant in Early and Middle Mississippian strata of England, Ireland, and eastern and western parts of North America (Webster 2003). This suggests that they migrated from Europe or North America into the Paleo-Tethyan realm in the Late Mississippian, apparently becoming extinct in Europe and North America. Only one genus (*Actinocrinites*) survived into the Permian where it has been reported from Timor (Wanner 1924, 1937), Western Australia (Webster 1987), and Queensland (Etheridge 1892). An indeterminate actinocrinitid was recently reported from the Early Permian of southern Nevada (Webster and Lane 2007), suggesting that the actinocrinitids migrated back into North America or unrecognized actinocrinitids existed in the Pennsylvanian of North America.

In some of the Paleo-Tethys faunas, one or more genera may have multiple species recognized. Hashimoto (2001) identified two species of *Actinocrinites* in the Morrowan fauna of Japan. Webster et al. (2004) recognized two species each of *Aacocrinus* and *Pimlicocrinus* Wright 1943 in the Algerian faunas. Thus, the diversity of these faunas is slightly greater at the species level than suggested at the generic level.

Pennsylvanian Disparids, Flexibles, and Articulates

Eleven genera of disparid crinoids are known from the Pennsylvanian. *Synbathocrinus* is a cosmopolitan Paleozoic crinoid. In the Pennsylvanian, it is known from China, Australia, Russia, and the United States. *Glaukosocrinus* is the lone Pennsylvanian genus of the family Cincinnaticrinidae. It is restricted to the United States. *Belskayacrinus* is a poorly constrained genus restricted to the Moscovian of Russia. The other 8 disparid genera belong to the Allagecrinoidea, which contains the catillocrinids and the allagecrinids. The three catillocrinid genera, *Rhepocatillocrinus*, *Metacatillocrinus*, and *Catillocrinus*, are endemic to China and North America, respectively. In contrast the allagecrinids are more geographically dispersed. Because most of the allagecrinid genera are microcrinoids, the current record probably underpresents their true temporal and geographic distribution.

Flexible crinoids are typically low diversity members of Paleozoic crinoid communities (Lane and Webster 1980). Nine genera of flexible crinoids are known from the Pennsylvanian. Five of these genera are geographically restricted to North America and one is endemic to Russia. *Cibolocrinus* is recorded from Algeria as well as North America, but *Aexitrophocrinus* and *Amphicrinus* have a much longer stratigraphic range and geographic distribution.

Nine genera of the family Ampelocrinidae occur in the Pennsylvanian. The taxonomic assignment of the family seems to be in dispute. Webster and Jell (1999) assigned the family to the subclass Articulata, whereas Kammer and Springer (2008) assigned it to the cladid superfamily Agassizocrinoidea. With the exception of *Ampelocrinus*, all members of the family are endemic to North America in the Pennsylvanian. Similarly the family Calceolispongiidae was assigned to the Articulata by Webster and Jell (1999). One genus, *Allosocrinus*, from this predominantly Permian family of crinoids is found in the Pennsylvanian of North America.

Pennsylvanian Blastoids

Waters (1990) provided a comprehensive review of the paleobiogeography of the blastoids; Macurda and Mapes (1982) provided the last survey of Pennsylvanian blastoids. The evolutionary history of the blastoids can be viewed as three phases of diversification, local dominance of echinoderm communities, and extinction (Waters 1988, 1990; Waters et al. 2003). Phase one consisted of the initial radiation from eastern North America in the Late Ordovician, followed by an increase in diversity and geographic range in the Silurian and Devonian, and culminating in a Middle/Late Devonian (Givetian/Frasnian) extinction. Phase two began with a Famennian (Late Devonian) radiation primarily in China, diversification and numerical dominance of local echinoderm communities in North America and Europe, and ended in the decline in blastoid diversity and abundance in the Serpukhovian (Late Mississippian). Although Mississippian blastoids had a worldwide distribution, the biogeographic center of the blastoid occurrences had shifted eastward toward the Tethyan seaway and away from North America and Europe by the Serpukhovian. Pennsylvanian blastoids tended to be small, were rare, and occurred in localized faunas, but they set the stage for the final radiation of blastoids in the Permian of Southeast Asia.

The following genera of blastoids are known from the Pennsylvanian: *Pentremites, Pentablastus, Angioblastus, Malchiblastus*, and a new genus of schizoblastid, *Orbitremites*, found at Xinxu, Guizhou Province, China. It has a range of Mississippian (Brigantian) to Permian, although Pennsylvanian specimens have not been encountered. The six blastoid genera from the Pennsylvanian are both long-ranged genera (*Pentremites, Angioblastus*, and *Orbitremites*, average generic longevity 50 million years) and short-lived genera (*Pentablastus*, the new blastoid from Xinxu, and *Malchiblastus*, average generic longevity 6 million years).

Pentremites was a widely distributed, abundant blastoid that originated in the late Tournaisian (Early Mississippian) and dominated many Visean (Middle Mississippian) echinoderm faunas in North America (Maples and Waters 1991). *Pentremites* declined to a single species in the Bashkirian (Early Pennsylvanian) and was geographically restricted to the Arkansas and Oklahoma area (Katz 1978). The youngest known specimen of *Pentremites* was collected from the Stanton Formation (Late Pennsylvanian; Kasimovian) from Kansas. The decline and extinction of *Pentremites* in the Pennsylvanian was probably the result of the continuing expansion of advanced cladid crinoids and the transition from the Middle Paleozoic to the Late Paleozoic Crinoid Macroevolutionary Fauna (Maples and Waters 1991; Kammer and Ausich 2006).

Orbitremites originated in the Brigantian (Late Mississippian) of England and Ireland where it was locally very abundant and coeval beds in Russia (Arendt 1960). Waters (1990) mistakenly ascribed a Moscovian (Pennsylvanian) age to the Russian occurrence. Although *Orbitremites* has not been found in Pennsylvanian-aged rocks, it is found in the Permian echinoderm communities from Timor (Wanner 1924) and follows the trend noted earlier of European Carboniferous genera migrating into the Tethyan realm in the Permian.

Angioblastus is the only other Carboniferous blastoid known to occur in the Permian. Breimer and Macurda (1972) synonymized the North American Pennsylvanian blastoid *Agmoblastus* with *Angioblastus*, a decision strongly disputed by Strimple and Mapes (1977). We have followed Macurda (1983) in upholding the synonymy of the two genera. Pennsylvanian occurrences of *Angioblastus* are primarily North American (Ellesmere Island in the Canadian Artic, Oklahoma, and Texas), but Permian occurrences (Bolivia, the Urals, and Indonesia) are primarily Tethyan (Macurda 1983).

Among the short-lived Pennsylvanian blastoid genera, *Pentablastus* has been found exclusively in the Bashkirian (Early Pennsylvanian) Rabanal Limestone of Spain. Specimens of a new, undescribed schizoblastid have been collected from the Luocheng Formation in Guangxi Province, China. Although previous papers (e.g., Waters et al. 2008) have given the age of this formation as Late Mississippian, the formation straddles the mid-Carboniferous boundary and the biostratigraphy is not well defined. The Luocheng Formation is a black shale with sideritic carbonate nodules and well-developed, hard ground horizons. Although crinoids and microcrinoids are present, the fauna is dominated by a very abundant molluscan assemblage. Overall this occurrence is very similar to the occurrences of *Angioblastus* reported by Strimple and Mapes (1977). *Malchiblastus* is known from the Moscovian (Middle Pennsylvanian) of Australia where it occurs in the fine lithic sandstones of the Neerkol Formation (McKellar 1966). *Malchiblastus* has also been described from coeval formations in Chubut Province, Argentina (Sabattini and Castillo 1989). Both blastoid occurrences are found in the *Levipustula levis* Brachiopod Biozone and are associated with fauna rich in bryozoans and brachiopods.

Arendt et al. (1968) described a blastoid fauna from the lower Namurian of north Kazakhstan consisting of four fissiculate species, *Dolichoblastus shimanski*, *Kazachstanoblastus carinatus*, *Mastoblastus ornatus*, and *Phaenoschisma rossica*. Although Macurda and Mapes (1982) included this fauna in their discussion of Pennsylvanian blastoids, the fauna is Late Mississippian in age based on the associated ammonoids. The fauna is interesting in the context of the present discussion, however, because it further demonstrates the shift of Carboniferous blastoid biogeography toward the Tethyan realm. Secondly, the blastoids locally are very common forming blastoid limestones, which are common in many Mississippian occurrences and in the Timor fauna but rare in Pennsylvanian blastoid occurrences. Third, the blastoids are associated with very abundant, diverse cephalopod faunas. As Arendt et al. (1968) point out, this association is rare in the Mississippian, but is more common in the Pennsylvanian.

Implications

Kammer and Ausich (2006) interpreted the Mississippian biodiversity spike in crinoid diversity as a function of changing carbonate sedimentological regimes that allowed camerate crinoids to flourish and later allowed pinnulate cladids to diversify. These conditions may have triggered the transition from the Middle Paleozoic Crinoid Macroevolutionary Fauna to the Late Paleozoic Crinoid Macroevolutionary Fauna. The most diverse Mississippian crinoid communities, particularly those in North American and Europe, were living on tropical passive margins or cratonic carbonate ramps. Although Kammer and Ausich (2006) concluded that the predominance of crinoids in the Mississippian ended because of Gondwanan glaciation and sea level change, crinoid faunas such as those found in the LaSalle Limestone show that the Late Paleozoic crinoid evolutionary fauna persisted into the Pennsylvanian in North America. Mississippian and Pennsylvanian crinoid faunas from the Bechar basin in western Algeria reported by Webster et al. (2004) occur in depositional settings very similar to those found in North America, but were located near the western margin of the West African craton (Hoepffner et al. 2006).

In contrast, many of the Pennsylvanian camerate crinoid and blastoid occurrences, both North American and Tethyan, occur in other depositional settings, which often are clastic dominated. The occurrence of *Malchiblastus* in fine-grained sandstones is one example as discussed above. The Gene Autry Shale in Oklahoma is another example. Abundant, well-preserved disarticulated blastoids have been found in beds that have also yielded the youngest carpoids (Kolata et al. 1991). The echinoderms are found in shales containing a coral thicket and an abundant, diverse molluscan assemblage. Macurda and Mapes (1982) discussed this fauna stating that three new blastoid species were found therein. As stated above, the well-preserved specimens are disarticulated. Complete specimens are rare, poorly preserved, and crushed. This fauna has not been formally described to date.

The crinoid assemblage from the Qiajiagou Formation described by Lane et al. (1996) was recovered from volcanoclastic sediments on a tectonically active margin. The echinoderms, which include unidentifiable blastoid radials, were living in ephemeral carbonate environments in a tropical volcanic island arc setting, likely similar to the modern-day Japan or the Philippines.

In contrast, the Pennsylvanian crinoid assemblage described by Hashimoto (2001, 2005) from the Akiyoshi Limestone is found in a mid-oceanic atoll sequence that records sedimentation on a seamount (Sano and Kanmera 1988). Paleoclimatic analysis of sedimentation in the Akiyoshi Limestone by Sano (2006) suggests the crinoids described by Hashimoto lived during a time of transition from warmer climates to cooler climates that began in the Middle Pennsylvanian.

We suggest that the Paleo-Tethys was a refugium for some camerate genera, especially periechocrinoids and some hexacrinitids, and blastoids during the Pennsylvanian. Reasons for the restriction of the camerates and blastoids to the Paleo-Tethys are uncertain. Large-scale reorganization of Paleozoic echinoderm communities and changing sedimentological regimes caused by Gondwanan glaciation and the emergence of the Appalachian orogen were likely contributing factors. One additional possibility is that after establishment in the Paleo-Tethys during the Mississippian these taxa may not have been able to migrate back into North America or Europe as the time duration before settlement and attachment of the larval stages precluded their dispersal across wide oceanic expanses.

References

Arendt YA (1960) Novyii podvid blastoidey iz podmoskovnogo karbona. Moskov [In Russian: New subspecies of blastoid from the Submoscovian Carboniferous]. Byulleten Moskovskogo Obshchestva Ispyitatelei Prirodyi, new series 65, Otdel Geologii. [Bulletin Societe Imperial Naturalists Moscou, Geologii Section] 35(4):149–150

Arendt YA (2002) Early Carboniferous echinoderms of the Moscow region. Paleontol J 36(Suppl 2):115–184

Arendt YA, Stupachenko AV (1983) Novyi dannye ob Akrokrinidakh [New data on the Acrocrinids]. Paleontologicheskii Zhurnal 2:69–80

Arendt YA, Breimer A, Macurda DB Jr (1968) A new blastoid fauna from the lower Namurian of north Kazachstan (USSR). Proc Koninklijke Nederlandse Akademie van Wetenschappen. Series B: Palaeont Geol Phys Chem Anthropol 71:159–174

Ausich WI, Kammer TW (2001) The study of crinoids during the 20th century and the challenges of the 21st century. J Paleontol 75:1161–1173

Ausich WI, Kammer TW, Baumiller TK (1994) Demise of the middle Paleozoic crinoid fauna: a single extinction event or rapid faunal turnover? Paleobiology 20:345–361

Bassler RS, Moodey MW (1943) Bibliographic and faunal index of Paleozoic pelmatozoan echinoderms. Geol Soc Am Spec Pap 45:734

Baumiller TK (1994) Patterns of dominance and extinction in the record of Paleozoic crinoids. In: David B, Guille A, Feral JP, Roux M (eds) Echinoderms through time. Balkema, Rotterdam, pp 193–198

Breimer A, Macurda DB Jr (1972) The phylogeny of the fissiculate blastoids Verhangelingen der Koninklijke Nederlandsche Akademie van Wetenschappen, Afdeeling Natuurkunde Eerste Reeks, Deel 26:1–390

Bowsher AL, Strimple HL (1986). *Platycrinites* and associated crinoids from Pennsylvanian rocks of the Sacramento Mountains, New Mexico. New Mex Bureau Mines Miner Resour Circular 197:1–37

de Marez Oyens FAHW (1940) Neue Permische Krinoiden von Timor, mit Bemerkungenber deren Vorkommen in Basleogebiet. Geological Expedition to the Lesser Sunda Island under leadership of HA Brouwer 1:285–348

Etheridge R Jr (1892) The organic remains of the Permo-Carboniferous System (Chapter 22). The organic remains of the Rolling Downs Formation (Chapter 32). The organic remains of the Desert Sandstone Formation (Chapter 34). Palaeontology of New Guinea (Chapter 39). In: Jack RL, Etheridge RJ (eds) The geology and palaeontology of Queensland and New Guinea. JC Beal, Brisbane, 2 vols, 768p

Golonka J (2002) Phanerozoic reef patterns. In: Kiessling W, Flügel E, Golonka J (eds) Plate-tectonic maps of the phanerozoic. SEPM (Society for Sedimentary Geology) Special Publications 72, pp. 21–75

Hashimoto K (2001) Pennsylvanian camerate crinoids from the Akiyoshi Limestone Group, southwest Japan. Bull Akiyoshi-dai Mus Nat Hist 36:1–16

Hashimoto K (2005) Pennsylvanian (Missourian) camerate crinoid from limestone blocks in the Tsunemori Formation. Bull Mine City Mus Yamaguchi Prefecture, Japan 20:1–15

Hayashaka I (1924) On the fauna of the Anthracolithic Limestone of Omi-Mura in the western part of Echigo. Sci Reports of Tohoku Imperial Univ, Sendai, Japan, ser. 2, Geology 8:1–82

Hoepffner C, Houari MR, Bouabdelli M (2006) Tectonics of the North African variscides (Morocco, Western Algeria): an outline. Comptes Rendus Geosci 338:25–40

Holtedahl O (1911) Zur Kenntnis der Karbonablagerungen des Westlichen Spitzbergens I. Eine Fauna der Moskauer Stufe. Kristiania, Videnskabsselskabets Skrifter I. Matematisk–Naturvidenskapelig Klasse 10:1–46

Itano WM, Webster GD, Houck KJ, Bateman WD (2003) The first Pennsylvanian batocrinid and other new echinoderms from the Minturn Formation of central Colorado. The Mountain Geologist 40:83–100

Kammer TW, Ausich WI (2006) The age of Crinoids: a Mississippian biodiversity spike coincident with widespread carbonate ramps. Palaios 21:238–248

Kammer TW, Springer GS (2008) Biostratigraphy of crinoids from the Wymps Gap Member of the Greenbrier Formation (Mississippian, Chesterian) in northern West Virginia. J Paleontol 82:1182–1189

Katz SG (1978) Revision of the Morrowan (Lower Pennsylvanian) *Pentremites* from Oklahoma and Arkansas. J Paleontol 52:675–682

Kolata DR, Frest TJ, Mapes RH (1991) The youngest carpoid: occurrence, affiities, and life mode of a Pennsylvanian (Morrowan) mitrate from Oklahoma. J Paleontol 65:844–855

Lane NG, Sevastopulo GD (1982) Microcrinoids from the Middle Pennsylvanian of Indiana. J Paleontol 56:103–115

Lane NG, Sevastopulo GD (1990) Biogeography of Lower Carboniferous crinoids. In: McKerrow WSS, Scotese SR (eds) Palaeozoic palaeogeography and biogeography. Geol Soc London Mem 12:333–338

Lane NG, Webster GD (1980) Crinoidea. In: Broadhead TW, Waters JA (eds) Echinoderms, Notes for a Short Course. Univ Tennessee, Stud Geol 3:144–157

Lane NG, Waters JA, Maples CG, Marcus SA, Liao ZT (1996) A camerate-rich Late Carboniferous (Moscovian) crinoid fauna from volcanic conglomerate, Xinjiang, Peoples Republic of China. J Paleontol 70:117–128

Macurda DB Jr (1983) Systematics of the ssiculate Blastoidea. Museum of Paleontology. Univ Michigan Pap Paleontol 22:291p

Macurda DB Jr, Mapes RH (1982) The enigma of Pennsylvanian blastoids. Proc North Am Paleontol Convention 3:343–345

Maples CG, Waters JA (1991) Mississippian pelmatozoan community reorganization; a predation-mediated faunal change. Paleobiology 17:400–410

McKellar RG (1966) A revision of the blastoids "*Mesoblastus*? *australis*", "*Granatocrinus*? *wachsmuthi*", and "*Tricoelocrinus*? *carpenteri*", described by Etheridge (1892) from the Carboniferous of Queensland. Mem Queensland Mus 14:191–197

Moore RC, Strimple HL (1969) Explosive evolutionary differentiation of unique group of Mississippian–Pennsylvanian camerate crinoids (Acrocrinidae). Univ Kansas Paleontol Contribs, Pap 39:1–44

Moore RC, Strimple HL (1973) Lower Pennsylvanian (Morrowan) crinoids from Arkansas, Oklahoma, and Texas. Univ Kansas Paleontol Contribs, Article 60, Echinodermata 12:1–84

Moore RC, Teichert C (1978) Treatise on Invertebrate Paleontology, Part T, Echinodermata 2, Crinoidea: 3 vols. Geological Society of America and University of Kansas, 1026p

Pabian RK, Strimple HL (1980) Some Late Pennsylvanian (Virgilian) crinoids from southeastern Nebraska and southwestern Iowa. Proc Iowa Acad Sci 87:1–19

Sabattini N, Castillo E (1989) Equinodermo blastoideo en el Carbonifero de las Sierras de Tepuel y Languineo, Provincia de Chubut, Argentina. Ameghiniana 26:219–224

Sano H (2006) Impact of long-term climate change and sea-level uctuation on Mississippian to Permian mid-oceanic atoll sedimentation (Akiyoshi Limestone Group, Japan). Palaeogeogr Palaeoclimatol Palaeoecol 236:169–189

Sano H, Kanmera K (1988) Paleogeographic reconstruction of accreted oceanic rocks, Akiyoshi, southwest Japan. Geology 16:600–603

Sepkoski JJ (2002) A compendium of fossil marine animal genera. Bull Am Paleontol 363:1–560

Sevastopulo GD (2005) The early ontogeny of blastoids. Geol J 40:351–362

Sevastopulo GD (2008) Chapter 3. Paleobiology of Carboniferous microcrinoids. In: Ausich WI, Webster GD (eds) Echinoderm paleobiology. Indiana University Press, Bloomington, pp 55–69, 456

Strimple HL (1975) Middle Pennsylvanian (Atokan) crinoids from Oklahoma and Missouri. Univ Kansas Paleontol Contrib, Pap 76:1–30

Strimple HL, Heckel PH (1978) A significant acrocrinid (Crinoidea: Camerata) from the Ladore Shale (Missourian, Upper Pennsylvanian) in Eastern Kansas. Kansas Geol Surv, Bull 211:5–9

Strimple HL, Mapes RH (1977) A new Upper Pennsylvanian fissiculate blastoid from Texas. J Paleontol 51:357–362

Strimple HL, Moore RC (1971) Crinoids of the LaSalle Limestone (Pennsylvanian) of Illinois. Univ Kansas Paleontol Contrib, Article 55, Echinodermata 11:1–48

Strimple HL, Watkins WT (1969) Carboniferous crinoids of Texas with stratigraphic implications. Palaeontogr Am 6(40):139–275

Termier G, Termier H (1950) Paleontologie Marocaine II. Invertebres de l'Ere Primaire. 4. Annelides, Arthropodes, Echinodermes, Conularides et Graptolithes. Service Carte Gologique Morocco, Notes et Memoires 79(4):1–279

Wanner J (1916) Die Permischen echinodermen von Timor, I. Teil. Palaontologie von Timor 11:1–329

Wanner J (1924) Die permischen Krinoiden von Timor. Mijnw Nederl Oost-Indie Verhandel 1921(3):348 p, 22 pl

Wanner J (1937) Neue Beiträge zur Kenntnis der permischen Echinodermen von Timor VIII–XIII. Palaeontographica Supp Bd 4, Abt 4 (2):57–212, pl 5–14

Waters JA (1988) The evolutionary palaeoecology of the Blastoidea. In: Paul CRC, Smith AB (eds) Echinoderm phylogeny and evolutionary biology. Clarendon Press, Oxford, UK, pp 215–233

Waters JA (1990) The palaeobiogeography of the Blastoidea (Echinodermata). In: McKerrow WSS, Scotese CR (eds) Palaeozoic palaeogeography and biogeography. Mem Geol Soc London 12:339–352

Waters JA, Maples CG, Lane NG, Marcus S, Liao ZT, Liu L, Hou HF, Wang JX (2003) A quadrupling of Famennian pelmatozoan diversity; new Late Devonian blastoids and crinoids from Northwest China. J Paleontol 77: 922–948

Waters JA, Marcus SA, Maples CG, Lane NG, Hou HF, Liao ZT, Wang JX, Liu LJ (2008) An overview of paleozoic stemmed echinoderms from China. In: Ausich WI, Webster GD (eds) Echinoderm paleobiology. Indiana University Press, Bloomington, pp 347–367, 456 p

Webster GD (1981) New Crinoids from the Naco Formation (Middle Pennsylvanian) of Arizona and a Revision of the Family Cromyocrinidae. J Paleontol 55:1176–1199

Webster GD (1987) Permian crinoids from the type-section of the Callytharra Formation, Callytharra Springs, Western Australia. Alcheringa 11:95–135

Webster GD (1998) Palaeobiogeography of Tethys Permian crinoids. Proc R Soc Victoria 110:289–308

Webster GD (2003) Bibliography and index of Paleozoic crinoids, coronates, and hemistreptocrinoids. 1758–1999. Geol Soc Am, Spec Pap 363:2335. GSA website: http://crinoid.gsajournals.org/crinoidmod/

Webster GD, Jell PA (1999) New Carboniferous crinoids from eastern Australia. Mem Queensland Mus 43:237–278

Webster GD, Kues BS (2006) Pennsylvanian crinoids of New Mexico. New Mexico Geol 28:1–37

Webster GD, Lane NG (1970) Carboniferous echinoderms from the southwestern United States. J Paleontol 44:276–296

Webster GD, Lane NG (2007) New Permian crinoids from the Battleship Wash patch 5 reef in southern Nevada. J Paleontol 81:951–965

Webster GD, Maples CG (2008) Cladid crinoid radial facets, brachials, and arm appendages: a terminology solution for studies of lineage, classification, and paleoenvironment. In: Ausich WI, Webster GC (eds) Echinoderm paleobiology. Indiana University Press, Bloomington, pp 196–226, 456

Webster GD, Yazdi M, Dastanpour M, Maples CG (2001) Preliminary analysis of Devonian and Carboniferous crinoids and blastoids from Iran. Travaux de Institut Scientique, Rabat, Serie Geologie et Geographie Physique (2000) 20:108–115

Webster GD, Maples CG, Sevastopulo GD, Frest T, Waters JA (2004) Carboniferous (Visean–Moscovian) echinoderms from the Bechar Basin area of western Algeria. Bull Am Paleontol 368:1–98

Webster GD, Waters JA, Liao ZT, Maples CG (2009) New Pennsylvanian (Moscovican) Echinoderms from Xinjiang–Uyghur Autonomous Region, Western China. Palaeoworld 18:241–250

Weller JM (1930) A group of larviform crinoids from Lower Pennsylvanian strata of the eastern Interior Basin. Illinois Geol Surv, Report 21:1–43

Yabe H, Sugiyama T (1934) An upper Paleozoic crinoid from Japan. Jpn J Geol Geogr 11:349–351

Yakovlev NN, Ivanov AP (1956). Morskie lilii i blastoidei kamennougolnykh i permskikh otlozheniy SSR.TrudyVsesoy Nauchno-Issledov Geol Inst 11:142 p, 21 pl

Jurassic and Early Cretaceous Ostracods from Western Australia: What They Reveal About Evolution of the Indian Ocean

Michelle Guzel

Abstract

Using a combination of previous studies, some reinterpreted, and material from the Barremian of the Southern Carnarvon Platform and the middle–late Albian of the Exmouth Plateau, the distribution of ostracods from Jurassic and Early Cretaceous sediments of Western Australia is used for analyzing evolution of the Indian Ocean. The Early Jurassic ostracod fauna of Western Australia is Tethyan, similar to those from northwestern Europe. By Middle Jurassic it was still primarily Tethyan, with strong links to an East Tethys Province (northwest India, Madagascar, Tanzania, and northern Somalia) and weaker links with a South Tethys Province (Saudi Arabia, Egypt, Jordan, and Israel). *Paradoxorhyncha* accords with migration along the marine shelf of southern Gondwana between west-central Argentina and Western Australia. During the Late Jurassic to early Valanginian, the Western Australian ostracods belonged to a distinctive Indian Ocean ostracod fauna characterized by Gondwanan genera: *Majungaella* and *Arculicythere*. By late Valanginian–Hauterivian, this fauna had strong links with South Africa and southern South America reflecting breaching of physical barriers between the Indian Ocean and the Pacific. By the Barremian, occurrences of *Rostrocytheridea* in Western Australia, South Africa and Patagonia indicate inception of a new Austral Province. During the late Aptian–Albian, the ostracod fauna of Western Australia belonged to a discrete Austral Province (Madagascar, southern India, southern Argentina, the Falkland Plateau, Agulhas Bank, and South Africa) with species distributed parallel to a cool West Wind Drift current encircling Antarctica, Australia, and Papua New Guinea. Characteristic species include *Arculicythere tumida* Dingle, 1971, *Isocythereis sealensis* Dingle, 1971, and *Cytherura? oertli* Dingle, 1984.

M. Guzel (✉)
School of Life and Environmental Sciences, Deakin University, Burwood, Vic 3125, Australia
e-mail: mguz@deakin.edu.au

J.A. Talent (ed.), *Earth and Life*, International Year of Planet Earth,
DOI 10.1007/978-90-481-3428-1_30,

Keywords
Ostracoda • Palaeobiogeography • Western Australia • Jurassic-early Cretaceous • Indian Ocean • Tethys • Gondwana

Introduction

Ostracods are microcrustaceans enclosed in a bivalved carapace of low-Mg calcium carbonate; it moults as they grow. Most ostracods and their larvae are benthic; they usually swim above, crawl on, or burrow into organic-rich muds or fine sands. Its distribution depends on the environmental factors, including water temperature, salinity, chemistry (e.g. oxygen concentration), and depth, as well as the substrate type, the physico-chemical equilibrium at the sediment–water interface, the hydrodynamic level, and the availability and competition for food resources (synthesis in Whatley 1983; papers in De Deckker et al. 1988).

Ostracods are particularly useful for plate-tectonic reconstructions and discriminating the opening of new seaways. The aim of this chapter is to use the geographic distribution of Western Australian Jurassic and Early Cretaceous ostracods to document the fragmentation of Gondwana and, especially, the genesis of the Indian Ocean.

Based on the striking similarities of Early Cretaceous benthic foraminifers and ostracods from Australia, South Africa, New Zealand, South America, peninsular India and Madagascar, Scheibnerová (1973: 408) proposed an Austral Bioprovince, a "non-tropical Southern Hemisphere climatic zone". Scheibnerová (1973: 409) pointed out that the Early Mesozoic (until Jurassic) foraminiferal faunas had shown quite strong cosmopolitanism, reflecting the incoming of climatic zonation.

Krömmelbein (1975a, 1976) was the first to note that *Majungaella* was a Gondwanan genus with occurrences in India, Madagascar, Tanzania, South Africa, Western Australia, Queensland, Argentina, and Brazil during the Middle Jurassic and Cretaceous. Subsequently, Dingle (1982: 377) noted that South African ostracod faunas had all the "essential faunal characteristics" of those from the Callovian to Cenomanian of Argentina, the Falkland Plateau, Tanzania, Madagascar, India, and Western Australia; they formed a "well-defined ostracod faunal province" based on a "relatively large number of species and genera that are regionally endemic". Regionally endemic genera listed by Dingle (1982) are *Songadella*, *Majungaella*, *Arculicythere*, *Pirileberis*, and *Rostrocytheridea*, as well as other genera including *Progonocythere/Amicytheridea* (species subsequently assigned to *Fastigatocythere* by Whatley and Ballent 1996), *Acrocythere*, and *Procytherura*. He called this the South Gondwana ostracod faunal province.

In 1988, Dingle extended the range of his South Gondwana Fauna down to the Bajocian and, for South Africa, divided it into Faunas A (Bajocian–Aptian) and B (Albian–Cenomanian) based on waning of the progonocytherids (with *Fastigatocythere* disappearing and *Majungaella* generally less abundant) and appearance of trachyleberids (including *Isocythereis sealensis* Dingle, 1971). Fauna A was subdivided into western (Argentina and South Africa) and eastern populations (Tanzania, Madagascar and India with links to Western Australia) (Dingle 1988). By the Albian, faunal diversity increased, the range of *Arculicythere tumida* Dingle, 1971 expanded, and the distinction between eastern and western populations collapsed (Dingle 1988: 846).

Babinot and Colin (1988, 1992) presented "a palaeobiogeographic synthesis of the ostracods of the Tethys margins including Australia during the Cretaceous" (Babinot and Colin 1988: 823). They defined subprovinces based on "a variable number of endemic species and a higher level of ubiquitous genera" (Babinot and Colin 1992: 284). During the Valanginian to Aptian, the occurrence of *Majungaella*, *Pirileberis*, and *Arculicythere* in Madagascar, India, South Africa, and Argentina was used to define a Southern Hemisphere Bioprovince at "the extreme border of the Tethyan realm" (Babinot and Colin 1988: 833), equivalent to Dingle's (1988) South Gondwana Fauna A. During the Albian, the occurrence of *Arculicythere* in India, Madagascar, South Africa, and Australia was used to define an Austral realm (Babinot and Colin 1988), equivalent to Dingle's (1988) South Gondwana Fauna B.

Further investigations led Babinot and Colin (1992) to define an Austral Bioprovince for the south-southeast of the Tethyan Realm during the

Cretaceous; it included Australia, India, Madagascar, and South Africa; it equates with the South Temperate Realm of Kauffman (1973) and the Pan Gondwana Fauna (Turonian–Maastrichtian) of Dingle (1988). The Austral Bioprovince was subsequently differentiated into Indian, Australian, and South African subprovinces (Babinot and Colin 1992).

This chapter focuses on the Jurassic to Early Cretaceous marine ostracod faunas of Western Australia, their temporal changes, and their links with other, primarily Gondwanan, ostracod faunas. Interpretations of these data illuminate the chronology of fragmentation of Gondwana and the development of new seaways and help significantly in developing a scenario for the overall evolution of the Indian Ocean.

Data

Most of the data for this study are based on an extensive review and, where necessary, reinterpretation of previous studies on the Jurassic and Early Cretaceous ostracods from Western Australia, Argentina, southern Chile, the Falkland Plateau, the Agulhas Bank, South Africa, the Mozambique Ridge, India, Madagascar, Tanzania, northern Somalia, Saudi Arabia, Jordan, Israel, and Egypt. All reinterpretations, including instances of synonymy, are also discussed below.

Unpublished data consist of a diverse middle–late Albian marine ostracod fauna from three petroleum exploration wells (Zeewulf-1, Scarborough-1, and Eendracht-1) on the Exmouth Plateau of offshore western Australia, and 10 previously unidentified species from the Barremian Muderong Shale of the Southern Carnarvon Platform of onshore Western Australia; the latter material was generously made available by Barry Taylor of the University of Western Australia.

Palaeobiogeographic Results and Discussion

Early Jurassic

"During the Late Triassic and Early Jurassic there was clear similarity between the facies and associated foraminiferal and ostracod microfauna of the northern Exmouth Plateau/Rowley Terrace area and those of other southern Tethyan margins, including the Northern Calcareous Alps" (Colwell et al. 1994: 11). Arias and Whatley (2005) noted that in the Pliensbachian (Early Jurassic), the northwestern and central European ostracod faunas of Denmark and Germany are very similar to the Tethyan faunas of northwest Africa; this indicates that Tethys had an open marine connection to the northern epicontinental sea. Lord et al. (1993: 118) also identified an Early Jurassic ostracod fauna of mostly poorly preserved and juvenile forms from NW Australia whose "basic assemblage composition is essentially similar to that found in northwestern Europe". This similarity was based on healdiid and cytheroid taxa with subordinate bairdioid, platycopine, and cladocopine forms. The NW Australian ostracod fauna, however, has also several apparently endemic species (Lord et al. 1993). Arias (2006) subsequently listed 17 species occurring in both Australia and the Northern Hemisphere, three of which occur also in the Pliensbachian of west-central Argentina. Such similarities indicate that the extensive Late Triassic and Early Jurassic marine shelf along northern Gondwana displayed little variation in its depositional pattern (Lord et al. 1993; Colwell et al. 1994).

The Australian ostracod fauna shared several healdiid and *Polycope* species – an Early Jurassic cosmopolitan element (Lord 1988) – with NW and central Europe, including the Paris Basin (cf. Arias and Whatley 2005, table 6). The healdiid *Hermiella* appears, however, to have originated from the E Himalaya–NW Australia–Timor area. Its first appearance in that region was ?Hettangian) rather than the Pliensbachian (Lord 1988; Lord et al. 1993; Arias 2006); this was much earlier than in North Africa and Europe. Other Australian Early Jurassic ostracod species occur also in Europe or are similar to European species (Arias 2006, table 2). Once the healdiids (and the whole Suborder Metacopina) became extinct in the earliest Toarcian (Arias and Whatley 2005), the cytherids rapidly diversified and occupied vacated niches in epicontinental seas (Ballent and Whatley 2000), the Early Jurassic epicontinental seas of NW Europe being thought to be "the locus of evolution" for the Cytheroidea (Lord 1988: 865). Although some Australian ostracods may be endemic, there is a strong European bias, both localities belonging to the eastern and western ends of Tethys, respectively. Refinement of Early Jurassic Tethyan palaeobiogeography awaits improved biostratigraphic resolution of the eastern end of Tethys and more information from the extensive marine shelf along northern Gondwana.

"Progonocytherinae" genus A, represented by a juvenile carapace and a juvenile valve from the Early Jurassic (Sinemurian–Pliensbachian) open marine deposits of Gandara-1, northwest Australia, was not illustrated by Lord et al. (1993), but is said to resemble *Progonocythere* of Bajocian–Kimmeridgian age according to Whatley and Ballent (1996); it may be ancestral to *Majungaella* of late Bathonian–early Callovian (Neale and Singh 1986; Khosla et al. 1997; Khosla and Jakhar 1999) to late Pliocene age (Whatley et al. 2005), "which is strictly gondwanine in its distribution" (Ballent et al. 1998: 45). Whatley and Ballent (1996: 925) note that *Progonocythere* "has a geographical range largely restricted to the Northern Hemisphere, particularly Britain and Europe". Exceptions are *Progonocythere sadharaensis* Khosla et al. (2003) and *P. jaisalmerensis* Khosla et al. (2003) from the Bajocian–Bathonian of northwest India (Khosla et al. 2005, 2006), *P. kutchensis* Guha, 1976 from the Bathonian of northwest India (Guha 1976), *P. laeviscula* Lyubimova and Mohan, 1960 from the late Bathonian–early Callovian of northwest India (Guha 1976; Kulshreshtha et al. 1985; Neale and Singh 1986; Khosla et al. 1997, 2003, 2004, 2005; Khosla and Jakhar 1999) and Madagascar (Grékoff 1963), and a *Progonocythere* sp. from the Kimmeridgian of northwest India (Khosla et al. 1997). Another *Progonocythere* sp. is reported in the early–middle Callovian of northwest India (Khosla et al. 2003) but is neither illustrated nor described. *Progonocythere* spp., including *P. laeviscula* Lyubimova and Mohan, 1960, also occurs in the Bajocian–Oxfordian of the Middle East (Saudi Arabia, Egypt, and Israel) (Dépêche et al. 1987; Rosenfeld et al. 1987a; Rosenfeld and Honigstein 1991).

"Progonocytherinae" genus B, a single specimen from the Early/Middle Jurassic (late Toarcian or Aalenian) open/restricted marine deposits of Brigadier-1, northwest Australia, is illustrated (Lord et al. 1993, fig. 6D) and is tentatively assigned here to *Fastigatocythere*. *Fastigatocythere grossepunctata* (Chapman, 1904), the earliest *Fastigatocythere* recorded by Whatley and Ballent (1996), is the most frequent species in the ostracod assemblage obtained from the Middle Jurassic of southwest Australia (i.e. the early Bajocian Newmarracarra Limestone) (Malz and Oertli 1993; Lord et al. 2006). This species is thought to be ancestral to *F. befotakaensis* (Grékoff, 1963) (Malz and Oertli 1993); it occurs in the Bajocian–Callovian of northwest India (Guha 1976; Neale and Singh 1986; Khosla et al. 1997, 2003, 2004, 2005, 2006; Khosla and Jakhar 1999), and the late Bathonian–early Callovian of Madagascar (Grékoff 1963; Mette 2004). Other *Fastigatocythere* species have been recorded in the Bajocian–Callovian of Madagascar (Grékoff 1963; Mette 2004), northwest India (Guha 1976; Neale and Singh 1986; Khosla et al. 1997, 2003, 2004, 2005, 2006; Khosla and Jakhar 1999), Saudi Arabia (Dépêche et al. 1987), and Egypt (Rosenfeld et al. 1987a), the Bathonian of Jordan (Basha 1980), the Callovian–Oxfordian of Israel (Rosenfeld and Honigstein 1991), the early Callovian of northern Somalia (Mette 1993), the middle Callovian of Tanzania (Bate 1975), the Oxfordian of Madagascar (Rafara 1990), the Oxfordian–early Kimmeridgian of Saudi Arabia (Dépêche et al. 1987), the Oxfordian–Tithonian of northwest India (Guha 1976; Neale and Singh 1986), the middle–late Kimmeridgian of Tanzania (Bate 1975), and the late Albian–Turonian of northwest India (Singh 1997; Andreu et al. 2007). This genus has also been recorded from the Bathonian–Kimmeridgian of offshore eastern Canada, Britain, and Europe, and possibly the middle Callovian of Argentina (Whatley and Ballent 1996).

If "Progonocytherinae" genus A and "Progonocythrinae" genus B from the Early Jurassic of Western Australia are indeed *Progonocythere* and *Fastigatocythere*, then they represent the earliest-known examples of these two genera and of the Family Progonocytheridae. In addition, the punctate juveniles identified as *Camptocythere/Praeschuleridea?* (figs. 7.17 and 7.18, Lord et al. 1993) in the Early Jurassic (Sinemurian–Pliensbachian) open marine deposits of Gandara-1, may be juvenile progonocytherids. Consequently, this may be the region from which the progonocytherids originated and dispersed, as did the Early Jurassic healdiid *Hermiella*.

Middle Jurassic

During the Bajocian, after the onset of rifting between Madagascar and East Africa at the Early/Middle Jurassic boundary (Geiger et al. 2004; Papini and Benvenuti 2008), East Gondwana (Australia, Antarctica, Greater India, Madagascar, and the

Seychelles) and West Gondwana (Africa and South America) began to drift apart, forming the west Indian Ocean between the east coast of Africa and the west coast of Madagascar/Greater India. Farther south, the Falkland Plateau rifted away from Gondwana (Stone et al. 2008). These plate movements impacted on the geographic distribution of marine ostracods. A Southern Hemisphere faunal element began to develop, characterized by occurrence of *Paradoxorhyncha*, *Gondwanacythere*, *Mandawacythere*, and Indeterminate Genus *australis*; this is illustrated in Fig. 1.

Table 1 records the distribution of ostracod genera from the Middle Jurassic of southwest Australia identified by Malz and Oertli (1993) and Lord et al. (2006); it illustrates that in the Middle Jurassic, Western Australia had links with Europe (including offshore eastern Canada), the Middle East, East Africa, Madagascar, and India via Tethys and Argentina via the southern margin of Gondwana. The strongest links were with west-central Argentina and northwest India, Madagascar, and Tanzania, all of which belong to Dingle's (1988) South Gondwana Province. There was a weaker link with Saudi Arabia and weak linkage with India, Madagascar, Tanzania, and northern Somalia (an East Tethys Province). During the Middle Jurassic, Saudi Arabia shared several species with Tanzania, Madagascar, and India Mette (2004, fig. 4); these, together with Israel, Egypt, and Jordan formed part of a South Tethys Province, isolated from the North Tethys by an oceanic barrier (Dépêche et al. 1987).

Malz and Oertli (1993) and Lord et al. (2006) studied material from the early Bajocian Newmarracarra Limestone at Bringo, Fossil Hill, and Mount Hill, Western Australia. Malz and Oertli (1993) revised and re-illustrated Chapman's (1904) ostracod types from the Newmarracarra Limestone. This ostracod assemblage is peculiar; it consists entirely of cytheroids (Lord et al. 2006: 191).

In addition to *F. grossepunctata* (Chapman 1904) – the most frequent species of the ostracod assemblage – Lord et al. (2006) identified *Amicytheridea* sp., a genus which, according to Whatley and Ballent (1996), is synonymous with *Fastigatocythere*.

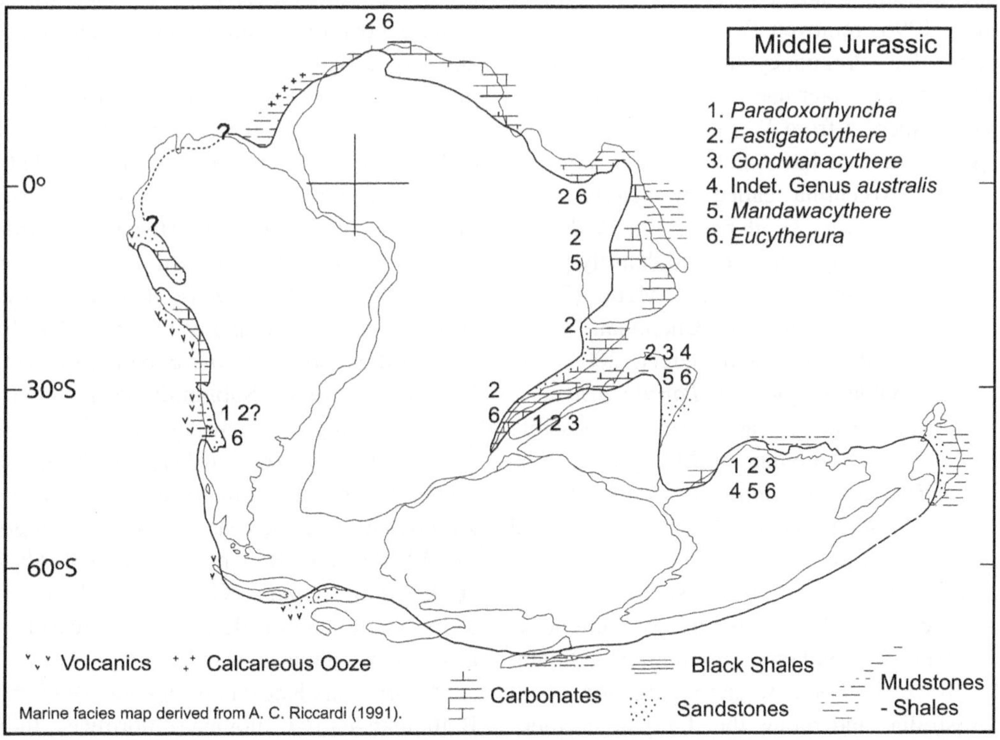

Fig. 1 Distribution of some Middle Jurassic ostracod genera listed in Table 1, illustrating emergence of a Southern Hemisphere faunal element characterized by *Paradoxorhyncha*, *Gondwanacythere*, *Mandawacythere*, and Indeterminate Genus *australis*

Table 1 Distribution of Western Australian Middle Jurassic ostracod genera

Western Australia	Argentina	India	Madagascar	Tanzania	Northern Somalia	Saudi Arabia	Jordan	Egypt	Israel	Europe (inc. offshore E Canada)
Fastigatocythere	?	*	*	*	*	*	*	*	*	*
Eucytherura	*	*		*				*		*
Paradoxorhyncha	*		*							
Gondwanacythere		*	*							
Mandawacythere		*				*				
Indeterminate Genus *australis*		*								
Aitkenicythere										
Bringocythere										
Fistulosacythere										
"*Kirtonella*"										
"*Hekistocythere*"										
"*Procytheridea*" ?										
"*Pleurocythere*" ?										
Gen. et sp. indet. sp. 2										

Paradoxorhyncha foveolata Chapman, 1904, *P. jurassica* (Chapman, 1904), *P? australiensis* (Chapman, 1904), misspelt as *P? australiense*, and *P?* sp. were identified by Malz and Oertli (1993). Ballent and Whatley (1996) removed *Paradoxorhyncha* from Progonocytheridae, Progonocytherinae, placing it in Cytheruridae, Cytheropterinae. Besides these four early Bajocian species from Western Australia, there is also a species from the late Aalenian–early Bajocian of west-central Argentina called *P. neuquenensis* (Ballent, 1991); it is very similar to *P? australiensis* (Chapman, 1904) in shape, outline, and densely pitted ornament, although smaller in size. Mette (2004) also recorded *P? australiensis* (Chapman, 1904) with *P. "malgachica"* Mette and Geiger, 2004, and another unidentifiable species of *Paradoxorhyncha* from the Bajocian of Madagascar, as well as a possible *P.* species from the Bathonian of Madagascar. The distribution of *Paradoxorhyncha*, "an endemic faunal element in the Southern Hemisphere" (Ballent and Whatley 1996: 315), indicates that a migration route existed along the marine shelf of the southern margin of Gondwana enabling faunas, including ammonoids (Riccardi 1991) and benthonic foraminifers (Ballent 1987), to move between western Argentina and Western Australia, and along the Tethyan northern margin of Gondwana between Western Australia and Madagascar.

However, based on new material, Lord et al. (2006: 197) are no longer convinced that the Western Australian species, first described by Chapman (1904), are indeed congeneric but are "uncertain as to their generic position". Ballent and Whatley (2009: 213) also noted that "Gen. indet. B of Bate et al. (1984, pl. 5, figs. 12, 14, 15)" from the Pliensbachian of DSDP Site 547 off North Africa "strongly resembles species of *Paradoxorhyncha*". More work is needed to unravel the relationships of this genus and to determine if any migrations occurred via the Hispanic Corridor.

Lord et al. (2006) defined three new genera, two of which, *Fistulosacythere* and *Bringocythere*, are endemic, monotypic and of uncertain familial affinity. The third, *Gondwanacythere*, belonging to the Family Progonocytheridae, Subfamily Progonocytherinae, is represented by the new species *G. decipiens* Lord et al. 2006, and *G. prolongata* (Malz and Oertli, 1993), described originally as *Strictocythere prolongata* (Chapman, 1904) and referred to *Progonocythere* by Whatley and Ballent (1996). According to Lord et al. (2006), *P. retusa* Grékoff, 1963 from the middle to late Bathonian of Madagascar (Grékoff, 1963) may also belong to this genus.

P. haboensis Khosla and Jakhar, 1997 from the late Bathonian–early Callovian of northwest India (Khosla et al. 1997), resembles *P. retusa* Grékoff, 1963; it should probably be assigned to *Gondwanacythere*.

This species is described as having an elongate-subovate outline, a rather entomodont hinge and straight marginal pore canals (Khosla et al. 1997: 22); illustrations (Ibid. pl. 6, figs. 1–2) demonstrate that the ventral inflation is not strong, and there are no marginal rims. Whereas *Gondwanacythere decipiens* Lord et al. (2006: 198, pl. 4, figs. 2–10 and pl. 5, figs. 1–8) and *Progonocythere retusa* Grékoff, 1963 (Grékoff 1963: 1737–1738, pl. 3, figs. 81–87) have chevron-shaped ridges, and *G. prolongata* (Malz and Oertli 1993) (Lord et al. 2006: 198, pl. 1, fig. 3 and pl. 6, figs. 1–8) has irregular punctae, *P. haboensis* Khosla and Jakhar, 1997 has indistinct pits as surface ornament. *P. tharensis* Khosla and Jakhar, 2006 from the Bajocian–Bathonian of northwest India (Khosla et al. 2006) should also probably be assigned to *Gondwanacythere*. This species is described as having an elongate-subquadrate outline, entomodont hinge, and a smooth surface with an indistinct anterodorsal furrow (Khosla et al. 2006: 20, 22); the illustrations (pl. 3, figs. 14–16) demonstrate that the ventral inflation is not strong.

Lord et al. (2006) are of the opinion that *Strictocythere* should not be considered synonymous with *Progonocythere*. The genus *Strictocythere* "is characterised by an elongate-oval shape in lateral outline versus the typical rectangular outline in *Progonocythere* with its strong ventral overhang" (Malz and Oertli 1993: 130). A similar distinction exists between the Southern Hemisphere genera *Gondwanacythere* and *Majungaella* (Lord et al. 2006).

Lord et al. (2006: 199, pl. 9, figs. 6–10) have also assigned *Procytherura*? sp. of Malz and Oertli (1993) to a new "species of unknown generic affiliation within the Superfamily Cytheroidea", that is, Indeterminate Genus *australis*. *Protobuntonia* sp. from the middle to late Callovian of northwest India (Khosla et al. 2006: 27, pl. 5, figs. 19–20 and pl. 6, fig. 1) should also probably be assigned to the same genus. They are both subtriangular in lateral outline, biconvex in dorsal view, with slightly inflated ventral regions and with greatest height at the anterior cardinal angle. The anterior margin is broadly rounded with a weak marginal rim; the dorsal and ventral margins taper to form a narrow, rounded posterior margin. The surface of the Australian species is "coarsely punctate anteriorly becoming finer in posterior half of valve" (Lord et al. 2006: 199), whereas that of the Indian species is smooth with the exception of "a faint downwardly curved rib near ventral margin giving it an angulated look" (Khosla et al. 2006: 27). The internal characters of the Indian species are not known, but the Australian species has "a weakly developed merodont" hinge and well-developed duplicature (Lord et al. 2006: 199). In contrast, the *Protobuntonia* has an amphidont hinge and an acutely pointed posterior end (Van Morkhoven 1963: 218–220).

Although *Mandawacythere* sp. is rare in the early Bajocian Newmarracarra Limestone of Western Australia (Malz and Oertli 1993; Lord et al. 2006), it is a very distinctive element of the Southern Hemisphere fauna. Malz and Oertli (1993: 134) note that *Mandawacythere* sp. has "a close relationship with the type species", *M. striata* Bate, 1975, from the Kimmeridgian (Late Jurassic) of Madagascar and Tanzania (Grékoff 1963; Bate 1975). Several species have been described from the Bajocian–Callovian (Middle Jurassic) of northwest India (Neale and Singh 1986; Khosla et al. 2003, 2004, 2005, 2006) and the middle Callovian of Saudi Arabia (Dépêche et al. 1987). *Mandawacythere* also occurs in the Oxfordian (Late Jurassic) of Madagascar (Rafara 1990; Mette 2004). Khosla et al. (2005, 2006) subsequently assigned this genus to the Family Protocytheridae, rather than the Family Trachyleberididae.

Gen. et sp. indeterminate sp. 1 (Lord et al. 2006: 201) is said to resemble the Early Jurassic genus *Pleurifera*. But comparison of the illustration (Lord et al. 2006, pl. 11, fig. 4) with *Aitkenicythere gracilis* (Bate 1975) from the middle/late Kimmeridgian (Late Jurassic) of Tanzania (Bate 1975: 202–203, pl. 10, figs. 11–14 and pl. 11, figs. 1–4) reveals a remarkable similarity, particularly in arrangement of the ridges, and as a consequence this species is assigned tentatively to *Aitkenicythere* cf. *gracilis* (Bate 1975); it also occurs in the Oxfordian (Late Jurassic) of Israel (Rosenfeld et al. 1987b).

Malz and Oertli (1993) compare *Renicytherura* sp. to "*Cytheropteron*" sp. A from the early Toarcian of southwest Germany, and Toarcian species from England formerly included in *Eucytherura*; they note differences in aspects of the caudal process. Lord et al. (2006: 200) are of the opinion that this assignment is provisional, the real generic affiliation being uncertain. However, *Renicytherura* sp. is very similar to *Eucytherura argentina* Ballent, 1991 from the Aalenian–Bajocian boundary of west-central Argentina (Ballent 1990, 1991; Ballent and Whatley

2009) with a similar shape and pattern of reticulation, a short caudal process, an eye-tubercle beneath the anterior cardinal angle, a short ocular rib extending antero-ventrally, a ventrolateral rib, and a faint dorsal rib. The main difference appears to be ornamental, i.e., in the type of fossae and degree of celation. The fossae of *E. argentina* Ballent, 1991 are subquadrate and, in some instances, subrounded due to celation (Ballent and Whatley 2009: 198, pl. 1, figs. 10–11). *Renicytherura* sp. is smoothly reticulate due to additional celation having reduced the fossae to narrow slits; it also has an almost vertical posterior rib (Malz and Oertli 1993: 134, figs. 6.37 and 6.38; Lord et al. 2006: 200–201, pl. 11, figs. 14–19). Consequently, *Renicytherura* sp. is here assigned to *Eucytherura.* It is noted by Ballent and Whatley (2009: 201) that the genus *Rutlandella* is regarded as "one of the many junior synonyms of *Eucytherura*", and consequently "*Rutlandella*" ? sp., which Lord et al. (2006: 201) notes "bears some resemblance in its small size and presence of eyespot to *R. mimica* Bate and Coleman 1975, described by those authors from the Toarcian of England", is also a species of *Eucytherura*.

Ballent and Whatley (2009) described several species of *Eucytherura* from the Early Jurassic (late Pliensbachian) and Middle Jurassic (late Aalenian–early Bajocian and middle–late Callovian) of west-central Argentina, with some of these species also recorded in the early Sinemurian of offshore southwest Ireland, the Sinemurian–early Toarcian of Wales, the late Sinemurian–early Toarcian of southwest Germany, the late Sinemurian–late Pliensbachian of the Danish Embayment and Sweden, the Pliensbachian of DSDP Site 547 off North Africa, the early–middle Toarcian of England, the Toarcian–Aalenian of west-central Europe, the Bajocian of Egypt, the late Bathonian of northwest Germany, and possibly the Oxfordian (Late Jurassic) of Israel. In the Southern Hemisphere, other species of *Eucytherura* have been recorded in the Bajocian-early Callovian of northwest India (Khosla et al. 2005, 2006), the middle Callovian of Tanzania (Bate 1975), and the Early Cretaceous of west-central Argentina and South Africa (Ballent and Whatley 2009; Brenner and Oertli 1976; Valicenti and Stephens 1984). Ballent and Whatley (2009) also noted that the specimens assigned to *Renicytherura* in the Sinemurian–Pliensbachian and late Toarcian–Aalenian of northwestern Australia (Lord et al. 1993) may belong to *Eucytherura michelseni* (Finger 1983), known from the Pliensbachian of the Danish Embayment, Sweden, and southwest Germany and the late Aalenian–early Bajocian of west-central Argentina.

In west-central Argentina, the shallow marine siltstones around the Aalenian–Bajocian boundary were dominated by a poorly diverse cytheroid fauna consisting mainly of Cytheruridae (Ballent 1990, 1991; Ballent and Whatley 1997a; Ballent and Whatley 2009) such as *Paradoxorhyncha neuquenensis* (Ballent 1991), eight species of *Eucytherura* including *E. argentina* Ballent, 1991 and *E. michelseni* (Finger, 1983), five species of *Procytherura*, *Kangarina* sp., and *Cytheropteron* sp. Two of the *Procytherura* species of west-central Argentina were previously identified in the middle Toarcian–Aalenian of the British Isles (Ballent and Whatley 2000). Also, in the preceding late Pliensbachian of west-central Argentina, two *Procytherura* specimens were obtained, one only tentatively assigned to the genus (Ballent and Whatley 2000). *Procytherura* first appears in the Sinemurian of DSDP Site 547B off North Africa (Bate et al. 1984); it is thought to have migrated from there into the epicontinental seas of northwest Europe and through the Hispanic Corridor to west-central Argentina (Ballent and Whatley 2000). In the Early Jurassic, Argentinian ostracod faunas were clearly differentiated from those of Australia, but shared numerous taxa with Western Europe via the Hispanic Corridor (Arias 2006).

Following Lord et al. (2006: 199) viewing *Procytherura*? sp. of Malz and Oertli (1993) as a new species of the "Indeterminate Genus *australis*", the genus *Procytherura* is construed as having made its first appearance in the western Indian Ocean in the Callovian of northern Somalia (Mette 1993), Tanzania (Bate 1975), and northwest India (Kulshreshtha et al. 1985). Elsewhere, in the Southern Hemisphere, a species of *Procytherura* was identified in the mid-Callovian of west-central Argentina (Ballent and Whatley 2000, 2009).

Lord et al. (2006: 200, pl. 9, figs. 16–17) note that "*Kirtonella*" sp., a provisional identification of uncertain affiliation, resembles a Bajocian species from northern England. This species is also similar to species of the genus *Afrocytheridea* (Whatley and Ballent 2004: 103–104, pl. 2, figs. 9–14), except that the hinge is antimerodont, rather than entomodont, and the dorsal margin is not sinuous. Species of *Afrocytheridea* have been described from the

Callovian (Middle Jurassic) of Tunisia, Syria, Israel, Saudi Arabia, Tanzania, northwest India, and possibly Madagascar and the Late Jurassic of Israel, Saudi Arabia, and northern Somalia (Whatley and Ballent 2004).

The generic affiliations of "*Procytheridea*" ? sp. (Lord et al. 2006: 200, pl. 11, figs. 1–3), "*Pleurocythere*" ? sp. (Lord et al. 2006: 200, pl. 11, figs. 6–8), and "*Hekistocythere*" sp. (Lord et al. 2006: 200, pl. 11, figs. 10–13) were all provisional. These three species and Gen. et sp. indeterminate sp. 2 (Lord et al. 2006: 201, pl. 11, fig. 5) have been compared with Northern Hemisphere genera but "further material is required to determine the relationships of these species and to throw light on their biogeographical distribution patterns, be they genuinely world-wide or endemic" (Lord et al. 2006: 200). Making comparison of these forms, together with species of *Trichordis* commonly occurring in the Middle and Upper Jurassic (Bajocian–Kimmeridgian) of Madagascar, Tanzania, India and Saudi Arabia (Whatley and Ballent 2004: 94), including the new species identified by Khosla et al. (2005, 2006), and the species of *Neurocythere* described from the Bajocian–Callovian of northwest India (Khosla et al. 1997, 2003, 2004, 2005, 2006; Khosla and Jakhar, 1999) and the Callovian of Madagascar (Grékoff 1963) may reveal interesting relationships. *Neurocythere* also ranges and has a much greater diversity, within the Bajocian to Tithonian? of Britain, northwest Europe, the Volga region of the former USSR, and North America (Whatley and Ballent 2004).

Like *Fastigatocythere* and *Gondwanacythere*, the genera *Afrocytheridea*, *Trichordis*, and *Neurocythere* belong to the Progonocytheridae. But in contrast to *Fastigatocythere* and *Gondwanacythere*, both belonging to the Subfamily Progonocytherinae, *Afrocytheridea*, *Trichordis*, and *Neurocythere* belong to the Subfamily Neurocytherinae. Of these genera, only *Gondwanacythere* is restricted to the Southern Hemisphere.

Late Jurassic

By the Late Jurassic, the western Indian Ocean extended discontinuously as far south as the Rocas Verdes rift basin of southern South America (Calderón et al. 2007); this was due to sea-floor spreading in the Weddell Sea between Antarctica and West Gondwana (Hathway 2000).

On the northwest margin of Australia, the onset of rifting in the Callovian (Middle Jurassic) is indicated by regional uplift and volcanic activity on the Wombat/Exmouth Plateau, Rowley Terrace, and Scott Plateau (Heine and Müller 2005). In the Oxfordian (Late Jurassic), sea-floor spreading began in the Argo Abyssal Plain (Heine and Müller 2005), but not the northern Gascoyne Abyssal Plain (Robb et al. 2005; Direen et al. 2008), with the West Burma Block drifting away from the northwest margin of Australia to form the east Indian Ocean. The presence of early Oxfordian, shallow marine, dinoflagellate cysts in the Cauvery Basin of southeast India, similar to those encountered in the Krishna–Godavari Basin of east India, the Papuan Basin and western Australian Basins, indicates that the eastern Indian Ocean extended as far south as southeast India (Narsimha 2006).

Also, during the Callovian–middle Oxfordian, Rais et al. (2007) note there is sedimentary evidence for strong current activity globally, as well as intense equatorial upwelling in the deeper Western Tethys; this coincided with cold global temperatures. This was followed by climate warming in the middle Oxfordian (Rais et al. 2007).

From the Argo Abyssal Plain, Oertli (1974: 947) listed and illustrated seven species, "each with only one specimen ... in a sample at Site 261, Core 33 (Oxfordian)". He suggested this sample might be from a nearshore deposit. "Though the few specimens of the lowermost core (Late Oxfordian) of Site 261 are of uncertain taxonomic position, their general shape and character suggest a shallow-water environment" (Oertli 1974: 949). The seven micro-ostracods are badly corroded, making identification difficult, but comparison of plate 7, figs. 6–12 with published illustrations of other Middle–Late Jurassic ostracods indicate a link with ostracods from northern Somalia, India, Madagascar, and especially Tanzania (an East Tethys Province) during the Late Jurassic, and a weak link with Saudi Arabia and Israel (part of a South Tethys Province). During the Oxfordian, northern Somalia shares numerous species with the South Tethys Province to its north, but during the Kimmeridgian it shares a few species with Tanzania to its south (Mette 2004, fig. 4). The new identifications are discussed below, as are links based on occurrences of similar genera in other Gondwanan

Table 2 Gondwanan distribution of Western Australian Late Jurassic ostracod genera

Western Australia	India	Madagascar	Tanzania	Northern Somalia	Saudi Arabia	Israel
Majungaella	*	*	*	*		
Arculicythere		*	*			
Eucytherura						*
Procytherura?	*		*			
Pontocyprella/Bythocypris			*	*		
Oligocythereis?	*	?			*	*

Late Jurassic localities (cf. Table 2). By the Oxfordian–Kimmeridgian a distinctive Indian Ocean fauna has developed; it is represented by the Gondwanan ostracod genera *Majungaella* and *Arculicythere* (illustrated in Fig. 2). Similar trends have been deciphered in the ammonoids and belemnites (Riccardi 1991), in the benthic foraminifers (Talib and Faisal 2006), and in the bivalves (Fürsich and Pandey 2003).

Challinor (in Grant-Mackie et al. 2000) noted that in the Late Jurassic, based on belemnites, northwestern Australia belonged to an Indo-Tethyan subprovince of the Tethyan Province; this included eastern Indonesia and Papua New Guinea and was distinct from the South Pacific Belemnite Province. The last extended along the southern margin of Gondwana from South America to New Caledonia. However, in the preceding Middle Jurassic, New Caledonia consisted of a mixture of Tethyan and South Pacific belemnites (Challinor in Grant-Mackie et al. 2000). In a similar vein, the Argo Abyssal Plain ostracod fauna of the Late Jurassic no longer displays any links with Argentina. This change is probably a consequence of a reorganization of ocean current patterns. Rais et al. (2007: 543) note that during the transition from Middle to Late Jurassic "the most important change in the ocean morphology is the deepening of the Hispanic Corridor separating Laurasia and

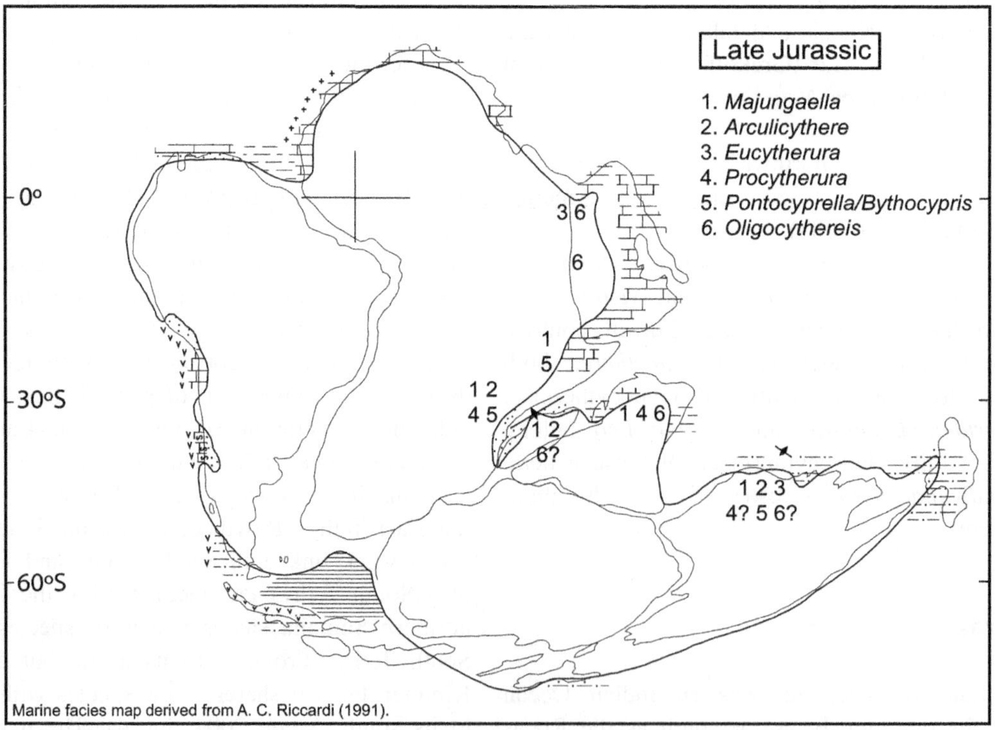

Fig. 2 Distribution of the Late Jurassic ostracod genera listed in Table 2, illustrating the distinctive Indian Ocean fauna which has developed and is characterized by the Gondwanan genera *Majungaella* and *Arculicythere*. See Fig. 1 for legend

Gondwana". This enabled establishment of a warm circum-equatorial current.

Oertli (1974) thought that the ostracod illustrated in pl. 7, fig. 11 might be a "*Metacytheropteron*", but this is very similar to the male right valve of *Majungaella oxfordiana* identified by Bate (1975, pl. 5, fig. 4) in the late Oxfordian of the Wami River Area, Tanzania. It is subtriangular in lateral view, ventro-laterally tumid, postero-dorsally upturned, tapers into a narrow posterior end, has an eye tubercle and shallow post-ocular sulcus, and is not significantly ornamented – just like *Majungaella oxfordiana* Bate, 1975. The genus *Majungaella* appeared first in the late Bathonian–early Callovian in the Kachchh region of northwest India (Neale and Singh 1986; Khosla et al. 1997; Khosla and Jakhar 1999) and in the early Callovian of Madagascar (Mette 2004), the Jaisalmer region of northwest India (Khosla et al. 2006), and the middle Callovian of Tanzania (Bate 1975). By the early/middle Oxfordian, it had spread to northern Somalia (Mette 1993) and by the late Oxfordian to Western Australia. The previous earliest-known occurrence of this genus in Australia was recorded by Damotte (1992) in the Berriasian to early Valanginian Barrow Group of Exmouth Plateau. By the end of the Kimmeridgian, it was in the Himalayan region of India (Jain and Mannikeri 1975) and by the late Valanginian it appeared on the Mozambique Ridge (Sigal 1974), in South Africa (Dingle 1969; Brenner and Oertli 1976; McLachlan et al. 1976a; Valicenti and Stephens 1984), southern Chile (Sigal et al. 1970), southern Argentina (Kielbowicz et al. 1983; Ballent et al. 1998) and Israel (Rosenfeld and Raab 1984). By the late Aptian, it appeared in Egypt (Bassiouni 2002) and by the middle Albian in the Eromanga Basin of Australia (Krömmelbein 1975b; Scheibnerová 1978). By the Campanian, it reached Brazil (Krömmelbein 1975a) and the Antarctic Peninsula (Fauth et al. 2003). The final resting place of *Majungaella* was in Antarctica, with the genus reported from the pre-glacial Eocene (Szczechura 2001), glacial Early Oligocene (Dingle and Majoran 2001), and interglacial Pliocene (Szczechura and Blaszyk 1996).

Oertli (1974) thought that the ostracod illustrated as his Plate 7, fig. 8 might be a "*Schuleridea*", but it is similar to the female left valve of *Trichordis triangula* identified by Bate (1975, pl. 6, fig. 7) in the late Oxfordian of the Wami River area, Tanzania; it differs from the other specimens of this species illustrated by Bate (1975, pl. 6). He identified this female left valve as *Trichordis triangular*; it is really an *Arculicythere* sp. with characteristic features of this genus such as three rounded ridges converging posteriorly, a broadly and asymmetrically rounded anterior margin, a strong and fairly straight dorsal margin inclined at an angle of about 20–30° towards a triangular posterior margin, a ventral margin overhung by a curved ventrolateral ridge, and a small eyespot. The ventral ridge is broadly convex and thick, the dorsal ridge convex but not as thick; the median longitudinal ridge is short and straight. It may be an ancestor of *Arculicythere tumida* Dingle, 1971, from the Hauterivian of Madagascar (Rafara 1990), the Aptian–Albian of the Tamil Nadu region of southern India (Jain 1976a, 1978; where *Pseudobythocythere kallakkudiensis* Jain, 1976 was not recognized as being *Arculicythere tumida* Dingle, 1971), and the Albian of Madagascar (Collignon et al. 1979, where *Arculicythere* DS 1 was not recognized as *A. tumida* Dingle, 1971), the Agulhas Bank off South Africa (Dingle 1971), the Falkland Plateau (Dingle 1984), southern Argentina (Ballent 1998), and Western Australia (Oertli 1974). In *A. tumida* Dingle, 1971, the median ridge is fairly tumid and begins with a large, subrounded subcentral tubercle with a shallow sulcus behind it. In *A. modica* Grékoff, 1963, from the Portlandian of Madagascar, and *A. defluxa* Grékoff, 1963, from the Valanginian, Hauterivian, and Albian of Madagascar (Ballent and Whatley 2006), the median ridge is distinctly divided into two by a sulcus, but in *A. kroemmelbeini* Rosenfeld and Raab, 1984, from the Neocomian of Israel, the median ridge is short with a weak subcentral tubercle and no sulcus. The *Arculicythere* sp. identified here in the late Oxfordian of Tanzania also lacks a sulcus and, as a consequence, there is a rounded depression between the dorsal and median ridges and an elongate, curved depression between the median and ventral ridges; compared with *Arculicythere tumida* Dingle, 1971, it has weak pits superimposed upon the ridges only, rather than being covered, for the most part, in small pits and hollows. Based on the similarity in outline in lateral view and the presence of a rounded depression between the median ridge and the dorsal margin and an elongate, curved depression between the median ridge and the ventral margin, the ostracod illustrated by Oertli (1974, pl. 7, fig. 8) is interpreted here to possibly be *Arculicythere* sp. The fact that the three ridges are not so prominent is attributed to the specimen being a juvenile. Illustrations of *A*? sp. A. (Oertli 1974, pl. 4, figs. 1–11) – subsequently identified as *A. tumida* Dingle, 1971 by Dingle (1984) – illustrates how the three ridges gradually become more prominent as the ostracod matured. The identification here of *Arculicythere* sp. in the late Oxfordian of Tanzania and possibly Western Australia makes these the earliest-known occurrences of the genus. The earliest report of this genus in Australia was by Damotte (1992) from the Berriasian to early Valanginian Barrow Group on the Exmouth Plateau. By the Portlandian, it appeared in Madagascar (Grékoff 1963), by the Valanginian, Israel (Rosenfeld and Raab 1984), by the Aptian, the Orange Basin on the southwest coast of Africa (McMillan 1990) and the Tamil Nadu region of southern India (Jain 1976a, 1978), and by the Albian, South Africa (Dingle 1971), the Falkland Plateau (Dingle 1984), and southern Argentina (Ballent 1998), and by the Cenomanian–Turonian, the Jaisalmer region of northwest India (Singh 1997) where *A. semilunata* Singh, 1997, occurs. The distribution of the *Arculicythere* is Gondwanan (Ballent and Whatley 2006), like that of *Majungaella*, except that in terms of time and space it is more restricted. Ballent and Whatley (2006) demonstrated that the genus *Arculicythere* ranged from late Tithonian to Albian, but this has now been extended from late Oxfordian to Cenomanian–Turonian. In Australia, the demise of the genus seems to have been at the Cenomanian–Turonian boundary because the author has not encountered *A. tumida* Dingle, 1971 beyond sediments of this age in the three petroleum exploration wells on the Exmouth Plateau of Western Australia.

Oertli (1974) thought that the ostracod illustrated as plate 7, fig. 6 might be an "*Acrocythere*"; Dingle (1984) suggested, based largely on the presence of three longitudinal carinae, that it was possibly synonymous with *Aitkenicythere*? sp. 327/18 from the middle Albian of the Falkland Plateau. However, the median carina of the ostracod illustrated by Dingle (1984, fig. 35A) is, however, longitudinal rather than oblique in orientation, so this synonymization is rejected here. Instead, as noted by Ballent (1991), the ostracod illustrated by Oertli (1974, pl. 7, fig. 6) is very similar to *Eucytherura pichia* (Ballent 1991) from the Aalenian–Bajocian of western Argentina, except that it has a more prominent dorsal ridge.

Oertli (1974) thought that the ostracod illustrated in his pl. 7, fig. 7 might be a "*Bythocypris*" or "*Pontocyprella*"; this ostracod is similar to *Bythocypris* sp. B of Bate, (1975, pl. 2, fig. 7) from the late Oxfordian of the Wami River area, Tanzania, except that it is more elongate. Bate (1975) reported both *Bythocypris* and *Pontocyprella* in the late Oxfordian-middle/late Kimmeridgian of Tanzania. *Bythocypris* has been recorded in the Aalenian–Bajocian of western Argentina (Ballent 1991), the Bajocian–Bathonian of northwest India (Khosla et al. 2006), the late Oxfordian of northern Somalia (Mette 1993), the Valanginian of southern Argentina (Kielbowicz et al. 1983), and the Hauterivian of Madagascar (Rafara 1990). *Pontocyprella* has been recorded in the early Callovian of northern Somalia (Mette 1993) and the middle Callovian of Madagascar (Grékoff 1963). In South Africa, both *Bythocypris* and *Pontocyprella* made their first appearance in the late Valanginian–Hauterivian of the Algoa Basin (Brenner and Oertli 1976). By the Aptian, *Bythocypris* appeared in Israel (Rosenfeld and Raab 1984) and by the Aptian/Albian in the nearby Sinai region of Egypt (Bassiouni 2002).

Oertli (1974) was unable to identify the ostracod illustrated in his pl. 7, fig. 9. It is suggested here that this may be a juvenile *Oligocythereis*. This identification was prompted by an illustration of *Oligocythereis minuta* by Khosla et al. (2006, pl. 2, fig. 8) from the early Callovian of the Jaisalmer region of northwest India; it is based primarily on shape and a similar length to height ratio (though the specimen is smaller than those described by Khosla et al. 2006) and, from the Callovian–Oxfordian of the same region (Kulshreshtha et al. 1985). Other species and possible species of *Oligocythereis* have been identified in the Bathonian–early Oxfordian of Israel (Maync 1965; Rosenfeld and Honigstein 1991), Egypt (Rosenfeld et al. 1987a) and Saudi Arabia (Dépêche et al. 1987), the early Callovian of northern Somalia (Mette 1993), the Bathonian–early Hauterivian of Madagascar (Grékoff 1963; Rafara 1990; Mette 2004), and the Valanginian to Hautervian of South Africa (Dingle and Klinger 1972; McLachlan et al. 1976a; Brenner and Oertli 1976).

The remaining two ostracods illustrated by Oertli (1974, pl. 7, figs. 10, 12) from the late Oxfordian of Western Australia are more problematic. Oertli (1974) was unable to identify them; the following identification is primarily based on shape. It is suggested here that they are possibly specimens of *Procytherura*, a genus previously described from the Early–Middle

Jurassic of west-central Argentina (Ballent 1990, 1991; Ballent and Whatley 2000, 2009) and the Middle Jurassic of northwest India (Kulshreshtha et al. 1985), Tanzania (Bate 1975), and northern Somalia (Mette 1993). In the Late Jurassic, *Procytherura* is recorded from the Callovian/Oxfordian of northwest India (Kulshreshtha et al. 1985) and the Kimmeridgian of Tanzania (Bate 1975). In the Early Cretaceous, *Procytherura* appears in the Neocomian of Madagascar (Rafara 1990), the Mozambique Ridge (Sigal 1974), South Africa (McLachlan et al. 1976a, b; Brenner and Oertli 1976; Valicenti and Stephens 1984) and southern Argentina (Musacchio 1978, 1979, 1981; Ballent and Whatley 2000, 2009), and the early–middle Albian of the Falkland Plateau (Dingle 1984). Ballent and Whatley (2000: 229) noted that *Procytherura* "was more or less equally diverse and widely distributed in both hemispheres during most of the Early and Middle Jurassic", but that in the Late Jurassic and Early Cretaceous it became gradually restricted to the Southern Hemisphere".

Early Cretaceous–Berriasian to Early Valanginian

The first known Cretaceous glaciation occurred in the Berriasian–early Valanginian on the southwest margin of the Eromanga Basin in South Australia. Alley and Frakes (2003) suggested that it was probably connected with upwelling of cold abyssal waters along the eastern margin of Australia, a mechanism which Jones et al. (2006) suggested for the formation of ice in eastern Australia during the Late Permian.

On the northwestern margin of Australia, deposition of the regressive Barrow Group and its equivalents during the Berriasian to early Valanginian was connected with the onset of rifting in the Gascoyne and Cuvier Abyssal Plains, when Greater India began to rift from Australia/Antarctica (Romine et al. 1997; Longley et al. 2002; Robb et al. 2005). Uplift along the continent–continent Cape Range transform fault, between the Cuvier and Gascoyne areas, shed sediment northwards, forming the Barrow Delta (Veevers 2006). Damotte (1992: 825) recorded an ostracod fauna from the Barrow Group in ODP 763C on the Exmouth Plateau of Western Australia; it is characteristic of "a shallow-marine, infralittoral to intertidal environment" with sediments "apparently carried into deeper water by currents and mass transport" (Exon et al. 1992: 246). These ostracods and their links to other Gondwanan localities (with references and associated ages) are listed in Table 3; their Berriasian–early Valanginian distribution is illustrated in Fig. 3. Because of the occurrence of the Gondwanan *Majungaella* and *Arculicythere*, the fauna is still distinctively of Indian Ocean type – developed in the Late Jurassic and with strong links with Madagascar.

The Berriasian–early Valanginian Barrow Group ostracod fauna shares several species and genera with the regressive marine sediments of Madagascar: *Majungaella nematis* Grékoff, 1963 and the genera *Majungaella*, *Oligocythereis?*, and *Arculicythere*. All of these also occur in the subsequent late Valanginian–Hauterivian transgressive sediments along with *Eocytheropteron corrosum* (Grékoff, 1963) and the genus *Cytheropteron*. The species "*Bythocypris*" *stroggylae* Brenner and Oertli, 1976 and the genus *Procytherura* also occur in the Hauterivian of Madagascar; *Majungaella nematis* Grékoff, 1963 and *Eocytheropteron corrosum* (Grékoff, 1963) also occur in northwest India in Kimmeridgian, Tithonian, and Callovian sediments respectively. In the Himalayas, *Majungaella spitiensis* (Jain and Mannikeri 1975) was obtained from the latest Kimmeridgian–Portlandian Chidamu Stage (Jain and Mannikeri 1975).

Marine Berriasian to early Valanginian sediments of west-central South America are also regressive (Riccardi 1991) with a microfauna characteristic of a "marginal marine environment, with warm-temperate to subtropical waters, normal salinity levels" and a "low energy regime" (Ballent and Ronchi 1999: 243); near-shore marine sediments occur interbedded with freshwater sediments off the coast of South Africa (Kirkwood Formation in McLachlan et al. 1976a: 350). Sediments of this age are not known from the Mozambique Ridge which, along with the Falkland Plateau, Agulhas Bank and Dronning Maud Land, formed a physical barrier to interchange between E Tethys and the SE Pacific (Riccardi 1991). *Oligocythereis?* is the only ostracod genus which the Barrow Group shares with the early Valanginian sediments of South Africa. There are, however, several Barrow Group ostracod genera and species appearing subsequently in the late Valanginian to Hauterivian transgressive marine and shallow-water sandstones and clays of South Africa: *Majungaella nematis* Grékoff, 1963 (also found in silty

Table 3 Jurassic –Cretaceous Gondwanan distribution of Western Australian Berriasian–early Valanginian ostracods

Western Australia	Argentina and Southern Chile	Falkland Plateau/Agulhas Bank	Brazil and Gabon	West Antarctica	South Africa
Majungaella nematis	Valanginian-Hauterivian (Ballent et al. 1998); Valanginian (Sigal et al. 1970)				Late Valanginian–Hauterivian (McLachlan et al. 1976a; Brenner and Oertli 1976)
Other *Majungaella* spp. (*in* Damotte 1992, sp. 20R and sp. 40R) Also occurs in the middle Albian–early Cenomanian of the Eromanga Basin, Queensland, and South Australia (Krömmelbein 1975b; Scheibnerová 1978)	Valanginian–Hauterivian (Kielbowicz et al. 1983; Ballent et al. 1998); Albian (Rossi de García *in* Malumián et al. 1972; Rossi de García 1979; Ronchi and Angelozzi 1994; Ballent and Whatley 1997b; Ballent 1998; Ballent et al. 1998); Santonian-Campanian (Ballent et al. 1998); late Campanian-early Maastrichtian (Rossi de García and Proserpio 1980; Ballent et al. 1998); mid Maastrichtian (Bertels 1975)		Campanian– Maastrichtian (Krömmelbein 1975a)	Middle–late Campanian (Fauth et al. 2003)	Late Valanginian–Hauterivian (Dingle 1969; McLachlan et al. 1976a; Brenner and Oertli 1976; Valicenti and Stephens 1984); late Aptian–early Cenomanian (Dingle 1971, 1984); Albian-Santonian (McMillan 1990)
Arculicythere spp. (*in* Damotte 1992, sp. 17R and sp. 46R)	Albian (Ballent 1998)	Early–late Albian (Dingle 1971, 1984)			Aptian–Albian (McMillan 1990)
Mandawacythere spp. (*in* Damotte 1992, sp. 763C)					
Procytherura spp. (*in* Damotte 1992, sp. and sp. 16R)	Late Pliensbachian, late Aalenian–early Bajocian and mid-Callovian (Ballent 1990, 1991; Ballent and Whatley 2000, 2009); late Valanginian-Hauterivian (Musacchio 1978, 1979, 1981; Ballent and Whatley 2000, 2009)	Early–middle Albian (Dingle 1984)	Early Cenomanian–Turonian (Grosdidier 1979)		Late Valanginian–Hauterivian (McLachlan et al. 1976a, b; Brenner and Oertli 1976; Valicenti and Stephens 1984); late Aptian (Dingle 1984)
Western Australia	Mozambique Ridge	Tanzania and Northern Somalia	Saudi Arabia, Egypt, and Israel	Madagascar	India
Majungaella nematis	Neocomian (Sigal 1974)			Portlandian–Valanginian (Grékoff 1963); Valanginian–Hauterivian (Rafara 1990)	Kimmeridgian–Tithonian (Guha 1976)

Table 3 (continued)

Western Australia	Argentina and Southern Chile	Falkland Plateau/Agulhas Bank	Brazil and Gabon	West Antarctica	South Africa
Other *Majungaella* spp. (*in* Damotte 1992, sp. 20R and sp. 40R). Also occurs in the middle Albian–early Cenomanian of the Eromanga Basin, Queensland and South Australia (Krömmelbein 1975b; Scheibnerová 1978)		Middle Callovian–Tithonian (Bate 1975; Mette 1993); Albian (Bate and Bayliss 1969)	Valanginian–Hauterivian (Rosenfeld and Raab 1984)	Early Callovian–early Valanginian (Grékoff 1963; Rafara 1990; Mette 2004); Hauterivian (Rafara 1990); middle to early late Albian (Collignon et al. 1979)	Late Bathonian–Kimmeridgian (Guha 1976; Kulshreshtha et al. 1985; Neale and Singh 1986; Khosla et al. 1997, 2003, 2004, 2005, 2006; Khosla and Jakhar 1999); latest Kimmeridgian–Portlandian (Jain and Mannikeri 1975); Cenomanian–Turonian (Singh 1997)
Arculicythere spp. (*in* Damotte 1992, sp. 17R and sp. 46R)		Late Oxfordian (Bate 1975)	Valanginian-Barremian (Rosenfeld and Raab 1984)	Portlandian–Valanginian (Grékoff 1963); Hauterivian (Rafara 1990); middle Albian–Vraconnian (Collignon et al. 1979)	Aptian–Albian (Jain 1976a, 1978); Cenomanian-Turonian (Singh 1997)
Mandawacythere spp. (*in* Damotte 1992, sp. 763C)		Middle/late Kimmeridgian (Bate 1975)	Middle Callovian (Dépêche et al. 1987)	Oxfordian (Rafara 1990; Mette 2004); Kimmeridgian (Grékoff 1963)	Bajocian–Callovian (Neale and Singh 1986; Khosla et al. 2003, 2004, 2005, 2006)
Procytherura spp. (*in* Damotte 1992, sp. and sp. 16R)	Neocomian (Sigal 1974)	Early-middle Callovian and early-middle/late Kimmeridgian (Bate 1975; Mette 1993)		Hauterivian (Rafara 1990)	Callovian-Oxfordian (Kulshreshtha et al. 1985)
Western Australia	Argentina and Southern Chile	Falkland Plateau/Agulhas Bank	Brazil and Gabon	East Antarctica	South Africa
Oligocythereis? spp. (*in* Damotte 1992, sp. 122)					Early Valanginian (McLachlan et al. 1976a); late Valanginian–Hauterivian (Dingle and Klinger 1972; McLachlan et al. 1976a; Brenner and Oertli 1976)
"*Bythocypris*" cf. *stroggylae*					Late Valanginian–Hauterivian (McLachlan et al. 1976a; Brenner and Oertli 1976; Valicenti and Stephens 1984)

Table 3 (continued)

Western Australia	Argentina and Southern Chile	Falkland Plateau/Agulhas Bank	Brazil and Gabon	West Antarctica	South Africa
Bythocypris? cf. *nodosa*					Hauterivian (Brenner and Oertli 1976; Valicenti and Stephens 1984)
Eocytheropteron cf. *corrosum*					Late Valanginian (McLachlan et al. 1976a)
Cytheropteron spp. (*in* Damotte 1992, sp. 30R)	Late Aalenian-early Bajocian (Ballent and Whatley 2009); Hauterivian (Musacchio 1979); late Campanian–early Maastrichtian (Rossi de García and Proserpio 1980)	Early–late Albian (Dingle 1971, 1984); late Maastrichtian (Majoran et al. 1997, 1998)		Maastrichtian (Majoran et al. 1997; Majoran and Widmark 1998)	Late Valanginian–Hauterivian (McLachlan et al. 1976a; Brenner and Oertli 1976; Valicenti and Stephens 1984); Maastrichtian (Dingle 1981)
Western Australia	Mozambique Ridge	Tanzania and Northern Somalia	Saudi Arabia, Egypt, and Israel	Madagascar	India
Oligocythereis? spp. (*in* Damotte 1992, sp. 122)		Early Callovian (Mette 1993)	Bathonian–early Oxfordian (Maync 1965; Dépêche et al. 1987; Rosenfeld et al. 1987a; Rosenfeld and Honigstein 1991)	Bathonian–middle Callovian and Portlandian–Valanginian (Grékoff 1963; Mette 2004); Hauterivian (Rafara 1990)	Callovian–Oxfordian (Kulshreshtha et al. 1985; Khosla et al. 2006)
"*Bythocypris*" cf. *stroggylae*				Hauterivian (Rafara 1990)	
Bythocypris? cf. *nodosa*					
Eocytheropteron cf. *corrosum*				Portlandian (Grékoff 1963)	Callovian–Oxfordian (Guha 1976)
Cytheropteron spp. (*in* Damotte 1992, sp. 30R)		Late Oxfordian and middle/late Kimmeridgian (Bate 1975); Albian (Bate and Bayliss 1969)	Aptian–Albian (Rosenfeld and Raab 1984)	Late Bathonian–middle Callovian and late Oxfordian–Portlandian (Grékoff 1963; Mette 2004)	Bajocian–Oxfordian (Kulshreshtha et al. 1985; Neale and Singh 1986; Khosla et al. 2003, 2004, 2005, 2006); middle Coniacian (Banerji 1970); late Campanian–Maastrichtian (Govindan 1969; Jain 1975; Sugumaran et al. 1997)

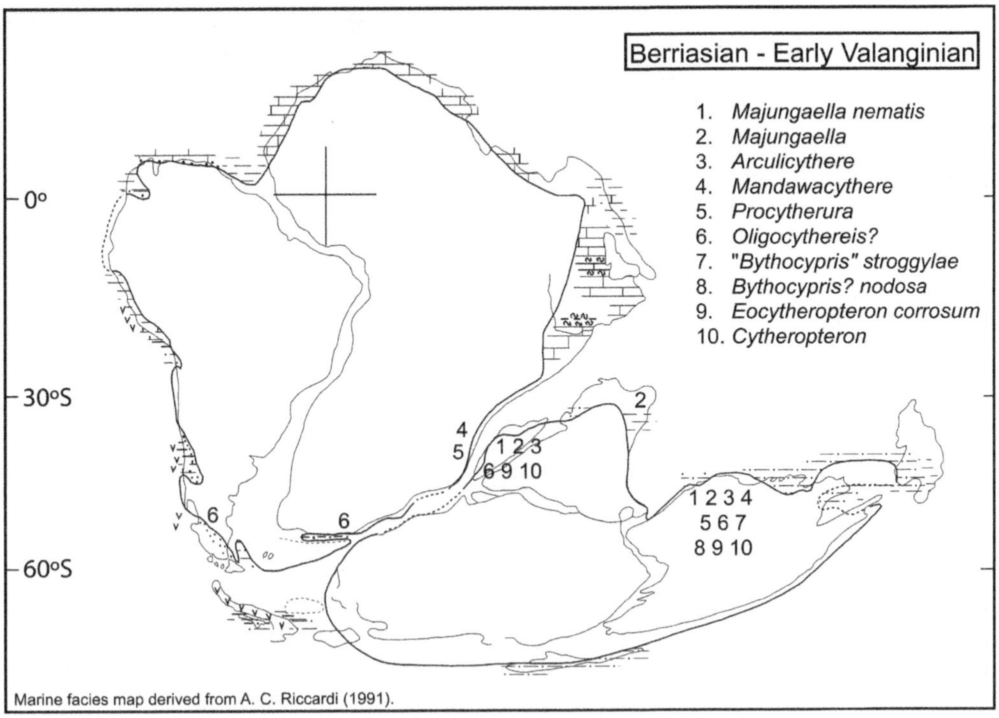

Fig. 3 Distribution of the Berriasian–early Valanginian ostracod genera and species listed as occurring in Western Australia in Table 3. It illustrates the distinctive Indian Ocean fauna characterized by the Gondwanan genera *Majungaella* and *Arculicythere*. See Fig. 1 for legend

claystones on the Mozambique Ridge), "*Bythocypris*" *stroggylae* Brenner and Oertli, 1976, *B? nodosa* Brenner and Oertli, 1976, *Eocytheropteron corrosum* (Grékoff 1963), and the genera *Majungaella, Procytherura* (also occurs on the Mozambique Ridge), *Oligocythereis?* and *Cytheropteron*. During the Valanginian–Hauterivian, *M. nematis* Grékoff, 1963, also appears in southern Argentina and southern Chile, whereas *Majungaella, Procytherura* (earliest appearance in Gondwana was in the late Pliensbachian of west-central Argentina), and *Cytheropteron* (also appeared in southern Argentina). Consequently, interchange between the E Tethys and the SE Pacific, via the W Indian Ocean, had occurred once the physical barriers between the two regions were breached in the late Valanginian.

Early Cretaceous – Late Valanginian to Earliest Aptian

Ogg et al. (2004) note that the base of the late Valanginian is associated with the onset of a major marine transgression; this breached the physical barriers between the Indian Ocean and the SE Pacific; it can be identified by migration of several ostracod genera and species between Madagascar, South Africa, and southern South America. This W Indian Ocean migratory pathway (Fig. 4) is based on Table 4 listing late Valanginian–earliest Aptian genera and species occurring in Western Australia during that interval as well as those previously present during the Berriasian–early Valanginian and which subsequently migrated between the Indian Ocean and the SE Pacific.

Majungaella, and in particular the species *M. nematis* Grékoff, 1963, demonstrates that ostracods migrated between the Indian Ocean and the SE Pacific because of its occurrence upon the Mozambique Ridge, and in South Africa and southern South America, as of the late Valanginian. *Eocytheropteron corrosum* (Grékoff, 1963), "*Bythocypris*" *stroggylae* Brenner and Oertli, 1976 and *B.*? *nodosa* Brenner and Oertli, 1976 also used this west Indian Ocean migratory pathway to reach South Africa.

By the late Valanginian–early Hauterivian, the genus *Procytherura* was confined mainly to the Southern Hemisphere (Ballent and Whatley 2000). *P. aerodynamica* Bate, 1975, originally described from

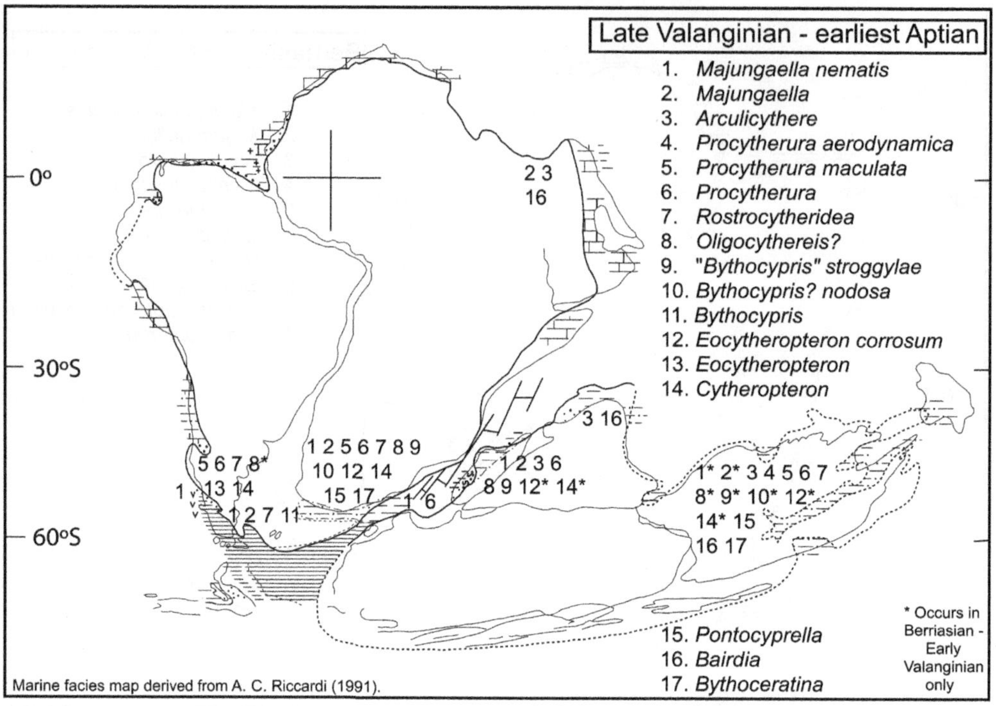

Fig. 4 Distribution of the late Valanginian–earliest Aptian ostracod genera and species listed in Table 4. It illustrates that the distinctive Indian Ocean fauna, characterized by the Gondwanan genera *Majungaella* and *Arculicythere*, has strong links with South Africa and southern South America due to breaching of the physical barriers that lay between the Indian Ocean and the southeast Pacific. By the Barremian, a new Austral Province was beginning to develop, characterized by the *Rostrocytheridea*. See Fig. 1 for legend

Table 4 Late Valanginian–Barremian Gondwanan distribution of Western Australian Neocomian and earliest Aptian ostracod genera and species

Western Australia	Argentina and Chile	South Africa and Mozambique Ridge	Madagascar	Israel
Majungaella nematis	*	*	*	
Majungaella	*	*	*	*
Arculicythere			*	*
Procytherura aerodynamica				
Procytherura maculata	*	*		
Procytherura	*	*	*	
Oligocythereis		*	*	
"*Bythocypris*" *stroggylae*		*	*	
Bythocypris? *nodosa*		*		
Bythocypris	*			
Eocytheropteron corrosum		*		
Cytheropteron	*	*		
Pontocyprella				
Bairdia				*
Rostrocytheridea	*	*		
Bythoceratina		*		

the early–middle/late Kimmeridgian of Tanzania (Bate 1975, pl. 11, figs. 7–16 and pl. 12, figs. 1–3), occurs at Site 263, in the SE corner of the Cuvier Abyssal Plain, Western Australia (as Indeterminate sp. E in Oertli 1974, pl. 7, fig. 5). *P. tintinnabulum* Valicenti and Stephens (1984) occurs in South Africa, *P. dinglei* Brenner and Oertli, 1976, in Madagascar (Rafara 1990), on the Mozambique Ridge (Sigal, 1974) and in South Africa (Brenner and Oertli 1976). *P. beerae* Brenner and Oertli (1976) occurs in Madagascar (Rafara 1990) and South Africa (Brenner and Oertli 1976), whereas *P. brenneri* Valicenti and Stephens (1984) occurs in South Africa and southern Argentina (Ballent and Whatley 2009). *P. maculata* Brenner and Oertli, 1976, occurs in South Africa (Brenner and Oertli 1976; McLachlan et al. 1976a, b), southern Argentina (Musacchio 1979, 1981; Ballent and Whatley 2000) and the Southern Carnarvon Platform of Western Australia (along with *P.* sp.). *P. kroemmelbeini* Musacchio, 1979 and *P. amygdala* Ballent and Whatley, 2009 occur in southern Argentina (Musacchio 1978, 1979, 1981; Ballent and Whatley 2000, 2009), whereas *P brenneri* Valicenti and Stephens (1984), known from the late Valanginian–Hauterivian of South Africa and southern Argentina, is considered to be an intermediary between the Indian Ocean *P. aerodynamica* Bate, 1975, and *P. maculata* Brenner and Oertli, 1976, from the late Valanginian–Hauterivian of South Africa and southern Argentina (McLachlan et al. 1976b; Valicenti and Stephens 1984; Ballent and Whatley 2000) and the Barremian of Western Australia. These data accord with migration between the Indian Ocean and the SE Pacific.

The late Valanginian transgression, and the cool late Valanginian–early Hauterivian climate, which McArthur et al. (2007) postulated had a substantial amount of polar ice associated with it, also resulted in the Gondwanan genera *Majungaella* and *Arculicythere* extending their range into the South Tethys Province; *Majungaella* has been recorded from the Valanginian–Hauterivian of Israel and *Arculicythere* from the Valanginian–Barremian of Israel (Rosenfeld and Raab 1984).

Prior to the late Valanginian transgression, there was a southward ridge-jump in the Argo Abyssal Plain, an anticlockwise change in spreading direction, volcanic activity on Joey Rise, and regional uplift terminating deposition of the deltaic Barrow Group (Heine and Müller 2005). The Gascoyne Volcanic Margin (Eldholm and Coffin 2000) formed just prior to the late Valanginian separation of Greater India from Australia/Antarctica (Romine et al. 1997; Longley et al. 2002; Robb et al. 2005) which, in the wake of sea-floor spreading in the Gascoyne and Cuvier Abyssal Plains, extended the E Indian Ocean southwards. Evidence of this extension is seen on the east coast of India where the Neocomian ferruginous sandstones and claystones (Gollapalli Formation) of the Krishna–Godavari Basin were deposited in a paralic to shallow marine environment; this conclusion is based on occurrence of arenaceous foraminifers and dinoflagellates (Prasad and Pundir 1999).

From the northern end of the Carnarvon Terrace (= the eastern margin of the Cuvier Abyssal Plain), Burger (1994) identified a shallow marine Valanginian–Hauterivian interval in three dredge samples. At Site 263, in the SE corner of the Cuvier Abyssal Plain, "rapid rifting associated with emplacement of exceptionally large volumes of magma" resulted in "a small, semi-enclosed oceanic basin (approximately 200 × 400 km), bounded by igneous ridges to the south and west and the Exmouth Plateau to the north" (Holbourn and Kaminski 1997: 11). Holbourn and Kaminski (1995, 1997) identified an agglutinated benthic foraminiferal fauna whose unique composition suggested "strong faunal differentiation or endemism within the Cuvier Basin" (Holbourn and Kaminski 1997: 25). The benthic foraminifers indicate that deposition occurred within a "relatively shallow-water environment, probably under deltaic influence, which gradually deepened as the rifted Indian Ocean margin subsided" (Holbourn and Kaminski 1996: 86). Oertli (1974) identified five ostracod species in the quartz-rich, slightly calcareous, silty claystones at the base of an interval of black organic-rich claystones interpreted as having been deposited in an outer shelf to upper slope environment, with "terrigenous silt-size grains of quartz, feldspar, muscovite and heavy minerals" derived from a nearby continental source (Veevers et al. 1974). These sediments are interpreted by the author to belong to Mutterlose's (1992) *Tegumentum striatum* Zone, which is predominantly late Valanginian in age rather than the late Albian/Aptian or older noted by Oertli (1974). Alternating dark and light laminations in the claystones indicate significant fluctuations in oxygenation and/or total organic carbon (TOC) (Holbourn and Kaminski 1995). The ostracods include *Procytherura aerodynamica* Bate, 1975, *Arculicythere?* sp. B, *Bairdia* sp. A, *Pontocyprella* sp. A, and Indeterminate sp. D. These ostracods have similarities with those from Tanzania, Madagascar, and South Africa; their distribution is presented in Fig. 4.

As mentioned previously, *Procytherura aerodynamica* Bate, 1975 originated in the early to middle/late Kimmeridgian sediments of Tanzania (Bate 1975, pl. 11, figs. 7–16 and pl. 12, figs. 1–3). Ballent and Whatley (2000) noted that the Indeterminate sp. E illustrated by Oertli (1974, pl. 7, fig, 5) is *P. aerodynamica* Bate, 1975. *P.* cf. *aerodynamica* Bate, 1975 was subsequently recorded from the late Aptian of Zululand, South Africa (Dingle 1984); it has been obtained by the author from the middle–late Albian of Scarborough-1 on the Exmouth Plateau.

Arculicythere? sp. B (Oertli 1974, pl. 4, fig. 13) is similar to *A. defluxa* Grékoff, 1963 from the Valanginian–Hauterivian of Madagascar (Grékoff 1963, pl. 7, figs. 182–187; Rafara 1990, pl. 1, figs. 14–16), except that the surface is reticulate and not smooth. Both species have a compressed anterior margin, a strong subcentral tubercle separated from a short horizontal median ridge by a narrow sulcus, a curved ventrolateral ridge overhanging the ventral margin, a weak bulge on the dorsal margin immediately behind a dorso-median groove, and a small eyespot.

Though *Bairdia s.l.* ranges from the Ordovician to Recent (Moore 1961) and has been construed as a remarkable survivor, there are only very rare occurrences in the Early Jurassic open marine sediments of Western Australia (Lord et al. 1993), the Middle Jurassic of Tanzania (Bate 1975), Madagascar (Mette 2004), northwest India (Neale and Singh 1986; Khosla et al. 2003, 2004, 2005, 2006), Saudi Arabia (Dépêche et al. 1987) and Israel (Maync 1965), and in the Late Jurassic of Tanzania and Madagascar (Bate 1975; Mette 2004). There are rare occurrences in the Valanginian–Albian of Israel (Rosenfeld and Raab 1984), the late Valanginian and early Aptian of Site 263 in the SE corner of the Cuvier Abyssal Plain (Oertli 1974), and in the Aptian of the Exmouth Plateau (Damotte 1992), South Africa (Dingle 1984), southern India (Bhatia and Jain 1969; Jain 1976a, b, 1978), and Egypt (Bassiouni 2002).

However, the true renaissance occurs in the Albian, with this genus dominating ostracod faunas on the Falkland Plateau (Dingle 1984), the Exmouth Plateau (Damotte 1992 and the author's study area), and Site 260 on the Gascoyne Abyssal Plain (Oertli 1974), diversifying its form at Site 259 on the Perth Abyssal Plain (Oertli 1974) and in southern India (Bhatia and Jain 1969; Jain 1976a, b, 1978), and reappearing in Madagascar (Collignon et al. 1979) and Tanzania (Bate and Bayliss 1969).

Separation of South America and Africa also occurred in the late Valanginian Stone et al. (2008), and in the wake of sea-floor spreading the southern South Atlantic Ocean was formed. Mutterlose and Wise (1990: 337–338) noted that on the Antarctic margin "dissolved oxygen content of the bottom waters was rising gradually from Valanginian to Aptian times", due to the sills that had been restricting deep oceanic circulation into the basins "becoming increasingly fragmented and ineffective as barriers" as sea-floor spreading continued. In the Valanginian of Site 692B, black shales were deposited on the continental slope of East Antarctica off Dronning Maud Land, due to Madagascar obstructing deep-water circulation from east Tethys into the newly formed anoxic Weddell Basin (Mutterlose and Wise 1990). During the late Valanginian to early Hauterivian, shallow marine sediments accumulated in the Patagonian region of west-central Argentina, but in S Argentina sediments accumulated in pelagic and euxinic environments (Riccardi 1991). There was no marine sedimentation on the Falkland Plateau as this had been uplifted in the Jurassic (Mutterlose and Wise 1990) and, since it was located adjacent to southeast South Africa, it blocked deep-water circulation into the southern South Atlantic (Dingle 1996).

The early Hauterivian is characterized by "a mixture of continental extension, magmatism, and continental fragments" (Robb et al. 2005) on the Exmouth Plateau's western margin, the South American Paraná and African Etendeka continental flood basalts, and the volcanic margins which formed both sides of the southern South Atlantic Ocean (Eldholm and Coffin 2000).

In the late Hauterivian, the climate was warm (McArthur et al. 2007); sea-floor spreading commenced in the Perth Abyssal Plain (Müller et al. 1998) between Antarctica and India (Veevers 2006; Subrahmanyam and Chand 2006), whereas the Bunbury Basalt was erupting in the SW corner of Australia (Frey et al. 2003). However, during the late Hauterivian to early Barremian, no ostracods were recorded in the Mardie Greensand, a low-energy middle- to outer-shelf deposit with little terrigenous influx (Hocking et al. 1988), nor the dark grey, glauconitic, outer-shelf claystones of the Muderong Shale Equivalent on the Exmouth Plateau. Based on the occurrence of "sparse, low diversity benthic foraminiferal assemblages" on the Exmouth Plateau, Holbourn and Kaminski (1997: 28) suggested an environment characterized by restricted circulation and low oxygenation. After a minor ridge jump to the west in the Cuvier Abyssal Plain, which Romine et al. (1997) records at the Hauterivian-Barremian boundary, and "which occurred at the same time as plate uncoupling along the Cape Range transform" (Boyd et al. 1992, page 57), ventilation improved. This resulted in diversification of the benthic foraminiferal assemblages in the overlying Barremian shelfal clays (Holbourn and Kaminski 1997). Taylor and Haig (2001: 136) note that the benthic foraminiferal assemblage on the Exmouth Plateau "is characteristic of the middle to outer neritic *Marsonella* Association" (Haig 1979; Haig and Lynch 1993). The Barremian to lowest Aptian claystones of the Muderong Shale Equivalent on the Exmouth Plateau are more calcareous and less pyritic than the dark, organic-rich, pyritic siltstones and claystones of the Exmouth-Barrow margin and Southern Carnarvon Platform. The former represents a moderately reducing, sea-floor environment characterized by low sedimentation rates (Exon et al. 1992); the latter is characterized by a moderately diverse *Ammobaculites* Association of siliceous organic-cemented agglutinated foraminifers which "inhabited dysaerobic seafloor muds in the upper offshore zone (10–50 m water depth) in a sea with a very gentle sea-floor gradient deepening slightly toward the north" (Taylor and Haig 2001, p. 140). The only ostracods recorded, but not identified, in the Barremian to lowest Aptian of the Muderong Shale and its equivalents occur in Barrabiddy-1 on the Southern Carnarvon Platform (Taylor and Haig 2001).

Barry Taylor provided the author with SEM images of 10 ostracod specimens he had collected from Barrabiddy-1. Some of the specimens are pyritized, but the author was able to identify *Procytherura maculata* Brenner and Oertli, 1976 (3 specimens), *P.* sp. (2 specimens), *Arculicythere* sp., *Rostrocytheridea* sp., and *Bythoceratina* sp. Two of the ostracods had damaged posterior margins and could not be adequately identified, but it is interesting to note that they are reminiscent of a reticulate, neocytherid-shaped, indeterminate specimen obtained by Dingle 1984: sp. 327/16B, p. 182) from the middle Albian of the Falkland Plateau.

As mentioned previously, *P. maculata* Brenner and Oertli, 1976, occurs in the Hauterivian of South Africa and southern Argentina (McLachlan et al. 1976a; Valicenti and Stephens 1984; Ballent and Whatley 2000), prior to its first-known occurrence in Western Australia in the Barremian.

The *Rostrocytheridea* sp. obtained from the Barremian of Western Australia is a closed carapace, so its internal features are not visible; it is very similar to *R.* sp. found by McLachlan et al. (1976a) in the late Valanginian of South Africa; it has a well-arched dorsal margin and a posterior margin which is not inclined downwards. In terms of Australia, the previous earliest-known occurrence of this genus was reported by Krömmelbein (1975b) in the Eromanga Basin middle Albian to early Cenomanian.

R. opisthorhynchus Ballent and Whatley, 2007 from the late Tithonian-Berriasian of the Patagonian region of Argentina is the oldest known occurrence of the genus; this particular species can be distinguished from other species of *Rostrocytheridea* based on its "rounded subtriangular outline in lateral view and the posterior "beak" mainly pointing backwards" (Ballent and Whatley 2007: 117). It is possible that this may be the point of origin for this genus. During the Valanginian–Hauterivian, *Rostrocytheridea* is restricted to the Patagonian region of Argentina (Musacchio 1978, 1979, 1981; Kielbowicz et al. 1983; Ballent and Whatley 2007) and South Africa (Dingle 1969; Brenner and Oertli 1976; McLachlan et al. 1976a; Valicenti and Stephens 1984; Dingle 1996); it is an early indicator of the inception of the Austral Province. It also indicates that migration occurred between the Indian Ocean and the southeast Pacific.

In Australia, *Rostrocytheridea* also occurs in the Santonian (Late Cretaceous) of the Carnarvon Basin (Bate 1972) and the Perth Basin (Neale 1975). It also occurs in the Campanian (Late Cretaceous) of the Antarctic Peninsula (Fauth et al. 2003; Ballent and Whatley 2007) and the Austral Basin of southern Argentina (Ballent and Whatley 2007). The youngest occurrence of *Rostrocytheridea* is a few metres below the Maastrichtian/Tertiary boundary of New Zealand (Dingle, in press).

The caudal process of the *Bythoceratina* sp. from the Barremian of Western Australia is broken off, but other characteristic features of this genus, such as its subtrapezoidal shape with a straight hinge-line, strong inflations separated by a dorso-median sulcus, and a strong ventrolateral spine in the posterior half, are obvious. *Bythoceratina* is also reported from the late Valanginian of South Africa (Valicenti and Stephens 1984) and the Campanian (Late Cretaceous) of the Carnarvon Basin of Australia (Bate 1972).

Based on dinoflagellates (Prasad and Pundir 1999), deposition of the shallow marine, inner to middle shelf Raghavapurum Shale in the Krishna–Godavari Basin of the east coast of India, began in the Barremian and ended in the early Aptian to early Albian. This, the first major marine transgression into that basin, was connected with sea-floor spreading between India and Antarctica, and the easterly tilt of the east coast of India (Prasad and Pundir 1999). There has been no record of ostracods from the Raghavapurum Shale.

In the late Hauterivian, marine connection between E Tethys, Antarctica, and S Argentina became less restricted; by the Barremian "an exclusively austral ammonoid fauna" had developed in the southern Argentinian, South African, and Madagascan regions (Riccardi 1991: 169). However, during the Barremian–early Aptian, southern Argentina experienced restricted marine conditions (Riccardi 1991). Black shales accumulated in the tectonically isolated and restricted Weddell and Mozambique Basins (Mutterlose and Wise 1990) – these were separated by the Mozambique Ridge. On the rapidly subsiding Falkland Plateau, Barremian to ?earliest Aptian, outer shelf to upper bathyal, anoxic, organic-rich claystones (Jeletzky 1983) were deposited prior to deposition of early Aptian regressive, fine-grained, calcareous sandstones derived from a well-aerated, inner neritic to littoral environment in a restricted marine environment with fluctuating water salinity, "presumably under the influence of a delta" (Jeletzky 1983). This transition from organic-rich claystones to regressive mudstones and calcareous chalks/claystones was probably due to the early Aptian marine regression associated with the final separation of Greater India from Antarctica. Early Aptian, organic-rich shales were also deposited in the Cape Basin off South Africa and on the Mozambique Ridge Mutterlose and Wise (1990). There was a narrow and shallow strait between the Falkland Plateau and the southern tip of South Africa – which received marine waters from the south (Riccardi 1991). The extensive salt deposits of the northern South Atlantic at that time is attributed to shallow marine waters intermittently washing over the Walvis Ridge–Rio Grande Rise (Riccardi 1991). There are no records of ostracods from any of these Barremian–early Aptian deposits.

However, the early Aptian black claystone in Sample 17-CC from Site 263, in the SE corner of the Cuvier Abyssal Plain, failed to produce even a single specimen of *Bairdia* (Oertli 1974). By the early Aptian, similar dark claystones were being deposited on the eastern edge of the Perth Abyssal Plain, in the SE Wharton Basin to the W, and on the Naturaliste Plateau, at the SW corner of Australia. Curiously, even in the more calcareous, green-grey sediments connected with the early Aptian marine regression, associated with a major ridge jump (Romine et al. 1997), the final separation of Greater India from Antarctica (Longley et al. 2002), and the beginning of the break-up of Australia from Antarctica (Frey et al. 2003), no ostracods have been found.

Early Cretaceous – Aptian

Frakes et al. (1987) noted that an Aptian transgression resulted in extensive marine flooding of the Australian continent during the Aptian–Albian (Fig. 5). This transgression was due to establishment of open oceanic conditions connected with the separation of Greater India from Antarctica (Longley et al. 2002). Formation of this circum-India seaway produced a new system of more oxygenated marine currents.

Upwelling of deep oceanic water enriched in silica resulted in deposition of the Windalia Radiolarite in the Carnarvon Basin of Western Australia (Ellis 1987), a "light to dark-grey radiolarite and claystone with varying amounts of limestone, greensand and chert", rare foraminifers, ammonites, belemnites, bivalves, ostracods, fish teeth, dinoflagellates, sponge spicules and trace fossils (Haig 1991: 34). Although Playford et al. (1975) recorded ostracods from the Windalia Radiolarite, Ellis (1987) did not report any in outcrops (Haig 1991). Lack of calcareous taxa and reduced numbers of benthic fossils is attributed to dysaerobic–anaerobic bottom waters (Haig 1991) in water depths no greater than upper bathyal (Ellis 1987).

Calcareous nannofossils in samples 38X-1, 19–20 cm to 41X-1, 30–31 cm in ODP 763B on the Exmouth Plateau are interpreted by the author as belonging to the latest early Aptian to late Aptian

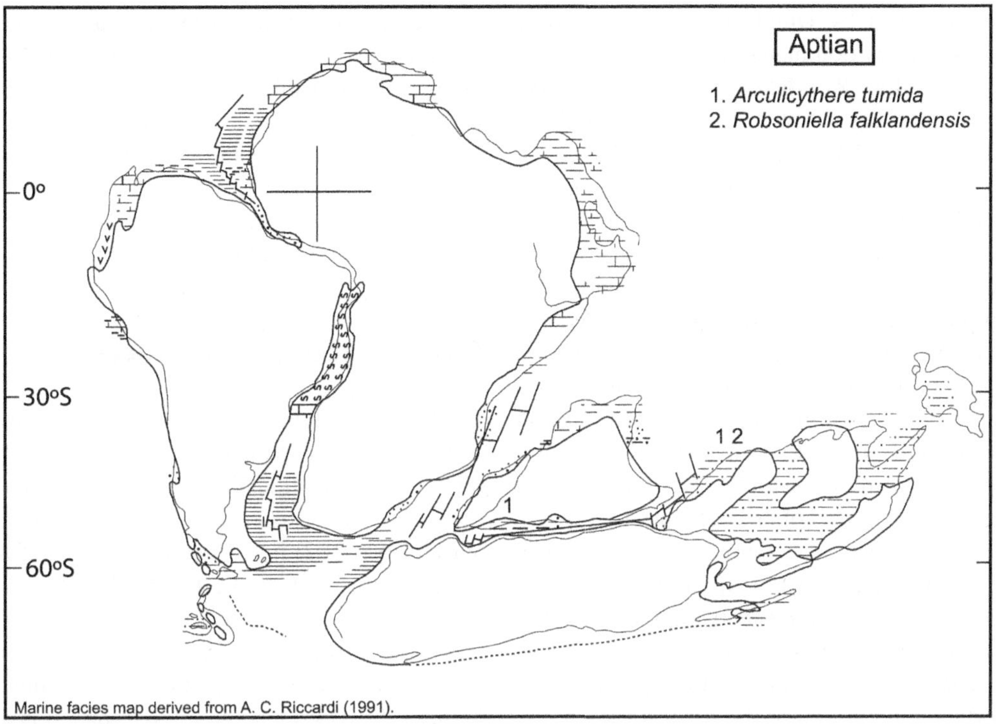

Fig. 5 Distribution of the Aptian ostracod species of Western Australia reflecting the extensive marine flooding of Australia and formation of the circum-India seaway that occurred in the Aptian. See Fig. 1 for legend

CC7b Subzone, and consequently equivalent of the Windalia Radiolarite. This interval consists of "very fine-grained hard carbonate layers that are interbedded with claystones and calcareous claystones" (Haq et al. 1990: 302). These greenish-grey, hemipelagic marls were deposited in an open marine, bathyal environment (Haq et al. 1990). The lower part of this interval is characterized by dark greenish-grey claystones, and "parallel laminations highlighted by dark detrital material" indicating that bioturbation was not extensive (Haq et al. 1990: 302). The calcareous content gradually increases upwards concurrent with glauconite content decreasing (Haq et al. 1990) as pelagic sedimentation overwhelmed the contribution of continental clays. At the top of this interval, in samples 38X-1, 38–41 cm and 39X-1, 88–90 cm, Damotte (1992) recorded the occurrence of the ostracods *Arculicythere tumida* Dingle, 1971 (Damotte 1992, pl. 3, figs. 1, 2) and *Robsoniella falklandensis* Dingle, 1984 (Damotte 1992, pl. 1, fig. 8), as well as the cosmopolitan *Cytherella*.

In Western Australia, this is the earliest-known record of *Arculicythere tumida* Dingle, 1971, a species whose earliest record is in the Hauterivian of Madagascar (Rafara 1990, pl. 1, figs. 17–18; pl. 2, figs. 1–2). The species identified by Jain (1976a, fig. 1H–J; 1978), in the late Aptian–Albian of the Cauvery Basin of southern India as *Pseudobythocythere kallakkudiensis* Jain, 1976, is here assigned to *Arculicythere tumida* Dingle, 1971. Consequently, following the early Aptian break-up of Greater India from Antarctica, which resulted in formation of the circum-India seaway and, ultimately, terminated restrictive marine conditions, *A. tumida* Dingle, 1971 migrated into the new seaway between India, Antarctica, and Western Australia.

Robsoniella falklandensis Dingle, 1984, a bairdioid, was originally described by Dingle (1984, fig. 8C–F) from early to late Albian nannofossil chalks and claystones of the Falkland Plateau; he also identified Indeterminate sp. A of Oertli (1974 pl. 6, figs. 1–11) from Site 260 on the Gascoyne Abyssal Plain of Western Australia, as being synonymous with it. Oertli (1974) attributed these ostracods to the middle–late Albian, but this interval of nannofossil chalks and claystones is reinterpreted by the author as early Albian in age. It was noted by Dingle (1984) that at both localities this species dominates the ostracod faunas. Because the bairdiid *R. falklandensis* Dingle, 1984 occurs in the latest early Aptian to late Aptian CC7b Subzone of ODP 763B on the Exmouth Plateau, it seems that it may have originated in Western Australia and migrated to the Falkland Plateau at the beginning of the Albian. Riccardi (1991) noted that during the Aptian there was a decrease in ammonoid similarity between southern Argentina and South Africa–Madagascar; this may be an evolutionary reflection of differing marine currents. Those of South Africa and Madagascar were influenced by a warm current flowing southwards from E Tethys, whereas southern Argentina, Antarctica, and the Falkland Plateau were under the influence of a cool current flowing around Australia, Papua New Guinea, and Antarctica. Thus the migratory pathways of *Arculicythere tumida* Dingle, 1971 and *R. falklandensis* parallel a cool West Wind Drift current (cf. Fig. 6) encircling Antarctica, Australia, and Papua New Guinea; the two pathways were important in formation of the Austral Province.

The late Aptian to earliest Albian Southern Ocean *Rhagodiscus angustus* Zone of Wise (1988), correlating with Subzone CC7b, has also been intersected in dark organic-rich claystones (Mutterlose and Wise 1990) on the continental slope of East Antarctica off Dronning Maud Land, on the Falkland Plateau and in the Cape Basin, off South Africa. There is no record of ostracods from any of these deposits.

Early Cretaceous – Early Albian

By the early Albian, Western Australia shared several ostracod species with the Austral Basin of southern Argentina, the Falkland Plateau, and the Agulhas Bank off South Africa (listed in Table 5; illustrated in Fig. 6). Wise (1988: 162) noted that at the close of the Aptian "chalk deposition on the Falkland Plateau began in a series of pulses" connected with "deep water ventilation of the South Atlantic when the Falkland Plateau moved past the tip of South Africa". Mutterlose and Wise (1990: 336) also noted that this "transition from

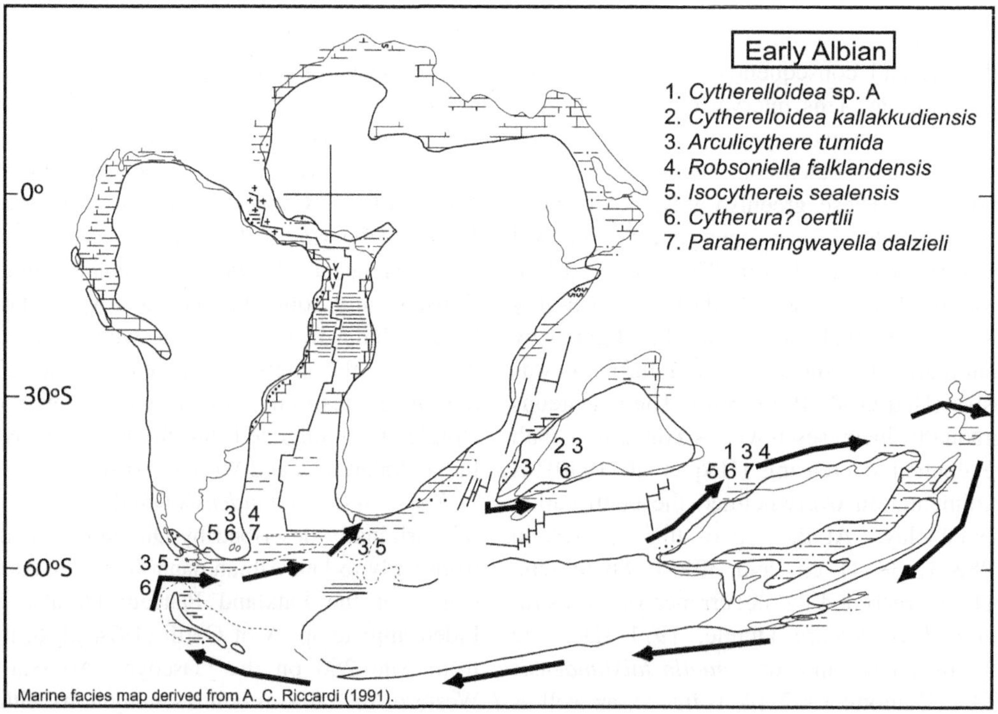

Fig. 6 Distribution of early Albian ostracod species of Western Australia listed in Table 5. *Arrows* illustrate the cool West Wind Drift current which resulted in formation of an Austral Province characterized by the distribution of *Arculicythere tumida* Dingle, 1971. See Fig. 1 for legend

Table 5 The Austral distribution of Western Australian early Albian ostracod species

Western Australia	Falkland Plateau	Agulhas Bank	Southern India	Madagascar	Southern Argentina
Cytherelloidea sp. A			*Cytherelloidea kallakkudiensis*		
Arculicythere tumida	*	*	*	*	*
Robsoniella falklandensis					
Isocythereis sealensis	*	*	Relation?		*
Cytherura? *oertlii*	*		*		*
Parahemingwayella dalzieli	*				

predominantly anoxic to truly oxic depositional conditions" also occurred at the top of this zone in the Cape Basin and on the continental slope of East Antarctica where the change at Site 693 to radiolarian diatomite signifies upwelling and "development of a reasonably strong oceanic circulation system within the basin". In addition, Gheerbrant and Rage (2006) indicate that the final separation between Africa and South America occurred in the Aptian/Albian. Based on ammonites, Owen and Mutterlose (2006: 725) noted that "some epicontinental sea connection probably occurred" between the North and South Atlantic during the Albian.

It is interesting to note that Western Australia shared several species with southern India (listed Table 5; illustrated Fig. 6); this is viewed as being connected with formation of the circum-India seaway. The similarities and synonymies identified by the author indicate that the late Aptian–Albian material from the Cauvery Basin of southern India (Jain 1976a, b, 1978) might be usefully re-studied.

Hart et al. (2001: 221) noted that the planktonic foraminifers of the earliest Albian to late Turonian Karai Formation of the Cauvery Basin were "morphologically quite distinctive" with many having "more pronounced keels and other ornamentation" than normal – just like those from the Northwest shelf of Australia and the Exmouth Plateau. This is indicative of a "distinctive palaeo-biogeographical province . . . restricted to the east coast basins of India and the Exmouth Plateau/N.W. Australia Shelf".

The author identified the early Albian CC8a Subzone in the nannofossil claystones and clayey chalks of the Haycock Marl on the Exmouth Plateau in ODP 762C and ODP 763B, and in similar sediments at Site 260 on the Gascoyne Abyssal Plain and Site 259 on the Perth Abyssal Plain. In this subzone, ostracods were identified by Damotte (1992) from the Exmouth Plateau and by Oertli (1974) from the Gascoyne and Perth Abyssal Plains. These included the cosmopolitan genus *Cytherella*, *Arculicythere tumida* Dingle, 1971, *Robsoniella falklandensis* Dingle, 1984, numerous forms of *Bairdia* as well as *Cytherelloidea* sp. A (which supports the onset of gradual warming of surface waters – indicated by Huber et al. (1995) as Albian to Cenomanian), *Isocythereis sealensis* Dingle, 1971 (identified here as synonymous with the *Cythereis*? aff. *brevicosta* Bate, 1972 of Oertli 1974), *Cytherura*? *oertlii* Dingle, 1984, and an indeterminate species from the Gascoyne Abyssal Plain identified by the author as *Parahemingwayella dalzieli* Dingle, 1984.

P. dalzieli Dingle, 1984, when originally described by Dingle (1984, fig. 14D–F, fig. 15A), was reported in the early to middle Albian nannofossil chalks and claystones of the Falkland Plateau. The Indeterminate sp. C of Oertli (1974, pl. 2, fig. 7) from the early Albian of Site 260 on the Gascoyne Abyssal Plain is identified here as *Parahemingwayella dalzieli* Dingle, 1984. Though the ornament is badly worn, it is still possible to see the subcentral process, two prominent converging longitudinal ribs, and even one of the coarse spines of the anterior margin – diagnostic of this species.

By the early Albian, *Arculicythere tumida* Dingle, 1971 expanded its territory and had a distribution paralleling the cool West Wind Drift current with occurrences in Madagascar (Collignon et al. 1979, pl. 4, fig. 12; with *Arculicythere* DS 1 considered here to be synonymous with *A. tumida* Dingle, 1971), southern India (Jain 1976a, fig. 1H–J; 1978), Western Australia (Oertli 1974, pl. 4, figs. 1–11, pl. 5, figs. 1–12; Damotte 1992, pl. 3, figs. 1, 2), the Agulhas Bank off South Africa (Dingle 1971, fig. 5), the Falkland Plateau (Dingle 1984, fig. 22A–F, fig. 23A–D), and the Austral Basin of southern Argentina (Ballent 1998, pl.

1, figs. 9–12). It was very abundant and its distribution is characteristic of the Austral Province.

Cytherura? *oertlii* Dingle, 1984, originally described by Dingle (1984, fig. 9E, F; fig. 10) from the middle to late Albian nannofossil chalks and claystones of the Falkland Plateau, and Indeterminate sp. B of Oertli (1974, pl. 7, fig. 1), from Site 260 on the Gascoyne Abyssal Plain of Western Australia, are regarded synonymous. This species has also been identified in the Albian of the Austral Basin of southern Argentina (Ballent 1998, pl. 1, fig. 7); the species identified by Jain (1978, fig. 1), in the late Albian of the Cauvery Basin of southern India as *Acrocythere* cf. *striata* Kaye, 1965 is here assigned to *Cytherura*? *oertlii* Dingle, 1984. *C*? *oertlii* Dingle, 1984 has a similar distribution to *Arculicythere tumida* Dingle, 1971, except that it does not occur in Madagascar.

Isocythereis sealensis, originally described by Dingle (1971, fig. 14) from the middle to late Albian on the Agulhas Bank off South Africa, was subsequently noted as occurring in the early to late Albian of the Falkland Plateau (Dingle 1984, fig. 27A–F, fig. 28A–D). The *Cythereis*? aff. *brevicosta* Bate, 1972 illustrated by Oertli (1974, pl. 4, fig. 12) from the early Albian of Site 259 on the Perth Abyssal Plain, is identified here as *Isocythereis sealensis* Dingle, 1971. Although the ornament is badly worn, it is still possible to see the stubs of the anterior and posterior marginal spines, the three longitudinal ribs, the stub of the posterior spine of the dorsal rib, the subdued subcentral tubercle of the right valve, and the prominent, narrow anterior marginal rim diagnostic of this species. This species was blind; there is considerable variability in its ornament. Dingle (1984) noted that in the Agulhas Bank specimens there is delicate reticulate ornament, whereas in the more calcified specimens of the Falkland Plateau the ornament tends to be coarser. Given this degree of variability, Ballent (1998, pl. 3, figs. 3–6) assigned only specimens from the Austral Basin of southern Argentina to *Isocythereis* sp., while still noting their strong similarity to *I. sealensis* Dingle, 1971, especially those from the Falkland Plateau. The species identified by Jain (1976a, fig. 2G), in the late Aptian–Albian of the Cauvery Basin of southern India, as "*Cythereis*" sp. B is possibly related to *I. sealensis* Dingle, 1971. The ostracod illustrated by Jain (1976a) in his fig. 2G is very similar to that illustrated by Dingle (1984, fig. 27D) except that it is much more elongate, and hence probably a male, has a small eyespot and has more strongly developed marginal denticles. Consequently, *Isocythereis sealensis* Dingle, 1971 has a similar distribution to *Arculicythere tumida* Dingle, 1971, except that it does not occur in Madagascar.

Cytherelloidea sp. A (Oertli 1974, pl. 1, figs. 3–4) has a shallow subcentral depression and ornament consisting of longitudinal striae becoming more reticulate in the anterior, dorsal, and posterior regions. Although, there are no SEMs of *Cytherella kallakkudiensis* Jain, 1976 from the mostly late Albian sediments of the Cauvery Basin of southern India (Jain 1976b, fig. 3A–B), that species is also characterized by a shallow subcentral depression and ornament of longitudinal striae.

Based on calcareous nannofossils, the early Albian Subzone CC8a has been identified at the base of the Gearle Siltstone (Haig et al. 1996) in the Giralia Anticline on the Southern Carnarvon Platform of Western Australia. Abundant calcareous fossils have been found only at the base of the Gearle Siltstone; this interval has produced nannofossils as well as belemnites and predominantly benthic foraminifers. Poorly preserved, presumably post-Albian calcareous nannofossils were reported by Shafik (1990: 18) from a "soft dark siltstone" at the top of the Gearle Siltstone.

The calcareous fossils at the base of the Gearle Siltstone may represent "a marine transgressive pulse that reactivated bottom-water circulation and facilitated a significant increase in the calcareous biogenic component of the sediment" (Haig et al. 1996: 41). The subsequent disappearance of the basal calcareous fossils as well as the siliceous agglutinated benthonic foraminifers "may reflect stagnant, anaerobic conditions on the sea floor" during deposition of the remainder of the Gearle Siltstone, resulting in "complete dissolution of calcareous planktonic and benthonic biogenic material" (Haig et al. 1996, page 65). Radiolarians, with their siliceous test, although poorly preserved, remained diverse and abundant; this implies that there were nutrient-rich, well-oxygenated surface waters, probably due to extensive upwelling along the continental margin (Haig et al. 1996). Radiolarian blooms generated substantial quantities of organic matter, decomposition of which resulted in anoxic conditions on the underlying sea-floor.

Early Cretaceous–Middle to Late Albian

By middle to late Albian, as India drifted northwards away from the Austral Province, the Western Australian ostracod fauna developed particularly strong links with the Falkland Plateau, whereas *Arculicythere tumida* Dingle, 1971, maintained its presence on the Agulhas Bank (Dingle 1971, 1984) and in the Austral Basin of southern Argentina (Ballent 1998). This is illustrated in Fig. 7 – based on co-occurrences of species between Australia, South Africa, the Falkland Plateau and Agulhas Bank, and southern Argentina (listed in Table 6).

Of the 20 ostracod species obtained by the author (from Zeewulf-1, Scarborough-1 and Eendracht-1) and by Damotte in 1992 (from ODP 762C and ODP 763B) from the middle to late Albian of the Exmouth Plateau, Western Australia, 7 have also been obtained from the Falkland Plateau by Dingle (1984): these (Table 6) are *Cytherella bensoni* Dingle, 1984, *Robsoniella falklandensis* Dingle, 1984, *Bairdia* sp. A, *Arculicythere tumida* Dingle, 1971, *A? dubia* Dingle, 1984, *Aitkenicythere? striosulcata* Dingle, 1984, and *Collisarboris? stanleyensis* Dingle, 1984. Two other species known from Western Australia were recorded in Zululand, South Africa, by Dingle (1984): *Pirileberis* aff. *mkuzensis* Dingle, 1984, reported by Damotte (1992) in ODP 763B, and *Procytherura* cf. *aerodynamica* Bate, 1975, obtained by the author from Scarborough-1.

Conclusions

The Jurassic and Early Cretaceous ostracods from Western Australia reflect palaeobiogeographic changes associated with fragmentation of Gondwana and formation of the Indian Ocean. Similarities in the Early Jurassic ostracod faunas of northwestern Europe – at the western end of Tethys – and the northwest coast of Australia at the eastern end of southern Tethys indicate there was little variation in depositional conditions along the extensive northern Gondwana marine shelf.

During the Middle Jurassic, Gondwana began to rift into East Gondwana (Australia, Antarctica, Greater India, Madagascar, and the Seychelles)

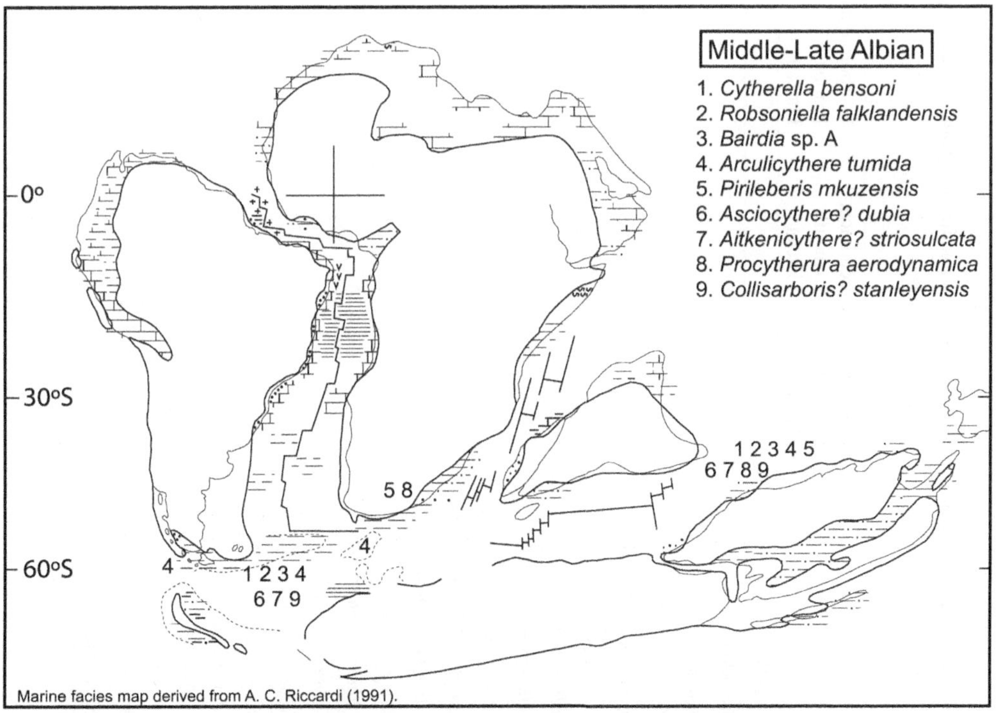

Fig. 7 Distribution of middle–late Albian ostracod species occurring in more than one locality of the Austral Province, as listed in Table 6. It illustrates that the fauna was still Austral in character – based on occurrence of *Arculicythere tumida* Dingle, 1971 – with strong faunal similarities between the Exmouth and Falkland Plateaus. See Fig. 1 for legend

Table 6 Austral distribution of Western Australian middle–late Albian ostracod species

	Western Australia					South Atlantic		South Africa	Southern Argentina
	Zeewulf-1 Exmouth Plateau	Eendracht-1 Exmouth Plateau	Scarborough-1 Exmouth Plateau	ODP 762 Exmouth Plateau	ODP 763 Exmouth Plateau	Falkland Plateau	Agulhas Bank	Zululand	Austral Basin
Cytherella bensoni		*		*		*			
Cytherella sp. E26		*							
Cytherella sp.					*				
Cytherelloidea sp. S79			*						
Robsoniella falklandensis	*	*	*	*	*	*			
Bairdia sp. A	*	*	*			*			
Bairdia sp. D		*	*		*				
Bairdia sp. E		*							
Bairdia sp. nov.		*	*						
Bythocypris?/Pontocyprella?			*						*
Arculicythere tumida		*	*	*		*	*		*
Arculicythere sp. 17R	*								
Eocytheropteron sp. nov.		*							
Pirileberis aff. *mkuzensis*					*			*	
Pirileberis sp.nov.		*	*						
Asciocythere? dubia		*				*			
Aitkenicythere? striosulcata		*	*			*			
Procytherura cf. *aerodynamica*			*					*	
Collisarboris? stanleyensis					*	*			
Indeterminate sp. S31			*						

and West Gondwana (Africa and South America), with the west Indian Ocean forming between the two fragments. The Western Australian Middle Jurassic ostracod fauna consisted of genera distributed along the extensive marine shelf of northern Gondwana (e.g. *Eucytherura* and *Fastigatocythere*), with a Southern Hemisphere element characterized by *Paradoxorhyncha*, *Gondwanacythere*, *Mandawacythere*, and "Indeterminate Genus" *australis*, plus endemic genera (e.g. *Bringocythere* and *Fistulosacythere*). The ostracod fauna of Western Australia had particularly strong links with northwest India, Madagascar, and Tanzania (an East Tethys Province) and weaker links with Saudi Arabia (part of a South Tethys Province). Another strong link is identified with west-central Argentina, based on the genus *Paradoxorhyncha* – which demonstrates that migration had also occurred along the marine shelf of southern Gondwana.

During the Late Jurassic, when the West Burma Block drifted away from the northwest margin of Australia and formed the eastern Indian Ocean, a distinctive Indian Ocean ostracod fauna developed characterized by the Gondwanan genera *Majungaella* and *Arculicythere*. This fauna had strong links with the East Tethys Province, including northern Somalia, and a weak link with Saudi Arabia and Israel (part of the South Tethys Province).

Sedimentation was regressive during the Berriasian to early Valanginian; the ostracods of Madagascar and Western Australia still belonged to a distinctive Indian Ocean ostracod fauna characterized by the Gondwanan genera *Majungaella* and *Arculicythere*.

During the late Valanginian, when Greater India drifted away from Australia–Antarctica and extended the eastern Indian Ocean southwards, there was a major marine transgression. As a consequence, interchange between the Indian Ocean and the southeast Pacific became possible during the late Valanginian to Hauterivian due to breaching of physical barriers between the two regions. The Indian Ocean ostracods of Western Australia thus came to display links with Madagascar, the Mozambique Ridge, South Africa, and southern South America. By the Barremian, the occurrence of *Rostrocytheridea* in Western Australia, South Africa, and southern Argentina indicated that a new province, the Austral Province, had been initiated.

The early Aptian break-up of Greater India from Antarctica resulted in formation of a circum-India seaway; it ultimately brought an end to the restricted environments between southern Argentina, South Africa, and east Antarctica, and along the southwest coast of Australia. By late Aptian, *Arculicythere tumida* Dingle, 1971 had migrated into the new seaway between India, Antarctica, and Western Australia; associated with it in Western Australia is the earliest-known occurrence of *Robsoniella falklandensis* Dingle, 1984. In the early Albian, these two species migrated to the Falkland Plateau via the marine shelf of southern Gondwana. Both migratory pathways paralleled a cool West Wind Drift current encircling Antarctica, Australia, and Papua New Guinea that resulted in formation of an Austral Province. By Albian times, the Exmouth Plateau had also developed links with the Agulhas Bank off South Africa, and with southern Argentina.

Acknowledgements The author is grateful to Mark Warne and Sheri McGrath and to the three reviewers, Sara C. Ballent, Richard V. Dingle, and Alan R. Lord, for their helpful suggestions to improve the manuscript. Barry Taylor, University of Western Australia, allowed identification from SEMs of ostracods from the Muderong Shale (Barremian). Financial support came from a Deakin University Postgraduate Research Scholarship. The School of Life and Environmental Sciences Conference Support (Deakin University) funded attendance and initial presentation of these results at the Second International Palaeontological Congress (Beijing) in June 2006.

References

Alley NF, Frakes LA (2003) First known Cretaceous glaciation: livingstone tillite member of the cadna-owie formation, South Australia. Austr J Earth Sci 50:139–144

Andreu B, Colin JP, Singh J (2007) Cretaceous (Albian to Coniacian) ostracodes from the subsurface of the Jaisalmer Basin, Rajasthan, India. Micropaleont 53:345–370

Arias C (2006) Northern and Southern Hemispheres ostracod palaeobiogeography during the Early Jurassic: possible migration routes. Palaeogeogr Palaeoclimatol Palaeoecol 233:63–95

Arias C, Whatley RC (2005) Palaeozoogeography of western European Lower Jurassic (Pliensbachian and Toarcian) Ostracoda. Geobios 38:697–724

Babinot JF, Colin JP (1988) Paleobiogeography of Tethyan Cretaceous marine ostracodes. In: Hanai T, Ikeya N, Ishizaki K (eds) Evolutionary biology of ostracoda, its fundamentals and applications. Proceedings of the 9th international symposium on ostracoda, Shizuoka, Japan, 1985. Developments

in Paleontology and Stratigraphy 11. Kodansha, Tokyo and Elsevier, Amsterdam, pp 823–839
Babinot JF, Colin JP (1992) Marine ostracod provincialism in the Late Cretaceous of the Tethyan Realm and the Austral Province. Palaeogeogr Palaeoclimatol Palaeoecol 92:283–293
Ballent SC (1987) Foraminiferos y Ostrácodos del Jurásico inferior de Argentina. Revista del Museo de la Plata, La Plata 9:43–130
Ballent SC (1990) Lower and Middle Jurassic Ostracoda from Argentina. In: Whatley RC, Maybury CA (eds) Ostracoda and global events. Chapman and Hall, London, pp 211–220
Ballent SC (1991) Ostrácodos del Jurásico medio (Límite Aaleniano-Bayociano) en la provincia del Neuquén, centro-oeste de Argentina. Rev Española Micropaleontología 23(3):21–56
Ballent SC (1998) Ostrácodos del Albiano de la cuenca Austral, Argentina. Ameghiniana 35:355–377
Ballent SC, Ronchi DI (1999) Platycopina (Ostracoda) del Neocomiano en el sector oriental de la cuenca Neuquina, Argentina. Ameghiniana 36:243–258
Ballent SC, Whatley RC (1996) The Middle Jurassic ostracod genus Paradoxorhyncha Chapman from Australia and Argentina: a Gondwanan zoogeographical enigma. Ameghiniana 33:315–318
Ballent SC, Whatley RC (1997a) The Ostracoda and environmental evidence in the marine Jurassic of Argentina. Proceedings of the 13th Colloquium of African Micropalaeontology, Yaounde, Cameroon, 1997. [Abstract]
Ballent SC, Whatley RC (1997b) Ostrácodos del Albiano de la Perforación SC-1 (Santa Cruz), Provincia de Santa Cruz, Argentina. Ameghiniana 34:251–253
Ballent SC, Whatley RC (2000) The distribution of the Mesozoic ostracod genus Procytherura Whatley: palaeogeographical implications with special reference to Argentina. Alcheringa 24:229–242
Ballent SC, Whatley RC (2006) The Mesozoic ostracod genus Arculicythere Grékoff: further evidence for the southern Gondwana seaway. Cretaceous Res 27:728–734
Ballent SC, Whatley RC (2007) The distribution of the gondwanine ostracod Rostrocytheridea Dingle: palaeozoogeographical implications. Geol J 42:113–125
Ballent SC, Whatley RC (2009) Taxonomy and zoogeography of the Mesozoic cytherurid Ostracoda from west-central Argentina. Palaeontology 52:193–218
Ballent SC, Ronchi DI, Whatley RC (1998) The ostracod genus Majungaella Grékoff in Argentina. Rev Geológica Chile 25:45–55
Banerji RK (1970) Marine Upper Cretaceous Ostracoda from the Lower Ariyalur Stage of Vridhachalam, South India. J Palaeontol Soc India 14:28–46
Basha SH (1980) Ostracoda from the Jurassic System of Jordan: including a stratigraphical outline. Rev Española Micropaleontol 12:231–254
Bassiouni MAA (2002) Mid-Cretaceous (Aptian-Early Turonian) Ostracoda from Sinai, Egypt. Neue Paläontologische Abhandlungen 5:123
Bate RH (1972) Upper Cretaceous Ostracoda from the Carnarvon Basin, Western Australia. Spec Pap Palaeontol 10: 85, 27 pls
Bate RH (1975) Ostracods from Callovian to Tithonian sediments of Tanzania, East Africa. Bull Br Mus (Nat Hist) Geol 26:165–220
Bate RH, Bayliss DD (1969) An outline of the Cretaceous and Tertiary Foraminifera and of the Cretaceous ostracods of Tanzania. Proceedings of the 3rd African Micropaleontological Colloquium, Cairo, Egypt, 1968. National Information and Document Centre, Cairo, pp 113–164
Bate RH, Lord AR, Riegraf W (1984). Jurassic Ostracoda from Leg 79, Site 547. Initial Reports of the Deep Sea Drilling Project 79:703–710
Bertels A (1975) Upper Cretaceous (middle Maastrichtian) ostracodes of Argentina. Micropaleontology 21:97–130
Bhatia SB, Jain SP (1969) Dalmiapuram formation: a new Lower Cretaceous horizon in South India. Bull Indian Geol Assoc 2:105–108
Boyd R, Williamson P, Haq BU (1992) Seismic stratigraphy and passive margin evolution of the southern Exmouth Plateau. Proc Ocean Drilling Program, Sci Res 122:39–59
Brenner P, Oertli HJ (1976) Lower Cretaceous ostracodes (Valanginian to Hauterivian) from the Sundays River Formation, Algoa Basin, South Africa. Bull Centre Recherche Pau – SNPA 10:471–533
Burger D (1994) Palynology of Mesozoic dredge samples from the North West Shelf. AGSO J Austr Geol Geophys 15(1):89–100
Calderón M, Fildani A, Hervé F, Fanning CM, Weislogel A, Cordani U (2007) Late Jurassic bimodal magmatism in the northern sea-floor remnant of the Rocas Verdes Basin, southern Patagonian Andes. J Geol Soc London 164:1011–1022
Chapman F (1904) On some Foraminifera and Ostracoda from Jurassic (Lower Oolite) strata, near Geraldton, Western Australia. Proc R Soc Victoria, NS 16:185–206
Collignon M, Sigal J, Grékoff N (1979) L'Albien et le Cénomanien du sondage de Diégo (Madagascar) et ses faunes d'ammonites, foraminifères planctoniques et ostracodes. Eclog Geol Helvetiae 72:215–249
Colwell JB, Röhl U, von Rad U, Kristan-Tollmann E (1994) Mesozoic sedimentary and volcaniclastic rocks dredged from the northern Exmouth Plateau and Rowley Terrace, offshore northwest Australia. AGSO J Austr Geol Geophys 15:11–42
Damotte R (1992) Data Report: Cretaceous ostracodes from Holes 761B and 764A (Wombat Plateau) and Holes 762C, 763B, and 763C (Exmouth Plateau). Proc Ocean Drilling Program, Sci Res 122:819–834
De Deckker PB, Colin JP, Peypouquet JP (1988) Ostracoda and the Earth Sciences. Elsevier, Amsterdam
Dépêche F, Le Nindre Y, Manivit J, Vaslet D (1987) Les ostracodes du Jurassique d'Arabie Saoudite Centrale: Systématique, répartition stratigraphique et paléogéographique. Geobios, Mém Spéc 9:221–275
Dingle RV (1969) Marine Neocomian Ostracoda from South Africa. Trans R Soc S Africa 38:139–165
Dingle RV (1971) Some Cretaceous ostracodal assemblages from the Agulhas Bank (South African Continental Margin). Trans R Soc S Africa 39:393–419
Dingle RV (1981) The Campanian and Maastrichtian Ostracoda of south-east Africa. Ann S Afr Mus 85:181

Dingle RV (1982) Some aspects of Cretaceous ostracod biostratigraphy of South Africa and relationships with other Gondwanide localities. Cretaceous Res 3:367–389

Dingle RV (1984) Mid-Cretaceous Ostracoda from southern Africa and the Falkland Plateau. Ann S Afr Mus 93(3):97–211

Dingle RV (1988) Marine ostracod distributions during the early break-up of Southern Gondwana. In: Hanai T, Ikeya N, Ishizaki K (eds) Evolutionary biology of ostracoda, its fundamentals and applications. Proceedings of the 9th international symposium on ostracoda, Shizuoka, Japan, 1985. Developments in Paleontology and Stratigraphy 11. Kodansha, Tokyo and Elsevier, Amsterdam, pp 841–854

Dingle RV (1996) Cretaceous Ostracoda of the SE Atlantic and SW Indian Ocean: a Stratigraphical Review and Atlas. Actes Colloques: Géologie de l'Afrique et de l'Atlantique Sud (1994), pp 1–17

Dingle RV (in press) Palaeozoogeographical implications for high latitude Gondwanide marine Ostracoda of new Upper Cretaceous faunas from New Zealand and the Antarctica Peninsula. Rev Española Micropleontología

Dingle RV, Klinger HC (1972) The stratigraphy and ostracod fauna of the Upper Jurassic sediments from Brenton, in the Knysna Outlier, Cape Province. Trans R Soc S Africa 40(4):279–302

Dingle RV, Majoran S (2001) Palaeo-climatic and – biogeographical implications of Oligocene Ostracoda from CRP-2/2A and CRP-3 drillholes, Victoria Land Basin, Antarctica. Terra Antarctica 8:369–382

Direen NG, Stagg HMJ, Symonds PA, Colwell JB (2008) Architecture of volcanic rifted margins: new insights from the Exmouth-Gascoyne margin, Western Australia. Aust J Earth Sci 55:341–363

Eldholm O, Coffin MF (2000) Large igneous provinces and plate tectonics. In: Gordon RG, van der Hilst RD, Richards MA (eds) The history and dynamics of global plate tectonics. American Geophysical Union, Geophysical Monograph Series, vol 121. pp 309–326

Ellis G (1987) Lower Cretaceous radiolarian biostratigraphy and depositional environment of the Windalia Radiolarite, Carnarvon Basin, Western Australia. BSc (Hons) thesis, University of Western Australia, Perth, Australia

Exon NF, Borella PE, Ito M (1992) Sedimentology of marine Cretaceous sequences in the central Exmouth Plateau (northwest Australia). Proc Ocean Drilling Program, Sci Res 122:233–257

Fauth G, Seeling J, Luther A (2003) Campanian (Upper Cretaceous) ostracods from southern James Ross Island, Antarctica. Micropaleontology 49:95–107

Finger KL (1983) Observations on the Lower Cretaceous ostracode zones of Alberta. Bull Canadian Petrol Geol 31: 326–337

Frakes LA, Burger D, Apthorpe M, Wiseman J, Dettmann M, Alley N, Flint R, Gravestock D, Ludbrook N, Backhouse J, Skwarko S, Scheibnerová V, McMinn A, Moore PS, Bolton BR, Douglas JG, Christ R, Wade M, Molnar RE, McGowran B, Balme BE, Day RA (1987) Australian Cretaceous shorelines, stage by stage. Palaeogeogr Palaeoclimatol Palaeoecol 59:31–48

Frey FA, Coffin MF, Wallace PJ, Weis D (2003) Leg 183 Synthesis: Kerguelen Plateau–Broken Ridgea large igneous province. Proc Ocean Drilling Program, Sci Res 183:1–48

Fürsich FT, Pandey DK (2003) Sequence stratigraphic significance of sedimentary cycles and shell concentrations in the Upper Jurassic–Lower Cretaceous of Kachchh, western India. Palaeogeogr Palaeoclimatol Palaeoecol 193: 285–309

Geiger M, Clark DN, Mette W (2004) Reappraisal of the timing of the breakup of Gondwana based on sedimentological and seismic evidence from the Morondava Basin, Madagascar. J Afr Earth Sci 38:363–381

Gheerbrant E, Rage J-C (2006) Paleobiogeography of Africa: how distinct from Gondwana and Laurasia? Palaeogeogr Palaeoclimatol Palaeoecol 241:224–246

Govindan A (1969) A preliminary note on the occurrence of ostracodes from the Upper Cretaceous rocks of Vridhachalam, South India. Bull Geol Soc India 6(4): 127–129

Grant-Mackie JA, Aita Y, Balme BE, Campbell HJ, Challinor AB, MacFarlan DAB, Molnar RE, Stevens GR, Thulborn RA (2000) Jurassic palaeobiogeography of Australasia. Mem Assoc Aust Palaeontols 23:311–353

Grékoff N (1963) Contribution à l'étude des ostracodes du Mésozoïque moyen (Bathonien–Valanginien) du Bassin de Majunga, Madagascar. Rev Inst Français Pétrole 18:1709–1783

Grosdidier E (1979) Principaux ostracodes marins de l'intervalle Aptien–Turonien du Gabon (Afrique Occidentale). Bull Centres Recherches Exploration-Production Elf Aquitaine 3:1–35

Guha DK (1976) On some Mesozoic Ostracoda from subcrops of Banni, Rann of Kutch, India. Proceedings of the 6th Indian colloquium on micropalaeontology and stratigraphy, pp 84–90. Banaras Hindu University, Varanasi

Haig DW (1979) Global distribution patterns for mid-Cretaceous foraminiferids. J Foraminiferal Res 9:29–40

Haig DW (1991) Guide to pre-symposium Excursion A3 – northern Perth Basin and southern Carnarvon Basin (Perth, June 1991). 11th International Symposium on Ostracoda, Warrnambool, Australia.

Haig DW, Lynch DA (1993) A late early Albian marine transgressive pulse over northeastern Australia, precursor to epeiric basin anoxia: foraminiferal evidence. Marine Micropaleontol 22:311–362

Haig DW, Watkins DK, Ellis G (1996) Mid-Cretaceous calcareous and siliceous microfossils from the basal Gearle Siltstone, Giralia Anticline, southern Carnarvon Basin. Alcheringa 20:41–68

Haq BU, von Rad U et al (1990) Proceedings of the Ocean Drilling Program, Initial reports, vol 122. Ocean Drilling Program, College Station, Texas. doi:10.2973/odp.proc.ir.122.1990

Hart MB, Joshi A, Watkinson MP (2001) Mid-Late Cretaceous stratigraphy of the Cauvery Basin and the development of the eastern Indian Ocean. J Geol Soc India 58:217–229

Hathway B (2000) Continental rift to back-arc basin: Jurassic-Cretaceous stratigraphical and structural evolution of the Larsen Basin, Antarctic Peninsula. J Geol Soc London 157:417–432

Heine C, Müller RD (2005) Late Jurassic rifting along the Australian North West Shelf: margin geometry and spreading ridge configuration. Aust J Earth Sci 52:27–39

Hocking RM, Voon JWK, Collins LB (1988) Stratigraphy and sedimentology of the basal Winning Group, northern

Carnarvon Basin. In: Purcell PG, Purcell RR (eds) The North West Shelf, Australia. Petroleum Exploration Society of Australia, Perth, pp 203–224

Holbourn AEL, Kaminski MA (1995) Lower Cretaceous benthic foraminifera from DSDP Site 263: micropalaeontological constraints for the early evolution of the Indian Ocean. Mar Micropaleontol 26:425–460

Holbourn AEL, Kaminski MA (1996) Lower Cretaceous benthic foraminifera from DSDP and ODP sites of the Indian Ocean: a review and synthesis. In: Moguilevsky A, Whatley RC (eds) Microfossils and oceanic environments. Proceedings of the symposium on the ocean drilling program and the marine biosphere, Aberystwyth, Wales, April 1994. Micropalaeontology Research Group, Institute of Earth Sciences, University of Wales, Aberystwyth, and British Micropalaeontological Society, London, pp 77–90

Holbourn AEL, Kaminski MA (1997) Lower Cretaceous deep-water benthic foraminifera of the Indian Ocean. Grzybowski Foundation, Spec Pub 4:172p. Jagiellonian University, Krakow, Poland

Huber BT, Hodell DA, Hamilton CP (1995) Middle–Late Cretaceous climate of the southern high latitudes: stable isotope evidence for minimal equator-to-pole thermal gradients. Geol Soc Am Bull 107:1164–1191

Jain SP (1975) Cytheracea (Ostracoda) from the Ariyalur Formation (Upper Cretaceous) of South India. Bull Indian Geol Assoc 8:41–70

Jain SP (1976a). Ostracoda from the Dalmiapuram Formation (Lower Cretaceous), Tiruchirapalli District, Tamil Nadu. Bull Indian Geol Assoc 9(1):33–42

Jain SP (1976b) Ostracoda from the Kallakkudi Limestone (Albian), Tiruchirapalli District, Tamil Nadu: Families Cytherellidae, Bairdiidae, Macrocyprididae and Paracyprididae. Indian J Earth Sci 3(2):199–207

Jain SP (1978) Further Ostracoda from the Kallakkudi Limestone (Albian), Tiruchirapalli District, Tamil Nadu, India. N Jb Geol Paläont Monat 8:502–512

Jain SP, Mannikeri MS (1975) Ostracoda from the Spiti Shales, Spiti Valley. Himalayan Geol 5:133–139

Jeletzky JA (1983) Macroinvertebrate paleontology, biochronology, and paleoenvironments of Lower Cretaceous and Upper Jurassic rocks, Deep Sea Drilling Hole 511, eastern Falkland Plateau. Initial Rep Deep Sea Drilling Project 71: 951–967

Jones AT, Frank TD, Fielding CR (2006) Cold climate in the eastern Australian mid to late Permian may reflect cold upwelling waters. Palaeogeogr Palaeoclimatol Palaeoecol 237:370–377

Kauffman EG (1973) Cretaceous Bivalvia. In: Hallam A (ed) Atlas of palaeobiogeography. Elsevier, Amsterdam, pp 353–382

Kaye P (1965) Some new British Albian Ostracoda. Bull Br Mus (Nat Hist) Geol 11:217–253

Khosla SC, Jakhar SR (1999) A note on the ostracode fauna from the Jurassic beds of Jumara Dome, Kachchh. J Geol Soc India 54:43–49

Khosla SC, Jakhar SR, Kumari M, Dubey S (2006) Middle Jurassic Ostracoda from the Jaisalmer Formation, Jaisalmer District, Rajasthan, India. J Palaeont Soc India 51:1–31

Khosla SC, Jakhar SR, Mohammed MH (1997) Ostracodes from the Jurassic beds of Habo Hill, Kachchh, Gujarat, India. Micropaleontology 43:1–39

Khosla SC, Jakhar SR, Mohammed MH (2004) Ostracodes from the Jurassic beds of Jhura Hill, Kachchh, Gujarat. J Geol Soc India 63:15–28

Khosla SC, Jakhar SR, Nagori ML, Felix AD, Kumari M (2003) Jurassic ostracode fauna from the type section of Khadir Formation, Kachchh and Lophocythere vertipolycostata n. sp. J Geol Soc India 61:61–74

Khosla SC, Kumari M, Felix AD, Jakhar SR, Nagori ML (2005) Middle Jurassic Ostracoda from the Northern Island Belt, Rann of Kachchh, Gujarat, India. J Palaeont Soc India 50:17–64

Kielbowicz AA, Ronchi D, Stach NH (1983) Foraminiferos y Ostrácodos Valanginianos de le Formación Springhill, Patagonia Austral. Rev Asoc Geol Argentina 38:313–339

Krömmelbein K (1975a) Remarks on marine Cretaceous ostracodes of Gondwanic distribution. In: Proceedings of the 5th African colloquium on micropalaeontology, Addis-Ababa, Ethiopia, 1972. Enadimsa, Madrid, pp 539–551

Krömmelbein K (1975b) Ostracoden aus der Kreide des Great Artesian Basin, Queensland, Australien. Senckenbergiana Lethaea 55:455–483

Krömmelbein K (1976) African Cretaceous ostracodes and their relations to surrounding continents. In: Gray J, Boucot AJ (eds) Historical biogeography, plate tectonics and the changing environment. Oregon State University, Corvallis, pp 305–310

Kulshreshtha SK, Singh RY, Tewari BS (1985) Ostracode fauna from Kuldhar Beds, Jaisalmer, Rajasthan, vol 1. Centre Advanced Study in Geology, Punjab University, Chandigarh, pp 123–154

Longley IM, Buessenschuett C, Clydsdale L, Cubitt CJ, Davis RC, Johnson MK, Marshall NM, Murray AP, Somerville R, Spry TB, Thompson NB (2002) The North West Shelf of Australiaa Woodside perspective. In: Keep M, Moss S (eds) The sedimentary basins of Western Australia 3. Petroleum Exploration Society of Australia, Perth, pp 27–88

Lord A (1988) Ostracoda of the Early Jurassic Tethyan Ocean. In: Hanai T, Ikeya N, Ishizaki K (eds) Evolutionary biology of ostracoda, its fundamentals and applications. Proceedings of the 9th international symposium on Ostracoda, Shizuoka, Japan, 1985. Developments in Paleontology and Stratigraphy 11. Kodansha, Tokyo and Elsevier, Amsterdam, pp 855–868

Lord A, Malz H, Apthorpe M (1993) Lower Jurassic Ostracoda from off Western Australia. In: McKenzie KG, Jones P (eds) Ostracoda in the earth and life sciences. Proceedings of the 11th international symposium on ostracoda, Warrnambool, VIC, Australia, 1991. AA Balkema, Rotterdam, pp 109–121

Lord AR, Malz H, Whittaker JE (2006) Bajocian Ostracoda of Western Australia and their faunal affinities. Senckenbergiana Lethaea 86:191–227

Majoran S, Widmark JGV (1998) Response of deep-sea ostracod assemblages to Late Cretaceous palaeoceanographic changes: ODP Site 689 in the Southern Ocean. Cretaceous Res 19:843–872

Majoran S, Kucera M, Widmark JGV (1998) Maastrichtian deep-sea ostracods from DSDP/ODP Sites 327, 356, 525, 527, 528, 529 and 698 in the South Atlantic. Revista Española Micropaleont 30(3):59–73

Majoran S, Widmark JGV, Kucera M (1997) Palaeoecological preferences and geographical distribution of Late

Maastrichtian deep-sea ostracods in the South Atlantic. Lethaia 30:3–64
Malumián N, Masiuk V, Rossi de García E (1972) Microfosiles del Cretacico superior de la perforacion SC-1 Provincia de Santa Cruz, Argentina. Rev Asoc Geol Argentina 27:265–272
Malz H, Oertli HJ (1993) Middle Jurassic Ostracoda from Western Australia. In: McKenzie KG, Jones P (eds) Ostracoda in the earth and life sciences. Proceedings of the 11th international symposium on ostracoda, Warrnambool, VIC, Australia, 1991. AA Balkema, Rotterdam, pp 123–140
Maync W (1965) Microbiostratigraphy of the Jurassic of Israel. Geol Surv Israel Bull 40:1–56 [Not seen]
McArthur JM, Janssen NMM, Reboulet S, Leng MJ, Thirlwall MF, van de Schootbrugge B (2007) Palaeotemperatures, polar ice volume, and isotope stratigraphy (Mg/Ca, $\delta^{18}O$, $\delta^{13}C$, $^{87}Sr/^{86}Sr$): The Early Cretaceous (Berriasian, Valanginian, Hauterivian). Palaeogeogr Palaeoclimatol Palaeoecol 248:391–430
McLachlan IR, Brenner PW, McMillan IK (1976a) The stratigraphy and micropalaeontology of the Cretaceous Brenton Formation and the PB-A/1 Well, near Knysna, Cape Province. Trans Geol Soc S Africa 79:341–370
McLachlan IR, Brenner PW, McMillan IK (1976b) Micropalaeontological study of the Cretaceous beds at Mbotyi and Mngazana, Transkei, South Africa. Trans Geol Soc S Africa 79:321–340
McMillan IK (1990) Foraminiferal biostratigraphy of the Barremian to Miocene rocks of the Kudu 9A-1, 9A-2 and 9A-3 boreholes. Geol Surv Namibia Comm 6:23–29. [Not seen]
Mette W (1993) Stratigraphie und fazies des Jura von Nordsomalia. Berliner Geowiss Abh A 149:1–125
Mette W (2004) Middle to Upper Jurassic sedimentary sequences and marine biota of the early Indian Ocean (southwest Madagascar): some biostratigraphic, palaeoecologic and palaeobiogeographic conclusions. J Afr Earth Sci 38:331–342
Mette W, Geiger M (2004) Bajocian and Bathonian ostracods and depositional environments in Madagascar (Morondava Basin and southern Majunga Basin). Beringeria 34. [Not seen]
Moore RC (1961) Treatise on invertebrate paleontology: part Q arthropoda 3 crustacea ostracoda. Geological Society of America and University of Kansas, Lawrence
Müller RD, Mihut D, Baldwin S (1998) A new kinematic model for the formation and evolution of the west and northwest Australian margin. In: Purcell PG, Purcell RR (eds) The sedimentary basins of Western Australia 2. Petroleum Exploration Society of Australia, Perth, pp 55–72
Musacchio EA (1978) Ostracodos del Cretacico Inferior en el Grupo Mendoza, cuenca del Neuquén, Argentina. Acta del Septimo Congreso Geologico Argentino, Neuquén, Argentina, 1978. Libart, Buenos Aires, pp 459–473
Musacchio EA (1979) Datos paleobiogeograficos de algunas asociaciones de foraminiferos, ostracodos y carofitas del Jurasico Medio y el Cretacico Inferior de Argentina. Ameghiniana 16:247–271
Musacchio EA (1981) South American Jurassic and Cretaceous Foraminifera, Ostracoda and Charophyta of Andean and Sub-Andean regions. Cuencas sedimentarias del Jurásico y Cretácico de América del Sur, vol 2. Comité Sudamericano del Jurásico y Cretácico, Buenos Aires, pp 461–498
Mutterlose J (1992) Lower Cretaceous nannofossil biostratigraphy off northwestern Australia (Leg 123). Proc Ocean Drilling Program, Sci Res 123:343–368
Mutterlose J, Wise SW Jr (1990) Lower Cretaceous nannofossil biostratigraphy of ODP Leg 113 Holes 692B and 693A, continental slope off East Antarctica, Weddell Sea. Proc Ocean Drilling Program, Sci Res 113:325–351
Narsimha K (2006) A record of Late Jurassic dinoflagellate cysts from Ariyalur-Pondicherry Sub-basin of Cauvery Basin, Tamil Nadu. J Geol Soc India 68:597–604
Neale JW (1975) The Ostracod Fauna from the Santonian Chalk (Upper Cretaceous) of Gingin, Western Australia. Special Papers in Palaeontology 16, 131p
Neale JW, Singh P (1986) Jurassic Ostracoda from the Banni Well No. 2, Kutch, India. Rev Española de Micropaleontología 17:347–372
Oertli HJ (1974) Lower Cretaceous and Jurassic ostracods from DSDP Leg 27a preliminary account. Initial Reports of the Deep Sea Drilling Project 27:947–965
Ogg JG, Agterberg FP, Gradstein FM (2004) The cretaceous period. In: Gradstein FM, Ogg JG, Smith AG (eds) A geologic time scale 2004. Cambridge University Press, Cambridge, pp 344–383
Owen HG, Mutterlose J (2006) Late Albian ammonites from offshore Suriname: implications for biostratigraphy and palaeobiogeography. Cretaceous Res 27:717–727
Papini M, Benvenuti M (2008) The Toarcian–Bathonian succession of the Antsiranana Basin (NW Madagascar): facies analysis and tectono-sedimentary history in the development of the East Africa–Madagascar conjugate margins. J Afr Earth Sci 51:21–38
Playford PE, Cope RN, Cockbain AE, Low GH, Lowry DC (1975) Phanerozoic. In The Geology of Western Australia. Mem Geol Surv Western Australia 2:223–433
Prasad B, Pundir BS (1999) Biostratigraphy of the exposed Gondwana and Cretaceous rocks of Krishna–Godavari Basin, India. J Palaeontol Soc India 44:91–117
Rafara AH (1990) Les ostracodes du Jurassique moyen–Néocomien du Bassin de Majunga (Madagascar)comparaison avec les autres faunes Gondwaniennes. Geobios 23:415–443
Rais P, Louis-Schmid B, Bernasconi SM, Weissert H (2007) Palaeoeanographic and palaeoclimatic reorganization around the Middle-Late Jurassic transition. Palaeogeogr Palaeoclimatol Palaeoecol 251:527–546
Riccardi AC (1991) Jurassic and Cretaceous marine connections between the southeast Pacific and Tethys. Palaeogeogr Palaeoclimatol Palaeoecol 87:155–189
Robb MS, Taylor B, Goodliffe AM (2005) Re-examination of the magnetic lineations of the Gascoyne and Cuvier Abyssal Plains, off NW Australia. Geophys J Int 163:42–55
Romine KK, Durrant JM, Cathro DL, Bernardel G (1997) Petroleum play element prediction for the Cretaceous–Tertiary basin phase, northern Carnarvon Basin. APPEA J 37(1):315–339
Ronchi D, Angelozzi G (1994) Bioestratigrafía del Cretácico–Terciario de dos pozos ubicados al oeste de le Cuenca Austral. Boletín de Informaciones Petroleras, Tercera Epoca 10(39):65–76

Rosenfeld A, Honigstein A (1991) Callovian-Oxfordian ostracodes from the Hamakhtesh Hagadol section, southern Israel. Rev Española Micropaleontología 23(3):133–148

Rosenfeld A, Raab M (1984) Lower Cretaceous ostracodes from Israel and Sinai. Geol Surv Israel Spec Pub 4. Israel J Earth Sci 33(3):85–134

Rosenfeld A, Gerry E, Honigstein A (1987a) Jurassic ostracodes from Gebel Maghara, Sinai, Egypt. Rev Española Micropaleontología 19:251–280

Rosenfeld A, Oertli H, Honigstein A, Gerry E (1987b) Oxfordian ostracodes from the Kidod Formation of the Majdal Shams area, Mount Hermon, Golan Heights. Bull Centres de Recherches Exploration-Production Elf Aquitaine 11(2):233–248. [Not seen]

Rossi de García E (1979) El genero Novocythere (Ostracoda) (Perforacion SC-1) Santa Cruz, Republica Argentina. Ameghiniana 14(1–4):117–121

Rossi de García E, Proserpio C (1980) Ostrácodos del Cretácico Superior de Patagonia, República Argentina (Hoja 44e, Valle General Racedo), Chubut, vol 2. Actas del Segundo Congresso Argentino de Paleontologia y Bioestratigrafia y Primer Congresso Latinoamericano de Paleontologia, Buenos Aires, Argentina, 1978, pp 15–34

Scheibnerová V (1973) A comparison of the austral and boreal Lower Cretaceous foraminiferal and ostracodal assemblages. In: Proceedings of the international symposium on the boreal lower cretaceous (1972). Seel House, Liverpool, pp 407–414

Scheibnerová V (1978) Comparative foraminiferal biostratigraphy of the Santos Oodnadatta No. 1 section, South Australia. Geol Surv New South Wales Rec 19:81–139

Shafik S (1990) Late Cretaceous nannofossil biostratigraphy and biogeography of the Australian western margin. Bur Min Res, Geol Geophys, Rep 295:1–164

Sigal J (1974) Comments on Leg 25 sites in relation to the Cretaceous and Paleogene stratigraphy in the eastern and southeastern African and Madagascar regional setting. Initial Reports of the Deep Sea Drilling Project 25:687–723

Sigal J, Grékoff N, Cañón A, Ernst M (1970) Sur l'âge et les affinitiés "gondwaniennes" de microfaunes (foraminifères et ostracodes) malgaches, indiennes et chiliennes au sommet du Jurassique et à la base du Crétacé. CR Acad Sci, Ser D 271:24–27

Singh P (1997) Ostracoda from the subsurface Cretaceous strata of Manhera Tibba Well-1 and Shahgarh Well-B, Jaisalmer Basin, Rajasthan, India with special remarks on foraminiferids. Geosci J 18:1–57

Stone P, Richards PC, Kimbell GS, Esser RP, Reeves D (2008) Cretaceous dykes discovered in the Falkland Islands: implications for regional tectonics in the South Atlantic. J Geol Soc London 165:1–4

Subrahmanyam C, Chand S (2006) Evolution of the passive continental margins of Indiaa geophysical appraisal. Gondwana Res 10:167–178

Sugumaran S, Nagaraj HM, Mallikarjuna UB (1997) Ostracode fauna from the Patti Formation (Late Cretaceous) of Vridhachalam area, Tamil Nadu, India. Palaeobotanist 46(1–2):133–140

Szczechura J (2001) Ostracods from the Eocene of Seymour Island, Antarctic Peninsula. Palaeontol Polonica 60:157–181

Szczechura J, Blaszyk J (1996) Ostracods from the Pecten Conglomerate (Pliocene) of Cockburn Island, Antarctic Peninsula. Palaeontol Polonica 55:175–186

Talib A, Faisal SMS (2006) A preliminary note on the occurrence of foraminifera in the Middle to Upper Jurassic sediments of Fakirwari, Kutch Mainland, Gujarat. J Geol Soc India 68:963–966

Taylor BA, Haig DW (2001) Barremian foraminifera from the Muderong Shale, oldest marine sequence in the Cretaceous of the southern Carnarvon Basin, Western Australia. Micropaleontology 47(2):125–143

Valicenti VH, Stephens JM (1984) Ostracods from the Upper Valanginian and Upper Hauterivian of the Sundays River Formation, Algoa Basin, South Africa. Rev Española Micropaleontología 16:171–239

Van Morkhoven FPCM (1963) Post-palaeozoic ostracodatheir morphology, taxonomy and economic use. Volume IIgeneric descriptions. Elsevier, Amsterdam

Veevers JJ (2006) Updated Gondwana (Permian-Cretaceous) earth history of Australia. Gondwana Res 9:231–260

Veevers JJ, Heirtzler JR et al (1974) Site reports: sites 259–263. Initial Reports of the Deep Sea Drilling Project 27:15–335. US Government Printing Office, Washington, DC

Whatley RC (1983) The application of Ostracoda to palaeoenvironmental analysis. In: Maddocks RF (ed) Applications of ostracoda. University of Houston, Geoscience, Houston, pp 51–77

Whatley RC, Ballent SC (1996) A review of the Mesozoic ostracod genus Progonocythere and its close allies. Palaeontology 39:919–939

Whatley RC, Ballent SC (2004) A review of the Mesozoic ostracod genus Lophocythere and its close allies. Palaeontology 47:81–108

Whatley RC, Ballent SC, Szczechura J (2005) Antarctic tertiary progonocytheridae: the last refuge of Majungaella, the survivor of a long lineage, a geographical and physiological migration from low to high latitudes. Palaeogeogr Palaeoclimatol Palaeoecol 225:302–316

Wise SW Jr (1988) Mesozoic–Cenozoic history of calcareous nannofossils in the region of the Southern Ocean. Palaeogeogr Palaeoclimatol Palaeoecol 67:57–179

Cretaceous Continental Bridges, Insularity, and Vicariance in the Southern Hemisphere: Which Route Did Dinosaurs Take?

Federico Fanti

Abstract

The history of dispersal of biota during the Mesozoic in the southern hemisphere is much debated. Clearly, the sequential break-up of Gondwana produced physical barriers that must, variously, have hampered dispersal events, and therefore impacted the phylogenetic hypotheses. Did various terrestrial organisms undergo dispersal via recently claimed continental bridges? When were the geographic and environmental conditions at an optimum for diffusion during Mesozoic times? Dinosaurs are arguably the most relevant group for illuminating the biogeography of the southern landmasses during the Cretaceous. Their vast stratigraphic and geographic occurrence is intimately linked to the evolution of Gondwana. Recent discoveries from all southern landmasses challenge several vicariant models. This study summarizses the most significant geologic, palaeogeographic, palaeontologic, and phylogenetic data on Cretaceous Gondwanan dinosaur evolution, with particular emphasis on the ephemeral land bridges that bulk large in recently developed biogeographic models. Comparison between different datasets accords with a complex and sequential mix of vicariance and dispersal patterns characterizing the fabric of dinosaurian faunas at that time. This study probes the significance of ephemeral intercontinental connections with regard to the biotic dispersal in the Late Cretaceous. An earlier peak in dinosaur diversity and dispersals seems likely.

Keywords

Gondwana • Cretaceous • Dinosaurs • Biogeography • Dispersal • Continental bridges • Vicariance

Introduction

The southern hemisphere has traditionally been considered as having a fundamental vicariant history in the Mesozoic: the sequential break-up of Gondwana that began in the Jurassic resulted in a successive division of ancestral areale and biota. Under this

F. Fanti (✉)
Dipartimento di Scienze della Terra e Geologico-Ambientali, Alma Mater Studiorum Università di Bologna, 40126 Bologna, Italy
e-mail: federico.fanti@unibo.it

J.A. Talent (ed.), *Earth and Life*, International Year of Planet Earth,
DOI 10.1007/978-90-481-3428-1_31, © Springer Science+Business Media B.V. 2012

scenario, continental break-up and the formation of intra-continental seas created barriers for terrestrial taxa and new ways of dispersion for marine species. In particular, the fragmentation of continents resulted in more opportunities for allopatric speciation, and therefore peaks in biological diversity (Valentine and Moores 1970; Tiffney and Niklas 1990; Sanmartin and Ronquist 2004; Walter 2004; Lieberman 2005). Recent paleontological discoveries combined with new paleogeographic reconstructions have greatly challenged this assumption. Dispersal events linked to the fragmentation of Gondwana have been proposed for several groups of vertebrates and invertebrates, as well as for selected floras (Krause et al. 1997, 1998, 2006; Forster 1999; Waters et al. 2000; Rana and Wilson 2003; Sanmartin and Ronquist 2004; Vences et al. 2004; Gheerbrant and Rage 2006).

A valuable contribution to this debate has been provided by the study of the Late Cretaceous Gondwanan dinosaurs. In the last decades dinosaur remains have been reported at an increasing rate from all Gondwanan landmasses, documenting the presence of a complex and heterogeneous assemblage of distinct faunas. They were surprisingly diverse, widespread across all southern landmasses from warm sub-tropical regions to sub-polar latitudes, and all were stratigraphically long lived. It is not surprising, therefore, that dinosaurs from the southern hemisphere represent an ideal case study in Mesozoic biogeography. All most recent hypothesis on Gondwanan dinosaurian biogeography during the Late Cretaceous imply the presence of geographically discrete areas that intermittently linked different landmasses (Fig. 1). In order to explain the occurrence of same family-level taxa in India, Madagascar, and South America during the latest Cretaceous, ephemeral land bridges connecting at different times Africa, Antarctica, Indo-Madagascar, and South America have been hypothesized. Vicariance- and dispersal-based models led to the development of two competing biogeographic models, named "Africa-First" (*sensu* Sereno et al. 2004 – see also Krause et al. 1997, 1999, 2006; O'Connor et al. 2006) and "Pan-Gondwana" (Sereno et al. 2004; Sereno and Brusatte 2008). The "Africa-First" hypothesis postulates that after the complete isolation of continental Africa in the Early Cretaceous, subaerial connections between Indo-Madagascar, Antarctica, and South America allowed dispersal of terrestrial taxa across southern landmasses until the end of the Cretaceous. The "Pan-Gondwana" hypothesis postulates subaerial connection between southern landmasses during the Early Cretaceous and the severing of one or more land connections starting with the late Cenomanian. Both hypotheses include a large amount of paleontological and geological data to which authors apparently have been giving different weight. Many studies, in fact, have been primarily based on the interpretation of the fossil record or exclusively on phylogenetic analyses, whereas very little data have been presented on the very innovative concept of these hypotheses: the existence and durability of land bridges. Postulation of the presence of one "ad hoc" ephemeral bridge has been often a simplification of global biological and geological dynamics.

The present study consists of three parts:

1. An overview of concepts and terminology currently used in biogeographic analyses;
2. Revision of geological data and paleogeographic reconstructions for hypothesized Cretaceous intercontinental connections; and
3. Revision of the geographic and stratigraphic occurrence of four predominantly Gondwanan dinosaur families (Titanosauria, Abelisauroidea, Carcharodontosauridae, and Spinosauridae).

Data presented herein dispute recent biogeographic hypotheses of the presence of a "southern way" responsible for dinosaur dispersal during the latest Cretaceous. In the specific case of the late Cretaceous Gondwanan faunas, such an integrative approach is particularly important. The question is whether they represent a neondemic, paleoendemic, or immigrant faunas.

Biogeography: Which Model?

A valid biogeographic hypothesis must incorporate all factors that may influence the distribution of a population, and it should not be supported solely in the light of evolutionary relationships between taxa. Three important and intimately linked parameters control the reconstruction of spatial and temporal dynamics of a population: *phylogeny*, *rate of evolution*, and *areal dynamics*. Such parameters can be confidentially used as reference point for extant biota as well as for extinct plants and animals. Phylogeny shows similarities that can link a derived population to its ancestors; the rate of evolution, and consequently relationships within a phylogenetic tree, is the basis to understand how fast

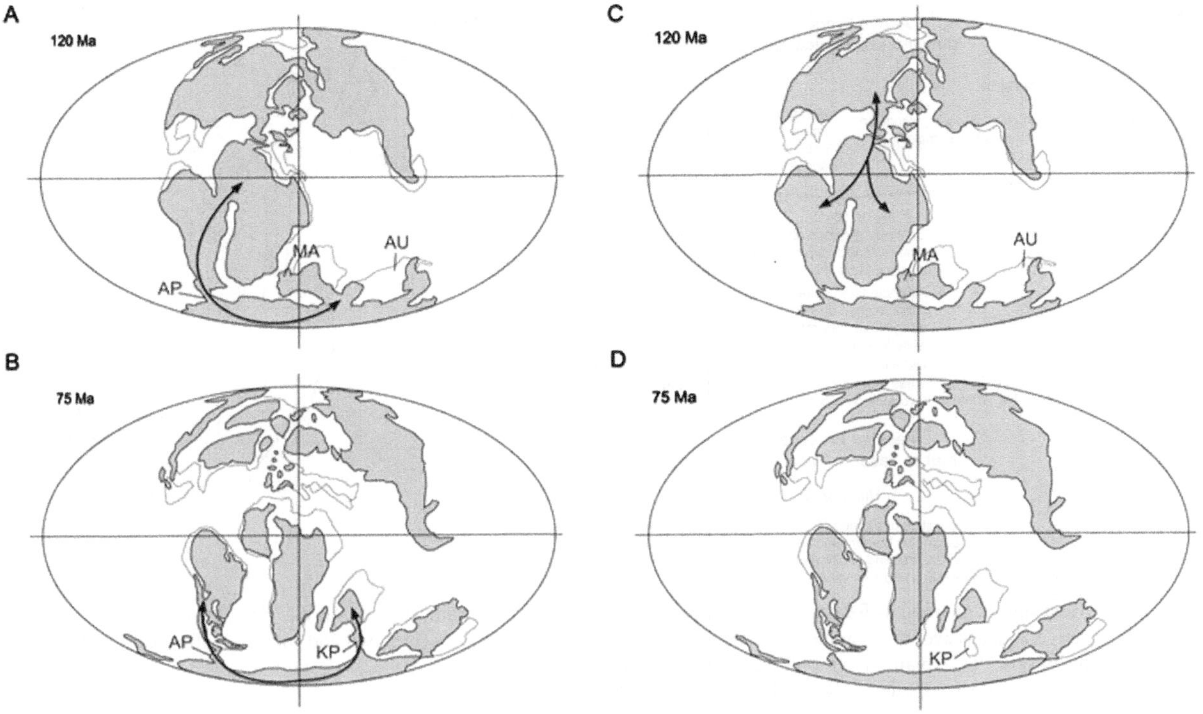

Fig. 1 Paleogeographic reconstructions of the two hypothesized scenarios for the Cretaceous period. (**a**, **b**) Laurasia and Gondwana 120 and 70 Ma (modified after Hay et al. 1999) showing continental land bridges linking Antarctica with South America and Indo-Madagascar. Arrows indicate dispersal routes proposed in the "Africa-First" and "Pan-Gondwana" models. (**c**, **d**) Paleogeographic reconstructions based on available data on the Kerguelen Plateau, Antarctic Peninsula, and Atlantic margin of South America. This model is the basis of proposed biogeographic analyses (see the text for discussion); arrows indicate hypothetical dispersal ways in the middle Cretaceous. AP, Antarctic Peninsula; AU, Australia; MA, Madagascar; KP, Kerguelen Plateau

and through which steps a population (or part of a population) evolved into a new taxon. Lastly, the study of areal dynamics introduces to all basal concepts of the two models that are currently at the base of all biogeographic hypothesis and models: dispersion and vicariance.

The presence of discontinuous filter routes such as continental bridges has been postulated in the literature since 1911 (Scharff 1911) to explain the presence of similar taxa in contiguous geographic regions. Within these active processes, it is important to distinguish between two commonly misused terms, *dispersion* and *dispersal*. Referred to a generic population, the first indicates a gradual demographic increase and consequent extension of the areale in the complete absence of natural constraints. On the contrary, the latter refers to a partial and non-gradual fragmentation or extension of both the areale and the population, thus implying a geographical or environmental variation (Raffi and Serpagli 1999; Zunino and Zullini 1999). Active distribution of terrestrial organisms may take place through three different ways: corridors, bridges (or filters), and occasional (sweepstakes) routes (Fig. 2).

The expression *continental bridge*, widely and sometimes improperly used in most recent biogeographic reconstructions, indicates ephemeral communication routes between separate landmasses: such routes represent occasional and short-lived dispersion ways for terrestrial taxa (often motivated by favorable environmental conditions) and intermittent barriers for marine species. This model implies a dispersal pattern with an active, rapid, and primarily unidirectional trend in population dynamics. The presence of ephemeral land bridges has been proposed at various stages of continental break-up during the Mesozoic, particularly between North and South America (Bonaparte 1986; Lucas and Hunt 1989), Europe and Africa (Rage 1981; Buffetaut et al. 1988; Le Loeuff and Buffetaut 1991; Buffetaut and Rage 1993), Africa and South America

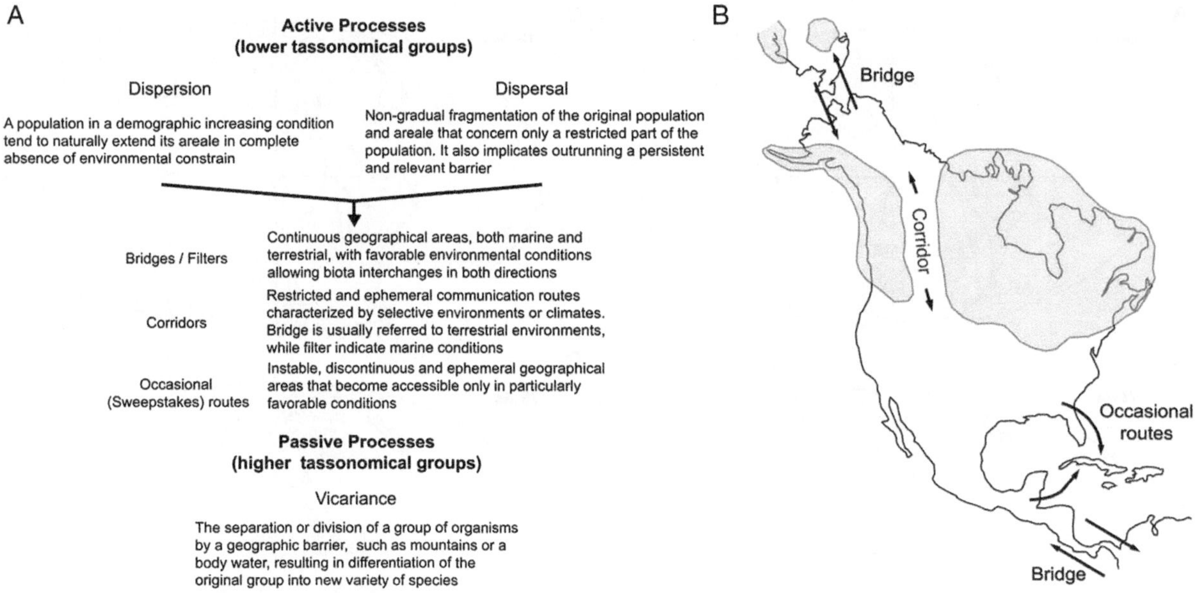

Fig. 2 (**a**) Most significant models for organism diffusion. Both processes, active and passive, are commonly used in dinosaur biogeography in order to explain geographic and phylogenetic affinities among taxa. (**b**) Case study of Pleistocene north American migration as example for active dispersion models. Ice caps in *gray* (redrawn and modified after Raffi and Serpagli 1999)

(Jacobs 1993; Goldblatt 1993; Gayet 2001; Sereno and Brusatte 2008), and Antarctica and Indo-Madagascar (Krause et al. 1997; Hay et al. 1999; Sparks and Smith 2004; Noonan and Chippindale 2006).

Corridors have been taken to represent continuous geographical areas with homogeneous and advantageous environments (and consequently a non-casual trend in population dynamics) that would allow biota to migrate in both directions for longer intervals. On the contrary, many studies have also discussed the role of vicariance as a dominant factor for Gondwanan dinosaur distribution, linking the dispersal of terrestrial organisms to the ongoing fragmentation of the southern landmasses (Milner and Molnar 1984; Russell 1993; Sampson et al. 1998; Forster 1999; Upchurch et al. 2002; Sanmartin and Ronquist 2004). The formation of vast epicontinental seas would have been responsible for major barriers to the dispersal of terrestrial organism, whereas marine taxa were able to rapidly spread across the southern hemisphere.

Insularity

The fragmentation of Gondwana resulted in major continental blocks such as Africa, South America, and Antarctica, as well as vast islands such as Australia, Madagascar, and India. It is also fundamental to remark that continental bridges originated as scattered islands that during intervals of marine low-stand or tectonic/volcanic uplift formed larger emerged areas surrounded by deep water environment. Therefore, beside the effective size of such islands, to discuss the distribution of Late Cretaceous dinosaur faunas and the effects of ephemeral bridges on a generic population, a last significant parameter is *insularity*.

Analysis of insularity is helpful to approach important aspects such as colonization rate, number of species, turnover and extinction rates, and relationships between insular faunas and their closest continental counterparts (also representing a possible area source for immigrants and/or competitors). The most significant practical aspects of insular conditions were exposed by Macarthur and Wilson (1967) in the homonymous theory. The number of taxa populating an island

1. tends to constant value,
2. decreases proportionally to the distance of the area source, and
3. is proportional to the dimension of the island.

An important consequence of the progressive reduction and stabilization of taxa number is a remarkable reduction in diversity. In a few cases the number of species can increase, whereas higher taxonomic groups (i.e., genus or family) are fewer in number (Karol et al. 2001; Williamson et al. 2001; Anderson et al. 2004; Badano et al. 2005).

In the specific case of Gondwanan dinosaurs, the presence of an active corridor (such as a region characterized by ongoing rifting events, i.e., Africa and South America in the Early Cretaceous), supports significant terrestrial faunal interchanges. In the case of insular setting (as in the case of the Mediterranean region) the presence of physical (occasional shallow seaways, size, and morphology of the island) and ecological barriers (dramatic reduction in food and particular fresh water supply, size of several terrestrial taxa) limited and strongly influenced dinosaur dispersal. Despite the fact that the fossil record still represents a small portion of the entire Late Cretaceous biodiversity, it can be assumed that dinosaur taxa recovered from Madagascar and India follow under a generic insularity model. Therefore, considering the high potential for new discoveries, the simplest assumption is to expect evidence for such parameters in the fossil record.

Cretaceous Continental Bridges

Madagascar Role in Gondwana Break-Up

Recent discoveries in the Late Cretaceous deposits of northern Africa, South America, and particularly from the Maevarano Formation of northwestern Madagascar, greatly emphasized the role of dinosaurs in Gondwana biogeography. In fact, both the "Africa-First" and "Pan-Gondwana" hypotheses were primarily developed after important finds of crocodyliforms, dinosaurs, and mammals from the island of Madagascar (Sues and Taquet 1979; Buckley and Brochu 1996, 1999, 2001; Forster et al. 1996; Krause and Hartman 1996; Sampson et al. 1996, 1998, 2001; Buckley et al. 2000; Curry-Rogers and Forster 2001; Vences et al. 2001; Carrano et al. 2002; Curry-Rogers 2002; Maganuco et al. 2005, 2006, 2007; Fanti and Therrien 2007; Sampson and Witmer 2007; Evans et al. 2008). Of paramount importance in this debate, the closest known relatives of many of the latest Cretaceous Malagasy taxa are coeval forms from South America, which separated from Madagascar in the Early Cretaceous (see also Krause et al. (2006) for a detailed revision of Malagasy taxa). Such gaps in the fossil record and discrepancies between available paleogeographic reconstructions led to the postulation of Late Cretaceous continental bridges linking southern landmasses.

Originally situated in central Gondwana, adjacent to present-day Kenya and Tanzania, the Indo-Madagascar block started its separation from continental Africa at 160 Ma (Fig. 3). This process led to opening of the Davie Ridge and the Mozambique Channel, where permanent marine conditions were established at 140 Ma (Coffin and Rabinowitz 1988; Storey et al. 1995; Krause et al. 1997; Marks and Tikku 2001; Tikku et al. 2002; Geiger et al. 2004; Winberry

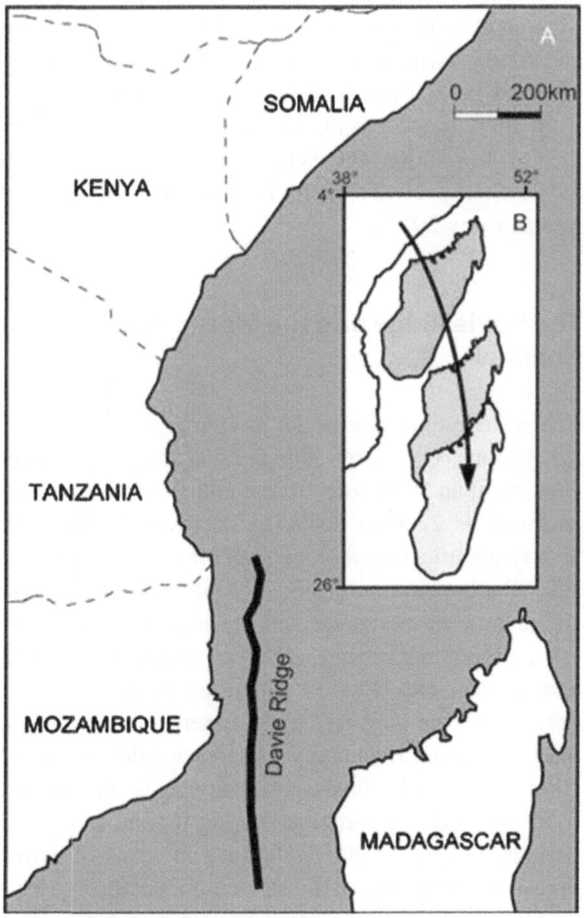

Fig. 3 (a) Present-day Mozambique Channel with the highlighted Davie Ridge. (b) Rotation of the island of Madagascar from its original position adjacent to present-day Kenya and Somalia (redrawn and modified after Rabinowitz et al. 1983)

and Anandakrishnan 2004). Separation between southern Africa and Antarctica around 150 Ma (Gaina et al. 2007) resulted in the southern oceans. Plate reconstruction models for the Indian Ocean indicate that seafloor spreading between the Indo-Madagascar block and Antarctica started at 135 Ma resulting in the isolation of the sub-continent from all other landmasses at 120 Ma (Scotese et al. 1988; Ramana et al. 2001; Frey et al. 2003; Jokat et al. 2003). The subduction of Greater India beneath Eurasia moved the Indian sub-continent, still joined to Madagascar, northwards. At about 88 Ma the spreading Indian Ridge system was captured by the Marion Hot Spot south of Madagascar that triggered the propagation of the nascent Mascarene Ridge to the north, opening the Mascarene Basin between Madagascar and India (Embleton and McElhinny 1975; Torsvik et al. 1998; Meert and Tamrat 2006). The separation between India and Madagascar was accompanied by an outpouring of flood basalts on the island of Madagascar: dates from these flood basalts cluster in age around 87.6 ± 0.6 Ma (Norton and Sclater 1979; Mahoney et al. 1991; Storey et al. 1995; Briggs 2003; Raval and Veeraswamy 2003).

The Davie Ridge and the Mozambique Channel

Recent investigations on the morphology and geological history of the area indicate that the ridge region consists of an alternation of seamounts that rise locally as much as 2200 m above the seafloor depths with an asymmetric morphology, steeping on its western side (Rabinowitz and Woods 2006). Thus, the Davie Ridge may have represented an area of great interest for biota interchanges between continental Africa and Madagascar (Fig. 3). However, to date no evidences indicate the presence of emerged areas during the Cretaceous. Eventually the area would have constituted of small islands or archipelagos surrounded by deep-water environments during marine low-stand periods. Despite its little relevance in terms of active dinosaur dispersion, it must be noticed that geodynamic events along the rifted margin may have influenced Madagascar's biota during the Late Cretaceous. The western coast of Madagascar was in fact intersected by repeated fluctuations of the sea level linked to the geodynamic evolution of the island (particularly during the Oxfordian–Callovian, Aptian, Coniacian, and latest Maastrichtian). Furthermore, these variations were not in phase along the entire length of the western Malagasy coast (Papini and Benvenuti 1998; Razafindrazaka et al. 1999). Hence, the following models proposed for the Late Cretaceous of North America by Lehman (1987, 1997, 2001), repeated sea-level oscillations could have reduced the adaptive pressure for biological specialization and led to rapid colonization of the region by a few opportunistic or generalist taxa.

The Kerguelen Plateau

Since 1999, when Hay et al. proposed the hypothesis on late Cretaceous connections between Indo-Madagascar and Antarctica through the Kerguelen Plateau, a variety of papers have dealt with the "Malagasy question": relationships between latest Cretaceous dinosaur faunas from Madagascar, India, and South America. However, given the "datum" of a subaerial connection, much has been discussed on vertebrate biogeography and very little on the Kerguelen Plateau.

Together with the Broken Ridge, the Kerguelen Plateau forms a large igneous province in the southern Indian Ocean (Coffin et al. 2002; Duncan 2002; Kent et al. 2002; Frey et al. 2003; Ali and Aitchison 2009). Virtually unknown 20 years ago, oceanic plateaus like the Kerguelen have been intensively explored by the Ocean Drilling Project (ODP) activities (Fig. 4). Data recovered, combined with plate reconstruction for the Indian and southern oceans, indicate that seafloor spreading between India and Western Australia started at 133 Ma, whereas the age of separation between India and Antarctica is still under examination and is now estimated between 165 and 132 Ma (Besse and Courtillot 1988; Briggs 2003; Frey et al. 2003; Jokat et al. 2003). The latest region of the Indo-Malagasy sub-continent that separated from Antarctica is represented by the actual Sri Lanka: originally confined between western Antarctica and the southern part of India, it separated from Antarctica no later than 130 Ma, as supported by Valanginian oceanic crust in the Enderby Basin (Ramana et al. 2001). Between 120 and 110 Ma, the Kerguelen hotspot produced a major magmatic event responsible for a rapid extension of the submarine plateau: at the end of this

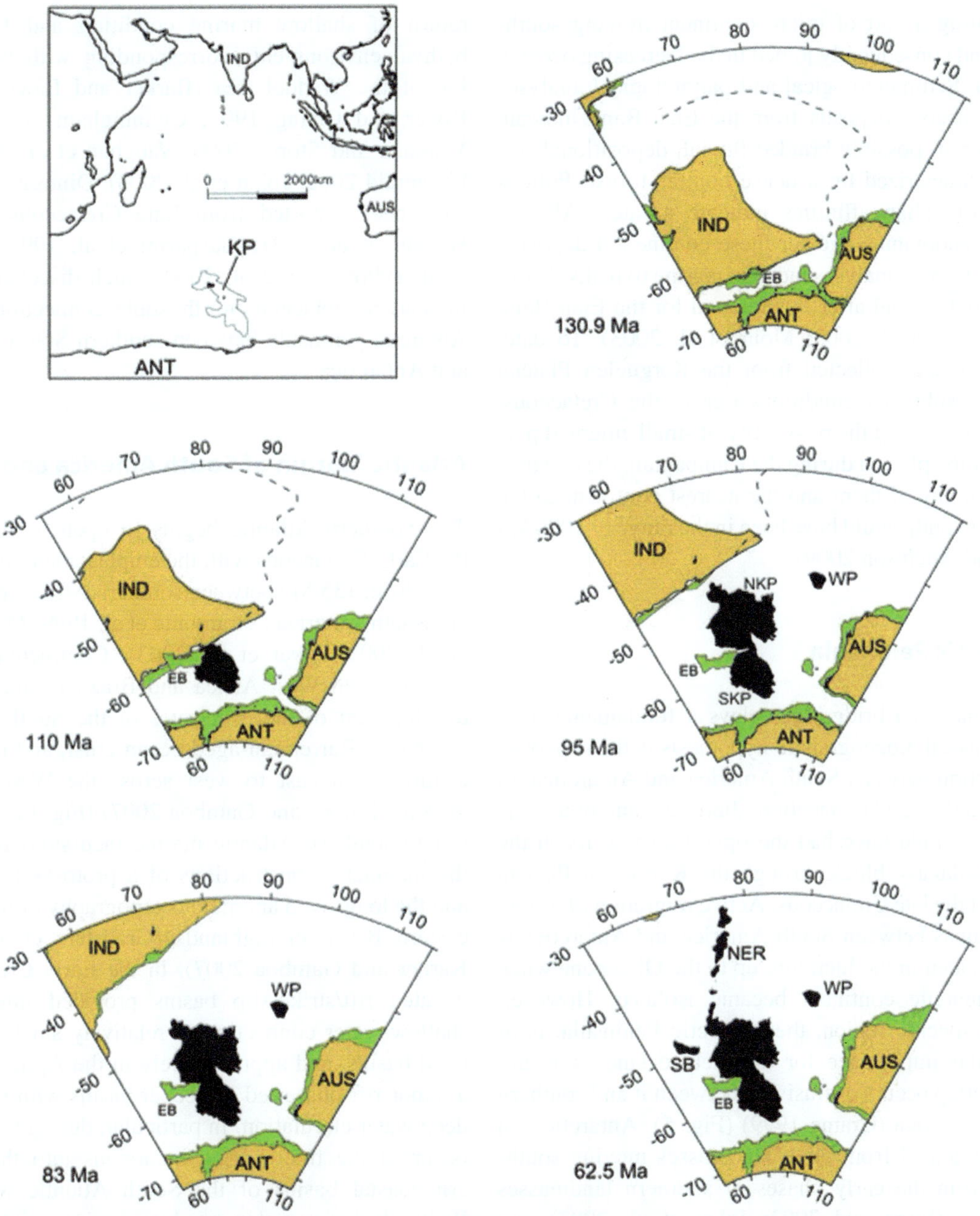

Fig. 4 Plate reconstructions of the southern Indian Ocean region showing the evolution of the Kerguelen Plateau and the Elan Bank during the Cretaceous. AUS, Australia; ANT, Antarctica; EB, Elan Bank; KP, present-day extension of the Kerguelen Plateau; IND, India; NKP, Northern Kerguelen Plateau; SKP, Southern Kerguelen Plateau; WP, Wallaby Plateau (redrawn and modified after Frey et al. 2003)

event, the easternmost regions of India were separated from Australia and Antarctica by approximately 2000 km (Frey et al. 2003; Gaina et al. 2007). Data obtained by the ODP activities indicate that only a restricted portion of the Kerguelen Plateau, the Elan Bank (Ingle et al. 2002; Borissova et al. 2003, and references therein), did actually emerge during the Cretaceous (Fig. 4). Today included in the Central Kerguelen Plateau, the Elan Bank separated from India in the early break-up between India and Antarctica, originating a sort of micro continent moving southward and consequently joined in the increasing oceanic plateau. Sedimentological and palynological analyses on Cretaceous deposits from the Elan Bank indicate a subaerial, possibly braided fluvial, depositional setting characterized by a dense conifer forest. Pollens with Australian affinities indicate a latest Albian–early Cenomanian age for these continental deposits. However, core analyses indicate marine to bathyal condition before and after this interval for the Elan Bank (Borissova et al. 2003; Mohr et al. 2003). To date, no other data collected from the Kerguelen Plateau indicate subaerial conditions during the Cretaceous. Even considering the possibility of small emerged portions of the plateau during the Campanian, the distance between any of them and the nearest continent to the north or south would have been in the range of 1000 km (Ali and Aitchison 2009).

Antarctic Peninsula

A second land bridge that plays a fundamental role in discussed biogeographic hypothesis is the supposed connection between South America and Antarctica; in fact, via this land connection, dinosaurs and other vertebrates would have had the opportunity to reach the Indo-Malagasy block through the Kerguelen Plateau up until the late Cretaceous. Active migration of terrestrial faunas between South America and Antarctica is documented in the literature up to the Oligocene when the Antarctic continent became isolated. However, only a specific region, the Antarctic Peninsula, is of particular importance for this debate since substantial overlap occurs exclusively between it and southern South America (Crame 1999) (Fig. 5). Antarctica, in fact, separated from other landmasses moving southward from the early phases of southern landmasses break-up (Fitzgerald 2002; Jokat et al. 2003), but probably part of the Antarctic Peninsula remained, at times linked to the Andean region. To date, there is not a widely accepted paleogeographic reconstruction for this region and speculations have been accumulating. Some authors indicate a continuous connection between the two continents from the middle Cretaceous to the Paleocene (Hay et al. 1999), whereas others suggested that south America and the Antarctic peninsula separated during the Campanian–Maastrichtian (Crame 1999), or at 131 Ma, with the return of shallow marine condition and following bathyal environments corresponding with the opening of the Weddel Sea (Barker and Lawver 1988; Royer and Chang 1991; Cunningham et al. 1995; Vaughan and Storey 2000; Vaughan et al. 2002a, b; Fitzgerald 2002; Jokat et al. 2003). Dinosaur remains have been reported from Late Cretaceous beds of Hooker et al. 1991; Gasparini et al. 1996; Molnar et al. 1996; Case et al. 2000). Such discoveries may indicate a Cretaceous north–south connection via the Antarctic peninsula between southern South America and Antarctica.

Atlantic Margin of South America and Africa

The Southern Atlantic begins to open northward in the Early Cretaceous with the emplacement of oceanic crust from 135 Ma between southern Africa and southern South America (Viramonte et al. 1999; Macdonald et al. 2003; Jokat et al. 2003). Continental extension between West Africa and Brazil resulted in the development of the evaporites of the South Atlantic salt basin, Barremian-aged ocean crust, and rift propagation from east to west across the West African margin (Karner and Gamboa 2007) (Fig. 6). The barrier to southern Atlantic marine incursions relates to the magmatic constructions of a proto-Walvis Ridge and the long-lived anomalous topography of the southeastern Brazilian highlands (Gradstein et al. 2004; Karner and Gamboa 2007). In the Early Cretaceous, isolated rift/strike-slip basins provided intermittent shallow-water connection in relatively small and isolated basins, and approximately in the Aptian–Albian a major rift phase led to deeper basins with restricted deep water circulation. In particular, during the Aptian occurred the first marine incursions into the northern coastal basins of the South Atlantic when the Walvis Ridge complex was bypassed: consequently, a mixed carbonate–silicoclastic platform system developed in a shallow epicontinental sea from late Aptian to mid–Albian (Koutsoukos 2001). Recent publications on the west African margin deposits also indicate the presence of marine deposits starting from the Aptian–Albian (Pletsch et al. 2001; Macdonald et al. 2003). No later than the early Cenomanian, a mature spreading stage allowed continuous and permanent deep-water connection between southern and northern Atlantic interrupting possible terrestrial connection

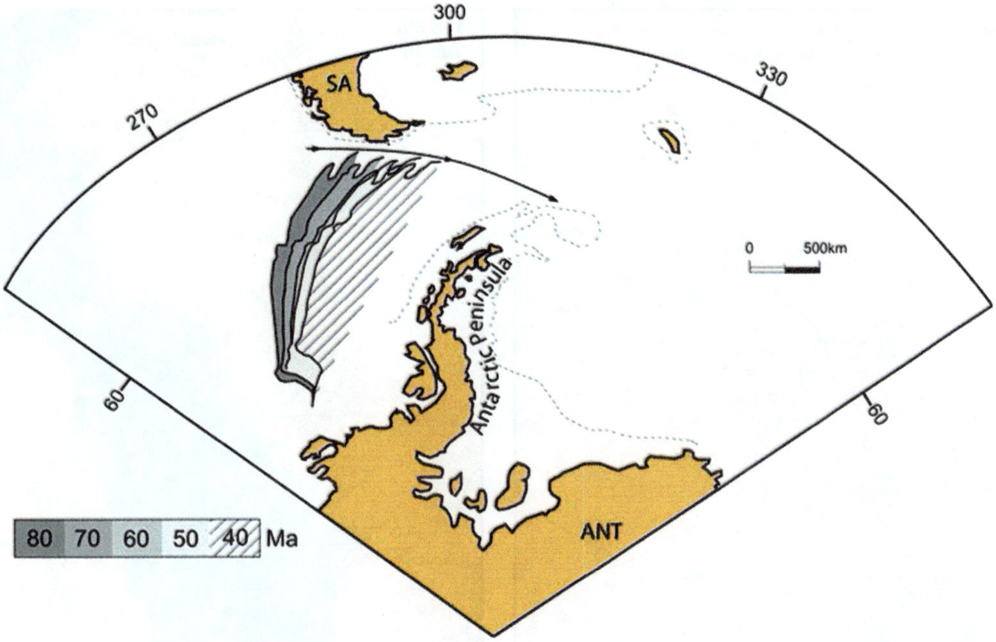

Fig. 5 Antarctic Peninsula motion relative to a fixed southern South America from 80 to 40 Ma (after Cunningham et al. 1995)

and faunal interchanges between south America and Africa (Petri 1987; Reyment and Dingle 1987; Pletsch et al. 2001; Tello Saenz et al. 2003; Guedes et al. 2005). Considering the geographical extension of the rifted margin, the presence of continental bridges, such as the Walvis Ridge, and occasional routes instead of a widely extended corridor between Africa and South America have to be considered as established probably from the Aptian (Fig. 6). Cretaceous faunal interchanges between western Africa and South America are widely documented in the dinosaur fossil record (Forster 1999; Candeiro et al. 2004, 2006; Sereno et al. 1994, 1996, 1998, 1999, 2004; Holtz et al. 2004; Makovicky et al. 2005; Coria and Currie 2006; Krause et al. 2006; Wilson 2006; Sereno and Brusatte 2008).

Mediterranean Region

Faunal interchanges between the proto-European region and Afro-Arabia, North America, and Asia are largely supported by the fossil record (Buffetaut and Le Loeuff 1991; Pérez-Moreno et al. 1994; Barrett and Upchurch 1995; Mateus and Antunes 2000; Dalla Vecchia 2002, 2005, 2008; Gierlinski and Sabath 2002; Dal Sasso 2003; Osi 2004; Evans et al. 2004; Sereno et al. 2004; Conti et al. 2005; Fejfar et al. 2005; Mateus 2006; Mateus et al. 2006, 2009; Prieto-Marquez et al. 2006; Sander et al. 2006; Wilson 2006) demonstrating the primary role of the circum-Mediterranean region in the Cretaceous paleobiogeography. Nevertheless, plate tectonic dynamics of the Mediterranean region has been controversial for decades with consequently uncertain reconstructions. Paleogeographic maps for the eastern Mediterranean area (Fig. 7) indicate a unique insular condition that persisted from the Late Jurassic to the latest Cretaceous characterized by large carbonate platforms (insular and peninsular conditions) occasionally linked to other landmasses (Stampfli et al. 2001; Bosellini 2002; Stampfli and Borel 2002; Vlahovic et al. 2005): however, due to geological gaps, reconstructions for the Turonian – early Campanian interval are largely incomplete (Philip and Floquet 2000). Geological and paleontological data available for the Cretaceous, according also to Dalla Vecchia (2002) indicate a temporal faunal diversification related to

1. geodynamics of the eastern Mediterranean plates,
2. carbonate platforms evolution, and

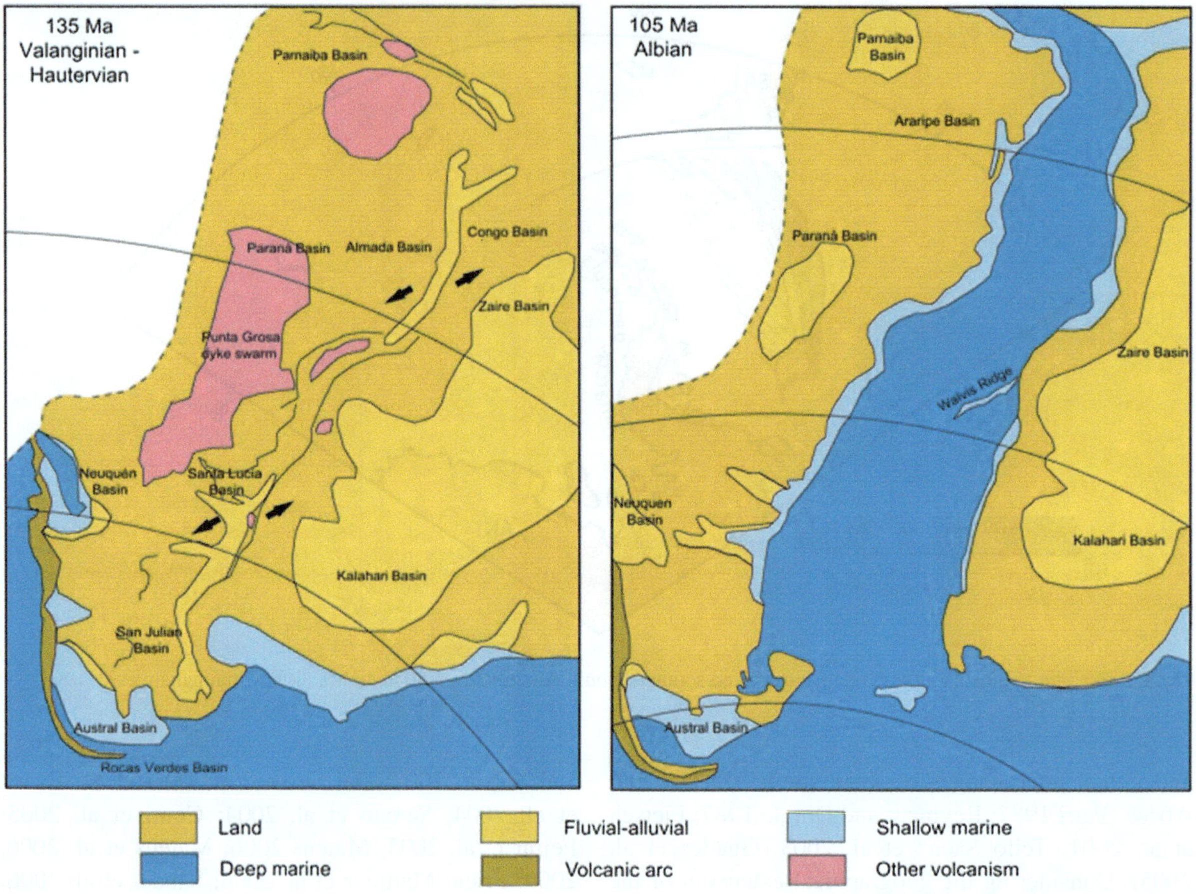

Fig. 6 Reconstruction of the South Atlantic region at approximately 135 and 105 Ma showing the progressive northward extension of the southern Atlantic Ocean (redrawn and modified after MacDonald et al. 2003). The Walvis Ridge has been proposed as a potential land bridge between Africa and South America. Major faunal exchange of terrestrial faunas between the two landmasses was interrupted at approximately 100 Ma

3. relative fluctuations of the sea level responsible for drowning or espousing of large areas. The mid-Cretaceous fragmentation of the areale and the presence of an insular environment, led to a higher adaptive pressure on terrestrial biota: as a consequence, local Cretaceous taxa display increasing adaptation to insular conditions (Dal Sasso and Signore 1998; Dalla Vecchia 2002, 2008; Dal Sasso 2003; Sander et al. 2006). Furthermore, hadrosaur, iguanodont, and nodosaur remains from the late Cretaceous of Europe suggest an Asian influence in the Mediterranean fauna (Dal Sasso 2003; Osi 2004; Fejfar et al. 2005; Prieto-Marquez et al. 2006). Thus, significant faunal turnover within the trans-Mediterranean region are suggested for the Bathonian–Callovian (end of north–south direct connection), for the Kimmerdgian–Tithonian (insular conditions), and for the middle Cretaceous (separation of Euro-Asiatic and African faunas).

Central America

During the late Jurassic (Callovian–Oxfordian), the proto-Caribbean Sea opened between North and South America creating a continuous seaway between the Pacific and the Atlantic margin, conditions that persisted during the lower Cretaceous with the progressive extension of the oceanic corridor. In the Albian, eastward subduction of the pacific plate led to the formation of a large shelf area in the westernmost part of the Caribbean plate (Meschede and Frisch 1998). Finally, from the Campanian, the relative motion between

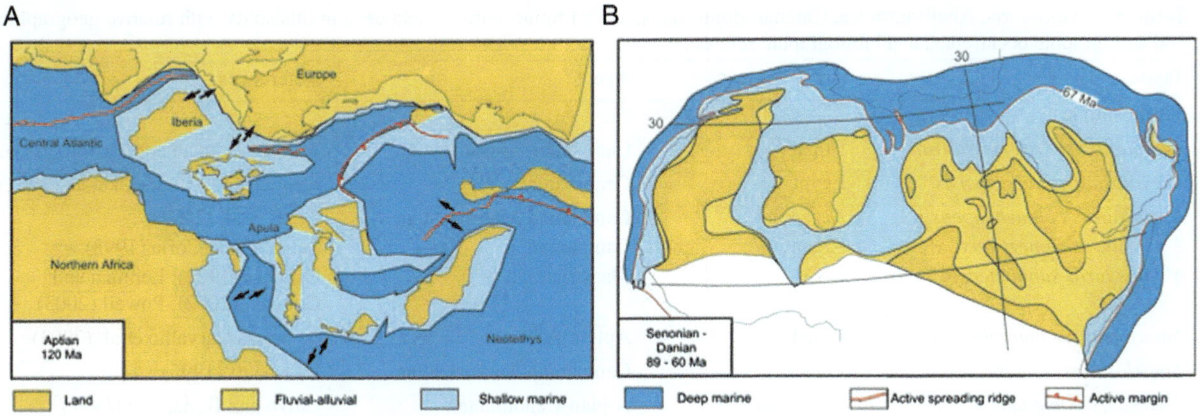

Fig. 7 Reconstruction of the North African and Mediterranean region. (**a**) Aptian (approx 120 Ma). (**b**) Senonian 89 Ma to Danian 60 Ma (modified after Stampfli et al. 2001). Recent discoveries from the Early Cretaceous of Portugal strongly support faunal connections between North America, Iberia, and northern Africa

North and South America became convergent (Pindell and Barrett 1990; Pindell 1993). Following recent paleogeographic reconstructions, only ephemeral island represented emerged land in the Caribbean area until the Miocene (Meschede and Frisch 1998): lacking further data on significant sea-level fluctuation that could have extended subaerial territories, a north–south dispersion pattern is not supported during the Cretaceous.

The geodynamic evolution of the Caribbean plate is discussed in order to test the presence of bridges or occasional routes between South America and North America during the late Jurassic and Cretaceous. In fact, reports of titanosaurs from the Campanian–Maastrichtian of North America (Gilmore 1922; McCord 1997; Lehman and Coulson 2002; Powell 2003) suggest the presence of a north–south dispersion pattern across the Caribbean area. However, *Venenosaurus dicrocei* from the Aptian of Utah (Tidwell et al. 2001) indicates an early dispersion of older members of the family.

Case Study: Dinosaur

During the Cretaceous, Gondwanan terrestrial faunas were composed by an enormous variety of taxa, including, crocodilians, turtles, lizards, snakes, dinosaurs, and mammals. Such groups of animals have various biogeographic history, resulting from different reactions to ecological changes and dispersal capabilities. As pointed out by Forster (1999), Holtz et al. (2004), and Krause et al. (2006) dispersal capabilities likely varied between dinosaur taxa as well. In order to test the most recent hypothesis on the Gondwanan dinosaur biogeography during the Cretaceous, a second perspective is employed. The geographical and stratigraphic occurrence of four predominantly Gondwanan dinosaur families (Titanosauria, Abelisauroidea, Carcharodontosauridae, and Spinosauridae) for a total of 98 taxa have been analyzed (Table 1). Their pattern of distribution appears to be strongly linked not only to the sequential break-up of Gondwana, but to all hypothesized Cretaceous land bridges discussed herein.

Titanosauria

Titanosaurs, the last surviving group of the giant sauropod dinosaurs, attained a nearly global distribution during the Cretaceous period (Table 1). Considered for a long time as the most significant endemic group of Gondwanan dinosaurs, findings of these large animals have been recorded throughout the world, including Africa, Asia, India, Europe, Madagascar, North and South America, and recently Australia (Wilson 2006) (Fig. 8a). This suggested the presence of persistent geographic connection between virtually all landmasses in the Cretaceous, particularly in the Campanian–Maastrichtian interval when phylogenetic analysis indicates close relationships among taxa from different continents.

The oldest titanosaur is *Janenschia robusta* from the Kimmeridgian–Berriasian deposits of Malawi (Wild

Table 1 Titanosauria, Abelisauroidea, Carcharodontosauridae, and Spinosauridae taxa used in this study with relative geographic and stratigraphic occurrence, and bibliographic reference

Taxon	Area	Age	References
Titanosauria			
Adamantinasaurus mezzalirai	Brazil	Campanian-Maastrichtian	Miloni-Santucci and Bertini (2006)
Aegyptosaurus baharijensis	Egypt	Cenomanian	Stromer (1932); Smith et al. (2001)
Aeolosaurus colhuehuapensis	Argentina	Campanian-Maastrichtian	Casal et al. (2007)
Aeolosaurus rionegrinus	Argentina	Campanian	Salgado and Coria (1993)
Alamosaurus sanjuanensis	USA	Maastrichtian	Gilmore (1922); Lehman and Coulson (2002); Powell (2003)
Amazonsaurus maranhensis	Brazil	Aptian-Albian	de Souza Carvalho et al. (2003)
Ampelosaurus atacis	France	Campanian-Maastrichtian	Le Loeuff (1995)
Andesaurus delgadoi	Argentina	Albian-Cenomanian	Calvo and Bonaparte (1991)
Antarctosaurus wichmannianus	Argentina	Campanian-Maastrichtian	Huene (1929); Powell (2003)
Argentinosaurus huinculensis	Argentina	Albian – Cenomanian	Bonaparte and Coria (1993)
Argyrosaurus superbus	Argentina	Campanian	Lydekker (1893); Powell (2003)
Austrosaurus mckillopi	Australia	Albian	Longman (1933); Salisbury et al. (2006)
Baurutitan britoi	Brazil	Maastricthian	Kellner et al. (2005)
Bonatitan reigi	Argentina	Campanian-Maastrichtian	Martinelli and Forasiepi (2004)
Bonitasaura salgadoi	Argentina	Santonian	Apesteguía (2004)
Borealosaurus wimani	China	Cenomanian-Turonian	You et al. (2004); Wilson (2005)
Daxiatitan binglingi	China	Lower Cretaceous	You et al. (2008)
Epachtosaurus sciuttoi	Argentina	Cenomanian-Turonian	Powell (1990); Martínez et al. (2004)
Erketu ellisoni	Mongolia	Aptian-Albian	Ksepka and Norell (2006)
Fusuisaurus zhaoi	China	Aptian- Albian	Mo et al. (2006)
Futalognkosaurus dukei	Argentina	Turonian-Coniacian	Calvo et al. (2007)
Gondwanatitan faustoi	Brazil	Campanian-Maas.	Kellner and de Azevedo (1999)
Huanghetitan liujiaxiaensis	China	Aptian- Albian	You et al. (2006)
Isisaurus colberti	India	Maastrichtian	Jain and Bandyopadhyay (1997)
Janenschia robusta	Malawi	Kimmeridgian-Berriasian	Fraas (1908); Jacobs et al. (1993)
Laplatasaurus araukanicus	Argentina	Maastrichtian	Huene (1929); Powell (2003)
Ligabuesaurus leanzai	Argentina	Aptian	Bonaparte et al. (2006)
Lirainosaurus astibiae	Spain	Campanian	Sanz et al. (1999); Powell (2003)
Macrurosaurus semmus	England	Valanginiano-Barremian	Seeley (1876); Powell (2003)
Magyarosaurus dacus	Romania	Maastrichtian	Powell (2003)
Malagasy taxon B	Madagascar	Campanian-Maastrichtian	Curry-Rogers (2002)
Malarguesaurus florenciae	Argentina	Turonian-Coniacian	González Riga et al. (2009)
Malawisaurus dixeyi	Malawi	Lower Cretaceous	Jacobs et al. (1993)
Maxakalisaurus topai	Brazil	Campanian	Kellner et al. (2006)
Maxakalisaurus topai	Brazil	Campanian-Maas.	Kellner et al. (2005)
Mendozasaurus neguyelap	Argentina	Turonian-Coniacian	González Riga (2003, 2005)
Muyelensaurus pecheni	Argentina	Turonian-Coniacian	Calvo et al. (2007)
Nemegtosaurus mongoliensis	Mongolia	Maastrichtian	Nowinski (1971); Wilson (2005)
Neuquensaurus australis	Argentina	Campanian	Salgado et al. (2005); Powell (2003)
Opisthocoelicaudia skarzynskii	Mongolia	Maastrichtian	Borsuk-Bialynicka (1977); Wilson (2005)

Table 1 (continued)

Taxon	Area	Age	References
Paralititan stromeri	Egypt	Cenomanian	Smith et al. (2001)
Pellegrinisaurus powelli	Argentina	Maastrichtian	Salgado (1996); Powell (2003)
Phuwiangosaurus sirindhornae	Thailand	Berriasian	Martin et al. (1994); Wilson (2005)
Puertasaurus reuili	Argentina	Maastricthian	Novas et al. (2005)
Quaesitosaurus orientalis	Mongolia	Santonian-Campanian	Kurzanov and Bannikov (1983); Wilson (2005)
Rapetosaurus krausei	Madagascar	Campanian-Maastrichtian	Curry-Rogers and Forster (2001)
Rinconsaurus caudamimus	Argentina	Upper Cretaceous	Calvo and González Riga (2003)
Rocasaurus muniozi	Argentina	Campanian-Maastrichtian	Salgado and Azpilicueta (2000)
Saltasaurus loricatus	Argentina	Campanian-Maastrichtian	Bonaparte and Powell (1980); Powell (1992); Powell (2003)
Titanos. Indet.	USA	Campanian	McCord (1997)
Titanos. Indet.	Mexico	Maastrichtian	Montellano-Ballesteros (2003); Jacobs et al. (1993)
Titanos. Indet.	Mali	Albian	O'Leary et al. (2004)
Titanos. Indet.	Morocco	Maastrichtian	Suberbiola et al. (2004)
Titanos. Indet.	Arabian Peninsula	Maastrichtian	Wilson et al. (2006)
Titanos. Indet.	India	Maastrichtian	Wilson and Mohabey (2006)
Titanos. Indet.	France	Maastrichtian	Le Loeuff et al. (1994)
Titanos. Indet.	France	Cenomanian	Buffetaut (1989)
Trigonosaurus pricei	Brazil	Maastricthian	Campos et al. (2005)
Uberabatitan riberoi	Brazil	Maastricthian	Salgado and De Souza Carvalho (2008)
Venenosaurus dicrocei (inc.sed.)	USA	Aptian	Tidwell et al. (2001)
Abelisauroidea			
Abelisauridae ge. Sp. Indet.	Argentina	Cenomanian-Turonian	Lamanna et al. (2002)
Abelisaurus comahuensis	Argentina	Campanian	Bonaparte and Novas (1985)
Aucasaurus garridoi	Argentina	Campanian	Coria et al. (2002)
Carnotaurus sastrei	Argentina	Campanian-Maastrichtian	Bonaparte (1985)
Ekrixinatosaurus novasi	Argentina	Cenomanian	Calvo et al. (2004a, b)
Ilokelesia aguadagrandensis	Argentina	Cenomanian	Coria and Salgado (1998)
Ligabueino andesi	Argentina	Hautervian	Bonaparte (1996)
MCF-PVPH-237 *(Bayosaurus)*	Argentina	Turonian-Coniacian	Coria et al. (2006)
Noasaurus leali	Argentina	Campanian-Maastrichtian	Bonaparte and Powell (1980)
Velocisaurus unicus	Argentina	Santonian	Bonaparte (1991)
Xenotarsosaurus bonapartei	Argentina	Cenomanian	Martínez et al. (1986); Coria and Rodríguez (1993)
Pycnonemosaurus nevesi	Brazil	Turonian-Santonian	Kellner and Campos (2002)
Indosaurus matleyi	India	Maastrichtian	Huene and Matley (1933); Chatterjee (1978)
Indosuchus raptorius	India	Maastrichtian	Huene and Matley (1933); Chatterjee (1978)
Rajasaurus narmadensis	India	Maastrichtian	Wilson et al. (2003)
Abelisauridae ge. Sp. Indet.	Morocco	Cenomanian	Mahler (2005)

Table 1 (continued)

Taxon	Area	Age	References
Deltadromeus agilis	Morocco	Aptian-Albian	Sereno et al. (1996)
Berberosaurus liassicus	Morocco	Pliensbachian-Toarcian	Allain et al. (2007)
Abelisauridae Indet.	Egitto	Campanian-Maastrichtian	Smith (2005)
Rugops primus	Niger	Cenomanian	Sereno et al. (2004)
Kryptops palaios	Nger	Aptian-Albian	Sereno and Brusatte (2008)
Majungasaurus crenatissimus	Madagascar	Campanian	Depéret (1896); Carrano (2007); O'Connor (2007); Sampson and Witmer (2007)
Masiakasaurus knopfleri	Madagascar	Campanian-Maastrichtian	Sampson et al. (2001)
Genusaurus sisteronis	France	Albian	Accarie et al. (1995); Buffetaut et al. (1995)
Tarascosaurus salluvicus	France	early Campanian	Le Loeuff and Buffetaut (1991)
Carcharodontosauridae			
Giganotosaurus carolinii	Argentina	Albian-Cenomanian	Coria and Salgado (1995)
Mapusaurus roseae	Argentina	Cenomanian	Coria and Currie (2006)
Tyrannotitan chubutensis	Argentina	Aptian	Novas et al. (2005b)
Bahariasaurus ingens	Egitto	early Cenomanian	Stromer (1934); Smith et al. (2006a, b)
Carcharodontosaurus saharicus	North Africa	Aptian-Turonian	Depéret and Savornin (1927); Sereno et al. (1996)
Eocarcharia dinops	Niger	Aptian-Albian	Sereno and Brusatte (2008)
Neovenator salerii	England	Barremian	Hutt et al. (1996)
Acrocanthosaurus atokensis	USA	Aptian-Albian	Stovall and Langston (1950); Currie and Carpenter (2000)
Spinosauridae			
Irritator challengeri	Brazil	Aptian	Martill et al. (1996); Sues et al. (2002)
Baryonychinae Indet.	Niger	Aptian	Taquet and Russell (1998)
Suchomimus tenerensis	Niger	Albian	Sereno et al. (1998)
Baryonyx walkeri	England	Barremian	Charig and Milner (1986)
Spinosaurus aegyptiacus	Egypt	Cenomanian	Stromer (1915); Buffetaut and Ouaja (2002);
			Machado and Kellner (2005); Dal Sasso et al. (2005)

1991). Chronologically, *Phuwiangosaurus sirindhornae* (Martin et al. 1994; Berriasian of Thailand) is the oldest titanosaur in Asia, *Macrurosaurus semmus* (Seeley 1876; Valanginian-Barremian of England) in Europe, *Venenosaurus dicrocei* (Tidwell et al. 2001; Aptian? of Utah) in North America, *Amazonsaurus maranhensis* (de Souza Carvalho et al. 2003, Aptian of Brazil), Gondwanatitan faustoi (Kellner and de Azevedo 1999; Late Cretaceous of Brazil), and *Ligabuesaurus leanzai* (Bonaparte et al. 2006; Aptian of Argentina) in South America, and *Austrosaurus mckillopi* (Longman 1933; Albian of Queensland) in Australia. To date, only one taxon is known from the Jurassic, 14 from the Early Cretaceous (Berriasian–Albian), and 45 from the Late Cretaceous (Cenomanian–Maastrichtian) indicating an increasing evolutionary success of this group (Fig. 8b). On the contrary, regional record in both Madagascar and India presents tremendous gaps, with titanosaurs reported only from Campanian and Maastrichtian deposits. Titanosaurs have been recently reported from the Late Cretaceous of Kenya (Sertich et al. 2006), Angola (Jacobs et al. 2006), and the Arabian Peninsula (Wilson et al. 2006), providing further evidences

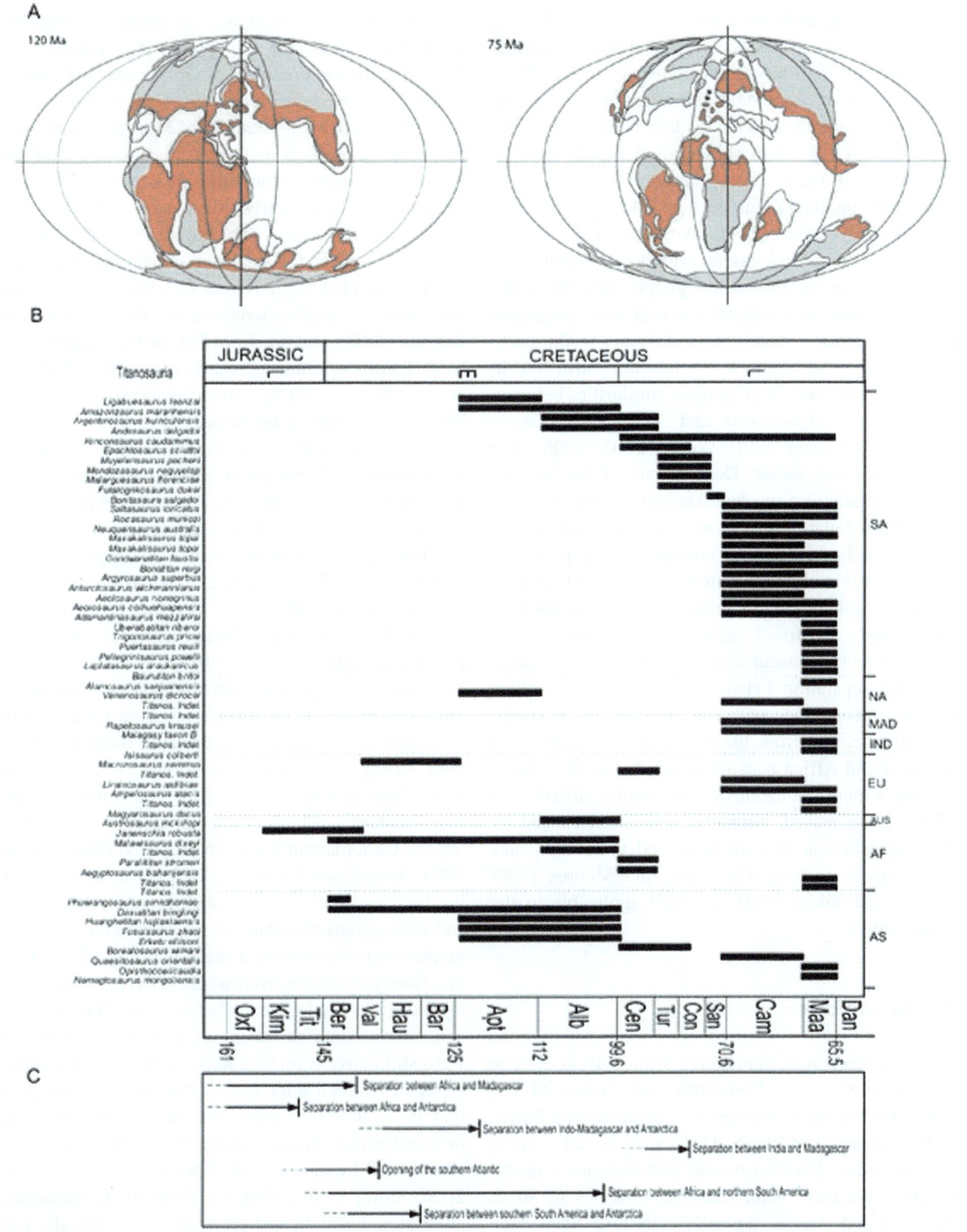

Fig. 8 (**a**) Hypothesized paleogeographic distribution of titanosaurs at 120 and 75 Ma (see the text for discussion). A northern route in the Early Cretaceous extended from North America to Asia supports phylogenetic analogies between contemporaneous faunas. Late Cretaceous continental break-up led to a forced fragmentation of distinct population with the only significant exception of South America. (**b**) Stratigraphic and geographic distribution of Titanosauria. (**c**) Major Gondwanan break-up events

for a Pan-African distribution of this family. Lastly, although one taxon has been reported from Australia (Molnar and Salisbury 2005; Salisbury et al. 2006), it represents an extremely important datum in the paleobiogeographic analysis of this family. Available data suggest that from the original areale sited in central Africa, titanosaurs spread rapidly into nearest continents, including Europe (and consequently Asia and North America), South America, and, through Antarctica, Indo-Madagascar prior to its isolation at 120 Ma. Sedimentological and geophysical data from the Caribbean area support an east-west migration for titanosaurus in northern landmasses in the early Cretaceous rather than a north–south diffusion in the late Cretaceous. Phylogenetic similarities between *Alamosaurus sanjuanensis* and Asian taxa (Curry-Rogers and Forster 2001; Wilson 2002, 2006) also support this hypothesis. Descendants of the original African populations reached Australia before its separation from Antarctica (also see Salisbury et al. 2006). Middle and late Cretaceous continental fragmentation led to a worldwide distribution of this family characterized by a significant increase in the species number. Interestingly, the fossil record displays peaks in the occurrence of titanosaur taxa in the Barremian–Aptian after the separation between Africa and Antarctica and in concomitance with the opening of the southern Atlantic, and from the Cenomanian when South America and Africa became isolated (Fig. 8c). Thus, lacking a more homogeneous geographic distribution of the fossil record, allopatric speciation caused by the fragmentation of continents and therefore peaks in biological diversity (Valentine and Moores 1970; Tiffney and Niklas 1990) are well applicable to the Titanosauria.

Abelisauroidea

It has now been more than 20 years since Abelisauridae was established by Bonaparte and Novas (1985) when *Abelisaurus comahuensis* was described. Novas (1991) coined the name Abelisauria to encompass Abelisauridae, *Noasaurus leali* and their most recent common ancestor. After recent discoveries in South America and Africa, nowadays 25 taxa have been included in the theropod superfamily Abelisauroidea. The current fossil record suggests that Abelisauroids underwent a significant evolutionary radiation starting with the middle-late Cretaceous colonizing southern landmasses (with the exception of Australia) and becoming the top predators of Gondwana (Fig. 9a).

To date, *Berberosaurus liassicus* (Allain et al. 2007; Pliensbachian–Toarcian of Morocco) is the oldest taxon classified within Ablisauroidea and the sole taxa recovered from Jurassic deposits (Table 1). Interestingly, the early Cretaceous (Hautervian–Albian) record includes one taxon from South America (*Ligabueino andesi*, Hautervian, Argentina; Bonaparte 1996), two taxa from Africa (*Kryptops palaios* and *Deltadromeus agilis*; Sereno et al. 1996; Sereno and Brusatte 2008), and one from Europe (*Genusaurus sisteronis*, Albian, France; Accarie et al. 1995; Buffetaut et al. 1995) suggesting a rapid dispersion before the break-up of southern landmasses (Fig. 9b).

According to the most recent analyses proposed by Carrano and Sampson (2008) and Sereno and Brusatte (2008), this group of dinosaur is particularly problematic in terms of geographic and stratigraphic relationships between taxa which share phylogenetic relationships. For instance, *Deltadromeus agilis* (Aptian–Albian, Morocco; Sereno et al. 1996), *N. leali* (Campanian–Maastrichtian, Argentina; Bonaparte and Powell 1980), and *Masiakasaurus knopfleri* (Campanian–Maastrichtian, Madagascar; Sampson et al. 2001) form a separate group within Abelisauroidea named Noasaurids. In order to support their strong phylogenetic relationships a "southern way" (linking Africa, South America, Antarctica, and Indo-Madagascar via continental bridges) active during the Cenomanian has been proposed (Sereno et al. 1994; Sereno and Brusatte 2008). However, considering both well-documented phylogenetic relationships and paleogeographic data, the most congruent biogeographic hypothesis would place the original areale of this family in Africa from where different taxa spread toward Indo-Madagascar (before 140 Ma), Europe, and South America when intercontinental corridors were still active (Fig. 9c). Reports of isolated material ascribed to abelisaurs from Northern Africa suggest that these theropod dinosaurs were a diverse group in continental Africa during the Early Cretaceous (Smith and Lamanna 2006; Smith and Dalla Vecchia 2006; Smith et al. 2006a, b). From the Cenomanian, abelisaurs fossil record increases progressively with 9 taxa in the Cenomanian–Santonian interval and 11 in the Campanian–Maastrichtian. Interestingly, as observed in Titanosauria, the number of abelisauroid

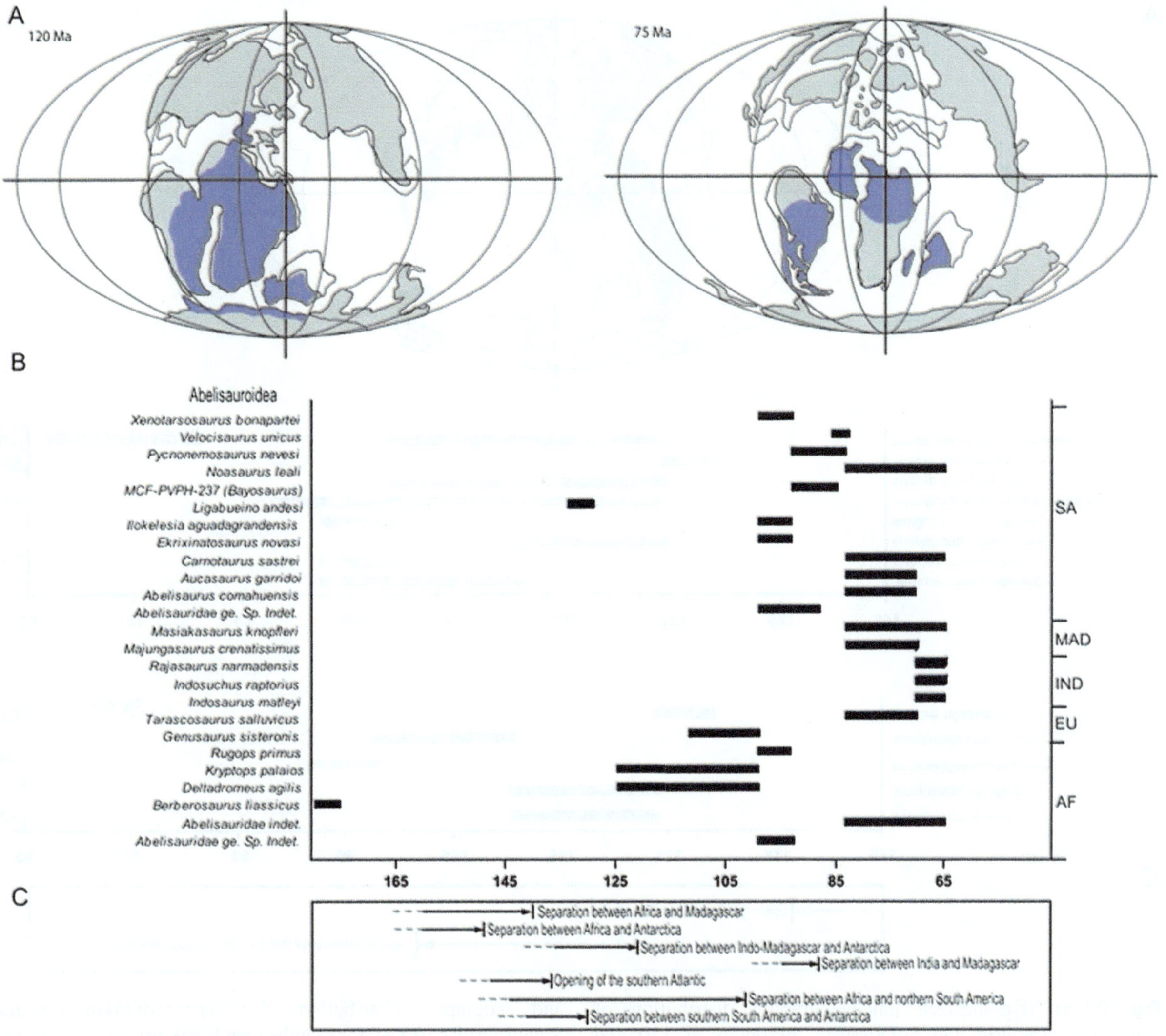

Fig. 9 (**a**) hypothesized paleogeographic distribution of Abelisauroid theropods at 120 and 75 Ma. Note a significant gap in the fossil record for the Late Cretaceous of central and southern Africa. (**b**) Stratigraphic and geographic distribution of Abelisauroidea. (**c**) Major Gondwanan break-up events

taxa significantly increases in concomitance and after the separation of Africa and northern South America.

Carcharodontosauridae and Spinosauridae

Despite the fact that carcharodontosaurids haven't been recovered in all southern continents, their remains have been reported from Europe, Africa, North, and South America. Thus their wide geographic distribution involves both the Mediterranean region and the Africa–South America margin where continental bridges have been proposed (Fig. 10a). *Acrocanthosaurus atokensis* (Stovall and Langston 1950; Currie and Carpenter 2000) from the Aptian–Albian of North America is nearly coeval with other carcharodontosaurids from Europe (*Neovenator salerii*, Hutt et al. 1996), Africa (*Carcharodontosaurus saharicus*, Depéret and Savornin (1927) and *Eocarcharia dinops*, Sereno and Brusatte 2008), and South America (*Tyrannotitan chubutensis*, Novas et al. 2005b) (see also Table 1) indicating rapid distribution of these large theropods in the Early Cretaceous. In the light of recent discoveries, spinosaurids display similarities with carcharodontosaurids in geographic and stratigraphic distribution.

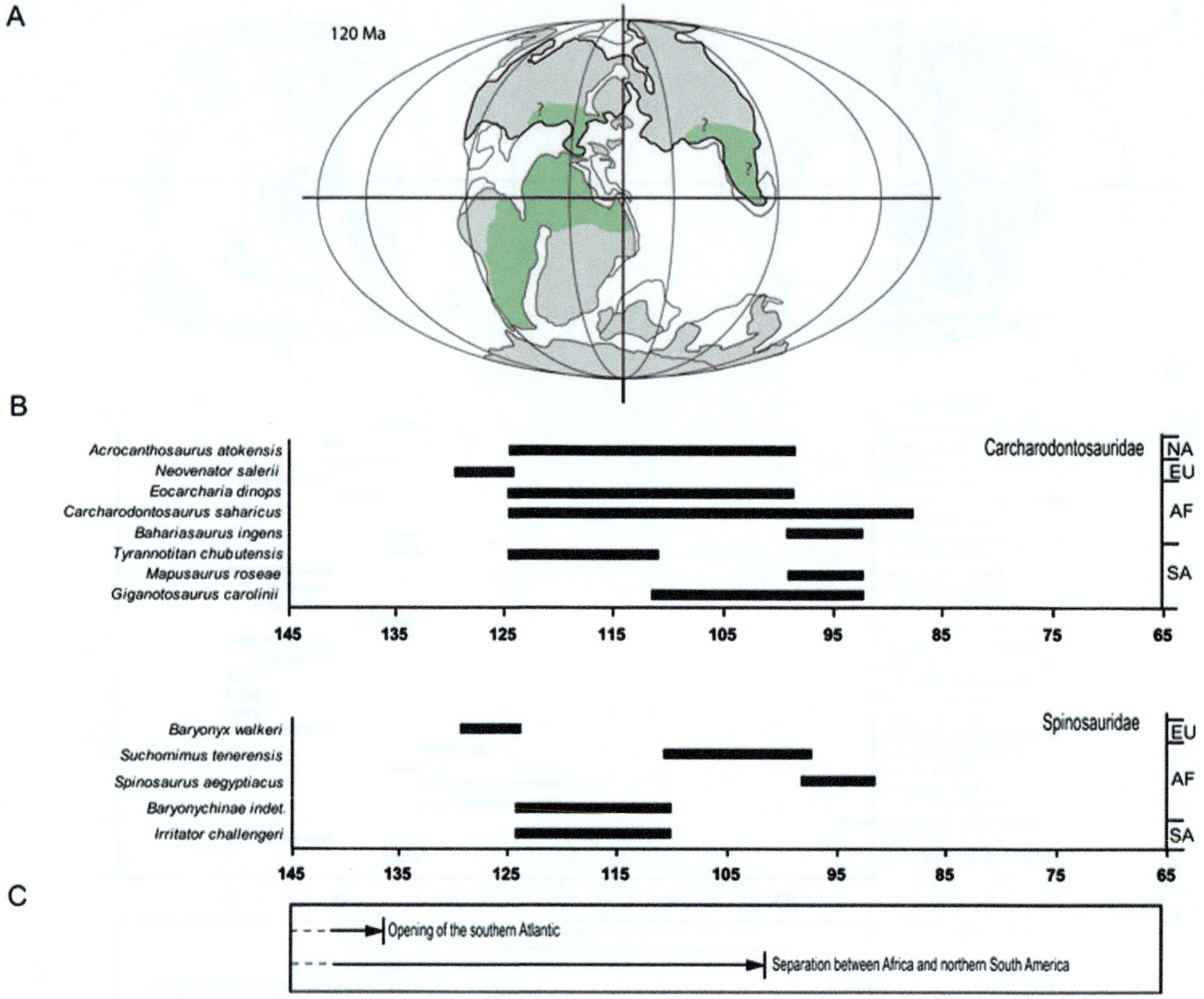

Fig. 10 (**a**) Hypothesized paleogeographic distribution of Carcharodontosaurid and Spinosaurid theropods at 120 and 75 Ma (see the text for discussion), also in the light of recent reports of spinosaurs from Asia. (**b**) Stratigraphic and geographic distribution of Carcharodontosauridae and Spinosauridae. (**c**) Major Gondwanan break-up events linked to the geographic distribution of the two theropod families

The oldest taxa of the family are *Baryonyx walkeri* (Barremian of England, Charig and Milner 1986), *Irritator challengeri* (Aptian of Brazil, Martill et al. 1996), and Baryonychinae indet (Aptian of Niger, Taquet and Russell 1998). Recently, isolated teeth and postcranial element referred to spinosaurid theropods have been reported from Japan, China, and Thailand (Buffetaut and Suteethorn 1998; Buffetaut and Ouaja 2002; Buffetaut et al. 2008). Pending further discoveries in these localities, such discoveries would remarkably extend the geographical range of Asian spinosaurs (Fig. 10b).

Nine taxa have, to date, been included in Carcharodontosauridae and 5 in Spinosauridae (Table 1): according to the fossil record, the majority of the taxa ascribed to these families appears geographically confined to central and northern Gondwana, and stratigraphically confined to the Barremian–Turonian interval (Fig. 10c). A possible explanation for the disappearance of both families are the global transgressive events in Cenomanian: in particular, northern Africa and South America were interested by widespread inland seas and short-termed climatic cooling (Voigt and Wiese 2000; Gale et al. 2002; Voigt et al. 2004; Jaimes and de Freitas 2006; Zagrarni et al. 2008).

Synthesis and Discussion

The synthesis of geological and paleogeographic data in this study enables fundamental events of the break-up of Gondwana to be placed sequentially in a robust chronostratigraphic framework (Fig. 11). For example, Africa separated from Madagascar starting at 160 Ma, from Antarctica at 150 Ma, from southern South America at 140 Ma, and from northern South America at 100 Ma. The Indo-Madagascar block became fully separated from Africa no later than 140 Ma, and from Antarctica between 135 and 120 Ma. India and Madagascar split at approximately 88 Ma. Lastly, southern South America and Antarctica separated at approximately 130 Ma. The Mediterranean region was characterized by insular conditions during most of the Cretaceous, whereas an ephemeral link between North and South America was established possibly in the latest Cretaceous. During the break-up of Gondwana, however, geographically restricted areas represented active corridors for terrestrial faunas: for example, a north-south connection through the Mediterranean region during the Late Jurassic and possibly for restricted intervals in the Early Cretaceous and the well-documented link between Africa and northern South America in the Early Cretaceous. On the contrary, geophysical and sedimentological data do not support the existence of a "southern way" postulated on the basis of paleontological and biological evidences and in particular the presence of continental bridges between Antarctica and Indo-Madagascar in the middle–late Cretaceous. Recent investigation on the Kerguelen Plateau and the Elan Bank (the sole section of the plateau that emerged in the Cretaceous) indicates that the Elan Bank

1. presented favorable conditions for dinosaur dispersal exclusively at 100 Ma, and not in the late Cretaceous as previously supposed, and
2. moved southward increasing progressively its distance from the Indo-Malagasy block.

Some authors indicate alternative bridges in order to link late Cretaceous fauna of Madagascar, India, South America, and Asia, for instance, a southern way through the Gunnerus Ridge (Case 2002; Noonan and Chippindale 2006), and a northern way that linked Madagascar with other landmasses through the Seychelles area and eventually to Great India (Rage 2003). To date, such hypotheses lack fundamental geological and paleontological support. As a consequence, the complete insulation of the Indo-Madagascar sub-continent and its dinosaurian faunas has to be confidentially placed at 120 Ma (see also Ramana et al. 2001; Briggs 2003; Jokat et al. 2003).

The presence of seamounts in the Mozambique Channel and in the Seychelles area as well as the vast submarine plateau between Antarctica and Indo-Madagascar probably represented fundamental corridors for marine taxa during the entire Cretaceous period. The South America–Antarctica connection via the Antarctic Peninsula is supported by the dinosaur fossil record but is still not universally accepted. Plate reconstructions proposed by Crame (1992, 1999) indicate the presence of a corridor between the two continental blocks characterized by a chain of emergent volcanic islands separated by shallow seas. Such environments suggest a minor role in terrestrial dispersion, whereas they represent fundamental dispersal ways for plants and marine taxa. As pointed out by Crame (1999), by the Late Cretaceous the area was located in the southern high latitudes and therefore affected by climatic variations, and by the rise of the proto-circum-Antarctic current.

With the dislocation of Gondwana and consequent high sea levels and the consequent inundation of lands by epicontinental seas, the Late Cretaceous is the most complex period for dinosaur biogeography. Remarkably, based on the available data, it is possible to observe an increase in diversity in both the Titanosauria and Abelisauroidea after the complete separation between Africa and South America at the end of the Albian stage. This trend continues with an increasing rate up to the Campanian and the Maastrichtian, when the fragmentation and consequent isolation of southern landmasses reaches its maximum. Such dynamics seem to reflect pulses in response to major environmental variations rather than intermittent, intra-continental migrations.

There is a clear imbalance in terms of collected and adequately described taxa with very few radiometric data available, between the Early and Late Cretaceous as well as between Africa and other southern landmasses. Similarly, the fossil record from Antarctica, Australia, and New Zealand is extremely poor relative to faunas recovered from other Gondwanan landmasses. Since the late 1970s, Australia has found from Aptian and Campanian–Maastrichtian deposits,

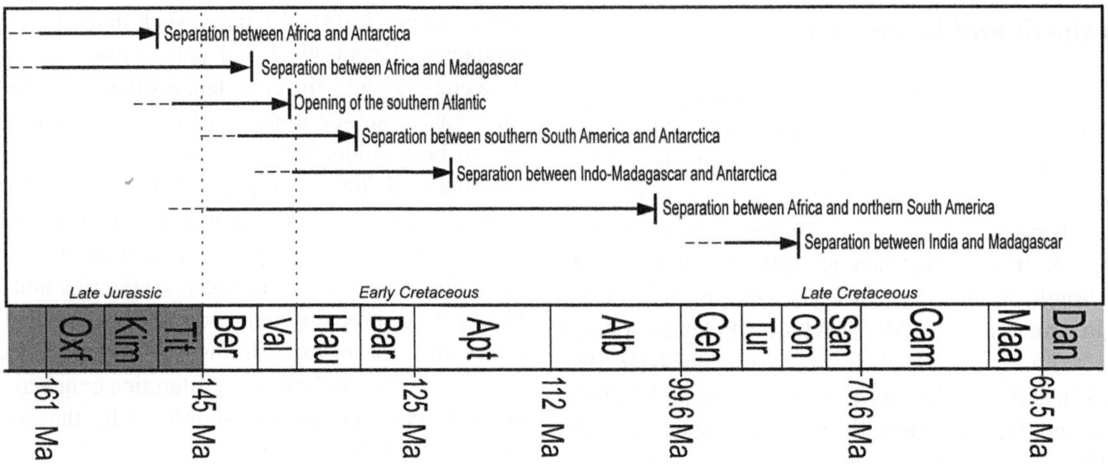

Fig. 11 Timing of the establishment of permanent marine conditions between southern landmasses (see text for discussion). In light of the available geological and paleontological data, the Late Jurassic and earliest Cretaceous (Berriasian–Valanginian) time represent the most likely interval for dinosaur dispersion across Gondwana

nearly all major dinosaur groups known elsewhere in Gondwanan landmasses, including ankylosaurs, iguanodonts, theropods, and titanosaurs (Longman 1933; Exon and Vine 1970; Molnar 1980; Molnar et al. 1981, 1985; Bartholomai and Molnar 1981; Hocknull et al. 2009; see also Weishampel et al. 2004 for a detailed record of Australasian taxa).

Recent phylogenetic analyses have clarified relationships between taxa separated both stratigraphically and geographically: however, proposed timing of dinosaur radiation displays several discrepancies with the available paleogeographic data. Beside Titanosauria, Abelisauroidea, Carcharodontosauridae, and Spinosauridae, maniraptoran dinosaurs represent another important element for Gondwana paleobiogeography. Recent claims that maniraptorans spread into Gondwanan continents before the Late Jurassic is now supported by several important finds (Rauhut and Werner 1995; Forster et al. 1996; Forster 1999; Krause et al. 1999; Calvo et al. 2004a, b, 2007; Novas and Agnolin 2004; Makovicky et al. 2005; Novas and Pol 2005; Fanti and Therrien 2007; Novas et al. 2008). Together with other Gondwanan dinosaurs discussed in this study, maniraptorans support a Late Jurassic to Early Cretaceous peak in dinosaur diffusion.

The two datasets included in this study suggest that Gondwanan dinosaurs may be overall considered immigrant during the Late Jurassic and locally in the Early Cretaceous when intercontinental connection represented active ways of dispersal for terrestrial taxa. After the fragmentation of Gondwana, primarily neoendemic taxa inhabited all southern continents.

Conclusions

In spite of important gaps in fossil geographic and stratigraphic distributions, this study indicates complex interaction between Gondwana fragmentation and dinosaur vicariance and dispersal during the Cretaceous. Geological data do not support the presence of hypothesized subaerial connections between Antarctica, South America, and Madagascar that constitute the bulk of most recent paleobiogeographic models for the Late Cretaceous Gondwanan dinosaurs, including both the "Africa-First" and "Pan-Gondwana" models.

To the contrary, paleontological and paleogeographic data discussed here support earlier theories for a Gondwanan distribution of several dinosaur groups prior to the complete separation of Laurasia and Gondwana and the break-up of southern landmasses. In particular, ancestors of the Late Cretaceous Malagasy and Indian theropods and sauropods (abelisauroids and titanosaurs) must have reached the Indo-Madagascar landmass before its separation from continental Africa around 145 Ma. The survival of taxa long after geographic separation from the original "population" is explained by early dispersal and following vicariance rather than through a late immigration. Interchanges of terrestrial faunas via corridors appear limited to the Late Jurassic with the sole exception of the Africa-northern South America connection before 100 Ma, whereas continental bridges *sensu stricto* may have occasionally linked Europe and Africa through the Mediterranean area and South America and Antarctica through the Antarctic Peninsula.

Although this study clarifies some aspects of Gondwana biogeography, the many questions still remaining will be answered by further discoveries from critical regions (such as continental Africa, Australia, and Antarctica). Paleontological data from the virtually unknown Jurassic deposits of Madagascar, India, and southern Africa will support present or new biogeographic scenarios.

Acknowledgments This study was conducted at the Dipartimento di Scienze della Terra e Geologico-Ambientali (Alma Mater Studiorum, Università di Bologna, Italy) and at the Royal Tyrrell Museum of Palaeontology (Drumheller, Canada). The fundamental suggestions of John Talent and Tom Rich greatly improved the manuscript. The author wishes to thank Matthew Lamanna, Andrea Cau, Simone Maganuco, and Tetsuto Miyashita for enlightening discussions and constructive comments at various stages of this research. The study was financially supported by the Rotary International Foundation (Chicago, USA) and the Museo Geologico Giovanni Capellini (Bologna, Italy).

References

Accarie H, Beaudoin B, Dejax J, Fries G, Michard J-C, Taquet P (1995) Decouverte d'un dinosaure theropode nouveau (Genusaurus sisteronis n. g., n. sp.) dans l'Albien marin de Sisteron (Alpes de Haute-Provence, France) et extension au Cretacé inferieur de la lignee ceratosaurienne. C R Hebd Seance Acad Sci Paris 320(2):327–334

Allain R, Tykoski R, Aquesbi N, Jalil N, Monbaron M, Russell D, Taquet P (2007) A basal abelisauroid from the late Early Jurassic of the High Atlas Mountains, Morocco, and the radiation of ceratosaurs. J Vertebr Paleontol 27:610–624

Ali J, Aitchison J (2009) Kerguelen Plateau and the Late Cretaceous southern-continent bioconnection hypothesis: tales from a topographic ocean. J Biogeogr 36:1778–1784

Anderson M, McNaughton S, Ritchie M (2004) Scale-dependent relationship between the spatial distribution of a limiting resource and plant species diversity in an African grassland ecosystem. Oecologia 139:277–287

Apesteguía S (2004) Bonitasaura salgadoi: a beaked sauropod in the Late Cretaceous of Gondwana. Naturwissenschaften 91:493–497

Badano E, Regidor H, Nùnez H, Acosta R, Gianoli E (2005) Species richness and structure of ant communities in a dynamic archipelago: effects of island area and age. J Biogeogr 32:221–227

Barker P, Lawver L (1988) South American–Antarctic plate motion over the past 50 Myr, and the evolution of the South American–Antarctic ridge. Geophys J R Astron Soc 94:377–386

Barrett P, Upchurch P (1995) Regnosaurus northamptoni, a stegosaurian dinosaur from the Lower Cretaceous of southern England. Geol Mag 132:213–222

Bartholomai A, Molnar R (1981) Muttaburrasaurus, a new iguanodontid (Ornithischian: Ornithopoda) dinosaur from the Lower Cretaceous of Queensland. Mem Queensland Mus 20:319–349

Besse J, Courtillot V (1988) Paleogeographic maps of the continents bordering the Indian Ocean since the Early Jurassic. J Geophys Res 93:11791–11808

Bonaparte J (1985) A horned Cretaceous carnosaur from Patagonia. Natl Geog Res 1:150–152

Bonaparte J (1986) History of the terrestrial Cretaceous vertebrates of Gondwana. Actas IC Congreso Argentino Paleontologia Bioestratigrafia 4. Actas 2:63–95

Bonaparte J (1991) Los vertebrados fósiles de la Formación Rio Colorado, de la Ciudad de Neuquén y Cercanías, Cretácico Superior, Argentina [The vertebrate fossils of the Rio Colorado Formation, from the city of Neuquén and surrounding areas, Upper Cretaceous, Argentina]. Rev Mus Argentino Cienc Nat "Bernardino Rivadavia" Inst Nac Invest Cienc Nat Paleontol 4:17–123

Bonaparte J (1996) Cretaceous tetrapods of Argentina. Münchner Geowissenschaft Abh A 30:73–130

Bonaparte J, Coria R (1993) Un nuevo y gigantesco sauropodo titanosaurio de la Formacion Rıo Limay (Albiano–Cenomaniano) de la provincia del Neuquen, Argentina. Ameghiniana 30:271–282

Bonaparte J, Powell J (1980) A continental assemblage of tetrapods from the Upper Cretaceous beds of El Brete, Northwestern Argentina (Sauropoda-Coelurosaria-Carnosauria-Aves). Mém Soc Géol France NS 139:19–28

Bonaparte J, Novas F (1985) Abelisaurus comahuensis, n g, n sp, Carnosauria from the Late Cretaceous of Patagonia. Ameghiniana 21:259–265

Bonaparte J, Gonzales-Riga B, Apesteguia S (2006) Ligabuesaurus leanzai gen. et sp. nov. (Dinosauria, Sauropoda), a new titanosaur from the Lohan Cura formation (Aptian, Lower Cretaceous) of Neuquén, Patagonia, Argentina. Cretaceous Res 27:364–376

Borissova I, Coffin M, Charvis P, Operto S (2003) Structure and development of a microcontinent: Elan Bank in the southern Indian Ocean. Geochem Geochim Geosyst 4:doi:10.1029/2003GC000535

Borsuk-Bialynicka M (1977). A new camarasaurid Opisthocoelicaudia gen. n. sp. n. from the Upper Cretaceous of Mongolia. Palaeontol Pol 37:5–64

Bosellini A (2002) Dinosaurs "re-write" the geodynamics of the eastern Mediterranean and the paleogeography of the Apulia Platform. Earth Sci Rev 59:211–234

Briggs J (2003) The biogeographic and tectonic history of India. J Biogeogr 30:381–388

Buckley G, Brochu C (1996) Campanian (Upper Cretaceous) crocodyliforms from Madagascar and their biogeographic implication. J Vertebr Paleontol 16:24A

Buckley G, Brochu C (1999) An enigmatic new crocodile from the Upper Cretaceous of Madagascar. Spec Pap Palaeontol 60:149–175

Buckley G, Brochu C (2001) A skull of Mahajangasuchus insignis (crocodiliformes) from the Upper Cretaceous of Madagascar. J Vertebr Paleontol 21:36A

Buckley G, Brochu C, Krause D, Pol D (2000) A pug-nosed crocodyliform from the Late Cretaceous of Madagascar. Nature 405:941–944

Buffetaut E (1989) Les Dinosaures du Crétacé terminal du Sud de la France: données tirées des restes osseux In La limite Crétacé–Tertiaire dans le bassin de l'Arc (Sud-est France). Cahiers Réserve Géol Digne 1:17–22

Buffetaut E, Le Loeuff J (1991) Late Cretaceous dinosaur faunas of Europe: some correlation problems. Cretaceous Res 12:159–176

Buffetaut E, Ouaja M (2002) A new specimen of Spinosaurus (Dinosauria, Theropoda) from the Lower Cretaceous of Tunisia, with remarks on the evolutionary history of the Spinosauridae. Bull Soc Geol France 5:415–421

Buffetaut E, Rage J (1993) Fossil amphibians and reptiles and the Africa–South America connection. In: George W, Lavocat R (eds) The Africa–South America connection. Clarendon Press, Oxford, pp 87–99

Buffetaut E, Suteethorn V (1998) The biogeographical significance of the Mesozoic vertebrates from Thailand. In: Hall R, Holloway D (eds) Biogeography and geological evolution of SE Asia. Backhuys Publisher, Leiden, pp 83–90

Buffetaut E, Mechin P, Mechin-Salessy A (1988) Un dinosaure theropode d'affinites gondwaniennes dans le Cretace superieur de Provence. C R Acad Sci Paris Ser II 306:153–158

Buffetaut E, Le Loeuff J, Accarie H, Beaudoin B, Dejax J, Fries G, Michard J, Taquet P (1995) Discovery of a new theropod dinosaur (Genusaurus sisteronis n. g., n. sp.) in the marine Albian of Sisteron (Alpes-de-Haute Provence, France) and the extension into the Lower Cretaceous of the ceratosaur lineage; reply. C R Acad Sci Ser II Sci Terre Planetes 321:79–83

Buffetaut E, Suteethorn V, Tong H, Amiot R (2008) An early Cretaceous spinosaurid theropod from southern China. Geol Mag 145:745–748

Calvo J, Bonaparte J (1991) Andesaurus delgadoi gen. et sp. nov. (Saurischia–Sauropoda) Dinosaurio Titanosauridae de la Formaci6n Rio Lirnay (Albiano–Cenomaniano) Neuqutn, Argentina. Ameghiniana 28:303–310

Calvo J, González Riga B (2003) Rinconsaurus caudamirus gen. et sp nov., a new titanosaurid (Dinosauria, Sauropoda) from the Late Cretaceous of Patagonia, Argentina. Rev Geol Chile 30:333–353

Calvo, J, Porfiri J, Kellner A (2004a) On a new maniraptoran dinosaur (Theropoda) from the Upper Cretaceous of Neuquen, Patagonia, Argentina. Rev Mus Argentino Cienc Nat NS 62:549–566

Calvo J, Porfiri J, Veralli C, Novas F, Poblete F (2004b) Phylogenetic status of Megaraptor namunhuaiquii Novas based on a new specimen from Neuquén, Patagonia, Argentina. Ameghiniana 41:565–575

Calvo J, Porfiri J, González-Riga B, Kellner A (2007) A new Cretaceous terrestrial ecosystem from Gondwana with the description of a new sauropod dinosaur. Anais Acad Brasil Ciências 79:529–541

Campos D, Kellner A, Bertini R, Santucci R (2005) On a titanosaurid (Dinosauria, Sauropoda) vertebral column from the Bauru Group, Late Cretaceous of Brazil. Arq Mus Nac 63:565–593

Candeiro C, Abranches C, Abrantes E, Avilla L, Martins V, Moreira A, Torres S, Bergqvist L (2004) Dinosaur remains from western Sao Paulo state, Brazil (Bauru Basin, Adamantina Formation, Upper Cretaceous). J African Earth Sci 18:1–10

Candeiro C, Martinelli A, Avilla L, Rich T (2006) Tetrapods from the upper Cretaceous (Turonian–Maastrichtian) Bauru group of Brazil: a reappraisal. Cretaceous Res 27: 923–946

Carrano M (2007) The appendicular skeleton of Majungasaurus crenatissimus (Theropoda:Abelisauridae) from the Late Cretaceous of Madagascar. Mem Soc Vertebr Paleontol 8, J Vertebr Paleontol 27:163–179

Carrano M, Sampson S (2008) The phylogeny of Ceratosauria (Dinosauria: Theropoda). J Syst Pal 6:183–236

Carrano M, Sampson S, Forster C (2002) The osteology of Masiakasaurus knopfleri, a small abelisauroid (Dinosauria: Theropoda) from the Late Cretaceous of Madagascar. J Vertebr Paleontol 22:510–534

Casal G, Martínez R, Luna M, Sciutto J, Lamanna M (2007) Aeolosaurus colhuehuapensis sp. nov. (Sauropoda, Titanosauria) de la Formacion Bajo Barreal, Cretacico superior de Argentina. Rev Bras Paleontol 10:53–62

Case J (2002) A new biogeographical model for dispersal of late Cretaceous vertebrates into Madagascar and India. J Vertebr Paleontol 22:42A

Case J, Martin J, Chaney D, Reguero M, Marenssi S, Santillana S, Woodburne M (2000) The first duck-billed dinosaur (family hadrosauridae) from Antarctica. J Vertebr Paleontol 20:612–614

Charig A, Milner A (1986) Baryonyx, a remarkable new theropod dinosaur. Nature 324:359–361

Chatterjee S (1978) Indosuchus and Indosaurus, Cretaceous Carnosaurus from India. J Paleontol 52:570–580

Coffin M, Rabinowitz P (1988) Evolution of the conjugate East African–Madagascan margins and the western Somali Basin. Geol Soc Am Spec Pap 225:1–78

Coffin M, Pringle M, Duncan R, Gladczenko T, Storey M, Műller R, Gahagan L (2002) Kerguelen hotspot magma output since 130 Ma. J Petrol 43:1121–1139

Conti M, Morsilli M, Nicosia U, Sacchi E, SavinoV, Wagensommer A, Di Maggio L, Gianolla P (2005) Jurassic dinosaur footprints from Southern Italy: footprints as indicators of constraints in paleogeographic interpretation. Palaios 20:534–550

Coria R, Currie P (2006) A new carcharodontosaurid (Dinosauria, Theropoda) from the Upper Cretaceous of Argentina. Geodiversitas 28:71–118

Coria R, Rodríguez J (1993) Sobre Xenatosaurus bonapartei Martinéz et al. 1986; un problem tico Neocaretosauria (Novas,1989) del Cretacico del Chubut. Ameghiniana 30:326–327

Coria R, Salgado L (1995) A new giant carnivorous dinosaur from the Cretaceous of Patagonia. Nature 377:224–226

Coria R, Salgado L (1998) A basal Abelisauria Novas, 1992 (Theropoda:Ceratosauria) from the Cretaceous of Patagonia, Argentina. Gaia 15:89–102

Coria R, Chiappe L, Dingus L (2002) A new close relative of Carnotaurus sastrei Bonaparte 1985 (Theropoda:

Abelisauridae) from the Late Cretaceous of Patagonia. J Vertebr Paleontol 22:460–465

Coria R, Currie P, Carabajal A (2006) A new abelisauroid theropod from northwestern Patagonia. Can J Earth Sci 43:1283–1289

Crame J (1992) Late Cretaceous paleoenvironments and biotas: an Antarctic perspective. Antarctic Sci 4:371–382

Crame J (1999) An evolutionary perspective on marine faunal connections between southernmost South America and Antarctica. Sci Marina 63:1–14

Cunningham W, Dalziel I, Lee T, Lawver L (1995) Southernmost South America–Antarctic Peninsula relative plate motions since 84 Ma: implications for the tectonic evolution of the Scotia Arc region. J Geophys Res 100:8257–8266

Currie P, Carpenter K (2000) A new specimen of Acrocanthosaurus atokensis (Theropoda, Dinosauria) from the Lower Cretaceous Antlers formation (Lower Cretaceous, Aptian) of Oklahoma, USA. Geodiversitas 22:207–246

Curry-Rogers K (2002) Titanosaur tails tell two tales: a second Malagasy titanosaur. J Vertebr Paleontol 22:47–48A

Curry-Rogers K, Forster C (2001) The last dinosaur titans: a new sauropod from Madagascar. Nature 412:530–534

Dal Sasso C (2003) Dinosaurs of Italy. C R Paleovol 2:46–66

Dal Sasso C, Signore M (1998) Exceptional soft-tissue preservation in a theropod dinosaur from Italy. Nature 392:383–387

Dal Sasso C, Maganuco S, Buffetaut E, Mendez MA (2005) New information on the skull of the enigmatic theropod *Spinosaurus*, with remarks on its sizes and affinities. J Vert Paleont 25:888–896

Dalla Vecchia M (2002) Cretaceous dinosaurs in the Adriatic–Dinaric carbonate platform (Italy and Croatia): paleoenvironmental implications and paleogeographical hypotheses. Mem Soc Geol Italiana 57:89–100

Dalla Vecchia M (2005) Between Gondwana and Laurasia: Cretaceous sauropods in an intraoceanic carbonate platform. In: Tidwell V, Carpenter K (eds) Thunder-Lizards, the sauropodomorph dinosaurs. Indiana University Press, Bloomington, IN, pp 395–429

Dalla Vecchia M (2008) Vertebrati fossili del Friuli – 450 milioni di anni di evoluzione, vol 50. Museo Friulano di Storia Naturale, Udine, 304pp

De Souza Carvalho I, dos Santos Avilla L, Salgado L (2003) Amazonsaurus maranhensis gen. et sp. nov. (Sauropoda, Diplodocoidea) from the Lower Cretaceous (Aptian–Albian) of Brazil. Cretaceous Res 24:697–713

Depéret C (1896) Sur l'existence de Dinosauriens, Sauropodes et Théropodes, dans le Crétacé supérieur de Madagascar. C R Acad Sci Paris 122:483–485

Depéret C, Savornin J (1927) Sur la decouverte d'una faune de vertebras albiens à Timimoun (Sahara occidental). C R Acad Sci Paris 181:1108–1111

Duncan R (2002) A time from for construction of the Kerguelen Plateau and Broken Ridge. J Petrol 43:1109–1119

Embleton B, McElhinny M (1975) The palaeoposition of Madagascar: palaeomagnetic evidence from the Isalo Group. Earth Planet Sci Lett 27:329–341

Evans S, Raia P, Barbera C (2004) New lizards and rhynchocephalians from the Lower Cretaceous of southern Italy. Acta Palaeontol Pol 49:393–408

Evans S, Jones M, Krause D (2008) A giant frog with South American affinities from the Late Cretaceous of Madagascar. Proc Natl Acad Sci USA 105:2951–2956

Exon N, Vine R (1970) Revised nomenclature of the "Blythsdale" sequence. Queensland Gov Mineral J 71:48–52

Fanti F, Therrien F (2007) Theropod tooth assemblages from the Late Cretaceous Maevarano formation and the possible presence of dromaeosaurid in Madagascar. Acta Palaeontol Pol 52:155–166

Fejfar O, Kostak M, Kvacek M, Mazuch J, Moucka M (2005) First Cenomanian dinosaur from Central Europe (Czech Republic). Acta Palaeontol Pol 50:295–300

Fitzgerald P (2002) Tectonics and landscape evolution of the Antarctic plate since the breakup of Gondwana, with an emphasis on the West Antarctic Rift System and the Transantarctic Mountains. R Soc N Z Bull 35:453–469

Forster C (1999) Gondwanan dinosaur evolution and biogeographic analysis. J African Earth Sci 28:169–185

Forster C, Chiappe L, Krause D, Sampson S (1996) The first Cretaceous bird from Madagascar. Nature 382:532–534

Fraas E (1908) Ostafrikanische Dinosaufler. Palaeontographica 55:105–144

Frey F, Coffin M, Wallace P, Weis D (2003) Leg 183 synthesis: Kerguelen Plateau–Broken Ridge – a large igneous province. In: Frey F, Coffin M, Wallace P, Quilty P (eds) Proceedings of Ocean Drilling Program, Initial Report 183. College Station, Texas

Gaina C, Dietmar Muller R, Brown B, Ishihara T, Ivanov S (2007) Breakup and early seafloor spreading between India and Antarctica. Geophys J Int 170:151–169

Gale A, Hardenbol J, Hathway B, Kennedy W, Young J, Phansalkar V (2002) Global correlation of Cenomanian (Upper Cretaceous) sequences: evidence for Milankovitch control on sea level. Geology 30:291–294

Gasparini Z, Pereda-Suberbiola X, Molnar R (1996) New datano the abkylosaurian dinosaur from the Late Cretaceous of the Antarctic Peninsula. In: Novas F, Molnar R (eds) Proceedings of the Gondwanan Dinosaur symposium. Mem Queensland Mus 39:583–594

Gayet M (2001) A review of some problems associated with the occurrences of fossil vertebrates in South America. J South Am Earth Sci 14:131–145

Geiger M, Clark D, Mette W (2004) Reappraisal of the timing of the breakup of Gondwana based on sedimentological and seismic evidence from the Morondova Basin, Madagascar. J African Earth Sci 38:363–381

Gheerbrant E, Rage J (2006) Paleobiogeography of Africa: how distinct from Gondwana and Laurasia. Palaeogeogr Palaeoclimatol Palaeoecol 212:224–246

Gierlinski G, Sabath K (2002) A probable stegosaurian track from the Late Jurassic of Poland. Acta Palaeontol Pol 47:561–564

Gilmore C (1922) A new sauropod dinosaur from the Ojo Alamo formation of New Mexico. Smithsonian Misc Collect 72:1–9

Goldblatt P (1993) Biological relationships between Africa and South America. Yale University Press, New Haven, CT, 630pp

González Riga B (2003) A new titanosaur (Dinosauria, Sauropoda) from the Upper Cretaceous of Mendoza Province, Argentina. Ameghiniana 40:155–172

González Riga B (2005) New fossil remains of Mendozasaurus neguyelap (Sauropoda, Titanosauria) from the Late Cretaceous of Mendoza, Argentina. Ameghiniana 42:535–548

González Riga B, Previtera E, Pirrone C (2009) Malarguesaurus florenciae gen. et sp. nov., a new titanosauriform (Dinosauria, Sauropoda) from the Upper Cretaceous of Mendoza, Argentina. Cretaceous Res 30:135–148

Gradstein F, Ogg J, Smith A (2004) A geologic time scale 2004. Cambridge University Press, Cambridge

Guedes E, Heilbron M, Vasconcelos P, Valeriano C, de Almeida J, Teixeira W, Filho A (2005) K-Ar and 40Ar/39Ar ages of dikes emplaced in the onshore basement of the Santos Basin, Resende Area, SE Brazil: implications for the south Atlantic opening and the Tertiary reactivation. J South Am Earth Sci 18:371–382

Hay W, De Conto R, Wold C, Wilson K, Voigt S, Schulz M, Wold A, Dullo W-C, Ronov A, Balukhovsky A, Soding E (1999) Alternative global Cretaceous paleogeography. In: Barrera E, Johnson C (eds) Evolution of the Cretaceous ocean-climate system. Geological Society of America Special Paper, vol 332. Geological Society of America, Boulder, CO, pp 1–47

Hocknull S, White M, Tischler T, Cook A, Calleja N, Sloan T, Elliott D, Hooker J, Milner A (2009) New mid-Cretaceous (Latest Albian) Dinosaurs from Winton, Queensland, Australia. J Vertebr Paleontol 14:480–519 (Published on-line: PloS ONE 4(7):e6190)

Holtz T, Chapman R, Lamanna M (2004) Mesozoic biogeography of Dinosauria. In: Weishampel D, Dodson P, Osmolska H (eds) The Dinosauria, 2nd edn. University of California Press, Berkeley, CA, pp 71–110

Hooker JJ, Milner AC, Sequeira SEK (1991) An ornithopod dinosaur from the Late Cretaceous of West Antarctica. Antarctic Science 3:330–331

Huene F von (1929) Los Saurisquios y Ornithisquios de Cretaceo Argentino. An Mus La Plata ser 2, 3:1–196

Huene F von, Matley C (1933) The Cretaceous saurischia and ornithischia of the Central Provinces of India. Mem Geol Surv India 21:1–72

Hutt S, Martill D, Barker M (1996) The first European allosaurid dinosaur (Lower Cretaceous, Wealden Group, England). N Jb Geol Paläont Mh 10:635–644

Ingle S, Weis D, Frey F (2002) Indian continental crust recovered from Elan Bank, Kerguelen Plateau (ODP Leg 183, site 1137). J Petrol 36:523–534

Jacobs L (1993) Quest for the African Dinosaurs: ancient roots of the modern world. Villard Books, New York, 314pp

Jacobs L, Winkler D, Downs R, Gomani E (1993) New material of an Early Cretaceous titanosaurid sauropod dinosaur from Malawi. Palaeontology 36:523–534

Jacobs L, Mateus O, Polcyn M, Schulp A, Antunes M, Morais M, Tavares T (2006) The occurrence and geological setting of Cretaceous dinosaurs, mosasaurs, plesiosaurs, and turtles from Angola. J Paleontol Soc Korea 22:91–110

Jaimes E, de Freitas M (2006) An Albian–Cenomanian unconformity in the northern Andes: Evidence and tectonic significance. J South Am Earth Sci 21:466–492

Jain S, Bandyopadhyay S (1997) A new Titanosaurid (Dinosauria: Sauropoda) from the Late Cretaceous of Centra India. J Vertebr Paleontol 17:114–136

Jokat W, Boebel T, Konig M, Meyer U (2003) Timing and geometry of early Gondwana breakup. J Geophys Res 108 (B9), 2428:1–15

Karner G, Gamboa L (2007) Timing and origin of the South Atlantic pre-salt sag basins and their capping evaporites. Geol Soc Spec Publ 285:15–35

Karol K, McCourt R, Cimino M, Delwiche C (2001) The closest living relatives of land plants. Science 294:2351–2353

Kellner A, Campos D (2002) On a theropod dinosaur (Abelisauria) from the continental Cretaceous of Brazil. Arq Mus Nac Rio de Janeiro 60:163–170

Kellner A, Campos D, Trotta M (2005) Description of a Titanosaurid caudal series from the Buru Group, Late Cretaceous of Brazil. Arq Mus Nac Rio de Janeiro 63:529–564

Kellner A, Campos D, Azevedo S, Trotta M, Henriques D, Craik M, Silva H (2006) On a new titanosaur sauropod from the Bauru Group, Late Cretaceous of Brazil. Bolet Mus Nac (Geol) 74:1–31

Kellner AWA, de Azevedo SAK (1999) A new sauropod dinosaur (Titanosauria) from the Late Cretaceous of Brazil. In: Tomida Y, Rich TH, Vickers-Rich P (eds), Proceedings of the Second Gondwanan Dinosaur Symposium. Austr Nat Sci Mus Monog 15:111–142

Kent R, Pringle M,Műller R, Saunders A, Ghose N (2002) ^{40}Ar/^{39}Ar geochronology of the Rajimahal basalts, India, and their relationship to the Kergueln Plateau. J Petrol 43:1141–1153

Koutsoukos E (2001) Foraminiferal assemblages and the early open marine record (upper Aptian–Albian) of northeastern Brazil. Abstr Prog Geol Soc Am 33:218p

Krause D, Hartman J (1996) Late Cretaceous fossils form Madagascar and their implication for biogeographic relationship with the Indian subcontinent. In: Sahni A (ed) Cretaceous stratigraphy and palaeoenvironments. Mem Geol Soc India 37:135–154

Krause D, Prasad G, von Koenigswald W, Sahni A, Grine F (1997) Cosmopolitanism among Gondwanan Late Cretaceous mammals. Nature 390:504–507

Krause D, Asher R, Buckley G, Gottfried M, LaDuke T (1998) Biogeographic origins of the non-dinosaurian vertebrate fauna of Madagascar: new evidence from the Late Cretaceous. J Vertebr Paleontol 18:57A

Krause D, Rogers R, Forster A, Hartman J, Buckley G, Sampson S (1999) The Late Cretaceous vertebrate fauna of Madagascar: implications for Gondwanan paleobiogeography. GSA Today 9:1–7

Krause D, O'Connor P, Curry Rogers K, Sampson S, Buckley G, Rogers R (2006) Late Cretaceous terrestrial vertebrates rom Madagascar: implications for Latin American biogeography. Ann Missouri Bot Gard 93:178–208

Ksepka D, Norell M (2006) Erketu ellisoni, a long-necked sauropod from Bor Guvé (Dornogov Aimag, Mongolia). Am Mus Novitates 3508:1–16

Kurzanov S, Bannikov A (1983) A new sauropod from the Upper Cretaceous of Mongolia. Paleontol Z 1983:91–97

Lamanna M, Ruben D, Martínez R, Smith J (2002) A definitive abelisaurid theropod dinosaur from the Early Late Cretaceous of Patagonia. J Vertebr Paleontol 22:58–69

Le Loeuff J (1995) Ampelosaurus atacis (nov. gen., nov. sp.), un nuveau Titanosauridae (Dinosauria, Sauropoda) du Cretace

Superieur de la haute vallee de l'Aude (France). C R Acad Sci Paris 321(2):693–696

Le Loeuff J, Buffetaut E (1991) *Tarascosaurus salluvicus*, new genus new species, a theropod dinosaur from the Upper Cretaceous of southern France. Geobios 24:585–594

Le Loeuff J, Buffetaut E, Cavin L, Martin M, Martin V, Tong H (1994) An Armoured titanosaurid sauropod from the Late Cretaceous of southern France and the occurrence of osteoderms in the Titanosauridae. Gaia 10:155–159

Lehman T (1987) Late Maastrichtian paleoenvironments and dinosaur biogeography in the Western Interior of North America. Palaeogeogr Palaeoclimatol Palaeoecol 60:189–217

Lehman T (1997) Late Campanian dinosaur biogeography in the western interior of North America. In: Wolberg D, Stump E, Rosenberg G (eds) Dinofest international. Academy of Natural Sciences, Philadelphia, PA, pp 223–240

Lehman T (2001) Late Cretaceous dinosaur provinciality. In: Tanke D, Carpenter K (eds) Mesozoic vertebrate life: new research inspired by the paleontology of Philip J. Currie. Indiana Press University, Bloomington, IN, pp 310–328

Lehman T, Coulson A (2002) A juvenile specimen of the sauropod Alamosaurus sanjuanensis from the Upper Cretaceous of Big Bend National Park, Texas. J Palaeontol 76:156–172

Lieberman B (2005) Geobiology and paleobiogeography: traching the coevolution of the Earth and its biota. Palaeogeogr Palaeoclimatol Palaeoecol 219:23–33

Longman H (1933) A new dinosaur from the Queensland Cretaceous. Mem Queensland Mus 10:131–144

Lucas S, Hunt A (1989) Alamosaurus and the sauropod hiatus in the Cretaceous of the North American Western Interior. Geol Soc Am Spec Pap 238:75–85

Lydekker R (1893) Contributions to the study of the fossil vertebrates of Argentina. I. The dinosaurs of Patagonia. Anal Mus La Plata Paleontol 2:1–14

MacArthur R, Wilson EO (1967) The theory of island biogeography. Princeton University Press, Princeton, NJ

Macdonald D, Gomez-Pereza I, Franzese J, Spalletti L, Lawverc L, Gahaganc L, Dalzielc I, Thomasd C, Trewind N, Holed M, Paton D (2003) Mesozoic break-up of SW Gondwana: implications for regional hydrocarbon potential of the southern South Atlantic. Mar Petrol Geol 20:287–308

Machado E, Kellner A (2005) Notas sobre Spinosauridae (Theropoda, Dinosauria). Ann Inst Geosci UFRJ 28:158–173

Maganuco S, Cau A, Pasini G (2005) First description of theropod remains from the Middle Jurassic (Bathonian) of Madagascar. Atti della Società Italiana di Scienze Naturali e del Museo Civico di Storia Naturale in Milano 146(II):165–202

Maganuco S, Dal Sasso C, Pasini G (2006) A new large predatory archosaur from the Middle Jurassic of Madagascar. Atti Soc Ital Sci Nat Mus Civ Stor Nat Milano 147:19–51 (Razanandrongobe sakalavae)

Maganuco S, Cau A, Dal Sasso C, Pasini G (2007) Evidence of large theropods from the Middle Jurassic of the Mahajanga Basin, NW Madagascar, with implications for ceratosaurian pedal ungual evolution. Atti Soc Ital Sci Nat Mus Civ Stor Nat Milano 148(II):261–271

Mahler L (2005) Record of Abelisauridae (Dinosauria: Theropoda) from the Cenomanian of Morocco. J Vertebr Paleontol 25:236–239

Mahoney J, Nicollet C, Dupuy C (1991) Madagascar basalts; tracking oceanic and continental sources. Earth Planet Sci Lett 104:350–363

Makovicky P, Apesteguia S, Agnolin F (2005) The earliest dromaeosaurid theropod from South America. Nature 437:1007–1011

Marks K, Tikku A (2001) Cretaceous reconstructions of East Antarctica, Africa and Madagascar. Earth Planet Sci Lett 186:479–495

Martill D, Cruickshank A, Frey E, Small P, Clarke M (1996) A new crested maniraptoran dinosaur from the Santana Formation (Lower Cretaceous) of Brazil. J Geol Soc London 153:5–8

Martin V, Buffetaut E, Suteethorn V (1994) A new genus of sauropod dinosaur from the Sao Khua formation (Late Jurassic or Early Cretaceous) of north-eastern Thailand. C R Acad Sci Paris 319(II):1085–1092

Martinelli A, Forasiepi A (2004) Late Cretaceous vertebrates from Bajo de Santa Rosa (Allen Formation), Rio Negro province, Argentina, with the description of a new sauropod dinosaur (Titanosauridae). Revista del Museo Argentino de Ciencias Naturales Nueva Serie 6:257–305

Martínez R, Ginmenez O, Rodriguez J, Bochatey G (1986) Xenotosaurus bonapartei gen. et. sp. nov. (Carnosauria, Abelisauridae), un nuevo Theropoda de la Formación Bajo Barreal, Chubut, Argentina. IV Congr. Argentino Paleontol Bioestratigr Actas 2:23–31

Martínez R, Gimenez O, Rodriguez J, Luna M, Lamanna M (2004) An articulated specimen of the basal titanosaurian (Dinosauria: Sauropoda) Epachthosaurus sciuttoi from the early Late Cretaceous Bajo Barreal Formation of Chubut Province, Argentina. J Vertebr Paleontol 24:107–120

Mateus O (2006) Late Jurassic dinosaurs from the Morrison Formation (USA), the Lourinha and Alcobaça formations (Portugal), and the Tendaguru Beds (Tanzania): a comparison. New Mexico Mus Nat Hist Sci Bull 36:223–231

Mateus O, Antunes M (2000) Torvosaurus sp. (Dinosauria: Theropoda) in the Late Jurassic of Portugal. I Congresso Ibérico de Paleontologia/XVI Jornadas de la Sociedad Española de Paleontología, pp 115–118

Mateus O, Walen A, Antunes M (2006) The large theropod fauna of the Lourinha Formation (Portugal) and its similarity to the Morrison Formation, with a description of a new species of Allosaurus. New Mexico Mus Nat Hist Sci Bull 36:123–129

Mateus O, Maidment CR, Christiansen NA (2009) A new long-necked 'sauropod-mimic' stegosaur and the evolution of the plated dinosaurs. Proc Roy Soc B 276:1815–1821

McCord R (1997) An Arizona titanosaurid sauropod and revision of the late Cretaceous Adobe Canion fauna. J Vertebr Paleontol 17:620–622

Meert J, Tamrat E (2006) Paleomagnetic evidence for a stationary Marion hotspot: additional paleomagnetic data from Madagascar. Gondwana Res 10:340–348

Meschede M, Frisch W (1998) A plate-tectonic model for the Mesozoic and Early Cenozoic history of the Caribbean plate. Tectonophysics 296:269–291

Milner A, Molnar D (1984) The biogeography of advanced ornithopod dinosaurs (Archosauria, Ornithischia) – a cladistic-vicariance model. In: Reif W, Westphal F (eds) Third symposium Mesozoic Terrestrial Ecosystem, pp 146–151

Miloni-Santucci R, Bertini R (2006) A new titanosaur from western São Paolo State, Upper Cretaceous Bauru Group, south-east Brazil. Palaeontology 49:171–185
Mo J, Wang W, Huang Z, Huang X, Xu X (2006) A Basal Titanosauriform from the Early Cretaceous of Guangxi, China. Acta Geologica Sinica 80:486–489
Mohr B, Wahnert V, Lazarus D (2003) Mid-Cretaceous paleobotany and palynology of the Central Kerguelen Plateau, southern Indian Ocean (ODP Leg 183, Site 1138). In: Frey F, Coffin M, Wallace P, Quilty P (eds) Proceedings of the ODP scientific results, vol 183, pp 1–39
Molnar R (1980) An ankylosaur (Ornithischia: Reptilia) from the Lower Cretaceous of southern Queensland. Mem Queensland Mus 20:77–87
Molnar R, Salisbury S (2005) Observation on Cretaceous Sauropods from Australia. In: Tidwell V, Carpenter K (eds) Thunder lizards: the sauropodomorph dinosaurs. University Indiana Press, Bloomington, pp 454–465
Molnar R, Flannery T, Rich T (1981) An allosaurid theropod dinosaur from the Early Cretaceous of Victoria, Australia. Alcheringa 5:141–146
Molnar R, Flannery T, Rich T (1985) Aussie Allosaurus after All. J Paleontol 59:1511–1513
Molnar R, Angriman A, Gasperini Z (1996) An Antarctic Cretaceous theropod. In: Novas F, Molnar R (eds) Proceeding of the Gondwanan Dinosaur symposium. Mem Queensland Mus 39:669–674
Montellano-Ballesteros M (2003) A titanosaurid sauropod from the Upper Cretaceous of Chihuahua, Mexico. Rev Mex Cienc Geol 20:160–164
Noonan B, Chippindale P (2006) Vicariant origin of Malagasy reptiles supports late Cretaceous Antarctic land bridge. Am Naturalist 168:730–741
Norton I, Sclater J (1979) Model for the evolution of the Indian Ocean and the breakup of Gondwanaland. J Geophys Res 84:6803–6830
Novas F (1991) Relaciones filogeneticas de los dinosaurios Teropodos Ceratosaurios. Ameghiniana 28(3–4):410
Novas F, Agnolin F (2004) Unquillosaurus cebali Powell, a giant maniraptoran (Dinosauria, Theropoda) from the Late Cretaceous of Argentina. Rev Mus Argentino Cienc Nat NS 6:61–66
Novas F, Pol D (2005) New evidence on deinonychosaurian dinosaur from the Late Cretaceous of Patagonia. Nature 433:858–861
Novas F, Salgado L, Calvo J, Agnolin F (2005a) Giant titanosaur (Dinosauria, Sauropoda) from the Late Cretaceous of Patagonia. Rev Mus Argentino Cienc Nat NS 7:37–42
Novas F, de Valas S, Vickers-Rich P, Rich T (2005b) A large Cretaceous theropod from Patagonia, Argentina, and the evolution of carcharodontosaurids. Naturwissenschaften 92:226–230
Novas F, Pol D, Canale J, Porfidi J, Calvo J (2008) A bizarre Cretaceous theropod dinosaur from Patagonia and the evolution of Gondwanan dromaeosaurids. Proc R Soc Lond B. doi:10.1098/rspb.2008.1554
Nowinski A (1971) Nemegtosaurus mongoliensis n. gen., n. sp. (Sauropoda) from the uppermost Cretaceous of Mongolia. Palaeontol Polon 25:57–81
O'Connor P (2007) The postcranial axial skeleton of Majungasaurus crenatissimus (Theropoda: Abelisauridae) from the late Cretaceous of Madagascar. Mem Soc Vertebr Paleontol 8, J Vertebr Paleontol 27:127–162
O'Connor P, Gottfried M, Stevens N, Roberts E, Ngasala S, Kapilima S, Chami R (2006) A new vertebrate fauna from the Cretaceous Red Sandstone Group, Rukwa Rift Basin, Southwestern Tanzania. J African Earth Sci 44:277–288
O'Leary M, Roberts E, Head J, Sissoko F, Bouare M (2004) Titanosaurian (Dinosauria: Sauropoda) remains from the "Continental Interclare" of Mali. J Vertebr Paleontol 24:923–930
Osi A (2004) The first dinosaur remains from the Upper Cretaceous of Hungary (Csehbanya Formation, Bakony Mts). Geobios 37:749–753
Papini M, Benvenuti M (1998) Lithostratigrphy, sedimentology and facies architecture of the Late Cretaceous successions in the central Mahajanga Basin, Madagascar. J African Earth Sci 2:229–247
Pérez-Moreno B, Sanz J, Buscalioni A, Moratalla J, Ortega F, Rasskin-Gutman D (1994) A unique multitoothed ornithomimosaur dinosaur from the Lower Cretaceous of Spain. Nature 370:363–367
Petri S (1987) Cretaceous paleogeographic maps of Brazil. Palaeogeogr Palaeoclimatol Palaeoecol 59:117–168
Philip J, Floquet M (2000) Early Campanian (83–80.5 Ma). In: Dercourt J, Gaetani M, Vrielynck B, Barrier E, Biju-Duval B, Brunet M, Cadet J, Crasquin S, Sandulescu M (eds) Atlas Peri-Tethys palaeogeographical maps – explanatory notes. CCGM/CGMW, Paris, pp 137–144
Pindell J (1993) Regional synopsis of the Gulf of Mexico and Caribbean evolution. GCSSEOM Foundation 13th Annual Research Conference, Houston, Texas, pp 251–274
Pindell J, Barrett S (1990) Geological evolution of the Caribbean region; a plate-tectonic perspective. In: Dengo G, Case J (eds) The Caribbean Region (The Geology of North America, vol H). Geological Society of America, Boulder, CO, pp 339–374
Pletsch T, Erbacher J, Holbourn A, Kuhnt W, Moullade M, Oboh-Ikuenobede F, Soding E, Wagner T (2001) Cretaceous separation of Africa and South America: the view from the West African margin (ODP Leg 159). J South Am Earth Sci 14:147–174
Powell J (1990) Epachthosaurus sciuttoi (gen. et sp. nov.) a sauropod dinosaur from the Cretaceous of Patagonia (Chubut Province). Actas Congr Argentino Paleontol Bioestrat 5:125–128
Powell J (1992) Osteologia de Saltasaurus loricatus (Sauropoda-Titanosauridae) del Cretacico Superior del Noroeste argentino. In: Sanz J, Buscalioni A (eds) Los dinosaurios y su entorno biotico, Serie Actas Academicas, vol 4. Instituto Juan de Valdes, Cuenca, Spain, pp 165–230
Powell J (2003) Revision of South American Titanosaurid dinosaurs: palaeobiological, palaeobiogeographical and phylogenetic aspects. Rec Queen Victoria Mus 111:1–94
Prieto-Marquez A, Gaete R, Rivas G, Galobart A, Boada M (2006) Hadrosauroid dinosaurs from the Late Cretaceous of Spain: Pararhabdodon isonensis revisited and Koutalisaurus kohlerorum, gen. et sp. nov. J Vertebr Paleontol 26: 929–943
Rabinowitz P, Woods S (2006) The Africa–Madagascar connection and mammalian migration. J African Earth Sci 44:270–276

Rabinowitz P, Coffin M, Falvey D (1983) The separation of Madagascar and Africa. Science 220:67–69

Raffi S, Serpagli E (1999) Introduzione alla paleontologia. Scienze della Terra, UTET Torino

Rage J (1981) Les continents péri-atlantiques au Crétacé Superieur: migrations des faunes continentales et problemes paleogeographiques. Cretaceous Res 2:65–84

Rage J (2003) Relationship of the Malagasy fauna during the Late Cretaceous: Northern or Southern routes? Acta Paleontol Pol 48:661–662

Ramana M, Ramprasad, T, Desa M (2001) Seafloor spreading magnetic anomalies in the Enderby Basin, East Antarctica. Earth Planet Sci Lett 191:241–255

Rana R, Wilson G (2003) New Late Cretaceous mammals from the Intertrappen bed of Rangapur, India and paleobiogeographic framework. Acta Palaeont Pol 48:331–348

Rauhut O, Werner C (1995) First record of the family Dromaeosauridae (Dinosauria: Theropoda) in the Cretaceous of Gondwana (Wadi Milk Formation, Northern Sudan). Paläontol Zeitsch 69:475–489

Raval U, Veeraswamy K (2003) India–Madagascar Separation: breakup along a pre-existing mobile belt and chipping of the craton. Gondwana Res 6:467–485

Razafindrazaka Y, Randriamanajara T, Piqué A, Thouin C, Laville E, Malod J, Réhault J (1999) Extension et sedimentation au Paleozoique terminal et au Mesozoique dans le basin de Majunga (nord-ouest Madagascar). J African Earth Sci 34:101–108

Reyment R, Dingle R (1987) Palaeogeography of Africa during the Cretaceous period. Palaeogeogr Palaeoclimatol Palaeoecol 59:93 116

Royer J, Chang T (1991) Evidence for relative motions between the Indian and Australian plates during the last 20 my from plate tectonic reconstructions; implications for the deformation of the Indo-Australian Plate. J Geophys Res 96:11779–11802

Russell D (1993) The role of central Asia in dinosaurian biogeography. Can J Earth Sci 30:2002–2012

Salisbury S, Molnar R, Lamanna M (2006) A new titanosauriform sauropod from the mid-Cretaceous (Albian–Cenomanian) Winton Formation of central-western Queensland, Australia. J Vertebr Paleontol 26:118A

Salgado L (1996) Pellegrinisaurus powelli nov. gen. et sp. (Sauropoda, Titanosauridae) from the Upper Cretaceous of lago Pelligrini, northwestern Patagonia, Argentina. Ameghiniana 33:355–365

Salgado L, Azpilicueta C (2000) Un nuevo saltasaurino (Sauropoda, Titanosauridae) de la provincia de Río Negro (Formacion Allen, Cretácico superior), Patagonia, Argentina. Ameginiana 37:259–264

Salgado L, Coria R (1993) Un nuevo titanosaurino (Sauropoda-Ttanosauridae) de la Fm. Allen (Campaniano–Maastrichtiano) de la Provincia de Rio Negro, Argentina. Ameghiniana 30:119–128

Salgado L, De Souza Carvalho I (2008) Uberabatitan riberoi, a new titanosaur from the Marília Formation (Bauru Group, Upper Cretaceous), Minas Gerais, Brazil. Palaeontology 51:881–901

Salgado L, Apesteguía S, Heredia S (2005) A new specimen of Neuquensaurus australis, a Late Cretaceous saltasaurine titanosaur from North Patagonia. J Vertebr Paleontol 25: 623–634

Sampson S, Witmer L (2007) Craniofacial anatomy of Majungasaurus crenatissimus (Theropoda: Abelisauridae) from the Late Cretaceous of Madagascar. Mem Soc Vert Pal 8, J Vertebr Paleontol 27:32–102

Sampson S, Krause D, Dodson P, Forster C (1996) The premaxilla of Majungasaurus (Theropoda, Ceratosauria), with implications for Gondwanan palaeobiogeography. J Vertebr Paleontol 16:601–605

Sampson S, Witmer L, Forster C, Krause D, O'Connor P, Dodson P, Ravoavy F (1998) Predatory dinosaur remains from Madagascar: implications for the Cretaceous biogeography of Gondwana. Science 280:1048–1051

Sampson S, Carrano M, Forster C (2001) A bizarre predatory dinosaur from the Late Cretaceous of Madagascar. Nature 409:504–506

Sander P, Mateus O, Laven T, Knotschke N (2006) Bone histology indicates insular dwarfism in a new Late Jurassic sauropod dinosaur. Nature 441:739–741

Sanmartin I, Ronquist F (2004) Southern hemisphere biogeography inferred by event-based models: plant versus animal patterns. Syst Biol 53:216–243

Sanz J, Powell E, Le Loueff J, Martínez R, Pereda-Suberbiola X (1999) Sauropod remains from the Upper Cretaceous of Lano (Northcentral Spain). Titanosaur phylogenetic relationships. Mus Cienc Nat Alava 14:235–255

Scharff R (1911) Distribution and origin of life in America. Constable, London, 497pp

Scotese C, Gahagan L, Larson R (1988) Plate tectonic reconstructions of the Cretaceous and Cenozoic ocean basins. Tectonophysics 155:27–48

Seeley H (1876) On Macrurosaurus semnus (Seeley), a long-tailed animal with procoelous vertebrae from the Cambridge Upper Greensand, preserved in the Woodwardian Museum of the University of Cambridge. Q J Geol Soc London 32:440–444

Sereno P, Brusatte S (2008) Basal abelisaurid and carcharodontosaurid theropods from the Lower Cretaceous Elrhaz Formation of Niger. Acta Palaeontol Pol 53:15–46

Sereno P, Wilson J, Larsson H, Dutheil D, Sues H-D (1994) Early Cretaceous dinosaurs from the Sahara. Science 266:267–271

Sereno P, Dutheil D, Iarochene M, Larsson H, Lyon G, Magwene P, Sidor C, Varricchio D, Wilson J (1996) Predatory dinosaurs from the Sahara and Late Cretaceous faunal differentiation. Science 272:986–991

Sereno P, Beck A, Dutheil D, Gado B, Larsson H, Lyon G, Marcot J, Rauhut O, Sadleir R, Sidor C, Varricchio D, Wilson G, Wilson J (1998) A long-snouted predatory dinosaur from Africa and the evolution of Spinosaurids. Science 282:1298–1302

Sereno P, Beck A, Dutheil D, Larsson H, Lyon G, Moussa B, Sadleir R, Sidor C, Varricchio D, Wilson G, Wilson J (1999) Cretaceous sauropods from the Sahara and the uneven rate of skeletal evolution among dinosaurs. Science 286:1342–1347

Sereno P, Wilson J, Conrad J (2004) New dinosaurs link southern landmasses in the Mid-Cretaceous. Proc R Soc Lond B, published online, 6 pp

Sertich J, Manthi F, Sampson S, Loewen M, Getty M (2006) Rift valley dinosaurs: a new Late Cretaceous vertebrate fauna from Kenya. J Vertebr Paleontol 26:124A

Smith JB (2005) Heterodonty in Tyrannosaurus rex: implications for the taxonomic and systematic utility of theropod dentitions. J Vertebr Paleontol 25:865–887
Smith J, Dalla Vecchia F (2006) An abelisaurid (Dinosauria: Theropoda) tooth from the Lower Cretaceous Chicla Formation of Libya. J African Earth Sci 46:240–244
Smith J, Lamanna M (2006) An abelisaurid from the Late Cretaceous of Egypt: implications for theropod biogeography. Naturwissenschaften 93:242–245
Smith J, Lamanna M, Lacoara K, Dodson P, Smith J, Poole J, Giegengack R, Attia Y (2001) A giant sauropod from an Upper Cretaceous mangrove deposit in Egypt. Science 292:1704–1706
Smith J, Tshakreen S, Rasmussen S, Lamanna M (2006a) New dinosaur discoveries from the Early Cretaceous of Libya. J Vertebr Paleontol 26:126A
Smith J, Lamanna M, Mayr H, Lacovara K (2006b) New information regarding the holotype of Spinosaurus aegyptiacus Stromer, 1915. J Paleontol 80:400–406
Sparks J, Smith W (2004) Phylogeny and biogeography of the Malagasy and Australasian rainbowfishes (Teleostei: Melanotaenioidei): Gondwanan vicariance and evolution in freshwater. Mol Phylogen Evol 33:719–734
Stampfli G, Borel G (2002) A plate tectonic model for the Paleozoic and Mesozoic constrained by dynamic plate boundaries and restored synthetic oceanic isochrons. Earth Planet Sci Lett 196:17–33
Stampfli G, Borel G, Cavazza W, Mosar J, Ziegler P (2001) The Paleotectonic Atlas of the PeriTethyan Domain. Episodes 222–228, Plus CD-ROM. European Geophysical Union, Brussels
Storey M, Mahoney J, Saunders A, Duncan R, Kelly S, Coffin M (1995) Timing of hot spot-related volcanism and the breakup of Madagascar and India. Science 267:852–855
Stovall J, Langston W (1950) Acrocanthosaurus atokensis, a new genus and species of Lower Cretaceous Theropoda from Oklahoma. Am Midland Nat 43:696–728
Stromer E (1915) Wirbelfier-Reste der Baharije-Stufe (unterstes Cenoman). 3. Das Original des Theropoden Spinosaurus aegyptiacus nov. gen. nov. spec. Abh. K. Bayer. Akad. Wissenscli. Math Phys Kl 28:1–32
Stromer E (1932) Wirbeltier-Reste der Baharije-Stufe (unterstes Cenoman). 11. Sauropoda. Abhandl Bayer Akad Wiss Math-naturwiss Abt 10:1–21
Stromer E (1934) Wirbeltierreste der Baharije-Stufe (unterstes Cenoman). 13. Dinosauria. Abh Bayer Akad Wissensch Math-naturwiss Abt 22:1–79
Suberbiola X, Bardet N, Iarochène M, Bouya B, Amaghzaz M (2004) The first record of a sauropod dinosaur from the Late Cretaceous phosphates of Morocco. J African Earth Sci 40:81–88
Sues H-D, Taquet P (1979) A pachycephalosaurid dinosaur from Madagascar and a Laurasia–Gondwanaland connection in the Cretaceous. Nature 279:633–635
Sues H-D, Frey E, Martill D, Scott D (2002) Irritator challengeri, a spinosaurid (Dinosauria; Theropoda) from the Lower Cretaceous of Brazil. J Vertebr Paleontol 22:535–547
Taquet P, Russell D (1998) New data on spinosaurid dinosaurs from the early Cretaceous of the Sahara. C R Acad Sci Paris 327:347–353
Tello Saenz C, Hackspacher P, Hadler Neto J, Iunes P, Guedes S, Ribeiro L, Paulo S (2003) Recognition of Cretaceous, Paleocene, and Neogene tectonic reactivation through apatite fission-track analysis in Precambrian areas of southeast Brazil: association with the opening of south Atlantic Ocean. J South Am Earth Sci 15:765–774
Tidwell V, Carpenter K, Meyer S (2001) New Titanosauriform (Sauropoda) from the Poison Strip Member of the Cedar Mountain Formation (Lower Cretaceous), Utah. In: Tanke D, Carpenter K (eds) Mesozoic vertebrate life: new research inspired by the paleontology of Philip J. Currie. Indiana University Press, Bloomington, IN, pp 139–165
Tiffney B, Niklas K (1990) Continental area, dispersion, latitudinal distribution and topographic variety; a test of correlation with terrestrial plant diversity. In: Ross R, Allmon W (eds) Causes of evolution; a paleontological perspective. University of Chicago Press, Chicago, IL, pp 76–102
Tikku A, Marks K, Kovacs L (2002) An Early Cretaceous extinct spreading center in the northern Natal Valley. Tectonophysics 374:87–108
Torsvik T, Tucker R, Ashwal L, Eide E, Rakotosolofo N, deWit M (1998) Late Cretaceous magmatism in Madagascar; palaeomagnetic evidence for a stationary Marion hotspot. Earth Planet Sci Lett 164:221–232
Upchurch P, Hunn C, Norman D (2002) An analysis of dinosaurian biogeography: evidence for the existence of vicariance and dispersal patterns caused by geological events. Proc R Soc Lond B 269:613–621
Valentine J, Moores E (1970) Plate tectonic regulation of faunal diversity and sea level: a model. Nature 228:657–659
Vaughan A, Storey B (2000) The eastern Palmer Land shear zone: a new terrane accretion model for the Mesozoic development of the Antarctic Peninsula. J Geol Soc 157:1243–1256
Vaughan A, Kelley S, Storey B (2002a) Mid-Cretaceous ductile deformation on the Eastern Palmer Land Shear Zone, Antarctica, and implications for timing of Mesozoic terrane collision. Geol Mag 139:465–471
Vaughan A, Pankhurst R, Fanning C (2002b) A mid-Cretaceous age for the Palmer Land event, Antarctic Peninsula: implications for terrane accretion timing and Gondwana palaeolatitudes. J Geol Soc 159:113–116
Vences M, Freyhof J, Sonnemberg R, Kosuch J, Veith M (2001) Reconciling fossils and molecules: Cenozoic divergence of Cichlid fishes and the biogeography of Madagascar. J Biogeogr 28:1091–1099
Vences M, Kosuch J, Rödel M, Lötters S, Channing A, Glaw F, Böhme W (2004) Phylogeography of Ptychadena mascareniensis suggests transoceanic dispersal in a widespread African–Malagasy frog lineage. J Biogeogr 31:593–601
Viramonte J, Kay S, Becchio R, Escayola M, Novitski I (1999) Cretaceous rift related magmatism in central-western South America. J South Am Earth Sci 12:109–121
Vlahovic I, Tisljar J, Velic I, Maticec D (2005) Evolution of the Adriatic Carbonate platform: palaeogeography, main events and depositional dynamics. Palaeogeogr Palaeoclimatol Palaeoecol 220:333–360
Voigt S, Wiese F (2000) Oxygen isotope variations and palaeobiogeographic changes in western and central Europe – evidence for late Turonian (Late Cretaceous) climate cooling. J Geol Soc 157:737–744
Voigt S, Gale A, Flögel S (2004) Midlatitude shelf seas in the Cenomanian–Turonian greenhouse world: temperature

evolution and North Atlantic circulation. Paleoceanography 19:PA4020, doi:10.1029/2004PA001015

Walter H (2004) The mismeasure of islands: implications for biogeographical theory and the conservation of nature. J Biogeogr 31:177–197

Waters J, Dijkstra L, Wallis G (2000) Biogeography of a southern hemisphere freshwater fish: how important is marine dispersal? Mol Ecol 9:1815–1821

Weishampel D, Barret P, Coria R, Le Loeuff J, Xing X, Xijin Z, Sahni A, Gomani E, Noto C (2004) Dinosaur distribution. In: Weishampel D, Dodson P, Osmolska H (eds) Dinosauria, 2nd edn. University of California Press, Berkley, CA, pp 517–606

Wild R (1991) Janenschia n. g. robusta (E. Frass 1908) pro Tornieria robusta (E. Fraas 1908) (Reptilia, Saurischia, Sauropodomorpha). Stuttgarteer Beitrage Naturkude Ser B (Geol und Paleontol) 173:1–4

Williamson M, Gaston K, Lonsdale M (2001) The species-area relationship does not have an asymptote! J Biogeogr 28:827–830

Wilson J (2002) Sauropod dinosaur phylogeny: critique and cladistic analysis. Zool J Linnean Soc 136:217–276

Wilson J (2005) Redescription of the Mongolian sauropod Nemegtosaurus mongoliensis Nowinski (Dinosauria: Saurischia) and comments on Late Cretaceous sauropod diversity. J Syst Palaeontol 3:283–318

Wilson J (2006) An overview of Titanosaur evolution and phylogeny. In: Colectivo Arqueològico-Paleontològico Salense (ed) Actas de las III Jornades sobre Dinosaurios y su Entorno. Salas de los Infantes, Burgos, Espana, pp 169–190

Wilson J, Mohabey D (2006) A titanosauriform (Dinosauira: sauropoda) axis from the Lameta Formation (Upper Cretaceous: Maastrichtian) of Nand, Central India. J Vertebr Paleontol 26:471–479

Wilson J, Mustafa H, Zalmout I (2006) Latest Cretaceous reptiles from the Hashemite kingdom of Jordan. J Vertebr Paleontol 26:140A

Wilson JA, Sereno PC, Srivastava S, Bhatt DK, Khosla A, Sahni A (2003) A new abelisaurid (Dinosauria, Theropoda) from the Lameta Formation (Cretaceous, Maastrichtian) of India. Contrib Mus Paleontol Univ Michigan 31(1):1–42

Winberry J, Anandakrishnan S (2004) Crustal structure of the West Antarctic rift system and Marie Byrd Land Hotspot. Geology 32:977–980

You H, Ji Q, Lamanna M, Li J, Li Y (2004) A titanosaurian sauropod dinosaur with opisthocoelous caudal vertebrae from the Early Late Cretaceous of Liaoning Province, China. Acta Geol Sinica 78:907–911

You H, Li D, Zhou L, Ji Q (2006) Huanghetitan liujiaxiaensis, a new sauropod dinosaur from the Lower Cretaceous Hekou Group of Lanzhou Basin, Gansu Province, China. Geol Rev 52:668–674

You H, Li D, Zhou L, Ji Q (2008) Daxiatitan binglingi: a giant sauropod dinosaur from the Early Cretaceous of China. Gansu Geol 17:1–10

Zagrarni M, Negra M, Hanini A (2008) Cenomanian–Turonian facies and sequence stratigraphy, Bahloul Formation, Tunisia. Sediment Geol 204:18–35

Zunino M, Zullini A (1999) Biogeografia – La dimensione spaziale dell'evoluzione. Casa Editrice Ambrosiana, Milano, 310pp

Palaeobiogeography of Mesozoic Mammals – Revisited

Thomas H. Rich and Patricia Vickers-Rich

Abstract

The fossil record of mammals in the Mesozoic is decidedly meagre in comparison to that of the Cainozoic, but some useful generalisations can be drawn about the biogeographic history of this group during the Mesozoic. Compared with the Jurassic, when cosmopolitanism was frequent among the mammalian families, regionalism became more pronounced in the Cretaceous, particularly the Late Cretaceous, probably reflecting the continental aggregation that produced Pangea and, subsequently Gondwana as the Mesozoic progressed. The conventional hypothesis that therians arose on the northern continents and dispersed to the southern continents reflects the poor Mesozoic mammalian record. Recent discoveries in Africa, South America and Australia suggest that caution is warranted before accepting the conventional hypothesis.

Keywords

Mammalia • Mesozoic • Metatherians • Eutherians • Monotremes • Marsupials • Placentals • Origins • Palaeobiogeography • Dispersals • Interchanges • Regionalism

Introduction

This review is an update of Rich (2008) which covered the literature through 2005.

In 1947, George Gaylord Simpson published an iconic analysis of the distribution of mammals in North America and Eurasia during the Cainozoic (Simpson 1947) (Fig. 1). He pioneered the use of quantitative methods for assessing the degree of interchange between the two broad regions. His methodology should be extended into the Mesozoic, but several factors make this impractical. First, there are far fewer mammalian taxa in the Mesozoic than the Cainozoic. Were Simpson writing his paper today using presently available data, there would be about 3,500 non-volant terrestrial mammalian genera available for analysis. This contrasts with the 300 mammalian genera known from the Mesozoic. In other words, there are 50 genera for every million years for the Cainozoic, but only 2 per million years for the Mesozoic.

There are large temporal gaps in the Mesozoic record (Fig. 2); the number and distribution of

T.H. Rich (✉)
Museum Victoria, Melbourne, Vic 3001, Australia

School of Geosciences, Monash University, Vic 3800, Australia
e-mail: trich@museum.vic.gov.au

J.A. Talent (ed.), *Earth and Life*, International Year of Planet Earth,
DOI 10.1007/978-90-481-3428-1_32, © Springer Science+Business Media B.V. 2012

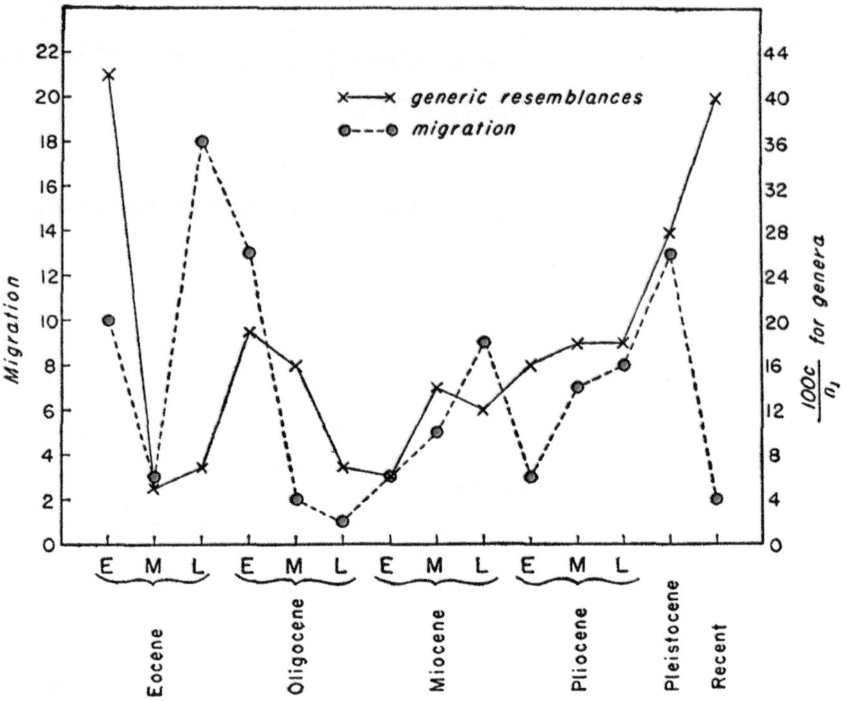

Fig. 1 Figure 4 in Simpson (1947). Number of genera in common between North America and Eurasia during the Cainozoic (*solid line*) and amount of migration (*dashed line*) between the two landmasses

Mesozoic mammal sites is uneven (Fig. 3). North America has Late Cretaceous sites comparable in number to sites in the Palaeocene of that continent, whereas Australia has only four Mesozoic sites, all late Early Cretaceous, and none from the rest of the Mesozoic. Generally, the Gondwanan continents have far fewer sites than Laurasia, making evaluation of biogeographic patterns hazardous.

Methods

Most of the distribution data for Mesozoic mammals in this analysis comes from Kielan-Jaworowska et al. (2004). It was generously supplied in an electronically readable form by Dr. Richard Cifelli. These data are updated in Figs. 7 and 8 with citation of additional information on family distributions indicated in brackets in the captions.

Discussion

The depth of our ignorance about the distribution of Mesozoic mammals is well illustrated by the multituberculates. A map of this group drawn in 1990 would show their distribution to have been restricted to the Laurasian continents where their remains were rather abundant (Fig. 4). Since then, multituberculates have been found in Africa, Australia (Rich et al. 2009), and South America (Fig. 5), though these reports are based on very few specimens. Does this imply that the group was abundant in Laurasia but sparse, with few species, in the Gondwanan terranes? Given that few specimens of mammals of any kind occur in the latter, such an interpretation would be unwarranted. Though Gondwanan multituberculates are rare as fossils, they must have been elements of palaeocommunities, even abundant ones. The paucity of available material calls for caution.

The Mesozoic palaeobiogeography of mammals can be conveniently divided into the Late Cretaceous and the pre-Late Cretaceous because, (1) during the Late Cretaceous, productive fossil mammal sites (and hence the record) are much better, (2) ancestors of the present-day metatherians and eutherians were a significant part of the Late Cretaceous mammalian assemblages – molecular techniques applied to their living descendants can provide useful insights regarding their Late Cretaceous ancestors.

Because the Morganucodontidae are regarded as being among the most primitive mammals (or

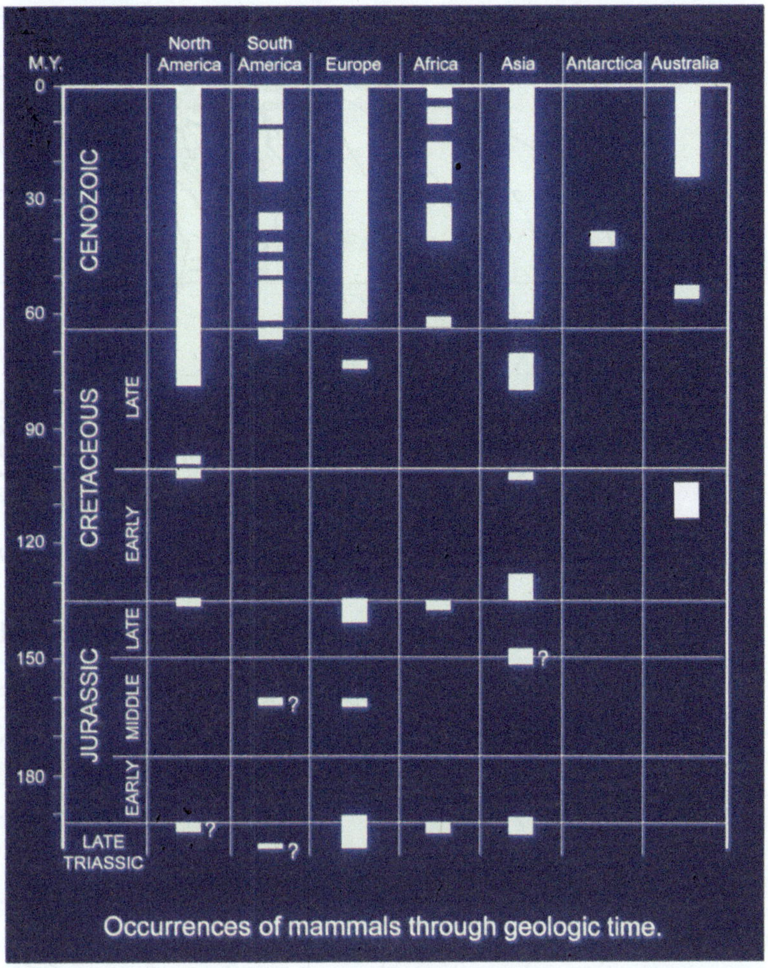

Fig. 2 Periods of time during the Mesozoic and Cainozoic when fossil mammals are known on the various landmasses (modified from Lillegraven et al. 1979)

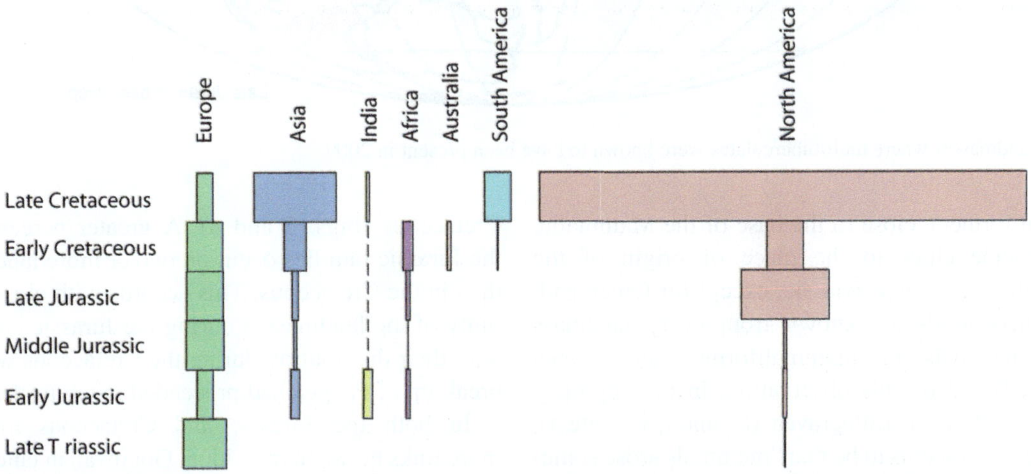

Fig. 3 Relative numbers of Mesozoic mammal occurrences by continents and age. "Occurrence" refers to a record at lowest identified taxonomic level within a locality or local fauna. For most pre-Late Cretaceous occurrences and virtually all occurrences outside of North America, the totals reflect all taxa from all known sites and in many cases reflect most known individual specimens. Lumping occurrences by local faunas results in under-representation for the North American Late Cretaceous, which nonetheless includes a disproportionately large number of occurrences (Kielan-Jaworowska et al. 2004, p. 108)

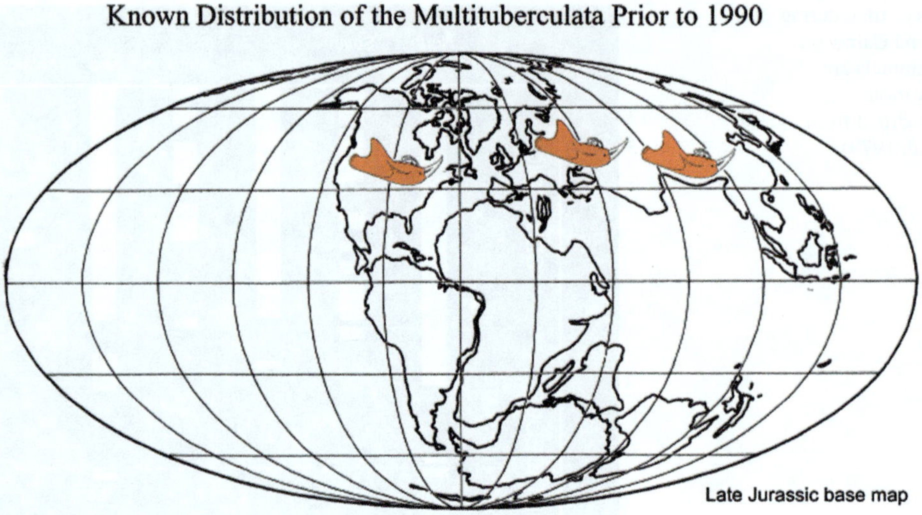

Fig. 4 Landmasses where multituberculates were known to have been present in 1990

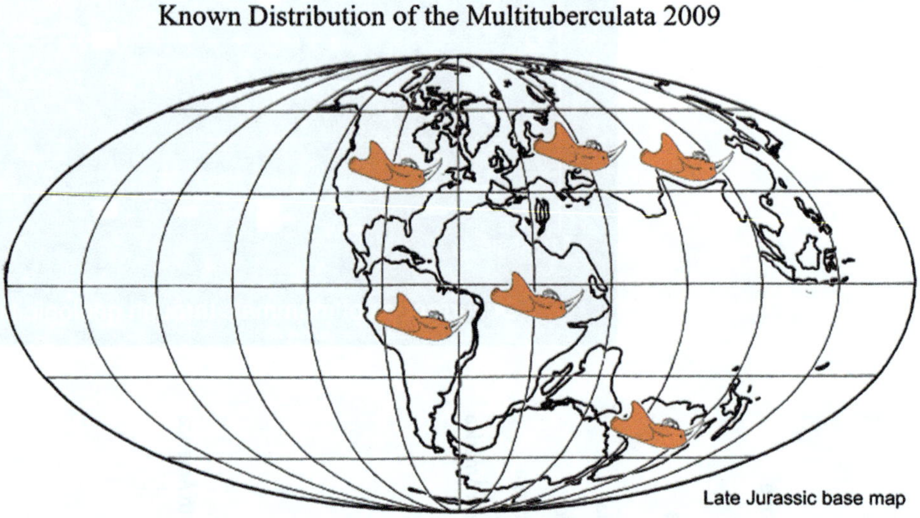

Fig. 5 Landmasses where multituberculates were known to have been present in 2009

mammaliformes), close to the base of the Mammalia, they provide clues to the place of origin of the Mammalia. In the late Triassic, except for Greenland, morganucodontids are known from every landmass where mammals or mammaliformes are known (Fig. 6). Based on this observation, in the appropriate words of Jason Lillegraven (in another context), the conclusion seems to be that "mammals arose somewhere on the Earth's dry surface".

The distribution of mammalian families in the Jurassic contrasts with their distribution in the Cretaceous (Figs. 7 and 8). A greater percentage of the Jurassic families occur on two or more landmasses than in the Cretaceous. This accords with closer proximity of the landmasses during the Jurassic compared with their distribution during the Cretaceous when the break-up of Pangea had proceeded much further.

In both the Jurassic and Cretaceous there are more links between individual Gondwanan landmasses and Laurasian ones than there are between pairs of Gondwanan landmasses. This is presumably connected with many more families being known on

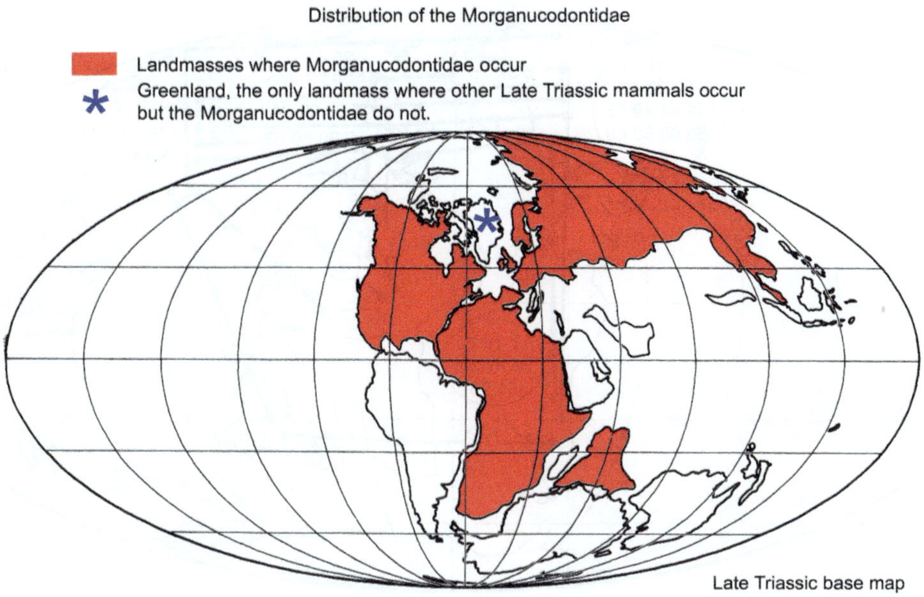

Fig. 6 Known distribution of the Morganucodontidae. Base map Late Triassic

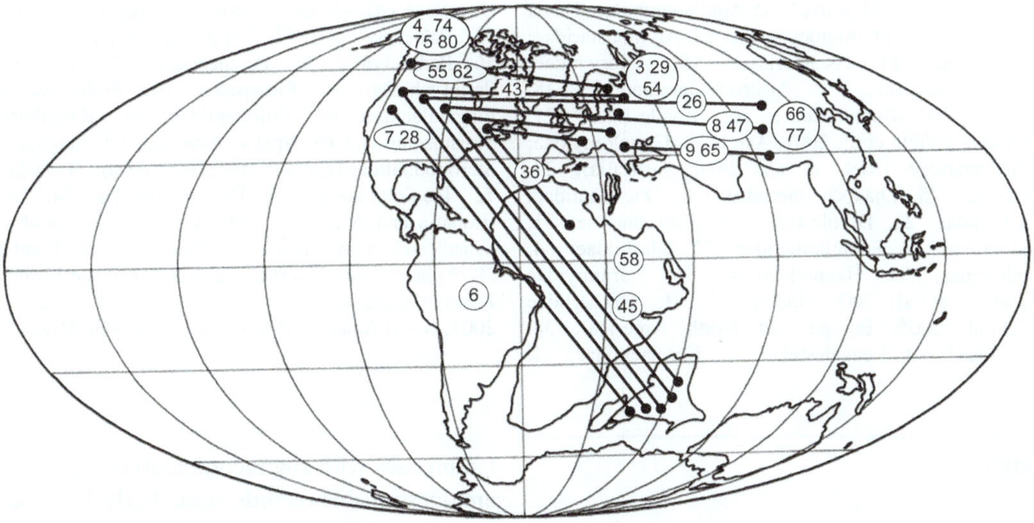

Fig. 7 Records of Jurassic mammalian families on a Callovian basemap. The same families found on two or more landmasses are linked together by a *solid line* with a *dot* on each landmass where the families occur. 3. Albionbaataridae; 4. Allodontidae; 6. Ameghinichnidae; 7. Amphidontidae; 8. "Amphilestidae"; 9. Amphitheriidae (Lopatin and Averianov 2007, Asia); 26. Docodontidae (Prasad and Manhas 2007, India); 28. Dryolestidae (Prasad et al. 2006, India); 29. Eleutherodontidae; 36. Haramiyidae; 43. Kuehneotheriidae; 45. Megazostrodontidae; 47. Morganucodontidae; 54. Paulchoffatiidae; 55. Paurodontidae; 58. Peramuridae; 62. Plagiaulacidae; 65. Shuotheriidae; 66. Sinoconodontidae; 74. Tinodontidae (Lopatin et al. 2005, Asia); 75. Triconodontidae; 77. Volaticotheriidae (Meng et al. 2006, Asia); 80. Zofiabaataridae

the Laurasian landmasses. What this broad brush-stroke data cannot do is provide evidence for the direction of movement of mammals between land masses (cf. dinosaurs, Fanti this volume). That would require far more information, especially well-dated sites with more extensive fossil assemblages and, in the case of the Gondwanan landmasses, many more sites.

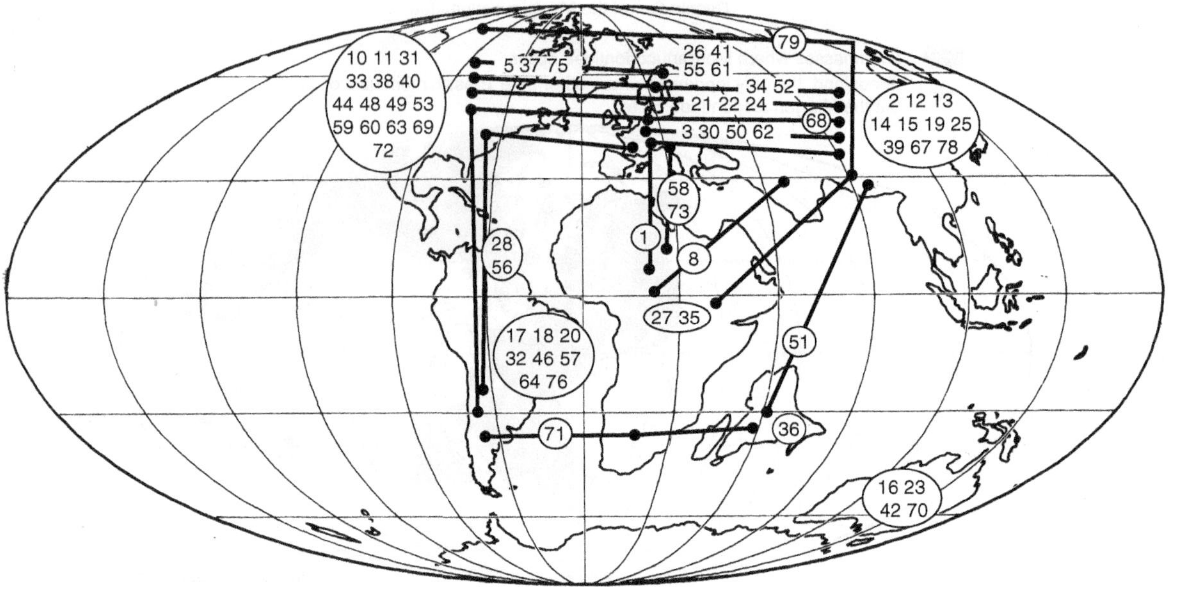

Fig. 8 Records of Cretaceous mammalian families on an Albian basemap. For explanation see caption for Fig. 1. Aegialodontidae; 2. Aguitheriidae; 3. Albionbaataridae; 5. Alphadontidae; 8. "Amphilestidae"; 10. Aquiladelphidae (Davis 2007, North America); 11. Arctocyonidae; 12. Arginbaataridae; 13. Arguimuridae; 14. Asiatheriidae; 15. Asioryctidae; 16. Ausktribosphenidae; 17. Austrotriconodontidae; 18. Barbereniidae; 19. Bobolestidae; 20. Bondesiidae; 21. Cimolodontidae (Wible et al. 2007, Asia); 22. Cimolomyidae; 23, Corriebaataridae (Rich et al. 2009, Australia); 24, Deltatheridiidae; 25. Djadochtatheriidae; 26. Docodontidae; 27. Donodontidae; 28. Dryolestidae; 30. Eobaataridae; 31. Eucosmodontidae; 32. Ferugliotheridae; 33. Glasbiidae; 34. Gobiconodontidae; 35. Hahnodontidae; 36. Haramiyidae (Anantharaman et al. 2006, India); 37. Herpetotheriidae (Martin et al. 2005, Europe and North America); 38. Hyopsodontidae?; 39. Kennalestidae; 40. Kermackiidae; 41. Kogiaononidae; 42. Kollikodontidae; 44. Leptictidae; 46. Mesungulatidae; 48. Nanocuridae (Fox et al. 2007, North America); 49: Neoplagiaulacidae; 50. Nyctitheriidae; 51. Otlestidae (Khosla et al. 2004 and Prasad et al. 2007, India); 52. Palaeoryctidae; 53. Pappotheriidae; 55. Paurodontidae; 56. Pediomyidae; 57. Peradectidae?; 58. Peramuridae; 59. Periptychidae; 60. Picopsidae; 61. Pinheirodontidae; 62. Plagiaulacidae; 63. Ptilodontidae; 64. Reigitheriidae; 67. Sloanbaataridae; 68. Spalacotheriidae; 69. Stagodontidae; 70. Steropodontidae (Lopatin et al. 2005, Asia); 71. Sudamericidae; 72. Taeniolabididae; 73. Thereuodontidae; 74. Tinodontidae (Lopatin and Averianov 2007 Asia); 75. Triconodontidae; 76. Vincelestidae; 78. Zalambdalestidae (Archibald and Averianov 2003 place *Kulbeckia*, the sole member of the Kulbeckiidae, in Zalambdalestidae); 79. Zhelestidae (Archibald and Averianov 2001, North America; Averianov et al. 2003, Africa)

The Data

Unquestioned docodontids, known only from Laurasia, occur in Middle Jurassic to Early Cretaceous sequences. Two specimens that may be docodonts are a single Late Triassic tooth from France assigned to *Delsatia* (Sigogneau-Russell and Godefroit 1997) and a jaw fragment with three teeth, *Reigitherium*, from the Late Cretaceous Los Alamitos Formation of Argentina; the latter, regarded by its describer as a docodont (Bonaparte 1990), is now regarded as Mammalia *incertae sedis* (Kielan-Jaworowska et al. 2004). The eutriconodonts are much more frequent and diverse in Asia, Europe and North America than in South America and Africa, the two Gondwanan landmasses with any record at all of this group. There are questionable records in the Early Jurassic of North America and India. By the Middle Jurassic, the eutriconodonts were well established in Asia and, in the Late Jurassic, they occur in North America, Asia, and Africa. They were most widely distributed in the Early Cretaceous, persisting into the Late Cretaceous in North and possibly South America. In Laurasia, the available record is consistent with some families not being present over that entire landmass. For example, whereas the Amphilestidae are known from North America and Asia including India, the Triconodontidae are known only from North America and Europe, the Gobiconodontidae are known only from Asia and North America.

The Haramiyidae are first and most widely known in the Late Triassic of Europe. Subsequent records are from single sites in the Early Jurassic of North America and the Late Jurassic of Africa. On the sparse evidence to date, the group would seem to have dispersed from Europe to North America and Africa.

The most diverse group of Mesozoic mammals is the Multituberculata. Only four of its 19 recognised families occur on more than one continent, all in Laurasia. This pattern contrasts with that of their modern analogues, the much more widely spread rodents. Of the 19 recent rodent families, 13 have records on two or more continents.

Although not a highly diverse group, the Symmetrodonts are one of the most widespread of Mesozoic mammals between the Late Triassic and mid Cretaceous. This diversity persisted from the time of Pangea in the Late Triassic to when the continents had split into Gondwana and Laurasia and these landmasses in turn had begun to split apart by the mid-Cretaceous. Five of the eight symmetrodont families are known from more than one continent; two of the three that are restricted to one continent are known from the Late Cretaceous Los Alamitos fauna of Patagonia; their protracted survival in South America may have been due to isolation of that continent. Contrastingly, the Spalacotheriidae are known from Africa, Asia, Europe, South America, and North America in the Early to mid-Cretaceous.

Monotremes are now known from the Cretaceous and Cainozoic of Australia, the Late Cainozoic of New Guinea, and the Early Cainozoic of South America. The strongest evidence supporting origination in Australia is their absence from the diverse mammalian assemblage from the Campanian Los Alamitos fauna of Patagonia. Absence of a monotreme is consistent with that group not having reached South America at a time, though they had been in Australia for at least 30 million years. Given that monotremes are generally regarded as quite primitive mammals, and hence presumably a distinct lineage that originated in the early Mesozoic (Rich et al. 2005), it is odd, given the configuration of the continents, that they occur nowhere else. However, given the meagre record of Jurassic terrestrial vertebrates from Australia – one temnospondyl (Warren and Hutchinson 1983) and one sauropod (Longman 1927) – monotremes may nevertheless have thrived in Australia during the Jurassic.

Three of the eight families of eupantotheres, each with more than one genus, are geographically widespread. The other five families each contain only a single genus and are restricted to one continent. In addition to these, there are two African eupantotherid taxa not assigned to a genus, and several European specimens that cannot be assigned with certainty to a family. Eupantotheres are most diverse in the Late Jurassic to Early Cretaceous; they occur on all the landmasses of Laurasia and South America. Two South American families and species are present in the Late Cretaceous Alamitos fauna.

The oldest marsupials occur in Laurasia – *Sinodelphis* from China dated at 125 myBP (Luo et al. 2003) and *Kokopellia* from North America at 100 myBP (Cifelli and De Muizon 1997). The diverse Late Cretaceous Los Alamitos local fauna of Patagonia has various archaic mammals similar to those of Jurassic localities elsewhere but lack therians of any kind (Bonaparte 1990). The early Palaeocene Tiupampa of Bolivia has both marsupials and placentals (Marshall et al. 1995). This accords with the conventional idea that marsupials arose in Laurasia, spread to South America, reached Antarctica no later than the Eocene (Woodburne and Zinsmeister 1982), and finally entered Australia in the Palaeocene or Eocene (Godthelp et al. 1992, 1999). What is unclear is whether there was a single marsupial dispersal event between South America and Australia, or multiple dispersals.

Extant marsupials can be divided quite sharply into Ameridelphia and Australodelphia. The Ameridelphia are restricted to the Americas whereas, with one exception, the Australodelphia are confined to Australasia. This taxonomic division was first recognised on the basis of foot structure (Szalay 1982), subsequently supported by molecular data (Nilsson et al. 2004). The one exception noted above is the South American Microbiotheridae. They appear in the fossil record in the Palaeocene and have one living australodelphian representative, *Dromiciops australis* from Chile. As *Dromiciops* appears to have been derived within the australadelphians, either its ancestors returned to South America after the dasyuromorphs plus peramelamorphs, on the one hand, and diprotodontians on the other differentiated in Australia, or alternatively, differentiation of the australodelphians into those two major groups occurred in South America and they reached Australia independently.

The Mesozoic palaeobiogeography of the eutherians remains uncertain, particularly with regard to the placentals. There are two schools of thought: the "classical" school bases its hypotheses primarily on the fossil record, the "molecular" school uses analysis of DNA and RNA sequences to frame their hypotheses.

The classical school holds that the eutherians arose in Laurasia and subsequently dispersed to Gondwana. This accords well with the vast bulk of the known mammalian fossil record, the oldest eutherian being either *Eomaia scansoria* (Ji et al. 2002) or *Acristatherium yanensis* (Hu et al. 2009) from the Early Cretaceous of China. Luo et al. (2011) recently described *Juramaia sinensis*, a eutherian from the Middle or Late Jurassic of China. The view that despite its unevenness, the fossil record is adequate to accurately locate these events in time and space is defended by Foote et al. (1999) and Wible et al. (2007). Using statistical arguments regarding the completeness of the fossil record, they suggest that the appearance of the eutherians took place in the Early Cretaceous of Laurasia. Likewise, they regard the placentals as having arisen close to the time of their appearance in the fossil record; i.e., in the aftermath of the KT boundary extinction event, probably as a consequence of opening up ecological space following demise of the non-avian

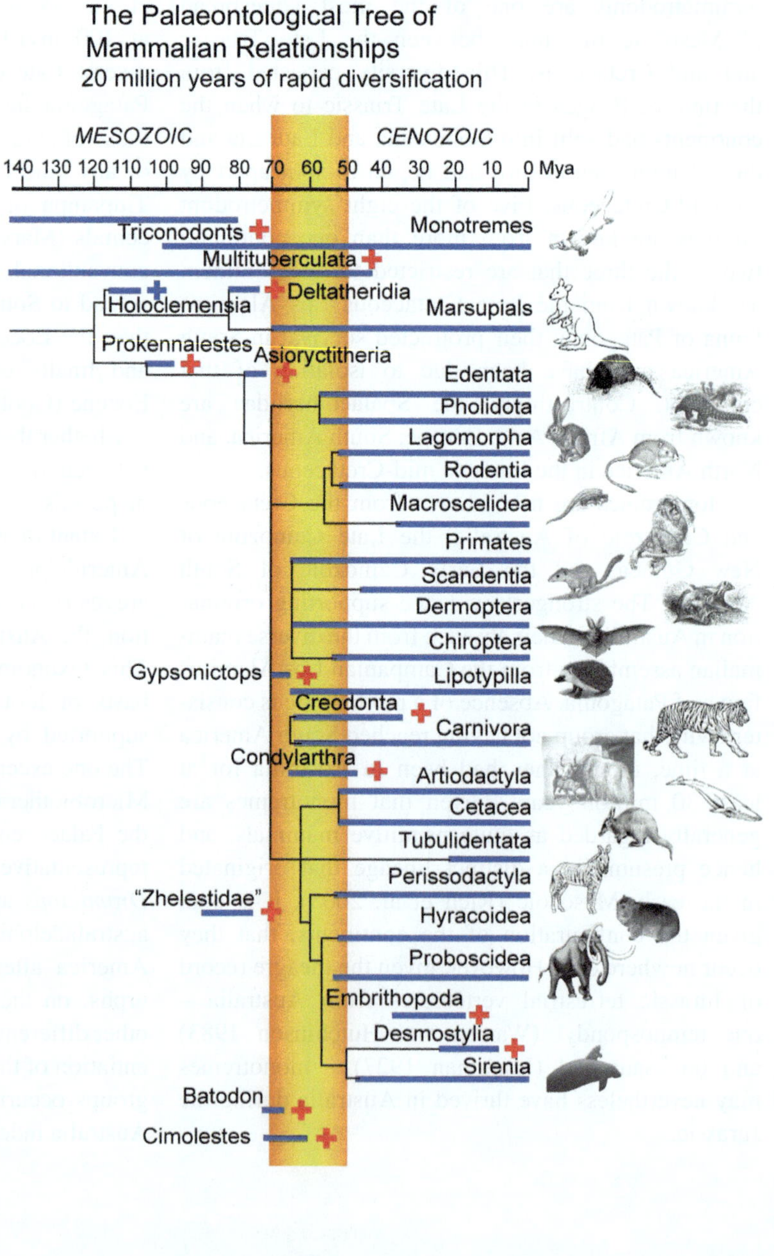

Fig. 9 Phylogeny of the Mammalia as seen by the majority of palaeontologists

dinosaurs, a view supported by Wible et al. (2007) and others (Fig. 9).

The molecular school is epitomised by Murphy et al. (2001) and Bininda-Emonds et al. (2007). Based on analysis of the structure of genetic material of modern species, four major clades of placentals are recognised: Afrotheria (Africa), Xenarthra (South America), Laurasiatheria, and Euarchontoglires. Laurasiatheria and Euarchontoglires are equivalent to the Boreutheria (Laurasia) (Murphy et al. 2001). This group suggested that the separation of Afrotheria from the Xenarthra about 110 million years ago was related to the separation of Africa and South America with formation of the South Atlantic Ocean at that time. Then the Boreutheria, constituting the bulk of the placentals, reached North America and spread from there to Europe and Asia. The molecular data have been interpreted to suggest that the majority of the modern placental orders arose 10–40 million years before their fossils are found in the geological record. If correct, this constrains the timing of movement of these placental groups to the Late Cretaceous when the afrotheres would have given rise to the Xenarthra which subsequently moved across the South Atlantic about 103 million years ago (Murphy et al. 2001). The Xenarthra gave rise to the boreosphenidians which then moved into Laurasia (Fig. 10).

The major weakness of the molecular school hypothesis concerns the calibration points used to determine the age in years of the splitting of the various placental clades from one another and thus the age of the clades themselves. The major weakness of the classic school hypothesis is the incompleteness of the fossil record.

Since 1997, a few tribosphenic mammal specimens have been discovered in the Middle Jurassic and Early Cretaceous of Gondwana. If they are not only indeed tribosphenic mammals, but also eutherians, this does not accord with the classical school of thought. The fossils in question include the Middle Jurassic *Ambondro* (Flynn et al. 1999), based on a single lower-jaw fragment from Madagascar; the Middle Jurassic *Asfaltomylos* (Rauhut et al. 2002), based on a single lower-jaw fragment and *Henosferus* (Rougier et al. 2007), based on three lower jaws and a possible isolated upper premolar from Argentina, and the Early Cretaceous Ausktribosphenos (Rich et al. 1997) and *Bishops* (Rich et al. 2001), based on about thirty

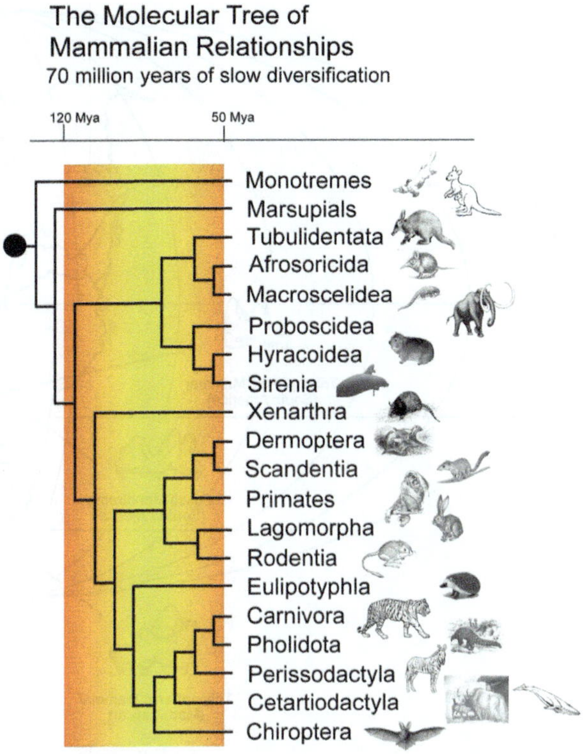

Fig. 10 Phylogeny of the Mammalia as seen by the majority of molecular biologists

lower-jaw fragments from Australia (Fig. 11). These forms have been variously interpreted. On the basis of their apparent tribosphenic dental morphology and dental formula, they have been referred to as eutherians (Woodburne et al. 2003). If this is the case, on the present evidence, eutherians arose earlier in Gondwana than in Laurasia. This accords with the conclusions based on molecular studies that placentals arose in Gondwana and subsequently spread to Laurasia.

Alternatively, these Mesozoic Gondwanan tribosphenic forms have been united with the monotremes under the hypothesis that a separate radiation of mammals with therian-like, but not true tribosphenic dentitions on a structurally primitive jaw took place in Gondwana (Fig. 12) (Luo et al. 2001, 2002, 2007). The primitive nature of the jaw is suggested based on the presence of an internal mandibular groove. This group was erected as the Australosphenida and the Laurasian eutherians were designated the Boreosphenida. A cladogram of the Mammalia was then constructed grouping all these australosphenidans together and placing them well

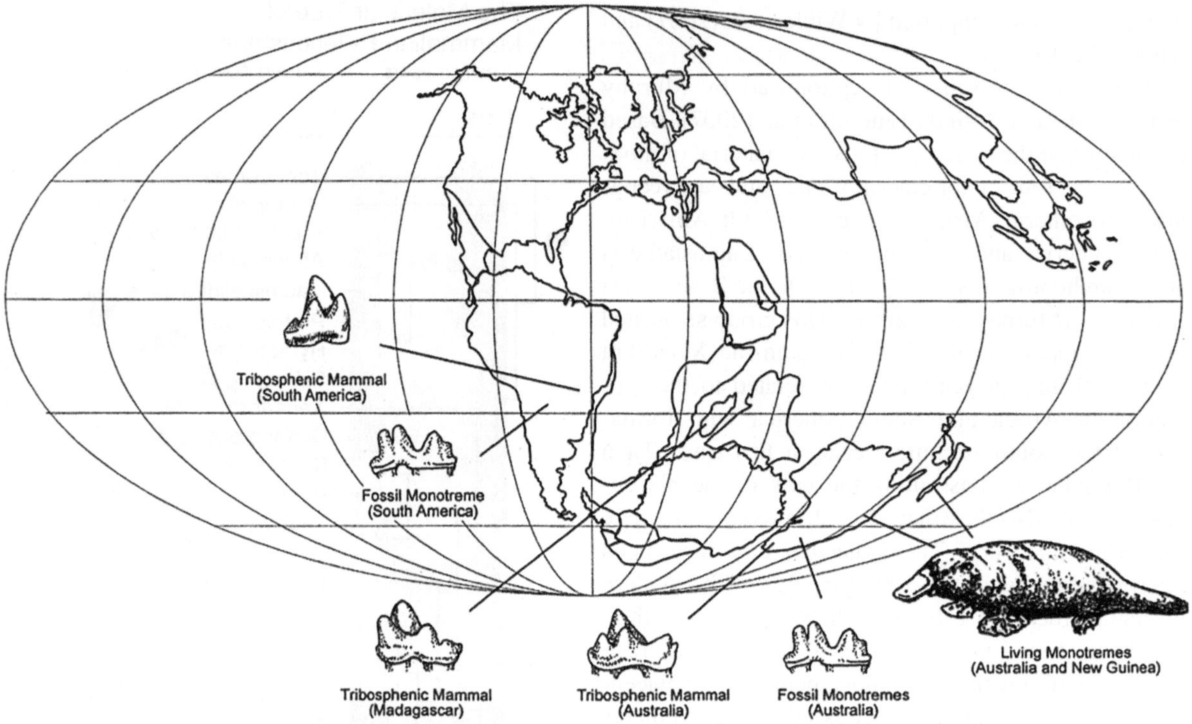

Fig. 11 Geographic distribution of the Australosphenida of Luo et al. (2001) (modified with permission from Zhe-Xi Luo. After Press Release of the Carnegie Museum of Natural History)

away from the Boreosphenida (Fig. 13) (Luo et al. 2001, 2002, 2007).

According to the alternative hypothesis of Luo et al. (2001, 2002, 2007), the southern hemisphere Australosphenida are closely allied with the northern hemisphere Jurassic Yinotheria, family Shuotheriidae. The shuotheriids are distinguished from both the Boreosphenida and Australosphenida by their pseudotribosphenic rather than tribosphenic molars (Fig. 14); this accords with them being only remotely related to either group (Woodburne et al. 2003; Rougier et al. 2007).

Examination of the data matrix used for construction of the cladogram of Luo et al. (2001, 2002, 2007) (Fig. 13) indicates that some important characters impinging on this hypothesis could be interpreted quite differently (Fig. 15) (Woodburne et al. 2003). Amongst other things this analysis separated the monotremes from other taxa referred to the Australosphenida and combined the non-monotreme Australosphenida [= tribosphenic Australosphenida] with the Boreosphenida into the Placentalia (the "traditional view" in Fig. 12). The basis for separating the monotremes specifically from the Ausktribosphenidae within the tribosphenic australosphenidians was given in Rich et al. (2002).

The essential aspects of the tribosphenic molar pattern are that, first, a cusp on the upper molar, the protocone, acts as a pestle in a basin or mortar formed on the lower molar by the talonid; and second, shearing occurs by successive upper molars abrading against the triangular pillar or trigonid of the lower molars, forming vertical or near-vertical facets (Fig. 14). The molars of the monotremes do not have the pattern of wear typical of a tribosphenic mammal (Pascual et al. 2002, Fig. 16 here, see especially D2, Woodburne 2003). There is no talonid on the lower molars into which a protocone occludes. Likewise, no near-vertical wear facets are present. Most marsupial and placental mammals, in fact, do not have a tribosphenic dentition although they are clearly descended from ancestors that did. It could very well be that monotremes are

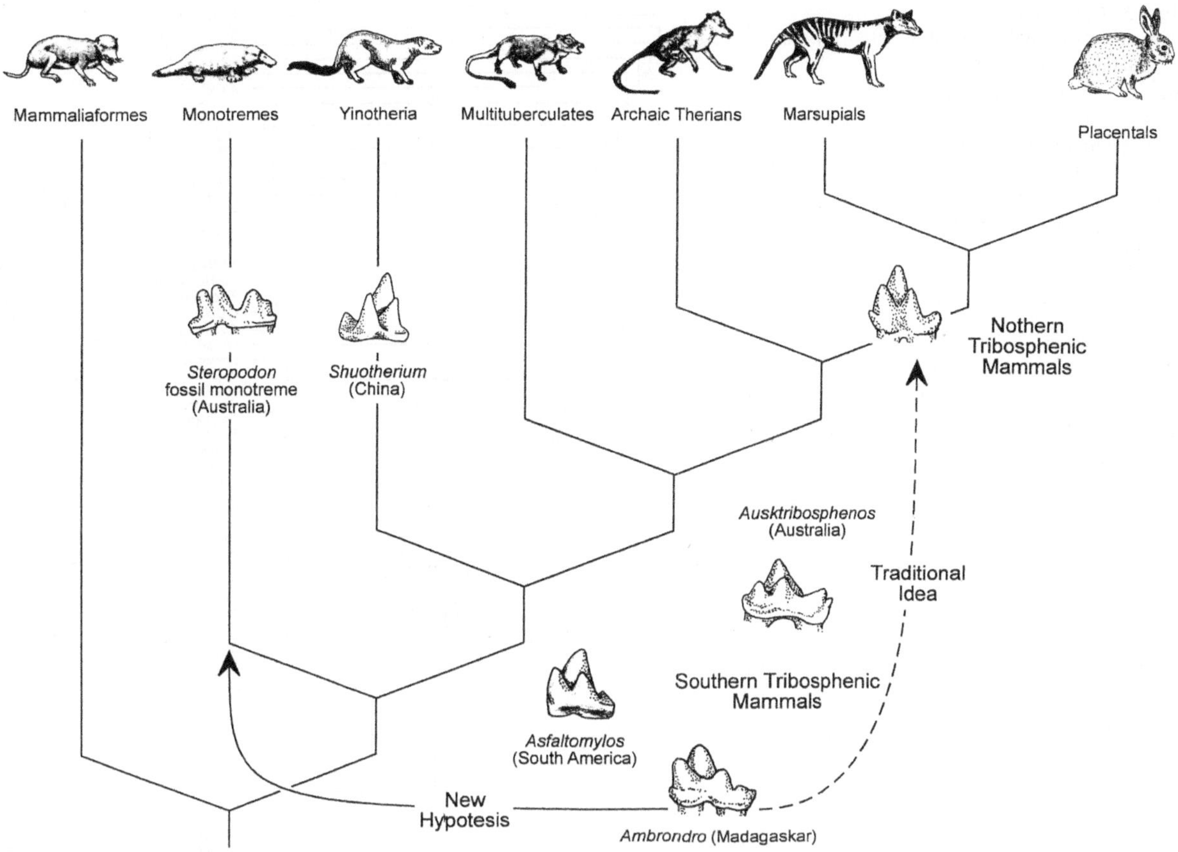

Fig. 12 Alternative phylogenies of the tribosphenic members of the Australosphenida of Luo et al. (2001) (modified with permission from Zhe-Xi Luo. After Press Release of the Carnegie Museum of Natural History)

descended from an ancestor with tribosphenic dentition. But, if so, the modifications that the teeth have undergone are so profound that there is no morphological trace of them necessarily having had a tribosphenic ancestor. In any case, the morphology of these teeth does not add evidence allying the monotremes with the tribosphenic australodelphians (Rich et al. 2002). For that matter, the dental pattern of the monotremes is as close to the boreosphenidians as to the tribosphenic australosphenidans.

The tribosphenic australosphenidans are characterised as having an advanced tribosphenic dentition on a primitive mandible (Luo et al. 2002), the primitive feature being the presence of an internal mandibular trough. This character of the jaw is quite variable in the different taxa regarded as australosphenidans. In the monotreme *Teinolophos* (Fig. 17), the groove is deep whereas in the monotreme *Steropodon*, it is shallow. In the ausktribosphenidan *Ausktribosphenos* (Fig. 18), the groove is shallow whereas in the other known ausktribosphenid, *Bishops*, there is no groove present (Fig. 19). The yinothere *Shuotherium* (Fig. 20) possesses a shallow groove as does the unquestioned boreosphenidian *Prokennalestes* (Kielan-Jaworowska and Dashzeveg 1989) (Fig. 21). Thus, it seems that this feature of the jaw is not consistent within the australosphenidans and is present in one unquestioned boreosphenidian. The presence of an internal mandibular trough is a plesiomorphic character and thus of little value in establishing relationships.

Two other characteristics used to discriminate australosphenidans from boreosphenidians are the height of the condyle and orientation of the angle on the mandible (Luo et al. 2002). Whereas this holds for the

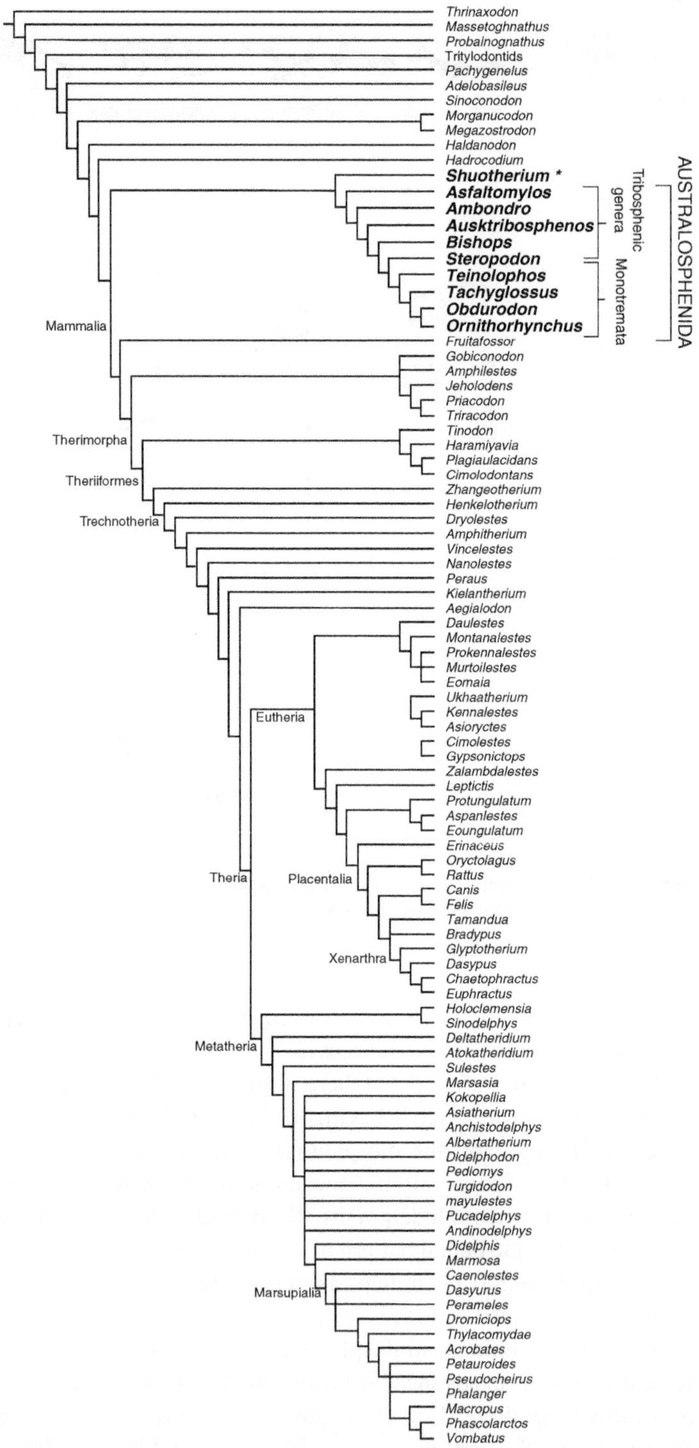

Fig. 13 Cladogram of the relationships of the Mammalia in Luo and Wible (2005). Note that the members of the Australosphenida are all clustered together, and the yinotheres (Chow and Rich 1982), emphasised by an *asterisk*, are closely allied with them

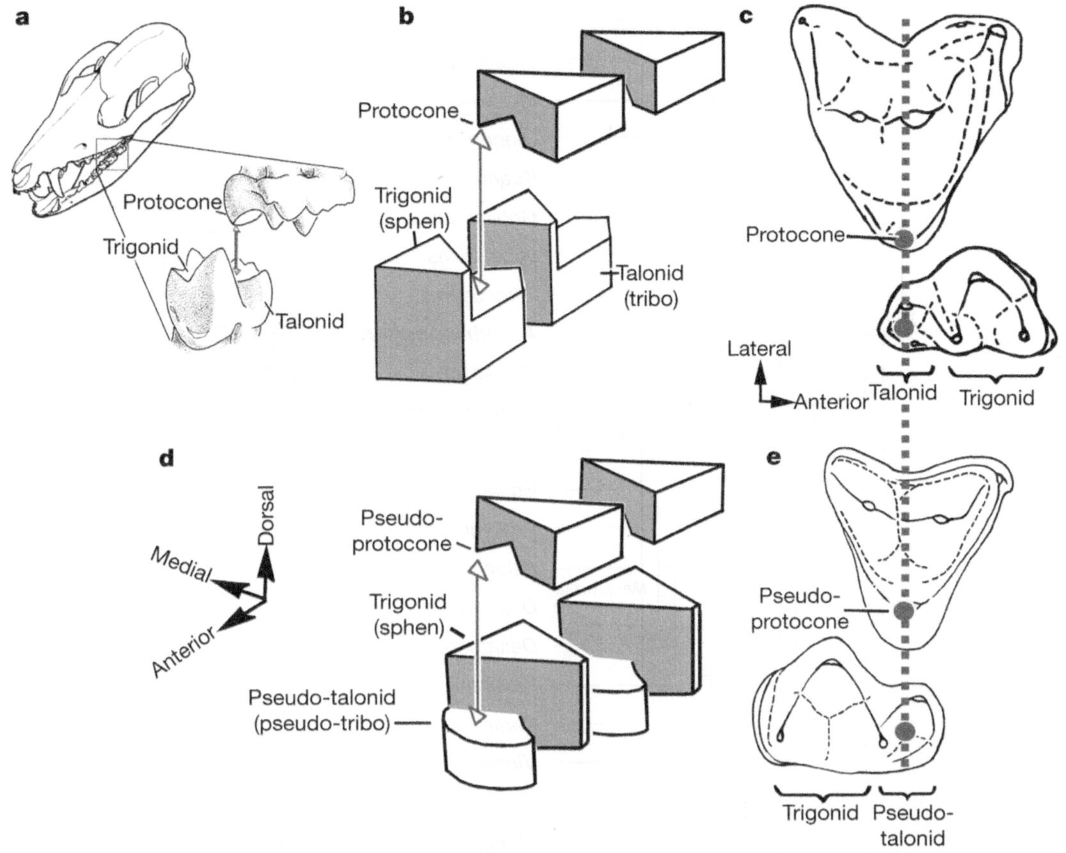

Fig. 14 Comparison of tribosphenic and pseudotribosphenic molars. (**a**) Skull of the marsupial *Didelphis virginiana* with its "true" tribosphenic molars. (**b**) Schematic illustration of the upper–lower occlusion of tribosphenic molars (trigonid anterior to talonid). (**c**) *Kielantherium* as an example of the ancestral tribosphenic pattern. (**d**) Schematic illustration of the upper–lower occlusion of pseudotribosphenic molars (trigonid posterior to pseudo-talonid). (**e**) Shuotheriids as an example of pseudotribosphenic molars (based on *Pseudotribos*). *Dashed line* shows the protocone-talonid occlusion and the pseudoprotocone-pseudotalonid occlusion. From Luo et al. (2007)

specimens shown in the left-hand column of Fig. 22 (Fig. 5 in Luo et al. 2002), additional erinaceids shown in the right-hand column (various figures from Butler 1948), have the orientation of these structures similar to that observed in the tribosphenic australosphenidans. Thus, these characteristics do not distinguish the tribosphenic australosphenidans from the boreosphenidians as Luo et al. (2002) have suggested.

Another character suggested by Luo et al. (2002) to distinguish australosphenidans from boreosphenidians was the nature of a cingulum in the vicinity of the paraconid on the lower molars. In the australosphenidans a shelf-like cingulum was interpreted by them as consistently wrapping around the paraconid in australosphenidans in contrast to the boreosphenidians where only a variable cuspule was present in the vicinity of the paraconid. Unfortunately, the view of the lower molar of *Teinolophos* shown in Fig. 3D of Luo et al. (2002) (Fig. 23d here) is lingual, whereas all the other teeth in that figure are in labial view. The actual labial view lies to the right of Fig. 23d.

When all three character differences are incorporated into the data matrix of Luo et al. (2002), the tribosphenic australosphenidans clustered with the eutherians whereas the monotremes and *Shuotherium* were quite remote from that group (Fig. 15) (Woodburne et al. 2003).

Because of the above, it is possible to conclude that the eutherians arose in Gondwana and subsequently spread to Laurasia. However, because the Jurassic context in which *Ambondro*, *Asfaltomylos* and *Henosferus* lived is still extremely poorly known – on the basis

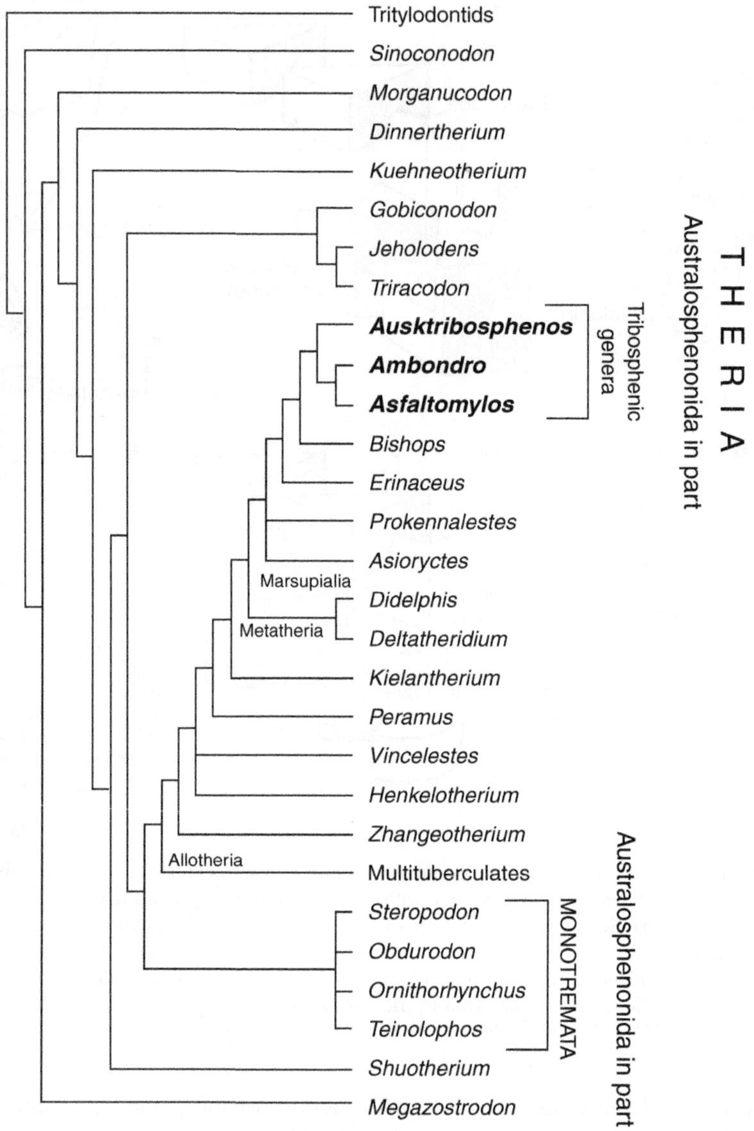

Fig. 15 Alternative cladogram of the relationships of the Mammalia (Fig. 2 in Woodburne et al. 2003). The characters analysed in this cladogram were virtually the same as those in Fig. 13. That the cladograms are not the same results from differences in assignment of character states. Whereas the australosphenidans were recognised by Luo et al. (2002) as a single entity, the taxa they included in that group are divided into the two quite distinct groups by Woodburne et al. (2003), eutherians and monotremes

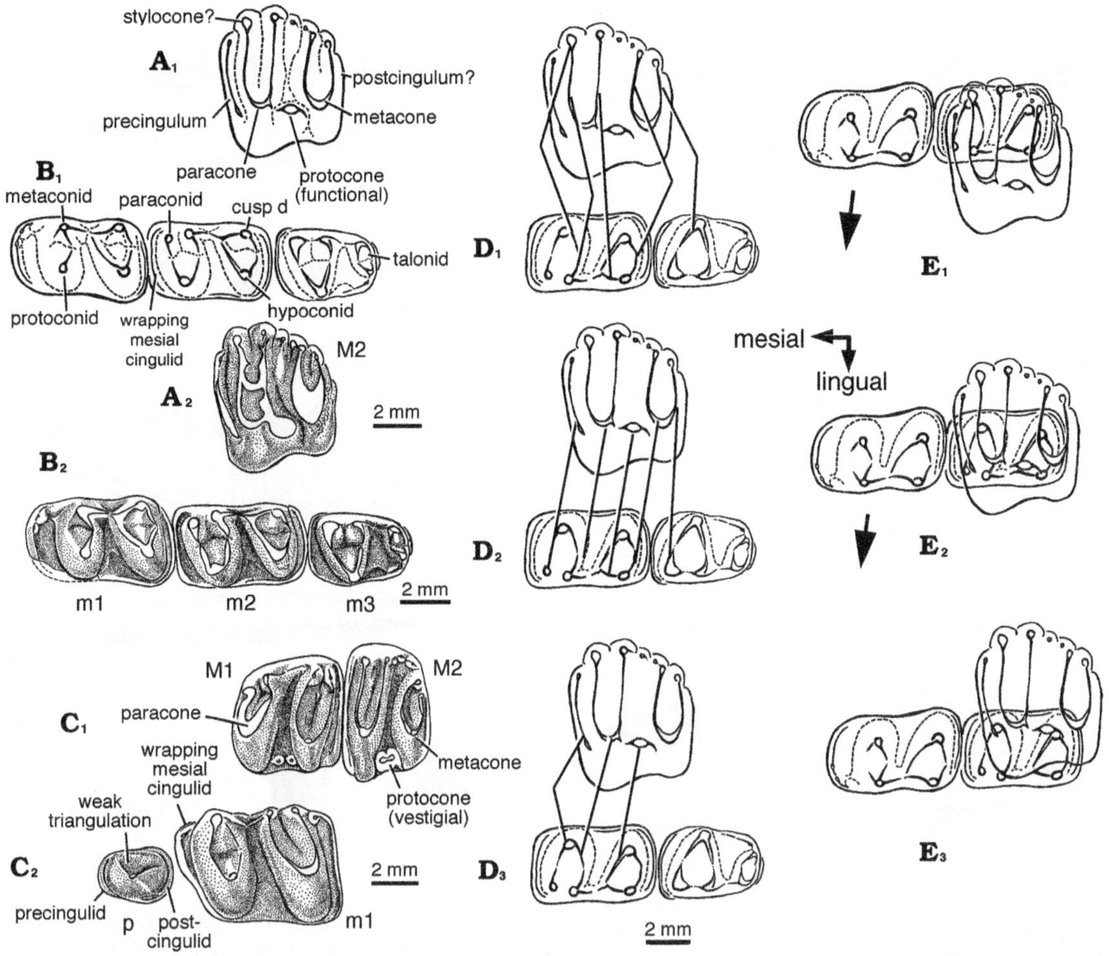

Fig. 16 Diagrammatic monotreme occlusal patterns. (**a1–2**) Upper molar of *Monotrematum sudamericanum*. (**b1–2**) Lower molars of *Steropodon galmani*. (**c1**) Upper molars of *Obdurodon dicksoni*. (**c2**) Lower premolar and molar of *O. dicksoni*. (**d**) Hypothetical occlusal relationships between upper and lower monotreme molars. (**d1**) Beginning of occlusal cycle. (**d2**) Middle of occlusal cycle which shows a functional protocone that is interpreted as having had a mortar and pestle action against cusp d. Even if this occlusal relationship did exist, it is analogous, not homologous to the tribosphenic pattern. (**d3**) End of occlusal cycle. (**e1–3**) Superimposed relationship of upper and lower monotreme molars, each numerical stage corresponding to the equivalent number in **d1–3**. After Luo et al. (2002)

of the fossils alone – Lillegraven's conclusion, "somewhere on the Earth's dry surface" is probably the most realistic conclusion to come to regarding the place of origin of the eutherians.

Molecular studies suggest the majority of the living orders of placentals arose well before the KT boundary (Bininda-Emonds et al. 2007). This contrasts with conclusions, drawn from the fossil record, suggesting only a few pre-Tertiary orders (Wible et al. 2007). If the molecular interpretation is correct, with many extant placental orders originating in the Late Cretaceous, there is an important lacuna in the fossil record. If that conclusion is correct, the unknown record may well be in areas that until now have been poorly sampled. This is the Garden of Eden hypothesis of Foote et al. (1999).

Foote et al. (1999) and Archibald and Deutschman (2001) have presented actuarial arguments that the fossil record should be taken as given. They suggested that it is reliable enough to accept the dates of first appearance of groups as close to their actual

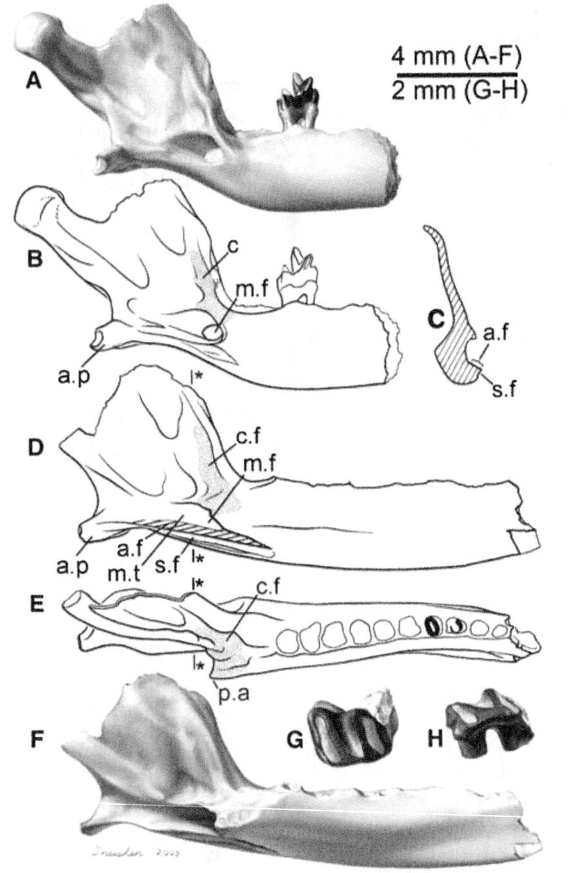

Fig. 17 (**a**) Medial view of holotype of *Teinolophos trusleri*, NMV P208231. (**b**) Diagrammatic medial view of NMV P208231; the stippled area indicates the position of the fused coronoid bone. (**c**) Cross section of mandible of referred specimen of *T. trusleri*, NMV P212933; position of cross section indicated in Figs. **d** and **e**, by lines terminated with *stars*. (**d**) Diagrammatic medial view of NMV P212933. The stippled area indicates the position of the contact facet for the coronoid bone. Traces of roots of a molar can be seen in alveoli three and four. (**e**) Diagrammatic dorsal view of NMV P212933. Traces of roots of a molar can be seen in alveoli three and four. (**f**) Medial view of NMV P212933, rotated slightly medially towards the viewer. (**g**) Occlusal and (**h**) medial views of isolated lower molar associated with dentary, NMV P212933. *Abbreviations*: a.f, angular facet; a.p, angular process; c, coronoid; c.f, coronoid facet; m.f, mandibular foramen; m.t, mandibular trough [= internal mandibular trough]; p.a, posterointernal angle; s.f, splenial facet. (From Rich et al. 2005, illustration by Peter Trusler)

dates of origin. Where this approach is weak is when evaluating the record of landmasses with no Late Cretaceous mammal sites, i.e., the Gondwana continents of Antarctica, Africa, and Australia.

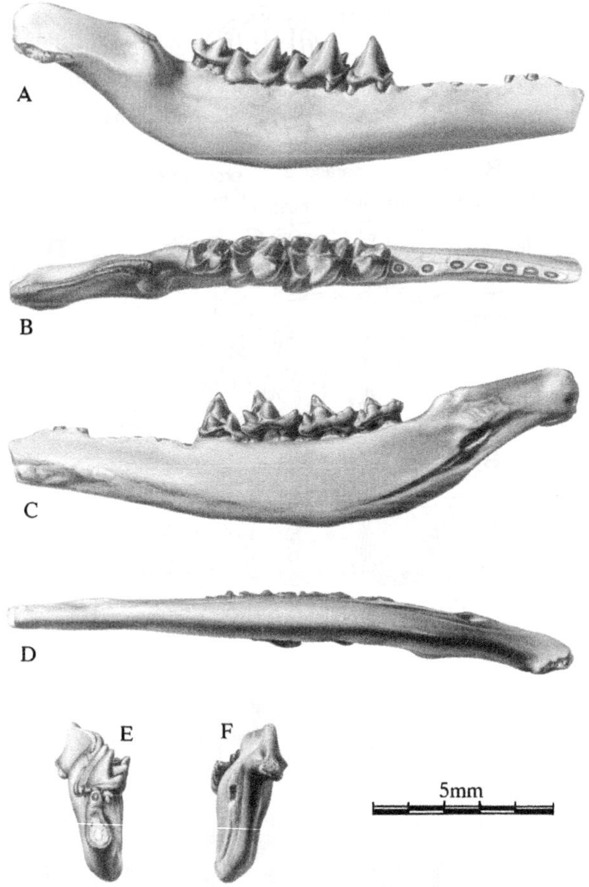

Fig. 18 Right mandible of the tribosphenic mammal *Ausktribosphenos nyktos* (Rich et al. 1997) (Fig. 2). From the Aptian Strzelecki Group, Flat Rocks, Victoria, Australia. Note presence of an internal mandibular trough, mt, in the middle image, the jaw in lingual view. Illustration by Peter Trusler

Hunter and Janis (2006) hypothesised that the placentals arose in the northern hemisphere. This was based in part on two parsimony arguments. The first was minimisation of the amount of missing evolutionary history (Foote et al. 1999). While this may be the best procedure for handling available data, given the uneven record of fossil placentals on various continents, whether it is even close enough to the actual events to be a useful guide rather than misleading is not clear. Second, it is not certain that a model which has the fewest number of continental interchanges for the various placental mammalian orders is the best estimate of their places of origin.

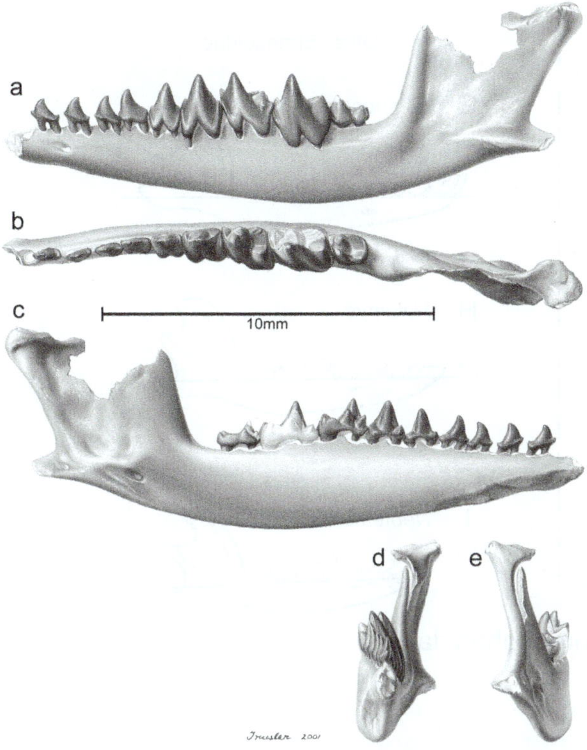

Fig. 19 Left mandible of the tribosphenic mammal *Bishops whitmorei* (Rich et al. 2001) (Fig. 1). From the Aptian Strzelecki Group, Flat Rocks, Victoria, Australia. Note the absence of an internal mandibular trough, mt, in the medial view of the jaw, Fig. **c**. Illustration by Peter Trusler

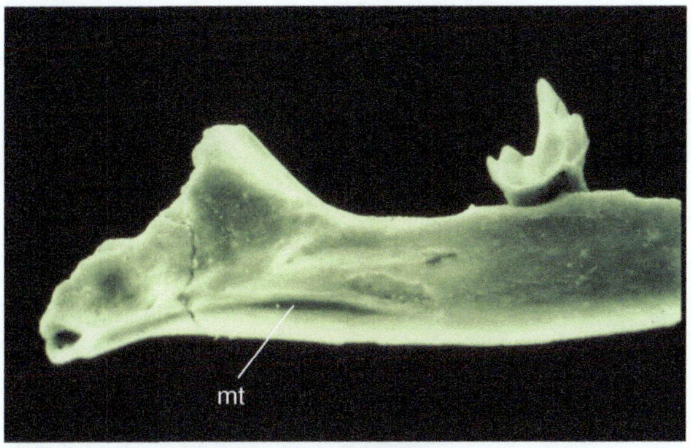

Fig. 21 Left mandible of the placental *Prokennalestes minor* Kielan-Jaworowska and Dashzeveg (1989) (Fig. 23). Note the presence of an internal mandibular trough in this medial view of the lower jaw

In this instance, parsimony is regarded as a trustworthy guide for deciding between one geographic model and another because intercontinental interchanges are considered unlikely (McKenna 1973).

We question whether such interchanges are not always rare. For example, between North and South America in the Pliocene Great American Interchange, 46 genera moved from one continent to the other (Webb 1985). This is far more than the ordinal interchanges analysed by Hunter and Janis (2006). If exchanges are possible at all, does their number provide a meaningful measure of the likelihood? Given the smaller numbers in the models compared by Hunter and Janis (2006), one can only wonder at just how meaningful the significance of the numerical differences might be.

An alternative hypothesis between the views of Luo et al. (2001, 2003) (Fig. 13) and Woodburne et al. (2003) (Fig. 15) is to be found in Rougier et al. (2007) (Fig. 24). In their cladogram, although the Australosphenida still include the monotremes and continue to be regarded as quite separate from the Boreosphenida, the shuotheriids are suggested to be only remotely related to the former.

Luo (2007), in reviewing the evolution of mammals during the Mesozoic, made the point that, at that time, there were many "evolutionary experiments" which came to dead ends. That is, there were several

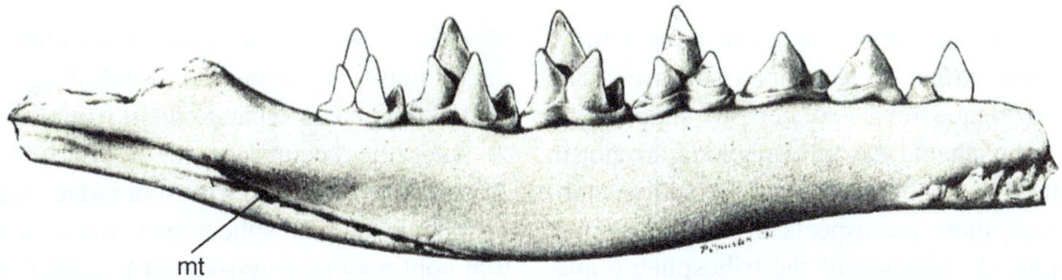

Fig. 20 Left mandible of the yinothere *Shuotherium dongi* Chow and Rich (1982) (Fig. 5B). Note the presence of an internal mandibular trough, mt, in this medial view of the lower jaw. Illustration by Peter Trusler

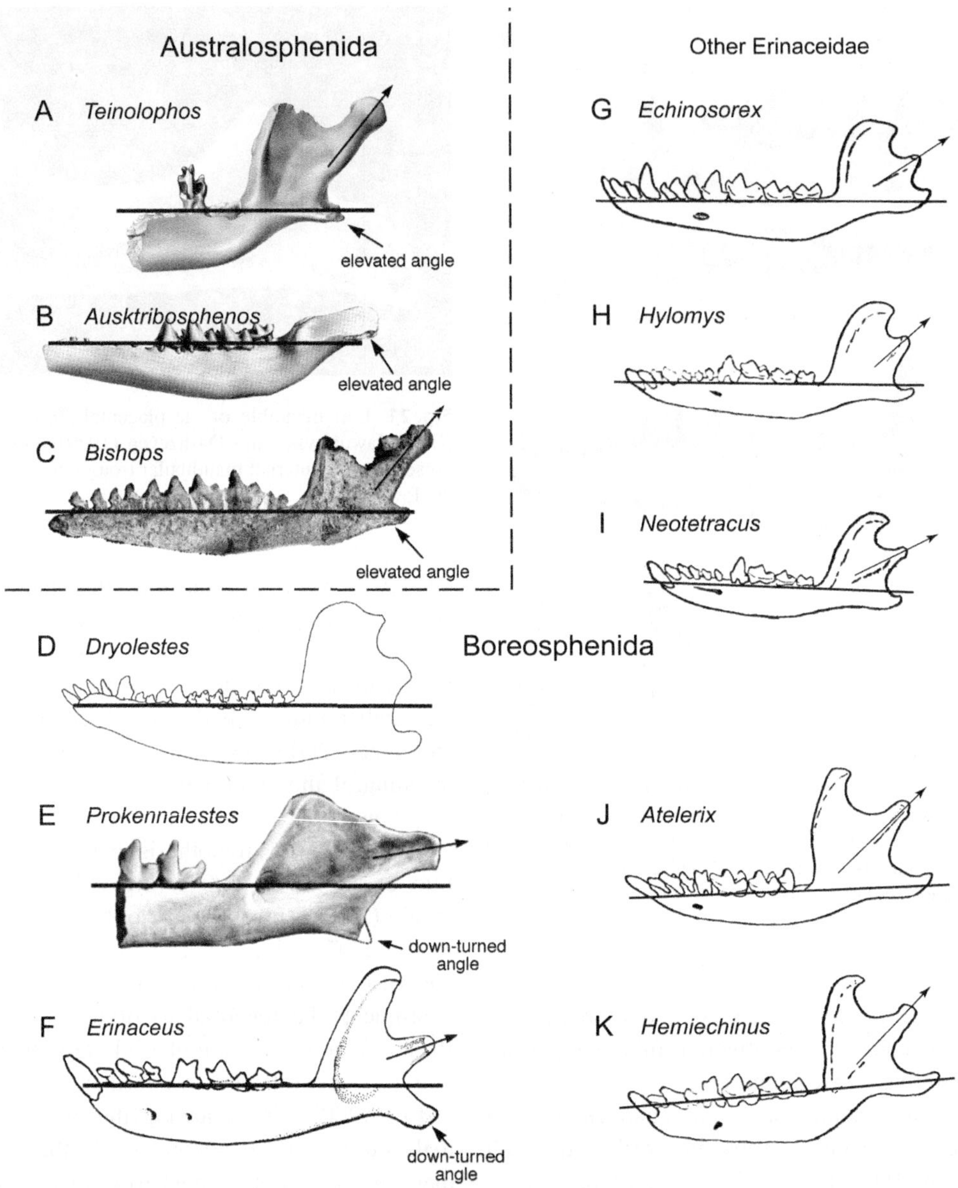

Fig. 22 Comparison of the height of the condyle and orientation of the condyle on the mandible of boreosphenidians and australosphenidans. See text for discussion. From Butler (1948) and Luo et al. (2002)

mammalian groups that arose, flourished for a time, and then became extinct without apparent issue. The morphologic evidence reviewed here to support the idea that the tribosphenic Australosphenida are not in fact part of the Placentalia is not unequivocally established. However, there are aspects of the geographic and stratigraphic distribution of the tribosphenic australosphenidans suggesting they may well have been one of the failed evolutionary experiments of the Mesozoic Mammalia *sensu* (Luo 2007) despite the uncertainties regarding the morphologic evidence presented thus far to separate them from the Eutheria.

After the occurrence in the Early Cretaceous of Australia of the Ausktribosphenidae, regarded as tribosphenic australosphenidans, there is no record on that continent of a non-volant placental mammal until the Pliocene with one possible exception. The exception is *Tingamarra poterorum*, a taxon based on a

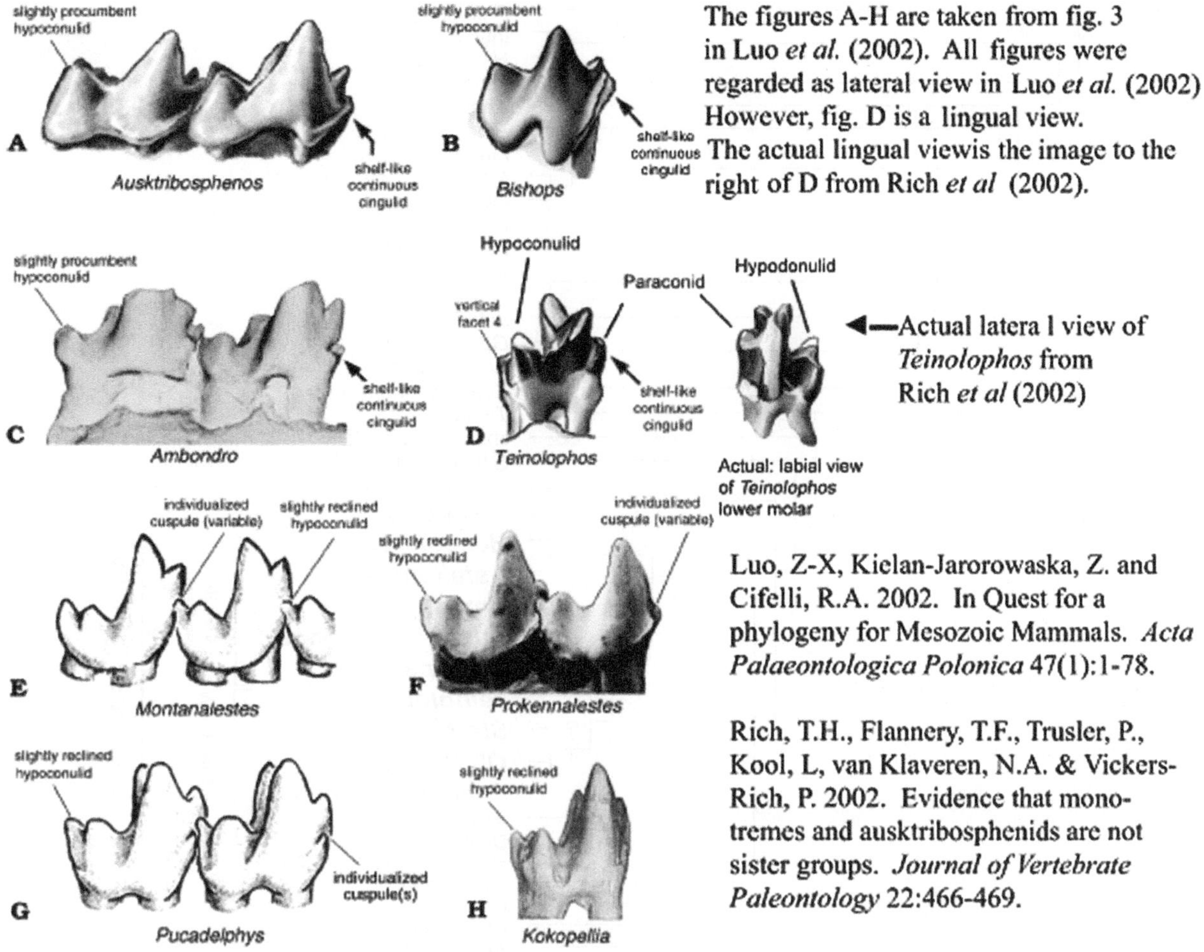

Fig. 23 Comparison of the dentitions of australosphenidans and boreosphenidians. See text for discussion. From Luo et al. (2002)

single lower molar. The holotype is from the Early Eocene at Murgon in southeastern Queensland. It was originally regarded as a placental, possibly a condylarth, when described by Godthelp et al. (1992). This determination was disputed by Woodburne and Case (1996) who regarded the holotype of *T. porterorum* as similar to the marsupial *Bobbschaefferia flumensis* from the Itaborai locality of Brazil. The taxonomy of this species remains unresolved.

As mentioned above, the Late Cretaceous Los Alamitos fauna of South America lacks any therians, although the mammals from this rich fossil site are diverse (Bonaparte 1990). As with Australia, it seems remarkable if the Asfaltomylidae, another family placed in the tribosphenic Australosphenida, were allied with placentals, that these Jurassic forms had no descendants represented in the single site in the Cretaceous of South America prior to the latest part of that period, where they would be reasonably expected if they were living on that continent.

After the Middle Jurassic record of the tribosphenic australosphenidan *Ambondro* in Madagascar (Flynn et al. 1999), there are no known mammals in Africa clearly allied with that genus. A single partial molar of a possible placental mammal from the Late Cretaceous/Maastrichtian of Madagascar (Averianov et al. 2003) was tentatively regarded by Averianov, Archibald, and Martin as a zhlestid ungulatomorph, a group widespread in the Late Cretaceous of Asia, Europe, and North America and thus possibly representing a group that entered Africa from Laurasia, rather than being a close relative of *Ambondro*.

So, all the three Gondwanan landmasses which have what can be regarded as tribosphenic

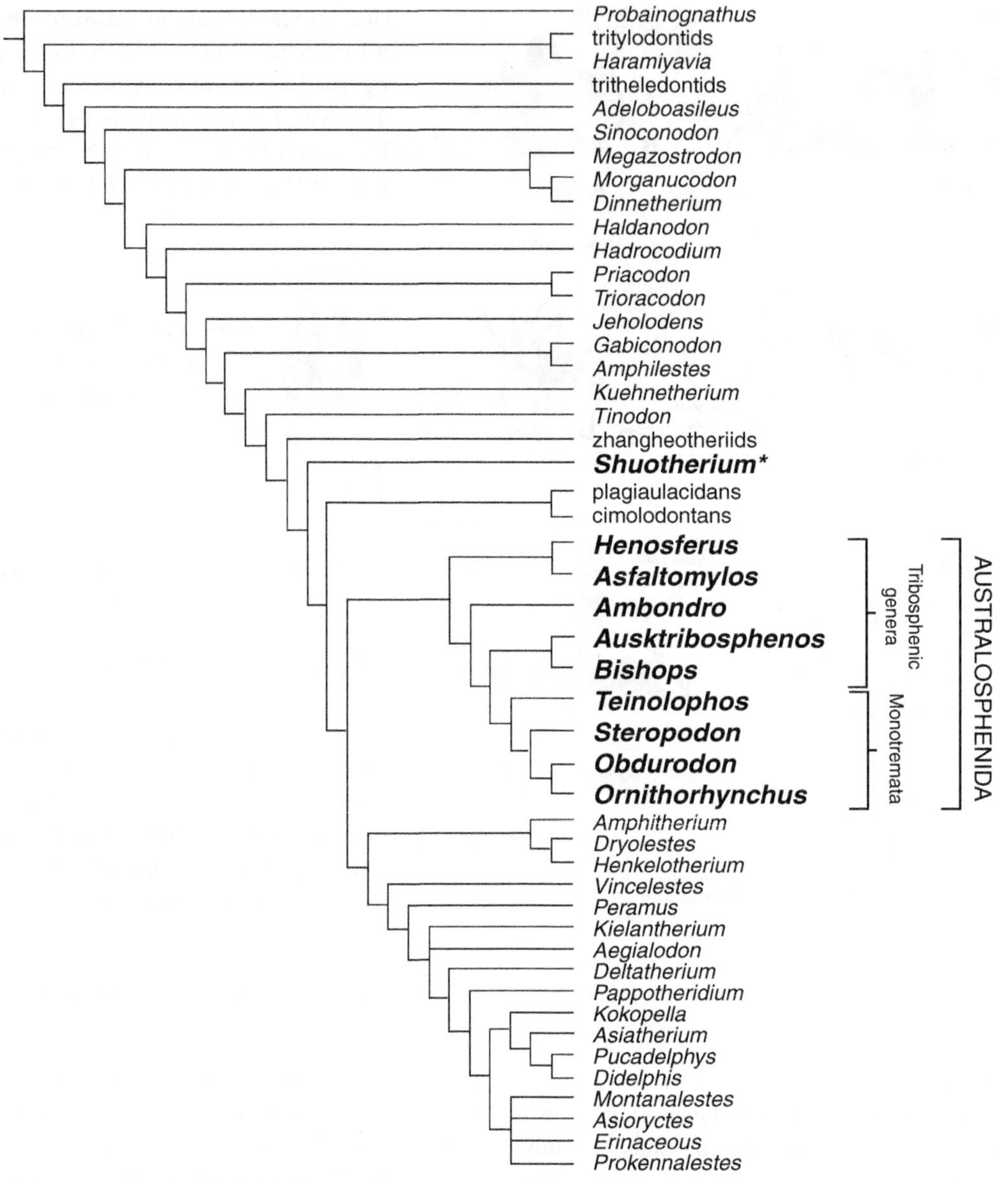

Fig. 24 Cladogram of the relationships of the Mammalia in Rougier et al. (2007). Note that like Fig. 15 and unlike Fig. 13, the Australosphenida are quite separate from the yinotheres which are emphasised by an asterisk. However, in agreement with Fig. 13 and unlike Fig. 15, the Australosphenida includes both monotremes and tribosphenic forms, the latter quite separate from the Placentalia

australosphenidans lack a record of any obvious descendants following their brief appearances in the geological record. Although this is hardly conclusive evidence that tribosphenic australosphenidans had no role in the evolution of the later placentals, this weak negative evidence of absence is consistent with that view.

At present, with the record in hand, particularly on the Gondwana continents, it seems prudent to follow the sage advice of Thomas Chamberlin (1890) and entertain multiple working hypotheses about this topic until such time as further material makes the interpretation of the evidence less equivocal.

Conclusion

Knowledge of the palaeobiogeography of Mesozoic mammals is extremely meagre and uneven both in time and space. Reconciliation of the interpretations of the fossil and molecular evidence relating to this problem is as central to future progress about this question as the discovery of additional fossils.

Acknowledgements Four colleagues greatly facilitated production of this updated understanding of the biogeography of Mesozoic mammals. We thank Richard Cifelli for generously supplying distribution data (in machine-readable format) for Mesozoic mammals that he compiled for Kielan-Jaworowska et al. (2004); Zhe-Xi Luo for kindly giving permission to use the Carnegie Museum web-site material, modified to produce Figs. 11 and 12; Draga Gelt for creating some of the illustrations and modifying others; and Mary Walters for proof-reading the manuscript.

References

Anantharaman S, Wilson GP, Das Armai DC, Clemens WA (2006) A possible Late Cretaceous "Haramiyidan" from India. J Vertebr Paleontol 26:488–490

Archibald JD, Averianov AO (2001) Paranyctoides and allies from the Late Cretaceous of North America and Asia. Acta Palaeontol Pol 46:533–551

Archibald JD, Averianov AO (2003) The Late Cretaceous placental mammal *Kulbeckia*. J Vertebr Paleontol 23:404–419

Archibald JD, Deutschman DH (2001) Quantitative analysis of the timing of the origin and diversification of extant placental orders. J Mamm Evol 8:107–124

Averianov AO, Archibald JD, Martin T (2003) Placental nature of the alleged marsupial from the Cretaceous of Madagascar. Acta Palaeontol Pol 48:149–151

Bininda-Emonds ORP, Cardillo M, Jones KE, Macphee RDE, Beck RMD, Grenyer R, Price SA, Vos RA Gittleman JL, Purvis A (2007) The delayed rise of present-day mammals. Nature 446:507–512

Bonaparte JF (1990) New Late Cretaceous mammals from the Los Alamitos Formation, northern Patagonia. Nat Geol Res 6:63–93

Butler PM (1948) On the evolution of the skull and teeth in the Erinaceidae, with special reference to fossil material in the British Museum. Proc Zool Soc Lond 118:446–500

Chamberlin TC (1890) The method of multiple working hypotheses. Science 15:92–96. Republished as Chamberlin TC (1965) The method of multiple working hypotheses. With this method the dangers of parental affection for a favourite theory can be circumvented. Science 148:754–759

Chow M, Rich TH (1982) Shuotherium dongi, gen.et sp. nov., a therian with pseudotribosphenic molars from the Jurassic of Sichuan, China. Aust Mammal 5:127–142

Cifelli RC, De Muizon C (1997) Dentition and jaw of Kokopellia juddi, a primitive marsupial or near-marsupial from the medial Cretaceous of Utah. J Mamm Evol 4:241–258

Davis BM (2007) A revision of "pediomyid" marsupials from the Late Cretaceous of North America. Acta Palaeontol Pol 52:217–256

Flynn JJ, Parrish JM; Rakotosamimanana B, Simpson WF, Wyss AE (1999) A middle Jurassic mammal from Madagascar. Nature 401:57–60

Foote M, Hunter JP, Janis CM, Sepkoski JR Jr (1999) Evolutionary and preservational constraints on origins of biologic groups: divergence times of Eutherian mammals. Science 283:1310–1314

Fox RC, Scott CS, Bryant HN (2007) A new, unusual therian mammal from the Upper Cretaceous of Saskatchewan, Canada. Cretaceous Res 28:821–829

Godthelp H, Archer M, Cifelli R, Hand SJ, Gilkeson CF (1992) Earliest known Australian Tertiary mammal fauna. Nature 356:514–516

Godthelp H, Wroe S, Archer M (1999) A new marsupial from the Early Eocene Tingamarra local fauna of Murgon, Southeastern Australia: a prototypical Australian marsupial? J Mamm Evol 6:289–313

Hu Y, Meng J, Li C, Wang Y (2009) New basal eutherian mammal from the Early Cretaceous Jehol biota, Liaoning, China. Proc R Soc B 277(1679):229–236

Hunter JP, Janis CM (2006) Spiny Norman in the Garden of Eden? Dispersal and early biogeography of Placentalia. J Mamm Evol 13:89–123

Ji Q, Luo Z-X, Yuan C-X, Wible JR, Zhang JP, George JA (2002) The earliest known Eutherian mammal. Nature 416:816–822

Khosla A, Prasad GVR, Verma O, Jain AK, Sahni A (2004) Discovery of a micromammal-yielding Deccan intertrappean site near Kisalpuri, Dindori District, Madhya Pradesh. Curr Sci 380:380–383

Kielan-Jaworowska Z, Dashzeveg D (1989) Eutherian mammals from the Early Cretaceous of Mongolia. Zool Scr 18: 347–355

Kielan-Jaworowska Z, Cifelli R, Luo Z-X (2004) Mammals in the age of dinosaurs. Columbia University Press, New York, 630p

Lillegraven J; Kielan-Jaworowka Z, Clemens WA (1979) Mesozoic Mammals: the first two-thirds of mammalian history. University of California, Berkeley, CA, 311 pp

Longman HA (1927) A giant dinosaur Rhoetosaurus brownie. Mem Qld Mus 9:1–18

Lopatin AV, Averianov AO (2007) The Earliest Asiatic Pretribosphenic Mammal (Cladotheria, Amphitheriidae) from the Middle Jurassic of Siberia. Dokl Akad Nauk (Biol Nauk) 417:432–434

Lopatin AV, Maschenko EN, Averianov AO, Rezvyi AS, Skutschas PP, Leshchinskiy SV (2005) Early Cretaceous Mammals from Western Siberia: 1. Tinodontidae. Paleont Zh 39:523–534

Luo Z-X (2007) Transformation and diversification in early mammal evolution. Nature 450:1011–1019

Luo Z-X, Wible JR (2005) A Late Jurassic digging mammal and early mammalian diversification. Science 308:103–107

Luo Z-X, Cifelli R, Kielan-Jaworowska Z (2001) Dual origin of tribosphenic mammals. Nature 409:53–57

Luo Z-X, Kielan-Jaworowska Z, Cifelli RL (2002) In quest for a phylogeny of Mesozoic mammals. Acta Palaeont Polonica 47:1–78

Luo Z-X, Ji Q, Wible JR, Yuan C-X (2003) An Early Cretaceous tribosphenic mammal and metatherian evolution. Science 302:1934–1940

Luo Z-X, Chen P, Li G, Chen M (2007) A new eutriconodont mammal and evolutionary development in early mammals. Nature 446:288–293

Luo Z-X, Yuan C-X, Meng Q-J, Qiang J (2011) A Jurassic eutherrian mammal and divergence of marsupials and placentals. Nature 476:442–445

Marshall LG, Muizon C, Sigogneau-Russell D (1995) Part 1: The locality of Tiupammpa: age, taphonomy and mammalian fauna. In: Muizon C (ed) *Pucadelphys andinus* (Marsupialia, Mammalia) from the early Paleocene of Bolivia. Mém Mus Nat Hist Nat 165: 11–20

Martin JE, Case JA, Jagt JWM, Schulp AS, Mulder EWA (2005) A new European marsupial indicates a Late Cretaceous high-latitude Transatlantic dispersal route. J Mamm Evol 12: 495–511

McKenna MC (1973) Sweepstakes, filters, corridors, Noah's arks, and beached Viking funeral shops in paleogeography. In: Tarling DH, Runcorn SK (eds) Implications of continental drift to the Earth sciences. Academic Press, London, New York, pp 21–46

Meng J, Hu Y, Wang Y, Wang X, Li C (2006) A Mesozoic gliding mammal from northeastern China. Nature 444:889–893

Murphy WJ, Eizirik E, O'Brien SJ, Madsen O, Scally M, Douady CJ, Teeling E, Ryder OA, Stanhope MJ, Jong WW, Springer MS (2001) Resolution of the early placental mammal radiation using Baysian phylogenetics. Science 294:2348–2351+Sup Material

Nilsson MA, Amason U, Spenser PBS, Janke A (2004) Marsupial relationships and a timeline. The palaeobiogeography of Mesozoic mammals: a review 249. Arq Mus Nac Rio de Janeiro 66(1):231–249 (jan./mar.2008 for marsupial radiation in South Gondwana). Gene 340:189–196

Pascual R, Goin FJ, Balarino L, Sauthier DEU (2002) New data on the Paleocene monotreme Monotrematum sudamericanum, and the convergent evolution of triangulate molars. Palaeontol Pol 47:487–492

Prasad GVR, Manhas BK (2007) A new docodont mammal from the Jurassic Kota Formation of India. Palaeontol Elect 10(2):7A, 11p

Prasad GVR, Verma O, Parmar V (2006) An overview of the Mesozoic mammalian fossil record of India. In: Barrett PM, Evans SE (eds) IX International symposium on Mesozoic terrestrial ecosystems and biota. Natural History Museum, Cambridge Publications, Manchester, London, UK, pp 101–104

Prasad GVR, Verma O, Sahni A, Parmar V, Khosla A (2007) A Cretaceous Hoofed mammal from India. Science 318:937

Rauhut OWM, Martin T, Ortiz-Jaureguizar E, Puerta P (2002) A Jurassic mammal from South America. Nature 416: 165–168

Rich TH (2008) The palaeobiogeography of Mesozoic mammals: a review. Arq Mus Nac Rio de Janeiro 66:231–249

Rich TH, Vickers-Rich P, Constantine A, Flannery TF, Kool L, van Klaveren N (1997) A tribosphenic mammal from the Mesozoic of Australia. Science 278:1438–1442

Rich TH, Flannery TF, Trusler P, Kool L, Van Klaveren NA, Vickers-Rich P (2001) A second tribosphenic mammal from the Mesozoic of Australia. Rec Queen Victoria Mus 110:1–9

Rich TH, Flannery TF, Trusler P, Kool L, van Klaveren NA, Vickers-Rich P (2002) Evidence that monotremes and ausktribosphenids are not sistergroups. J Vertebr Palaeontol 22:466–469

Rich TH, Hopson JA, Musser AM, Flannery TF, Vickers-Rich P (2005) Independent origins of middle ear bones in monotremes and therians. Science 307:910–914

Rich TH, Vickers-Rich P, Flannery TF, KearBP, Cantrill D, Komarower P, Kool L, Pickering D, Trusler P, Morton S, vanKlaveren N, Fitzgerald EMG (2009) An Australian multituberculate and its palaeobiogeographic implications. Acta Paleontol Pol 54:1–5

Rougier GW, Martinelli AG, Forasiepi AM, Novacek MJ (2007) New Jurassic mammals from Patagonia, Argentina: a reappraisal of australosphenidan morphology and interrelationships. Am Mus Novitates 3566:1–54

Sigogneau-Russell D, Godefroit P (1997) A primitive docodont (Mammalia) from the Upper Triassic of France and the possible therian affinities of the order. C R Acad Sci 324:135–140

Simpson GG (1947) Holarctic mammalian faunas and continental relationships during the Cenozoic. Bull Geol Soc Am 68:613–688

Szalay FS (1982) A new appraisal of marsupial phylogeny and classification. In: Archer M (ed) Carnivorous marsupials. Surrey Beatty & Sons and Royal Zoological Society of NSW, Mosman, vol 2, pp 621–640

Warren AH, Hutchinson MN (1983) The last labyrinthodont? A new brachyopoid (Amphibia, Temnospondyli) from the Early Jurassic Evergreen Formation of Queensland, Australia. Philos Trans R Soc Ser B 303:1–62

Webb SD (1985) Late Cenozoic mammal dispersals between the Americas. In: Stehli FG, Webb SD (eds) The Great American biotic interchange. Plenum, New York, pp 357–386

Wible JR, Rougier GW, Novacek MJ, Asher RJ (2007) Cretaceous eutherians and Laurasian origin for placental mammals near the K/T boundary. Nature 447:1003–1006

Woodburne MO (2003) Monotremes as pretribosphenic mammals. J Mamm Evol 10:195–248

Woodburne MO, Case JA (1996) Dispersal, vicariance, and the Late Cretaceous to early Tertiary land mammal biogeography from South America to Australia. J Mamm Evol 3:121–161

Woodburne MO, Zinsmeister WJ (1982) Fossil land mammal from Antarctica. Science 218:284–286

Woodburne MO, Rich TH, Springer M (2003) The evolution of tribospheny in Mesozoic mammals. Mol Phylogenet Evol 28:360–385

Part V
Cenozoic Era

Cenozoic Environmental Shifts and Foraminiferal Evolution

Brian McGowran

Abstract

The dense record of Cenozoic foraminifera simultaneously supplies a mosaic of biostratigraphy, a rich field for evolutionary studies and the vehicles for geochemical environmental proxies. Four groups are discussed: the larger foraminifera on the warm-water shelves and platforms, the planktonics, the deep-sea faunas and the southern-extratropical benthics. The environmental trajectory from greenhouse in the later Cretaceous and earlier Paleogene to icehouse in the Neogene is not smooth but punctuated, and there are two particularly critical intervals, later Eocene and early-middle Miocene. The foraminiferal record is not smooth but chunky at 10^7 years' scale. There are several good examples of two powerful synchroneities, one being between the faunas of the different realms and the other between the fossil record and the physical-environmental record.

Keywords

Late Cretaceous • Cenozoic • Paleogene • Neogene • Biostratigraphy • Environments • Foraminifers • Planktonics • Benthics • Paleoclimate

Introduction

Naturalists have been aware of the foraminifera since the late seventeenth century. Advances in microscopy rapidly expanded general appreciation of biodiversity during the great Enlightenment projects of discovery and classification. The foraminifera displayed a striking range of variety in their shells in marginal-marine and shallow-water sediments, soon to be confirmed in a fossil record to match. They contributed little to the building of the geological time scale, which was accomplished by using the macrofossil record.

Background of Foraminifera

Despite the above, the large shells of the genus *Nummulites* ("coin-like", also identified as petrified lentils by the Greek chronicler Herodotus) were the basis for the "Nummulitique" as illustrated by Lyell (1871, Frontispiece, also McGowran 2005, Fig. 1). The Nummulitique became the Paleogene Period of the Cenozoic Era (Gignoux 1955).

B. McGowran (✉)
Earth and Environmental Sciences, The University of Adelaide, Adelaide, SA 5005, Australia
e-mail: brian.mcgowran@adelaide.edu.au

J.A. Talent (ed.), *Earth and Life*, International Year of Planet Earth,
DOI 10.1007/978-90-481-3428-1_33, © Springer Science+Business Media B.V. 2012

The d'Orbigny Legacy

Even when the succession of life upon which this time-scale-building achievement was realized was explained by Darwin's *On the origin...* (1859), these abundant and beautiful shells remained in limbo, thanks largely to the attacks by the "English school" of the mid-nineteenth century on Alcide d'Orbigny, the prolific French paleontologist and stratigrapher with a "catastrophist" world view, and their influence on Darwin (Lipps 1981; Cifelli 1990). I have already quoted the following from Lipps (McGowran 2005) but it bears repeating:

> Although d'Orbigny is now recognized as a generally good descriptive worker, he was ridiculed by English micropaleontologists for proliferating species and his theory of creations was resoundingly beaten by evolutionists. Thus, the founder of micropaleontology made no contribution to the use of microfossils in evolutionary studies. In fact, it seems that the English foraminiferologists were so incensed by d'Orbigny's attitudes that they too destroyed the potential of foraminifera in this type of work. Had d'Orbigny not clung so tenaciously to his version of successive creations and had the English micropaleontologists not later damned him so thoroughly, foraminifera (and micropaleontology) might have made sound contributions to stratigraphy and evolutionary paleontology very early on....
>
> Carpenter (Carpenter et al. 1862, pp. x–xi) arrived at a set of conclusions that dominated the study of foraminifera until the rise of industrial micropaleontology. Chief among his conclusions were the following:
>
> I. There was a great range of variation in foraminifera.
> II. The ordinary notion of species, as assemblages of individuals marked out from each other by definite characters have been genetically transmitted from original prototypes similarly distinguished, is quite "inapplicable to this group".
> III. The only natural classification of the vast aggregate of diversified forms which this group contains will be one that ranges them according to their direction and degree of divergence from a small number of principal family-types.
> IV. The evidence in regard to the genetic continuity between the foraminifera of successive geological periods, and between those of the later of these periods and the existing inhabitants of our seas, is as complete as the nature of the case admits.
> V. "There is no evidence of any fundamental modification or advance in the Foraminiferous type from the Palaeozoic period to the present time." In short, "...there is no indication of any tendency to elevation towards a higher type."
>
> Carpenter's views, especially regarding the longevity of species, were passed to Darwin who, in the *Origin*, was forced to consider Carpenter's assertions that there had been no change among the foraminifera since the earliest geologic period. Darwin reasoned that organisms should show continued advancement through time and, although he knew it was not a necessary outcome of evolution, it seemed at that time generally to be true. Foraminifera were one exception, and Darwin, accepting Carpenter's opinion, strained to reconcile such a record in the fourth edition of the *Origin...*: "It is not an insuperable difficulty that Foraminifera have not, as insisted on by Dr Carpenter, progressed in organization since even the Laurentian epoch; for some organisms would have to remain fitted for simple conditions of life, and what could be better fitted for this end than these lowly organized Protozoa?" Thus, Darwin expressed the two most common and destructive views regarding foraminifera—that they ranged long without evolving, and that they were "simple" organisms that we perhaps should expect not to evolve at all.
>
> "Thus micropaleontology, as a technology that was able to assist geology, was essentially killed by Carpenter's and Williamson's attacks upon d'Orbigny, and the acceptance of Carpenter's views, as with Darwin, eliminated foraminifera from serious consideration in evolutionary studies."

Carpenter's opinion, that the foraminifera had not evolved since the Precambrian, was not observational, reflecting instead the tradition of the times that fossil specimens be assigned to previously described extant taxa (Cifelli 1990). Although the stratigraphic importance of identified morphotypes was pointed out clearly in 1865 by A.E. Reuss, it was not generally appreciated for a long time afterwards (Glaessner 1945).

The HMS *Challenger* expedition of the 1870s provided the first real insight into the diversity and distribution of the foraminifera on the continental shelves and slopes and in the deep ocean. The rise of sedimentary basin analysis in petroleum exploration and development stimulated the study of foraminifera as paleoenvironmental indicators (*biofacies analysis*) and in correlation and age determination (*biostratigraphy*), all depending on morphology and systematics in the age dominated by Joseph A Cushman (e.g., 1940).

But it was not until the 1930s, six decades after the first reconstructions of evolutionary lineages in the horses, that we saw such lineages for the foraminifera—in this case, the late Cretaceous and Cenozoic planktonic foraminifera, rapidly proving themselves biostratigraphically in deconstructing the tectonic complexities of the Alpine fold belts and adjacent basins. Two areas of exploration in the 1930s stand out—the Caucasus and surrounds (especially

Glaessner 1937; Subbotina 1953) and the Caribbean (e.g., Cushman and Stainforth 1945; Bolli 1957; Blow 1959).

Evolution and the Cenozoic Fossil Record: Foraminifera, the Negative Ledger

Concerning the fossil record, easily the outstanding point about the succession of life, pre-Darwin, was the biostratigraphic construction of the geological time scale. Sandwiched between the theory and fact of extinction (Cuvier in the 1790s) and the theory and fact of speciation (Darwin and Wallace in the 1850s), the rapid sketching of a succession of organisms and a history of life through geological time was a—or perhaps the—major scientific achievement of the early nineteenth century. The historical components of evolution were accepted readily enough in the first Darwinian revolution of the 1860s–1870s. Fossil foraminifera played no part, nor in the ensuing decades of rancor and apathy over the status and significance of evolution in biology. In the second Darwinian revolution of the 1930s–1940s, the so-called synthetic theory incorporating population genetics and natural selection (e.g., Mayr and Provine 1980; Smocovitis 1996), foraminifera were again absent. Intrinsic to this ferment were the rise of evolutionary taxonomy and the various evolution-centered species concepts, followed in due course by numerical phenetics and cladistics (Hull 1988). Occasional seeps from all this could be detected in foraminiferology but there was virtually no actual contribution from foraminiferology.

Evolution and the Cenozoic Fossil Record: Foraminifera, the Positive Ledger

But change was afoot in foraminiferology and the discipline was deeply involved in the following succession of technological and scientific advances down the decades: (i) an economic drive, stimulating drilling and coring in petroleum exploration and production (1930s onward); (ii) a further petroleum-technological drive, offshore drilling leading to deep-ocean drilling during the plate-tectonic revolution (1960s onward); (iii) ultra-microscopic technology, especially the scanning electron microscope (1960s); (iv) an environmental drive in the rise of paleoceanography and the development of chemical proxies from ancient oceans carried in the mineralized skeleton (1950s, especially 1970s); and (v) a biological drive in the form of evolutionary genetics, which is challenging both species discrimination and phyletic configurations (1980s).

Inherent properties of foraminifera underpinned those advances—their small size (attaining >1 mm diameter in very few lineages down the mega-anna), their robust skeletons recovered in huge numbers in small subsurface and subsea samples, and their inhabiting the neritic, marginal-marine and pelagic environmental realms. In the pelagial their habitats are benthic, down to the abyssal plains and the trenches, and planktonic, in and below the photic zone and mixed layer.

The Cenozoic Environmental Panorama

It is commonplace that the global environment has shifted extensively from the Cretaceous "greenhouse" state to the Neogene "icehouse" state. In the former state, sea levels were higher and continental floodings were more extensive. In the very early days of paleoceanography Lipps (1970) developed a far-sighted scenario of plankton evolution in terms of two end-member model oceans with various environmental and paleobiological implications (Fig. 1).

The Cenozoic trajectory of the greenhouse-icehouse transition is now "described" (if that is the word) by the oxygen and carbon isotope records compiled from numerous oceanic sections drilled during the Deep Sea Drilling Project and the Ocean Drilling Program (Zachos et al. 2001) (Figs. 2 and 3). Note that these readings come from the shells of deep-benthic foraminifera, a reminder that the entire water mass of the global ocean is involved, not merely the skin of the mixed layer above the thermocline.

Given certain caveats concerning postmortem changes to the isotopic composition of the original shells, there is fair consensus on the stepwise shifts from the early Paleogene to the late Neogene (Figs. 2 and 3). At the coarsest time scale (10^{8-7} years) we see an overall progressive cooling of the global ocean pointing to polar cooling through the Cenozoic Era. This trajectory is actually the last half, approximately, of the downslope from the highpoint in the middle

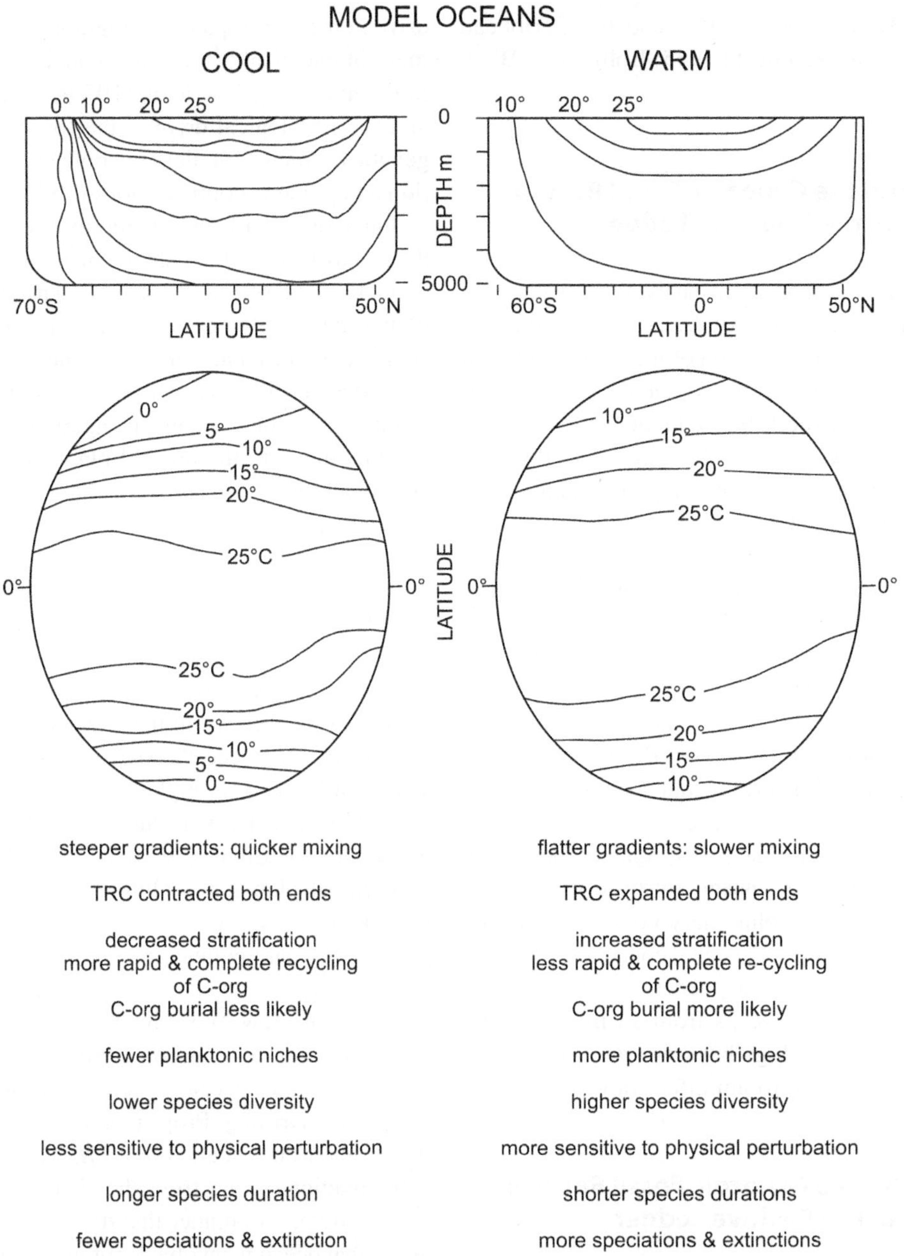

Fig. 1 Model oceans, in an heuristic pioneering attempt to contrast greenhouse with icehouse oceans (Lipps 1970); the generalizations added on environments and biotas (McGowran 2005, Fig. 6.25) are still useful and broadly meaningful. TRC, trophic resource continuum (Hallock 1987); C-org, organic carbon. Reprinted with permission of Cambridge University Press

Cretaceous Period of the later Phanerozoic supercycle (seen in Fig. 4). The history—geohistory and biohistory—of the Cenozoic Era is a twofold configuration. The Paleogene Period (Paleocene-Oligocene) was more greenhouse-like with higher (perhaps much higher?) levels of pCO_2 (Fig. 3). The Neogene Period (Miocene-Holocene) was more icehouse-like with, strikingly, pCO_2 levels seemingly refractory to changes in global temperature, sea level, or the waxing-waning of ice sheets. The Oligocene Epoch was a kind of interregnum between the bulk of the Paleogene and the Neogene.

It has long been argued whether stepwise patterns in geohistory and biohistory were real or merely

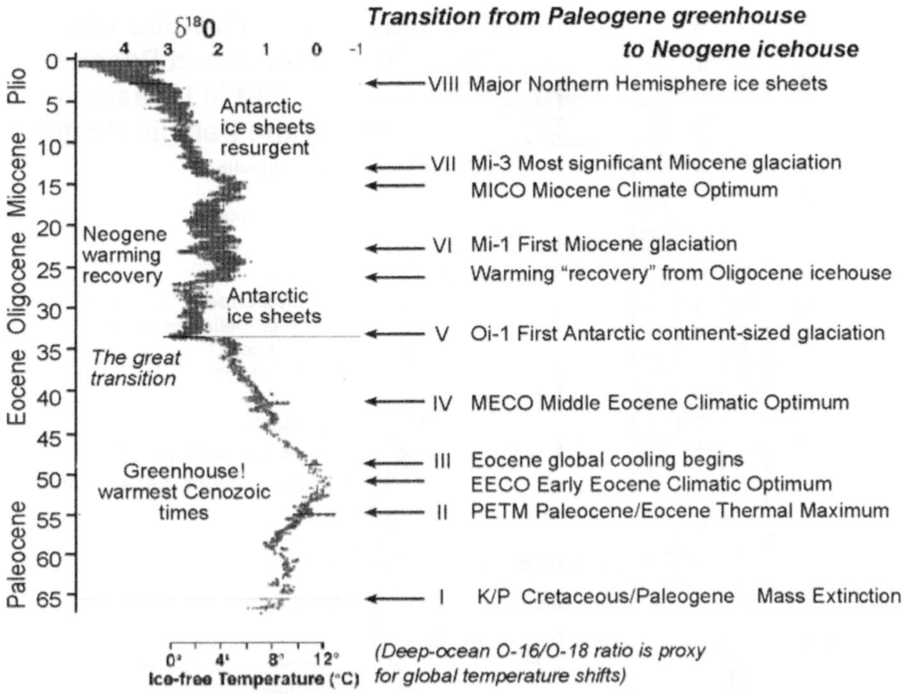

Fig. 2 Steps in the global climatic transformation during the Cenozoic Era. Oxygen isotope ratios (δ^{18}O) from Zachos et al. (2001, 2008) are from deep-ocean, benthic foraminiferal shells. Significant environmental shifts or transient events from numerous sources, including four coolings at levels III, V, VII, and VIII. Lines I–VIII from McGowran (2005, Fig. 6.12). See also Fig. 4

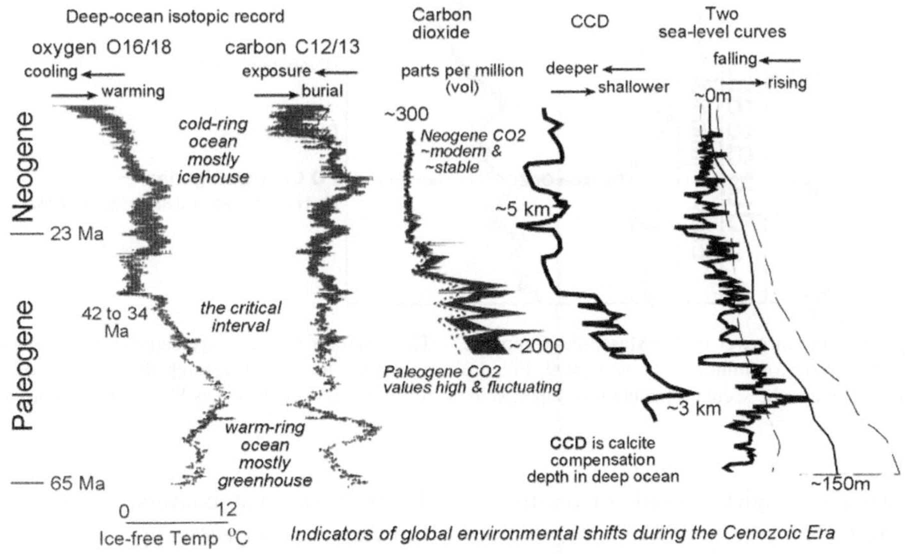

Fig. 3 Physical framework for Cenozoic history. Oxygen and carbon isotope ratios, from Zachos et al. (2001, 2008). Atmospheric carbon dioxide estimates, from Pagani et al. (2005). Calcite compensation depth, eastern equatorial Pacific, from Lyle et al. (2010). Sea level curves, from MA Kominz, Oceanography, 18, 152 (2005)

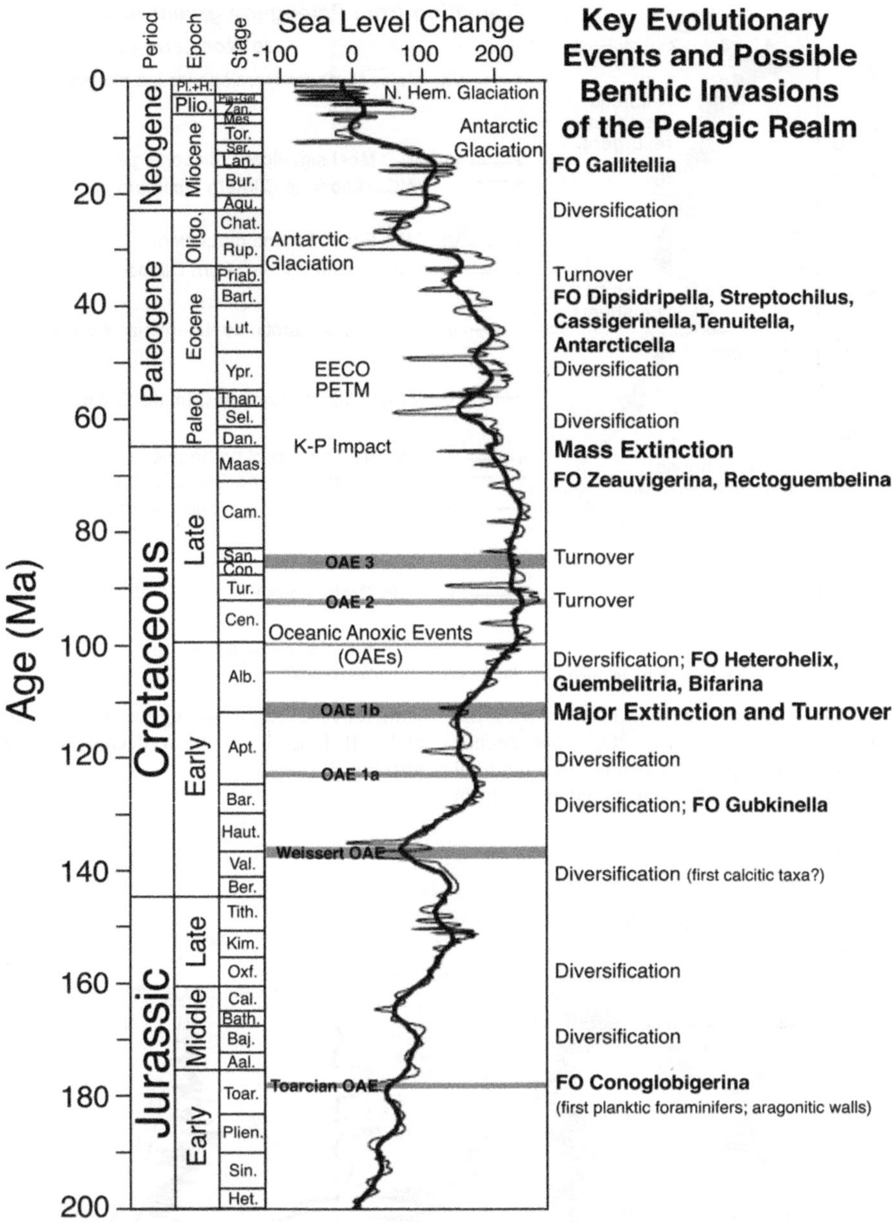

Fig. 4 Diversification and turnover in the Mesozoic–Cenozoic evolution of the planktonic foraminifera (Leckie 2009, Fig. 1). The sea level curve displays the second Phanerozoic supercycle. The genera in bold are suspected to be closely related to benthic ancestors. Reprinted through the courtesy of Mark Leckie and with permission of the US National Academy of Sciences

artefacts of highly incomplete records of fossils and strata. The more our timescales and time-correlations improve in refinement and accuracy, the more clearly we see that the stepwise patterns are not artefacts. Thus, at the next level down (time scale at 10^{7-6} years) we see four coolings, labeled III, V, VII, and VIII. Each of the four great coolings is stronger than the previous, and conversely the ensuing warming recoveries are successively weaker. The first cooling terminates the warmest times of the era, the Early Eocene climatic optimum; the second terminates a still warm and wet but transitional time with (it is inferred) full-scale Antarctic ice sheets; the third terminates the latest Paleogene-early Neogene warming and more specifically the Miocene climatic optimum with the reestablishing of the West Antarctic ice sheet; and the fourth terminates Early Pliocene warming with the rapid expansion of northern ice sheets. These so-called chills comprise four of the eight levels I–VIII identified as significant environmental shifts (Figs. 2 and 3). Others include the Cretaceous/Paleogene mass extinction, the extreme and transient spike known as the Paleocene/Eocene thermal maximum, the later Eocene times of change and transition ushered in by the Khirthar transgression, and the late Oligocene warming that reversed the Oligocene ice times.

Foraminifera in the Cenozoic Era

Most species of foraminifera are benthic in their habitat and neritic, and diversities are highest in the outer neritic and upper slope. Some species tend to be epifaunal and others infaunal, and an infaunal/epifaunal ratio can be a rough but useful indication of nutrient availability. However, the habitat has to be inferred in the fossil record and the infaunal/epifaunal ecological distinction is blurred at best in the vicinity of the sediment/waterface in soft sediments. Diversities tend to decrease along gradients of increasing stress (salinity, oxygen, physical fluctuations), but more relevant here is a geographic component. Buzas and Culver (1999) summarized extensive documentation of benthic foraminifera around the continental margins of North America and extracted the following generalizations: there are five geographical regions—Pacific, Arctic, Atlantic, Gulf of Mexico, and Caribbean, and species occurrences in each fit a log series distribution. Species richness is highest in the Caribbean and lowest in the Arctic, but differences are due to rare species with few occurrences each, most with no fossil record and most endemic species rare.

Buzas et al. (2002) compared changes in diversity during the Neogene in the tropical Central American Isthmus and the temperate Central Atlantic Coastal Plain. The within-habitat diversity measure, Fisher's alpha, increased 2.1 times or 106% in the tropics and 1.4 times or 40% in the temperate area. These contrasting diversity increases generated a more pronounced Holocene gradient (164%) than existed in the Miocene (80%)—seemingly an outcome of global cooling.

Four environmental groups are selected for special mention:

- the deep-ocean benthics,
- the neritic and benthic larger foraminifera employing photosymbiosis,
- the planktonics, and
- the neritic benthics.

(For the fourth but actually central group, I refer to our studies on the neritic faunas of southern Australia.)

Deep-Ocean Benthic Foraminifera

Diverse assemblages with numerous cosmopolitan species inhabit this, the largest habitat on earth. Before the plate-tectonic revolution it was regarded, somewhat vaguely, as very old and environmentally very stable. Along with the realization of much younger seafloor ages came the discovery of strong fluctuations in the lysocline and calcite compensation depth (CCD) with obvious implications for foraminiferal biocalcification and preservation, oxygen levels, and for reconstructing or modeling CO_2 levels in the past. The CCD was generally shallow during the greenhouse world, plunged at the kilometer-scale from Eocene to Oligocene, and stayed deep in the icehouse world except when shoaling at the Miocene optimum (van Andel et al. 1975). Advances in stratigraphic resolution and age determination have clarified the later Eocene transition, where Tripati et al. (2005) could reconstruct CCD fluctuations to become proxies for transient glaciations, of which there are four in the initial study of the transition. Lyle et al. (2010) extended that work to produce the CCD profile through time in Fig. 3. In broad terms there has been a shelf/basin fractionation through time: more extensive neritic carbonate bodies and a shallow deep-oceanic CCD in the Paleogene, vice versa in the Neogene.

Thomas (2007) magisterially reviewed the Cenozoic record of the deep benthics and the following (including statements supported there, but not here) is largely taken from her paper. The Cenozoic faunas were emplaced in broad terms during the "Mesozoic marine revolution" with many species persisting from the Campanian into the Paleocene. Culver (2003) concluded from a review of the smaller benthic foraminiferal record—neritic to deep ocean—across the K/P boundary that they did not suffer mass extinction, whatever the depth of water.

The Cenozoic bathyal and abyssal succession was sorted into assemblage zones (Berggren and Miller

1989; Miller et al. 1992), and three (later, four) turnovers were distinguished as the basis for four (now five) faunas. A "Cretaceous fauna" persisted into the Paleocene until its abrupt extinction at the end of that epoch, indeed, the severest extinction of the entire Cenozoic (Thomas 1998). A "Paleogene fauna" persisted for ~20 myr until a gradual but severe turnover during the Late Eocene and earliest Oligocene (Corliss 1981; Thomas 1992; Thomas and Gooday 1996). A "Transitional fauna" was almost as long-lived, undergoing a turnover in the Early to Middle Miocene. Thomas and Vincent (1987) demonstrated that this turnover began before the global cooling, when carbon-isotopic excursions imply global changes in production/burial. The "Modern fauna" includes a major transition within the Middle Pleistocene (Hayward 2001, 2002; Hayward et al. 2007).

Thomas contrasted the abrupt event at the PETM with the three later turnovers which were recognized as gradual. The latter three had pronounced cooling in common in their paleoceanographic contexts, two occurring when polar ice sheets were sharply expanding (but triggered before that expansion) and the Pleistocene event when northern hemisphere glaciation intensified. In all three there was loss of "cylindrical" species, mostly uniserial or elongate-biserial and belonging to the stilostomellids, pleurostomellids or uniserial lagenids (some of the latter surviving), and their ecology has to be inferred. By analogy with extant forms with which their abundances correlated (*Bulimina*, *Bolivina*, *Uvigerina*), these hard-hit forms may have lived infaunally in high-food and/or low-oxygen situations. Thomas emphasized the concept of *bentho-pelagic coupling*, whereby phytoplankton, produced in the photic zone by photosynthesis, are the main food source for the benthic foraminifera, so that events affecting the former will strongly influence the latter. "Such phytodetritus-using species bloom opportunistically when fresh, labile organic material reaches the seafloor, and they rapidly increased in abundance during the E/O turnover" (Thomas 2007, p. 7). Thus it is hypothesized that bentho-pelagic coupling originated in the fundamental greenhouse-icehouse transition of the Priabonian-Rupelian—circulation intensified, diatoms became much more significant as primary producers, and fresh phytodetritus was delivered in large seasonal pulses. The E/O turnover marked the birth of the modern.

In contrast, the early Paleogene deep-sea foraminiferal assemblages may have lacked the phytodetritus niche. Their members were more commonly deposit feeders relying on bacteria and archaea. Perhaps the much warmer deep ocean (10–12°C warmer than today's) increased bacterial metabolic rates. That bentho-pelagic coupling was weak or absent before the Priabonian is suggested by the decoupling of surface water and deep-water extinctions (Thomas 2007). At the K/P boundary the spectacular extinctions were in the plankton, the calcareous nannoplankton, and the planktonic foraminifera, while the benthic foraminifera did not suffer mass extinction. It may be that delivery of food from the photic zone resumed more rapidly after the catastrophe than is generally supposed. At the PETM it was the other way round: the calcareous nannoplankton and planktonic foraminifera were surely affected but the mass extinction was in the deep sea. It is linked tightly to the hyperthermal event and global warming may well be the central factor; but factors such as oxygen supply, productivity, and carbonate corrosion with a pronounced shoaling of the CCD do not explain the global event, with no refugia available.

In summary,

(i) the Cenozoic record of deep-ocean benthic foraminifera is punctuated by turnovers distinguishing faunal blocks of 10^7 years' duration.

(ii) The Priabonian-Rupelian, greenhouse-icehouse transition is fundamental in triggering either the onset or the pronounced strengthening of the bentho-pelagic coupling.

Molecular evidence (Pawlowski et al. 2007) indicates that three common deep-sea species, *Epistominella exigua*, *Cibicides wuellerstorfi*, and *Oridorsalis umbonatus*, possess high similarity in their nuclear ribosomal RNA genes among Arctic and Antarctic populations up to 17,000 km apart. This is in strong contrast to the high levels of cryptic diversity usual in neritic and planktonic species and implies modest levels of global diversity in the global deep ocean.

The Planktonic Foraminifera

The planktonic foraminifera acquired their first notoriety from the discovery of the blanketing "*Globigerina* ooze" on the voyage of *HMS Challenger* (1872–1976). Notions of the global diversity gradient (tropics to poles) and latitudinal provinciality go back to

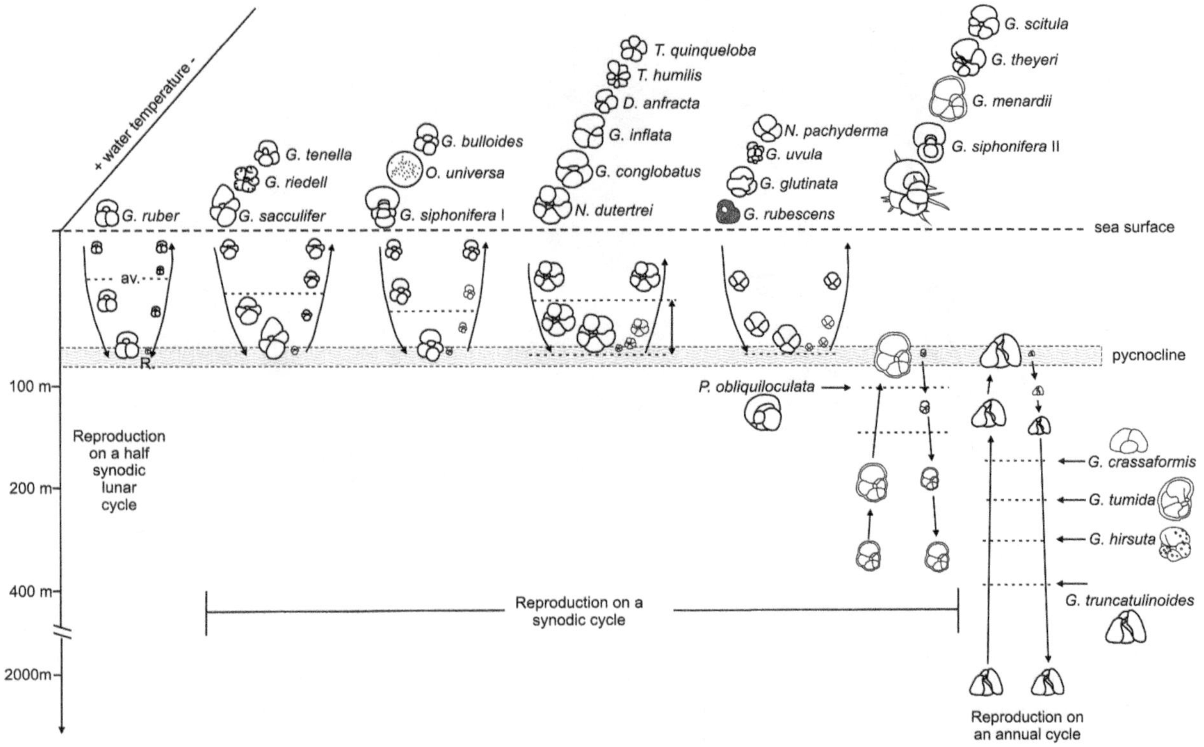

Fig. 5 Schiebel and Hemleben (2005, Fig. 2) summarized the global oceanic habitat of planktonic foraminifera. The reproductive cycles are shown in seven life cycles and the species in six cycles are arranged in order of cooling temperature; in the seventh, in depth (average is stippled). Courtesy of Ralf Schiebel and with kind permission from Springer Science+Business Media B.V.

Alfred Russel Wallace and other nineteenth-century biogeographers but were boosted by the mapping of these pelagic microorganisms, popularized by John Murray (in Murray and Hjort 1912) (see also McGowran 2005, Fig. 4.4). Modern distributions are summarized in Fig. 5.

Petroleum exploration in the 1920s–1930s accomplished progress in systematics, stratigraphic distribution and biozonation, and phylogenetic reconstructions. The pattern that emerged in subsequent decades is repeated expansion and contraction in both diversity and range of morphologies in the preserved shell, or test. Thus the angular and keeled form, at one time lumped into the genus *Globorotalia*, in due course became distinguished as *Rotalipora* (middle Cretaceous), *Globotruncana* (late Cretaceous), *Morozovella* (Paleogene), and *Globorotalia* (Neogene), each signifying a clade independent of but allochronously paralleling the others. Iterative evolution is found in several other shell morphologies, for example planispiral coiling, biseriality, and triseriality, or a globular shape. Each expansion is an evolutionary radiation that can be matched plausibly with physical patterns, oxygen isotopes, and sequence-stratigraphic supercycles (Fig. 6). As in any repeated bio- or geohistorical configuration, the iterations have both common and distinctive characteristics.

Also iterative is the adopting of photosymbiosis, widely observed to be fundamental in conserving and recycling nutrients in low-nutrient environments. Photosymbiosis is inferred isotopically in extinct clades in the fossil record, the empirical basis being that asymbiotic and photosymbiotic taxa occupy distinct fields in oxygen-carbon crossplots (e.g., Norris 1996; Pearson 1998; McGowran 2005). Photosymbiosis was not adopted in the mid-Cretaceous radiation even though there was clear depth-stratification (Norris and Wilson 1998); it was adopted exuberantly in the late Cretaceous but all the participating clades went extinct (d'Hondt and Zachos 1998); and it was reinvented in the Paleocene (Fig. 7).

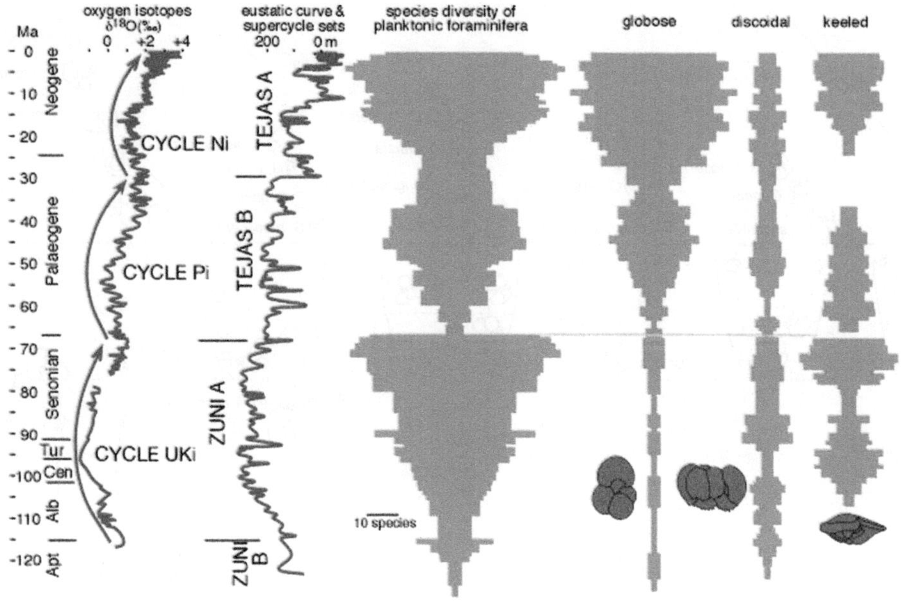

Fig. 6 Three radiations congruent with three natural environmental divisions of the past 100 myr of biogeohistory, from McGowran (2005, Fig. 6.2). The three-part division of the physical environment is seen independently in oceanic oxygen isotopes as a proxy for water-temperature change, and a eustatic curve based on oceanic spillage across the continental margins (Abreu et al. 1998). The Senonian, Palaeogene, and Neogene radiations in planktonic foraminifera are expressed simply on species diversity, also broken down to recurring gross test forms, globose, discoidal, and keeled (Norris 1991). The latter morphogroup is clearly quadripartite through time and polyphyletic. Reprinted with permission of Cambridge University Press

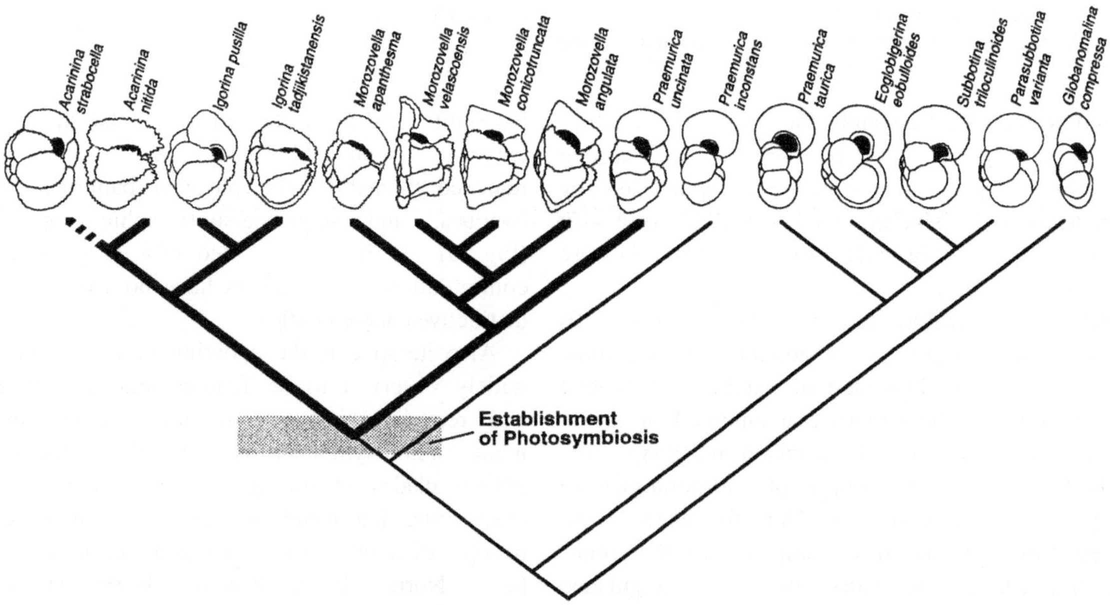

Fig. 7 Paleocene trochospiral planktonic foraminifera: cladogram illustrating the suggested phylogenetic relationships (McGowran 2005, Fig. 4.12 after Norris 1996, Fig. 3). The *shaded* area indicates the position at which photosymbiosis becomes an important ecological strategy in the early Paleogene radiation. Heavy lines: species inferred isotopically (and independently of the cladogram) to possess that strategy. Reprinted with permission of Cambridge University Press

Planktonic Foraminifera and Molecular Systematics

There are some 47 morphospecies extant in some 21 genera which fall into three superfamilies within the order Globigerinina: the Heterohelicoidea (micro-perforate non-spinose), the Globorotaloidea (macro-perforate non-spinose), and Globigerinoidea (spinose) (Schiebel and Hemleben 2005). As of the major review by Darling and Wade (2008), 19 morphospecies had been searched for their small subunit (SSU) ribosomal RNA (rRNA) genotypes/subtypes (with species from the three superfamilies), and the 19 morphospecies yielded 54 genotypes. Thus morphospecies may underestimate cryptic species by a factor of ~3. Darling and Wade listed eight morphospecies as having been studied in enough detail to allow biogeographic interpretation of the data. The status and integrity of the three superfamilies are supported by this work.

It is clear that many cryptic species await discovery. It is clear too that there will be major difficulties in matching morphospecies with cryptic, even in the modern biota, where intraspecific phenotypic variation can be substantial, as for instance in the well-known *Globigerina bulloides*. Darling and Wade (2008) suggested that cryptic clusters within morphospecies be raised to the level of genus. I am more attracted to the notion of the *planktonic super-species* of De Vargas et al. (2004), who perceive significance in the very conservative nature of the morphological species or phenotype, which in fact is a monophyletic cluster of sibling species which split several million years ago (on molecular clock calculations) and still maintain slight, subtle, far from obvious morphological divergence between the ecotypes/genotypes after all that time. They propose that this distinctive evolutionary mode characterizes most marine planktonic taxa, metazoans as well as protists.

The efflorescence of documented genotypes has ecological and biogeographic—*phylogeographic*, in a felicitous neologism—implications. On a local and ecological scale different genotypes in some cases can be shown to be different ecotypes (nutrient, water mixing characteristics) but found in the same regional water mass. On a biogeographic scale the ecotype/genotype can have a highly disjunct distribution. Thinking on speciation models is strongly influenced by the possibility—or not—of pelagic populations being geographically isolated (water mass fronts, tectonic barriers) long enough for divergence and isolating mechanisms to operate. The models accordingly tend to be allopatric and vicariant on the one hand or sympatric and parapatric on the other. On evidence of rapid and far-reaching dispersal by *Globorotalia truncatulinoides*, and more general evidence of strong gene flow at oceanic scale, Sexton and Norris (2008) were sceptical of effective geographic isolation. On evidence from the high-latitude species *Neogloboquadrina pachyderma*, Darling et al. (2007) acknowledged the power of gene flow over enormous distances but found effective Arctic geographic isolation—north Pacific populations originated in the southern hemisphere while the Arctic was in genetic continuum with the Atlantic.

The Tychopelagic Mode and Polyphyly: Bridging the Benthic/Planktonic Divide

"Tychopelagic" refers to diatoms that live as benthos but can also survive in the plankton, even in large numbers well offshore. Darling et al. (2009) demonstrated that *Bolivina variabilis* in the neritic infauna and *Streptochilus globigerus* in the surface waters of the open ocean are one and the same species—the first tychopelagic species recognized among the foraminifera. Meanwhile, Ujiié et al. (2008) argued on SSU rRNA evidence that the triserial planktonic species *Gallitellia vivans* is a separate lineage of planktonics, branching close to the benthic rotaliids *Stainforthia* and *Virgulinella* and diverging in the early Miocene, much later than other triserial planktonics.

As reviewed elegantly by Leckie (2009) (Fig. 4) these are modern items refueling an old suspicion—that the planktonic foraminiferal order Globigerinina is not the neatly monophyletic package arising in the Jurassic. Attention is focused on one of the three superfamilies, the Heterohelicoidea, the non-spinose microperforates. In the Paleogene especially (Huber et al. 2006) but also in the late Cretaceous and Neogene, the microperforates are a motley collection of taxa presenting more problems than do the other superfamilies in examples of disjunct stratigraphic distributions, mysterious phylogenetic connections, and uncertainties as to benthic or planktonic habitat.

Li and McGowran (1996) observed that two Paleogene groups, the earliest Paleocene *Parvulorugoglobigerina* and the late Eocene *Tenuitella*, acquired their strong similarities—peculiarly elongated aperture and microperforate walls—through evolutionary convergence not direct relationship. They suggested further that this convergence was driven by eutrophic conditions and perhaps dysoxia. Most of the biostratigraphically important events in the late Eocene and earliest Oligocene of southern Australia are microperforates (McGowran 2009), sometimes in neritic facies almost bereft of normal planktonics but, regionally and beyond, during the high-nutrient times of the Priabonian in contrast to the Bartonian below and the Rupelian above.

Leckie (2009) developed a scenario for the outer shelf and upper slope, where organic matter from terrestrial sources and from the photic zone provided optimal conditions for the benthos. In addition, the thermocline, the chlorophyll maximum zone, and the oxygen minimum zone (OMZ) may intersect the benthic habitats here, and to complete the ingredients this sector is sensitive to sea-level fluctuations and OMZ intensification. Leckie proposed that the shelf-slope break accordingly makes an ideal "jumping off" platform for benthics to spring into the planktonic realm—presumably, on the correlation with eutrophy, finding their niches during an intensified OMZ. Presumably too the removal of competition makes the jump attractive and more readily accomplished. In the longer perspective of the Mesozoic-Cenozoic (Fig. 4), there are numerous suspects and numerous environmental punctuations, especially the oceanic anoxic events of the greenhouse times.

Trends Through Cenozoic Time

Radiation and iteration give a sense of repetition and recurrence, of "time's cycle," which is true enough. But "time's arrow" is also true, and is exemplified by a trend in the sizes of planktonic foraminiferal tests (Fig. 8). With cooling, there was biogeographic differentiation and size differentiation, apparent in the

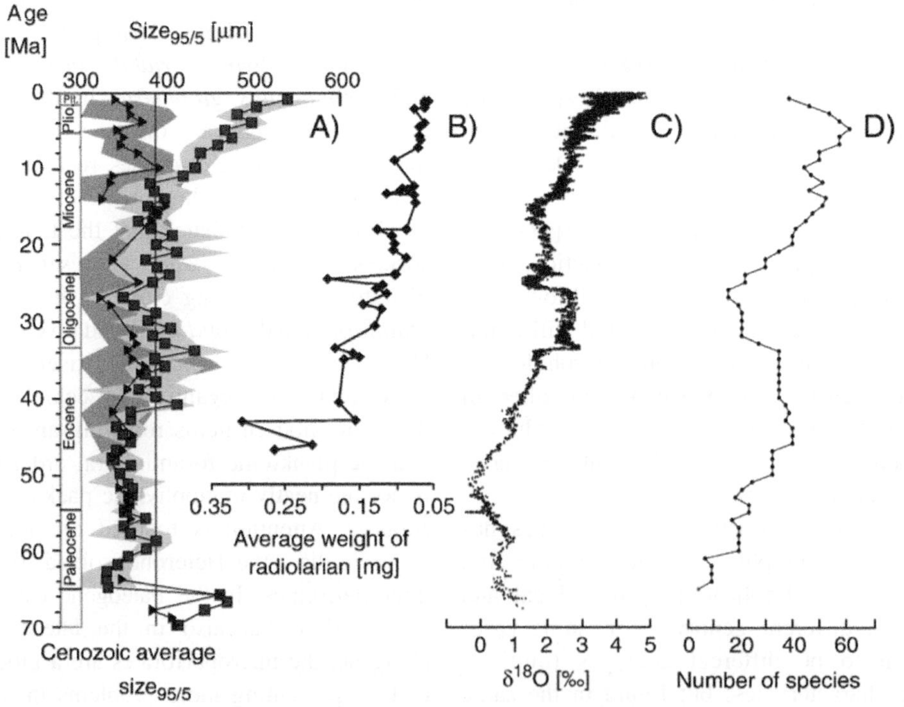

Fig. 8 Cenozoic changes in test sizes in planktonic foraminifera (Schmidt et al. 2006, Fig. 25). (**a**) test sizes as 95-percentile, 1–Myr means, tropical-subtropical (*squares*) and temperate-subpolar (*triangles*) (*vertical line*, Cenozoic average). (**b**) average weight of radiolarians' opaline tests. (**c**) deep-ocean temperature record, as in Fig. 2. (**d**) morphospecies of planktonic foraminifera, per 1–Myr interval. Reprinted with kind permission of Elsevier B.V.

late Paleogene but clearest from the Middle Miocene onward.

Benthic Larger Foraminifera

The larger foraminifera attracted attention as prominent components of neritic carbonate masses—they were and are major contributors in the so-called carbonate factory of lower latitudes and warmer waters (Hottinger 1987). Anatomically they are characterized by large size (extreme forms up to 15 cm diameter) and internal skeletal complexity. Ecologically they are characterized in two ways which boil down to the same thing—as ultra-specialized, extreme *K*-strategists (Hottinger 1982, 1996) and as inhabiting the oligotrophic end of the expanded trophic resource continuum (Hallock 1987; Hallock et al. 1991). Some of the main biological characteristics of the larger foraminifera, in contrast to asymbiotic smaller benthics, are (Hottinger 1983) (i) long life cycles, up to two years in *Marginopora*, implying an extensive ontogeny constructing complex shell structures and attaining exceptionally large body sizes with maximized surface area/volume ratio; (ii) refined reproduction cycles, often including protection of juveniles; relatively suppressed sexual phase; dimorphic form with larger megalospheric tests; (iii) symbiosis protected by shells; thinned upper walls, numerous spaces formed by structural complexity; (iv) protoplasmic differentiation in canaliferous foraminifera; (v) abundance and diversity in regions of low primary productivity where these organisms and other oligotrophic calcifiers generate carbonates in warm-neritic environments. The abundance of larger foraminifera in fossil assemblages, often to the point of rock-forming density, reflects living conditions: adequate light and trace-nutrients are not density-limited to the degree that filtering or foraging for food would be. The diversity of larger foraminifera reflects a refined partitioning of niches. More than other physical factors, light intensity and quality would appear to be the prime control: different wavelengths penetrating to different water depths are exploited by different, host-specific symbionts.

In the modern ocean they are limited to a global climatic belt described by winter minimum isotherms of 15–20°C (Langer and Hottinger 2000). There is a low-diversity Caribbean realm and a high-diversity area comprising an inner Central Pacific province centered on the Pacific Warm Pool and within a broader Central Indo-Pacific realm, and the Western Indian Ocean including the Red Sea and Persian Gulf.

When AM Davies published his *Tertiary faunas* in 1934, he ranked the larger benthic foraminifera as taking "a good second place" to the Mammalia in the Tertiary era in approaching the ideal zone fossil—"a species that can spread over the whole earth in a time which is negligible compared with its duration as a species, though that in turn is very short on the geological time scale" (quoted in McGowran 2005). Davies made no mention of other foraminifera or microfossils. On the paleobiological side lineage studies have long been attractive, if always to that small and dedicated group of practitioners willing and able to master the arcane disciplines of the anatomy and systematics (McGowran 2005, Chs. 4 and 6). For the flavor, consult Reichel (1937), Tan Sin Hok (1939), van der Vlerk (1959), Hottinger (1960), Adams (1983), and Drooger (1993). In recent years Hottinger has preeminently sustained enquiry into the deeper meaning and potential of these unique protists in time and space. For the modern large photosymbiotics and their habitats see especially Reiss and Hottinger (1984), Hallock (1999), Hohenegger (2000), and Renema (2002).

The term "larger foraminifera" lumps numerous lineages arising independently and iteratively in several higher taxa of foraminifera. Hottinger (1982, 1997) described "phyla" (clades) as trajectories of $r \rightarrow K$ strategy (Fig. 9) which invented and reinvented large, discoidal, and cigar-shaped forms. His ideas on larger foraminiferal evolution, based initially in the rich Paleogene successions of nummulitids and alveolinids in western and central Tethys, have developed at different levels. At the species level less emphasis is placed on distinctly bounded spatiotemporal entities and more on segments of closely related and gradually evolving lineages displaying extensive parallelism (Hottinger 1961, Fig. 1, 1981, Fig. 1) (McGowran 2005, Fig. 4.30) (Fig. 10). One does not deconstruct and trace a lineage in isolation but as one strand in a cluster of sympatric lineages.

At a higher level Hottinger (1998, 2001) developed a form of cycle in *K*-strategist foraminifera known as the *global community maturation* (GCM) cycle (Table 1, Fig. 11). There are two primary strands in his thinking here—first, the notion of direct environmental action in which genetic information is lost, and second a "focus on the recovery of the genetic

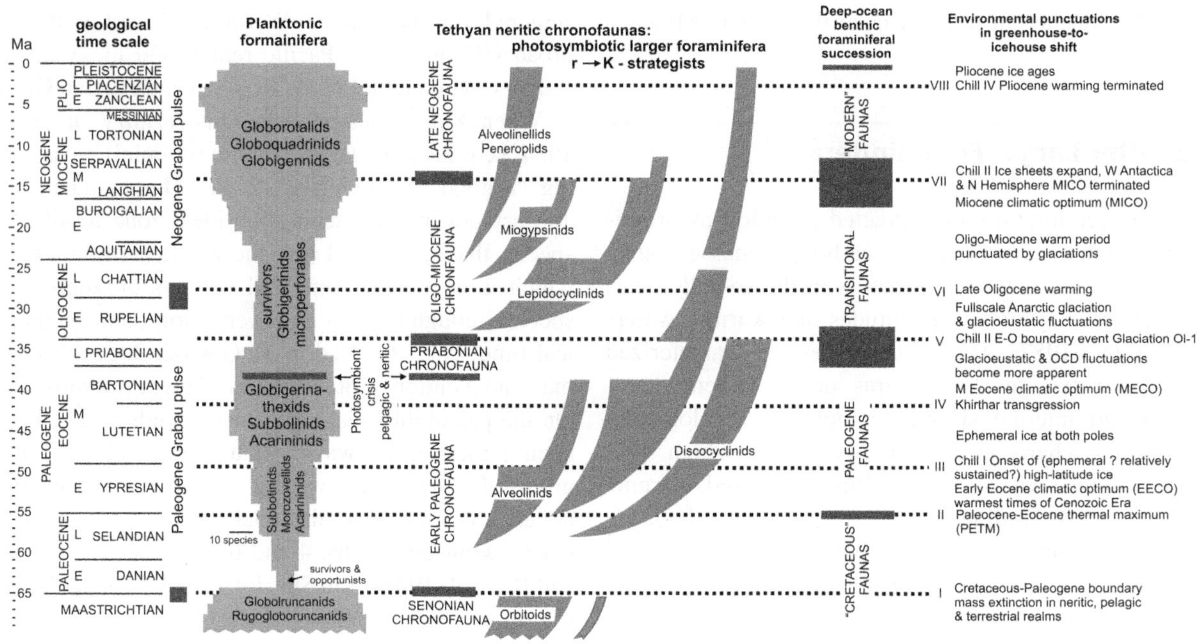

Fig. 9 Cenozoic events and foraminiferal events. Environmental punctuations: many sources; lines I–VIII from Fig. 2. Planktonic foraminifera, from Fig. 5 with taxa from many sources. Neritic large-foraminiferal chronofaunas, from McGowran and Li (2000). $r \rightarrow K$ trajectories of taxa, from McGowran and Li (2000) after Hottinger (1983). Deep-ocean faunas, from Thomas (2007). Bartonian photosymbiotic crisis, from Fig. 13 and text

complexity necessary to reinstall *K*-strategists in the ecosystem over geological time spans, as reflected by the succession of phenotypes in geologic times and over oceanwide extension, within the frame of benthic foraminifera" (2001, p. 472). The recovery of the genetic complexity occurs in a sequence of five typical phases (numbers as in Table 1). (1) There was a long, slow recovery from comprehensive extinction. (2) Diversification occurred in disparate communities: this was a time of smallish, monospecific pioneers in the "generic" radiation—the genera *Discocyclina, Ranikothalia, Glomalveolina, Fallotella,* and *Broeckinella.* (3) In this phase only a few possessors of each body plan begin to dominate—three porcellanous families, respectively fusiform, spherical-concentric, and annular-concentric, and three lamellar-perforate, respectively planispiral-involute nummulitids, annular-concentric with lateral chamberlets in othophragminids, and trochospiral rotaliids. The prominently iterative adult dimorphism is an adaptation accelerating the growth of cloned generations by increasing the individual's initial biomass and, by a large inheritance of symbionts, significantly reducing the number of symbiont fissions needed to fully stock the greenhouse of the adult host (Hottinger 2000). (4) These dominances are played out in parallel evolution, the formation of "odd pairs" (Hottinger 1999) and very large sizes. (5) In this stage progressive morpho-differentiations and progressive geographic differentiations (endemism) reach a point of the largest forms disappearing.

The global community maturation cycle (GCM) is a period of continuous, gradual biotic change between two discontinuities involving major genetic revolutions (2001, p. 474). Hottinger is well aware of the importance of physical perturbations on a flourishing but vulnerable fauna of *K*-strategists, that vulnerability being particularly to an excess of nutrient, at which point bryozoans and "algae" will outcompete them. But Hottinger also engages with high-level ecological notions of communities becoming self-organizing and self-developing—such as where a minor revolution "seems to be unrelated to major environmental change and may be exclusively the consequence of biotic processes, either by failing to keep more successful immigrants out or by leaving open

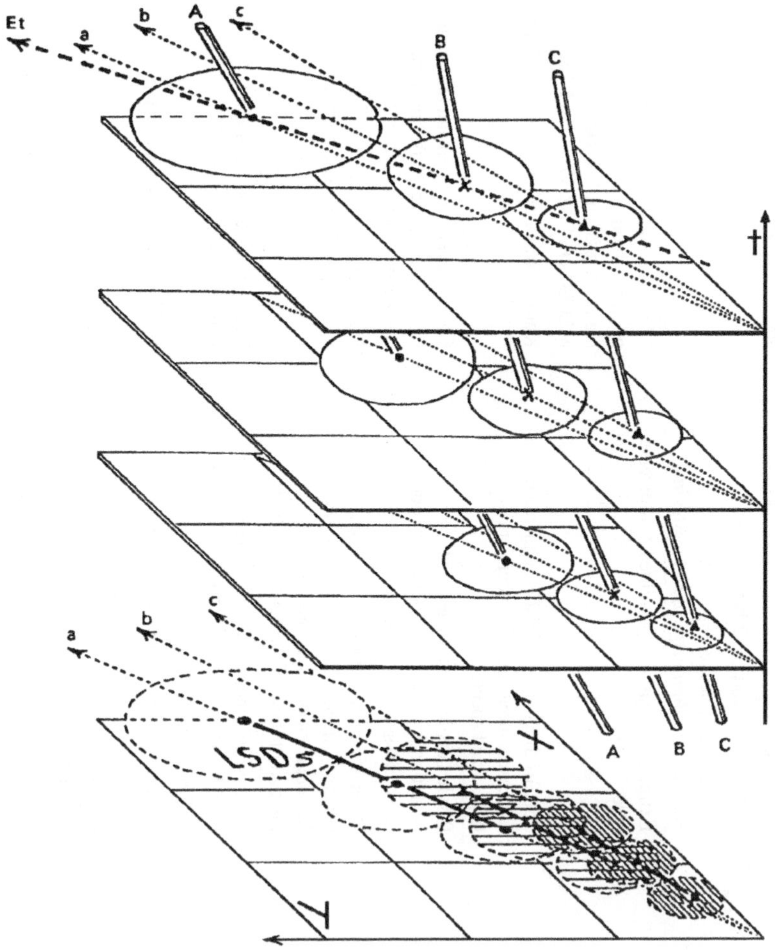

Fig. 10 Phylogenetic trends and limits of biostratigraphic resolution in larger foraminifera, based especially but not exclusively in the alveolinids (McGowran 2005, Fig. 4.30 after Hottinger 1981). Three horizontal planes are three timeslices (vertical axis t = time). Horizontal axes X and Y represent morphological characteristics changing quantitatively and measurably with time. Data from the three timeslices are projected onto the plane at bottom. A, B, and C are mean values of successive species interpreted as phylogenetic *lineages*, and *circles* define specific variation at the given time. Phylogenetic *trends* extracted from the lineage reconstruction are denoted by a, b, and c, different for each lineage. Et marks trends independent of time, usually reflecting ecological gradients. "The similar directions of phylogenetic and ecological trends correspond in reality in many groups of larger foraminifera; they are not fortuitous and have a functional significance" (Hottinger 1981). In the projection plane, LSD is the least significance difference for successional species in a lineage. There is clear discrimination here between (i) neighboring species within each timeslice (between-lineage), and (ii) successional species between timeslices (within-lineage). Two implications are (i) species can be separated and identified only when the timeslices can be separated, and (ii) the spacing of these timeslices represents the limit of biostratigraphic resolution (Fig. 17). Reprinted with permission of Cambridge University Press and courtesy of Lukas Hottinger

environmental niches after having genetic trouble by too many cloned generations in the symbionts that have to reproduce asexually within their host over its full reproduction cycle during many years" (2001, p. 474). However, recent progress on critical stratigraphic levels strengthens a case for environmental forcing (see following sections).

The Neritic and Benthic Foraminifera in Southern Australian Neogene

A Late Oligocene-Late Miocene section in southern Australia was searched for indications of neritic faunal succession and, perhaps, changes that might be called benthic foraminiferal evolution (Li and McGowran 1997, 2000; McGowran and Li 1996). Detailed faunal

Table 1 Hottinger's (1998, 2001) Global community maturation (= GCM) cycles, based on the Paleocene–Eocene of Central and Western Tethys

E/O boundary			*GCM boundary*	Extinction of orthophragminids	Chronofauna terminated
Priabonian	3(?)	SBZ 20 SBZ 19	Success of the competition	New genera (*Spiroclypeus*?) together with Pellatispirids and Orthphragminids	*Priabonian Chronofauna*
Late Bartonian	2(?)	SBZ 18	Progress of competition	Insurgence of new competitors: *Pellatispira, Biplanispira, Heterostegina*	
(most of) Bartonian		SBZ 17		Extinction of large nummulites, last diverse alveolinids	*Chronofauna terminated:* SBZ 17/18 photosymbiotic crisis ~coeval with plankton crisis at E 13/14, low C17 MECO is in SBZ 17 and in C18r
Latest Lutetian	5	SBZ 16 C19r	*GCM boundary (?)* Removal of competitors	Species diversity diminishes, endemism increases, monospecific endemism of large-sized species	Khirthar transgression is here at C19r
Lutetian Cuisian Late Ilerdian	4	SBZ 15 to SBZ 9	Size increase	Parallel evolution of diversified evolutionary lines, development of odd pairs, classes of over 10 cm attained	Cooling event end-Cuisian terminates EECO immediately after maximum species diversity in SBZ 12
Early Ilerdian	3	SBZ 8 to SBZ 5	Complete success and expansion through neritic depth zones	Diversification *within* the successful genera—*Alveolina, Orbitolites, Assilina, Nummulites*, orthophragminids First adult dimorphism and large shells	*EECO at maximum* *LFT is synchronous with PETM*
Late Paleocene	2	SBZ 4 SBZ 2	Experimenting with new life strategies	Increase in generic diversity and reduction of endemism; species of smaller size and complex structure	Developmental phase of *Early Paleogene chronofauna*
Early Paleocene	1	SBZ 1	Prepare for K-strategy	No larger form; no complex structure (?)	No K-strategists in recovery phase
K/P boundary			*GCM boundary*	K-strategists extinct; small neritic benthics affected but not a mass extinction	Mass extinctions except in deep sea

Hottinger's stages tentatively ascribed as reverting to 2(?) and 3(?) are here referred to the new and short-lived Priabonian Chronofauna following the photosymbiotic crisis in the late Bartonian.

Abbreviations: E/O Eocene/Oligocene, *K/P* Cretaceous/Paleogene, *SBZ* Shallow benthic zone, *GCM* Global community maturation, *LFT* Large foraminifera turnover (and diversity expansion), *PETM* Paleocene/Eocene thermal maximum, *MECO* Middle Eocene climatic optimum, *EECO* Early Eocene climatic optimum, *C19r* geomagnetic Chron C19reversed

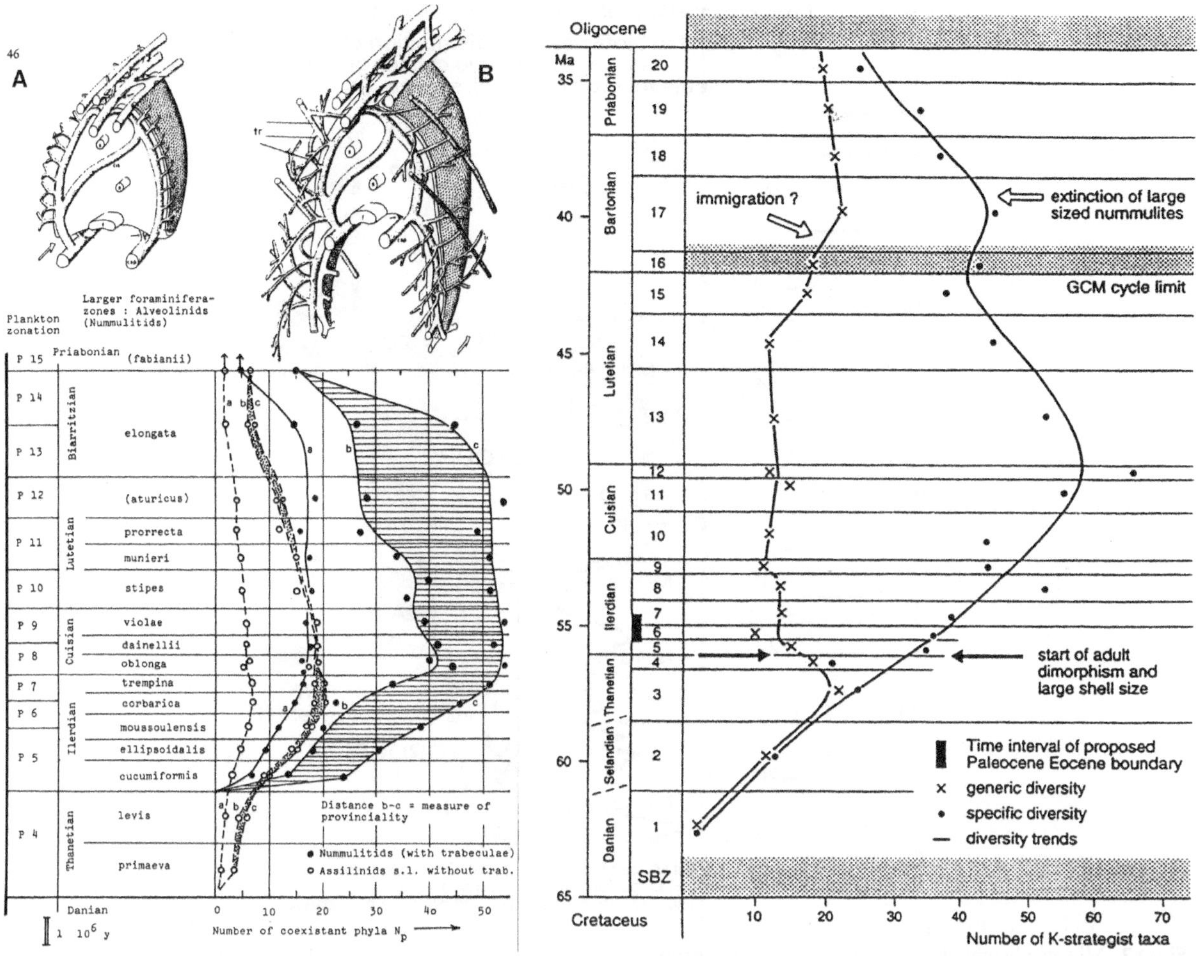

Fig. 11 Hottinger's *Assilina–Nummulites* diversity and global community maturation (GCM) cycles. *Left* (1988, Fig. 6), Canal system in relatively simple (*Assilina A*) and complex (*Nummulites B*) single chambers. Line (**a**) diversity of phyla (species clades) zone by zone, which, (**b**) are weighted by 1, 2, or 3, according to number of provinces they occur in. Distance **b–c** is a measure of provinciality, (**c**) being 3.a to give positive values. Thus diversity reflects both ecological adaptations (earlier) and production of bioprovinces (in due course). *Right* (1998, Fig. 1), Generic and specific diversity of larger foraminiferal K-strategists in Paleocene and Eocene. Reprinted through courtesy of Lukas Hottinger

analysis might be split into two parts, "time's cycle" at the third order (sequence-stratigraphic) and "time's arrow" at the second order. By this I mean separating recurrence from secular change. A quantified biofacies change developed by Qianyu Li is based on a ratio of inner-neritic to outer neritic benthics, giving a proxy for shifts in sea level which correlate plausibly with third-order sequence boundaries (Fig. 12). This is a pattern of recurrence, or time's cycle. Cluster analysis of the reconstructed benthic foraminiferal record revealed three clusters, groups 1 to 3, which are successively more tightly clustered (McGowran and Li 1996), perhaps implying ecological progress in Neogene recovery from the Oligocene crises. Perhaps more significantly, the clustered groups 1–3 are strongly and cleanly successional (with only 3 group 1C samples out of position in the entire data set). It is a pattern of moving forward, time's arrow, rather than a clustering of cooler vis-à-vis warmer, or upwelling vis-à-vis oligotrophic (not shown here but discussed in the above references). The clean breaks are correlated respectively with the base of the Burdigalian stage and the upper Serravallian. The former can be correlated with a major

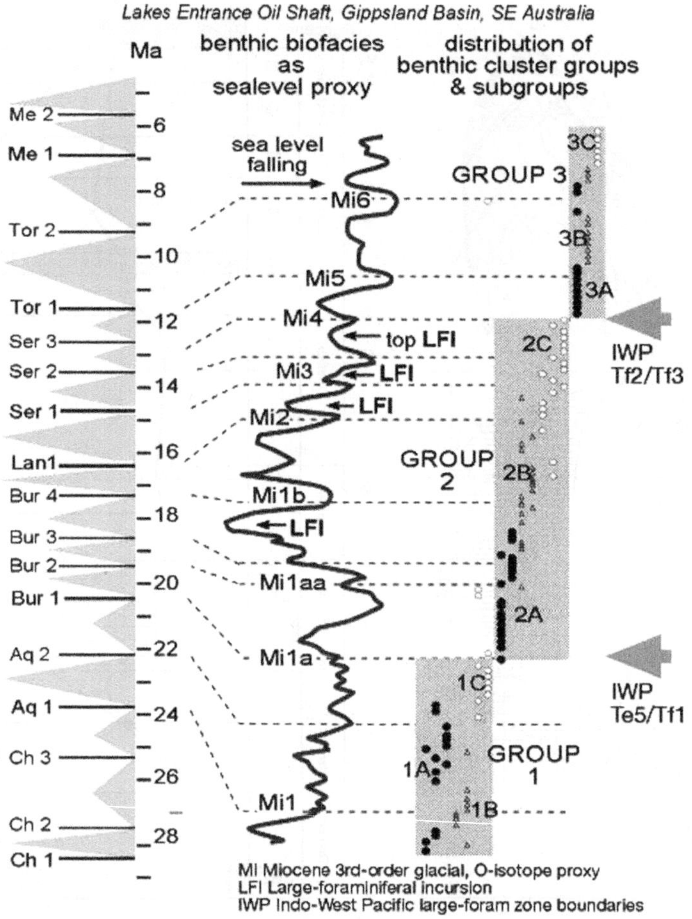

Fig. 12 Miocene benthic foraminifera faunas in the extratropical neritic of southern Australia display both "time's cycle" and "time's arrow" components at 10^6 years' and 10^7 years' scale respectively. The unscaled paleodepth curve (Li and McGowran 1997) is based on proportions of inner, middle, and outer neritic biofacies of benthic foraminifera. This reversible or quasi-cyclical pattern encourages estimates of where global sequence boundaries (Hardenbol et al. 1998) fall. (Global stage names abbreviated from Chattian, Aquitanian, Burdigalian, Langhian, Serravallian, Tortonian, Messianian.) It also suggests correlations with the third-order Mi glacials Mi1 to Mi6, adapted from a composite deep-oceanic curve of $\delta^{18}O$ *Cibicidoides* (Wright and Miller 1993). Although more rigor in correlation is desirable, the biostratigraphic matches with the global curves are plausible using biostratigraphic events for correlation. The larger foraminiferal horizons imply episodes of warming interspersed with the third-order glaciations. The same biofacies data are clustered into groups and subgroups (Li and McGowran 2000, modified from McGowran and Li 1996). The three clustered groups 1–3 ("second-order") are almost perfectly successional, giving very sharp breaks at the Aquitanian/Burdigalian boundary and in the late Serravallian. Within the main groups (i.e. at "third-order") there is very little coherence among samples from similar environments, such as glacials or upwellings or warm intervals (see McGowran and Li 1996). Instead of communities cohering as they retreat in the face of environmental pressure and recurring (i.e. returning) when times are right again, there is a very strong sense here of ongoing faunal rehash and reassembly. The two breaks in the southern extratropical reaches of the Indo-West Pacific Province correlate strongly with the two major breaks in the tropical IWP, evident on inspection of Fig. 15 but taking the Te_5/Tf_1 boundary at base-Burdigalian from Renema (2007 and personal communication 2009)

interface in the Indo-West Pacific succession (but see following sections). The latter correlates with the second such interface and is within the major climatic transformation from the mid-Neogene warm times including MICO (Fig. 2) to the late Neogene icehouse conditions.

Chronofaunas and the Significance of Trans-Realm Parallels

The GCMs would seem to fit the notion of chronofauna among the neritic larger foraminifera. The concept of chronofauna and chronoflora recognized the persistence through time by community types. "Time" is in the long term, at 10^6–10^7 years' scale. Networks of species were held together through local variants of the overall environment; those species could be replaced by others in more or less the same role without destroying the structure of the chronofauna. There was stability and coadaptation while evolution proceeded. In the Cenozoic fossil record, chronofaunas have been recognized mostly in the North America land mammal ages (NALMA), exemplified by Webb's (1984) study of the Clarendonian chronofauna in the late Miocene. He described the development of an essentially self-contained entity, based in large ungulates, held together by an elaborate set of positive and negative feedback loops, all moving in response to a climatic tendency toward cooling and drying at temperate latitudes. McGowran and Li (2000) and McGowran (2005) discussed Cenozoic chronofaunas.

The following section discusses parallels in terms of the punctuated greenhouse-to-icehouse shift outlined in Fig. 2.

I Cretaceous/Paleogene Boundary

The mass extinction, terrestrial, neritic, pelagic, at the end of the Cretaceous has been discussed extensively, now at the level of integrated, ecosystem thinking (e.g., d'Hondt 2005). The event affected, drastically, the rich calcareous planktonics, the coccoliths and their relatives, and the planktonic foraminifera. The macro-benthos also suffered; the loss of K-strategist foraminifera was almost total and none contributed to Paleocene restocking, which took several million years to get underway. Benthic smaller foraminifera did not suffer mass extinction (Culver 2003) and the "Cretaceous faunas" of the deep ocean (Thomas 2007) continued to the end of the Paleocene.

II Paleocene/Eocene Thermal Maximum

The PETM also displays simultaneous effects among neritic, terrestrial, and pelagic chronofaunas. The North American land mammals experienced a major chronofaunal turnover (Paleocene/Eocene chronofaunal transition) at the Clarkforkian-early Wasatchian transition in the biggest immigrational wave of the Cenozoic (Woodburne and Swisher 1995). The planktonic foraminifera genera *Morozovella* and *Acarinina* speciated very rapidly to produce short-lived taxa, possibly facilitated by enhanced photosymbiosis during the oligotrophic conditions of this extreme and ephemeral event (Kelly et al. 1996). The deep-benthic extinction event was the strongest of the entire Cenozoic (Thomas 2007). Hottinger (1998, 2001) expressed doubt that a pronounced turnover in large benthic foraminifera (LFT) was to be correlated with the PETM (Fig. 11) and we suggested (McGowran and Li 2000) that the seeming lack of impact at this horizon may have been because K-strategy had yet to attain mature and metastable form. However, tight correlations have now been demonstrated between the deep-benthic event, the LFT, and the carbon-isotopic excursion signaling the PETM (Fig. 13). Therefore, one might postulate that the benthic photosymbionts were stimulated evolutionarily by the same extreme oligotrophy as were the planktonics.

III Cooling, Terminating Early Eocene Optimum

This was the first major shift away from the greenhouse world, the oceanic response being a steepening of gradients and more physical differentiation among water masses, and the beginnings of contraction of thermophilic plankton to lower latitudes. The recent monograph of the planktonic foraminifera (Pearson et al. 2006a, b) shows considerable turnover in the vicinity among the acarininids and globigerinids and a sharp iteration and transfer from *Morozovella* to the extraordinarily similar but independently derived *Morozovelloides*. Maximum specific diversity in the large benthics occurs in SBZ12 at end-Cuisian (=end-Ypresian). The turnaround in specific diversity follows at the cooling (Table 1).

IV to V Khirthar Transgression to Eocene/Oligocene Boundary: The Auversian Facies Shift

Hottinger placed the GCM boundary here in SBZ 16 correlated with C19r (Fig. 11), latest Lutetian (not earliest Bartonian). We suggested instead (McGowran and Li 2000) that the highly mature phase of the early Palaeogene chronofauna was brought on by the Khirthar transgression, which increased neritic living

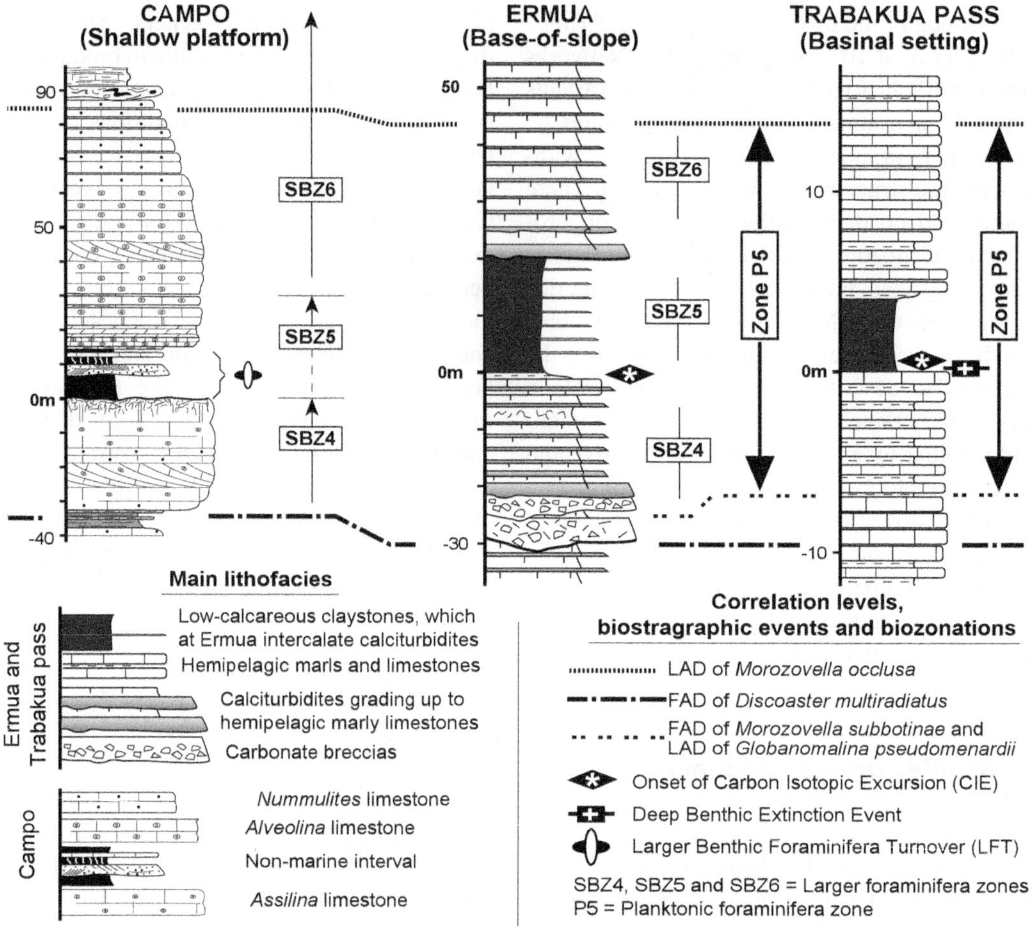

Fig. 13 Three sections in Pyrenean Spain demonstrating synchroneity of the deep-benthic extinction event with the larger benthic foraminiferal turnover, and of both with the onset of the carbon isotopic excursion which signals the Paleocene–Eocene thermal maximum. Figure 2B in Pujalte et al. (2009). Reprinted through courtesy of Victoriano Pujalte Navarro and with permission of Geologica Acta

space, encouraged latitudinal expansion of tropical or pantropical biotas, and brought on the strong but ephemeral Middle Eocene climatic optimum (MECO). This is the climactic interval in the greenhouse-to-icehouse shift, for which the name Auversian Facies Shift is urged (Berger and Wefer 1996; McGowran and Li 2000; McGowran 2009) (Fig. 14). For our present purposes the AFS is twofold. Its earlier part is the Bartonian Stage, rich in neritic carbonate facies and more oligotrophic environments, tropical and extra-tropical, and spanning MECO. Its later part is the Priabonian Stage, distinctly cooler and distinctly more eutrophic in neritic environments. This is seen in central and western Tethys where large photosymbionts contract very markedly, only the small *Nummulites* and orthophragminids surviving into the Priabonian (Nebelsick et al. 2005; McGowran 2009, Fig. 3).

The truly critical transition within this critical interval is in the late Bartonian, ~38 Ma, around the geomagnetic C18/C17 boundary. This is the horizon of extinction of the large nummulitids and large alveolinids at the SBZ17/18 boundary, leaving the field largely to the surviving orthophragminids and the newly radiating pellatispirids. Most significantly, there is a parallel crisis in the photosymbiotic plankton where acarininids (almost all) and *Morozovelloides* are extinguished. On the available evidence the two events are slightly offset, but undoubtedly will be shown to be coeval. The crisis is due to a surfeit of nutrient on the shelves in Tethys, the Indo-Pacific

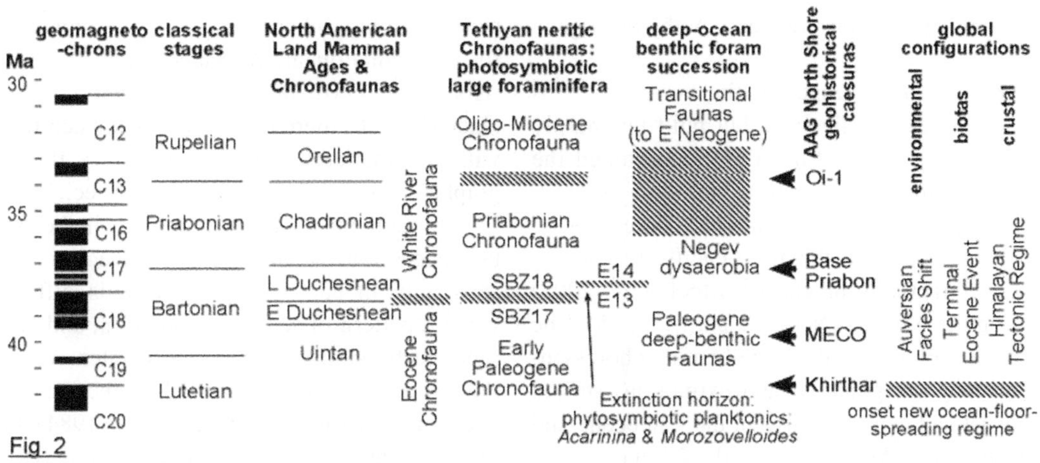

Fig. 14 Correlations in the late Lutetian–early Rupelian interval from the terrestrial, neritic, and pelagic realms (modified from McGowran 2009, Fig. 2) drawn to impart (i) the sense of long-term stability punctuated by short-term turnover, and (ii) chronological parallels between the three environmental realms. The "fauna" or "chronofauna" in the mammals, neritic foraminifera and deep-ocean foraminifera each persisted throughout the Eocene until late Bartonian. Underlying environmental and biotic shifts are a new tectonic regime exemplified by invigorated ocean floor spreading and the major stratigraphic event, the Khirthar transgression and onset of extensive neritic carbonate accumulations. Mammals, Woodburne (2004, 2007) and chronofaunal boundary, Prothero (2004). Large foraminifera, Hottinger (1997), McGowran and Li (2000); SBZ17, 18, Shallow benthic zones 17, 18 (Serra-Kiel et al. 1998). Planktonics E14/E13 boundary, planktonic foraminiferal zones and extinction horizon, Berggren and Pearson (2005), Sexton et al. (2006). Successional benthic foraminiferal faunas (chronofaunas), Thomas (2007); Negev dysaerobia, Barbieri et al. (2003). AAG North Shore, southern Australian stratigraphic record display on main turning points, or caesuras, from McGowran (2009)

region and the Australo-Antarctic Gulf, linked in some way to the cooling after MECO (Bohaty et al. 2009). The crisis extended into the warm water photic zone and, at the same time, the major change occurred in the deep-benthic environment at global-chronofaunal level, when the above-mentioned bentho-pelagic coupling was established. In the North American mammal succession the long-lived Eocene chronofauna was succeeded by the equally long-lived White River Chronofauna, the transition occurring in the high turnovers of the Duchesnean "Age."

Thus, we suggest that both the ultra-mature phase of the Early Paleogene large-foraminiferal chronofauna and, eventually, the chronofaunal demise were forced environmentally in concert with changes in all environmental realms.

Glaciation Oi-1 marks the first modern-sized Antarctic ice growth. Disappearing abruptly from the plankton were the characteristically Eocene genera *Hantkenina* and *Turborotalia*, and from the neritic, the orthophragminids and pellatispirids. Both events were due to the climatic change but in different ways—the plankton losing their nutrient-rich warm habitat, the benthos disrupted glacioeustatically. In southern Australia the major overturn in the neritic benthics is exactly at this level (Moss and McGowran 2003) and this too is a habitat change characterized by reduced delivery of nutrient.

V to VI The Oligocene Interregnum

In the Oligocene the world was at its coolest since the Gondwanaland ice age, sandwiched between the Eocene greenhouse and the last greenhouse hurrah of the Miocene. In its fossil record the Oligocene is transitional in every sense, from the Paleogene, essentially Eocene in early nineteenth century terms with its "archaic" biotas, to the Neogene, essentially Miocene but which "modern" world we still inhabit.

The foraminifera illustrate this transitional character in all their realms—diversities are low and morphological ranges are contracted. The onset of Neogene warming (Fig. 2) (actually beginning in the late Oligocene) and redevelopment of benign neritic environments favorable for rediscovery of photosymbiosis were the stimulus for the Neogene radiations and their numerous convergences with Paleogene antecedents.

Large K-strategist lineages in the neritic benthos were few in the early Oligocene but built up sharply in the Late Oligocene. Hottinger (1997) detailed the lists of genera tracing the photosymbiotic communities' spreading into the middle and outer neritic (deeper photic zone). In the Indo-Pacific region the "letter stages" (actually biochrons: McGowran 2005) are based on these forms (Neogene part: Fig. 15). The main locus of evolution of the large-benthic photosymbiotics (diversity, origination, and extinction) shifted from the west and central Tethys in the Paleocene and Eocene to the Indo-West Pacific in the Oligocene and Neogene (Renema 2007).

A major concern of anatomists (e.g., van der Vlerk 1959; Adams 1983; Drooger 1993) has been the development through time of the embryonic apparatus—surely the same selection pressures for large biomass and a healthy inheritance of symbionts as inferred for the radiation at the Paleocene/Eocene boundary (Hottinger 2000). In the *Miogypsinoides-Miogypsina* clade (Fig. 16) there is a "long-lasting central lineage" with a remarkable parallelism across three biogeographic provinces—many of the species names for lineage segments names are the same in the cosmopolitan succession. Meanwhile side lineages are produced separately in each province. Details differ, but the general pattern of recovery, radiation, and community seem to be similar to the (more intensively studied) Paleogene of Tethys. Assembling the succession has been more difficult (Adams 1970, 1984) but documentation and systematics are proceeding (Boudagher-Fadel and Lokier 2005; Renema 2002, 2007).

The general pattern is the same for the planktonics. Small microperforates and globigerinids survived the turnover and diversification and morphological range increased toward the Miocene.

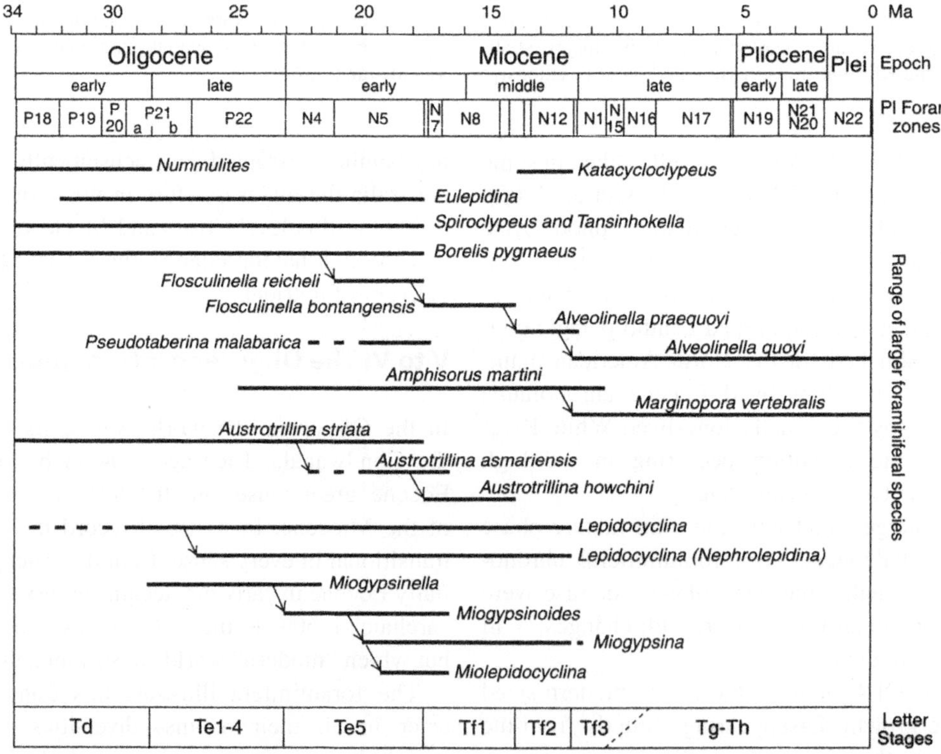

Fig. 15 The Indo-Pacific "letter stages" of the tropical and subtropical neritic are defined by the ranges of key larger benthic foraminifera and correlated with planktonic foraminiferal zones (Li et al. 2009, Fig. 3.13, modified after Adams (1970), and Boudagher-Fadel and Banner 1999). Courtesy of Qianyu Li and with permission of Springer Science+Business Media B.V.

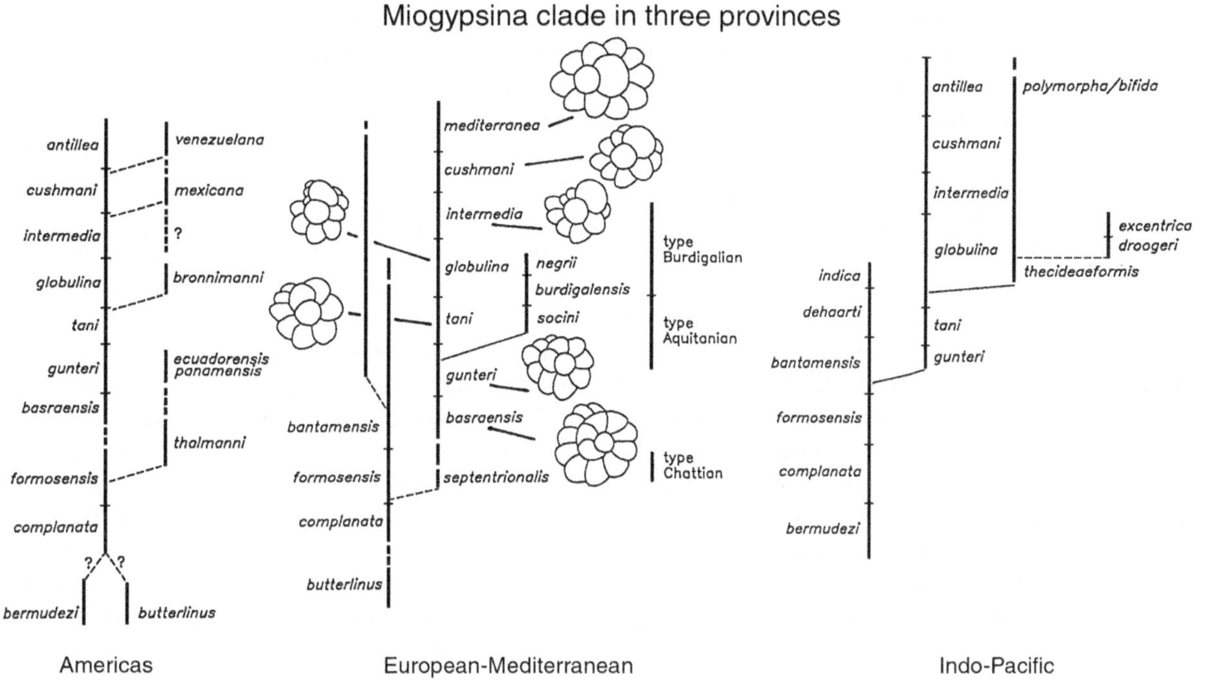

Fig. 16 The *Miogypsinoides–Miogypsina* clade in three provinces—Americas, western Tethys (European–Mediterranean), and Indo-Pacific (McGowran 2005, Fig. 4.28, collated from Drooger 1993, Figs. 56, 61, and 64). Sketches were added (Drooger's Fig. 59) of embryonic apparatus for seven successive Mediterranean species. There is a "long-lasting central lineage" in which nepionic acceleration was rapid and the average duration of ten successive morphometrically discriminated species was of the order of one million years per species. Side-lineages developed separately in the three provinces. Reprinted with permission of Cambridge University Press

VII to VIII MICO Terminated; Bipolar Ice Gets a Grip

Warming and transgression that began in the late Oligocene reached their zeniths in the late-early and early-middle Miocene, to be truncated at glaciation Mi 3b centered about 13.74 Ma (Abels et al. 2005). Planktonic species diversity rose strongly in the early to middle Miocene then dropped slightly (Figs. 8 and 9). Benthic K-strategist diversity also rose during the late Oligocene to Middle Miocene but dropped sharply (e.g. McGowran 1979, Fig. 5). These major shifts occurred in an aura of nutrient instability implied by the Monterey carbon excursion which would have perturbed the oligotrophic communities. Indeed, the prominent intra-Burdigalian biochronal boundary, Te5/Tf1, is recognized on the extinction of the *Eulepidina* and *Spiroclypeus/Tansinhokella* clades and the anagenetic end of *Borelis* (Fig. 15) and this is closely linked in time with the start of a eutrophying event signaled by the Monterey carbon excursion (Fig. 2). (However, this major boundary may be at the base of the Burdigalian: Fig. 14.)

The main changes in the Indo-West Pacific photosymbiotic faunas are losses in several lineages and extinctions of several, especially the prominent Lepidocyclinids and Miogypsinids after *Austrotrillina* and the anagenetic end of *Flosculinella.* This is the termination of the great Indo-West Pacific *K*-strategic journey which began in the ruins of the cold late Paleogene. But the conclusion spanned the ~2 myr of "Stage" Tf2, during which there were more glacials after Mi3b.

The deep-ocean transition occurred gradually but at that time. In the North American mammal faunas there was turnover during MICO before a settling into the durable Clarendonian chronofauna which began at this time.

IX Termination of Pliocene Warming

Dramatic as it was, the rise of the modern ice age did not affect the foraminiferal or other biotas to anywhere near the same extent as the Paleogene or Middle Miocene events (McGowran 2005; Aubry et al. 2009) did. The modern world is still the Neogene world. The lack of impact of the late Neogene climate on biotas was dramatically confirmed by a high-resolution, cyclostratigraphically analyzed study of turnover in small mammals (van Dam et al. 2006).

Evolution: Species Taxonomy and Biostratigraphy

The development of modern planktonic protistan biostratigraphy is an ascending conceptual series—planktonic foraminiferal zones have evolved from assemblage zones to oppelzones and range zones, to phylozones, to chronozones, to datums along with datums from other microfossil groups, all of which are integrated with geophysical and geochemical signals in the "age model" of paleoceanography (McGowran 2005, Fig. 2.2). Defining events are given with magnetochronological calibration and estimated age down to 10^4 years.

Serra-Kiel et al. (1998) assembled 20 SBZ zones on the Tethyan Paleocene-Eocene larger foraminifera, spanning about 30 myr. For the same time span two successions from the pelagic realm respectively contained 23 zones/subzones (calcareous nannoplankton) and 25 zones/subzones (planktonic foraminifera). These numbers are similar, encouraging a comparison of the different approaches to the organic evolution which supports the geo- and biohistorical enterprise (Fig. 17).

In reviewing the taxonomic philosophies (such as they are) and methods of foraminiferology, McGowran (2005) contrasted the sheer density of the fossil record of planktonic protists with intensively studied groups such as dinosaurs and horses. This is a major reason why cladistic methods have been applied but never flourished in the morphotaxonomy of the fossil record. Pearson et al. (2006a, b) strongly defended the continued application of a form of stratophenetics to a group, Eocene planktonics, almost all of whose member

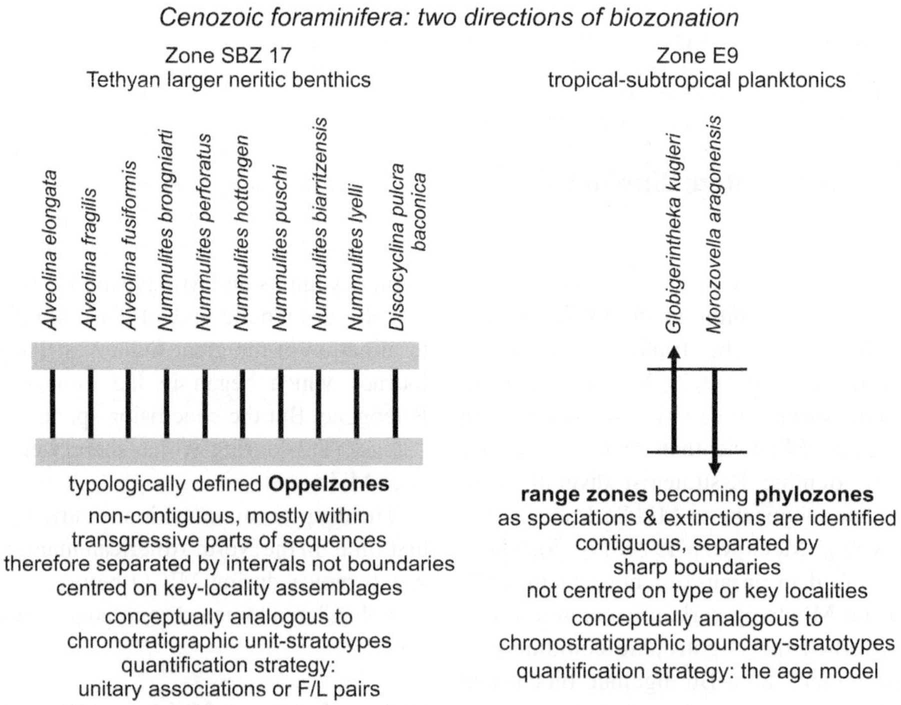

Fig. 17 Two kinds of Eocene foraminiferal biozone exemplifying two biostratigraphic traditions, plankton in the pelagic realm and larger benthics in the neritic realm (McGowran and Li 2007, Fig. 4). In the Tethyan neritic, shallow benthic zone SBZ 17 is characterized by a key-locality assemblage of ten species, none of which chararacterize SBZ 16 or SBZ 18 (Serra-Kiel et al. 1998). The planktonic zone is defined by two events (Berggren and Pearson 2005). The two kinds of biozone are conceptually analogous to chronostratigraphic unit-stratotypes and boundary-stratotypes, respectively. Unitary associations, Guex (1991); F/L pairs, Alroy (1994). Reprinted courtesy of Micropaleontology Press

morphospecies have well-tested stratigraphic ranges and more-than-plausible positions in a reconstructed phylogeny.

Conclusions

1. We have scenarios for environmental change during the Cenozoic Era—descriptions of how the world transformed from the greenhouse times of the dinosaurs to the icehouse times of today. The main change is cooling, the underlying drive is the replacement by continental drift of the "warm ring" (Tethys) by the "cold ring" (the circum-Antarctic Southern Ocean). Changes occurred in steps, each cooling stronger, each warming reversal weaker. Correlation and age determination enable stronger ties between the neritic, terrestrial, and pelagic realms. Eight environmental shifts or transient events are outlined (Figs. 1 and 2) as a template. The two critical intervals are the Auversian Facies shift (Bartonian to early Rupelian) and late Burdigalian to Serravallian—both more profound environmentally than the late Neogene ice ages, both shaping the modern biosphere more than has the Plio-Pleistocene.
2. Three groups of foraminifera are discussed, one simply environmental (the deep ocean faunas), one nontaxonomic but strongly iterative (the photosymbiotic, benthic larger foraminifera all following the $r \rightarrow K$ ecological strategy on oligotrophic shelves and platforms), and one ecological group that once was thought to be monophyletic but no more (the planktonic foraminifera).
3. Major changes or turnovers in the deep-ocean and large-photosymbiotic groups justify the concept of chronofaunas of 10^7 years' duration. There are five in the deep ocean and four in the neritic. The planktonic equivalent is a two-part Cenozoic record, Paleogene, recovering from the end-Cretaceous catastrophe, and Neogene, recovering from the Eocene extinctions.
4. The construction of a chronofauna is partly opportunist in initial response to opportunity, partly evolutionary in the classical sense of adaptive radiation, and partly environmental in being shaped by external events. Parallel events in the different realms indicate global factors.
5. The end-Cretaceous event was profoundly influential—except on the deep-sea foraminifera. Recent corrections to chronology and correlation show that the end-Paleocene hyperthermal affected all realms. The End-Ypresian or Cuisian cooling was influential but not as well known. The later Eocene was turbulent but two events stand out—the late Bartonian photosymbiotic crisis and the end-Priabonian ice age Oi-1. Neogene recovery began in the late Oligocene. The Monterey carbon excursion and abrupt terminating of the Miocene warm period affected all realms.

Ackowledgments I thank Qianyu Li and Paul Pearson for critically reading the manuscript and the editor, John Talent, for initiating the project, inviting the contribution, and steadfastly displaying patience and encouragement.

References

Abels HA, Hilgen FJ, Krijgsman W, Kruk RW, Raffi I, Turco E, Zachariasse WJ (2005) Long-period orbital control on middle Miocene global cooling: integrated stratigraphy and astronomical tuning of the Blue Clay Formation on Malta. Paleoceanography 20:PA4012. doi:10.1029/2004PA001129
Abreu VS, Hardenbol J, Haddad GA, Baum GR, Droxler AW, Vail PR (1998) Oxygen isotope synthesis: a Cretaceous icehouse? In:de Graciansky P-C, Hardenbol J, Jacquin T, Vail PR (eds) Mesozoic and Cenozoic sequence stratigraphy of European basins. SEPM (Soc Sedimentary Geol), Spec Pub No. 60, pp 75–80
Adams CG (1970) A reconsideration of the East Indies Letter Classification of the Tertiary. Bull Br Mus (Nat Hist), London 19:87–137
Adams CG (1983) Speciation, phylogenesis, tectonism, climate and eustasy: factors in the evolution of Cenozoic larger foraminiferal bioprovinces. In: Sims RW, Price JH, Whalley PES (eds) Evolution time and space: the emergence of the biosphere. Systematics Assoc Spec, vol 23. Academic, London, pp 255–289
Adams, CG (1984) Neogene larger foraminifera, evolutionary and geological events in the context of datum planes. In: Ikebe N, Tsuchi R (eds) Pacific Neogene datum planes. University of Tokyo Press, Tokyo, pp 47–67
Alroy J (1994) Appearance event ordination: a new biochronological method. Paleobiology 20:191–207
Aubry M-P, Berggren WA, Van Couvering JA, McGowran B, Hilgen F, Steininger F, Lourens L (2009) The Neogene and Quaternary: chronostratigraphic compromise or non-overlapping magisteria? Stratigraphy 6:1–16
Barbieri R, Benjamini C, Monechi S, Reale V (2003) Stratigraphy and benthic foraminiferal events across the Middle–Late Eocene transition in Western Negev, Israel. In: Prothero DR, Ivany LC, Nesbitt EA (eds) From greenhouse

to icehouse: the marine Eocene–Oligocene transition. Ch 26. Columbia University Press, New York, pp 453–470

Berger WH, Wefer G (1996) Expeditions into the past: paleoceanographic studies in the South Atlantic. In: Wefer G, Berger WH, Siedler G, Webb DJ (eds), The South Atlantic: present and past circulation. Springer-Verlag, Berlin, pp 363–410

Berggren WA, Miller KG (1989) Cenozoic bathyal and abyssal calcareous benthic foraminiferal zonation. Micropaleontology 35:308–320

Berggren WA, Pearson PN (2005) A revised tropical to subtropical Paleogene planktonic foraminiferal zonation. J Foraminiferal Res 35:279–298

Blow WH (1959) Age, correlation and biostratigraphy of the Upper Tocuyo (San Lorenzo) and Pozón formations eastern Falcón, Venezuela. Bull Am Paleontol 39:1–251

Bohaty SM, Zachos JC, Florindo F, Delaney ML (2009) Coupled greenhouse warming and deep-sea acidification in the middle Eocene. Paleoceanography 24:PA2207. doi:10.1029/2008PA001676

Bolli HM (1957) The genera *Globigerina* and *Globorotalia* in the Paleocene–lower Eocene Lizard Springs Formation of Trinidad, B.W.I. United States. Nat Mus, Bull 215:51–81

Boudagher-Fadel MK, Banner FT (1999) Revision of the stratigraphic significance of the Oligo-Miocene "letter-stages". Revue de Micropaléontologie 42:93–97

Boudagher-Fadel MK, Lokier SW (2005) Significant Miocene larger foraminifera from south central Java. Revue de Paléobiologie 24:291–309

Buzas MA, Culver SJ (1999) Understanding regional species diversity through the log series distribution of occurrences. Diversity Distribution 8:187–195

Buzas MA, Collins LS, Culver SJ (2002) Latitudinal difference in biodiversity caused by higher tropical rate of increase. Proc Natl Acad Sci USA 99:7841–7843

Carpenter WB, Parker WK, Jones TR (1862) Introduction to the study of the foraminifera. The Ray Society, London

Cifelli R (1990) A history of the classification of the foraminifera (1826–1933). Part I, Foraminiferal classification from d'Orbigny to Galloway. Cushman Foundation Foraminiferal Res, Spec Pub 27:1–88

Corliss BH (1981) Deep-sea benthic foraminiferal faunal turnover near the Eocene/Oligocene boundary. Mar Micropaleontol 6:367–384. doi:10.1016/0377-8398(81)90007-4

Culver SJ (2003 Benthic foraminifera across the Cretaceous–Tertiary (K–T) boundary: a review. Mar Micropaleontol 47:177–226

Cushman JA (1940) Foraminifera, their classification and economic use. Harvard University Press, Boston

Cushman JA, Stainforth RM (1945) The foraminifera of the Cipero Marl Formation of Trinidad, British West Indies. Cushman Lab Foraminiferal Res, Spec Pub 14:1–74

Darling K, Wade CM (2008) The genetic diversity of planktic foraminifera and the global distribution of ribosomal RNA genotypes. Mar Micropaleontol 67:216–238

Darling KF, Kucera M, Wade CM (2007) Global molecular phylogeography reveals persistent Arctic circumpolar isolation in a marine planktonic protist. PNAS 104:5003–5004

Darling KF, Thomas E, Kasemann SA, Seears HA, Smart CW, Wade CM (2009) Surviving mass extinction by bridging the benthic/planktic divide. PNAS 106:12629–12633

Darwin C (1859) On the origin of species [Facsimile of first edition, 1859, ed. E. Mayr (1964)]. Harvard University Press, Cambridge, MA

Davies AM (1934) Tertiary faunas: a text-book for oilfield palaeontologists and students of geology. Thomas Murby, London, 1934–35 [v 1, 1935], v 2

De Vargas C, Saez AG, Medlin LK, Thierstein HR (2004) Superspecies in the calcareous plankton. In: Thierstein HR, Young JR (eds) Coccolithophores: from molecular processes to global impact. Springer, Berlin, pp 271–298

d'Hondt S (2005) Consequence of the Cretaceous/Paleogene mass extinction for marine ecosystems. Annu Rev Ecol Syst 36:295–317

d'Hondt S, Zachos JC (1998) Cretaceous foraminifera and the evolutionary history of planktic photosymbiosis. Paleobiology 24:512–523

Drooger CW (1993) Radial foraminifera: morphometrics and evolution. North-Holland, Amsterdam, New York, 242p

Gignoux M (1955) Stratigraphic geology (English translation by GG Woodford of the 1950 French edition of "Géologie Stratigraphique"). WH Freeman, San Francisco, 682pp

Glaessner MF (1937) Planktonforaminiferen aus der Kreide und dem Eozän und ihre stratigraphische Bedeutung. Studies in Micropaleontol 1 (Moscow, 1937):27–52

Glaessner MF (1945) Principles of micropalaeontology. Melbourne University Press, Melbourne, 296p

Guex, J (1991) Biochronological correlations. Springer-Verlag, Berlin

Hallock P (1987) Fluctuations in the trophic resource continuum: a factor in global diversity cycles? Paleoceanography 2:457–471

Hallock P (1999) Symbiont-bearing foraminifera. In: Sen Gupta BK (ed) Modern foraminifera. Kluwer Academic, Dordrecht, pp 123–139

Hallock P, Premoli Silva I, Boersma A (1991) Similarities between planktonic and larger foraminiferal evolutionary trends through Paleogene paleoceanographic changes. Palaeogeogr Palaeoclimatol Palaeoecol 83:49–64

Hardenbol J, Thierry J, Farley MB, Jacquin T, de Graciansky P-C, Vail PR (1998) Mesozoic and Cenozoic sequence chronostratigraphic framework of European basins. In: de Graciansky P-C, Hardenbol J, Jacquin T, Vail PR (eds), Mesozoic and Cenozoic sequence stratigraphy of European basins. SEPM (Society of Sedimentary Geology) Special Publication 60:3–13, Tulsa

Hayward BW (2001) Global deep-sea extinctions during the Pleistocene ice ages. Geology 29:599–602. doi:10.1130/00917613(2001)029<0599:GDSEDT>2.0.CO;2

Hayward BW (2002) Late Pliocene to Middle Pleistocene extinctions of deep-sea benthic foraminifera ("*Stilostomella* extinction") in the southwest Pacific. J Foraminiferal Res 32:274–307. doi:10.2113/32.3.274

Hayward BW, Kawagata S, Grenfell HR, Sabaa AT, O'Neill T (2007) Last global extinction in the deep sea during the mid-Pleistocene climate transition. Paleoceanography 22:PA3103. doi:10.1029/2007PA001424

Hohenegger J (2000) Coenoclines of larger foraminifera. Micropaleontology 46(Suppl 1):127–151

Hok TS (1939) The results of phylomorphogenetic studies of some larger Foraminifera (a review). Mijnbouw en Geologie 6:93–97

Hottinger L (1960) Recherches sur les Alvéolines du Paléocène et de l'Eocène. Mém. Suisses Paléontol 75/76:1–243

Hottinger L (1981) The resolution power of the biostratigraphic clock based on evolution and its limits. International Symposium on Concepts and Methods in Palaeontology, Universidad Barcelona, Barcelona, pp 233–242

Hottinger L (1982) Larger foraminifera, giant cells with a historical background. Naturwissenschaft 69:361–371

Hottinger L (1983) Processes determining the distribution of larger foraminifera in space and time. Utrecht Micropaleontol Bull 30:239–253

Hottinger L (1987) Conditions for generating carbonate platforms. Mem Soc Geol Italy 40:265–271

Hottinger L (1988) Significance of diversity in shallow benthic foraminifera. Atti del Quarto Simposio Ecologia e Paleoecologia Communità Benthonice, Sorrento, 1-5 November 1988. Museo Regionale di Scienze Natural—Torino, pp 35–51

Hottinger L (1996) Sels nutritifs et biosédimentation. Mémoir de la Societé géologique de France, n.s., 169:99–107

Hottinger L (1997) Shallow benthic foraminiferal assemblages as signals for depth of their deposition and their limitations. Bull de la Societé géologique de France, t 168(4): 491–505

Hottinger L (1998) Shallow benthic foraminifera at the Paleocene–Eocene boundary. Strata 9:61–64

Hottinger L (1999) "Odd partnership", a particular size relationship between close species of larger foraminifera, with an emendation of an outstandingly odd partner, *Glomalveolina delicatissima* (Smout, 1954), Middle Eocene. Eclogae geologicae Helvetiiae 92:385–393

Hottinger L (2000) Adaptations of the foraminiferal cell to life in shallow carbonate environments. Acad Naz Sci Lett Arti di Modena, Collana di Studi 21:135–140

Hottinger L (2001) Learning from the past. In: Levi-Montalcini L (ed) Frontiers of life. Discovery and spoliation of the biosphere, vol 4(2). Academic, London and San Diego, pp 449–477

Huber BT, Olsson RK, Pearson PN (2006) Chapter 16: Taxonomy, biostratigraphy and phylogeny of Eocene microperforate planktonic foraminifera (*Jenkinsina, Cassigerinelloita, Chiloguembelina, Streptochilus, Zeauvigerina, Tenuitella,* and *Cassigerinella*) and problematica (*Dipsidripella*). In: Pearson PN, Olsson RK, Huber BT, Hemleben C, Berggren WA (eds) Atlas of Eocene planktonic foraminifera. Cushman Foundation Special Publications, vol 41, Washington, DC, pp 461–508

Hull D (1988) Science as a process: an evolutionary account of the social and conceptual development of science. University of Chicago Press, Chicago, 583pp

Kelly DC, Bralower TJ, Zachos JC, Premoli Silva I, Thomas E (1996) Rapid diversification of planktonic foraminifera in the tropical Pacific (ODP Site 865) during the late Paleocene thermal maximum. Geology 24:423–426

Langer MR, Hottinger L (2000) Biogeography of selected "larger" foraminifera. Micropaleontology 46(Suppl 1): 105–126

Leckie RM (2009) Seeking a better life in the plankton. PNAS 106:14183–14184. doi: 10.1073/pnas.0907091106

Li Q, McGowran B (1996) Early Paleocene *Parvularugoglobigerina* and late Eocene *Praetenuitella*: evolutionary convergence implies similar habitat? J Micropalaeontol 14:119–134

Li Q, McGowran B (1997) Miocene climatic oscillations recorded in the Lakes Entrance oil shaft, southeastern Australia: benthic foraminiferal response on a mid-latitude margin. Micropaleontology 43:149–164

Li Q, McGowran B (2000) The Miocene foraminifera from Lakes Entrance Oil Shaft, southeastern Australia. Assoc Aust Palaeontols, Mem 22:142

Li Q, Zhong G, Tian J (2009) Chapter 3: Stratigraphy and sea level changes. In: Wang P, Li Q (eds) The South China Sea. Developments in paleoenvironmental Research, vol 13, pp 75–170. doi:10.1007/978-1-4020-9745-4_3

Lipps JH (1970) Plankton evolution. Evolution 24:1–22

Lipps JH (1981) What, if anything, is micropaleontology? Paleobiology 7:167–199

Lyell C (1871) Students elements of geology. John Murray, London

Lyle M, Pälicke H, Nishi H, Raffi I, Gamage K, Klaus A, the IODP Expeditions 320/321 Scientific Party (2010) The Pacific equatorial age transect, IODP Expeditions 320 and 321: building a 50-million-year-long environmental record of the equatorial Pacific. Sci Drilling 9:4–15. doi:10.2204/iodp.sd.9.01.2010

Mayr E, Provine WB (1980) The evolutionary synthesis. Harvard University Press, Cambridge

McGowran B (1979) The Australian Tertiary: foraminiferal overview. Mar Micropaleontol 4:235–264

McGowran B (2005) Biostratigraphy: microfossils and geological time. Cambridge University Press, Cambridge, 459pp

McGowran B (2009) The Australo-Antarctic Gulf and the Auversian facies shift. In: Koeberl C, Montanari A (eds) The Late Eocene Earth—hothouse, icehouse, and impacts. Geological Society of American Special Paper 452, Ch 14, Boulder, CO, pp 215–240. doi:10.1130/2009.2452(14)

McGowran B, Li Q (1996) Ecostratigraphy and sequence biostratigraphy, with a neritic foraminiferal example from the Miocene in southern Australia. Hist Biol 11: 137–169

McGowran B, Li Q (2000) Evolutionary palaeoecology of Cainozoic foraminifera: Tethys, IndoPacific, southern Australasia. Hist Biol 15:3–28

McGowran B, Li Q (2007) Stratigraphy: gateway to geohistory and biohistory. Stratigraphy 4:173–185

Miller KG, Katz ME, Berggren WA (1992) Cenozoic deep-sea benthic foraminifera: a tale of three turnovers. In: Takayanagi Y, Saito T (eds) Studies in benthic foraminifera: proceedings of the fourth international symposium on Benthic Foraminifera, Sendai, 1990 (Benthos '90), Tokai University Press, Tokyo, pp 245–248

Moss G, McGowran B (2003) Oligocene neritic foraminifera in Southern Australia: spatiotemporal biotic patterns reflect sequence-stratigraphic environmental patterns. In: Olson H, Leckie M (eds) Paleobiological, geochemical, and other proxies of sea level change. SEPM (Society of Sedimentary Geology) special volume 75. Tulsa, pp 117–138

Murray J, Hjort J (1912) The ocean: a general account of the science of the sea. Williams & Norgate, London

Nebelsick JH, Rasser MW, Bassi D (2005) Facies dynamics in Eocene to Oligocene circumalpine carbonates. Facies 51:197–216

Norris RD (1991) Parallel evolution in the keel structure of planktonic foraminifera. J Foraminiferal Res 21:319–331
Norris RD (1996) Symbiosis as an evolutionary innovation in the radiation of Paleocene planktonic foraminifera. Paleobiology 22:461–480
Norris RD, Wilson PA (1998) Low-latitude sea-surface temperatures for the mid-Cretaceous and the evolution of planktic foraminifera. Geology 26:823–826
Pagani M, Zachos JC, Freeman KH, Tipple B, Bohaty S (2005) Marked decline in atmospheric carbon dioxide concentrations during the Palaeogene. Science 309:600–603
Pawlowski J, Fahrnis J, Lecroq B, Longet G, Cornelius N, Excoffier L, Cedhagen T, Gooday AJ (2007) Bipolar gene flow in deep-sea benthic foraminifera. Mol Ecol 16:4089–4096
Pearson PN (1998) Stable isotopes and the study of evolution in planktonic foraminifera. Paleontol Soc Pap 4:138–178
Pearson PN, Olsson RK, Huber BT, Hemleben C, Berggren WA (eds) (2006a) Atlas of Eocene planktonic foraminifera: Fredericksburg, Cushman Foundation Foraminiferal Res, Spec Publ No. 41, 514pp
Pearson PN, Olsson RK, Huber BT, Hemleben C, Berggren WA, Coxall HK (2006b) Chapter 1: Overview of Eocene planktonic foraminiferal taxonomy, paleoecology, phylogeny, and biostratigraphy. In: Pearson et al. (2006a), pp 11–28
Prothero DR (2004) Did impacts, volcanic eruptions, or climate change affect mammalian evolution? Palaeogeogr Palaeoclimatol Palaeoecol 214:283–294
Pujalte V, Schmitz B, Baceta JI, Orue-Etxebarria X, Bernaola G, Dinares-Turell J, Payros A, Apellaniz E, Caballero F (2009) Correlation of the Thanetian–Ilerdian turnover of larger foraminifera and the Paleocene–Eocene thermal maximum: confirming evidence form the Campo area (Pyrenees, Spain). Geol Acta 7:161–175. doi:10.1344/105.000000276
Reichel M (1937) Étude sur les Alvéolines. Mém Suisses Paléontol 57 et 59:1–147
Reiss Z, Hottinger L (1984) The Gulf of Aquaba: ecological micropaleontology. Springer, New York
Renema W (2002) Larger foraminifera as marine environmental indicators. Scr Geol 124:263pp
Renema W (2007) Faunal development of larger benthic foraminifera in the Cenozoic of southeast Asia. In: Renema W (ed) Biogeography, time, and place: distributions, barriers, and islands. Topics in geobiology, vol 29, Ch 6. Springer, Netherlands, pp 179–215
Schiebel R, Hemleben C (2005) Modern planktic foraminifera. Paläontologische Zeitschrift 79:135–148
Schmidt DN, Lazarus D, Young JR, Kucera M (2006) Biogeography and body size in marine plankton. Earth-Sci Rev 78:239–266
Serra-Kiel J, Hottinger L, Caus E, Drobne K, Ferràndez C, Jauhri AK, Less G, Pavlovec R, Pignatti J, Samso JM, Schaub H, Sire E., Strougo, Tambareau Y, Tosquella J, Zakrevskaya E (1998) Larger foraminiferal biostratigraphy of the Tethyan Paleocene and Eocene. Bull Société Géologique France 169:281–299
Sexton PF, Norris RD (2008) Dispersal and biogeography of marine plankton: long-distance dispersal of the foraminifer. Geology 36:899–902. doi:10.1130/G25232A.1
Sexton PF, Wilson PA, Pearson PN (2006) Palaeoecology of late middle Eocene planktic foraminifera and evolutionary implications. Mar Micropaleontol 60:1–16
Smocovitis VB (1996) Unifying biology: the evolutionary synthesis and evolutionary biology. Princeton University Press, Princeton, 248pp
Subbotina NN (1953) [Globigerinidae, Hantkeninidae, and Globorotaliidae. Fossil foraminifera of the U.S.S.R.] Vses. Neft. Nauchno-Issled. Geol-Razved Inst (VNIGRI), Trudy, ns 6:1–296
Thomas E (1992) Middle Eocene-late Oligocene bathyal benthic foraminifera (Weddell Sea): faunal changes and implications for ocean circulation. In: Prothero DR, Berggren WA (eds) Eocene–Oligocene climatic and biotic evolution. Princeton Univ Press, Princeton, pp 245–271
Thomas E (1998) The biogeography of the late Paleocene benthic foraminiferal extinction. In: Aubry M-P, Lucas S, Berggren WA (eds) Late Paleocene–Early Eocene climatic and biotic events in the marine and terrestrial records. Columbia University Press, New York, pp 214–243
Thomas E (2007) Cenozoic mass extinctions in the deep sea: what perturbs the largest habitat on Earth? Geol Soc Am, Spec Pap 424:1–23. doi:10.11.30/2007.2424(01)
Thomas E, Gooday AJ (1996) Deep-sea benthic foraminifera: tracers for Cenozoic changes in oceanic productivity? Geology 24:355–358. doi:10.11.30/0091–7613(1996)024<0355:CDSBFT>2.3.co;2
Thomas E, Vincent E (1987) Equatorial Pacific deep-sea benthic foraminifera: faunal changes before the Middle Miocene polar cooling. Geology 15:1035–1039
Tripati A, Backman J, Elderfield H, Ferretti P (2005) Eocene bipolar glaciation associated with global carbon cycle changes. Nature 436:341–346
Ujiié Y, Kimoto K, Pawlowski J (2008) Molecular evidence for an independent origin of modern triserial planktonic foraminifera from benthic ancestors. Mar Micropaleontol 69:334–340
van Andel TJH, Heath GR, Moore TC Jr (1975) Cenozoic history and paleoceanography of the central Pacific Ocean. Geol Soc Am Mem 143:1–134
van Dam JA, Aziz HA, Sierra MAA, Hilgen FJ, van den Hoek Ostende LW, Lourens, LJ, Mein P, van der Meulen AJ, Pelaez-Campomanes P (2006) Long-period astronomical forcing of mammal turnover. Nature 443:687–691
Van der Vlerk IM (1959) Problems and principles of Tertiary and Quaternary stratigraphy. Q J Geol Soc London 115:49–63
Webb SD (1984) On two kinds of rapid faunal turnover. In: Berggren WA, van Couvering JA (eds) Catastrophes and earth history. Princeton University Press, Princeton, pp 417–436
Woodburne MO (ed) (2004) Late Cretaceous and Cenozoic mammals of North America: biostratigraphy and geochronology. Columbia University Press, New York, 391pp
Woodburne MO (2007) Mammal ages. Stratigraphy 3(2006):229–261
Woodburne M, Swisher C (1995) Mammal high-resolution geochronology, intercontinental overland dispersals, sea level, climate and vicariance. In: Berggren WA, Kent DV, Aubry M-P, Hardenbol J (eds) Geochronology time scales and global stratigraphic correlation. SEPM (Society of Sedimentary Geology) Special Publications, vol 54. Tulsa, pp 335–364

Wright JD, Miller KG (1993) Southern Ocean influences on late Eocene to Miocene deepwater circulation. In: Kennett JP, Warnke DA (eds), The Antarctic Paleoenvironment: a perspective on global change. Antarctic Research Series, Part Two, American Geophysical Union, Washington, DC, pp 601–625

Zachos JC, Pagani M, Sloan L, Thomas E, Billups K (2001) Trends, rhythms, and aberrations in global climate 65 Ma to Present. Science 292:686–693

Zachos JC, Dickens GR, Zeebe RE (2008) An early Cenozoic perspective on greenhouse warming and carbon-cycle dynamics. Nature 451:279–283. doi:10.1038/nature06588

Miocene Asteroid Impacts: Proposed Effects on the Biogeography and Extinction Patterns of Eastern North American Gastropods

Edward J. Petuch

Abstract

New information on the pattern of Early Neogene asteroid impacts and interplanetary dust clouds appear to parallel extinction events seen in the eastern North American Transmarian Province. It is hypothesized that the extraterrestrial events of the Miocene were sufficiently severe to account for the three main pulses of extinction and speciation during this time. Gastropod faunal composition is used to test this theory.

Keywords

Eastern North America • Asteroid impacts • Transmarian Province • Serravallian asteroid impacts and hypothesized extinctions • Interplanetary dust clouds • Climatic optimum speciation pulses

Introduction

The marine climates of Miocene North America are now known to have fluctuated wildly throughout the epoch with extensive warm periods punctuated by rapid-onset cooling events (Petuch 1993, 1995; Farley et al. 2006; Oleinik et al. 2008). In North America, no place is better suited for studying these extreme climatic fluctuations than the eastern coast of the United States. Centered on Chesapeake Bay (Maryland, Virginia, and Delaware), this region is tectonically stable with no major epeirogenic events having distorted the fossil record. This area contains some of the richest and best-preserved fossil shell beds found anywhere on the continent. The fossiliferous strata of the Chesapeake Miocene are also temporally extensive, covering most of the epoch from the early Burdigalian to the Messinian.

Biogeographically, the eastern North American molluscan faunas are also situated in an appropriate range of latitudes to give maximum insight into the ecological effects of Miocene temperature fluctuations. Although normally occurring in temperate water conditions and containing temperate endemic gastropods, the Chesapeake molluscan faunas became inundated with tropical-subtropical migrants during warm times and, conversely, with boreal taxa during cool times (Petuch 1988, 2004; Petuch and Drolshagen 2009). These faunal shifts were accompanied by dramatic extinctions and extirpations reflecting the rapidly changing marine climates. The molluscan faunas used for studying these extinction patterns are from four of

E.J. Petuch (✉)
Department of Geosciences, Florida Atlantic University, Boca Raton, FL 33431, USA
e-mail: epetuch@fau.edu

J.A. Talent (ed.), *Earth and Life*, International Year of Planet Earth,
DOI 10.1007/978-90-481-3428-1_34, © Springer Science+Business Media B.V. 2012

the formations of the Chesapeake Group; the Calvert (Langhian), Choptank (Serravallian), St. Mary's (late Serravallian–early Tortonian), and Eastover (late Tortonian–Messinian). A generalized stratigraphic column showing the subjacent and suprajacent units bounding these four Miocene formations is shown in Fig. 1.

Based on gastropod faunal composition, three long warm intervals can be seen to have occurred in eastern North America during the Miocene, one extending from late Burdigalian to the late Langhian (approximately 16–14.8 Ma and corresponding to the Middle Miocene Climatic Optimum (MMCO) of Oleinik et al. (2008)), one extending from latest Serravallian to early Tortonian (approximately 11.4–10 Ma), and one extending throughout the Messinian (approximately 7–5 Ma). These warm intervals correspond to depositions of the Calvert, St. Mary's and Eastover formations respectively. Bracketing these three major warm intervals were four cold intervals, one extending from Aquitanian to late Burdigalian (approximately 23–16 Ma), one during the Serravallian (approximately 14.5–11.5 Ma), one during the late Tortonian (approximately 8 Ma), and one during the latest Messinian (5.3 Ma). These correspond to depositions of the Old Church (poorly preserved and rarely exposed) and Choptank formations, the unconformity between the St. Mary's and Eastover formations, and the unconformity between the Eastover and Yorktown formations (Miocene–Pliocene boundary), respectively. These warm and cool climatic episodes are discussed in more detail in the following sections.

The geochronology of astroblemes, flood basalts, and cosmological events (e.g. Shoemaker and Chao 1961; Gorter and Bayford 2000; Glikson 2005; Farley et al. 2006) has shown that several major asteroid impacts occurred around the world during the Miocene. These include the Ries-Steinheim Binary Impact (Ries and Steinheim craters, Bavaria, 14.5 Ma), the Ewing Impact (Ewing Structure, abyssal plain, equatorial Pacific, 11.7 Ma), the Puffin Impact (Puffin Structure, northwestern Australia, 12 Ma), and the Kara-Kul Impact (Kara-Kul Crater, Tajikistan, 5 Ma). Of special interest is the cold interval during the middle Tortonian, apparently caused by solar obscuration due to a dust cloud that formed by colliding asteroids

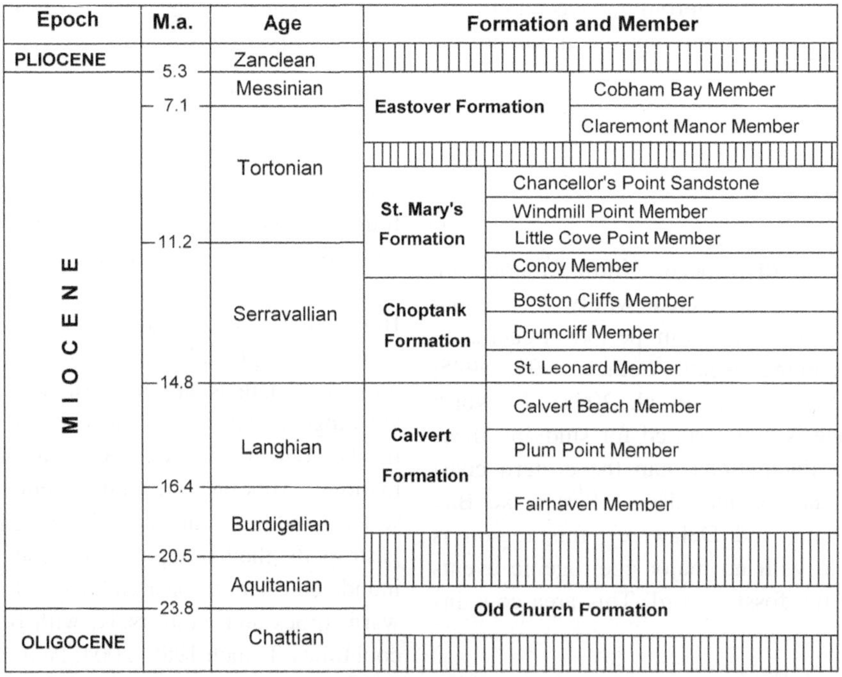

Fig. 1 Generalized stratigraphic column for the Chesapeake Bay region of Maryland, showing the five formations included within the Miocene component of the Chesapeake Group. Pliocene formations are absent in Maryland. Farther south in Virginia and northern North Carolina, the Yorktown Formation spans the Zanclean and Piacenzian Pliocene and unconformably overlays the Eastover Formation (taken from Petuch and Drolshagen 2009)

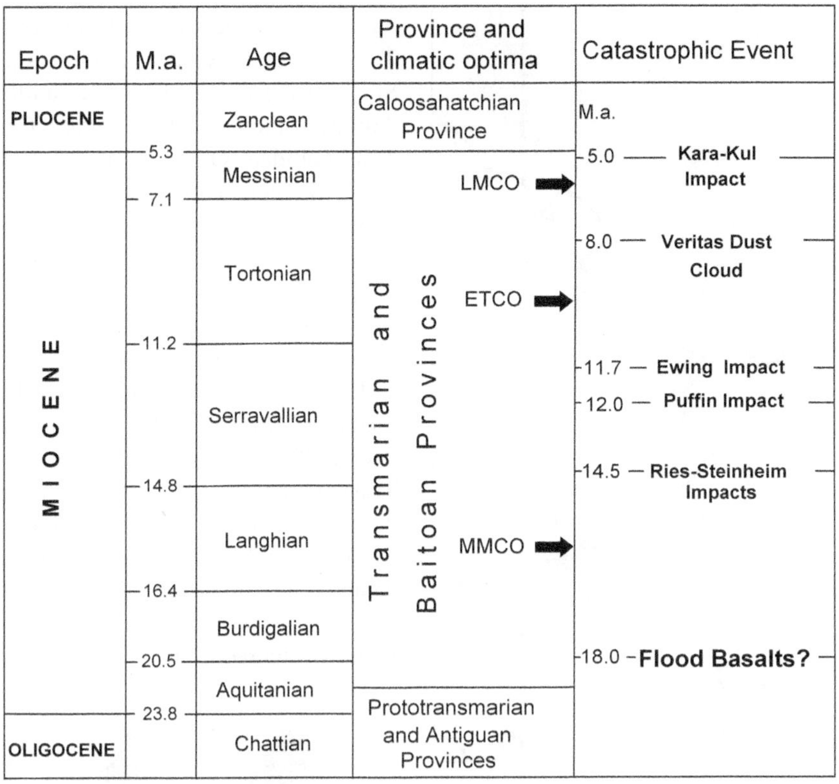

Fig. 2 Chronologic chart showing the timeline of Miocene asteroid impacts and interplanetary dust clouds in relation to Miocene eastern North American marine molluscan faunal provinces. The three main Miocene climatic optima, MMCO, the ETCO, and the LMCO are shown in relation to cool times that are possibly linked to extraterrestrial catastrophic events. The Ries, Steinheim, Puffin, and Ewing Impacts can be seen to be clustered together in the Serravallian cool time. The Kara-Kul Impact can be seen to correlate with the demise of the Transmarian Province

in the main asteroid belt (Farley et al. 2006). This non-impact catastrophic event correlates with extinction of the gastropod fauna of the St. Mary's Formation, which is discussed in detail later in this chapter. A chronologic arrangement of Miocene extraterrestrially derived catastrophic events is shown on Fig. 2.

Miocene Marine Biogeography of Eastern North America

As defined by the 50% rule of mutual exclusivity of species (Valentine 1973; Briggs 1974, 1995; Petuch 2004; Petuch and Drolshagen 2009), two molluscan faunal provinces extended down the eastern coast of the United States during the Miocene: the Transmarian Province (named for the St. Mary's River of Maryland; Petuch 1988, 2004) extended from the Gulf of Maine south to northern North Carolina; the Baitoan Province (named for the Baitoa Formation of the Dominican Republic; Petuch 2004) extended from southern North Carolina to northern Brazil (Fig. 3). The Baitoan Province, with nine subprovinces, embraced eutropical faunas composed of classic tropical Tethyan gastropod families and genera. Two Baitoan subprovinces extended across the southeastern United States: the Onslowian Subprovince along the Carolinas, Georgia, and northeastern Florida, and the Chipolan Subprovince along southeastern, southern, and western Florida, and along the northern coast of the Gulf of Mexico (Fig. 3). The Transmarian Province, with three subprovinces (Sankatian, Calvertian, and Pungoian), had stenothermal paratropical faunas with only a few cold-adapted tropically derived taxa associated with a large endemic temperate fauna and some representatives of warm-adapted boreal groups (Petuch 2004).

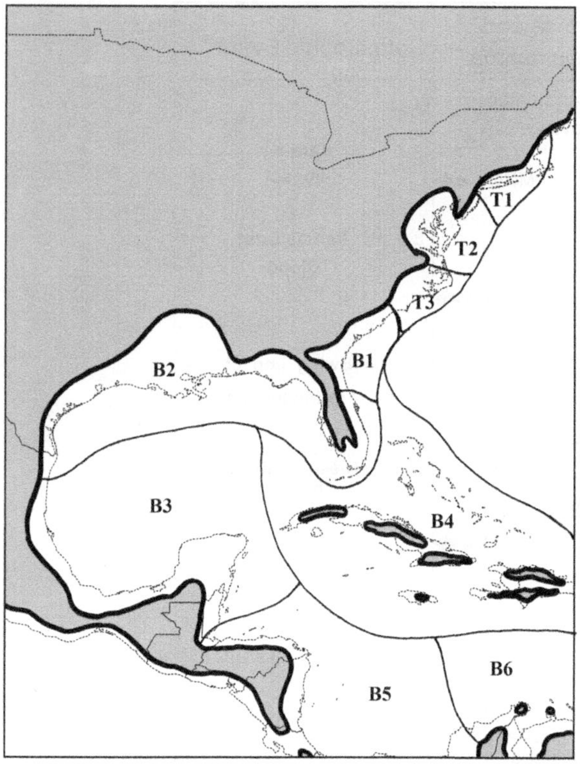

Fig. 3 Biogeographical map of the eastern coast of North America during the Miocene, showing the limits of the Transmarian (T) and Baitoan (B) Provinces. The Transmarian subprovinces include T1 = Sankatian Subprovince, T2 = Calvertian Subprovince (centered within the Salisbury Sea), and T3 = Pungoian Subprovince (centered within the Albemarle Sea). The Baitoan subprovinces include B1 = Onslowian Subprovince, B2 = Chipolan Subprovince, B3 = Agueguexquitean Subprovince, B4 = Anguillan Subprovince, B5 = Culebran Subprovince, and B6 = Cantaurean Subprovince. The approximate configuration of the Miocene coastline (*shaded area*) is shown superimposed upon the outline of Recent North America (*stippled line*) (taken from Petuch 2004)

The Baitoan Province had a classic West Tethyan gastropod fauna with species radiations and index species in the key eutropical families Cypraeidae, Ovulidae, Strombidae, Potamididae, Cerithiidae, Modulidae, Melongenidae, Turbinellidae (Vasinae and Turbinellinae), the volutid subfamilies Lyriinae and Volutinae, and the fasciolariid subfamily Peristerniinae. These high tropical groups are well represented in the fossil records of the Baitoa Formation (Hispaniola), the Cantaure Formation (Venezuela), the Chipola Formation (northwestern Florida), and the Pirabas Formation (Brazil), but are missing from the Transmarian region. A few eutropical groups did however manage to invade the cooler waters of the Transmarian Province, giving it a paratropical status. Some of these were from the families Xenophoridae, Architectonicidae, Personiidae, Cassidae, Tonnidae, Marginellidae, Olividae, and Conidae (Petuch 2004: 30–43; Petuch and Drolshagen 2009: 15–20). Because of its intermediate position between the low-latitude eutropical Baitoan Province and the high-latitude Boreal Province, the Transmarian Province had stenothermal faunas that were conspicuously sensitive to latitudinal climatic shifts.

Endemism in the Transmarian Province was high, reaching almost 100% at the species level and over 50% at the generic level. Some of the classic Transmarian endemic genera, shown in Figs. 4, 5, 6, 7, 8, 9, and 10, are included in Table 1. Other endemic components include a species radiation of cone shells centered around *Gradiconus deluvianus* (Green 1840) (Fig. 6), a large endemic radiation of scaphelline volutes centered around *Volutifusus mutabilis* (Conrad

Fig. 4 Eutropical genera found in the Transmarian Province during the MMCO (Langhian Miocene). (**a**) *Carinorbis* (*C. dalli* (Whitfield 1894), length 13 mm); (**b**) *Oliva* (*O. harrisi* Martin (1904), length 24 mm); (**c**) *Cirsotrema* (*C. calvertensis* (Martin 1904), length 28 mm); (**d**) *Ficus* (*Ficus harrisi* (Martin 1904), length 29 mm); (**e**) *Amaea* (*A. prunicola* (Martin 1904), length 20 mm); (**f**) *Calvertitella* (*C. indenta* (Conrad 1841), length 41 mm); (**g**, **h**) *Xenophora* (*X. lindae* Petuch and Drolshagen 2009, width 31 mm)

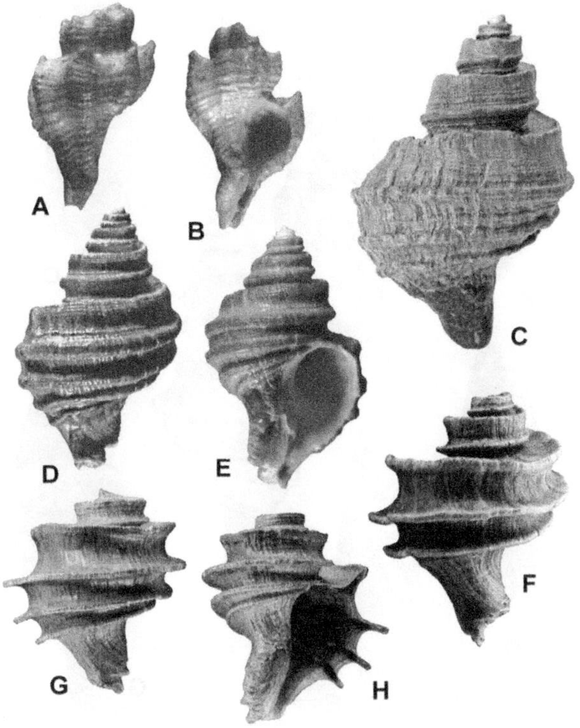

Fig. 5 Transmarian endemic taxa that became extinct after the Ries, Steinheim, Puffin, and Ewing Impacts (Serravallian Miocene). (**a, b**) *Stephanosalpinx* (*S. candelabra* Petuch (1988), length 18 mm); (**c**) *Ecphorosycon* (*E. kalyx* (Petuch 1988), length 68 mm); (**d, e**) *Chesathais* (*C. lindae* Petuch (1988) length 59 mm); (**f**) *Trisecphora* (*T. patuxentia* (Petuch 1988), length 37 mm); (**g, h**) *Ecphora (Planecphora)* (*E. (P). delicata* Petuch (1989), length 39 mm)

Fig. 6 Eutropical genera found in the Transmarian Province during the ETCO (early Tortonian Miocene). (**a**) *Conus (Gradiconus)* (*C. (G.) deluvianus* (Green 1830), length 58 mm); (**b**) *Conus (Gradiconus)* (*C. (G.) sanctaemariae* Petuch (1988), length 40 mm); (**c**) *Stenorhytis* (*S. expansa* (Conrad 1842), length 23 mm); (**d**) *Hesperisternia* (*H. petuchi* Vermeij (2006), length 23 mm); (**e**) *Eudolium* (*E. marylandicum* Petuch (1988), length 44 mm); (**f**) *Cymatosyrinx* (*C. limatula* (Conrad 1830), length 24 mm); (**g**) *Conus (Gradiconus)* (*C. (G.) asheri* Petuch (1988), length 42 mm); (**h**) *Laevityphis* (*L. acuticosta* (Conrad 1830), length 22 mm)

1834), and an endemic radiation of the busyconid genus *Turrifulgur* (Petuch 1988) (Fig. 7).

After the Kara-Kul Impact at the end of the Messinian, the marine climate of the Transmarian region became much colder; sea level dropped precipitously, exposing the shallow water neritic environments to subaerial conditions (the "Messinian Event" and "Transmarian Extinction" (Petuch 2004: 261–262). This double-edged environmental catastrophism obliterated the biogeographic boundaries and, by the latest Messinian, both the Baitoan and Transmarian Provinces ceased to exist. During the early Zanclean Pliocene, the faunal remnants of the Onslowian, Chipolan, and Agueguexquitean subprovinces of the Baitoan Province and the Transmarian Province combined to produce a new biogeographic region, the Caloosahatchian Province (named for the Caloosahatchee Formation of southern Florida (Petuch 1982, 2004) (Fig. 11). The combined remnants of the other six Baitoan subprovinces gave rise to the new Gatunian Province (named for the Gatun Formation of Panama (Petuch 1982, 2004) (Fig. 11). Six endemic Transmarian genera managed to survive the terminal Messinian extinction, persisting into the basal member of the Yorktown Formation (Sunken Meadow Member, Zanclean Pliocene). These included the muricids *Scalaspira*, *Lirosoma*, and *Chesatrophon*, the volutids *Megaptygma* and *Scaphellopsis*, and the fasciolariid *Mariafusus*. Occurring only in the Yorktownian Subprovince of the Caloosahatchian Province (Fig. 11), these Transmarian relicts became extinct before the mid-Piacenzian Pliocene.

Fig. 7 Transmarian endemic genera that became extinct during the time of the Veritas Dust Cloud (mid-Tortonian Miocene). (**a**) *Pseudaptyxis* (*P. sanctaemariae* Petuch (1988), length 26 mm); (**b**) *Turrifulgur* (*T. fusiforme* (Conrad 1840), length 69 mm); (**c**) *Conradconfusus* (*C. parilis* (Conrad 1832), length 106 mm); (**d**) *Sycopsis* (*S. tuberculatum* (Conrad 1840), length 58 mm); (**e**) *Mariasveltia* (*M. lunata* (Conrad 1830), length 14 mm; (**f**) *Marianarona* (*M. marylandica* Petuch (1988), length 12 mm); (**g**) *Poliniciella* (*P. marylandica* Petuch (1988), length 20 mm)

Fig. 8 Transmarian endemic turrid gastropod genera (Family Turridae) that became extinct during the time of the Veritas Dust Cloud. (**a**) *Mariadrillia* (*M. parvoidea* (Martin 1904), length 10 mm); (**b**) *Nodisurculina* (*N. engonata* (Conrad 1862), length 14 mm); (**c**) *Sediliopsis* (*S. incilifera* (Conrad 1834), length 16 mm); (**d**) *Chesaclava* (*C. dissimilis* (Conrad 1830), length 16 mm); (**e**) *Hemipleurotoma* (*H. communis* (Conrad 1830), length 23 mm); (**f**) *Chesasyrinx* (*C. rotifera* (Conrad 1830), length 23 mm; (**g**) *Mariaturricula* (*M. biscatenaria* (Conrad 1830), length 36 mm)

The Transmarian Province During the Middle Miocene Climatic Optimum

Following the climatic cooling caused by massive flood basalts in the early Burdigalian (Glikson 2005), marine climates warmed quickly during the late Burdigalian and Langhian allowing warm-temperate to subtropical conditions to exist as far north as present-day New Jersey (Petuch 2004: 90–91). This worldwide warm interval, the mid-Miocene Climatic Optimum (MMCO), persisted for almost 2 Ma (Oleinik et al. 2008). At this time, the Transmarian Province became established as a distinct biogeographic entity, evolving numerous endemic taxa and species complexes that persisted until the end of the Miocene. The largest macrogastropod fauna ever known (in the Transmarian Province—over 160 species in the Calvert Formation) developed during the time span of the MMCO.

During the MMCO, the Transmarian region was invaded by numerous Baitoan tropical groups and, over time, developed its own unique paratropical fauna. Some of the more important tropical migrants included the genera *Cerithiopsis* (Cerithiopsidae), *Seila* (Seilidae), *Amaea* and *Cirsotrema* (both Epitoniidae), *Niso* (Melanellidae), *Architectonica* (Architectonicidae), *Erato* (Eratoidae), *Ficus* (Ficidae), *Sassia* (Personiidae), *Metula* (Buccinidae), *Oliva* (Olividae), *Persicula* (Marginellidae), and *Glyphostoma* (Clathurellidae) (some illustrated in Fig. 4). None of these tropical taxa had been found as far north as this since the Langhian. The tropical migrants were sympatric with some of the first members of endemic Transmarian taxa such

Fig. 9 Eutropical genera found in the Transmarian Province during the LMCO (Messinian Miocene). (**a**) *Conus (Gradiconus)* (*C. (G.) spenceri* Ward (1992), length 33 mm); (**b**) *Oliva* (*O. idonea* Conrad (1839), length 25 mm); (**c**) *Petaloconchus (Anguinella)* (*P. (A.) virginica* (Conrad 1839), largest individual 22 mm, with 3 individuals attached to a fragment of the scallop *Chesapecten middlesexensis*); (**d**, **e**) *Crepidula (Bostrycapulus)* (*C. (B.) bonnieae* Petuch and Drolshagen (2009), length 23 mm); (**f**) *Serpulorbis* (*S. granifera* (Say 1824), length 37 mm, attached to the oyster *Mansfieldostrea geraldjohnsoni*)

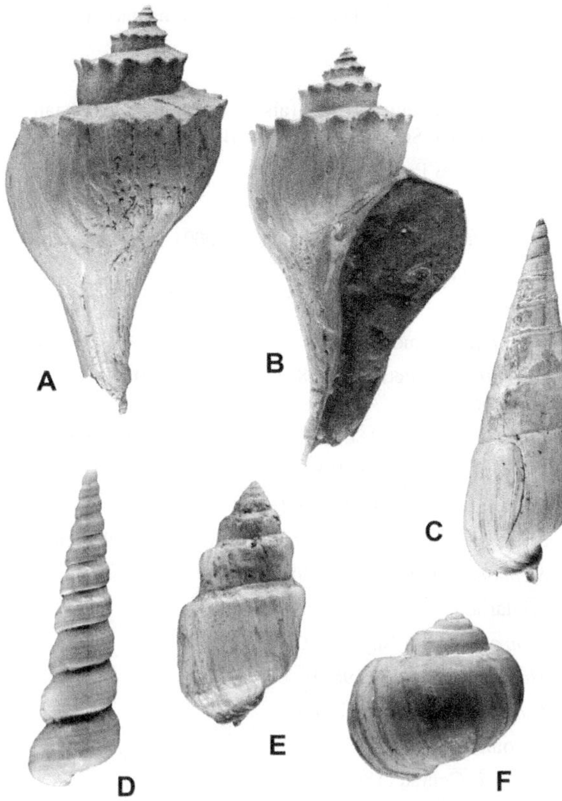

Fig. 10 Transmarian endemic genera that became extinct after the Kara-Kul Impact (latest Messinian Miocene). (**a**, **b**) *Coronafulgur* (*C. kendrewi* Petuch (2004), length 122 mm); (**c**) *Laevihastula* (*L.* cf. *simplex* (Conrad 1830), length 36 mm); (**d**) *Mariacolpus* (*M. plebeia* (Say 1824), length 51 mm); (**e**) *Bulliopsis* (*B. bowlerensis* Allmon (1990), length 24 mm); (**f**) *Euspirella* (*E. permeliae* Petuch and Drolshagen (2009), length 15 mm)

as *Chesathais* and *Trisecphora* (both Muricidae), *Conradconfusus* (Fasciolariidae), *Scaphellopsis* (Volutidae), and *Calverturris* and *Transmariaturris* (both Turridae). A complete list of Langhian tropical genera and their Transmarian offshoot species is given in Table 2.

Although containing a large number of tropically derived southern taxa, the Transmarian region during the MMCO lacked many of the key eutropical index taxa. Some of these missing groups included the Cerithiidae, Modulidae, Cypraeidae, Ovulidae, Strombidae, Tonnidae, Cassidae, Turbinellidae, and Conidae. Later, during the early Tortonian warm time (the ETCO, discussed later in this chapter), the families Tonnidae, Cassidae, and Conidae did manage to invade the Transmarian Province and evolved their own endemic species complexes. The conspicuous absence of these tropical taxa during the MMCO may have been more due to ecological competition with established Transmarian taxa than actual climatic controls. The Cassidae and Tonnidae would have been in direct competition with Calvert *Ficus* and *Sassia* species and the Conidae would have been in direct competition with an endemic fauna of large turrids such as *Transmariaturris*, *Mariaturricula*, and *Calverturris* (see Petuch and Drolshagen 2009).

Extinctions Correlated with the Ries, Steinheim, Puffin, and Ewing Impacts

In contrast to the warm Langhian Age, the subsequent Serravallian was characterized as a cold time with wildly fluctuating sea levels (Petuch 2004; Petuch

Table 1 Listing of classic endemic Transmarian mollusk genera

Turritellidae	
	Calvertitella Petuch (1988) (also found in the Onslowian and Chipolan Subprovinces of the Baitoan Province)
	Mariacolpus Petuch (1988)
Naticidae	
	Euspirella Petuch and Drolshagen (2009)
	Poliniciella Petuch (1988)
Muricidae	
	Chesathais Petuch (1988)
	Chesatrophon Petuch (1988) (relict in the earliest Pliocene)
	Lirosoma Conrad (1862) (relict in the earliest Pliocene)
	Mariasalpinx Petuch (1988)
	Patuxentrophon Petuch (1992)
	Scalaspira Conrad (1862) (relict in the earliest Pliocene)
	Stephanosalpinx Petuch (1988)
Fasciolariidae	
	Conradconfusus Snyder (2000)
	Mariafusus Petuch (1988)
	Pseudaptyxis Petuch (1988)
Busyconidae	
	Sycopsis Conrad (1867)
Buccinidae	
	Bulliopsis Conrad (1862)
Volutidae	
	Megaptygma Conrad (1863)
	Scaphellopsis Petuch (1988)
Cancellariidae	
	Cancellariella Martin (1904)
	Marianarona Petuch (1988)
	Mariasveltia Petuch (1988)
Terebridae	
	Laevihastula Petuch (1988)
Turridae	
	Calverturris Petuch (1992)
	Chesaclava Petuch (1988)
	Chesasyrinx Petuch (1988)
	Mariadrillia Petuch (1988)
	Mariaturricula Petuch (1988)
	Nodisurculina Petuch (1988)
	Sediliopsis Petuch (1988)
	Transmariaturris Petuch (1992)

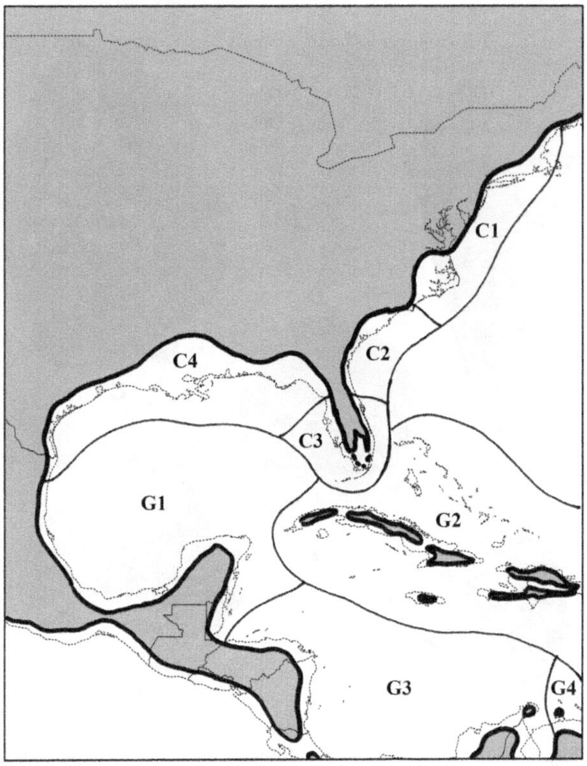

Fig. 11 Biogeographical map of the eastern coast of North America during the Pliocene, showing the limits of the Caloosahatchian (C) and Gatunian (G) Provinces. The Caloosahatchian subprovinces include C1 = Yorktownian Subprovince, C2 = Duplinian Subprovince, C3 = Buckinghamian Subprovince, C4 = Jacksonbluffian Subprovince. The Gatunian subprovinces include G1 = Veracruzan Subprovince, G2 = Guraban Subprovince, G3 = Limonian Subprovince, and G4 = Puntagavilanian Subprovince. The approximate configuration of the Pliocene coastline (*shaded area*) is shown superimposed upon the outline of Recent North America (*stippled line*) (taken from Petuch 2004)

and Drolshagen 2009). In total, the destructive effects of three distinct cooling episodes can be seen in the late Langhian-Serravallian fossil record of Maryland. The first of these, at around 14.5–14.8 Ma, represents the terminal Langhian event and the extinction and extirpation of a large part of the Calvert Formation fauna. Out of the 65 macrogastropod genera found in the Calvert Formation, 19 (29%) (mostly tropical taxa) disappeared from the Transmarian Province at this time. This major extinction correlates exactly with the Ries-Steinheim Binary Impact in Bavaria (Shoemaker and Chao 1961). The albedo effect from this double impact was apparently strong enough to plunge the northern hemisphere into a severe cold time that resulted in the extinction and extirpation of all tropical taxa.

Because of the lingering effects of the Ries-Steinheim Impact, the Serravallian Transmarian fauna was much less species-rich than that of the older

Table 2 Tropical genera in the Transmarian Province during the Middle Miocene climatic optimum (found in the plum point member of the Calvert formation)

Fissurellidae
Emarginula (*E. marylandica* Martin 1904)
Cerithiopsidae
Cerithiopsis (*C. calvertensis* Martin 1904)
Seilidae
Seila (*S. clavulus* (H.C. Lea 1843))
Fossaridae
Carinorbis (*C. dalli* (Whitfield 1894))
Epitoniidae
Amaea (*A. prunicola* (Martin 1904) and *A. reticulata* (Martin 1904))
Cirsotrema (*C. calvertensis* (Martin 1904))
Stenorhytis (*S. pachypleura* (Conrad 1841))
Melanellidae
Niso (*N. lineata* (Conrad 1841))
Architectonicidae
Architectonica (*A. trilineatum* (Conrad 1841))
Eratoidae
Erato (*E. perexigua* (Conrad 1841))
Xenophoridae
Xenophora (*X. lindae* Petuch and Drolshagen 2009)
Ficidae
Ficus (*F. harrisi* (Martin 1904))
Personiidae
Sassia (*S. centrosum* (Conrad 1863))
Muricidae
Murexiella (*M. shilohensis* (Heilprin 1887))
Buccinidae
Metula (*M. calvertana* (Martin 1904))
Olividae
Oliva (*O. harrisi* Martin 1904)
Marginellidae
Dentimargo (*D. calvertensis* (Martin 1904))
Persicula (*P. conulus* (H.C. Lea 1843)
Clathurellidae
Glyphostoma (*G. obtusa* (Martin 1904))
Turridae
Cymatosyrinx (*C.* sp.)

Langhian fauna (65 genera in the Calvert Formation and only 25 in the subsequent Choptank Formation). Typically, the Serravallian faunas were dominated by cold-tolerant endemic taxa such as *Mariacolpus* (Turritellidae), *Euspirella* (Naticidae), *Chesathais* and *Trisecphora* (both Muricidae), *Sycopsis* (Busyconidae), *Cancellariella* and *Marianarona* (both Cancellariidae), and *Calverturris*, *Mariaturricula*, and *Transmariaturris* (all Turridae).

The boreal genus *Euspira* (Naticidae) was abundant and dominant in the Choptank faunas, attesting to the colder marine climate of the early Serravallian. This mid-Miocene cooling was further exacerbated by a third major asteroid impact at 12 Ma; the Puffin Impact off northwestern Australia (Gorter and Bayford 2000). Being a continental shelf impact, a large amount of ejecta would have been thrown into the stratosphere, again increasing the earth's albedo.

A fourth asteroid impact occurred soon after the Puffin Impact, at 11.7 Ma, on the Pacific sea floor near the equator southeast of Hawaii. Referred to as the Ewing Impact, this was the largest of the four mid-Miocene impacts and produced the most widespread effects. As pointed out by Abbott et al. (2003) and Leung et al. (2004), silicon carbide spherules traced to the impact were found as far away as Arizona. This indicates that a tremendous amount of dust and debris was blasted into the stratosphere or higher, increasing the earth's albedo and lowering worldwide temperatures. Evidence of such a massive impact and the resultant climatic degeneration would readily explain the large extinction at the end of the Serravallian (the last of Choptank deposition).

The Transmarian Province During the Early Tortonian Climatic Optimum

By the Serravallian-Tortonian boundary, water temperatures warmed and a more stable marine climate was again established within the Transmarian Province. This warming initiated the early Tortonian Climatic Optimum (ETCO). Based on the high percentage of resident eutropical taxa, the early Tortonian and the ETCO had the warmest marine climate of the Transmarian time span. For the first time during the Miocene, high tropical taxa such as *Gradiconus* (Conidae), *Mitra (Nebularia)* (Mitridae), *Phalium* (Cassidae), *Eudolium* (Tonnidae), and *Hesperisternia* (Buccinidae) migrated northward from Florida and the Carolinas and invaded the Salisbury Sea (Fig. 6). Interestingly, the earlier Langhian eutropical taxa, which had invaded the Calvert Subsea during the MMCO (i.e. *Xenophora*, *Ficus*, *Seila*, *Architectonica*, *Amaea*, *Metula*, and *Glyphostoma*) did not reinvade the Maryland region along with the other tropical taxa. This exclusion may have been due to ecological competition with established Transmarian taxa rather than differences in marine climate.

Of the 20 eutropical gastropod genera that occurred in the Transmarian Province during the Langhian (listed under the MMCO section), only four reinvaded the Salisbury Sea during the ETCO. These included *Stenorhytis* (Epitoniidae), *Laevityphis* (Muricidae), *Dentimargo* (Marginellidae), and *Cymatosyrinx* (Turridae). The classic tropical genus *Gradiconus* underwent a species radiation during the early Tortonian, producing at least three species (*G. deluvianus* (Green 1840), *G. sanctaemariae* (Petuch 1988), and *G. asheri* (Petuch 1988); see Fig. 6. These three cone species were part of the second-richest macrogastropod fauna ever found in the Chesapeake Miocene (Windmill Point Member of the St. Mary's Formation, with 97 species (Petuch and Drolshagen 2009), attesting to the warm, stable marine climate that existed during the early Tortonian.

Of special interest during the ETCO was the appearance of the richest fauna of vermivorous turrid gastropods ever found in the Chesapeake Miocene (examples shown in Fig. 8). Several classic Langhian taxa, such as *Chesaclava*, *Hemipleurotoma*, *Mariaturricula*, and *Sediliopsis*, managed to survive the Serravallian extinctions and evolved new species complexes (Petuch and Drolshagen 2009). These Transmarian endemics occurred together with the tropical genus *Cymatosyrinx*, which had also invaded the region during the Langhian. Three new endemic genera also appeared at this time, including *Nodisurculina*, *Chesasyrinx*, and *Mariadrillia*, and these short-lived taxa existed only during the ETCO.

Extinctions Correlated with the Veritas Interplanetary Dust Cloud

At the end of the mid-Tortonian (approximately 8.3 Ma), the Transmarian ecosystems underwent abrupt collapses, resulting in the extinction of almost one-third of the macrogastropod fauna (26.6%; with 16 genera extinct out of a total of 60). This is readily seen by comparing the rich early Tortonian St. Mary's Formation fauna with that of the impoverished late Tortonian Claremont Manor Member fauna of the subsequent Eastover Formation (Petuch and Drolshagen 2009). Particularly hard-hit were the families Fasciolariidae (with the extinction of *Conradconfusus* and *Pseudaptyxis*; see Fig. 7), Busyconidae (with the extinction of *Sycopsis* and *Turrifulgur*; see Fig. 7), and Turridae (with the extinction of *Chesaclava*, *Chesasyrinx*, *Hemipleurotoma*, *Mariaturricula*, *Nodisurculina*, *Sediliopsis*, and *Mariadrillia*; see Fig. 8).

This major extinction and extirpation event, which essentially destroyed the Transmarian Province, correlates directly with a newly discovered catastrophic drop in solar intensity caused by an interplanetary dust cloud. Based on meteoric dust samples, Farley et al. (2006) demonstrated that two or more members of the Veritas Family of asteroids (in the main asteroid belt) collided and disintegrated into an immense dust cloud at 8.3 Ma. This thick dust cloud settled into an earth-crossing orbit, obscuring the sun, and greatly diminishing the amount of incident solar radiation. The lowered insolation resulted in severe cooling events worldwide.

The Veritas Dust Cloud solar obscuration led to an almost instantaneous cool time, plunging the warm temperate-to-subtropical Transmarian Province into cold temperate conditions. The complete collapse of the rich early Tortonian St. Mary's fauna, with the abrupt disappearance of tropical groups such as *Phalium*, *Eudolium*, *Laevityphis*, *Hesperisternia*, *Mitra*, and *Ventrilia*, points to a rapid-onset cooling event. The quickness of the cooling event did not allow enough time for these temperature-sensitive groups to migrate southward. The tropically derived taxa, and stenothermal Transmarian endemics such as *Conradconfusus*, *Pseudaptyxis*, and *Sycopsis*, were trapped in a marine climate that was cooling faster than they could adapt, resulting in their extinction.

The Transmarian Province During the Late Miocene Climatic Optimum

The eventual dispersal of the Veritas Dust Cloud allowed marine climates to return to warmer conditions during the Messinian Age (the Late Miocene Climatic Optimum; referred to here as the LMCO). For nearly 2 Ma, the Transmarian region contained warm temperate-to-subtropical conditions and several tropical taxa reinvaded the province from the south. Principal among these were the conid *Gradiconus* and the olivid *Oliva* (Fig. 9), the latter having reappeared after an absence during the Serravallian and Tortonian Ages. By this time, however, the spatial configuration of the Transmarian region had changed completely, with the sedimentary infilling of the Salisbury Sea basin and the partial infilling of the remaining Albemarle Sea (Petuch 2004: 123–127; Petuch Drolshagen 2009: 11–15) (Fig. 3). The reduced number of biotopes and habitats resulted in a relatively impoverished molluscan fauna, even though the LMCO marine climate was one of the warmest of the entire Miocene.

The Messinian Transmarian fauna, exemplified by the mollusks of the Cobham Bay Member of the Eastover Formation (Ward and Blackwelder 1980; Ward 1992; Petuch and Drolshagen 2009), contained a number of newly evolved gastropod groups that have persisted into the Recent in tropical and subtropical areas. Appearing for the first time in the fossil record was the spiny crepidulid subgenus *Bostrycapulus* (Fig. 9), a group that is abundant in the Recent Caribbean Province. Representatives of the high tropical groups *Gradiconus*, *Oliva*, and *Crepidula (Bostrycapulus)* occurred together with surviving remnants of the rich early Tortonian St. Mary's fauna such as *Coronafulgur* (Busyconidae), *Bulliopsis* (Buccinidae), *Mariacolpus* (Turritellidae), *Laevihastula* (Terebridae), and *Euspirella* (Naticidae) (Fig. 10). Although constrained by the limited number of biotopes (mostly sand and mud bottoms), the Messinian Transmarian neritic gastropod fauna was vibrant, as evidenced by the evolution of new species of *Torcula* (Turritellidae), *Volutifusus* (Volutidae), *Ecphora* and *Urosalpinx* (both Muricidae), *Ilyanassa* (Nassariidae), and a species radiation of the naticid *Neverita* (Petuch and Drolshagen 2009: 101–113). By the late Messinian, the reconfigured Transmarian Province was rebounding and evolving a new fauna for the third time since the Langhian.

Extinctions Correlated with the Kara-Kul Impact

At the end of the Miocene, one of the most significant marine fauna extinctions of the entire Neogene occurred worldwide (Petuch 1993). During this time, the northern hemisphere quickly cooled and ice sheets began to expand, resulting in a major glacioeustatic low. This is readily seen in the large unconformity between the Miocene Eastover Formation and the Pliocene Yorktown Formation (Petuch and Drolshagen 2009: 22–23). Worldwide current patterns were also disrupted, as exemplified by the changes in the Pacific North Equatorial Countercurrent detected in the latest Miocene of Panama (Collins et al. 1996). Along eastern North America, this extreme eustatic low, coupled with severely lowered water temperatures, quickly annihilated the newly rebounded Transmarian faunas, resulting in the "Transmarian Extinction" (Petuch 2004: 261–262). At this time, classic Transmarian genera such as *Mariacolpus*, *Euspirella*, *Mariasalpinx*, *Coronafulgur*, *Bulliopsis*, and *Laevihastula* (Fig. 10) became extinct, the Transmarian ecosystems collapsed, and the province, as a biogeographical entity, ceased to exist.

The late Messinian Transmarian Extinction, at 5 Ma, correlates exactly with the formation of the large Kara-Kul impact structure in the Pamir Mountains of Tajikistan (Gurov et al. 1993; Gurov and Gurova 1993). Being a central continental impact, Kara-Kul would have produced immense amounts of ejecta and aerosols which, in turn, would have greatly increased the earth's albedo. The resultant lowered temperatures were severe enough to have cooled the northwestern Atlantic to nearly boreal conditions and would have dropped sea level by at least 100 m. This would have been sufficient to decimate the stenothermal, shallow neritic Transmarian ecosystems. Chronostratigraphically, the effects of the Kara-Kul Impact were sufficient to have caused both the extinction of many marine faunas around the world and the alteration of sedimentation patterns. This chronologic marker was recognized by early workers as the Miocene-Pliocene boundary.

Summary and Conclusions

In light of new data on early Neogene asteroid impacts and interplanetary dust clouds, it is here conjectured that the patterns of speciation and extinction seen in the eastern North American Transmarian Province may have been governed by extraterrestrial events. Improved dating of many large Neogene impact structures around the world has now shown that some impacts were exactly correlative with major extinction events within the Transmarian region. This implies that these impacts were severe enough to have produced worldwide climatic degeneration. The Transmarian Province, being a paratropical warm temperate region, would have been perfectly situated latitudinally to record and preserve any evidence of these climatic fluctuations.

The Transmarian Province underwent three main pulses of speciation; one during the late Burdigalian-Langhian, one during the early Tortonian, and one during the Messinian. These speciation events correspond to three major Miocene warm times; the Middle Miocene Climatic Optimum (MMCO), the early Tortonian Climatic Optimum (ETCO), and the Late Miocene Climatic Optimum (LMCO). During these times, eutropical molluscan taxa such as *Architectonica*, *Gradiconus*, *Oliva*, *Phalium*, *Xenophora*, *Mitra*, and *Persicula* migrated northward and invaded the Transmarian region. Here, they evolved their own endemic species complexes and were sympatric with species radiations of temperate-water Transmarian endemic genera such as *Chesathais*, *Ecphorosycon*, *Conradconfusus*, *Pseudaptyxis*, *Sycopsis*, *Coronafulgur*, *Bulliopsis*, *Calverturris*, and *Transmariaturris*. In the Chesapeake Group of formations, the MMCO corresponds to the deposition of the Calvert Formation, the ETCO corresponds to the deposition of the rich fossil beds of the St. Mary's Formation, and the LMCO corresponds to the deposition of the upper beds of the Eastover Formation. A list of all Transmarian macrogastropod genera, and their last appearances in the fossil record, is given in Table 3.

Separating the three warm intervals were three cold times: one during the Serravallian, one during the mid-to-late Tortonian, and one at the end of the Messinian. Within the stratigraphic record of the Chesapeake Group, these cold times are easily seen as sets of unconformities between formations and as extinctions of large percentages of the molluscan faunas occurring in the four main Miocene formations. The most complex cooling event occurred during deposition of the Choptank Formation in the Serravallian, when sequential extinction events effectively split the Transmarian Province into two faunal groupings. These included a Langhian-Serravallian temporal component dominated by genera such as *Trisecphora*, *Chesathais*, and *Ecphorosycon*, and multiple species offshoots of *Ecphora* (all Muricidae), and a Tortonian-Messinian temporal

Table 3 Last appearances of Transmarian macrogastropod genera in the fossil record

Fissurellidae
Emarginula R
Glyphis R
Trochidae
Calliostoma R
Leiotrochus Y
Turritellidae
Calvertitella CA
Mariacolpus E
Torcula R
Vermetidae
Petaloconchus (Anguinella) E
Serpulorbis R
Cerithiopsidae
Cerithiopsis R
Seilidae
Seila R
Fossaridae
Carinorbis Y
Crepidulidae
Calyptraea R
Crepidula R
Crepidula (Bostrycapulus) R
Crucibulum R
Crucibulum (Dispotaea) R
Epitoniidae
Amaea R
Cirsotrema R
Epitonium R
Stenorhytis R
Melanellidae
Niso R
Architectonicidae
Architectonica R
Eratoidae
Erato R
Xenophoridae
Xenophora R
Naticidae
Euspira R
Euspirella E
Neverita R
Polinicíella SM
Sinum R

Table 3 (continued)

Tonnidae
Eudolium R
Cassidae
Phalium (Mariacassis) SM (subgenus only)
Ficidae
Ficus R
Personiidae
Sassia R
Muricidae
Chesathais CH
Chesatrophon Y
Ecphora Y
Ecphora (Planecphora) Y
Ecphorosycon CH
Laevityphis R
Lirosoma Y
Mariasalpinx E
Murexiella R
Patuxentrophon CH
Scalaspira Y
Stephanosalpinx CH
Trisecphora CH
Urosalpinx R
Fasciolariidae
Conradconfusus SM
Mariafusus Y
Pseudaptyxis SM
Busyconidae
Busycotypus R
Coronafulgur E
Sycopsis SM
Turrifulgur SM
Buccinidae
Bulliopsis E
Celatoconus Y
Hesperisternia R
Metula R
Ptychosalpinx R
Columbellidae
Mitrella R
Nassariidae
Ilyanassa R
Volutidae
Megaptygma Y
Scaphellopsis Y
Volutifusus R

Table 3 (continued)

Mitridae
Mitra (Nebularia) R
Marginellidae
Dentimargo R
Persicula R
Olividae
Oliva (Strephona) R
Cancellariidae
Admete R
Cancellariella CH
Marianarona SM
Mariasveltia SM
Ventrilia R
Conidae
Conus (Gradiconus) R
Terebridae
Laevihastula E
Myurellina R
Strioterebrum R
Clathurellidae
Glyphostoma R
Turridae
Calverturris CH
Chesaclava SM
Chesasyrinx SM
Cymatosyrinx R
Hemipleurotoma SM
Mariadrillia SM
Mariadrillia SM
Mariaturricula SM
Nodisurculina SM
Oenopota R
Sediliopsis SM
Transmariaturris CH
Acteonidae
Acteon R

CA at end of Calvert Formation, *CH* at end of Choptank Formation, *SM* at end of St. Mary's Formation, *E* at end of Eastover Formation, *Y* survived as relicts in Pliocene Yorktown Formation only, *R* extant in Recent

component dominated by genera such as *Gradiconus*, *Coronafulgur*, *Urosalpinx*, and *Mariasalpinx*. Fully one-third of the older Transmarian genera became extinct, sequentially, during the deposition of the three members of the Serravallian Choptank Formation.

The Serravallian cooling event and accompanying extinctions, essentially the paradigm shift in the composition of Transmarian molluscan assemblages, now appear to have resulted from four separate asteroid impacts in quick succession. The first, the Ries and Steinheim Impacts occurred simultaneously at 14.5 Ma and produced a double crater feature in Bavaria. This first Serravallian catastrophic event, causing global cooling, destroyed the late Langhian Transmarian molluscan communities, leading to the extinction of stratigraphically important genera such as *Calvertitella* and the extirpation of tropical groups such as *Xenophora, Amaea, Architectonica, Murexiella, Ficus, Metula*, and *Persicula*. Later in the Serravallian, two more asteroid impacts, the Puffin Impact (at 12 Ma) and the Ewing Impact (at 11.7 Ma), sustained the global cooling produced by the Ries-Steinheim Binary Impact. This led to the faunal impoverishment seen in the Choptank Formation and the sequential extinctions of the Serravallian index genera.

The late Tortonian cooling event cannot be correlated with any known asteroid impact, but does correlate directly with the creation of the immense interplanetary Veritas Dust Cloud. Formed by the collision of asteroids of the Veritas Family in the main asteroid belt and dated at 8.3 Ma, this dust cloud produced a rapid-onset global cooling event. The rich paratropical faunas of the St. Mary's Formation were immediately affected by the Veritas solar obscuration, rapidly succumbing to the marine climatic degeneration. Such classic late Transmarian genera as *Poliniciella, Turrifulgur, Sycopsis, Conradconfusus, Pseudaptyxis, Mariadrillia, Nodisurculina*, and *Chesasyrinx* disappeared abruptly from the fossil record during this time. Other extant tropical groups such as *Stenorhytis, Phalium, Eudolium, Mitra (Nebularia)*, and *Dentimargo* also disappeared from the fossil record of Maryland, never again being found that far north.

The post-Veritas marine climate warmed appreciably and the Messinian Transmarian region was again invaded by tropical faunas from the south. Some of these tropical genera included *Gradiconus* and *Oliva*, groups that do not range farther north than Cape Hatteras in the Recent. Although much impoverished by the Veritas extinction, some remnants of the Transmarian faunas, such as the endemic genera *Mariacolpus, Euspirella, Coronafulgur, Bulliopsis*, and *Laevihastula*, did manage to survive into the late Messinian (Cobham Bay Member of the Eastover Formation) and preserved the faunal integrity of the Transmarian Province. This final pulse of Transmarian evolution was quickly destroyed by the global cooling event caused by the Kara-Kul Impact at 5 Ma, the terminal event that ended the Miocene. The Transmarian Province, as a whole, can now be seen to have been temporally confined to the Miocene, with its oldest elements appearing in the Aquitanian (Old Church Formation) and its last remnants disappearing at the Messinian-Zanclean (Pliocene) boundary (Eastover Formation).

If the closely spaced Serravallian asteroid impacts and extinctions are considered to be a single, rapidly sequential event, then the obliteration of the Transmarian Province can be seen to have been essentially two-staged. As pointed out in an earlier work (Petuch 1995), the collective Serravallian extinction produced a "wounding factor," where a large number of endemic Transmarian genera disappeared and the local ecosystems were impoverished to the verge of collapse. This was exacerbated by the Veritas Dust Cloud interval, which further weakened the Transmarian marine communities. As evidenced by the first appearance of newly evolved molluscan taxa such as the gastropod *Crepidula (Bostrycapulus)* and the bivalves *Carolinapecten, Conradostrea*, and *Mansfieldostrea*, the Messinian Transmarian Province was rapidly rebounding, producing new endemic ecosystems. This quickly ended with the coup de grace event, in the form of the terminal Miocene Kara-Kul Impact. The Transmarian faunas simply did not have enough time during the Messinian to evolve ecosystems strong and complex enough to survive the Kara-Kul coup de grace. The severe global cooling, coupled with sedimentary infilling and spatial loss in the neritic zone, led to the demise of an entire faunal province.

During the post-catastrophic Zanclean Pliocene, the Virginia and northern North Carolina coasts represented an area of geographical heterochrony (Petuch 1982), where relictual taxa from an ancestral province persisted into a later time and became components of a new province. Here, and only in the lower members of the Yorktown Formation, classic Transmarian genera such as *Scalaspira, Lirosoma, Chesatrophon, Mariafusus, Megaptygma*, and *Scaphellopsis* were sympatric with newly evolved endemic species complexes of the Caloosahatchian Province. By the early Piacenzian Pliocene, however, these regionally confined (Secondary Relict Pocket) Transmarian relicts were extinct. A few typical Transmarian index genera such as *Busycotypus*, *Ptychosalpinx*, and *Volutifusus* did manage to survive the terminal Miocene extinction and have persisted on into the Recent. These Transmarian derivatives were prominent components of the subsequent Plio-Pleistocene Caloosahatchian Province and later the Pleistocene and Recent Carolinian Province (Petuch 2004).

Acknowledgments It is a pleasure to thank Anton E. Oleinik of Florida Atlantic University for reviewing the manuscript, helping lay out the plates and for drafting Figs. 1 and 2. Mardie Drolshagen of Black Diamond Studios, Delray Beach, Florida, took excellent digital photos of specimens and helped draft the biogeographic maps. Thomas Darragh of Museum Victoria gave a helpful review of the manuscript. John Talent encouraged me to submit this manuscript for the International Year of Planet Earth volume.

References

Abbott DH, Nunes AA, Leung IS, Hagstrum JT (2003) The Ewing impact structure progress report. Abstract #P52A-0472, Fall Meeting 2003. American Geophysical Union

Briggs JC (1974) Marine zoogeography. McGraw-Hill, New York, 270p

Briggs JC (1995) Global biogeography. Elsevier, Amsterdam, 452p

Collins LS, Coates AG, Berggren WA, Aubry M-P, Zhang J (1996) The late Miocene Panama isthmian strait. Geology 24:687–690

Farley KA, Vokrouhlicky D, Bottke W, Nesvorny D (2006) A late Miocene dust shower from the break-up of an asteroid in the main belt. Nature 439:295–297

Glikson A (2005) Asteroid/comet impact clusters, flood basalts and mass extinctions: significance of isotopic age overlaps. Earth Planet Sci Lett 236:933–937

Gorter JD, Bayford SW (2000) Possible impact origin for the middle Miocene (Serravallian) Puffin Structure, Ashmore Platform, northwest Australia. Aust J Earth Sci 47:707–714

Gurov EP, Gurova HP (1993) The Kara-Kul Depression in the Pamirs – a reliable impact structure. Geol Zh 6:53–64

Gurov EP, Gurova HP, Rakitskaya RB, Yamnichenko AY (1993) The Karakul Depression in Pamirs – the first impact structure in central Asia (abstract). Lunar Planet Sci XXIV:591–592

Leung IS, Winston R, Abbott DH (2004) Silicon carbide from the Canyon Diablo meteorite and the Ewing Impact Structure. Abstract #P33C-10, Spring Meeting 2004, American Geophysical Union

Oleinik AE, Marinkovich L, Barinov KB, Swart PK (2008) Magnitude of middle Miocene warming in north Pacific high latitudes: stable isotope evidence from *Kaneharaia* (Bivalvia, Dosiniinae). Bull Geol Surv Jpn 59:339–353

Petuch EJ (1982) Geographical heterochrony: contemporaneous co-existence of Neogene and Recent molluscan faunas in the Americas. Palaeogeogr Palaeoclimatol Palaeoecol 37:277–312

Petuch EJ (1988) Neogene history of tropical American Mollusks: biogeography and evolutionary patterns of tropical Western Atlantic Mollusca. The Coastal Education and Research Foundation, Charlottesville, Virginia, 217p

Petuch EJ (1993) Patterns of diversity and extinction in Transmarian Muricacean, Buccinacean, and Conacean Gastropods. The Nautilus 106(4):155–173

Petuch EJ (1995) Molluscan diversity in the Late Neogene of Florida: evidence for a two-staged mass extinction. Science 270:275–277

Petuch EJ (2004) Cenozoic Seas: the View from Eastern North America. CRC Press, Boca Raton, London, New York, 308p

Petuch EJ, Drolshagen M (2009) Molluscan Paleontology of the Chesapeake Miocene. CRC Press, Boca Raton, London, New York, 152p

Shoemaker EM, Chao ECT (1961) New Evidence for the impact origin of the Ries Basin, Bavaria, Germany. J Geophys Res 66:3371–3378

Valentine JW (1973) Evolutionary paleoecology of the marine biosphere. Prentice-Hall, Englewood, NJ, 511p

Ward LW (1992) Molluscan biostratigraphy of the Miocene, Middle Atlantic Coastal Plain of North America. Virginia Mus Nat Hist Mem 2:159

Ward LW, Blackwelder BW (1980) Stratigraphic revision of Upper Miocene and Lower Pliocene Beds of the Chesapeake Group, Middle Atlantic Coastal Plain. US Geol Surv Bull 1482-D:61

The Rise of Australian Marsupials: A Synopsis of Biostratigraphic, Phylogenetic, Palaeoecologic and Palaeobiogeographic Understanding

Karen H. Black, Michael Archer, Suzanne J. Hand, and Henk Godthelp

Abstract

The origins, evolution and palaeodiversity of Australia's unique marsupial fauna are reviewed. Australia's marsupial fauna is both taxonomically and ecologically diverse comprising four extant orders (Dasyuromorphia, Peramelemorphia, Notoryctemorphia and Diprotodontia) and one extinct order (Yalkaparidontia). Molecular divergence dates estimate a Palaeocene origin for the Australian marsupial orders yet ordinal differentiation is obscured by significant gaps in the fossil record with a single terrestrial mammal-bearing deposit known between the late Cretaceous and the late Oligocene. This deposit, the 55 million-year-old early Eocene Tingamarra Local Fauna of southeastern Queensland, contains Australia's oldest marsupial (Superorder Australidelphia) as well as taxa tentatively interpreted to represent South American groups (Order Polydolopimorphia). Palaeobiogeographic hypotheses regarding the distribution and interordinal relationships of Australian and South American marsupials are discussed. Dasyuromorphia and Peramelemorphia were possibly also present in the early Eocene, Diprotodontia in at least the late Oligocene and Notoryctemorphia and Yalkaparidontia in the early Miocene. Palaeobiodiversity was highest during the early to middle Miocene as evidenced by a spectacular array of marsupial groups in the rainforest assemblages of the Riversleigh World Heritage Area. The onset of icehouse conditions during the middle Miocene saw significant faunal turnover with loss of many archaic groups and the emergence of a range of modern lineages. Few deposits of late Miocene age are known. Development of Australia's first grasslands and arid habitats occurred in the Pliocene, accompanied by an explosive radiation of grazing kangaroos. The Pleistocene was characterised by severe and unpredictable climatic conditions and the extinction of the Australian megafauna. Lowered sea levels allowed faunal interchange between mainland Australia and neighbouring New Guinea as well as the arrival of the

K.H. Black (✉)
Evolution of Earth and Life Sciences Research Group, School of Biological, Earth and Environmental Sciences, University of New South Wales, Sydney, NSW 2052, Australia
e-mail: k.black@unsw.edu.au

J.A. Talent (ed.), *Earth and Life*, International Year of Planet Earth,
DOI 10.1007/978-90-481-3428-1_35,

first humans. Resolution of the role of humans and/or climate change in megafaunal extinction requires more precise dating of late Pleistocene deposits. We reflect on the predictive power of the fossil record to enhance understanding of the effects of climate change and humans on the future of the Australian marsupial fauna.

Keywords

Australia • Endemic marsupials • Metatherians • Phylogenetic affinities • Molecular divergence dating • 'Icehouse' (late Oligocene) • 'Greenhouse' (early Miocene) • Riversleigh faunas • Megafaunal extinction

Introduction

Australian marsupials are the product of a 125 million-year-long evolutionary marathon that began in China before sledging on through Europe, North America, South America, Antarctica and, by at least 55 million years ago (Ma), finally Australia – the last piece of land on the then remnant of Gondwana. Australia subsequently drifted free from Antarctica by at least 35 Ma to become an island laboratory where some of the world's most unusual marsupials evolved. Throughout their globe-spanning journey, these marsupials survived brutal bouts of climate change both hot and cold, global catastrophes in the form of giant meteorite impacts and super volcanoes, life and death struggles with hostile competitors including dinosaurs, predatory birds and placental mammals, and finally the many challenges that arrived with humans. Australia's marsupials are the seasoned survivors of the worst challenges Earth could throw at any group of mammals – and they survived and thrived while many other groups that began the same race fell out along the way.

To begin, we introduce the major groups of Australian participants, living and extinct. The modern Australasian marsupial fauna comprises four orders (Aplin and Archer 1987): Dasyuromorphia, Peramelemorphia, Diprotodontia and Notoryctemorphia. The carnivorous/insectivorous dasyuromorphians (Fig. 1) have long been regarded as ecological correlates of placental insectivores, cats, badgers, wolves and anteaters. Three families (~65 spp) are recognised: Dasyuridae (marsupial insectivores, badgers and cats), Thylacinidae (†, marsupial wolves) and Myrmecobiidae (the numbat; a marsupial anteater). The omnivorous peramelemorphians (Fig. 2) are placed in three families (~20 spp): Thylacomyidae (bilbies); Peramelidae (ordinary bandicoots); and Peroryctidae (forest bandicoots). Extinct peramelemorphian families include yaralids (very plesiomorphic bandicoots) and possibly the bizarre numbigilgids (marsupial plagiomenids), although the latter is more commonly regarded to be Marsupialia *incertae sedis* (Beck et al. 2008a; Section 'The Pliocene Marsupial Fauna'). The order Diprotodontia (Fig. 3; koalas, wombats, possums and kangaroos) is the most diverse, ecologically as well as taxonomically, of all marsupial orders. It contains 21 families placed in two suborders (Aplin and Archer 1987): Vombatiformes (koalas and wombats) and Phalangerida (possums and kangaroos). Extinct diprotodontian families include the following: Thylacoleonidae (marsupial lions); Diprotodontidae (marsupial sheep, hippos and rhinos), Palorchestidae (marsupial tapirs); Ilariidae (marsupial anthracotheres); Wynyardiidae (no analogue among placentals); Maradidae (overall form uncertain); Ektopodontidae (marsupial rodents); Miralinidae (pugilistic possums); Pilkipildridae (marsupial carpolestid primates); and Balbaridae (archaic galloping kangaroos). Although mostly herbivores (including specialist browsers, grazers, folivores, frugivores, nectarivores and graminivores), diprotodontians also include large carnivores (e.g. marsupial lions and some propleopine hypsiprymnodontid kangaroos) and omnivore/insectivores (e.g. burramyids, acrobatids and petaurids). In contrast, notoryctemorphians (marsupial moles) contain only a single family, Notoryctidae, with two extant species, *Notoryctes typhlops* (Fig. 4) and *N. caurinus*. Representatives of all extant orders are recorded in the fossil record from the late Oligocene except notoryctemorphians, which are so far only known from the early Miocene to Holocene, and peramelemorphians and dasyuromorphians, which are possibly represented as far back as the early Eocene (Archer et al. 1999; Long et al. 2002). However, the

Fig. 1 Representatives of extant (and recently extinct †) dasyuromorphian families (Table 1): (**a**, **b**) Dasyuridae; (**c**) Thylacinidae; (**d**) Myrmecobiidae. (**a**) Kimberley Planigale *Planigale subtilissima* (M. Archer); (**b**) Tiger Quoll *Dasyurus maculatus* (G. Settle); (**c**) Thylacine *Thylacinus cynocephalus*† (Tasmanian Museum and Art Gallery); and (**d**) Numbat *Myrmecobius fasciatus* (D. Whitford). Most of Australia's carnivores (e.g., **b**–**d**) are now either extinct or endangered

Eocene dasyuromorphian-like taxon (*Djarthia murgonensis*) exhibits no undoubted synapomorphies of that group (Beck et al. 2008b) and there is also some doubt about the peramelemorphian nature of the teeth that may represent this group. A fifth order, Yalkaparidontia, was established by Archer et al. (1988) and contains the single extinct family Yalkaparidontidae (marsupial woodpeckers; Beck 2009) represented by two species unique to the early to middle Miocene deposits of Riversleigh, northwestern Queensland. A sixth order, Microbiotheria, which is otherwise only known from the Cenozoic records of Antarctica and South America, with one living representative in southern Argentina and Chile, may be represented by undescribed taxa in an early Eocene deposit near Murgon in southeastern Queensland (Archer et al. 1999; McGowran et al. 2000).

Australia's distinctive and endemic marsupial fauna is the result of a long period of relative isolation of the Australian continent since its separation from Antarctica about 35 million years ago. The modern fauna has evolved to occupy a wide variety of ecological niches in habitats ranging from wet tropical rainforests to deserts. Understanding the history of this radiation has been hindered by significant gaps in the fossil record between the late Cretaceous and early

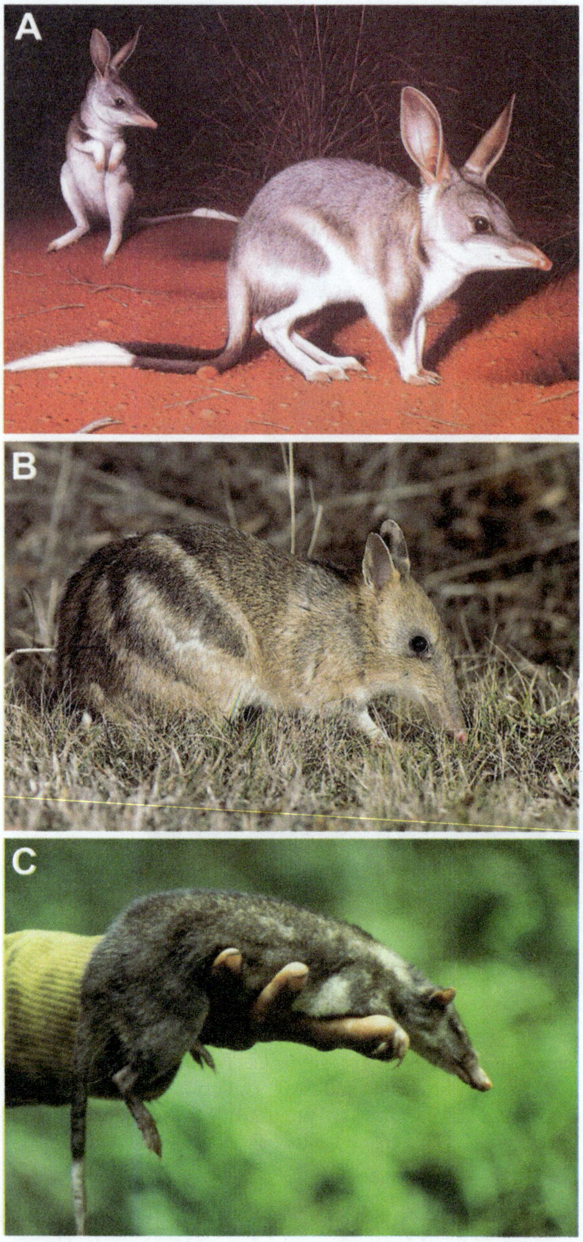

Fig. 2 Representative extant members of peramelemorphian families (Table 1). (**a**) Thylacomyidae, Greater Bilby *Macrotis lagotis* (R. Scott); (**b**) Peramelidae, Eastern Barred Bandicoot *Perameles gunni* (Wikimedia Commons, Noodle Snacks); and (**c**) Peroryctidae *Peroryctes longicaudata* (T. Flannery). These families proliferated only after rainfall in Australia began to steeply decline in the late Miocene

Eocene when marsupials first entered Australia, and between the early Eocene and late Oligocene when all modern Australasian marsupial orders appear to have differentiated. What follows is a synopsis of our current understanding of the origins, biogeography and palaeodiversity of the Australian marsupial radiation. Key events in the history of the Australian marsupial radiation are summarised in Fig. 5.

Throughout this work, higher-level systematic nomenclature (Table 1) follows Aplin and Archer (1987) except that the superfamily Macropodoidea is divided here into four rather than two families following Kear and Cooke (2001): Macropodidae; Potoroidae; Balbaridae (extinct); and Hypsiprymnodontidae. Further, an additional vombatomorphian family has been recognised, Maradidae, following Black (2007), and the superfamily Polydolopoidea is given ordinal status (Polydolopomorphia) following Kirsch et al. (1997) and Goin et al. (2003). In general, when we refer to 'Australian marsupials', we are including marsupials known from Irian Jaya and Papua New Guinea as well as surrounding islands.

Common abbreviations: Ma = million years; Fm = Formation; LF = Local Fauna (i.e., a faunal assemblage from a specific locality or series of geographically and stratigraphically related localities that have been interpreted to represent the same palaeocommunity); N.S.W., New South Wales; N.T., Northern Territory; Qld, Queensland; S.A., South Australia; Tas., Tasmania; Vic., Victoria; W.A., Western Australia.

Origins, Relationships and Biogeography

Metatherian Origins and Dispersal

Metatherians (crown group Marsupialia plus their stem relatives; Luo et al. 2003; Rougier et al. 1998) evidently originated in Laurasia sometime prior to 125 Ma, the age of the oldest metatherian, the mouse-sized *Sinodelphys szalayi* from the Yixian Fm of Liaoning Province, China (Luo et al. 2003). *Holoclemensia texana*, from the 115 Ma Butler Farm locality, Texas, is the sister taxon to *S. szalayi* (Luo et al. 2007) making it one of the oldest and most plesiomorphic metatherians from North America. The dispersal route of metatherians from Asia to North America either involved eastward dispersal via Beringia (Bering

Fig. 3 Representatives of extant members of diprotodontian families (Table 1). (**a**) Phascolarctidae, *Phascolarctos cinereus* (Wikimedia Commons, Sanjay ach); (**b**) Vombatidae, *Vombatus ursinus* (Wikimedia Commons, Materialscientist); (**c**) Pseudocheiridae, *Pseudocheirus archeri* (M. Archer); (**d**) Macropodidae, *Macropus parryi* (M. Archer); (**e**) Phalangeridae, *Spilocuscus maculatus* (Wikimedia Commons, Aviceda); (**f**) Acrobatidae, *Acrobates pygmaeus* (D. Andrew); and (**g**) Burramyidae, *Cercartetus nanus* (D. Whitford). All of these families have representatives extending back to late Oligocene and some (e.g., macropodids and pseudocheirids) have massively increased in diversity since the late Miocene

land bridge) or potentially, following discovery of a plesiomorphic marsupial (*Arcantiodelphys marchandi*) from Cenomanian (99.6–93.5 Ma) deposits in southwestern France (Vullo et al. 2009), westward dispersal via Europe and Greenland.

The timing as well as number of metatherian dispersal events from North America to South America is controversial with suggestions spanning the late Cretaceous to early Paleocene, a window of ~24 Ma (Beck 2008b; Fig. 6). The 75 Ma (Campanian) Los Alamitos fauna of South America is devoid of metatherians as well as eutherians (Bonaparte 1990) suggesting dispersal occurred after this time. The oldest undoubted South American metatherians come from the early Paleocene (64.5–63.0 Ma) Tiupampa Fauna of Bolivia with at least 11 species representing five families and four marsupial orders (Wroe and Archer 2006; de Muizon 1998). The relatively high diversity of Tiupampian metatherians is regarded by some authors (e.g. de Muizon and Cifelli 2001) as support for a late Cretaceous (Maastrichtian; 70.6–65.5 Ma) dispersal event.

Fig. 4 A modern representative of the order Notoryctemorphia containing the sole family Notoryctidae, the Southern Marsupial Mole *Notoryctes typhlops* (M. Gillam). While today only known from deserts in central Australia, their ancestors appear to have first evolved in lush rainforests

Extant South American marsupials comprise three orders and three families (Aplin and Archer 1987): Didelphimorphia (didelphidae and caluromyids; opposums such as *Marmosa*); Paucituberculata (caenolestids: rat opossums such as *Caenolestes*); and Microbiotheria (microbiotheriids; microbiotheres such as *Dromiciops*; Fig. 7). The fossil record is far more diverse and includes 22 families (Rose 2006) and one or two additional orders, the Sparassodonta (borhyaenoids) and Polydolopimorphia (Kirsch et al. 1997; Goin et al. 2003), although the ordinal status and monophyly of the latter are controversial (Beck et al. 2008a). Aplin and Archer (1987), for example, placed the superfamily Polydolopoidea within Paucituberculata.

Antarctic marsupials known from the middle Eocene (~50 Ma) La Meseta Fm of Seymour Island include didelphimorphians, polydolopids and microbiotheres (Reguero et al. 2002). On balance, these middle Eocene Antarctic marsupials are distinctly South American in affinity although the presence of at least two different microbiotheres also suggests palaeobiogeographic affinity to Australian marsupials such as the early Eocene *Djarthia murgonensis* (see below).

Extant Australian marsupials have been placed, as summarised above, in four orders: Dasyuromorphia, Peramelemorphia, Notoryctemorphia and Diprotodontia. The first known occurrence of marsupials in Australia is as part of the earliest Eocene Tingamarra LF (Fig. 8) from southeastern Qld (Godthelp et al. 1992). This assemblage includes a variety of small insectivorous to omnivorous marsupials (see below) some of which are not readily allocated to extant or extinct orders. Of particular interest in this assemblage is the first (and only) appearance of an apparent terrestrial (non-volant) placental group in Australia represented by *Tingamarra porterorum*, a small condylarth-like taxon known from teeth (Godthelp et al. 1992) and a petrosal (Beck et al. in prep.). Details of the post-early Eocene Australian radiation are unknown until the late Oligocene when an additional order of marsupials, the Yalkaparidontia, makes an appearance among a vast array of other taxa which are all referable to still-living Australian orders.

Competing palaeobiogeographic hypotheses to account for the relationships between, and palaeogeography of, the South American and Australian marsupials include the following: (1) microbiotheriids evolved first and persisted in South America but at least one lineage dispersed across Antarctica to Australia by the early Eocene to trigger evolution of the Australian orders; (2) South America, Antarctica and Australia shared a similar early Cenozoic metatherian fauna with differential extinctions resulting in a disjunct distribution of microbiotheriids in South America and Australia by the middle Eocene; and (3) a metatherian group dispersed across Antarctica to Australia by the early Eocene to found Microbiotheria which subsequently evolved into the other Australian orders with one microbiothere dispersing to South America by the middle Eocene to found the South American microbiothere radiation (Beck et al. 2008b).

On balance, South American marsupials, although better documented for all periods of the early to mid Cenozoic, exhibit less high-level as well as overall diversity than Australian marsupials. This apparent difference reflects the fact that, with the exception of a couple of small, species-depauperate groups such as argyrolagids and groeberiids, South American marsupials were largely restricted to insectivorous, carnivorous and omnivorous kinds while a diversity of placental orders filled the herbivorous as well as some of the insectivorous, carnivorous and omnivorous roles. In contrast, in Australia, apart from bats, a seemingly narrow range of monotremes, one short-lived lineage of early Miocene insectivorous/omnivorous

Eon	Era	Period		Epoch	Climate, Geology, Habitat	Fauna
PHANEROZOIC	CENOZOIC	QUATERNARY		Holocene 10,000 yBP to present	Interglacial conditions: warmer, wetter climate as at present	Europeans arrive by 1606 Dingos introduced 5,000-4,000 yBP
PHANEROZOIC	CENOZOIC	QUATERNARY		Pleistocene 1.8 Ma to 10,000 yBP	Period of glacial/interglacial cycles Dunes and desert features develop by 1 Ma	Mass extinctions 68,000-35,000 yBP Gigantism in marsupial lineages- "Megafauna" Humans arrive 100,000-60,000 yBP
PHANEROZOIC	CENOZOIC	TERTIARY	NEOGENE	Pliocene 5.3 to 1.8 Ma	First grasslands and arid habitats develop Short early Pliocene greenhouse event	Evolution of first grazing marsupials First Australian rodents First representatives of modern bandicoot families
PHANEROZOIC	CENOZOIC	TERTIARY	NEOGENE	Miocene: L. Mio. 10.4 to 5.3 Ma	Icehouse conditions peak 6-5 Ma: low sea-levels result in emergent land-bridges between Australia and New Guinea Dry, seasonal climates spread by 6 Ma	Faunal interchange between Australia and New Guinea Extensive radiations of modern marsupial groups e.g. dasyurids and macropodine kangaroos
PHANEROZOIC	CENOZOIC	TERTIARY	NEOGENE	Miocene: M. Mio. 16.4 to 10.4 Ma	Emergence of New Guinea highlands Start of major icehouse event 15 Ma: formation of permanent Antarctic ice-cap Retraction of rainforests and spread of sclerophyll forest and woodland Miocene climatic optimum 16-15 Ma: rainfall, sea-levels and biodiversity at record high levels	Significant faunal turnover with extinction of archaic groups and emergence of ancestors of modern lineages
PHANEROZOIC	CENOZOIC	TERTIARY	NEOGENE	Miocene: E. Mio. 23.0 to 16.4 Ma	Widespread rainforest Beginning of a major greenhouse event 23 Ma	Highly diverse fossil assemblages from Riversleigh Diversification of 'possum' groups with first appearance of acrobatids, phalangerids and dactylopsilines and spectacular radiation of pseudocheirids
PHANEROZOIC	CENOZOIC	TERTIARY	PALAEOGENE	Oligocene 33.9 to 23.0 Ma	Icehouse phase: cool, seasonal climates Open forest and woodland assemblages	Earliest appearance of modern Australian marsupial orders in the fossil record ~25.5 Ma
PHANEROZOIC	CENOZOIC	TERTIARY	PALAEOGENE	Eocene 55.8 to 33.9 Ma	Warm, wet climate with broad-leaved rainforest assemblages Rapid sea-floor spreading resulting in final separation of Australia from Antarctica 35 Ma	Subordinal split of vombatiform and phalangeridan diprotodontians ~54 Ma Oldest Australian marsupials from Tingamarra LF, Qld, 54.6 Ma, including the oldest Australidelphian *Djarthia murgonensis*
PHANEROZOIC	CENOZOIC	TERTIARY	PALAEOGENE	Paleocene 65.5 to 55.8 Ma	Warm, wet climate with Gondwanan forest Tasman Sea widens, further separating Australia from New Zealand	All four modern Australian marsupial orders had diverged, yet no known Australian Paleocene vertebrate fossil record Oldest Sth American marsupials from Tiupampa Fauna, Bolivia 64.5-63 Ma
PHANEROZOIC	MESOZOIC	CRETACEOUS		Late Cret. 99.6 to 65.5 Ma	Asteroid collision Diversification and spread of angiosperms throughout Gondwana Start of separation of Australia and New Zealand > 82 Ma	Mass extinctions 65 Ma Marsupials disperse from North America to South America 70.6-65.5 Ma
PHANEROZOIC	MESOZOIC	CRETACEOUS		Early Cret. 145.5 to 99.6 Ma	Australia/Antarctica rifting commences 100 Ma First Australian angiosperms Sea-levels high: Australia inundated	Dinosaurs dominate First monotremes from Flat Rocks, Victoria 115 Ma Oldest metatherian, *Sinodelphys szalayi*, from Yixian Fm, China 125 Ma

Fig. 5 Key events in the evolution of Australian marsupials. In contrast to the relatively rich record of events and evolutionary outcomes for marsupials spanning the last 26 Ma, no marsupial fossils have been found in Australia older than those in the 55 Ma Tingamarra LF, and none is known from that time until 26 Ma. (*Abbreviations*: Ma, million years ago; yBP, years before present)

Table 1 Marsupial classification used in this study

Marsupialia
Cohort AMERIDELPHIA
Order Didelphimorphia
Order Paucituberculata
Order Sparassodonta†
Order Polydolopimorphia†
Cohort AUSTRALIDELPHIA
Order Microbiotheria
Family Microbiotheriidae
Order Dasyuromorphia
Family Dasyuridae
Subfamilies: Dasyurinae, Phascogalinae, Sminthopsinae, Barinyainae†
Family Thylacinidae†
Family Myrmecobiidae
Order Peramelemorphia
Family Peramelidae
Family Peroryctidae
Family Thylacomyidae
Family Yaralidae†
Order Notoryctemorphia
Family Notoryctidae
Order Diprotodontia
Suborder Vombatiformes
Infraorder Phascolarctomorphia
Family Phascolarctidae
Infraorder Vombatomorphia
Family Vombatidae
Family Ilariidae†
Family Wynyardiidae†
Family Thylacoleonidae†
Family Maradidae†
Superfamily Diprotodontoidea
Family Diprotodontidae†
Subfamilies: Zygomaturinae†, Diprotodontinae†
Family Palorchestidae†
Suborder Phalangerida
Superfamily Phalangeroidea
Family Phalangeridae
Family Ektopodontidae†
Family Miralinidae†
Superfamily Burramyoidea
Family Burramyidae
Superfamily Petauroidea
Family Petauridae
Family Pseudocheiridae

Table 1 (continued)

Superfamily Tarsipedoidea
Family Tarsipedidae
Family Acrobatidae
Superfamily Macropodoidea
Family Hypsiprymnodontidae
Subfamilies: Hypsiprymnodontinae, Propleopinae†
Family Potoroidae
Family Balbaridae†
Family Macropodidae
Subfamilies: Bulungamayinae†, Sthenurinae, Macropodinae
Phalangerida superfamily *incertae sedis*
Family Pilkipildridae†
Order Yalkaparidontia†
Family Yalkaparidontidae†
Marsupialia cohort *incertae sedis*
Family Yingabalinaridae†
Family Numbigilgidae†

placentals, and rodents in the late Cenozoic, marsupials filled the majority of herbivorous as well as all of the insectivorous, carnivorous and omnivorous mammalian roles.

Phylogenetic Affinities of the Australian Marsupial Radiation

The world's seven modern marsupial orders are commonly divided into two cohorts (Table 1, Fig. 9): Ameridelphia containing Didelphimorphia and Paucituberculata and Australidelphia containing the four extant Australasian marsupial orders plus Microbiotheria, although the latter is represented today by only a single South American species (Aplin and Archer 1987). These cohorts were originally proposed by Szalay (1982) based primarily on tarsal (ankle bone) morphology. Szalay (1982) argued that australidelphians were united by the synapomorphic possession of a continuous lower ankle joint between the calcaneum and astragulus (CLAJP) while ameridelphians were characterised by the plesiomorphic condition of a separate lower ankle joint pattern (SLAJP).

Australidelphia has since found strong support from studies based on a wide range of character systems, both molecular and morphological (e.g.

Fig. 6 A tiny bright *red dot* in a vast landscape of the past. Henk Godthelp (UNSW) searches in the late Cretaceous (*below* him) to Paleogene sediments (*above* him) in the Grand Baranca, Patagonia, Argentina, for traces of the first marsupials that reached South America (M. Archer)

Amrine-Madsen et al. 2003; Horovitz and Sánchez-Villagra 2003; Asher et al. 2004; Cardillo et al. 2004; Nilsson et al. 2004; Drummond et al. 2006; Phillips et al. 2006; Sánchez-Villagra et al. 2007; Beck 2008a; Beck et al. 2008b; Meredith et al. 2008a; Phillips and Pratt 2008; Meredith et al. 2009).

Fig. 7 Sole living representative of the australidelphian order Microbiotheria, *Dromiciops gliroides*. Microbiotheriids, which only survive today in South America, are a vital evolutionary link between the marsupials of South America, Antarctica and Australia (J.A.W. Kirsch)

Although australidelphian monophyly is widely accepted, the precise phylogenetic relationships of microbiotherians to other australidelphians are unresolved. As noted by numerous authors (e.g. Woodburne and Case 1996; Springer et al. 1998; Amrine-Madsen et al. 2003; Beck 2008a; Beck et al. 2008b), clarifying this relationship is important for deconstructing the palaeobiogeographic history of the Australian marsupial radiation. Two alternate phylogenetic schemes for australidelphians have been proposed: (1) microbiotheres are the sister-group to all Australian marsupials (e.g. Amrine-Madsen et al. 2003; Phillips et al. 2006; Beck 2008a; Meredith et al. 2008a; Meredith et al. 2009; Fig. 9); and (2) microbiotheres nest within the Australian marsupial radiation (e.g. Kirsch et al. 1997; Burk et al. 1999; Asher et al. 2004; Cardillo et al. 2004; Nilsson et al. 2004; Beck et al. 2008b; Horovitz et al. 2009).

Australia's Oldest Marsupials

The oldest known Australian marsupials are from the early Eocene Tingamarra LF from Murgon (Fig. 8),

Fig. 8 Site of the early Eocene Tingamarra LF, near Murgon, southeastern Queensland (K. Black; *inset* H. Godthelp). From this 'pit in a paddock', in 55 myo sediments that accumulated long before Australia had finally separated from Antarctica 35 Ma, have come Australia's oldest marsupials such as *Djarthia* as well as the world's first songbirds and possibly the oldest bat

southeastern Qld. Of these, *Djarthia murgonensis* Godthelp et al., 1999 represents the oldest australidelphian marsupial known (see below) while *Chulpasia jimthorselli* Sigé et al., 2009 (Fig. 10) and *Thylacotinga bartholomaii* Archer et al., 1993 have been tentatively referred to the South American order Polydolopimorphia (Sigé et al. 2009).

The Tingamarra LF has been radiometrically dated (Godthelp et al. 1992) using K/Ar ratios at 54.6 Ma. Woodburne and Case (1996) challenged this interpretation and suggested instead a late Oligocene age. However, presence in the Tingamarra LF of graculavid birds (Boles 1999) which are globally only known from the early Eocene, madtsoiid snakes most similar to late Cretaceous taxa known from South America (Scanlon 2005), an archaeonycteridid bat (Hand et al. 1994; Sigé et al. 2004) of the kind known globally from the early Eocene, and a species of *Chulpasia*, a genus otherwise known only from the late Paleocene of Peru (Sigé et al. 2009), do not provide support for the late Oligocene age proposed by Woodburne and Case (1996). They do, however, provide additional support for the original interpretation of Godthelp et al. (Archer et al. 1999; Beck et al. 2008b).

Djarthia murgonensis was originally described on the basis of its dentition. Because there were no undoubted dental synapomorphies distinguishing australidelphian from ameridelphian marsupials, it was not allocated at the time to either clade (Godthelp et al. 1999). But now recently described petrosals and tarsal bones (Beck et al. 2008b; Fig. 11) indicate that *D. murgonensis* belongs in Australidelphia as the sister taxon to all other extant Australasian marsupials. Because of its plesiomorphic position within Australidelphia and the extremely plesiomorphic nature of its dental, cranial and postcranial morphology, Beck et al. (2008b) suggested that *D. murgonensis* may represent the ancestral morphotype of all Australian marsupials and possibly of all australidelphians.

Building this possibility and the apparent lack of any older australidelphian known from South America, Beck et al. (2008b) supported the hypothesis noted above that the current distribution of *Dromiciops* within South America was the result of back dispersal of microbiotheres from eastern Gondwana. Two possibly older microbiotherian australidelphians have been described: the early or middle Eocene *Khasia cordillerensis* (Marshall and de Muizon 1988; de Muizon 1991; from Tiupampa, Bolivia) and the late Paleocene *Mirandatherium alipioi* (de Paula Couto 1962; from Itaboraí, Brazil). However, their referral to Microbiotheria is questioned by Beck et al. (2008b) because it is based on features of the dentition that have been shown to be homoplastic (Szalay 1994; Wroe

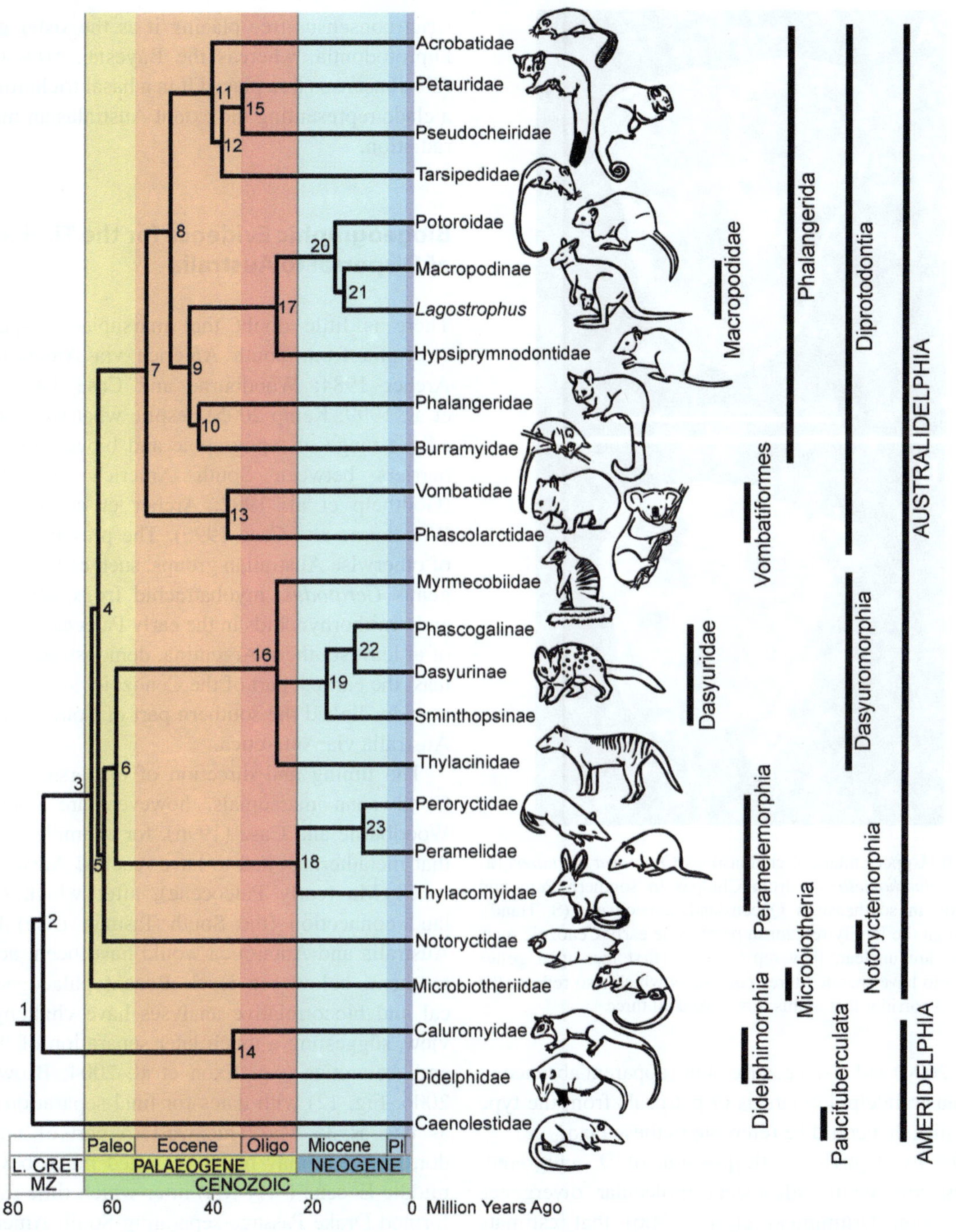

Fig. 9 Phylogeny and divergence estimates for modern marsupial groups based on nuclear gene sequence data of Meredith et al. (2009). Node numbers and divergence dates are listed in Table 2. *Abbreviations*: L.CRET, late Cretaceous; MZ, Mesozoic; Oligo, Oligocene; Paleo, Paleocene; Pl, Pliocene. The basic dichotomy between the 'ameridelphian' (South America) and australidelphian marsupials (Australian plus ubiquitous microbiotheres) demonstrates higher familial diversity of marsupials in Australia. Despite the calculated antiquity based on molecular divergence, apart from microbiotheriids none of the other australidelphian families are known as fossils prior to the late Oligocene

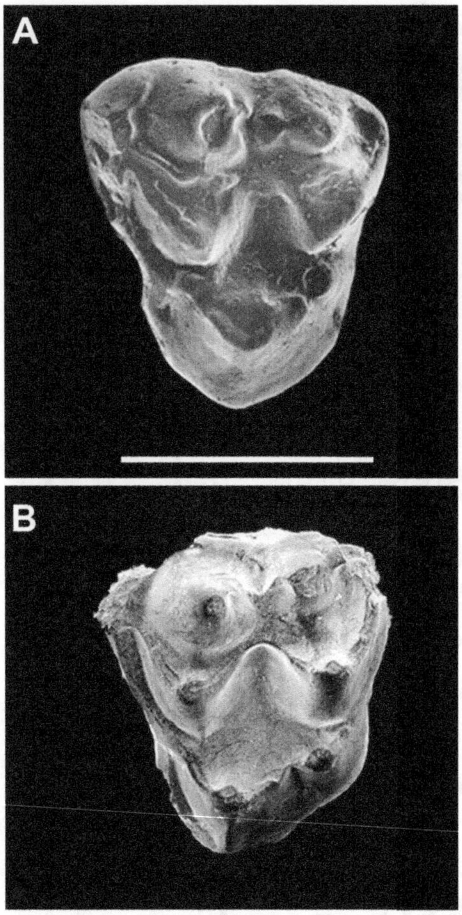

Fig. 10 Upper molars of chulpasiines *Chulpasia mattaueri* (**a**) and *C. jimthorselli* (**b**) from Chulpas in southern Peru and Murgon in southeastern Queensland respectively (S. Hand). Although the family relationships of these early Cenozoic marsupials are unclear, they represent the first marsupial genus known to have spanned three continents which also reflects the forested corridor that at this time linked all three

et al. 2000) and also because of the apparent absence of any australidelphian tarsals or petrosals from the type localities that could be referable to these taxa.

The basal phylogenetic position of *D. murgonensis* is congruent with recent molecular divergence dates (e.g. Drummond et al. 2006) that estimate Australidelphia to have originated during the late Cretaceous or very early Paleocene (76.5±4.7 Ma, Beck 2008a; 65.0–75.1 Ma, Drummond et al. 2006), and modern australidelphian orders to have diverged from each other soon after (65.2±3.2 Ma, Beck 2008a; 56.9–65.5 Ma, Drummond et al. 2006).

The phylogenetic position of *Dromiciops* within Beck's (2008a) analysis was unresolved, with the strict consensus tree placing it as the sister group to Diprotodontia, whereas the Bayesian 50% majority rule consensus tree placed it in a basal trichotomy with a clade representing the extant Australasian marsupial radiation.

Biogeographic Evidence for the Timing of Dispersal to Australia

There is little doubt that marsupials dispersed to Australia from South America via Antarctica (e.g. Archer 1984; Woodburne and Case 1996; Archer et al. 1999; Kemp 2005) despite what were undoubtedly a range of topographic and biogeographic filter barriers between South America and Australia (Godthelp et al. 1992; Archer et al. 1993, 1999; Woodburne and Case 1996). The presence of a suite of otherwise Australian groups, such as lungfish in the genus *Ceratodus*, myobatrachid frogs, chelid turtles and ornithorhynchids in the early Paleocene (~63 Ma) of at least southern Argentina, demonstrates that by at least the earliest part of the Cenozoic, a biogeographic corridor linked the southern part of South America to Australia via Antarctica.

The timing and direction of dispersal(s) of these Gondwanan marsupials, however, are contentious. Woodburne and Case (1996), for example, suggested that metatherians had to have reached Australia prior to 64 Ma (early Paleocene), after which time the land connection (the South Tasman Rise) between Australia and Antarctica would have been, according to them, below sea level. Recent palaeogeographical and biocorrelative analyses have challenged this view, suggesting a much later separation of Australia and Antarctica (e.g. Exon et al. 2004; Brown et al. 2006; Fig. 12) with dates for final separation ranging as late as 35 Ma. This western part of this corridor, however, may not have lasted much beyond the middle Eocene (~41 Ma) after which time the newly formed Drake Passage separating South America and Antarctica (Scher and Martin 2006) would have made dispersal increasingly unlikely.

Recent palaeontological discoveries further constrain speculation about the timing of dispersal of marsupials into Australia. The presence of marsupials in the early Eocene (54.6 Ma) Tingamarra LF of Qld (Godthelp et al. 1999; Archer et al. 1993; Beck et al. 2008b) sets a minimum age for arrival from East

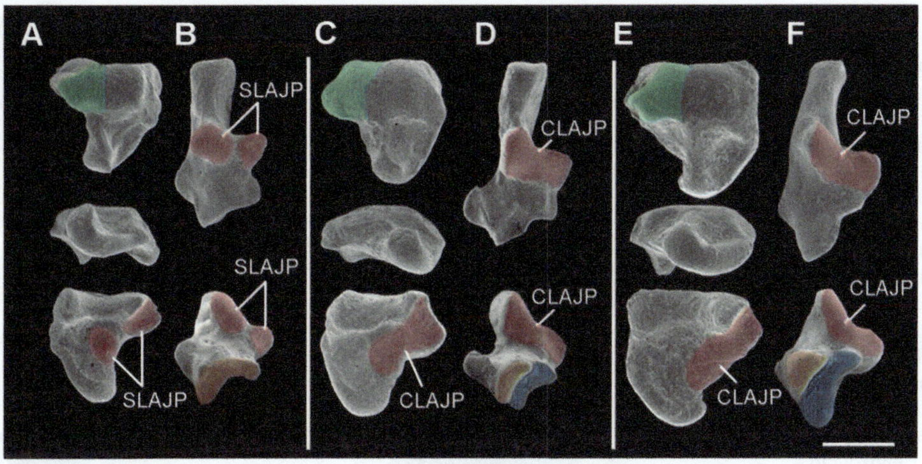

Fig. 11 Comparison of isolated anklebones of the Australian early Eocene *Djarthia murgonensis* with an extant australidelphian marsupial and an extant ameridelphian marsupial. Astragali (**a**, **c**, **e**) and calcanea (**b**, **d**, **f**) of the 'ameridelphian' didelphid *Thylamys elegans* (**a–b**), *Djarthia murgonensis* (**c–d**) and the australidelphian microbiotheriid *Dromiciops australis* (**e–f**). Astragali illustrated in dorsal (*top*), distal (*middle*) and ventral (*bottom*) views. Calcanea illustrated in dorsal (*top*) and distal (*bottom*) views. Scale bar, 1 mm. Presence of the continuous lower ankle joint pattern (CLAPJ) and subdivision of the calcaneocuboid facet into three distinct facets are australidelphian synapomorphies (after Beck et al. 2008b). The differences between the ankle bones of these two great groups of marsupials are also reflected in molecular and other system differences

Antarctica. Further, close resemblance of a Tingamarra species of *Chulpasia* (Polydolopimorphia) to another in Paleocene deposits of Peru (Sigé et al. 2009; see below) argues for a Paleocene connection between Australia and South America.

Interordinal Relationships of Australasian Marsupials

The interrelationships of the Australasian marsupial orders are largely unresolved with most conflict surrounding the relative positions of Peramelemorphia (bandicoots and bilbies) and Notoryctemorphia (marsupial moles).

Peramelemorphians have been variously placed as the sister group to Diprotodontia (collectively forming the taxon Syndactyla) based on pedal morphology (Szalay 1982); the sister group to Dasyuromorphia (Fig. 9) based on comparative serology (e.g. Kirsch 1977a), nuclear genes (Amrine-Madsen et al. 2003; Springer et al. 2009) and mitochondrial genes (Nilsson et al. 2004; Phillips et al. 2006); and the sister group to Notoryctemorphia based on postcranial, craniodental and soft tissue anatomy (e.g. Horovitz and Sánchez-Villagra 2003). It has alternatively been suggested that peramelemorphians may lie outside the rest of the australidelphian radiation based on dental and basicranial morphology (Archer 1976b, c), DNA hybridisation (Kirsch et al. 1997) and mitochondrial RNA gene sequences (Burk et al. 1999). Kirsch et al. (1997) went one step further by placing them in a monotypic cohort (Perametatheria) to further emphasise their distinction from all other australidelphian orders (which he placed in Cohort Eometatheria). Recent analyses using 'super' matrices constructed from combined data sources also produce conflicting results. The combined character analysis of Asher et al. (2004), which included skeletal, dental, soft tissue and nuclear and mitochondrial gene sequence data, also supported monophyly of Eometatheria (*sensu* Kirsch et al. 1997) to the exclusion of peramelemorphians. The phylogenetic 'supertree' of marsupial relationships proposed by Cardillo et al. (2004) based on 158 source trees (generated from previous phylogenetic analyses) supported inclusion of Peramelemorphia within Australidelphia as its most basal clade. In contrast Beck's (2008a) phylogenetic analysis based on a 20.1-kilobase molecular supermatrix that combined molecular data from three previous phylogenetic analyses (Amrine-Madsen et al. 2003; Asher et al. 2004; Nilsson et al. 2004) found Peramelemorphia to be the sister group of a combined Dasyuromorphia/Notoryctemorphia clade.

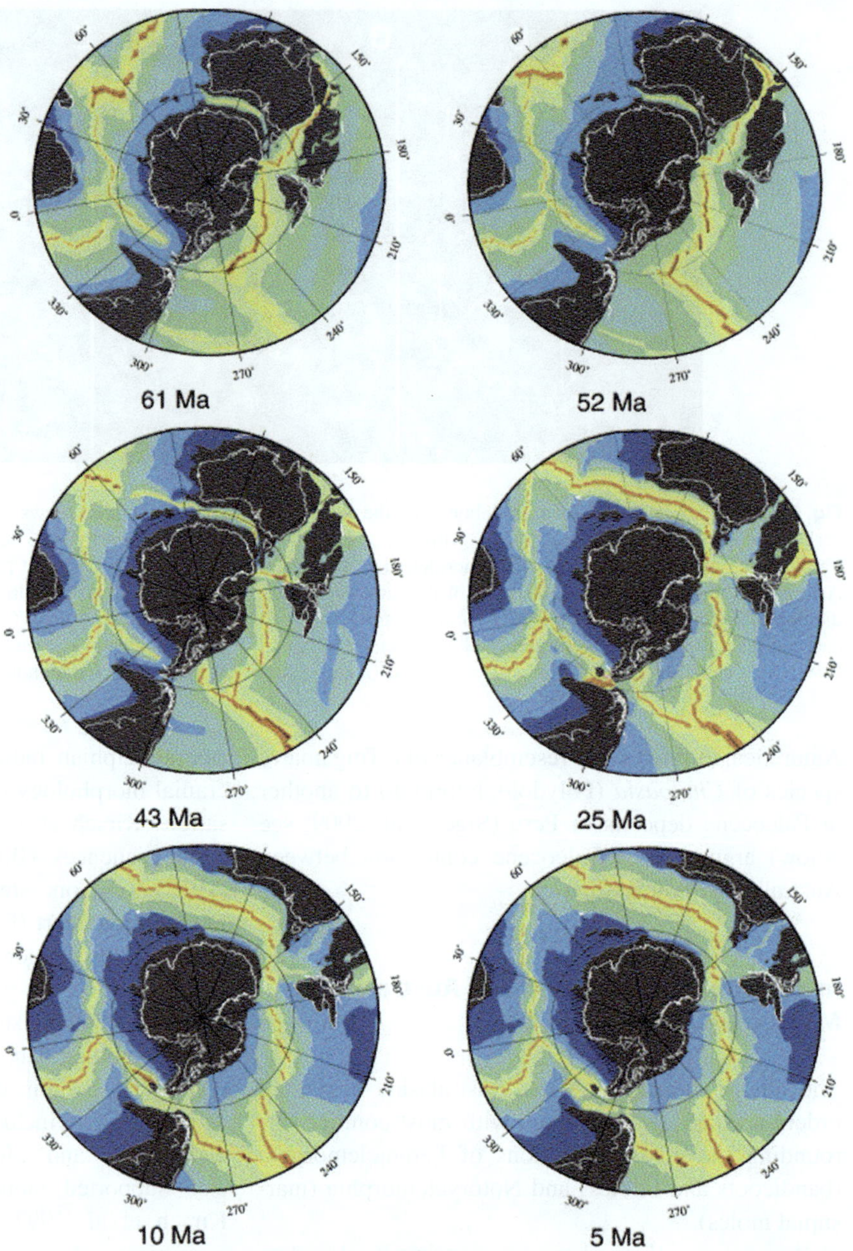

Fig. 12 Palaeogeographic reconstruction of the break-up of Gondwana from 61 to 5 Ma. Australia's separation from Antarctica began in the west in the Cretaceous but final separation in the east did not occur until the late Eocene about 35 million years ago. Reprinted from Palaeogeography, Palaeoclimatology, Palaeoecology, 231(1–2), B. Brown, C. Gaina, R. Dietmar Müller, Circum-Antarctic palaeobathymetry: Illustrated examples from Cenozoic to recent times, pp 158–168, Copyright (2006), with permission from Elsevier

The affinities of Notoryctemorphia within Australidelphia are no less uncertain despite ample morphological and molecular data (Beck 2008a). The lack until recently of any fossil taxa attributable to this highly autapomorphic group has made it that much more difficult to work out its relationships. In terms of both dental morphology (zalambdodonty) and skeletal morphology, marsupial moles are strikingly convergent on placental golden moles (chrysochlorids). Archer (1976c) suggested possible affinities to diprotodontians based on some similarities in middle ear construction and soft anatomical features. Description of the first known extinct notoryctid from early Miocene sediments of the Riversleigh World Heritage fossil deposits of Qld (Archer et al. 2011) may help shed light on the ordinal affinities of the group. Beck (2008a) notes potential synapomorphies in tarsal morphology between this plesiomorphic notoryctid and peramelemorphians, although most recent molecular studies of the extant species support a dasyuromorphian affinity (e.g. Amrine-Madsen et al. 2003; Nilsson et al. 2004; Phillips et al. 2006; Meredith et al. 2008a).

Despite conflicting interpretations about the interrelationships of the Australasian orders, there is resounding support from both morphological and molecular data for the monophyly of each individual order (Amrine-Madsen et al. 2003; Aplin and Archer 1987; Beck 2008a; Beck et al. 2008b; Horovitz and Sánchez-Villagra 2003; Kirsch et al. 1997; Meredith et al. 2008a, 2009; Phillips et al. 2006; Phillips and Pratt 2008).

Interordinal Molecular Divergence Dates

In light of recent advances in molecular sequencing and phylogenetic data analysis, and in the face of a relatively poor early Cenozoic record for Australian marsupials, analyses of higher-level marsupial systematics based on molecular data have outnumbered those based on morphology in recent years. When combined with fossil data that is available, these analyses have provided useful frameworks for interpreting the probable time of origin of the major Australasian marsupial clades. At the forefront of molecular divergence dating methods is the application of relaxed molecular clocks that allow for variation in evolutionary rates across different phylogenetic groups. This algorithm is regarded to be more biologically realistic (Lepage et al. 2007) compared with strict molecular clock methods which assume homogeneity in rates of change (Drummond et al. 2006). Recent proponents of the relaxed method include Drummond et al. (2006), Beck (2008a), and Meredith et al. (2008a, b, c, 2009).

Beck (2008a), using a 20.1-kilobase molecular supermatrix (7 nuclear and 15 mitochondrial genes) and a Bayesian relaxed molecular clock with multiple fossil constraints, found that the first major split within the extant Australasian marsupial radiation was the divergence of Diprotodontia from the 'polyprotodont' orders (i.e., Dasyuromorphia, Peramelemorphia and Notoryctemorphia) at 65.2 Ma (± 3.7 Ma) (Table 2). He suggests this divergence probably occurred within Australia, as evidenced by both the monophyly of the Australian marsupial radiation and the absence of any Australian marsupial orders in South American or Antarctic fossil deposits. Slightly younger dates for this split were estimated by Springer et al. (2009; 63.9 Ma) and Meredith et al. (2008a; 62.2 Ma).

According to these analyses, peramelemorphians appear to have diverged from a dasyuromorphian/notoryctemorphian clade at 63.4±3.7 Ma, with the latter orders diverging from each other at 60.1±3.6 Ma (Beck 2008a). Comparable dates were also found by Meredith et al. (2008a; 60.9 Ma and 59.1 Ma respectively). Meredith et al. (2009) and Springer et al. (2009) found an alternate tree topology (Fig. 9) with notoryctemorphians diverging from a dasyuromorphian/peramelemorphian clade at 62.7 Ma, and dasyuromorphians and peramelemorphians last sharing a common ancestor around 60.7 Ma. Clearly, if these dates are correct, all four modern Australasian marsupial orders had differentiated by the earliest Eocene (the age of the Tingamarra LF).

Curiously, no Tingamarran taxon discovered to date can be unambiguously assigned to any of these Australian orders with the possible exception of a putative peramelemorphian (Archer et al. 1999; Long et al. 2002). This could mean one or more of four possibilities:

1. the Tingamarra LF is older than early Eocene (but this flies in the face of too much globally based biocorrelation as well as the radiometric date);
2. the Tingamarran taxa are unrepresentative of the diversity of marsupials elsewhere on the continent (unlikely given the hundreds of specimens found to date);
3. the Tingamarran taxa may be members of these Australian orders but not yet have developed the otherwise distinctive dental morphology that otherwise characterises these groups, as suggested by Godthelp et al. (1999) and Beck (2008a) (possible but surprising that at least one dental synapomorphy would not have shown up by this time); or
4. the relaxed clock algorithm for estimating divergence dates is too relaxed.

Australian Marsupial Palaeodiversity

In the sections that follow, we present overviews of the evolving Cenozoic diversity of Australia's marsupials and their palaeoenvironments. Unfortunately, despite the obviously important events of the early

Table 2 Divergence times (Ma) and confidence intervals (in parentheses) for marsupials based on recent relaxed molecular clock analyses

Node/clade		Divergence estimates (Ma) Meredith et al. (2009)[a]	Meredith et al. (2011)[b]	Beck (2008a)[c]	Meredith et al. (2008a)[d]
1	Marsupialia	75.5 (65.2–86.2)	78.1 (68–87)	80.6 (70.7–90.0)	78.5 (67.8–89.4)
2	Didelphimorphia + Australidelphia	72.5 (65.2–86.2)	–	–	–
3	Australidelphia	63.0 (55.6–70.0)	64.8 (57–70)	67.4 (59.8–73.6)	62.9 (55.3–70.0)
4	Australasian taxa	62.2 (54.8–69.2)	63.9 (56–70)	65.2 (57.9–70.9)	62.2 (54.7–69.3)
5	Notoryctemorphia + Dasyuromorphia + Peramelemorphia	61.1 (53.9–68.1)	62.7 (55–68)	63.4 (56.3–69.2)	60.9 (53.5–68.1)
6	Peramelemorphia + Dasyuromorphia	59.2 (52.2–66.4)	60.7 (54–67)	–	–
7	Diprotodontia (Phalangerida + Vombatiformes)	53.3 (46.6–60.0)	55.3 (48–61)	57.1 (50.5–62.9)	54.1 (47.3–60.9)
8	Phalangerida	48.4 (42.1–55.1)	49.9 (44–56)	51.1 (44.9–57.0)	47.6 (41.4–54.1)
9	Macropodoidea + Phalangeridae + burramyidae	45.3 (39.0–52.0)	46.1 (40–52)	47.1 (41.1–53.0)	44.9 (38.9–51.3)
10	Phalangeridae + burramyidae	42.8 (36.6–49.4)	42.8 (37–49)	40.0 (33.0–47.4)	41.9 (36.0–48.2)
11	Acrobatidae + Tarsipedidae + Petauridae + Pseudocheiridae	40.5 (34.5–46.8)	41.6 (36–48)	–	–
12	Tarsipedidae +Petauridae + Pseudocheiridae	38.2 (32.4–44.4)	39.1 (33–45)	–	–
13	Vombatiformes	36.8 (30.8–43.3)	39.0 (33–45)	40.7 (34.4–47.0)	37.3 (31.4–43.6)
14	Didelphimorphia	32.5 (26.3–39.2)	37.4 (30–45)	–	37.5 (30.4–45.6)
15	Petauroidea (Petauridae + Pseudocheiridae)	34.7 (29.2–40.6)	36.8 (31–43)	32.4 (26.6–38.7)	33.1 (27.7–38.9)
16	Dasyuromorphia	27.3 (24.8–31.7)[e]	27.3 (25–31)[e]	26.0 (23.2–1.4)[f]	–
17	Macropodoidea	26.8 (22.0–32.3)	26.4 (22–31)	–	–
18	Peramelemorphia	23.2 (19.2–27.8)	23.8 (20–28)	–	–
19	Dasyuridae	16.9 (14.2–20.4)	–	–	15.0 (12.3–18.3)
20	Potoroidae + Macropodidae	15.9 (12.7–19.7)	14.2 (12–18)	–	13.8 (12.1–17.0)
21	Macropodidae	13.8 (10.9–17.5)	–	–	–
22	Phascogalinae + Dasyurinae	11.9 (9.6–14.8)	–	11.6 (8.8–15.2)	10.2 (8.1–12.8)
23	Peramelidae + Peroryctidae	10.0 (7.8–12.8)	–	11.7 (9.1–15.1)	9.1 (7.3–11.3)

Node numbers refer to Fig. 9
[a]Data from *Multidivtime* partitioned analysis of Meredith et al. (2009) taken from supplementary data
[b]As cited in Springer et al. (2009)
[c]Data (point estimates and 95% confidence intervals) from *Multidivtime* analysis of Beck (2008a) using all sequences and all constraints (table 3, p. 182)
[d]Data from *Multidivtime* partitioned analysis (diprotodontian maximum = 65 Ma; table 5, p. 22) of Meredith et al. (2008a)
[e]Dasyuridae + Myrmecobiidae: Thylacinidae not included in analysis of Meredith et al. (2009, 2010)
[f]Dasyuridae + Thylacinidae: Myrmecobiidae not included in analysis of Beck (2008a)

Cenozoic, only a single assemblage, the early Eocene Tingamarra LF, provides any information about what happened to Australia's mammals between the early Cretaceous at about 110 Ma until the late Oligocene at 26 Ma (Fig. 13). These two 'dark ages' on either side of the Tingamarra LF, lasting 55 Ma before and 29 Ma after, obscure virtually all of the fascinating events that must have occurred during the early differentiation of Australia's modern orders.

The Palaeogene Marsupials of Australia: The Early Eocene

As noted above, the Australian fossil mammal record is almost entirely unknown between the late Cretaceous and late Oligocene. The sole exception spanning this period is the early Eocene Tingamarra LF near Murgon, southeastern Qld. At least seven mammal species are represented in the fauna (Archer

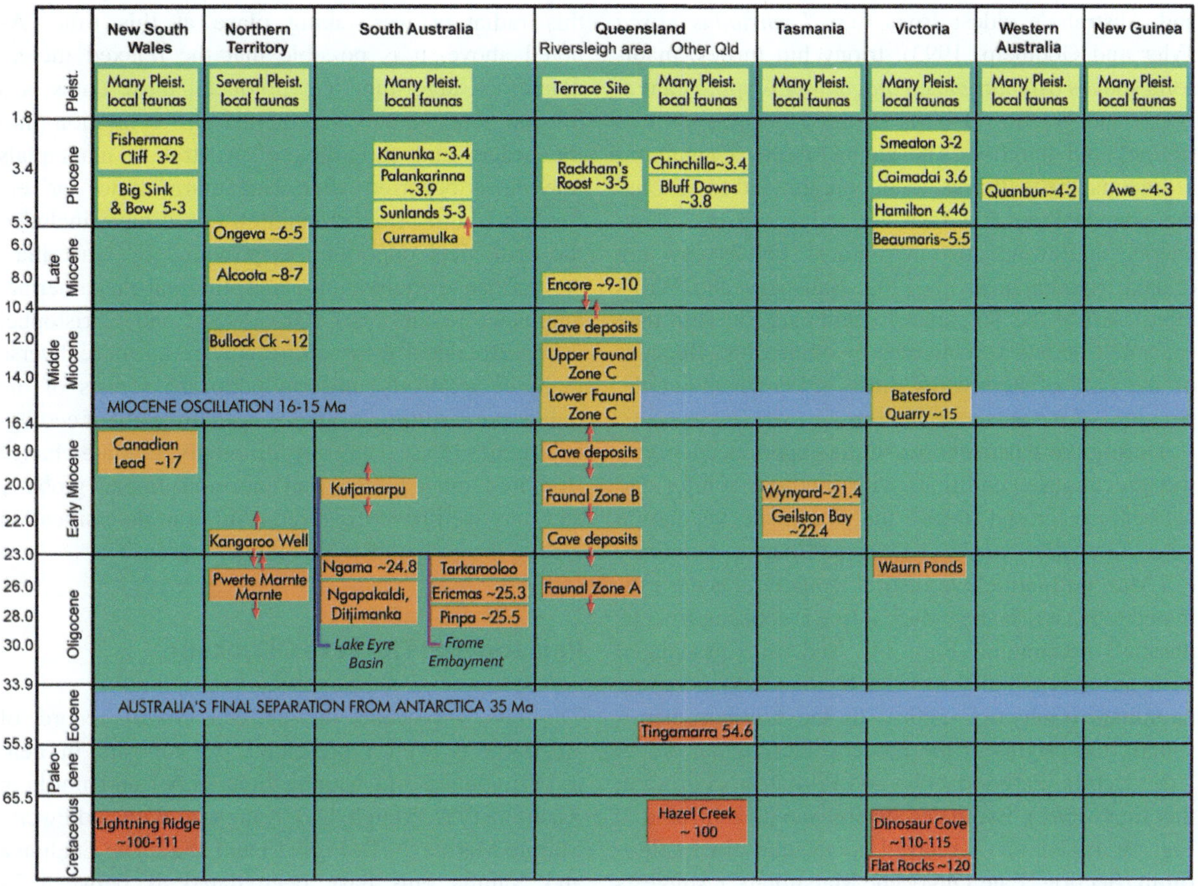

Fig. 13 Correlation of Australasia's Cretaceous to Pleistocene terrestrial fossil mammal assemblages, with age of local faunas given in millions of years. 'Holes' in the record where nothing is known are the late Cretaceous to early Eocene and the middle Eocene to late Oligocene. However, very little is also known from the late Miocene in part because, as a critical interval of palaeoclimatic change, Australia's increasing dryness appears to have resulted in few sites suitable for fossilisation

et al. 1993), five of which have been described and include the condylarth-like placental *Tingamarra porterorum* (Godthelp et al. 1992), the oldest bat in the Southern Hemisphere *Australonycteris clarkae* (Hand et al. 1994) and the marsupials *D. murgonensis* (Godthelp et al. 1999; Beck et al. 2008b), *Thylacotinga bartholomaii* (Archer et al. 1993) and *Chulpasia jimthorselli* Sigé et al., 2009. The latter two taxa are placed in the subfamily Chulpasiinae along with *Chulpasia mattaueri* Crochet and Sigé, 1993 from the Laguna Umayo red mudstone unit at Chulpas, southern Peru, interpreted to be late Paleocene–early Eocene in age (Sigé et al. 2004). Although its familial affinities are unresolved, chulpasiines were referred to the South American order Polydolopimorphia as the first evidence that generic-level groups of therian mammals once spanned Gondwana (Sigé et al. 2009). This Gondwanan intimacy is also evidenced among ornithorhynchids with the Paleocene *Monotrematum* from Argentina almost certainly being a junior synonym of the Oligo-Miocene *Obdurodon* from Australia and the same species of freshwater lungfish, *Neoceratodus fosteri*, being present in Paleocene deposits of Argentina as well as spanning the late Cretaceous to Holocene record of Australia. The phylogenetic relationships and biogeographic significance of *D. murgonensis* are discussed in Section 'Australia's Oldest Marsupials'. Undescribed Tingamarran marsupials include a *Mirandatherium*-like microbiotheriid and a possible plesiomorphic peramelemorphian (Archer et al. 1999; McGowran et al. 2000). Other vertebrates of the Tingamarran faunal assemblage include the world's oldest mekosuchine crocodiles (*Kambara murgonensis* Willis et al., 1993; *Kambara implexidens* Salisbury and Willis, 1996) and oldest passerine birds (Boles 1999) and Australia's oldest frogs (e.g. *Lechriodus casca* Tyler and Godthelp, 1993), trionychid turtles, madtsoiid snakes (*Alamitophis*, *Patagoniophis* spp; Scanlon 2005) and graculavid birds (Boles 1999), as well as teleosts and lungfish. The nature of the Tingamarra sediments, the well-preserved, non-abraded fossils and the abundant aquatic taxa are indicative of a low-energy, shallow aquatic environment such as a swamp or lake (Godthelp et al. 1992; Salisbury and Willis 1996). All of the Tingamarra mammals found to date are small-bodied insectivores or omnivores. Because of the lack of herbivores, little is known about the structure of the palaeovegetation.

During the Paleocene Australia experienced warm, wet greenhouse conditions with southern conifer and *Nothofagus* closed forest the dominant vegetation type. Globally, at the Paleocene/Eocene boundary (55 Ma), mid-latitude terrestrial temperatures are estimated to have risen by ~4–5°C (the Paleocene/Eocene thermal maximum; Fig. 14) and are associated with an increase in terrestrial biodiversity including mammals (Gingerich 2006). In the northern hemisphere, the origins of the modern mammalian orders Artiodactyla, Perissodactyla and Primates occurred at this time (ibid.). Estimated divergence dates (Table 2, Fig. 9) based on a relaxed molecular clock algorithm (Section 'Late Oligocene Marsupials'), suggests that diprotodontians underwent rapid diversification during the early to middle Eocene including the subordinal split of Vombatiformes and Phalangerida. However, there is no palaeontological evidence that this radiation was taking place at this time. As noted above, it is possible that the relaxed molecular clock algorithm that produced these dates is a bit too relaxed and that actual differentiation into the orders and suborders of Australian marsupials may have occurred after Tingamarra time. Drivers for this round of differentiation could have included the ecological opportunities provided by Australia's early Eocene multistratal, mesothermal-megathermal angiosperm rainforests (Greenwood and Christophel 2005). Botanically, these forests were far more diverse than the relatively uniform podocarp forests of the Paleocene resulting in an increased diversity of ecological niches (Case 1989). Additionally, climate change from mid to late Eocene had enormous impact globally and may well have triggered a surge in diversification among the early Australian marsupials.

Palaeogene: The Late Oligocene

The late Oligocene (26–23 Ma) fossil record of Australia is well represented by deposits in central (Etadunna and Namba Fms, S.A.) and northern Australia (Riversleigh, Qld, Kangaroo Well and Pwerte Marnte Marnte, N.T.; Figs. 13 and 15a). The Etadunna and Namba Fms have been dated as between 24 and 26 Ma using radiometry, magnetostratigraphy and biocorrelation of forams and pollens (Lindsay 1987; Woodburne et al. 1985, 1993). Riversleigh's Faunal Zone A vertebrate assemblages (= System A of Archer

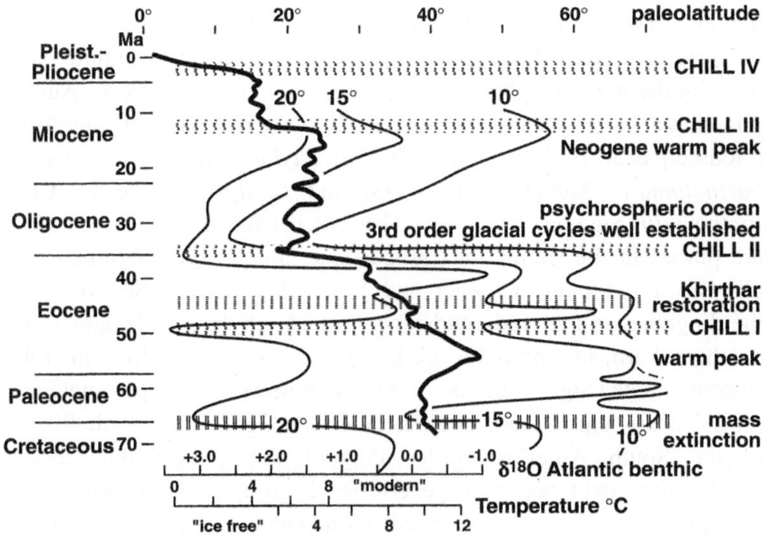

Fig. 14 From evidence of high-latitude excursions and extinction episodes for large forams in southern Australian waters, McGowran et al. (2000) calculated changes in temperature that suggest Australia endured at least four long-term cycles of climate change from greenhouse to icehouse conditions. Changes in marsupial diversity over the same period of time provide support for this hypothesis

Fig. 15 Map of Australasian fossil mammal-bearing deposits. (**a**) Tertiary deposits; (**b**) Pleistocene deposits. Conspicuous challenges are the almost total lack of Tertiary deposits from Western Australia and the single site known from New Guinea. The richest, longest records for the Tertiary come from Queensland, South Australia and the Northern Territory. Pleistocene sites, however, are most abundant in Victoria, New South Wales and South Australia

Fig. 15 (continued)

et al. 1989, 1991) are interpreted as late Oligocene in age (Fig. 13) on the basis of biocorrelation with those of the Etadunna Fm (Archer et al. 1989, 1991, 1997; Travouillon et al. 2006). Similarly, through biocorrelation, the Kangaroo Well LF from the N.T. (Fig. 16a) is regarded to be slightly younger than the Ngama LF (Zone D; Fig. 16b) of the Etadunna Fm (Woodburne et al. 1993; Megirian et al. 2004), and the Pwerte Marnte Marnte LF is regarded to be of a similar age as, if not older than, the Ngapakaldi LF (Zone B–C) of the Etadunna Fm (Murray and Megirian 2006a).

Palaeoenvironment

Sometime before the end of the Eocene (33.9 Ma; ICS 2008) Australia finally severed all land ties with Antarctica and began drifting northwards towards southeastern Asia as an island continent (Fig. 12). Because South America also separated from Antarctica during the late Eocene with consequent development of the Drake Passage, the cold Antarctic circumpolar current began to churn its way around the southern oceans (Scher and Martin 2006). Global oceanic circulation patterns began to change, affecting climates worldwide. Glaciers began to form in Antarctica and global sea temperatures cooled, which had a major impact on climate in the southern hemisphere (White 2006). Australia had entered an 'icehouse' phase resulting in low temperatures, reduced rainfall and consequent declining biodiversity (Frakes et al. 1987; McGowran and Li 1994). Late Oligocene forests were less diverse than those of the early Eocene and Oligocene and were dominated by casuarinaceans, myrtaceans and species of *Nothofagus* with araucariaceans and podocarpaceans prominent in southeastern Australia. With the advent of greater seasonality in the Oligocene, sclerophyll taxa such as species of

Fig. 16 Late Oligocene fossil deposits from central Australia. (**a**) Kangaroo Well N.T. (S. Hand); (**b**) Lake Palankarinna, S.A. (S. Williams). Fossil marsupials, which are hard to find in the limestone exposures of Kangaroo Well but relatively abundant in the Etadunna Formation exposures (*white* and *greenish* sediments in the photo), are taxonomically similar. At least two species from the Etadunna Formation have also been found in the late Oligocene deposits of Riversleigh

Acacia first appeared and became more widespread (McGowran et al. 2000).

Palaeoecological analyses of Riversleigh's late Oligocene Faunal Zone A mammalian assemblages (e.g. White Hunter Site), using cenogram and body mass distribution analyses (Travouillon et al. 2009), suggest the possibility of open forest habitats. Fossil wood from Riversleigh's late Oligocene Dunsinane LF identified (J. O'Brien 1996, 'personal communication') as fagacian (?*Nothofagus* sp.) shows growth rings suggestive of cool conditions with seasonality or at least episodic changes in growth rates. Guerin and Hill (2006) identified other plant taxa from the Dunsinane LF which they suggest provide no evidence for rainforest (although most of the genera they identified including *Rubus*, *Capparis* and *Alectryon* do have some modern species in rainforests). Guerin and Hill (2006) misattributed the Dunsinane assemblage to the late Oligocene 'Riversleigh Fauna' apparently assuming there was just one faunal assemblage from Riversleigh when there are in fact over 200 spanning the late Oligocene to early, mid and late Miocene, Pliocene and Pleistocene (Fig. 13). A Burdekin Plum (*Pleiogynium* sp. cf. *P. timorense*), represented by a well-preserved seed pod from Dunsinane Site (see also Arena 1997), represents a species that today occurs in rainforests, open forests and vine scrubs. On balance, the Dunsinane plants suggest cool, possibly seasonal, open forest and woodland in the Riversleigh area in contrast to closed forests which appear to have characterised the early to mid Miocene of this region (see below).

The Late Oligocene Marsupial Fauna

Three (Dasyuromorphia, Peramelemorphia, Diprotodontia) of the five Australasian marsupial orders have fossil representatives dating back to the late Oligocene (Table 3) but all five had probably differentiated by this time (see Section 'Interordinal Molecular Divergence Dates').

Dasyuromorphians

Dasyuromorphians include the vast majority of carnivores, insectivores and many of the omnivores of Australia. Dasyuromorphians are classified into three families: Dasyuridae (64 living species); Myrmecobiidae (1 species); and the Thylacinidae (1 recently extinct species). Although its interordinal affinities are in doubt (see Section 'Interordinal Relationships of Australasian Marsupials'), the monophyly of Dasyuromorphia is well supported by morphological (e.g. Wroe et al. 2000) and molecular data (e.g. Beck 2008a; Meredith et al. 2008a; Springer et al. 2009). Regardless of whether Notoryctemorphia or Peramelemorphia is the sister group to Dasyuromorphia, most molecular dates estimate that this order originated in the late Paleocene sometime between 63 and 56 Ma (Beck 2008a; Meredith et al. 2008a; Springer et al. 2009; Table 2).

The relationships of dasyuromorphian families to each other are unclear (Fig. 9). Krajewski et al. (2000a, b) found weak to moderate support for a sister group relationship between Dasyuridae and Thylacinidae to the exclusion of Myrmecobiidae based on mitochondrial *cytochrome b* and *12S rRNA* and protamine P1 gene sequences. The marsupial supertree analysis of Cardillo et al. (2004) produced varied results depending on the phylogenetic method used, although their unweighted supertree found strong support for a dasyurid-myrmecobiid clade to the exclusion of thylacinids. Morphology-based analyses have produced similarly equivocal results with Wroe's (1997a) cladistic analyses of craniodental data favouring neither sister group relationship over the other.

Contributing to this uncertainty is the absence of any pre-Pleistocene fossil record of myrmecobiids and the absence of undoubted dasyurids in late Oligocene deposits as well as the rarity of dasyurids in early to middle Miocene deposits Wroe, 1999. The latter appears to challenge the long-held view that dasyurids were structurally ancestral to the Australian marsupial radiation (Bensley 1903; Ride 1964; Szalay 1994). Wroe (1999) concludes that the oldest undoubted dasyurid, as determined by cranial synapomorphies, is *Barinya wangala* Wroe, 1999 (Fig. 17) from the early Miocene Neville's Garden LF (Faunal Zone B) of Riversleigh. This taxon was assigned to Dasyuridae on the basis of four unequivocal basicranial synapomorphies. Wroe (1999) further indicated a range of synapomorphic features of the auditory region shared by *B. wangala* and modern dasyurids that would preclude a plesiomorphic position for the family within Dasyuromorphia. Murray and Megirian (2006b) indicate three of these putative dasyurid synapomorphies are present in the Miocene thylacinid *Mutpuracinus archibaldi* Murray and Megirian, 2000. The results of Murray and Megirian's (2006b) cladistic analysis conflict with those of Wroe (1999) by suggesting

Table 3 Generic and species-level diversity (genus/species) of Australasian marsupial families from the late Oligocene to recent

	Late Oligo.	Early Mio.	Mid Mio.	Late Mio.	Plioc.	Pleist.	Holo.
Dasyuromorphia							
Dasyuridae		1/1	1/1	1/1	7/7	11/25	21/66
Thylacinidae	3/3	4/4	5/6	2/4	1/1	1/1	1/1
Myrmecobiidae							1/1
Dasyuromorphia *i.s.*	4/4	3/3	1/1	1/1			
Peramelemorphia							
Yaralidae	1/1	1/1	1/1				
Peramelidae					1/4	3/4	3/8
Peroryctidae					1/1		4/11
Thylacomyidae					1/1	2/2	2/2
Peramelemorphia *i.s.*	1/1	4/5	5/6				
Notoryctemorphia							
Notoryctidae		1/1					1/2
Yalkaparidontia							
Yalkaparidontidae		1/1	1/1				
Diprotodontia							
VOMBATIFORMES							
Phascolarctidae	4/7	3/4	2/3	1/1	1/3	2/4	1/1
Vombatidae	1/1	2/2	2/2	2/2	2/3	6/8	2/3
Ilariidae	3/4						
Wynyardiidae	3/5	2/4					
Maradidae	1/1						
Diprotodontidae	5/8	2/2	2/5	5/6	7/9	4/4	
Palorchestidae	1/1	1/1	1/1	1/2	1/3	1/3	
Thylacoleonidae	2/2	4/5	2/3	1/2	1/2	1/2	
Vombatiformes *i.s.*					1/2		
PHALANGERIDA							
Acrobatidae		2/2	1/1			1/1	2/2
Tarsipedidae						1/1	1/1
Petauridae		1/1	1/1		2/3	3/6	3/11
Pseudocheiridae	2/4	4/8	4/14	3/3	6/7	3/5	6/17
Petauroidea *i.s.*		3/3	1/1	1/1			
Burramyidae	1/2	2/3	2/2	1/1	2/2	2/3	2/5
Phalangeridae		3/5	2/5	1/1	2/2	1/1	6/23
Miralinidae	2/5	1/2	1/2				
Ektopodontidae	2/4	3/4			2/2	1/1	
Pilkipildridae	2/4	1/1	1/1				
Balbaridae	5/11+	4/13	3/5				
Hypsiprymnodontidae	1/1	2/3	2/4	1/2	2/2	1/3	1/1
Potoroidae	1/1	1/1	1/1		3/3	4/9	4/9
Macropodidae	3/5	4/4	4/6	5/6	15/33	18/69	12/60
Macropodoidea *i.s.*	3/3						
Marsupialia i.s.							
Yingabalinaridae		1/1					
Numbigilgidae					1/1		

Updated from Archer et al. (1999)
i.s. incertae sedis

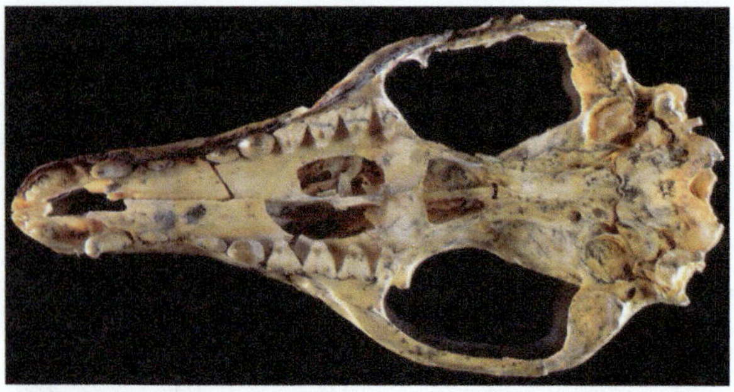

Fig. 17 Well-preserved skull of the oldest-known undoubted dasyurid *Barinya wangala* from Riversleigh (R. Arnett). Because they have been preserved in freshwater limestone, preservation of many of the Oligocene and Miocene skulls from Riversleigh is unusually good

instead that *B. wangala* with extant dasyurids form the plesiomorphic sister group to Thylacinidae.

Thylacinids (with the thylacoleonids) appear to be the dominant marsupial carnivores of the late Oligocene and much of the Miocene, with numerous fossil taxa being described in the last two decades (Archer 1982b; Muirhead and Archer 1990; Muirhead 1992, 1997; Muirhead and Wroe 1998; Wroe 1996b, 2001; Wroe and Musser 2001; Murray and Megirian 2000, 2006b). The oldest thylacinids from late Oligocene deposits include *Badjicius turnbulli* Muirhead and Wroe, 1998 from White Hunter LF, Faunal Zone A, Riversleigh; *Nimbacinus dicksoni* Muirhead and Archer, 1990 from Faunal Zones A-C, Riversleigh (Fig. 18); *Nimbacinus* sp. cf *N. dicksoni*

Fig. 18 *Nimbacinus dicksoni* skull and dentary emerging from a partially processed limestone block from Riversleigh's AL90 deposit (A. Gillespie). This fox-sized thylacinid was found as an intact skeleton curled up in the relaxed posture of an animal that had died in its sleep about 15 Ma. It lay adjacent to a similarly well-preserved skeleton of an archaic kangaroo

(Kangaroo Well LF, N.T.; Megirian et al. 2004); and an un-named new genus and species from the Pwerte Marnte Marnte LF of the N.T. (Murray and Megirian 2006a).

Beck (2008a) estimated that thylacinids diverged from dasyurids around 26 Ma. Krajewski et al. (2000b), on the basis of mitochondrial *cytochrome b* and *12S rRNA* sequences, estimated myrmecobiids diverged from a dasyurid/thylacinid clade between 25.4 and 27.8 Ma. Both estimates are congruent with the known fossil record. However, as noted by Beck (2008a), these dates imply a relatively rapid rate of cladogenesis within Thylacinidae. The presence of numerous derived thylacinids in late Oligocene to early Miocene deposits prompted Murray and Megirian (2006b, Fig. 17) to suggest an early Oligocene divergence for the thylacinid and dasyurid lineages.

As a consequence of the apparent absence of dental synapomorphies to define Dasyuridae, Wroe (1996b, 1997a) allocated several central Australian late Oligocene taxa, originally described as dasyurids on the basis of dental evidence, to Dasyuromorphia *incertae sedis*. The mouse-sized *Ankotarinja tirarensis* Archer, 1976a, from the Ditjimanka LF of S.A. has been variously argued to be a dasyurid, peradectid or microbiotheriid (Long et al. 2002). Other taxa include *Apoktesis cuspis* Campbell, 1976 from the Ngapakaldi LF; *Keeuna woodburnei* Archer, 1976a from the Ditjimanka LF; and *Dasylurinja kokuminola* Archer, 1982a from the Ngama LF. Murray and Megirian (2006b) have challenged Wroe's (1996b, 1997a) removal of the above taxa from Dasyuridae. However, certain attribution to a particular family (or potentially certainty that they do not represent any known dasyuromorphian family) requires discovery of basicranial remains.

Peramelemorphians

Bandicoots and bilbies (Fig. 2) are the modern representatives of an enigmatic, potentially endemic group of Australian marsupials with unclear ordinal affinities. Peramelemorphia has been variously interpreted at one time or another as the sister group to each of the four extant Australasian marsupial orders as well as sister group to all other australidelphians or potentially even part of the paraphyletic 'ameridelphian' radiation (see Section 'Interordinal Relationships of Australasian Marsupials'). Discovery of a potential bandicoot upper molar in the early Eocene Tingamarra

LF of Qld could be taken as support for the argument that peramelemorphians differentiated before all of the other Australian orders. Woodburne and Case (1996) suggested that the bunodont *T. bartholomaii*, from the Tingamarra LF, was a peramelemorphian. However there is no credible basis for this suggestion given that it shares no synapomorphies with peramelemorphians (see Section 'The Palaeogene Marsupials of Australia: The Early Eocene').

Undoubted peramelemorphians are currently divided into two superfamilies: Perameloidea, containing the three extant families Thylacomylidae, Peroryctidae and Peramelidae; and Yaraloidea, containing the extinct family Yaralidae (Fig. 19). If numbigilgids (containing *Numbigilga ernielundeliusi*; Fig. 20) are peramelemorphians (a possibility discussed by Beck et al. 2008a), their relationships to one or the other of these two superfamilies is unclear. Molecular divergence dates suggest a late Oligocene to early Miocene (20–29 Ma) divergence of Thylacomyidae from Peramelidae (Meredith et al. 2008c). Meredith et al. (2008c, 2009) further estimated Peroryctidae (their Peroryctinae + Echimyperinae) and Peramelidae (their Peramelinae) diverged during the late Miocene (7–13 Ma), with similar dates obtained by Beck (2008a; 11.7±1.5 my).

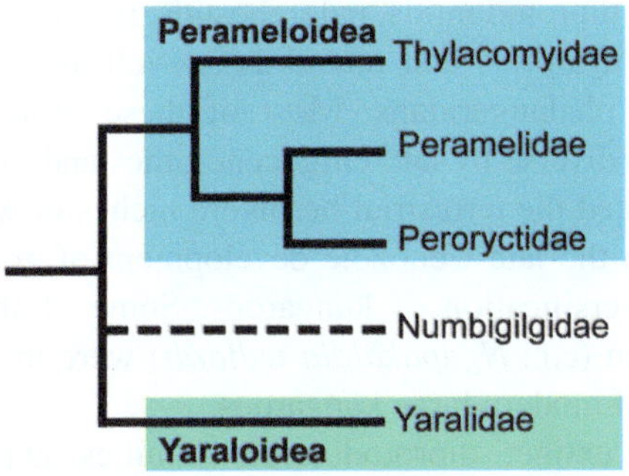

Fig. 19 Phylogenetic interrelationships of peramelemorphians including extinct groups based on dental morphology. The *dashed line* indicates tentative relationships of the enigmatic numbigilgids to this order of marsupials (Fig. 20). Yaraloids, all of which vanished before the Pleistocene, are more archaic than any of the living groups most of which do not have a fossil record until the Pliocene. However, these almost certainly began to diversify during the relatively poorly documented late Miocene

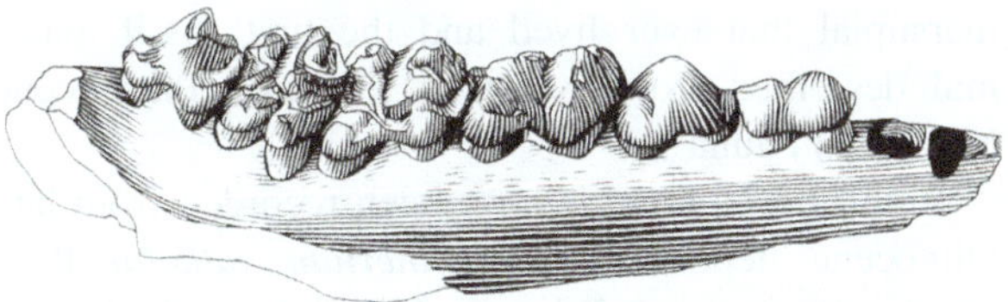

Fig. 20 The partial right dentary of a very strange numbigilgid marsupial (drawing J. Muirhead). The relationships of this group are a mystery; they may be peramelemorphians (Fig. 19) or they may represent an entirely different order. However there is nothing like this taxon in any of the other known marsupial orders. Strange features include the complex extra cusps and blades on the molars and the two-cusped second premolar

Superficially, the peramelemorphian fossil record appears relatively depauperate with only six named species. However, numerous undescribed taxa (Fig. 21) are recorded from late Oligocene to middle Miocene deposits at Riversleigh (Archer et al. 1994, 1997) and central Australia (Woodburne et al. 1993). Further, Schwartz (2006a) indicated that bandicoots are the most abundant faunal element in the late Oligocene Kangaroo Well LF of the N.T., and Murray and Megirian (2006a) indicate the presence of a new perameloid family in the late Oligocene Pwerte Marnte Marnte LF (N.T.).

Yaralids are a geographically widespread group of plesiomorphic, short-snouted bandicoots (Long et al. 2002) including one of the oldest known (with the possible exception of the early Eocene Tingamarran taxon discussed above) peramelemorphians, *Yarala kida* Schwartz, 2006a from the late Oligocene Kangaroo Well LF. *Yarala kida* is more plesiomorphic and possibly ancestral to the type species, *Yarala burchfieldi* Muirhead and Filan, 1995 from early to middle Miocene deposits of Riversleigh (Schwartz 2006a). Peramelemorphians in the late Oligocene Ditjimanka

Fig. 21 Yaralid bandicoots are among the most common marsupials found in the rich early to mid Miocene fossil deposits of Riversleigh (K. Atkinson). Sometimes, as in this case where the specimen has been partially exposed by dissolution of limestone in dilute acetic acid, skeletons are preserved with associated elements found next to each other

LF (S.A.) include small yaralids (*fide* MA) but these have yet to be formally described. The feeding preferences for species of *Yarala* have been interpreted, by analogy with extant dasyurids, to have been insectivorous/carnivorous, as opposed to insectivorous/omnivorous preferences exhibited by modern bandicoots (Muirhead and Filan 1995; Schwartz 2006a). During the late Oligocene to middle Miocene, bandicoots appear to have dominated the insectivorous/carnivorous niche later occupied by the smaller dasyurids during the latter half of the Neogene (Muirhead 1994; Wroe 1999; Schwartz 2006a). It is possible that as dasyurids differentiated in the late Miocene, they out-competed comparably sized peramelemorphians and in particular the previously diverse yaraloids which were extinct by the mid Pliocene.

Diprotodontians

By the late Oligocene, diprotodontians were already highly diverse with at least 17 of the 21 families already in existence (Table 3). Molecular divergence dates (Drummond et al. 2006; Beck 2008a; Meredith et al. 2008a, 2009) indicate early to middle Eocene origins for most of the major diprotodontian groups (Table 2). Beck (2008a) estimated that the diprotodontian suborders Vombatiformes (koalas and vombatomorphians) and Phalangerida (possums and kangaroos) diverged from each another approximately 57.1 Ma (± 3.4 Ma; 50.5–62.9, 95% confidence interval). He recognised, however, that the absence of any diprotodontian taxon from the Tingamarra LF (54.6 Ma) may be compelling evidence for a younger date within the 95% confidence limits. Divergence dates for the phalangeridan/vombatiform split younger than the Tingamarran LF age were proposed by Drummond et al. (2006, 48 Ma), Meredith et al. (2008a, 54.1 Ma) and Meredith et al. (2009, 53.3 Ma; Fig. 9).

Vombatiforms

Calculated dates for the phascolarctomorphian (koala-like)/vombatomorphian (wombat-like) split range from 40 to 35 Ma (Beck 2008a; Meredith et al. 2008a, 2009; Springer et al. 2009). Seven koala species (Table 3), some of questionable specific distinction, are recorded from late Oligocene deposits in central Australia, although no more than two coexisted in any given faunal assemblage: *Madakoala devisi* Woodburne et al., 1987a from Pinpa, Ericmas and Tarkarooloo LFs, S.A.; *M. wellsi* Woodburne et al., 1987a from Ericmas LF, S.A.; *Madakoala* sp. (Wadikali LF, S.A.); *Perikoala palankarinnica* Stirton, 1957b from Ditjimanka LF, S.A.; *P. robustus* Woodburne et al., 1987a from Ditjimanka and Palankarinna South LFs, S.A.; *Litokoala kutjamarpensis* Stirton et al., 1967 includes *L. kanunkaensis* Springer, 1987 from Kanunka North LF, S.A.; and *Nimiokoala* sp. Black and Archer, 1997b from South Prospect B LF, S.A. Koalas are relatively rare elements in all faunal assemblages in which they occur, being more abundant today than they have been at any previous time spanning the last 25 million years (Black 1999).

The spread of open forest and woodland habitats during the late Oligocene appears to have particularly benefited large-bodied vombatomorphian browsers such as diprotodontids, palorchestids, ilariids, vombatids, wynyardiids and maradids as well as kangaroos among phalangeridans. Most of these groups were already diverse by late Oligocene times and may have dominated the terrestrial herbivore niches in Australia prior to the late Cenozoic development of grasslands and diversification of kangaroos. Some of the most common (e.g. *Ngapakaldia tedfordi*) were in the size range of modern large kangaroos.

The extinct diprotodontids (families Diprotodontidae and Palorchestidae) are a geographically and temporally diverse group of quadrupedal lophodont browsers. Their abundance in late Oligocene to late Pleistocene fossil deposits has made them arguably the most useful marsupial group for biocorrelating Australia's Tertiary fossil assemblages (Stirton et al. 1967; Black 1997b; Murray et al. 2000a, b; Piper 2006). Among their ranks was the Pleistocene *Diprotodon optatum* Owen, 1838 (Fig. 22), the largest marsupial that ever lived and the first fossil mammal described from Australia (Section 'Pleistocene Marsupial Fauna').

Eight diprotodontids have been recorded from late Oligocene deposits: *Raemeotherium yatkolai* Rich et al., 1978 from Ericmas LF, S.A.; *Silvabestius johnnilandi* Black and Archer, 1997a, *S. michaelbirti* Black and Archer, 1997a and *S.* sp. Black and Archer, 1997a from Faunal Zone A, Riversleigh, Qld; *Ngapakaldia tedfordi* Stirton, 1967a and *Pitikantia dailyi* Stirton, 1967a from Ngapakaldi LF Faunal Zone C, S.A.; *N. bonythoni* Stirton, 1967a from

Fig. 22 The largest marsupial known, the three-tonne diprotodontid *Diprotodon optatum* from the Australian Pleistocene (P. Murray). While this dry-country megafaunal species may have looked wombat-like, diprotodontids were only distantly related to the much smaller, fossorial vombatids

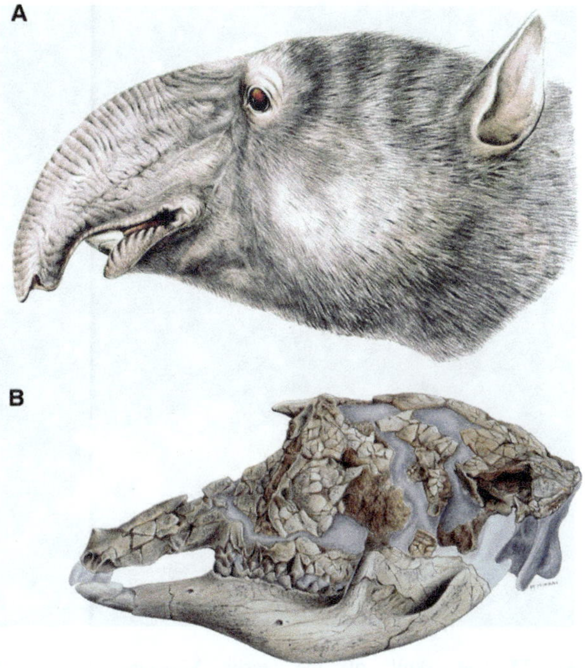

Fig. 23 The late Miocene palorchestid *Palorchestes painei* from Alcoota in the Northern Territory. (**a**) Reconstruction showing the characteristic tapir-like trunk of palorchestids (P. Murray). (**b**) Rendering of the skull that formed the basis for the cranial reconstruction, showing at the *top* the tapir-like position of the tips of the nasal bones (P. Murray)

Ngapakaldi LF, Faunal Zone C, S.A.; Faunal Zones A-B, Riversleigh; and *Neohelos tirarensis* Stirton, 1967b from Kutjamarpu LF, S.A.; Faunal Zones A-C, Riversleigh. Most late Oligocene diprotodontids were sheep-sized, with *N. tirarensis* being calf-sized (though increasing in size throughout Faunal Zones A–C) and the largest species, *N. bonythoni*, comparable in size to a domestic cow. *Ngapakaldia bonythoni* was the most abundant diprotodontid during the late Oligocene at Riversleigh, known from 20 individuals from 14 LFs Black 2010). In contrast, its smaller congener, *N. tedfordi*, was the most abundant diprotodontid in the late Oligocene deposits of central Australia (Rich and Rich 1987).

Compared with diprotodontids, palorchestids are rare components of Australia's fossil assemblages which may indicate a solitary lifestyle (Flannery and Archer 1985). Commonly referred to as 'marsupial tapirs' (e.g. Bartholomai 1978a; Murray 1986, 1990a), palorchestids display modifications in their crania (i.e. retraction of the nasal bones and presence of a large infraorbital foramen on the rostrum) that suggest the presence of a large, well-innervated, mobile trunk (Fig. 23). One late Oligocene species is known: *Propalorchestes ponticulus* Murray, 1990a from Riversleigh's Faunal Zone A assemblages.

Wynyardiids are dog-sized browsers with semi-lophodont/sub-selenodont teeth (Fig. 24). They are not uncommon in late Oligocene to early Miocene assemblages. The family was originally described on the basis of an edentulous skull and partial skeleton found at the base of a cliff in the coastal town of Wynyard, Tasmania. Spencer (1901) interpreted *Wynyardia bassiana* to be structurally intermediate between polyprotodont and diprotodont marsupials, but later authors recognised diprotodontian affinities but specifically affinities with phalangeroids (e.g. Jones 1930, Haight and Murray 1981). Basicranial anatomy led Aplin (1987) and Aplin and Archer (1987) to include this family within the suborder Vombatiformes (Fig. 25). Five late Oligocene wynyardiid species are known: *Namilamadeta snideri* Rich and Archer, 1979 from Tarkarooloo LF, S.A.; *N. albivenator* Pledge, 2005 from Faunal Zone A, Riversleigh, Qld; *Ayekaye jaredi* Megirian et al., 2004 from Kangaroo Well LF, N.T.; *Muramura williamsi* Pledge, 1987a from Ditjimanka LF, Etadunna Fm, S.A.; and *M. pinpensis* Pledge, 2003 from Pinpa LF, Namba Fm, S.A. *Muramura* species are known from relatively complete skeletons which show adaptations to a possible semi-fossorial lifestyle (Pledge 2003).

Ilariids were quadrupedal, dog- to calf-sized, short-faced browsers with selenodont dentitions (Fig. 26). Two genera and three species are currently named and all are restricted to late Oligocene deposits: *Ilaria illumidens* Tedford and Woodburne, 1987 from Pinpa

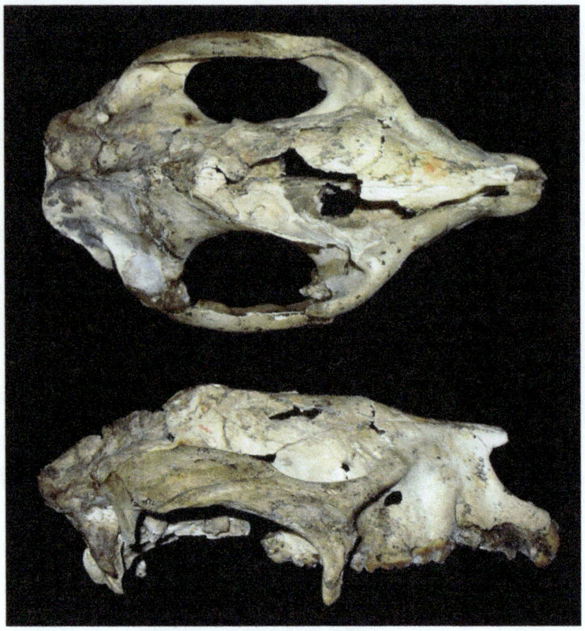

Fig. 24 The skull of the early Miocene wynyardiid *Namilamadeta superior* from Upper Site at Riversleigh (K. Black). Wynyardiids were first known from a Tasmanian specimen that lacked teeth. Until discovery of dentitions in central Australia and Riversleigh their now undoubted relationship to vombatiform marsupials was unclear. All of them were extinct by the end of the middle Miocene

Fig. 26 Semi-selenodont dentition of the ilariid *Kuterintja ngama* from the late Oligocene White Hunter Site at Riversleigh (T. Myers). These vombatiforms, although fox-size, had teeth that were similar to those of the extinct placental anthracotheres. The shapes of teeth indicate probable diet and in this case they almost certainly ate soft leaves possibly of trees

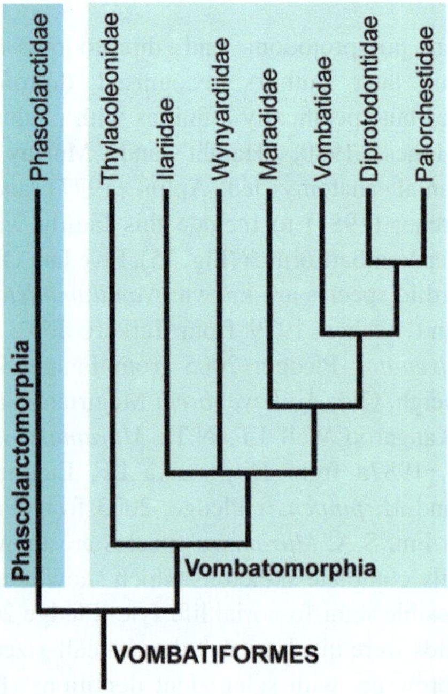

Fig. 25 Relationships of extinct families within the suborder Vombatiformes (after Black 2008). This suborder in itself is as or more diverse than the placental order Perissodactyla and included forms that ranged from tapir-like herbivores (palorchestids) to the most specialised of mammalian carnivores – the marsupial lions (Thylacoleonidae)

LF, S.A.; *Ilaria lawsoni* Tedford and Woodburne, 1987 from Ditjimanka LF, S.A.; and *Kuterintja ngama* Pledge, 1987b from Ngama LF, S.A. and White Hunter LF, Faunal Zone A, Riversleigh, Qld (Myers and Archer 1997). An as yet unnamed new genus and species has also been found in the late Oligocene Pwerte Marnte Marnte LF of the N.T., and may represent the plesiomorphic sister group or structural precursor to all other ilariids (Murray and Megirian 2006a). Ilariids have been variously interpreted as the plesiomorphic sister group to a wynyardiid-vombatid-diprotodontoid clade (e.g. Marshall et al. 1990; Murray 1998), or, on the basis of postcranial morphology, the sister group of vombatids (Munson 1992). Myers et al. (1999) expressed their unresolved status within Vombatomorphia as a trichotomy with vombatoids and a wynyardiid-diprotodontoid clade.

Another enigmatic vombatomorphian group is the recently established family Maradidae which contains the single species *Marada arcanum* Black, 2007. Black (2007) erected the family on the basis of a single mandible (Fig. 27) from the Riversleigh Faunal Zone A Hiatus LF. The mandible reveals a unique

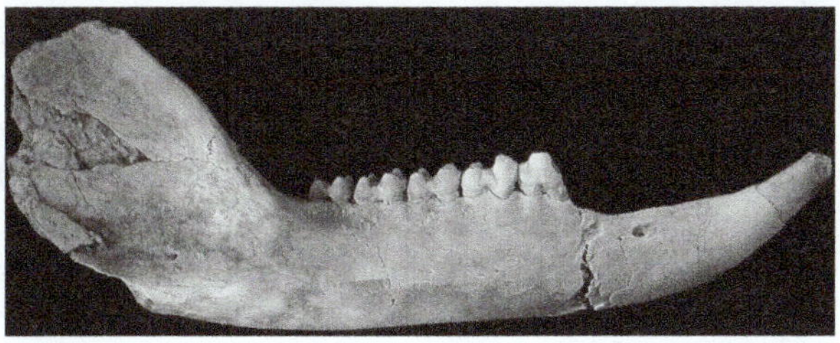

Fig. 27 The only known jaw (and Holotype) of the enigmatic vombatomorphian family Maradidae, *Marada arcanum*, from the late Oligocene Hiatus Site at Riversleigh (K. Black). In the absence of more information, the ecological role of this marsupial remains a mystery

(among vombatiforms) bunolophodont dentition and a combination of plesiomorphic and autapomorphic features unlike those of any other vombatomorphian family. Subsequent analysis (KHB) of a cast of a lower molar (LM_3) from the late Oligocene Pwerte Marnte Marnte LF from the N.T., noted by Murray and Megirian (2006a, Fig. 10), suggests this tooth may represent the same species. Maradidae is the sister group of a clade containing vombatids and diprotodontoids (Black 2008; Fig. 25).

The oldest, most plesiomorphic wombat is *Rhizophascolonus crowcrofti* Stirton et al., 1967 from the Kutjamarpu LF, S.A. and Bone Reef LF, Faunal Zone A, Riversleigh (Archer et al. 2006). Unlike all other vombatids which possess ever-growing, open-rooted (hypselodont) teeth, *R. crowcrofti* possessed rooted, albeit hypsodont teeth (Fig. 28) that may have been an adaptation for eating rhizophagus and/or abrasive vegetation such as bark or stems (Brewer et al. 2008).

Thylacoleonids (marsupial lions) are unique among vombatiforms in being the only carnivorous members of the suborder. Their dentition is characterised by progressive elaboration and hypertrophy of the posterior premolar (P3) into an elongate, shearing blade and coincident reduction of the molar row from the posterior end forward. Three genera are currently recognised (*Priscileo*, *Wakaleo*, *Thylacoleo*) but a re-evaluation of these taxa is currently underway along with description of new taxa from the Oligo-Miocene deposits of Riversleigh (Gillespie 2007). Two species are known from late Oligocene deposits: the cat-sized *Priscileo pitikantensis* Rauscher, 1987 from Ngapakaldi LF, S.A.; and *Wakaleo* n. sp. 1 from Faunal Zones A–B (Archer et al. 2006). Marsupial lions are generally rare elements in Tertiary fossil assemblages which is no doubt a reflection of their ecological roles as carnivores (Gillespie 1999).

Overall, survival of a single vombatomorphian family today (wombats) belies the highly diverse nature

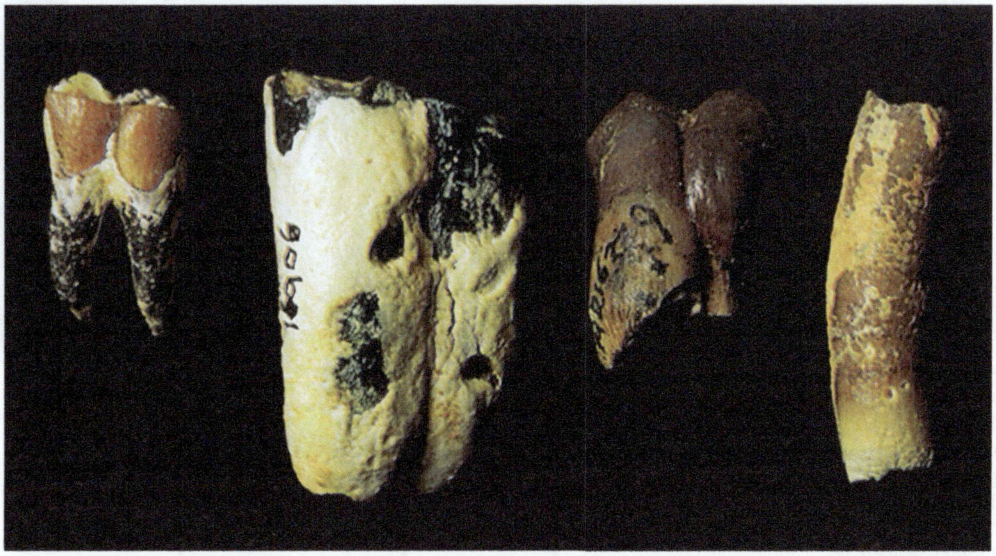

Fig. 28 Teeth of primitive wombats, all from Riversleigh (M. Archer). They demonstrate, moving left to right from the left tooth, which is early Miocene in age, a steady increase in the height of the crown until by late Miocene time (tooth at for right) they had lost their roots altogether and the teeth erupted continuously throughout life. This evolutionary transition suggests adaptation to an increasingly abrasive diet within 10 million years

of the vombatomorphian radiation. However, at least three families (maradids, ilariids and wynyardiids) do not have a fossil record beyond the early Miocene (Table 3); hence family-level diversity in this suborder appears to have been declining in steps through the latter part of the Cenozoic. It has been suggested (Archer 1984; Black 2007) that coincident increases in generic diversity in the kangaroo (macropodoid) radiation, particularly from the late Miocene to Pleistocene, may indicate long-term competition between the two groups.

Phalangeridans

Four of the five phalangeridan (possums and kangaroos) superfamilies and all but four families (Phalangeridae, Tarsipedidae, Acrobatidae, Pilkipildridae) are represented in late Oligocene faunal assemblages (Table 3). Divergence dates (Table 2) suggest the basal split within Phalangerida occurred during the early Eocene (~50–51 Ma; Beck 2008a; Meredith et al. 2009; Springer et al. 2009). The analysis of five nuclear gene sequences by both Meredith et al. (2009) and Springer et al. (2009) suggests a basal split between a macropodoid/phalangeroid/burramyoid clade and a clade containing petauroids, tarsipedids and acrobatids (Fig. 9). The molecular results do not support the monophyly of Tarsipedoidea (Tarsipedidae + Acrobatidae) suggested by Aplin and Archer (1987) on the basis of morphological evidence, but support the inclusion of tarsipedids and acrobatids within Petauroidea. Beck (2008a) did not include tarsipedids or acrobatids in his analysis and found that the basal split occurred between Petauroidea and a Macropodoidea/Burramyoidea/Phalangeroidea clade. Beck (2008a) and Phillips and Pratt (2008) suggested members of the latter clade were united by the synapomorphic possession of a hypertrophied, serrated third premolar. Meredith et al. (2009) accordingly proposed the name Australoplagiaulacoida for this clade. Point estimates for the subsequent divergence of Macropodoidea from a Phalangeroidea–Burramyoidea clade range from 42 to 48 Ma (Beck 2008a; Meredith et al. 2008a, 2009; Springer et al. 2009).

Burramyoids The superfamily Burramyoidea, containing the single family Burramyidae (Table 3, Fig. 29), was established by Aplin and Archer (1987) in recognition of the degree of morphological

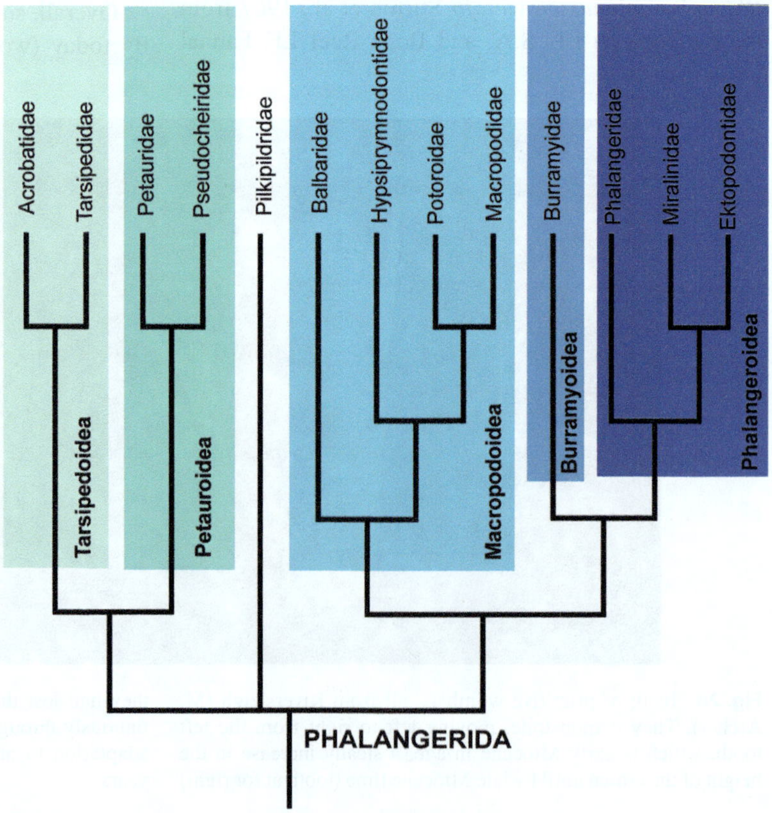

Fig. 29 Relationships of extinct families within the suborder Phalangerida (after Archer et al. 1999). The relationships of the five superfamilies reflect growing understanding about molecular biology as well as cranial and dental morphology. However, the relationships of the diminutive extinct pilkipildrid possums remain in doubt

Fig. 30 The extant mountain pygmy-possum *Burramys parvus*, one of the largest of the burramyids (H. Bates). Having first been discovered and named as a fossil in the 1800s before being discovered alive in 1966 in an alpine ski lodge, this highly specialised possum is now in danger of becoming an extinct fossil again because of growing climate-change-related threats to its alpine habitat

distinction of the members of this family from all other phalangeridans. Modern burramyids include the Mountain Pygmy Possum (*Burramys parvus*, Fig. 30) and four species of *Cercartetus* (Fig. 3g), although the latter are sometimes placed in three genera (*Cercartetus*, *Dromiciola*, *Eudromicia*) based primarily on differences in premolar morphology and degree of reduction at the posterior end of the molar row.

Burramyoidea is estimated to have diverged from Phalangeroidea between 40 and 43 Ma (Beck 2008a; Meredith et al. 2008a, 2009; Springer et al. 2009). The oldest burramyids are *Burramys wakefieldi* Pledge, 1987c from the late Oligocene Ngama LF, S.A., and *Burramys brutyi* Brammall and Archer, 1997 from the late Oligocene of Riversleigh. *Burramys brutyi* has been recorded from all four faunal zones (A–D) at Riversleigh giving it one of the longest temporal spans (late Oligocene–early late Miocene) of any marsupial taxon. It is generally abundant in all Riversleigh sites in which it occurs although considerably less abundant in Faunal Zone A (Brammall and Archer 1997). It appears that the spread of lowland rainforest during the early Miocene was beneficial to species of *Burramys*. An additional species (*Burramys triradiatus* Turnbull, Rich and Lundelius, 1987) has been recorded from the Pliocene rainforest assemblage of the Hamilton LF, Victoria. The modern species, *B. parvus*, is restricted to alpine and subalpine habitats in two small areas (20 × 10, 30 × 10 km, respectively) in the Snowy Mountains of Vic. and N.S.W. (Strahan 1995), a habitat under threat from imminent global warming. The abundance of *Burramys* in rainforest palaeohabitats has led Archer and Hand (2006) to suggest that extant lowland temperate rainforests may provide a suitable alternative habitat for the modern species.

As yet, no extinct species of *Cercartetus* have been described, although Brammall (1998) and Archer et al. (2006) indicate the presence of new species in early to middle Miocene sediments at Riversleigh.

Phalangeroids Phalangeroidea contains three families, two of which are extinct: Phalangeridae, Miralinidae (extinct; with two subfamilies, Miralininae and Durudawirinae) and Ektopodontidae (extinct). A fourth family, Pilkipildridae (extinct) may belong to this superfamily. Although affinity with miralinids has been suggested Aplin and Archer (1987) noted other reasons for possible affinity to petauroids. Acknowledging the uncertainty, they formally classified Pilkipildridae as Phalangerida *incertae sedis* (Table 1).

Within Phalangeroidea, there is a general trend of increasing molar complexity from phalangerids, through miralinids to ektopodontids. Aplin and Archer (1987) and Woodburne et al. (1987b) regarded miralinids to be the sister group to ektopodontids with phalangerids the sister group to a miralinid–ektopodontid clade (Fig. 29). Both miralinids and ektopodontids have been described from late Oligocene deposits of central Australia. However, undoubted phalangerids do not make an appearance in the fossil record until the early Miocene.

Late Oligocene miralinids include *Miralina doylei* Woodburne, Pledge and Archer, 1987b and *M. minor* Woodburne, Pledge and Archer, 1987b, both from the late Oligocene Ditjimanka LF of S.A.; and a possible third and fourth species from the late Oligocene Tarkarooloo LF (*Miralina* sp.), and the Yanda and Wadikali LFs (*Miralina* sp. cf. *M. minor)* of South Australia. Schwartz (2006b) recently described a new genus, *Barguru* and three new miralinine species from the late Oligocene Kangaroo Well and middle Miocene Bullock Creek LFs of the N.T., indicating a greater diversity and temporal range for the family than previously known.

Ektopodontids are an enigmatic group of possums with complex, highly serrated dentitions (Fig. 31)

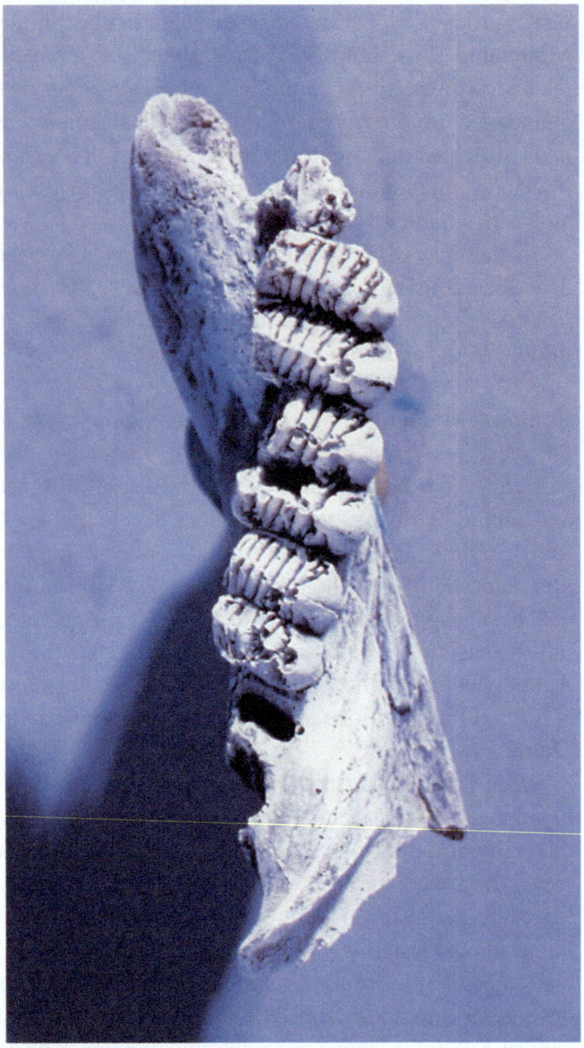

Fig. 31 The strikingly complex, somewhat rodent-like lower teeth of *Ektopodon stirtoni*, from Mammalon Hill, S.A., are typical of all ektopodontid possums (N. Pledge). These similarities may have led to competition with native rodents and extinction prior to the late Pleistocene

that have been likened to those of multituberculates (Crosby et al. 2004) and rodents (Archer 1984; Archer and Hand 2006). Pledge (1982) speculated about feeding strategies which he suggested may have ranged from being granivorous, to folivorous and/or insectivorous to omnivorous. The oldest ektopodontids are known from late Oligocene deposits in central Australia and the N.T. and include *Chunia illuminata* Woodburne and Clemens, 1986 from Ditjimanka LF, S.A.; *Chunia omega* Woodburne and Clemens, 1986 from Tarkarooloo LF, S.A.; *Ektopodon stirtoni* Pledge, 1986 from Ngama LF, S.A.; and *Ektopodon ulta* Megirian et al., 2004 from Kangaroo Well LF, N.T. Most species are known from isolated teeth or dentitions except *E. stirtoni* which is also known from a partial palate preserving a portion of the zygomatic arch. Pledge (1982) noted the short face and large anteriorly orientated orbits of this specimen which indicate enhanced stereoscopic vision similar to that seen in arboreal primates (Crosby et al. 2004).

Petauroids Modern petauroids are divided between two families: Petauridae which includes most of smaller gliding petauroids, striped possums and Leadbeater's Possum; and Pseudocheiridae, the ringtail possums and greater gliders. Although circum-continental in distribution, the highest diversity of pseudocheirids occurs in high-altitude tropical rainforests in northeastern Queensland and New Guinea. Petaurids occupy diverse habitats. The gliding petaurids (species of *Petaurus*) are commonly found in open forest/woodland environments while the striped possums are restricted to rainforest (Strahan 1995). These families are estimated on the basis of molecular data to have diverged from each another during the late Eocene or earliest Oligocene (Fig. 9, 29–35 Ma; Beck 2008a; Meredith et al. 2008a, 2010). This relatively late divergence (compared with other possum families) may explain the low diversity of petaurids and pseudocheirids in late Oligocene deposits. It was not until the onset of greenhouse conditions in the early Miocene that these groups underwent significant diversification.

Overall, few fossil petaurids have been described and none that are known are from the late Oligocene. Late Oligocene pseudocheirids comprise two genera (*Marlu* and *Pildra*) which are known from the Tirari Desert and Frome Basin of S.A., and from the Kangaroo Well LF of the N.T. Travouillon et al. (2009) also note the presence of three sympatric species in the late Oligocene White Hunter LF of Riversleigh. Species of *Paljara* were previously interpreted (Bassarova and Archer 1999; Bassarova et al. 2001; Crosby et al. 2004) to be present at Riversleigh during the late Oligocene. However, reassessment of the age of the deposits in which they are found suggests species of this genus may be restricted to the Miocene throughout Australia (Roberts et al. 2008).

Macropodoids Kear and Cooke (2001) most recently revised kangaroo higher-level systematics. They recognise four kangaroo families: Balbaridae (extinct), Hypsiprymnodontidae, Potoroidae and Macropodidae (Fig. 29). This contrasts with the more traditional two family subdivision of Macropodoidea (e.g. Archer 1984; Aplin and Archer 1987; Flannery 1989) into Potoroidae (subfamilies Potoroinae, Palaeopotoroinae, Hypsiprymnodontinae, Bulungamayinae) and Macropodidae (subfamilies Macropodinae, Sthenurinae, Balbarinae). A subsequent comprehensive review of kangaroo relationships by Cooke (2006) confirms the basic results of Kear and Cooke (2001). In these more recent classifications (Kear and Cooke 2001), Balbaridae is regarded to be the plesiomorphic sister group of all other macropodoids, Hypsiprymnodontidae is the sister group of a Potoroidae/Macropodidae clade and Bulungamayinae is a stem group within Macropodidae (cf. Owen 1874).

Recent molecular analyses of extant taxa support Hypsiprymnodontidae as the sister group to all other macropodoids with divergence dates estimated to be between 26 and 35 Ma (Meredith et al. 2008b, 2009; Springer et al. 2009). Hypsiprymnodontidae is commonly (e.g. Archer et al. 1999) divided into the subfamilies Hypsiprymnodontinae, containing the single extant species musky rat-kangaroo *H. moschatus* and the extinct form *H. bartholomaii* Flannery and Archer, 1987b and Propleopinae (containing the extinct giant rat-kangaroo genera *Ekaltadeta*, *Jackmahoneya* and *Propleopus*). However, Wroe (1997b), Wroe et al. (1998) and Cooke and Kear (1999) have suggested that propleopines may share a closer relationship with balbarids than with hypsiprymnodontines.

A single propleopine species, *Ekaltadeta ima* Archer and Flannery, 1985 has been recorded from late Oligocene deposits at Riversleigh (Faunal Zone A Hiatus and White Hunter LFs) as well as numerous early to middle Miocene sites (Archer et al. 2006). Since its original description there has been much debate about the palaeobiology of this enigmatic kangaroo (Fig. 32). *Ekaltadeta ima* was the size of a female Grey Kangaroo (*Macropus giganteus*) with bunodont molars yet striking, enlarged plagiaulacoid premolars and marks on the enamel of the cheekteeth that have led some authors (e.g. Archer and Flannery 1985; Ride et al. 1997) to suggest this species (and potentially some later Cenozoic propleopines such as *Propleopus chillagoensis* Archer et al., 1978) was carnivorous. Wroe et al. (1998), in describing an adult skull of *E. ima*, noted the capacity of the premolars for powerful vertical shear and numerous features of the skull that could indeed be construed as carnivorous adaptations, and suggested instead that it might have been a bear-like omnivore.

Burk et al. (1998) estimated the divergence of potoroids from macropodids to have occurred 30 Ma on the basis of mitochondrial genome sequences. Meredith et al. (2008b, 2009) suggest divergence occurred between 20 and 24 Ma or between 13 and 20 Ma, respectively, but these dates are inconsistent with the fossil record given that potoroines are well known from the late Oligocene including

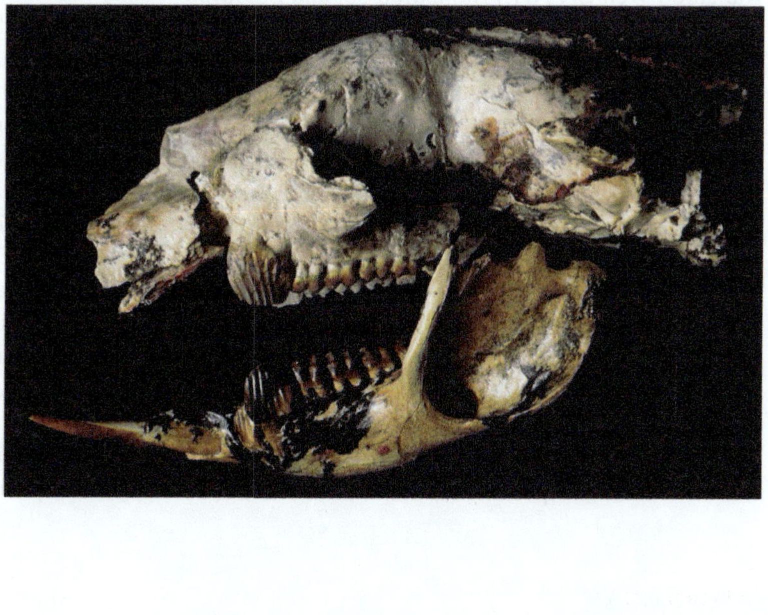

Fig. 32 Skull and lower jaw of the carnivorous kangaroo *Ekaltadeta ima* from Camel Sputum Site, Riversleigh (R. Arnett). These decidedly 'un-kangaroo-like' propleopine kangaroos were almost certainly quadrupedal with powerful forelimbs. From marks on their teeth, it appears that they were capable of using their dagger-like lower incisors to stab and their bizarre premolars and powerful molars to dismantle and consume small to medium-sized prey

Gumardee pascuali Flannery et al., 1983 from Site D, Faunal Zone A of Riversleigh and *Purtia mosaicus* Case, 1984 and *Kyeema mahoneyi* from localities in the Etadunna Fm, S.A. (Woodburne et al. 1993). However, Long et al. (2002) question placement of *P. mosaicus* and *G. pascuali* within Potoroinae, suggesting they are better regarded as Macropodoidea *incertae sedis*, and Meredith et al. (2008b) have suggested that *K. mahoneyi* may be a stem taxon basal to the macropodid-potoroid clade. Even so, the diversity of bulungamayine kangaroos in the late Oligocene deposits of Riversleigh and central Australia suggest that a hypothetical early Miocene divergence date for the potoroid-macropodid split is too recent. Late Oligocene bulungamayines from Riversleigh include *Bulungamaya delicata* Flannery et al., 1983 and *Wabularoo naughtoni* Archer, 1979. Further, Woodburne et al. (1993) note the presence of a species referable to *Bulungamaya* in late Oligocene Faunal Zones D and E of the Etadunna Formation.

The relatively plesiomorphic family Balbaridae was highly diverse by late Oligocene times (Table 3). Five genera and at least 11 species have been described from faunal assemblages of this age (Fig. 33). These include (from Faunal Zone A deposits at Riversleigh) *Balbaroo fangaroo* Cooke, 2000, *B. gregoriensis* Flannery et al., 1983, *Galanarla tessellate* Flannery et al., 1983, *Ganawamaya aediculis* Cooke, 1992, *Nambaroo couperi* Cooke, 1997a and *Wururoo dayamayi* Cooke, 1997a; (from the Tarkarooloo LF, S.A.) *Nambaroo tarrinyeri* Flannery and Rich, 1986, *Nambaroo saltavus* Flannery and Rich, 1986 and *Nambaroo novus* Flannery and Rich, 1986; (from the Kangaroo Well LF, N.T.) *Nambaroo* sp. and *Balbaroo* species. Cooke (2006) indicates that at least seven additional species of *Nambaroo* have been found in the Oligo-Miocene assemblages of Riversleigh and Woodburne et al. (1993) note three additional undescribed species of *Nambaroo* (sp. A–C) from the Etadunna Fm. The extent to which these Riversleigh and central Australian taxa may possibly be conspecific is not known.

Balbarids were traditionally included as a subfamily (Balbarinae) within Macropodidae but on the basis of subsequent cladistic analyses (e.g. Cooke 1997a; Cooke and Kear 1999; Kear 1998; Wroe et al. 1998) were given familial status (Cooke and Kear 1999; Kear and Cooke 2001) (Table 1, Fig. 6). Recent description of cranial and postcranial material of *Nambaroo* *gillespiae* Kear et al., 2007 from late Oligocene to early Miocene sediments of Riversleigh, has confirmed the monophyly of Balbaridae and its plesiomorphic position within Macropodoidea. Kear et al. (2007) noted similarities in postcranial morphology (e.g. robust forelimbs, lateral flexibility of the pes and opposable first digit of the pes) to *H. moschatus* and species of *Dendrolagus* (tree kangaroos) that suggested a similar quadrupedal gait and perhaps some capacity to climb. Consequently, as suggested by Cooke and Kear (1999) and Kear et al. (2007), evolution of the bipedal hopping gait characteristic of the majority of modern kangaroos, may not have occurred prior to the late Oligocene/early Miocene interval.

Fig. 33 Reconstruction of an undescribed primitive balbarid kangaroo from the late Oligocene of central Australia (R. Scott). Balbarids were among the most common of the archaic kangaroos in the forests of the late Oligocene to middle Miocene. Although in the main herbivorous, some would have been omnivorous and others carnivorous (Fig. 32). Flamingos, crocs and fish were abundant in the freshwater lakes that dotted inland Australia at this time

Phalangerida *incertae sedis*: Pilkipildrids The phylogenetic position of pilkipildrids within Phalangerida is currently uncertain (Fig. 29), hence Aplin and Archer's (1987) placement of the family as Phalangerida *incertae sedis*. Discovery of cranial material may be required before their interfamilial relationships can be resolved (Archer and Hand 2006). At present, the group is known from teeth and jaws. The bunodont molars and large burramyid-like premolar of all species suggest an omnivorous diet specializing in fruit or seeds (Archer et al. 1987). There are some striking resemblances between the dentitions of these marsupials and those of carpolestid primates (e.g. Paleocene species of *Carpolestes* and *Carpoadaptes*) which may also have been frugivores, granivores and/or insectivores (Biknevicius 1983). The following species are known from late Oligocene deposits: *Pilkipildra handae* Archer et al., 1987 from the Billeroo Creek LF of S.A.; *Pilkipildra taylorae* Archer et al., 1987 from the Ditjimanka LF of S.A.; and *Djilgaringa thompsonae* Archer et al., 1987 from the Yanda LF of South Australia. An additional, as yet, unnamed species has been recorded from the Kangaroo Well LF of the N.T. (Megirian et al. 2004).

Neogene

The Early Miocene (23–16.3 Ma)

Well-documented early Miocene terrestrial mammal assemblages are known from Queensland (Faunal Zone B of Riversleigh; Fig. 34) and South Australia (Kutjamarpu LF). Riversleigh's Faunal Zone B deposits have traditionally been dated using marsupial stage-of-evolution biocorrelation but more recently (Archer et al. in review) their early Miocene age has been confirmed by radiometric dating of U/Pb isotopes in speleothem. The age of the Kutjamarpu LF has been the subject of some debate. The Wipajiri Fm which contains this LF lies unconformably on the Etadunna Fm but was nevertheless interpreted by Woodburne et al. (1993) to be late Oligocene in age. Archer et al. (1997), on the basis of biocorrelation with Riversleigh faunas, suggested instead that it was either early or middle Miocene in age. Taxa shared by the Kutjamarpu LF and Riversleigh's Faunal Zone B assemblages include the kangaroo *Wakiewakie lawsoni* Woodburne, 1984 wombat *R. crowcrofti* Stirton et al., 1967 and diprotodontid *N. tirarensis* Stirton, 1967b. The latter two are also known from Riversleigh's Faunal Zone A and C deposits making them less informative. Black (1997b), however, notes that *N. tirarensis* from Riversleigh's Faunal Zone B most closely resembles the *N. tirarensis* type material from the Kutjamarpu LF. Taxa shared between the Kutjamarpu LF and Riversleigh's Faunal Zone C assemblages include: *Wakaleo oldfieldi*; *Litokoala kutjamarpensis*; *Marlu kutjamarpensis* Woodburne et al., 1987c; *Marlu syke* Roberts et al., 2009; and *Marlu ampelos* Roberts et al., 2009. Multivariate analysis (Travouillon et al. 2006) did not resolve which of Riversleigh's Faunal Zones, B or C, grouped more closely to the Kutjamarpu Local Fauna.

Other early Miocene deposits containing marsupials include the Geilston Bay and Wynyard LFs of respectively southeastern and northwestern Tasmania (Fig. 15a). The Geilston Bay LF is located on the eastern shore of the Derwent River near Hobart. Dating of overlying basalts indicates a minimum age of 22.4±0.5 Ma for this faunal assemblage (Tedford and Kemp 1998) which contains fragmentary remains of two indeterminate species of petauroid, a dasyurid and a burramyid. The predominantly marine faunal assemblage at Fossil Bluff, Wynyard, contains a single mammal species, the vombatomorphian *Wynyardia bassiana*. It has been dated using forams and by biocorrelation to be 21.4 Ma (Macphail 1996).

Palaeoenvironment

During the early Miocene, Australia was undergoing a 'greenhouse' climate phase characterised by rising sea-levels (as polar ice caps melted), warm, humid conditions and high biological diversity (Frakes et al. 1987; McGowran and Li 1994). Rainforest was widespread across northern and eastern Australia. Plant macrofossils from Vic., N.S.W. and Qld indicate species most closely related to those of the modern-day wet tropical rainforests of northeastern Qld (Greenwood and Christophel 2005).

Riversleigh's northwestern Qld early Miocene faunal assemblages (Faunal Zone B) include many taxa (Fig. 35) that are congeneric with living rainforest taxa (including the musky rat-kangaroo, cuscuses, striped possums, logrunners and lyrebirds) as well as highly diverse sympatric arboreal folivorous marsupials (Archer et al. 1997; Roberts et al. 2007) and hipposiderid bats (Hand 1997; Hand and Archer 2005) of the kind that characterise modern subtropical/tropical

Fig. 34 The Riversleigh World Heritage area of northwestern Queensland. (**a**) Tertiary freshwater limestone jutting up out of a semi-arid woodland (K. Black). (**b**) Excavation work at the early Miocene PIR Site (A. Nelson). The blocks of limestone have been carved using light explosives into blocks that can be more easily transported to UNSW for acid-processing

closed forest communities. Further, Faunal Zone B frog communities, which consist primarily of tiny to small bubble-nesting mybatrachids, resemble modern communities that exist in permanently wet, cool environments in northern Australia (M. Tyler 2000, 'personal communication', in McGowran et al. 2000). Taken together, the evidence suggests widespread (as opposed to riparian) rainforest assemblages which contradicts the conclusion reached by Megirian (1992) based on inadequate data and flawed arguments that the early Miocene deposits of Riversleigh could only have accumulated under dry or semi-arid conditions (Archer

Fig. 35 Reconstruction of part of an early Miocene palaeoenvironment based on fossils from Upper Site, Riversleigh (D. Dunphy). These perpetually wet, cool to warm closed forests were far more biodiverse than any of Australia's modern rainforests and more similar in this respect to the modern lowland rainforests of Borneo

et al. 1995; McGowran et al. 2000). In terms of overall mammal diversity alone, the early Miocene (Faunal Zone B) assemblages of Riversleigh exhibit significantly greater diversity than that found in contemporary Queensland wet tropical rainforest communities (Archer et al. 1997) making them more comparable to contemporary mid-montane rainforests of New Guinea (Archer et al. 1995) or the lowland rainforests of Borneo or the Amazon (McGowran et al. 2000).

Martin (2006) indicates drier, more seasonal conditions for central Australia during the early Miocene compared with northern and eastern Australia, although central Australia was still significantly wetter than it is today. Sclerophyll communities with species of *Acacia, Eucalyptus* and *Casuarina* bordered freshwater lakes and ephemeral swamps. Less common rainforest species were probably restricted to wetter habitats within this region (Martin 2006).

In higher palaeolatitudes, temperate rainforests with lower diversity were dominant. The palaeoflora of the Geilston Bay LF, Tas. (located at 50 degrees south during the early Miocene) indicates a mixture of podocarp and araucarian conifers, species of *Nothofagus*, myrtaceans, proteaceans and lauraceans (Tedford and Kemp 1998). At Fossil Bluff, Wynyard, *Nothofagus* dominated the rainforest canopy while araucarian and podocarp conifers and casuarinaceans formed a minor component. The low-diversity understorey was dominated by cyatheacean and dicksoniacean tree ferns (Macphail 1996).

The Early Miocene Marsupial Fauna

The best-documented early Miocene faunal assemblages are those of Faunal Zone B from the Riversleigh World Heritage Area, northwestern Qld (Archer et al. 1989, 1994, 1995, 1997, 1999). Within these highly diverse faunal assemblages, alongside more archaic groups some of which were in decline, we see the first appearance of many modern groups such as notoryctids, dasyurids, phalangerids and acrobatids as well as extinct families such as yingabalinarids and yalkaparidontids that are so far unique to Riversleigh (Table 3).

Dasyuromorphians

Dasyuromorphians were moderately diverse during the early Miocene with both thylacinids and dasyurids represented. Four species of thylacinid are known: *Nimbacinus dicksoni Wabulacinus ridei* Muirhead, 1997; *Ngamalacinus timmulvaneyi* Muirhead, 1997; and *Thylacinus macknessi* Muirhead, 1992. All four species are known from Faunal Zone B deposits at Riversleigh and at least two species, *W. ridei* and *N. timmulvaneyi*, if not all, coexisted during this period. Muirhead (1997) suggests that high thylacinid diversity is a reflection of the capacity for greater niche-partitioning in the biologically diverse rainforests of Riversleigh. This high diversity may also in part complement the relatively low diversity of large dasyurids. In this regard, this early Miocene balance of species is the reverse of that seen in the Pliocene to Holocene.

Four subfamilies are currently recognised within Dasyuridae (Long et al. 2002): the extinct Barinyainae; the extant Dasyurinae (devils, kowaris, mulgaras, kalutas, quolls, pseudantechinuses); the Phascogalinae (antechinuses, phascogales); and the Sminthopsinae (planigales, dunnarts, ningauis, kultarrs). Molecular divergence dates generated by the analysis of Krajewski et al. (2000b) of cytochrome *b*, 12S rRNA and protamine P1 gene sequences suggest that the ancestors of all major extant dasyurid lineages had differentiated by 20 Ma. This date may be consistent with the fossil record of the oldest representative of Barinyainae, *B. wangala* (Wroe 1999) (see Section 'Late Oligocene Marsupials'; Fig. 17) and the oldest (yet to be described) probable phascogalines and dasyurines recorded in Faunal Zone B sediments at Riversleigh (Archer et al. 1994; Wroe 1999). Tedford and Kemp (1998) recorded a dasyurid in the Geilston Bay LF of Tasmania represented by a single femur fragment said to be 'modern looking' and similar to species of *Dasyurus*.

Peramelemorphians

Although only a single bandicoot, *Yarala burchfieldi*, is described from the early Miocene, bandicoot diversity had increased markedly since the late Oligocene with at least five new genera and seven new species of peramelemorphian recorded from Faunal Zone B deposits at Riversleigh (Archer et al. 2006). Description of these taxa is presently underway and not one is referrable to any modern bandicoot family (K. Travouillon 2010, 'personal communication'). A further five species ('Peramelid' sp. D–H, Woodburne et al. 1993, table 4, p. 508) have been recorded from the Kutjamarpu LF, South Australia. However, it is not clear to which family these actually belong or whether any are conspecific with yaralids known from Riversleigh.

Notoryctemorphians

Until recently the order was known from a single extant genus containing two species: the northern marsupial mole, *Notoryctes caurinus*, which is restricted to the sandy deserts of W.A.; and the southern marsupial mole, *Notoryctes typhlops* (Fig. 4), which is widely distributed across Australia's arid interior and a small region of Western Australia. Little is known about the ecology of these enigmatic, highly specialised marsupials, in part because they spend most of their lives underground. Subterranean specialisations include reduction of the eyes to vestigial subcutaneous lenses; absence of external ears; development of a horny shield on the snout; and well-developed forelimbs with spade-like claws (Strahan 1995). The uncertain phylogenetic affinities of the order (see Section 'Interordinal Relationships of Australasian Marsupials'), combined with the highly specialised nature of the modern genus, implies a long period of independent evolution which is congruent with recent molecular divergence dates that suggest notoryctemorphians originated sometime between 63 and 56 Ma (Beck 2008a; Meredith et al. 2008a; Springer et al. 2009; see Section 'Interordinal Molecular Divergence Dates'). The recognition of a generically distinct fossil species in the early Miocene rainforest assemblages of Riversleigh was unexpected considering the ecological niche of the modern genus (Archer 1989). Analysis of postcranial remains indicate the forelimbs of the fossil species were also specialised for digging, possibly an adaptation for burrowing through leaf litter or moss on the rainforest floor (Archer et al. 1994; Long et al. 2002; Archer et al. 2011). This adaptation may then have become a preadaptation for burrowing through other kinds of friable substrates including sand. A comparable range of habitats spanning swamps to deserts characterises chrysochlorid moles (placentals) of Africa all of which have similar skeletal adaptations for burrowing. Clearly, arid substrates are not required for pre-moles to evolve into moles.

Yalkaparidontians

Archer, Hand and Godthelp (1988) erected the order Yalkaparidontia to encompass an extraordinarily unusual new family (Yalkaparidontidae) and two new species (*Yalkaparidon coheni* and *Yalkaparidon jonesi*) of marsupials from early to middle Miocene sediments at Riversleigh. *Yalkaparidon coheni* is known from a relatively complete skull and lower jaw (Fig. 36). Despite a relatively plesiomorphic basicranium, it possessed greatly enlarged, ever-growing incisors and extremely zalambdodont (V-shaped) cheekteeth which led Archer et al. (1988) to hypothesise a diet possibly consisting of insect larvae, worms, egg content or some other food that could be processed without any of the horizontal shearing component normal for most other plesiomorphic mammals. Beck (2009) noted numerous similarities between the cranium of *Y. coheni* and the highly specialised crania of striped possums (species of *Dactylopsila* and *Dactylonax*) as well as

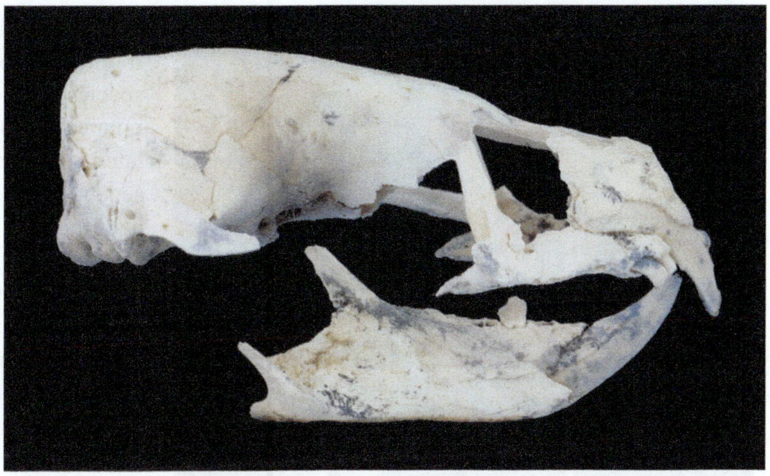

Fig. 36 The skull and lower jaw of the yalkaparidontian *Yalkaparidon coheni* (R. Arnett). This is one of the most bizarre Australian mammals known representing a unique order only found in the fossil deposits of Riversleigh. Speculation about what it did for a living ranges from egg-eating to being a mammalian version of a woodpecker

the Madagascan primate *Daubentonia madagascariensis*. Both of these extant species fed on xylophagus (wood-boring) larvae, using their enlarged incisors to gouge holes in the wood and an elongate digit of the hand to extract larvae. Elsewhere in the world this ecological niche is filled by picid birds (woodpeckers). In Australia and Madagascar, where woodpeckers are absent, the niche appears to have been filled by species of *Dactylopsila*, *Dactylonax* and *Daubentonia* (Cartmill 1974). Following this pattern, Beck (2009) has predicted that more complete specimens of yalkaparidontids will show a similarly elongate digit. This interpretation may be complicated by the fact that at Riversleigh species of dactylopsiline petaurids are sympatric with yalkaparidontids. If both groups were in fact in this, what was for mammals, extraordinary niche, they probably would have competed with one another. If that were the case, it may explain why only one of these groups, the dactylopsilines, survived beyond the middle Miocene.

Because of the highly specialised skull and dentition (Fig. 36) of species of *Yalkaparidon*, the phylogenetic affinity of this group to other australidelphians remains unclear.

Diprotodontians

The structurally and floristically diverse rainforest environments of the early Miocene appear to have triggered increases in diversity in phalangeridans. As might be expected, arboreal possum families in particular show a significant increase in generic and species level diversity (Table 3). At least 18 genera and 24 species of possums are recorded from early Miocene deposits compared with only 11 genera and 20 species during the late Oligocene. In direct contrast is an associated decrease in diversity in the predominantly terrestrial, large-bodied vombatiforms with the apparent extinction of two families (Ilariidae and Maradidae) and a reduction in generic and species level diversity in the remaining families. At least 20 genera and 29 species of vombatiform were present during the late Oligocene, yet only 13 genera and 18 species are recorded from the early Miocene. Macropodoid diversity appears to have remained relatively unchanged from the moderate diversity displayed during the late Oligocene (Table 3).

Vombatiforms

Phascolarctomorphians A reduction in koala diversity is evident from four genera comprising seven relatively rare species in the late Oligocene to three genera comprising four rare species during the early Miocene. However, as in the late Oligocene, commonly one or two albeit uncommon species are present in any of the larger, better-sampled faunal assemblages. From late Miocene times onwards the infraorder is represented by only one genus, *Phascolarctos*, although by the Quaternary the species *Phascolarctos cinereus* had become common in open forest communities, far more so than any other phascolarctid had been prior to that time. Early Miocene taxa include *Litokoala kutjamarpensis* (Stirton et al. 1967) from the Kutjamarpu LF, S.A.; *Litokoala garyjohnstoni* (Louys et al. 2007a) from the Outasite LF, Faunal Zone B, Riversleigh; *Nimiokoala greystanesi* (Black and Archer 1997b) from the Faunal Zones B and C,

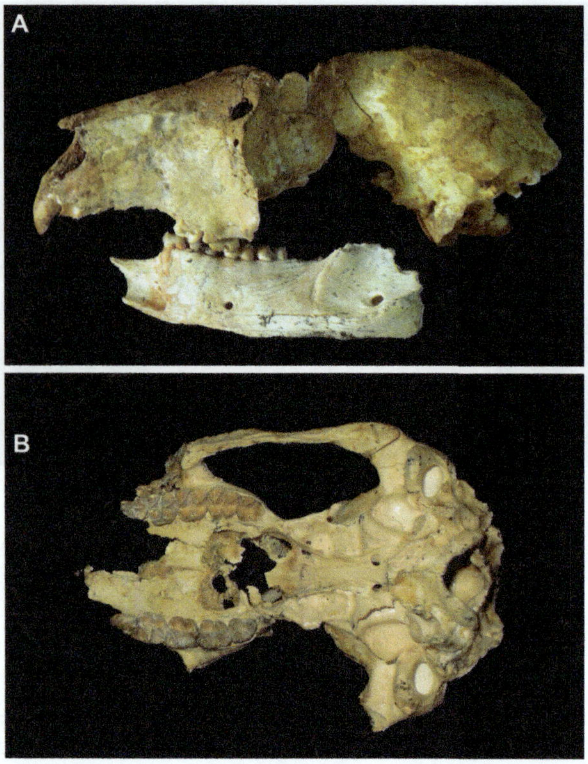

Fig. 37 Fossil koalas (phascolarctids) from Riversleigh. (**a**) *Nimiokoala greystanesi* from the Miocene had very odd, highly crenulated teeth. (**b**) A partial skull of *Litokoala kutjamarpensis* showing enlarged tympanic bullae (K. Black) that suggest it was very sensitive to low-frequency sounds of the kind that modern koalas utter when communicating across large distances

Riversleigh; Fig. 37a); and an unnamed genus and species from Riversleigh's Faunal Zone B sediments that may well be the most plesiomorphic phascolarctid known (Black 1999; Black et al. in press). *Litokoala kutjamarpensis* was originally described on the basis of a single upper molar from the Wipajiri Fm (source of the Kutjamarpu LF) which is a measure of the relative rarity of this taxon in central Australia. However, more complete cranial (Louys et al. 2007a, 2009) and dental material (Black and Archer 1997b) has since been found in Riversleigh's Faunal Zone C deposits (Fig. 37b), allowing a re-evaluation of species boundaries within the genus. Consequently, *Litokoala kanunkaensis* is now regarded to be a junior synonym of *L. kutjamarpensis* (Louys et al. 2007a).

Louys et al. (2009) noted interesting similarities in the auditory region of species of both *Litokoala* and *Nimiokoala* to the modern species, notably the enlarged auditory bullae (Fig. 37b). In the living koala, enlarged bullae function to increase perception of low-frequency sounds. The relatively sedentary modern species uses frequent low-pitched bellows to communicate long distances. Louys et al. (2009) speculated that these characteristic behavioural traits (i.e., sedentism and loud, frequent vocalisations) may have had their origins in the early Miocene with rainforest koalas utilising low-frequency vocalisations to achieve greater directionality of sound in a closed forest environment. It may also have enabled long-distance communication necessary because of lower species densities within those closed forests.

Early to middle Miocene phascolarctids are notably smaller in body size than at any other period in their evolutionary history. This may be a reflection of more nutrient-rich rainforest plants compared with those available in the late Oligocene and late Miocene to Holocene.

Vombatomorphians Five Vombatomorphian genera are recorded from early Miocene deposits with Ilariidae and Maradidae not surviving beyond the late Oligocene. Of families that continued into the early Miocene, diprotodontids show a significant decline in diversity compared with those of the late Oligocene, while palorchestids and wynyardiids maintained the relatively low to moderate late Oligocene levels of generic and species level diversity (Table 3). The carnivorous thylacoleonids and the herbivorous vombatids are the only vombatomorphian families to have increased in diversity since the late Oligocene.

For thylacoleonids, like thylacinids, this may have been a response to increased prey diversity and niche availability in the rainforest strata. A greater range of body sizes is displayed by early Miocene thylacoleonids, with the evolution of smaller forms, that may have taken advantage of the abundant small prey such as possums and yaraloid bandicoots. Early Miocene species include *W. oldfieldi* Clemens and Plane, 1974 (Kutjamarpu LF, Wipajiri Fm; Riversleigh Faunal Zone C); *Wakaleo* n. sp 1 (Riversleigh Faunal Zones A–B); *Prisciléo roskellyae* (Riversleigh Faunal Zones A–C; Fig. 38); and two new unnamed genera and species of 'micro-lions' from Riversleigh's Faunal Zone B (Gillespie 2007).

Early Miocene vombatids include *R. crowcrofti* from the Kutjamarpu LF, S.A. and two new genera and species from Riversleigh's Faunal Zone B deposits (Archer et al. 2006).

Fig. 38 Skull of a small species of marsupial lion (thylacoleonid), *Priscileo roskellyae*, from the early Miocene Upper Site of Riversleigh (M. Archer). Although barely cat-size, this at least omnivorous vombatiform marsupial was soon replaced by larger and larger, ever more specialised descendents until by the late Pleistocene marsupial lions were lioness-size

Wynyardiid generic and species level diversity remained relatively constant between the late Oligocene and early Miocene although some faunal turnover is evident within the family. No species of *Ayekaye* or *Muramura* survived the Oligocene while *Wynyardia bassiana* appears in the Fossil Bluff LF, Tasmania. Within *Namilamadeta*, *Namilamadeta snideri* remains confined to the Oligocene, two new species (*Namilamadeta crassirostrum* Pledge, 2005; *Namilamadeta superior* Pledge, 2005) appear in Faunal Zone B deposits of Riversleigh while *Namilamadeta albivenator* spans the late Oligocene to early Miocene of Riversleigh.

Diprotodontid diversity declines from five genera and eight species during the late Oligocene to just two genera and two species (*N. bonythoni*, *N. tirarensis*) during the early Miocene: the lowest number of taxa for any period of the Tertiary. Interestingly, the largest of the late Oligocene diprotodontids were those that survived into the early Miocene. Black (2008) speculated that extinction or low abundance of the smaller forms may be associated with increased competition for resources from an increased diversity and abundance of similar-sized macropodoids (Cooke 1997c) and wynyardiids (Pledge 2005).

Phalangeridans

The early Miocene sees the first appearance of some families including acrobatids (feather-tailed and pen-tailed possums; Fig. 3f) and phalangerids (brush-tailed possums, scaly-tailed possums cuscuses), an increase in diversity of burramyids and petaurids and a spectacular radiation of pseudocheirids that continues on into the middle Miocene. Among possum families, only tarsipedids (honey possums) are not represented in the early Miocene – or, for that matter, at any time until the late Quaternary.

Phalangeroids Phalangerids are the most speciose and geographically diverse of the extant possum families comprising six genera and 23 species including cuscuses (*Phalanger*, *Spilocuscus*, *Strigocuscus*, *Ailurops*); brush-tailed possums (*Trichosurus*); and scaly-tailed possums (*Wyulda*). Nineteen species are distributed throughout New Guinea and surrounding islands, two of which also occur in Australia. Another four species occur only in Australia. The oldest recorded phalangerid is an unnamed taxon from the Geilston Bay LF from Tasmania that was originally reported by Tedford et al. (1975) to be a phalangerid. Tedford and Kemp (1998) reinterpret this specimen as a petauroid with uncertain relationships to other petauroids. However, Crosby et al. (2001) support its original reference to Phalangeridae. Woodburne et al. (1993) indicate the presence of a phalangerid in the Kutjamarpu LF, S.A., but it too has yet to be formally described. Crosby et al. (2004) indicate the presence of at least two unnamed genera from the early to middle Miocene of Riversleigh. The only formally described early Miocene phalangerids are from Riversleigh's Faunal Zone B assemblages. These include: *Onirocuscus silvicultrix* Crosby, 2007; *Onirocuscus inversus* Crosby, 2007; and *Wyulda asherjoeli* Crosby et al., 2001. All three species as well as one undescribed taxon coexisted during the early Miocene. Differences in body size and tooth morphology suggest niche partitioning on the basis of size and diet. Both species of *Onirocuscus* are estimated to have been medium-sized (2000–3000 g) phalangerids (Crosby 2007). In distinct contrast, *W. asherjoeli* is reported to be the smallest (660 g) phalangerid known, living or extinct (Crosby et al. 2001). Like its modern descendant *Wyulda squamicaudata* which nests in rock crevices, Crosby et al. (2001) suggest that this small possum probably utilised as dens the numerous rock fissures and cavities that would have been present in the limestone substrate that underlay the Riversleigh rainforests.

Corresponding with the appearance of phalangerids in the fossil record is a decline in diversity of morphologically similar but rarer miralinid possums. Miralinids are represented by a single early Miocene genus and two species (*Durudawiri inusitatus* Crosby and Archer, 2000; *Durudawiri anfractus* Crosby, 2002) compared with the two genera and five species present during the late Oligocene. Both species of *Durudawiri* belong to the subfamily Durudawirinae which is restricted to Faunal Zone B deposits at Riversleigh. This reduction in diversity, however, may be more apparent than real given the rarity of miralinids in most of the assemblages in which they occur. Certainly, the presence of the miralinine genus *Barguru* in the late Oligocene (Kangaroo Well LF) and middle Miocene (Bullock Creek LF) deposits of the N.T. indicate that another lineage was present during the early Miocene. Crosby and Archer (2000) interpret the rarity of durudawirines in the Riversleigh rainforest assemblages to mean that they were specializing on a resource that was limited in that environment. On the other hand, the extraordinary degree of thegosis evident in the teeth of at least species of *Miralina* suggests that this group of possums may have exhibited high levels of intraspecific aggression which could have resulted in reduced number of individuals present in any one location. Thegotic behaviour in mammals often reflects levels of aggression in addition to the need to restore the sharpness of cutting blades in the dentition (Every 1970). It has been previously noted to be a conspicuous feature of the teeth of some extinct fossil mammals from Riversleigh (Archer et al. 1990).

Ektopodontids maintain a moderate diversity into the early Miocene with three genera and four species: *Ektopodon serratus* Stirton et al., 1967 from the Kutjamarpu LF, S.A. and Faunal Zone B, Riversleigh (Archer et al. 2006); *Ektopodon litolophus* Pledge et al., 1999 from the Kutjamarpu LF, S.A.; *Chunia* sp. from Faunal Zone B, Riversleigh (Archer et al. 2006); and a new ektopodontid genus and species, Faunal Zone B, Riversleigh (ibid.). However, apart from *E. serratus* which appears to have been reasonably common judging by the plethora of isolated molars in the Kutjamarpu LF, the other taxa noted here are rare elements in their faunal assemblages. Given the very strange, multituberculate/rodent-like molars, it is possible that this group of possums was specialising on some resource that was rare in at least the rainforests of Riversleigh. While the family persisted until its apparent extinction after the early Pleistocene, it was never again as common as it was in the early Miocene of central Australia.

Petauroids Brammall (1998) and Brammall and Archer (1997, 1999) indicate the presence of a species of *Dactylopsila* in early to middle Miocene sediments of Riversleigh, which represents the only pre-Pleistocene taxon that can be confidently assigned to the subfamily Dactylopsilinae. Prior to its discovery, it was generally accepted that dactylopsilines had evolved in New Guinea and dispersed into Australia sometime during the Pleistocene (McGowran et al. 2000). In view of the Riversleigh discoveries, it now seems at least as likely that the reverse direction of dispersal must have been the case. Having originated by the early Miocene in rainforest assemblages of Riversleigh, the group must have subsequently dispersed to New Guinea (which was in the main below sea level until about 15 million years ago), probably during the late Miocene to early Pliocene when emergent land-bridges would have facilitated dispersal (Hall 2001), and then possibly back-dispersed in the Pliocene/Pleistocene, via land emergent across Torres Strait, to establish populations of *Dactylopsila trivirgata* in northeastern Qld. This scenario is congruent with molecular divergence dates which indicate the dactylopsiline lineage diverged from petaurines approximately 20–18 Ma (Meredith et al. 2010) and that the modern species, *Dactylopsila palpator* and *D. trivirgata*, radiated during the late Miocene to earliest Pliocene (10.17–5.23 Ma; Meredith et al. 2010).

Other early Miocene non-pseudocheirid petauroids include *Djaludjangi yadjana* Brammall, 1998 from

Faunal Zone B and C assemblages of Riversleigh. This relatively plesiomorphic taxon lacks key petaurid synapomorphies leading Brammall (1998) to speculate that it may in fact represent a new family. Presently it is placed as Petauroidea *incertae sedis* as are two indeterminate genera from the Geilston Bay LF of Tasmania (Tedford and Kemp 1998).

In contrast to petaurids, pseudocheirids represent one of the most successful marsupial radiations of the early middle Miocene. As many as six sympatric species have been reported from the Upper Site LF of Faunal Zone B, Riversleigh (Roberts et al. 2007; Travouillon et al. 2006) and five sympatric species are known from the Kutjamarpu LF (Woodburne et al. 1993). Much of this material (Fig. 39) is currently in the process of being described (Roberts et al. 2009). Published taxa from the Kutjamarpu LF include *Pildra tertius* Woodburne et al., 1987c; *M. kutjamarpensis* (ibid.); *M. ampelos* Roberts et al., 2009; *M. syke* Roberts et al., 2009; and *Paljara tirarensae* Woodburne et al., 1987c. The latter two species are also known from Faunal Zone B, Riversleigh (Bassarova et al. 2001; Roberts et al. 2009). Additional published species from Faunal Zone B, Riversleigh, include *Gawinga aranaea* Roberts et al., 2007; *Paljara maxbourkei* Bassarova et al., 2001; and *Paljara nancyhawardae* Bassarova et al., 2001. This high diversity of ringtail possums almost certainly reflects a high diversity of rainforest plants on which they fed. However, not all were necessarily obligate folivores; some may also have eaten mosses and other arboreal plants as do some of the smaller living pseudocheirids of New Guinea. The smallest of the extinct species (e.g. species of *Pildra* and *Paljara*) often exhibit significant reduction in tooth size near the end of the molar row similarly suggesting that these forms may have focused on plant foods other than tree leaves.

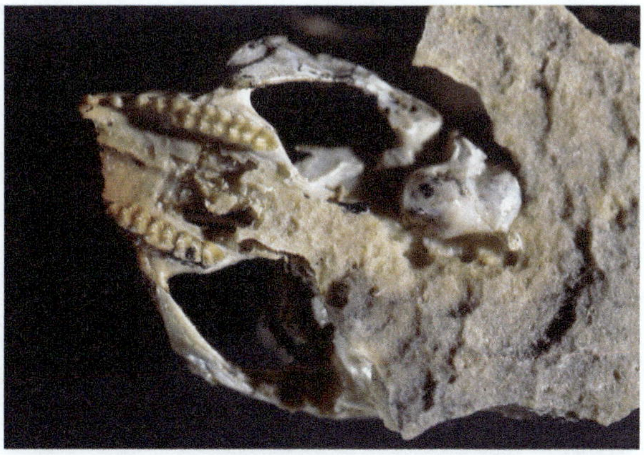

Fig. 39 Fossil ringtail possum (pseudocheirid) skull from the early Miocene Neville's Garden Site, Riversleigh (Photo R. Murphy). Although still being acid-etched from the freshwater limestone in which this skull was found, the leaf-eating teeth and wide space for the powerful chewing muscles of a folivore are clear

Tarsipedoids Molecular divergence dates suggest that acrobatids, tarsipedids petaurids and pseudocheirids last shared a common ancestor (Fig. 9) around 40 Ma (Meredith et al. 2009) with crown group acrobatids last sharing a common ancestor during the late Oligocene approximately 25.1 Ma (Meredith et al. 2009). Although some molecular studies (e.g. Meredith et al. 2009; Phillips and Pratt 2008; Kavanagh et al. 2004) challenge the interpretation of Aplin and Archer (1987) that of these four families, tarsipedids and acrobatids are the sister groups of each other, we tentatively follow this morphology-based view here (Table 1, Fig. 29).

The phylogenetic position of tarsipedids has been difficult to work out, in part because of the highly specialised nature of its sole extant representative, the omnivorous/nectivorous honey possum, *Tarsipes rostratus* (Fig. 40), which has spicule-like remnants for teeth, and the fact that there are no pre-Pleistocene tarsipedids known, or at least none that have been recognised. Given the vestigial nature of the teeth of the living species, it is possible that some of the more unusual isolated teeth unique to Riversleigh (e.g. yingabalanarids) might represent this group without us having any idea that this is the case.

Acrobatid material from the early Miocene of Riversleigh represents at least two species (Brammall and Archer 1997) currently under study (by MA).

Burramyoids Burramyids are represented by three species known from the early Miocene. Faunal Zone B specimens of *B. brutyi* (Brammall and Archer 1997) of this age are significantly more abundant than specimens of the same species in late Oligocene deposits at Riversleigh. The early Miocene marks the first appearance of species of the genus *Cercartetus* although the fossil record for this group at this time is relatively poor. *Cercartetus* sp. has been recorded from three Faunal Zone B deposits at Riversleigh (Archer et al. 1997, 2006; Brammall 1998; Brammall and Archer

Fig. 40 Tarsipedids, or honey possums, now only known from W.A., have no pre-Pleistocene fossil record (J. Gould). Having spent millions of years becoming increasingly nectarivorous, their cheekteeth are degenerate. From molecular biology, they appear to be at least distantly related to feather-tail gliders (acrobatids)

1997). Tedford and Kemp (1998) report the presence of an unnamed burramyoid in the Geilston Bay LF which they suggest is most similar to (albeit larger than) modern species of *Cercartetus*.

Macropodoids As in the late Oligocene, the most diverse macropodoids of the early Miocene are balbarids. Some in this family have ranges that extend from the late Oligocene to early Miocene: *Balbaroo gregoriensis* Flannery et al., 1983; *Nambaroo gillespieae* Kear et al., 2007 (= *Nambaroo* sp. 3 of Archer et al. 2006); and *Nambaroo* sp. 5 (Archer et al. 2006). Balbarids that first appear in the early Miocene Faunal Zone B assemblages of Riversleigh include *Balbaroo fangaroo* Cooke, 2000; four unnamed species of *Nambaroo* (sp. 2, sp.4, sp. 6, sp. 7; Archer et al. 2006); three species of *Ganawamaya* (*Ganawamaya ornata* Cooke, 1992; *Ganawamaya acris* Cooke, 1992; *Ganawamaya* sp. 4); and two unnamed species of *Wururoo* (sp. 2, sp. 3; Archer et al. 2006). Kear et al., (2007) described an almost complete skeleton of *N. gillespieae*, one of the few skeletons known for these primitive kangaroos. He concluded that the very short feet, well-developed opposable first toe on the hind foot, powerful forelimbs and other skeletal features indicated that this animal galloped rather than hopped and may have been able to climb. Taken together, it would not have looked or behaved like modern kangaroos. The closest any living kangaroo comes to this is the bandicoot-sized musky rat-kangaroo (Hypsiprymnodontidae; *H. moschatus*) which gallops rather than hops (see below) but has relatively less well-built forelimbs.

Of special mention, and unique among all kangaroos, is the unusual condition found in *B. fangaroo* (Fig. 41) of extraordinary hypertrophied upper canines (Cooke 2000). Enlarged canines that are sexually dimorphic in size are found in modern artiodactyls such as tragulids and moschids. These small-to-tiny deer occupy dense vegetation and the males use their large canines as close-range visual cues of sexual identity, agonistic displays and for defence. By analogy, Cooke (2000, 2006) proposed that the enormous (for a kangaroo) canines (Fig. 41) of *B. fangaroo* may also be a sexually dimorphic feature in a species that occupied the dense understorey of Riversleigh's early Miocene rainforests. Cooke (2006) further suggested that like *N. gillespiae, B. fangaroo* may have been quadrupedal and hence less able to use its forearms and hind legs in fights in contrast to living kangaroos. Consequently, its large canines may have also been used, as they sometimes are in moschids and tragulids, as a defensive weapon. Interestingly, modern tragulids and moschids also have relatively short legs compared with other deer.

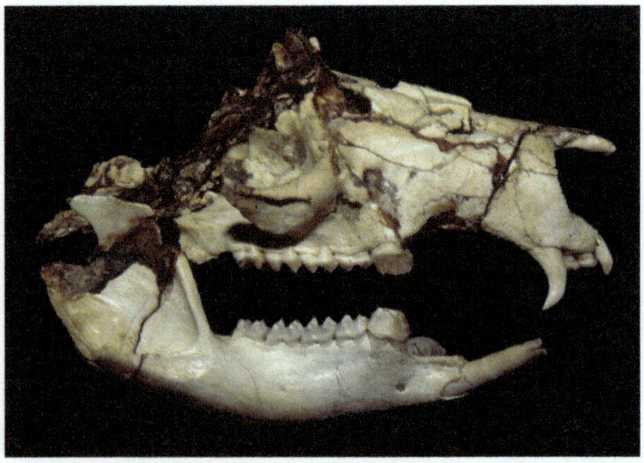

Fig. 41 Skull of the early Miocene balbarid kangaroo *Balbaroo fangaroo* from Riversleigh (A. Gillespie). While the cheekteeth indicate that it was primarily herbivorous, the enormous, recurved and sharp upper canines suggest it was capable of putting up a good fight to avoid becoming the dinner of carnivorous kangaroos and other predators

Australia's modern musky rat-kangaroo, *H. moschatus*, of north Qld rainforests is the smallest and most plesiomorphic of all living kangaroos. Unlike other living macropodoids, it has a quadrupedal gait and, at high speed, gallops rather than hops. Although terrestrial, it has a mobile first digit on the pes that allows it to climb on inclined or fallen branches on the rainforest floor (Strahan 1995). Other plesiomorphic features include its prehensile tail which it uses to carry nesting materials, its similar length forelimbs and hindlimbs, an omnivorous diet of insects, fruits and seeds, a non-sacculated stomach and normally giving birth to two rather than one young at a time (Burk et al. 1998). The genus makes its first appearance in the early Miocene assemblages of Riversleigh where it is represented by two species: *H. bartholomaii* Flannery and Archer, 1987b; and *Hypsiprymnodon* sp. (Wroe and Archer 1995; Archer et al. 2006). Dental morphology has remained relatively conservative within the genus with a limited number of differences between the extinct and the living species. Cooke (2006) attributed these similarities, as well as the numerous plesiomorphic behavioural and anatomical attributes of the modern species, to the retention of its ancestral rainforest habitat.

Less diverse during the early Miocene are propleopine hypsiprymnodontids which are represented by a single, albeit relatively common species, *Ekaltadeta ima* Archer and Flannery, 1985 from Faunal Zone B of Riversleigh. This taxon (Fig. 32), unique to Riversleigh, has been interpreted on the basis of microwear on the teeth to have been at least partly carnivorous and has been commonly referred to as the 'Carnivorous Kangaroo'. This propleopine hypsiprymnodontid genus survived into the late Miocene (see below). A related and possibly descendent lineage of more bear-like species referable to the genus *Propleopus* is represented in Pliocene and Pleistocene deposits.

Moderate diversity is displayed by bulungamayine kangaroos during the early Miocene with an increase from two to four genera since late Oligocene times. *Bulungamaya delicata* Flannery et al., 1983, *Bulungamaya* sp. A and *W. naughtoni* Archer, 1979 span both the late Oligocene and early Miocene while two new genera and species, *Ganguroo bilamina* Cooke, 1997b and *Nowidgee matrix* Cooke, 1997b, first appear in Faunal Zone B assemblages of Riversleigh. Woodburne et al. (1993) also note the presence of *Bulungamaya* sp. B in the Kutjamarpu LF of South Australia. Discovery of dentitions of *N. matrix* and *G. bilamina* has played a significant role in current understanding of macropodoid phylogeny and classification. Cooke (1997b) showed how the molar morphologies of *N. matrix*, *B. delicata* and *G. bilamina* represent stages in the transition from an omnivorous, bunolophodont kangaroo into an herbivorous, completely lophodont form. A similar transition to lophodonty was previously noted for balbarine kangaroos (albeit via a different morphological pathway) which led Flannery (1989) to propose that balbarine kangaroos gave rise to later macropodids. It is now apparent that lophodonty has convergently developed at least twice within kangaroos. Cooke (1997b, 1999) further noted similarities in premolar and molar morphology between derived bulungamayines (e.g. *G. bilamina*) and late Miocene macropodids (e.g. *Hadronomus puckridgi* Woodburne, 1967) and, on this basis, suggested bulungamayines were ancestral to modern kangaroos. This hypothesis is reflected in current classifications wherein Bulungamayinae is basal within Macropodidae and Balbaridae has been given family-level status (Kear and Cooke 2001).

Potoroid diversity remains low throughout the Miocene, possibly as a result of competition with diversifying, small macropodoid groups such as hypsiprymnodontines and bulungamayines. The family is represented by a single species, *Wakiewakie lawsoni* Woodburne, 1984 from the Kutjamarpu LF of S.A., and the Upper Site LF of Riversleigh. It has an enormously long third premolar similar to but far larger than that tooth in species of *Bettongia*. The reason for such hypertrophy of this sectorial blade is a complete mystery.

Marsupialia *incertae sedis*

Archer et al. (1990) established the family Yingabalanaridae for the enigmatic, mouse-sized species *Yingabalanara richardsoni* Archer et al., 1990, named on the basis of a single lower molar from the Upper Site LF, Riversleigh. The holotype has well-defined wear facettes on the tooth indicating that it interacted with precision with a well-matched tooth in the upper molar row. Given the bizarre form of this lower molar, an upper molar would be instantly recognisable as comparably distinct.

Although making the point that the molar was unlike that of any other Australian clade warranting erection of a new family, they were unable to find a single synapomorphy that would unambiguously relate the family to any known marsupial, placental or pre-eutherian group. Uncertainty even about the orientation of the tooth within the lower jaw resulted in Archer et al. (1990) proposing six alternative hypotheses of dental structure and homology. Potential similarities to such diverse groups as symmetrodonts, tribotheres (e.g. *Potamotelses*), zalambdodont marsupials, adapid primates and even phyllostomoid bats were noted. Although yingabalanarids are currently regarded as Mammalia *incertae sedis* (Table 1), Long et al. (2002) suggested that some (albeit probably convergent) similarities to the zalambdodont yalkaparidontids may be evidence for inclusion within Marsupialia. Despite years of collecting since the original description, only one other molar representing this taxon has been found. This must have been a very rare taxon in the Faunal Zone B assemblages of Riversleigh.

Middle Miocene (16.3–10.4 Ma)

Australia's middle Miocene marsupial fauna is best represented in the Faunal Zone C deposits of Riversleigh and the Bullock Creek LF (Fig. 42) of N.T. An additional mammal-bearing deposit, the lower middle Miocene Batesford Limestone at Batesford Quarry, near Geelong in Victoria (Fig. 15a), is relatively depauperate containing a single indeterminate diprotodontid partial dentary (Piper et al. 2006).

The Bullock Creek LF has been dated at approximately 12 Ma on the basis of stage-of-evolution biocorrelation of its marsupial fauna and specifically its zygomaturine diprotodontids (e.g. *N. stirtoni*, *Nimbadon lavarackorum*), palorchestids (e.g. *Propalorchestes novaculacephalus*) and thylacoleonids (*Wakaleo vanderleueri*) (Woodburne et al. 1985; Murray et al. 2000a, b; Black and Hand 2010). These species are also represented in Faunal Zone C deposits at Riversleigh suggesting that they are similar in age (Archer et al. 1994, 1997; Murray et al. 2000a, b; Black 1997b, 2006). At Riversleigh, Faunal Zone C assemblages are known from a relatively thick sequence of horizontally bedded freshwater carbonates suggesting that they represent a longer span of time than the stratigraphically limited Camfield Beds at Bullock Creek. For this reason, some of Riversleigh's upper Faunal Zone C assemblages, although also interpreted here as middle Miocene, may postdate the Bullock Creek LF. For example, the presence of a highly derived species of *Neohelos* (*Neohelos* sp. C; Murray et al. 2000a) in the high Faunal Zone C Jaw Junction LF, suggests this deposit is younger than the Bullock Creek LF yet older than the late Miocene Encore LF of Riversleigh and the Alcoota LF of the N.T.

Palaeoenvironment

Globally the middle Miocene was a period of pronounced climatic change with a shift occurring from warm, wet, greenhouse conditions of the early middle Miocene to cooler, drier conditions towards the end of the middle Miocene, anticipating even more significant climate change in the late Miocene (see below). The equable greenhouse conditions of the early Miocene culminated in the middle Miocene climatic optimum at 16–15 Ma (Fig. 14), when rainfall, sea levels and biotic diversity were at record high levels (McGowran and Li 1994). What followed was a period of increasingly cooler, drier conditions triggered at least in part by development of a permanent Antarctic ice cap. The progressive movement of the Australian landmass northward and its subsequent collision with South East Asia effectively closed the deep sea passage between Australia and Indonesia, severing a major connection between the Indian and Pacific Oceans (Kennett 1977). This change in oceanic circulation is correlated with expansion of the east Antarctic ice sheet between 14 and 11 Ma (Lawver and Gahagan 2003) resulting in a further lowering of sea level and declines in temperature and rainfall as well, probably, as environmental predictability. The drying of the Australian continent was further compounded by the emergence of the New Guinea Highlands which created a rain-shadow effect across Australia's north and the increasing height of the eastern Dividing Range which similarly captured on its eastern flank an increasing amount of rain brought in by winds from the Pacific Ocean. The net effect was a steady, profound change in the vegetation of the interior of the continent with retraction of rainforests to continental margins and their gradual replacement in the interior by open forests and woodlands. However, widespread grasslands did not become a feature of the continent until mid to late Pliocene time (see below).

Within Riversleigh's middle Miocene Faunal Zone C assemblages, there is a distinct change in taxonomic diversity between LFs from lower (e.g. Dwornamor

Fig. 42 The source for some of Australia's Miocene marsupials. (**a**) Excavating the middle Miocene Bullock Creek limestone of the southern N.T. which may have been a shallow freshwater lake (M. Archer). (**b**) Quarrying AL90 Site of Riversleigh (A. Gillespie) which was a limestone cave that had trapped thousands of forest animals spanning the middle to late Miocene

LF) and higher stratigraphic units (e.g. Henk's Hollow LF, Jaw Junction LF) within the depositional sequence (Archer et al. 1994, 1995; Creaser 1997). The early middle Miocene assemblages (e.g. the Dwornamor LF), like Faunal Zone B assemblages, are species-rich, with abundant sympatric arboreal folivores and frog assemblages suggestive of warm, wet, aseasonal, floristically diverse rainforest environments. For example, 15 species of arboreal folivores, 10 of which are pseudocheirids, have been recorded from the Dwornamor LF alone (Archer et al. 2006). Higher Faunal Zone C assemblages, however, have fewer arboreal taxa (e.g. seven species in Henk's Hollow LF), a greater diversity of large terrestrial herbivores and fewer frog species, suggestive of an opening up of the canopy and possibly the beginning of seasonal variations in conditions.

The palaeoenvironment of the late middle Miocene Bullock Creek LF has been interpreted, on the basis of its abundant aquatic and stream bank fauna, as a large meandering river and surrounding floodplain with seasonal fluctuations in water supply. Murray and Megirian (1992) interpret the proximal vegetation to be a mix of shrubs and sedges. Unlike Riversleigh's Faunal Zone C assemblages, small terrestrial animals and arboreal taxa are scarce (Murray and Megirian 1992; Schwartz and Megirian 2004) and may suggest that species-rich forests were distal to the depositional area. Murray and Vickers-Rich (2004) suggest that the rarer elements of the Bullock Creek LF, such as the phalangeroids, small kangaroos and small carnivores, may have been sourced from a gallery forest community some distance upstream.

Middle Miocene marsupial assemblages from central Australia are not known. However, palynological evidence indicates drier, more seasonal conditions with a further retreat of rainforest (Greenwood and Christophel 2005) accompanied by the spread of sclerophyll forests and woodlands (White 2006). Grass pollen has been recorded in low quantities during the Miocene but was by no means abundant until at least the early to middle Pliocene and then primarily in the northwest to central parts of the continent (Martin 1990; Archer et al. 1995). Earlier claims about extensive grasslands in central Australia during the middle Miocene were based on misidentification of restionacean (swamp plants) pollens for grass pollens.

The Middle Miocene Marsupial Fauna

The middle Miocene was a period of significant faunal turnover characterised by a decline in overall diversity. By the end of the middle Miocene we see the extinction of entire marsupial families (e.g. wynyardiids, balbarids, pilkipildrids, miralinids, yalkaparidontids, yaralids, yingabalinarids) accompanied by the appearance of more derived taxa including the ancestors of many new lineages (e.g. species of *Phascolarctos* and *Bettongia*). Molecular divergence dates indicate that the majority of terrestrial marsupial groups within Dasyuromorphia, Peramelemorphia, Macropodoidea and Vombatidae radiated during the middle Miocene in response to the progressive opening up of the rainforest canopy (Meredith et al. 2009).

Dasyuromorphians

Thylacinids continued their reign as the dominant marsupial carnivores, increasing in diversity with five genera and seven species recorded from middle Miocene deposits. The temporal range of both *N. dicksoni* (Riversleigh and Bullock Creek) and *T. macknessi* (Riversleigh) extends into the middle Miocene and an additional three genera and four species appear in the fossil record. *Muribacinus gadiyuli* Wroe, 1996b and *Maximucinus muirheadae* Wroe, 2001 are recorded from Faunal Zone C of Riversleigh and *Nimbacinus richi* Murray and Megirian, 2000 and *Mutpuracinus archibaldi* Murray and Megirian, 2000 are recorded from the Bullock Creek LF, N.T. Material comparable with the latter species has also been found in the Jaw Junction LF of Riversleigh (Archer et al. 2006). There is direct evidence of at least two sympatric species recorded in three different middle Miocene faunal assemblages: *N. richi* plus *M. archibaldi* from Top Site at Bullock Creek; *M. gadiyuli* plus *T. macknessi* from the Dwornamor LF of Riversleigh; and *N. dicksoni* plus *M. gadiyuli* from the Henk's Hollow LF of Riversleigh. It is probable that all five species coexisted during the middle Miocene. Wroe (2001) found extensive differences in body weight estimates for the above taxa suggesting considerable trophic diversity. *Maximucinus muirheadae* is estimated to have weighed 18 kg which is comparable to a small adult *Thylacinus cynocephalus*. In contrast, *Mutpuracinus archibaldi* is the smallest thylacinid presently known with a body weight ($\sim$1 kg) similar to that of the modern Eastern Quoll (*Dasyurus viverrinus*). *Muribacinus gadiyuli* is estimated to have weighed $\sim$1.5 kg which is intermediate in size between *D. viverrinus* and the larger Tiger Quoll, *Dasyurus maculatus* while *N. dicksoni* and *N. richi* were equivalent in size ($\sim$5 kg) to a domestic cat (Wroe 2001). The diversity of thylacinids and other large carnivores (e.g. thylacoleonids and propleopine kangaroos) in Miocene deposits has led Wroe (1999, 2001) to challenge the traditionally held view (e.g. Archer and Bartholomai 1978; Flannery 1991, 1994) that reptiles dominated the large terrestrial carnivore niches of Australia.

The dasyurid *B. wangala* (Fig. 17) is recorded from two middle Miocene assemblages: Jim's Jaw and Henk's Hollow LFs of Riversleigh. The paucity of middle Miocene dasyurids is in stark contrast to the diversity of thylacinids of this age. Although the record of the late Miocene is poor, it is evident that the major radiation of this group must have occurred sometime between the middle Miocene and Pliocene because by Pliocene time many dasyurid groups are in evidence.

Other dasyuromorphians include the plesiomorphic *Joculusium muizoni* Wroe, 2001 and a new as yet undescribed dasyuromorphian genus and species (Archer et al. 2006), both from the Dwornamor LF of Riversleigh. *Joculusium muizoni* is known from limited dental material and, as a consequence, cannot be confidently placed within any known dasyuromorphian family. Another undescribed middle Miocene genus and species of highly specialised dasyuromorphian has turned up in the Rick's Rusty Rocks LF of Riversleigh (Arena et al. in preparation). This genus also occurs in the late Miocene Encore LF of Riversleigh.

Peramelemorphians

Bandicoots are as diverse and abundant during the middle Miocene as they were in the early Miocene. *Yarala burchfieldi* as well as four undescribed genera and five species of bandicoot that first appeared during the early Miocene continue on into the middle Miocene. We also see the first appearance of an additional new genus and species (Peramelemorphia new genus 5 sp. 1) in the Dwornamor LF (Archer et al. 2006). Many of these new taxa are referrable to Yaralidae (Travouillon 2010, 'personal communication'). Murray and Megirian (1992) also note the presence of 'V.D. bandicoots' (= yaraloid bandicoots) in the Bullock Creek LF. However, unlike those found at Riversleigh, these appear to represent a relatively minor component of the fauna. Given the abundance of yaralid specimens in the Ditjimanka LF of central Australia, it may be that their apparent rarity in the Bullock Creek LF is the result of taphonomic factors.

Yalkaparidontians

The middle Miocene marks the disappearance of the relatively common *Y. coheni* (recorded from nine Faunal Zone B deposits) and the appearance of the less abundant *Yalkaparidon jonesi* Archer et al., 1988 known from only two Faunal Zone C deposits (Dwornamor and Last Minute LFs) at Riversleigh. *Yalkaparidon jonesi* differs from its cogener in being slightly larger and in having an additional lower premolar (Archer et al. 1988). It was the last-known representative of the order Yalkaparidontia. Whatever niche these curious animals occupied in the early to middle Miocene, marsupial 'woodpeckers' (see the early Miocene) or other, it has either disappeared with them or remained vacant since the middle Miocene. If they were competing with dactylopsiline petaurids, this was the point at which the latter won.

Diprotodontians

The vegetation changes that occurred during the second half of the middle Miocene are most directly reflected in changes in the relative diversity of the various herbivorous diprotodontian groups. Possum species form a large component of the marsupial fauna within early middle Miocene rainforest assemblages whereas the large terrestrial diprotodontids began to diversify towards the end of the middle Miocene coincident with progressive opening up of the closed forests. Many groups demonstrate an increase in body size, probably in response to the increasingly poor quality of the most abundant food plants.

Vombatiforms

Phascolarctomorphians A continued decline in koala diversity is evident with only two genera (*Litokoala* and *Nimiokoala*) and three species (*L. kutjamarpensis*, *Litokoala* n. sp 1 and *N. greystanesi*) known from the middle Miocene. *Nimiokoala greystanesi* (Fig. 37a) is considerably less abundant being recorded from an isolated specimen from a single Faunal Zone C deposit (compared with 12 Faunal Zone B assemblages) at Riversleigh. Coincident with low species diversity, all species were also uncommon in the assemblages in which they occur.

Litokoala is the dominant genus in Faunal Zone C (Archer et al. 2006) which is congruent with the phylogenetic interpretation of Black and Archer (1997b) that it is the sister group of the modern genus *Phascolarctos*. At present, koalas are not known from the Bullock Creek LF of the N.T. but this may reflect taphonomic factors as well as the rarity of Miocene phascolarctids.

Vombatomorphians With the apparent demise of Wynyardiidae, only four vombatomorphian families (Thylacoleonidae, Diprotodontidae, Palorchestidae and Vombatidae) are represented in middle Miocene deposits (Table 3), a family-level diversity that was maintained until the late Pleistocene.

Thylacoleonid diversity declined to two genera and three species with the loss of the 'micro-lions' of the early Miocene. *Priscileo roskellyae* and *W. oldfieldi* continue into the middle Miocene and a new species, *Wakaleo vanderleueri* Clemens and Plane, 1974) (Fig. 43), is recorded from both the Bullock Creek LF and the middle to late Miocene of Riversleigh. Like a number of Miocene vombatomorphian lineages (see below), species of *Wakaleo* exhibit a chronological morphocline as evidenced by the gradual increase in size accompanied by an increase in the proportional length of the posterior premolar within the tooth row and a progressive reduction in molar number (Murray and Megirian 1990; Gillespie 1999). This reflects an evolutionary focus on the premolar/molar junction as a particularly important functional region of the dentition. This interface between the last premolar and first molar has also been an important area of evolutionary

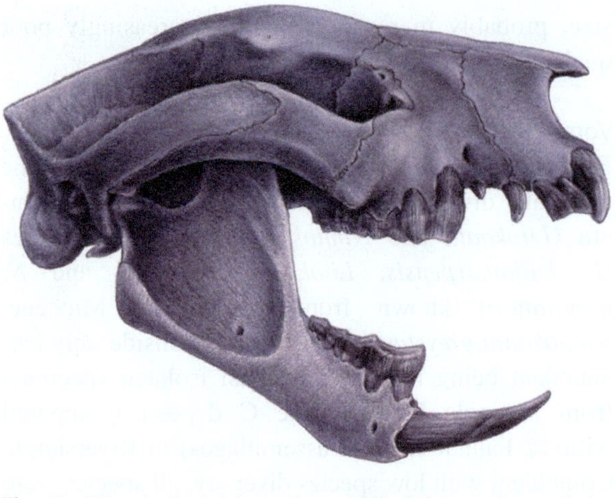

Fig. 43 The skull and jaw of a middle Miocene leopard-size thylacoleonid, *Wakaleo vanderleuri* from the Bullock Creek LF, N.T. (P. Murray). Larger than *Priscjleo* but smaller than *Thylacoleo*, the species of this genus of marsupial carnivores already have vicious stabbing lower incisors and increasingly elongate shearing premolars

change in other groups such as the phalangeridan families Pilkipildridae, Burramyidae, Phalangeridae, Miralinidae and all macropodoid families.

Diprotodontoids are the only vombatomorphian group to have increased in diversity during the middle Miocene almost certainly in response to the spread of more open forest environments. Three genera and six species are represented with a maximum of three species recorded from any given faunal assemblage. *Propalorchestes novaculacephalus* Murray, 1986, *Nimbadon lavarackorum* Hand et al., 1993 and *Neohelos stirtoni* Murray et al., 2000b are known from both the Henk's Hollow (Riversleigh) and Bullock Creek LFs. The genus *Neohelos*, represented by a single species (*N. tirarensis*) during the late Oligocene to early Miocene, underwent a significant middle Miocene radiation resulting in three new species (*N. stirtoni*, *Neohelos* sp. A, *Neohelos* sp. C). As is the case for species of *Wakaleo*. Black (1997b) and Murray et al. (2000a) have documented a gradual evolution in cranial and dental morphology accompanied by an increase in body size (with the exception of *Neohelos* sp. A) throughout the early to middle Tertiary. *Neohelos tirarensis*, the smallest (~100 kg), least-derived species of the morphocline, is most common in Faunal Zone B, and is succeeded by the larger (~300 kg), more-derived *N. stirtoni* common to Faunal Zone C assemblages of Riversleigh and the Bullock Creek LF. *Neohelos* sp. C is unique to Riversleigh's Jaw Junction LF and is the largest and most-derived member of the genus, displaying a number of features that are structurally antecedent to late Miocene diprotodontids (e.g. species of *Kolopsis*) from the Alcoota LF, N.T. The Jaw Junction LF is situated at the top of the Faunal Zone C stratigraphic sequence (Creaser 1997) and may postdate the Bullock Creek LF.

Several diprotodontid species (e.g. *N. lavarackorum*, *Neohelos* sp. A, *N. stirtoni*) exhibit high abundance in many of the deposits in which they are found which suggests that each of these species moved together as large 'mobs' (Black 1997b; Black and Hand 2010; Murray et al. 2000a, b). Of note is an exceptionally well-preserved *N. lavarackorum* population that includes cranial (Fig. 44) and postcranial material (Fig. 45) of pouch-young to mature adults, from AL90 Site at Riversleigh. This material has allowed benchmark studies of intraspecific (e.g. Black and Hand 2010) and ontogenetic (e.g. Black et al. 2010) variation within a fossil marsupial population. As a consequence of the former study, *Nimbadon whitelawi* Hand et al., 1993 known from a single specimen from the Bullock Creek LF, has since been synonymised with *N. lavarackorum* (Black and Hand 2010). *Nimbadon lavarackorum* is small (approx. 75 kg) compared to other mid-Miocene diprotodontids which seems to buck the general trend for increasing body size through time. Black and Hand (2010) suggest this plesiomorphic species may represent a relict taxon with phylogenetic relationships to even smaller

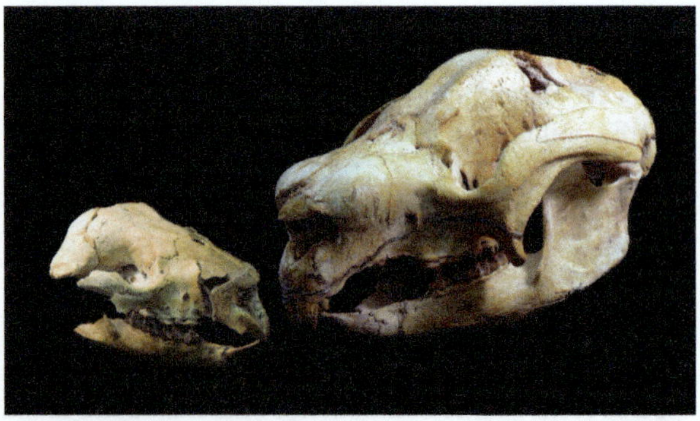

Fig. 44 Skulls of an adult and juvenile diprotodontid *Nimbadon lavarackorum* from the middle Miocene AL90 Site of Riversleigh (K. Black). Dozens of skulls representing all developmental stages of this sheep-sized herbivore have come from the AL90 cave deposit

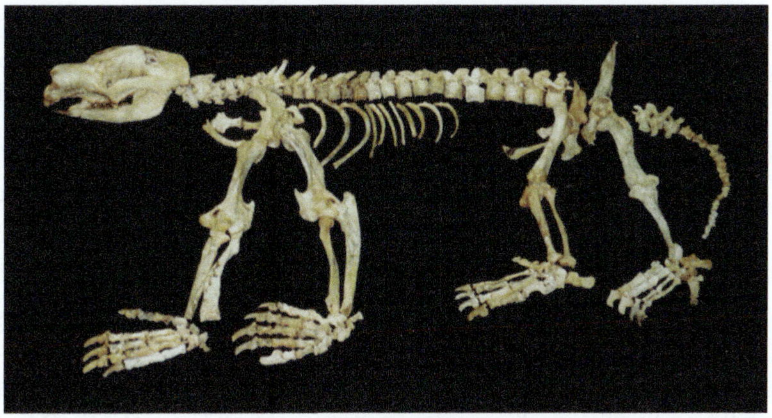

Fig. 45 Composite skeleton of *Nimbadon lavarackorum* from the middle Miocene AL90 Site, Riversleigh (K. Black). In addition to the many skulls and isolated limb elements, two articulated skeletons, one evidently with a pouch young in the pelvic region, are currently being prepared. Curious features include the long, powerful forelimbs and large, narrow, recurved claws

late Oligocene taxa such as the relatively rarer species of *Silvabestius*.

Phalangeridans

Phalangeroids Until recently, the middle Miocene phalangeroid fossil record was relatively depauperate. Miralinids and ektopodontids were not known from middle Miocene deposits and phalangerids were represented by only two species, both of which had been referred to modern genera (*Strigocuscus* and *Trichosurus*). Two species of miralinine miralinids have since been recorded from the Bullock Creek LF (*Barguru kayir* Schwartz, 2006b and *Barguru maru* Schwartz, 2006b). Schwartz (2006b) notes that these species of *Barguru* lack the characteristic molar crenulations that typify other miralinids. She suggests that in this respect they may be transitional between phalangerids and miralinids.

Ektopodontids are so far missing from middle Miocene deposits which may be a reflection of their rarity within most palaeocommunities. One genus and species has been recorded in Plio-Pleistocene assemblages in Victoria clearly indicating that they must have existed somewhere in the middle Miocene if not in the N.T. or Riversleigh areas.

Previously, the modern phalangerid genera *Strigocuscus* and *Trichosurus* were thought to have had their origins in the middle Miocene. Both *Strigocuscus reidi* Flannery and Archer, 1987a and *Trichosurus dicksoni* Flannery and Archer, 1987a had been described from Faunal Zone C deposits of Riversleigh. However, following an analysis of trichosurine periotic morphology (Crosby and Norris 2003) and a reassessment of cranial material assigned to *S. reidi*, Crosby (2007) reassigned *reidi* Flannery and Archer to a new genus, *Onirocuscus*. Crosby and Norris (2003) and Crosby (2007) also suggested that *dicksoni* Flannery and Archer represented a new as yet undescribed genus of phalangerids. Other middle Miocene phalangerids from Riversleigh include *Onirocuscus rupina* Crosby, 2007; *Onirocuscus inversus*; and an additional undescribed species referable to the same new genus as '*Trichosurus*' *dicksoni* (Fig. 46).

Burramyoids Both *B. brutyi* and *Cercartetus* sp. extend their temporal range into the middle Miocene with levels of abundance similar to those in the early Miocene. Given the rainforest palaeoenvironments of

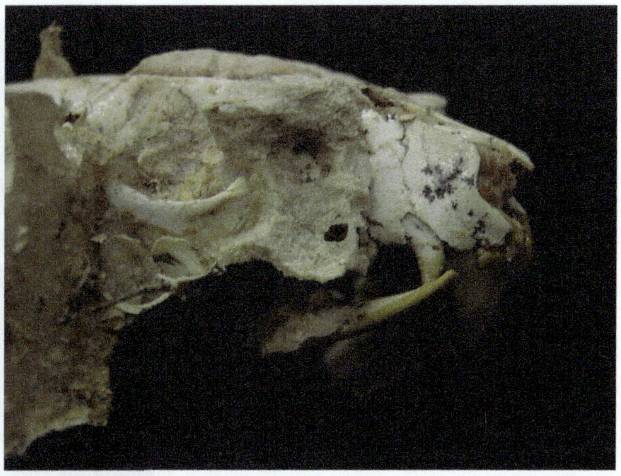

Fig. 46 Partially processed skull and jaws of a middle Miocene phalangerid from Dome Site South at Riversleigh (K. Black). The exquisite preservation of Riversleigh's fossils may reflect the speed with which they were encased by limestone after falling into carbonate-rich freshwater pools. In some cases, even soft tissues have been fossilised here because of the globally unusual process of bacteria-mediated phosphatisation

species of *Burramys* in the late Oligocene to early Pliocene, it seems probable that middle Miocene populations of *B. brutyi* occupied remnant rainforest patches within the Riversleigh region. The very similar species of this genus, starting with *B. wakefieldi* in lowland scrubby rainforests of the late Oligocene, *B. brutyi* in the early to middle Miocene lowland rainforests of Riversleigh and *B. triradiatus* in the early Pliocene lowland rainforest of Victoria suggest that this lineage of highly specialised possums has occupied lowland rainforests for most of the last 24 million years. This has provided the incentive to consider experimentally acclimatizing climate-change threatened alpine individuals of *B. parvus* for translocation to more resilient lowland rainforest release sites (Archer and Beal 2004).

On the basis of mitochondrial DNA sequence data, extant species of *Cercartetus* were estimated (Osborne and Christidis 2002) to have diverged from one another sometime between 27 and 16 Ma. However, the apparent absence of any modern species in the early to middle Miocene fossil record suggests that this estimate of divergence time is too old. The phylogenetic analysis of Osborne and Christidis (2002) indicates that the more plesiomorphic species (e.g. *C. caudatus*) are rainforest-adapted taxa, whereas the more derived living species are found in sclerophyll forest (*C. nanus*) or mallee and heathlands (*C. lepidus* and *C. concinnus*). They suggest that the fragmentation of closed and spread of open forests during the Miocene may have facilitated speciation within the genus.

Petauroids The middle Miocene petauroid fauna consists of *Dactylopsila* sp. and *D. yadjana* and a spectacular radiation of ringtail possums (Fig. 47). Fourteen pseudocheirid species have been reported from Riversleigh's middle Miocene assemblages (Archer 1992; Archer et al. 1999; Bassarova et al. 2001; Archer et al. 2006; Roberts et al. 2008, 2009) including members of *Pildra* (4 undescribed species); *Marlu* (*M. kutjamarpensis, M. syke, M. ampelos, M. karya* Roberts et al. 2009); *Paljara* (*P. nancyhawardae, P. tirarensae, P. hushae* Roberts et al. 2008); and an unnamed genus and species. Two *Pseudochirops* sp. had previously been reported from Faunal Zone C deposits (Archer et al. 1997, 2006; Roberts et al. 2007) but have since been identified as a new genus that awaits description. Ten sympatric species are recorded in Riversleigh's lower Faunal Zone C Dwornamor LF (Archer et al. 2006) whereas only three species are known from the upper Faunal Zone C Henk's Hollow LF. Pseudocheirids are not known from the Bullock Creek LF, although a bias towards large terrestrial species and aquatic taxa has been demonstrated (Murray and Megirian 1992). Be that as it may, modern woodland/open forest

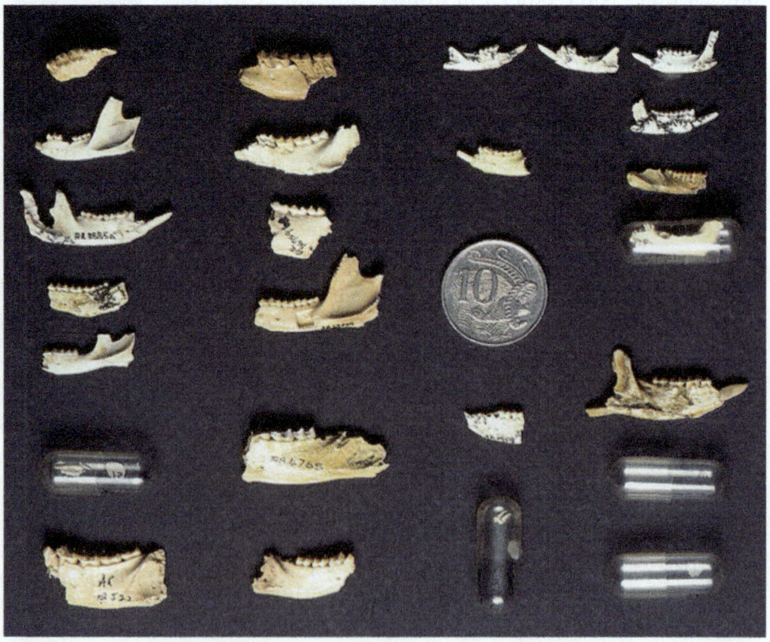

Fig. 47 A range of specimens indicating the high diversity of ringtail possums (pseudocheirids) from the early to middle Miocene deposits of Riversleigh (M. Archer). While many new species have been described, none appear to represent any of the living genera of ringtail possums all of which appear to have evolved in the late Miocene or later

habitats of the kind presumed by Murray and Megirian (1992) to have characterised the Bullock Creek palaeoenvironment support fewer arboreal folivores than more floristically diverse rainforests. The decline of more archaic pseudocheirids (e.g. *Marlu*, *Pildra* and *Paljara*) appears to coincide with the arrival of relatively more modern taxa. Meredith et al. (2010) suggested that, on the basis of biomolecular differences, diversification within the crown group subfamilies of Pseudocheiridae (Hemibelidinae, Pseudocheirinae and Pseudocheiropsinae) began in the middle to late Miocene between 12 and 9 Ma.

Phalangerida *incertae sedis* A single pilkipildrid, *Djilgaringa gillespieae* Archer et al., 1987, is known from the middle Miocene Last Minute and Dwornamor LFs of Riversleigh. This species was a small, omnivorous possum with short, deep jaws with powerful premolars adapted for cutting through hard surfaces such as chitinous insects, seed coats or possibly even small vertebrates. *Djilgaringa gillespieae* appears to be the last of its kind, the whole family vanishing after the middle Miocene.

Macropodoids Although macropodoid generic and species level diversity remained relatively unchanged from the late Oligocene, the middle Miocene marks a profound change in faunal composition with a rapid decline of the archaic balbarids and a corresponding rise of modern macropodids. Molecular divergence dates suggest (Meredith et al. 2008b) that an extensive diversification of macropodids occurred around 12 Ma, a time that coincides with the opening up of rainforest and the continued cooling and drying of the continent. Middle Miocene bulungamayine macropodids show numerous adaptations to more open-forested environments including the development of completely lophodont browsing dentitions (e.g. *G. bilamina*; Cooke 1997b) as well as more bipedal hopping gaits (e.g. *Wanburoo* sp.; Kear et al. 2001).

Three genera and five species of balbarids are known from middle Miocene deposits: *Balbaroo camfieldensis* Flannery et al., 1983 and *Nambaroo bullockensis* Schwartz and Megirian, 2004. Both are recorded from the Bullock Creek LF while *B. fangaroo* (Fig. 41), *Balbaroo* sp. 4 and *Wururoo* sp. 2 are recorded from Riversleigh's Faunal Zone C deposits (Archer et al. 2006). Bulungamayine kangaroos include three new species (*Wanburoo hilarus* Cooke, 1999, *Wanburoo* sp. and *Ganguroo* sp.) as well as the persistent early Miocene taxa *G. bilamina*, *N. matrix* and *B. delicata*.

Hypsiprymnodontid diversity remains relatively constant with *H. bartholomai*, *Hypsiprymnodon* sp. and *E. ima* (Fig. 32) extending their range into the middle Miocene. A new propleopine species, *Ekaltadeta jamiemulvaneyi* Wroe, 1996a has been recorded from the Cleft Of Ages LF (and late Miocene Faunal Zone D Encore LF) of Riversleigh. *Ekaltadeta jamiemulvaneyi* had a dentary 50% larger than *E. ima* but probably occupied a more omnivorous niche based on its relatively reduced premolar size and less of a gradient in molar size reduction towards the posterior end of the toothrow. In this regard *E. jamiemulvaneyi* shares more affinities with the Pleistocene *Propleopus oscillans* and may be more appropriately placed within that genus. However, further material is required to test these possibilities (Wroe 1996a).

Potoroid diversity remains as low as it was in the early Miocene with a single middle Miocene species described and an additional unnamed potoroid identified from the Bullock Creek LF (L. Schwartz 'personal communication' to J. Case in Westerman et al. 2002). *Bettongia moyesi* Flannery and Archer, 1987c from the Two Trees LF of Riversleigh is known from an exceptionally well-preserved skull and associated dentaries (Fig. 48). This faunal assemblage is the only one at Riversleigh that has produced this taxon. Morphologically, this potoroid is broadly similar to the living burrowing bettong, *Bettongia lesueur*

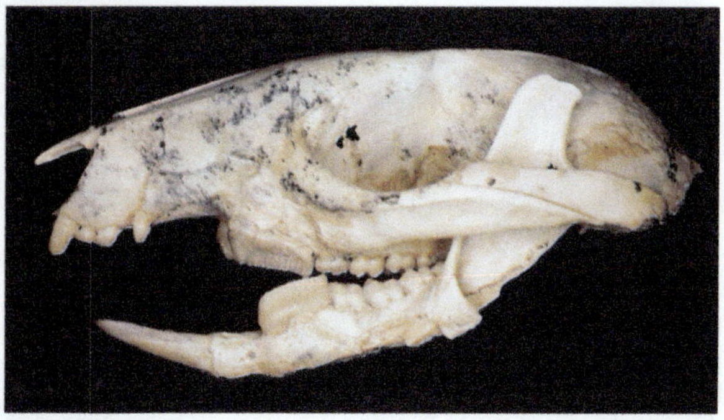

Fig. 48 The skull and dentary of the middle Miocene potoroid *Bettongia moyesi*, from Two Tree Site at Riversleigh (D. Maitland). While other kinds of archaic kangaroos were common in these ancient forests, potoroids (rat-kangaroos) were extremely rare. However, by Quaternary time they were common in all habitats of Australia from rainforests to deserts

(Flannery and Archer 1987c). Recent molecular studies (e.g. Westerman et al. 2004) support a plesiomorphic position for *B. lesueur* relative to all other modern *Bettongia* species. *Bettongia moyesi* represents the first occurrence of the modern genus in the fossil record and is congruent with recent molecular divergence dates (e.g. Westerman et al. 2004) that suggest a middle to late Miocene origin for crown group *Bettongia*. Some authors (e.g. Cooke and Kear 1999; Kear and Cooke 2001) have raised doubts about the generic and even subfamilial position of *B. moyesi*, as well as *W. lawsoni*, suggesting these taxa may be basal bulungamayines. Clearly this area of macropodid phylogeny is in need of major review.

Late Miocene (10.4–5.2 Ma)

Australian late Miocene terrestrial faunas are rare and relatively depauperate, a factor that has been attributed to the increasingly dry climatic conditions of the time which were less conducive to preservation of fossils (McGowran and Li 1994). The best-known deposits are the Alcoota (Fig. 49) and Ongeva LFs of the N.T. and the Encore LF (Faunal Zone D) of Riversleigh, although there are other much less diverse assemblages such as the Beaumaris LF of Victoria (Figs. 13 and 15a).

Comprehensive studies of the palaeontology and geology of the Alcoota locality (Alcoota and Ongeva LFs) have been undertaken by Woodburne (1967), Murray and Megirian (1992), Murray et al. (1993), Megirian et al. (1996) and Murray and Vickers-Rich (2004). Age estimates for these deposits are based on stage-of-evolution biocorrelation of zygomaturine diprotodontids (Murray et al. 1993; Megirian et al. 1996). The Ongeva LF, which comes from a stratigraphic unit in the Waite Fm that overlies that which produces the Alcoota LF, contains zygomaturines (e.g. *Zygomaturus gilli*; Megirian et al. 1996) at an equivalent stage of evolution to zygomaturines from the Beaumaris LF of Victoria. The Beaumaris LF has been variably interpreted as latest late Miocene or earliest Pliocene (i.e., Cheltenhamian) in age (Rich et al. 1991). Strontium dating of molluscs indicates the deposit is between 4.5 and 5.5 Ma (Dickinson et al. 2002). The Alcoota LF has been traditionally interpreted as late Miocene (Mitchellian) in age, somewhat older than the Ongeva LF. However, recognition of *Kolopsis torus* in both the Alcoota and Ongeva LFs (Megirian et al. 1996) and *Kolopsis* sp. cf. *K. torus* (Rich et al. 2003) in the Beaumaris LF has led Rich et al. to suggest the Alcoota LF is also Cheltanhamian in age.

The Encore LF of Riversleigh is interpreted to be early late Miocene (~9–10 myo) in age on the basis

Fig. 49 Applying preservatives to exposed bones in the late Miocene Alcoota deposit of the N.T. (M. Archer). Bones here are so common that it is hard to isolate sections of bone bed in preparation for plaster-jacketing and transportation

of biocorrelation of its marsupial fauna (Myers et al. 2001) including palorchestids (Black 1997a), vombatids (Brewer et al. 2007), thylacoleonids (Gillespie 2007), propleopine kangaroos (Wroe 1996a), koalas (Archer et al. 1995) and dasyuromorphians (Wroe 1997a). The Encore LF contains the oldest representatives of characteristic Plio-Pleistocene genera (e.g. *Phascolarctos*, *Palorchestes*, *Warendja*) that are structurally intermediate between middle Miocene species and closely related taxa from the Alcoota LF, as well as dasyuromorphians that show affinities to extant dasyurines and sminthopsines (Wroe 1997a, 1998).

Palaeoenvironment

Icehouse conditions increased in severity throughout the late Miocene resulting in further cooling and drying of the continent. By the end of the Miocene (5–6 Ma), the Antarctic ice cap had greatly increased in volume (Kennett 1977) resulting in a significant drop in sea level and a dramatic decrease in southern sea temperatures (Zachos et al. 2001). Rainforest disappeared or became further fragmented persisting only in small patches in southern and southeastern Australia with Auracarian forests present on the northeastern coast of Queensland. Wet sclerophyll forest with an open *Eucalyptus* canopy and rainforest understorey gradually replaced rainforests (Martin 1994). In central Australia dry sclerophyll forests and woodlands rose to dominance with abundant casuarinaceans, species of *Eucalyptus*, other myrtaceans and herbaceous groups such as asteraceans, restionaceans and poaceans (Martin 1998). Although palynological evidence indicates the presence of some grasses there is no evidence in the late Miocene of extensive grasslands (Martin 2006).

Continued tectonic activity involving adjacent areas of the Indo-Australian and Eurasian and Pacific plates, combined with the shallowing of the passage between Australia and Indonesia (10–9 Ma) because of a drop in sea level, resulted in the emergence of some of the modern-day Moluccan islands. By the end of the Miocene, the 25 million-year-long period of virtually total isolation of the Australian continent began to come to an end because these islands provided increasingly effective dispersal routes for both plants and animals into and out of Australia (Aplin 2006). Today, the Moluccan Island faunas are characterised by a blend of Asian and Australasian groups with a high level of endemism. The continued rise of the New Guinea Highlands during the late Miocene may have provided ecological refuges for rainforest-adapted taxa that had dispersed from Australia during the late Miocene.

The early late Miocene Encore LF is interpreted to represent an albeit open forest palaeocommunity (Myers et al. 2001; Travouillon et al. 2009) because of the presence of at least six arboreal folivores, a possible representative of the modern phascolarctid genus *Phascolarctos*, and the low-crowned browsing dentitions of its diprotodontoids and macropodoids.

The palaeoenvironment of the Alcoota region has been interpreted to be fluvio-lacustrine with a large, shallow, spring-fed, ephemeral lake (Murray and Megirian 1992). The reason for the accumulation of fossils at Alcoota, which are dominated by large diprotodontids, has been argued by Murray and Megirian, to be waterhole-tethering that occurred during a period of prolonged drought. Alcoota fossils collected to date consist of hundreds of densely packed skull and skeletal elements representing more than 150 medium- to large-sized partial skeletons of diprotodontoids (6 spp.), thylacinids (2 spp.), macropodoids (2 spp.) and thylacoleonids (1 spp.), as well as giant flightless dromornithid and large casuariid birds, crocodiles, varanids and a range of smaller animals (Murray 1997a). Material found so far represents only a minor portion of the death assemblage. Murray (1997a) estimates there to be at least 2500 individuals in the deposit. Reptiles include a species of crocodile (*Baru* sp.) and possibly a giant goanna referable to *Megalania* (Murray and Megirian 1992). A single arboreal species, a ringtail possum referable to the modern genus *Pseudochirops*, has been found. Murray and Megirian (1992) interpret the proximal palaeovegetation as subtropical savanna, with sedge and tussock grasses along the lake shore and riparian forest distal to the depositional environment. Rainfall was interpreted to have been highly seasonal with the climate oscillating between subhumid and semi-arid conditions. Archer et al. (1995) have questioned the suggestion of semi-arid conditions suggesting that the relatively low-crowned, thin-enamelled, browsing dentitions of the Alcoota diprotodontoids and macropodids would not be suited for processing abrasive semi-arid plants but would be suited to processing the softer vegetation found in sclerophyll forests. Further, Martin (2006) concluded, based on Australia's pollen record, that there were no grasslands as such in Australia prior to the Pliocene (Section 'Palaeoenvironment'). Murray and Megirian (1992) interpret the paucity of arboreal folivores to be the result of taphonomic processes such as fluviatile sorting of smaller elements before deposition.

The Ongeva LF is interpreted to have accumulated via natural attrition around a permanent or semi-permanent water hole (Megirian et al. 1996). It is faunistically less diverse than the Alcoota LF. The marsupial fauna consists of two zygomaturine diprotodontids, *K. torus* and *Zygomaturus gilli* (= *Kolopsis yperus*), a macropodid species probably referrable to the genus *Dorcopsoides* (Megirian et al. 1996) and the thylacinid *Thylacinus megiriani* Murray, 1997b. Other vertebrates include a crocodile of the genus *Quinkana* and two giant flightless birds (*Dromornis* sp. cf *Dromornis stirtoni* and a species of *Ilbandornis*). Little is known about the proximal vegetation of the Ongeva LF. However, coprolites attributed to diprotodontids and/or dromornithids have provided a window into aspects of the palaeoenvironment. These coprolites are well rounded with concentric ripples which, by analogy with modern domesticated bovid excreta, suggest conditions were favourable for herbivores with comparatively moist vegetation and an adequate supply of water (Megirian et al. 1996). It seems probable to us, given the basic similarity of the mammalian herbivores, that this was also the case for the Alcoota palaeoenvironment.

The Late Miocene Marsupial Fauna

The paucity of late Miocene fossil deposits has meant that very little is known about the fauna of this interval. In particular, smaller mammals are very poorly known with only seven arboreal species recorded (Table 3). The degree to which this is a reflection of sampling/preservational bias or limited niche diversity resulting from the retraction of rainforests and the expansion of open forest (McGowran et al. 2000) is difficult to determine. Certainly there are many families (e.g. perameloids, burramyids, petaurids, phalangerids, ektopodontids, potoroids) that are represented in both middle Miocene and Pliocene deposits that do not have a late Miocene record. This suggests that either their absence from the late Miocene assemblages can be attributed to preservational factors or these groups survived in more coastal habitats having become scarce or regionally extinct in central Australia and northwestern Qld. Whatever the case, known late Miocene taxa indicate a significant change in the structure of mammal communities during this interval. These changes included (Archer and Hand 2006) extinction of older groups that once dominated the early to middle Miocene rainforest environments (e.g. yalkaparidontids, balbarid kangaroos, yaralid bandicoots) and appearance of more modern forms (e.g. sthenurine and macropodine kangaroos, hypselodont vombatids). This phenomenon is supported by molecular evidence (e.g. Meredith et al. 2008a, b, c, 2009; Beck 2008a) which indicates many modern marsupial groups (e.g. macropodine kangaroos, dasyurids, peramelids) underwent extensive radiations during the late Miocene, presumably in response to climate change.

Dasyuromorphians

Four thylacinids are known from late Miocene deposits: *Thylacinus* sp. cf *T. macknessi* from the Encore LF; *Thylacinus potens* (Figs. 50 and 51) and *Tyarrpecinus rothi* Murray and Megirian, 2000 from the Alcoota LF; and *Thylacinus megiriani* from the Ongeva LF. In keeping with its late occurrence in the Miocene, *T. megiriani* is the largest thylacine species known with body estimates of around 57 kg (Wroe 2001), significantly greater than the recently extinct *T. cynocephalus* (15–35 kg; Strahan 1995). *Thylacinus potens* was similarly large (~30–45 kg; Murray 1997a; Wroe 2001) while *T. macknessi* and *Ty. rothi* were 9 and 5 kg, respectively (Wroe 2001). Murray and Megirian (2000) suggest that *Tyarrpecinus* is the sister group to the *Thylacinus* crown group with *T. megiriani* the sister group to *T. cynocephalus*. Subsequent to the late Miocene, no more than a single species of thylacinid (*T. cynocephalus*) has been found in any faunal assemblage. The once morphologically and ecologically diverse Thylacinidae had entered a period of dramatic decline. A similar decline (and subsequent extinction) is noted for the yaralid bandicoots. Wroe (1996b, 1999) and Muirhead and Filan (1995) speculated this demise may have been the result of competition with diversifying dasyurids and was inextricably linked to climate change.

The late Miocene dasyurid fossil record is correspondingly poor, with only a single species recorded (*Ganbulanyi djadjinguli* Wroe, 1998). However, molecular studies (e.g. Krajewski et al. 1997, 2000b; Beck 2008a) suggest that the modern

Fig. 50 Palate of the giant late Miocene thylacinid *Thylacinus potens* from the Alcoota LF of the N.T. (L. Meier). This wolf-sized carnivorous marsupial was the largest thylacinid ever found. Its large size reflects the increasing size of the late Miocene herbivores, a trend that culminated in the megafaunal giants of the Pleistocene

Fig. 51 Reconstruction of the largest known marsupial thylacine, *Thylacinus potens* from the late Miocene Alcoota LF, NT (courtesy S. Wroe). In keeping with its role as the king of mammalian beasts in the late Miocene, it is very rare in the Alcoota deposit

dasyurid subfamilies (Dasyurinae, Phascogalinae and Sminthopsinae) originated during the middle Miocene and underwent a period of active cladogenesis during the late Miocene. Beck (2008a) estimated that Phascogalinae and Dasyurinae diverged from each other around 11.6 Ma which corresponds with the middle late Miocene boundary, a time of significant global cooling and drying. This date is also congruent with the occurrence of the oldest dasyurine, *G. djadjinguli*, in the early late Miocene of Riversleigh. Whether the dasyurid radiation was the result of competitive exclusion of thylacinids and yaralids or whether the demise of these groups resulted in the availability of vacant carnivorous and insectivorous niches in which dasyurids subsequently diversified, cannot be determined. Wroe (1999) hypothesised that unique adaptations in the middle ear of *B. wangala* (Fig. 17) and modern dasyurids may have been an adaptive advantage over smaller thylacinids as Australia became increasingly drier from the late Miocene to the present.

Ganbulanyi djadjinguli from the Encore LF (Riversleigh Faunal Zone D) has been described by Wroe (1998) as the 'smallest mammalian bone cracker' displaying unusual blunt-cusped premolars and molars similar to those of the modern Tasmanian devil (*Sarcophilus harrisii*). An alternative interpretation of a molluscivorous diet was presented by Arena et al. (1998). An additional species from the Encore LF, *Mayigriphus orbus* Wroe, 1997a, is regarded as Dasyuromorphia *incertae sedis*, sharing synapomorphies with both dasyurids (specifically *Planigale*) and thylacinids, making its placement within either family problematic (Wroe 1997a).

Peramelemorphians

No bandicoots have been described from late Miocene deposits. However, Myers et al. (2001) and Murray and Megirian (1992) note the presence of an indeterminate genus and species of bandicoot in the Encore and the Alcoota LFs, respectively. Muirhead (2000) suggests this is the result of a post middle Miocene to pre-Pliocene bottleneck for peramelemorphians with the extinction of the archaic yaralids at some stage during the middle to late Miocene and the radiation of modern bandicoot groups from the late Miocene to early Pliocene. At present, the oldest crown group bandicoots are recorded from Pliocene deposits. The oldest peroryctid is from the Hamilton LF (Vic.; Turnbull et al. 2003) and the oldest peramelids are from the Bow (N.S.W.) and Bluff Downs (northeastern Qld) LFs. Beck (2008a) estimated the divergence of Peramelidae and Peroryctidae to have occurred around 11.7±1.5 Ma with similar estimates (7–13 Ma) obtained by Meredith et al. (Peramelinae and Peroryctinae under their classification system; 2008c). Consequently, a greater diversity of crown-group bandicoots in late Miocene deposits should be expected to turn up in the future.

Diprotodontians

The predominantly herbivorous diprotodontians were the most abundant forms in late Miocene fossil assemblages with 10 families represented. Of these, diprotodontids and macropodids are the most diverse and abundant with possum families relatively rare and restricted to three families (Phalangeridae, Burramyidae and Pseudocheiridae). All herbivores are browsing grade and characterised by relatively low crowned molars (Archer et al. 1995; Murray 1997a) with the exception of a single vombatid species from the Encore LF of Riversleigh. Many diprotodontian lineages (e.g. diprotodontoids, thylacoleonids and macropodids) had begun to show the early stages of gigantism, a response in herbivores to the need to consume greater amounts of poorer quality browse.

Vombatiforms

Marked changes in vombatiform composition are evident with larger, more derived taxa. Relatively common and speciose middle Miocene genera such as *Litokoala*, *Neohelos*, *Propalorchestes* and *Rhizophascolonus* were replaced by more derived forms such as, respectively, *Phascolarctos*, *Kolopsis*, *Palorchestes* and *Warendja* (Fig. 52).

The late Miocene saw an adaptive radiation of large-bodied diprotodontoids in response to more widespread open forests. Eight new species in six new genera (*Palorchestes*, *Kolopsis*, *Plaisiodon*, *Alkwertatherium*, *Zygomaturus*, *Pyramios*) have been described, two of which (*Palorchestes*, *Zygomaturus*) survived into the late Pleistocene. Two new diprotodontoid species are recorded from the Encore LF (*Palorchestes anulus* Black, 1997a and an indeterminate zygomaturine), five new species from the Alcoota LF (*K. torus* Woodburne, 1967, *Plaisiodon centralis* Woodburne, 1967, *Alkwertatherium webbi* Murray, 1990b, *Pyramios alcootensis* Woodburne, 1967 and *Palorchestes painei* Woodburne, 1967), two species from the Ongeva LF (*K. torus* and *Z. gilli* Stirton, 1967c), and four species from the Beaumaris LF (*Z. gilli*, *Kolopsis* sp. cf. *K. torus*, *Kolopsis* sp. and *Palorchestes* sp.; Piper 2006). An additional species, *Kolopsis yperus* Murray and Megirian, 1992 from the Ongeva LF has since been synonymised with *Z. gilli* (Megirian et al. 1996).

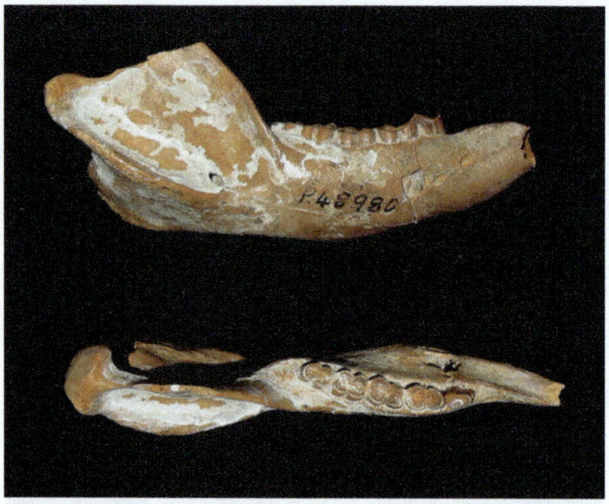

Fig. 52 Dentary (two views of the holotype) of the Pleistocene wombat *Warendja wakefieldi* from McEacherns Cave, Vic. (K. Black). As in all vombatids that have ever-growing, rootless teeth, the premolars and molars constantly erupt to replace tooth material worn away by abrasion. The cheekteeth of early to middle Miocene forest-dwelling vombatids from Riversleigh had roots and crowns. Neither became irrelevant until the climate change of the late Miocene

Kolopsis torus is the smallest (125–250 kg; Murray 1997a) and most abundant late Miocene diprotodontid. Assuming the Beaumaris material is referrable to this species, it is also temporally the longest-lasting and one of the most geographically widespread diprotodontid. While *P. centralis* was relatively common at Alcoota, the similar-sized (300–500 kg) *Pyramios alcootense* and slightly smaller (350 kg) *A. webbi* are rare suggesting they may have belonged to a distal community (Murray and Megirian 1992; Murray 1997a).

Many of the Alcoota taxa appear to have structural antecedents within the upper Faunal Zone C and

Encore LFs of Riversleigh and the middle Miocene Bullock Creek LF, N.T. *Palorchestes anulus* from the Encore LF, the most plesiomorphic member of the genus, is proportionately and morphologically intermediate between *Propalorchestes novaculacephalus* (Faunal Zone C, Riversleigh; Bullock Creek LF) and *Palorchestes painei* (Fig. 23) from the Alcoota LF (Black 1997a). The structural transition from *Neohelos* through *Kolopsis* to *Zygomaturus* is well documented (Stirton et al. 1967; Murray and Megirian 1992; Murray et al. 2000a, b) and has been fundamental in establishing a biocorrelative framework for Australian Tertiary fossil deposits. A similar chronological morphocline is evident in *Wakaleo* spp. with *W. vanderleuri* (Bullock Creek LF; Encore LF) being intermediate in size and morphology between *W. oldfieldi* (Riversleigh Faunal Zone C; Kutjamarpu LF) and *Wakaleo alcootaensis* Archer and Rich, 1982 (Alcoota LF; Murray and Megirian 1990).

Warendja encorensis Brewer et al., 2007 from the Encore LF provides the earliest record of a hypselodont wombat in the Australian fossil record (Brewer et al. 2007). All modern and Plio-Pleistocene vombatids are hypselodont, characterised by evergrowing, rootless teeth, an adaptation to consuming highly abrasive vegetation. The transition in vombatids from rooted, brachyodont teeth (e.g. *R. crowcrofti*) to unrooted hypselodont teeth during the late Miocene signals a marked change in the abrasiveness of at least some component of the vegetation. However, given that even middle Miocene vombatids (*R. crowcrofti*) were exhibiting increasingly higher crowns with diminishing roots, this group of marsupials must have already begun to specialise on some perhaps uniquely abrasive food types, such as roots or stems, within the otherwise relatively luxurious early to middle Miocene forests of Riversleigh and presumably central Australia.

Phalangeridans

Of the possum families, pseudocheirids, phalangerids and burramyids are the only groups recorded in late Miocene deposits and all are exceedingly rare. *Trichosurus* sp., the temporally extensive *B. brutyi* and a new petauroid genus and species are recorded from the Encore LF as well as three sympatric pseudocheirids including a species of the modern genus *Pseudochirops*, a species of *Pildra* and *P. tirarensae* (Roberts et al. 2008; Archer et al. 2006). Archer and Bartholomai (1978) noted a species of *Pseudochirops* in the Alcoota LF. Today, four of the five *Pseudochirops* species are restricted to New Guinea with a single Australian species (*Pseudochirops archeri*) found in northern Queensland. The appearance of *Pseudochirops* within late Miocene fossil deposits is congruent with the molecular divergence dates of Meredith et al. (2010) who found that speciation within the subfamily Pseudochiropsinae commenced 12–10 Ma on mainland Australia, with subsequent dispersal into New Guinea during the late Miocene prior to 6.5 Ma. However, it is also possible that the Alcoota taxon represents a species of *Petropseudes* the only living species of which, *Petropseudes dahli*, the rock-haunting Ringtail Possum, is closely related to species of *Pseudochirops*. Unlike species of *Pseudochirops*, *P. dahli* spends as much of its time on the ground as in trees, in seasonally dry vine forests and woodlands of northern Australia.

Macropodoids Hypsiprymnodontids are represented in the Encore LF by the propleopine kangaroos *E. ima* and *E. jamiemulvaneyi*. The early late Miocene is the last record of this long-ranging genus which is succeeded by *Propleopus* spp. during the Pliocene and Pleistocene. Similarly, bulungamayine kangaroos are not recorded from deposits younger than the early late Miocene which coincides with the first appearance of representatives of the modern macropodid subfamilies Sthenurinae and Macropodinae. Three bulungamayine species are recorded from the Encore LF: *Wanburoo hilaris*, *Wanburoo* sp. and *Ganguroo* sp. (Archer et al. 2006).

Sthenurines are a group of (mostly) large to giant-sized, short-faced kangaroos (e.g. *Procoptodon*, *Sthenurus*, *Simosthenurus*) that were prolific during the Pliocene and Pleistocene (Prideaux 2004). The living, endangered Banded Hare-wallaby *Lagostrophus fasciatus* (Fig. 53) may be the sole living representative of this once diverse subfamily. This possibility was first noted by Bartholomai (1967) when making comparisons with the teeth of species of the Plio-Pleistocene species of the sthenurine genus *Troposodon*, then later further supported by Archer (1981) and Flannery (1983, 1989). This possibility is supported by recent molecular analyses (e.g. Westerman et al. 2002; Meredith et al. 2008b) which found *L. fasciatus* to be a genetically distinct lineage within Macropodidae

Fig. 53 A possible surviving sthenurine kangaroo, the banded hare-wallaby *Lagostrophus fasciatus* is an endangered species whose range is now restricted to two islands in Shark Bay, W.A. (J. Gould). While some studies suggest they are too 'primitive' to be part of the sthenurine radiation, most researchers agree that they are surviving members of the group that contained species of the extinct genus *Troposodon* and, as such, unlike any of the other living kangaroos

and the sister group to Macropodinae. Estimates that have been given for the time of divergence of the *L. fasciatus* lineage from Macropodinae are, however, early Miocene (16–23 Ma) in age (Westerman et al. 2002; Meredith et al. 2008b). This significantly predates the oldest fossil record for both groups Prideaux and Warburton's (2010) analysis of osteological data found strong support for the monophyly of *L. fasciatus* with species of *Troposodon*, leading them to establish the subfamily Lagostrophinae. However, affinity of lagostrophines with sthenurines was not supported, members of the former being distinctly more plesiomorphic than the latter, which could be interpreted as support for an early Miocene divergence.

Rhizosthenurus flanneryi Kear, 2002 is the oldest sthenurine known from postcranial elements occurring in the Encore LF. This material was originally attributed by Kear et al. (2001) to the bulungamayine genus *Wanburoo*, contrary to the results of their phylogenetic analysis which showed clear affinities with sthenurines. To address the subsequent polyphyly of Bulungamayinae, Kear (2002) reassigned the material to a new sthenurine genus and species, *R. flanneryi*. Modifications evident in the tarsus of *R. flanneryi* indicated a shift in the distribution of body weight to pedal digit IV, a condition antecedent to the highly derived, functionally monodactylus pes of later sthenurines such as species of *Procoptodon* (Kear et al. 2001). Consequently, *R. flanneryi* was found to represent the most plesiomorphic member of a clade containing *Hadronomas puckridgi* Woodburne, 1967 and crown group sthenurines (Kear et al. 2001; Kear 2002).

Hadronomas puckridgi from the Alcoota LF was originally described on the basis of dental material. Woodburne (1967) assigned this taxon to Potoroinae. Subsequent discovery of cranial and postcranial remains has indicated sthenurine affinities first noted by Murray (1991, 1995). *Hadronomus puckridgi* was equivalent in size to a modern red kangaroo and showed, apart from its size, numerous adaptations for high-browse eating, including long forelimbs with elongate curved claws on digits III and IV of the manus allowing tall branches to be reached (Murray and Vickers-Rich 2004); this is similar to the way the long-armed species of *Procoptodon* (Fig. 54) are presumed to have functioned.

The oldest unequivocal macropodine, *Dorcopsoides fossilis* Woodburne, 1967 is recorded from the Alcoota LF and possibly the Ongeva LF. *Dorcopsoides fossilis*

Fig. 54 Reconstruction of the largest known kangaroo, the sthenurine macropodid *Procoptodon goliah*. It stood 2.5 m tall and is known from many Pleistocene sites in Australia (R. Scott). By the time it was common across much of inland as well as coastal areas of the continent, grasslands and deserts were a feature of the interior. A wisp of smoke in the distance controversially suggests that humans may have arrived in time to see some of these now extinct megafaunal giants

is the plesiomorphic sister taxon to all other macropodines which are united, to the exclusion of *D. fossilis*, in their reduction of the lateral tuberosity on metatarsal V (Kear 2002).

The paucity of diverse kangaroos so far known from the late Miocene is striking compared with the diversity in the late Oligocene to middle Miocene and then again from the Pliocene to Holocene. Clearly much of the record of kangaroo evolution during the late Miocene must be hiding somewhere, awaiting discovery.

Pliocene

Pliocene (5.3 to 1.8 Ma) mammal deposits are well represented in Australia, occurring in all states except Tasmania and the Northern Territory. The oldest records for marsupials in New Guinea are also Pliocene in age. Pliocene marsupial-bearing deposits (Figs. 13 and 15a) include Queensland's Rackham's Roost (Fig. 55), Bluff Downs and Chinchilla LFs; the Bow and Big Sink LFs of central N.S.W.; Quanbun LF of W.A. (Archer et al. 1984; Rich et al. 1991); the Awe LF of P.N.G. (Fig. 56); and numerous localities in the Lake Eyre, Frome, Murray, Otway and Port Philip Basins of southeastern Australia such as Palankarinna, Kanunka, Sunlands, Fisherman's Cliff, Hamilton (Woodburne 1972; Fig. 57) and Dog Rocks LFs (reviewed by Tedford et al. 2006).

Some of these Pliocene assemblages are radiometrically dated by basalts associated with, capping or sandwiching the fossil-bearing sediments, including Hamilton LF (4.46±0.1 Ma; Turnbull and Lundelius 1970; Turnbull et al. 2003; Tedford et al. 2006) and Bluff Downs LF (3.6±0.5 Ma; Rich et al. 1991; Mackness et al. 2000; Mackness and Archer 2001). Other terrestrial fossil mammal deposits, mostly in southeastern Australia, occur between tongues of well-dated marine sediments (e.g. Coimadai LF; Piper et al. 2006). The ages of other Australian Pliocene mammal assemblages have been estimated by biocorrelation of the marsupial species they contain (e.g. Rackham's Roost LF, Archer et al. 1991; Godthelp et al. 1999; Big Sink LF, Dawson et al. 1999; Quanban LF, Rich et al. 1991). The Curramulka LF from Corra Lynne Cave, Yorke Peninsula, S.A., is regarded to be late Miocene to Pliocene on the basis of biocorrelation (Pledge 1992; Archer et al. 1997; Prideaux 2004).

Palaeoenvironment

During the Pliocene the global climate fluctuated markedly with the expansion and contraction of northern and southern hemisphere ice sheets (Gallagher et al. 2003). Following the cool and dry conditions that characterised the late Miocene, in the early Pliocene there was a brief respite of warmer, wetter conditions before Australia began to dry once more. During this brief milder greenhouse phase, *Nothofagus* and other

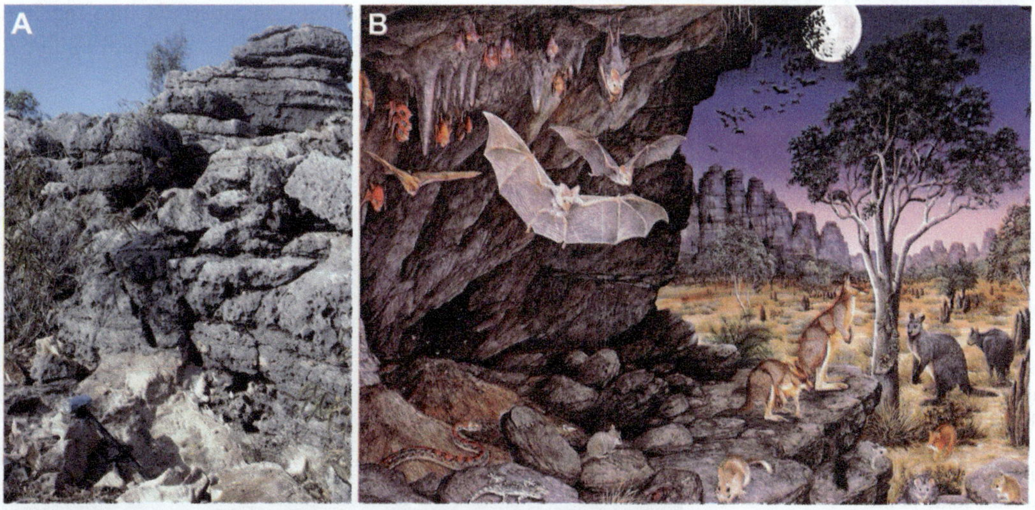

Fig. 55 Riversleigh's only known Pliocene deposit. (**a**) Early Pliocene Rackham's Roost site (S. Hand) with the pale Pliocene deposit lapping up against the base of the former walls of the cave; (**b**) reconstruction of Rackham's Roost faunal assemblage (D. Dunphy). By this time, the forests of northern and inland Australia had opened up, the first rodents had arrived and the first grazing kangaroos had evolved

Fig. 56 Early Pliocene Otibanda Formation, the source of the Awe Local Fauna, New Guinea (P. Woolley). The diprotodontids and kangaroos in this deposit appear to be most similar to those of the Alcoota LF of the NT suggesting that Australian marsupials first colonised NG sometime during the late Miocene

rainforest species that had survived during the late Miocene in small habitats in the eastern highlands expanded their range (Martin 2006). Significant climatic/vegetation gradients occurred across Australia (McGowran et al. 2000), with wet sclerophyll forests common in peripheral regions, including biodiverse rainforests near the coastal margins of the east, southeast and northwest (Hill 1992, 1994) but drier forests and woodlands further inland (McGowran et al. 2000; Martin 2006). These palaeoenvironmental gradients are reflected in early Pliocene marsupial assemblages, such as in the rainforest-type assemblages of Hamilton and Sunlands LFs of coastal Victoria and South Australia which are dominated by arboreal and browsing taxa, compared to those of (e.g.) the Palankarinna and Bluff Downs LFs which appear to reflect more open forest environments.

In the later Pliocene, rainforest remained a significant component of the southeastern Australian landscape but the tropical floristic elements had gone from this region (Gallagher et al. 2003). At this time, the rise of herbaceous taxa (e.g. Asteraceae, Chenopodiaceae, Poaceae, Restionaceae), which are well represented in today's open-canopy sclerophyll vegetation in eastern Australia, reflects a shift from non-seasonal to seasonal rainfall patterns (Greenwood and Christophel 2005).

The oldest palaeobotanical evidence of grasslands occurs in the later Pliocene (Martin 2006). Reflecting this fact, mid to late Pliocene grazing mammals, in particular kangaroos and wombats, underwent significant diversification that continued on into the Pleistocene (see Section 'The Pliocene Marsupial Fauna').

Fig. 57 Source of the early Pliocene Hamilton LF, Vic. (M. Archer). The massive basalt flows overran and buried a bone-rich deposit that outcrops near the feet of the person in the centre of the image. At least some of the extinct Hamilton marsupials appear to have lived in rainforest or wet sclerophyll forest

Overall, despite the increasingly drier conditions, much of Australia remained forested throughout the Pliocene. The vegetation appears to have been mainly sclerophyll/woodland, with araucarian rainforest and casuarinaceous sclerophyll forests in northeastern Australia appearing relatively unchanged from those of the late Miocene, but with arid shrubs and grasslands emerging in northwestern central Australia later in the epoch (Martin 2006).

The Pliocene Marsupial Fauna

During the Pliocene the Australian fauna also began to assume its distinctive modern appearance and some living marsupials made their first appearances in the fossil record (e.g. Agile Wallaby *Macropus agilis*). Also during this epoch the first undoubted grazing species, suggestive of dietary specialisation on grasses, appear. Australian Pliocene marsupial assemblages lack many of the distinctive 'older' groups that characterise Oligo-Miocene assemblages (e.g. ilariids, wynyardiids, miralinids, bulungamayines) and include the first representatives of many modern marsupial groups such as dasyurines, sminthopsines, thylacomyids and peramelids as well as many genera including *Dasyurus, Planigale, Sminthopsis, Perameles, Protemnodon, Sthenurus, Prionotemnus, Macropus* and *Zygomaturus* that became diverse during the Quaternary. Some generic-level continuity spanning the Miocene/Holocene interval is also evident in the Pliocene including *Burramys*, *Strigocuscus*, *Trichosurus*, *Palorchestes, Warendja* and *Thylacinus* (McGowran et al. 2000; Brewer et al. 2007).

However, some early Pliocene marsupial assemblages, such as the Hamilton LF of Victoria (Fig. 57), represent rainforest communities that retain several Oligo-Miocene lineages (e.g. ektopodontids). Still other early Pliocene assemblages (e.g. the Bluff Downs LF) include some evidently archaic lineages surprisingly not encountered in the older fossil record (i.e., numbigilgids; see below). These seemingly anachronistic taxa, as well as the seemingly abrupt first appearance of other lineages in the Pliocene such as *Macropus, Sthenurus* and *Troposodon*, suggest that these lineages will eventually prove to have at least late Miocene representation when fossil faunas of this age are better known.

Throughout the rest of the Pliocene, depending to some extent on the location of the particular fossil deposit, assemblages of this age include a mix of marsupial genera some of which (based on living relatives) indicate drier conditions (e.g. those from central Australia) and others moderately moist conditions (those from southern, northeastern and eastern Australia). By the late Pliocene some contained genera that today have species only in semi-arid to arid regions (Tedford et al. 2006).

There is much debate concerning the origins of New Guinea's marsupial fauna (particularly its now extinct diprotodontids; see below). Following the middle Miocene collision of the Australian plate with the island arcs of south east Asia, the mountainous spine of New Guinea began to build and small islands emerged at the edge of the subducting Australian Plate. Orogeny was most rapid between 8 and 4 Ma, but has continued until the present day to produce elevations of 5 km in the Central Ranges of New Guinea and establishing its current topography (Cloos et al. 2005). The increasingly high altitudinal forests of New Guinea appear to have provided a refugium for some of Australian rainforest-adapted marsupial lineages (such as tree kangaroos, cuscuses and pseudocheirid possums) as the continent itself dried out. Exchange across Torres Strait may have first occurred during periods of lower sea level in the late Miocene (but see below) and this seems to be reflected in New Guinea's Pliocene Awe LF which shares many similarities with early Pliocene and late Miocene Australian marsupial assemblages (e.g. congeneric forms of *Protemnodon* and *Thylacinus*). Lineages characteristic of earlier Oligo-Miocene Australian assemblages (such as ilariids, wynyardiids, etc.) are absent from the New Guinea record, as are many extant Australian lineages with ancient pedigrees missing from the modern New Guinea fauna (e.g. yaralids, numbigilgids, thylacomyids, hypsiprymnodontids, sthenurines, potoroids, tarsipedids, petaurines, phascolarctids, vombatids).

Dasyuromorphians

By the Pliocene only one thylacinid species – the extant Tasmanian Tiger *T. cynocephalus* – appears to be represented. It is widely distributed, occurring in the Pliocene fossil record of both Australia (e.g. Chinchilla LF) and New Guinea (e.g. Awe LF), but is not common.

Many of the modern dasyurid genera had also emerged by the early Pliocene (e.g. *Antechinus* and *Planigale*). Other early Pliocene dasyurine taxa include: *Archerium chinchillaensis* Wroe and

Mackness, 2000 from the Chinchilla LF; *Dasyurus dunmalli* Bartholomai, 1971 from Chinchilla, Bluff Downs and Bow LFs (Wroe and Mackness 1998); '*Dasycercus*' *worboysi* Dawson et al., 1999 from Big Sink LF; *Glaucodon ballaratensis* Stirton, 1957a, a dasyurine from the Smeaton LF, Vic.; and *Myoictis* sp. from the Awe LF, P.N.G. (Archer 1982a).

Drying out of the continent in the late Miocene probably resulted in a wave of extinctions with phasing out of older forms as new, relatively more modern groups originated and diversified. This is also probably reflected in the independent evolution of arid-adapted features such as highly fenestrated palates and hypertrophied alisphenoid hypotympanic sinuses in the middle ear, the latter being an adaptation to increasing sensitivity to more distant, low-frequency sounds which is relatively important for inhabitants of more open environments (Louys et al. 2009).

Fig. 58 The pig-footed bandicoot, *Chaeropus ecaudatus*, a recently extinct peramelid that once lived in central Australia (J. Gould). From studies of gut contents in preserved specimens, it appears to have been the only grazing bandicoot and one of the smallest mammalian grazers known

Peramelemorphians

The radiation of modern kinds of bandicoots and bilbies probably began in the late Miocene and proliferated during the Pliocene. Today they occupy most Australian habitats from rainforest to desert. Although the divergence time of modern bandicoot lineages has been estimated to have occurred as early as 13 Ma (e.g. Meredith et al. 2008c; see above), the first fossil representatives of all three modern bandicoot families, Peramelidae, Peroryctidae and Thylacomyidae, first appear in the early Pliocene. Pliocene peramelids (ordinary bandicoots) include *Perameles allinghamensis* from the Bluff Downs LF, Qld (Archer 1976d; 3.62 Ma, Mackness et al. 2000; Mackness and Archer 2001) and *Perameles bowensis* from the Bow LF, N.S.W. (c. 3.62 Ma; Muirhead et al. 1997). They also include *Chaeropus* sp. from a Quaternary deposit near Chillagoe (Muirhead and Godthelp 1996), a close relative of *Chaeropus ecaudatus* (the Quaternary pig-footed bandicoot). The latter, the only known grazing bandicoot (Fig. 58), became extinct relatively recently as a result of European activities such as land clearing and the introduction of cats, foxes, sheep and cattle. The oldest peroryctid (forest bandicoots) is *Peroryctes tedfordi* from the Hamilton LF, Vic. (4.46 Ma; Turnbull et al. 2003). The oldest thylacomyid (bilbies or rabbit bandicoots), and in fact the first-named Tertiary peramelemorphian, is *Ischnodon australis* Stirton, 1955 from the Palankarinna LF in the Tirari Desert, central Australia.

Although *Numbigilga ernielundeliusi* Beck et al. 2008a (Fig. 20) is formally regarded as Marsupialia *incertae sedis*, one of us (Mike Archer) is nevertheless of the view that this highly autapomorphic taxon is most probably a highly derived lineage of peramelemorphians. Its appearance in the early Pliocene without obvious ancestral relatives in the Miocene is a mystery and suggests rapid evolutionary change in one of the already known mid to late Cenozoic marsupial lineages such as yaralids or another of the diverse peramelemorphian groups that appear to have been the result of a late Miocene proliferation within this order.

Diprotodontians

As in the late Miocene, diprotodontians are the most common taxa in Australian Pliocene fossil assemblages, with 14 families recorded (Table 3). Kangaroo species (38 spp) far outnumber diprotodontoids (12 spp) for the first time in the record and a tendency towards gigantism is common in both groups. Diprotodontians include for the first time a growing number of grazing species (e.g. species of *Macropus*) with dentitions that appear to reflect the challenges of dining on abrasive grasses.

Vombatiforms

Phascolarctomorphians From the early Pliocene Chinchilla and Bluff Downs LFs of Queensland, two species of the poorly understood genus *Koobor* (Archer 1976d) are recorded: *notabilis* from Chinchilla and *jimbarratti* from Bluff Downs (Archer 1976d). Both have selenodont upper molars suggesting that they were eating tree leaves. They were regarded by Archer (1976d) to be members of the koala family Phascolarctidae although Myers and Archer (1997) were at that time uncertain about the familial affinities of this genus. Aplin and Archer (1987), Pledge (1987b), Black (1999) and Marshall et al. (1990) have suggested potential alternative links to ilariids. Unfortunately at present there is insufficient material to resolve the affinity of these rare folivores. The modern koala genus *Phascolarctos* is represented by three Pliocene species, *Phascolarctos yorkensis*, *Phascolarctos maris* and *Phascolarctos stirtoni*, from Curramulka (S.A.), Sunlands (S.A.) and Gore Cement Mills (Queensland) LFs respectively (Pledge 1987d, 1992; Price et al. 2009). The age of the Gore Cement Mills LF is in doubt but may in fact be Pleistocene in age (Price et al. 2009). Further, *P. stirtoni* may be conspecific with *Phascolarctos cinereus* given that there are few if any significant morphological differences between the two, the former differing mainly in its slightly larger size. *Cundokoala* (Pledge 1992) (type species *Cundokoala yorkensis* Pledge, 1992) is now regarded to be a junior synonym of *Phascolarctos* (Black 1999).

Vombatomorphians Wombats are rare in Pliocene deposits but include the very large *Ramsayia lemleyi* Archer, 1976d from the Bluff Downs LF of Qld, an early representative of the gigantic *Phascolonus gigas* Owen, 1858 in the Quanban LF of W.A. and possibly the Bow LF of N.S.W. This giant fossil wombat, which was relatively common in the Pleistocene, was about 1.6 m long, stood about 1 m at the shoulder and weighed up to 200 kg (Murray 1991). Its skeleton suggests that this wombat was not as well adapted for burrowing as modern wombats. But if it did burrow, it must have created some substantial excavations and stumbling places for other large, contemporaneous, terrestrial marsupials such as the three-tonne *D. optatum* (Fig. 22). Given the presence of species of the plesiomorphic vombatid genus *Warendja* (Fig. 52) in both the late Miocene (*W. encorensis* Brewer et al., 2007, from the Encore LF, Qld) and the Pleistocene (*W. wakefieldi* Hope and Wilkinson, 1982, from McEacherns Cave, Vic.), this lineage must also have been present during the Pliocene despite its apparent absence from known Pliocene assemblages.

In Australia, Pliocene marsupial lions (thylacoleonids) are widely distributed, with the larger *Thylacoleo crassidentatus* Bartholomai, 1962 represented in the Bow, Chinchilla and Bluff Downs LFs of N.S.W. and Qld, and the much smaller *Thylacoleo hilli* Pledge, 1977 in the Bow LF and Curramulka LF from N.S.W. and S. A. Among Pliocene assemblages, only the Bow LF of N.S.W. includes two sympatric thylacoleonids – *T. crassidentatus* and *T. hilli*.

Diprotodontids are represented by at least seven genera and nine species in the Pliocene, this being the highest diversity recorded in the family's long history. The spread of open forests were evidently beneficial for this now-extinct family. Notable Pliocene diprotodontids include the zygomaturines *Kolopsis rotundus* Plane, 1967 and *Kolopsoides cultridens* Plane, 1967 from the Awe LF of N.G. and *Zygomaturus keanei* Stirton, 1967c from the Palankarinna LF of S. A. Representative diprotodontines include *Euowenia grata* De Vis, 1887 (Fig. 59) from the Chinchilla LF, Qld and from Pliocene sediments of the Tirari Formation, S. A. (Camens and

Fig. 59 The palate and mandible of the Pliocene diprotodontine *Euowenia grata* from the Pliocene Chinchilla LF, Qld (K. Black). This cow-sized diprotodontid was highly unusual in appearing to have, like vombatids, only a single pair of upper incisors. The very worn teeth indicate that whatever plant material it was eating was highly abrasive

Wells 2010), and *Euryzygoma dunense* De Vis, 1888a from the Chinchilla LF, Qld; *Meniscolophus mawsoni* Stirton, 1955 from the Palankarinna LF, S.A.; and *Nototherium watutense* Anderson, 1937 from the Awe LF of New Guinea.

The Plio-Pleistocene New Guinean zygomaturine radiation includes four younger species: *Hulitherium tomasettii* Flannery and Plane, 1986; *Maokopia ronaldi* Flannery, 1992; *N. watutense*; and *Zygomaturus nimborensia* Hardjasasmita, 1985. Recently, the generic assignment and specific distinction of many of these taxa have been questioned (e.g. Murray 1992; Flannery 1994). Murray (1992) suggested *K. rotundus*, *M. ronaldi* and *H. tomasettii* are all referable to *Zygomaturus*. Flannery (1994) suggested *Z. nimborensia* is a junior synonym of *N. watutense* and that *N. watutense* may be more appropriately placed in the otherwise late Miocene genus *Kolopsis*. Clearly a holistic revision of these small zygomaturines is needed.

Two theories regarding the timing and origins of the New Guinean zygomaturine radiation have been proposed: one by Flannery and Plane (1986) and Flannery (1988) in which *H. tomasettii*, *M. ronaldi* and *K. rotundus* represent relict, plesiomorphic species derived from late Miocene zygomaturines such as *K. torus* and *P. centralis*; the second, by Murray (1992), in which all of the New Guinean zygomaturines (except *K. cultridens*) are derived members of the *Zygomaturus* lineage that speciated in New Guinea during the late Pliocene.

The New Guinean zygomaturines appear to represent a unique and diverse radiation of small-bodied forms that almost certainly occupied very different habitats than those occupied by zygomaturines in Australia. Flannery and Plane (1986) suggested *H. tomasettii* possessed a unique hindlimb morphology that allowed for a higher range of movement indicating that the species was not graviportal like other diprotodontids. *Hulitherium tomasettii* was the largest of the New Guinean diprotodontids, weighing between 75 and 200 kg. It occupied a mid-montane rainforest habitat, an ecological niche similar to that of China's extant Giant Panda *Ailuropoda melanoleuca* and South America's Spectacled Bear *Tremarctos ornatus* (Flannery and Plane 1986). Flannery and Plane (1986) interpreted *H. tomasettii*'s numerous cranial and skeletal specialisations as an indication that these must have evolved over a significant period of time following a late Tertiary origin for this lineage.

Pliocene palorchestids (trunked, 'marsupial tapirs') include *Palorchestes pickeringi* Piper, 2006 from the Plio-Pleistocene Nelson Bay and Hamilton LFs of Victoria; *Palorchestes selestiae* Mackness, 1995 from the Bluff Downs LF, northeastern Qld; and *P. parvus* De Vis, 1895 from the Chinchilla, Bow and Big Sink LFs of Qld and N.S.W.

Phalangeridans

The diversity of possums in the Pliocene is greater than that so far known from the late Miocene with five families (Pseudocheiridae, Phalangeridae, Burramyidae, Petauridae and Ektopodontidae) and 16 species represented. This increase in arboreal taxa probably in part reflects the respite in cooling and drying that occurred during the early Pliocene.

One Pliocene burramyid is known: *B. triradiatus* Turnbull et al., 1987 from the early Pliocene Hamilton LF of Victoria.

Pliocene phalangerids include *Strigocuscus notialis* Flannery et al., 1987 (reassigned to *Onirocuscus* by Crosby 2007), and *Trichosurus hamiltonensis* Flannery et al., 1987. Undescribed phalangerids have also been reported from the Chinchilla and Bow LFs (Archer and Hand 1987). One is also known from the Bluff Downs LF, Qld. (B. Mackness 2010, 'personal communication').

Ektopodontids make their last stand in the Plio-Pleistocene of southeastern Australia. *Darcius duggani* Rich, 1986 was described originally from the Hamilton LF but its range is now known to extend into the early Pleistocene with two isolated teeth described from Nelson Bay LF by Rich et al. (2006). *Ektopodon paucicristata* Rich et al., 2006 has been described from the Childers Cove LF and Dutton Way LF, Victoria.

The oldest-known crown group pseudocheirids are from the Hamilton and Bluff Downs LFs of Vic. and northeastern Qld respectively. The *Hemibelideus-Petauroides* lineage appears in the Hamilton LF (*Petauroides stirtoni* and *Petauroides marshalli*, which were originally described as species of *Pseudocheirus* by Turnbull and Lundelius 1970). *Pseudocheirops winteri* is from the Bluff Downs LF (Mackness and Archer 2001). *Petropseudes* is recorded from the early Pliocene of Riversleigh (Archer et al. 1994; Archer et al. 2006) but more material may be required to confirm its presence (Meredith et al. 2010).

Relaxed molecular clock dates for radiation within the New Guinea *Pseudochirops* clade range from 9.7 to 4.3 Ma and are generally coincident with the uplift of the central cordillera and other highlands of New Guinea (Meredith et al. 2010). These authors suggest that diversification within the New Guinean *Pseudochirulus* clade occurred later, in the last 2.5 Ma, hence postdating the origin and establishment of the central cordillera and other highlands in New Guinea and indicating that *Pseudochirulus* species diversified into habitats with pre-existing, altitudinal differentiation (Meredith et al. 2010).

Plio-Pleistocene species of *Pseudokoala* superficially resemble koalas in their selenodont teeth, and are nearly as big as the living koala, but they were in fact giant ringtail possums: the Pliocene *Pseudokoala erlita* Turnbull and Lundelius, 1970 is known from the Hamilton LF, Vic. and *Pseudokoala curramulkensis* Pledge, 1992 from the Curramulka LF, South Australia.

The petaurid fossil record is relatively poor. In regard to species of *Petaurus*, this may be because open forest (the current habitat of the gliding petaurids) was not widespread before the early Pliocene. Until the forests had opened up, there would have been little selection pressure on arboreal possums to develop the ability to glide (Archer and Hand 2006). The first occurrence of the modern genus *Petaurus* (Fig. 60) is from the Hamilton LF, Vic., with Turnbull et al. (1987) reporting *Petaurus* sp. cf. *P. norfolcensis* and *P.* sp. cf. *P. australis* from that site. Turnbull et al. (2003) suggest as many as three species may be present (species similar to *P. norfolcensis*, *P. breviceps* and *P. australis*) but the material may alternatively represent a single highly variable species.

Fig. 60 Petaurids like these living Sugar Gliders, *Petaurus breviceps*, were probably stimulated into evolving gliding adaptations by the opening of forests from the late Miocene (A. Easton, Queensland Museum). Like all three groups of gliding marsupials, which occupy open forests to woodlands, its closest relatives (in this case Leadbeater's Possum) are non-gliding possums found in lush forests

Macropodoids Species of *Hypsiprymnodon* have been reported from the Hamilton LF, Vic. and Bow LF, N.S.W. (Cooke 2006).

The Pliocene propleopine hypsiprymnodontid *Jackmahoneyia toxoniensis* Ride, 1993 from the Bow LF, N.S.W., is structurally intermediate between Riversleigh species of the Miocene *Ekaltadeta* and species of the Plio/Pleistocene *Propleopus* in size, morphology and broadly age, although *Propleopus chillagoensis* Archer et al., 1978 may also be Pliocene in age.

Pliocene potoroids include *Milliyowi bungantitj* Flannery et al., 1992 from the Hamilton LF, Victoria and *Potorous* sp. from the Curramulka LF, S.A. which may, however, be late Miocene rather than early Pliocene (Pledge 1992) in age.

Macropodids By the start of Pliocene time, the low-crowned browsing kangaroo lineages of the early to mid Miocene (balbarids and bulungamayines) had been replaced by an explosive adaptive radiation of more modern groups of grazing kangaroos (macropodines and sthenurines), responding to the spread of grasslands across Australia.

Grazing (macropodine) kangaroos are characterised by high-crowned (hypsodont) teeth with well-developed transverse molar crests and molar progression whereby the more heavily worn anterior molars and premolars move forward and eventually fall out. As this happens, the more posterior molars gradually move forward to replace the worn and eventually lost teeth. This molar progression better enables these grazing kangaroos to survive the heavy dental wear caused by tough, fibrous, siliceous grasses (Cooke 2006). They are also characterised by their bipedal bounding (saltatorial) gate (Cooke 2006). There are 14 extant genera comprising 50 species ranging in size from tiny pademelons (*Thylogale* spp) to the Red Kangaroo (*Macropus rufus*) occupying a diverse range of habitats from rainforest to open forest, arid inland plains and rocky escarpments. There are two tribes:

the Dendrolagini which include the rainforest-adapted arboreal tree kangaroos that are now restricted to north Queensland and New Guinea; and the Macropodini which include the widespread 'ordinary' terrestrial kangaroos and wallabies.

Extinct tree kangaroos (cf. *Dendrolagus* sp. 1992) are known from early Pliocene rainforests of the Hamilton LF, Victoria. The plesiomorphic forest wallaby *Dorcopsis wintercookorum* Flannery et al., 1992 also comes from the Hamilton LF. An unnamed species of *Dorcopsis* is also recorded from the Awe LF of Papua

Other Pliocene macropodine taxa include the first representatives of the extant Agile Wallaby *Macropus agilis* (Gould 1842) and *Macropus pavana* Bartholomai, 1978b from the Bluff Downs LF, Qld; *Macropus pan* De Vis, 1895 from the Chinchilla LF, Qld; *Macropus* sp. (Flannery et al. 1992) and *Kurrabi pelchenorum* Flannery et al., 1992 from the Hamilton LF, Vic.; *Prionotemnus palankarinnicus* Stirton, 1955 from the Palankarinna LF, S. A.; *Kurrabi mahoneyi* and *Kurrabi merriwaensis* Flannery and Archer, 1984 from the Bow LF, N.S.W.; *Protemnodon bandharr* Dawson et al., 1999 from the Big Sink LF, N.S.W.; *Protemnodon chinchillaensis* Bartholomai, 1978b and *Protemnodon devisi* Bartholomai, 1978b from the Chinchilla LF, Qld; *Protemnodon snewini* Bartholomai, 1978b from the Bluff Downs LF, Qld; *Protemnodon otibandus* Plane, 1967, *Protemnodon buloloensis* Plane, 1967 and *Watutia novaeguineae* Flannery et al., 1989 from the Awe LF, Papua; and the extinct pademelon *Thylogale ignis* Flannery et al., 1992 from the Hamilton LF, Victoria.

Sthenurines are the largest and most bizarre of the late Cenozoic kangaroos. Pliocene sthenurines include the genera *Sthenurus, Simosthenurus, Archaeosimos* and *Troposodon* (Fig. 61). *Lagostrophus fasciatus*, the Banded Hare-wallaby (Fig. 53), an endangered species whose range is now restricted to two islands in Shark Bay, W.A. (Cooke 2006) has been claimed (e.g. Flannery 1983) to be the only living sthenurine. Molecular data lend support to this hypothesis (e.g. Meredith et al. 2008b) but overviews of the interrelationships of all kangaroos (e.g. Prideaux and Warburton 2010) suggest it is unlikely. Pliocene sthenurines include *Sthenurus andersoni* Marcus, 1962 and *Sthenurus notabilis* (Bartholomai 1963) both from the Chinchilla LF, Qld (Prideaux 2004); *Simosthenurus antiquus* (Bartholomai 1963), Chinchilla LF, Qld;

Fig. 61 Reconstruction of a Pliocene propleopine kangaroo eating the juvenile of an extinct relative (*Troposodon* sp.) of the living banded hare-wallaby *Lagostrophus fasciatus* (R. Scott). Propleopines include the carnivorous kangaroos of the early to middle Miocene (*Ekaltadeta* spp) as well as slender, somewhat bear-like species (*Propleopus wellingtonensis*) that survived into the late Pleistocene before the subfamily went extinct along with other megafaunal giants

Simosthenurus tirarensis Prideaux, 2004, Camel Swamp Yard, Tirari Formation, S.A.; *Simosthenurus* sp. Bow LF, N.S.W. (Flannery and Archer 1984); *Simosthenurus* sp. Hamilton LF, Vic. (Flannery et al. 1992); *Archaeosimos correlli* Prideaux, 2004 and *Archaeosimos cegsai* (Pledge, 1992), Curramulka LF, S.A.; *Troposodon minor* (Owen 1877) Bluff Downs, Chinchilla and Hamilton LFs; *Troposodon bluffensis* Bartholomai, 1978b Bluff Downs LF, Qld; *Troposodon bowensis* Flannery and Archer, 1983 Bow LF, N.S.W.; and *T. kenti* Campbell, 1973 Kanunka LF, S. A.

Marsupialia *incertae sedis*

Numbigilgids

N. ernielundeliusi is an enigmatic member of the Australian marsupial fauna (Beck et al. 2008a)

(Fig. 20) that cannot be referred with certainty to any of the known orders although, as noted above, the overall appearance of this taxon suggests it may be a highly autapomorphic group of perameliforms. It is known from a fragmentary right mandible preserving p2–m4 and a single left upper molar, both specimens from the Bluff Downs LF of northeastern Qld. However, regardless of its precise phylogenetic affinities, *Numbigilga* represents an important addition to the known dental diversity of Australian marsupials. If its highly distinctive features are the result of a long period of evolutionary specialisation, it is surprising that no members of this group or probable antecedents for it have been identified in any of the highly diverse Miocene fossil assemblages known from Queensland. This suggests it is a group that differentiated rapidly, perhaps during the late Miocene given that very little is known about the evolutionary history of most of the smaller mammals during this interval.

Pleistocene

The Pleistocene epoch (1.8 Ma to 10,000 yBP) was marked by rapid and severe changes in climate and megafaunal extinctions. Like the rest of the world, Australia experienced extreme climatic conditions as the Arctic and Antarctic polar ice caps repeatedly grew and retreated, resulting in a constantly changing pattern of forests, grasslands and deserts because of consequent fluctuations in rainfall. Australian vertebrates responded to the challenges of these rapid environmental changes in various ways including retreat to refugia, adaptation, gigantism and extinction. Temporary land bridges were exposed between mainland Australia and Tasmania, and Australia and New Guinea, during periods of lowered sea level (by some 130 m during the last glacial maximum; Pillans and Bourman 2001) facilitating the dispersal of mammals, including humans, when large areas of the Australian continental shelf were exposed. Ocean circulation and weather patterns were also profoundly affected. For example, McGowran et al. (1997) have pointed out that the warm Leeuwin Current, which sweeps down the west coast and into the Great Australian Bight, was effectively shut down during periods of low sea level resulting in colder water in the Bight which contributed to a lowering of onshore rainfall in southern Australia (Pillans and Bourman 2001).

Australia has many Pleistocene marsupial-bearing deposits in all states and territories, from the interior to today's coastal fringe (Fig. 15b). Many have been poorly or unreliably dated but the increasingly routine use of rapidly improving techniques to date fossiliferous sediments or the fossils themselves – such as uranium series dating, OSL (optically stimulated luminescence), ESR (electron spin resonance), AAR (amino acid racemisation) and next generation C14 dating – is steadily redressing this shortcoming. Relatively well-dated mammal assemblages include the Early Pleistocene Nelson Bay LF of southern Vic. (Piper et al. 2006); Middle and Late Pleistocene Mount Etna LFs of eastern central Qld (Hocknull et al. 2007); Middle to Late Pleistocene Naracoorte Caves LFs (Fig. 62) of eastern S.A. (Moriarty et al. 2000; Grün et al. 2008, Pillans 2003); Middle Pleistocene Thylacoleo Caves, W.A. (Prideaux et al. 2007a); and Late Pleistocene assemblages such as the Katapiri LF, S.A. (Tedford et al. 2006), Russenden Cave LF, Qld (Price et al. 2009), Cuddie Springs LF, N.S.W. (Field 2004) and Kelangurr Cave, Irian Jaya (Flannery 1992) (Fig. 15b).

Palaeoenvironment

In the northern hemisphere, Arctic ice sheets up to a kilometre thick advanced south during glacial periods, covering much of North America and Eurasia, and retreated during warmer interglacials as many as 17 times. During at least the last one million years each glacial phase lasted on average 100,000 years and interglacials 10,000 years (Berger et al. 1992), although 1 to 3 million years ago these glacial (Milankovitch) cycles may have occurred more frequently at 40,000-year intervals (Hays et al. 1976). This change in cyclicity is referred to as the Mid-Pleistocene Transition and its timing and effects appear to have differed in the northern and southern hemispheres.

In Australia, no major continental ice sheets have been detected, with minor glaciations restricted to Tasmania and the Southern Highlands (White 2006). Rather, glacial periods were generally characterised by dry, cold conditions and interglacial periods by warmer, wetter climates. Fluctuations in vegetation types broadly reflect those cycles with expansion of forests and woodlands dominating during interglacials and expansion of open shrublands and grasslands during glacials.

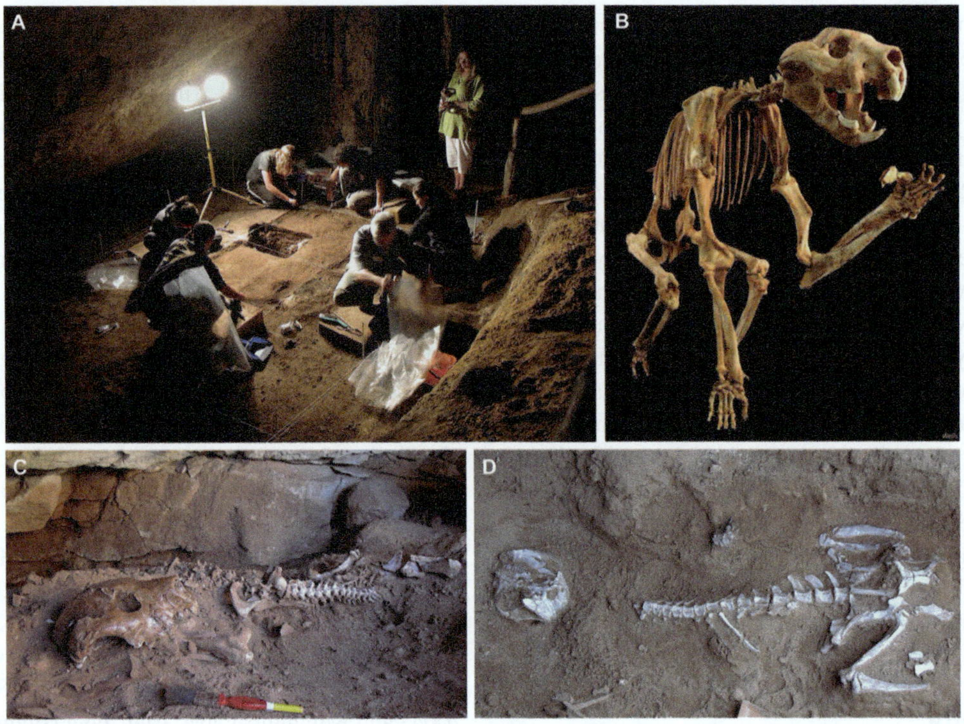

Fig. 62 The Naracoorte Caves, all 27 of them, have acted as pitfall traps, collecting animals for at least 500,000 years including many megafaunal species. (**a**) The excavation process; (**b**) reconstructed *Thylacoleo carnifex* skeleton; (**c**) *T. carnifex* skull and other skeletal remains; (**d**) articulated *Simosthenurus occidentalis* skeleton (S. Bourne). The Naracoorte Caves are part of the Fossil Mammals of Australia World Heritage listing alongside Riversleigh

Despite the cyclical fluctuations, the overall trend throughout the Pleistocene was towards reduced precipitation (Martin 2006). At the beginning of the epoch, Australia's climate was wetter than now and interglacials throughout the Pleistocene were generally wetter. Interglacial monsoon-fed megalakes occurred in northwestern and southeastern Australia (Gregory and Bungunnia Lakes respectively; Bowler et al. 2001; Chen and Barton 1991). A major shift to aridity occurred prior to 780,000 yBP in northwestern and central Australia, where rainfall is linked to the strength of the Asian monsoon (Kershaw et al. 1993, 2003a; Rhodes et al. 2005). In central Australia, reduced vegetation cover facilitated development of extensive longitudinal dune fields around 1 million years ago, with Lake Amadeus providing evidence of an arid shift from lacustrine sediments to evaporites and dunes (Bowler 1981; Chen and Barton 1991; Fujioka et al. 2005 earliest Tertiary therian mammal). Fully arid conditions were established in some areas of the continent about 500,000 yBP as evidenced for example by gypsum-dominated sediments in southwestern Australia (Zheng et al. 1998, 2002; Rhodes et al. 2005). In southern Australia, Lake Bungunnia underwent a shift from lacustrine sediments to saline clays and dunes about 600,000 yBP (Jansen et al. 1986; Stephenson 1986). Pillans and Bourman (2001) report a change to aridity 600 to 500,000 yBP characterised by oxide-weathering zones changing to carbonate-weathering zones in weathered dune systems at Sellicks Beach and Kangaroo Island, South Australia.

Australia-wide, progressively more arid conditions were in evidence by the Middle Pleistocene 430,000 to 300,000 yBP, at the time of the global mid-Brunhes Climate Event (Jansen et al. 1986), with records of increase in terrigenous dust in drill cores in the Tasman and Coral Seas and Indian Ocean (Hesse 1994; Kawahata 2002; Hesse and McTainsh 2003). Pollen records indicate a progressive increase in open-arid vegetation in northern Australia from 300,000 yBP and intensifying since 180,000 yBP (Kawamura et al.

2006). In the case of the Gregory lake system of northwestern Australia, over the last 300,000 yBP each megalake (at approximately 300, 200 and 100,000 yBP) was smaller than the previous one, reflecting the decreasing influence of the Asian monsoon (Bowler et al. 2001). In northeastern Australia, araucarian rainforest and sclerophyll casuarinaceae forests persisted (albeit cyclically shrinking and expanding; Kershaw et al. 2003b; Moss and Kershaw 2000, 2007), but there is evidence of a complete collapse of rainforest in central eastern Queensland about 280,000 yBP (Hocknull et al. 2007, see below).

During the last 100,000 yBP, aridity has increased, with an increase in ENSO activity beginning around 60,000 yBP (Kershaw et al. 2003a). Nevertheless, between 60,000 and 40,000 yBP, by which time the first people had arrived in Australia, wetter conditions in southern and eastern Australia maintained well-watered corridors that facilitated rapid colonisation of the continent (Archer et al. 1998). The last glacial period, about 17,000 to 23,000 yBP, was particularly dry and since then, even during the current interglacial, rainfall has not reached pre-glacial levels (Ayliffe et al. 1998; Hesse et al. 2004; Martin 2006).

Pleistocene Marsupial Fauna

The extreme and rapid climatic oscillations of the Pleistocene appear to have triggered cycles of range restriction and expansion in many Australian marsupial lineages as well as localised and continental extinctions (see below).

However, it is apparent that Pleistocene climate change influenced regional faunas in different ways. For example, Pleistocene records from southern Australia show remarkable faunal stability over the last 500,000 yBP (spanning several glacial cycles) for non-megafaunal marsupial taxa (Prideaux et al. 2007b). On the other hand, fossiliferous cave deposits from Mount Etna, central eastern Qld, document a major faunal turnover between 280,000 and 205,000 yBP, when a diverse rainforest assemblage was replaced by a xeric-adapted fauna (Hocknull et al. 2007). In southern Australia, stability in marsupial faunas appears to reflect long-term adaptation to increasing aridity which began 400 to 700,000 yBP (see above). In northern Australia, palaeoclimate proxies indicate that the climatic shift to aridity occurred later at around 300,000 yBP (Hocknull et al. 2007).

Middle Pleistocene connectivity between the faunas of northern Australia and New Guinea is demonstrated by Mount Etna cave deposits (Hocknull et al. 2007). Congeneric taxa include possums (species of *Dactylopsila*, *Pseudochirulus*, *'Petauroides'*, *Cercartetus*), pademelons (*Thylogale*) and tree kangaroos (*Dendrolagus*) that are more closely related to living species in New Guinea's montane rainforests than to species in Australia's wet tropics.

In the Pleistocene record, all four modern Australian marsupial orders are well represented. Major radiations are restricted to some groups (e.g. dasyurids, macropodine and sthenurine kangaroos, potoroids, vombatids), with extinctions in marsupial lineages more typically characterizing this epoch.

Dasyuromorphians

Among dasyuromorphians, dasyurids increased in diversity, with Wroe and Muirhead (1999) noting 25 Pleistocene dasyurids, all of which are assigned to modern genera. These include *Dasyuroides achilpatna* Archer, 1982a (sister taxon to the modern Kowari *Dasyuroides byrnei*) and the devil *Sarcophilus moornaensis* Crabb, 1982 both from Fishermans Cliff LF, N.S.W., a dunnart *Sminthopsis floravillensis* Archer, 1982a from the Floraville LF and species of *Antechinus* including *Antechinus puteus* Van Dyck, 1982 from the Russenden Cave LF, Qld, and *Antechinus yammal* Cramb and Hocknull, 2010 and *Antechinus yuna* Cramb and Hocknull, 2010, both from the Mount Etna mid Pleistocene deposits, Qld. Although far more diverse than in earlier epochs, known Pleistocene dasyurid taxa represent less than half the number of genera and species known from the Holocene (Table 3).

As in the Pliocene, only *Thylacinus cynocephalus* is represented in the Pleistocene, albeit by some very large individuals from, for example, Wellington Caves, N.S.W. (Fig. 63). Previous views that these large Pleistocene individuals represented a distinct species, *Thylacinus spelaeus* Owen, 1845, were challenged by Ride (1964) who demonstrated that they were not specifically distinct from the modern species. They appear instead to be an example of the gigantism that occurred in many Australian marsupial lineages in the Pleistocene (see below). That said, some Pleistocene individuals that are otherwise indistinguishable from the modern Thylacine were 13% smaller than modern Tasmanian individuals, hence it appears that

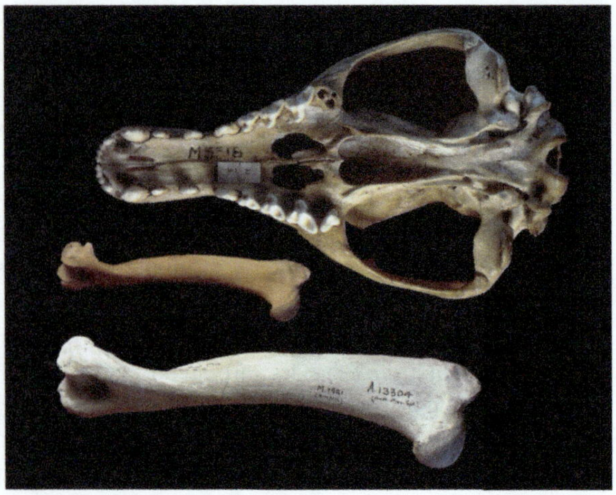

Fig. 63 Skull of the Recent Thylacine, *Thylacinus cynocephalus*, alongside a humerus from the Kimberley region of northwestern W.A., and a much larger Pleistocene humerus from Wellington Caves, central N.S.W. (M. Archer). None of the mainland Thylacines appears to have survived much beyond 4,000 yBP after which time they were restricted to Tasmania until Europeans drove them to extinction by 1936

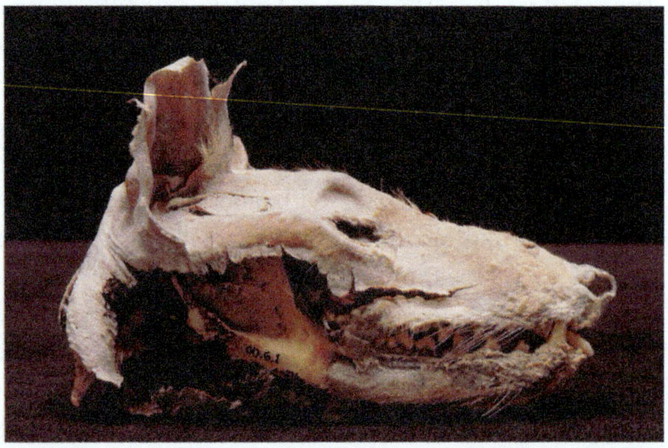

Fig. 64 *Thylacinus cynocephalus* mummified head from the surface of a cave on the Western Australian section of the Nullarbor Plain (John Long and WAM). These individuals are among the last survivors of the mainland Thylacine. Introduction and probably domestication of the comparably sized Dingo (*Canis lupus*) by unknown people about 4,000 years ago may have led to extinction of the Thylacine on the mainland. Dingos never reached Tasmania

there was considerable palaeogeographic variation in size. Exceptionally preserved mummified remains of *T. cynocephalus* have been found in caves in the Nullabor region (Fig. 64).

The family Myrmecobiidae is known from a single extant, endangered species, the numbat or banded anteater, *Myrmecobius fasciatus*, which is today restricted to small areas of southwestern Western Australia. It has no pre-Pleistocene fossil record and all Pleistocene fossils represent the modern species (Fig. 1c). It is represented in Pleistocene cave deposits in Western Australia (e.g. Madura Cave), in caves on the Nullarbor Plain and in lacustrine deposits around Lake Menindee, N.S.W. (Archer et al. 1984). There remains confusion about the phylogenetic relationships of numbats to other dasyuromorphians (see Section 'Late Oligocene Marsupials') due in part to the vestigial dentition of the living species reflecting its diet of termites and ants.

Peramelemorphians

Pleistocene records of extant and recently extinct species include seven peramelids ('ordinary' bandicoots) and two thylacomyids (bilbies) (e.g. species of *Isoodon* and *Perameles*: *Isoodon obesulus*, *Isoodon macrourus*, *Isoodon auratus*, *Perameles bougainville*, *Perameles gunnii*, *Perameles nasuata*; Archer et al. 1984). A relative of the pig-footed bandicoot (Fig. 58), *Chaeropus* sp. cf. *C. ecaudatus*, is reported by Muirhead and Godthelp (1996) from Chillagoe, north Qld. Quaternary thylacomyids, the extant bilby *Macrotis lagotis* (Fig. 2a) and recently extinct lesser bilby *Macrotis leucura*, are known from various deposits across Australia including Wellington Caves and Lake Menindee in N.S.W. and Madura Cave in Western Australia (Archer et al. 1984).

Notoryctemorphians

No Pleistocene fossils of marsupial moles are yet known, but it is likely that the desert-dwelling habits of Australia's two living species became established in the Pleistocene as aridification gripped the heart of the continent and sandy deserts spread in central Australia. Nevertheless, it is clear that the characteristic morphological and behavioural adaptations of marsupial moles for burrowing long pre-date the Quaternary (as shown by Riversleigh Miocene fossils; Section 'The Early Miocene Marsupial Fauna') and probably first evolved in the mossy-floored rainforests of Australia's early Cenozoic.

Diprotodontians

The overall diversity of diprotodontians remained similar with the exception of sthenurine and macropodine kangaroos which, in the case of the latter group in

Australia, doubled in genera and species during the Pleistocene (Table 3).

Vombatiforms

Among vombatiforms, diprotodontid diversity decreased as vombatid diversity increased and, among phalangeridans, pseudocheirid diversity decreased as petaurid diversity increased (see below).

Phascolarctomorphians At least four koala species are known from the Pleistocene, all but one being referable to the modern genus *Phascolarctos*: *P. yorkensis* (Pledge 1992) from the Phosphate Mine, Wellington Caves; *P. stirtoni* (Bartholomai 1968) from sites such as Cement Mills (Gore) and Nelson Bay LFs; and the extant species *P. cinereus* (Goldfuss 1817) from many sites including Mammoth Cave, W.A., Victoria Fossil Cave, S.A., Lake Victoria and Wellington Caves, N.S.W. and Cement Mills, Qld. The rainforest species *Invictokoala monticola* Price and Hocknull, 2011, from Mt Etna, Qld, represents an archaic 'ghost' lineage of koalas with affinities to late Oligocene species of *Madakoala*.

Previously, doubt has been expressed about the specific distinction of *P. stirtoni* from the modern koala *P. cinereus*, with suggestions that the modern *cinereus* is a dwarf descendent of the Pleistocene giant koala *stirtoni* (e.g. Archer and Hand 1987; Long et al. 2002). Price (2008) has reviewed the Pleistocene record for koalas and concluded that the two species were not parts of a chronocline involving a single evolving lineage, arguing that the two species are dentally distinct and overlapped temporally during at least the Late and Middle Pleistocene, and were sympatric in at least the Naracoorte region of South Australia during the Middle Pleistocene. A consequence of accepting specific distinction for *P. stirtoni* is that it then represents yet another casualty of Late Pleistocene extinction processes (see below).

Vombatomorphians Wombats were more diverse in the Pleistocene than at any other time before or since, with as many as eight different kinds in five genera. A number of faunal assemblages include at least two species (e.g. Victoria Fossil Cave and Fishermans Creek LFs; Tedford et al. 2006). Taxa include extinct genera (*Phascolonus*, *Ramsayia* and *Warendja*), extinct species of the modern genera *Vombatus* and *Lasiorhinus*, and all three extant species (i.e., *Vombatus ursinus*, *Lasiorhinus krefftii* and *Lasiorhinus latifrons*). The giant wombat, *P. gigas*, which was relatively common in Australian Pleistocene faunas, particularly in eastern Australia, was three times the size of the largest modern species but possibly did not burrow like modern wombats '. . .due to the physical and physiological limitations of burrows' (Woolnough and Steele 2001). Species of the Plio-Pleistocene genus *Ramsayia* Tate, 1951 were also large, with the Pleistocene *Ramsayia magna* (Owen, 1872) from Wellington Caves being twice the size of the modern common wombat, *V. ursinus*. Ironically, the currently endangered north Qld hairy-nosed wombat, *L. krefftii*, was first described as a presumed extinct taxon from Wellington Caves, N.S.W. before it was found to be still living in Qld and northern N.S.W..

The Pleistocene marsupial lion *Thylacoleo carnifex* Owen, 1859 (Fig. 62b–c) is the largest mammalian carnivore known from Australia. At nearly 2 m in length and with an estimated weight of 90–160 kg (Wroe et al. 2003), *T. carnifex* was the size of a lioness and twice the size of a panther. It was also a hyper-specialised carnivore with greatly enlarged carnassial (meat-eating) premolars that formed secateur-like cutting blades (at the expense of its greatly reduced molar row), powerful canine-like lower incisors well-suited for stabbing and hypertrophied thumb claws probably used to suppress or dismember prey. Van Valen (1969) suggested the degree of carnivorous specialisation in terms of the hypertrophy of its carnassials and hypotrophy of its molar teeth was unmatched by any other mammalian carnivore. *Thylacoleo carnifex* occurs in numerous Pleistocene deposits throughout N.S.W., Qld, Vic., W.A., S.A. and Tasmania, including Kanunka, Wellington Caves, Thylacoleo Caves, Victoria Fossil Cave, Beginners Luck Cave and Kings Creek LFs. Extremely large individuals are known from some of the deposits in southeastern Australia while the smallest are known from caves on the Nullarbor caves and from the Balladonia deposit of Western Australia.

Diprotodontoids Arguably the most famous fossil marsupial described from Australia is *Diprotodon optatum* Owen, 1838 (Fig. 22). It is the largest marsupial known, standing 2 m at the shoulder, 3 m in length and weighing an estimated 2–3 tonnes (Wroe et al. 2003, 2004). It was also the first fossil marsupial described from Australia by the renowned English anatomist Sir Richard Owen in 1838, on the basis of dental remains from Wellington Caves, N.S.W. *Diprotodon* is the best represented diprotodontine genus in terms of fossil material. It is known from complete crania and postcranial skeletons, as well as soft anatomical impressions and trackways, and has been recorded from all Australian states except the Northern Territory and Tasmania. At least seven *Diprotodon* species have previously been identified but only two of these are commonly accepted as valid: the type species *D. optatum* and a smaller species *Diprotodon minor* Huxley, 1862. However, a recent review of *Diprotodon* taxonomy (Price 2008) suggests that morphological differences between *D. optatum* and *D. minor* may be due to sexual dimorphism and a high degree of individual variation implying that there is in fact only one highly variable species, *D. optatum*.

It has been suggested (Rich 1983) that *D. optatum* was adapted more to the open woodlands and grasslands in contrast to species of *Zygomaturus* which appear to have been adapted to forested areas. This would explain the occurrence of species of *Zygomaturus* in central Australia in the Pliocene and the appearance and spread of *D. optatum* only in the Pleistocene, a spread coincident with the establishment of arid/semi-arid conditions throughout central Australia (Archer and Hand 2006).

Pleistocene zygomaturine diprotodontids are also known from New Guinea: *H. tomasettii* Flannery and Plane, 1986 from the Pureni LF; *M. ronaldi* Flannery, 1992 from cave deposits; and *Z. nimborensia* Hardjasasmita, 1985 from Nimboran, Irian Jaya. *Maokopia ronaldi* was described on the basis of a partial cranium and partial dentary from Kelangurr Cave, West Baliem Valley, Irian Jaya, which has been dated at 41,500±1,500 yBP. Abundant material, including elements of the postcranial skeleton, has since been recovered, suggesting this small (50–100 kg) zygomaturine was relatively common in the subalpine herbfields of the region during the late Pleistocene (Flannery 1992). The dentition of *M. ronaldi* is unique among zygomaturines in having relatively well-developed longitudinal links between the lophs, interpreted by Flannery (1992) as an adaptation to an abrasive diet of siliceous ferns and grasses. Flannery (1994) suggested that like *H. tomasettii*, *M. ronaldi* may have also resembled a panda, but on a much smaller scale. Analysis of the cranium suggests a short-faced animal with a high forehead and forward-facing eyes. In contrast to Flannery and Plane (1986), Murray (1992) has suggested that both *H. tomasettii* and *M. ronaldi* are derived members of the *Zygomaturus* clade, their small size and shortened crania being the result of relatively rapid habitat-induced insular dwarfing during the Pleistocene.

The Pleistocene *Palorchestes azael* is the most widely distributed palorchestid and as such appears to have been able to adapt to a wide range of habitats and environmental conditions (Davis and Archer 1997). Postcranial material suggests the cow-sized *P. azael* had powerful forelimbs and large sharp claws. Its ecological niche has long been a mystery. It may have been capable of climbing or even felling trees, stripping bark and uprooting shrubs, possibly for tubers (Flannery and Archer 1985; Flannery 1994).

Phalangeridans

In the Australian Pleistocene, pseudocheirid diversity appears to have decreased in contrast to levels noted in the Pliocene and Holocene and seems more comparable with the diversity evident in the late Miocene (Table 1.3). Petaurids, on the other hand, appear in general to have increased in diversity during the Pleistocene. Both trends in Australia probably reflect a response to the opening up and reduction in extent of the forests in Australia. However, in the more resilient highland rainforests of New Guinea, there appears to have been a late radiation of *Pseudochirulus* ringtail possums (Meredith et al. 2010; see Section 'The Pliocene Marsupial Fauna').

The last surviving ektopodontid has been found in the Early Pleistocene Nelson Bay LF, Victoria. There is also a marked decline in the apparent number of phalangerid taxa known from the Pleistocene (e.g. *Trichosurus* sp. recorded from deposits such as Malkuni and Nelson Bay LFs) in the lead up to a marked rise in diversity in the Holocene (23 species in six genera being known from Australia and New Guinea; Table 3).

The oldest-known honey possums (tarsipedids) occur in the Pleistocene record of southwestern Australia (Balme et al. 1978). These appear to represent the extant species *T. rostratus* (Fig. 40). The evolutionary relationships of honey possums remain unclear, in part because of the vestigial dentition of the living species, a tiny, nocturnal nectar-, pollen- and blossom-feeding possum that is about half the size of a mouse (see also Section 'The Early Miocene Marsupial Fauna'). Today honey possums occur only in coastal heathlands in the southwestern corner of Western Australia.

Macropodoids *Hypsiprymnodon* is not recorded in Pleistocene deposits (including Mount Etna's Middle Pleistocene rainforest assemblages) but its presence is inferred from records before and after the Pleistocene.

Pleistocene propleopine species include *Propleopus oscillans* (De Vis, 1888b) from numerous sites in Qld, N.S.W., S.A. and Vic., as well as (e.g.) *P. chillagoensis* Archer et al., 1978 and *P. wellingtonensis* Archer and Flannery, 1985. *Propleopus oscillans* was as large as a living Eastern Grey Kangaroo (70 kg) and interpreted to be an omnivore or opportunistic carnivore (Ride et al. 1997; Wroe et al. 1998).

There are many Pleistocene macropodine taxa. One of the most common is *Macropus giganteus titan* Owen, 1838 (see Section 'Gigantism and Megafaunal Extinction') known from many sites in eastern Australia, a taxon about 30% larger than its descendent *M. giganteus giganteus*, the Eastern Grey Kangaroo. Others include a giant tree kangaroo *Bohra paulae* Flannery and Szalay, 1982 from Wellington Caves LF; *Baringa nelsonensis* Flannery and Hann, 1984 from the Nelson Bay LF; and *Congruus congruous* McNamara, 1994 from a cave fill near Naracoorte, South Australia. The many species of the large, wallaby-like species of *Protemnodon* represent the last stand for this genus. Many genera (including the latter three) disappeared during the Pleistocene but the genus *Macropus* continued to radiate and today numbers 14 species that between them occupy most Australian habitats except rainforests which in terms of macropodines is occupied today by pademelons (*Thylogale* spp), rock-wallabies (*Petrogale* spp), tree-kangaroos (*Dendrolagus* spp) and, in New Guinea, by forest wallabies (*Dorcopsis* and *Dorcopsulus* species). All of the rainforest macropodines are dominantly browsers rather than grazers (the species of *Macropus*) which occupy the more open habitats.

The Pleistocene also witnessed a significant radiation and last stand of the large to gigantic, short-faced sthenurine kangaroos (e.g. species of *Procoptodon*, *Sthenurus*, *Simosthenurus*, Metasthenurus; Fig. 65). In the sthenurine hind foot, all digits except IV were reduced or lost, making some (e.g. species of *Procoptodon*) monodactyl and in this respect technically similar to late Cenozoic horses. The most primitive members of the group were adapted to life in open forests or woodlands. Widely distributed from Qld to S.A., *Procoptodon goliah* Owen, 1874 at 2.5 m tall and weighing 200 to 300 kg (Fig. 54) is the largest kangaroo that ever lived, but other *Procoptodon* species (*rapha*, *pusio*) were also enormous. *Procoptodon*'s long arms and long-clawed fingers may have been used to reach browse in high tree branches, while its high-crowned teeth with thickened, crenulated enamel suggest that it was processing highly abrasive foods (Prideaux 2004). Other Pleistocene sthenurines include *Simosthenurus occidentalis* Glauert, 1910 with at least 10 other species of that genus, *Sthenurus atlas* (Owen, 1838) with at least five other congeners, and the monotypic *Metasthenurus newtonae* (Prideaux 2004). The last *Troposodon* (or giant forest wallaby) also occurred in the Pleistocene, e.g. *T. minor* (Owen 1877) from Wellington Caves LF. The relationships of species of *Troposodon* to other macropodine kangaroos have been previously considered and, while normally regarded as sthenurines, they may in fact represent a more plesiomorphic group of macropodoids (Prideaux and Warburton 2010).

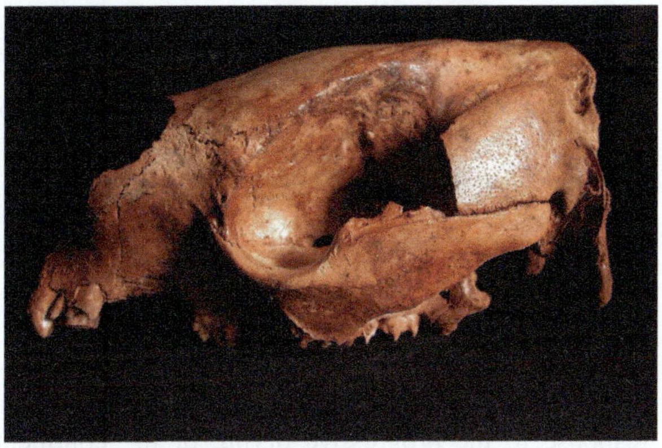

Fig. 65 Skull of the short-faced sthenurine macropodid kangaroo *Simosthenrus browneorum* from Pleistocene deposits in Naracoorte Caves, S.A. and Mammoth Cave, W.A. (S. Bourne). This giant, boof-headed kangaroo was one of about 55 megafaunal species alive in the late Pleistocene. Most but not all appear to have been driven to extinction by climate change prior to the arrival of humans

Gigantism and Megafaunal Extinction

The gigantism that commenced in the late Miocene and Pliocene in Australian marsupial lineages culminated in the Pleistocene in a suite of large animals (generally >40 kg) known as megafauna. Gigantism was a global phenomenon seen in many terrestrial vertebrate lineages. In Australia, the megafauna included not only marsupials but also gigantic birds (*Genyornis*), goannas (*Megalania*), crocodiles (*Quinkana, Pallimnarchus*), horned turtles (*Meiolania*), snakes (*Wonambi*) and echidnas (*'Zaglossus' hacketti*).

Increasingly large body size in Australian marsupials in the later Cenozoic is thought to represent a response to the progressive drying of the continent since the late Miocene. All other things being equal, drier conditions meant poorer quality vegetation, with larger animals better able to cope because they are able to consume more. Gigantism in carnivores would follow the increased size in their herbivorous prey. Increasing size in carnivores would also have conferred competitive advantages in terms of obtaining and maintaining access to carcasses. It has also been suggested that a tendency to increase in size over time, often called Cope's Rule (e.g. Van Valkenburgh et al. 2004), may be an intrinsic part of evolutionary processes in vertebrate lineages. Correlated with increasing body size however, there are a number of potential disadvantages (in terms of environmental challenges of the Australian Pleistocene): water requirement increases; gestational period increases and litter size decreases potentially resulting in overall reduction in population size. As precipitation decreased during the Pleistocene, with free-standing water harder to find and increasing unpredictability of good seasons, larger animals would have been relatively disadvantaged in terms of their ability to hydrate their larger fermentation 'vats' and successfully raise their young (Archer et al. 1998; Cardillo et al. 2005). The decline in predictability of resources would have been among the most stressful aspects of the Pleistocene for the biota – flora as well as fauna (Archer et al. 1998).

As noted above (Section 'Pleistocene Marsupial Fauna'), the Pleistocene produced the largest marsupial that ever lived (*D. optatum*; Fig. 22), the most specialised and in Australia largest known mammalian carnivore (*T. carnifex*; Fig. 62b–c), the largest kangaroo (*Procoptodon goliah*; Fig. 54), the largest devil (*Sarcophilus laniarius*) and many other extremes. Gigantism also extended to arboreal marsupials including koalas (*P. Yorkensis*) and ringtail possums (*Pseudokoala cathysantamaria*). Some 60 Australian megafaunal species are generally recognised (e.g. Flood 1995).

Most of these gigantic mammals, birds and reptiles became extinct during the Late Pleistocene. The reasons for their demise are controversial, with views divided between anthropogenic causes (e.g. Jones 1968; Merrilees 1968; Flannery 1990, 1994; Miller et al. 1999, 2005; Roberts et al. 2001; Prideaux et al. 2007b; Turney et al. 2008) and climate/vegetation change alone or in combination with human pressures (e.g. Horton 1980; Head 1995; Bowman and Choquenot 1999; Lundelius and Graham 1999; Archer et al. 1998; Barnosky et al. 2004; Burney and Flannery 2005; Wroe and Field 2006).

A central issue for both sides of the debate is timing – that of the arrival of people in Australia, and that of megafaunal extinctions. Neither is well resolved. Dates based on evidence, however substantial, for the arrival of people in Australia range from 40,000 to 120,000 yBP. For example, O'Connell and Allen (2004) provide evidence for modern human arrival at 45,000 to 43,000 yBP, Turney et al. (2001) at 50,000 yBP, Bowler et al. (2003) 50,000 to 46,000 yBP, and Roberts et al. (1994) 53,000 to 60,000 yBP. Other dates based on interpretation of peaks in charcoal dust as a signal of human firing of the bush have been up to 120,000 yBP (references cited in Bradstock et al. 2002). Although humans have been in the Indonesian Archipelago for at least 800,000 years (Morwood et al. 1998), it has been argued (e.g. Finlayson 2005) that colonisation of Australia before the penultimate glacial maximum (e.g. 130,000 yBP, Singh et al. 1981) would not have been by modern humans. Wroe and Field (2006) note that 65% of extinct Australian Pleistocene megafauna are absent from deposits younger than 130,000 yBP and only eight species were clearly present at the earliest point for which there is broad acceptance for human arrival (*D. optatum, P. goliah, Sthenurus andersoni, S. occidentalis, Simosthenurus brownei, Protemnodon roechus, T. carnifex, P. gigas, V. hacketti*).

Interactions between Australia's first people and Late Pleistocene marsupials have been noted by many (e.g. Archer et al. 1980; Kiernan et al. 1983; Furby et al. 1993; Flannery 1994; Dodson et al. 1993; Miller et al. 1999, 2005; Roberts et al. 2001; Field 2004). From Mammoth Cave, WA, Archer et al. (1980) report Late Pleistocene *Zygomaturus* and *Sthenurus* bones that appear to have been interfered with by people.

Field et al. (2001) document the overlap of megafaunal remains and stone tools between 36 and 30,000 yBP at Cuddie Springs, N.S.W., as well-cut marks on the tibia of a Red Kangaroo (*Macropus rufus*) and evidence of burning of a sthenurine femur. Butchering of modern species of kangaroos in Kutikina Cave, Tasmania during the Late Pleistocene has been described by, among others, Kiernan et al. (1983) and Garvey (2010).

Arguments and evidence for human-mediated extinction of Australian megafauna include the partial concurrence of extinctions and human colonisation not correlated with significant climate change (e.g. Miller et al. 1999; Roberts et al. 2001); the climate of the last two glacial cycles no more severe than previous cycles (e.g. Turney et al. 2008); attenuation of megafauna in pre-human record due to differential sampling probabilities rather than disappearance (e.g. Signor–Lipps effect, Kidwell and Holland 2002); humans critically altering habitats by the use of fire and other land-use practices (e.g. Jones 1968; Horton 1984; Miller et al. 2005); and humans entering Australia encountering a naive fauna that was easily predated and rapidly (in as little as 750 yrs) wiped out (e.g. overkill or blitzkrieg models; Merrilees 1968; Flannery 1994; Brook and Bowman 2004). Roberts et al. (2001) report the human-mediated continent-wide extinction of many megafaunal marsupials at approximately 46,000 yBP in the absence of acute climate change but could not rule out the effects of other climatic impacts, or an extended period of anthropogenic ecosystem disruption.

Contemporary arguments and evidence from the other side of the debate – the role of environmental change (driven by climate) in Australian megafaunal extinctions – include staggered but continuous megafaunal extinctions over several glacial cycles in the absence of humans (e.g. Price 2008); significant long-term climate deterioration particularly during the last 500,000 yBP (see Section 'Palaeoenvironment'); total absence of megafaunal kill sites of the kind that have been found in some places in North America (e.g. Horton 1984); lack of specialised technology or tools in the 'kits' of the first humans entering Australia for killing megafauna (e.g. Flood 1995; Johnson and Wroe 2003); absence of any individual megafaunal species having been killed (rather than potentially scavenged) by humans; absence of evidence for human-mediated firing of the bush (e.g. Kershaw et al. 2003a, b; Lynch et al. 2007); evidence that routine firing of the bush at least in N.S.W. did not occur prior to about 6,000 yBP (Black et al. 2007); evidence that humans moving through Indonesia on their way to Australia did not exterminate the megafauna on the islands they passed through (Louys et al. 2007b); evidence that only hunter/gatherer strategies were being used to obtain food rather than agriculture which has led elsewhere to significant alteration of the habitat. Price and Sobbe (2005) found no evidence of human occupation until long after the loss of megafauna on the Darling Downs, southeastern Queensland, and Price (2008) argues that the response of the biota to climate change, and impact of humans, may have been very different across Australia due to the differential timing of major climate-induced habitat modifications.

The role of humans in the demise of small island megafaunas in the southwestern Pacific is reasonably well documented (e.g. Holdaway and Jacomb 2000; White et al. 2010). Human-correlated extinctions of megafauna occurred on two islands, Madagascar and New Zealand. But in contrast to the situation with respect to Australia, this occurred very late in the Holocene, impacted relatively much smaller landmasses and was precipitated by essentially contemporary people who arrived with agriculture, killing tools and domesticated carnivores. This was not the situation that unfolded in Australia and, until the arrival of Europeans on this continent, the species that have gone extinct in the last 200 years because of European agricultural practices and other non-Australian activities were doing fine. Indigenous Australians have long practised, almost certainly since their arrival whenever that was, sustainable wild-harvesting strategies for obtaining food. This does not lead to extinctions; European-style degradation of natural ecosystems and its replacement for agriculture focused on production of surpluses of introduced species does (Archer and Beal 2004).

Debate continues as to whether or not megafauna survived in Tasmania later than elsewhere in Australia and, if so, whether it was because people arrived there less than 43,000 yBP (e.g. Cosgrove and Allen 2001; Turney et al. 2008). Between 43,000 and 39,000 yBP (Lambeck and Chappell 2001) a land bridge was established between Tasmania and the mainland which brought the first humans to the island. The known Tasmanian megafauna, consisting of six marsupials and one echidna species, has been dated as older than

127,000 yBP; those species would have made the crossing into Tasmania during one of the glacial maxima. Turney et al. (2008) argue that human involvement in their extinction by 41,000 yBP is implicated, although no megafaunal remains have been found in Tasmanian human occupation sites, the oldest of which are interpreted to be 43,000 to 40,000 yBP.

Most Australian Pleistocene megafauna (some 60 species) became extinct during the Late Pleistocene but some species survived into the Holocene. These include the Red Kangaroo, Eastern Grey Kangaroo, Western Grey Kangaroo, the Euro, the Black Kangaroo and the Antilopine Kangaroo. The living Eastern Grey Kangaroo *M. giganteus giganteus* is approximately 30% smaller than its Pleistocene relative *M. giganteus titan*, and it has generally been accepted (Marshall and Corruccini 1978) that similar 'dwarfing' of this kind has occurred in several other marsupial species lineages that span the late Pleistocene to Holocene. Other examples may include the giant mainland Tasmanian Devils previously known as *S. laniarius* and their conspecific modern descendant *Sarocphilus harrisii* and the giant Thylacines previously known as *Thylacinus spelaeus* and its conspecific descendant *T. cynocephalus*. However, Price (2008) questions the reality of dwarfing at least in the *M. giganteus* lineage, suggesting that better sampling and dating of Pleistocene assemblages, combined with additional morphological and phylogenetic analyses, may well show that the two forms represent two different species which overlapped both temporally and geographically, as he has demonstrated to be the case in the Pleistocene koalas. Recognition of a range of gigantic ('predwarf') forms as distinct species would substantially increase the known number of Late Pleistocene extinct megafaunal species. It would also, as Price (2008) points out, invite a rethink about some of the proposed drivers for putative dwarfing such as responses in Late Pleistocene lineages to climate-driven resource limitations (Marshall and Corruccini 1978) and/or selection for smaller forms because of human hunting of the larger individuals (Johnson and Wroe 2003).

No matter on which side of the Australian megafaunal extinction debate, researchers agree that resolution will require more precise dating for both the arrival of humans and megafaunal extinctions, as well as a better understanding of human interactions with the broader Australian biota during early occupation (Choquenot and Bowman 1998; Roberts et al. 2001; Barnosky et al. 2004; Burney and Flannery 2005; Wroe and Field 2006).

Retrospective and Prospective – 'Big' Events and Their Consequences for the Evolution of Marsupials

Looking back over this compilation, there are a range of transcendent messages that can be drawn. Broadly, these could be placed in two categories:

1. Clusters of evolutionary outcomes some of which appear to have abiogenic correlates while others appear to be synergistic outcomes of complex chains of events, some with biological components; and
2. Multi-million-year trends in the relative fortunes of lineages that have survived these events.

Considering the first, in an ideal world we should be able to learn from the history of the relationships between the triggers and their consequences about what we might expect from current and future events such as climate change (see below).

Considering the second, we have argued elsewhere (Archer et al. 1991; Archer and Beal 2004) that by carefully documenting these changes in the lead up to the modern world, we should be able to improve our ability to determine the relative health of individual lineages which should in turn inform efforts to more effectively conserve as much of the global genome as possible for the future. Much of this documentation remains to be done in sufficient detail to provide this kind of overview, and that research is underway.

The first relevant event was the early Cretaceous trigger that resulted in the metatherian/marsupial origins in the first place. While it is generally assumed to have occurred about 125 million years ago in Asia with the evolution of *Sinodelphys* (see earlier section), there is still some uncertainty about this interpretation (e.g. Averianov et al. 2010). But it is clear that the early Cretaceous was a vibrant time for the early radiation of mammals of all kinds throughout the world, including Australia. Drivers for this period of rapid diversification almost certainly included the warm climates of the early Cretaceous which in turn may have been triggered by the increase in CO_2 resulting from increased tectonic activity during this period. Rising CO_2 levels would have triggered and supported greenhouse conditions at least until the end of the Cretaceous. The evolution of angiosperms and the co-evolutionary expansion of insects at this time may also have added incentives for the diversification of many new kinds of small, insectivorous and nectarivorous mammals including the first metatherians. The fact that much of Gondwana was also being torn asunder at this time, with Africa and India hiving off and heading north creating the South Atlantic and Indian Oceans, and the fact that sea levels were rising and inundating many continents including Australia which itself had been turned into three large islands separated by inland seas, would have contributed to the creation of wetter, warmer climates on all continents. In this churning maelstrom of greenhouse conditions, the first metatherians arose and thrived.

While we do not yet know what happened with respect to Australian marsupials prior to the early Eocene, the fate of metatherians in the rest of the world is better known. Clearly they had spread to all northern continents and possibly Africa by at least 100 Ma where they continued to diversify and, by 70 Ma, were more diverse than eutherians. But gradually the cooling climates of the late Cretaceous began to spell disaster for metatherians on the northern continents. With the bolide impact at 65 Ma, mass volcanism soon thereafter in southern Asia and global mass extinctions, metatherian diversity in the north began to decline and eutherian mammals began to dominate. While this could mean that for some reason, in contrast to early eutherians, early metatherians differentially prospered under greenhouse conditions and declined when these gave way to cooler, drier climates, the ability of Cenozoic marsupials to thrive and diversify in tropical as well as arid zones and even seasonally sub-zero alpine areas in Australia suggests there would have been no inherent limit to their relative ability to adapt to a wide range of palaeoclimates.

As detailed above, by 55 Ma metatherians had become established and diversified on all continents of the Earth, including Antarctica. The movement of these early metatherians and subsequently early marsupials from their probable source in Asia to, eventually, Australia, was clearly accomplished by dispersal rather than vicariant events. Hence continental movements following the breakup of Gondwana had little impact on the radiation of early marsupials. Starting about 70 Ma and until at least 55 Ma, hence right through the environmental catastrophes at the K/T boundary seemingly without missing a beat, there was a rapidly dispersing front pushing its way southward out of North America to South America and then eastward across Antarctica and into Australia. All of these southern lands were dominated at the time by specialised but relatively plesiomorphic groups of mammals such as dryolestoids, gondwanatheres and monotremes. Perhaps the global collapse of ecosystems that followed the catastrophe at the end of the Cretaceous enabled the migrating marsupials to take advantage of thousands of vacant niches that had suddenly become available.

Curiously, in Australia at least, the same opportunities do not appear to have been open to the earliest placental immigrants. While it is clear from the presence of *Tingamarra* in the 55 Myo deposit at Murgon, Qld that terrestrial placental mammals were travelling with the early marsupial peregrines and that both groups were easily able to insinuate into and diversify within these southern ecosystems, these land-bound (in contrast to bats which were also present in the Murgon deposit) placentals soon vanished and marsupials overwhelmed Australia. We have, above, explored possible reasons for why this may have happened including the fact that marsupials have greater control over their reproductive investment, but the real answer is that we do not know why this very different outcome occurred in Australia in contrast to what happened in the northern continents. The fact that a different outcome again occurred in South America where the first marsupials became the mammalian carnivores and the first placentals became herbivores and the marsupials' dinner, suggests that the outcomes of continental competitions of this kind may well be no more or less predictable than a throw of the dice (Kirsch 1977b).

Unfortunately, with only a single early Tertiary mammal-bearing fossil site known from Australia, we have just one peek at what was happening here between 110 and 26 Ma, and that was at 55 Ma, the age of the Tingamarra assemblage from Queensland. Although as noted above this fauna contains the not unexpected indications of South American links, it also contains no specialised herbivores or carnivores. While this surprised us at first, it is in fact probably attributable to a global stage of ecosystem restoration following the massive destruction of ecosystems 10 Ma before this time. All of the Tingamarra mammals are either insectivores or omnivores with no clear indication of any undoubted representative of the Diprotodontia, the largest order of Australian marsupials, most of which are herbivores. Of course it is possible, given the molecular divergence date for diprotodontians and the other orders of Australian marsupials of 65 to 62 Ma, that they may be among the Tingamarran taxa but not yet as herbivores. Although greenhouse conditions predominated in the early Eocene over much of the world, this relative paucity of obligate herbivores is a feature of most of the ecosystems of the time throughout the world. Without evidence suggesting any other explanation, it is possible that most of the Paleocene to early Eocene mammal communities, including those from Australia, had not yet rebalanced in terms of niche specialists and that as a result omnivores and insectivores remained in control. This is apparent from North American Paleocene mammals which are dominantly insectivores and omnivores with very few exceptions. Specialised herbivores do not begin to become a significant feature of these communities until well into the Eocene. Because we have not found any fossil mammal communities in Australia representing the interval between the earliest Eocene and latest Oligocene, we do not know how long it took here for specialised herbivores to appear and once again proliferate. Certainly by the latest Oligocene, there are many specialised herbivorous marsupials known from Australia. The moral of the story appears to be that in times of extreme ecological crisis and for at least 10 million years afterwards, whether you are a placental or a marsupial, it pays to be either an insectivore or an omnivore.

Today, as we speed headlong into Earth's sixth mass extinction event (e.g. Woinarski et al. 2010; Fitzsimons et al. 2010; Steffen et al. 2010), perhaps the same thing can be seen occurring again. If we consider what has been happening to Australian marsupials since arrival of Europeans with their cultural predisposition for replacing diverse native ecosystems with monocultures of introduced species, 100% of Australia's six marsupial carnivores are now either extinct or uncomfortably close to the brink as are about 30% of the 78 species of herbivores. In contrast, only about 10% of the 49 species of insectivores and 10% of the 18 species of omnivores have become extinct or endangered. Further, the marsupial that appears to be in the least danger following our environmentally unfortunate ministrations is a classic opportunistic omnivore – the common brushtailed possum. So perhaps these are the first signs of an unfolding Armageddon similar to the one that occurred 65 million years ago which took nature more than 10 million years to repair.

Moving on, by the time we see the late Oligocene communities in Australia, which functioned within an icehouse, cooler and drier world than today, it is clear that ecosystem balance had been restored with herbivore numbers outnumbering carnivores and omnivores by a factor of at least 10. However, there are some striking differences from the contemporary situation. Dasyurids of any size were rare while bandicoots appeared to be doing most of the things that dasyurids do today. Carnivore niches for mammals had become dominated by two groups: thylacines and marsupial lions, a dominion maintained into the late Pleistocene. In the relatively dense forest vegetation, kangaroos primarily galloped rather than hopped. As these communities transformed into those of the early Miocene, we see once again a return to greenhouse conditions and marsupial biodiversity leaps accordingly. It is clear looking backward and forward from this point, that mammal biodiversity in Australia peaks during greenhouse conditions and slumps during icehouse conditions. These climatic factors controlling or influencing mammalian diversity would appear to be similar all over the world. Certainly the global greenhouse event of the early Eocene (Fig. 14) saw high levels of biodiversity spring up through Europe and North America. The main difference with respect to Australia, besides the obvious one of the unique marsupial groups involved, relates to its continental isolation. No matter how cold the world became, how much ice accumulated at the poles or how far sea levels fell, Australia remained an island. On other continents with the exception of Antarctica, regular changes in sea level either blocked or drove intercontinental dispersals which then had complex consequences for both the invaders and the invaded. Australia has remained for the last 35 million years an isolated natural laboratory. This is why what happened here during pre-modern cycles of climate change may be the best way to study intrinsic responses by mammal communities to climate change.

The shift to greenhouse conditions in the early Miocene, detailed previously, had a huge impact on marsupial biodiversity. When we first confronted the very high diversity of marsupials emerging from the freshwater limestone deposits of Riversleigh, some colleagues refused to believe that mammalian biodiversity could have been this high in Australia and that it must be an artifact of mixed faunas. Many research projects and Ph.D. theses later, it is evident that this paleobiodiversity, which can reach 64 sympatric marsupials which is higher than the number for the whole of the Wet Tropics World Heritage area, is natural and not the result of mixing. This diversity of terrestrial mammals is more akin to that found in the lowland rainforests of Borneo or the Amazon. What now appears to need explaining instead is why Australian marsupial communities of the Holocene, which have previously been taken to indicate the 'normal' levels of marsupial biodiversity for the continent, are relatively impoverished. The main answer would appear to be the late Cenozoic decline in rainfall with concomitant increased seasonality and the consequent collapse and coastward retreat of the vast wet forests that previously dominated the continent. But it provides evidence that Australia has no inherent attributes that limit its ability to produce levels of mammalian biodiversity that are as high as anywhere else in the world whether our 'tired' soils are nutrient deficient or not.

The over-riding challenge is to understand exactly what happened during the subsequent mid to late Miocene. As noted above, the mid Miocene climatic oscillation with the steep drop in rainfall from 15 my marked the start of a decline in biodiversity and, unfortunately, a decline in the number of Australian fossil sites that document changes in the biota from this time until the early Pliocene about 4.5 my. Although sites such as Alcoota in the N.T. and Encore Site at Riversleigh are providing insights into what happened during the seemingly critical late Miocene, we are now intensifying our search for more late Miocene sites to try to better understand what happened during this period of increasing climatic stress. Certainly it is clear from differences between the marsupial assemblages of the mid Miocene and those of the early Pliocene that during this brief 8 Ma interval there were major transformations in kind and in ecological role-playing with respect to which groups dominated the new, expanding, relatively less well-watered niches, particularly in central Australia. But at this point, with limited actual data, much of what is currently presumed to have happened is speculative deduction based on differences between the mid Miocene and early Pliocene biotas. The late Miocene was also a period of massive change in most other continents. Among other events with huge consequences that occurred during this period, ancestral humans diverged from their chimp relatives to end up triggering the next global mass extinction.

Another mystery of late Miocene vintage is what happened and when in the developing relationship between the biotas of the newly emergent island of New Guinea and those of the Australian mainland. While we have noted above the probable origins and timing for the dispersal of late Miocene diprotodontoids to New Guinea, this sidesteps questions about every other group of marsupials that must have had a similar potential for dispersal. Why are there no potoroids, hypsiprymnodontids or any other relatively plesiomorphic kangaroo groups present somewhere in New Guinea? Part of the answer may be that they were there but have since become extinct. Given the discovery that many groups once thought to have had their evolutionary furnace in New Guinea, such as striped possums, tree kangaroos and cuscuses, are now known from older Tertiary deposits in Australia, it now seems equally likely that they originated in Australia and subsequently dispersed to New Guinea at sometime during the late Miocene to Pliocene. Unfortunately, because the Tertiary marsupial record of New Guinea is unknown prior to the Pliocene, we cannot test competing hypotheses about areas of origin or direction of subsequent dispersal. In fact the New Guinea record is only just better than that for the whole of Western Australia, which, with just two biotically impoverished, poorly dated sites (Quanbun in the Kimberley, Jandakot in the southwest; Figs. 13 and 15a), is abysmal. Hence the timing of arrival and early history of marsupials in New Guinea are areas of vast ignorance and significant challenge.

While under normal circumstances one might expect the record of biotic events to steadily improve the closer we get to the present, the curse of declining rainfall into the latter part of the Cenozoic has left us with exactly the opposite situation. Compared with the record for the late Oligocene to mid Miocene, the Pliocene record is only just better than that of the late Miocene. Hence the details of what happened in the brief flip back to greenhouse conditions in the early Pliocene are largely a mystery and an impetus to find more Pliocene sites, particularly in northern Australia. The ongoing mystery of what was happening in New Guinea persists into and through the Pliocene with just one significant site (Awe in Papua) to represent the whole of the Pliocene.

One of the key but imprecisely dated events of the latest Miocene or early Pliocene was the arrival of rodents into Australia. Given the tantalizing discovery of dendromurine ('tree mice') rodents along with many other diverse groups in the Rackham's Roost deposit at Riversleigh (Fig. 55), it is increasingly clear that rodent origins in Australia may have had a very complex beginning, one that probably had significant consequences for early Pliocene marsupial groups. Is it possible, for example, that these invaders negatively impacted on the once-common rodent-like ektopodontid possums which only just survived into the early Pleistocene?

The three key and most controversial events of the Pleistocene in terms of marsupial evolution and fortunes were decline of the megafauna, the arrival and impact of humans and the severity of contemporary climate change. In each case, understanding about what actually happened into the late Pleistocene, and the potential interactions between these three, has been critically constrained by a lack of sufficient data. Understanding about climate change during the late Pleistocene is perhaps more advanced than the other two, but the severity of the last period of aridity in terms of its potential role as a cause of extinction of megafaunal species remains controversial. That said, the most recent research into the potential correlation of climate change intensity and extent of extinction of megafaunal species on a global scale (e.g. Nogués - Bravo et al. 2010) has concluded that these events are indeed correlated on most continents but not necessarily globally coeval. The other increasingly recognised fact is that most of the megafauna on the larger continents (e.g. Europe, North America and Australia) had begun to decline and in some cases vanished well before human arrival on those continents (e.g. Cosgrove et al. 2010), although most of these studies seem reluctant to exclude the possibility that humans may have exacerbated the negative impact of climate change. In other words, while the idea that newly arrived humans 'blitzkrieged' the megafauna has no substantial support, the relative importance of human behaviour, climate change and potentially other factors such as disease and ecosystem destabilisation in extinction events have yet to be clarified. For this situation to change, we need a lot more primary data from more sites where these factors can be distinguished and credibly assessed. As data steadily accumulates, the direct role of early humans in the extinction events continues to appear to diminish. For example, recent research into the role of anthropogenic fire (e.g. Black et al. 2007; and work in progress by Scott Mooney et al. 2010, 'personal communication'), once argued to have been a major factor in the Pleistocene extinctions through habitat alteration, demonstrates that this was a relatively trivial factor compared with climate change and may not have been a factor of any significance prior to about 6,000 years ago, long after megafaunal diversity had collapsed.

It is ironic that while the capacity of Europeans to cause extinctions through mismanagement of the land is accepted, on the eve of impending crises about to be visited on Australia's marsupials as a result of accelerated climate change (Steffen et al. 2010), any scientist would challenge the capacity of climate change to alter pre-modern environments, destabilise ecosystems and cause mass extinctions. Given also the evidence for the impact of climate change on Australia's marsupials over the last 55 million years detailed above, it seems probable that the next time the history of these unique marsupials is reviewed there will be considerably less scepticism about the capacity of climate change to drive extinctions.

Acknowledgements We first thank John Talent for inviting us to compile this overview for the volume. We thank the ARC and other funding bodies that have contributed over the years to enable us and our colleagues to explore Australia's unique fossil record to develop an understanding of palaeobiodiversity, trends through time and how these have led to the situation all of us are trying to conserve for the future. Phil Creaser has helped immeasurably by establishing the CREATE Funds to assist palaeoscience research projects conducted at UNSW. We also thank Queensland National Parks, the World Heritage Unit in Canberra, Xstrata Community Partnership Program North Queensland, Mount Isa City Council, Lawn Hill Riversleigh Pastoral Holding Company, Riversleigh Interpretive Centre, Adele's Grove and the Waanyi People for permission to investigate the rocks at Riversleigh. Colleagues who have assisted with information, ideas, access to work in progress and helped us obtain the materials that resulted in our current understanding include Ken Aplin, Rick Arena, Mina Bassarova, Alan Bartholomai, Hayley Bates, Bob Beale, Robin Beck, Jan Bimrose, Steve Bourne, Jenni Brammall, Pip Brewer, Linda Broome, Lizard Cannell, Bill Clemens, Bernie Cooke, Phil Creaser, Kirsten Crosby, Judith Field, Gillian Garvey, Anna Gillespie, Francisco Goin, Glen Graham, Ian Graham, Yamila Gurovich, Bob Jones, Scott Hocknull, Christine Janis, Jenny Jones, John Kirsch, Chris Larkin, Julien Louys, Rod and Michelle Low Mow, Dirk Megirian, Alexis Meyer, Jeanette Muirhead, Peter Murray, Frank Nissen, Rosendo Pascual, Neville Pledge, Elizabeth Price, Gilbert Price, Gavin

Prideaux, Alan Rackham, Liz Reed, Tom Rich, David Ride, Karen Roberts, John Scanlon, Bernard Sigé, Dick Tedford, Kenny Travouillon, Steve Van Dyck, Rod Wells, Arthur White, Karen White, Steph Williams, Paul Willis, Vera Weisbecker, Mike Woodburne, Jon Woodhead, Trevor Worthy, Steve Wroe and Jian-xin Zhao.

References

Amrine-Madsen H, Scally M, Westerman M, Stanhope MJ, Krajewski C, Springer MS (2003) Nuclear gene sequences provide evidence for the monophyly of australidelphian marsupials. Mol Phylogen Evol 28:186–196

Anderson C (1937) Palaeontological notes. No. IV. Fossil marsupials from New Guinea. Rec Aust Mus 20:73–78

Aplin KP (1987) Basicranial anatomy of the early Miocene Diprotodontian *Wynyardia bassiana* (Marsupialia: Wynyardiidae) and its implications for wynyardiid phylogeny and classification. In: Archer M (ed) Possums and opossums: studies in evolution. Surrey Beatty and Sons Pty Ltd, Chipping Norton, pp 369–39

Aplin KP (2006) Ten million years of rodent evolution in Australasia: phylogenetic evidence and a speculative historical biogeography. In: Merrick JR, Archer M, Hickey GM, Lee MSY (eds) Evolution and biogeography of Australasian vertebrates. Auscipub Pty Ltd, Sydney, pp 707–744

Aplin KP, Archer M (1987) Recent advances in marsupial systematics with a new syncretic classification. In: Archer M (ed) Possums and opossums: studies in evolution. Surrey Beatty and Sons Pty Ltd, Chipping Norton, pp xv–lxxii

Archer M (1976a) Miocene marsupicarnivores (Marsupialia) from central south Australia, *Ankotarinja tirarensis* gen. et sp. nov., *Keeuna woodnei* gen. et sp. nov., and their significance in terms of early marsupial radiations. Trans R Soc South Aust 100:53–73

Archer M (1976b) The dasyurid dentition and its relationships to that of didelphids, thylacinids, borhyaenids (Marsupicarnivora) and peramelids (Peramelina: Marsupialia). Aust J Zool Suppl Ser 39:1–34

Archer M (1976c) The basicranial region of marsupicarnivores (Marsupialia), inter-relationships of the carnivorous marsupials, and affinities of the insectivorous marsupial peramelids. Zool J Linn Soc 59:217–322

Archer M (1976d) Results of the Ray E. Lemley expeditions, Part 1. The Allingham formation and a new Pliocene vertebrate fauna from northern Queensland. Mem Queensland Mus 17:379–397

Archer M (1979) *Wabularoo naughtoni* gen. et sp. nov., an enigmatic kangaroo (Marsupialia) from the middle Tertiary Carl Creek Limestone of northwestern Queensland. Results of the Ray E. Lemley expeditions, part 4. Mem Queensland Mus 19:299–307

Archer M (1981) A review of the origins and radiations of Australian mammals. In: Keast A (ed) Ecological biogeography of Australia. Junk, The Hague, pp 1437–1488.

Archer M (1982a) Review of the dasyurid (Marsupialia) fossil record, integration of data bearing on phylogenetic interpretation, and suprageneric classification. In: Archer M (ed) Carnivorous marsupials, Surrey Beatty and Sons, Chipping Norton, vol 2, pp 397–443

Archer M (1982b) A review of Miocene thylacinids (Thylacinidae, Marsupialia), the phylogenetic position of the Thylacinidae and the problem of apriorisms in character analysis. In: Archer M (ed) Carnivorous marsupials, Surrey Beatty and Sons, Chipping Norton, vol 2, pp 445–476

Archer M (1984) The Australian marsupial radiation. In: Archer M, Clayton G (eds) Vertebrate zoogeography and evolution in Australasia (Animals in space and time). Hesperian Press, Carlisle, WA, pp 633–808

Archer M (1989) The science of being wrong. Aust Nat Hist 23:170–171

Archer M (1992) Ringtail possums (Pseudocheiridae, Marsupialia) from the Tertiary deposits of Riversleigh. The Beagle, Records of the Northern Territory Mus Arts Sci 9:257–258

Archer M, Bartholomai A (1978) Tertiary mammals of Australia: a synoptic review. Alcheringa: An Australas J Palaeontol 2:1–19

Archer M, Beal B (2004) Going native. Hodder and Headline, Sydney

Archer M, Flannery TF (1985) Revision of the extinct gigantic rat kangaroos (Potoroidae: Marsupialia) with a description of a new Miocene genus and species and a new Pleistocene species of *Propleopus*. J Paleontol 59:1331–1349

Archer M, Hand SJ (1987) Evolutionary considerations. In: Cronin L (ed) Koala: Australia's endearing marsupial. Reed Books, Sydney, pp 79–106

Archer M, Hand SJ (2006) The Australian marsupial radiation. In: Merrick JR, Archer M, Hickey GM, Lee MSY (eds) Evolution and biogeography of Australasian vertebrates. Auscipub Pty Ltd, Sydney, pp 575–646

Archer M, Rich TH (1982) Results of the Ray E. Lemley expeditions. *Wakaleo alcootaensis* n. sp. (Thylacoleonidae, Marsupialia), a new marsupial lion from the Miocene of the Northern Territory with a consideration of early radiation in the family. In: Archer M (ed) Carnivorous marsupials. Surrey Beatty and Sons, Chipping Norton, vol 2, pp 495–502

Archer M, Bartholomai A, Marshall LG (1978) *Propleopus chillagoensis*, a new north Queensland species of extinct giant rat-kangaroo (Macropodidae: Potoroinae). Mem Natl Mus Victoria 39:55–60

Archer M, Crawford IM, Merrilees D (1980) Incisions, breakages and charring, some probably man-made, in fossil bones from Mammoth Cave, Western Australia. Alcheringa 4:115–131

Archer M, Clayton G, Hand SJ (1984) A checklist of Australasian fossil mammals. In: Archer, M, Clayton G (eds) Vertebrate zoogeography and evolution in Australasia (Animals in space and time). Hesperian Press, Carlisle, WA, pp 1027–1087

Archer M, Tedford RH, Rich TH (1987) The Pilkipildridae, a new family and four new species of ?Petauroid possums (Marsupialia: Phalangerida) from the Australian Miocene. In: Archer M (ed) Possums and opossums: studies in evolution. Surrey Beatty and Sons Pty Ltd, Chipping Norton, pp 607–627

Archer M, Hand S, Godthelp H (1988) A new order of Tertiary zalambdodont marsupials. Science 239:1528–1531

Archer M, Godthelp HJ, Hand SJ, Megirian D (1989) Fossil mammals of Riversleigh, northwestern Queensland: preliminary overview of biostratigraphy, correlation and environmental change. Australian Zool 25:29–65

Archer M, Every RG, Godthelp HJ, Hand SJ, Scally KB (1990) Yingabalanaridae, a new family of enigmatic marsupials from Tertiary deposits of Riversleigh, northwestern Queensland. Mem Queensland Mus 28:193–202

Archer M, Hand SJ, Godthelp HJ (1991) Riversleigh: the story of animals in ancient rainforests of inland Australia, 1st edn. Reed Books, Balgowlah, NSW, pp 1–264

Archer M, Godthelp H, Hand SJ (1993) Early Eocene marsupial from Australia. In: Schrenk F, Ernst K (eds) Kaupia: Darmstadter beitrage zur naturgeschichte monument grube messel-perspectives and relationships part 2. Hessisches Landesmuseum Darmstadt, Darmstadt, Germany, pp 193–200

Archer M, Hand SJ, Godthelp HJ (1994) Riversleigh: the story of animals in ancient rainforests of inland Australia. Reprint with revisions, Reed Books, Chatswood, NSW, pp 1–264

Archer M, Hand SJ, Godthelp HJ (1995) Tertiary environmental and biotic change in Australia. In: Vrba E, Denton GH, Partridge TC, Burckle LH (eds) Palaeoclimate and evolution, with emphasis on human origins. Yale University Press, New Haven, pp 77–90

Archer M, Hand SJ, Godthelp H, Creaser P (1997) Correlation of the Cainozoic sediments of the Riversleigh World Heritage fossil property, Queensland, Australia. In: Aguilar J–P, Legendre S, Michaux J (eds) Actes du Congrès BiochroM'97, Mémoires et Travaux de l'E.P.H.E., Institut de Montpellier, vol 21, pp 131–152

Archer M, Burnley, I, Dodson J, Harding R, Head L, Murphy A (1998) From plesiosaurs to people: 100 million years of Australian environmental history. State of the Environment Technical Paper Series (Portrait of Australia), Department of the Environment, Canberra

Archer M, Arena DA, Bassarova M, Black K, Brammall J, Cooke BN, Creaser P, Crosby K, Gillespie A, Godthelp H, Gott M, Hand SJ, Kear BP, Krikmann A, Mackness B, Muirhead J, Musser A, Myers T, Pledge NS, Wang Y, Wroe S (1999) The evolutionary history and diversity of Australian mammals. Aust Mammal 21:1–45

Archer M, Arena DA, Bassarova M, Beck RMD, Black K, Boles WE, Brewer P, Cooke BN, Crosby K, Gillespie A, Godthelp H, Hand SJ, Kear BP, Louys J, Morrell A, Muirhead J, Roberts KK, Scanlon JD, Travouillon KJ, Wroe S (2006) Current status of species-level representation in faunas from selected fossil localities in the Riversleigh World Heritage Area, northwestern Queensland. Alcheringa: An Australas J Palaeontol 30:1–17

Archer M, Beck RMD, Gott M, Hand SJ, Black KH (2011) Australia's first fossil marsupial mole (Notoryctemorphia) resolves controversy about the evolution of zalambdodonty in marsupials. Proc R Soc B 278:1498–1506

Arena DA (1997) The palaeontology and geology of Dunsinane Site, Riversleigh. Mem Queensland Mus 41:171–179

Arena R, Wroe S, Archer M (1998) Additional Riversleigh material referred to the dasyurid *Ganbulanyi djadjinguli*: phylogenetic and palaeobiological implications. Riversleigh Symposium 1998 Abstracts. The University of New South Wales, Sydney, December 3–4, pp 1–2

Asher RJ, Horovitz I, Sánchez-Villagra MR (2004) First combined cladistic analysis of marsupial mammal interrelationships. Mol Phylogen Evol 33:240–250

Averianov AO, Archibald JD, Ekdale EG (2010) New material of the Late Cretaceous deltatheroidan mammal *Sulestes* from Uzbekistan and phylogenetic reassessment of the metatherian-eutherian dichotomy. J Syst Palaeontol 8: 301–330

Ayliffe LK, Marianelli PC, Moriarty KC, Wells RT, McCulloch MT, Mortimer GE, Hellstrom JC (1998) 500 ka precipitation record from southeastern Australia: evidence for inter-glacial relative aridity. Geology 26:147–150

Balme JM, Merrilees D, Porter JK (1978) Late Quaternary mammal remains, spanning about 30 000 years, from excavations in Devil's Lair, Western Australia. J R Soc West Aust 61:33–65

Barnosky AD, Koch PL, Feranec RS, Wing SL, Shabel AB (2004) Assessing the causes of Late Pleistocene extinctions on the continents. Science 306:70–75

Bartholomai A (1962) A new species of *Thylacoleo* and notes on some caudal vertebrae of *Palorchestes azael*. Mem Queensland Mus 14:33–40

Bartholomai A (1963) Revision of the extinct macropodid genus *Sthenurus* Owen in Queensland. Mem Queensland Mus 16:19–26

Bartholomai A (1967) *Troposodon*, a new genus of fossil Macropodinae (Marsupialia). Mem Queensland Mus 15:21–33

Bartholomai A (1968) A new fossil koala from Queensland and a reassessment of the taxonomic position of the problematic species, *Koalemus ingens* DeVis. Mem Queensland Mus 15:65–71

Bartholomai A (1971) *Dasyurus dunmalli*, a new species of fossil marsupial (Dasyuridae) in the upper Cainozoic deposits of Queensland. Mem Queensland Mus 16:19–26

Bartholomai A (1978a) The rostrum in *Palorchestes* Owen (Marsupialia: Diprotodontidae). Results of the Ray E. Lemley expedition, part 3. Mem Queensland Mus 18:145–149

Bartholomai A (1978b) The Macropodidae (Marsupialia) from the Allingham Formation, northern Queensland. Results of the Ray E. Lemley expeditions, part 2. Mem Queensland Mus 18:127–142

Bassarova M, Archer M (1999) Living and extinct pseudocheirids (Marsupialia, Pseudocheiridae): phylogenetic relationships and changes in diversity through time. Aust Mammal 21:25–27

Bassarova M, Archer M, Hand S (2001) New Oligo-Miocene pseudocheirids (Marsupialia) of the genus *Paljara* from Riversleigh, northwestern Queensland. Mem Assoc Australas Palaeontol 25:61–76

Beck R (2008a) A dated phylogeny of marsupials using a molecular supermatrix and multiple fossil constraints. J Mammal 89:175–189

Beck RMD (2008b) Form, function, phylogeny and biogeography of enigmatic Australian metatherians. University of New South Wales, Sydney, NSW

Beck RMD (2009) Was the Oligo-Miocene Australian metatherian *Yalkaparidon* a 'mammalian woodpecker'? Biol J Linnean Soc 97:1–17

Beck R, Archer M, Godthelp H, Mackness B, Hand S, Muirhead J (2008a) A bizarre new family of Marsupialia (*incertae sedis*) from the early Pliocene of northeastern Australia: implications for the phylogeny of bunodont marsupials. J Paleontol 82:749–762

Beck RMD, Godthelp H, Weisbecker V, Archer M, Hand SJ (2008b) Australia's oldest marsupial fossils and their biogeographical implications. PLoS One 3(3):e1858. doi:10.1371/journal.pone.0001858

Bensley BA (1903) On the evolution of the Australian Marsupialia; with remarks on the relationships of the marsupials in general. Trans Linnean Soc London, Zool 9:83–217

Berger A, Loutre MF, Laskar J (1992) Stability of the astronomical frequencies over the Earth's History for paleoclimate studies. Science 255:560–566

Biknevicius AR (1983) Dental function and diet in the Carpolestidae (primates, Plesiadapiformes). Am J Phys Anthropol 71:157–171

Black K (1997a) A new species of Palorchestidae (Marsupialia) from the late middle to early late Miocene Encore Local Fauna, Riversleigh, northwestern Queensland. Mem Queensland Mus 41:181–186

Black K (1997b) Diversity and biostratigraphy of the Diprotodontoidea of Riversleigh, northwestern Queensland. Mem Queensland Mus 41:187–192

Black K (1999) Diversity and relationships of living and extinct koalas (Phascolarctidae, Marsupialia). Aust Mammal 21:16–17, 34–45

Black K (2006) Description of new material for *Propalorchestes novaculacephalus* (Marsupialia: Palorchestidae) from the mid Miocene of Riversleigh, northwestern Queensland. Alcheringa: An Australas J Palaeontol 30:351–361

Black K (2007) Maradidae: a new family of vombatomorphian marsupial from the late Oligocene of Riversleigh, northwestern Queensland. Alcheringa: An Australas J Palaeontol 31:17–32

Black K (2008) Diversity, phylogeny and biostratigraphy of diprotodontoids (Marsupialia: Diprotodontidae, Palorchestidae) from the Riversleigh World Heritage Area. Doctor of Philosophy, University of New South Wales, Sydney, NSW

Black KH (2010) *Ngapakaldia bonythoni* (Marsupialia, Diprotodontidae): new material from Riversleigh, northwestern Queensland, and a reassessment of the genus *Bematherium*. Alcheringa: An Australas J Palaeontol 34:471–492

Black K, Archer M (1997a) *Silvabestius*, a new genus and two new species of primitive zygomaturines (Marsupialia, Diprotodontidae) from Riversleigh, northwestern Queensland. Mem Queensland Mus 41:181–208

Black K, Archer M (1997b) *Nimiokoala* gen. nov. (Marsupialia, Phascolarctidae) from Riversleigh, northwestern Queensland, with a revision of *Litokoala*. Mem Queensland Mus 41:209–228

Black KH, Hand SJ (2010) First crania and assessment of species boundaries in *Nimbadon* (Marsupialia: Diprotodontidae) from the middle Miocene of Australia. Am Mus Novitates 3678:1–60

Black KH, Archer M, Hand SJ, Godthelp HJ (2010) First comprehensive analysis of cranial ontogeny in a fossil marsupial-from a 15-million-year-old cave deposit in northern Australia. J Vertebrate Paleontol 30:993–1011

Black KH, Archer M, Hand SJ (in press) New Tertiary koala (Marsupialia: Phascolarctidae) from Riversleigh, Australia with a revisionof phascolarctid phylogenetics, paleoecology and paleobiodiversity. J Vertebrate Paleontol

Black MP, Mooney SD, Haberle SG (2007) The fire, human and climate nexus in the Sydney Basin, eastern Australia. The Holocene 17:465–478

Boles WE (1999) Early Eocene shorebirds (Aves: Charadriiformes) from the Tingamarra Local Fauna, Murgon, Queensland, Australia. Rec West Aust Mus, Suppl 57:229–238

Bonaparte JF (1990) New Late Cretaceous mammals from the Los Alamitos Formation, northern Patagonia. Natl Geogr Res 6:63–93

Bowler JM (1981) Australian salt lakes: a palaeohydrological approach. Hydrobiologia 82:431–444

Bowler JM, Wyrwoll KH, Lu Y (2001) Variations of the northwest Australian summer monsoon over the last 300,000 years: the paleohydrological record of the Gregory (Mulan) lakes system. Quaternary Int 83–85:63–80

Bowler JM, Johnston H, Olley JM, Prescott JR, Roberts RG, Shawcross W, Spooner NA (2003) New ages for human occupation and climatic change at Lake Mungo, Australia. Nature 421:837–840

Bowman DMJS, Choquenot D (1999) The operational details of marsupial megafaunal overkill. Australian Biol 12:1–5

Bradstock RA, Williams JE, Gill AM (eds) (2002) Flammable Australia: the fire regimes and biodiversity of a continent. Cambridge University Press, Cambridge

Brammall JR (1998) A new petauroid possum from the Oligo-Miocene of Riversleigh, northwestern Queensland. Alcheringa: An Australas J Palaeontol 23:31–50

Brammall J, Archer M (1997) A new Oligocene–Miocene species of Burramys (Marsupialia. Burramyidae) from Riversleigh, northwestern Queensland. Mem Queensland Mus 41:247–268

Brammall J, Archer M (1999) Living and extinct petaurids, acrobatids, tarsipedids and Burramyids (Marsupialia): relationships and diversity through time. Australian Mammal 21:24–25

Brewer P, Archer M, Hand S, Godthelp H (2007) A new species of the wombat *Warendja* from late Miocene deposits at Riversleigh, northwest Queensland, Australia. Palaeontology 50:811–828

Brewer P, Archer M, Hand S (2008) Additional specimens of the oldest wombat *Rhizophascolonus crowcrofti* (Vombatidae; Marsupialia) from the Wipajiri Formation, South Australia: an intermediate morphology? J Vertebrate Paleontol 28:1144–1148

Brook BW, Bowman DMJS (2004) The uncertain blitzkrieg of Pleistocene megafauna. J Biogeogr 31:517–523

Brown B, Gaina C, Müller RD (2006) Circum-Antarctic palaeobathymetry: illustrated examples from Cenozoic to recent times. Palaeogeogr Palaeoclimatol Palaeoecol 231: 158–168

Burk A, Westerman M, Springer M (1998) The phylogenetic position of the musky rat-kangaroo and the evolution of bipedal hopping in kangaroos (Macropodidae: Diprotodontia). Syst Biol 47:457–474

Burk A, Westerman M, Kao DJ, Kavanagh JR, Springer MS (1999) An analysis of marsupial interordinal relationships based on 12S rRNA, tRNA Valine, 16S rRNA, and Cytochrome *b* sequences. J Mammalian Evol 6:317–334

Burney DA, Flannery TF (2005) Fifty millennia of catastrophic extinctions after human contact. Trends Ecol Evol 20: 395–401

Camens AB, Wells RT (2010) Palaeobiology of *Euowenia grata* (Marsupialia: Diprotodontinae) and it's presence in Northern South Australia. J Mammal Evol 17:3–19

Campbell CR (1973) A new species of *Troposodon* Bartholomai; from the early Pleistocene Kanunka Fauna, South Australia (Macropodinae: Marsupialia). Rec South Aust Mus 16:1–18

Campbell C (1976) Tertiary Dasyuridae and Peramelidae from the Tirari Desert, South Australia. Dissertation Abstr Int B Sci Eng 37:4375

Cardillo M, Bininda-Emonds O, Boakes E, Purvis A (2004) A species-level phylogenetic supertree of marsupials. J Zool 264:11–31

Cardillo M, Mace GM, Jones KE, Bielby J, Bininda-Emonds ORP, Sechrest W, Orme CDL, Purvis A (2005) Multiple causes of high extinction risk in large mammal species. Science 309:1239–1241

Cartmill M (1974) *Daubentonia, Dactylopsila*, woodpeckers, and klinorhynchy. In: Martin RD, Doyle GA, Walker AC (eds) Prosimian biology. Gerald Duckworth and Co. Ltd, London, pp 655–670

Case JA (1984) A new genus of Potoroinae (Marsupialia: Macropodidae) from the Miocene Ngapakaldi Local Fauna, South Australia, and a definition of the Potoroinae. J Paleontol 58:1074–1086

Case JA (1989) Antarctica: the effect of high latitude heterochroneity on the origin of the Australian marsupials. In: Crame JA (ed) Origins and evolution of the Antarctic Biota, vol 47. Geological Society, London, Special Publication, pp 217–226

Chen XY, Barton CE (1991) Onset of aridity and dune-building in Central Australia: sedimentological and magnetostratigraphic evidence from Lake Amadeus. Palaeogeogr Palaeoclimatol Palaeoecol 84:55–71

Choquenot D, Bowman DMJS (1998) Marsupial Megafauna, Aborigines and the overkill hypothesis: application of predator-prey models to the question of Pleistocene extinction in Australia. Global Ecol Biogeogr Lett 7:167–180

Clemens WA, Plane MD (1974) Mid-Tertiary Thylacoleonidae (Marsupialia, Mammalia). J Paleontol 48:652–660

Cloos M, Sapiie B, van Ufford AQ, Weiland RJ, Warren PQ, McMahon TP (2005) Collisional delamination in New Guinea: the geotectonics of subducting slab breakoff. Geol Soc Am Spec Pap 400:1–51

Cooke B (1992) Primitive macropodids from Riversleigh, north-western Queensland. Alcheringa: An Australasian J Palaeontol 16:201–217

Cooke BN (1997a) Two new balbarine kangaroos and lower molar evolution within the subfamily. Mem Queensland Mus 41:269–280

Cooke B (1997b) New Miocene bulungamayine kangaroos (Marsupialia, Potoroidae) from Riversleigh, northwestern Queensland. Mem Queensland Mus 41:281–294

Cooke B (1997c) Biostratigraphic implications of fossil kangaroos at Riversleigh, northwestern Queensland. Mem Queensland Mus 41:295–302

Cooke B (1999) *Wanburoo hilarus* gen. et. Sp. nov., a lophodont bulungamayine kangaroo (Marsupialia: Macropodoidea: Bulungamayinae) from the Miocene deposits of Riversleigh, northwestern Queensland. Rec West Aust Mus Suppl 57:239–253

Cooke BN (2000) Cranial remains of a new species of balbarine kangaroo (Marsupialia: Macropodoidea) from the Oligo-Miocene freshwater limestone deposits of Riversleigh World Heritage Area, northern Australia. J Paleontol 74:317–326

Cooke BN (2006) Kangaroos. In: Merrick JR, Archer M, Hickey GM, Lee MSY (eds) Evolution and biogeography of Australasian vertebrates. Auscipub Pty Ltd, Sydney, pp 647–672

Cooke BN, Kear BP (1999) Evolution and diversity of kangaroos (Macropodoidea: Marsupialia). Aust Mammal 21:27–29, 34–45

Cosgrove R, Allen J (2001) Prey choice and hunting strategies in the Late Pleistocene: evidence from Southwest Tasmania. In: Anderson A, O'Connor S, Lilley I (eds) Histories of old ages: essays in honour of Rhys Jones. Coombs Academic Publishing, Australian National University, Canberra, pp 397–429

Cosgrove R, Field J, Garvey J, Brenner-Coltrain J, Goede A, Charles B, Wroe S, Pike-Tay A, Grün R, Aubert M, Lees W, O'Connell J (2010) Overdone overkill – the archaeological perspective on Tasmanian megafaunal extinctions. J Archaeol Sci 37:2486–2503

Crabb PL (1982) Pleistocene dasyurids (Marsupialia) from southwestern New South Wales. In: Archer M (ed) Carnivorous marsupials, vol 2. Surrey Beatty and Sons, Chipping Norton, pp 511–516

Cramb J, Hocknull S (2010) Two new species of *Antechinus* Macleay (Dasyuridae: Marsupialia) from mid-Pleistocene cave deposits in eastern central Queensland. Aust Mammal 32:127–144

Creaser P (1997) Oligocene–Miocene sediments of Riversleigh: the potential significance of topography. Mem Queensland Mus 41:303–314

Crochet J-Y, Sigé B (1993) Les mammiferes de Chulpas (Formation Umayo, transition Crétacé–Tertiaire, Pérou): Données préliminaires. Documents des Laboratoires de Géologie de Lyon 125:97–107

Crosby K (2002) A second species of the possum *Durudawiri* (Marsupialia: Miralinidae) from the early Miocene of Riversleigh, northwestern Queensland. Alcheringa: An Australas J Palaeontol 26:333–340

Crosby K (2007) Rediagnosis of the fossil species assigned to *Strigocuscus* (Marsupialia, Phalangeridae), with description of a new genus and three new species. Alcheringa: An Australas J Palaeontol 31:33–58

Crosby K, Archer M (2000) Durudawirines, a new group of phalangeroid marsupials from the Miocene of Riversleigh, northwestern Queensland. J Paleontol 74:327–335

Crosby K, Norris CA (2003) Periotic morphology in the trichosurin possums *Strigocuscus celebensis* and *Wyulda squamicaudata* (Diprotodontia, Phalangeridae) and a revised diagnosis of the tribe Trichosurini. Am Mus Novitates 3414:1–14

Crosby K, Bassarova M, Archer M, Carbery K (2004) Fossil possums in Australasia discovery, diversity and evolution. In: Goldingay RL, Jackson SM (eds) The biology of Australian possums and gliders. Surrey Beatty and Sons, Chipping Norton, NSW, pp 161–176

Crosby K, Nagy M, Archer M (2001) *Wyulda asherjoeli*, a new phalangerid (Diprotodontia: Marsupialia) from the early Miocene of Riversleigh, northwestern Queensland. Mem Assoc Australas Palaeontol 25:77–82

Davis AC, Archer M (1997) *Palorchestes azael* (Mammalia, Palorchestidae) from the late Pleistocene Terrace Site Local Fauna, Riversleigh, northwestern Queensland. Mem Queensland Mus 41:315–320

Dawson L, Muirhead J, Wroe S (1999) The Big Sink Local Fauna: a lower Pliocene mammalian fauna from the Wellington Caves complex, Wellington, New South Wales. Rec West Aust Mus Suppl 57:265–290

de Muizon C (ed) (1991) La fauna de mamíferos de Tiupampa (Paleoceno inferior, Formación Santa Lucia), Bolivia. Fósiles y facies de Bolivia. 1 Vertebrados. Revista Técnica de Yacimientos Petroleros Fiscales de Bolivia, Santa Cruz, Bolivia

de Muizon C (1998) *Mayulestes ferox*, a borhyaenoid (Metatheria, Mammalia) from the early Palaeocene of Bolivia. Phylogenetic and paleobiologic implications. Geodiversitas 20:19–142

de Muizon C, Cifelli RL (2001) A new basal "didelphoid" (Marsupialia, Mammalia) from the early Paleocene of Tiupampa (Bolivia). J Vertebrate Paleontol 21: 87–97

de Paula Couto C (1962) Didelfídeos fósiles del Paleoceno de Brasil. Revista del Museo Argentino de Ciencias Naturales "B Rivadavia", Zoologìa 112:135–166

De Vis CW (ed) (1887) Anonymous. The Brisbane Courier No. 9224:6

De Vis CW (ed) (1888a) Anonymous. On a supposed new species of *Nototherium*. Abstract of the Proceedings on the Linnean Society of New South Wales for December 28th 1887:v

De Vis CW (1888b) On an extinct genus of the marsupials allied to *Hypsiprymnodon*. Proc Linnean Soc New South Wales 3:5–8

De Vis CW (1895) A review of the fossil jaws of the Macropodidae in the Queensland museum. Proc Linnean Soc New South Wales 10:75–133

Dickinson JA, Wallace MW, Holdgate GR, Gallagher SJ, Thomas L (2002) Origin and timing of the Miocene–Pliocene unconformity in southeast Australia. J Sedimentary Res 72:288–303

Dodson JR, Fullagar R, Furby JH, Prosser IP (1993) Humans and megafauna in a late Pleistocene environment at Cuddie Springs, northwestern New South Wales. Archaeol Oceania 28:93–99

Drummond AJ, Ho SYW, Phillips MJ, Rambaut A (2006) Relaxed Phylogenetics and dating with confidence. PLoS Biol 4(5):e88

Every RG (1970) Sharpness of teeth in man and other primates. Postilla 143:1–30

Exon NF, Kennett JP, Malone MJ (2004) Leg 189 synthesis: Cretaceous–Holocene history of the Tasmanian Gateway. In: Exon NF, Kennett JP, Malone MJ (eds) Proceedings of the ocean drilling program, scientific results, vol 189. Ocean Drilling Program, Texas, pp 1–38

Field J (2004) The archaeology of Late Pleistocene faunal extinctions in Australia: a view from Cuddie Springs. In: Murray T (ed) Archaeology from Australia. Australian Scholarly Publishing, Melbourne, pp 18–35

Field J, Fullagar R, Lord J (2001) A large area archaeological excavation at Cuddie Springs. Antiquity 75:695–702

Finlayson C (2005) Biogeography and evolution of the genus *Homo*. Trends Ecol Evol 20:457–463

Fitzsimons J, Legge S, Traill B, Woinarski J (2010) Into oblivion? The disappearing native mammals of northern Australia. Report Prepared for the Nature Conservancy, pp 1–20

Flannery TF (1983) Revision in the macropodids subfamily Sthenurinae (Marsupialia: Macropodoidea) and the relationships of the species of *Troposodon* and *Lagostrophus*. Aust Mammal 6:15–28

Flannery TF (1988) Origins of the Australo–Pacific land mammal fauna. Aust Zool Rev 1:15–24

Flannery TF (1989) Phylogeny of the Macropodoidea; a study in convergence. In: Grigg GJ, Jarman PJ, Hume ID (eds) Kangaroos, wallabies and rat-kangaroos, vol 1. Surrey Beatty and Sons, Sydney, pp 1–46

Flannery TF (1990) Pleistocene faunal loss: implications of the aftershock for Australia's past and future. Archaeol Oceania 25:45–67

Flannery TF (1991) The mystery of the meganesian meat-eaters. Aust Nat Hist 3:722–729

Flannery TF (1992) New Pleistocene marsupials (Macropodidae, Diprotodontidae) from subalpine habitats in Irian Jaya, Indonesia. Alcheringa 16:321–331

Flannery TF (1994) The future eaters. An ecological history of the Australasian lands and people. Reed Books, Sydney

Flannery TF, Archer M (1983) Revision of the genus *Troposodon* Bartholomai (Macropodidae: Marsupialia). Alcheringa 7:263–279

Flannery TF, Archer M (1984) The macropodoids (Marsupialia) of the early Pliocene Bow Local Fauna, central eastern New South Wales. Aust Zool 21:357–383

Flannery TF, Archer M (1985) *Palorchestes* Owen, 1874. Large and small palorchestids. In: Rich PV, van Tets GF (eds) Kadimakara. Extinct vertebrates of Australia. Pioneer Design Studio, Lilydale, pp 234–239

Flannery TF, Archer M (1987a) *Strigocuscus reidi* and *Trichosurus dicksoni*, two new fossil phalangerids (Marsupialia: Phalangeridae) from the Miocene of northwestern Queensland. In: Archer M (ed) Possums and opossums: studies in evolution. Surrey Beatty and Sons Pty Ltd, Chipping Norton, pp 527–536

Flannery TF, Archer M (1987b) *Hypsiprymnodon bartholomaii* (Potoroidae: Marsupialia), a new species from the Miocene Dwornamor Local Fauna and a reassessment of the phylogenetic position of *H. moschatus*. In: Archer M (ed) Possums and opossums: studies in evolution. Surrey Beatty and Sons Pty Ltd, Chipping Norton, pp 749–758

Flannery TF, Archer M (1987c) *Bettongia moyesi*, a new and plesiomorphic kangaroo (Marsupialia: Potoroidae) from Miocene sediments of northwestern Queensland. In: Archer M (ed) Possums and opossums: studies in evolution. Surrey Beatty and Sons Pty Ltd, Chipping Norton, pp 759–767

Flannery TF, Hann LM (1984). A new macropodine genus and species (Marsupialia: Macropodidae) from the early Pleistocene of southwestern Victoria. Australian Mammal 7:193–204

Flannery TF, Plane MD (1986) A new late Pleistocene diprotodontid (Marsupialia) from Pureni, Southern Highlands Province, Papua New Guinea. Bur Miner Resour J Aust, Geol Geophys 10:65–76

Flannery TF, Rich TH (1986) Macropodoids from the middle Miocene Namba Formation, South Australia, and the homology of some dental structures in kangaroos. J Paleontol 60:418–447

Flannery TF, Szalay FS (1982) *Bohra paulae*, a new giant fossil tree kangaroo (Marsupialia: Macropodidae) from New South Wales, Australia. Aust Mammal 5:83–94

Flannery TF, Archer M, Plane MD (1983) Middle Miocene kangaroos (Macropodoidea: Marsupialia) from three localities in northern Australia, with a description of two new subfamilies. Bur Miner Res J Aust, Geol Geophys 7:287–302

Flannery TF, Turnbull WD, Rich TH, Lundelius EL Jr (1987) The phalangerids (Marsupialia: Phalangeridae) of the early Pliocene Hamilton Local Fauna, southwestern Victoria. In: Archer M (ed) Possums and opossums: studies in evolution. Surrey Beatty and Sons Pty Ltd, Chipping Norton, pp 537–546

Flannery TF, Hoch E, Aplin KP (1989) Macropodines from the Pliocene Otibanda Formation, Papua New Guinea. Alcheringa 13:145–152

Flannery TF, Rich TH, Turnbull WD, Lundelius EL (1992) The Macropodoidea (Marsupialia) of the early Pliocene Hamilton Local Fauna, Victoria, Australia. Fieldiana Geol 25: 1–37

Flood J (1995) Archaeology of the Dreamtime: the story of prehistoric Australia and its people. Angus and Robertson, Sydney

Frakes LA, McGowran B, Bowler JM (1987) Evolution of Australian environments. In: Dyne GR, Walton DW (eds) Fauna of Australia, vol 1A. Australian Government Publishing Service, Canberra, pp 1–16

Fujioka T, Chappell J, Honda M, Yatsevich I, Fifield K, Fabel D (2005) Global cooling initiated stony deserts in central Australia 2–4 Ma, dated by cosmogenic ^{21}Ne–^{10}Be. Geology 33:993–996

Furby J, Fullagar R, Dodson J, Prosser I (1993) The Cuddie Springs bone bed revisited, 1991. In: Smith MA, Spriggs M, Fankhauser BL (eds) Sahul in review: Pleistocene archaeology in Australia, New Guinea and Island Melanesia. Department of Prehistory, Research School of Pacific Studies, Australian National University, Canberra, pp 204–210

Gallagher SJ, Greenwood DR, Taylor D, Smith AJ, Wallace MW, Holdgate GR (2003) The Pliocene climatic and environmental evolution of southeastern Australia: evidence from the marine and terrestrial realm. Palaeogeogr Palaeoclimatol Palaeoecol 193:349–382

Garvey J (2010) Economic anatomy of the Bennett's wallaby (*Macropus rufogriseus*): implications for understanding human hunting strategies in late Pleistocene Tasmania. Quaternary Int 211:144–156

Gillespie AK (1999) Diversity and evolutionary relationships of marsupial lions. Aust Mammal 21:21–22, 34–45

Gillespie AK (2007) Diversity and systematics of marsupial lions from the Riversleigh World Heritage Area and the evolution of the Thylacoleonidae. Doctor of Philosophy, University of New South Wales, Sydney

Gingerich PD (2006) Environment and evolution through the Paleocene–Eocene thermal maximum. Trends Ecol Evol 21:246–253

Glauert L (1910) *Sthenurus occidentalis* (Glauert). Bull Geol Surv West Aust 36:53–70

Godthelp H, Archer M, Cifelli R, Hand SJ, Gilkeson CF (1992) Earliest known Australian Tertiary mammal fauna. Nature 356:514–516

Godthelp H, Wroe S, Archer M (1999) A new marsupial from the early Eocene Tingamarra Local Fauna of Murgon, Southeastern Queensland: a prototypical Australian marsupial? J Mamm Evol 6:289–313

Goin FJ, Candela AM, de Muizon C (2003) The affinities of *Roberthoffstetteria nationalgeographica* (Marsupialia) and the origin of the polydolopine molar pattern. J Vertebrate Paleontol 23:869–876

Goldfuss GA (1817) Die Säugethiere, Abbildungen nach der Natur, mit Beschreibungen von Schreber JCD (ed). Fortgesetzt von A. Goldfuss. 65e cahier

Gould J (1842) Descriptions of four new species of kangaroos. Proc Zool Soc London 1841:80–83

Greenwood DR, Christophel DC (2005) The origins and Tertiary history of Australian "tropical" rainforests. In: Bermingham E, Dick C, Moritz C (eds) Tropical rainforests: past, present, future. University of Chicago Press, Chicago, pp 336–373

Grün R, Wells RT, Eggins S, Spooner N, Aubert M, Brown L, Rhodes E, (2008) Electron spin resonance dating of South Australian megafauna sites. Aust J Earth Sci 55:917–935

Guerin G, Hill R (2006) Plant macrofossil evidence for the environment associated with the Riversleigh fauna. Aust J Bot 54:717–731

Haight JR, Murray PF (1981) The cranial endocast of the early Miocene marsupial, *Wynyardia bassiana*: an assessment of taxonomic relationships based upon comparisons with recent forms. Brain, Behav Evol 19:17–36

Hall R (2001) Cenozoic reconstructions of SE Asia and the SW Pacific: changing patterns of land and sea. In: Metcalfe I, Smith JMB, Morwood M, Davidson I (eds) Faunal and floral migrations and evolution in SE Asia–Australasia. Swets and Zeitlinger b. v., Lisse, pp 35–56

Hand SJ (1997) *Hipposideros bernardsigei*, a new hipposiderid (Microchiroptera) from the Miocene of Australia and a reconsideration of the monophyly of related species groups. Müchner geowissenschaftliche Abhandlungen, A 34:73–92

Hand S, Archer M (2005) A new hipposiderid genus (Microchiroptera) from an early Miocene bat community in Australia. Palaeontology 48:371–383

Hand S, Archer M, Rich T, Pledge N (1993) *Nimbadon*, a new genus and three species of Tertiary zygomaturines (Marsupialia, Diprotodontidae) from northern Australia, with a reassessment of *Neohelos*. Mem Queensland Mus 33:193–210

Hand SJ, Novacek MJ, Godthelp H, Archer M (1994) First Eocene bat from Australia. J Vertebrate Paleontol 14:375–381

Hardjasasmita HS (1985) Fosil diprotodontid: *Zygomaturus* Owen 1859 dari Nimboran, Irian Jaya. Pertemuan Ilmiah Arkeologi 3:999–1004

Hays JD, Imbrie J, Shackleton NJ (1976) Variations in the Earth's orbit: pacemaker of the Ice Ages. Science 194:1121–1132

Head L (1995) Meganesian barbecue: reply to George Seddon. Meanjin 54:702–709

Hesse PP (1994) The record of continental dust from Australia in Tasman Sea sediments. Quaternary Sci Rev 13:257–272

Hesse PP, McTainsh GH (2003) Australian dust deposits: modern processes and the Quaternary record. Quaternary Sci Rev 22:2007–2035

Hesse PP, Magee JW, van der Kaars S (2004) Late Quaternary climates of the Australian arid zone: a review. Quaternary Int 118–119:87–102

Hill RS (1992) *Nothofagus*: evolution from a southern perspective. Trends Ecol Evol 7:190–194

Hill RS (1994) The history of selected Australian taxa. In: Hill RS (ed) History of the Australian vegetation: Cretaceous to recent. Cambridge University Press, Cambridge, pp 390–419

Hocknull SA, Zhao J-X, FengY-X, Webb GE (2007) Responses of Quaternary rainforest vertebrates to climate change in Australia. Earth Planet Sci Lett 264:317–331

Holdaway R, Jacomb C (2000) Rapid extinction of the Moas (Aves: Dinornithiformes): model, test, and implications. Science 287:2250–2254

Hope JH, Wilkinson HE (1982) *Warendja wakefieldi*, a new genus of wombat (Marsupialia, Vombatidae) from Pleistocene sediments in McEacherns Cave, western Victoria. Mem Natl Mus Victoria 43:109–120

Horovitz I, Sánchez-Villagra MR (2003) A morphological analysis of marsupial mammal higher-level phylogenetic relationships. Cladistics 19:181–212

Horovitz I, Martin T, Bloch J, Ladevèze S, Kurz C, Sánchez-Villagra MR (2009) Cranial anatomy of the earliest marsupials and the origin of opossums. PLoS One 4(12):e8278

Horton DR (1980) A review of the extinction question: man, climate and megafauna. Archaeol Phys Anthropol Oceania 15:86–97

Horton D (1984) Red kangaroos: last of the Australian megafauna. In: Martin PS, Klein RG (eds) Quaternary extinctions: a prehistoric revolution. University of Arizona Press, Tucson, pp 639–680

Huxley TH (1862) On the premolar teeth of *Diprotodon* and on a new species of that genus. Q J Geol Soc London 18:422–427

International Commission on Stratigraphy (ICS) (2008) http://essayweb.net/geology/timeline/images/ICS.gif

Jansen JHF, Kuijpers A, Troelstra SR (1986) A mid-brunhes climatic event: long-term changes in global atmosphere and ocean circulation. Science 232:619–622

Johnson CN, Wroe S (2003) Causes of extinction of vertebrates during the Holocene of mainland Australia: arrival of the dingo, or human impact? The Holocene 13:941–948

Jones FW (1930) A re-examination of the skeletal characters of *Wynyardia bassiana*, an extinct Tasmanian marsupial. Pap Proc R Soc Tasmania 1930:96–115

Jones R (1968) The geographical background to the arrival of man in Australia and Tasmania. Archaeol Phys Anthropol Oceania 3:186–215

Kavanagh JR, Burk-Herrick A, Westerman M, Springer MS (2004) Relationships among families of Diprotodontia (Marsupialia) and the phylogenetic position of the autapomorphic honey possum (*Tarsipes rostratus*). J Mamm Evol 11:207–222

Kawahata H (2002) Shifts in oceanic and atmospheric boundaries in the Tasman Sea (Southwest Pacific) during the Late Pleistocene: evidence from organic carbon and lithogenic fluxes. Palaeogeogr Palaeoclimatol Palaeoecol 184:225–249

Kawamura H, Holbourn A, Kuhnt W (2006) Climate variability and land–ocean interactions in the Indo PacificWarm Pool: a 460-ka palynological and organic geochemical record from the Timor Sea. Mar Micropaleontol 59:1–14

Kear B (1998) Postcranial morphology and phylogenetics of Oligo-Miocene kangaroos (Marsupialia: Macropodoidea) from Riversleigh, northwestern Queensland. University of New South Wales, Sydney, NSW

Kear BP (2002) Phylogenetic implications of macropodid (Marsupialia: Macropodoidea) postcranial remains from Miocene deposits of Riversleigh, northwestern Queensland. Alcheringa: An Australas J Palaeontol 26:299–318

Kear BP, Cooke BN (2001) A review of macropodoid (Marsupialia) systematics with the inclusion of a new family. Assoc Australas Palaeontol Mem 25:83–102

Kear BP, Archer M, Flannery TF (2001) Bulungamayine (Marsupialia: Macropodidae) postcranial elements from the late Miocene of Riversleigh northwestern Queensland. Assoc Australas Palaeontol Mem 25:103–122

Kear BP, Cooke BN, Archer M, Flannery TF (2007) Implications of a new species of the Oligo-Miocene kangaroo (Marsupialia: Macropodoidea) *Nambaroo*, from the Riversleigh World Heritage Area, Queensland. J Paleontol 81:1147–1167

Kemp TS (2005) The Origin and evolution of mammals. Oxford University Press, Oxford

Kennett JP (1977) Cenozoic evolution of Antarctic glaciation, the Circum-Antarctic Ocean, and their impact on global paleoceanography. J Geophys Res 82:3843–3860

Kershaw AP, McKenzie GM, McMinn A (1993) A Quaternary vegetation history of northeastern Queensland from pollen analysis of ODP site 820. Proc Ocean Drilling Program Sci Results 133:107–114

Kershaw AP, van der Kaars S, Moss PT (2003a) Late Quaternary Milankovitch-scale climate change and variability and its impact on monsoonal Australasia. Mar Geol 201:81–85

Kershaw P, Moss P, van der Kaars S (2003b) Causes and consequences of long-term climatic variability on the Australian continent. Freshwater Biol 48:1274–1283

Kidwell SM, Holland SM (2002) Quality of the fossil record: implications for evolutionary biology. Annu Rev Ecol Syst 33:561–588

Kiernan K, Jones R, Ranson D (1983) New evidence from Fraser cave for glacial age man in Southwest Tasmania. Nature 301:28–32

Kirsch JAW (1977a) The comparative serology of Marsupialia, and a classification of marsupials. Aust J Zool, Suppl Ser 52:1–152

Kirsch JAW (1977b) The six-percent solution: second thoughts on the adaptedness of the Marsupialia. Am Sci 65:276–288

Kirsch J, Lapointe F, Springer M (1997) DNA-hybridisation studies of marsupials and their implications for metatherian classification. Aust J Zool 45:211–280

Krajewski C, Blacket M, Buckley L, Westerman M (1997) A multigene assessment of phylogenetic relationships within the dasyurid marsupial subfamily Sminthopsinae. Mol Phylogen Evol 8:236–248

Krajewski C, Blacket MJ, Westerman M (2000a) DNA sequence analysis of familial relationships among dasyuromorphian marsupials. J Mamm Evol 7:95–108

Krajewski C, Wroe S, Westerman M (2000b) Molecular evidence for the pattern and timing of cladogenesis in dasyurid marsupials. Zool J Linnean Soc 130:375–404

Lambeck K, Chappell J (2001) Sea level change through the last glacial cycle. Science 292:679–686

Lawver LA, Gahagan LM (2003) Evolution of Cenozoic seaways in the circum-Antarctic region. Palaeogeogr Palaeoclimatol Palaeoecol 198:11–37

Lepage T, Bryant D, Philippe H, Lartillot N (2007) A general comparison of relaxed molecular clock models. Mol Biol Evol 24:2669–2680

Lindsay, JM (1987) Age and habitat of a monospecific foraminiferal fauna from near-type Etadunna Formation, Lake Palankarinna, Lake Eyre Basin. Department of Mines and Energy South Australia Report Book 87/93 (unpublished)

Long J, Archer M, Flannery T, Hand S (2002) Prehistoric mammals of Australia and New Guinea: one hundred million years of evolution. University of New South Wales Press, Sydney, NSW

Louys J, Black K, Archer M, Hand SJ, Godthelp H (2007a) Descriptions of koala fossils from the Miocene of Riversleigh, northwestern Queensland and implications for *Litokoala* (Marsupialia, Phascolarctidae). Alcheringa: An Australasian J Palaeontol 31:99–110

Louys J, Curnoe D, Tong H (2007b) Characteristics of Pleistocene megafauna extinctions in Southeast Asia. Palaeogeogr, Palaeoclimatol, Palaeoecol 243:152–173

Louys J, Aplin K, Beck RMD, Archer M (2009) Cranial anatomy of Oligo-Miocene koalas (Diprotodontia: Phascolarctidae): stages in the evolution of an extreme leaf-eating specialization. J Vertebrate Paleontol 29:981–992

Lundelius EL Jr, Graham R (1999) The weather changed: shifting climate dissolved ancient animal alliances. Discovering Archaeol 1:48–50, 53

Luo Z-X, Ji Q, Wible JR, Yuan C-X (2003) An Early Cretaceous tribosphenic mammal and metatherian evolution. Science 302:1934–1940

Luo Z-X, Chen P, Li G, Chen M (2007) A new eutriconodont mammal and evolutionary development in early mammals. Nature 446:288–293

Lynch AH, Abramson D, Görgen K, Beringer J, Uotila P (2007) Influence of savanna fire on Australian monsoon season precipitation and circulation as simulated using a distributed computing environment. Geophys Res Lett 34:L20801

Mackness BS (1995) *Palorchestes selestiae*, a new species of palorchestid marsupial from the early Pliocene Bluff Downs Local Fauna, northeastern Queensland. Mem Queensland Mus 38:603–609

Mackness BS, Archer M (2001) A new petauroid possum (Marsupialia, Pseudocheiridae) from the Pliocene Bluff Downs Local Fauna, northern Queensland. Alcheringa 25:439–444

Mackness BS, Whitehead P, McNamara G (2000) A new potassium-argon date in relation to the Pliocene Bluff Downs Local Fauna, northern Australia. Australian J Earth Sci 47:807–811

Macphail MK (1996) A habitat for the enigmatic *Wynyardia bassiana* Spencer, 1901 Australia's first described Tertiary land mammal? Alcheringa: An Australas J Palaeontol 20:227–243

Marcus LF (1962) A new species of *Sthenurus* (Marsupialia, Macropodidae) from the Pleistocene of New South Wales. Rec Aust Mus 15:299–304

Marshall LG, Corruccini RS (1978) Variability, evolutionary rates, and allometry in dwarfing lineages. Paleobiology 4:101–119

Marshall LG, de Muizon C (1988) The dawn of the age of mammals in South America. Natl Geogr Res 4:23–55

Marshall LG, Case JA, Woodburne MO (1990) Phylogenetic relationships of the families of marsupials. In: Genoways HH (ed) Current mammalogy, vol 2. Plenum Press, New York, pp 433–502

Martin HA (1990) The palynology of the Namba Formation in the Wooltana-1 bore, Callabonna Basin (Lake Frome), South Australia, and its relevance to Miocene grasslands in central Australia. Alcheringa: An Australas J Palaeontol 14:247–255

Martin HA (1994) Australian Tertiary phytogeography. In: Hill R (ed) History of the Australian vegetation. Cambridge University Press, Cambridge, pp 104–142

Martin HA (1998) Tertiary climatic evolution and the development of aridity in Australia. Proc Linnean Soc New South Wales 119:115–136

Martin HA (2006) Cenozoic climatic change and the development of the arid vegetation in Australia. J Arid Environ 66:533–563

McGowran B, Li Q (1994) The Miocene oscillation in southern Australia. Rec South Aust Mus 27:197–212

McGowran B, Li Q, Cann J, Padley D, McKirdy DM, Shafik S (1997) Biogeographic impact of the Leeuwin Current in southern Australia since the late middle Eocene. Palaeogeogr Palaeoclimatol Palaeoecol 136:19–40

McGowran B, Archer M, Bock P, Darragh TA, Godthelp H, Hageman S, Hand SJ, Hill R, Li Q, Maxwell PA, McNamara KJ, Macphail M, Mildenhall D, Partridge AD, Richardson J, Shafik S, Truswell EM, Warne M (2000) Australasian palaeobiogeography: the Palaeogene and Neogene record. Mem Assoc Australas Palaeontol 23:405–470

McNamara JA (1994) A new fossil wallaby (Marsupialia: Macropodidae) from the south east of South Australia. Rec South Aust Mus 27:111–115

Megirian D (1992) Interpretation of the Miocene Carl Creek Limestone, northwestern Queensland. The Beagle, Records Northern Territory Mus Arts Sci 9:219–248

Megirian D, Murray P, Wells R (1996) The late Miocene Ongeva Local Fauna of central Australia. The Beagle, Records Northern Territory Mus Arts Sci 13:9–38

Megirian D, Murray P, Schwartz L, Borch CVD (2004) Late Oligocene Kangaroo Well Local Fauna from the Ulta Limestone (new name), and climate of the Miocene oscillation across central Australia. Aust J Earth Sci 51: 701–741

Meredith R, Westerman M, Case J, Springer M (2008a) A phylogeny and timescale for marsupial evolution based on sequences for five nuclear genes. J Mammalian Evol 15:1–36

Meredith RW, Westerman M, Springer MS (2008b) A phylogeny and timescale for the living genera of kangaroos and kin (Macropodiformes: Marsupialia) based on nuclear DNA sequences. Australian J Zool 56:395–410

Meredith RW, Westerman M, Springer MS (2008c) A timescale and phylogeny for "Bandicoots" (Peramelemorphia: Marsupialia) based on sequences for five nuclear genes. Mole Phylogen Evol 47:1–20

Meredith RW, Westerman M, Springer MS (2009) A phylogeny of Diprotodontia (Marsupialia) based on sequences for five nuclear genes. Mol Phylogen Evol 51:554–571

Meredith R, Mendoza M, Roberts K, Westerman M, Springer M (2010) A phylogeny and timescale for the evolution of Pseudocheiridae (Marsupialia: Diprotodontia) in Australia and New Guinea. J Mamm Evol 17:75–99

Meredith RW, Krajewski C, Westerman M, Springer MS (2011) Relationships and divergence times among the orders and families of Marsupialia. Mus Northern Arizona, Bul 65: 383–406

Merrilees D (1968) Man the destroyer: late Quaternary changes in Australian marsupial fauna. J R Soc West Aust 51:1–24

Miller GH, Magee JW, Johnson BJ, Fogel ML, Spooner NA, McCulloch MT, Ayliffe LK (1999) Pleistocene extinction of *Genyornis newtoni*: human impact on Australian megafauna. Science 283:205–208

Miller GH, Fogel ML, Magee JW, Gagan MK, Clarke SJ, Johnson BJ (2005) Ecosystem collapse in Pleistocene Australia and a human role in megafaunal extinction. Science 309:287–290

Moriarty KC, McCulloch M, Wells RT, McDowell MC (2000) Mid-Pleistocene cave fills, megafaunal remains and climate change at Naracoorte, South Australia: towards a predictive model using U/Th dating of speleothems. Palaeogeogr Palaeoclimatol Palaeoecol 159:113–143

Morwood MJ, O'Sullivan PB, Aziz F, Raza A (1998) Fission–track ages of stone tools and fossils on the east Indonesian island of Flores. Nature 392:173–176

Moss PT, Kershaw AP (2000) The last glacial cycle from the humid tropics of northeastern Australia: comparison of a terrestrial and a marine record. Palaeogeogr Palaeoclimatol Palaeoecol 155:155–176

Moss PT, Kershaw AP (2007) A late Quaternary marine palynological record (Oxygen Isotope Stages 1 to 7) for the Humid Tropics of northeastern Australia based on ODP Site 820. Palaeogeogr, Palaeoclimatol, Palaeoecol 251:4–22

Muirhead J (1992) A specialised thylacinid, *Thylacinus macknessi* (Marsupialia: Thylacinidae) from Miocene deposits of Riversleigh, northwestern Queensland. Aust Mammal 15:67–76

Muirhead J (1994) Systematics, evolution, and palaeobiology of recent and fossil bandicoots (Peramelemorphia, Marsupialia). Doctor of Philosophy, University of New South Wales, Sydney

Muirhead J (1997) Two new early Miocene thylacines from Riversleigh, northwestern Queensland. Mem Queensland Mus 41:367–377

Muirhead J (2000) Yaraloidea (Marsupialia, Peramelemorphia), a new superfamily of marsupial and a description and analysis of the cranium of the Miocene *Yarala Burfieldi*. J Paleontol 74:512–523

Muirhead J, Archer M (1990) *Nimbacinus dicksoni*, a plesiomorphic thylacine (Marsupialia: Thylacinidae) from Tertiary deposits of Queensland and the Northern Territory. Mem Queensland Mus 28:203–221

Muirhead J, Filan S (1995) *Yarala burchfieldi*, a plesiomorphic bandicoot (Marsupialia, Peramelemorphia) from Oligo-Miocene deposits of Riversleigh, northwestern Queensland. J Paleontol 69:127–134

Muirhead J, Godthelp HJ (1996) Fossil bandicoots of Chillagoe (northeastern Queensland) and the first known specimens of the Pig–footed Bandicoot *Chaeropus* Ogilby, 1838 from Queensland. Aust Mammal 19:73–76

Muirhead J, Wroe S (1998) A new genus and species, *Badjjcinus turnbulli* (Thylacinidae: Marsupialia), from the late Oligocene of Riversleigh, northern Australia, and an investigation of thylacinid phylogeny. J Vertebrate Paleontol 18:612–626

Muirhead J, Dawson L, Archer M (1997) *Perameles bowensis*, a new species of *Perameles* (Peramelemorphia, Marsupialia) from Pliocene faunas of Bow and Wellington Caves, New South Wales. Proc Linnean Soc New South Wales 117:163–173

Munson CJ (1992) Postcranial descriptions of *Ilaria* and *Ngapakaldia* (Vombatiformes, Marsupialia) and the phylogeny of the vombatiforms based on postcranial morphology, vol 125. University of California Publications in Zoology. University of California Press, Berkeley, CA

Murray P (1986) *Propalorchestes novaculacephalus* gen. et sp. nov., a new palorchestid (Diprotodontoidea: Marsupialia) from the middle Miocene Camfield Beds, Northern Territory, Australia. The Beagle, Records Northern Territory Mus Arts Sci 3:195–211

Murray P (1990a) Primitive marsupial tapirs (*Propalorchestes novaculacephalus* Murray and *P. ponticulus* [Marsupialia: Palorchestidae] sp. nov.) from the mid-Miocene of north Australia. The Beagle, Records Northern Territory Mus Arts Sci 7:39–51

Murray R (1990b) *Alkwertatherium webbi*, a new zygomaturine genus and species from the late Miocene Alcoota Local Fauna, Northern Territory (Marsupialia: Diprotodontidae). The Beagle, Records Northern Territory Mus Arts Sci 7:53–80

Murray P (1991) The sthenurine affinity of the late Miocene kangaroo, *Hadronomas puckridgi* Woodburne (Marsupialia, Macropodidae). Alcheringa: An Australas J Palaeontol 15:255–283

Murray PF (1992) The smallest New Guinea zygomaturines—derived dwarfs or relict plesiomorphs? The Beagle, Records Northern Territory Mus Arts Sci 9:89–110

Murray PF (1995) The postcranial skeleton of the Miocene kangaroo, *Hadronomas puckridgi* Woodburne (Marsupialia, Macropodidae). Alcheringa: An Australas J Palaeontol 19:119–170

Murray P (1997a) Alcoota: a snapshot of the Australian late Miocene. Riversleigh Notes 35:2–7

Murray PF (1997b) *Thylacinus megiriani*, a new species of thylacine (Marsupialia: Thylacinidae) from the Ongeva Local Fauna of central Australia. Rec South Aust Mus 30:43–61

Murray P (1998) Palaeontology and palaeobiology of wombats. In: Wells R, Pridmore PA (eds) Wombats. Surrey Beatty and Sons, Sydney, pp 1–33

Murray P, Megirian D (1990) Further observations on the morphology of *Wakaleo vanderleuri* (Marsupialia: Thylacoleonidae) from the mid-Miocene Camfield Beds, Northern Territory. The Beagle, Records Northern Territory Mus Arts Sci 7:91–102

Murray P, Megirian D (1992) Continuity and contrast in middle and late Miocene vertebrate communities from the Northern Territory. The Beagle, Records Northern Territory Mus Arts Sci 9:195–218

Murray P, Megirian D (2000) Two new genera and three new species of Thylacinidae (Marsupialia) from the Miocene of the Northern Territory, Australia. The Beagle, Records Northern Territory Mus Arts Sci 16:145–162

Murray PF, Megirian D (2006a) The Pwerte Marnte Marnte Local Fauna: a new vertebrate assemblage of presumed Oligocene age from the Northern Territory of Australia. Alcheringa Special Issue 1:211–228

Murray PF, Megirian D (2006b) Cranial morphology of the Miocene thylacinid *Mutpuracinus archibaldi* (Thylacinidae, Marsupialia) and relationships within the Dasyuromorphia. Alcheringa Special Issue 1:229–276

Murray PF, Vickers-Rich P (2004) Magnificent Mihirungs: the colossal flightless birds of the Australian dreamtime. Indiana University Press, Bloomington, IN

Murray P, Megirian D, Wells R (1993) *Kolopsis yperus* sp. nov. (Zygomaturinae, Marsupialia) from the Ongeva Local Fauna: new evidence for the age of the Alcoota fossil beds of central Australia. The Beagle, Records Northern Territory Mus Arts Sci 10:155–172

Murray P, Megirian D, Rich T, Plane M, Black K, Archer M, Hand S, Vickers-Rich P (2000a) Morphology, systematics, and evolution of the marsupial genus *Neohelos* Stirton (Diprotodontidae, Zygomaturinae). Museums and Art Galleries of the Northern Territory Research Report No. 6, pp 1–141

Murray P, Megirian D, Rich T, Plane M, Vickers-Rich P (2000b) *Neohelos stirtoni*, a new species of Zygomaturinae (Diprotodontidae: Marsupialia) from the mid-Tertiary of northern Australia. Mem Queen Victoria Mus 105:1–47

Myers T, Archer M (1997) *Kuterintja ngama* (Marsupialia, Ilariidae): a revised systematic analysis based on material from the late Oligocene of Riversleigh, northwestern Queensland. Mem Queensland Mus 41:379–392

Myers T, Archer M, Krikmann A, Pledge N (1999) Diversity and evolutionary relationships of ilariids, wynyardiids, vombatids and related groups of marsupials. Aust Mammal 21:18–19, 34–45

Myers T, Crosby K, Archer M, Tyler M (2001) The Encore Local Fauna, a late Miocene assemblage from Riversleigh, northwestern Queensland. Mem Assoc Australas Palaeontol 25:147–154

Nilsson MA, Arnason U, Spencer PBS, Janke A (2004) Marsupial relationships and a timeline for marsupial radiation in South Gondwana. Gene 340:189–196

Nogués-Bravo D, Ohlemüller R, Batra P, Araú MB (2010) Climate predictors of late Quaternary extinctions. Evolution 64:2442–2449

O'Connell JF, Allen J (2004) Dating the colonization of Sahul (Pleistocene Australia–New Guinea): a review of recent research. J Archaeol Sci 31:835–853

Osborne MJ, Christidis L (2002) Systematics and biogeography of pygmy possums (Burramyidae:*Cercartetus*). Aust J Zool 50:25–37

Owen R (1838) Fossil remains from Wellington Valley, Australia. Marsupialia. In: Appendix to Mitchell TL (ed) Three expeditions into the interior of eastern Australia, with descriptions of the recently explored region of Australia felix, and of the present colony of New South Wales, vol 2. T. and W. Boone, London, pp 359–369

Owen R (1845) Descriptive and illustrated catalogue of the Fossil organic remains of Mammalia and Aves contained in the museum of the Royal College of Surgeons of England. Taylor, London

Owen R (1858) Odontology. Teeth of mammals. Encyclopaedia Britannica, or dictionary of arts, sciences, and general literature, 8th edn, vol 16, Encyclopaedia Britannica Inc, Edinburgh, p447

Owen R (1859) On the classification and geographical distribution of the Mammalia. Lecture on Sir Robert Reade's Foundation, delivered before the University of Cambridge in the Senate-House, May 10, 1859. London, John W. Parker and Son, 103pp

Owen R (1872) On the fossil mammals of Australia. Part VII. Genus *Phascolomys*. Philos Trans R Soc 162:172–196, 241–258

Owen R (1874) On the fossil mammals of Australia. Part IX. Family Macropodidae. Genera *Macropus*, *Pachysiagon*, *Leptosiagon*, *Procoptodon* and *Palorchestes*. Philos Trans R Soc 164:783–803

Owen R (1877) On a new species of *Sthenurus*, with remarks on the relation of the genus to *Dorcopsis*, Müller. Proc Sci Meetings Zool Soc London 1877:352–361

Phillips MJ, Pratt RC (2008) Family-level relationships among the Australasian marsupial "herbivores" (Diprotodontia: koala, wombats, kangaroos and possums). Mol Phylogen Evol 46:594–605

Phillips MJ, McLenachan PA, Down C, Gibb GC, Penny D (2006) Combined mitochondrial and nuclear DNA sequences resolve the interrelations of the major Australasian marsupial radiations. Syst Biol 55:122–137

Pillans B (2003) Subdividing the Pleistocene using the Matuyama–Brunhes boundary (MBB): an Australasian perspective. Quaternary Sci Rev 22:1569–1577

Pillans B, Bourman R (2001) Mid Pleistocene arid shift in southern Australia, dated by magnetostratigraphy. Aust J Soil Res 39:89–98

Piper K (2006) A new species of Palorchestidae (Marsupialia) from the Pliocene and early Pleistocene of Victoria. Alcheringa: An Australas J Palaeontol 30(Issue S1): 281–294

Piper KJ, Fitzgerald EMG, Rich TH (2006) Mesozoic to early Quaternary mammal faunas of Victoria, south-east Australia. Palaeontology 49:1237–1262

Plane MD (1967) Two new diprotodontids from the Pliocene Otibanda Formation, New Guinea. Bull Bureau Mineral Resour Geol Geophys Aust 85:105–128

Pledge NS (1977) A new species of *Thylacoleo* (Marsupialia: Thylacoleonidae) with notes on the occurrences and distribution of Thylacoleonidae in South Australia. Rec South Aust Mus 17:277–283

Pledge NS (1982) Enigmatic *Ektopodon*: a case history of palaeontological interpretation. In: Rich PV, Thompson EM (eds) The fossil vertebrates of Australasia. Monash University Offset Printing Unit, Clayton, pp 477–488

Pledge N (1986) A new species of *Ektopodon* (Marsupialia; Phalangeroidea) from the Miocene of South Australia. University of California Publications in Geological Sciences, vol 131, pp 43–67

Pledge N (1987a) *Muramura williamsi*, a new genus and species of ?wynyardiid (Marsupialia: Vombatoidea) from the middle Miocene Etadunna Formation of South Australia. In: Archer M (ed) Possums and opossums: studies in evolution, Surrey Beatty and Sons Pty Ltd, Chipping Norton, pp 393–400

Pledge NS (1987b) *Kuterintja ngama*, a new genus and species of primitive vombatoid marsupial from the medial Miocene Ngama Local Fauna of South Australia. In: Archer M (ed) Possums and opossums: studies in evolution. Surrey Beatty and Sons Pty Ltd, Chipping Norton, pp 419–422

Pledge NS (1987c) A new species of *Burramys* Broom (Marsupialia: Burramyidae) from the middle Miocene of South Australia. In: Archer M (ed) Possums and opossums: studies in evolution. Surrey Beatty and Sons Pty Ltd, Chipping Norton, pp 725–728

Pledge NS (1987d) *Phascolarctos maris*, a new species of koala (Marsupialia: Phascolarctidae) from the early Pliocene of South Australia. In: Archer M (ed) Possums and opossums: studies in evolution. Surrey Beatty and Sons Pty Ltd, Chipping Norton, pp 327–330

Pledge NS (1992) The Curramulka local fauna: a new late Tertiary fossil assemblage from Yorke Peninsula, South Australia. The Beagle, Records Northern Territory Mus Arts Sci 9:115–142

Pledge NS (2003) A new species of *Muramura* Pledge (Wynyardiidae: Marsupialia) from the middle Tertiary of the Callabonna Basin, northeastern South Australia. Bull Am Mus Nat Hist 279:541–555

Pledge N (2005) The Riversleigh wynyardiids. Mem Queensland Mus 51:135–169

Pledge N, Archer M, Hand SJ, Godthelp H (1999) Additions to knowledge about ektopodontids (Marsupialia: Ektopodontidae): including a new species *E. litolophus*. Rec West Aust Mus, Suppl 57:255–264

Price GJ (2008) Is the modern koala (*Phascolarctos cinereus*) a derived dwarf of a Pleistocene giant? Implications for testing megafauna extinction hypotheses. Quaternary Sci Rev 27:2516–2521

Price GJ, Hocknull SA (2011) *Invictokoala monticola* gen. et sp. nov. (Phascolarctidae, Marsupialia), a Pleistocene koala holdover from Oligocene ancestors. J Syst Palaeontol 9:327–335

Price GJ, Sobbe IH (2005) Pleistocene palaeoecology and environmental change on the Darling Downs, southeastern Queensland, Australia. Mem Queensland Mus 51:171–201

Price GJ, Zhao J-X, Feng Y-X, Hocknull SA (2009) New U/Th ages for Pleistocene megafauna deposits of southeastern Queensland, Australia. J Asian Earth Sci 34:190–197

Prideaux GJ (2004) Systematics and evolution of the sthenurine kangaroos. UC Publications in Geological Sciences, University of California Publications in Geological Sciences, vol 146, pp 1–622

Prideaux GJ, Warburton NM (2010) An osteology based appraisal of the phylogeny and evolution of kangaroos and wallabies (Macropodidae, Marsupialia). Zool J Linnean Soc 159:954–987

Prideaux GJ, Long JA, Ayliffe LK, Hellstrom JC, Pillans B, Boles WE, Hutchinson MN, Roberts RG, Cupper ML, Arnold LJ, Devine PD, Warton NM (2007a) An arid-adapted middle Pleistocene vertebrate fauna from south-central Australia. Nature 445:422–425

Prideaux GJ, Roberts RG, Megirian D, Westaway KE, Hellstrom JG, Olley JM (2007b) Mammalian responses to Pleistocene climate change in southeastern Australia. Geology 35: 33–36

Rauscher B (1987) *Priscilieo pitikantensis*, a new genus and species of thylacoleonid marsupial (Marsupialia: Thylacoleonidae) from the Miocene Etadunna Formations, South Australia. In: Archer M (ed) Possums and opossums: studies in evolution. Surrey Beatty and Sons Pty Ltd, Chipping Norton, pp 423–432

Reguero MA, Marenssi SA, Santillana SN (2002) Antarctic Peninsula and South America (Patagonia) Paleogene terrestrial faunas and environments: biogeographic relationships. Palaeogeogr Palaeoclimatol Palaeoecol 179:189–210

Rhodes E, Chappell J, Fujioka T, Fitzsimmons K, Magee J, Aubert M, Hewitt D (2005) The history of aridity in Australia: chronological developments. In: Roach IC (ed) Regolith 2005 – ten years of CRC LEME. Cooperative Research Centre for Landscape Environments and Mineral Exploration, Bentley, WA, pp 265–268

Rich TH (1983) The largest marsupial. *Diprotodon optatum*. In: Quirk S, Archer M (eds) Prehistoric animals of Australia. Australian Museum, Sydney, pp 62–63

Rich TH (1986) *Darcius duggani*, a new ektopodontid (Marsupialia; Phalangeroidea) from the early Pliocene Hamilton Local Fauna, Australia. University of California Publications in Geological Sciences, Berkeley, vol 131, pp 68–74

Rich TH, Archer M (1979) *Namilamadeta snideri*, a new diprotodontan (Marsupialia, Vombatoidea) from the medial Miocene of South Australia. Alcheringa: An Australas J Palaeontol 3:197–208

Rich TH, Rich PV (1987) New specimens of *Ngapakaldia* (Marsupialia: Diprotodontoidea) and taxonomic diversity in medial Miocene palorchestids. In: Archer M (ed) Possums and opossums: studies in evolution. Surrey Beatty and Sons Pty Ltd, Chipping Norton, pp 467–476

Rich TH, Archer M, Tedford RH (1978) *Raemeotherium yatkolai*, gen. et sp. nov., a primitive diprotodontid from the medial Miocene of South Australia. Mem Natl Mus Victoria 39:85–91

Rich TH, Archer M, Hand SJ, Godthelp HJ, Muirhead J, Pledge NS, Flannery TF, Woodburne MO, Case JA, Tedford RH, Turnbull WD, Lundelius ELJ, Rich LSV, Whitelaw MJ, Kemp A, Rich PV (1991) Australian Mesozoic and Tertiary terrestrial mammal localities. In: Vickers-Rich P, Monaghan JM, Baird RF, Rich TH (eds) Vertebrate palaeontology of Australasia. Pioneer Design Studio and Monash University Publications Committee, Melbourne, pp 1005–1058

Rich TH, Darragh TA, Vickers-Rich P (2003) The strange case of the wandering fossil. Bull Am Mus Nat Hist 279: 556–567

Rich TH, Piper KJ, Pickering D, Wright S (2006) Further Ektopodontidae (Phalangeroidea, Mammalia) from south-western Victoria. Alcheringa: An Australas J Palaeontol 30:133–140

Ride WDL (1964) A review of Australian fossil marsupials. J R Soc West Aust 47:97–131

Ride WDL (1993) *Jackmahoneya* gen. nov. and the genesis of the macropodiform molar. Mem Assoc Australas Palaeontol 15:441–459

Ride WDL, Pridmore PA, Barwick RE, Wells RT, Heady RD (1997) Towards a biology of *Propleopus oscillans* (Marsupialia: Propleopinae, Hypsiprymnodontidae). Proc Linnean Soc New South Wales 117:243–328

Roberts RG, Jones R, Spooner NA, Head MJ, Murray AS, Smith MA (1994) The human colonisation of Australia: optical dates of 53,000 and 60,000 years bracket human arrival at Deaf Adder Gorge, Northern Territory. Quaternary Sci Rev 13:575–583

Roberts RG, Flannery TF, Ayliffe LK, Yoshida H, Olley JM, Prideaux GJ, Laslett GM, Baynes A, Smith MA, Jones R, Smith BL (2001) New ages for the last Australian Megafauna: continent-wide extinction about 46,000 years ago. Science 292:1888–1892

Roberts KK, Bassarova M, Archer M (2008) Oligo-Miocene ringtail possums of the genus *Paljara* (Pseudocheiridae: Marsupialia) from Queensland, Australia. Geobios 41:833–844

Roberts KK, Archer M, Hand SJ, Godthelp H (2007) New genus and species of extinct Miocene ringtail possums (Marsupialia, Pseudocheiridae). Am Mus Novitates 3560: 1–15

Roberts KK, Archer M, Hand SJ, Godthelp H (2009) New Australian Oligocene to Miocene ringtail possums (Pseudocheiridae) and revision of the genus *Marlu*. Palaeontology 52:441–456

Rose KD (2006) The beginning of the age of mammals. John Hopkins University Press, Baltimore

Rougier GW, Wible JR, Novacek MJ (1998) Implications of *Deltatheridium* specimens for early marsupial history. Nature 396:459–463

Salisbury SW, Willis PMA (1996) A new crocodylian from the early Eocene of southeastern Queensland and a preliminary investigation into the phylogenetic relationships of crocodyloids. Alcheringa: An Australas J Palaeontol 20:179–226

Sánchez-Villagra M, Ladevèze S, Horovitz I, Argot C, Hooker JJ, Macrini TE, Martin T, Moore-Fay S, de Muizon C, Schmelzle T, Asher RJ (2007) Exceptionally preserved North American Paleogene metatherians: adaptations and discovery of a major gap in the opossum fossil record. Biol Lett 3:318–322

Scanlon JD (2005) Australia's oldest known snakes: *Patagoniophis, Alamitophis*, and cf. *Madtsoia* (Squamata: Madtsoiidae) from the Eocene of Queensland. Mem Queensland Mus 51:215–235

Scher HD, Martin EE (2006) Timing and climatic consequences of the opening of Drake Passage. Science 312:428–430

Schwartz L (2006a) A new species of bandicoot from the Oligocene of northern Australia and implications of bandicoots for correlating Australian Tertiary mammal faunas. Palaeontology 49:991–998

Schwartz LRS (2006b) Miralinidae (Marsupialia: Phalangeroidea) from northern Australia, including the youngest occurrence of the family. Alcheringa: An Australasian J Palaeontol 30:343–350

Schwartz LRS, Megirian D (2004) A New Species of *Nambaroo* (Marsupialia; Macropodoidea) from the Miocene Camfield Beds of Northern Australia with observations on the phylogeny of the Balbarinae. J Vertebrate Paleontol 24: 668–675

Sigé B, Sempere T, Butler Robert F, Marshall LG, Crochet J-Y (2004) Age and stratigraphic reassessment of the fossil-bearing Laguna Umayo red mudstone unit, SE Peru, from regional stratigraphy, fossil record, and paleomagnetism. Geobios 37:771–794

Sigé B, Archer M, Crochet J-Y, Godthelp H, Hand S, Beck R (2009) *Chulpasia* and *Thylacotinga*, late Paleocene–earliest Eocene trans-Antarctic Gondwanan bunodont marsupials: new data from Australia. Geobios 42:813–823

Singh G, Kershaw AP, Clark R (1981) Quaternary vegetation and fire history in Australia. In: Gill AM, Groves RA, Noble JC (eds) Fire and the Australian biota. Australian Academy of Science, Canberra, pp 23–54

Spencer WB (1901) A description of *Wynyardia bassiana*, a fossil marsupial from the Tertiary beds of Table Cape, Tasmania. Proc Zool Soc London 1900:776–794

Springer MS (1987) Lower molars of *Litokoala* (Marsupialia: Phascolarctidae) and their bearing on phascolarctid evolution. In: Archer M (ed) Possums and opossums: studies in evolution. Surrey Beatty and Sons Pty Ltd, Chipping Norton, pp 319–325

Springer MS, Westerman M, Kavanagh JR, Burk A, Woodburne MO, Kao DJ, Krajewski C (1998) The origin of the Australasian marsupial fauna and the phylogenetic affinities of the enigmatic monito del monte and marsupial mole. Proc R Soc B 265:2381–2386

Springer MS, Krajewski C, Meredith RW (2009) Metatheria. In: Hedges SB, Kumar S (eds) The timetree of life. Oxford University Press, New York, pp 466–470

Steffen W, Burbidge A, Hughes L, Kitching R, Lindenmayer D, Musgrave W, Smith MS, Werner PA (2010) Australia's biodiversity and climate change. CSIRO Publishing, Collingwood

Stephenson AE (1986) Lake Bungunnia—a Plio-Pleistocene megalake in southern Australia. Palaeogeogr Palaeoclimatol Palaeoecol 57:137–156

Stirton RA (1955) Late Tertiary marsupials from South Australia. Rec South Aust Mus 11:247–268

Stirton RA (1957a) Tertiary marsupials from Victoria, Australia. Mem Natl Mus Victoria 21:121–134

Stirton RA (1957b) A new koala from the Pliocene Palankarinna fauna of South Australia. Rec South Aust Mus 13: 71–81

Stirton R (1967a) The Diprotodontidae from the Ngapakaldi Fauna, South Australia. Bull Bureau Mineral Resour, Geol Geophys Aust 85:1–44

Stirton R (1967b) A diprotodontid from the Miocene Kutjamarpu Fauna, South Australia. Bull Bureau Mineral Resour, Geol Geophys Aust 85:45–51

Stirton RA (1967c) New species of *Zygomaturus* and additional observations on *Meniscolophus*, Pliocene Palankarinna

Fauna, South Australia. Bull Bureau Mineral Resour, Geol Geophys 85:129–147

Stirton RA, Tedford RH, Woodburn MO (1967) A new Tertiary formation and fauna from the Tirari Desert, South Australia. Rec South Aust Mus 15:427–462

Strahan R (ed) (1995) The mammals of Australia. Reed Books, Sydney

Szalay FS (1982) A new appraisal of marsupial phylogeny and classification. In: Archer M (ed) Carnivorous marsupials, vol 2. Royal Zoological Society of NSW, Sydney, pp 621–640

Szalay F (1994) Evolutionary history of the marsupials and an analysis of osteological characters. Cambridge University Press, Cambridge

Tate GHH (1951) The wombats (Marsupialia, Phascolomyidae). Am Mus Novitates 1521:1–18

Tedford RH, Kemp NR (1998) Oligocene marsupials of the Geilston Bay Local Fauna, Tasmania. Am Mus Novitates 3244:1–22

Tedford RH, Woodburne MO (1987) The Ilariidae, a new family of vombatiform marsupials from Miocene strata of South Australia and an evaluation of the homology of molar cusps in the Diprotodontia. In: Archer M (ed) Possums and opossums: studies in evolution. Surrey Beatty and Sons Pty Ltd, Chipping Norton, pp 401–418

Tedford RH, Banks MR, Kemp NR, McDougall I, Sutherland FL (1975) Recognition of the oldest known fossil marsupials from Australia. Nature 225:141–142

Tedford RH, Wells R, Prideaux G (2006) Pliocene and earlier Pleistocene marsupial evolution in southeastern Australia. Alcheringa 30:313–322

Travouillon KJ, Archer M, Hand SJ, Godthelp H (2006) Multivariate analyses of Cenozoic mammalian faunas from Riversleigh, northwestern Queensland. Alcheringa, Special Issue 1:323–349

Travouillon KJ, Legendre S, Archer M, Hand SJ (2009) Palaeoecological analyses of Riversleigh's Oligo-Miocene sites: implications for Oligo-Miocene climate change in Australia. Palaeogeogr Palaeoclimatol Palaeoecol 276: 24–37

Turnbull WD, Lundelius EL Jr (1970) The Hamilton fauna. A late Pliocene mammalian fauna from the Grange Burn, Victoria, Australia. Fieldiana: Geology 19:1–163

Turnbull WD, Rich TH, Lundelius EL Jr (1987) Burramyids (Marsupialia: Burramyidae) of the early Pliocene Hamilton Local Fauna, southwestern Victoria. In: Archer M (ed) Possums and opossums: studies in evolution. Surrey Beatty and Sons Pty Ltd, Chipping Norton, pp 729–739

Turnbull WD, Lundelius EL Jr, Archer M (2003) Dasyurids, perameloids, phalangeroids, and vombatoids from the Early Pliocene Hamilton Fauna, Victoria, Australia. Bull Am Mus Nat Hist 279:513–540

Turney CSM, Bird MI, Fifield LK, Roberts RG, Smith M, Dortch CE, Gr¨un R, Lawson E, Ayliffe LK, Miller GH, Dortch J, Cresswell RG (2001) Early human occupation at Devil's Lair, southwestern Australia 50,000 years ago. Quaternary Res 55:3–13

Turney C, Flannery TF, Roberts RG, Reid C, Fifield LK, Higham TFG, Jacobs Z, Kemp N, Colhoun EA, Kalin RM, Ogle N (2008) Late-surviving megafauna in Tasmania, Australia, implicate human involvement in their extinction. Proc Natl Acad Sci USA 105:12150–12153

Tyler MJ, Godthelp H (1993) A new species of *Lechriodus* Boulenger (Anura: Leptodactylidae) from the early Eocene of Queensland. Trans R Soc South Aust 117:187–189

Van Dyck S (1982) *Antechinus puteus* (Marsupialia: Dasyuridae), a new fossil species from the Texas Caves, southeastern Queensland. Aust Mammal 5:59–68

Van Valen L (1969) Variation genetics of extinct animals. Am Nat 103:193–224

Van Valkenburgh B, Wang X, Damuth J (2004) Cope's Rule, hypercarnivory and extinction in North American canids. Science 306:101

Vullo R, Gheerbrant E, de Muizon C, Néraudeau D (2009) The oldest modern therian mammal from Europe and its bearing on stem marsupial paleobiogeography. Proc Natl Acad Sci USA 106:19910–19915

Westerman M, Burk A, Amrine-Madsen HM, Prideaux GJ, Case JA, Springer MS (2002) Molecular evidence for the last survivor of an ancient kangaroo lineage. J Mamm Evol 9:209–223

Westerman M, Loke S, Springer MS (2004) Molecular phylogenetic relationships of two extinct potoroid marsupials, *Potorous platyops* and *Caloprymnus campestris* (Potoroinae: Marsupialia). Mol Phylogen Evol 31:476–485

White ME (2006) Environments of the geological past. In: Merrick JR, Archer M, Hickey GM, Lee MSY (eds) Evolution and biogeography of Australasian vertebrates. Auscipub Pty Ltd, Sydney, pp 17–50

White AW, Worthy TH, Hawkins S, Bedford S, Spriggs M (2010) Megafaunal meiolaniid horned turtles survived until early human settlement in Vanuatu, Southwest Pacific. Proc Natl Acad Sci USA 107(35):15512–15516

Willis PMA, Molnar RE, Scanlon JD (1993) An early Eocene crocodylian from Murgon, southeastern Queensland. Kaupia Darmstädter Beiträge zur Naturgeschichte 3:25–32

Woinarski JCZ, Armstrong M, Brennan K, Fisher A, Griffiths AD, Hill B, Milne DJ, Palmer C, Ward S, Watson M, Winderlich S, Young S (2010) Monitoring indicates rapid and severe decline of native small mammals in Kakadu National Park, northern Australia. Wildlife Res 37: 116–126

Woodburne MO (1967) The Alcoota Fauna, central Australia. An integrated palaeontological and geological study. Bull Bureau Mineral Resour Geophys Aust Geol 87:1–187

Woodburne M (1972) The Hamilton Fauna-A late Pliocene mammalian fauna from the Grange Burn, Victoria, Australia. J Paleontol 46:930

Woodburne MO (1984) *Wakiewakie lawsoni*, a new genus and species of Potoroinae (Marsupialia: Macropodidae) of medial Miocene age, South Australia. J Paleontol 58: 1062–1073

Woodburne MO, Case JA (1996) Dispersal, vicariance, and the late Cretaceous to early Tertiary land mammal biogeography from South America to Australia. J Mamm Evol 3:121–161

Woodburne MO, Clemens WA (1986) Revision of the Ektopodontidae (Mammalia, Marsupialia, Phalangeroidea) of the Australian Neogene. Univer California Pub Geol Sci v. 131. University of California Press, Berkeley

Woodburne MO, Tedford RH, Archer M, Turnbull WD, Plane MD, Lundelius EL Jr (1985) Biochronology of the continental mammal record of Australia and New Guinea. South Australian Dep Mines Energy, Spec Pub 5:347–364

Woodburne MO, Tedford R, Archer M, Pledge N (1987a) *Madakoala*, a new genus and two species of Miocene koalas (Marsupialia: Phascolarctidae) from South Australia and a new species of *Perikoala*. In: Archer M (ed) Possums and opossums: studies in evolution. Surrey Beatty and Sons Pty Ltd, Chipping Norton, pp 293–317

Woodburne MO, Pledge NS, Archer M (1987b) The Miralinidae, a new family and two new species of phalangeroid marsupials from Miocene strata of South Australia. In: Archer M (ed) Possums and opossums: studies in evolution. Surrey Beatty and Sons Pty Ltd, Chipping Norton, pp 581–602

Woodburne MO, Tedford RH, Archer M (1987c) New Miocene ringtail possums (Marsupialia: Pseudocheiridae) from South Australia. In: Archer M (ed) Possums and opossums: studies in evolution. Surrey Beatty and Sons Pty Ltd, Chipping Norton, pp 639–679

Woodburne MO, MacFadden BJ, Case JA, Springer MS, Springer KB (1993) Land mammal biostratigraphy and magnetostratigraphy of the Etadunna Formation (Late Oligocene) of South Australia. J Vertebrate Paleontol 13:483–515

Woolnough AP, Steele VR (2001) The palaeoecology of the Vombatidae: did giant wombats burrrow? Mammal Rev 31:33–45

Wroe S (1996a) An investigation of phylogeny in the giant extinct rat kangaroo *Ekaltadeta* (Propleopinae, Potoroidae, Marsupialia). J Paleontol 70:681–690

Wroe S (1996b) *Muribacinus gadiyuli* (Thylacinidae; Marsupialia), a very plesiomorphic thylacinid from the Miocene of Riversleigh, northwestern Queensland, and the problem of paraphyly for the Dasyuridae (Marsupialia). J Paleontol 70:1032–1044

Wroe S (1997a) *Mayigriphus orbus* gen. et sp. nov., a Miocene dasyuromorphian from Riversleigh, northwestern Queensland. Mem Queensland Mus 41:439–448

Wroe S (1997b) Stratigraphy and phylogeny of the giant extinct rat kangaroos (Propleopinae, Hypsiprymnodontidae, Marsupialia). Mem Queensland Mus 41:449–456

Wroe S (1998) A new 'bone-cracking' dasyurid (Marsupialia), from the Miocene of Riversleigh, northwestern Queensland. Alcheringa: An Australas J Palaeontol 22:277–284

Wroe S (1999) The geologically oldest dasyurid (Marsupialia), from the Miocene Riversleigh, northwestern Queensland. Palaeontology 42:1–27

Wroe S (2001) *Maximucinus muirheadae*, gen. et sp. nov. (Thylacinidiae, Marsupialia), from the Miocene of Riversleigh, northwestern Queensland, with estimates of body weights for fossil thylacinids. Aust J Zool 49:603–614

Wroe S, Archer M (1995) Extraordinary diphyodonty-related change in dental function for a tooth of the extinct marsupial *Ekaltadeta ima* (Propleopinae, Hypsiprymnodontidae). Archs oral Biol 40:597–603

Wroe S, Archer M (2006) Origins and early radiations of marsupials. In: Merrick JR, Archer M, Hickey GM, Lee MSY (eds) Evolution and biogeography of Australasian vertebrates. Auscipub Pty Ltd, Sydney, pp 551–574

Wroe S, Field J (2006) A review of the evidence for a human role in the extinction of Australian megafauna and an alternative interpretation. Quaternary Sci Rev 25:2692–2703

Wroe S, Mackness BS (1998) Revision of the Pliocene dasyurid *Dasyurus dunmalli* (Dasyuridae, Marsupialia). Mem Queensland Mus 42:605–612

Wroe S, Mackness B (2000) A new genus and species of dasyurid from the Pliocene Chinchilla Local Fauna of south-eastern Queensland. Alcheringa: An Australasian J Palaeontol 24:319–326

Wroe S, Muirhead J (1999) Evolution of Australia's marsupicarnivores: Dasyuridae, Thylacinidae, Myrmecobiidae, Dasyuromorphia *incertae sedis* and Marsupialia *incertae sedis*. Aust Mammal 21:10–11, 34–45

Wroe S, Musser A (2001) The skull of *Nimbacinus dicksoni* (Thylacinidae: Marsupialia). Aust J Zool 49: 487–514

Wroe S, Brammall JR, Cooke BN (1998) The skull of *Ekaltadeta ima* (Marsupialia, Hypsiprymnodontidae?): and analysis of some marsupial cranial features and a re-investigation of propleopine phylogeny, with notes on the inference of carnivory in mammals. J Paleontol 72:738–751

Wroe S, Crowther M, Dortch J, Chong J (2004) The size of the largest marsupial and why it matters. Proc R Soc London B 271:S34–S36

Wroe S, Ebach M, Ahyong S, Muizon C, Muirhead J (2000) Cladistic analysis of dasyuromorphians (Marsupialia) phylogeny using cranial and dental characters. J Mammal 81:1008–1024

Wroe S, Myers TJ, Seebacher F, Kear B, Gillespie A, Crowther M, Salisbury S (2003) An alternative method for predicting body mass: the case of the Pleistocene marsupial lion. Paleobiology 29:404–412

Zachos J, Pagani M, Sloan L, Thomas E, Billups K (2001) Trends, rhythms, and aberrations in global climate 65 Ma to present. Science 292:686–693

Zheng H, Wyrwoll K-H, Li Z, Powell C McA (1998) Onset of aridity in southern Western Australia—a preliminary appraisal. Global Planet Change 18:175–187

Zheng H, Powell CMcA, Zhao H (2002) Eolian and lacustrine evidence of late Quaternary palaeoenvironmental changes in southwestern Australia. Global Planet Change 35: 75–92

Epilogue – A Perspective

John A. Talent

Abstract

Earth and biologic (including climatic) scientists are in a pivotal position for reading the 'library' of Earth's past and present as backdrop to global modelling of past as well as future patterns (including cyclicities) of gross environmental change—marine, estuarine, terrestrial and climatic. They have the tools for deciphering the implications of climate modeling, and the globally decreasing biodiversity and carrying capacity, and to discriminate scientifically credible information from the aggressive rhetoric and willful disinformation of professional optimists. Ultimately, it is only scientists who can accurately decipher quantitatively the anthropogenic overprint (the human impact) that presently undermines the global support systems. This requires vastly improved knowledge of extinction patterns (at all scales), causes of land and sea (including deep-ocean) extinctions, reef dynamics, changes in gross ocean circulation and ecosphere collapse. Nature may display remarkable capacity for rebound but conceivably with dramatically depleted biodiversity; it need not necessarily include *Homo sapiens* among its constituents. Better understanding of the natural world is imperative for our longer-term survival.

Keywords

Science literacy • Interconnectedness of Earth's systems • Extinctions in the deep past • Depletion of biodiversity • Anthropogenic overprint • Delaying ecosphere collapse

By definition, someone who is science-literate understands the essential principles of the sciences including biologic and climatic systems. He or she knows how to assess scientifically credible information and is able to make informed and responsible decisions. Scientific observations and the results of climate modeling indicate that human activities are now the primary cause of the continuing increase in Earth's globally averaged surface temperature.

Regardless of the pronouncements of the world's professional optimists (there are many of these with no idea how the biosphere works), there is strong evidence of these phenomena becoming rapidly and increasingly serious, pointing towards a tipping point for ecosphere collapse on a scale as large as (and potentially larger than) the seven or so major extinction events of the deep past (there have been dozens of lesser ones; Walliser 1990), this time bringing with it global economic decline and even collapse on a scale

J.A. Talent (✉)
Earth and Planetary Sciences, Macquarie University 2109, Australia
e-mail: jatalent32@gmail.com

J.A. Talent (ed.), *Earth and Life*, International Year of Planet Earth,
DOI 10.1007/978-90-481-3428-1, © Springer Science+Business Media B.V. 2012

never before experienced or imagined. In such circumstances, it is too easy to ignore scientific facts and logic, accepting baseless assertions from those who loudly deny the facts of science built up by countless thousands of scientists – earth scientists including palaeontologists, climate scientists and biologists – particularly in the past 150 years or so, especially since the appearance in 1859 of Charles Darwin's *The Origin of Species*. A systems-thinking approach is imperative, i.e. an ability to understand complex interconnections among all components of the sciences, including climate and biologic systems. Aggressive rhetoric, wilful disinformation, 'bald assertions' and glib rejection of the enormous database available to us are unhelpful as we confront enormous changes that will affect almost every aspect of human society. The general public, lacking knowledge of compelling data, is easy prey for such attacks, especially if there is a hint that attempts at rectification may require funding. As someone once said, the most sensitive nerve in the human body is the hip-pocket nerve!

The United Nations 'Millennium Ecosystem Assessment' (MA) is now widely accepted. It has stated succinctly that 'Unwittingly, the collective impact of human activity is undermining the support systems of life on Earth. Yet we remain deeply dependent on the world's natural and managed ecosystems for our survival'.

Life – from microbes to plants and animals (including humanity) – is the major driver of the global carbon cycle and can influence global climate by modifying the chemical makeup of the atmosphere. This can be tracked through the geologic (especially sedimentary and isotopic) record. Growing evidence shows that changes in many physical and biological systems are linked to changes resulting from human activities, decreasing the environment's capacity to support various biological communities and individual species with the net result that biodiversity and ecological resilience have been substantially reduced. Natural processes driving our planet's long-term climate do not explain the rapid climate change of the last 150 years or so. Average temperatures have increased markedly in the past 50 years, especially in polar regions. The only explanation consistent with all available evidence is that human activities are primarily responsible, causing increases in greenhouse gas concentrations by burning fossil fuels. Natural processes removing carbon dioxide from the atmosphere are slow compared to the processes now adding it to the atmosphere. Thus, carbon dioxide introduced into the atmosphere today may remain there a century or more. Other greenhouse gases, principally methane and nitrous oxide, may remain in the atmosphere for much longer, even thousands of years. The abundance of carbon in the atmosphere is reduced through seafloor accumulation of marine sediments and accumulation of plant biomass; it is increased by deforestation and the burning of fossil fuels. The abundance of greenhouse gases in the atmosphere is controlled by cycles that continually move these components between ocean, land, life and atmosphere reservoirs. No matter what we do about it, the warming influence of greenhouse gases is expected to persist well into the next century and beyond – see Dodson (2010) for further information on the complex interaction and dependencies of the solar, terrestrial, atmospheric and living components that make up our Earth's systems. The interconnectedness of Earth's systems means that a significant change in any one component of the climate system can significantly influence the equilibrium of some or all of the biosphere.

However, significant Cenozoic events pre-dated the appearance of *Homo sapiens* and cannot therefore be attributed in any way to human activities. Ed Petuch, in an earlier chapter, has shown how a pattern of extinction events (not grand scale, but nonetheless significant) reduced the diversity of Western Atlantic molluscan faunas during the Miocene, long before humankind appeared in that part of the globe. Marine and non-marine events associated with the glacial and inter-glacial patterns occurred in even later Cenozoic times. Bruce Hayward and colleagues, incidentally, have demonstrated from numerous deep ocean bores that major extinction in the bottom-dwelling microfaunas, specifically foraminifers, occurred during Late Pliocene to Middle Pleistocene times (3 million to 12,000 years ago) (Hayward 2002; O'Neill et al. 2007). There is no way this extinction pattern can be connected with the advent of *H. sapiens*. Can humankind be expected to rectify such gross phenomena that are not of our making? Doubtless not, because the scale is too large. But we nevertheless need to know more about extinction patterns, extinction optima and the causes of extinction on land and in the shallow seas and deep ocean environments, especially for the past million years or so, if we are to compellingly decipher what may be called the anthropogenic overprint – the contribution to environmental devastation, globally, brought about by the activities of *H. sapiens* – and to be able to demonstrate this quantitatively.

The pattern of these events is part of the backdrop against which modeling the pattern of future environmental change needs to be evaluated. The drivers of evolution and the dramatic set-backs which have occurred in the deep past suggest that there will be future global events that are too large for humankind to contend with. Nevertheless, deciphering such patterns and processes increases in relevance for the human species over the last few hundred thousand years of Cenozoic (viewed as including Quaternary) time. One of the drivers for the background pattern is climatic cyclicities, interpreted from the sedimentary and palaeontologic records and, seemingly to some extent, predicted from the astronomical record. Not only have these events affected terrestrial, lacustrine and estuarine ecosystems, locally and globally, but also have affected continental-shelf ecosystems (from which most relevant data have been obtained) and, as noted elsewhere in this chapter, deep-sea ecosystems as well.

Because of their focus on life in the recent and deep past, palaeontologists, in association with other earth and biological scientists, are in a pivotal position for 'reading the library' of Earth's past history, even deciphering the tiny 'time-capsules' referred to earlier that unlock the secrets of past environmental change (see Tasch 1973). This is the backdrop against which modeling the pattern of future environmental change comes into play. Extension of such models into the future is a major challenge. Experiments in recolonizing devastated ecosystems with predators and prey are interesting (e.g., Berger et al. 2001) but, sadly, trivial compared to the scale of devastation already wrought globally by humankind, and the scale of any major attempt that would need to be mounted to substantially rectify this.

Most scientific and popular attention has focused on modification of extant terrestrial and estuarine ecosystems and extinction of individual larger biota (mammals, birds, reptiles, fish), but serious changes are also occurring in marine ecosystems, seemingly driven by global climatic change. In eastern Australia, for instance, an increasing number of tropical crustaceans, fish, molluscs and reef-building corals are intruding southwards from the tropical Indo-Pacific Province into the gradually warming temperate waters of the Peronian Province. The gross pattern of provinciality for shallow marine organisms, based in the first instance on molluscs, is thus starting to undergo modification. This is exemplified by the gradual (though still minimal) southward shift of the Peronian-Maugean provincial boundary in southeastern Australia.

The oceans, covering 70% of the Earth's surface, exert a major control on climate by dominating Earth's energy and water cycles. They have the capacity to absorb large amounts of solar energy. Heat and water vapour are redistributed globally through density-driven ocean currents and atmospheric circulation (Ivanova 2009 for a recent synthesis; Dodson 2010). Changes in ocean circulation caused by large influxes of fresh water from melting ice can lead to significant and even abrupt changes in gross climate, locally and globally. Tectonic movements, opening or closing gateways to oceanic circulation, as has happened repeatedly in the deep past, can lead to gross climatic change as well as major changes in oceanic circulation, but these are commonly on a scale of a million years and more. For example, there is great potential for using palaeontologic data (coupled with isotopic data) for improving understanding of the history and climatic impact of change in the Gulf Stream and its counter-current, and the history of marine circulation in the northern Pacific (Beringia–Japan), most obviously the impact of opening and closing of 'gateways' such as the Bering Straits. Quantitative palaeontologic data from offshore marine deposits has potential for improving understanding of monsoonal systems – as background for improved prediction of their frequency, cyclicity and intensity. This, however, is beyond the scope of the present volume.

Reef systems (presently under threat globally) are pivotal in man's survival as carbon sinks and as contributors to regulating the sort of air we breathe and thus, indirectly, to food supplies. The dynamics of reef systems in the geologically recent past provide a specific initiative where traditional palaeontologic investigations may be utilised – curiously, there has been no attempt to present the dynamics of an entire reef system since Odum and Odum (1955). Excellent target areas for parallel/integrated analyses of contemporary reefs and the dynamics of Late Cenozoic reefs through the last 35 million years or so are available in the southwest Pacific, e.g., on the Loyalty Islands and on Mangaia in the Cook Islands. These have elegantly exposed raised reef systems that call for team-research by carbonate sedimentologists interacting with experts on scleractinians, calcareous algae, foraminifers and ostracods – groups on which palaeontologists have long been major contributors – in elucidating reef dynamics through this long period of time. This would help with providing a portion of the baseline for modelling the future of our threatened reef systems but, is there sufficient time? The foremost student of living corals, 'Charlie' Veron, has insisted (Veron 2008) that the time-frame for saving the reefs of the world, including the Great Barrier Reef, is now less than 20 years. It is a massive task that beggars imagination and may be impossible to turn around.

The palaeontologic database and its interpretation for the past million years or so, and especially the last few hundred thousand years (building up to the present, establishing cylicities), impact on the future of the Earth's environments and the possible place of *H. sapiens* (or descendants of that species) within them. The importance of having a comprehensive knowledge of this 'background' pattern, mentioned at the beginning of the introductory chapter to this volume, increases in relevance as we approach the present day where over-exploitation of key species, declining water quality, global warming, ocean acidification and the decline of terrestrial and marine faunas and floras (including fish, reptile, bird and mammal faunas), and the die-back of entire reef communities, have become features of the modern world.

In coming decades, scientists expect climate change to have an increasing impact on human and natural systems, impacting on accessibility to raw materials (including water), energy, biodiversity and, by implication, food and economic stability. The resolve of individuals, communities and countries to identify and implement effective management strategies for critical institutional and natural (including biological) resources will be necessary to ensure the stability of both human and natural systems as temperatures rise.

Extension of models of environmental change into the future is a major challenge. For this we need the best possible databases of living and extinct species. As the dominant species on the planet, we need to understand the natural world for our longer-term survival. Putting it another way, the biosphere's survival depends on our knowledge of it.

Good science, and I focus here on refer here on biologic and palaeontologic sciences, is based on good taxonomy, not what one might refer to as superficial 'nuts and bolts' taxonomy, naming things solely for the pleasure of doing so. Museum Studies guru Andrew Simpson has pointed out to me that, as a species, *H. sapiens* is 'hard-wired' to classify the natural world – because we have always needed to do so for survival. All human cultures do this to know what can be eaten, what should be avoided and what changes in our part of the natural world signify about the environment. All human cultures, he says, are compelled to collect and classify as a way of understanding.

Anthropologist Claude Lévi-Strauss insisted that 'The world began without the human race and will certainly end without it'. An opinion given in the New York Times a couple of years ago, suggested there was no doubt that '. . .if [humankind] were to disappear, Nature would quickly reclaim every place that *Homo sapiens* called home', but that 'reclamation' would be from a world with dramatically depleted biodiversity and loss of numerous entire ecosystems. Experiments in recolonizing devastated ecosystems with predators and prey, as noted earlier, are interesting but, sadly trivial compared to the scale of devastation already wrought by humankind.

These views are not unduly pessimistic. In terms of billions of years, Lévi-Strauss is certainly right. The New York Times scenario, on a much shorter time-frame, is plausible. It is nevertheless incumbent on us to take every possible step to prevent global catastrophe occurring or, at the very least, to modify or delay this for as long as reasonably possible.

Acknowledgements This chapter draws on short essays authored by John A. Talent and Bettina Reichenbacher in E. Derbyshire (Ed.), Planet Earth: International Year, in issues of WebMags, published in 2008 and 2009 for UNESCO by Boston-Hannah

References

Berger J, Swenson JE, Persson I-L (2001) Recolonizing carnivores and naïve prey: conservation lessons from Pleistocene extinctions. Science 291:1036–1039

Dodson J (ed) (2010) Changing climates, Earth systems and society. Springer, Dordrecht, 244pp

Hayward BW (2002) Pliocene to middle Pleistocene extinction of deep-sea benthic Foraminifera "*Stilostomella* extinction" in the southwest Pacific. J Foraminiferal Res 32:274–307

Ivanova EV (2009) The global thermohaline paleocirculation. Springer, Dordrecht, 314pp
Odum HT, Odum EP (1955) Trophic structure and productivity of a windward coral community on Eniwetok Atoll. Ecol Monogr 35:291–320
O'Neill TA, Hayward BW, Kawagata S, Sabaa AT, Grenfell HR (2007) Pleistocene extinction of deep-sea benthic Forminifera. Palaeontology 50:1073–1102
Tasch P (1973) Paleobiology of the invertebrates: data retrieval from the Fossil record. Wiley, New York, NY, 946pp
Veron JEN (2008) A reef in time: the Great Barrier Reef from beginning to end. Harvard University, Cambridge, 304pp
Walliser OH (1990) Extinction events in Earth history. Springer, Berlin, 333pp

General Index

Note: The letters 'f' and 't' following the locators refer to figures and tables respectively.

D

F

G

H

Taxonomic Index (Genera and Species)

Note: The letters 'f' and 't' following the locators refer to figures and tables respectively.

GPSR Compliance

The European Union's (EU) General Product Safety Regulation (GPSR) is a set of rules that requires consumer products to be safe and our obligations to ensure this.

If you have any concerns about our products, you can contact us on ProductSafety@springernature.com

In case Publisher is established outside the EU, the EU authorized representative is:

Springer Nature Customer Service Center GmbH
Europaplatz 3
69115 Heidelberg, Germany

Batch number: 10770416

Printed by Printforce, the Netherlands